Physics for Scientists and Engineers: Foundations and Connections

To Zak and Jeff
with love.

Physics for Scientists and Engineers: Foundations and Connections

Advance Edition, Volume I
Not for Redistribution

Professor Debora M. Katz
United States Naval Academy

Australia • Brazil • Mexico • Singapore • United Kingdom • United States

***Physics for Scientists and Engineers: Foundations and Connections*, Advance Edition, Volume 1**
Not for Redistribution
Debora M. Katz

Product Director: Mary Finch
Product Manager: Charlie Hartford
Managing Developer: Peter McGahey
Product Development Specialist: Nicole Hurst
Senior Content Developer: Susan Dust Pashos
Product Assistant: Chris Robinson
Media Developer: Andrew Coppola
Content Project Managers: Cathy Brooks and Alison Eigel Zade
Art Director: Bruce Bond
Manufacturing Planner: Sandee Milewski
Rights Acquisition Specialist: Shalice Shah-Caldwell
Production Service and Compositor: Graphic World Inc
Photo Researcher: Sharon Donahue
Cover Designer: Bruce Bond
Cover Image: United Launch Alliance

Library of Congress Control Number: 2013952110

Student Edition:

ISBN-13: 978-1-305-07798-0

ISBN-10: 1-305-07798-9

Cengage Learning
200 First Stamford Place, 4th Floor
Stamford, CT 06902
USA

Printed in the United States of America
1 2 3 4 5 6 7 17 16 15 14 13

Physics for Scientists and Engineers

Beyond the Quantitative

Debora M. Katz | *United States Naval Academy*

The main goal of *Physics for Scientists and Engineers: Foundations and Connections* is to offer a calculus-based introductory physics textbook designed to assist you in taking your students "beyond the quantitative." Debora Katz leverages physics education research (PER) best practices and her extensive classroom experience to motivate her readers and address the areas where students struggle the most—bridging the gap between abstract language and application, overcoming common preconceptions, and connecting mathematical formalism and physics concepts.

Students often view physics as a series of unrelated facts, concepts, and equations that have little or no bearing on their everyday lives. This text is designed to change these impressions.

1. **Case studies to motivate students and make abstract concepts concrete.** This text uses case studies to draw readers into the story of physics. In this text, case studies are introduced and revisited throughout the chapters in pedagogy such as the concept exercises, examples, and end-of-chapter problems.

   Some case studies are based on students' contemporary "real-world" experiences, including events they may read about in the news. These case studies make abstract physics concepts understandable and help bridge the gaps between the key concepts, the formal language, and the mathematics of physics.

   For example, the case study in Chapter 5 illustrates Newton's laws of motion by examining an actual train collision described in news articles. The author introduces free-body diagrams and Newton's second law, asking students to determine why backward-facing passengers received less bodily harm than did the forward-facing passengers. Students apply physics "tools"—Newton's laws and free-body diagrams—to get to the bottom of this real-world example.

   Other case studies are based on historical events. For example, the case study in Chapter 24 (Volume 2) discusses Benjamin Franklin's study of lightning and his subsequent invention of the lightning rod. Although many students know that Franklin flew a kite during a storm, many do not know he was trying to test one of his scientific hypotheses or that he subsequently invented the lightning rod. The shape of lightning rods was of great debate—Franklin's design included a pointed tip, but a rival of Franklin had argued that the end of a lightning rod must be rounded. As students work through the debate, they are motivated to calculate the electric field produced by charged conductors of various shapes.

2. **Student dialogues to address preconceptions.** Students come to the classroom with preconceptions (what some call misconceptions). By acknowledging and addressing these preconceptions, we can transform them into building blocks toward proper understanding. Often, students simply need a connection between their preconceptions and the true physical principles.

   For example, many students believe that when a truck and a car collide, the force of the truck on the car is greater than the force of the car on the truck. This can be viewed in one of two ways—either as a misconception about Newton's third law or as a misapplied resource in need of additional knowledge. Students know that the car's "experience" in the collision is different from the truck's experience. They need to connect this preconception to the acceleration of the car instead of to the force exerted on the car by the truck. Once students make that connection, it is much easier for them to understand that the force on each vehicle must have an equal magnitude. Thus, they are able to incorporate Newton's third law into their own newly unified worldview.

   In this text, preconceptions are primarily addressed by **dialogues between fictional students.** These student dialogues allow the readers to discover their preconceptions without a sense of failure. Dialogues may be incorporated into case studies, concept exercises, examples, and end-of-chapter problems. For example, a student dialogue discussing the train collision case study (Chapter 5) helps students address their preconceptions around Newton's laws.

3 **Two-column format for examples and derivations to connect mathematical formalism and physics concepts.** Research shows that students struggle to make these connections. By presenting many of the **examples and derivations in two columns (what an expert problem-solver thinks on the left and what that expert would write on the board on the right),** students can make the connections between the concept being taught and the mathematical steps to follow. It is like having an instructor within the text: in one column he or she explains the concept, and in the other he or she shows the mathematical steps to follow, just as an instructor would verbally explain a problem in class while simultaneously solving the problem on the board.

A Note to Instructors from the Author

Before I began this project, I was not aware of just how many people are involved in writing a textbook. After all, the cover only lists the authors. Now I know that each chapter is read by many colleagues, like you, from around the country. Comments from colleagues not only allow me to hone the book into something I hope you will find useful in your classrooms, but also provide an opportunity to have a dialogue on topics that I love—physics and teaching. I have learned a great deal, and I thank all my colleagues for being part of this dialogue.

My Background

When I decided to pursue physics and astrophysics, I didn't think about teaching; I just knew I was interested in the subjects. As a graduate student at the University of Minnesota, I won a fellowship that included training in physics education by well-known figures in physics education research (PER). I learned to take teaching seriously, and when it was time to find a job, I knew I wanted to work at an institution committed to undergraduate education.

Fortunately, I found that job at the United States Naval Academy. I have had the opportunity to teach a year-long calculus-based course not just to physics majors but also to a varied audience with a diverse set of abilities, interests, learning styles, and preconceptions. To engage my students in the process of learning physics, I use many of the techniques that have come out of PER, and over the past couple of decades, my students have taught me how to teach them. I have tried my best to integrate the lessons learned from my students into the pages of this text.

My Book

Although PER offers many ways to approach teaching, all of these approaches require students to take an active part in their education while I act as a coach, urging them to take the necessary steps. There is a wide range of pedagogy in introductory physics, but one thing is always true: Students do better if they read their textbook.

Many textbooks were originally written for previous generations of students. Although these textbooks have been updated for content, today's students don't find them readable. (My students easily read 100 pages of history but find it difficult to get through five pages of physics.) My primary goal in writing this book was to make it readable without compromising content, because I believe that our students are willing and able to read a book they find engaging.

Students in required courses often tell us that they don't understand how physics fits into their lives. So to make physics engaging, I use case studies in my classroom and in my book. Case studies relate interesting topics to the concepts, principles, and tools of physics. (In my class, students write their own case studies, relating such topics as sports, movies, and history to physics concepts.)

Case studies also help students realize how the preconceptions they have developed over decades of observation tie into the formalism of physics. For example, students know that seat belts keep passengers in cars, but they are not usually aware of the connection between that fact and Newton's first law. To make and reinforce these types of connections, some case studies use a dialogue between fictional students to highlight and clarify commonly held misidentified preconceptions.

Of course we know (as did our teachers) that physics is not learned passively. We expect our students to learn physics by solving problems. Recent editions of traditional textbooks have begun to include problem-solving strategies. However, many steps in example problems are still not well explained and often seem mysterious to students. I have found that students understand the examples better when I use a two-column format. The column on the right contains the formalism you would typically find in a textbook or write on the classroom board. The left column includes the words you might use to explain each step.

I hope you'll find this textbook program to be a better learning tool for your students than what has previously been available.

Thank you for taking the time to consider this Advance Edition for your own classroom use. If you have any comments or questions, feel free to contact me at deborakatz@yahoo.com.

Sincerely,
Debora

A Note to Students from the Publisher

We want to receive student feedback regarding the author's approach and the use of specific pedagogical devices. To facilitate this process, we have incorporated QR codes within selected chapters that link to brief online surveys. Randomly selected participants will be awarded a $25 American Express gift card at the end of each semester to thank them for their time and feedback. Your input is important to us and could influence the development of future editions.

To take the survey scan or visit **www.cengage.com/community/katz**

Contents

Part I Classical Mechanics

Chapter 1 Getting Started 1

1-1 Physics 1
1-2 How Are Laws of Physics Found? 2
1-3 A Guide to Learning Physics 3
Useful Features of this Book 3
Advice from the Author 4
First Case Study 4
1-4 Solving Problems in Physics 5
1-5 Systems of Units 6
Time 6
Length 7
Mass 7
Scientific Notation and Common Prefixes for SI Units 7
Converting Units 7
1-6 Dimensional Analysis 9
1-7 Error and Significant Figures 10
1-8 Order-of-Magnitude Estimates 12
Welcome to the Beginning of Your Adventure 17

Chapter 2 One-Dimensional Motion 22

2-1 What Is One-Dimensional Translational Kinematics? 23
2-2 Motion Diagrams 23
2-3 Coordinate Systems and Position 24
Introduction to Vectors 24
2-4 Position-Versus-Time Graphs 25
Position-Versus-Time Graph 26
2-5 Displacement and Distance Traveled 27
Displacement (Change in Position) 28
Displacement, Translation, and the Particle Model 28
Distance Traveled 29
2-6 Average Velocity and Speed 30
Average Velocity 31
Average Speed 31
Average Velocity from a Position-Versus-Time Graph 32
2-7 Instantaneous Velocity and Speed 34
Instantaneous Speed 35
Displacement from a Velocity-Versus-Time Graph 36
2-8 Average and Instantaneous Acceleration 37
Average Acceleration 37
Instantaneous Acceleration 37
2-9 Special Case: Constant Acceleration 40
Velocity as a Function of Time for Constant Acceleration 40
Displacement as a Function of Velocity and Time 40
The Five Kinematic Equations for Constant Acceleration 41
Using Integral Calculus 41
2-10 A Special Case of Constant Acceleration: Free Fall 45
Graphical Solutions 49

Chapter 3 Vectors 59

3-1 Geometric Treatment of Vectors 60
Drawing Vectors 60
Adding Vectors Geometrically 61
Multiplying a Vector by a Scalar 61
Subtracting Vectors Geometrically 62
Using a Scale 64
3-2 Cartesian Coordinate Systems 65
Axes and Coordinates 65
Unit Vectors 67
Right-Handed Coordinate Systems 67
3-3 Components of a Vector 68
Vector Components and Scalar Components 68
Resolving a Vector into Components 69
Vector Magnitude and Direction 72
Angles, Inverse Trigonometric Functions, and Vector Directions 72
3-4 Combining Vectors by Components 76
Vector Components of Motion Variables 77

Chapter 4 Two- and Three-Dimensional Motion 85

4-1 What Is Multidimensional Motion? 86
4-2 Motion Diagrams for Multidimensional Motion 87
4-3 Position and Displacement 88
4-4 Velocity and Acceleration 90
4-5 Special Case of Projectile Motion 94
The Equations for Projectile Motion 94
Range Equation 96
4-6 Special Case of Uniform Circular Motion 99
Polar Coordinate System 100
Linear and Angular Speed 100
Centripetal Acceleration 101
4-7 Relative Motion in One Dimension 104
4-8 Relative Motion in Two Dimensions 106
A Word About Air Resistance 110

Chapter 5 Newton's Laws of Motion 119

5-1 Our Experience With Dynamics 120
5-2 Newton's First Law 121

5-3 Force 122
Contact Versus Field Forces 123
Internal Versus External Forces 123
5-4 Inertial Mass 124
5-5 Inertial Reference Frames 124
5-6 Newton's Second Law 126
5-7 Some Specific Forces 130
Gravity Near the Earth's Surface 130
Spring Force 132
Normal Force 132
Tension Force 133
Kinetic Friction 133
5-8 Free-Body Diagrams 134
5-9 Newton's Third Law 141
5-10 Fundamental Forces 146

Chapter 6 Applications of Newton's Law of Motion 153
6-1 Newton's Laws in a Messy World 153
6-2 Friction and the Normal Force Revisited 155
6-3 A Model for Static Friction 156
6-4 Kinetic and Rolling Friction 160
Model for Kinetic Friction 161
Rolling Friction 163
6-5 Drag and Terminal Speed 163
Terminal Speed 166
6-6 Centripetal Force 168
Nonuniform Circular Motion 174

Chapter 7 Gravity 184
7-1 A Knowable Universe 185
7-2 Kepler's Laws of Planetary Motion 187
Kepler's First Law 187
Kepler's Second Law 188
Kepler's Third Law 188
7-3 Newton's Law of Universal Gravity 190
Gravitational and Inertial Mass 195
7-4 The Gravitational Field 196
7-5 Variations in the Earth's Gravitational Field 203
The Earth as a Noninertial Reference Frame 204

Chapter 8 Conservation of Energy 213
8-1 Another Approach to Newtonian Mechanics 214
8-2 Energy 215
Kinetic Energy 215
Potential Energy 217
8-3 Gravitational Potential Energy Near the Earth 218
Reference Configuration 219
Path Independence 219
8-4 Universal Gravitational Potential Energy 221
Reference Configuration for Universal Gravity 222
8-5 Elastic Potential Energy 223
Reference Configuration for Spring Potential Energy 224
8-6 Conservation of Mechanical Energy 225
8-7 Applying the Conservation of Mechanical Energy 228
8-8 Energy Graphs 232
Force Approach Versus Conservation Approach 234
8-9 Special Case: Orbital Energies 235
Elliptical Orbits 236

Chapter 9 Energy in Nonisolated Systems 247
9-1 Energy Transfer to and from the Environment 248
9-2 Work Done by a Constant Force 248
9-3 Dot Product 252
9-4 Work Done by a Nonconstant Force 254
9-5 Conservation and Nonconservative Forces 256
9-6 Particles, Objects, and Systems 259
Work and Mechanical Energy 260
Center of Gravity and Center of Mass 260
Zero-Work Forces 261
9-7 Thermal Energy 261
Globular Cluster Analogy 262
Change in Thermal Energy due to Moving Friction 262
9-8 Work–Energy Theorem 264
9-9 Power 269

Chapter 10 Systems of Particles and Conservation of Momentum 281
10-1 A Second Conservation Principle 282
10-2 Momentum of a Particle 283
10-3 Center of Mass Revisited 284
10-4 Systems of Particles 287
Momentum of a System of Particles 288
10-5 Conservation of Momentum 289
10-6 Case Study: Rockets 292
10-7 Rocket Thrust: An Open System (Optional) 296

Chapter 11 Collisions 306
11-1 What is a Collision? 307
11-2 Impulse 307
11-3 Conservation During a Collision 310
Conservation of Momentum During a Collision 310
Conservation of Kinetic Energy During a Collision 311
11-4 Special Case: One-Dimensional Inelastic Collisions 312
11-5 One-Dimensional Elastic Collisions 315
Stationary Target in an Elastic Collision: Special Cases 316
11-6 Two-Dimensional Collisions 320
Elastic Two-Dimensional Two-Particle Collision 321
Completely Inelastic Two-Dimensional Two-Particle Collision 322

Chapter 12 Rotation I: Kinematics and Dynamics 331

12-1 Rotation Versus Translation **331**
12-2 Rotational Kinematics **333**
Angular Position and Angular Displacement **334**
Angular Velocity **335**
Angular Acceleration **336**
12-3 Special Case of Constant Angular Acceleration **338**
12-4 The Connection Between Rotation and Circular Motion **340**
Distance Traveled by a Point on a Rotating Object **340**
Translational Velocity of a Point on a Rotating Object **340**
Tangential Acceleration **341**
Centripetal Acceleration **341**
Rotational Versus Translational Parameters **341**
12-5 Torque **344**
12-6 Cross Product **346**
12-7 Rotational Dynamics **347**
Newton's Second Law in Rotational Form **348**
Levers **351**

Chapter 13 Rotation II: A Conservation Approach 360

13-1 Conservation Approach **360**
13-2 Rotational Inertia **362**
Rotation Axis Must Be Specified **363**
Rotational Inertia of Continuous Objects **365**
13-3 Rotational Kinetic Energy **367**
13-4 Special Case of Rolling Motion **369**
13-5 Work and Power **373**
Application: Waterwheels **374**
13-6 Angular Momentum **376**
Angular Momentum of a Particle **376**
13-7 Conservation of Angular Momentum **379**

Part II Mechanics of Complex Systems

Chapter 14 Static Equilibrium, Elasticity, and Fracture 390

14-1 What is Static Equilibrium? **391**
Types of Static Equilibrium **391**
14-2 Conditions for Equilibrium **392**
Cross Product Revisited **393**
14-3 Examples of Static Equilibrium **394**
14-4 Elasticity and Fracture **404**
Stress **404**
Strain **405**
Tensile Deformation **405**
Compressive Deformation **406**
Shear Deformation **406**

Chapter 15 Fluids 418

15-1 What Is a Fluid? **418**
Fluid Model **419**
15-2 Static Fluid on the Earth **420**
15-3 Pressure **420**
Definition and Units of Pressure **421**
Pressure Variation with Depth in a Static Fluid **421**
Change in an Object's Volume with Pressure **422**
15-4 Archimedes's Principle **423**
The Buoyant Force **424**
15-5 Measuring Pressure **428**
Pascal's Principle **428**
Manometers, Gauge Pressure, and Absolute Pressure **430**
Barometers **430**
15-6 Ideal Fluid Flow **432**
15-7 The Continuity Equation **433**
15-8 Bernoulli's Equation **436**
Pressure in a Moving Fluid **437**
A Final Note **443**

Chapter 16 Oscillations 450

16-1 Picturing Harmonic Motion **451**
16-2 Kinematic Equations of Simple Harmonic Motion **453**
Position Versus Time in Simple Harmonic Motion **454**
Velocity Versus Time in Simple Harmonic Motion **455**
Acceleration Versus Time in Simple Harmonic Motion **455**
16-3 Connection With Circular Motion **456**
16-4 Dynamics of Simple Harmonic Motion **459**
16-5 Special Case: Object–Spring Oscillator **461**
16-6 Special Case: Simple Pendulum **463**
16-7 Special Case: Physical Pendulum **466**
16-8 Special Case: Torsion Pendulum **468**
16-9 Energy in Simple Harmonic Motion **470**
Energy of an Object–Spring Oscillator **471**
16-10 Damped Harmonic Motion **473**
Describing Damped Oscillations: The Time Constant **474**
Frequency of Damped Harmonic Motion **475**
16-11 Driven Oscillators **476**

Chapter 17 Traveling Waves 486

17-1 Introducing Mechanical Waves **487**
17-2 Pulses **487**
Wave Function for a Particular Pulse **488**
17-3 Harmonic Waves **490**
Transverse Harmonic Waves **491**
Longitudinal Harmonic Waves **493**
Speed of a Harmonic Wave **494**
17-4 Special Case: Transverse Wave on a Rope **494**
17-5 Sound: Special Case of a Traveling Longitudinal Wave **497**
Speed of Sound **498**
Pressure Waves **498**

17-6 Energy Transport in Waves 500
Energy and Power of a Transverse Harmonic Wave 500
Energy and Power of Sound 501
17-7 Two- and Three-Dimensional Waves 502
Intensity and Loudness 502
17-8 Refraction and Diffraction 506
Refraction 506
Diffraction 507
17-9 The Doppler Shift 507
Stationary Source, Moving Observer 508
Moving Source, Stationary Observer 509
Source and Observer Both Moving 510
Shock Waves 513
17-10 The Wave Equation 513

Chapter 18 Superposition and Standing Waves 522
18-1 Superposition 522
18-2 Reflection 524
Fixed-End Reflection 524
Free-End Reflection 524
Law of Reflection 525
18-3 Interference 526
Interference in Pulses and One-Dimensional Waves 527
Two- and Three-Dimensional Interference 528
18-4 Standing Waves 531
Producing a Standing Wave 531
Wave Function of a Standing Wave 531
Position of Nodes and Antinodes 532
Standing Waves in Musical Instruments 533
18-5 Guitar: Resonance on a String Fixed at Both Ends 534
18-6 Flute: Resonance in a Tube Open at Both Ends 537
18-7 Clarinet: Resonance in a Tube Closed at One End and Open at the Other End 540
18-8 Beats 541
18-9 Fourier's Theorem 543

Chapter 19 Temperature, Thermal Expansion, and Gas Laws 552
19-1 Thermodynamics and Temperature 553
Units of Temperature 553
19-2 Zeroth Law of Thermodynamics 555
19-3 Thermal Expansion 555
Microscopic Model of Thermal Expansion 556
Macroscopic Observation of Thermal Expansion 557
19-4 Thermal Stress 560
Thermal Expansion of Water 562
19-5 Gas Laws 564
19-6 Ideal Gas Law 566
Avogadro's Number 567
19-7 Temperature Standards 570
Case Study: Constant-Volume Gas Thermometer 572

Chapter 20 Kinetic Theory of Gases 580
20-1 What Is the Kinetic Theory? 581
20-2 Average and Root-Mean-Square Quantities 582
20-3 The Kinetic Theory Applied to Gas Temperature and Pressure 584
20-4 Maxwell-Boltzmann Distribution Function 588
Maxwell-Boltzmann Distribution 589
20-5 Mean Free Path 591
Diffusion 593
20-6 Real Gases: The Van der Waals Equation of State 596
Comparing Van der Waals and the Ideal Gas Equations of State 597
20-7 Phase Changes 599
20-8 Evaporation 601
Humidity 603

Chapter 21 Heat and the First Law of Thermodynamics 611
21-1 What Is Heat? 612
21-2 How Does Heat Fit Into the Conservation of Energy? 613
Thermal Energy, Work, and Heat 613
21-3 The First Law of Thermodynamics 615
21-4 Heat Capacity and Specific Heat 616
21-5 Latent Heat 620
21-6 Work in Thermodynamic Processes 623
21-7 Specific Thermodynamic Processes 625
Adiabatic Process ($Q = 0$) 625
Isothermal Process ($\Delta T = 0$) 626
Constant-Volume Process ($\Delta V = 0$) 627
Constant-Pressure Process ($\Delta P = 0$) 628
Cyclic Process ($\Delta E_{th} = 0$) 628
Free Expansion ($Q = W = 0$) 629
21-8 Equipartition of Energy 631
Molar Specific Heat of Gases 631
Degrees of Freedom 634
21-9 Adiabatic Processes Revisited 637
21-10 Conduction, Convection, and Radiation 638
Conduction 639
Convection 641
Radiation 641
Power Absorbed from Sunlight 642

Chapter 22 Entropy and the Second Law of Thermodynamics 651
22-1 Second Law of Thermodynamics, Clausius Statement 652
22-2 Heat Engines 652
22-3 Second Law of Thermodynamics, Kelvin-Planck Statement 656
22-4 The Most Efficient Engine 656
Reversible and Irreversible Processes 656
Carnot Engine 657
Third Law of Thermodynamics 659
22-5 Case Study: Refrigerators 664
22-6 Entropy 665
22-7 Second Law of Thermodynamics, General Statements 670
The Arrow of Time 671

22-8 Order and Disorder **672**
22-9 Entropy, Probability, and the Second Law **673**

Appendix A Mathematics 683
A-1 Algebra and Geometry **683**
A-2 Trigonometry **684**
A-3 Calculus **685**
Derivatives **685**
Integrals **686**
A-4 Propagation of Uncertainty **687**
Sums and Differences **687**
Products, Quotients, and Powers **688**
Multiplication by an Exact Number **688**

Appendix B Reference Tables 689
B-1 Symbols and Units **689**
B-2 Conversion Factors **691**
B-3 Some Astronomical Data **692**
B-4 Rough Magnitudes and Scales **693**

Periodic Table of the Elements 696

Answers to Concept Exercises and Odd-Numbered Problems A-1–A-24

Index I-1

LIST OF CASE STUDIES, VOLUME I

Page numbers list initial mention only. Case studies may have multiple parts and are often revisited in worked examples, concept exercises, and end-of-chapter problems within a chapter or in multiple chapters.

Chapter	Topic	Page
1	Raisin	5
2	An experiment in motion	25
3	Skydiving	60
4	Skateboarding	86
5	Train collision	120
6	Skydiving in formation	154
7	Dark matter and MOND	186
8	Comet Halley	214
9	A meteor shower	248
10	Rockets	282
11	Train collision revisited	307
12	Ancient megaliths	332
13	Laboratory challenge: Accounting for all the energy	361
14	Designing an air and space exhibit	392
15	Flying airplanes	419
16	Experiment: Bouncy vs. not very bouncy	452
17	Rolling stones and kidney stones	487
18	A big-band concert	523
19	Inventing the thermometer	554
20	The Earth's atmosphere	581
21	You	616
22	Engines and refrigerators	652

PART ONE
Classical Mechanics

1 Getting Started

Underlying Principle

Physics is the fundamental natural, experimental science.

Major Concepts

1. Theory
2. Scientific evidence
3. Standard unit
4. The SI system
5. Uncertainty and error
6. Significant figures
7. Mass density

Tools

1. Conversion factor
2. Dimensional analysis
3. Order-of-magnitude estimate

Key Question

What is physics, and how can I best study it?

1-1 Physics 1

1-2 How are laws of physics found? 2

1-3 A guide to learning physics 3

1-4 Solving problems in physics 5

1-5 Systems of units 6

1-6 Dimensional analysis 9

1-7 Error and significant figures 10

1-8 Order-of-magnitude estimates 12

How does your brain tell your body to run? Will the Universe expand forever, or will it collapse down to a single point? How do bicycles, airplanes, and rockets work? Why is the sky blue? How did the Earth form? Why can a cockroach survive a fall off a refrigerator? Why do ballerinas and basketball players seem to hover in midair? These questions are just a few of the ones that physics can answer (Fig. 1.1).

1-1 Physics

Physics is a natural science; that is, it deals with natural phenomena, as opposed to a social or political science that deals with human society or human governments. Not all artificial creations are beyond the realm of physics, though. Physicists build spacecraft, fight cancer, and design better bicycles. Physics is also called a physical science. This wording may seem redundant, but it signifies that the goal of physics is to discover the laws governing the *physical* Universe. These laws are not invented by people. For example, not being allowed to drive your car at 100 miles per hour (mph) on an open highway is a law invented by people, but not being able to pedal your ordinary bicycle at 100 mph is determined by the laws of the physical Universe.

There are other natural sciences, such as astronomy, biology, chemistry, and geology. Physics is *the* fundamental natural science because it examines the principles that apply to *all* parts of the physical world, whereas other sciences focus on a more limited part of the physical world. Biology, for example, focuses on living organisms.

Geir Mogen/NTNU Info

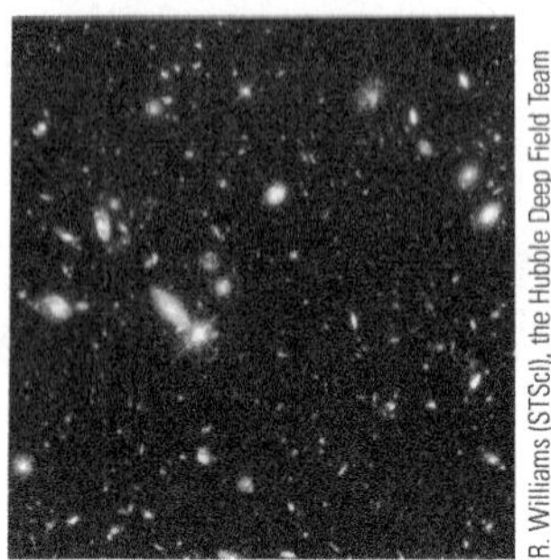

R. Williams (STScI), the Hubble Deep Field Team and NASA

Ayakovlev/Shutterstock.com

FIGURE 1.1 **A.** How does your brain tell your body to run? **B.** Will the Universe expand forever or recollapse to a single point? **C.** Why do ballerinas seem to hover in the air?

Because physics is such a broad topic, it is helpful to break it into several branches or fields. Each field roughly corresponds to one or two *principles* or *laws* of nature. A **physical principle** or **law** is a rule that governs some behavior or property of the physical Universe. Traditionally, there is a slight distinction between them: Laws are principles that have withstood many experiments and observations. Some principles, however, earn the title of *law* and are later found to have some limitations, but they do not get demoted back to *principle.* Throughout this book, we use the terms *principle* and *law* interchangeably. Understanding the laws of physics is one of your major goals.

You will also learn many *concepts.* A **concept** is an idea that makes it possible to describe the physical world clearly. For example, acceleration (Chapter 2) is a *concept*, but Newton's second law of motion (Chapter 5) is a *principle* that explains how an object accelerates.

This book is divided into six major parts, with each part focusing on just a few principles.

Part I focuses on describing and explaining how objects move. This branch of physics is known as **classical mechanics**. Classical mechanics is governed by Newton's laws of motion.

Part II continues the study of motion. The focus in this part is on more complicated systems such as fluids and more complicated motion such as oscillations and waves. The laws of **thermodynamics** provide the basis for studying complicated systems such as gases.

Part III focuses on **electricity** (how charged particles interact). Electricity is governed by Coulomb's law and Gauss's law.

Part IV shifts to **magnetism**, the study of the interaction between moving charged particles. Magnetism is based on Ampère's law and is so closely related to electricity that physicists consider these two phenomena as one (**electromagnetism**). Faraday's law provides part of the connection between electricity and magnetism.

Part V shows that the connection between electricity and magnetism explains **light** or **radiation**. Maxwell's equations govern electricity, magnetism, and radiation.

Part VI briefly describes two important branches of physics discovered in the 20th century, including **relativity** and **quantum mechanics**. Relativity is divided into two principles. Special relativity describes the motion of objects moving at very high constant speeds. General relativity expands the principles of relativity to include acceleration. Quantum mechanics is important to understanding very small objects, such as **atoms**. Both quantum mechanics and relativity are important to **nuclear physics** and **cosmology** (study of the Universe), and both of these 20th century principles challenge the principles of classical mechanics.

1-2 How Are Laws of Physics Found?

THEORY Major Concept

Physical principles are discovered (not invented) by people. A principle starts off as someone's idea or theory. How do you know when one of your theories is a law of the Universe? A scientific **theory** makes testable predictions. For example, in the early 1500s, Nicolaus Copernicus, a church canon, economist, and physician, theorized that the planets move in circular orbits around the Sun. Copernicus used his theory to make several predictions about the location, brightness, and phase of the planets at certain times. The Copernican theory roughly matched observations of the planets, but it failed to predict their location precisely (Fig. 1.2). This failure means that the theory of circular orbits is not a natural law. In fact, the German mathematician and astronomer Johannes Kepler (1571–1630) showed that elliptical orbits work much better at predicting the location of the planets than do circular orbits (Chapter 7).

Copernicus's model was not original. He revived an ancient Greek theory established by Aristarchus of Samos (310–230 BCE).

This example shows how a scientific theory can be refuted, but not how it can be proved. In fact, no scientific theory can be *proved.* Scientific theories are tested and retested. A theory that holds up under much testing is eventually accepted as a law. Later scientific evidence, however, may refute a theory, even one that was once considered a law.

SCIENTIFIC EVIDENCE Major Concept

Scientific evidence consists of measured observations. In the case of the Copernican theory of planetary orbits, the measured observation is the position of a planet with respect to the background stars (Fig. 1.2). In many branches of science, the observations are made in a laboratory. In that case, a carefully planned and controlled experiment is conducted with the purpose of producing evidence to support or refute a scientific theory. Because theories of physics are often tested in a laboratory, physics is called an *experimental science.*

Once a physical law has been discovered and tested by science, practical applications are often produced by engineers. In the 19th century, physicists discovered the principles of electricity and magnetism. Today, electrical engineers use those principles to design computers, communication devices, and personal music players, just to name a few applications. As we develop an understanding of physics, we will learn how some common and even a few exotic devices work.

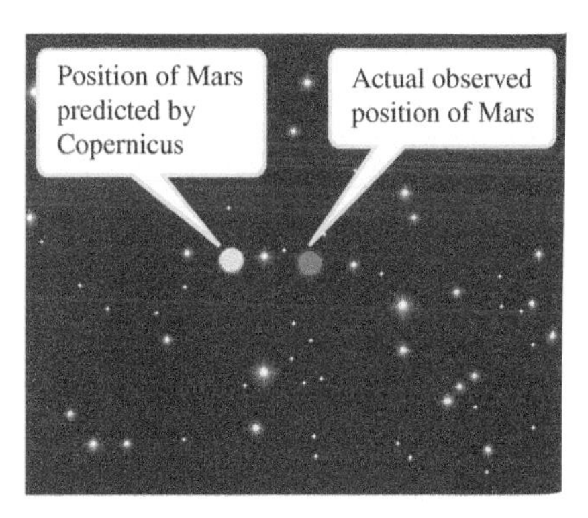

FIGURE 1.2 The Copernican theory of circular planetary orbits was refuted because it failed to predict the position of a planet such as Mars on a particular night and time. Tests of theories are repeated many times to allow for the possibility of observational errors. (The white dots represent fixed background stars.)

1-3 A Guide to Learning Physics

Learning physics is a lot like learning to drive a car. When you learn to drive a car, you learn *concepts* such as *indicating your intention to turn*. Then, you learn traffic *laws* that tell you when to indicate your turn. You *practice* those traffic laws when you drive on the road. The car is usually equipped with *tools* (such as a turn signal indicator) to help you. In physics, you will learn concepts and principles. You will practice physics when you solve problems and perform observations and experiments in a laboratory. The tool that will help you is mathematics.

Your everyday experience helped you master driving, and it will make learning physics easier too. Everyday experience is usually very complicated, however, and it is not easy to apply the principles of physics. Take the common experience of walking. What causes you to walk? What principles of physics are involved? Your experience tells you that your muscles cause you to walk, but that doesn't tell you why it is more difficult to walk on some surfaces (sand or ice) than on other surfaces (carpeted floors) or why people cannot walk on water. Part of learning physics is expanding on your experience and connecting that experience to physics principles and to mathematical tools. This book has features that will help you make those connections and use those tools.

Useful Features of this Book

Become familiar with the various useful features of this book.

Concept overview: Every chapter begins with an overview of the concepts, an outline listing the physical principles, new major concepts, and the key questions addressed in the chapter (page 1). The overview also tells you which new mathematical tools you will pick up. Finally, it will tell you if any special cases of physical principles or concepts are covered in the chapter. For example, in Chapter 2's overview, acceleration is listed as major concept, and *constant* acceleration is listed as a special case.

Summaries: Every chapter ends with a summary. It is similar to the overview at the beginning of the chapter, but it briefly describes each principle, major concept, mathematical tool, and special case. The summary provides references to particular equations in the main part of the chapter to help you organize your knowledge before a quiz, test, or exam. The summaries are not a substitute for your own notes. The best summaries are made by individual students. Together, the concept overview at the beginning of a chapter and the summary at the end help you *organize* all the new material you will learn. The details come from your careful reading of the chapter.

Two-column worked examples: In the main body of each chapter are worked examples. Watching an expert solve a physics problem often seems baffling because the expert has trouble explaining what she is thinking. The left column of the worked examples provides those often-unstated thoughts. The column on the right is what the expert would likely write on the board. When you encounter an example, first try to solve the problem without looking at the solution. When you have made your best attempt, read both columns to see how you did. Together, the two columns should help you connect principles and concepts to their mathematical expression. It is okay if you do not get very far on your own. You will still learn a lot by carefully reading the solution. There will be many other opportunities for you to practice your problem-solving skills.

Concept exercises: Physics textbooks are known for being dense, with a lot of information on every page. Throughout each chapter, you will find several concept exercises. Their purpose is to give you a chance to digest some of what you just read. Take a few moments when you encounter a concept exercise to answer the questions. You may find it helpful to work on scratch paper, in your notebook, or even directly in the book. Answers are provided near the end of the book along with those for odd-numbered homework problems. It is very important to read the textbook before the material is discussed in class, and the concept exercises will help make sure your reading is effective.

Case studies: Each chapter has a CASE STUDY to help connect physics principles and concepts to everyday or laboratory experience. Many of the case studies involve discussions between fictional students based on things that real students have said. Use these discussions to uncover your own ideas about physics. You may find that you agree with one or more fictional students.

Some case studies are based on experiments that students like you have performed. Their data help make the explanations throughout the chapter more concrete. Still other case studies are based on historical experiments and discoveries. These help you connect your learning process to the greater scientific endeavor.

Advice from the Author

There are many ways to learn physics and succeed in class. Take some time to think about how you plan to study. If you are a very successful student, you may have never needed to do this kind of planning; university-level physics, however, often requires more effort than other subjects, so it is worth coming up with a plan. Here are some steps that have led many other students to success in physics.

Steps to success in physics

Step 1: On the first day of class, your instructor will probably give you a course outline or syllabus listing each reading assignment. Read each assignment *before* the material is covered in class. Not everything will make sense to you on this first reading. Your goals for the first reading are to (a) become familiar with the material and (b) identify questions or problem areas.

Step 2: Take notes on your reading and in class. Use the concept overview at the beginning of the chapter to help you organize your reading notes. Jot down questions you have and be sure to get your questions answered by your professor, another student, or your teaching assistant.

As you are reading, you may write notes in the textbook itself, on note cards, or in a notebook. Each method has advantages. If you make notes in the textbook, you are likely to find them again even years later. Yes, you should keep your textbook! If you plan to be a scientist or engineer, you need to build up your own personal library of books. That is part of what it means to get a college education. You will refer to this textbook for decades. However, using note cards has the advantage that they can easily be used as flashcards when you review for a test. Finally, if you write your reading notes in a notebook, then leave room on each page for your classroom notes so all your notes are kept in one neat place.

Step 3: Working problems is an important part of any physics class. Even if you are not required to turn in homework problems for credit, you should still work several problems after class. After class, reread the material in the textbook. Keep your notes available and make any additions or changes as you reread. Then, work problems. In the next section, we will discuss a problem-solving strategy.

First Case Study

In subsequent chapters, the CASE STUDY often appears near the beginning of the chapter. The case study is referred to throughout the chapter and sometimes even in later chapters. After you read a case study for the first time, think about how you might approach the problem on your own.

CASE STUDY Raisin

Physicists take pride in being able to estimate the value of important quantities. In this case study, we will estimate the *density* of a single raisin from the photo (Fig. 1.3). We will return to this problem several times as we learn about mass, volume, density, and estimation.

FIGURE 1.3 A box of raisins, surrounded by everyday objects (for scale).

1-4 Solving Problems in Physics

Solving problems in physics, whether for homework or on a test, is often the hardest part of the class. Many students do not see the point of solving problems. They believe that they understand everything they have read in the book or have heard in class; they just cannot solve the problems. Many professors believe that problem solving is the primary way to learn physics and certainly one of the best ways to test a student's knowledge of the subject.

Physics professors have been known to say, "Physics is not a spectator sport." They mean that in a physics class you are expected to *do* physics, just as in a physical education class you are expected to *play* the game. By contrast, in your history course you are not expected to *make* history. You learn about history, you discuss history, and you write essays about history. In physics, you are expected to work problems and do experiments in a laboratory using the same sort of skills as a professional physicist.

Our goal is to teach you to think like a physicist. This goal is challenging for you, your instructor, and this textbook's author. Problem solving is an important part of meeting that challenge because when we solve problems our minds are active. Keeping with the sports analogy, problem solving is analogous to practicing. You cannot become a good athlete by watching sports on TV; you must practice on the field.

When you begin a sport, you are given detailed instructions on how to position and move your body. As you become a better player, you often modify what you have learned. In fact, after playing for a long time, an athlete often has trouble remembering what it was like to be a novice. The same is true for solving problems in physics.

To help students develop their problem-solving skills, some expert solvers have developed detailed guidelines, although the experts tend to stray from those guidelines when they work problems. The guidelines we use in this textbook are streamlined and mimic expert methods. Experts devise a strategy, execute a plan, and challenge their results, so in this book, we break problem solving into three procedures: (1) **INTERPRET and ANTICIPATE,** (2) **SOLVE,** and (3) **CHECK and THINK.**

Problem-solving steps

Procedure 1: INTERPRET and ANTICIPATE

The goal of this step is to come up with a **strategy** or plan based on knowing which physical principles are important to the question we must answer. There are two parts to meeting this goal.

The first is to **interpret** the question. What have we been asked to find? What physical principles are relevant to this question? It may be helpful to restate the question in your own words or to ask yourself if you have seen a similar problem before.

The second part is to **anticipate** the result. There are many questions we can ask to develop an expectation, such as the following:

a. Should the result be numerical or algebraic?

b. What are the proper units or what are the expected dimensions? (Units and dimensions are discussed in Sections 1-5 and 1-6.)

c. Can we make an educated guess at an approximate value or order of magnitude (Section 1-8) for the desired quantity?

Often, the interpretation of a problem involves creating some sort of visualization such as a sketch. Students often find **INTERPRET and ANTICIPATE** to be the hardest of the three steps. To help, we'll provide steps and visualization tools (such as diagrams and sketches) in some chapters for specific types of problems.

To take the survey scan or visit **www.cengage.com/community/katz**

Procedure 2: SOLVE
In procedure 1, we come up with a strategy or plan. In procedure 2, we **execute** this plan and arrive at an answer. Often, this part involves using mathematical tools. Again, as we learn more physics we'll add some specific steps to this procedure.

Procedure 3: CHECK and THINK
This final procedure is crucial and the most interesting of the three. We must **check** our results. Do we believe that our answers are correct? We "check" by seeing if our expectations from procedure 1 match our results found in procedure 2. If they agree, we may be confident that our answer is correct. Of course, it is possible that we have made mistakes in both procedures, but this comparison is usually a good test.

Once we believe that our answers are correct, we need to stop and think about what we have just learned. What are the **implications** of our results? Often, students are so happy to get to the end of a problem that they forget to reap the benefits of all their hard work. For experts, this part is the most pleasing because at this point major discoveries are made. Examples are *The truck driver was exceeding the speed limit, My results predict the existence of neutron stars,* or *It should be possible to send wireless radio signals.*

Just as it is helpful to watch skilled athletes play when you are learning a new sport, it is helpful to work through expertly solved problems when you are learning physics. It is likely that your physics instructor will demonstrate his or her problem-solving techniques. You will also find many worked examples throughout this book written in a two-column format, using these three procedures. As you solve your own problems, be aware of your own thoughts. Jot down some notes to yourself to help you become aware of what you are thinking.

1-5 Systems of Units

Physics relies on quantitative measurements. For example, if you are interested in learning about gravity near the surface of the Earth, you might drop a cannonball off the Leaning Tower of Pisa (Fig. 1.4) and measure the time it takes to land. To record your measurement, you need to decide what *standard time unit* to use. A **standard unit** is a precisely defined quantity to which measurements are compared. It is widely believed that the experimenter Galileo Galilei (1564–1642) sometimes used his own heartbeat as a standard time unit. If you use your own heartbeat, you would probably find that it takes roughly four heartbeats for the cannonball to land. Your heartbeat is not a good standard time unit, however, because (1) if anyone else wanted to compare the results of their experiment with yours, they would need to borrow your heart; and (2) your heartbeat is not regular (consistent). If you are sleeping, your heartbeat slows down, and if you exert yourself, it speeds up. A standard unit must be universal (reproducible by all observers) and consistent.

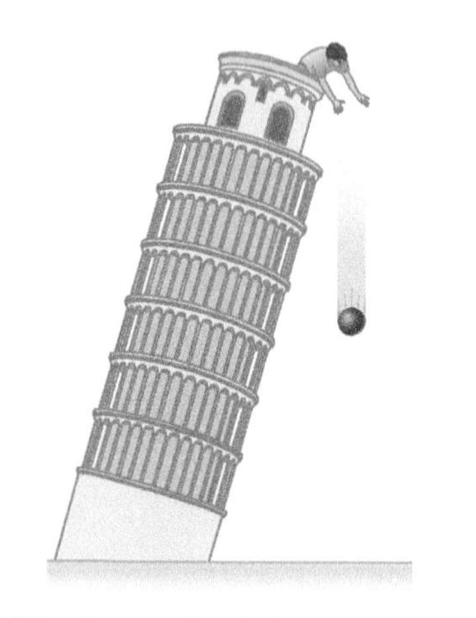

FIGURE 1.4 Galileo is famous for studying the effect of gravity through experimentation. You might be inspired to try your own experiment of dropping a cannonball off a building.

To develop a good system of standard units, an international committee agreed on a set of definitions and standards for comparison of measurements. The resulting unit system, known as the **SI system** (Système International d'Unités), is used worldwide in the scientific community and in this textbook. A complete list of the SI base units is provided in Appendix B. In this section, we consider the SI units for time, length, and mass, three of the fundamental quantities in physics.

THE SI SYSTEM Major Concept

Time

The SI unit of time is the **second**, abbreviated with a lowercase "s." Units are written in usual plain (roman) type, not in *italics*. The unit of time was formerly based on the rotation period of the Earth, but that standard is not accurate enough for today's experiments. The time standard is now determined by an atomic clock.

Atomic clocks measure the radiation from a particular atom. Atomic radiation has certain characteristic periods. The second is defined in terms of a certain characteristic period of cesium-133's radiation:

> 1 second is the duration of 9,192,631,770 periods of the radiation (corresponding to the transition between hyperfine levels of the ground state) of the cesium-133 atom.

Cesium clocks are kept at several locations, such as the U.S. Naval Observatory in Washington, D.C., and the National Institute of Standards and Technology (NIST) in Boulder, Colorado (Fig. 1.5). The time is broadcast on the World Wide Web and by radio signal, so it is possible to access these clocks from remote locations. This time standard is much more precise than we will need in this textbook, and a typical stopwatch will do when we need to measure time intervals.

Courtesy, NIST

FIGURE 1.5 Cesium clock at the National Institute of Standards and Technology in Boulder, Colorado.

Length

The SI unit of length is the **meter** (m). The meter is defined in terms of the speed of light in a vacuum, $c \equiv 299{,}792{,}458\ \text{m/s}$:

1 meter is the distance light travels through empty space in 1/299,792,458 second.

Again, this standard is more precise than we will need, and a meterstick will serve our purposes for measuring length or distance.

Mass

The SI unit of mass is the **kilogram** (kg). The standard is a specific platinum–iridium alloy cylinder kept at the International Bureau of Weights and Measures in Sèvres, France (Fig. 1.6). The mass of that cylinder is defined as 1 kg. In Chapter 5, we discuss the difference between weight and mass. For now, we point out that the **pound** is the U.S. customary unit for weight, not mass.

Photograph courtesy of the BIPM

FIGURE 1.6 The mass standard is a platinum–iridium cylinder kept at the International Bureau of Weights and Measures in Sèvres, France.

Scientific Notation and Common Prefixes for SI Units

Many numbers in science are very large or very small. For example, the mass of the Earth is 5,980,000,000,000,000,000,000,000 kg, and the mass of a proton is 0.00000000000000000000000000167262158 kg. In both cases, a long series of zeros is needed as placeholders. All those zeros make the numbers difficult to read. Scientists use **scientific notation** to write numbers in a compact form in terms of powers of 10. So, the mass of the Earth is written as 5.98×10^{24} kg, and the mass of a proton is $1.67262158 \times 10^{-27}$ kg.

The power of 10 used to express a value in scientific notation may also be expressed with a prefix in front of the unit. The most common prefixes are in Table 1.1, and a more extensive list is in Appendix B. For example, $\ell = 9.3 \times 10^{-2}$ m may be expressed as $\ell = 9.3$ cm, where cm stands for centimeters.

TABLE 1.1 Commonly used power-of-10 prefixes.

Power of 10	Prefix	Abbreviation
10^{-12}	pico	Lowercase p
10^{-9}	nano	Lowercase n
10^{-6}	micro	Lowercase Greek μ
10^{-3}	milli	Lowercase m
10^{-2}	centi	Lowercase c
10^{3}	kilo	Lowercase k
10^{6}	mega	Uppercase M
10^{9}	giga	Uppercase G
10^{12}	tera	Uppercase T

CONCEPT EXERCISE 1.1

To practice using the prefixes in Table 1.1, complete the following puns.

a. What do you call 10^6 phones?
b. What do you call 10^{-12} lo?
c. What do you call 2000 mockingbirds?
d. What do you call 0.000001 fish?
e. What do you call 1,000,000,000,000 pins?

Converting Units

Although SI units are the primary standard used in science and engineering, other systems of units are in popular use.

When we know a quantity measured in some other system of units, we often need to convert that measurement to SI units. For example, in U.S. customary units, speed is measured in miles per hour (abbreviated as mph or mi/h). Suppose a speed limit on a highway is 65 mph and we need to convert that speed to SI units (meters per second, abbreviated m/s). To do so, we need a *conversion factor*.

A **conversion factor** comes from writing equal quantities in terms of a fraction equal to unity. For example, we know that 1 hour equals 3600 seconds:

$$1\ \text{h} = 3600\ \text{s}$$

The commonly used metric system is similar to SI, often using the prefixes in Table 1.1, but time may be expressed in hours or minutes.

CONVERSION FACTOR Tool

We find a conversion factor between hours and seconds if we divide each side by 3600 s:

$$\frac{1\text{ h}}{3600\text{ s}} = \frac{3600\text{ s}}{3600\text{ s}}$$

(We could have also come up with a conversion factor by dividing both sides by 1 h instead of 3600 s.) The units may be treated algebraically so that "seconds" cancel out on the right, and the conversion factor is

$$\frac{1\text{ h}}{3600\text{ s}} = 1$$

Because multiplying any number by unity does not change that number's value, we can multiply a measurement by a conversion factor without changing its value, only its units.

To convert 65 mph to the SI unit m/s, we need another conversion factor to convert from miles to meters. From Appendix B,

$$1\text{ mi} = 1609\text{ m}$$

Choose the numerator and the denominator of a conversion factor so that the original units cancel out and the desired units remain after you complete the multiplication.

So, the conversion factor is

$$\frac{1609\text{ m}}{1\text{ mi}}$$

We use both conversion factors to convert 65 mph to SI units:

$$65\ \cancel{\text{mi}}/\cancel{\text{h}}\left(\frac{1\ \cancel{\text{h}}}{3600\text{ s}}\right)\left(\frac{1609\text{ m}}{1\ \cancel{\text{mi}}}\right) = 29\text{ m/s}$$

CONCEPT EXERCISE 1.2

CASE STUDY Mass of a Box of Raisins

Find the mass in SI units of the raisins shown in the box in Figure 1.3.

EXAMPLE 1.1 How Far Is a Light-Year?

A light-year (ly) is a unit of distance (length) commonly used in astronomy. A light-year is the distance light travels through a vacuum in 1 year. The speed of light in a vacuum is $c \equiv 299{,}792{,}458\text{ m/s}$. Find the number of meters in 1 light-year.

INTERPRET and ANTICIPATE

This example tells us how far light travels through a vacuum in 1 s, and we need the distance light travels in 1 yr. First, we convert 1 yr into seconds. Then, we multiply our answer by the distance light travels in 1 s to find the distance light travels in 1 yr. We expect a numerical result in the form 1 ly = _____ m.

SOLVE

There are 365 days in 1 year, 24 hours in 1 day, and 3600 s in 1 hour.	$1\ \cancel{\text{yr}}\left(\frac{365\ \cancel{\text{day}}}{1\ \cancel{\text{yr}}}\right)\left(\frac{24\ \cancel{\text{h}}}{1\ \cancel{\text{day}}}\right)\left(\frac{3600\text{ s}}{1\ \cancel{\text{h}}}\right) = 3.154 \times 10^7\text{ s}$
Because light travels 299,729,458 m in 1 second, we multiply to find the distance light travels in 1 year.	$(3.154 \times 10^7\text{ s})(299{,}729{,}458\text{ m/s}) = 9.45 \times 10^{15}\text{ m}$ $1\text{ ly} = 9.45 \times 10^{15}\text{ m} \quad (1)$

CHECK and THINK

Our answer has the form that we expected. In the process of solving this example, we found that $1\text{ yr} = 3.154 \times 10^7\text{ s}$, which is an important conversion factor. It is easy to remember if you think of it as $1\text{ yr} \approx \pi \times 10^7\text{ s}$. Of course, $\pi = 3.14159\ldots$, which is a little smaller than 3.154, but usually that slight difference is insignificant.

1-6 Dimensional Analysis

There are seven fundamental quantities that are mutually independent. In Part I, we encounter three of these fundamental quantities: time, length, and mass. (We'll learn about the other four in later chapters.) Quantities such as speed, volume, and density are *derived* from fundamental quantities and are known as **derived quantities**. For instance, to derive speed, we divide a length by a time; to derive volume; we multiply three lengths; and to derive a *mass density* (also known simply as *density*), we divide a mass by a volume. To get started and make this section less abstract, let us think about mass density. Density is symbolized by the lowercase Greek letter "rho" ρ. (For a list of Greek letters, see Appendix B.) Mathematically, **density** is expressed as

$$\rho = \frac{m}{V} \tag{1.1}$$

where m is the mass of an object and V is its volume.

We will encounter two meanings of the word *dimension* in physics. One definition of **dimension** is the type or category of a measured quantity. (We won't encounter the other definition until Chapter 2.) For example, distance may be measured in feet, meters, or miles, but all these measurements are units of length. So, the dimension of distance is length. We use the nonitalic uppercase symbols T, L, and M for the fundamental dimensions time, length, and mass, respectively. The dimensions of all the derived quantities in Part I can be written in terms of these three fundamental dimensions.

DIMENSIONAL ANALYSIS Tool

Dimensional analysis is a method in which the dimensions of a quantity rather than its value or other properties are used to tackle a problem. Physicists often use dimensional analysis to check a result. If your result does not have the expected dimensions, you have made a mistake. (Checking the dimensions of a result is just one way to test your work.) Dimensional analysis may also be used in the first problem-solving step to help interpret a problem.

We use the symbol $[\![Q]\!]$ to mean the dimensions (or units) of quantity Q. For example, the phrase *dimensions of density* may be written as $[\![\rho]\!]$.

Some quantities have no dimensions. We say that these quantities are *dimensionless*. For example, the ratio of two lengths is a dimensionless quantity.

Dimensional analysis is based on a few rules:

Rule 1: Dimensions may be treated as algebraic symbols. For example, to find the volume of a rectangular box such as the one in Figure 1.7, we must multiply its length by its width by its height:

$$V = \ell wh$$

The dimensions of volume are L^3:

$$[\![V]\!] = [\![\ell]\!][\![w]\!][\![h]\!] = (\text{L})(\text{L})(\text{L}) = \text{L}^3$$

Rule 2: Quantities can be added or subtracted only if they have the same dimensions.

Rule 3: The terms on both sides of an equation must have the same dimensions.

Rule 4: Trigonometric functions such as sine, cosine, and tangent apply only to (dimensionless) angular quantities, those measured in degrees or radians.

Rule 5: Special functions such as logarithms and exponential functions apply only to dimensionless quantities.

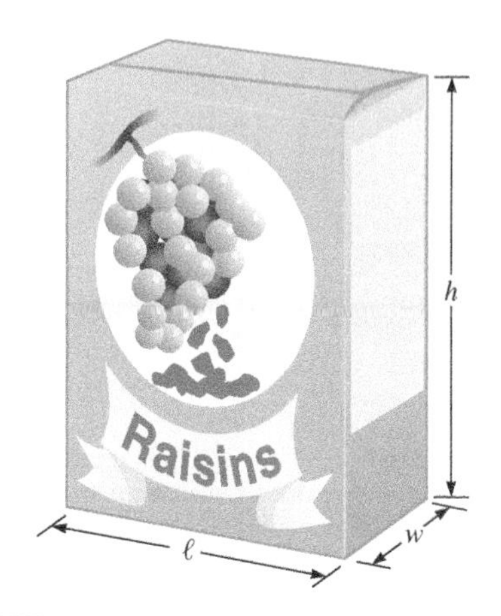

FIGURE 1.7 The volume of the box is $V = \ell wh$.

CONCEPT EXERCISE 1.3

In Einstein's famous equation $E = mc^2$, m stands for mass and c stands for the speed of light. Use this equation to find the dimensions of energy E.

CONCEPT EXERCISE 1.4

Use your results from Concept Exercise 1.3 to find which of the following quantities Q may represent energy. In these expressions, r is the radius of a circle and v is speed.

a. $Q = mv$ **b.** $Q = \frac{1}{2}mv^2$ **c.** $Q = m\frac{v^2}{r}$ **d.** $Q = mvr$

CONCEPT EXERCISE 1.5

CASE STUDY SI Units of Density

The goal of our case study is to find the density of a single raisin. Use dimensional analysis to find the SI units of density. Knowing these units will help us check the answer to the case study.

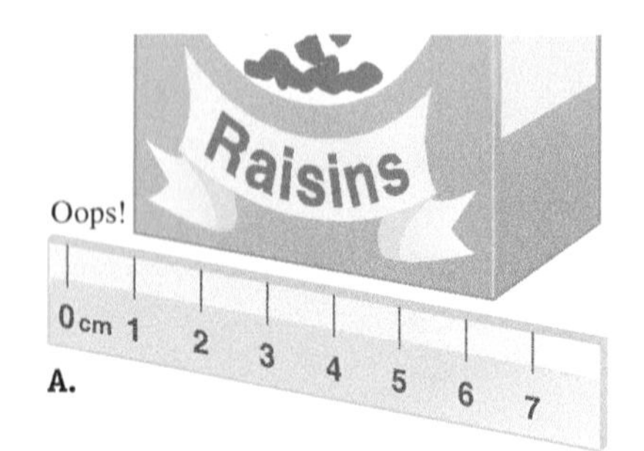

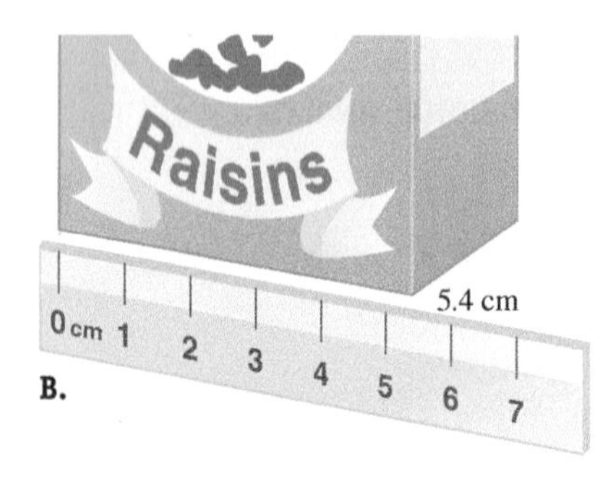

FIGURE 1.8 A. This is a mistake. The edge of the box does not line up with the zero mark on the ruler. **B.** A coarse ruler (few tick marks) leads to a large uncertainty in the measurement. **C.** A finer ruler (many tick marks) reduces the uncertainty in the measurement.

1-7 Error and Significant Figures

Because physics is an experimental science, physicists make a lot of measurements. No measurement is perfect. The terms **uncertainty** or **error** describe the imperfection of measurements. Normally, we think that an "error" is a mistake or fault, such as when a computer message says "Error: Document Failed to Print." A measurement error is not a mistake but rather is a natural result of the measurement process. Mistakes can be fixed. For example, if you are measuring the length ℓ of the raisin box (Fig. 1.8A), you might forget to align the edge of the box with the "0" on the ruler. You can correct that mistake by being careful and paying attention. On the other hand, measurement errors can be reduced but not completely eliminated. When reporting measured quantities, you must estimate the uncertainty of your measurements.

For example, if you are pulled over by a police officer who used a radar gun to measure your speed, the uncertainty in the measurement might make a difference in whether or not you get a ticket. If the speed limit is 65 mph and the radar detector measured your speed at 69 mph with an uncertainty of 1 mph, it is very likely that you were exceeding the speed limit. On the other hand, if the uncertainty in the measurement is 5 mph, the range of the measurement is between 64 and 74 mph, and it is possible that you were driving within the speed limit.

Let's see how to estimate uncertainty. Suppose you try to measure the length of the raisin box with a ruler that only has tick marks for centimeters as shown in Figure 1.8B. You find that the length is between the 5-cm and 6-cm tick marks. You estimate that the length is $\ell = 5.4\,\text{cm}$, but it may be anywhere in the range of 5.3 to 5.5 cm. You estimate the uncertainty in your measurement to be ± 0.1 cm, and so you report your measurement as $\ell = (5.4 \pm 0.1)$ cm.

You can reduce the error in your measurement by using a ruler that has millimeter tick marks (Fig. 1.8C). You see that the length is between the 5.3-cm and 5.4-cm tick marks. You estimate that the length is $\ell = 5.35\,\text{cm}$. This time, the uncertainty in your measurement is ± 0.05 cm. You report your measurement as $\ell = (5.35 \pm 0.05)$ cm.

By using the coarse ruler in Figure 1.8B, you can measure length to the nearest tenths place (one place to the right of the decimal), and by using the finer ruler in Figure 1.8C, you can measure the length to the nearest hundredths place. When you report a measurement, the number of digits you use to express the measurement indicates how precisely the measurement is known even if you do not explicitly report the uncertainty. In other words, the number of reported digits, known as the number of **significant figures**, implicitly expresses the uncertainty of the measurement. The last digit you report is uncertain. As another example, if a time is reported as $t = 12.89$ s, the measurement has four significant figures and the last digit is uncertain.

SIGNIFICANT FIGURES

✪ Major Concept

The number of significant figures used to express some quantity must be reported correctly. When using a calculator, it is tempting to write down all the digits returned by the calculator display. This practice, however, is incorrect because it represents our results as being much more certain than they really are. A few rules about significant figures help when making calculations with measured or reported quantities.

There is some disagreement about how to interpret a measurement that is reported without explicitly stating the uncertainty. For example, if a length is reported as 36 cm, it is possible that both digits are known and that the measurement is between 35.5 and 36.5 cm. In this textbook, we will assume the last digit, in this example the "6", is uncertain.

Rule 1: When multiplying or dividing, report the result with the same number of significant figures as the least certain value. For example,

$$\frac{12.3}{4.6} = 2.7$$

because 4.6 has only two significant figures.

Rule 2: When adding or subtracting, the number of decimal places in the result should equal the smallest number of decimal places in any of the given terms. For example,

$$12.34 + 2.006 - 8.9 = 5.4$$

because 8.9 has only one decimal place.

Rule 3: Numbers that are not measured may be considered exact. Irrational numbers such as π and e are known to many significant figures and do not limit your results. For example,

$$\frac{1}{3}(4.56\pi) = 4.78$$

is reported to three significant figures because neither $\frac{1}{3}$ nor π is measured, and our answer is limited only by the three significant figures of 4.56.

Rule 4: It is best to use scientific notation because a zero that acts as a placeholder is not necessarily a significant figure. For example, $m = 390$ kg may have two or three significant figures. To avoid that ambiguity, you may add a decimal point; for example, $m = 390.$ kg has three significant figures. A better way to clarify the number of significant figures is to use scientific notation: $m = 3.90 \times 10^2$ kg has three significant figures, and $m = 3.9 \times 10^2$ kg has two significant figures.

Rule 5: You should keep extra significant figures in intermediate steps when making a calculation, but you should round the final answer to the correct number of significant figures. The extra significant figures in an intermediate result help avoid introducing an error due to rounding a number up or down. This step is particularly important if an intermediate result is a number ending in 5.

Rule 6: When your answer begins with a 1, it is okay to keep one extra significant figure (as long as none of the operands has a leading 1). For example,

$$\frac{5}{4.3} = 1.2$$

CONCEPT EXERCISE 1.6

How many significant figures does each number have? If the number is exact or if the number of significant figures is ambiguous, explain.

a. $\frac{1}{2}$ in the formula $r = \frac{1}{2}d$, where r is radius and d is diameter

b. 105

c. 150

d. 1.50×10^2

e. 1.5×10^2

f. 0.15×10^3

CONCEPT EXERCISE 1.7

Complete the following arithmetic. Report your answers using scientific notation and the correct number of significant figures.

a. $\frac{1}{2}(199) =$

b. $(9.81)\dfrac{4.5}{1.23 \times 10^{-3}} =$

c. $6.789 - 14.1 =$

d. $\dfrac{39.1}{7.75}(0.456 - 1.23) =$

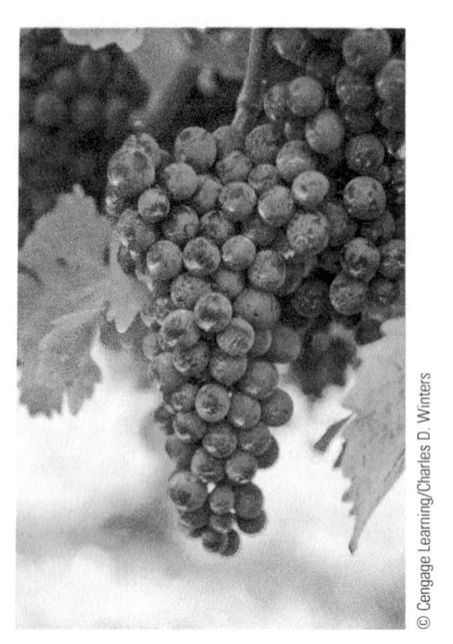
© Cengage Learning/Charles D. Winters

FIGURE 1.9 It takes longer to eat a nice bunch of grapes than to eat a small box of raisins, but the calorie content is the same.

1-8 Order-of-Magnitude Estimates

At lunch with a group of physicists, many topics are discussed: current events, politics, and personal matters. In many ways, the conversation sounds like any lunchtime conversation, but one difference you may notice is that physicists pride themselves on being able to estimate many numerical quantities and will frequently do so at informal gatherings. An **estimate** is not a guess; it is a calculation based on a few roughly known values. For example, if the physicists' conversation turns to the long lines in the cafeteria, it is likely that at least one of them will estimate the total time he will spend in cafeteria lines over the course of his lifetime. His estimate is based on knowing the time he typically waits in line, the number of years he has eaten in cafeterias, and the number of years he expects to eat in cafeterias. Returning to the subject of our CASE STUDY, when comparing their lunches one physicist may remark that it is more efficient to eat raisins instead of grapes because raisins are denser than grapes and provide the same caloric intake. A small box of raisins has 130 calories, which is equivalent to a medium bunch of grapes (Fig. 1.9). Perhaps the physicists will estimate the density of a single raisin and compare it with the density of a single grape.

As a student, part of your job is to learn to think like a physicist, so you, too, must learn to make estimates. Some homework problems will specifically ask you to make an estimate. Estimation is important even if you are expected to find an exact result. Your estimate is another way for you to check your numerical answer.

You may wish to commit some roughly known numbers to memory. Some of these numbers come from common experience, and you may already know them. If remembering such numbers in U.S. customary units is easier for you, it is also worth memorizing a few common conversion factors. To help you think about the sort of values that are worth knowing, consider three categories: your body, your daily experience, and your education.

Your body: You should know several facts about your body besides your height and weight. For example, how long is your thumb? What is the area of your palm? How long is your walking stride? How fast can you run? What is your pulse rate? Table 1.2 provides some typical answers, but it would be handy for you to know these values for your own body.

TABLE 1.2 Typical values associated with a human body.

Quantity	U.S. Customary Units	SI and Metric Units
Length of thumb	2 in.	5 cm
Area of palm	6 in.^{2}	40 cm^2
Height	5–6 ft	1.5–2 m
Weight	110–200 lb	500–1000 N
Mass	4–7 slugs*	50–100 kg
Average stride	1 yd	1 m
Resting heartbeat	60–80 per min	1–1.25 per second
Running speed	6-minute mile (10 mph)	16 km/h

*U.S. customary of mass; see Appendix B.

Your daily experience: You should know facts about the things you experience on a regular basis. How much does your car or bicycle weigh? How tall is one story of a typical building? How much does your physics book weigh? What is the area of your cell phone? What is the time between rings on your phone? How long is a matchstick? How long do you spend eating breakfast? What is the area of your campus? Table 1.3 provides some typical numbers, but you should try to make the list specific to your experience, and you should also try to expand the list. Because these facts come from your daily life, you will find them helpful in making estimates for the rest of your life no matter how much or how little physics you do in the future.

TABLE 1.3 Typical values associated with common objects and events.

Quantity	U.S. Customary Units	SI and Metric Units
Top speed of car	120 mph	200 km/h
Top speed of *typical* bicycle	30 mph	50 km/h
Weight of car	1–2 tons	10,000–20,000 N
Mass of car	70–140 slugs*	1000–2000 kg
Weight of physics book	10 lb	50 N
Mass of physics book	0.4 slug	5 kg
Area of cell phone	5 in.^{2}	30 cm^2
Length of matchstick	2 in.	5 cm
Length of a housefly	0.2 in.	0.5 cm
Weight of a quarter	0.2 oz	6×10^{-2} N
Mass of a quarter	4×10^{-4} slug	6 g
Area of a dollar bill	17 in.^{2}	100 cm^2
Height of typical story	10 ft	3 m
Area of U.S. Naval Academy	6.5 mi^2	15 km^2
Density of water	62 lb/ft^3	1000 kg/m^3
Density of ice	57 lb/ft^3	9.17×10^2 kg/m^3

*U.S. customary unit of mass; see Appendix B.

Your education: Many calculations you will make in physics involve objects and phenomena outside the range of daily experience. In your school or professional career, however, you may often use certain facts that you learned in an academic setting. Table 1.4 and Appendix B provide some facts to help build up your intuition. As you solve more physics problems, take the time to add new facts to your list.

TABLE 1.4 Approximate values of selected interesting facts.

Quantity	Value in Convenient Units
Age of the Universe	14 billion years
Age of the Earth	4.5 billion years
Time for light to travel from the Sun to the Earth	8 minutes
Time for light to cross the diameter of a proton	3.3×10^{-24} s
Diameter of the Milky Way galaxy	10^5 ly
Size of the smallest visible dust particle	0.1 mm
Size of a living cell	10 μm
Diameter of a hydrogen atom	10^{-10} m
Diameter of a proton	10^{-15} m
Mass of the Milky Way galaxy	10^{42} kg
Mass of an elephant	5×10^3 kg
Mass of a frog	100 g

An estimate is based on roughly known values. To indicate that a result is approximate, we use the mathematical symbol "≈" instead of an equals sign. Suppose you want to know your height in terms of the length of your thumb. If you are 5.5 feet tall,

$$h = 5.5 \text{ ft}\left(\frac{12 \text{ in.}}{1 \text{ ft}}\right) = 66 \text{ in.}$$

$$\left(\frac{h}{\ell_{\text{thumb}}}\right) = \frac{66 \text{ in.}}{2 \text{ in.}} = 33 \approx 3 \times 10$$

Our final estimate is reported to one significant figure because the length of your thumb is only known to one significant figure. So, your height is roughly 30 thumb lengths.

When only the correct power of 10 is known, we say that the value has zero significant figures and the resulting estimate is only expected to provide the correct power of 10. **Order of magnitude** is another term for the power of 10; an estimate based on values with zero significant figures is sometimes referred to as an **order-of-magnitude calculation** or **order-of-magnitude estimate**. The mathematical symbol "~" is used instead of an equals sign to indicate that a result came from an order-of-magnitude estimate. If we model the Milky Way galaxy as a disk, we can use the diameter provided in Table 1.4 to find an order-of-magnitude estimate for its area:

ORDER-OF-MAGNITUDE ESTIMATE

Tool

$$A = \pi r^2 = \pi\left(\frac{10^5 \text{ ly}}{2}\right)^2 = 8 \times 10^9 \text{ ly}^2 \approx 10 \times 10^9 \text{ ly}^2$$

$$A \sim 10^{10} \text{ ly}^2$$

In this case, because 8 is close to 10, the order-of-magnitude estimate is 10^{10} ly^2, not 10^9 ly^2.

EXAMPLE 1.2 A Great Way to Save Money

A Estimate the money you spend in a school year on your favorite drink purchased from a vending machine or cafe.

FIGURE 1.10 How much money do you spend on your favorite drink in a school year?

wdeon/iStockphoto1.com

INTERPRET and ANTICIPATE

You probably know roughly how much money you spend on your favorite drink on a typical school day. After taking into account vacation days and weekends, you can come up with a weekly average. Multiply by the number of weeks you are in school each year to find the money you spend during that time on your favorite drink. Your answer will depend on your particular habits. Our calculation here is based on the habits of an actual physics professor who drinks half a dozen diet sodas during a workday. At home on the weekends, he does not buy soda from vending machines, but during vacations such as spring break, he works five days per week in his office and drinks 8 sodas per weekday. At his university, winter break is 3 weeks long, spring break is 1 week long, each semester is 16 weeks long, and there are two semesters in a school year. We expect a numerical answer of the form $____.

SOLVE

N_{work} is the number of sodas consumed in a typical work week.	$N_{\text{work}} = (6 \text{ sodas/weekday})(5 \text{ weekdays})$ $N_{\text{work}} = 30 \text{ sodas}$
If P is the price of a single soda, $X = PN_{\text{work}}$ is the amount spent on soda in a work week. This particular professor spends \$1.50 on each soda from the vending machine.	$P = \$1.50/\text{soda}$ $X = PN_{\text{work}} = (\$1.50/\text{soda})(30 \text{ sodas})$ $X = \$45$ (work week)
Follow a similar procedure to find the amount Y spent on soda during a vacation week.	$N_{\text{vaca}} = (8 \text{ sodas/weekday})(5 \text{ weekdays}) = 40 \text{ sodas}$ $Y = PN_{\text{vaca}} = (\$1.50/\text{soda})(40 \text{ sodas})$ $Y = \$60$ (vacation week)
We must find how many weeks this professor spends X and Y on soda during the school year. (Summers don't count because the problem specified "the school year.") Let Z = total amount spent on diet soda.	$Z = (\text{number of work weeks})X + (\text{number of vacation weeks})Y$ (1)
In our professor's case, each semester is 16 weeks long, and there are two semesters in a school year.	$\text{number of work weeks} = (2 \text{ semesters})\left(\frac{16 \text{ weeks}}{\text{semester}}\right)$ $\text{number of work weeks} = 32$

At the professor's school, winter break is 3 weeks long and spring break is 1 week long.	number of vacation weeks = 3 + 1 = 4
Substitute the relevant numbers into Equation (1).	$Z = (32)(\$45) + 4(\$60)$ $Z = \$1680$
Although the cost of a single soda from the vending machine is known to three significant figures, the number of sodas consumed per week is only known to one significant figure. The answer has a leading "1", but none of the operands does, so we keep one extra significant figure.	$Z \approx \$1700$

CHECK and THINK

Our result is in the form we expected. We will think more about the implications after we finish part B.

B About how much money would be saved in one school year if the beverages in part A were purchased from a grocery store?

INTERPRET and ANTICIPATE

We assume the number of sodas consumed per school year remains unchanged, but the cost per soda is lower if the sodas are purchased in a grocery store. We need some everyday experience with grocery store soda prices. There should be a positive savings of the form $_____.

SOLVE Find the number n of sodas the professor in part A consumes in a school year. We said he spent \$1.50 per soda for a total of \$1680 per year.	$n = \frac{\$1680}{\$1.50/\text{soda}}$ $n = 1120 \text{ sodas}$
A package of 12 cans of soda costs around \$4 in a grocery store. To find the total cost z if the professor purchased his sodas at the store, multiply the number of sodas by the cost of a single soda.	$z = 1120 \text{ sodas}\left(\frac{\$4}{12 \text{ sodas}}\right)$ $z = \$373$ $z \approx \$400$
His yearly savings S is found from the difference between Z (vending machine cost) and z (grocery store cost).	$S = Z - z$ $S \approx \$1700 - \$400$ $S \approx \$1300$

CHECK and THINK

As expected, purchasing sodas from the grocery store saves money, but just how much might seem surprising. The cost of one soda purchased in a grocery store is approximately \$4/12 = \$0.33. So, the savings on just a single soda that costs \$1.50 from a vending machine is more than a dollar (\$1.17). That is about a 78% savings, and 78% of \$1700 is \$1300, consistent with our results. The professor has made this calculation and now buys his soda from grocery stores.

EXAMPLE 1.3 A Grape Way to Spend Your Time

Estimate the density of a grape.

INTERPRET and ANTICIPATE

Make this estimate by drawing on everyday experience. For example, you might estimate the number of grapes in a kilogram, about 2.2 pounds. We will approach the problem by estimating the volume and mass of a single grape. Our answer should be numerical, in the form ____ kg/m^3. A grape is mostly water, so we expect our result to be close to the density of water (1000 kg/m^3; see Table 1.3).

Example continues on page 16 ▶

SOLVE Assume the grape is essentially spherical. The diameter of a grape is about half the length of a person's thumb. Use this fact and Table 1.2 to find the radius of a grape.	$r = \frac{2.5\text{ cm}}{2} = 1.25\text{ cm}$ $r = 1.2 \times 10^{-2}\text{ m}$
Find the volume V. (See Appendix A for the volume of a sphere.)	$V = \frac{4}{3}\pi r^3 = \frac{4}{3}\pi(1.2 \times 10^{-2}\text{ m})^3$ $V \approx 7 \times 10^{-6}\text{ m}^3$
Estimate the mass of a grape by holding a grape in the palm of one hand and a quarter in the palm of your other hand. Their weights are about the same, and so are their masses. (Table 1.3 lists the mass of a quarter.)	$m = 6\text{ g} = 6 \times 10^{-3}\text{ kg}$
Use Equation 1.1 to find the density of a grape to one significant figure.	$\rho = \frac{m}{V} = \frac{6 \times 10^{-3}\text{ kg}}{7 \times 10^{-6}\text{ m}^3}$ $\rho \approx 9 \times 10^2\text{ kg/m}^3$

CHECK and THINK

Our result has the expected form and is the same order of magnitude as the density of water. This estimate will be important for the next example. We expect a raisin to be denser than a grape.

EXAMPLE 1.4 CASE STUDY Raisin Your Estimation Skills

Return to the case study (page 5 and Fig. 1.3) and estimate the density of a raisin. Use the mass of the box of raisins found in Concept Exercise 1.2.

INTERPRET and ANTICIPATE

Our job is to gather data from Figure 1.3 to estimate the density of a single raisin. We know the mass of the box (4.25×10^{-2} kg). We need to find the volume of the raisins. If the box is packed tightly, the volume of the box equals the volume of the raisins. We must estimate the volume of the box. We expect to find a numerical solution of the form _____ kg/m^3. We expect the density of a raisin to be greater than the density of a grape.

SOLVE Several objects in the photo provide a reference for finding the size of the raisin box. We have decided to use the box of matches, but you may use the cell phone or the battery. Estimate the length ℓ, width w, and height h in terms of the length of a matchstick.	$\ell \approx 1$ match-length $w \approx \frac{1}{3}$ match-length $h \approx 1.3$ match-lengths
According to Table 1.3, one matchstick is about 5 cm or 0.05 m long. Use this estimate to convert ℓ, w, and h to meters.	$\ell \approx 1\,(0.05\text{ m}) \approx 5 \times 10^{-2}\text{ m}$ $w \approx \frac{1}{3}\,(0.05\text{ m}) \approx 1.7 \times 10^{-2}\text{ m}$ $h \approx 1.3\,(0.05\text{ m}) \approx 6.5 \times 10^{-2}\text{ m}$
Calculate the estimated volume. Because the length of a matchstick is only known to one significant figure, the volume should have only one significant figure. We keep an extra significant figure until we have calculated the density.	$V = \ell wh$ $V \approx (5 \times 10^{-2}\text{ m})(1.7 \times 10^{-2}\text{ m})(6.5 \times 10^{-2}\text{ m})$ $V \approx 5.5 \times 10^{-5}\text{ m}^3$
Find the density from Equation 1.1 and the mass, 4.25×10^{-2} kg. Because the volume only has one significant figure, the density estimate is reported to one significant figure.	$\rho = \frac{m}{V}$ (1.1) $\rho \approx \frac{4.25 \times 10^{-2}\text{ kg}}{5.5 \times 10^{-5}\text{ m}^3} \approx 8 \times 10^2\text{ kg/m}^3$

CHECK and THINK

We have a numerical answer in the correct SI units, but our estimated density is lower than we expected. We thought that the density should be greater than the density of a grape. What went wrong? Our estimate was based on the assumption that the box was tightly packed full of raisins, but the manufacturer actually packs the boxes loosely. The raisins in Figure 1.11 have been tightly packed down. (No raisins were removed from the box.) Now we can see that the volume of raisins is about half the volume of the box.

FIGURE 1.11 A box with tightly packed raisins.

Dr. Jenny Spinner, St. Joseph's University

SOLVE

Let's now assume the volume of the box should be half of the old volume estimate.

$$V_{\text{new}} \approx \frac{5.5 \times 10^{-5}\text{ m}^3}{2} \approx 2.8 \times 10^{-5}\text{ m}^3$$

Use the new volume estimate to find a new estimate for the density of a raisin. By rule 6 on page 13, it is okay to keep one extra significant figure when your answer begins with a 1.

$$\rho_{\text{new}} \approx \frac{4.25 \times 10^{-2}\text{ kg}}{2.8 \times 10^{-5}\text{ m}^3} \approx 1.5 \times 10^3\text{ kg/m}^3$$

$$\rho_{\text{new}} \approx 1.5 \times 10^3\text{ kg/m}^3$$

CHECK and THINK

The new estimate fits our expectations: A raisin is denser than a grape.

Welcome to the Beginning of Your Adventure

Physicists love to study all aspects of nature, from the details of subatomic particles to the Universe as a whole. Discoveries made by physicists have been harnessed by engineers who have designed devices and machines that you use every day. Studying physics is an adventurous journey full of uphill struggles with surprising twists and turns. Along the way are beautiful views to see and interesting people to meet. It is an awesome adventure; enjoy it. Many older physicists envy you because you are at the beginning of the road.

SUMMARY

Underlying Principle: Physics is the fundamental natural, experimental science.

Major Concepts

1. Scientific **theories** make testable predictions.
2. **Scientific evidence** comes from measured observations.
3. A **standard unit** is a precisely defined quantity to which measurements are compared.
4. The **SI system** (Système International d'Unités) is a set of standard units used worldwide in the scientific community and in this textbook.
5. The terms **uncertainty** and **error** are used to describe the imperfection of a measurement.
6. The number of reported digits—known as the number of **significant figures**—implicitly expresses the uncertainty of the measurement.
7. **Mass density** is

$$\rho = \frac{m}{V} \tag{1.1}$$

Tools

1. A **conversion factor** comes from writing equal quantities in terms of a fraction equal to unity. Multiplying a quantity by a conversion factor does not change its value, only its units.
2. **Dimensional analysis** is a method in which the dimensions of a quantity rather than its value or other properties are used to attack a problem. It is a good tool for checking or anticipating a result.
3. An **order-of-magnitude estimate** is a calculation based on values with no significant figures. Only the power of 10 is known.

PROBLEM-SOLVING STRATEGY

INTERPRET and ANTICIPATE

Often it is best to start with a sketch. Your goal is to come up with a plan. Ask yourself:

1. Can I restate the problem in my own words?
2. What concepts and physical principles are relevant?
3. Should the result be algebraic or numerical?
4. What should the units or dimensions be?
5. What is the approximate value or order of magnitude?

SOLVE

Execute your plan.

CHECK AND THINK

Examine your result by asking:

1. Does my result match my expectation?
2. What is implied by my result?

PROBLEMS AND QUESTIONS

A = **algebraic** C = **conceptual** E = **estimation** G = **graphical** N = **numerical**

1-5 Systems of Units

1. N The average life expectancy in Japan is 81 years. What is this time in SI units?
2. N If you live in the United States, you probably know your height in feet and inches. In other countries, metric units are commonly used for measuring such quantities. First, find your height in inches. Then determine your height in **a.** centimeters and **b.** meters
3. N The age of the Earth is 4.5 billion years. What is the age of the Earth in the appropriate SI units?
4. N How many cubic centimeters (cm^3) are in one cubic meter (m^3)?
5. N In a laboratory, it is often convenient to make measurements in centimeters and grams, but SI units are needed for calculations. Convert the following measurements to SI units.
   a. 0.53 cm **b.** 128.92 g **c.** 35.7 cm^3 **d.** 65.7 g/cm^3
6. C What do *all* conversion factors have in common?
7. N A certain pure 0.9999 gold bullion bar with a mass of 311 g has a volume of 16.1 cm^3. What is the density of gold in appropriate SI units?
8. N In Jules Verne's novel, *Twenty Thousand Leagues Under the Sea*, Captain Nemo and his passengers undergo many adventures as they travel the Earth's oceans. **a.** If 1.00 league equals 3.500 km, find the depth in meters to which the crew traveled if they actually went 2.000×10^4 leagues below the ocean surface. **b.** Find the difference between your answer to part (a) and the radius of the Earth, 6.38×10^6 m. (Incidentally, author Jules Verne meant that the total distance traveled, and not the depth, was 20,000 leagues.)
9. N The distance to the Sun is 93 million miles. What is the distance to the Sun in the appropriate SI units?
10. E, N A popular unit of measure in the ancient world was the cubit (approximately the distance between a person's elbow and the end of the middle finger, when outstretched). Estimate the length of 1 cubit in centimeters. Given that there are about 1.609×10^3 m in 1 mile, how many cubits are there in 1 mile?
11. N CASE STUDY On planet Betatron, mass is measured in bloobits and length in bots. You are the Earth representative on the interplanetary commission for unit conversions and find that 1 kg = 0.23 bloobits and 1 m = 1.41 bots. Express the density of a raisin (2×10^3 kg/m^3) in Betatron units.
12. N Use your weight in pounds to find your mass in kilograms. (On the Earth, 1 kg weighs roughly 2.2 lb.)
13. N A garden snail named Archie, owned by Carl Branhorn of Pott Row, England, covered a 33-cm course in 2.0 min at the 1995 World Snail Racing Championships, held in Longhan, England (Fig. P1.13). Determine Archie's average speed $S_{av} = d/t$ in the appropriate SI units.
14. C As part of a biology field trip, you have taken an equal-arm balance (Fig. P1.14) to the beach. Your plan was to measure the masses of various mol-

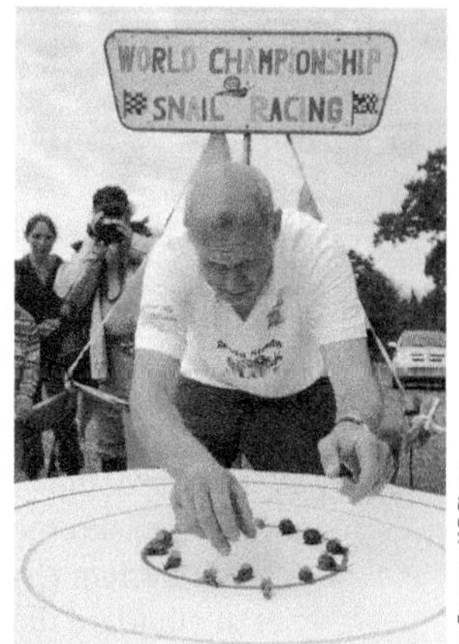

Rex Features/AP Photo

FIGURE P1.13

lusks, but you forgot to bring along your set of standard gram masses. You notice that the beach is full of pebbles. Although there are variations in color, texture, and shape, you wonder whether you can somehow use the pebbles as a standard mass set. Develop a procedure for assembling a standard mass set from the pebbles on the beach. Describe your procedure step by step so that someone else could follow it.

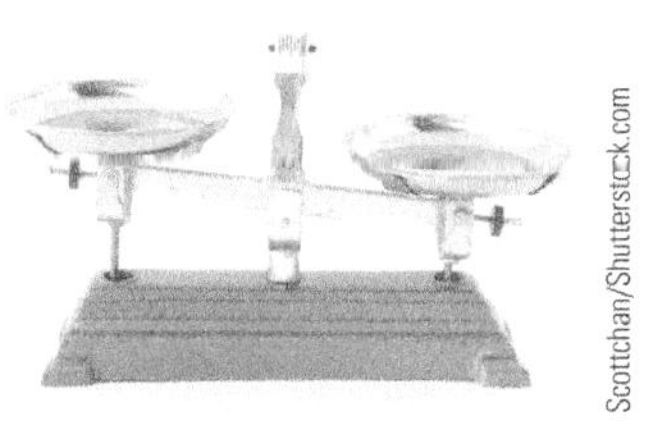

FIGURE P1.14

Scottchan/Shutterstock.com

15. N An Olympic-sized swimming pool, with a 660,000-gallon capacity, is filled with a large hose in 10.0 hours. **a.** What is the rate at which the pool is filled, in gallons per minute and **b.** in liters per second? **c.** At this rate, how long would it take to fill a typical backyard pool, with a volume of 40.0 cubic meters?

1-6 Dimensional Analysis

16. C Which of the following exaggerated statements nevertheless uses the term "light-year" correctly? **a.** "It will be light-years before I graduate from college." **b.** "Aaron drives so quickly; he must be a dozen light-years in front of us." Explain your answer.

17. N The kilogram standard is a circular cylinder whose height and diameter both equal 39.17 mm (Fig. 1.7). What is the density of the alloy used in the standard kilogram?

Problems 18 and 19 are paired.

18. C Acceleration a has the dimensions of length per time squared, speed v has the dimensions of length per time, and radius r has the dimension of length. Which of the following expressions may be correct? Explain your answer in each case.
a. $a = vr$ **b.** $a = v/r$ **c.** $a = v^2/r$ **d.** $a = v/r^2$

19. C See Problem 18. What are the dimensions of the following combinations of quantities?
a. ma **b.** mv^2/r **c.** mv **d.** mvr

20. A In the study of electricity later in this book, you will encounter the intriguing idea that energy can be stored by an electric field present in some region of space. In this case the energy stored is often described in terms of *energy density*, symbolized as u, which is the stored energy per unit volume. **a.** Determine the dimensions of u. *Hint*: Refer back to your results from Concept Exercise 1.3 to find the dimensions of energy. **b.** What are the SI units for u?

21. A In subsequent chapters, two different types of a physical quantity called *energy* will be defined. The kinetic energy K of an object is $K = \frac{1}{2}mv^2$. The gravitational potential energy U associated with an object–Earth system can sometimes be expressed as $U = mgy$. In these expressions, m stands for mass, g is the gravitational acceleration with dimensions of length per time squared, v is a quantity called speed that has the dimensions of length per time, and y is a distance with units of length. Show that the two different expressions for energy are consistent in that they have the same dimensions.

22. C The symbols for volume flow rate, area, and speed are, respectively, R, A, and v. The SI units of volume flow rate, area, and speed are, respectively, cubic meters per second, square meters, and meters per second. Could $v = RA$? Check by using dimensional analysis.

23. A Later in this book, you will study oscillating motion. Such motion is repetitive. Each complete repetition is called a cycle, and the frequency f is the number of cycles per unit time. The SI unit of frequency is s^{-1}. For a block oscillating at the end of a spring, the frequency depends on the mass m of the block, and the stiffness (spring constant) k of the spring. The dimensions of k are mass per time squared. Use dimensional analysis to find the mathematical dependence of f on k and m.

24. C What problem might arise if we were to multiply or divide any two numbers that have the same dimension (such as length) but different units (such as meters and feet), while working within a formula or string of calculations, as opposed to multiplying or dividing two numbers that have the same unit of measure?

25. A Force is the central concept in classical mechanics. It is measured using the derived SI unit called the newton, represented by N. **a.** Use the relationship $a = F/m$ to determine the dimensions of the newton in terms of mass, length, and time. Here, m stands for mass, a is the acceleration with dimensions of length per time squared, and F is the force. **b.** What is the "fundamental SI unit" of force? (That is, express 1 N in terms of kg, m, and s.) **c.** The force exerted by a spring depends on the length x by which it is stretched or compressed: $F = kx$. The constant of proportionality k, known as the spring constant, depends on the stiffness of the spring and is the force the spring exerts per unit length. Use your results from part (b) to find the fundamental SI units for k.

1-7 Error and Significant Figures

26. N Convert 13.7 billion years (the age of the Universe) to the appropriate SI unit. Be sure to report your answer with the correct number of significant figures.

27. C How many significant figures does 0.00130 m have?

28. C A distance with two significant figures divided by a time interval with three significant figures equals an average speed. How many significant figures does the average speed have?

29. C What is the number of significant figures in each of these numbers?
a. 7.913×10^{11} **b.** 0.00643 **c.** 4.1×10^{-4} **d.** 615 ± 3

Problems 30 and 34 are paired.

30. C In a laboratory, a researcher measures time using a stopwatch and finds that the time is between 10.53 s and 10.56 s. What should he report for the measured time and its error?

31. N Perform the following arithmetic operations, keeping the correct number of significant figures in your answer.
a. The product 56.2×0.154 **b.** The sum $9.8 + 43.4 + 124$
c. The quotient $81.340/\pi$

32. N Calculate the result for each of the following cases using the correct number of significant figures.
a. $3.07670 - 10.988$ **b.** $1.0093 \times 10^5 - 9.98 \times 10^4$
c. $\dfrac{5.4423 \times 10^6}{4.008 \times 10^3}$

33. N Calculate the result for the following operations using the correct number of significant figures.

$$3.07670 - 10.988 + \frac{\left(\dfrac{5.4423 \times 10^6}{4.008 \times 10^3}\right)}{(1.0093 \times 10^5 - 9.98 \times 10^4)}$$

34. C In a laboratory, a researcher fails to start a stopwatch at the beginning of an experiment. Instead he starts it a little late and finds that the time is between 17.89 s and 17.92 s. What should he report for the measurement of the time?

35. N Complete the following calculations and report your answer using scientific notation, the correct number of significant figures, and SI units. **a.** Model the Earth as a sphere with a radius of 6378.1 km. Find its volume. **b.** The mass of the Earth is 5.98×10^{24} kg. Find the density of the Earth.

36. C A pendulum consists of a bob at the end of a string (Fig. P1.36) In one cycle the bob swings out and back to its original position. The time for one cycle is called "the period," T. Johanna measures the period and finds it is (0.9 ± 0.1) s. Jimmy replaces the bob of the pendulum with a bob that has twice the mass, but is otherwise identical. He finds the period is now (0.95 ± 0.01) s. Jimmy claims that the heavier bob slowed down the pendulum and increased its period. Johanna argues that their measurements are consistent with one another and the period has not changed. Whom do you agree with? Explain.

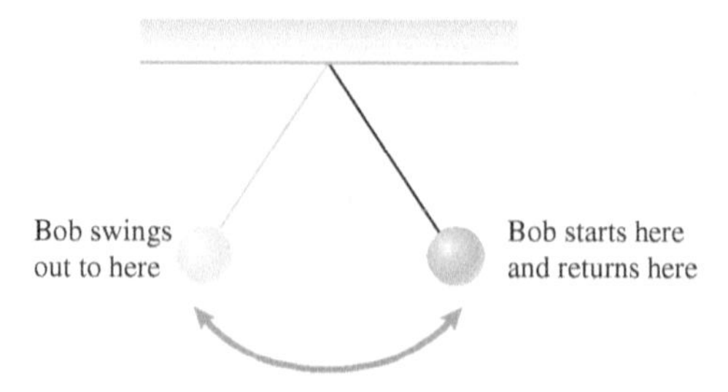

FIGURE P1.36 Problems 36, 37, and 58.

37. C In each case (a) through (d), your lab partner has measured the mass of a pendulum bob (Fig. P1.36) and reported the results as shown. Identify and correct the flaw in each case, if one exists. If no correction is possible, explain why you cannot correct the flaw. **a.** $m = 1.23 \pm 0.1$ g **b.** $m = 10.64 \pm 0.03$ g **c.** $m = 18.70 \pm 0.01$ g **d.** $m = 7.6 \pm 0.01$ g

1-8 Order-of-Magnitude Estimates

38. E Estimate the area of your room's floor. Measure the area and compare your results.

39. E Estimate both the number of hours you will spend studying physics during the current term and the total number of hours the entire class will spend studying physics this term. Show your work and explain your assumptions.

40. E Measure the distance from your outstretched thumb to your outstretched pinky finger. Use that information to estimate (a) the top surface area and (b) the volume of your dresser or desk. (Indicate your choice of furniture.)

41. E Use your height in SI units to estimate the height of your shower stall or length of your bathtub.

42. E Estimate the volume of your physics classroom. Explain your process.

43. E Roughly speaking, what are the volume of your lungs and the volume of your stomach? Explain your work.

44. E Approximately how much money do you spend each year on entertainment?

45. E Estimate the number of living cells in a tiger.

46. E How many protons are there in a standard deck of 52 cards?

47. E Estimate the time it would take you to hike from Boston to Miami.

48. E In 2011, artist Hans-Peter Feldmann covered the walls of a gallery at the New York Guggenheim Museum with 100,000 one-dollar bills (Fig. P1.48). Approximately how much would it cost you to wallpaper your room in one-dollar bills, assuming the bills do not overlap? Consider the cost of the bills alone, not other supplies or labor costs.

FIGURE P1.48 Courtesy Solomon R. Guggenheim Foundation, New York. Photo by David Heald

General Problems

49. N The mass of one electron is 9.11×10^{-31} kg, and the mass of one proton is 1.67×10^{-27} kg. How many electrons would it take to equal the mass of one proton?

50. N Convert the following distances into SI units. Which is larger?

$$d_1 = 1.02 \text{ km} \qquad d_2 = 102 \times 10^{-5} \text{ Mm}$$

51. N A 350-seat rectangular concert hall has a width of 60.0 ft, length of 81.0 ft, and height of 26.0 ft. The density of air is 0.0755 lb/ft^3. **a.** What is the volume of the concert hall in cubic meters? **b.** What is the weight of the air in the concert hall in newtons?

52. G Later in this book, you will learn that sound is a wave. The wavelength λ and frequency f of a wave are related by $\lambda f = v$ where v is the speed of the wave. Musicians refer to these different wavelengths or frequencies by their notes (A–G). Use the information in the following table to plot the frequency on the vertical axis and $1/\lambda$ on the horizontal axis. Give a conceptual interpretation and numerical value of the slope on your graph.

Pitch	Wavelength λ (m)	Frequency f (s^{-1})
A	0.7800	440.0
B	0.6949	493.9
C	0.6559	523.2
D	0.5843	587.3
E	0.5206	659.3
F	0.4914	698.5
G	0.4378	784.0

53. N Two decorative spheres are carved from the same slab of marble. The radius of the first sphere, which has four times the mass of the second sphere, is 14.5 in. What is the radius of the second sphere in inches?

54. A Newton's law of universal gravity $F = G\dfrac{m_1 m_2}{r^2}$ gives the magnitude of the force exerted by two particles with masses m_1 and m_2 on each other. The SI units of force are kg · m/s^2; G is the universal gravitational constant and r is the distance separating the two particles. What are the SI units of G?

55. A Two different expressions for finding the magnitude of a quantity called force are $F = ma$ and $F = \dfrac{p_f - p_i}{t_f - t_i}$. The acceleration a has dimensions of length per time squared. Momentum p is equal to mv, where v is speed with the dimensions of length per time. The subscripts f and i refer to the *final* and *initial* values of those quantities, respectively. Also, m stands for mass and t stands for time. Show that the two different expressions for force are consistent in that they result in a quantity with the same dimensions.

56. The size of a cell in the human body is determined by a need to optimize the exchange rates between the cell's interior and exterior environments. This process requires a suitable surface area-to-volume ratio.

a. A Using a sphere of radius r, find an expression for the ratio between the surface area and the volume of a sphere.

b. G Cells are typically on the order of micrometers in radius. Plot the ratio A/V as a function of r from part A for values of r ranging from 1 μm to 6 μm.

57. N During a visit to New York City, Lil decides to estimate the height of the Empire State Building (Fig. P1.57). She measures the angle θ of elevation of the spire atop the building as 20°. After walking 9.0×10^2 ft closer to the iconic building, she finds the angle to be 25°. Use Lil's data to estimate the height h of the Empire State Building.

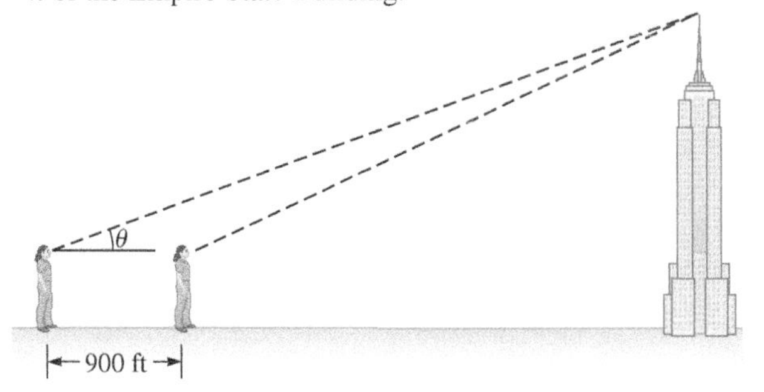

FIGURE P1.57 (Not to scale.)

58. C, G You hypothesize that the period of a pendulum is proportional to the length of the string (Fig P1.36). The period is defined as the time for one complete cycle. You conduct some experiments to test your hypothesis. In the first trial, you measure the period to be 0.40 ± 0.08 s. In the second trial, the length of the string is doubled, and you measure the period to be 0.65 ± 0.08 s. All other variables are kept the same in the two trials.

Discuss the extent to which the experimental evidence can be used to support or refute your belief. Are the data consistent with the notion that period is proportional to length? Can the data be used to rule out this possibility? Draw a diagram with a number line to support your answer.

59. N You are part of a team in an engineering class that is working on a scale model of a new design for a life vest. You have been asked to find the mass of a piece of foam that will be used for flotation. Because the piece is too bulky to fit on your balance, you break it into two parts. You measure the mass of the first part as 128.3 ± 0.3 g and the second part as 77.0 ± 0.3 g. **a.** What are the maximum and minimum values for the total mass you might reasonably report? **b.** What is the best estimate for the total mass of the foam?

Hint: Propagation of uncertainty is described in Appendix A.

60. A Model the human body as three cylinders: a large one for the torso and two small ones for the legs. The height of each cylinder is half the height h of the person. The radius of the large cylinder is r, and the radii of the smaller cylinders are $r/2$. The body has a mass m. Assume the average density of the body is about the same as water, $\rho = 1000\ \text{kg/m}^3$, and the radius of each cylinder is small compared with its height. Show that the body's surface area in meters squared is roughly given by $A = 0.129m^{0.5}h^{0.5}$, where m is in kilograms and h is in meters. *Note*: An empirical fit produces a more accurate estimate of the body's surface area: $A = 0.202m^{0.425}h^{0.725}$.

61. N A unit of distance used in astronomy is the parsec (pc): 1 pc = 3.26 ly. The distance to the Earth's next-nearest star, α-Centauri, is 1.3 pc. Find the distance d to α-Centauri in light-years and in meters.

2 One-Dimensional Motion

Key Question
How do physicists describe one-dimensional motion using concepts and mathematics?

2-1 **What is one-dimensional translational kinematics?** 23
2-2 **Motion diagrams** 23
2-3 **Coordinate systems and position** 24
2-4 **Position-versus-time graphs** 25
2-5 **Displacement and distance traveled** 27
2-6 **Average velocity and speed** 30
2-7 **Instantaneous velocity and speed** 34
2-8 **Average and instantaneous acceleration** 37
2-9 **Special case: constant acceleration** 40
2-10 **A special case of constant acceleration: free fall** 45

Underlying Principles

Kinematics and one-dimensional motion

Major Concepts

1. Translational motion
2. Particle
3. Position
4. Displacement
5. Distance traveled
6. Average and instantaneous velocity
7. Average and instantaneous speed
8. Average and instantaneous acceleration

Special Cases

1. Constant acceleration
2. Free fall

Tools

1. Motion diagrams
2. Vector and scalar quantities
3. Coordinate systems
4. Graphs of position, velocity, and acceleration versus time

Scientific progress is often most exciting when a serendipitous observation leads to a major discovery, which is what happened in 1915 when V. M. Slipher surprised himself and the astronomy community. Measuring the speeds of 15 "spirals" thought to be infant solar systems, Slipher showed that these spirals were moving up to 100 times faster than expected. His observations supported an emerging theory of the Universe. According to this theory, these spirals are whole galaxies outside our own Milky Way galaxy, all moving at high speeds away from one another as the entire Universe expands.

Slipher's experience is not unique. In physics, the most important progress is often made through the careful observation of motion. The study of motion has led to our understanding of many fundamental principles of physics, such as Einstein's theory of relativity and Newton's laws of motion. Newton's laws provide the underpinning for most of the concepts in this book. So, to grasp physics, we must develop a clear and precise description of motion.

Galileo Galilei—whose scientific research during the Italian Renaissance laid the groundwork for Newton's discoveries—divided the study of motion into two parts: kinematics and dynamics. **Kinematics**, the topic of this chapter, is the careful description of motion independent of its cause. **Dynamics** deals with the cause of motion and is the focus of Chapter 5.

KINEMATICS and ONE-DIMENSIONAL MOTION
Underlying Principles

2-1 What Is One-Dimensional Translational Kinematics?

TRANSLATIONAL MOTION and PARTICLE
Major Concepts

The goal of this chapter is to describe motion using the concepts and equations of kinematics. To make this goal easier to achieve, we restrict the motion described in this chapter in two ways. In general, motion can be rotational, vibrational, or translational (Fig. 2.1). **Rotational** motion is spinning motion, such as the motion of a carousel. The back-and-forth motion of a guitar string is an example of **vibrational** motion. The descent of a landing airplane is an example of **translational** motion.

In Chapters 2–11, we restrict our study to objects that are purely translating without rotating or vibrating. A purely translating object may be modeled as a particle. A **particle** is an idealized point that has no spatial extent, no shape, and no internal structure. It has no distinct top or bottom, front or back, or left or right side and therefore cannot rotate and cannot vibrate. Although no real physical object is an ideal particle, many real objects can be approximated, or *modeled*, as particles even if they are not microscopically small or lacking in structure.

Here in Chapter 2, we restrict our study to one-dimensional motion, such as that of a ball falling vertically or a train traveling on a straight track. One-dimensional motion is mathematically simpler than two- and three-dimensional motion. Most of the concepts of kinematics—such as position, displacement, velocity, and acceleration—are mathematically described by vectors. **Vector** quantities ("vectors") have both a magnitude, such as 35 m/s, and a direction, such as northeast. In one-dimensional motion, the direction of any vector is limited to two possibilities, such as east or west, north or south, up or down, or left or right. Mathematically, we can describe these two choices as positive or negative. Our study of physics will also include **scalar** quantities ("scalars") that lack direction; scalars have only magnitude. Examples of scalars are mass, temperature, and time.

A. Rotational motion

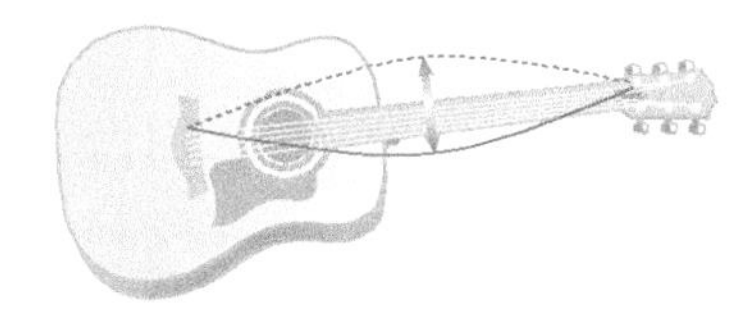

B. Vibrational motion

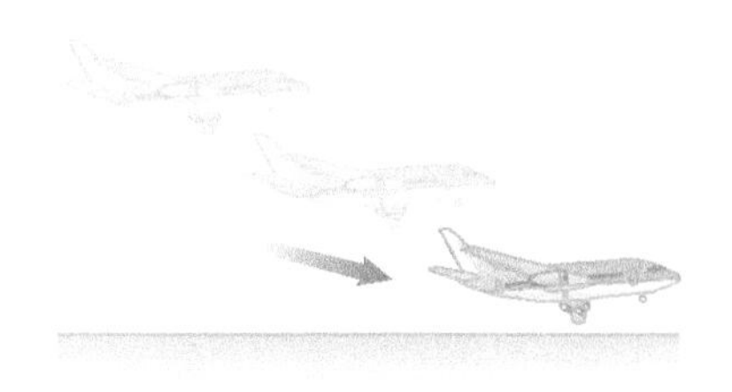

C. Translational motion

FIGURE 2.1 Three types of motion. **A.** The carousel is rotating. **B.** The guitar string is vibrating. **C.** The airplane is translating.

2-2 Motion Diagrams

We observe and describe motion every day. Runners would like to achieve a "4-minute mile," for instance, and the 1968 Ford GT40 can race "from 0 to 100 mph in 8 seconds." Our everyday language is often vague, however. Can a Ford GT40 reach 200 mph in 16 seconds? Although such ambiguity may be okay in some situations, physics requires a carefully defined vocabulary to describe motion. Kinematics can be expressed using words, mathematics, diagrams, and graphs. In this section, we will start with a visual description known as a *motion diagram*.

MOTION DIAGRAM Tool

A **motion diagram** is any illustration that shows the location of a particle at regular time intervals. For example, imagine observing the motion of Mars in the night sky. You might start by sketching the region of the night sky around Mars at a particular time. You model Mars as a particle, representing it on your sketch by a dot and labeling its initial location with an A (Fig. 2.2). If you then observed Mars's position at ten-night intervals for a few weeks and recorded and labeled those positions on your sketch, you would end up with a motion diagram of Mars. (In a laboratory course, you may use a camera or motion sensor to make a motion diagram.)

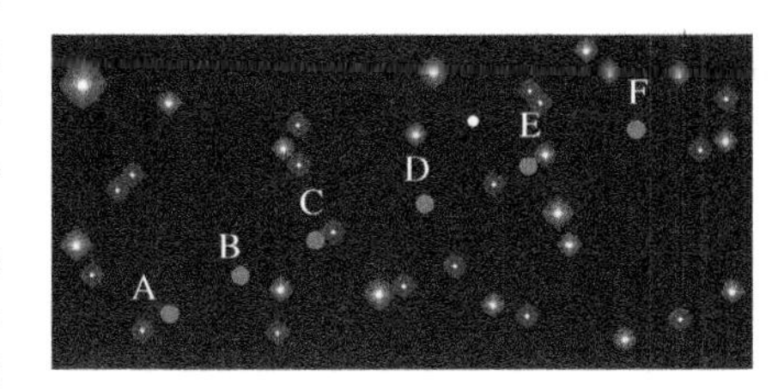

FIGURE 2.2 In this motion diagram, Mars moves from A to F. The time interval between each measurement is ten nights.

A motion diagram contains a lot of information. First, we can see the shape of the particle's path in the plane of the diagram. In Figure 2.2, it is clear that Mars is moving up and to the right in the plane of the sky. Second, because the images are taken at regular time intervals, we can infer something about the particle's speed. A particle at rest is represented by only one dot on a motion diagram because the particle does not move during the time intervals. If a particle maintains its speed, the spacing

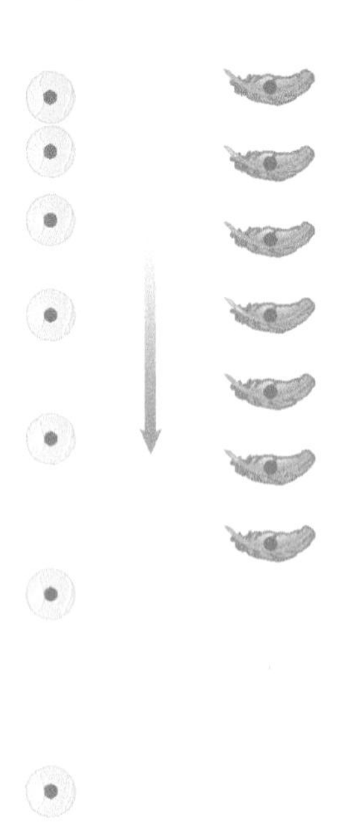

FIGURE 2.3 Motion diagrams of a falling ball and of a falling feather.

between dots is uniform. If the particle is slowing down, the dots get closer together. If the particle is speeding up, the dots get farther apart.

Figure 2.3 shows the motion diagrams for a ball and a feather dropped at the same time. The feather's images are equally spaced, but the ball's images are closer together at the top near the beginning of its fall. The difference in these motion diagrams tells us the feather maintained its speed but the ball sped up as it fell. Because we are using the particle model, it is convenient to represent the object as a dot on a motion diagram as shown in Figure 2.3.

CONCEPT EXERCISE 2.1

In each of the five motion diagrams shown in Figure 2.4, a particle moves in space from position A to position E. For each diagram, describe the motion of the particle as maintaining speed, speeding up, slowing down, or remaining at rest.

FIGURE 2.4

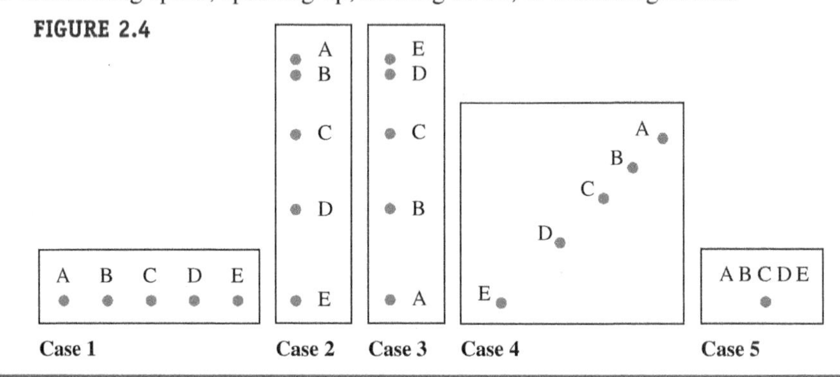

2-3 Coordinate Systems and Position

COORDINATE SYSTEM ◉ Tool

Motion diagrams alone are not enough to enable us to study kinematics. A precise description of motion requires us to know the location of a particle at particular times. We use a coordinate system to measure location. In this chapter, we need only a one-dimensional **coordinate system** (Fig. 2.5) with two elements: a reference point, called the **origin**, indicated by a zero; and a line passing through the origin called either the **coordinate axis** or just the **axis**, usually labeled x, y, or z. Although the label may be chosen arbitrarily, often a horizontal axis is labeled x and a vertical axis y. One direction along the coordinate axis is chosen to be positive, indicated by an arrow. The opposite direction is negative.

POSITION ✪ Major Concept

The **position** of a particle is its location with respect to the chosen coordinate system. Position is a vector whose direction in one dimension is indicated by a positive or negative sign, depending on whether it is on the positive or negative side of the origin. A particle at the origin has a position of zero.

Introduction to Vectors

VECTORS ◉ Tool

The overhead arrow points to the right as shown no matter in which direction the vector actually points.

We will study four vector quantities in this chapter: position, displacement, velocity, and acceleration. An arrow $\vec{}$ above a symbol indicates that the symbol represents a vector quantity. No arrow appears above a symbol representing a scalar quantity.

For one-dimensional motion, vectors point in either the positive or negative direction along a single x, y, or z axis. When writing a vector, we must indicate both its

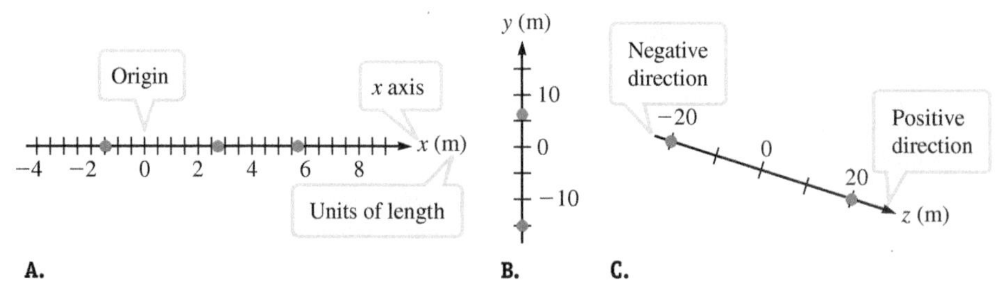

FIGURE 2.5 Three examples of a one-dimensional coordinate system. Each one has an origin; an x, y, or z axis; the distance measured in meters; and an arrow to indicate the positive direction. The labels x, y, and z are an arbitrary choice. In all three cases, the blue particles are at positive positions and the red at negative positions.

magnitude and direction. Mathematically, we use unit vectors to indicate direction. **Unit vectors** always have a magnitude of one and point in the positive direction along a single coordinate axis. The unit vectors $\hat{\imath}$, $\hat{\jmath}$, and $\hat{k}$ point in the positive x, y, and z directions, respectively. Suppose we wish to write a mathematical expression for a one-dimensional vector $\vec{Q}$ pointing in the negative y direction. We use the unit vector $\hat{\jmath}$ to indicate the y axis:

Use a hat ^ to indicate a unit vector.

$$\vec{Q} = -Q\hat{\jmath}$$

where the negative sign indicates that the vector points in the negative direction. The magnitude of the vector is $|-Q| = Q$.

The factor that appears before the unit vector is called the **scalar component**. The scalar component of $\vec{Q} = -Q\hat{\jmath}$ is $Q_y = -Q$. The subscript y indicates a scalar component along the y axis, and so $\vec{Q}$ may be written as

$$\vec{Q} = Q_y\hat{\jmath}$$

The magnitude of a vector is always positive, but the scalar component of a vector may be positive or negative. The scalar component multiplied by the unit vector is called the **vector component**. The vector component of $\vec{Q}$ is $\vec{Q}_y = Q_y\hat{\jmath} = -Q\hat{\jmath}$. In one dimension, the vector component is identical to the vector itself.

Because position is a vector, we must indicate its direction and magnitude. The symbol for position is $\vec{r}$. The vector components of position are written as $\vec{x}$, $\vec{y}$, and $\vec{z}$ and their scalar components as x, y, and z, respectively.

In this book, position is the only vector that does not use subscripts to indicate vector or scalar components.

In one dimension, position has just one vector component and is written in terms of any one of the three vector components $\vec{x}$, $\vec{y}$, or $\vec{z}$:

$$\vec{x} = x\hat{\imath}, \quad \vec{y} = y\hat{\jmath}, \quad \text{or } \vec{z} = z\hat{k}$$

For example, $\vec{r} = -15\hat{\jmath}$ m gives the position of a particle at $y = -15$ m (Fig. 2.5B). Notice that we indicate the SI unit for position (meter) at the end, after the other symbols. The vector component of this position vector is $\vec{y} = -15\hat{\jmath}$ m, and its scalar component is $y = -15$ m. The magnitude of this position is $r = |y| = |-15|$ m $= 15$ m.

CONCEPT EXERCISE 2.2

For each of the following, give the vector component, the scalar component, and the magnitude.

a. $\vec{r} = -2.4\hat{\imath}$ m **b.** $\vec{r} = -2.4\hat{k}$ m **c.** $\vec{z} = -2.4\hat{k}$ m **d.** $x = -2.4$ m

2-4 Position-Versus-Time Graphs

The world does not come with an attached coordinate system; we must *choose* a system that works for a particular problem. There may be more than one good choice of coordinate systems, and choosing a good system takes practice. We consider a **CASE STUDY** to see how coordinate systems are used. This case study is based on experiments two students performed in their physics laboratory.

CASE STUDY An Experiment in Motion

Two students (Crall and Whipple) set up an experiment to study the motion of a small cart along a straight north–south track (Fig. 2.6A). Crall chooses a coordinate system such that the *positive x* direction points north. Crall and Whipple recorded the position of the cart at 0.4-s time intervals in meters from the origin along Crall's axis. Their data are recorded in Table 2.1.

They use these data to make the motion diagram on the left side of Figure 2.6B. We can describe the cart's motion in words as follows: The cart remained at rest at position $\vec{x}_B = 1.64\hat{\imath}$ m for at least the first 0.4 s. It then moved in the negative x direction at a nearly constant speed to position $\vec{x}_P = 0.33\hat{\imath}$ m (point P).

TABLE 2.1 Position data of an experimental cart at 16 times, using Crall's choice of coordinates.

Label	t (s)	x (m)
A	0.0	1.64
B	0.4	1.64
C	0.8	1.55
D	1.2	1.45
E	1.6	1.35
F	2.0	1.25
G	2.4	1.16
H	2.8	1.06
I	3.2	0.97
J	3.6	0.87
K	4.0	0.78
L	4.4	0.69
M	4.8	0.60
N	5.2	0.51
O	5.6	0.42
P	6.0	0.33

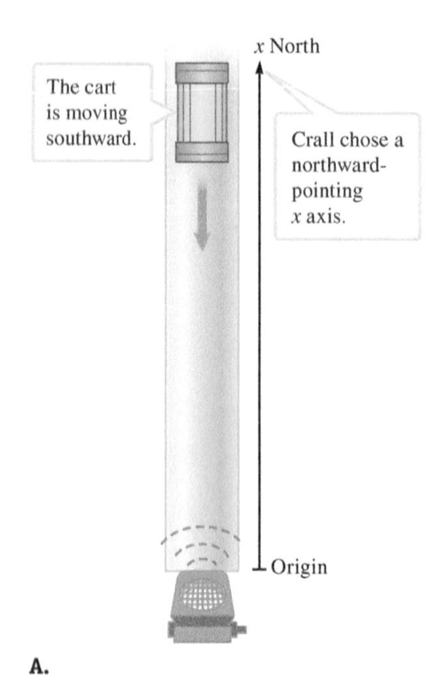

A.

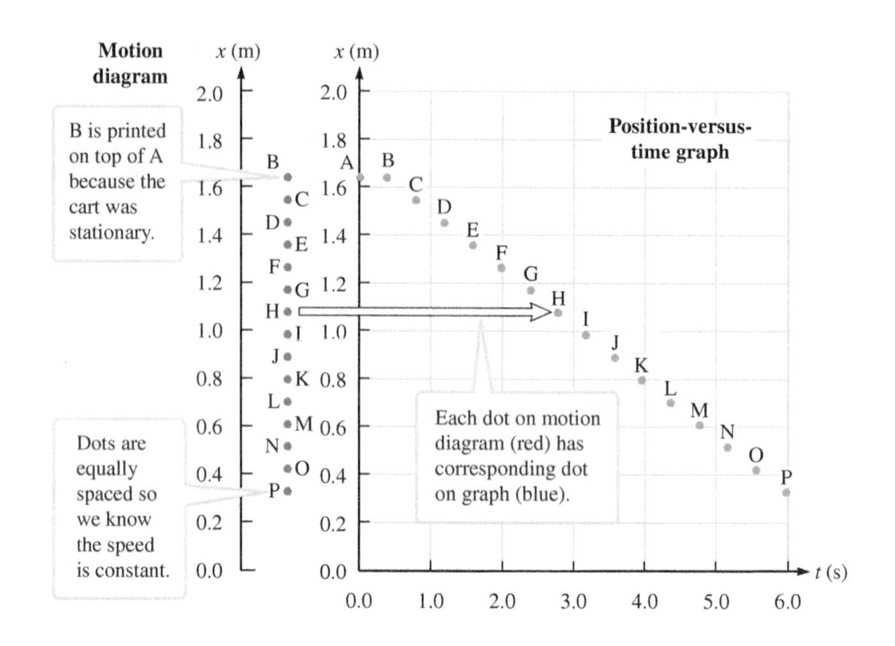

B.

FIGURE 2.6 A. Overhead view of Crall and Whipple's experimental setup. A sonic ranger (a motion sensor) detects the position of a cart moving along a straight, horizontal track by sending an ultrasonic pulse that echoes back. **B.** The motion diagram (left side) shows the location of the cart every 0.4 s. The position-versus-time graph (right side) is a two-dimensional representation of the same data. Position is plotted on the vertical axis, and time is plotted on the horizontal axis.

Position-Versus-Time Graph

Even with a coordinate system for position imposed on it, a motion diagram is not enough to describe the cart's motion completely. For example, we cannot determine the speed of the cart if we do not know the time interval between each position on the motion diagram. So, when measuring motion, we must also measure the time. We begin by setting an initial time to zero; other times are measured with respect to the initial time. Time is a scalar quantity, and it can be positive or negative. Events that occur *after* the initial time are assigned positive times, and events that occur *before* the initial time are assigned negative times. Like choosing a coordinate system, choosing an initial time takes some care. It is often convenient (but not always practical) to set the initial time to the moment the motion begins.

Crall and Whipple's data in Table 2.1 can be arranged in the position-versus-time graph on the right side of Figure 2.6B. In such a graph, the position's scalar component is plotted on the vertical axis, and time is plotted on the horizontal axis. A complete **position-versus-time graph** has the following features:

POSITION-VERSUS-TIME GRAPH

Tool

1. A vertical position axis labeled x, y, or z corresponding to the chosen coordinate system, with position increasing upward
2. A horizontal time axis labeled t, with time increasing to the right
3. Labeled units on both axes (typically "m" for meters and "s" for seconds)
4. Tick marks labeled along both axes

Let's compare a position-versus-time graph to a motion diagram. First, notice that Crall and Whipple's position-versus-time graph (Fig. 2.6B, right) looks like a stretched-out version of the motion diagram (Fig. 2.6B, left). Each dot in a position-versus-time graph represents two pieces of information, the position on the vertical axis and the time on the horizontal axis. In contrast, a motion diagram represents only location.

A position-versus-time graph differs from a motion diagram in one other way. A motion diagram shows the path of the particle, but a position-versus-time graph does not. For example, the position-versus-time graph on the right in Figure 2.6B looks like a particle moving down an incline, but we know that the cart really moved along a horizontal track.

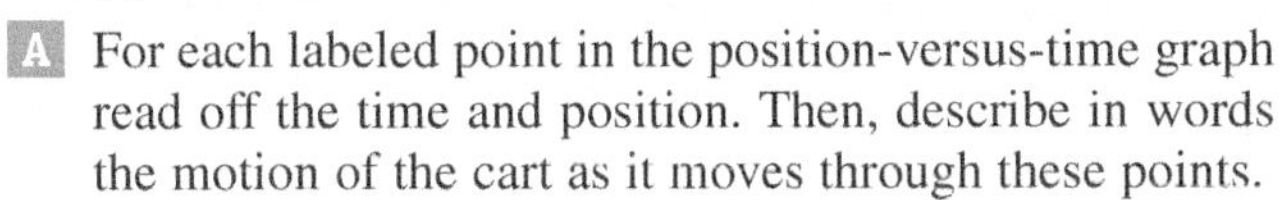

CASE STUDY A Laboratory Cart

Figure 2.7 shows the position-versus-time graph for a cart moving along a straight track like the one used by Crall and Whipple in Figure 2.6A.

A For each labeled point in the position-versus-time graph read off the time and position. Then, describe in words the motion of the cart as it moves through these points.

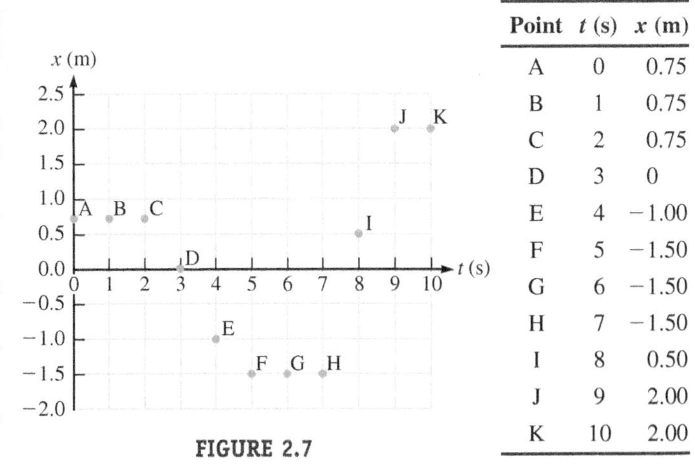

FIGURE 2.7

Point	t (s)	x (m)
A	0	0.75
B	1	0.75
C	2	0.75
D	3	0
E	4	−1.00
F	5	−1.50
G	6	−1.50
H	7	−1.50
I	8	0.50
J	9	2.00
K	10	2.00

INTERPRET and ANTICIPATE

Work your way from A to K on the position-versus-time graph, reading values off the graph as needed.

SOLVE

For each point, read off the position and time. These data are best displayed in a table as at the far right.

Points on graph	*Motion of cart*
Points A through C all have the same position: $\vec{x}_A = \vec{x}_B = \vec{x}_C = 0.75\hat{\imath}$ m.	The cart is at rest at the positive position $\vec{x} = 0.75\hat{\imath}$ m.
Points C through F are successively lower on the x axis.	The cart moves in the negative x direction.
Points F through H are all at the same position: $-1.50\hat{\imath}$ m.	The cart is at rest at $\vec{x} = -1.50\hat{\imath}$ m.
Points H though J are successively higher on the x axis.	The cart moves in the positive x direction.
Points J and K have the same position: $\vec{x}_J = \vec{x}_K = 2.00\hat{\imath}$ m.	The cart is at rest at the positive position $x = 2.00i$ m.

B Create a motion diagram from the position-versus-time graph of Figure 2.7.

INTERPRET and ANTICIPATE

Crall and Whipple's position-versus-time graph looks like a stretched-out motion diagram (Fig. 2.6B). So, we can create a motion diagram from a position-versus-time graph by sliding each point on the graph horizontally onto the position axis, preserving the labeling of the dots.

SOLVE

Figure 2.8 shows the motion diagram on the left and the position-versus-time graph on the right as at the far right.

CHECK and THINK

Compare the motion diagram in Figure 2.8 with the verbal description in part A. Three groups of points (A through C, F through H, and J and K) overlap in the motion diagram, indicating that the cart is at rest, consistent with our previous description. The motion diagram indicates that the cart moves in the negative x direction from C to F and in the positive x direction from H to J, also consistent with our description in part A.

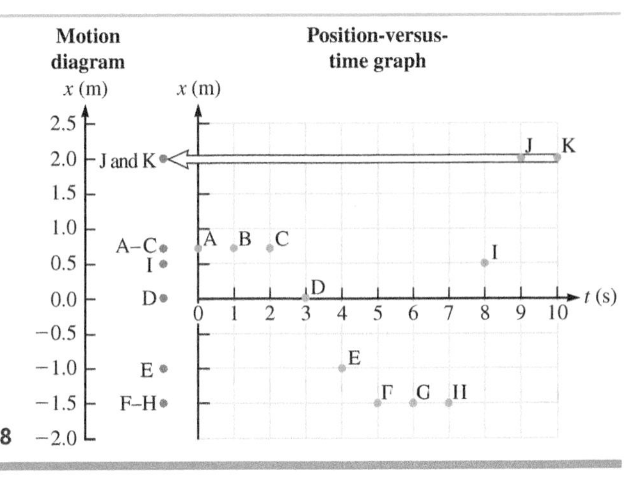

FIGURE 2.8

2-5 Displacement and Distance Traveled

In many ways, kinematics is the study of how certain quantities *change*. We use the language of mathematics to describe kinematics, so we need a mathematical symbol to indicate change. The symbol for change is an uppercase Greek letter delta Δ; the

DISPLACEMENT ✪ Major Concept

The symbol ≡ does not mean *equals*; instead, it means *is defined to be.*

notation ΔQ means "the change in the quantity Q." This notation is shorthand for writing $Q_f - Q_i$, where the subscripts f and i stand for *f*inal and *i*nitial, respectively. When describing motion, the initial and final times must be *chosen*. The initial time is not necessarily set to zero, and the initial and final times are not necessarily the first and last times the particle moves. They are two relevant moments.

Displacement (Change in Position)

The change in position ($\Delta\vec{r}$) comes up so frequently in kinematics that it is given its own term, ***displacement.*** For one-dimensional motion, the vector component of displacement is

$$\Delta\vec{x} \equiv (x_f - x_i)\hat{\imath}, \qquad \Delta\vec{y} \equiv (y_f - y_i)\hat{\jmath}, \qquad \text{or } \Delta\vec{z} \equiv (z_f - z_i)\hat{k} \qquad (2.1)$$

depending on the choice of the coordinate axis.

Like position, *change* in position is also a vector quantity. In one dimension, the direction of a displacement vector is indicated by a positive or negative sign. The sign of the displacement, and therefore its direction, are determined by the subtraction in Equation 2.1. Looking back at Crall's coordinate system (Table 2.2), we see that the cart started at position $\vec{x}_i = 1.64\hat{\imath}$ m and ended at $\vec{x}_f = 0.33\hat{\imath}$ m. The cart's displacement between its initial and final positions is therefore

$$\Delta\vec{x} = (x_f - x_i)\hat{\imath} = (0.33 - 1.64)\hat{\imath} = -1.31\hat{\imath}\text{ m}$$

The displacement is negative, meaning that the cart moved in the negative x direction, which in this case is southward toward the origin (Fig. 2.6).

Position depends on the choice of coordinate systems, but displacement does not. As shown in Figure 2.9, Whipple has not chosen the same coordinate system as Crall. Whipple's and Crall's measured positions do not agree (Table 2.2). Crall and Whipple do agree on the cart's displacement between its initial and final positions, however. Using Whipple's coordinate system, we have

$$\Delta\vec{x} = (x_f - x_i)\hat{\imath} = (-1.31 - 0)\hat{\imath} = -1.31\hat{\imath}\text{ m}$$

So, in both coordinate systems, the displacement is the same: 1.31 m southward.

TABLE 2.2 Crall's and Whipple's data

Label	t (s)	Crall's x (m)	Whipple's x (m)
A	0.0	1.64	0
B	0.4	1.64	0
C	0.8	1.55	−0.09
D	1.2	1.45	−0.19
E	1.6	1.35	−0.29
F	2.0	1.25	−0.39
G	2.4	1.16	−0.48
H	2.8	1.06	−0.58
I	3.2	0.97	−0.67
J	3.6	0.87	−0.77
K	4.0	0.78	−0.86
L	4.4	0.69	−0.95
M	4.8	0.60	−1.04
N	5.2	0.51	−1.13
O	5.6	0.42	−1.22
P	6.0	0.33	−1.31

Displacement, Translation, and the Particle Model

An object can be modeled as a particle undergoing purely translational motion if every point on the object undergoes exactly the same displacement as every other point on the object. Figure 2.10A shows the motion of a ball in a sled sliding down a slippery slope; each point on the ball has the same displacement. Figure 2.10B shows the same ball rolling down a hill; in this case, the displacement of one eye is greater than the displacement of the other. The sliding ball may be modeled as a particle undergoing pure translational motion, whereas the rolling ball is both rotating and translating, and cannot be modeled as a particle.

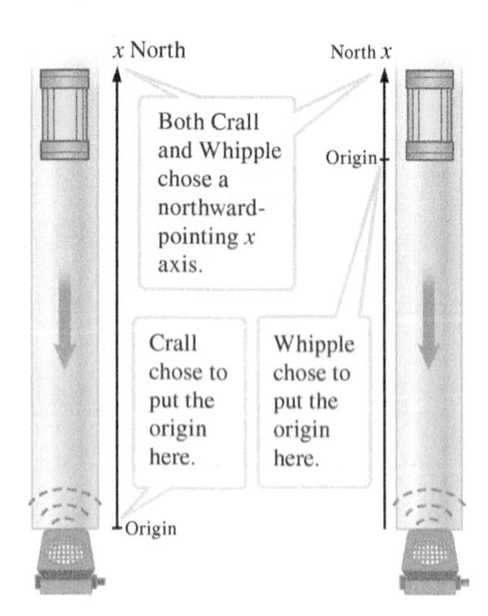

FIGURE 2.9 Position depends on the choice of coordinate system, but displacement does not. Crall chose to place the origin of the coordinate system at the motion sensor, and Whipple placed the origin at the cart's initial position. Crall's and Whipple's position measurements do not agree, but they both measure the same displacement.

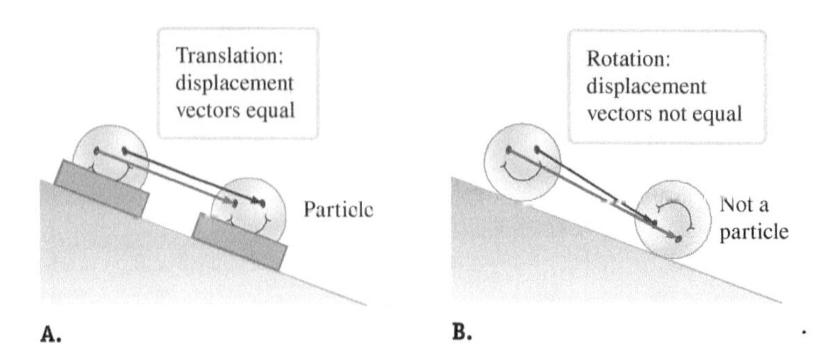

FIGURE 2.10 **A.** A ball in a sled slides down an incline. The black arrow is the displacement vector for one eye and the red arrow is the displacement vector for the other. **B.** The ball rolls down the incline. If the displacement of every point on an object is the same, the object may be modeled as a particle undergoing pure translational motion. The displacement of the two eyes in part B is not the same because the object is rotating in addition to translating.

CONCEPT EXERCISE 2.3

Figure 2.11 shows the motion of various objects:

Case 1. A person on a moving Ferris wheel
Case 2. A person on a loop-the-loop roller coaster
Case 3. The tire on a moving bicycle
Case 4. The person on the bicycle

Consider the displacement of points on each object, and decide whether the object is undergoing pure translational motion and therefore may be modeled as a particle.

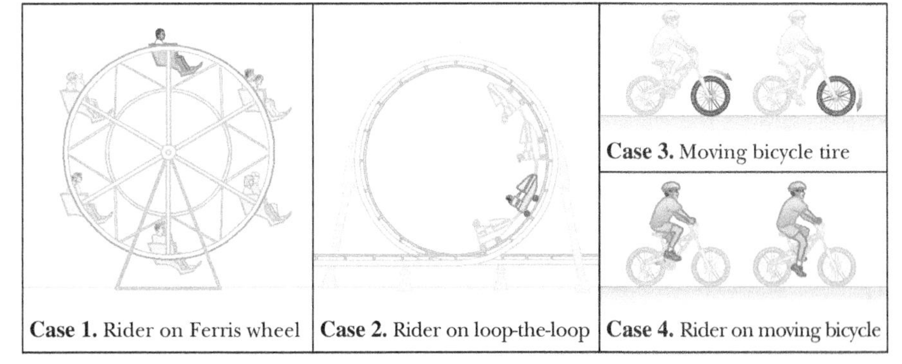

FIGURE 2.11

Distance Traveled

Displacement is not the same thing as distance traveled. The **distance traveled** d by any moving particle is the length of the whole path the particle covers. Distance traveled is a scalar quantity and therefore has no direction. A tiger in a zoo may pace the length of a 20-m enclosure 100 times in a day. Her total distance traveled would be $100 \times 20\text{ m} = 2000\text{ m}$, but if she ends at the same position at which she started, her displacement is zero. Distance traveled and the magnitude of displacement are equal if, and only if, the path is a straight line in one direction.

DISTANCE TRAVELED
Major Concept

EXAMPLE 2.2 Block and Spring: Displacement and Distance Traveled

A block on a nearly frictionless table is attached to a spring (Fig. 2.12). The other end of the spring is fixed to a wall. When the block is pulled to the right and then released, it moves back and forth for a long time before coming to rest. A coordinate axis, arbitrarily labeled z, lies along the table; the origin is at the original position of the block, and the positive direction is to the right. The block's position is measured every 0.25 s as given in Table 2.3.

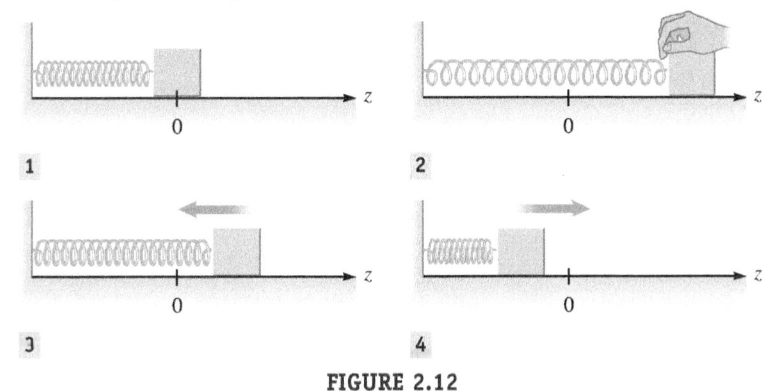

FIGURE 2.12

TABLE 2.3 Position and time data

z (m)	t (s)
0.65	0
0.63	0.25
0.57	0.50
0.48	0.75
0.35	1.00
0.20	1.25
0.05	1.50
−0.12	1.75
−0.27	2.00
−0.41	2.25
−0.52	2.50
−0.60	2.75
−0.64	3.00
−0.65	3.25
−0.61	3.50
−0.53	3.75
−0.42	4.00
−0.29	4.25
−0.14	4.50
0.02	4.75
0.18	5.00
0.33	5.25
0.46	5.50
0.56	5.75
0.62	6.00
0.65	6.25

A Create a position-versus-time graph from these data.

INTERPRET and ANTICIPATE

The block moves back and forth, so we expect our position-versus-time graph to show a repeating curve. The initial time ($t_i = 0$) has already been chosen as the time of the first recorded position (Table 2.3). When working with your own laboratory data, you may make a similarly convenient choice.

Example continues on page 30 ▶

SOLVE
Make a position-versus-time graph with z on the vertical axis and t on the horizontal axis (Fig. 2.13). (It may be helpful to use a spreadsheet program.)

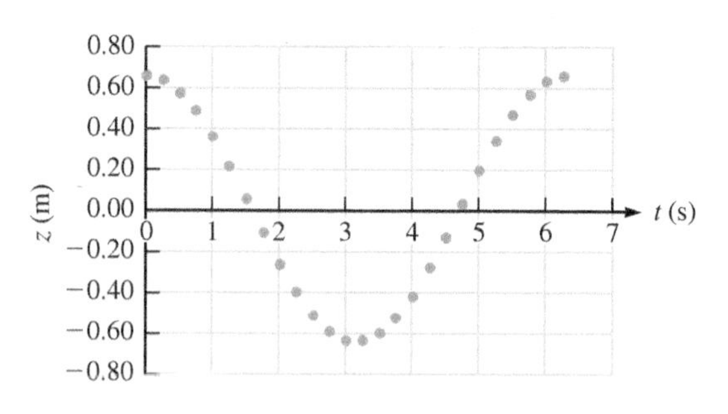

FIGURE 2.13

CHECK and THINK
As expected, Figure 2.13 shows a repeating pattern for the motion.

B What is the displacement of the block during the entire 6.25 s?

INTERPRET and ANTICIPATE
In 6.25 s, the block returned to the original position, so we expect the displacement to be zero.

SOLVE
The initial and final positions are at the same height on the graph. Find the displacement by subtracting.

$$\vec{z}_f = \vec{z}_i = 0.65\hat{k}\text{ m}$$

$$\Delta\vec{z} = \vec{z}_f - \vec{z}_i = (0.65 - 0.65)\hat{k}\text{ m} = 0 \quad \text{as expected.}$$

C What is the displacement of the block during the first 3.25 s?

SOLVE
For this time interval, the final position is negative and the initial position is positive. Find the displacement by subtracting.

$$\vec{z}_f = -0.65\hat{k}\text{ m}$$

$$\vec{z}_i = 0.65\hat{k}\text{ m}$$

$$\Delta\vec{z} = (z_f - z_i)\hat{k} = (-0.65 - 0.65)\hat{k}\text{ m} = -1.30\hat{k}\text{ m}$$

CHECK and THINK
The block moved in the negative z direction during this time interval.

D What is the total distance traveled for the recorded measurement interval?

INTERPRET and ANTICIPATE
Distance is not a vector, so we are not concerned with the algebraic signs of the positions. All we need is the length of the path covered in the recorded 6.25 s.

SOLVE
From part C, we know that the block moved 1.30 m in the first 3.25 s. In the next 3 s, it returned to its original position, so it must have traveled another 1.30 m. The total distance traveled is thus the sum of these two distances.

$$d = (1.30 + 1.30)\text{ m} = 2.60\text{ m}$$

CHECK and THINK
The block's back-and-forth motion is much like a tiger pacing in her cage. When the block returns to its initial position, its displacement is zero. The distance traveled, however, is not zero!

2-6 Average Velocity and Speed

There are several ways to measure how fast a particle is moving. This section and the next will cover four such measures: average velocity, average speed, instantaneous velocity, and instantaneous speed.

Average Velocity

AVERAGE VELOCITY

Major Concept

Every marathon runner who completes the race undergoes the same displacement, but the one who does it in the shortest time wins. The **average velocity** is the displacement divided by the change in time; for displacement in the x direction,

$$\vec{v}_{\text{av},x} \equiv \frac{\Delta\vec{x}}{\Delta t} = \frac{x_f - x_i}{t_f - t_i}\hat{\imath} \tag{2.2}$$

Similar equations can also be written for displacement in the y and z directions; the unit vectors then become $\hat{\jmath}$ and $\hat{k}$, respectively.

where the subscript "av" means that the velocity is averaged over the time interval Δt.

Average velocity is a vector quantity whose magnitude in SI units is measured in meters per second. The change in time in the denominator of Equation 2.2 must always be positive because the final time is always after the initial time. In other words, the sign of the average velocity is determined by the sign of the displacement in the numerator. Therefore, the average velocity points in the same direction as the displacement.

Average Speed

AVERAGE SPEED Major Concept

Sometimes, we are more interested in knowing how fast an object moves without regard to the direction of that motion. **Average speed** S_{av} depends on total distance traveled (not on displacement) such that

$$S_{\text{av}} \equiv \frac{d}{\Delta t} \tag{2.3}$$

where d is the total distance traveled in the time interval Δt. Average speed is always positive. Like average velocity, average speed in SI units is measured in meters per second. Because average speed is not a vector quantity, no subscript x, y, or z is needed.

Average speed is important in our everyday experience and is sometimes more meaningful than average velocity. As illustrated in Figure 2.14, the start-to-finish displacement is not necessarily the same for different marathon courses, but all marathon courses are 26.2 miles long. A runner's average *speed* may be (nearly) the same on every course, but the runner's average *velocity* will vary from course to course because the displacement varies. Therefore, the average speed is a better measure of the runner's ability than average velocity, which is more a measure of the shape of the course. Average velocity is a measure of how quickly the displacement changes. The runner who completes the marathon in the shortest time interval has the highest average speed and wins the race.

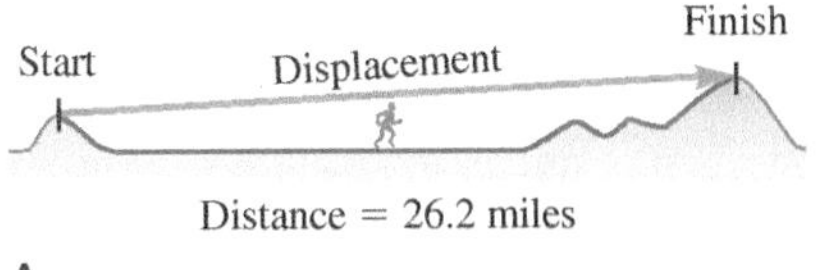

A.

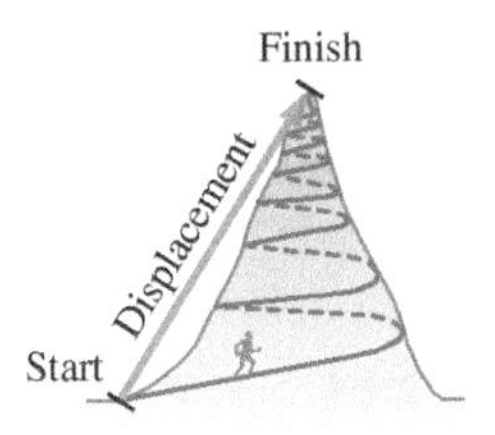

B.

FIGURE 2.14 Two marathon courses. The distance run on each course is 26.2 miles, but the magnitude of the displacement of a runner on course B is less than on course A.

CONCEPT EXERCISE 2.4

The top marathon runners complete the race in around 2 hours, so such an elite marathoner's average speed is around 13.1 mph. To calculate a runner's average velocity in a marathon, we would have to know the displacement, which means knowing the particular course. If the magnitude of thd runner's displacement is 20 mi over course A and 12 mi over course B, find the magnitude of the average velocity for each course (Fig. 2.14) and compare your results to the elite marathon average speed.

CONCEPT EXERCISE 2.5

In our everyday experience, we sometimes use the phrase "as the crow flies." Does this expression refer to total distance traveled, or does it refer to magnitude of the displacement?

CONCEPT EXERCISE 2.6

A tiger paces the length of his 20-m enclosure 100 times in 2 hours. He then stops at his initial position. Find his average speed and average velocity over this period. How do they compare with the average speed and average velocity of a tiger that lies asleep in one place for 2 hours?

EXAMPLE 2.3 Motion of a Particle

A The average velocity of a particle measured over a 5.4-s interval is $\vec{v}_{\text{av},x} = 4.9\hat{\imath}$ m/s. Find its displacement during this time interval.

INTERPRET and ANTICIPATE

Because the average velocity points in the positive x direction, we expect the displacement to be in the positive x direction.

SOLVE

Solve Equation 2.2 for displacement.

$$\vec{v}_{\text{av},x} = \frac{\Delta\vec{x}}{\Delta t} \tag{2.2}$$

$$\Delta\vec{x} = \vec{v}_{\text{av},x}\,\Delta t$$

Make the appropriate substitutions.

$$\Delta\vec{x} = (4.9\hat{\imath}\text{ m/s})(5.4\text{ s}) = 2.6 \times 10^1\hat{\imath}\text{ m}$$

CHECK and THINK

The displacement is 26 m in the positive x direction as expected.

B The average velocity of a particle measured over a 5.4-s interval is $\vec{v}_{\text{av},x} = -6.9\hat{\imath}$ m/s. Find its displacement over this time interval.

INTERPRET and ANTICIPATE

This part is similar to part A except now we expect the displacement to be in the negative x direction.

SOLVE

As in part A, solve for displacement and make the appropriate substitutions.

$$\Delta\vec{x} = \vec{v}_{\text{av},x}\,\Delta t$$

$$\Delta\vec{x} = (-6.9\hat{\imath}\text{ m/s})(5.4\text{ s}) = -3.7 \times 10^1\hat{\imath}\text{ m}$$

CHECK and THINK

The displacement is 37 m in the negative x direction as predicted.

Average Velocity from a Position-Versus-Time Graph

Graphs are particularly useful tools for presenting data in a visual form. Average velocity may be found from a position-versus-time graph as shown in the following experiment.

EXAMPLE 2.4 CASE STUDY Another Experiment

In a new experiment, Crall and Whipple attached a fan to the cart (Fig. 2.15). They placed the cart near the bottom of an inclined ramp and gave it a push to start it moving up the ramp. Their position-versus-time graph is shown in Figure 2.16. The cart started at $\vec{x} = 0.5\hat{\imath}$ m. It traveled in the positive x direction to a peak position of 1.8 m. Then it traveled in the negative x direction back to a final position of 0.5 m. For clarity on their graph, Crall and Whipple marked the initial, peak, and final positions.

Crall believes that the cart moved faster going up the ramp than it moved going down. Use Crall and Whipple's data from Figure 2.16 to calculate the average velocity for the trip up the ramp and the average velocity for the trip down the ramp. In the **CHECK and THINK** step, compare the magnitude of the average velocity up the ramp to the magnitude of the average velocity down the ramp to decide if Crall is correct.

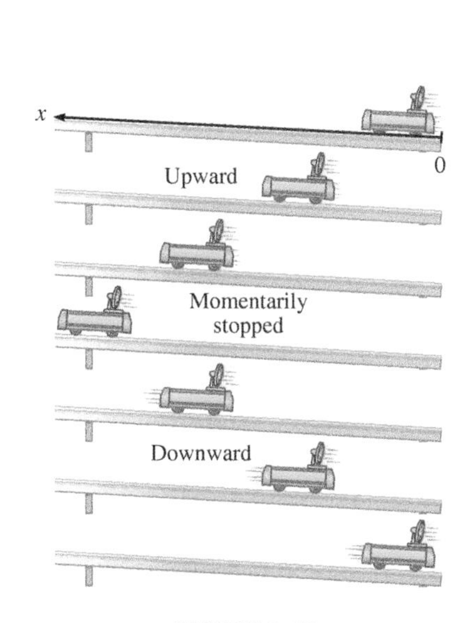

FIGURE 2.15

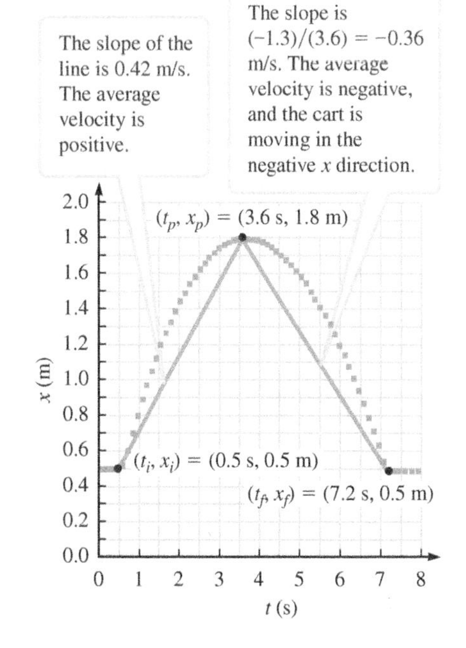

FIGURE 2.16

INTERPRET and ANTICIPATE

In Crall and Whipple's coordinate system, the x axis points up the ramp. So, the cart's displacement is positive when it moves up the ramp and negative when it moves down. We expect the average velocity to be positive when the cart goes up the ramp and negative when it goes down.

SOLVE

Let's find the average velocity for the trip up the ramp (Eq. 2.2). We use the subscript "*ip*" to indicate the average velocity over the time interval $t_p - t_i$ when the cart moved from its *i*nitial position to its *p*eak position and drop the subscript "*x*" for simplicity.

$$\vec{v}_{\text{av},\,ip} = \frac{\Delta \vec{x}_{ip}}{\Delta t} = \frac{x_p - x_i}{t_p - t_i}\hat{\imath}$$

$$\vec{v}_{\text{av},\,ip} = \frac{(1.8 - 0.5)\text{m}}{(3.6 - 0.5)\text{s}}\hat{\imath} \tag{1}$$

$$\vec{v}_{\text{av},\,ip} = 0.42\hat{\imath}\ \text{m/s}\ \ \text{(up the ramp)}$$

We apply a similar process to find the average velocity on the way down the ramp.

$$\vec{v}_{\text{av},\,pf} = \frac{\Delta \vec{x}_{pf}}{\Delta t} = \frac{x_f - x_p}{t_f - t_p}\hat{\imath} = \frac{(0.5 - 1.8)\text{m}}{(7.2 - 3.6)\text{s}}\hat{\imath}$$

$$\vec{v}_{\text{av},\,pf} = -0.36\hat{\imath}\ \text{m/s}\ \ \text{(down the ramp)}$$

CHECK and THINK

As expected, the average velocity up the ramp is positive and down the ramp is negative.

Crall is interested in whether the cart is faster going up the ramp or down the ramp; to address his interest, we need to compare the magnitude of these average velocities. When we do so, we find that Crall is correct: The cart was faster on the way up the ramp.

$$|\vec{v}_{\text{av},\,ip}| = 0.42\ \text{m/s up the ramp}$$

$$|\vec{v}_{\text{av},\,pf}| = 0.36\ \text{m/s down the ramp}$$

$$|\vec{v}_{\text{av},\,ip}| > |\vec{v}_{\text{av},\,pf}|$$

Let's take another look at how we found the average velocity for the cart in the previous example. The magnitude of the average velocity of the cart on the way up is given by (Eq. 1)

$$\vec{v}_{\text{av},\,ip} = \left[\frac{(1.8 - 0.5)\text{m}}{(3.6 - 0.5)\text{s}}\right]\hat{\imath}$$

Recall that the slope of a line is its rise divided by its run.

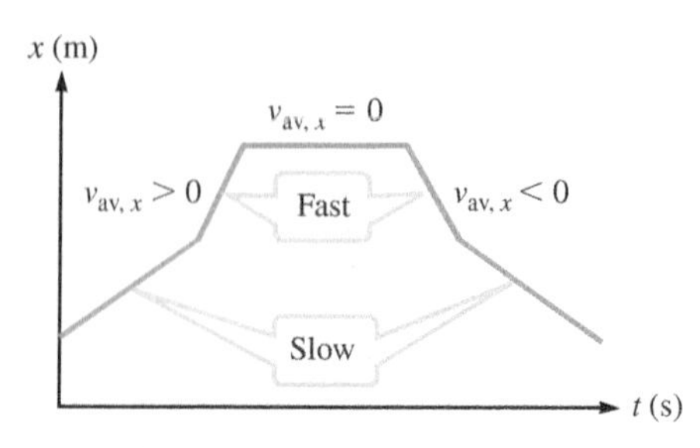

FIGURE 2.17 Finding average velocity from a position-versus-time graph. Steep slope implies a high average velocity. The sign of the slope gives the sign and the direction of the average velocity. A horizontal line implies that the particle is at rest for the time interval during which the position does not change.

The ratio in brackets, [ ], is the same as the slope of the upward-sloping orange line drawn between the initial point and the maximum point in Figure 2.16. On this graph, the rise has units of length (m), the run has units of time (s), and the slope of the orange line is 0.42 m/s.

So, another way to find the average velocity is from the position-versus-time graph. *The slope of the straight line joining two points in a position-versus-time graph is the average velocity over that time interval.* In practice, we inspect a position-versus-time graph to estimate or compare average velocities over different time intervals (Fig. 2.17) using these facts:

1. A steeper slope means a higher average velocity (faster motion).
2. The sign of the slope gives the sign of the average velocity and therefore the direction of motion.
3. A horizontal line has zero rise and therefore zero slope. So, a horizontal line means that the average velocity is zero.

INSTANTANEOUS VELOCITY

Major Concept

2-7 Instantaneous Velocity and Speed

If a particle does not speed up, slow down, or change direction, we say that it is moving at *constant* or *uniform* velocity. In this case, the particle's average velocity is equal to its velocity at every instant in time. The velocity at each instant is called the **instantaneous velocity**, usually referred to simply as **velocity**. Figure 2.18A shows a position-versus-time graph for a particle moving at constant velocity. Because the average and instantaneous velocities are equal in this case, the slope of the line gives both. Further, the slope of the line does not depend on the time interval and is the same between points a and b, b and c, or any pair of points on the position-versus-time graph:

$$\frac{\Delta \vec{x}_{ab}}{\Delta t_{ab}} = \frac{\Delta \vec{x}_{bc}}{\Delta t_{bc}} = \frac{\Delta \vec{x}_{ac}}{\Delta t_{ac}}$$

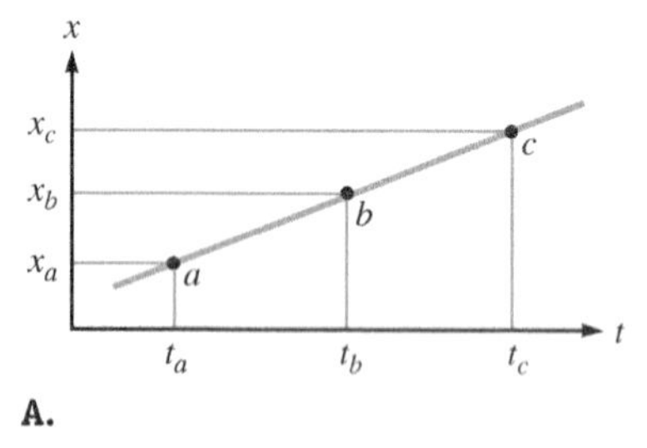

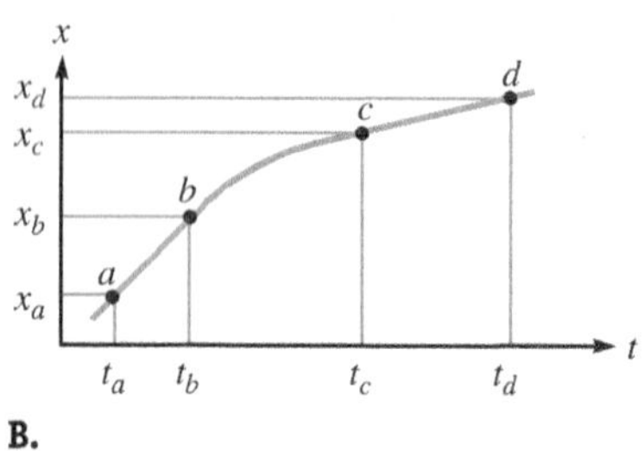

FIGURE 2.18 A. The position-versus-time graph has a constant slope, so the velocity is uniform and the average velocity does not depend on the time interval. **B.** The decreasing slope means the particle is slowing down.

If the speed of the particle changes (Fig. 2.18B), the average velocity depends on the time interval during which it is measured. The slope of the line between points a and b is steeper than the slope between c and d. Therefore, the particle had a higher average velocity over the first time interval Δt_{ab} than it did over the second time interval Δt_{cd}:

$$\frac{\Delta x_{ab}}{\Delta t_{ab}} > \frac{\Delta x_{cd}}{\Delta t_{cd}}$$

Instantaneous velocity at one particular time t is obtained in much the same way as average velocity over some time interval Δt, but with a shrinking Δt. Figure 2.19 shows the position-versus-time graph for a particle whose velocity is continuously changing. As the time interval gets shorter, the dashed slanted line in Figure 2.19 becomes a better and better approximation to the velocity at point a (coordinates t_a, x_a). Also, as the time interval approaches zero, the dashed

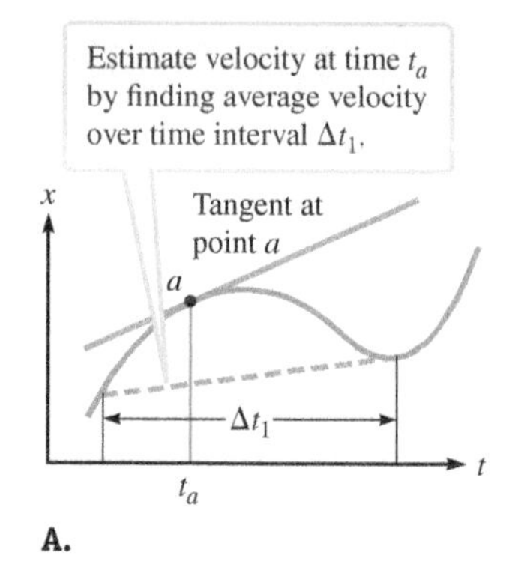

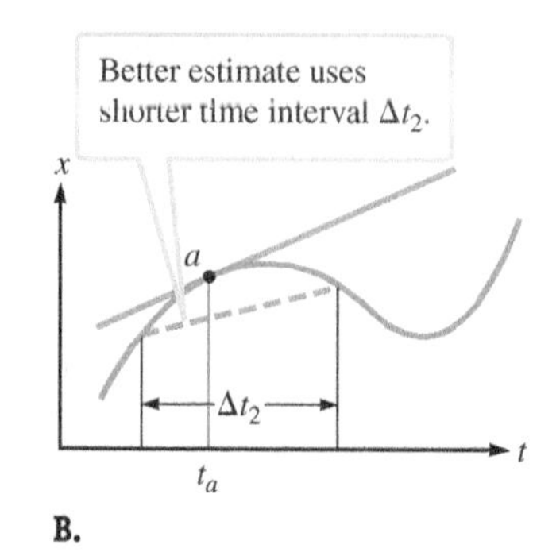

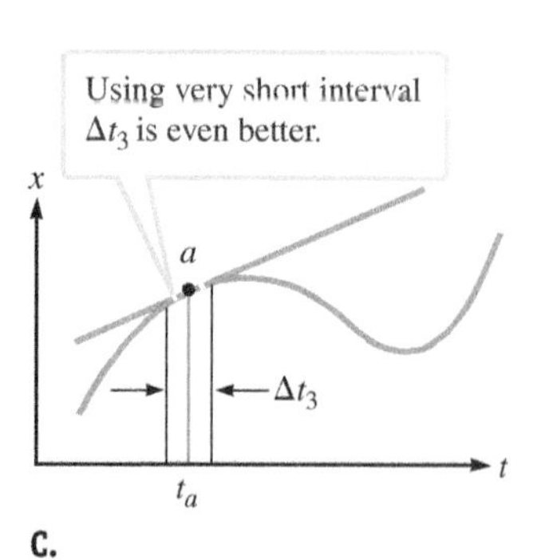

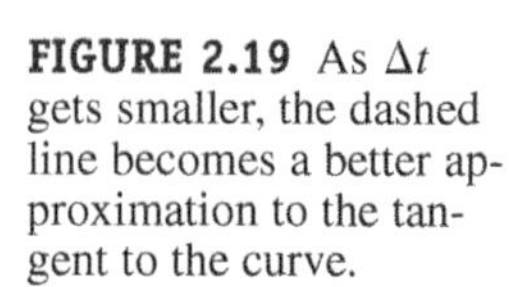

FIGURE 2.19 As Δt gets smaller, the dashed line becomes a better approximation to the tangent to the curve.

slanted line touches the curve at only one point a and is identical to the tangent to the curve at that point (Fig. 2.19C). *The slope of the tangent line at point a is the instantaneous velocity at time* t_a.

Figure 2.19 shows that as Δt approaches zero, the average velocity approaches the instantaneous velocity. So, we find expressions for the instantaneous velocity (in the x direction) by taking the limit of $\vec{v}_{\text{av},x} = \Delta\vec{x}/\Delta t$ (Eq. 2.2) as Δt approaches zero:

$$\vec{v}_x = \lim_{\Delta t \to 0} \frac{\Delta \vec{x}}{\Delta t} = \frac{d\vec{x}}{dt} = \frac{dx}{dt}\hat{\imath} \tag{2.4}$$

Like average velocity, instantaneous velocity is a vector quantity. To distinguish the average velocity from instantaneous velocity, we use the additional subscript "av" for the average velocity and no additional subscript for the instantaneous velocity.

Similar equations can also be written for velocity in the y and z directions.

Instantaneous Speed

The **instantaneous speed** (or simply the **speed**) is the magnitude of the instantaneous velocity. Speed is a scalar quantity and has no direction. The symbol for speed is v without an arrow.

INSTANTANEOUS SPEED

Major Concept

The definition of *instantaneous* speed is very different from the definition of *average* speed. Both speed and average speed are scalars. Speed is just the magnitude of the velocity vector, but average speed is not simply the magnitude of the average velocity. Average speed is uniquely defined by $S_{\text{av}} \equiv d/\Delta t$ (Eq. 2.3).

In everyday language, we often use speed and velocity interchangeably, but not in physics.

Block and Spring: Velocity and Speed

In this example, we take another look at the block attached to the spring (Fig. 2.12). The block's position as a function of time is given by $\vec{z}(t) = (0.65 \cos \omega t)\hat{k}$, where $\omega = 1.0$ rad/s is a constant. The symbol $\vec{z}(t)$ means "z as a function of t" and not "z multiplied by t," and *rad* is the abbreviation for radians.

A Find the average velocity and average speed over the 6.25-s interval.

INTERPRET and ANTICIPATE

To find the average velocity, we divide the displacement by the time interval. To find the average speed, we must first find the total distance traveled. We expect to find two numerical answers of the form $\vec{v}_{\text{av},z} = \_\_\_\_\hat{k}$ m/s and $S_{\text{av}} = \_\_\_\_$ m/s.

SOLVE

Average velocity is given by Equation 2.2 written in terms of z instead of x.	$\vec{v}_{\text{av},z} = \dfrac{\Delta\vec{z}}{\Delta t}$ (2.2)
We already found the displacement was zero in part B of Example 2.2. Therefore the average velocity is zero.	$\Delta\vec{z} = 0$ $\vec{v}_{\text{av},z} = \dfrac{0}{6.25\text{ s}} = 0$
Use Equation 2.3 for average speed.	$S_{\text{av}} = \dfrac{d}{\Delta t}$ (2.3)
We found the total distance traveled in part D of Example 2.2. Now we divide by the total time, 6.25 s.	$d = 2.60$ m $S_{\text{av}} = \dfrac{2.60\text{ m}}{6.25\text{ s}} = 0.416$ m/s

CHECK and THINK

Because the displacement is zero, the average velocity is zero. The distance traveled by the block is not zero, and so its average speed is not zero.

Example continues on page 36 ▶

B Find the velocity and speed at $t = 2.00$ s.

INTERPRET and ANTICIPATE
In part A, we found the *average* velocity and *average* speed. Here, we are asked for the *instantaneous* velocity and *instantaneous* speed. We see from the position-versus-time graph (Fig. 2.20) that the slope of the tangent line at $t = 2.00$ s is negative, so we expect the velocity to be negative. Instantaneous speed is the magnitude of velocity and will be positive.

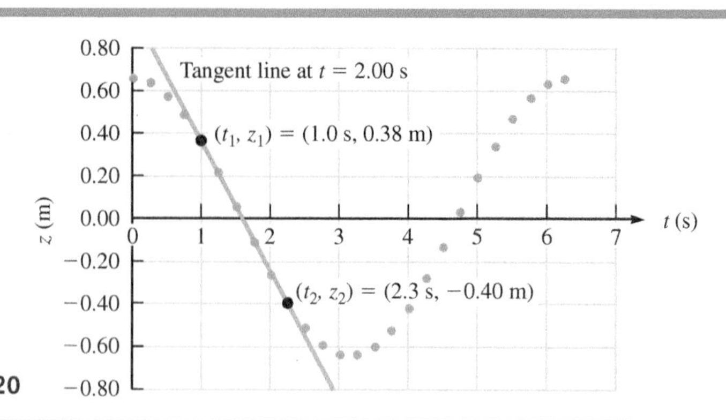

FIGURE 2.20

Let's further anticipate our results by estimating the slope of the tangent line. Choose two points on the tangent line and find the slope from these two points. For example, $(t_1, z_1) = (1.0 \text{ s}, 0.38 \text{ m})$ and $(t_2, z_2) = (2.3 \text{ s}, -0.40 \text{ m})$.

$$\text{slope} = \frac{z_2 - z_1}{t_2 - t_1} = \frac{(-0.40 - 0.38) \text{ m}}{(2.3 - 1.0) \text{ s}}$$

$$\text{slope} = -0.6 \text{ m/s}$$

SOLVE
Equation 2.4 (written in terms of z) gives the velocity $\vec{v}_z(t)$ as the time derivative of $\vec{z}(t)$.

$$\vec{v}_z(t) = \frac{d\vec{z}}{dt} = \frac{d(0.65 \cos \omega t)\hat{k}}{dt} = 0.65 \frac{d(\cos \omega t)}{dt}\hat{k}$$

We need the chain rule (Appendix A).

$$\vec{v}_z(t) = 0.65 \frac{d(\omega t)}{dt} \frac{d(\cos \omega t)}{d(\omega t)}\hat{k}$$

Use $d(\cos \theta)/d\theta = -\sin \theta$ to arrive at the velocity at any time t.

$$\vec{v}_z(t) = -(0.65\omega \sin \omega t)\hat{k}$$

We need the velocity at $t = 2.00$ s. Substitute for t and ω. Notice that the "rad" has disappeared from our final answer. Because a radian is defined as the ratio of two lengths (circumference of a circle to its diameter) and is therefore dimensionless, we drop "rad" from our expression whenever it is possible. Here, the velocity's units are simply expressed as meters per second.

$$\vec{v}_z(2.00 \text{ s}) = -(0.65 \text{ m})(1 \text{ rad/s}) \sin [(1 \text{ rad/s})(2.00 \text{ s})]\hat{k}$$

$$\vec{v}_z(2.00 \text{ s}) = -0.59\hat{k} \text{ m/s}$$

The (instantaneous) speed is just the magnitude of the (instantaneous) velocity.

$$|v_z(2.00 \text{ s})| = 0.59 \text{ m/s}$$

CHECK and THINK
As expected, the estimated slope (-0.6 m/s) of the tangent line is close to the velocity we found, and the absolute value of the slope is close to the speed we found.

Displacement from a Velocity-Versus-Time Graph

VELOCITY-VERSUS-TIME GRAPH

Tool

In much the same way that we made a position-versus-time graph, we can make a **velocity-versus-time graph** by plotting instantaneous velocity on the vertical axis and time on the horizontal axis. A velocity-versus-time graph is useful for showing when the particle is speeding up, slowing down, or maintaining a constant velocity.

We can find the displacement of a particle from its velocity-versus-time graph. For example, Figure 2.21 shows velocity-versus-time graphs for two particles. For both particles, the velocity is constant, so the graphs are horizontal lines with $\vec{v}_x = \vec{v}_{\text{av},x}$. Let's find the displacement of each particle during the first 5.00 seconds. According to Equation 2.2, the displacement is

$$\Delta\vec{x} = \vec{v}_{\text{av},x}\Delta t = (v_{\text{av},x}\Delta t)\hat{\imath} \tag{2.5}$$

Notice that the magnitude of the displacement of each particle over the first 5 seconds is the area of the corresponding rectangle. The region between the curve and the horizontal t axis is called "the area under the curve," so the displacement is

$$\Delta\vec{x} = (\text{area under the curve})\hat{\imath}$$

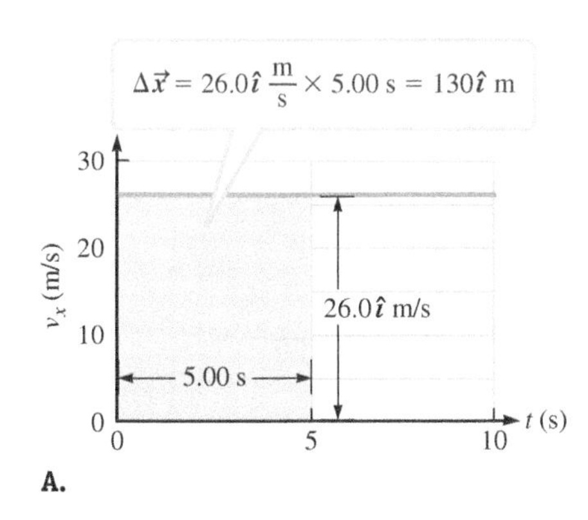

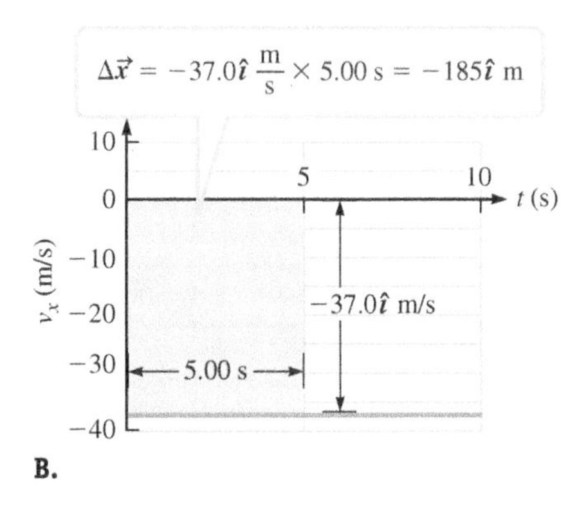

FIGURE 2.21 A. Velocity-versus-time graph for a particle moving at constant velocity $\vec{v}_x = 26.0\hat{\imath}$ m/s. The area of the blue rectangle equals the displacement in the first 5.00-s time interval. The sign of this area and the displacement are positive. **B.** Velocity-versus-time graph for a particle moving at constant velocity $\vec{v}_x = -37.0\hat{\imath}$ m/s. The sign of this area (blue rectangle) and the displacement are negative.

The area of the rectangles in Fig. 2.21 has the units (m/s) × s = m, not m^2. The sign of the displacement comes from the location of the rectangle. If the rectangle is above the t axis, the displacement is positive (Fig. 2.21A). If the rectangle is below the t axis, the displacement is negative (Fig. 2.21B).

2-8 Average and Instantaneous Acceleration

AVERAGE and INSTANTANEOUS ACCELERATION

Major Concepts

Sections 2-3 through 2-7 described motion mathematically using the concepts of position, displacement, speed, and velocity. To complete this mathematical description, we need to explore another concept, acceleration. A particle that is speeding up, slowing down, or changing direction is accelerating.

Average Acceleration

Average acceleration is the change in a particle's velocity over some time interval:

$$\vec{a}_{av,\,x} \equiv \frac{\Delta\vec{v}_x}{\Delta t} = \frac{\Delta v_x}{\Delta t}\hat{\imath} = \frac{v_x(t_f) - v_x(t_i)}{t_f - t_i}\hat{\imath} = \frac{v_{fx} - v_{ix}}{t_f - t_i}\hat{\imath} \tag{2.6}$$

Similar equations can also be written in the y and z directions. The notation v_{fx} means the x component of velocity at the *final* time, and $v_x(t_f) = v_{fx}$, not velocity multiplied by time.

Average acceleration is a vector whose direction is determined by the subtraction in the numerator of Equation 2.6.

Instantaneous Acceleration

The **instantaneous acceleration**, also known as simply the **acceleration**, is derived from the average acceleration in the same way that instantaneous velocity is derived from average velocity:

$$\vec{a}_x \equiv \lim_{\Delta t \to 0}\frac{\Delta\vec{v}_x}{\Delta t} = \frac{d\vec{v}_x}{dt} = \frac{dv_x}{dt}\hat{\imath} \tag{2.7}$$

In SI units, both average and instantaneous acceleration are measured in meters per second per second, which is usually expressed as meters per second squared (m/s^2). Acceleration is often used in everyday language to describe something that is speeding up or sometimes even something that is increasing, such as accelerating unemployment. In physics, acceleration does not mean that a particle is moving quickly or that the speed is necessarily increasing. Rather, in physics, acceleration refers only to the rate at which the particle's velocity changes. In other words, any particle that is either speeding up or slowing down is accelerating. Furthermore, because velocity is a vector, any particle that changes direction is also accelerating, even if the particle's speed remains constant as the direction of motion changes. This last fact will be very important when we study motion in two and three dimensions.

For one-dimensional motion, a particle's acceleration at any given instant is either in the same direction as its velocity or in the opposite direction (Fig. 2.22). If the acceleration and the velocity are in the same direction, the particle is speeding up. If the acceleration and velocity are in opposite directions, the particle is slowing down, a situation sometimes called **deceleration**. Deceleration does not necessarily indicate motion in the negative direction. Deceleration is really shorthand for the phrase *acceleration in the direction opposite the velocity.*

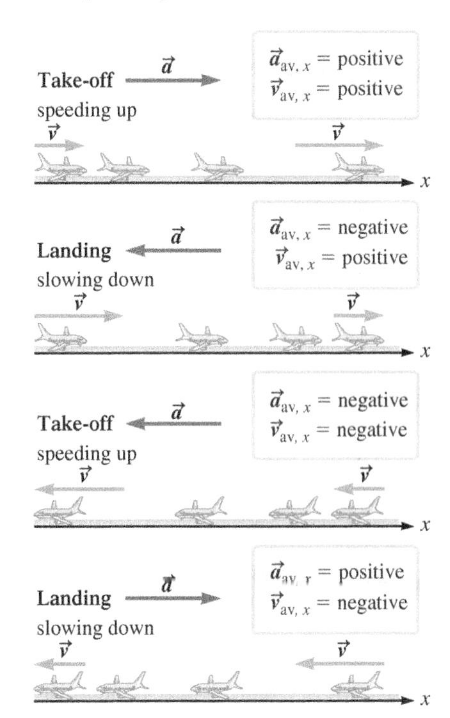

FIGURE 2.22 When the acceleration is in the same direction as the velocity, the plane speeds up. When the acceleration is in the opposite direction to the velocity, the plane slows down.

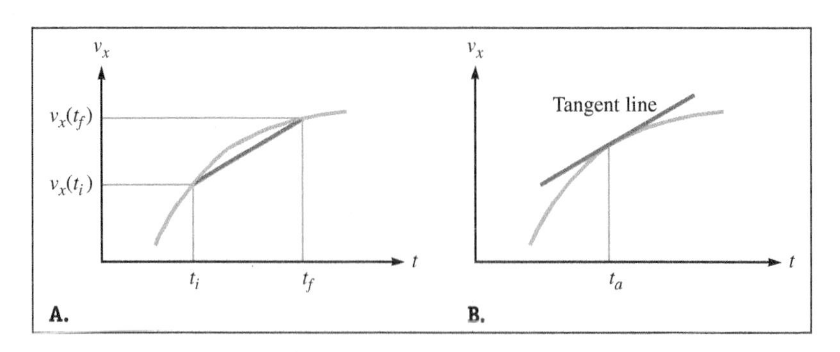

FIGURE 2.23 A. The slope of the line between two points gives the average acceleration over the time interval $t_f - t_i$. **B.** The slope of the tangent line gives the instantaneous acceleration at time t_a.

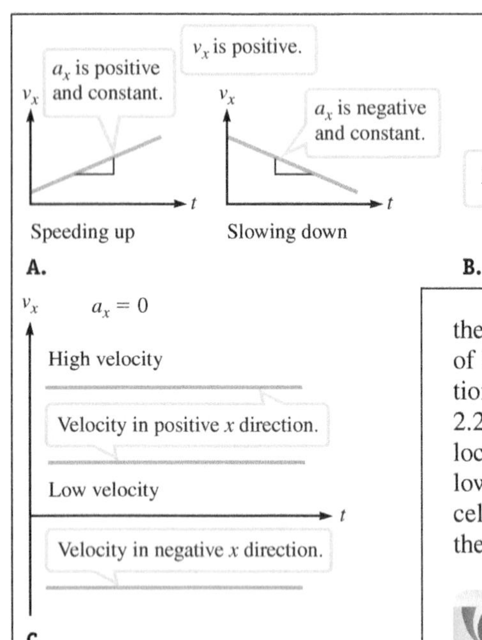

FIGURE 2.24 A. The sign of the slope gives the sign and direction of the acceleration. In both cases, the velocity is positive. **B.** The steeper slope means a greater acceleration. **C.** A horizontal line on a velocity-versus-time graph means that the acceleration is zero and the velocity is constant.

Just as average and instantaneous velocity can be found from slopes on a position-versus-time graph, average and instantaneous acceleration can be found from slopes on a velocity-versus-time graph. The slope of the line joining two points on a velocity-versus-time graph is the average acceleration over that time interval (Fig. 2.23A). The slope of the tangent line at a particular point is the instantaneous acceleration at that point (Fig. 2.23B). The slope has dimensions of length per time squared. The sign of the slope gives the direction of the acceleration. If the velocity-versus-time curve is linear, the acceleration is constant (Fig. 2.24A). A steeper slope means a greater acceleration, but not necessarily a high velocity. Figure 2.24B shows velocity versus time for a particle that starts off with a low velocity and a large acceleration but later has a high velocity and a small acceleration. Finally, a horizontal line on a velocity-versus-time graph indicates that the velocity is constant and the acceleration is zero (Fig. 2.24C).

ACCELERATION-VERSUS-TIME GRAPH

Tool

CONCEPT EXERCISE 2.7

Kinematics graphs are great for showing how a quantity depends on time. For example, an **acceleration-versus-time graph** shows how acceleration depends on time.

a. Based on your experience with two other kinematics graphs (position- and velocity-versus-time), describe how to plot an acceleration-versus-time graph.

b. Describe how all three kinematics graphs are connected by slopes of tangent lines (derivatives).

c. If a particle's acceleration is positive and constant, what is the shape of the curve in each of the three kinematics graphs?

EXAMPLE 2.6 Launching a Rocket

Launching a rocket into space is complicated. In this example, we will consider a very simplified scenario, ignoring the effects of the Earth's motion, the effects of the atmosphere, and the details of the rocket's design.

A rocket launched from the surface of the Earth at 11.2 km/s travels along a one-dimensional path. With the origin of the coordinate system at the center of the Earth, the rocket's position for the first few minutes after takeoff is approximately given by

$$y(t) = A + Bt - Ct^2 + Dt^3 \tag{2.8}$$

where A, B, C, and D are constants and

$$A = 6.38 \times 10^6\ \text{m} \qquad C = 4.90\ \text{m/s}^2$$
$$B = 1.12 \times 10^4\ \text{m/s} \qquad D = 5.73 \times 10^{-3}\ \text{m/s}^3$$

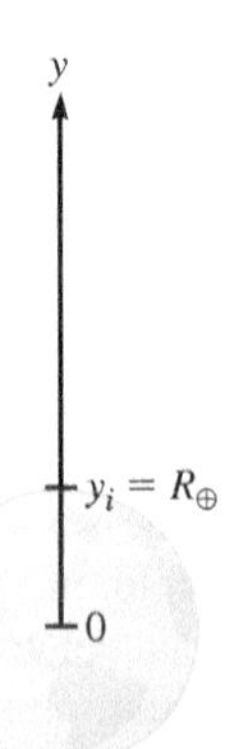

FIGURE 2.25

A Find the initial position y_i of the rocket at $t = 0$ and then find an expression for its displacement $\Delta\vec{y}(t)$.

INTERPRET and ANTICIPATE

To help visualize this problem, start with a simple sketch, representing the Earth with a circle (Fig. 2.25). The origin of the y axis is at the center of the Earth, and positive is upward (outward from the center).

SOLVE

To find the rocket's initial position, substitute $t = 0$ into Equation 2.8.

$$y(t) = A + Bt - Ct^2 + Dt^3$$
$$y(0) = A + (B)(0) - (C)(0)^2 + (D)(0)^3$$
$$y_i = A \quad = 6.38 \times 10^6 \text{ m}$$

CHECK and THINK

The rocket is initially on the surface of the Earth, so $y_i = R_\oplus$, where $R_\oplus$ is the radius of the Earth. As listed in reference material (inside cover, Appendix B, or Internet), $R_\oplus = 6.38 \times 10^6$ m, as we found.

SOLVE

The magnitude of the displacement comes from Equation 2.1 written in terms of the vector component of y.

$$\Delta\vec{y}(t) = \vec{y}(t) - \vec{y}_i \quad (2.1)$$
$$\Delta\vec{y}(t) = [(A + Bt - Ct^2 + Dt^3) - (A)]\hat{\jmath}$$
$$\Delta\vec{y}(t) = (Bt - Ct^2 + Dt^3)\hat{\jmath} \quad (1)$$

CHECK and THINK

Let's check that our expression has the correct dimensions.

First, find the dimensions of the constants from their SI units.

$$[A] = \text{L} \quad [B] = \frac{\text{L}}{\text{T}} \quad [C] = \frac{\text{L}}{\text{T}^2} \quad [D] = \frac{\text{L}}{\text{T}^3}$$

Second, substitute the dimensions for each constant and variable in Equation (1). As expected, the expression for displacement has the dimensions of length.

$$[\Delta y] = \frac{\text{L}}{\text{T}}\text{T} - \frac{\text{L}}{\text{T}^2}\text{T}^2 + \frac{\text{L}}{\text{T}^3}\text{T}^3$$
$$[\Delta y] = \text{L}$$

B Derive expressions for $\vec{v}_y(t)$ and $\vec{a}_y(t)$.

SOLVE

Use the same coordinate system (Fig. 2.25). To find an expression for the velocity, take the time derivative of position (Eq. 2.4 written in terms of y).

$$\vec{v}_y = \frac{d\vec{y}}{dt} \quad (2.4)$$
$$\vec{v}_y = \frac{d(A + Bt - Ct^2 + Dt^3)\hat{\jmath}}{dt}$$
$$\vec{v}_y(t) = (B - 2Ct + 3Dt^2)\hat{\jmath} \quad (2)$$

To find an expression for the acceleration, take the time derivative of velocity (Eq. 2.7).

$$\vec{a}_y = \frac{d\vec{v}_y}{dt} \quad (2.7)$$
$$\vec{a}_y = \frac{d(B - 2Ct + 3Dt^2)\hat{\jmath}}{dt}$$
$$\vec{a}_y(t) = (-2C + 6Dt)\hat{\jmath} \quad (3)$$

CHECK and THINK

Use dimensions of the constants we found in part A and substitute the dimensions for each constant and variable in Equation (2). As expected, the expression for velocity has the dimensions of length per time.

$$[v_y] = \left(\frac{\text{L}}{\text{T}} - \frac{\text{L}}{\text{T}^2}\text{T} + \frac{\text{L}}{\text{T}^3}\text{T}^2\right)$$
$$[v_y] = \frac{\text{L}}{\text{T}}$$

Substitute the dimensions for each constant and variable in Equation (3). As expected, the expression for acceleration has the dimensions of length per time squared.

$$[a_y] = \frac{\text{L}}{\text{T}^2} + \frac{\text{L}}{\text{T}^3}\text{T}$$
$$[a_y] = \frac{\text{L}}{\text{T}^2}$$

CONSTANT ACCELERATION

Special Case

2-9 Special Case: Constant Acceleration

When we want to describe the motion of some particle, there are three quantities—position, velocity, and acceleration—that we would like to know at all times. In the most general situations, acceleration is *not* a constant (such as the rocket in Example 2.6). The resulting motion can be difficult to describe mathematically, often requiring computer assistance. When the *acceleration is constant*, our task is simplified because the acceleration has the same value at all times. We can describe the position and velocity of the particle at a time t in terms of the particle's position and velocity at an earlier time t_0. Let's assume the particle's motion is one-dimensional along the x axis. Typically, we set $t_0 = 0$. Then, there are five (or six) motion variables:

Usually, the initial and final positions are replaced by the displacement $\Delta x = x(t) - x_0$.

1. t is the time elapsed since $t_0 = 0$.
2A. $x(t)$ is the position at t.
2B. x_0 is the position at $t_0 = 0$.
3. $v_x(t)$ is the velocity at t.
4. v_{0x} is the velocity at $t_0 = 0$.
5. a_x is the constant acceleration, with the same value at all times.

Recall that position depends on the choice of coordinate system, and in any situation, there are many good choices for a coordinate system. The motion of the particle, however, cannot depend on our choice of coordinate system, so kinematic equations cannot depend on that choice. It works out that the mathematical description of motion is in terms of the particle's displacement ($\Delta x = x - x_0$), which does not depend on our choice of the coordinate system. So, for constant acceleration, there are really just five variables: t, Δx, v_x, v_{0x}, and a_x.

We can derive five equations corresponding to these five variables that can be used to solve all constant-acceleration problems. Throughout these derivations, we will work with the scalar components of the motion variables along an x axis, so direction is indicated simply by the sign of the scalar component. Similar derivations can be done for any other axis.

Velocity as a Function of Time for Constant Acceleration

From Figure 2.24A, we know that the velocity-versus-time graph is linear in the special case of constant acceleration. The tangent at any point on the line coincides with the line itself. Therefore, in the case of constant acceleration a_x, the average acceleration $a_{av,x}$ over any time interval and the instantaneous acceleration $a_x(t)$ at all particular times are equal:

Here, we use the symbol a_x without (t) to indicate that the acceleration is a constant at all times.

$$a_x(t) = a_x = a_{av,x} = \frac{v_x(t_f) - v_x(t_i)}{t_f - t_i}$$

It is conventional practice to set the initial time to zero and replace t_f with t. The initial velocity $v_x(0)$ is written v_{0x}, and the acceleration is

$$a_x = \frac{v_x(t) - v_{0x}}{t}$$

Solve this equation for $v_x(t)$:

$$v_x(t) = v_{0x} + a_x t \quad (2.9)$$

Equation 2.9 is the first of the five equations and gives velocity as a function of time for constant acceleration. Equation 2.9 is linear, in the form $y = mx + b$, describing either of the lines in Figure 2.24A with a_x equal to the slope (m) and v_{0x} equal to the y intercept (b).

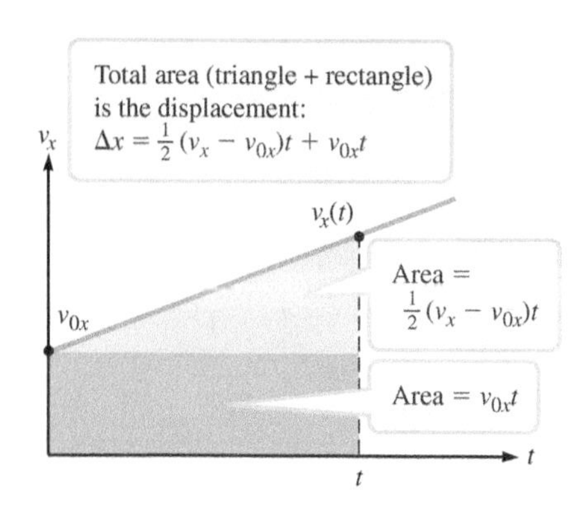

FIGURE 2.26 Velocity-versus-time graph for the special case of constant acceleration. The area under the curve is the displacement in time t.

Displacement as a Function of Velocity and Time

In Figure 2.21, we found the displacement from the area under a velocity-versus-time curve in the case of zero acceleration. When there is a nonzero acceleration, we can still find the displacement as the area under the curve. We can break this area into two shapes, the light blue triangle and the dark blue rectangle (Fig. 2.26). The sum of these two areas is the displacement in the time interval t,

$$\Delta x = \tfrac{1}{2}(v_{0x} + v_x)t \quad (2.10)$$

where we've replaced $v_x(t)$ with v_x to keep the notation compact. This equation is the second of the five.

Before deriving further equations, let's take a closer look at Equation 2.10. Consider the definition of average velocity (Eq. 2.2),

$$v_{\text{av},x} = \frac{\Delta x}{\Delta t} = \frac{\Delta x}{t}$$

if $t_i = 0$ and $t_f = t$. Solving this expression for Δx, we get

$$\Delta x = v_{\text{av},x}\, t$$

Then, substituting this result into $\Delta x = \frac{1}{2}(v_{0x} + v_x)t$ (Eq. 2.10), we find

$$v_{\text{av},x} = \tfrac{1}{2}(v_{0x} + v_x)$$

In words, the average velocity is the arithmetic mean of the initial and final velocities but *only* in the special case of constant acceleration.

The Five Kinematic Equations for Constant Acceleration

Notice that each of the equations we've derived so far only involves four of the five kinematics variables (Δx, t, a_x, v_{0x} and v_x); $v_x(t) = v_{0x} + a_x t$ (Eq. 2.9) does not involve displacement, and $\Delta x = \frac{1}{2}(v_{0x} + v_x)t$ (Eq. 2.10) does not involve acceleration. We can use this information to help us solve problems. For example, if we know acceleration, initial velocity, and final velocity, we can use Equation 2.9 to solve for the time.

If we know acceleration, initial velocity, and time and we need to know the displacement, however, we have to use both Equations 2.9 and 2.10 simultaneously. A more efficient method would be to develop a new equation by eliminating final velocity from the two equations. To do so, substitute Equation 2.9 for v_x in Equation 2.10 and simplify:

$$\Delta x = \tfrac{1}{2}(v_{0x} + v_{0x} + a_x t)t$$

$$\Delta x = v_{0x}t + \tfrac{1}{2}a_x t^2 \qquad (2.11)$$

Equation 2.11 is the third of the five kinematic equations we need for one-dimensional motion with constant acceleration. A similar procedure is used to eliminate either time or initial velocity to obtain

$$\Delta x = v_x t - \tfrac{1}{2}a_x t^2 \qquad (2.12)$$

$$v_x^2 = v_{0x}^2 + 2a_x\Delta x \qquad (2.13)$$

Table 2.4 summarizes these five equations. The derivation of the last two equations—Equation 2.12 in which v_{0x} is missing and Equation 2.13 in which t is missing—is left for homework.

TABLE 2.4 Kinematic equations of motion for constant acceleration.

Equation		Eliminated Variable
$v_x = v_{0x} + a_x t$	(2.9)	Displacement Δx
$\Delta x = \frac{1}{2}(v_{0x} + v_x)t$	(2.10)	Acceleration a_x
$\Delta x = v_{0x}t + \frac{1}{2}a_x t^2$	(2.11)	Final velocity v_x
$\Delta x = v_x t - \frac{1}{2}a_x t^2$	(2.12)	Initial velocity v_{0x}
$v_x^2 = v_{0x}^2 + 2a_x\Delta x$	(2.13)	Final time t

Note: The equations are written in terms of scalar components of the motion variables along the x axis. Similar equations can also be written in the y and z directions.

Using Integral Calculus

Calculus is *the* mathematical tool most widely used by physicists today. Probably one of the first uses of calculus was in the study of motion. You can gain insight into Equation 2.9 by rederiving it using integral calculus. If you have not learned integral calculus, proceed to the **Problem-Solving Strategy: Constant Acceleration** and return to this derivation when you are ready.

(ALTERNATIVE) DERIVATION Equation 2.9

Use integral calculus to derive Equation 2.9, $v_x(t) = v_{0x} + a_x t$ in the special case of constant acceleration.

Start with Equation 2.7 for acceleration in the x direction, then rewrite as a differential equation.	$a_x = \dfrac{dv_x}{dt}$ (2.7) $a_x\,dt = dv_x$

Derivation continues on page 42 ▶

Integrate both sides from an initial time t_i to a final time t_f.	$\int_{t_i}^{t_f} a_x dt = \int_{v_{ix}}^{v_{fx}} dv_x$	(1)
Rewrite using the usual conventions of setting $t_i = 0$, $t_f = t$, $v_{ix} = v_{0x}$, and $v_{fx} = v_x$. Because a_x is constant, it can be pulled outside the integral. (That is why constant acceleration makes the mathematics simpler.)	$a_x \int_0^t dt = \int_{v_{0x}}^{v_x} dv_x$	
Carry out the integration and evaluate over the limits of integration.	$a_x t \Big\vert_0^t = v_x \Big\vert_{v_{0x}}^{v_x}$ $a_x t = v_x - v_{0x}$	
Solve for the final velocity.	$v_x = v_{0x} + a_x t$ ✓	(2.9)

COMMENTS

In many situations, the acceleration is not constant. According to Equation (1), however, you can always integrate $a(t)$ over time to find the speed. Sometimes, it is difficult to do by hand, and a computer may be needed.

 **PROBLEM-SOLVING STRATEGY**

Constant Acceleration

As part of the **INTERPRET and ANTICIPATE** procedure, draw a **sketch** that includes a **coordinate system.**

There are four parts to the **SOLVE** procedure:

Step 1 Start by **listing** the six kinematic variables (initial position, final position, initial velocity, final velocity, acceleration, and time). Often, the initial and final positions are combined as the displacement, so your list typically includes five variables. Be sure to include the sign of each value. Write "need" next to any variable you need to solve for and write "not needed" next to any variable that you don't know and don't need to find.

Step 2 Once you have listed the parameters, you may use the constant acceleration equations **(Table 2.4)**. You can avoid doing algebra by **choosing the equation** that does not include the "not needed" variable.

Step 3 Do **algebra** before substituting values.

Step 4 Substitute **values** if appropriate.

EXAMPLE 2.7 Breaking the Sound Speed Barrier

On October 15, 1997, driver Andy Green drove a vehicle dubbed the ThrustSuperSonicCar (ThrustSSC) faster than the speed of sound, officially breaking the sound speed barrier in a land vehicle. The course he drove is 13 miles long. The first 6 miles are used to get the car up to speed. The next mile is timed, and Green's speed over the timed mile broke the record. The last 6 miles are used to slow the car back down. According to the rules, Green had to make two runs within 1 hour. His speed averaged over the two runs of the timed mile was reported as 763.035 mph (1233.704 km/h).

Assume the acceleration was constant over the first and last 6 miles and the velocity was constant over the timed mile. Report your answers to a conservative three significant figures (rather than seven!).

A Find the acceleration for the first 6 miles.

INTERPRET and ANTICIPATE

Make a **simple sketch including a coordinate system**. This sketch applies to all parts of the problem (Fig. 2.27). Because the car is speeding up for the first 6 miles, we expect the acceleration to point in the same direction as the velocity. So, we expect a numerical expression for the acceleration in the form $\vec{a}_x = +\_\_\_\_\_ \hat{\imath}\ \text{m/s}^2$.

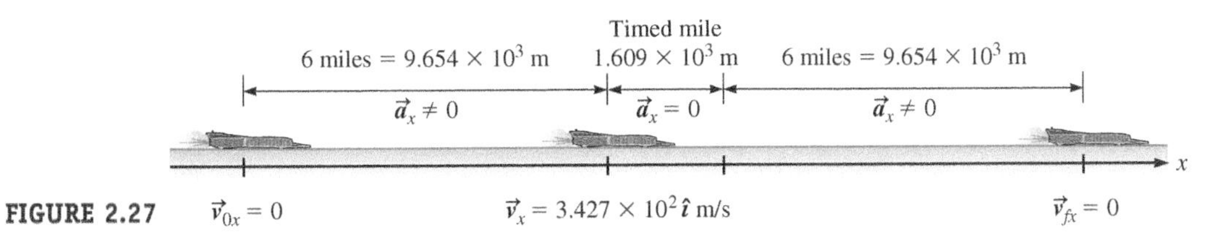

FIGURE 2.27

SOLVE

Convert any values not given in SI units to SI units. It is helpful to write these results on your sketch (Fig. 2.27).	$6\,\text{mi}\left(\frac{1.609 \times 10^3\,\text{m}}{1\,\text{mi}}\right) = 9.654 \times 10^3\,\text{m}$ $1233.704\,\text{km/h}\left(\frac{0.2778\,\text{m/s}}{1\,\text{km/h}}\right) = 3.427 \times 10^2\,\text{m/s}$
Step 1 **List** the known and unknown variables. Be sure to account for all five.	$\Delta\vec{x} = 9.654 \times 10^3 \hat{\imath}\,\text{m}$ $\vec{v}_{0x} = 0$ $\vec{v}_x = 3.427 \times 10^2 \hat{\imath}\,\text{m/s}$ $\vec{a}_x$ needed t not needed
Step 2 Because time is unknown and not needed (at this point), it is best to **choose the equation** in **Table 2.4** that does not include time.	$v_x^2 = v_{0x}^2 + 2a_x\Delta x$ (2.13)
Step 3 Do **algebra.** Solve for a_x.	$a_x = \frac{v_x^2 - v_{0x}^2}{2\Delta x}$
Step 4 Substitute the given **values.**	$a_x = \frac{(3.427 \times 10^2\,\text{m/s})^2 - 0^2}{2(9.654 \times 10^3\,\text{m})} = 6.08\,\text{m/s}^2$ $\vec{a}_x = 6.08\hat{\imath}\,\text{m/s}^2$ Our result has the form we expected.

B How long did the ThrustSSC take to cover the timed mile?

SOLVE

Step 1 Again, **list** the known and unknown variables in SI units. Because we assume the velocity was constant for this part, $a_x = 0$ and $v_{0x} = v_x$. We therefore know four of the five kinematic variables.	$\Delta\vec{x} = 1\hat{\imath}\,\text{mi} = 1.609 \times 10^3 \hat{\imath}\,\text{m}$ $\vec{v}_{0x} = \vec{v}_x = 3.427 \times 10^2 \hat{\imath}\,\text{m/s}$ $\vec{a}_x = 0$ t needed
Steps 2 and 3 Because the acceleration is zero, the equations in **Table 2.4** are greatly simplified. Equations 2.9 and 2.13 become $v_{0x} = v_x$, and the other three reduce to $\Delta x = v_x t$. Solve for time.	$\Delta x = v_x t$ $t = \frac{\Delta x}{v_x}$
Step 4 Substitute the **values.**	$t = \frac{1.609 \times 10^3\,\text{m}}{3.427 \times 10^2\,\text{m/s}} = 4.70\,\text{s}$

Example continues on page 44 ▶

CHECK and THINK

To check this result, compare it to the time needed for a car traveling at highway speed to cover 1 mile. Normal highway speed is around 60 mph. A car at 60 mph covers 1 mile in 1 minute, taking 12 to 13 times longer than the ThrustSSC took to cover the same distance. This difference makes sense because the ThrustSSC's record speed was about 12 to 13 times faster than that of a typical car on a highway.

C What was the acceleration for the last 6 miles?

SOLVE	
The magnitude of the acceleration has to be the same as for the first 6 miles, but it must point in the negative x direction.	$\vec{a}_x = -6.08\hat{\imath}\ \text{m/s}^2$

CHECK and THINK

The ThrustSSC's velocity is always in the positive x direction, but its acceleration switches direction. When the car is speeding up over the first 6 miles, its acceleration is in the same direction as its velocity. Over the last 6 miles, however, the car is slowing down, so its acceleration is opposite in direction to its velocity.

EXAMPLE 2.8 High-Speed Electrons

Because so much modern technology depends on electricity, moving electrons play an important role in our everyday lives. One way to accelerate an electron is with oppositely charged plates known as electrodes, where one plate carries a positive charge and the other carries a negative charge (Fig. 2.28). We will model the electron as a particle moving with constant acceleration.

An electron initially at rest is released from the negative electrode. The electrodes are 7.5 mm apart. The electron accelerates at $1.5 \times 10^{17}\ \text{m/s}^2$ toward the positive electrode.

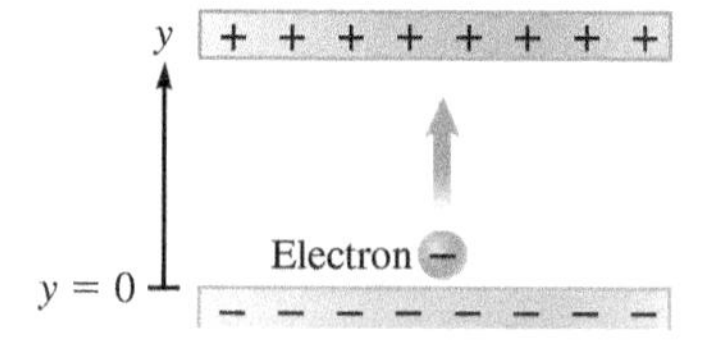

FIGURE 2.28

A Find the velocity of the electron as it strikes the positive electrode.

INTERPRET and ANTICIPATE

A sketch was provided. Because we need to find a vector quantity (velocity), it is important to choose a coordinate system before doing anything else. One good choice in this case is a vertical y axis with its origin at the negative electrode as shown in Figure 2.28. Because the acceleration is constant, we may apply the constant-acceleration equations in Table 2.4. We expect a numerical result in the form $\vec{v}_y = +\_\_\_\_\hat{\jmath}\ \text{m/s}$.

SOLVE		
Step 1 As in the previous example, **list** the known and unknown variables. Convert them to SI units.	$\Delta\vec{y} = 7.5 \times 10^{-3}\hat{\jmath}\ \text{m}$ $\vec{v}_{0y} = 0$ $\vec{v}_y$ needed	$\vec{a}_y = 1.5 \times 10^{17}\hat{\jmath}\ \text{m/s}^2$ t not needed
Step 2 **Choose an equation.** Equation 2.13 does not involve time and therefore is the best choice to use here. Replace x with y because we chose to describe the motion using a y axis.	$v_y^2 = v_{0y}^2 + 2a_y\Delta y$	
Step 3 Do **algebra.** Solve for velocity.	$v_y = \pm(\sqrt{2a_y\Delta y})$	
Step 4 Substitute the **values** and choose the positive root because the electron moves in the positive direction of our chosen axis.	$v_y = \sqrt{2(1.5 \times 10^{17}\ \text{m/s}^2)(7.5 \times 10^{-3}\ \text{m})}$ $\vec{v}_y = 4.7 \times 10^7\hat{\jmath}\ \text{m/s}$	

CHECK and THINK
The answer is in the form we predicted, but the electron's final speed is five orders of magnitude greater than the ThrustSSC's highest recorded speed, which seems very fast. Consider the electron's very great acceleration, 1.5×10^{17} m/s^2 (typical in the cathode-ray tube of an older television). If the electron accelerated at this rate for just 1 second, its speed would be 1.5×10^{17} m/s. That is nine orders of magnitude greater than the speed of light in a vacuum (3×10^8 m/s). According to Einstein, nothing can exceed the speed of light. So, as we will find in part B, the time for an electron to get to the positive electrode must be much shorter than 1 second.

B Find the time the electron takes to go from the negative to the positive electrode.

INTERPRET and ANTICIPATE
As we concluded in part A, we expect to find $t \ll 1$ s.

SOLVE	
Step 1 As always, **list** the known and unknown kinematic variables in SI units. Now we know four of the five variables.	$\Delta\vec{y} = 7.5 \times 10^{-3}\hat{\jmath}$ m $\vec{a}_y = 1.5 \times 10^{17}\hat{\jmath}$ m/s^2 $\vec{v}_{0y} = 0$ t needed $\vec{v}_y = 4.7 \times 10^7\hat{\jmath}$ m/s
Step 2 Any equation in **Table 2.4** that involves time will work. We arbitrarily **choose** Equation 2.9.	$v_y = v_{0y} + a_y t$
Step 3 Do **algebra.** Solve for t.	$t = \dfrac{v_y - v_{0y}}{a_y}$
Step 4 Substitute the **values.**	$t = \dfrac{(4.7 \times 10^7 \text{ m/s}) - 0}{1.5 \times 10^{17} \text{ m/s}^2} = 3.2 \times 10^{-10}$ s $t = 0.32$ ns

CHECK and THINK
As expected, the electron was accelerated for a very short time.

2-10 A Special Case of Constant Acceleration: Free Fall

After the rocket engines in Example 2.6 shut off, the rocket slows down because of the Earth's gravity. When an object accelerates solely under the influence of gravity, it is said to be in **free fall**. The term *free fall* seems to imply that the object must be moving downward, but that is not the case. The rocket, for example, still moves upward after engine shutoff, and the term *free fall* still applies. As long as the object is affected only by gravity, free fall describes its motion after it has been thrown, launched, or dropped. The "fall" refers to the direction of the acceleration. Because this acceleration is due to gravity, it is directed downward toward the center of the Earth.

FREE FALL Special Case

Free fall is an idealization. In real situations, such factors as a planet's atmosphere affect the motion of the object. Often, these effects are much weaker than the effect of gravity, and in these cases, free fall is a good approximation.

When an object rises far above the Earth's surface, its acceleration may drop appreciably. For objects that stay close to the Earth's surface, however, acceleration is nearly constant in free fall, and the magnitude is given by

$$g = 9.81 \text{ m/s}^2$$

Do **not** refer to "g" as "gravity."

The name for "g" is somewhat controversial (more about that in Chapters 5 and 7). For now, we will use the term *free-fall acceleration*. Although we will assume g is given by the constant above, it is really an approximation. The free-fall acceleration at the Earth's surface varies slightly with altitude and latitude. For example, at the poles, $g \approx 9.83\ \text{m/s}^2$, and at the top of Mount Everest, $g \approx 9.77\ \text{m/s}^2$.

PROBLEM-SOLVING STRATEGY

Free Fall

Free fall near the surface of the Earth is a special case of constant acceleration, and the **Problem-Solving Strategy: *Constant Acceleration*** (p. 42) is applicable. When listing the kinematic variables in step 1, you know that the acceleration is $\pm g$, depending on your choice of coordinate system. (It is often convenient to choose an upward-pointing y axis, and for this choice, $\vec{a}_y = -g\hat{\jmath}$.)

EXAMPLE 2.9 Welcome to the Model Rocket Club

A model rocket is launched straight up at 11.2 m/s. Assume the engine fires very briefly and then burns out. What peak height does the rocket reach?

INTERPRET and ANTICIPATE

This example is similar to Example 2.6, so we can use the same sketch, and by using the same upward-pointing y axis, we can easily compare the two problems. The major difference between them is that the model rocket's initial speed is three orders of magnitude lower than the full-size rocket's initial speed, so we expect the model rocket will not go nearly as high.

SOLVE Step 1 **List** the known and unknown kinematic variables. At the peak height, the model rocket's velocity is momentarily zero.	$\Delta\vec{y} = \text{needed}$ $\vec{a}_y = -g\hat{\jmath} = -9.81\hat{\jmath}\ \text{m/s}^2$ $\vec{v}_{0y} = 11.2\hat{\jmath}\ \text{m/s}$ t not needed $\vec{v}_y = 0$
Step 2 We **choose an equation** that does not involve time (Eq. 2.13).	$v_y^2 = v_{0y}^2 + 2a_y\Delta y$ (2.13)
Step 3 Do **algebra.** Solve for displacement.	$\Delta y = \dfrac{v_y^2 - v_{0y}^2}{2a_y}$
Step 4 Substitute **values.**	$\Delta y = \dfrac{0^2 - (11.2\ \text{m/s})^2}{2(-9.81\ \text{m/s}^2)}$ $\Delta\vec{y} = 6.39\hat{\jmath}\ \text{m}$

CHECK and THINK

As expected, the model rocket stays near the surface of the Earth. Its displacement is about the height of a two-story building. (Often when checking an answer, it is helpful to consult Appendix B's tables of approximate values.) Think about how this model rocket example differs from the full-size rocket in Example 2.6. The model rocket stays near the surface of the Earth. Its acceleration is essentially constant for its entire flight. Over the course of the full-size rocket's flight, its acceleration decreases according to $\vec{a}_y(t) = (-2C + 6Dt)\hat{\jmath}$. We cannot apply the constant acceleration equations to the full-size rocket, but we can apply them to the model rocket.

EXAMPLE 2.10 Hard Hats Required!

A construction worker is riding an exterior elevator to the top of a skyscraper. The elevator's speed is 4.3 m/s. When the elevator is 72.4 m above the ground, a screwdriver falls off.

A What is the velocity of the screwdriver as it hits the ground?

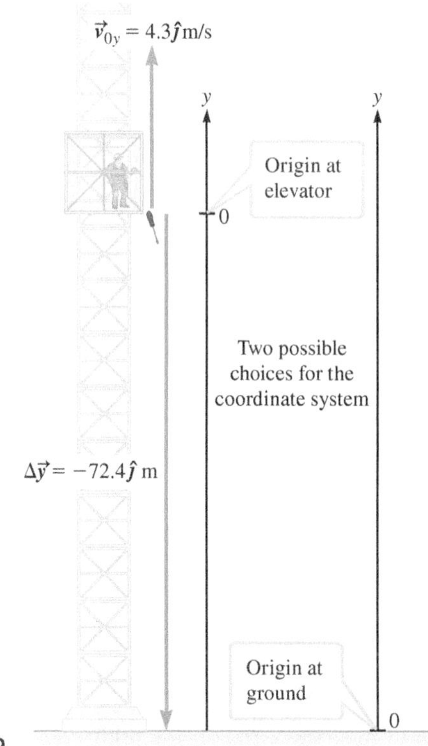

FIGURE 2.29

INTERPRET and ANTICIPATE

If we **choose an upward-pointing *y* axis** as in Example 2.9, we expect the velocity of the screwdriver to be in the negative *y* direction when it hits the ground. Your choice for the origin does not affect your results. Figure 2.29 shows two possible choices.

SOLVE

Step 1 **List** the known and unknown variables. The screwdriver's initial velocity is the same as the velocity of the elevator, so the screwdriver will actually continue *upward* for a short time. Notice that the displacement does not depend on which origin you choose.	$\Delta\vec{y} = -72.4\hat{\jmath}$ m $\vec{v}_{0y} = +4.3\hat{\jmath}$ m/s $\vec{v}_y$ needed $\vec{a}_y = -g\hat{\jmath} = -9.81\hat{\jmath}$ m/s^2 t not needed
Step 2 Of the equations in **Table 2.4**, Equation 2.13 is the one that contains no time term.	$v_y^2 = v_{0y}^2 + 2a_y\Delta y$ (2.13)
Step 3 Do **algebra.** Solve for v_y. Choose the negative root because as the screwdriver hits the ground, its velocity is downward (in the negative *y* direction in the coordinate system we have chosen).	$v_y = \pm\sqrt{v_{0y}^2 + 2a_y\Delta y}$
Step 4: Substitute the **values.**	$\vec{v}_y = -\sqrt{(4.3 \text{ m/s})^2 + 2(-9.81 \text{ m/s}^2)(-72.4 \text{ m})}\hat{\jmath}$ $\vec{v}_y = -37.9\hat{\jmath}$ m/s

CHECK and THINK

The velocity is about 85 mph, or a little faster than typical highway speed.

B How long does it take the screwdriver to reach the ground?

SOLVE

Step 1 As in previous examples, we now know four of the five kinematic variables. We need time. Any of Equations 2.9 through 2.13 will do; we **choose** Equation 2.9.	$v_y = v_{0y} + a_y t$
Step 2 Do **algebra** Solve for time.	$t = \dfrac{v_y - v_{0y}}{a_y}$
Step 3 Substitute the **values.**	$t = \dfrac{-37.9 \text{ m/s} - 4.3 \text{ m/s}}{-9.81 \text{ m/s}^2} = 4.3$ s

Example continues on page 48 ▶

CHECK and THINK

The model rocket (Example 2.9) and this falling screwdriver problem are very similar. Both required a solution for time after finding some other kinematic variable. In this example, we solved first for v_y and then for t. What if we tried to solve for t first? Our list of known variables would be the same, but we would label t "needed" and v_y "not needed." In that case, choose Equation 2.11, $\Delta y = v_{0y}t + \frac{1}{2}a_y t^2$. Solving for t, we find

$$t = \frac{-v_{0y} \pm \sqrt{v_{0y}^2 + 2a_y\Delta y}}{a_y}$$

We are left with a decision to make. How do we choose between the positive and negative roots of the square-root term? To answer this question, we notice that the expression under the square-root sign is the right side of Equation 2.13, so we can write

$$t = \frac{-v_{0y} \pm \sqrt{v_{0y}^2 + 2a_y\Delta y}}{a_y} = \frac{-v_{0y} \pm \sqrt{v_y^2}}{a_y}$$

As in part A, the direction of $\vec{v}_y$ determines whether we choose the positive root or the negative one. In this example, the correct choice is the negative root, indicating motion in the downward direction.

EXAMPLE 2.11 Movie Stunt

In many movies, a hero jumps off a structure and lands on a moving vehicle. The hero must, of course, watch the moving vehicle and figure out just the right moment to jump.

Superwoman stands on a bridge as a flatbed truck drives toward the bridge at a constant speed of 18 m/s. She steps off the bridge and falls straight down, landing 10.0 m below in the bed of the truck. How far from the bridge was the truck when she stepped off the bridge?

INTERPRET and ANTICIPATE

Make a **simple sketch** that includes two **coordinate axes,** one for Superwoman and one for the truck (Fig. 2.30). The key to solving this problem is to realize that the time it takes Superwoman to fall 10.0 m must be equal to the time it takes the truck to get to the bridge. Once you know that time, use the truck's speed to find the truck's displacement during that time.

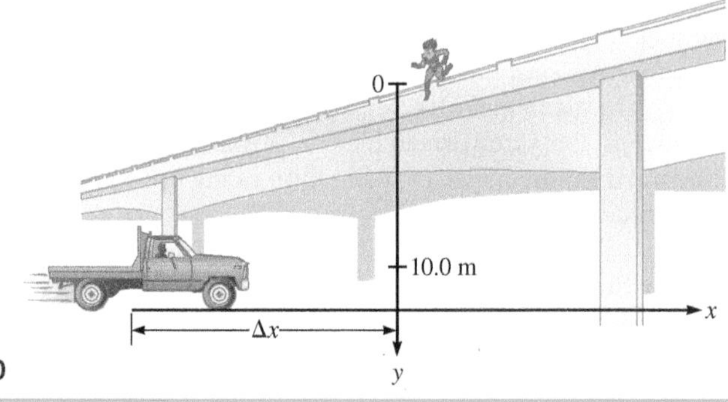

FIGURE 2.30

SOLVE

Step 1 **List** the known and unknown variables for Superwoman. We have chosen a downward-pointing y axis and for this choice, $\vec{a}_y = g\hat{\jmath}$.	*Superwoman:* $\Delta\vec{y} = 10.0\hat{\jmath}$ m $\vec{a}_y = g\hat{\jmath} = 9.81\hat{\jmath}$ m/s^2 $\vec{v}_{0y} = 0$ t must be same as for truck $\vec{v}_y$ not needed
List the known and unknown variables for the truck.	*Truck:* $\Delta\vec{x}$ = needed $\vec{a}_x = 0$ $\vec{v}_{0x} = \vec{v}_x = 18\hat{\imath}$ m/s (constant) t, same as for Superwoman
Step 2 We **choose an equation** that can be used to find the time it takes Superwoman to fall the 10.0 m. Because we do not know and do not need her final velocity v_y, our best choice is Equation 2.11.	$\Delta y = v_{0y}t + \frac{1}{2}a_y t^2 = 0 + \frac{1}{2}gt^2$ $t = \sqrt{\frac{2\Delta y}{g}}$ (1)
Next, we select an equation for Δx for the truck, assuming t is known from Equation (1). Again our **choice** is Equation 2.11.	$\Delta x = v_{0x}t + \frac{1}{2}a_x t^2 = v_{0x}t + 0$ $\Delta x = v_{0x}t$

Steps 3 and 4: Do **algebra**. Substitute t from Equation (1) and the given numerical **values** from the problem statement.

$$\Delta x = v_{0x}\sqrt{\frac{2\Delta y}{g}} = (18\ \text{m/s})\sqrt{\frac{2(10.0\ \text{m})}{9.81\ \text{m/s}^2}}$$

$$\Delta \vec{x} = 26\hat{\imath}\ \text{m}$$

CHECK and THINK

The truck was 26 m or nearly 90 ft from the bridge, or about the distance between adjacent bases in a baseball diamond, when Superwoman stepped off the bridge.

Graphical Solutions

We have seen that graphs of position, velocity, or acceleration versus time may be used in the **INTERPRET and ANTICIPATE** procedure and then in the **CHECK and THINK** procedure. Sometimes, these graphs may be used in the **SOLVE** procedure as shown the next example.

EXAMPLE 2.12 Circus Rehearsal

Two circus performers practice one segment of their act. Horatio stands on a platform 8.0 m above the ground and drops a ball straight down. At the same moment, Amelia uses a spring-loaded device on the ground to launch a dart straight up toward the ball. The dart is launched at 11.5 m/s. How far above the ground and how long after launch does the dart hit the ball? Solve this problem graphically.

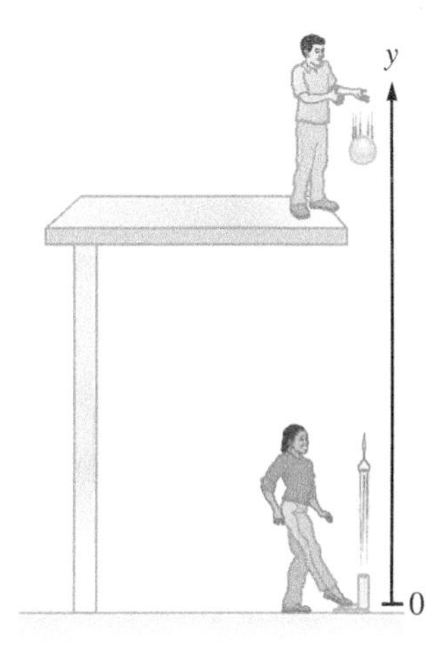

FIGURE 2.31

INTERPRET and ANTICIPATE

Make a **sketch** and **choose** an upward-pointing y axis with its origin at ground level (Fig. 2.31). To solve this problem graphically, make position-versus-time graphs for the ball and the dart on the same set of axes. To do so, we need position as a function of time for the ball and for the dart. We expect the dart to hit the ball at some point between the ground and the platform.

SOLVE

List the known and unknown variables for the ball. Because we need position and not displacement, we must list initial and final positions separately. The subscript "B" stands for ball.

Ball:

y_{Bf} need equation	v_{Bfy} not needed
$y_{Bi} = 8.0$ m	$\vec{a}_y = -g\hat{\jmath} = -9.81\hat{\jmath}\ \text{m/s}^2$
$v_{Biy} = 0$	t must be part of equation

Because the final velocity is not needed, use Equation 2.11.

$$\Delta y_B = v_{Biy} + \tfrac{1}{2}a_y t^2 \tag{2.11}$$

Use the definition of displacement to convert this equation to one for final position of the ball.

$$\Delta y_B = y_{Bf} - y_{Bi}$$
$$y_{Bf} = y_{Bi} + v_{Biy}t + \tfrac{1}{2}a_y t^2$$

Substitute known values.

$$y_{Bf} = \left(8.0 - \frac{9.81}{2}t^2\right)\ \text{m} \tag{1}$$

Use the same y axis and follow the same procedure for the dart. The subscript "D" stands for dart.

Dart:

y_{Df} need equation	v_{Dfy} unknown
$y_{Di} = 0$	$\vec{a}_y = -g\hat{\jmath} = -9.81\hat{\jmath}\ \text{m/s}^2$
$v_{Diy} = 11.5$ m/s	t must be part of equation

As in the case of the ball, the dart's position comes from Equation 2.11.

$$y_{Df} = y_{Di} + v_{Diy}t + \tfrac{1}{2}a_y t^2$$

$$y_{Df} = \left(11.5t - \frac{9.81}{2}t^2\right)\ \text{m} \tag{2}$$

Example continues on page 50 ▶

Plot Equations (1) and (2), the position as a function of time for each object, on the same set of axes (Fig. 2.32). A spreadsheet program is helpful.

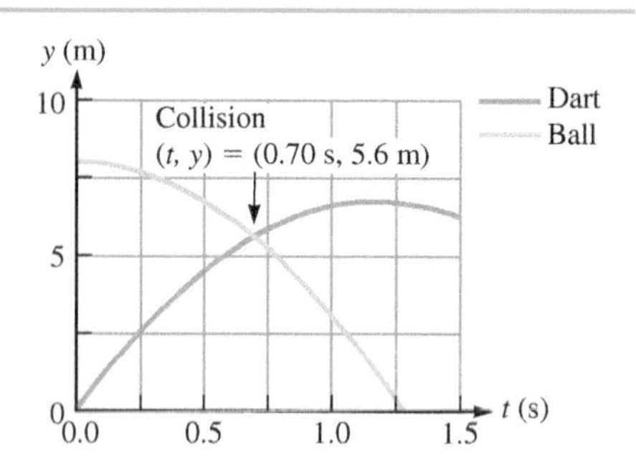

FIGURE 2.32

A close inspection reveals that these curves intersect at $(t, y_f) = (0.70 \text{ s}, 5.6 \text{ m})$.

The dart hits the ball 5.6 m above the ground 0.70 s after the dart is launched.

CHECK and THINK

As expected, the point of intersection $y = 5.6$ m is between the ground $(y = 0)$ and the platform $(y = 8.0 \text{ m})$. We solved this problem graphically, but we could have solved Equations (1) and (2) simultaneously (Problem 88).

SUMMARY

Underlying Principles

1. **Kinematics:** description of motion independent of its cause.
2. **One-dimensional motion:** motion along a single straight line.

Major Concepts

1. **Translational motion:** an object can be treated as a particle undergoing purely translational motion if every point on the object undergoes exactly the same displacement as every other such point.
2. **Particle:** an idealized point with no spatial extent, no shape, and no internal structure.
3. **Position** $\vec{r}$ of a particle: its location with respect to a chosen coordinate system. The direction of position in one dimension is indicated by a positive or a negative sign.
4. **Displacement:** the change in position $(\Delta\vec{r})$. The vector component of displacement for each coordinate axis is

$$\Delta\vec{x} \equiv (x_f - x_i)\hat{\imath} \quad \Delta\vec{y} \equiv (y_f - y_i)\hat{\jmath} \quad \Delta\vec{z} \equiv (z_f - z_i)\hat{k} \tag{2.1}$$

5. **Distance traveled** (d): the length of the whole path covered by a moving particle; a scalar quantity.

6. **Average velocity:** the ratio of displacement to change in time.

$$\vec{v}_{\text{av},x} \equiv \frac{\Delta\vec{x}}{\Delta t} = \frac{x_f - x_i}{t_f - t_i}\hat{\imath} \quad \vec{v}_{\text{av},y} \equiv \frac{\Delta\vec{y}}{\Delta t} = \frac{y_f - y_i}{t_f - t_i}\hat{\jmath} \quad \vec{v}_{\text{av},z} \equiv \frac{\Delta\vec{z}}{\Delta t} = \frac{z_f - z_i}{t_f - t_i}\hat{k} \tag{2.2}$$

Instantaneous velocity (also known as **velocity**): the average velocity as Δt approaches zero.

$$\vec{v}_x = \lim_{\Delta t \to 0} \frac{\Delta\vec{x}}{\Delta t} = \frac{d\vec{x}}{dt} = \frac{dx}{dt}\hat{\imath} \quad \vec{v}_y = \lim_{\Delta t \to 0} \frac{\Delta\vec{y}}{\Delta t} = \frac{dy}{dt}\hat{\jmath} \quad \vec{v}_z = \lim_{\Delta t \to 0} \frac{\Delta\vec{z}}{\Delta t} = \frac{dz}{dt}\hat{k} \tag{2.4}$$

7. **Average speed:** depends on total distance traveled such that

$$S_{\text{av}} \equiv \frac{d}{\Delta t} \tag{2.3}$$

Instantaneous speed (or simply the **speed**): magnitude of the velocity.

Major Concepts—cont'd

8. **Average acceleration:** change in a particle's velocity over some time interval.

$$\vec{a}_{\text{av},x} \equiv \frac{\Delta \vec{v}_x}{\Delta t} = \frac{\Delta v_x}{\Delta t}\hat{\imath} = \frac{v_{xf} - v_{xi}}{t_f - t_i}\hat{\imath}$$
$$\vec{a}_{\text{av},y} \equiv \frac{\Delta v_y}{\Delta t}\hat{\jmath} \quad \vec{a}_{\text{av},z} \equiv \frac{\Delta v_z}{\Delta t}\hat{k} \tag{2.6}$$

Instantaneous acceleration (also known as simply **acceleration**): the average acceleration as Δt approaches zero.

$$\vec{a}_x \equiv \lim_{\Delta t \to 0} \frac{\Delta \vec{v}_x}{\Delta t} = \frac{d\vec{v}_x}{dt} = \frac{dv_x}{dt}\hat{\imath}$$
$$\vec{a}_y \equiv \frac{dv_y}{dt}\hat{\jmath} \quad \vec{a}_z \equiv \frac{dv_z}{dt}\hat{k} \tag{2.7}$$

Special Cases

1. In the special case of **constant acceleration** the acceleration has the same magnitude and direction at all times, greatly simplifying the mathematical analysis.
2. An object accelerating solely under the influence of gravity is said to be in **free fall**. An object in free fall near the surface of the Earth has acceleration of magnitude $g = 9.81\ \text{m/s}^2$.

Tools

1. **Motion diagrams** show the location of a particle at regular time intervals.
2. **Vector** quantities have both magnitude and direction. In one-dimensional motion, a vector can have one of two directions. A **scalar** quantity has a magnitude but not a direction.
3. **Coordinate systems** have two elements in one dimension, (1) a reference point called the **origin** indicated by a zero and (2) a **coordinate axis** (or just **axis**) passing through the origin and usually labeled x, y, or z.
4. A **position-versus-time graph** is made by plotting position on the vertical axis and time on the horizontal axis. **Velocity-versus-time** and **acceleration-versus-time graphs** are similar to position-versus-time graphs except that velocity or acceleration instead of position is plotted on the vertical axis.

GENERAL PROBLEM-SOLVING STRATEGIES

Three visualization tools may help in the **INTERPRET and ANTICIPATE** procedure:

1. Motion diagram
2. Sketch with a coordinate system
3. Position-, velocity-, or acceleration-versus-time graphs

Not all problems involve constant acceleration. When the acceleration changes, you must use calculus (not Table 2.4) to solve the problem.

PROBLEM-SOLVING STRATEGY

SPECIAL CASES: CONSTANT ACCELERATION AND FREE FALL

As part of the **INTERPRET and ANTICIPATE** procedure: Draw a **sketch** that includes a **coordinate system**.

There are four parts to the **SOLVE** procedure:

1. Start by **listing** the six kinematic variables (initial position, final position, initial velocity, final velocity, acceleration, and time). Often, the initial and final positions are combined as the displacement, so your list typically includes five variables. Include the sign of each value. Write "need" next to any variable you need to solve for and write "not needed" next to any variable that you don't know and don't need to find. If the problem involves free fall, remember that the acceleration has a magnitude of $g = 9.81\ \text{m/s}^2$ and points downward toward the center of the Earth.
2. Once you have listed the parameters, you may use the constant acceleration equations (**Table 2.4**). You can avoid doing algebra by choosing the equation that does not include the "not needed" variable.
3. Do **algebra** before substituting values.
4. Substitute **values** if appropriate.

PROBLEMS AND QUESTIONS

A = algebraic C = conceptual E = estimation G = graphical N = numerical

2-1 What Is One-Dimensional Translational Kinematics?

1. C Is the Moon's motion around the Earth one-dimensional? Explain your answer.

2-2 Motion Diagrams

Problems 2 and 4 are paired.

2. C An animal's tracks are frozen in the snow (Fig. P2.2). Can these tracks be used to make a motion diagram? If so, what are the shortcomings of a motion diagram made from these data? If not, why not?

Christopher P. Grant/Shutterstock.com

FIGURE P2.2 Problems 2 and 4.

2-3 Coordinate Systems and Position

Problems 3 and 12 are paired.

3. G A particle moves from position A to position C as shown in Figure P2.3. Two different coordinate systems have been chosen. Write expressions for the particle at all three positions using **a.** the top coordinate system, and **b.** the bottom coordinate system.

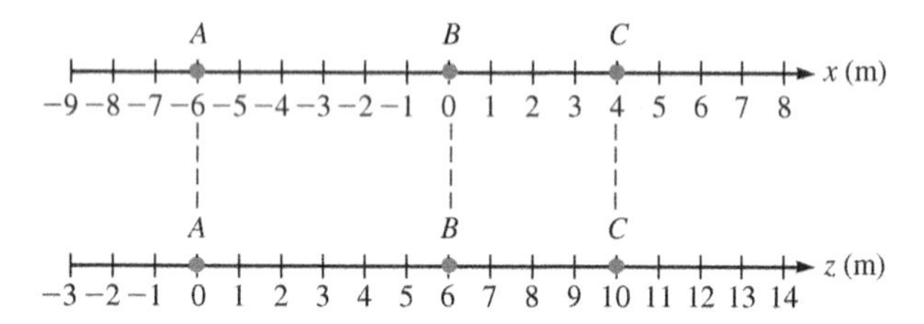

FIGURE P2.3 Problems 3 and 12.

2-4 Position-Versus-Time Graphs

4. C An animal's tracks are frozen in the snow (see Fig. P2.2). Can these tracks be used to make a position-versus-time graph? Explain.
5. N For each of the following velocity vectors, give the vector component, the scalar component, and the magnitude.
   a. $\vec{v} = 35.0\hat{\jmath}$ m/s **c.** $v_z = -3.50$ m/s
   b. $v_x = 53.0$ m/s **d.** $\vec{v} = -5.30\hat{\imath}$ m/s
6. In the traditional Hansel and Gretel fable, the children drop crumbs of bread on the ground to mark their path through the woods. Unfortunately, the crumbs are eaten by birds, and the children cannot find their way home. In this modern-day problem, the children use a device that releases a drop of food dye once per minute. As long as it does not rain, they can find their way home. As an extra bonus, they make a motion diagram as shown in Figure P2.6.
   a. C Describe the motion of the children in words.
   b. G Using the coordinate system in Figure P2.6, make a position-versus-time graph. Note any ambiguities you encounter.
   c. C Is your position-versus-time graph consistent with your description in part (a)? Explain.
7. G After a long and grueling race, two cadets, A and B, are coming into the finish line at the Marine Corps marathon. They move in the same direction along a straight path; the position-versus-time graphs for the runners are shown in Figure P2.7 for a minute near the end of the race. **a.** At time t_1 is the speed of cadet B greater than, less than, or equal to the speed of cadet A? **b.** At time t_2, is cadet B speeding up, slowing down, or moving with constant speed? **c.** From time $t = 0$ to time $t = 60$ s, is the average speed of cadet B greater than, less than, or equal to the average speed of cadet A? Explain.

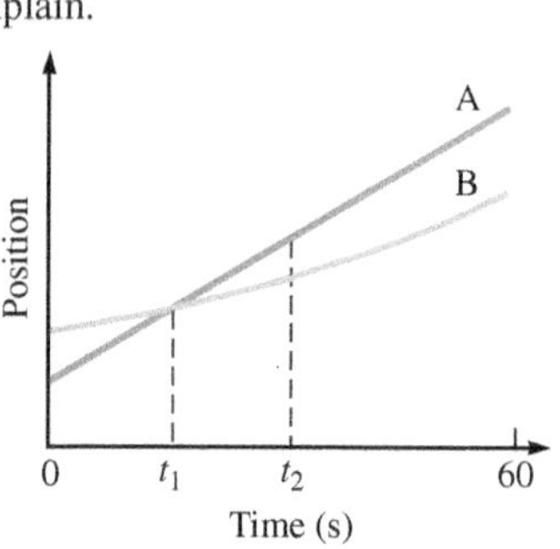

FIGURE P2.7

8. G Cassidi Reese (who graduated with a degree in physics) enjoys skydiving in her free time. She volunteered to wear a device during a jump that automatically measures her altitude as a function of time. Assume she maintains a nearly straight path from the plane to her target on the ground during a jump. (We will learn in Chapter 4 that in the absence of air resistance an objected dropped from a plane does not fall along a straight path.) Her data are presented in a position-versus-time graph (Fig. P2.8.) The device automatically stops taking data at an altitude of approximately 500 m. **a.** Draw a sketch of this problem. The position-versus-time graph implies a particular choice of coordinate system. Include this coordinate system on your sketch. **b.** Describe Reese's motion in words. Your description should include when she sped up, slowed down, or maintained

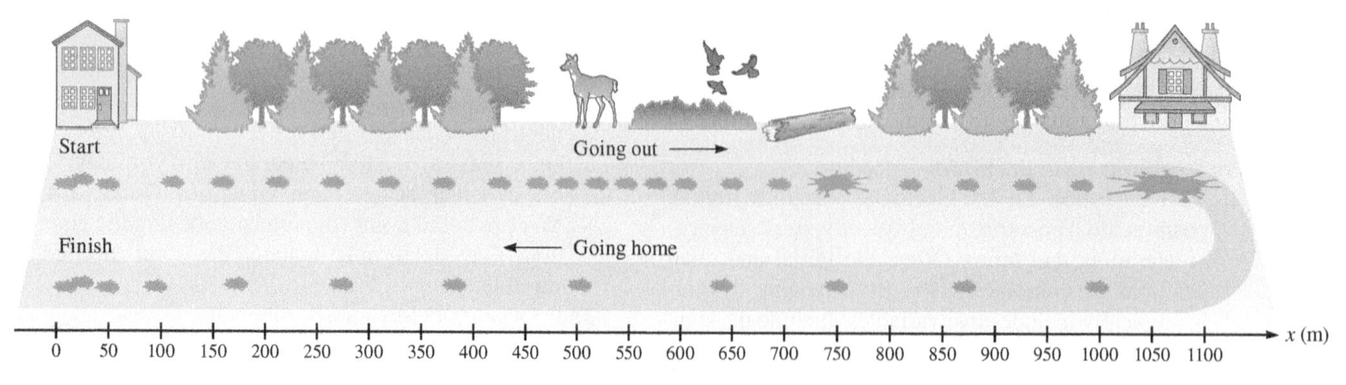

FIGURE P2.6

speed. **c.** For convenience, her altitude was written on the graph every 10 s starting at $t = 0$. Use these data to create a motion diagram for Reese. **d.** Does your motion diagram match your description in part (b)?

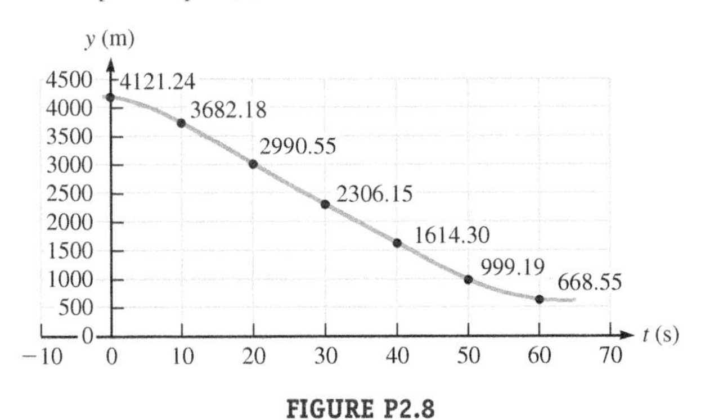

FIGURE P2.8

9. Elisha Graves Otis invented the elevator brake in the mid-1800s, making it possible to build tall skyscrapers with fast elevators. Today's skyscrapers are a large fraction of a mile tall; for example, Taipei 101 in Taiwan has 101 stories and is 515 m (0.32 miles) tall. The top speed of the elevator in the Taipei 101 tower is roughly three times greater than the ascent rate of a commercial jet airplane. The position and time data in the table are based on such an elevator.

a. G Working in SI units, make a position-versus-time graph for the elevator. (You may wish to use a spreadsheet program.)

b. C Describe the motion of the elevator in words.

c. N Find the highest speed of the elevator. When is the elevator going at this speed?

d. C What sort of considerations would the engineers need to make to ensure the comfort of the passengers?

Position (stories)	Time (s)	Position (stories)	Time (s)	Position (stories)	Time (s)
0.0	0	32.7	13	72.0	25
0.3	1	36.0	14	75.3	26
1.1	2	39.3	15	78.5	27
2.5	3	42.5	16	81.6	28
4.4	4	45.8	17	84.2	29
6.8	5	49.1	18	86.5	30
9.8	6	52.4	19	88.2	31
13.1	7	55.6	20	89.6	32
16.4	8	58.9	21	90.0	33
19.6	9	62.2	22	90.0	34
22.9	10	65.5	23	90.0	35
26.2	11	68.7	24	90.0	36
29.5	12				

10. CASE STUDY As shown in Figure 2.9, Whipple chose a coordinate system that was different from Crall's. There are many possible coordinate systems that can be used to analyze the motion of the cart. For example, a third student (Yoon) chose to place the origin of her coordinate system at position A (the cart's initial position) and use a southward-pointing y axis as her positive axis.

a. N Use the distance information given in Table 2.2 to make a new table that gives the cart's position at each of the 16 times according to Yoon's coordinate system. *Hint*: A sketch similar to Figure 2.9 is a good place to start.

b. G Use the data in your table to make a new position-versus-time graph. *Hint*: **CHECK and THINK** about your graph by comparing it with Figure 2.6.

2-5 Displacement and Distance Traveled

11. C When is the distance traveled by a particle along one straight line less than the magnitude of its displacement?

12. G A particle moves from position A to position C as shown in Figure P2.3. Two different coordinate systems have been chosen. Write expressions for the particle's displacement from A to C using **a.** the top coordinate system and **b.** the bottom coordinate system. **c.** Compare your results. Do they make sense? Explain.

13. N A race car travels 825 km around a circular sprint track of radius 1.313 km. How many times did it go around the track?

14. A woman lives on the fourth floor of an apartment building. She works in a high-rise office building 8.5 blocks away from her apartment on the same street. Her office is on the 12th floor. Assume each story of her apartment building is 4.0 m, each story of her office building is 5.5 m, and a block is 146.6 m long.

a. C Sketch her path.

b. E Estimate the distance she travels to work.

c. N Find the magnitude of her displacement.

15. N A train leaving Albuquerque travels 293 miles, due east, to Amarillo. The train spends a couple of days at the station in Amarillo and then heads back west 107 miles where it stops in Tucumcari. Suppose the positive x direction points to the east and Albuquerque is at the origin of this axis. **a.** What is the total distance traveled by the train from Albuquerque to Tucumcari? **b.** What is the displacement of the train for the entire journey? Give both answers in appropriate SI units.

16. E Milwaukee, Wisconsin, and Grand Rapids, Michigan, are on opposite sides of Lake Michigan. Assume they are at the same elevation. The best highway route between the two cities is 266.9 miles long. Use a map to estimate the displacement of a car that travels from Milwaukee to Grand Rapids. Give your answer in miles and in meters. Be sure to include an estimate of the direction.

Problems 17, 18, 19, and 40 are grouped.

17. N The position of a particle attached to a vertical spring is given by $\vec{y} = (y_0 \cos \omega t)\hat{\jmath}$. The y axis points upward, $y_0 = 14.5$ cm, and $\omega = 18.85$ rad/s. Find the position of the particle at **a.** $t = 0$ and **b.** $t = 9.0$ s. Give your answers in centimeters.

18. A particle is attached to a vertical spring. The particle is pulled down and released, and then it oscillates up and down. Using an upward-pointing y axis, the position of the particle is given by

$$\vec{y} = \left(y_0 \cos \frac{2\pi t}{T}\right)\hat{\jmath}$$

Both y_0 and T are constants in time. The amplitude (y_0) is usually given in meters, and the period (T) is usually in seconds.

a. G Draw a sketch of this situation. Include the y axis.

b. N What is the position of the particle when $t = 0, \frac{1}{2}T, T, \frac{3}{2}T, 2T$ and $\frac{5}{2}T$? Add these positions with clear labels to your sketch.

19. Return to the description of the particle attached to the spring in Problem 17.

a. N Find the displacement of the particle during the time interval from $t = 0$ to $t = 9.0$ s. What does your answer mean?

b. N Find the distance (in centimeters) the particle traveled during this time interval. How can this distance be larger than the magnitude of the displacement?

c. C Many physical systems are modeled by a particle attached to a spring. List some examples of systems that may be modeled by springs. It may be helpful to use the index of this book or the Internet.

2-6 Average Velocity and Speed

20. C A particle is constrained such that it can only move in one dimension to the right or to the left. During some fixed interval of time, its average speed was twice the magnitude of its average velocity. Was the particle always moving during this time interval, or was there necessarily a moment when it was at rest? Explain your answer.

21. N During a relay race, you run the first leg of the race, a distance of 2.0×10^2 m to the north, in 22.23 s. You then run the same distance back to the south in 24.15 s in the second leg of the race. Suppose the positive y axis points to the north. What is your average velocity **a.** for the first leg of the relay race and **b.** for the entire race?

22. C When is the average speed of a particle moving along one straight line less than the magnitude of its average velocity over the same time interval?

Problems 23 through 25 are grouped.

23. Light can be described as a wave or as a particle known as a photon. The speed of light c is 3.00×10^8 m/s.

a. E Imagine flipping a switch on the wall that turns on a lamp in the middle of the room. Estimate the time it takes a photon to reach your eye.

b. N The Earth–Sun distance is 1.50×10^{11} m. Find the time it takes a photon leaving the surface of the Sun to reach us.

c. C Why is it difficult to measure the speed of a photon? Contrast it to measuring the speed of a jogger.

24. N Light can be described as a wave or as a particle known as a photon. The speed of light c is 3.00×10^8 m/s. **a.** Sirius is the brightest star in the night sky, 8.18×10^{16} mfrom the Earth. Find the time it takes a photon to reach us from Sirius. Give your answer in years. **b.** A light-year (ly) is the distance that light travels in 1 year. How far from the Earth is Sirius in light-years?

25. During a thunderstorm, a frightened child is soothed by learning to estimate the distance to a lightning strike by counting the time between seeing the lightning and hearing the thunder (Fig. P2.25). The speed v_s of sound in air depends on the air temperature, but assume the value is 343 m/s. The speed of light c is 3.00×10^8 m/s.

a. E A child sees the lightning and then counts to eight slowly before hearing the thunder. Assume the light travel time is negligible. Estimate the distance to the lightning strike.

b. N Using your estimate in part (a), find the light travel time. Is it fair to neglect the light travel time?

c. C Think about how time was measured in this problem. Is it fair to neglect the difference between the speed of sound in cold air (v_s at 0°C = 331.4 m/s) and the speed of sound in very warm air (v_s at 40°C = 355.4 m/s)?

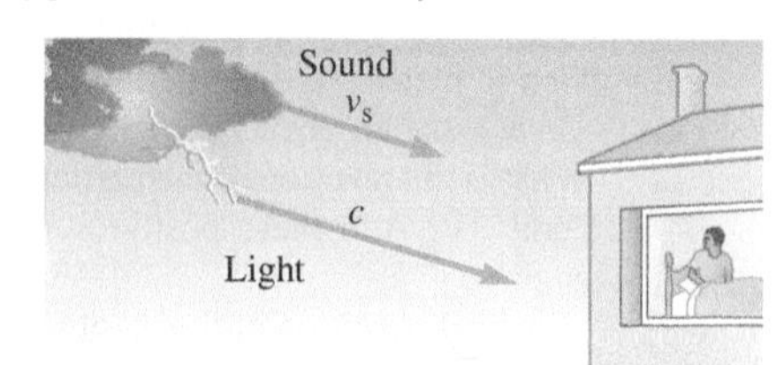

FIGURE P2.25

26. Scientists and engineers must interpret problems from various sources. We can practice this skill anytime we read a newspaper or magazine or browse the Internet. Consider the "Rocket Car" urban legend that can be found on many Internet sites, in which the Arizona Highway Patrol allegedly found the vaporized wreckage of an automobile. The story goes that after some analysis and investigation, it was believed that a former Air Force sergeant attached solid-fuel rockets to his 1967 Chevy Impala and ignited the rockets approximately 3.9 miles from the crash site. The vehicle quickly reached a speed of approximately 275 mph. It continued at this speed for 20 to 25 seconds. The car remained on the highway for 2.6 miles before the driver applied the brakes. The brakes melted and the tires blew out, causing the vehicle to become airborne. It traveled through the air for 1.3 miles before it hit a cliff face 125 feet above the road. Of course, this story was debunked. It is physically implausible, but it can still provide an opportunity to practice analyzing a problem.

a. C Draw a sketch of the situation.

b. C For the constant-velocity part of the car's motion, identify initial and final positions, the velocity, and the time interval.

c. N Calculate the displacement using the position data and then again using the velocity and time data. Are your results consistent?

d. C If your results are not consistent, reread the legend and identify possible sources of the discrepancy.

27. N The Hawaiian Islands are being formed by the 4.0-in/yr motion of the Pacific Plate over an undersea hot spot in the Earth's crust. New volcanoes are formed as the plate moves, and the age of each volcano can be determined by measuring its distance from Kilauea, the Big Island volcano currently atop the hot spot. Assuming the plate moves at constant speed, what is the age of the Kauai volcano if that island is currently 519 km from Kilauea?

28. When you hear a noise, you usually know the direction from which it came even if you cannot see the source. This ability is partly because you have hearing in two ears. Imagine a noise from a source that is directly to your right. The sound reaches your right ear before it reaches your left ear. Your brain interprets this extra travel time (Δt) to your left ear and identifies the source as being directly to your right. In this simple model, the extra travel time is maximal for a source located directly to your right or left ($\Delta t = \Delta t_{max}$). A source directly behind or in front of you has equal travel time to each ear, so $\Delta t = 0$. Sources at other locations have intermediate extra travel times ($0 \le \Delta t \le \Delta t_{max}$). Assume a source is directly to your right.

a. E If the speed of sound in air at room temperature is v_s = 343 m/s, find Δt_{max}.

b. E Find Δt_{max} if instead you and the source are in seawater at the same temperature, where v_s = 1531 m/s.

c. C Why is it difficult to locate the source of a noise when you are under water?

29. A In attempting to break one of his many swimming records, Michael Phelps swims the length L of a swimming pool in time t_1 and returns to his starting point in time t_2, completing the lap in world record time. Assume his first lap is in the positive y-direction and use the symbols L, t_1, and t_2. What is Phelps's average velocity during **a.** the first half of this lap and **b.** the second half of the lap? What is his **c.** average velocity and **d.** average speed for the entire lap?

2-7 Instantaneous Velocity and Speed

30. A The instantaneous speed of a particle moving along one straight line is $v(t) = ate^{-5t}$, where the speed v is measured in meters per second, the time t is measured in seconds, and the magnitude of the constant a is measured in meters per second squared. What is its maximum speed, expressed as a multiple of a?

31. A particle's velocity is given by $v_y(t) = -at$, where a = 0.758 m/s² is a constant.

a. C Describe the particle's motion. In particular, is it speeding up, slowing down, or maintaining constant speed?

b. N Find the particle's velocity at $t = 0$, $t = 10.0$ s, and $t = 5.00$ min.

c. N Find the particle's speed at $t = 0$, $t = 10.0$ s, and t = 5.00 min.

32. C An object initially traveling in the positive x direction undergoes a change in velocity so that, after a finite amount of time passes, it ends up traveling in the negative x direction. Sketch and describe the slope of the position-versus-time graph for this object's motion.

33. N Figure P2.33 shows the y-position (in blue) of a particle versus time. **a.** What is the average velocity of the particle during the time interval $t = 1.00$ s to $t = 3.50$ s? **b.** Using the tangent to the curve (shown as the orange line in the figure), what is the instantaneous velocity of the particle at $t = 1.50$ s? **c.** At what time is the velocity of the particle equal to zero?

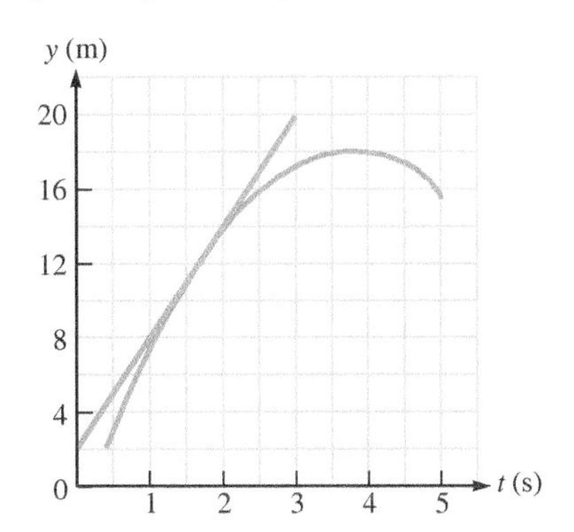

FIGURE P2.33

Problems 34 and 35 are paired.

34. A particle's position is given by $z(t) = -(7.50\text{ m/s}^2)\,t^2$ for $t \geq 0$.
   a. A Find an expression for the particle's velocity as a function of time.
   b. C Is the particle speeding up, slowing down, or maintaining a constant speed?
   c. N What are the particle's position, velocity, and speed at $t = 6.50$ min?

35. N A particle's position is given by $z(t) = -(7.50\text{ m/s}^2)\,t^2$ for $t \geq 0$. **a.** Find the particle's velocity at $t = 1.50$ s and $t = 3.50$ s. **b.** What is the particle's average velocity during the time interval from $t = 1.50$ s to $t = 3.50$ s?

36. C Two sprinters start a race along a straight track at the same time and cross the finish line at the same time. **a.** Are their average velocities necessarily equal? Explain. **b.** Are their instantaneous velocities necessarily *always* equal? Explain. **c.** Are their final velocities necessarily equal? Explain.

2-8 Average and Instantaneous Acceleration

37. N An electronic line judge camera captures the impact of a 57.0-g tennis ball traveling at 33.0 m/s with the side line of a tennis court (Fig. P2.37). The ball rebounds with a speed of 20.0 m/s and is seen to be in contact with the ground for 4.00 ms. What is the magnitude of the average acceleration of the ball during the time it is in contact with the ground? Assume one-dimensional motion.

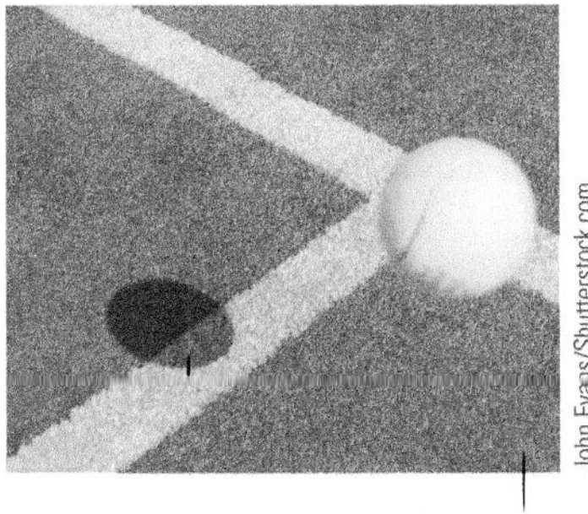

John Evans/Shutterstock.com

FIGURE P2.37

38. C During a bungee jump, a student (i) initially moves downward with increasing speed and then (ii) moves downward with decreasing speed, (iii) reaches the turnaround point (that is, the low point of the motion), (iv) moves upward with increasing speed, and finally (v) moves upward with decreasing speed. For each situation (i) through (v), state the direction of the bungee jumper's velocity and acceleration (that is, upward, downward, or zero). *Note*: (i), (ii), (iv), and (v) refer to intervals of motion, whereas (iii) refers to an instant.

39. C While studying for your physics test, a friend reminds you that the velocity of an object is zero when the slope on a position-versus-time graph for the object is equal to zero. Your friend also says that the acceleration must also therefore be zero at these times because the slope of the velocity-versus-time graph would also have to be equal to zero. **a.** Has your friend given you good advice for the exam? Explain why or why not. **b.** Construct and describe an example, or case, that would illustrate your response to part (a).

Problems 40 and 17 are paired.

40. As in Problem 17, a particle is attached to a vertical spring. The particle is pulled down and released, and then it oscillates up and down. Using an upward-pointing y axis, the position of the particle is given by

$$\vec{y} = \left(y_0 \cos\frac{2\pi t}{T}\right)\hat{\jmath}$$

Both y_0 and T are constants in time. The amplitude (y_0) is usually given in meters, and the period (T) is usually in seconds.
   a. G Draw a sketch of this problem. Include the y axis.
   b. G Plot position versus time, extending your graph at least to $t = 2T$.
   c. A, G Find $\vec{v}_y(t)$ and plot velocity versus time for the same time interval as in part (b).
   d. A, G Find $\vec{a}_y(t)$ and plot acceleration versus time for the same time interval as in part (b).
   e. C At what times is speed at a maximum? Where is the particle at these times? Label the locations on a sketch of the physical situation.
   f. C At what times is the magnitude of the acceleration at a maximum? Where is the particle at these times? Label the locations on the sketch from part (e).

41. C The graph of the scalar component of the position versus time for a particle moving horizontally is a parabola (Fig. P2.41). What can be said about the particle's acceleration?

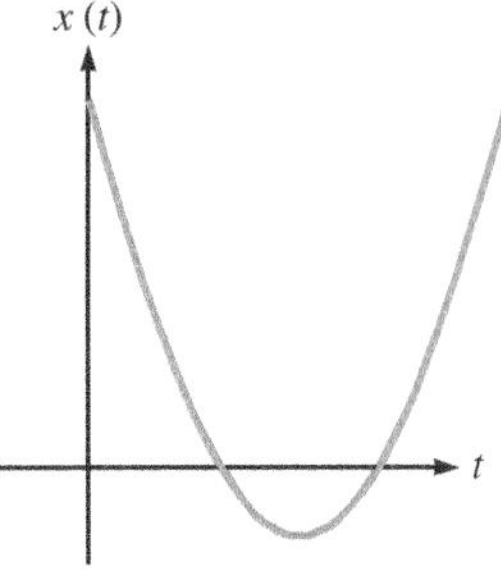

FIGURE P2.41

42. CASE STUDY Back in the laboratory, Crall gives a cart a quick push up an incline. The cart's speed just after the cart leaves Crall's hand is v_0. The cart rolls up the incline, reaches its highest point, and then rolls back down. When the cart returns to Crall's hand, its speed is once again v_0.
   a. G Sketch a velocity-versus-time graph for the cart's motion from just after the initial push to just before it returns. Use a coordinate system in which the positive x direction is upward and parallel to the incline.
   b. C At the highest point of the cart's motion, is the acceleration of the cart positive, negative, or zero? Explain.

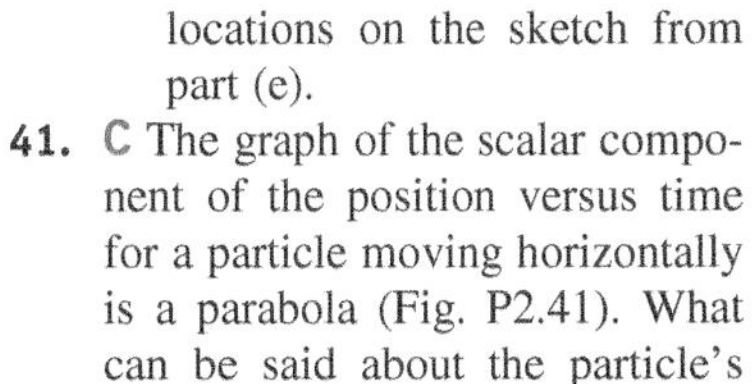

Problems 43 and 44 are paired.

43. In general, when an object moves through a medium such as water or air, the medium affects the object's motion. If nothing else (like gravity or a motor) acts to counter the effect of the medium, the object will decelerate. Above a certain speed threshold, the speed of the object is given by

$$v_x(t) = \frac{v_0}{bt + C}$$

along an arbitrary x axis. Both b and C are constants in time. The units of b are s^{-1}, and C is unitless. Finally, v_0 is a constant with units of meters per second.

a. C How do you know that the direction of the acceleration is opposite that of the velocity?

b. A Show that the magnitude of its acceleration is given by $a_x = (b/v_0)v_x^2$.

c. C Is the acceleration constant? Can the equations in Table 2.4 be applied to this scenario?

d. C How can a submarine move through the water at constant speed?

44. Consider an object moving through a medium as in Problem 43. Below a certain speed threshold, the speed of the object is given by $v_x(t) = v_{0x}e^{-bt}$ along an arbitrary x axis. The constant b is given in s^{-1}.

a. G Make a graph of velocity versus time.

b. C What is the direction of the acceleration? Explain.

c. A, G Find and graph acceleration as a function of time.

d. A Show that $\vec{a}_x = -b\vec{v}_x$.

45. A computer system, using a preset coordinate system, begins tracking the motion of a high-speed train. The computer system determines the position of the train in that coordinate system, starting at time $t = 0$, and models the motion via the equation

$$\vec{z}(t) = \left(129.1 \text{ m} - \frac{246.3 \text{ m} \cdot \text{s}}{t + 2.0 \text{ s}}\right)\hat{k}$$

a. A Find an expression for the acceleration of the train as a function of time.

b. C Is the train slowing down or speeding up? Explain your answer.

c. C Given the function supplied by the computer system, does the train ever turn around and move in the opposite direction? Explain your answer.

46. In Example 2.6, we considered a simple model for a rocket launched from the surface of the Earth. A better expression for the rocket's position measured from the center of the Earth is given by

$$\vec{y}(t) = \left(R_\oplus^{3/2} + 3\sqrt{\frac{g}{2}}R_\oplus t\right)^{2/3}\hat{\jmath}$$

where $R_\oplus$ is the radius of the Earth (6.38×10^6 m) and g is the constant acceleration of an object in free fall near the Earth's surface (9.81 m/s^2).

a. A Derive expressions for $\vec{v}_y(t)$ and $\vec{a}_y(t)$.

b. G Plot $y(t)$, $v_y(t)$, and $a_y(t)$. (A spreadsheet program would be helpful.)

c. N When will the rocket be at $y = 4R_\oplus$?

d. N What are $\vec{v}_y$ and $\vec{a}_y$ when $y = 4R_\oplus$?

2-9 Special Case: Constant Acceleration

47. N A uniformly accelerating rocket is found to have a velocity of 15.0 m/s when its height is 5.00 m above the ground, and 1.50 s later the rocket is at a height of 58.0 m. What is the magnitude of its acceleration?

48. C A piece of debris is in space, far from any planets or stars, and is initially moving. The debris is suddenly subject to a constant acceleration. **a.** Is it *possible* the debris is slowing down? **b.** Is it *possible* the debris could ever reverse its motion? Explain your answers.

49. N A driver uniformly accelerates his car such that $\vec{a} = 6.851\hat{\imath}$ m/s^2. **a.** Assuming he starts from rest, find the velocity of the car after it has accelerated for 4.55 s. **b.** If immediately after that 4.55 s the driver lays off the accelerator, slams on the brakes, and comes to a stop in the subsequent 5.62 s, what is the acceleration he experiences during that time, assuming the acceleration is constant?

50. G Car A and car B travel in the same direction along a straight section of the interstate highway. For the entire interval shown on the velocity-versus-time graph (Fig. P2.50), car A is ahead of car B. **a.** At time t_3, is the magnitude of the acceleration of car A greater than, less than, or equal to that of car B? Explain. **b.** From time t_1 to time t_2, does the distance between cars A and B increase, decrease, or remain constant? Explain.

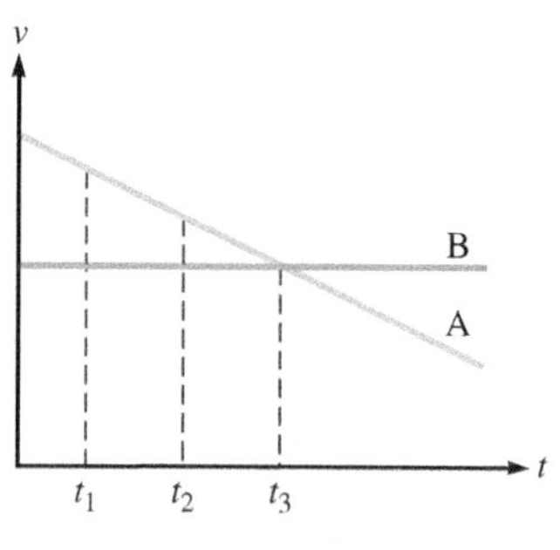

FIGURE P2.50

51. N Accelerating uniformly to overtake a slow-moving truck, a car moving initially at 24.0 m/s covers 68.0 m in 2.50 s. **a.** What is the final speed of the car? **b.** What is the magnitude of the car's acceleration?

52. An object that moves in one dimension has the velocity-versus-time graph shown in Figure P2.52. At time $t = 0$, the object has position $x = 0$.

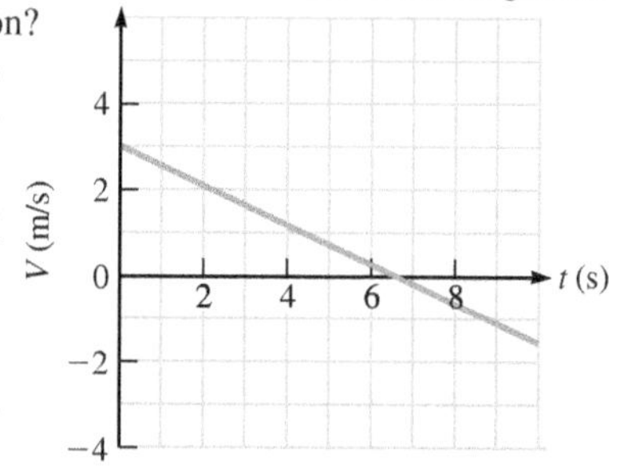

FIGURE P2.52

a. G At time $t = 5$ s, is the acceleration of the object positive, negative, or zero? Explain.

b. G At time $t = 8$ s, is the object speeding up, slowing down, or moving with constant speed? Explain.

c. A Write an expression for the position of the object as a function of time. Explain how you use the graph to obtain your answer.

d. N Use your expression from part (c) to determine the time (if any) at which the object reaches its maximum position. Check your results by examining the graph. *Hint*: To get started with finding the maximum of a function, take the derivative and set it equal to zero.

53. N A particle moves along the positive x axis with a constant acceleration of 3.00 m/s^2 and over time reaches a final speed of 15.0 m/s. **a.** If the particle's initial velocity was $2.00\hat{\imath}$ m/s, what is its displacement during this time, once it reaches its final speed? **b.** What is the distance the particle travels during this time? **c.** If the particle's initial velocity was instead $-2.00\hat{\imath}$ m/s, what is its displacement during this time? **d.** What is the total distance it travels given the initial velocity in part (c)?

54. CASE STUDY Crall and Whipple attached a fan to a cart placed on a level track and then released the cart. They made a position-versus-time graph (Fig. P2.54) and fit a curve to these data such that

$$x = 0.036 \text{ m} + (0.0080 \text{ m/s})t + (0.10 \text{ m/s}^2)t^2$$

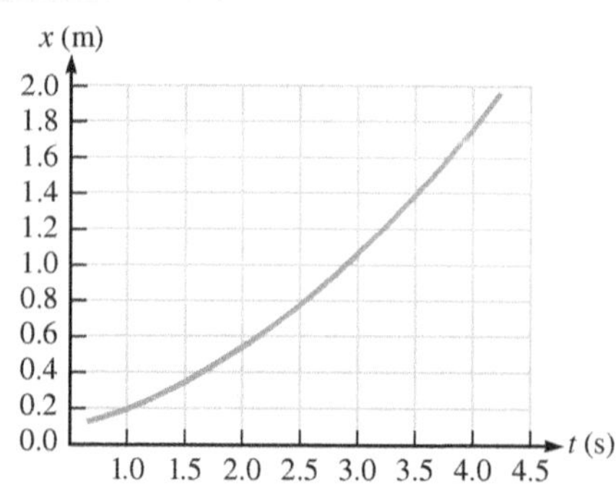

FIGURE P2.54

a. A, G Find and graph the velocity as a function of time.

b. C What is the shape of the velocity-versus-time graph? What do you expect the acceleration-versus-time graph to look like? Explain.

c. A, G Find and graph the acceleration as a function of time.

55. N A vehicle moves along the x axis according to the equation $x = 35.0 + 20.0t + 12.0t^2$, where x is in feet and t is in seconds. What is the **a.** position, **b.** velocity, and **c.** acceleration of the vehicle when $t = 4.00$ s?
Give your answer in U.S. customary units.

56. N The engineer of an intercity train observes a rock slide blocking the train's path 225.0 m ahead and activates the train's emergency brakes. The train decelerates uniformly at 1.8 m/s^2 for 12.70 s before reaching the rock slide. What is the speed with which the train reaches the rock slide?

2-10 A Special Case of Constant Acceleration: Free Fall

57. N A pebble is thrown downward from a 44.0-m-high cliff with an initial speed of 7.70 m/s. How long does it take the pebble to reach the ground?

58. N In a cartoon program, Peter tosses his baby, Stewie, up into the air to keep the child entertained. Stewie reaches a maximum height of 0.873 m above the release point. Suppose the positive y axis points upward. **a.** With what initial velocity was Stewie thrown? **b.** How much time did it take Stewie to reach the peak height?

59. N Tadeh launches a model rocket straight up from his backyard that takes 4.50 s to reach its maximum altitude. (After launch, the rocket's motion is only influenced by gravity.) **a.** What is the rocket's initial velocity? **b.** What is the maximum altitude reached by the rocket?

60. According to several newspapers, on January 25, 2000, an elevator in the Empire State Building fell 40 stories just after a second passenger boarded on the 44th floor.

a. C Draw a sketch and include a coordinate system.

b. E Use the data in Appendix B to estimate the displacement of the elevator in SI units.

c. N According to at least one newspaper, the fall took 4 s. Find the acceleration, assuming it is constant.

d. C Compare your answer to the acceleration of free fall. Does your answer make sense? Explain your reasoning. If needed, explain any sources of discrepancy.

61. N In the movie *Star Wars: The Empire Strikes Back*, after being told that Darth Vader is his father, Luke Skywalker falls from a ledge in Cloud City (not on the Earth, so the magnitude of the free-fall acceleration is not necessarily 9.81 m/s^2). Suppose he falls a distance of 28.5 m in 2.5 s. Assuming he starts from rest, answer the following questions. **a.** What is Luke's velocity 2.5 s after he starts to fall? **b.** What is the constant acceleration due to gravity, experienced by Luke, on Cloud City? Use an upward-pointing y axis.

62. N A worker tosses bricks one by one to a coworker on a scaffold 3.00 m above his location. Each brick is in flight for 1.80 s. **a.** What is the initial velocity with which the bricks are thrown? **b.** What is the velocity of the bricks just as they are caught?

Problems 63 and 64 are paired.

63. N A rock is thrown straight up into the air with an initial speed of 24 m/s at time $t = 0$. Ignore air resistance in this problem. At what times does it move with a speed of 12 m/s? *Note*: There are two answers to this problem.

64. A For the rock in Problem 63, determine a symbolic expression in terms of g for the time t when the rock is moving with a speed of $v/2$. *Note*: There are two answers to this problem.

65. N A sounding rocket, launched vertically upward with an initial speed of 75.0 m/s, accelerates away from the launch pad at 5.50 m/s^2. The rocket exhausts its fuel, and its engine shuts down at an altitude of 1.20 km, after which it falls freely under the influence of gravity. **a.** How long is the rocket in the air? **b.** What is the maximum altitude reached by the rocket? **c.** What is the velocity of the rocket just before it strikes the ground?

General Problems

66. N An object decelerates at a constant rate from an initial velocity of $-7.00\hat{\imath}$ m/s to a final velocity of $10.0\hat{\imath}$ m/s. **a.** What is the object's acceleration if its displacement is $15.0\hat{\imath}$ m? **b.** What would be the object's acceleration if the total distance it traveled were 15.0 m?

67. N While strolling downtown on a Saturday afternoon, you stumble across an old car show. As you are walking along an alley toward a main street, you glimpse a particularly stylish Alpha Romeo pass by. Tall buildings on either side of the alley obscure your view, so you see the car only as it passes between the buildings. Thinking back to your physics class, you realize that you can calculate the car's acceleration. You estimate the width of the alleyway between the two buildings to be 4 m. The car was in view for 0.5 s. You also heard the engine rev when the car started from a red light, so you know the Alpha Romeo started from rest 2 s before you first saw it. Find the magnitude of its acceleration.

68. A particle is attached to a vertical spring. The particle is pulled down and released, and then it oscillates up and down. Using an upward-pointing y axis, the position of the particle is given by $\vec{y} = (y_0 \cos \omega t)\hat{\jmath}$. Both y_0 and ω are constants in time.

a. A Show that $\vec{a}_y = -\omega^2\vec{y}$.

b. C Is the acceleration constant?

c. C Can the equations in Table 2.4 be applied to a particle on a spring? Explain.

69. N A trooper is moving due south along the freeway at a speed of 21 m/s. At time $t = 0$, a red car passes the trooper. The red car moves with constant velocity of 28 m/s southward. At the instant the trooper's car is passed, the trooper begins to speed up at a constant rate of 2.0 m/s^2. What is the maximum distance ahead of the trooper that is reached by the red car?

70. A dancer moves in one dimension back and forth across the stage. If the end of the stage nearest to her is considered to be the origin of an x axis that runs parallel to the stage, her position, as a function of time, is given by

$$\vec{x}(t) = [(0.02 \text{ m/s}^3)t^3 - (0.35 \text{ m/s}^2)t^2 + (1.75 \text{ m/s})t - 2.00 \text{ m}]\hat{\imath}$$

a. A Find an expression for the dancer's velocity as a function of time.

b. G Graph the velocity as a function of time for the 14 s over which the dancer performs (the dancer begins when $t = 0$) and use the graph to determine when the dancer's velocity is equal to 0 m/s.

71. E The electrical impulse initiated by the nerves in Lina's hand, signaling she has touched a hot stove, travels to her brain as fast as 200 m/s. At this speed, estimate the travel time of this impulse.

72. C Two cars leave Seattle at the same time en route to Boston on Interstate 90. The first car moves uniformly the whole way, with constant speed v. The second car travels with constant speed $(v + 1)$ mph for the first half of the distance and travels with constant speed $(v - 1)$ mph for the second half of the distance. Which car gets to Boston first (or is it a tie)? Explain your reasoning.

73. N At 2.00 s, an object begins to move along the y axis and its position is given by the equation $y = 6t^2 - 5t - 2$, with y in meters and t in seconds. **a.** What is the position of the object when it changes its direction? **b.** What is the object's velocity when it returns to its original position at $t = 0$?

74. C An object starts from rest, traversing a distance d in a time T while moving with constant acceleration along a straight-line path. **a.** After a time $T/2$ has elapsed, has the object traveled a distance greater than, less than, or equal to $d/2$? Explain. **b.** Let v_{av} represent the average velocity of the object. When the object has traveled a distance $d/2$, is its instantaneous velocity greater than, less than, or equal to v_{av}? Explain.

75. N The initial velocity of a military jet is 205 m/s eastward. The pilot ignites the afterburners, and the jet accelerates eastward at a constant rate for 1.75 s. The final velocity of the jet is 315 m/s eastward. What was the jet's displacement during the time it was accelerating?

76. Two carts are set in motion at $t = 0$ on a frictionless track in a physics laboratory. The first cart is launched from an initial position of $x = 18.0$ cm with an initial velocity of $11.8\hat{\imath}$ cm/s and a constant acceleration of $-3.40\hat{\imath}$ cm/s^2. The second cart is launched from $x = 20.0$ cm with a constant velocity of $4.30\hat{\imath}$ cm/s.

a. N What are the times for which the two carts have equal speeds?

b. N What are the speeds of the carts at that time?

c. N What are the locations and times at which the carts pass each other?

d. C What is the difference between what is asked in parts (a) and (c) of this problem with regard to the times you found?

77. N The motion of a spacecraft in the outer solar system is described by the equation $x = 4.00t^2 - 3.00t + 5.00$, where x is in astronomical units (AU) and t is in years. **a.** What is the average speed of the spacecraft between $t = 1.00$ yr and $t = 3.00$ yr? **b.** What is the instantaneous speed of the spacecraft at $t = 1.00$ yr and at $t = 3.00$ yr? **c.** For what time t is the speed of the spacecraft zero?
Report answers in AU/year and years.

78. C Cars A and B each move to the right with constant acceleration along a straight road. The velocity vectors of each car are shown in Figure P2.78 for several times separated by equal time intervals. For the entire interval from time t_1 to time t_4, car B is ahead of car A (that is, car B is to the right of car A). **a.** Is the acceleration of car B to the left, to the right, or zero? Explain. **b.** Is the magnitude of the acceleration of car A greater than, less than, or equal to the magnitude of the acceleration of car B? Explain your reasoning. **c.** Is the distance between car A and car B at time t_3 greater than, less than, or equal to the distance between car A and car B at time t_2? Explain.

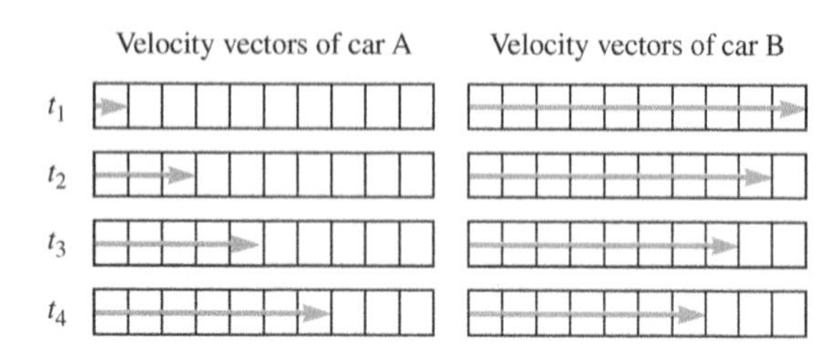

FIGURE P2.78

79. N A hydrogen ion moves along the z axis of a bubble chamber according to the equation $z = 5.00t^2 + 4.00t$ where z is in meters and t is in seconds. What is its average velocity during the time interval **a.** $t = 2.50$ to 3.50 s and **b.** $t = 3.50$ to 3.60 s?

80. N Trying to determine its depth, a rock climber drops a pebble into a chasm and hears the pebble strike the ground 3.20 s later. **a.** If the speed of sound in air is 343 m/s at the rock climber's location, what is the depth of the chasm? **b.** What is the percentage of error that would result from assuming the speed of sound is infinite?

81. G CASE STUDY Similar to that shown in Figure 2.6, Crall and Whipple set up three other experiments (Fig. P2.81). In experiment A, a small fan is attached to a cart. The cart is released from rest and moves along the horizontal track. In experiment B, the cart starts at rest at the bottom of the incline. It is then given a push and moves up the incline. In experiment C, a rubber disk is hung from a spring. The rubber disk is pulled down a few centimeters and then released. The spring causes the disk to oscillate up and down. Pick coordinate systems for each experiment. Be sure to choose an origin and a positive direction. Challenge yourself to think of other possible choices and pick the best coordinate system.

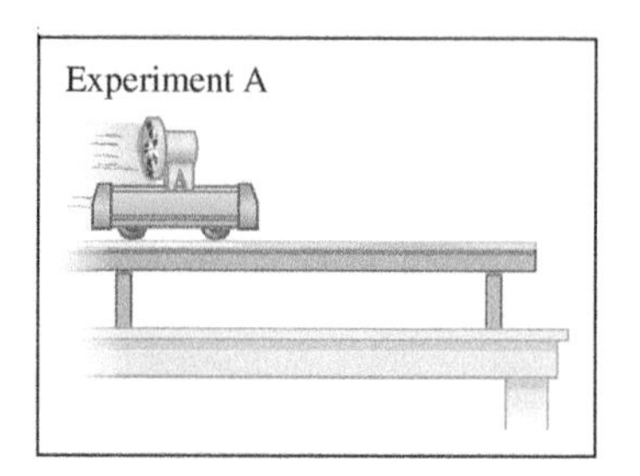

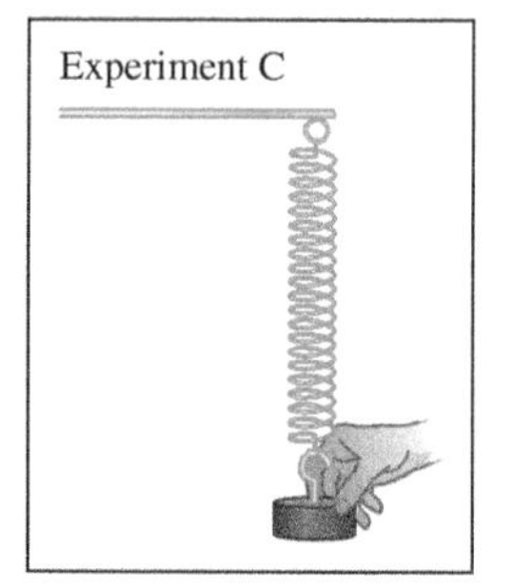

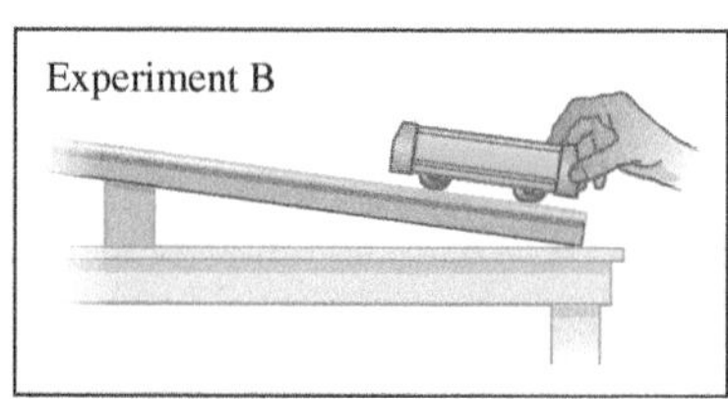

FIGURE P2.81

Problems 82 through 85 are grouped.

82. A Write expressions for the average velocity in the y and z directions.

83. A Write expressions for the instantaneous velocity in the y and z directions.

84. A Write expressions for the average acceleration in the y and z directions.

85. A Write expressions for the instantaneous acceleration in the y and z directions.

86. G Return to Example 2.6 and calculate the displacement $\Delta\vec{y}(t)$, velocity $\vec{v}_y(t)$, and acceleration $\vec{a}_y(t)$ for $t = 0$, 0.5, 1, 1.5, 2, 2.5, and 3 min. Make graphs of displacement, velocity, and acceleration as functions of time.

87. N In 1898, the world land speed record was set by Gaston Chasseloup-Laubat driving a car named Jeantaud. His speed was 39.24 mph (62.78 km/h), much lower than the limit on our interstate highways today. Repeat the calculations of Example 2.7 (acceleration for first 6 miles, time of timed mile, acceleration for last 6 miles) for the Jeantaud car. Compare the results of the ThrustSSC to Jeantaud.

88. N In Example 2.12, two circus performers rehearse a trick in which a ball and a dart collide. We found the height and time of the collision graphically. Return to that example, and find height and time by simultaneously solving the equations for the ball and the dart.

89. A Use integral calculus to show $\Delta x = v_{0x}t + \frac{1}{2}a_xt^2$ (Eq. 2.11) in the case of constant acceleration.

Vectors

3

Underlying Principles

No new physical principles. This chapter describes mathematical tools and concepts.

Major Concepts

1. Vector and scalar components
2. Resolving a vector into components
3. Vector magnitude and direction
4. Vector addition and subtraction
5. Commutative
6. Associative

Tools

1. Vector manipulation
   a. Parallelogram addition and subtraction
   b. Head-to-tail addition and subtraction
   c. Tail-to-tail subtraction
2. Coordinate systems
   a. Cartesian coordinate system
   b. Right-handed coordinate systems

Key Question

How do you add and subtract vectors?

3-1 Geometric treatment of vectors 60

3-2 Cartesian coordinate systems 65

3-3 Components of a vector 68

3-4 Combining vectors by components 76

Each day, we speak our native language, creating sentences that have never been uttered before without thinking about the grammatical rules that govern our speech. Our expectations about speech are based on our native language's grammar. Much of our struggle with learning another language is due to developing a new set of expectations.

Developing new mathematical skills is like learning a language. We have combined scalar quantities using arithmetic and algebraic rules for so long that we often are barely aware of these rules. Now we are going to learn the "language" of **vectors** (quantities with magnitude and direction), including the rules for combining them. Some of these rules will seem familiar from our experience with scalars, whereas other rules will seem new and foreign. The more experience you have with vectors, the more intuitive working with them will become. Vectors such as position, velocity, and acceleration appear frequently in physics.

3-1 Geometric Treatment of Vectors

In Chapter 2, we learned that **vectors** have both a magnitude and a direction, whereas **scalars** have only magnitude. Many quantities in physics—such as position, displacement, average and instantaneous velocity, and average and instantaneous acceleration—are vectors. If motion is one-dimensional, we can use these quantities without learning all the rules of vectors. Vector rules are needed, however, for the description of two-dimensional and three-dimensional motion.

We already know the rules for combining scalars: how to add, subtract, multiply, and divide them. Using these rules allows us to solve scalar equations. Because physical laws are mathematically expressed as equations that involve both vectors and scalars, we must learn to make calculations with vectors. There are two ways to calculate with vectors: geometrically and algebraically. We start with the geometric technique.

The geometric technique will allow you to *anticipate* the result of a problem and *check* your result. Geometric treatment of vectors will also help you think more like a physicist. Recognizing these vectors and coordinate systems is often the first step in *interpreting* a problem.

Because this chapter is dedicated to developing mathematical tools that are used for all vectors, we will often write the rules and relationships in terms of generic vectors designated by uppercase letters such as $\vec{A}$, $\vec{B}$, $\vec{C}$, $\vec{D}$, and $\vec{R}$. To keep the chapter from being too abstract, we will sometimes work with concepts from Chapter 2 such as position, displacement, and average velocity. These vector quantities appear in a number of examples and in the following case study.

CASE STUDY Skydiving

Cassidi Reese, the author's former student who graduated with a degree in physics, likes to skydive and collect data on her dives. Just 10 months after learning to skydive, she celebrated her 300th jump. Achieving such a goal is not easy, and some of her jumps were especially challenging.

On the morning of her 136th jump, Reese boarded a plane in Williamstown, New Jersey, with a group of other skydivers. She happened to be the last jumper out of the plane. The other skydivers were dropped over an empty field, but Reese saw that she was heading for a forest and knew that she had to do something. She took two important measures to avoid landing in a tree. First, she set her parachute in a configuration called "brakes." In this configuration, both her descent velocity and her horizontal velocity were minimized. Second, she opened her parachute when she was at an altitude of 1200 m. (Normally, she would have opened her parachute at an altitude of roughly 750 m.) Reese landed safely in the clearing. Throughout this chapter, we will return to this case study to see what would have happened had Reese deployed her parachute at 750 m. The distance data are given in Figure 3.1. (Reese's motion is influenced by the force exerted by the Earth's gravity as well as by the force exerted by the air, as described in Chapters 5 and 6. In this chapter, we use this case study to practice working with vectors and therefore won't concern ourselves with causes of her motion.)

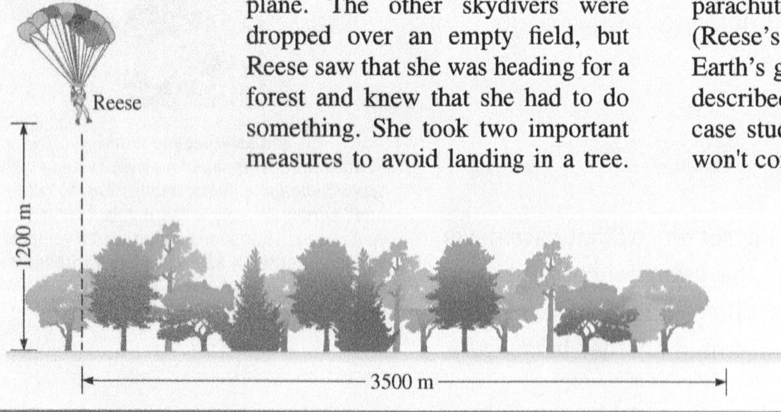

FIGURE 3.1 To avoid landing in a forest, Reese set her parachute to "brakes" to slow her travel and she deployed her parachute at 1200 m instead of 750 m. The clearing was 3500 m away from the point just below her when she opened her parachute. She made it to the clearing unharmed, but what would have happened had she not opened her parachute early? (Reese, her parachute, and the trees are not shown to scale.)

Drawing Vectors

Any representation of a vector must reflect both the vector's magnitude and its direction (Fig. 3.2). The length of the arrow represents the magnitude of the vector, and the arrowhead indicates its direction. Two vectors are equal to each other if their magnitudes are equal and they point in the same direction, no matter where the vectors are located. The three vectors in Figure 3.2 are all equal.

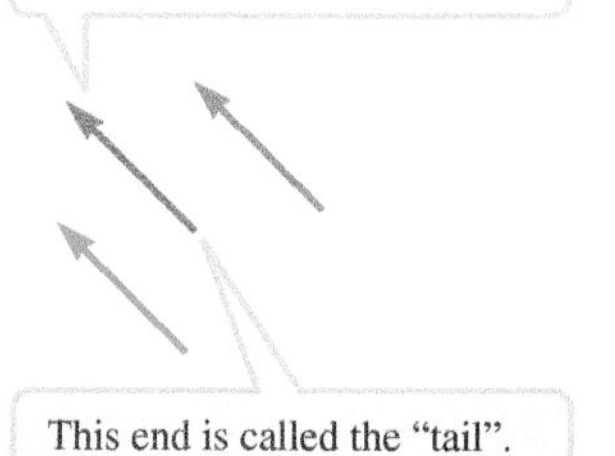

FIGURE 3.2 A vector may be represented pictorially by an arrow. All three vectors shown are equal.

A vector can be shifted to a new location as long as its magnitude and direction remain unchanged. We will use this property of vectors when we want to combine them geometrically.

Adding Vectors Geometrically

When adding two or more vectors, we must take into account both the magnitude and the direction of each vector. Suppose we need to find the vector $\vec{R}$ resulting from the addition of two other vectors $\vec{A}$ and $\vec{B}$:

$$\vec{R} = \vec{A} + \vec{B} \tag{3.1}$$

The vector $\vec{R}$ is called the ***resultant vector*** or simply the ***resultant***.

PARALLELOGRAM VECTOR ADDITION; HEAD-TO-TAIL VECTOR ADDITION

Tools

The resultant can be found geometrically by either of two methods. The first method is known as ***parallelogram addition*** and is illustrated in Figure 3.3A. The second method of geometric vector addition is known as ***head-to-tail addition*** and is illustrated in Figure 3.3B. Two copies of each vector are used in the parallelogram method, and only one copy of each vector is used in the head-to-tail method.

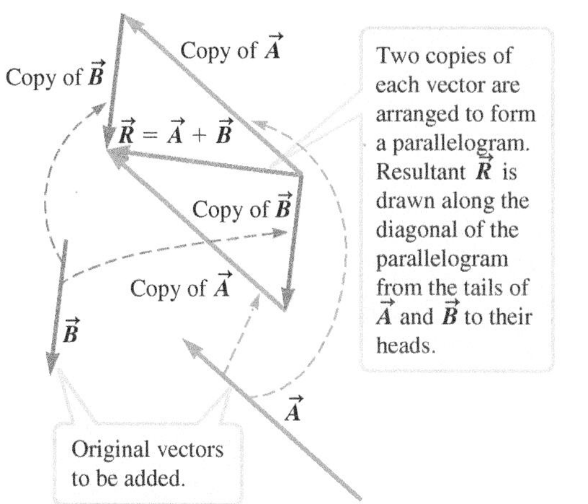

A. Parallelogram method

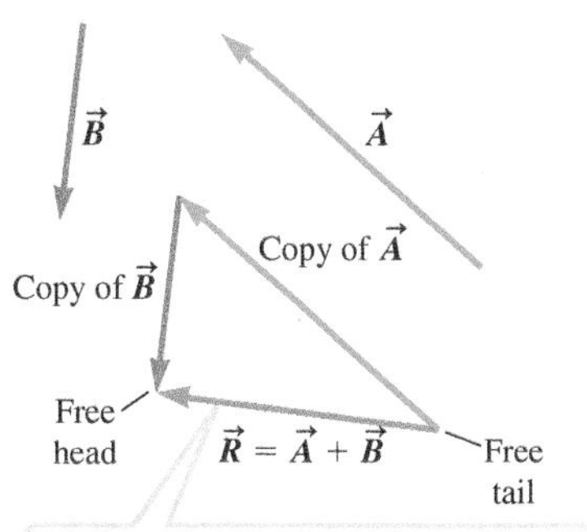

B. Head-to-tail method

FIGURE 3.3 Two methods for adding vectors geometrically.

COMMUTATIVE; ASSOCIATIVE

Major Concepts

Vector addition is **commutative**, meaning that the order of addition does not matter:

$$\vec{A} + \vec{B} = \vec{B} + \vec{A} \tag{3.2}$$

The two triangles in Figure 3.4 show the head-to-tail addition of vector $\vec{A}$ plus vector $\vec{B}$. The triangle on the right was formed by placing the head of vector $\vec{A}$ at the tail of vector $\vec{B}$. The third leg of the triangle is the resultant vector $\vec{A} + \vec{B}$. In the triangle on the left, the head of vector $\vec{B}$ was placed at the tail of vector $\vec{A}$. The third leg in this triangle is the vector $\vec{B} + \vec{A}$.

We can learn two things from Figure 3.4. First, the vectors $\vec{A} + \vec{B}$ and $\vec{B} + \vec{A}$ are the same length and point in the same direction; therefore, these two vectors are equal, demonstrating that vector addition is commutative. Second, we can see how parallelogram addition is related to the commutative property. Each triangle in Figure 3.4 is half of the parallelogram that we would draw to add vectors $\vec{A}$ and $\vec{B}$ in either order using the parallelogram addition method. Therefore, the vector drawn along the diagonal of the parallelogram from the tails to the heads of these vectors must be equal to both $\vec{A} + \vec{B}$ and $\vec{B} + \vec{A}$.

As you will show in Problem 12, vector addition is also **associative**, meaning that if there are more than two vectors to add, they can be grouped in any order:

$$(\vec{A} + \vec{B}) + \vec{C} = \vec{A} + (\vec{B} + \vec{C}) \tag{3.3}$$

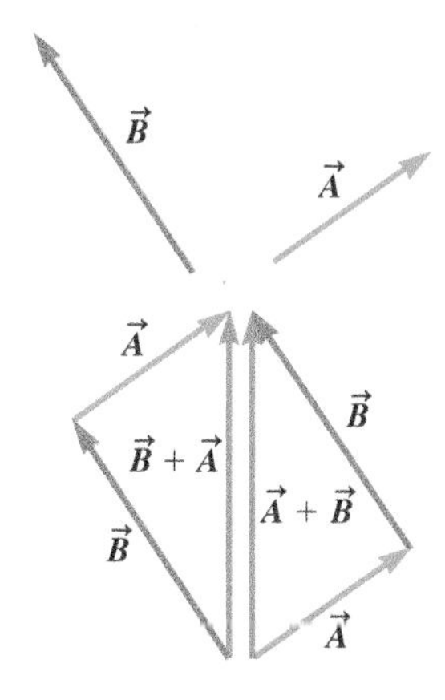

FIGURE 3.4 Vector addition is commutative, which means that vectors can be added in any order: $\vec{A} + \vec{B} = \vec{B} + \vec{A}$. This property of vector addition allows us to use parallelogram addition.

Multiplying a Vector by a Scalar

When a vector $\vec{A}$ is multiplied or divided by a positive scalar s, the resultant $\vec{R}$ is a vector that points in the same direction as $\vec{A}$ but differs in magnitude from that of $\vec{A}$ by the factor s:

$$\vec{R} = s\vec{A} \tag{3.4}$$

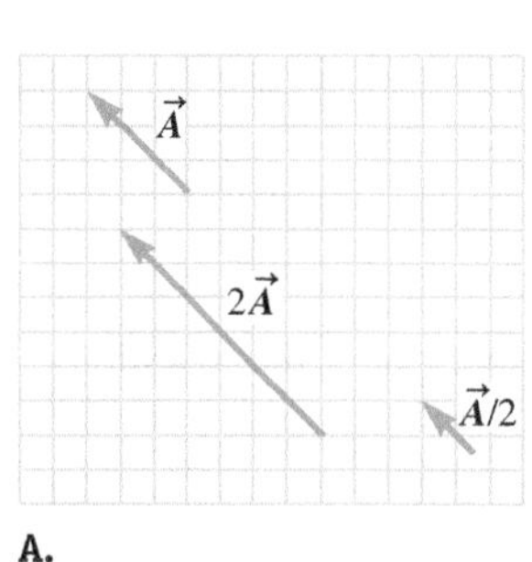

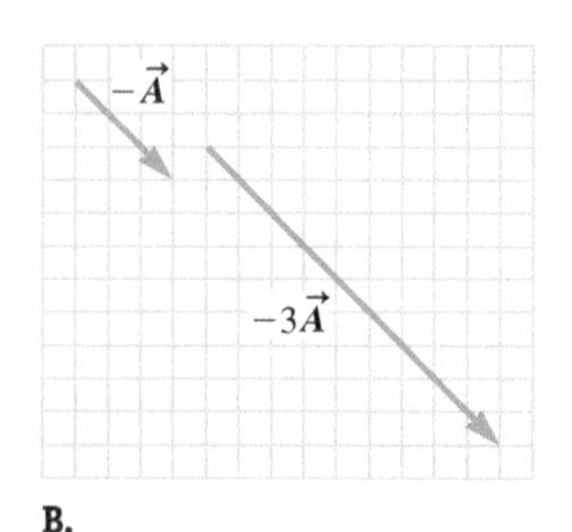

FIGURE 3.5 **A.** When a vector is multiplied or divided by a positive scalar, the resulting vector points in the same direction as the original vector. **B.** When a vector is multiplied (or divided) by a negative scalar, the resulting vector points in the opposite direction.

Figure 3.5A shows vector $\vec{A}$ both multiplied and divided by 2. Both resultant vectors point in the same direction as $\vec{A}$, but the vector $2\vec{A}$ is twice as long as $\vec{A}$ and the vector $\vec{A}/2$ is half as long as $\vec{A}$.

If a vector is multiplied or divided by a negative scalar, its direction reverses. Figure 3.5B shows the same vector $\vec{A}$ multiplied by -1 and -3 to obtain $-\vec{A}$ and $-3\vec{A}$. Both resultant vectors point in the direction opposite to the direction of $\vec{A}$. However, vector $-\vec{A}$ is the same length as $\vec{A}$, so the magnitudes of these two vectors are equal.

Subtracting Vectors Geometrically

Vector subtraction is like vector addition except that one vector is negative: $\vec{R} = \vec{A} - \vec{B}$. We can rewrite the subtraction in the form of an addition:

$$\vec{R} = \vec{A} - \vec{B} = \vec{A} + (-\vec{B}) \tag{3.5}$$

The vector $-\vec{B}$ has the same magnitude as $\vec{B}$ but points in the opposite direction. The vector $-\vec{B}$ is derived from $\vec{B}$ by drawing a vector of the same length but pointing in the opposite direction. Once you have drawn $-\vec{B}$, you can then add $-\vec{B}$ to $\vec{A}$ using either head-to-tail addition or parallelogram addition (Fig. 3.6).

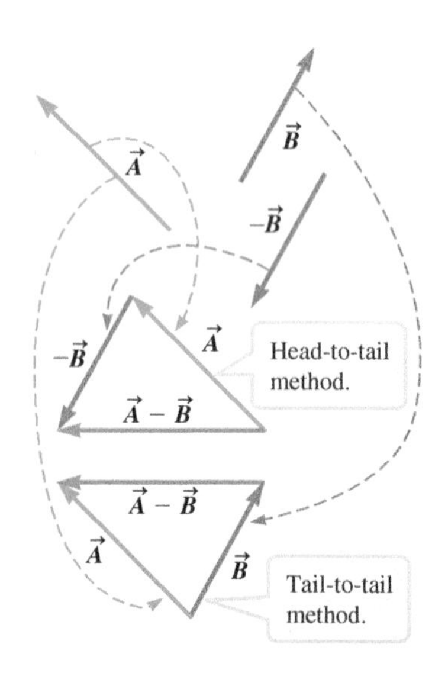

FIGURE 3.6 There are several ways to find $\vec{A} - \vec{B}$ geometrically. As shown in the upper triangle, we may multiply $\vec{B}$ by -1 and then add the result to $\vec{A}$. In the lower triangle, we place $\vec{A}$ and $\vec{B}$ tail to tail, and the resulting vector points from the head of $\vec{B}$ to the head of $\vec{A}$.

An alternative method for geometrically subtracting two vectors $\vec{R} = \vec{A} - \vec{B}$ is **tail-to-tail subtraction.** As shown in the lower triangle in Figure 3.6, the two original vectors $\vec{A}$ and $\vec{B}$ are placed tail to tail, and the resultant vector points from the head of vector $\vec{B}$ to the head of vector $\vec{A}$. This method is often most useful when finding displacement (Example 3.2).

TAIL-TO-TAIL VECTOR SUBTRACTION

◉ Tool

CONCEPT EXERCISE 3.1

The three vectors $\vec{A}$, $\vec{B}$, and $\vec{C}$ in Figure 3.7 all have the same magnitude. Which of these combinations results in a vector of zero magnitude? (More than one choice may be correct.)

a. $\vec{A} - \vec{B}$ **b.** $\vec{B} - \vec{A}$ **c.** $\vec{A} - \vec{C}$ **d.** $\vec{C} - \vec{A}$ **e.** $\vec{A} + \vec{C}$

$\vec{A}$ $\vec{B}$ $\vec{C}$

FIGURE 3.7

CONCEPT EXERCISE 3.2

The three vectors $\vec{A}$, $\vec{B}$, and $\vec{C}$ in Figure 3.7 all have the same magnitude.

a. Does $\vec{A} + \vec{C}$ equal $\vec{C} + \vec{A}$? **b.** Does $\vec{A} - \vec{C}$ equal $\vec{C} - \vec{A}$?
c. Is vector subtraction commutative?

EXAMPLE 3.1 Adding Vectors Head to Tail

Three vectors $\vec{A}$, $\vec{B}$, and $\vec{C}$ are shown in Figure 3.8. Find the vector $\vec{R} = \vec{A} + \frac{1}{2}\vec{B} - 2\vec{C}$.

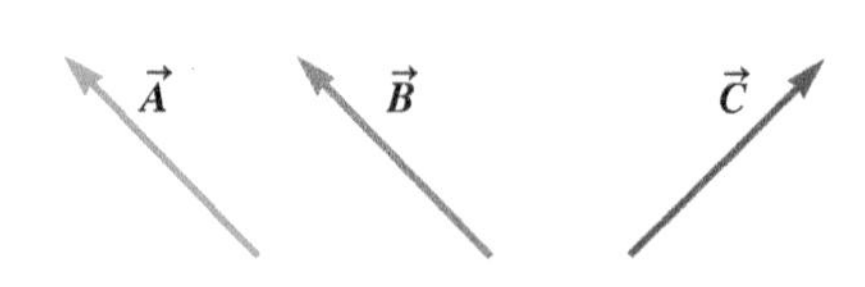

FIGURE 3.8

INTERPRET and ANTICIPATE

As discussed, there are several ways to add and subtract vectors. Because each method produces the same result, the choice is a matter of personal taste. In this example, we choose the head-to-tail method.

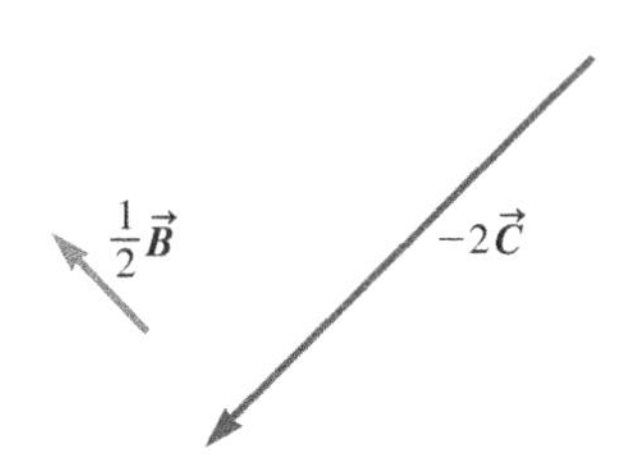

FIGURE 3.9

SOLVE

Regardless of which addition method we choose, our first step must be to multiply vectors $\vec{B}$ and $\vec{C}$ by the appropriate scalars. In Figure 3.9, $\frac{1}{2}\vec{B}$ points in the same direction as $\vec{B}$ but is only half as long. Similarly, $-2\vec{C}$ points in the direction opposite that of $\vec{C}$ and is twice as long as $\vec{C}$.

We arbitrarily choose to add by the head-to-tail method. First, add $\vec{A} + \frac{1}{2}\vec{B}$ by placing the tail of $\frac{1}{2}\vec{B}$ next to the head of $\vec{A}$ (Fig. 3.10). In this problem, the vectors $\vec{A}$, $\frac{1}{2}\vec{B}$, and $(\vec{A} + \frac{1}{2}\vec{B})$ all point in the same direction.

Finally, we use head-to-tail addition to combine $-2\vec{C}$ and $(\vec{A} + \frac{1}{2}\vec{B})$ as shown in Fig. 3.11.

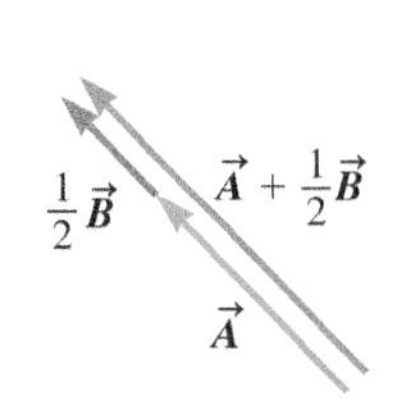

FIGURE 3.10

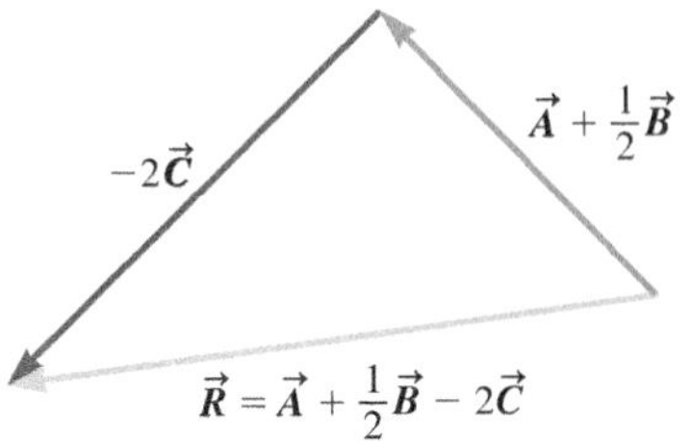

FIGURE 3.11

CHECK and THINK

The resultant vector points to the left and downward. As a check, you can combine the vectors in another order. For example, you could first combine $\frac{1}{2}\vec{B} - 2\vec{C}$ and then add $\vec{A}$. Try it. The resultant should be the same, verifying that vector addition is associative.

Vector addition and subtraction are common in physics. For example, the displacement $\Delta\vec{r}$ (Eq. 2.1) of a particle is found by subtracting its initial position $\vec{r}_i$ from its final position $\vec{r}_f$:

$$\Delta\vec{r} = \vec{r}_f - \vec{r}_i \tag{3.6}$$

The initial and final positions $\vec{r}_i$ and $\vec{r}_f$ need to be measured from a common reference point, the origin of a coordinate system (Section 2-3).

EXAMPLE 3.2 CASE STUDY Cassidi Reese's Displacement

As described previously, Reese opened her parachute when she was directly above a grove of trees, but she landed safely in a nearby field. In Figure 3.12, a reference point has been chosen, and her position $\vec{r}_i$ when she opened the parachute and her position $\vec{r}_f$ when she landed are measured from this reference point. Draw the vector that represents her displacement during this time interval.

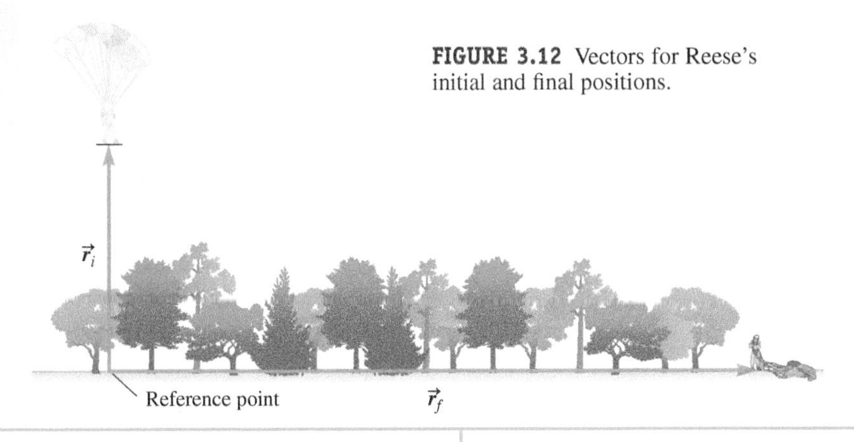

FIGURE 3.12 Vectors for Reese's initial and final positions.

INTERPRET and ANTICIPATE

According to Equation 3.6, to find her displacement, we subtract her initial position vector $\vec{r}_i$ from her final position vector $\vec{r}_f$. There are two ways to do this step geometrically.

$$\Delta\vec{r} = \vec{r}_f - \vec{r}_i$$

Example continues on page 64 ▶

SOLVE
Because the two vectors are already tail to tail, the most convenient subtraction method is tail-to-tail subtraction, in which the displacement vector $\Delta\vec{r}$ points from the head of $\vec{r}_i$ to the head of $\vec{r}_f$ (Fig. 3.13).

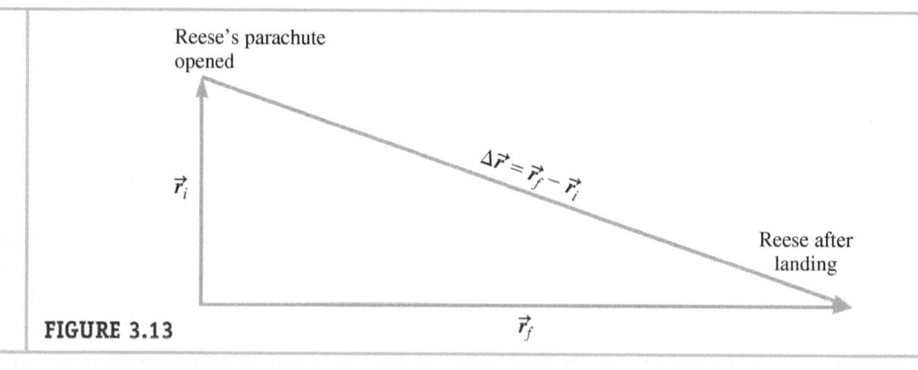

FIGURE 3.13

CHECK and THINK
Displacement is the change in position, so it makes sense that the displacement vector is an arrow that runs from Reese's initial position above the trees to her final position on the ground.

EXAMPLE 3.3 Finding an Acceleration Direction Geometrically

A moon orbits a planet in a circular path. At one instant, the moon's velocity is $\vec{v}_1$; at a later time, its velocity is $\vec{v}_2$ (Fig. 3.14). The average acceleration is given by Equation 2.6:

$$\vec{a}_{\text{av}} = \frac{\Delta\vec{v}}{\Delta t}$$

Find the direction of the average acceleration.

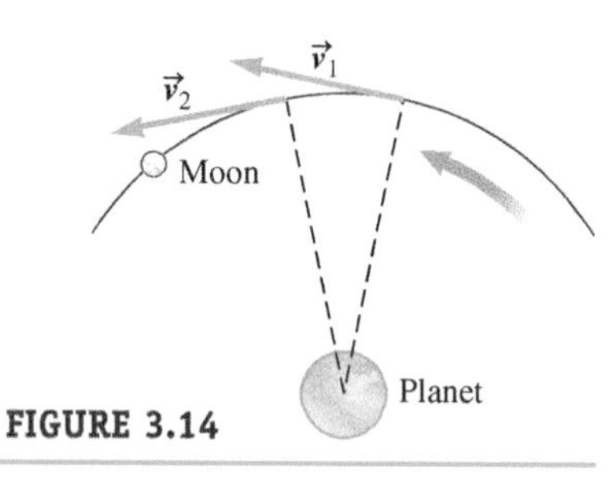

FIGURE 3.14

INTERPRET and ANTICIPATE
The time interval is a scalar. Therefore, the average acceleration must point in the same direction as the change in velocity. If we know the direction of $\Delta\vec{v}$, we also know the direction of $\vec{a}_{\text{av}}$.

SOLVE
To determine the direction of $\Delta\vec{v}$, subtract $\vec{v}_2 - \vec{v}_1$ geometrically. We could use tail-to-tail subtraction, but we have arbitrarily chosen to multiply $\vec{v}_1$ by -1 and add the result to $\vec{v}_2$. The resultant vector $\Delta\vec{v}$ runs from the tail of $\vec{v}_2$ to the head of $-\vec{v}_1$. The average acceleration vector is proportional to the change in velocity and therefore points in the same direction.

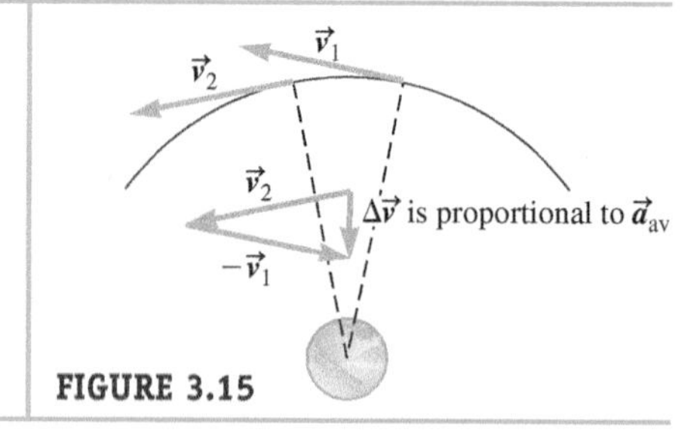

FIGURE 3.15

CHECK and THINK
The two velocity vectors have the same length, so the moon's speed is the same at both instants: $v_1 = v_2$. It might seem surprising that we found a nonzero acceleration, but our answer is correct. In this case, the nonzero acceleration means that the *direction* of the moon's velocity is changing: $\vec{v}_1 \neq \vec{v}_2$.

Using a Scale

Many physical quantities (such as position, displacement, velocity, and acceleration) are vectors that can be represented as arrows and manipulated geometrically. The length of the arrow must be proportional to the magnitude of the vector, but it is often either impossible or not practical to draw an arrow whose length actually equals the magnitude. In such cases, we must use a scale, much like the scale on a map.

For instance, a cartographer may set a scale such that 3 cm on a map equals 10 miles on the Earth, and we write this scale as 3 cm → 10 mi. The procedure for drawing position or displacement vectors is similar to drawing lines on a map. For example, in the CASE STUDY (Fig. 3.1), Reese opened her parachute when her initial position was $r_i = 1200$ m above the ground. By setting a scale such that 1 cm $\rightarrow$ 400 m, we represented her initial position vector $\vec{r}_i$ by an arrow that is only 3 cm long in Figure 3.12.

Physics involves many vector quantities that do not have the dimensions of length. The scale we use to represent such vectors must take into account the dimensions of the vector. If we want to draw a velocity vector, we need to set a scale so that a unit of length on the paper represents a velocity. For example, we might set a scale such that 1 cm on the paper represents 5 m/s: 1 cm $\rightarrow$ 5 m/s.

In practice, we draw vectors so as to *anticipate* a result that we want to calculate algebraically. An approximate scale rather than an exact one may be useful. For example, if you need to draw two velocity vectors representing 10 m/s and 5 m/s, you may not need an exact scale, but you should draw one vector approximately twice as long as the other vector.

CONCEPT EXERCISE 3.3

a. You wish to represent free-fall acceleration using a scale such that 1 cm $\rightarrow$ 1 m/s^2. How long would your vector be?
b. If instead you use a scale such that 0.5 cm $\rightarrow$ 1 m/s^2, how long would the acceleration vector be? (Give answers to three significant figures.)

3-2 Cartesian Coordinate Systems

In Chapter 2, we combined vector components algebraically and defined a one-dimensional coordinate system for some particular physical situations. By building on those skills, we can study two- and three-dimensional motion.

Axes and Coordinates

A one-dimensional coordinate system requires only one axis, usually labeled x, y, or z. A two-dimensional coordinate system requires two axes (usually x and y), and a three-dimensional coordinate system requires all three axes (x, y, and z). In a **Cartesian coordinate system**, the axes are at right angles to one another. Figure 3.16 shows two two-dimensional coordinate systems. The axes of each system cross at right angles but do not need to be horizontal and vertical. The two axes of each system lie in the same plane and define a two-dimensional space. As before, an arrow on the end of an axis indicates the positive direction.

CARTESIAN COORDINATE SYSTEM
Tool

Two numbers known as **coordinates** specify the location of a point in a two-dimensional system. We write these two coordinates as (x, y). The axes cross at their mutual origin $(x, y) = (0, 0)$. The axes divide up the two-dimensional space into four quadrants labeled with roman numerals. Figure 3.16 shows the relationship between the sign of the coordinates and the four quadrants.

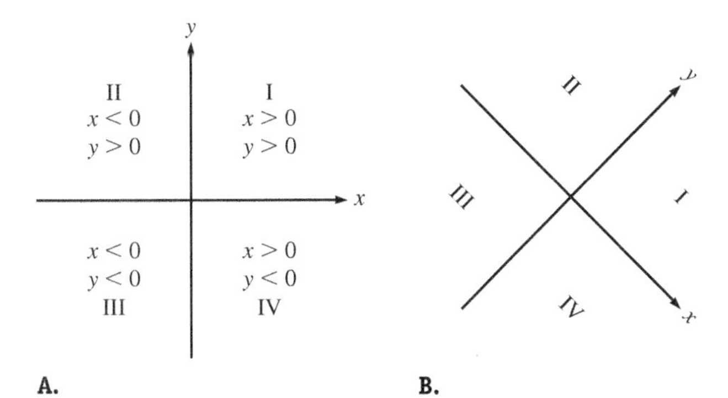

FIGURE 3.16 Two-dimensional Cartesian coordinate systems. **A.** The intersection of the axes divides space into four quadrants. The signs of the x and y coordinates depend on the quadrant as shown. **B.** An x axis might not be horizontal and a y axis might not be vertical, but the axes must be perpendicular to each other.

EXAMPLE 3.4 Reading Coordinates

A pool cue lies on a pool table. The two-dimensional coordinate system shown in Figure 3.17 has been chosen with the origin at the center of the table.

A What are the coordinates of the two ends of the cue?

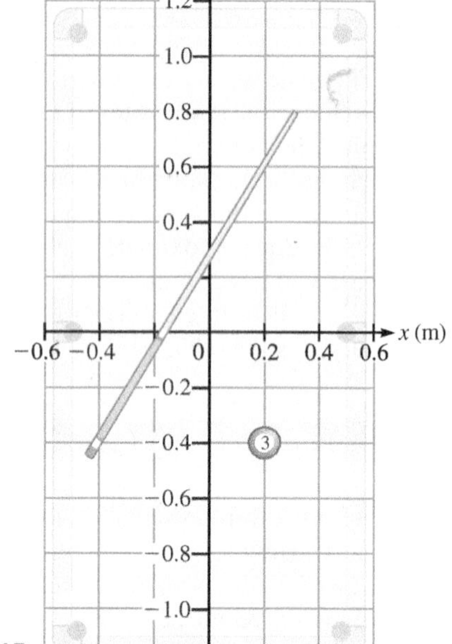

FIGURE 3.17

INTERPRET and ANTICIPATE

In this coordinate system, the x axis is horizontal with the positive direction pointing to the right. The y axis is vertical with the positive direction pointing upward. Both axes are labeled in meters. Figure 3.16 may be used to check the sign of our answer.

SOLVE

Read off the coordinates for each end of the cue to two significant figures. We estimate that the tip (narrow end) is half way between $x = 0.2$ and 0.4 m.	$(x_t, y_t) = (0.30 \text{ m}, 0.80 \text{ m})$
The bumper (thick end) does not lie on a grid line. So, we must estimate both its x and y coordinates.	$(x_b, y_b) = (-0.45 \text{ m}, -0.45 \text{ m})$

CHECK and THINK

Because the tip is in quadrant I, both coordinates are positive. The bumper is in quadrant III, so both coordinates are negative.

B How long is the cue?

INTERPRET and ANTICIPATE

Draw two distances labeled Δx and Δy to form a right triangle that has the cue as its hypotenuse (Fig. 3.18). Then, use the Pythagorean theorem to find the length of the cue.

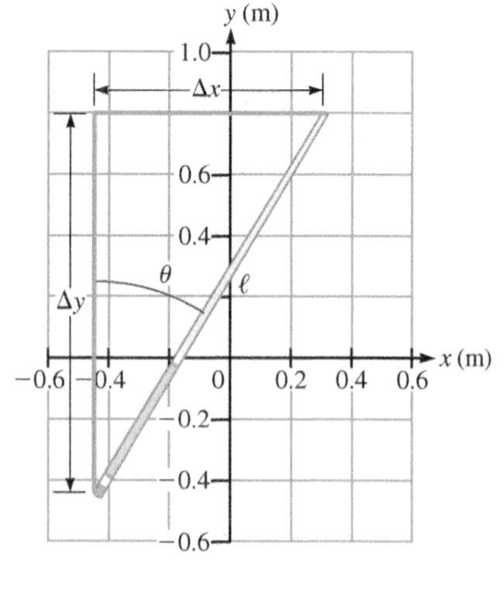

FIGURE 3.18

SOLVE

First, we find the lengths Δx and Δy from the coordinates of the tip and bumper found in part B. Because we plan to square Δx and Δy, it does not matter in which order we do the subtraction.	$\Delta x = x_t - x_b$ $\Delta x = 0.30 - (-0.45)$ $\Delta x = 0.75$ m $\Delta y = y_t - y_b$ $\Delta y = 0.80 - (-0.45)$ $\Delta y = 1.25$ m
To get the length of the cue, use the Pythagorean theorem.	$\ell^2 = \Delta x^2 + \Delta y^2$ $\ell = \sqrt{\Delta x^2 + \Delta y^2}$ $\ell = \sqrt{(0.75 \text{ m})^2 + (1.25 \text{ m})^2}$ $\ell = 1.46$ m

CHECK and THINK

If you are familiar with pool, you know that a cue is somewhat shorter than an adult. So, this answer seems reasonable.

C Figure 3.19 shows a different coordinate system imposed on the same pool table we have been working with. Which of the answers above change? Which remain the same? Use this new coordinate system to find the positions of the ends of the cue as well as its length.

INTERPRET and ANTICIPATE

Any coordinate system is *artificially* imposed on a physical situation, meaning that the origin can be anywhere. Because the location of the origin determines the coordinates of any object in the system, the numerical coordinates of an object change as the location of the origin changes. The length of an object cannot depend on the choice of coordinate system. Therefore, the length of the cue must still be 1.46 m.

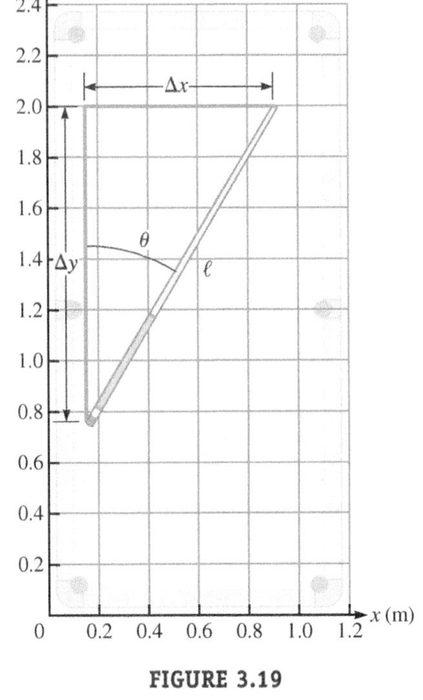

FIGURE 3.19

SOLVE

The new coordinates are read off the figure. In this case, all the coordinates are positive because both ends of the cue are in quadrant I.

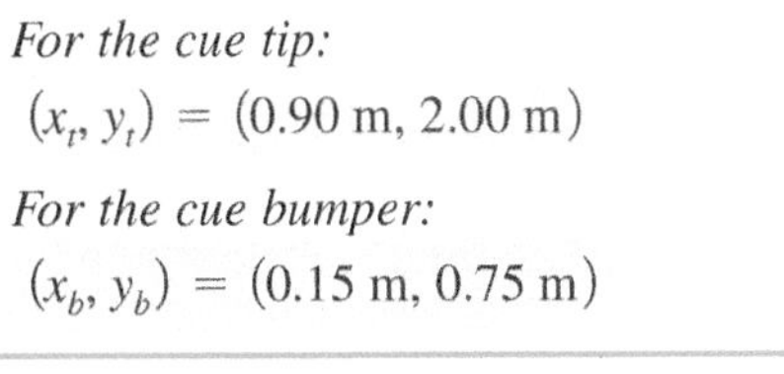

For the cue tip:

$(x_t, y_t) = (0.90\text{ m}, 2.00\text{ m})$

For the cue bumper:

$(x_b, y_b) = (0.15\text{ m}, 0.75\text{ m})$

Recalculate the length of the cue using the same procedure as in part C but using the new coordinate system.

$$\Delta x = x_t - x_b$$
$$\Delta x = 0.90 - 0.15$$
$$\Delta x = 0.75\text{ m}$$

$$\Delta y = y_t - y_b$$
$$\Delta y = 2.00 - 0.75$$
$$\Delta y = 1.25\text{ m}$$

$$\ell = \sqrt{(0.75\text{ m})^2 + (1.25\text{ m})^2}$$
$$\ell = 1.46\text{ m}$$

CHECK and THINK

As expected, the coordinates depend on the choice of coordinate system, but the length of the pool cue does not depend on the coordinate system.

Unit Vectors

As in one dimension, unit vectors specify the direction of vector quantities in two and three dimensions. In the case of one-dimensional vectors, we needed to use only one unit vector ($\hat{\imath}$, $\hat{\jmath}$, or $\hat{k}$) for a given vector quantity. When working with two-dimensional vectors, however, we will need to use two unit vectors simultaneously; and with three-dimensional vectors, all three unit vectors are needed. All facts about unit vectors from Chapter 2 still hold. Unit vectors always have a magnitude of 1. The three unit vectors $\hat{\imath}$, $\hat{\jmath}$, and $\hat{k}$ point along the positive x, y, and z axes, respectively. Like any other vector, a unit vector may be moved to any location without changing it as long as it remains parallel to its original direction.

Right-Handed Coordinate Systems

Figure 3.20 shows a three-dimensional Cartesian coordinate system along with the three unit vectors $\hat{\imath}$, $\hat{\jmath}$, and $\hat{k}$. Because it is a Cartesian system, all three axes cross at right angles. Three-dimensional objects are difficult to represent on a two-dimensional piece of paper. In Figure 3.20, the x and y axes, meet at right angles; both lie in the plane of the page, just as in the two-dimensional case shown in Figure 3.16. Now imagine poking a pencil through the page, at the point where the x and y axes intersect. When held perpendicular to the page, the pencil represents the z axis.

When drawing three-dimensional situations, it is often better to represent an axis or vector that is perpendicular to the page with either a circled dot ⊙ or a circled cross ⊗. An axis or vector pointing out of the page is represented by ⊙, and one pointing into the page is represented by ⊗. The coordinate system shown in Figure 3.21A is equivalent to the one in Figure 3.20. One way to remember the difference is to picture an archery arrow with tail feathers. If the arrow were pointing at you, you would see the tip as represented by the ⊙ symbol. If the arrow were pointing away from you, you would see the cross made by the tail feathers, resembling the ⊗.

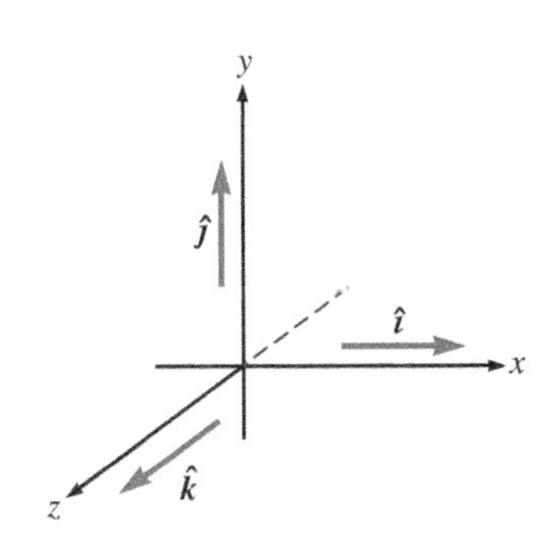

FIGURE 3.20 A three-dimensional coordinate system has x and y axes in the plane of the page; the z axis is perpendicular to the page. The three unit vectors point along the positive x, y, and z axes, respectively.

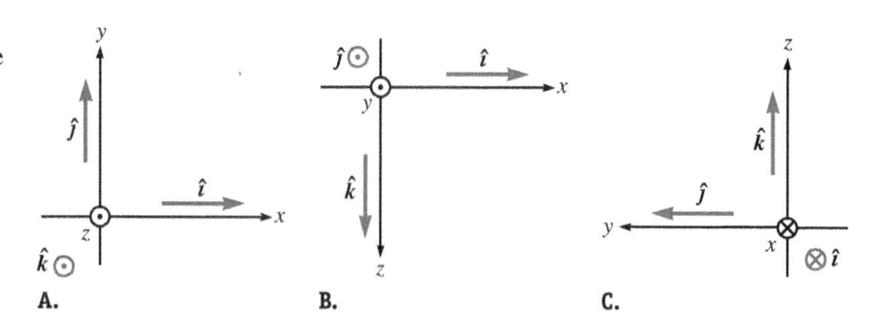

FIGURE 3.21 A. A right-handed coordinate system equivalent to the one in Figure 3.20. **B.** A right-handed coordinate system with the y axis pointing out of the page. **C.** A right-handed coordinate system with the x axis pointing into the page.

RIGHT-HANDED COORDINATE SYSTEM Tool

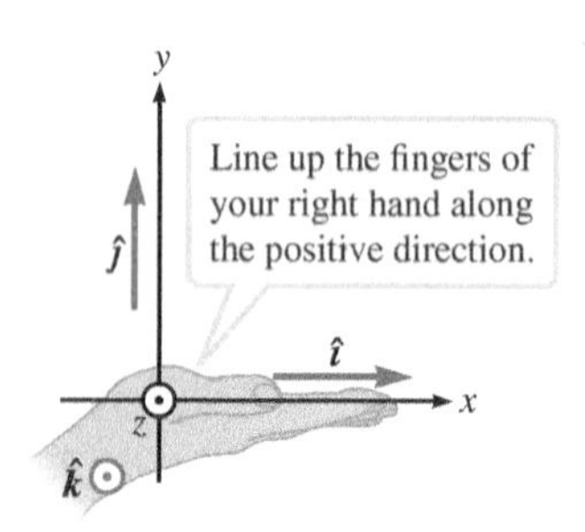

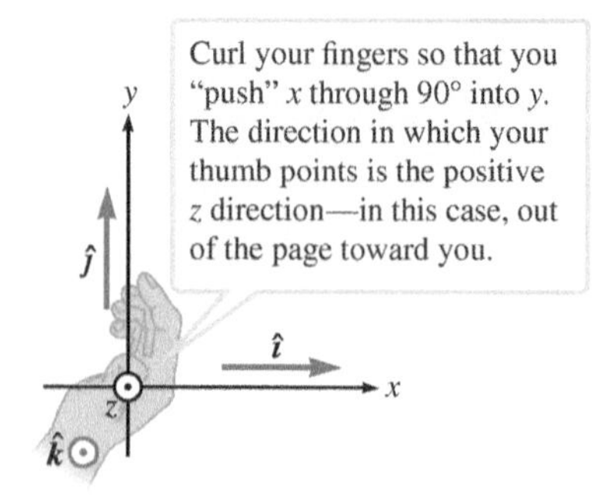

FIGURE 3.22 Finding the z direction in a right-handed coordinate system.

All the coordinate systems used in this book are **right-handed coordinate systems.** Figure 3.21 shows three right-handed x–y–z coordinate systems on a two-dimensional page. In each case, the direction of the z axis (and the unit vector $\hat{k}$) is determined by what is called a *right-hand convention.* Figure 3.22 shows how to use this convention; and it shows that for this coordinate system z points out of the page as it does in Figure 3.21A. Verify for yourself that parts B and C of Figure 3.21 also follow the right-hand convention.

Because the majority of the problems in the first part of this textbook can be analyzed with a two-dimensional Cartesian coordinate system, the rest of this chapter concentrates on two-dimensional vectors.

CONCEPT EXERCISE 3.4

Which coordinate system in Figure 3.23 is right-handed? (More than one choice may be correct.)

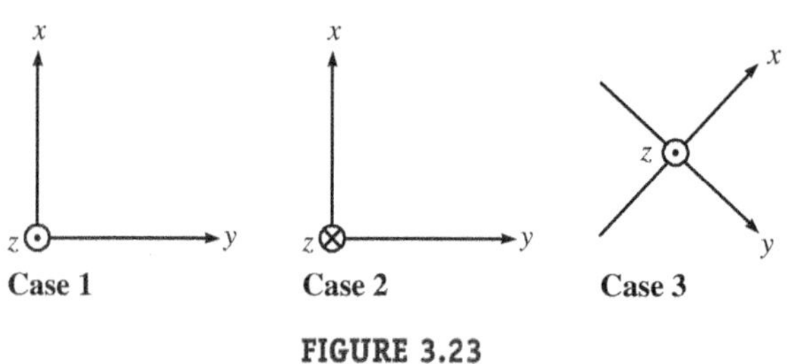

FIGURE 3.23

3-3 Components of a Vector

Vector Components and Scalar Components

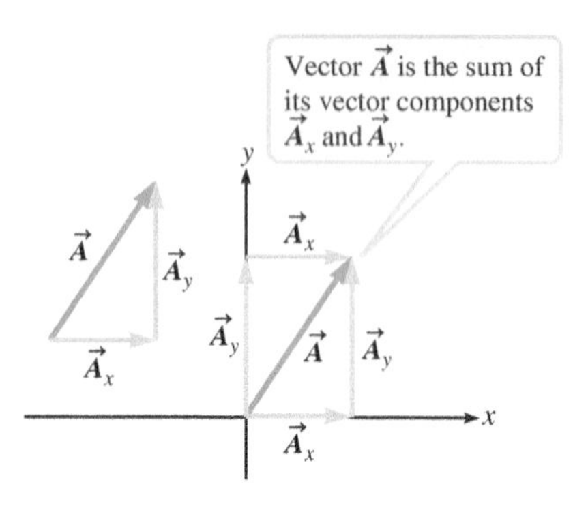

FIGURE 3.24 We may slide the vector $\vec{A}$ and its vector components $\vec{A}_x$ and $\vec{A}_y$ to any location as long as they remain parallel to their original directions.

Vector quantities can be represented numerically using their components. Vector $\vec{A}$ is shown in a two-dimensional coordinate system (Fig. 3.24). Any vector can be written as the sum of other vectors. In this case, $\vec{A} = \vec{A}_x + \vec{A}_y$, where the vectors $\vec{A}_x$ and $\vec{A}_y$ are the x and y **vector components** of $\vec{A}$. Vector components are either parallel or antiparallel to the coordinate axes. (A vector component antiparallel to a coordinate axis is aligned with the coordinate axis but points in the negative direction.) As is true for all vectors, sliding the vector components to new locations does not change them as long as they remain parallel to their original directions (Fig. 3.24).

VECTOR COMPONENTS; SCALAR COMPONENTS

Major Concepts

Because vector components are always either parallel or antiparallel to the coordinate axes, they can easily be written in terms of unit vectors. For example, in a three-dimensional coordinate system, the vector components of $\vec{B}$ are

$$\vec{B}_x = B_x\hat{\imath}, \quad \vec{B}_y = B_y\hat{\jmath}, \quad \vec{B}_z = B_z\hat{k} \tag{3.7}$$

where B_x, B_y, and B_z are the **scalar components** of $\vec{B}$. Like all scalars, the scalar components of a vector do not have direction. They may be positive or negative. A negative scalar component means that the vector component is pointing in the negative direction along its particular axis.

Any vector is completely specified by either its vector components or its scalar components. For example, the vector $\vec{B}$ may be expressed as a sum of its vector components:

$$\vec{B} = \vec{B}_x + \vec{B}_y + \vec{B}_z$$

Vector $\vec{B}$ can also be written in **component form**, meaning that it is written in terms of its scalar components multiplied by unit vectors:

$$\vec{B} = B_x\hat{\imath} + B_y\hat{\jmath} + B_z\hat{k} \tag{3.8}$$

Vector and scalar components are closely related; if we know one, we can easily figure out the other. For example, if we know the vector component $\vec{A}_x = -3.5\hat{\imath}$, we also know that the scalar component is $A_x = -3.5$. If we know the scalar component $A_y = -7.0$, we know that the vector component is $\vec{A}_y = -7.0\hat{\jmath}$. In this textbook, the term *components* means *scalar* components. Vector components will be specified explicitly when necessary.

Resolving a Vector into Components

RESOLVING A VECTOR INTO COMPONENTS

Major Concept

The process of finding a vector's (vector or scalar) components is known as *resolving the vector into components*. Before learning the mathematical process for resolving a vector, let's build a conceptual understanding. Imagine a vector $\vec{A}$ represented by a real arrow sticking out of the floor at an angle. A vertical screen is placed near the arrow as shown in Figure 3.25. A two-dimensional x–y coordinate system is imposed on the floor and on the screen. We use a light source to find the vector components of the arrow "vector". When we place the light source above the arrow, we see a shadow of the arrow on the floor along the x axis (Fig. 3.25A). This shadow is the x vector component $\vec{A}_x$. When we place the light source to the right of the arrow, we see a shadow of the arrow on the screen along the y axis (Fig. 3.25B). This shadow is the y vector component $\vec{A}_y$.

We can find vector $\vec{A}$'s scalar components mathematically. Figure 3.26A shows vector $\vec{A}$ and its vector components using the conventional arrow representation. The vector components along the x and y axes are the "shadows" of vector $\vec{A}$ projected perpendicular to those axes. These vector components have been assembled to form a right triangle with $\vec{A}$ as the hypotenuse. We can use trigonometry to find the scalar components A_x and A_y in terms of θ and A, where θ is the angle the vector makes with the x axis. For the scalar component A_x, we have

$$\cos\theta = \frac{\text{adjacent}}{\text{hypotenuse}} = \frac{A_x}{A}$$

$$A_x = A\cos\theta \tag{3.9}$$

With a little more trigonometry, we can find the scalar component A_y:

$$\sin\theta = \frac{\text{opposite}}{\text{hypotenuse}} = \frac{A_y}{A}$$

$$A_y = A\sin\theta \tag{3.10}$$

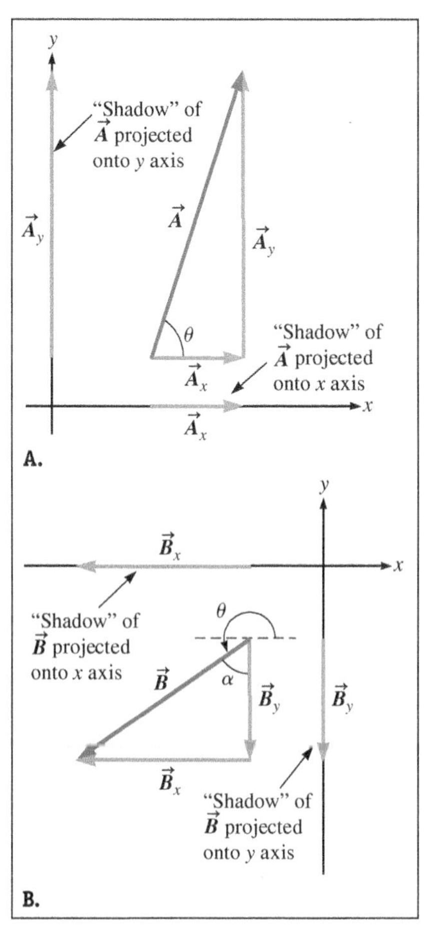

FIGURE 3.26 **A.** The vector components $\vec{A}_x$ and $\vec{A}_y$ are the legs of a right triangle with $\vec{A}$ as the hypotenuse. **B.** The vector components $\vec{B}_x$ and $\vec{B}_y$ are both negative.

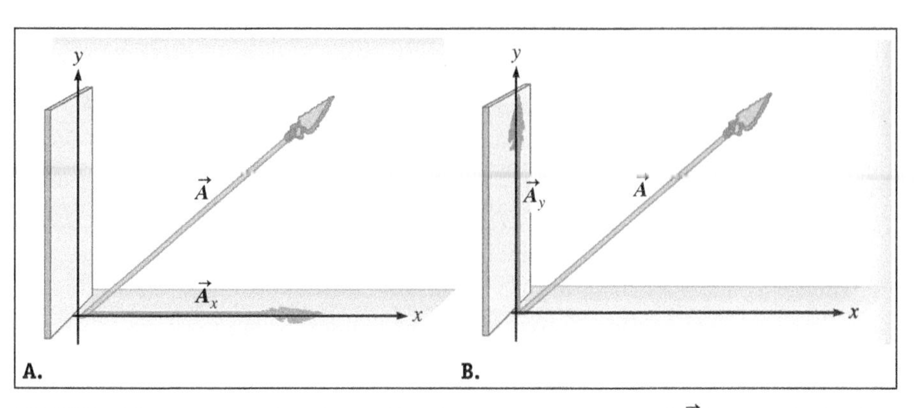

FIGURE 3.25 One way to imagine finding the vector components of $\vec{A}$ is by projecting its shadow onto two perpendicular surfaces. **A.** The x component is the shadow projected onto the horizontal ground. **B.** The y component is the shadow projected onto a vertical screen.

We use these scalar components A_x and A_y to write $\vec{A}$ in component form (Eq. 3.8):

$$\vec{A} = A\cos\theta\,\hat{\imath} + A\sin\theta\,\hat{\jmath} \tag{3.11}$$

Vector $\vec{A}$ points up and to the right. Both of its vector components point in the positive direction along their respective axes. As long as the angle θ is measured counterclockwise from the x axis, Equation 3.11 still holds, even if a vector has one or more components that point in the negative direction. For example, the vector components of $\vec{B}$ shown in Figure 3.26B both point in negative directions, and $\vec{B}$ is expressed as

$$\vec{B} = B\cos\theta\,\hat{\imath} + B\sin\theta\,\hat{\jmath}$$

The angle θ is between 180° and 270°, so both $\cos\theta$ and $\sin\theta$ are negative.

Although Equation 3.11 can always be used to express a vector in component form, it is often not practical because angles are not always measured counterclockwise from the x axis. It is more common for angles to be expressed as an acute angle measured either from the horizontal or vertical such as the angle α in Figure 3.26B. Start by constructing a right triangle; the legs are the vector components, and the hypotenuse is the vector. Then use trigonometry to express vector $\vec{B}$ in terms of the angle α. Because the angle α is less than 90°, both $\sin\alpha$ and $\cos\alpha$ are positive. In this case, both vector components are negative, so we must insert a negative sign for each component:

$$\vec{B} = -B\sin\alpha\,\hat{\imath} - B\cos\alpha\,\hat{\jmath}$$

CONCEPT EXERCISE 3.5

Show that if the angle $\theta = 215°$ in Figure 3.26B, then

a. $\alpha = 55°$
b. $\vec{B} = B\cos\theta\,\hat{\imath} + B\sin\theta\,\hat{\jmath} = -B\sin\alpha\,\hat{\imath} - B\cos\alpha\,\hat{\jmath}$

EXAMPLE 3.5 Expressing Vectors in Component Form, with a Twist

Three vectors and a two-dimensional coordinate system are shown in Figure 3.27. The magnitudes of these vectors are $A = 3.6$ m, $B = 2.7$ m, and $C = 3.3$ m.

A Write these three vectors in component form.

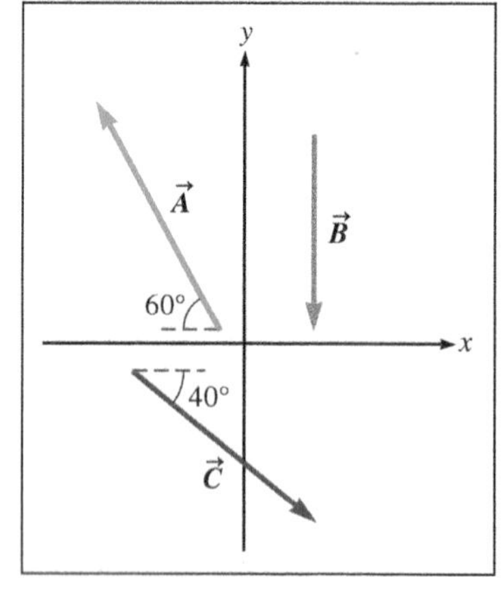

FIGURE 3.27

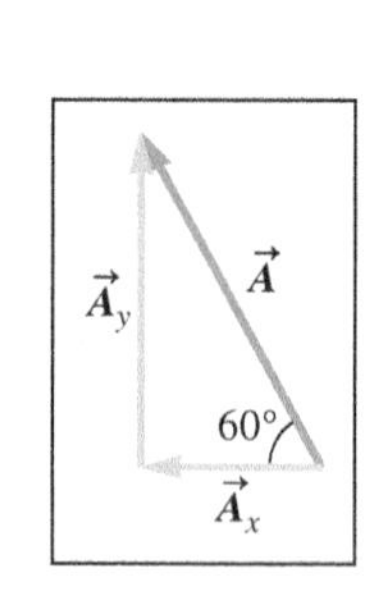

FIGURE 3.28

INTERPRET and ANTICIPATE

To write a vector in component form, we must first find its scalar components and then place the scalar components in front of the appropriate unit vectors.

SOLVE

Draw $\vec{A}_x$ and $\vec{A}_y$ so that together they form a right triangle with $\vec{A}$ as the hypotenuse. The vector components $\vec{A}_x$ and $\vec{A}_y$ are parallel to the x and y axes, respectively (Fig 3.28).

Use trigonometry to find the magnitude of each scalar component. The scalar component A_x must be negative because $\vec{A}_x$ points in the negative x direction.	$A_x = -A\cos 60°$ $A_x = -3.6\cos 60° = -1.8$ m
We can similarly reason that A_y is positive.	$A_y = A\sin 60°$ $A_y = 3.6\sin 60° = 3.1$ m
Place these scalar components in front of the appropriate unit vectors and add the expressions. The dimensional unit appears outside the parentheses because both scalar components are measured in meters.	$\vec{A} = (-1.8\hat{\imath} + 3.1\hat{\jmath})$ m

We see that $\vec{B}$ is antiparallel to the y axis, meaning that it points in the negative y direction. Therefore, $\vec{B}$ is equal to its vector component $\vec{B}_y$ because $\vec{B}_x$ is zero.

$$\vec{B}_x = 0$$
$$\vec{B}_y = -2.7\hat{\jmath}\ \text{m}$$
$$\vec{B} = (0\hat{\imath} - 2.7\hat{\jmath})\ \text{m}$$
$$\vec{B} = -2.7\hat{\jmath}\ \text{m}$$

To write $\vec{C}$ in component form, start with a sketch similar to that for $\vec{A}$ (Fig. 3.29).

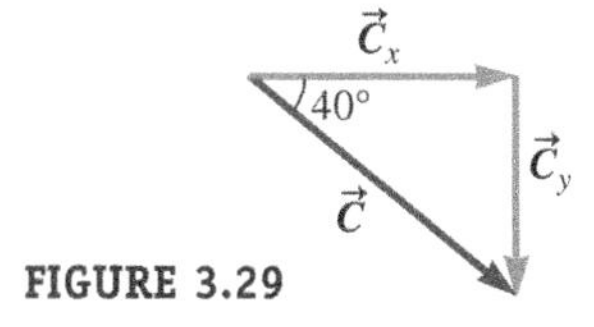

FIGURE 3.29

In this case, C_x is positive and C_y is negative.

$$C_x = C\cos 40° = 3.3\cos 40° = 2.5\ \text{m}$$
$$C_y = -C\sin 40° = -3.3\sin 40° = -2.1\ \text{m}$$
$$\vec{C} = (2.5\hat{\imath} - 2.1\hat{\jmath})\ \text{m}$$

B Figure 3.30 here shows vector $\vec{B}$ in a new coordinate system with the axes labeled x' and y'. This x'–y' coordinate system is turned 30° counterclockwise with respect to the x–y coordinate system. Write $\vec{B}$ in component form in the x'–y' coordinate system.

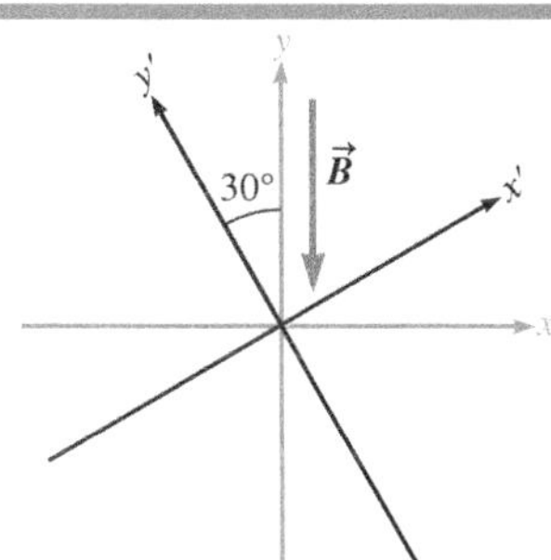

FIGURE 3.30

INTERPRET and ANTICIPATE

In this new coordinate system, $\vec{B}$ is not parallel to an axis, so we must use a procedure similar to that used in finding the components of $\vec{A}$ and $\vec{C}$ in part A. We must find the angle that $\vec{B}$ makes with the new axes to find its components.

SOLVE

Slide $\vec{B}$ leftward and down so that its tail sits on the common origin of the two coordinate systems. This change shows that $\vec{B}$ makes an angle of 30° with the y' axis. Sketch the vector components (Fig. 3.31).

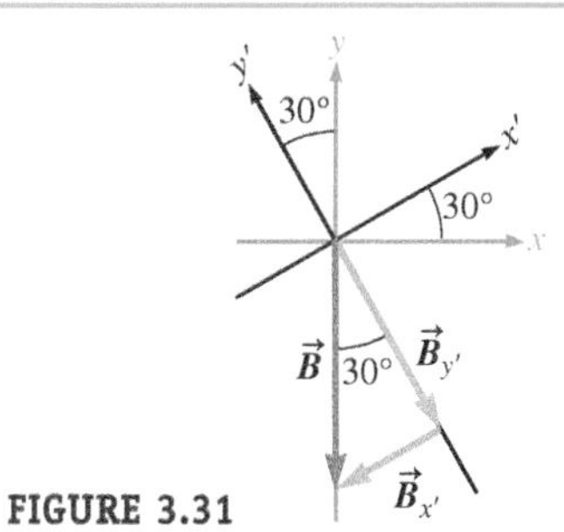

FIGURE 3.31

Use trigonometry to find the scalar components $B_{x'}$ and $B_{y'}$ (both negative).

$$B_{x'} = -B\sin 30° = -2.7\sin 30° = -1.4\ \text{m}$$
$$B_{y'} = -B\cos 30° = -2.7\cos 30° = -2.3\ \text{m}$$
$$\vec{B} = (-1.4\hat{\imath} - 2.3\hat{\jmath})\ \text{m}\quad \text{(primed coordinates)}$$

CHECK and THINK

Many beginners think that the x component of a vector is always found by taking the cosine of some angle and that the y component comes from taking the sine of that angle. In this example, we needed the sine function to find the x component and the cosine function to find the y component. It is best to draw a right triangle and use the general definition of the sine and cosine trigonometric functions (Appendix A) to find vector components. This procedure will give the correct components for any coordinate system, no matter how the angles are drawn.

EXAMPLE 3.6 Using Vector Information to Make a Graphical Representation

A vector $\vec{D}$ is given by $\vec{D} = -3.5\hat{\imath} + 4.0\hat{\jmath}$. Draw a representation of $\vec{D}$ as an arrow on an appropriate coordinate system.

INTERPRET and ANTICIPATE

Because the unit vectors $\hat{\imath}$ and $\hat{\jmath}$ are used, we need a two-dimensional x–y coordinate system.

Example continues on page 72 ▶

SOLVE
Identify the vector components.

$\vec{D}_x = -3.5\hat{\imath}$ $\vec{D}_y = 4.0\hat{\jmath}$

Draw a two-dimensional coordinate system with a suitable scale. The given vector is unitless, so do not show any units on the axes. Draw the vector components $\vec{D}_x$ and $\vec{D}_y$ on the coordinate system, placing the tail of $\vec{D}_y$ at the head of $\vec{D}_x$ (Fig. 3.32). $\vec{D}_x$ is negative, pointing along the negative x axis. $\vec{D}_y$ is positive, pointing parallel to the positive y axis.

To find $\vec{D}$, add $\vec{D}_x + \vec{D}_y$ using the head-to-tail method. The resultant vector is $\vec{D}$, as shown.

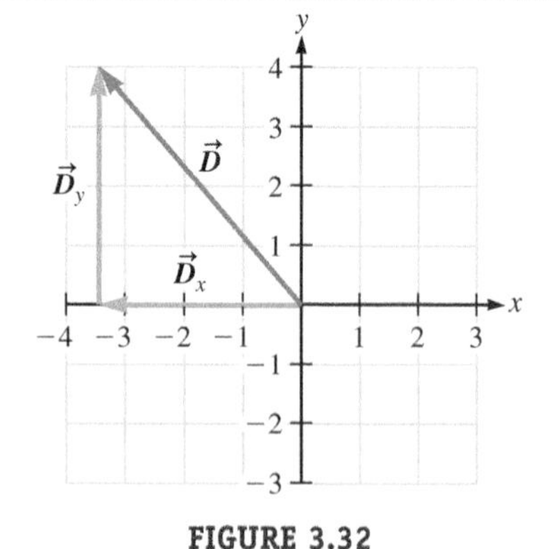

FIGURE 3.32

CHECK and THINK
It makes sense that vector $\vec{D}$ points to the left and up because its x component is negative (left) and its y component is positive (up).

VECTOR MAGNITUDE; VECTOR DIRECTION
Major Concepts

Vector Magnitude and Direction

We have represented vector quantities pictorially and in terms of components. A final way to describe a vector is in terms of its magnitude and direction.

Direction is usually given as an angle in some coordinate system. In the everyday world, this coordinate system is usually the cardinal system used on maps. For example, the phrase "Hurricane Isabelle is heading northeast" means that the hurricane's velocity vector points 45° to the east of north. In physics, the coordinate system used to describe the magnitude and direction of a vector quantity is often a two-dimensional x–y coordinate system. In such a system, we indicate direction as an angle measured from the x axis. The usual convention is that an angle measured counterclockwise from the x axis is considered positive. In Figure 3.33A, the angle between $\vec{A}$ and the x axis is θ, so we say "the direction of $\vec{A}$ with respect to x is θ." If a direction is given as a negative angle, the angle is measured *clockwise* from the x axis. In Figure 3.33B, the direction of $\vec{B}$ is θ measured counterclockwise from the x axis so that $270° < \theta < 360°$. The angle $\alpha = \theta - 360°$ is a negative number.

We can use the scalar components of a vector to find its magnitude and direction. Because $\vec{A}_x$, $\vec{A}_y$, and $\vec{A}$ form a right triangle as in Figure 3.26A, we can use the Pythagorean theorem to find the magnitude A:

$$A^2 = A_x^2 + A_y^2$$

$$A = \sqrt{A_x^2 + A_y^2} \tag{3.12}$$

Because A_x and A_y are squared, their signs do not affect A. Because A is a magnitude, it cannot be negative; therefore, we always choose the positive square root.

The process for finding the magnitude of a three-dimensional vector is similar:

$$A = \sqrt{A_x^2 + A_y^2 + A_z^2} \tag{3.13}$$

We use trigonometry to find the direction θ in terms of A_x and A_y. From Figure 3.26A, we see that

$$\tan\theta = \frac{A_y}{A_x}$$

$$\theta = \tan^{-1}\frac{A_y}{A_x} \tag{3.14}$$

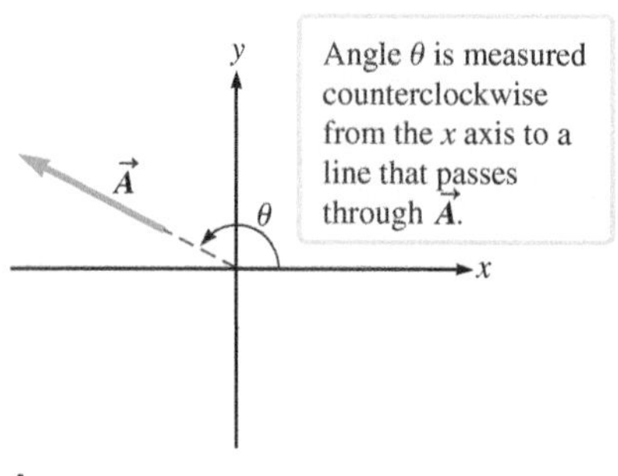

A.

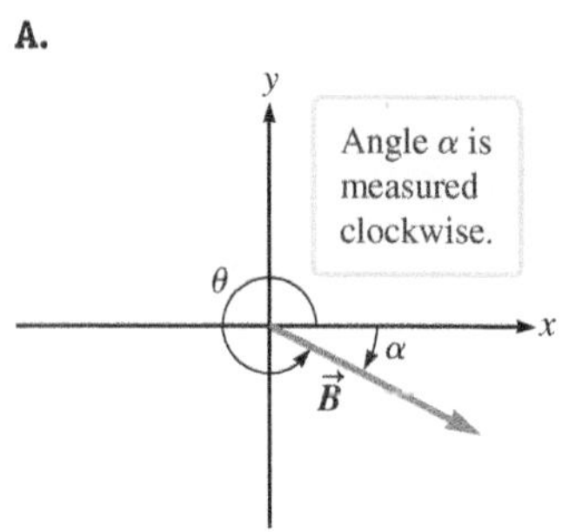

B.

FIGURE 3.33 **A.** The direction of a two-dimensional vector is given by the positive angle θ measured counterclockwise from the x axis. **B.** The direction may also be given by the negative angle $\alpha = \theta - 360°$ measured clockwise from the x axis.

Angles, Inverse Trigonometric Functions, and Vector Directions

You must be careful when using a calculator to take the inverse of a trigonometric function. Because trigonometric functions repeat (are periodic), the inverse of a trigonometric function has an infinite number of solutions. For example, $\tan^{-1} 0 = 0$, 180°, 360°, . . . even though a calculator usually gives only one solution, $\tan^{-1} 0 = 0$.

In a given physics problem, the solution given by a calculator may not be physically reasonable.

The best way to make sure that your answer for a vector's direction is correct is to draw a sketch and estimate the angle. If the angle shown on your calculator is very different from your estimate, you probably need to correct for the geometry of the problem and apply the convention of positive angles measured counterclockwise from the x axis. This convention has been programmed into most calculators.

For example, let us find the direction θ_B of the vector $\vec{B}$ in Figure 3.34. The vector points into quadrant II. So, the angle θ_B must be between 90° and 180°; we estimate that it is about 150°. Reading off $B_x = -3.0$ m and $B_y = 1.5$ m and entering

$$\tan^{-1}\left(-\frac{1.5}{3}\right)$$

into a calculator yields $\tan^{-1}\left(-\frac{1}{2}\right) = -27°$, which is nowhere near our estimate. This answer would mean that the vector points into quadrant IV.

In Figure 3.35, we see that the tangent function repeats every 180°, but a calculator gives only a single solution between −90° and 90°. In other words, the "inverse tangent button" on a calculator cannot tell which of B_x or B_y is negative. To find θ_B, we must take the solution from the calculator and add 180° to return the vector $\vec{B}$ to the proper quadrant (quadrant II; see Fig. 3.16):

$$\theta_B = -27° + 180° = 153°$$

which is very close to our estimate.

The inverse tangent button on a calculator will similarly confuse a vector in quadrant III with one in quadrant I. You must apply the appropriate correction for the geometry of the situation.

One way around this ambiguity produced by our calculators is to sketch the vector along with the angle θ measured from the x axis. If $|\theta| > 90°$, draw an acute angle α between the vector and the x axis as in Figure 3.34. Use the inverse tangent button on your calculator to find α from the magnitudes of the vector components. For the vector in Figure 3.34,

$$\alpha = \tan^{-1}\left(\frac{1.5}{3}\right) = 27°$$

Then use your sketch to find θ. In this case, $\theta_B = 180° - 27° = 153°$.

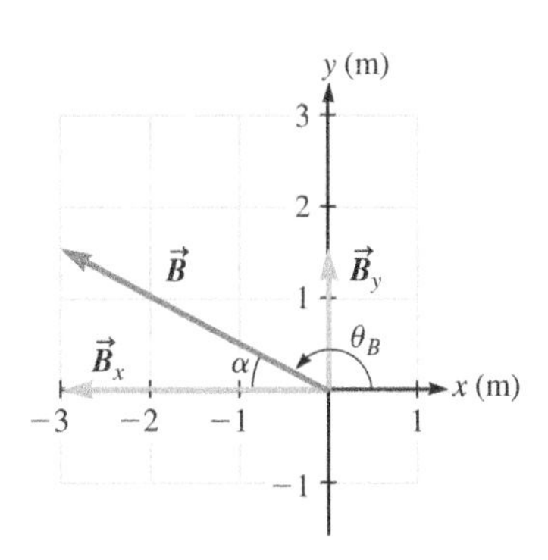

FIGURE 3.34 The direction of a vector is given by the angle measured counterclockwise from the x axis. The Pythagorean theorem and trigonometric functions are used to relate vector components to the magnitude and direction of a vector.

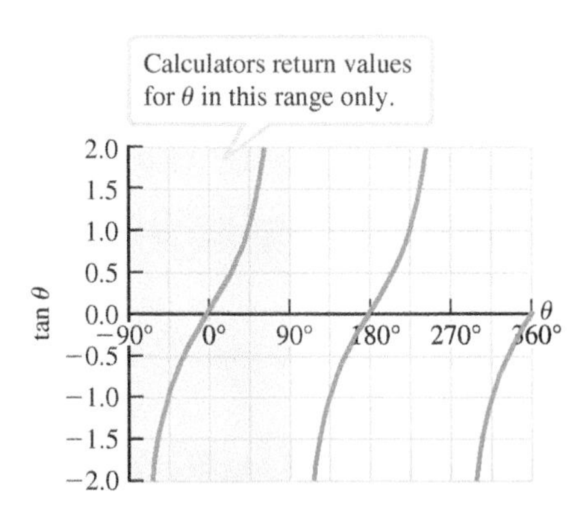

FIGURE 3.35 The tangent function repeats. When you use a calculator to find the inverse tangent, it will only give angles between −90° and 90°.

EXAMPLE 3.7 Translating Unit-Vector Information to Magnitude and Direction

Find the magnitude and direction of the vector $\vec{D} = (-3.5\hat{\imath} + 4.0\hat{\jmath})$ from Example 3.6.

INTERPRET and ANTICIPATE

Our work in Example 3.6 helps us anticipate the answer here. According to our sketch (Fig. 3.32), $\vec{D}$ vector points into quadrant II, so we expect that $90° < \theta < 180°$.

SOLVE The scalar components are the coefficients in front of the unit vectors.	$D_x = -3.5 \quad D_y = 4.0$
Find the magnitude from Equation 3.12.	$D = \sqrt{D_x^2 + D_y^2}$ $D = \sqrt{(-3.5)^2 + 4.0^2}$ $D = 5.3$
Use Equation 3.14 and a calculator to find the direction.	$\tan^{-1}\frac{4.0}{-3.5} = -49°$
CHECK and THINK This result does not meet our expectation that $90° < \theta < 180°$. So, we add 180° to correct the answer returned by the calculator because $\vec{D}$ should point into quadrant II.	$\theta = -49° + 180° = 131°$

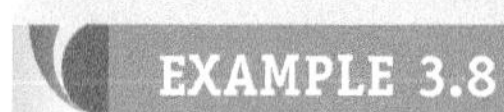

EXAMPLE 3.8 CASE STUDY Magnitude and Direction to Miss the Trees

In Example 3.2, we drew a representation of Reese's displacement from the time she set her parachute in the brakes configuration to the moment she landed safely in the field. In this example, we will find the magnitude and direction of her displacement assuming she just barely missed the trees as shown in Figure 3.1. In the next section, we will determine whether she would have survived the jump had she opened her parachute at the usual 750 m instead of at 1200 m.

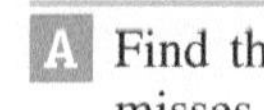

A Find the magnitude and direction of Reese's displacement if she just barely misses the trees. Use the distance data given in Figure 3.1, page 60.

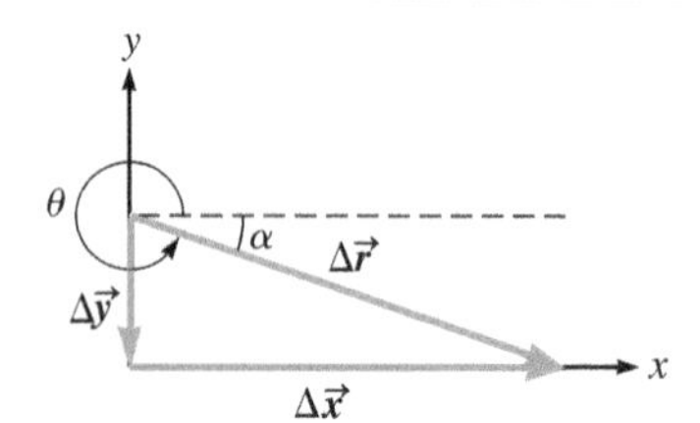

FIGURE 3.36

INTERPRET and ANTICIPATE

Roughly sketch Reese's displacement $\Delta\vec{r}$ in a two-dimensional coordinate system and show the vector components $\Delta\vec{x}$ and $\Delta\vec{y}$ of $\Delta\vec{r}$. Also show the angles θ and α, either of which describes the direction of the displacement vector (Fig. 3.36).

SOLVE

We are given Δx and Δy. Assign Δy a negative value (downward).	$\Delta x = 3500\text{ m} \qquad \Delta y = -1200\text{ m}$
Modify $A = \sqrt{A_x^2 + A_y^2}$ (Eq. 3.12) to find the magnitude of the displacement.	$\Delta r = \sqrt{\Delta x^2 + \Delta y^2}$ (3.12) $\Delta r = \sqrt{(3500\text{ m})^2 + (-1200\text{ m})^2}$ $\Delta r = 3.7 \times 10^3\text{ m}$
Modify $\theta = \tan^{-1} A_y/A_x$ (Eq. 3.14) to find the direction.	$\tan^{-1}\frac{\Delta y}{\Delta x} = \tan^{-1}\frac{-1200}{3500} = -19°$
From our drawing, we see that we have just found α. To find θ, we need to add 360° to our α value.	$\alpha = -19°$ $\theta = 360° + \alpha = 360° - 19°$ $\theta = 341°$

CHECK and THINK

The answer expressed as $\alpha = -19°$ means that Reese fell through an angle of 19° from the horizontal after she set her parachute in the brakes configuration at an altitude of 1200 m.

B Find the magnitude and direction of Reese's displacement if she opens her parachute at an altitude of 750 m instead of 1200 m. Assume she still covers the same horizontal distance as in part A.

INTERPRET and ANTICIPATE

With her parachute closed, Reese falls straight down so that when she opens her parachute the point directly below her is still 3500 m from the edge of the clearing. Her altitude is only 750 m, however.

SOLVE

This situation is like part A except that the y component of displacement has changed.	$\Delta x = 3500\text{ m} \qquad \Delta y = -750\text{ m}$
As before, modify Equation 3.12 to find the magnitude of the displacement.	$\Delta r = \sqrt{\Delta x^2 + \Delta y^2} = \sqrt{(3500\text{ m})^2 + (-750\text{ m})^2}$ $\Delta r = 3.6 \times 10^3\text{ m}$
Modify Equation 3.14 to find the direction in terms of α.	$\alpha = \tan^{-1}\frac{\Delta y}{\Delta x} = \tan^{-1}\left(\frac{-750}{3500}\right) = -12°$

To find θ we need to add 360° as in part A.	$\theta = 360° + \alpha = 360° - 12°$ $\theta = 348°$

CHECK and THINK

When we return to this CASE STUDY in the next section, we will determine whether or not Reese would have cleared the trees with her chute opening at 750 m.

EXAMPLE 3.9 Translating Magnitude and Direction into Component Form

Jai alai (pronounced "high-lie") is a court game in which players use a long hand-shaped basket strapped to their wrist to propel a ball (Fig. 3.37). The object of the game is to bounce the ball against the wall in such a way that an opponent cannot catch and return the ball. Suppose a jai alai ball collides with a wall at a speed of 69 m/s and a 55° angle from the vertical as shown in Figure 3.38. Choose an x–y coordinate system and write the ball's velocity in component form.

FIGURE 3.37 A jai alai player in action.

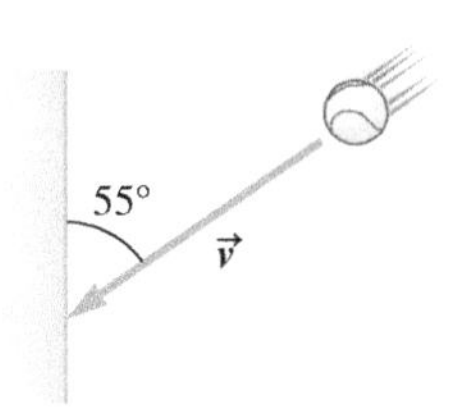

FIGURE 3.38

INTERPRET and ANTICIPATE

Choose a coordinate system and draw a sketch (Fig 3.39). For convenience, place the tail of $\vec{v}$ at the origin and draw the vector components of $\vec{v}$. In anticipation of using $\theta = \tan^{-1}(A_y/A_x)$ (Eq. 3.14), we also show the angle θ measured counterclockwise from the x axis. The velocity vector points in the negative x and y directions, so we expect a numerical result of the form $\vec{v} = (-\_\_\_\_\_\ \hat{\imath} - \_\_\_\_\_\ \hat{\jmath})$ m/s.

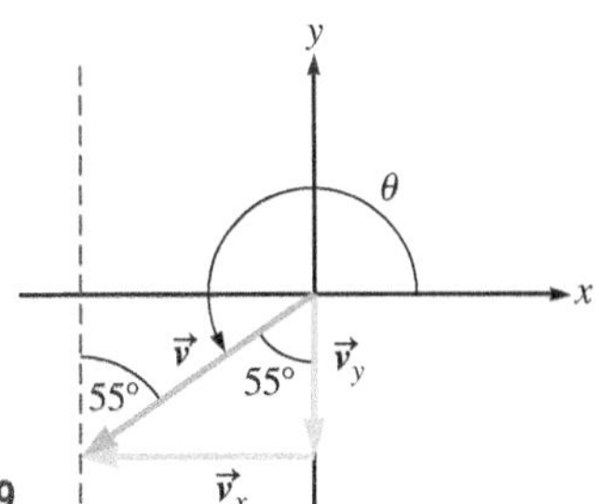

FIGURE 3.39

SOLVE Solve for θ.	$\theta + 55° = 270°$ $\theta = 270° - 55° = 215°$
Trigonometry (Eqs. 3.9 and 3.10) gives the scalar components.	$v_x = v\cos\theta = (69\text{ m/s})\cos 215° = -57\text{ m/s}$ $v_y = v\sin\theta = (69\text{ m/s})\sin 215° = -40\text{ m/s}$ (two significant figures)
To write the ball's velocity in component form, place the scalar components in front of the appropriate unit vectors.	$\vec{v} = (-57\hat{\imath} - 40\hat{\jmath})\text{ m/s}$
CHECK and THINK Both components are negative as we expected. Breaking a velocity into its horizontal and vertical scalar components is an important part of analyzing two-dimensional motion. So, let's use the acute 55° angle and trigonometry to find the components a second time. The velocity vector is the hypotenuse of a right triangle; the x component is opposite the angle, and the y component is adjacent. We can use the sine and cosine functions to find the magnitude of the vector components.	$\sin 55° = \frac{v_x}{v}$ $v_x = v\sin 55°$ $v_x = (69\text{ m/s})\sin 55° = 57\text{ m/s}$ $\cos 55° = \frac{v_y}{v}$ $v_y = v\cos 55°$ $v_y = (69\text{ m/s})\cos 55° = 40\text{ m/s}$

The negative signs do not come automatically using this method. Instead, we must use our sketch to reason that both components are negative.

3-4 Combining Vectors by Components

In Sections 3-1 and 3-3, we represented vectors pictorially, in component form, and in terms of magnitude and direction. When solving a problem, we often use two or three of these representations. The pictorial representation helps most when planning how to solve a problem, making estimates, and checking a solution. A problem often gives a vector in terms of magnitude and direction, or the final answer may require that form. Complicated calculations are easiest to do using component form, which is the subject of this section. There are three things you need to know about manipulating components.

First, if two vectors are equal to each other, their components are equal. For example, if $\vec{A} = \vec{B}$, then

$$A_x = B_x \tag{3.15}$$

$$A_y = B_y \tag{3.16}$$

These equalities ensure that their magnitudes and directions are equal. To show that the magnitudes are equal, start by finding the magnitude of $\vec{A}$ (Eq. 3.12):

$$A = \sqrt{A_x^2 + A_y^2}$$

Now substitute Equations 3.15 and 3.16 for each component and show that their magnitudes are equal:

$$A = \sqrt{B_x^2 + B_y^2} = B$$

Next, let's find θ_A (the direction of $\vec{A}$) from Equation 3.14:

$$\theta_A = \tan^{-1}\frac{A_y}{A_x}$$

Finally, substitute Equations 3.15 and 3.16 for each component and show that their directions are the same:

$$\theta_A = \tan^{-1}\frac{B_y}{B_x} = \theta_B$$

The second thing you need to know is that when a vector is multiplied (or divided) by a scalar, each of the vector's components is multiplied (or divided) by that scalar. For example, if $\vec{R}$ is the resultant of multiplying $\vec{A}$ by a scalar s, then

$$\vec{R} = s\vec{A} = sA_x\hat{\imath} + sA_y\hat{\jmath} \tag{3.17}$$

The resultant's magnitude is changed by the factor s:

$$R = \sqrt{(sA_x)^2 + (sA_y)^2}$$
$$R = s\sqrt{A_x^2 + A_y^2} = sA$$

but the direction remains unchanged:

$$\tan\theta_R = \frac{sA_y}{sA_x} = \frac{A_y}{A_x} = \tan\theta_A$$

If s is negative, however, the direction of the original vector is reversed.

VECTOR ADDITION
Major Concept

Finally, you need to know that to add or subtract two vectors. you must add or subtract each of their components. For example, if

$$\vec{R} = \vec{A} + \vec{B}$$

the components of $\vec{R}$ are found by adding the components of $\vec{A}$ and $\vec{B}$:

$$R_x = A_x + B_x$$
$$R_y = A_y + B_y$$

The resultant $\vec{R}$ is expressed in component form as

$$\vec{R} = (A_x + B_x)\hat{\imath} + (A_y + B_y)\hat{\jmath} \tag{3.18}$$

VECTOR SUBTRACTION
Major Concept

As an example of the subtraction of two vectors, if

$$\vec{Q} = \vec{A} - \vec{B}$$

the components of $\vec{Q}$ are found by subtracting the components of $\vec{B}$ from those of $\vec{A}$, and $\vec{Q}$ is expressed in component form as

$$\vec{Q} = (A_x - B_x)\hat{\imath} + (A_y - B_y)\hat{\jmath} \tag{3.19}$$

Vector Components of Motion Variables

We are now ready to connect the kinematics we learned in Chapter 2 to the vector mathematics we have learned here. Most of the equations for one-dimensional motion in Chapter 2 were written in terms of vector components. The advantage of using vector components is that it is relatively easy to expand these relationships to three-dimensional motion. Let's do that now for position, displacement, and average velocity.

In Chapter 2, we wrote the position vector $\vec{r}$ in terms of one of three vector components ($\vec{x}$, $\vec{y}$, or $\vec{z}$). The position vector of an object in three-dimensional space is the vector sum of these three vector components:

$$\vec{r} = \vec{x} + \vec{y} + \vec{z} \tag{3.20}$$

By using unit vectors, we write the position vector in terms of its scalar components:

$$\vec{r} = x\hat{\imath} + y\hat{\jmath} + z\hat{k} \tag{3.21}$$

The displacement vector is found by subtracting the initial position vector from the final position vector $\Delta\vec{r} = \vec{r}_f - \vec{r}_i$. This step is done by subtracting each initial component from the corresponding final component (Eq. 3.19):

$$\Delta\vec{r} = (x_f - x_i)\hat{\imath} + (y_f - y_i)\hat{\jmath} + (z_f - z_i)\hat{k} \tag{3.22}$$

Each component in Equation 3.22 is the displacement in one direction (Eq. 2.1):

$$\Delta\vec{r} = \Delta x\hat{\imath} + \Delta y\hat{\jmath} + \Delta z\hat{k} \tag{3.23}$$

or

$$\Delta\vec{r} = \Delta\vec{x} + \Delta\vec{y} + \Delta\vec{z} \tag{3.24}$$

To find the average velocity, divide the displacement vector $\Delta\vec{r}$ by the time interval Δt. The time interval is a scalar, so we just need to divide each component of $\Delta\vec{r}$ (Eq. 3.23) by Δt:

$$\vec{v}_{\text{av}} = \frac{\Delta\vec{r}}{\Delta t} = \frac{\Delta x}{\Delta t}\hat{\imath} + \frac{\Delta y}{\Delta t}\hat{\jmath} + \frac{\Delta z}{\Delta t}\hat{k} \tag{3.25}$$

Each component is the average velocity in that particular direction (Eq. 2.2):

$$\vec{v}_{\text{av}} = v_{\text{av},x}\hat{\imath} + v_{\text{av},y}\hat{\jmath} + v_{\text{av},z}\hat{k} \tag{3.26}$$

EXAMPLE 3.10 Adding Vectors

Three vectors $\vec{A} = 2\hat{\imath} - 3\hat{\jmath}$, $\vec{B} = 4\hat{\imath} - 2\hat{\jmath}$, and $\vec{C} = -5\hat{\imath} - 7\hat{\jmath}$ are dimensionless. Find vector $\vec{R} = \vec{A} + \vec{B} + \vec{C}$ in component form. Also give the magnitude and direction of $\vec{R}$.

INTERPRET and ANTICIPATE

The best way to anticipate the result of adding three vectors is to sketch the vectors and add them geometrically (Fig. 3.40). The three vectors are drawn head to tail. The resultant vector $\vec{R}$ is drawn from the tail of $\vec{A}$ to the head of $\vec{C}$.

From our sketch, we see that the x component of $\vec{R}$ is positive and small and that the y component is negative and large. We see that angle θ is between 270° and 360°.

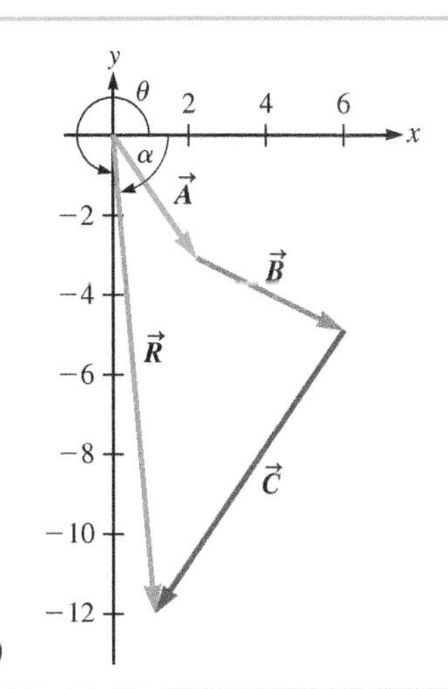

FIGURE 3.40

Example continues on page 78 ▶

SOLVE According to Equation 3.18, to find the vector sum, we add the x components together and the y components together.	$R_x = A_x + B_x + C_x = 2 + 4 - 5 = 1$ $R_y = A_y + B_y + C_y = -3 - 2 - 7 = -12$
Writing $\vec{R}$ in component form means adding the vector components of $\vec{R}$.	$\vec{R} = R_x\hat{\imath} + R_y\hat{\jmath}$ $\vec{R} = 1\hat{\imath} - 12\hat{\jmath}$

CHECK and THINK
As expected, the x component is positive and small, and the y component is negative and large.

SOLVE The magnitude of the resultant comes from modifying Equation 3.12.	$R = \sqrt{R_x^2 + R_y^2} = \sqrt{1^2 + (-12)^2}$ (3.12) $R = 12$ (two significant figures)
The direction comes from Equation 3.14.	$\alpha = \tan^{-1}\frac{R_y}{R_x} = \tan^{-1}\frac{-12}{1} = -85°$ (3.14)
Our calculator is giving us angle α instead of angle θ, so we must add 360° to find θ. That is one reason our sketch is so important.	$\theta - \alpha = 360°$ $\theta = \alpha + 360° = -85° + 360°$ $\theta = 275°$

CHECK and THINK
Because the y component is 12 times greater than the x component, the x component has little effect on the magnitude. From our sketch, we would estimate that the magnitude is near 12. Also, we can see that θ is close to 270°, which matches our algebraic results.

EXAMPLE 3.11 CASE STUDY Clearing the Trees in the Brakes Configuration

When Reese saw that she was in danger of hitting the trees, she set her parachute in the brakes configuration (case study, page 60). In this configuration, her velocity was constant at $\vec{v} = (6.0\hat{\imath} - 1.5\hat{\jmath})$ m/s. Had she deployed her parachute at 750 m, would she have cleared the trees?

FIGURE 3.41 Suppose Reese releases her parachute at 750 m instead of at 1200 m. Compare with Figures 3.1 and 3.12.

INTERPRET and ANTICIPATE
Our plan is to use Reese's velocity to find her horizontal displacement during the descent. If her horizontal displacement is at least 3500 m ($\Delta x \geq 3500$ m), she would safely clear the trees.

SOLVE Because Reese's velocity is constant, her average velocity and her instantaneous velocity are equal.	$\vec{v}_{av} = \vec{v} = (6.0\hat{\imath} - 1.5\hat{\jmath})$ m/s $v_{av,x} = 6.0$ m/s $v_{av,y} = -1.5$ m/s
If she deploys her parachute at 750 m, then $\Delta y = -750$ m for the jump. Her horizontal displacement Δx depends on time Δt it takes her to descend straight down. Find Δt from the known $v_{av,y}$ and Δy using Equation 3.25.	$v_{av,y} = \frac{\Delta y}{\Delta t}$ (3.25) $\Delta t = \frac{\Delta y}{v_{av,y}} = \frac{-750\text{ m}}{-1.5\text{ m/s}} = 500$ s

Use this Δt to find Δx because the x component of velocity is also constant.	$v_{av,x} = \frac{\Delta x}{\Delta t}$ (3.25) $\Delta x = v_{av,x}\Delta t = (6.0\ m/s)(500\ s)$ $\Delta x = 3000\ m$ No, she would have hit the trees.

CHECK and THINK

We conclude that Reese would have landed in the trees had she waited and opened her parachute at 750 m. The advantage of deploying her parachute at a higher altitude is that the longer she spends in the air, the larger the value of Δx, assuming that she has the parachute open in the brakes configuration.

SUMMARY

Underlying Principles

No new physical concepts were introduced in the chapter. Instead, we focused on mathematical tools. We have not finished adding all the tools needed to work with vectors, but the rest will be added as we need them. The tools we now have will take care of our needs for the next seven chapters.

Major Concepts

1. Any vector can be written as the sum of its **vector components**: $\vec{A} = \vec{A}_x + \vec{A}_y$. Alternatively, a vector can be written in **component form**, which means that the vector is written in terms of its **scalar components** and the appropriate unit vectors: $\vec{A} = A_x\hat{\imath} + A_y\hat{\jmath}$.
2. The process of finding a vector's (vector or scalar) components is known as **resolving the vector into components**. Trigonometry is used to resolve a vector into components.
3. The **magnitude** of a two-dimensional vector $\vec{A}$ is
$$A = \sqrt{A_x^2 + A_y^2} \tag{3.12}$$
The **direction** of a two-dimensional vector measured counterclockwise from the x axis may be written in terms of its scalar components as
$$\theta = \tan^{-1}\frac{A_y}{A_x} \tag{3.14}$$
4. Vectors may be **added**, $\vec{R} = \vec{A} + \vec{B}$, in terms of their scalar components:
$$\vec{R} = (A_x + B_x)\hat{\imath} + (A_y + B_y)\hat{\jmath} \tag{3.18}$$
Vectors may be **subtracted**, $\vec{Q} = \vec{A} - \vec{B}$, in terms of their scalar components:
$$\vec{Q} = (A_x - B_x)\hat{\imath} + (A_y - B_y)\hat{\jmath} \tag{3.19}$$
5. Vector addition is **commutative**, which means that the order of addition does not matter:
$$\vec{A} + \vec{B} = \vec{B} + \vec{A} \tag{3.2}$$
6. Vector addition is **associative**, which means that if there are more than two vectors to add, they can be grouped in any order:
$$(\vec{A} + \vec{B}) + \vec{C} = \vec{A} + (\vec{B} + \vec{C}) \tag{3.3}$$

Tools

1. Vector manipulation
   a. **Parallelogram addition** is a geometric method for adding vectors $\vec{R} = \vec{A} + \vec{B}$. In this method, two copies of each vector $\vec{A}$ and $\vec{B}$ are used to form a parallelogram. The resultant vector $\vec{R}$ is drawn along the diagonal of the parallelogram from the tails of the two vectors $\vec{A}$ and $\vec{B}$ to their heads.
   b. **Head-to-tail addition** is another geometric method for adding vectors $\vec{R} = \vec{A} + \vec{B}$. In this method, the head of one of the vectors is placed at the tail of the other vector. The resultant vector is drawn from the free tail of the first vector $\vec{A}$ to the free head of the second vector $\vec{B}$.

Tools—cont'd

Geometric **subtraction** $\vec{R} = \vec{A} - \vec{B}$ may be performed by first multiplying $\vec{B}$ by -1 and then applying either the parallelogram or head-to-tail method.

c. **Tail-to-tail subtraction** is an alternative geometric method for subtracting vectors $\vec{R} = \vec{A} - \vec{B}$. The two vectors $\vec{A}$ and $\vec{B}$ are placed tail to tail, and the resultant vector points from the head of the second vector $\vec{B}$ to the head of the first vector $\vec{A}$.

2. Coordinate systems
   a. In a **Cartesian coordinate system**, two or more axes cross at right angles to each other.
   b. All the coordinate systems used in this book are **right-handed coordinate systems.** In a right-handed coordinate system, the direction of $\hat{k}$ is determined by the *right-hand convention* (Fig. 3.22).

PROBLEMS AND QUESTIONS

A = algebraic C = conceptual E = estimation G = graphical N = numerical

3-1 Geometric Treatment of Vectors

1. A velocity vector has a magnitude of 720 m/s. Two students draw arrows representing this vector. Clarisse chooses a scale such that 1 cm → 100 m/s.
   a. N What is the length of the arrow that Clarisse draws?
   b. N Francois's arrow is half as long as Clarisse's. What is Francois's scale?
   c. C Is one student's choice better than the other? If so, what makes it a better scale?
2. A young boy throws a baseball through a window.
   a. C Sketch the problem and pick a reference point.
   b. G Use this reference point to draw a vector representing the initial and final position of the ball.
   c. G Draw the displacement vector.
   d. C Which of these vectors change if you pick a different reference point?
3. G Vectors $\vec{A}$ and $\vec{B}$ are perpendicular and have the same nonzero magnitude ($A = B$). If $\vec{C} = \vec{A} + \vec{B}$, what is C, the magnitude of $\vec{C}$? *Hint*: Sketch these vectors.
4. G Vectors $\vec{A}$ and $\vec{B}$ have the same nonzero magnitude ($A = B$), where $\vec{C} = \vec{A} + \vec{B}$ and $\vec{D} = \vec{A} - \vec{B}$. If the magnitudes of $\vec{C}$ and $\vec{D}$ are the same, how are the directions of $\vec{A}$ and $\vec{B}$ related? *Hint*: Sketch these vectors.
5. N Vector $\vec{A}$, with a magnitude of 18 units, points in the positive x direction. Adding vector $\vec{B}$ to vector $\vec{A}$ yields a resultant vector that points in the negative x direction with a magnitude of 6 units. What are the magnitude and direction of vector $\vec{B}$?

Problems 6, 42, and 64 are grouped.

6. G Figure P3.6 shows three vectors. Copy this figure on to your own paper and find $\vec{R} = \vec{F}_1 + \vec{F}_2 + \vec{F}_3$ geometrically.
7. C A student makes a mistake in finding displacement $\Delta\vec{r}$ using head-to-tail addition. Instead of drawing $-\vec{r}_i$, the student draws $-\vec{r}_f$. What, if anything, is wrong with the resulting displacement vector's magnitude or direction?

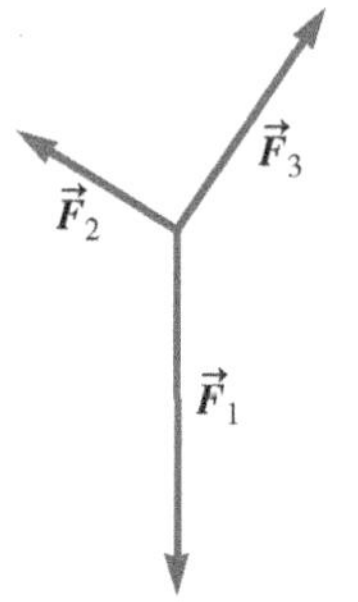

FIGURE P3.6 Problems 6 and 64.

8. The layout of the town of Popperville is a perfectly square grid, with blocks 100 feet long (Fig. P3.8). A cat leaves her house and travels to Mike Mulligan's place 4 blocks west and 3 blocks north of its starting point. After lapping up the milk that Mike always leaves out, the cat travels an additional 1 block north and 3 blocks east to Mrs. McGillicutty's house.
   a. G On your paper, carefully copy the grid of streets that contains the cat's house, Mike Mulligan's house, and Mrs. McGillicutty's house. Then draw and label three displacement vectors on your sketch: $\Delta\vec{r}_1$, the cat's displacement from her house to Mike Mulligan's; $\Delta\vec{r}_2$, the cat's displacement from Mike Mulligan's to Mrs. McGillicutty's; and $\Delta\vec{r}_{tot}$, the cat's total displacement.
   b. A Write an equation to relate the three displacement vectors you have drawn.
   c. C Is the distance the cat traveled greater than, less than, or equal to the magnitude of the cat's total displacement vector? Explain.

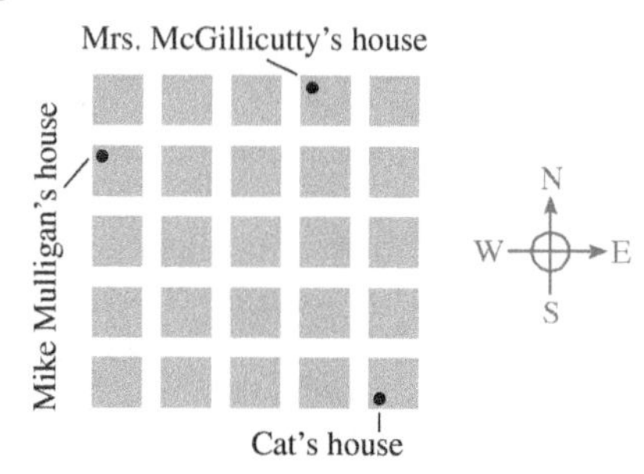

FIGURE P3.8

9. C The magnitude of average velocity is miscalculated by dividing the displacement Δr by $2\Delta t$ instead of Δt. The mistake is carried forward, and an average velocity vector is drawn to scale. What, if anything, is wrong with this vector's magnitude or direction?
10. G An Olympic runner races through the final turn of the track on her way to the finish line. For the following questions, assume the curved section is a semicircle and that the runner moves clockwise around the track as watched from above. **a.** On your paper, draw a sketch of the curved section of the track as seen from above. Mark two points on the track, one to represent the position of the runner at time t_i, just after she has entered the turn, and one to represent her position at time t_f, part of the way through the turn. (You have freedom in your choices.) **b.** Using a point in the interior of the track as an origin, draw vectors to represent the position of the runner at time t_i (vector $\vec{r}_i$) and the position of the runner at time t_f (vector $\vec{r}_f$). **c.** Use your position vectors to find the displacement of the runner.
11. A Three vectors $\vec{A}$, $\vec{B}$, and $\vec{C}$ are related to one another such that $2\vec{A} + 3\vec{B} = 18\vec{C}$. A fourth vector $\vec{D}$ exists such that $6\vec{C} + \vec{D} = \vec{A}$. Determine an expression for $\vec{D}$ in terms of the vectors $\vec{A}$ and $\vec{B}$.

12. G Draw three arbitrary vectors $\vec{A}$, $\vec{B}$, and $\vec{C}$. Use these three vectors to demonstrate that vector addition is associative (Eq. 3.3).

13. C In Chapter 5, you will study a very important vector, *force*. Each case in Figure P3.13 shows an example of force vectors exerted on an object. These forces are all of the same magnitude F_0. Assume that the forces lie in the plane of the paper. Rank the cases from greatest to smallest according to the magnitude of the total force. *Note*: The total force is the vector sum of the individual forces exerted on the object.

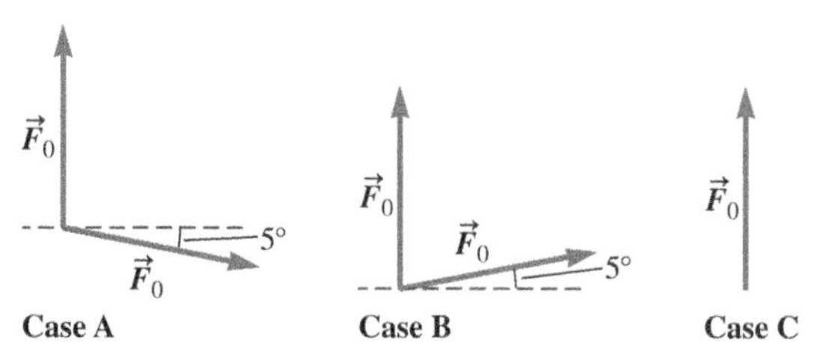

FIGURE P3.13

Problems 14 and 15 are paired.

14. G For this problem, you will need a ruler. A classroom clock has a small magnifying glass embedded near the end of the minute hand. The magnifying glass may be modeled as a particle undergoing translational motion along a circular path. Class begins at 7:55 and ends at 8:50 (Fig. P3.14) **a.** Suppose the minute hand of the clock has an actual length of 0.300 m. Use your ruler to find the scale we used when drawing Figure P3.14. Use that scale to draw the displacement vector of the magnifying glass during the 55-min class period. In the **Check and Think** step, be sure to comment on the direction of the displacement. **b.** Use your ruler to find the magnitude of the displacement vector.

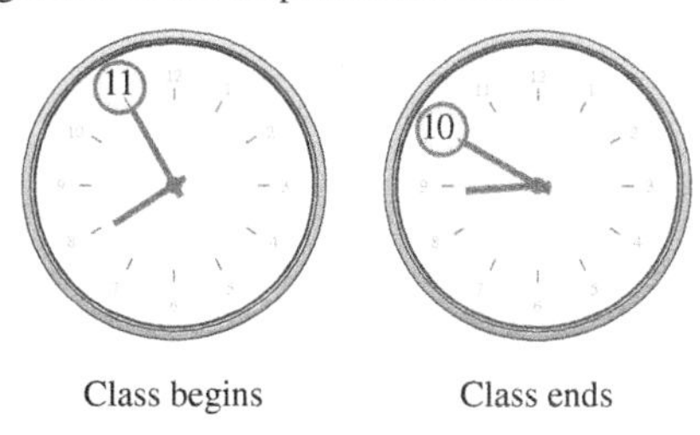

FIGURE P3.14 Problems 14 and 15.

15. Refer to the situation described in Problem 14. Average velocity is the displacement divided by the time interval:

$$\vec{v}_{av} = \frac{\Delta \vec{r}}{\Delta t}$$

a. N Graphically find the magnitude of the average velocity of the magnifying glass during the 55-min class period.

b. C Describe the direction of the average velocity vector.

16. G Vector $\vec{A}$ has a magnitude of 4.50 m and makes an angle of 64.0° with the positive x axis. Vector $\vec{B}$, with a magnitude equal to that of vector $\vec{A}$, points along the negative y axis (Fig. P3.16). Graphically find the magnitude and direction of the resultant vector **a.** $\vec{A} + \vec{B}$, **b.** $\vec{A} - \vec{B}$, **c.** $\vec{B} - \vec{A}$, and **d.** $2\vec{A} - \vec{B}$.

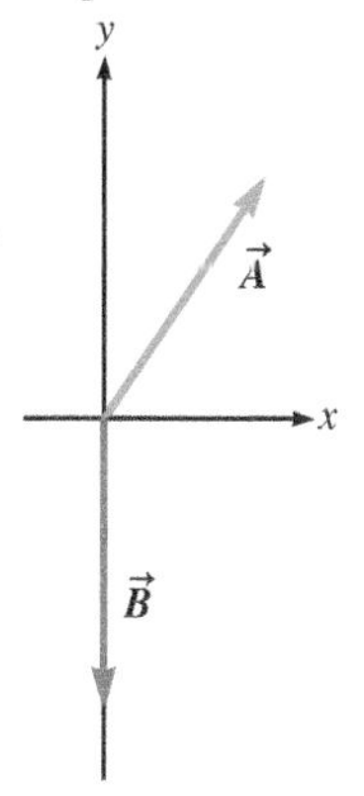

FIGURE P3.16

17. Miguel, an Ultimate Frisbee player, is running three drills (Fig. P3.17). In the first drill, Miguel runs a distance d straight down the field and then makes a 90° turn to the right, running an additional distance d. In the second drill, Miguel runs a distance d straight down the field and makes a 95° turn to the right before running an additional distance d. In the final drill, Miguel runs a distance d straight down the field and makes an 85° turn to the right before running an additional distance d.

a. C In which case did Miguel end up farthest from his starting point? In which case did he end up closest to his starting point?

b. C In which case(s) does Miguel end up farther away from his starting point than distance d?

c. N At what angle will his distance away be exactly d?

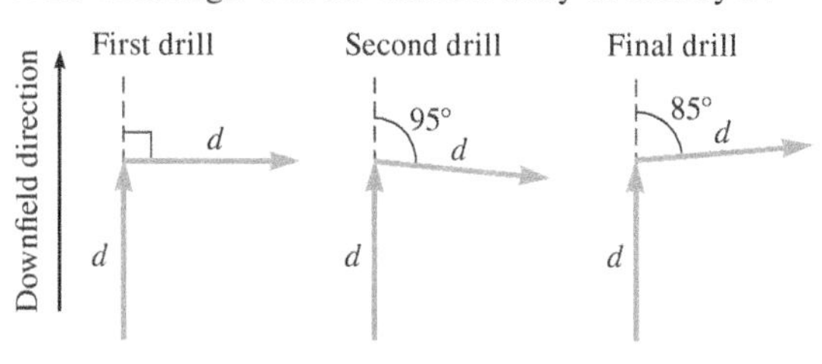

FIGURE P3.17

3-2 Cartesian Coordinate Systems

18. N A baseball diamond consists of four plates arranged in a square. Each side of the square is 90 ft (27.43 m) long. Use an $x-y$ coordinate system with the origin at the center of the diamond as shown in Figure P3.18. **a.** What is the position of each plate in this system? **b.** What is the distance from home plate to second base?

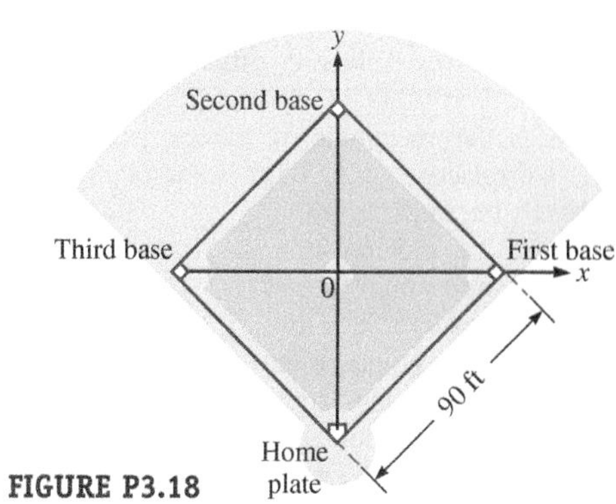

FIGURE P3.18

19. N A museum curator is setting up a new display. The show has a particularly large painting that is 6.6 m long. The floor plan for the gallery is shown in Figure P3.19. Can the curator display this large painting in this gallery? If so, on which walls can it hang?

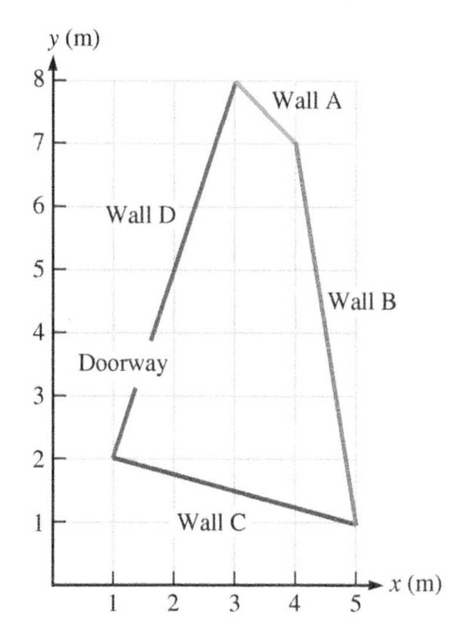

FIGURE P3.19

20. A football is thrown along an arc. The ball is released at a height of 6.54 ft (1.99 m) above the ground and, after traveling a horizontal distance of 87.8 ft (26.8 m), is caught by a receiver at a point that is 4.22 ft (1.29 m) above the ground.

a. G Sketch the throw from the point of release to the moment the ball is caught (as viewed by an observer from the sideline), placing an x–y coordinate system in the picture with the origin at ground level, directly below where the ball is released.

b. N Determine the coordinates of the ball at the point it is released and where it is caught.

c. N What is the magnitude of the displacement between the initial and final points in the ball's flight?

21. N Two aircraft approaching an aircraft carrier are detected by radar. The first aircraft is at a horizontal distance of 34.3 km in the direction 42.0° north of east and has an altitude of 2.40 km. The second aircraft, at half this altitude, is at a horizontal distance of 42.6 km in the direction 37.0° north of east. What is the distance separating the two aircraft? *Hint*: Use a coordinate system with the x axis pointing east, the y axis pointing north, and the z axis in the vertical direction.

22. C Because right-handed people hold the pencil in their right hand, they often mistakenly use their free left hand to draw a coordinate system. In effect, they draw a left-handed coordinate system. Suppose that such a person has drawn the x axis pointing to the right and the y axis pointing toward the top of the page. Which way does the z axis point in a left-handed system?

23. N A truck driver delivering office supplies downtown travels 5.00 blocks north, 3.00 blocks east, and finally 2.00 blocks south. **a.** What is the magnitude (in blocks) and the direction of the driver's displacement? **b.** What is the total distance traveled (in blocks) by the truck driver?

24. C Instead of closing their right hand so that their fingers push the x axis into the y axis and their thumb points in the direction of the z axis, some people like to hold their right index finger, middle finger, and thumb at right angles to one another as shown in Figure P3.24. Suppose that the z axis is associated with the thumb. With which axis is the index finger associated? With which axis is the middle finger associated?

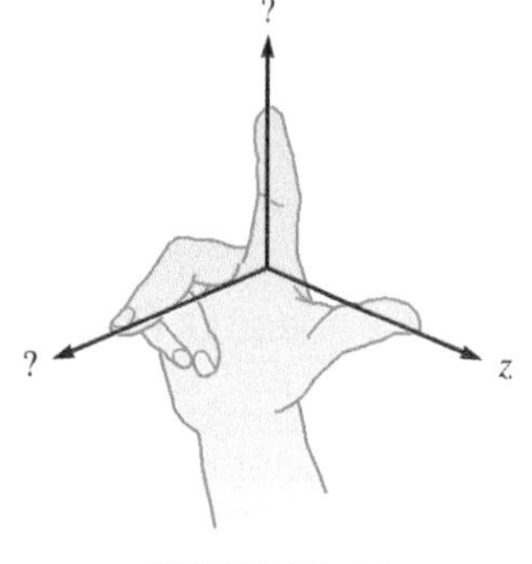

FIGURE P3.24

3-3 Components of a Vector

25. N Carolyn rides her bike 40.0° south of west for 5.40 miles as illustrated in Figure P3.25. What is the distance she would have to ride due south and due west to reach the same location?

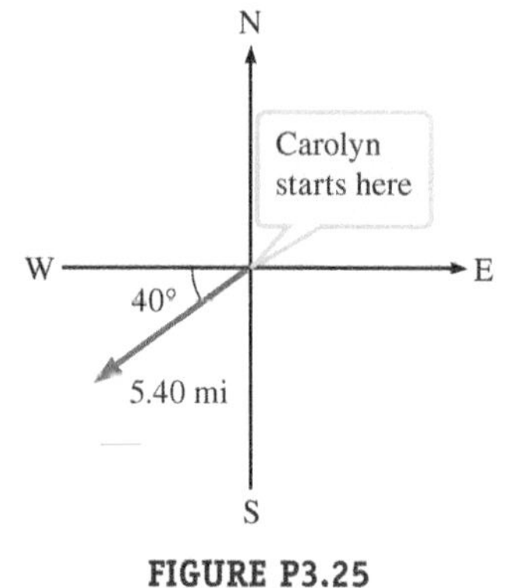

FIGURE P3.25

Problems 26 and 27 are paired.

26. G Draw the vector $\vec{A} = 16.5\hat{\imath} - 33.0\hat{\jmath}$ on a coordinate system.

27. N Find the magnitude and direction of the vector $\vec{A} = 16.5\hat{\imath} - 33.0\hat{\jmath}$.

28. N A vector's x component is twice its y component. What is the acute angle between the two vector components?

Problems 29 and 30 are paired.

29. N Vector $\vec{B}$ has a magnitude of 19.45. Its direction measured counterclockwise from the x axis is 127.3°. Find its vector components.

30. G Draw the vector in Problem 29 on a coordinate system.

31. N A vector points into the first quadrant, and its x and y components are both positive. If its magnitude is equal to twice the magnitude of its x component, what is the angle between the vector and the positive x axis?

32. N Find the magnitude of the vector $\vec{B} = -2.00\hat{\imath} + 3.00\hat{\jmath} - 4.00\hat{k}$.

33. N Vector $\vec{A}$ has scalar components of 4.00 units in the negative x direction and 2.00 units in the negative y direction. **a.** Write the vector $\vec{A}$ in component form. **b.** What are the magnitude and direction of $\vec{A}$? **c.** Find the vector $\vec{B}$ that when added to $\vec{A}$ yields a resultant vector with an x component 6.00 units in the negative x direction and no y component.

34. A soccer player starts at one end of the field and runs along a straight line making an angle θ with the goal line. She is able to move d meters closer to the opposing goal line before going out of bounds.

a. G Make a sketch of this situation that includes the soccer player's displacement vector $\Delta\vec{r}$. Label d and θ on your sketch.

b. A Write an expression for the distance the soccer player ran in terms of θ and d.

35. N A firecracker explodes into four equal pieces (Fig. P3.35). Given the magnitude and direction of the velocity for each piece and the coordinate system shown, determine the x and y velocity components for each piece of the firecracker.

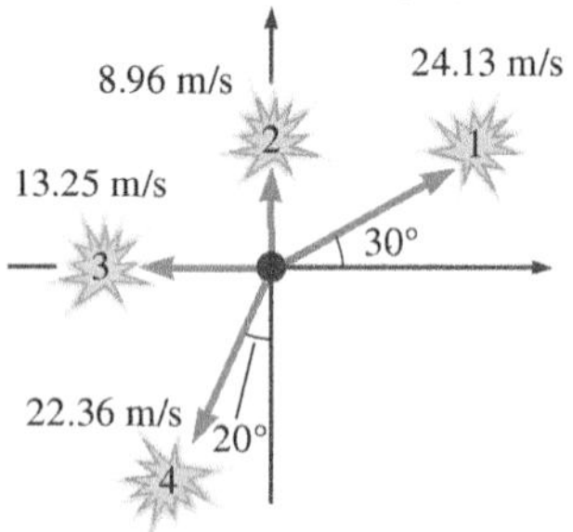

8.96 m/s, 24.13 m/s, 13.25 m/s, 30°, 22.36 m/s, 20°

FIGURE P3.35

36. A In a soccer match, a breakaway forward is streaking toward the goal line with speed v_0. A defender closes from the side as shown in the motion diagram in Figure P3.36. Notice that the "downfield positions" of the forward and the defender are the same at each instant in time. If the defender moves in a direction making an angle θ with respect to the downfield direction, what is his speed? Answer in terms of the given quantities.

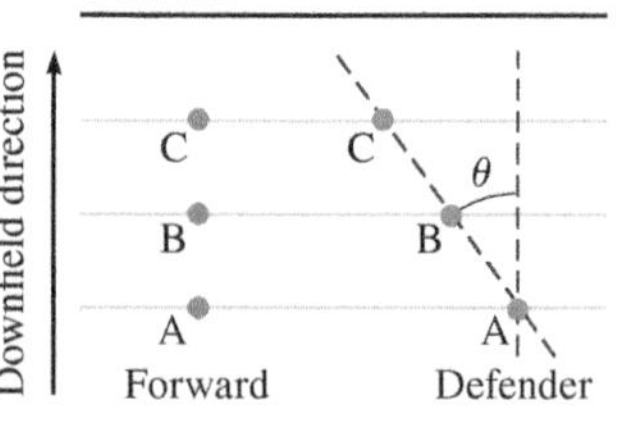

FIGURE P3.36

37. N Ezri, a greyhound dog, is running 30.00° west of north with a constant speed of 17.88 m/s. If the positive x axis points to the east and the positive y axis points to the north, determine Ezri's displacement (in component form) during a time interval of 4.000 s.

38. N Consider the vector $\vec{A} = 3.00\hat{\imath} + 2.00\hat{\jmath} + 5.00\hat{k}$. **a.** What are the magnitudes of the scalar components of $\vec{A}$? **b.** What is the magnitude of the vector $\vec{A}$? **c.** What are the angles the vector $\vec{A}$ makes with the x, y, and z axes?

39. N Vector $\vec{A}$ has an x component of 15.0 m, a y component of −6.00 m, and a z component of −3.00 m. **a.** Write the vector $\vec{A}$ in component form. **b.** Find the vector $\vec{B}$, pointing in the same direction as vector $\vec{A}$, but having one-third its length. **c.** Find the vector $\vec{C}$, pointing in the opposite direction of vector $\vec{A}$, but with three times its length.

3-4 Combining Vectors by Components

40. G Figure P3.40 shows a map of Grand Canyon National Park in Arizona. You need a ruler and protractor for this problem. **a.** Paul hikes from Cape Royale to Point Sublime. Find the magnitude and direction of his displacement, ignoring any difference in altitude between the two points. **b.** Lil hikes from Point Sublime to Cape Royale. Find the magnitude and direction of her displacement. Compare your answer with that of part (a).

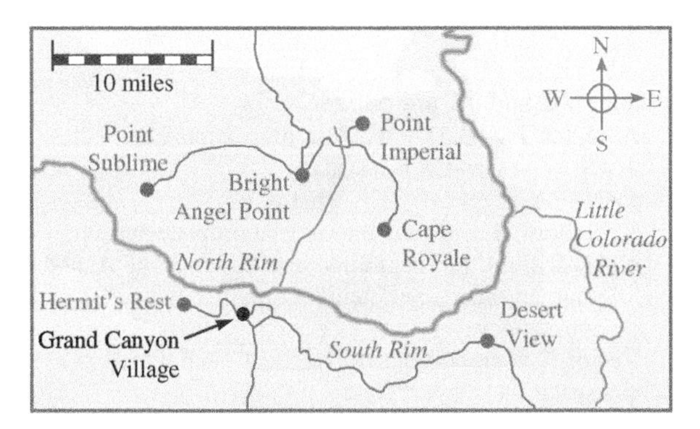

FIGURE P3.40

41. N CASE STUDY In the case study, Reese opened her parachute at 1200 m. **a.** Find α_{safe}, the angle below the horizontal at which she would just barely miss the grove of trees and would land safely in the field. **b.** Use the data in Example 3.11 to find the direction of her actual displacement. Comment on her relative safety. (Was her landing too close for comfort, or did she land far from the grove of trees?) Your comment should make reference to your results in part (a).

Problems 6, 42 and 64 are grouped.

42. The same vectors that are shown in Figure P3.6 are shown in Figure P3.42. The magnitudes are $F_1 = 1.90f$, $F_2 = f$, and $F_3 = 1.4f$, where f is a constant.

a. N Use the coordinate system shown in Figure P3.42 to find $\vec{R} = \vec{F}_1 + \vec{F}_2 + \vec{F}_3$ in component form in terms of f.

b. N If $R_x = 0.33$, what is R_y?

c. C Check your result by comparing your answer to that of Problem 6.

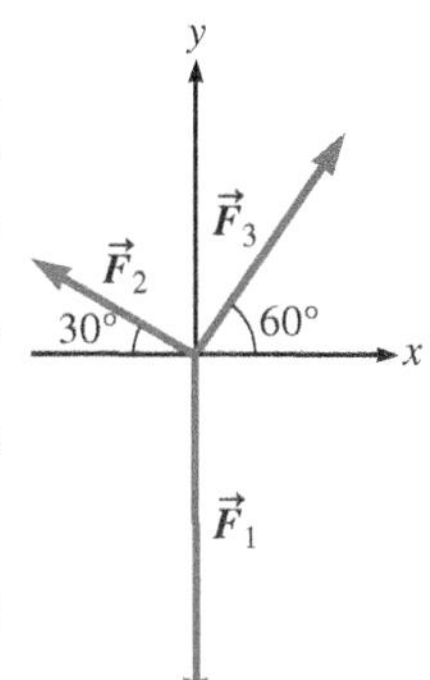

FIGURE P3.42

43. N A supertanker begins in Homer, Alaska, sails 125 km west, and then sails 275 km south. In which direction would it need to sail to return in a straight line to where it began?

44. A Three vectors are shown in Figure P3.44, but they are not drawn to scale. The sum of the three vectors is $\vec{R} = \vec{F}_1 + \vec{F}_2 + \vec{F}_3$. If $R_y = 0$ and $F_2 = 0.5F_3$, find R_x in terms of F_1.

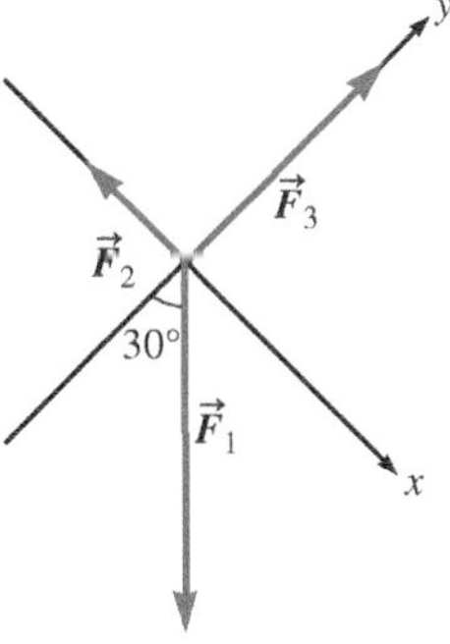

FIGURE P3.44

45. N A vector $\vec{A} = (5.20\hat{\imath} - 3.70\hat{\jmath})$ m and a vector $\vec{B} = (1.04\hat{\imath} + B_y\hat{\jmath})$ m are related by a scalar quantity such that $\vec{A} = s\vec{B}$. Determine the value of the y component of the vector $\vec{B}$.

46. N Frustrated with her physics class, Janet leaves campus and walks into town to acquire some refreshments. She continues on to a friend's house, where she happens to meet a physics tutor. The tutor agrees to help her, and Janet and the tutor walk to the nearby library to study before the upcoming exam. Janet's three displacements can be described by the following vectors: $\Delta\vec{r}_1 = (625\hat{\imath} - 271\hat{\jmath})$ m, $\Delta\vec{r}_2 = (415\hat{\jmath})$ m, and $\Delta\vec{r}_3 = (-296\hat{\imath} + 181\hat{\jmath})$ m. What is Janet's total displacement?

Problems 47 and 48 are paired.

47. N Consider the vectors $\vec{A} = 16.5\hat{\imath} - 33.0\hat{\jmath}$ and $\vec{B} = -2.00\hat{\imath} + 3.00\hat{\jmath} - 4.00\hat{k}$. Find $\vec{R} = \vec{B} - \vec{A}$ in component form.

48. N Consider vectors $\vec{A} = 16.5\hat{\imath} - 33.0\hat{\jmath}$, $\vec{B} = -2.00\hat{\imath} + 3.00\hat{\jmath} - 4.00\hat{k}$, and $\vec{R} = \vec{B} - \vec{A}$. Find the magnitude of $\vec{R}$.

49. A An airplane leaves city A and flies a distance d_1 due north, landing at city B. The next day, the airplane leaves city B and flies a distance d_2 in a direction making an angle θ to the east of due north, landing at city C. Determine the distance between city A and city C. Express your answer in terms of d_1, d_2, and θ.

50. N An aircraft undergoes two displacements. If the first displacement is directed at an angle of 225° with the positive x axis and has magnitude 30.0 miles, and if the resultant displacement is directed at an angle of 75.0° with the positive x axis and magnitude 22.0 miles, what are the magnitude and direction of the second displacement?

51. N The resultant vector $\vec{R} = 2\vec{A} - \vec{B} - 2\vec{C}$ has zero magnitude. Vector $\vec{A}$ has an x component of 4.60 m and a y component of -12.1 m, and vector $\vec{B}$ has an x component of -3.00 m and a y component of -4.00 m. What are the x and y components of vector $\vec{C}$? (All of these are two-dimensional vectors.)

52. A Three vectors all have the same magnitude. The symbol for the magnitude of each of these vectors is M. The first vector $\vec{A}$ points in the positive x direction. The second vector $\vec{B}$ points in the negative y direction. The third vector $\vec{C}$ points in the positive z direction. These three vectors added together are equal to a fourth vector $\vec{D}$. What is the magnitude of the fourth vector?

53. N The two-dimensional vectors $\vec{A}$ and $\vec{B}$ both have magnitudes of 7.00 m. If $\vec{A} + \vec{B} = 4\hat{\imath}$, what is the angle between and $\vec{A}$ and $\vec{B}$?

Problems 54 and 55 are paired.

54. A Two birds begin next to each other and then fly through the air at the same elevation above level ground at the same speed v measured in meters per second. One flies northeast, and the other flies northwest. Northeast is exactly halfway between north and east, and northwest is exactly halfway between north and west. After flying for a time t measured in seconds, what is the distance d measured in meters between them? Ignore the curvature of the Earth.

55. N Two birds begin next to each other and then fly through the air at the same elevation above level ground at 22.5 m/s. One flies northeast, and the other flies northwest. After flying for 10.5 s, what is the distance between them? Ignore the curvature of the Earth.

56. N Vector $\vec{A} = -\hat{\imath} + 2\hat{\jmath} - 5\hat{k}$ and vector $\vec{B} = -5\hat{\imath} - 3\hat{\jmath} - 2\hat{k}$. **a.** What are the scalar components of vector $\vec{R} = \vec{A} + \vec{B}$? What is the magnitude of vector $\vec{R}$? **b.** What are the scalar components of vector $\vec{S} = 3\vec{A} - 2\vec{B}$? What is the magnitude of vector $\vec{S}$?

General Problems

57. G A spider undergoes the displacement represented by the vector $\Delta\vec{r}_1$ in Figure P3.57. The spider rests for a while and then undergoes the displacement represented by the vector $\Delta\vec{r}_2$. Each of these

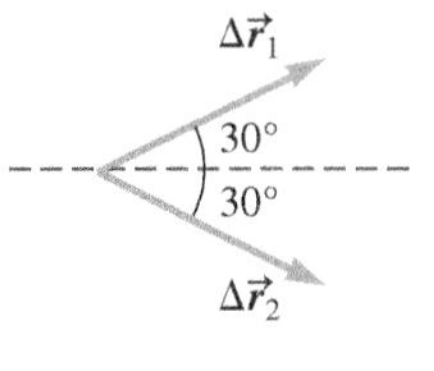

FIGURE P3.57

displacements has magnitude d. After making these two consecutive displacements, is the spider's distance from its original starting point greater than $2d$, equal to $2d$, between d and $2d$, equal to d, or less than d? Explain your reasoning.

Problems 58 and 59 are paired.

58. Peter throws a baseball through a house's window. He stands 3.4 m from the window. The ball starts in his hand 1.1 m above the ground and goes through the window at a point 2.4 m off the ground.
a. G Sketch the problem, including an x–y coordinate system with the origin on the ground directly below the initial position of the ball, the y axis pointing upward, and the x axis pointing toward the house.
b. N What are the initial and final coordinates of the ball?

59. N After finding the initial and final coordinates of the ball in Problem 58, find the displacement of the ball in component form.

60. N Consider the two vectors $\vec{A} = -4\hat{\imath} + 3\hat{\jmath}$ and $\vec{B} = -2\hat{\imath} - 5\hat{\jmath}$. Find the vectors **a.** $\vec{A} + \vec{B}$ and **b.** $\vec{A} - \vec{B}$. Find the quantities **c.** $|\vec{A} + \vec{B}|$ and **d.** $|\vec{A} - \vec{B}|$. **e.** Find the directions of the vectors $\vec{A} + \vec{B}$ and $\vec{A} - \vec{B}$.

61. N Vector $\vec{R}$, with a magnitude of 22.5 units, is directed at an angle of 73.0° below the positive x axis. What are its x and y components?

62. N A glider aircraft initially traveling due west at 85.0 km/h encounters a sudden gust of wind at 35.0 km/h directed toward the northeast (Fig. P3.62). What are the speed and direction of the glider relative to the ground during the wind gust? (The velocity of the glider with respect to the ground is the velocity of the gilder with respect to the wind plus the velocity of the wind with respect to the ground.)

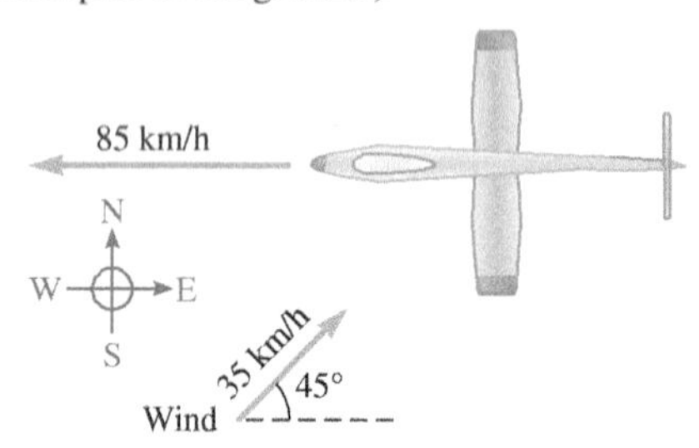

FIGURE P3.62

63. N What are the magnitude and direction of a vector that has an x component of -33.0 units and a y component of -45.0 units?

Problems 6, 42 and 64 are grouped.

64. G Three vectors are shown in Figure P3.6. Copy this figure onto your paper and find $\vec{R} = \vec{F}_2 - \vec{F}_3 - \frac{1}{2}\vec{F}_1$ geometrically.

65. C Vector $\vec{A}$ is in the positive x direction and has magnitude 10. Vector $\vec{B}$ makes an angle of $+\theta$ (measured counterclockwise) from the positive x axis, whereas vector $\vec{C}$ makes an angle of $-\theta$ (measured counterclockwise) from the positive x axis. Vectors $\vec{B}$ and $\vec{C}$ have equal magnitudes. The three vectors $\vec{A}$, $\vec{B}$, and $\vec{C}$ are related according to the equation $\vec{A} = \vec{B} + \vec{C}$. Are the magnitudes of vectors $\vec{B}$ and $\vec{C}$ greater than 5, less than 5, or equal to 5? Explain. If your answer depends on the specific value of θ state that explicitly. *Hint*: Sketch these vectors, including θ.

66. A Using the rules of vector addition, prove that $A = \sqrt{A_x^2 + A_y^2 + A_z^2}$ represents the magnitude of a three-dimensional vector.

67. C A vector $\vec{A}$ and a vector $\vec{B}$ each have the same magnitude but point in different directions. Under what circumstances will the sum of these two vectors result in a vector $\vec{C}$ with the same magnitude as vectors $\vec{A}$ and $\vec{B}$? *Hint*: Draw a sketch.

68. Draw two unit vectors $\hat{\imath}$ and $\hat{\jmath}$.
a. G Geometrically find the sum of these two vectors.
b. N What are the magnitude and direction of the resultant vector?
c. N Write the resultant in component form.

69. C Vector $\vec{A}$ and vector $\vec{B}$ point in different directions. **a.** Under what circumstances are the two vectors related by a scalar quantity as in $\vec{B} = s\vec{A}$? **b.** What must be true about the scalar quantity?

Problems 70 and 71 are paired.

70. A vector $\vec{F} = 3.3\hat{\imath} + 6.3\hat{\jmath}$ is proportional to vector $\vec{A}$ such that $\vec{F} = m\vec{A}$ and m is a scalar.
a. N If $\vec{A} = 1.1\hat{\imath} + 2.1\hat{\jmath}$, what is m?
b. G Draw $\vec{A}$ and $\vec{F}$ on the same coordinate system.
c. N, C Find the magnitude and direction of $\vec{A}$ and $\vec{F}$. Are your answers consistent? Explain how you checked for consistency.

71. Vector $\vec{F}$ is proportional to vector $\vec{A}$ such that $\vec{F} = m\vec{A}$ and m is a scalar.
a. N If $\vec{A} = 2.4\hat{\imath} + 3.0\hat{\jmath}$ and $\vec{F} = 4.0\hat{\imath} + 5.0\hat{\jmath}$, what is m?
b. C Why is it impossible to have $\vec{A} = 2.4\hat{\imath} + 3.0\hat{\jmath}$ and $\vec{F} = 40.0\hat{\imath} + 0.50\hat{\jmath}$, given the relationship between the two vectors?

72. N A chef drops a chopstick on a countertop that is made of square tiles that form a regular grid. She immediately notices that the chopstick spans almost exactly two grids in the horizontal direction. From years of experience, the chef knows that the chopstick has a length of 7.3 grid units. What is the orientation of the chopstick with respect to the horizontal? Answer by specifying an angle in degrees.

73. N A function is given as $f(x) = 3x$, where $x > 0$. Plot this function. What is the angle between the graph of this function and the positive x axis?

Problems 74 and 75 are paired.

74. N A classroom clock has a small magnifying glass embedded near the end of the minute hand. The magnifying glass may be modeled as a particle. Class begins at 7:55 and ends at 8:50. The length of the minute hand is 0.300 m. **a.** Find the average velocity of the magnifying glass at the end of the minute hand using the coordinate system shown in Figure P3.74. Give your answer in component form. **b.** Find the magnitude and direction of the average velocity. **c.** Find the average speed and in the **Check and Think** step, compare to the average velocity.

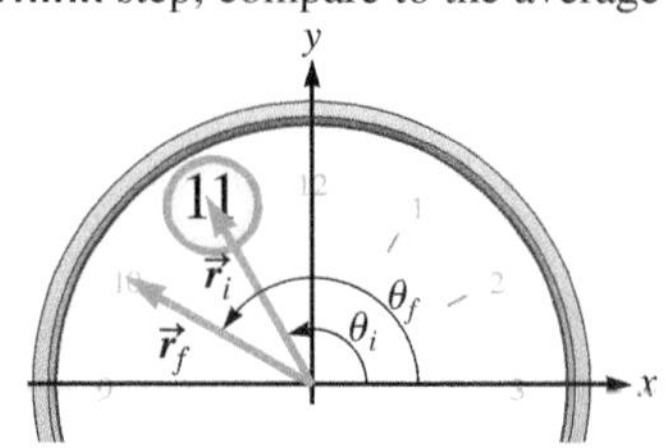

FIGURE P3.74 Problems 74 and 75.

75. C In Problem 74, we found the direction of the average velocity of a magnifying glass at the end of the minute hand on a clock. Now consider the direction of the *instantaneous* velocity when the minute hand points to the 10 and when it points to the 11. If you did Problem 74, compare the direction of the instantaneous velocity vectors to the direction of the average velocity.

Two- and Three-Dimensional Motion

4

Underlying Principles

Two- and three-dimensional kinematics

Major Concepts

1. Position
2. Displacement
3. Average and instantaneous velocity
4. Average and instantaneous acceleration
5. Relative motion and reference frames

Special Cases

1. Projectile motion
   a. Range
   b. Maximum range
2. Uniform circular motion
3. Centripetal acceleration

Key Question

How can the principles of kinematics and the mathematics of vectors be used to describe two- and three-dimensional motion?

4-1 What is multidimensional motion? 86

4-2 Motion diagrams for multidimensional motion 87

4-3 Position and displacement 88

4-4 Velocity and acceleration 90

4-5 Special case of projectile motion 94

4-6 Special case of uniform circular motion 99

4-7 Relative motion in one dimension 104

4-8 Relative motion in two dimensions 106

Imagine the thrill of zigzagging down a ski slope in the Alps, cliff diving into the crystal blue waters of Hawaii, or playing a game of Ultimate Frisbee in front of the physics building on campus. All these thrills and many more are possible because we live in a three-dimensional world.

Why stop at three dimensions? An even more exciting prospect is that, according to some theories in physics, the Universe may have as many as 11 dimensions! These models are beyond the scope of this book, of course, but it is fun to think that what we are studying here is the basis for such mind-boggling ideas about the Universe.

It would be a lot easier to ace a physics class if our Universe were only one-dimensional because we would have already learned all the kinematics we need in Chapter 2. In such a one-dimensional world, however, we would only be able to move back and forth along a single straight line. Fortunately, our world is not that dull.

TWO- AND THREE-DIMENSIONAL KINEMATICS

Underlying Principle

4-1 What Is Multidimensional Motion?

Except for relative motion, the concepts in this chapter are similar to those we looked at in Chapter 2. This chapter combines the kinematics concepts of Chapter 2 with the mathematical tools of Chapter 3 to study two- and three-dimensional motion. We only study objects that can be modeled as particles, but our particles are no longer restricted to motion along a straight line.

One-dimensional motion occurs along a single straight line, whereas multidimensional motion may be along a curved path or along a path made up of more than one straight line. Figure 4.1 shows motion in one, two, and three dimensions. Notice that the number of axes required corresponds to the number of dimensions.

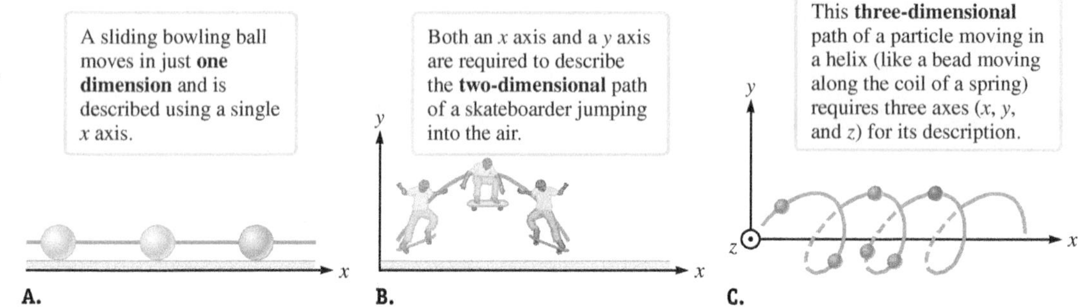

FIGURE 4.1 Examples of motion. **A.** One dimension. **B.** Two dimensions. **C.** Three dimensions. In part C, recall that the symbol ⊙ means out of the page; here, the z axis points out of the page.

Because many of the situations we study in this book are two-dimensional, we will emphasize two-dimensional motion in this chapter. Many equations are written in the more general three-dimensional form, however, and we will work out several three-dimensional examples.

A powerful way to study multidimensional motion is to break the vector quantities (position, displacement, velocity, and acceleration) into components that describe the motion. Each component can then be manipulated and analyzed separately. This idea is illustrated throughout the chapter as we work on different aspects of the following case study.

CASE STUDY Skateboarding

If you have ever skateboarded or watched skateboarders, you might have noticed that many skateboarding tricks are based on a move known as the *ollie*. As shown in Figure 4.2A, both rider and skateboard fly through the air, with the rider's feet never losing contact with the board. This move, invented by Alan "Ollie" Gelfand of Florida in the late 1970s, was originally used to get over an obstacle.

Before Gelfand invented the ollie, a popular trick at skateboarding competitions was the high jump, now known as the hippy jump. Like the ollie, the hippy involves an obstacle, but instead of both rider and skateboard going over the obstacle, only the rider goes over. The skateboard goes under the obstacle, and the rider lands back on the board on the other side (Fig. 4.2B). This case study focuses on the hippy jump.

FIGURE 4.2 A. The ollie is a skateboarding trick in which both the rider and the board pass above an obstacle. **B.** In the hippy jump trick, the skateboard passes under the obstacle, and the skater goes over it.

The motion of the skater in either trick is more complicated than simple translational motion. Nevertheless, we treat the skater as a particle. Two students, Avi and Cameron, are asked to model this trick using a small cart on a track (Fig. 4.3A). The cart is equipped with a spring gun that is used to launch a ball. The cart models the skateboard, and the ball models the rider. An obstacle is placed above the track and does not impede the cart's motion. The students can set the launch angle θ and they are to determine at what value of θ the ball must be launched so that it clears the obstacle and lands back in the cart. As we proceed, we will develop the tools needed to figure out which student has the right idea. Here is part of their discussion.

Avi: We need to point the launcher forward. I would guess that θ needs to be less than 90°, maybe around 45°.

Cameron: I think we need to point the launcher straight up.

Avi: That won't work because the ball has to travel a greater distance than the cart. It has to go up in an arc and travel back down. The cart goes only in a straight line. We need to figure out where the cart will be when the ball lands and then aim for that place (Fig. 4.3B). Just imagine that there is a target where the cart will be on the other side of the obstacle. We need to aim for that imaginary target. The motion of the cart doesn't really affect how we aim the launcher.

Cameron: I don't think it works that way. I think it is more like playing tennis on a cruise ship. If you bounce the ball straight up off the face of your racket, it lands right back on your racket (Fig. 4.3C). You don't have to worry about the speed or direction of the ship when you're playing. You just play like you always do.

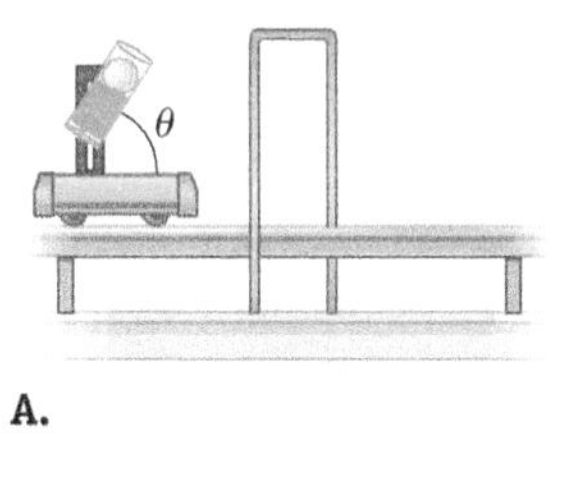

A.

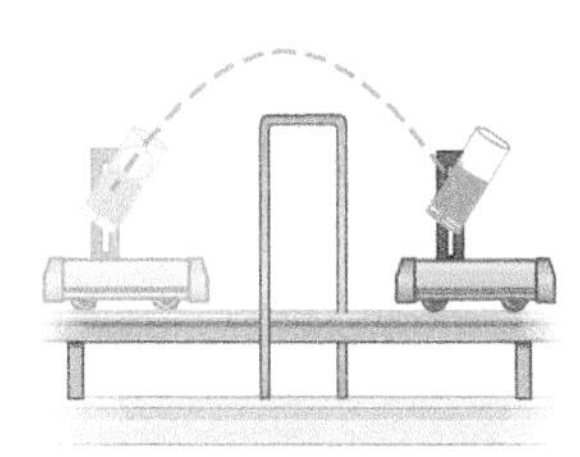
B.

C.

FIGURE 4.3 **A.** Avi and Cameron must model the hippy jump with a cart and spring gun launcher. **B.** Avi suggests launching at an angle smaller than 90°, arguing that the ball must be aimed at a target on the track where the cart will soon be. Essentially, Avi argues to ignore the motion of the cart. **C.** Cameron mentions that if a person on a ship hits a tennis ball straight up, it will land back on the racket. Cameron uses this argument to say that the ball should be launched straight up.

CONCEPT EXERCISE 4.1

In each case, determine whether the object is moving in one, two, or three dimensions.

a. The rider in the ollie (Fig. 4.2A).
b. The skateboard in the ollie (Fig. 4.2A).
c. The rider in the hippy (Fig. 4.2B).
d. The skateboard in the hippy (Fig. 4.2B).

4-2 Motion Diagrams for Multidimensional Motion

A motion diagram is an illustration that shows the location of a particle at regular time intervals (Chapter 2). We often use a dot to represent the particle in a motion diagram. As in one-dimensional motion, the spacing between the dots in a diagram for two- or three-dimensional motion indicates the relative speed of the particle. A particle at rest is represented by a single dot. If a particle maintains its speed, the dot spacing is uniform. If the particle is slowing down, the dots get closer together. If the particle is speeding up, the dots get farther apart.

Figure 4.4 shows the motion of three particles. We use the term **nonuniform** to describe the motion of a particle whose speed changes and the term **uniform** to describe the motion when the speed does not change. Whether a particle maintains its speed or not is independent of its path as shown by the examples of one- and two-dimensional motion in Figure 4.4.

We need to choose a coordinate system to analyze motion. Figure 4.4 shows a possible coordinate choice for each motion diagram. (Remember, though, that there are many other good choices that are not shown here.) As usual, we are free to choose both the location of the origin and which direction on each axis is to be the positive direction. In Figure 4.4B, for instance, we have chosen to place the origin at A, but in Figure 4.4C, we have chosen to place the origin at the center of the circular path. One thing we are *not* free to choose about our coordinate system is the minimum number of axes; this number is determined by the dimensionality of the motion.

Dot spacing decreases; particle is **slowing down**.
L, M, and N dots overlap **(particle stops)**.
A.

Dot spacing from G to L increases; particle is **speeding up**.
Dot spacing from A through F decreases; particle is **slowing down**.
B.

Uniform dot spacing; particle's **speed is constant**.
C.

FIGURE 4.4 Motion diagram of a particle moving along **A.** a one-dimensional path, **B.** a two-dimensional parabolic path, and **C.** a two-dimensional circular path. A possible choice of coordinate system for each motion diagram is shown in all three cases.

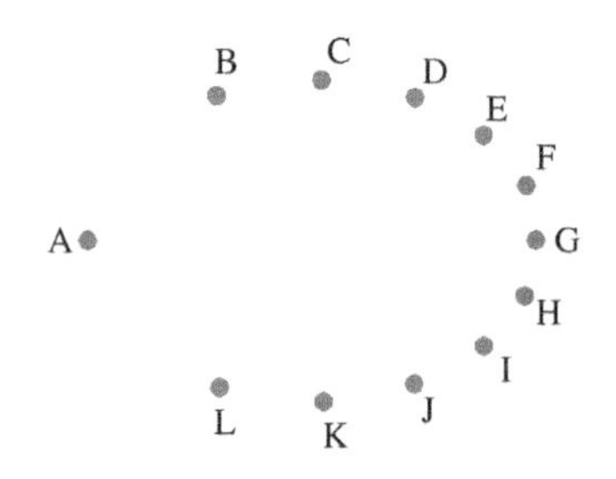

FIGURE 4.5 All dots are in the plane of the page.

CONCEPT EXERCISE 4.2

Based on the particle's motion diagram in Figure 4.5:

a. Is this path one-, two-, or three-dimensional?

b. Is the particle speeding up, slowing down, or maintaining a constant speed? Does your answer change depending on the range of dots you consider? If so, be sure you include this information in your description.

4-3 Position and Displacement

A position vector is used to locate a particle. For example, Figure 4.6 shows the position vectors $\vec{r}_C$ and $\vec{r}_G$ for each motion diagram in Figure 4.4. In each case, the tail of the position vector is at the origin of the chosen coordinate system, and the head is at dot C or dot G. The displacement vector $\Delta\vec{r}_{CG}$ from C to G is also shown in each case. The displacement vector is found by subtracting $\vec{r}_C$ from $\vec{r}_G$ geometrically.

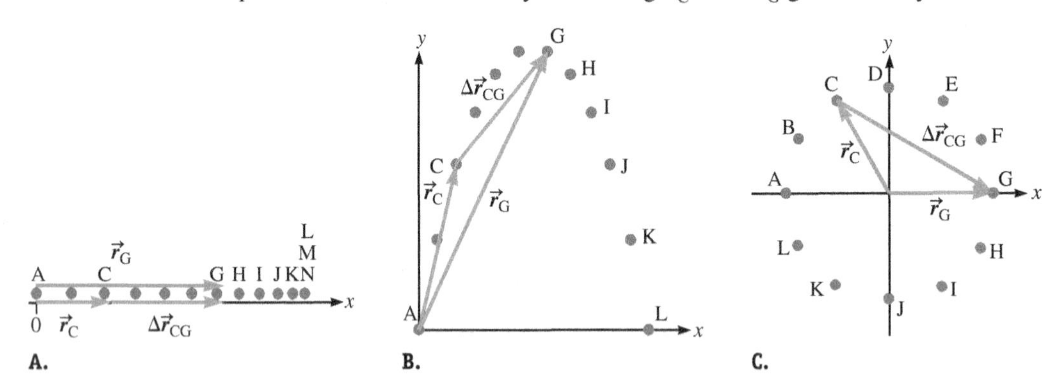

FIGURE 4.6 The same motion diagrams from Figure 4.4, showing position vectors $\vec{r}_C$ and $\vec{r}_G$ for a particle moving along **A.** a one-dimensional path, **B.** a two-dimensional parabolic path, and **C.** a two-dimensional circular path.

Often, the best way to work with multidimensional vectors is in component form. (Eq. 3.8). The position vector is written

POSITION ✪ Major Concept

$$\vec{r} = x\hat{\imath} + y\hat{\jmath} + z\hat{k} \quad (4.1)$$

where x, y, and z are the scalar components.

The displacement $\Delta\vec{r}$ is found by subtracting the initial position vector from the final one:

DISPLACEMENT ✪ Major Concept

$$\Delta\vec{r} = \vec{r}_f - \vec{r}_i = \Delta x\hat{\imath} + \Delta y\hat{\jmath} + \Delta z\hat{k} \quad (4.2)$$

Part of the power of working with components is that each component can be calculated independently. The next example illustrates this procedure.

EXAMPLE 4.1 Mathematical Description of a Circular Path

A circle is an important example of a two-dimensional path. For a particle moving at constant speed in a circular path, the particle's position is given by the components

$$x = A\cos(\omega t) \qquad y = A\sin(\omega t) \quad (4.3)$$

where the numerical values of A and ω are constant for a given circle, but may differ from one circle to another. Suppose their values are $A = 2.0$ m and $\omega = 0.50\pi$ rad/s.

A Show that Equation 4.3 describes a circular path.

INTERPRET and ANTICIPATE

There are at least two ways to find the shape of the path defined by the functions $x(t)$ and $y(t)$. Method 1 is to make a graph of x versus y. Method 2 is to eliminate t between the functions, creating one function of the form $y(x)$. As an illustration, we will use both methods here. Method 1 should produce a circular graph, and method 2 should result in the equation of a circle (Appendix A).

SOLVE

Method 1. We get numerical values (coordinates) for the points to plot by substituting a range of time values in Equation 4.3 to find values for x and y simultaneously. We show just four times here, but you need many more to draw an adequate circle.

t (s)	x (m)	y (m)
0	2.0	0
1.0	0	2.0
2.0	−2.0	0
3.0	0	−2.0

Plot these points on an x–y coordinate system. Connect them with a smooth curve as in Figure 4.7. (The red dots indicate points that are in the table above.) Each tick mark in Figure 4.7 represents 0.5 m along either the x or y axis.

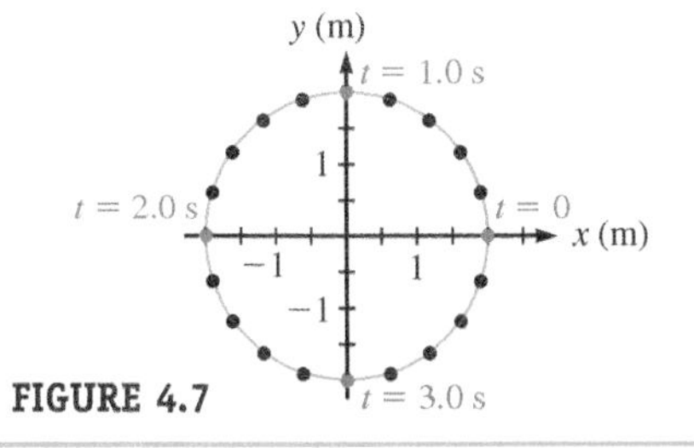

FIGURE 4.7

Method 2. Because the expressions for x and y both involve trigonometric functions, use the trigonometric identity $\cos^2\theta + \sin^2\theta = 1$ to combine the two equations. Square the expressions for x and y.

$$x^2 = A^2\cos^2(\omega t)$$
$$y^2 = A^2\sin^2(\omega t)$$

Add these expressions together.

$$x^2 + y^2 = A^2[\cos^2(\omega t) + \sin^2(\omega t)]$$

The term in square brackets equals 1.

$$x^2 + y^2 = A^2$$

CHECK and THINK

The expression we just obtained is the equation of a circle with radius A (Appendix A). So, the constant A in Equation 4.3 is the radius of the circular path. In this case, the radius $A = 2.0$ m, which we can also see on the figure we drew by using method 1.

B Find the position in component form at an initial time $t_i = 0$ and at a final time $t_f = 5.0$ s.

INTERPRET and ANTICIPATE

We expect to find two numerical answers of the form $\vec{r} = (x\hat{\imath} + y\hat{\jmath})$ m. We find the x and y components directly from Equation 4.3. Our sketch can be used to check our answer.

SOLVE

Find the components at $t_i = 0$ by substitution.

$$x_i = A\cos(\omega t_i) = 2.0\text{ m}\cos(0) = 2.0\text{ m}$$
$$y_i = A\sin(\omega t_i) = 2.0\text{ m}\sin(0) = 0$$

Combine these scalar components to find the initial position (Eq. 4.1).

$$\vec{r}_i = x_i\hat{\imath} + y_i\hat{\jmath} = (2.0\hat{\imath} + 0\hat{\jmath})\text{ m}$$
$$\vec{r}_i = 2.0\hat{\imath}\text{ m}$$

Repeat this procedure to find the position at $t_f = 5.0$ s.

$$\omega t_f = (0.50\pi\text{ rad/s})(5.0\text{ s}) = 2.5\pi\text{ rad}$$
$$x_f = (2.0\text{ m})\cos(2.5\pi\text{ rad}) = 0$$
$$y_f = (2.0\text{ m})\sin(2.5\pi\text{ rad}) = 2.0\text{ m}$$
$$\vec{r}_f = x_f\hat{\imath} + y_f\hat{\jmath}$$
$$\vec{r}_f = 2.0\hat{\jmath}\text{ m}$$

CHECK and THINK

It makes sense that the magnitudes r_i and r_f have the same value because as the particle moves along a circular path, its distance from the origin does not change. So, the magnitude of the position vector is always equal to the radius of the circle. Also, our drawing in part A confirms our result for $\vec{r}_i$.

Example continues on page 90 ▶

C What is the particle's displacement over this 5-s interval?

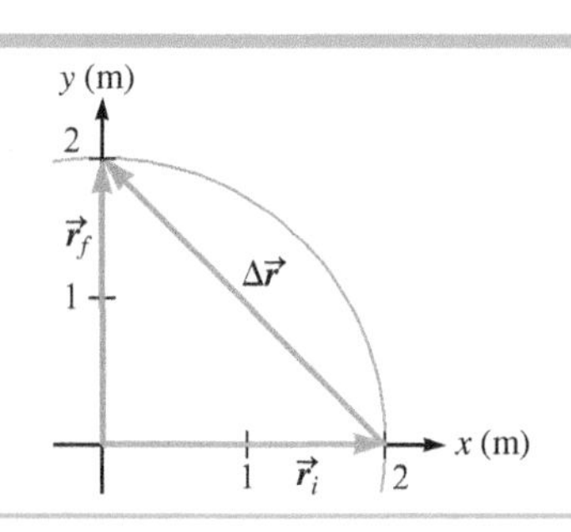

FIGURE 4.8

INTERPRET and ANTICIPATE

We start by getting a rough idea of the displacement geometrically. Because we know both the direction and the magnitude of the two position vectors, we draw these vectors on a coordinate system as in Figure 4.8. Using tail-to-tail subtraction, we find the direction of the displacement vector $\Delta\vec{r} = \vec{r}_f - \vec{r}_i$.

SOLVE

Next, we find an algebraic solution, using Equation 4.2 to calculate the displacement.

$$\Delta\vec{r} = \Delta x\hat{\imath} + \Delta y\hat{\jmath} = (x_f - x_i)\hat{\imath} + (y_f - y_i)\hat{\jmath}$$

$$\Delta\vec{r} = [(0 - 2.0)\hat{\imath} + (2.0 - 0)\hat{\jmath}]\text{ m}$$

$$\Delta\vec{r} = (-2.0\hat{\imath} + 2.0\hat{\jmath})\text{ m}$$

CHECK and THINK

Our result is consistent with $\Delta\vec{r}$ found geometrically. Both show that $\Delta r_x < 0$ and $\Delta r_y > 0$.

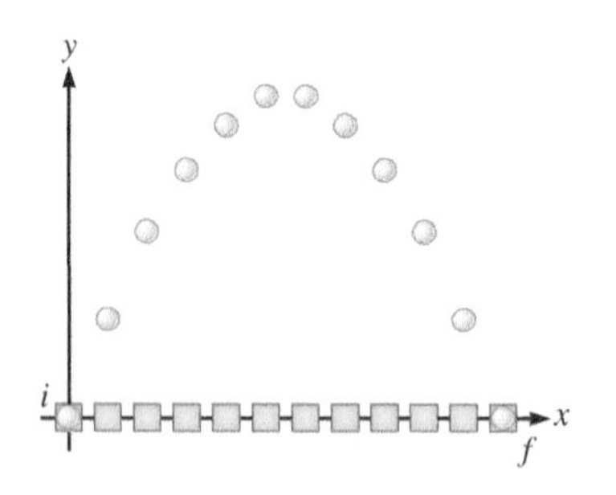

FIGURE 4.9 Avi and Cameron's motion diagram for the ball and cart in the case study. The ball is represented by the circles, and the cart is represented by the squares.

CONCEPT EXERCISE 4.3

CASE STUDY Ball's Displacement

Avi and Cameron have come up with a motion diagram in Figure 4.9 for the ball and cart in the case study, along with their choice of coordinate system. They define the initial position of both objects to be their common location at the moment the ball leaves the cart and the final position to be the location where the ball lands in the cart. If the displacement of the cart is $\Delta\vec{r} = 0.75\hat{\imath}$ m, what is the displacement of the ball?

4-4 Velocity and Acceleration

We saw in Section 2-4 that to find the average velocity of a particle, divide the particle's displacement $\Delta\vec{r}$ by the time interval Δt over which that displacement occurred. So,

AVERAGE VELOCITY Major Concept

$$\vec{v}_{av} = \frac{\Delta\vec{r}}{\Delta t} \quad (4.4)$$

It is often best to write the average velocity in terms of its components:

$$\vec{v}_{av} = v_{av,x}\hat{\imath} + v_{av,y}\hat{\jmath} + v_{av,z}\hat{k} \quad (4.5)$$

Each component is the displacement in one particular direction divided by the time interval (Eq. 2.2). So, the average velocity is written as

$$\vec{v}_{av} = \frac{\Delta x}{\Delta t}\hat{\imath} + \frac{\Delta y}{\Delta t}\hat{\jmath} + \frac{\Delta z}{\Delta t}\hat{k} \quad (4.6)$$

or as

$$\vec{v}_{av} = \frac{x_f - x_i}{t_f - t_i}\hat{\imath} + \frac{y_f - y_i}{t_f - t_i}\hat{\jmath} + \frac{z_f - z_i}{t_f - t_i}\hat{k} \quad (4.7)$$

To take the survey scan or visit **www.cengage.com/community/katz**

We also know that instantaneous velocity (usually referred to as velocity) comes from taking the limit of the average velocity as Δt approaches zero (Section 2-7). In other words, we find the velocity $\vec{v}$ from the time derivative of $\vec{r}$:

VELOCITY Major Concept

$$\vec{v} = \frac{d\vec{r}}{dt} \quad (4.8)$$

By taking the limit of each component in Equation 4.6, we find an expression for the velocity in component form:

$$\vec{v} = \frac{dx}{dt}\hat{\imath} + \frac{dy}{dt}\hat{\jmath} + \frac{dz}{dt}\hat{k} \tag{4.9}$$

Each term is the corresponding velocity component:

$$\vec{v} = v_x\hat{\imath} + v_y\hat{\jmath} + v_z\hat{k} \tag{4.10}$$

The average acceleration and the instantaneous acceleration are found in a similar manner (Section 2-8). The average acceleration comes from dividing the change in velocity by the time interval:

AVERAGE ACCELERATION
Major Concept

$$\vec{a}_{av} = \frac{\Delta \vec{v}}{\Delta t} \tag{4.11}$$

The average acceleration is written in component form as

$$\vec{a}_{av} = \frac{\Delta v_x}{\Delta t}\hat{\imath} + \frac{\Delta v_y}{\Delta t}\hat{\jmath} + \frac{\Delta v_z}{\Delta t}\hat{k} \tag{4.12}$$

or as

$$\vec{a}_{av} = a_{av,x}\hat{\imath} + a_{av,y}\hat{\jmath} + a_{av,z}\hat{k} \tag{4.13}$$

The instantaneous acceleration (usually referred to as acceleration) is the time derivative of the velocity:

ACCELERATION
Major Concept

$$\vec{a} = \frac{d\vec{v}}{dt} \tag{4.14}$$

Each component of the acceleration is the time derivative of the velocity component:

$$\vec{a} = \frac{dv_x}{dt}\hat{\imath} + \frac{dv_y}{dt}\hat{\jmath} + \frac{dv_z}{dt}\hat{k} \tag{4.15}$$

So, each term is the corresponding acceleration component:

$$\vec{a} = a_x\hat{\imath} + a_y\hat{\jmath} + a_z\hat{k} \tag{4.16}$$

Because the velocity is the time derivative of position, we can write the acceleration as the second derivative of position:

$$\vec{a} = \frac{d^2x}{dt^2}\hat{\imath} + \frac{d^2y}{dt^2}\hat{\jmath} + \frac{d^2z}{dt^2}\hat{k} \tag{4.17}$$

Working with these quantities in component form is a powerful tool for analyzing multidimensional motion.

EXAMPLE 4.2 Helical Motion

A particle travels along the helical path in Figure 4.10. It is initially at $\vec{r}_i = (3.0\hat{\imath} + 4.0\hat{\jmath} + 2.3\hat{k})$ m, and after 3.0 s its position is $\vec{r}_f = (3.0\hat{\imath} + 4.0\hat{\jmath} + 5.0\hat{k})$ m. The position and displacement vectors are shown. What is the average velocity of the particle during this time interval?

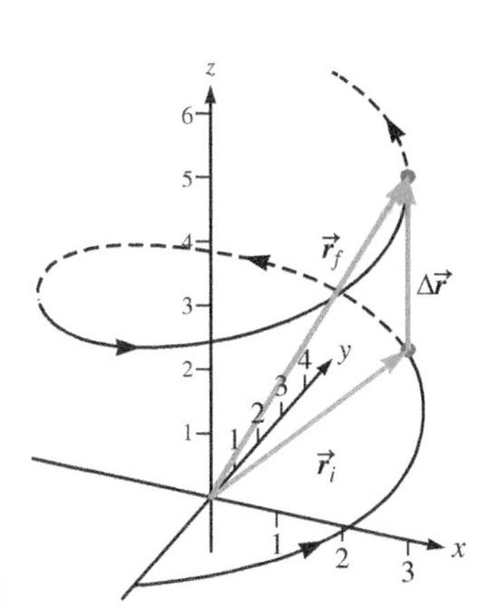

FIGURE 4.10

INTERPRET and ANTICIPATE

From Chapter 2, the average velocity points in the same direction as the displacement, so we expect to find a numerical answer in the form $\vec{v}_{av}$ = _____ $\hat{k}$ m/s because the displacement is parallel to the z axis.

Example continues on page 92 ▶

SOLVE
Substitute values into Equation 4.7, which is the average velocity in component form.

$$\vec{v}_{av} = \frac{x_f - x_i}{t_f - t_i}\hat{\imath} + \frac{y_f - y_i}{t_f - t_i}\hat{\jmath} + \frac{z_f - z_i}{t_f - t_i}\hat{k} \quad (4.7)$$

$$\vec{v}_{av} = \frac{(3.0 - 3.0)\text{ m}}{3.0\text{ s}}\hat{\imath} + \frac{(4.0 - 4.0)\text{ m}}{3.0\text{ s}}\hat{\jmath} + \frac{(5.0 - 2.3)\text{ m}}{3.0\text{ s}}\hat{k}$$

$$\vec{v}_{av} = 0\hat{\imath} + 0\hat{\jmath} + 0.90\hat{k}\text{ m/s}$$

$$\vec{v}_{av} = 0.90\hat{k}\text{ m/s}$$

CHECK and THINK
As expected, the displacement and the average velocity point in the same direction: straight up along the z axis.

EXAMPLE 4.3 Velocity and Speed in Circular Motion

From Example 4.1, the motion of a particle traveling at a constant speed around a circular path is described by

$$x = A\cos\omega t \qquad y = A\sin\omega t$$

where the radius A and factor ω have specific constant values for a given circular path.

A Find an expression for the velocity of a particle moving along this circular path.

INTERPRET and ANTICIPATE
The expressions for x and y tell us the components of the particle's position vector. According to Equation 4.9, we need to take the time derivative of each component to find velocity.

SOLVE
Start with the x component. The constant A does not depend on time, so we pull it outside the derivative. The derivative of the cosine function is negative one multiplied by the sine function (Appendix A).

$$\frac{dx}{dt} = \frac{d(A\cos\omega t)}{dt} = A\frac{d(\cos\omega t)}{dt}$$

$$\frac{dx}{dt} = A(-\omega\sin\omega t) = -A\omega\sin\omega t = v_x$$

Now we do the same thing for the y component. The derivative of the sine function is the cosine function.

$$\frac{dy}{dt} = \frac{d(A\sin\omega t)}{dt} = A\frac{d(\sin\omega t)}{dt}$$

$$\frac{dy}{dt} = A(\omega\cos\omega t) = A\omega\cos\omega t = v_y$$

Combining these scalar components gives us the velocity in component form.

$$\vec{v}(t) = (-A\omega\sin\omega t)\hat{\imath} + (A\omega\cos\omega t)\hat{\jmath} \quad (4.18)$$

B Find an expression for the speed of this particle.

INTERPRET and ANTICIPATE
The speed is the magnitude of velocity.

SOLVE
Use the Pythagorean theorem to find the magnitude of a vector (Eq. 3.12). Choose the positive root because magnitude is always positive.

$$|\vec{v}| = \sqrt{v_x^2 + v_y^2} = \sqrt{(-A\omega\sin\omega t)^2 + (A\omega\cos\omega t)^2}$$

$$|\vec{v}| = \sqrt{(A\omega)^2(\sin^2\omega t + \cos^2\omega t)} = \sqrt{(A\omega)^2}$$

$$|\vec{v}| = A\omega$$

CHECK and THINK
This result is very important. Because both A and ω have a constant value for any given circle, we just found that the speed of this particle moving in a circular path is also constant (does not depend on time). By contrast, our results in part A show that the velocity vector *does* depend on time and is therefore not constant. We conclude that only the direction of the velocity changes as a particle travels in uniform circular motion.

EXAMPLE 4.4 Acceleration and Velocity in Circular Motion

Return to the particle traveling at a constant speed around a circular path as described in Example 4.3.

A Find an expression for the particle's acceleration.

INTERPRET and ANTICIPATE
Take the time derivative of the velocity to find acceleration (Eq. 4.15).

SOLVE We found the velocity in Equation 4.18. Take the time derivative of v_x to find the x component of acceleration.	$\frac{dv_x}{dt} = \frac{d(-A\omega \sin \omega t)}{dt} = -A\omega \frac{d(\sin \omega t)}{dt}$ $a_x = -A\omega^2 \cos \omega t$
Take the time derivative of v_y to find the y component of acceleration.	$\frac{dv_y}{dt} = \frac{d(A\omega \cos \omega t)}{dt} = A\omega \frac{d(\cos \omega t)}{dt}$ $a_y = -A\omega^2 \sin \omega t$
The acceleration in component form is found by combining these scalar components (Eq. 4.15).	$\vec{a}(t) = (-A\omega^2 \cos \omega t)\hat{\imath} - (A\omega^2 \sin \omega t)\hat{\jmath}$ $\vec{a}(t) = -A\omega^2(\cos \omega t\, \hat{\imath} + \sin \omega t\, \hat{\jmath})$ (4.19)

CHECK and THINK
We will consider the implications of this result in part B. Let's just check the units here. Because there are no units associated with trigonometric functions, the units are determined by the coefficient $A\omega^2$ and are $\text{m}(\text{rad/s})^2 = \text{m/s}^2$ as expected for acceleration. (Recall that a radian is defined as the ratio of two lengths, an arc length and the radius of a circle. So, a radian is really dimensionless.)

B In Example 4.1, with $A = 2.0$ m and $\omega = 0.50\pi$ rad/s, we found that the particle's positions at $t = 0$ and 5.0 s are $\vec{r}(0) = 2.0\hat{\imath}$ m and $\vec{r}(5.0\text{ s}) = 2.0\hat{\jmath}$ m, respectively. Find the velocity and acceleration at these times. Make a sketch showing the relative directions of the position, velocity, and acceleration vectors at $t = 0$ and at $t = 5.0$ s.

INTERPRET and ANTICIPATE
To find the velocity at these two times, substitute values into the expression for $\vec{v}$ (Eq. 4.18). Find acceleration in a similar manner, substituting into the expression for $\vec{a}$ (Eq. 4.19).

SOLVE It is convenient to find a value for the constant factor $A\omega$, since it appears in both velocity components.	$A\omega = (2.0\text{ m})(0.50\pi\text{ rad/s}) = 3.1\text{ m/s}$
Substitute $t = 0$ into Equation 4.18 for velocity.	$\vec{v}(t) = A\omega(-\sin \omega t\, \hat{\imath} + \cos \omega t\, \hat{\jmath})$ $\vec{v}(0) = (3.1\text{ m/s})(-\sin 0\, \hat{\imath} + \cos 0\, \hat{\jmath}) = 3.1\hat{\jmath}\text{ m/s}$
Repeat for $t = 5.0$ s.	$\vec{v}(5.0\text{ s}) = (3.1\text{ m/s})(-\sin 2.5\pi\, \hat{\imath} + \cos 2.5\pi\, \hat{\jmath})$ $\vec{v}(5.0\text{ s}) = -3.1\hat{\imath}\text{ m/s}$
The factor $A\omega^2$ appears in both acceleration components.	$A\omega^2 = (2.0\text{ m})(0.50\pi\text{ rad/s})^2 = 4.9\text{ m/s}^2$
Substitute $t = 0$ into Equation 4.19 for acceleration.	$\vec{a}(t) = -A\omega^2(\cos \omega t\, \hat{\imath} + \sin \omega t\, \hat{\jmath})$ $\vec{a}(0) = -4.9\text{ m/s}^2(\hat{\imath} + 0\, \hat{\jmath}) = -4.9\hat{\imath}\text{ m/s}^2$
Repeat for $t = 5.0$ s.	$\vec{a}(5.0\text{ s}) = -4.9\text{ m/s}^2(\cos 2.5\pi\hat{\imath} + \sin 2.5\pi\, \hat{\jmath})$ $\vec{a}(5.0\text{ s}) = -4.9\hat{\jmath}\text{ m/s}^2$

Example continues on page 94 ▶

Sketch these six vectors on a two-dimensional coordinate system. There is no need to set a scale for the vectors because we are only interested in their relative directions.

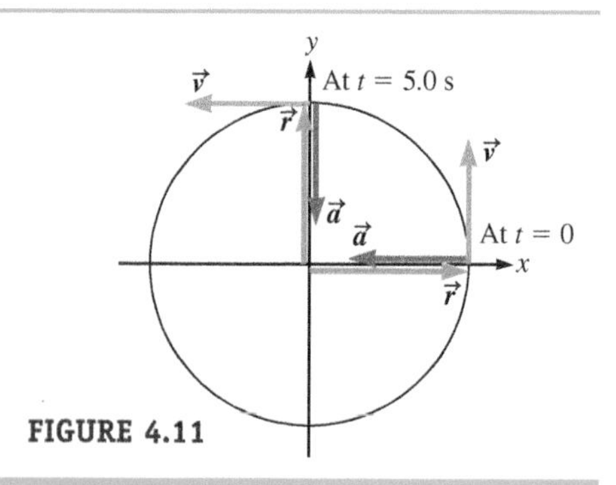

FIGURE 4.11

CHECK and THINK

Our sketch shows that the velocity is perpendicular to the acceleration in both instances, as is *always* the case for uniform circular motion (circular motion at constant speed). In fact, whenever the velocity's direction changes while the speed remains constant, the acceleration must be perpendicular to the velocity. We will look more closely at uniform circular motion in Section 4-6.

4-5 Special Case of Projectile Motion

A bullet shot out of a gun, a water balloon dropped out a third-story dormitory window, and a skateboarder doing a hippy are all projectiles. A **projectile** is any particle that is launched or *projected*. The two defining properties of a projectile are (1) that the launching gives it an initial velocity $\vec{v}_0$ and (2) that after it is launched its motion is influenced only by gravity and the projectile is therefore in free fall. For example, once the ball has left a basketball player's hand, the ball is a projectile (Fig. 4.12). The player gives the ball an initial velocity $\vec{v}_0 = \vec{v}_{0x} + \vec{v}_{0y}$, and after that *only the vertical vector component* $\vec{v}_y$ *is altered by gravity*. The horizontal vector component $\vec{v}_x = \vec{v}_{0x}$ is a constant and therefore never changes from its initial value.

Projectile motion is a special case in which (1) there is no acceleration in the x (horizontal) direction and (2) the acceleration in y (vertical) direction is the constant free-fall acceleration g. When the initial velocity's x component is zero ($\vec{v}_{0x} = 0$), the projectile's path is along a straight, vertical line. When the initial velocity's x component is nonzero ($\vec{v}_{0x} \neq 0$), the projectile's path is a parabola in the xy plane.

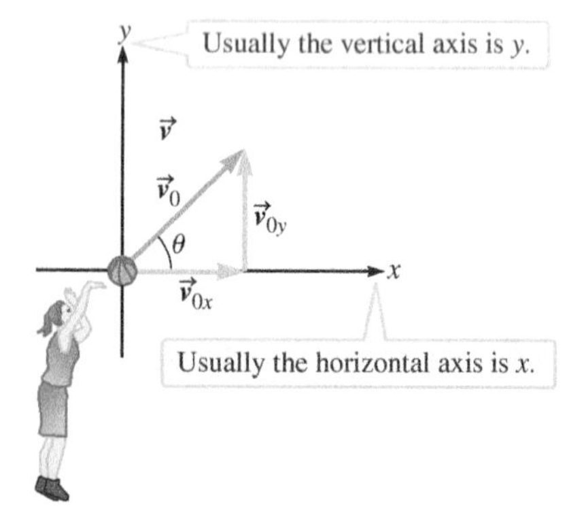

FIGURE 4.12 A basketball player shoots a ball. After the ball leaves her hand, the ball is a projectile.

The Equations for Projectile Motion

A mathematical description of projectile motion means finding equations for the position, velocity, and acceleration as functions of time. We'll continue to use the coordinate system shown in Figure 4.12. We already know the equation for the acceleration because the projectile is in free fall: $\vec{a} = -g\hat{\jmath}$.

Before we get into the mathematical description, let's create a visual description. Figure 4.13 shows the motion diagram of a projectile such as a basketball in a free

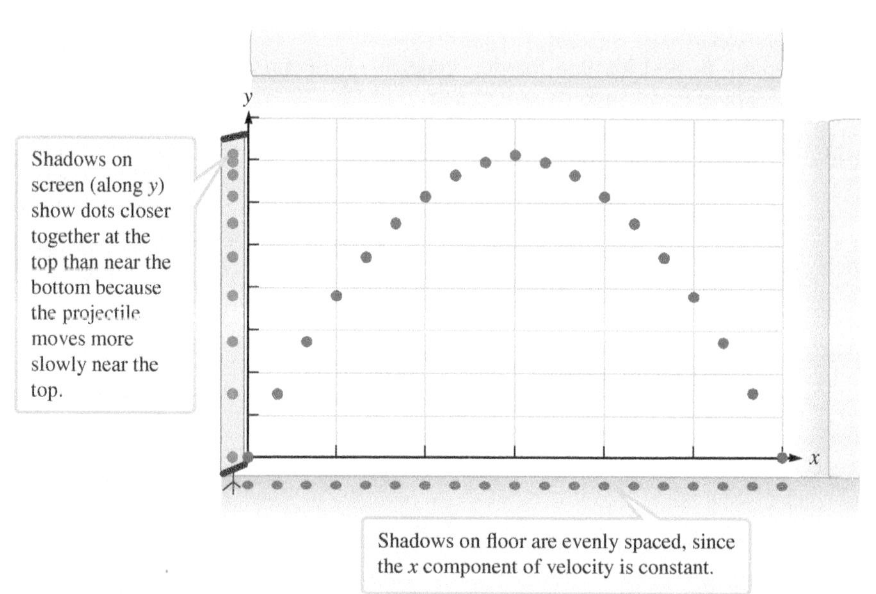

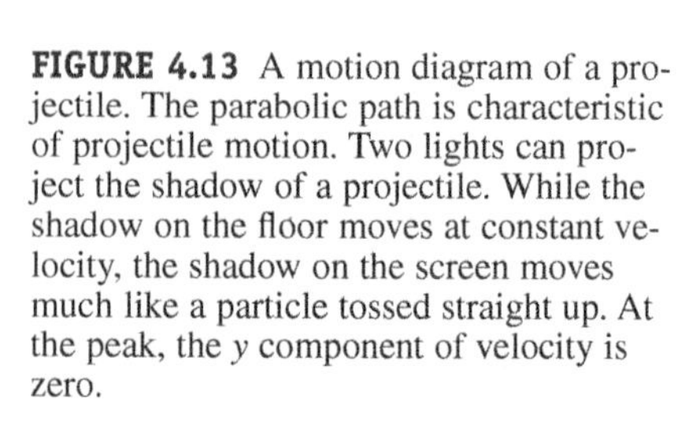
FIGURE 4.13 A motion diagram of a projectile. The parabolic path is characteristic of projectile motion. Two lights can project the shadow of a projectile. While the shadow on the floor moves at constant velocity, the shadow on the screen moves much like a particle tossed straight up. At the peak, the y component of velocity is zero.

throw or a skateboarder doing an ollie. The projectile is modeled as a particle and is represented by a dot. Imagine using two light sources to break this two-dimensional motion into motion in the x direction and motion in the y direction. One light source is overhead, emitting rays perpendicular to the floor. The other light source is placed in front of the projectile, producing rays perpendicular to a vertical screen behind the projectile. The shadows on the floor are evenly spaced, indicating uniform motion in the x direction. The shadows on the screen are closer together near the top, indicating that the particle moves more slowly near the top. In fact, these shadows look exactly like the motion diagram of a particle in purely vertical free fall.

The shadows in Figure 4.13 show us how to connect the motion of a projectile to its mathematical description. First, because the acceleration is constant in either direction, the five constant acceleration equations (Table 2.4, repeated here) may be applied in both directions. Second, in the x direction, because $a_x = 0$, the five constant equations collapse to two:

Projectile motion is a combination of uniform motion in the horizontal direction and free fall in the vertical direction.

$$v_x = v_{0x} \quad (4.20)$$

$$x = x_0 + v_{0x}t \quad (4.21)$$

Third, in the y direction, $\vec{a} = -g\hat{j}$, and so the velocity and position are given by

$$v_y = v_{0y} - gt \quad (2.9)$$

$$y = y_0 + v_{0y}t - \tfrac{1}{2}gt^2 \quad (2.11)$$

Typically, we know the initial velocity $\vec{v}_0$, and we must break it into its components. For example, in Figure 4.12, we have

$$v_{0x} = v_0 \cos\theta \qquad v_{0y} = v_0 \sin\theta \quad (4.22)$$

where the launch angle θ is measured counterclockwise from the x axis.

We combine the above equations to write vector equations for the projectile's motion:

$$\vec{a} = -g\hat{j} \quad (4.23)$$

$$\vec{v} = (v_0 \cos\theta)\hat{\imath} + (v_0 \sin\theta - gt)\hat{j} \quad (4.24)$$

$$\vec{r} = [v_0 t \cos\theta + x_0]\hat{\imath} + [(v_0 t \sin\theta - \tfrac{1}{2}gt^2) + y_0]\hat{j} \quad (4.25)$$

We often choose to put the origin of the coordinate system at the initial position so that $(x_0, y_0) = (0, 0)$. The projectile's position is then

$$\vec{r} = (v_0 t \cos\theta)\hat{\imath} + (v_0 t \sin\theta - \tfrac{1}{2}gt^2)\hat{j} \quad (4.26)$$

and we don't need to use Equation 4.25 for this choice.

TABLE 2.4 Kinematic equations of motion for constant acceleration

Equation	
$v_x = v_{0x} + a_x t$	(2.9)
$\Delta x = \frac{1}{2}(v_{0x} + v_x)t$	(2.10)
$\Delta x = v_{0x}t + \frac{1}{2}a_x t^2$	(2.11)
$\Delta x = v_x t - \frac{1}{2}a_x t^2$	(2.12)
$v_x^2 = v_{0x}^2 + 2a_x \Delta x$	(2.13)

Notes: The equations are written in terms of scalar components of the motion variables along the x axis. Similar equations can also be written in the y and z directions.

EXAMPLE 4.5 CASE STUDY How Do the Shadows Move?

In Concept Exercise 4.3, Avi and Cameron came up with a motion diagram (Fig. 4.9) for the cart and ball in the case study. Imagine placing two light sources and a screen in this drawing so that the moving ball casts shadows on horizontal and vertical surfaces (similar to Fig. 4.13). Describe the path of the shadows.

SOLVE

If we used lights to create shadows of the ball, the shadows would look much like those in Figure 4.13. The vertical shadows would be close together near the top and farther apart near the bottom, indicating the change in velocity along this vertical axis. Because there is no horizontal acceleration in projectile motion, the horizontal component of the ball's velocity is constant. Thus, the horizontal shadows would be evenly spaced. Because the (constant) *horizontal* velocity of the ball is the same as the (constant) velocity of the cart, the ball's horizontal shadows would, at every moment, be projected onto the cart as the cart moves along the track. So, the ball is always directly above the cart. We will need this important conclusion as we wrap up our case-study analysis in Section 4-8.

EXAMPLE 4.6 Parabolic Path

Use Equation 4.26, $\vec{r} = (v_0 t \cos\theta)\hat{\imath} + (v_0 t \sin\theta - \frac{1}{2}gt^2)\hat{\jmath}$, to show that the path of a projectile is a parabola.

INTERPRET and ANTICIPATE

Finding a path means finding y as a function of x. Equation 4.26 provides x and y as functions of time t. From method 2 in Example 4.1, one way to find the path is to eliminate t from the x and y equations.

SOLVE

Use the x component of the projectile's position vector (Eq. 4.26) to obtain an expression for time.

$$x = v_0 t \cos\theta$$

$$t = \frac{x}{v_0 \cos\theta}$$

Substituting this result into the expression for the y component of position allows us to eliminate t between the equations.

$$y = v_0 t \sin\theta - \tfrac{1}{2}gt^2 = \frac{v_0 x \sin\theta}{v_0\cos\theta} - \tfrac{1}{2}g\left(\frac{x}{v_0\cos\theta}\right)^2$$

$$y = x\tan\theta - \left(\frac{g}{2v_0^2\cos^2\theta}\right)x^2 \tag{4.27}$$

CHECK and THINK

The coefficients a, b, and c determine the geometry—height, width, and center—of a given parabola. Because g is constant, our result tells us that the geometry of a projectile's parabolic path depends only on the initial velocity of the projectile.

This result is the equation of a parabola in the form $y = ax^2 + bx + c$ with

$$a = -\frac{g}{2v_0^2\cos^2\theta} \qquad b = \tan\theta \qquad c = 0$$

RANGE EQUATION Special Case

Maximum range R_{max} occurs at launch angle of 45°.

R

0° 15° 30° 45° 60° 75° 90°

θ

Two different angles give the same range.

FIGURE 4.14 Range of a projectile depends on launch angle θ. The maximum range occurs for $\theta = 45°$. For all other ranges, there is a low angle θ_L and a high angle θ_H that both result in the same R. Notice that $\theta_L + \theta_H = 90°$.

Range Equation

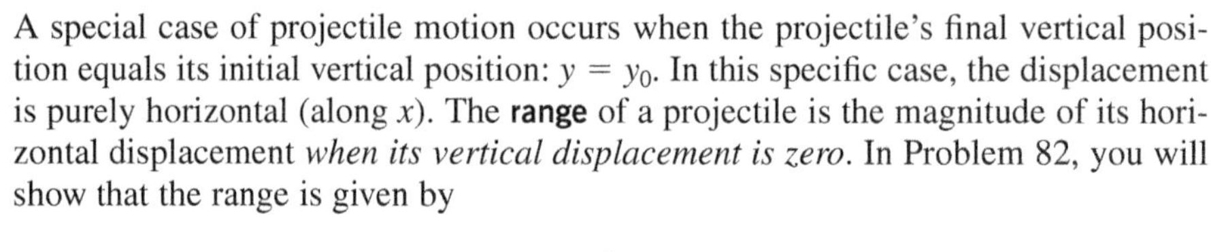

A special case of projectile motion occurs when the projectile's final vertical position equals its initial vertical position: $y = y_0$. In this specific case, the displacement is purely horizontal (along x). The **range** of a projectile is the magnitude of its horizontal displacement *when its vertical displacement is zero*. In Problem 82, you will show that the range is given by

$$R = \frac{v_0^2}{g}\sin 2\theta \tag{4.28}$$

Equation 4.28 (the range equation) is used to find the magnitude of the displacement *only if the displacement along y is zero*: $R = \Delta r$ only if $\Delta y = 0$. (Because the vertical displacement is zero, the horizontal displacement is the total displacement.)

According to the range equation, the range of a projectile launched on the Earth depends on two factors: initial speed v_0 and launch angle θ. Suppose a launcher always gives the projectile the same initial speed v_0. The range of the projectile is then controlled only by the launch angle θ as shown in Figure 4.14. We see that (1) the longest range occurs at $\theta = 45°$ and (2) two possible angles can be used to achieve any range value smaller than the maximum.

Whenever a manufacturer specifies the range for a launch device, the range is understood to mean the maximum value, which is the value at which $\theta = 45°$ (and $\sin 2\theta = 1$):

MAXIMUM RANGE Special Case

$$R_{max} = \frac{v_0^2}{g} \tag{4.29}$$

PROBLEM-SOLVING STRATEGY

Two-Dimensional Projectile Motion

There are two parts to the **INTERPRET and ANTICIPATE** procedure:

1. Start with a **sketch** that includes a **coordinate system**. Often, it is best to put the origin of the coordinate system at the position where the projectile is launched.
2. Ask yourself if the vertical displacement is zero ($\Delta y = 0$). If your answer is no, you **cannot** apply the **range equation** (Eq. 4.28).

There are also two parts to the **SOLVE** procedure:

1. Start by **listing the six kinematic variables** (initial position, final position, initial velocity, final velocity, acceleration, and time) for x and y separately. The time t is the same in both lists. Assuming you have chosen a horizontal x axis and a vertical y axis, there are a few things about these parameters that are always true:
   a. $a_x = 0$
   b. $v_x = v_{0x}$
   c. $a_y = \pm g$ (Choose the minus sign if your y axis points upward.)
   d. Trigonometry (Eq. 4.22) may be used to find the initial velocity components when the launch angle θ is known.
2. Once you have listed the parameters, you may use the **constant-acceleration equations** (Table 2.4). In practice, Equations 4.24 through 4.26 (page 111) were derived from the constant-acceleration equations for the special case of two-dimensional projectile motion.

CONCEPT EXERCISE 4.4

A manufacturer claims to have a three-person water-balloon launcher with a 300-yd range (Fig. 4.15), which is a distance equal to the length of three football fields placed end to end!

a. What is the displacement of a water balloon launched at $\theta = 45°$ if it is launched from the ground and returns to the ground? Give your answer in yards.

b. If a water balloon is launched off a building at $\theta = 45°$ and lands on the ground 50 ft below, is its horizontal displacement greater than, less than, or equal to the range?

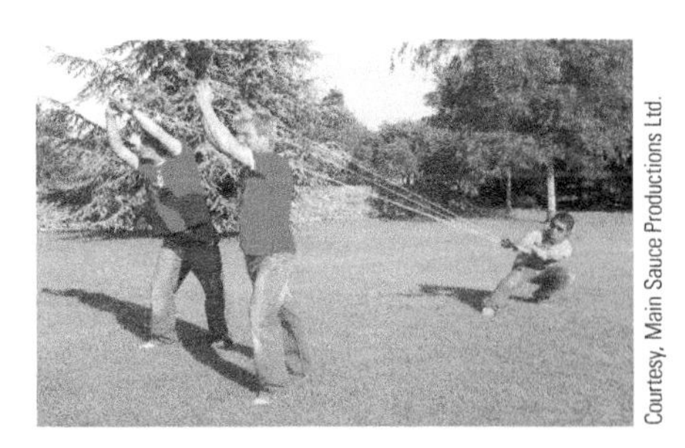

Courtesy, Main Sauce Productions Ltd.

FIGURE 4.15

To take the survey scan or visit **www.cengage.com/community/katz**

CONCEPT EXERCISE 4.5

Does the range of a launcher change if it is used on the Moon? If so, is the range longer or shorter on the Moon than on the Earth?

EXAMPLE 4.7 **A Training Exercise**

A pilot learns how to hit a target by dropping mock bombs on an abandoned building as seen in Figure 4.16. The bomber flies quickly (505 km/h) at a low altitude (1.83×10^3 m). Assume air resistance is negligible.

A How far from the abandoned building should the bomber drop its mock bomb?

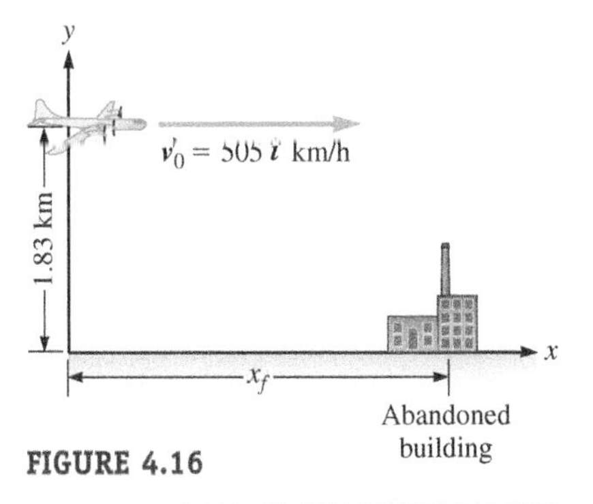

FIGURE 4.16

INTERPRET and ANTICIPATE

Steps 1 and 2 The problem statement provides a **sketch with a coordinate system**, so we don't need to draw our own. We might be tempted to use the range equation, but we cannot because $\Delta y \neq 0$. Instead, we need to think about the motion of the mock bomb along x and y separately.

Example continues on page 98 ▶

SOLVE

Step 1 Start by **listing the six kinematic** quantities for x and y separately. The mock bomb starts inside the plane. Therefore, its initial velocity is the same as the velocity of the plane. In this list, all the variables pertain to the mock bomb. Notice that the initial and final values for both x and y are a function of the coordinate system provided in the problem statement.

$x_i = 0$	$y_i = 1.83 \times 10^3\,\text{m}$
$x_f =$ needed	$y_f = 0$
$v_x = v_{0x} = 505\,\text{km/h} = 1.40 \times 10^2\,\text{m/s}$	$v_{0y} = 0$
$a_x = 0$	v_y not needed
t unknown	$a_y = -g$
	t unknown

Step 2 We start with the displacement along y and solve for the time t the mock bomb takes to fall to the target. The displacement along y comes from modifying Equation 2.11.

$$\Delta y = v_{0y}t - \tfrac{1}{2}gt^2 = 0 - \tfrac{1}{2}gt^2$$

$$t = \sqrt{\frac{2\Delta y}{-g}} = \sqrt{\frac{2(-1.83 \times 10^3\,\text{m})}{-9.81\,\text{m/s}}}$$

$$t = 19.3\,\text{s} \tag{1}$$

The x component of position comes from the first term in Equation 4.26. This term involves the time of flight (which we just found above) and the x component of initial velocity, which we have in Equation 4.22.

At impact, $x_f = v_0 t \cos\theta = (v_0 \cos\theta)t$

$$v_{0x} = v_0 \cos\theta \tag{4.22}$$

$$x_f = v_{0x}t$$

Substitute values, including the time found in Equation (1).

$$x_f = (1.40 \times 10^2\,\text{m/s})(19.3\,\text{s})$$

$$x_f = 2.70 \times 10^3\,\text{m} = 2.70\,\text{km}$$

CHECK and THINK

It might seem incredible that the mock bomb drops when the plane is 2.70 km (1.7 mi) away from the target. To get a better feel for this seemingly strange fact, think about the shadow of the plane on the ground. As in Figure 4.13, suppose a light directly above the plane gives off rays perpendicular to the ground, projecting the plane's shadow. The mock bomb leaving the plane would always be directly above the plane's shadow. So, the mock bomb and the plane's shadow would be on the abandoned building at the same time.

B Many bomber pilots are trained to fly upward as they drop their bomb. This maneuver allows them to be farther from the explosion. If $\theta = 20.0°$ in Figure 4.17, at what distance from the abandoned building should the mock bomb be dropped?

FIGURE 4.17

INTERPRET and ANTICIPATE

The only difference between this question and part A is the mock bomb's initial velocity. It now has a nonzero y component. Use part A to take some shortcuts.

SOLVE

Find the new x and y components of the initial velocity (Eq. 4.24). The other parameters are the same as in part A.

$$v_{0x} = v_0 \cos\theta = (1.40 \times 10^2\,\text{m/s}) \cos 20.0°$$

$$v_{0x} = 1.32 \times 10^2\,\text{m/s}$$

$$v_{0y} = v_0 \sin\theta = (1.40 \times 10^2\,\text{m/s}) \sin 20.0°$$

$$v_{0y} = 47.9\,\text{m/s}$$

As before, find the time it takes the mock bomb to fall from the plane to the target from the displacement along y. Now, however, we must use the quadratic formula to solve for t because the v_{0y} term in our equation is not zero.

$\Delta y = v_{0y}t - \tfrac{1}{2}gt^2$; rearrange to

$$(-\tfrac{1}{2}g)t^2 + (v_{0y})t - \Delta y = 0$$

$$t = \frac{-v_{0y} \pm \sqrt{v_{0y}^2 - 2g\Delta y}}{-g}$$

Substitute values.	$t = \dfrac{-47.9 \text{ m/s} \pm \sqrt{(47.9 \text{ m/s})^2 - 2(9.81 \text{ m/s}^2)(-1.83 \times 10^3 \text{ m})}}{-9.81 \text{ m/s}^2}$ $t = -15.0 \text{ s or } 24.8 \text{ s}$
Choose the positive solution for the time and substitute it in the expression for the x component of the mock bomb's displacement. The negative solution occurs before the mock bomb is dropped and is not a physically reasonable solution.	$x = v_{0x}t = (1.32 \times 10^2 \text{ m/s})(24.8 \text{ s})$ $x = 3.27 \times 10^3 \text{ m} = 3.27 \text{ km}$

CHECK and THINK

Compared with part A, the mock bomb is dropped when the plane is even farther from the target (a little more than 2 mi) because the mock bomb's initial velocity now has an upward component. So, the mock bomb travels above its original altitude and as a result takes longer to fall back to the Earth. In addition, its horizontal velocity is less than in part A. The mock bomb must be dropped sooner to compensate.

4-6 Special Case of Uniform Circular Motion

UNIFORM CIRCULAR MOTION

Special Case

A particle moving along a circular path at constant speed is said to undergo uniform circular motion, as the term suggests. To ancient mathematicians, uniform circular motion was considered perfect and divine. They believed that only heavenly bodies such as the Moon, planets, and stars could exhibit such perfect motion. Today, we do not associate uniform circular motion with perfection, yet it is still an important special case of two-dimensional motion requiring a mathematical description.

Let's start by finding the speed of a particle moving in a circle of radius r. The particle completes one revolution around the circle in a time T known as the **period**. In one period, the distance traveled is the circumference of the circle, $2\pi r$. Because the speed is uniform, it is given by

$$v = \frac{2\pi r}{T} \tag{4.30}$$

Equation 4.30 gives the particle's "translational" or "linear" speed.

Next, let's consider the particle's position as it moves. At one particular instant (Fig. 4.18) the particle's position is

$$\vec{r} = (r \cos \theta)\hat{\imath} + (r \sin \theta)\hat{\jmath} \tag{4.31}$$

where θ is measured counterclockwise from the positive x axis. Notice that the radius r of the circle is also the magnitude of the position vector $\vec{r}$. As the particle travels around the circle, only the direction of $\vec{r}$ changes. Figure 4.19 shows that only θ changes as the particle moves in a circle. So, once we know the radius of the circle, we need only θ to locate the particle.

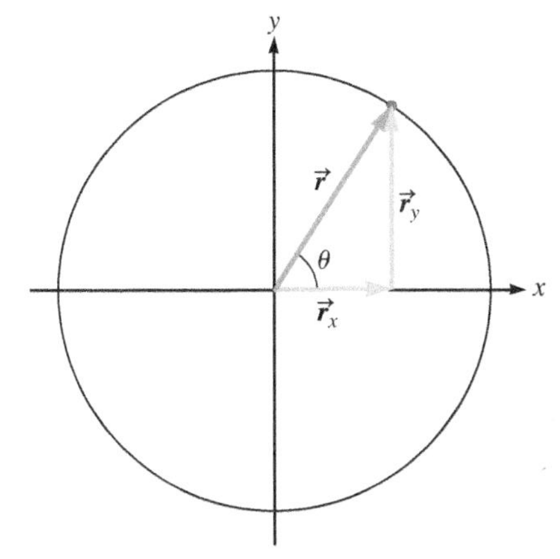

FIGURE 4.18 A particle moves uniformly in a circle. Its position vector at one instant is $\vec{r}$.

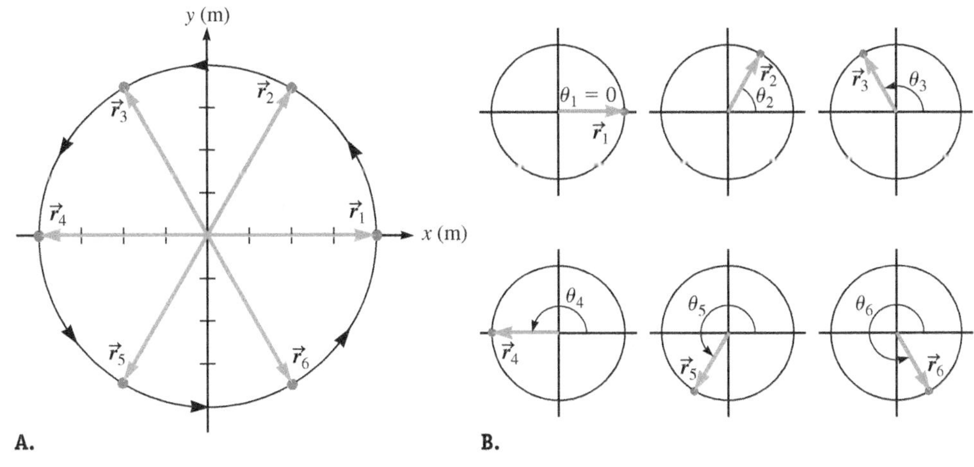

FIGURE 4.19 A. A motion diagram of a particle in uniform circular motion. **B.** The same positions shown on separate diagrams. Only the direction θ changes as the particle moves around the circle.

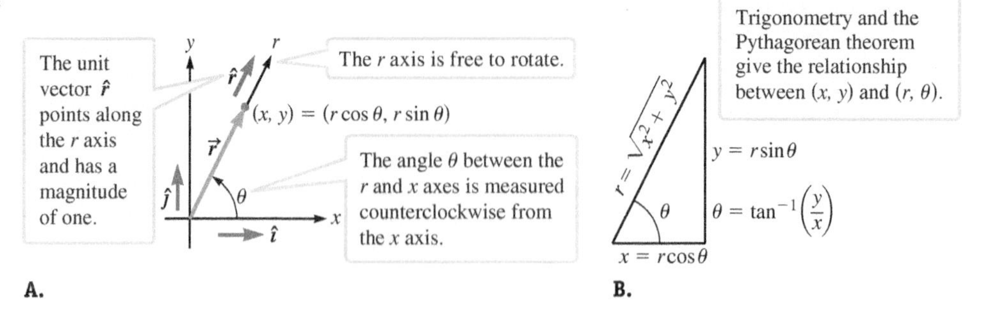

FIGURE 4.20 Plane polar coordinate system.

Polar Coordinate System

We now introduce a coordinate system that will help us exploit the key feature of uniform circular motion—as a particle moves at constant speed along a circular path, only the direction of its position vector changes. As we have seen, two coordinates are needed to describe the position of a particle that moves in two dimensions. In a Cartesian coordinate system, the two coordinates are x and y. As shown in Figure 4.20A, a particle's position can also be given by the magnitude of its position vector $\vec{r}$ and the angle θ. These two coordinates r and θ are called **polar coordinates**, an alternative to Cartesian coordinates. In a polar coordinate system, the r axis is free to rotate. The angle θ is measured counterclockwise from a fixed x axis to the r axis.

By counterclockwise, we mean in the direction that goes from quadrant I to quadrant II to quadrant III to quadrant IV.

The two coordinate systems are related by trigonometry (Fig. 4.20B). To transform from x, y coordinates to r, θ coordinates, use the relationships

$$r = \sqrt{x^2 + y^2} \qquad \text{and} \qquad \tan\theta = \frac{y}{x}$$

and to get from (r, θ) to (x, y), use the relationships

$$x = r\cos\theta \qquad \text{and} \qquad y = r\sin\theta$$

Finally, the polar coordinate system involves two unit vectors, but for the work in this book, we only need to know about one of them. The unit vector $\hat{r}$ points radially outward from the origin as shown in Figure 4.20A.

Linear and Angular Speed

Once we know the radius of the circular path, only the θ component is needed to determine the particle's position at any given moment. We can describe how quickly the particle moves in terms of the change in θ with time. Speed measured in this way is called **angular speed** and is symbolized by the Greek letter omega, ω:

Equation 4.32 gives the same ω described in Example 4.1.

$$\omega = \frac{d\theta}{dt} \tag{4.32}$$

The angle θ is measured in radians, so ω is in radians per second. In one period, the position vector sweeps through an angle $\theta = 2\pi$ rad. Therefore, for a particle in uniform circular motion,

$$\omega = \frac{2\pi}{T} \tag{4.33}$$

We sometimes refer to the instantaneous speed v (Section 2-7) as the **linear** or **translational speed** to distinguish it from the angular speed ω.

Figure 4.21 shows the initial and final positions $\vec{r}_i$ and $\vec{r}_f$ of a particle traveling in uniform circular motion. The position vector swept out an angle θ, and the particle has traveled a distance s along the circle such that

$$s = \theta r \tag{4.34}$$

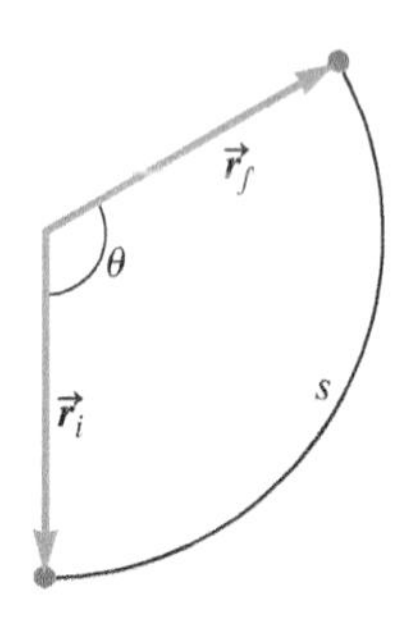

FIGURE 4.21 A particle travels along a portion of a circular path.

Taking the time derivative of s in Equation 4.34 gives the translational speed of a particle in uniform circular motion.

$$v = \frac{ds}{dt} = \frac{d(\theta r)}{dt}$$

Because the radius of the circle r does not change, we can take it outside the derivative. Doing so leaves us with the derivative defining angular speed.

$$\frac{ds}{dt} = r\frac{d\theta}{dt}$$

$$v = r\omega \tag{4.35}$$

Because the radius is measured in meters and the angular speed in rad/s, you might be tempted to report the translational speed v in m·rad/s. The "rad" is dropped, though, because a radian is a ratio of two lengths (Eq. 4.34), and is therefore dimensionless.

CONCEPT EXERCISE 4.6

A particle travels at a uniform linear speed around a circle of radius $r = 0.5$ m, completing one revolution in 12 s.

a. What is the value of the linear speed v?
b. Find the particle's angular speed.
c. What are the angle θ and arc length s swept out by the particle in 4 s?

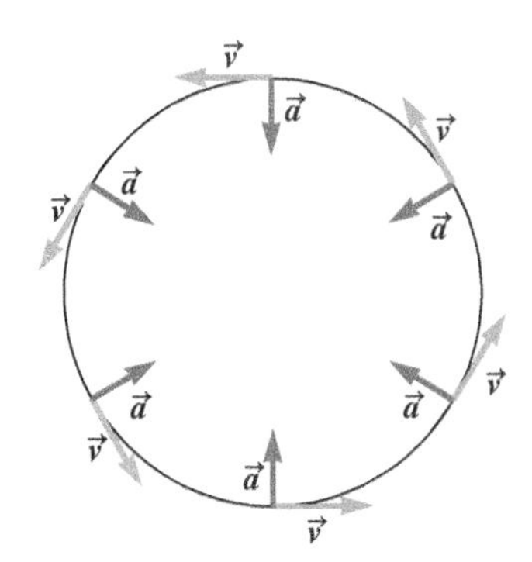

FIGURE 4.22 A particle undergoes uniform circular motion. The particle's velocity is tangent to the circle. Its acceleration points to the center of the circle.

Centripetal Acceleration

The velocity of a particle in uniform circular motion is always tangent to its path (Fig. 4.22). For a particle's velocity to change direction while its speed remains constant, the acceleration must be perpendicular to the velocity (Example 4.4). So, in the case of uniform circular motion, the acceleration always points toward the center of the path and is known as **centripetal acceleration** $\vec{a}_c$. The term *centripetal* comes from a Latin word meaning "center-seeking."

CENTRIPETAL ACCELERATION
Special Case

DERIVATION Centripetal Acceleration

In uniform circular motion, the magnitude of the centripetal acceleration is constant and depends on the particle's speed and the path's radius. We will show that for a particle undergoing uniform circular motion centered on the origin as in Figure 4.23, the centripetal acceleration is given by

$$\vec{a}_c = -\frac{v^2}{r}\hat{r} \tag{4.36}$$

and

$$\vec{a}_c = -\omega^2 r\hat{r} \tag{4.37}$$

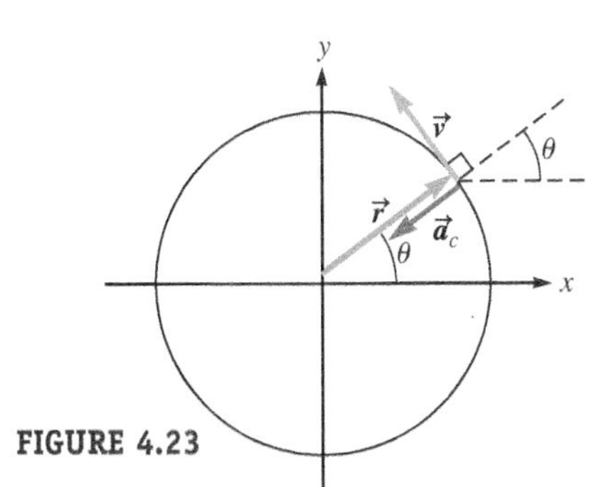

FIGURE 4.23

Consider the arbitrary moment shown in Figure 4.23. Sketch the velocity vector and its components at that moment (Fig. 4.24).	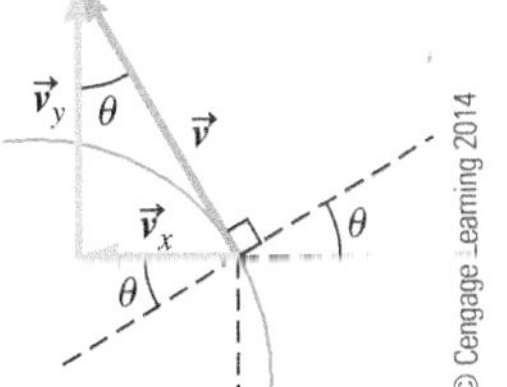 **FIGURE 4.24**
Write the velocity in component form.	$\vec{v} = (-v\sin\theta)\hat{\imath} + (v\cos\theta)\hat{\jmath}$
Take the time derivative (Eq. 4.14) to find the centripetal acceleration. The speed v is constant, but θ is a function of time.	$\vec{a}_c = \frac{d\vec{v}}{dt} = \frac{d(-v\sin\theta)}{dt}\hat{\imath} + \frac{d(v\cos\theta)}{dt}\hat{\jmath}$ $\vec{a}_c = -v\frac{d(\sin\theta)}{dt}\hat{\imath} + v\frac{d(\cos\theta)}{dt}\hat{\jmath}$

Derivation continues on page 102 ▶

Use the chain rule (Appendix A) to differentiate, then substitute $\omega = d\theta/dt$ (Eq. 4.32).	$\vec{a}_c = -v\frac{d(\sin\theta)}{d\theta}\frac{d\theta}{dt}\hat{\imath} + v\frac{d(\cos\theta)}{d\theta}\frac{d\theta}{dt}\hat{\jmath}$ $\vec{a}_c = -(v\cos\theta)\left(\frac{d\theta}{dt}\right)\hat{\imath} - (v\sin\theta)\left(\frac{d\theta}{dt}\right)\hat{\jmath}$ $\vec{a}_c = -v\omega\cos\theta\,\hat{\imath} - v\omega\sin\theta\,\hat{\jmath}$
Replace the angular speed ω using $v = r\omega$ (Eq. 4.35).	$\vec{a}_c = -\frac{v^2}{r}\cos\theta\,\hat{\imath} - \frac{v^2}{r}\sin\theta\,\hat{\jmath}$
Ah! At last. The magnitude of the centripetal acceleration is constant and is obtained from the Pythagorean theorem applied to the components of the acceleration.	$a_c = \sqrt{\left(-\frac{v^2}{r}\cos\theta\right)^2 + \left(-\frac{v^2}{r}\sin\theta\right)^2}$ $a_c = \frac{v^2}{r}\sqrt{(\cos\theta)^2 + (\sin\theta)^2}$ $a_c = \frac{v^2}{r}$ (4.38)
The centripetal acceleration always points to the center of the circle. So, using polar coordinates, the centripetal acceleration is in the negative $\hat{r}$ direction. Use $v = \omega r$ to replace v with ω.	$\vec{a}_c = -\frac{v^2}{r}\hat{r}$ ✓ (4.36) $\vec{a}_c = -\omega^2 r\hat{r}$ ✓ (4.37)

EXAMPLE 4.8 Ferris Wheel

Estimate the angular speed ω and the linear speed v of a passenger on a Ferris wheel like the one in Figure 4.25.

INTERPRET and ANTICIPATE

To find the angular speed, we need to know the period of revolution, and we need the radius of the wheel to find the linear speed. The problem statement gives us none of this information, and so we must come up with our own values based on experience.

SOLVE

To estimate the period, recall from experience that the Ferris wheel is a slow ride. If a ride lasts 2 minutes after all the passengers are seated, a given cart might get to the top four to six times. Therefore, let us assume a passenger makes five revolutions in 2 minutes and from this estimate calculate the period in seconds.	$T = \frac{(2\text{ min})(60\text{ s/min})}{5\text{ rev}} = \frac{120\text{ s}}{5}$ $T \approx 24\text{ s}$

FIGURE 4.25

Courtesy, RampantScotland.com

To estimate the angular speed, use Equation 4.33, $\omega = 2\pi/T$	$\omega = \frac{2\pi}{T} = \frac{2\pi}{24\text{ s}}$ $\omega = 0.26\text{ rad/s}$
Because our period value is an estimate, we should not be surprised to find that the period is actually not 24 s but is instead 30 s or even as low as 10 s. Therefore we should report our estimate for ω to one significant figure.	$\omega \approx 0.3\text{ rad/s}$
Now we need to estimate the radius of the Ferris wheel. Figure 4.25 shows a four-story building just behind the Ferris wheel. The building's height is about the same as the radius of the wheel. One story in a typical building is between 2.5 and 4 m high. From the photo, it looks like a medium-size building, so we will assume 3 m per story.	$r = (4\text{ stories})(3\text{ m/story})$ $r = 12\text{ m}$
Now we can find the linear speed. To avoid rounding error, we use two significant figures, but our final estimate should have only one significant figure.	$v = r\omega = (12\text{ m})(0.26\text{ rad/s})$ $v \approx 3\text{ m/s}$

CHECK and THINK

Ferris wheels come in a great range of sizes and speeds. The first Ferris wheel was built for the Chicago World's Fair in 1893. It held 36 cars, each capable of carrying 60 passengers. The wheel's diameter was 250 ft (which is more than 76 m) and the wheel had a period of 9 min (540 s). Passengers experienced a linear speed of only 0.4 m/s, which is slower than walking speed. Still, the ride was so overwhelming that there were reports of a man who became suicidal and tried to get out before the car reached the bottom. The wheel was built as the American response to the Eiffel Tower, which had been built for the 1889 World's Fair in Paris. Unfortunately, after the Chicago World's Fair, the wheel lost money and so was dynamited.

EXAMPLE 4.9 Centrifuge

NASA Amesbury

FIGURE 4.26

Centrifuges are used to accelerate a variety of objects, including medical specimens, wet laundry, and even human test subjects. After the centrifuge reaches a set speed, the subject moves in uniform circular motion so that the acceleration is centripetal. A centrifuge used by NASA (Fig. 4.26) to measure the effects of great acceleration on human subjects has a revolution rate of 33.5 revolutions per minute (rpm) and a radius of 8.84 m. Find the centripetal acceleration of the human subject using a polar coordinate system with the origin at the center of the circle. For the **CHECK and THINK** procedure, consider that acceleration around $4g$ to $6g$ can cause a person to have tunnel vision or even lose consciousness in just a few seconds. Would this subject experience such effects?

INTERPRET and ANTICIPATE

Because the centripetal acceleration always points to the center of the circle, we should write our final answer in the form $\vec{a}_c = -(\underline{\qquad})\hat{r}\ \text{m/s}^2$ in plane polar coordinates.

SOLVE First, convert the revolution rate to revolutions per second.	$33.5\ \text{rpm}\left(\frac{1\ \text{min}}{60\ \text{s}}\right) = 0.558\ \text{rev/s}$
Use the revolution rate to find the angular speed. The subject sweeps through 2π rad in each revolution.	$\omega = 2\pi\ \text{rad}(0.558\ \text{rev/s}) = 3.51\ \text{rad/s}$
Substitute ω and r into Equation 4.37. The direction is always toward the center of the circle, which is the negative $\hat{r}$ direction.	$\vec{a}_c = -\omega^2 r\hat{r}$ (4.37) $\vec{a}_c = -(3.51\ \text{rad/s})^2(8.84\ \text{m})\hat{r}$ $\vec{a}_c = -109\hat{r}\ \text{m/s}^2$

CHECK and THINK

Our expression has the form that we expected. It is common for acceleration to be expressed in terms of the free-fall acceleration—that is, as a number n times the free-fall acceleration: $a_c = ng$. We find n by dividing

$$a_c = \frac{109\ \text{m/s}^2}{9.81\ \text{m/s}^2}g \approx 11g.$$

So, the subject would likely experience tunnel vision or even lose consciousness.

4-7 Relative Motion in One Dimension

RELATIVE MOTION

Major Concept

The word *relative* has so many meanings. Think of such statements as "My uncle is my favorite relative," and "Compare the relative merits of democracy and fascism." A physicist uses the phrase **relative motion** when motion is observed (and measured) from different perspectives. To a physicist, an observer's perspective depends only on the observer's motion: at rest, moving at constant nonzero velocity, or accelerating.

You have probably observed motion from different perspectives. For instance, imagine that you are standing by the side of the road. Your friend—a passenger in a car—passes by at 30 mph and waves. You barely have time to see that he is waving to you because he appears to be moving so quickly. Now imagine riding along on a multilane highway at 60 mph and spotting your friend riding in the next lane, also traveling at 60 mph. This time, he does not appear to speed past you; instead, he seems to be stationary. If your windows were open, you could (aside from a little shouting) converse with him as you would with any passengers in your car.

REFERENCE FRAME

Major Concept

Measurements of motion depend on the observer's perspective. A **reference frame** or simply a **frame** is a coordinate system attached to an observer's particular perspective; in other words, an observer selects a reference frame by choosing a coordinate system that—from her perspective—is at rest. The observer's measurements are made *relative to* or *with respect to* that frame. (These two phrases mean the same thing, and we will use them interchangeably through this discussion.)

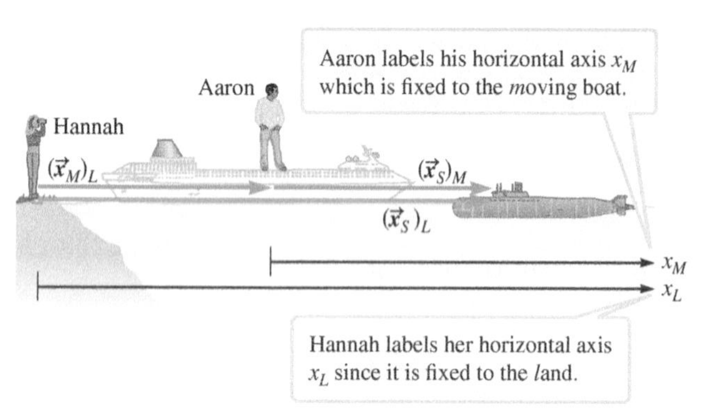

FIGURE 4.27 A submarine is observed from two frames.

The relationships between position, velocity, and acceleration measured in two frames are the foundation for our later study of Einstein's theory of relativity (Chapter 39). Let's use a specific one-dimensional example to derive these relationships. In Figure 4.27, a submarine is observed both by Hannah on the shore and by Aaron on a boat moving at constant velocity. Each of the two observers measures the position of the submarine relative to his or her own frame, and each observer chooses an x axis pointing to the right. Each observer is at rest at the origin of his or her own coordinate system. Hannah's coordinate system is stationary with respect to the shore. Aaron's coordinate system is fixed to the boat and therefore moves with respect to the shore. To distinguish between the two reference frames, Aaron uses an M subscript for *m*oving boat and Hannah uses an L for *l*and. (The choice of M and L will become clear in the next section.) Therefore, the position of the *s*ubmarine relative to Hannah on the *l*and is $(\vec{x}_S)_L$ and relative to Aaron is $(\vec{x}_S)_M$. According to Hannah on the *l*and, the position of Aaron on the *m*oving boat is $(\vec{x}_M)_L$.

The first subscript tells you which object is observed, and the second subscript tells you from which frame. When you are writing, it is common to drop the parentheses and just write a double subscript.

Figure 4.27 shows that these three vectors are related by the expression

$$(\vec{x}_S)_L = (\vec{x}_S)_M + (\vec{x}_M)_L \tag{4.39}$$

We take the time derivative of Equation 4.39 to find the relationship between the submarine velocity measured by Hannah $(\vec{v}_S)_L$ and that measured by Aaron $(\vec{v}_S)_M$:

$$\frac{d(\vec{x}_S)_L}{dt} = \frac{d(\vec{x}_S)_M}{dt} + \frac{d(\vec{x}_M)_L}{dt}$$

$$(\vec{v}_S)_L = (\vec{v}_S)_M + (\vec{v}_M)_L \tag{4.40}$$

where $(\vec{v}_M)_L$ is Aaron's velocity as measured by Hannah on the land.

We take another time derivative to find the acceleration:

$$\frac{d(\vec{v}_S)_L}{dt} = \frac{d(\vec{v}_S)_M}{dt} + \frac{d(\vec{v}_M)_L}{dt}$$

Because Aaron's velocity $(\vec{v}_M)_L$ (measured by Hannah) is constant, its time derivative is zero, so both observers measure the same acceleration for the submarine:

$$(\vec{a}_S)_L = (\vec{a}_S)_M \tag{4.41}$$

EXAMPLE 4.10 Here Comes the Pilot

A pilot is in a boat moving at constant velocity as in Figure 4.27, on his way to a submarine to help the vessel's captain navigate through a difficult channel. A harbormaster stationed on shore observes that the submarine's velocity is $5.3\hat{\imath}$ m/s and that the pilot boat's velocity is $6.5\hat{\imath}$ m/s. According to the pilot, the submarine is initially 5.65×10^2 m away. How long will it take the pilot boat to reach the submarine?

INTERPRET and ANTICIPATE

Use the coordinate system shown in Figure 4.27. We know the velocities of the *s*ubmarine $(\vec{v}_S)_L$ and the pilot boat $(\vec{v}_M)_L$ relative to the harbormaster on the *l*and. If we want to know how long the pilot boat must travel to reach the sub, our first task is to find the velocity $(\vec{v}_S)_M$ of the submarine relative to the pilot boat.

SOLVE

Start with Equation 4.40 and solve for the velocity of the submarine with respect to the pilot boat's frame $(\vec{v}_S)_M$.

$$(\vec{v}_S)_L = (\vec{v}_S)_M + (\vec{v}_M)_L \tag{4.40}$$

$$(\vec{v}_S)_M = (\vec{v}_S)_L - (\vec{v}_M)_L$$

$$(\vec{v}_S)_M = (5.3\hat{\imath} - 6.5\hat{\imath})\text{ m/s} = -1.2\hat{\imath}\text{ m/s}$$

CHECK and THINK

The negative sign means that, according to the pilot, the submarine is moving in the negative x direction. In other words, to the pilot it looks like the submarine is moving toward him. If that troubles you, think about the situation from the harbormaster's point of view. To her, both the pilot boat and the submarine are moving away from shore. Because the pilot boat has a higher velocity, the harbormaster sees the gap between the pilot boat and the submarine getting smaller.

SOLVE (continued)

We use the submarine's velocity $(\vec{v}_S)_M$ and its initial position $(\vec{x}_S)_M$ measured by the pilot on the boat to find the time it takes the pilot's boat to meet the submarine. To do so, we use $\Delta x = v_{0x}t + \frac{1}{2}a_x t^2$ (Eq. 2.11) with $a = 0$.

When the pilot's boat reaches the submarine, the submarine will be at the origin of his coordinate system $(x_S)_M = 0$. (The time t measured by the harbormaster and by the pilot must be the same.)

$$(\vec{v}_S)_M = \frac{(\Delta\vec{x}_S)_M}{t} = \frac{0 - (x_S)_M}{t}\hat{\imath}$$

$$t = \frac{-(x_S)_M}{(v_S)_M} = \frac{-5.65 \times 10^2\text{ m}}{-1.2\text{ m/s}}$$

$$t = 4.7 \times 10^2\text{ s} = 7.8\text{ min}$$

EXAMPLE 4.11 Does the Speeder Get a Ticket?

A local officer and a state trooper each measure the velocity of a speeding car. The local officer is parked by the side of the road, and the state trooper is pursuing the speeder. According to the local officer, the speeder is going 95.0 mph (42.4 m/s), whereas according to the state trooper, the speeder is going 15.1 mph (6.70 m/s).

A What is the velocity of the trooper relative to the local officer?

INTERPRET and ANTICIPATE

Because this question involves relative motion, we must use two coordinate systems. We use the subscript L for the *l*ocal officer and M for the *m*oving trooper. Figure 4.28 applies to all three parts of the problem.

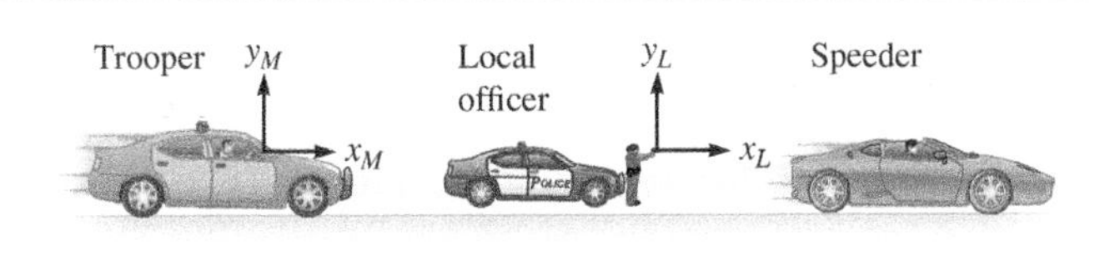

FIGURE 4.28

Example continues on page 106 ▶

SOLVE

We know $(\vec{v}_S)_L$, the speeder's velocity measured by the stationary *local* officer (42.4 m/s), and $(\vec{v}_S)_M$, the speeder's velocity measured by the moving trooper (6.70 m/s). We can therefore use Equation 4.40 to find $(\vec{v}_M)_L$, the velocity of the trooper measured by the local officer.

$$(\vec{v}_S)_L = (\vec{v}_S)_M + (\vec{v}_M)_L \quad (4.40)$$

$$(\vec{v}_M)_L = (\vec{v}_S)_L - (\vec{v}_S)_M$$

$$(\vec{v}_M)_L = (42.4\hat{\imath} - 6.70\hat{\imath})\ \text{m/s}$$

$$(\vec{v}_M)_L = 35.7\hat{\imath}\ \text{m/s} = 79.8\hat{\imath}\ \text{mph}$$

CHECK and THINK

The stationary officer sees both cars traveling in the same direction. Because the trooper's speed is lower (79.8 mph) than that of the speeder (95.0 mph), the local officer sees the gap between them widen. (Don't worry if you worked in US customary units and found a slight difference in your answer due to a rounding error.)

B The speeder sees the local officer in his rearview mirror and slows down such that 2.5 s after he hits the brakes his speed is, according to the local officer, 60.0 mph (26.8 m/s). What new velocity does the state trooper measure?

SOLVE

We now know $(\vec{v}_M)_L$, velocity of state trooper measured by local officer, from part A. We were just given $(\vec{v}_S)_L$, new velocity of speeder measured by the local officer. Use Equation 4.40 to find $(\vec{v}_S)_M$, velocity of speeder measured by the state trooper.

$$(\vec{v}_S)_L = (\vec{v}_S)_M + (\vec{v}_M)_L \quad (4.40)$$

$$(\vec{v}_S)_M = (\vec{v}_S)_L - (\vec{v}_M)_L$$

$$(\vec{v}_S)_M = (26.8\hat{\imath} - 35.7\hat{\imath})\ \text{m/s}$$

$$(\vec{v}_S)_M = -8.93\hat{\imath}\ \text{m/s} = -20.0\hat{\imath}\ \text{mph}$$

CHECK and THINK

The state trooper now sees the speeder going in the negative x direction; in other words, the speeder seems to be moving toward the state trooper. The local officer still sees both cars moving away, but the gap between them is getting smaller.

C Find the acceleration of the speeder measured by each observer during this 2.5-s interval.

SOLVE

We can use $v_x = v_{0x} + a_x t$ (Eq. 2.9) for either observer. Let's start with the acceleration observed by the local officer. According to $(\vec{a}_S)_L = (\vec{a}_S)_M$ (Eq. 4.41), it should be the same as the acceleration measured by the state trooper.

$$v_x = v_{0x} + a_x t \quad (2.9)$$

$$a_x = \frac{(v_x - v_{0x})_L}{t} = \frac{(26.8 - 42.4)\ \text{m/s}}{2.5\ \text{s}}$$

$$(\vec{a}_S)_L = -6.25\hat{\imath}\ \text{m/s}^2$$

CHECK and THINK

To check, we apply Equation 2.9 to the state trooper's observation and find that the two observers agree on the acceleration.

$$(a_S)_M = \frac{(v_x - v_{0x})_M}{t} = \frac{(-8.93 - 6.70)\ \text{m/s}}{2.5\ \text{s}}$$

$$(\vec{a}_S)_M = -6.25\hat{\imath}\ \text{m/s}^2$$

4-8 Relative Motion in Two Dimensions

RELATIVE MOTION ✪ Major Concept

We are now ready to consider two-dimensional motion observed from two reference frames. After this discussion, we will be equipped to wrap up the skateboarding case study.

To make the extension to two-dimensional relative motion, we return to the submarine. This time, though, the submarine is in a holding pattern just offshore, and the boat's approach is not along the line joining Hannah to the submarine. Figure 4.29 shows this scene from above. Now each observer must use a two-dimensional coordinate system, with the corresponding axes of the coordinate systems parallel to each other. For us to be able to work out the relationship between these two coordinate systems, they must maintain their relative orientation. In other words, as the boat moves, it may translate, but is not allowed to rotate.

Hannah, on *l*and, measures the position vectors of Aaron on the *m*oving boat $(\vec{r}_M)_L$ and of the *s*ubmarine $(\vec{r}_S)_L$. Aaron measures the position $(\vec{r}_S)_M$ of the submarine, and the three position vectors in Figure 4.29 are related by

$$(\vec{r}_S)_L = (\vec{r}_S)_M + (\vec{r}_M)_L \tag{4.42}$$

As before, the first time derivative gives the velocity:

$$\frac{d(\vec{r}_S)_L}{dt} = \frac{d(\vec{r}_S)_M}{dt} + \frac{d(\vec{r}_M)_L}{dt}$$

$$(\vec{v}_S)_L = (\vec{v}_S)_M + (\vec{v}_M)_L \tag{4.43}$$

We take the second time derivative to find the acceleration:

$$\frac{d(\vec{v}_S)_L}{dt} = \frac{d(\vec{v}_S)_M}{dt} + \frac{d(\vec{v}_M)_L}{dt}$$

As before, the boat is moving at constant velocity, so the time derivative of $(\vec{v}_M)_L$ is zero, and both observers measure the same acceleration:

$$(\vec{a}_S)_L = (\vec{a}_S)_M \tag{4.44}$$

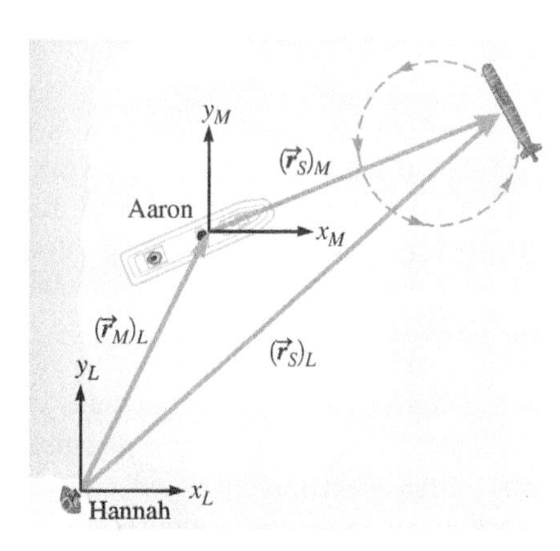

FIGURE 4.29 A submarine is observed from two frames. One frame is attached to the land, and the other frame is attached to the boat and moves relative to the land. Each observer uses a two-dimensional coordinate system. The axes of the two systems remain parallel to each other as the boat moves toward the submarine.

These three equations (4.42, 4.43, and 4.44) are two-dimensional vector equations, and as is often the case, it is helpful to write them in terms of their scalar components:

$$(x_S)_L = (x_S)_M + (x_M)_L \qquad (y_S)_L = (y_S)_M + (y_M)_L \tag{4.45}$$

$$(v_{Sx})_L = (v_{Sx})_M + (v_{Mx})_L \qquad (v_{Sy})_L = (v_{Sy})_M + (v_{My})_L \tag{4.46}$$

$$(a_{Sx})_L = (a_{Sx})_M \qquad (a_{Sy})_L = (a_{Sy})_M \tag{4.47}$$

Don't be intimidated by all the subscripts. The *x* or *y* indicates the component. Think of *S* as the *s*ubject being observed from the *l*ab (*L*) frame and from the *m*oving (*M*) frame with respect to the lab.

We have developed all our equations for relative motion with two observers in mind. It is not necessary for there to be two actual human observers, however. In other words, there can be two reference frames (lab and moving) that are not occupied by human observers. For example, in an airplane, there may be two velocity gauges, ground velocity and air velocity. The ground velocity gauge gives the airplane's velocity relative to the ground, and the air velocity gauge gives its velocity relative to the wind. Although there are likely actual observers on the ground (in the control tower), there is (probably) no observer traveling with the wind. We can still define two reference frames, one attached to the ground ("the lab") and another "attached" to the (moving) wind (Fig. 4.30). The speed of the airplane relative to the wind is called the airspeed, and its speed relative to the ground is its groundspeed.

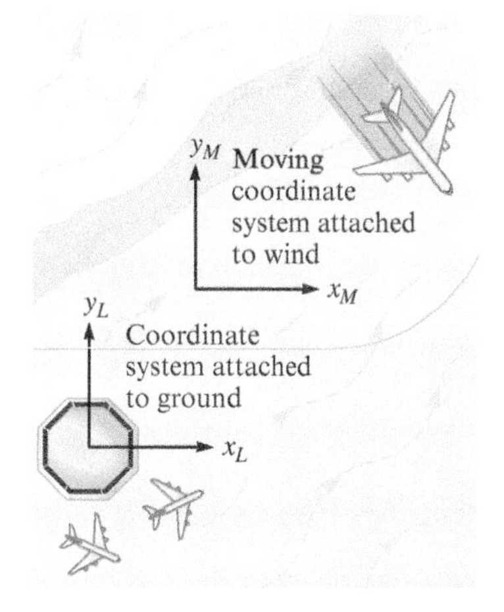

FIGURE 4.30 The speed of the airplane relative to the wind (the airspeed) and its speed relative to the ground (groundspeed) define two reference frames: "the lab" attached to the ground and "moving" attached to the wind.

EXAMPLE 4.12 Groundspeed

An airplane flies at an airspeed of 809 km/h at an angle $\alpha = 20°$ (exactly) south of east. The wind blows at 85.0 km/h toward the northeast with respect to the ground. What are the plane's groundspeed and direction?

Example continues on page 108 ▶

INTERPRET and ANTICIPATE

Airspeed tells us how fast the plane is moving relative to the 85.0-km/h wind; *groundspeed* is the plane's speed as measured by someone standing on the ground. Because the plane's motion is not aligned with the wind's direction, we must use our two-dimensional relative-motion equations.

Make a sketch (Fig. 4.31) to get an estimate for our final results, aligning the coordinate systems with the cardinal (compass) directions. The term *northeast* means exactly 45° north of east, so the direction of the wind's velocity is $\theta = 45°$ counterclockwise from the x axis. Our sketch shows a rough geometric addition of $(\vec{v}_S)_L = (\vec{v}_S)_M + (\vec{v}_M)_L$ as a guide. The vector $(\vec{v}_S)_L$ is slightly longer than $(\vec{v}_S)_M$ (groundspeed a little faster than airspeed), and $(\vec{v}_S)_L$ points more easterly than $(\vec{v}_S)_M$.

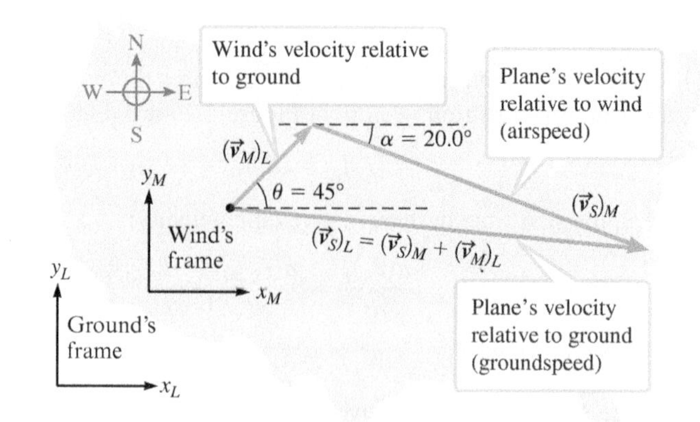

FIGURE 4.31

SOLVE

Convert the given speeds to SI units.

$$(v_M)_L = (85.0\ \text{km/h})\left(\frac{0.278\ \text{m/s}}{1\ \text{km/h}}\right) = 23.6\ \text{m/s}$$

$$(v_S)_M = (809\ \text{km/h})\left(\frac{0.278\ \text{m/s}}{1\ \text{km/h}}\right) = 225\ \text{m/s}$$

Write the velocities $(\vec{v}_M)_L$ and $(\vec{v}_S)_M$ in component form.

$$(\vec{v}_M)_L = (v_{ML}\cos\theta)\hat{\imath} + (v_{ML}\sin\theta)\hat{\jmath}$$

$$v_{ML}\cos 45° = v_{ML}\sin 45° = (23.6\ \text{m/s})\left(\frac{\sqrt{2}}{2}\right) = 16.7\ \text{m/s}$$

$$(\vec{v}_M)_L = (16.7\hat{\imath} + 16.7\hat{\jmath})\ \text{m/s}$$

$$(\vec{v}_S)_M = (v_{SM}\cos\alpha)\hat{\imath} - (v_{SM}\sin\alpha)\hat{\jmath}$$

$$(\vec{v}_S)_M = (211\hat{\imath} - 76.9\hat{\jmath})\ \text{m/s}$$

We find the plane's groundspeed components by adding the wind's velocity and the airspeed velocity.

$$(v_{Sx})_L = (v_{Sx})_M + (v_{Mx})_L \tag{4.46}$$

$$(v_{Sx})_L = (211 + 16.7)\ \text{m/s} = 227.7\ \text{m/s}$$

$$(v_{Sy})_L = (v_{Sy})_M + (v_{My})_L \tag{4.46}$$

$$(v_{Sy})_L = (-76.9 + 16.7)\ \text{m/s} = -60.2\ \text{m/s}$$

The groundspeed and direction are found from these components using $v = \sqrt{v_x^2 + v_y^2}$ and $\theta = \tan^{-1}(v_y/v_x)$ (Eqs. 3.12 and 3.14).

$$(v_S)_L = \sqrt{(227.7\ \text{m/s})^2 + (-60.2\ \text{m/s})^2}$$

$$(v_S)_L = 236\ \text{m/s}$$

$$\tan^{-1}\left(\frac{-60.2\ \text{m/s}}{227.7\ \text{m/s}}\right) = -14.8°$$

CHECK and THINK

The airplane is heading 14.8° south of east, which meets our expectation of being slightly more easterly than the 20° south-of-east angle seen by our imaginary wind observer. Also, as expected from our sketch, the groundspeed of 236 m/s is a little higher than the airspeed of 225 m/s.

CONCEPT EXERCISE 4.7

Which speed, groundspeed or airspeed, is needed to determine the flight time of an airplane that flies from city A to city B?

EXAMPLE 4.13 CASE STUDY Who's Right?

Reread the case study (pages 86–87) and determine which student is correct: Cameron, who wants to use a 90° launch angle; or Avi, who wants to use a smaller angle.

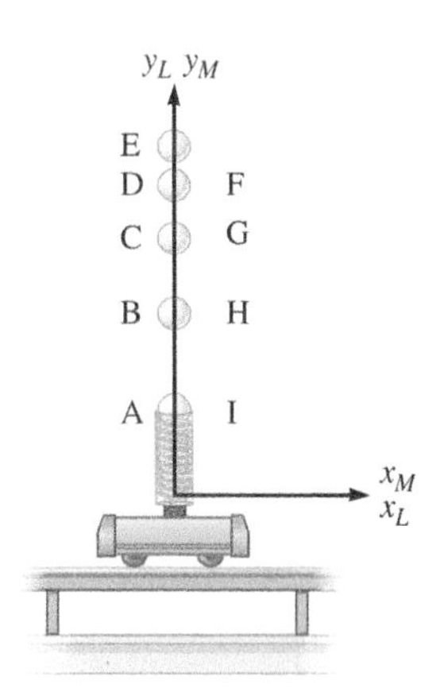

FIGURE 4.32

INTERPRET and ANTICIPATE

The solution we present here is based on viewing the situation from two reference frames. One is the laboratory frame L, the frame from which Cameron and Avi observe the experiment. The other frame M is attached to the cart. These two coordinate systems are shown in Figure 4.32.

If the cart is not in motion, there is no difference between the two reference frames. If the ball is launched straight up, observers in both frames see the same motion diagram. The ball moves along a single axis (y_L or y_M) from point A to point E and back down to point I.

SOLVE

If the cart is **not** in motion and the ball is launched at an angle smaller than 90° as in Figure 4.33, observers in both frames see the ball move along a parabolic path. They agree that the ball lands in front of the cart at point I.

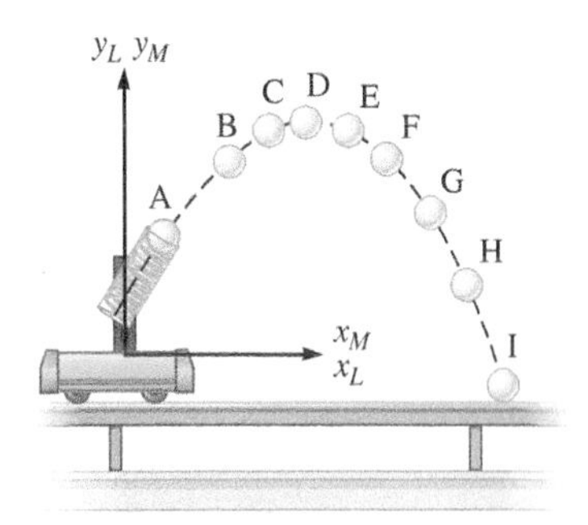

FIGURE 4.33

What changes when the cart (with its frame) is already in motion at constant velocity at the instant the ball is launched? For the observer in the moving cart frame, the answer is simple: Nothing changes. To this observer, it does not matter whether the cart is stationary as the ball is launched or moving with constant velocity during the launch. Either way, the path of the ball in the cart frame is along a straight, vertical line if the launch angle is 90° and along a parabola if the launch angle is smaller than 90°. A straight-up launch brings the ball back to the cart, and a launch at $\theta < 90°$ means that the ball lands at some point in front of the cart.

The ball can land only in one place, of course. So, what the observer on the cart sees must also be what the laboratory observer sees: a 90° launch angle lands the ball in the cart, and a launch angle smaller than 90° lands the ball on the track in front of the cart.

So, Cameron is correct: The ball must be launched at a 90° angle.

CHECK and THINK

Consider what the laboratory observer sees for a 90° launch from the moving cart. To land back in the cart, the ball must move along a straight, vertical path in the cart frame, yet the ball's path in the laboratory frame cannot be along a straight, vertical line.

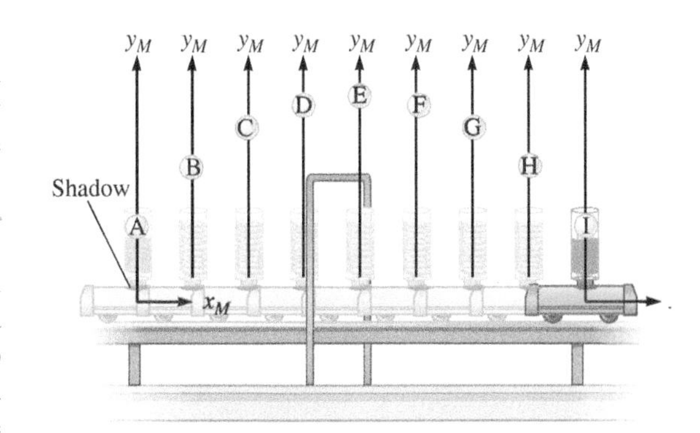

FIGURE 4.34

Figure 4.34 shows how this apparent contradiction is reconciled. The cart moves at constant velocity along the x axis while, from the perspective of the laboratory observer, the ball moves along a two-dimensional parabolic path. If we use an overhead light to project the ball's shadow (Section 4-5) along the x axis, we would find that this shadow is always superimposed on the cart. When the ball reaches point I in its parabolic path, the cart is also at that point, and the ball falls into the cart.

Now consider what the laboratory observer sees when the cart is moving and the launch angle is less than 90°. The motion diagram shows that the ball still moves in a parabolic path, but its velocity along the x axis is now greater than the velocity of the cart (Fig. 4.35). The result of this greater horizontal velocity is that the parabola is wider. The ball's shadow along the x axis is still

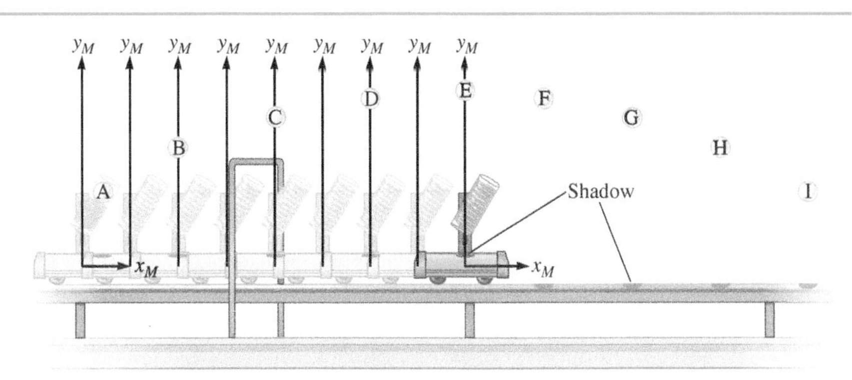

FIGURE 4.35

Example continues on page 110 ▶

moving at a constant velocity, but the magnitude of that velocity is greater than the magnitude of the cart's velocity. The ball's shadow is no longer superimposed on the cart; instead, the shadow is in front of the cart, and the gap between the shadow and the cart widens. When the ball lands on the track, the cart is still far behind.

According to this model, the skateboarder executing the hippy must jump straight up to land on his skateboard. While he is in the air, his shadow will be on top of the skateboard just as the ball's shadow is on the cart.

A Word About Air Resistance

We have just seen that Cameron is correct: The ball must be launched straight up if it is to land back in the cart, whether the cart is moving or not. Still, there is something about Avi's belief that the launch has to be aimed not straight up but rather along the track that agrees with our intuition, which is usually based on our experiences in the everyday world. In this case, our intuition may be based on what we know about throwing lightweight objects such as balloons, foam balls, and feathers. In this chapter, we ignored air, the medium through which all these projectiles move.

In many situations, ignoring air is acceptable, but experience tells us that sometimes the air greatly affects the motion of objects. For example, imagine replacing the metal ball in the case study with a very light foam ball. The air in the laboratory opposes the motion of the foam ball, with the result that its maximum height and range are noticeably shorter than if it were moving in a vacuum. So, like Avi, we might try to compensate for the air's resistance by aiming the launcher at a lower angle. (The air actually opposes the motion of both balls, but the effect is more noticeable in the case of the foam ball.)

In physics, we often think of how things work under ideal conditions—in this case, projectile motion in a vacuum. In the laboratory, we do our best to design experiments and equipment that are a close approximation to ideal conditions. For example, we use objects whose physical dimensions minimize air resistance, and in some laboratories, we may even evacuate a chamber. So, under laboratory conditions, Cameron is correct. In some situations, however, we may need to factor in air resistance, as we will in Chapter 6.

SUMMARY

Underlying Principles: Two- and three-dimensional kinematics

One-dimensional motion is along a single straight line, whereas multidimensional motion may be along a curved path or along a path consisting of two or more straight lines.

Major Concepts

Summary of one- and three-dimensional kinematic quantities.

Quantity	Three Dimensions		One Dimension (along x axis)
1. **Position** is a vector used to locate a particle with respect to the origin of a particular coordinate system.	$\vec{r} = x\hat{\imath} + y\hat{\jmath} + z\hat{k}$	(4.1)	$\vec{r} = x\hat{\imath}$
2. **Displacement** $\Delta\vec{r}$ is found by subtracting the initial position vector from the final position vector.	$\Delta\vec{r} = \vec{r}_f - \vec{r}_i = \Delta x\hat{\imath} + \Delta y\hat{\jmath} + \Delta z\hat{k}$ $\Delta x \equiv (x_f - x_i)$ $\Delta y \equiv (y_f - y_i)$ $\Delta z \equiv (z_f - z_i)$	(4.2)	$\Delta\vec{r} = (x_f - x_i)\hat{\imath}$

Major Concepts—cont'd

Quantity	Three Dimensions		One Dimension
3a. Average velocity is the displacement divided by the time interval.	$\vec{v}_{av} = \dfrac{\Delta\vec{r}}{\Delta t}$	(4.4)	$\vec{v}_{av} = \dfrac{\Delta x}{\Delta t}\hat{\imath}$
	$\vec{v}_{av} = \dfrac{\Delta x}{\Delta t}\hat{\imath} + \dfrac{\Delta y}{\Delta t}\hat{\jmath} + \dfrac{\Delta z}{\Delta t}\hat{k}$	(4.6)	
3b. Velocity is the time derivative of $\vec{r}$.	$\vec{v} = \dfrac{d\vec{r}}{dt}$	(4.8)	$\vec{v} = \dfrac{dx}{dt}\hat{\imath}$
	$\vec{v} = \dfrac{dx}{dt}\hat{\imath} + \dfrac{dy}{dt}\hat{\jmath} + \dfrac{dz}{dt}\hat{k}$	(4.9)	
4a. Average acceleration comes from dividing the change in velocity by the time interval.	$\vec{a}_{av} = \dfrac{\Delta\vec{v}}{\Delta t}$	(4.11)	$\vec{a}_{av} = \dfrac{\Delta v_x}{\Delta t}\hat{\imath}$
	$\vec{a}_{av} = \dfrac{\Delta v_x}{\Delta t}\hat{\imath} + \dfrac{\Delta v_y}{\Delta t}\hat{\jmath} + \dfrac{\Delta v_z}{\Delta t}\hat{k}$	(4.12)	
4b. Acceleration is the time derivative of the velocity.	$\vec{a} = \dfrac{d\vec{v}}{dt}$	(4.14)	$\vec{a} = \dfrac{dv_x}{dt}\hat{\imath}$
	$\vec{a} = \dfrac{dv_x}{dt}\hat{\imath} + \dfrac{dv_y}{dt}\hat{\jmath} + \dfrac{dv_z}{dt}\hat{k}$	(4.15)	

5. **Relative motion** refers to motion observed from different perspectives. An observer's perspective depends only on the observer's motion.

A **reference frame** or simply a **frame** is a coordinate system attached to an observer's particular perspective.

When a relative-motion situation involves two frames, the one that is stationary relative to the Earth is usually called the *l*aboratory frame, and the other is the *m*oving frame. For relative motion in two dimensions, we write these relationships in terms of their scalar components:

$$(x_S)_L = (x_S)_M + (x_M)_L \qquad (4.45)$$
$$(y_S)_L = (y_S)_M + (y_M)_L$$
$$(v_{Sx})_L = (v_{Sx})_M + (v_{Mx})_L \qquad (4.46)$$
$$(v_{Sy})_L = (v_{Sy})_M + (v_{My})_L$$
$$(a_{Sx})_L = (a_{Sx})_M$$
$$(a_{Sy})_L = (a_{Sy})_M \qquad (4.47)$$

Special Cases

1. **Two-dimensional projectile motion** is a special case in which (a) there is no acceleration in the x (horizontal) direction and (b) the acceleration in the y (vertical) direction is the constant free-fall acceleration g. A projectile's path is parabolic. The equations of motion for a projectile are

$$\vec{a} = -g\hat{\jmath} \qquad (4.23)$$
$$\vec{v} = (v_0\cos\theta)\hat{\imath} + (v_0\sin\theta - gt)\hat{\jmath} \qquad (4.24)$$
$$\vec{r} = [v_0 t\cos\theta + x_0]\hat{\imath} + [(v_0 t\sin\theta - \tfrac{1}{2}gt^2) + y_0]\hat{\jmath} \qquad (4.25)$$

a. The **range** of a projectile is the magnitude of its horizontal displacement when its vertical displacement is zero:

$$R = \frac{v_0^2}{g}\sin 2\theta \qquad (4.28)$$

b. The **maximum range** occurs when $\theta = 45°$:

$$R_{max} = v_0^2/g \qquad (4.29)$$

2. **Uniform circular motion** is a particular case of two-dimensional motion in which the path of the particle is a circle and its speed is constant. The magnitude of the position vector is the radius r of the circle.

The particle completes one revolution around the circle in a time T known as the **period**. In one period, the distance traveled is the circumference of the circle, $2\pi r$. So, its speed is given by

$$v = \frac{2\pi r}{T} \qquad (4.30)$$

and the centripetal acceleration is

$$a_c = \frac{v^2}{r} = r\omega^2 \text{ toward the center} \qquad (4.36 \text{ and } 4.37)$$

The rate at which θ changes is the **angular speed** of the motion:

$$\omega = \frac{d\theta}{dt} \qquad (4.32)$$

For a particle moving in uniform circular motion,

$$\omega = \frac{2\pi}{T} \qquad (4.33)$$

The speed and angular speed are related by

$$v = r\omega \qquad (4.35)$$

PROBLEM-SOLVING STRATEGY Two-Dimensional Projectile Motion

There are two parts to the **INTERPRET and ANTICIPATE** procedure:

1. Start with a **sketch** that includes a **coordinate system**.
2. If the vertical displacement is zero ($\Delta y = 0$), you can apply the range equation $R = (v_0^2/g)\sin 2\theta$ (Eq. 4.28).

There are also two parts to the **SOLVE** procedure:

1. **List six kinematic variables** for x and y separately.
2. You may use the **constant acceleration equations** (Table 2.4 or Equations 4.23 through 4.25).

PROBLEMS AND QUESTIONS

A = **algebraic** C = **conceptual** E = **estimation** G = **graphical** N = **numerical**

4-1 What Is Multidimensional Motion?

1. C In each case, determine if the train is moving in one or two dimensions. Explain your answers. **a.** A train moves in one direction along a flat, straight track. **b.** A train moves in one direction along a flat, straight track and then reverses direction, moving in the opposite direction along the same track. **c.** A train moves in one direction along a straight, downhill track.

2. C In each case, determine whether the object is moving in one, two, or three dimensions. Explain. **a.** The car driving up the hill in Figure P4.2A **b.** The car winding down the mountain in Figure P4.2B

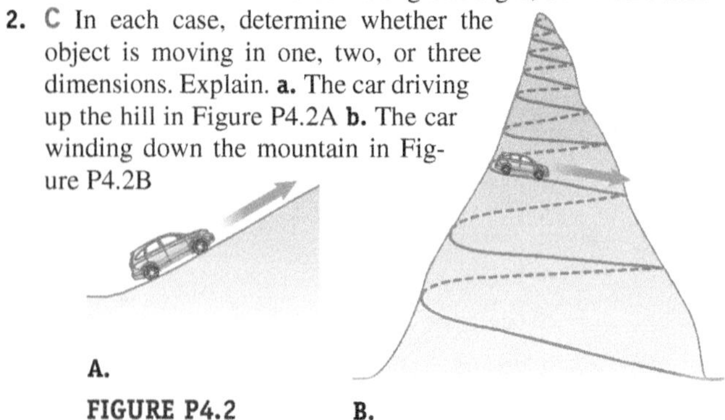

FIGURE P4.2

3. C CASE STUDY Imagine an indoor tennis court on a cruise ship moving at constant velocity. Assume the court is oriented so that one player faces the bow (forward) and the other the stern (backward). Does one player have an advantage over the other? If so, which one? Explain your answers.

4-2 Motion Diagrams for Multidimensional Motion

4. G A basketball player dribbles the ball while running at a constant speed straight across the court. Think about the velocity and acceleration of the ball and then sketch a motion diagram for the ball. Explain how you arrived at your sketch.

5. C A motion diagram of a bouncing ball is shown in Figure P4.5. **a.** Does the ball move in one, two, or three dimensions? **b.** For which points is the ball's speed the highest? **c.** Where is the ball at its lowest speeds?

A B C D E F G H I J K L M N O

FIGURE P4.5 All dots are in the plane of the page.

6. A ball hangs from a string. The string is kept taut as the ball is displaced to one side and released. The ball swings freely back and forth. This is an example of a simple pendulum.
a. G Use the data in the accompanying table to create a motion diagram for this ball.
b. C Does the ball maintain a constant speed? If not, where does it speed up or slow down?

t (s)	x (cm)	y (cm)	t (s)	x (cm)	y (cm)
0	30.90	4.89	11	−29.44	4.43
1	29.44	4.43	12	−25.14	3.21
2	25.14	3.21	13	−18.36	1.70
3	18.36	1.70	14	−9.69	0.47
4	9.69	0.47	15	0.00	0.00
5	0.00	0.00	16	9.69	0.47
6	−9.69	0.47	17	18.36	1.70
7	−18.36	1.70	18	25.14	3.21
8	−25.14	3.21	19	29.44	4.43
9	−29.44	4.43	20	30.90	4.89
10	−30.90	4.89			

4-3 Position and Displacement

7. An ice skater moves along a circular path at constant speed with $\omega = \pi$ rad/s. With the origin located at the center of the circular path, her coordinates at time $t = 0$ are $x = 5.00$ m and $y = 0$.
a. C Sketch the circular path and include the coordinate system, marking the location of the origin and the initial position of the skater. What is the radius of the circular path?
b. N Write an expression for the x coordinate of the skater as a function of time.
c. G Plot the function you found in part (b) with time on the horizontal axis and position x along the vertical axis for all times between $t = 0$ and $t = 6.0$ s.
d. C After the first time the skater goes around the path, she is then repeating her motion as she goes around again and again. Describe how this repetitive behavior is illustrated in your graph from part (c).

Problems 8 and 26 are paired.

8. Figure P4.8 shows the motion diagram of two balls, one on the left and one on the right. Each ball starts at a point labeled i. The ball on the left is released and falls straight down. At the same time, the ball on the right is launched horizontally and follows the path shown.

a. C Use the given coordinate system to write the position of points i, C, E, G, and K in component form for each ball.

b. N Find the displacement of each ball from i to points C, E, G, and K.

c. C Compare your answers for the two balls in part (b). What similarities do you notice?

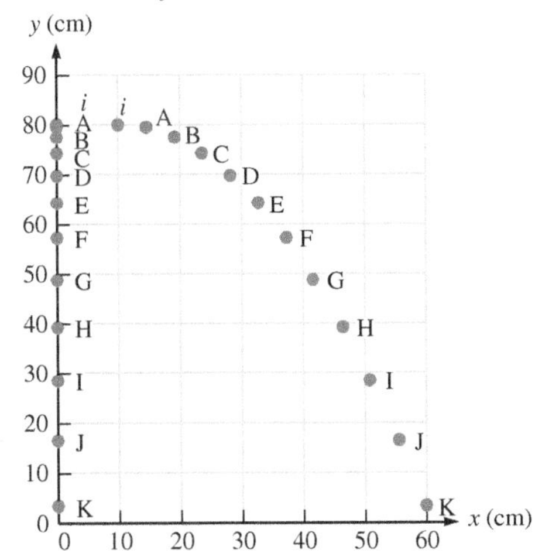

FIGURE P4.8 Problems 8 and 26.

9. A particle moves at constant speed in a circular path, centered about the origin, such that $\omega = 2.0$ rad/s. At some instant, its position is $(x, y) = (3.56, 0.44)$ m.

a. N What is the radius of the circle?

b. C How does your answer change if $\omega = 4.0$ rad/s?

4-4 Velocity and Acceleration

10. N A glider dives toward the ground at a constant velocity of 4.50 m/s and at an angle of 56.0° below the horizontal. If the Sun is directly overhead, what is the speed of the glider's shadow on the level ground below?

11. N An object moves with an initial velocity $\vec{v}_i = 3.00\hat{\jmath}$ m/s and an acceleration $\vec{a} = 2.50\hat{\imath}$ m/s^2. Assume the object is initially at the origin. **a.** What is the position vector of the object as a function of time? **b.** What is the velocity vector of the object as a function of time? **c.** What is the position of the object at time $t = 3.00$ s? **d.** What is the speed of the object at time $t = 3.00$ s?

Problems 12 and 13 are paired.

12. C If a particle's speed is always increasing, what are the possible angles between the particle's velocity and acceleration?

13. C If a particle's speed is always decreasing, what are the possible angles between the particle's velocity and acceleration?

14. N An aircraft flies at constant altitude (with respect to sea level) over the South Rim of the Grand Canyon (Fig. P3.40, page 83). Consider a coordinate system such that the positive x axis points to the east, and the positive y axis points north. The aircraft's initial position and velocity are 1350 m at an angle of 145° and 60.0 m/s at an angle of 55.0° where both angles are measured counterclockwise with respect to the positive x axis. The aircraft's acceleration is 4.0 m/s^2 at an angle of 195° with respect to the positive x axis. **a.** What is the velocity of the aircraft after 7.50 s have elapsed? **b.** What is the position vector of the aircraft after 7.50 s have elapsed?

15. N A glider is initially moving at a constant height of 3.59 m. It is suddenly subject to a wind such that its velocity at a later time t can be described by the equation $\vec{v}(t) = 15.72\hat{\imath} - 7.88(1 + t)\hat{\jmath} + 0.79t^3\hat{k}$, where $\vec{v}$ and its components are in meters per second, t is in seconds, and the z axis is perpendicular to the level ground. **a.** What was the initial velocity of the glider? **b.** Write an expression for the acceleration of the glider in component form, when $t = 2.15$ s.

Problems 16 and 17 are paired.

16. N If the vector components of the position of a particle moving in the xy plane as a function of time are $\vec{x} = (2.5\text{ m/s}^2)t^2\hat{\imath}$ and $\vec{y} = (5.0\text{ m/s}^3)t^3\hat{\jmath}$, at what time t is the angle between the particle's velocity and the x axis equal to 45°?

17. A If the vector components of a particle's position moving in the xy plane as a function of time are $\vec{x} = bt^2\hat{\imath}$ and $\vec{y} = ct^3\hat{\jmath}$, where b and c are positive constants with the appropriate dimensions such that the components will be in meters, at what time t is the angle between the particle's velocity and the x axis equal to 45°?

18. C An object is subject to a constant acceleration. Under what circumstances does the object travel (a) in a straight line, (b) along a circular path, or (c) along a curved noncircular path?

19. A The spiral is an example of a mathematical form appearing in nature, from the visible construction of seashells, pinecones, and galaxies to the movement behavior of certain animals. The position of a hungry animal that moves outward along a spiral path, searching for food, can be written as $\vec{r}(t) = A\omega t\cos(\omega t)\hat{\imath} + A\omega t\sin(\omega t)\hat{\jmath}$. Write an expression for the velocity of the animal in component form.

4-5 Special Case of Projectile Motion

20. N A circus performer stands on a platform and throws an apple from a height of 45 m above the ground with an initial velocity $\vec{v}_0$ as shown in Figure P4.20. A second, blindfolded performer must catch the apple. If $v_0 = 26$ m/s, how far from the end of the platform should the second performer stand?

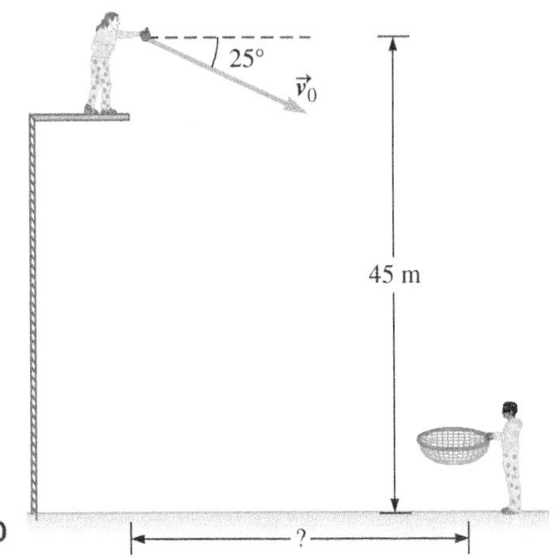

FIGURE P4.20

21. N Anthony carelessly rolls his toy car off a 74.0-cm-high table. The car strikes the floor a horizontal distance of 97.0 cm from the edge of the table. **a.** What was the velocity with which the car left the table? **b.** What was the angle of the car's velocity with respect to the floor just prior to impact?

22. C A physics student stands on a second-story balcony and uses a potato gun to launch a potato horizontally with speed v. The potato has flight time t and lands on the ground a horizontal distance d from the balcony. **a.** If the launch speed of the potato were doubled, would the time of flight increase, decrease, or stay the same? If the flight time changes, would it double or be halved? Explain. **b.** If the launch speed of the potato were doubled, would the horizontal distance increase, decrease, or stay the same? If the horizontal distance changes, would it double or be halved? Explain.

Problems 23 and 24 are paired.

23. N During the battle of Bunker Hill, Colonel William Prescott ordered the American Army to bombard the British Army camped near Boston. The projectiles had an initial velocity of 45 m/s at 35° above the horizon and an initial position that was 35 m higher than where they hit the ground. How far did the projectiles move horizontally before they hit the ground? Ignore air resistance.

24. A During the battle of Bunker Hill, Colonel William Prescott ordered the American Army to bombard the British Army camped near Boston. The projectiles had an initial velocity of v measured in meters per second at an angle θ above the horizon and an initial position that was h higher than where they hit the ground. How far did the projectiles move horizontally before they hit the ground? Ignore air resistance.

25. N A softball is hit with an initial velocity of 29.0 m/s at an angle of 60.0° above the horizontal and impacts the top of the outfield fence 5.00 s later. Assuming the initial height of the softball was 0.500 m above (level) ground, what are the ball's horizontal and vertical displacements?

26. Figure P4.8 shows the motion diagram of two balls. The time interval between images is 0.036 s. Each ball starts at point i. The ball on the left is released and falls straight down. At the same time, the ball on the right is launched horizontally and follows the path shown.

a. A Write the velocity and acceleration of each ball as a function of time.

b. N What is the velocity and acceleration of each ball at i, G, and K?

c. C Compare your answers for the two balls in part (b). What similarities do you notice?

27. A circus performer throws an apple toward a hoop held by a performer on a platform (Fig. P4.27). The thrower aims for the hoop and throws with a speed of 24 m/s. At the exact moment the thrower releases the apple, the other performer drops the hoop. The hoop falls straight down.

a. N At what height above the ground does the apple go through the hoop?

b. C If the performer on the platform did not drop the hoop, would the apple pass through it?

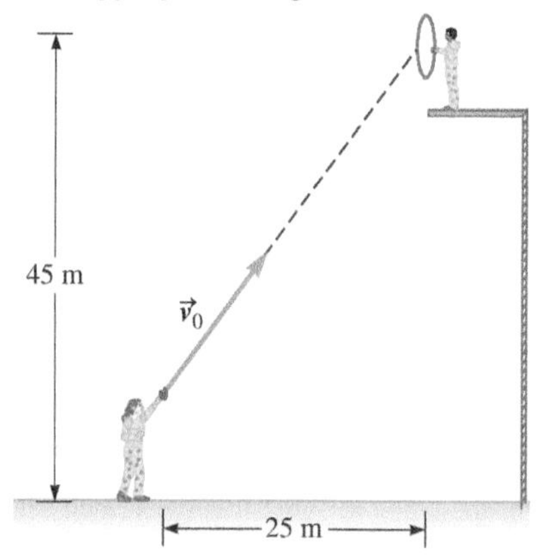

FIGURE P4.27

28. A An arrow is fired with initial velocity v_0 at an angle θ from the top of battlements, a height h above the ground. **a.** In terms of h, v_0, θ, and g, what is the time at which the arrow reaches its maximum height? **b.** In terms of h, v_0, θ, and g, what is the maximum height above the ground reached by the arrow?

29. N A rock is thrown horizontally off a 56.0-m-high cliff overlooking the ocean, and the sound of the splash is heard 3.60 s later. If the speed of sound in air at this location is 343 m/s, what was the initial velocity of the rock?

30. A A projectile is launched up and to the right over flat, level ground. If air resistance is ignored, its maximum range occurs when the angle between its initial velocity and the ground is 45°. Which angles would result in the range being equal to half the maximum?

31. N Sienna tosses a ball from the window of her high-rise apartment building with an initial velocity of 6.50 m/s at 25.0° above the horizontal. The ball strikes the ground 4.50 s later. **a.** What is the horizontal distance from the base of the building to the point where the ball strikes the ground? **b.** What is the height from which the ball is thrown? **c.** What is the time it takes for the ball to reach a point 15.0 m below the window where it was thrown?

32. N Some cats can be trained to jump from one location to another and perform other tricks. Kit the cat is going to jump through a hoop. He begins on a wicker cabinet at a height of 1.750 m above the floor and jumps through the center of a vertical hoop, reaching a peak height 3.125 m above the floor. **a.** With what initial velocity did Kit leave the cabinet if the hoop is at a horizontal distance of 1.544 m from the cabinet? **b.** If Kit lands on a bed at a horizontal distance of 3.587 m from the cabinet, how high above the ground is the bed?

33. N Dock diving is a great form of athletic competition for dogs of all shapes and sizes (Fig. P4.33). Sheba, the American Pit Bull Terrier, runs and jumps off the dock with an initial speed of 9.02 m/s at an angle of 25° with respect to the surface of the water. **a.** If Sheba begins at a height of 0.84 m above the surface of the water, through what horizontal distance does she travel before hitting the surface of the water? **b.** Write an expression for the velocity of Sheba, in component form, the instant before she hits the water. **c.** Determine the peak height above the water reached by Sheba during her jump.

Brian Cahn/Zuma Press/Corbis

FIGURE P4.33

4-6 Special Case of Uniform Circular Motion

34. C A graduate student discovers that the only centrifuge in the laboratory is malfunctioning. The angular speed ω was supposed to be adjustable, but now it is stuck on a high angular speed, and the student is running an experiment that requires a low translational speed. Fortunately, the arm is adjustable. What can the student do to the length of the arm to get the correct translational speed?

35. N The bola is a traditional weapon used for tripping up or grounding an animal (Fig. P4.35). Once it is set into motion, each ball at the end of the bola can be thought of as a single object in uniform circular motion. Suppose it takes the bola 0.3250 s to traverse a circular path with a radius of 0.8661 m. What is the magnitude of the centripetal acceleration experienced by either ball at the end of the bola?

FIGURE P4.35 A bola is spun in a circle above the hunter, eventually being released and thrown forward.

36. C In three different driving tests, a car moves with constant speed v_0. In case 1, the car passes over a mark painted on a horizontal, straight section of road. In case 2, the car passes over a mark painted at the crest of a small hill. In case 3, the car passes over a mark painted at the bottom of a small dip. The hill and the dip are circular in profile, with the same radius (Fig. P4.36). Rank the cases from greatest to least according to the magnitude of the acceleration of the car when it passes the mark. Explain.

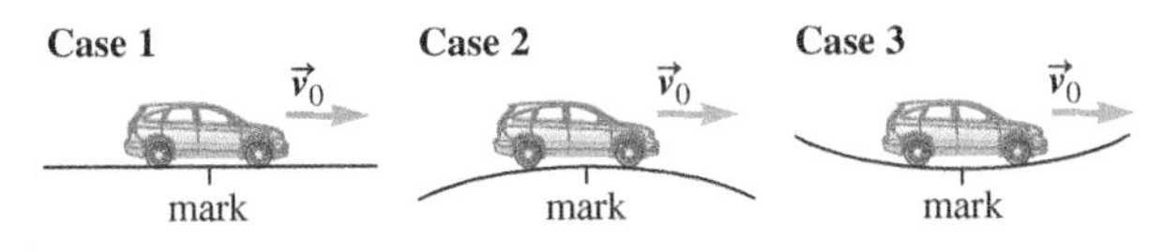

FIGURE P4.36

37. A child swings a tennis ball attached to a 0.750-m string in a horizontal circle above his head at a rate of 5.00 rev/s.

a. N What is the centripetal acceleration of the tennis ball?

b. C The child now increases the length of the string to 1.00 m but has to decrease the rate of rotation to 4.00 rev/s. Is the speed of the ball greater now or when the string was shorter?

c. N What is the centripetal acceleration of the tennis ball when the string is 1.00 m in length?

Problems 38, 39, and 40 are grouped.

38. A Two particles A and B move at a constant speed in circular paths at the same angular speed ω. Particle A's circle has a radius that is twice the length of particle B's circle. What is the ratio T_A/T_B of their periods?

39. A For particles A and B in Problem 38, what is the ratio v_A/v_B of their translational speeds?

40. A For particles A and B in Problem 38, what is the ratio a_A/a_B of (the magnitude of) their centripetal accelerations?

41. N Approaching one of the many sharp horizontal turns in the Monaco Grand Prix, an experienced Formula-1 driver slows down from 135 km/h to 55.0 km/h while rounding the bend in 10.0 s. If the driver continues to decelerate at this same rate and the radius of the curve is 15.0 m, what is the acceleration of the car the moment that its speed reaches 55.0 km/h?

42. A pendulum constructed with a bowling ball at the end of a cable 4.00 m in length has an acceleration vector $\vec{a} = (-8.50\hat{\imath} + 24.3\hat{\jmath})\,\text{m/s}^2$ when it is 24.7° past the lowest point in its swing.

a. G Sketch the scalar components of the acceleration vector in a vector diagram at this point in the pendulum's motion.

b. N What is the magnitude of the radial acceleration of the pendulum at this point?

c. N What are the speed and the velocity of the bowling ball at this point?

Problems 43 and 44 are paired.

43. N The Moon's orbit around the Earth is nearly circular and has a period of approximately 28 days. Assume the Moon is moving in uniform circular motion. **a.** Find the angular speed of the Moon. **b.** What is its centripetal acceleration?

44. The Earth's orbit around the Sun is nearly circular. Assume the Earth is moving in uniform circular motion.

a. N Find the angular speed of the Earth.

b. N What is its centripetal acceleration?

c. C Compare your answers to parts (a) and (b) with those in Problem 43 for the Moon. Do your answers make sense? Explain.

4-7 Relative Motion in One Dimension

45. N Pete and Sue, two reckless teenage drivers, are racing eastward along a straight stretch of highway. Pete is traveling at 98.0 km/h, and Sue is chasing him at 125 km/h. **a.** What is Pete's velocity with respect to Sue? **b.** What is Sue's velocity with respect to Pete? **c.** If Sue is initially 325 m behind Pete, how long will it take her to catch up to him?

46. A state trooper parked near the side of the road sees a car pass a truck. According to the trooper, the car's velocity is 75 mph and the truck's velocity is 58 mph.

a. G Draw a motion diagram for these two vehicles as seen by the trooper.

b. G Draw a motion diagram of the truck as seen by the car's driver.

c. N What is the velocity of the truck according to the car's driver?

d. C Are your answers to parts (b) and (c) consistent? Explain.

47. C A person might use the relative motion of a train, car, or similar vehicle to perform the apparently superhuman act of throwing a 200-mph fastball. **a.** Explain at least one way to do this trick. **b.** An observer in what reference frame would see this act as superhuman? Could a baseball pitcher ever observe his or her own fastball moving at 200 mph?

48. C A brother and sister, Alan and Beth, have just adopted a new dog, Sparky. Alan and Beth walk directly toward each other, each moving with constant speed along a straight-line path. Sparky leaves Alan and runs to Beth. Is the magnitude of Sparky's displacement with respect to Alan greater than, less than, or equal to the magnitude of Sparky's displacement with respect to Beth? Explain your reasoning.

49. N A man paddles a canoe in a long, straight section of a river. The canoe moves downstream with constant speed 3 m/s relative to the water. The river has a steady current of 1 m/s relative to the bank. The man's hat falls into the river. Five minutes later, he notices that his hat is missing and immediately turns the canoe around, paddling upriver with the same constant speed of 3 m/s relative to the water. How long does it take the man to row back upriver to reclaim his hat?

50. A trooper drives her car with a constant speed of 20.0 m/s to the east along a straight road. A speeder drives his car to the east on the same road while slowing down at a constant rate. The speeder moves at 16.0 m/s at time $t = 0$, and 8.00 m/s at time $t = 4$ s. For the entire interval from $t = 0$ to $t = 4$ s, the speeder is ahead of the trooper (that is, the speeder is located to the east of the trooper).

a. N In the frame of the trooper, what is the speeder's velocity at $t = 2$ s?

b. C In the reference frame of the trooper, is the speeder's car speeding up, slowing down, or moving with a constant speed at $t = 2$ s? Explain.

c. C The trooper has a laser ranging device that can determine the distance between the two cars. She finds that the distance is d at time $t = 2$ s and $0.9d$ at time $t = 3$ s. At time $t = 4$ s, will the distance between the cars be greater than, less than, or equal to $0.8d$? Explain.

4-8 Relative Motion in Two Dimensions

51. N CASE STUDY In the case study, we saw that a skateboarder should jump straight up at an angle $\theta = 90°$ to do a high (hippy) jump over an obstacle and land back on the board. We modeled this trick on a level track. If instead a skateboarder must perform a high jump down a 20.0° incline, what should be the launch angle relative to the board?

52. An ant and a spider each move with constant velocity on a horizontal table. The velocity vectors and positions of the ant and the spider (with respect to the table) at time $t = 0$ s are shown in Figure P4.52.

a. G Draw the velocity vector of the ant in the frame of the spider.

b. C Is there a time at which the ant and the spider will have the same position? Explain.

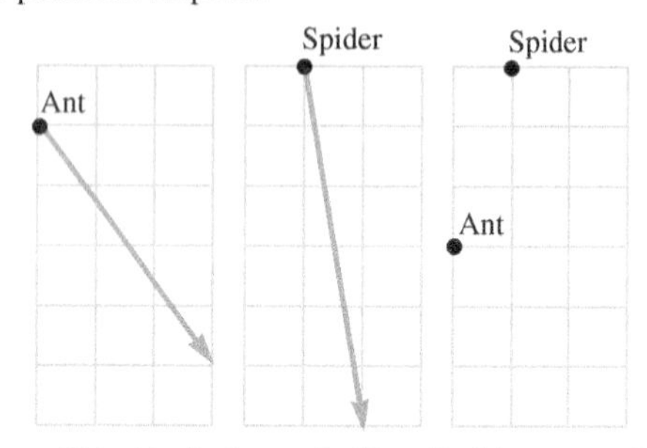

FIGURE P4.52 Velocities in frame of table (1 unit = 1 cm/s) Positions at $t = 0$ (1 unit = 1 cm)

Problems 53 and 54 are paired.

53. N Suppose at one point along the Nile River a ferryboat must travel straight across a 10.3-mile stretch from west to east. At this location, the river flows from south to north with a speed of 2.41 m/s. The ferryboat has a motor that can move the boat forward at a constant speed of 20.0 mph in still water. In what direction should the ferry captain direct the boat so as to travel directly across the river?

54. N Return to the ferryboat scenario in Problem 53. If the captain points the boat directly at the target location on the east bank of the river, how far downstream will she be from the target when she lands on the east bank?

55. N A jetliner travels with a constant speed of 710.0 km/h relative to the air to a city 880.0 km due south. **a.** What is the time interval required to complete the trip if the jetliner is experiencing a headwind of 55.0 km/h toward the north? **b.** What is the time interval required to complete the trip if the jetliner now experiences a tailwind of the same speed? **c.** What is the time interval required to complete the trip if the jetliner now experiences a crosswind of the same speed relative to the ground towards the west?

56. C Avi and Cameron are back in physics class and are working on the following problem:

Car A moves to the east along a straight road as shown in the motion diagram in Figure P4.56. A traffic cone is at rest on the road at the location shown. Car B, not shown on the diagram, is located due south of the traffic cone and is moving to the south, directly away from the cone. Their task is to describe the approximate direction ("south," "southeast," and so forth) of the velocity of car B relative to car A at the following instants: instant 1, when car A is west of the cone; instant 2, when car A is next to the cone; and instant 3, when car A is east of the cone. Consider the discussion that Avi and Cameron have about this problem.

Avi: The velocity of car B relative to car A is to the southeast at instant 1, due south at instant 2, and southwest at instant 3. If car A is picked as an origin, the velocity would follow the line from car A to car B.

Cameron: I agree with your answer for the instant at which car A is next to the cone. At this time, the velocity of car B is due south relative to car A because no east–west movement is seen. The velocity, though, should be southwest at both of the other times. The west–east separation between the cars is decreasing at 1 and increasing the other way at 3.

State whether you agree or disagree with each statement. If you disagree, describe what specifically is incorrect and how it could be corrected.

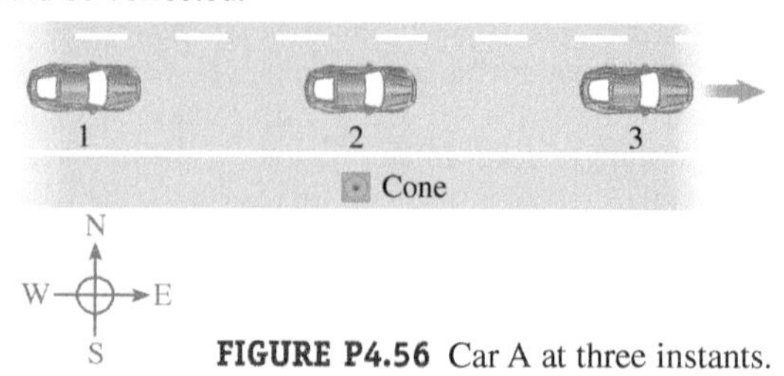

FIGURE P4.56 Car A at three instants.

57. N A pair of sunglasses balanced on a car's rearview mirror suddenly drops down as the car accelerates eastward at 2.33 m/s^2. **a.** What is the magnitude and direction of the sunglasses' acceleration as seen by an observer inside the car? **b.** What is the acceleration of the sunglasses as seen by an observer standing outside the car?

58. N Two bicyclists in a sprint race begin from rest and accelerate away from the origin of an x–y coordinate system. Miguel's acceleration is given by $(-0.700\hat{\imath} + 1.00\hat{\jmath})$ m/s^2, and Lance's acceleration is given by $(1.20\hat{\imath} + 0.300\hat{\jmath})$ m/s^2. **a.** What is Miguel's acceleration with respect to Lance? **b.** What is Miguel's speed with respect to Lance after 4.50 s have elapsed? **c.** What is the distance separating Miguel and Lance after 4.50 s have elapsed?

General Problems

59. C A particle has a nonzero acceleration and a nonzero constant speed at all times. What is the angle between the particle's velocity and acceleration?

60. N A golfer hits his approach shot at an angle of 50.0°, giving the ball an initial speed of 38.2 m/s (Fig. P4.60). The ball lands on the elevated green, 5.50 m above the initial position near the hole, and stops immediately. **a.** How much time passed while the ball was in the air? **b.** How far did the ball travel horizontally before landing? **c.** What was the peak height reached by the ball?

FIGURE P4.60

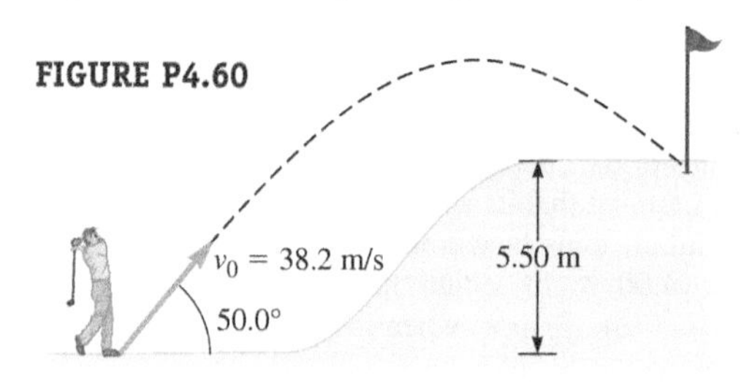

61. A You are watching a friend practice archery when he misses the target completely, and the arrow sticks into the ground. Discouraged, your friend asks whether you can help by estimating the speed with which the arrow left the bow. Remembering your physics class, you realize that you can do so by measuring its height above the ground when it was launched, if you assume the arrow was launched horizontally and that the ground is level. For your analysis, you let h represent the arrow's height above the ground when it was launched, v_0 represent the launch speed, and θ represent the angle the arrow makes with the horizontal when it is stuck into the ground. Find an expression for v_0 in terms of h and θ.

Problems 62 and 63 are paired.

62. C A ball starts from rest at the left end of the track shown in Figure P4.62. The ball flies off the right end of the track and follows the parabolic trajectory shown. At each instant 1 and 2, is the direction of the acceleration of the ball upward, downward, to the right, to the left, or is the acceleration zero? Explain.

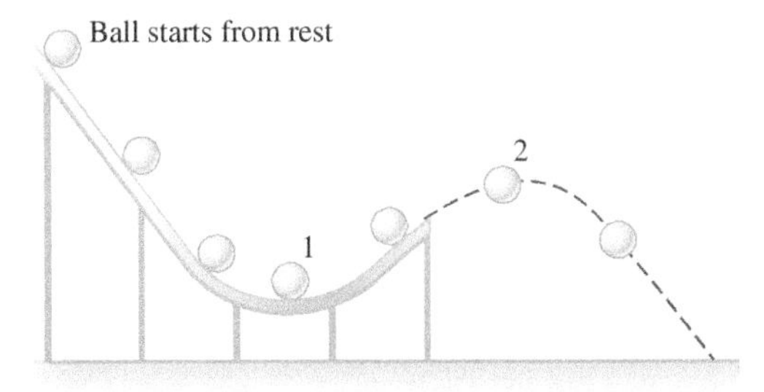

FIGURE P4.62 Problems 62 and 63.

63. N A ball starts from rest at the left end of the track shown in Figure P4.62. The ball flies off the right end of the track and follows the parabolic trajectory shown. Suppose the right end of the track is 1.0 m above the floor and makes an angle of 30.0° with the horizontal. Also suppose the ball leaves the track with a speed of 4.0 m/s. Determine where the ball hits the floor.

64. N David Beckham has lined up for one of his famous free kicks from a point 25.0 m from the goal. He kicks the soccer ball with a speed of 22.0 m/s at 28.0° to the horizontal. The height of the goal's crossbar is 2.44 m. **a.** What is the distance from the crossbar with which the ball will go into the goal or sail over? **b.** Does the soccer ball reach the goal on its way up or on its way down?

Problems 65, 66, and 67 are grouped.

65. G Suppose a particle's position is given by

$$x = At \qquad \text{and} \qquad y = Bt - Ct^2$$

where A, B, and C are constants. What is the shape of the particle's path?

66. For the particle in Problem 65, $A = 2$ m/s, $B = 4$ m/s, and $C = 4$ m/s^2.
    a. G Sketch the particle's path.
    b. N Complete the accompanying table.
    c. C On your sketch of the particle's path, draw vector components for the velocity and acceleration for the times in the accompanying table. Indicate any components that equal zero.

t (s)	x (m)	y (m)	v_x (m/s)	v_y (m/s)	a_x (m/s^2)	a_y (m/s^2)
0	0			4		
0.5		1			0	
1.0			2			−8

67. A Return to the particle in Problem 65 where A, B, and C are constants. Find expressions for the particle's velocity and acceleration.

68. N Frequently, a weapon must be fired at a target that is closer than the weapon's maximum range. To hit such a target, a weapon has two possible launch angles (Fig. P4.68A): one higher than 45° (θ_H) and one lower than 45° (θ_L). Although the displacement of the projectile is the same for the two angles, a projectile launched at θ_H has a longer flight time and a higher peak position than one launched at θ_L. Usually, some tactical situation makes one angle preferable to the other. For example, if the projectile must go over some nearby object such as a grove of trees, the higher angle may be desirable. A shorter flight time and therefore θ_L are preferable if the target is mobile.

In practice, many weapons are designed to operate either at angles lower than 45° or at angles higher than 45°, but not both. Tanks, for example, often must face mobile targets; to minimize the time the target has to move, tanks fire at low angles. Grenades, on the other hand, are launched at high angles because a soldier launching a grenade is often close to the target, but has no armor plating for protection. The high launch angle allows the soldier to stay out of sight by hiding behind some obstacle, and the longer flight time may make it possible for the soldier to move farther from the exploding grenade.

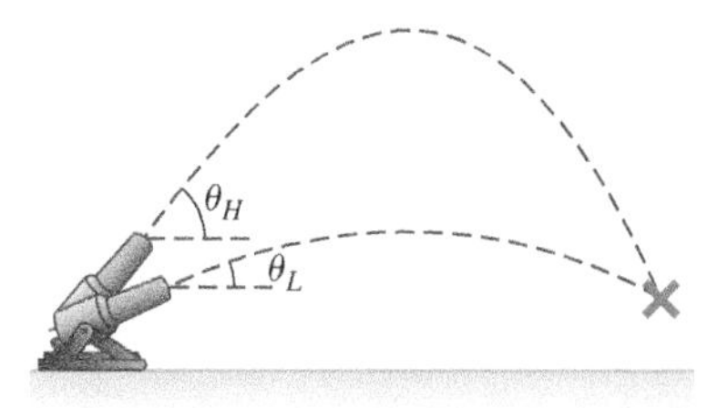

A.

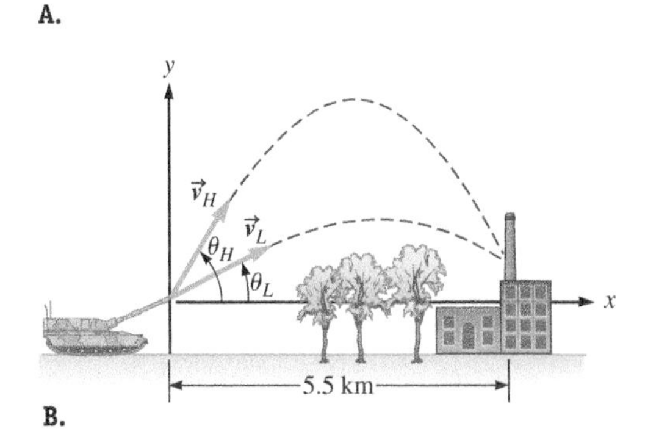

B.

FIGURE P4.68

Imagine an unusual scenario in which a large gun mounted on a vehicle is required to hit an explosives factory (Fig. P4.68B). A huge explosion is expected, and there must be time for the gunner to retreat. A grove of trees provides cover. The maximum range of the gun is 17.6 km, and the maximum speed of the vehicle is 80.0 km/h. **a.** What is the muzzle speed v_0? (Muzzle speed is the speed at which the projectile leaves the barrel of the gun.) **b.** The target is 5.5 km away. Find the low angle θ_L and the high angle θ_H at which the gunner may aim so as to hit the target. **c.** Find the time the projectile takes to hit the target for both angles. **d.** Assume the vehicle retreats at its maximum speed (80.0 km/h) to be as far from the ensuing explosion as possible. How far is it from the factory at the time of the explosion for each launch angle?

Problems 69 and 70 are paired.

69. N A projectile is launched up and to the right over flat, level ground. Its range is 177 m, and its maximum elevation above the ground is 354 m. What was the angle between its initial velocity and the ground? Ignore air resistance.

70. A A projectile is launched up and to the right over flat, level ground. Its range is equal to half of its maximum elevation above the ground. What was the angle between its initial velocity and the ground? Ignore air resistance.

71. N A World War II–era dive bomber is being used to drop food and supplies to outposts on remote mountaintops. During one such drop, the bomber descends with velocity v directed at an angle of 37.0° below the horizontal. The cargo is released at an altitude of 2450 m and reaches its intended drop zone with a displacement Δr of 3850 m. What is the speed of the bomber when it releases its cargo?

72. C An observer sitting on a park bench watches a person walking behind a runner. Figure P4.72A is the motion diagram representing what this observer sees. To better reveal the changing distance between runner and walker, five observations (A through E) are shown on five separate lines in Figure P4.72B. To the observer on the bench, both the runner and the walker move to the right, and the gap between them widens. Draw the motion diagram of the runner from the reference frame of the walker.

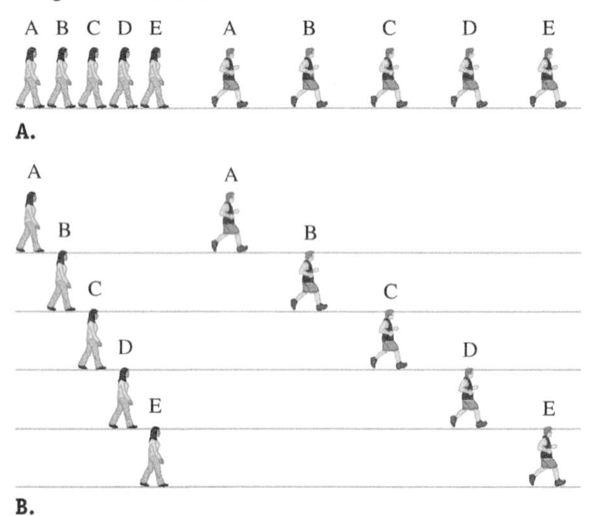

FIGURE P4.72

73. N In a dramatic physics demonstration, an apple is suspended 3.45 m above the floor. A pellet gun on the floor is 4.62 m from the point just below the apple. The gun is aimed at the apple; the gun's muzzle speed is 38.3 m/s. At the moment the pellet is shot from the gun, the apple is released. At what height above the floor does the pellet hit the apple?

Problems 74 and 75 are paired.

74. G Figure P4.74 is a motion diagram of Mars from positions A to T. The time interval between "dots" is 10 days. Describe the motion of Mars, including changes in Mars's velocity. Also note the fastest and slowest points in Mars's motion. (White dots represent the positions of fixed stars.)

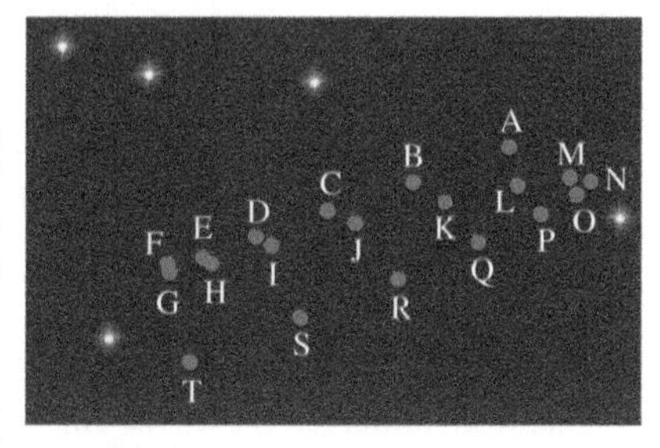

FIGURE P4.74 Problems 74 and 75.

75. C Figure P4.74 is a motion diagram of Mars from positions A to T. The time interval between "dots" is 10 days. The diagram is made with respect to an observer on the Earth. Although Mars appears to stop near points F and N, it does not actually stop in its orbit around the Sun. Why does it appear to stop in this motion diagram?

76. A riverboat with a speed in still water of 20 knots (10.3 m/s) travels on a river that has a constant speed of 0.650 m/s.

a. N What is the time interval required for the riverboat to travel a distance of 4.00 km upstream and return to its starting point?

b. N What would be the time interval required for the same trip in still water?

c. C Why does the boat trip take longer when there is a river current?

77. A An object located at the origin at $t = 0$ moves with an initial velocity of $(2.00\hat{\imath} + 7.00\hat{\jmath})$ m/s and a nonconstant acceleration of $\vec{a} = (\sqrt{3}t\hat{\imath} - t\hat{\jmath})$ m/s^2. **a.** What is the velocity of the particle as a function of time? **b.** What is the position of the particle as a function of time?

Hint: Integration is required.

78. Stretching his arm to a height of 1.05 m above the ground, Arman mischievously fires his Nerf dart gun at Irene. The dart leaves the gun with initial speed v_0 at an angle of 42.0° above the horizontal. Irene eludes the dart, and it harmlessly strikes the ground.

a. A What is the horizontal displacement x of the dart as a function of the initial speed v_i, written as $x(v_i)$, just as the dart strikes the floor?

b. N What is x if $v_0 = 0.300$ m/s?

c. N What is x if $v_0 = 30.0$ m/s?

d. A For small but nonzero values of v_0, show that $x(v_0)$ simplifies because one of its terms dominates.

e. A What is the form of $x(v_i)$ for large values of v_0?

79. N A circus cat has been trained to leap off a 12-m-high platform and land on a pillow. The cat leaps off at $v_0 = 3.5$ m/s and an angle $\theta = 25°$ (Fig P4.79). **a.** Where should the trainer place the pillow so that the cat lands safely? **b.** What is the cat's velocity as she lands in the pillow?

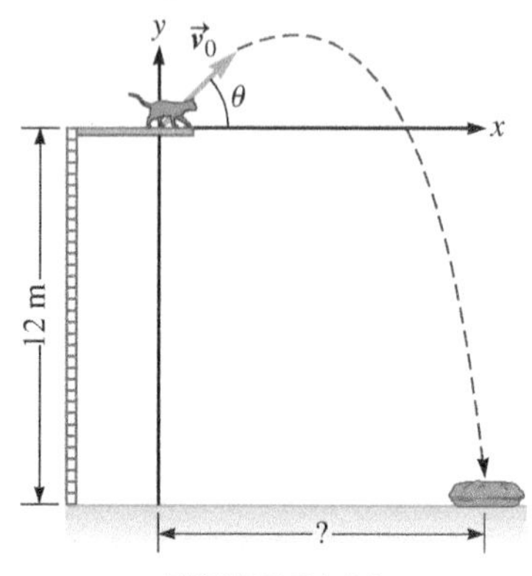

FIGURE P4.79

Problems 80 and 81 are paired.

80. Cosmic rays are high-speed charged particles from space. They are harmful to people, other organisms, and equipment, but fortunately the Earth's upper atmosphere and magnetic field act to shield us from these particles. The trajectory of charged particles trapped by the Earth's atmosphere is complicated. In this problem, we will explore a simpler scenario in which a single charged particle is trapped by a magnetic field and travels in a helical path. The position of this particle is given by

$$\vec{r}(t) = A\cos\omega t\hat{\imath} + A\sin\omega t\hat{\jmath} + Bt\hat{k}$$

where A, B, and ω are constants.

a. G Substitute values for t to confirm graphically that the path of the particle is a helix.

b. A Find $\vec{v}(t)$ and $\vec{a}(t)$.

c. C Describe the z component of $\vec{v}(t)$ and $\vec{a}(t)$.

d. C Do $\vec{r}(t)$, $\vec{v}(t)$, and $\vec{a}(t)$ in this problem apply only to charged particles in a magnetic field, or do they apply to any particle moving in a helical path? Explain.

81. An experimentalist in a laboratory finds that a particle has a helical path. The position of this particle in the laboratory frame is given by

$$\vec{r}(t) = R\cos\omega t\hat{\imath} + R\sin\omega t\hat{\jmath} + v_z t\hat{k}$$

where R, v_z, and ω are constants. A moving frame has velocity $(\vec{v}_M)_L = v_z\hat{k}$ relative to the laboratory frame.

a. C What is the path of the particle in the moving frame?

b. A What is the velocity of the particle as a function of time relative to the moving frame?

c. A What is the acceleration of the particle in each frame?

d. C How should the acceleration in each frame be related? Does your answer to part (c) make sense? Explain.

82. A Derive the range equation $R = \frac{v_0^2}{g}\sin 2\theta$ (Eq. 4.28). *Hint*: What is the vertical displacement when the horizontal displacement equals the range?

Newton's Laws of Motion

5

Underlying Principles

Newton's three laws of motion

Major Concepts

1. Dynamics
2. Force
3. System
4. Inertia
5. Inertial reference frames

Special Cases: Specific Forces

1. Gravity
2. Hooke's law (spring force)
3. Normal force
4. Tension force
5. Kinetic friction

Tools

Free-body diagrams

Key Questions

How is constant velocity maintained?

What causes acceleration?

How do two objects interact?

5-1 Our experience with dynamics 120

5-2 Newton's first law 121

5-3 Force 122

5-4 Inertial mass 124

5-5 Inertial reference frames 124

5-6 Newton's second law 126

5-7 Some specific forces 130

5-8 Free-body diagrams 134

5-9 Newton's third law 141

5-10 Fundamental forces 146

When the greatest minds in physics today grab a cup of coffee together, they might argue about an 11-dimensional model of our Universe, about the requirements of sending a crewed mission to Mars, or about the way signals are transmitted in the human brain. Scientific argument was also taking place over three centuries ago—although on vastly different topics, of course—in letters exchanged between such great scientists as Isaac Newton, Edmond Halley, and Robert Hooke. The crux of those scientific arguments was likely to be motion and its causes, such as what kept the Moon in its orbit or why apples and coconuts falling from trees have the same acceleration. The research resulting from these 17th-century arguments led to such practical applications as the manufacture of accurate clocks and watches.

17th-century science may seem mundane because clocks just do not seem as exciting as, say, nanomachines. You might have trouble believing that motion was a hot topic then. In fact, the debate among some of those early scientists was bitter and hard fought. Imagine losing your temper over the trajectory of an apple!

The 17th-century study of motion laid the foundation for science today not just because of actual discoveries but also because of the scientific process that developed as these discoveries were made. Without that earlier work, we could not even *begin* to imagine an 11-dimensional Universe, a mission to Mars, or detailed studies of the human brain.

5-1 Our Experience with Dynamics

In Chapters 2 and 4 (kinematics), we learned how to define and describe motion both in words and mathematically. Relative to an observer, objects may be at rest, may move with constant velocity, may accelerate by speeding up or slowing down, or may accelerate without changing speed at all. We did not discuss what *causes* the motion (or lack thereof), however. Why are some objects at rest? Why do others move at constant velocity and still others accelerate? In this chapter, we investigate the answers to these questions as we look at the branch of mechanics known as **dynamics** (the cause of motion).

DYNAMICS 

For centuries, dynamics was on the cutting edge of scientific thought in much the same way that certain questions in cosmology and human genetics are today. In 1687, English scientist Isaac Newton (1642–1727) published the *Principia,*[1] a report on his scientific studies in which he explained the causes of motion. Today, his explanations, having been debated and tested innumerable times over the intervening centuries, are largely accepted and are considered laws of nature.[2]

Principia is pronounced prin-kip-pee-ah.

Newton's work in dynamics is summarized in his three laws of motion. Before these three laws were formulated by him and then accepted by the scientific community, many great thinkers reasoned that a cause or a force was required for an object to maintain its motion, even if that motion was constant velocity. This belief seems reasonable, as you can easily see by considering just one simple example: If you push a book across your desktop, the book will come to rest soon after you stop pushing. If you want it to keep moving, you need to keep pushing.

Such common experiences in our everyday lives have helped shape our intuition concerning motion. We observe that to keep something moving, it must be pushed or pulled in some way. Newton's first law of motion, however, says that this belief is not true. It may seem frustrating to find that the laws of physics run counter to our intuition, but we can become more comfortable with physics by seeing how it fits into our experiences. Therefore, an important step in mastering these laws is becoming aware of our intuition and our current conceptions about motion and about forces. We begin this chapter with a case study to help us gain this awareness.

FIGURE 5.1 Aerial view of emergency workers helping the injured from a train crash near Los Angeles (April 23, 2002).

CASE STUDY Train Collision

On April 23, 2002, a passenger train about 35 miles outside of Los Angeles was hit by a freight train (Fig. 5.1). The accident killed two people and injured more than 260, with all the injured being on the passenger train. Witnesses reported that those people who were seated facing backward suffered little or no injury. News reports said that the passenger train came to a quick stop before the collision and that the impact with the freight train pushed the passenger train 370 ft backward (Fig. 5.2).

One of the most controversial parts of the early reports was how fast the freight train was going at the moment of impact. In Chapter 11, we will reconstruct the ac-

[1] The full title is *Philosophiae Naturalis Principia Mathematica,* Latin for *Mathematical Principles of Natural Philosophy.*

[2] In later chapters, we will discuss situations in which the laws of Newtonian mechanics are known to break down.

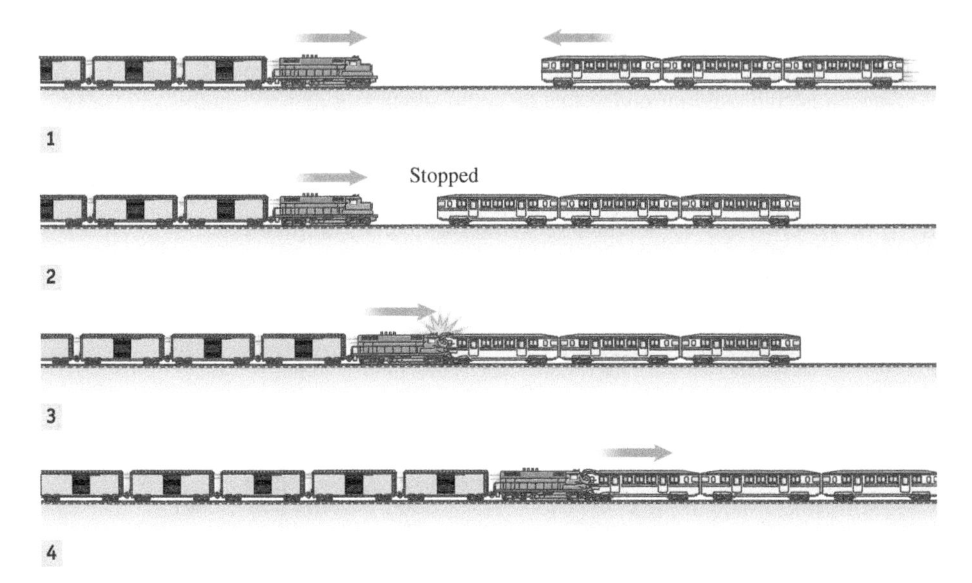

FIGURE 5.2 **1** A freight train and a passenger train move toward each other. **2** The passenger train stops. **3** The freight train continues and collides with the passenger train. **4** The passenger train is shoved backward.

cident and estimate the speed of the freight train upon impact. In this chapter, we are concerned only with the following questions:

1. Why did passengers seated facing backward fare better than those who were either standing or seated facing forward?
2. The passenger train was at rest before the collision and was pushed backward. What do those facts tell us about how hard the freight train pushed on the passenger train? Did the passenger train push on the freight train? If so, how hard did it push?

5-2 Newton's First Law

In the years since Newton wrote his laws of motion in Latin, his words have been translated and interpreted many times. Although physicists generally agree on the principles of dynamics, they often do not agree on the best way to state Newton's laws. For this reason, we will state it three times in this chapter and discuss the subtle differences in each statement.

A close translation of the Latin in which **Newton's first law** was originally written is

1ST STATEMENT OF NEWTON'S FIRST LAW **Underlying Principle**

Every body continues in its state of rest, or of uniform motion in a straight line, unless it is compelled to change that state by forces impressed on it.

Newton's first law says that there is no essential difference between rest and moving in a straight line at constant speed. Both states will continue unchanged unless a force acts on the object. This law is in sharp contrast to the idea that a cause or a force is required to maintain constant velocity.

A book set on a level tabletop will not start sliding until you apply a force to it with your hand. Why does the moving book stop, though? Once it is moving and you have removed your hand, the book has a certain velocity. According to Newton's first law, the book should continue in a straight line at the same speed. Because it slows down and stops, a force must have acted on it and caused this change in motion. The force of friction between the book and the table causes the book to decelerate and eventually come to rest. Because of friction, we might conclude (mistakenly) that rest is the natural state of an object because moving objects eventually stop.

We will study friction in detail in Section 5-7 and in Chapter 6. For now, we will imagine ways to reduce its effects. For example, instead of sliding a book across a clean, dry tabletop, suppose you slide it across a tabletop coated in a slippery wax. If you push the same book with the same force across this slippery surface, the book will slide much farther before coming to rest because the wax has reduced the force of friction.

Newton is credited with saying that he could see farther than most other people because he stood on the shoulders of giants, by which he meant the people who did scientific research before him. One such giant was Galileo Galilei (1564–1642), an Italian scientist whose insight was to imagine a world without friction. Galileo reasoned that if we could reduce friction to nothing, an object set in motion would continue to move at constant velocity forever. This insight led to Newton's first law.

CONCEPT EXERCISE 5.1

Because Newton's first law is counterintuitive, it is important to take some time to think about what the law says and about how and why it differs from our intuition.

a. Why did the unavoidable presence of friction make it difficult for earlier scientists to come to the conclusion expressed in Newton's first law?
b. What is the natural state of an object?
c. How much force does it take to keep an object moving at constant velocity?

CONCEPT EXERCISE 5.2

CASE STUDY Train Collision and Newton's First Law

A group of college students discusses the train collision case study. Use Newton's first law to decide which underlined statements are correct and which are false. Explain your answers.

To take the survey scan or visit **www.cengage.com/community/katz**

Shannon: This newspaper says that the people who got really hurt were either standing up or sitting in a forward-facing seat. Those people got thrown forward when the train stopped.

Avi: That's why there are seat belts in cars. If you get into a crash, the force can throw you through the windshield.

Cameron: There is no force that throws you through the windshield. You fly through the windshield because you are already moving and it would take a force to stop you from going forward. *That's* why there's a seat belt.

Avi: That doesn't make sense. Because then you would need a force to stop you from flying through the windshield even when you just stop slowly at a red light.

Cameron: That's right, but when you slow down slowly, you don't need such a big force and the car seat can take care of it.

Shannon: The seat? I don't think a seat can exert a force. It can't move on its own or hold you. That's why the people who were sitting forward on the train were hurt. The people who were sitting backward had the back of the seat to block them.

5-3 Force

In our everyday language, force has many meanings and usages, such as *he forced me to do my homework, the force of evil,* or *an armed force*. As with most terminology, the term *force* is more specific in physics than it is in our everyday usage. We can derive an operational definition of force from Newton's first law.

When Newton wrote the *Principia*, the language and mathematics of physics were not as well defined as they are today. Newton's own statement of his first law (page 121) is somewhat vague for a modern reader. Another way we can grasp Newton's first law is by restating it using the term *accelerate* that we carefully defined in Chapter 2:

2ND STATEMENT OF NEWTON'S FIRST LAW **Underlying Principle**

If no force acts on an object, then the object cannot accelerate.

This statement means that an object's velocity cannot change in either magnitude or direction unless a force is applied. From Newton's first law, we can reason that a **force** is a push or pull that is required to make an object accelerate. Force is a vector quantity; it has both magnitude and direction. Force must therefore be manipulated using the vector algebra we discussed in Chapter 3.

FORCE ✪ Major Concept

Contact Versus Field Forces

In classical mechanics, it is convenient to divide forces into two types. **Contact forces** are forces that require the source to touch the subject. If you want to accelerate your textbook with your hand, you must touch the book and exert a contact force on it. **Field forces** are those that do not require contact between source and subject; instead, field forces can act through empty space. One of the first forces we learn about as infants is gravity. Gravity is a field force; an apple falls from a baby's hand to the ground because of the Earth's gravitational pull on the apple, but the Earth is not in contact with the apple. Another example of a field force is the electric force (Chapter 23). The source of the electric force is charged particles, such as protons and electrons. This force is responsible for such phenomena as lightning and for keeping our electronic equipment running. A force results from an *interaction* between two objects, such as the interaction between your hand and a textbook or the interaction between the Earth and an apple. We may arbitrarily think of one member of the pair involved in the interaction—that is, your hand or the Earth—as the **source** of the force. The other member—that is, the textbook or the apple—is then the **subject** on which the force acts.

Internal Versus External Forces

In addition to distinguishing between contact forces and field forces, we must also distinguish between internal and external forces. Any collection of two or more objects is known as a **system.** An **internal force** is any force that acts *inside a system*, that is, any force exerted by one object in a system on another object in the system. To determine which forces are internal to the system and which are external, we must first decide what we want to call the "system." For example, is the system in Figure 5.3 the whole car? Only some part of it? One or more occupants? Suppose we decide the whole car and its occupants are the system as indicated by the red loop. Then, any force exerted on the car (the system) by anything *outside* the system—such as the road, the Earth, or another car—is an **external force.** Any forces that are inside the system, such as the force exerted by the driver on the steering wheel or the force exerted by one child on the other child—are internal forces. In this chapter, we continue to use the particle model for objects and systems, so we represent the whole car and its occupants as a particle.

SYSTEM ✪ Major Concept

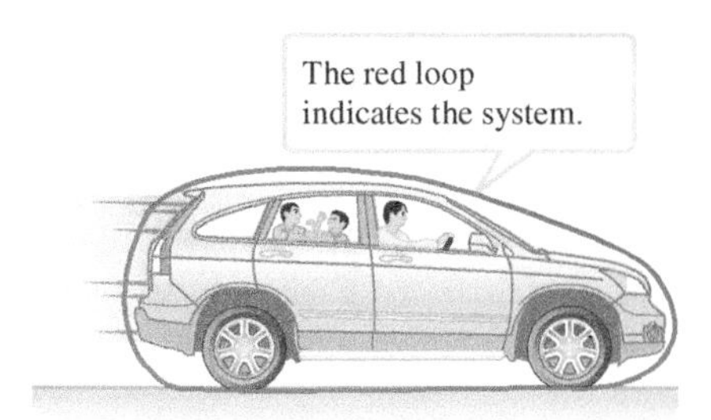

FIGURE 5.3 The people and the car have been chosen as the system (encircled in red). External forces due to the road or the Earth can accelerate the system. Internal forces such as the two children pushing on each other cannot.

CONCEPT EXERCISE 5.3

Shown in Figure 5.4 are four situations in which a force acts on a subject. The subject is labeled in each case.

Case 1. A baseball glove stops a vertically falling baseball.
Case 2. CASE STUDY A freight train collides with a stopped passenger train.
Case 3. A satellite orbits the Earth.

For each case, identify the source of the force and the direction of the force. Then state whether a contact force or a field force is involved.

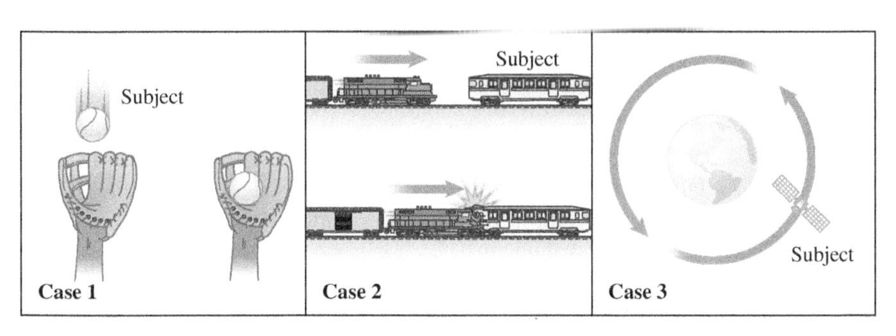

FIGURE 5.4

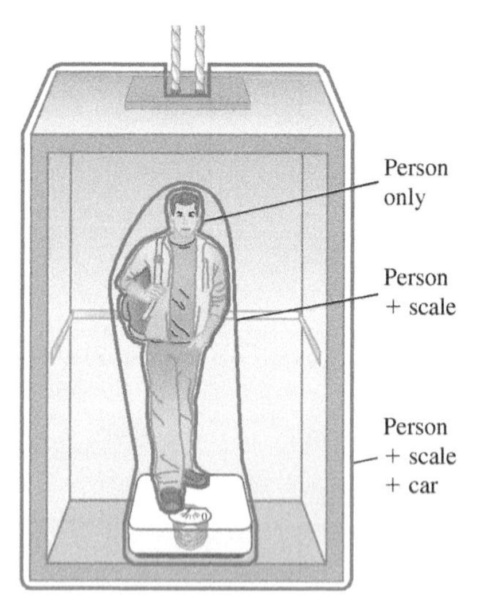

FIGURE 5.5

CONCEPT EXERCISE 5.4

A person stands on a spring scale in an elevator car as shown in Figure 5.5. Which of these sources—the Earth, spring scale, elevator car, and cable—exert an external force if the system consists of:

a. Only the person?
b. The person and the spring scale?
c. The person, the spring scale, and the elevator car?

5-4 Inertial Mass

INERTIA Major Concept

Newton's first law represents a revolutionary change in human understanding. It implies that what is natural about motion is an object's tendency to maintain constant velocity. This idea is so important that a new word, *inertia*, is required to express the concept. **Inertia** is the tendency of an object to maintain its constant velocity (to resist changes in velocity). Consequently, Newton's first law is often referred to as the *law of inertia*.

Not all objects have the same inertia. It is, for instance, difficult to stop a runaway truck but easy to flip a quarter into the air. It is difficult to change the truck's velocity because the truck has a lot of inertia. The quarter, on the other hand, has little inertia, and it is therefore easy to change its velocity with just your thumb. The truck and quarter are different from each other in many ways—shape, size, and composition, to name just a few—but their difference in mass is what counts when we try to accelerate them. **Mass,** also known as **inertial mass,** is an intrinsic scalar property of any object. Mass measures the object's inertia. In SI units, mass is measured in kilograms. In the U.S. customary system, the unit of mass is the slug, a term we almost never use in everyday conversation.

The more mass an object has, the more difficult it is to accelerate that object. If we apply the same force to objects having different masses, we find that the acceleration of each object is inversely proportional to its mass:

$$a \propto \frac{1}{m}$$

Mass is sometimes confused with weight. Weight is a measure of the gravitational force acting on an object (Section 5-7). The weight of an object can change from place to place, but mass does not. For example, an astronaut weighs less on the Moon than she does on the Earth, but her mass is the same no matter where she is.

CONCEPT EXERCISE 5.5

Often, words that are precisely defined in physics are used loosely in everyday language. A person who cannot seem to get things done is described as having a lot of *inertia*.

a. How does that metaphor fit with the physics definition of inertia?
b. How can that metaphor be misleading?
c. How is the term *massive* used in everyday language?

5-5 Inertial Reference Frames

An object can be observed from a number of different perspectives, which are called *reference frames* in physics. The most important difference between reference frames is their relative motion (Sections 4-7 and 4-8).

Newton's first law is not valid in all reference frames. As an example, consider a cruise ship equipped with a skating rink. As hockey and figure-skating fans know, a skating rink needs a Zamboni machine to smooth out the ice surface every now and then. As shown in Figure 5.6A, a Zamboni has been parked on the ice of our cruise ship

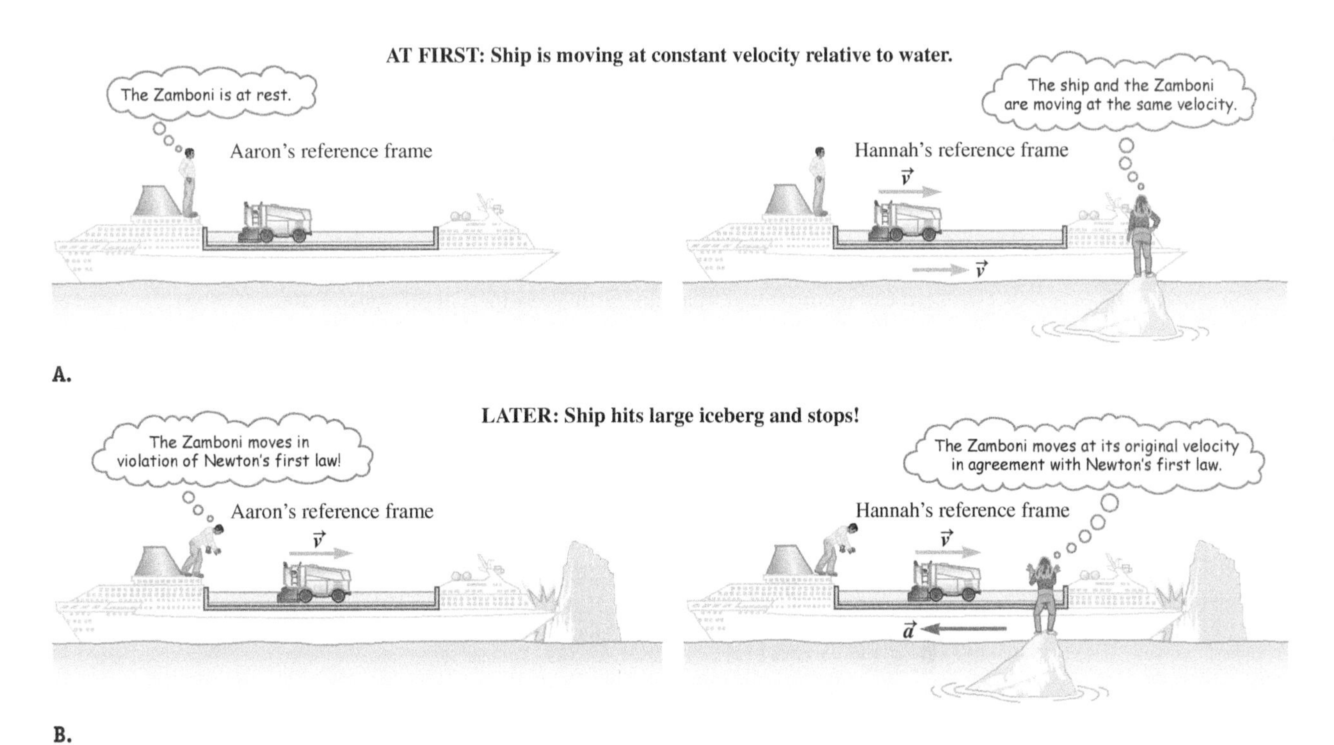

FIGURE 5.6 Two observers see a Zamboni machine on an ice rink aboard a cruise ship. **A.** At first, Aaron on the ship sees the Zamboni at rest. Hannah is on an iceberg, and she sees the Zamboni moving at the same velocity $\vec{v}$ as the ship. **B.** When the ship suddenly stops, Aaron sees the Zamboni move across the ice, and he cannot identify a force that caused the Zamboni's acceleration. Hannah on the iceberg notices that as the ship stops (accelerates to the left), the Zamboni continues moving rightward at its original velocity $\vec{v}$ until it reaches the wall (consistent with Newton's first law).

rink and is at rest relative to the ship. Assume friction between the Zamboni and ice is negligible. Aaron, on the ship looking at the rink, would report that the Zamboni has zero velocity. According to Hannah on a nearby iceberg, however, both the ship and the Zamboni are moving at constant velocity to the right relative to the water.

Now suppose the ship stops suddenly (Fig. 5.6B). Hannah notes that, as the ship's speed decreases to zero, the ship has an acceleration to the left relative to the water. According to Hannah, however, the Zamboni continues to move to the right at its original velocity relative to the water. Hannah can account for the Zamboni's motion in terms of Newton's first law. To her, while the ship is slowing down, the Zamboni continues moving at constant velocity relative to the water because no force acts to change its velocity (until it reaches the wall).

To Aaron, the Zamboni starts with zero velocity but then accelerates toward the rink wall. He cannot identify any force that could cause this acceleration. To him, the Zamboni moves in violation of Newton's first law. According to this law, the Zamboni should remain at rest unless acted upon by a force.

The observers' explanations differ due to their different reference frames. Aaron is in an *accelerating reference frame* (the frame is the ship, which is slowing down), but Hannah is in a stationary reference frame.

INERTIAL REFERENCE FRAMES

Major Concept

Reference frames in which Newton's first law is valid are called **inertial reference frames.** Inertial reference frames may move with constant velocity, but they do not accelerate. Accelerating frames are known as **noninertial reference frames,** and Newton's first law is not valid in them. In our example, the shipboard observer is in a noninertial reference frame once the ship starts slowing down.

Before the ship began decelerating, both observers were in inertial reference frames. Neither of them observed the Zamboni violate Newton's first law. The Zamboni was at rest from the point of view of the shipboard observer, and moving with a nonzero constant velocity according to the point of view of the observer on the iceberg.

As long as no external force acts on an object, it is always possible to find an inertial reference frame in which the velocity of the object is zero. In other words, there is nothing special or "natural" about a state of rest.

If there had been another reference frame moving at constant velocity—another ship, for example—it would also be an inertial reference frame. Any frame moving at constant velocity with respect to an inertial reference frame is also an inertial reference frame.

There is a subtle point about all reference frames on the Earth. The Earth is spinning on its axis and at the same time orbiting the Sun, which is orbiting the center of our galaxy. From Chapter 4, spinning and orbital motion mean that the Earth is accelerating. For most situations in this book, however, the Earth's acceleration is small enough that we can ignore it and treat the Earth as an inertial reference frame.

CONCEPT EXERCISE 5.6

Which of the following reference frames are inertial frames?

a. An airplane cruising in a straight path at constant speed
b. An airplane taking off
c. A car taking a sharp turn

5-6 Newton's Second Law

Before stating Newton's second law, let us review the consequences of Newton's first law:

1. Rest is not a particularly special state; it is merely a special case of constant velocity.
2. A force is required to change an object's velocity.
3. Inertia is the tendency of an object to maintain its constant velocity. An object's inertia is determined by its mass. The more mass (inertia) an object has, the more difficult it is to accelerate that object.

Newton's second law relates the acceleration of an object to the vector sum of all the forces exerted on that object. This vector sum of all those external forces is called either the **total force** or the **net force** on the object. Although Newton did not write his second law using mathematical symbols, today we combine the concepts he presented into one equation:

NEWTON'S SECOND LAW
Underlying Principle

$$\vec{F}_{\text{tot}} \equiv \sum \vec{F} = m\vec{a} \quad (5.1)$$

where $\sum \vec{F}$ is the total external force acting on an object of mass m and $\vec{a}$ is the acceleration of that object. The symbol $\sum$ (a summation sign, represented by an oversized uppercase Greek letter sigma) indicates that we must take the sum of all the forces acting on the object. According to Newton's second law, the total external force acting on a system can accelerate the system, but forces internal to a system cannot.

As discussed in Chapter 3, multiplying the vector $\vec{a}$ by the scalar m results in a vector that points in the same direction as $\vec{a}$. Applying this rule to Newton's second law (Eq. 5.1), shows that the total force on an object points in the same direction as the object's acceleration.

If only one force acts on an object, the object must accelerate in the direction of that force. If more than one force acts on an object, it is possible for the total force to be zero, in which case the acceleration is also zero, which leads to the third way to state Newton's first law:

3RD STATEMENT OF NEWTON'S FIRST LAW
Underlying Principle

The net force on an object is zero if and only if the acceleration of the object is also zero.

Like all other vector equations, Newton's second law can be written in terms of scalar equations with one equation for each scalar component. For a three-dimensional Cartesian coordinate system, Newton's second law is therefore written

$$\sum F_x = ma_x \qquad \sum F_y = ma_y \qquad \sum F_z = ma_z \quad (5.2)$$

Because $\sum \vec{F}$ is the vector sum of all the forces acting on the object, $\sum F_x$ is the sum of all x components of force acting on the object. Further, $\sum F_x$ can cause acceleration only in the x direction. Similar statements can be made for $\sum F_y$ and $\sum F_z$.

Newton's second law is used to define a unit of force. If a single force is applied to a standard mass of 1 kg such that the mass accelerates at 1 m/s^2, the applied force is defined to be 1 newton (1 N):

$$1 \text{ N} \equiv (1 \text{ kg})(1 \text{ m/s}^2) = 1 \text{ kg} \cdot \text{m/s}^2 \quad (5.3)$$

In U.S. customary units, force is measured in pounds:

$$1 \text{ lb} \equiv (1 \text{ slug})(1 \text{ ft/s}^2) = 4.45 \text{ N}$$

Newton's second law is a fundamental principle of classical mechanics. Although it might seem straightforward, its full meaning and subtlety are brought out when it is applied to problems; therefore, much of the rest of this chapter and the next are devoted to studying applications of this law.

CONCEPT EXERCISE 5.7

a. Take a moment to be sure that you understand the distinction between Newton's first two laws. How are they different from each other?

b. According to Newton's second law, what is the acceleration of an object if there are no forces acting on it? Is your answer consistent with Newton's first law?

EXAMPLE 5.1 Electrodes

In Example 2.8 (page 44), an electron initially at rest is released from a negative electrode positioned 7.5 mm away from a positive electrode (Fig. 5.7). Together the two electrodes produce an electric force that gives the electron an acceleration of 1.5×10^{17} m/s^2 toward the positive electrode. No other force is exerted on the electron (mass $m_e = 9.109 \times 10^{-31}$ kg).

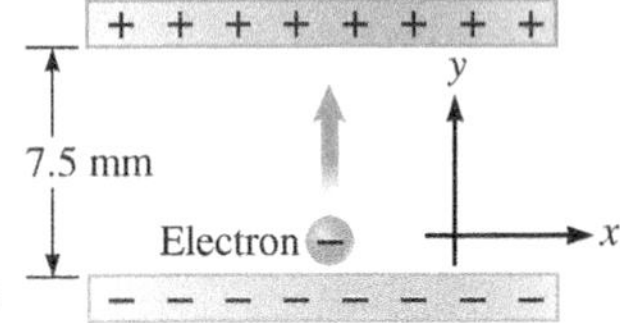

FIGURE 5.7

A Find the electric force $\vec{F}_E$ acting on the electron.

INTERPRET and ANTICIPATE

Treat the electron as just another particle with mass m_e, accelerating upward. We can use Newton's second law to find the force on a particle whose mass and acceleration we know. Force is a vector quantity, so our answer should have both a magnitude and a direction. The total force and the acceleration point in the same direction. Because only one force (the electric force) acts on the electron, that force must point upward because the acceleration is upward toward the positive electrode. Our answer should be in the vector form $\vec{F} =$ ____ $\hat{\jmath}$ N.

SOLVE

Using Newton's second law in component form, find the scalar component of the electric force. The mass of the particle is the mass of an electron. The only force exerted on the electron is the electric force, so $\sum F_y = F_E$.	$\sum F_y = ma_y$ $m = m_e$ $F_E = m_e a_y$
Substitute values for the mass of the electron and the given acceleration.	$F_E = (9.109 \times 10^{-31} \text{ kg})(1.5 \times 10^{17} \text{ m/s}^2)$
Using the definition 1 N = 1 kg · m/s^2, express this value for the electric force ($\vec{F}_E$) in newtons.	$\vec{F}_E = 1.4 \times 10^{-13}\, \hat{\jmath}$ N

CHECK and THINK

The calculated force might seem small, but only a small force is required for the given acceleration because the electron's inertial mass is so small.

Example continues on page 128 ▶

B If the electron is replaced by a proton (mass $m_p = 1.673 \times 10^{-27}$ kg) initially at rest and released from the positive electrode, find the velocity of the proton as it strikes the negative electrode. Assume the electric force exerted on the proton has the same magnitude as the force exerted on the electron found in part A but is in the opposite direction (Fig. 5.8).

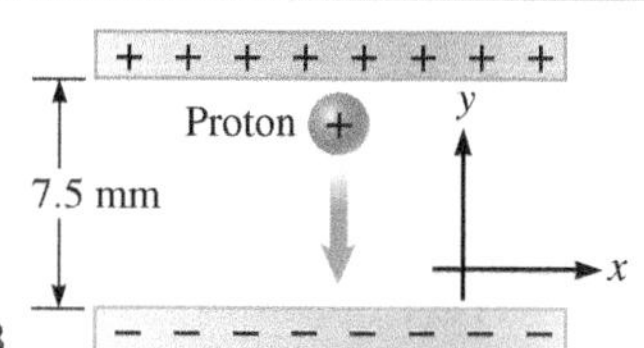

FIGURE 5.8

:• **INTERPRET and ANTICIPATE**

There are two differences from part A. First, the force on the particle and its velocity are now downward. Second, we are asked to find velocity instead of force. Our answer should be in the form $\vec{v} = -(\_\_\_)\hat{\jmath}$ m/s. We will use Newton's second law and our answer to part A to find the acceleration, and from that we will use kinematics to find the velocity.

:• **SOLVE**

The force on and acceleration of the proton are both in the negative y direction in the coordinate system chosen. The mass is the mass of a proton: $m = m_p$.

$$\sum F_y = ma_y = -F_E$$

$$a_y = \frac{-F_E}{m_p} \qquad (1)$$

The acceleration is constant, so we can use the constant acceleration problem-solving strategy (page 42) to find the proton's velocity. We used this strategy for a similar situation in Example 2.8 (pages 44–45).

$$v_y^2 = v_{0y}^2 + 2a_y\Delta y$$

$$v_{0y} = 0 \quad \text{(initially at rest)}$$

$$v_y = \pm\sqrt{2a_y\Delta y} \qquad (2)$$

Here we need the negative square root because the velocity is in the negative y direction. Substitute from Equation (1) for a_y.

$$v_y = -\sqrt{2\left(\frac{-F_E}{m_p}\right)\Delta y}$$

Substitute values, remembering that Δy for this proton is negative in our coordinate system.

$$v_y = -\sqrt{2\left(\frac{-1.4 \times 10^{-13}\ \text{N}}{1.673 \times 10^{-27}\ \text{kg}}\right)(-7.5 \times 10^{-3}\ \text{m})}$$

$$\vec{v} = -1.1 \times 10^6\,\hat{\jmath}\ \text{m/s}$$

:• **CHECK and THINK**

A force of the same magnitude acts on the proton, which is 1800 times more massive than the electron. According to Newton's second law, the proton's acceleration should be about 1800 times smaller. Because the particle's speed depends on the square root of a_y (Eq. 2), the proton's final speed should be about $\frac{1}{\sqrt{1800}}$ times the electron's speed. In Example 2.8, we found the electron's speed is $4.7 \times 10^7\,\hat{\jmath}$ m/s. So, our answer seems reasonable.

$$\frac{v_p}{v_e} = \frac{1.1 \times 10^6}{4.7 \times 10^7} \approx 0.023 \approx \frac{1}{\sqrt{1800}}\ ✓$$

EXAMPLE 5.2 More Electrodes

Two pairs of electrodes are oriented as shown in Figure 5.9. An electron is at rest in the center of the configuration. When the electrodes are turned on, each pair produces a force on the electron such that

$$\vec{F}_E = (1.8 \times 10^{-13}\,\hat{\imath} + 7.2 \times 10^{-14}\,\hat{\jmath})\ \text{N}$$

Find the acceleration of the electron.

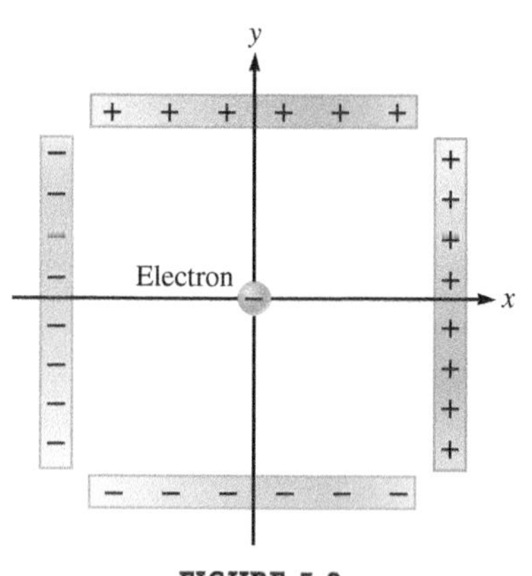

FIGURE 5.9

:• **INTERPRET and ANTICIPATE**

From Example 5.1, we can simply treat the electron as a particle on which a two-component force acts. We need to find a magnitude and direction for the electron's acceleration. Because there are forces in both the positive x direction and the positive y direction, we expect the acceleration to have positive x and y components as in $\vec{a} = (\_\_\_\,\hat{\imath} + \_\_\_\,\hat{\jmath})$ m/s^2.

SOLVE Follow a procedure similar to that used in Example 5.1, but now use Newton's second law twice, once for each component of the force.	$\sum F_x = ma_x$ $\sum F_y = ma_y$
The mass is the electron's mass. Because there is only one force in each direction, the sums are reduced to a single term each.	$F_x = m_e a_x$ $F_y = m_e a_y$
Solve both equations for acceleration.	$a_x = \dfrac{F_x}{m_e}$ and $a_y = \dfrac{F_y}{m_e}$
Substitute the values. The electron mass is 9.109×10^{-31} kg.	$a_x = \dfrac{1.8 \times 10^{-13}\text{ N}}{9.109 \times 10^{-31}\text{ kg}} = 2.0 \times 10^{17}\text{ m/s}^2$ $a_y = \dfrac{7.2 \times 10^{-14}\text{ N}}{9.109 \times 10^{-31}\text{ kg}} = 7.9 \times 10^{16}\text{ m/s}^2$
Write the solution in component form.	$\vec{a} = (2.0\,\hat{\imath} + 0.79\,\hat{\jmath}) \times 10^{17}\text{ m/s}^2$

CHECK and THINK
The final answer is in the form we expected and roughly the same order of magnitude as the acceleration given in Example 5.1.

EXAMPLE 5.3 Three Balanced Forces

Two ropes are attached to a ring and exert forces as shown (Fig. 5.10). The magnitude of these forces is given by $F_1 = 2F_2$ and $f \equiv F_2 = 22.0$ N. A third force is applied by a rope so that the ring's acceleration is zero. What is the magnitude and direction of the force applied by the third rope?

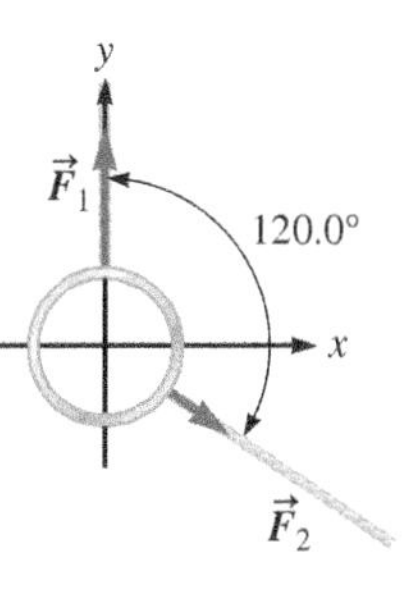

FIGURE 5.10

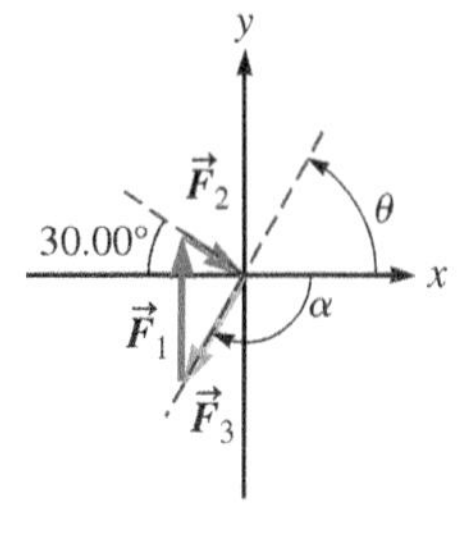

FIGURE 5.11

INTERPRET and ANTICIPATE
In order for the acceleration to be zero the sum of the three forces must be zero. To anticipate the result, we geometrically add $\vec{F}_1$ and $\vec{F}_2$. In order for the net force to be zero, $\vec{F}_3$ must point from the tip of $\vec{F}_2$ to the tail of $\vec{F}_1$. Our goal is find the magnitude of $\vec{F}_3$ and the angle α; from our sketch (Fig. 5.11) we expect the magnitude of $\vec{F}_3$ is similar to that of $\vec{F}_1$ and absolute value of α is greater than 90°.

SOLVE The ring is not accelerating, so the net force in both the x and y direction is zero.	$\sum F_x = ma_x = 0$ $\sum F_y = ma_y = 0$
Use Figure 5.11 to resolve the three forces into their components. Set the sum of the forces in each direction equal to zero.	$\sum F_x = F_2 \cos 30.00° - F_3 \cos\theta = 0$ $\sum F_y = F_1 - F_2 \sin 30.00° - F_3 \sin\theta = 0$
We have two equations and two unknowns (F_3 and θ).	$F_2 \cos 30.00° = F_3 \cos\theta$ (1) $F_1 - F_2 \sin 30.00° = F_3 \sin\theta$ (2)
Eliminate F_3 by dividing Equation (2) by Equation (1). Use $F_1 = 2F_2$ and $f \equiv F_2$ to simplify. Solve for θ by using $\tan\theta = \sin\theta/\cos\theta$.	$\dfrac{\sin\theta}{\cos\theta} = \dfrac{2f - f\sin 30.00°}{f\cos 30.00°}$ $\theta = \tan^{-1}\left(\dfrac{2 - \sin 30.00°}{\cos 30.00°}\right) = 60.00°$
Solve either Equation (1) or (2) for F_3 and substitute values. (We choose Equation 1.)	$F_3 = \dfrac{F_2 \cos 30.00°}{\cos\theta} = \dfrac{(22.0\text{ N}) \cos 30.00°}{\cos 60.00°}$ $F_3 = 38.1$ N

Example continues on page 130 ▶

To find the direction α, we use Figure 5.11. We see $\alpha + \theta = 180.0°$, and because the direction is measured clockwise from the x axis, it must be negative.	$\alpha = -(180.0° - 60.00°)$ $\alpha = -120.0°$

CHECK and THINK

The magnitude $F_1 = 2f = 44.0$ N. So, as expected the magnitude of F_3 is similar to that of F_1. Also as expected the absolute value of α is greater than 90°.

5-7 Some Specific Forces

As you study physics, you will encounter many forces. You should expect to learn three things about each one: its source, its magnitude, and its direction. This section presents a partial list that will make our study of Newton's laws less abstract.

Gravity Near the Earth's Surface

GRAVITY ▶ Special Case

Gravity, or the **gravitational force** (the two terms are synonymous), is the field force that keeps us on the surface of the Earth and keeps the planets in orbit around the Sun. In general, any two objects that have mass exert an attractive gravitational force on each other. The closer the objects are together, the stronger their gravitational attraction. The general gravitational force between any two bodies is described in Chapter 7. For now, let's focus on the gravitational force exerted by the Earth on objects located near its surface.

From Chapter 2, when an object is in free fall, the only force acting on it is gravity. Experiments show that in the absence of air resistance, all objects in free fall near the surface of the Earth have the same downward acceleration g. Because the Earth's gravitational force $\vec{F}_g$ is the only force acting on such an object, Newton's second law for this situation is simply

$$\vec{F}_{tot} = \vec{F}_g = m\vec{a} \tag{5.4}$$

If we choose an upward-pointing y axis, we can represent the acceleration as

$$\vec{a} = -g\hat{\jmath} \tag{5.5}$$

Substitute Equation 5.5 into Equation 5.4 to find the gravitational force acting on an object of mass m near the Earth's surface:

$$\vec{F}_g = -mg\hat{\jmath} \tag{5.6}$$

The gravitational force is always attractive; that is, all objects are always pulled toward the center of the Earth. Because of its role in Equation 5.6, g is also known as the **local acceleration due to gravity.**

There is an intimate connection between gravity and what we call *weight*. In this book, we define the **weight** w of any object to be the magnitude of the gravitational force acting on the object:

There are other definitions of weight. Sometimes, weight is defined as a vector quantity, and other definitions are operational in that they involve measuring weight with a scale.

$$w \equiv F_g = mg \tag{5.7}$$

Because the local acceleration due to gravity g is greater at sea level than at high elevations, the weight of an object is greater near sea level.

We can also talk about the weight of an object on another planet. In this case, g is not the local gravitational acceleration near the Earth's surface (9.81 m/s^2), but rather the local gravitational acceleration near the surface of that planet. *Mass*, however, does *not* depend on the local acceleration due to gravity, which means that the mass of an object is the same at sea level, on top of a mountain, or even on another planet.

Because weight is the magnitude of a force, the SI unit for weight is the newton. The U.S. customary unit for weight is the familiar pound. The conversion factors between newtons and pounds are 1 N = 0.225 lb and 1 lb = 4.45 N. Note that the kilogram is *not* a unit of weight; it is the SI unit for mass. A translation can be made, however, between pounds and kilograms: 1 kg = 2.2 lb. This translation is not a true conversion because one side of the equation is a mass and the other side is a weight.

The translation 1 kg = 2.2 lb should only be applied for objects near sea level on Earth, but it is commonly used at other elevations.

An object does not need to be in free fall for gravity to act on it. For instance, even though gravity is acting on you right now, your acceleration is (probably) zero because another force (due to the chair) is also acting on you.

You may be reading your textbook in a moving vehicle, in which case your acceleration is not necessarily zero.

Equation 5.6 gives the gravitational force on any object near the surface of a planet such as the Earth whether the object is in free fall, stationary, or moving with some acceleration other than the free-fall acceleration. If the object is not in free fall, another force besides gravity must be acting on it, and its acceleration must be determined by applying Newton's second law.

EXAMPLE 5.4 A Heavy Space Suit

On the Earth, an astronaut's space suit weighs 1.36×10^3 N (about 300 lb).

A What is the mass of the space suit?

FIGURE 5.12 Edwin E. (Buzz) Aldrin on the Moon, July 1969.

INTERPRET and ANTICIPATE

The weight is a force, so its SI units are newtons as given. We expect to find the mass in kilograms.

SOLVE

Find the mass of the space suit from Equation 5.7.

$$w = mg$$

$$m = \frac{w}{g} = \frac{1.36 \times 10^3 \text{ N}}{9.81 \text{ m/s}^2}$$

$$m = 1.39 \times 10^2 \text{ kg}$$

CHECK and THINK

These units are what we expected.

B The free-fall acceleration near the surface of the Moon is $g_M = 0.16g$. What does the space suit weigh on the Moon?

INTERPRET and ANTICIPATE

We expect the space suit to weigh less on the Moon than it does on the Earth.

SOLVE

The weight of the suit on the Moon is the magnitude of the Moon's gravitational force on it.

$$w_M = mg_M = (1.39 \times 10^2 \text{ kg})(0.16 \times 9.81 \text{ m/s}^2)$$

$$w_M = 2.18 \times 10^2 \text{ N}$$

CHECK AND THINK

The weight of the space suit on the Moon is about one-sixth of its weight on the Earth: $1360 \text{ N}/218 \text{ N} \approx 6$. So would the space suit seem heavy to the astronaut on the Moon? To answer this, imagine walking around on the Earth in your heaviest winter clothes and boots. How much do you think they weigh? Perhaps 5 to 10 lb. Let's say 10 lb. Since you can manage wearing about 10 lb of clothing, let's assume that if the space suit weighs no more than 10 lb on the Moon, the astronaut won't consider it too heavy. Using 1 lb = 4.45 N, we find that 10 lb is about 45 N. The space suit's weight on the Moon is about 4 times heavier (or about 50 lb). So, the suit would feel heavy to the astronaut. (In fact, lower mass suits have been developed for today's astronauts.)

HOOKE'S LAW Special Case

Spring Force

Figure 5.13 shows a spring fixed on one end and attached to a block on the other end, which is free to move. A *relaxed spring*—that is, one that is neither compressed nor stretched—does not exert a force. When either compressed or stretched from the relaxed state, however, the spring exerts a contact force on the block. Depending on whether the spring is compressed or stretched, the contact force can be either a push or a pull.

The force due to the spring is a **restoring force**; that is, it is directed so as to return ("restore") the spring to its relaxed state. The farther the spring is from being in its relaxed state, the stronger the restoring force. For many springs, the force exerted is proportional to how far the attached block is displaced from the relaxed position. These springs are said to obey **Hooke's law**. For a spring lying along the x axis, Hooke's law is mathematically expressed as

$$\vec{F}_H = -k\Delta\vec{x} \tag{5.8}$$

where the subscript H stands for *H*ooke. The negative sign means that the restoring force is in the direction opposite the direction of the displacement. Similar equations are written for a spring oriented along a y or z axis. The constant k is known as the **spring constant**, a scalar property that differs from one spring to another; it measures the stiffness of the spring. The SI unit of k is newtons per meter (N/m).

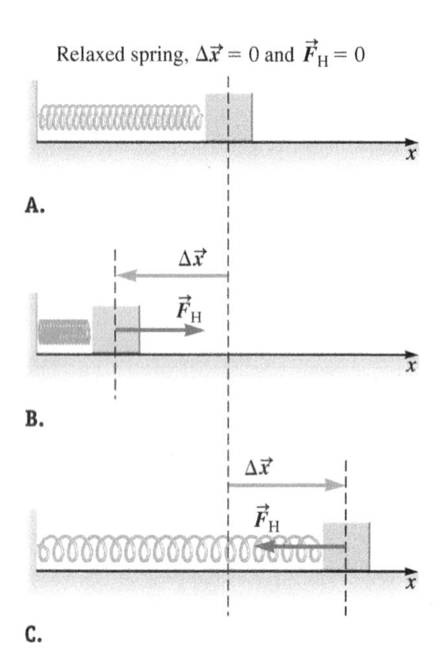

FIGURE 5.13 **A.** The spring is relaxed and does not exert a force. **B.** A compressed spring pushes on the block (contact force). **C.** A stretched spring pulls on the block (contact force).

Springs are used in many mechanical systems, including such common objects as retractable pens, mattresses, and jewelry clasps. In addition, many forces exerted by agents other than springs can be modeled as spring forces. We will return to springs many times throughout this book.

Spring scales are often used to measure the weight of an object. An object may be hung from a spring scale (Fig. 5.14A) or placed on top (Fig. 5.14B). We define an object's **apparent weight** as the reading of a scale displaying the magnitude of the spring force on that object. Why does the dial reading give the magnitude of the spring force? The force F_H exerted by the spring is proportional to how far the spring is stretched or compressed. The scale manufacturer knows the k value of the spring and designs the scale in such a way that the extension or compression of the spring moves a pointer on the dial to a number that is equal to the magnitude of the force exerted by the spring. We will show in Example 5.8 that the apparent weight (scale reading) does not necessarily equal the true weight of the object (magnitude of the gravitational force exerted on it).

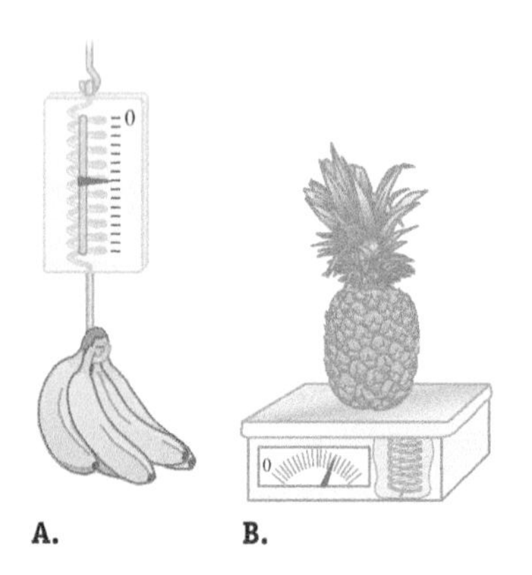

FIGURE 5.14 Spring scales are used to measure the apparent weight of an object. **A.** An object may hang from the spring. **B.** A pan scale employs a main spring (cutaway) connected to a system of levers (not shown).

CONCEPT EXERCISE 5.8

Imagine weighing the same bunch of bananas with two different spring scales like the one in Figure 5.14A. Scale 1 has a spring with spring constant k_1. When the bananas are hung from scale 1, the spring stretches Δy_1 from its relaxed position. Therefore, the apparent weight of the bananas is

$$w_{app} = k_1 \Delta y_1$$

Scale 2 has a stiffer spring; its spring constant is $k_2 = 3k_1$. If the same bunch of bananas is hung from scale 2, what is the displacement of its spring Δy_2 in terms of Δy_1?

NORMAL FORCE Special Case

Normal Force

Whenever any object is in contact with a surface, the surface exerts a contact force on the object. The component of this contact force that is perpendicular to the surface is called the **normal force**. (*Normal* is another word for *perpendicular*.) The normal force gets its name because its direction is always perpendicular to the surface.

The magnitude of the normal force varies depending on the situation. The harder the object presses against the surface, the greater the normal force exerted by the surface. Because our intuition tells us that only living organisms can respond to being touched, it is counterintuitive to imagine that an inanimate object like a tabletop can alter the force it exerts. The force exerted by an inanimate object can indeed change, though, and to see how, consider placing a bowling ball on a mattress.

A good spring mattress contains hundreds of springs. If you place a bowling ball on the mattress, the springs compress downward (Fig. 5.15A), and the spring force on the bowling ball is consequently upward. Because the bowling ball is not accelerating, we reason that the spring force on the ball exactly balances the gravitational force on the ball. How far the springs compress depends on how stiff they are. If the mattress were made of springs much stiffer than those shown in Figure 5.15A (that is, if k were higher), the springs, according to $\vec{F}_H = -k\,\Delta\vec{x}$ (Eq. 5.8), would not compress as far to exert the same spring force. If the mattress springs were very, very stiff, you might not even notice the compression, but they would still be slightly compressed and would exert the same upward force on the bowling ball.

Now imagine placing the bowling ball on a table (Fig. 5.15B). The table is made of molecules held together by molecular bonds that act like very stiff springs. They are slightly compressed by the bowling ball, and as a result they exert an upward force on the ball.

This principle is true of any surface. If a force pushes on a surface, the molecular bonds between the atoms in that surface are compressed, and the surface pushes back with a normal force. If the bowling ball is replaced by a Ping-Pong ball, on either the mattress with its springs or the tabletop with its molecular "springs," the compression will, of course, not be as great. As a result, the normal force exerted by either the mattress or the tabletop on the Ping-Pong ball will be smaller than the normal force on the bowling ball. In other words, the surface can alter the amount of force it exerts, counterintuitive though that may seem.

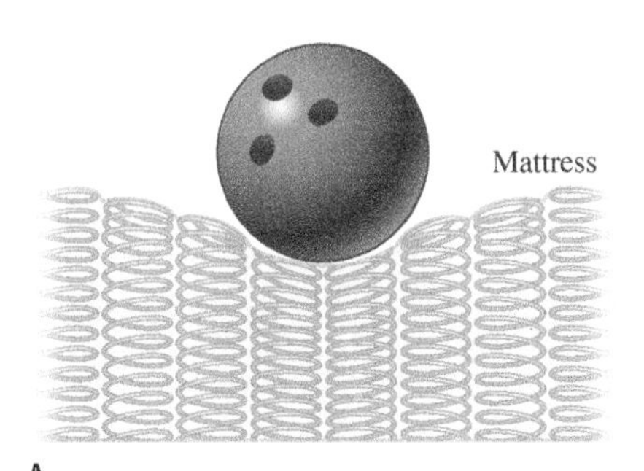

Table

B.

FIGURE 5.15 A. The bowling ball is supported by the springs in a mattress. If stiffer (higher k) springs are used, a smaller compression results in the same force on the bowling ball. **B.** The molecular bonds in the table act like very, very stiff springs in a mattress.

Tension Force

A **tension force** $\vec{F}_T$ is exerted by a taut rope, cable, or similar cord such as the rope pulling on the bananas in Figure 5.16. The rope must be in contact with the object to exert a tension force. Because the rope pulls on the object, the tension force is directed along the rope. In Figure 5.16, the rope pulls upward on the bananas. The magnitude of the tension force is called the **tension**.

Like the normal force, the tension force can be modeled as a spring force exerted by molecular bonds between molecules (close-up, Fig. 5.16). In the case of the tension force, a taut rope stretches the molecular bonds. Like very stiff springs, these molecular bonds pull back. If the "springs" are very stiff, the rope does not stretch perceptibly, and we say the rope is *unstretchable*.

In this textbook, we will consider only ropes that are unstretchable and massless, where by "massless" we mean that the mass of any rope in our discussion is so small that we can ignore the effect of gravity on the rope. For such an ideal rope, the tension is the same everywhere in it. Thus, if the rope connects two objects, the magnitude of the tension force on each object is the same.

TENSION FORCE Special Case

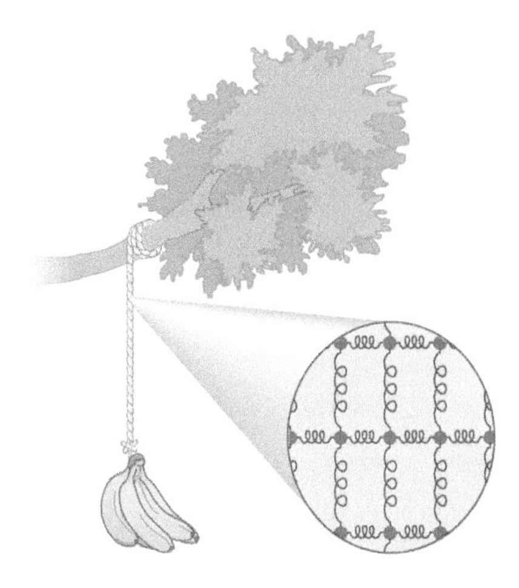

FIGURE 5.16 Bananas hang from a rope. The tension force on the bananas is upward. Molecular bonds in the rope act like very, very stiff springs (close-up). When those bonds are stretched, they pull back.

CONCEPT EXERCISE 5.9

For all three situations, find the magnitude and direction of the tension force(s) exerted on Rochelle. If not enough information is given, say so. Explain your answers.

a. Rochelle and Buddy pull on opposite ends of rope. The tension force exerted on Buddy is 15 N and is directed toward Rochelle.

b. Now, one end of the rope is tied to a sturdy pole and Rochelle pulls on the other end. The tension force on the pole is 15 N directed toward Rochelle.

c. Finally, Rochelle holds one rope in her left hand, and the other end of that rope is pulled by Joe. She holds another rope in her right hand, and the other end of that rope is pulled by Buddy. Both Buddy and Joe experience a 15-N force directed toward Rochelle.

Kinetic Friction

As stated previously, when an object is in contact with a surface, the surface exerts a force on the object. The component of that force perpendicular to the surface is the normal force. The force exerted by the surface may also have a parallel component, which is known as the frictional force. Experiments have shown that friction is a function of the relative motion of the object and the surface. Whenever an object is sliding across a surface, **kinetic friction** is the parallel component of the contact force exerted by the surface on the object. **Static friction** may exert a force on an object that is at rest with respect to the surface. Finally, if an object rolls along a surface, such as a tire on a road, **rolling friction** acts on the object.

KINETIC FRICTION Special Case

In Chapter 6, we will discuss all three forms of friction in detail. In this chapter, we will mainly consider kinetic friction, which causes a sliding object to slow down and finally stop. The direction of kinetic friction is always parallel to the surface and opposite the direction in which the object is moving relative to the surface.

The magnitude of kinetic friction is proportional to the normal force acting on the object:

$$F_k = \mu_k F_N \tag{5.9}$$

where the constant of proportionality μ_k is known as the **coefficient of kinetic friction.** (The symbol μ is the lowercase Greek letter mu.)

Table 5.1 is a summary of the five forces we have just described. The table mentions the source, magnitude, and direction for each force. Each time you study a new force, be sure that you can identify these three parameters for it.

TABLE 5.1 Summary of the five forces examined in this chapter.

Force	Source	Magnitude	Direction
Gravity ($\vec{F}_g$)	Earth	"Weight," mg	Toward center of the Earth
Hooke's law ($\vec{F}_H$)	Spring	$k\,\Delta x$	Push or pull along spring
Normal ($\vec{F}_N$)	Surface	Varies	Perpendicular to surface
Tension ($\vec{F}_T$)	Rope	"Tension," the same all along the rope	Pull along rope
Kinetic friction ($\vec{F}_k$)	Surface	$\mu_k F_N$	Parallel to surface, opposite to object's $\vec{v}$ relative to surface

5-8 Free-Body Diagrams

This section describes an important technique for applying Newton's second law to solve problems. A good way to begin analyzing such problems is with a drawing. Although a nicely drawn, artistic rendering of the problem may be aesthetically pleasing, it is not only unnecessary but can actually be distracting. Instead, physicists use a type of drawing called a **free-body diagram,** having the following elements:

FREE-BODY DIAGRAMS 

Element 1. A simple representation of a subject. We will continue to use the particle model for all objects until Chapter 13. Until then, represent any object (or system) by a dot.

Element 2. Clearly labeled vector representations of all the external forces exerted on the subject. In most cases, the tails of the vectors should be on the subject. To make sure that you don't miss any force exerted on the subject, consult Table 5.1 and decide which of the five forces are present. Make your decision based on whether or not the source of a particular force is present in the situation.

Element 3. A clearly labeled **coordinate system.** Picking a good coordinate system comes with practice. One helpful tip is to choose one axis to be parallel to the subject's acceleration.

Element 4. An indication of the acceleration. If relevant, draw a vector arrow indicating the direction of the subject's acceleration, but visually different from the arrows representing the force vectors. For example, you might use a dashed line or a different color. If the subject's acceleration is zero, note that on the diagram.

Drawing free-body diagrams takes practice. You'll get that practice solving problems that involve Newton's second law. In the following problem-solving strategy, drawing a free-body diagram is the primary tool used in the **INTERPRET and ANTICIPATE** procedure.

PROBLEM-SOLVING STRATEGY

Applying Newton's Second Law

:• INTERPRET and ANTICIPATE

Identify the system (often, a single object) that is subject to external force(s). Draw a free-body diagram for that system, making sure that the diagram has all four elements given above. Once the free-body diagram is complete, there are three steps that help in the **SOLVE** procedure when an algebraic or numerical result is required.

:• SOLVE

Step 1 Apply Newton's second law in component form. Use the coordinate system on the free-body diagram to apply Equation 5.2. You will have one equation for each direction in which there is at least one force.

$$\sum F_x = ma_x \quad \sum F_y = ma_y \quad \sum F_z = ma_z$$

Step 2 Write down any other equations that are relevant to the forces involved. Table 5.1 lists magnitudes of the gravitational force (weight), the spring force, and kinetic friction. If the situation involves one or more of these forces, write down the appropriate equation.

Step 3 Do algebra before substitution. Review your equations. Which parameters are known? Which are unknown? Which do you need to solve for? You may have more equations than you need. Find an algebraic expression for the parameter you need before you substitute any numerical values. This practice makes it easier to find a mistake if you make one and makes it easier for another person to understand your work.

EXAMPLE 5.5 Sheldon's Morning Routine

Sheldon, of mass 60.4 kg, stands on his bathroom scale. What is the normal force exerted by the scale on Sheldon? What does the scale read (in newtons)?

:• INTERPRET and ANTICIPATE

Sheldon is the subject, represented by a dot on our free-body diagram (Fig. 5.17). Consider the five forces listed in Table 5.1. There is no rope in this problem, and Sheldon is not moving; so, there is no tension force or kinetic friction exerted on him. He does experience the downward pull of gravity $\vec{F}_g$, and the scale exerts an upward normal force $\vec{F}_N$ on him.

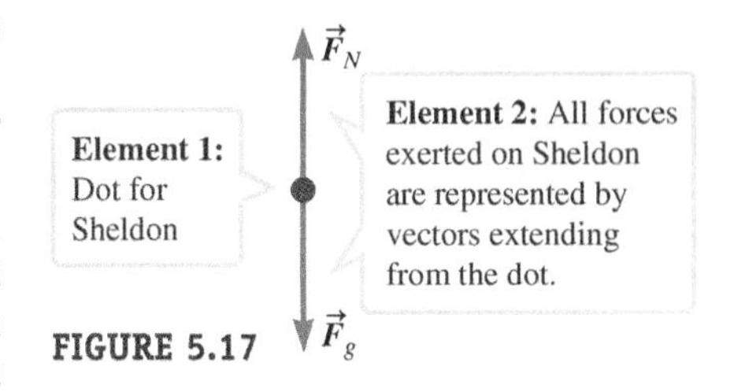

FIGURE 5.17

The two forces are vertical, so we only need a one-dimensional coordinate system (Fig. 5.18). We choose an upward-pointing y axis. Sheldon is at rest, so his acceleration is zero as indicated.

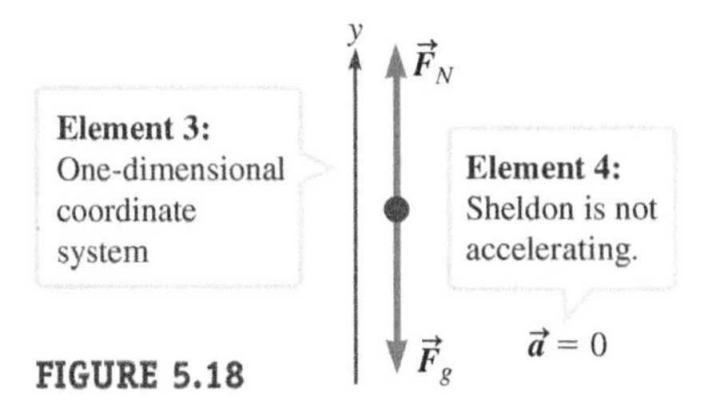

FIGURE 5.18

The reading on the scale equals the magnitude of the force that the scale exerts. So, in this case, the magnitude of the normal force is the reading on the scale. We expect $\vec{F}_N = +(\_\_\_\_\_\hat{\jmath})$ N because the normal force is upward.

:• SOLVE

Step 1 **Apply Newton's second law in component form.** We only need the y-component portion of Newton's second law (Eq. 5.2). In this case, the normal force is in the positive direction, and gravity is in the negative direction. The sum of the two forces is zero because Sheldon is not accelerating.

$$\sum F_y = ma_y \quad (5.2)$$

$$F_N - F_g = 0 \quad (1)$$

Step 2 **Write relevant force equations.** Table 5.1 lists the magnitude of the gravitational force. There is nothing more we can write about the normal force.

$$w = F_g = mg \quad (2)$$

Step 3 **Do algebra.** Substitute Equation (2) into Equation (1) and solve for the scalar component of the normal force.

$$F_N - mg = 0$$

$$F_N = mg$$

Example continues on page 136 ▶

The normal force is in the positive y direction.	$\vec{F}_N = mg\hat{\jmath}$
Substitute values. Remember that g is a positive number.	$\vec{F}_N = (60.4\text{ kg})(9.81\text{ m/s}^2)\hat{\jmath}$ $\vec{F}_N = 593\hat{\jmath}\text{ N}$
The reading on the scale is the magnitude of the normal force.	$F_N = 593\text{ N}$

CHECK and THINK

Our answers have the form we expected. It is interesting that Sheldon's weight in U.S. customary units is about 130 lb, which is reasonable. Many physics problems involve the reading on a scale. Keep in mind that the scale's readout is the magnitude of the force it exerts on the object. However, in your laboratory you are likely to find a scale that reports mass in kilograms rather than weight in newtons.

EXAMPLE 5.6 Christa's Backpack

Christa's backpack is full of physics books and hangs from a single strap that loops around a hook as shown in Figure 5.19. The combined mass of the backpack and its contents is 3.14 kg, and the mass of the strap is very much less than this combined mass. Use a free-body diagram and Newton's second law to find the tension F_T in the strap and the total tension force $(\vec{F}_T)_{\text{tot}}$ exerted by the strap on the backpack.

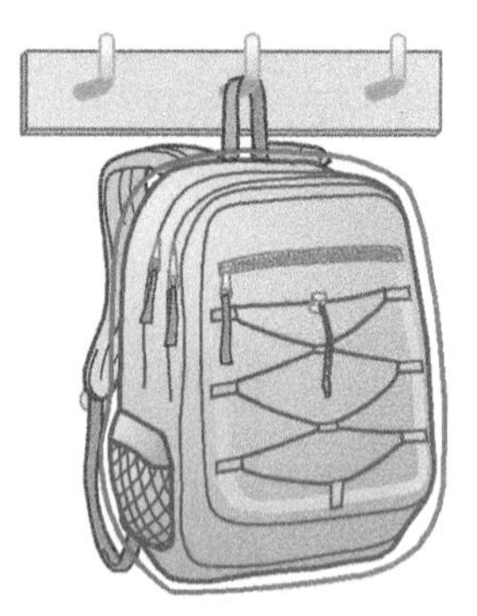

FIGURE 5.19

INTERPRET and ANTICIPATE

Start by identifying the system for which the free-body diagram will be drawn. Because we must find a force exerted on the backpack, the backpack and its contents are the system. We must consider the strap to be external to the system to find the force it exerts.

Review the four other forces in Table 5.1 to determine which are exerted on the system.

No surface in this problem → no friction or normal force.

No springs → no spring force.

The strap exerts a tension force at each point of contact.

Gravity is acting.

Only gravity and the tension force act on the backpack.

Draw a free-body diagram, including all four elements (Figs. 5.20 and 5.21).

1. **Represent the system**—the backpack and its contents—**by a dot.**

2. **Draw and label the forces.** The tail of each force is on the dot. Gravity points to the center of the Earth (straight down in this problem). The strap attaches vertically to the backpack at two points, so it exerts an upward tension force at both of these points (Fig. 5.20).

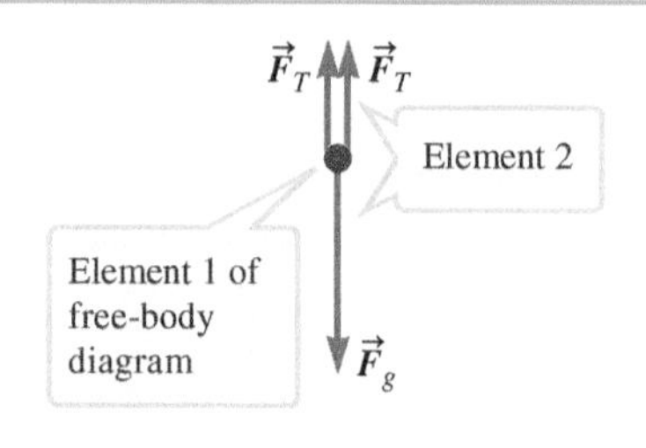

FIGURE 5.20

3. Choose a **coordinate system.** A vertical y axis was chosen because all the forces lie along this axis. An x and a z axis are implied but not drawn because they are not used in this one-dimensional problem (Fig. 5.21).
4. Because the backpack hangs from the hook and does not move, its **acceleration is zero as indicated.**

We expect both numerical answers to have the SI units of newtons. Also, the total tension force points up, so we expect to find an answer in the form $(\vec{F}_T)_{\text{tot}} = \underline{\quad\quad}\,\hat{\jmath}\text{ N}$.

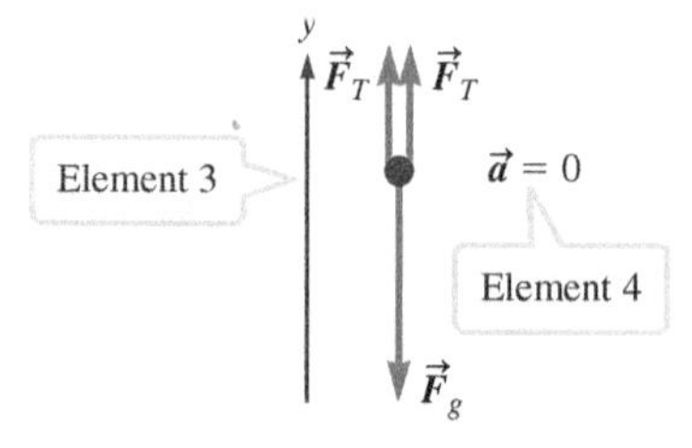

FIGURE 5.21

SOLVE

Step 1 Apply Newton's second law in component form. From our free-body diagram, all forces are along the y axis, so we use just the y-component portion of Newton's second law (Eq. 5.2).

$$\sum F_y = ma_y$$

Because the acceleration is zero, the ma_y term is zero.

$$\sum F_y = 0 \quad (1)$$

Now, write each force in the free-body diagram in component form and substitute them into Equation (1). The magnitude of the gravitational force is the weight w. The tension has the same magnitude F_T throughout the single strap. The strap attaches to the backpack in two places, however, so there are two tension forces of magnitude F_T in the positive y direction. The total tension force is the vector sum of these two (Eq. 2).

$$\vec{F}_g = -w\hat{\jmath}$$

$$(\vec{F}_T)_{tot} = 2\vec{F}_T = 2F_T\hat{\jmath} \quad (2)$$

$$\sum F_y = 2F_T - w = 0 \quad (3)$$

Step 2 Write relevant force equations. The weight w of the backpack is the magnitude mg of the Earth's gravitational force on it.

$$w = mg \quad (4)$$

Step 3 Do algebra. Solve Equations (3) and (4) for the tension F_T by eliminating w from the two equations and isolating F_T.

$$\sum F_y = 2F_T - mg = 0$$

$$F_T = \tfrac{1}{2}mg$$

Substitute values to find the tension in the strap. Notice that the tension is half the weight because the strap is attached at two points.

$$F_T = \tfrac{1}{2}mg = \tfrac{1}{2}(3.14\text{ kg})(9.81\text{ m/s}^2)$$

$$F_T = 15.4\text{ N}$$

The total tension force exerted on the backpack comes from Equation (2).

$$(\vec{F}_T)_{tot} = 2F_T\hat{\jmath} = 2(15.4\text{ N})\hat{\jmath}$$

$$(\vec{F}_T)_{tot} = 30.8\hat{\jmath}\text{ N}$$

CHECK and THINK

We just found that the magnitude of the total tension force is equal to the weight of the backpack and its contents, which makes sense because the backpack is not accelerating and so the net force must be zero.

EXAMPLE 5.7 CASE STUDY Train Collision

When the passenger train of our case study came to a sudden stop before the collision, passengers who were in forward-facing seats were thrown forward, and passengers in backward-facing seats were not (Fig. 5.22). Friction due to the seat bottoms is exerted on both passengers. (For the backward-facing passenger, it is static friction; for the forward-facing passenger, it is kinetic friction.) For both passengers, friction is parallel to the seat bottom and points to the right in this case. Draw two free-body diagrams: one for a forward-facing passenger and one for a backward-facing passenger. Assume both passengers have roughly the same mass. Use Newton's laws to explain why the forward-facing passenger fell forward but the backward-facing passenger did not.

FIGURE 5.22

INTERPRET and ANTICIPATE

First, determine which of the five forces discussed in Section 5-7 are exerted on each passenger.

Example continues on page 138 ▶

No tension or spring forces.

Gravity $\vec{F}_g$ acts on both passengers (equal masses assumed, so equal gravitational force on each passenger).

Normal force:

1. $\vec{F}_N$ due to seat bottoms acts on both passengers.
2. Passenger seated facing backward is acted on by additional normal force $\vec{f}_N$ due to the back of her seat.

Friction $\vec{F}_f$ (static or kinetic) due to the seat bottoms points to the right for both passengers.

The two free-body diagrams each contain all four elements (Fig. 5.23). Each passenger is **represented as a dot**. The **accelerations** are indicated. A **coordinate system** has been chosen so that the x axis is parallel to the direction of the accelerations (often a convenient choice). All the **forces** exerted on each passenger have been **drawn and labeled**. There are three forces exerted on the forward-facing passenger and four forces on the backward-facing passenger. Notice that friction due to the seat bottoms points in the same direction as the acceleration (the x direction).

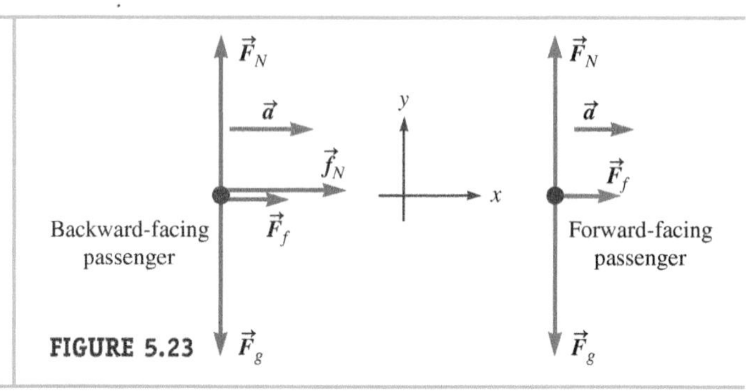

FIGURE 5.23

SOLVE

We can tell why one passenger fell and the other didn't by examining the free-body diagrams. Before the train decelerated suddenly, both passengers were moving at constant velocity in the negative x direction. According to Newton's first law, a force is required to slow these passengers down (to give them the same acceleration as the train). This force must be in the same direction as the train's acceleration, the positive x direction. Only the frictional force due to the seats and the normal force due to the back of the backward-facing passenger's seat point in the x direction. Friction alone is too weak to provide the required acceleration in this case. Therefore, the forward-facing passenger continues to move forward at a speed somewhat less than his original speed. The backward-facing passenger accelerates with the train and remains in her seat because of the additional normal force on her back.

CHECK and THINK

This example explains why seat belts are so important. If you are in a car that decelerates suddenly, friction due to the seat is not strong enough to keep you accelerating along with the car. Inertia will carry you through the windshield if you rely on that friction alone. The seat belt provides the additional force needed to slow you down and keep you in the car.

EXAMPLE 5.8 Astronaut on an Elevator

We hear of "weightless" astronauts in orbiters like the International Space Station. The gravitational force exerted by the Earth on such orbiters is not zero, so why do the occupants feel weightless? The answer is that orbiting astronauts experience weightlessness not because there is no gravitational pull on them but because they are in a noninertial reference frame. In this example, we will use an elevator on the Earth to see how the perception of "weight" works.

In Figure 5.24, an astronaut stands on a spring scale in an elevator car on the Earth. The elevator moves downward with a constant acceleration of magnitude $a < g$. From Section 5-7, the astronaut's apparent weight w_{app} is the reading on the scale.

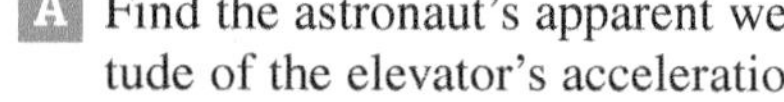

A Find the astronaut's apparent weight in terms of her true weight w, the magnitude of the elevator's acceleration a, and the free-fall acceleration g.

FIGURE 5.24

INTERPRET and ANTICIPATE

Our free-body diagram for the astronaut (Fig. 5.25) includes all four elements. The **astronaut is represented as a dot.** Only two forces are exerted on the astronaut: gravity $\vec{F}_g$ and the contact force from the spring scale. In this case, the contact force is provided by the spring so that $\vec{F}_N = \vec{F}_H$. These **forces are labeled** on the free-body diagram, and are vertical. We choose an upward-pointing y axis for our **coordinate system**. Finally, the astronaut accelerates downward with the elevator, which is indicated by the downward-pointing **acceleration** vector.

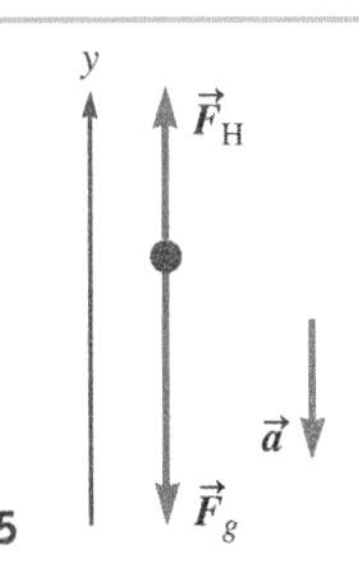

FIGURE 5.25

The reading on the scale dial is the magnitude of the spring force. This reading is also the astronaut's apparent weight, so we can equate these two quantities. We seek an algebraic expression for the apparent weight, which is a positive scalar.

$$w_{app} = F_H \tag{1}$$

SOLVE

Step 1 Apply Newton's second law in component form. Both forces lie along the y axis, so only one equation is needed.

$$\sum F_y = F_H - F_g = ma_y \tag{2}$$

The acceleration is downward as indicated on the free-body diagram.

$$\vec{a}_y = -a\hat{\jmath} \tag{3}$$

Write Newton's second law explicitly in terms of the magnitude of the elevator's acceleration by substituting Equation (3) into Equation (2).

$$\sum F_y = F_H - F_g = -ma \tag{4}$$

Step 2 Write other relevant force equations. The astronaut's true weight is the magnitude of the gravitational force on her (Eq. 5.7), and her mass can be written in terms of this (true) weight. (There is no need to write $F_H = k\,\Delta y$ because we are not interested in k or Δy.

$$w = F_g = mg \tag{5}$$

$$m = \frac{w}{g} \tag{6}$$

Step 3 Do algebra. Substitute Equations (1) (w_{app} for F_H), (5) (w for F_g), and (6) (w/g for m) into Equation (4), and then solve for w_{app}.

$$w_{app} - w = -\frac{w}{g}a$$

$$w_{app} = w\left(1 - \frac{a}{g}\right) \tag{7}$$

CHECK and THINK

According to Equation (7), as long as the elevator is accelerating downward, the astronaut's apparent weight is less than her true weight. (For an elevator accelerating upward, the minus sign becomes a plus sign and the apparent weight is greater than her true weight.)

How does a person experience "apparent weight"? You do not actually feel the pull of gravity downward; instead, you feel the normal force on your body due to a chair, the floor, or whatever surface is preventing you from falling to the center of the Earth. In this case, the spring scale is acting on the astronaut in this way. She feels lighter because the force due to the scale has decreased.

Because the elevator is accelerating, the astronaut is in a noninertial reference frame. Just as the observer on the ship in Figure 5.6 cannot identify a force that caused the Zamboni to accelerate, the astronaut cannot identify a force that caused her weight to appear to decrease.

B What happens to the astronaut's apparent weight if the elevator is in free fall as seen in Figure 5.26?

FIGURE 5.26

INTERPRET and ANTICIPATE

We expect that something special must happen if the elevator is in free fall. The free-body diagram for the astronaut is now as shown in Figure 5.27. The spring force F_H is no longer acting on the astronaut because she is not in contact with the scale.

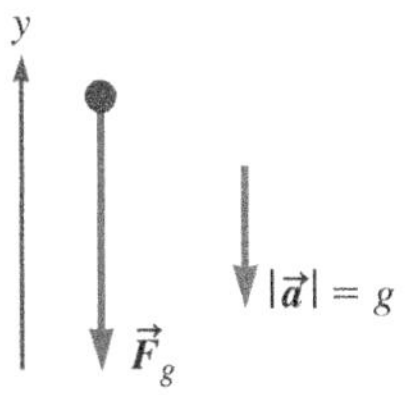

FIGURE 5.27

Example continues on page 140 ▶

SOLVE

If the elevator is in free fall, its acceleration is exactly g. Substitute $a = g$ into Equation (7).

$$w_{\text{app}} = w\left(1 - \frac{g}{g}\right) = 0$$

CHECK and THINK

If the elevator is in free fall, the astronaut's apparent weight is zero, and we say that she is "weightless." She does not experience any normal force and therefore does not feel her own weight. The force of the Earth's gravity is not zero, but the force of the spring scale is zero because the astronaut's feet hover above the scale. Therefore, the spring is in its relaxed state, with the pointer at zero on the dial. Similarly, an astronaut in orbit around the Earth is also in free fall, so her apparent weight is zero. Both astronauts feel "weightless."

EXAMPLE 5.9 Pulleys Are Everywhere

Pulleys are found on the end of cranes, on sailboats, and even in the weight room of a workout gym. Ideal pulleys are considered to be massless and frictionless. In the common laboratory setup shown in Figure 5.28, the hanging cylinder of mass m is released from rest. If the friction between the cart of mass M and the track is negligible, find the acceleration of the cart and the tension in the rope.

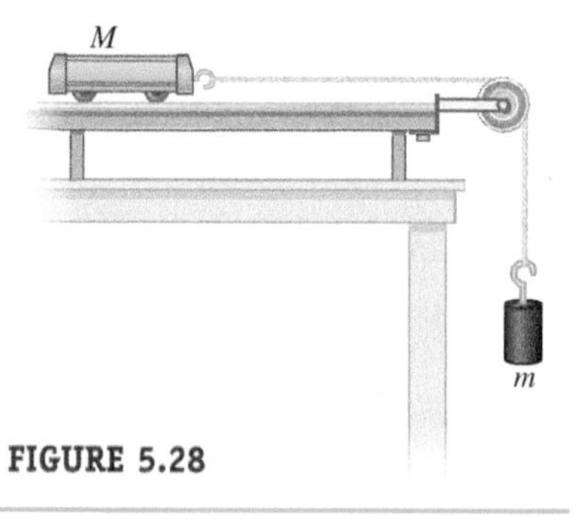

FIGURE 5.28

INTERPRET and ANTICIPATE

Draw one free-body diagram for the cart and another for the hanging cylinder as in Figure 5.29.

1. Each subject is **represented by a dot** (labeled with the mass) in Figure 5.29.
2. **Forces are drawn and labeled on each object.**
   No springs or friction.
   Two forces on the cylinder: gravity $\vec{F}_g$ and the tension force $\vec{F}_T$.
   Three forces on the cart: gravity $\vec{F}_G$, the normal force $\vec{F}_N$, and the tension force $\vec{F}_T$.
   For an ideal pulley, the tension is the same throughout the rope (same symbol $\vec{F}_T$ in both diagrams). We chose an uppercase subscript for gravitational force on the cart and a lowercase subscript for gravitational force on the hanging cylinder.

You can add elements 3 and 4 in any order. Here we'll add element 4 first.

4. The **acceleration** of each subject is indicated. The cart accelerates to the right when the cylinder accelerates downward.
3. **Coordinate systems** were picked to make calculations easy. For the hanging cylinder, the choice may seem unconventional but the taut rope ensures that the acceleration of both the cylinder and the cart has the same magnitude. This choice of coordinates means that the acceleration of both objects is along a positive x axis (to the right for the cart; downward for the cylinder).

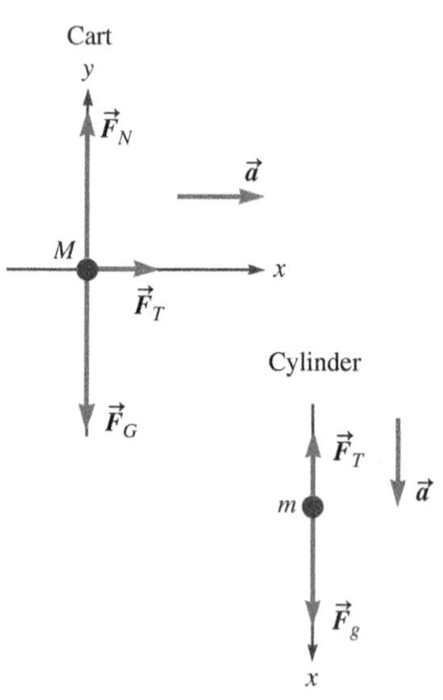

FIGURE 5.29

We expect to find the same algebraic expression for the acceleration of both objects in the form $\vec{a} = (\_\_\_\_)\hat{\imath}$ and an additional algebraic scalar expression for the tension.

SOLVE

Step 1 **Apply Newton's second law in component form.** We must write two equations for the cart because there are forces along both x and y directions. There is only one equation for the cylinder. The acceleration of the cart along x must equal the acceleration of the cylinder, so we have used the same symbol a for both. The cart has no acceleration in the y direction.

Cart:

$$\sum F_x = F_T = Ma \tag{1}$$

$$\sum F_y = F_N - F_G = 0$$

Hanging cylinder:

$$\sum F_x = F_g - F_T = ma \tag{2}$$

Step 2 **Write other relevant force equations.** Use Equation 5.7 to write expressions for the weight of each subject.	$F_g = mg$ (3) $F_G = Mg$
Step 3 **Do algebra.** The y equation for the cart gives no information about the acceleration, so we don't need it. Equations (1) and (2) both have two unknowns, F_T and a. We use these equations along with Equation (3) to solve for the unknowns. Substitute Equations (3) and (1) into Equation (2), and then solve for a.	$mg - Ma = ma$ $mg = Ma + ma = a(M + m)$ $a = \frac{mg}{M + m}$ (4)
Write the acceleration in vector notation. Find the tension in the rope by substituting Equation (4) into Equation (1). Remember tension is a scalar.	$\vec{a} = \left(\frac{mg}{M + m}\right)\hat{\imath}$ $\quad F_T = \frac{Mmg}{M + m}$ (5)

CHECK and THINK

Our expressions for tension and acceleration are in the forms we expected as we can verify by checking their SI units.

Finally, the $\hat{\imath}$ in Equation (5) means that the cart accelerates to the right while the hanging cylinder accelerates downward.

$$[a] = \left[\frac{mg}{M + m}\right] = \frac{(\text{kg})(\text{m/s}^2)}{\text{kg}} = \text{m/s}^2$$

and

$$[F_T] = \left[\frac{Mmg}{M + m}\right] = \frac{(\text{kg})(\text{N})}{\text{kg}} = \text{N}$$

5-9 Newton's Third Law

Unlike Newton's first two laws, which deal with forces acting on a single subject, Newton's third law is concerned with the forces between two objects that interact with each other. Conceptually, Newton's third law states that if two objects A and B interact, the force exerted by A on B is equal in magnitude to the force exerted by B on A but in the opposite direction.

We experience Newton's third law all the time. When you are sitting in a chair, your body presses down on the chair, and you can feel the chair pressing back on you. (If the chair did not exert this force on you, you would end up falling through the chair and sitting on the floor. Then you and the *floor* would exert a pair of forces on each other.) As another example, when your arm lifts a heavy backpack upward, you can feel the backpack pulling your arm downward.

We can write Newton's third law mathematically. First, we define $\vec{F}_{[\text{B on A}]}$ as the force exerted *by* object B *on* object A. In the language of Section 5-3, B is the source of the force $\vec{F}_{[\text{B on A}]}$, and A is the subject on which the force acts. Because B exerts a force on object A, it must be true that A also exerts a force on B, and we denote this force by $\vec{F}_{[\text{A on B}]}$. By Newton's third law, these two forces are a pair that must be equal in magnitude and opposite in direction. Mathematically, this idea is expressed as

It is common to drop the word *on* from the subscript so that $\vec{F}_{[\text{B on A}]}$ is written as $\vec{F}_{\text{BA}}$.

$$\vec{F}_{[\text{B on A}]} = -\vec{F}_{[\text{A on B}]} \tag{5.10}$$

NEWTON'S THIRD LAW

Underlying Principle

CONCEPT EXERCISE 5.10

In Concept Exercise 5.3, we identified the source and subject of a force in three situations. Return to those situations and draw a free-body diagram for each subject. In each diagram, draw a loop around the vector arrow representing the force that is part of a third-law pair associated with the source. Label this with the [source on subject] subscript notation.

Sometimes it is hard to see the difference between Newton's second and third laws. There are two important distinctions, however. (1) The third law always applies to *a pair of* interacting objects and to *a pair of* forces; one force acts on one object of the pair, and the other force acts on the other object of the pair. The second law involves all the forces acting *on a single object.* (2) The third law says nothing about acceleration, whereas the second law relates total force exerted on an object to its acceleration.

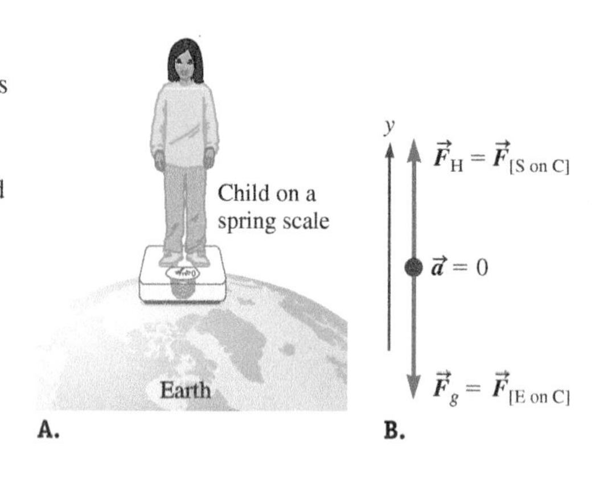

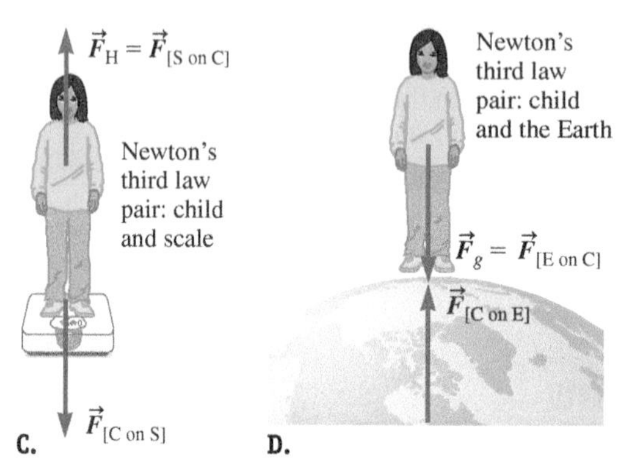

FIGURE 5.30 **A.** A child stands on a spring scale. **B.** A free-body diagram shows all the forces that act on the child, but not the forces that she exerts. According to Newton's second law and because the child is not accelerating, $\vec{F}_{[\text{S on C}]} = -\vec{F}_{[\text{E on C}]}$. **C.** The child interacts with the spring scale. According to Newton's third law, $\vec{F}_{[\text{S on C}]} = -\vec{F}_{[\text{C on S}]}$. **D.** The child also interacts with the Earth, and, by the third law, $\vec{F}_{[\text{E on C}]} = -\vec{F}_{[\text{C on E}]}$.

To further illuminate the distinction between the second and third laws, consider a child standing on a spring scale (Fig. 5.30A). When applying Newton's second law, we use a free-body diagram to represent all the forces acting on the child (Fig. 5.30B). *The forces for which the child is a source are never drawn in the free-body diagram for the child.* The free-body diagram shows two forces acting on her: the spring force $\vec{F}_{\text{H}}$, which in our $\vec{F}_{[\text{source on subject}]}$ notation becomes $\vec{F}_{[\text{S on C}]}$; and gravity $\vec{F}_g$, which in our third-law notation is $\vec{F}_{[\text{E on C}]}$. The subscripts stand for *S*pring, *C*hild, and *E*arth.

Parts C and D of Figure 5.30 help illustrate Newton's third law. They are *not* free-body diagrams, differing from a free-body diagram in three ways:

1. Each figure has two objects instead of one.
2. In addition to being acted on by external forces, each object is also the source of a force.
3. Most important, not all the forces exerted on each object are drawn. Instead, each figure shows only interacting force pairs.

In Figure 5.30C, the source of $\vec{F}_{[\text{S on C}]}$ is the spring, and the subject is the child. This force forms an interaction pair with $\vec{F}_{[\text{C on S}]}$, the force for which the child is the source and the spring is the subject. Similarly, Figure 5.30D shows the interaction force pair between the Earth and the child, $\vec{F}_{[\text{E on C}]}$ and $\vec{F}_{[\text{C on E}]}$. Because the child is not the subject of $\vec{F}_{[\text{C on S}]}$ or $\vec{F}_{[\text{C on E}]}$, these two forces do not appear in the free-body diagram for the child (Fig. 5.30B).

In some situations, the distinction between the second and third laws is further blurred because all the forces have the same magnitude. This child on the spring scale is such a case: $F_{[\text{S on C}]} = F_{[\text{E on C}]} = F_{[\text{C on S}]} = F_{[\text{C on E}]}$, and we must look at both the second law and the third law to see why. According to Newton's second law and because the child is not accelerating (Fig. 5.30B), the two forces acting on her are equal in magnitude:

$$\sum F_y = F_{[\text{S on C}]} - F_{[\text{E on C}]} = ma_y = 0$$
$$F_{[\text{S on C}]} = F_{[\text{E on C}]}$$

Notice that the two forces have different sources: the spring scale for $F_{[\text{S on C}]}$ and the Earth for $F_{[\text{E on C}]}$. Both of these forces are acting on the same subject (the child) as denoted by "on C" in the subscript.

We use Newton's third law to relate the magnitudes of forces that act on a pair of interacting objects (Fig. 5.30C). For the child-spring pair, we have

$$F_{[\text{S on C}]} = F_{[\text{C on S}]}$$

and for the child–Earth pair (Fig. 5.30D), we have

$$F_{[\text{E on C}]} = F_{[\text{C on E}]}$$

The wording of the subscripts should help you keep track of a third-law force pair. The subscripts should form a pair involving only two objects; for example, E and C in the case of [E on C] and [C on E].

EXAMPLE 5.10 Pushy Skaters

Two ice skaters initially at rest push off from each other as shown in Figure 5.31. The coefficient of kinetic friction between their skates and the ice is $\mu_k = 0.05$. Wilma weighs 6.45×10^2 N, and Mark has a mass of 74.8 kg. At the moment of the push-off, the magnitude of his acceleration is 4.58 m/s^2.

A Find the force with which Wilma pushed Mark.

FIGURE 5.31

INTERPRET and ANTICIPATE

To find the force responsible for Mark's 4.58-m/s^2 acceleration, start with a free-body diagram for him, **represented as usual by a dot** (Fig. 5.32). We have included a two-dimensional **coordinate system** and indicated the direction of his **acceleration**. All the **forces** exerted on him have been **drawn and labeled**. To help keep track of our symbols, we'll use lowercase f for all the forces acting on Mark and uppercase F for those acting on Wilma.

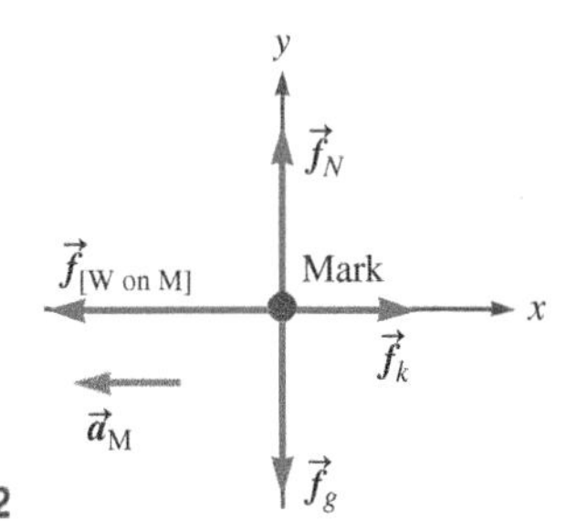

FIGURE 5.32

From our free-body diagram, we expect $\vec{f}_{[\text{W on M}]}$, the force exerted by Wilma on Mark, to be in the negative x direction. We anticipate a numerical answer of the form $\vec{f}_{[\text{W on M}]} = -\_\_\_\_\ \hat{\imath}$ N.

SOLVE

Step 1 **Apply Newton's second law in component form.** There are forces in both the x and y directions, so there are two equations.	$\sum F_x = f_k - f_{[\text{W on M}]} = ma_x$ (1) $\sum F_y = f_N - f_g = ma_y$ (2)
Mark is not accelerating in the direction perpendicular to the ice (the y direction), so $a_y = 0$ in Equation (2).	$f_N - f_g = 0$ $f_N = f_g$ (3)
Mark's acceleration (magnitude 4.58 m/s^2) must be entirely in the negative x direction. Write Equation (1) in terms of his acceleration a_{Mx}.	$\vec{a}_x = -a_{Mx}\hat{\imath} = -4.58\,\hat{\imath}$ m/s^2 $f_k - f_{[\text{W on M}]} = -ma_{Mx}$ (4)
Step 2 **Write other force equations.** Use Equation 5.7 to express Mark's weight.	$f_g = mg$ (5)
We also need Equation 5.9 for kinetic friction.	$f_k = \mu_k f_N$ (6)
Step 3 **Do algebra.** To find $\vec{f}_{[\text{W on M}]}$, start by eliminating f_N from Equation (6) with Equation (3).	$f_k = \mu_k f_g$ (7)
Substitute Equation (5) into Equation (7) to write the frictional force in terms of Mark's weight.	$f_k = \mu_k mg$ (8)
Substitute Equation (8) into Equation (4) and solve for force exerted by Wilma on Mark.	$\mu_k mg - f_{[\text{W on M}]} = -ma_{Mx}$ $f_{[\text{W on M}]} = ma_{Mx} + \mu_k mg$ $f_{[\text{W on M}]} = m(a_{Mx} + \mu_k g)$
Insert numerical values and write the final answer in component form.	$f_{[\text{W on M}]} = (74.8\text{ kg})[4.58\text{ m/s}^2 + (0.05)(9.81\text{ m/s}^2)]$ $\vec{f}_{[\text{W on M}]} = -3.79 \times 10^2\,\hat{\imath}$ N

CHECK and THINK

Our answer is in the form we expected.

Example continues on page 144 ▶

B Find Wilma's acceleration.

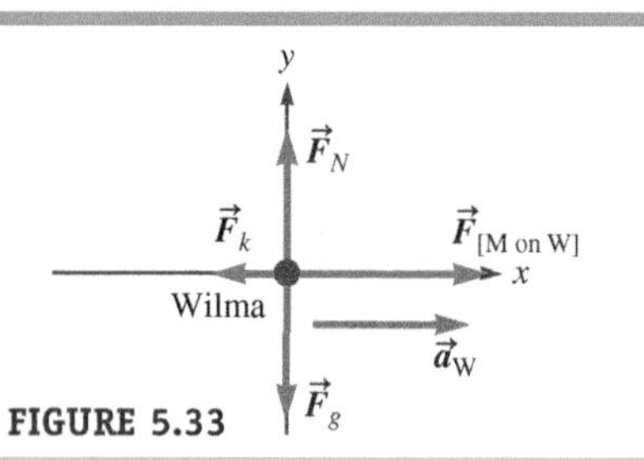

FIGURE 5.33

INTERPRET and ANTICIPATE

To find her acceleration, we need a free-body diagram for Wilma (Fig. 5.33). Her free-body diagram contains all four elements and is similar to Mark's diagram.

SOLVE

Step 1 **Apply Newton's second law in component form.** The results are similar to Equations (1) and (2) for Mark.

$$\sum F_x = F_{[M\text{ on }W]} - F_k = Ma_{Wx} \quad (9)$$

$$\sum F_y = F_N - F_g = Ma_{Wy} \quad (10)$$

Like Mark, Wilma accelerates only in the x direction, so $a_{Wy} = 0$ in Equation (10).

$$F_N - F_g = 0$$

$$F_N = F_g \quad (11)$$

Step 2 **Write other force equations.** Use Equation 5.7 to express Wilma's weight. We also need Equation 5.9 for kinetic friction.

$$F_g = Mg \quad (12)$$

$$F_k = \mu_k F_N \quad (13)$$

According to Newton's third law, the force exerted by Mark on Wilma has the same magnitude as the force she exerts on him, but in the opposite direction. We found the force she exerts on him in part A.

$$\vec{F}_{[M\text{ on }W]} = -\vec{f}_{[W\text{ on }M]} = 3.79 \times 10^2 \hat{\imath}\text{ N} \quad (14)$$

Step 3 **Do algebra.** Combine Equations (11) and (13) to find an expression for kinetic friction in terms of Wilma's weight.

$$F_k = \mu_k F_g \quad (15)$$

Substitute Equation (15) into Equation (9), and solve for her acceleration a_{Wx}.

$$F_{[M\text{ on }W]} - \mu_k F_g = Ma_{Wx}$$

$$a_{Wx} = \frac{1}{M}(F_{[M\text{ on }W]} - \mu_k F_g)$$

We were given Wilma's weight, but not her mass. Use Equation (12) to express her mass in terms of her weight: $M = F_g/g$.

$$a_{Wx} = \frac{g}{F_g}(F_{[M\text{ on }W]} - \mu_k F_g)$$

$$a_{Wx} = g\left(\frac{F_{[M\text{ on }W]}}{F_g} - \mu_k\right)$$

Substitute numerical values, including Equation (14) for the force Mark exerts on Wilma.

$$a_{Wx} = (9.81\text{ m/s}^2)\left(\frac{3.79 \times 10^2\text{ N}}{6.45 \times 10^2\text{ N}} - 0.05\right)$$

$$\vec{a}_W = 5.27\hat{\imath}\text{ m/s}^2$$

CHECK and THINK

According to Newton's *third* law, each skater exerted the same magnitude of force on the other skater. According to Newton's *second* law, the skater with the smaller mass had a higher acceleration. Because the direction of the force on Wilma was opposite the direction of the force on Mark, her acceleration was in the opposite direction, too.

EXAMPLE 5.11 Superman of the Insect World

In 2003, Malcolm Burrows of the University of Cambridge, England, reported that the 6-mm froghopper insect could leap up 70 cm, equivalent to a person leaping up 70 stories.[3] Burrows reported that the best jumps have an acceleration of $408g$ at an angle of $(58 \pm 2.6)°$. The mass of his froghopper was (12.3 ± 0.7) mg.

A Identify the source of the force that propels the froghopper upward.

[3]Malcom Burrows, "Froghopper Insects Leap to New Heights," *Nature* 424:509 (2003).

SOLVE

To jump, the froghopper must push downward on the ground. By Newton's third law, the ground must therefore push upward on the froghopper, and in the opposite direction as shown in Figure 5.34. The vertical component of the force exerted by the ground is the normal force acting on the insect; the horizontal component is due to static friction, which we will study in detail in Chapter 6.

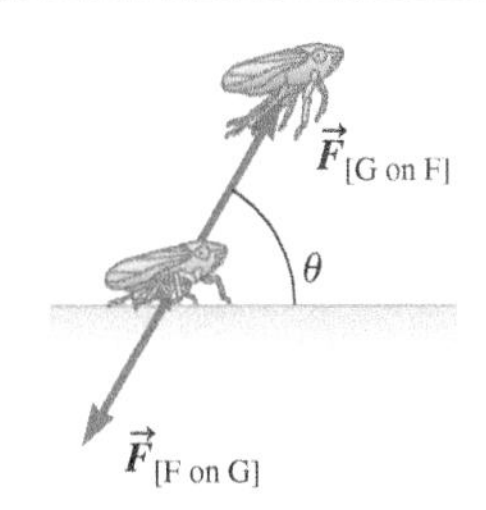

FIGURE 5.34

B Assume a 12.3-mg froghopper leaps at a 60° angle with the ground and accelerates at $408g$. What is the magnitude of the force the froghopper exerts on the ground? *Hint*: Make a simplification.

INTERPRET and ANTICIPATE

Start by drawing a free-body diagram for the froghopper including all four elements as in Figure 5.35. The froghopper is **represented by a dot**. The **forces are drawn and labeled**. There are two forces acting on the froghopper: $\vec{F}_{[G\ on\ F]}$ due to the ground and $\vec{F}_g$ due to gravity. The **acceleration** has been indicated, and a **coordinate system** has been chosen so that the y axis lines up with the acceleration (often a good choice).

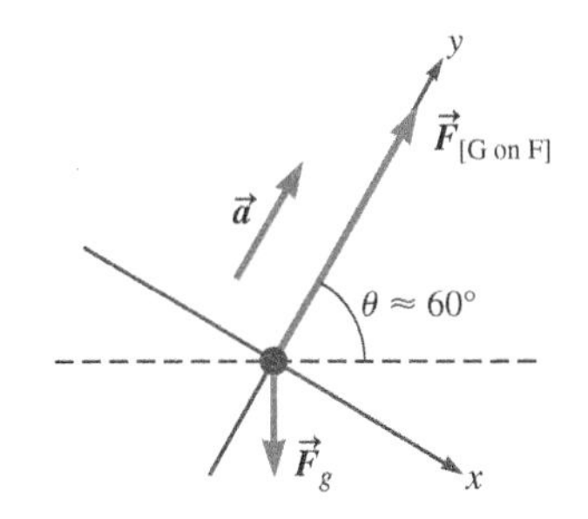

FIGURE 5.35

SOLVE

Step 1 **Apply Newton's second law in component form.** We could solve this problem exactly by applying Newton's second law in component form along both the x and y axes, but that step is not necessary. Because the froghopper is so light, the gravitational force on it is much smaller than the force exerted by the ground. Let's assume we can ignore the gravitational force for now, and find the force exerted by the ground on the froghopper. If the magnitude of that force is comparable to the froghopper's weight, we will rework the problem and include gravity.

Because we are ignoring gravity, we simply apply Newton's second law in the y direction.

$$\sum F_y = ma_y$$
$$F_{[G\ on\ F]} = ma_y$$

Step 2 **Write down relevant force equations.** There are no other relevant force equations.

Step 3 **Do algebra.** The force we are looking for is already isolated on the left side of the equation. We only need to substitute the given value of the acceleration.

$$F_{[G\ on\ F]} = ma_y = 408mg \quad (1)$$
$$F_{[G\ on\ F]} = (408)(12.3 \times 10^{-6}\,\text{kg})(9.81\ \text{m/s}^2)$$
$$F_{[G\ on\ F]} = 4.92 \times 10^{-2}\,\text{N}$$

CHECK and THINK

According to Equation (1), the force exerted by the ground on the insect is more than 400 times its weight, so it is reasonable to neglect gravity. By Newton's third law, the force exerted by the froghopper on the ground ($F_{[F\ on\ G]}$) is equal in magnitude to this force but in the opposite direction. In other words, the force exerted by the froghopper is more than 400 times its weight. Imagine exerting more than 60,000 pounds with your legs!

CONCEPT EXERCISE 5.11

A child jumping off the monkey bars at a playground accelerates toward the ground because of the gravitational force exerted on him by the Earth. From Newton's third law, the force exerted by the Earth on the child is equal in magnitude to the force exerted by the child on the Earth. Does the Earth accelerate? Explain.

CONCEPT EXERCISE 5.12

CASE STUDY Train Collision and Newton's Third Law

Avi, Cameron, and Shannon continue to discuss the train collision case study. Use Newton's third law to decide which student is correct. Explain your answer.

Cameron: The passenger train got shoved backward. That means that the freight train did all the pushing. The passenger train couldn't push on the freight train because the passenger train wasn't moving.

Shannon: No. It's not like the freight train exerted all the force. They both exerted a force on each other.

Avi: Shannon's right. Newton's third law says that they have to push on each other. The freight train just pushed harder than the passenger train so that the net force was in the direction the freight train was moving.

Shannon: The force is the same on each train. The passenger train got shoved backward because it was lighter than the freight train.

Avi: I think the freight train pushed harder because it was heavy. Look, the freight train suffered almost no damage and the passenger train was a wreck. The passenger train had to have been hit harder.

5-10 Fundamental Forces

In this chapter, we studied five specific forces: gravity, the spring force, the normal force, the tension force, and kinetic friction. Although we will encounter many other forces throughout this text, the modern theory of physics states that there are only four fundamental forces: gravitational, electromagnetic, strong, and weak. The gravitational force was introduced previously and will be revisited often.

The electromagnetic force is discussed in Parts III and IV of this book. It is the force between electrical charges and is an important part of our daily lives, governing computers, motors, and even human nerve function and thought. In Section 5-7, we found that tension and the normal force can be understood in terms of the molecular bonds between atoms. These bonds are due to electromagnetic forces. In fact, most of the forces we study in this book are really a consequence of the electromagnetic force.

The strong force and the weak force are not studied either in classical mechanics or in electricity and magnetism, so they are briefly mentioned in this book. The strong force keeps protons (and neutrons) together in the nucleus of an atom. Without the strong force, the electromagnetic force between any two protons would cause them to repel each other (with the result that atoms could not form and the Universe would be a very different place). The weak force acts in the nucleus of atoms. Neutrons are unstable outside the nucleus, and the weak force governs how they decay as a result of this instability. The weak force also plays a role in fusion reactions that power the Sun.

SUMMARY

❶ Underlying Principles: Newton's three laws of motion

1. Close translation of **Newton's first law**:
   Every body continues in its state of rest, or of uniform motion in a straight line, unless it is compelled to change that state by forces impressed on it.

   More contemporary statement of Newton's first law:
   If no force acts on an object, then the object cannot accelerate.

   Newton's second law contains his first law:
   The net force on an object is zero if and only if the acceleration of the object is also zero.

Underlying Principles: Newton's three laws of motion—cont'd

2. **Newton's second law**: The acceleration of an object is proportional to the total force acting on it:
$$\vec{F}_{\text{tot}} \equiv \sum \vec{F} = m\vec{a} \qquad (5.1)$$
Newton's second law is often applied in component form:
$$\sum F_x = ma_x \quad \sum F_y = ma_y \quad \sum F_z = ma_z \quad (5.2)$$

3. **Newton's third law**: If two objects A and B interact, the force exerted by A on B is equal in magnitude to the force exerted by B on A but in the opposite direction. Mathematically,
$$\vec{F}_{[\text{B on A}]} = -\vec{F}_{[\text{A on B}]} \qquad (5.10)$$

Major Concepts

1. **Dynamics:** a branch of mechanics that focuses on the causes of motion.
2. **Force:** a push or pull; required to make an object accelerate. A force has three properties:
   a. Force does not exist in isolation; it must act on a "subject."
   b. If a force is exerted on a subject, there must be some "source" responsible for that force.
   c. Force is a vector quantity having both magnitude and direction.
3. **System**: any collection of objects.
4. **Inertia**: the tendency of an object to maintain its constant velocity. **Mass**, also known as **inertial mass**, is the intrinsic property of an object that determines its inertia.
5. **Inertial reference frame**: one in which Newton's first law is valid. Inertial reference frames may move with constant velocity, but they do not accelerate. Accelerating frames are known as **noninertial** reference frames, and Newton's first law is not valid in them.

Special Cases

Table 5.1 on page 134 summarizes the five specific forces examined in this chapter.

Tools

Four elements of a **free-body diagram**:

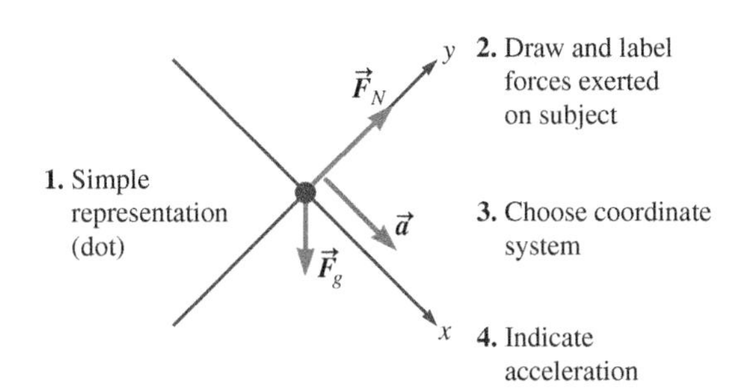

PROBLEM-SOLVING STRATEGY

Applying Newton's Second Law, $\vec{F}_{\text{tot}} \equiv \sum \vec{F} = m\vec{a}$

INTERPRET and ANTICIPATE

Draw a free-body diagram.

SOLVE

1. Apply Newton's second law in component form. Write one equation for each direction that has at least one force.
2. Write any other relevant force equations.
3. Do algebra before substituting values.

PROBLEMS AND QUESTIONS

A = algebraic C = conceptual E = estimation G = graphical N = numerical

5-1 Our Experience with Dynamics

1. C Why is it easier to lift a very large beach ball than a small lead ball?
2. C A heavy bowling ball and a light soccer ball are both rolling toward you at the same speed. Which object is easier to stop with your foot? Which is more likely to break your toe?
3. C Imagine pushing two blocks on ice. The light block slides farther than the heavy block. Why?

5-2 Newton's First Law

4. C When Julia Child would cook an omelet, she would rapidly jostle the pan back and forth (Fig. P5.4). The egg would slosh

back and forth in the pan as it cooked. Use Newton's laws to explain the egg's motion.

FIGURE P5.4

5. C In a science-fiction movie, the hero must get from one spaceship to another, which is several kilometers away. The hero (wearing the appropriate boots) stands on his ship and runs along its surface toward the other ship. When he reaches the edge of his ship, he jumps toward the other ship. Assuming he is far from any major sources of gravity such as planets and friction is negligible. Does he arrive with a lower speed, higher speed or about the same speed at which he left his ship? Explain.
6. C A car is traveling directly east for 90 minutes, yet there is no net force on the car during this time. Explain how this is possible.
7. C Determine whether at least one force is necessary for the following motions. A particle moves **a.** in a straight line at constant speed, **b.** in a straight line while slowing down and **c.** along an L-shaped path at constant speed.

5-3 Force

8. C A bar magnet is used to lift keys that have fallen into a sewer. The keys fly up a few millimeters to meet the bar magnet. Is the magnetic force a contact force or a field force? Explain.
9. N Two forces act on an object with $\vec{F}_1 = (7.263\hat{\imath} + 8.889\hat{\jmath})$N and $\vec{F}_2 = (-13.452\hat{\imath} + 7.991\hat{\jmath})$N. What is the magnitude of the net force experienced by the object?
10. C An object is subject to a single constant force. Under what circumstances does the object travel in a straight line, along a curved path, or along a circular path?
11. N Three forces act on an object with $\vec{F}_1 = (6.03\hat{\imath} - 10.64\hat{\jmath})$N and $\vec{F}_2 = (-3.71\hat{\imath} - 12.93\hat{\jmath})$N. If the net force on the object is zero, what is the unknown force, $\vec{F}_3$?
12. C You blow a small piece of paper through the air. Is the force on the paper a contact force or a field force? Explain.
13. C A fish swims in the ocean. Does the fish exert a force on the water? If so, is it a field force or a contact force? Explain.

5-4 Inertial Mass

14. C Read the following (fictional) dialogue between two people in 2066. Find at least two statements that are incorrect. Explain why they are incorrect.

Europa: I am making Thanksgiving dinner for 12 people. So, what weight turkey do I need?

Dallas: I would suggest about 1 kilogram per person. So, get a 12-kilogram turkey.

Europa: Oh, I forgot to mention that this dinner is for people on the Moon. I plan to shop in the Moon market. How do I find the turkey's weight on the Moon?

Dallas: That's easy. It still weighs 12 kilograms on the Moon.

15. C What has more inertia, a bowling ball at rest or the same ball when it is rolling? Explain.
16. C Do you have more inertia when you are sitting or when you are running? Explain.
17. C Who has more inertia, a child sitting on the sofa or a heavy man jogging?

5-5 Inertial Reference Frames

Problems 18, 19, and 42 are grouped.

18. C A ball hanging from a light string or rod can be used as an accelerometer (a device that measures acceleration) as shown in Figure P5.18. What force causes the deflection of the ball? Is the cart in the lower part of the photo an inertial reference frame? How can the ball's deflection be used to find the cart's acceleration? In which direction is the cart accelerating? Explain your answers.

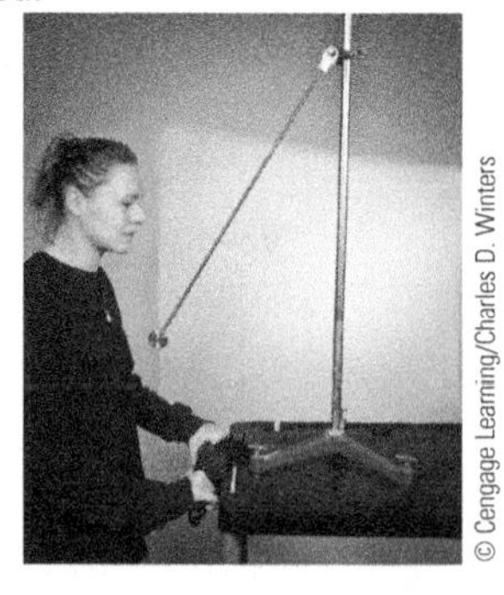

FIGURE P5.18 Problems 18, 19, and 42.

19. C Suppose the cart in Figure P5.18 is moving at constant velocity in a strong wind. What force causes the deflection of the ball? Is the cart an inertial reference frame? Explain.
20. C You are riding a luxury bus. In front of you is a cup of tea resting on the seat-back tray. Which of the following events may lead to spilled tea in your lap? The bus **a.** remains at rest, **b.** moves at constant velocity, **c.** speeds up or **d.** slows down. Don't worry about other circumstances such as a person knocking your cup over. (More than one choice may be correct.) Explain your answers.

5-6 Newton's Second Law

21. N A particle's acceleration is $\vec{a} = (3.45\hat{\imath} - 1.84\hat{\jmath})$ m/s^2, and the total force exerted on the particle is $\vec{F}_{tot} = (10.8\hat{\imath} - 5.76\hat{\jmath})$ N. What is the particle's mass?
22. N A particle with mass $m = 4.00$ kg accelerates according to $\vec{a} = (-3.00\hat{\imath} + 2.00\hat{\jmath})$ m/s^2. **a.** What is the net force acting on the particle? **b.** What is the magnitude of this force?
23. N The x and y coordinates of a 4.00-kg particle moving in the xy plane under the influence of a net force F are given by $x = t^4 - 6t$ and $y = 4t^2 + 1$, with x and y in meters and t in seconds. What is the magnitude of the force F at $t = 4.00$ s?
24. C In the movie *Garden State*, one of the characters uses a bow to launch an arrow straight up into the air. The arrow travels straight up, stops momentarily, and then comes straight down along the same path. The graph of its position versus time is a parabola. What can be said about the sum of the forces on the arrow?
25. N The starship *Enterprise* has its tractor beam locked onto some valuable debris and is trying to pull it toward the ship. A Klingon battle cruiser and a Romulan warbird are also trying to recover the item by pulling the debris with their tractor beams as shown in Figure P5.25. **a.** Given the following magnitudes of the tractor beam forces, find the net force experienced by the debris: $F_{Ent} = 7.59 \times 10^6$ N, $F_{Rom} = 2.53 \times 10^6$ N, and $F_{Kling} = 8.97 \times 10^5$ N. **b.** If the debris has a mass of 2549 kg, what is the net acceleration of the debris?

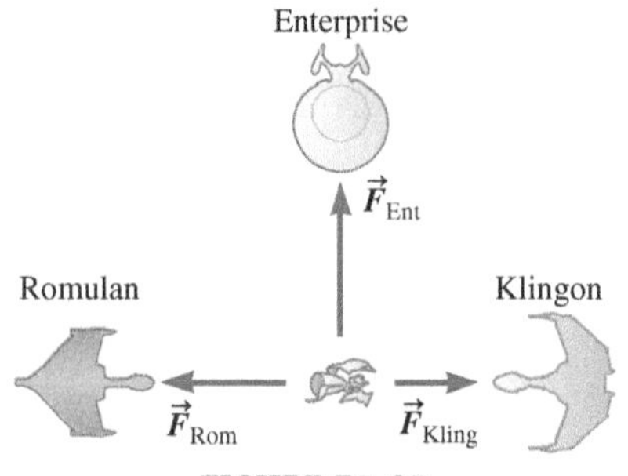

FIGURE P5.25

26. C A race car is moving around a circular track at a constant speed of 65 m/s. It continues moving in this manner at this speed for 15 minutes. During this time, what is the direction of the net force exerted on the race car?

27. N A particle of mass m_1 accelerates at 4.25 m/s^2 when a force F is applied. A second particle of mass m_2 experiences an acceleration of only 1.25 m/s^2 under the influence of this same force F. **a.** What is the ratio of m_1 to m_2? **b.** If the two particles are combined into one particle with mass $m_1 + m_2$, what is the acceleration of this particle under the influence of the force F?

28. N Jim Bob needs to tow his truck home using his neighbor's SUV. The SUV pulls the truck by exerting a horizontal force of 3133 N. If this force alone causes the truck to accelerate at a rate of 2.91 m/s^2, what is the mass of Jim Bob's truck?

29. N Two forces $\vec{F}_1 = (62.98\hat{\imath} - 15.80\hat{\jmath})$N and $\vec{F}_2 = (23.66\hat{\imath} - 78.05\hat{\jmath})$N are exerted on a particle. The particle's mass is 14.23 kg. **a.** Find the particle's acceleration in component form. **b.** What are the magnitude and direction of the acceleration?

30. N Three forces $\vec{F}_1 = (62.98\hat{\imath} - 15.80\hat{\jmath})$N, $\vec{F}_2 = (23.66\hat{\imath} - 78.05\hat{\jmath})$N, and $\vec{F}_3 = (-86.64\hat{\imath} + 233.4\hat{\jmath})$N are exerted on a particle. The particle's mass is 14.23 kg. Find the particle's acceleration.

31. N A hockey stick pushes a 0.160-kg puck with constant force across the frictionless surface of an ice rink. During this motion, the puck's velocity changes from $4.00\hat{\imath}$ m/s to $(6.00\hat{\imath} + 12.00\hat{\jmath})$ m/s in 4.00 s. **a.** What are the scalar components of the force acting on the puck? **b.** What is the magnitude of the force acting on the puck?

Problems 32 and 33 are paired.

32. N If the vector components of the position of a particle moving in the xy plane as a function of time are $\vec{x}(t) = \left(2.5\ \frac{\text{m}}{\text{s}^2}\right)t^2\hat{\imath}$ and $\vec{y}(t) = \left(5.0\ \frac{\text{m}}{\text{s}^3}\right)t^3\hat{\jmath}$, when is the angle between the net force on the particle and the x axis equal to 45°?

33. A If the vector components of the position of a particle moving in the xy plane as a function of time are $\vec{x}(t) = bt^2\hat{\imath}$ and $\vec{y}(t) = ct^3\hat{\jmath}$, where b and c are positive constants, b has dimensions of length per time squared, and c has dimensions of length per time cubed, when is the angle between the net force on the particle and the x axis equal to 45°?

5-7 Some Specific Forces

34. N A 15.0-kg object is in free fall near the surface of the Earth. What is its weight? What is its acceleration? What is the direction of the gravitational force exerted on it? How do your answers change if the same object is at rest on the surface of the Earth?

35. N A black widow spider hangs motionless from a web that extends vertically from the ceiling above. If the spider has a mass of 1.5 g, what is the tension in the web?

36. C Determine whether each of the following statements is true or false. **a.** An object's weight is always equal to its mass. **b.** The force of tension always pushes. **c.** The magnitude of the sum of the forces on an object is never greater than its weight. Explain.

37. N You place tomatoes in the pan of a hanging spring scale and find that they weigh 2.5 lb. You measure the downward displacement of the scale's pan to be 1.5 in. What is the spring constant? Give your answer in lb/in. and in SI units.

38. C Kinetic friction is proportional to the normal force (Eq. 5.9). Why should there be an intimate connection between these two forces?

39. N A student takes the elevator up to the fourth floor to see her favorite physics instructor. She stands on the floor of the elevator, which is horizontal. Both the student and the elevator are solid objects, and they both accelerate upward at 5.19 m/s^2. This acceleration only occurs briefly at the beginning of the ride up. Her mass is 80.0 kg. What is the normal force exerted by the floor of the elevator on the student during her brief acceleration?

40. A sleigh is being pulled horizontally by a train of horses at a constant speed of 8.05 m/s. The magnitude of the normal force exerted by the snow-covered ground on the sleigh is 6.37×10^3 N.
    a. N If the coefficient of kinetic friction between the sleigh and the ground is 0.23, what is the magnitude of the kinetic friction force experienced by the sleigh?
    b. C If the only other horizontal force exerted on the sleigh is due to the horses pulling the sleigh, what must be the magnitude of this force?

5-8 Free-Body Diagrams

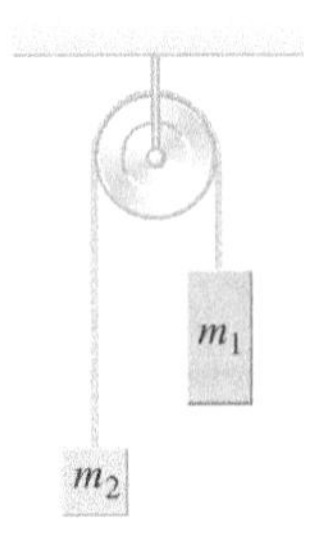

FIGURE P5.41

41. N Two blocks are connected by a rope that passes over a massless and frictionless pulley as shown in Figure P5.41. Given that m_1 = 15.93 kg and m_2 = 10.45 kg, determine the magnitudes of the tension in the rope and the blocks' acceleration.

42. A Find an expression for the cart's acceleration in Figure P5.18 in terms of the ball's mass m and the angle θ.

43. A A woman uses a rope to pull a block of mass m across a level floor at a constant velocity. The coefficient of kinetic friction between the block and the floor is μ_k. The rope makes an angle θ with the floor. Find an algebraic expression for the tension in the rope in terms of the parameters listed in the problem and any constants.

44. N A student working on a school project modeled a trampoline as a spring obeying Hooke's law and measured the spring constant of a certain trampoline as 4617 N/m. If a child of mass 27.0 kg compresses the trampoline vertically by a maximum of 0.25 m, while bouncing up and down, what is the child's acceleration at the moment of maximum compression?

45. N One great form of athletic competition for bulldogs, American pit bull terriers, huskies, and many other breeds is the weight pull. Many of these dogs can pull weights two orders of magnitude greater than their own weight! Railsplitter, an American pit bull terrier, can pull a sled that has a total weight of 2215 lb along a horizontal surface. Suppose he pulls with a horizontal force of 3.50×10^3 N while friction is also working against the sled. If the magnitude of the net acceleration of the sled is 0.152 m/s^2, find the value of the coefficient of kinetic friction, μ_k, between the sled and the pulling surface. (The sled is normally on a set of rails.)

46. A heavy crate of mass 50.0 kg is pulled at constant speed by a dockworker who pulls with a 345-N force at an angle θ with the horizontal (Fig. P5.46). The magnitude of the friction force between the crate and the pavement is 212 N.
    a. G Draw a free-body diagram of the forces acting on the crate.
    b. N What is the angle θ of the rope with the horizontal?
    c. N What is the magnitude of the normal force exerted by the pavement on the crate?

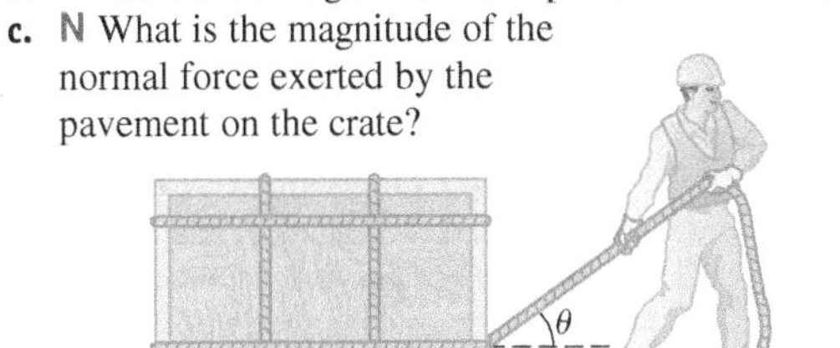

FIGURE P5.46

47. N A block with mass m_1 hangs from a rope that is extended over an ideal pulley and attached to a second block with mass m_2 that sits on a ledge. The second block is also connected to a third block with mass m_3 by a second rope that hangs over a second ideal pulley as shown in Figure P5.47. If the friction between the ledge and the second block is negligible, $m_1 = 3.00$ kg, $m_2 = 5.00$ kg, and $m_3 = 8.00$ kg, find the magnitude of the tension in each rope and the acceleration of each block.

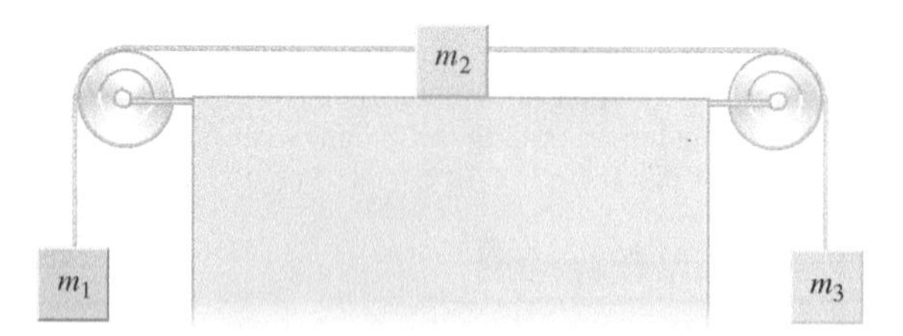

FIGURE P5.47

48. N To get in shape, you head to the local gym to exercise by lifting weights. Using a lat machine (Fig. P5.48), you notice that the wire connected to the bar is run around three different pulleys and is connected to a vertical stack of weights on the other end. The weights move up and down as you pull down on the bar. Suppose you set the machine so that the weight to be lifted is 120.0 lb. When you pull down on the bar, exerting a constant force, the acceleration experienced by the weights is 0.250 m/s^2. What tension must exist in the wire connecting the bar and weights under these conditions? Assume the pulleys are massless and frictionless.

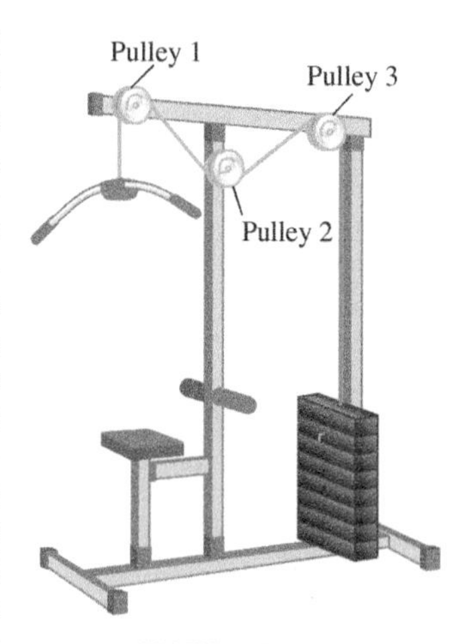

FIGURE P5.48

Problems 49 and 50 are paired.

49. N A block with mass m_1 hangs from a rope that is extended over an ideal pulley and attached to a second block with mass m_2 that sits on a ledge slanted at an angle of 20° (Fig. P5.49). Suppose the system of blocks is initially motionless and held still, and then it is released. If $m_1 = 7.00$ kg and $m_2 = 2.00$ kg, find the magnitude of the acceleration of the blocks, assuming there is no friction between the second block and the ledge.

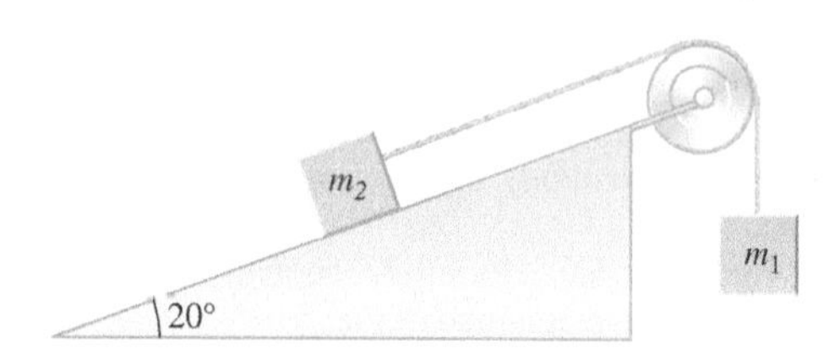

FIGURE P5.49 Problems 49 and 50.

50. N Suppose the system of blocks in Problem 49 is initially held motionless and, when released, begins to accelerate. **a.** If $m_1 =$ 7.00 kg, $m_2 = 2.00$ kg, and the magnitude of the acceleration of the blocks is 0.134 m/s^2, find the magnitude of the kinetic frictional force between the second block and the ledge. **b.** What is the value of the coefficient of kinetic friction between the block and the ledge?

51. Two objects, $m_1 = 3.00$ kg and $m_2 = 8.50$ kg, are attached by a massless cord passing over a frictionless pulley as shown in Figure P5.51. Assume the horizontal surface is frictionless.
a. G Draw a free-body diagram for each of the two objects.
b. N What is the tension in the cord?
c. N What is the magnitude of the acceleration of the two objects?

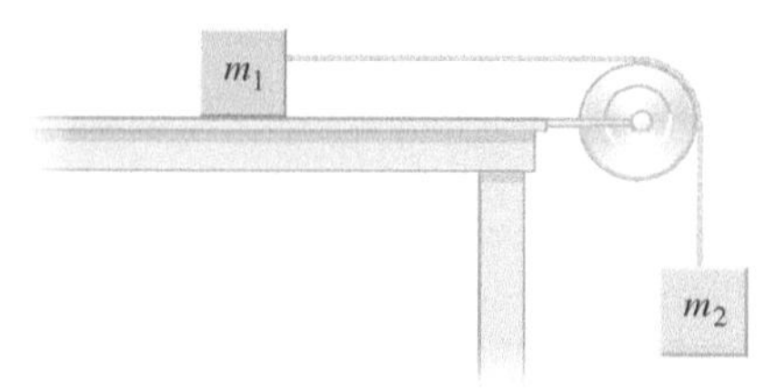

FIGURE P5.51 Problems 51 and 65.

52. A runaway piano starts from rest and slides down a 20.0° frictionless incline 5.00 m in length.
a. G Draw a free-body diagram of the piano.
b. N What is the acceleration of the piano?
c. N What is the speed of the piano at the bottom of the incline?

5-9 Newton's Third Law

53. C Does the ground need to exert a force on you for you to jump off the ground, or do you need to exert a force on the ground? If the ground must exert a force on you, is that force greater than the force you exert on the ground?

54. C A boxer breaks his hand by punching another boxer's jaw. How is it possible to break your hand when you exert such a force?

55. Paul places a toy block of mass m on a hockey puck of mass M and sets the two in motion at constant velocity on the surface of a frozen lake.
a. G Draw a free-body diagram of the block, identifying each of the forces acting on the block.
b. G Draw a free-body diagram of the hockey puck, identifying each of the forces acting on the puck.
c. C Which are the action and reaction pairs of forces in the block–puck–frozen lake system?

56. C A textbook rests on a movable wooden plank that is initially parallel to the ground. **a.** How does the normal force on the book compare to the gravitational force on the book as it rests in the horizontal position? **b.** If you push down on the book, what happens to the magnitude of the normal force as it rests in the horizontal position? **c.** The normal force on the book is part of a third-law interaction pair. Describe the third-law partner of this normal force.

57. An astronaut is stationary while floating in space, a short distance from the safety of her spacecraft. The mass of the astronaut including all her gear is 106.4 kg. Her tether to the craft becomes disconnected, and she needs to get back before she runs out of air. She removes her backpack unit, which has a mass of 30.1 kg and, putting herself between the backpack and the craft, pushes the pack with a force of 212 N directly away from the craft.
a. C Explain how this action returns the astronaut to the craft.
b. N What is the magnitude of the acceleration experienced by the backpack while the force is applied?
c. N What is the magnitude of the acceleration experienced by the astronaut while the force is applied?

58. C A talking horse is attached to a wagon. The driver says, "Giddy-up, boy. Let's go!" The horse says, "I can't. When I pull on the wagon, the wagon pulls back on me with a force of the same magnitude and in the opposite direction. The two forces cancel, so I cannot accelerate the cart." The horse is very clever, but doesn't really understand Newton's laws. Explain where the horse has gone wrong.

59. C When you dive off a cliff into the ocean, the Earth's gravity pulls down on your body, and you accelerate downward. According to Newton's third law, your body's gravity exerts a force on the Earth. In which direction is this force? Why doesn't the Earth accelerate in this direction at 9.81 m/s^2 as a result?

60. N A worker is attempting to lift a 55.0-kg palette of bricks resting on the ground by means of a rope attached to a pulley. **a.** Before the worker pulls on the rope, what is the force exerted by the ground on the palette? **b.** The worker exerts a force of 295 N downward on his end of the rope. What is the force exerted by the ground on the palette? **c.** If the worker doubles the downward force, what is the force exerted by the ground on the palette?

61. C CASE STUDY Return to the dialogue in Concept Exercise 5.12 and review the train collision. The passenger train was stopped on the tracks when it was hit by the freight train. Imagine instead that both trains were moving and both accelerating toward each other. Assume the freight train's acceleration is greater than that of the passenger train. In this case, what can you say about the relative force each train exerts on the other when they collide?

5-10 Fundamental Forces

62. C A concept map is a visual representation of concepts and their connections. Create a concept map including the five specific forces listed in Table 5.1 and the four forces discussed in Section 5.10. Justify the placement of each force.

General Problems

63. N A 75.0-g arrow, fired at a speed of 110 m/s to the left, impacts a tree, which it penetrates to a depth of 12.5 cm before coming to a stop. Assuming the force of friction exerted by the tree is constant, what are the magnitude and direction of the friction force acting on the arrow?

64. C An airplane is moving to the south with a speed of 125 m/s while maintaining a constant altitude. It continues moving in this direction at this speed for 15 seconds. During this time, what is the direction of the net force exerted on the airplane?

65. N A box with mass $m_1 = 6.00$ kg sliding on a rough table with a coefficient of kinetic friction of 0.220 is connected by a massless cord strung over a massless, frictionless pulley to a second box of mass $m_2 = 12.0$ kg hanging from the side of the table (Fig. P5.51). What is the tension in the cord connecting the boxes?

66. C One hundred and twenty-one forces act on an object that is at rest and stays at rest. If one of these forces is removed and stops acting on the object and all the others continue acting on the object with the same magnitudes and in the same directions, what would happen to the object?

67. N A cosmic ray muon with mass $m_\mu = 1.88 \times 10^{-28}$ kg impacting the Earth's atmosphere slows down in proportion to the amount of matter it passes through. One such particle, initially traveling at 2.50×10^8 m/s in a straight line, decreases in speed to 1.50×10^8 m/s over a distance of 1.20 km. **a.** What is the magnitude of the force experienced by the muon? **b.** How does this force compare to the weight of the muon?

68. N A boulder of mass 80.0 kg rests directly on a spring with a spring constant of 6850 N/m. **a.** What is the compression of the spring? **b.** Now, instead, the boulder and the spring are in an elevator accelerating upward at 5.19 m/s^2. What is the compression of the spring in this case?

69. C Only four forces act on an object. They all have the same magnitude. Their directions are north, south, northwest, and northeast. In which direction does the object accelerate?

70. N A 1.50-kg particle initially at rest and at the origin of an x–y coordinate system is subjected to a time-dependent force of $\vec{F}(t) = (3.00t\hat{\imath} - 6.00\hat{\jmath})$ N with t in seconds. **a.** At what time t will the particle's speed be 15.0 m/s? **b.** How far from the origin will the particle be when its velocity is 15.0 m/s? **c.** What is the particle's total displacement at this time?

Problems 71 and 72 are paired.

71. N A block of ice ($m = 15.0$ kg) with an attached rope is at rest on a frictionless surface. You pull the block with a horizontal force of 95.0 N for 1.54 s. **a.** Determine the magnitude of each force acting on the block of ice while you are pulling. **b.** With what speed is the ice moving after you are finished pulling?

72. Repeat Problem 71, but this time you pull on the block at an angle of 20.0°.

73. A Ezra is pulling a sled, filled with snow, by pulling on a rope attached to the sled. The rope makes an angle θ with respect to the horizontal ground, and the sled is being pulled at a constant speed. If the sled and snow have a total mass of m, the acceleration due to gravity is g, the magnitude of the normal force is F_N, and the coefficient of kinetic friction between the sled and the ground is μ_k, what is the angle θ that the rope makes with the ground in terms of these five quantities?

74. N Starting from rest, a rectangular toy block with mass 300 g slides in 1.30 s all the way across a table 1.20 m in length that Zak has tilted at an angle of 42.0° to the horizontal. **a.** What is the magnitude of the acceleration of the toy block? **b.** What is the coefficient of kinetic friction between the block and the table? **c.** What are the magnitude and direction of the friction force acting on the block? **d.** What is the speed of the block when it is at the end of the table, having slid a distance of 1.20 m?

75. When a 1.50-kg dress hangs midway from a taut clothesline stretched between two poles planted 7.50 m apart, the clothesline is seen to sag 0.0500 m.

a. G Draw a free-body diagram of the dress.

b. N What is the tension produced on the clothesline by the dress? Assume the clothesline is massless.

76. N Jamal and Dayo are lifting a large chest, weighing 207 lb, by using the two rope handles attached to either side. As they lift and hold it up so that it is motionless, each handle makes a different angle with respect to the vertical side of the chest (Fig. P5.76). If the angle between Jamal's handle and the vertical side is 25.0° and the angle between Dayo's handle and the vertical side of the chest is 30.0°, what are the tensions in each handle?

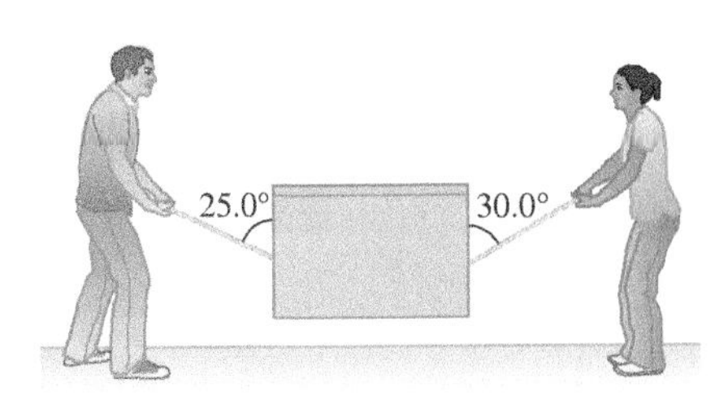

FIGURE P5.76

77. N A heavy chandelier with mass 125 kg is hung by chains in equilibrium from the ceiling of a concert hall as shown in Figure P5.77, with $\theta_1 = 37.0°$ and $\theta_2 = 64.0°$. Assuming the chains are massless, what are the tensions F_{T1}, F_{T2}, and F_{T3} in the three chains?

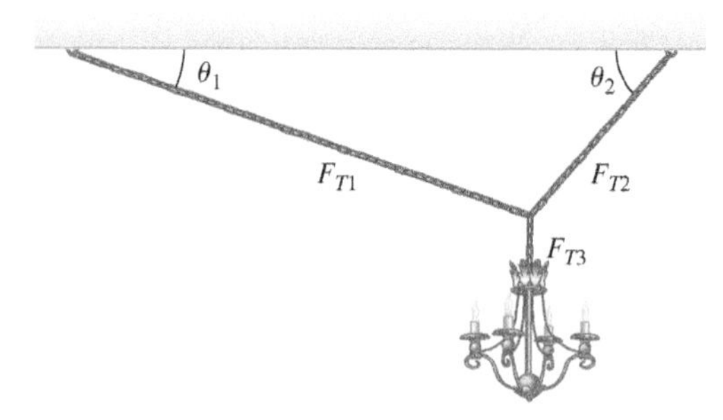

FIGURE P5.77

78. Two children, Raffi and John, sitting on sleds tied together with a massless rope, are being dragged across a frozen river by their playful Siberian husky Rex who is supplying a 112-N horizontal force (Fig. P5.78). The coefficient of friction between the sleds and the ice is 0.08, the combined mass of Raffi and his sled is 42.0 kg, and the combined mass of John and his sled is 51.0 kg.

a. G Draw a free-body diagram for each of the child–sled systems.
b. N What is the acceleration of the system?
c. N What is the tension F_T in the rope connecting the two sleds?

FIGURE P5.78

79. Two boxes with masses $m_1 = 4.00$ kg and $m_2 = 10.0$ kg are attached by a massless cord passing over a frictionless pulley as shown in Figure P5.79. The incline is frictionless, and $\theta = 30.0°$.

m_2 m_1 θ

FIGURE P5.79

a. G Draw a free-body diagram for each of the boxes.
b. N What is the magnitude of the acceleration of the boxes?
c. N What is the tension in the cord connecting the boxes?
d. N What is the speed of each of the boxes 3.00 s after the system is released from rest?

80. Two blocks of mass $m_1 = 1.50$ kg and $m_2 = 5.00$ kg are connected by a massless cord passing over a frictionless pulley as shown in Figure P5.80, with $\theta = 35.0°$. The coefficient of kinetic friction between block 1 and the horizontal surface is 0.400, and the coefficient of kinetic friction between block 2 and the inclined surface is 0.330.

a. G Draw a free-body diagram for each of the two blocks.
b. N What is the acceleration of the system when it is released from rest?
c. N What is the tension in the cord connecting the blocks?

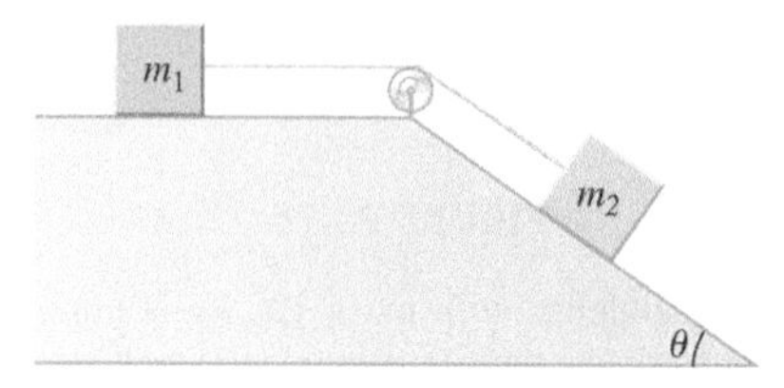

FIGURE P5.80

81. N An aerial demonstration aircraft dives at an angle θ, accelerating from 95.0 m/s to 145 m/s in 7.00 s (Fig. P5.81). The pilot has hung a 0.150-kg keepsake at the end of a string from the hinge on the aircraft's canopy. During the dive, the string of the keepsake is seen to remain perpendicular to the top of the canopy. **a.** What is the angle θ of the aircraft's dive? **b.** What is the tension in the string?

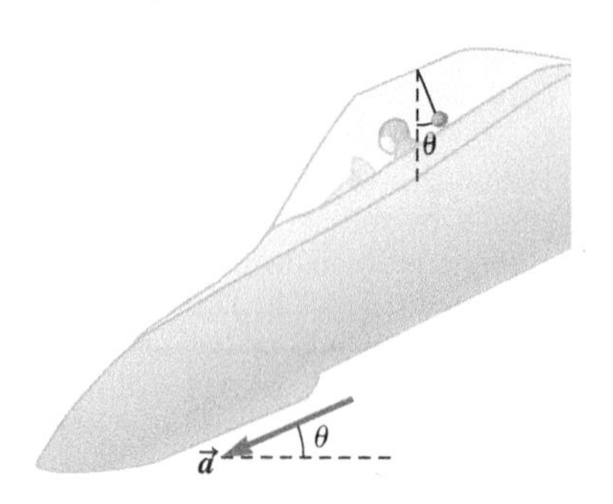

FIGURE P5.81

82. N A painter sits on a scaffold that is connected to a rope passing over a pulley. The other end of the rope rests in the hands of the painter who wants to lift the scaffold. She plans to pull downward on the loose end of the rope, thinking that the scaffold will then rise vertically with her along for the ride. The scaffold has a mass of 52 kg, and her mass is 63 kg. The painter pulls downward on the rope with a force of 600.0 N, while she and the scaffold are hanging from the other end above the ground. **a.** What is the net acceleration on the system consisting of the painter and the scaffold? **b.** What is the magnitude of the normal force exerted on the painter by the scaffold?

83. N Three crates with masses $m_1 = 5.45$ kg, $m_2 = 7.88$ kg, and $m_3 = 4.89$ kg are in contact on a frictionless surface. A horizontal force $F = 205$ N is applied to the third crate as shown in Figure P5.83. **a.** What is the magnitude of the contact force between crates 1 and 2? **b.** What is the magnitude of the contact force between crates 2 and 3?

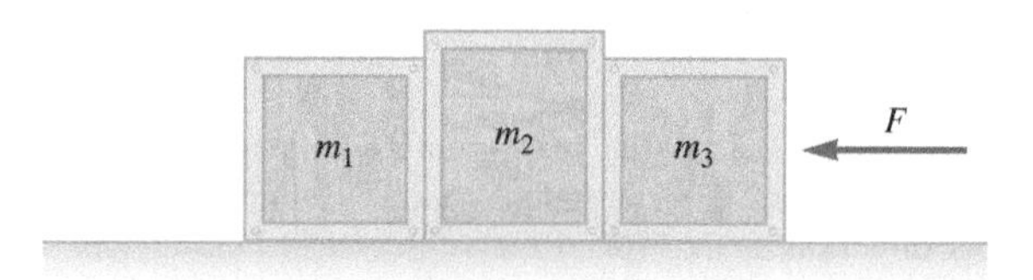

FIGURE P5.83

84. A A small block with mass m is set on the top of an upside-down hemispherical bowl. If the coefficient of static friction between the block and the bowl is μ_s and the block is slowly repositioned at different points down the surface of the bowl, at what angle measured from the vertical will the block begin to slide? Write your answer in terms of the mass, m; the gravitational acceleration on Earth, g; and the coefficient of static friction, μ_s.

Applications of Newton's Laws of Motion

6

Underlying Principle

Newton's laws of motion

Major Concepts

No new major concepts

Special Cases

1. Static friction (revisited)
2. Kinetic friction (revisited)
3. Rolling friction
4. Drag force
5. Terminal speed
6. Centripetal force
7. Nonuniform circular motion

Key Questions

How are Newton's laws of motion applied to complicated situations?

How do we handle friction and air resistance in real situations?

6-1 Newton's laws in a messy world 153

6-2 Friction and the normal force revisited 155

6-3 A model for static friction 156

6-4 Kinetic and rolling friction 160

6-5 Drag and terminal speed 163

6-6 Centripetal force 168

Newton's laws of motion help us understand many different situations. We can use them to figure out how skydivers fly in formation, to set a speed limit on a curved road, and to pick out a good pair of running shoes. Newton's three laws are at the heart of classical mechanics, and we will continue to use their power throughout the next three parts of this book.

In this chapter, we introduce no new principles or major concepts. Instead, we digest, refine, and synthesize the previously introduced principles as we practice applying them to practical situations.

6-1 Newton's Laws in a Messy World

Kinematics enables us to describe motion, and Newton's laws govern the *dynamics* (causes) of motion. In this chapter, we (1) practice using Newton's laws, (2) study a few new forces, and (3) weave kinematics and dynamics together. As a particularly important example of this last point, we revisit circular motion.

In this chapter, we tackle more complicated, realistic, and interesting problems that involve resistive forces such as friction and drag. Part of the reason we focus on resistive forces is that our intuition is based on their presence in our daily lives. We always need to reconcile our intuition with the laws of physics. The following CASE STUDY will help you sharpen your intuition. Later, you can compare that intuition with what the laws of physics say about the situation.

CASE STUDY Skydiving in Formation

A group of college students is working on physics homework in the study lounge. An inspirational poster hangs on the wall. It declares "teamwork" under a picture of skydivers, their parachutes not yet deployed, holding hands to form a large circle (Fig. 6.1). Avi notices the picture.

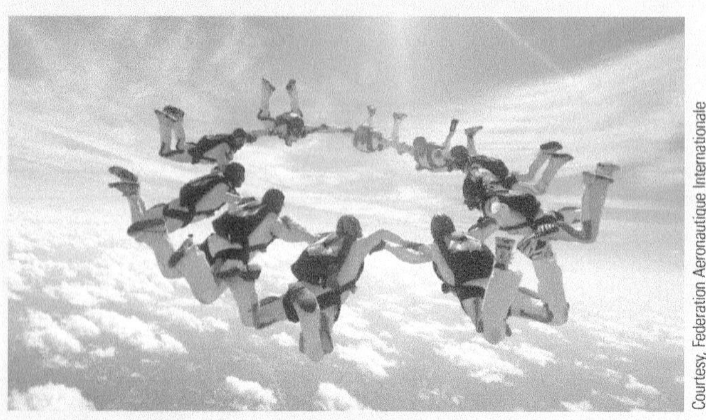

Courtesy, Federation Aeronautique Internationale

FIGURE 6.1 A poster shows skydivers flying in a formation like the one shown here. Avi wonders how this formation is possible.

Avi: Check out that picture. I don't get it. There are about 10 people holding hands. Did they all jump out of the plane together?

Shannon: I don't see how 10 people can jump out of a plane at the same time. Why do you ask?

Avi: Well, if they didn't jump out together, they can't all be together now. Think about it. Their parachutes are closed, which means that they are all in free fall. They each start with the initial velocity of the plane, and then they leave the plane. After that, only gravity acts on them. Each skydiver travels in her or his own parabolic trajectory (Fig. 6.2A). How can they possibly catch up to one another? They should each fall separately and land on the ground in the same order they jumped out of the plane.

Cameron: The problem is that what we learn in class has almost nothing to do with what happens in the real world. Physics works only in ideal conditions, like in our imagination or maybe in a perfect lab. The real world is just too messy for physics. The prof even said that real problems have to be solved on a computer.

Shannon: Okay. I see what Avi means. If all the skydivers are in free fall and if they left the plane separately, they cannot get back together while they are in the air. But, Cameron, just because we study physics under ideal conditions doesn't mean that everything outside of class happens by magic. Physics still holds. We just have to figure out what else is going on that makes a situation more complicated, like maybe the plane was flying at a downward angle instead of flying horizontally (Fig. 6.2B). That way, the second skydiver does not leave the plane until the door is at the level of the first skydiver, who's in free fall. The plane then continues at a slant so that it is always at the same level as the first and second skydivers, and then the third skydiver exits, and so on.

Avi: Shannon, I think you hit on something when you mentioned free fall, but I think it's the opposite. The divers are *not* in free fall. Instead, the air exerts a force on them. Remember that we learned that the range of a projectile is shorter if we take air into account.

Cameron: Okay. I see what you're both saying. The problem is harder than the stuff we did in class where the plane flies level and we ignore air, but the reason we do it that way is because the real world is too hard to study with physics. There is no way to figure it all out. All we do in physics is pretend that everything is a particle, even with two huge trains that are stuck together. Maybe not everything can be modeled as a particle. People jumping out of airplanes are not particles, so maybe we cannot pretend that they are.

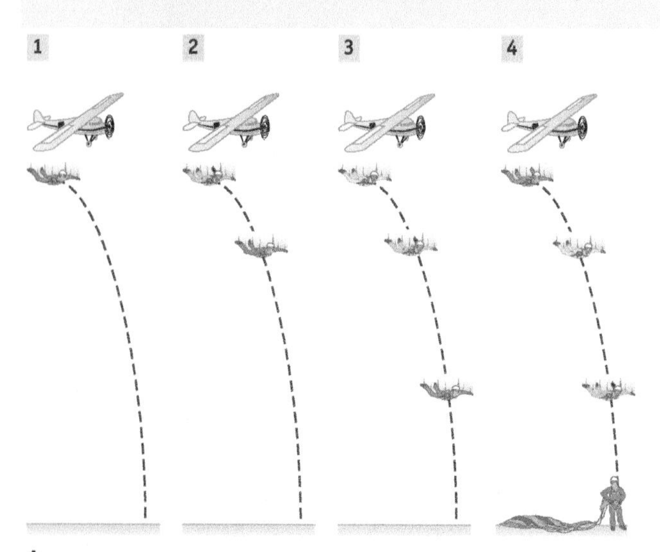

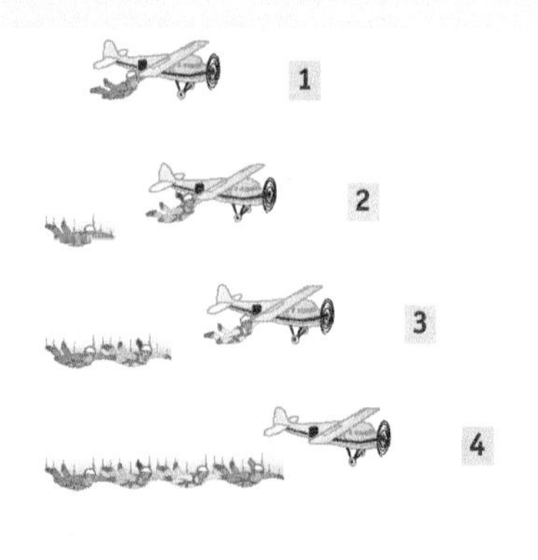

FIGURE 6.2 A. Avi explains that each skydiver falls along a separate parabolic path. **1** Only one skydiver has left the plane. **4** The last skydiver has left the plane, and the first one is already on the ground. Avi argues the skydivers never are close enough to hold hands in a circle. **B.** Shannon's explanation of how the skydivers can hold hands is that the plane loses altitude as it flies, remaining close to the skydivers as they exit the plane.

CONCEPT EXERCISE 6.1

CASE STUDY **Skydiving Arguments**

Take a moment to think about each student's argument.

a. Avi argues that the air exerts a force on the skydivers. Does that argument seem possible? If so, how could air help the skydivers create the formation we see in Figure 6.1? As you formulate your answers to these questions, think about this key question: Why do skydivers use parachutes in the first place?

b. According to Shannon's suggestion, the plane's altitude is constantly decreasing so that it is always alongside the divers. Does this situation mean that the plane must be accelerating? If so, what is that acceleration, and what sensation would a person on the plane experience? If not, what can you say about the motion of the plane?

c. Cameron argues that the physics we study in the classroom is good only for ideal conditions and that the world is more complicated than any situation we see in class. Cameron says people jumping out of airplanes are not particles and cannot be treated like particles. Do you agree or disagree? Explain your answer. It may be helpful to return to Chapter 2 to see how a particle was defined.

6-2 Friction and the Normal Force Revisited

Whenever two surfaces rub against each other, a friction force acts on both of them. Friction can be reduced but never truly eliminated. Our intuition about the nature of motion, an intuition that tells us that a force is required to keep things moving, is based on the presence of friction in most real-world situations. Early researchers studying motion believed that a force or cause was required to keep an object moving, even at constant velocity, because they often observed moving objects slowing down and then stopping. The presence of kinetic friction probably delayed the discovery of Newton's law of inertia. Like Galileo and Newton, we often try to imagine a world without friction. We develop theories based on this ideal condition, and then we test these theories in laboratory experiments designed to minimize friction.

It might sound like friction is some ugly fact of our Universe that we wish would just go away. That is far from the truth, however, for our world would be very weird if friction simply vanished. Without friction, we could not walk, drive a car, pet a dog, or strike a match.

In Figure 6.3A, a person does not support the box from below, but instead presses it against a wall. If there were no friction, the free-body diagram for the box would look like Figure 6.3B. The person pushes with $\vec{F}_p$, and the wall exerts a normal force $\vec{F}_N$. The box does not accelerate in the x direction, so by Newton's second law these two forces have equal magnitudes. The Earth's gravity pulls the box downward, and in the absence of friction, the box would consequently accelerate along the negative y direction. The box is *not* accelerating, however. Therefore, another force must also be acting on the box, one with the same magnitude as the gravitational force but in the positive y direction. It is the static friction force (Fig. 6.3C).

The wall is the source of the normal force and of static friction, so the wall actually exerts a single force up and to the left (Fig. 6.3D). For convenience, we break this force into two components. The component perpendicular to the wall is called the normal force, and the component parallel to the wall is called friction (in this case, static friction). Because the wall really exerts only one force on the box, you may think that as one component increases, the other would, too. In fact, that is true. Friction (the parallel force) is proportional to the normal force.

We observe the large-scale, or macroscopic, effect of the stationary box against the wall, but the normal force and friction underlying this macroscopic observation are due to many microscopic, or small-scale, interactions. These interactions are the electromagnetic force exerted by the molecules in each surface. The number of interactions depends on how much contact the two surfaces have with each other. Even

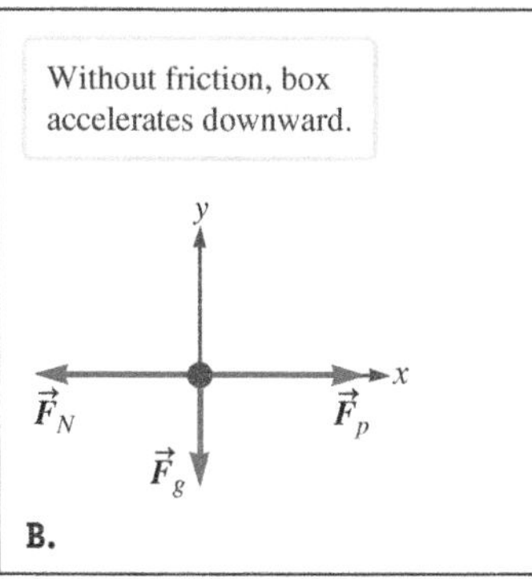

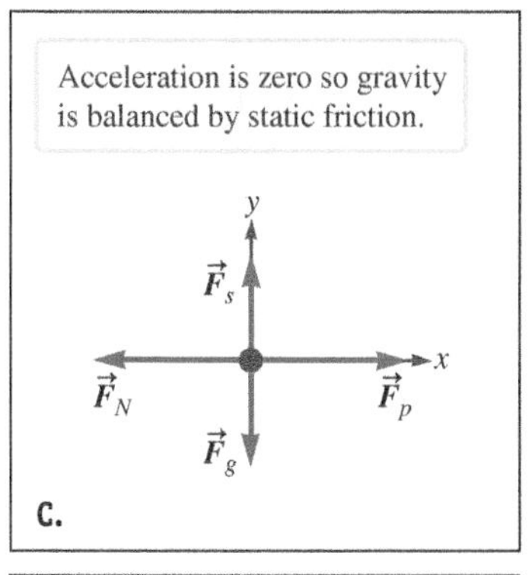

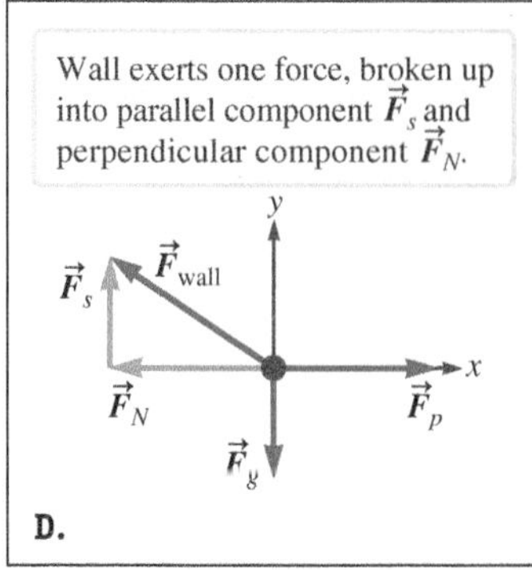

FIGURE 6.3 **A.** A person holds a box against a wall. **B.** If there were no friction, the box would accelerate downward. **C.** The box does not accelerate because static friction supports the box against gravity. **D.** The wall actually exerts one force that is broken into its parallel and perpendicular components for convenience.

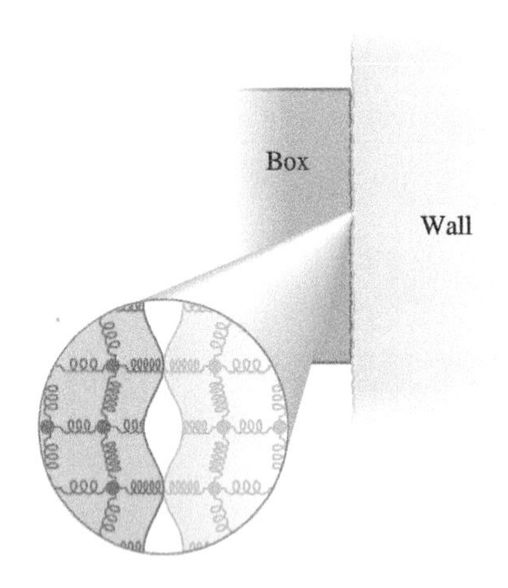

FIGURE 6.4 A close-up of the wall and box show that they touch each other at certain points. In between, there are pockets of no contact. Molecular bonds form at the points of contact. We model these bonds as stiff springs.

surfaces that seem smooth on the macroscopic scale are jagged on the microscopic scale. So, whenever two surfaces come into contact with each other, the surfaces touch only at certain points. As a result, there are, in addition to the contact points, numerous pockets of no contact (Fig. 6.4). So, the *real* contact area is smaller than the *apparent* contact area. The real contact area depends on the normal force. A large normal force means a large real contact area, which means a strong frictional force.

It might seem counterintuitive, but polishing two surfaces increases the magnitude of the friction force between them because polishing increases their real contact area. For example, imagine two highly polished pieces of metal that are kept in a vacuum chamber to make sure that they stay clean. When they are brought into contact with each other, there are very few pockets; consequently, a very high proportion of their surfaces are in contact. It is therefore possible for these two metals to bond, or *cold-weld*, together. We can think of cold-welding as an extreme case of friction. Further, if you want to reduce friction between two highly polished surfaces, it is a good idea to sprinkle the surfaces with small particles such as powder so that their real contact area is reduced.

How can we possibly expect to measure and calculate the electromagnetic forces between all the molecules that interact on the microscopic scale? There are two answers. (1) Usually, we don't. Instead of making all the necessary calculations on the microscopic scale, we run experiments on the macroscopic scale and measure the effects of friction. (2) When we need to think more deeply about friction on the microscopic scale, we use computers to make the large number of required calculations. Of course, the computer calculations involve simulated data as opposed to observed laboratory data. The results of the computer simulations are a deep understanding of how friction works on the microscopic scale and a framework for the macroscopic scale.

Part of what the macroscopic laboratory experiments have shown is that the friction force acts differently depending on the relative motion of the two surfaces: whether they are at rest, sliding past each other, or one rolling across the other. One goal in this chapter is to deepen our understanding of friction in all three of these cases. To help, we use a model of the microscopic interaction between surfaces. When two surfaces come into contact with each other, molecules in one surface form electromagnetic bonds with molecules in the other surface. Through each bond, the surfaces exert a force on one another. So, the more bonds formed between the two surfaces, the stronger the force. The bonds form wherever the surfaces touch each other (Fig. 6.4). As in Chapter 5, we model these microscopic bonds by stiff springs.

Modeling, which is used throughout this book, is an important technique used in physics to understand complex systems. Models are used in other fields as well. For example, a globe is a model for the Earth. Whenever you use a model to understand a complex system, you must keep in mind that the model is simpler than the real systems and so has certain limitations. For example, a globe is usually a smooth spherical model of the Earth. You may find it useful to consult a globe to compare the relative sizes of India and China, but you would not consult a globe to compare the relative climates of the two countries. With this caveat in mind, we model friction.

6-3 A Model for Static Friction

Let's begin with two surfaces at rest with respect to each other such as the box and wall (Fig. 6.3A). When an object is at rest with respect to a surface with which it is in contact, the surface may exert a force on the object that is parallel to the surface. This parallel force is **static friction**. If the object is not accelerating, the static friction force is balanced by the vector sum of any other forces that are applied to the object.

To learn more about static friction, let us combine the results of a macroscopic experiment with a microscopic picture (Fig. 6.5). (This experiment is similar to Example 5.9, so we use the same coordinate systems here.) At first, a box rests on a steel surface, with no string or hook attached. A free-body diagram for the box shows just two forces: gravity $\vec{F}_g$ due to the Earth and the normal force $\vec{F}_N$ due to the surface (Fig. 6.5A). On the microscopic scale, we imagine that the normal force is due to molecular bonds in the steel surface that act like very stiff springs (Section 5-7). The box compresses these bonds slightly, and, like springs, they exert an upward force on the box (Fig. 6.5A, close-up).

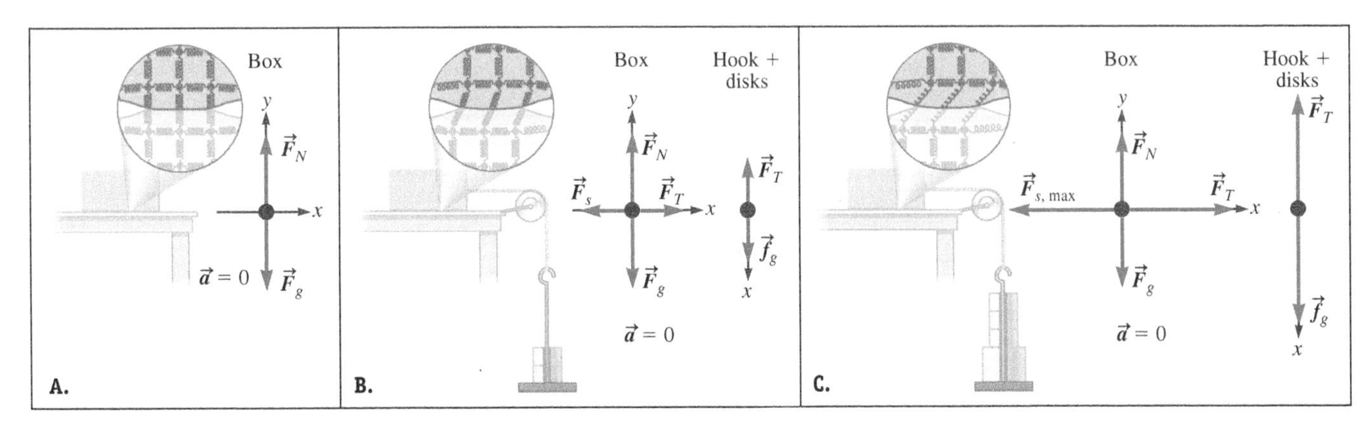

FIGURE 6.5 In this experimental setup, lead disks added to a hook attached by a pulley to a box on a steel surface increase the tension applied to the box. As long as the box does not accelerate, static friction balances this tension. **A.** When two surfaces are in contact, bonds form. These bonds are modeled as springs. **B.** As a force is applied to one of the objects, these bonds are stretched. Like stretched springs, the bonds apply a restoring force. **C.** When the applied force is increased to $\vec{F}_{s,\,max}$, these bonds are stretched to their limit. An increase in the applied force snaps the bonds.

Next, we attach a string and hook to the box. First, we put only a few lead disks on the hook so that the box remains at rest. There are now four forces acting on the box: gravity, the normal force, a tension force $\vec{F}_T$, and static friction force $\vec{F}_s$, (Fig. 6.5B). The normal force and gravity are the same as before (Fig. 6.5A). Because the box is not accelerating, we know from Newton's second law that $F_s = F_T$ (and $F_N = F_g$).

We also draw a free-body diagram for the system made up of the hook and lead disks. There are two forces acting on this system: gravity $\vec{f}_g$ and a tension force. Because the system is not accelerating, its weight must be equal to the tension in the string. So, by adding more disks to the hook, we can increase the tension force exerted on the box.

We find that we can add more lead disks to the hook without accelerating the box. The only way to do so is if the force of static friction F_s increases as the tension F_T increases. Our microscopic model of molecular bonds forming between molecules in the box surface and molecules in the table surface shows how this situation is possible. When there is tension force on the box, the molecular bonds are skewed (Fig. 6.5B, close-up). The bonds pull back (that is, leftward) on the box just like stretched springs would. This pulling by the microscopic bonds is responsible for the static friction force we observe on the macroscopic scale. A stronger tension force exerted on the box means that the bonds stretch farther and, according to Hooke's law, pull back with a greater force. On the macroscopic scale, this increase is described as being an increase in static friction.

Our experience tells us that there must be a limit to static friction. As we add more lead disks, the tension increases and so does static friction. With enough disks added, the box starts to slide. Just before sliding begins, static friction is at its maximum value $F_{s,\,max}$ (Fig. 6.5C). On the microscopic level, this maximum friction level corresponds to the molecular bonds being stretched as far as they can go without breaking (Fig. 6.5C, close-up). Just a little more tension in the string breaks these bonds, and then the box slides.

We reason that the maximum force of static friction must be related to the number of molecular bonds that form between the two surfaces. We already know that the number of bonds between the surfaces depends on how hard the surfaces press on each other (i.e., on the magnitude of the normal force they exert on each other). Mathematically the maximum force of static friction is proportional to the normal force exerted by the two surfaces on each other:

$$F_{s,\,max} = \mu_s F_N \tag{6.1}$$

The constant of proportionality μ_s is known as the **coefficient of static friction** and depends on the composition and smoothness of the two surfaces. Highly polished metal on metal, for instance, will have more points of contact than wood on wood, so the coefficient of static friction is higher for metal on metal.

STATIC FRICTION Special Case

Notice that Equation 6.1 is not a vector equation. *Static friction is always parallel to the two contacting surfaces, whereas the normal force is perpendicular to the surfaces.* For Equation 6.1 to be a vector equation, these two forces would need to be in the same direction. We infer the direction of the static friction force from a free-body diagram.

The coefficient of static friction is a positive, unitless scalar that is usually determined by experiments. Some values for μ_s are found in Table 6.1.

TABLE 6.1 Coefficients of static, kinetic, and rolling friction.

Surfaces	Coefficient of Static Friction μ_s	Coefficient of Kinetic Friction μ_k	Coefficient of Rolling Friction μ_r
Aluminum on steel	0.56 ± 0.08	0.47	
Brass on steel	0.45 ± 0.07	0.42 ± 0.03	
Copper on cast iron	1.08 ± 0.03	0.30 ± 0.01	
Copper on steel	0.53	0.36	
Glass on glass	0.92 ± 0.02	0.4	
Ice on ice	0.10	0.03	
Rubber on dry concrete	1.0	0.8	0.01–0.02
Rubber on wet concrete	0.4 ± 0.2	0.3 ± 0.1	
Steel on steel (dry)	0.72 ± 0.06	0.58 ± 0.02	0.001–0.002
Steel on steel (lubricated)	0.10	0.05	
Synovial joints in humans	0.01	0.006 ± 0.005	
Teflon on Teflon	0.04	0.04	
Teflon on steel	0.04	0.04	
Wood on wood	0.4 ± 0.1	0.2	
Wood on snow	0.11 ± 0.01	0.06	
Waxed ski on snow	0.1	0.04 ± 0.01	

These coefficients are found experimentally. Where possible, an estimate of the margin of error is given along with the actual value. These values are reasonable for classroom use, but if your work requires more precision, you should do your own literature search or experimentation.

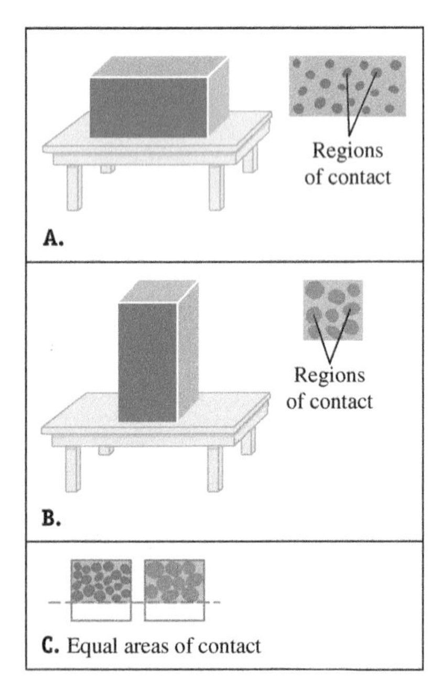

FIGURE 6.6 A. When the large side touches the table, there are many small regions of contact. **B.** When the small side touches the table, there are fewer, but larger areas of contact. **C.** Total effective areas of contact are equal.

Experiments have also shown that the magnitude of the *apparent* contact area between the two surfaces does not significantly affect the magnitude of static friction force because the *effective* contact area depends only on the normal force. For example, when the largest side of a rectangular block rests on a table, there are many small regions of contact between the block and the table (Fig. 6.6A). If the block is rotated so that one of its smallest sides rests on the table, there are fewer, but larger regions of contact (Fig. 6.6B). The total effective contact areas (and the normal force) in the two cases are equal (Fig. 6.6C), so friction between the block and the table is the same, too.

We saw that before the static friction force acting on the box in Figure 6.5 reaches its maximum value, the magnitude of that force is equal to the tension in the string. Notice that if no force parallel to the contacting surfaces acts on the box, there is no static friction force acting on the box either (Fig. 6.5A). So, you might guess that the magnitude of the static friction force is equal to the vector sum of all the other forces applied parallel to the contacting surfaces. Although that statement is often true, it is *not* always the case. For example, imagine that our box is at rest on the bed of a flatbed truck (Fig. 6.7A). When the truck is at rest, the only forces acting on the box are gravity and the normal force due to the truck bed. Now imagine that the truck accelerates to the right as shown. If the box remains at rest with respect to the truck, the acceleration of the box relative to the ground must be the same as the acceleration of the truck relative to the ground (Fig. 6.7B). We know from Newton's first law that there must be a force responsible for the box's acceleration, and here that force is static friction (Fig. 6.7C). In this case, there are no other forces acting either parallel or antiparallel to the

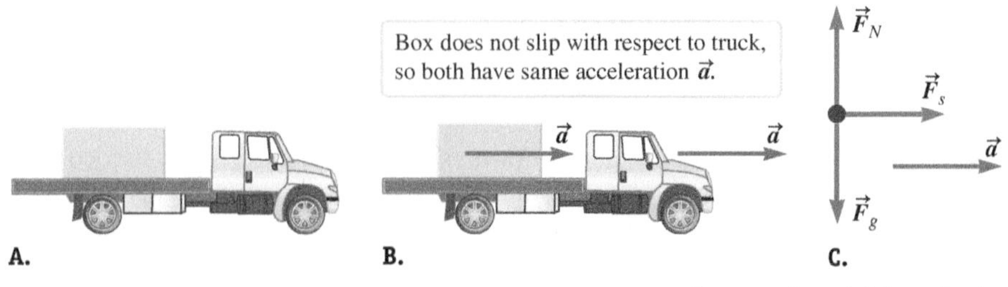

FIGURE 6.7 A. A box rests on a flatbed truck. **B.** When the truck accelerates, the box remains at rest with respect to the truck. So, the box must have the same acceleration as the truck with respect to the ground. **C.** A free-body diagram for the box shows that static friction must accelerate the box.

bed surface; thus, it would be incorrect to describe the friction force as being a force opposing the vector sum of other forces acting along the bed's surface.

CONCEPT EXERCISE 6.2

A box rests on a steel surface. Four sides of the box are made of aluminum, and the other two sides are coated with Teflon. Each panel in Figure 6.8 shows a pair of experiments. In each pair, decide which experiment requires the larger tension force to move the box and therefore has the higher maximum static friction force.

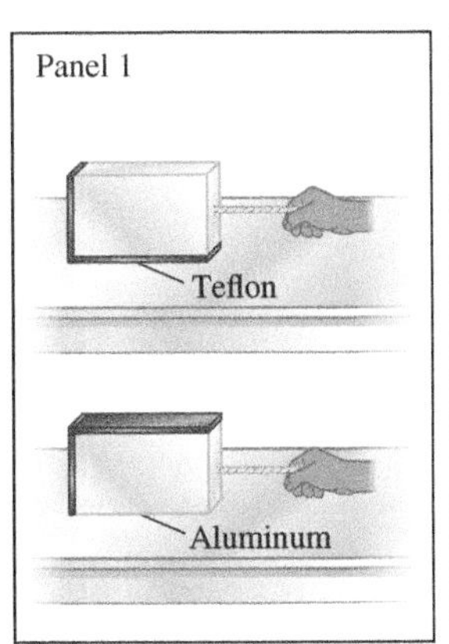

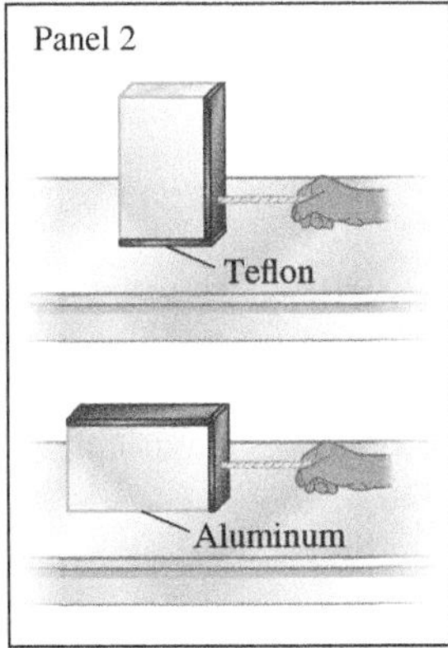

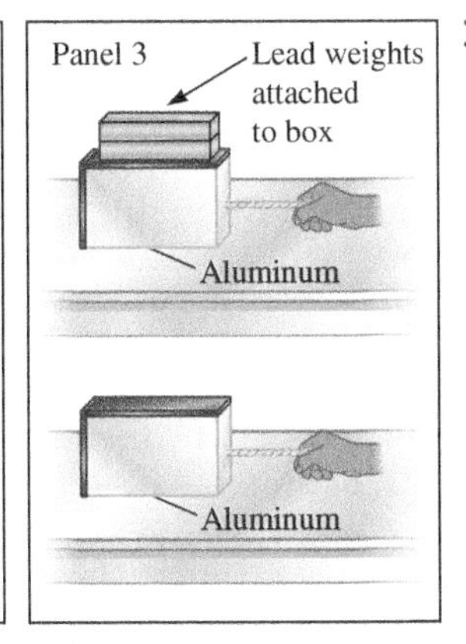

FIGURE 6.8

CONCEPT EXERCISE 6.3

For a particular object and a particular tabletop, the maximum value possible for the magnitude of the static friction force exerted on the object is 15 N. The object and table are at rest, and a free-body diagram for this object is shown in Figure 6.9.

a. If the tension force on the object is zero, what is F_s?
b. If the tension on the object is $F_T = 5$ N, what is F_s?
c. If the tension is increased to $F_T = 15$ N, what is F_s?
d. What happens if the tension is increased to $F_T = 25$ N?

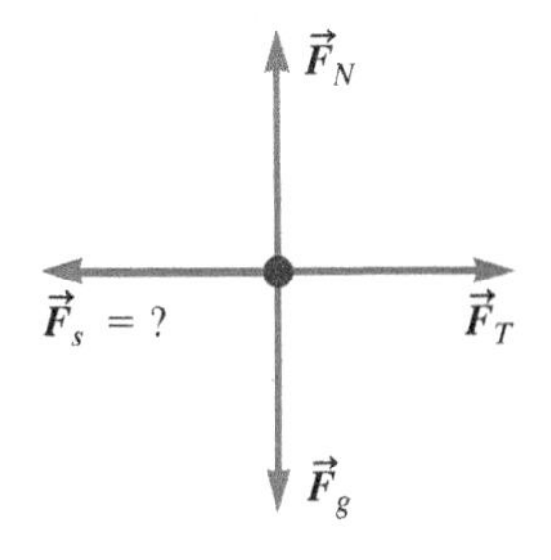

FIGURE 6.9

EXAMPLE 6.1 Testing Running Shoes

The coefficients of friction are usually found experimentally. In this example, we explore an experiment that requires only a protractor to find the coefficient of static friction between two surfaces.

Lisa, a runner, has purchased new shoes and wishes to measure the coefficient of static friction between her shoes and the running track. A higher coefficient of static friction means a faster run (see Concept Exercise 6.7). The coach has extra tiles of track and has given her a piece for her experiment.

She places a shoe on the tile and then slowly tilts the tile, measuring the angle θ at which the shoe just begins to move (Fig. 6.10).

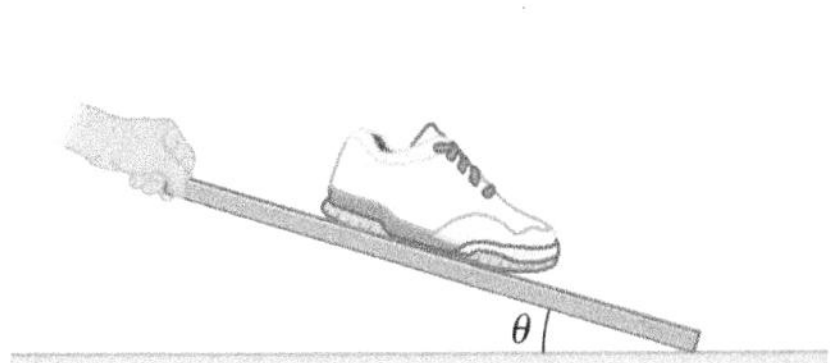

FIGURE 6.10

A Find an expression she can use to determine μ_s in terms of θ.

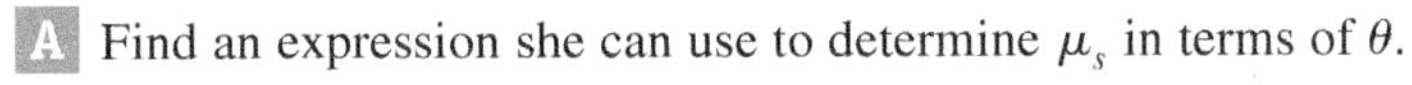

INTERPRET and ANTICIPATE

Start by drawing a free-body diagram for the shoe. It is often best to choose a coordinate system that has one axis parallel to the incline (Fig. 6.11). We are interested in the moment just before the shoe slips, at which time the acceleration is zero.

Our answer for μ_s should be unitless and algebraic. We expect that the coefficient of static friction depends on the tilt angle. Imagine if Lisa used Velcro between her shoe and the tile; she probably could stand the tile straight up without the shoe slipping.

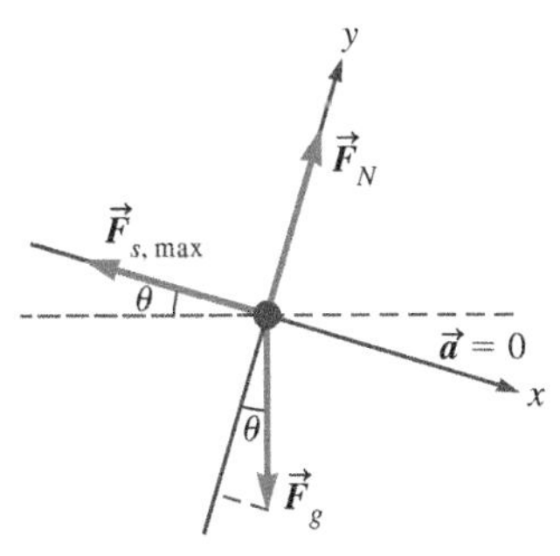

FIGURE 6.11

Example continues on page 160 ▶

SOLVE

Next, apply Newton's second law with the acceleration equal to zero. The gravitational force must be broken into components parallel and perpendicular to the chosen x axis.

$$\sum F_x = F_g \sin\theta - F_{s,\,\max} = 0 \quad (1)$$

$$\sum F_y = F_N - F_g \cos\theta = 0 \quad (2)$$

Rewrite $F_{s,\,\max}$ using $F_{s,\,\max} = \mu_s F_N$ (Eq. 6.1) and substitute into Equation (1).

$$F_g \sin\theta - \mu_s F_N = 0 \quad (3)$$

Solve Equations (2) and (3) for μ_s simultaneously. Often, when two equations involve sine and cosine functions, it is helpful to divide one equation by the other.

$$\mu_s F_N = F_g \sin\theta$$

$$F_N = F_g \cos\theta$$

$$\frac{\mu_s F_N}{F_N} = \frac{F_g \sin\theta}{F_g \cos\theta} \qquad \mu_s = \frac{\sin\theta}{\cos\theta}$$

$$\mu_s = \tan\theta$$

CHECK and THINK

Our answer is in the form we expected; as θ increases from 0 to 90°, μ_s increases. Thus, a protractor is the only tool needed to find the coefficient of static friction between two materials.

B If Lisa finds that $\theta = 52°$ at the instant the shoe begins to slip, what is μ_s for her shoes on the track?

We just need to substitute Lisa's measurement into our solution.

$$\mu_s = \tan\theta = \tan 52° = 1.3$$

CHECK and THINK

This value is a bit higher than μ_s for rubber on dry concrete (Table 6.1), which seems about right.

6-4 Kinetic and Rolling Friction

Let us now move to the kinetic friction between two objects, once again imagining a box on a surface (Fig. 6.12). Once the molecular bonds between the two surfaces are broken and the box begins to move, static friction no longer plays a role, but there is still friction between the two surfaces. Whenever an object is in contact with some surface and slides with respect to that surface, kinetic friction works to stop that motion. The direction of kinetic friction is always parallel to the surface and in the direction opposite the direction of the motion with respect to the surface (Fig. 6.12B).

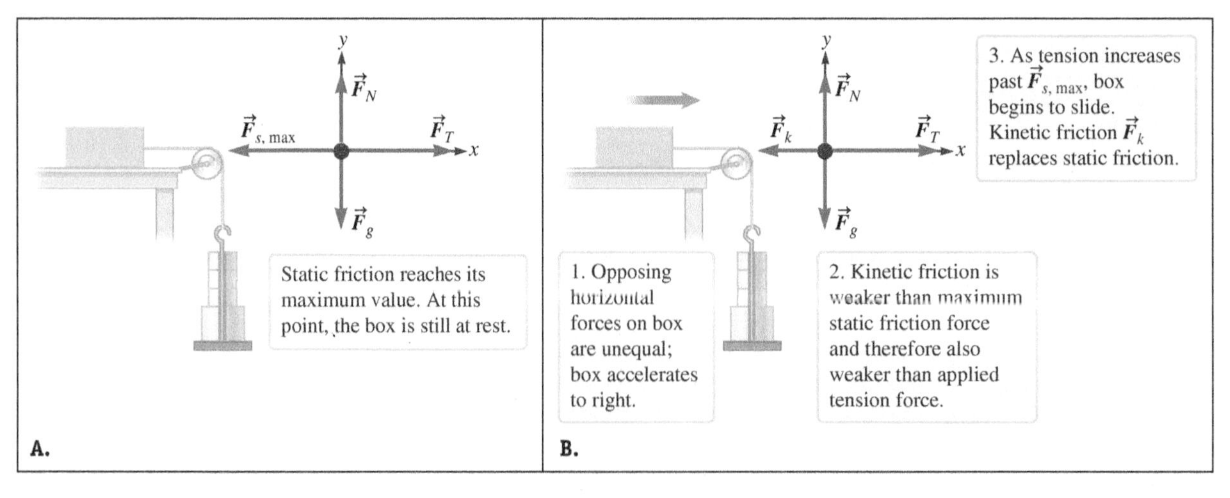

FIGURE 6.12 Experimental setup corresponding to Figure 6.5, but with increased force on the box. **A.** As long as the box does not accelerate, static friction balances the tension in the string. **B.** Once tension exceeds the maximum static friction force, kinetic friction takes over, and the box accelerates to the right.

Model for Kinetic Friction

To learn more about kinetic friction, let us return to our microscopic model. As the sliding surfaces rub against each other, points of contact form weak bonds momentarily. Individually, these bonds are relatively easy to break, but added together they are responsible for kinetic friction on the macroscopic scale. Because it is easier to break the bonds once the surfaces are in motion, the kinetic friction between any two surfaces is always weaker than the maximum value of the static friction force between the surfaces.

Like static friction, kinetic friction must be related to the number of molecular bonds that form between the two surfaces, which depends, in turn, on the composition of the surfaces and on the magnitude of the normal force they exert on each other. According to Equation 5.9, the magnitude of kinetic friction is proportional to the normal force acting on the object:

$$F_k = \mu_k F_N \quad (6.2)$$

KINETIC FRICTION Special Case

where the constant of proportionality μ_k is known as the coefficient of kinetic friction. Like the coefficient of static friction, μ_k is a unitless scalar that depends on the composition of both surfaces and is found experimentally (Table 6.1). Experiments have shown that the speed of an object experiencing kinetic friction does not significantly affect the magnitude of this force. In fact, unlike static friction, kinetic friction is relatively constant.

Like Equation 6.1, Equation 6.2 is *not* a vector equation because $\vec{F}_N$ is perpendicular to the surface and $\vec{F}_k$ is parallel to the surface and in the direction opposite the motion. Because kinetic friction is weaker than the maximum value of static friction, we expect $\mu_s > \mu_k$ for each pair of surfaces in Table 6.1.

Figure 6.12 shows two free-body diagrams for the box as the tension increases. The box remains at rest as long as the static friction force balances the tension in the string. When a tiny bit more weight on the hook increases the tension such that $F_T > F_{s,\max}$, however, the box begins to move (Fig. 6.12B). At that moment, static friction is replaced by kinetic friction. Because $F_k < F_{s,\max}$, we know that $F_k < F_T$. So, the box is accelerated to the right:

$$\sum F_x = ma_x = F_T - F_k$$

$$a_x = \frac{F_T - F_k}{m} > 0$$

Because both F_T and F_k are constant, the acceleration of the box is constant. This information is helpful because it means that we can use the constant-acceleration kinematics equations (Table 2.4).

CONCEPT EXERCISE 6.4

Imagine trying to push a heavy sofa across the room. At first you have difficulty getting the sofa to budge, but once it starts to move your task becomes easier.

a. Does the sofa accelerate for a moment after you get it moving?

b. Use what you now know about static and kinetic friction to explain what is going on. Why won't the sofa move at first? Why do you have to apply a stronger and stronger force to get it to move? Why is the force you must apply to keep the sofa moving less than the force you applied just at the instant motion began?

Sometimes when one surface slides over another, a few strong molecular bonds form at the points of contact. When that happens, there are two possible results. One is that the bonds break almost as soon as they are formed. In this case, the moving object stops only momentarily and then begins sliding again. When this making and breaking of bonds happens repeatedly, the motion is jerky. This jerky motion can lead to squeaky, screeching sounds, such as when a car stops suddenly.

The other possibility is that the strong molecular bonds do not break. Instead, weaker bonds break so that some bits of one surface are left embedded in the other surface, such as tire marks left on the road (Fig. 6.13). Wear and tear on the tire surfaces causes tires to go bald, and roads eventually need to be resurfaced.

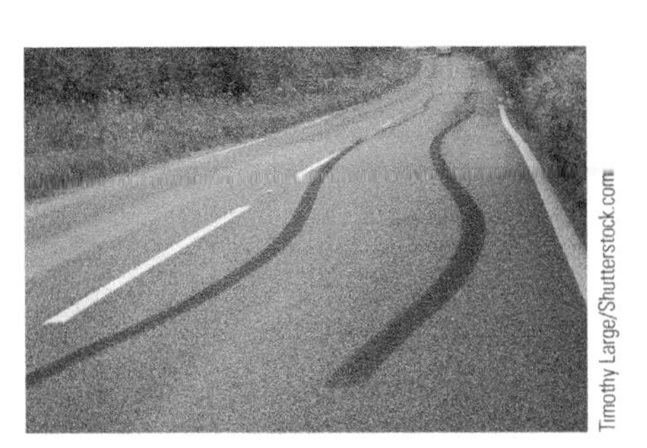

FIGURE 6.13 When the wheels locked and slid without rolling, small pieces of the car tires became embedded in the road, leaving skid marks.

CONCEPT EXERCISE 6.5

Why is it important to lubricate such things as car engines, bicycle chains, and pocket knives?

EXAMPLE 6.2 Fun in the Snow

Children are sledding on a hill as illustrated in Figure 6.14. Zak gives Sophia a push to start her off with a speed of 1.7 m/s at a vertical height of 4.6 m above the bottom of the hill. They are using a low-friction sled, and we will assume this sled sliding on snow is like waxed skis on snow. Find the velocity of the sled at the bottom of the hill.

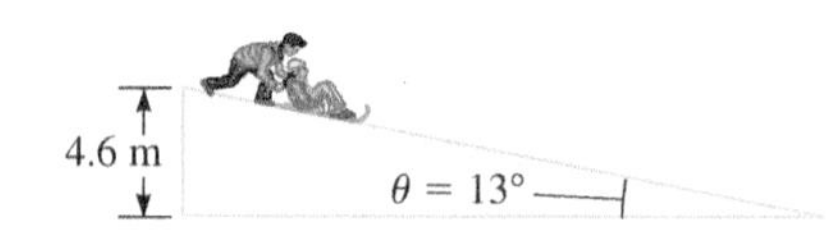

FIGURE 6.14

INTERPRET and ANTICIPATE

This problem requires the synthesis of dynamics and kinematics. We need dynamics to find the acceleration of the sled and kinematics to find the final velocity. We will use the initial velocity of the sled imparted by Zak when we work on the kinematics. Our problem begins just after Zak has lost contact with Sophia.

Start with a free-body diagram for the sled, which has three forces acting on it: gravity, the normal force, and kinetic friction (Fig. 6.15). The acceleration is downhill. We have decided to use an x axis that points in the same direction as the acceleration, as is often best. With this coordinate system our answer should be in the form $\vec{v} =$ ____$\hat{\imath}$ m/s.

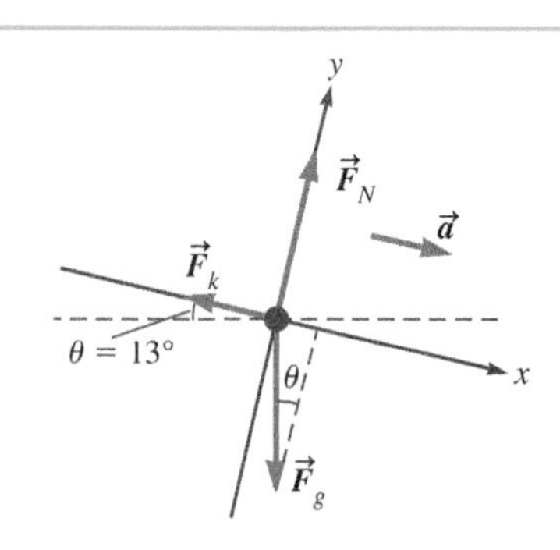

FIGURE 6.15

SOLVE

To find the acceleration, use Newton's second law.	$\sum F_x = F_g \sin\theta - F_k = ma_x$ $\sum F_y = F_N - F_g \cos\theta = 0$
Substitute $F_k = \mu_k F_N$ (Eq. 6.2) for F_k and mg for F_g.	$mg\sin\theta - \mu_k F_N = ma_x$ $F_N - mg\cos\theta = 0$
Solve these two equations simultaneously for a_x, eliminating F_N. Notice that mass m cancels out.	$ma_x = mg\sin\theta - \mu_k mg\cos\theta$ $a_x = g(\sin\theta - \mu_k\cos\theta)$
Substitute values. From Figure 6.14, $\theta = 13°$. Look up μ_k for waxed skies on snow from Table 6.1.	$a_x = 9.81\text{ m/s}^2(\sin 13° - 0.04\cos 13°)$ $a_x = 1.82\text{ m/s}^2$

CHECK and THINK

As a quick check, it makes sense that the sled's acceleration is less than the free-fall acceleration $g = 9.81\text{ m/s}^2$. Also, we expect the acceleration to be less than it would be if there were no kinetic friction (which would be $a_x = g\sin\theta = 2.2\text{ m/s}^2$), and it is.

SOLVE

Because the acceleration is constant, we can use Equation 2.9 to find the speed of the sled at the bottom of the hill. Apply trigonometry to Figure 6.15, and find $\Delta x = 4.6/\sin(13)$.

$$v_x^2 = v_{0x}^2 + 2a_x\Delta x \quad (2.9)$$

$$v_x^2 = (1.7\text{ m/s})^2 + 2(1.82\text{ m/s}^2)\left(\frac{4.6\text{ m}}{\sin 13°}\right) = 77.3\text{ (m/s)}^2$$

$$v_x = \pm 8.8\text{m/s}$$

$$\vec{v} = 8.8\,\hat{\imath}\text{ m/s}$$

CHECK and THINK

It seems reasonable that the sled is going about five times faster at the bottom of the hill than at the top. This speed is roughly 20 mph, however, which is probably a bit faster than a typical sled. Notice that our answer did not depend on the mass of Sophia or that of the sled.

Rolling Friction

It is probably obvious that the invention of the wheel made life a lot easier. Like objects sliding on a horizontal surface, however, rolling ones are also subject to friction that causes them to slow down and stop.

The friction associated with the rolling motion of one object against a surface is called **rolling friction**, and it is weaker than both kinetic and static friction. That is why it is easier to move a cart with wheels than to move one without wheels.

On the microscopic level, rolling friction is similar to both static and kinetic friction. Consider a ball rolling along a horizontal floor. Where the ball is in contact with the floor, molecular bonds form and then immediately break as the ball rolls over the floor. New bonds form and break repeatedly. The number of bonds that forms depends on the real contact area.

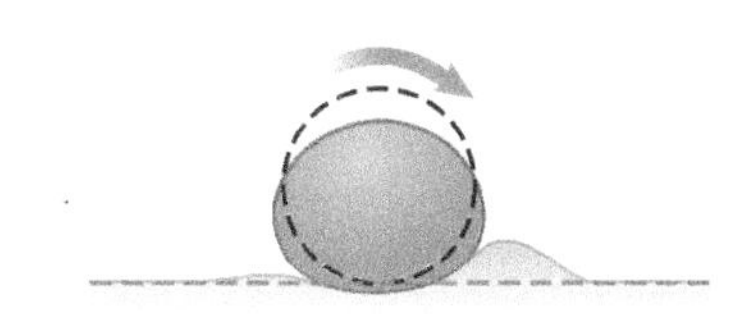

FIGURE 6.16 A ball rolling on a floor causes the ball to flatten and the floor to deform.

The portion of the ball in contact with the pavement at any instant is slightly flattened, and the portion of the floor in contact with the ball is slightly deformed as well (Fig. 6.16). The more the ball and floor deform, the greater their real contact area. All other things being equal, a greater normal force causes a greater deformation. So, rolling friction depends on the normal force between the two surfaces:

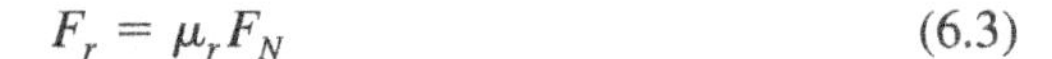

$$F_r = \mu_r F_N \quad (6.3)$$

ROLLING FRICTION  Special Case

Like the other two coefficients of friction, the **coefficient of rolling friction** μ_r is a unitless scalar that is found experimentally (Table 6.1). Like kinetic friction, the direction of the rolling friction force is opposite the direction of motion with respect to the surface. Because the object moves with respect to the surface, the term ***moving friction*** is use to include both kinetic and rolling friction, but not static friction.

CONCEPT EXERCISE 6.6

A car is on an incline at an angle θ to the horizontal (Fig. 6.17). Clearly indicating particular frictional forces, draw a free-body diagram (no need to indicate $\vec{a}$) for the following:

a. The car while parked on the incline
b. The car rolling downhill
c. The car sliding downhill when the surface is icy (so the car's wheels are not rolling)

FIGURE 6.17

CONCEPT EXERCISE 6.7

What forces act on you as you walk across a room? Draw a free-body diagram showing all of them. Which force or forces propel you forward? Why is it more difficult to walk on a slippery surface than on a nonslippery one? Explain how you use Newton's third law to control your motion.

6-5 Drag and Terminal Speed

When an object moves in a *fluid medium* such as air or water, the medium exerts a resistive **drag force** on the object. If you parachute out of an airplane, the drag force exerted by the air on the chute keeps you from free falling to the Earth. Moving friction and drag are called *resistive forces* because they work to resist the motion of an object relative to the source (the surface or medium).

DRAG FORCE Special Case

Like friction, the drag force is a macroscopic effect generated by many microscopic interactions. Drag results from many collisions that take place between molecules in the medium and the object. When an object is moving with respect to a medium, there are more collisions to the object's front side than to the object's back side, and the object tends to slow down. The drag force on an object depends on the density (mass per unit volume) of the medium and also on the cross-sectional area of the object perpendicular to the direction of motion.

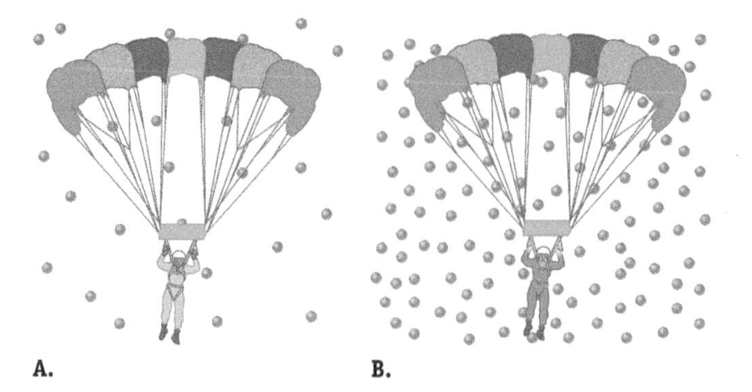

FIGURE 6.18 **A.** In a diffuse (low-density) medium, the drag force on the parachute is lower because relatively few molecules of the medium collide with the chute. **B.** In a denser medium, the drag force is greater because here there are more molecules of the medium colliding with the parachute. (Differences in air density have been exaggerated.)

To illustrate density dependence, let us look at two skydivers who have opened their parachutes at different altitudes (Fig. 6.18). The density of air decreases with increasing altitude. So, the skydiver at the higher altitude is surrounded by lower-density air. The skydiver in the more diffuse medium (Fig. 6.18A) experiences fewer collisions with molecules in the medium than the skydiver in the denser medium (Fig. 6.18B). As a result of the higher number of collisions, the skydiver in the denser medium has a larger drag force acting on her.

Now consider how drag force depends on the area of the object. The cross-sectional area of importance is the area perpendicular to the velocity. We can use shadow projections to show how the size of this cross-sectional area is related to drag force magnitude. Imagine that the Sun is directly above a plane when a jump takes place so that its rays are perpendicular to the ground below the skydiver. Figure 6.19 shows that the area of the shadow of a skydiver falling without an open parachute is smaller than the area of a skydiver falling with an open parachute. The latter skydiver experiences a larger drag force because the number of air molecules his parachute encounters is greater than the number encountered by the skydiver who has not yet opened her chute.

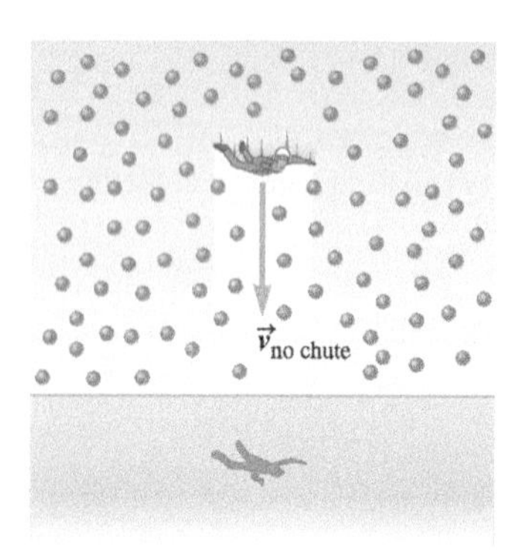

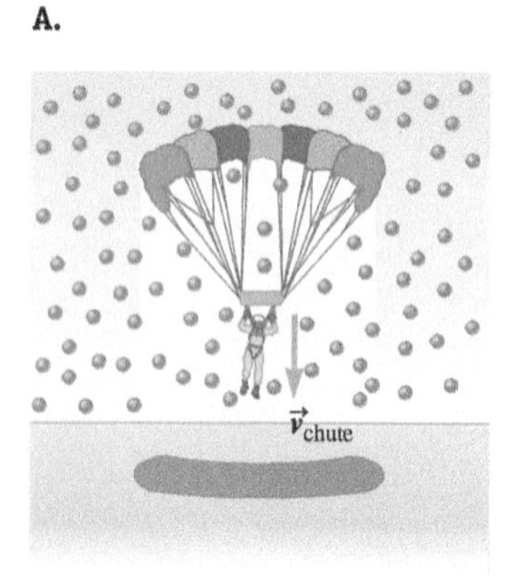

FIGURE 6.19 The cross-sectional area of an object is the area of its shadow projected on a plane that is perpendicular to its velocity. **A.** A skydiver has not yet opened her parachute. Her cross-sectional area is small. **B.** A skydiver has opened his parachute. The larger cross-sectional area of his parachute means that he (and his chute) experience a greater drag force than the other skydiver.

CONCEPT EXERCISE 6.8

Figure 6.20 shows four objects moving downward. Find the cross-sectional area of each object for the surface perpendicular to the direction of its motion.

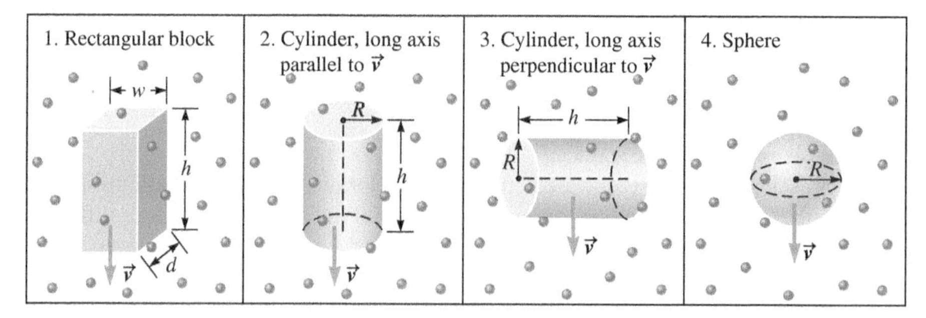

FIGURE 6.20

We are building a mathematical description of drag. So far, we know that the magnitude of the drag force F_D acting on any object is proportional to the density of the medium ρ and the cross-sectional area A of the object:

$$F_D \propto \rho A$$

This magnitude also depends on the relative motion between the object and the medium because a faster relative motion means that each molecule–object collision is stronger. (Think of the difference between getting hit by a fast-moving baseball and a slow-moving one.)

Relative motion between the object and medium is necessary if an object is to experience a drag force, but the object may be at rest with respect to the ground. So, you do not have to jump out of an airplane to experience a drag force; instead, you can just sit on a park bench on a windy day.

The faster an object moves through a medium (or, in our park-bench case, the faster the medium moves past the object), the stronger the drag force on the object. Expressed as a proportionality, we have

$$F_D \propto v^n$$

The constant n is found experimentally; it depends on both the medium and the object. Experiments have shown that $n = 1$ both for small objects (dust particles in air, for instance) moving at low speeds and for large objects moving at typical speeds through liquids (such as a boat in water). In these cases, the drag force is given by

$$\vec{F}_D = -b\vec{v} \tag{6.4}$$

where b is a constant found experimentally with the dimensions of force per speed or mass per time. The negative sign indicates that the direction of the drag force is opposite the velocity's direction.

In this book, we are mainly interested in the drag force exerted on blunt objects moving through air at high speeds. For all such objects (airplanes, skydivers, baseballs, cars), $n = 2$.

One additional factor—the object's shape—affects the drag force on an object moving through air at high speed. For example, car designers who wish to reduce drag use wind tunnels to study the effect of shape on drag. These studies give rise to the tapered, smooth design of energy-efficient automobiles.

We write all these factors—object's shape, density of the medium ρ, object's perpendicular cross-sectional area A, and speed v— into one neat equation:

$$F_D = \tfrac{1}{2}C\rho A v^2 \tag{6.5}$$

The **drag coefficient** C is a unitless number that is found experimentally to be between 0.4 and 1.0. It depends primarily on the shape of the object.

The drag coefficient is not necessarily a constant, but we ignore such complications in this textbook.

Equation 6.5 is not a vector equation, so we cannot use it to determine the direction of the drag force. Instead, we simply state without proof that the direction of the drag force is opposite the direction of the relative velocity.

EXAMPLE 6.3 Ball Tossed

A baseball is tossed straight up (Fig. 6.21). Draw a free-body diagram for the ball at points A, B, and C.

INTERPRET and ANTICIPATE

Gravity always points to the center of the Earth, and drag points in the direction opposite the direction of the relative velocity. At A and B, the velocity is upward, so the drag force is downward.

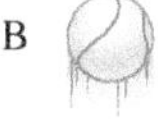

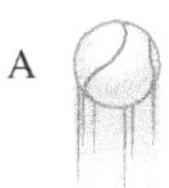

FIGURE 6.21

SOLVE

The force of gravity is the same at all three points, so we draw three arrows all of the same length (Fig. 6.22).

The drag force on the ball depends on the ball's speed. Because of the downward force of gravity, the ball's speed decreases as it moves from A to C. As a result, the drag force also decreases, so we draw our $\vec{F}_D$ arrow shorter in the free-body diagram at B than in the diagram at A.

At the top of the ball's trajectory, its speed is zero, so there is no drag on the ball at the top (C).

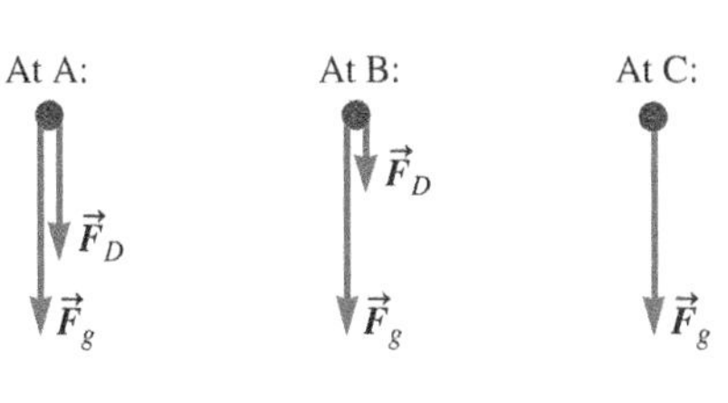

FIGURE 6.22

CHECK and THINK

Let us compare this ball's trajectory with the trajectory of a ball under the ideal free-fall condition (no drag). A ball in free fall without drag has only gravity acting on it. The time for the velocity to reach zero will be longer and the ball's height at the top will be greater if the ball is thrown in a vacuum than if it is thrown in a medium such as air.

TERMINAL SPEED ▶ Special Case

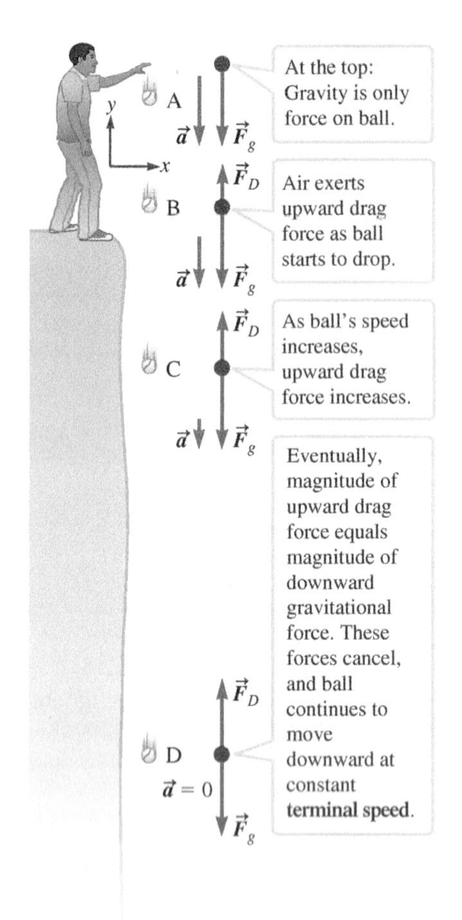

FIGURE 6.23 A ball falls from a great height, reaching terminal speed at position D.

Terminal Speed

When an object falls through the air, the Earth's gravity pulls the object downward, and at the same time the drag force due to the air pushes the object upward. Imagine standing at the edge of a very high cliff (Fig. 6.23). You drop a baseball and then watch as it falls to the base of the cliff far below. As the ball's speed increases, the upward drag force increases. Eventually, the magnitude of the upward drag force equals the ball's weight, and the ball stops accelerating. The ball continues to move downward at a constant speed known as the **terminal speed**. The terminal speed is the highest speed the ball reaches. (If the ball fell in a vacuum instead of through air, no drag force would act on it, and its speed would continue to increase until it hit the ground.)

We use Newton's second law to find an expression for the terminal speed:

$$\sum F_y = F_D - F_g = ma_y$$

At the terminal speed, the acceleration is zero (position D in Fig. 6.23), which means that this expression becomes

$$F_D - F_g = 0$$
$$F_D = F_g$$

Once we substitute for F_D from Equation 6.5 and rearrange to solve for the speed, we end up with

$$v_t = \sqrt{\frac{2mg}{C\rho A}} \tag{6.6}$$

Equation 6.6 confirms our intuition that heavier objects fall faster than light ones. Our intuition is based on observing objects moving not in a vacuum but rather through air. For example, imagine dropping a golf ball and a Ping-Pong ball off the Leaning Tower of Pisa. The two balls are nearly the same shape and size, and they fall through the same air, which means that the density of the medium is the same for both. The only factor that is different is their masses. According to Equation 6.6, the more massive golf ball will reach a higher terminal speed than the Ping-Pong ball and will therefore reach the ground more quickly. This result matches our experience, but what is hard to accept is what happens in a vacuum.

In a vacuum, the only force acting on either ball is gravity. Then, according to Newton's second law,

$$\sum F_y = -F_g = ma_y$$
$$-mg = ma_y$$

The two balls have the same acceleration $a_y = -g$ independent of mass so that if they are both in free fall in a vacuum, they land on the ground at the same time and are moving at the same speed.

CONCEPT EXERCISE 6.9

Do all objects falling through the air reach terminal speed? Explain.

CONCEPT EXERCISE 6.10

a. An object falls through an evacuated space. Which of the three graphs in Figure 6.24 best represents the object's acceleration as a function of time? Which of the three graphs in Figure 6.24 best represents its *speed* as a function of time?

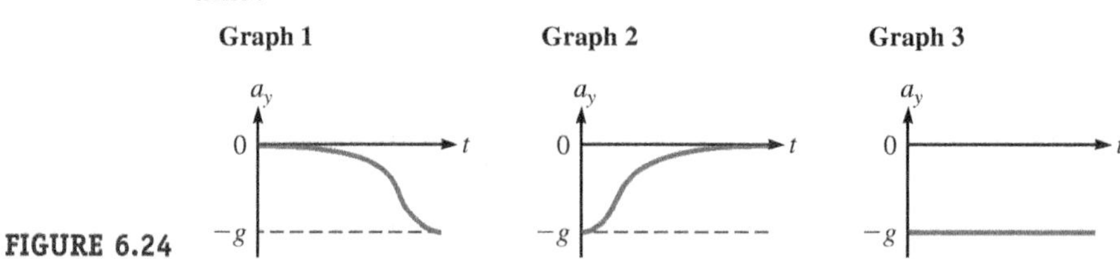

FIGURE 6.24

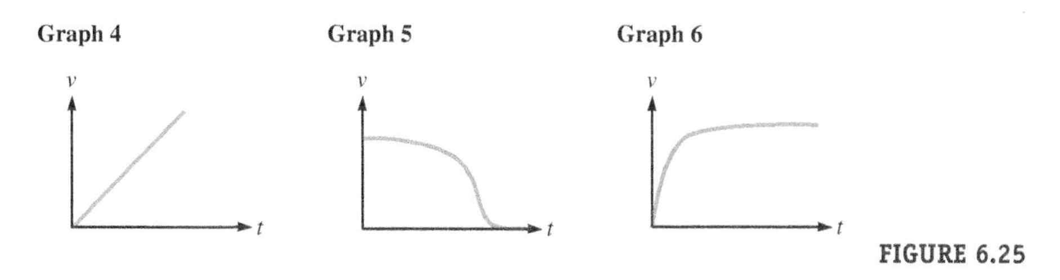

FIGURE 6.25

b. Repeat part (a) for an object that falls through air and at some point reaches its terminal speed.

EXAMPLE 6.4 CASE STUDY Data From a Skydiving Physicist

Cassidi Reese—the skydiver from Chapter 3—agreed to collect jump data for us so that we can study terminal speed. Her data will help us figure out the case study in this chapter. On her data-collecting jump, she varied her speed in the interval between about $t = 6$ s and $t = 65$ s as shown in Figure 6.26.

With her parachute closed, Reese was able to vary her speed by holding her body in different positions as she fell through the air. To increase her speed, she positioned her head downward and held her arms straight back by her sides. To slow down, she oriented herself belly-to-Earth, with her body stretched out and acting almost like a parachute.

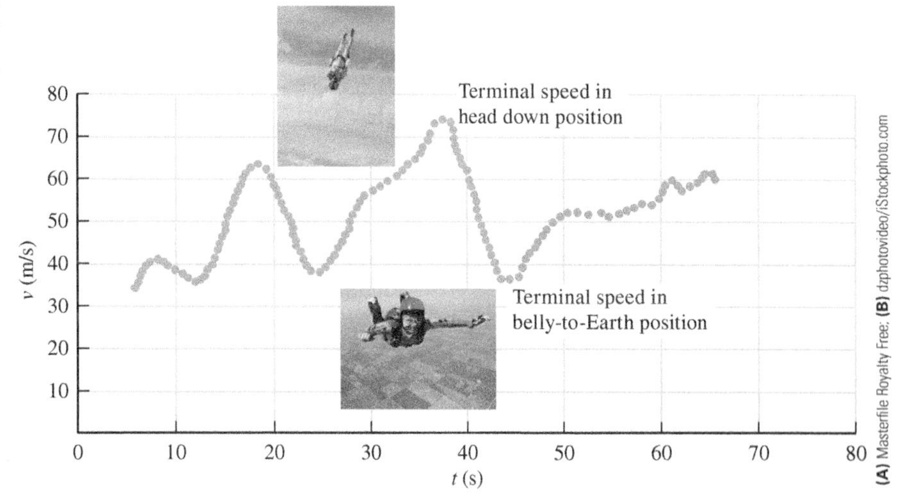

FIGURE 6.26 Reese's speed versus time during a jump in which she varied her body's orientation as shown.

A Explain how these two orientations can account for Reese's variations in speed.

With her head pointing downward and her arms at her side, Reese's cross-sectional area (which is equal to the area of her shadow cast on the ground with the Sun directly overhead) is as small as she can make it. In addition, this orientation is tapered and smooth, like a well-designed car. Therefore, this position minimizes both C and A in $v_t = \sqrt{2mg/C\rho A}$ (Eq. 6.6). Because both these factors appear in the denominator, minimizing them maximizes her terminal speed.

In the belly-to-Earth position, Reese's cross-sectional area is large, and her shape is almost like a parachute. In this position, she maximizes C and A, and her terminal speed is consequently minimized.

B Reese and her gear weighed 667 N on the day she collected data. The air density is 1.15 kg/m^3. Use her data to estimate her **effective cross-sectional area** CA at her highest and lowest speeds.

INTERPRET and ANTICIPATE

From our reasoning in part A, we expect that the effective cross-sectional area is largest when Reese is at a lowest terminal speed.

SOLVE

Our first step is to solve Equation 6.6 for CA because this term is the one we are after.

$$v_t = \sqrt{\frac{2mg}{C\rho A}} \quad (6.6)$$

$$CA = \frac{2mg}{\rho v_t^2}$$

Example continues on page 168 ▶

Of the factors on the right in this equation, the only one we do not have a numerical value for is v_t. Because our task is to determine CA at two speeds, we read off Reese's highest and lowest speeds from her data graph.	$v_{t,\,max} \approx 73$ m/s $v_{t,\,min} \approx 36$ m/s
We substitute values.	$(CA)_{min} = \frac{2(667\text{ N})}{(1.15\text{ kg/m}^3)(73\text{ m/s})^2} = 0.22\text{ m}^2$ $(CA)_{max} = \frac{2(667\text{ N})}{(1.15\text{ kg/m}^3)(36\text{ m/s})^2} = 0.90\text{ m}^2$

CHECK and THINK

Reese can change her effective area by about a factor of four, which allows her to control her speed even before she opens her parachute.

EXAMPLE 6.5 CASE STUDY Flying in Formation

Return to the case study in Section 6-1. Explain how the skydivers in the poster (Fig. 6.1) ended up falling together.

Avi was on to the correct idea. If the skydivers were free-falling, they could not all catch up with one another in flight. Therefore, to all be at the same altitude at the same moment, they would have to step out of the plane together. Because of the drag force exerted by the air, they weren't in free-fall. The drag exerted by the air allows them to join up after exiting the plane individually.

We learned in Example 6.4 that a skydiver can control her speed by changing her effective area. A skydiving team could use this physical fact to join up in the air. Skydivers exiting the plane early on orient their bodies to have a large effective area CA and consequently a low terminal speed. The later skydivers then catch up by holding their bodies in a tapered orientation so as to have a high terminal speed.

A refinement of this technique is to order the skydivers by weight. Equation 6.6, $v_t = \sqrt{2mg/C\rho A}$ tells us that, for a given body orientation, the lighter skydivers have the lower terminal speeds and so should exit the plane before the heavier ones.

To take the survey scan or visit **www.cengage.com/community/katz**

Shannon's idea that the plane must fly at a downward slant is not necessary. The plane flies level while the divers jump out of it, one after the other.

Cameron believes that problems like this one are too complicated to be solved using physics alone. In a sense, that is correct. In many cases, our numerical results are based on parameters that usually are determined experimentally, such as the coefficients of friction μ_s, μ_k, and μ_r and the drag coefficient C. There is no analytical basis for giving exact values for these parameters. Once such values are found in a controlled environment, however, they can be used to analyze complicated problems outside the laboratory.

6-6 Centripetal Force

Let us now return to uniform circular motion to show the connection between the kinematics and dynamics of this special case of two-dimensional motion. From Chapter 4, a particle moving in uniform circular motion has a centripetal acceleration. This acceleration points to the center of the circle, and the particle's velocity is tangent to the circle. The velocity and acceleration vectors are perpendicular to each other, which means that the speed remains constant but the direction of the velocity changes.

The constant magnitude of the centripetal acceleration is $a_c = v^2/r$ (Eq. 4.38). Newton's second law tells us that there must be a net force on the object that is responsible for this acceleration. This net force points to the center of the circle and is known as the **centripetal force**. Its magnitude is given by

CENTRIPETAL FORCE Special Case

$$F_c = m\frac{v^2}{r} \tag{6.7}$$

If the origin of a polar coordinate system is at the center of the circle, the centripetal force is written as

$$\vec{F}_c = -m\frac{v^2}{r}\hat{r} \tag{6.8}$$

The centripetal force is *not* a *new* force. It is not *generated by* the circular motion of a particle; instead, it is a *requirement of* circular motion. Some physical force (or forces)—gravity, a spring force, the normal force, a tension force, static friction—must act on an object in uniform circular motion in such a way that the net force on the object is perpendicular to the velocity and points to the center of the circular path. Neither drag nor moving friction can generate a centripetal force because they are always directed opposite the velocity.

In the case of uniform circular motion, the net force is the centripetal force, which is always perpendicular to the velocity. So, imagine that the source of the centripetal force were suddenly removed such that there was no net force exerted on the object. Then, according to Newton's first law, the object would continue at the same speed but in a straight line tangent to the point where the object was when the force suddenly vanished.

PROBLEM-SOLVING STRATEGY

When Centripetal Force Is Present

Problems that involve centripetal force are no different from other problems that require us to apply Newton's second law. So, the strategy developed in Section 5-8 works here. Two modifications, however, may be helpful.

:• INTERPRET and ANTICIPATE

As always, indicate the direction of the object's acceleration. In the case of uniform circular motion, the acceleration is directed toward the center of the circle and is the centripetal acceleration. As in Chapter 5, it is often best to choose to align an axis with the acceleration.

Modification 1 Draw a free-body diagram for one particular instant and **circle the centripetal force(s)**. A free-body diagram should only include forces that are exerted by particular sources. The source of the centripetal force is due to one or more of these sources; there is no separate source of the centripetal force. So, when you draw a free-body diagram, do *not* draw a separate vector for the centripetal force. Instead, circle the force or forces that are parallel (or antiparallel) to the centripetal acceleration. In some cases, a force may have a component that is parallel to the centripetal acceleration and a component that is perpendicular. In that case, circle the force and indicate the parallel component with a short phrase.

:• SOLVE

Modification 2 The vector sum of the force or forces that you have circled is the centripetal force. So, when you apply Newton's second law, you can expect to set the **sum of the forces equal to the centripetal force** whose magnitude is given by $F_c = m(v^2/r)$ (Eq. 6.7).

CONCEPT EXERCISE 6.11

The following objects are moving in uniform circular motion. Draw a free-body diagram for each object and identify the force responsible for the centripetal acceleration.

Object 1. A person riding on the barrel-of-fun ride (Fig. 6.27, top)
Object 2. The lead object in the laboratory set-up (Fig. 6.27, center)
Object 3. A jogger running on a circular track (Fig. 6.27, bottom)

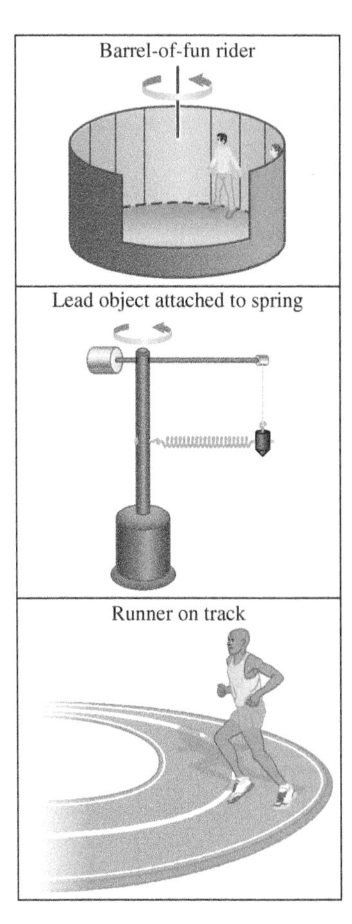

FIGURE 6.27

EXAMPLE 6.6 Particle Accelerator

Particle accelerators are used to study subatomic particles, fundamental building blocks of the Universe. In the type of accelerator known as a *cyclotron*, the particles are accelerated around a circular path, constrained to that path by a magnetic force. The first cyclotron particle accelerator was developed in the early 1930s by American physicists Ernest O. Lawrence and M. Stanley Livingston. In this example, we estimate the strength of the magnetic force in the original cyclotron.

For the purpose of our estimate, we assume a proton is moving in uniform circular motion with a speed of 3.9×10^6 m/s. The diameter of the first cyclotron was about 4.5 in. (1.1×10^{-1} m).

A Find the magnitude of the magnetic force required to keep a proton in uniform circular motion at the specified speed and radius.

INTERPRET and ANTICIPATE

It might seem intimidating to be asked to find a *magnetic* force when we know almost nothing about magnetism. What we need to keep in mind, though, is that this question is really asking us to find the magnitude of the centripetal force. We do not need to know anything about the agent responsible for the centripetal force. Without this force, the proton would travel in a straight line.

SOLVE

Look up the mass of the proton.	$m_p = 1.67 \times 10^{-27}$ kg	
We were given the cyclotron diameter, but Equation 6.7 is given in terms of the radius of the motion, which means that we need to calculate the radius and substitute.	$F = m_p \dfrac{v^2}{r}$	(6.7)
	$F = (1.67 \times 10^{-27}\text{ kg}) \dfrac{(3.9 \times 10^6\text{ m/s})^2}{5.5 \times 10^{-2}\text{ m}}$	
	$F = 4.6 \times 10^{-13}$ N	

CHECK and THINK

As we expect for any force when we work in SI units, our answer is in newtons.

B Using the same magnetic force, how could you design a cyclotron for a faster proton?

INTERPRET and ANTICIPATE

This question is asking us to examine the relationships among the parameters in $F_c = m(v^2/r)$ (Eq. 6.7). We must find an algebraic result and think about how changing one or more parameters increases the proton's speed.

SOLVE

Because we need to use the same magnetic force and the particle is a proton, both F and m_p are constants. The only other variable in Equation 6.7 that affects particle speed is r. Solve for r in terms of v to see how these two parameters are related.	$F = m_p \dfrac{v^2}{r}$	(6.7)
	$r = \left(\dfrac{m_p}{F}\right) v^2$	

CHECK and THINK

If F and m_p are constants, a faster proton travels in a larger radius, and a larger cyclotron is therefore required. To achieve great speeds, contemporary (synchrotron) particle accelerators use other techniques that are more efficient than the original cyclotrons, and still these accelerators are miles in diameter.

EXAMPLE 6.7 Flat Turn

An engineer is designing a road with a curve in it as in Figure 6.28. The curve is a circular arc of radius 43.3 m, and the road surface is to be concrete. The engineer must figure out the maximum speed limit for this part of the road. So, she must assume the road will be wet sometimes and a driver will attempt to maintain the maximum speed all along the road. Find the maximum speed limit under the wet-road condition.

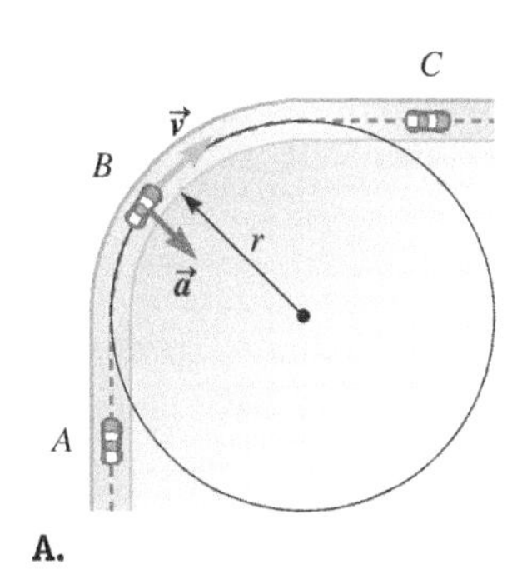

FIGURE 6.28

INTERPRET and ANTICIPATE

This example combines dynamics and kinematics. When the car is in the curve, it is in uniform circular motion. Its velocity is tangent to the curve, and its acceleration is directed toward the center of the circular arc. We need to identify the force or forces that are acting as the centripetal force and then use Equation 6.7, $F_c = m(v^2/r)$, to find the maximum speed at which the car can stay on the road and not fly off (nearly) tangent to the curve. This speed will be the speed limit posted for the bend.

Consider all the forces you know about. Identify which forces act on the car and then which of them may contribute to the centripetal force.

Spring force: There is no spring and so no spring force.
Tension force: There is no rope and therefore no tension force.
Gravity: Because the downward force of gravity has no component directed toward the center of this circle, it does not contribute to the centripetal force.
Normal force: This upward force has no component directed toward the center of the circle and so cannot contribute to the centripetal force.
Rolling friction: Rolling friction always opposes the motion, which in this case means that rolling friction is along the circular path in the counterclockwise direction. Because rolling friction is not directed toward the center, it cannot contribute to the centripetal force. The rolling friction between the tires and the road acts to slow down the car, but the engine compensates for it. Therefore, we ignore rolling friction in this example.
Drag force: The air acts to slow down the car, but the engine compensates for it, meaning that here is another force we can ignore.
Static friction: It may seem surprising, but static friction is the only force that can and does produce the required centripetal force. It may be helpful to think back to the situation with the box in the flatbed truck (Fig. 6.7), where the box was accelerating. We reasoned that because the box was at rest with respect to the flatbed truck, static friction caused the acceleration in the horizontal direction.

You might think that because the car is moving with respect to the road (unlike the box on the flatbed), there can be no static friction. The tires are rolling along the road, and we expect there to be rolling friction. How can static friction also act in this case?

Static friction is parallel to the road surface and is directed toward the center of the circle. Figure 6.28B shows the car at point B, well into the turn. In this figure, the car is heading into the page. The center of the circle is to the right, which is the direction of the centripetal acceleration. The tires do not roll or slip to the right (or left), and static friction is directed to the right.

We are ready to **draw a free-body** diagram for the car **when it is at point *B*** from the perspective of Figure 6.28B. The free-body diagram (Fig. 6.29) includes three forces; there is **no** separate vector drawn for the centripetal force. **Static friction has been circled** because it is parallel to the centripetal acceleration.

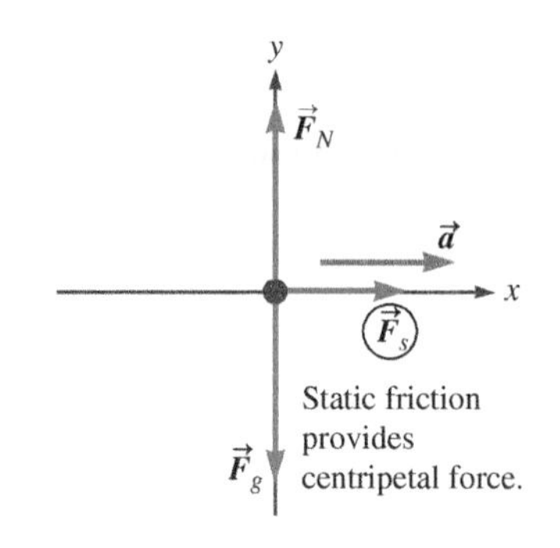

FIGURE 6.29

Example continues on page 172 ▶

SOLVE

Apply Newton's second law. The **total force in the *x* direction equals the centripetal force** $F_c = m(v^2/r)$, Equation 6.7. Because all three forms of friction are proportional to the normal force (Eqs. 6.1–6.3), it is often helpful to solve for F_N when friction is involved.

$$\sum F_x = ma_x$$

$$F_s = m\frac{v^2}{r}$$

$$\sum F_y = F_N - F_g = ma_y = 0$$

$$F_N = F_g = mg$$

To find the maximum speed, consider the maximum possible centripetal force. That is when static friction is at its maximum. Substitute $F_{s,\max} = \mu_s F_N$ (Eq. 6.1).

$$F_{s,\max} = m\frac{v_{\max}^2}{r}$$

$$\mu_s F_N = m\frac{v_{\max}^2}{r}$$

$$\mu_s mg = m\frac{v_{\max}^2}{r}$$

$$v_{\max} = \sqrt{\mu_s gr}$$

Consult Table 6.1 for μ_s for rubber on wet concrete and substitute values.

$$v_{\max} = \sqrt{\mu_s gr}$$

$$v_{\max} = \sqrt{(0.4)(9.81\ \text{m/s}^2)(43.3\ \text{m})}$$

$$v_{\max} = 13.0\,\text{m/s} \approx 29\,\text{mph}$$

CHECK and THINK

There are several important points to be learned from this result. (1) The maximum speed limit does not depend on the mass of the vehicle. There are no separate speed limits by weight class. (2) We found the maximum speed of a vehicle in this turn. If a vehicle is going slower than this speed, the static friction force required to maintain circular motion is smaller than its maximum, and the vehicle safely negotiates the turn. (3) If the car exceeds the speed limit, static friction cannot supply the required centripetal force. As a result, the car slips out of the turn. Kinetic friction takes over, but it is weaker than static friction. With the weaker force acting on it, the car cannot complete the turn and careens off the road. (4) If the road is not wet, $\mu_s = 1.0$, and the car can safely take the turn at $v_{\max} = \sqrt{(1.0)(9.81\ \text{m/s}^2)(43.3\ \text{m})} = 20.6\ \text{m/s} \approx 46$ mph. It makes sense that this speed is higher than the speed limit posted.

EXAMPLE 6.8 Banked Turn

An engineer needs to design the curve in Example 6.7 for a higher speed limit. One way to design a road that allows for higher speeds in a curve is to bank the road as shown in Figure 6.30.

We consider the role of the same forces as in Example 6.7: gravity, the normal force, and static friction. Because of the banking, it is possible to take this curve in the absence of static friction. To do so, however, the car must be traveling at one particular speed v_N. Whenever a car takes the curve at any speed other than v_N, static friction comes into play.

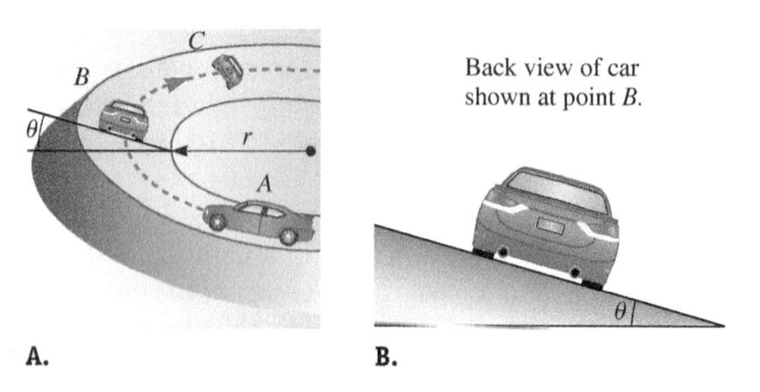

FIGURE 6.30

A Find an expression for v_N. What is v_N if $\theta = 7.5°$?

INTERPRET and ANTICIPATE

There is no static friction in this case from the way v_N is defined above. So, we need only to concern ourselves with gravity and the normal force.

The car does not slide up or down the banked roadway, so the car's circular path is in a horizontal plane. Therefore, the direction of the centripetal acceleration is in that same horizontal plane. It is helpful to choose a coordinate system in which one axis is along the acceleration, such as the x axis in Figure 6.31.

The x component of the normal force is parallel to the centripetal acceleration, so **we circle the normal force and write a phrase near the circle.**

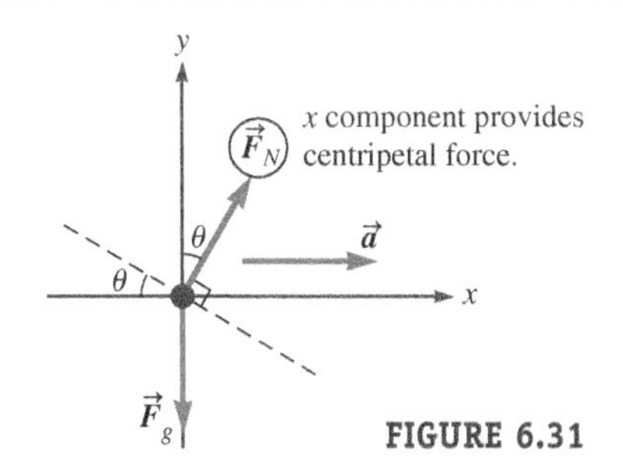

FIGURE 6.31

SOLVE

Apply Newton's second law. **The x component of the normal force acts as the centripetal force.** There is no acceleration in the y direction.

$$\sum F_x = ma_x \qquad \sum F_y = F_N \cos\theta - F_g = ma_y = 0$$

$$F_N \sin\theta = \frac{mv_N^2}{r} \text{ in } x \text{ direction} \qquad (1)$$

$$F_N \cos\theta = F_g \text{ in } y \text{ direction} \qquad (2)$$

To solve Equations (1) and (2) simultaneously to find v_N, first substitute $F_g = mg$ into Equation (2) to produce Equation (3). Eliminate F_N by dividing Equation (1) by Equation (3). Notice that the mass m cancels.

$$F_N \cos\theta = mg \qquad (3)$$

$$\frac{\sin\theta}{\cos\theta} = \frac{\cancel{m}v_N^2}{\cancel{m}gr} = \tan\theta$$

$$v_N = \sqrt{gr\tan\theta}$$

Substitute values.

$$v_N = \sqrt{(9.81 \text{ m/s}^2)(43.3 \text{ m}) \tan 7.5°}$$

$$v_N = 7.5 \text{ m/s} \approx 17 \text{ mph}$$

CHECK and THINK

This speed is less than the maximum speed limit we found for the unbanked turn in Example 6.7. This result seems surprising because our experience tells us that banking a turn should allow the car to go faster. It does, once we take static friction into account (part B).

B If the car exceeds v_N, what is the direction of F_s?

INTERPRET and ANTICIPATE

As in Example 6.7, static friction cannot be in the direction in which the car is moving. Static friction must be parallel to the surface, which in this case means the embankment. Thus, static friction may be directed either up or down the embankment. Because the car's speed is greater than v_N, the x component of the normal force alone is not enough to supply the required centripetal force. Static friction directed down the embankment adds to centripetal force as shown in the free-body diagram (Fig. 6.32). So, the **x component of the static friction and the normal force both contribute to the centripetal force.** If friction were directed up the embankment, the centripetal force would be smaller.

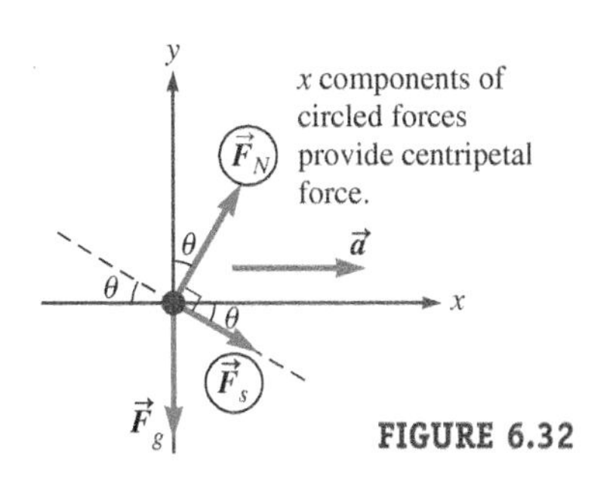

FIGURE 6.32

C What is the speed limit of this banked road if $\theta = 7.5°$? (As before, assume the speed limit is set for a wet road and the road surface is concrete.)

SOLVE

Apply Newton's second law to the forces in the free-body diagram. Because we need the highest possible speed, we set static friction to its maximum.

$$\sum F_x = ma_x$$

$$F_N \sin\theta + F_{s,\max}\cos\theta = m\frac{v_{\max}^2}{r}$$

$$\sum F_y = ma_y$$

$$F_N \cos\theta - F_{s,\max}\sin\theta - F_g = 0$$

Example continues on page 174 ▶

In both equations, substitute $\mu_s F_N$ (Eq. 6.1) for static friction $F_{s,\,max}$.

$$F_N \sin\theta + \mu_s F_N \cos\theta = m\frac{v_{max}^2}{r} \quad (1)$$

$$F_N \cos\theta - \mu_s F_N \sin\theta = F_g \quad (2)$$

Substitute $F_g = mg$ in Equation (2), and factor F_N in Equation (1). Divide Equation (4) by Equation (3) to eliminate F_N. Notice that mass m cancels out.

$$F_N = \frac{mg}{\cos\theta - \mu_s \sin\theta} \quad (3)$$

$$F_N(\sin\theta + \mu_s\cos\theta) = m\frac{v_{max}^2}{r} \quad (4)$$

$$\frac{mg(\sin\theta + \mu_s\cos\theta)}{\cos\theta - \mu_s\sin\theta} = m\frac{v_{max}^2}{r}$$

$$v_{max} = \sqrt{\frac{gr(\sin\theta + \mu_s\cos\theta)}{\cos\theta - \mu_s\sin\theta}}$$

$$v_{max} = \sqrt{\frac{(9.81\text{ m/s}^2)(43.3\text{ m})(\sin 7.5^\circ + 0.4\cos 7.5^\circ)}{\cos 7.5^\circ - 0.4\sin 7.5^\circ}}$$

$$v_{max} = 15\text{ m/s} \approx 34\text{ mph}$$

As expected, the speed limit on this banked turn is higher than the limit on the unbanked turn (29 mph). The banking angle results in a normal force with a component directed toward the center of the circular path, resulting in a centripetal force that is stronger than the centripetal force due to static friction alone. If the car exceeds this speed limit, it cannot maintain the circular path and will move off the road.

D What is the direction of static friction if $v < v_N$?

If $v < v_N$, the x component of the normal force is greater than the required centripetal force. In this case, static friction reduces the centripetal force. To do so, it must be directed up the embankment as shown in the free-body diagram (Fig. 6.33).

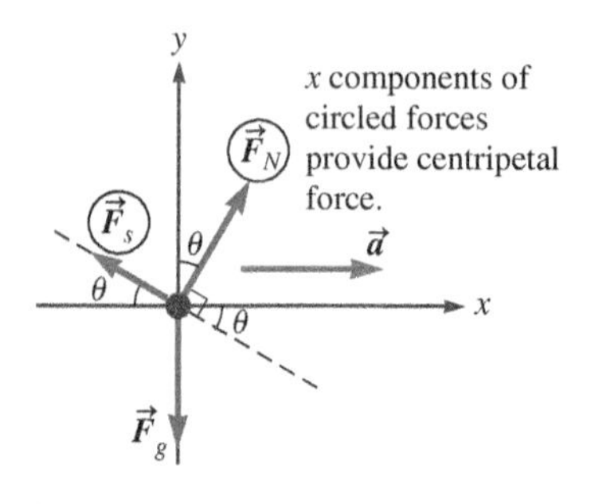

FIGURE 6.33

This answer may seem counterintuitive, but think of the extreme case where $v = 0$ so that the car is at rest. Without static friction, gravity would pull the car down the embankment. In fact, if the embankment is very steep, $F_{s,\,max}$ may be too small to keep the car from slipping downward, which is true even if the car is moving at some nonzero speed $v < v_N$. So, engineers cannot increase the embankment angle beyond a practical limit.

NONUNIFORM CIRCULAR MOTION

Special Case

Nonuniform Circular Motion

We have seen that if the total force on a particle is constant and perpendicular to its velocity, the particle will travel in uniform circular motion. In this case, the particle's speed is constant, its velocity is tangent to the circle, and its acceleration is directed to the center of the circle.

If a particle's path is circular and the total force is not perpendicular to the velocity, the particle speed changes (Fig. 6.34). This type of circular motion, in which both the magnitude and direction of the velocity change, is called **nonuniform circular motion**. The speed of a particle in nonuniform circular motion changes because the total force has a component parallel to the velocity. (Of course, to maintain a circular path, the total force must also have a component that is centripetal.)

As we have seen, a powerful way to analyze vectors is to break them into components. In this case, instead of using the usual x and y components, we break $\vec{F}_{tot}$ (or $\vec{a}$)

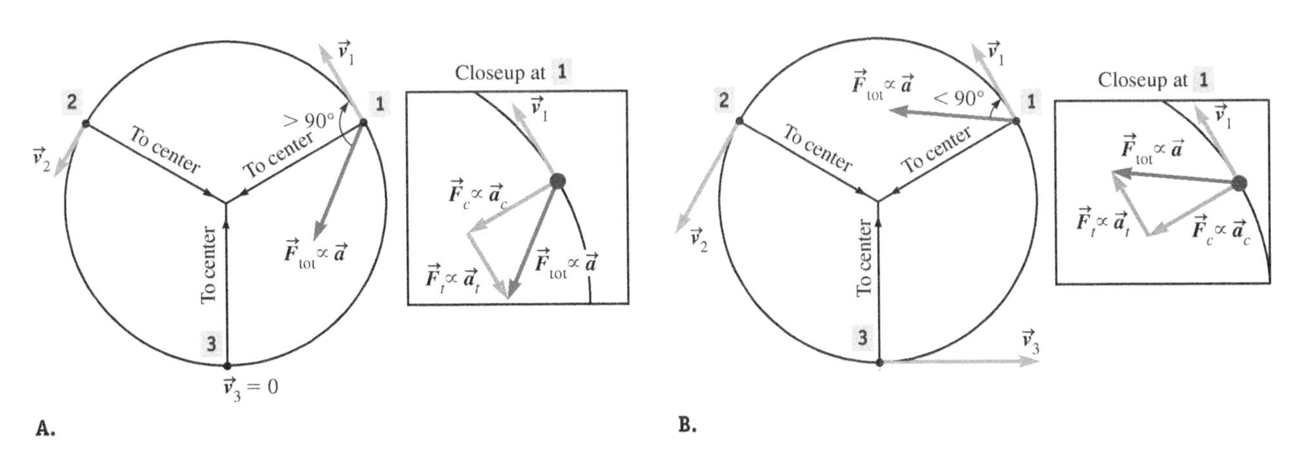

FIGURE 6.34 **A.** A particle travels in a circular path. The net force on it makes an angle greater than 90° with its velocity. The result is that the particle slows down and momentarily stops at **3**. **B.** A particle travels in a circular path. The net force on it makes an angle less than 90° with its velocity. The result is that the particle speeds up indefinitely.

into components that are parallel and perpendicular to the velocity vector as shown in the insets of Figure 6.34. The parallel component $\vec{F}_t$ is referred to as the **tangential component** because it is tangent to the circle, and the perpendicular component $\vec{F}_c$ is called the **centripetal component** because it acts toward the center of the motion. The tangential component of the force is responsible for changing the speed of the particle. If this component points in the direction opposite the direction of the velocity as in Figure 6.34A, the particle slows down. If the tangential component points in the direction of the velocity as in Figure 6.34B, the particle speeds up.

The magnitude of the centripetal force F_c and centripetal acceleration a_c are the same as in uniform circular motion:

$$F_c = m\frac{v^2}{r} \quad \text{and} \quad a_c = \frac{v^2}{r}$$

The difference is that in nonuniform circular motion these two quantities are not constant. As the speed changes, the centripetal force and acceleration must also change.

EXAMPLE 6.9 Loop-the-Loop Ride

The loop-the-loop ride is a roller coaster in which the cars travel on the inside of a vertical circular track (Fig. 6.35). After the cars are launched at the bottom B, no motor or engine pulls them along the track.

A Treat each car as a particle and assume rolling friction and drag are negligible. Draw a free-body diagram for a car at points (*b*ottom) B, (*r*ight) R, (*t*op) T, and (*l*eft) L. Indicate the acceleration in each case.

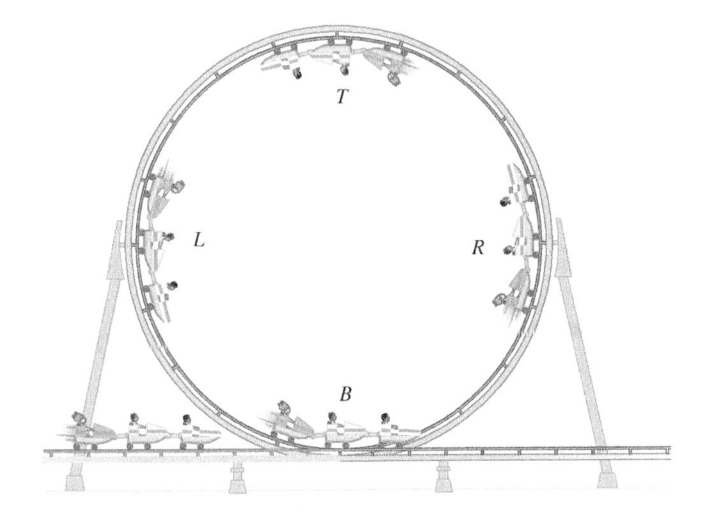

FIGURE 6.35

Example continues on page 176 ▶

Only two forces act on the car: gravity and the normal force due to the track. (Unlike the banked turn in Example 6.8, static friction does not act on a car in the vertical loop-the-loop.) The gravitational force on a car is the same at all four positions. The normal force changes in both magnitude and direction (Fig. 6.36). The direction of the acceleration is estimated by adding the two forces geometrically.

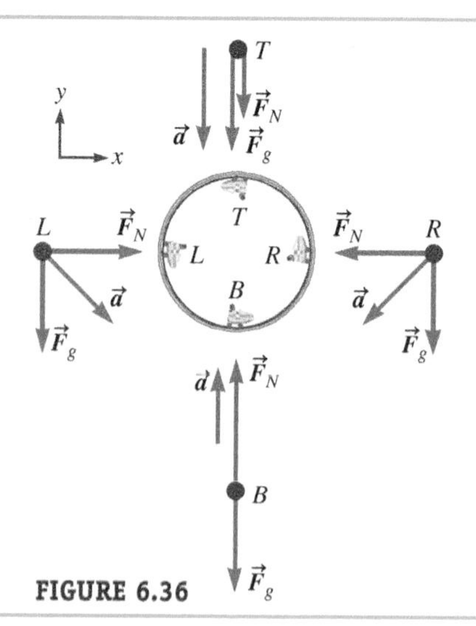

FIGURE 6.36

B Is this uniform circular motion? Explain.

No. Uniform circular motion requires that the velocity and acceleration be perpendicular to each other. In our diagrams, the acceleration is perpendicular to the velocity only at T and B. At all other positions, the acceleration has a tangential component. At R and all other positions on the right half of the track, the tangential component opposes the velocity. Therefore, cars slow down as they move from B to T. On the left side of the track, the tangential acceleration is in the same direction as the velocity. Consequently, cars speed up as they move from T back down to B.

C Write an expression for the magnitude of the centripetal force F_c at each of the four positions in terms of the car's mass m, radius of the loop r, and the speed at the four positions.

INTERPRET and ANTICIPATE

At each position, we must identify the force or forces that point toward the center of the circle. If more than one force points in this direction, we must add them to find the magnitude of the total centripetal force at that point. The top and bottom points (T and B) are the simplest because for those two points both forces are radial.

SOLVE

At T, both the normal force and gravity point to the center of the circle, so they both contribute to the centripetal force.

For the top T:

$$F_c = F_N + F_g = m\frac{v_T^2}{r}$$

At B, both the normal force and gravity affect the centripetal force just as at T. Now, however, gravity points away from the center of the circle, so it is negative.

For the bottom B:

$$F_c = F_N - F_g = m\frac{v_B^2}{r}$$

Gravity is tangent to the circle at R and L. Therefore, only the normal force contributes to the centripetal force at these two points.

For the left L and right R:

$$F_c = F_N = m\frac{v_{L \text{ or } R}^2}{r}$$

CHECK and THINK

The loop-the-loop ride might seem very dangerous because our intuition tells us that a car should fall from T to B. Our analysis seems to confirm our intuition. The net force exerted on a car at point T points straight down. If the velocity were zero at T, the car would indeed fall straight down to B. Because the velocity is tangent to the track at T, however, a car accelerates downward at T, but that does not mean that its motion is straight down. Instead, its motion is downward and *along the track.*

D If the cars are just barely in contact with the track at T, find an expression for their speed at that position.

INTERPRET and ANTICIPATE

If the cars are just barely in contact with the track, the normal force exerted by the track on the cars is approximately zero. Only gravity is responsible for the centripetal force.

$$F_c = F_g$$

$$m\frac{v_T^2}{r} = mg$$

We solve for v_T.

$$v_T = \sqrt{gr}$$

CHECK and THINK

If the cars are moving at any speed higher than $\sqrt{gr}$ when they get to the top, the normal force must contribute to the centripetal force. A problem arises if the cars are moving at a speed less than $\sqrt{gr}$ at the top. If $v = 0$ at the top, the car would fall straight down. If $0 < v < \sqrt{gr}$, the car would follow a parabolic path as it fell to the ground. Next time you plan to board such a ride, estimate the speed of the cars and be sure it greater than $\sqrt{gr}$.

SUMMARY

Underlying Principles: Newton's laws of motion

Special Cases

1. When an object is in contact with a surface and is at rest relative that surface, the **static friction** force on the object is parallel to the surface. If the object is not accelerating, the static friction force is balanced by the vector sum of any other forces that are applied to the object parallel to the surface.

   The maximum force of static friction is given by

   $$F_{s,\max} = \mu_s F_N \quad (6.1)$$

   The **coefficient of static friction** μ_s is a positive, unitless scalar that is usually found through experimentation (Table 6.1).
2. Whenever two surfaces are in contact and sliding relative to each other, **kinetic friction** works against that motion. The direction of kinetic friction on the object is always parallel to the contacting surfaces and in the direction opposite the direction of the object's motion relative to the other surface. The magnitude of kinetic friction is given by

   $$F_k = \mu_k F_N \quad (6.2)$$

   where μ_k is the **coefficient of kinetic friction** and is found experimentally (Table 6.1).
3. The friction force associated with the rolling motion of an object along a surface is called **rolling friction**. The direction of the rolling friction force is opposite the motion of the object relative to the surface. The magnitude of the rolling friction force is given by

   $$F_r = \mu_r F_N \quad (6.3)$$

   The **coefficient of rolling friction** μ_r is a unitless scalar that is found experimentally (Table 6.1).
4. There is a **drag force** on an object when it moves relative to a fluid medium. The direction opposes the object's relative velocity. For small objects moving slowly through air or larger objects moving through water,

   $$\vec{F}_D = -b\vec{v} \quad (6.4)$$

   For blunt objects moving at high speed through air, the magnitude of the drag force is

   $$F_D = \frac{1}{2}C\rho A v^2 \quad (6.5)$$
5. When an object falling through the air stops accelerating, it continues to move downward at a constant speed known as its **terminal speed**:

   $$v_t = \sqrt{\frac{2mg}{C\rho A}} \quad (6.6)$$
6. The net force exerted on object in uniform circular motion points to the center of the circle and is known as the **centripetal force**. Its magnitude is given by

   $$F_c = m\frac{v^2}{r} \quad (6.7)$$

PROBLEM-SOLVING STRATEGY When Centripetal Force Is Present

Problems that involve centripetal force are no different from other problems that require us to apply Newton's second law. Two modifications, however, may be helpful.

INTERPRET and ANTICIPATE

Modification 1 Draw a free-body diagram for one particular instant and **circle the centripetal force(s)**.

SOLVE

Modification 2 The **sum of the forces that are parallel or antiparallel to the centripetal acceleration equals the centripetal force** (Eq. 6.7).

PROBLEMS AND QUESTIONS

A = algebraic C = conceptual E = estimation G = graphical N = numerical

6-1 Newton's Laws in a Messy World

1. C In many textbook problems, we ignore certain complications such as friction and drag. The problems contain key words that indicate such a simplification is being used. For example, if a surface is described as "slippery," it means that we can ignore friction. Look at the previous chapters' problem sets. Find five uses of these key words and explain how to interpret each case.
2. C CASE STUDY On page 154, Cameron says, "the real world is too hard to study with physics." Do you agree with that statement? If so, explain why. If not, come up with a better statement and explain why it is better.
3. C CASE STUDY Are skydivers with their parachutes closed in free fall? Explain.

6-2 Friction and the Normal Force Revisited

4. C We often need to model a problem, such as when we model molecular bonds as springs or people as particles. For which circumstances are these examples good models? Under what conditions do they break down?

Problems 5 and 9 are paired.

5. C In Figure 6.3, a man holds a box against the wall. The box is at rest. **a.** Compare the normal force exerted by the wall to the force exerted by the man. **b.** If the man reduces his force on the box, what will happen to the normal force? **c.** If the man reduces his force on the box, what will happen to static friction?

FIGURE P6.6

6. G Draw a free-body diagram for the burglar, who is shown at rest while sneaking through a chimney in Figure P6.6.
7. C The shower curtain rod in Figure P6.7 is called a tension rod. The rod is not attached to the wall with screws, nails, or glue, but is pressed into the wall instead. Explain why the rod remains at rest, supporting the curtain. Explain why the name is misleading and come up with a better name.

Courtesy, Signature Hardware and Gatco Inc.

FIGURE P6.7

6-3 A Model for Static Friction

8. N, C A rectangular block has a length that is five times its width and a height that is three times its width. The block's surfaces are all identical except for size. When the block is placed on a horizontal tabletop so that the area in contact with the table is length × width, it is found that a horizontal force of 10.0 N applied to the block is just sufficient to overcome the static friction force and cause the block to move. The block is then knocked over so that the area in contact with the table is length × height. Now, what minimum horizontal force will cause the block to move? Explain.
9. N A man exerts a force of 16.7 N horizontally on a box so that it is at rest in contact with a wall as in Figure 6.3. The box weighs 6.52 N. **a.** Find the static friction force exerted on the box, given the forces being applied. **b.** If the coefficient of static friction between the wall and the box is 0.50, find the maximum static friction force that may be exerted on the box. Comment on your results.

Problems 10 and 11 are paired.

10. C A makeshift sign hangs by a wire that is extended over an ideal pulley and is wrapped around a large potted plant on the roof as shown in Figure P6.10. When first set up by the shopkeeper on a sunny and dry day, the sign and the pot are in equilibrium. Is it possible that the sign falls to the ground during a rainstorm while still remaining connected to the pot? What would have to be true for that to be possible?

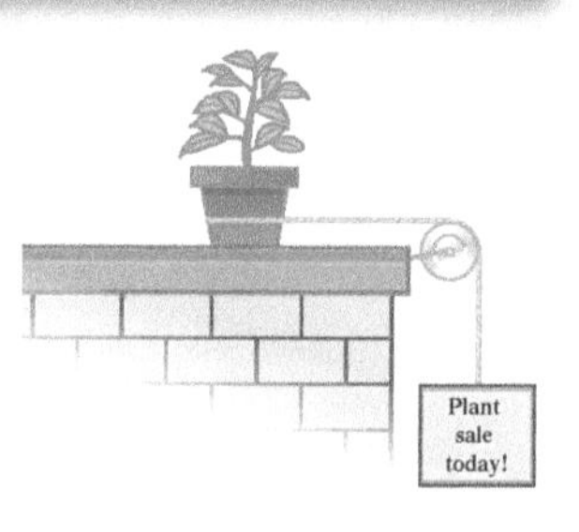

FIGURE P6.10 Problems 10 and 11.

11. N In Problem 10, the mass of the sign is 25.4 kg, and the mass of the potted plant is 66.7 kg. **a.** Assuming the objects are in equilibrium, determine the magnitude of the static friction force experienced by the potted plant. **b.** What is the maximum value of the static friction force if the coefficient of static friction between the pot and the roof is 0.572?
12. C A heavy box of tools rests on a tabletop. You push on the box with a horizontal force of 100 N and observe that the box does not move. You then push on the box with a horizontal force of 200 N and observe that the box does not move. How does the magnitude of the static friction force compare to the force with which you push in each case?
13. N A motorcyclist is traveling at 55.0 mph on a flat stretch of highway during a sudden rainstorm. The rain has reduced the coefficient of static friction between the motorcycle's tires and the road to 0.080 when the motorcyclist slams on the brakes, locking the tires in place. **a.** What is the minimum distance required to bring the motorcycle to a complete stop? **b.** What would be the stopping distance if it were not raining and the coefficient of static friction were 0.630?

Problems 14 and 15 are paired.

14. A small steel I-beam (Fig. P6.14) is at rest with respect to the steel surface of a truck. The truck is accelerating with respect to the road. The mass of the I-beam is 5.8×10^3 kg.
    a. G Draw a free-body diagram for the I-beam.
    b. C What force or forces accelerate the I-beam with respect to the ground?
    c. N The I-beam must remain at rest with respect to the truck. What is the maximum acceleration of the truck? Evaluate your answer.
    d. G On the highway, the truck moves with a constant velocity. Draw a free-body diagram for the I-beam. Compare it with your diagram in part (a).

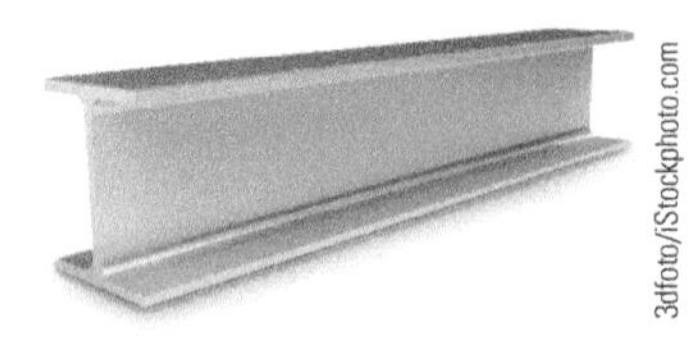

3dfoto/iStockphoto.com

FIGURE P6.14

15. A box is at rest with respect to the surface of a flatbed truck. The coefficient of static friction between the box and the surface is μ_S.
    a. A Find an expression for the maximum acceleration of the truck so that the box remains at rest with respect to the truck. Your expression should be in terms of μ_S and g.
    b. C How does your answer change if the mass of the box is doubled?

Problems 16 and 17 are paired.

16. A A filled treasure chest of mass m with a long rope tied around its center lies in the middle of a room. Dirk wishes to drag the chest, but there is friction between the chest and the floor with a coefficient of static friction μ_s. If the angle between the rope and the floor is θ, what is the magnitude of the tension required to just get the chest moving? Express your answer in terms of m, μ_s, θ, and g.

17. N A filled treasure chest ($m = 375$ kg) with a long rope tied around its center lies in the middle of a room. Dirk wishes to drag the chest, but there is friction between the chest and the floor with $\mu_s = 0.52$. If the angle between the rope and the floor is 30.0°, what is the magnitude of the tension required to just get the chest moving?

18. N Rochelle holds her 2.80-kg physics textbook by pressing horizontally against both sides of the textbook with the palms of her hands. If the coefficient of static friction between her hands and the textbook is 0.500, what is the minimum force with which she must compress the textbook with each hand to keep it from slipping out of her hands?

Problems 19, 20, and 21 are grouped.

19. A A sled and rider have a total mass M. They are at rest on a snowy hill. The coefficient of static friction between the sled and the snow is μ_S. Find an expression for the maximum angle of the hill's slope measured upward from the horizontal in terms of μ_S. How does your answer change if the rider is not on the sled?

20. N A sled and rider have a total mass 56.8 kg. They are at rest on a snowy hill. The coefficient of static friction between the sled and the snow is 0.225. What is the maximum angle of the hill's slope measured upward from the horizontal?

21. C, N A sled and rider have a total mass 56.8 kg. They are on a snowy hill. The coefficient of static friction between the sled and the snow is 0.225. The angle of the hill's slope measured upward from the horizontal is 19.5°. Are they at rest?

6-4 Kinetic and Rolling Friction

22. N You need to design an experiment to measure the free-fall acceleration. You have an inclined plane made of steel and several objects that you may place on the plane. The plane is fixed at 45°. You have a Teflon-coated block with a mass of 0.5 kg, a cart with steel wheels and a mass of 1.5 kg, and a steel box with a mass of 0.75 kg. You have a sonar device to measure the acceleration. You are to place one or more objects on the plane and measure their acceleration along the plane. Does it matter which object or objects you choose? If not, why not? If so, which would you use? Explain your answer.

23. N A block with mass $M = 2.00$ kg is placed on an inclined plane. The plane makes an angle of 30° with the horizontal, and the coefficients of kinetic and static friction between the block and the plane are $\mu_k = 0.400$ and $\mu_s = 0.600$, respectively. Will the block slide down the plane, or will it remain motionless? Justify your answer.

24. C Lisa measured the coefficient of static friction between two pairs of running shoes and the track in Example 6.1 (page 159). If she wants to have an advantage in a race, which shoes should she wear, the ones with a high coefficient or the ones with the low coefficient of static friction? Explain.

25. N An ice cube with a mass of 0.0507 kg is placed at the midpoint of a 1.00-m-long wooden board that is propped up at a 50° angle. The coefficient of kinetic friction between the ice and the wood is 0.133. **a.** How much time does it take for the ice cube to slide to the lower end of the board? **b.** If the ice cube is replaced with a 0.0507-kg wooden block, where the coefficient of kinetic friction between the block and the board is 0.275, at what angle should the board be placed so that the block takes the same amount of time to slide to the lower end as the ice cube does? You may find a spreadsheet program helpful in answering this question.

26. C In a pinewood derby race, children release small cars they have constructed from the top of a ramp, record the time it takes to reach the finish line, and rank the winners through several trials (Fig. P6.26). Juan proclaims to his friend, Antonio, that his car will reach the finish line first because he has adjusted his wheels to be more narrow. This adjustment means that the area in contact with the racetrack will be less than normal. Juan suggests that this adjustment will make the magnitude of the friction between his wheels and the track less than that for Antonio's car. Assuming the masses of the cars are the same and air resistance can be ignored, is Juan correct, or does Antonio have a valid counterargument?

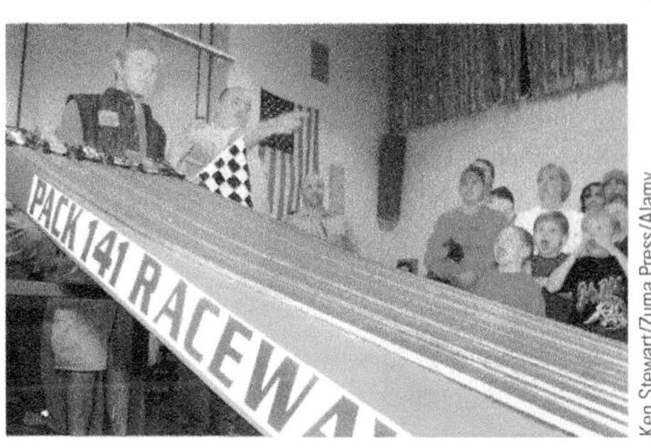

FIGURE P6.26

27. A Curling is a game similar to lawn bowling except it is played on ice and instead of rolling balls on the lawn, stones are slid along ice. A curler slides a stone across a sheet of ice with an initial speed v_i in the positive x direction. The coefficient of kinetic friction between the stone and the curling lane is μ_k. Express your answers in terms of v_i, μ_k, and g only. **a.** What is the acceleration of the stone as it slides down the lane? **b.** What distance does the curling stone travel?

28. C Janelle and Jordan need to push their car to the mechanic. Janelle pushes the car while Jordan steers with the car in the neutral gear. (It is free to roll.) Initially, Janelle is pushing the car downhill, later on level ground, and finally uphill to get it into the mechanic's lot. She finds it much easier to push the car downhill than uphill. Assume the angles of elevation θ of the downhill and uphill slopes are the same with respect to the level part of the road. **a.** If the coefficient of rolling friction, μ_r, between the tires and the road is the same on each part of the journey, how does the magnitude of the friction force compare on each part? **b.** Is the friction force always working against Janelle's pushing force on the car? Explain why it is easier to push the car downhill.

Problems 29, 30, and 31 are grouped.

29. N A sled and rider have a total mass of 56.8 kg. They are on a snowy hill. The coefficient of kinetic friction between the sled and the snow is 0.195. The angle of the hill's slope measured upward from the horizontal is 19.5°. What is the acceleration of the rider? Is the acceleration greater, less than, or equal to your result if a more massive rider uses the same sled on the same hill? Explain.

30. N A sled and rider have a total mass of 56.8 kg. They are on a snowy hill accelerating at $0.7g$. The coefficient of kinetic friction between the sled and the snow is 0.18. What is the angle of the hill's slope measured upward from the horizontal? You may find a spreadsheet program helpful in answering this question.

31. N A cart and rider have a total mass of 56.8 kg. The cart is rolling down a hill and accelerating at $0.5g$. The coefficients of kinetic friction and of rolling friction between the cart's wheels and the road are 0.20 and 0.020 respectively. What is the angle of the hill's slope measured upward from the horizontal? You may find a spreadsheet program helpful in answering this question.

6-5 Drag and Terminal Speed

32. C Why do you crumple up a sheet of paper before you toss it toward the garbage can?

33. N On a particularly hot day at the racetrack, Dale Earnhardt's famous No. 88 Chevy Impala ($m = 1600$ kg), speeding at 200.0 mph, blows a gasket in the engine and begins to decelerate. The car is shifted into neutral and coasts toward the pit lane. Neglecting all other sources of friction and assuming the drag coefficient is 0.330 and the car's frontal area is 2.76 m^2, what is the initial deceleration of the Impala? Report your answer to three significant figures.

34. C An intrepid experimenter with two identical objects stands at the edge of a cliff, looking over a canyon. She throws one of the objects directly upward with speed v while holding her hand out over the edge. She throws the second object directly downward with speed v from the same height. Neither object reaches terminal velocity, yet they each have the same speed just before they strike the canyon floor below. Explain how this could be true, considering the drag force on the object during each object's motion.

35. N A small sphere of mass $m = 0.500$ kg is dropped from rest into a viscous liquid in which the resistive force on the sphere can be expressed as $\vec{F}_D = -b\vec{v}$, and reaches one-fourth its terminal speed in 3.45 s. **a.** What is the terminal speed of the sphere? **b.** What is the distance traveled by the sphere in 3.45 s?

Problems 36 and 37 are paired.

36. A racquetball with a radius of 0.0285 m is thrown vertically into the air with an initial speed of 3.07 m/s. The drag coefficient of the ball is 0.47, and the density of the air is 1.29 kg/m^3.

a. N What is the cross-sectional area of the ball?

b. N What is the magnitude of the drag force on the ball the instant it is thrown?

c. N What is the magnitude of the drag force on the ball when it reaches the peak?

d. G Make a sketch of the magnitude of the drag force versus time from the moment the ball is thrown until it reaches the peak.

37. N A racquetball has a radius of 0.0285 m. The drag coefficient of the ball is 0.47, and the density of the air is 1.29 kg/m^3. What would be the terminal speed for the racquetball if it were dropped from a very high cliff, assuming it has a mass of 0.0450 kg?

38. C The terminal speed of a penny in air has been found to be between 11 m/s and 32 m/s. A man tosses a penny in the air and fails to catch it. **a.** Does the penny reach its terminal speed before landing at the man's feet? **b.** If the man is on top of the Empire State Building in Manhattan and the penny falls to the ground (roughly 1000 ft below), does it reach its terminal speed? **c.** If a penny falls off a large building, would it bore a hole into the sidewalk below? Explain.

39. CASE STUDY Let's take another look at the skydivers flying in formation (Example 6.5, page 168). Assume there are three skydivers of different body types. Let's model each one as a cylinder. Skydiver A's cylinder is 2.00 m tall, has a radius of 0.25 m, and has a total weight of 670 N; skydiver B's cylinder is 1.75 m tall, has a radius of 0.33 m, and has a total weight of 675 N; and skydiver C's cylinder is 1.95 m tall, has a radius of 0.43 m, and has a total weight of 907 N. Assume the drag coefficient is 0.5 when the skydivers are pointing head down and the drag coefficient is 1.0 when the skydivers are oriented belly-to-Earth. The density of air is 1.29 kg/m^3.

a. N Find the maximum and minimum terminal speed for each skydiver.

b. C If the skydivers wish to fly in formation as in Figure 6.1, in which order should they leave the plane?

c. N Imagine that the plane flies in a circular path so that skydivers can exit over the same spot. Suppose that the time between skydivers exiting the plane is 30 s. Assume the skydivers hit terminal speed immediately (not true; actually takes about 10 s). Find the time that the first skydiver must wait after exiting the plane until the last skydiver joins the formation. How would your answer change if you took into account that each diver accelerates for about 10 s just after exiting the plane?

40. In this problem, we compare the drag force to rolling friction on a typical car. The density of air is 1.29 kg/m^3.

a. E Estimate the weight and effective area of a typical car. Assume the car has been designed to reduce drag.

b. G Use your estimate to plot v^2 versus F_D. Does the shape of your graph make sense? Explain.

c. E Estimate rolling friction on this car. Add your estimate to the graph.

d. G At what speed does rolling friction equal the drag on the car? For what range of speeds is it okay to ignore drag?

e. C How do your answers change if you take into account that at slow speeds drag is proportional to v, not v^2?

41. N An inflated spherical beach ball with a radius of 0.3573 m and average density of 10.65 kg/m^3 is being held under water in a pool by Janelle. The density of the water in the pool is 1000.0 kg/m^3. When Janelle releases the ball, it begins to rise to the surface. If the drag coefficient of the ball in the water is 0.470 and the constant upward force on the ball is 1875 N, what will be the terminal speed of the ball as it rises? Ignore the effects of gravity on the ball.

42. E CASE STUDY In the train collision case study (Chapter 5, page 119), we ignored the drag force on the trains. Estimate the drag on the trains and compare it to the kinetic friction on them. Is it okay to ignore drag? Explain.

Problems 43 and 44 are paired.

43. N Your sailboat has capsized! Fortunately, you are no longer aboard the boat. Instead, you are hanging onto the end of a long rope, the other end of which is attached to a Coast Guard helicopter. Model yourself as a particle of mass $M = 55.0$ kg with a diameter equal to 0.500 m. The density of the air is $\rho = 1.29$ kg/m^3. Assume the drag coefficient between you and the air is $C = 0.500$. **a.** First, ignore the drag force due to the air. If the helicopter is flying at a constant speed $v_0 = 35.0$ m/s, what angle will the rope make with the vertical? **b.** Now, consider the drag force due to the air. What angle does the rope make with the vertical given the information in part (a)?

44. You are hanging onto the end of a long rope, the other end of which is attached to a Coast Guard helicopter. Model yourself as a particle of mass $M = 55.0$ kg with a diameter equal to 0.500 m. The density of the air is $\rho = 1.29$ kg/m^3. Assume the drag coefficient between you and the air is $C = 0.500$. Now, at time $t = 0$, the helicopter begins to accelerate horizontally with an acceleration $a = 3.00$ m/s^2.

a. N If we ignore the effect of the air drag force, what is the angle of the rope with the vertical?

b. A Suppose in this case where the helicopter accelerates we also consider the drag force due to the air. Derive an expression for the angle made by the rope with the vertical as a function of time.

c. C Neither in this problem nor Problem 43 has a terminal speed been calculated. Is terminal speed a meaningful concept for this situation? Why or why not?

d. C What happens to the tension if the helicopter continues to accelerate? What would be the result for a real rope in this situation?

45. C The **drag coefficient** C in $F_D = \frac{1}{2}C\rho Av^2$ (Eq. 6.5) depends primarily on the shape of the object. You already have developed an intuition about what shapes correspond to a low C by observing the shapes of aerodynamic cars, boats, and even bullets. Which object, a sphere or a cube, would have a larger drag coefficient, assuming they are nearly the same size? Explain your reasoning. What aspect of an object most determines its drag coefficient?

6-6 Centripetal Force

46. N Amir (m = 56 kg) begins from rest and starts to jog around a circular track with a radius of 32.0 m. After a brief 1.56 s, Amir reaches his top speed of 3.20 m/s. What is the magnitude of the centripetal force responsible for keeping Amir on the circular path?
47. N The speed of a 100-g toy car at the bottom of a vertical circular portion of track is 8.00 m/s. If the radius of curvature of this portion of the track is 60.0 cm, what are the magnitude and direction of the force the track exerts on the car? Report your answer to three significant figures.
48. You are riding in a car around a turn. You do not move with respect to the seat. The turn is flat.
    a. G Draw a free-body diagram for you.
    b. C Which force or forces are responsible for the centripetal force on you?
    c. C What happens to you if the driver takes the turn too quickly?
49. N Artificial gravity is produced in a space station by rotating it, so it is a noninertial reference frame. The rotation means that there must be a centripetal force exerted on the occupants; this centripetal force is exerted by the walls of the station. The space station in Arthur C. Clarke's *2001: A Space Odyssey* is in the shape of a four-spoked wheel with a diameter of 155 m. If the space station rotates at a rate of 2.40 revolutions per minute, what is the magnitude of the artificial gravitational acceleration provided to a space tourist walking on the inner wall of the station?
50. N Escaping from a tomb raid gone wrong, Lara Croft (m = 57.0 kg) swings across an alligator-infested river from a 9.00-m-long vine. If her speed at the bottom of the swing is 7.50 m/s and she makes it safely across the river, what is the minimum breaking strength of the vine?

Problems 51 and 52 are paired.

51. N Harry Potter decides to take Pottery 101 as an elective to satisfy his arts requirement at Hogwarts. He sets some clay (m = 3.25 kg) on the edge of a pottery wheel (r = 0.600 m), which is initially motionless. He then begins to rotate the wheel with a uniform acceleration, reaching a final angular speed of 2.400 rev/s in 3.00 s. **a.** What is the speed of the clay when the initial 3.00 s has passed? **b.** What is the centripetal acceleration of the clay initially and when the initial 3.00 s has passed? **c.** What is the magnitude of the constant tangential acceleration responsible for starting the clay in circular motion?
52. C Harry sets some clay (m = 3.25 kg) on the edge of a pottery wheel (r = 0.600 m), which is initially motionless. He then begins to rotate the wheel with a uniform acceleration, reaching a final angular speed of 2.400 rev/s in 3.00 s, while not touching the clay. As a result, the clay is subject to a tangential and centripetal acceleration while it sits on the edge of the wheel. **a.** What force is responsible for the tangential acceleration, and what force is responsible for the centripetal acceleration? **b.** Which of these two forces, tangential or centripetal, will necessarily fail to keep the clay in place on the wheel first? Why?

Problems 53 and 54 are paired.

53. A A small disk of mass m is attached by a rope to a block with a larger mass M through a hole in a table as shown in Figure P6.53. The disk moves in circle of radius R at constant speed, and the block is at rest. Assume friction between the disk and the table is negligible. Find an expression for its speed v_{ucm} in terms of the parameters given and the tension F_T in the rope.

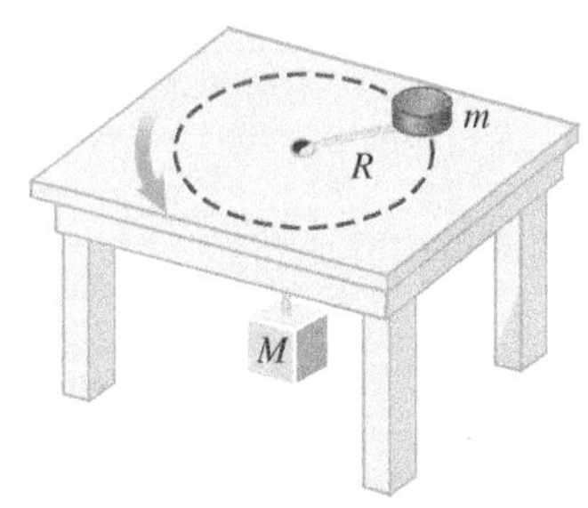

FIGURE P6.53 Problems 53 and 54.

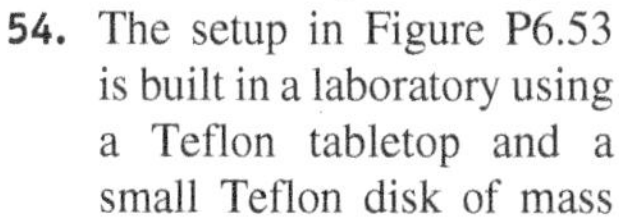

54. The setup in Figure P6.53 is built in a laboratory using a Teflon tabletop and a small Teflon disk of mass m. The disk is attached by a rope to a block with a larger mass M. The small disk starts with a speed v_{ucm} (from Problem 53 in which friction is negligible) and travels in a circular path.
    a. G Draw two free-body diagrams for the disk. One should be from an overhead perspective; the other should be from behind.
    b. C Describe the path of the disk. Explain.
    c. C What happens to the block? Explain.
55. N A cyclist experiences a total horizontal force of 85.0 N when he rounds a horizontal curve on a track at a constant speed of 8.00 m/s. What are the magnitude and direction of the total horizontal force he would experience after he doubled his speed through the curve?

Problems 56, 57, and 58 are grouped.

56. E The Earth's gravity is responsible for the centripetal force on the Moon. Assume the Moon's orbit represents uniform circular motion. Estimate the centripetal force on the Moon.
57. N When a star dies, much of its mass may collapse into a single point known as a black hole. The gravitational force of a black hole on surrounding astronomical objects can be very great. Astronomers estimate the strength of this force by observing the orbits of such objects around a black hole. What is the gravitational force exerted by a black hole on a 1-solar-mass star whose orbit has a 1.4×10^{10} m radius and a period of 5.6 days?
58. N A satellite of mass 16.7 kg in geosynchronous orbit at an altitude of 3.58×10^4 km above the Earth's surface remains above the same spot on the Earth. Assume its orbit is circular. Find the magnitude of the gravitational force exerted by the Earth on the satellite. *Hint*: The answer is not 163 N.
59. N Banked curves are designed so that the radial component of the normal force on the car rounding the curve provides the centripetal force required to execute uniform circular motion and safely negotiate the curve. A car rounds a banked curve with banking angle $\theta = 22.0°$ and radius of curvature 150 m. **a.** If the coefficient of static friction between the car's tires and the road is $\mu_s = 0.400$, what is the range of speeds for which the car can safely negotiate the turn without slipping? **b.** What is the minimum value of μ_s for which the car's minimum safe speed is zero? Note that friction points up the incline here.

General Problems

60. C A block lies motionless on a horizontal tabletop. You apply a force $\vec{F}_{app}$ horizontally to the block, but it does not move. What can you say about the relative sizes and magnitudes of $\vec{F}_{app}$, the static friction force between the block and the table, and the kinetic friction force between the block and the table?
61. N A car with a mass of 1453 kg is rolling along a flat stretch of road and eventually comes to a stop due to rolling friction. If the car begins with a speed of 10.0 m/s and the car comes to a stop in 6.88 s, what is the coefficient of rolling friction between the tires and the road?

62. To improve health and train physically, some people choose to run in a pool of water rather than outside or on a treadmill in air. Typical values for the density of air and water are 1.29 kg/m^3 and 1.00×10^3 kg/m^3, respectively. Consider a runner who is moving at 2.57 m/s in air.
   a. E, N Assuming the drag coefficient for the human body is approximately 1.00, estimate the cross-sectional area of the person running and determine the resulting drag force on the person while running in air.
   b. E, N Again, using your estimate of the cross-sectional area and the same drag coefficient, determine the resulting drag force on the person running at the same speed in water.
   c. C Does the speed in part (b) seem achievable while running in water? Explain your answer.

63. N A force of 87.0 N in the horizontal direction is required to set a 30.0-kg box, initially at rest on a flat surface, in motion. After the box begins to move, a horizontal force of 64.0 N is required to maintain a constant speed. **a.** What is the coefficient of static friction between the box and the surface? **b.** What is the coefficient of kinetic friction between the box and the surface?

64. A box rests on a surface (Fig. P6.64). A force $\vec{F}_{app}$ is applied to the box in two different ways. In both cases, $\vec{F}_{app}$ has the same magnitude, but in case 1 the force is directed below the horizontal, whereas in case 2 it is directed above the horizontal.
   a. G Draw a free-body diagram for both cases.
   b. C Now F_{app} is increased in both cases until the box just barely remains at rest. Compare $F_{s,max}$ for each free-body diagram.
   c. C Use your answer to part (b) to find a best way to move a heavy desk. Describe and explain your solution.

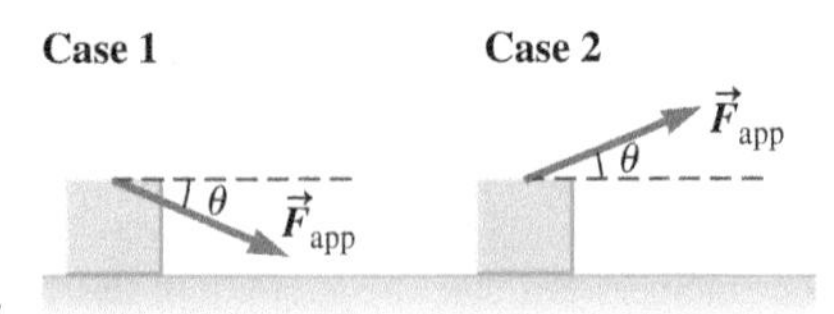

FIGURE P6.64

65. A A box of mass m rests on a rough, horizontal surface with a coefficient of static friction μ_s. If a force $\vec{F}_p$ is applied to the box at an angle θ as shown, what is the minimum value of θ for which the box will not move regardless of the magnitude of $\vec{F}_p$?

FIGURE P6.65

66. C A cylinder of mass M at rest on the end of a string causes a tension F_{T1} in the string. If the cylinder is allowed to swing, the tension in the string when it is vertical is F_{T2}. How do the two tensions F_{T1} and F_{T2} compare in magnitude?

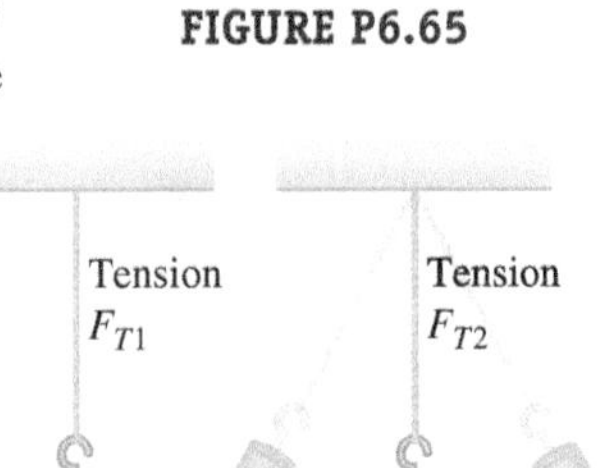

FIGURE P6.66

Problems 67, 70, 71, and 72 are grouped.

67. A A block of mass M is placed on a frictionless plane. The plane is inclined at an angle θ, and the block is a distance d from its end. Of course, we would expect the block to slip down the plane. Suppose we revolve the incline around the vertical axis shown in Figure P6.67 instead. At what period of revolution will the block remain in place on the plane?

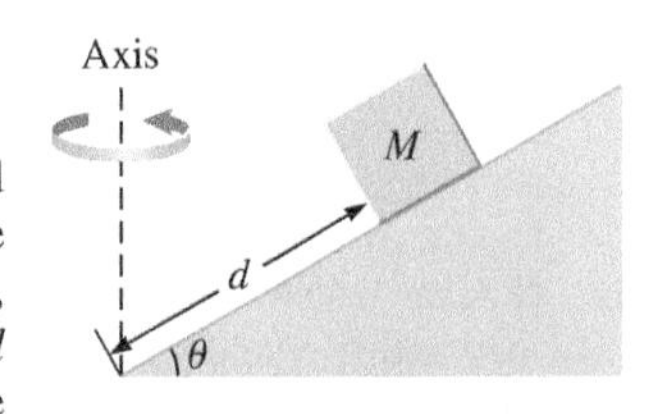

FIGURE P6.67 Problems 67, 71, and 72.

68. N Instead of moving back and forth, a conical pendulum moves in a circle at constant speed as its string traces out a cone (Fig. P6.68). One such pendulum is constructed with a string of length $L = 12.0$ cm and bob of mass 0.210 kg. The string makes an angle $\theta = 7.00°$ with the vertical. **a.** What is the radial acceleration of the bob? **b.** What are the horizontal and vertical components of the tension force exerted by the string on the bob?

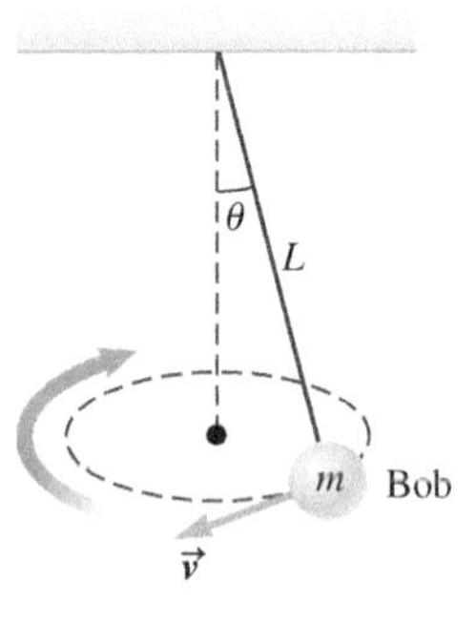

FIGURE P6.68

69. N A child is swinging a 325-g ball at the end of a 74.0-cm-long string in a vertical circle. The string can withstand a tension of 12.0 N before breaking. **a.** What is the tension in the string when the ball is at the top of the circle if its speed at that point is 3.40 m/s? **b.** What is the maximum speed the ball can have at the bottom of the circle if the string does not break?

70. A Suppose you place a block of mass M on a plane inclined at an angle θ. Show that the mass will slide down the plane if $\mu_s < \tan\theta$, where μ_s is the coefficient of static friction between the block and the plane.

71. A A block of mass M is placed on a plane, where μ_s is the coefficient of static friction between the block and the plane. The plane is inclined at an angle θ. When the plane is stationary, the block slips down the plane. We can keep the block in place on the incline by revolving the plane around the vertical axis shown in Figure P6.67. In this case, what is the *maximum* period of revolution (that is, the slowest rotation) that will keep the block on the incline a distance d from the lower end?

72. A Return to the situation in Problem 71, where the plane rotates around the axis as shown in Figure P6.67. What is the plane's *minimum* rotation period if the block is to remain stationary? *Hint*: Use the solution to Problem 71.

Problems 73 and 74 are paired.

73. N A car is driving around a flat, circularly curved road with a radius of 5.00×10^2 m. The mass of the car is 1500 kg, and the coefficient of static friction between its tires and the road is 0.30. **a.** Using the work of Example 6.7 (pages 202–203), what is the maximum speed the car can have without slipping? **b.** What is the magnitude of the centripetal force on the car if it were to travel at the maximum speed around the track?

74. A car is driving around a banked, circularly curved road with a radius of 5.00×10^2 m. The mass of the car is 1500 kg, and the road is banked at an angle of 20°.
   a. N Using the results of Example 6.8 (pages 204–205) and ignoring friction, what is the maximum speed the car can have without slipping?
   b. N What is the magnitude of the centripetal force on the car if it were to travel at the maximum speed around the track?
   c. C Would including friction in this problem increase the maximum speed allowed or decrease it? Explain your answer.

75. N Two children, with masses $m_1 = 35.0$ kg and $m_2 = 43.0$ kg, are swinging on a tire swing attached to a tree overhanging a pond. The mass of the tire is negligible. At the lowest point of the swinging motion, the tension in each of the three vertical 4.00-m-long chains supporting the swing is 275 N. **a.** What is the speed of the children at the lowest point of the swinging motion? **b.** What is the force exerted on each child by the tire at the lowest point of the swinging motion?

76. C Chris, a recent physics major, wanted to design and carry out an experiment to show that an object's mass determines its inertia. He used an ultrasound device to measure acceleration of a low-friction cart attached to a hanging block to provide the same force on the cart during each run (Fig. P6.76A). Chris varied the mass of the cart by varying the number of lead rods placed in it.

Chris used Newton's second law $\sum F_x = F_T = Ma_x$ to predict his results. He reasoned that because $\vec{F}_T$ is the same for each run, the cart's acceleration should be inversely proportional to its mass:

$$a_x = \frac{F_T}{M} = \frac{\text{constant}}{M} \quad (1)$$

Chris's goal was to show that his data fit Equation (1). He decided to analyze his results by plotting a_x as a function of $1/M$; Equation (1) predicted that he should get a straight line, passing through the origin with a slope equal to the tension (red line in Fig. P6.76B):

Chris ran several trials for each run, averaged his results and estimated the error. He then plotted his data (green line in Fig. P6.76B). Chris was excited to see that he correctly predicted that the data fell along a straight line:

$$a_x = (0.27\ \text{N})\frac{1}{M} - (0.048\ \text{m/s}^2)$$

According to the straight-line fit to the data, the slope of the line is 0.27 N, which was close to the weight of the hanging mass and therefore close to the tension in the string. Chris, though, was disappointed to see that the line had a negative intercept. Mathematically, as $M \to \infty$, $\frac{1}{M} \to 0$. Chris was confused because he believed that as the mass increased, the cart's acceleration should approach zero. He was quite sure that he did not discover some new property of inertia or mass. After convincing himself that he was not being careless in the laboratory and that his data were correct, he started to search for an explanation for the discrepancy between his prediction and his data. Help Chris find an explanation.

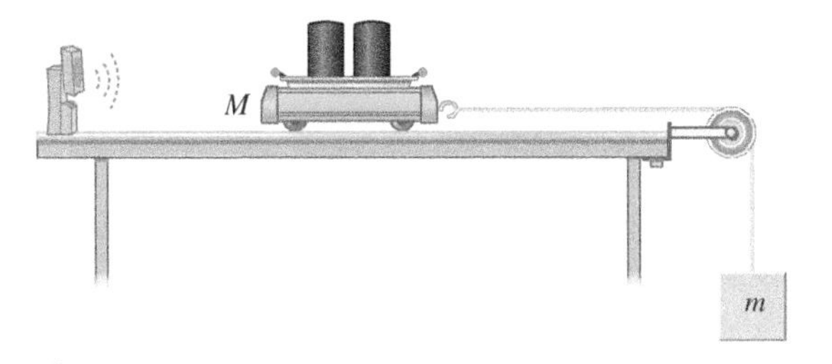

A.

B.

FIGURE P6.76 A. Chris's experimental apparatus. **B.** Chris's prediction (red line) and experimental results (green line).

77. A Having just crossed the finish line in the 200-km Finland ice marathon with a final straight-line sprint, a skater stands to full height and coasts to a stop by the resistive force of air, which is proportional to the square of his speed v and is given by $F_D = -kmv^2$, where m is the skater's mass and k is a constant. What is the skater's speed at time t after he begins to coast?

78. An aircraft carrier approaching home port cuts its engines and coasts to a full stop at the dock. Upon cutting engine power, the aircraft carrier's speed is seen to be decreasing exponentially, given by $v = v_i e^{-0.050t}$, where t is in seconds.

a. N If the aircraft carrier's initial speed is 22 knots (11.3 m/s), what is its speed at $t = 30.0$ s after cutting its engines?

b. A What is the acceleration of the aircraft carrier as a function of time?

79. N The radius of circular electron orbits in the Bohr model of the hydrogen atom are given by $(5.29 \times 10^{-11}\ \text{m})n^2$, where n is the electron's energy level (Fig. P6.79). The speed of the electron in each energy level is $(c/137n)$, where $c = 3 \times 10^8$ m/s is the speed of light in vacuum. **a.** What is the centripetal acceleration of an electron in the ground state ($n = 1$) of the Bohr hydrogen atom? **b.** What are the magnitude and direction of the centripetal force acting on an electron in the ground state? **c.** What are the magnitude and direction of the centripetal force acting on an electron in the $n = 2$ excited state?

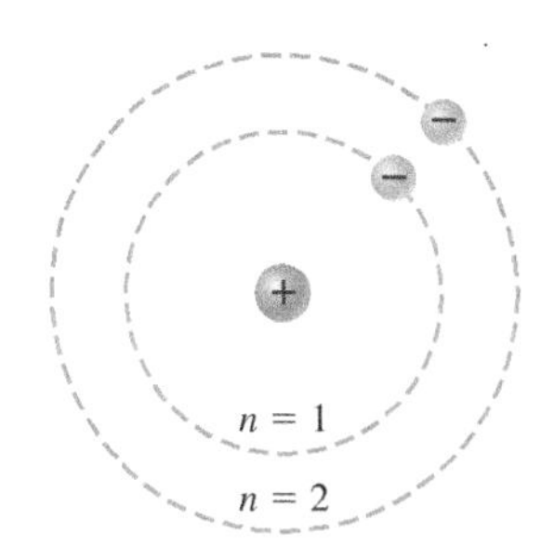

FIGURE P6.79

80. N A particle of dust lands 45.0 mm from the center of a compact disc (CD) that is 120 mm in diameter. The CD speeds up from rest, and the dust particle is ejected when the CD is rotating at 90.0 revolutions per minute. What is the coefficient of static friction between the particle and the surface of the CD?

81. N Since March 2006, NASA's Mars Reconnaissance Orbiter (MRO) has been in a circular orbit at an altitude of 316 km around Mars (Fig. P6.81). The acceleration due to gravity on the surface of the planet Mars is $0.376g$, and its radius is 3.40×10^3 km. Assume the acceleration due to gravity at the satellite is the same as on the planet's surface. **a.** What is MRO's orbital speed? **B.** What is the period of the spacecraft's orbit?

FIGURE P6.81

7 Gravity

Key Questions

How is Newton's law of universal gravity applied in various situations?

What is the scientific process that leads to important developments such as Newton's law of universal gravity?

7-1 A knowable universe 185

7-2 Kepler's laws of planetary motion 187

7-3 Newton's law of universal gravity 190

7-4 The gravitational field 196

7-5 Variations in the Earth's gravitational field 203

Underlying Principles

Newton's law of universal gravity

Major Concepts

1. Geocentric model
2. Heliocentric model
3. Empirical laws
4. Inertial mass
5. Gravitational mass
6. Gravitational field

Special Cases

1. Kepler's first law
2. Kepler's second law
3. Kepler's third law

Tools

1. Ellipse
2. Inverse-square law

Most of us cannot remember a time in our lives before we heard about gravity. It seems perfectly logical to us that gravity keeps our feet on the ground and the planets in their orbit. When these notions of gravity—which we take for granted—were discovered, however, they overthrew two millennia of thoughts on the nature of the Universe, and most people found gravity incomprehensible. The powerful effect of studying gravity does not stop there. When Albert Einstein studied gravity, he developed theories that took physics in a whole new direction. Even today, gravity is the most elusive of the fundamental forces.

This chapter has two goals: (1) to study the physical properties of gravity, and (2) to think about the process that led to its discovery. As any historian will say, we learn a lot about the present day by studying the past. So, as we study the history of gravity, let us reflect on the way science operates today and the lessons we can learn from the past.

7-1 A Knowable Universe

Scientists are very much a part of the society in which they live. That society influences the work scientists do; in turn, discoveries made by scientists influence society. Technology, wars, national poverty or affluence, and religious beliefs are among the many factors that impact scientific progress. For one scientist's discovery to become accepted as part of our collective picture of the Universe, that discovery must be supported by other scientists. Scientists, like all other members of society, may have trouble seeing the truth of a discovery because they have personal biases, and sometimes a discovery acceptable in the final analysis is initially rejected.

Today, we take for granted many scientific discoveries that were once highly controversial. We teach our children that the Earth is a planet, for instance, and that, like the other planets in the solar system, it orbits the Sun. When the idea that the Sun (and not the Earth) is the center of the solar system was first put forth, however, it ran counter to societal beliefs at the time. The scientists who put forth this and other equally controversial theories were bold, courageous, and insightful. Once such theories were supported by observation and experimentation, they were accepted and changed society. Newton's law of universal gravity is one such important theory.

This chapter is devoted to one force, gravity, one of the four fundamental forces (Section 5-10).

A closer look at gravity will be a good foundation for our study of the electric force in Chapters 23-29.

By focusing on gravity, Einstein formulated his theory of general relativity (Chapter 39). So, to appreciate general relativity, we must have a good understanding of gravity.

Newton's theory of gravity marked the birth of modern science; accepting his findings required an important shift in how scientists think, what questions they ask, and how they answer them. Part of any engineer's or scientist's job is to develop the critical thinking skills necessary to evaluate new theories, and understanding the process that led to Newton's understanding of gravity will help you hone these skills.

FIGURE 7.1 Tycho Brahe's observatory. He made the best-possible naked-eye observations of the sky.

Although we think of it as Newton's work alone, formulating the law of universal gravity involved many scholars and required the collection of accurate data, intellectual honesty, and bold insight in interpreting those data. The data came from astronomical observations. Tycho Brahe (1546–1601) spent two decades, from 1577 until he left Denmark in 1597, making the best-possible naked-eye observations of the sky (Fig. 7.1). At the time Tycho was collecting his data, ideas about the Universe and how it "works" were heavily influenced by early Greek philosophers as well as by Christian theology. Many scholars still accepted the **geocentric model** (*geo-*, "Earth"; *-centric*, "centered") of the solar system, a model described in detail by Ptolemy, a Greek astronomer and mathematician who lived in the second century CE. According to this model, the Sun, the Moon, the stars, and the planets orbited the Earth. Even before Ptolemy's time, however, a few scholars supported a **heliocentric model** (*helio-*, "Sun") of the Universe, a model in which the Earth and other planets orbit the Sun. This heliocentric model was then redeveloped by Nicolaus Copernicus (1473–1543). Both of these early models assumed astronomical bodies move in uniform circular motion. To account for astronomical observations, both models included elaborate combinations of circular motion (Fig. 7.2).

GEOCENTRIC MODEL AND HELIOCENTRIC MODEL

Major Concept

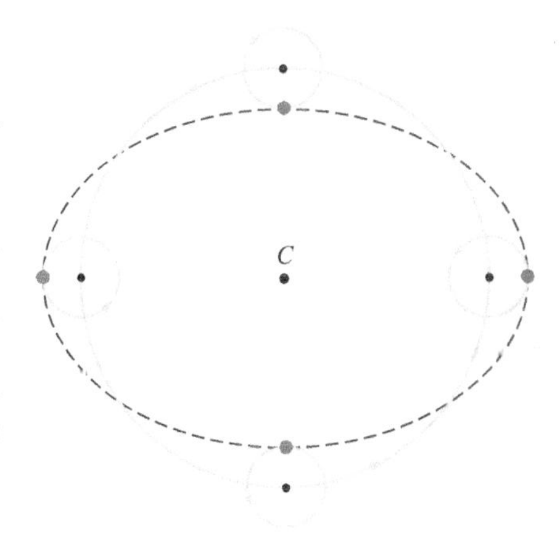

FIGURE 7.2 The Ptolemaic model of the solar system was based on the theory that the planets move in uniform circular motion around the Earth located near the center at C. Today, we know that planetary motion is not uniform and not circular.

In his own day, Tycho became famous for his discovery of what he thought was the birth of a new star. (Today, we know that what he observed was actually a star *dying*, but that's another story.) In gratitude for his findings, King Frederick of Denmark gave Tycho an estate and money to establish an astronomical observatory. At the time, most people who studied nature did not believe that quantitative observations were necessary. Tycho, unlike his contemporaries, strived to make the most accurate measurements possible. He observed as often as he could and estimated the magnitude of his measurement *uncertainties,* also called *errors*. He examined his methods so as to find sources of errors and make improvements. His hard work paid off, and his measurements are the best ever made without binoculars or a telescope.

Tycho collected data on the naked-eye planets (Mercury, Venus, Mars, Jupiter, and Saturn), tracking movements over periods of days, months, and years. He then used his data to develop a model of the solar system. Needing the help of a skilled mathematician, he hired Johannes Kepler (1571–1630) to develop a new model, but the two disagreed about it. Tycho's model was a combination of geocentric and heliocentric, and not until Tycho died did Kepler develop a fully heliocentric model.

CONCEPT EXERCISE 7.1

What important experimental skills can we learn from Tycho Brahe?

CASE STUDY Dark Matter and MOND

Today, the field of astrophysics extends far beyond our solar system. We have models of our galaxy, other galaxies, and the Universe as a whole.

There is still much to learn, however. This case study focuses on one controversial issue: the existence of *dark matter* (matter that we have not observed directly). If we haven't observed it, how do we know that dark matter exists? The answer is that we have observed dark matter's influences on ordinary objects such as stars and galaxies. The term *luminous matter* describes objects we can detect directly with our eyes, a telescope or other instrument. So far, the existence of dark matter has only been deduced *indirectly* by observing its interaction with luminous matter.

To get a handle on one way the existence of dark matter is deduced, let's look at the story of the star Sirius, the brightest star in the night sky. In the early 1800s, Sirius was observed to wander back and forth in the sky (Fig. 7.3). From Newton's laws of motion, a force must act on this star to produce this nonuniform motion, and there must be a source of this force. By observing the motion of Sirius, scientists in the early 1800s *indirectly* detected an unseen object.

In 1862, Alvan Clark, an American astronomer and telescope maker, directly observed that Sirius is actually made up of two stars, one much brighter than the other. The brighter one is now called Sirius A; it was the one that was seen to wander back and forth. The fainter one is called Sirius B; it could only be detected directly by Clark's powerful telescope. Although the fainter Sirius B is hard to observe, it provides the gravitational force responsible for the wandering motion observed in Sirius A, the brighter star.

Both Sirius A and B are ordinary matter, not dark matter. The original story of Sirius A and B before Clark's observation, however, illustrates one way that dark matter is indirectly detected. Just as in the case of Sirius B, the existence of dark matter is deduced from the acceleration of luminous matter. According to Newton's second law, a force must be responsible for accelerating the luminous matter. The claim is that the dark matter supplies the necessary gravitational force.

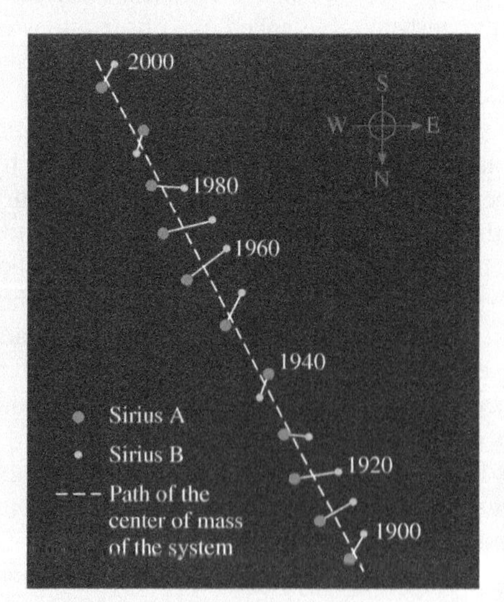

FIGURE 7.3 Observations of Sirius A made at 10-year intervals show that it appears to wobble in the night sky. Initially, astronomers speculated that the gravity of an unseen object caused the wobble. Today, we are able to observe a small secondary star named Sirius B.

There is, however, a major difference between faint objects such as Sirius B and dark matter. Sirius B (before Clark's observation) is an example of *"non-luminous" ordinary matter* that happens to be very faint and hard to detect directly. Such ordinary matter is made up of familiar particles such as protons, neutrons, and electrons that have all been detected directly in laboratories. **Dark matter** is *extraordinary* matter made up of some other types of particles. So far, no laboratory experiments have provided direct evidence for the existence of dark matter, but because of the great deal of indirect evidence, many scientists today are certain that dark matter exists.

Among the many astronomical observations providing evidence for dark matter, we will consider only one greatly simplified piece of evidence, the motion of stars in our own galaxy. Our Milky Way galaxy is a spiral galaxy that from the side looks like a disk with a spherical bulge (Fig. 7.4). The galaxy consists of roughly 200 billion to 400 billion stars and 10^{40} kg of gas. These stars and gas orbit the central bulge much as the eight major planets of our solar system orbit the Sun. For the stars and gas far from the center of the Milky Way, the observed centripetal acceleration is greater than can be accounted for by the gravitational force exerted by the ordinary matter in the galaxy. Scientists hypothesize that dark matter must also be present to account for the

Spitzer/NASA

FIGURE 7.4 The Andromeda galaxy is a spiral galaxy believed to look much like the Milky Way galaxy.

gravitational force required. In this case study, we will use Newton's law of gravity to estimate the dark matter content in our own galaxy.

Because dark matter has yet to be directly observed, some physicists are skeptical that dark matter exists. Instead, they say that Newton's laws must be incorrect. While scientists agree that these laws are valid in many situations, some scientists argue that Newton's laws must be modified to explain phenomena that naturally occur only in astronomical situations. For example, theoretical astrophysicist Mordehai Milgrom has developed a theory of *mo*dified *N*ewtonian *d*ynamics (MOND), a very controversial theory that has not been widely accepted and the details of which are beyond the scope of this book. In this chapter, however, we consider MOND as an alternative to the theory of dark matter as we think about the scientific method.

7-2 Kepler's Laws of Planetary Motion

Kepler began his work on planetary orbits by analyzing the information Tycho had collected on Mars. Some of the data gave a circular orbit for this planet, in agreement with the belief at the time that all heavenly bodies moved in circular orbits. Two of Tycho's Mars measurements, however, could not be fit to a circular orbit. At that time, most mathematicians might have ignored the discrepancy, arguing that all measurements are subject to experimental uncertainty. Kepler, however, realized that even taking into account the uncertainty Tycho estimated for his measurements, these data did not fit a circular model. Intellectually honest in his search for the truth, Kepler boldly tossed out the 2000-year-old assumption of circular orbits. This action was the critical step in his subsequent discovery of his three laws of planetary motion.

Kepler's First Law

KEPLER'S FIRST LAW

Special Case

To understand Kepler's first law, you need to refresh what you know about ellipses. One way to see the difference between an ellipse and a circle is to draw them both. Figure 7.5A shows how to draw a circle. A tack holds a piece of string on one end. The length of the string is r. A pencil is attached to the other end. The pencil traces out a circle because each point on the circle is the same distance r from the tack.

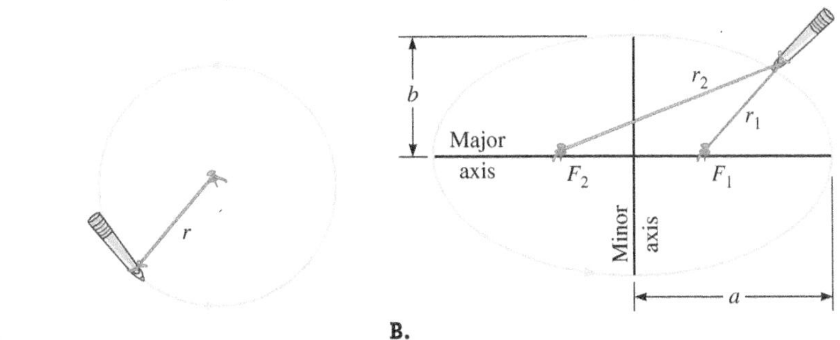

FIGURE 7.5 A. A circle is drawn by attaching a pencil to a string held down by a single tack. Every point on the circle is the same distance r from the center of the circle. **B.** An ellipse is drawn by sliding a pencil against a string that is fixed by two tacks located at the two foci F_1 and F_2. The total distance $r_1 + r_2$ is constant. A circle is a special case of an ellipse.

Two tacks are needed to draw an ellipse. One end of the string is tacked at point F_1, and the other end is tacked at point F_2 (Fig. 7.5B). The points F_1 and F_2 are known as the **foci** (plural of **focus**) of the ellipse. The length of the string $r_1 + r_2$ is greater than the distance between the two tacks. A pencil keeps the string taut. The distances between the pencil and the points F_1 and F_2 are r_1 and r_2, respectively. As the pencil traces an ellipse, r_1 and r_2 change, but the length of the string $r_1 + r_2$ remains constant. So, an **ellipse** is the locus of points for which the sum of the distances from the two foci F_1 and F_2 to the perimeter of the ellipse is constant. If the tacks are brought closer together, the ellipse becomes more like a circle. In fact, if one tack is driven into the top of the other, the ellipse is a circle of radius $(r_1 + r_2)/2$.

ELLIPSE Tool

Figure 7.5B shows the major and minor axes of the ellipse. The **major axis** passes through both foci. The **minor axis** is the perpendicular bisector of the major axis. Either half of the major axis is called a *semimajor axis*, and either half of the minor axis is called a *semiminor axis*. The lengths of the semimajor and semiminor axes are a and b, respectively.

Kepler's first law, illustrated in Figure 7.6, states that **planetary orbits are ellipses with the Sun at one focus**. The ellipse in Figure 7.6 is obviously not a circle, but the elliptical orbits of most planets are very nearly circular. Also, notice that there is nothing at the other focus.

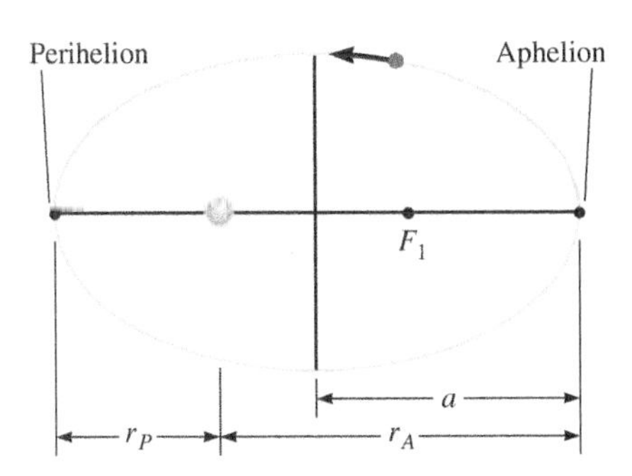

FIGURE 7.6 According to Kepler's first law, planetary orbits are elliptical. The Sun is at one focus, and there is nothing at the other focus.

The closest position of a planet to the Sun is called the planet's **perihelion** (*peri-*, "around, near"; *helios*, "Sun"). The farthest position is the planet's **aphelion** (*api-*, "away from"). The length of the semimajor axis a is the average distance the planet is from the Sun, and this distance can be found from the perihelion and aphelion distances r_P and r_A:

$$a = \frac{r_P + r_A}{2} \tag{7.1}$$

For a circular orbit, $r_A = r_P = a = r$. The Earth's orbit is only slightly elliptical; its perihelion distance, $r_P = 1.48 \times 10^{11}$ m, is nearly equal to its aphelion distance, $r_A = 1.52 \times 10^{11}$ m. The semimajor axis of the Earth's orbit is

$$a = \frac{r_P + r_A}{2} = \frac{1.48 \times 10^{11}\text{ m} + 1.52 \times 10^{11}\text{ m}}{2} = 1.50 \times 10^{11}\text{ m}$$

This average Earth–Sun distance is used to define the **astronomical unit (AU)**: 1 AU $= 1.50 \times 10^{11}$ m. Distances in the solar system are usually measured in astronomical units.

KEPLER'S SECOND AND THIRD LAWS

Special Cases

Kepler's Second Law

For nearly two millennia, scholars believed that the motion of astronomical objects was based on uniform circular motion. Kepler's first law, published in 1609, eliminated the circular part, and Kepler's second law, published at the same time, removed the uniform part. According to **Kepler's second law, a line joining a planet to the Sun sweeps out equal areas in equal time intervals.**

Put simply, Kepler's second law says that a planet's speed is highest when it is near the Sun.

Figure 7.7 illustrates Kepler's second law for a planet shown at 12 positions. Because it is a motion diagram, the time intervals between adjacent positions are equal all around the ellipse. According to Kepler's second law, all the pie-shaped wedges have the same area. For that to be true, the planet's speed must change. The planet moves more slowly near aphelion than it does near perihelion.

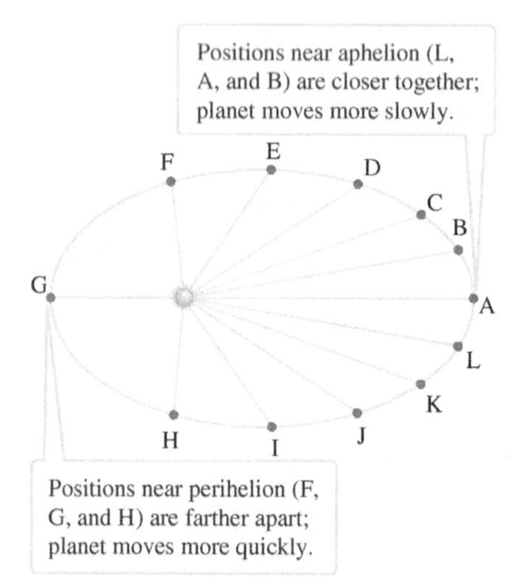

FIGURE 7.7 A motion diagram for a planet orbiting the Sun. According to Kepler's second law, all the pie-shaped wedges have the same area.

Kepler's Third Law

Unlike Kepler's first two laws, which are about the orbit of a single planet, his third law involves all the planets. The time it takes a planet to complete one orbit is known as the planet's **period**. Kepler found that a planet's period depends on its average distance from the Sun. Planets that are closer to the Sun have shorter periods than planets that are farther from the Sun. Mercury, the planet closest to the Sun, has a period of less than three months; Earth's is, of course, one year; and Pluto, a very distant body, has a period of nearly two and a half centuries.

Kepler's third law states that **the square of a planet's period T is proportional to the cube of its semi-major axis,** or $T^2 = Ca^3$. The constant C is the same for every object orbiting the Sun. We can find C easily from the Earth's orbital data. For the Earth, $T = 1$ yr and $a = 1$ AU; therefore,

$$C = \frac{T^2}{a^3} = \frac{(1\text{ yr})^2}{(1\text{ AU})^3} = 1\text{ yr}^2/\text{AU}^3$$

Therefore, if the period T of any object is measured in years and its semimajor axis a in astronomical units, Kepler's third law is expressed mathematically as

$$T^2 = (1\text{ yr}^2/\text{AU}^3)a^3$$

$$T^2_{[\text{yr}]} = a^3_{[\text{AU}]} \tag{7.2}$$

where the subscripts are a reminder that this form of Kepler's third law holds only when these two particular units are used. The constant C with its units is implied in Equation 7.2, so the equation does not imply that time is somehow equivalent to distance. Figure 7.8 shows the elliptical orbit, period, and average distance from the Sun for each major planets and Pluto.

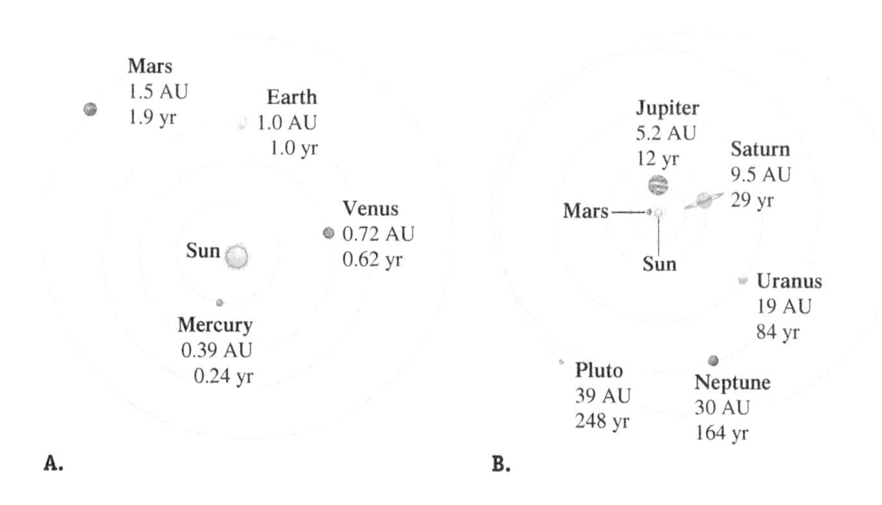

FIGURE 7.8 A. The period and semimajor axis of each planet in the inner solar system. **B.** The period and semimajor axis of each planet in the outer solar system and Pluto. (Scales are different in parts A and B.)

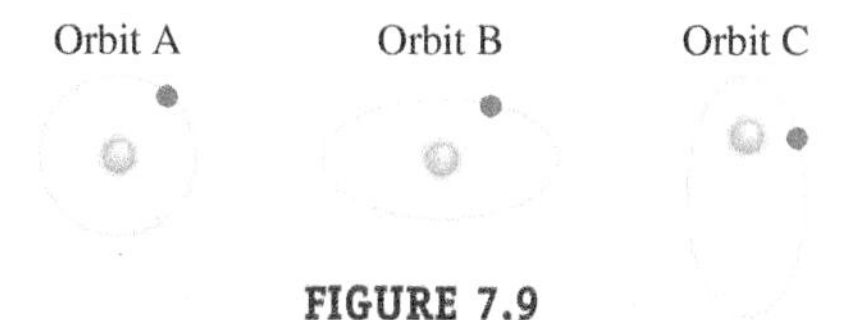

FIGURE 7.9

CONCEPT EXERCISE 7.2

Three possible planetary orbits are shown in Figure 7.9. According to Kepler's first law, which orbits are possible and which are not? Explain.

CONCEPT EXERCISE 7.3

Planet A's orbit is circular with the Sun at the center. Planet B's orbit is elliptical with the Sun at one focus (Fig. 7.10). The radius of planet A's orbit is equal to the semimajor axis of planet B's orbit: $r = a$.

a. Compare the periods of the two planets.
b. Which planet reaches the higher speed? Explain.

FIGURE 7.10

EXAMPLE 7.1 The Largest Planet in the Solar System

Use Kepler's third law to find Jupiter's period from its semimajor axis $a = 5.2$ AU.

INTERPRET and ANTICIPATE

An object with a semimajor axis greater than 1 AU should have a period longer than 1 year (Fig. 7.8).

SOLVE

Solve Equation 7.2 for period T.	$T^2_{[\text{yr}]} = a^3_{[\text{AU}]}$
Substitute the semimajor axis value.	$T = (5.2\text{ AU})^{3/2} = 12\text{ yr}$

CHECK and THINK

The period is greater than 1 year as expected.

EXAMPLE 7.2 The Closest Planet to the Sun

Mercury's period is 0.24 yr. Find Mercury's average distance from the Sun.

INTERPRET and ANTICIPATE

An object with a period shorter than 1 year should have a semimajor shorter than 1 AU (Fig. 7.8).

Example continues on page 190 ▶

SOLVE Solve Equation 7.2 for a.	$T^2_{[\text{yr}]} = a^3_{[\text{AU}]}$ $a_{[\text{AU}]} = T^{2/3}_{[\text{yr}]}$
Substitute the given value for the period.	$a = (0.24\ \text{yr})^{2/3} = 0.39\ \text{AU}$

CHECK and THINK
As expected, our result is less than 1 AU.

EMPIRICAL LAWS Major Concept

Kepler's three laws of planetary motion are often described as **empirical laws**. Empirical laws are formulated based solely on data, without any hypothesis presented first, which was certainly true of Kepler's first law. Kepler had no theoretical reason to try fitting an ellipse to Tycho's observations of Mars, and he had not observed any mechanism that causes elliptical motion in the heavens. Kepler's contemporaries theorized that astronomical objects were "perfect" and therefore would exhibit perfect motion, which to them was uniform and circular. People often find it difficult to give up their theories, and such researchers would probably have never tried fitting an ellipse to these data.

It is not entirely correct to say that Kepler's second and third laws are strictly empirical. Kepler theorized that the Sun influences the motion of the planets. He reasoned that the closer a planet is to the Sun, the faster the planet would move. It is true that he did not need this theoretical basis to derive his second or third laws, but it seems likely that his theory helped him discover these laws more easily. It still remains true that Kepler had no single theory that accounts for all three laws; as far as he knew, a different mechanism may have been responsible for each law. It was Newton who developed a theoretical basis for all three laws.

CONCEPT EXERCISE 7.4

Imagine a comet whose period is 500 years.

a. What is its average distance from the Sun?
b. If its perihelion distance is 0.5 AU, what is its aphelion distance?
c. Where does the comet spend most of its time: in the inner solar system, where it is < 2 AU from the Sun, or in the outer solar system, where it is > 2AU from the Sun?

CONCEPT EXERCISE 7.5

Today's employees are rewarded for *thinking outside the box*. In what ways did Kepler think outside the box?

7-3 Newton's Law of Universal Gravity

Because Galileo supported the idea of a heliocentric solar system, the Roman Catholic Church placed him under house arrest.The conflict between the Roman Catholic Church and Galileo was complicated. His house arrest was likely influenced by his personality and by global factors such as the Protestant Reformation. In the 1990s, Pope John Paul II expressed regret for the church's response to Galileo and acquitted him.

Galileo Galilei (1564–1642) used a telescope for astronomical observations. It was the first known time a scientist used a device to explore the heavens beyond the capability of his own senses. Galileo's observations of Venus provided important supporting evidence for Kepler's first law. Galileo found that, like the Earth's moon, Venus goes through phases (new, crescent, quarter, gibbous, and full). (In the geocentric model, Venus is always in a region of space that is between the Earth and the Sun; therefore, the geocentric model predicts that Venus would never go through quarter, gibbous, or full phases.) For an observer on the Earth to see these phases of Venus, it must orbit the Sun and not the Earth.

Isaac Newton (1642–1727) was born the year that Galileo died. By the time he studied Kepler's laws, Newton could assume they were solid. Galileo's observations supported the heliocentric model, and Kepler's laws fit observations of planetary orbits. Newton was thus able to focus on finding a general theory to explain all three laws. One of his remarkable insights was his hypothesis that terrestrial and celestial physical laws are the same. Starting from this idea, Newton theorized that gravity is not limited to objects near the Earth but instead applies to all objects in the Universe. Today, this theory is known as **Newton's law of universal gravity.**

Like Kepler, Newton theorized that the Sun exerts a force on the planets and that this force is responsible for their kinematics. Newton identified this force as gravity. He used Kepler's second and third laws to derive two properties of gravity. From Kepler's second law, Newton found that the force of gravity must be directed straight toward the Sun. Using Kepler's third law, he mathematically reasoned that the magnitude of the gravitational force F_G felt by any planet depends on the distance r between the Sun and the planet. In particular, Newton found that gravity obeys an *inverse-square law*, which means that the gravitational force on a planet diminishes with the inverse square of its distance from the Sun:

UNIVERSAL GRAVITATIONAL FORCE
Underlying Principle

INVERSE-SQUARE LAW Tool

$$F_G \propto \frac{1}{r^2} \tag{7.3}$$

We will see in later chapters that many other quantities in physics diminish with the inverse square of distance.

Having figured out these two properties of gravity, Newton's next step was an important test of the validity of his thinking. He used his laws of motion and these two properties of gravity that he had inferred from Kepler's second and third laws to derive Kepler's first law. His success in deriving Kepler's first law showed that all three of Kepler's planetary laws stemmed from one physical theory: The Sun exerts a gravitational force on all the planets.

Newton's law of universal gravity has three features:

1. Every particle with nonzero mass in the Universe exerts a gravitational force on every other particle with nonzero mass in the Universe.
2. The gravitational force is attractive. For example, the Sun pulls the Earth toward the Sun, and the Earth pulls the Sun toward the Earth.
3. The magnitude of the gravitational force between two particles is directly proportional to the product of their masses and inversely proportional to the square of their separation.

Mathematically, the law of universal gravity is summarized as

$$F_G = G\frac{m_1 m_2}{r^2} \tag{7.4}$$

where G is the **universal gravitational constant** found through experimentation to be

$$G = 6.673 \times 10^{-11}\,\text{N} \cdot \text{m}^2/\text{kg}^2$$

Figure 7.11A shows two particles separated by a distance r. First, consider particle 1 as the source of the gravitational force. Then, particle 1 exerts a gravitational force $\vec{F}_{[1 \text{ on } 2]}$ on the subject, particle 2. According to Newton's third law, we could, instead, consider particle 2 to be the source. Then, particle 2 exerts a gravitational force $\vec{F}_{[2 \text{ on } 1]}$ on the subject, particle 1. Further, from Newton's third law, the magnitude of $\vec{F}_{[1 \text{ on } 2]}$ is equal to the magnitude of $\vec{F}_{[2 \text{ on } 1]}$. It doesn't matter how unequal the two masses are; the magnitude of the forces is the same. Particle 1's pull on particle 2 is just as strong as particle 2's pull on particle 1. We can see this equality from Equation 7.4 because the universal gravitational constant G, the separation r, and the product of the two masses $m_1 m_2$ are the same no matter which is considered the source and which the subject.

All the features of universal gravity can be expressed as one vector equation using the polar coordinate system introduced in Section 4-6. Particle 1 is placed at the origin of a polar coordinate system, and $\hat{r}$ points from particle 1 to particle 2 (Fig. 7.11B). With this choice, the gravitational force exerted by particle 1 on particle 2 is

$$\vec{F}_{[1 \text{ on } 2]} = -F_G\hat{r}$$

$$\vec{F}_{[1 \text{ on } 2]} = -G\frac{m_1 m_2}{r^2}\hat{r}$$

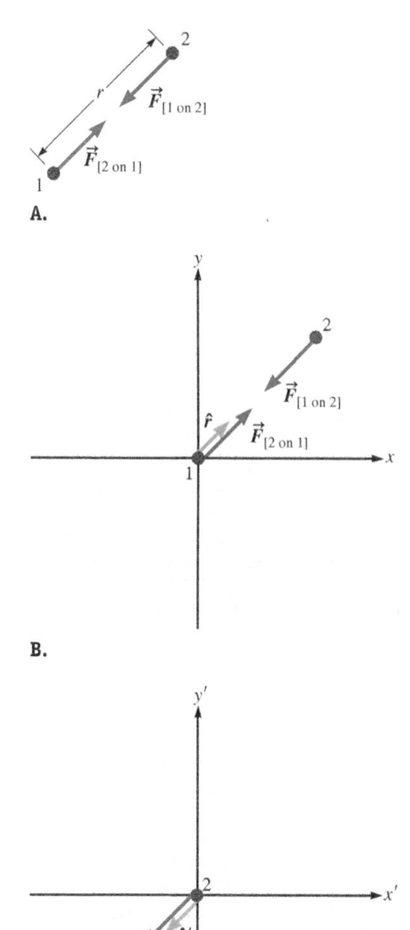

FIGURE 7.11 **A.** The gravitational force particle 1 exerts on particle 2 is equal to the magnitude of the gravitational force exerted by particle 2 on particle 1. **B.** With particle 1 located at the origin of a coordinate system, the unit vector $\hat{r}$ points from particle 1 toward particle 2. **C.** Another choice of coordinate system places particle 2 at the origin. Now unit vector $\hat{r}'$ points from particle 2 toward particle 1.

and the force exerted by particle 2 on particle 1 is

$$\vec{F}_{[2 \text{ on } 1]} = F_G\hat{r} = G\frac{m_1 m_2}{r^2}\hat{r}$$

Of course, it is possible to put particle 2 at the origin (Fig. 7.11C). With this coordinate system, we have

$$\vec{F}_{[1 \text{ on } 2]} = F_G\hat{r}' = G\frac{m_1 m_2}{r^2}\hat{r}'$$

$$\vec{F}_{[2 \text{ on } 1]} = -F_G\hat{r}' = -G\frac{m_1 m_2}{r^2}\hat{r}'$$

where $\hat{r}'$ is the unit vector that points from particle 2 toward particle 1.

The law of universal gravity is stated in terms of particles. Is it possible to treat a large object such as a planet as a particle? The answer depends on both the shape of the source, and distance between the source and the subject. A uniform spherical source may be treated as a particle as long as the subject it acts on is outside the sphere. For example, if we are interested in the gravitational force exerted by the Earth on a comet, we treat the Earth as a particle because it is (nearly) uniform and spherical and the comet is outside the Earth. For a subject outside a sphere, the gravitational force acts as though all the sphere's mass were concentrated at its center. Then, r in Equation 7.4 is measured from the center of the sphere. (This property is due to the inverse-square nature of the law of universal gravity, a feature shared by the electric force.) If, on the other hand, we are interested in a comet that has fallen inside the gaseous Sun, we cannot model the Sun as a particle.

Nonspherical objects such as galaxies consisting of hundreds of billions of stars may be modeled as particles if the center-to-center distance between them is much greater than their size. The distance between the Earth and any object located on the surface of the Earth is approximated as being the radius of the Earth $r = R_\oplus$ (Fig. 7.12). (Recall that the symbol ⊕ stands for "Earth.") We can model objects located on the surface on the Earth as particles as far as gravity is concerned because the distance ($R_\oplus = 6.37 \times 10^6$ m) between the Earth's center and anything on its surface is far greater than the size of the object.

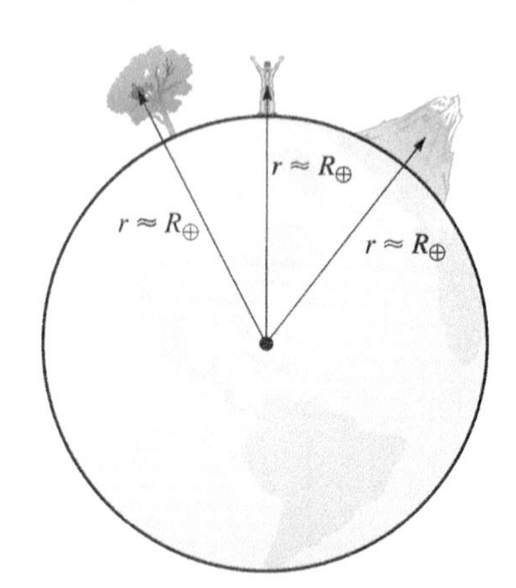

FIGURE 7.12 We can model objects such as apples, people, and mountains on the surface of the Earth as particles.

EXAMPLE 7.3 Cavendish's Experiment

With any two nonastronomical objects, the gravitational force is weak, which makes it difficult to measure the universal gravitational constant G. In fact, this constant is the least well determined of all the physical constants.

In 1798, Henry Cavendish used a device designed by John Michell to measure the density of the Earth. His results could have been used to measure G. Today, many physics students use a tabletop apparatus based on Michell's design to do just that.

Figure 7.13 is a schematic of a contemporary "Cavendish" apparatus. Two large spheres of mass M are attached to the frame of the apparatus. Two smaller spheres of mass m are attached to a lightweight rod that is free to rotate. A laser beam is reflected from a mirror attached to the rod, first with the large spheres in one position and then with the large spheres moved to a second position. By measuring the change in the reflected light's path, a student infers the motion of the two small spheres caused by the gravitational force exerted on them by the large spheres. This device is a torsion balance (Chapter 16).

The masses of the spheres are 1.50 kg and 38.3 g. What is the magnitude of the gravitational force between one large sphere and one small one when they are separated by 46.5 mm?

Cavendish apparatus, top view

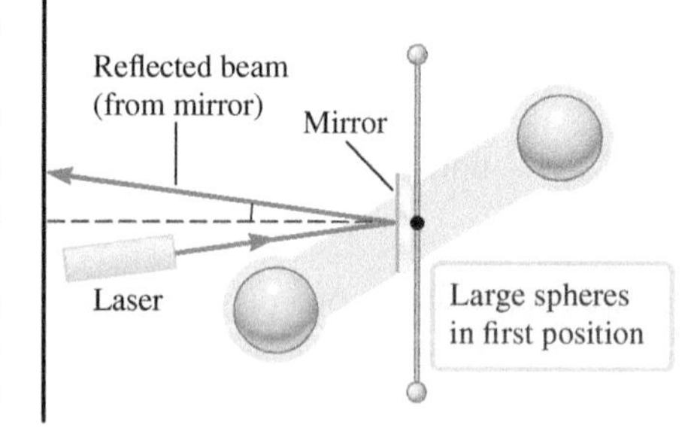

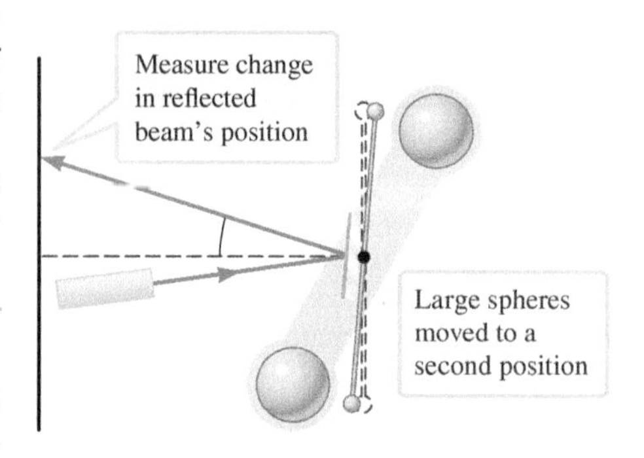

FIGURE 7.13

INTERPRET and ANTICIPATE

This example is a straightforward application of Newton's law of universal gravity (Eq. 7.4). The key is to realize that the separation distance given is actually the center-to-center distance.

Remember to work in SI units.	$F_G = G\frac{m_1 m_2}{r^2} = (6.67 \times 10^{-11}\,\text{N}\cdot\text{m}^2/\text{kg}^2)\frac{(1.50\,\text{kg})(38.3 \times 10^{-3}\,\text{kg})}{(46.5 \times 10^{-3}\,\text{m})^2}$ $F_G = 1.77 \times 10^{-9}\,\text{N}$

CHECK and THINK

Gravity is weak. This result is about seven orders of magnitude lower than the weight of a single grain of rice.

EXAMPLE 7.4 A Cluster of Galaxies

Three galaxies are at the vertices of an equilateral triangle, with the distance d between any two of them being much, much greater than their diameters (Fig. 7.14). Use the relative masses indicated to find an expression for the gravitational force on C due to galaxies A and B.

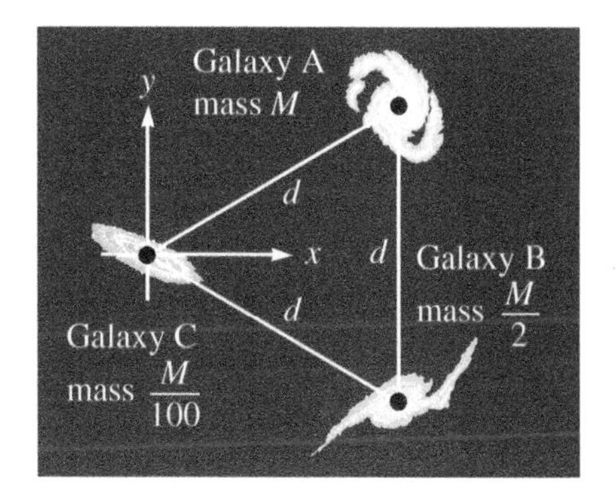

FIGURE 7.14 Not to scale.

INTERPRET and ANTICIPATE

Because the distance between galaxies is much greater than their sizes, we can treat them as particles. A free-body diagram helps interpret the situation.

We choose to put C at the origin of a coordinate system. As shown on our free-body diagram (Fig. 7.15), two gravitational forces act on C: the force exerted by A on C, $\vec{F}_{[\text{A on C}]}$, and the force exerted by B on C, $\vec{F}_{[\text{B on C}]}$. We have sketched the triangular configuration of the galaxies to help find the angles between these forces and the x axis. Because the interior angles of an equilateral triangle are 60°, the angle between each force and the x axis is 30° as shown. We do not need the acceleration, so we do not indicate it on the diagram.

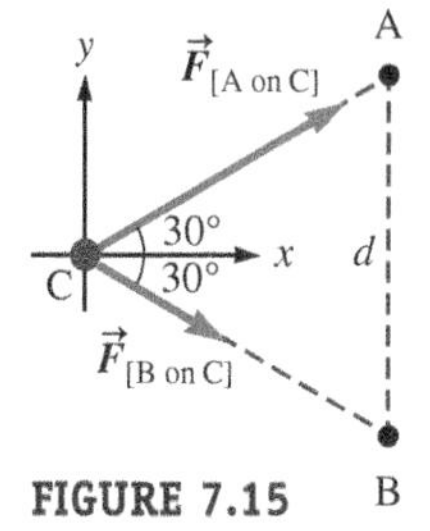

FIGURE 7.15

SOLVE

Start by breaking $\vec{F}_{[\text{A on C}]}$ and $\vec{F}_{[\text{B on C}]}$ into components.	$(F_{[\text{A on C}]})_x = F_{[\text{A on C}]}\cos 30° = \frac{\sqrt{3}}{2}F_{[\text{A on C}]}$ $(F_{[\text{A on C}]})_y = F_{[\text{A on C}]}\sin 30° = \frac{1}{2}F_{[\text{A on C}]}$ $(F_{[\text{B on C}]})_x = F_{[\text{B on C}]}\cos 30° = \frac{\sqrt{3}}{2}F_{[\text{B on C}]}$ $(F_{[\text{B on C}]})_y = -F_{[\text{B on C}]}\sin 30° = -\frac{1}{2}F_{[\text{B on C}]}$
Then, find the total force on C in component form.	$\vec{F}_{\text{tot}} = (\frac{\sqrt{3}}{2}F_{[\text{A on C}]} + \frac{\sqrt{3}}{2}F_{[\text{B on C}]})\hat{\imath} + (\frac{1}{2}F_{[\text{A on C}]} - \frac{1}{2}F_{[\text{B on C}]})\hat{\jmath}$ (1)
We now take into account that $\vec{F}_{[\text{A on C}]}$ and $\vec{F}_{[\text{B on C}]}$ are gravitational forces by applying $F_G = G(m_1 m_2/r^2)$ (Eq. 7.4) to find their magnitudes.	$F_{[\text{A on C}]} = G\frac{M(M/100)}{d^2} = \frac{GM^2}{100d^2}$ (2) $F_{[\text{B on C}]} = G\frac{(M/2)(M/100)}{d^2} = \frac{1}{2}F_{[\text{A on C}]}$
Simplify Equation (1) for $\vec{F}_{\text{tot}}$ by substituting $F_{[\text{B on C}]} = \frac{1}{2}F_{[\text{A on C}]}$.	$\vec{F}_{\text{tot}} = (\frac{\sqrt{3}}{2}F_{[\text{A on C}]} + (\frac{\sqrt{3}}{2})\frac{1}{2}F_{[\text{A on C}]})\hat{\imath} + (\frac{1}{2}F_{[\text{A on C}]} - (\frac{1}{2})\frac{1}{2}F_{[\text{A on C}]})\hat{\jmath}$ $\vec{F}_{\text{tot}} = \frac{3\sqrt{3}}{4}F_{[\text{A on C}]}\hat{\imath} + \frac{1}{4}F_{[\text{A on C}]}\hat{\jmath}$
Next, substitute Equation (2) for $F_{[\text{A on C}]}$.	$\vec{F}_{\text{tot}} = \frac{3\sqrt{3}}{4}\left(\frac{GM^2}{100d^2}\right)\hat{\imath} + \frac{1}{4}\left(\frac{GM^2}{100d^2}\right)\hat{\jmath} = \left(\frac{3\sqrt{3}GM^2}{400d^2}\right)\hat{\imath} + \left(\frac{GM^2}{400d^2}\right)\hat{\jmath}$

CHECK and THINK

To check our expression, we first consider the dimensions. In each term, we have GM^2/d^2, which has the dimensions of force.

$$\left[\frac{GM^2}{d^2}\right] = \left(\frac{\mathrm{F \cdot L^2}}{\mathrm{M^2}}\right)(\mathrm{M^2})\left(\frac{1}{\mathrm{L^2}}\right) = \mathrm{F}$$

Next, we see from our free-body diagram that the x components of the forces add together and the y components partially cancel each other. We therefore expect the term in x to be greater than the term in y, as we found.

$$\left(\frac{3\sqrt{3}GM^2}{400d^2}\right) > \left(\frac{GM^2}{400d^2}\right)$$

DERIVATION Kepler's Third Law

Newton derived the law of universal gravity from Kepler's second (equal areas) and third ($T^2 \propto a^3$) laws. In Section 13-7, we will derive Kepler's second law. Here, we derive Kepler's third law for an object in a circular orbit around the Sun,

$$T^2 = \left(\frac{4\pi^2}{GM_\odot}\right)r^3 \tag{7.5}$$

GENERAL FORM OF KEPLER'S THIRD LAW

Special Case

from Newton's laws. We then extend this equation to the general form of Kepler's third law of an object in an elliptical orbit with another object of mass M at one focus:

$$T^2 = \left(\frac{4\pi^2}{GM}\right)a^3 \tag{7.6}$$

Consider a planet in a circular orbit around the Sun. Draw a free-body diagram for this planet at a particular moment (Fig. 7.16). We have chosen a polar coordinate system with $\hat{r}$ directed from the planet to the Sun. The only force acting on it is the gravitational force due to the Sun.

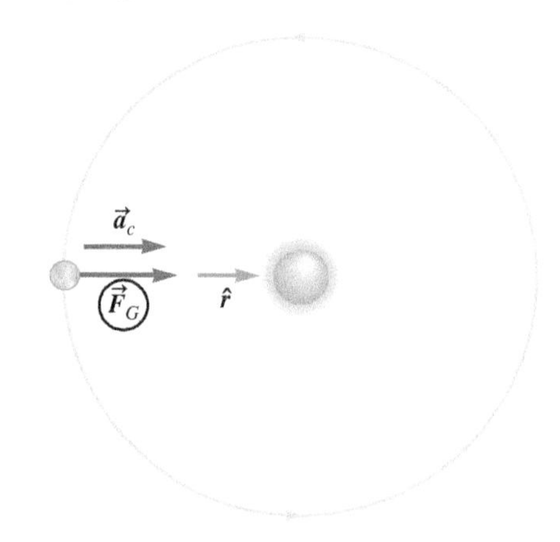

FIGURE 7.16

Gravity depends on the planet's distance to the Sun r, and the centripetal force required depends on the planet's period T. The key to finding the relationship between T and r is that the gravitational force of the Sun provides the centripetal force (circled in the free-body diagram).

Applying Newton's second law shows that the net force exerted on the planet is gravity and that it equals the centripetal force. Substitute Equation 6.7, $F_c = m(v^2/r)$, for F_C and Equation 7.4, $F_G = G(m_1m_2/r^2)$, for F_G.

$$\sum F_r = M_P a_c$$

$$F_G = F_C$$

$$G\frac{M_\odot M_P}{r^2} = M_P\frac{v^2}{r}$$

The planet mass M_P cancels out of this equation, and so does one power of r.

$$v^2 = G\frac{M_\odot}{r} \tag{7.7}$$

We find the planet's speed from Equation 4.30.

$$v = \frac{2\pi r}{T} \tag{4.30}$$

Square that speed and set it equal to Equation 7.7.

$$v^2 = \frac{4\pi^2 r^2}{T^2} = \frac{GM_\odot}{r}$$

Rearrange this equation to arrive at Kepler's third law for a planet in a circular orbit around the Sun.

$$T^2 = \left(\frac{4\pi^2}{GM_\odot}\right)r^3 \; \checkmark \tag{7.5}$$

COMMENTS

A more general derivation yields Equation 7.6 which holds when the orbit is elliptical with semimajor axis a and an object of mass M is at one focus. If G and M are expressed in SI units, then T and a are in seconds and meters, respectively.

$$T^2 = \left(\frac{4\pi^2}{GM}\right)a^3 \tag{7.6}$$

Next, let's compare the first equation for Kepler's third law (Eq. 7.2) with this general expression. The term $(4\pi^2/GM)$ is the same for each planet in an elliptical orbit around the Sun, and we arbitrarily use the letter C to represent that term.	$T^2 = Cr^3$
For the Earth, $T = 1$ yr and $r = 1$ AU. Therefore, if T is measured in years and r is measured in astronomical units, then $C = 1\ \text{yr}^2/\text{AU}^3$. So, Equation 7.2 is consistent with Equation 7.6 when time is measured in years and distance in astronomical units.	$T^2_{[\text{yr}]} = a^3_{[\text{AU}]}$ (7.2)

Gravitational and Inertial Mass

It may seem obvious that the gravitational force between any two objects depends on the objects' masses, but there is no theoretical basis for this. That is, no one knows why the gravitational force depends on inertia (mass) and not some other property of the objects, such as volume, density, or composition.

The mass in the law of universal gravity (Eq. 7.4) is referred to as the object's **gravitational mass**. It is the property of the particles that creates a gravitational force between them. As we know, the mass in Newton's second law ($\sum \vec{F}_{\text{tot}} = m\vec{a}$) is the *inertial mass* of an object. Experimental evidence supports the idea that the gravitational mass of any object equals its inertial mass.

GRAVITATIONAL MASS AND INERTIAL MASS

Major Concept

Let us see how this reasoning works. We *assume* the gravitational mass m_{grav} is identical to the inertial mass m_{inert} and then see if this assumption matches experimental evidence. From Equation 7.4, we know that the magnitude of the gravitational force on an object of gravitational mass m_{grav} near the Earth's surface is

$$F_G = G\frac{M_\oplus m_{\text{grav}}}{R_\oplus^2}$$

Suppose our object of inertial mass m_{inert} is in free fall near the Earth's surface. If it is, the only force acting on it is gravity, and we use Newton's second law to find the object's free-fall acceleration g:

$$F_G = m_{\text{inert}}g$$

$$G\frac{M_\oplus m_{\text{grav}}}{R_\oplus^2} = m_{\text{inert}}g$$

Using our assumption that inertial mass and the gravitational mass are identical,

$$m_{\text{inert}} = m_{\text{grav}} \equiv m$$

This assumption leads to

$$G\frac{M_\oplus m}{R_\oplus^2} = mg$$

$$\frac{GM_\oplus}{R_\oplus^2} = g \qquad (7.8)$$

Equation 7.8 predicts that g is a constant because $GM_\oplus/R_\oplus^2$ is a constant. We have found a testable prediction for the assumption that $m_{\text{inert}} = m_{\text{grav}} \equiv m$: If the assumption is correct, the gravitational acceleration near the surface of the Earth g has the same value for all objects and does not depend on any property such as size or shape of the object in free fall.

We know from countless experiments carried out over centuries that in the absence of air resistance, all objects near the surface of the Earth fall at the same acceleration g no matter what physical properties they have. Therefore, we infer that the gravitational mass of any object is equal to its inertial mass from the observation that the acceleration due to gravity is constant near the surface of the Earth.

The idea that gravitational mass and inertial mass are identical may seem trivial, but this idea is actually very important and led Einstein to discover his theory of *general relativity* (Chapter 39). As a last thought on this matter, imagine that the gravitational mass did *not* equal the inertial mass. If that were the case, you might

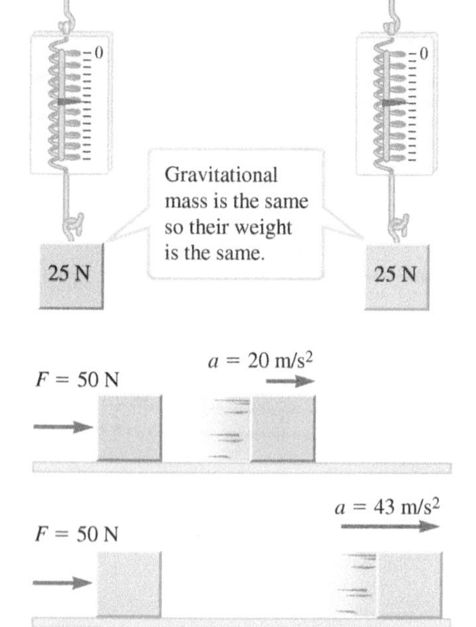

FIGURE 7.17 If gravitational mass and inertial mass were not the same, two objects that have the same weight may not acquire the same acceleration when the same net force is exerted on each.

weigh two objects on a spring scale and find that they have identical weights, but when you pushed each object horizontally along a frictionless surface, the same force would *not* produce the same acceleration (Fig. 7.17).

7-4 The Gravitational Field

Imagine the task of putting several spacecraft with different masses into orbit at some distance r from the center of the Earth (Fig. 7.18). One thing you would need to know in such a project is the Earth's gravitational force exerted on each spacecraft when it is in orbit. We find that force from Newton's law of universal gravity (Eq. 7.4), expressed in vector form:

$$\vec{F}_G(r) = -G\frac{M_\oplus m_{SC}}{r^2}\hat{r} \tag{7.9}$$

where m_{SC} is the mass of a particular spacecraft and the unit vector $\hat{r}$ points outward from the Earth toward a spacecraft. You would use Equation 7.9 three times for the three different spacecraft to obtain three different values for $\vec{F}_G$. If the three spacecraft were all at the same distance r from the Earth, the difference in the gravitational force exerted on a particular spacecraft would depend only on the mass of that spacecraft. In Figure 7.18, $\vec{F}_G$ is strongest for the middle spacecraft because it has the greatest mass and is weakest for the rightmost spacecraft because it has the smallest mass.

In this situation, where the Earth is present in all three cases but there are three different spacecraft, it is convenient to think of the Earth as the source of the gravitational force and each spacecraft as a subject of that force. We can rewrite Equation 7.9 as

$$\vec{F}_G(r) = m_{SC}\left(-G\frac{M_\oplus}{r^2}\hat{r}\right) \tag{7.10}$$

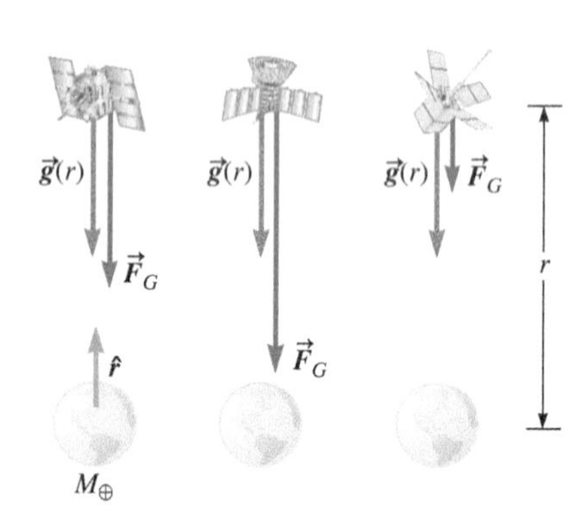

FIGURE 7.18 The gravitational force $\vec{F}_G$ on a spacecraft at a position r depends on the mass of the spacecraft that is actually there. The gravitational field $\vec{g}$ at r is the same no matter which spacecraft is there. In fact, the gravitational field is the same even if nothing is there.

The mass m_{SC} is a property of the subject (each spacecraft). The term in parentheses is the same in all three cases independent of the particular spacecraft; it depends on the source (the Earth). For convenience, let's give this term its own mathematical symbol:

$$\vec{g}(r) \equiv -\frac{GM_\oplus}{r^2}\hat{r} \tag{7.11}$$

where $\vec{g}(r)$ depends only on the source. Then, the gravitational force is written in terms of two pieces, one that depends on the subject (m_{SC}) and the other that depends on the source $\vec{g}(r)$:

$$\vec{F}_G(r) = m_{SC}\vec{g}(r)$$

Breaking a field force such as gravity (or the electric force) into pieces—one piece that depends on the subject and the other that depends on the source—is a powerful technique that is used in many situations, and it introduces another concept, the field. The general term *field* refers to a region under the influence of some physical source. So, in this case, the source is the Earth, and the region is the environment around the Earth. Every position in a field is assigned a particular value of some quantity that is influenced by the source. If the source's influence is a vector quantity, the field is a **vector field**, and each position in the field is associated with a particular magnitude and direction. The Earth's gravitational field is a vector field, and Equation 7.11 assigns a magnitude and direction to every position in that field (Fig. 7.19). The term *field* also refers to the mathematical description of the source's influence; in this case, $\vec{g}(r)$. So, Equation 7.11 is known as the *gravitational vector field* of the Earth.

Although the terms *force* and *field* sound alike, they have very different meanings. Unlike the gravitational force, the gravitational field depends only on a source (the Earth) and not on any subject (spacecraft) in the field. The Earth's gravitational field $\vec{g}(r)$ is a measure of how the Earth influences its environment and is a function only of distance from the center as indicated by the (r) in the notation. In Figure 7.18, the gravitational field $\vec{g}(r)$ is the same at the location of each spacecraft. In fact, even if all the spacecraft vanished, $\vec{g}(r)$ would still have the same magnitude and direction because it depends only on the mass $M_\oplus$ of the Earth and the distance r from the center of the Earth.

The word *vector* is often dropped when referring to specific vector fields, so we refer to a *gravitational field*, not a *gravitational* ***vector*** *field.*

This case of the various spacecraft and the Earth illustrates the following important concepts behind a vector field.

1. The field has a **source** that influences its surroundings. We are interested in knowing the gravitational force on a variety of spacecraft, so we choose the Earth as the source of the gravitational field.
2. At each position r, the magnitude and direction of the Earth's gravitational field $\vec{g}(r)$ **only depend on properties of the source**, which is the Earth's mass in this case.
3. **Subjects** are items that may be placed in the source's field. The force exerted on a subject depends on properties of the source and the subject. In this case, the gravitational force exerted on the spacecraft depends on the mass of the spacecraft and the Earth's gravitational field at the spacecraft's location.

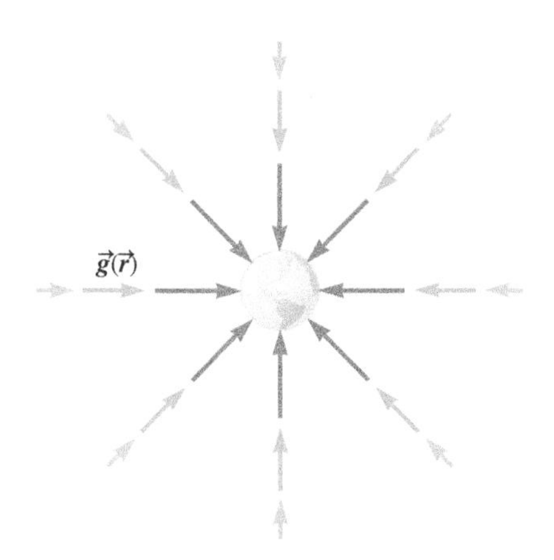

FIGURE 7.19 The Earth's gravitational field points to the center of the Earth at each point. It is strongest near the surface of the Earth, where the vectors are the longest. The gravitational field is equivalent to the local free-fall acceleration at each point.

The dimensions of the gravitational field are force per mass, so it has SI units of newtons per kilogram. Also notice that the gravitational field has the dimensions of acceleration

$$[\![g]\!] = \frac{\text{F}}{\text{M}} = \frac{\text{M} \cdot \text{L/T}^2}{\text{M}} = \frac{\text{L}}{\text{T}^2}$$

We can think of the Earth's gravitational field at a particular point as being the local free-fall acceleration (which is not true for other fields, such as the electric field). Figure 7.19 shows that the Earth's gravitational *field* is strongest near the surface. So, the free-fall acceleration is higher near the surface than it is at any point out in space. In fact, near the surface of the Earth, the gravitational field has the value of the free-fall acceleration we encountered in Chapter 2 and Equation 7.8:

$$\vec{g}(R_\oplus) = -\frac{GM_\oplus}{R_\oplus^2}\hat{r} \approx -9.8\hat{r}\ \text{m/s}^2 \tag{7.12}$$

We use only two significant figures here because the free-fall acceleration $g = 9.81\ \text{m/s}^2$ we have been using is actually slightly less than the value calculated when we substitute into Equation 7.12. The reason for the discrepancy is discussed in Section 7.5.

EXAMPLE 7.5 Apples and the Moon

One of Newton's important contributions to science was the idea that the laws of physics are universal; there is no difference between terrestrial and celestial laws. This idea underlies the law of universal gravity. Newton said that the same gravitational pull by the Earth that causes an apple to fall to the ground is what keeps the Moon in its orbit. Earlier scientists did not necessarily think that the Earth could influence the Moon and would not have used properties of the Moon to better understand the Earth. Newton, however, used the period of the Moon's orbit to estimate the magnitude of the Earth's gravitational *field* at the Moon's distance. This example is based on his estimation.

A Use the period of the Moon's orbit to find the magnitude of the Moon's centripetal acceleration, assuming the orbit is circular.

INTERPRET and ANTICIPATE

This part of the problem is a straightforward application of the concept of uniform circular motion from Chapter 4. We expect a numerical result of the form $a_c =$ ____ m/s^2.

SOLVE

Because our ultimate task is to find the value of something at the location of the Moon, let us label the Earth–Moon distance $r_{[\text{at Moon}]}$. Apply $\vec{a}_c = -(v^2/r)\hat{r}$ (Eq. 4.36) for centripetal acceleration.	$a_c = \dfrac{v^2}{r_{[\text{at Moon}]}}$ (1)
The period of the Moon's orbit is roughly a month. Find the Moon's speed from its period using $v = 2\pi r/T$ (Eq. 4.30).	$v = \dfrac{2\pi r_{[\text{at Moon}]}}{T}$
Substitute into Equation (1). Notice that one power of $r_{[\text{at Moon}]}$ cancels out.	$a_c = \dfrac{4\pi^2 r_{[\text{at Moon}]}}{T^2}$

Example continues on page 198 ▶

The Earth–Moon distance is $r_{[\text{at Moon}]} = 3.84 \times 10^8$ m, and the period of the Moon's orbit is $T = 27.3$ d $= 2.36 \times 10^6$ s.	$a_c = \dfrac{4\pi^2(3.84 \times 10^8 \text{ m})}{(2.36 \times 10^6 \text{ s})^2}$ $a_c = 2.72 \times 10^{-3} \text{ m/s}^2$

CHECK and THINK

Our result is in the form we expected.

B Now use the mass of the Earth and the Earth–Moon distance to find the magnitude of the Earth's gravitational *field* at the Moon, $g(r_{[\text{at Moon}]})$.

INTERPRET and ANTICIPATE

Let us make the approximation that the only force acting on the Moon is the gravitational force of the Earth. Therefore, (1) the Moon is in free fall, and (2) the Earth's gravity alone provides the centripetal force (Section 6-6) that keeps the Moon in its orbit. So, we expect the centripetal acceleration of the Moon to equal the Earth's gravitational *field* at distance $r_{[\text{at Moon}]}$.

SOLVE Use Equation 7.11, omitting the unit vector $\hat{r}$ because we are interested in magnitude only.	$g(r_{[\text{at Moon}]}) = \dfrac{GM_{\oplus}}{r^2_{[\text{at Moon}]}}$ (7.11) $g(r_{[\text{at Moon}]}) = \dfrac{(6.67 \times 10^{-11} \text{N} \cdot \text{m}^2/\text{kg}^2)(5.98 \times 10^{24} \text{ kg})}{(3.84 \times 10^8 \text{ m})^2}$ $g(r_{[\text{at Moon}]}) = 2.70 \times 10^{-3} \text{ N/kg} = 2.70 \times 10^{-3} \text{ m/s}^2$

CHECK and THINK

We found essentially what we expected to find: $a_c \approx g(r_{[\text{at moon}]})$. There is a slight discrepancy (less than 1%) due to our assumptions that the Moon's orbit is a circle centered on the center of the Earth and that only the Earth provides a gravitational force on it. Other objects, such as the Sun, also exert a gravitational force on it.

EXAMPLE 7.6 Your TV Satellite

A *geosynchronous* satellite is one that orbits with the same 24-hour period as the Earth's rotation. Thus, such a satellite is always above the same point on the Earth. Find the altitude of a geosynchronous satellite in a circular orbit.

INTERPRET and ANTICIPATE

As in Example 7.5, we assume the only force acting on the satellite is the Earth's gravity, which must provide a centripetal force.

SOLVE The magnitude of the centripetal acceleration is equal to the magnitude of the gravitational field at the satellite's position.	$a_c = g(r)$ $\dfrac{v^2}{r} = \dfrac{GM_{\oplus}}{r^2}$ $v^2 = \dfrac{GM_{\oplus}}{r}$ (1)
For a geosynchronous satellite to remain always above the same place on the Earth, its period must be 24 hours. Use Equation 4.30 to find the satellite's speed.	$v = \dfrac{2\pi r}{T}$ (4.30)
We eliminate v from Equations (1) and 4.30 and solve for r. Notice that our result looks like Kepler's third law (Eq. 7.5) with the Sun's mass replaced by the Earth's mass.	$\dfrac{4\pi^2 r^2}{T^2} = \dfrac{GM_{\oplus}}{r}$ $r = \left(\dfrac{GM_{\oplus} T^2}{4\pi^2}\right)^{1/3}$

Substitute numerical values.	$r = \left[\frac{(6.67 \times 10^{-11}\ \text{N}\cdot\text{m}^2/\text{kg}^2)(5.98 \times 10^{24}\ \text{kg})(8.64 \times 10^4\ \text{s})^2}{4\pi^2}\right]^{1/3}$ $r = 4.23 \times 10^7\ \text{m}$
This result is the distance from the center of the Earth to the satellite. To find the altitude h of the satellite above the ground, subtract the Earth's radius.	$h = r - R_\oplus$ $h = 4.23 \times 10^7\ \text{m} - 6.37 \times 10^6\ \text{m}$ $h = 3.59 \times 10^7\ \text{m}$

CHECK and THINK

We are often asked to find a numerical result that is not easy to confirm, but in this case it is easy to find the actual altitudes of geosynchronous orbits. For example, according to the National Aeronautics and Space Administration (NASA) and the *Encyclopedia Britannica*, the altitude of a satellite's geosynchronous orbit is about 36,000 km.

So far, we have focused on the Earth as a source of a gravitational field, but, of course, anything with mass creates a gravitational field. We can find the gravitational field of any source that may be modeled as a particle of mass M using

$$\vec{g}(r) \equiv -\frac{GM}{r^2}\hat{r} \tag{7.13}$$

GRAVITATIONAL FIELD OF A PARTICLE

Major Concept

where $\hat{r}$ is a unit vector pointing away from the particle. Equation 7.13 also applies at the surface and outside any spherical source, such as a planet or star; in this case, $\hat{r}$ points away from the center of the spherical source. Equation 7.13 also applies to sources that are not spherical—galaxies, for instance, which might be disk-shaped—when the distance from the source is much greater than the size of the source.

If more than one particle is the source of a gravitational field, the resulting field at any point is found by adding the gravitational fields due to each particle individually. We will learn how to find the gravitational field due to more complicated sources when we study another field force, electricity, in Chapter 24.

EXAMPLE 7.7 SOHO: *So*lar and *H*eli*o*spheric Satellite

SOHO is a space observatory used to study the Sun, orbiting the Sun with the same 1-yr period as the Earth. As it does so, the observatory maintains its position, known as the first Lagrange or L1 point, between the Earth and the Sun. In this example, we will find the distance between the Earth and the L1 point.

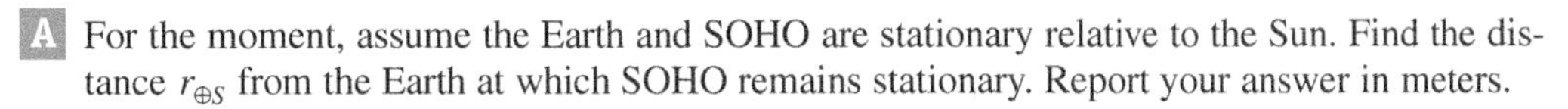

A For the moment, assume the Earth and SOHO are stationary relative to the Sun. Find the distance $r_{\oplus S}$ from the Earth at which SOHO remains stationary. Report your answer in meters.

INTERPRET and ANTICIPATE

A sketch as in Figure 7.20 helps interpret this problem. The Earth and SOHO are assumed to be stationary. SOHO is located at a point S between the Earth and the Sun, and S must be the point at which the vector sum of the Earth's gravitational field and the Sun's gravitational field is zero. Because the Sun is much more massive than the Earth, we expect S to be closer to the Earth than to the Sun.

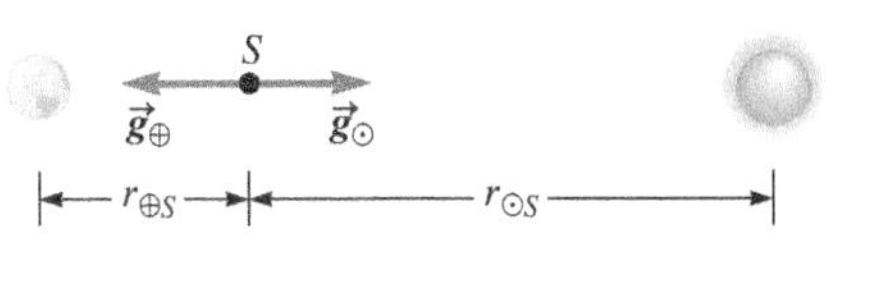

FIGURE 7.20

SOLVE The gravitational fields due to the Earth and the Sun are both given by $\vec{g}(r) \equiv -\frac{GM}{r^2}\hat{r}$ (Eq. 7.13). At S, the magnitudes of the two fields are equal. Let's call the Sun–SOHO distance $r_{\odot S}$ and the Earth–SOHO distance $r_{\oplus S}$.	$\frac{GM_\odot}{r_{\odot S}^2} = \frac{GM_\oplus}{r_{\oplus S}^2}$ (1)

Example continues on page 200 ▶

Because we have two unknowns, $r_{\odot S}$ and $r_{\oplus S}$, we need another equation. This second equation comes from the average distance between the Earth and the Sun, 1 AU.

$$r_{\odot S} + r_{\oplus S} = 1\text{ AU} \quad (2)$$

SOLVE

Solve Equations (1) and (2) simultaneously for $r_{\oplus S}$. Start by simplifying Equation (1).

$$\frac{M_\odot}{r_{\odot S}^2} = \frac{M_\oplus}{r_{\oplus S}^2} \qquad \frac{r_{\odot S}}{r_{\oplus S}} = \sqrt{\frac{M_\odot}{M_\oplus}}$$

$$r_{\odot S} = r_{\oplus S}\sqrt{\frac{M_\odot}{M_\oplus}}$$

Substitute the result into Equation (2).

$$r_{\oplus S}\sqrt{\frac{M_\odot}{M_\oplus}} + r_{\oplus S} = 1\text{ AU}$$

$$r_{\oplus S} = \frac{1\text{ AU}}{1 + \sqrt{M_\odot/M_\oplus}}$$

Substitute values and convert the answer to meters. Recall 1 AU = 1.50×10^{11} m.

$$r_{\oplus S} = \frac{1\text{ AU}}{1 + \sqrt{1.99 \times 10^{30}\text{ kg}/5.98 \times 10^{24}\text{ kg}}}$$

$$r_{\oplus S} = 1.73 \times 10^{-3}\text{ AU} = 2.60 \times 10^{8}\text{ m}$$

CHECK and THINK

As expected, point S is much closer to the Earth (1.73×10^{-3} AU) than to the Sun ($1 - 1.73 \times 10^{-3}\text{ AU} = 0.998$ AU).

B Now assume SOHO stays aligned with the Earth as both bodies orbit the Sun. The L1 point lies along the imaginary straight line connecting the Earth and the Sun, and this point is the location of SOHO as it orbits. Is $r_{\oplus L}$, the distance between the Earth and L1, the same as the distance $r_{\oplus S}$ found in part A? If not, is the distance $r_{\oplus L}$ longer or shorter than the value for $r_{\oplus S}$ found in part A?

INTERPRET and SOLVE

Because it orbits the Sun, SOHO is accelerating. Its centripetal acceleration is the net gravitational field at its position L1; therefore, this field cannot be zero. The centripetal acceleration must point toward the Sun; therefore, the net gravitational field at L1 must point toward the Sun (Fig. 7.21).

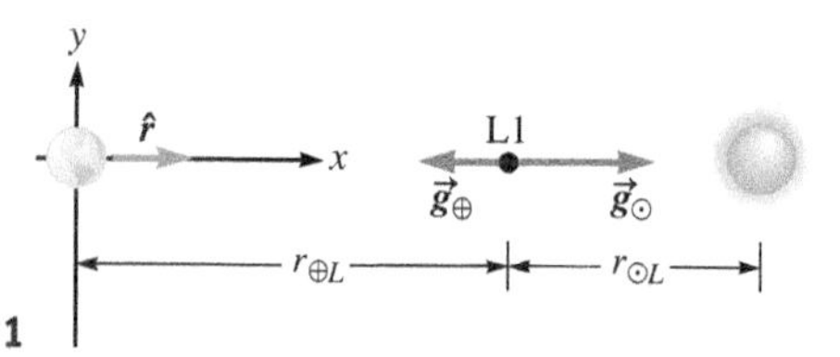

FIGURE 7.21

For the total gravitational field at L1 to point toward the Sun, $g_\odot > g_\oplus$. So, L1 must be closer to the Sun than S. Therefore, $r_{\oplus L} > r_{\oplus S}$.

C According to NASA, the L1 point for SOHO is $r_{\oplus L} = 1.55 \times 10^{9}$ m. Confirm this value.

INTERPRET and ANTICIPATE

We must develop an appropriate expression and substitute values to show that $r_{\oplus L} = 1.55 \times 10^{9}$ m is correct. The key to finding this appropriate expression comes from realizing that the centripetal acceleration of SOHO must equal the total gravitational field at L1.

SOLVE

SOHO's period is same as the Earth's: 1 year. Write the centripetal acceleration in terms of the period.

$$v = \frac{2\pi r_{\odot L}}{T}$$

$$a_c = \frac{v^2}{r_{\odot L}} = \frac{4\pi^2 r_{\odot L}}{T^2} \quad (1)$$

The total gravitational field $(\vec{g}_L)_{\text{tot}}$ at L1 comes from adding the gravitational fields due to the Earth and the Sun at L1. For this vector addition, we choose a coordinate system centered on the Earth as shown in Figure 7.21.

$$(\vec{g}_L)_{\text{tot}} = \frac{GM_\odot}{r_{\odot L}^2}\hat{r} - \frac{GM_\oplus}{r_{\oplus L}^2}\hat{r} \quad (2)$$

As in part A, the average distance between the Earth and the Sun is 1 AU.	$r_{\oplus L} + r_{\odot L} = 1\text{ AU}$ $r_{\odot L} = 1\text{ AU} - r_{\oplus L}$
Eliminate $r_{\odot L}$ from Equations (1) and (2).	$a_c = \dfrac{4\pi^2(1\text{ AU} - r_{\oplus L})}{T^2}$ (3) $(\vec{g}_L)_{\text{tot}} = \dfrac{GM_\odot}{(1\text{ AU} - r_{\oplus L})^2}\hat{r} - \dfrac{GM_\oplus}{r_{\oplus L}^2}\hat{r}$ (4)
Now substitute values into Equations (3) and (4). If we find that they are equal, we will have confirmed SOHO's distance from the Earth.	$r_{\oplus L} = 1.55 \times 10^9\text{ m}$ $r_{\odot L} = 1\text{ AU} - r_{\oplus L} = 1.48 \times 10^{11}\text{ m}$
Start with Equation (3) for the centripetal acceleration.	$a_c = \dfrac{4\pi^2(1.48 \times 10^{11}\text{ m})}{(3.15 \times 10^7\text{ s})^2} = 5.98 \times 10^{-3}\text{ m/s}^2$ (5)
Next, find the magnitude of the total gravitational field at the L1 point from Equation (4).	$(g_L)_{\text{tot}} = (6.67 \times 10^{-11}\text{ N}\cdot\text{m}^2/\text{kg}^2)\left[\dfrac{1.99 \times 10^{30}\text{ kg}}{(1.48 \times 10^{11}\text{ m})^2} - \dfrac{5.98 \times 10^{24}\text{ kg}}{(1.55 \times 10^9\text{ m})^2}\right]$ $(g_L)_{\text{tot}} = 5.98 \times 10^{-3}\text{ m/s}^2$ (6)
Compare Equations (5) and (6).	$(g_L)_{\text{tot}} = a_c = 5.98 \times 10^{-3}\text{ m/s}^2$ ✓

CHECK and THINK

Our work confirms that the L1 point is six times farther from the Earth than S, the (fictitious) stationary position. (Notice that Figures 7.20 and 7.21 are not drawn to scale.)

EXAMPLE 7.8 CASE STUDY Dark Matter in Our Galaxy

To estimate the amount of dark matter in the Milky Way galaxy, let us apply a simple model that describes the galaxy in terms of two components: a central spherical component known as the bulge and a flat component known as the disk in which the arms of the galaxy are embedded (Fig. 7.22). Stars in the disk orbit the bulge. Consider the circular orbit of a star far from the center of the galaxy on the outer edge of the disk, with orbital radius $r = 7.7 \times 10^{20}$ m and linear speed $v = 2.2 \times 10^5$ m/s. Assume we can treat all the matter in the bulge and disk as a particle located at the galactic center. An estimate of the amount of ordinary matter (both luminous matter and nonluminous)—such as stars, gas, and planets—in the galaxy is $M_{\text{ord}} = 1.8 \times 10^{41}$ kg.

Estimate the mass of the dark matter in the Milky Way galaxy.

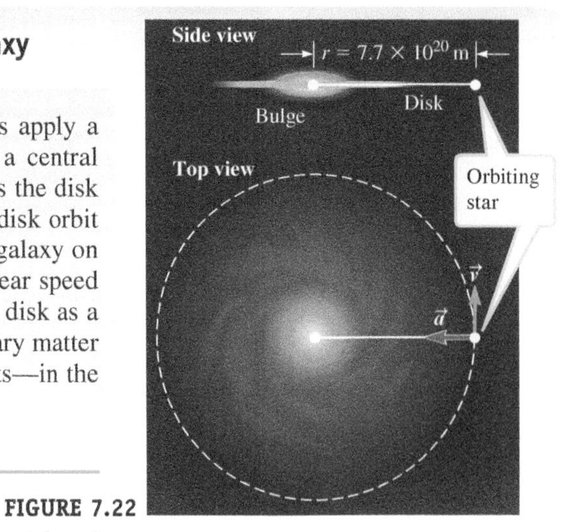

FIGURE 7.22

INTERPRET and ANTICIPATE

As in Examples 7.6 and 7.7, the key to solving this one is to realize that the centripetal force on the star at the edge of the Milky Way is provided by a gravitational force. In this case, that gravitational force is due to all the mass of the bulge and disk of the Milky Way. Once we have found the mass of the Milky Way, we will need to subtract the mass of the ordinary matter to find a numerical estimate for the mass of the dark matter.

SOLVE Start with the magnitude of the centripetal acceleration of the orbiting star.	$a_c = \dfrac{v^2}{r}$ (4.38)
Because the gravitational force of the Milky Way's bulge and disk provides the centripetal force on the orbiting star, the star's centripetal acceleration must equal the gravitation *field* of the Milky Way at the star's location. Because we are modeling the matter in the Milky Way as a particle at the center, we use Equation 7.13, $\vec{g}(r) \equiv -(GM/r^2)\hat{r}$. The mass M_{MW} is the mass of both dark and ordinary matter.	$g(r) = \dfrac{GM_{MW}}{r^2}$

Example continues on page 202 ▶

Set the magnitude of the gravitational field equal to the magnitude of the centripetal acceleration and solve for M_{MW}.	$\frac{v^2}{r} = \frac{GM_{MW}}{r^2} \qquad M_{MW} = \frac{v^2 r}{G}$
Substitute values.	$M_{MW} = \frac{(2.2 \times 10^5\,\text{m/s})^2(7.7 \times 10^{20}\,\text{m})}{6.67 \times 10^{-11}\,\text{N} \cdot \text{m}^2/\text{kg}^2}$ $M_{MW} = 5.6 \times 10^{41}$ kg
To find the mass of the dark matter, we must subtract the mass of the ordinary matter.	$M_{DM} = M_{MW} - M_{\text{ord}} = 5.6 \times 10^{41}\,\text{kg} - 1.8 \times 10^{41}\,\text{kg}$ $M_{DM} = 3.8 \times 10^{41}\,\text{kg}$

CHECK and THINK

This result means that $(3.8/5.6) \times 100 = 68\%$ of the Milky Way's matter is dark matter. We just found that there is more dark matter than all the ordinary matter in the Milky Way. To put this value in perspective, imagine standing in a room with 100 people, 68 of whom are invisible. The room might look pretty empty with only 32 people in it, but if you started to walk around, you would bump into people and find that it is actually quite crowded.

Many other indirect observations support the theory that dark matter exists. Like the evidence here, these observations show that much of the Universe is made up of extraordinary dark matter. It seems hard to believe that the kind of ordinary matter (protons, neutrons, and electrons) we are used to seeing everywhere is actually just a small component of the Universe.

EXAMPLE 7.9 CASE STUDY MOND in Our Galaxy

According to Example 7.8, 68% of our own galaxy is made up of extraordinary particles—dark matter—that have never been observed on the Earth. There is no laboratory evidence for the existence of dark matter. Some scientists—Milgrom among them—argue that there is little or no dark matter. So, how do these scientists account for the high speeds of luminous objects such as stars observed in our own Milky Way galaxy? Milgrom argues that Newton's law must be modified to take these observations into account and has developed a new theory, called MOND (*mo*dified *N*ewtonian *d*ynamics). According to this theory, Newton's laws as presented in Chapter 5 are good approximations for objects moving with relatively great acceleration, but the modified version of Newton's laws is necessary for accelerations smaller than 10^{-10} m/s^2. Is the centripetal acceleration of the star in Example 7.8 small enough that MOND (as proposed by Milgrom) plays an important role?

INTERPRET and ANTICIPATE

All the necessary concepts are covered in Chapter 4 about centripetal acceleration.

SOLVE Find the centripetal acceleration of the star (modeled as a particle).	$a_c = \frac{v^2}{r} = \frac{(2.2 \times 10^5\,\text{m/s})^2}{7.7 \times 10^{20}\,\text{m}}$ $a_c = 6.3 \times 10^{-11}\,\text{m/s}^2$

Yes, the resulting acceleration is below the MOND limit.

CHECK and THINK

Let's think about the dark-matter problem in the Milky Way galaxy. The problem stems from the observation of stars orbiting on the galaxy's outer edge. The centripetal acceleration of such stars requires a gravitational force that is too great to be supplied by the amount of luminous matter we observe. Even if we take faint objects such as planets and dim stars into account, nonluminous ordinary matter cannot make up the difference.

There are at least two possible solutions. One is that there must be some type of extraordinary matter (dark matter). Another solution is that Newton's laws do not hold up for the motion of the stars on the galaxy's outer edge.

To sort out which solution, if either, is correct, we must test each one. Tests for dark matter and MOND are beyond the scope of this book, but we can take a moment to think about tests for the solar system models in Galileo's time. When Galileo observed the sky through a telescope, the geocentric model of the solar system was widely accepted. His observations of Venus tested that geocentric model. Because the geocentric model could not account for the phases of Venus he observed, the model failed the test, and the heliocentric model gained acceptance. When there are two different competing theories in science, observations and experiments are used to eliminate one or both of them.

7-5 Variations in the Earth's Gravitational Field

Because the Earth is not a perfect sphere and because its density is not uniform, the gravitational field is not the same everywhere on the planet's surface. Roughly speaking, the Earth's gravitational field is slightly stronger at sea level than at higher elevations. The situation is made more complicated because the Earth is always changing as a result of earthquakes and volcanic eruptions. Such changes mean that the Earth's gravitational field changes over time. Space-based instruments measure these variations in the Earth's gravitational field near its surface.

The gravitational field varies both with elevation and over time, which means that the free-fall acceleration near the Earth's surface depends both on location and on time, but the variation is fairly small. Even though they are small, these variations can affect sensitive experiments. For most of the work we do in this textbook, however, we do not need to worry about these slight variations.

EXAMPLE 7.10 How Much Do You Weigh on Mount Everest?

To see just how much the gravitational field varies on the surface of the Earth, let us calculate it at sea level and on top of the highest mountain, Everest.

Find $\Delta g_{\text{alt}} = g_{\text{sea level}} - g_{\text{Everest}} = g(R_\oplus) - g(R_\oplus + h)$, where the subscript "alt" stands for altitude. To find a precise value, work to four significant figures and use the polar radius of the Earth, $R_{\oplus,\,\text{pole}} = 6.365 \times 10^6\,\text{m}$, for your sea-level calculation. The height of Mount Everest is $h = 8.848 \times 10^3\,\text{m}$ above the average radius of the Earth, $R_\oplus = 6.376 \times 10^6\,\text{m}$.

INTERPRET and ANTICIPATE

We expect to find a relatively small numerical result for Δg_{alt}.

SOLVE

Use Equation 7.13, $\vec{g}(r) \equiv -(GM/r^2)\hat{r}$, to find the magnitude of the gravitational field at each location.	$g(R_{\oplus,\,\text{pole}}) = \dfrac{GM_\oplus}{R^2_{\oplus,\,\text{pole}}}$ $\quad g(R_\oplus + h) = \dfrac{GM_\oplus}{(R_\oplus + h)^2}$
The numerator is the same in both equations.	$GM_\oplus = (6.673 \times 10^{-11}\,\text{N}\cdot\text{m}^2/\text{kg}^2)(5.976 \times 10^{24}\,\text{kg})$ $GM_\oplus = 3.9878 \times 10^{14}\,\text{N}\cdot\text{m}^2/\text{kg}$
Substitute values. Use $1\,\text{N/kg} = 1\,\text{m/s}^2$ to express g in familiar units.	$g(R_{\oplus,\,\text{pole}}) = \dfrac{3.9878 \times 10^{14}\,\text{N}\cdot\text{m}^2/\text{kg}}{(6.365 \times 10^6\,\text{m})^2}$ $g(R_{\oplus,\,\text{pole}}) = 9.843\,\text{m/s}^2$ $g(R_\oplus + h) = \dfrac{3.9878 \times 10^{14}\,\text{N}\cdot\text{m}^2/\text{kg}}{(6.376 \times 10^6 + 8.848 \times 10^3\,\text{m})^2}$ $g(R_\oplus + h) = 9.782\,\text{m/s}^2$
Find the difference in the gravitational field.	$\Delta g_{\text{alt}} = g(R_{\oplus,\,\text{pole}}) - g(R_\oplus + h) = (9.843 - 9.782)\,\text{m/s}^2$ $\Delta g_{\text{alt}} = 0.061\,\text{m/s}^2$

CHECK and THINK

This difference is small but easily measurable. It means that a 50-kg scientist weighs 492.2 N (110.7 lb) at sea level but only 489.1 N (110.0 lb) on top of Mount Everest.

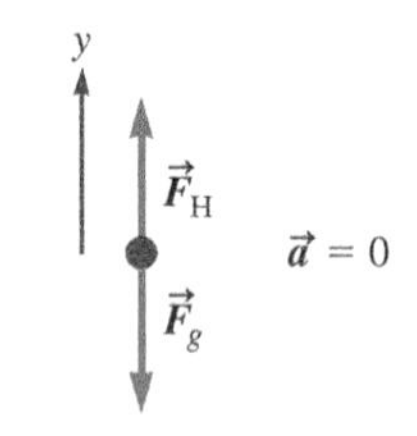

FIGURE 7.23 A. Aaron on the Earth weighs a turtle at two locations on Earth. Aaron is in a *non*inertial reference frame. **B.** The free-body diagram drawn by Aaron is the same in both locations.

Here are a few warnings about notation. First, do not confuse G and g. Uppercase G is the universal gravitational constant. Lowercase g without any other symbols is the free-fall acceleration value near the surface of the Earth. The National Institute of Standards and Technology (NIST) defined g to be $9.80665\,\text{m/s}^2$; unless other information is given, you may assume this value. Second, $\vec{g}(r)$ is the notation used for the gravitational field produced by any source. It is a vector quantity, and its SI units are newtons per kilogram. It is numerically equivalent to the local free-fall acceleration when that acceleration is expressed in meters per second squared. Its magnitude is written as $g(r)$, and only near the surface of the Earth is $\vec{g}(R_\oplus) \approx -g\hat{r}$.

The Earth as a Noninertial Reference Frame

There is one more thing we must consider regarding the gravitational field near the Earth's surface. The Earth is a *non*inertial frame. Recall from Chapter 5 that an inertial reference frame is one that is not accelerating. Because the Earth rotates on its axis and orbits the Sun, it is accelerating and is therefore a noninertial frame. Observers in the Earth's noninertial frame may see a particle violate Newton's first law. To Earth-bound observers (at the same altitude), it seems that at the Earth's surface the gravitational field varies with latitude. We refer to the magnitude of the gravitational field measured by an Earth-bound observer as the *effective* gravitational field g_{eff}.

To see how latitude affects our observations of the Earth's gravitational field, let us investigate g_{eff} at the equator and at the North Pole. Imagine Aaron weighing a turtle on a spring scale first in the Galapagos Islands, located at the Earth's equator, and then at the North Pole (Fig. 7.23A). Aaron, being in the noninertial frame of the Earth, draws the same free-body diagram for the turtle at both locations (Figure 7.23B). He identifies two forces, gravity $\vec{F}_g$ and Hooke's spring force $\vec{F}_H$, and he does not observe the turtle accelerating in either location. Applying Newton's second law to the turtle, he says that

$$\sum F_y = F_H - F_g = ma_y = 0$$

$$F_H = F_g$$

$$w_{\text{app}} = w = mg$$

For those who might wonder if turtles are found near the poles: The leatherneck turtle can withstand extreme temperatures and can be found anywhere from the equator to polar regions.

Aaron reasons that the apparent weight w_{app} of the turtle (the value read off the scale) equals the turtle's weight w (magnitude of the Earth's gravitational force on the turtle), and he expects the weight to be the same in the Galapagos as it is at the North Pole. The scale reads slightly less in the Galapagos, however, meaning that the turtle's apparent weight there is slightly less than its apparent weight at the North Pole. To Aaron on the Earth, it therefore appears that the Earth's gravitational field is weaker at the *e*quator than at the *p*ole, or $g_e < g_p$. This apparent difference in the Earth's effective gravitational field results from the observer being in a noninertial frame.

To find a mathematical expression for the effective gravitational field at the equator g_e, we must imagine another observer—Hannah—in an inertial frame hovering above the North Pole watching Aaron weigh the turtle (Fig. 7.24). Her free-body diagram is essentially the same as Aaron's at the North Pole (Fig. 7.23B). (There is one slight difference in that she would see the turtle rotating in place like a record on a turntable. For our purposes here, however, we can ignore this effect and continue to treat the turtle as a particle.) So, both Hannah hovering above the Earth and Aaron standing on the ground find that at the North Pole,

$$w_{\text{app}} = w = mg_p \tag{7.14}$$

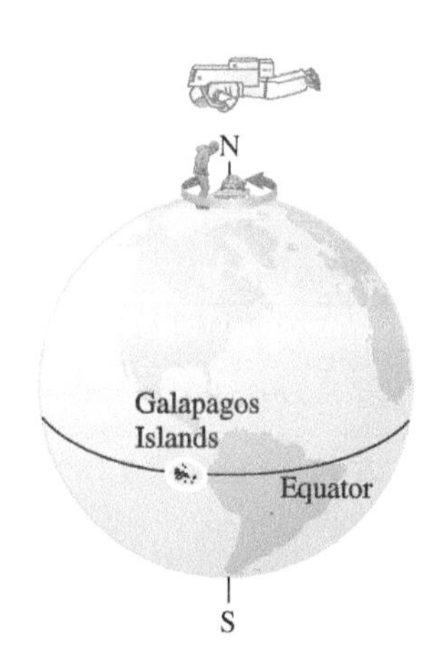

FIGURE 7.24 Hannah in an inertial reference frame above the Earth's North Pole.

The major difference between the noninertial and inertial frames comes from observing the turtle in the Galapagos. From Hannah's view above the North Pole, she sees both Aaron and the turtle moving in a circle around the Earth's axis (Fig. 7.25A). Like Aaron, Hannah identifies two forces acting on the turtle, but from her vantage point in an inertial frame, these forces must account for the turtle's centripetal acceleration (Fig. 7.25B). When Hannah applies Newton's second law, she finds that the turtle's apparent weight at the equator equals its weight minus the centripetal force, $m(v^2/R_\oplus)$:

$$\sum F_y = F_H - F_g = ma_y = -m\frac{v^2}{R_\oplus}$$

$$F_H = F_g - m\frac{v^2}{R_\oplus}$$

$$w_{app} = w - m\frac{v^2}{R_\oplus} \quad (7.15)$$

We can now use Equation 7.15 to find an expression for g_e, the magnitude of the Earth's effective gravitational field at the equator. The weight w is the magnitude of the gravitational force, which must be the same at the North Pole and in the Galapagos:

$$w = mg_p \quad (7.14)$$

The apparent weight read off the scale by Aaron on the equator is written in term of the gravitational field g_e at the equator:

$$w_{app} = mg_e \quad (7.16)$$

Substitute Equations 7.14 and 7.16 into Equation 7.15:

$$mg_e = mg_p - m\frac{v^2}{R_\oplus}$$

Removing the common factor m, we see that to Aaron on the equator, the gravitational field appears to be

$$g_e = g_p - \frac{v^2}{R_\oplus} \quad (7.17)$$

To him, it appears that the Earth's gravitational field at the equator is weaker than at the poles, $g_e < g_p$. To Hannah in an inertial frame (hovering above the North Pole), the gravitational field at the equator is still g, and the reason $w_{app} < w$ is because of the turtle's centripetal acceleration at the equator.

We have found the effective gravitational field g_{eff} at two extremes—the North Pole and the equator. An object at the equator has the highest centripetal acceleration and therefore the lowest effective gravitational field $(g_{eff})_{min} = g_e$ (Eq. 7.17). An object at the North Pole has no centripetal acceleration and therefore the highest effective gravitational field $(g_{eff})_{max} = g_p$.

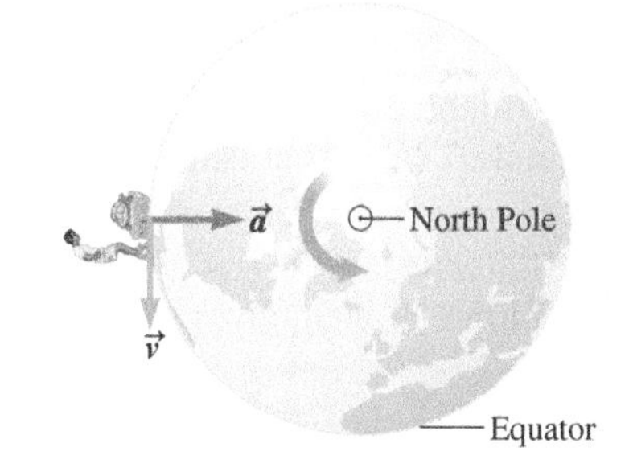

A.

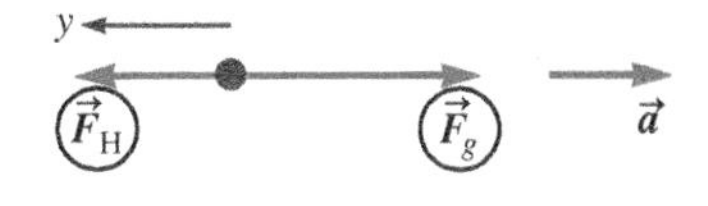

B.

FIGURE 7.25 A. Hannah sees that both Aaron and the turtle in the Galapagos are moving in a circle. Therefore, they are both are accelerating. Aaron in the Galapagos is in a noninertial reference frame. **B.** Free-body diagram for a turtle as drawn by Hannah in an inertial reference frame off the Earth. Hannah sees that the turtle has a centripetal acceleration, but Aaron does not see this centripetal acceleration.

EXAMPLE 7.11 Ignoring the Earth's Rotation

A In Chapter 5, we said that we can usually ignore that we are observers in a noninertial frame. To see how good this approximation actually is, find the numerical value of the difference between the effective gravitational field at either pole and at the Earth's equator, $\Delta g_{lat} = g_p - g_e$.

INTERPRET and ANTICIPATE

As in Example 7.10, we expect to find a small numerical result for Δg_{lat}.

Start with Equation 7.17 to find an expression for Δg_{lat}. Rearrange this equation to get the two g terms together.	$g_e = g_p - \dfrac{v^2}{R_\oplus}$ $\Delta g_{lat} = g_p - g_e = \dfrac{v^2}{R_\oplus}$ (1)
The right side of this equation is the centripetal acceleration from Equation 4.38, $a_c = v^2/r$. We do not know the speed v, but $v = 2\pi r/T$ allows us to eliminate v and write this expression in terms of the period T, which we know to be 24 h for the Earth. For $R_\oplus$, we use the Earth's equatorial radius.	$\dfrac{v^2}{R_\oplus} = \left(\dfrac{1}{R_\oplus}\right)\left(\dfrac{4\pi^2 R_\oplus^2}{T^2}\right)$ $\dfrac{v^2}{R_\oplus} = \dfrac{4\pi^2 R_\oplus}{T^2} = \dfrac{4\pi^2(6.387 \times 10^6 \text{ m})}{(8.64 \times 10^4 \text{ s})^2} = 3.38 \times 10^{-2} \text{ m/s}^2$

Example continues on page 206 ▶

According to Equation (1), the centripetal acceleration is Δg_{lat}.	$\Delta g_{lat} = 3.38 \times 10^{-2}\ \text{m/s}^2 \text{or N/kg}$

B Compare this value for Δg_{lat} with the variation in the gravitational field due to altitude Δg_{alt} (Example 7.10). What is the ratio $\Delta g_{alt}/\Delta g_{lat}$?

INTERPRET and ANTICIPATE

This part of the problem gives us an opportunity to think about our result in part A.

SOLVE

In Example 7.10, we found the variation in the Earth's gravitational field due to the altitude difference between the North Pole and the top of Mount Everest.

$$\Delta g_{alt} = 0.061\ \text{N/kg or m/s}^2$$

$$\frac{\Delta g_{alt}}{\Delta g_{lat}} = \frac{0.061\ \text{N/kg}}{0.0338\ \text{N/kg}} = 1.8$$

CHECK and THINK

Although the latitude variation Δg_{lat} is only about half the altitude variation, it is easily observable. In fact, this variation is the reason planets are wider at their equators than at their poles.

SUMMARY

Underlying Principles: Newton's law of universal gravity

Newton theorized that gravity is not limited to objects near the Earth, but instead applies to all objects in the Universe. The gravitational force is attractive, and its magnitude is

$$F_G = G\frac{m_1 m_2}{r^2} \quad (7.4)$$

Major Concepts

1. A **geocentric model** was described by Ptolemy, a Greek astronomer and mathematician who lived in the second century CE. According to this model, the Sun, the Moon, the stars, and the planets orbited the Earth.
2. In the **heliocentric model**, the Earth and other planets orbit the Sun.
3. **Empirical laws** are laws formulated based solely on data, without any hypothesis presented first.
4. The mass in Newton's second law ($\Sigma\vec{F}_{tot} = m\vec{a}$) is the **inertial mass** of an object. It is a measure of the object's resistance to change its velocity.
5. **Gravitational mass** is the property of the particles that creates a gravitational force between them. Experimental evidence supports the idea that the gravitational mass of any object equals its inertial mass.
6. *Field* is a general term referring to a region under the influence of some physical source. A **gravitational field** $\vec{g}(r)$ depends only on a source and not on any subject in the field. The value of the gravitational field at a particular point is equal to the local free-fall acceleration at that point. Usually, the SI units for the gravitational field are newtons per kilogram (N/kg), but they may also be expressed as meters per second squared (m/s^2). The gravitational field due to any particle with mass M is given by

$$\vec{g}(r) \equiv -\frac{GM}{r^2}\hat{r} \quad (7.13)$$

where $\hat{r}$ is a unit vector pointing away from the particle.

Special Cases

1. **Kepler's first law:** planetary orbits are ellipses with the Sun at one focus.
2. **Kepler's second law:** a line joining a planet to the Sun sweeps out equal areas in equal times.
3. If period T is measured in years and the semimajor axis a is measured in astronomical units (AU),

Kepler's third law is expressed mathematically as

$$T^2_{[yr]} = a^3_{[AU]} \quad (7.2)$$

The general form of Kepler's third law is

$$T^2 = \left(\frac{4\pi^2}{GM}\right)a^3 \quad (7.6)$$

Tools

1. An **ellipse** is the locus of points for which the sum of the distances from the two foci F_1 and F_2 is constant (Fig. 7.5B). The **major axis** passes through both foci. The **minor axis** is the perpendicular bisector of the major axis. The length of the semimajor and semiminor axis are a and b, respectively.
2. **Inverse-square law:** Newton found that gravity obeys an *inverse-square law*, which means that the gravitational force on a planet varies as the inverse square of its distance from the Sun:

$$F_G \propto \frac{1}{r^2} \tag{7.3}$$

Any quantity that diminishes with the inverse square of distance is said to obey an inverse-square law.

PROBLEMS AND QUESTIONS

A = **algebraic** C = **conceptual** E = **estimation** G = **graphical** N = **numerical**

7-1 A Knowable Universe

1. C We use the terms *sunset* and *sunrise*. In what way are these terms misleading?
2. C Briefly describe a contemporary scientific endeavor. How is that scientific project influenced by society? How does it affect society?

7-2 Kepler's Laws of Planetary Motion

3. N For many years, astronomer Percival Lowell searched for a "Planet X" that might explain some of the perturbations observed in the orbit of Uranus. These perturbations were later explained when the masses of the outer planets and planetoids, particularly Neptune, became better measured (*Voyager 2*). At the time, however, Lowell had proposed the existence of a Planet X that orbited the Sun with a mean distance of 43 AU. With what period would this Planet X orbit the Sun?
4. C Many scientific projects are carried out by teams rather than by individuals. Tycho was a great observer, and Kepler was a great mathematician. They had difficulty working together. It is likely that you will be expected to work with others. Come up with strategies for working with people whom you might find difficult.
5. N You are given a string of length 10 cm, two tacks, and a pencil. You are to use these to draw an ellipse as in Figure 7.5B. How far apart must you place the tacks to ensure that the semimajor axis of the ellipse is 5 cm? Explain your answer. If possible, try it yourself.
6. N Io and Europa are two of Jupiter's many moons. The mean distance of Europa from Jupiter is about twice as far as that for Io and Jupiter. By what factor is the period of Europa's orbit longer than that of Io's?

Problems 7 and 8 are paired.

7. Model the Moon's orbit around the Earth as an ellipse with the Earth at one focus. The Moon's farthest distance (apogee) from the center of the Earth is $r_A = 4.05 \times 10^8$ m, and its closest distance (perigee) is $r_P = 3.63 \times 10^8$ m.
   - **a.** N Calculate the semimajor axis of the Moon's orbit.
   - **b.** N How far is the Earth from the center of the Moon's elliptical orbit?
   - **c.** G Use a scale such as 1 cm $\rightarrow 10^8$ m to sketch the Earth–Moon system at apogee and at perigee and the Moon's orbit. (The semiminor axis of the Moon's orbit is roughly $b = 3.84 \times 10^8$ m.)
8. Deimos is one of Mars's two moons (Fig. P7.8). Deimos's farthest distance from the center of Mars is $r_A = 2.359 \times 10^7$ m, and its closest distance is $r_P = 2.345 \times 10^7$ m.
   - **a.** G Use a scale such that 1 cm $\rightarrow 10^7$ m to sketch the Mars–Deimos system and include the orbit of Deimos.
   - **b.** N Calculate the semimajor axis of Deimos's orbit.
   - **c.** N How far is Mars from the center of Deimos's elliptical orbit? Give your answer in meters and in terms of Mars's radius.

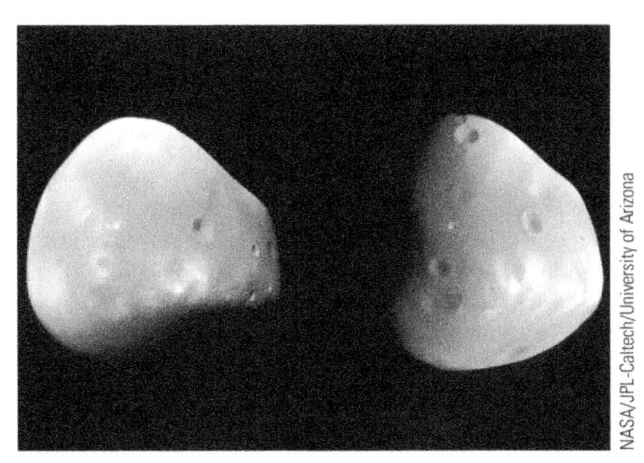

FIGURE P7.8

9. N Figure P7.9 is a scaled representation of a planet's orbit. with a semimajor axis of 1.524 AU. **a.** Find the ratio of the aphelion-to-perihelion distance. **b.** Find the perihelion and aphelion distances in astronomical units. **c.** Find the distance the Sun is from the center of the orbit.

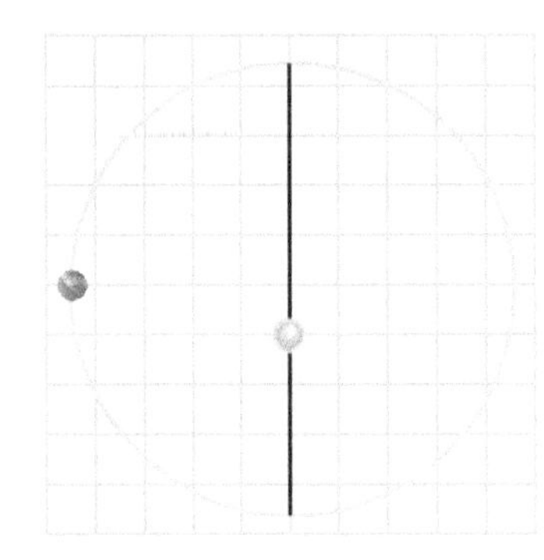

FIGURE P7.9 Problems 9 and 63.

10. C Students often incorrectly assert that the Earth's seasons are caused by its elliptical orbit around the Sun. Why is that assertion incorrect?

11. N One method for estimating the travel time from the Earth to the Moon is to use Kepler's third law to find half the period of an orbit whose perigee is at the Earth and whose apogee is at the Moon. What is the travel time to the Moon, a distance of 3.84×10^5 km from the Earth's center, for a spacecraft initially at an altitude of 200 km in the Earth's orbit, assuming it does not use any means of propulsion?

12. Before Kepler, the prevailing model for the shape of a planetary orbit was that it was circular. Kepler's first law of planetary motion challenges this model, stating planetary orbits are ellipses with the Sun at one focus. Of the eight major planets, Mercury's orbit is the one that differs most from a circular orbit. It has an aphelion distance from the Sun (r_A) of 0.46670 AU and a perihelion distance from the Sun (r_P) of 0.30750 AU.
    a. N Compute the length of the semimajor axis of Mercury.
    b. G The semiminor axis b of the orbit is 0.37883 AU. Using this and the information from part (a), make an accurate sketch of the orbit of Mercury. The sketch should contain not only Mercury's orbit, but also labels for perihelion distance, aphelion distance, major axis, and minor axis. For this sketch, you will need to position the Sun accurately.
    c. C Does the orbit you sketched in part (b) look very different from a circle? Was Copernicus a poor scientist for trying to fit Mercury's orbit with a circle?
    d. C Does the orbit you sketched have the Sun near the center as Copernicus would have thought when he tried to model planetary orbits as circles centered on the Sun?

13. N A massive black hole is believed to exist at the center of our galaxy (and most other spiral galaxies). Since the 1990s, astronomers have been tracking the motions of several dozen stars in rapid motion around the center. Their motions give a clue to the size of this black hole. **a.** One of these stars is believed to be in an approximately circular orbit with a radius of about 1.50×10^3 AU and a period of approximately 30 yr. Use these numbers to determine the mass of the black hole around which this star is orbiting. **b.** What is the speed of this star, and how does it compare with the speed of the Earth in its orbit? How does it compare with the speed of light?

14. N Since 1995, hundreds of extrasolar planets have been discovered. There is the exciting possibility that there is life on one or more of these planets. To support life similar to that on the Earth, the planet must have liquid water. For an Earth-like planet orbiting a star like the Sun, this requirement means that the planet must be within a habitable zone of 0.9 AU to 1.4 AU from the star. The semimajor axis of an extrasolar planet is inferred from its period. What range in periods corresponds to the habitable zone for an Earth-like planet orbiting a Sun-like star?

15. N When Sedna was discovered in 2003, it was the most distant object known to orbit the Sun. Currently, it is moving toward the inner solar system. Its period is 10,500 years. Its perihelion distance is 75 AU. **a.** What is its semimajor axis in astronomical units? **b.** What is its aphelion distance?

7-3 Newton's Law of Universal Gravity

16. C Which is greater, the gravitational force of the Sun on an asteroid or the gravitational force of an asteroid on the Sun? Clearly explain your reasoning without using any formulas. A clever student should be able to answer this question *without* citing Newton's law of universal gravity.

17. N The mass of the Earth is approximately 5.98×10^{24} kg, and the mass of the Moon is approximately 7.35×10^{22} kg. The Moon and the Earth are separated by about 3.84×10^8 m. **a.** What is the magnitude of the gravitational force that the Moon exerts on the Earth? **b.** If Serena is on the Moon and her mass is 25 kg, what is the magnitude of the gravitational force on Serena due to the Moon? The radius of the Moon is approximately 1.74×10^6 m.

Problems 18 and 27 are paired.

18. C Saturn's beautiful rings are composed of millions of particles ranging in size from boulders to tiny dust particles. Does a particle nearer Saturn orbit with the same period as one in the outer part of the rings? Why or why not? Does a particle with a greater mass than a second particle orbit with a different period if they are the same distance from Saturn? Why or why not?

19. N Maria, with mass $m = 48.0$ kg, is in the second row of seats, a distance of 3.00 m from Prof. Karen ($M = 55.0$ kg). What is the magnitude of the gravitational force between Maria and Prof. Karen?

20. A black hole is an object with mass, but no spatial extent. It truly is a particle. A black hole may form from a dead star. Such a black hole has a mass several times the mass of the Sun. Imagine a black hole whose mass is ten times the mass of the Sun.
    a. C Would you expect the period of an object orbiting the black hole with a semimajor axis of 1 AU to have a period greater than, less than, or equal to 1 yr? Explain your reasoning.
    b. N Use Equation 7.6 to calculate this period.

21. N Two spheres of mass $M_1 = 750$ kg and $M_2 = 350$ kg are placed 5.00 m apart. A particle of mass $m = 10.0$ kg is now placed midway between the two spheres. **a.** What is the net gravitational force on the particle due to the two spheres? **b.** At what position between the two spheres should the particle be placed so that the net gravitational force on the particle is zero?

22. E Estimate the magnitude of the gravitational force between the electron and proton in a hydrogen atom.

23. N The Lunar Reconnaissance Orbiter (LRO), with mass $m =$ 1850 kg, maps the surface of the Moon from an orbital altitude of 50.0 km. What are the magnitude and direction of **a.** the force the LRO experiences due to the Moon's gravity and **b.** the force exerted by the LRO on the Moon?

Problems 24 and 25 are paired.

24. A Suppose a planet with mass m is orbiting star with mass M and the mean distance between the planet and the star is a. Using Newton's law of universal gravity, derive an algebraic expression for the speed of the planet when it is at the mean distance from the star.

25. N Suppose a planet with a mass of 2.44×10^{25} kg is orbiting a star with a mass of 3.65×10^{31} kg and the mean distance between the planet and the star is 1.12×10^{12} m. Using Newton's law of universal gravity, determine the speed of the planet when it is at the mean distance from the star.

Problems 26 and 41 are paired.

26. N Three billiard balls, the two-ball, the four-ball, and the eight-ball, are arranged on a pool table as shown in Figure P7.26. Given the coordinate system shown and that the mass of each ball is 0.150 kg, determine the gravitational force on the eight-ball due to the other two balls.

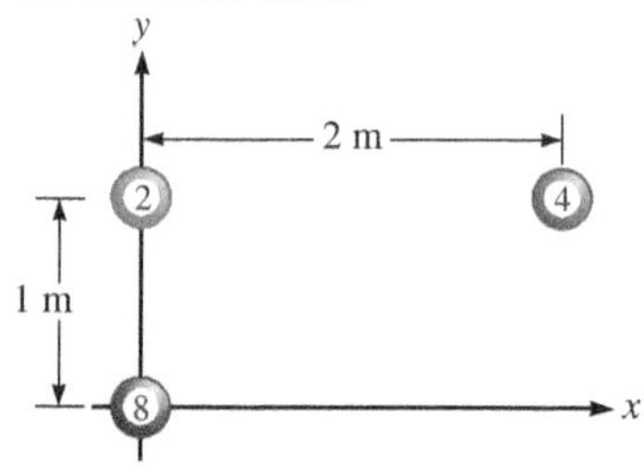

FIGURE P7.26 Problems 26 and 41.

27. N Saturn's ring system forms a relatively thin, circular disk in the equatorial plane of the planet. The inner radius of the ring system is approximately 92,000 km from the center of the planet, and the outer edge is about 137,000 km from the center of the planet. The mass of Saturn itself is 5.68×10^{26} kg. **a.** What is the period of a particle in the outer edge compared with the period of a particle in the inner edge? **b.** How long does it take a particle in the inner edge to move once around Saturn? **c.** While this inner-edge particle is completing one orbit abound Saturn, how far around Saturn does a particle on the outer edge move?

28. N Three spheres are arranged in the xy plane as in Figure P7.28. The first sphere, of mass $m_1 = 12.5$ kg, is located at the origin; the second sphere, of mass $m_2 = 4.50$ kg is located at $(-6.00, 0.00)$ m; and the third sphere, of mass $m_3 = 8.00$ kg is located at $(0.00, 5.00)$ m. Assuming an isolated system, what is the net gravitational force on the sphere located at the origin?

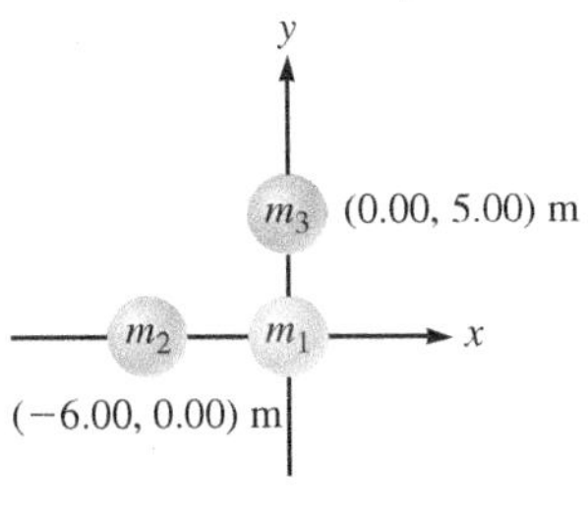

FIGURE P7.28

Problems 29 and 30 are paired.

29. N Find the magnitude of the Sun's gravitational force on the Earth when the Earth is at perihelion and at aphelion. The perihelion and aphelion distances are 1.48×10^{11} m and 1.52×10^{11} m, respectively.

30. Andromeda is a spiral galaxy much like the Milky Way. Andromeda's mass is about 2.8×10^{12} times the mass of the Sun and is 2.42×10^{22} m away. The mass of the Milky Way is 2.1×10^{12} times the mass of the Sun.
   a. E Estimate the magnitude of the gravitational force between the Milky Way and Andromeda.
   b. N Assuming no other forces act on the Milky Way, find its acceleration.

7-4 The Gravitational Field

31. N When first detected, near-Earth asteroid 2011 MD was at its closest approach of only 12,000 km above the Earth's surface. What was the asteroid's acceleration due to the Earth's gravity at this point in its trajectory?

32. C CASE STUDY To estimate the mass of the dark matter in our galaxy, we need to adopt Newton's hypothesis that there is no difference between celestial and terrestrial laws of physics. Explain why this idea is a necessary part of our dark-matter estimate and of astrophysical study in general.

33. A Three particles, each with mass m, are located at coordinates $(0, L)$, $(L, 0)$, and (L, L) as shown in Figure P7.33. What are the magnitude and direction of the gravitational field at the origin due to the three particles?

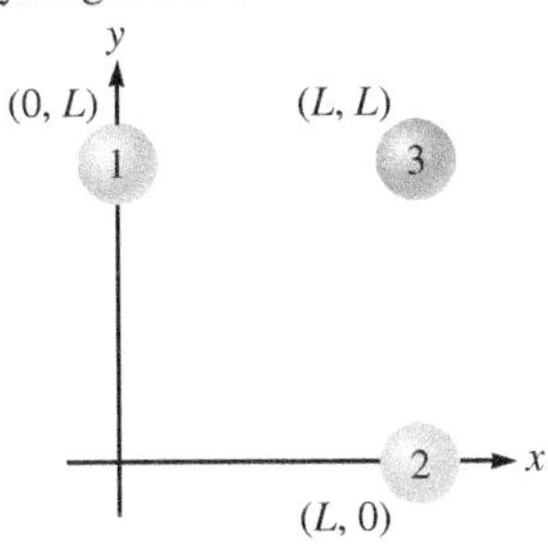

FIGURE P7.33

Problems 34, 35, and 36 are grouped.

34. C It is common for students to confuse the terms *force* and *field*. Let us find a way to keep them straight. **a.** What are the SI units of a gravitational force? Is it a vector or a scalar? **b.** What are the SI units of a gravitational field? Give at least two answers. Is it a vector or a scalar?

35. C You sit on your chair. You then get off and let your dog sit on your chair. **a.** Is the Earth's gravitational force on you equal to the Earth's gravitational force on your dog? **b.** What happens to the Earth's gravitational field when you are replaced by your dog on the chair? Does it increase, decrease, or stay the same. Explain.

36. C In your own words, describe the difference between the terms *gravitational force* and *gravitational field*.

Problems 37, 38, and 39 are grouped.

37. The Sun has a mass of approximately 1.99×10^{30} kg.
   a. N Given that the Earth is on average about 1.50×10^{11} m from the Sun, what is the magnitude of the Sun's gravitational field at this distance?
   b. G Sketch the magnitude of the gravitational field due to the Sun as a function of distance from the Sun. Indicate the Earth's position on your graph. Assume the radius of the Sun is 7.00×10^{8} m and begin the graph there.
   c. N Given that the mass of the Earth is 5.97×10^{24} kg, what is the magnitude of the gravitational force on the Earth due to the Sun?

38. The Earth has a mass of approximately 5.97×10^{24} kg.
   a. N Given that the Earth is on average about 1.50×10^{11} m from the Sun, what is the magnitude of the Earth's gravitational field at the location that is occupied by the Sun?
   b. G Sketch the magnitude of the gravitational field due to the Earth as a function of distance from the Earth.
   c. N Given that the mass of the Sun is 1.99×10^{30} kg, what is the magnitude of the gravitational force on the Sun due to the Earth?

39. C There are gravitational fields associated with the Earth and the Sun. The magnitude of the gravitational field due to the Earth at the location of the Sun is significantly less than the magnitude of the gravitational field due to the Sun at the location of the Earth. This seems to violate Newton's third law. Explain how Newton's third law is verified when comparing the gravitational force each body exerts on the other.

40. C CASE STUDY **MOND** According to Example 7.8 (page 201), 68% of our own galaxy is made up of extraordinary particles—dark matter—that have never been observed on the Earth. In fact, there is no laboratory evidence for the existence of dark matter. So, some scientists say that there is little or no dark matter; instead, they argue that Newton's law must be modified to take these observations into account. This theory is called MOND, for *mo*dified *N*ewtonian *d*ynamics. **a.** If the estimated amount of dark matter were much smaller, say a few percent instead of 68% (or more), do you believe that the MOND picture of the Universe would generally be supported? **b.** In what way are MOND proponents like Kepler? Does your answer depend on whether MOND or dark matter is supported by future experiments?

41. N Three billiard balls, the two-ball, the four-ball, and the eight-ball, are arranged on a pool table as shown in Figure P7.26. Given the coordinate system shown and that the mass of each ball is 0.150 kg, determine the gravitational field at $x = 2$ m, $y = 0$.

42. C Example 7.7 discusses the Lagrange point, L1, located between the Earth and the Sun. Suppose we placed an object in orbit *around* this point. Do you suppose we could obtain a circular orbit? Why or why not? *Hint*: At the Earth's distance from the Sun, the Sun's gravitational pull is just sufficient to pull it into an almost circular orbit. If you launch an object into a nearby orbit around the Earth, the object feels a gravitational field due almost entirely to the Earth. What would be the comparable situation at L1?

43. C Example 7.6 discusses *geosynchronous* satellites, which remain above a fixed point on the Earth's surface. Suppose you want to put a satellite in orbit so that it always remains above New York City. Is it possible to do so? Why or why not?

44. C In the 2009 motion picture *Star Trek*, the antagonist uses a device to convert a planet into a *black hole*, an object with mass

but no spatial extent. Assume the only fictional aspect of this event is the existence of a device capable of turning a planet's mass into a black hole; otherwise, the laws of physics described in this chapter hold. What happens to the orbit of that planet around its star? Carefully explain your reasoning.

45. Figure P7.45 shows a picture of American astronaut Clay Anderson experiencing weightlessness on board the International Space Station.

a. N Most people have the misconception that a person in a spacecraft is weightless because he or she is no longer affected by gravity. Show that this premise cannot be true by computing the gravitational field of the Earth at an altitude of 200 km, the typical altitude of a spacecraft in orbit. Compare this result with the gravitational field on the surface of the Earth.

b. C Why would astronauts in orbit experience weightlessness even if they are experiencing a gravitational field (and therefore a gravitational force)?

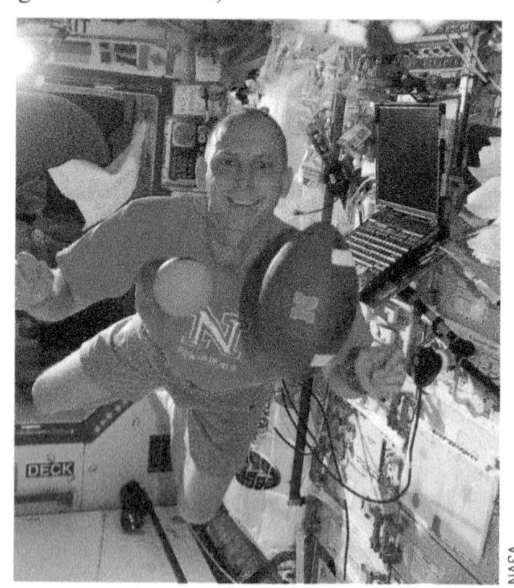

FIGURE P7.45

7-5 Variations in the Earth's Gravitational Field

46. G A simple plumb line may be constructed by hanging a small, dense object such as a lead ball from a lightweight string. It is used to determine a vertical line by holding the open end of the string such that the lead ball hovers just above the ground. Imagine observing three plumb lines from an inertial reference frame. For each location listed, make a sketch and determine if the plumb line is pointing to the center of the Earth. If it is not, indicate on your sketch where it is pointing. Explain your answers. **a.** North Pole **b.** Equator **c.** 45° north latitude

47. The Moon rotates on its axis as it orbits the Earth. The same side of the Moon always faces the Earth.

a. G Sketch a motion diagram for the Moon. Include the Earth on your sketch.

b. N Find the Moon's rotational period.

c. N Compare (by finding $w_{\text{pole}} - w_{\text{equator}}$) the apparent weight of a 1-kg object on the Moon's equator with its apparent weight at one of the Moon's poles. Assume the same elevation.

d. N How does your answer compare with what you would expect to find for a 1-kg object's apparent weight measured on the Earth's equator and at one of its poles? Explain your results.

48. E Suppose you dropped a small object such as a golf ball from a very tall building such as the Willis Tower in Chicago. Ignoring air drag, estimate the direction and horizontal position of where the ball lands with respect to the point just below where it was dropped.

Problems 49 and 50 are paired.

49. N In 1851, Leon Foucault built a small pendulum in his basement that confirmed the theory that the Earth rotates. Today, you can see his much larger pendulum at the Pantheon in Paris, France (Fig. P7.49). To an inertial observer, the pendulum appears to swing back and forth along a single plane. To an observer on the Earth, this plane appears to rotate. **a.** Imagine that such a pendulum is at the North Pole. Through how many degrees would the plane appear to rotate in a single day? **b.** Now imagine that such a pendulum is on the equator. Through how many degrees would the plane appear to rotate in a single day?

FIGURE P7.49 Problems 49 and 50. Close-up of the bob of the Foucault pendulum.

50. N Consider the pendulum shown in Figure P7.49. To an inertial observer, the pendulum appears to swing back and forth along a single plane. To an observer on the Earth, this plane appears to rotate. The apparent rotation angle θ of the plane in a single day depends on the pendulum's latitude φ: $\theta = 360° \sin \varphi$. The latitude of Paris is 48° 48'. **a.** Confirm that this equation holds at the poles and at the equator. **b.** Find θ for the pendulum in Paris.

General Problems

51. N The International Space Station (ISS) experiences an acceleration due to the Earth's gravity of 8.83 m/s^2. What is the orbital period of the ISS?

Problems 52, 55, 56, 57, and 58 are grouped.

52. If you were to calculate the pull of the Sun on the Earth and the pull of the Moon on the Earth, you would undoubtedly find that the Sun's pull is much stronger than that of the Moon, yet the Moon's pull is the primary cause of tides on the Earth. Tides exist because of the *difference* in the gravitational pull of a body (Sun or Moon) on opposite sides of the Earth. Even though the Sun's pull is stronger, the difference between the pull on the near and far sides is greater for the Moon.

a. A Let $F(r)$ be the gravitational force exerted on one mass by a second mass a distance r away. Calculate $dF(r)/dr$ to show how F changes as r is changed.

b. N Evaluate this expression for $dF(r)/dr$ for the force of the Sun at the Earth's center and for the Moon at the Earth's center.

c. N Suppose the Earth–Moon distance remains the same, but the Earth is moved closer to the Sun. Is there any point where $dF(r)/dr$ for the two forces has the same value?

53. N Two black holes (the remains of exploded stars), separated by a distance of 10.0 AU (1 AU = 1.50×10^{11} m), attract one another with a gravitational force of 8.90×10^{25} N. The combined mass of the two black holes is 4.00×10^{30} kg. What is the mass of each black hole?

54. C Quito is a city in Ecuador located at the equator but in the very high Andes Mountains. How do you expect the apparent weight of a turtle measured in the Galapagos Islands to differ from that measured in Quito?

55. Tides on the Earth are caused by the difference in the Moon's gravitational field at the far side and at the near side of the Earth. The side of the Earth facing the Moon experiences a greater gravitational force than the side facing away from the Moon. The distance from the Earth's center to the Moon's center is 3.84×10^8 m, and the radius of the Earth is 6.37×10^6 m. The mass of the Moon is 7.36×10^{22} kg.

a. N Calculate the Moon's gravitational field at the side of the Earth facing the Moon.

b. N Calculate the Moon's gravitational field at the side of the Earth facing away from the Moon.

c. N Calculate the Moon's gravitational field at the center of the Earth.

d. G Sketch the Earth and include the three vectors from parts (a) through (c).

e. C Qualitatively explain why there are two tides a day on most places on the Earth due to the Moon. (Because of friction, the high tides do not exactly line up with the Moon.)

56. Consider the Earth and the Moon as a two-particle system.

a. A Find an expression for the gravitational field $\vec{g}$ of this two-particle system as a function of the distance r from the center of the Earth. (Do not worry about points inside either the Earth or the Moon.)

b. G Plot the scalar component of $\vec{g}$ as a function of distance from the center of the Earth.

57. N Consider the Earth and the Moon as a two-particle system. **a.** Find the gravitational field of this two-particle system at the point that is exactly halfway between the Earth and the Moon. **b.** An asteroid of mass 6.69×10^{15} kg is at the point exactly halfway between the Earth and the Moon. What is the magnitude of the gravitational force on it?

58. N Consider the Earth and the Moon as a two-particle system. **a.** How far from the center of the Earth is the gravitational field of this two-particle system zero? **b.** Sketch gravitational field vectors $\vec{g}$ along the line joining the Earth and the Moon. Indicate the point at which $\vec{g} = 0$. (Do not consider positions inside either object.)

59. N Many science-fiction spacecraft are shown with cylindrical modules that rotate to provide the crew with artificial gravity—a centripetal acceleration that is comparable to the gravitational acceleration on the Earth, 9.81 m/s^2. If one such module is 500 m in diameter, what is the speed of rotation that would provide an artificial gravity equal to that on the Earth's surface? (Report your answer to 2 significant figures in revolutions per minute and radians per second.)

60. You are a planetary scientist studying the atmosphere of Jupiter through a large telescope when you observe an asteroid approaching the planet. This asteroid is large, so you know it is held together by gravity rather than the cohesive forces that hold a large rock together. If the asteroid gets too close to Jupiter, the massive tidal forces will tear it apart, scattering small particles that will add to the ring system. You have calculated the closest distance the asteroid will come to Jupiter. How do you know if the asteroid will survive?

a. N A measure of the cohesive gravitational force holding such an asteroid together is the gravitational field on the surface due to the mass of the asteroid. This field is independent of the distance of the asteroid from Jupiter. Calculate the gravitational field at the surface of the asteroid due only to the mass of the asteroid. Assume the asteroid has a diameter of 10,000 km and a density of 1300 kg/m^3.

b. G Tidal forces from Jupiter tend to disrupt the asteroid by pulling it apart. The tidal forces depend on the distance between Jupiter and the asteroid. There is a distance between Jupiter and the asteroid known as the *Roche limit* where the tidal forces are balanced by the asteroid's own cohesive gravitational force. If the asteroid is within the Roche limit, it will be torn apart. Figure P7.60 shows Jupiter's gravitational field as a function of distance from its center. By looking at this graph, can you determine an approximate value for the Roche limit for this asteroid in the vicinity of this planet?

c. C What will happen to the Roche limit if we consider an asteroid of lower density?

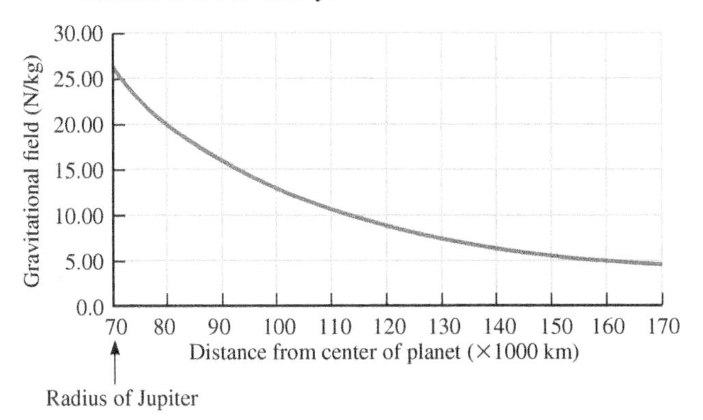

FIGURE P7.60

61 C Is it possible to put a geosynchronous satellite above London by placing it in an orbit parallel to the Earth's equator and moving eastward (the same direction as the Earth's rotation)? Why or why not?

62. **CASE STUDY** According to MOND, Newton's second law ($\vec{F}_{tot} = m\vec{a}$) must be modified when we are dealing with very slowly accelerating objects such as stars on the outer edge of the Milky Way. In MOND, if the acceleration of an object is very small (less than about 10^{-10} m/s^2), the net force applied to the object is proportional not to the acceleration as in Newton's second law, but rather to the square of the acceleration: $F_{tot} \propto a^2$.

a. C Why has the MOND theory not been tested in a laboratory on the Earth? Explain.

b. N Can the theory be tested in a space station? Explain.

63. Planetary orbits are often approximated as uniform circular motion. Figure P7.9 is a scaled representation of a planet's orbit with a semimajor axis of 1.524 AU.

a. G Use Figure P7.9 to find the ratio of the Sun's maximum gravitational field to its minimum gravitational field on the planet's orbit.

b. N What is the ratio of the planet's maximum speed to its minimum speed?

c. C Comment on the validity of approximating this orbit as uniform circular motion.

64. Kepler's second law of planetary motion can be used to relate the ratio of perihelion to aphelion speeds of a planet to the ratio of its perihelion to aphelion distances.

a. G Start by sketching the area swept out by the line joining the planet to the Sun during a short period (say 1 day) when the Sun is near aphelion and another area for when the Sun is near perihelion. Based on your sketch, argue that it would be safe to approximate the shapes of those areas as triangles.

b. A Write the expression for the "triangular" area swept out by the line connecting the planet to the Sun at aphelion in terms of r_A, v_A, and time t. Write a similar expression for the "triangular" area swept out at perihelion in terms of r_P, v_P, and t. *Note*: The speeds of the planet at aphelion and perihelion are v_A, and v_P, respectively.

c. A Show that the ratio of aphelion to perihelion distances, r_A/r_P, is expressible in terms of a ratio of v_A and v_P.

d. N The Earth varies in speed during its orbit. The Earth's lowest orbital speed is 29.29 km/s. What is the Earth's highest orbital speed?

65. Newton's version of Kepler's third law offers one of the few ways of determining the mass of objects in space. One form of this relationship is expressed in Equation 7.6. To derive it, however, we assumed the orbit of the less massive object is circular and is centered on the more massive object, which is not completely realistic. In reality, both objects move in an orbit in response to the gravitational force between them! A careful derivation shows that for two masses M_1 and M_2 in orbit around one another with semimajor axis a, the proper expression of Newton's version of Kepler's third law is

$$T^2 = \left[\frac{4\pi^2}{G(M_1 + M_2)}\right]a^3$$

Equation 7.6 differs from this "proper" expression in that we've replaced the mass of a single object with the sum of the masses of both objects.

a. N Consider Jupiter, the second most massive object in the solar system. Compute the orbital period of Jupiter around the Sun using this "proper" expression of Newton's version of Kepler's third law and compare the result with the answer you find using Equation 7.6. Do they differ significantly given $M_\odot = 1.98892 \times 10^{30}$ kg, $M_J = 1.8987 \times 10^{27}$ kg, and $a = 7.7857 \times 10^{11}$ m?

b. C Would you expect Equation 7.6 to hold for the Moon orbiting the Earth? Clearly explain your reasoning.

c. C Would you expect Equation 7.6 to be accurate for two asteroids orbiting each other? Clearly explain your reasoning.

66. E Some people insist on believing that the positions of the planets at the time of your birth can have an influence on your life, a subject sometimes referred to as *astrology*. Some astrologers even insist that there is a physical basis in astrology because the gravitational pull of the various planets affects a newborn's body. Estimate the magnitude of the gravitational force on a newborn from the planet Jupiter and compare it with the magnitude of the gravitational force due to the pediatrician who happens to be present during the birth. Which is more likely to have a gravitational "influence" on a newborn child? Is this method a valid scientific test of the gravitational basis of astrology?

67. A One intriguing bit of trivia is that for almost any planet made of solid matter, the orbital period of a satellite in a low orbit is about the same, on the order of 90 minutes! Here, a low orbit is one that has a semimajor axis on the order of the planet's radius in size. (Obviously, the orbit must be slightly bigger than the radius of the planet; otherwise, the satellite would hit the ground.) Show that for any planet with a density comparable to the density of the Earth, the orbital period for a low orbit is the same by solving for this period in terms of the planet's density. For simplicity, assume these planets are spherical and of uniform density.

68. E In his book *Death from the Skies!*, astronomer Phil Plait describes what would happen if a very massive, very compact object (a *black hole*) were to get near enough to the Earth to destroy it. Estimate at what distance a 5-solar-mass black hole would literally start ripping the surface of the Earth apart. *Note*: Plait estimates the likelihood of this sort of interaction to be on the order of one in one trillion in your lifetime, so it's not really worth losing much sleep over.

69. In his book *A Brief History of Time*, physicist Stephen Hawking popularized the term *Spahettification* to describe what happens to someone who falls feet first into a small but highly massive object. In essence, the gravitational field at the person's feet is sufficiently higher than the gravitational field at the head, and the person gets stretched out like a spaghetti noodle (a phenomenon that would likely be fatal). Imagine that you are 1000 km from an object with a mass of 1 solar mass and you are approximately 1.5 m tall. You want to figure out if you are in danger of being "spaghettified."

a. A As a first step, we need to come up with an expression that will allow us to compute the change in gravitational field, $\Delta\vec{g}(d)$, over a small distance ℓ such as a person's height at a distance d from the source of the gravitational field. Show that when $d \gg \ell$, you can write this change as

$$\Delta\vec{g}(d) = \vec{g}\left(d + \frac{\ell}{2}\right) - \vec{g}\left(d - \frac{\ell}{2}\right) \approx -GM\left(\frac{2\ell}{d^3}\right)\hat{r}$$

b. N Calculate the difference between the gravitational field of the black hole at your feet and your head if you are falling feet first into the black hole.

c. C Is this difference in gravitational field large enough to "spaghettify" you? Clearly explain your reasoning.

70. N **CASE STUDY** If the only force exerted on a star far from the center of the Galaxy ($r = 7.7 \times 10^{20}$ m) is the gravitational force exerted by the ordinary matter ($M_{\text{ord}} = 1.8 \times 10^{41}$ kg), find the speed of the star. Assume a circular orbit and assume all the Galaxy's matter is concentrated at the center.

Conservation of Energy

8

Underlying Principle

Conservation of mechanical energy

Major Concepts

1. Energy
2. Kinetic energy
3. Potential energy
4. Isolated system

Special Cases

1. Potential energy
   a. Gravitational potential energy near Earth's surface of the Earth
   b. Universal gravitational potential energy
   c. Elastic potential energy
2. Orbital energies
   a. Circular orbit
   b. Elliptical orbit

Tools

1. Bar chart
2. Energy graph

Key Questions

What is energy?

How does the principle of conservation of mechanical energy simplify complicated situations?

8-1 Another approach to Newtonian mechanics 214

8-2 Energy 215

8-3 Gravitational potential energy near the Earth 218

8-4 Universal gravitational potential energy 221

8-5 Elastic potential energy 223

8-6 Conservation of mechanical energy 225

8-7 Applying the conservation of mechanical energy 228

8-8 Energy graphs 232

8-9 Special cases Orbital energies 235

It seems as if nothing stays the same. You go home on a break from college only to find that your room is now a home gym, your favorite science teacher has retired, and your younger brother is taller than you. Change is not unique to the human condition. Our solar system formed out of a cloud of gas, radioactive elements below the Earth's surface decay, and the Sun will eventually die.

So, with everything apparently changing all the time, it is remarkable that philosopher René Descartes (1596–1650) believed that some quantities in the Universe remain constant. He took a very different approach to physics than Newton. Whereas Newton searched to explain changes in motion, Descartes said that the *motion* of the Universe as a whole is conserved, or constant. Today, we describe Descartes's original idea more precisely and say that the *energy* and *momentum* of the Universe are conserved. The notion that some quantities remain constant throughout a process—like the formation of the solar system—is a new approach for us.

8-1 Another Approach to Newtonian Mechanics

In Chapter 5, we introduced the underlying principle of dynamics in the form of Newton's laws of motion. It might seem that all we need to do is continue learning about other specific forces, and then we could study the physics of any phenomenon. In principle, this idea is true because, according to Newton's second law, if we know all the forces acting on a particle, we can find its acceleration. Once we know its acceleration, kinematics (the mathematical description of motion) tells us how to find the particle's velocity and position.

There are, however, many practical situations in which applying Newton's second law does not easily lead to a description of the motion. Fortunately, there is an alternative approach that does not replace Newton's laws (which still govern mechanics); instead, it complements Newton's laws. Some problems are more easily solved using this alternative approach (the conservation approach). The following case study is such a problem.

CASE STUDY Comet Halley

Comet Halley has a period of 76 years. Its first recorded sighting dates from 240 BCE. All its subsequent 30 trips through the inner solar system have been observed from the Earth. *Giotto*, a European Space Agency spacecraft, made history by taking the first close-up pictures of a comet nucleus in 1986.

Since then, other missions have been flown to observe comets as they pass near the Earth. Such missions take careful planning. Because the spacecraft needs to get close to the comet without crashing into it, mission planners must know the comet's orbit precisely (Fig. 8.1A). In this case study, we focus on just one aspect of the Comet Halley's orbit, its speed when *Giotto* flew by it.

Comet Halley had already passed perihelion (closest point to the Sun) and was on its way back out of the inner solar system when *Giotto* made its rendezvous. The spacecraft encountered the comet on March 13, 1986, at a distance $r_G = 0.89$ AU from the Sun and 0.98 AU from the Earth. Our goal is to find Comet Halley's speed when *Giotto* made its rendezvous.

Figure 8.1B illustrates the difficulty in using the force approach. At all positions, the gravitational force exerted on the comet is directed toward the Sun, but the magnitude of that force depends on distance. Because acceleration is proportional to this force, Comet Halley's acceleration depends on position. Also, the relative angle between the acceleration and the velocity depends on position. Using the force approach to find the speed of an object is very complicated when the force, acceleration, and velocity are all changing in both magnitude and direction. The conservation approach, however, greatly simplifies this problem and readily yields the desired quantity.

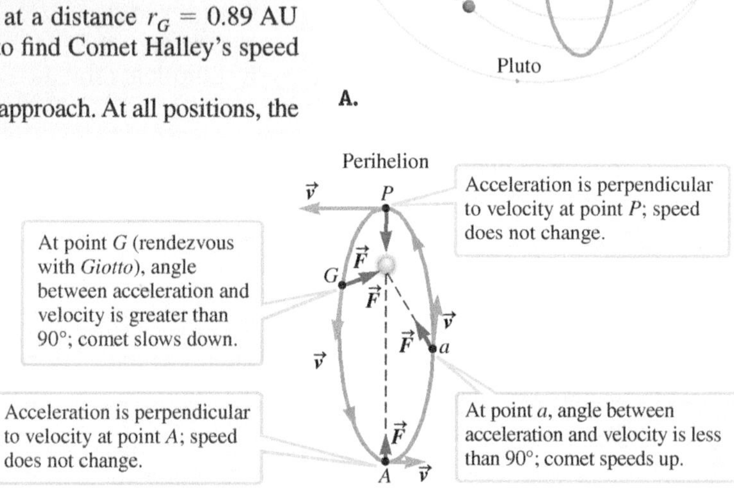

FIGURE 8.1 **A.** Comet Halley's orbit. (Orbits not drawn to scale.) **B.** Gravitational force on the comet and orbital velocity vectors at various points along its orbit.

CONCEPT EXERCISE 8.1

CASE STUDY Comet Halley's Orbital Parameters

Figure 8.1 shows Comet Halley's elliptical orbit.

a. Use the information given in the case study above and Kepler's third law from Chapter 7 to find the semimajor axis a of Comet Halley's orbit.

b. The comet's perihelion distance is $r_P = 0.59$ AU. Find its aphelion distance (farthest point from the Sun).

8-2 Energy

Our first approach to studying mechanics focused on understanding forces and applying Newton's laws of motion (the *force approach*). Our second approach to mechanics is the *conservation approach*, the basic idea of which is that certain quantities remain unchanged during a particular process or activity. In this chapter, we focus on the *conservation of mechanical energy*, so we must start by defining energy.

Newton's first law provides a working definition of force. There is no such law to define energy. *Energy*, like so many terms in physics, also has many everyday usages, some of which can help us to develop our intuition. For example, we might say, "My physics professor has a lot of energy." In this context, *energy* describes the teacher's style: High energy is associated with a lively style, and a low-energy professor has a dull style. This usage fits well into the physics meaning of the word: In physics, the term **energy** describes the state of a particle, object, or system. So, energy is *not* a material substance; instead, it is a scalar quantity that describes the state—in particular, the motion and the configuration—of a system.

ENERGY ✪ **Major Concept**

For brevity, we use the term *system* to stand for a particle, an object, or a collection of particles and objects.

Kinetic Energy

Kinetic energy K describes the motion of a system. The kinetic energy of a single particle depends on its mass m and speed v:

$$K = \tfrac{1}{2}mv^2 \tag{8.1}$$

KINETIC ENERGY ✪ **Major Concept**

To find the kinetic energy of a system that consists of more than one particle, we add up the kinetic energy of each one:

$$K = \frac{1}{2}m_1v_1^2 + \frac{1}{2}m_2v_2^2 + \frac{1}{2}m_3v_3^2 \ldots + \frac{1}{2}m_nv_n^2 = \sum_{i=1}^{n}\frac{1}{2}m_iv_i^2 \tag{8.2}$$

The SI unit for energy is named for British physicist James Joule. The joule is abbreviated as J and is given by

$$1\,\text{J} = 1\,\text{kg}\cdot\text{m}^2/\text{s}^2 = 1\,\text{N}\cdot\text{m}$$

Because the speed of a particle depends on the reference frame in which speed is measured, the particle's kinetic energy also depends on the choice of reference frame. For example, if a spider sits on the dashboard of a moving car, the spider's kinetic energy is zero according to the driver of the car, but not according to an observer parked on the side of the road.

CONCEPT EXERCISE 8.2

A ball is tossed straight up. What is its kinetic energy at the top of its flight?

CONCEPT EXERCISE 8.3

Can scalar quantities be negative? Can kinetic energy be negative? Explain your answers.

EXAMPLE 8.1 **CASE STUDY** **Comet Halley's Kinetic Energy**

Suppose a comet has the same period (76 years) and mass (1.7×10^{15} kg) as Comet Halley, but it moves in uniform circular motion around the Sun. Find its kinetic energy.

INTERPRET and ANTICIPATE

Knowing the period of the comet, we use Kepler's third law to find the comet's speed. We can then find a numerical value for the comet's kinetic energy. Kinetic energy must always be positive (Concept Exercise 8.3). Always check to see that any kinetic energy you have calculated is positive.

Example continues on page 216 ▶

SOLVE

From Kepler's third law, if the period of the comet in the circular orbit has the same period as that of Comet Halley, its orbital radius must equal the semimajor axis of Comet Halley's orbit.

$$r = a$$

Find the comet's speed from Equation 4.30, with $r = a$.

$$v = \frac{2\pi r}{T} = \frac{2\pi a}{T} \quad (4.30)$$

Substitute this result into $K = \frac{1}{2}mv^2$ (Eq. 8.1).

$$K = \frac{1}{2}mv^2 = \frac{1}{2}m\left(\frac{2\pi a}{T}\right)^2 = \frac{2\pi^2 ma^2}{T^2}$$

We used Kepler's third law to find a in Concept Exercise 8.1.

$$a = 18 \text{ AU} = 2.7 \times 10^{12}\text{ m}$$

Convert 76 years to seconds and substitute.

$$K = \frac{2\pi^2(1.7 \times 10^{15}\text{ kg})(2.7 \times 10^{12}\text{ m})^2}{(2.4 \times 10^9\text{ s})^2}$$

$$K = 4.2 \times 10^{22}\text{ J}$$

CHECK and THINK

Our kinetic energy result is positive and in the right units. More than 10^{22} J seems like a lot of energy. To check the magnitude of our result, let us calculate the Earth's kinetic energy. The orbital speed of the Earth is about 30 km/s, and its mass is 6×10^{24} kg, so the Earth's kinetic energy is $K_\oplus \approx \frac{1}{2}(6 \times 10^{24}\text{ kg})(30 \times 10^3\text{ m/s})^2 \approx 3 \times 10^{33}$ J. The comet is much less massive than the Earth, so it is reasonable that it has much less kinetic energy.

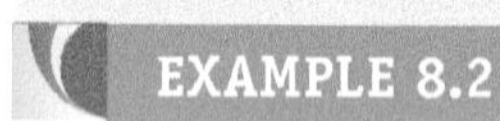

EXAMPLE 8.2 Tossing a Ball Revisited

A ball is tossed upward, leaving the person's hand at speed v_0. It returns to the same height and is caught moving at speed v_0. The ball has mass m. Sketch the ball's kinetic energy as a function of time.

INTERPRET and ANTICIPATE

The ball is in free fall, so we can easily find an expression for its speed as a function of time. Once we know that expression, we can find kinetic energy.

SOLVE

The ball's speed v is given by Equation 2.9. We take upward to be positive. The acceleration is then $a = -g$.

$$v = v_0 + at \quad (2.9)$$

$$v = v_0 - gt \quad (1)$$

To find an expression for kinetic energy as a function of time, we substitute Equation (1) into Equation 8.1.

$$K = \tfrac{1}{2}mv^2 = \tfrac{1}{2}m(v_0 - gt)^2 \quad (2)$$

$$K(t) = \tfrac{1}{2}m(v_0^2 - 2v_0gt + g^2t^2) \quad (3)$$

Our job now is to sketch Equation (3). There are three *positive* constants: m, v_0, and g. The t^2 term indicates that we have a parabola. We know that at the top of the motion, the velocity is zero, so by Equation (1) $v_0 = gt_{top}$; this is the maximum speed of the ball. From Equation (2), the ball's kinetic energy is at its maximum as it leaves the hand at $t = 0$ and again when it returns at $t = 2t_{top}$ (Fig. 8.2).

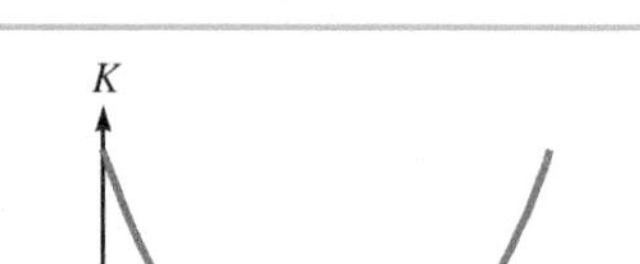

FIGURE 8.2 Kinetic energy versus time

CHECK and THINK

As expected, the kinetic energy is never negative. It is zero when the ball momentarily stops at the top (Fig. 8.2). Contrast this figure to the velocity-versus-time graph, which is positive until the ball reaches the top and is negative when it falls back down (Fig. 8.3).

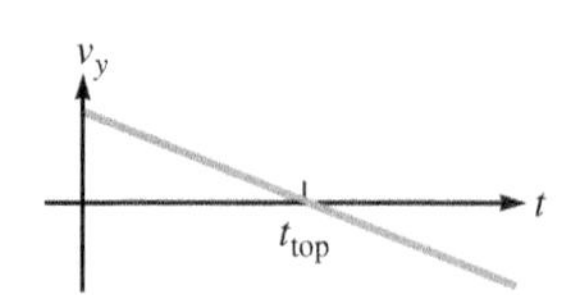

FIGURE 8.3 Velocity versus time

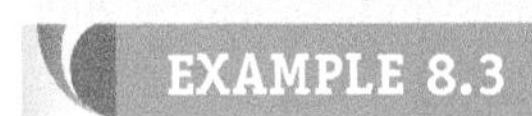

EXAMPLE 8.3 Crall and Whipple's Carts Collide

Crall and Whipple (CASE STUDY Chapter 2) place two carts on a track and set them on a collision course (Fig. 8.4). Cart A has a mass of 0.831 kg, and cart B has a mass of 1.336 kg. Just before they collide, their velocities are $\vec{v}_A = 1.23\hat{\imath}\,\text{m/s}$ and $\vec{v}_B = -0.91\hat{\imath}\,\text{m/s}$ (measured in a frame that is at rest with respect to the laboratory).

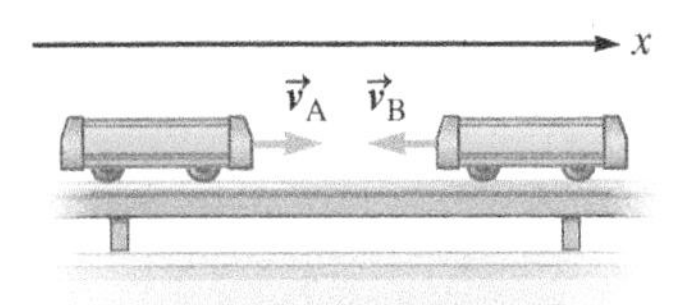

FIGURE 8.4 Two carts are set on a collision course.

A Crall chooses a system that consists of both carts. According to Crall, what is the kinetic energy of the system?

INTERPRET and ANTICIPATE

Because the system consists of two particles, add the kinetic energy of each one to find the kinetic energy of the system.

SOLVE

According to Crall, each cart has nonzero kinetic energy. We add the kinetic energy of each cart.

$$K = \tfrac{1}{2}m_A v_A^2 + \tfrac{1}{2}m_B v_B^2$$

$$K = \tfrac{1}{2}(0.831\text{ kg})(1.23\text{ m/s})^2 + \tfrac{1}{2}(1.336\text{ kg})(-0.91\text{ m/s})^2$$

$$K = 1.18\text{ J} \quad \text{in Crall's (lab's) frame}$$

CHECK and THINK

Because kinetic energy only depends on speed, the direction in which the carts are moving does not matter. The kinetic energy of the system according to Crall would be the same whether the carts were moving toward each other, away from each other, or in the same direction.

B Whipple decides to calculate the kinetic energy of cart B as measured by an imaginary observer moving at the same velocity as cart A. What is the kinetic energy measured by this observer?

INTERPRET and ANTICIPATE

This part is similar to part A in that the system has two particles. What is different is the speed of carts in Whipple's chosen frame. In this frame, cart A is at rest, and only cart B is in motion. So, only cart B has nonzero kinetic energy.

SOLVE

We must find the velocity of cart B (the subject) in Whipple's *m*oving reference frame by using Equation 4.46. The relative velocity between the two frames $(v_M)_L$ is the velocity of cart A as observed in the *l*aboratory (Crall's) frame.

$$(\vec{v}_B)_L = (\vec{v}_B)_M + (\vec{v}_M)_L \qquad (4.46)$$

$$(\vec{v}_B)_M = (\vec{v}_B)_L - (\vec{v}_M)_L$$

$$(\vec{v}_B)_M = -0.91\hat{\imath}\text{ m/s} - 1.23\hat{\imath}\text{ m/s} = -2.14\hat{\imath}\text{ m/s}$$

Use $K = \frac{1}{2}mv^2$ (Eq. 8.1) to find the kinetic energy, substituting for v the speed seen by the moving observer.

$$K = \tfrac{1}{2}m_B[(v_B)_M]^2 \qquad (8.1)$$

$$K = \tfrac{1}{2}(1.336\text{ kg})(-2.14\text{ m/s})^2 = 3.06\text{ J}$$

CHECK and THINK

The kinetic energy measured in the moving frame (Whipple's choice) does not equal the kinetic energy measured in the laboratory frame (Crall's choice). The lesson here is that once you choose a particular reference frame, you must stick with it throughout all relevant calculations. (It is particularly important when working with potential energy, our next topic.)

Potential Energy

POTENTIAL ENERGY
Major Concept

Kinetic energy describes the system's motion, but even a system that has no motion may still have energy associated with the arrangement of the particles in it. This energy, known as **potential energy**, depends on the configuration of a system. Potential energy is only used to describe a system consisting of two or more particles that interact with each other via one or more *internal* forces. That is, forces acting on subjects in the system come only from sources within that system.

Not all types of internal forces can be associated with potential energy. Potential energy can only be associated with internal **conservative** forces. We will define *conservative* and *nonconservative* forces completely (and learn about a third classification) in Chapter 9. Of the forces we have studied so far, only gravity and the spring force (Hooke's law) are conservative.

Potential energy cannot be used to describe a single particle because a single (isolated) particle cannot experience internal forces. To associate potential energy with a system, we must define the system so that it includes both the source(s) and subject(s) of a conservative force or forces.

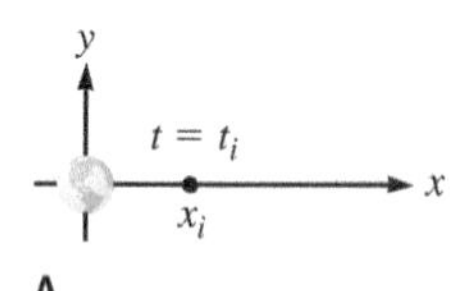

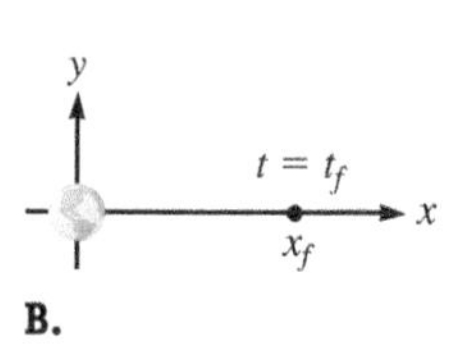

FIGURE 8.5 A two-particle system such as the Earth and a spacecraft (represented by a black dot). **A.** The spacecraft starts at an initial position x_i. **B.** The spacecraft has moved to a final position x_f.

Consider a two-particle system such as the Earth and a small spacecraft (Fig. 8.5). A coordinate system has been chosen with its origin is at the center of the Earth. Initially, at time t_i, the spacecraft is at $\vec{r}_i = x_i\hat{\imath}$ (Fig. 8.5A). At a later time, $t = t_f$, the spacecraft is at a new position, $\vec{r}_f = x_f\hat{\imath}$, and the two particles are farther apart in this new configuration (Fig. 8.5B). Because the system's configuration has changed, there is a change in potential energy U. The change in potential energy is defined by

$$\Delta U = U_f - U_i \equiv -\int_{x_i}^{x_f} F_x dx \tag{8.3}$$

where F_x is the x component of the force exerted by the Earth on the spacecraft. In this case, that force is gravity, directed along the x axis. The spacecraft's displacement is also along the x axis. Equation 8.3 may be rewritten for displacement along any axis (x, y, z, r) as long as the component of the force along that axis is used.

We find the units of potential energy from Equation 8.3:

$$1\ \text{N} \cdot \text{m} = 1\ \text{kg} \cdot \text{m/s}^2\ \text{m} = 1\ \text{kg} \cdot \text{m}^2/\text{s}^2$$
$$1\ \text{N} \cdot \text{m} = 1\ \text{J}$$

So, the SI unit of energy—both potential and kinetic—is the joule. In the next three sections, we will use Equation 8.3 to derive expressions for the potential energy associated with gravity and the spring force.

GRAVITATIONAL POTENTIAL ENERGY NEAR EARTH'S SURFACE

Special Case

8-3 Gravitational Potential Energy Near the Earth

A very common system you will encounter consists of the Earth and an object near the Earth. In this section, we'll use Equation 8.3 to find an easily-remembered expression for the change in the system's gravitational potential energy.

DERIVATION Potential Energy Near the Surface of a Planet

We show here that the change in a system's potential energy near the surface of a planet such as the Earth is given by

$$\Delta U = mgy_f - mgy_i \tag{8.4}$$

To calculate the gravitational potential energy of a system involving the Earth, imagine a two-particle system in which one particle is the Earth (Fig. 8.6). For the sake of argument, imagine that the other particle is an apple of mass m. We have chosen to place the origin of the coordinate system on the surface of the Earth. The apple (represented as a dot) moves from y_i to y_f. (The distance shown in Figure 8.6 is exaggerated.)

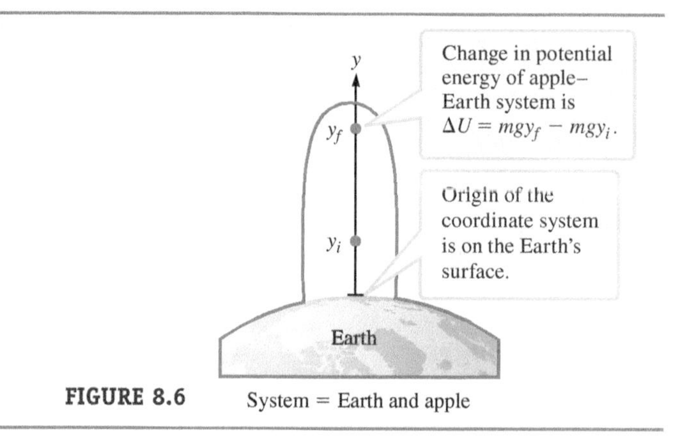

FIGURE 8.6 System = Earth and apple

The Earth's gravitational force on an object near its surface does not depend on distance. So, as the apple is raised from its initial to its final position, the gravitational force on it does not change.	$\vec{F}_g = -mg\hat{\jmath}$	
Modify Equation 8.3 to find the change in the gravitational potential energy of the Earth–apple system due to the apple's displacement along the y axis from y_i to y_f.	$\Delta U = -\int_{y_i}^{y_f} F_y dy$	(8.3)
Substitute $F_y = -mg$.	$\Delta U = -\int_{y_i}^{y_f} (-mg)dy$	
Because $-mg$ does not depend on y, we pull it outside the integral and integrate.	$\Delta U = mg\int_{y_i}^{y_f} dy = mgy\Big\vert_{y_i}^{y_f}$	
Substitute the limits of integration.	$\Delta U = U_f - U_i = mgy_f - mgy_i$ ✓	(8.4)

COMMENTS

Only changes in potential energy ΔU are physically important. The initial (or final) potential energy is set to a convenient value. Choosing the zero point for the potential energy (or equivalently assigning the reference configuration) is an important step in solving problems as discussed below.

Reference Configuration

To make calculations easier, we typically choose a particular configuration to be the *reference configuration* and assign it a potential energy of zero. The potential energy of all other configurations can then be found from a change in potential energy relative to the reference configuration. Choosing the reference configuration in a situation that involves the Earth and a particle near its surface means setting the origin of the coordinate system to a convenient place. For example, if we place the origin of the coordinate system in Figure 8.7 at the initial position of the apple, it becomes the reference configuration, with a potential energy of zero:

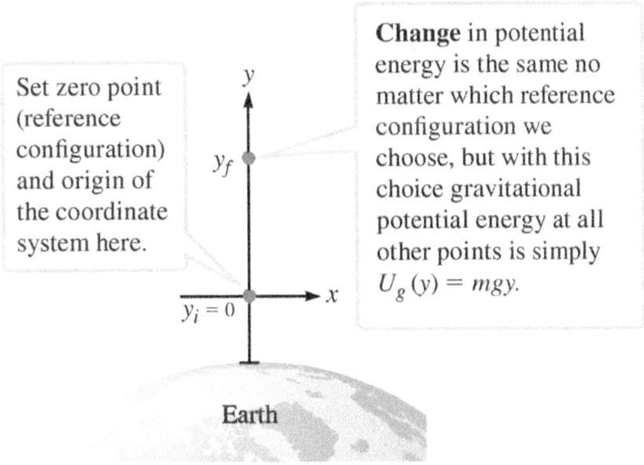

FIGURE 8.7 As in Figure 8.6 the system consists of the apple and the Earth, but here the choice of the origin and reference configuration is more convenient.

$$\Delta U = U_f - 0 = mgy_f - 0$$

The **gravitational potential energy** of any other configuration may now be written as

$$U_g(y) = mgy \quad (8.5)$$

So, if the apple is at the origin, the potential energy of the Earth–apple system is zero. If the potential energy of the system is greater than zero, the apple is above the origin. If the potential energy is negative, the apple is below the origin. Potential energy is a scalar, so the signs do not indicate direction; rather, they indicate a relative change in potential energy. If the apple is above the origin, the system has *more* energy than if the apple is below the origin.

Path Independence

Only a vertical change in the particle's position results in a change in the potential energy of the Earth–particle system. If the particle moves upward, the gravitational potential energy of the system increases. If the particle's displacement is downward, the potential energy of the system decreases. If the particle's displacement is horizontal, the potential energy remains the same. Furthermore, the change in potential energy only depends on the final and initial configurations of the system, no matter

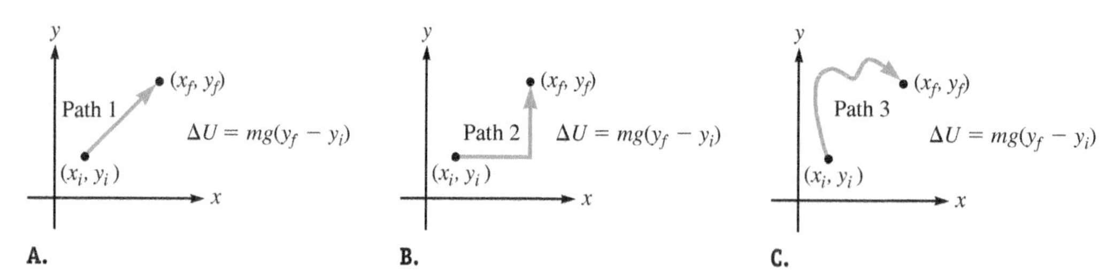

FIGURE 8.8 The change in gravitational potential energy does not depend on the path. The change in potential energy is the same for all three paths because the initial and final positions are the same.

what configurations the system takes in between. We say that the change in potential energy of the system is *path independent.* Figure 8.8 shows three possible paths a particle may take getting from (x_i, y_i) to (x_f, y_f); in all three cases, the change in potential energy is the same: $\Delta U = mg(y_f - y_i)$.

EXAMPLE 8.4 Look Out Below!

A construction worker is repairing the roof of a house. His hammer of mass 0.55 kg slips from his hand and falls through a basketball hoop to the driveway below. The basketball hoop is 3.0 m above the ground and 5.7 m below the roof where the hammer left the worker's hand.

A To calculate potential energy, what must be in the system?

The system must have at least two objects that interact with each other through one of the conservative forces we know, gravity or the spring force.	The system includes the hammer and the Earth. The force between them is gravity.

B Place the origin of an upward-pointing y axis on the driveway. What is the change in the potential energy of the system?

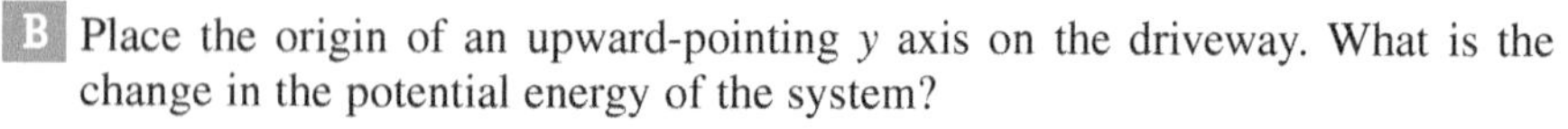

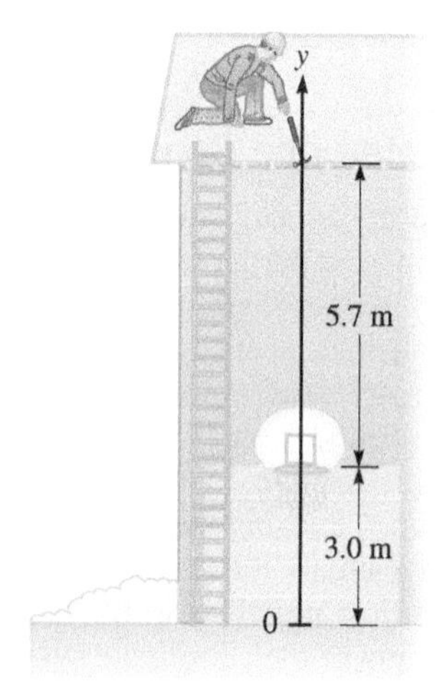

FIGURE 8.9

INTERPRET and ANTICIPATE

Following the directions, we draw an upward-pointing y axis on our sketch (Fig. 8.9).

SOLVE

With this choice for the origin, the reference point for the potential energy of the system is at the driveway. When the hammer is on the *d*riveway, the potential energy of the system is zero.	$U_D = 0$
The potential energy of the system when the hammer is at the *r*oof is given by Equation 8.5.	$U_R = mgy_R$ (8.5)
Find the change in the system's gravitational potential energy due the hammer falling from the roof to the driveway.	$\Delta U = U_D - U_R = 0 - mgy_R = -mgy_R$
Figure 8.9 shows that $y_R = 3.0\text{ m} + 5.7\text{ m} = 8.7\text{ m}$.	$\Delta U = -(0.55\text{ kg})(9.81\text{ m/s}^2)(8.7\text{ m}) = -47\text{ kg m}^2/\text{s}^2$ $\Delta U = -47\text{ J}$

CHECK and THINK

The negative sign means that the Earth–hammer system lost potential energy due to the hammer's fall.

C Now use the basketball hoop as the origin of the y axis and find the change in potential energy.

INTERPRET and ANTICIPATE

In Figure 8.10, we use an upward-pointing y axis with the origin at the basketball hoop. The roof is at a positive position, and the driveway at a negative position. With this choice, the reference point for the potential energy of the system is at the basketball hoop. When the hammer is at the roof, the potential energy of the system is positive, and when the hammer is on the driveway, the potential energy is negative. We expect to find the same change in potential energy.

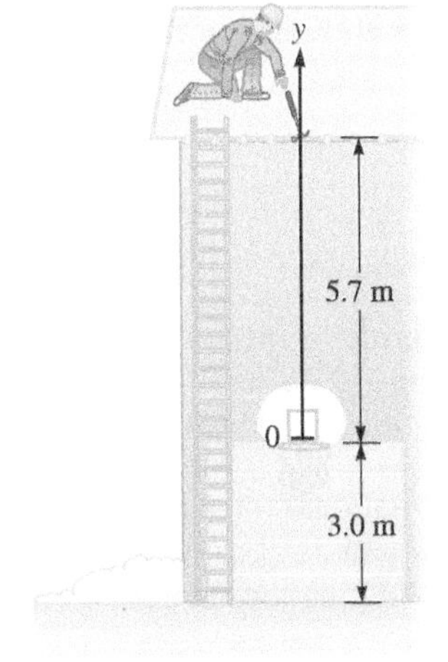

FIGURE 8.10

SOLVE Use Equation 8.5 to find the potential energy of the system for both configurations: hammer at roof and hammer on driveway.	$U_R = mgy_R$ $U_D = mgy_D$
Find the change in the system's gravitational potential energy.	$\Delta U = U_D - U_R = mgy_D - mgy_R$ $\Delta U = mg(y_D - y_R)$
Our sketch shows that the roof's position is positive and the driveway's is negative: $y_R = +5.7$ m and $y_D = -3.0$ m.	$\Delta U = (0.55\text{ kg})(9.81\text{ m/s}^2)(-3.0\text{ m} - 5.7\text{ m})$ $\Delta U = (0.55\text{ kg})(9.81\text{ m/s}^2)(-8.7\text{ m}) = -47\text{ J}$

CHECK and THINK

There are no surprises here; this result is exactly what we found in part B. With this choice for the origin, neither the final (nor the initial) configuration was set as the reference configuration. Therefore, neither U_D nor U_R was zero making our calculation slightly more complicated. It is helpful to choose either the initial or final configuration as the reference configuration, but it is not necessary.

CONCEPT EXERCISE 8.4

In Figure 8.11, a person launches a ball off of a building three times.

Case 1. He drops the ball and it lands at the base of the building.
Case 2. He uses a slingshot to launch the ball in a parabolic path. The ball lands on the ground at the same level as the base of the building.
Case 3. He launches the ball. It flies in a parabolic path and lands on the roof of a building at the same height at which the ball was launched.

Compare the change in potential energy of the Earth–ball system for each case. Explain your answer.

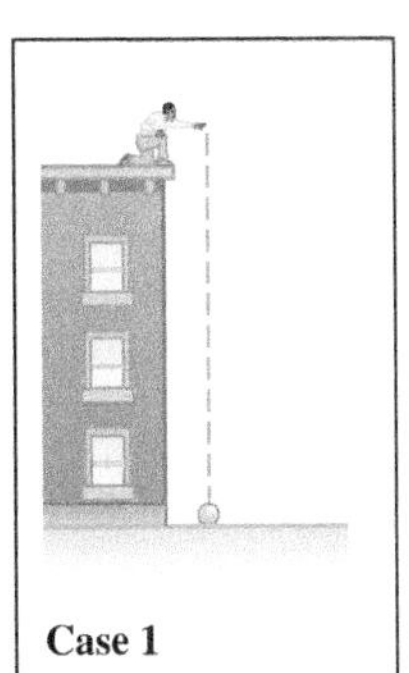
Case 1

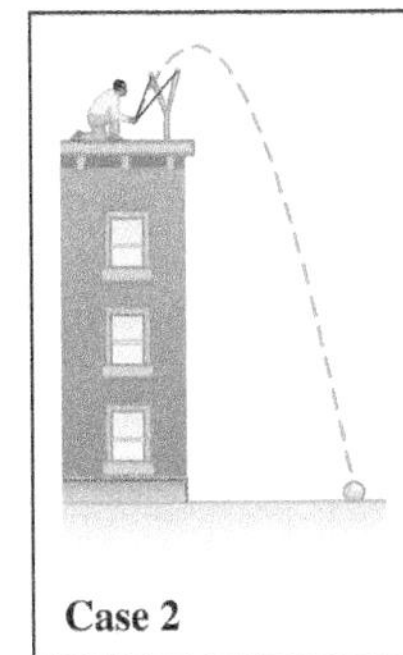
Case 2

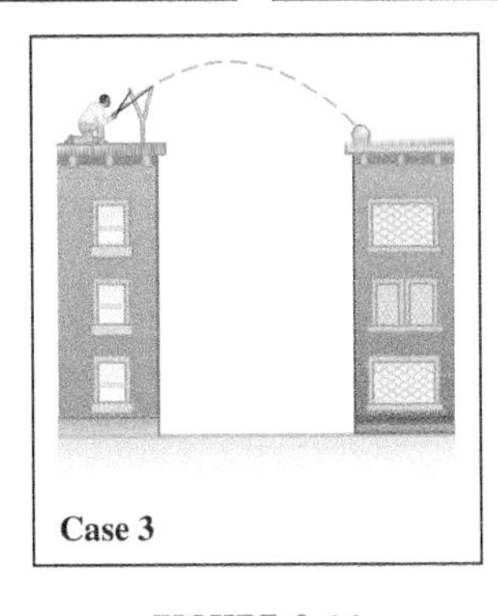
Case 3

FIGURE 8.11

8-4 Universal Gravitational Potential Energy

Equation 8.5, $U_g(y) = mgy$, can be used to find the gravitational potential energy of any system that consists of a particle near the surface of a large object such as the Earth, the Moon, or another planet. In fact, it can be used whenever the gravitational force exerted on the particle is roughly constant. If it isn't (approximately) constant such as when a rocket is launched from the Earth, however, we need to use an expression for *universal gravitational potential energy*.

DERIVATION Universal Gravitational Potential Energy

Starting with Equation 8.3, we will derive the expression

$$\Delta U = \frac{GMm}{r_i} - \frac{GMm}{r_f} \quad (8.6)$$

for the change in the gravitational potential energy of any system that consists of two particles such as the Earth (of mass M) and a spacecraft (of mass m).

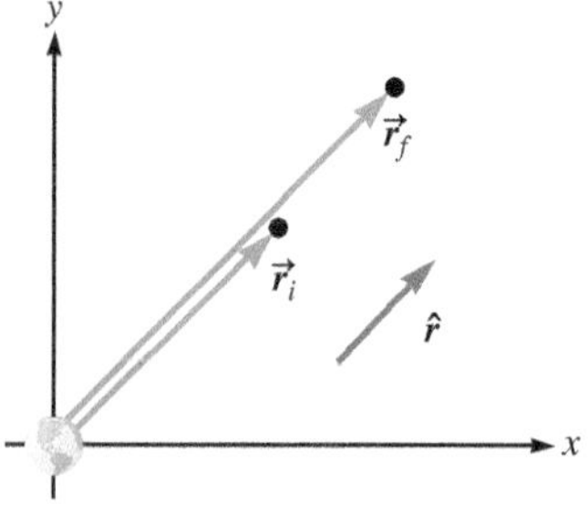

FIGURE 8.12

Let's consider a spacecraft (represented by a dot) that moves away from the Earth from $\vec{r}_i$ to $\vec{r}_f$. We have chosen a unit vector $\hat{r}$ that points away from the Earth in the direction of the spacecraft (Fig. 8.12).

The gravitational force of the Earth on the spacecraft is given by Newton's law of universal gravity (Eq. 7.4). The subscript r indicates that the force is along the radial axis joining the Earth to the spacecraft.	$\vec{F}_G = F_r\hat{r} = -G\frac{Mm}{r^2}\hat{r}$
Rewrite Equation 8.3 in terms of this radial axis.	$\Delta U = -\int_{r_i}^{r_f} F_r dr$ (8.3)
Substitute for F_r.	$\Delta U = -\int_{r_i}^{r_f}\left(-G\frac{Mm}{r^2}\right)dr$
Pull the constants outside the integral and integrate. Notice that the negative signs cancel out.	$\Delta U = GMm\int_{r_i}^{r_f}\left(\frac{1}{r^2}\right)dr = GMm\left(-\frac{1}{r}\right)\Big\|_{r_i}^{r_f}$
Substitute the limits of integration.	$\Delta U = -\frac{GMm}{r_f} - \left(-\frac{GMm}{r_i}\right) = \frac{GMm}{r_i} - \frac{GMm}{r_f}$ ✓ (8.6)

COMMENTS

We showed in the previous section that if a particle moves upward from the surface of the Earth, the gravitational potential energy increases. So we expect $\Delta U > 0$ in this case because the spacecraft moves upward (Fig. 8.12).

The initial position is closer to the origin than the final position.	$r_i < r_f$
Because r_i is in the denominator of the first term and r_f is in the denominator of the second term in Equation 8.6, the first term is larger than the second term.	$\frac{GMm}{r_i} > \frac{GMm}{r_f}$
So, the change in potential energy is positive as predicted.	$\Delta U = \frac{GMm}{r_i} - \frac{GMm}{r_f} > 0$

Reference Configuration for Universal Gravity

If the spacecraft moves away from the Earth, the potential energy of the system increases (Fig. 8.13A). If the spacecraft moves toward the Earth, the system's potential energy decreases (Fig. 8.13B). If spacecraft ends up at the same distance from the Earth at which it started such as on part of a circular orbit, there is no change in the system's gravitational potential energy (Fig. 8.13C).

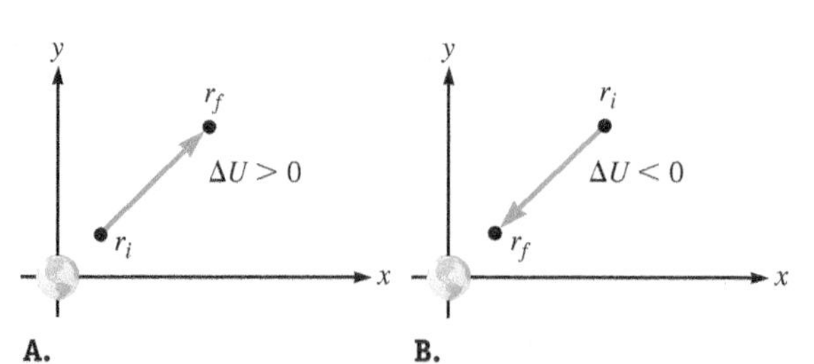

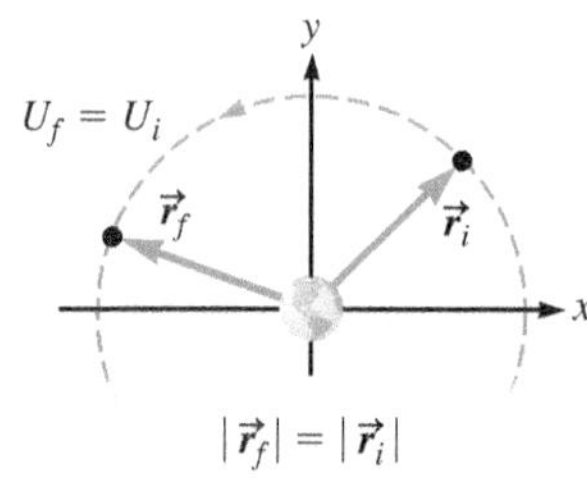

FIGURE 8.13 A system consists of a spacecraft and the Earth.

In principle, the reference configuration can be set to any arbitrary choice such as the spacecraft on the surface of the Earth or in a circular geosynchronous orbit. In practice, because position appears in the denominator, it is best and therefore *usual* to set the reference configuration such that when the two particles are infinitely far apart, the potential energy is zero. So, for a system that consists of two particles 1 and 2 with the reference configuration chosen such that $\lim_{r\to\infty} U = 0$, the gravitational potential energy is

$$U_G(r) = -G\frac{m_1 m_2}{r} \quad (8.7)$$

UNIVERSAL GRAVITATIONAL POTENTIAL ENERGY

Special Case

In any particular problem, you may need $U_g = mgy$ (Eq. 8.5) *or* $U_G = -G(m_1m_2/r)$ (Eq. 8.7), *but not both.* Use $U_g = mgy$ only if the system consists of a large object such as a planet and another particle such as an apple that stays near the surface of the large object.

The subscript G is a reminder that Equation 8.7 is good for particles that interact through (universal) gravity.

EXAMPLE 8.5 Plotting Universal Gravitational Potential Energy

Plot $U_G(r)$ versus r and show with a sketch how it connects to a spacecraft moving directly away from the Earth. Should the change in the system's gravitational potential energy be positive, negative, or zero? Does the graph give the predicted results for ΔU_G?

INTERPRET and ANTICIPATE

We put r on the horizontal axis and $U_G(r)$ on the vertical axis (Fig. 8.14). It may be helpful to use a spreadsheet or similar program. Our sketch shows the Earth and the spacecraft at two different positions $\vec{r}_1$ and $\vec{r}_2$. Because the spacecraft moves away from the Earth, we expect that the system's gravitational potential energy will increase: $\Delta U_G > 0$.

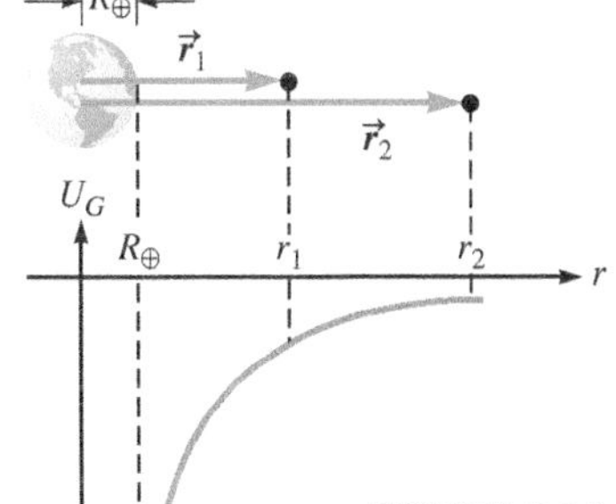

FIGURE 8.14

SOLVE

From the graph, $U_G(r_2) > U_G(r_1)$. Therefore the change in potential energy if the spacecraft moves away from the Earth from $\vec{r}_1$ to $\vec{r}_2$ is positive, which means that the gravitational potential energy increases as the spacecraft moves away.

$$\Delta U_G = U_G(r_2) - U_G(r_1)$$
$$U_G(r_2) > U_G(r_1)$$
$$\Delta U_G > 0$$

CHECK and THINK

The graph shows that as $r \to \infty$, the potential energy $U_G \to 0$ as required by our choice of the usual reference configuration.

8-5 Elastic Potential Energy

ELASTIC POTENTIAL ENERGY

Special Case

The other conservative force we study in mechanics is exerted by a spring that obeys Hooke's law. In this section, we derive an expression for the potential energy stored by such a spring, which is known as **elastic potential energy**.

DERIVATION Elastic Potential Energy

Figure 8.15 shows a block on a frictionless, horizontal surface attached to a spring. The other end of the spring is attached to a fixed wall. The system consists of the block, the spring, and the wall. For brevity, we omit the wall and refer to the system as a spring–block system. The block is on a frictionless surface. The block starts at position x_i (Fig. 8.15A). It moves to the left to position x_f, and the spring compresses. The force exerted by the spring on the block (and the wall) is internal to the system, is conservative, and is given by Hooke's law $\vec{F}_H = -k\Delta\vec{x}$ (Eq. 5.8) where $\Delta\vec{x}$ is the displacement of the block from the relaxed (or equilibrium) position. We will show that the change in the system's spring potential energy is given by

$$\Delta U = \tfrac{1}{2}kx_f^2 - \tfrac{1}{2}kx_i^2 \quad (8.8)$$

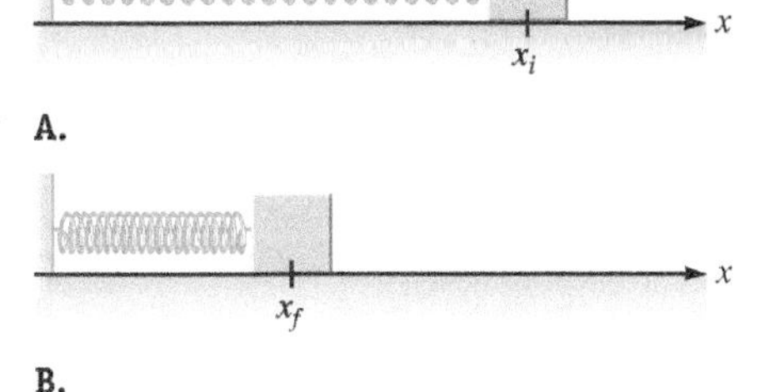

FIGURE 8.15 A system consists of a block, a wall, and a spring. **A.** The block is initially at x_i. **B.** The block moves to x_f.

Derivation continues on page 224 ▶

To find the potential energy of this system, we take the relaxed position to be at the origin and substitute the resulting expression for force into $\Delta U = -\int_{x_i}^{x_f} F_x dx$ (Eq. 8.3).	Hooke's law with origin at relaxed position: $F_x = -kx$ $\Delta U = -\int_{x_i}^{x_f} -kx\,dx$
Simplify and integrate.	$\Delta U = k\int_{x_i}^{x_f} x\,dx = k\frac{x^2}{2}\Big\|_{x_i}^{x_f}$
Substitute the limits of integration.	$\Delta U = \frac{1}{2}kx_f^2 - \frac{1}{2}kx_i^2$ ✓ (8.8)

COMMENTS

We derived this expression for a horizontal spring lying along the x axis. Our derivation would be the same, however, if the spring were vertical and parallel to the y axis as is common in many problems. (See Example 8.6.)

Reference Configuration for Spring Potential Energy

We usually place the origin of the coordinate system at the relaxed (or equilibrium) position and set this position to the reference configuration. Then, the **elastic potential energy** of the spring–block system is given by

$$U_e = \tfrac{1}{2}kx^2 \tag{8.9}$$

where the spring lies along the x axis. If the spring lies along a different axis such as the y axis, then

$$U_e = \tfrac{1}{2}ky^2 \tag{8.10}$$

where e stands for *e*lastic. Because the component of position is squared, the potential energy associated with any configuration (except the reference configuration) is positive. If the block is at the relaxed position (Fig. 8.16A), the potential energy of the system is zero. If the spring is compressed (Fig. 8.16B) or stretched (Fig. 8.16C), the potential energy is positive. In other words, if the block moves from the relaxed position to any other position, the potential energy increases. However, this statement does *not* mean that all changes in potential energy are positive.

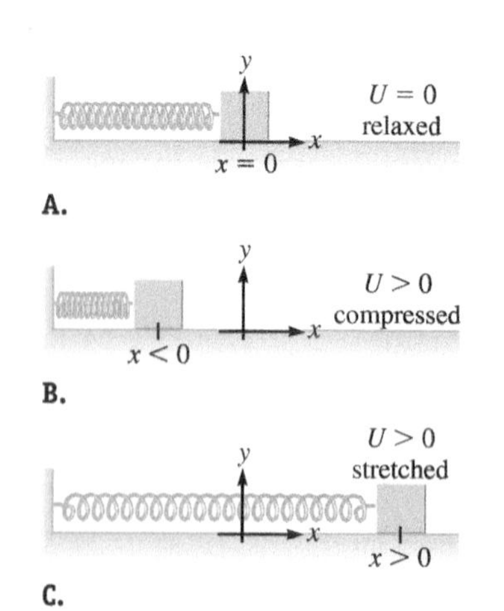

FIGURE 8.16 A. The reference configuration and the origin are set to the position of the block when the spring is relaxed. If the spring is then compressed **B.** or stretched **C.**, the elastic potential energy of the system increases.

EXAMPLE 8.6 A Vertical Spring

In your laboratory, you are likely to encounter a vertical rather than a horizontal spring because then there is no need to create a frictionless surface. You will probably hang a ball of mass m from the spring of spring constant k. The experiment will begin when the ball is displaced straight downward by a distance h (Fig. 8.17).

A To find the change in the system's potential energy, what must be included in the system? (Be sure that there are no external forces exerted on your system.)

It is probably obvious that the system must include the ball and the spring (and the ceiling). It might not be as obvious that the Earth must also be in the system. The Earth must be included, however, because as the ball moves along the vertical axis there is a change in gravitational potential energy.

B Find an expression for the change in the system's potential energy from the equilibrium position (after the ball has been attached to the spring as in Fig. 8.17, middle) to a distance h below equilibrium (Fig. 8.17, right).

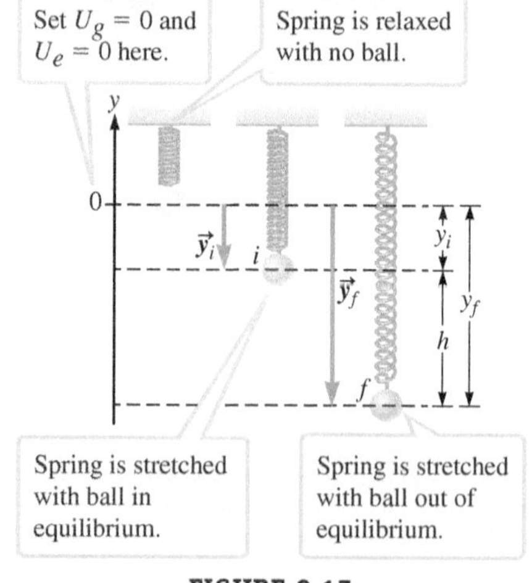

FIGURE 8.17

INTERPRET and ANTICIPATE

Two conservative forces—gravity and the spring force—are internal to this system. It is convenient to set the reference configuration to the point where the spring is *relaxed* (before the ball is hung from it), so $y = 0$ at the bottom end of the spring (Fig. 8.17, left). The elastic and gravitational potential energies are then both zero when the spring is relaxed. At the *initial* position with the ball in equilibrium attached to the stretched spring (Fig. 8.17, middle), the gravitational potential energy is negative, and the elastic potential energy is positive.

SOLVE

Use $U_g(y) = mgy$ (Eq. 8.5) to find the initial gravitational potential energy and $U_e = \frac{1}{2}ky^2$ (Eq. 8.10) to find the initial elastic potential energy at $\vec{y}_i$.	$U_{gi} = -mgy_i$ $U_{ei} = \frac{1}{2}ky_i^2$	(1) (2)
Next, find the potential energies at the final position. According to Figure 8.17, right the final position is given by $\vec{y}_f = -(h + y_i)\hat{\jmath}$.	$U_{gf} = mg(-y_f) = -mg(h + y_i)$ $U_{gf} = -mgh - mgy_i$ $U_{ef} = \frac{1}{2}k(-y_f)^2 = \frac{1}{2}k(h + y_i)^2$ $U_{ef} = \frac{1}{2}kh^2 + khy_i + \frac{1}{2}ky_i^2$	(3) (4)
Find the change in gravitational potential energy by subtracting Equation (1) from Equation (3).	$\Delta U_g = (-mgh - mgy_i) - (-mgy_i)$ $\Delta U_g = -mgh$	(5)
Find the change in elastic potential energy by subtracting Equation (2) from Equation (4).	$\Delta U_e = \left(\frac{1}{2}kh^2 + khy_i + \frac{1}{2}ky_i^2\right) - \frac{1}{2}ky_i^2$ $\Delta U_e = \frac{1}{2}kh^2 + khy_i$	(6)
Add Equations (5) and (6) to find the change in the system's potential energy (both gravitational and elastic).	$\Delta U = \Delta U_g + \Delta U_e$ $\Delta U = -mgh + \frac{1}{2}kh^2 + khy_i$	(7)

The first term, $-mgh$, is due to change in gravitational potential energy, and the second term, $\frac{1}{2}kh^2$, is due to the change in elastic potential energy. The system's potential energy was not zero at the initial (equilibrium) position, however, so a third term, khy_i, arises. Let's use Newton's second law to deal with this third term. When the ball is in equilibrium (Fig. 8.17, middle), two forces, the spring force and gravity, are exerted on it, and its acceleration is zero (Fig. 8.18).

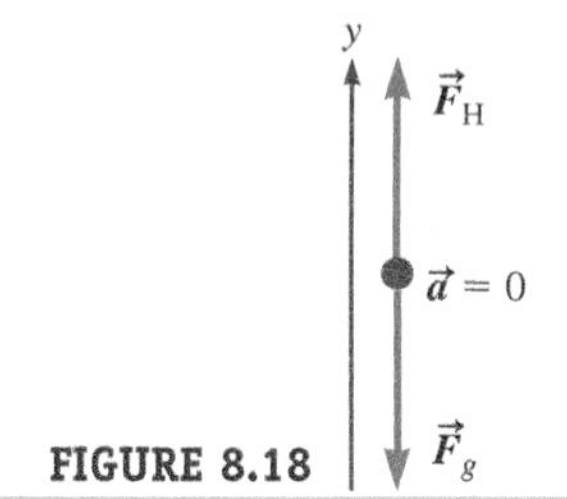

FIGURE 8.18

Applying Newton's second law, we find that the spring force is equal in magnitude to the ball's weight.	$\sum F_y = F_H - F_g = 0$ $F_g = F_H$ $mg = ky_i$	(8)
Substitute Equation (8) into Equation (7).	$\Delta U = -mgh + \frac{1}{2}kh^2 + mgh$ $\Delta U = \frac{1}{2}kh^2$	(9)

CHECK and THINK

Our final expression, Equation (9), does not appear to depend on gravitational potential energy, which is very helpful in the laboratory. If you measure the displacement h of the ball from its equilibrium position, you apparently can ignore changes in gravitational potential energy. Actually, these changes balance out because the system has potential energy in its equilibrium configuration.

8-6 Conservation of Mechanical Energy

Because potential energy can only be associated with a system that consists of two or more particles, the conservation approach requires that we choose a system of two or more particles. In this chapter, we choose systems that include only particles and objects that can be modeled as particles; in the next chapter, we consider more complicated systems.

ISOLATED SYSTEM ✪ **Major Concept**

In this chapter, we also make the special choice of an *isolated system*, one that does not interact with its environment. An **isolated system** does not exert a force on its environment, and the environment does not exert a force on it. For example, if a situation involves an asteroid falling to the Earth, we would not choose the asteroid alone to be the system (Fig. 8.19A). Instead, we should choose a system that includes both the asteroid and the Earth so that gravity is internal to the system (Fig. 8.19B).

An isolated system is an idealization in that almost no real system can be truly isolated. For example, in the Earth–asteroid system in Figure 8.19B, the environment is not truly empty space. The Moon, the Sun, and other planets are in the environment, and they exert gravitational forces on the system. Also, as the asteroid passes through the Earth's atmosphere, molecules interact with the system through the drag force. We can often ignore such environmental effects and consider the system to be isolated, however.

Once we have decided what is in the system, we cannot change that decision partway through our analysis. We then apply the conservation approach to our chosen system. Part of the conservation approach comes from the *conservation of mechanical energy.* The sum of a system's kinetic and potential energy is its **mechanical energy,**

$$E = K + U \tag{8.11}$$

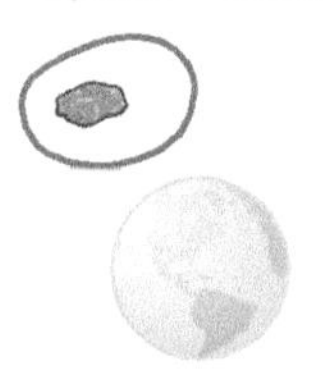

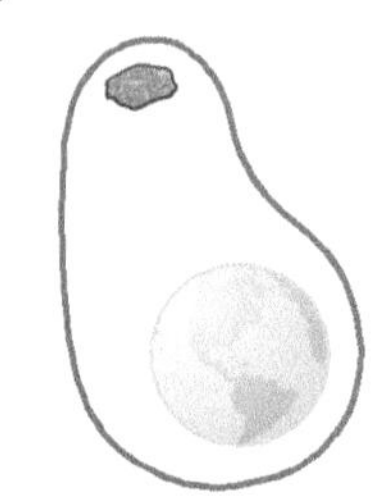

FIGURE 8.19 Two possible system choices. **A.** Only the asteroid is in the system. The system is **not isolated** because the Earth exerts a force on it. **B.** Both the Earth and the asteroid are in the system. The environment is considered to be empty and cannot exert a force on the **isolated** Earth–asteroid system.

According to the **principle of conservation of mechanical energy,** *if only conservative forces are acting, the mechanical energy of an isolated system is conserved.* Put another way, the kinetic energy $K(t)$ and the potential energy $U(t)$ change in time, but the mechanical energy E is a constant that does *not* depend on time. So, it is possible for the system's energy to change forms from kinetic to potential and from potential to kinetic, but the sum of the two is a constant.

The conservation of mechanical energy may be expressed in terms of an initial time and a final time:

$$E_i = E_f \tag{8.12}$$

Another way to express the conservation of mechanical energy is

$$\Delta E = 0 \tag{8.13}$$

Because mechanical energy is the sum of the kinetic and potential energy of the system, we have

$$K_i + U_i = K_f + U_f \tag{8.14}$$

CONSERVATION OF MECHANICAL ENERGY ❶ **Underlying Principle**

By rearranging terms, we can write Equation 8.14 in terms of the change in energy:

$$(K_f - K_i) + (U_f - U_i) = 0$$

$$\Delta K + \Delta U = 0 \tag{8.15}$$

Equations 8.12 through 8.15 are all mathematical expressions for the conservation of mechanical energy principle. To illustrate the conservation of mechanical energy approach, let us begin with an example we have already considered many times.

EXAMPLE 8.7 A Falling Ball Revisited

A ball of mass m falls straight down and lands in a person's hand (Fig. 8.20). The ball's initial speed is v_i, and it falls with constant acceleration $\vec{a}_y = -g\hat{\jmath}$. Starting with the conservation of mechanical energy, show that

$$v_f^2 = v_i^2 - 2g\Delta y$$

in agreement with our work in Chapter 2.

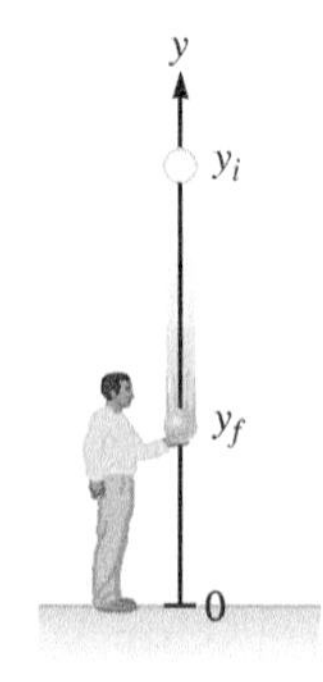

FIGURE 8.20

INTERPRET and ANTICIPATE

The system consists of the ball and the Earth. If we ignore the surrounding air, the system is essentially isolated, and mechanical energy is conserved.

SOLVE

Find an expression for the initial mechanical energy. Use $K = \frac{1}{2}mv^2$ (Eq. 8.1) and $U_g(y) = mgy$ (Eq. 8.5).	$E_i = K_i + U_i = \frac{1}{2}mv_i^2 + mgy_i$ (8.11)
Find a similar expression for the final mechanical energy.	$E_f = K_f + U_f = \frac{1}{2}mv_f^2 + mgy_f$
Mechanical energy is conserved, so set the initial and final mechanical energies equal to each other.	$E_i = E_f$ $\frac{1}{2}mv_i^2 + mgy_i = \frac{1}{2}mv_f^2 + mgy_f$
The mass drops out of this equation. Solve for the square of the final speed.	$v_i^2 + 2gy_i = v_f^2 + 2gy_f$ $v_f^2 = v_i^2 + 2gy_i - 2gy_f = v_i^2 - 2g(y_f - y_i)$ $v_f^2 = v_i^2 - 2g\Delta y$

CHECK and THINK

This result is exactly what we would find if we started with kinematic Equation 2.13 for constant acceleration. The conservation approach does not replace the kinematics and dynamics we learned previously, but it gives us other tools for solving complicated problems.

DERIVATION Principle of Conservation of Mechanical Energy

We started with this familiar example of a free-falling particle to show that conservation of mechanical energy provides an alternative and consistent approach to mechanics. To strengthen that connection, we will use Newton's second law to derive Equation 8.13, $\Delta E = 0$, for any isolated system in which only conservative forces are exerted between the system's particles.

Start with Newton's second law in one dimension along the x axis to make the derivation mathematically simpler.	$F_x = ma_x$
Write the acceleration in terms of its time derivative (Eq. 2.7).	$F_x = m\frac{dv_x}{dt}$ (1)
To eliminate time, use the chain rule and Equation 2.4.	$\frac{dv_x}{dt} = \frac{dx}{dt}\frac{dv_x}{dx} = v_x\frac{dv_x}{dx}$ (2)
Rewrite Newton's second law by substituting (2) into (1).	$F_x = mv_x\frac{dv_x}{dx}$
Solve this differential equation by moving dx to the left side and then integrating.	$F_x dx = mv_x dv_x$ $\int F_x dx = \int mv_x dv_x$
We must make sure that the limits of integration match. The integration is from the initial configuration and speed to the final configuration and speed. At the initial time, the particle's position and speed are x_i and v_i, respectively. Its final position and speed are x_f and v_f.	$\int_{x_i}^{x_f} F_x dx = \int_{v_i}^{v_f} mv_x dv_x$
Integrate and substitute the limits on the right side.	$\int_{x_i}^{x_f} F_x dx = \frac{1}{2}mv^2\Big\vert_{v_i}^{v_f} = \frac{1}{2}mv_f^2 - \frac{1}{2}mv_i^2$
The right side of the equation is the change in kinetic energy (Eq. 8.1). The left side is the negative of the change in potential energy (Eq. 8.3).	$-\Delta U = \Delta K$

Derivation continues on page 228 ▶

Upon rearranging, we find that we have derived Equation 8.13, the conservation of mechanical energy.	$\Delta K + \Delta U = 0$ $\Delta E = 0$ ✓

COMMENTS

This derivation shows that the conservation of mechanical energy may be derived from Newton's second law. So, the conservation approach does not replace the force approach; rather, they complement one another.

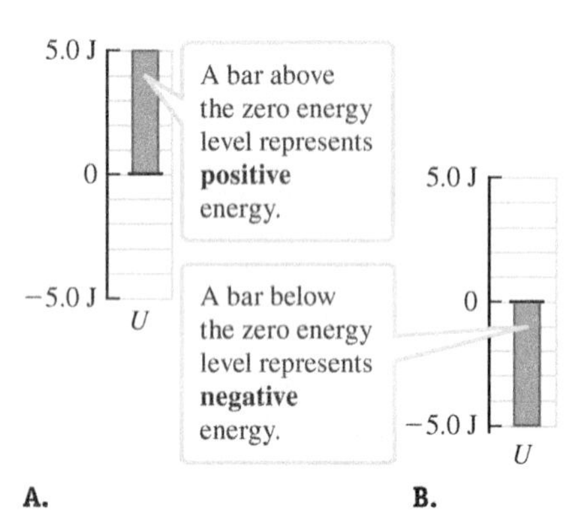

FIGURE 8.21 Bar charts for potential energy.

BAR CHART Tool

8-7 Applying the Conservation of Mechanical Energy

Applying conservation of mechanical energy is a good approach when you are interested in motion at two separate times, and the conservation approach is particularly powerful when the acceleration is *not* constant. In this section, we learn a strategy for applying the conservation of mechanical energy.

When we use the force approach, we choose a coordinate system and draw a free-body diagram. Likewise, using the conservation approach involves choices and diagrams. We must decide what to include in the system and what is in the outside environment. Some choices may make a problem easier to analyze, but the answer cannot depend on our choice.

Two kinds of diagrams help us visualize conservation of mechanical energy. The first is a **bar chart** (Fig. 8.21). (The second is discussed in the next section.) The vertical height of a bar represents the amount of energy. If there is no energy, write "zero" on the horizontal blank. Kinetic energy can never be negative, so it cannot be represented by a bar below the zero energy level. The mechanical energy bar's height (E) must equal the sum of the heights of the kinetic (K) and potential (U) energy bars (Fig. 8.22). If mechanical energy ($E = K + U$) is conserved, the height of its bar must remain unchanged.

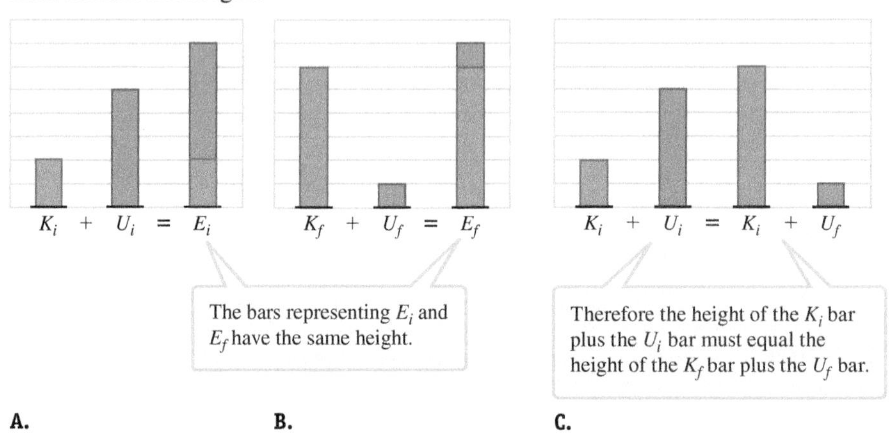

FIGURE 8.22 A. The initial mechanical energy bar is the sum of the initial kinetic energy and potential energy bars. **B.** The final kinetic energy and potential energy bars add up to the final mechanical energy bar. **C.** We can combine the initial and final bar charts. The sum of the initial energy bars equals the sum of the final energy bars.

CONCEPT EXERCISE 8.5

Each chart shown in Figure 8.23 has one unknown energy. Use the conservation of mechanical energy to find the missing energy and draw in the appropriate bar. If it is zero, write the word *zero* in place of the bar.

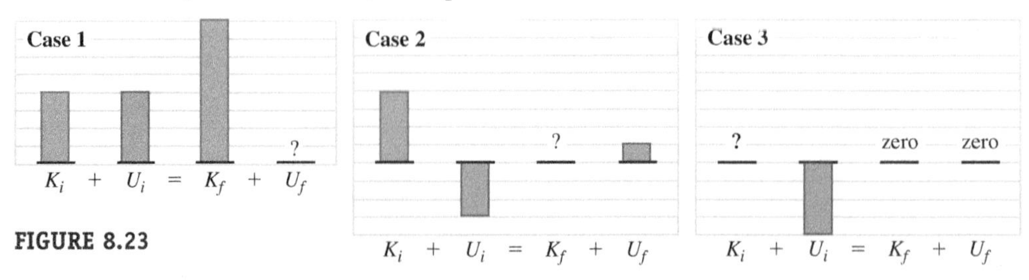

FIGURE 8.23

To solve problems using the principle of conservation of mechanical energy, it is useful to rewrite $E_i = E_f$ (Eq. 8.12). First, the potential energy is associated with conservative forces. Right now, we only know of two conservative forces, gravity and the Hooke's law spring force. So, we write the potential energy as the sum of these two:

$$U = U_g + U_e \quad \text{or} \quad U = U_G + U_e$$

Second, we write mechanical energy in terms of the kinetic and potential energies:

$$K_i + U_{gi} + U_{ei} = K_f + U_{gf} + U_{ef} \tag{8.16}$$

Each term in Equation 8.16 is represented by a bar in a bar chart, and so when applying Equation 8.16 expect three bars on each side of the equal sign.

Remember that we have two expressions for gravitational potential energy, $U_g = mgy$ and $U_G = -G(m_1m_2/r)$ (Eqs. 8.5 and 8.7), and we need only one for any situation. We have used U_g for simplicity in Equation 8.16, but the choice is still implied.

PROBLEM-SOLVING STRATEGY

Applying the Principle of Conservation of Mechanical Energy

INTERPRET and ANTICIPATE

Step 1 Choose an **isolated system** (Section 8-6).

Step 2 Pick a **coordinate system and the reference configuration** (Section 8-6).

Step 3 Identify the configuration of the system at an **initial time and at a final time**. (These times do not necessarily come at the beginning and end of the motion.)

Step 4 Draw a **bar chart** with three bars on each side of the equal sign to visualize the conservation of mechanical energy.

SOLVE

Step 5 Apply the principle of **conservation of mechanical energy** $K_i + U_{gi} + U_{ei} = K_f + U_{gf} + U_{ef}$ (Eq. 8.16).

Step 6 Write down any **other equations** (Eqs. 8.1, 8.5, 8.7, 8.9, and 8.10) that are relevant to the kinetic and potential energies involved.

Step 7 **Do algebra** before substitution.

EXAMPLE 8.8 Escape Speed

To launch a spacecraft so that it never returns, its initial speed must at least be equal to the escape speed of the planet. If a spacecraft is launched at exactly the escape speed, its kinetic energy is zero when it is infinitely far from the planet. Because a spacecraft can never truly be infinitely far away, we assume *very far away* is a good approximation to *infinitely* far away. Find an expression for the escape speed of a planet of mass M and radius R.

INTERPRET and ANTICIPATE

This example is an ideal candidate for solving using the principle of conservation of mechanical energy because we are interested in the spacecraft's motion at two different times: when it is launched and when it is very far away. (It would be difficult to use the force approach because the force exerted on the spacecraft gets weaker as the spacecraft moves away from the Earth.)

Step 1 A suitable **isolated system** consists of the spacecraft and the planet. (As long as any other objects are very far from the system so that their gravitational force is negligible, the system may be considered isolated.)

Step 2 Place the **coordinate system** at the center of the planet as shown in Figure 8.24. The unit vector $\hat{r}$ points outward. Set the **reference configuration** to when the spacecraft is infinitely far away from the planet.

Step 3 Set the **initial time** to when the spacecraft is still on planet and the **final time** to when the spacecraft is very (infinitely) far away.

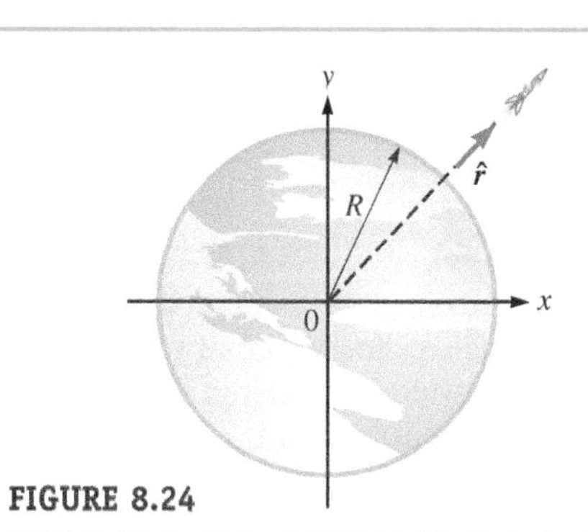

FIGURE 8.24

Example continues on page 230 ▶

Step 4 **Draw a bar chart.** Initially, the system has positive kinetic energy and negative gravitational potential energy. At the final time, the spacecraft is very far away, and there is no kinetic energy (the spacecraft *just* barely escaped) and no potential energy. (The final gravitational potential energy is zero because infinite separation is the reference configuration.) There is no spring involved in this example, so there can be no elastic potential energy (Fig. 8.25).

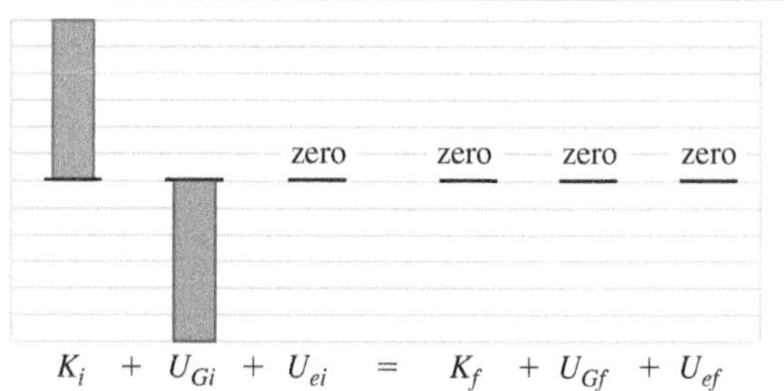

FIGURE 8.25

SOLVE

Step 5 Apply the **conservation of mechanical energy** (Eq. 8.16). The bar chart helps us see which terms are zero. We find that the sum of the initial kinetic and potential energies must equal zero.

$$K_i + U_{Gi} + U_{ei} = K_f + U_{Gf} + U_{ef} \quad (8.16)$$

$$K_i + U_{Gi} + 0 = 0 + 0 + 0$$

$$K_i + U_{Gi} = 0$$

Step 6 **Write relevant energy equations.** For launch at escape speed, the initial kinetic energy is found from Equation 8.1.

$$K_i = \tfrac{1}{2}mv_{esc}^2 \quad (8.1)$$

The initial gravitational potential energy of the system is given by Equation 8.7. The masses involved are M (mass of the planet) and m (mass of the spacecraft). At launch, the spacecraft begins on the planet's surface at a distance R from the origin. You might be tempted to use $U_g(y) = mgy$ (Eq. 8.5) for the initial gravitational potential energy, but this equation will not be valid over the entire path of the escaping spacecraft. We must use the universal gravitational potential energy.

$$U_{Gi} = -G\frac{Mm}{R} \quad (8.7)$$

Step 7 **Do algebra** to find the escape speed.

$$\tfrac{1}{2}mv_{esc}^2 - G\frac{Mm}{R} = 0$$

$$v_{esc} = \sqrt{\frac{2GM}{R}} \quad (8.17)$$

CHECK and THINK

Notice that the escape speed of a planet does not depend on the mass of the spacecraft. (The force required to accelerate the spacecraft up to the escape speed does depend on its mass, however.)

Finally, it is convenient to write our result in terms of the free-fall acceleration near the surface of the planet using Equation 7.13, $\vec{g}(r) = -\frac{GM}{r^2}\hat{r}$.

$$g = \frac{GM}{R^2}$$

$$v_{esc} = \sqrt{2gR} \quad (8.18)$$

EXAMPLE 8.9 Dart Gun

Figure 8.26 shows a spring-loaded dart gun used to launch a small ball of mass m straight up. When the spring is relaxed, it comes to the end of the gun barrel. If the spring constant is k and the spring is compressed by y_c, find an expression for the maximum height y_{max} of the ball above the end of the gun barrel.

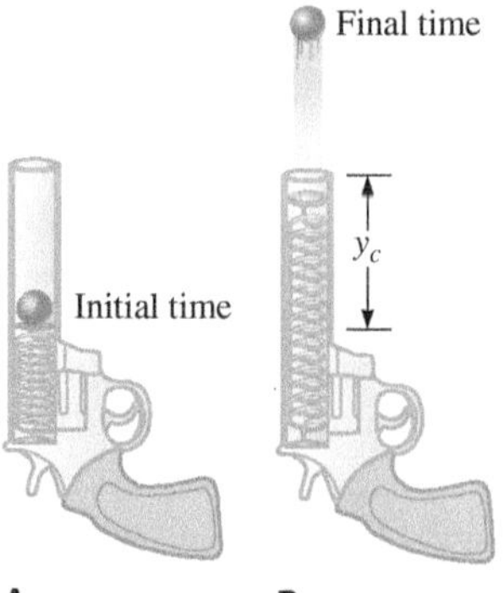

FIGURE 8.26

INTERPRET and ANTICIPATE

Step 1 We choose to include the ball, the spring, the gun, and the Earth in the system. As long as the drag force of the air and the friction due to the gun barrel may be ignored, the **system is isolated.**

Step 2 Place the origin of the **coordinate system** at the end of the gun barrel as shown in Figure 8.27.

The **reference configuration** is the one in which the ball is at the end of the gun barrel. The spring is relaxed at that point. The elastic and gravitational potential energies of the system are zero in that configuration.

Step 3 We set the **initial time** to when the ball is on top of a fully compressed spring and the **final time** to when the ball has reached its maximum height.

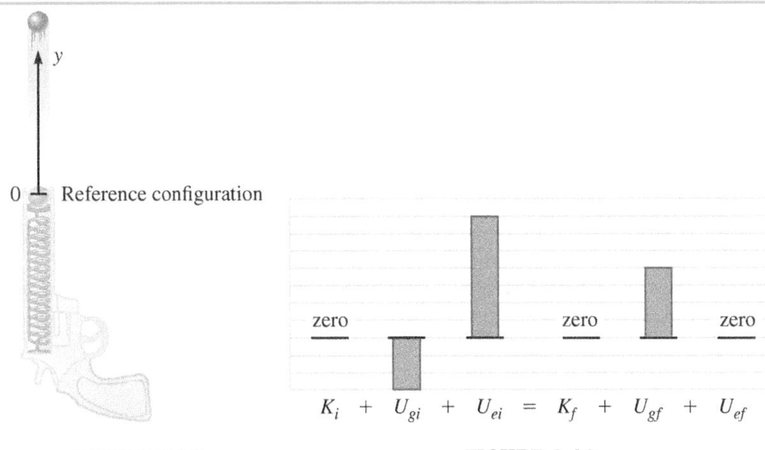

FIGURE 8.27

FIGURE 8.28

Step 4 **Draw a bar chart.** The initial kinetic energy is zero, but the initial gravitational potential energy is negative, and the initial spring potential energy is positive (Fig. 8.28). At the final time, the spring is relaxed and not touching the ball, so the elastic potential energy is zero. The final kinetic energy is zero because the ball momentarily stops at the top of its path. The gravitational potential energy is positive because the ball is above the reference level.

SOLVE

Step 5 Apply the **conservation of mechanical energy** (Eq. 8.16). The bar chart helps us see which terms are zero, and our expression is very simple.

$$U_{gi} + U_{ei} = U_{gf} \quad (1)$$

Step 6 **Write relevant energy equations.** The gravitational potential energy terms come from $U_g(y) = mgy$ (Eq. 8.5), and the elastic potential energy term comes from $U_e = \frac{1}{2}ky^2$ (Eq. 8.10).

$$U_{gi} = -mgy_c$$
$$U_{gf} = mgy_{\max}$$
$$U_{ei} = \tfrac{1}{2}ky_c^2$$

Step 7 **Do algebra.** Substitute these energies into Equation (1) and solve for $y_{\max}$.

$$-mgy_c + \tfrac{1}{2}ky_c^2 = mgy_{\max}$$
$$y_{\max} = \frac{k}{2mg}y_c^2 - y_c$$

CHECK and THINK

To see if this expression makes sense, check the dimensions. We expect to find that $[\![y_{\max}]\!] = \text{L}$.

$$\left[\!\left[\frac{k}{2mg}y_c^2\right]\!\right] - [\![y_c]\!] = \left[\frac{(\text{F/L})}{\text{F}}\text{L}^2\right] - \text{L} = \text{L}$$

As a further check, think about how $y_{\max}$ should depend on the various parameters. A stiffer spring (larger k) and a greater compression (greater y_c) should give a greater maximum height. A more massive ball means a lower maximum height. Our result is consistent with all of these.

EXAMPLE 8.10 Roller Coaster

The first roller coaster built in the United States was the Gravity Pleasure Switchback Railway at Coney Island in Brooklyn, New York (Fig. 8.29). The Gravity Pleasure consisted of two identical towers connected by a roller-coaster track. The ride reached a top speed of 6 mph. The cars had to be manually towed to the top of the hills at the beginning of both tracks. The passengers dismounted at the end of the first track, climbed the stairs up the second tower, and then got back in the cars for the return ride.

Most of today's roller coasters have a closed-loop track, and passengers do not dismount until after the ride is completed. Steel Force is a roller coaster located in Dorney Park in Allentown, Pennsylvania. Its tallest hill is 205 ft. Assume rolling friction between the wheels of the car and the track is negligible. Find the maximum speed of a car on Steel Force. For **CHECK and THINK:** How does that speed compare with that of the Gravity Pleasure?

The Granger Collection, New York

FIGURE 8.29 The Gravity Pleasure Switchback Railway (1884).

Example continues on page 232 ▶

INTERPRET and ANTICIPATE

We expect the maximum speed of the car to be at the bottom of the hill.

Step 1 The **isolated system** includes the Earth and the roller-coaster car in the system. As long as rolling friction and drag are negligible, mechanical energy is conserved.

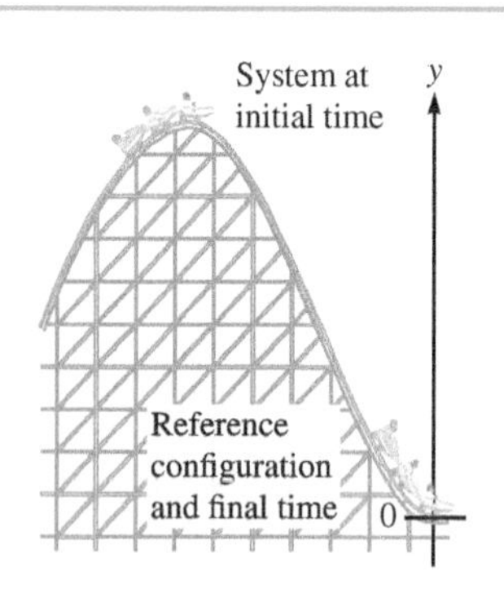

FIGURE 8.30

Step 2 It is convenient to place the origin of the **coordinate system** at the bottom of the track as shown in Figure 8.30. Then the **reference configuration** is the one in which the car is at the bottom of the track. The gravitational potential energy is zero for this configuration.

Step 3 Set the **initial time** to when the car is at the top of the hill and the **final time** to when the car is at the bottom of the hill.

Step 4 **Draw a bar chart** (Fig. 8.31). We can assume the initial kinetic energy is zero because the car moves very slowly at the top of the hill before it begins its descent. The initial gravitational potential energy is positive. The final kinetic energy is positive. The final gravitational potential energy is zero (at the reference configuration).

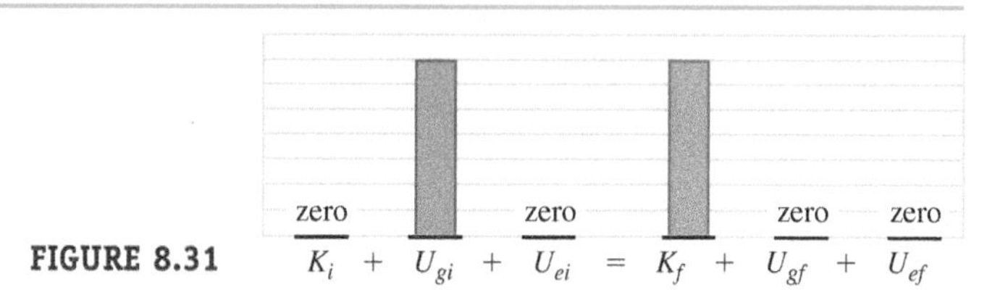

FIGURE 8.31

SOLVE

Step 5 Apply the **conservation of mechanical energy** (Eq. 8.16). Again the bar chart shows that the equation we need is very simple.

$$U_{gi} = K_f \quad (1)$$

Step 6 **Write relevant energy equations.** In the expression for kinetic energy, v_{max} represents the maximum speed.

$$K_f = \tfrac{1}{2}mv_{max}^2 \quad (8.1)$$

$$U_{gi} = mgy_i \quad (8.5)$$

Step 7 **Do algebra.** Substitute these energies into Equation (1) and solve for v_{max}. Convert 205 ft to 62.5 meters and substitute.

$$mgy_i = \tfrac{1}{2}mv_{max}^2$$

$$v_{max} = \sqrt{2gy_i} = \sqrt{2(9.81\ \text{m/s}^2)(62.5\ \text{m})}$$

$$v_{max} = 35.0\ \text{m/s}$$

CHECK and THINK

This speed is about 78 mph, or about 13 times the maximum speed attained on the Gravity Pleasure. The Dorney Park website claims that the Steel Force is somewhat slower than 78 mph, which may be because rolling friction and drag play a role in dissipating mechanical energy. In the next chapter, we will learn how to apply the conservation approach to situations that involve nonconservative forces such as friction.

8-8 Energy Graphs

In the previous section, we used energy bar charts to give us a pictorial representation of the conservation of mechanical energy. Another pictorial representation comes from plotting the potential energy. Once a coordinate system and reference configuration have been established, it is possible to write an expression for the potential energy in terms of the relative position between particles in the system (Eqs. 8.5, 8.7, 8.9, and 8.10).

Graphs of potential energy versus relative position, also known as **potential energy curves** (or U curves), give us a way to visualize potential energy (Fig. 8.32). A potential energy curve alone is not enough to visualize the conservation of mechanical energy, however. Instead, we must create an **energy graph** to represent a system's mechanical energy E, potential energy U, and kinetic energy K on a single set of axes.

ENERGY GRAPH Tool

The energy graph starts with a U curve to represent the potential energy. Then, because we are only considering systems that have constant mechanical energy, the me-

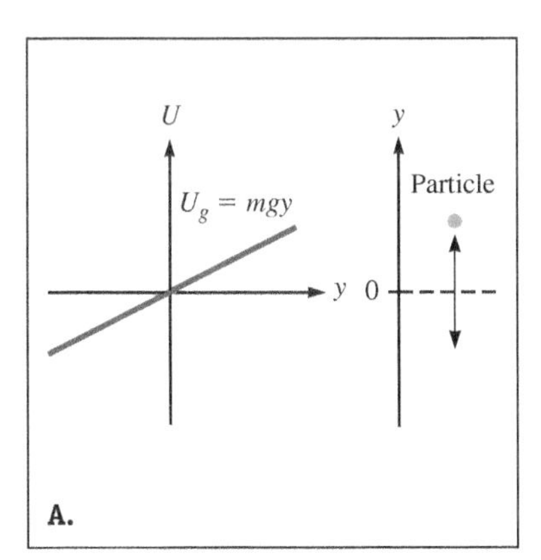

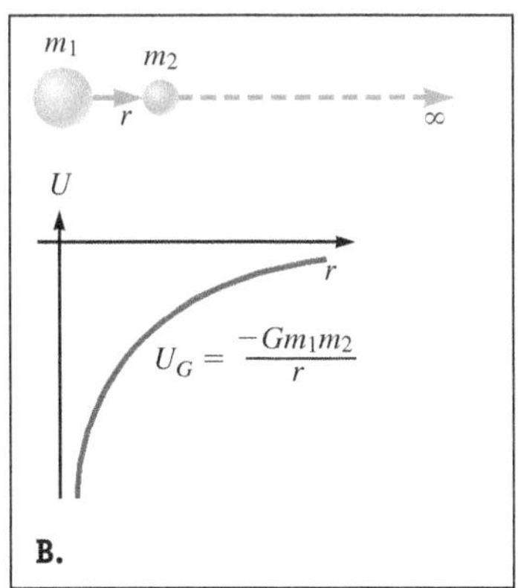

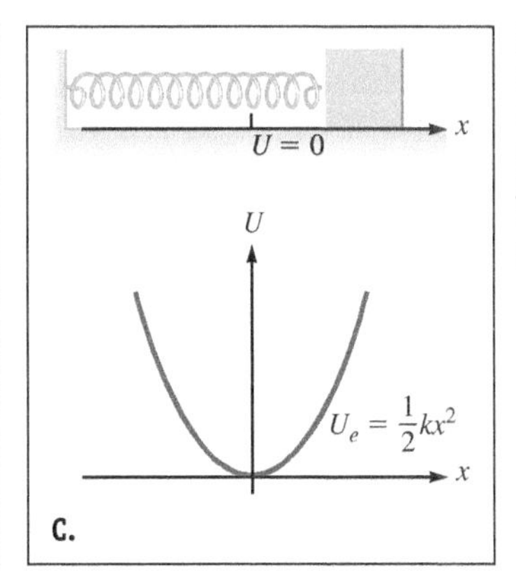

FIGURE 8.32 Potential energy curves for **A.** gravity near the surface of a large object such as the Earth, **B.** universal gravity, and **C.** Hooke's law.

chanical energy does not depend on the relative position of the particles in the system. So, E is represented by a horizontal line on an energy graph ("the horizontal E line").

We do not actually plot kinetic energy on the energy graph, but we find it in the following way. The kinetic energy is given by (Eq. 8.11)

$$K = E - U$$

So, on the energy graph, kinetic energy is represented by the vertical space between the horizontal E line and the potential energy curve.

To see how to use an energy graph, consider a familiar physical situation: a ball thrown straight upward from the top of a tall building (Fig. 8.33A) with a certain initial velocity. The (nearly) isolated system consists of the ball and the Earth, and we assume mechanical energy of the system is conserved. We start by drawing a potential energy curve. For this situation, it looks like the one in Figure 8.32A. The mechanical energy of the system is constant and is therefore represented by the horizontal line shown in Figure 8.33B. To help see the relationship between the bar charts and the energy graph, we have overlaid the bar charts on the energy graph. Usually, an energy graph does not have such bar charts drawn on it.

To better understand Figure 8.33B, consider two facts about kinetic energy. First, kinetic energy cannot be negative; second, kinetic energy is given by $K = E - U$. Taken together, we reason that a system *cannot* be in a configuration that would have $U > E$ because then the kinetic energy would have to be negative. Figure 8.33B shows that point B corresponds to the highest position of the ball because, for higher positions $y > y_{max}$, the potential energy would be greater than the mechanical energy, $U > E$. In other words, the ball cannot rise higher than the corresponding position y_{max} because that situation would require more mechanical energy than the system possesses. We say that some configurations are *forbidden* to the system. In this case, any position y to the right of $y = y_{max}$ in Fig. 8.33B is not allowed for the ball's given initial velocity.

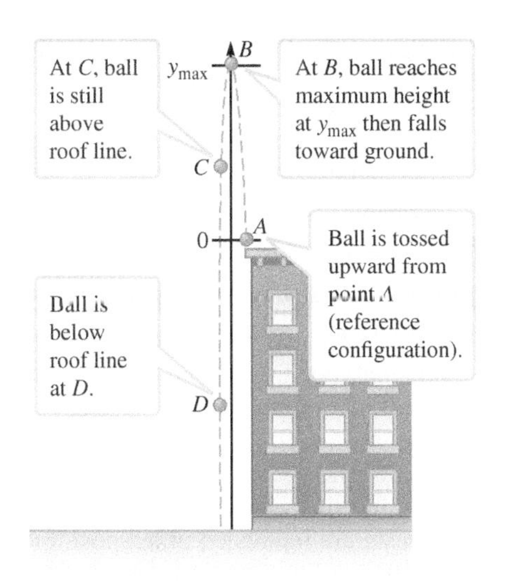

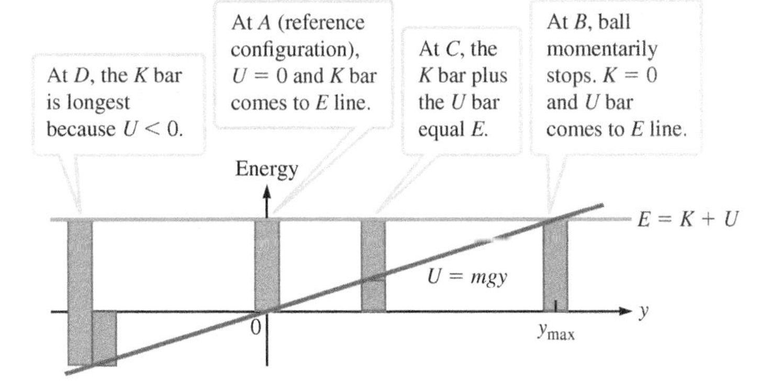

FIGURE 8.33 A. A system consists of a ball and the Earth. The ball is tossed straight up off a tall building. Four points along the ball's path are labeled A through D. **B.** The energy graph is overlaid by four bar charts corresponding to the four points labeled in part A. In all cases, the U bar touches the U curve, while the K bar fills the vertical space between the horizontal E line and the U curve.

EXAMPLE 8.11 Horizontal Spring

A block is attached to a spring and is supported by a frictionless, horizontal surface (Fig. 8.34). The block is displaced such that the spring stretches from its relaxed position to x_s and is then released. Draw a sketch, an energy graph, and bar charts for this example. Then describe the block's allowed positions and motion, connecting your description to the three visual aids.

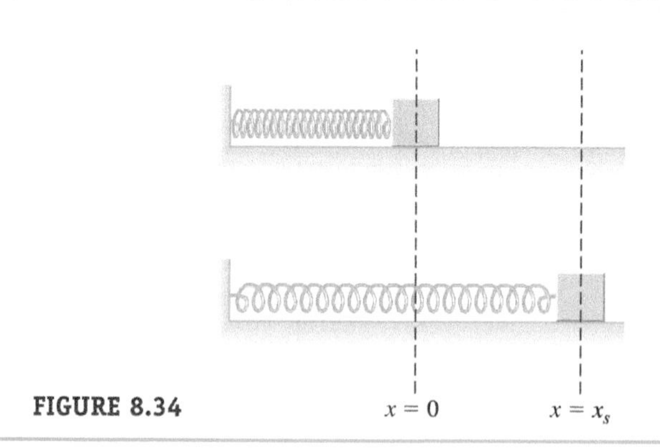

FIGURE 8.34

SOLVE

The potential energy curve is a parabola as in Figure 8.32C. The mechanical energy E is represented by a horizontal line that crosses the potential energy curve in two places, at x_s (representing maximum stretch of the spring) and x_c (representing maximum compression of the spring; Fig. 8.35). When the block is at x_s or x_c, the kinetic energy of the spring–block system is zero. We know that the only allowable configurations of a system have $E \geq U$, so the block must always satisfy $x_c \leq x \leq x_s$. (The system does not have enough mechanical energy to be outside this range.)

Figure 8.35 shows that the block moves from x_s to x_c and back again repeatedly; the bar charts indicate the kinetic and potential energies in three configurations. At x_s, the spring is fully stretched, and the block is momentarily at rest. The kinetic energy is zero. At x_c, K is again zero; now U has the same value that it does at x_s, but the spring is fully compressed. At $x = 0$, the potential energy of the system is zero. All the mechanical energy is in the form of kinetic energy $E = K$. So, the block is at its maximum speed when it passes through the origin.

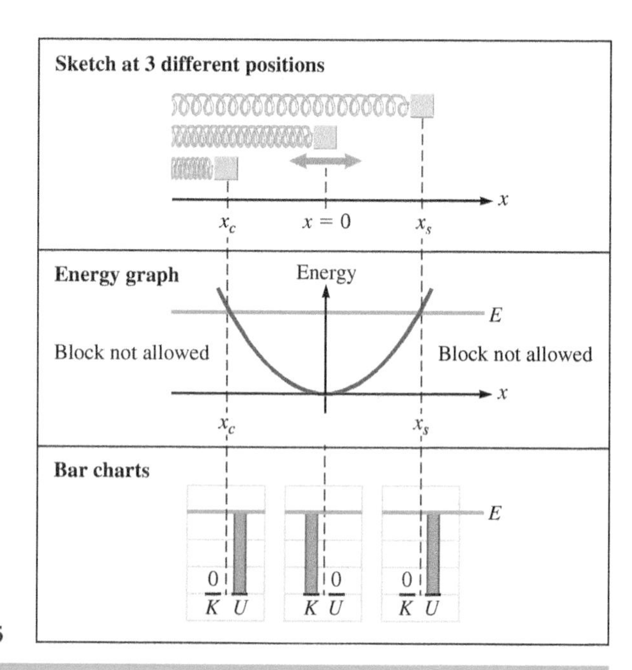

FIGURE 8.35

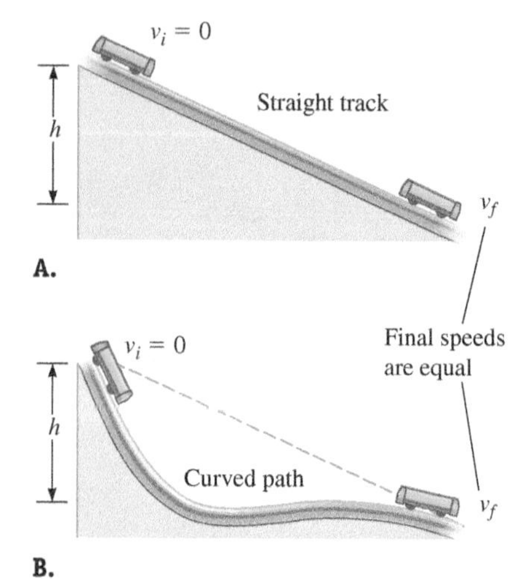

FIGURE 8.36 Two possible paths for a cart to take. In both cases, the cart starts from rest and undergoes the same displacement. The final speed does not depend on the path, but the time elapsed does.

Force Approach Versus Conservation Approach

Let us take a moment to compare the force and conservation approaches. The force approach involves vectors: force, acceleration, velocity, displacement, and position. The conservation of mechanical energy approach involves scalars: kinetic energy, potential energy, the magnitude of velocity (speed), and the scalar components of position (x, y, z, and r). We use free-body diagrams and graphical vector manipulation to help us solve problems visually using the force approach. To visualize the conservation of energy approach, we use bar charts and energy graphs.

Newton's laws of motion govern the force approach. The conservation approach involves the conservation of mechanical energy. We have shown that in the case of an isolated particle, the conservation approach is equivalent to Newton's second law. (In the next chapter, we will show how to expand this approach to include nonisolated systems and systems that cannot be modeled as a collection of particles.)

Why do we need two approaches? Can we just pick our favorite and ignore the other approach? The answer is that you must learn both approaches because some problems are better solved with one approach than the other, and many advanced problems are best solved by combining the two approaches.

You also must be able to identify the sort of problems that are best suited for each approach. The conservation approach is good when we are interested in the position and speed of particles in a system at two different times. For example, Figure 8.36 shows two possible paths a small cart may take. In both cases the cart starts at the same position from rest and ends at the same position. If we want to compare the final speed of the cart in the two cases, the conservation approach works well. We know that the Earth–cart system has the same change in gravitational potential energy, so we conclude that the cart has the same final speed in both cases.

If, however, we want to know which is the quicker path, the conservation approach cannot help. The conservation approach does not help us figure out the time elapsed. So, if we are interested in knowing the elapsed time, we must use the force approach. Because the path in Figure 8.36B is curved, the net force on the cart is not a constant, and it is difficult to find the net force on the cart at every moment. So, applying the force approach would be difficult, but with the help of a computer, we could find the answer.

8-9 Special Case: Orbital Energies

ORBITAL ENERGIES Special Case

In this section, we take a closer look at the energy associated with a particle in orbit around another, much more massive particle at rest such as a comet in orbit around the Sun. We consider this two-particle system to be isolated with only conservative forces acting so that mechanical energy is conserved.

DERIVATION Mechanical Energy for Circular Orbit System

Let us start with the circular orbit of a particle (of mass m) such as a comet around a much more massive particle (of mass M) such as the Sun (Fig. 8.37). The comet maintains a constant distance r from the Sun and a constant speed v_c, where the subscript c stands for circular. We will show that the mechanical energy of the system is given by

$$E = -\frac{1}{2}\left(\frac{GMm}{r}\right) \quad (8.19)$$

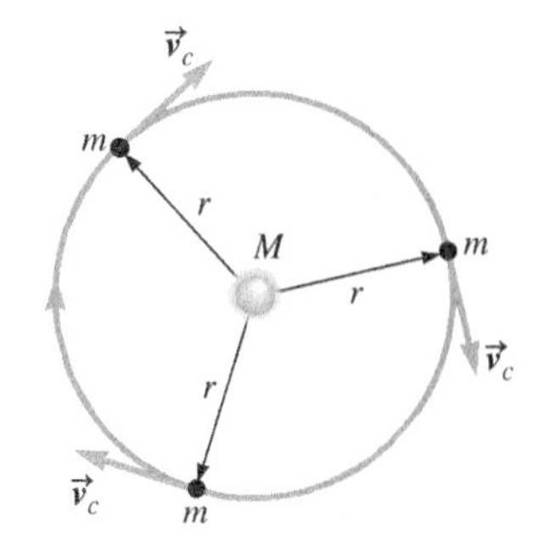

FIGURE 8.37 A two-particle system. A particle of mass m orbits a particle of mass M at constant speed v_c in a circular path of radius r.

For the purpose of this derivation, we neglect the motion of the more massive particle (the Sun). Then, the system's kinetic energy is given by the comet's motion, and K is constant because the comet's mass and speed are constant.	$K = \frac{1}{2}mv_c^2$
The system's potential energy is constant because the masses are constant, and so is the separation r.	$U_G = -\frac{GMm}{r}$ (8.7)
Because the kinetic and potential energies are constant, the mechanical energy is constant, too.	$E = K + U_G = \frac{1}{2}mv_c^2 - \frac{GMm}{r}$ (1)
The centripetal acceleration of the comet is equal to the gravitational field of the Sun at r. So, $\vec{a}_c = -\frac{v^2}{r}\hat{r}$ (Eq. 4.36) equals $\vec{g}(r) = -\frac{GM}{r^2}\hat{r}$ (Eq. 7.13)	$a = g(r)$ $\frac{v_c^2}{r} = \frac{GM}{r^2}$
Cancel out one r.	$v_c^2 = \frac{GM}{r}$ (8.20)
Substitute Equation 8.20 into Equation (1) and simplify to find the mechanical energy of the system.	$E = \frac{1}{2}m\left(\frac{GM}{r}\right) - \frac{GMm}{r}$ $E = -\frac{1}{2}\left(\frac{GMm}{r}\right)$ ✓ (8.19)

COMMENTS

We can take this result a step further by substituting Equation 8.7 for the term in parentheses.

$$E = \tfrac{1}{2}U_G \quad (8.21)$$

What we found is important and remarkable: For a system in which one particle orbits the other in a circular orbit, the mechanical energy is a constant equal to one-half the potential energy. In Problem 59, you will be asked to show that $K = -E$. If these relationships among the potential, kinetic, and mechanical energy don't hold for an orbiting system, the orbit is not circular.

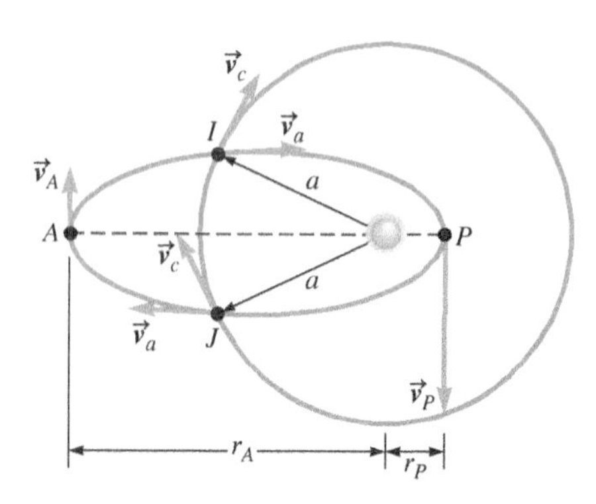

FIGURE 8.38 Comparison of a circular and an elliptical orbit. The semimajor axis of the ellipse is the radius of the circular orbit. At points I and J on the ellipse, the particle's speed is the same as it would be on the circle. At perihelion P, the particle's speed is too high to maintain a circular orbit, and at aphelion A, it is too low.

Elliptical Orbits

Let's start by comparing two systems. In one system, a comet has an elliptical orbit around the Sun; in the other, the comet's orbit is circular (Fig. 8.38). For the purpose of our comparison, the radius of the circular orbit is equal to the semimajor axis of the elliptical orbit ($r = a$), and both systems have the same mechanical energy (Eq. 8.19):

$$E = -\frac{1}{2}\frac{GMm}{r} = -\frac{1}{2}\frac{GMm}{a} \quad (8.22)$$

The kinetic and potential energies are constant for a circular orbit, but they are not constant for an elliptical orbit. We can, however, easily find the kinetic and potential energies of the elliptical orbit system when the comet is at point I or point J (Fig. 8.38). At these locations, the circular orbit crosses the elliptical orbit and $r = a$.

So, at I and J, the potential energy of the system is

$$U_G = -\frac{GMm}{a} \quad (8.23)$$

The kinetic energy at I or J is found from the conservation of mechanical energy (with Eqs. 8.22 and 8.23):

$$K = E - U_G = -\frac{1}{2}\frac{GMm}{a} - \left(-\frac{GMm}{a}\right)$$

$$K = \frac{1}{2}\frac{GMm}{a} \quad (8.24)$$

In the case of an elliptical orbit, only the mechanical energy is constant. The equations for potential (Eq. 8.23) and kinetic (Eq. 8.24) energies only hold at I and J where $r = a$. We can gain insight into how these energies change by examining the comet's speed as it moves in the elliptical orbit.

First, we find the comet's speed v_a at point I or J from Equation 8.24:

$$K = \frac{1}{2}\frac{GMm}{a} = \frac{1}{2}mv_a^2$$

$$v_a = \sqrt{\frac{GM}{a}} = \sqrt{\frac{GM}{r}} \quad (8.25)$$

where we substituted $r = a$ at these two points. By comparing Equation 8.25 with Equation 8.20, we find that at I or J (where the two orbits cross), the comet's speed on the elliptical orbit equals the speed it would have on the circular orbit, $v_a = v_c$. We know that the speed v_c of the comet in a circular orbit is constant. From Kepler's second law, however, we know that the speed of the comet in the elliptical orbit varies.

When the comet is at I or J, $v_a = v_c$, but it does not maintain a circular orbit because its velocity is not perpendicular to the centripetal force provided by the Sun's gravity. So, for a comet at I (analogous to point a in Fig. 8.1B, p. 214), the Sun's gravity causes the comet to speed up, and for a comet at J, gravity causes it to slow down. The comet's velocity is perpendicular to its acceleration only at perihelion and aphelion.

Even at those points, though, the comet cannot maintain a circular orbit. For a comet to maintain a circular orbit of radius r, its speed must be given by $v_c = \sqrt{GM/r}$

(Eq. 8.20). The comet in an elliptical orbit moves faster near perihelion than it does in a circular orbit ($v_P > v_c$), so the comet moves away from the Sun. Similarly, when the comet is at aphelion, its speed is too low ($v_A < v_c$) to maintain a circular orbit, so the comet moves toward the Sun. In the next example, we connect the comet's changing speed to the system's changing kinetic and potential energies.

EXAMPLE 8.12 Bar Charts for an Elliptical Orbit

Sketch an elliptical orbit for a comet around the Sun, including points A, P, I, and J as in Figure 8.38. For each of these four points, sketch a bar chart.

INTERPRET and ANTICIPATE

At points I and J, the comet's kinetic and potential energies are equal to what they would be if the comet were in a circular orbit. At perihelion P, the comet is moving faster than v_c, so it has more kinetic energy. Likewise, at aphelion A, the comet is moving slower than v_c, so it has less kinetic energy. The mechanical energy is constant through the entire orbit, so we can find the potential energy at P and A.

SOLVE

The orbit closely resembles Figure 8.38.

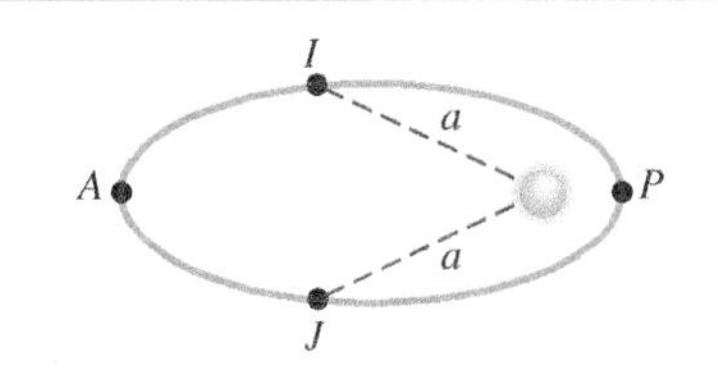

FIGURE 8.39

Bar charts are shown in Figure 8.40. Mechanical energy is constant, so E bars are the same length at all points (A, I, P, and J).

At aphelion A, the comet moves too slowly to maintain a circular orbit. The K bar is shorter than the E bar. The comet is farthest from the Sun and the U_G bar is shorter than in other configurations (less negative U_G).

At points I and J, the bar chart is the same as for a circular orbit.

At perihelion P, the comet moves too fast to maintain a circular orbit. The K bar is longer than the E bar. The comet is closer to the Sun than at I or J and U_G is less, so it is shown as a longer (negative) bar.

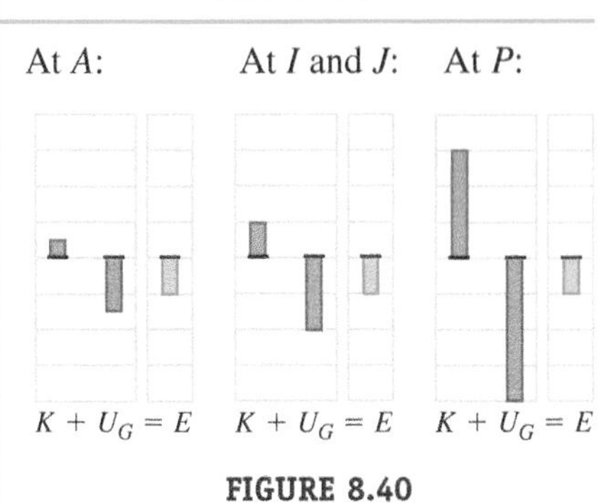

FIGURE 8.40

CHECK and THINK

When working with universal gravity, remember that when the reference configuration is set so that $U_G(\infty) = 0$, the potential energy becomes more negative as the particles move closer together as represented by the long negative U bar at perihelion in Figure 8.40. This change is nonetheless a *decrease* in potential energy, so the kinetic energy must *increase*. (The comet's highest speed is near perihelion.)

EXAMPLE 8.13 CASE STUDY Halley's Speed

Find the speed of Comet Halley when *Giotto* took its pictures. At that time, the comet was at $r_G = 1.33 \times 10^{11}$ m. To check your result, find the comet's speed at perihelion $r_P = 8.83 \times 10^{10}$ m, at aphelion $r_A = 5.27 \times 10^{12}$ m, and at $a = 2.678 \times 10^{12}$ m. (There is some variation in the observed values of the comet's orbital parameters. The values in this example come from the average of a small sample of such measurements. Work with all significant figures given and round your final answer in kilometers per second to two significant figures.)

Example continues on page 238 ▶

INTERPRET and ANTICIPATE

The ellipse in Figure 8.38 works as a sketch of Comet Halley's orbit. The bar charts are similar to Figure 8.40 in Example 8.12. Figure 8.41 is our energy graph. It might seem that we still need the mass of Comet Halley to find its speed, but we will show that the speed of a comet does not depend on its mass.

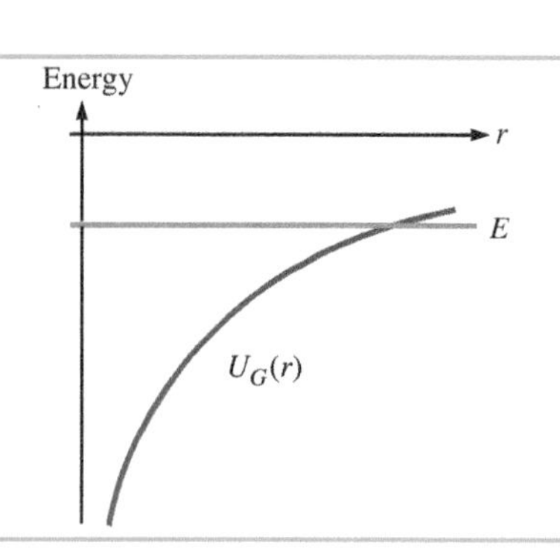

FIGURE 8.41

SOLVE

We need an expression for Comet Halley's speed. We start by writing an expression for its mechanical energy.

$$E = -G\frac{M_{\odot}m}{2a} \quad (8.22)$$

Now we need to write down the potential energy of the system at some arbitrary position.

$$U_G = -G\frac{M_{\odot}m}{r} \quad (8.7)$$

We also need an expression for the kinetic energy.

$$K = \tfrac{1}{2}mv^2$$

We know that the potential energy plus the kinetic energy equals the mechanical energy.

$$E = K + U$$

$$-G\frac{M_{\odot}m}{2a} = \frac{1}{2}mv^2 - G\frac{M_{\odot}m}{r}$$

Solve this equation for speed. The mass of the comet falls out of the equation.

$$-G\frac{M_{\odot}}{2a} = \frac{1}{2}v^2 - G\frac{M_{\odot}}{r}$$

$$v^2 = 2\left(\frac{GM_{\odot}}{r} - \frac{GM_{\odot}}{2a}\right)$$

$$v = \sqrt{2GM_{\odot}\left(\frac{1}{r} - \frac{1}{2a}\right)}$$

Because we need to evaluate this expression for four different values of r, it is useful to substitute for all the other quantities (G, $M_{\odot}$, and a).

$$v = \sqrt{2(6.673 \times 10^{-11}\,\text{N}\cdot\text{m}^2/\text{kg}^2)(1.989 \times 10^{30}\,\text{kg})}$$

$$\times \sqrt{\left[\frac{1}{r} - \frac{1}{2(2.678 \times 10^{12}\,\text{m})}\right]}$$

$$v(r) = \sqrt{(2.655 \times 10^{20}\,\text{m}^3/\text{s}^2)\left(\frac{1}{r} - \frac{1}{5.356 \times 10^{12}\,\text{m}}\right)}$$

Now convert AU to m and substitute the four values of r. We have shown only the explicit calculation at r_G; you may verify the other results. We make our calculations with three significant figures and report our final answers to two significant figures.

$$v(r_G) = \sqrt{(2.655 \times 10^{20}\,\text{m}^3/\text{s}^2)\left(\frac{1}{1.33 \times 10^{11}\,\text{m}} - \frac{1}{5.356 \times 10^{12}\,\text{m}}\right)}$$

$$v(r_G) = 4.41 \times 10^4\,\text{m/s} = 44\,\text{km/s}$$

$$v(r_P) = 54\,\text{km/s}$$

$$v(r_A) = 0.92\,\text{km/s}$$

$$v(a) = 7.0\,\text{km/s}$$

CHECK and THINK

These speeds are consistent with the energy graph and bar charts, in Example 8.12, from which we expect $K_P > K_G > K_a > K_A$. So, we expect $v(r_P) > v(r_G) > v(a) > v(r_A)$, just as we found. Let's now think about the value we found; 44 km/s is about 100,000 mph. Imagine trying to meet up with a friend moving at that speed. If you were running just 10 minutes behind schedule, your friend would be nearly 20,000 miles from your rendezvous point at the time you got there. Clearly, a lot of precise calculations must go into trying to rendezvous with a small, fast-moving object like a comet.

To take the survey scan or visit **www.cengage.com/community/katz**

SUMMARY

Underlying Principles: Conservation of mechanical energy

If only conservative forces are acting, the mechanical energy ($E = K + U$) of an isolated system is constant:

$$K_i + U_i = K_f + U_f$$

Major Concepts

1. No energy is exchanged between an **isolated system** and its environment.
2. **Energy** describes the motion and configuration of a system.
3. **Kinetic energy** K describes the motion of a system. It depends both on the mass and speed of the system:

$$K = \tfrac{1}{2}mv^2 \quad (8.1)$$

4. **Potential energy** depends on the configuration of the system. For a system that consists of two particles, the change in potential energy is given by

$$\Delta U = U_f - U_i = -\int_{x_i}^{x_f} F_x\,dx \quad (8.3)$$

when one particle moves from x_i to x_f.

Special Cases

1. Potential energy
   a. **Gravity near surface of the Earth:** Gravitational potential energy of the system is given by

   $$U_g(y) = mgy \quad (8.5)$$

   where the reference configuration is at $y = 0$.

   b. For a system that consists of two particles 1 and 2 with the reference configuration chosen such that $\lim_{r\to\infty} U = 0$, the **universal gravitational potential energy** is

   $$U_G(r) = -G\frac{m_1 m_2}{r} \quad (8.7)$$

   c. If the origin of a coordinate system is at the relaxed position of a spring, which is set to the reference configuration, the **elastic potential energy** of the particle–spring system is given by

   $$U_s = \tfrac{1}{2}kx^2 \quad (8.9)$$

2. Conservation of mechanical energy and orbital motion
   a. For a **circular orbit**, there is a special relationship between the mechanical energy, the potential energy, and the kinetic energy of the system, with $E = \frac{1}{2}U_G$ (Eq. 8.21) and $E = -K$.

   b. For an **elliptical orbit**, the mechanical energy is

   $$E = -\frac{1}{2}\frac{GMm}{a} \quad (8.22)$$

Tools

1. A **bar chart** helps visualize the conservation of mechanical energy. The vertical height of the bar represents the energy. A horizontal line is drawn to represent the zero point of the energy. If the energy is positive, the bar is drawn above that horizontal line, and if the energy is negative, the bar is below the line.
2. Another pictorial representation is an **energy graph**, which displays a potential energy curve and a horizontal E line representing the constant mechanical energy. The kinetic energy is not displayed, but is represented by the space between the horizontal E line and the potential energy curve.

PROBLEM-SOLVING STRATEGY Applying the Principle of Conservation of Mechanical Energy

INTERPRET and ANTICIPATE

1. Choose an **isolated system** (Section 8-6).
2. Pick a **coordinate system and the reference configuration.**
3. Identify an **initial and final time.**
4. Draw a **bar chart.**

SOLVE

5. Apply the principle of **conservation of mechanical energy.**
6. Write down any **other energy equations.**
7. Do **algebra** before substitution.

PROBLEMS AND QUESTIONS

A = algebraic C = conceptual E = estimation G = graphical N = numerical

8-1 Another Approach to Newtonian Mechanics

1. C CASE STUDY From Figure 8.1B for Comet Halley, is the comet speeding up, slowing down, or maintaining its speed just before it reaches aphelion? What about just after it passes aphelion? Explain.

8-2 Energy

2. E Estimate the kinetic energy of the following: **a.** An ant walking across the kitchen floor **b.** A baseball thrown by a professional pitcher **c.** A car on the highway **d.** A large truck on the highway
3. N A 70-kg man runs by a seated woman at 4.5 km/h. The woman's mass is 55 kg. **a.** According to the woman, what is the man's kinetic energy? **b.** According to the man, what is the woman's kinetic energy?
4. N An astronaut and his gear (m_A = 90.0 kg) are at rest with respect to his shuttlecraft (m_S = 12,500 kg). The astronaut and the shuttlecraft are initially separated by a distance of 10.0 m. If the astronaut moves farther away from the shuttlecraft to a distance of 20.0 m, what is the change in gravitational potential energy for the astronaut–shuttlecraft system? (*Hint*: Write an expression for the gravitational force between the astronaut and the shuttle using Eq. 8.3.)
5. N A 0.430-kg soccer ball is kicked at an initial speed of 34.0 m/s at an angle of 35.0° to the horizontal. What is the kinetic energy of the soccer ball when it has reached the apex of its trajectory?
6. Both a car (m_c = 1550 kg) and a truck (m_t = 8150 kg) are initially at rest and are each sped up to a final speed of 30 mph.
   a. N What is the final kinetic energy of the car?
   b. N What is the final kinetic energy of the truck?
   c. C Are we able to determine which vehicle was subject to a greater acceleration? If so, explain how. If not, what would we need to know to make that determination?

Problems 7 and 8 are paired.

7. N According to a seated woman, a 67.7-kg man runs with his 32.4-kg dog toward a 20.2-kg child who is running toward the man. The man and dog's speed is 3.50 km/h, and the child's speed is 1.25 km/h. **a.** According to the woman, what is the kinetic energy of the man–dog–child system? **b.** According to the man, what is the kinetic energy of the man–dog–child system?
8. According to a seated woman, a 67.7-kg man runs toward a 20.2-kg child who is running toward the man. The man's speed is 3.50 km/h, and the child's speed is 1.25 km/h.
   a. N According to the woman, what is the kinetic energy of the child–man system?
   b. C Is it possible to find a reference frame in which the child–man system's kinetic energy is zero? If so, describe the reference frame. If not, explain why not.
   c. C If the child and man were running in the same direction, would the kinetic energy of the system increase or would it decrease according to the woman?
9. N Lottery balls are thrown into a hopper and are set into motion by air forced through the container. Each ball has a mass of 0.0050 kg. At one instant in time, ball 1 is moving upward with a speed of 1.5 m/s, and ball 2 is moving downward with a speed of 1.75 m/s. If the kinetic energy of the three-ball system at this instant is 0.0260 J, what must be the speed of ball 3?
10. C If you model the Earth as a particle, can you estimate its potential energy? Why or why not?

8-3 Gravitational Potential Energy Near the Earth

11. N A ball (m = 2.5 kg) is given an initial velocity upwards of 15 m/s. What is the change in the gravitational potential energy of the ball from its initial height to when the ball is at the peak of its motion?

Problems 12, 13, and 14 are grouped.

12. C When you are asked to find the change in potential energy, the contents of the system are implied by the question. For example, in Figure P8.12 several situations are shown in which a ball moves from point A to point B. If you are asked to find the change in gravitational potential energy, what object or objects must be included in the system in each case?

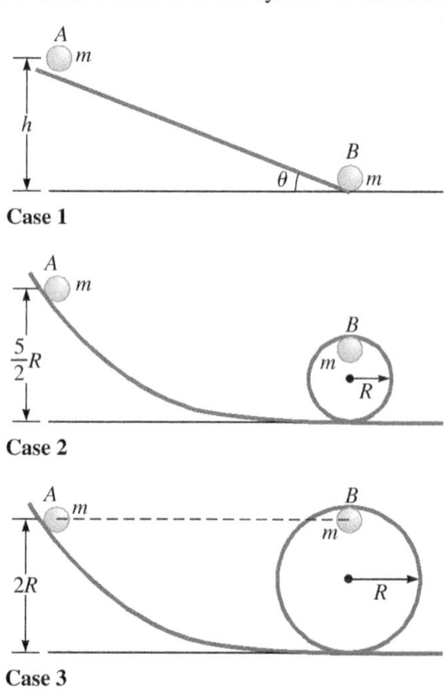

FIGURE P8.12 Problems 12, 13, and 14.

13. A In each situation shown in Figure P8.12, a ball moves from point A to point B. For each case, find an algebraic expression for the change in potential energy in terms of the parameters given in the figure. You can assume that the radius of the ball is negligible.
14. N In each situation shown in Figure P8.12, a ball moves from point A to point B. Use the following data to find the change in the gravitational potential energy in each case. You can assume that the radius of the ball is negligible.
   a. h = 1.35 m, θ = 25°, and m = 0.65 kg
   b. R = 33.5 m and m = 756 kg
   c. R = 33.5 m and m = 756 kg
15. N A wooden block (m = 1.75 kg) slides down an incline from a height 1.50 m above a tabletop and eventually comes to a stop due to friction between the block and the table. What is the change in the block-Earth system's gravitational potential energy?

16. A Two blocks of masses m_1 and m_2 are connected by a cord that passes over a pulley as shown in Figure P8.16. They start at the same height, and the hanging block is lowered through a distance h, causing the block on the incline to rise. Calculate the change in gravitational potential energy for each of the blocks.

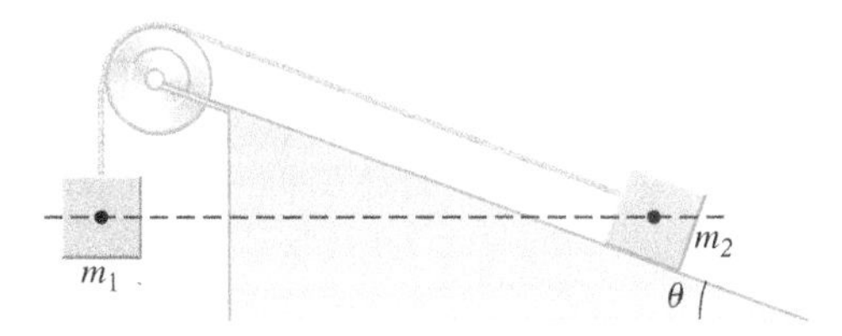

FIGURE P8.16

17. N A playground slide is 8.00 ft long and makes an angle of 33.0° with the horizontal. A 55.0-kg child, initially at the top, slides all the way down to the bottom of the slide. **a.** Choosing the bottom of the slide as the reference configuration, what is the system's potential energy when the child is at the top and at the bottom of the slide? What is the change in potential energy as the child slides from the top to the bottom of the slide? **b.** Repeat part (a), choosing the top of the slide as the reference configuration.
18. N In Example 8.4, we calculated the change in potential energy of the Earth–hammer system when the reference configuration was set at the driveway and at the basketball hoop. Repeat this calculation, this time setting the reference configuration to the roof at the point of the hammer's release.
19. N A ball of mass 0.40 kg hangs straight down on a string of length 15 cm. It is then swung upward, keeping the string taut, until the string makes an angle of 65° with respect to the vertical. Find the change in the gravitational potential energy of the Earth-ball system.
20. N A 34.0-cm-long pendulum with a 245-g bob swings through an angle of 180°. Choose the reference configuration as the point where the pendulum string is vertical. Assuming that the pendulum string remains taut, what is the gravitational potential energy of the pendulum bob–Earth system when the pendulum string makes an angle of **a.** 90° with the vertical? and **b.** 40.5° with the vertical? **c.** What is the gravitational potential energy of the pendulum bob–Earth system at the lowest point of the pendulum's motion?

8-4 Universal Gravitational Potential Energy

21. The average distance between the Earth and the Sun is about 1.5×10^{11} m, whereas the average distance between the Earth and Mars is about 2.3×10^{11} m. Assume the usual convention that the gravitational potential energy referenced to the Earth is zero when infinitely far away.
    a. N What is the gravitational potential energy of the Earth–Sun system?
    b. N What is the gravitational potential energy of the Earth–Mars system?
    c. C Describe how these values indicate that the Sun has a greater effect on the Earth's motion than Mars does. What is the most dominant factor that causes the Sun to have the greater effect?
22. E Imagine that an object the size of a car is raised to the top of a skyscraper. Estimate the change in the Earth–car system's potential energy. If the car were raised into geosynchronous orbit instead, estimate the change in the system's potential energy. *Hint*: The altitude of geosynchronous orbit is $h = 3.59 \times 10^7$ m; see Example 7.6.

8-5 Elastic Potential Energy

23. One type of toy car contains a spring that is compressed as the wheels are rolled backward along a surface. The spring remains compressed until the wheels are freed and the car is allowed to roll forward. Jose learns that if he rolls the car backward for a greater distance (up to a certain point), the car will go faster when he releases it. The spring compresses 1.00 cm for every 10.0 cm the car is rolled backward.
    a. N Assuming the spring constant is 150.0 N/m, what is the elastic potential energy stored in the spring when Jose rolls the car backward 20.0 cm?
    b. N What is the elastic potential energy stored in the spring when he rolls the car backward 30.0 cm?
    c. C Explain the correlation between the results for parts (a) and (b) and Jose's observations of different speeds.

Problems 24 and 26 are paired.

24. A block is placed on top of a vertical spring, and the spring compresses. Figure P8.24 depicts a moment in time when the spring is compressed by an amount h.
    a. C To calculate the change in the gravitational and elastic potential energies, what must be included in the system?
    b. A Find an expression for the change in the system's potential energy in terms of the parameters shown in Figure P8.24.
    c. N If $m = 0.865$ kg and $k = 125$ N/m, find the change in the system's potential energy when the block's displacement is $h = 0.0650$ m, relative to its initial position.

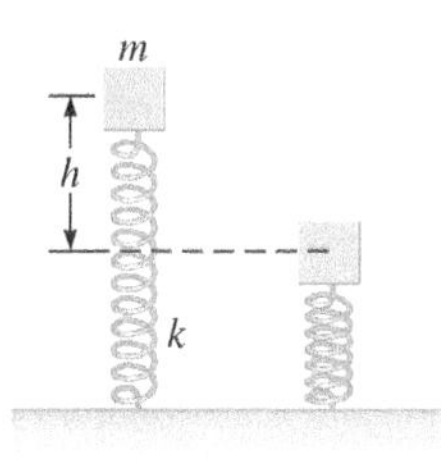

FIGURE P8.24

25. N Rubber tends to be nonlinear as an elastic material. Suppose a particular rubber band exerts a restoring force given by $F_x(x) = -Ax - Bx^2$, where the empirical constants are $A = 14$ N/m and $B = 3.3$ N/m^2 so that F_x is in newtons when x is in meters. Calculate the change in elastic potential energy of the rubber band when an external force stretches it from $x = 0$ to $x = 0.20$ m.
26. N A block is hung from a vertical spring. The spring stretches ($h = 0.0650$ m) as shown for a particular instant in time in Figure P8.26. Consider the Earth, spring, and block to be in the system. If $m = 0.865$ kg and $k = 125$ N/m, find the change in the system's potential energy between the two times depicted in the figure.

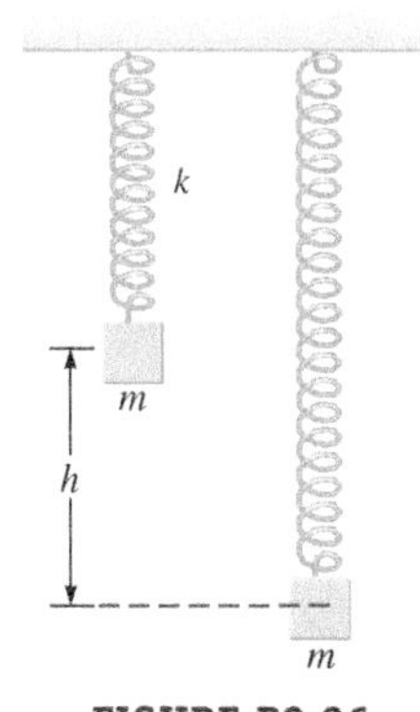

FIGURE P8.26

27. A A spring of spring constant k lies along an incline as shown in Figure P8.27. A block of mass m is attached to the spring. The spring compresses, and the block comes to rest as shown. Find an expression for the change in the Earth–block–spring system's potential energy in terms of the parameters given in the figure.

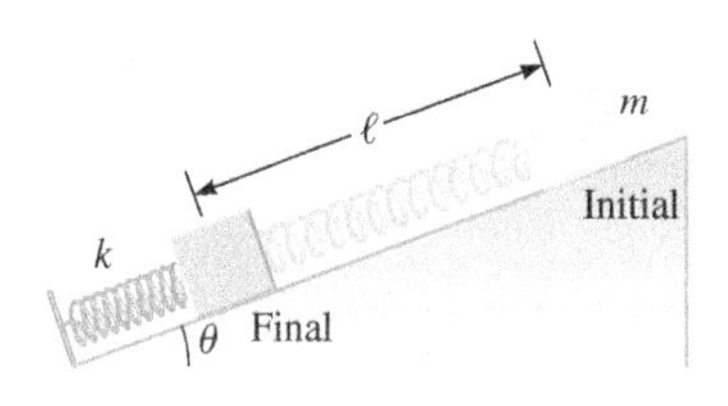

FIGURE P8.27

28. A block on a frictionless, horizontal surface is attached to two springs as shown in Figure P8.28. The block is displaced, compressing one spring and stretching the other.

a. A Find an expression for the change in the block–springs system's potential energy in terms of the parameters given in the figure.

b. C Is it possible to displace the block in such a way that the system's potential energy does not change?

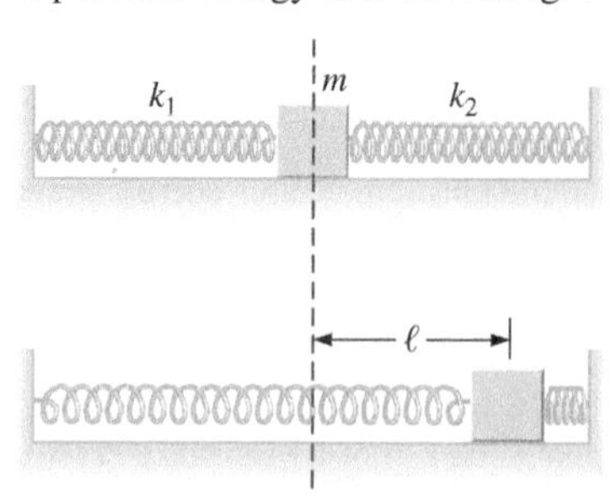

FIGURE P8.28

8-6 Conservation of Mechanical Energy

29. N A falcon is soaring over a prairie, flying at a height of 45.0 m with a speed of 12.9 m/s. The falcon spots a mouse running along the ground and dives to catch its dinner. Ignoring air resistance, and assuming the falcon is only subject to the gravitational force as it dives, how fast will the falcon be moving the instant it is 5.00 m above the ground?

30. A stellar black hole may form when a massive star dies. The star collapses down to a single point. Our Sun will not become a black hole because it does not have enough mass, so let us imagine a black hole with eight times the mass of the Sun. A single point has no radius, but a black hole has an effective radius known as the *Schwarzschild radius*. If an object gets inside the Schwarzschild radius, it never escapes. The Schwarzschild radius of an eight-solar-mass black hole is about 24 km.

a. N Find the escape speed using the Schwarzschild radius.

b. C Compare your answer to the speed of light and speculate on how these objects got the name "black hole."

31. N A newly established colony on the Moon launches a capsule vertically with an initial speed of 1.50 km/s. Ignoring the rotation of the Moon, what is the maximum height reached by the capsule?

32. The Flybar high-tech pogo stick is advertised as being capable of launching jumpers up to 6 ft. The ad says that the minimum weight of a jumper is 120 lb and the maximum weight is 250 lb. It also says that the pogo stick uses "a patented system of elastometric rubber springs that provides up to 1200 lbs of thrust, something common helical spring sticks simply cannot achieve (rubber has 10 times the energy storing capability of steel)."

FIGURE P8.32

a. E Use Figure P8.32 to estimate the maximum compression of the pogo stick's spring. Include the uncertainty in your estimate.

b. N What is the effective spring constant of the elastometric rubber springs? Comment on the claim that rubber has 10 times the energy-storing capability of steel.

c. N Check the ad's claim that the maximum height a jumper can achieve is 6 ft.

33. N An uncrewed mission to the nearest star, Proxima Centauri, is launched from the Earth's surface as a projectile with an initial speed of 43.0 km/s, just enough for the spacecraft to escape the Earth's gravity and leave the solar system. Ignoring air resistance and the Earth's rotation, what is the speed of the spacecraft when it is more than halfway to the star? Assume we are ignoring the effect of the Sun on the spacecraft.

34. C A small ball is tied to a string and hung as shown in Figure P8.34. It is released from rest at position 1, and, during the swing, the string meets a fixed peg as shown. Explain why position 2 at which the ball comes momentarily to rest must be at the same height as position 1.

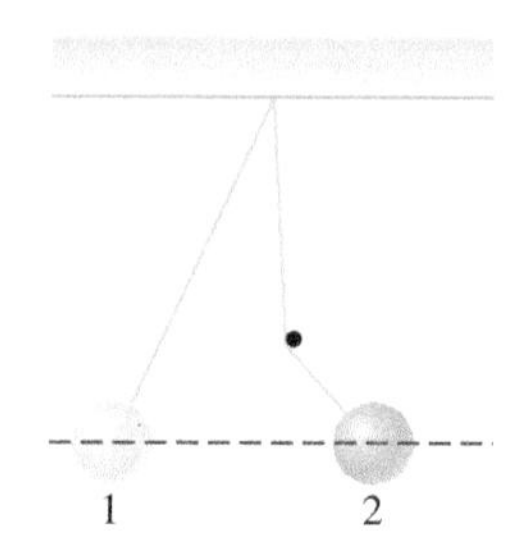

FIGURE P8.34

35. C On a late-night TV talk show, the host tosses watermelons from the top of a building with the sole purpose of watching them splatter as they hit the sidewalk several stories below. If the host tosses a watermelon up, down, or horizontally with the same initial speed, it will hit the sidewalk with the same final speed. Explain why.

36. Suppose an astronaut jumps straight up in his full gear, first on the Earth and later on the Moon. Assume his take-off speed is the same in each case.

a. C **INTERPRET and ANTICIPATE:** Do you expect him to jump higher on the Earth or on the Moon?

b. N Find the ratio of his maximum positions, $y_{\max \oplus}/y_{\max \text{ Moon}}$. Does your result match your expectations? Explain.

37. N In the Marvel comic series *X-Men*, Colossus would sometimes throw Wolverine toward an enemy in what was called a *fastball special*. Suppose Colossus throws Wolverine at an angle of 30.0° with respect to the ground (Fig. P8.37). Wolverine is 2.15 m above the ground when he is released, and he leaves Colossus's hands with a speed of 20.0 m/s. **a.** Using conservation of energy and the components of the initial velocity, find the maximum height attained by Wolverine during the flight. **b.** Using conservation of energy, what is Wolverine's speed the instant before he hits the ground?

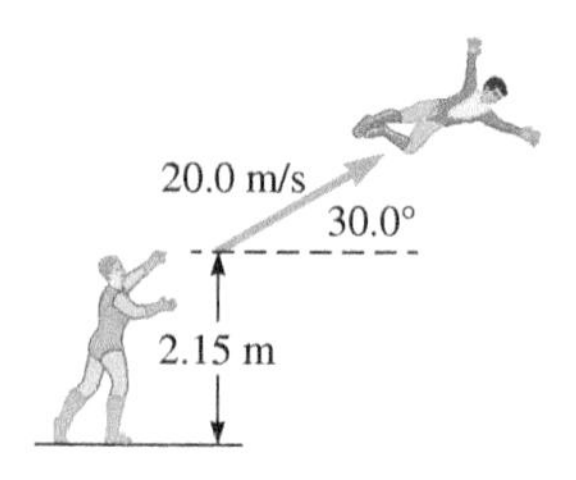

FIGURE P8.37

8-7 Applying the Conservation of Mechanical Energy

38. N A game booth at a carnival invites contestants to strike a softball affixed to the bottom end of a lightweight stick 1.25 m long whose opposite end is attached to a frictionless, horizontal axle in such a way that the ball and stick are able to swing around in a full circle. If the ball and stick are initially at rest and hanging straight down, what is the minimum speed the softball must have after it is struck for the contestant to win?

39. N Figure P8.39 shows two bar charts. In each, the final kinetic energy is unknown. **a.** Find K_f. **b.** If $m = 2.5$ kg, find v_f.

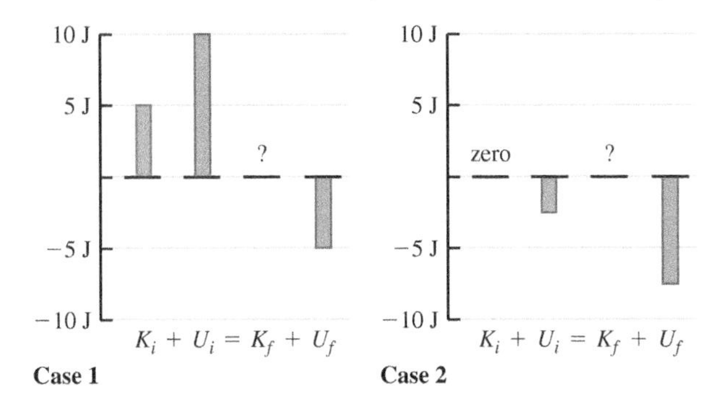

FIGURE P8.39

Problems 40 and 41 are paired.

40. N What is the escape speed of the Earth and of the Moon?

41. N If a spacecraft is launched from the Moon at the escape speed of the Earth, how fast will the spacecraft be going when it is very far away from the Moon, ignoring the effects of other celestial bodies?

42. N A 1.50-kg box rests atop a massless vertical spring with $k = 4250$ N/m that has been compressed by 15.0 cm from its equilibrium position. The box is released and leaves the spring when it reaches its equilibrium position. What is the maximum height the box reaches above its original position?

43. N A man unloads a 5.0-kg box from a moving van by giving it an initial speed of 1.3 m/s down a ramp that makes an angle of 20° with the horizontal. It slides with negligible friction for 2.7 m along the ramp before reaching the man's helper. What is the speed of the box as the helper catches it?

44. N Starting at rest, Tina slides down a frictionless waterslide with a horizontal section at the bottom that is 4.00 ft above the surface of the swimming pool and strikes the water a distance of 15.0 ft away from the end of the slide. Using conservation of energy, what is Tina's initial height on the waterslide?

Problems 45 and 46 are paired.

45. N Karen and Randy are playing with a toy car and track. They set up the track on the floor as shown in Figure P8.45, where the apparatus on the far left is used to launch the car forward by pressing down on the top portion. After the car is launched, it follows the track and continues upward, leaving the track. If the car has a mass of 130 g and it reaches a maximum vertical height of 1.2 m above the floor, what was the speed of the car while it was moving along the floor? Ignore the effects of friction and air resistance.

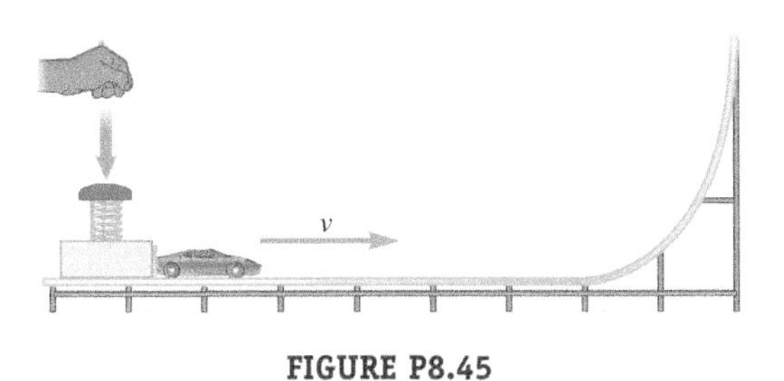

FIGURE P8.45

46. N Karen and Randy are playing with a toy car and track. They set up the track on the floor as shown in Figure P8.46, where the apparatus on the far left is used to launch the car forward by pressing down on the top portion. After the car is launched, it follows the track and continues, leaving the track at an angle of 35° with respect to the floor, at a height of 0.35 m above the floor. If the car has a mass of 130 g and it leaves the track with a speed of 3.85 m/s, what is the maximum height reached by the car, relative to the floor? Use conservation of energy to solve this problem. Ignore the effects of friction and air resistance.

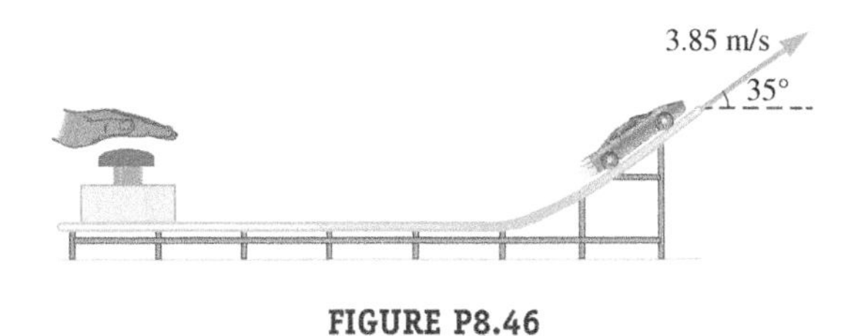

FIGURE P8.46

47. N An intrepid physics student decides to try bungee jumping. She obtains a cord that is 9.00 m long and has a spring constant of 5.00×10^2 N/m. When fully suited, she has a mass of 70.0 kg. She looks for a bridge to which she can tie the cord and step off. Determine the minimum height of the bridge that will allow her to stay dry (that is, so that she stops just before hitting the water below). Assume air resistance is negligible.

48. N A block of mass $m = 1.50$ kg attached to a horizontal spring with force constant $k = 6.00 \times 10^2$ N/m that is secured to a wall is stretched a distance of 5.00 cm beyond the spring's relaxed position and released from rest. What is the elastic potential energy of the block–spring system **a.** just before the block is released and **b.** when the block passes through the spring's relaxed position? What is the speed of the block **c.** as it passes through the spring's relaxed position and **d.** when it has compressed the spring 2.50 cm beyond its relaxed position?

49. N A roller-coaster track is being designed so that the roller-coaster car can safely make it around the circular vertical loop of radius $R = 10.0$ m on the frictionless track. The loop is immediately after the highest point in the track, which is a height h above the bottom of the loop. What is the minimum value of h for which the roller-coaster car will barely make it around the vertical loop?

8-8 Energy Graphs

50. G A jack-in-the-box is actually a system that consists of an object attached to the top of a vertical spring (Fig. P8.50). **a.** Sketch the energy graph for the potential energy and the total energy of the spring–object system as a function of compression distance x from $x = -x_{max}$ to $x = 0$, where $-x_{max}$ is the maximum amount of compression of the spring. Ignore the change in gravitational potential energy. **b.** Sketch the kinetic energy of the system between these points—the two distances in part (a)—on the same graph (using a different color).

FIGURE P8.50 Problems 50 and 79.

51. G A side view of a half-pipe at a skateboard park is shown in Figure P8.51. Sketch a graph of the gravitational potential energy of the skateboarder–Earth system as a function of position for a skateboarder who travels from the left side of the half-pipe to the right side. Let the leftmost point be where $x = 0$ and the lowest point in the half-pipe be where $U = 0$.

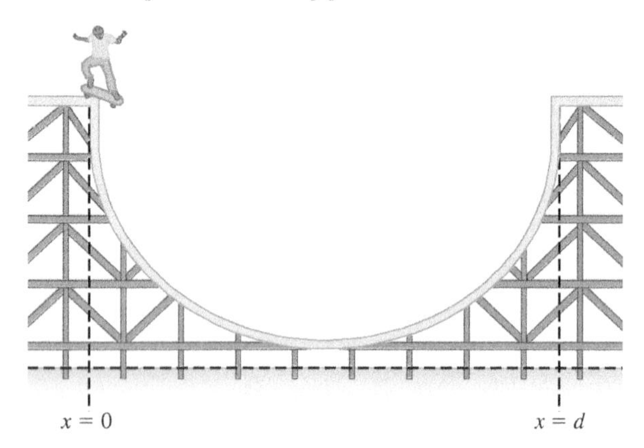

FIGURE P8.51

Problems 52, 53, and 54 are grouped.

52. G For this problem, you may wish to use a spreadsheet or similar program to make the following potential energy graphs. Make two graphs of potential energy versus position, where $U = mgy$, on the same set of axes, where you first set $mg = 1$ N and then $mg = 2$ N. Describe your results.

53. G For this problem, you may wish to use a spreadsheet or similar program to make the following potential energy graphs. Make two graphs of potential energy versus position, where $U = -GMm/r$, on the same set of axes, where you first set $GMm = 1$ J · m and then $GMm = 2$ J · m. Describe your results.

54. G For this problem, you may wish to use a spreadsheet or similar program to make the following potential energy graphs. Make two graphs of potential energy versus position, where $U = \frac{1}{2}kx^2$, on the same set of axes, where you first set $k = 1$ N/m and then $k = 2$ N/m. Describe your results.

55. A particle moves in one dimension under the action of a conservative force. The potential energy of the system is given by the graph in Figure P8.55. Suppose the particle is given a total energy E, which is shown as a horizontal line on the graph.

a. G Sketch bar charts of the kinetic and potential energies at points $x = 0$, $x = x_1$, and $x = x_2$.

b. C At which location is the particle moving the fastest?

c. C What can be said about the speed of the particle at $x = x_3$?

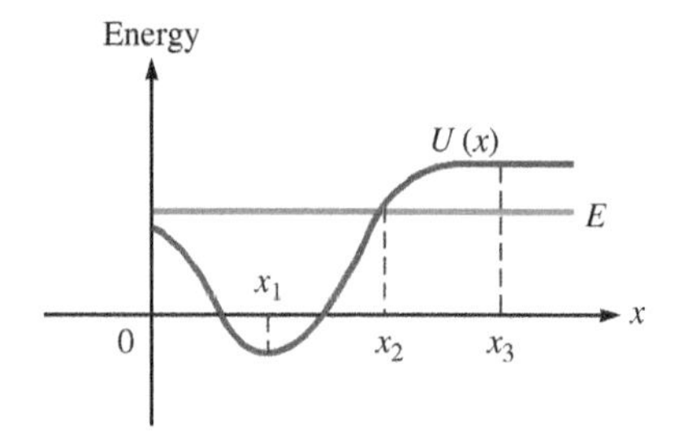

FIGURE P8.55

56. The Sun is powered by nuclear fusion. In each chain reaction, four hydrogen nuclei fuse together to become a single helium nucleus. The total combined mass of the individual hydrogen nuclei before the reaction is greater than the mass of the helium nucleus. The mass lost as a result of this chain reaction is turned into energy according to Einstein's famous equation, $E = mc^2$ (Chapter 39). Consider the first part of the chain reaction in which two hydrogen nuclei (each a single proton) fuse. Figure P8.56A shows the potential energy curve of a two-proton system. Energy in this graph is given in keV, where "eV" is the abbreviation for the unit electron-volt and 1 eV $= 1.6 \times 10^{-19}$ J. When the protons are more than about 2 fm (2×10^{-15} m) apart, the potential energy is positive. When the protons are close together, the potential energy is negative, and the protons are fused. Figure P8.56B shows a close-up of the potential energy curve when the protons are far apart.

a. G What is the minimum mechanical energy required for the two protons to fuse? Give your answer in keV.

b. G The mechanical energy of an average two-proton system in the Sun is around 2 keV. What is the minimum distance between the protons in such a system?

c. C Explain why it seems impossible for fusion to take place in the Sun. *Note*: Of course, fusion must take place in the Sun. Although it seems impossible according to classical mechanics, quantum mechanics (Part VI) makes fusion possible. According to quantum mechanics, the seemingly impossible is merely improbable, and even events with a low probability of occurrence may still happen. Because there are so many protons in the Sun, there is a good chance that at any moment many of them will fuse.

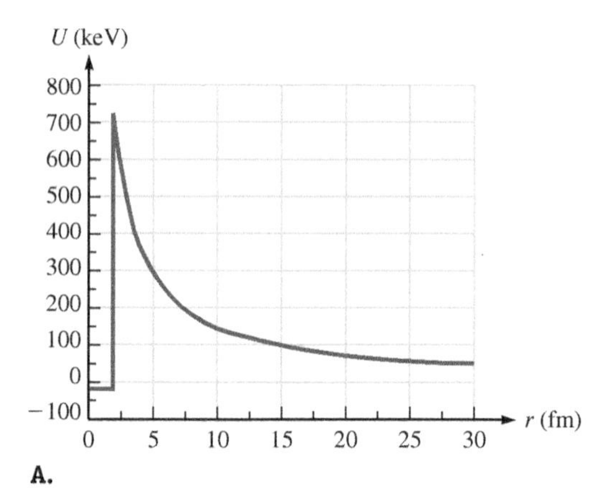

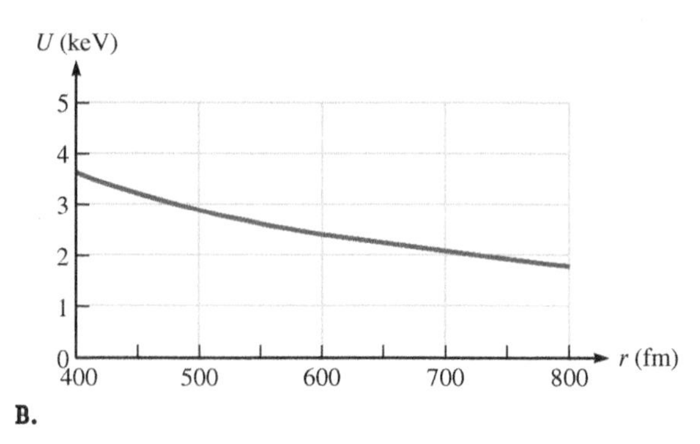

FIGURE P8.56

57. Figure P8.57A shows the potential energy curve for a two-particle system. Particle 1 remains at rest at the origin, and particle 2 ($m = 2.75$ kg) is initially at $\vec{x}_i = 24.0\hat{\imath}$ m with velocity $\vec{v}_i = -3.00\hat{\imath}$ m/s as shown in Figure P8.57B. Report numerical answers to three significant figures.

a. N What is the mechanical energy of this system?

b. N What is the kinetic energy of the system when particle 2 is at $\vec{x} = 9.00\hat{\imath}$ m?

c. C Will the two particles ever touch? Explain why or why not.

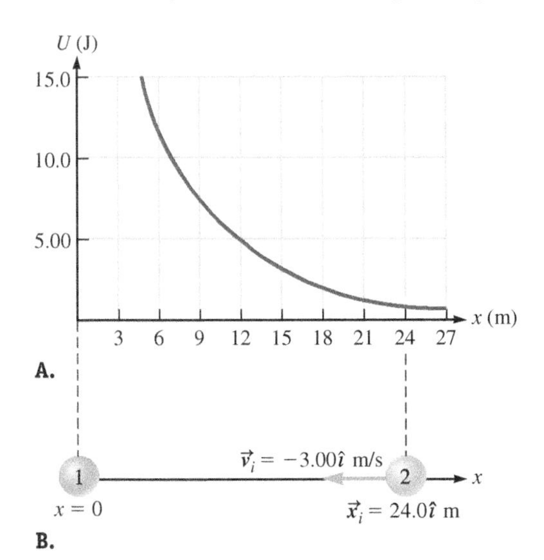

FIGURE P8.57

8-9 Special Case: Orbital Energies

58. G Make a bar chart for a two-particle system in which one particle is in a circular orbit around the other as in Figure 8.37 (page 235).

59. A Show that for the case of a circular orbit, $E = -K$. *Hint:* First show that $K = -\frac{1}{2}U_G$.

60. N Much of the mass of our Milky Way galaxy is concentrated in a central sphere of radius $r = 2$ kpc, where "pc" is the abbreviation for the unit "parsec"; 1 pc = 3.26 ly. Assume the Sun is in a circular orbit of radius $r = 8.0$ kpc around the central sphere of the Milky Way. The Sun's orbital speed is approximately 220 km/s; assume the central sphere is at rest.

a. N Estimate the mass in the inner Milky Way. Report your answer in kilograms and in solar masses.

b. E What is the escape speed of the Milky Way?

c. C **CHECK and THINK:** Do you believe that stars in the Milky Way have been observed to have speeds of 500 km/s? Explain.

61. A stellar black hole may form when a massive star dies. The mass of the star collapses down to a single point. Imagine an astronaut orbiting a black hole having eight times the mass of the Sun. Assume the orbit is circular.

a. N Find the speed of the astronaut if his orbital radius is $r = 1$ AU.

b. N Find his speed if his orbital radius is $r = 11.8$ km.

c. C **CHECK and THINK:** Compare your answers to the speed of light in a vacuum. What would the astronaut's orbital speed be if his orbital radius were smaller than 11.8 km?

62. C Astronomers are searching for Planet X in the plane of our solar system. It is expected to be beyond Pluto. Is there more mechanical energy in the Pluto–Sun system or in the Planet X–Sun system? Explain.

63. N CASE STUDY Comet Hale-Bopp's closest approach to the Sun is 0.914 AU, and its aphelion, or maximum distance from the Sun, is 371 AU in an elliptical orbit. (The average distance of the Earth from the Sun is 1 AU = 1.496×10^{11} m.) **a.** What is the eccentricity of Comet Hale-Bopp's orbit? **b.** What is the period of this comet's orbit? **c.** The comet's mass is estimated to be 1.30×10^{16} kg. What is the potential energy of the Comet Hale-Bopp–Sun system when it is at its farthest distance from the Sun?

64. G CASE STUDY Comet Hale-Bopp's elliptical orbit is described in Problem 63. Draw an energy graph for the Sun–comet system. For points *A*, *P*, *I*, and *J* (Fig. 8.38, page 236), superimpose a bar chart on the energy graph.

General Problems

65. A 50.0-g toy car is released from rest on a frictionless track with a vertical loop of radius *R*. The initial height of the car is $h = 4.00R$.

a. A What is the speed of the car at the top of the vertical loop?

b. N What is the magnitude of the normal force acting on the car at the top of the vertical loop?

66. A In Chapter 23, you will encounter the electric force that exists between two electrically charged objects, given by the formula $F_E = kq_1q_2/r^2$, where k is a constant called Coulomb's constant, each q represents the amount of excess charge on each object, and r is the distance between the objects. Find the change in electric potential energy between two charges that are initially a distance r_1 apart and later a distance r_2 apart.

67. N The Earth's perihelion distance (closest approach to the Sun) is $r_P = 1.48 \times 10^{11}$ m, and its aphelion distance (farthest point) is $r_A = 1.52 \times 10^{11}$ m. What is the change in the Sun–Earth's gravitational potential energy as the Earth moves from aphelion to perihelion? What is the change in its gravitational potential energy from perihelion to aphelion?

68. E, N After ripping the padding off a chair you are refurbishing, you notice that there are six springs beneath, which are intended to contribute equally in supporting your weight when you sit. You find a tag that indicates that the springs are identical and that each has a spring constant of 1.5×10^3 N/m. What would be the elastic potential energy stored in the six-spring system if you were to sit on the chair?

69. A In a classic laboratory experiment, a cart of mass m_1 on a horizontal air track is attached via a string over an ideal pulley to a hanging cylinder of mass m_2. The system is released from rest, and the motion is measured. Find the speed of the cart after the cylinder has descended a distance *H*.

70. G A block is attached to a spring, and the block makes contact with a frictionless surface. Sketch a graph of the potential energy of the block–spring system as a function of position along with the corresponding system as in Figure 8.16 (page 224). Use this sketch to show how the block can move to the left from a positive position to a negative position and decrease the elastic potential energy of the system.

Problems 71 and 72 are paired.

71. N At the start of a basketball game, a referee tosses a basketball straight into the air by giving it some initial speed. After being given that speed, the ball reaches a maximum height of 4.25 m above where it started. Using conservation of energy, find **a.** the ball's initial speed and **b.** the height of the ball when it has a speed of 2.5 m/s.

72. A At the start of a basketball game, a referee tosses a basketball straight into the air by giving it some initial speed. After being given that speed, the ball reaches a maximum height y_{max} above where it started. Using conservation of energy, find expressions for **a.** the ball's initial speed in terms of the gravitational acceleration g and the maximum height y_{max} and **b.** the height of the ball when it has speed v in terms of its current height y, the gravitational acceleration g, and the maximum height y_{max}.

73. N A rocket carrying a new 950-kg satellite into orbit misfires and places the satellite in an orbit with an altitude of 125 km, well below its operational altitude in low-Earth orbit. **a.** What would be the height of the satellite's orbit if its total energy were 500 MJ greater? What would be the difference in the system's **b.** kinetic energy and **c.** potential energy?

74. N In the far future, astronauts travel to the planet Saturn and land on Mimas, one of its 62 moons. Mimas is small compared with the Earth's moon, with mass $M_m = 3.75 \times 10^{19}$ kg and radius $R_m = 1.98 \times 10^5$ m, giving it a free-fall acceleration of $g = 0.0636$ m/s^2. One astronaut, being a baseball fan and having a strong arm, decides to see how high she can throw a ball in this reduced gravity. She throws the ball straight up from the surface of Mimas at a speed of 40 m/s (about 90 mph, the speed of a good major league fastball). **a.** Predict the maximum height of the ball assuming g is constant and using energy conservation. Mimas has no atmosphere, so there is no air resistance. **b.** Now calculate the maximum height using universal gravitation. **c.** How far off is your estimate of part (a)? Express your answer as a percent difference and indicate if the estimate is too high or too low.

75. N At 220 m, the bungee jump at the Verzasca Dam in Locarno, Switzerland, is one of the highest jumps on record. The length of the elastic cord, which can be modeled as having negligible mass and obeying Hooke's law, has to be precisely tailored to each jumper because the margin of error at the bottom of the dam is less than 10.0 m. Kristin prepares for her jump by first hanging at rest from a 10.0-m length of the cord and is observed to stretch the rope to a total length of 12.5 m. **a.** What length of cord should Kristin use for her jump to be exactly 220 m? **b.** What is the maximum acceleration she will experience during her jump?

76. N Two 67.0-g arrows are fired in quick succession with an initial speed of 80.0 m/s. The first arrow makes an initial angle of 33.0° with the horizontal, and the second arrow is fired straight upward. Assume an isolated system and choose the reference configuration at the initial position of the arrows. **a.** What is the maximum height of each of the arrows? **b.** What is the total mechanical energy of the arrow–Earth system for each of the arrows at their maximum height?

77. N A block of mass $m_1 = 4.00$ kg initially at rest on top of a frictionless, horizontal table is attached by a lightweight string to a second block of mass $m_2 = 3.00$ kg hanging vertically from the edge of the table and a distance $h = 0.450$ m above the floor (Fig. P8.77). If the edge of the table is assumed to be frictionless, what is the speed with which the first block leaves the edge of the table?

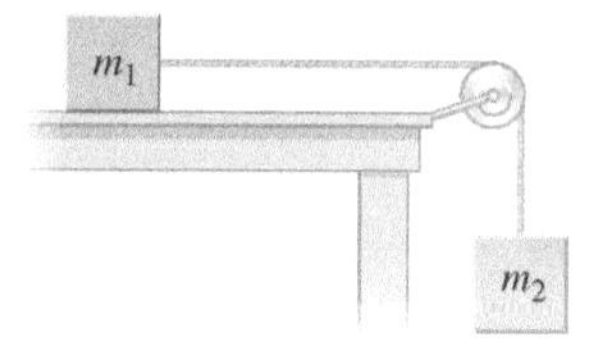

FIGURE P8.77

78. A Eric is twirling a ball of mass $m = 0.150$ kg attached to a lightweight string in a vertical circle. If the total energy of the system is conserved, by what factor does the tension on the string increase at the bottom compared with the top of the circle?

79. N A jack-in-the-box with a box of mass M and clown of mass $m = 200.0$ g rests on a horizontal table (Fig. P8.50). The clown is attached to the box by means of a lightweight, vertical spring with force constant $k = 1.00 \times 10^2$ N/m. Zara pushes down on the clown with an additional force of 4 mg, further compressing the spring. When Zara removes her hand, the clown jumps upward, and the spring lifts the box off the table. What is the maximum value for the mass M of the box for this action to occur?

Problems 80 and 81 are paired.

80. C A system consists of an object and the source of a conservative force. No other forces are exerted on the object. Figure P8.80 shows a graph of the potential energy of the system as a function of the object's position. At what location could the object be in equilibrium? Explain your answer. (The location is an example of what is sometimes called a *stable equilibrium*; explain why.) *Hint*: Considering Eq. 8.3, the force F would be equal to the negative of the derivative of U as a function of position. Use this fact to help answer the question.

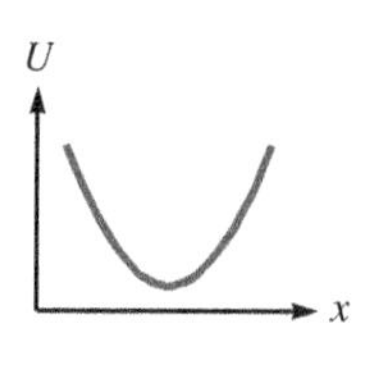

FIGURE P8.80

81. C A system consists of an object and the source of a conservative force. No other forces are exerted on the object. Figure P8.81 shows a graph of the potential energy of the system as a function of the object's position. At what location could the object be in equilibrium? Explain your answer. (The location is an example of what is sometimes called an *unstable equilibrium*; explain why.) *Hint*: Considering Eq. 8.3, the force F would be equal to the negative of the derivative of U as a function of position. Use this fact to help answer the question.

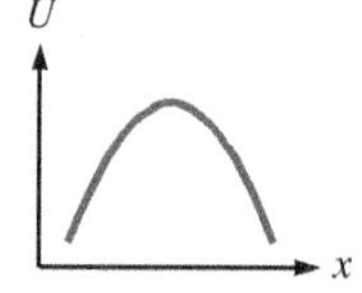

FIGURE P8.81

82. N A cluster of grapes is removed from a frictionless, hemispherical bowl 44.0 cm in diameter, leaving behind a single spherical grape of mass 3.00 g initially at rest at the upper edge of the bowl along its horizontal diameter. Choose the bottom of the bowl as the reference configuration where $h = 0$, and answer the following questions as the grape slides to the bottom of the bowl. **a.** What is the gravitational potential energy of the grape–Earth system at the grape's initial position? **b.** What is the kinetic energy of the grape when it reaches the bottom of the bowl? **c.** What is the speed of the grape when it reaches the bottom of the bowl? **d.** What are the potential and kinetic energies of the grape when it reaches a point that is a height $h = 15.0$ cm above the bottom of the bowl?

83. N A planet ($m = 5.542 \times 10^{26}$ kg) is in orbit around a star ($M = 1.056 \times 10^{31}$ kg) in another part of our galaxy. **a.** What is the total energy of the planet if its orbit is circular with a radius of 6.35×10^{12} m? **b.** What is the total energy of the planet if its orbit is instead elliptical and if its maximum distance is 7.21×10^{12} m from the star?

84. N As of 2012, all of the shuttles in NASA's space shuttle program have been retired. When they once launched, a space shuttle (mass = 2.03×10^6 kg) would lift off as the thrust of its three main engines was suddenly augmented by the firing of the solid rocket boosters at $t = 0$ s. Suppose the shuttle's acceleration is given by $a = 3.00t + 0.330t^2 - 0.100t^3$, where a is in meters per second squared and t is in seconds, in the first few seconds after liftoff. What is the change in the kinetic energy of the space shuttle during the first 3.50 s after liftoff in this scenario?

Energy in Nonisolated Systems

9

Underlying Principles

1. Work–kinetic energy theorem
2. Work–mechanical energy theorem
3. Work–energy theorem

Major Concepts

1. Work
2. General expression for work done
3. General expression for change in potential energy
4. Conservative (and nonconservative) force
5. Zero-work force
6. Center of mass
7. Internal energy
8. Dissipative forces
9. Power

Special Cases: Internal Energy

1. Thermal energy
2. Chemical energy

Tools

Dot product (scalar product)

Key Questions

How can the conservation approach be extended to account for energy transfer to and from nonisolated systems?

How are dissipative forces such as moving friction and drag accounted for in the conservation approach?

9-1 Energy transfer to and from the environment 248

9-2 Work done by a constant force 248

9-3 Dot product 252

9-4 Work done by a nonconstant force 254

9-5 Conservative and nonconservative forces 256

9-6 Particles, objects, and systems 259

9-7 Thermal energy 261

9-8 Work–energy theorem 264

9-9 Power 269

The Sun is the source of energy that supports our lives. Deep in the Sun's core, nuclear reactions between protons release energy. This energy does not stay in the Sun; instead, it is radiated away in the form of light and other electromagnetic waves (Part V). The Sun's energy lights and heats the solar system, driving the process of photosynthesis that captures that energy in the form of food we eventually eat. Therefore, the Sun is a nonisolated system, transferring energy to its environment. If the Sun's energy were not transferred to the environment, the Earth could not support life at all.

Unlike the conversion of energy from one form to another form—say from potential to kinetic energy—the transfer of energy to or from a system means a net gain or loss of energy. The Sun is actually losing energy, and in about 4.5 billion years, it will no longer be able to produce sufficient nuclear reactions in its core to sustain its present structure. The Sun will then become a white dwarf, the compact remains of a dead star.

This chapter is the second in a trilogy with the same goal of developing a conservation approach to mechanics. In Chapter 8, we focused on the conservation of mechanical energy in isolated systems. In this chapter, we adapt the idea of conservation of energy to account for energy transfer in nonisolated systems.

9-1 Energy Transfer to and from the Environment

For the isolated systems in Chapter 8, no energy is transferred and mechanical energy is conserved. Any real system is not completely isolated, however. Therefore, energy is transferred between the system and its environment. The transfer of energy can lead to spectacular events such as described in the following case study.

CASE STUDY A Meteor Shower

Comet Halley's orbit brings it into the inner solar system every 76 years. Each time a comet comes into the inner solar system, the Sun heats it up and small particles are left behind, forming a debris stream. When the Earth passes through a debris stream, these small particles, known as meteors, interact with the atmosphere. The Earth's atmosphere heats the meteors, which give off light before they evaporate. Meteors are observable without the use of any optical aid and are commonly called *shooting stars*. When the Earth passes through a large debris stream, it is possible to see several shooting stars per second in what is known as a *meteor shower* (Fig. 9.1). Meteor showers are named after the constellation from which the shooting stars seem to originate.

One evening in mid-August, three physics professors—Black, Noir, and Kuro—observed the Perseid meteor shower. On their drive home, Black asked if either of the other two professors knew the size of a piece of the debris. Neither of the other two professors knew the answer, but between the three of them, they knew enough information to estimate an answer. They worked in the dark without a calculator and came up with a good estimate using a conservation approach. In fact, they tried two different approaches and found the same estimate to within a factor of two. By the end of this chapter, we will make our first estimate, and we will make another estimate in Chapter 34.

The basis of their first estimate and ours is that the meteor is the system and the Earth's atmosphere is the environment. The meteor starts with a certain amount of energy that is transferred to the atmosphere. We observe this transfer as a shooting star.

FIGURE 9.1 Three meteors are visible in this photo taken during a meteor shower. The meteors appear to come from the same direction in the sky.

Let us compare the Comet Halley case study (Example 8.13) with this meteor case study. In the Halley case study, we included the Sun and comet in the system, which we considered to be isolated. An isolated system's energy is only *converted* or *changed* from one form to another, but no energy is *transferred* into or out of the system. In the meteor case study, the meteor is the system, and the atmosphere is outside the system. The meteor transfers some of its energy to the atmosphere. The meteor loses energy, and the atmosphere gains energy.

9-2 Work Done by a Constant Force

WORK ✪ Major Concept

One way for a system to gain or lose energy is to be acted on by an external force. If the force changes the energy of the system, we say that the force *does work* on the system. **Work** is the amount of energy transferred to (or from) the system by an external force. Work has the dimensions of energy usually expressed as newton meters (N · m), but recall that 1 N · m = 1 J. The use of the word *work* in physics is very different from our common usages such as "the work done by Michelangelo covers the ceiling of the Sistine Chapel" or "I would like to go to a movie, but I have too much work to do." Instead of looking to our intuition for the meaning of "work" in physics, let us turn to three scenarios and develop a mathematical and conceptual sense.

In each case, we will consider a simple system consisting of one object that is modeled as a particle. The system is not isolated. Energy is transferred to or from the system through some external force.

Figure 9.2A shows a hockey player pushing a puck (the system) to his coach, who then stops it. We can think of this action as two separate interactions: first the player and the puck and then the coach and the puck. Because the puck is on ice, we will ignore friction. Because the puck is modeled as a single particle, we cannot associate potential energy with the system. (Recall that potential energy is only associated with a system that consists of two or more particles.) Therefore, any energy gained or lost by the system must appear as a change in the puck's kinetic energy. We will assume in each case that the player or the coach exerts a constant force on the puck as long as his or her stick is in contact with the puck. (Keeping a constant force on the puck is actually quite difficult to do, but the assumption makes our task much easier.)

Parts B and C of Figure 9.2 show free-body diagrams for the puck: one for when the player's stick is in contact with the puck and the other for when the coach's stick is in contact with it. The normal force and the gravitational force are the same in both free-body diagrams, but the *p*layer's stick exerts a force $\vec{F}_P$ in the positive x direction, and the *c*oach's stick exerts a force $\vec{F}_C$ in the negative x direction.

The conservation approach stems from Newton's laws of motion. From both free-body diagrams, there is no acceleration in the y direction because the gravitational force and the normal force cancel out. The sticks, though, cause acceleration in the x direction. We need to look at each stick–puck interaction separately. Our goal is to find the change in kinetic energy in each case and associate it with the work done by the stick.

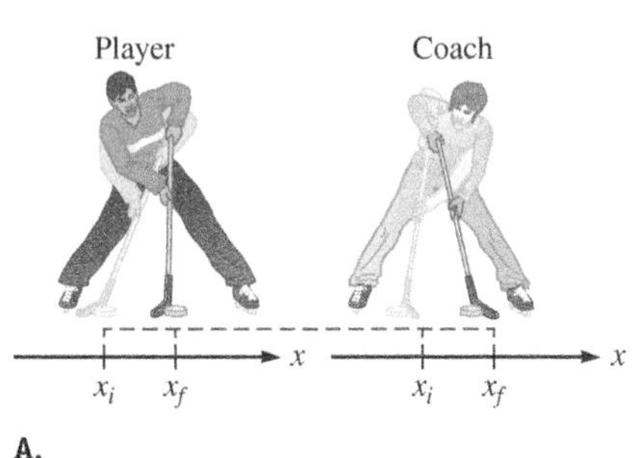

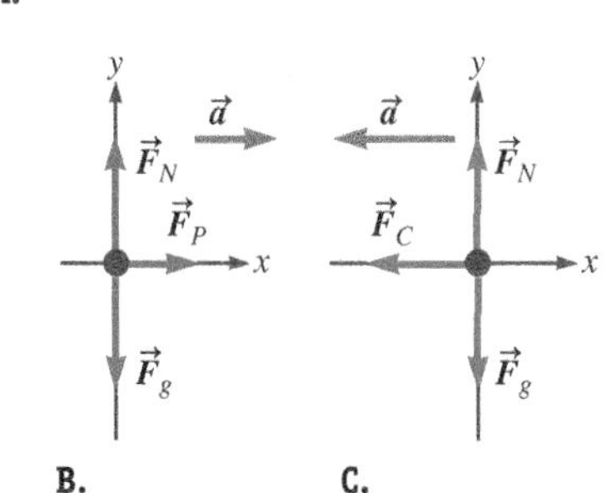

FIGURE 9.2 A. A hockey player passes a puck to his coach. **B.** Free-body diagrams show the forces acting on the puck when the player's stick is in contact and then **C.** when the coach's stick is in contact with the puck.

DERIVATION Work Done by a Constant Force in Same Direction as Displacement

We will use the specific example of the player's stick pressing on the puck to show that the work done by a constant force in the x direction is given by

$$W = F_x \Delta x \tag{9.1}$$

where the puck's displacement while in contact with the stick is Δx.

Using the free-body diagram in Figure 9.2B, apply Newton's second law in component form.	$\sum F_x = ma_x$ $F_P = ma_x \quad (1)$
The stick is in contact with the puck from x_i to x_f, and the puck accelerates as it moves in the x direction between these two points. We can write a differential equation describing the puck's motion during the time it is contact with the stick. Substitute $a_x = dv_x/dt$ (Eq. 2.7) into Equation (1), and use the chain rule.	$F_P = ma_x = m\frac{dv_x}{dt}$ $F_P = m\frac{dv_x}{dx}\frac{dx}{dt}$
Use $v_x = dx/dt$ (Eq. 2.4).	$F_P = m\frac{dv_x}{dx}v_x$
This differential equation is solved by bringing dx to the left side and integrating from the moment the stick first makes contact with the puck at x_i to when the puck loses contact at x_f. The limits on the left are simply x_i to x_f, but the integral on the right is over dv_x. The limits on the right must be the x component of velocity at x_i and x_f, which we call v_{xi} to v_{xf}.	$F_P dx = mv_x dv_x$ $\int_{x_i}^{x_f} F_P dx = \int_{v_{xi}}^{v_{xf}} mv_x dv_x \quad (9.2)$
The mass of the puck is constant, and we have assumed the stick exerts a constant force on the puck. Therefore, we can pull m and F_P outside the integrals.	$F_P\int_{x_i}^{x_f} dx = m\int_{v_{xi}}^{v_{xf}} v_x dv_x$ $F_P x\Big\vert_{x_i}^{x_f} = \frac{1}{2}mv_x^2\Big\vert_{v_{xi}}^{v_{xf}}$ $F_P(x_f - x_i) = \frac{1}{2}mv_{xf}^2 - \frac{1}{2}mv_{xi}^2 \quad (9.3)$

Derivation continues on page 250 ▶

We use $\Delta x = (x_f - x_i)$ and $K = \frac{1}{2}mv^2$ (Eqs. 2.1 and 8.1) to rewrite Equation 9.3 in terms of displacement Δx and the change in kinetic energy ΔK.	$F_P \Delta x = \Delta K$	(9.4)
Equation 9.4 relates the change in the puck's kinetic energy to the force applied by the player's stick. In the language of Newton's second law, we say that the constant force exerted by the stick on the puck accelerated the puck. In the language of the conservation approach, we say that the **work done by the stick on the puck** changes (increases) the puck's kinetic energy. To write a general expression, we have dropped subscript P (for player) and added a subscript x as a reminder that the stick's force is in the x direction. (Work is a scalar, and no such direction subscript is needed.)	$W_P = \Delta K = F_P \Delta x$ $W = F_x \Delta x$ ✓	(9.1)

COMMENTS

Remember that Equation 9.1 was derived in the special case of a constant force exerted on a particle whose displacement is parallel to that force; in fact, it is just our first step in finding a general expression for work.

We now turn our attention to the work done by the coach (Fig 9.2C). We do not need to repeat the derivation we did for the player's interaction with the puck. Let's just see if Equation 9.1 gives a reasonable result for the work done by the coach's stick on the puck. The puck has a positive x component of velocity when it makes contact with the coach's stick, and the stick decelerates the puck. Therefore, the *change* in the puck's kinetic energy is negative ($\Delta K < 0$) as a result of this interaction.

Because the puck loses energy, we expect that the work done by the coach's stick on the puck must be negative. The force exerted by the coach is in the negative x direction: $\vec{F}_C = -F_C\hat{\imath}$. The puck's displacement Δx is in the positive x direction while it is in contact with the coach's stick. So, the work done by the coach's stick on the puck is negative (Eq. 9.1):

$$W_C = -F_C \Delta x$$

which is consistent with what we reasoned: The work done by the coach's stick is negative, which results in a loss in kinetic energy.

WORK–KINETIC ENERGY THEOREM

Underlying Principle

In the two scenarios above, only one force (due either to the hockey player's stick or the coach's stick) does work on the puck. If more than one external force does work on the particle at one time, the total work done is found from the arithmetic sum of the work done by each force. The change in the system's kinetic energy is a result of the total work done. These facts are summarized in the **work–kinetic energy theorem**, which says that the total work done on a particle by all external forces equals the change in the particle's kinetic energy. Mathematically, this theorem is written as

$$W_{tot} = \Delta K \tag{9.5}$$

Now consider the third scenario, where a constant force is applied to a particle whose displacement is not parallel or antiparallel to the force; our goal is to derive a somewhat more general expression for work. Figure 9.3A shows Paul pulling a sled along the ice. As in the case of the puck, we consider the sled (and its contents) to be a one-particle system and friction between the ice and the sled to be negligible. Paul exerts a constant force $\vec{F}$ at an angle θ with respect to the x axis (Fig. 9.3B). The sled's displacement Δx is purely along the x axis (Fig. 9.3C). According to Equation 9.1, we need the x component of Paul's force, $F_x = F\cos\theta$. Then, the work done by him on the sled is

$$W = F_x \Delta x = (F\cos\theta)\Delta x$$

This example of Paul pulling the sled gives us an idea of how to write a more general expression for the work done by a constant force $\vec{F}$ exerted on a particle whose displacement is $\Delta\vec{r}$:

$$W = F\Delta r \cos\theta \tag{9.6}$$

where θ is the angle between $\vec{F}$ and $\Delta\vec{r}$, and F and Δr are (positive) magnitudes.

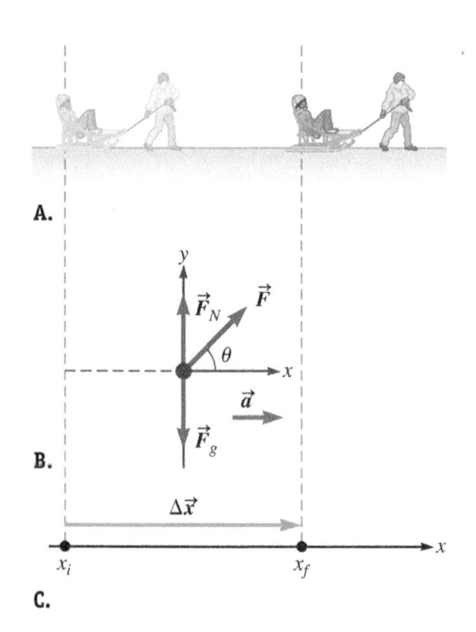

FIGURE 9.3 A. Paul pulls a sled. **B.** The free-body diagram shows that the force exerted by Paul makes an angle θ with the x axis. **C.** The displacement of the sled is along the x axis from x_i to x_f.

EXAMPLE 9.1 Work Done on a Laboratory Cart

In the common laboratory experiment shown in Figure 9.4, a small cart of mass $m = 0.78$ kg is pulled up a ramp. The experiment is designed so that friction is negligible. Find the work done by each force shown acting on the cart as it is displaced 1.5 m up the ramp. The hanging object weighs 17.8 N, and the tension in the rope is 6.0 N.

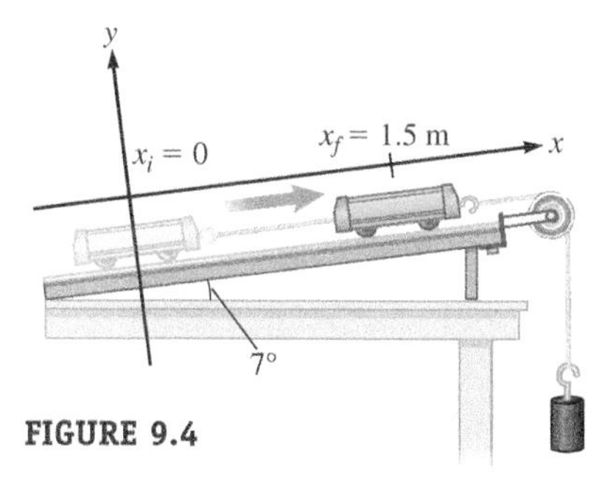

FIGURE 9.4

INTERPRET and ANTICIPATE

The problem statement implies that the system consists of the cart alone modeled as a particle. Our job is to identify each force acting on the cart and then apply $W = F\Delta r\cos\theta$ (Eq. 9.6) once for each force. We expect to find numerical results with SI units of newton meters. Because the weight of the hanging mass is greater than the tension, we expect that the hanging mass accelerates downward and the cart accelerates up the ramp. So, we expect that the kinetic energy of the cart increases and the total work done by all the forces is positive.

Start with a free-body diagram identifying each force acting on the cart. They are gravity, the normal force, and the tension force. When calculating the work done on an object, it is often convenient to indicate the displacement of the object as we have done in Figure 9.5.

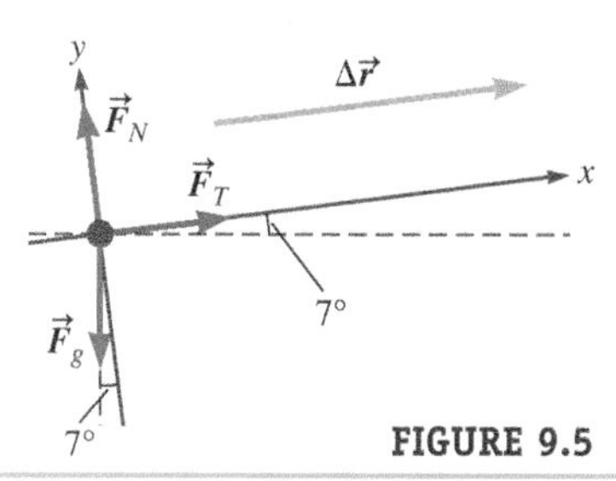

FIGURE 9.5

SOLVE

Apply $W = F\Delta r\cos\theta$ three times. The displacement of the cart is the same ($\Delta\vec{r} = 1.5\hat{\imath}$ m) each time, but the angle between the force and the displacement changes for each force. The tension force on the cart is parallel to the cart's displacement, so the angle between $\Delta\vec{r}$ and $\vec{F}_T$ is $\theta_T = 0$.

$$W_T = F_T\Delta r\cos\theta_T = F_T\Delta r\cos 0$$
$$W_T = F_T\Delta r = (6.0\,\text{N})(1.5\,\text{m})$$
$$W_T = 9.0\,\text{N}\cdot\text{m}$$

CHECK and THINK

The tension force and the displacement are parallel, and the work done by the tension on the cart is positive.

SOLVE

The angle between the normal force $\vec{F}_N$ and the cart's displacement $\Delta\vec{r}$ is $\theta_N = 90°$.

$$W_N = F_N\Delta r\cos\theta_N$$
$$W_N = F_N\Delta r\cos 90° = 0$$

CHECK and THINK

The work done by the normal force on the cart is zero because the normal force is perpendicular to the cart's displacement. The cosine part of $W = F\Delta r\cos\theta$ (Eq. 9.6) ensures that the work done by any force that is perpendicular to the displacement is zero.

SOLVE

The angle between the gravitational force $\vec{F}_g$ and the displacement $\Delta\vec{r}$ is $\theta_g = 97°$ (Fig. 9.5).

$$W_g = F_g\Delta r\cos\theta_g = mg\,\Delta r\cos 97°$$
$$W_g = (0.78\,\text{kg})(9.81\,\text{m/s}^2)(1.5\,\text{m})\cos 97°$$
$$W_g = -1.4\,\text{N}\cdot\text{m}$$

CHECK and THINK

The gravitational force on the cart has both x and y components. Like the normal force, the y component of the gravitational force $\vec{F}_g$ is perpendicular to the displacement and therefore does **no** work on the cart. The x component of the gravitational force is antiparallel (opposite direction) to the displacement and therefore does negative work on the cart. The cosine function in $W = F\Delta r\cos\theta$ selects the vector component of the force that is parallel (or antiparallel) to the displacement. This fact is important in the next section.

To check our work, we find the total work done on the cart by all three external forces. As expected, the total work is positive. According to the work–kinetic energy theorem $W_{\text{tot}} = \Delta K$, the change in kinetic energy is positive, which makes sense because the cart's speed increases as it is pulled up the ramp.

$$W_{\text{tot}} = W_T + W_N + W_g$$
$$W_{\text{tot}} = 9.0\,\text{N}\cdot\text{m} + 0 - 1.4\,\text{N}\cdot\text{m}$$
$$W_{\text{tot}} = 7.6\,\text{N}\cdot\text{m}$$

DOT PRODUCT Tool

9-3 Dot Product

From Equation 9.6, $W = F\Delta r\cos\theta$, work involves the multiplication of two vectors: force and displacement. In Section 9-2, to build a basic understanding of work, we avoided the mathematics behind multiplying two vectors. To deal with more general cases involving nonconstant forces in up to three dimensions, we now present the necessary mathematics.

There are three types of multiplication involving vectors. The first involves the multiplication of a vector by a scalar. As discussed in Chapter 3, this type of multiplication results in new vector that is parallel or antiparallel to the original vector. The subject of this section is the second type, in which the multiplication of two vectors results in a scalar. This type of vector multiplication is called the **dot product** or **scalar product**. In this section, we will work with generic vectors $\vec{A}$ and $\vec{B}$.

Figure 9.6A shows two vectors $\vec{A}$ and $\vec{B}$ that lie in a plane; φ is the smaller angle between them when they are placed tail to tail. The dot product or scalar product of these two vectors is defined to be

$$D = \vec{A}\cdot\vec{B} \equiv AB\cos\varphi \tag{9.7}$$

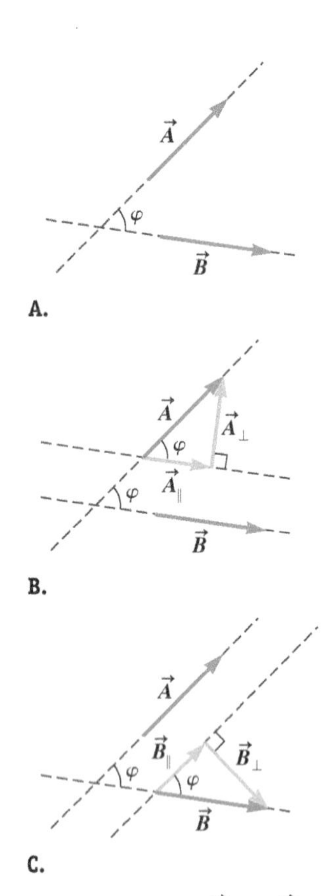

FIGURE 9.6 A. Vectors $\vec{A}$ and $\vec{B}$ line in a plane; the angle between them is φ. **B.** Vector $\vec{A}$ is broken into two components, one parallel to $\vec{B}$ and the other perpendicular to $\vec{B}$. **C.** Now vector $\vec{B}$ is broken into components that are parallel and perpendicular to $\vec{A}$.

We can see why both names fit this multiplication. First, the symbol for the dot product is a dot •. (It is *not* correct to write $\vec{A}\vec{B}$.) Second, the result of taking a dot product is a scalar quantity D. The dot product is found by multiplying the magnitudes of the two vectors and the cosine of the angle φ. Because A and B are magnitudes, they must be positive, so the cosine term determines the sign of D. If $\varphi < 90°$, then $D > 0$, and if $90° < \varphi < 180°$, then $D < 0$.

Because D is a scalar quantity, we cannot represent the result of a dot product D with an arrow on a piece of paper, but pictorial representations of the two constituent vectors $\vec{A}$ and $\vec{B}$ help us interpret the role of the dot product. In Figure 9.6B, we have broken vector $\vec{A}$ into two vector components: one perpendicular ($\vec{A}_\perp$) and the other parallel ($\vec{A}_\parallel$) to vector $\vec{B}$. Trigonometry gives us the scalar component $A_\parallel = A\cos\varphi$. Rewriting Equation 9.7 in terms of this parallel scalar component gives

$$D = \vec{A}\cdot\vec{B} = A_\parallel B \tag{9.8}$$

We use a similar procedure to rewrite Equation 9.7 in terms of the component of $\vec{B}$ that is parallel to $\vec{A}$. From Figure 9.6C, we see that $B_\parallel = B\cos\varphi$, and we rewrite Equation 9.7 as:

$$D = \vec{A}\cdot\vec{B} = AB_\parallel \tag{9.9}$$

According to Equations 9.8 and 9.9 the dot product can be thought of either as B multiplied by the component of $\vec{A}$ that is parallel to $\vec{B}$ (Eq. 9.8) or as A multiplied by the component of $\vec{B}$ that is parallel to $\vec{A}$ (Eq. 9.9).

Another way to find the dot product is by using a coordinate system and breaking each vector into components (Fig. 9.7). Then, the dot product of these two vectors is found by multiplying their like components and then adding: $D = A_xB_x + A_yB_y$ (Problem 13). In three dimensions, the dot product may be found using

$$D = \vec{A}\cdot\vec{B} = A_xB_x + A_yB_y + A_zB_z \tag{9.10}$$

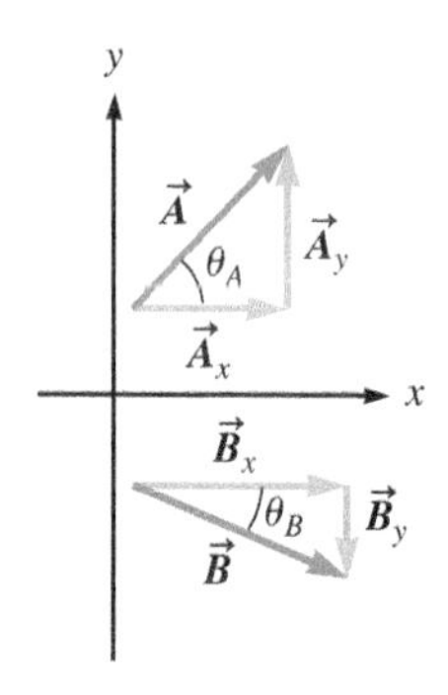

FIGURE 9.7 Once a coordinate system has been chosen, both vectors may be broken into their components.

In addition to knowing how to calculate a dot product, there are three important properties of the dot product. (For homework, you may be asked to prove these properties.)

1. The dot product of a vector with itself gives the square of the magnitude of that vector:

$$\vec{A}\cdot\vec{A} = A^2 \tag{9.11}$$

2. The dot product is *commutative*:

$$\vec{A}\cdot\vec{B} = \vec{B}\cdot\vec{A} \tag{9.12}$$

3. The dot product is *associative*:

$$\vec{A}\cdot(\vec{B} + \vec{C}) = \vec{A}\cdot\vec{B} + \vec{A}\cdot\vec{C} \tag{9.13}$$

We are ready to go back to *work* (pun intended). By comparing work $W = F\,\Delta r\cos\theta$ (Eq. 9.6) to the dot product $\vec{A}\cdot\vec{B} \equiv AB\cos\varphi$ (Eq. 9.7), we see that the **work done by a constant force** is the dot product of the force and the displacement:

$$W = \vec{F}\cdot\Delta\vec{r} \tag{9.14}$$

We use Equations 9.8 through 9.10 to write work in three other convenient forms:

$$W = F_{\parallel}\,\Delta r \tag{9.15}$$

$$W = F\,\Delta r_{\parallel} \tag{9.16}$$

$$W = F_x\Delta x + F_y\Delta y + F_z\Delta z \tag{9.17}$$

EXAMPLE 9.2 Laboratory Cart Revisited

Return to the cart in Example 9.1. Using each of Equations 9.15 through 9.17, find the work done by gravity. Compare your results.

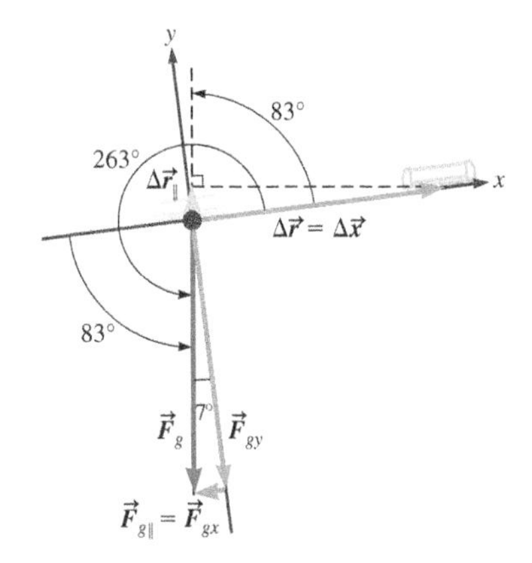

FIGURE 9.8

INTERPRET and ANTICIPATE

In Figure 9.8, we have drawn vectors $\vec{F}_g$ and $\Delta\vec{r}$ on a common coordinate system. This figure will help us find different components of the two vectors as we work each part of the example.

A Use Equation 9.15 to find the work done by gravity.

SOLVE

In all three parts, it is helpful to know the weight of the cart.

$$w = mg = (0.78\,\text{kg})(9.81\,\text{m/s}^2)$$
$$w = 7.65\,\text{N}$$

We need to find $F_{\parallel}$. Figure 9.8 shows that $\vec{F}_{g\parallel}$ is antiparallel to $\Delta\vec{r}$, so $\vec{F}_{g\parallel}$ is negative. Find the magnitude of $\vec{F}_{g\parallel}$ using trigonometry.

$$\vec{F}_{g\parallel} = -w\sin 7° = -(7.65\,\text{N})\sin 7°$$
$$\vec{F}_{g\parallel} = -0.93$$

Apply $W = F_{\parallel}\,\Delta r$ (Eq. 9.15).

$$W = \vec{F}_{g\parallel}\,\Delta r = (-0.93\,\text{N})(1.5\,\text{m})$$
$$W = -1.4\,\text{N}\cdot\text{m}$$

B Use Equation 9.16 to find the work done by gravity.

SOLVE

We need to find $\Delta r_{\parallel}$. Figure 9.8 shows that $\Delta\vec{r}_{\parallel}$ is antiparallel to $\vec{F}_g$. So, $\Delta r_{\parallel}$ is negative, and its magnitude comes from trigonometry.

$$\Delta r_{\parallel} = -\Delta r\cos 83°$$
$$\Delta r_{\parallel} = -(1.5\,\text{m})\cos 83° = -0.18\,\text{m}$$

Use $W = F\Delta r_{\parallel}$ (Eq. 9.16) to find work.

$$W = w\,\Delta r_{\parallel} = (7.65\,\text{N})(-0.18\,\text{m})$$
$$W = -1.4\,\text{N}\cdot\text{m}$$

C Use Equation 9.17 to find the work done by gravity.

SOLVE

We must find the x and y components for both vectors $\vec{F}_g$ and $\Delta\vec{r}$. By inspecting Figure 9.8, we notice that $\vec{F}_{g\parallel} = \vec{F}_{gx}$ and $\Delta\vec{r} = \Delta\vec{x}$.

$$F_{gx} = w\cos(263°) = (7.65\,\text{N})\cos(263°) = -0.932\text{N}$$
$$F_{gy} = w\sin(263°) = (7.65\,\text{N})\sin(263°) = -7.59\text{N}$$
$$\Delta\vec{r} = \Delta\vec{x} = 1.5\hat{\imath}\,\text{m}$$

Example continues on page 254 ▶

Using $W = F_x\Delta x + F_y\Delta y + F_z\Delta z$ (Eq. 9.17), we find work one last time.	$W = F_x\Delta x + F_y\Delta y + F_z\Delta z = F_{gx}\Delta x$ $W = (-0.932\ \text{N})(1.5\ \text{m}) = -1.4\ \text{N}\cdot\text{m}$

CHECK and THINK

We have now found the work done by gravity on the cart in four different ways, three here and one in Example 9.1. All four times, we found the same result.

9-4 Work Done by a Nonconstant Force

We derived our expressions for work by thinking about hockey sticks and a puck. Our assumption was that the hockey sticks exerted a constant force on the puck as long as the sticks were in contact with the puck. Because that assumption was the basis for our derivation, $W = \vec{F}\cdot\Delta\vec{r}$ (Eq. 9.14) holds only if the force is constant. We need to find an expression for work that applies even when the force is *not* constant.

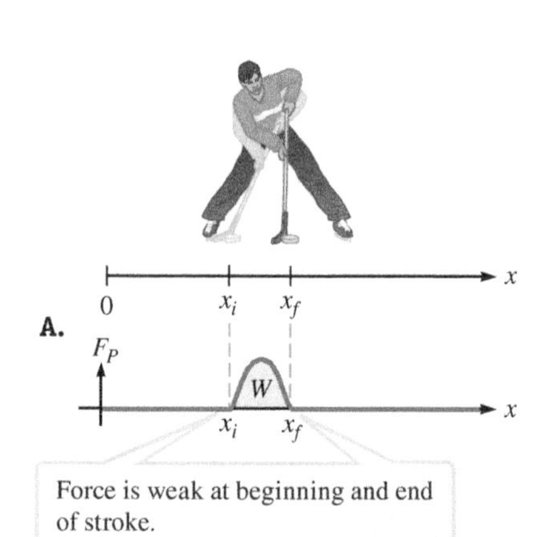

FIGURE 9.9 A. A hockey player exerts a nonconstant force on a puck. While the player's stick is in contact with it, the puck is displaced from x_i to x_f. **B.** A plot of the player's force F_P on the puck as a function of the puck's position x looks like a hill because the player's force is strongest in the middle of his stroke. The work done by the player on the puck is the area under the curve.

Let us return to the hockey player and the puck (Fig. 9.9A), but this time we will assume the magnitude of the force exerted by the player's stick is a function of position. As before, the stick is in contact with the puck from x_i to x_f, and it exerts a force of varying magnitude on the puck in the positive x direction (Fig. 9.9B).

In our earlier derivation of work (page 249), we assumed the stick exerted a constant force on the puck, so when we got to Equation 9.2, we pulled F_p out of the integral. In this case, F_x is not constant and cannot be pulled out of the integral. We must conclude that work is given by

$$W = \int_{x_i}^{x_f} F_x dx \tag{9.18}$$

We interpret this integral as the area under the force-versus-position graph (Fig. 9.9B).

The hockey player's force is parallel to the puck's displacement, so Equation 9.18 is only good for the special case of a nonconstant force that is parallel (or antiparallel) to the displacement of the particle. A more general situation is shown in Figure 9.10, where a nonconstant force acts on an amusement park car that is spiraling up a steep hill. The motion of the car is three-dimensional, so a three-dimensional coordinate system is used. At any instant, the car's infinitesimal displacement $d\vec{r}$ is tangent to its path at that point. As shown in Figure 9.10, the force $\vec{F}$ at point A is neither parallel or antiparallel to the car's displacement $d\vec{r}$.

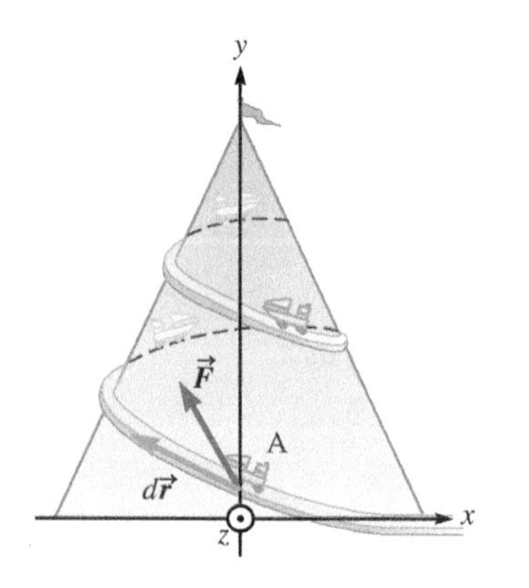

FIGURE 9.10 A nonconstant force acts on an amusement park car that spirals up a steep hill.

In general, the force and infinitesimal displacement has three components, and we write $\vec{F}$ and $d\vec{r}$ in terms of their components:

$$\vec{F} = F_x\hat{\imath} + F_y\hat{\jmath} + F_z\hat{k} \tag{9.19}$$

$$d\vec{r} = dx\hat{\imath} + dy\hat{\jmath} + dz\hat{k}$$

The work dW done by the external force that results in the car's infinitesimal displacement is the dot product $\vec{F}\cdot d\vec{r}$. Because we know the vector components, we use Equation 9.10:

$$dW = \vec{F}\cdot d\vec{r} = F_x dx + F_y dy + F_z dz \tag{9.20}$$

To find the work done by the force over some finite displacement $\Delta\vec{r}$, we integrate (add up) the work dW done over each infinitesimal displacement $d\vec{r}$:

GENERAL EXPRESSION FOR WORK DONE Major Concept

$$W = \int_{r_i}^{r_f} \vec{F}\cdot d\vec{r} = \int_{r_i}^{r_f} (F_x dx + F_y dy + F_z dz) \tag{9.21}$$

Equation 9.21 is the general expression for work. The integral in this equation is sometimes called a *path integral*, and we say that the force is integrated over the path from r_i to r_f.

CONCEPT EXERCISE 9.1

In the three cases shown in Figure 9.11, a force acts on a particle, and the particle is displaced from an initial position to a final position. From the position-versus-force graphs, determine if the force is parallel or antiparallel to the displacement in each case. Explain.

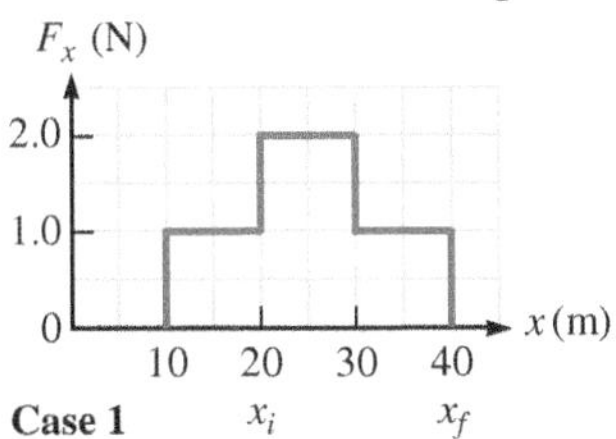

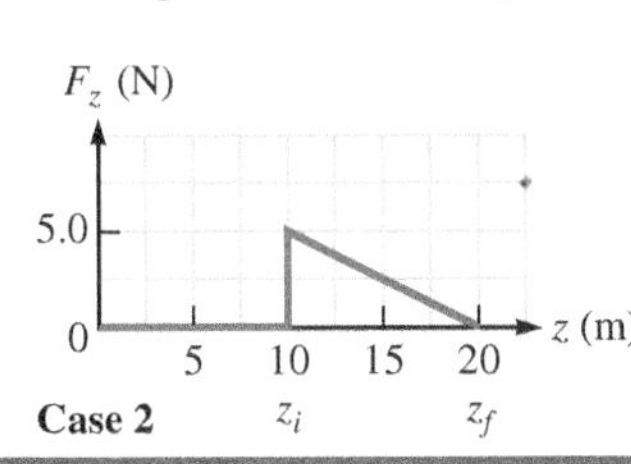

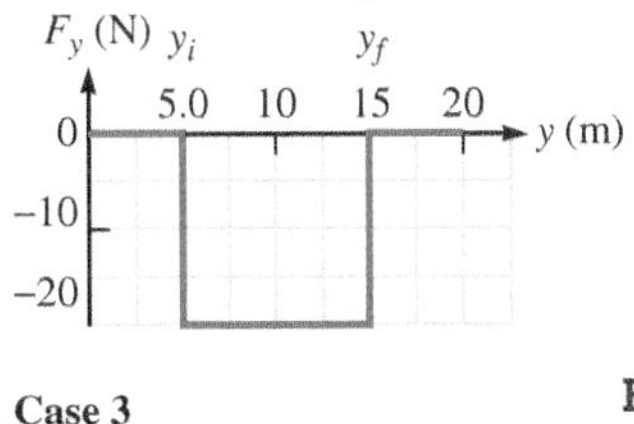

FIGURE 9.11

EXAMPLE 9.3 Work Done by a General Force

A force $\vec{F} = (3x\hat{\imath} - 8y\hat{\jmath})$ N acts on a particle. Find the work done by this force as the particle moves in an xy plane from $(x_i, y_i) = (1, 0)$ m to $(x_f, y_f) = (-3, 8)$ m.

INTERPRET and ANTICIPATE

The force depends on both x and y, so it is two-dimensional and not constant. We expect to find a numerical answer in the form $W =$ _____ N · m.

SOLVE

Write Equation 9.21 in component form. We have two integrals, one for x and one for y.

$$W = \int_{r_i}^{r_f} \vec{F} \cdot d\vec{r} = \int_{x_i}^{x_f} F_x dx + \int_{y_i}^{y_f} F_y dy$$

Substitute for the force components. The limits on the integrals come from the initial and final positions of the particle.

$$W = \int_1^{-3} 3x\,dx + \int_0^8 (-\,8y)dy$$

We dropped the units from the integral, but the integrand is in newtons and the limits are in meters.

$$W = \tfrac{3}{2}x^2\Big|_1^{-3} - \tfrac{8}{2}y^2\Big|_0^8 = \tfrac{3}{2}[(-3)^2 - 1^2] - \tfrac{8}{2}[8^2 - 0]$$

$$W = -244\,\text{N} \cdot \text{m}$$

CHECK and THINK

Our answer is in the form we expected; the negative sign means that the particle loses energy. Why can't we use a simpler expression for work such as $W = \vec{F} \cdot \Delta\vec{r}$ (Eq. 9.14)? The reason is that as the particle moves, the force acting on it changes. In fact, the x component of the force initially points in the positive direction, but by the end of the path, F_x is negative. Also, the magnitude of the y component of force varies from 0 to 64 N.

EXAMPLE 9.4 CASE STUDY Work Done by Gravity on a Meteor

Model a meteor as a single particle of mass m (Fig. 9.12). External forces such as gravity do work on the particle. Suppose the particle starts very far from a large planet (such as the Earth) with mass M. Find the work done by the gravitational force of the planet on the meteor as the meteor falls along a radial path to a position r measured from the planet's center.

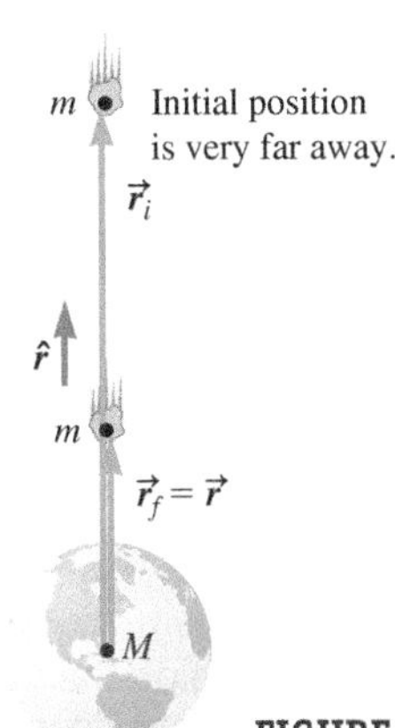

FIGURE 9.12

INTERPRET and ANTICIPATE

Whenever we are asked to find the work done by some force, a choice has been made for us; the source of the force must be external to the system. So, here the system is just the meteor and not the planet. We will take the initial position "very far" to mean infinitely far away, $r_i \to \infty$. As the meteor falls to the position $\vec{r}_f = \vec{r}$, gravity does work on it. Because the gravitational force gets stronger as the meteor gets closer to the planet, we must integrate. We expect to find an algebraic expression.

Example continues on page 256 ▶

SOLVE

Using the unit vector $\hat{r}$ shown, write an expression for the gravitational force exerted on the meteor. This force is toward the planet in the negative $\hat{r}$ direction.

$$\vec{F}_G = F_G\hat{r} = -G\frac{Mm}{r^2}\hat{r}$$

Substitute this expression for force into Equation 9.21. We have left the limits of integration in their general form for now.

$$W = \int_{r_i}^{r_f} \vec{F} \cdot d\vec{r} = \int_{r_i}^{r_f} F_G(\hat{r} \cdot d\vec{r})$$

$$W = \int_{r_i}^{r_f} F_G\,dr = \int_{r_i}^{r_f}\left(-G\frac{Mm}{r^2}\right)dr$$

Integrate and substitute limits. The final position is simply r, and the initial position is infinitely far away. (Recall that $1/\infty \to 0$.)

$$W = -GMm\left[-\frac{1}{r}\right]\Bigg|_{r_i}^{r_f} = GMm\left[\frac{1}{r_f} - \frac{1}{r_i}\right]$$

$$W = GMm\left[\frac{1}{r} - \frac{1}{\infty}\right] = \frac{GMm}{r} \quad (9.22)$$

CHECK and THINK

We have found an algebraic expression as expected. Equation 9.22 is important and requires further discussion. (1) It is the work done by any large object's gravitational force on a particle that moves toward the large object from very far away. (2) This expression for work is always positive as expected because the force and the displacement of the particle are in the same direction: toward the center of the large object. (3) Equation 9.22 is very similar to Equation 8.7 for the gravitational potential energy of a two-particle system, $U_G(r) = -G(Mm/r)$. The two expressions only differ by a negative sign. We will discuss this difference at length in the next section. (4) Finally, remember that other external forces may do work on the particle.

9-5 Conservative and Nonconservative Forces

To take the survey scan or visit **www.cengage.com/community/katz**

From Example 9.4, the expression for the gravitational potential energy of a planet–particle system $U_G(r) = -G(Mm/r)$ differs from the work done by the planet on the particle $W = G(Mm/r)$ only by a negative sign. (This relationship between potential energy and work is true for any conservative force, not just gravity.) The difference comes from deciding what is included in the system. When we choose the contents of a system, our choice affects which forces are internal and which are external. This choice subsequently determines which forces can do work on the system and which forces determine the potential energy of a system. In this section, we see how our choices affect our analysis. Rest assured that our choices cannot change the actual results. It is only our analysis that changes.

For example, suppose a (nonconstant) force is exerted on some particle, and the particle's displacement is parallel to the force such as when a spring exerts a force on a block. If we decide to include the source of that force (the spring) inside the system, then the system may experience changes in its potential energy. In one dimension, the change in potential energy is given by Equation 8.3:

$$\Delta U = U_f - U_i \equiv -\int_{x_i}^{x_f} F_x\,dx$$

However, if we decide that the source of the force is external to the system, then the source (the spring) does work (Eq. 9.18):

$$W = \int_{x_i}^{x_f} F_x\,dx$$

When a source of a force is included in a system, there may be a change in the system's potential energy ΔU. If the source is outside the system, however, the force may do work on the system.

When we compare these two choices for what is in the system, we find a general rule: The amount of work done if the source is outside the system equals the negative of the change in energy if the source is included in the system, or

$$W_{\text{ext}} = -\Delta U_{\text{int}} \quad (9.23)$$

where the subscript "ext" is a reminder that to do work the source must be *ext*ernal to the system and the subscript "int" is a reminder that to change the potential energy the source must be *int*ernal. (Often these subscripts are dropped.)

In Chapter 8, we only considered a one-dimension equation for change in potential energy. The most general expression for change in potential energy comes from multiplying Equation 9.21 by -1:

GENERAL EXPRESSION FOR CHANGE IN POTENTIAL ENERGY

Major Concept

$$\Delta U = -\int_{r_i}^{r_f} \vec{F} \cdot d\vec{r} \tag{9.24}$$

We also said in Chapter 8 that potential energy could only be associated with conservative forces such as gravity and the spring force obeying Hooke's law. We can now define a **conservative force** as one that has two properties:

CONSERVATIVE FORCE

Major Concept

1. A change in a conservative force depends only on the change in the (relative) position of the subject.
2. The work done by a conservative force on a subject does not depend on the particular path of the subject.

Let's look at each property separately using universal gravity between the Earth (the source) and a moving meteor (the subject) as a specific example of a conservative force.

The first property might seem puzzling because universal gravity depends not only on the relative position of the source and the subject but also on their masses: $F_G = GMm/r^2$ (Eq. 7.4). Does this dependence on mass mean that gravity isn't a conservative force? The answer is no because as the position r of the subject changes, neither its mass m nor the source's mass M changes. Any change in the gravitational force between them is due only to a change in their relative position.

By contrast, a change in a nonconservative force may depend on some other factors such as the relative velocity of the source and the subject. Moving (kinetic and rolling) friction and drag are **nonconservative** forces because these forces depend on the relative motion between the source and the subject. Take the case of the drag force due to air on an accelerating meteor. As the meteor's speed increases, the drag force increases. So, the change in the drag force does not depend (solely) on the position of the subject; therefore, drag is a nonconservative force.

NONCONSERVATIVE FORCE

Major Concept

The second property says that if a subject moves from an initial position to a final position, the work done on the subject by a conservative force does not depend on the path the subject takes in getting from $\vec{r}_i$ to $\vec{r}_f$. For instance, if a meteor falls along a straight path toward the Earth from $\vec{r}_i$ to $\vec{r}_f$ or takes some roundabout path, the work done by the gravity is the same in either case.

There are two ways to test a force to see if it has this second (path-independent) property. The first test involves calculating the work done by the force on a subject along several different paths. If the work done is not the same along each path, the force is nonconservative. The alternate test involves calculating the work done by the force on a particle along a closed path. If the work done along a closed path does not equal zero, the force is nonconservative.

EXAMPLE 9.5 Gravity Is a Conservative Force

A particle of mass m moves from $\vec{r}_i$ to $\vec{r}_f$ and is acted on by the gravitational force exerted by a planet of mass M as shown in Figure 9.13. Show that the work done by gravity as the particle moves along path 1 (green) equals the work done along path 2 (red) and that gravity (so far) passes the first test of a conservative force.

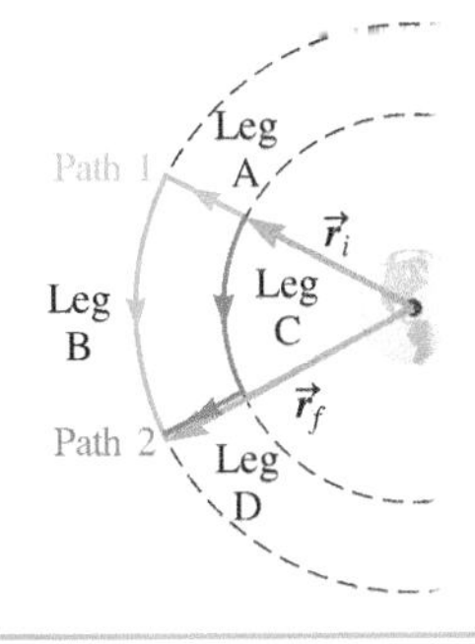

FIGURE 9.13 Two paths from $\vec{r}_i$ to $\vec{r}_f$. Path 1 is in green and path 2 is in red.

INTERPRET and ANTICIPATE

Each path has a straight part and a curved part. So, path 1 is made of leg A (straight part) and leg B (curved part), and path 2 is made of leg C (curved) and leg D (straight). Finding the work done along each whole path means finding the work done along each leg of the path and then adding them together.

Example continues on page 258 ▶

SOLVE

Finding the work done along leg A is similar to our procedure in finding the work done on the meteor falling toward the Earth (Example 9.4). The difference here is that the particle moves directly away from the planet. So, $r_f > r_i$.

$$W = \int_{r_i}^{r_f} \vec{F} \cdot d\vec{r}$$

$$W_A = \int_{r_i}^{r_f} \vec{F}_G \cdot d\vec{r} = \int_{r_i}^{r_f} -F_G \hat{r} \cdot d\vec{r} = -\int_{r_i}^{r_f} F_G dr$$

$$W_A = -\int_{r_i}^{r_f} \left(G\frac{Mm}{r^2} \right) dr$$

$$W_A = GMm\left[\frac{1}{r_f} - \frac{1}{r_i}\right]$$

The work done by gravity as the particle moves along leg B is zero because the force (toward the center) and the displacement (along the circular arc) are perpendicular all along this part of the path. The dot product of perpendicular vectors is zero.

$$W = \int_{r_i}^{r_f} \vec{F} \cdot d\vec{r}$$

$$W_B = 0$$

The total work done by gravity along path 1 is the sum of the work done along each leg.

$$W_1 = W_A + W_B = GMm\left[\frac{1}{r_f} - \frac{1}{r_i}\right] \quad (1)$$

The work done along path 2 is found in a similar way. First, the work along leg C is zero for the same reason that the work done along leg B is zero.

$$W_C = 0$$

Finding the work done by gravity along leg D is just like finding the work along leg A.

$$W_D = GMm\left[\frac{1}{r_f} - \frac{1}{r_i}\right]$$

So, the total work done along path 2 is the sum of the work done along each leg.

$$W_2 = W_C + W_D = GMm\left[\frac{1}{r_f} - \frac{1}{r_i}\right] \quad (2)$$

Compare Equation (1) with Equation (2) to find that the work done by gravity along path 1 equals the work done along path 2.

$$W_2 = W_1$$

CHECK and THINK

Because the work done along each path is the same, we can conclude that gravity *has not failed* this test of a conservative force. We cannot possibly test every path from $\vec{r}_i$ to $\vec{r}_f$ because there are an infinite number of them. So, passing this test is a requirement, but not a sufficient condition, to prove that gravity is a conservative force. Even the second test—that zero work is done around a closed path—is a necessary, but not sufficient, condition. A more sophisticated test exists, but it is beyond the scope of this book. Instead, we state that gravity, the spring force obeying Hooke's law, and the electrostatic force (Chapter 23) all are conservative forces.

EXAMPLE 9.6 A Nonconservative Force

To illustrate the second test for a conservative force, let us return to Paul and his sled to examine two forces that are *not* conservative. The sled is on the ice, and Paul pulls a rope that is attached to the sled (Fig. 9.14A). The tension F_P in Paul's rope is constant. (This constant tension is, of course, difficult for Paul, but it makes our calculations simpler.) The sled is constrained to move in a circle because a second rope attached to it is fixed to a stake in the ice. We assume friction between the sled and ice is low enough to be ignored. As a test, we calculate the work done by the tension force in Paul's rope in one circuit of the sled.

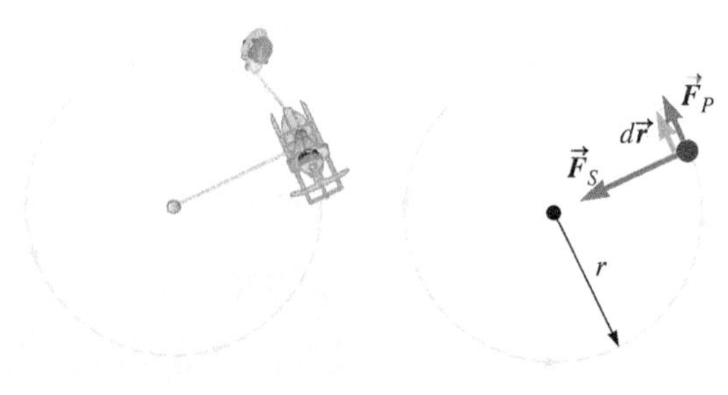

FIGURE 9.14 **A. Overhead view** **B. Free-body diagram**

INTERPRET and ANTICIPATE

We are interested in finding the work done by Paul's rope around the closed circular path, so Paul and his rope are outside the system. The system consists only of the sled and its contents.

SOLVE

Start with the general expression for work (Eq. 9.21). At every point on the circle as shown in Fig. 9.14B, the tension force $\vec{F}_P$ exerted by Paul and the infinitesimal displacement $d\vec{r}$ are parallel ($\varphi = 0$), so the cosine of the angle between them is 1.

$$W = \int_{r_i}^{r_f} \vec{F}_P \cdot d\vec{r} \qquad (9.21)$$

$$W = \int_{r_i}^{r_f} F_P dr \cos\varphi$$

$$W = \int_{r_i}^{r_f} F_P dr \cos 0 = \int_{r_i}^{r_f} F_P dr$$

Because the tension is constant over the entire path, we can pull it outside the integral.

$$W = F_P \int_{r_i}^{r_f} dr$$

The integral is a path integral over a single circular path. The result of such an integral is the length of the path; in this case, it is the circumference of the circle.

$$W = F_P 2\pi r$$

CHECK and THINK

We just found that the work done by the tension force in Paul's rope is *not* zero around a closed path. We conclude that this tension force is nonconservative. Also notice that because the work is positive, the sled's kinetic energy increases. This means that the sled and Paul must be accelerating; a feat that would be difficult for Paul to maintain for an extended period of time.

In the scenario in Example 9.6 in which Paul pulls a sled around a closed, circular path, the sled is constrained to move in a circle by a rope attached to a stake. Figure 9.14B shows the sled at one particular moment along with its infinitesimal displacement and the tension forces exerted on it. We want to calculate the work done by the rope attached to the stake as the sled moves on its circular path. From the figure, we do not need to make any actual calculation because the force $\vec{F}_S$ exerted by the stake's rope is always perpendicular to $d\vec{r}$. Therefore, the work done by the tension force in the stake's rope is always zero. We say that the force $\vec{F}_S$ is a zero-work force. A **zero-work force** may accelerate a particle without doing any work on it. Any centripetal force is a zero-work force. A force may be a zero-work force in some situations and not in others. For instance, gravity—a conservative force—can be a zero-work force when it keeps a particle in a circular orbit. Other examples of zero-work forces are static friction and the normal force. To see why, we need to take another look at the particle model.

ZERO-WORK FORCE

Major Concept

9-6 Particles, Objects, and Systems

So far in this book, we have modeled all sorts of real objects—such as people, cars, and planets—as particles. This model does not work when we are interested in the energy associated with some situations that involve friction and the normal force.

Recall from Chapter 2 that a **particle** is an idealized point that has no spatial extent, no shape, and no internal structure. An object can be treated as a particle undergoing purely translational motion if every point on the object undergoes exactly the same displacement as every other point on the object.

Now let us look at a simple object that cannot be modeled as a single particle. A chain of five metal links is held by an end link and then dropped (Fig. 9.15). The chain falls straight down so that the bottom link lands first. The displacement of the bottom link is less than the displacement of the top link. Because the displacement of each link of the chain is not exactly the same, we say that the chain is a **deformable object** or a **deformable system**, and it cannot be modeled as a particle in this situation.

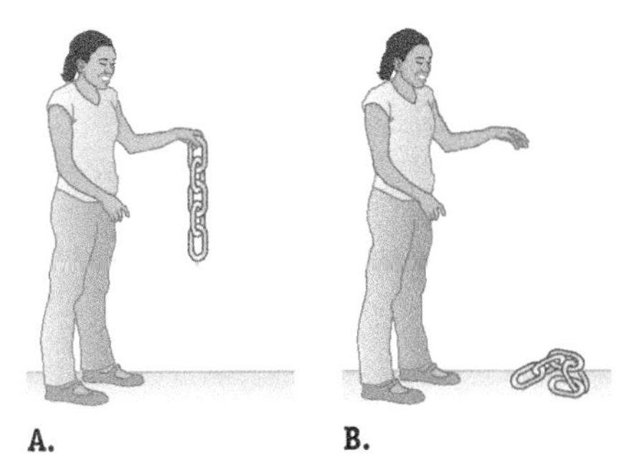

FIGURE 9.15 **A.** A chain consisting of five links is held by an end link and then dropped. **B.** Each link lands on the ground, but each link has a different displacement. For example, the bottom link has a smaller displacement than the top link.

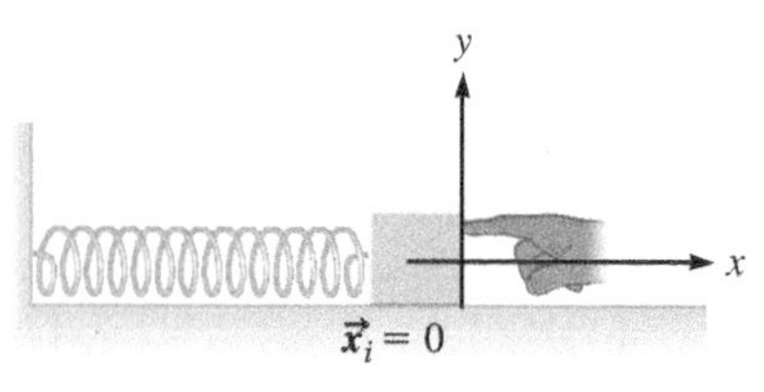

A.

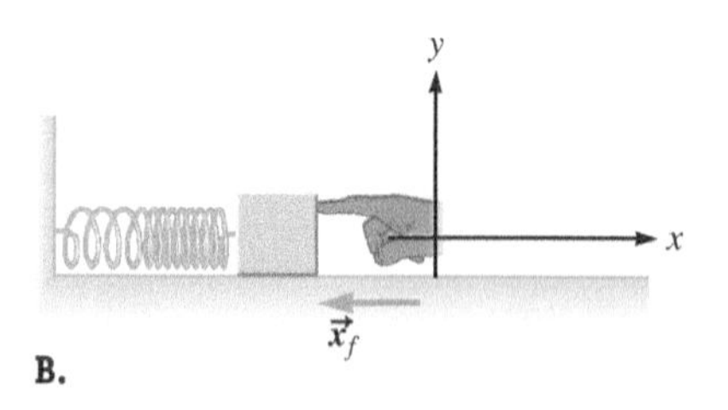

B.

FIGURE 9.16 A. A person's hand pushes on the block, and the spring compresses. **B.** The person's hand is outside the block–spring system, so it does work on the system. That work increases the system's potential energy.

Work and Mechanical Energy

Our goal is to incorporate deformable systems into the conservation approach, leading us to extend the work–*kinetic* energy theorem into a work–*mechanical* energy theorem. We start by finding an expression for the work done by a person's hand when compressing a spring (Fig. 9.16). In this case, the deformable system consists of the spring, the wall, and the block.

When the person pushes on the block, the coil attached to the block is displaced, but because the spring is deformable, the coil attached to the wall is not displaced. So, what displacement should we use to calculate the work done by the person on the spring–block system? In this case, we must use the displacement of the *point of application* of the force. The **point of application** is the place in the system where the force is exerted. For a system modeled as a particle, it is simply the location of the particle, but that is not the case for a deformable system. The hand exerts a contact force, so the point of application is the place where the hand is touching the block. The hand has displacement $\Delta\vec{x} = -|x_f - x_i|\hat{\imath}$ (Fig. 9.16).

We also need to know the force exerted by the person to find the work done on the system. According to Newton's second law, as long as the block is not accelerating, the magnitude of the force exerted by the person must be given by Hooke's law, $\vec{F}_H = -|kx|\hat{\imath}$. Because the force is not constant, we must integrate to find the work done by the person. Because the force and the displacement point in the same direction, the angle between them is zero, and the work done by the hand is positive:

$$W = \int_{x_i}^{x_f} (F_H dx)(\cos 0) = \int_{x_i}^{x_f} kx\,dx$$

$$W = k\int_{x_i}^{x_f} x\,dx = \tfrac{1}{2}kx^2\Big|_{x_i}^{x_f}$$

$$W = \tfrac{1}{2}k(x_f^2 - x_i^2), \text{ done by person} \quad (9.25)$$

In this example, there is no friction between the block and the surface, and the block is at rest at both its initial and final positions; so, its kinetic energy does not change. Therefore, the energy transferred from the person's hand to the block–spring system did not go into the system's kinetic energy. Instead, the work done on the system went into changing the system's configuration; the spring was initially relaxed, and in its final configuration it is compressed. So, in this case, the positive work done increased the system's potential energy. In fact, if you compare work done by the external person (Eq. 9.25) to the change in elastic potential energy of spring–block system (Eq. 8.8), you will find that they are equal: $W_{ext} = \Delta U_e$.

In a more general case, the work done can change both the kinetic energy and the potential energy. For example, if the block was initially at rest but moving at the final time, the system's kinetic energy and potential energy would have increased. So, in general, the total work done on a system by external forces in the absence of resistive forces (such as friction and drag) changes the system's mechanical energy; this **work–mechanical energy theorem** is

WORK–MECHANICAL ENERGY THEOREM

Underlying Principle

$$W_{tot} = \Delta E$$
$$W_{tot} = \Delta K + \Delta U \quad (9.26)$$

Center of Gravity and Center of Mass

For a deformable system, we need to know the displacement of the point of application to find the work done by an external force. For the case of a contact force such as a person pushing on a spring–block system, the point of application is the place where the hand makes contact with the system.

In the case of the falling chain (Fig. 9.15), however, the external force is gravity, and gravity is a field force. Gravity pulls each component (each link) of the system downward. So, where is the point of application in that case? The answer is that we define a point called the **center of gravity**, and total gravitational force exerted on the system is modeled as acting on the center of gravity.

When the gravitational field is constant (or nearly so) as it is near the surface of the Earth, the center of gravity is at the same point as the *center of mass*. The **center of mass (CM)** is a point associated with any system such that when a net external force is exerted on the system, the center of mass accelerates according to Newton's second law $\vec{F}_{tot} = m\vec{a}_{CM}$, where m is the system's mass. When that net external force is due to a uniform gravitational field near the surface of the Earth, the center of mass (and equivalently the center of gravity) accelerates downward. (In general, not every component has the same acceleration, so it is possible that only the center of mass has the acceleration $\vec{a}_{CM}$.) Put in terms of energy, the kinetic energy of the center of mass depends on the system's mass and its center-of-mass speed: $K_{CM} = \frac{1}{2}mv_{CM}^2$.

CENTER OF MASS Major Concept

In the next chapter, we will learn how to calculate the center-of-mass position. For now, you can think of the center of mass as the average location of the system's mass, and you can find it from symmetry. For example, the center of mass of a sphere with uniform density is at the center of the sphere. There does not need to be any matter at the center of mass. In the case of the chain, the center of mass is in the center of the middle link, where there is no metal.

Zero-Work Forces

With the deformable system model in mind, let's take another look at zero-work forces. As an example, consider the simple act of walking across a level floor. To walk, you swing one leg forward and then place your foot down and press against the floor. So, when you walk, you change your shape (deform the system), allowing yourself to press on the floor. By Newton's third law, the floor exerts a force back on you, and a horizontal component accelerates you forward. This horizontal force exerted by the floor is static friction, and its point of application is on the bottom of your shoe. By the definition of *static* friction, however, that point of application has no displacement. So, static friction does zero work on you.

Static friction is not always a zero-work force. For example, when a box sits in a flatbed truck, static friction accelerates the box, matching the truck's acceleration (Fig. 9.17). In this case, the point of application is displaced in the same direction as the static friction force, so we conclude that static friction does positive work on the box.

The box does not slip with respect to the truck, so both have the same acceleration $\vec{a}$.

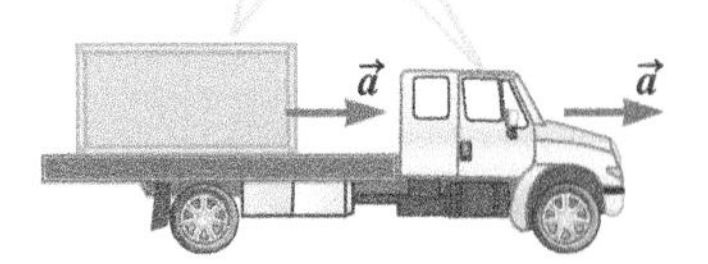

FIGURE 9.17 The point of application of static friction moves in the same direction as the box's displacement, so static friction does positive work in this case.

CONCEPT EXERCISE 9.2

For a person to jump straight up, she must crouch down and quickly extend her legs.

a. What force accelerates her upward?
b. Does this force do work on her? Explain.

CONCEPT EXERCISE 9.3

A man is accelerated upward in an elevator.

a. What force causes him to accelerate?
b. Does this force do work on him? Explain.

9-7 Thermal Energy

So far in our conservation approach to mechanics, we have avoided situations that involve moving friction and drag. In these next two sections, we learn how to include these nonconservative forces.

We experience thermal energy whenever we rub our hands together to warm them up. **Thermal energy** is the internal energy associated with the temperature of a system. It is not really a new form of energy, though; it's just a way to describe the kinetic energy (and potential energy) of the many particles in a system.

THERMAL ENERGY Special Case

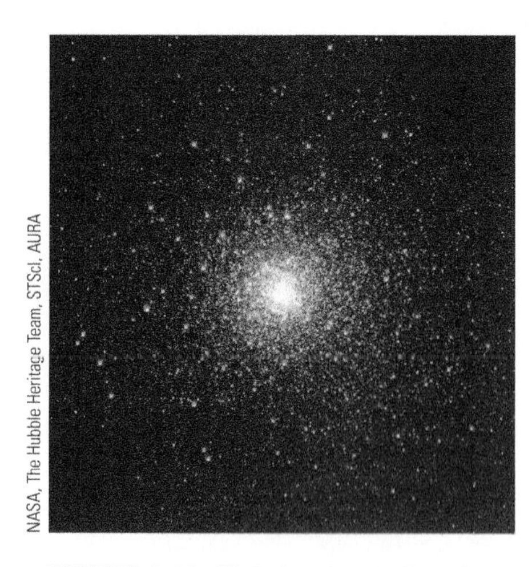
NASA, The Hubble Heritage Team, STScI, AURA

FIGURE 9.18 Globular clusters have hundreds of thousands of stars. The center of mass of a globular cluster orbits the Milky Way much as a planet orbits the Sun. The stars in the globular cluster whiz around the cluster like a swarm of bees.

Globular Cluster Analogy

We can develop our intuition about thermal energy by looking at an analogy (developed by Richard Feynman, an American Nobel Prize–winning physicist). A globular cluster is a roughly spherical group of hundreds of thousands of stars (Fig. 9.18). The center of mass of a globular cluster orbits the center of the Milky Way galaxy much like a planet orbits the Sun. In addition, the stars in the cluster zip around inside the cluster on erratic paths.

First, we model the globular cluster as a system of particles and imagine that some external force (due perhaps to the gravitational attraction of another very large star cluster) does work on our globular cluster. The work W done by this external force is likely to do three things: (1) change the shape of the cluster, (2) accelerate its center of mass, and (3) accelerate the individual stars moving inside the cluster. Each of these changes corresponds to some change in the system's energy as follows: (1) Because potential energy depends on the configuration of the system, a change in the cluster's shape means that its potential energy U changes. (2) The acceleration of the cluster's center of mass means that the cluster's center-of-mass kinetic energy K_{CM} changes. (3) Similarly, the acceleration of the individual stars means that the sum of their kinetic energy K_{stars} changes. So, the work done by the external force equals the change the system's potential energy plus the change in the center-of-mass kinetic energy plus the change in kinetic energy of all the individual stars:

$$W = \Delta U + \Delta K_{CM} + \Delta K_{stars} \tag{9.27}$$

Equation 9.27 is actually just the work–mechanical energy theorem $W_{tot} = \Delta E$ (Eq. 9.26) for this cluster.

Now imagine observing this cluster from a great distance so that we cannot see the individual stars and thus can no longer model the cluster as a system of particles. Instead, we model the cluster as a single deformable object. There are no other changes to the situation; an external force still does work on the cluster. Even from this great distance, we can still observe the cluster deform and its center of mass accelerate. There is a major difference, though. Because we cannot see the individual stars, we cannot observe the change in the individual stars' kinetic energy. It seems that some of the work done on the cluster is missing because $W > \Delta U + \Delta K_{CM}$.

We know, though, that energy cannot be destroyed, so we assume the missing energy is really internal to the system and account for the missing energy this way:

$$W = \Delta U + \Delta K_{CM} + \Delta E_{int} \tag{9.28}$$

INTERNAL ENERGY ✪ Major Concept

By comparing Equations 9.27 and 9.28, we equate the change in internal energy ΔE_{int} with the change in kinetic energy of the individual stars ΔK_{stars}. **Internal energy** is a manifestation of potential or kinetic energy on a small scale (compared with the whole system) involving a large number of particles. Imagine trying to add up the kinetic energy of the hundred thousand stars in the cluster; the concept of internal energy makes it possible for us to avoid making all those individual calculations and still apply the conservation approach.

Internal energy comes in several forms. For now, the form of internal energy we are concerned with is thermal energy, so we write the change in internal energy as $\Delta E_{int} = \Delta E_{th}$. In our analogy, the motion of the individual stars is thermal energy. (This wording may seem silly because that motion has nothing to do with the star's temperature, but astronomers have been known to use the phrase *thermal energy* in this way.)

Change in Thermal Energy due to Moving Friction

Suppose we need to calculate the change in thermal energy due to moving friction. Consider a particular example in which a student slides a book across a table from point i to point f along the straight path S (Fig. 9.19). To make our argument simple and clear, we ignore the momentary acceleration at i and the momentary deceleration at f. Four forces—gravity, the normal force, the push of the person, and kinetic friction—act on the book. The person pushes with a constant force $\vec{F}_P$ equal in magnitude to kinetic friction on the book so that the book moves with constant velocity.

At first, let's assume we can treat the book as a particle and find the work done on it by the four forces. Gravity and the normal force do no work because they are perpendicular to the book's displacement. The work done by the person on the book is positive and is given by $W_P = F_P \Delta r$. Now, we might be tempted to argue that the work done by kinetic friction on the book is $W_k = -F_k \Delta r$ because the magnitude of kinetic friction is the same as that of the person's applied force, but in the opposite direction as the book's displacement. Then, the total work W_{tot} done on the book would be zero.

There are two problems with that temptation. First, the book is not a particle, and we know that sliding the book across the table increases the book's thermal energy. (If we touch the bottom of the book, it feels warmer.) So, the total work done on the book cannot be zero[1]. The second problem is that the table's thermal energy also increases. It would be very difficult to separate the book's increase in thermal energy from the table's increase. We avoid this difficulty by cleverly choosing to include both the book and the tabletop in the system. Then, we identify $F_k \Delta r$ as the change in the system's thermal energy. It doesn't matter how much of that thermal energy is in the book versus how much is in the table.

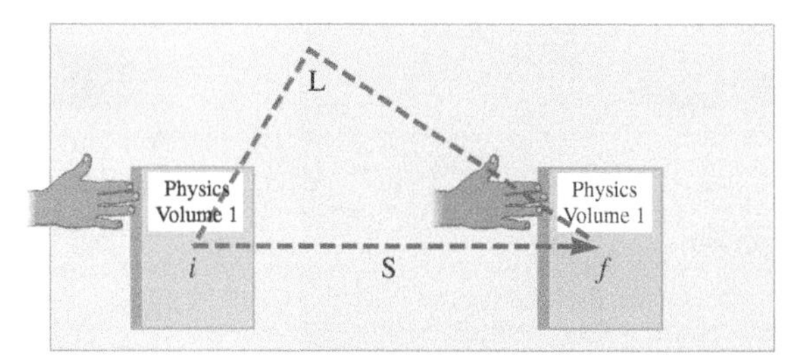

A. Top view

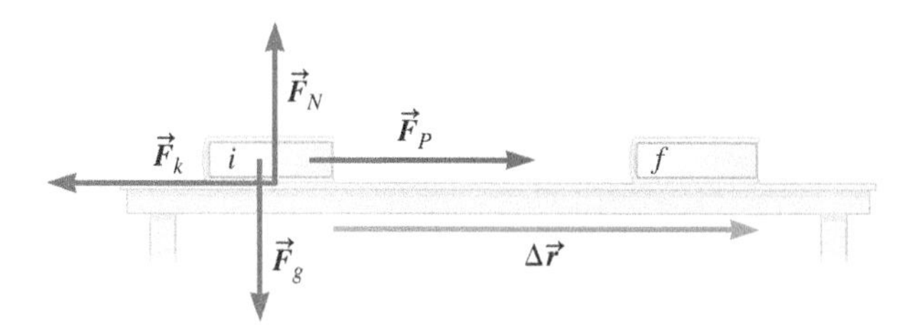

B. Side view

FIGURE 9.19 A. A student slides a book across a table along the straight path S. Later we consider a student sliding the book along the L–shaped path L. **B.** Four forces act on the book. The book does not accelerate, so $F_P = F_k$.

So, with this clever choice for the system, the person does positive work on the book–tabletop system. This work does *not* go into the kinetic energy of the system's center of mass. (The book moves at constant velocity.) The work does not go into the potential energy either. (There are no conservative forces in the system.) Instead, the work done by the person goes into the thermal energy of the system. (Both the tabletop and the book get warmer.) Comparing this system to the globular cluster, the increased temperature is analogous to the change in the stars' individual speeds due to (part) of the work done by the external force on the cluster. For the book–tabletop system, the work done by the person equals the change in thermal energy $W_P = \Delta E_{th}$.

THERMAL ENERGY
Special Case

To find the change in thermal energy, you need to know one more thing: that it depends on the path taken by objects in a system. In the book–tabletop system, if the student slides the book along the L-shaped path instead of the straight path S (Fig. 9.19A), the change in the system's thermal energy is greater because path L is longer than path S. The change in thermal energy depends on the total path length s such that

$$\Delta E_{th} = F_k s \quad (9.29)$$

Equation 9.29 is always positive. So, if kinetic (or rolling) friction acts in some system, it always increases the system's thermal energy. Unlike changes in potential or kinetic energy, this increase in thermal energy cannot be converted back to mechanical energy. The increase in thermal energy is at the expense of mechanical energy, and we say that mechanical energy has been dissipated. The thermal energy will eventually leave the system entirely and increase the energy of the environment (Chapter 21). Nonconservative forces (such as moving friction and drag) are known as **dissipative forces** because they reduce the mechanical energy of the system, and this energy is eventually lost by the system.

DISSIPATIVE FORCE
Major Concept

CONCEPT EXERCISE 9.4

Avi passes a book up to Cameron who's on the upper bunk, and Cameron then passes the book back down to Avi. Consider the Earth and the book to be the system. In this round trip, does the system's gravitational energy increase, decrease, or remain the same? Explain.

[1]This argument assumes the only way to transfer energy to a system is through work. In Chapter 21, we will learn about another way to transfer energy to a system. Our argument here is not affected because work is the only energy transfer mechanism possible in this case.

CONCEPT EXERCISE 9.5

Avi and Cameron sit on opposite sides of the table as in Figure 9.19. Avi pushes a book to Cameron, Cameron pushes the book back toward Avi, and they repeat this rapid pushing for several rounds before the book is returned to Avi. Consider the book and the tabletop to be the system. At the end of these rounds, does the system's thermal energy increase, decrease, or remain the same? Explain. Compare your answer here with your answer to Concept Exercise 9.4.

9-8 Work–Energy Theorem

Now we are prepared to expand the work–*mechanical energy* theorem to a work–energy theorem that includes energy dissipation. To do so, we use the familiar example of a block sliding on an inclined plane (Fig. 9.20A). First, suppose no friction acts, only the Earth and the block are in the system, and no external forces do work on the system. (Set the reference configuration to the bottom of the inclined plane.) As illustrated by the bar chart (Fig. 9.20B), the mechanical energy is conserved in this isolated system.

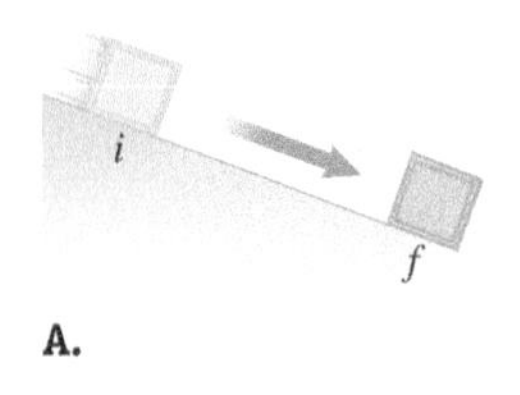

A.

No friction (isolated system consists of Earth + block)

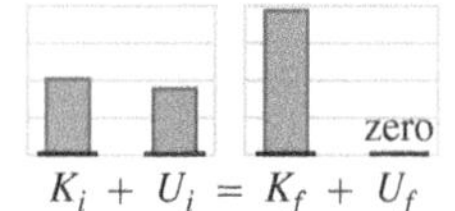

B.

Kinetic friction present (system consists of Earth + block + inclined plane)

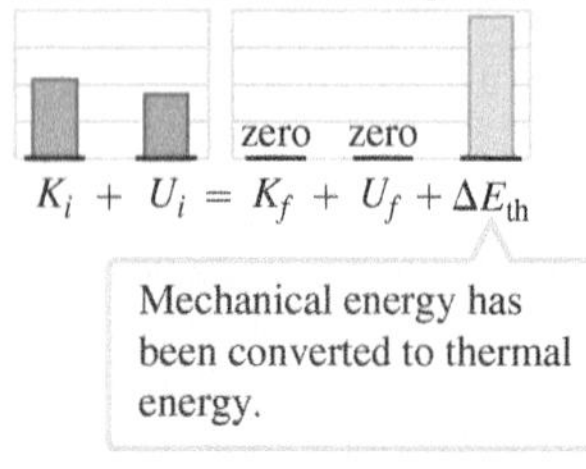

C.

FIGURE 9.20 A. A block slides down a plane. **B.** If kinetic friction is negligible, mechanical energy is conserved. **C.** If kinetic friction causes the block to stop, mechanical energy is not conserved and is instead converted to thermal energy.

Now assume kinetic friction acts between the block and the inclined plane. The new system consists of the block, the Earth, and the inclined plane. As before, the system starts with positive kinetic energy and potential energy, but this time when the system reaches its final configuration, the block has stopped. In the final configuration, the system has no kinetic or potential energy. All the mechanical energy has been converted to thermal energy. A new bar must be included in the bar chart for thermal energy ΔE_{th} (Fig. 9.20C). Mechanical energy $(K + U)$ has been dissipated, but the overall energy is conserved in this isolated system.

Now let us add one more complication: that of an external force that does work on the system. Now, the block–Earth–inclined plane system is not isolated. Specifically, work is done by a person who pulls a rope attached to the block (Fig. 9.21A). (The person and the rope are outside the system.) So, we must add a new bar to the chart (Fig. 9.21B). All the energies are the same as they were previously except that the kinetic energy in the final configuration is not zero because the tension in the rope did positive work on the nonisolated system.

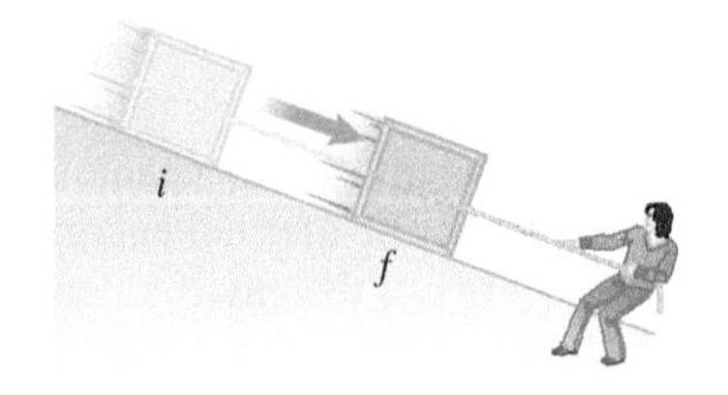

A. B.

FIGURE 9.21 A. A person uses a rope to pull a block down an inclined plane. Kinetic friction acts between the block and the plane. We consider the block, the Earth, and the plane to be inside the system; the person and the rope are outside the system. The rope does work on the system, and kinetic friction increases the thermal energy of the system. **B.** The total height of the three bars on the left equals the total height of the two bars on the right.

We use this bar chart (Fig. 9.21B) to write a general work–energy theorem for a nonisolated system including internal dissipative forces. According to that bar chart, the initial mechanical energy (potential energy plus kinetic energy) plus the energy transferred to or from the system equals the final mechanical energy plus the change in internal energy (thermal energy, in this case.) Mathematically,

$$\begin{pmatrix}\text{initial}\\ \text{mechanical}\\ \text{energy}\end{pmatrix} + \begin{pmatrix}\text{total work,}\\ \text{which is}\\ \text{energy transferred}\end{pmatrix} = \begin{pmatrix}\text{final}\\ \text{mechanical}\\ \text{energy}\end{pmatrix} + \begin{pmatrix}\text{change in}\\ \text{internal}\\ \text{energy}\end{pmatrix}$$

$$E_i + W_{tot} = E_f + \Delta E_{int} \tag{9.30}$$

We write the mechanical energy in terms of the kinetic energy and potential energy:

$$K_i + U_i + W_{tot} = K_f + U_f + \Delta E_{int} \quad (9.31)$$

Equation 9.31 is a very general expression for the **work–energy theorem**. Purely for convenience, we rewrite Equation 9.31 explicitly for the two forms of potential energy—gravitational (Eq. 8.5 and 8.7) and elastic (Eq. 8.9)—we are most concerned with in this part of the book. So,

$$K_i + U_{gi} + U_{ei} + W_{tot} = K_f + U_{gf} + U_{ef} + \Delta E_{int} \quad (9.32)$$

The work–energy theorem involved includes a change in internal energy ΔE_{int}. Internal energy comes in several forms, including chemical energy. Chemical energy is important when we consider certain deformable objects and forces that do no work. For example, a froghopper insect leaps upward by pushing on the ground, and by Newton's third law, the ground pushes back on the froghopper, accelerating the froghopper (Example 5.11). Consider the froghopper and the Earth as the system. The components of the force exerted by the ground are the normal force and static friction. The point of application of these forces has no displacement, so these forces do no work on the froghopper (Section 9.6). No external forces do work on the system, yet the system's mechanical energy increases (Fig. 9.22A). Where does this energy come from? The answer is that the increase in mechanical energy must come from the froghopper's internal energy. The froghopper is a deformable object: To push against the ground and leap into the air, it compresses and quickly releases muscles. The ultimate source of energy that allows an animal to control its muscles is food. Chemical reactions (the digestive process) release energy, and the animal uses that energy (in part) to control its muscles. So, this form of energy is known as *chemical energy*.

Living organisms are not the only objects that use chemical energy to gain mechanical energy. For example, a car runs on fossil fuel. Like food in animals, fossil fuel is the source of chemical energy that is converted to mechanical energy. When the insect leaps, the change in its chemical ΔE_{chem} is negative (Fig. 9.22B). So, the source of chemical energy (gasoline or pizza) must be replenished. Table 9.1 lists the rough energy content of some common fuels.

WORK–ENERGY THEOREM **Underlying Principle**

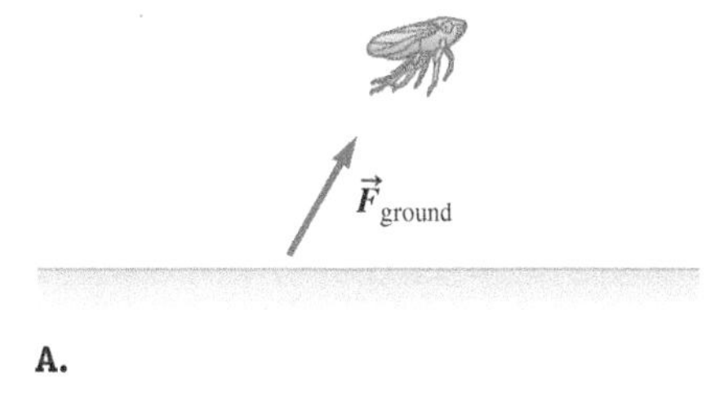

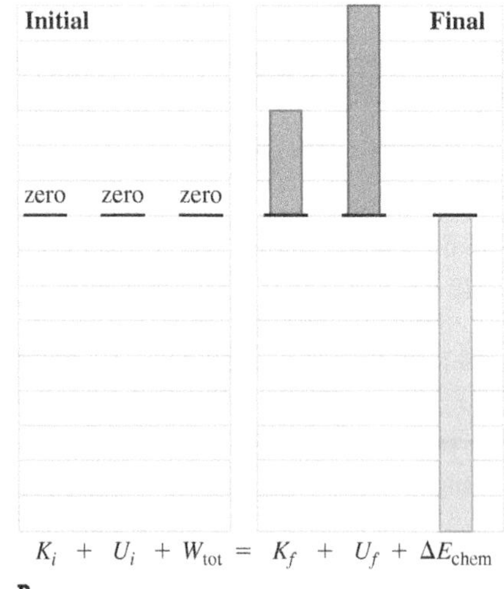

FIGURE 9.22 **A.** A froghopper starts on the ground at rest. The froghopper leaps up, and mechanical energy of the system increases. **B.** The mechanical energy comes from the decrease in internal chemical energy.

CHEMICAL ENERGY **Special Case**

TABLE 9.1 Energy Content in Joules.

Astronomical sources	Joules (J)
Typical supernova (exploding star)	10^{43}
Milky Way's radiation in 1 s	10^{38}
Sun's radiation in 1 year	10^{34}

Terrestrial sources	Joules (J)
Earth's rotation	10^{29}
Volcanic detonation	10^{19}
Largest recorded earthquake	10^{18}
San Francisco earthquake of 1906	10^{17}
Tornado or thunderstorm	10^{15}
Lightning flash	10^{10}

Human needs and everyday phenomena	Joules (J)
1000-MW power station in 1 year	10^{16}
1 lb of uranium-235	3.7×10^{13}
Energy to put space shuttle into orbit	10^{13}
Crossing Atlantic Ocean in a jet plane	10^{12}
1 U.S. gallon of gasoline	1.3×10^{8}
1 lb of coal	1.6×10^{7}
Two-ton truck traveling at highway speed	10^{6}
1 lb of TNT	10^{6}
1 can lighter fluid	10^{6}
1 candy bar, order of french fries, or slice of pizza	10^{3}
1 AA battery	10^{3}
1 tablespoon sugar, 1 apple, or 5 crackers	10^{2}
Major league pitch	10^{2}
Striking a typewriter key	10^{-2}

Adapted from Howard Keller, "Energy Yield of Various Sources," *Physics Teacher* 30:455 (1992).

Remember that we have two expressions for gravitational potential energy, $U_g = mgy$ and $U_G = -G(m_1m_2/r)$, and we need only one for any situation.

A more detailed study of chemical energy is beyond the scope of this book; instead, here we are primarily interested in thermal energy. So, for convenience in problem solving, we write the work–energy equation in terms of one specific internal energy, thermal energy:

$$K_i + U_{gi} + U_{ei} + W_{\text{tot}} = K_f + U_{gf} + U_{ef} + \Delta E_{\text{th}} \tag{9.33}$$

PROBLEM-SOLVING STRATEGY

Applying the Work–Energy Theorem

The problem-solving strategy for applying the work–energy theorem (Eq. 9.33) is nearly the same as it is when applying the conservation of energy principle. However, now the system no longer needs to be isolated. Also, you need four bars on each side of the chart to account for the four terms on each side of the work–energy theorem.

:• INTERPRET and ANTICIPATE

Step 1 Choose a **system**. Here are some tips for choosing a system:

- The source of a conservative force (gravity or spring force) is best included in the system. Then, associate potential energy with these forces.
- Dissipative forces such as moving friction increase the thermal energy of two objects; include both objects in the system.
- Sources of other nonconservative forces such as a rope are often best left out of the system. Then, the work done by the force is easily calculated, and it accounts for the transfer of energy from the environment to the system.

Step 2 Pick a **coordinate system and the reference configuration.**

Step 3 Identify an **initial time and a final time.**

Step 4 Draw a **bar chart.**

:• SOLVE

Step 5 Apply the **work–energy theorem** (Eq. 9.32 or 9.33).

Step 6 Write down any **other equations.**

Step 7 **Do algebra** before substitution.

EXAMPLE 9.7 My Kitten Is Stuck in a Tree!

By examining the same problem three times, this example shows how choosing a system affects our analysis. A girl uses a rope to lower a basket of kittens down from a tree at a constant velocity (Fig. 9.23). She wears leather gloves to protect her hands, and the rope slips through them. Analyze the problem by making bar charts using three different systems: the basket of kittens; the basket of kittens and the Earth; and the basket of kittens, the Earth, the rope, and the girl's leather gloves. To keep the charts simple, leave out the bars for elastic potential energy because there are no springs in this problem.

FIGURE 9.23

:• INTERPRET and ANTICIPATE

For each system choice, consider whether a force is internal or external. If a force is internal to the system, we need to consider whether it changes the system's potential or thermal energy. If it is external, calculate the work done by the force as it acts on the system. We expect that the total height of the bars on the left should equal the total height of the bars on the right in each case, no matter which system choice has been made. In all three cases, the kinetic energy of the system is unchanged because the basket is lowered at constant velocity. So, the initial kinetic energy bar has the same height as the final kinetic energy bar in all three cases.

A Make a bar chart using only the basket of kittens as the system.

:• SOLVE

Two external forces act on the basket of kittens: gravity exerted by the Earth and the tension in the rope. We model the basket and its contents as a single particle. A particle has no internal structure, so there can be no potential energy or internal energy. Both gravity and tension do work on the particle.

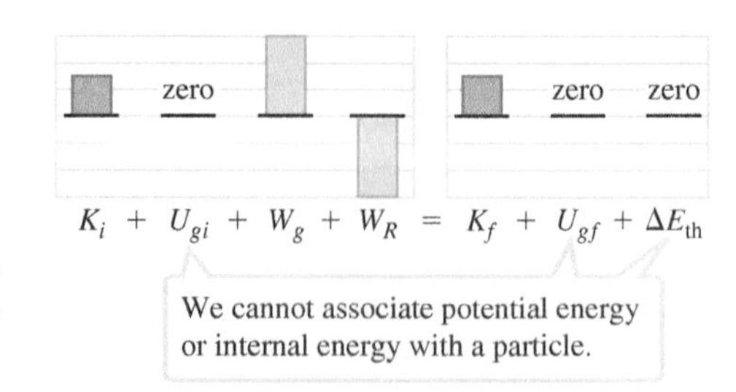

FIGURE 9.24

The work W_g done by gravity is positive because the force and the displacement are in the same direction. The work W_R done by the tension in the rope is negative because the force and the displacement are in opposite directions. For the total height of the bars on the left of Figure 9.24 to equal the total height of the bars on the right, the bars for W_g and W_R must be the same length.

CHECK and THINK

According to the work–kinetic energy theorem for a particle (Eq. 9.5), the total work done on a particle equals the change in its kinetic energy. Because the basket is lowered at constant velocity, its kinetic energy does not change. Therefore, the total work done by gravity plus tension must equal zero.

B Make a bar chart using the basket of kittens and the Earth as the system.

SOLVE

The rope is still outside the system, so tension still does work on the system. With the Earth inside the system, however, gravity no longer does work. Instead, the gravitational potential energy of the system changes, so we must choose a reference configuration. We choose the reference configuration so that it corresponds to the final configuration of the system just before the basket makes contact with the ground.

The work done by the rope is the same as it was in part A. Earth cannot do work because it is *in* the system. The gravitational potential energy is initially positive, and is zero in the final configuration (Fig. 9.25).

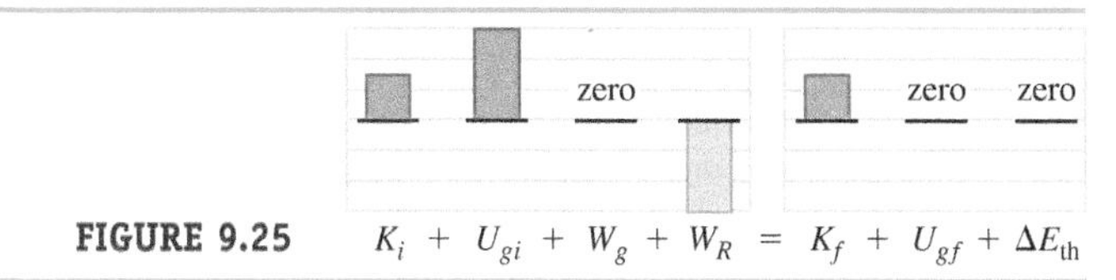

FIGURE 9.25

CHECK and THINK

The total height of the bars on the left equals the total height of the bars on the right. This situation is a special case of the work–mechanical energy theorem (Eq. 9.26) in which the change in kinetic energy is zero, so we have $W_{tot} = \Delta U$. In this particular case, the only work done is by the rope W_R, and the potential energy is gravitational U_g. We have $W_R = U_{gf} - U_{gi} = -U_{gi}$ as seen in the bar chart.

C Make a bar chart using the basket of kittens, the Earth, the rope, and the girl's leather gloves as the system.

SOLVE

With this situation, there are no forces left outside the system to do work. The rope slides through the leather gloves, and kinetic friction acts on both the rope and the gloves. The internal thermal energy of the system increases.

The gravitational potential energy is the same as it was in part B. The rope is in the system, so it cannot do work on the system. Instead, friction between the rope and the leather gloves increases the system's thermal energy (Fig. 9.26).

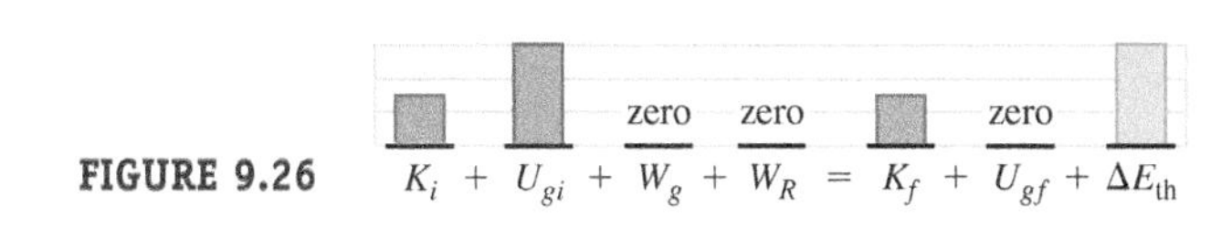

FIGURE 9.26

CHECK and THINK

If there were no friction, we would expect mechanical energy to be conserved, and the decrease in gravitational potential energy would be converted to an increase in kinetic energy. This system includes friction, however, and the decrease in gravitational potential energy goes instead into increasing the thermal energy.

EXAMPLE 9.8 Dart Gun Revisited

In Example 8.9 (page 230), we found an expression for the maximum height of a small ball launched straight up with a spring-loaded dart gun. In that example, we ignored the friction between the ball and the barrel of the gun, and we found that the maximum height of the ball is given by

$$y_{max} = \frac{k}{2mg} y_c^2 - y_c$$

where y_c is the distance by which the spring is initially compressed. Now assume kinetic friction F_k is significant and find an expression for the maximum height of the ball.

Example continues on page 268 ▶

INTERPRET and ANTICIPATE

Step 1 In solving Example 8.9, our (isolated) system consisted of the Earth, the ball, and the spring. Now that we need to take kinetic friction into account, it is best to include the gun barrel in the system. The system is still isolated, but kinetic friction dissipates some mechanical energy and increases the thermal energy. We expect the ball's maximum height to be lower than it was when friction was negligible.

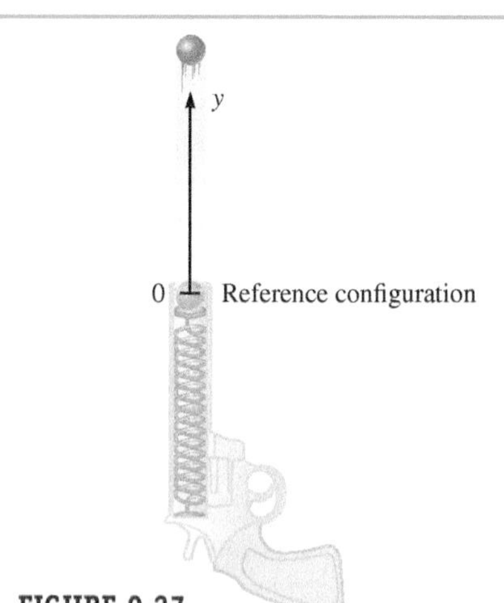

FIGURE 9.27

Step 2 As before, place the origin of the **coordinate system** at the end of the gun barrel in Figure 9.27. The **reference configuration** is the one in which the ball is at the end of the gun barrel. The spring is relaxed at that point. The elastic and gravitational potential energies of the system are zero in that configuration.

Step 3 Also as before, we set the **initial time** to when the ball is on top of a fully compressed spring and the **final time** to when the ball has reached its maximum height.

Step 4 Draw a **bar chart.** The initial and final kinetic energies are both zero. The initial gravitational potential energy is negative, and the initial elastic potential energy is positive. When the system is in its final configuration, the spring is relaxed and has lost contact with the ball, and the gravitational potential energy is positive. The difference between the bar chart in Figure 9.28 and the one in Example 8.9 is that here there is a bar for the increase in the thermal energy of the system.

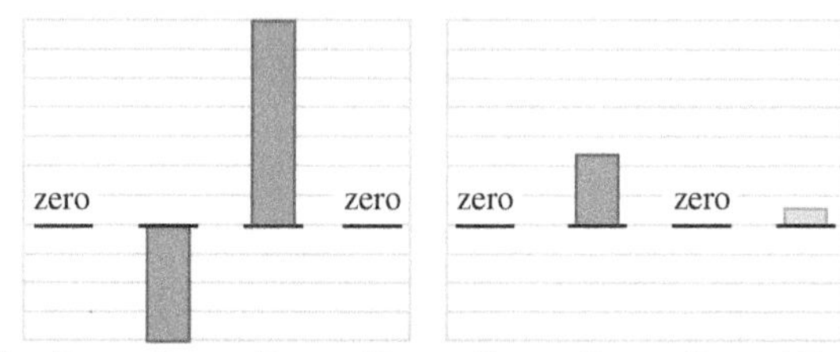

FIGURE 9.28 $K_i + U_{gi} + U_{ei} + W = K_f + U_{gf} + U_{ef} + \Delta E_{th}$

SOLVE

Step 5 Apply the **work–energy theorem** (Eq. 9.33) to write an algebraic expression, substituting zeros as indicated by the bar chart.

$$K_i + U_{gi} + U_{ei} + W_{tot} = K_f + U_{gf} + U_{ef} + \Delta E_{th}$$
$$0 + U_{gi} + U_{ei} + 0 = 0 + U_{gf} + 0 + \Delta E_{th}$$
$$U_{gi} + U_{ei} = U_{gf} + \Delta E_{th} \qquad (1)$$

Step 6 Write down **other equations.** Write expressions for potential energy (Eqs. 8.5 and 8.10) and thermal energy (Eq. 9.29). The path length over which friction acts is $s = y_c$.

$$U_{gi} = -mgy_c \qquad U_{ei} = \tfrac{1}{2}ky_c^2$$
$$U_{gf} = mgy_{max} \qquad \Delta E_{th} = F_k y_c$$

Step 7 Do **algebra.** Substitute for these energies in Equation (1) and solve for y_{max}.

$$-mgy_c + \tfrac{1}{2}ky_c^2 = mgy_{max} + F_k y_c$$
$$y_{max} = \frac{k}{2mg}y_c^2 - y_c - \frac{F_k}{mg}y_c$$

CHECK and THINK

The first two terms are the same as what we found for the maximum height of the ball in the absence of friction. Because $F_k y_c/mg$ is subtracted from these terms, the maximum height y_{max} reached by the ball is lower in this case, as expected.

EXAMPLE 9.9 Snow Tubing

At a winter recreation resort, snow tubers at the bottom of the hill hook their tubes to a tow rope. A motor pulls the rope so that tubers move at constant velocity to the top of the hill. (Ignore the momentary acceleration of the tuber when he first attaches his tube to the rope.) The coefficient of kinetic friction between the tube and the snow is $\mu_k = 0.16$. A boy and his tube with a total weight of 415 N are pulled a distance of 255 m up the 18° incline (Fig. 9.29). Consider the Earth, the boy, his tube, and the snow along his path to make up the system. Find the energy transferred from the motor to the system via the work done by the rope.

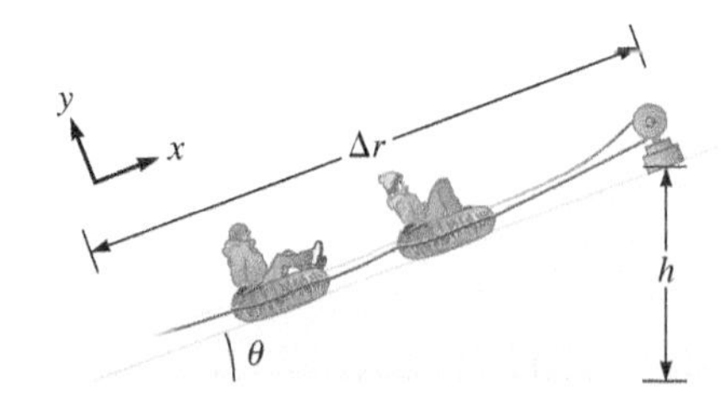

FIGURE 9.29

INTERPRET and ANTICIPATE

Step 1 A **system** has been chosen.

Step 2 A **coordinate system** has been chosen, and we set the **reference configuration** to when the boy is at the bottom of the hill.

Step 3 The **initial time** is when the boy is at the reference configuration, and the **final time** is when he is at the top of the hill.

Step 4 Draw a **bar chart.** The kinetic energy K of the system is unchanged because the tuber moves up at constant velocity. The thermal energy of the system increases due to friction between the tube and the snow. The initial gravitational potential energy is zero. We expect that the motor transfers energy to the system so that the rope must do positive work W_R on the system (Fig. 9.30).

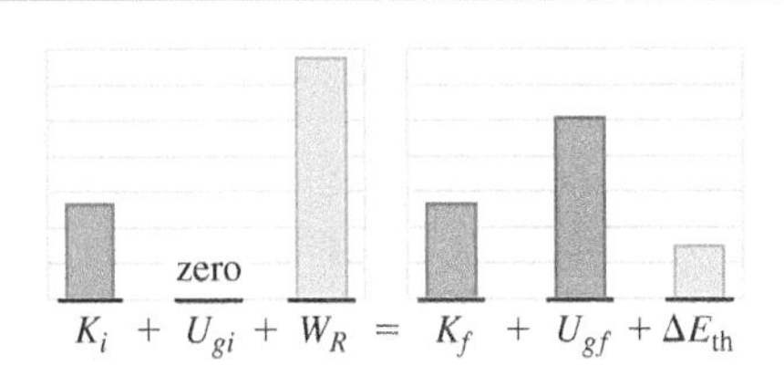

FIGURE 9.30

SOLVE

Step 5 Use the bar chart to apply the **work–energy theorem** (Eq. 9.33). The kinetic energy is constant and cancels out.

$$K_i + U_{gi} + U_{ei} + W_{tot} = K_f + U_{gf} + U_{ef} + \Delta E_{th}$$
$$K + W_R = K + U_{gf} + \Delta E_{th}$$
$$W_R = U_{gf} + \Delta E_{th} \quad (1)$$

Step 6 Write down **other equations.** Equation 8.5 gives the gravitational potential energy of the tuber, whose vertical displacement is h. Using trigonometry, rewrite h in terms of the tuber's given displacement Δr along the slope.

$$U_{gf} = mgh = mg(\Delta r \sin\theta) \quad (2)$$

The change in thermal energy comes from Equation 9.29. Kinetic friction is related to the normal force by Equation 5.9.

$$\Delta E_{th} = F_k \Delta r = \mu_k F_N \Delta r \quad (3)$$

Find the normal force using a free-body diagram (Fig. 9.31).

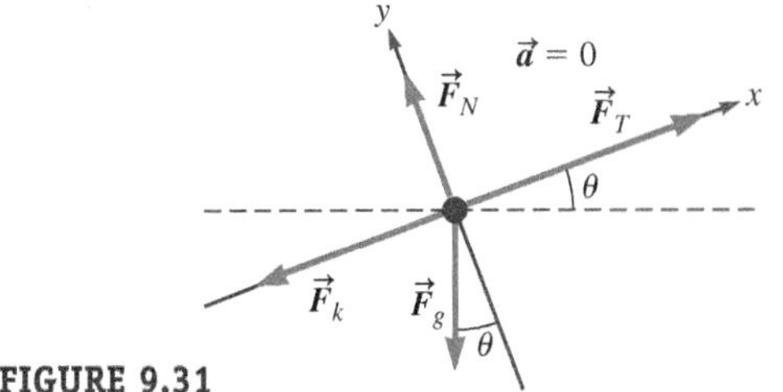

FIGURE 9.31

From Newton's second law and because the tuber is not accelerating, we find the normal force in terms of the weight mg of the tuber and his tube.

$$\sum F_y = F_N - F_g \cos\theta = 0$$
$$F_N = F_g \cos\theta = mg\cos\theta \quad (4)$$

Step 7 **Do algebra.** Substitute Equation (4) into Equation (3).

$$\Delta E_{th} = \mu_k mg\,\Delta r \cos\theta \quad (5)$$

Substitute Equations (2) and (5) into Equation (1).

$$W_R = mg\,\Delta r \sin\theta + \mu_k mg\,\Delta r\cos\theta$$
$$W_R = mg\,\Delta r(\sin\theta + \mu_k\cos\theta)$$
$$W_R = (415\,\text{N})(255\,\text{m})[\sin 18° + (0.16)\cos 18°]$$
$$W_R = 4.9 \times 10^4\,\text{J}$$

CHECK and THINK

As expected, the motor transfers energy to the system, or, in other words, the rope does positive work on the system. It might seem that more than 10,000 J is a lot of energy just to pull a boy up a hill, but it is roughly the energy content of a few french fries (Table 9.1).

9-9 Power

In Example 9.9, we found the energy transferred from a motor to the snow tuber's system. The work done by the rope does not depend on the time it takes the motor to get the tuber up the hill; the same amount of energy is transferred whether the uphill trip takes a long time or a short time. The rate of energy transfer, however, is often important. We would not want a motor that transferred energy so slowly that it took all afternoon to get the tuber up the hill or one that transferred energy so quickly that the trip became unsafe.

POWER ✪ **Major Concept**

Power is the rate at which energy is transferred into or out of a system or the rate at which energy is converted from one form to another within a system. The word *rate* implies a change in a quantity per unit time. Just as velocity is the rate of change of position, power is the rate of change of energy. We defined the average velocity as the displacement divided by the time interval for that displacement. Similarly, the average power P_{av} transferred to a system is the work W done on the system divided by the time interval Δt over which the energy was transferred:

$$P_{av} = \frac{W}{\Delta t} \tag{9.34}$$

When energy is converted from one form to another within a system, the average power for the conversion is found in a similar way. For example, if we are interested in finding the average power dissipated by friction, we replace W by the change in thermal energy ΔE_{th}:

$$P_{av} = \frac{\Delta E_{th}}{\Delta t} \tag{9.35}$$

Keeping with our velocity analogy, just as instantaneous velocity is simply called *velocity* and is given by the time derivative of position, so instantaneous power is simply called *power* and is given by the time derivative of energy transferred or converted. So, for power transferred to or from a system through work, the instantaneous power is

$$P = \frac{dW}{dt} \tag{9.36}$$

If we are interested in finding the power in converting from one form of energy to another, we take the time derivative of that energy converted. Again for the conversion of mechanical energy to thermal energy, the power dissipated is

$$P = \frac{dE_{th}}{dt} \tag{9.37}$$

Both *work* and *watt* are symbolized by the same letter. Keep them distinct in your mind by remembering that italic uppercase *W* stands for work, and regular (roman, or nonitalic) uppercase W stands for watts.

Power is a scalar with the dimensions of energy per time. Its SI units are joules per second or newton meters per second. This combination of units is called the *watt*, abbreviated "W," where

$$1\ \text{W} = 1\ \text{N} \cdot \text{m/s} = 1\ \text{J/s}$$

We are familiar with the watt as a unit of power. We buy 100-W lightbulbs, the exercise machine reports that we are converting chemical energy to mechanical energy (commonly called *burning* fat) at 400 W, and our regional power plant supplies 10^9 W. In U.S. customary units, power is measured in horsepower (hp), where

$$1\ \text{hp} = 746\ \text{W}$$

In the case of power transferred to or from a system through work, it is often convenient to rewrite Equation 9.36 in terms of external force $\vec{F}$ applied. To do so, substitute $dW = \vec{F} \cdot d\vec{r}$ (Eq. 9.20):

$$P = \frac{\vec{F} \cdot d\vec{r}}{dt} = \vec{F} \cdot \left(\frac{d\vec{r}}{dt}\right)$$

The term in parentheses is velocity (Eq. 4.8), so

$$P = \vec{F} \cdot \vec{v} \tag{9.38}$$

where $\vec{v}$ is the velocity of the point at which the force is applied.

CONCEPT EXERCISE 9.6

Electric companies charge consumers for the amount of energy used in each billing cycle. The companies often express energy in units of kilowatt-hours (kWh), where

$$1 \text{ kWh} = (1000 \text{ W})(3600 \text{ s}) = 3.6 \times 10^6 \text{ J}$$

In Maryland, a major electric supplier charges 4.72 cents per kilowatt-hour. If a Maryland consumer accidentally leaves a 100-W lightbulb on continuously for a 30-day billing cycle, how much would she owe the electric company for her mistake?

EXAMPLE 9.10 Let's Take the Elevator

The Taipei 101 tower shown in Figure 9.32 has the fastest elevators in the world today. The elevator's ascent is so fast that a special pressure-control system was installed. After the express elevator is loaded, it quickly reaches a constant speed of 16.6 m/s. Assume the elevator and its passengers have a mass of 4800 kg and there is a counterweight that reduces the effective weight of the elevator by 40%. Also assume there is a constant 5000-N kinetic friction force acting between the elevator and the shaft. Find the power supplied by the elevator's motor.

FIGURE 9.32

INTERPRET and ANTICIPATE

This example is similar to Example 9.9 in that a rope is used to transfer energy from a motor to an object that is pulled upward. We include the elevator, the passengers, the Earth, and the elevator shaft in the system. Unlike Example 9.9, however, here we are asked to find the rate at which work is transferred from the motor to the system. We expect to find a numerical result in the form $P = +$____W.

SOLVE

The external force is due to the tension in the rope. So, we need to find the tension force on the elevator. As usual, draw a free-body diagram (Fig. 9.33).

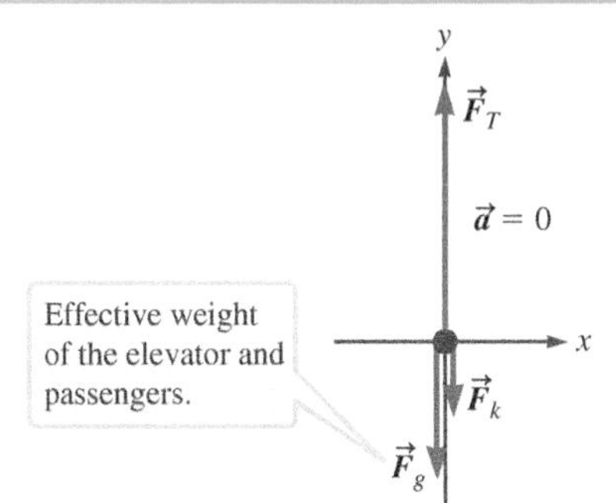

FIGURE 9.33

Apply Newton's second law to find the tension. The elevator moves at constant velocity, so the acceleration is zero. The effective weight is $F_g = 0.60mg$.

$$\sum F_y = F_T - F_g - F_k = 0$$
$$F_T = F_g + F_k = 0.60mg + F_k$$
$$F_T = (0.60)(4800 \text{ kg})(9.81 \text{ m/s}^2) + 5000 \text{ N}$$
$$\vec{F}_T = 3.3 \times 10^4 \hat{\jmath} \text{ N}$$

Power is found using Equation 9.38. Both the velocity and the tension force are upward, so the angle between them is zero.

$$P = \vec{F} \cdot \vec{v} = (3.3 \times 10^4 \text{ N})(16.6 \text{ m/s}) \cos 0° \quad (9.38)$$
$$P = 5.5 \times 10^5 \text{ W}$$

CHECK and THINK

Our result has the form we expected. It is sometimes helpful to convert the power output of a motor into horsepower. In this case, the elevator motor supplies more than 700 hp. This result may seem high until we compare it with something we know, such as a typical car with a 130-hp engine. The elevator's motor puts out as much power as roughly eight car engines. The speed of the Taipei 101 elevator is comparable to the speed limit on a small road, but the elevator is designed to hold five to six times more passengers than a typical car. So, it seems reasonable that the elevator motor would be several times as powerful as a typical car's engine.

EXAMPLE 9.11 Cruising Down the Highway

A car manufacturer reports that when a particular model is driven on a level highway at a constant speed of 60 mph, it has an energy efficiency (gasoline mileage) of 30 miles to 1 gallon of gasoline. One gallon of gasoline yields 1.3×10^8 J. What is the power requirement of the engine for the car to maintain its highway cruising speed? Comment on your results.

INTERPRET and ANTICIPATE

This example is much like converting a measurement from one system of units to another. The interesting part is thinking about our results.

SOLVE

Using the car's speed of 60 mph and its gasoline mileage of 30 miles per gallon, find the rate at which the car consumes gasoline.

$$(60\,\text{mi/h})\left(\frac{1\text{ gal}}{30\text{ mi}}\right) = 2\text{ gal/h}$$

Using the equivalence of 1 gallon of gasoline to 1.3×10^8 J, find the power supplied by the engine.

$$P = (2\text{ gal/h})(1.3 \times 10^8\,\text{J/gal})(1\text{ h}/3600\,\text{s})$$

$$P = 7.2 \times 10^4\,\text{W} \approx 97\text{ hp}$$

CHECK and THINK

We just found that the car's engine must supply about 100 hp just to maintain a constant speed of 60 mph on a level highway. This power is within the capability of a typical car (about 130 hp). The example statement asks us to comment on our results.

What is amazing is that this power is not changing the car's mechanical energy. The car maintains constant speed, so its kinetic energy does not change, and the highway is level, so there is no change in gravitational potential energy of the Earth–car system. The car's engine converts chemical energy, and none of it goes into mechanical energy. Where does it go? The converted energy is dissipated by both rolling friction between the car's tires and the road and by the drag force of the air on the body of the car. Energy is also dissipated by kinetic friction between moving parts inside the car's drivetrain, engine, and other systems. In fact, most of the power supplied by the engine never makes it to the wheels of the car, and is instead dissipated in the car's internal workings. Of course, even more power is required if the system's mechanical energy is increasing, such as when the car accelerates or climbs a hill.

EXAMPLE 9.12 CASE STUDY Size of a Meteor

In the meteor case study, three physics professors estimated the size of a meteor. They worked in the dark without a calculator, no steps were written down, and their estimate was made to one significant figure. One professor had read an article stating that the light given off by a meteor is equivalent to turning on a 100-W lightbulb for 2.5 s. The professors knew that a shooting star results when a meteor travels through the Earth's atmosphere, heating both the meteor and the atmosphere. The meteor gives off light and is vaporized. Three other facts they knew came in handy: $1\text{ AU} = 1.5 \times 10^{11}\text{m}$, $1\text{ yr} \approx \pi \times 10^7\text{ s}$, and the density of water is $\rho = 1000\,\text{kg/m}^3$. Estimate the size (radius) of a meteor.

INTERPRET and ANTICIPATE

The professors decided to include only the meteor in the system. The meteor had some mechanical energy when it entered the Earth's atmosphere, and all that energy was lost in 2.5 s, at which time the meteor was completely vaporized. The professors started with three assumptions. First, the meteor is small, so little energy is needed to vaporize it. They also reasoned that if the meteor has only a small mass, the amount of work done on it by the Earth's gravity during the 2.5-s time interval must also be negligible. Second, they assumed the half of the meteor's energy goes into heating the atmosphere and the other half is radiated away in the form of light. Third, they assumed the meteor's initial speed is twice the speed of the Earth around the Sun. Finally, they assumed the meteor was about twice as dense as water. (Don't worry if you wouldn't come up with all these.)

The bar chart in Figure 9.34 summarizes these assumptions. The meteor starts with kinetic energy K_i. Half of that kinetic energy is transferred in the form of light E_L, and the other half goes into thermal energy E_{th} of the atmosphere.

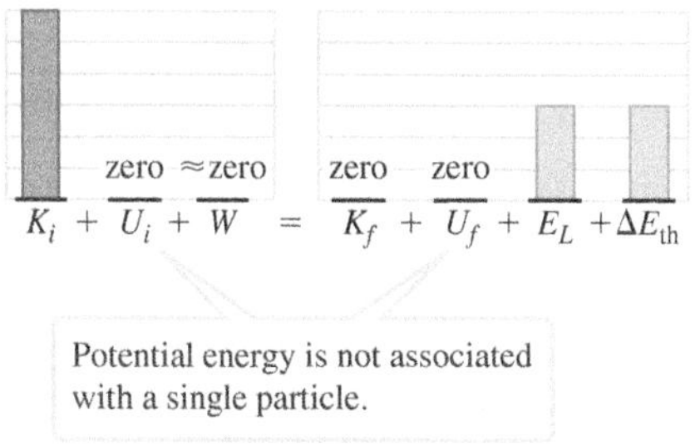

$$K_i + U_i + W = K_f + U_f + E_L + \Delta E_{th}$$

zero ≈zero zero zero

Potential energy is not associated with a single particle.

FIGURE 9.34

SOLVE

Our professors estimated the energy that goes into light from the article that one professor read.

$$E_L = P\Delta t = (100\,\text{W})(2.5\,\text{s}) = 250\,\text{J}$$

They assumed the energy that is radiated away in the form of light E_L is half the kinetic energy.

$$E_L = \tfrac{1}{2}K_i = \tfrac{1}{2}\left(\tfrac{1}{2}mv_i^2\right)$$

They solved for the mass of the meteor.

$$m = \frac{4E_L}{v_i^2}$$

The speed of the Earth is its orbital circumference divided by 1 year. It is very handy to memorize the number of seconds in 1 year this way because the π's cancel.

$$v_\oplus = \frac{2\pi a}{t} \approx \frac{2\pi(1.5\times10^{11}\,\text{m})}{\pi\times10^7\,\text{s}}$$

$$v_\oplus \approx \frac{3\times10^{11}\,\text{m}}{10^7\,\text{s}} = 3\times10^4\,\text{m/s}$$

The professors assumed the meteor's speed was twice the speed of the Earth, giving a (very approximate) mass of the meteor.

$$m = \frac{4E_L}{(2v_\oplus)^2} \approx \frac{4(250\,\text{J})}{4(3\times10^4\,\text{m/s})^2}$$

$$m \approx \frac{(300\,\text{J})}{(3\times10^4\,\text{m/s})^2} = \frac{1}{3}\times10^{-6}\,\text{kg} \approx 3\times10^{-7}\,\text{kg}$$

If the meteor is roughly spherical, we can find its size if we know its density.

$$m = \rho V = \rho\frac{4\pi R^3}{3} \approx \rho 4R^3$$

$$R \approx \left(\frac{m}{4\rho}\right)^{1/3}$$

The professors assumed the meteor had twice the density of water.

$$R \approx \left[\frac{3\times10^{-7}\,\text{kg}}{4(2000\,\text{kg/m}^3)}\right]^{1/3} = \left(\frac{300\times10^{-9}\,\text{kg}}{8\times10^3\,\text{kg/m}^3}\right)^{1/3}$$

$$R \approx (30\times10^{-12}\,\text{m}^3)^{1/3}$$

$$R \approx 3\times10^{-4}\,\text{m} = 0.3\,\text{mm}$$

CHECK and THINK

The radius of a meteor is roughly 0.3 mm, about the size of a grain of sand, and this result is about the correct size. A second estimate in Chapter 34 confirms this result.

This estimate shows the power of the conservation approach. We do not need to know the detailed physics of how the meteor vaporizes, gives off light, and heats the atmosphere. To find a reasonably good estimate of a meteor's size, all we need to do is account for the transfer and conversion of energy.

SUMMARY

Underlying Principles

1. **Work–kinetic energy theorem:** The total work done on a particle by all external forces equals the change in the particle's kinetic energy:

$$W_{tot} = \Delta K \qquad (9.5)$$

2. **Work–mechanical energy theorem:** The total work done on a system by external forces in the absence of resistive forces changes the system's mechanical energy:

$$W_{tot} = \Delta K + \Delta U \qquad (9.26)$$

3. **Work–energy theorem:** The equation

$$K_i + U_i + W_{tot} = K_f + U_f + \Delta E_{int} \qquad (9.31)$$

is an extension of the work–mechanical energy theorem that holds when there are internal dissipative forces.

Major Concepts

1. **Work** is the amount of energy transferred to (or from) the system by an external force.
2. A **general expression for work done** is

$$W = \int_{r_i}^{r_f} \vec{F} \cdot d\vec{r} = \int_{r_i}^{r_f} (F_x dx + F_y dy + F_z dz) \qquad (9.21)$$

3. A **general expression for change in potential energy** is

$$\Delta U = -\int_{r_i}^{r_f} \vec{F} \cdot d\vec{r} \qquad (9.24)$$

4. Conservative (and nonconservative) force: A **conservative force** depends only on the relative position of the source and object. The work done by a conservative force on a particle does *not* depend on the particle's path, and the work done by a **nonconservative force** *does* depend on the particle's path.
5. A **zero-work force** may accelerate a particle without doing any work on it.
6. The **center of mass** is a point associated with any object that moves as though all the mass were concentrated there and all external forces were applied there.
7. **Internal energy** within a system or deformable object is a manifestation of potential or kinetic energy that is on a scale that is too small for us to calculate directly.
8. When **dissipative forces** (nonconservative forces) act in a system, the system's thermal energy increases, and its mechanical energy decreases.
9. **Power** is the rate at which energy is transferred into or out of a system or the rate at which energy is converted from one form to another within a system. The instantaneous power is

$$P = \frac{dW}{dt} \quad (9.36) \qquad \text{and} \qquad P = \vec{F} \cdot \vec{v} \quad (9.38)$$

Special Cases: Internal Energy

1. **Thermal energy** is internal energy associated with the temperature of the system or deformable object. A change in thermal energy due to moving friction is given by

$$\Delta E_{th} = F_k s \qquad (9.29)$$

2. **Chemical energy** is another form of internal energy governed by chemical reactions within the system or deformable object.

Tools

If two vectors $\vec{A}$ and $\vec{B}$ lie in a plane at an angle φ, the **dot product** or **scalar product** of these two vectors is given by

$$D = \vec{A} \cdot \vec{B} \equiv AB \cos \varphi \quad (9.7) \qquad \text{and} \qquad D = \vec{A} \cdot \vec{B} = A_x B_x + A_y B_y + A_z B_z \quad (9.10)$$

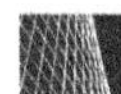

PROBLEM-SOLVING STRATEGY Applying the Work–Energy Theorem

INTERPRET and ANTICIPATE

1. Choose a **system**. Here are some tips for choosing a system:
   - Include the source of a conservative force (gravity or spring force).
   - Be sure that dissipative forces are internal.
   - Sources of other nonconservative forces should be external.
2. Pick a **coordinate system and the reference configuration**.
3. Identify an **initial time** and a **final time**.
4. Draw a **bar chart**.

SOLVE

5. Apply the **work–energy theorem** (Eq. 9.32 or 9.33).
6. Write down any **other equations**.
7. **Do algebra** before substitution.

PROBLEMS AND QUESTIONS

A = algebraic C = conceptual E = estimation G = graphical N = numerical

9-1 Energy Transfer to and from the Environment

1. C Pick an isolated system for the following scenarios while including the fewest number of objects as possible. **a.** A satellite in orbit around the Earth **b.** An airplane in flight **c.** A truck driving along the road **d.** A person jumping

9-2 Work Done by a Constant Force

2. E Estimate the work done by the air drag applied to a parachute when a paratrooper descends from an initial height of 3.00×10^3 ft above the ground.
3. C Riders on a merry-go-round experience a force accelerating them toward the center while they are in circular motion. Does this force do any work? Explain your answer.

Problems 4 and 5 are paired.

4. During practice, a hockey player passes the puck to his coach who is 6.5 m away as shown in Figure 9.2. The puck, which has a mass of 2.0 kg, was initially at rest. The hockey player exerts a constant 47.4-N force as his stick pushes the puck 0.25 m. Include only the puck in the system and assume the friction between the ice and the puck is negligible.
   a. C Draw a free-body diagram for the puck while it is in contact with the player's stick.
   b. N Find the work done by all the forces in your free-body diagram.
   c. N What is the speed of the puck as it leaves the player's stick?
   d. C In an all-star hockey game, the puck reaches speeds of 100 mph. Use that information to check your results to parts (a) through (c).
5. N In Problem 4, what work must the coach do on the puck to stop it?
6. A team of dockworkers does 5.25×10^3 J of total work while pushing a large 435-kg container by applying a constant horizontal force. The container is displaced through a horizontal distance of 15.0 m.
   a. N What is the magnitude of the force F applied by the dockworkers?
   b. C How would the motion of the crate change if the dockworkers applied a total force greater than F?
   c. C How would the motion of the crate change if the dockworkers applied a total force smaller than F?
7. N Kerry is pulling a 154-kg sled along a snowy, horizontal path with a 615-N force directed at an angle of 30.0° above the ground. If he pulls the sled over a distance of 30.0 m, how much work has Kerry performed on the sled?
8. N A 537-kg trailer is hitched to a truck. Find the work done by the truck on the trailer in each of the following cases. Assume rolling friction is negligible. **a.** The trailer is pulled at constant speed along a level road for 2.30 km. **b.** The trailer is accelerated from rest to a speed of 88.8 km/h. **c.** The trailer is pulled at constant speed along a road inclined at 12.5° for 2.30 km.
9. N Consider two objects, one that is moving to the east with a speed of 25 m/s and a second that is initially motionless. If each object has an identical mass and each object is to have a final speed of 25 m/s moving to the west, compare and contrast the necessary net amount of work that must be done on each object. Does one object require a greater amount of work? If so, which one?
10. N A helicopter rescues a trapped person of mass $m = 65.0$ kg from a flooded river by lifting the person vertically upward using a winch and rope. The person is pulled 12.0 m into the helicopter with a constant force that is 15% greater than the person's weight. **a.** Find the work done by each of the forces acting on the person. **b.** Assuming the survivor starts from rest, determine his speed upon reaching the helicopter.

9-3 Dot Product

11. N Find the dot product of the vectors $\vec{A} = 7.12\hat{\imath} + 2.00\hat{\jmath} - 3.90\hat{k}$ and $\vec{B} = 4.10\hat{\imath} - 11.00\hat{\jmath}$.
12. N An object is subject to a force $\vec{F} = (512\hat{\imath} - 134\hat{\jmath})$ N such that 10,125 J of work is performed on the object. If the object travels 25.0 m in the positive x direction while this work is performed, what must be the displacement of the object in the y direction?
13. A Show that the dot product of $\vec{A}$ and $\vec{B}$ is given by $D = A_xB_x + A_yB_y$. *Hint*: Refer to Figure 9.7.

Problems 14 and 15 are paired.

14. Vector $\vec{A} = (A\cos\alpha)\hat{\imath} + (A\sin\alpha)\hat{\jmath}$ and vector $\vec{B} = (B\cos\alpha)\hat{\imath} + (B\sin\alpha)\hat{\jmath}$.
    a. A What is the angle between vectors $\vec{A}$ and $\vec{B}$?
    b. G To check your answer, sketch the vectors on an x–y coordinate system.
15. Vector $\vec{A} = (A\cos 15°)\hat{\imath} + (A\sin 15°)\hat{\jmath}$ and vector $\vec{B} = (B\cos 15°)\hat{\imath} + (B\sin 15°)\hat{\jmath}$.
    a. N What is the angle between vectors $\vec{A}$ and $\vec{B}$?
    b. G To check your answer, sketch the vectors on an x–y coordinate system.

16. N In Figure P9.16, the magnitude of the vectors are $A = 6.00$ and $B = 3.00$. Find $\vec{A}\cdot\vec{B}$ in each case.

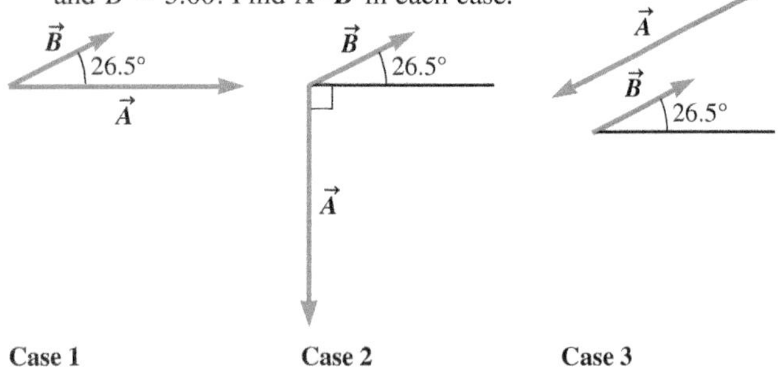

FIGURE P9.16

17. N A cart of mass $m = 1.30$ kg travels 2.4 m down a frictionless ramp inclined at 15° relative to a horizontal plane. Find the work done by gravity using **a.** Equation 9.15, **b.** Equation 9.16, and **c.** Equation 9.17.
18. A Show that the dot product of a vector with itself gives the square of the magnitude of that vector: $\vec{A}\cdot\vec{A} = A^2$ (Eq. 9.11).
19. A Show that the dot product is *commutative*: $\vec{A}\cdot\vec{B} = \vec{B}\cdot\vec{A}$ (Eq. 9.12).
20. A Show that the dot product is *associative*: $\vec{A}\cdot(\vec{B}+\vec{C}) = \vec{A}\cdot\vec{B} + \vec{A}\cdot\vec{C}$ (Eq. 9.13).

Problems 21 and 22 are paired.

21. A Figure P9.21 shows two vectors $\vec{A}$ and $\vec{B}$. Find the dot product $D = \vec{A}\cdot\vec{B}$ using Equation 9.8.

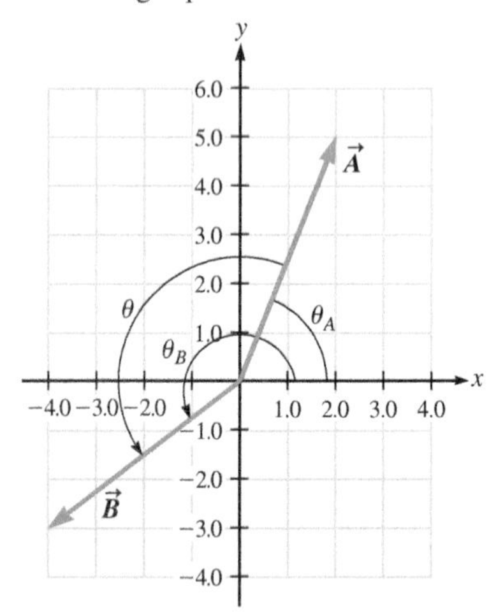

FIGURE P9.21 Problems 21 and 22.

22. A Figure P9.21 shows two vectors $\vec{A}$ and $\vec{B}$. Find the dot product $D = \vec{A}\cdot\vec{B}$ using Equation 9.10. If you did Problem 21, compare your results.
23. N A constant force of magnitude 4.75 N is exerted on an object. The force's direction is 60.0° counterclockwise from the positive x axis in the xy plane, and the object's displacement is $\Delta\vec{r} = (4.2\hat{\imath} - 2.1\hat{\jmath} + 1.6\hat{k})$ m. Calculate the work done by this force.

9-4 Work Done by a Nonconstant Force

24. C In three cases, a force acts on a particle, and the particle is displaced from an initial position to a final position. Figure 9.11 (page 255) shows the position-versus-force graphs, indicating the initial and final positions of the particle in each case. Find the work done by the force on the particle and sketch the force and displacement vectors along with the appropriate axis in each case.
25. N An object of mass $m = 5.8$ kg moves under the influence of one force. That force causes the object to move along a path given by $x = 6.0 + 5.0t + 2.0t^2$, where x is in meters and t is in seconds. Calculate the work done by the force on the object from $t = 2.0$ s to $t = 7.0$ s.

Problems 26 and 27 are paired.

26. N A nonconstant force is exerted on a particle as it moves in the positive direction along the x axis. Figure P9.26 shows a graph of this force F_x versus the particle's position x. Find the work done by this force on the particle as the particle moves as follows. **a.** From $x_i = 0$ to $x_f = 10.0$ m **b.** From $x_i = 10.0$ to $x_f = 20.0$ m **c.** From $x_i = 0$ to $x_f = 20.0$ m

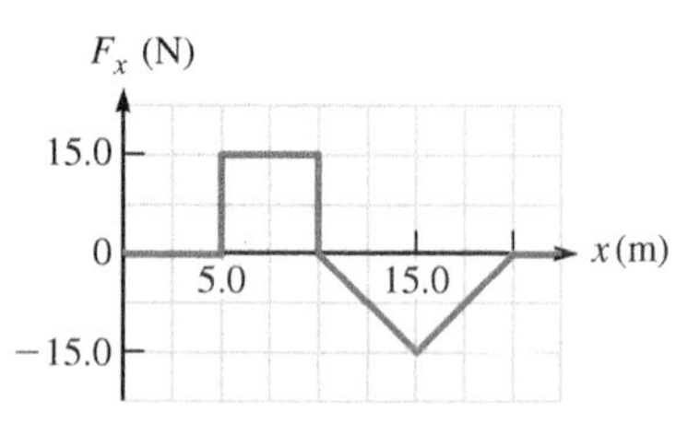

FIGURE P9.26 Problems 26 and 27.

27. N If the force in Figure P9.26 is the net force on the particle, find the change in the particle's kinetic energy as it moves as follows. **a.** From $x_i = 0$ to $x_f = 10.0$ m **b.** From $x_i = 10.0$ to $x_f = 20.0$ m **c.** From $x_i = 0$ to $x_f = 20.0$ m
28. A block on a frictionless table is attached to a horizontal spring of spring constant $k = 345$ N/m.
    a. G Sketch a graph of the spring force on the block F versus the block's position x. Use the conventional choice of placing the origin at the relaxed position of the spring.
    b. N Use your graph to estimate the work done by the spring on the block as the block moves from the relaxed position of the spring, $x_i = 0$, to $x_f = 35.0$ m.
    c. N Use Equation 9.18 to find the work done by the spring on the block as the block moves from $x_i = 0$ to $x_f = 35.0$ m.
29. N A force $\vec{F} = (4.000x^2\hat{\imath} - 6.000y\hat{\jmath})$ N acts on an object that moves 550.0 m along the direction pointing 22.5° clockwise from the positive y axis. Find the work done by the force on the object as it moves along the path.
30. A particle moves in the xy plane (Fig. P9.30) from the origin to a point having coordinates $x = 7.00$ m and $y = 4.00$ m under the influence of a force given by $\vec{F} = 3y^2\hat{\imath} + x\hat{\jmath}$.

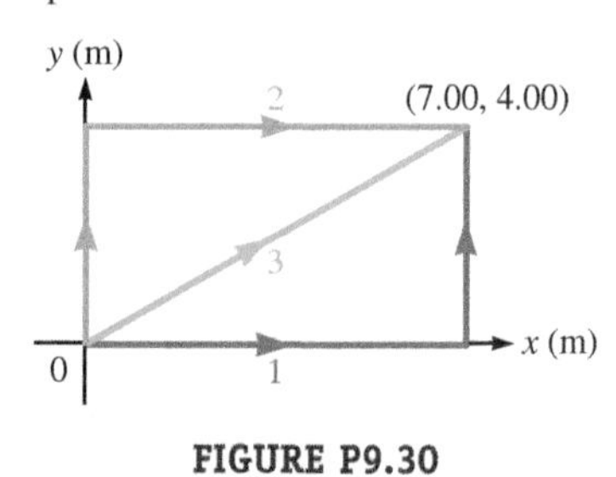

FIGURE P9.30

    a. N What is the work done on the particle by the force F if it moves along path 1 (shown in red)?
    b. N What is the work done on the particle by the force F if it moves along path 2 (shown in blue)?
    c. N What is the work done on the particle by the force F if it moves along path 3 (shown in green)?
    d. C Is the force F conservative or nonconservative? Explain.
31. A A small object is attached to two springs of the same length ℓ, but with different spring constants k_1 and k_2 as shown in Figure P9.31. Initially, both springs are relaxed. The object is then displaced straight along the x axis from x_i to x_f. Find an expression for the work done by the springs on the object.

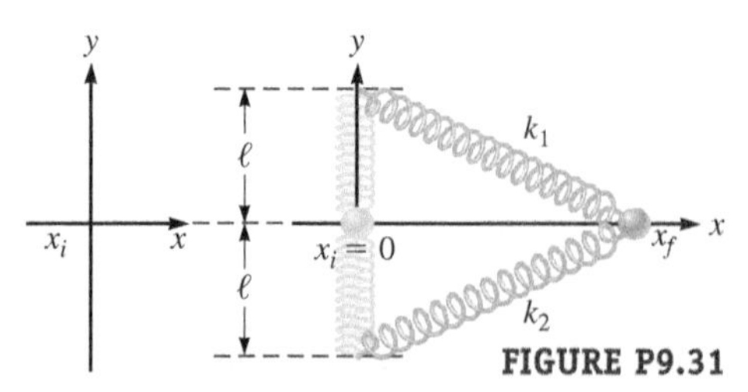

FIGURE P9.31

32. The force acting on a particle is given by $\vec{F}(r) = (A/r^2)\hat{r}$, where A is a positive constant.
   a. C What are the dimensions of A? Can the force be negative?
   b. G Sketch F versus r.
   c. A Find the work done by the force on the particle as it moves from $r_i = r_0$ to $r_f = 2r_0$.
   d. C Check your answer by comparing it with Example 9.5.

9-5 Conservative, Nonconservative, and Zero-Work Forces

33. A Show that the total work done by universal gravity on an object around a closed path is zero.
34. C In each of the situations described, determine whether or not the work done by the man is positive, negative, or zero and briefly explain your reasoning. **a.** A man pushes a child in a stroller, causing it to move horizontally across an ordinary floor, where friction is present. **b.** A man moves a child in a stroller across a frictionless floor so that the man and child glide together at a constant speed. **c.** A man prevents a child in a stroller from accelerating as it travels downhill. **d.** A man pushes the Great Wall of China, yet it remains motionless.
35. C **Review** In Example 9.6, Paul exerts a constant force on a sled that is constrained to move along a circular path. Because the sled is on ice, friction between the sled and the ice is negligible. Is the sled speeding up, slowing down, or maintaining its speed? Is the sled accelerating? If so, draw an overhead sketch and show the sled's acceleration, direction of motion along the path, and forces that are easily drawn from this perspective. If the sled isn't accelerating, show that the sum of the forces exerted on it is zero.

9-6 Particles, Objects, and Systems

36. N A 2.15-g hailstone, which can be modeled as a particle, falls a vertical distance of 145 m at constant speed. What is the work done on the hailstone by **a.** gravity and **b.** air resistance?

Problems 37 and 38 are paired.

37. A 6.0-lb dry concrete block is on a rubber conveyor belt in a factory. The block remains stationary on the conveyor belt as it goes 2.45 m along a 12.0° incline. The conveyor belt maintains a constant speed of 0.95 m/s.
   a. C Draw a free-body diagram for the concrete block.
   b. N Consider the Earth and the block to be the system. Find the work done by any external force on the system.
38. A 6.0-lb dry concrete block is on a rubber conveyor belt in a factory. The block remains stationary on the conveyor belt as it moves 2.45 m horizontally. The conveyor belt maintains a constant speed of 0.95 m/s.
   a. C Draw a free-body diagram for the concrete block.
   b. N Consider the Earth and the block to be the system. Find the work done by any external force on the system.
   c. C Compare your result with your result from Problem 37.
39. N A shopper weighs 3.00 kg of apples on a supermarket scale whose spring obeys Hooke's law and notes that the spring stretches a distance of 3.00 cm. **a.** What will the spring's extension be if 5.00 kg of oranges are weighed instead? **b.** What is the total amount of work that the shopper must do to stretch this spring a total distance of 7.00 cm beyond its relaxed position?
40. A In Figure 9.15, a chain of five links is dropped to the ground. Show that the work done on the whole chain by the Earth's gravity is

$$W = Mg\,\Delta y_{CM}$$

where M is the mass of the whole chain and Δy_{CM} is the displacement of the chain's center of mass. *Hint*: Start by treating each link in the chain as a particle of mass m. (You may wonder if it is okay to model each link as a particle. Each part of a single link has nearly the same displacement. For our discussion here, the slight variation in displacement of a single link is not important, so it is okay to model each individual link as a particle.) Work done on the whole chain is found by adding up the work done on the individual links.

9-7 Thermal Energy

41. A 4.50-kg wooden block slides 2.35 m down a wooden incline at constant velocity. (The coefficient of static friction is 0.400, and the coefficient of kinetic friction is 0.200.)
   a. N What is the angle of the incline?
   b. C When we are considering energy, why is it useful to include the surface of the incline in the system?
   c. N Calculate the increase in the system's thermal energy.
42. C The term *internal energy* could use improvement. In some sense, all energies are internal. Potential energy is the energy associated with the relative position of the particles *in* the system, and kinetic energy is associated with the *motion* of the particles in the system. Come up with a better term for internal energy and explain why it is better.
43. N CASE STUDY Consider a problem similar to the case study, but brought down to the Earth. A 6.0×10^3 kg bus traveling at 90.0 km/h skids to a halt on a wet (horizontal) road, where $\mu_k = 0.30$. How much thermal energy was generated by the friction?
44. C In calculating the work done on a deformable object, we must use the displacement of the point at which the force is applied. Some people choose a system in which the source of friction is external, which is not the choice we use in this book. Explain why it is especially complicated in the case of kinetic friction to find the displacement of the point of application.

Problems 45 and 46 are paired.

45. A A bullet flying horizontally hits a wooden block that is initially at rest on a frictionless, horizontal surface. The bullet gets stuck in the block, and the bullet–block system has a final speed v_f. Find the final speed of the bullet–block system in terms of the mass of the bullet m_b, the speed of the bullet before the collision v_b, the mass of the block m_{wb}, and the amount of thermal energy generated during the collision E_{th}.
46. A A bullet flying horizontally hits a wooden block that is moving before the collision in the same direction as the bullet with an initial speed v_{wb}. The bullet gets stuck in the block, and the bullet–block system has a final speed v_f. Find the final speed of the bullet–block system in terms of the mass of the bullet m_b, the speed of the bullet before the collision v_b, the mass of the block m_{wb}, the speed of the block before the collision v_{wb}, and the amount of thermal energy generated during the collision.
47. N You wish to make a simple amusement park ride in which a steel-wheeled roller-coaster car travels down one long slope, where rolling friction is negligible, and later slows to a stop through kinetic friction between the roller coaster's locked wheels sliding along a horizontal plastic (polystyrene) track. Assume the roller-coaster car (filled with passengers) has a mass of 750.0 kg and starts 80.7 m above the ground. **a. Review** Calculate how fast the car is going when it reaches the bottom of the hill. **b.** How much does the thermal energy of the system change during the stopping motion of the car? **c.** If the car stops in 230.66 m, what is the coefficient of kinetic friction between the wheels and the plastic stopping track?

9-8 Work–Energy Theorem

48. A Show that $K_i + U_{gi} + U_{ei} + W_{tot} = K_f + U_{gf} + U_{ef} + \Delta E_{th}$ (Eq. 9.33) is the same as $K_i + U_i = K_f + U_f$ (Eq. 8.14) in the

case of an isolated system in which only conservative forces act.

49. A Show that $K_i + U_{gi} + U_{ei} + W_{tot} = K_f + U_{gf} + U_{ef} + \Delta E_{th}$ (Eq. 9.33) is the same as $W_{tot} = \Delta K + \Delta U$ (Eq. 9.26) in the case of a particle.

Problems 50 and 51 are paired.

50. A small 0.65-kg box is launched from rest by a horizontal spring as shown in Figure P9.50. The block slides on a track down a hill and comes to rest at a distance d from the base of the hill. Kinetic friction between the box and the track is negligible on the hill, but the coefficient of kinetic friction between the box and the horizontal parts of track is 0.35. The spring has a spring constant of 345 N/m, and is compressed 30.0 cm with the box attached. The block remains on the track at all times.

a. C What would you include in the system? Explain your choice.

b. N Calculate d.

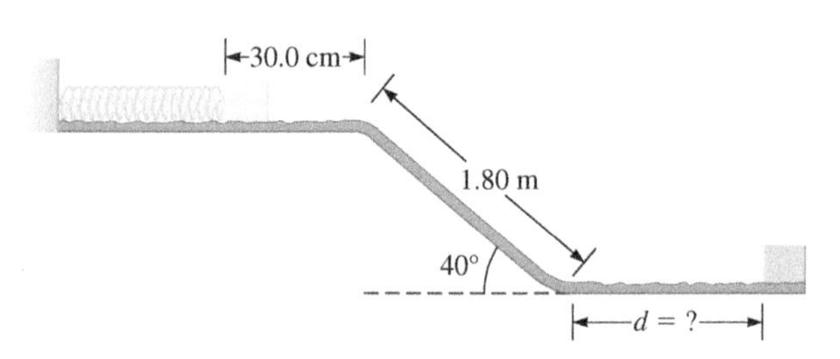

FIGURE P9.50 Problems 50 and 51. (Not to scale.)

51. A small 0.65-kg box is launched from rest by a horizontal spring as shown in Figure P9.50. The block slides on a track down a hill and comes to rest at a distance d from the base of the hill. The coefficient of kinetic friction between the box and the track is 0.35 along the entire track. The spring has a spring constant of 345 N/m, and is compressed 30.0 cm with the box attached.

a. C What would you include in the system? Explain your choice.

b. N Calculate d.

c. C Compare your answer with your answer to Problem 50 if you did that problem.

52. N A horizontal spring with force constant $k = 625$ N/m is attached to a wall at one end and to a block of mass $m = 3.00$ kg at the other end that rests on a horizontal surface. The block is released from rest from a position 3.50 cm beyond the spring's equilibrium position. **a.** If the surface is frictionless, what is the speed of the block as it passes through the equilibrium position? **b.** If the surface is rough and the coefficient of kinetic friction between the box and the surface is $\mu_k = 0.280$, what is the speed of the block as it passes through the equilibrium position?

53. N A box of mass $m = 2.00$ kg is dropped from rest onto a massless, vertical spring with spring constant $k = 2.40 \times 10^2$ N/m that is initially at its natural length. How far is the spring compressed by the box if the initial height of the box is 1.75 m above the top of the spring?

54. A person jumps straight upwards.

a. E Estimate the change in the person's internal energy.

b. C What is the form of the internal energy? What is the ultimate source of that energy?

55. N Return to Example 9.9 and use the result to find the tension in the rope.

Problems 56 and 57 are paired.

56. A Crall and Whipple design a loop-the-loop track for a small toy car (Fig. P9.56). The car starts at height y_i above the bottom of the loop, goes through the loop of radius R, and then travels along a flat, horizontal track. Rolling friction is negligible. What is the minimum height y_i from which the car can be released so that the car just barely makes it around the loop?

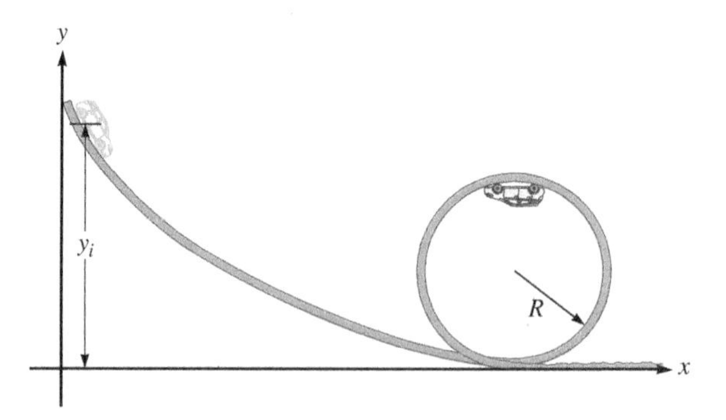

FIGURE P9.56 Problems 56 and 57.

57. N Crall and Whipple design a loop-the-loop track for a small toy car. The car starts at height y_i above the bottom of the loop, goes through the loop of radius R, and then travels along a flat, horizontal track before coming to rest. Rolling friction between the horizontal track and the car is significant, but it is negligible along the rest of the track. Assume Crall and Whipple release the car from the minimum height y_i (found in Problem 56). If the coefficient of rolling friction between the car and the horizontal track is 0.30 and the radius of the loop is 0.45 m, how far does the car travel along the horizontal track from the base of the loop before coming to rest?

58. C A pendulum is constructed by hanging a 0.500 kg-ball at the end of a 1.00-m-long lightweight, strong string. The other end of the string is attached to a support in the ceiling so that the ball is free to swing back and forth within a plane perpendicular to the ceiling. If this pendulum is displaced from the equilibrium position by 10.0° (with respect to vertical) and allowed to move freely, do you expect the pendulum to swing back and forth, reaching this angle again? Explain your answer.

59. N Calculate the force required to pull a stuffed toy duck (mass $m = 1.25$ kg) at a constant velocity of 3.6 m/s horizontally across the floor if the string is 50.0° above the horizontal. The coefficient of kinetic friction between the duck and the floor is 0.70.

9-9 Power

60. N What is the power output of a 75.0-kg student that climbs a knotted vertical rope 9.50 m in height at his high school gym at constant speed in 10.0 s?

61. N Return to Example 9.10 and the elevators in the Taipei 101 tower. What power is dissipated due to friction between the shaft and the elevator?

62. E Estimate the average power required for you to climb two floors in an office building in 10 s.

63. N An elevator motor moves a car with six people upward at a constant speed of 2.50 m/s. The mass of the elevator is 8.00×10^2 kg, and the average mass of a person on the elevator is about 80.0 kg. Calculate the electric power that must be delivered to lift the elevator car, assuming half of the necessary power delivered goes into thermal energy.

64. N A pail in a water well is hoisted by means of a frictionless winch, which consists of a spool and a hand crank. When Jill turns the winch at her fastest water-fetching rate, she can lift the pail the 25.0 m to the top in 12.2 s. Calculate the average power supplied by Jill's muscles during the upward ascent. Assume the pail of water when full has a mass of 6.82 kg.

65. N Figure P9.65A shows a crate attached to a rope that is extended over an ideal pulley. Boris pulls on the other end of the rope with a constant force until the crate has risen a total distance of 6.53 m (Fig. P9.65B). If the crate has a mass of 81.36 kg, what is the average power exerted by Boris, assuming he accomplishes the task in 5.33 s?

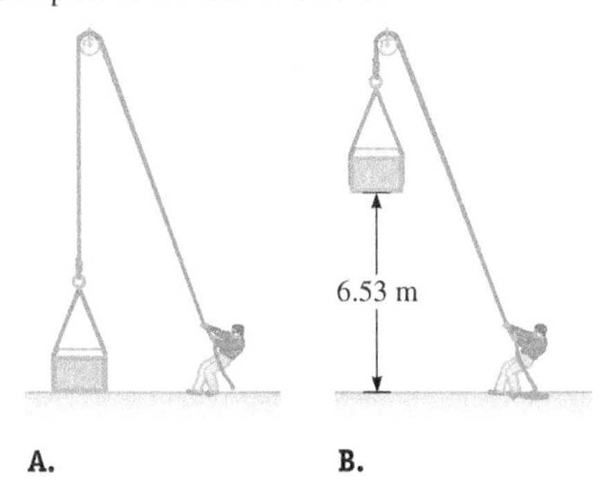

FIGURE P9.65

66. E, N While playing an automobile simulation computer program, you are test-driving a 2011 Ford F-150 pickup truck. The program allows several players to choose a vehicle and drive around in a virtual world, using real physical parameters and characteristics for the simulation. While waiting at a red light, another player pulls up in a 2011 Chevrolet Malibu and challenges you to a race. The race will be over as soon as someone reaches 60 mph. By estimating the average power of each engine in watts and the mass of each vehicle, determine who wins the race. (You may find it helpful to search for information about the engine of each vehicle on the Internet.)

General Problems

67. N Use the definition of the scalar dot product to determine the angle between each pair of vectors.
   a. $\vec{A} = -3\hat{\imath} - \hat{\jmath} + 4\hat{k}$ and $\vec{B} = 2\hat{\imath} + 2\hat{\jmath} + 2\hat{k}$
   b. $\vec{A} = \hat{\imath} + 2\hat{\jmath}$ and $\vec{B} = -2\hat{\jmath} - 3\hat{k}$
   c. $\vec{A} = 4\hat{\imath} + 2\hat{k}$ and $\vec{B} = -\hat{\imath} + 5\hat{\jmath} + 3\hat{k}$
68. C A physics professor keeps a yo-yo moving in a horizontal circle by its string as it is brought up from a canyon floor to the edge of a tall cliff by an elevator. Determine whether or not the work done on the yo-yo by the each of the following forces is positive, negative, or zero and briefly explain your reasoning. **a.** The elevator (contact force) **b.** The Earth's gravity **c.** The yo-yo string's tension
69. N A constant horizontal force of $F = 200.0$ N pushes a crate of mass $m = 25.0$ kg, initially at rest, along a rough, horizontal surface. The crate travels a total distance of 8.00 m, and the coefficient of kinetic friction between the crate and the surface is 0.250. **a.** What is the work done on the crate by the force F? **b.** How much does friction increase the internal energy of the crate–surface system? **c.** What is the change in kinetic energy of the crate? **d.** What is the final speed of the crate?
70. N Given the vectors $\vec{A} = \hat{\imath} + \hat{\jmath} + \hat{k}$, $\vec{B} = 4\hat{\imath} - 3\hat{\jmath} - 2\hat{k}$, and $\vec{C} = 2\hat{\imath} + 2\hat{\jmath}$, calculate $\vec{A}\cdot(\vec{C} - \vec{B})$.
71. N An object is subject to a nonconstant force $\vec{F} = (6x^3 - 2x)\hat{\imath}$ such that the force is in newtons when x is in meters. Determine the work done on the object as a result of this force as the object moves from $x = 0$ to $x = 1.00 \times 10^2$ m.
72. E Estimate the power required for a boxer to jump rope.
73. N A mother gently lifts her 4.55-kg baby from the crib at constant speed through a vertical distance of 85.0 cm.
   a. What is the work done on the baby by the mother?
   b. What is the total force exerted on the mother by the baby?
74. Kerry is pulling a 154-kg sled along a snowy, horizontal path with a 615-N force directed at an angle of 30.0° above the ground. He pulls the sled over a distance of 30.0 m, and the coefficient of kinetic friction between the sled and the ground is 0.0612.
   a. C Define the system to be used to account for the change in thermal energy as the sled is moved across the ground.
   b. N How much work is performed by Kerry as he pulls the sled over this distance?
   c. N What is the increase in thermal energy experienced by the system during the motion?
75. N A particle of mass $m = 2.50$ kg moving along the x axis from $x = 0$ to $x = 10.0$ m experiences a net conservative force in an isolated system given by $F = 3x - 5$, where F is in newtons and x is in meters. **a.** What is the work done on the particle by the force F? **b.** What is the change in the potential energy of the system during this motion? **c.** If the speed of the particle at the origin is 1.25 m/s, what is its kinetic energy at $x = 10.0$ m?
76. N A particle with mass $m = 750$ g is found to be moving with velocity $\vec{v} = (-4.00\hat{\imath} + 3.00\hat{\jmath})$ m/s. From the definition of the scalar product, $v^2 = \vec{v}\cdot\vec{v}$. **a.** What is the particle's kinetic energy at this time? **b.** If the particle's velocity changes to $\vec{v} = (5.00\hat{\imath} - 4.00\hat{\jmath})$ m/s, what is the net work done on the particle?
77. N Maria sets up a simple track for her toy block ($m = 0.25$ kg) as shown in Figure P9.77. She holds the block at the top of the track, 0.54 m above the bottom, and releases it from rest. **a.** Neglecting friction, what is the speed of the block when it reaches the bottom of the curve (the beginning of the horizontal section of track)? **b.** If friction is present on the horizontal section of track and the block comes to a stop after traveling 0.75 m along the bottom, what is the magnitude of the friction force acting on the block?

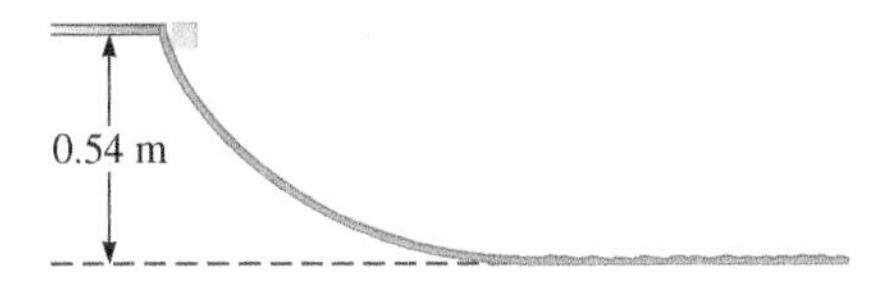

FIGURE P9.77

78. A A particle in an isolated system experiences a net conservative force given by $\vec{F} = (Ay - By^2)\hat{\jmath}$, where $\vec{F}$ is in newtons, y is in meters, and A and B are constants, with $U(y = 0) = 0$. **a.** What is the potential energy function $U(y)$ of the system? **b.** What is the change in potential energy of the system if the particle moves from $y = 1.00$ m to $y = 4.00$ m? **c.** What is the change in kinetic energy of the system if the particle moves from $y = 1.00$ m to $y = 4.00$ m?
79. N A snowmobile is pulling a sled ($m = 375$ kg) across the snowy tundra (Fig. P9.79). The snowmobile applies a pulling force of 2511 N at an angle of 15° with respect to the ground and pulls the sled over a distance of 50.0 m. **a.** How much work is done by the snowmobile? **b.** If the sled is also being acted on by a constant kinetic friction force of 253 N and it started from rest, what is the final speed of the sled?

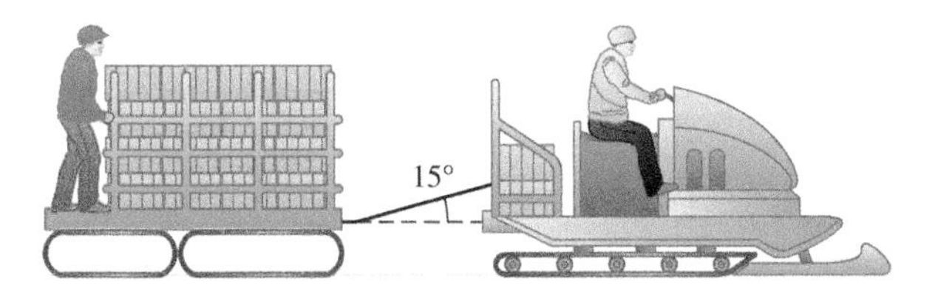

FIGURE P9.79

80. N A block of mass $m = 0.250$ kg is pressed against a spring resting on the bottom of a plane inclined an angle $\theta = 45.0°$ to the horizontal. The spring, which has a force constant of 955 N/m, is compressed a distance of 8.00 cm, and the block is released from rest. Consider the total energy of the spring–block–Earth system. **a.** What is the total distance the block moves from its initial position if the incline is frictionless? **b.** What is the total distance the block moves from its initial position if the coefficient of kinetic friction between the incline and the block is 0.330?

81. N On a movie set, an alien spacecraft is to be lifted to a height of 30.0 m for use in a scene. The 200.0-kg spacecraft is attached by ropes to a massless pulley on a crane, and four members of the film's construction crew lift the prop at constant speed by delivering 185 W of power each. If 20.0% of the mechanical energy delivered to the pulley is lost to friction, what is the time interval required to lift the spacecraft to the specified height?

82. N The elevators in the Empire State Building travel the 101 stories of the structure in less than 1 minute. In doing so, they accelerate from rest for the first 5.00 s of the upward trip to reach their cruising speed of 6.10 m/s. An elevator full of passengers has a mass of 1155 kg. What is the average power of each elevator's motor during **a.** the acceleration and **b.** the cruising phase of its motion?

83. N A spring-loaded toy gun is aimed vertically and fired after a spherical projectile of mass $m = 30.0$ g is loaded by compressing the spring with force constant $k = 825$ N/m a distance of 9.00 cm. As the projectile travels a total distance of 59.0 cm along the barrel of the gun, it experiences a friction force of 1.75 N. What is the height reached by the projectile above its initial location with the spring compressed?

84. N Shawn ($m = 45.0$ kg) rides his skateboard at a local skate park. He starts from rest at the top of the track as seen in Figure P9.84 and begins a descent down the track, always maintaining contact with the surface. The mass of the skateboard is negligible, as is friction except where noted. **a.** What is Shawn's speed when he reaches the bottom of the initial dip, 12.0 m below the starting point? **b.** He then ascends the other side of the dip to the top of a hill, 8.0 m above the ground. What is his speed when he reaches this point? **c.** As he begins to descend again, down a straight, 18.0-m-long slope, he slows his skateboard down by using friction on the tail of the board. He is able to produce a friction force with a magnitude of 120.0 N. What is the change in thermal energy of the board–rider–track system as he descends the 18.0-m length of track? **d.** What is his speed when he reaches the bottom (the end of the 18.0-m length of track)?

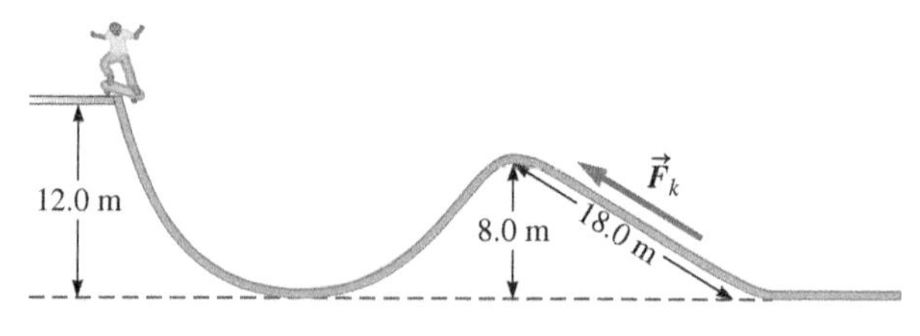

FIGURE P9.84

85. The motion of a box of mass $m = 2.00$ kg along the x axis can be described by the function $x = 4.00 + 3.00t^2 + 2.00t^3$, where x is in meters and t is in seconds.

a. A What is the kinetic energy of the box as a function of time?

b. A What are the acceleration of the box and the force acting on the box as a function of time?

c. A What is the power delivered to the box as a function of time?

d. N What is the work performed on the particle during the time interval $t = 1.00$ s to $t = 3.00$ s?

86. C A hardware store clerk cuts a 10.0-m length of a uniform chain. The clerk lays out the chain on a high horizontal shelf so that 4.00 m of its length is hanging over the edge of the shelf. If the coefficient of static friction between the shelf and the chain is 0.450, will the chain slide off the table? Why?

87. A The dot product can be used to find the scalar components of a vector. Show that $\vec{A} \cdot \hat{\imath} = A_x$ where vector $\vec{A}$ is three-dimensional.

88. E CASE STUDY Return to Example 9.12. Check the professor's assumption that work done by gravity on the meteor is negligible.

Systems of Particles and Conservation of Momentum

10

Underlying Principles

1. General form of Newton's second law
2. Conservation of momentum
3. Newton's second law applied to a system of particles and to one of constant mass

Major Concepts

1. Momentum (particle, system)
2. Center of mass
3. Closed and open systems

Special Cases: Rockets

First and second rocket equations

Key Questions

What is momentum, and when is it conserved?

What can be learned by applying the conservation of momentum that cannot be learned from the conservation of energy?

10-1 A second conservation principle 282

10-2 Momentum of a particle 283

10-3 Center of mass revisited 284

10-4 Systems of particles 287

10-5 Conservation of momentum 289

10-6 Case study: Rockets 292

10-7 Rocket thrust: An open system (optional) 296

In Chapters 8 and 9, we introduced a very powerful concept: the conservation of energy principle. With this concept, we showed how to calculate the escape speed of a planet, the size of a meteor, and the power requirement of the world's fastest elevator.

Energy is a scalar quantity, however, so the conservation of energy principle only helps us find *magnitudes*, but not *directions*, of quantities. For example, imagine that we have the task of programming a robot to play pool. We might program the robot to use conservation of energy to find the *speed* of a particular ball after the cue ball hits it. Whether the robot makes the shot, however, depends mostly on the *direction* of the ball's velocity, not on its speed.

Fortunately, energy is not the only conserved quantity. The vector quantity known as *momentum* (the product of mass and velocity) is also conserved. Because momentum is a vector, conservation of momentum may be used to find directional information. To play a winning game of pool, our robot must also be programmed using the principle of conservation of momentum. Of course, this principle applies to much more than just a game of pool. In this chapter, we apply conservation of momentum to problems ranging in scale from rockets to atomic nuclei. In Chapter 11, we will focus on a single application: collisions like those of billiard balls in the game of pool.

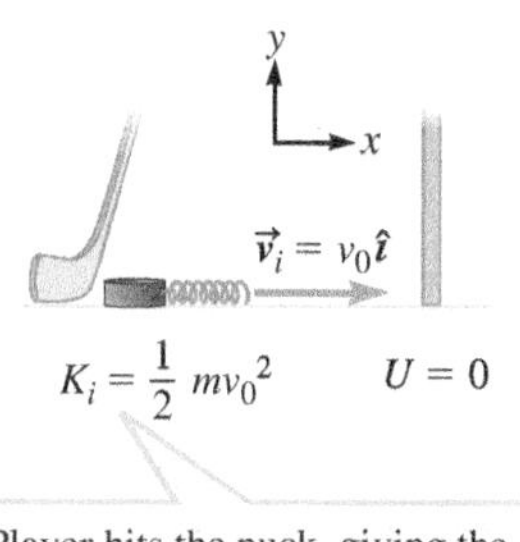

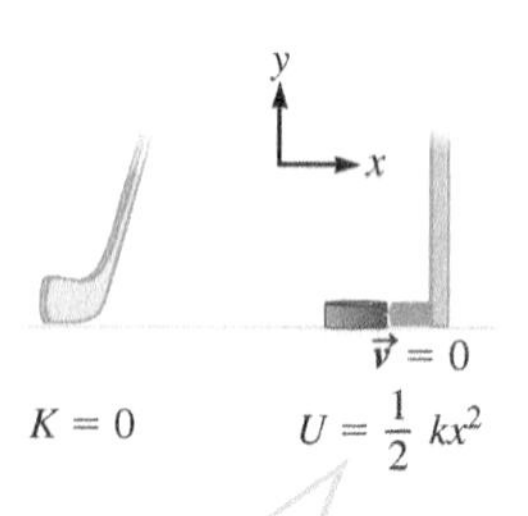

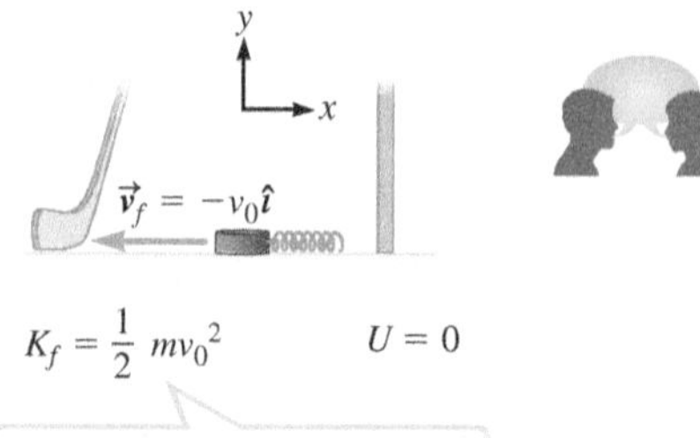

FIGURE 10.1 **A.** A hockey player sends a puck with a spring attached toward a wall. **B.** The puck–spring momentarily stops at the wall. **C.** The puck–spring returns to the player. Nothing in the conservation of energy principle can account for the change in the puck–spring system's direction.

10-1 A Second Conservation Principle

Newton's laws of motion are the foundation of mechanics. Working directly with these laws means taking a force approach to mechanics problems. The force approach involves vectors (such as force, acceleration, and velocity). The conservation of energy approach, however, involves only scalars (such as work, kinetic energy, potential energy, and speed), so directional information is lost.

Take, for example, a hockey player practicing with a puck that has a small spring attached to it (Fig. 10.1). The puck–spring system's kinetic energy is converted to potential energy as the spring compresses against the wall. As the spring stretches out again, the system's potential energy is converted to kinetic energy. As long as the puck has the same speed when it returns to the player, mechanical energy is conserved.

We see that the puck–spring system reverses direction, but nothing in the conservation of energy principle can account for the change in direction. We must expand the conservation approach to include a vector quantity: *momentum*.The expanded version of the conservation approach allows us to analyze complicated situations such as the following case study about rockets in space.

CASE STUDY Rockets

In 1919, Robert Goddard, an American rocketry pioneer, published a paper describing a method for accelerating a rocket into space. On January 13, 1920, the *New York Times* printed an editorial claiming that Goddard's rocket wouldn't work because, according to Newton's third law, the rocket would need something other than a vacuum to react against. The editorial says that Goddard "lacks the knowledge ladled out daily in high schools." After Apollo 11 was launched at the Moon, the *New York Times* printed an apology (July 17, 1969). A group of college students discusses the original 1920 editorial.

Shannon: Actually, the editorial makes sense to me. It seems obvious that chemical energy in the rocket is converted to kinetic energy, and any basketball player can do that. But here's the problem. If you jump up, you need the floor to push off of. If you are in space all by yourself, there's nothing to push against. You can contract your leg muscles and get nowhere. So, I figure that a rocket on takeoff pushes against the ground and later the air, but I can't figure out how it moves in the vacuum of space.

Avi: I think the rocket does not accelerate once it gets into space. Gravity is very weak out there because the rocket escaped the Earth. Say you wanted to send a rocket to the Moon. After the rocket escapes the Earth's atmosphere, it just coasts to the Moon at constant velocity until it gets close enough for the Moon to pull it in.

Cameron: There's more to it than that. When they want to send a spacecraft to Saturn, they use the gravity of an inner planet to slingshot the thing out that far.

Shannon: Avi, spacecrafts have to accelerate in space, and they can't just count on gravity to provide the necessary force. Think about when they do some kind of space dock. They have to guide the spacecraft in just like when you drive a car into a garage. You need to be able to turn the spacecraft, speed up, or slow down.

Cameron: Yeah, but that is all done by using gravity to line things up.

Avi: In fact, when spacecraft dock together, their gravity pulls them together. Then, when you need to separate them, they just push off each other.

Shannon: If that is the case, why did the *Times* have to apologize in 1969?

CONCEPT EXERCISE 10.1

CASE STUDY What Do You Already Know About Rockets?

Think about how spacecraft change course once they are in space. Do spacecraft use rocket thrusters to maneuver in space, or are rockets only used to launch the spacecraft?

10-2 Momentum of a Particle

To start thinking about momentum, consider a simple object that can be modeled as a particle: a hockey puck of mass m and velocity $\vec{v}$. The **momentum** $\vec{p}$ of a particle is defined as the product of its mass and its velocity:

MOMENTUM OF A PARTICLE

Major Concept

$$\vec{p} \equiv m\vec{v} \tag{10.1}$$

Momentum is a vector pointing in the same direction as the particle's velocity. Momentum has the dimensions of mass multiplied by speed, and there are no special SI units for momentum; it is reported in kg · m/s. Because the velocity of a particle depends on the reference frame, momentum also depends on the reference frame.

The momentum given by Equation 10.1 is sometimes referred to as **linear** or **translational momentum** to distinguish it from *angular* momentum defined in Chapter 13. The plural of *momentum* is *momenta*.

The definition of momentum gives us another way to write Newton's second law. We start with Newton's second law for a particle and use $\vec{a} = d\vec{v}/dt$ (Eq. 4.14):

$$\vec{F}_{\text{tot}} \equiv \sum \vec{F} = m\vec{a} = m\frac{d\vec{v}}{dt}$$

We are considering a particle whose mass does not change, so bringing m inside the derivative makes no difference, and

$$\sum \vec{F} = \frac{d(m\vec{v})}{dt}$$

According to Equation 10.1, $m\vec{v}$ is the momentum of the particle. We have just found that the sum of the forces acting on a particle is the rate of change of the particle's momentum.

GENERAL FORM OF NEWTON'S SECOND LAW

Underlying Principle

$$\sum \vec{F} = \frac{d\vec{p}}{dt} \tag{10.2}$$

Not only is Equation 10.2 another way to express Newton's second law, it is also a more *general* way. This form of Newton's second law is important when we study a system of particles (as in the next section). Equation 10.2, $\vec{F}_{\text{tot}} = d\vec{p}/dt$, holds even for systems whose mass changes, and $\vec{F}_{\text{tot}} = m\vec{a}$ is only good in the special case of a system with constant mass. Also, $\vec{F}_{\text{tot}} = d\vec{p}/dt$ gives us a fourth way to express Newton's *first* law (the law of inertia): *A* (net) *force is required to change a particle's momentum.*

You may hear physicists say that Newton expressed his second law as Equation 10.2, but history of science scholars don't believe it. This discrepancy is probably due to physicists' beliefs that $\vec{F}_{\text{tot}} = d\vec{p}/dt$ is more fundamental than $\vec{F}_{\text{tot}} = m\vec{a}$.

EXAMPLE 10.1 "Look Out Below!"

Two balls are dropped from the top of a building at the same time. One ball has a mass of 2.5 kg, and the other has a mass of 5.0 kg. Drag on the balls is negligible, and they land on the ground 5.3 s after they were released. Find the momentum of each ball just as it reaches the ground.

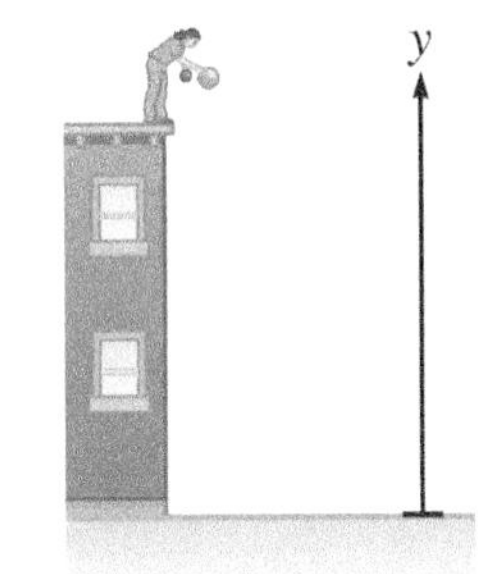

FIGURE 10.2

INTERPRET and ANTICIPATE

Momentum is a vector, so we must define a coordinate system to specify its direction as in Figure 10.2. We expect to find two numerical solutions of the form $\vec{p} = -(\_\_\_\_)\,\hat{\jmath}$ kg · m/s.

There are at least two ways to solve for the momentum of each ball. Method 1 uses $\vec{p} \equiv m\vec{v}$ (Eq. 10.1), requiring kinematics to find the velocity of a ball just before it lands. Method 2 involves integrating $\sum \vec{F} = d\vec{p}/dt$ (Eq. 10.2). We start with method 1 because we are more familiar with it; after we have used method 2, we will compare our results. (A third method is used in Problem 12.)

SOLVE

Method 1

In Equation 2.9, the initial speed is zero, and the acceleration is the free-fall acceleration. We find the final velocity of each ball as it lands.

$$v_y = v_{y0} + a_y t = -gt \tag{2.9}$$

$$\vec{v}_f = -gt\hat{\jmath} = -(9.81\ \text{m/s}^2)(5.3\ \text{s})\hat{\jmath} = -52\hat{\jmath}\ \text{m/s}$$

Example continues on page 283 ▶

Substitute this result into Equation 10.1 to find each ball's momentum, using subscript 1 for the 2.5-kg ball and subscript 2 for the 5.0-kg ball. Both results have the form we expected.	$\vec{p}_f = m\vec{v}_f$ (10.1) $\vec{p}_{1f} = -(2.5\,\text{kg})(52\,\text{m/s})\hat{\jmath} = -1.3 \times 10^2\hat{\jmath}\,\text{kg}\cdot\text{m/s}$ $\vec{p}_{2f} = -(5.0\,\text{kg})(52\,\text{m/s})\hat{\jmath} = -2.6 \times 10^2\hat{\jmath}\,\text{kg}\cdot\text{m/s}$
SOLVE Method 2 Equation 10.2 is simplified because only one force—the Earth's gravity—acts on either ball.	$\Sigma\vec{F} = \vec{F}_g = \dfrac{d\vec{p}}{dt}$ (10.2) $-mg\hat{\jmath} = \dfrac{d\vec{p}}{dt}$
Multiply each side by dt and integrate from the instant the ball is released ($t_i = 0$) to the instant the ball lands ($t_f = t$). Because the ball was released from rest, its initial momentum is $\vec{p}_i = 0$. Because $-mg\hat{\jmath}$ is constant, it is pulled outside the integral.	$-(mg\,dt)\hat{\jmath} = d\vec{p}$ $\int_0^t -(mg\,dt)\hat{\jmath} = -(mg)\hat{\jmath}\int_0^t dt = \int_0^{\vec{p}_f} d\vec{p}$ $-(mgt)\hat{\jmath} = \vec{p}_f$
It takes one ball 5.3 s to reach the ground, which we use to find the momentum of each.	$\vec{p}_{1f} = -(2.5\,\text{kg})(9.81\,\text{m/s}^2)(5.3\,\text{s})\hat{\jmath} = -1.3 \times 10^2\hat{\jmath}\,\text{kg}\cdot\text{m/s}$ $\vec{p}_{2f} = -(5.0\,\text{kg})(9.81\,\text{m/s}^2)(5.3\,\text{s})\hat{\jmath} = -2.6 \times 10^2\hat{\jmath}\,\text{kg}\cdot\text{m/s}$

CHECK and THINK

We found the same results using two different methods. As we have come to expect, both balls land with the same velocity. The ball with twice the mass, however, has twice as much momentum. A commonly held belief is that when two different objects are dropped from a great height (in the absence of drag), the heavier one is faster. These sorts of mistaken ideas are usually based on experience that is not identified with the correct physical interpretation. Imagine catching these two balls; clearly, the heavier one is harder to stop. It is incorrect to connect this experience to the speed of the balls. Instead, we should say that it is harder to stop the heavier ball because it has a greater momentum.

10-3 Center of Mass Revisited

When you think about the case of a falling ball, the ideas of momentum and the general form of Newton's second law (Eq. 10.2) don't seem very important. (We studied problems like that in Chapter 2.) The real power of the concept of momentum comes out when we analyze a system of two or more particles. To do so, we need to know how to find a system's center of mass.

The **center of mass (CM)** is a point associated with any system such that when a net external force is exerted on the system, the center of mass accelerates according to Newton's second law (Section 9-6). When a system may be modeled as a particle, the center of mass is just the position of the particle, and there is no need to distinguish the center-of-mass motion from the motion of the rest of the system. For example, we often are able to model a walking person as a particle. However, when a person walks, her arms and legs do not move with the same velocity and acceleration as her center of mass. If we need to know how much energy a person needs to walk a great distance, we must take into account the motion of her arms and legs, as well as her center-of-mass motion.

To locate the center of mass, we start by considering a simple two-particle system (Figure 10.3A).We choose a coordinate system with the origin on particle 1. The x component of particle 2's position is then x_2. For this coordinate system, the x component of the center-of-mass position is defined to be

$$x_{CM} \equiv \left(\frac{m_2}{m_1 + m_2}\right)x_2$$

Because we placed the origin on particle 1, x_2 is the distance between the two particles. The factor $m_2/(m_1 + m_2)$ is a unitless fraction between 0 and 1. If the two particles have equal mass, the fraction is $\frac{1}{2}$, and the center of mass is halfway between the two particles. If $m_2 > m_1$, the fraction is greater than $\frac{1}{2}$, and the center of mass is closer to particle 2. Of course, if $m_2 < m_1$, the center of mass is closer to particle 1.

Sometimes, it is not convenient to place the origin on one of the particles. A change in coordinate system, of course, cannot change the physical location of the center of mass, but it changes that point's coordinates relative to the new origin. Using the coordinate system in Figure 10.3B, the center of mass of the same two-particle system is now

$$x_{\mathrm{CM}} = \frac{m_1 x_1 + m_2 x_2}{m_1 + m_2}$$

If there are more than two particles, we find the x component of the center-of-mass position in a similar way. For n particles, we have

$$x_{\mathrm{CM}} = \frac{m_1 x_1 + m_2 x_2 + m_3 x_3 + \cdots + m_n x_n}{m_1 + m_2 + m_3 + \cdots + m_n}$$

The sum in the denominator is the total mass M of the system. Using summation notation, we can write a compact expression for the x component of the center-of-mass position:

$$x_{\mathrm{CM}} = \frac{1}{M}\sum_{j=1}^{n} m_j x_j \tag{10.3}$$

According to this expression, the center of mass is a *weighted average of position*. The term *weighted average* does not have anything to do with the gravitational force; rather, it is a mathematical description meaning that the average takes into account the relative importance of the quantities being averaged. To understand this term, imagine students standing in a line along one wall of a classroom. You choose an x axis that runs parallel to the wall and whose origin is at one corner, and then you measure the position of each student. The average position of the students is found by adding up the position of each student and dividing by the total number of students. In this case, each student's position is given equal importance, or *weight*, in your calculation. You could instead find a weighted average based on some property of the students such as their physics grades. You would assign "weight" to each student's position by multiplying each position by his or her grade and then find the average of these *weighted* positions. The students with the best grades would then count more when you find the average position. The center of mass is weighted based on the relative mass of each particle in the system; mathematically, the weighting is done by multiplying each term by m_j/M. So, the more massive particles are more important when finding the center-of-mass position.

We started by thinking about two particles lying along a single x axis. A more complicated system may have particles located in two or three dimensions as in Figure 10.3C. Similar expressions are written for the y and z components of the center-of-mass position:

$$y_{\mathrm{CM}} = \frac{1}{M}\sum_{j=1}^{n} m_j y_j \qquad z_{\mathrm{CM}} = \frac{1}{M}\sum_{j=1}^{n} m_j z_j$$

The position vector of the center of mass is then

$$\vec{r}_{\mathrm{CM}} = x_{\mathrm{CM}}\hat{\imath} + y_{\mathrm{CM}}\hat{\jmath} + z_{\mathrm{CM}}\hat{k} \tag{10.4}$$

which is written in compact form as

$$\vec{r}_{\mathrm{CM}} = \frac{1}{M}\sum_{j=1}^{n} m_j \vec{r}_j \tag{10.5}$$

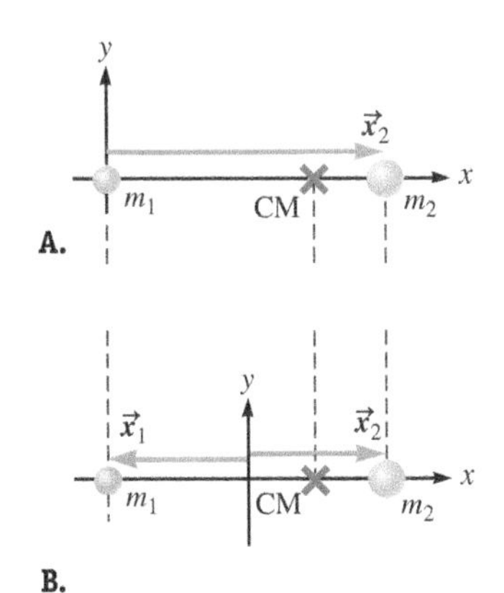

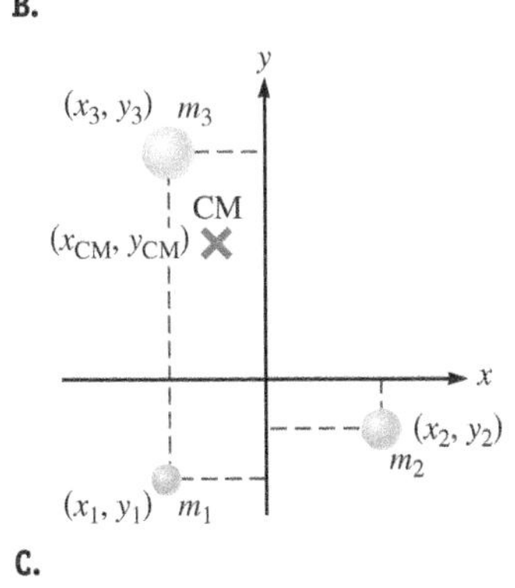

FIGURE 10.3 A. A two-particle system with the origin on particle 1. **B.** A two-particle system with the origin between the two particles. **C.** A three-particle system with the origin arbitrarily placed.

$x_{1 \text{ or } 2}$ is the scalar x component of the particle's position, which may be positive or negative. With the origin as shown in Figure 10.3B, x_1 is negative and x_2 is positive.

CENTER OF MASS Major Concept

CONCEPT EXERCISE 10.2

We often think of the Earth and the Moon as a two-particle system with the center of mass located at the center of the Earth. Find the center of mass of the Earth–Moon system and comment on the validity of the usual assumption that it is at the center of the Earth. *Hint*: The Earth and the Moon's masses are 5.98×10^{24} kg and 7.36×10^{22} kg, respectively; their center-to-center distance is 3.84×10^{8} m; and the Earth's radius is 6.37×10^{6} m.

EXAMPLE 10.2 A Strange Game of Pool

Two billiard balls of mass 0.170 kg and one bowling ball of mass 4.54 kg are on a pool table. Use the coordinate system in Figure 10.4 to find the center-of-mass position of the three-ball system. Report your answer to two significant figures.

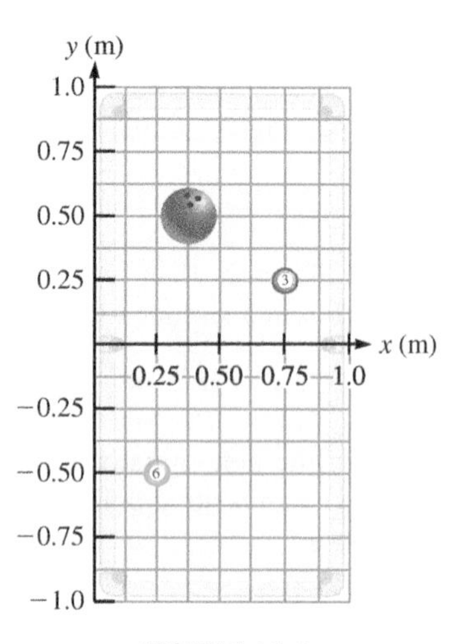

FIGURE 10.4

INTERPRET and ANTICIPATE

The bowling ball is much more massive than the billiard balls, so we expect the center of mass to be close to the bowling ball. Our result should be in the form $\vec{r}_{CM} = (\_\_\_\_\hat{\imath} + \_\_\_\_\hat{\jmath})\text{ m}$.

SOLVE

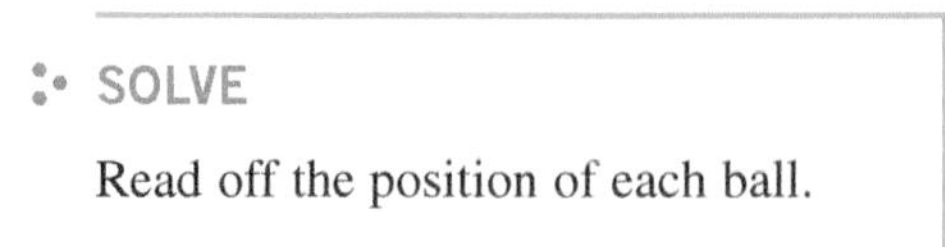

Read off the position of each ball.

$$\vec{r}_3 = (0.75\hat{\imath} + 0.25\hat{\jmath})\text{ m}$$
$$\vec{r}_6 = (0.25\hat{\imath} - 0.50\hat{\jmath})\text{ m}$$
$$\vec{r}_b = (0.375\hat{\imath} + 0.50\hat{\jmath})\text{ m}$$

Find the total mass of the system.

$$M = 2 \times (0.170\text{ kg}) + 4.54\text{ kg}$$
$$M = 4.88\text{ kg}$$

Use Equation 10.3 to find the x component of the center-of-mass position. Then, find the y component in a similar manner.

$$x_{CM} = \frac{1}{M}\sum_{j=1}^{n} m_j x_j \quad (10.3)$$

$$x_{CM} = \frac{(0.170\text{ kg})(0.75\text{ m}) + (0.170\text{ kg})(0.25\text{ m}) + (4.54\text{ kg})(0.375\text{ m})}{4.88\text{ kg}} = 0.41\text{ m}$$

$$y_{CM} = \frac{1}{M}\sum_{i=1}^{n} m_i y_i$$

$$y_{CM} = \frac{(0.170\text{ kg})(0.25\text{ m}) + (0.170\text{ kg})(-0.50\text{ m}) + (4.54\text{ kg})(0.50\text{ m})}{4.88\text{ kg}} = 0.46\text{ m}$$

Write the center-of-mass position in component form using Equation 10.4.

$$\vec{r} = x_{CM}\hat{\imath} + y_{CM}\hat{\jmath} \quad (10.4)$$
$$\vec{r}_{CM} = (0.41\hat{\imath} + 0.46\hat{\jmath})\text{ m}$$

CHECK and THINK

As expected, the center of mass is near the bowling ball.

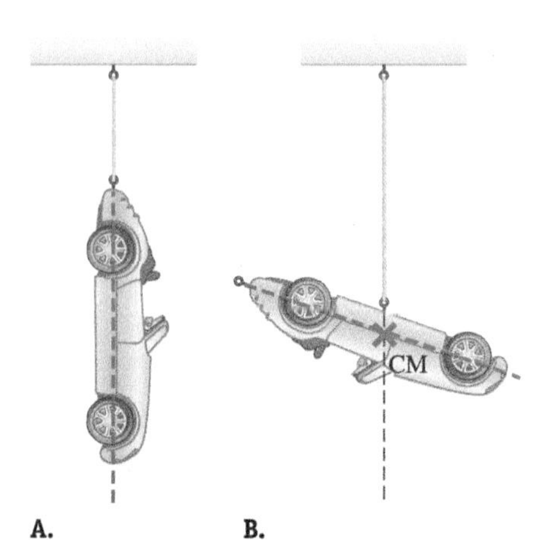

FIGURE 10.5 Finding the car's center of mass by suspending it from two separate points.

Equations 10.3 through 10.5 are useful for finding the center of mass for a system of a small number of particles, but how do we find the center of mass for an extended object such as a car? In that case, we imagine slicing the object into a great number of very small pieces. Each piece is modeled as a particle of mass dm. Then, when we find the weighted average over this great number of slices, the sums become integrals:

$$x_{CM} = \frac{1}{M}\int x\,dm \qquad y_{CM} = \frac{1}{M}\int y\,dm \qquad z_{CM} = \frac{1}{M}\int z\,dm$$

Often, there is no need to actually do these integrals. For an object with uniform density and a regular shape, we exploit the object's symmetry. For example, for a sphere of uniform density, the center of mass is at the center of the sphere. For a donut shape of uniform density, the center of mass is in the center of the hole.

To find the center of mass of an irregularly shaped object or one with nonuniform density, we perform an experiment. External forces applied to an extended object act as though all the mass of the object were concentrated at the center of mass. So, if we suspend an object from a rope, two forces—gravity and the tension force—act on the object. The center of mass must be somewhere along the line that runs through the length of the rope (Problem 14.63).

For example, Figure 10.5A shows a car suspended by a rope; the car's center of mass must be somewhere on the line shown. To find the exact location of the center of mass, we need to suspend the car from a different point (Fig. 10.5B); now we know that the center of mass must be somewhere on the new line. The two lines intersect at one point, and this point must be the center of mass of the car. There is nothing unique or special about the two points from which we suspended the car; any two well-separated points work.

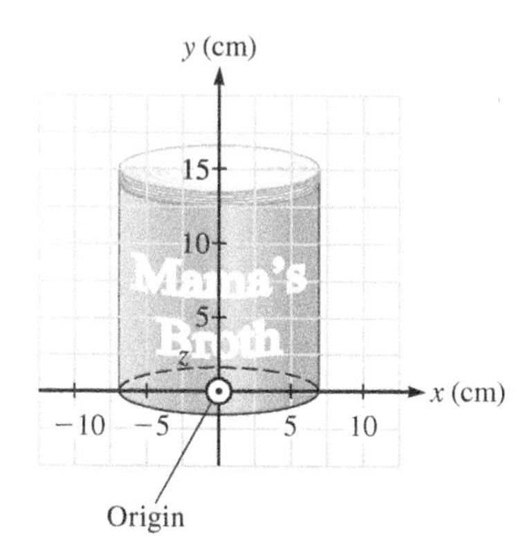

FIGURE 10.6 The z axis points out of the page, directly at the viewer.

CONCEPT EXERCISE 10.3

The car's trunk in Figure 10.5 is empty. What would happen to the center-of-mass position if instead the trunk were full of heavy rocks?

CONCEPT EXERCISE 10.4

Unlike a can of vegetable soup, a can of broth has uniform density. Find the center-of-mass position (x_{CM}, y_{CM}, z_{CM}) for the can of broth in Figure 10.6 assuming it is vacuum sealed so that there is no air bubble. The origin of the coordinate system is in the center of the can's circular base.

10-4 Systems of Particles

Now that we know how to locate the center of mass, we are ready to apply Newton's second law to a system of particles. If an external force is applied to the system, the center of mass accelerates as if all the mass of the system were concentrated at that point. Newton's second law for the system is written as

NEWTON'S SECOND LAW APPLIED TO A SYSTEM WITH CONSTANT MASS

Underlying Principle

$$\sum \vec{F}_{ext} = M\vec{a}_{CM} \tag{10.6}$$

There are two subtle but important facts about Eq. 10.6:

1. It is possible that the net external force may be zero, in which case the center-of-mass acceleration $\vec{a}_{CM}$ is zero. In fact, choosing a system such that the net external force *is* zero is an important part of applying the conservation of momentum principle.
2. The equation $\sum \vec{F}_{ext} = M\vec{a}_{CM}$ involves the center-of-mass acceleration $\vec{a}_{CM}$, which is generally *not* the same as the acceleration of any particle within the system. So, for example, the center of mass may have a constant velocity, but the particles within the system may be accelerating.

EXAMPLE 10.3 Back to the Drawing Board

Zak, a novice artillery enthusiast, decides to launch a projectile at an angle $\theta = 85°$. The total mass of the projectile is $M = 0.45$ kg. Fortunately, he has taken the precaution of launching into an empty farm field, but unfortunately, there is a malfunction. At the peak of the projectile's trajectory, the projectile splits in two sections. Zak finds the back section ($m_1 = 0.15$ kg) 155 m from the launch point (Fig. 10.7). If the launch speed was $v_0 = 125$ m/s and drag is negligible, where should he look for the front section ($m_2 = 0.30$ kg)?

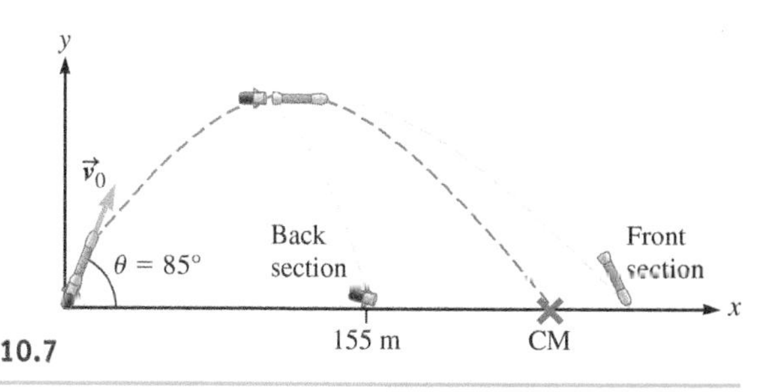

FIGURE 10.7

INTERPRET and ANTICIPATE

If drag is negligible, after launching only gravity acts on the center of mass, which acts as a simple projectile. We can easily find where the center of mass lands, although there will not be any physical material there. Because we know where the back section landed, we can calculate where the front section landed. We expect to find a numerical result of the form $\vec{x}_2 =$ _____ $\hat{\imath}$, and we expect that the front section is farther from the origin than either the center of mass or the back section.

Example continues on page 288 ▶

SOLVE
Find the final center-of-mass position from the range of the projectile (Eq. 4.28).

$$R = \frac{v_0^2}{g}\sin 2\theta \qquad (4.28)$$

$$R = \frac{(125\text{ m/s})^2}{9.81\text{ m/s}^2}\sin[2(85)°] = 2.76 \times 10^2\text{ m}$$

$$\vec{x}_{CM} = 2.76 \times 10^2\hat{\imath}\text{ m}$$

Model the projectile as a two-particle system. Particle 1 is the back section. The front section is particle 2. Solve Equation 10.3 for $\vec{x}_2$, particle 2's position.

$$x_{CM} = \frac{1}{M}(m_1x_1 + m_2x_2) \qquad (10.3)$$

$$m_2x_2 = Mx_{CM} - m_1x_1$$

$$x_2 = \frac{Mx_{CM} - m_1x_1}{m_2}$$

$$x_2 = \frac{(0.45\text{ kg})(276\text{ m}) - (0.15\text{ kg})(155\text{ m})}{0.30\text{ kg}}$$

$$\vec{x}_2 = 3.4 \times 10^2\hat{\imath}\text{ m}$$

CHECK and THINK
Our answer matches our expectations. It might seem bizarre that if Zak were to go to where the center of mass of the system lands ($\vec{x}_{CM} = 2.76 \times 10^2\hat{\imath}$ m), he would not find any part of the projectile there. Even if there is no physical matter at the center of mass, it still follows Newton's laws of motion. The front section travels farther if the back section falls off than if the entire projectile remains intact and follows the center-of-mass trajectory. This fact will be helpful when we return to the case study.

Momentum of a System of Particles

MOMENTUM OF A SYSTEM OF PARTICLES
Major Concept

The concept of center of mass is also important when we consider the momentum of a system of particles. For a system of n particles, the total momentum is the vector sum of the momentum of each particle and is given by

$$\vec{p}_{tot} = \sum_{j=1}^{n}\vec{p}_j \qquad (10.7)$$

In Problem 81 you will show that the total momentum of a system of particles equals the momentum of the center of mass:

$$\vec{p}_{tot} = \sum_{j=1}^{n}\vec{p}_j = \vec{p}_{CM} \qquad (10.8)$$

If the center of mass is not accelerating, there is an inertial reference frame in which the center of mass is at rest ($\vec{v}_{CM} = 0$) known as the *center-of-mass frame*. In the center-of-mass frame, $\vec{p}_{CM} = 0$; therefore, the total momentum of the system is zero ($\vec{p}_{tot} = 0$).

EXAMPLE 10.4 Skating Partners

Consider a two-particle system consisting of two ice skaters gliding on the ice. Paul's mass is 76.0 kg, and his speed is $v_P = 7.30$ m/s; Lil's mass is 52.0 kg, and her speed is $v_L = 6.50$ m/s. The direction of each skater's velocity is shown in Figure 10.8. Friction and drag are negligible, so the skaters' velocities are constant.

A Use the coordinate system shown in Figure 10.8 to find the total momentum of the system with respect to the ice.

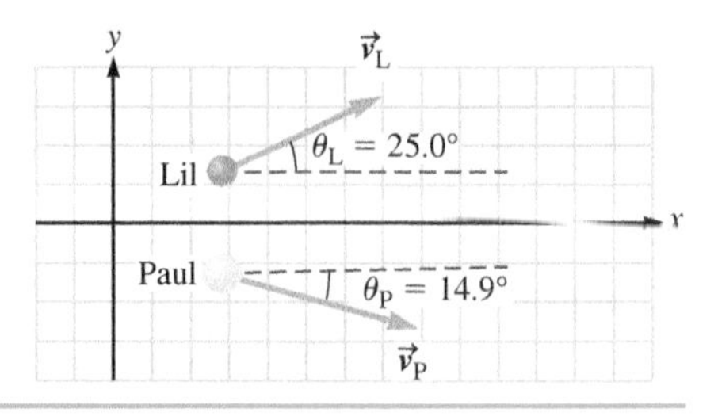

FIGURE 10.8 Velocity vectors for Paul and Lil.

INTERPRET and ANTICIPATE
We will find the momentum of each skater with respect to the ice and then add our results. We can develop our expectations graphically.

Find the magnitude of each skater's momentum by taking the magnitude of $\vec{p} \equiv m\vec{v}$ (Eq. 10.1). The subscript L stands for *L*il, and the subscript P stands for *P*aul.

$$p_L = m_L v_L = (52.0\ \text{kg})(6.50\ \text{m/s}) = 338\ \text{kg}\cdot\text{m/s}$$
$$p_P = m_P v_P = (76.0\ \text{kg})(7.30\ \text{m/s}) = 555\ \text{kg}\cdot\text{m/s}$$

Momentum is parallel to velocity. Sketch the two momentum vectors on the coordinate grid (Fig. 10.9). The y components of momentum cancel, so we expect the total momentum to be in the form $\vec{p}_{tot}$ = _____ $\hat{\imath}$ kg · m/s.

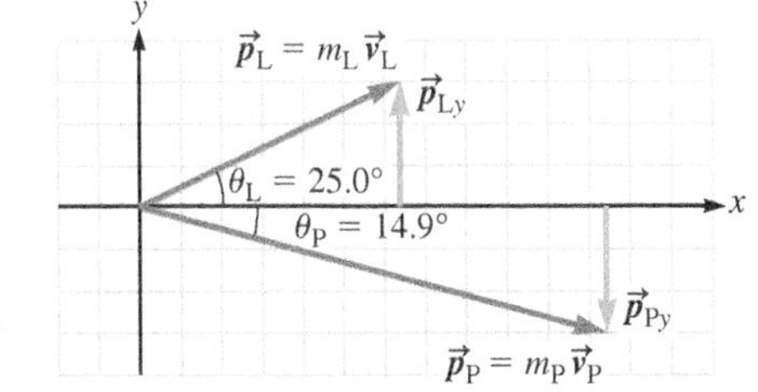

FIGURE 10.9 Momentum vectors for Paul and Lil.

:• SOLVE

Use the magnitudes found above and the directions given in the figures to write the momentum of each skater in component form.

$$\vec{p}_L = (p_L\cos\theta_L)\hat{\imath} + (p_L\sin\theta_L)\hat{\jmath}$$
$$\vec{p}_L = (338\ \text{kg}\cdot\text{m/s})(\cos 25.0°)\hat{\imath} + (338\ \text{kg}\cdot\text{m/s})(\sin 25.0°)\hat{\jmath}$$
$$\vec{p}_L = (306\hat{\imath} + 143\hat{\jmath})\ \text{kg}\cdot\text{m/s}$$
$$\vec{p}_P = (p_P\cos\theta_P)\hat{\imath} - (p_P\sin\theta_P)\hat{\jmath}$$
$$\vec{p}_P = (555\ \text{kg}\cdot\text{m/s})(\cos 14.9°)\hat{\imath} - (555\ \text{kg}\cdot\text{m/s})(\sin 14.9°)\hat{\jmath}$$
$$\vec{p}_P = (536\hat{\imath} - 143\hat{\jmath})\ \text{kg}\cdot\text{m/s}$$

Now add the two momentum vectors to find the total momentum.

$$\vec{p}_{tot} = (306\ \text{kg}\cdot\text{m/s} + 536\ \text{kg}\cdot\text{m/s})\hat{\imath} + (143\ \text{kg}\cdot\text{m/s} - 143\ \text{kg}\cdot\text{m/s})\hat{\jmath}$$
$$\vec{p}_{tot} = 842\hat{\imath}\ \text{kg}\cdot\text{m/s}$$

:• CHECK and THINK

As expected, the total momentum points in the positive x direction.

B What is the velocity of the center of mass with respect to the ice?

:• INTERPRET and ANTICIPATE

The total momentum of the system and the momentum of the center of mass are the same. Therefore, the velocity of the center of mass should point in the same direction as the total momentum. So, we expect to find $\vec{v}_{CM}$ = _____ $\hat{\imath}$ m/s.

:• SOLVE

Divide the total momentum by the total mass of the system to find the center-of-mass velocity.

$$\vec{p}_{CM} = \vec{p}_{tot} = M\vec{v}_{CM} \qquad (10.8)$$
$$\vec{v}_{CM} = \frac{\vec{p}_{tot}}{M} = \frac{842\hat{\imath}\ \text{kg}\cdot\text{m/s}}{(52.0\ \text{kg} + 76.0\ \text{kg})}$$
$$\vec{v}_{CM} = 6.58\hat{\imath}\ \text{m/s}$$

:• CHECK and THINK

Our result is in the form we expected. Think about the total momentum in the center-of-mass frame. In this frame, the velocity and the momentum of the center of mass are zero. Therefore, the total momentum of the system *relative* to the center of mass is zero. This does not mean that the skaters are at rest in the center-of-mass frame. They are still moving in this frame, but their momenta cancel out.

10-5 Conservation of Momentum

In Chapters 8 and 9, we used the conservation of energy approach to tackle a number of difficult problems. Momentum conservation is an equally powerful concept. Although we can think of momentum transfer from one system to another, often the most powerful and useful way to consider momentum is to define a system for which momentum is conserved. So, we must first figure out what conditions are required for momentum to be conserved.

Unlike energy, momentum does not come in different forms, so we will not be concerned with momentum converting from one form to another.

First, conservation of momentum means that the *total* momentum does not change in time. Second, when any quantity is conserved, the time derivative of that quantity is zero. So, when momentum is conserved,

$$\frac{d\vec{p}_{\text{tot}}}{dt} = 0$$

Third, the total momentum is the same as the momentum of the center of mass $\vec{p}_{\text{tot}} = \vec{p}_{\text{CM}}$ (Eq. 10.8). So, when momentum is conserved, the time derivative of the center of mass momentum is zero:

$$\frac{d\vec{p}_{\text{CM}}}{dt} = 0$$

Next, write the center-of-mass momentum in terms of the system's total mass and the center-of-mass velocity:

$$\frac{d\vec{p}_{\text{CM}}}{dt} = \frac{d(M\vec{v}_{\text{CM}})}{dt} = 0$$

If the total mass of the system is constant, we can pull M outside the derivative:

$$\frac{d\vec{p}_{\text{CM}}}{dt} = M\frac{d(\vec{v}_{\text{CM}})}{dt} = M\vec{a}_{\text{CM}} = 0$$

Finally, when we compare this expression with $\sum\vec{F}_{\text{ext}} = M\vec{a}_{\text{CM}}$ (Eq. 10.6), we find that **when momentum is conserved, the total external force must be zero**:

$$\frac{d\vec{p}_{\text{CM}}}{dt} = M\vec{a}_{\text{CM}} = \sum\vec{F}_{\text{ext}} = 0$$

This last expression is an important finding. The total momentum of a system is conserved if two conditions are met: (1) there is *no* net external force acting on the system, and (2) the mass of the system is constant. Condition 1 does *not* require the system to be isolated. Forces may act on the system as long as the total (net) external force is zero. Of course, the net force on an isolated system is zero, so an isolated system meets condition 1. Condition 2 means that no mass may enter or leave the system. A system that does not lose or gain mass is called a **closed system**. So, we must choose a system that meets both conditions 1 and 2 if we want to apply the conservation of momentum.

CLOSED SYSTEM
Major Concept

Conservation of momentum is best expressed in practice by thinking about the momentum of the system at two times, usually referred to as the *initial* and *final* times. Using this convention for a system for which momentum is conserved, we have

CONSERVATION OF MOMENTUM
Underlying Principle

$$\vec{p}_{i,\text{tot}} = \vec{p}_{f,\text{tot}} \tag{10.9}$$

The initial and final times are chosen to be convenient and do not necessary refer to the beginning or end of motion.

Before we try some examples, we notice that if the total external force on a system is not zero, momentum is not conserved. In such cases, the change in momentum is given by Newton's second law:

NEWTON'S SECOND LAW APPLIED TO A SYSTEM OF PARTICLES
Underlying Principle

$$\sum\vec{F}_{\text{ext}} = \frac{d\vec{p}_{\text{CM}}}{dt} = \frac{d\vec{p}_{\text{tot}}}{dt} \tag{10.10}$$

CONCEPT EXERCISE 10.5

What is the purpose of the ropes attached to the cannon in Figure 10.10?

FIGURE 10.10

Because momentum is a vector, it may be conserved in one, two, or three dimensions. If the vector sum of the external forces on a closed system is zero, momentum is conserved in all three dimensions, and Equation 10.9 may be written in component form as

$$\sum p_{ix} = \sum p_{fx} \qquad \sum p_{iy} = \sum p_{fy} \qquad \sum p_{iz} = \sum p_{fz} \tag{10.11}$$

The summation symbol indicates that we must add the momentum components of all particles in the system to arrive at the total momentum component in each direction.

In some cases, the total external force is zero in only one or two dimensions. In those cases, momentum is only conserved along the directions for which the total external force is zero. For example, in the absence of drag, the only force acting on a projectile near the Earth's surface is gravity along a vertical y axis. Therefore, the momentum of the projectile is not conserved in the y direction, but it is conserved in the x direction. So, for a projectile (moving in the xy plane), $\sum p_{ix} = \sum p_{fx}$, but $\sum p_{iy} \neq \sum p_{fy}$.

EXAMPLE 10.5 Nuclear Decay

An unstable nucleus decays into three particles. Particles 1 and 2 each have mass $2m$, and particle 3 has mass m. We wish to use two detectors to observe particles 1 and 2. In the center-of-mass frame, the nucleus initially has no kinetic energy. When the nucleus decays, the three particles have the velocities shown in Figure 10.11 and total kinetic energy $K = \frac{17}{8}mv^2$. Particles 1 and 2 have the same speed v_α, and particle 3's speed is $\sqrt{2}v$. Find the angles θ_1 and θ_2.

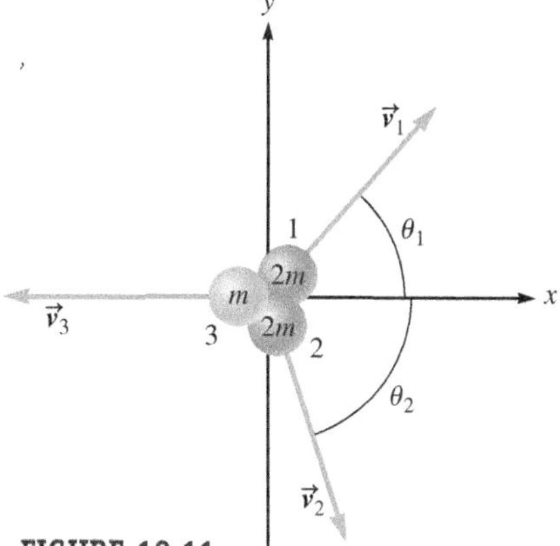

FIGURE 10.11

INTERPRET and ANTICIPATE

If we consider all three particles to make up the system, the system is closed, and no external forces act on it. Therefore, momentum is conserved. In the center-of-mass frame, the system's momentum is zero. The particles travel in the xy plane, so total momentum is conserved in two dimensions. Our result should be numerical values for angles θ_1 and θ_2.

SOLVE

Find the speed v_α of particle 1 in terms of v from the expression for kinetic energy $K = \sum_{i=1}^{n} \frac{1}{2}m_i v_i^2$ (Eq. 8.2).

$$K = \frac{1}{2}m_1v_1^2 + \frac{1}{2}m_2v_2^2 + \frac{1}{2}m_3v_3^2 = \frac{17}{8}mv^2$$

$$K = \frac{1}{2}(2m)v_\alpha^2 + \frac{1}{2}(2m)v_\alpha^2 + \frac{1}{2}m(\sqrt{2}v)^2 = \frac{17}{8}mv^2$$

$$2mv_\alpha^2 = \frac{17}{8}mv^2 - mv^2 = \frac{9}{8}mv^2$$

$$v_\alpha^2 = \frac{9}{16}v^2$$

$$v_\alpha = \frac{3}{4}v \tag{1}$$

Momentum is conserved in both directions. Let's consider momentum in the y direction first. Initially, the y component of the nucleus's momentum is zero. After the decay, only particles 1 and 2 have momentum in the y direction. For momentum to be conserved, the sum of the two y components of momentum must be zero. We find $\theta_1 = \theta_2$. For convenience, we define this angle as θ.

$$\sum p_{iy} = 0$$

$$\sum p_{fy} = 2mv_\alpha \sin\theta_1 - 2mv_\alpha \sin\theta_2$$

$$\sum p_{iy} = \sum p_{fy}$$

$$0 = 2mv_\alpha \sin\theta_1 - 2mv_\alpha \sin\theta_2$$

$$\theta_1 = \theta_2 \equiv \theta$$

Next, consider momentum in the x direction. Initially, there is no momentum in the x direction, so the final momentum must also be zero.

$$\sum p_{ix} = 0$$

$$\sum p_{fx} = 2(2mv_\alpha \cos\theta) - \sqrt{2}mv$$

$$\sum p_{ix} = \sum p_{fx}$$

$$0 = 4mv_\alpha \cos\theta - \sqrt{2}mv$$

Substitute $v_\alpha = \frac{3}{4}v$ (Eq. 1).

$$4m\left(\frac{3}{4}v\right)\cos\theta = \sqrt{2}mv$$

$$3\cos\theta = \sqrt{2}$$

$$\theta = \cos^{-1}\left(\frac{\sqrt{2}}{3}\right) = 62°$$

$$\theta_1 = \theta_2 = 62°$$

Example continues on page 292 ▶

CHECK and THINK

We found that in Figure 10.11, the angles θ_1 and θ_2 should be equal. In the center-of-mass frame of this closed, isolated system, the total momentum must be zero before and after the decay. The center of mass must remain at rest. The kinetic energy, however, actually increases from zero before the decay to $K = \frac{17}{8}mv^2$. Kinetic energy is not a conserved quantity in this case. Instead, internal energy of the system is converted to kinetic energy. Because this process involves only internal forces, momentum is conserved. As a result, parts of the system move while its center of mass remains at rest.

EXAMPLE 10.6 "Here I Come to Save the Day!"

Imagine that a dastardly villain has planted dynamite on a runaway abandoned mine cart of mass $m_c = 93$ kg. The cart is moving quickly at 45 mph ($v_c = 20$ m/s) along a straight, level track (Fig. 10.12). The hero, who weighs an incredible 750 lbs ($m_h = 340$ kg), jumps off a bridge and falls straight down into the cart. What is the speed of the cart after the hero has made his landing? Assume rolling friction and drag are negligible.

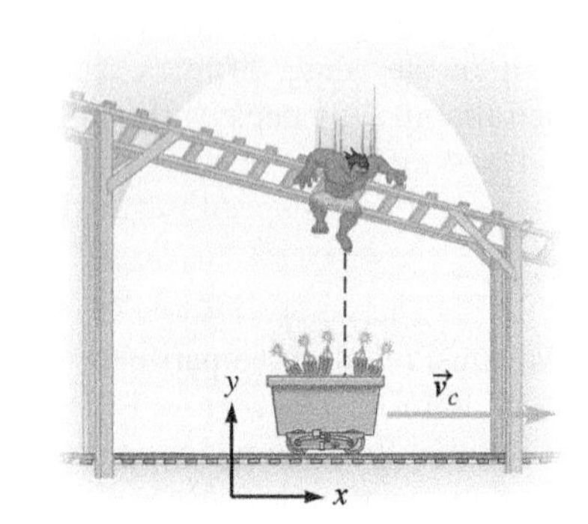

FIGURE 10.12

INTERPRET and ANTICIPATE

Let us consider the cart and the hero to make up the system. The total external force on the system is *not* zero. Before the hero falls, there is a net force due to gravity in the y direction. So, momentum is *not* conserved in the y direction. Because friction and drag are negligible, there is no net force in the x direction. Therefore, momentum is conserved in the x direction. We expect to find a numerical result for the speed.

SOLVE

Before the hero lands, only the cart has momentum in the x direction.

$$p_{ix} = m_c v_c$$

After the hero lands in the cart, both he and the cart move with the same velocity $\vec{v}_f$ (in the x direction).

$$p_{fx} = (m_c + m_h)v_f$$

Use the information that momentum is conserved in the x direction to find the final speed of the cart–hero system.

$$m_c v_c = (m_c + m_h)v_f$$

$$v_f = \frac{m_c v_c}{m_c + m_h} = \frac{(93 \text{ kg})(20 \text{ m/s})}{93 \text{ kg} + 340 \text{ kg}} = 4.3 \text{ m/s}$$

CHECK and THINK

We just found that the cart slows down by nearly a factor of five when the hero falls into it. This finding makes sense because initially the hero has no momentum in the x direction, but the relatively lightweight cart has momentum in the x direction. When the hero lands, he acquires an x component of momentum. Because the hero is so massive, the cart must slow down considerably when he lands for momentum in the x direction to be conserved. This example also illustrates collision, the subject of Chapter 11.

10-6 Case Study: Rockets

When you think of a rocket, you probably imagine a spacecraft being launched. Here we start with an Earth-bound craft that illustrates the physics of a rocket. Imagine Sophia on a light raft in a still pond. She uses a slingshot launcher to shoot large water balloons horizontally off the back of her raft (Fig. 10.13A). To think about conservation of momentum, we must choose a closed system. We include Sophia, the slingshot, the raft, and the water balloons in the system. So, even after she launches a balloon, we will still consider it part of the system. For now, we will also assume no net external force acts on the system, so momentum of the system is conserved. (In Section 10-7, we consider the raft and its contents to be its own separate system distinct from the launched water balloons, and we will also consider the effects of external forces. For now, let's accept this somewhat unrealistic situation in which water does *not* exert a drag force on the raft.) The goal of this section is to derive an expression—known as the *first rocket equation*—for the velocity of the raft.

Imagine that the raft is initially at rest with respect to the pond so that the system's momentum is zero in the pond's frame (*not* depicted in Fig. 10.13). Now imagine that Sophia launches a water balloon in the negative x direction (Fig. 10.13B). The balloon's momentum is in the negative x direction, and for the total momentum of the system to remain zero, the rest of the system (raft and contents) must have momentum in the positive x direction. If Sophia continues to launch her water balloons, her speed will continue to increase. This raft example has the essence of a rocket. The water balloons are the fuel, the launcher is the exhaust system, the raft is the body of the rocket, and Sophia is the astronaut.

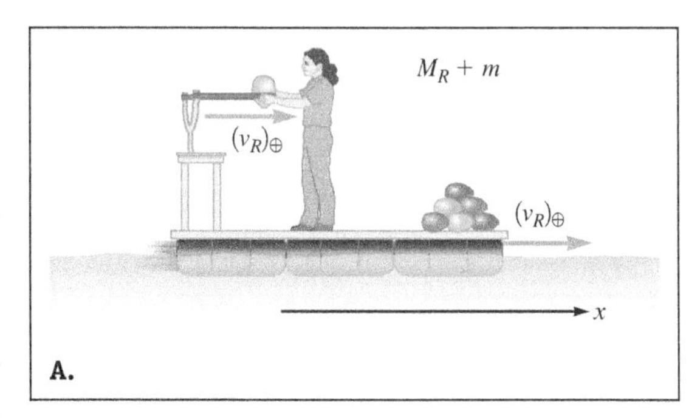

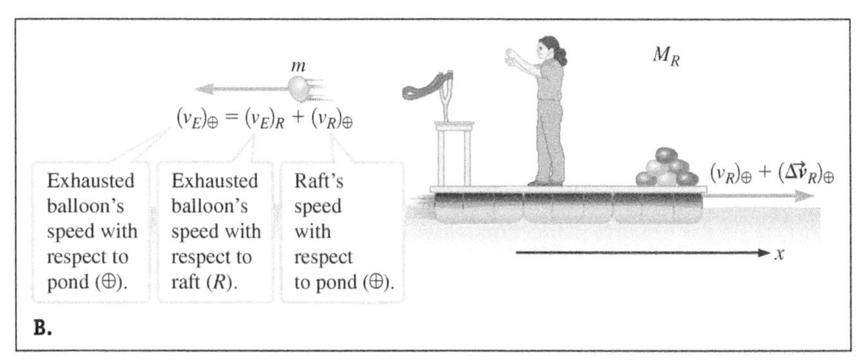

FIGURE 10.13 A. Sophia on a raft with water balloons and a launcher. **B.** Sophia launches a water balloon in the negative x direction. To conserve momentum, the raft's velocity in the positive x direction must increase.

Engineers need to know how the change in a rocket's velocity relates to the velocity of its exhausted fuel and to the masses involved (rocket and fuel). This relation is contained in the **first rocket equation**.

DERIVATION First Rocket Equation

Let $(\vec{v}_E)_R$ be the velocity of the *e*xhausted fuel with respect to the *r*ocket. Let $\Delta\vec{v}_R$ be the change in the rocket's velocity with respect to the Earth, ⊕. Here, M_{Ri} is the mass of the rocket (plus the remaining fuel) at some initial time, and M_{Rf} is its mass at some final time. We derive the first rocket equation:

$$\Delta\vec{v}_R = (\vec{v}_E)_R \ln\left(\frac{M_{Rf}}{M_{Ri}}\right) \quad (10.12)$$

FIRST ROCKET EQUATION
Special Case

Let's start with Sophia's rocket. At the initial time (Fig. 10.13A), Sophia has already launched one or more balloons; the *r*ocket has velocity $(\vec{v}_R)_\oplus$ relative to the Earth. She is about to launch a balloon of mass m, so initially the mass of the rocket and its contents is $M_R + m$. The final time occurs just after she has launched the balloon with velocity $(\vec{v}_E)_R$ with respect to the rocket (Fig. 10.13B).

One way to model this system is as just two particles. One particle is the balloon that has just been launched. The other particle is the remainder of the system: the rocket and its contents.

Momentum of this system is conserved.	$\sum\vec{p}_i = \sum\vec{p}_f$ (10.9)
There are two particles in the system: the exhausted balloon (E) and the rocket (R). Collect terms that involve the rocket on the right and terms that involve the balloon on the left.	$\vec{p}_{Ei} + \vec{p}_{Ri} = \vec{p}_{Ef} + \vec{p}_{Rf}$ $\vec{p}_{Ei} - \vec{p}_{Ef} = \vec{p}_{Rf} - \vec{p}_{Ri}$ $-(\vec{p}_{Ef} - \vec{p}_{Ei}) = \vec{p}_{Rf} - \vec{p}_{Ri}$
We find that the change in the rocket's momentum must be equal in magnitude and opposite in direction to the change in the balloon's momentum (with respect to the pond fixed to the Earth).	$-(\Delta\vec{p}_E)_\oplus = (\Delta\vec{p}_R)_\oplus$ (1) (exhaust) (rocket)
Before the balloon is launched, it moves along with the rocket at velocity $(\vec{v}_R)_\oplus$, so its initial momentum is $m(\vec{v}_R)_\oplus$. The balloon's final momentum is $m(\vec{v}_E)_\oplus$.	$(\Delta\vec{p}_E)_\oplus = m(\vec{v}_E)_\oplus - m(\vec{v}_R)_\oplus$ (2)

Derivation continues on page 294 ▶

So far, we have been working in terms of the Earth's reference frame. In the first rocket equation, the speed of the balloon is measured with respect to the rocket, so we switch to the rocket's reference frame (Eq. 4.20).

$$(\vec{v}_E)_\oplus = (\vec{v}_E)_R + (\vec{v}_R)_\oplus \tag{3}$$

Use Equation (3) to eliminate $(\vec{v}_E)_\oplus$ from Equation (2).

$$(\Delta\vec{p}_E)_\oplus = m[(\vec{v}_E)_R + (\vec{v}_R)_\oplus] - m(\vec{v}_R)_\oplus$$

$$(\Delta\vec{p}_E)_\oplus = m(\vec{v}_E)_R \tag{4}$$

Now, we find the change in the balloon's momentum in the rocket's frame. In the rocket's frame, the balloon initially is at rest and has no momentum. When the balloon is launched, its velocity is $(\vec{v}_E)_R$.

$$(\Delta\vec{p}_E)_R = m(\vec{v}_E)_R - 0$$

$$(\Delta\vec{p}_E)_R = m(\vec{v}_E)_R \tag{5}$$

Compare Equations (4) and (5). The balloon's change in momentum is the same in both frames.

$$(\Delta\vec{p}_E)_\oplus = (\Delta\vec{p}_E)_R \tag{10.13}$$

Substitute into Equation (1). We find that the change in the rocket's momentum $(\Delta\vec{p}_R)_\oplus$ (in the positive x direction *relative to the Earth*) is equal to the change in the balloon's momentum, $-(\Delta\vec{p}_E)_R$ (in the negative x direction *relative to the rocket*).

$$(\Delta\vec{p}_R)_\oplus = -(\Delta\vec{p}_E)_R \tag{10.14}$$

Now we need the rocket's change in momentum. Initially, the rocket has velocity $(\vec{v}_R)_\oplus$. After the balloon is launched, the rocket's velocity is $(\vec{v}_R)_\oplus + (\Delta\vec{v}_R)_\oplus$, where $(\Delta\vec{v}_R)_\oplus$ is the increase in the rocket's velocity (relative to the pond).

$$(\Delta\vec{p}_R)_\oplus = M_R[(\vec{v}_R)_\oplus + (\Delta\vec{v}_R)_\oplus] - M_R(\vec{v}_R)_\oplus$$

$$(\Delta\vec{p}_R)_\oplus = M_R(\Delta\vec{v}_R)_\oplus \tag{6}$$

Substitute Equations (5) and (6) into Equation 10.14.

$$M_R(\Delta\vec{v}_R)_\oplus = -m(\vec{v}_E)_R \tag{7}$$

Now we are ready to consider the continuous exhausting of water with the fire hose (Fig. 10.14). Just as in the case of the balloon, the water is exhausted from the hose at $(\vec{v}_E)_R$ with respect to the raft rocket. We imagine breaking up the steady stream of water coming out of the fire hose into small disks of mass dm. Each of these imaginary disks of water is like a very small water balloon. When a small mass of water dm is ejected from the rocket, the rocket's velocity changes by a small amount $(d\vec{v}_R)_\oplus$.

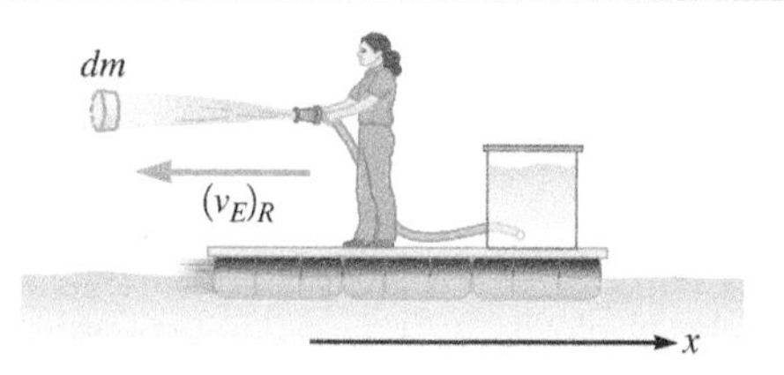

FIGURE 10.14

Rewrite Equation (7) in terms of these differential changes.

$$M_R(d\vec{v}_R)_\oplus = -(dm)(\vec{v}_E)_R \tag{10.15}$$

When a small mass of water dm is ejected, the mass of the rocket and its contents is reduced by dM_R. Therefore, $dM_R = -dm$.

$$M_R(d\vec{v}_R)_\oplus = -(-dM_R)(\vec{v}_E)_R$$

$$(d\vec{v}_R)_\oplus = (\vec{v}_E)_R\left(\frac{dM_R}{M_R}\right) \tag{8}$$

Integrate Equation (8) from some initial time when the rocket's velocity (with respect to the water) was $\vec{v}_{Ri}$ and its mass was M_{Ri} to a final time when the rocket's velocity (with respect to the Earth) and mass are $\vec{v}_{Rf}$ and M_{Rf}. The exhaust velocity of the water with respect to the rocket $(\vec{v}_E)_R$ is constant, so it is pulled outside the integral. The integral on the right results in the natural logarithm function (Appendix A).

$$\int_{\vec{v}_{Ri}}^{\vec{v}_{Rf}} (d\vec{v}_R)_\oplus = \int_{M_{Ri}}^{M_{Rf}} (\vec{v}_E)_R\left(\frac{dM_R}{M_R}\right)$$

$$\vec{v}_{Rf} - \vec{v}_{Ri} = (\vec{v}_E)_R \ln\left(\frac{M_{Rf}}{M_{Ri}}\right)$$

$$\Delta\vec{v}_R = (\vec{v}_E)_R \ln\left(\frac{M_{Rf}}{M_{Ri}}\right) \checkmark \tag{10.12}$$

COMMENTS

Equation 10.12 is the first of two rocket equations. It gives the change in the rocket velocities in terms of its mass. Because the final mass is less than the initial mass of the rocket ($M_{Rf} < M_{Ri}$) and Equation 10.12 involves the natural logarithm of a fraction less than 1, the result is a negative number. So, as expected, the change in the rocket's velocity $(\Delta\vec{v}_R)$ is in the opposite direction to the exhaust velocity $(\vec{v}_E)_R$.

EXAMPLE 10.7 CASE STUDY Rocket Science

Return to the rocket case study (page 282). How would you explain to Shannon that a rocket does not need anything to push against? A spacecraft usually has several small rocket thrusters (Fig. 10.15) that exhaust the fuel at several different angles. Explain how these thrusters might be used to maneuver the spacecraft.

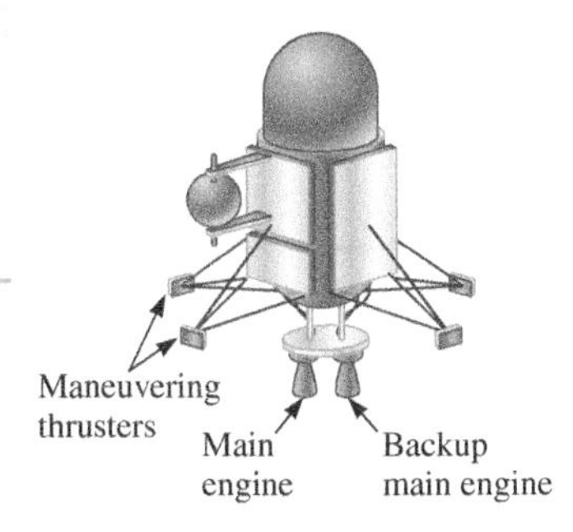

FIGURE 10.15

SOLVE

We assume the spacecraft is far from any massive objects so that gravity can be ignored and there are no other external forces.

If we consider the spacecraft and its exhausted fuel to make up a single system, momentum is conserved because no external forces act on the system. Therefore, to conserve momentum, any change in the exhausted fuel's momentum must result in a change in the spacecraft's momentum, equal in magnitude and opposite in direction: $(\Delta\vec{p}_R)_\oplus = -(\Delta\vec{p}_E)_R$ (Eq. 10.14).

The change in the rocket's velocity is always in the direction opposite to the exhausted fuel's velocity. The small thrusters are used to change the spacecraft's direction. Imagine using the hose on the raft in Figure 10.14 for steering. If Sophia wishes to turn the rocket into the page, she would need to direct the hose out of the page. If resistive forces are very small as they are in space, there is considerable risk of oversteering. Small maneuvering thrusters are about 100 times weaker than the main thrusters used to control the spacecraft.

EXAMPLE 10.8 "How Fast Will We Go?"

A team of engineering students has entered a competition. Each team is provided with a long, thin boat known as a crew shell. Two team members must be on board, and the team that achieves the fastest speed wins. The students may use any mechanical means to power the crew shell. They have decided to use a small commercial pump that ejects water at an impressive speed of 36 m/s. The students mount a 55-gal drum of water on the crew shell. The total mass of the crew shell, pump, empty water drum, and two team members is 157 kg. Ignore resistive forces and find the maximum possible speed of their crew shell on a still pond.

INTERPRET and ANTICIPATE

We include the crew shell and its contents even after the water is ejected in the system. There are no net external forces on the system. Therefore, momentum is conserved. The crew shell starts at rest with respect to the pond. As long as fuel is being exhausted, the shell's speed increases. The shell reaches its maximum speed just as all the fuel is exhausted. We expect to find a numerical result in the form v_{max} = ______ m/s.

SOLVE

Use the first rocket equation (Eq. 10.12) to find the maximum speed of the crew shell. The initial velocity is zero, and the final velocity is the maximum velocity of the shell.

$$\vec{v}_{Rf} - \vec{v}_{Ri} = (\vec{v}_E)_R \ln\left(\frac{M_{Rf}}{M_{Ri}}\right)$$

$$\vec{v}_{max} = (\vec{v}_E)_R \ln\left(\frac{M_{Rf}}{M_{Ri}}\right) \quad (1)$$

The initial mass includes the mass of the 55-gal drum of fuel. We convert from gallons to cubic meters and use the density of water to find the mass of this fuel.

$$V_w = 55\ \text{gal}\left(\frac{3.786 \times 10^{-3}\ \text{m}^3}{1\ \text{gal}}\right) = 0.21\ \text{m}^3$$

$$M_E = \rho_w V_w = (0.998 \times 10^3\ \text{kg/m}^3)(0.21\ \text{m}^3)$$

$$M_E = 2.1 \times 10^2\ \text{kg}$$

In Equation (1), the final mass M_{Rf} is the mass of the raft's contents not including the mass of the fuel because it has been ejected; the initial mass includes the mass of the fuel.

$$M_{Rf} = 157\ \text{kg}$$

$$M_{Ri} = M_{Rf} + M_E = 367\ \text{kg}$$

We are only interested in speed, so we take the absolute value of the velocity.

$$v_{max} = \left|(36\ \text{m/s}) \ln\left(\frac{157\ \text{kg}}{367\ \text{kg}}\right)\right|$$

$$v_{max} = 31.\ \text{m/s}$$

Example continues on page 296 ▶

CHECK and THINK

Our answer is in the form we expected, and 31 m/s is about 70 mph, which seems very fast. If we set up this situation as an experiment, the shell's maximum speed would be significantly lower than our answer. Drag plays a major role and cannot be ignored in determining the shell's motion. We will return to this situation in Example 10.10, after we have learned how to account for external forces acting on a rocket system.

10-7 Rocket Thrust: An Open System (Optional)

In Section 10-6, we considered the rocket and its exhausted fuel to be part of the system. The total mass of the system is a constant. If we decide to choose just the rocket and its contents to be the system, the mass of the rocket decreases as fuel is exhausted. We have then chosen an **open system**, that is, a system that either gains or loses mass.

OPEN SYSTEM Major Concept

The mass M_R in this open system is *not* constant; it is the mass of the rocket plus the fuel that has not yet been exhausted. The force exerted by the fuel is known as the **thrust**, and

$$\vec{F}_{\text{thrust}} = (\vec{v}_E)_R \frac{dM_R}{dt} \tag{10.16}$$

where $(\vec{v}_E)_R$ is the velocity of the exhausted fuel with respect to the rocket as defined in Section 10-6.

In Problem 83 you will show that for an open system such as a rocket, the **second rocket equation** is

SECOND ROCKET EQUATION Special Case

$$M_R(\vec{a}_R)_\oplus = \sum \vec{F}_{\text{ext}} + \vec{F}_{\text{thrust}} \tag{10.17}$$

where $(\vec{a}_R)_\oplus$ is the rocket's acceleration observed in the Earth's frame. Because the exhausted fuel is not considered to be in the system, the thrust exerted by the fuel on the rocket is an external force. The first term in Equation 10.17 represents the sum of the *other* external forces such as gravity and drag on the rocket, but not the thrust exerted by the fuel. If no other forces (such as gravity or drag) act on the rocket, the rocket's acceleration is due solely to the thrust exerted on it by the exhausted fuel.

EXAMPLE 10.9 CASE STUDY Rocket Science Revisited

Suppose Shannon is not convinced by the explanation made in Example 10.7. Try another explanation. This time, do *not* include the exhausted fuel in the system. How would you account for the rocket's acceleration?

SOLVE

As before, we assume the spacecraft is far from any massive objects, so gravity can be ignored, and there are no other external forces. If we consider the spacecraft and its contents, but not the exhausted fuel, to be the system, we can use Newton's third law to explain how the spacecraft is maneuvered. The spacecraft's thrusters exert a force on the fuel. By Newton's third law, the fuel must exert a force on the spacecraft of equal magnitude and in the opposite direction. Because there are essentially no resistive forces in space, the force exerted by the fuel on the spacecraft easily causes it to accelerate.

EXAMPLE 10.10 What a Drag!

Let's return to the crew shell competition in Example 10.8 and find a more realistic expectation for the team's maximum speed. The engineering students tested their pump and found that it ejects water at a rate of 23 gal/min. Further tests on the crew shell have shown that the drag force (Eq. 6.5) is given by $F_D = \frac{1}{2}C\rho A v^2 = (19.02 \text{ kg/m})v^2$. Find the terminal speed of the crew shell. (Don't worry about the pump running out of water.)

INTERPRET and ANTICIPATE

The water exhausted provides the thrust and the other external force on the shell is due to the drag force. At the terminal speed, the crew shell is no longer accelerating (Section 6-5), so $(\vec{a}_R)_\oplus = 0$. Therefore, according to the second rocket equation, the drag force must be balanced by the thrust. We expect a lower speed than the one found in Example 10.8.

SOLVE

Set $(\vec{a}_R)_\oplus = 0$ in Equation 10.17.

$$0 = \sum \vec{F}_{\text{ext}} + \vec{F}_{\text{thrust}}$$

$$\sum \vec{F}_{\text{ext}} = -\vec{F}_{\text{thrust}}$$

Because we only need to find the magnitude of the velocity, we only need to work with the magnitude of the forces.

$$F_D = F_{\text{thrust}} = (19.02 \text{ kg/m})v_t^2$$

$$v_t = \sqrt{\frac{F_{\text{thrust}}}{19.02 \text{ kg/m}}} \tag{1}$$

Mass is lost from the crew shell at a rate of 23 gal/min. Use the density of water and several conversion factors to express this rate in kilograms per second.

$$\frac{dM_R}{dt} = \left(\frac{23 \text{ gal}}{1 \text{ min}}\right)\left(\frac{1 \text{ min}}{60 \text{ s}}\right)\left(\frac{3.786 \times 10^{-3} \text{ m}^3}{1 \text{ gal}}\right)\left(\frac{998 \text{ kg}}{1 \text{ m}^3}\right)$$

$$\frac{dM_R}{dt} = 1.4 \text{ kg/s}$$

Use Equation 10.16 to find (the magnitude of) the thrust. The exhaust speed $(v_E)_R$ was given in Example 10.8 as 36 m/s.

$$F_{\text{thrust}} = \left(\frac{dM_R}{dt}\right)(v_E)_R = (1.4 \text{ kg/s})(36 \text{ m/s})$$

$$F_{\text{thrust}} = 50 \text{ N}$$

The terminal speed comes from substituting the thrust into Equation (1).

$$v_t = \sqrt{\frac{F_{\text{thrust}}}{19.02 \text{ kg/m}}} = \sqrt{\frac{50 \text{ N}}{19.02 \text{ kg/m}}} = 1.6 \text{ m/s}$$

CHECK and THINK

Once drag is taken into account, the crew shell is expected to have a much lower maximum speed. Drag slows down the shell by more than one order of magnitude, to just about 4 mph. This result seems reasonable because we know that a crew shell—which is normally rowed—does not get far once the rowers stop rowing.

EXAMPLE 10.11 CASE STUDY "The *Eagle* Has Landed!"

Our modern daily life relies heavily on artificial satellites, but crewed space missions are rare. It is hard for us to imagine that only in the 1960s did we plan such a mission to the Moon, a remarkable act of bravery and ingenuity.

On July 16, 1969, *Apollo 11* was launched from the Kennedy Space Center with three astronauts on board. In lunar orbit on Sunday morning, July 20, Neil Armstrong and Buzz Aldrin climbed into the lunar lander known as the *Eagle*. Michael Collins remained on board the command module, and the *Eagle* of total mass 15,065 kg was released.

Rocket thrusters on the *Eagle* were used to control its descent to the Moon. Although the *Eagle* had overshot the planned landing spot, Armstrong knew that he could still manage to land manually. Aldrin and Armstrong worked without a computer, with Aldrin providing calculations to Armstrong.

The descent (engine) rocket thruster had a maximum thrust of 45,000 N to control the *Eagle*'s deceleration. In addition, the *Eagle* was equipped with sets of smaller thrusters to control its direction (Fig. 10.16). The descent engine had 8212 kg of fuel. It was fired for 756.3 s, and it only had 60 s of fuel remaining when the *Eagle* landed.

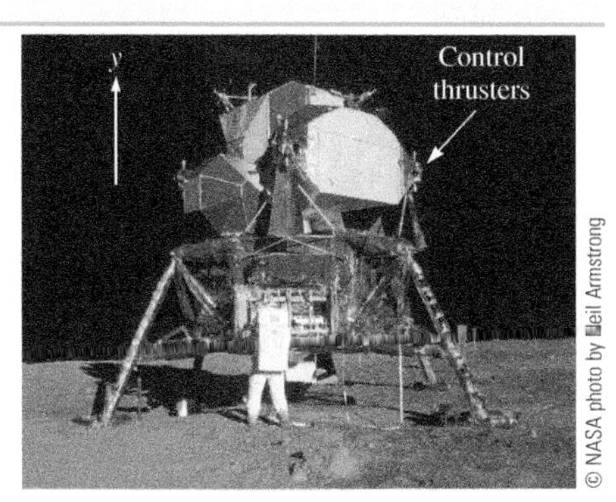

FIGURE 10.16 Control thrusters on the *Eagle* were grouped in sets of five (thrusting up, down, forward, backward, and sideways). The descent engine is on the underside of the spacecraft to Aldrin's right in this photo taken by Armstrong.

A Assume the velocity and rate at which mass was ejected from the descent engine were constant at maximum thrust. Use the given data to find dM_R/dt and $(\vec{v}_E)_R$ for the *Eagle*'s descent engine.

Example continues on page 298 ▶

INTERPRET and ANTICIPATE

To find dM_R/dt, calculate how long it would take to expel all the fuel. Assuming the fuel was ejected at a steady rate, divide the mass of the fuel by the time to expel it all to obtain the rate of mass loss. Once we have that result, we find the velocity of the fuel from the thrust.

SOLVE

The descent engine fired for 756.3 s and had 60 s of fuel remaining. Add these figures to find the total time to expel all the fuel.

$$\Delta t = 756.3\text{ s} + 60\text{ s} = 816.3\text{ s}$$

The initial mass of the *Eagle* is greater than its final mass; therefore, dM_R/dt is negative. In the time $\Delta t = 816.3$ s, the *Eagle* would have lost all the mass of the fuel.

$$\frac{dM_R}{dt} = \frac{M_{Rf} - M_{Ri}}{\Delta t} = \frac{-8212\text{ kg}}{816.3\text{ s}}$$

$$\frac{dM_R}{dt} = -10.06\text{ kg/s}$$

To find the velocity of the ejected fuel, use Equation 10.16. Choose a coordinate system with an upward-pointing y axis. Here, $\vec{F}_{\text{thrust}}$ is in the positive y direction (slowing the spacecraft).

$$\vec{F}_{\text{thrust}} = (\vec{v}_E)_R \frac{dM_R}{dt} \tag{10.16}$$

$$(\vec{v}_E)_R = \left(\frac{dM_R}{dt}\right)^{-1} \vec{F}_{\text{thrust}} = \left(-\frac{1}{10.06\text{ kg/s}}\right)(45000\hat{\jmath}\text{ N})$$

$$(\vec{v}_E)_R = -4.5 \times 10^3 \hat{\jmath}\text{ m/s} = -4.5\hat{\jmath}\text{ km/s}$$

CHECK and THINK

Our numerical results are reasonable. The direction for exhaust velocity makes sense. The thrust must point upward to decelerate the *Eagle*, which would otherwise be in free fall toward the Moon. Therefore, the exhaust must be ejected downward.

B What was the *Eagle*'s acceleration just before it landed on the Moon? Assume the descent engine was at maximum thrust. The local gravitational acceleration near the surface of the Moon is $\vec{g}_{\text{moon}} = -1.67\hat{\jmath}\text{ m/s}^2$.

INTERPRET and ANTICIPATE

Clearly, the Moon's gravity acted on the *Eagle*. Because the Moon has (almost) no atmosphere, there are no other external forces to consider. The engine slowed the descent, so we expect the *Eagle*'s acceleration to be upward.

SOLVE

Solve the second rocket equation (Eq. 10.17) for acceleration.

$$M_R(\vec{a}_R)_\oplus = \sum \vec{F}_{\text{ext}} + \vec{F}_{\text{thrust}} \tag{10.17}$$

$$(\vec{a}_R)_\oplus = \frac{1}{M_R}\left(\sum \vec{F}_{\text{ext}} + \vec{F}_{\text{thrust}}\right) = \frac{1}{M_R}(M_R\vec{g}_{\text{Moon}} + \vec{F}_{\text{thrust}})$$

$$(\vec{a}_R)_\oplus = \left(\vec{g}_{\text{Moon}} + \frac{\vec{F}_{\text{thrust}}}{M_R}\right) \tag{1}$$

The mass M_R needed in Equation (1) is the mass of the *Eagle* just before touching down. We find that information from the initial mass of the lunar lander minus the mass M_E of the fuel that was burned.

$$M_E = (756.3\text{ s})(10.06\text{ kg/s}) = 7608\text{ kg}$$

$$M_R = M_i - M_E = 15065\text{ kg} - 7608\text{ kg} = 7457\text{ kg}$$

$$(\vec{a}_R)_\oplus = \left(-1.67\hat{\jmath}\text{ m/s}^2 + \frac{4.5 \times 10^4\hat{\jmath}\text{ N}}{7457\text{ kg}}\right) = 4.4\hat{\jmath}\text{ m/s}^2$$

CHECK and THINK

As expected, the acceleration was upward. The acceleration just before landing was about 2.5 times the gravitational acceleration on the surface of the Moon, but less than half of the Earth's gravitational acceleration ($g = 9.81\text{ m/s}^2$). Astronauts are expected to experience about three g's when launched from the Earth's surface. Because the Moon's gravitational acceleration is about six times weaker than that of the Earth (measured at their surfaces), landing on and taking off from the Moon are much more pleasant rides than doing the same from the Earth.

SUMMARY

Underlying Principles

1. A **general form of Newton's second law** is $\vec{F}_{\text{tot}} = d\vec{p}/dt$. This form holds even for systems whose mass changes, and $\vec{F}_{\text{tot}} = m\vec{a}$ is only good in the special case of a system with constant mass.
2. The **total momentum of a system is conserved** if two conditions are met: (1) there is no net external force acting on it, with $\sum\vec{F}_{\text{ext}} = 0$; and (2) the system is **closed**. Mathematically, conservation of momentum is
$$\vec{p}_{i,\text{tot}} = \vec{p}_{f,\text{tot}} \quad (10.9)$$
3. **Newton's second law applied to a system of particles:**
$$\sum\vec{F}_{\text{ext}} = \frac{d\vec{p}_{\text{CM}}}{dt} = \frac{d\vec{p}_{\text{tot}}}{dt} \quad (10.10)$$
If the **system's mass is constant**, then
$$\sum\vec{F}_{\text{ext}} = M\vec{a}_{\text{CM}} \quad (10.6)$$

Major Concepts

1. **Momentum** is a vector that points in the same direction as a **particle's** velocity:
$$\vec{p} \equiv m\vec{v} \quad (10.1)$$
For a **system of *n* particles, the total momentum** is given by
$$\vec{p}_{\text{tot}} = \sum_{j=1}^{n}\vec{p}_j = \vec{p}_{\text{CM}} \quad (10.7 \text{ and } 10.8)$$
2. **Center of mass** is a point associated with an object or system that moves as though all of its mass was concentrated there and all the external forces were applied there. It is found from
$$\vec{r}_{\text{CM}} = \frac{1}{M}\sum_{j=1}^{n} m_j\vec{r}_j \quad (10.5)$$
The momentum of the center of mass is the total momentum of the system $\vec{p}_{\text{tot}} = \vec{p}_{\text{CM}}$ (Eq. 10.8).
3. A system that does not lose or gain mass is called a **closed system**. An **open system** is one that either gains or loses mass.

Special Cases: Rockets

1. The **first rocket equation** is
$$\Delta\vec{v}_R = (\vec{v}_E)_R \ln\left(\frac{M_{Rf}}{M_{Ri}}\right) \quad (10.12)$$
2. The **second rocket equation** is
$$M_R(\vec{a}_R)_{\oplus} = \sum\vec{F}_{\text{ext}} + \vec{F}_{\text{thrust}} \quad (10.17)$$
where the force exerted by the fuel is the thrust:
$$\vec{F}_{\text{thrust}} = (\vec{v}_E)_R\frac{dM_R}{dt} \quad (10.16)$$

PROBLEMS AND QUESTIONS

A = algebraic C = conceptual E = estimation G = graphical N = numerical

10-1 A Second Conservation Principle

1. C Which of the following questions cannot be answered by applying the conservation of energy principle? Explain your answers. **a.** A ball is tossed up at a given speed. What is the ball's maximum height? **b.** If a ball strikes a wall at a given angle, at what angle does it bounce back? **c.** An object initially at rest explodes into two pieces of equal mass. If the velocity of one piece is measured, what is the velocity of the other piece?
2. **Review (C)** Chapters 8 through 10 were about the conservation approach. List the general equations in those three chapters that are worth memorizing. Explain your choices.

10-2 Momentum of a Particle

3. N Find the momentum in each of the following cases.
   a. A 65-kg person is walking with a velocity of $1.2\hat{\imath}$ m/s.
   b. A 2.3-g bullet is traveling at 720 m/s in the $-\hat{\imath}$ direction.

4. A mother pushes her son in a stroller at a constant speed of 1.52 m/s. The boy tosses a 56.7-g tennis ball straight up at 1.75 m/s and catches it. The boy's father sits on a bench and watches.
   a. N According to the mother, what are the ball's initial and final momenta?
   b. N According to the father, what are the ball's initial and final momenta?
   c. C According to the mother, is the ball's momentum ever zero? If so, when? If not, why not?
   d. C According to the father, is the ball's momentum ever zero? If so, when? If not, why not?
5. N A pitcher throws a 145-g ball horizontally with velocity $\vec{v} = 32.0\,\hat{\imath}$ m/s, and it lands in the catcher's glove 0.450 s later. What is the momentum of the ball the instant before it lands in the glove? Ignore air resistance.
6. E Estimate the magnitude of the momentum of a car on the highway.
7. A An object moves with linear momentum of magnitude p. If the object's mass is m, show that its kinetic energy is given by $K = p^2/2m$.
8. E In comparing a thrown spear versus an arrow shot from a bow, you find that the spear's speed is approximately 35.0 m/s, whereas the arrow's speed is nearly 125 m/s. Treating the spear and the arrow as particles, estimate the mass of the spear and the arrow and, given the speeds discussed here, determine which object would have a greater momentum.
9. N What is the magnitude of the Earth's momentum associated with its orbital motion around the Sun?
10. The velocity of a 10-kg object is given by $\vec{v} = 5t^2\hat{\imath} - (7t + 2t^3)\hat{\jmath}$, where if t is in seconds, then v will be in meters per second.
    a. A What is the net force on the object as a function of time?
    b. N What is the momentum of the object when $t = 15.0$ s?
11. N A particle has a momentum of magnitude 40.0 kg · m/s and a kinetic energy of 3.40×10^2 J. **a.** What is the mass of the particle? **b.** What is the speed of the particle?
12. **Review (N)** In Example 10.1, two balls are dropped from the top of a building at the same time. One ball has a mass of 2.5 kg, and the other has a mass of 5.0 kg. Drag on the balls is negligible, and they land on the ground 5.3 s after they were released. Using the conservation of energy principle, find the momentum of each ball just as it reaches the ground.
13. N Latoya, sitting on a sled, is being pushed by Dewain on the horizontal surface of a frozen lake. Dewain slips and falls, giving the sled one final push, and the sled comes to rest 9.50 s later. The speed of the sled after the final push is 4.00 m/s, and the combined mass of the sled and Latoya is 32.5 kg. Using a momentum approach, determine the magnitude of the average friction force acting on the sled during this interval.
14. A A baseball is thrown vertically upward. The mass of the baseball is m, and its initial speed is v_i. **a.** What is the momentum of the baseball at its maximum height? **b.** What is the momentum of the baseball when it has traveled three-fourths of the total distance in its up-and-down motion?

10-3 Center of Mass Revisited

15. N Find the center of mass of a system with three particles of masses 1.0 kg, 2.0 kg, and 3.0 kg kept at the vertices of an equilateral triangle of side 1.0 m (Fig. P10.15).

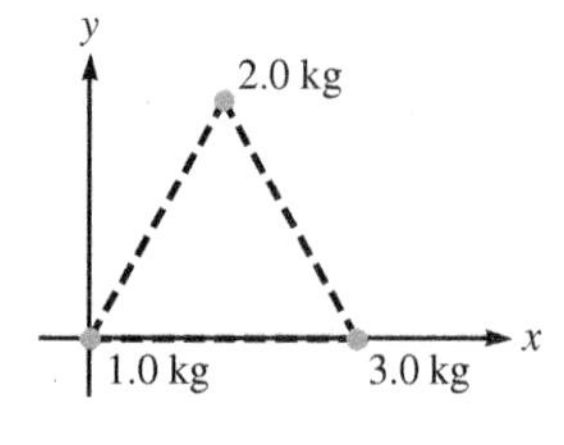

FIGURE P10.15

Problems 16 and 17 are paired.

16. N Fifteen students are lined up along a wall. An x axis has been chosen to run parallel to the wall with the origin in one corner. The positions and (fractional) physics grades of the students are given in Table P10.16. **a.** Find the average of the students' positions. **b.** Find the weighted average of their positions based on their grades. Comment on your results.

TABLE P10.16

x (m)	Grade	Mass (kg)	x (m)	Grade	Mass (kg)
0.25	0.92	54	4.27	0.69	61
0.78	0.95	55	4.65	0.84	56
1.16	0.88	49	5.32	0.75	54
2.04	0.88	69	5.67	0.89	84
2.56	0.78	63	6.10	0.83	78
3.20	0.55	67	6.31	0.82	76
3.62	0.65	65	6.55	0.98	96
3.95	0.46	53			

17. N Fifteen students are lined up along a wall. An x axis has been chosen to run parallel to the wall with the origin in one corner. The positions and masses of the students are given in Table P10.16. **a.** Find the average of the students' positions. **b.** Find the position of their center of mass. Comment on your results.
18. Two metersticks are connected at their ends as shown in Figure P10.18. The center of mass of each individual meterstick is at its midpoint, and the mass of each meterstick is m.
    a. N Where is the center of mass of the two-stick system as depicted in the figure, with the origin located at the intersection of the sticks?
    b. C Can the two-stick system be balanced on the end of your finger so that is remains lying flat in front of you in the orientation shown? Why or why not?

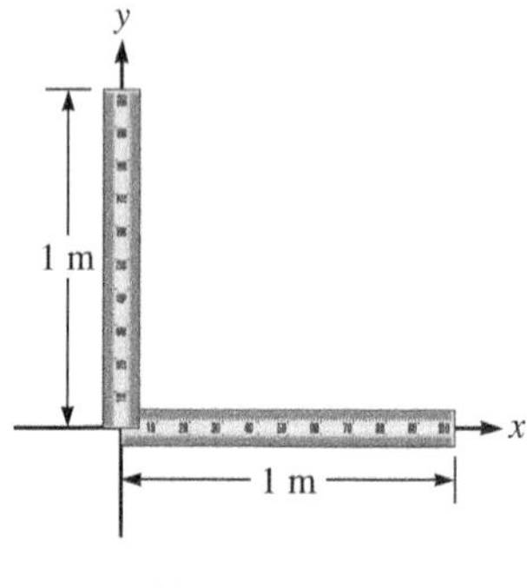

FIGURE P10.18

19. N A boy of mass 25.0 kg is sitting on one side of a seesaw while his older sister, who has a mass of 35.0 kg, is sitting on the other side. Each child is 1.50 m from the center of the seesaw. **a.** Where is the center of mass of the two children relative to the center of the seesaw? **b.** The older sister moves so that the center of mass of the two children is directly above the center of the seesaw. How far did she move and in which direction? Did she move closer to or further from the center?
20. E Estimate the position of your center of mass. Explain your assumptions.
21. N A water molecule, consisting of two hydrogen atoms and one oxygen atom, resides in the plane of an x–y coordinate system. The coordinates of the oxygen atom and the two hydrogen atoms are, respectively, (0, 0), (-7.57×10^{-11} m, 5.86×10^{-11} m) and (7.57×10^{-11} m, 5.86×10^{-11} m). Where is the center of mass of the water molecule located in this coordinate system?
22. The center of mass of a meterstick is at the 50-cm mark. (We are ignoring the thickness of the stick.) You know where it is from the stick's symmetry, and you can confirm it by trying to balance the stick at one point. (You can only get it to balance when you support it at the 50-cm mark.) In this problem, we

find the center of mass of a meterstick by integration. Because we know the answer, this problem is actually about learning the method. So, follow these steps.

a. G Draw the meterstick lying horizontally along an x axis. Place the origin at the 0 mark.

b. G Imagine slicing the stick into small pieces. Each piece has mass dm and length dx. Draw on your sketch one such piece at an arbitrary position.

c. C The linear mass density λ of the stick is its mass M divided by its length L:

$$\lambda = \frac{M}{L}$$

If you had the meterstick in the laboratory, describe what you would measure to find its linear mass density. What are the SI units of λ?

d. A Find an expression for dm in terms of λ and dx.

e. N The stick's center of mass is found by substituting your expression for dm into $x_{CM} = (1/M)\int x\, dm$ and integrating. Do so, being sure to find the appropriate limits for the integration. Check your answer.

Problems 23 and 24 are paired.

23. N When there is a solar eclipse, the Moon is between the Earth and the Sun. Find the center of mass of this three-particle system during a solar eclipse.

24. When the Moon's phase is a "quarter Moon," the Earth, the Moon, and the Sun are at the vertices of a right triangle.

a. N Find the center of mass of the Earth–Moon–Sun system during this phase of the Moon.

b. C If you did Problem 23, use it to check and think about your result.

10-4 Systems of Particles

25. N The distance between Jupiter and one of its moons, Io, is 4.22×10^8 m. What is the location of the center of mass of the Jupiter–Io system if Jupiter's mass is 1.9×10^{27} kg and Io's mass is 8.9×10^{22} kg?

26. A person of mass m stands on a rope ladder that is hanging from a freely floating balloon of mass M. The balloon is initially at rest with respect to the ground. (The buoyant force on the person–balloon system is countering the force of gravity.)

a. C In what direction will the balloon move if the person starts to climb the rope ladder at constant velocity $\vec{v}$ relative to the ladder?

b. A At what speed will the balloon move if the person starts to climb the rope ladder at constant velocity $\vec{v}$ relative to the ladder?

27. N A rugby player with a mass of 65.0 kg is running to the right at a speed of 6.00 m/s toward another player of mass 90.0 kg, who is running in the opposite direction at a speed of 5.00 m/s. What is the total momentum of the two players (both magnitude and direction)?

28. A Your goal here is to show that $\Sigma\vec{F}_{ext} = M\vec{a}_{CM}$ holds for a system of particles. The system's mass M is constant; no particles enter or leave the system. **a.** Show that the velocity of the center of mass is given by

$$\vec{v}_{CM} = (1/M)\sum_{j=1}^{n} m_j\vec{v}_j$$

where m_j and $\vec{v}_j$ are the mass and the velocity of the jth particle, respectively. **b.** Find an expression for the acceleration of the center of mass. Then show that

$$M\vec{a}_{CM} = \sum_{j=1}^{n} m_j\vec{a}_j = \sum_{j=1}^{n}\vec{F}_j$$

where $\vec{F}_j$ is the net force exerted on particle j. **c.** The forces exerted on any particle j include both internal and external forces. Use Newton's third law to argue that the sum of the internal forces exerted on all the particles is zero so that $\sum_{j=1}^{n}\vec{F}_j$ is the sum of only the external forces $\Sigma\vec{F}_{ext}$. Explain your answer.

29. N Two particles with masses 2.0 kg and 4.0 kg are approaching each other with accelerations of 1.0 m/s^2 and 2.0 m/s^2, respectively, on a smooth, horizontal surface (with negligible friction). Find the magnitude of the acceleration of the center of mass of the system.

30. N A billiard player sends the cue ball toward a group of three balls that are initially at rest and in contact with one another. After the cue ball strikes the group, the four balls scatter, each traveling in a different direction with different speeds as shown in Figure P10.30. If each ball has the same mass, 0.16 kg, determine the total momentum of the system consisting of the four balls immediately after the collision.

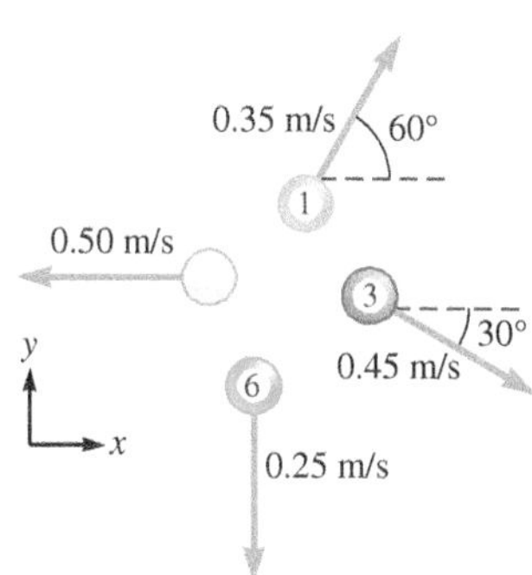

FIGURE P10.30

Problems 31 and 32 are paired.

31. C A crate of mass M is initially at rest on a frictionless, level table. A small block of mass m ($m < M$) moves toward the crate as shown in Figure P10.31. Later, the block and crate are stuck together and are moving with some final speed. The momentum of the block–crate system is the same both before and after the collision. Is the magnitude of the change in momentum of the crate greater than, less than, or equal to the magnitude of the change in the momentum of the block? Explain.

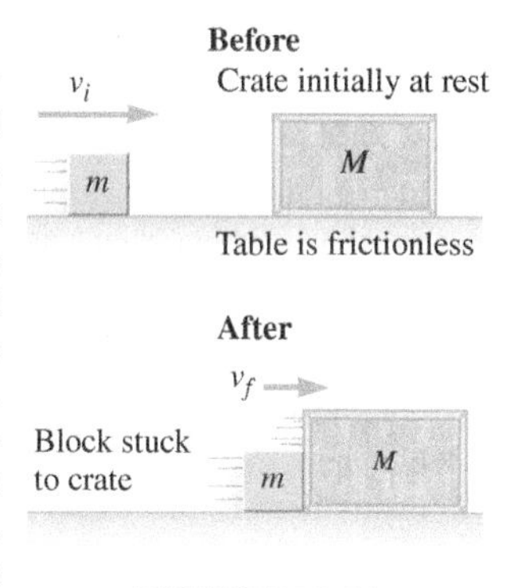

FIGURE P10.31

32. G Shown in the grid in Figure P10.32 are vectors representing the final momenta of the block and the crate in Figure P10.31. Reproduce the grid on your paper and draw the vector that represents the initial momentum of the block m in the space provided.

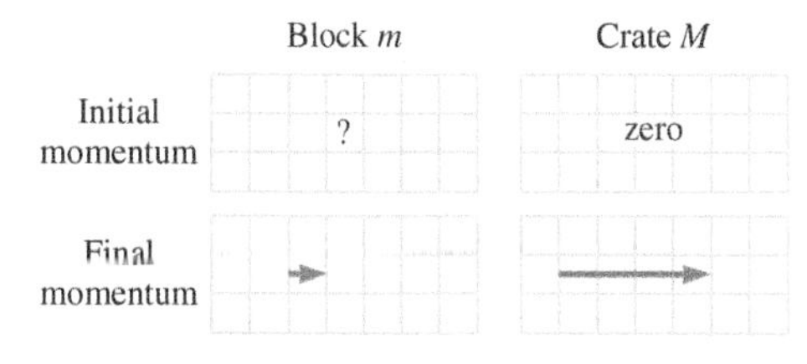

FIGURE P10.32

33. N A particle initially at rest is constrained to move on a smooth, horizontal surface. Another identical particle moving along the surface hits the stationary particle with a speed v. If the particles move together after the collision and their total momentum is the same as the initial momentum of the two-particle system, what is the speed of the combination just after the impact?

34. According to the National Academy of Sciences, the Earth's surface temperature has risen about 1°F since 1900. There is evidence that this *climate change* may be due to human activity. The organizers of World Jump Day argue that if the Earth were in a slightly larger orbit, we could avoid global warming and climate change. They propose that we move the Earth into this new orbit by jumping. The idea is to get people in a particular time zone to jump together. The hope is to have 600 million people jump in a 24-hour period. Let's see if it will work. Consider the Earth and its inhabitants to make up the system.

a. E Estimate the number of people in your time zone. Assume they all decide to jump at the same time; estimate the total mass of the jumpers.

b. C What is the net external force on the Earth–jumpers system?

c. N Assume the jumpers use high-tech Flybar pogo sticks (Fig. P8.32), which allow them to jump 6 ft. What is the displacement of the Earth as a result of their jump?

d. C What happens to the Earth when the jumpers land?

35. N Four metal spheres are placed along the x axis. The first sphere, with $m_1 = 1.00$ kg, is located at the origin; the second sphere, with mass $m_2 = 4.00$ kg, is located at $x = -4.00$ m; the third sphere, with mass $m_3 = 3.00$ kg, is located at $x = 2.00$ m; and the final sphere, with mass $m_4 = 6.00$ kg, is located at $x = 4.00$ m. What is the location of the center of mass of the system of the four spheres?

36. C In 1970, the carcass of a sperm whale beached near Florence, Oregon. The Oregon Highway Division had the task of removing it. They decided to use half a ton of dynamite to blow it up. The idea was that if the whale carcass were blown into smaller pieces, animals such as seagulls would eat the remains. A television news reporter caught the event on film. Pieces of whale blubber were thrown up to 0.25 mi from the blast site. One chunk of blubber even smashed a car. Include the dynamite and whale carcass in the system. **a.** What is the net external force on the system? **b.** Assuming the dynamite was placed near the whale carcass's center of mass, how was the blubber distributed after the blast? **c.** Would you expect to find the larger chunks of blubber near the blast site, far from the blast site, or randomly distributed? Explain your answer.

10-5 Conservation of Momentum

37. N Nicholas, with mass 75.0 kg, jumps vertically upward to a maximum height of 45.0 cm. With what speed does the Earth recoil because of his jump?

38. C Usually, we do not walk or even stand on a lightweight boat or raft because of the danger of falling into the water. If you have ever stepped off a small boat onto a dock, however, you have probably noticed that the boat moves away from the dock as you step toward the dock or out of the boat. A heavy dog running on a long lightweight raft presents a similar situation. At first, the raft and the dog are at rest with respect to the water (Fig. P10.38A) so that $\vec{v}_i = 0$. The dog then runs on top of the raft at $\vec{v}_d = v_d\hat{\imath}$ with respect to the water (Fig. P10.38B). The dog has twice the mass of the raft. Find an expression for the velocity of the raft after the dog began running.

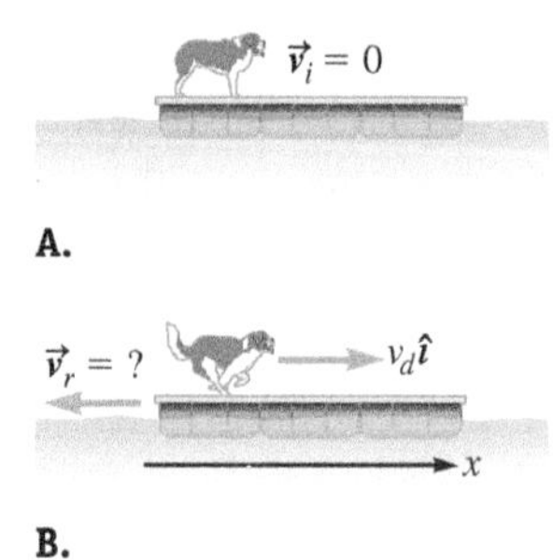

FIGURE P10.38

39. N A 175-g billiard ball is shot toward an identical ball at velocity $\vec{v}_i = 6.50\hat{\imath}$ m/s. The identical ball is initially at rest. After the balls hit, one of them travels with velocity $\vec{v}_{1,f} = (1.20\hat{\imath} + 2.52\hat{\jmath})$ m/s. What is the velocity of the second ball after the impact? Ignore effects of friction during this process.

Problems 40 and 41 are paired.

40. There is a compressed spring between two laboratory carts of masses m_1 and m_2. Initially, the carts are held at rest on a horizontal track (Fig. P10.40A). The carts are released, and the cart of mass m_1 has velocity $\vec{v}_1$ in the positive x direction (Fig. P10.40B). Assume rolling friction is negligible.

a. C What is the net external force on the two-cart spring system?

b. A Find an expression for the velocity of cart 2.

c. C Sometimes, mistakes are made in a laboratory. For example, what changes in parts (a) and (b) if the track is not level as shown in Figure P10.40C? Explain your answer.

41. N There is a compressed spring between two laboratory carts of masses $m_1 = 105$ g and $m_2 = 212$ g. Initially, the carts are held at rest on a horizontal track (Fig. P10.40A). The carts are released, and the cart of mass m_1 has velocity $\vec{v}_1 = 2.035\hat{\imath}$ m/s in the positive x direction (Fig. 10.40B). Assume rolling friction is negligible. **a.** What is the net external force on the two-cart system? **b.** Find the velocity of cart 2.

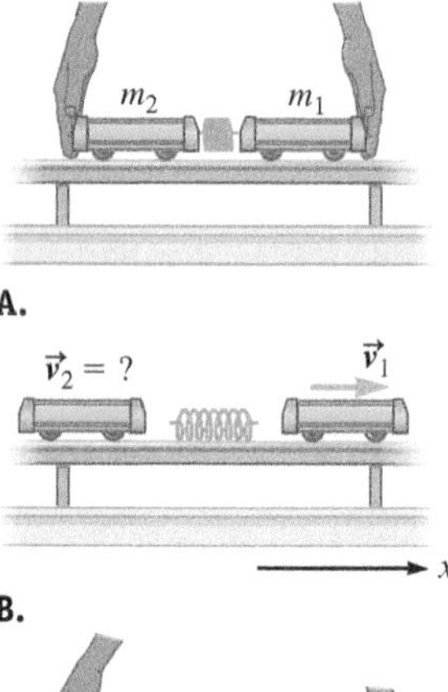

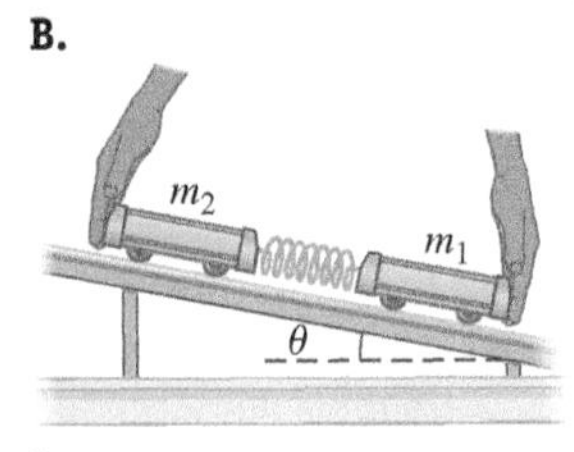

FIGURE P10.40 Problems 40 and 41.

42. N A submarine with a mass of 6.26×10^6 kg contains a torpedo with a mass of 354 kg. The submarine fires the torpedo at an angle of 25° with respect to the horizontal as shown in Figure P10.42. **a.** If the submarine and the torpedo were initially at rest and the torpedo left the submarine with a speed of 89.2 m/s, what is the recoil speed of the submarine? **b.** What is the direction of recoil of the submarine?

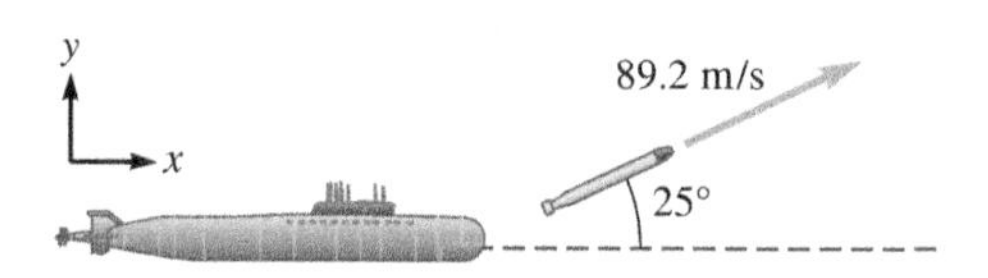

FIGURE P10.42

43. A 44.0-kg child finds himself trapped on the surface of a frozen lake, 10.0 m from the shore. The child slips with each step on the frictionless ice and remains the same distance from the shoreline. Egged on by his parents, he throws a 0.750-kg ball he is carrying toward the center of the lake with a horizontal speed of 1.50 m/s, in the direction opposite that of the shoreline.

a. N Does the act of throwing the ball cause the child to move? If so, what are the speed and the direction of his motion with respect to the Earth?

b. C What are the forces acting on the child when he throws the ball?

Problems 44 and 45 are paired.

44. C A model rocket is shot straight up. As it reaches the highest point in its trajectory, it explodes in midair into three pieces with velocities indicated by the arrows in Figure P10.44, as viewed from directly above the explosion. Rank the mass of each piece in order from smallest to largest and justify your answer.

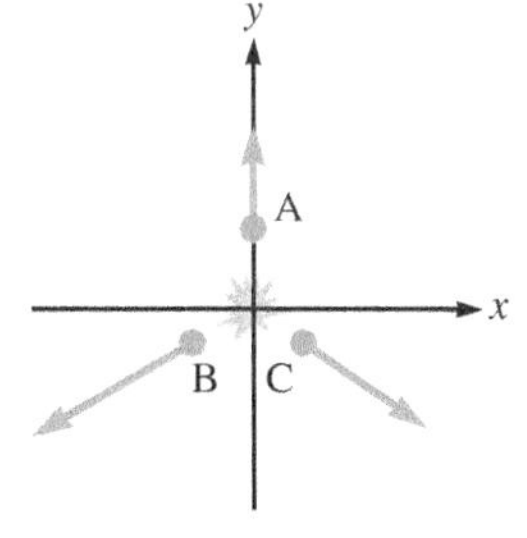

FIGURE P10.44 Problems 44 and 45.

45. N A model rocket is shot straight up and explodes at the top of its trajectory into three pieces as viewed from above and shown in Figure P10.44. The masses of the three pieces are $m_A = 100.0$ g, $m_B = 20.0$ g, and $m_C = 30.0$ g. Immediately after the explosion, piece A is traveling at 1.50 m/s, and piece B is traveling at 7.00 m/s in a direction 30° below the negative x axis as shown. What is the velocity of piece C?

Problems 46 and 47 are paired.

46. C, N An astronaut finds herself in a predicament in which she has become untethered from her shuttle. She figures that she could get back to her shuttle by throwing one of three objects she possesses in the opposite direction of the shuttle. The masses of the objects are 5.6 kg, 7.9 kg, and 10.3 kg, respectively. She is able to throw the first object with a speed of 15.00 m/s, the second with a speed of 10.0 m/s, and the third with a speed of 7.5 m/s. If the mass of the astronaut and her remaining gear is 75.0 kg, which object should she throw to get back to the shuttle as quickly as possible? Justify your choice.

47. N Using the information in Problem 46, determine the final speed of the astronaut with respect to the shuttle if she were to throw each object successively, starting with the least massive and ending with the most massive. Assume that the speeds described are those measured in the rest frame of the astronaut.

10-6 Case Study: Rockets

48. N A model rocket consisting of a 60.7-g chassis and a 30.5-g engine is taken on an interstellar mission and fired in outer space. The engine produces 4.95 N of thrust while expending 13.5 g of fuel at a constant rate during a 2.10-s burn. **a.** What is the average exhaust speed of this rocket's engine? **b.** If the rocket starts from rest, what is its final velocity?

Problems 49, 50 and 55 are grouped.

49. N A tennis ball machine holds 200 balls and has a mass of 38.5 kg when empty. Each tennis ball has a mass of 56.7 g. A coach has set the machine to launch balls horizontally with a speed of 52.0 m/s. He forgot to set the brakes on the machine. If rolling friction between the machine and the court is negligible and all 200 balls are launched, what is the final speed of the machine?

50. N A physics student has set the ball machine in Problem 49 to launch balls horizontally with a speed of 52.0 m/s. The brakes on the machine are set so that it does not move. The student has constructed a "land boat" out of a large grocery cart and a cardboard sail. Each ball is guided into the cart by the cardboard sail. The cart of mass 15.3 kg is initially at rest. If rolling friction between the cart and the floor is negligible and all 200 balls are launched, what is the final speed of the cart?

51. N The space shuttle uses its thrusters with an exhaust velocity of 4440 m/s. The shuttle is initially at rest in space and accelerates to a final speed of 1.00 km/s. **a.** What percentage of the initial mass of the shuttle (including the full fuel tank) must be ejected to reach that speed? **b.** If the mass of the shuttle and fuel is initially 1.85×10^6 kg, how much fuel is expelled?

52. N A shuttlecraft is initially moving at 94.5 m/s as it travels in deep space. The mass of the shuttle given its current fuel level is approximately 6.33×10^6 kg. When the shuttle fires its rockets to move faster, the speed of the exhaust leaving the rockets is 2750 m/s. What is the mass of the fuel necessary to get the shuttlecraft up to a final speed of 125 m/s?

Problems 53 and 54 are paired.

53. C CASE STUDY Shannon is not convinced by the argument in Example 10.7 (page 295), so the conversation continues. Avi and Shannon are sitting on the grass near the softball field.

Shannon: I just don't think that you can propel a rocket or even a boat by pushing something off of it. Catch this. (Shannon throws a softball to Avi). Why am I not moving backward?

Answer Shannon's question.

54. C CASE STUDY Shannon and Avi continue to explore the case study (page 282). They sit on skateboards facing each other. Initially, they are both at rest. Shannon, who is very strong, holds a heavy (medicine) ball.

Shannon: I'm going throw this ball to you. You catch it. If the system consists of you, me, and the ball, there are no external forces and momentum is conserved, right? So, I will move backward.

Avi: Yeah, you'll go backward, and so will I.

Shannon: That doesn't make sense. Momentum is already conserved by my going backward. If you go backward too, somehow the system would have to gain momentum. You got that from watching too many cartoons.

Shannon throws the ball to Avi, who catches it. Are either of them moving afterward? If so, who? Explain your answers.

10-7 Rocket Thrust: An Open System (Optional)

55. N A coach has set the machine in Problem 49 to launch balls horizontally with a speed of 52.0 m/s at a rate of one ball every 1.50 s. He forgot to set the brake on the machine. If the coefficient of rolling friction between the machine and the court is 0.00300 and all 200 balls are launched out, what is the maximum speed of the machine? (Compare your result with that of Problem 49.)

56. N The cryogenic main stage of a rocket has an exhaust speed of 4.21×10^3 m/s and burns liquid hydrogen and liquid oxygen at a combined rate of 317 kg/s. **a.** What is the thrust produced by the rocket's main engine? **b.** If the initial mass of the rocket is 1.10×10^5 kg, what is the initial acceleration of the rocket upon liftoff from the Earth?

57. N To lift off from the Moon, a 9.50×10^5 kg rocket needs a thrust larger than the force of gravity. If the exhaust velocity is 4.25×10^3 m/s, at what rate does the exhaust need to be expelled to provide sufficient thrust? The acceleration due to gravity on the Moon is 1.62 m/s^2.

General Problems

58. N A system consists of two particles. The first particle has mass $m_1 = 5.00$ kg and a velocity of $(-4.00\hat{\imath} + 2.00\hat{\jmath})$ m/s, and the second particle has mass $m_2 = 2.00$ kg and a velocity of $(2.00\hat{\imath} + 4.00\hat{\jmath})$ m/s. **a.** What is the velocity of the center of mass of this system? **b.** What is the total momentum of this system?

59. N A ball of mass $m = 450.0$ g traveling at a speed of 8.00 m/s impacts a vertical wall at an angle of $\theta_i = 45.0°$ below the horizontal (x axis) and bounces away at an angle of $\theta_f = 45.0°$ above the horizontal. What is the average force exerted by the wall on the ball if the ball is in contact with the wall for 250.0 ms?

60. C In the original concept, Superman reaches great heights by jumping. In later stories, Superman is able to fly. **a.** In the original concept, is Superman modeled as a rocket? Explain. **b.** Ignore the means by which he levitates in the later stories, and account for how he propels himself.

61. N Two balls with masses $m_1 = 2.0$ g and $m_2 = 3.0$ g move in a plane with velocities $\vec{v}_1 = 6.0\hat{\imath}$ m/s and $\vec{v}_2 = 4.0\hat{\jmath}$ m/s. What is the total momentum of the system?

62. N An astronaut out on a spacewalk to construct a new section of the International Space Station walks with a constant velocity of 2.00 m/s on a flat sheet of metal placed on a flat, frictionless, horizontal honeycomb surface linking the two parts of the station. The mass of the astronaut is 75.0 kg, and the mass of the sheet of metal is 245 kg. **a.** What is the velocity of the metal sheet relative to the honeycomb surface? **b.** What is the speed of the astronaut relative to the honeycomb surface?

63. N Two particles with masses 2.0 kg and 4.0 kg are moving in the same direction with speeds 2.0 m/s and 3.0 m/s, respectively, on a smooth, horizontal surface (with negligible friction). Find the speed of the center of mass of the system.

64. The large slingshot launcher discussed in Section 10-6 exerts a force on the water balloon that is given by the formula $F = 500(1 - t^2)$, where the time is measured in seconds and the calculated force is in newtons.

a. G The slingshot is released and begins exerting a force on the water balloon when $t = 0$. The balloon is considered to be launched when the force on the balloon is zero. Plot the force versus time for the event of launching the balloon. Indicate the time at which the balloon is launched.

b. A Write an expression for the magnitude of the momentum of the water balloon as a function of time during its launch.

c. G Plot the momentum versus time until the moment the balloon is launched.

65. N A racquetball of mass $m = 43.0$ g, initially moving at 30.0 m/s horizontally in the positive x direction, is struck by a racket. After being struck, the ball moves back in the opposite direction at an angle of 30.0° above the horizontal with a speed of 50.0 m/s. What is the average vector force exerted on the racket by the ball if they are in contact for 2.50 ms?

66. N A parcel moving in a horizontal direction with speed $v_0 = 12$ m/s breaks into two fragments of weights 1.0 N and 1.5 N, respectively. The speed of the larger piece remains horizontal immediately after the separation and increases to $v_{1.5} = 25$ m/s. Find the necessary speed and direction of the smaller piece immediately after the separation.

67. N The position of a particle of mass $m_1 = 1.00$ kg is described by the vector $\vec{r}_1 = (2t + 5t^2)\hat{\imath} + 4t\hat{\jmath}$, where t is in seconds and $\vec{r}$ is in meters. The motion of a second particle, of mass $m_2 = 2.50$ kg, is described by the vector $\vec{r}_2 = (3 - 5t)\hat{\imath} + (-t - 2t^2)\hat{\jmath}$. **a.** What is the vector position of the center of mass at $t = 3.00$ s? **b.** What is the velocity of the center of mass at $t = 3.00$ s? **c.** What is the total linear momentum of the system at $t = 3.00$ s?

68. Alika, with mass 45.0 kg, and Armon, with mass 55.0 kg, are both atop their skateboards at rest on a frictionless, horizontal surface. Alika pushes Armon, giving him a speed of 3.00 m/s to the right.

a. N What are the magnitude and the direction of Alika's speed after the push?

b. N What is the amount of potential energy converted by Alika into kinetic energy for the system?

c. C Is momentum conserved in the Alika–Armon system during the push?

d. C Explain your answer to part (c) given that there are large forces acting between Alika and Armon during the push and that their speeds are zero initially and nonzero after the push.

69. N A comet is traveling through space with speed 3.33×10^4 m/s when it encounters an asteroid that was at rest. The comet and the asteroid stick together, becoming a single object with a single velocity. If the mass of the comet is 1.11×10^{14} kg and the mass of the asteroid is 6.66×10^{20} kg, what is the final velocity of their combination?

70. N A ballistic pendulum is used to measure the speed of bullets. It comprises a heavy block of wood of mass M suspended by two long cords. A bullet of mass m is fired into the block horizontally. The block, with the bullet embedded in it, swings upward (Fig. P10.70). The center of mass of the combination rises through a vertical distance h before coming to rest momentarily. In a particular experiment, a bullet of mass 40.0 g is fired into a wooden block of mass 10.0 kg. The block–bullet combination is observed to rise to a maximum height of 20.0 cm above the block's initial height. **a.** What is the initial speed of the bullet? **b.** What is the fraction of initial kinetic energy lost after the bullet is embedded in the block?

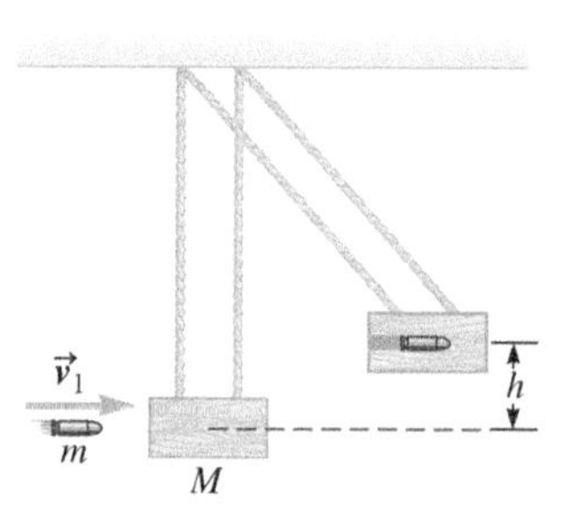

FIGURE P10.70

71. A Rochelle and Sheldon are two astronauts in space tethered together by a strong rope of length L. Initially, Sheldon is holding a large tool, which Rochelle needs. He then throws it to Rochelle, who catches it. Sheldon, Rochelle, and the tool all have the same mass m. Find an expression for the distance traveled by the tool relative to the center of mass of the system.

72. C Physics students at a prominent school were given an assessment test. Consider a winter service truck moving at constant velocity $\vec{v}$ along an icy road. The truck ejects sand out the back at a constant rate $\Delta m/\Delta t$. The sand falls straight down with zero horizontal momentum. The question is, what is the net force on the truck? No students got the correct answer. Most students said that the net force on the truck is zero. Why is this answer incorrect? What is the correct answer?

73. A Joe is teaching Buddy to ice-skate. They each hold an end of a long rope of length L. Initially, they are at rest. Buddy then pulls herself along the rope until she reaches Joe. Buddy's mass is three-fourths of Joe's mass. Find an expression for the distance traveled by Buddy.

74. C Figure P10.74 provides artists with human proportions. Notice that the center of mass moves lower in the body as the person grows. Explain this change. What does it tell you about human proportions as a person grows?

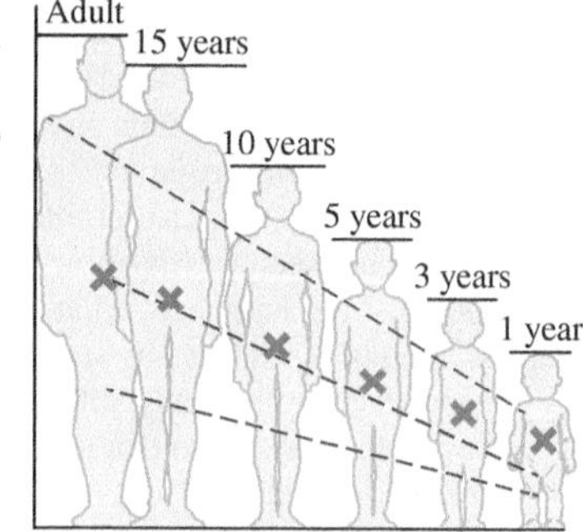

FIGURE P10.74

75. N During the final seconds of the Iditarod race, a sled's harness breaks, sending the team of dogs and the sled and driver careening away from one another on the icy surface. The sled, its driver, and his backpack, with combined mass 275 kg, move with a speed of 3.00 m/s in the positive x direction. In an attempt to catch the harness without slowing down, the driver throws his heavy backpack of mass 20.0 kg in the direction opposite the motion of the sled, giving it a final speed relative to the sled of 5.00 m/s in the opposite direction. **a.** What is the final speed of the sled and driver relative to the ground? **b.** What is the final speed of the backpack relative to the ground?

76. A single-stage rocket of mass 308 metric tons (not including fuel) carries a payload of 3150 kg to low-Earth orbit. The exhaust speed of the rocket's cryogenic propellant is 3.20×10^3 m/s.

a. N If the speed of the rocket as it enters orbit is 8.00 km/s, what is the mass of propellant used during the rocket's burn?

b. N The rocket is redesigned to boost its exhaust speed by a factor of two. What is the mass of propellant used in the redesigned rocket to carry the same payload to low-Earth orbit?

c. C Because the exhaust speed of the redesigned rocket is increased by a factor of two, why is the fuel consumption of the redesigned rocket not exactly half that of the original rocket?

77. C If you try standing on one leg, you will see that your body shifts (Fig. P10.77). Explain why this shift is necessary to prevent you from falling.

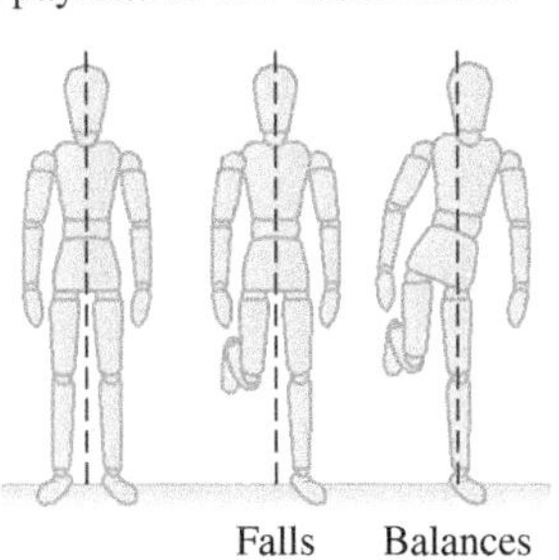

FIGURE P10.77

78. N A light spring is attached to a block of mass $4m$ at rest on a frictionless, horizontal table. A second block of mass m is now placed on the table, in contact with the free end of the spring, and the two blocks are pushed together (Fig. P10.78). When the blocks are released, the more massive block moves to the left at 2.50 m/s. **a.** What is the speed of the less massive block? **b.** If $m = 1.00$ kg, what is the elastic potential energy of the system before it is released from rest?

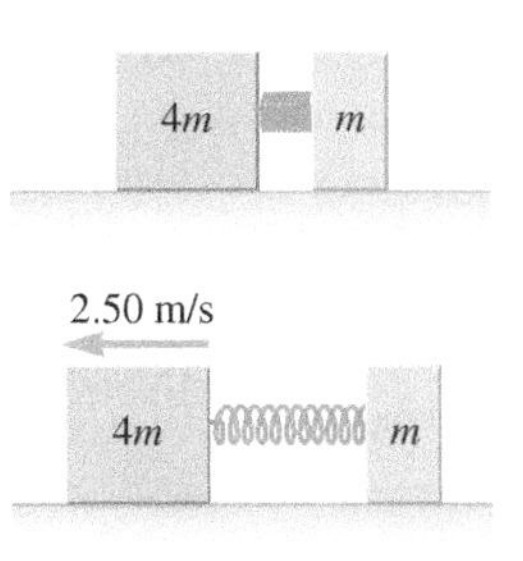

FIGURE P10.78

79. N Two blocks of mass 1.00 kg and 0.500 kg are pressed together on a frictionless, horizontal surface, compressing a light spring of force constant 8.1 N/m between them by 6.50 cm (similar to Fig. P10.78). The two blocks are released simultaneously from rest. What is the maximum speed of each block?

80. C In Figure P10.80A, a man shows a yoga pose known as the chair pose. Notice that the man leans forward so that he looks like a folding chair. In Figure P10.80B, a woman uses a wall so that she forms a straight-back chair. Why is the wall necessary to create this straight-back chair?

FIGURE P10.80

81. A Show that the total momentum of a system of particles equals the momentum of the center of mass:

$$\vec{p}_{\text{tot}} = \sum_{j=1}^{n} \vec{p}_j = \vec{p}_{\text{CM}} \quad (10.8)$$

82. N Your friend is making an animated short film, and he consults you on the physics of a scene. In the scene, a penguin and a dog are riding on a model train set. The penguin is in the front car, and the dog is in the second car (Fig. P10.82). The train is moving quickly at speed v_0 when the power is cut. The penguin then pulls the pin connecting its car to the rest of the train. The dog and the remaining cars continue at the same speed v_0 along the track. The penguin and the first car have about one-fifth of the total mass, and the dog and the remaining cars make of four-fifths of the mass. Your friend believes that the penguin will speed up because momentum is conserved and the penguin's part of the system has less mass than the other part of the system. Would such a scene violate laws of physics? To answer this question, find the penguin's speed v_p in terms of v_0 after the cars have been separated. You may ignore friction and drag.

FIGURE P10.82

83. A CASE STUDY Show that for an open system such as a rocket, the second rocket equation is

$$M_R(\vec{a}_R)_{\oplus} = \sum \vec{F}_{\text{ext}} + \vec{F}_{\text{thrust}} \quad (10.17)$$

11 Collisions

Key Question

How can the principle of conservation of momentum be applied to the complicated situation of colliding objects?

11-1 What is a collision? 307

11-2 Impulse 307

11-3 Conservation during a collision 310

11-4 Special case: One-dimensional inelastic collisions 312

11-5 One-dimensional elastic collisions 315

11-6 Two-dimensional collisions 320

Underlying Principles

1. Impulse–momentum theorem
2. Conservation of momentum (Chapter 10)

Major Concepts

1. Collision
2. Impulse approximation
3. Impulse
4. Inelastic collision
5. Completely inelastic collision
6. Elastic collision

Special Cases

One-dimensional and two-dimensional collisions

Collisions occur naturally on scales from subatomic particles to whole galaxies. In July 1994, Comet Shoemaker-Levy 9, consisting of more than 20 fragments, crashed into Jupiter. The larger comet fragments left scars in Jupiter's atmosphere that were larger than the Earth and took more than 6 months to heal. Not all collisions are as dramatic. Air molecules colliding with your skin keep you warm, and photons colliding with cells in your eyes allow you to see. Some collisions are deliberate, such as when a car manufacturer tests the safety of a new car or when a physicist studies subatomic particles in an accelerator. In a particularly spectacular example, a spacecraft was designed to collide with a comet so as to create a crater and observe the comet's interior (Fig. 11.1).

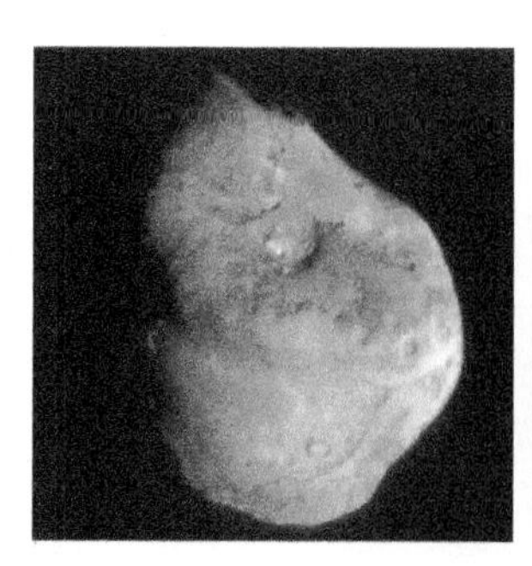

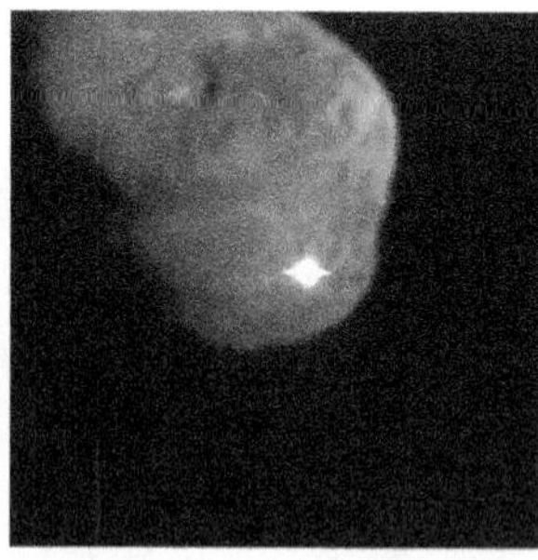

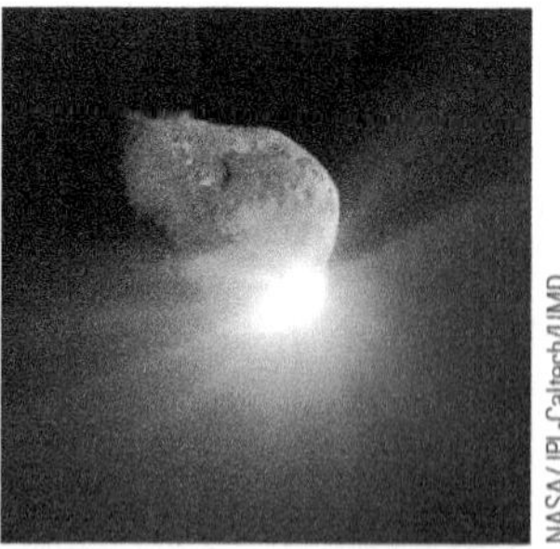

FIGURE 11.1 A spacecraft impacts Comet Tempel 1, throwing material into space.

11-1 What Is a Collision?

We have already analyzed a variety of situations using the principles of the conservation of energy and momentum. In this chapter, we apply the conservation approach to *collisions*. Some collisions involve direct contact between the colliding object, as when two cars collide. Other collisions involve "close encounters," as when a spacecraft is sent on a special type of collision course with a planet known as a *gravitational slingshot*, allowing the spacecraft to pick up kinetic energy as it passes near the planet. In physics, a **collision** is an isolated event in which two or more objects exert relatively strong forces on one another for a relatively short time. We usually assume the force between the colliding objects during the time of the collision is much stronger than any other forces exerted on either of them; this assumption is known as the **impulse approximation**. If the force of interaction is a field force such as gravity between a spacecraft and a planet, the objects do not need to touch for them to collide.

COLLISION ✪ Major Concept

IMPULSE APPROXIMATION ✪ Major Concept

CASE STUDY Train Collision Revisited

In Chapter 5, we studied a train collision that took place on April 23, 2002. A passenger train about 35 miles outside of Los Angeles was hit by a freight train (Fig. 5.1). The accident killed two people and injured more than 260. All the injured were on the passenger train. News reports said that the passenger train came to a quick stop before the collision and that the impact with the freight train pushed the passenger train 370 ft backward. One of the most controversial parts of the early reports was the speed of the freight train at the moment of impact. Using physics to reconstruct the events of such an accident is routine in an investigation. In this chapter, we reconstruct the accident and estimate the freight train's speed upon impact.

CONCEPT EXERCISE 11.1

CASE STUDY Forensic Science

Forensic science is the application of scientific techniques and methodology to discover the cause of a particular event such as an accident. Modern forensics have been used to solve historic crimes and mysteries such as the cause of death of King Tutankhamen of ancient Egypt and composer Ludwig van Beethoven in 1827. Forensic evidence is often admitted into a court case. For example, skid marks at an accident scene are used to find the initial speed of a car. Explain why skid marks are left by the car and how they may be used to find the car's initial speed.

11-2 Impulse

You might not usually think of it this way, but catching something can be modeled as a collision. Consider an old picnic game known as the *egg toss*. A team of two players tosses a raw egg back and forth. Each time a player catches the egg without breaking it, he must take a step backward. If the egg breaks, the team is out of the game. The team whose members end up the farthest apart wins. Players seem to know instinctively how to catch the egg gently; no player tries to catch it quickly. Instead, he tries to meet the egg with his fingertips, cupping his hands and moving them backward as the egg stops (Fig. 11.2). This gentle catch means that the egg takes a longer time to stop moving than would a tossed baseball.

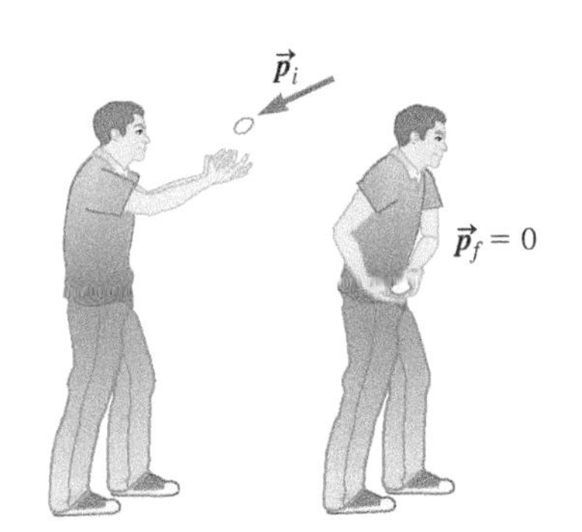

FIGURE 11.2 A player in an egg toss gently catches an egg by allowing it to move from his fingertips to his palm as he moves his hands backward and downward.

EXAMPLE 11.1 How to Win the Egg Toss

The length of time the player spends making the catch actually determines the magnitude of the force he must apply to the egg to make it stop. Show that the average force $\vec{F}_{p,\,\text{av}}$ exerted by a player (Fig. 11.2) is related to the time Δt he takes to stop the egg and the egg's initial momentum $\vec{p}_i$ by

$$\vec{F}_{p,\,\text{av}}\Delta t \approx -\vec{p}_i \tag{11.1}$$

INTERPRET and ANTICIPATE

Newton's second law written in terms of momentum (Eq. 10.2) relates change in momentum to force, so it is a good starting point.

$$\sum \vec{F}_{\text{ext}} = \frac{d\vec{p}}{dt}$$

SOLVE

Two forces act on the egg as the player catches it: gravity $\vec{F}_g$ and the player's applied force $\vec{F}_p$.

$$\vec{F}_g + \vec{F}_p = \frac{d\vec{p}}{dt}$$

Multiply each side by dt and integrate from the initial time when the player first makes contact with the egg to the final time when the egg has stopped. Because the gravitational force on the egg is constant, we can pull it outside the integral. The force exerted by the player is not constant, however.

$$\vec{F}_g dt + \vec{F}_p dt = d\vec{p}$$

$$\vec{F}_g \int_{t_i}^{t_f} dt + \int_{t_i}^{t_f} \vec{F}_p dt = \int_{p_i}^{p_f} d\vec{p}$$

The magnitude of the integral $\int_{t_i}^{t_f} \vec{F}_p dt$ is the area under the graph of F_p versus t (Fig. 11.3A). This area is equal to the area of the rectangle of height $F_{p,\,\text{av}}$ and width Δt, where $F_{p,\,\text{av}}$ is the magnitude of the average force exerted by the player on the egg during the time interval $\Delta t = t_f - t_i$ (Fig. 11.3B).

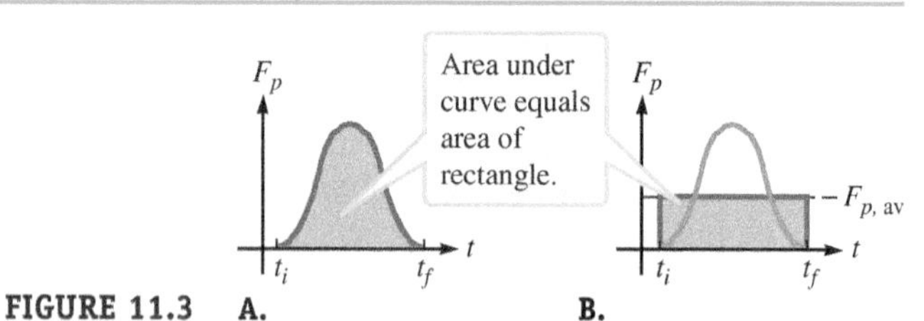

FIGURE 11.3 A. B.

Complete the integration.

$$\vec{F}_g \int_{t_i}^{t_f} dt + \vec{F}_{p,\,\text{av}} \int_{t_i}^{t_f} dt = \int_{p_i}^{p_f} d\vec{p}$$

$$\vec{F}_g(t_f - t_i) + \vec{F}_{p,\,\text{av}}(t_f - t_i) = \vec{p}_f - \vec{p}_i$$

$$\vec{F}_g\Delta t + \vec{F}_{p,\,\text{av}}\Delta t = \Delta\vec{p}$$

According to the impulse approximation, during the short time of the collision, the force exerted by the colliding objects on each other is much stronger than any other forces exerted on either of them. So, the force exerted by the player on the egg is much greater than gravity. In other words, the change in the egg's momentum is primarily due to the force exerted by the player.

$$\vec{F}_{p,\,\text{av}}\Delta t \approx \Delta\vec{p} = \vec{p}_f - \vec{p}_i \tag{11.2}$$

The egg has momentum $\vec{p}_i$ when it first makes contact with the player's fingers. The player's job is to stop the egg so that its final momentum $\vec{p}_f$ is zero.

$$\vec{F}_{p,\,\text{av}}\Delta t \approx 0 - \vec{p}_i = -\vec{p}_i \ ✓ \tag{11.1}$$

CHECK and THINK

Equation 11.1 shows how the player uses time to catch the egg gently. The egg has a nonzero initial momentum when it reaches the player's hand; that initial momentum is a result of gravity and how the egg was thrown by his teammate. According to Equation 11.1, the player catching the egg must apply an average force $\vec{F}_{p,\,\text{av}}$ over a time interval Δt such that combination $\vec{F}_{p,\,\text{av}}\Delta t$ equals the magnitude of the egg's initial momentum. The player cannot change the initial momentum of the egg, but he can control the force he exerts on the egg by controlling the time over which he makes the catch. If he takes a long time, the average force he exerts on the egg is weak, and the egg will not break. If he stops the egg too quickly, the average force he exerts on the egg will be strong, the egg will break, and he is out of the game.

The egg toss illustrates that the outcome of a collision depends in part on the time interval during which the collision occurs. This relationship between time interval and force is explicitly described by a quantity known as *impulse*. The **impulse** exerted on an object during a time interval Δt is a vector defined as

$$\vec{I} \equiv \int_{t_i}^{t_f} \vec{F}(t)dt = \vec{F}_{av}\Delta t \qquad (11.3)$$

IMPULSE Major Concept

The impulse points in the same direction as the force exerted on the object. Be careful, though; impulse is not the same as force.

Comparing $\vec{I} = \vec{F}_{av}\Delta t$ (Eq. 11.3) with $\vec{F}_{p,\,av}\Delta t = \Delta\vec{p}$ (Eq. 11.2), we can see that the impulse is also equal to the change in momentum. To put it generally, the total impulse exerted on an object equals the change in the object's momentum:

$$\vec{I}_{tot} = \sum\vec{I} = \Delta\vec{p} \qquad (11.4)$$

IMPULSE–MOMENTUM THEOREM Underlying Principle

Equation 11.4 is referred to as the **impulse–momentum theorem**, and it is another statement of Newton's second law (Problem 12). According to the impulse–momentum *theorem*, the total impulse exerted on an object is in the same direction as the *change* in the object's momentum. According to the impulse *approximation* during a collision, we can usually ignore all the forces except those exerted by one colliding object on the other. So, applying the impulse approximation means that Equation 11.4 reduces to just the impulses exerted by one colliding object on the other(s).

Impulse has both the dimensions of force multiplied by time (Eq. 11.3) and of momentum (Eq. 11.4). In SI units, impulse is given either in N · s or kg · m/s.

CONCEPT EXERCISE 11.2

Why does a coach instruct a gymnast to bend her knees when she lands on the ground?

EXAMPLE 11.2 "Heads Up!"

Two balls, one of mass 2.5 kg and the other of mass 5.0 kg, are dropped from a building at the same time. Drag on the balls is negligible, and they land on the ground 5.3 s after they are released. In Example 10.1, we found that the momentum of the two balls just as they reached the ground was $\vec{p}_1 = -1.3 \times 10^2\hat{\jmath}$ kg · m/s and $\vec{p}_2 = -2.6 \times 10^2\hat{\jmath}$ kg · m/s. Use the same coordinate system as in Figure 10.2 (page 283).

A Find the impulse exerted by the ground on the balls.

INTERPRET and ANTICIPATE

The momenta provided are *initial* momenta for the ball–ground collisions. The ground stops the balls, and their final momenta are zero. The only other force acting on the balls during their collision with the ground is gravity, which (according to the impulse approximation) is weak compared with the force exerted by the ground. We ignore gravity, so, according to the impulse–momentum theorem, the change in each ball's momentum equals the impulse exerted by the ground. We expect to find two numerical results of the form $\vec{I} =$ ______ $\hat{\jmath}$ kg · m/s.

SOLVE

During the collision, the only significant force on each ball is exerted by the ground (subscript "grd"), so Equation 11.4 reduces to one term on the left.

$$\vec{I}_{tot} = \sum\vec{I} = \Delta\vec{p} \qquad (11.4)$$

$$\vec{I}_{tot} = \vec{I}_{grd} = \Delta\vec{p}$$

Substitute for the momenta. The final momentum is zero because the balls eventually come to rest on the ground. The initial momenta were found in Example 10.1.

$$\vec{I}_{grd} = \Delta\vec{p} = \vec{p}_f - \vec{p}_i = 0 - \vec{p}_i = -\vec{p}_i$$

$$\vec{I}_{1\,grd} = -\vec{p}_{1\,i} = 1.3 \times 10^2\hat{\jmath} \text{ kg} \cdot \text{m/s}$$

$$\vec{I}_{2\,grd} = -\vec{p}_{2\,i} = 2.6 \times 10^2\hat{\jmath} \text{ kg} \cdot \text{m/s}$$

Example continues on page 310 ▶

CHECK and THINK

The impulse exerted by the ground on the balls points upward as we expected. Notice that the ground exerts twice as much impulse on the ball having twice the mass.

B If it takes 0.005 s for the ground to stop each ball, find the average force exerted by the ground on the 5.0-kg ball. Compare the magnitude of the force exerted by the ground with the ball's weight.

INTERPRET and ANTICIPATE

In part A, we found the impulse. Now that we know the time interval of the collisions, we can find the average force exerted by the ground. We can compare it with the weight of the ball to see if gravity can be ignored as we did in part A. Our results should be in the form $\vec{F}_{av} = \underline{\quad\quad}\hat{\jmath}$ N.

SOLVE

Solve $\vec{I} = \vec{F}_{av}\Delta t$ (Eq. 11.3) for the average force.	$\vec{F}_{av} = \dfrac{\vec{I}}{\Delta t}$
Use the impulse found in part A. Recall that $1\text{ N} = 1\text{ kg}\cdot\text{m/s}^2$. The subscript "grd" was omitted from the average force for simplicity.	$\vec{F}_{av} = \dfrac{\vec{I}_{grd}}{\Delta t} = \dfrac{1.3 \times 10^2\hat{\jmath}\text{ kg}\cdot\text{m/s}}{0.005\text{ s}}$ $\vec{F}_{av} = 2.6 \times 10^4\hat{\jmath}\text{ N}$
Now find the ball's weight.	$w = m_1 g = (2.5\text{ kg})(9.81\text{ m/s}^2) = 25\text{ N}$
To compare the magnitude of the average force exerted by the ground with the ball's weight, find the ratio F_{av}/w.	$\dfrac{F_{av}}{w} = \dfrac{2.6 \times 10^4\text{ N}}{25\text{ N}} = 1.0 \times 10^3$

CHECK and THINK

The results are in the form we expected. The force exerted by the ground is about 1000 times the ball's weight. Therefore, we are justified in neglecting gravity during these collisions.

11-3 Conservation During a Collision

To analyze a collision with the conservation approach, we choose a system that includes the colliding objects. For objects (modeled as particles) to collide, at least one of the objects must be moving relative to the other(s). So, before the collision, the system has both momentum and kinetic energy. For the collisions we study, usually momentum is conserved (or nearly so), and kinetic energy may or may not be conserved.

Conservation of Momentum During a Collision

If the system of colliding objects is closed (constant mass) and no net external force is exerted on the system, the system's momentum must be conserved (Section 10-5). According to the impulse approximation, the forces exerted by the colliding objects on each other are often much stronger than any external forces; so, it may be acceptable to apply conservation of momentum to a system if the net external force is relatively weak. For example, if we are interested in the collision of an arrow and a tin can, we may ignore the gravitational force exerted by the Earth on the arrow–can system and apply the conservation of momentum principle to the system during the time of the collision. Of course, we must take gravity into account before and after the collision. Throughout this textbook, we assume *the total linear momentum of a system of colliding objects remains constant during the collision.*

So, if the impulse approximation holds, the total momentum of the system before the collision must equal the total momentum of the system after the collision:

$$(\vec{p}_{tot})_i = (\vec{p}_{tot})_f \tag{11.5}$$

Sometimes a more convenient way to express conservation of momentum is in terms of the system's center of mass. The total momentum of a system is equal to the momentum of the system's center of mass (Eq. 10.8). So, conservation of momentum can be written as

CONSERVATION OF MOMENTUM
Underlying Principle

$$(\vec{p}_{CM})_i = (\vec{p}_{CM})_f \tag{11.6}$$

From this equation, we can show that during a collision the center-of-mass velocity is conserved:

$$M(\vec{v}_{CM})_i = M(\vec{v}_{CM})_f$$

$$(\vec{v}_{CM})_i = (\vec{v}_{CM})_f \tag{11.7}$$

Equation 11.7 is consistent with what we already know from Chapter 10. A net external force is required to accelerate a system's center of mass. So, if there is no net external force (or if the net external force is negligible) during the collision, the center-of-mass velocity is not changed. The implications of conservation of momentum during a collision are summarized in the concept map shown in Figure 11.4.

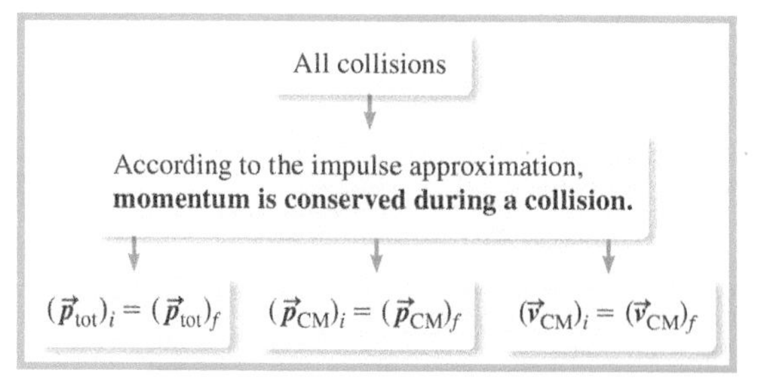

FIGURE 11.4 This concept map shows the implications of the conservation of momentum principle during a collision.

CONCEPT EXERCISE 11.3

When two objects collide, the impulse exerted on object 1 by object 2 is equal in magnitude and opposite and direction to the impulse exerted on object 2 by object 1:

$$\vec{I}_{[1\text{ on }2]} = -\vec{I}_{[2\text{ on }1]} \tag{11.8}$$

And the change in their momenta is given by:

$$\Delta\vec{p}_1 = \Delta\vec{p}_2 \tag{11.9}$$

Which of Newton's three laws justifies these two questions?

Conservation of Kinetic Energy During a Collision

You have probably noticed that when you drop a rubber ball it doesn't quite get back up to the original height after it bounces off the floor. Clearly the Earth–ball system loses some mechanical energy. Where does that energy go?

You may find the answer from another experience. If you have ever played a racket sport (such as squash) with a small rubber ball, you may have noticed that the ball gets very warm after it has been hit around for a few minutes. Some of the ball–wall system's kinetic energy has been converted into thermal energy. The system also loses energy to the surrounding air in the form of sound waves. During any collision, some or even all of the system's kinetic energy is converted into another form or is lost to the environment. In an **inelastic collision**, the system of colliding objects loses kinetic energy. So, the system's initial kinetic energy is greater than the kinetic energy after the collision:

INELASTIC COLLISION
Major Concept

$$K_{tot,\,i} > K_{tot,\,f}$$

The greatest loss of kinetic energy occurs when the colliding objects stick together in what is known as a **completely inelastic collision**. For example, a dart thrown into a dartboard is a completely inelastic collision. However, the objects do not need to be at rest after the collision. For example, if the legendary William Tell actually shot an apple off his son's head with an arrow, the arrow stuck in the apple, and the apple and the arrow moved together after the completely inelastic collision.

COMPLETELY INELASTIC COLLISION
Major Concept

An **elastic collision** is an ideal case in which the system of colliding objects conserves its total kinetic energy:

ELASTIC COLLISION
Major Concept

$$K_{tot,\,i} = K_{tot,\,f} \tag{11.10}$$

Although no actual collision is ever truly elastic, many collisions may be approximated as elastic.

So, although *linear momentum is always conserved* for a colliding system when external forces can be neglected, *kinetic energy is only conserved if the collision is elastic*. We apply conservation of momentum to both elastic and completely inelastic collisions, but we apply conservation of kinetic energy only to elastic collisions (Fig. 11.5).

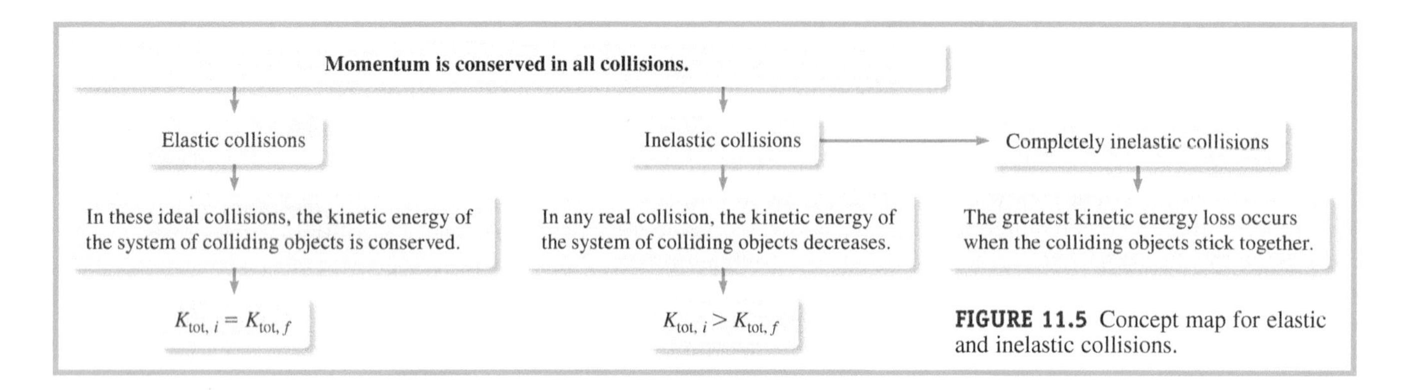

FIGURE 11.5 Concept map for elastic and inelastic collisions.

CONCEPT EXERCISE 11.4

CASE STUDY **Trains Stick Together**

Two trains collide head-on and stick together. Is the collision elastic, inelastic, or completely inelastic? Explain. Which of Equations 11.5 through 11.10 apply to the two-train system? Explain.

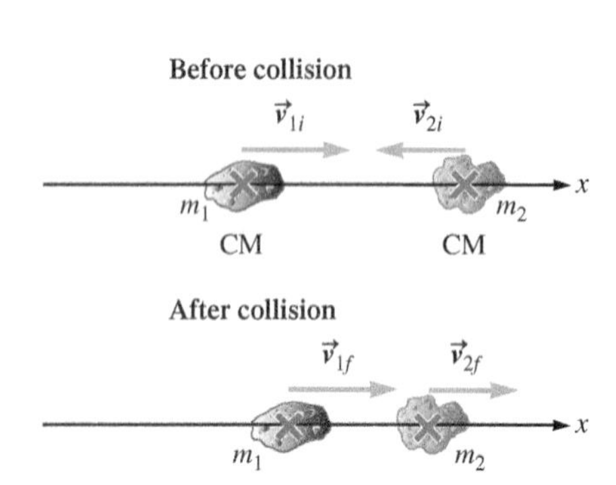

FIGURE 11.6 Two asteroids collide along the line joining their centers of mass. The collision is one-dimensional.

11-4 Special Case: One-Dimensional Inelastic Collisions

In a one-dimensional collision, the motion of the objects before and after the collision is along a single coordinate axis. So, the objects must collide along the line joining their centers of mass (Fig. 11.6).

Only momentum is conserved in an inelastic collision. For a one-dimensional inelastic collision between two objects along the x axis such as in Figure 11.6, we have

$$\sum \vec{p}_i = \sum \vec{p}_f$$

$$(p_{1i} + p_{2i})\hat{\imath} = (p_{1f} + p_{2f})\hat{\imath}$$

$$(m_1 v_{1i} + m_2 v_{2i})\hat{\imath} = (m_1 v_{1f} + m_2 v_{2f})\hat{\imath} \tag{11.11}$$

In many collisions, one of the objects is initially at rest. The stationary object is known as the *target*, and the moving object is called the *projectile*. If object 2 in Equation 11.11 is the target, then

$$v_{2i} = 0$$

$$m_1 v_{1i}\hat{\imath} = (m_1 v_{1f} + m_2 v_{2f})\hat{\imath} \tag{11.12}$$

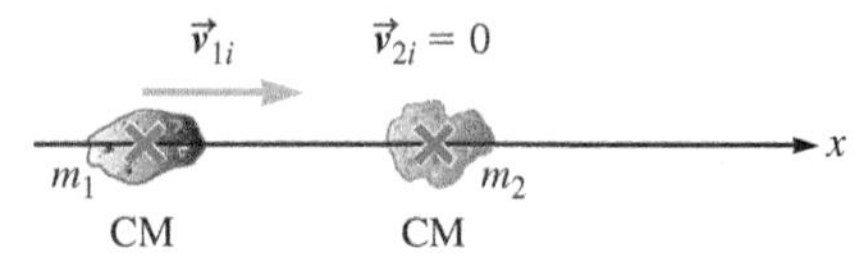

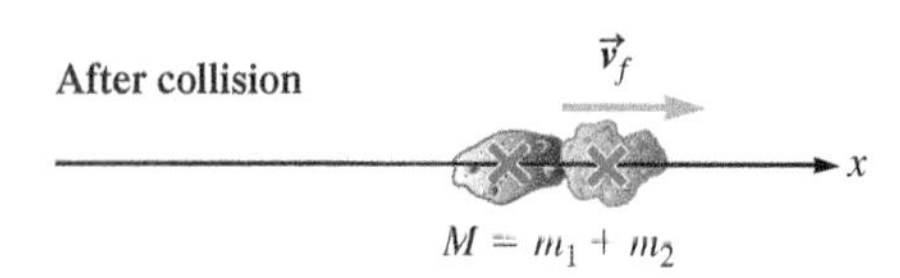

FIGURE 11.7 A completely inelastic, one-dimensional collision involving a stationary target.

If the collision is *completely* inelastic such as shown in Figure 11.7, the two objects stick together upon collision. After the collision, they act as one large particle of mass $M = m_1 + m_2$ moving at velocity $\vec{v}_f$. Equation 11.12 is simplified in the case of a completely inelastic collision:

COMPLETELY INELASTIC ONE-DIMENSIONAL COLLISION WITH A STATIONARY TARGET

Special Case

$$m_1 v_{1i}\hat{\imath} = (m_1 + m_2)v_f\hat{\imath} = Mv_f\hat{\imath} \tag{11.13}$$

Equations 11.12 and 11.13 are good for certain special cases (Fig. 11.8). When solving an inelastic collision problem, it is best to start with conservation of momentum in a general form (Eqs. 11.5–11.7).

Special case: One-dimensional collision in which only two objects are involved and the target (particle 2) is initially at rest. Momentum is conserved and simple to express.

$$m_1v_{1i}\hat{\imath} = (m_1v_{1f} + m_2v_{2f})\hat{\imath}$$

Very special case: One-dimensional completely **inelastic** collision (objects stick together); **only** momentum is conserved.

$$m_1v_{1i}\hat{\imath} = (m_1 + m_2)v_f\hat{\imath} = Mv_f\hat{\imath}$$

FIGURE 11.8 Concept map for conservation of momentum in the special case of a one-dimensional collision involving only two objects, one of which is initially at rest.

EXAMPLE 11.3 Two Colliding Particles

Figure 11.9 shows a head-on collision between two particles. Particle 1 has three times the mass of particle 2. Initially, particle 1's velocity is $\vec{v}_{1i} = v_0\hat{\imath}$, and particle 2's velocity is $\vec{v}_{2i} = -2v_0\hat{\imath}$. After the collision, particle 2's velocity is $\vec{v}_{2f} = v_0\hat{\imath}$. Find the change in the two-particle system's kinetic energy. In thinking about your result, answer this question: Is the collision elastic, inelastic, or completely inelastic?

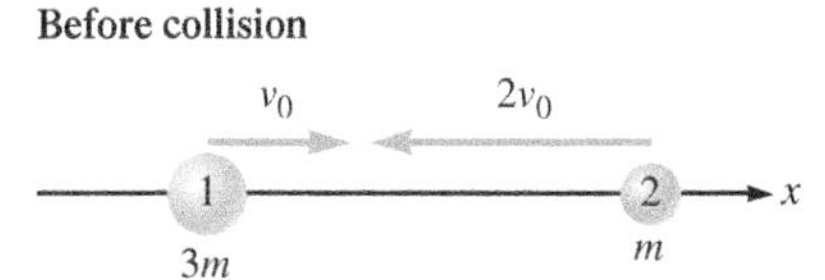

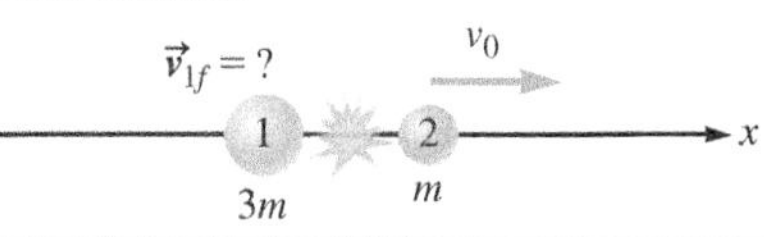

FIGURE 11.9

INTERPRET and ANTICIPATE

Both particles make up the system. Use conservation of momentum to find particle 1's velocity after the collision and then Equation 8.2 to find the kinetic energy of the system before and after the collision. Our result should be algebraic.

SOLVE

Use the coordinate system shown in Figure 11.9 to find the momentum of the system before and after the collision (Eq. 11.5). (It is reassuring to see that the third equation is the same as Eq. 11.11.)

$$\sum \vec{p}_i = \sum \vec{p}_f \qquad (11.5)$$

$$(p_{1i} + p_{2i})\hat{\imath} = (p_{1f} + p_{2f})\hat{\imath}$$

$$(m_1v_{1i} + m_2v_{2i})\hat{\imath} = (m_1v_{1f} + m_2v_{2f})\hat{\imath}$$

$$(3mv_0 - 2mv_0)\hat{\imath} = (3mv_{1f} + mv_0)\hat{\imath}$$

Solving for v_{1f} shows that particle 1 is at rest after the collision.

$$3v_0 - 2v_0 = 3v_{1f} + v_0$$

$$v_0 = 3v_{1f} + v_0$$

$$v_{1f} = 0$$

Find the total kinetic energy K before and after the collision.

$$K_i = \tfrac{1}{2}m_1v_{1i}^2 + \tfrac{1}{2}m_2v_{2i}^2 = \tfrac{1}{2}(3m)v_0^2 + \tfrac{1}{2}m(2v_0)^2$$

$$K_i = \tfrac{3}{2}mv_0^2 + \tfrac{4}{2}mv_0^2 = \tfrac{7}{2}mv_0^2$$

$$K_f = \tfrac{1}{2}m_1v_{1f}^2 + \tfrac{1}{2}m_2v_{2f}^2 = 0 + \tfrac{1}{2}mv_0^2 = \tfrac{1}{2}mv_0^2$$

The change in kinetic energy is negative, meaning that the system loses kinetic energy during the collision.

$$\Delta K = K_f - K_i = \tfrac{1}{2}mv_0^2 - \tfrac{7}{2}mv_0^2 = -\tfrac{6}{2}mv_0^2$$

$$\Delta K = -3mv_0^2$$

CHECK and THINK

Because the kinetic energy was not conserved during the collision, the collision must be *inelastic*. (The energy lost may go into the internal energy of each ball or may be lost to the environment through sound.) The collision is not completely inelastic because the particles have two different velocities after the collision, and so cannot be stuck together.

EXAMPLE 11.4 CASE STUDY How Fast Was the Train Going?

We return to the train collision case study from Chapter 5. One of the most important details—the speed of the freight train upon impact—was never reliably reported by witnesses. Initial reports indicated that the freight train was going only about 10 mph at the time of the collision. The goal of this example is to test this estimated train speed from the following reliably known information.

The freight train operator failed to slow down at a yellow signal. The freight train[1] of mass $m_1 = 5.221 \times 10^6$ kg approached a red signal at a speed of 48 mph. The operator applied the emergency brakes after the train passed the red signal. The passenger train of mass $m_2 = 4.1 \times 10^5$ kg was stopped on the tracks.[2] When the trains collided, they locked together and skidded 370 ft.

Two additional assumptions are needed. First, when the emergency brake locks the wheels, the train slides along the track, and kinetic friction (not rolling friction) slows the train. Second, the coefficient of kinetic friction for steel wheels on steel rails varies with weather conditions and lubrication, but we can assume a range of $0.03 \leq \mu_k \leq 0.4$.

With this information, find the speed of the freight train upon impact. *Hint*: The answer must be given as a range.

INTERPRET and ANTICIPATE

The system consists of the two trains. Figure 11.10 shows the system at three key times: (1) just before the collision, (2) just after the collision, and (3) when the trains finally come to rest. We want to find $\vec{v}_{1i}$ shown before the collision.

During the collision, the net external force on the system is due to kinetic friction. We assume friction is weak compared to the force exerted by the trains on each other; therefore, the momentum of the system is conserved. The two trains stick together, so the collision is completely inelastic.

We can use either the conservation approach or the force approach to find the velocity of the system during the period from just after the collision until the trains come to rest. We choose the conservation approach. (You may use the force approach in Problem 32). There are two major principles used in the solution: conservation of momentum and conservation of energy.

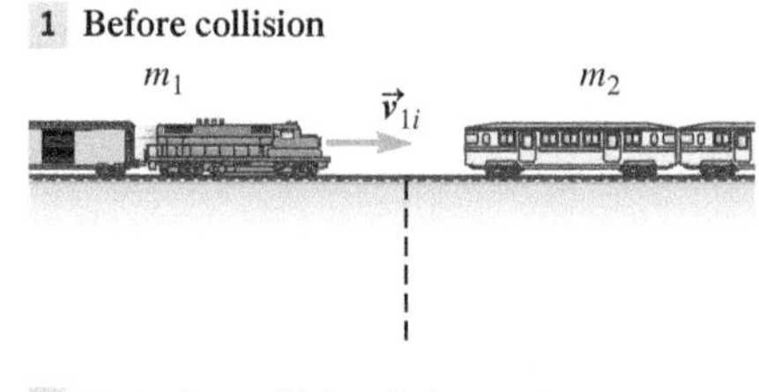

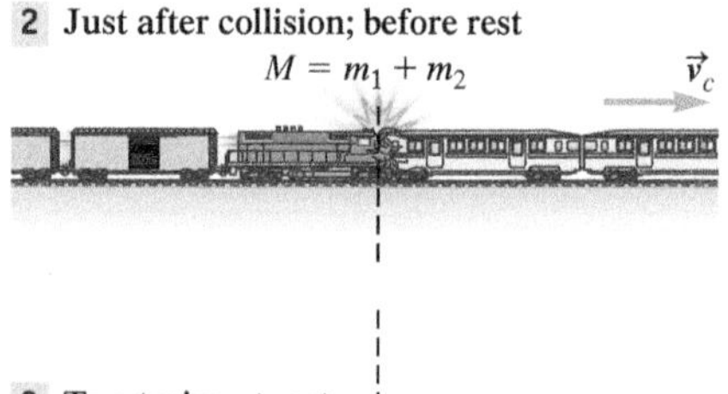

FIGURE 11.10

SOLVE

Let's first use conservation of momentum. Initially, only the freight train has momentum. After the collision, the two trains move together at $\vec{v}_c$. (It is reassuring to find that the penultimate equation is the same as Eq. 11.13.)

$$\sum \vec{p}_i = \sum \vec{p}_f$$

$$(m_1 v_{1i} + m_2 v_{2i})\hat{\imath} = (m_1 v_{1f} + m_2 v_{2f})\hat{\imath}$$

$$m_1 v_{1i}\hat{\imath} = (m_1 + m_2)v_c\hat{\imath}$$

$$m_1 v_{1i}\hat{\imath} = M v_c\hat{\imath}$$

$$v_{1i} = \frac{M}{m_1} v_c \qquad (1)$$

If we know the speed v_c of the two-train system just after the collision, we can use Equation (1) to find the initial velocity of the freight train. We now seek v_c by applying the work–energy theorem, using the 7-step problem-solving strategy on page 266.

Steps 1–4 The **system** consists of the two trains and the track. Our energy bar chart is shown in Figure 11.11. There are no external forces doing work on the system. For this part of the example, we set the **initial time** to the moment of impact, when both trains are moving in the positive x direction. Therefore, the system has kinetic energy initially. There is friction between the track and the trains, so this kinetic energy is converted to thermal energy, and at the **final time** the trains have stopped.

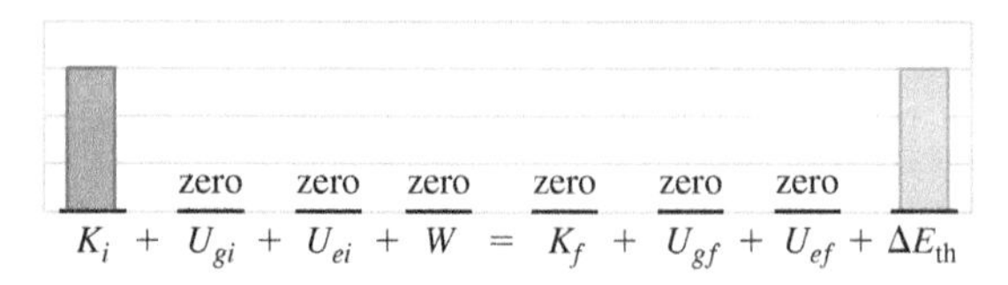

FIGURE 11.11 Energy bar chart for the system.

[1] The spokeswoman for the freight railroad reported the weight of the train to four significant figures.
[2] The passenger train company gave the average weight of a passenger train to two significant figures.

Step 5 Apply the **work–energy theorem** (Eq. 9.33) to write an algebraic expression, using the bar chart as a guide.	$K_i + U_{gi} + U_{ei} + W_{tot} = K_f + U_{gf} + U_{ef} + \Delta E_{th}$ (9.33) $K_i = \Delta E_{th}$ (2)
Step 6 Other equations. The initial kinetic energy is due to the motion of the combined trains moving at v_C. Kinetic friction increases the thermal energy over the stopping distance Δx. The track is level, and only gravity and the normal force are exerted on the system. Therefore, the magnitude of the normal force exerted on the trains equals their weight.	$K_i = \frac{1}{2}Mv_C^2$ (3) $\Delta E_{th} = F_k \Delta x = \mu_k F_N \Delta x$ $\Delta E_{th} = \mu_k Mg\,\Delta x$ (4)
Step 7 Do **algebra.** Find an expression for v_C by substituting Equations (3) and (4) into Equation (2).	$\frac{1}{2}Mv_C^2 = \mu_k Mg\,\Delta x$ $v_C = \sqrt{2\mu_k g\,\Delta x}$ (5)
Substitute Equation (5) in Equation (1) and solve for freight train's speed v_{1i} just prior to the collision.	$v_{1i} = \frac{M}{m_1}\sqrt{2\mu_k g \Delta x}$

To find the range of possible initial freight train speeds, use the extreme values of the kinetic friction coefficient. This estimate is limited to one significant figure. Use 370 ft = 110 m.

$$v_{1i}\,(\max) = \left(\frac{5.63 \times 10^6\,\text{kg}}{5.221 \times 10^6\,\text{kg}}\right)\sqrt{2(0.4)(9.81\text{ m/s}^2)(1.1 \times 10^2\text{ m})} = 30\text{ m/s}$$

$$v_{1i}\,(\min) = \left(\frac{5.63 \times 10^6\,\text{kg}}{5.221 \times 10^6\,\text{kg}}\right)\sqrt{2(0.03)(9.81\text{ m/s}^2)(1.1 \times 10^2\text{ m})} = 9\text{ m/s}$$

$$(9 \le v_{1i} \le 30)\text{ m/s}$$

According to the report, the freight train was going 48 mph (21 m/s) when it passed the red signal, so the upper limit we found should be replaced by this figure. (We have to assume the train did not speed up after passing the signal.)	$(9 \le v_{1i} \le 21)$ m/s $(20 \le v_{1i} \le 48)$ mph

CHECK and THINK

The lower limit of 20 mph tells us that the freight train was going at least twice as fast as the first reports indicated. Later reports stated that the freight train exceeded 20 mph at the moment of impact. In one report, it was suggested that the train may not have slowed even after passing the red signal.

11-5 One-Dimensional Elastic Collisions

In Section 11-4, we considered inelastic one-dimensional collisions; in such collisions, only momentum is conserved. If the collision is elastic, kinetic energy is also conserved. Although no collision is truly elastic, some may be modeled as elastic collisions.

In many problems involving the one-dimensional elastic collision of two objects, we often know the mass of each object and two velocities, leaving two velocities unknown. For instance, if we know the initial velocity of each particle $\vec{v}_{1i}$ and $\vec{v}_{2i}$, but the final velocity of each particle $\vec{v}_{1f}$ and $\vec{v}_{2f}$ is unknown (Fig. 11.12), we can use conservation of momentum and conservation of kinetic energy to solve for the two unknown velocities. The algebra leading to $\vec{v}_{1f}$ and $\vec{v}_{2f}$ is messy, but it involves nothing new. To help you solve such problems, we derive expressions for $\vec{v}_{1f}$ and $\vec{v}_{2f}$ here.

Before collision

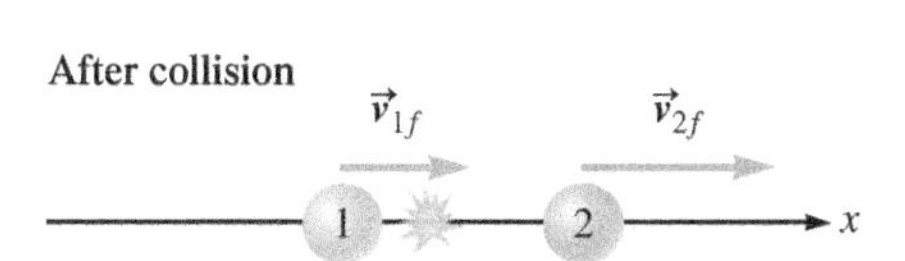

After collision

$\vec{v}_{1f}$ $\vec{v}_{2f}$

FIGURE 11.12 Two particles collide elastically along the line joining their centers of mass. The collision is one-dimensional.

DERIVATION Particle Velocities After One-Dimensional Elastic Collision

We will show that the velocity of the two particles in Figure 11.12 after their one-dimensional elastic collision is given by

$$\vec{v}_{1f} = \frac{m_1 - m_2}{M}\vec{v}_{1i} + \frac{2m_2}{M}\vec{v}_{2i} \tag{11.14}$$

$$\vec{v}_{2f} = \frac{2m_1}{M}\vec{v}_{1i} + \frac{m_2 - m_1}{M}\vec{v}_{2i} \tag{11.15}$$

FINAL VELOCITIES IN ONE-DIMENSIONAL ELASTIC COLLISION

Special Case

where m_i is the mass of the ith particle and M is the sum of their masses.

Derivation continues on page 316 ▶

Because the collision is elastic, kinetic energy is conserved. Write the kinetic energy of each particle (before and after the collision) in terms of its mass and speed.	$K_{1i} + K_{2i} = K_{1f} + K_{2f}$ $\frac{1}{2}m_1v_{1i}^2 + \frac{1}{2}m_2v_{2i}^2 = \frac{1}{2}m_1v_{1f}^2 + \frac{1}{2}m_2v_{2f}^2$
The factor of $\frac{1}{2}$ cancels out. Collect terms for particle 1 on the left and particle 2 on the right.	$m_1(v_{1i}^2 - v_{1f}^2) = m_2(v_{2f}^2 - v_{2i}^2)$
Factor the difference of the squares as $(a^2 - b^2) = (a - b)(a + b)$.	$m_1(v_{1i} - v_{1f})(v_{1i} + v_{1f}) = m_2(v_{2f} - v_{2i})(v_{2f} + v_{2i})$ (1)
Next, apply conservation of momentum. Both particles have momentum before and after the collision.	$m_1\vec{v}_{1i} + m_2\vec{v}_{2i} = m_1\vec{v}_{1f} + m_2\vec{v}_{2f}$
Again, collect terms for particle 1 on the left and particle 2 on the right.	$m_1\vec{v}_{1i} - m_1\vec{v}_{1f} = m_2\vec{v}_{2f} - m_2\vec{v}_{2i}$ $m_1(v_{1i} - v_{1f})\hat{\imath} = m_2(v_{2f} - v_{2i})\hat{\imath}$ $m_1(v_{1i} - v_{1f}) = m_2(v_{2f} - v_{2i})$ (2)
Use Equation (2) to eliminate $m_2(v_{2f} - v_{2i})$ from Equation (1).	$m_1(v_{1i} - v_{1f})(v_{1i} + v_{1f}) = m_1(v_{1i} - v_{1f})(v_{2i} + v_{2f})$
Divide by the common factor $m_1(v_{1i} - v_{1f})$ to obtain an expression for v_{2f}.	$v_{1i} + v_{1f} = v_{2i} + v_{2f}$ $v_{2f} = v_{1i} + v_{1f} - v_{2i}$ (3)
Solve for v_{1f} by substituting Equation (3) into Equation (2).	$m_1(v_{1i} - v_{1f}) = m_2(v_{1i} + v_{1f} - v_{2i} - v_{2i})$ $(m_1 - m_2)v_{1i} + 2m_2v_{2i} = (m_1 + m_2)v_{1f}$ $v_{1f} = \frac{m_1 - m_2}{m_1 + m_2}v_{1i} + \frac{2m_2}{m_1 + m_2}v_{2i}$ (4)
We find v_{2f} by substituting Equation (4) into Equation (3).	$v_{2f} = v_{1i} + \frac{m_1 - m_2}{m_1 + m_2}v_{1i} + \frac{2m_2}{m_1 + m_2}v_{2i} - v_{2i}$ $v_{2f} = \frac{2m_1}{m_1 + m_2}v_{1i} + \frac{m_2 - m_1}{m_1 + m_2}v_{2i}$ (5)
Finally, Equations (4) and (5) are rewritten in terms of the total mass of the two-particle system, $M = m_1 + m_2$, and as vector equations.	$\vec{v}_{1f} = \frac{m_1 - m_2}{M}\vec{v}_{1i} + \frac{2m_2}{M}\vec{v}_{2i}$ ✓ (11.14) $\vec{v}_{2f} = \frac{2m_1}{M}\vec{v}_{1i} + \frac{m_2 - m_1}{M}\vec{v}_{2i}$ ✓ (11.15)

COMMENTS

Usually, it is best to start any collision problem by writing down the conservation expressions that are valid. For an elastic collision, you would start by writing down an expression for conservation of momentum and another for conservation of kinetic energy. You now have two equations, so you could solve for two unknown quantities. As this derivation shows, however, the algebra involved in finding the two unknowns may be daunting. So, if the problem is a one-dimensional elastic collision, you may wish to start by using Equations 11.14 and 11.5 in that special case.

Stationary Target in an Elastic Collision: Special Cases

Our derivation of the final velocities of the two particles involved in one-dimensional elastic collision allows us to consider the special case of an initially stationary target. In this case, particle 2 (the target) is initially at rest ($v_{2i} = 0$), and Equations 11.14 and 11.15 become

$$\vec{v}_{1f} = \frac{m_1 - m_2}{M}\vec{v}_{1i} \quad (11.16)$$

$$\vec{v}_{2f} = \frac{2m_1}{M}\vec{v}_{1i} \quad (11.17)$$

According to Equation 11.16, particle 1 (the projectile) reverses direction if it has less mass than the target (particle 2) so that $m_1 < m_2$. To further explore how mass affects the final velocity of the particles, imagine a one-dimensional collision between two bumper cars at an amusement park (Fig. 11.13). In all three cases, car 1 is initially moving and car 2 is initially at rest. Their relative mass determines how the cars move after the collision. If the cars have equal mass, then after the collision car 1 is at rest and car 2 is moving at the same speed as car 1 originally had (Fig. 11.13A). If car 1 is much less massive than car 2, after the collision car 1 has a high speed in the reverse direction and car 2 moves slowly in the same direction as car 1's initial velocity (Fig. 11.13B). Finally, if car 1 is much more massive than car 2, after the collision car 1 continues moving with nearly its same velocity and car 2 moves in the same direction at nearly twice the speed as car 1 (Fig. 11.13C).

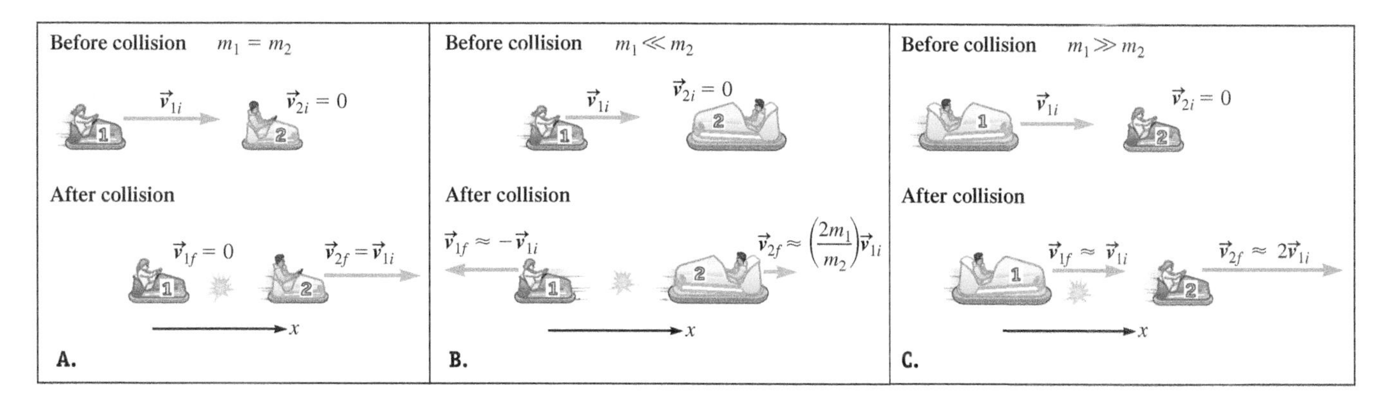

FIGURE 11.13 A one-dimensional collision between two bumper cars in which car 2 is initially at rest in all three cases. **A.** The cars have equal mass. **B.** Car 2 is much more massive than car 1. **C.** Car 1 is much more massive than car 2.

In this subsection, we considered the special case in which the target is initially stationary. As long as the target is not accelerating, it is always possible to find an inertial reference in which the target is initially at rest. As illustrated in the next two examples, using the target's frame is often the best way to tackle a problem.

EXAMPLE 11.5 One-Dimensional Collision in Two Reference Frames

Crall and Whipple set up a laboratory experiment to study a one-dimensional collision between two low-friction carts of mass $m_1 = 2m$ and $m_2 = m$. A small spring on cart 2 ensures that the collision is essentially elastic. Crall decides to work with a reference frame fixed to the laboratory (Fig. 11.14). In the laboratory frame, Crall measures the initial velocities of both carts before the collision: $\vec{v}_{1i} = 3.2\hat{\imath}$ m/s and $\vec{v}_{2i} = 0.8\hat{\imath}$ m/s.

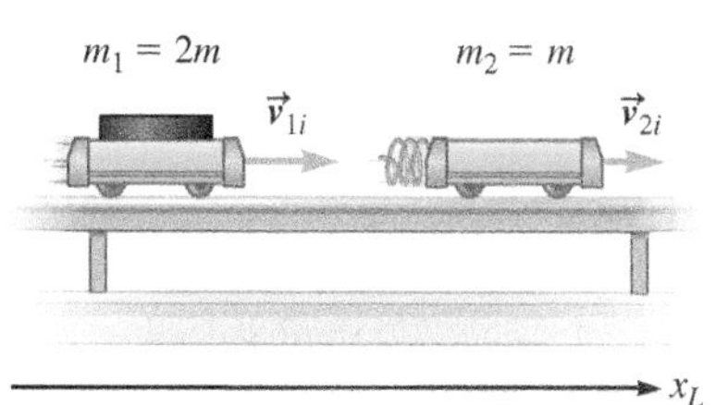

FIGURE 11.14 Crall's laboratory frame, before collision

A Use Crall's measurements to find the velocity of the two carts after the collision in the laboratory reference frame.

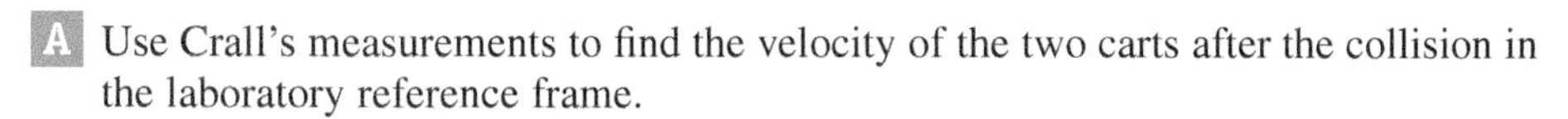

INTERPRET and ANTICIPATE

This example is a one-dimensional elastic collision between two objects. Neither object is initially at rest. We expect to find two numerical results of the form $\vec{v}_f =$ _____ $\hat{\imath}$ m/s.

SOLVE

Find the velocities from a straightforward substitution. For cart 1, use Equation 11.14.

$$v_{1f} = \frac{m_1 - m_2}{M}v_{1i} + \frac{2m_2}{M}v_{2i} \quad (11.14)$$

$$v_{1f} = \frac{2m - m}{3m}v_{1i} + \frac{2m}{3m}v_{2i}$$

$$v_{1f} = \tfrac{1}{3}v_{1i} + \tfrac{2}{3}v_{2i}$$

$$\vec{v}_{1f} = \tfrac{1}{3}(3.2\hat{\imath}\text{ m/s}) + \tfrac{2}{3}(0.8\hat{\imath}\text{ m/s})$$

$$\vec{v}_{1f} = 1.6\hat{\imath}\text{ m/s}$$

Example continues on page 318 ▶

For cart 2, use Equation 11.15.

$$v_{2f} = \frac{2m_1}{M}v_{1i} + \frac{m_2 - m_1}{M}v_{2i} \qquad (11.15)$$

$$v_{2f} = \frac{2(2m)}{3m}v_{1i} + \frac{m - 2m}{3m}v_{2i} = \tfrac{4}{3}v_{1i} - \tfrac{1}{3}v_{2i}$$

$$\vec{v}_{2f} = \tfrac{4}{3}(3.2\hat{\imath}\text{ m/s}) - \tfrac{1}{3}(0.8\hat{\imath}\text{ m/s}) = 4.0\hat{\imath}\text{ m/s}$$

CHECK and THINK

According to our results, the carts continue to move in the positive x direction after the collision. Before the collision, cart 1 was moving faster than cart 2. After the collision, cart 2 is moving faster than cart 1.

To check our results, we compare the change in the two cart's momenta. We don't know the value of m, so we leave our expressions in terms of m. As expected, $\Delta\vec{p}_1 = -\Delta\vec{p}_2$ (Eq. 11.9).

$$\Delta\vec{p}_1 = 2m(\vec{v}_{1f} - \vec{v}_{1i}) = 2m(1.6\hat{\imath} - 3.2\hat{\imath})\text{ m/s}$$

$$\Delta\vec{p}_1 = -(3.2\text{ m/s})m\hat{\imath}$$

$$\Delta\vec{p}_2 = m(\vec{v}_{2f} - \vec{v}_{2i}) = m(4.0\hat{\imath} - 0.8\hat{\imath})\text{ m/s}$$

$$\Delta\vec{p}_2 = (3.2\text{ m/s})m\hat{\imath}$$

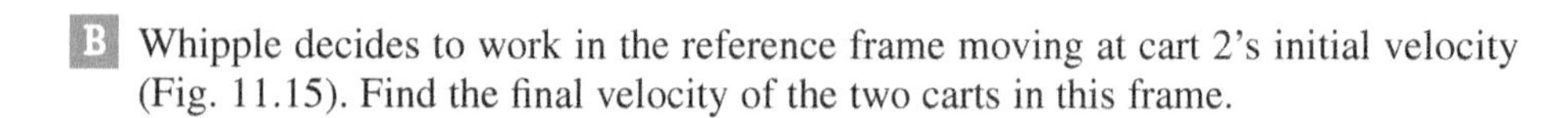

B Whipple decides to work in the reference frame moving at cart 2's initial velocity (Fig. 11.15). Find the final velocity of the two carts in this frame.

INTERPRET and ANTICIPATE

Whipple's chosen frame is an inertial one that moves at cart 2's *initial* velocity relative to the laboratory. So, cart 2 is initially at rest in Whipple's frame, but its speed is nonzero after the collision. We use the notation from Section 4-7 to indicate the two reference frames: L for *l*aboratory and M for *m*oving. Cart 2 is initially a stationary target in Whipple's frame like the targets in Figure 11.13, so we can use the equations for the special case of a stationary target. Our results should be consistent with part A.

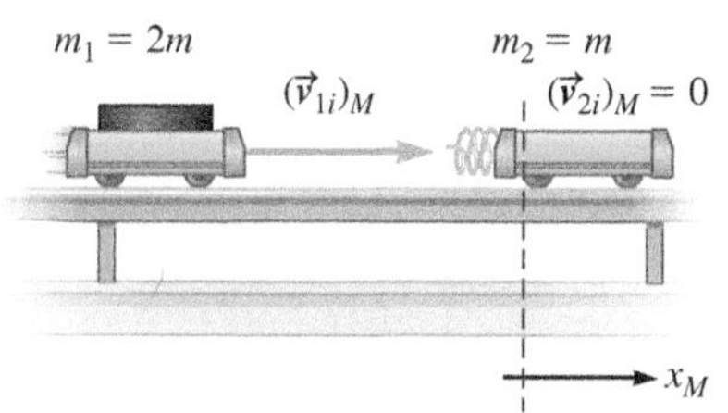

FIGURE 11.15 Whipple's moving frame, before collision

SOLVE

Solve for cart 1's initial velocity $(\vec{v}_{1i})_M$ in Whipple's *moving* frame (Eq. 4.40). The relative velocity between the two frames $(\vec{v}_M)_L$ is cart 2's initial velocity with respect to the laboratory, $\vec{v}_{2i} = 0.8\hat{\imath}$ m/s.

$$(\vec{v}_{1i})_L = (\vec{v}_{1i})_M + (\vec{v}_M)_L \qquad (4.40)$$

$$(\vec{v}_{1i})_M = (\vec{v}_{1i})_L - (\vec{v}_M)_L = 3.2\hat{\imath}\text{ m/s} - 0.8\hat{\imath}\text{ m/s}$$

$$(\vec{v}_{1i})_M = 2.4\hat{\imath}\text{ m/s}$$

Cart 1's final velocity in Whipple's reference frame comes from Equation 11.16 in this special case of an elastic one-dimensional collision with a stationary target.

$$(\vec{v}_{1f})_M = \frac{m_1 - m_2}{M}(\vec{v}_{1i})_M \qquad (11.16)$$

$$(\vec{v}_{1f})_M = \frac{2m - m}{3m}(\vec{v}_{1i})_M = \tfrac{1}{3}(\vec{v}_{1i})_M$$

$$(\vec{v}_{1f})_M = \tfrac{1}{3}(2.4\hat{\imath}\text{ m/s}) = 0.8\hat{\imath}\text{ m/s}$$

Similarly, cart 2's final velocity in Whipple's frame comes from Equation 11.17.

$$(\vec{v}_{2f})_M = \frac{2m_1}{M}(\vec{v}_{1i})_M = \frac{2(2m)}{3m}(\vec{v}_{1i})_M$$

$$(\vec{v}_{2f})_M = \tfrac{4}{3}(\vec{v}_{1i})_M = \tfrac{4}{3}(2.4\hat{\imath}\text{ m/s})$$

$$(\vec{v}_{2f})_M = 3.2\hat{\imath}\text{ m/s}$$

CHECK and THINK

Use Equation 4.40 to compare the results of parts A and B. The two sets of results are consistent.

$$(\vec{v}_{1f})_L = (\vec{v}_{1f})_M + (\vec{v}_M)_L$$

$$(\vec{v}_{1f})_L = 0.8\hat{\imath}\text{ m/s} + 0.8\hat{\imath}\text{ m/s} = 1.6\hat{\imath}\text{ m/s} \checkmark \quad \text{(lab frame)}$$

$$(\vec{v}_{2f})_L = (\vec{v}_{2f})_M + (\vec{v}_M)_L$$

$$(\vec{v}_{2f})_L = 3.2\hat{\imath}\text{ m/s} + 0.8\hat{\imath}\text{ m/s} = 4.0\hat{\imath}\text{ m/s} \checkmark \quad \text{(lab frame)}$$

Because the target is initially stationary in the Whipple's frame, our calculation using Equations 11.16 and 11.17 is simpler here than it was in part A. This problem-solving skill will help us solve our next example.

EXAMPLE 11.6 A Gravitational Slingshot

In the rocket case study in Chapter 10 (page 282), Cameron suggests that rockets are accelerated by the gravitational attraction of large objects such as planets and moons. It's true. Rocket scientists sometimes set spacecraft on collision courses in which the spacecraft does not crash on the planet's surface. The spacecraft and the planet interact through gravity, a field force that does not require the two objects to touch to collide. We require only that the gravitational force exerted by one object on the other be relatively strong for a relatively short time. When the spacecraft collides with a planet, the spacecraft is accelerated in a process known as a *gravitational slingshot*.

Consider an elastic collision between a spacecraft and a planet (Fig. 11.16A). The spacecraft is initially moving along the positive x axis, and the planet is moving in the negative x direction. During the collision, the spacecraft arcs around the planet so that after the collision, the spacecraft has reversed direction. Of course, the spacecraft's path around the planet is two-dimensional, but because the spacecraft's motion before and after the collision is along a single straight line, we consider it to be a one-dimensional collision. In the solar system's (fixed "laboratory") frame, the initial velocity of the spacecraft is $\vec{v}_{Si} = v_{Si}\hat{\imath}$. We also consider the planet's frame (Fig. 11.16B); the *p*lanet's velocity relative to the solar system frame is $\vec{v}_P = -v_P\hat{\imath}$. Find an expression for the spacecraft's final velocity $\vec{v}_{Sf}$ relative to the solar system's frame. Comment on the change in the spacecraft's speed.

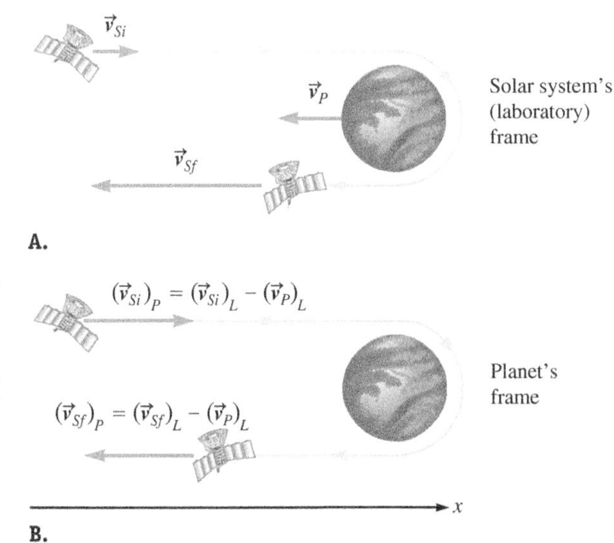

FIGURE 11.16

INTERPRET and ANTICIPATE

The planet is much more massive than the spacecraft, so the planet's velocity $\vec{v}_P$ is essentially unchanged by the collision. To verify that observation, we work in the solar system's (*l*aboratory) frame and use Equation 11.15 (because we have a one-dimensional elastic collision) with $m_1 \ll m_2$.

$$(\vec{v}_P)_f = \frac{2m_1}{M}\vec{v}_{Si} + \frac{m_2 - m_1}{M}(\vec{v}_P)_i \tag{11.15}$$

$$(\vec{v}_P)_f \approx 0 + \frac{m_2}{m_2}(\vec{v}_P)_i \approx (\vec{v}_P)_i$$

$$(\vec{v}_P)_L = -(v_P\hat{\imath})_L$$

is approximately constant in the laboratory (solar system) frame.

Now consider the planet's (moving) frame. Because the collision does not change the planet's velocity in the planet's own reference frame, its velocity before and after the collision is zero. If we make our calculation using the planet's frame, we have a special case of a one-dimensional elastic collision with a very massive ($m_1 \ll m_2$) stationary target. So, we can use $\vec{v}_{1f} \approx -\vec{v}_{1i}$ (Fig. 11.13B) to find the velocity of the spacecraft after the collision in the planet's frame. We will then transform our answer back to the solar system's frame.

SOLVE

Working in the *p*lanet's frame is very simple. According to Figure 11.13B, the spacecraft's speed is constant, and it simply reverses direction after the collision.

$$(\vec{v}_{Sf})_P \approx -(\vec{v}_{Si})_P \tag{1}$$

We must transform the spacecraft's initial velocity relative to the solar system frame to the planet's (moving) frame (Eq. 4.40). The subscript L stands for the *l*aboratory (solar system) frame, P stands for *p*lanet, and S stands for spacecraft (the subject).

$$(\vec{v}_{Si})_L = (\vec{v}_{Si})_P + (\vec{v}_P)_L \tag{4.40}$$

$$(\vec{v}_{Si})_P = (\vec{v}_{Si})_L - (\vec{v}_P)_L$$

$$(\vec{v}_{Si})_P = [v_{Si}\hat{\imath} - (-v_P\hat{\imath})]_L$$

$$(\vec{v}_{Si})_P = [(v_{Si} + v_P)\hat{\imath}]_L \tag{2}$$

Substitute Equation (2) into Equation (1). The result is the spacecraft's velocity after the collision with respect to the planet's reference frame.

$$(\vec{v}_{Sf})_P \approx -[(v_{Si} + v_P)\hat{\imath}]_L$$

Example continues on page 320 ▶

Now we must transform the spacecraft's final velocity back to the solar system (laboratory) frame (Eq. 4.40). For simplicity, we dropped the subscript L in our final expression.	$(\vec{v}_{Sf})_L = (\vec{v}_{Sf})_P + (\vec{v}_P)_L$ (4.40) $(\vec{v}_{Sf})_L = -[(v_{Si} + v_P)\hat{\imath}]_L + [-v_P\hat{\imath}]_L$ After the gravitational slingshot, the spacecraft's speed (in the solar system's frame) is: $\vec{v}_{Sf} = -(v_{Si} + 2v_P)\hat{\imath}$
CHECK and THINK Finally, find the change in the spacecraft's speed in the solar system's reference frame.	$\Delta v_S = \lvert\vec{v}_{Sf}\rvert - \lvert\vec{v}_{Si}\rvert = (v_{Si} + 2v_P) - v_{Si}$ $\Delta v_S = 2v_P$

The change in spacecraft's speed in the solar system's frame is twice the planet's speed. It seems like the planet-spacecraft's system gains kinetic energy. How is that possible if no net external force is exerted on the system (as it must be if momentum is conserved)? The explanation is that we have an approximation built into our work. The approximation is that the planet's speed and kinetic energy are unchanged by the collision. Although it is a very good approximation, the planet actually loses kinetic energy, which means a very small decrease in its speed (Problem 48).

Finally, in this scenario, the spacecraft reverses direction. Most gravitational slingshots are not simply 180° reversals. Instead, the spacecraft's path is deflected along some other angle, and rocket scientists may plan several slingshots along the trajectory of a single spacecraft.

CONCEPT EXERCISE 11.5

If a spacecraft is headed for the outer solar system, it may require several gravitational slingshots with planets in the inner solar system. If a spacecraft undergoes a head-on slingshot with Venus as in Example 11.6, find the spacecraft's change in speed Δv_S. *Hint*: Venus's orbital period is 1.94×10^7 s, and its average distance from the Sun is 1.08×10^{11} m.

11-6 Two-Dimensional Collisions

Figure 11.17 shows examples of two-dimensional collisions. A cue ball hits the eight-ball on its side, so the collision is not head-on. Afterward, neither ball's motion is along the x axis (Fig. 11.17A). In Figure 11.17B, a car moves in the positive x direction while a truck moves in the positive y direction before the collision. After the collision, the truck and the car are stuck together and move at an angle θ with respect to the x axis. In both situations, two coordinate axes are needed to describe the motion of each system before and after the collision.

The physics of two-dimensional collisions is the same as that for one-dimensional collisions. The momentum of the system is conserved, and kinetic energy is also conserved if the collision is elastic.

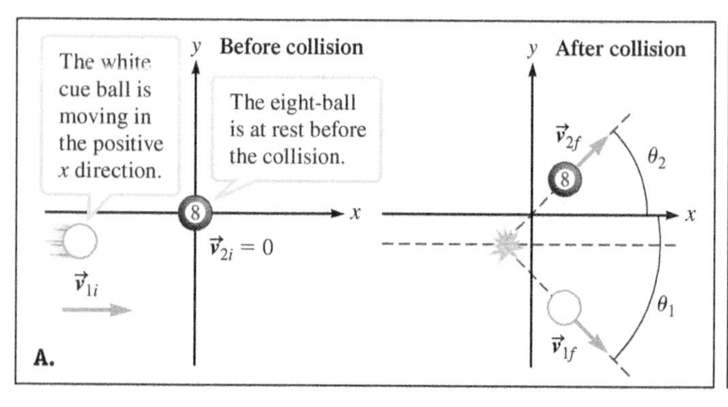

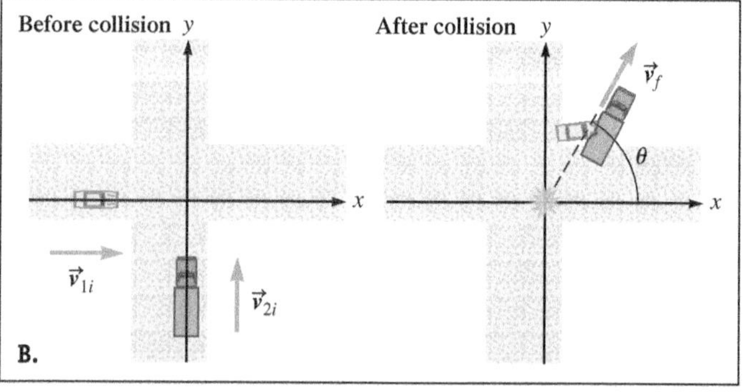

FIGURE 11.17 Two examples of two-dimensional collision. **A.** A white cue ball collides elastically with the side of the eight-ball in a game of pool (billiards). **B.** Two vehicles collide at an intersection. This collision is completely inelastic.

The real difference between one- and two-dimensional collisions is seen when we express conservation of momentum mathematically (Eq. 11.5):

$$\sum \vec{p}_i = \sum \vec{p}_f$$

For a two-dimensional collision, it is often convenient to write the conservation of momentum equation in terms of components:

$$\sum p_{ix} = \sum p_{fx} \qquad \sum p_{iy} = \sum p_{fy} \tag{11.18}$$

Because we have two conservation of momentum equations and not just one, the mathematics behind a two-dimensional elastic collision is more complicated. To keep things manageable, for the rest of this section, we'll focus our attention on two-dimensional collisions that only involve two particles such as those in Figure 11.17.

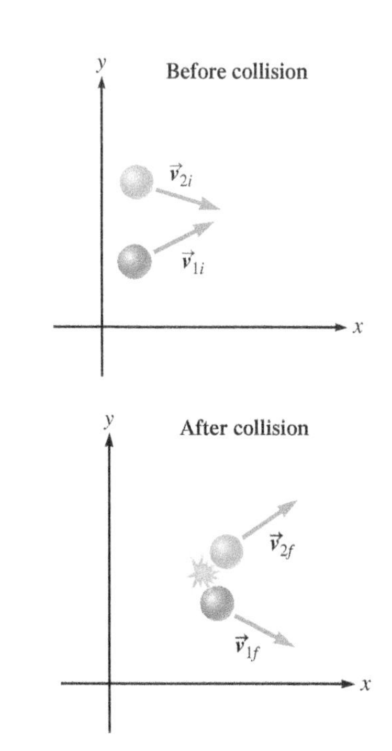

FIGURE 11.18 Two moving particles collide in two dimensions.

Elastic Two-Dimensional Two-Particle Collision

Figure 11.18 shows a two-dimensional elastic collision between two particles. Before and after the collision, both particles are moving. For such a collision, conservation of kinetic energy is rewritten in terms of the particle's speeds:

$$\tfrac{1}{2}m_1 v_{1i}^2 + \tfrac{1}{2}m_2 v_{2i}^2 = \tfrac{1}{2}m_1 v_{1f}^2 + \tfrac{1}{2}m_2 v_{2f}^2 \tag{11.19}$$

Equation 11.19 holds for any elastic collision involving two objects, regardless of the number of dimensions.

For collisions involving two objects, the conservation of momentum (Eq. 11.5) is

$$m_1\vec{v}_{1i} + m_2\vec{v}_{2i} = m_1\vec{v}_{1f} + m_2\vec{v}_{2f} \tag{11.20}$$

Equation 11.20 is good for any collision of any dimensionality involving two objects. It is helpful to rewrite conservation of momentum in terms of the scalar components. These velocity components can be found in the usual way from the angle each vector makes with the x axis. Using the coordinate system shown in Figure 11.18:

$$m_1 v_{1ix} + m_2 v_{2ix} = m_1 v_{1fx} + m_2 v_{2fx} \tag{11.21}$$

$$m_1 v_{1iy} + m_2 v_{2iy} = m_1 v_{1fy} + m_2 v_{2fy} \tag{11.22}$$

Although the physical principles (conservation of momentum and conservation of kinetic energy) of one- and two-dimensional collisions are the same, one thing that makes two-dimensional collisions more complicated is keeping track of all the subscripts. These subscripts account for the particle's label (1 or 2), the time (*i*nitial or *f*inal), and component (x or y).

When solving a problem involving a two-dimensional elastic collision of two objects, there are three independent equations—one from conservation of kinetic energy (Eq. 11.19) and two from conservation of momentum (Eqs. 11.21 and 11.22)—which means that you can solve for three unknown quantities. Often the problem is simplified because one object is initially at rest as is the case in a game of pool, the subject of the next example.

EXAMPLE 11.7 Playing Billiards

Figure 11.17A is an example of a somewhat simplified two-dimensional elastic collision. Often, but not always, a shot in a game of pool may be approximated as an elastic collision between two objects of equal mass $m_1 = m_2$. In addition, only one of the objects (the cue ball) is initially moving, so $\vec{v}_{2i} = 0$. Find the total angle $(\theta_1 + \theta_2)$ between the velocities of the two balls after the collision.

INTERPRET and ANTICIPATE

Both momentum and kinetic energy are conserved in this two-dimensional collision.

Example continues on page 322 ▶

SOLVE

Conservation of kinetic energy (Eq. 11.19) is simplified because the masses are equal ($m_1 = m_2$) and particle 2 is initially at rest ($\vec{v}_{2i} = 0$).

$$\tfrac{1}{2}m_1 v_{1i}^2 + \tfrac{1}{2}m_2 v_{2i}^2 = \tfrac{1}{2}m_1 v_{1f}^2 + \tfrac{1}{2}m_2 v_{2f}^2 \quad (11.19)$$

$$\tfrac{1}{2}m_1 v_{1i}^2 + 0 = \tfrac{1}{2}m_1 v_{1f}^2 + \tfrac{1}{2}m_1 v_{2f}^2$$

$$v_{1i}^2 = v_{1f}^2 + v_{2f}^2 \quad (1)$$

It is possible to apply conservation of momentum in component form, but here we use the vector form (Eq. 11.20). This expression becomes a simple vector addition of the velocities in the case of billiards.

$$m_1\vec{v}_{1i} + m_2\vec{v}_{2i} = m_1\vec{v}_{1f} + m_2\vec{v}_{2f} \quad (11.20)$$

$$m_1\vec{v}_{1i} + 0 = m_1\vec{v}_{1f} + m_1\vec{v}_{2f}$$

$$\vec{v}_{1i} = \vec{v}_{1f} + \vec{v}_{2f} \quad (2)$$

In Figure 11.19, we graphically represent the vector addition in Equation (2). Equation (1) is the Pythagorean theorem applied to the triangle shown. We conclude that $\vec{v}_{1i}$ is the hypotenuse of the triangle and that $\alpha = 90°$.

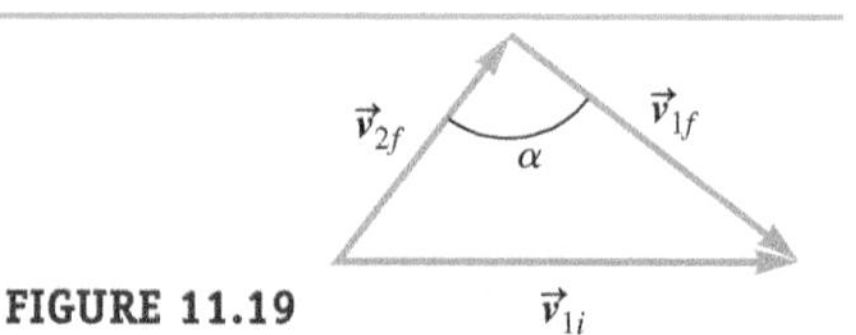

FIGURE 11.19

To compare this triangle to the pool balls after the collision, in Figure 11.20 we slide $\vec{v}_{1f}$ so that its tail touches the tail of $\vec{v}_{2f}$.

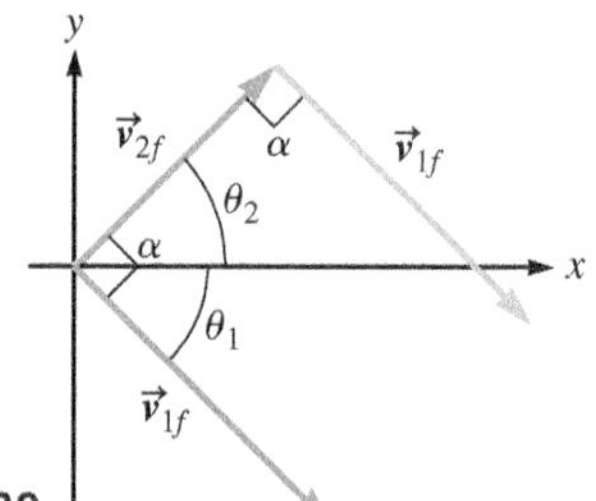

FIGURE 11.20

The angle α is the sum of $\theta_1 + \theta_2$, and it is the angle between $\vec{v}_{1f}$ and $\vec{v}_{2f}$.

$$\theta_1 + \theta_2 = 90° \quad (11.23)$$

CHECK and THINK

The angle $\alpha = 90°$ holds only for an elastic two-dimensional collision with $m_1 = m_2$ and $\vec{v}_{2i} = 0$. Because many collisions in pool are nearly elastic, $\alpha = 90°$ for many shots throughout the game, and players can predict both where the struck ball will go and where the cue ball will go after the collision. If the collision is head-on, it is a one-dimensional elastic collision between objects of equal mass. In this case, the cue ball stops, and the other ball continues along the same straight line. The angle between the two final velocities isn't well-defined in this case because the cue ball stops.

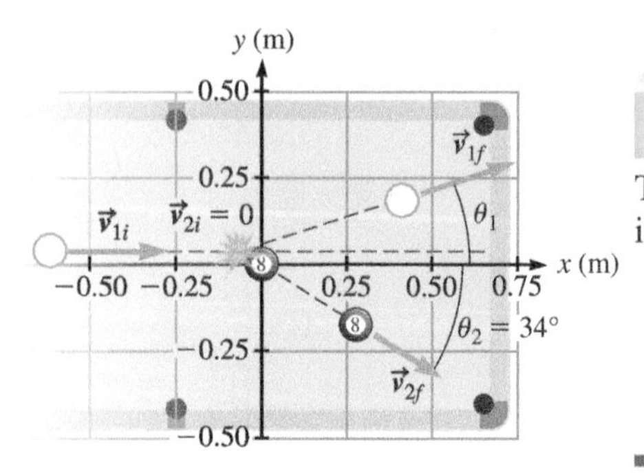

FIGURE 11.21

CONCEPT EXERCISE 11.6

The cue ball hits the eight-ball in a game of pool (Fig. 11.21). Assume the collision is elastic.

a. Will the eight-ball be sunk into a pocket if it is hit at the angle $\theta_2 = 34°$ shown in the figure?

b. If the cue ball is also sunk, the player is said to have scratched. If she scratches when trying to sink the eight-ball, she loses. Did the player lose?

Completely Inelastic Two-Dimensional Two-Particle Collision

In Figure 11.22, the two particles stick together in a completely inelastic collision. Kinetic energy is not conserved, so there is one fewer equation that can be applied. Because both objects stick together, however, they have the same final velocity $\vec{v}_f$, which reduces the number of possible unknown

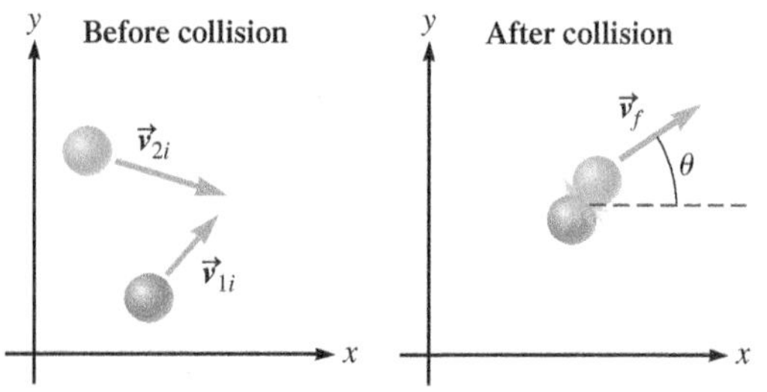

Before collision — After collision

FIGURE 11.22 A two-dimensional completely inelastic collision.

quantities. Further, the conservation of momentum equations are simplified. For the collision shown in Figure 11.22, conservation of momentum in component form is

$$m_1v_{1ix} + m_2v_{2ix} = (m_1 + m_2)v_{fx} = Mv_{fx} \quad (11.24)$$

$$m_1v_{1iy} + m_2v_{2iy} = (m_1 + m_2)v_{fy} = Mv_{fy} \quad (11.25)$$

Sometimes, it is convenient to rewrite Equations 11.24 and 11.25 in terms of their common final speed v_f and the angle θ:

$$m_1v_{1ix} + m_2v_{2ix} = Mv_f\cos\theta \quad (11.26)$$

$$m_1v_{1iy} + m_2v_{2iy} = Mv_f\sin\theta \quad (11.27)$$

EXAMPLE 11.8 Molecular Collision

A hydrogen molecule (H_2) of mass $m_1 = 3.35 \times 10^{-27}$ kg and an oxygen atom (O) of mass $m_2 = 2.66 \times 10^{-26}$ kg collide and form a water molecule (H_2O) as in Figure 11.23. If the initial speeds of the H_2 molecule and O atom are $v_{1i} = 550$ m/s and $v_{2i} = 378$ m/s, respectively, find the velocity of the H_2O molecule after the collision.

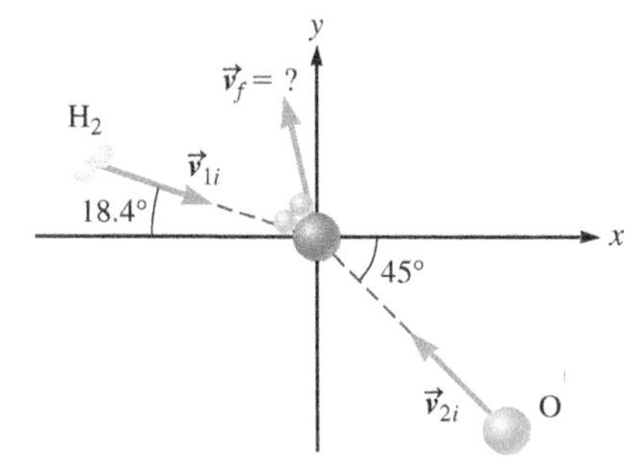

FIGURE 11.23

INTERPRET and ANTICIPATE

The two colliding objects stick together, so the collision is completely inelastic. Therefore, kinetic energy is not conserved, but momentum in each direction is conserved. Our result should be in the form $\vec{v}_f = (\_\_\_\_\,\hat{\imath} + \_\_\_\_\,\hat{\jmath})$ m/s.

SOLVE

Write each object's initial velocity in component form.	$\vec{v}_{1i} = (v_{1i}\cos 18.4°)\hat{\imath} - (v_{1i}\sin 18.4°)\hat{\jmath} = (522\hat{\imath} - 174\hat{\jmath})$ m/s $\vec{v}_{2i} = (-v_{2i}\cos 45°)\hat{\imath} + (v_{2i}\sin 45°)\hat{\jmath} = (-267\hat{\imath} + 267\hat{\jmath})$ m/s
Find the total mass of the two-object system.	$M = m_1 + m_2 = 3.35 \times 10^{-27}\text{ kg} + 2.66 \times 10^{-26}\text{ kg}$ $M = 2.99 \times 10^{-26}$ kg
The two particles stick together. We express conservation of momentum in terms of the particles' common final speed v_f and the angle θ (Eqs. 11.24 and 11.25).	$m_1v_{1ix} + m_2v_{2ix} = Mv_{fx}$ (11.24) $m_1v_{1iy} + m_2v_{2iy} = Mv_{fy}$ (11.25)
Solve Equations 11.24 and 11.25 for the components of the final velocity.	$v_{fx} = \dfrac{m_1v_{1ix} + m_2v_{2ix}}{M}$ $v_{fx} = \dfrac{(3.35 \times 10^{-27}\text{ kg})(522\text{ m/s}) + (2.66 \times 10^{-26}\text{ kg})(-267\text{ m/s})}{2.99 \times 10^{-26}\text{ kg}} = -179\text{ m/s}$ $v_{fy} = \dfrac{m_1v_{1iy} + m_2v_{2iy}}{M}$ $v_{fy} = \dfrac{(3.35 \times 10^{-27}\text{ kg})(-174\text{ m/s}) + (2.66 \times 10^{-26}\text{ kg})(267\text{ m/s})}{2.99 \times 10^{-26}\text{ kg}} = 218\text{ m/s}$
Write the result in component form.	$\vec{v}_f = (-179\hat{\imath} + 218\hat{\jmath})$ m/s

CHECK and THINK

Our answer has the form we expected. Our analysis shows that the water molecule moves up and to the *left*. Our result makes sense because even though both objects have comparable initial speeds, the oxygen atom is about eight times more massive than the hydrogen molecule. Therefore, the oxygen atom's momentum makes up most of the system's initial momentum. After the collision, the motion of the water molecule is nearly in the same direction as the oxygen atom's initial motion.

SUMMARY

Underlying Principles

1. Another statement of Newton's second law is the **impulse–momentum theorem**:
$$\vec{I}_{\text{tot}} = \sum \vec{I} = \Delta\vec{p} \quad (11.4)$$
2. **Conservation of momentum** (Chapter 10): If the impulse approximation holds, the total momentum of the system before the collision must equal the total momentum of the system after the collision:
$$(\vec{p}_{\text{tot}})_i = (\vec{p}_{\text{tot}})_f \quad (11.5)$$
Equation 11.5 can be written in terms of the center-of-mass momentum:
$$(\vec{p}_{\text{CM}})_i = (\vec{p}_{\text{CM}})_f \quad (11.6)$$

Major Concepts

1. A **collision** is an isolated event in which two or more objects exert relatively strong forces on one another for a relatively short time.
2. The **impulse approximation** assumes the force between the colliding objects during the time of the collision is much stronger than any other force exerted on either of them.
3. The **impulse** exerted on an object during a time interval Δt is a vector defined as
$$\vec{I} \equiv \int_{t_i}^{t_f} \vec{F}(t)dt = \vec{F}_{\text{av}}\Delta t \quad (11.3)$$
4. In an **inelastic collision**, the system of colliding objects loses kinetic energy.
5. In a **completely inelastic collision**, the system of colliding objects loses kinetic energy, and after the collision, the objects are stuck together.
6. An **elastic collision** is an ideal case in which the system of colliding objects conserves its total kinetic energy.

To take the survey scan or visit **www.cengage.com/community/katz**

Special Cases

1. Momentum in a completely **inelastic one-dimensional** collision with a stationary target:
$$m_1 v_{1i}\hat{\imath} = (m_1 + m_2)v_f\hat{\imath} = Mv_f\hat{\imath} \quad (11.13)$$
2. Final velocities in an **elastic one-dimensional** collision:
$$\vec{v}_{1f} = \frac{m_1 - m_2}{M}\vec{v}_{1i} + \frac{2m_2}{M}\vec{v}_{2i} \quad (11.14)$$
$$\vec{v}_{2f} = \frac{2m_1}{M}\vec{v}_{1i} + \frac{m_2 - m_1}{M}\vec{v}_{2i} \quad (11.15)$$
3. Conservation of momentum for two-object, two-**dimensional collision**:
$$m_1 v_{1ix} + m_2 v_{2ix} = m_1 v_{1fx} + m_2 v_{2fx} \quad (11.21)$$
$$m_1 v_{1iy} + m_2 v_{2iy} = m_1 v_{1fy} + m_2 v_{2fy} \quad (11.22)$$
4. Conservation of kinetic energy for two-object elastic collision:
$$\tfrac{1}{2}m_1 v_{1i}^2 + \tfrac{1}{2}m_2 v_{2i}^2 = \tfrac{1}{2}m_1 v_{1f}^2 + \tfrac{1}{2}m_2 v_{2f}^2 \quad (11.19)$$

PROBLEMS AND QUESTIONS

A = algebraic C = conceptual E = estimation G = graphical N = numerical

11-1 What Is a Collision?

1. C When a spacecraft collides with a planet, it is not necessary for them to actually touch each other, but when a car collides with a truck, the car and the truck must touch. Explain the difference between these two types of collisions.

11-2 Impulse

2. C When a person feels that he is about to fall, he will often put out his hand to try to "break the fall." Explain why this natural reaction usually leads to bruises or minor broken bones such as in the wrists instead of major broken bones such as the skull.
3. N A tall man walking at 1.25 m/s accidentally bumps his head of mass 3.10 kg on a steel doorjamb. His head stops in 0.010 s. What is the magnitude of the average force exerted by the doorjamb on the man's head? As part of the **CHECK and THINK** step, explain how padding the doorjamb can help reduce this force.
4. N A 35.0-kg child steps off a 4.0-ft-high diving board and executes a cannonball jump into a pool. (The child holds her body in a tight ball so that air resistance is negligible as she falls

downward.) What is the impulse exerted by the water on the child? In the **CHECK and THINK** step, explain why the child would be greatly injured if she were to land on the cement edge of the pool instead of in the water.

5. N A basketball of mass $m = 625$ g rolls off the hoop's rim, falls from a height of 3.05 m to the court's floor, and then bounces up to a height of 1.40 m. **a.** What are the magnitude and direction of the impulse delivered to the basketball by the floor? **b.** If the ball is in contact with the floor for 0.150 s, what is the average force exerted on the basketball by the floor?

6. E Two glass pickle jars fall off the same kitchen table and land on the floor. One jar is empty, and the other is full of (cucumber) pickles (and brine). Estimate the impulse exerted by the floor on each jar. As part of the **CHECK and THINK** step, answer the question: Which is more likely to break? Explain your answer.

7. N Sven hits a baseball ($m = 0.15$ kg). He applies an average force of 50.0 N. The ball had an initial velocity of 35.0 m/s to the right and a final velocity of 40.0 m/s to the left as viewed by a fan in the stands. **a.** What is the impulse delivered by Sven's bat to the baseball? **b.** How long is his bat in contact with the ball?

8. C A car's air bag helps protect the driver in the case of a collision. **a.** If a driver traveling at a particular speed hits an obstacle and comes to rest, what factors influence the total impulse the driver experiences while coming to rest? **b.** The air bag increases the time over which the collision takes place compared to a collision without an air bag present. How does this increase help protect the occupants of the car?

9. N A 65.0-kg driver of a vehicle traveling with a velocity of $11.5\hat{\imath}$ m/s in a parking lot hits a parked car and comes to rest in 0.150 s. **a.** Find the impulse experienced by the driver. **b.** Find the average force acting on the driver during the collision.

10. In a laboratory, a cart collides with a wall and bounces back. Figure P11.10 shows a graph of the force exerted by the wall versus time.
    a. N Find the impulse exerted by the wall on the cart.
    b. N What is the average force exerted by the wall on the cart?
    c. N If the cart has a mass of 0.448 kg, what is its change in velocity?
    d. C Make a sketch of the situation. Include a coordinate system and explain the significance of the signs in parts (a) through (c).

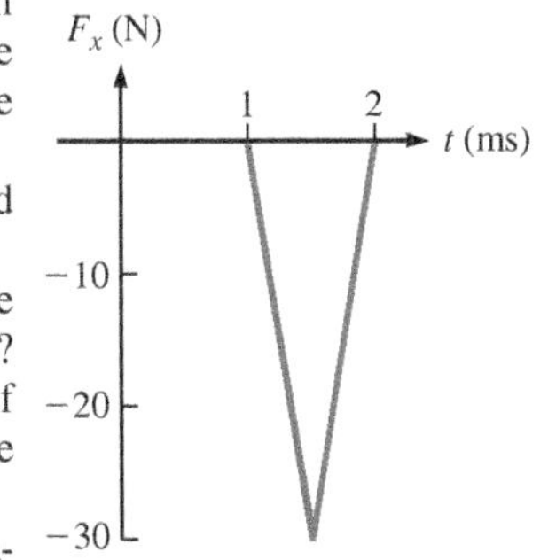

FIGURE P11.10

11. N An empty bucket is placed on a scale, and the scale is reset to read 0.00 N. Water falls from a faucet, high overhead, into the bucket without splashing at a rate of 350 mL/s. What is the reading on the scale 4.50 s after the water first hits the bottom of the empty bucket if at that time the falling water travels 1.25 m before hitting the water already in the bucket?

12. A Show that Equation 11.4 (the **impulse–momentum theorem**) is another statement of Newton's second law.

13. C A crate of mass M is initially at rest on a level, frictionless table. A small block of mass m ($m < M$) moves toward the crate as shown in Figure P11.13. After the collision, the block sticks to the crate. Is the magnitude of the impulse exerted on the crate by the block greater than, less than, or equal to the magnitude of the impulse exerted on the block by the crate during the collision process? Explain.

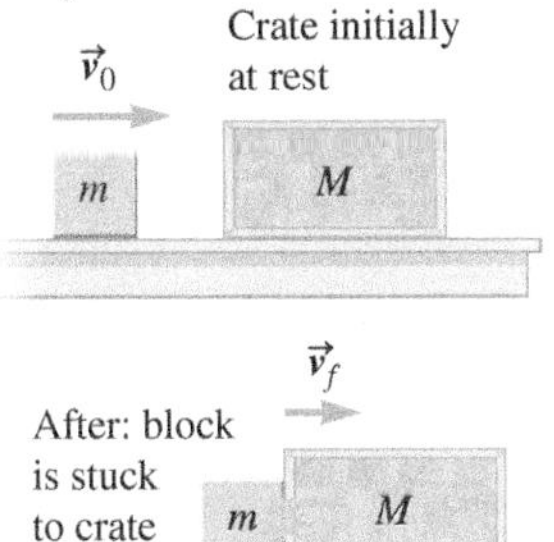

FIGURE P11.13

11-3 Conservation During a Collision

14. C Two students, Cameron and Avi, are riding in a small car and have the following discussion.

Avi: If I shoot a cue ball and hit another ball that is just resting there, the cue ball stops and the other ball is propelled at the same speed as the incoming cue ball. So, if a small car rear ends a parked truck, the car would stop, and the truck would be propelled forward at the same speed the car had been traveling.

Cameron: No way, and don't even think of trying that one. The truck has more mass, so it experiences a smaller impulse during the collision. The small car would experience a huge impulse.

Do you agree with either of these students? Explain.

15. N Two pucks in a laboratory are placed on an air table. Puck 1 has twice the mass of puck 2. They are pushed toward each other and strike in a head-on collision. Initially, puck 2 is twice as fast as puck 1. **a.** What is the total momentum before the collision? **b.** What is the center-of-mass velocity before the collision? **c.** If the pucks are initially 2.70 m apart, how far did puck 1 move before the collision?

16. C A truck collides with a small, empty parked car. Explain your answers to the parts below. **a.** Compare the force exerted by the truck on the car with the force exerted by the car on the truck. **b.** Compare the impulse exerted by the truck on the car with the impulse exerted by the car on the truck. **c.** Compare the change in the truck's momentum with the change in the car's momentum.

Problems 17 and 18 are paired.

17. N A comet is traveling through space with a speed of 3.33×10^4 m/s when it collides with an asteroid that was at rest. The comet and the asteroid stick together during the collision process. The mass of the comet is 1.11×10^{14} kg, and the mass of the asteroid is 6.66×10^{20} kg. **a.** What is the speed of the center of mass of the asteroid–comet system before the collision? **b.** What is the speed of the system's center of mass after the collision?

18. N For the comet and asteroid in Problem 17, **a.** what is the magnitude of the momentum of the system's center of mass before the collision? **b.** What is the magnitude of the momentum of the system's center of mass after the collision?

Problems 19 and 20 are paired.

19. A A skater of mass m standing on ice throws a stone of mass M with speed v in a horizontal direction. Find the distance over which the skater will move in the opposite direction if the coefficient of kinetic friction between the skater and the ice is μ_k.

20. N A skater of mass 45.0 kg standing on ice throws a stone of mass 7.65 kg with a speed of 20.9 m/s in a horizontal direction. Find the distance over which the skater will move in the opposite direction if the coefficient of kinetic friction between the skater and the ice is 0.03.

11-4 Special Case: One-Dimensional Inelastic Collisions

21. N An object of mass 2.0 kg moving with a velocity of 3.0 m/s collides with another object of mass 1.0 kg moving with a velocity of 4.0 m/s in the same direction. The two objects get stuck together in the collision. What is the velocity of the combination after the collision?

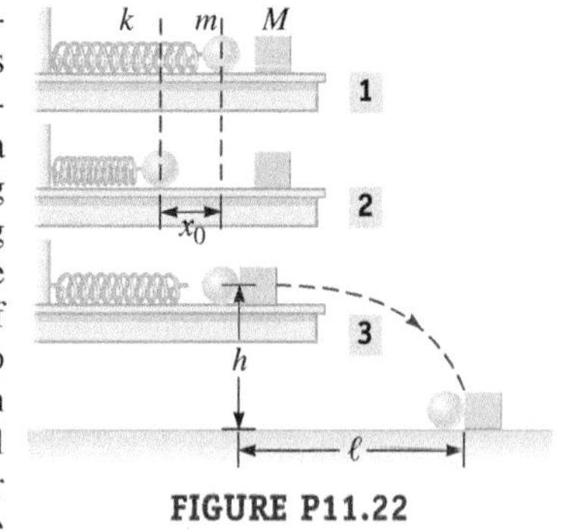

FIGURE P11.22

22. A In a laboratory experiment, **1** a block of mass M is placed on a frictionless table at the end of a relaxed spring of spring constant k. **2** The spring is compressed a distance x_0 and **3** a small ball of mass m is launched into the block as shown in Figure P11.22. The ball and block stick together and are projected off the table of height h. Find an expression for the horizontal displacement ℓ of the ball–block system from the end of the table until it hits the floor in terms of the parameters given.

23. N Ezra ($m = 25.0$ kg) has a tire swing and wants to swing as high as possible. He thinks that his best option is to run as fast as he can and jump onto the tire at full speed. The tire has a mass of 10.0 kg and hangs 3.75 m straight down from a tree branch. Ezra stands back 10.0 m and accelerates to a speed of 3.50 m/s before jumping onto the tire swing. **a.** How fast are Ezra and the tire moving immediately after he jumps onto the swing? **b.** How high does the tire travel above its initial height?

24. E A suspicious physics student watches a stunt performed at an ice show. In the stunt, a performer shoots an arrow into a bale of hay (Fig. P11.24). Another performer rides on the bale of hay like a cowboy. After the arrow enters the bale, the bale–arrow system slides roughly 5 m along the ice. Estimate the initial speed of the arrow. Is there a trick to this stunt?

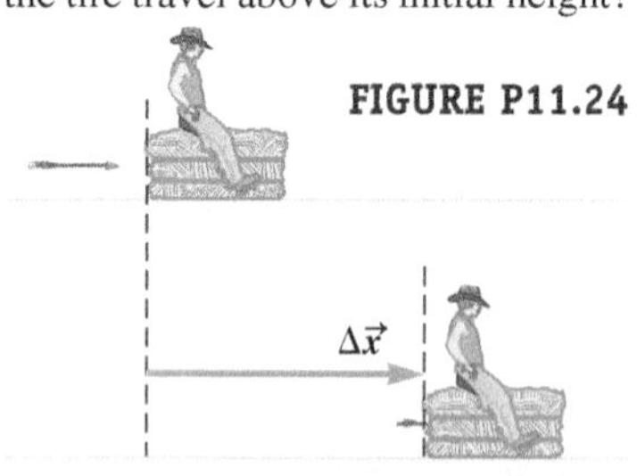

FIGURE P11.24

25. N A 2.45-kg ball is shot into a 0.450-kg box that is at rest on a frictionless, horizontal table (Fig. P11.25); after the collision, the ball is embedded in the box. The box is attached to a 0.30-m rope that is attached to the table on the other end. The ball's initial velocity is perpendicular to the rope as shown. If the ball's initial speed before impact is 13.5 m/s, what is the tension in the rope after the collision?

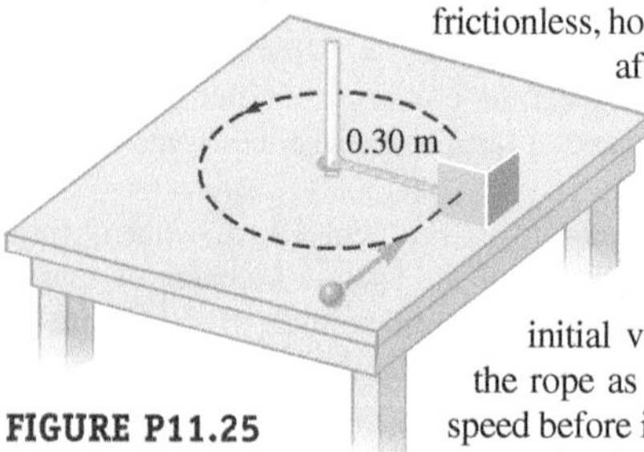

FIGURE P11.25

26. A In an attempt to overtake a slow truck of mass $3m$ moving with speed v_T, a car with mass m initially moving with speed v_C in the same direction instead collides and locks bumpers with the truck. **a.** What is the final speed of the car and the truck immediately after the collision? **b.** What is the change in kinetic energy of the car–truck system during the collision?

Problems 27 and 28 are paired.

27. N Jeff and Zak have worked out a stunt-show performance involving hovercrafts on a collision course. Just before the collision, Jeff and Zak will safely leap out of the hovercrafts, but they still have a little physics to work out. The two hovercrafts, each with a mass of 966 kg, are hovering at a constant height above the ground and are traveling toward each other as shown in Figure P11.27. The first hovercraft is traveling at 15 m/s to the right, and the second is traveling at 20 m/s to the left. **a.** If the hovercrafts undergo a completely inelastic collision, what is their final velocity after the collision? (Report to two significant figures.) **b.** How much kinetic energy is lost in the collision? (Report to three significant figures.)

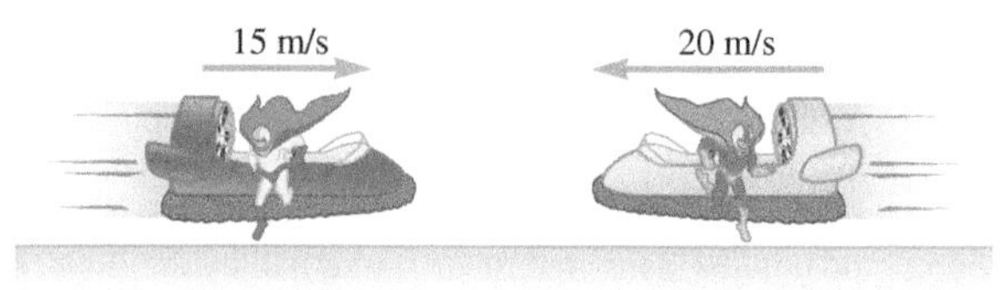

FIGURE P11.27 Problems 27 and 28.

28. C The two (unoccupied) hovercrafts, each with a mass of 966 kg, are hovering at a constant height above the ground and are traveling toward each other as shown in Figure P11.27. The first hovercraft is traveling at 15 m/s to the right, and the second is traveling at 20 m/s to the left. Suppose the two hovercrafts collide, but the collision is not completely inelastic. What bounds can you put on the kinetic energy lost in the collision process? Explain your answer.

Problems 29 and 30 are paired.

29. A A dart of mass m is fired at and sticks into a block of mass M that is initially at rest on a rough, horizontal surface. The coefficient of kinetic friction between the block and the surface is μ_k. After the collision, the dart and the block slide a distance D before coming to rest. If the dart were fired horizontally, what would its speed be immediately before impact with the block?

30. N A dart of mass $m = 10.0$ g is fired at and sticks into a block of mass $M = 85.0$ g that is initially at rest on a rough, horizontal surface. The coefficient of kinetic friction between the block and the surface is 0.400. After the collision, the dart and the block slide a distance of 6.00 m before coming to rest. If the dart were fired horizontally, what would its speed be immediately before impact with the block?

31. N A bullet of mass $m = 8.00$ g is fired into and embeds itself in a large 1.50-kg block of wood, initially at rest. What was the original speed of the bullet if that block with the embedded bullet were moving at a speed of 1.10 m/s immediately after the collision?

32. N **Review** CASE STUDY In Example 11.4, we used the conservation of energy principle to find the velocity of the two-train system just after impact. Now, use Newton's second law to find v_c, the velocity of the system just after the collision. Compare your result with that found in Example 11.4.

33. A A bullet of mass m is fired into a ballistic pendulum and embeds itself in the wooden bob of mass M (Fig. P11.33). After the collision, the pendulum reaches a maximum height h above its original position. **a.** Show that the kinetic energy of the system decreases by the factor $m/(m + M)$ immediately after the collision. **b.** What is the change in momentum of the bullet-bob system due to the collision?

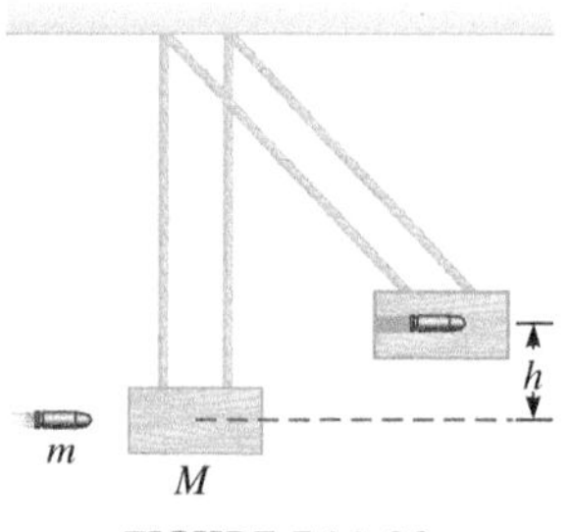

FIGURE P11.33

34. N In Examples 8.9 (page 230) and 9.8 (page 267), we found the maximum height of a small ball launched straight up from a spring-loaded dart gun. In this problem, we find the muzzle speed v_m of the dart gun. The dart gun is held horizontally such that its muzzle is very close to an open container of modeling dough (called *Fun Doh*) so that the ball hits the Fun Doh at the muzzle velocity. The ball of mass $m_1 = 2.7 \times 10^{-2}$ kg is fired into the packed Fun Doh container of total mass $m_2 = 0.15$ kg. The container is suspended from a string so that it can swing as a pendulum bob. The ball becomes embedded in the Fun Doh, and the center of mass of the ball–Fun Doh system rises to a maximum height $h = 0.26$ m. Assume dissipative forces may be ignored. Find the muzzle speed of the dart gun.

11-5 One-Dimensional Elastic Collisions

35. N One object ($m_1 = 0.200$ kg) is moving to the right with a speed of 2.00 m/s when it is struck from behind by another object ($m_2 = 0.300$ kg) that is moving to the right at 6.00 m/s. If friction is negligible and the collision between these objects is elastic, find the final velocity of each.

Problems 36, 37, and 38 are grouped.

36. C Particle 1 collides head-on with particle 2, which is initially at rest. After the collision, particle 1 has *reversed* direction and is moving at nearly the same speed it had before the collision. What can you say about the relative mass of the particles and particle 2's velocity after the collision? Explain.

37. C Particle 1 collides head-on with particle 2, which is initially at rest. After the collision, particle 1 *continues moving* at nearly its original velocity. What can you say about the relative mass of the particles and particle 2's velocity after the collision? Explain.

38. C Particle 1 collides head-on with particle 2, which is initially at rest. After the collision, particle 1 has *stopped*. What can you say about the relative mass of the particles and particle 2's velocity after the collision? Explain.

39. N Two objects collide head-on (Fig. P11.39). The first object is moving with an initial speed of 8.00 m/s, and the second object is moving with an initial speed of 10.00 m/s. Assuming the collision is elastic, $m_1 = 5.15$ kg, and $m_2 = 6.25$ kg, determine the final velocity of each object.

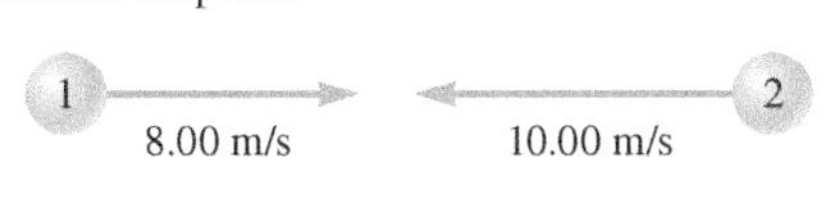

FIGURE P11.39

Problems 40 and 41 are paired.

40. A Initially, ball 1 rests on an incline of height h, and ball 2 rests on an incline of height $h/2$ as shown in Figure P11.40. They are released from rest simultaneously and collide in the trough of the track. If $m_2 = 4\,m_1$ and the collision is elastic, find an expression for the velocity of each ball immediately after the collision.

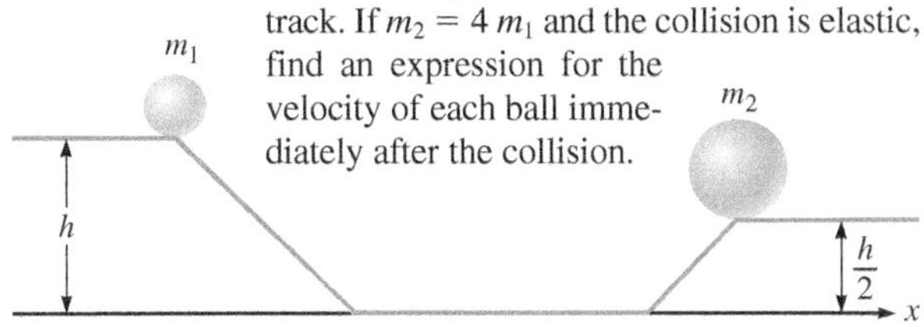

FIGURE P11.40 Problems 40 and 41.

41. N Initially, ball 1 rests on an incline of height h, and ball 2 rests on an incline of height $h/2$ as shown in Figure P11.40. They are released from rest simultaneously and collide elastically in the trough of the track. If $m_2 = 4\,m_1$, $m_1 = 0.045$ kg, and $h = 0.65$ m, what is the velocity of each ball after the collision?

42. N In an attempt to produce exotic new particles, a proton of mass $m_p = 1.67 \times 10^{-27}$ kg is accelerated to $0.99c$ ($c = 3.00 \times 10^8$ m/s is the speed of light) and crashed into a helium nucleus of mass $m_{He} = 6.64 \times 10^{-27}$ kg initially at rest. The collision is elastic. **a.** What is the kinetic energy of the helium nucleus after the collision? **b.** What is the kinetic energy of the proton after the collision? (In Chapter 39, we'll learn what Einstein says about making such calculations.)

43. Pendulum bob 1 has mass m_1. It is displaced to height h_1 and released. Pendulum bob 1 elastically collides with pendulum bob 2 of mass m_2 (Fig. P11.43).

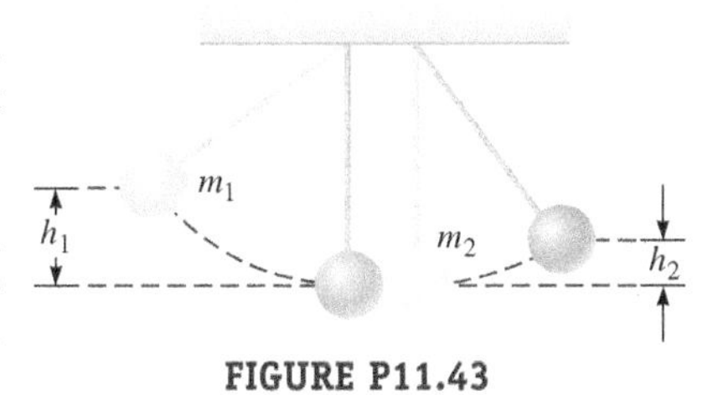

FIGURE P11.43

a. A Find an expression for the maximum height h_2 of pendulum bob 2.

b. N If $m_2 = 2.5m_1$ and $h_1 = 5.46$ m, what is h_2?

Problems 44, 45, and 46 are grouped.

44. N Crall and Whipple observe two carts collide elastically in a laboratory experiment. Cart 1 has a mass of 1.504 kg, and cart 2 has a mass of 1.040 kg. Crall chooses to work in the laboratory frame. Before the collision in his frame, the carts' speeds are $\vec{v}_{1i} = 0.353\hat{\imath}$ m/s and $\vec{v}_{2i} = -0.294\hat{\imath}$ m/s. According to Crall, what are the velocities of the carts after the collision?

45. N Refer to Crall and Whipple's experiment in Problem 44. Whipple chooses to do his calculations in a frame moving at cart 2's initial velocity. According to Whipple, what are the velocities of the carts before and after the collision?

46. N Refer to Crall and Whipple's experiment in Problem 44. Consider the two carts to make up a single system. What is that system's center-of-mass velocity before and after the collision?

47. N A roller-coaster car of mass $m_1 = 8.00 \times 10^2$ kg starts from rest and slides on a frictionless track from point A to point B at a height $h = 10.0$ m below point A, where it collides elastically with a second car of mass $m_2 = 2.00 \times 10^2$ kg, initially at rest. What is the maximum height that the second car rises above point B on the track?

48. E In Example 11.6, we found the change in a spacecraft's speed during a gravitational slingshot with a planet. In working that example, we assumed the planet does not lose any kinetic energy. In this problem, we test that assumption. In 1972, *Pioneer 10* was launched. It was the first spacecraft to take close-up images of Jupiter. It was launched from the Earth at a very high speed, fast enough to pass our Moon in just 11 hours. It approached Jupiter at a speed of 14 km/s. The gravitational slingshot with Jupiter increased its speed to 37 km/s. The mass of *Pioneer 10* is 270 kg, and Jupiter's mass is 1.9×10^{27} kg. Find the decrease in Jupiter's kinetic energy and speed as a result of this slingshot. Jupiter's speed is about 1.3×10^4 m/s. Comment on your findings.

49. N Two skateboarders, with masses $m_1 = 75.0$ kg and $m_2 = 65.0$ kg, simultaneously leave the opposite sides of a frictionless half-pipe at height $h = 4.00$ m as shown in Figure P11.49. Assume the skateboarders undergo a completely elastic head-on collision on the horizontal segment of the half-pipe. Treating the skateboarders as particles and assuming they don't fall off their skateboards, what is the height reached by each skateboarder after the collision?

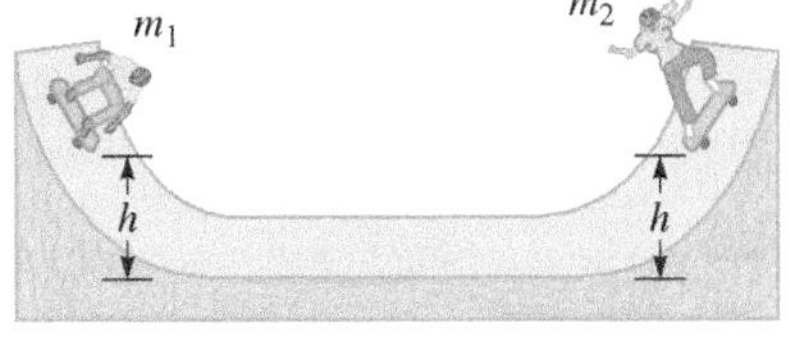

FIGURE P11.49

11-6 Two-Dimensional Collisions

50. In a laboratory experiment, an electron with a kinetic energy of 50.5 keV is shot toward another electron initially at rest (Fig. P11.50). (1 eV = 1.602×10^{-19} J) The collision is elastic. The initially moving electron is deflected by the collision.

a. C Is it possible for the initially stationary electron to remain at rest after the collision? Explain.

b. N The initially moving electron is detected at an angle of 40.0° from its original path. What is the speed of each electron after the collision?

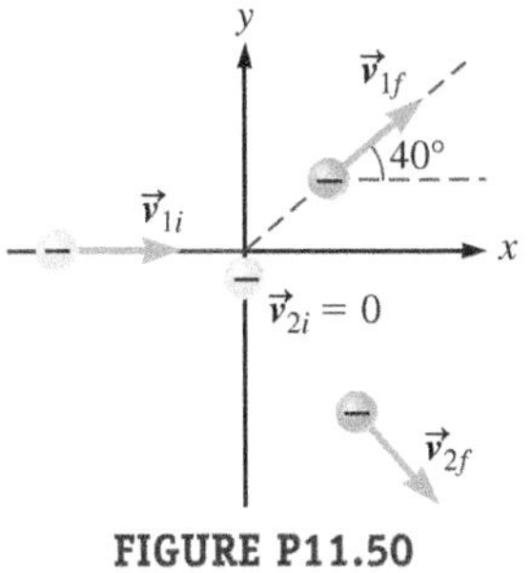

FIGURE P11.50

51. N In Figure P11.51, a cue ball is shot toward the eight-ball on a pool table. The cue ball is shot at the eight-ball with a speed of 8.00 m/s in a direction 30.0° from the y axis. Both balls have the same mass of 0.170 kg. After the balls undergo an elastic collision, the eight-ball travels in the negative x direction into the side pocket. What is the velocity of the cue ball after this collision?

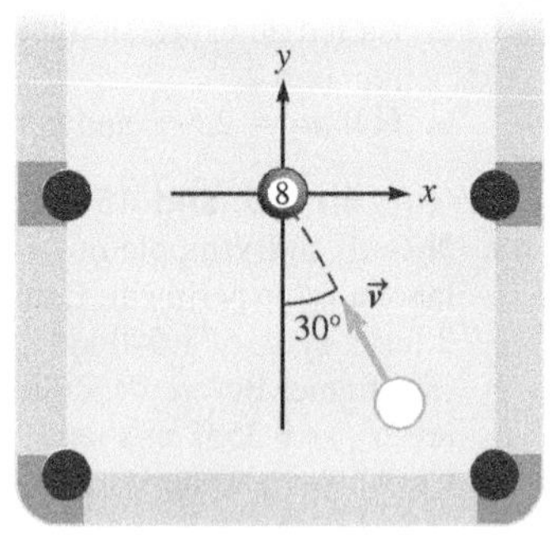

FIGURE P11.51

52. N A proton with an initial speed of 2.00×10^8 m/s in the x direction collides elastically with another proton initially at rest. The first proton's velocity after the collision is 1.60×10^8 m/s at an angle of 35.0° with the horizontal. What is the velocity of the second proton after the collision?

53. N A football player of mass 95 kg is running at a speed of 5.0 m/s down the field as shown in Figure P11.53. A second player of mass 140 kg, running at a speed of 2.5 m/s, tackles the first player so that they move together after the collision. What is the velocity of the two players immediately after the collision?

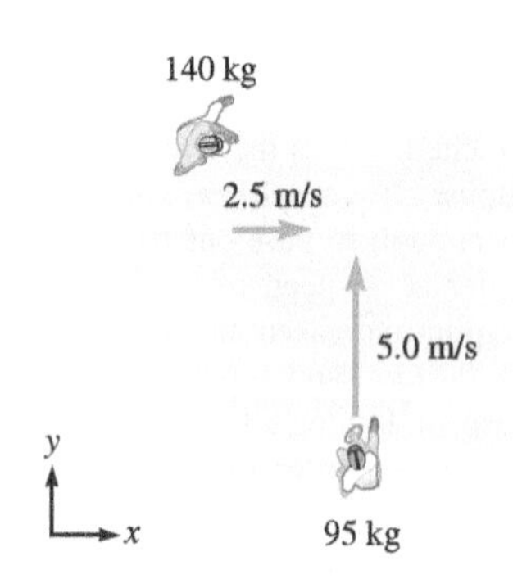

FIGURE P11.53

Problems 54 and 55 are paired.

54. N Two bumper cars at the county fair are sliding toward one another (Fig. P11.54). Initially, bumper car 1 is traveling to the east at 5.62 m/s, and bumper car 2 is traveling 60.0° south of west at 10.00 m/s. After they collide, bumper car 1 is observed to be traveling to the west with a speed of 3.14 m/s. Friction is negligible between the cars and the ground. **a.** If the masses of bumper cars 1 and 2 are 596 kg and 625 kg respectively, what is the velocity of bumper car 2 immediately after the collision? **b.** What is the kinetic energy lost in the collision?

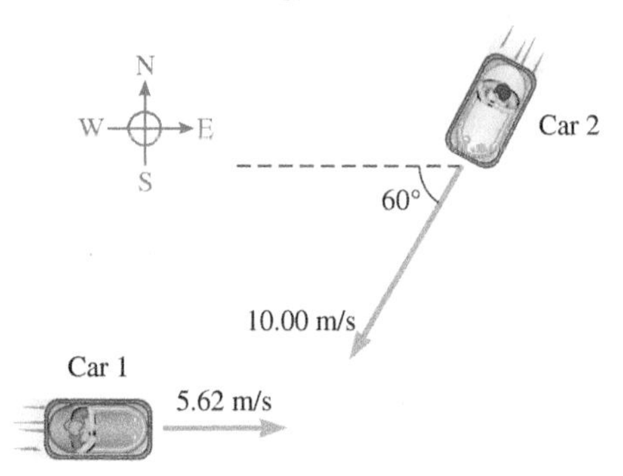

FIGURE P11.54 Problems 54 and 55.

55. Two bumper cars at the county fair are sliding toward one another (Fig. P11.54). Initially, bumper car 1 is traveling to the east at 5.62 m/s, and bumper car 2 is traveling 60.0° south of west at 10.00 m/s. They collide and stick together, as the driver of one car reaches out and grabs hold of the other driver. The two bumper cars move off together after the collision, and friction is negligible between the cars and the ground.

a. N If the masses of bumper cars 1 and 2 are 596 kg and 625 kg respectively, what is the velocity of the bumper cars immediately after the collision?

b. N What is the kinetic energy lost in the collision?

c. C Compare your answers to part (b) from this and Problem 54. Is one answer larger than the other? Discuss and explain any differences you find.

56. N A police officer is attempting to reconstruct an accident in which a car traveling southward with a speed of 23.0 mph collided with another car of equal mass traveling eastward at an unknown speed. After the collision, the two cars coupled and slid at an angle of 60.0° south of east. If the speed limit in that neighborhood is 25.0 mph, should the officer cite the second driver for speeding? Explain your answer.

57. N A bomb explodes into three pieces A, B, and C of equal mass. Piece A flies with a speed of 40.0 m/s, and piece B with a speed of 30.0 m/s at an angle of 90° relative to the direction of A as shown in Figure P11.57. Determine the speed of piece C and the direction of its velocity relative to the direction of piece A.

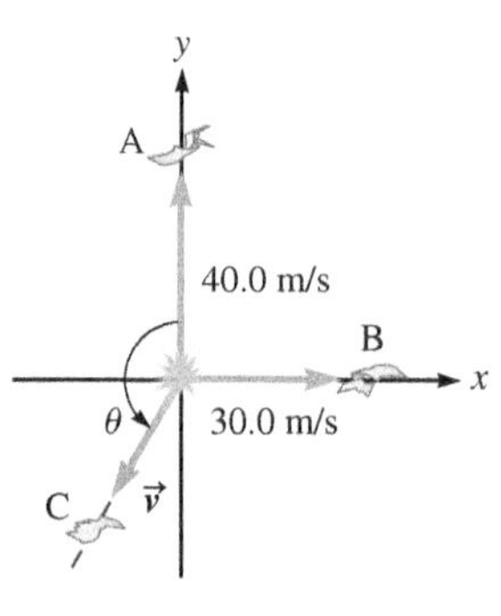

FIGURE P11.57

58. N The spontaneous decay of a heavy atomic nucleus of mass $M = 14.5 \times 10^{-27}$ kg that is initially at rest produces three particles. The first particle, with mass $m_1 = 6.64 \times 10^{-27}$ kg, is ejected in the positive x direction with a speed of 2.50×10^7 m/s. The second particle, with mass $m_2 = 3.00 \times 10^{-27}$ kg, is ejected in the negative y direction with a speed of 2.00×10^7 m/s. **a.** What is the velocity of the third particle? **b.** What is the increase in the kinetic energy of the system after the decay?

General Problems

59. N An object of mass $m = 4.00$ kg that is moving with a speed of 10.0 m/s collides head-on with another object, and the collision lasts 1.50 s. A graph showing the magnitude of the force during the collision versus time is shown in Figure P11.59, where the force is exerted in the direction opposite the initial velocity. Find the speed of the 4.00-kg mass after collision.

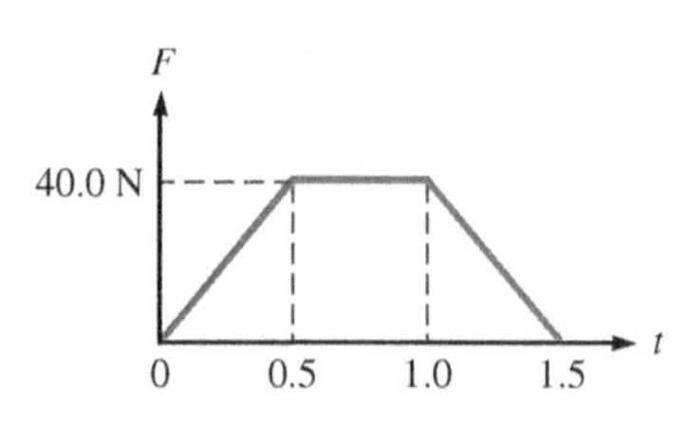

FIGURE P11.59

60. A A wooden block of mass M is initially at rest at the edge of a frictionless table at a height h above the ground. A bullet of mass m is fired horizontally into the block and embeds itself in the block. The block lands a distance d from the edge of the table. Find an expression for the speed of the bullet just before the collision.

61. N A particle of mass $m_1 = 1.50$ kg with an initial velocity $\vec{v}_1 = 4.50\hat{j}$ m/s has a completely inelastic collision with a second particle of mass $m_2 = 3.50$ kg with an initial velocity

$\vec{v}_2 = 3.00\hat{\imath}$. What is the velocity of the combined particles immediately after the collision?

62. A person jumping on a trampoline lands at time $t = 0$ and experiences a force given by $F = 280(1.8 - t)t$. The time t is measured in seconds, and the resulting force is in newtons. After 1.8 s, the force is zero, and the person is launched off of the trampoline.
a. G Sketch the magnitude of the force as a function of time.
b. N What is the magnitude of the total impulse in the time interval from 0 to 1.8 s?

63. N In an experiment designed to determine the velocity of a bullet fired by a certain gun, a wooden block of mass $M = 500.0$ g is supported only by its edges, and a bullet of mass $m = 8.00$ g is fired vertically upward into the block from close range below. After the bullet embeds itself in the block, the block and the bullet are measured to rise to a maximum height of 25.0 cm above the block's original position. What is the speed of the bullet just before impact?

64. From what might be a possible scene in the comic book *The X-Men*, the Juggernaut (m_J) is charging into Colossus (m_C) and the two collide. The initial speed of the Juggernaut is v_{Ji} and the initial speed of Colossus is v_{Ci}. After the collision, the final speed of the Juggernaut is v_{Jf} and the final speed of Colossus is v_{Cf} as they each bounce off of the other, heading in opposite directions.
a. A What is the impulse experienced by the Juggernaut?
b. A What is the impulse experienced by Colossus?
c. C In your own words, explain how these impulses must compare with each other and how they are related to the average force each superhero experiences during the collision.

65. N A 110-kg rugby player running east with a speed of 4.00 m/s tackles an 85.0-kg opponent running north with a speed of 3.50 m/s. Assume the tackle is a perfectly inelastic collision. **a.** What is the velocity of the players immediately after the tackle? **b.** What is the amount of mechanical energy lost during the collision?

Problems 66 and 67 are paired.

66. Two pucks in a laboratory are placed on an air table (Fig. P11.66). Puck 2 has four times the mass of puck 1 ($m_2 = 4m_1$). Initially, puck 1's speed is three times puck 2's speed ($v_{1i} = 3v_{2i}$), puck 1's position is $\vec{r}_{1i} = -x_{1i}\hat{\imath}$, and puck 2's position is $\vec{r}_{2i} = -y_{2i}\hat{\jmath}$. The pucks collide at the origin.

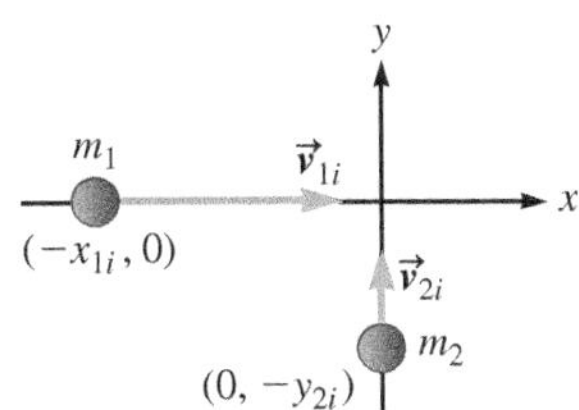

FIGURE P11.66 Problems 66 and 67.

a. G Copy Figure P11.66 and then add $\vec{v}_{CM}$ to your sketch.
b. C Does puck 2 travel a greater distance, lesser distance, or the same distance as puck 1?
c. A Find an expression for y_{2i} in terms of x_{1i}.
d. N If puck 1 moves 1.33 m, how far does puck 2 move before the collision?

67. N Assume the pucks in Figure P11.66 stick together after their collision at the origin. Puck 2 has four times the mass of puck 1 ($m_2 = 4m_1$). Initially, puck 1's speed is three times puck 2's speed ($v_{1i} = 3v_{2i}$), puck 1's position is $\vec{r}_{1i} = -x_{1i}\hat{\imath}$, and puck 2's position is $\vec{r}_{2i} = -y_{2i}\hat{\jmath}$. **a.** Find an expression for their velocity after the collision in terms of puck 1's initial velocity. **b.** What is the fraction K_f/K_i that remains in the system?

68. C Two objects travel toward each other, collide, and are motionless after the collision. Object 1 has a greater speed than object 2. **a.** Can you determine whether the collision was elastic or inelastic? **b.** Can you determine which object has more mass? **c.** Can you determine which object initially had a larger kinetic energy? Explain all your answers.

69. N An object of mass 4.0 kg collides elastically with a second object, initially at rest, and then continues to move in the original direction with one-half its original speed. What is the mass of the second object?

Problems 70 and 71 are paired.

70. N A ball of mass 50.0 g is dropped from a height of 10.0 m. It rebounds after losing 75% of its kinetic energy during the collision process. If the collision with the ground took 0.010 s, find the magnitude of the impulse experienced by the ball.

71. N A ball of mass 0.10 kg that is released from a height of 2.5 m above the ground rebounds to a height 0.65 m. The time of contact of the ball with the ground is 0.010 s. Find the magnitude of the average force exerted by the ball on the ground during the collision.

72. A A pendulum consists of a wooden bob of mass M suspended by a massless rod of length ℓ. A bullet of mass $m \ll M$ is fired horizontally with speed v at the bob and emerges from the bob with speed $v/3$ as shown in Figure P11.72. If the pendulum bob just barely reaches the highest point such that it is able to swing through one complete circle, find an expression for the speed v of the bullet before the collision.

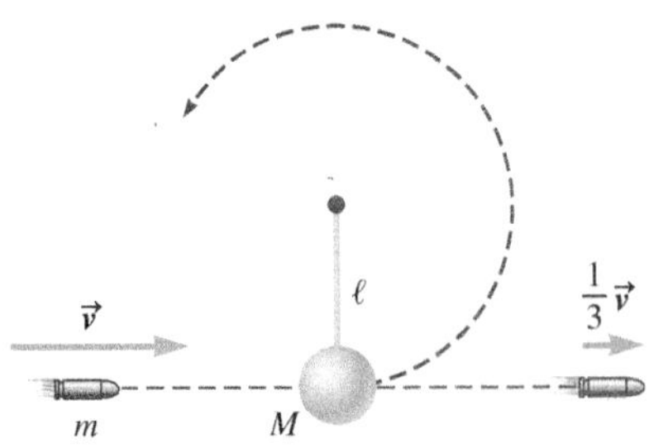

FIGURE P11.72

73. Three runaway train cars are moving on a frictionless, horizontal track in a railroad yard as shown in Figure P11.73. The first car, with mass $m_1 = 1.50 \times 10^3$ kg, is moving to the right with speed $v_1 = 10.0$ m/s; the second car, with mass $m_2 = 2.50 \times 10^3$ kg, is moving to the left with speed $v_2 = 5.00$ m/s, and the third car, with mass $m_3 = 1.20 \times 10^3$ kg, is moving to the left with speed $v_3 = 8.00$ m/s. The three railroad cars collide at the same instant and couple, forming a train of three cars.
a. N What is the final velocity of the train cars immediately after the collision?
b. C Would the answer to part (a) change if the three cars did not collide at the same instant? Explain.

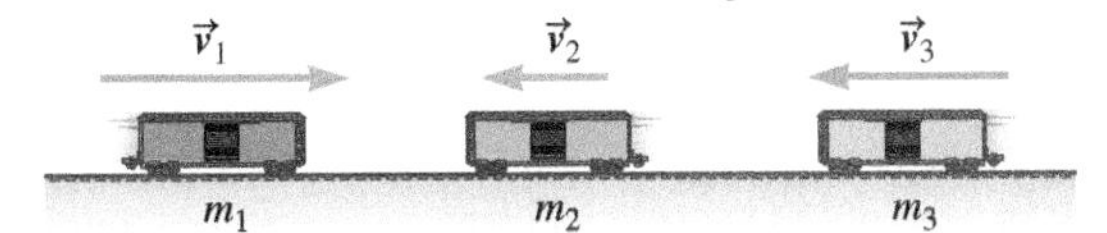

FIGURE P11.73

Problems 74 and 75 are paired.

74. In the early part of the 20th century, Sir Joseph J. Thomson (discoverer of the electron) proposed a "plum pudding" model of the atom. He believed that the positive charge of the atom was spread out like a pudding and that the negative charges (electrons) were embedded in the pudding like plums. His student Ernest Rutherford performed an experiment in 1911 that disproved the plum pudding model. Rutherford fired a beam of alpha particles (helium nuclei) at a thin metal sheet. Because alpha particles are positive, the plum pudding model predicted that they should only be slightly deflected by the positive pudding. Instead, Rutherford found that the alpha particles were greatly deflected, and some even reversed direction completely. Rutherford was surprised by his results and said that it was like firing a bullet at a tissue paper and seeing it bounce back. Rutherford concluded that the positive charge of the atom was not spread out like pudding, but rather was concentrated in the center or *nucleus* of the atom. The alpha particles used in the experiment had an initial speed of 2×10^7 m/s and a mass of 6.7×10^{-27} kg.

a. N Assuming the nucleus is initially at rest, the collision is head-on and elastic, and the nucleus is much more massive than the alpha particle, find the final speed of the alpha particle after it collides with the nucleus.

b. C Rutherford used gold in his experiment. Check the assumption that the nucleus is much more massive than the alpha particle by finding the speed of the nucleus after the collision.

75. N Rutherford fired a beam of alpha particles (helium nuclei) at a thin sheet of gold. An alpha particle was observed to be deflected by 90.0°; its speed was unchanged. The alpha particles used in the experiment had an initial speed of 2×10^7 m/s and a mass of 6.7×10^{-27} kg. Assume the alpha particle collided with a gold nucleus that was initially at rest. Find the speed of the nucleus after the collision.

Problems 76 and 77 are paired.

76. A dramatic (and perhaps unexpected) collision occurs if you hold a tennis ball on top of a basketball and drop them at the same time. After impacting the ground, the tennis ball can be launched much higher than the original height of the two balls. Assume this process can be modeled as an elastic collision of the basketball with the ground followed by a second elastic collision of the basketball with the tennis ball. The basketball (with mass m_b) and tennis ball (with mass m_t) each falls a distance h with an acceleration g before the basketball hits the ground.

a. A What is the velocity of each ball the instant before the basketball collides with the ground in terms of the variables specified?

b. A Assume the basketball first undergoes an elastic collision with the ground. How fast is the basketball moving after this collision?

c. A The basketball then collides elastically with the tennis ball. What is the final velocity of the tennis ball?

d. N Assume the basketball has a mass that is eight times that of the tennis ball. What is the ratio of the final speed of the tennis ball to the speed of the balls just before impact?

77. N Refer to the dramatic collision described in Problem 76. The basketball and the tennis ball have masses $m_b = 0.45$ kg and $m_t = 57$ g, respectively, and fall from a height of $h = 150$ cm. What is the maximum height reached by the tennis ball after the balls collide with the ground?

78. E February 3, 2009, was a very snowy day along Interstate 69 just outside of Indianapolis, Indiana. As a result of the slippery conditions and low visibility (50 yards or less), there was an enormous accident involving approximately 30 vehicles, including cars, tractor-trailers, and even a fire truck. Many witnesses said that people were driving too fast for the conditions and were too close together. In this problem, we explore two rules of thumb for driving in such conditions. The first is to drive at a speed that is half of what it would be in ideal conditions. The other is the "8-second" rule: Watch the vehicle in front of you as it passes some object such as a street sign, and you should pass that same object 8 seconds later. On a dry road, the 8-second rule is replaced by a 3-second rule. **a.** Assume vehicles on a slippery interstate highway follow both rules. What is the distance between the vehicles? **b.** If a driver followed the first rule of thumb, driving at a lower speed, but used the 3-second rule instead of the 8-second rule, what is the distance between the vehicles? How does that distance compare with the visibility on the day of the accident? **c.** Suppose drivers do not follow either rule of thumb for slippery conditions. What is the distance between vehicles? How does that distance compare with the visibility on that day? **d.** Suppose a driver was not obeying either rule of thumb when she sees a tractor-trailer that stopped on the highway. She presses on her brakes, locking the wheels, and her car crashes into the truck. Estimate the magnitude of the impulse exerted on her car. **e.** Estimate the impulse on the car in part (d) had the driver followed both rules of thumb for slippery conditions instead of ignoring them.

Problems 79 and 80 are paired.

79. N A cart filled with sand rolls at a speed of 1.0 m/s along a horizontal path without friction. A ball of mass $m = 2.0$ kg is thrown with a horizontal velocity of 8.0 m/s toward the cart as shown in Figure P11.79. The ball gets stuck in the sand. What is the velocity of the cart after the ball strikes it? The mass of the cart is 15 kg.

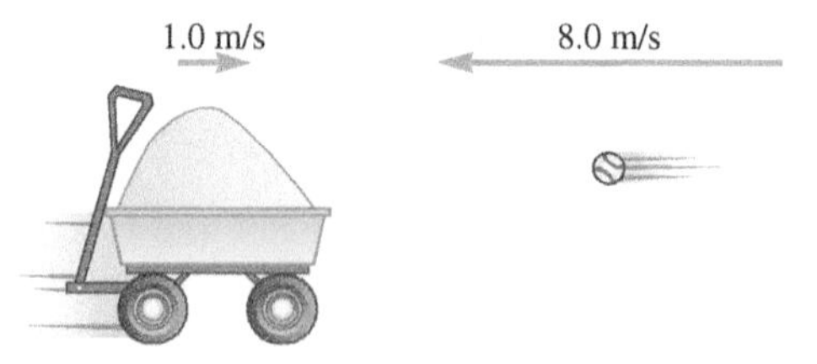

FIGURE P11.79 Problems 79 and 80.

80. Consider the cart, ball, and sand in Problem 79.

a. N What are the initial and final kinetic energies of the system? Is the energy conserved in this collision process?

b. C What type of collision process is it?

81. A ball of mass m moving in the positive x direction with speed v_i experiences an elastic glancing collision with a second ball of mass $3m$ moving in the opposite direction with the same speed as the first ball. After the collision, the first ball moves in the positive y direction.

a. A What are the speeds of the two balls in terms of their initial speeds?

b. N What is the angle θ at which the second ball travels after the collision, measured with respect to the original direction of the first ball (the positive x direction)?

82. A The force on a particle is given by $\vec{F} = F_{max} \cos(2\pi t/T)\hat{\imath}$, where F_{max} and T are constants. Find expressions for the impulse on the particle during each of the following intervals. **a.** $0 < t < T/4$ **b.** $T/4 < t < 3T/4$ **c.** $0 < t < T$

83. A Consider a system of two colliding objects, and derive the conservation of momentum (Equation 11.5) by considering the impulse on each object.

84. N Return to Example 11.8 (page 323), and find the system's center-of-mass velocity before and after the collision.

Rotation I: Kinematics and Dynamics

12

Underlying Principles

Newton's second law in rotational form

Major Concepts

1. Rigid object
2. Rotation axis
3. Fixed axis
4. Angular kinematic quantities
   a. Angular position
   b. Angular displacement
   c. Angular velocity
   d. Angular acceleration
5. Torque
6. Rotational inertia

Tools

1. Right-hand rule
2. Cross product (vector product)

Key Questions

How do we describe the kinematics of a rotating object?

How can we apply Newton's second law to a rotating object?

12-1 Rotation versus translation 331

12-2 Rotational kinematics 333

12-3 Special case of constant angular acceleration 338

12-4 The connection between rotation and circular motion 340

12-5 Torque 344

12-6 Cross product 346

12-7 Rotational dynamics 347

From the earliest times, rotational motion of wheels and other objects has played a major role in culture and technology. Rotating waterwheels may be used to transfer energy from a river, the rotation of gears on a bicycle allows a rider to pedal up steep mountains more easily (Fig. 12.1), and the rotation of a lever enables a person to lift very heavy objects. In this chapter, we explore **rotational kinematics** (the study of rotational motion independent of its cause) and **rotational dynamics** (the cause of rotational motion).

FIGURE 12.1 Wheels and rotational motion have played a major role in human endeavors since ancient times. Even without powered machinery, gears allow people to pedal easily up steep hills.

12-1 Rotation Versus Translation

In Chapter 2, we treated an object as a particle undergoing purely translational motion if every point of the object underwent exactly the same displacement. Figure 12.2A shows Zak riding a Ferris wheel from an initial position to a final position. As the Ferris wheel goes around, Zak's chair pivots so that he is always facing the same direction. All points associated with Zak have the same displacement as he moves between an initial position and a final position. Thus, Zak's motion is translational, and he may be approximated as a particle.

Compare his motion with that of Sophia on a witch's wheel in Figure 12.2B. Sophia's chair does not pivot, so when she is at the top of the wheel, she is upside

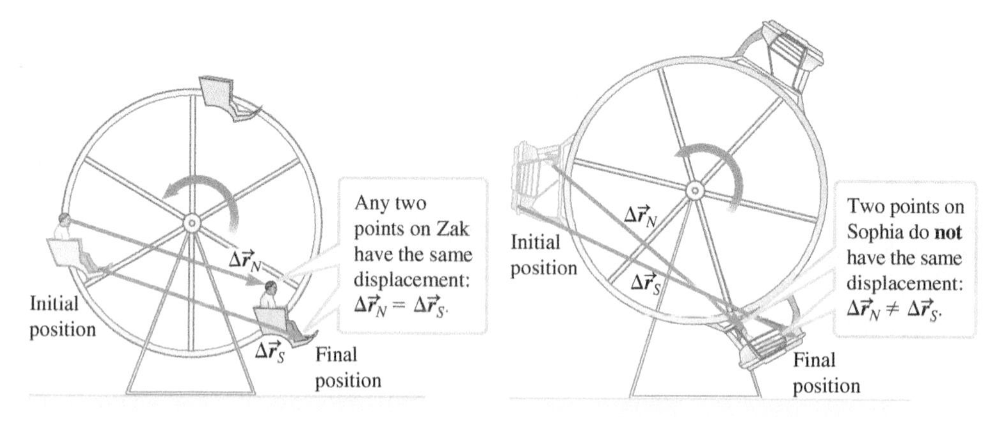

FIGURE 12.2 A. When Zak rides a Ferris wheel, the orientation of his body with respect to the ground remains the same; he is always facing the same direction. His motion is translational. In a given time interval, the displacement of his nose equals the displacement of his shoe: $\Delta\vec{r}_N = \Delta\vec{r}_S$. He can be modeled as a particle. **B.** When Sophia rides a witch's wheel, the orientation of her body with respect to the ground changes. Her motion is not translational; in a given time interval, the displacement of her nose does not equal the displacement of her shoe: $\Delta\vec{r}_N \neq \Delta\vec{r}_S$. She cannot be modeled as a particle.

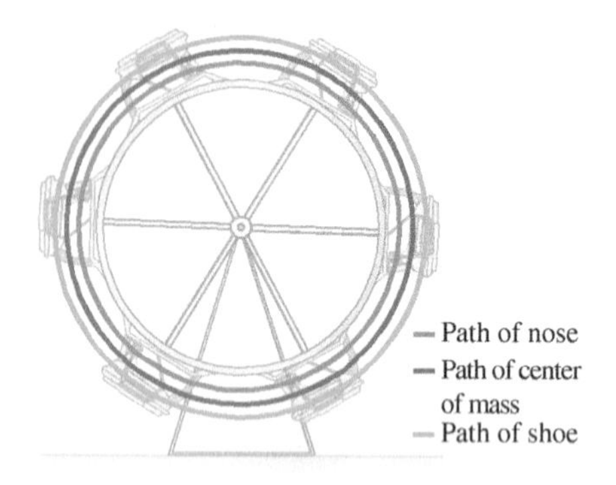

FIGURE 12.3 Sophia on the witch's wheel ride may be modeled as a rotating rigid object. Points such as her shoe, center of mass, and nose travel along arcs of concentric circles.

down. The displacements of her nose and shoe are unequal: $\Delta\vec{r}_N \neq \Delta\vec{r}_S$. Because *not* every point associated with Sophia has the same displacement, her motion is *not* translational, and we *cannot* approximate her as a particle. Instead, we consider the path of several points associated with her. Figure 12.3 shows the path of her nose, shoe, and center of mass. Each point is traveling along the arc of a circle, and all the circles have a common center at the center of the witch's wheel. We model Sophia as a *rotating rigid object.*

RIGID OBJECT; ROTATION AXIS; FIXED AXIS

Major Concepts

A **rigid object** is an object that moves with all its components locked together so that its shape does not change. When we are considering a real object, we need to decide if it can be modeled as a rigid object. A test-crash dummy on the witch's wheel is likely to maintain its shape and so is well modeled as a rigid object, but can we model a live woman moving her arms and legs as a rigid object? If the internal movement of her arms and legs is small compared with the overall motion, we may be safe in modeling her as a rigid object. A typical rider on an amusement park ride is usually well modeled as a rigid object. When an ice skater pulls her arms inward as she spins, however, the rigid object model fails to fit her motion.

If a rigid object is rotating, every particle in the object moves in a circular path centered on the same axis, known as the **rotation axis.** In the case of Sophia on the witch's wheel, the rotation axis runs through the center of the wheel and is perpendicular to the page. In this chapter, we limit our discussion to rotations around a **fixed axis,** that is, a rotation axis that does not move relative to the observer. Sophia's rotation axis is fixed, but the rotation axis of a bicycle wheel rolling down the road is not fixed. We discuss rolling motion in Chapter 13. An axis may be fixed in one reference frame but not in another. If we observe the bicycle wheel in a frame moving along with the bicycle, the axis of rotation is fixed.

For the object's motion to be considered a rotation, the particles in the object do not need to make a complete circle. The following case study illustrates such rotations.

CASE STUDY Ancient Megaliths

Ancient peoples built megalithic monuments of impressive size and mass with primitive technology. Three examples are shown in Figure 12.4.

- The great trilithons of Stonehenge in England were constructed approximately 4600 years ago (Fig. 12.4A). A large, upright stone in a trilithon weighs about 40 tons and stands about 25 ft above the ground.
- Queen Hatshepsut's obelisk was erected approximately 3500 years ago in Egypt (Fig. 12.4B). The largest Egyptian obelisks are made of a single block of granite; they weigh about 400 tons and stand 100 ft tall.
- Moai statues on Easter Island were created roughly 400 to 1000 years ago (Fig. 12.4C). There are nearly 900 standing Easter Island Moai, weighing an average of 14 tons and standing about 14 ft tall.

Today, it is hard to imagine creating any megalithic structure without the use of modern construction equipment. Because moving a 400-ton granite obelisk safely would be difficult even with all the technological advantages of today, some people believe that it was *impossible* for ancient people to accomplish such feats with their technology. It has even been suggested that extraterrestrial aliens visited ancient peoples and constructed the megaliths.[1]

Science usually proceeds by testing the simplest theories first. A claim that an extraterrestrial workforce created the ancient monuments is not supported by other evidence such as written records, and this assertion dismisses the much simpler explanation that ancient people built their own megaliths using tools and teamwork. Archeological evidence from artifacts and written records supports this simpler theory.

Many of the details of ancient construction were unrecorded, or perhaps the records have been lost. So, archeologists must develop theories and test them by using a team of contemporary workers to build replicas of ancient monuments using primitive technology. Of course, even when a modern workforce successfully builds a replica of an ancient monument, it does not prove how the ancient workforce built it. In this case study, we explore how ancient people may have used the Earth's gravity and levers to raise an obelisk or an upright stone in a trilithon and how levers may have been used to rotate a Moai.

A. B. C.

FIGURE 12.4 **A.** A trilithon is made of three large stones. Together they form a square arch. **B.** An obelisk is a single large stone. **C.** Easter Island has about 900 standing Moai such as the four intact statues shown here.

CONCEPT EXERCISE 12.1

Figure 12.5 shows two rotating objects. Indicate the rotation axis in each case and determine in which reference frame the rotation axis is fixed. Explain.

Case 1. The Earth's daily rotation on its axis.

Case 2. A bowling ball.

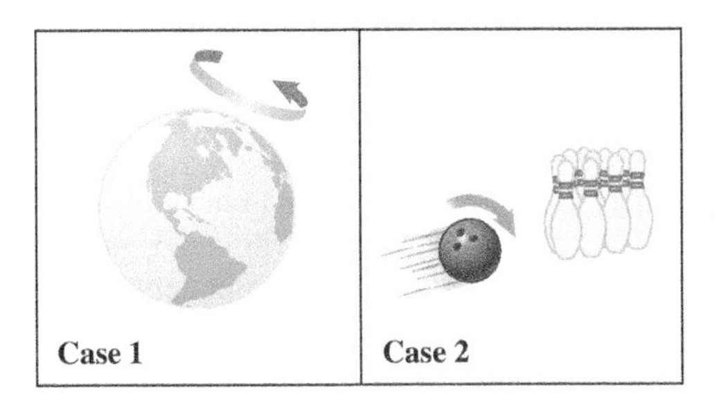

FIGURE 12.5

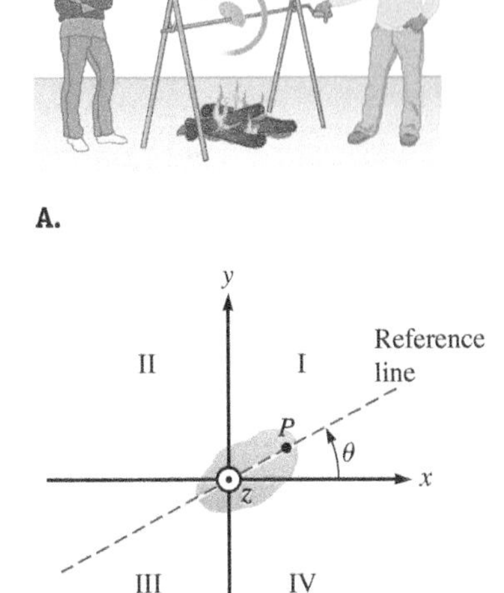

FIGURE 12.6 **A.** Aaron rotates a potato. **B.** The potato rotates counterclockwise in the xy plane.

12-2 Rotational Kinematics

As we did for translational motion, we begin our study of rotational motion with kinematics, that is, the mathematical description of rotational motion. We will use analogies with translational kinematics (Chapters 2 and 4) to help us.

Figure 12.6A shows a potato being cooked over an open fire. A stake has been driven through the potato, and it slowly rotates as the stake is turned. As in translational kinematics, we must choose a coordinate system (right-handed in this book). Figure 12.6B shows our choice. The z axis is the rotation axis, aligned with the stake. With this choice, the potato rotates in the xy plane. The coordinate system in Figure 12.6B is right-handed because it follows the rule from Chapter 3: Line up the fingers of your right hand along the positive x axis and curl your fingers so that you "push" the x axis through 90° into the positive y axis; then, the direction in which your thumb points is the direction of the positive z axis (Fig. 3.22). Although you may orient the rotation axis in any convenient way, in this book we will usually align the z axis with the rotation axis.

We will sometimes include the adjective *translational* for clarity when describing a quantity, but if no adjective is explicitly used, *translational* is implied.

[1]See, for example, Erich von Däniken's best-selling book *Chariots of the Gods?*

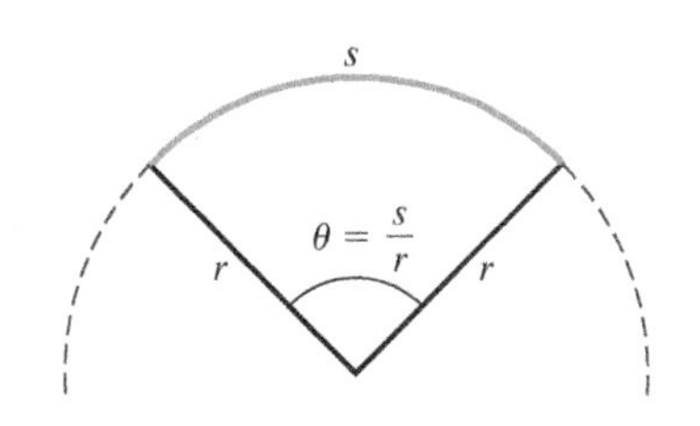

FIGURE 12.7 The arc length s and radius r are measured in the same length units (such as meters).

To describe the rotational motion of the potato, we need to choose a **reference line** fixed on the object and perpendicular to the rotation axis. If the rotation axis is aligned with the z axis, the reference line is parallel to the xy plane. The reference line rotates with the object. For the potato in Figure 12.6B, we have picked a reference line that passes through the rotation axis and a spot ("an eye") on the potato labeled P.

Angular Position and Angular Displacement

ANGULAR POSITION

Major Concept

Polar coordinates are used to describe rotational motion (Chapter 4). The **angular position** θ of the reference line is measured counterclockwise from the positive x axis as shown in Figure 12.6B. The angular position θ is negative if it is measured clockwise from the positive x axis.

In rotational kinematics, we specify the angular position in radians because it is simpler to work with radians than degrees or revolutions. Consider the angle θ shown in Figure 12.7. The circle has radius r, and s is the arc length subtended by θ; so, in radians, θ is

$$\theta = \frac{s}{r} \tag{12.1}$$

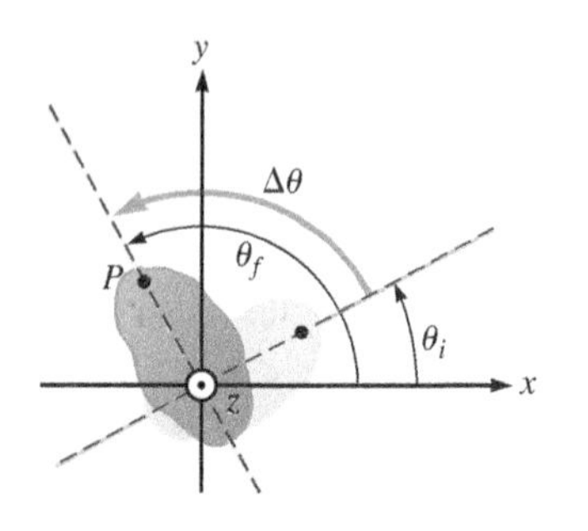

FIGURE 12.8 The angular displacement $\Delta\theta$ of a potato.

From Equation 12.1, an angle in radians is the ratio of two lengths and is therefore dimensionless. For example, if $s = 5.0$ m and $r = 2.5$ m, then $\theta = s/r = 5.0\text{ m}/2.5\text{ m} = 2.0$ according to Equation 12.1. We write $\theta = 2.0$ "rad" to clarify that the angle is measured in radians and *not* in degrees or revolutions. Equation 12.1 does not hold if θ is measured in degrees or revolutions; in rotational kinematics, we convert all angles to radians. Because a radian comes from a dimensionless ratio, we often drop the "rad" when radian measure is understood from the context.

The angular position $\theta(t)$is a function of time for a rotating object. If, for example, at $t = 0$ the reference line were aligned with the positive x axis, then $\theta(0) = 0$. When the potato completes one revolution so that the reference line is once again lined up with the positive x axis, the angular position is not zero, but 2π rad.

ANGULAR DISPLACEMENT

Major Concept

A change in the angular position is known as the **angular displacement**,

$$\Delta\theta = \theta_f - \theta_i \tag{12.2}$$

where (as usual) i and f stand for *i*nitial and *f*inal, respectively. Figure 12.8 shows the angular displacement of the potato. Equation 12.2 does not require angular displacement to be in radians, but subsequent kinematic equations will.

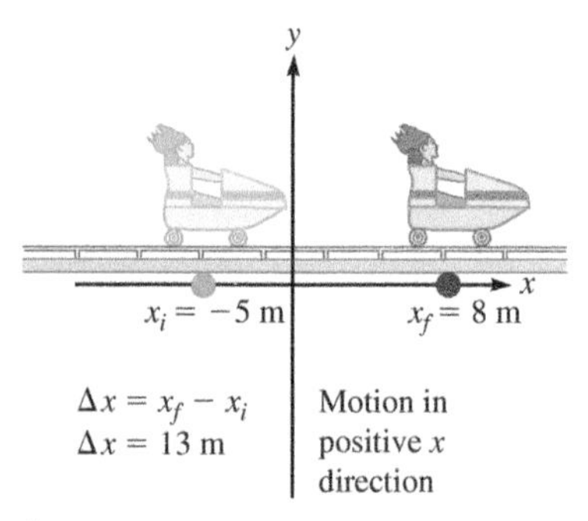

A.

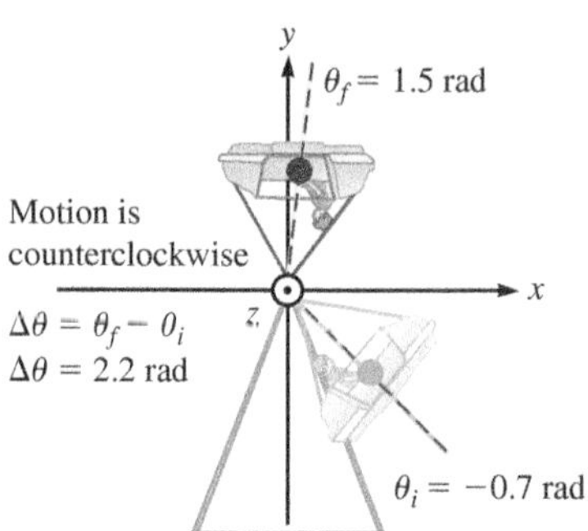

B.

FIGURE 12.9 An analogy between one-dimensional motion along an x axis and rotational motion around a z axis.

Although angular displacement has both magnitude and direction, it does not obey the commutative rule of vector addition. So, angular displacement is not a vector, and we should not represent its direction with a straight-line arrow. Instead, we refer to the direction of angular displacement with the terms *clockwise* and *counterclockwise*. These terms are consistent with the sign convention for angular position, so a counterclockwise angular displacement is positive. If $\theta_f > \theta_i$, the angular displacement is positive, and the net rotation of the object is counterclockwise as shown in Figure 12.8.

In this book, we focus on objects that rotate around a single rotation axis, similar to restricting translational kinematics to one dimension. A single axis makes the mathematics considerably simpler and allows us to concentrate on a conceptual understanding of rotation. When restricted to a single rotation axis, angular displacement *is* commutative. For example, an object may be rotated around a single axis by 1.5 rad and then by 0.75 rad, or it may be rotated by 0.75 rad and then by 1.5 rad with the same result.

Let's take a closer look at the analogy with one-dimensional translational kinematics. Figure 12.9A shows Charlotte riding in a car that is restricted by a track to move in one dimension while Figure 12.9B shows Jim riding in a gondola that is restricted by its axle to rotate about a single, fixed axis. We'll compare the choice of coordinate system, model, measured position, and displacement for the two different riders.

Coordinate system: For one-dimensional translational motion, we typically choose to align one coordinate axis with the direction of movement. If the motion is horizontal, we often choose the direction of motion to be the x axis as shown in Figure 12.9A. For motion around a single rotation axis, we typically align the z axis with the rotation axis. Then, the object's rotation is in the xy plane as shown in Figure 12.9B.

Model: Charlotte may be modeled as a particle and therefore may be represented by a single dot on the x axis. Jim is modeled as a rigid object, and a reference line is drawn on the coordinate system to represent him.

Measured position: Charlotte's position x is positive if she is to the right of the origin. Charlotte's initial position is shown as $x_i = -5$ m. By analogy, Jim's angular position θ is positive if θ is measured counterclockwise from the positive x axis. Measuring θ is different from measuring x. Once a coordinate system has been established, there is only one way to measure x, but θ may be measured either clockwise or counterclockwise. Jim's initial position is measured clockwise and is therefore negative: $\theta_i = -0.7$ rad. (Had it been measured counterclockwise, it would have been $2\pi - 0.7$ rad.) Linear or translational position x is analogous to angular position θ.

Displacement: Charlotte's motion may be either in the positive or the negative x direction. If $x_f > x_i$, her displacement is positive and her motion is in the positive x direction as shown in Figure 12.9A. Jim's rotation around the z axis may be either counterclockwise or clockwise in the xy plane. If $\theta_f > \theta_i$, his angular displacement is positive and his rotation is counterclockwise as shown in Figure 12.9B. Translational displacement Δx is analogous to angular displacement $\Delta\theta$.

The direction in which we measure angular positions is chosen to be consistent with direction of angular displacement.

CONCEPT EXERCISE 12.2

In Figure 12.10, the angle θ subtended is 15.5°, and the radius is 0.25 m. Find the arc length and explain why it would be simpler to find it if the angle were measured in radians instead of degrees.

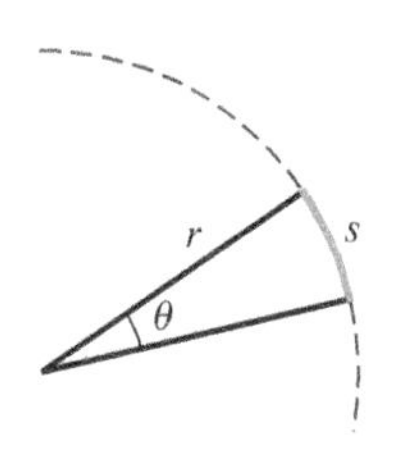

FIGURE 12.10

CONCEPT EXERCISE 12.3

Table 12.1 gives the angular position of a rotating bottle (Fig. 12.11) at five instants in time.

a. Sketch a motion diagram for the bottle along with a coordinate system. It is not necessary to draw the actual bottle; instead, draw the reference line at each time corresponding to positions A through E.

b. What is the angular displacement from A to E? Is the rotation clockwise or counterclockwise?

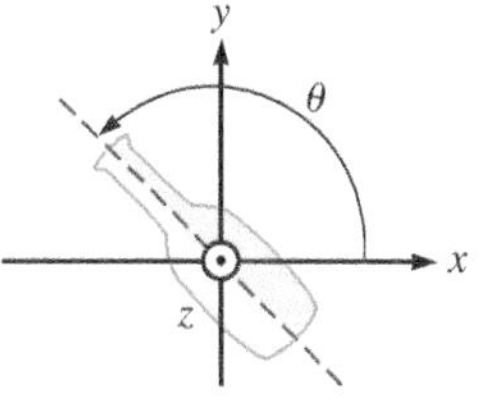

FIGURE 12.11

TABLE 12.1 Angular position of the rotating bottle at various time instants.

Position	t (s)	θ (rad)
A	0	0.75
B	2	3.45
C	4	5.95
D	6	8.25
E	8	10.05

Angular Velocity

The average angular speed and instantaneous angular speed are analogous to average and instantaneous translational speed. Average and instantaneous angular speed are represented by the lowercase Greek letter omega, ω. The **average angular speed** ω_{av} over a time interval Δt is the angular displacement divided by the time interval:

Neither angular displacement nor average angular speed is a vector.

$$\omega_{av} = \frac{\theta_f - \theta_i}{t_f - t_i} = \frac{\Delta\theta}{\Delta t} \quad (12.3)$$

The **instantaneous angular speed** comes from taking the limit of Equation 12.3 as the time interval goes to zero:

$$\omega = \lim_{\Delta t \to 0} \frac{\Delta\theta}{\Delta t} = \frac{d\theta}{dt} \quad (12.4)$$

If we measure θ in radians, the SI units of ω_{av} and ω are radians per second, or rad/s.

The instantaneous angular speed is the magnitude of the **instantaneous angular velocity** vector. (We usually drop the word "instantaneous") Just as the direction of $\vec{v}$ refers to the specific coordinate system we choose, so does the direction of $\vec{\omega}$.

ANGULAR VELOCITY
Major Concept

It might seem convenient to use the terms *clockwise* or *counterclockwise* to describe the direction of angular velocity, but those terms are too ambiguous as shown by the two observers of the rotating potato in Figure 12.12A. Aaron is in the background, and Hannah is in the foreground. To Hannah, the potato appears to be rotating counterclockwise, but to Aaron, the potato is rotating clockwise.

RIGHT-HAND RULE  Tool

There is an unambiguous way to describe the direction of the angular velocity. An observer uses the **right-hand rule** convention to find the direction of $\vec{\omega}$ by curling the fingers of his or her right hand in the same sense as the rotation. The observer's right thumb then points in the direction of $\vec{\omega}$ (Fig. 12.12B). Try it for yourself; you'll find that your right thumb points out of the page. Aaron and Hannah would also agree that $\vec{\omega}$ points out of the page.

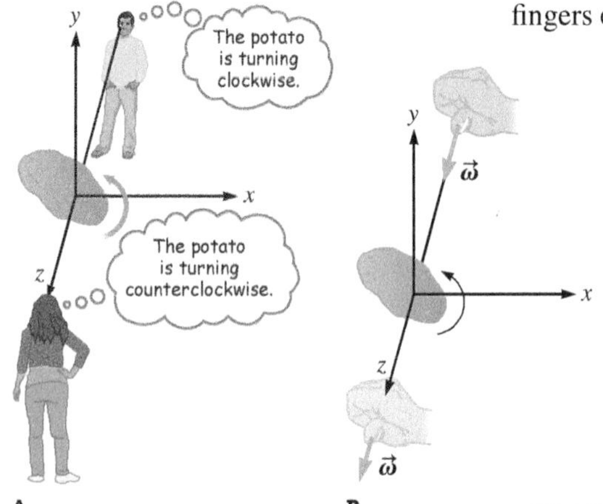

FIGURE 12.12 A. Aaron and Hannah watch a rotating potato. **B.** An observer uses the right-hand rule convention to find the direction of $\vec{\omega}$ by curling the fingers of his or her right hand in the same sense as the rotation. The observer's right thumb then points in the direction of $\vec{\omega}$.

We have chosen a particular coordinate system as shown in Figure 12.12. Because we have decided to align the rotation axis with the z axis, $\vec{\omega}$ will point in either the positive or negative z direction. We found that $\vec{\omega}$ points out of the page, which corresponds to the positive z direction in our coordinate system. The potato's motion is in the xy plane, but its angular velocity points in the positive z direction. Although the right-hand rule convention might seem counterintuitive, with practice it becomes second nature. The right-hand rule for angular velocity provides an unambiguous, consistent way to describe rotational motion.

CONCEPT EXERCISE 12.4

What is the average angular speed of the rotating bottle in Concept Exercise 12.3 over the 8-second time interval?

Angular Acceleration

If an object's translational speed changes, the object must be accelerating. Likewise, if an object's rotation speeds up or slows down, the object has a nonzero angular acceleration. The **average angular acceleration** $\vec{\alpha}_{av}$ is given by

$$\vec{\alpha}_{av} = \frac{\vec{\omega}_f - \vec{\omega}_i}{t_f - t_i} = \frac{\Delta\vec{\omega}}{\Delta t} \tag{12.5}$$

The **instantaneous angular acceleration** (also known simply as **angular acceleration**) $\vec{\alpha}$ is found by taking the limit of $\vec{\alpha}_{av}$ as Δt goes to zero:

ANGULAR ACCELERATION
Major Concept

$$\vec{\alpha} = \lim_{\Delta t \to 0} \frac{\Delta\vec{\omega}}{\Delta t} = \frac{d\vec{\omega}}{dt} \tag{12.6}$$

If an object's rotation is speeding up, then $\vec{\alpha}$ points in the same direction as $\vec{\omega}$. If the object is slowing down, then $\vec{\alpha}$ and $\vec{\omega}$ point in opposite directions. Because we are only studying rotations about a single stationary rotation axis, we are only concerned with angular accelerations that are parallel or antiparallel to the angular velocity.

Let us return to Charlotte riding on the track and Jim riding on the wheel (Fig. 12.9) to see how angular velocity and angular acceleration fit in the analogy between rotation and one-dimensional translation.

(Angular) velocity: In Figure 12.13, Charlotte is represented by a dot. Her motion and initial velocity $\vec{v}_i$ are in the positive x direction (Fig. 12.13A, left). Jim is represented by a reference line rotating counterclockwise about the z axis in the xy plane. According to the right-hand rule, Jim's initial angular velocity $\vec{\omega}_i$ is in the positive z direction as shown. Translational velocity $\vec{v}$ is analogous to angular velocity $\vec{\omega}$.

(Angular) acceleration: If Charlotte speeds up so that $v_f > v_i$ (Fig. 12.13B, left), her acceleration $\vec{a}$ must be in the same direction as her velocity. In this case,

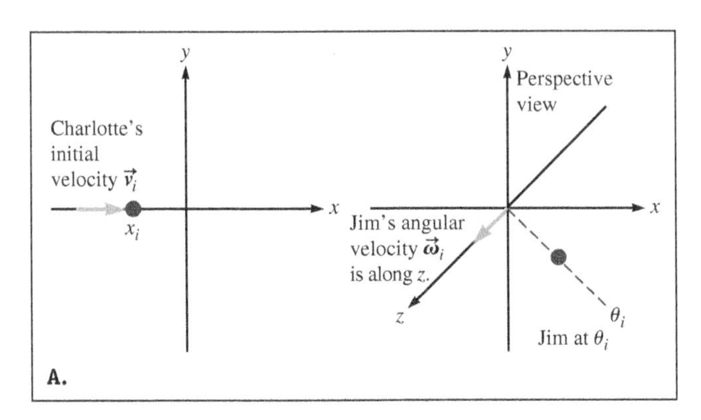

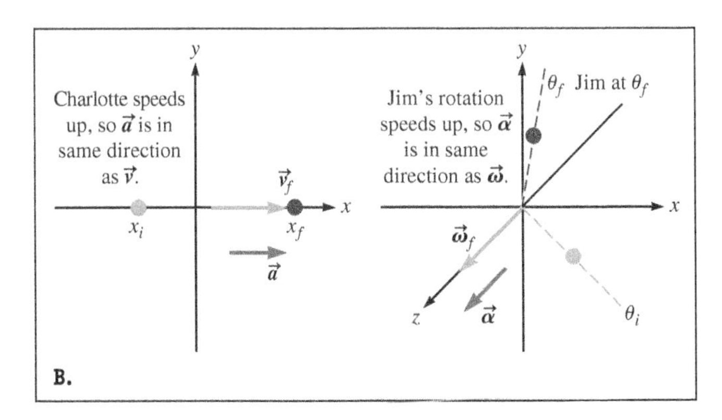

FIGURE 12.13 A. An analogy between translational velocity and angular velocity. **B.** An analogy between translational acceleration and angular acceleration. In both parts, Jim's situation is shown in perspective with the z axis extending from upper right to lower left.

$\vec{a}$ is in the positive x direction. Now imagine that Jim's rotation speeds up: $\omega_f > \omega_i$. The increase in his angular speed means that his angular acceleration $\vec{\alpha}$ is in the same direction as his angular velocity. In this case, $\vec{\alpha}$ is in the positive z direction. Translational acceleration $\vec{a}$ is analogous to angular acceleration $\vec{\alpha}$.

CONCEPT EXERCISE 12.5

What are the directions of $\vec{\omega}$ and $\vec{\alpha}$ for the rotating bottle in Concept Exercise 12.3?

EXAMPLE 12.1 "As the World Turns"

Treat the Earth as a rigid object rotating on a fixed axis and consider a time interval $\Delta t = 12$ h. Find the magnitude of the Earth's angular displacement $\Delta\theta$, angular speed ω, and angular acceleration α during that time interval. The Earth's rotation is very nearly uniform.

INTERPRET and ANTICIPATE

Every 24 hours, the Earth completes one revolution. We expect to find numerical answers in the form $\Delta\theta =$ _____ rad, $\omega =$ _____ rad/s, and $\alpha =$ _____ rad/s^2.

SOLVE

The uniform rotation of the Earth means that in half its period it completes half a revolution.

$$\Delta\theta = \tfrac{1}{2}\text{ rev} = \tfrac{1}{2}(2\pi\text{ rad}) = \pi\text{ rad}$$

The Earth's angular speed is constant. The instantaneous angular speed (Eq. 12.4) equals the average angular speed (Eq. 12.3).

$$\omega = \omega_{av} = \frac{\Delta\theta}{\Delta t} = \frac{\pi\text{ rad}}{12\text{ h}\cdot(3600\text{ s}/1\text{ h})} \quad (12.3)$$

$$\omega = \frac{\pi\text{ rad}}{4.3\times10^4\text{ s}} = 7.3\times10^{-5}\text{ rad/s}$$

The Earth's constant angular speed means that its angular acceleration is zero.

$$\alpha = 0$$

CHECK and THINK

Uniform rotation makes this example mathematically simple. Solving this example gives us a chance to develop our intuition for angular speed: Once-a-day rotation corresponds to $\omega = 7.3\times10^{-5}$ rad/s.

12-3 Special Case of Constant Angular Acceleration

In one-dimensional translational kinematics (Chapter 2), we considered the special case in which a particle moves with constant acceleration. The special case of $\vec{a}$ = constant for translational motion is analogous to the special case of $\vec{\alpha}$ = constant for rotational motion. The mathematics describing an object rotating at constant angular acceleration is exactly the same as that describing a particle moving along a line at constant acceleration. In Section 2-9, we solved $v = dx/dt$ and $\vec{a} = d\vec{v}/dt$ (Eqs. 2.4 and 2.7, respectively) in the special case of $\vec{a}$ = constant. It is mathematically equivalent to solving $\omega = d\theta/dt$ and $\vec{\alpha} = d\vec{\omega}/dt$ (Eqs. 12.4 and 12.6, respectively) in the special case of $\vec{\alpha}$ = constant, so we do not need to work out the solution again. Instead, all we need to do is rewrite the solutions from Chapter 2 in terms of the variables for rotational motion. The analogy we developed (Figs. 12.9 and 12.13) shows us how to make the conversion from translational variables to rotational variables: $\Delta x \rightarrow \Delta\theta$, $v_x \rightarrow \omega$, and $a \rightarrow \alpha$. Analogous pairs of kinematic equations are listed in Table 12.2. The two left columns of the table are from Chapter 2 and apply to translational motion, and the two right columns apply to rotational motion and come from converting the translational variables to rotational variables.

TABLE 12.2 Translational kinematic equations for constant acceleration (left) and rotational kinematic equations for constant angular acceleration (right).

Translation (along x axis) with $\vec{a}$ = constant		Rotation (single rotation axis) with $\vec{\alpha}$ = constant	
Equation	**Eliminated variable**	**Equation**	**Eliminated variable**
$v_x = v_{0x} + a_x t$ (2.9)	Displacement Δx	$\omega = \omega_0 + \alpha t$ (12.7)	Angular displacement $\Delta\theta$
$\Delta x = \frac{1}{2}(v_{0x} + v_x)t$ (2.10)	Acceleration a_x	$\Delta\theta = \frac{1}{2}(\omega_0 + \omega)t$ (12.8)	Angular acceleration α
$\Delta x = v_{0x}t + \frac{1}{2}a_x t^2$ (2.11)	Final velocity v_x	$\Delta\theta = \omega_0 t + \frac{1}{2}\alpha t^2$ (12.9)	Final angular velocity ω
$\Delta x = v_x t - \frac{1}{2}a_x t^2$ (2.12)	Initial velocity v_{0x}	$\Delta\theta = \omega t - \frac{1}{2}\alpha t^2$ (12.10)	Initial angular velocity ω_0
$v_x^2 = v_{0x}^2 + 2a_x\Delta x$ (2.13)	Final time t	$\omega^2 = \omega_0^2 + 2\alpha\Delta\theta$ (12.11)	Final time t

Recall from Chapter 2 that these five equations are not independent. Only two are independent; the other three are derived by algebraically combining the two independent equations. Writing out these five equations with a single variable eliminated from each often makes problem solving easier. Many physicists only memorize two of the equations (usually Eqs. 12.7 and 12.8) and then solve them simultaneously as needed for any particular situation. It is not necessary to memorize all five of them, but it is important to be able to derive any one of them as needed to solve a problem.

PROBLEM-SOLVING STRATEGY

Constant Angular Acceleration

We only need to slightly modify the steps for one-dimensional constant-acceleration problems. As part of the **INTERPRET and ANTICIPATE** procedure: Draw a sketch that includes a coordinate system. Keep in mind that you will likely need a three-dimensional coordinate system.

There are also two parts to the **SOLVE** procedure:

Step 1 Start by **listing** the six kinematic variables (initial angular position, final angular position, initial angular velocity, final angular velocity, angular acceleration, and time). Often, the initial and final angular positions are combined as the displacement, so your list typically includes five variables. Be sure to include the sign of each value. Write the word *need* next to any variable you need to solve for and write *not needed* next to any variable that you don't know and don't need to find.

Step 2 Once you have listed the parameters, you may use the constant angular acceleration equations **(Table 12.2)**. You can avoid doing algebra by choosing the equation that does not include the unneeded variable.

Step 3 Do **algebra** before substituting values.

Step 4 Substitute **values** if appropriate.

EXAMPLE 12.2 A Spinning Pulsar

Pulsars are astronomical objects that give off very regular pulses of radio waves with periods roughly between 1 millisecond and 1 second. When pulsars were first discovered in 1967 by Jocelyn Bell, their uniform pulses led her to think that she had discovered extraterrestrial intelligence. Today we know that pulsars are rotating compact objects about the size of a city. Two spots on a pulsar emit radio waves, one of them becoming visible to us periodically as the pulsar rotates (Fig. 12.14) much like a light spinning on top of a lighthouse. Astronomers measure the angular speed of a rotating pulsar by observing the period of its radio wave pulses.

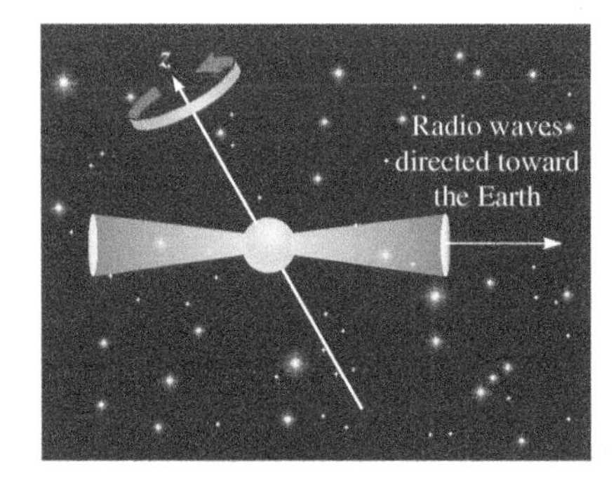

FIGURE 12.14

A pulsar may form when a star of medium mass dies. On July 4, 1054 CE, Chinese astronomers witnessed the death of a star that led to the formation of the pulsar in the Crab Nebula. Contemporary astronomers have measured the angular speed of the nebula's pulsar as $\omega = 1.89 \times 10^2$ rad/s and have found that the pulsar is slowing down such that $\alpha = 2.39 \times 10^{-9}$ rad/s^2. What was the initial angular speed of the Crab Nebula's pulsar in 1054 CE?

INTERPRET and ANTICIPATE

We expect a numerical result for ω_0 that is somewhat greater than ω. Because the angular acceleration is so low, however, we should expect that the pulsar's angular speed has not decreased very much since its formation; the exact amount of time that has elapsed since the pulsar was formed is not very significant. So, the date you are working this problem will not matter very much. To be specific, we have decided to work this problem for the year 2015. Work the problem for whatever year you'd like—from the year you were born to 50 years from now—and you will get roughly the same answer. A sketch with a coordinate system is provided as part of the problem statement.

SOLVE

Step 1 List variables. We have assumed the Crab Nebula's pulsar formed 961 years ago. In Figure 12.14, the angular velocity ω is aligned with the positive z axis, and because the pulsar is slowing down, α must be in the negative z direction.

$\Delta\theta =$ not needed $\qquad \omega_0 =$ needed

$$\omega = 1.89 \times 10^2 \text{ rad/s}$$
$$\alpha = -2.39 \times 10^{-9} \text{ rad/s}^2$$
$$t = 961 \text{ y} = 3.03 \times 10^{10} \text{ s}$$

Steps 2 and 3 Because $\Delta\theta$ is not needed, use Equation 12.7 to solve for ω_0.

$$\omega = \omega_0 + \alpha t \qquad (12.7)$$
$$\omega_0 = \omega - \alpha t$$

Step 4 Substitute.

$$\omega_0 = (1.89 \times 10^2 \text{ rad/s}) - (-2.39 \times 10^{-9} \text{ rad/s}^2)(3.03 \times 10^{10} \text{ s})$$
$$\omega_0 = 2.61 \times 10^2 \text{ rad/s}$$

CHECK and THINK

As expected, the Crab Nebula's pulsar had a greater angular speed in 1054 CE than it does today.

$$\frac{\omega_0 - \omega}{\omega_0} = \frac{2.61 \times 10^2 \text{ rad/s} - 1.89 \times 10^2 \text{ rad/s}}{2.61 \times 10^2 \text{ rad/s}} = 0.28$$

In the past millennium, the pulsar has only slowed down about 28%. A pulsar's rotation is so nearly uniform that some pulsars have been used to keep time. In fact, the value $\omega = 1.89 \times 10^2$ rad/s was measured in 2004, but even if you are working on this problem in 2019 and you assume that year is when ω was measured, your answer will be the same to three significant figures.

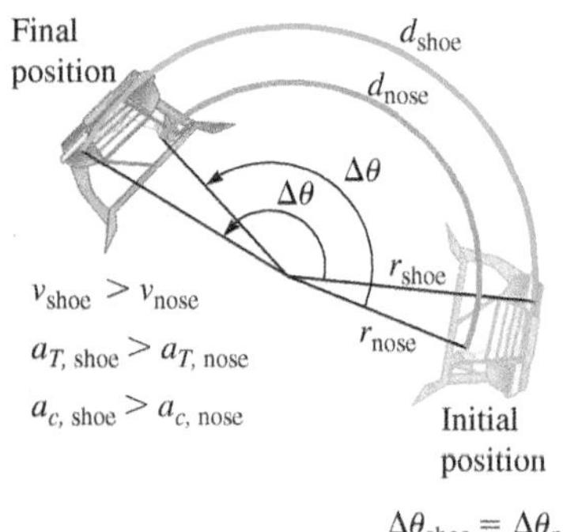

FIGURE 12.15 Sophia in rotation is modeled as a rigid object, so her nose and her shoe have equal angular displacements $\Delta\theta$, angular speeds ω, and angular accelerations α. Her shoe is farther from the axis of rotation than her nose is. Therefore, the translational parameters d, v, a_T, and a_c of her shoe are greater than those for her nose.

12-4 The Connection Between Rotation and Circular Motion

In Section 12-1 we modeled Sophia on a witch's wheel (Fig. 12.2B) as a rigid rotating object. Another way to model her is as a collection of particles, each undergoing circular motion around a common center. In this section, we explore how the rigid-object and collection-of-particles models are related.

Distance Traveled by a Point on a Rotating Object

Take another look at Sophia riding on the witch's wheel (Fig. 12.15). If we model her as a rigid object as she moves from some initial position to a final position, her nose and shoe have equal angular displacements $\Delta\theta$.

Now we instead model Sophia as a collection of particles, each in circular motion. For example, her nose travels along the arc of a circle of radius r_{nose} and her shoe along a circle of radius r_{shoe}. The distance her shoe travels along its circular path is d_{shoe}, and the distance traveled by her nose along its circular path is d_{nose}. From Equation 12.1,

$$\Delta\theta = \frac{d_{shoe}}{r_{shoe}} \quad \text{and} \quad \Delta\theta = \frac{d_{nose}}{r_{nose}}$$

Because the angular displacement $\Delta\theta$ is equal for all of her particles, we can equate these expressions:

$$\frac{d_{shoe}}{r_{shoe}} = \frac{d_{nose}}{r_{nose}}$$

and solve for d_{shoe}:

$$d_{shoe} = d_{nose}\frac{r_{shoe}}{r_{nose}}$$

Sophia's shoe travels along a larger circle than her nose: $r_{shoe} > r_{nose}$, so her shoe travels a longer distance than her nose: $d_{shoe} > d_{nose}$. This particular example illustrates a statement that is true for any rigid rotating object: *A particle far from the rotation axis of a rigid object travels a greater distance than a particle that is closer to the axis.* The distance traveled is directly proportional to r, the distance from the particle to the axis of rotation. Expressed mathematically,

$$d = (\Delta\theta)r \tag{12.12}$$

where $\Delta\theta$ is measured in radians. In SI units, both d and r are measured in meters, showing again that radians are dimensionless.

Translational Velocity of a Point on a Rotating Object

Now let us look at the translational and angular speeds of Sophia's nose and shoe. Sophia moves from her initial position to her final position in a time interval Δt. Her nose and shoe undergo the same angular displacement $\Delta\theta$ in the same time interval. Therefore, her nose and shoe have the same average angular speed $\omega_{av} = \Delta\theta/\Delta t$ (Eq. 12.3), and if we consider an infinitesimal angular displacement $d\theta$ in an infinitesimal time interval dt, her nose and shoe have identical instantaneous angular speeds $\omega = d\theta/dt$ (Eq. 12.4).

You already know that *all particles of a rotating rigid object have the same angular speed* ω, but you may be surprised to learn that not all the particles have the same *translational* speed. Sophia's shoe travels a greater distance than her nose in the same time interval, so her shoe must be moving at a higher average speed than her nose:

$$v_{av,\,shoe} = \frac{d_{shoe}}{\Delta t} \quad \text{and} \quad v_{av,\,nose} = \frac{d_{nose}}{\Delta t}$$

Using $d = r(\Delta\theta)$ (Eq. 12.12) for distance traveled,

$$v_{av,\,shoe} = \frac{r_{shoe}\Delta\theta}{\Delta t} \quad \text{and} \quad v_{av,\,nose} = \frac{r_{nose}\Delta\theta}{\Delta t}$$

but because $r_{\text{shoe}} > r_{\text{nose}}$,

$$v_{\text{av, shoe}} > v_{\text{av, nose}}$$

We extend this result to the more general case of any rotating rigid object. In each case, the average translational speed of a particle of the object is proportional to r, the particle's distance from the rotation axis. Therefore, a particle far from the rotation axis has a higher average translational speed than does a particle close to the axis.

We can extend this idea further to infinitesimal time intervals: The instantaneous translational speed of any particle in a rotating rigid object is proportional to r, the particle's distance from the rotation axis. *Therefore, a point far from the rotation axis has a higher instantaneous translational speed than does a point close to the axis.* Mathematically, we write

$$v = \left(\frac{d\theta}{dt}\right)r = \omega r \tag{12.13}$$

ω is measured in rad/s but v is in m/s, so we drop the "rad" when expressing the units of v.

Equation 12.13 gives the translational speed of any point in a rotating rigid object. The direction of the translational velocity of each point is *tangential* to the circular path of that point at each instant in time.

Tangential Acceleration

Imagine that as Sophia rotates her angular speed changes; perhaps the ride is slowing down. In that case, the *tangential* velocity of both her nose and her shoe must decrease. Because her nose and her shoe have identical angular speeds at each time instant, the nose and the shoe must also have identical *angular* accelerations α, but their *tangential* accelerations are not the same. We find an expression for *tangential* acceleration by differentiating Equation 12.13 with respect to time:

$$a_T = \frac{dv}{dt} = \left(\frac{d^2\theta}{dt^2}\right)r = \left(\frac{d\omega}{dt}\right)r$$

$$a_T = \alpha r \tag{12.14}$$

α is measured in rad/s^2 but we drop the "rad" when writing the units of a_T.

The subscript T stands for *tangent* to the circle. Conceptually, Equation 12.14 says that the tangential acceleration of a particle in a rotating rigid object is proportional to r, the particle's distance from the rotation axis. A particle far from the axis has a greater tangential acceleration than does a particle close to the rotation axis.

Centripetal Acceleration

Recall that when a particle moves in a circle, the direction of its velocity continually changes and the particle must therefore be accelerating toward the center of the circle. The same is true for any point of a rotating rigid object; each particle has a centripetal acceleration given by $a_c = v^2/r$ (Eq. 4.38). We use Equation 12.13 to rewrite v:

$$a_c = \frac{(\omega r)^2}{r} = \omega^2 r \tag{12.15}$$

Equation 12.15 says that *the centripetal acceleration of a particle in a rotating rigid object is proportional to the particle's distance from the rotation axis. A particle far from the axis has a higher centripetal acceleration than does a point close to the rotation axis.* So, at any instant, Sophia's shoe has a greater centripetal acceleration than does her nose.

Rotational Versus Translational Parameters

Equations 12.12 through 12.15 connect the translational kinematics of any particle in a rotating rigid object to the rotational kinematics of the object as a whole. The rotational parameters (angular displacement, angular velocity, and angular acceleration) are useful because every particle in a rotating rigid object has the same $\Delta\theta$, the same ω, and the same α. The translational parameters (d, v, a_T, and a_c) depend on the particular particle's distance r from the axis of rotation.

EXAMPLE 12.3 Two Cities

Treat the Earth as a rigid object rotating on a fixed axis and consider a time interval $\Delta t = 12$ h. Assume the Earth's angular speed is constant.

A Esmeraldas, Ecuador, is nearly on the equator. Find the translational distance traveled d, translational speed v, tangential acceleration a_T, and centripetal acceleration a_c of Esmeraldas during this time interval.

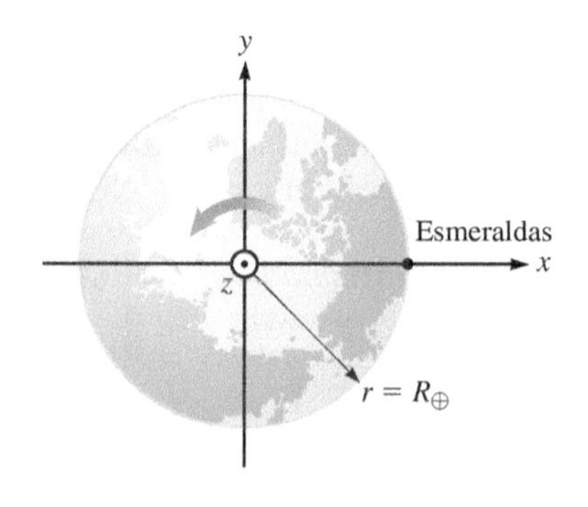

FIGURE 12.16 Looking down from the North Pole.

INTERPRET and ANTICIPATE

We are asked to use rotational parameters for the *entire* Earth to find the translational parameters of one particle, Esmeraldas. The results from Example 12.1 are helpful.

Figure 12.16 shows the Earth from above the north geographic pole. The z axis is aligned with the rotation axis. Esmeraldas is on the equator, so its distance from the rotation axis is equal to the radius of the Earth: $r = R_\oplus$.

SOLVE

The distance traveled comes from Equation 12.12 and the angular displacement found in Example 12.1. We drop "rad" when writing the units of d.

$$d = (\Delta\theta)r \tag{12.12}$$

$$d = (\Delta\theta)R_\oplus = (\pi \text{ rad})(6.37 \times 10^6 \text{ m})$$

$$d = 2.00 \times 10^7 \text{ m} = 20{,}000 \text{ km}$$

Use Equation 12.13 and the angular speed found in Example 12.1 to find Esmeraldas's translational speed.

$$v = \omega r \tag{12.13}$$

$$v = \omega R_\oplus = (7.3 \times 10^{-5} \text{ rad/s})(6.37 \times 10^6 \text{ m})$$

$$v = 4.7 \times 10^2 \text{ m/s} = 1700 \text{ km/h}$$

Because the Earth's angular acceleration is zero, Esmeraldas's tangential acceleration must be zero: $a_T = \alpha r$ (Eq. 12.14).

$$a_T = \alpha r = (0)R_\oplus$$

$$a_T = 0$$

Esmeraldas's centripetal acceleration is not zero and is given by Equation 12.15.

$$a_c = \omega^2 r \tag{12.15}$$

$$a_c = \omega^2 R_\oplus = (7.3 \times 10^{-5} \text{ rad/s})^2(6.37 \times 10^6 \text{ m})$$

$$a_c = 3.4 \times 10^{-2} \text{ m/s}^2 \approx g/300$$

CHECK and THINK

We continue to develop our intuition for rotational motion. A point on the Earth's equator travels a distance of 20,000 km in half a day or 40,000 km per day with a constant speed of 1700 km/h. The centripetal acceleration of a point on the equator is much less than gravitational acceleration at the surface of the Earth: $a_c \approx g/300$.

B Repeat part A for New Orleans, which is at roughly 30° north latitude.

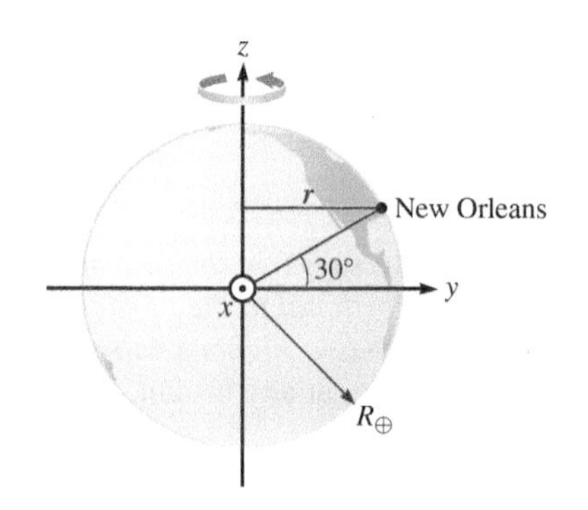

FIGURE 12.17 Looking directly at the equator.

INTERPRET and ANTICIPATE

The difference between this part of the problem and part A is that New Orleans is closer to the axis of rotation than is Esmeraldas.

Figure 12.17 shows the rotation axis aligned with z as in part A, but in this figure, z is up instead of out of the page. From this figure, $r = R_\oplus \cos 30°$. Because $r < R_\oplus$, we expect the distance traveled d, speed v, and centripetal acceleration a_c of New Orleans to have smaller magnitudes than they do for Esmeraldas.

SOLVE

Use the new value of r and follow the same procedure as in part A.

$$r = (6.37 \times 10^6 \text{ m}) \cos 30° = 5.52 \times 10^6 \text{ m}$$

Use Equation 12.12 to find the distance traveled by New Orleans in 12 h.	$d = (\Delta\theta)r$ (12.12) $d = (\pi \text{ rad})(5.52 \times 10^6 \text{ m}) = 1.73 \times 10^7 \text{ m}$
From Equation 12.13, find the translational speed for New Orleans.	$v = \omega r$ (12.13) $v = (7.3 \times 10^{-5} \text{ rad/s})(5.52 \times 10^6 \text{ m}) = 4.0 \times 10^2 \text{ m/s}$
Just as in the case of Esmeraldas, New Orleans's tangential acceleration must be zero (Eq. 12.14). Its centripetal acceleration, however, is nonzero and is given by Equation 12.15.	$a_T = 0$ $a_c = \omega^2 r = (7.3 \times 10^{-5} \text{ rad/s})^2(5.52 \times 10^6 \text{ m})$ (12.15) $a_c = 2.9 \times 10^{-2} \text{ m/s}^2$

CHECK and THINK

New Orleans's distance traveled, translational speed, and centripetal acceleration are all less than corresponding quantities for Esmeraldas.

Part B gives us a chance to see the difference between rotational and translational parameters. All the Earth's particles such as Esmeraldas and New Orleans have the same value for each rotational parameter—the same $\Delta\theta$, the same ω and the same α— but the values of the translational parameters—d, v, a_T, and a_c—depend on the particle's location. Because Esmeraldas is farther from the rotation axis than is New Orleans, Esmeraldas travels farther in the same time interval, at a higher speed, and with a greater centripetal acceleration than does New Orleans.

Finally, notice that the Earth's rotation is a special case. Because the Earth's rotation is uniform ($\alpha = 0$), all points on the Earth have zero tangential acceleration $a_T = 0$, but that is not the case for a rotating object with $\alpha \neq 0$.

EXAMPLE 12.4 Shift Gears When You Need To

Gears play major role in our everyday lives. Nearly all mechanical devices that move, such as the windshield wipers on your car and the antique clock in the tower on campus, have gears. Because the gears are often hidden inside the device, we cannot see them in action. A bicycle's gears, on the other hand, are easily observed.

A bicycle rider moves the bike by pedaling the front gear known as the *chainwheel*. The chainwheel is connected by a chain to the rear gear known as the *freewheel*, which is attached to the rear tire (Fig. 12.18). The angular speed of the rear tire determines the translational speed of the bicycle. A typical rider is most comfortable when he pedals the chainwheel at 80 rpm (revolutions per minute). Most modern bicycles are equipped with several chainwheels and freewheels of varying sizes. The gear ratio γ is defined to be the radius of the chainwheel divided by the radius of the freewheel r_c/r_f. A modern bicycle has many gear ratios available, and the rider may vary the angular speed of the rear tire to maintain his ideal pedaling rate by selecting a gear ratio.

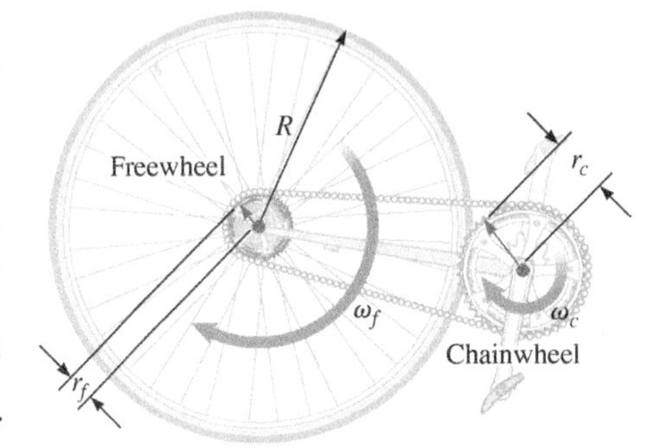

FIGURE 12.18

A certain bicycle has a rear tire 27 inches in diameter. Find the bicycle's translational speed if a typical rider selects a gear ratio of **A** $r_c/r_f = 4.82$ and **B** $r_c/r_f = 0.688$. Assume the rider maintains a pedal rate of 80 rpm and the translational speed of a point on the tire's rim equals the bicycle's translational speed.

INTERPRET and ANTICIPATE

The angular speed of the rear tire ω_{tire} determines the translational speed of the bicycle v_{bike}; the faster the tire rotates, the faster the bicycle moves. The freewheel gear is attached to the rear tire. So, the angular speed of the tire must equal the angular speed of the freewheel $\omega_{\text{tire}} = \omega_f$. Now consider the chainwheel, connected to the freewheel by a chain. All the links in the chain have the same translational speed, so the translational speed of a tooth on the freewheel must equal the translational speed of a tooth on the chainwheel: $v_{\text{chain}} = \omega_c r_c = \omega_f r_f$. The freewheel and chainwheel may have different radii, however, so their angular speeds are not necessarily equal. If the chainwheel is larger than the freewheel ($r_c > r_f$), the freewheel will have a greater angular speed ($\omega_f > \omega_c$). Therefore, a high gear ratio makes for a fast bicycle, and we expect that the bicycle's speed in part A is faster than its speed in part B.

Example continues on page 344 ▶

SOLVE

The bicycle tire rolls along the ground, so a point on the tire's rim is a distance R from the axis of rotation, where R is the tire's radius. Use $v = \omega r$ (Eq. 12.13) and $\omega_{\text{tire}} = \omega_f$ to write an expression for the bicycle's speed in terms of R.	$v_{\text{bike}} = \omega_f R$	(1)
As mentioned above, the translational speed of a tooth on the freewheel must equal the translational speed of a tooth on the chainwheel. This fact gives us an expression for the angular speed of the freewheel.	$v_{\text{chain}} = \omega_c r_c = \omega_f r_f$ $\omega_f = \omega_c \dfrac{r_c}{r_f}$	(2)
Substitute Equation (2) into Equation (1).	$v_{\text{bike}} = \omega_c R \dfrac{r_c}{r_f} = \omega_c R\gamma$	(3)
Both ω_c and R have the same values for both parts of the problem.	$\omega_c = 80\ \text{rpm}\left(\dfrac{2\pi\ \text{rad}}{1\ \text{rev}}\right)\left(\dfrac{1\ \text{min}}{60\ \text{s}}\right) = 8.4\ \text{rad/s}$ $R = \left(\dfrac{27\ \text{in.}}{2}\right)\left(\dfrac{0.0254\ \text{m}}{1\ \text{in.}}\right) = 0.34\ \text{m}$	
For part A, substitute $\gamma = r_c/r_f = 4.82$ into Equation (3).	$v_{\text{bike}} = \omega_c R\gamma = (8.4\ \text{rad/s})(0.34\ \text{m})(4.82)$ $v_{\text{bike}} = 14\ \text{m/s} = 31\ \text{mph}$	
For part B, substitute $\gamma = r_c/r_f = 0.688$.	$v_{\text{bike}} = \omega_c R\gamma = (8.4\ \text{rad/s})(0.34\ \text{m})(0.688)$ $v_{\text{bike}} = 2.0\ \text{m/s} = 4.4\ \text{mph}$	

CHECK and THINK

As expected, the bicycle's speed is faster for the higher gear ratio in part A. This example illustrates the usefulness of gears. The gears allow the rider to maintain a comfortable pedaling rate even as the terrain changes. More athletic riders often do not make full use of the gears available to them. Instead, they change their pedaling rate to maintain their speed despite any changes in the terrain. These riders are not making use of the mechanical advantage gears add to a bicycle. A story is told that the inventor of the multiple-ratio bicycle gear, Paul de Vivie (1853–1930), who was also known as Velocio, convinced bicycle racers to start using multiple gears by holding a race. A racing champion with a single-gear bicycle rode against an inexperienced rider who rode a bicycle with just three gear ratios. The inexperienced rider won the race.

TORQUE **Major Concept**

12-5 Torque

Imagine trying to loosen a bolt that has rusted into place. If your wrench has a short handle as shown in Figure 12.19A, you might find the task impossible. A longer-handled wrench as in Figure 12.19B makes the job much easier.

This common experience illustrates the factors that cause an object to rotate. The first factor is the magnitude F of the force applied to the handle. If one person cannot get the wrench to budge, he might ask a stronger person to give it a try.

The second factor is the distance r from the axis of rotation to the point of application of the force. In both cases shown in Figure 12.19, the same person applies the same force to rotate the bolt–wrench system. The axis of rotation is the same in both cases (along the axis of the bolt). In part B, the longer handle of the wrench allows the person to apply a force farther from the rotation axis than in part A, allowing him to rotate the bolt–wrench system.

The third factor is the direction in which the force is applied. If you want to rotate the bolt, you know from experience not to pull or push in a direction parallel to the wrench handle. In fact, it is best to apply a force perpendicular to the handle.

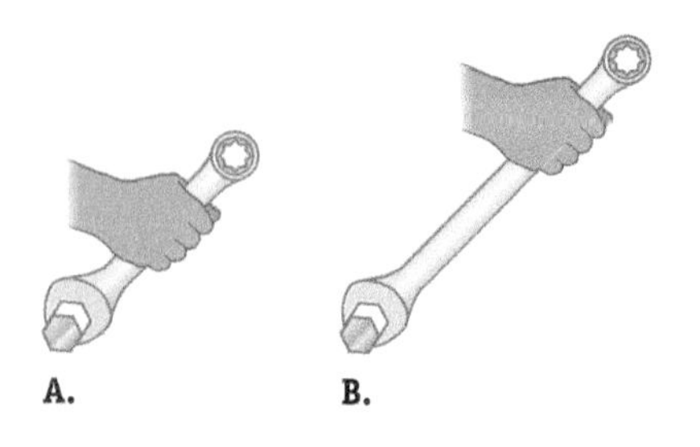

FIGURE 12.19 Wrenching a bolt.

We combine these three factors into one concept: *torque*, a word coming from the Greek verb meaning "to twist." A **torque** on a rotating object is analogous to a force on a translating object: Just as a single force causes the acceleration of a particle, a single torque causes the angular acceleration of a rigid object.

Figure 12.20 illustrates the similarity and difference between a force and a torque. Figure 12.20A shows a crane lifting a steel beam off the ground. The tension force accelerates the beam upward. Figure 12.20B shows a crane rotating a radio antenna into an upright position. The base of the radio antenna is fixed on one side, forming the rotation axis. The tension force is applied at point P, which is at position $\vec{r}$ from the rotation axis as shown. The angle between $\vec{F}$ and $\vec{r}$ is φ. The torque due to the crane's cable rotates the radio antenna upright.

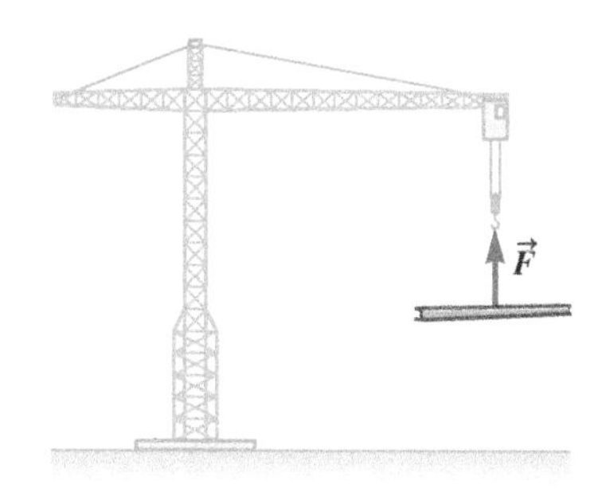

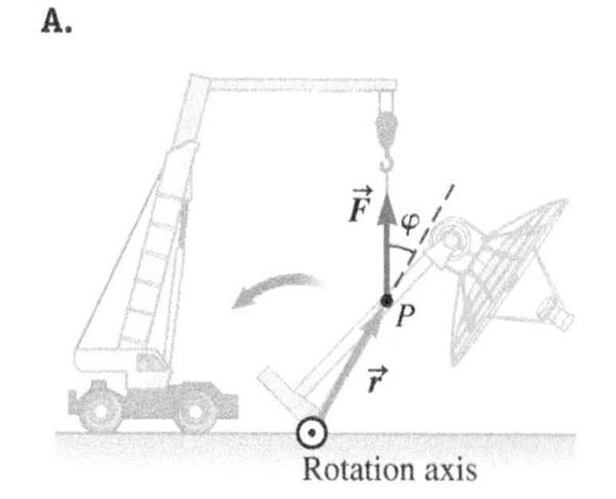

FIGURE 12.20 Analogy between force and torque.

Torque is symbolized by the lowercase Greek letter τ (pronounced "tau"). There are several mathematical expressions for torque. The expression

$$\tau = rF \sin \varphi \qquad (12.16)$$

best shows how the three factors F, r, and φ are combined in the concept of torque. As shown in Figure 12.20B, $\vec{r}$ points from the rotation axis to the point of application P, and the angle φ between $\vec{F}$ and $\vec{r}$ is found by extending $\vec{r}$. Only the magnitudes of force and position enter into Equation 12.16. The SI units of torque are newton meters $(\text{N} \cdot \text{m})$.

Equation 12.16 fits our experience of trying to turn a bolt–wrench system. A stronger person applying a larger force F will produce a greater torque on the system. A longer wrench handle r allows a person to apply a greater torque. If a person pulls parallel to the handle, then $\varphi = 0$; so, according to Equation 12.16, the torque is zero. The torque is also zero if the person pushes parallel to the handle because then $\varphi = 180°$ and $\tau = Fr \sin 180° = 0$. In fact, the torque is a maximum for $\varphi = 90°$, which confirms our experience that the best way to turn the bolt–wrench system is to apply a force perpendicular to the handle.

Let's use the example of rotating the radio antenna to look at two other ways to write Equation 12.16. Figure 12.21A shows $\vec{F}$, $\vec{r}$, and φ from Figure 12.20B. We can think of vector $\vec{r}$ as defining a coordinate axis. Vector $\vec{F}$ may then be broken into two components, one parallel to the r axis and the other perpendicular to it. The perpendicular component $\vec{F}_\perp$ is shown in Figure 12.21A. The magnitude of the perpendicular component is found using trigonometry:

$$F_\perp = F \sin \varphi \qquad (12.17)$$

Substituting Equation 12.17 into Equation 12.16, we find a new expression for torque:

$$\tau = rF_\perp \qquad (12.18)$$

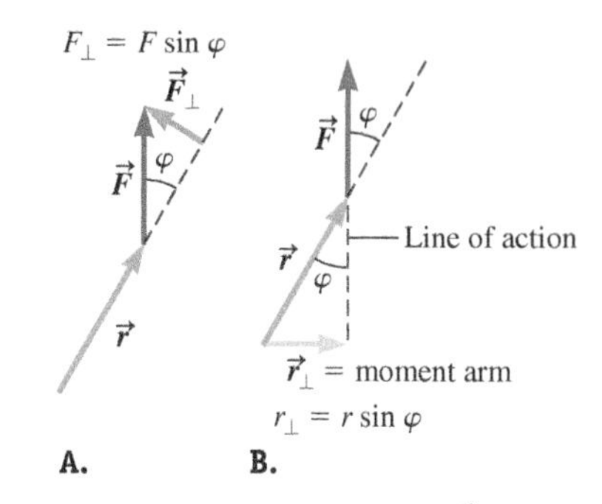

FIGURE 12.21 A. The vector $\vec{F}$ may be broken into two components: one parallel to the axis defined by the vector $\vec{r}$ and the other perpendicular to it. We've shown only $\vec{F}_\perp$ here. **B.** The line along $\vec{F}$ is known as the *line of action*. The moment arm is the component of $\vec{r}$ that is perpendicular to line of action.

Equation 12.18 says that torque only depends on the component of force that is applied perpendicular to $\vec{r}$ and not on the parallel component.

Another way to express torque comes from looking at the components of $\vec{r}$ relative to $\vec{F}$. As shown in Figure 12.21B, we draw a line along $\vec{F}$ known as the *line of action*. This time, we think of $\vec{F}$ as defining a coordinate axis, and we break $\vec{r}$ into components parallel and perpendicular to that axis. The **moment arm** is defined as the component of $\vec{r}$ that is perpendicular to the force $\vec{F}$. The magnitude of the moment arm is found from trigonometry:

$$r_\perp = r \sin \varphi \qquad (12.19)$$

We find a third expression for torque by substituting Equation 12.19 into Equation 12.16:

$$\tau = r_\perp F \qquad (12.20)$$

Equation 12.20 says that the magnitude of the torque is proportional to the moment arm. A longer moment arm produces a greater torque.

CONCEPT EXERCISE 12.6

For each exercise shown in Figure 12.22, how does the moment arm on the dumbbell compare to the length of the person's arm?

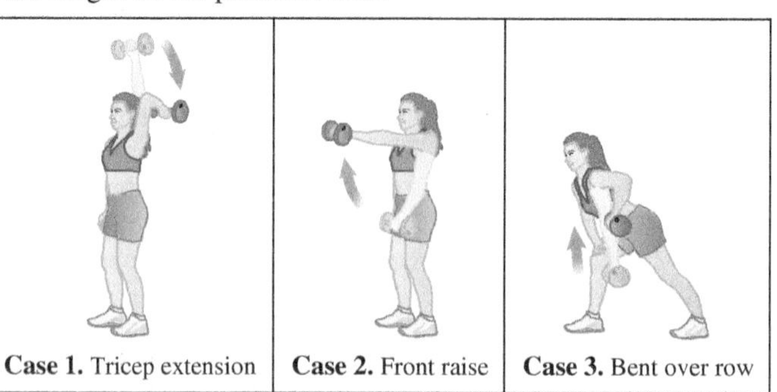

FIGURE 12.22

Recall the first way: A vector multiplied by a scalar results in a vector parallel to the original vector as in $\vec{F}_{tot} = m\vec{a}$. The second way is the dot product between two vectors, yielding a scalar as in $W = \vec{F}\cdot\Delta\vec{r}$.

12-6 Cross Product

Like force, torque is a vector. Equations 12.16, 12.18, and 12.20 give the magnitude of torque, but not its direction. To write a vector equation for torque, we must introduce the third and final way to multiply vectors.

The third type of vector multiplication is known as the *cross product* or *vector product.* We will begin with three generic vectors $\vec{A}$, $\vec{B}$, and $\vec{R}$. Figure 12.23A shows two vectors $\vec{A}$ and $\vec{B}$ that define a single plane. The **cross product** or **vector product** between these two vectors $\vec{A}$ and $\vec{B}$ results in a third vector $\vec{R}$:

VECTOR CROSS PRODUCT  Tool

$$\vec{R} = \vec{A} \times \vec{B} \quad (12.21)$$

The magnitude of $\vec{R}$ is given by

$$R = AB \sin \varphi \quad (12.22)$$

where A and B are magnitudes of the vectors $\vec{A}$ and $\vec{B}$, and φ is the angle between the two vectors (Fig. 12.23A); the angle φ must be between 0° and 180°. R is always positive because the magnitudes A and B must be positive, and the sine of an angle between 0 and 180° is positive. R may also be written as $R = A_{\perp}B = AB_{\perp}$, where $A_{\perp} = A \sin \varphi$ and $B_{\perp} = B \sin \varphi$.

The vector resulting from a cross product $\vec{R}$ is always perpendicular to the plane defined by the original two vectors $\vec{A}$ and $\vec{B}$. If $\vec{A}$ and $\vec{B}$ are in the plane of the page as in Figure 12.23, then $\vec{R}$ must either point into or out of the page. The direction of $\vec{R}$ comes from the **right-hand rule** for cross products, a procedure similar to that used in Section 3-2 to find the direction of the z axis in a right-handed coordinate system.

Line up the fingers of your right hand along the first vector in the cross product (in this case, vector $\vec{A}$) and then imagine closing your fingers so that you "push" $\vec{A}$ through the angle φ into the second vector (in this case, vector $\vec{B}$; see Fig. 12.23B). The direction in which your thumb points as you push $\vec{A}$ into $\vec{B}$ is the direction of the resulting vector $\vec{R}$, which in this case is out of the page. Figure 12.23C shows another way to use the right-hand rule. In this procedure, align your right index finger with the first vector $\vec{A}$ and align your right middle finger with the second vector $\vec{B}$. Your right thumb points in the direction of the resulting vector $\vec{R}$.

Switching the order of vectors in the cross product changes the direction of the result. For example, let us find the direction of $\vec{Q} = \vec{B} \times \vec{A}$ using the procedure illustrated in Figure 12.23B. Because $\vec{B}$ is the first vector, align the fingers of your right hand with $\vec{B}$. To close your hand such that you push vector $\vec{B}$ into vector $\vec{A}$, you must orient your hand

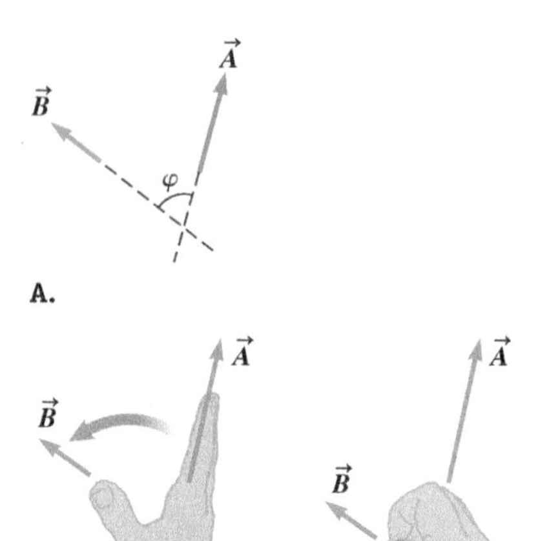

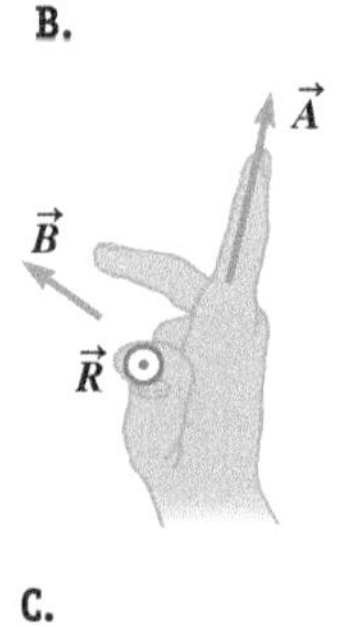

FIGURE 12.23 A. Two vectors and $\vec{A}$ and $\vec{B}$ are placed tail to tail and are separated by the angle φ. **B.** We find the direction of the cross product using the right-hand rule. Align the fingers of your right hand with the first vector $\vec{A}$. Then imagine closing your hand such that you push vector $\vec{A}$ through angle φ into vector $\vec{B}$. Your right thumb points in the direction of the resulting vector $\vec{R}$. **C.** Another way to use the right-hand rule is to align the index finger of your right hand with the first vector $\vec{A}$ and your middle finger with the second vector $\vec{B}$. Your right thumb points in the direction of the resulting vector $\vec{R}$.

such that your right thumb goes into the page. (Remember that $0° \le \varphi \le 180°$.) Therefore, $\vec{Q}$ points in the opposite direction to $\vec{R}$ and

$$\vec{A} \times \vec{B} = -\vec{B} \times \vec{A}$$

Thus, the cross product is not commutative and the two vectors must be written in the correct order or you will find the wrong direction for the result.

Now we are ready to write a vector equation for torque. By comparing Equations 12.16 and 12.22, we see that torque may be written in terms of a cross product:

$$\vec{\tau} = \vec{r} \times \vec{F} \tag{12.23}$$

The direction of the torque is found using the right-hand rule for cross products. For example in Figure 12.20B, if you align the fingers of your right hand with $\vec{r}$ and close your fingers into $\vec{F}$, your thumb points out of the page. So, the torque due to the crane on the radio antenna is out of the page.

The direction of torque (just like the direction of angular velocity or angular acceleration) is perpendicular to the plane in which the object rotates. As described in Section 12-2, we restrict our study to rotations around a single, fixed axis usually aligned with the z axis. In that case, a torque will either be in the positive or negative z direction.

EXAMPLE 12.5 Parallelograms and the Cross Product

Two vectors $\vec{B}$ and $\vec{C}$ make up a parallelogram as shown in Figure 12.24. Show that the area A of the parallelogram can be found from the magnitude of the cross product $\vec{A} = \vec{B} \times \vec{C}$.

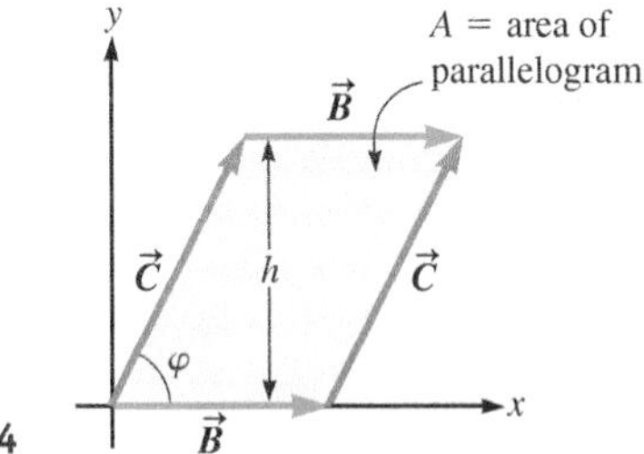

FIGURE 12.24

INTERPRET and ANTICIPATE

Our goal is to show that something we know, the formula for the area of a parallelogram, is equal to another expression, the cross product $\vec{A} = \vec{B} \times \vec{C}$. By working this simple proof, we gain insight into the cross product.

SOLVE

Start with the area of a parallelogram (Appendix A), where b is the length of the "base" and h is the "height" labeled in Figure 12.24.	$A = bh$
From the figure, $b = B$, the magnitude of $\vec{B}$, and $h = C \sin \varphi$. Make the substitutions.	$A = BC \sin \varphi$ (1)
Comparing Equation (1) with Equation 12.22, we see that the area A is the magnitude of $\vec{A} = \vec{B} \times \vec{C}$.	$\lvert\vec{A}\rvert = BC \sin \varphi$ ✓

12-7 Rotational Dynamics

Our goal in this section is to find an expression for Newton's second law that is convenient for rotational motion using our analogy with translational motion. In Section 12-5, we found that force is analogous to torque: $\vec{F} \rightarrow \vec{\tau}$. When we want to find the acceleration of an object, we must first find the net force acting on it. By analogy, when we want to find the angular acceleration of an object, we must first find the net torque acting on it.

If more than one torque acts on an object, we find the net torque on the object with vector addition. Restriction to a single, fixed axis greatly simplifies this vector addition. Let's find the total torque acting on the radio antenna in Figure 12.25. Three forces act on the radio antenna: the normal force $\vec{F}_N$ due to the ground, the tension force $\vec{F}_T$ due to the cable, and gravity $\vec{F}_g$ due to the Earth. The normal force acts where the ground touches the antenna along the rotation axis. Because the point of application is on the rotation axis, $\vec{r}_N = 0$, and according to Equation 12.23, the normal force produces zero torque on the antenna. That leaves just the torque due to the crane's cable and the torque due to gravity. The tension force exerted by the crane acts at point P. The vectors $\vec{r}_P$ and

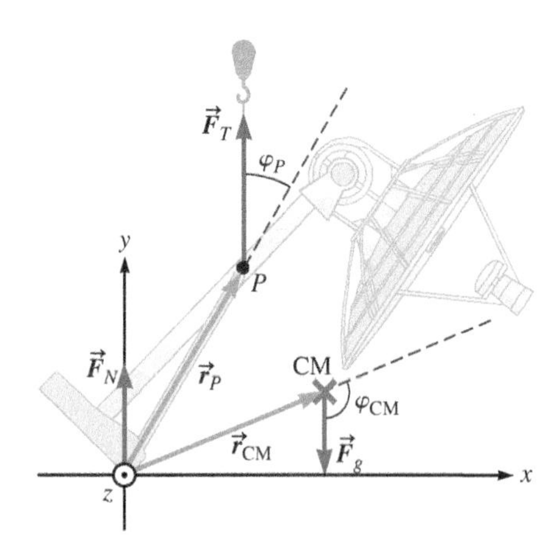

FIGURE 12.25 A crane rotating a radio antenna to a vertical position. The normal force does not exert a torque on the radio antenna because its point of application is on the rotation axis. The torque exerted by the crane is in the positive z direction, and the torque exerted by gravity is in the negative z direction.

$\vec{F}_T$ are shown in Figure 12.25. The torque due to the tension $\vec{\tau}_T$ points out of the page in the positive z direction. The Earth's gravity acts at the antenna's center of mass, which is slightly below the antenna. Applying the right-hand rule to the vectors $\vec{r}_{CM}$ and $\vec{F}_g$ shown in Figure 12.25, we find that the torque due to gravity points into the page (in the negative z direction). So, the total torque acting on the radio antenna is

$$\vec{\tau}_{tot} = \vec{\tau}_N + \vec{\tau}_T + \vec{\tau}_g = 0 + \tau_T\hat{k} - \tau_g\hat{k}$$

Equation 12.16 gives the magnitudes τ_T and τ_g in terms of the variables shown in Figure 12.25:

$$\vec{\tau}_{tot} = (r_P F_T \sin\varphi_P)\hat{k} - (r_{CM}F_g \sin\varphi_{CM})\hat{k}$$

The expression we just found for the total torque exerted on the antenna consists of two terms. The first term is the torque due to the crane, and the second is the torque due to gravity. The direction of the total torque is either in the positive or negative z direction depending on the relative magnitude of these two torques.

Newton's Second Law in Rotational Form

NEWTON'S SECOND LAW IN ROTATIONAL FORM

Underlying Principle

By Newton's second law, the total force acting on a particle is proportional to the particle's acceleration:

$$\vec{F}_{tot} \propto \vec{a}$$

Because the total force and acceleration vectors are proportional, the total force points in the same direction as the particle's acceleration. By analogy, the total torque acting on an object is proportional to the object's angular acceleration:

$$\vec{\tau}_{tot} \propto \vec{\alpha}$$

As in the translational case, because the two vectors are proportional, the total torque points in the same direction as the object's angular acceleration. If we restrict rotational motion to a single, fixed axis aligned with the z axis, the total torque and angular acceleration point in either the positive or negative z direction.

In the case of the antenna (Fig. 12.25), if the torque due to the crane is greater than the torque due to gravity, the angular acceleration is in the positive z direction. However, if the torque due to gravity is the greater one, the angular acceleration is in the negative z direction. If the two torques have the same magnitude, the total torque is zero, and the antenna has zero angular acceleration. It may be rotating at a constant angular speed or be at rest.

The proportionality constant in Newton's second law is inertial mass (Eq. 5.1):

$$\vec{F}_{tot} = m\vec{a}$$

ROTATIONAL INERTIA

Major Concept

Many people refer to rotational inertia as *moment of inertia.* We will not use that phrase in this book because it is too easy to confuse it with other terms.

Inertial mass is the measure of a particle's resistance to acceleration; in other words, mass is a measure of a particle's tendency to maintain its translational velocity. By analogy, **rotational inertia** I is a measure of an object's resistance to angular acceleration; in other words, rotational inertia is a measure of an object's tendency to maintain its angular velocity. Mathematically, Newton's second law for a rotating object is

$$\vec{\tau}_{tot} = I\vec{\alpha} \quad (12.24)$$

where I is the constant of proportionality just as m is in $\vec{F}_{tot} = m\vec{a}$. The SI units of I are found from the units of τ and α.

$$[I] = \frac{[\tau]}{[\alpha]} = \frac{\text{N}\cdot\text{m}}{\text{rad/s}^2} = \frac{\text{kg}\cdot\text{m/s}^2\cdot\text{m}}{\text{rad/s}^2}$$

$$[I] = \text{kg}\cdot\text{m}^2$$

In Chapter 13, we show how to calculate the rotational inertia I, but it is possible to find I experimentally. Imagine applying a single known torque (Eq. 12.23) to an object and measuring its resulting angular acceleration. The torque and angular acceleration vectors point in the same direction, and we find I from Equation 12.24:

$$I = \frac{\tau}{\alpha} \quad (12.25)$$

If more than one torque is applied to the object, then τ in Equation 12.25 is the total torque.

EXAMPLE 12.6 Crall and Whipple Set Things Spinning

Two student researchers, Crall and Whipple, used the laboratory setup shown in Figure 12.26 to measure the rotational inertia of a system consisting of two metal cylinders attached to a light crossbar. The crossbar is mounted on a light axle of radius $R = 0.50$ cm. A lead ball attached to a lightweight string wrapped around the axle rotates the system as the tension in the string provides a constant torque. The tension is measured to be 0.981 N. When the two cylinders are far apart as shown in Figure 12.26, Crall and Whipple measure a constant angular acceleration $\alpha = 0.331$ rad/s^2.

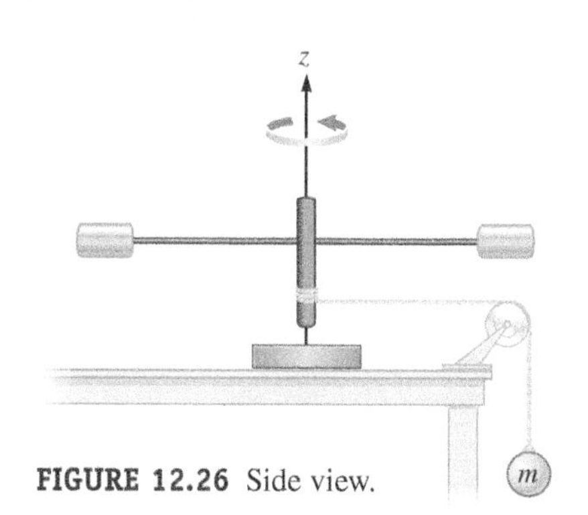

FIGURE 12.26 Side view.

A Find the rotational inertia of the system.

INTERPRET and ANTICIPATE

The only torque acting on the system is due to the tension in the string (Fig. 12.27). Therefore, both the (single) torque and the angular acceleration point in the same direction—the positive z direction—and we can use Equation 12.25 to find the rotational inertia once we find the magnitude of the torque. We should find a numerical result in the form $I =$ _____ kg · m^2.

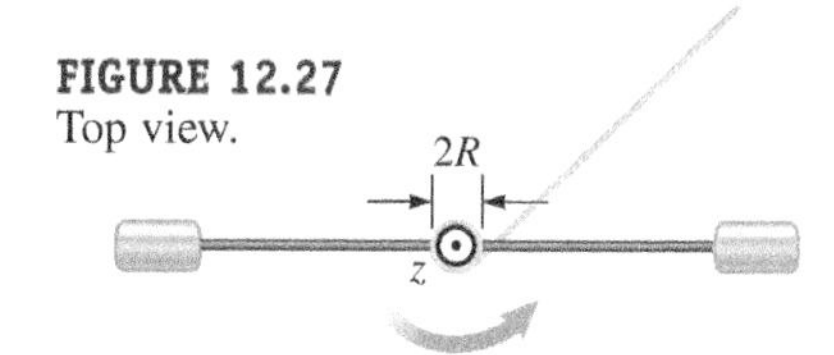

FIGURE 12.27 Top view.

SOLVE

As shown in Figure 12.27, the string is tangent to the axle. Therefore, the tension force in the string is perpendicular to the radius of the axle. This radius is the moment arm in Equation 12.20 with $r_\perp = R$.

$$\tau = r_\perp F$$

$$\tau = RF_T \quad (1)$$

Substitute Equation (1) into Equation 12.25 and the given angular acceleration to find the rotational inertia. We drop the "rad" notation when stating the units of I.

$$I = \frac{\tau}{\alpha} = \frac{RF_T}{\alpha}$$

$$I = \frac{(0.50 \times 10^{-2}\ \text{m})(0.981\ \text{N})}{0.331\ \text{rad/s}^2}$$

$$I = 1.5 \times 10^{-2}\ \text{kg} \cdot \text{m}^2$$

CHECK and THINK

The units are as we expected.

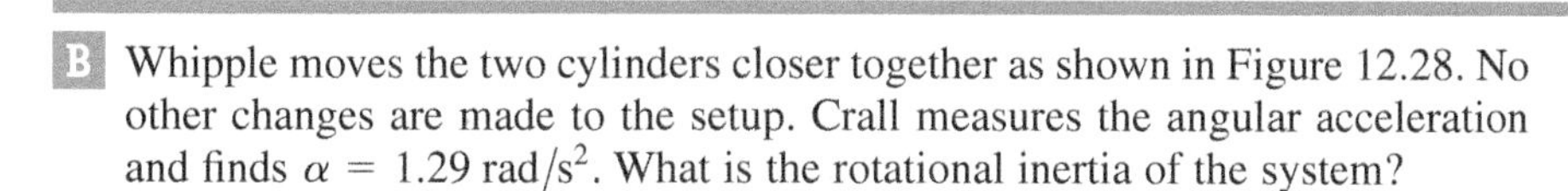
B Whipple moves the two cylinders closer together as shown in Figure 12.28. No other changes are made to the setup. Crall measures the angular acceleration and finds $\alpha = 1.29$ rad/s^2. What is the rotational inertia of the system?

INTERPRET and ANTICIPATE

Because the change to the setup only involves the cylinders, the torque is the same as it was in part A. All we need to do is use the new observed angular acceleration to find the new rotational inertia.

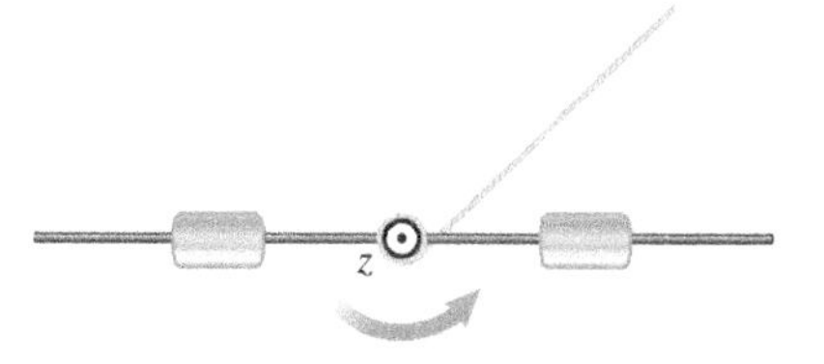

FIGURE 12.28 The cylinders have been moved closer together.

SOLVE

Substitute the new value of α into Equation 12.25.

$$I = \frac{\tau}{\alpha} = \frac{(0.50 \times 10^{-2}\text{m})(0.981\ \text{N})}{1.29\ \text{rad/s}^2}$$

$$I = 3.8 \times 10^{-3}\text{kg} \cdot \text{m}^2$$

CHECK and THINK

We just found that moving the two cylinders closer together reduced the rotational inertia of the system. Would it be easier to stop the rotating system shown in Figure 12.27 or the one shown in Figure 12.28? When the cylinders are moved closer together, the system has a smaller rotational inertia, so a weaker torque would result in the same angular acceleration, and it is easier to stop the system in Figure 12.28. The difference between the systems in Figures 12.27 and 12.28 is analogous to the difference between a bowling ball and a soccer ball. Imagine a bowling ball and a soccer ball moving with the same velocity. It is much easier to stop the moving soccer ball because it has much less inertial mass than does the bowling ball.

EXAMPLE 12.7 CASE STUDY Queen Hatshepsut's Obelisk

Our case study focused on how ancient people constructed megalithic monuments. For example, how did ancient Egyptians raise granite obelisks that were up to 100 ft tall and weighed 400 tons without damaging them? One possibility is that they used gravity to get an obelisk to a near-vertical position on its pedestal and then pulled it up from there. In a PBS *Nova* documentary[2], researchers are shown testing this idea with a 40-ft, 40-ton replica obelisk in a quarry in Massachusetts (Fig. 12.29A). To prevent damage, the obelisk's rotation is primarily controlled by sand, which required the construction of a large sandbox. To understand why the sand is required, let us imagine the same scenario but with no sand (or ropes) as shown in Figure 12.29B. Assume the center of mass is at the center of the obelisk; the obelisk hangs over a pit and pivots on an edge that is about 1 m to left of the center of mass. Find the magnitude of the obelisk's angular acceleration and the magnitude of the tangential acceleration of a point on its base at the moment the obelisk is released. The obelisk's rotational inertia is $I = 4.7 \times 10^5$ kg · m^2.

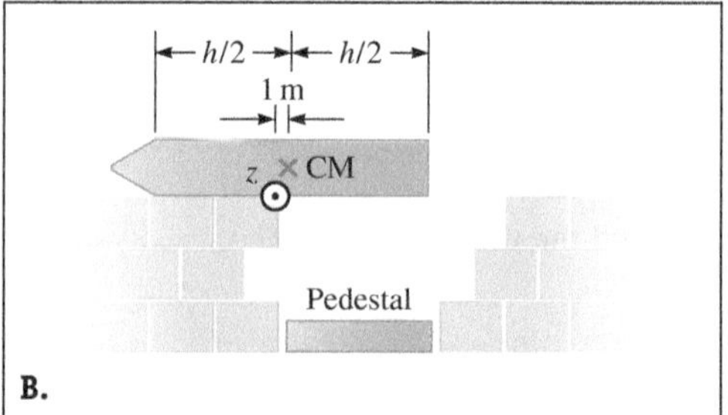

FIGURE 12.29 **A.** In the PBS *Nova* documentary, researchers try to raise a replica obelisk. **B.** Raising an obelisk without sand or ropes.

INTERPRET and ANTICIPATE

When the obelisk is released, the only torque acting on it is due to gravity $\vec{\tau}_g$. There is a normal force acting at the edge of the pit, but it does not contribute a torque because that force is applied along the axis of rotation. With only one torque acting on the obelisk, Equation 12.24 is simplified. Gravitational torque $\vec{\tau}_g$ and $\vec{\alpha}$ both point in the negative z direction. Our results should be numerical in the forms $\alpha =$ _____ rad/s^2 and $a_T =$ _____ m/s^2.

SOLVE

First convert the obelisk's mass and height to SI units.	$m = 3.6 \times 10^4$ kg $\quad h = 12$ m
Use Equation 12.20 to find the torque due to gravity.	$\tau_g = r_\perp F_g = r_\perp mg$
The moment arm $r_\perp = 1$ m. Substitute numerical values.	$\tau_g = (1\text{ m})(3.6 \times 10^4\text{ kg})(9.81\text{ m/s}^2) = 3.5 \times 10^5\text{ N}\cdot\text{m}$
The angular acceleration comes from $\vec{\tau}_{tot} = I\vec{\alpha}$ (Eq. 12.24).	$\tau_g = I\alpha$ $\alpha = \frac{\tau_g}{I} = \frac{3.5 \times 10^5\text{ N}\cdot\text{m}}{4.7 \times 10^5\text{ kg}\cdot\text{m}^2} = 0.74\text{ rad/s}^2$
The tangential acceleration comes from Equation 12.14, where $r = (h/2) + 1$ m.	$a_T = \alpha r$ $a_T = (0.74\text{ rad/s}^2)(7\text{m}) = 5.2\text{ m/s}^2 \approx \frac{1}{2}g$

CHECK and THINK

The results are in the form we expected. The tangential acceleration is about $\frac{1}{2}g$, which is rather high. Without the sand, the obelisk would rotate too quickly and would probably break when the base hit the pedestal. When damage to the base of the stones was not a concern as it was at Stonehenge (Example 12.8), it is believed that rotation was controlled without using sand.

[2]The PBS television show *Nova* organized teams to build replicas of several famous ancient monuments in two series of documentaries, *Secrets of Lost Empires* (1997) and *Secrets of Lost Empires II* (2000).

Levers

Archimedes (c. 287–212 BCE), an ancient Greek mathematician and inventor, discussed levers in his book *On the Equilibrium of Planes*. Archimedes understood that levers provide a mechanical advantage allowing a person to move very massive objects that would otherwise be impossible to move with just his or her own human strength. A famous quotation is attributed to Archimedes, "Give me a place to stand and I will move the Earth."[3]

A **lever** is a simple machine used to amplify the force exerted on an object. In Figure 12.30A, a load is placed on one end of the arm of the lever, and a force $\vec{F}_{app}$ is applied to the other end as shown. The net torque causes the arm to rotate around the fulcrum or pivot, moving the load.

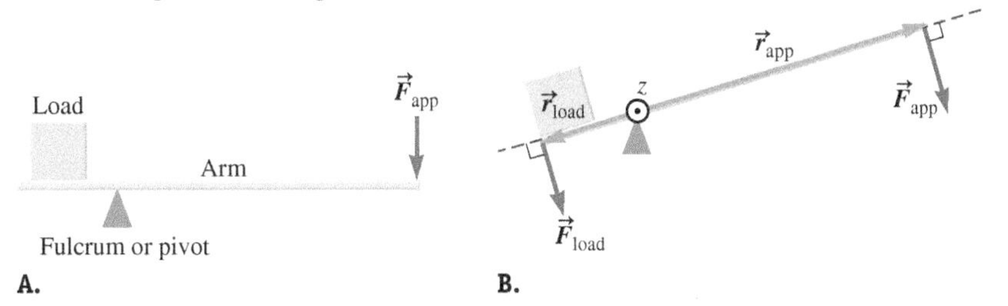

FIGURE 12.30 A. The arm of a lever may rotate on a fulcrum. **B.** Two torques act on the lever: one exerted by the load, the other by the applied force. (The normal force exerted by the fulcrum exerts no torque, so it is not included here.)

We can find an expression for $\vec{F}_{app}$ in terms of the parameters shown in Figure 12.30B. Assume the arm rotates at a constant angular speed so that $\vec{\alpha} = 0$. Then, according to Equation 12.24 (Newton's second law for rotation), the net torque is zero. Assume also the weight of the arm is negligible so that there are three forces acting on the arm: the normal force exerted by the fulcrum, the applied force $\vec{F}_{app}$, and the normal force $\vec{F}_{load}$ due to the load. (Remember that the normal force exerted by the load will not always equal its weight.) The rotation axis passes through the fulcrum, so the normal force exerted by the fulcrum does not exert a torque. We use Equation 12.23 (vector definition of torque) to find the torque $\vec{\tau}_{app}$ due to the applied force,

$$\vec{\tau}_{app} = -(r_{app}F_{app} \sin 90°)\hat{k} = -r_{app}F_{app}\hat{k} \tag{12.26}$$

and the torque $\vec{\tau}_{load}$ due to the load,

$$\vec{\tau}_{load} = (r_{load}F_{load} \sin 90°)\hat{k} = r_{load}F_{load}\hat{k} \tag{12.27}$$

Substitute Equations 12.26 and 12.27 into Equation 12.24 and set the expression equal to zero because $\vec{\alpha} = 0$:

$$\vec{\tau}_{tot} = \vec{\tau}_{app} + \vec{\tau}_{load} = I\vec{\alpha}$$

$$\vec{\tau}_{tot} = -r_{app}F_{app}\hat{k} + r_{load}F_{load}\hat{k} = 0$$

Solve for $\vec{F}_{app}$:

$$r_{app}F_{app}\hat{k} = r_{load}F_{load}\hat{k}$$

$$F_{app} = \frac{r_{load}}{r_{app}}F_{load}$$

By Newton's third law, the normal force $\vec{F}_{load}$ exerted by the load on the arm is equal in magnitude to the normal force $\vec{F}_N$ exerted by the arm on the load, so we have

$$F_{app} = \frac{r_{load}}{r_{app}}F_N \tag{12.28}$$

If $r_{app} > r_{load}$ as shown in Figure 12.30B, then $F_{app} < F_N$. In other words, the force F_N exerted by the lever on the load is greater than the applied force F_{app} by a factor of r_{app}/r_{load}. If a person applies a 500-N force and $r_{app} = 3r_{load}$, the force exerted on the load is 1500 N!

[3]Archimedes may never have actually made the statement, but many ancient authors such as Pappus of Alexandria (290–350 CE) and Plutarch (c. 45–120 CE) attribute such a comment to him.

EXAMPLE 12.8 CASE STUDY Building Stonehenge

The upright stones in the largest trilithon at Stonehenge are partially buried in the ground (Fig. 12.31A), unlike Queen Hatshepsut's obelisk, whose base is exposed. Therefore, Stonehenge's ancient builders in Britain did not need to be concerned about slightly damaging the base of the upright. The PBS *Nova* documentary presents a test of how the upright may have been raised. Archeologists believe that pits were first excavated for the upright stones (Fig. 12.31B) and that each upright was then rotated by a torque due to gravity so that the stone was nearly vertical as in Figure 12.31C. An A-frame lever then aided in bringing the stone to vertical (Fig. 12.31D). The A-frame lever is approximately parallel to the stone. The *Nova* team, consisting of 70 people, successfully raised a replica stone. Estimate the number of people that would be required if the A-frame lever were not used. Model the A-frame lever as a simple rod with the fulcrum at the ground and approximately parallel ropes attached at the top and at a distance r_{load} from the bottom, measured along the rod.

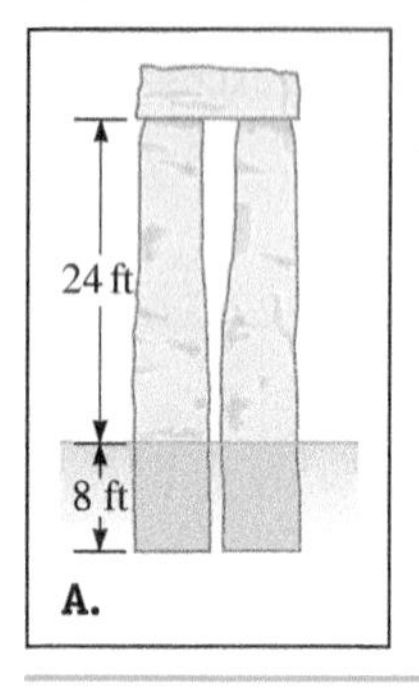

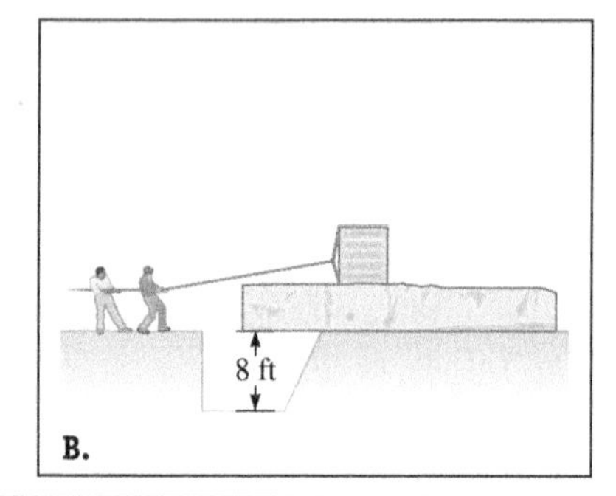

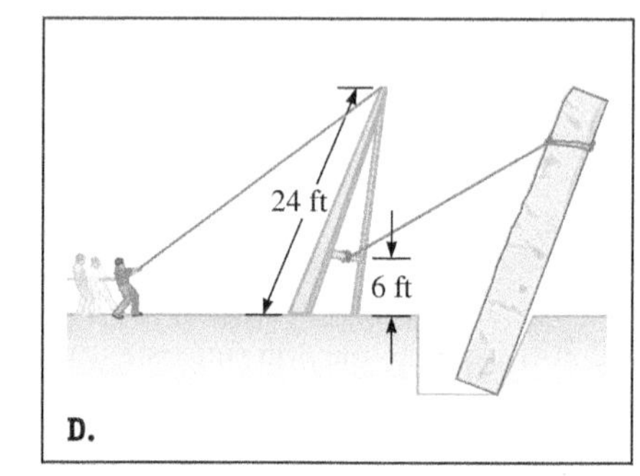

FIGURE 12.31 A. The largest stone in Stonehenge supports a trilithon. The stone is about 32 ft tall with 8 ft under ground. **B.** A large stone is slid into place partially over a pit. **C.** A smaller stone is slid across so that a torque due to gravity rotates the large stone into the pit. **D.** The large stone is brought to a vertical position with an A-frame lever.

INTERPRET and ANTICIPATE

We need to find the mechanical advantage the A-frame lever provides. In other words, we need to find by what factor the lever amplifies the force exerted by the people. Then, we assume the size of the team must be increased by that factor if no lever is used.

Start with a coordinate system and a sketch (Fig. 12.32) to identify the torques exerted on the rod representing the A-frame lever. The z axis is on the fulcrum and points out of the page. Two nonzero torques act on the rod: the torque $\vec{\tau}_{app}$ due to the force $\vec{F}_{app}$ exerted by the team and the torque $\vec{\tau}_{load}$ due to the force $\vec{F}_{load}$ exerted by the stone. The torque $\vec{\tau}_{app}$ points in the positive z direction, and the torque $\vec{\tau}_{load}$ points in the negative z direction. When the angular acceleration is zero (so that the stone is just on the verge of moving), the two torques have equal magnitudes.

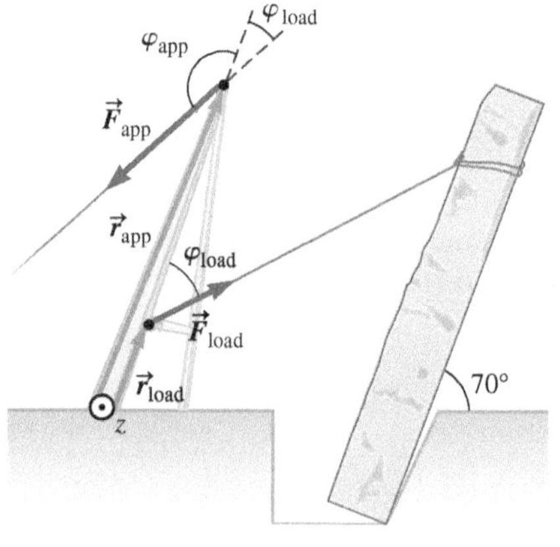

FIGURE 12.32

SOLVE

Assume the ropes are parallel to find the relationship between the angles φ_{app} and φ_{load} and the sine of these angles.

$$\varphi_{app} = 180° - \varphi_{load}$$
$$\sin \varphi_{app} = \sin(180° - \varphi_{load})$$
$$\sin \varphi_{app} = \sin \varphi_{load}$$

Because $\sin \varphi_{app} = \sin \varphi_{load}$, Equation 12.28 applies to this lever problem, but instead of a normal force F_N on the load, we have a tension force F_T. Solve for the ratio F_T/F_{app}. Because we are interested in finding a unitless ratio, we can work in the U.S. customary units of feet (Fig. 21.31D).

$$r_{load}F_T \sin \varphi_{load} = r_{app}F_{app} \sin \varphi_{load}$$
$$F_{app} = \frac{r_{load}}{r_{app}}F_T \quad \text{which resembles Eq. 12.28}$$
$$\frac{F_T}{F_{app}} = \frac{r_{app}}{r_{load}} = \frac{24\text{ ft}}{6\text{ ft}} = 4$$

The A-frame amplifies the force exerted by the team by a factor of four. If no A-frame lever were used, the team would need to be four times larger.

size of team using A-frame = 70 people

size of team without A-frame = 4 × 70 = 280 people

CHECK and THINK

This example illustrates the power of even simple mechanical tools. Levers fashioned from modest materials such as wood and rope were probably used by many ancient civilizations to construct monuments. No extraterrestrial workforce or tools were required.

We end this chapter by considering another example of ancient engineering. The Moai statues on Easter Island are counted among the seven wonders of the modern world. There is essentially no written record on how the Moai were constructed, transported, and erected, but many surviving partially constructed Moai provide important evidence. A volcanic crater served as the primary quarry (Fig. 12.33A). The Moai's fronts were carved while they lay inside the quarry. Just outside the quarry are incomplete Moai that have been erected facing the sea so that their backs could be carved (Fig. 12.33B). There are broken Moai along the transportation path to the sea. The final Moai stand near the sea facing inland (Fig. 12.33C).

Archeologists have developed and tested theories for each stage of the Moai construction processes. The Easter Islanders probably used stone tools to carve the statues and may have used wooden sleds to transport and erect them. Ramps made out of stones were probably used to raise the Moai to about 45°, and then levers were likely used to lift them to a vertical position. This method is similar to the tools and processes we discussed for raising obelisks in Egypt and stones at Stonehenge. One part of the Moai construction process is unique, however. All the complete Moai face inland, but all the partially complete Moai near the quarry face the sea. The Easter Islanders must have had a way to rotate their statues. One possibility is that a small workforce used levers to rotate the statues while they were attached to their sleds. Another recently described possibility for transport and rotation (thought not accepted by all experts) is that once a statue was nearly vertical, three teams of islanders "walked" the statue with a rocking motion using ropes attached to the statue's head.[4]

Although no experiment can be performed that conclusively refutes the extraterrestrial hypothesis, there is no evidence that uniquely supports the claim that extraterrestrials built the ancient monuments. There is, however, very strong evidence that ancient people used modest tools made out of readily available materials and basic physics to construct megaliths on their own.

A.

B.

C.

[4]Hannah Bloch, "The Riddle of Easter Island," *National Geographic* 222(1), 31-49, July 2012.

FIGURE 12.33 **A.** Stones are quarried near the top of a volcano. **B.** Moai are erected facing the sea so that their backs can be carved. **C.** Moai are transported to the sea, where they are erected facing inland.

SUMMARY

Underlying Principle

Newton's second law in rotational form: $\vec{\tau}_{\text{tot}} = I\vec{\alpha}$ (Eq. 12.24)

Major Concepts

1. A **rigid object** is an object that moves with all its components locked together so that its shape does not change.
2. Every particle in a rotating rigid object moves in a circular path around the **rotation axis**.
3. A **fixed axis** is a rotation axis that does not move relative to the observer.
4. Angular kinematic quantities
   a. The (positive) **angular position** θ of the reference line is measured counterclockwise from the positive x axis.
   b. A change in the angular position is known as the **angular displacement** and is given by Equation 12.2: $\Delta\theta = \theta_f - \theta_i$.

Major Concepts—cont'd

c. **Angular speed** is given by Equation 12.4: $\omega = d\theta/dt$. The direction of **angular velocity** is given by the right-hand rule.

d. **Angular acceleration** is given by Equation 12.6: $\vec{\alpha} = d\vec{\omega}/dt$. If an object's rotation is speeding up, then $\vec{\alpha}$ points in the same direction as $\vec{\omega}$. If the object is slowing down, then $\vec{\alpha}$ and $\vec{\omega}$ point in opposite directions.

5. A **torque** on a rotating object is analogous to a force on a translating object: A force causes the acceleration of a translating object; a torque causes the angular acceleration of a rotating object. Mathematically, torque is given by Equation 12.23: $\vec{\tau} = \vec{r} \times \vec{F}$. The torque's direction is found from the right-hand rule for vector cross products.
6. **Rotational inertia** I is a measure of an object's resistance to have an angular acceleration.

Tools

1. An observer uses the **right-hand rule** convention to find the direction of $\vec{\omega}$ by curling the fingers of his or her right hand in the same sense as the rotation. The observer's right thumb then points in the direction of $\vec{\omega}$. Because we have decided to use the convention that the rotation axis is aligned with the z axis, $\vec{\omega}$ will point in either the positive or negative z direction.
2. The **cross product** or **vector product** between two vectors $\vec{A}$ and $\vec{B}$ results in a third vector $\vec{R}$: $\vec{R} = \vec{A} \times \vec{B}$. The magnitude of R is given by Equation 12.22: $R = AB \sin \varphi$. The direction of $\vec{R}$ comes from the right-hand rule for cross products. Line up the fingers of your right hand along the first vector in the cross product (in this case, vector $\vec{A}$) and then imagine closing your fingers so that you "push" $\vec{A}$ through the angle φ into the second vector (in this case, vector $\vec{B}$); see Figure 12.23B. The direction in which your thumb points as you push $\vec{A}$ into $\vec{B}$ is the direction of the resulting vector $\vec{R}$.

PROBLEMS AND QUESTIONS

A = algebraic C = conceptual E = estimation G = graphical N = numerical

12-1 Rotation Versus Translation

Problems 1 and 2 are paired.

1. C Often, we model the Moon as a particle in a circular orbit around the Earth. The same side of the Moon always faces the Earth. Sketch the Moon in its orbit. Explain in what way the particle model is insufficient.
2. C Suppose a satellite orbits the Earth such that it is well modeled as a particle. Draw a sketch of it in its orbit. Explain how its motion is different from the Moon's motion around the Earth.

Problems 3 and 4 are paired.

3. C An ice skater spins on one foot. Is he rotating? If so, describe his axis of rotation. Explain your answer.
4. C A speed ice skater travels around a closed, circular track. Is she rotating? If so, describe her axis of rotation. Explain your answer.

12-2 Rotational Kinematics

5. N A ceiling fan is rotating counterclockwise with a constant angular acceleration of π rad/s^2 about a fixed axis perpendicular to its plane and through its center. Assume the fan starts from rest. **a.** What is the angular velocity of the fan after 4.0 s? **b.** What is the angular displacement of the fan after 4.0 s? **c.** How many revolutions has the fan gone through in 4.0 s?
6. N As seen from above the Earth's North Pole, the Moon's orbit is counterclockwise. Use a coordinate system with the positive z axis pointing north. Find the magnitude and direction of the Moon's angular velocity. *Hint*: Draw a sketch of the Moon's orbit from this perspective above the North Pole, including the coordinate system.

Problems 7 and 8 are paired.

7. N A rotating object's angular position is given by $\theta(t) = (1.54t^2 - 7.65t + 2.75)$ rad, where t is measured in seconds. Find **a.** the object's angular speed when $t = 3.50$ s and **b.** the magnitude of the angular acceleration when $t = 3.50$ s.
8. N A rotating object's angular position is given by $\theta(t) = (1.54t^2 - 7.65t + 2.75)$ rad, where t is measured in seconds. **a.** When is the object momentarily at rest? **b.** What is the magnitude of the angular acceleration at that time?
9. Jupiter rotates about its axis once every 9 hours 55 minutes.
   a. N What is Jupiter's angular speed of rotation?
   b. C What is the effect of this rapid rotation on the shape of the planet?
10. C The text of this chapter says that "angular displacement does not obey the commutative rule of vector addition." What does that statement mean? Try the following. Place a book on a table with the front cover facing upward and the spine toward you. Now make two rotations.

    First, perform a 90° rotation (clockwise as seen from above) around the axis perpendicular to the table and through the book. Next, perform a 90° rotation around the horizontal axis parallel to the table and through the book (clockwise as seen from the left side of the book). The book should now be upright and facing you.

Now, return the book to its original position and make these same two rotations, *but in the opposite order*. What result do you get this time? Does angular displacement obey the commutative rule of vector addition? Explain.

Problems 11 and 12 are paired.

11. N The long hand on a clock is known as the *minute hand*. It completes one clockwise rotation each hour. Choose a coordinate system such that a z axis points out of the clock. Consider a 15-minute time interval. Find the minute hand's **a.** angular displacement, **b.** angular velocity, and **c.** angular acceleration.

12. The long hand on a clock is known as the *minute hand*. Usually, it completes one clockwise rotation each hour. A particular clock, however, is slowing down at a constant rate of 2.909×10^{-7} rad/s^2. Choose a coordinate system such that a z axis points out of the clock.

a. C What is the direction of the minute hand's angular acceleration?

b. N Suppose the slow clock and a working clock both read noon at the same moment and at that moment their angular velocities are equal. What does the broken clock read when the working clock reads 1 PM?

13. E Assuming an automobile in Los Angeles is driven a total of 12,000 miles per year in commuting to work, what is the approximate number of revolutions the car's tires undergo each year?

Problems 14 and 15 are paired.

14. N Use the information about the Crab Nebula's pulsar from Example 12.2 (page 339). **a.** Determine the period of the pulsar's rotation **b.** Determine the number of seconds by which this period increases each second (dT/dt).

15. N Example 12.2 gives the rotation rate for the Crab Nebula's pulsar and also the (negative) angular acceleration of this rotation. The rotation rate of this pulsar as well as that of many other pulsars does not show a long, steady decrease, however. There are occasional abrupt *increases* in the rotation rate called *glitches* that are caused by mechanisms in the interior of the pulsar. For the glitch observed in the nebula's pulsar in June 2000, the fractional increase in the rotation rate was $\Delta\omega/\omega = (\omega_{\text{new}} - \omega_{\text{orig}})/\omega_{\text{orig}} = 2.4 \times 10^{-8}$. **a.** What was the rotation rate after the glitch? **b.** The angular acceleration (or spin-down rate, as it is called) also changed after the glitch: $\Delta\alpha/\alpha = 5 \times 10^{-3}$. What was the new angular acceleration? **c.** How long after the glitch did it take the pulsar to return to its pre-glitch rotation rate?

16. C A disk rolls up an inclined plane as shown in Figure P12.16, reaches point A, stops there momentarily, and then rolls down the inclined plane. Use the coordinate system shown to determine the direction of the angular velocity and the angular acceleration in each part of the motion as given below. If either one is zero, say so. Explain your answers. **a.** When the disk is going up the incline. **b.** At point A when the disk stops momentarily. **c.** When the disk is rolling down the incline

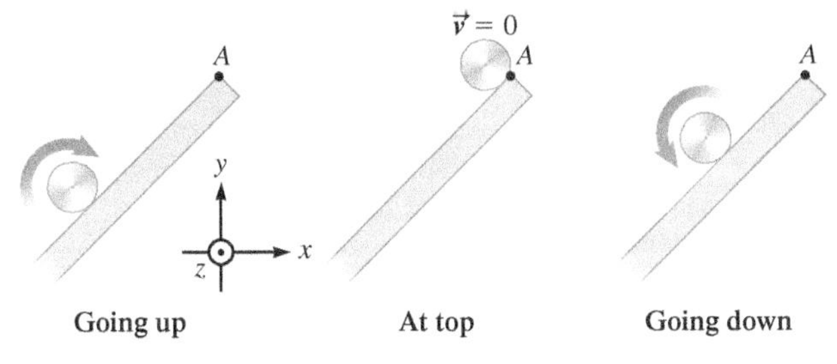

FIGURE P12.16

17. N Jeff, running outside to play, pushes on a swinging door, causing its motion to be briefly described by $\theta = t^2 + 0.800t + 2.00$, where t is in seconds and θ is in radians. At $t = 0$ and at $t = 1.50$ s, what are the **a.** angular position, **b.** angular speed, and **c.** angular acceleration of the door?

12-3 Special Case of Constant Angular Acceleration

18. N A potter's wheel rotating at 240 rev/min is switched off and rotates through 80.0 revolutions prior to coming to rest. What is the constant angular acceleration of the potter's wheel during this interval?

19. N Friction in an old clock causes it to lose 1 minute per hour. Assume the angular acceleration is constant, the positive z axis points out of the face of the clock, and at the beginning of the hour it was at its ideal angular speed. Find the angular acceleration of the minute hand (long hand) and its angular velocity at the end of the hour.

20. N A wheel starts from rest and in 12.65 s is rotating with an angular speed of 5.435 π rad/s. **a.** Find the magnitude of the constant angular acceleration of the wheel. **b.** Through what angle does the wheel move in 6.325 s?

21. N Some people are able to spin a basketball on the tip of a finger. It is often done by balancing the ball on the tip of the finger and brushing the other hand along the side of the ball to cause it to rotate. Suppose the ball begins from rest and reaches a final angular speed of 18.65 rad/s in 1.10 s. **a.** Assuming the ball is subject to a constant angular acceleration, what is the magnitude of the constant acceleration? **b.** Through how many revolutions does the ball rotate during the 1.10 s?

22. N Starting from rest, a wheel reaches an angular speed of 15.0 rad/s in 5.00 s. **a.** What is the magnitude of the constant angular acceleration of the wheel? **b.** Through what angle in radians does the wheel rotate during this time interval?

23. N A potter's wheel is rotating with an angular velocity of $\vec{\omega} = 24.2\hat{k}$ rad/s. The wheel is slowing down as it is subject to a constant angular acceleration and eventually comes to a stop. **a.** If the wheel goes through 21.0 revolutions as it is slowed to a stop, what is the constant angular acceleration applied to the wheel? **b.** How long does it take for the wheel to come to rest?

Problems 24 and 25 are paired.

24. The angular speed of a wheel is given by
$\omega(t) = 72.5 \text{ rad/s} + (9.34 \text{ rad/s}^2)t$.

a. C Is the wheel's angular acceleration constant? Explain.

b. A Find an expression for the angular acceleration.

25. The angular speed of a wheel is given by
$\omega(t) = 72.5 \text{ rad/s} + (9.34 \text{ rad/s}^2)t + (2.24 \text{ rad/s}^3)t^2$.

a. C Is the wheel's angular acceleration constant? Explain.

b. A Find an expression for the angular acceleration.

26. N A wheel is programmed to follow a specific cycle in which it accelerates from rest for 3.40 s and reaches an angular speed of 25.0 rev/s. Then, 3.00 s after reaching this speed, a braking mechanism is engaged, stopping the wheel smoothly in 15.0 s. What is the number of revolutions completed by the wheel during this cycle?

12-4 The Connection Between Rotation and Circular Motion

27. N An electric food processor comes with many attachments for blending and slicing food. Assume the motor maintains the same angular speed for all the various attachments. The largest attachment has a diameter of 12.0 cm, and the smallest has a diameter of 5.00 cm. Find the ratio of the translational speeds of points on the edges of each attachment.

28. C A truck's tire starts from rest and begins to rotate with a constant angular acceleration. Consider a point on the outer edge of the tire. **a.** Is the tangential acceleration component constant? Explain. **b.** Is the centripetal acceleration component constant? Explain.

29. N A bicyclist is testing a new racing bike on a circular track of radius 53.0 m. The bicyclist is able to maintain a constant speed of 24.0 m/s throughout the test run. **a.** What is the angular speed of the bike? **b.** What are the magnitude and direction of the bike's acceleration?

30. N Paul is listening to an old-fashioned vinyl record on a turntable. He likes to place coins on the record and watch them spin. A penny is 3.55 cm from the axis of rotation, and a nickel is 8.10 cm from the axis of rotation. The record rotates at 33 rpm (revolutions per minute). What is the tangential speed of each coin?

31. N A disk is initially at rest. A penny is placed on it at a distance of 1.0 m from the rotation axis. At time $t = 0$ s, the disk begins to rotate with a constant angular acceleration of 2.0 rad/s^2 around a fixed, vertical axis through its center and perpendicular to its plane. Find the magnitude of the net acceleration of the coin at $t = 1.5$ s.

Problems 32, 33, and 34 are grouped.

32. N The blades on modern wind turbines are typically 20 to 40 meters in length. Two different turbines have their blades attached with one end at the center of rotation. Consider two points, each at the outer end of a blade on each of the turbines: one with a blade length of 20.0 m, and the other with a blade length of 40.0 m. Each blade is rotating with a constant angular speed of 3.88 rad/s, and the rotation is considered for 10.0 s. **a.** What is the speed of each point in meters per second? **b.** What is the angular distance traveled by each point? **c.** What is the translational distance traveled by each point? **d.** What is the magnitude of the centripetal acceleration that would be experienced by an object located at each point?

33. C Consider again the two wind turbines in Problem 32. **a.** At what point along the 40-m blade would the angular speed of the point be the same as the angular speed at the end of the 20-m blade? Explain your answer. **b.** At what point along the 40-m blade would the translational speed of the point be the same as the speed at the end of the 20-m blade? Explain your answer.

34. C Consider again the two wind turbines in Problem 32. **a.** At what point along the 40-m blade would the angular distance traveled by the point be the same as the angular distance traveled by the point at the end of the 20-m blade? **b.** At what point along the 40-m blade would the translational distance traveled by the point be the same as the translational distance traveled by the point at the end of the 20-m blade?

35. N In testing an automobile tire for proper alignment, a technician marks a spot on the tire 0.200 m from the center. He then mounts the tire in a vertical plane and notes that the radius vector to the spot is at an angle of 35.0° with the horizontal. Starting from rest, the tire is spun rapidly with a constant angular acceleration of 3.00 rad/s^2. **a.** What is the angular speed of the wheel after 4.00 s? **b.** What is the tangential speed of the spot after 4.00 s? **c.** What is the magnitude of the total accleration of the spot after 4.00 s?" **d.** What is the angular position of the spot after 4.00 s?

Problems 36 and 37 are paired.

36. N Two children, each with a mass of 25.0 kg, are at fixed locations on a merry-go-round (a disk that spins about an axis perpendicular to the disk and through its center; Fig. P12.36). One child is 0.75 m from the center of the merry-go-round, and the other is near the outer edge, 3.00 m from the center. With the merry-go-round rotating at a constant angular speed, the child near the edge is moving with translational speed of 12.5 m/s. **a.** What is the angular speed of each child? **b.** Through what angular distance does each child move in 5.0 s? **c.** Through what distance in meters does each child move in 5.0 s? **d.** What is the centripetal force experienced by each child as he or she holds on? Which child has a more difficult time holding on?

FIGURE P12.36 Problems 36 and 37. A merry-go-round in a neighborhood park.

37. N A merry-go-round at a children's park begins at rest and is pushed by a parent while a child holds onto the ride at the outer edge, 3.00 m from the center. The parent pushes with a constant force for 2.54 s until the child is moving with a speed of 12.5 m/s. What is the magnitude of the net acceleration on the child the instant before the parent stops pushing the merry-go-round?

38. N A wheel rotating at a constant rate of 1850 rev/min has a diameter of 17.8 cm. **a.** What is the angular speed of the wheel in radians per second? **b.** What is the tangential speed of a point on the wheel's rim? **c.** What is the centripetal acceleration of a point on the wheel's rim? **d.** What is the total distance traversed by a point on the wheel's rim from $t = 0$ to $t = 3.00$ s?

12-5 Torque

39. C Why are doorknobs placed on the edge opposite the hinges?

40. C Why would a person with weak hands prefer to have a door handle such as the one shown in Figure P12.40 instead of a round doorknob?

FIGURE P12.40

41. N A student pulls with a 300.0-N force on the outer edge of a door of width 1.51 m that pivots about point *P* as shown in Figure P12.41. Find the magnitude of the torque applied to the door in each case, where the force is applied in a different direction as shown.

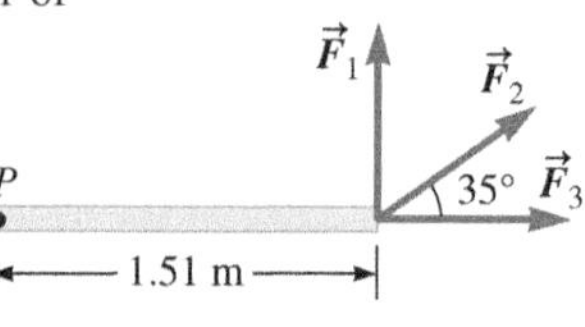

FIGURE P12.41

42. C You are working on your car and need to remove a cover in the engine compartment, which requires removing several screws. As you search through your tools, you find that you have two screwdrivers that will fit the screws, but each has a different handle diameter. One of the handles is very thick, whereas the other handle diameter is rather small by comparison. Which screwdriver should you choose to make the job easier for you to accomplish? Explain your choice and explain what quantity most directly applies in describing why the job is "easier" for you.

43. N A wheel of inner radius $r_1 = 15.0$ cm and outer radius $r_2 = 35.0$ cm shown in Figure P12.43 is free to rotate about the axle through the origin *O*. What is the magnitude of the net torque on the wheel due to the three forces shown?

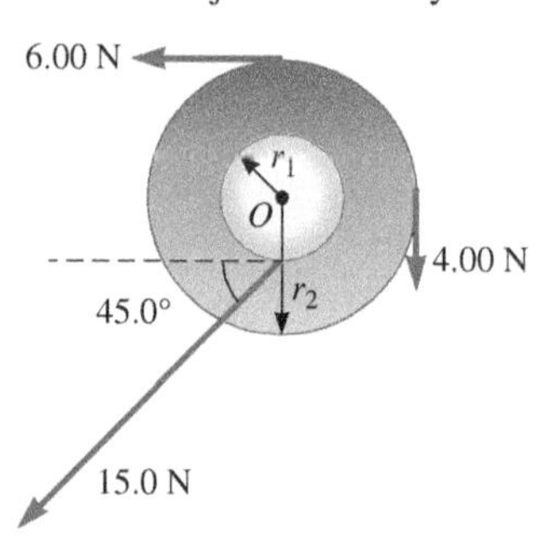

FIGURE P12.43

44. N A uniform plank 6.0 m long rests on two supports, 2.5 m apart (Fig. P12.44). The gravitational force on the plank is 100 N. The left end of the plank is 1.5 m to the left of the left support, so the plank is not centered on the supports. A person is standing on the plank half a meter to the right of the right support. The gravitational force on this person is 80.0 N. How far to the right can the person walk before the plank begins to tip?

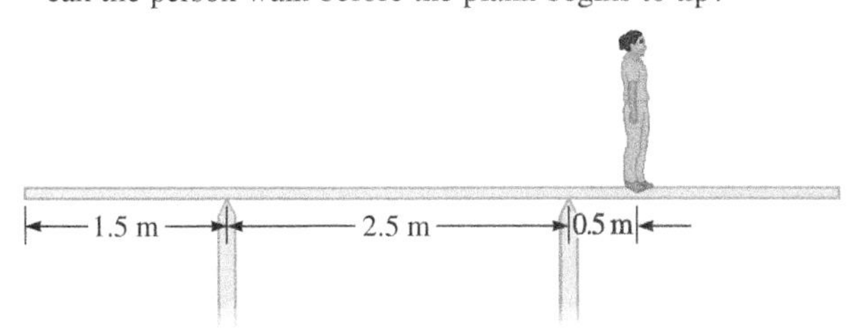

FIGURE P12.44

45. N Five forces of equal magnitude are applied to a rod of length L (Fig. P12.45). Rank in order from smallest to largest the magnitudes of the torques produced by these forces around the hinge shown.

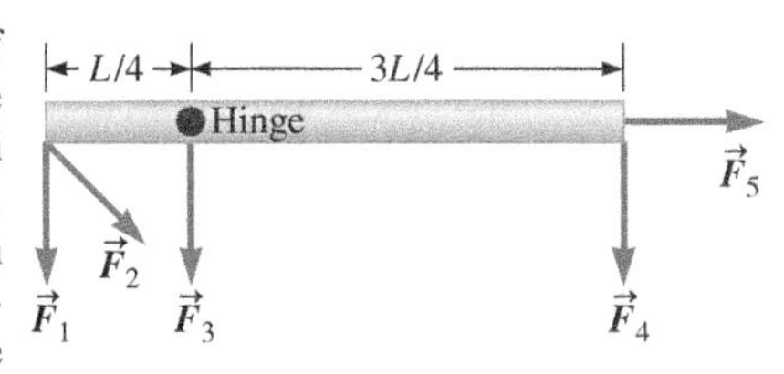

FIGURE P12.45

46. N At one point during its swing, a wrecking ball exerts a tension force of 9800 N on its cable, which makes an angle of 30.0° with the horizontal. The crane's 9.00-m-long boom is at an angle of 55.0° with the horizontal. What is the torque exerted by the wrecking ball on the crane about an axis perpendicular to the page and passing through point P shown in Figure P12.46?

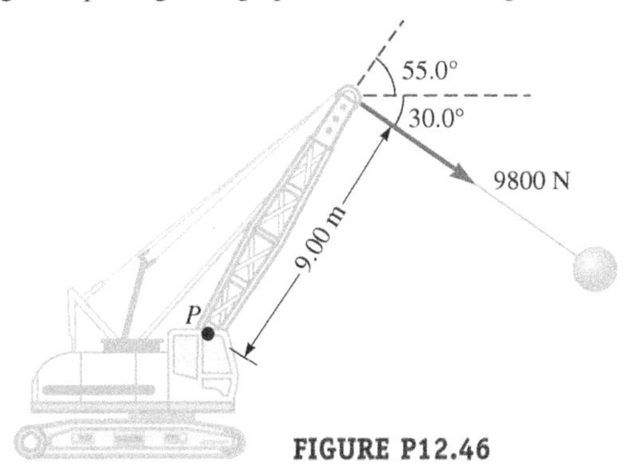

FIGURE P12.46

12-6 Cross Product

47. N Find the cross product $\vec{A} \times \vec{B}$ in each case: **a.** $\vec{A} = 15.0\hat{\imath}$ and $\vec{B} = 15.0\hat{\imath}$ **b.** $\vec{A} = 15.0\hat{\imath}$ and $\vec{B} = 15.0\hat{\jmath}$ **c.** $\vec{A} = 15.0\hat{\imath}$ and $\vec{B} = -15.0\hat{\jmath}$

Problems 48 and 49 are paired.

48. N The cross product $\vec{A} \times \vec{B} = 42.0\hat{k}$, and the magnitudes of the vectors are $A = 12.0$ and $B = 7.00$. What is the angle between the vectors?

49. N The cross product $\vec{A} \times \vec{B} = -42.0\hat{k}$, and the magnitudes of the vectors are $A = 12.0$ and $B = 7.00$. What is the angle between the vectors?

50. C Can the dot product and the cross product between two vectors both be zero? Explain.

51. N A force $\vec{F} = (2\hat{\imath} + 3\hat{\jmath} + 4\hat{k})$N is applied to a point with position vector $\vec{r} = (3\hat{\imath} + 2\hat{\jmath} + \hat{k})$m. Find the torque due to this force about the axis passing through the origin.

52. N Given a vector $\vec{A} = 4.5\hat{\imath} + 4.5\hat{\jmath}$ and a vector $\vec{B} = -4.5\hat{\imath} + 4.5\hat{\jmath}$, determine the magnitude of the cross product of these two vectors, $\vec{A} \times \vec{B}$. *Hint*: Make a sketch of both vectors including a coordinate system.

12-7 Rotational Dynamics

Problems 53 and 54 are paired.

53. N A square plate with sides 2.0 m in length can rotate around an axle passing through its center of mass (CM) and perpendicular to its surface (Fig. P12.53). There are four forces acting on the plate at different points. The rotational inertia of the plate is 24 kg · m². Use the values given in the figure to answer the following questions. **a.** What is the net torque acting on the plate? **b.** What is the angular acceleration of the plate?

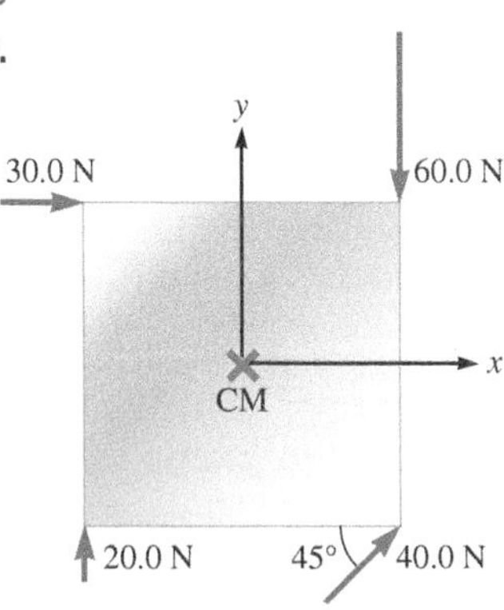

FIGURE P12.53
Problems 53 and 54.

54. N Assume the plate in Problem 53 (Fig. P12.53) starts from rest. **a.** What is the angular velocity of the plate after 4.0 s? **b.** Through how many revolutions will the plate rotate in 4.0 s?

55. A disk with a radius of 4.5 m has a 100-N force applied to its outer edge at two different angles (Fig. P12.55). The disk has a rotational inertia of 165 kg · m².
    a. N What is the magnitude of the torque applied to the disk in case 1?
    b. N What is the magnitude of the torque applied to the disk in case 2?
    c. N Assuming the force on the disk is constant in each case, what is the magnitude of the angular acceleration applied to the disk in each case?
    d. C Which case is a more effective way of spinning the disk? Describe which quantity you are using to determine "effectiveness" and why you chose that quantity.

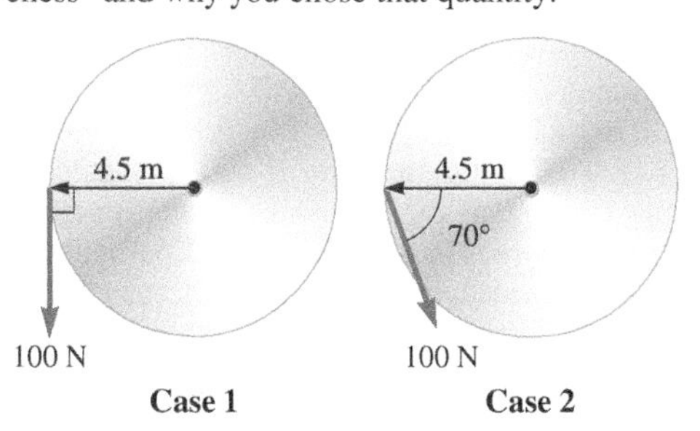

FIGURE P12.55

56. N Disc jockeys (DJs) use a turntable in applying their trade, often using their hand to speed up or slow down a disc record so as to produce a desired change in the sound (Fig. P12.56). Suppose DJ Trick wants to slow down a record initially rotating clockwise (as viewed from above) with an angular speed of 33.0 rpm to an angular speed of 22.0 rpm. The record has a rotational inertia of 0.012 kg · m² and a radius of 0.15 m. **a.** What angular acceleration is necessary if he wishes to accomplish this feat in exactly 0.65 s with a constant acceleration?

FIGURE P12.56

b. How many revolutions does the record go through during this change in speed? **c.** If DJ Trick applies a vertical force with his finger to the edge of the record, with what force must he push so as to slow the record in the above time? Assume the coefficient of kinetic friction between his finger and the record is 0.50, and ignore the mass of the finger.

57. N Frictional torque causes a disk to decelerate from an angular speed of 4.00 rad/s at $t = 0$ to 1.00 rad/s at $t = 5.00$ s. The equation describing the angular speed of the wheel during this time interval is given by $d\theta/dt = \omega_0 e^{-bt}$, where b and ω_0 are constants. **a.** What are the values of b and ω_0 during this time interval? **b.** What is the magnitude of the angular acceleration of the disk at $t = 5.00$ s? **c.** How many revolutions does the disk make during the interval $t = 0$ to $t = 5.00$ s?

General Problems

58. N A wheel rotates through 45.0 revolutions in 4.00 s, during which time it reaches a final angular speed of 66.0 rad/s. What is the constant angular acceleration of the wheel during this 4.00-s interval?

59. N A wheel initially rotating at 85.0 rev/min decelerates at the constant rate of 3.50 rad/s^2 when a braking mechanism is engaged. **a.** What is the time interval required to bring the wheel to a complete stop? **b.** What is the angle in radians through which the wheel rotates in this time interval?

60. A child's pinwheel rotates as the wind passes through it.

a. N If the pinwheel rotates from $\theta = 0°$ to $\theta = 90°$ in a time of 0.150 s, what is the average angular speed of the pinwheel?

b. N If the pinwheel rotates from $\theta = 0°$ to $\theta = 180°$ in a time of 0.300 s, what is the average angular speed of the pinwheel?

c. N If the pinwheel rotates from $\theta = 0°$ to $\theta = 270°$ in a time of 0.450 s, what is the average angular speed of the pinwheel?

d. N If the pinwheel rotates from $\theta = 0°$ through one revolution to $\theta = 360°$ in a time of 0.600 s, what is the average angular speed of the pinwheel?

e. C How is your answer to part (d) counterintuitive given your answers to parts (a), (b), and (c)?

61. N A centrifuge used for training astronauts rotating at 0.810 rad/s is spun up to 1.81 rad/s with an angular acceleration of 0.050 rad/s^2. **a.** What is the magnitude of the angular displacement that the centrifuge rotates through during this increase in speed? **b.** If the initial and final speeds of the centrifuge were tripled and the angular acceleration remained at 0.050 rad/s^2, what would be the factor by which the result in part (a) would change?

Problems 62 and 63 are paired.

62. C A disk is rotating around a fixed axis that passes through its center and is perpendicular to the face of the disk. Consider a point on the rim of the disk (point R) and another point halfway between the center and the rim (point H) at one particular instant. **a.** How does the angular speed ω of the disk at point H compare with the angular speed of the disk at point R? **b.** How does the tangential speed of the disk at point H compare with the tangential speed of the disk at point R? **c.** Suppose we pick a point H on the disk at random (by throwing a dart, for example), and we compare the speeds at that point with the speeds at point R. How will the answers to parts (a) and (b) be different? Explain.

63. C Consider the rotating disk in Problem 62 at one particular instant. **a.** How does the angular acceleration α of the disk at point H compare with the angular acceleration of the disk at point R? **b.** How does the tangential acceleration a_T of the disk at point H compare with the tangential acceleration of the disk at point R? **c.** Suppose we pick a point H on the disk at random (by throwing a dart, for example), and we compare the accelerations at that point with the accelerations at point R. How will the answers to parts (a) and (b) be different? Explain. **d.** Suppose the disk is rotating at a *constant rate*. Will the answers to parts (a) and (b) change? Explain.

64. A A potter's wheel rotates with an angular acceleration $\alpha = 4at^3 - 3bt^2$, where t is the time in seconds, a and b are constants, and α has units of radians per second squared. The initial position of the wheel is θ_0 and the initial angular velocity is ω_0. Obtain expressions for the angular velocity and the angular displacement of the wheel as a function of time.

65. N A racing motorcycle attempting to break a track record accelerates uniformly from rest and without slipping and tops 100 mph (44.7 m/s) in 4.99 s. The motorcycle's tires have a radius of 28.9 cm. **a.** What is the number of revolutions through which the motorcycle's tires turn during this time? **b.** What is the angular speed of the tires in revolutions per second after 4.99 s?

Problems 66 and 67 are paired.

66. N As seen from above the Earth's North Pole, the Earth's rotation is counterclockwise. In Example 12.1, we assumed the Earth's angular velocity is constant. Actually, the Earth is slowing down at a rate of roughly 0.002 s per century. Choose a coordinate system such that the positive z axis points north. Find the angular acceleration of the Earth, keeping at least eight significant figures in your calculations. Report your answer using two significant figures.

67. N As seen from above the Earth's North Pole, the Earth's rotation is counterclockwise. In Example 12.3, we assumed the Earth's angular velocity is constant. Actually, the Earth is slowing down at a rate of roughly 0.002 s per century. Find the magnitude of the tangential acceleration of Esmeraldas and New Orleans.

68. N Lara is running just outside the circumference of a carousel, looking for her favorite horse to ride, with a constant angular speed of 1.00 rad/s. Just as she spots the horse, one-fourth of the circumference ahead of her, the carousel begins to move, accelerating from rest at 0.050 rad/s^2. **a.** Taking the time when the carousel begins to move as $t = 0$, when will Lara catch up to the horse? **b.** Lara mistakenly passes the horse and keeps running at constant angular speed. If the carousel continues to accelerate at the same rate, when will the horse draw even with Lara again?

69. N The propeller of an aircraft accelerates from rest with an angular acceleration $\alpha = 4t + 6$, where α is in rad/s^2 and t is in seconds. What is the angle in radians through which the propeller rotates from $t = 1.00$ s to $t = 6.00$ s?

70. G A ball rolls to the left along a horizontal surface, up the slope, and then continues along a horizontal surface (Fig. P12.70). Sketch the angular speed ω and the magnitude of the angular acceleration α of the ball as functions of time.

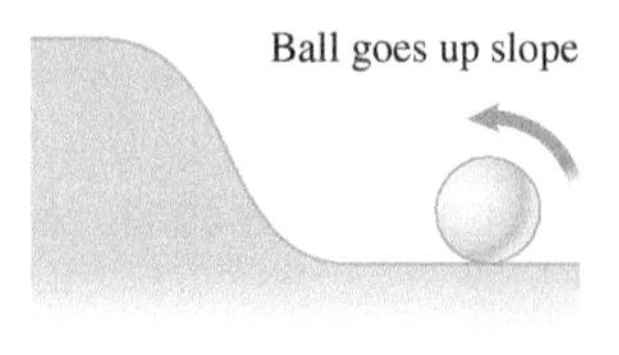

FIGURE P12.70

Problems 71 through 75 are grouped.

71. A Three forces are exerted on the disk shown in Figure P12.71, and their magnitudes are $F_3 = 2F_2 = 2F_1$. The disk's outer rim has radius R, and the inner rim has radius $R/2$. As shown in the figure, $\vec{F}_1$ and $\vec{F}_3$ are tangent to the outer rim of the disk, and $\vec{F}_2$ is tangent to the inner rim. $\vec{F}_3$ is parallel to the x axis, $\vec{F}_2$ is parallel to the y axis, and $\vec{F}_1$ makes a 45° angle with the negative x axis. Find expressions for the magnitude of each torque exerted around the center of the disk in terms of R and F_1.

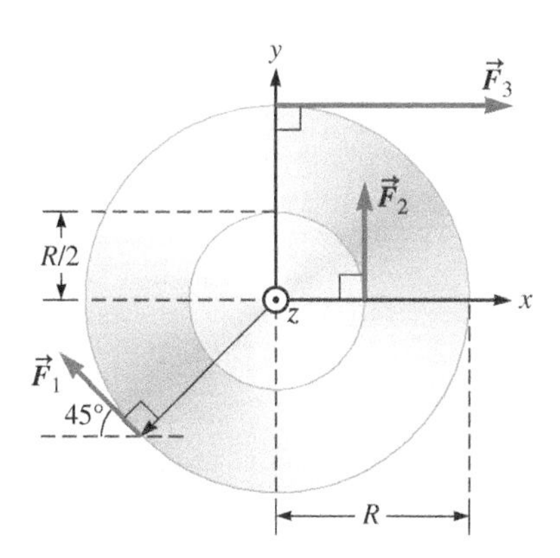

FIGURE P12.71 Problems 71–75.

72. N Consider the disk in Problem 71. The disk's outer rim has radius $R = 4.20$ m, and $F_1 = 10.5$ N. Find the magnitude of each torque exerted around the center of the disk.

73. A For the disk in Problem 71, find an expression for total torque exerted around the center of the disk in terms of R and F_1.

74. N Consider the disk in Problem 71. The disk's outer rim has radius $R = 4.20$ m, and $F_1 = 10.5$ N. The disk's rotational inertia is 0.545 kg · m². Find the angular acceleration of the disk.

75. N Consider the disk in Problem 71. Three forces are exerted on the disk (Fig. P12.71), with $F_2 = F_1 = 10.5$ N. The disk's outer rim has radius $R = 4.20$ m, and the inner rim has radius $R/2$. The angular acceleration is zero. What is the magnitude of $\vec{F}_3$?

76. Suppose the angular displacement of a rotating object is given by the equation $\Delta\theta(t) = (-5.3 \text{ rad/s}^3)t^3 + (7.0 \text{ rad/s}^2)t^2 + (8.1 \text{ rad/s})t$.

a. N Find a time when the object is momentarily not rotating.

b. C Determine whether the object is beginning to turn clockwise or counterclockwise. (Let clockwise be indicated by a negative angular velocity.)

77. N Starting from rest, an automobile accelerates with a uniform tangential acceleration of 2.00 m/s² while negotiating a semicircular unbanked curve. Three-fourths of the way into the curve, the car's tires lose traction with the road, causing the car to skid off the curve. What is the coefficient of static friction between the car's tires and the road?

78. N Consider a particle on the rim of a disk of radius R rotating around a fixed axis that passes through its center and is perpendicular to the face of the disk. The rim has a tangential acceleration that is constant in magnitude, a_T What is the direction of the total acceleration vector of this particle after it has made three complete rotations, beginning from rest? Give your answer to three significant figures.

Problems 79 through 81 are grouped.

79. N A uniform 4.55-kg horizontal rod is fixed on one end as shown in Figure P12.79. The rod is 1.75 m long. A rope is attached to the opposite end, and it makes a 30.0° angle with respect to the rod. What is the magnitude of the torque exerted by the tension in the rope around the fixed end of the rod?

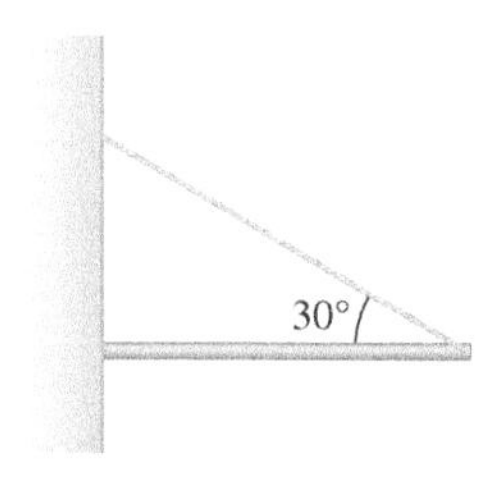

FIGURE P12.79 Problems 79–81.

80. N Consider the rod in Problem 79. What is the total torque around the fixed end of the rod? Could the rod be at rest?

81. N If the rod in Problem 79 rotates with a constant angular velocity, what is the tension in the rope at the instant shown?

82. N As a compact disc (CD) spins clockwise as seen from above, information is read from it, starting with the innermost ring and moving outward. When the information is being read from the innermost ring, the CD's angular speed is $\omega_0 = 52.4$ rad/s. The CD slows down so that when information is read from the outermost ring, $\omega = 20.9$ rad/s. It takes 74 min 33 s to read the music from a particular CD. Find the constant angular acceleration of the CD.

83. N A disk-shaped machine part has a diameter of 40.0 cm. Its angular position is given by $\theta = -1.20t^3 + 1.50t^2$, where t is in seconds and θ is in radians. **a.** What is the maximum angular speed of the part during this time interval? **b.** What is the maximum tangential speed of a point halfway to the rim of the part during this time interval? **c.** What is the time t for which the part reverses its direction of motion after $t = 0$? **d.** Through how many revolutions has the part turned when it reaches the time found in part (c)?

84. E CASE STUDY Wally Wallington is a retired carpenter who believes that he found a way for ancient people to have built Stonehenge. In fact, he has been reconstructing Stonehenge on his own using only simple tools such as wooden levers, stones, sand, water, ropes, and old paint drums. On his own, he has erected concrete columns weighing approximately 20,000 lb. He says that his passion for moving heavy objects with modest equipment started when he was working as a carpenter and had to remove 1200-lb concrete blocks. One solution would have been to break the blocks into smaller pieces and then use wheelbarrows to remove the pieces. Wally thought of a better way. He used levers to move the blocks; he found that the process became very easy with practice. Think of how you might reproduce Wally's original work. Assume he had some reasonably strong lengths of wood and a few old paint drums that he could fill with loose rock, sand, or water. Describe in detail how you might lift a 1200-lb block using a lever.

13 Rotation II: A Conservation Approach

Key Question

When and how can we apply the principles of conservation of energy and angular momentum to rotational motion?

Underlying Principles

1. Conservation of energy
2. Newton's second law
3. Conservation of angular momentum

Major Concepts

1. Rotational inertia
2. Parallel-axis theorem
3. Rotational kinetic energy
4. Work
5. Power
6. Angular momentum

Special Case

Pure rolling

13-1 Conservation approach 360

13-2 Rotational inertia 362

13-3 Rotational kinetic energy 367

13-4 Special case of rolling motion 369

13-5 Work and power 373

13-6 Angular momentum 376

13-7 Conservation of angular momentum 379

In Chapter 12, we began our study of rotational motion by considering the relatively uncomplicated situation of a rigid object rotating around a fixed axis. Rotational motion, however, can be much more complicated and exciting. First, objects such as Frisbees, bicycle wheels, and somersaulting divers rotate and translate at the same time. Because they translate, their rotation axis must also be in motion. Second, rotating objects are not necessarily rigid: for instance, collapsing stars, spinning figure skaters, and clay thrown on a potter's wheel all change shape as they rotate.

We've faced complicated motion before, and we have found that conservation principles are important tools for analyzing such situations. This chapter has two goals: (1) to find the conditions under which conservation principles may be applied to a rotating object or system and (2) to find the most convenient way to apply these principles.

13-1 Conservation Approach

So far, we have restricted our study of rotational motion to rotation around a fixed axis, analogous to translational motion restricted to one dimension. We also only considered rigid objects, analogous to studying translational motion of objects that can be approximated as particles. For more complicated translational motion such as the elliptical orbit of Comet Halley or the flight of a rocket, we needed the principles of conservation of energy and momentum. These same conservation principles hold for rotational motion, and they give us another way to analyze complicated problems.

CASE STUDY Accounting for All the Energy

The case study in this chapter is based on an experiment a student—Chris—performed in a physics laboratory. Chris applies the principle of conservation of energy to an object that both rotates and translates.

EXAMPLE 13.1 CASE STUDY Loop-the-Loop Laboratory Challenge

Chris's laboratory instructor gives each student an identical steel loop-the-loop setup as shown in Figure 13.1. The instructor has a solid steel ball that rolls on the track. The challenge is to calculate the minimum height h from which the ball can be released so as to just barely keep in contact with the track as it completes the loop. The instructor hands out the ball only when a student is ready to test his or her calculation. Each student will be given three opportunities. The student who has the smallest h in any one of his or her successful trials wins.

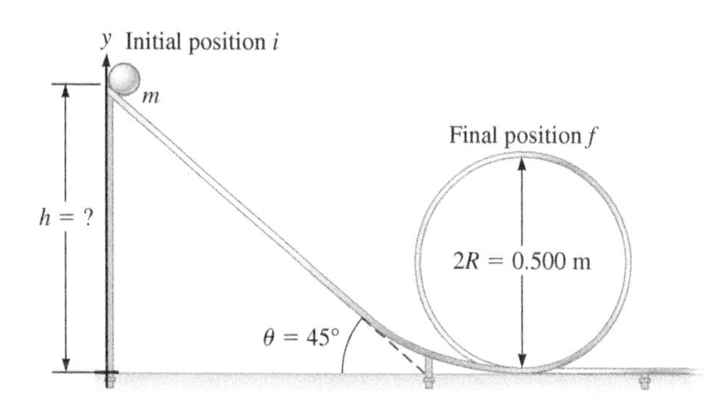

FIGURE 13.1 Experimental setup for a laboratory challenge.

INTERPRET and ANTICIPATE

Chris took a conservation of energy approach. Step 1: He included the Earth and the ball in the **system**. Step 2: He set the **reference configuration** and chose a **coordinate system**. Step 3: He set the **initial** position i to the place where the ball is released and the **final** position f to the top of the loop (Fig. 13.1), ignoring the small horizontal portion of the track. (Compare Chris's work here to Problem 9.56.)

SOLVE

Step 4 Many of the energy terms are zero as shown in Chris's **bar chart** (Fig. 13.2). The ball is released from rest, so $K_i = 0$. There are no external forces to do work on the system; therefore, $W_{tot} = 0$. Chris ignored internal thermal changes due to friction so that $\Delta E_{int} = 0$.

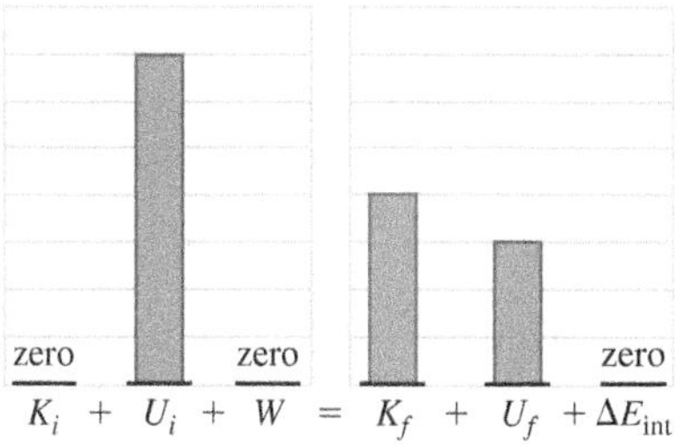

zero, zero, zero

$K_i + U_i + W = K_f + U_f + \Delta E_{int}$

FIGURE 13.2

Step 5 Chris wrote the **conservation of energy equation** using his bar chart.

$$0 + U_i + 0 = K_f + U_f + 0$$

Step 6 Then he used **other equations** to write the conservation of energy equation in terms of the variables shown in Figure 13.1.

$$mgh = \tfrac{1}{2}mv_f^2 + mg(2R) \qquad (13.1)$$

If the ball just barely makes contact with the track at the top of the loop, the only force acting on the ball at that moment is gravity. The ball's centripetal acceleration equals the acceleration due to gravity.

$$a_c = \frac{v_f^2}{R} = g \qquad (13.2)$$

Step 7 **Do algebra.** Chris substituted Equation 13.2 into Equation 13.1 and simplified to find h.

$$mgh = \tfrac{1}{2}mv_f^2 + 2mgR$$

$$mgh = \tfrac{1}{2}mgR + 2mgR = \tfrac{5}{2}mgR$$

$$h = \tfrac{5}{2}R = \tfrac{5}{2}(0.250\,\text{m})$$

$$h = 0.625\text{ m} = 62.5\text{ cm}$$

Example continues on page 362 ▶

CHECK and THINK

When Chris tested his calculation, the ball fell off the track. He figured the problem was that he had ignored rolling friction.

SOLVE

Step 4 Chris made a new bar chart (Fig. 13.3), taking into account changes in internal thermal energy: $\Delta E_{\text{int}} = \Delta E_{\text{th}}$.

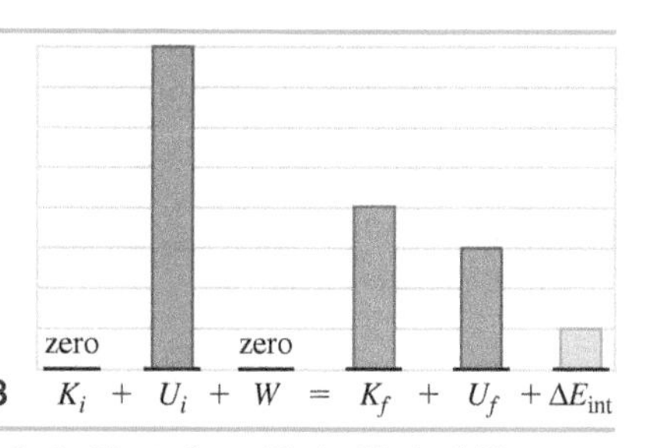

FIGURE 13.3

Step 5 He wrote the **conservation of energy equation.**

$$0 + U_i + 0 = K_f + U_f + \Delta E_{\text{th}}$$

Step 6 To find ΔE_{th}, he used $\Delta E_{\text{th}} = F_k s$ (Eq. 9.29) with rolling friction substituted for kinetic friction: $F_k \rightarrow F_r$. He assumes that the magnitude of the normal force is $F_N = F_g \cos\theta$.

$$\Delta E_{\text{th}} = F_r s = \mu_r F_N s$$
$$\Delta E_{\text{th}} = \mu_r (F_g \cos\theta) s$$
$$\Delta E_{\text{th}} = \mu_r (mg \cos\theta) s$$

The total distance s traveled from i to f is the distance $h/\sin\theta$ down the incline plus the distance πR around half the circular loop. The normal force is *not* constant as the ball travels over the circular loop. So, Chris's assumption that the normal force is $mg\cos\theta$ over the whole trip means that he is making an estimate.

$$\Delta E_{\text{th}} = \mu_r mg \cos\theta\left(\frac{h}{\sin\theta} + \pi R\right)$$
$$\Delta E_{\text{th}} = \mu_r mg \cos 45^\circ\left(\frac{h}{\sin 45^\circ} + \pi R\right)$$
$$\Delta E_{\text{th}} = \mu_r mg\left(h + \frac{\pi}{\sqrt{2}}R\right) \quad (13.3)$$

Step 7 **Do algebra.** Now, taking rolling friction into account, Chris substituted ΔE_{th} into the conservation of energy equation and solved for h. He chose the largest value for the coefficient of rolling friction for steel on steel from Table 6.1: $\mu_r = 0.002$.

$$mgh = \tfrac{1}{2}mv_f^2 + 2mgR + \mu_r mg[h + (\pi/\sqrt{2})R]$$
$$mgh = \tfrac{5}{2}mgR + \mu_r mg[h + (\pi/\sqrt{2})R]$$
$$h = \frac{[\tfrac{5}{2} + \mu_r(\pi/\sqrt{2})]R}{1 - \mu_r} = \frac{[\tfrac{5}{2} + (0.002)(\pi/\sqrt{2})](0.25\text{ m})}{1 - 0.002}$$
$$h = 0.627\text{ m} = 62.7\text{ cm}$$

CHECK and THINK

This result is only 2 mm higher than Chris's previous calculation, so apparently not much mechanical energy is lost to rolling friction in this case. Chris's instructor gave him the steel ball to test his second calculation, but again the ball fell off the track.

The trouble is that Chris has missed part of the ball's motion. We will return to Chris's experiment to find the missing piece later in this chapter. (His assumption that the normal force is a constant over the whole path does not account for the ball falling off the track.)

CONCEPT EXERCISE 13.1

CASE STUDY When Is Energy Conserved?

Under what conditions is a system's energy conserved? Review Chapter 8 if necessary. Was Chris wrong to think that energy is conserved in the apparatus in Figure 13.1?

ROTATIONAL INERTIA

Major Concept

13-2 Rotational Inertia

In Section 12-7, we introduced rotational inertia as a measure of an object's tendency to maintain its angular velocity. In translational motion, a more massive particle has more inertia, so we expect that a more massive object also has more rotational inertia. It turns out that mass is only one factor that determines an object's rotational inertia.

A clue to the second factor comes from Example 12.6 (page 349). Crall and Whipple measured the rotational inertia of a dumbbell-shaped object (Figs. 12.26–12.28; also Fig. 13.4) and found that when the lead cylinders were close together, the rotational inertia of the dumbbell was reduced. The two factors in an object's rotational inertia are (1) its mass and (2) how its mass is distributed with respect to the rotation axis. An object whose mass is far from the axis of rotation has a larger rotational inertia than an object with the same mass located closer to the rotation axis. Imagine starting to rotate Crall and Whipple's dumbbell; it is easier to rotate the dumbbell when the cylinders are closer to the axis of rotation (Fig. 13.4B).

The rotational inertia of an object that consists of n particles may be found mathematically from

$$I = \sum_{i=1}^{n} m_i r_i^2 \tag{13.4}$$

where m_i is the mass of the ith particle and r_i is that particle's distance from the rotation axis. The SI unit of rotational inertia is $\text{kg} \cdot \text{m}^2$.

EXAMPLE 13.2 Crall and Whipple's Dumbbells Revisited

Crall and Whipple measured the rotational inertia of a system of two lead cylinders (Example 12.6, page 349) in two different configurations. Each lead cylinder has a mass of 0.200 kg. For each of the configurations, use Crall and Whipple's experimental results (shown in Fig. 13.4) to determine the distance d between each cylinder's center of mass for the two different configurations. Assume the cross bar and axle are very light compared with the cylinders so that their contribution to the rotational inertia of the object may be ignored.

INTERPRET and ANTICIPATE

The object is modeled as two particles of mass $m = 0.200$ kg. Each particle is a distance r from the rotation axis. Use the rotational inertia experimentally determined by Crall and Whipple to solve for r in Equation 13.4. Our answer should match the configurations shown; that is, we expect $d_a > d_b$.

SOLVE

Determine the distance d_a in Figure 13.4A. Equation 13.4 takes a simple form because both particles have the same mass m and are the same distance r from the rotation axis.

$$I = \sum_{i=1}^{n} m_i r_i^2 = m_1 r_1^2 + m_2 r_2^2 \tag{13.4}$$

$$I = mr^2 + mr^2 = 2mr^2$$

$$r = \sqrt{\frac{I}{2m}} \tag{13.5}$$

Crall and Whipple found $I = 1.5 \times 10^{-2}\ \text{kg} \cdot \text{m}^2$ for the object in Figure 13.4A. The distance between the cylinders is twice their distance from the axis of rotation, $d_a = 2r_a$.

$$r_a = \sqrt{\frac{1.5 \times 10^{-2}\ \text{kg} \cdot \text{m}^2}{2(0.200\ \text{kg})}}$$

$$r_a = 0.194\ \text{m} = 19.4\ \text{m}$$

$$d_a = 39\ \text{cm}$$

Now determine the distance d_b. For Figure 13.4B, Crall and Whipple found $I = 3.8 \times 10^{-3}\ \text{kg} \cdot \text{m}^2$.

$$r_b = \sqrt{\frac{3.8 \times 10^{-3}\ \text{kg} \cdot \text{m}^2}{2(0.200\ \text{kg})}}$$

$$r_b = 9.75 \times 10^{-2}\ \text{m} = 9.75\ \text{cm}$$

$$d_b = 19\ \text{cm}$$

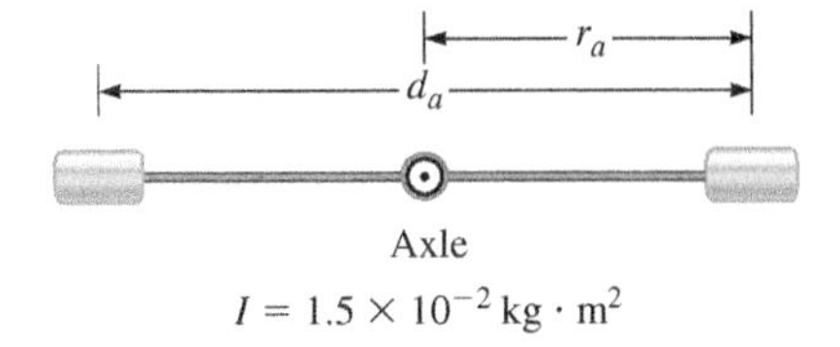

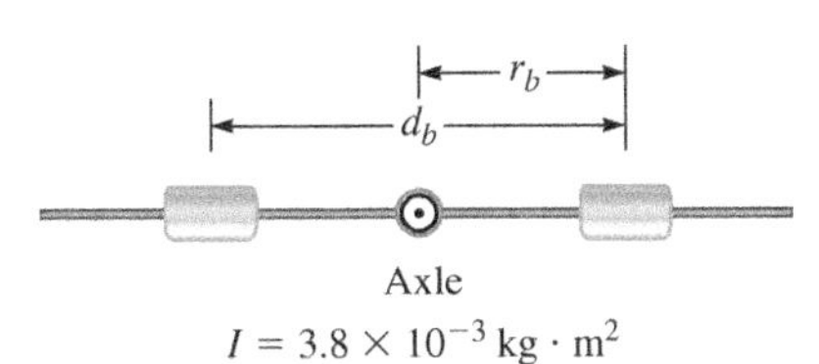

FIGURE 13.4 Crall and Whipple measured the rotational inertia of a dumbbell-shaped object. Rotational inertia depends both on mass and the location of the mass with respect to the axis of rotation.

CHECK and THINK

As expected, $r_a > r_b$ and $d_a > d_b$. In fact, $r_a \approx 2r_b$, and the rotational inertia of the larger dumbbell is about four times greater than that of the small dumbbell even though the two dumbbells are identical in mass. Clearly, the distribution of mass within an object is a major factor in determining the object's rotational inertia.

Rotation Axis Must Be Specified

Because the distribution of an object's mass around the rotation axis is important in determining its rotational inertia, the location of the rotation axis must be specified when reporting a value for I. In other words, it is not good enough to say the rotational inertia of the dumbbell in Figure 13.4A is $I = 1.5 \times 10^{-2}\ \text{kg} \cdot \text{m}^2$ without specifying that the rotation axis is along the axle at the center of the dumbbell (perpendicular to the rod joining the two cylinders). In fact, if a different rotation axis is used, the rotational inertia of the dumbbell is not equal to what Crall and Whipple found experimentally. In the following example, we find the rotational inertia of an object around two different rotation axes.

EXAMPLE 13.3 Rotational Inertia: Same Object, Different Axes

Figure 13.5 shows an object made up of three lead spheres attached to three lightweight, stiff rods. The mass of the rods is negligible compared with the mass of the spheres. Each of the three spheres has mass m, and each lies at the vertex of an equilateral triangle of side ℓ. Find expressions for the rotational inertia of this object around an axis that is perpendicular to the plane of the triangle, (1) first passing through the object's center of mass (Fig. 13.5A) and then (2) passing through a vertex (Fig. 13.5B).

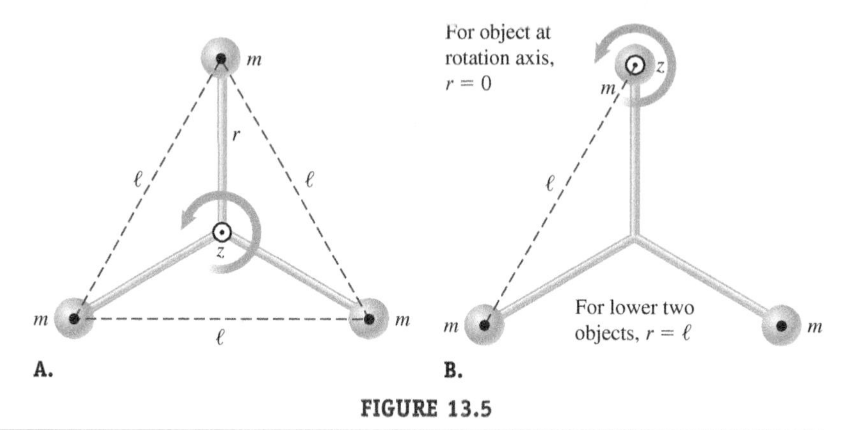

FIGURE 13.5

INTERPRET and ANTICIPATE

To find the rotational inertia around each axis, we need the distance r of each sphere with respect to the rotation axis. In Figure 13.5A, r is the same for each sphere; in Figure 13.5B, however, r is zero for one of the spheres. We expect to find two algebraic expressions for the rotational inertia with dimensions $M \cdot L^2$.

SOLVE

In Figure 13.5A, each mass is attached to a rod that runs to the center of mass of the three-sphere system. Using Figure 13.6, we can express the length of each rod in terms of the length ℓ of each side of the equilateral triangle as $r = \ell/\sqrt{3}$.

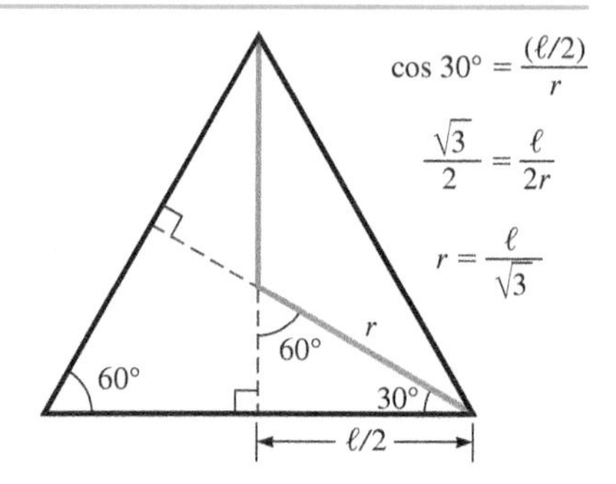

FIGURE 13.6 Geometry for finding r in Figure 13.5A.

The rotational inertia around the center of mass comes from Equation 13.4.

$$I = \sum_{i=1}^{n} m_i r_i^2 = \sum_{i=1}^{3} m_i r_i^2 \quad (13.4)$$

$$I = m\left(\frac{\ell}{\sqrt{3}}\right)^2 + m\left(\frac{\ell}{\sqrt{3}}\right)^2 + m\left(\frac{\ell}{\sqrt{3}}\right)^2$$

$$I_{CM} = m\ell^2 \quad (13.6)$$

In Figure 13.5B, now the axis of rotation passes through one of the three spheres. For that sphere, $r = 0$. For the other two spheres, $r = \ell$.

$$I = \sum_{i=1}^{n} m_i r_i^2 = \sum_{i=1}^{3} m_i r_i^2$$

$$I = m(0)^2 + m\ell^2 + m\ell^2$$

$$I_{\text{vertex}} = 2m\ell^2 \quad (13.7)$$

CHECK and THINK

The expressions we found both have the correct dimensions: $M \cdot L^2$. If the axis of rotation passes through a vertex of the triangle (Fig. 13.5B), the rotational inertia is twice the rotational inertia found when the axis of rotation passes through the center of mass (Fig. 13.5A): $I_{\text{vertex}} = 2I_{CM}$. This result makes some sense. When we rotate the system around the vertex, the mass is on average farther from the rotation axis than when we rotate the system around the center of mass. Just moving the axis of rotation changes the rotational inertia of the object. So, if you want the system to have a particular angular acceleration, it takes twice the torque to get that angular acceleration if the axis of rotation passes through a vertex instead of through the center of mass.

Rotational Inertia of Continuous Objects

The summation in Equation 13.4 is only practical if we need to calculate the rotational inertia of an object that can be modeled as a small collection of particles. To calculate the rotational inertia of a continuous object like a bicycle tire, a Frisbee, or baseball bat, we must model the continuous object as a large number of infinitesimal pieces of mass dm (Fig. 13.7). Each infinitesimal piece is a perpendicular distance r from the axis of rotation. The rotational inertia of each infinitesimal particle around the axis is

$$dI = r^2 dm$$

To find the rotational inertia of the whole object, we must add up the contribution of each infinitesimal particle using integration:

$$\int dI = \int r^2 dm$$

$$I = \int r^2 dm \tag{13.8}$$

If the mass of the object is uniformly distributed and the shape of the object is simple and symmetric such as a hoop, disk, or straight rod, Equation 13.8 may be solved with a pencil and paper in a few lines of calculation. Table 13.1 gives the rotational inertia of nine symmetric objects around axes that pass through the object's center of mass, found from Equation 13.8. If the object is more complicated, such as a hammer, gymnast, or airplane, it may be necessary to either model the object as a collection of simpler objects or solve Equation 13.8 numerically with the aid of a computer.

In some situations, we may be able to apply a very useful shortcut. Suppose we know the rotational inertia of some object (say a hammer; Fig. 13.8) around an axis through its center of mass, but we need to know the rotational inertia around a dif-

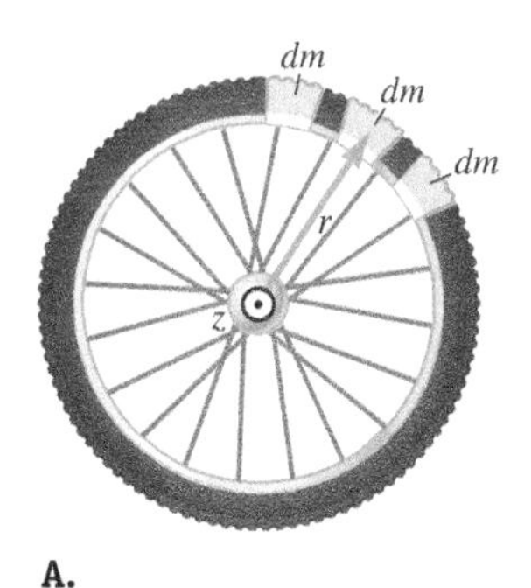

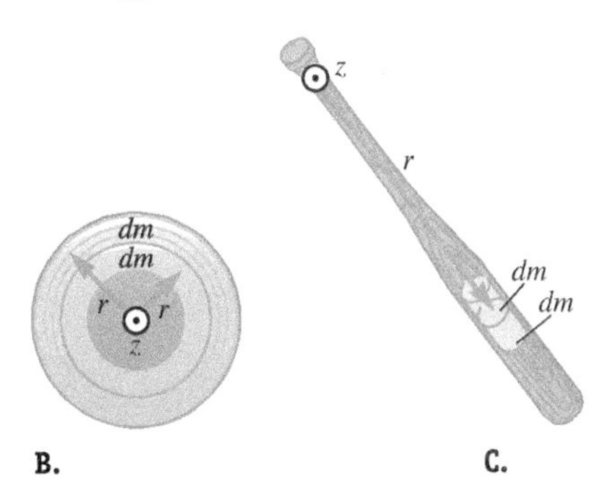

FIGURE 13.7 To find the rotational inertia, we model a continuous object as a large number of particles of mass dm, exploiting the symmetry of the object (if present).

TABLE 13.1 Rotational inertia of various objects.

Hoop around the central axis $I_{CM} = MR^2$ Rotation axis, R	Thin spherical shell around any diameter $I_{CM} = \frac{2}{3}MR^2$ Rotation axis, R	Thin rod around axis through center and perpendicular to rod $I_{CM} = \frac{1}{12}ML^2$ Rotation axis, L
Hollow cylinder around the central axis $I_{CM} = \frac{1}{2}M(R_1^2 + R_2^2)$ Rotation axis, R_1, R_2	Solid sphere around any diameter $I_{CM} = \frac{2}{5}MR^2$ Rotation axis, R	Solid cylinder or disk around a central diameter $I_{CM} = \frac{1}{4}MR^2 + \frac{1}{12}ML^2$ Rotation axis, R, L
Hoop around any diameter $I_{CM} = \frac{1}{2}MR^2$ Rotation axis, R	Thin rectangular solid around axis through center $I_{CM} = \frac{1}{12}M(a^2 + b^2)$ Rotation axis, a, b	Solid cylinder or disk around the central axis $I_{CM} = \frac{1}{2}MR^2$ Rotation axis, R, L

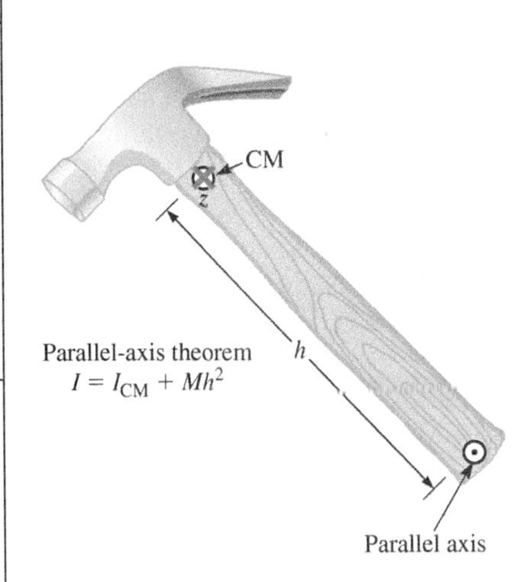

FIGURE 13.8 If you know the rotational inertia I_{CM} around an axis through an object's center of mass (CM), the parallel-axis theorem may be used to find the rotational inertia I around a new axis parallel to the original axis.

ferent axis. If that axis is parallel to the axis that passes through the center of mass, we can use the *parallel-axis theorem* to find the new rotational inertia. If M is the mass of the object, h is the perpendicular distance between the new axis and the axis through the center of mass, and I_{CM} is the rotational inertia around the axis passing through the center of mass, the rotational inertia around the new axis is

PARALLEL AXIS THEOREM

Major Concept

$$I = I_{CM} + Mh^2 \quad (13.9)$$

according to the **parallel-axis theorem**.

EXAMPLE 13.4 Rotational Inertia of a Bicycle Tire

Model a bicycle tire as a hoop of radius R. Use $I = \int r^2 dm$ (Eq. 13.8) to find the rotational inertia of the bicycle tire around the axis.

INTERPRET and ANTICIPATE

Model the object as a large number of infinitesimal pieces (Fig. 13.7A). We expect our results to be consistent with Table 13.1.

SOLVE

The distance from the axis of rotation to each infinitesimal piece equals the radius of the tire: $r = R$.	$I = \int r^2 dm = \int R^2 dm$
Because R is a constant for each piece of mass dm, we can bring R outside the integral.	$I = R^2 \int dm$
Integration over all pieces of mass dm equals the total mass M of the tire.	$I_{tire} = MR^2$

CHECK and THINK

The rotational inertia of the tire equals the rotational inertia of a hoop as shown in Table 13.1. We often need to know the rotational inertia of a hoop or disk around its center of mass. A disk of the same mass and radius has half the rotational inertia of a hoop because the disk's mass is distributed such that on average it is closer to the axis of rotation (Table 13.1, lower right).

EXAMPLE 13.5 Using the Parallel-Axis Theorem

In Example 13.3 (page 364), we found the rotational inertia of an object consisting of three spheres lying at the vertices of an equilateral triangle around an axis perpendicular to the plane of the triangle and passing through the object's center of mass (Fig. 13.5A) is $I_{CM} = m\ell^2$. Now use the parallel-axis theorem to once again find rotation inertia around an axis passing through a vertex (Fig. 13.5B).

INTERPRET and ANTICIPATE

Start with $I_{CM} = m\ell^2$ (Eq. 13.6) and use the parallel-axis theorem to find the rotational inertia around the axis shown in Figure 13.5B. We expect our result to be consistent with Example 13.3 ($I_{vertex} = 2m\ell^2$).

SOLVE

Figure 13.6 shows that the perpendicular distance between the two axes of rotation is the same as the distance r from one vertex to the center of mass.	$h = r = \frac{\ell}{\sqrt{3}}$
Substituting for I_{CM} and h into the parallel-axis theorem (Eq. 13.9) gives the rotational inertia around the axis through the vertex.	$I = I_{CM} + Mh^2 \quad (13.9)$ $I_{vertex} = m\ell^2 + 3m\left(\frac{\ell}{\sqrt{3}}\right)^2$ $I_{vertex} = m\ell^2 + 3m\left(\frac{\ell^2}{3}\right) = m\ell^2 + m\ell^2 = 2m\ell^2$

CHECK and THINK
As expected, this result is exactly what we found in Example 13.3 (Eq. 13.7). Often, the parallel axis theorem makes finding the rotational inertia much easier than using Equation 13.4 or Equation 13.8 directly.

EXAMPLE 13.6 "Swing, Batter!"

Model the baseball bat in Figure 13.7C as a thin rod of length L. Find the rotational inertia around an axis through one end of the bat.

INTERPRET and ANTICIPATE
Table 13.1 lists the rotational inertia of a thin rod around an axis through the center of the rod. We can use that information and the parallel-axis theorem to find the rotational inertia around an axis that runs through one end of the rod. Our result should be an algebraic expression with dimensions $M \cdot L^2$. Because more of the mass is farther from the new axis of rotation, we expect our expression will be greater than the rotational inertia given in Table 13.1 for an axis through the center.

SOLVE
The rotational inertia around the center of mass comes from Table 13.1. The distance h is half the length of the rod, $h = L/2$.

$$I = I_{CM} + Mh^2$$

$$I = \frac{1}{12}ML^2 + M\left(\frac{L}{2}\right)^2$$

$$I = \frac{1}{12}ML^2 + \frac{1}{4}ML^2 = \frac{1}{3}ML^2$$

CHECK and THINK
As expected, the rotational inertia is greater for an axis through the end of the bat: $I > I_{CM}$. Sometimes, a coach will instruct a weak batter to "choke up" on the bat, that is, to move his or her hands closer to the bat's center of mass. This change reduces the rotational inertia so that the bat is easier to rotate or "swing." (Of course, a real bat is more complicated because it is not a uniform rod.)

13-3 Rotational Kinetic Energy

All moving objects have kinetic energy. If the object may be modeled as a translating particle, its kinetic energy is easily found from $K = \frac{1}{2}mv^2$ (Eq. 8.1). If the object is rotating, however, Equation 8.1 is inadequate because different parts of the object move with different translational speeds v around the rotation axis.

ROTATIONAL KINETIC ENERGY
Major Concept

DERIVATION Rotational Kinetic Energy

For any object or system with rotational inertia I rotating at angular speed ω, the rotational kinetic energy K_r is given by

$$K_r = \frac{1}{2}I\omega^2 \quad (13.10)$$

To keep our derivation simple, we will show that this expression holds in the case of an object that consists of just two particles. For example, Figure 13.9 shows two small balls attached to a lightweight, stiff rod.

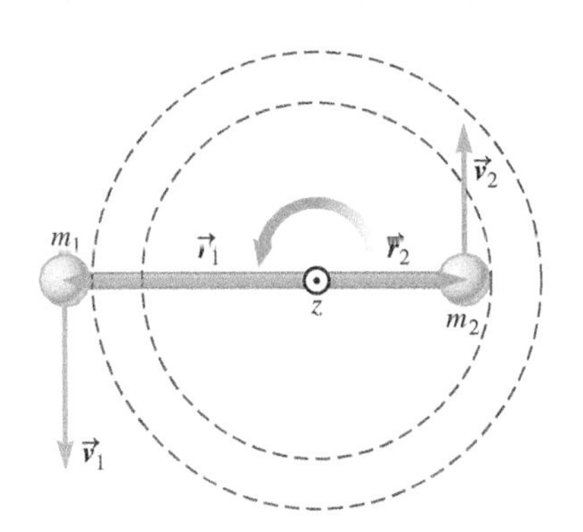

FIGURE 13.9 A dumbbell-shaped object is modeled as two particles moving in concentric circles.

Recall from Chapter 12 that one way to model a rotating object is as a collection of particles moving in concentric circles. So, each ball moves on a circular path around the z axis at the speed indicated in the figure. The total kinetic energy of this two-particle system comes from Equation 8.2.

$$K_r = \sum_{i=1}^{n} \frac{1}{2}m_i v_i^2$$

$$K_r = \frac{1}{2}m_1 v_1^2 + \frac{1}{2}m_2 v_2^2$$

Derivation continues on page 368 ▶

Both balls have the same angular speed ω. Using $v = \omega r$ (Eq. 12.13), we rewrite their translational speeds in terms of their angular speed.	$K_r = \frac{1}{2}m_1(\omega r_1)^2 + \frac{1}{2}m_2(\omega r_2)^2$ $K_r = \frac{1}{2}(m_1 r_1^2 + m_2 r_2^2)\omega^2$ (1)
The term in parentheses in Equation (1) is the rotational inertia of this two-particle object from Equation 13.4.	$I = \sum_{i=1}^{n} m_i r_i^2$ (13.4) $I = m_1 r_1^2 + m_2 r_2^2$ (2)
Substitute Equation (2) into Equation (1) to find the expression for rotational kinetic energy.	$K_r = \frac{1}{2}I\omega^2$ ✓ (13.10)

COMMENTS

We found the rotational kinetic energy of a simple two-particle system, but $K_r = \frac{1}{2}I\omega^2$ (Eq. 13.10) holds for any rotating object or system. It is easy to remember this expression because it is so similar to $K = \frac{1}{2}mv^2$; the inertia m is replaced by the *rotational* inertia I, and the speed v is replaced by the *angular* speed ω.

CONSERVATION OF ENERGY

Underlying Principle

The **conservation of energy** principle is slightly modified to include rotational motion. The initial mechanical energy (translational kinetic energy K, **rotational kinetic energy** K_r, and potential energy U) of the system plus the total work W done on the system by external forces must equal the final mechanical energy of the system plus changes in the system's internal (thermal) energy ΔE_{th}:

$$K_i + K_{ri} + U_i + W = K_f + K_{rf} + U_f + \Delta E_{th}$$

So, in practice, when we apply this modified version of the conservation of energy principle, we must include an extra bar for rotational kinetic energy in our bar chart as we do in the following example.

EXAMPLE 13.7 An Atwood Machine

Pulleys are found on the end of cranes, in the resistance machines in the gym, and in the hand reels on fishing poles. So far, we have only considered very light pulleys, ignoring their rotational inertia. (For instance, see Example 5.9, page 140.) Now, we consider an Atwood machine (Fig. 13.10). A light rope runs over a pulley and connects two objects of masses m and $2m$. The pulley is modeled as a solid disk of radius R and mass $2m$. When the lighter object is displaced upward by an amount Δy, the heavier object is displaced downward by the same distance. Therefore, it is convenient to use a coordinate axis that bends along the rope (similar to the coordinate system used in Example 5.9). The system starts at rest (Fig. 13.10A). Assume there are no energy losses. Find an expression for the angular speed of the pulley at a later time (Fig. 13.10B) in terms of Δy, R, and g.

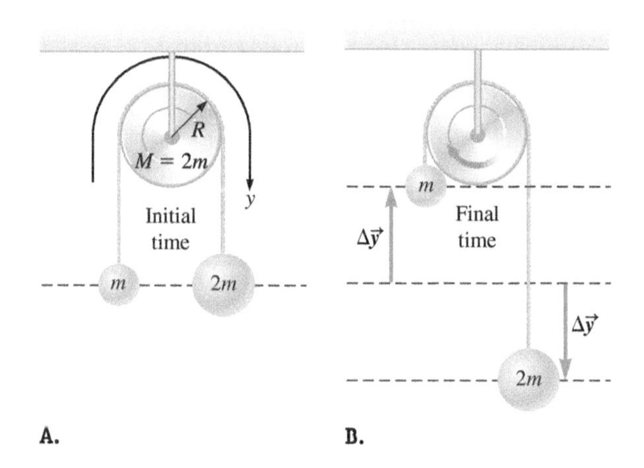

FIGURE 13.10 An Atwood machine.

INTERPRET and ANTICIPATE

Step 1 Using an energy approach, we choose a system including the hanging objects, the pulley, and the Earth. **Steps 2** (choosing a coordinate system) and **3** (indicating the initial and final times) were done for us.

Step 4 **Bar chart.** To save space, the bars for the various components of the system on the left have been combined (Fig. 13.11). On the right, however, we must consider the components (each hanging object and the pulley) separately. As shown in the bar chart, the change in system's thermal energy is zero. No external forces do work on the system. After the system is released, the lighter object rises, and the heavier object falls. The net result is that the system loses gravitational potential energy as demonstrated by adding up the two bars drawn for U_g. Both hanging objects have translational kinetic energy, and the pulley has rotational kinetic energy. Because the hanging masses are attached to each other by the rope, they have the same translational speed v

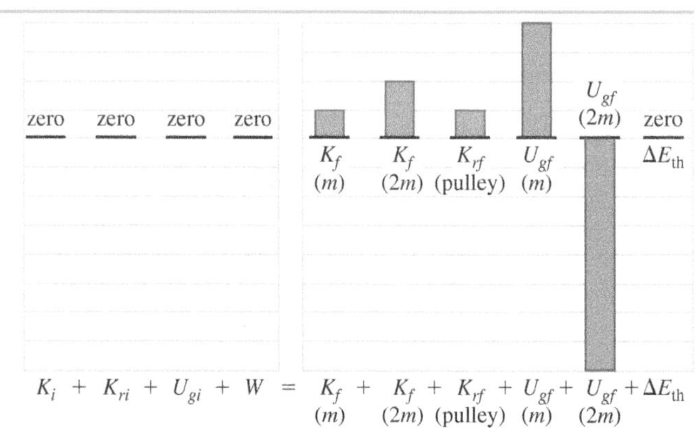

FIGURE 13.11

SOLVE

Step 5 Aided by the bar chart, write an **equation for the conservation of energy**.

$$0 = K_f + K_{rf} + U_{gf}$$

$$0 = \tfrac{1}{2}mv^2 + \tfrac{1}{2}(2m)v^2 + \tfrac{1}{2}I\omega^2 + mg\Delta y - 2mg\Delta y$$

$$0 = \tfrac{3}{2}mv^2 + \tfrac{1}{2}I\omega^2 - mg\Delta y \quad (1)$$

Step 6 **Other equations.** The rotational inertia of the pulley comes from Table 13.1 (lower right). The translational speed v of the hanging objects is proportional to the angular speed of the pulley (Eq. 12.13).

$$I = \tfrac{1}{2}MR^2 = \tfrac{1}{2}(2m)R^2 = mR^2$$

$$v = \omega R \quad (12.13)$$

Step 7 **Do algebra.** Substitute for v and I in Equation (1).

$$0 = \tfrac{3}{2}m\omega^2R^2 + \tfrac{1}{2}mR^2\omega^2 - mg\Delta y$$

$$mg\Delta y = 2m\omega^2R^2$$

$$\omega = \frac{1}{R}\sqrt{\frac{g\Delta y}{2}}$$

CHECK and THINK

Does the result make sense? It makes sense that ω depends directly on Δy because as Δy increases, more gravitational potential energy is converted into kinetic energy, increasing ω. Also, if Δy is zero, the hanging blocks are at rest, and the system has no kinetic energy. It also makes sense that ω is inversely proportional to R because it is more difficult to rotate a larger disk, so we expect a larger disk to have a lower angular speed. (As always, you should confirm that the dimensions are correct.)

13-4 Special Case of Rolling Motion

We have studied pure translational motion and pure rotational motion. Rolling motion is a combination of these two. Imagine, for instance, a game of Frisbee. The Frisbee's center of mass translates from the thrower to the catcher; meanwhile, the Frisbee also rotates around an axis passing through its center of mass. The axis is not fixed; rather, it moves at the center-of-mass velocity. There are many other examples of a combination of rotational and translational motion: a gymnast doing a backflip across a balance beam, a stunt plane corkscrewing across the sky, and a graduation hat being tossed high into the air. In this section, we consider one special case of combination motion, rolling without slipping.

Imagine an object such as an automobile or truck tire rolling along a flat, horizontal surface without slipping. If we attach a light to the tire rim, then a long-exposure photo of the tire as it rolls reveals a continuous-motion diagram (Fig. 13.12). The tire's axle is along the rotation axis passing through the tire's center of mass. The center of mass moves along a straight line, parallel to the images formed by the running light and taillight in Figure 13.12. By contrast, the path of the light attached at the rim forms a repeating curve. The shape of this curve comes from the combination of the tire's rotational and translational motion.

FIGURE 13.12 A light attached to the rim of a truck or automobile tire describes a repeating curved pattern as the tire rolls.

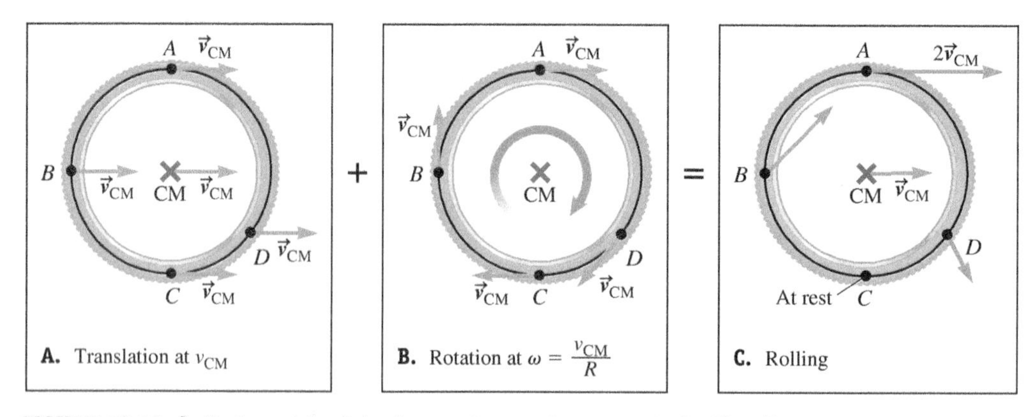

FIGURE 13.13 **A.** Each particle of the tire translates at the same velocity $\vec{v}_{CM}$. **B.** Each particle on the rim rotates around the center of mass at $\omega = v_{CM}/R$. **C.** Pure rolling can be modeled as translational motion plus rotational motion; so the velocity at each point is vector sum of the velocities from parts A and B. Notice that the top of the tire has the highest speed $2v_{CM}$, and the point of contact with the horizontal surface is momentarily at rest.

Figure 13.13 further illustrates pure rolling as a special case of motion that combines translation and rotation. Because rolling is a combination of translating (Fig. 13.13A) and rotating (Fig. 13.13B), the velocity of each point comes from adding up the velocity vectors for translating and rotating (Fig. 13.13C). The speed at the center of mass is v_{CM}. At the top of the tire in Figure 13.13C (point A), the speed is $2v_{CM}$, so point A is momentarily the fastest-moving point on the tire. At the bottom, where the tire is in contact with the road (point C), the speed is zero as the photograph in Figure 13.14 shows. Even though we know that the tire continues to roll, zero velocity at point C is another way to say that the tire rolls *without slipping*. If the tire slips as it rolls, the velocity of the point of contact would be nonzero.

FIGURE 13.14 The tire and spokes are blurry near the top and are clear near the bottom, showing that the point of contact between the tire and the ground is at rest.

Because the tire is rolling without slipping, it is relatively easy to connect the translational and rotational motions mathematically. Figure 13.15 shows how a string may be used to measure the center-of-mass displacement of a rolling tire. After one revolution, the magnitude of the center-of-mass position measured from the origin equals the length of the string. In this case, the string is as long as the circumference of the tire, so $x_{CM} = 2\pi R$.

What happens if the tire does not complete exactly one revolution, but instead rotates through an angle θ? In that case, the center-of-mass position does not equal the circumference of the tire; it equals the arc length s:

$$x_{CM} = s = \theta R \tag{13.11}$$

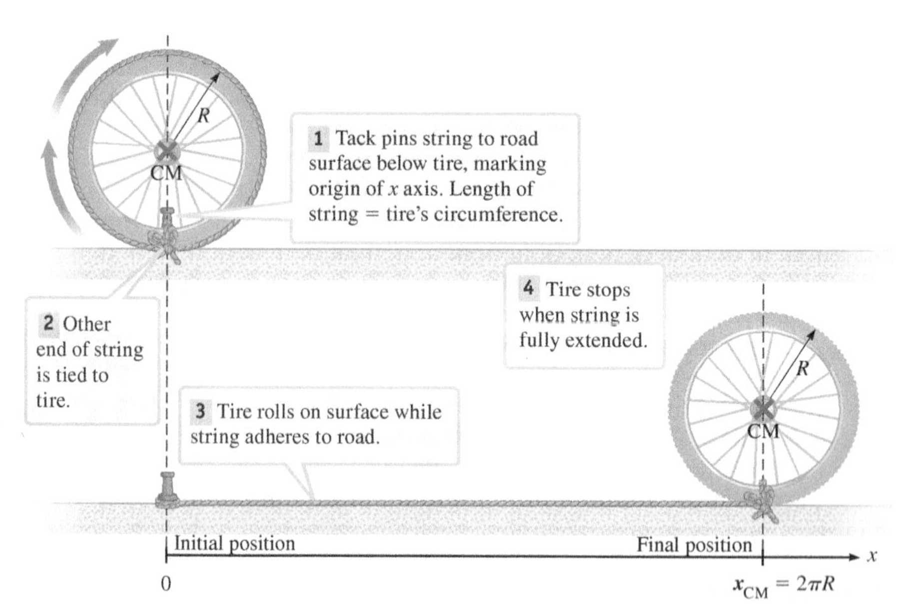

FIGURE 13.15 A string is tied to the tire and tacked to the surface. The tire rolls along the surface through one complete revolution. The tire's center-of-mass displacement equals its circumference, $x_{CM} = 2\pi R$.

The center-of-mass speed v_{CM} comes from taking the time derivative of the center-of-mass position:

$$v_{CM} = \frac{dx_{CM}}{dt} = \frac{d(\theta R)}{dt}$$

Of course, the tire's radius does not change as it rolls, so we bring R outside the derivative and recognize $d\theta/dt$ as the angular speed ω (Eq. 12.4):

$$v_{CM} = R\frac{d\theta}{dt}$$

$$v_{CM} = \omega R \tag{13.12}$$

PURE ROLLING Special Case

In Equation 13.12, v_{CM} is the translational speed of the center of mass parallel to the surface and ω is the angular speed around an axis through the center of mass.

If an object rolls without slipping, Equation 13.12 applies, and we describe the object's motion as **pure rolling**. According to this equation, pure rolling is a special combination of translational and rotational motion in which the center-of-mass speed v_{CM} is determined by the angular speed ω and the radius R of the object.

The magnitude of the center-of-mass acceleration a_{CM} comes from taking the time derivative of Equation 13.12:

$$a_{CM} = \frac{dv_{CM}}{dt} = \frac{d(\omega R)}{dt} = R\frac{d\omega}{dt}$$

where $d\omega/dt$ is the angular acceleration α (Eq. 12.6):

$$a_{CM} = \alpha R \tag{13.13}$$

If an object's motion is described by a combination of rotation around an axis through its center of mass and translation of the center of mass, its kinetic energy must be given by the sum of Equations 8.1 and 13.10:

$$K = K_{CM} + K_r = \tfrac{1}{2}mv_{CM}^2 + \tfrac{1}{2}I_{CM}\omega^2$$

In the special case of a pure rolling, the center-of-mass speed and the angular speed are connected by Equation 13.12.

EXAMPLE 13.8 CASE STUDY Loop-the-Loop Laboratory Challenge Revisited

Return to the loop-the-loop laboratory challenge in Example 13.1 on page 361 and help Chris find the winning value for h.

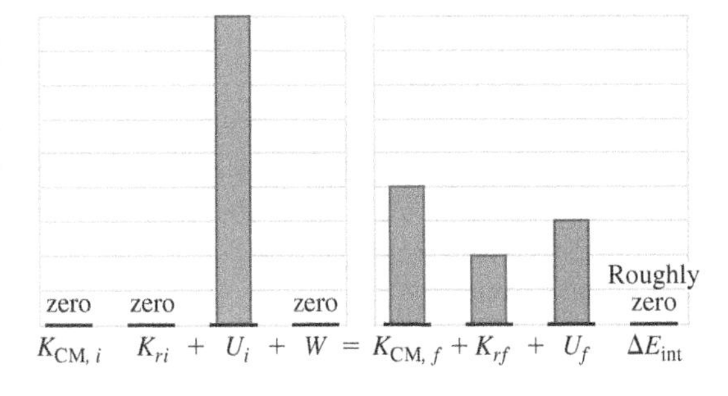

FIGURE 13.16

INTERPRET and ANTICIPATE

Chris failed to take the rotational kinetic energy of the ball into account. The ball is rolling, so it has translational and rotational kinetic energy (Fig. 13.16). Chris already found that rolling friction did not make much of a difference to his results. We will ignore rolling friction here. So, we start with Step 4, making a new **bar chart**.

SOLVE

Step 5 Chris should rewrite the conservation of energy equation using this new bar chart. Refer to Figure 13.1 for the initial and final positions of the ball.

$$K_i + U_i + W_{tot} = K_f + U_f + \Delta E_{int}$$

$$0 + U_i + 0 = K_{CM,f} + K_{rf} + U_f + 0$$

$$mgh = \tfrac{1}{2}mv_f^2 + \tfrac{1}{2}I_{CM}\omega_f^2 + 2mgR \tag{13.14}$$

Step 6 **Other equations.** The ball's rotational inertia is found in Table 13.1. Its angular speed at the top of the loop is given by $v_{CM} = \omega R$ (Eq. 13.12).

$$I_{CM} = \tfrac{2}{5}mR^2 \qquad \omega_f = \frac{v_f}{R}$$

Example continues on page 372 ▶

Step 7 **Do algebra.** Substitute into Equation 13.14 and solve for h.	$mgh = \frac{1}{2}mv_f^2 + \frac{1}{2}\left(\frac{2}{5}mR^2\right)\left(\frac{v_f}{R}\right)^2 + 2mgR$ $mgh = \frac{7}{10}mv_f^2 + 2mgR$ (1)
Now we need Equation 13.2, a result from Chris's earlier work. If the ball just barely makes contact with the track at the top of the loop, the ball's centripetal acceleration equals the acceleration due to gravity.	$a_c = \frac{v_f^2}{R} = g$ (13.2) $v_f^2 = gR$
Substitute for v_f^2 in Equation (1).	$mgh = \frac{7}{10}mgR + 2mgR = (2.7)mgR$ $h = 2.7R = (2.7)(0.250\text{ m})$ $h = 0.675\text{ m} = 67.5\text{ cm}$

CHECK and THINK

This new release height is about 5 cm higher than what Chris found previously. He needs to release the ball from this greater height so that the system starts with enough potential energy to complete the loop. Some of the potential energy is converted into the ball's rotational kinetic energy, and the ball successfully makes it around the loop. Without this extra initial potential energy, the ball's center-of-mass speed at the top of the loop is too low, and the ball cannot complete the loop.

Notice that when Chris added the effect of rolling friction to his first answer, he only found a height difference of 2 mm. Because the new difference we found is 25 times greater, we are safe to ignore mechanical energy losses due to friction.

EXAMPLE 13.9 "And the Winner Is . . ."

Imagine a downhill race between four objects each of mass M (Fig. 13.17). One object such as a skier is best modeled as a particle (shown as a small flat object) translating down the hill. The other objects are modeled as a solid cylinder, hoop, and solid sphere each of radius R rolling down the hill without slipping. All four objects start from rest at height h. Assume mechanical energy losses due to friction may be ignored and determine the order in which the objects cross the finish line.

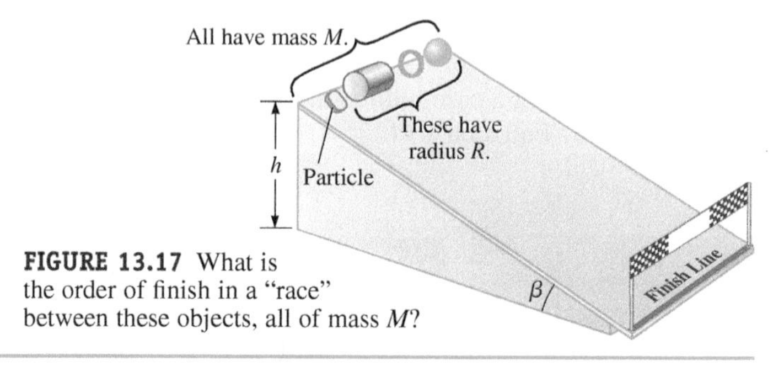

FIGURE 13.17 What is the order of finish in a "race" between these objects, all of mass M?

INTERPRET and ANTICIPATE

The object whose center of mass crosses the finish line in the shortest time interval wins the race. Because all objects start from rest and travel the same distance, the object with the fastest center-of-mass speed as it crosses the finish line wins. For Step 1, consider each object to be part of a **system** with the Earth and use conservation of mechanical energy to find the object's center-of-mass speed at the finish line. For Steps 2 and 3, set the **reference configuration** to the bottom of the incline, choose an upward-pointing y **axis**, set the **initial** time to when an object is at the top of the incline, and set the **final** time to when the object is at the bottom.

Step 4 Each **bar chart** (Fig. 13.18) shows the same initial energy. Because the particle does not rotate, its bar chart has zero rotational kinetic energy. The bar charts for the rolling objects have nonzero rotational kinetic energy.

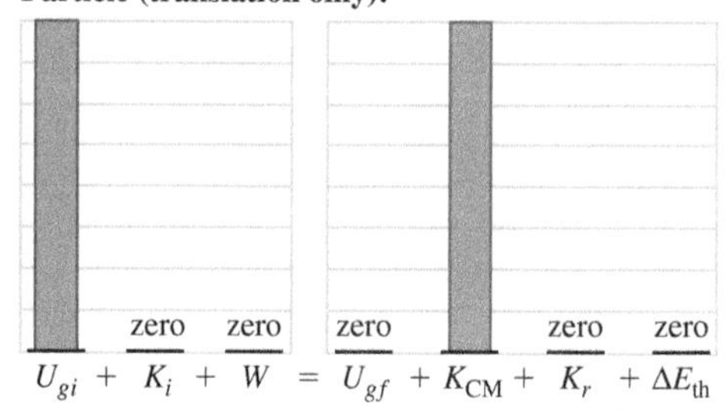

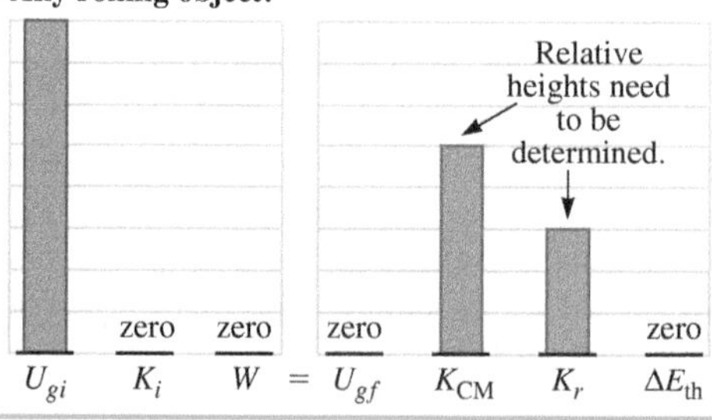

FIGURE 13.18

SOLVE

Step 5 **Conservation of energy** for the Earth–particle system produces a familiar result (Chapter 8).

$$U_{gi} = K_{CM}$$

Steps 6 and 7 **Other equations** and **do algebra**. Write the potential energy in terms of the particle's initial height and the kinetic energy in terms of the particle's final speed. Solve for the particle's final speed.	$Mgh = \frac{1}{2}Mv_{CM}^2$ $v_{CM,\,part} = \sqrt{2gh}$ (13.15)
Steps 5 and 6 **Conservation of energy** and **other equations**. There are three systems involving a rolling object and the Earth. Conservation of energy for these three systems has the same mathematical form.	$U_{gi} = K_{CM} + K_r$ $Mgh = \frac{1}{2}Mv_{CM}^2 + \frac{1}{2}I_{CM}\omega^2$
From Table 13.1, the rotational inertia of any of these objects may be written as $I_{CM} = (f)MR^2$, where f is a unitless fraction such as $\frac{2}{5}$ in the case of the solid sphere.	$Mgh = \frac{1}{2}Mv_{CM}^2 + \frac{1}{2}(f)MR^2\omega^2$
We use the condition for pure rolling motion (Eq. 13.12, $v_{CM} = \omega R$) to write the expression in terms of v_{CM} and eliminate ω and R.	$Mgh = \frac{1}{2}Mv_{CM}^2 + \frac{1}{2}(f)Mv_{CM}^2$ (13.16) $gh = \frac{1}{2}v_{CM}^2 + \frac{1}{2}(f)v_{CM}^2 = \frac{1}{2}(1+f)v_{CM}^2$ $v_{CM}^2 = \frac{2gh}{(1+f)}$ $v_{CM} = \sqrt{\frac{2gh}{1+f}}$ (13.17)
Find f in Table 13.1 and substitute into Equation 13.17. For a solid cylinder, $f = \frac{1}{2}$.	$v_{CM,\,cyl} = \sqrt{\frac{2gh}{3/2}} = \sqrt{\frac{4}{3}gh} = \sqrt{1.33gh}$
For a hoop, $f = 1$.	$v_{CM,\,hoop} = \sqrt{\frac{2gh}{2}} = \sqrt{gh}$
For a solid sphere, $f = \frac{2}{5}$.	$v_{CM,\,sph} = \sqrt{\frac{2gh}{7/5}} = \sqrt{\frac{10}{7}gh} = \sqrt{1.43gh}$
So, the order in which the objects cross the finish line from first to last is particle, solid sphere, solid cylinder, and hoop.	$v_{CM,\,part} = \sqrt{2gh}$ $v_{CM,\,sph} = \sqrt{1.43gh}$ $v_{CM,\,cyl} = \sqrt{1.33gh}$ $v_{CM,\,hoop} = \sqrt{gh}$ $v_{CM,\,part} > v_{CM,\,sph} > v_{CM,\,cyl} > v_{CM,\,hoop}$

CHECK and THINK

As a quick check, we see that all the expressions have the correct dimensions. A particle does not rotate, so none of the gravitational potential energy goes into rotation; instead, all the initial potential energy goes into translational motion, so the particle wins the race. For objects that rotate, the amount of gravitational potential energy that goes into rotational kinetic energy depends on how close the mass is to the rotation axis. If much of the mass is located farther away from the axis, more of the gravitational potential energy goes into rotational kinetic energy, and less is available for translation. So, the object's center-of-mass speed does *not* depend on its mass or radius; rather, it depends on f, which is determined by the geometry of the object.

13-5 Work and Power

On New Year's Day 1982, the movie *Conan the Barbarian*, starring Arnold Schwarzenegger, opened. His character, Conan, developed his muscles by pushing a very heavy wheel around a vertical axle. Today, *Conan wheels* are used as part of certain strength competitions (Fig. 13.19). Energy is transferred to the wheel by the work done by the athlete. If a system such as a Conan wheel is rotating, work is more conveniently expressed in terms of torque rather than force.

FIGURE 13.19 The contestant is pushing a heavy load around a rotation axis to the left in the photo.

DERIVATION Work in Terms of Torque

Figure 13.20 shows a Conan wheel from above. A force $\vec{F}$ is applied to the wheel's arm at point P, making an angle φ with respect to the arm. We will assume that the wheel moves counterclockwise as shown and derive work in terms of torque:

$$W = \int_{\theta_i}^{\theta_f} \tau \, d\theta \tag{13.18}$$

We will also show that the power is given by

$$P = \tau\omega \tag{13.19}$$

WORK; POWER ✪ Major Concepts

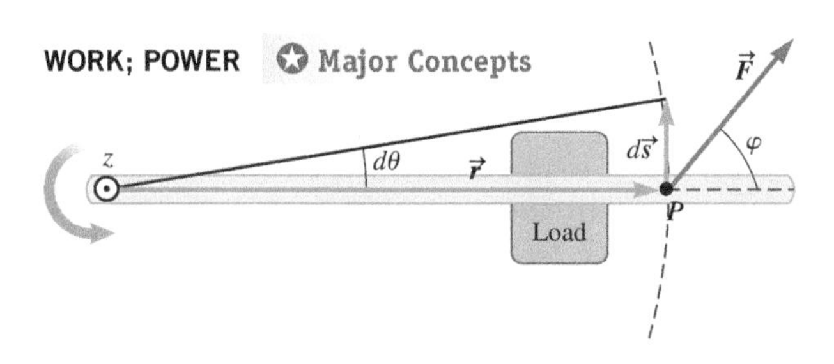

FIGURE 13.20 A person applies a torque to this wheel.

Consider an infinitesimal displacement $d\vec{s}$ of the point of contact P. The work done by the force on the Conan wheel during this infinitesimal displacement comes from Equation 9.20 written in terms of the dot product.	$dW = \vec{F} \cdot d\vec{s}$ (9.20) $dW = F\,ds \cos(90° - \varphi)$ $dW = F\,ds \sin\varphi$ (13.20)
Because the wheel is rotating, angular displacement is much more convenient to work with than translational displacement. The translational displacement ds can be written in terms of the infinitesimal angle $d\theta$ and distance r between point P and the rotation axis ($ds = r\,d\theta$, Eq. 12.12).	$ds = r d\theta$ (13.21)
Substitute Equation 13.21 into Equation 13.20.	$dW = (Fr \sin\varphi) d\theta$
The term $Fr\sin\varphi$ is the torque applied by the person to the Conan wheel ($\tau = Fr\sin\varphi$, Eq. 12.16).	$dW = \tau\, d\theta$ (13.22)
The work done by the person on the Conan wheel as the wheel rotates from some initial angular position θ_i to a final angular position θ_f comes from integrating Equation 13.22.	$W = \int_{\theta_i}^{\theta_f} \tau\, d\theta$ ✓ (13.18)

COMMENTS

In keeping with the analogy in Chapter 12 between rotation around a single axis and translation in one dimension, Equation 13.18 is analogous to $W = \int_{x_i}^{x_f} F_x dx$ (Eq. 9.18).

We often need to know the power (the rate at which energy is transferred to the system). To find the power, we divide Equation 13.22 by dt. The resulting Equation 13.19 is analogous to $P = \vec{F} \cdot \vec{v}$ (Eq. 9.38).	$P = \dfrac{dW}{dt} = \tau \dfrac{d\theta}{dt}$ $P = \tau\omega$ ✓ (13.19)

Applications: Waterwheels

One of the oldest technological innovations was the wheel. You probably think about how useful the wheel is for transportation, but it has another important use: transferring gravitational potential energy into (rotational) kinetic energy.

Falling water loses gravitational potential energy. That energy may be transferred to a waterwheel (Fig. 13.21) that is then used to operate machinery or generate an electric current. The earliest record of a waterwheel dates from 400 BCE. Ancient waterwheels were mainly used to grind grains into flour. Later waterwheels were used in other industries, for example to operate bellows and hammers in forging iron, to sharpen tools and weapons, and to make textiles. The basic principle of the waterwheel is still in use today in modern hydroelectric power stations (Chapter 32).

There are several major variations in the design of a waterwheel, and many details of these designs contribute to the overall efficiency of a particular device. For example, an efficient waterwheel will have buckets designed to hold the water until the bucket reaches the bottom (as in the next example).

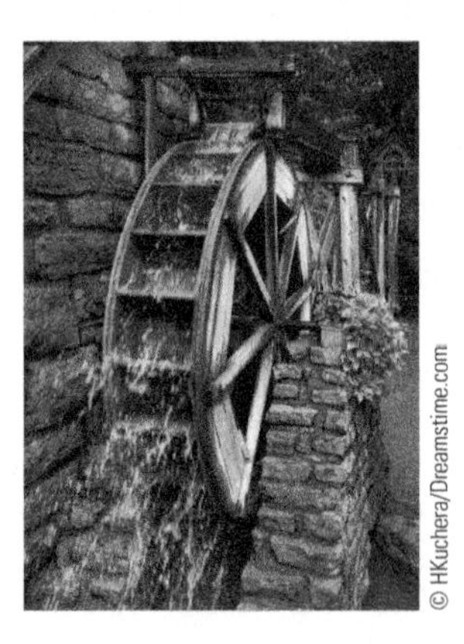

FIGURE 13.21 Waterwheels may be used to transfer energy from moving water to other systems.

EXAMPLE 13.10 What a Grind!

Consider a 12-ft-diameter waterwheel that may have been used to run a textile factory in the United States in the 19th century. The wheel has 24 buckets, each holding 30 gal of water. Assume (somewhat unrealistically) water fills a 30-gal bucket at 10° from the top of the wheel and the water remains in the bucket until it is at 100° from the top, at which point all the water is emptied completely out of the bucket. Friction acts on the waterwheel so that the net torque is zero and the angular speed is a constant 10 rpm. At what rate is energy transferred from the waterfall to the wheel? *Hint*: First find the work done by the water in a single bucket, ignoring the mass of the bucket.

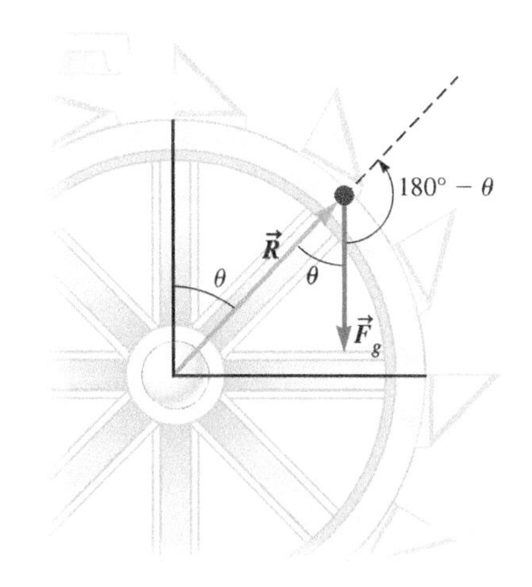

FIGURE 13.22 The black dot represents a bucket of water, where the water has mass m.

INTERPRET and ANTICIPATE

The gravitational force on the water in a full bucket exerts a torque on the waterwheel. A free-body diagram (Fig. 13.22) of a full bucket when its angular position is θ from the top of the waterwheel shows that the torque depends on θ. The torque is zero when the bucket is at the top ($\theta = 0$) and is maximized when the bucket is at $\theta = 90°$.

SOLVE

The magnitude of the torque comes from $\tau = Fr\sin\varphi$ (Eq. 12.16). Use $\sin(180° - \theta) = \sin\theta$.

$$\tau = F_g R\sin(180° - \theta)$$

$$\tau = mgR\sin\theta \quad (13.23)$$

Because the torque changes as the waterwheel rotates, we must integrate Equation 13.18. Substitute Equation 13.23 for the torque and integrate.

$$W = \int_{\theta_i}^{\theta_f} \tau\, d\theta \quad (13.18)$$

$$W = \int_{\theta_i}^{\theta_f} mgR\sin\theta\, d\theta = mgR\int_{\theta_i}^{\theta_f} \sin\theta\, d\theta$$

$$W = -mgR\cos\theta\Big|_{\theta_i}^{\theta_f} = mgR(\cos\theta_i - \cos\theta_f)$$

Before evaluating our expression, convert all the given values to SI units and then find the mass of water in the bucket from the density of water. We do not need to convert θ into radians because θ only appears as the input to a trigonometric function.

$$R = \frac{D}{2} = \frac{12\text{ ft}}{2} = 6\text{ft}\left(\frac{1\text{ m}}{3.28\text{ ft}}\right) = 1.83\text{ m}$$

$$30\text{ gal} \times \frac{1\text{ m}^3}{264\text{ gal}} = 0.114\text{ m}^3$$

$$m = \rho V = (1000\text{ kg/m}^3)(0.114\text{ m}^3)$$

$$m = 1.14 \times 10^2\text{ kg}$$

Substitute for m, R, θ_i, and θ_f.

$$W = mgR(\cos\theta_i - \cos\theta_f)$$

$$W = (1.14 \times 10^2\text{ kg})(9.81\text{ m/s}^2)(1.83\text{ m})(\cos 10° - \cos 100°) = 2.4 \times 10^3\text{ J}$$

During one revolution, a single bucket transfers this energy to the waterwheel. Because water is only in a bucket as it rotates through 90°, roughly one-fourth of the buckets have water at any given instant. Therefore, in a single revolution, energy is transferred from the water to the waterwheel by one-fourth of the buckets (six buckets).

$$W_{tot} = 6W = 6(2.4 \times 10^3\text{ J})$$

$$W_{tot} = 1.4 \times 10^4\text{ J}$$

The power transferred to the waterwheel is given by the total work done by the water in one revolution divided by the period. Find the period from the angular speed.

$$T = \frac{60\text{ s}}{10\text{ rev}} = 6\text{ s}$$

$$P = \frac{W_{tot}}{T} = \frac{1.4 \times 10^4\text{ J}}{6\text{ s}}$$

$$P = 2.4 \times 10^3\text{ W} = 2.4\text{ kW}$$

Example continues on page 376 ▶

CHECK and THINK

Our results provide an estimate of the capabilities of a historic waterwheel. A typical modern house requires between 4 kW and 7 kW. So, the waterwheel in this example would not have enough power for a modern house, but may be able to supply a small apartment. As long as the weather conditions provided water, the waterwheel could provide a steady source of power 24 hours a day, seven days a week. If a factory needed more power, additional waterwheels were added.

For more than two millennia, the waterwheel provided an important means to power machinery. The main alternatives were either human muscles or animal muscles. Also, this sort of waterwheel was a reliable and steady power source. By comparison, consider a person on a stationary exercise bicycle who transfers power from food he has eaten to the kinetic energy of the bicycle. A person in good condition can put out about 100 W for an hour or so but would then need to rest and eat more food. A waterwheel such as the one in Example 13.10 can deliver about 2000 W without needing a rest.

ANGULAR MOMENTUM

Major Concept

13-6 Angular Momentum

A Segway can be safely maneuvered through busy streets. A unicycle is steered without a steering wheel. Helicopters require two separate rotors. According to Kepler's second law, planets slow down when they are far away from the Sun and speed up when they are close to the Sun. To understand such phenomena, we need to introduce another concept: *angular momentum*. Like linear momentum, angular momentum helps when analyzing complicated motion. Conservation of angular momentum is yet another powerful tool for problem solving in physics.

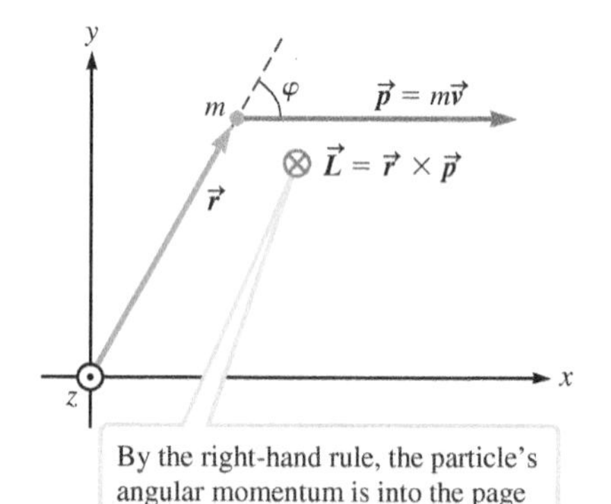

FIGURE 13.23 Definition of angular momentum. A particle does not need to be in orbit around the z axis to have angular momentum around that axis.

Angular Momentum of a Particle

Angular momentum $\vec{L}$ is defined for a particle of mass m moving with translational momentum $\vec{p} = m\vec{v}$:

$$\vec{L} \equiv \vec{r} \times \vec{p} \tag{13.24}$$

where $\vec{r}$ is the position vector of the particle pointing from the origin of the coordinate system to the particle (Fig. 13.23). Angular momentum is defined in terms of the motion of a single particle that does not need to be orbiting the origin of the coordinate system. In fact, the particle does not even need to be moving in a curved path.

Because angular momentum can be determined by using the vector cross product, its direction can be found by using the right-hand rule (Section 12-6). The vector $\vec{L}$ must be perpendicular to the plane containing the vectors $\vec{r}$ and $\vec{p}$. As with any cross product, the magnitude of angular momentum may be found from $R = AB \sin \varphi$ (Eq. 12.22):

$$L = rp \sin \varphi = r_\perp p = rp_\perp \tag{13.25}$$

where φ is the angle between $\vec{r}$ and $\vec{p}$, $r_\perp$ is the component of $\vec{r}$ that is perpendicular to $\vec{p}$, and $p_\perp$ is the component of $\vec{p}$ that is perpendicular to $\vec{r}$. The SI units of angular momentum are $\text{kg} \cdot \text{m}^2/\text{s}$.

CONCEPT EXERCISE 13.2

Figure 13.24 shows a particle with momentum $\vec{p}$. Using the coordinate systems shown, determine the direction of the angular momentum of the particle around the origin in each case, and write expressions for $\vec{L}$, using symbols defined in Figure 13.23.

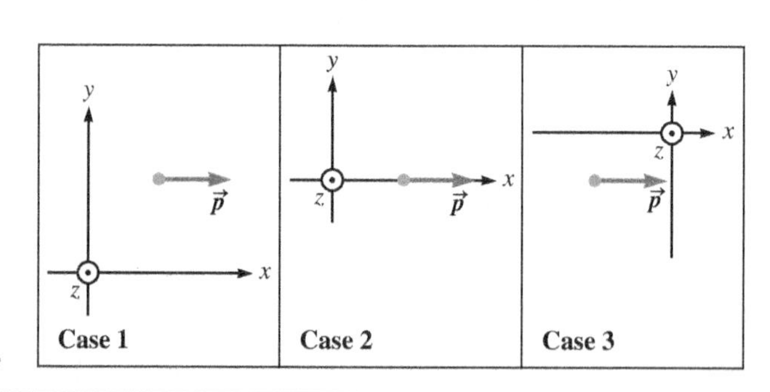

FIGURE 13.24

DERIVATION Angular Momentum of a Rotating Rigid Object

We can find an expression for the angular momentum of a rotating rigid object by considering it to be a collection of particles orbiting the common rotation axis. As an example, consider the dumbbell-shaped object in Figure 13.25. Modeling the dumbbell as two particles moving in concentric circles around the z axis, we will show that its angular momentum is given by

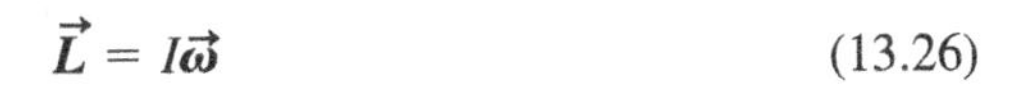

$$\vec{L} = I\vec{\omega} \qquad (13.26)$$

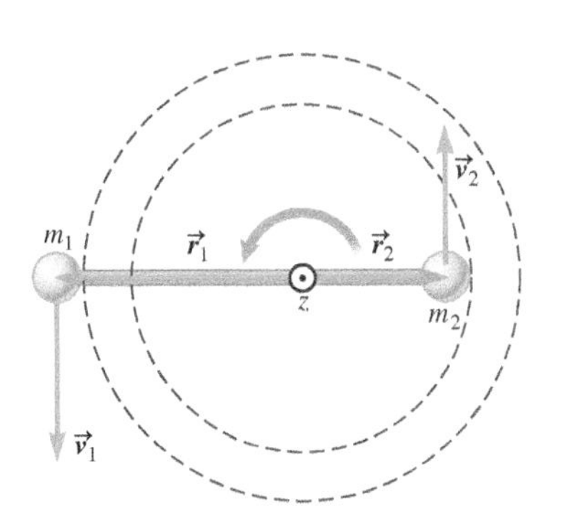

FIGURE 13.25

The angular momentum of each particle is given by $\vec{L} = \vec{r} \times \vec{p}$ (Eq. 13.24).	$\vec{L}_1 = \vec{r}_1 \times \vec{p}_1$ $\vec{L}_2 = \vec{r}_2 \times \vec{p}_2$
By the right-hand rule, both $\vec{L}_1$ and $\vec{L}_2$ point in the positive z direction. From Figure 13.25, $\vec{r}$ is perpendicular to $\vec{p}$ for each particle.	$\vec{L}_1 = r_1 p_1 \hat{k}$ $\vec{L}_2 = r_2 p_2 \hat{k}$
The angular momentum $\vec{L}$ of the entire object is the sum of $\vec{L}_1$ plus $\vec{L}_2$.	$\vec{L} = \vec{L}_1 + \vec{L}_2$ $\vec{L} = (r_1 p_1 + r_2 p_2)\hat{k}$
From $p = mv$ (Eq. 10.1), we can write $\vec{L}$ in terms of the mass and velocity of each particle.	$\vec{L} = (r_1 m_1 v_1 + r_2 m_2 v_2)\hat{k}$
Both particles have the same angular speed ω, related to their translational speeds by $v = \omega r$ (Eq. 12.13).	$\vec{L} = (r_1 m_1 \omega r_1 + r_2 m_2 \omega r_2)\hat{k}$ $\vec{L} = (m_1 r_1^2 + m_2 r_2^2)\omega\hat{k}$
The term in parentheses is the rotational inertia of the dumbbell-shaped object (Eq. 13.4).	$\vec{L} = I\omega\hat{k}$
The angular velocity in Figure 13.25 is in the z direction: $\vec{\omega} = \omega\hat{k}$. Therefore, the dumbbell-shaped object's angular momentum may be written as a vector.	$\vec{L} = I\vec{\omega}$ ✓ (13.26)

COMMENTS

We found Equation 13.26 by calculating the angular momentum around the rotational axis of the dumbbell. The relationship between angular momentum and angular velocity may be more complicated in some situations, but Equation 13.26 holds for any object if $\vec{L}$ is the component of the angular momentum along the rotation axis.[1]

EXAMPLE 13.11 CASE STUDY Comet Halley's Orbital Angular Momentum

Use the results of the case study in Chapter 8 to find the orbital angular momentum of Comet Halley around the Sun **A** when the comet is at perihelion ($r_P = 8.83 \times 10^{10}$ m) and **B** when it is at aphelion ($r_A = 5.27 \times 10^{12}$ m). In Example 8.13, we found the perihelion and aphelion speeds, $v(r_P) = 54.4$ km/s and $v(r_A) = 0.92$ km/s, respectively. Recall that the mass of Comet Halley is $m = 1.7 \times 10^{15}$ kg. (We have provided an extra significant figure here for $v(r_P)$. Use all the significant figures given and report your final results to two significant figures.)

Example continues on page 378 ▶

[1]According to Equation 13.26, $\vec{L}$ and $\vec{\omega}$ are parallel. Sometimes, though, the motion is more complicated, and the two vectors point in different directions. When $\vec{L}$ and $\vec{\omega}$ are not parallel, I is not a scalar, and Equation 13.26 does not hold. This sort of complicated motion is beyond the scope of this textbook.

INTERPRET and ANTICIPATE

The example asks for the *orbital* angular momentum around the Sun. In this case, we can treat Comet Halley as a particle, ignoring its rotational motion around its center of mass. At perihelion and aphelion, the position vector is perpendicular to the velocity vector (Fig. 13.26). Finding the angular momentum for these two positions is straightforward.

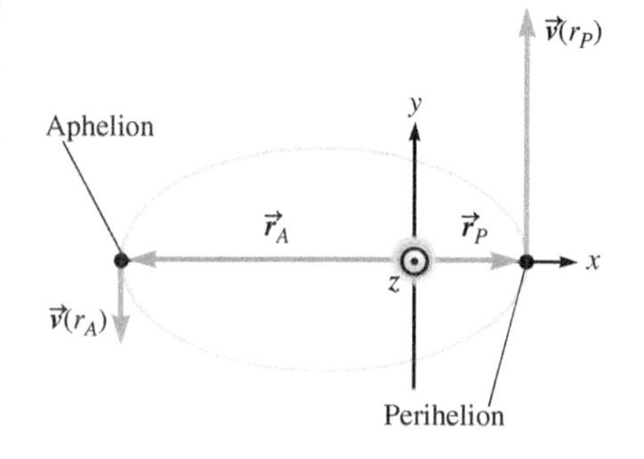

FIGURE 13.26

SOLVE

According to the right-hand rule, the angular momentum is in the positive z direction given the coordinate system shown in Figure 13.26.

$$\vec{L} = \vec{r} \times \vec{p} \quad (13.24)$$
$$\vec{L} = rp\hat{k}$$
$$\vec{L} = rmv\hat{k}$$

A Substitute the perihelion values.

$$\vec{L}(r_P) = r_P m v_P \hat{k}$$
$$\vec{L}(r_P) = (8.83 \times 10^{10}\,\text{m})(1.7 \times 10^{15}\,\text{kg})(5.44 \times 10^{4}\,\text{m/s})\hat{k}$$
$$\vec{L}(r_P) = (8.2 \times 10^{30})\hat{k}\,\text{kg} \cdot \text{m}^2/\text{s}$$

B Substitute the aphelion values.

$$\vec{L}(r_A) = r_A m v_A \hat{k}$$
$$\vec{L}(r_A) = (5.27 \times 10^{12}\,\text{m})(1.7 \times 10^{15}\,\text{kg})(9.2 \times 10^{2}\,\text{m/s})\hat{k}$$
$$\vec{L}(r_A) = (8.2 \times 10^{30})\hat{k}\,\text{kg} \cdot \text{m}^2/\text{s}$$

CHECK and THINK

As a quick check, notice that our results have the correct units. A more interesting point is that Comet Halley's angular momentum seems to be the same at perihelion and aphelion. It is *not* a coincidence, as we show in the next section.

EXAMPLE 13.12 The Spinning CD

Find the change in angular momentum of a CD (m = 16.3 g, R = 6.00 cm) as it is played from beginning to end. From Problem 12.82 (page 359), when information is being read from the innermost ring, the CD's angular speed is 52.4 rad/s. The CD slows down so that when information is read from the outermost ring, its angular speed is 20.9 rad/s.

INTERPRET and ANTICIPATE

Figure 13.27 shows the initial and final angular velocities of the CD. The change in angular velocity is in the positive z direction. Because $\vec{L}$ is parallel to $\vec{\omega}$, we expect the change in angular momentum to be in the positive z direction also.

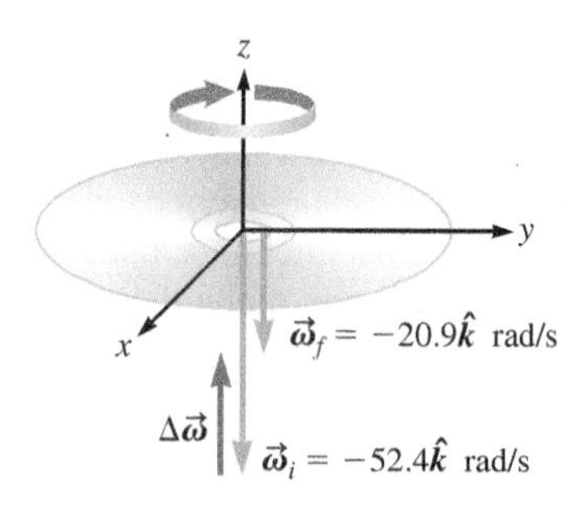

FIGURE 13.27

SOLVE

Model the CD as a solid disk and use Table 13.1 to find its rotational inertia.

$$I = \tfrac{1}{2}MR^2 = \tfrac{1}{2}(16.3 \times 10^{-3}\,\text{kg})(6.00 \times 10^{-2}\,\text{m})^2$$
$$I = 2.934 \times 10^{-5}\,\text{kg} \cdot \text{m}^2$$

The change in angular momentum is derived from $\vec{L} = I\vec{\omega}$ (Eq. 13.26). The rotational inertia does not change as the CD is played.

$$\Delta\vec{L} = \Delta(I\vec{\omega}) = I\Delta\vec{\omega}$$
$$\Delta\vec{L} = I(\vec{\omega}_f - \vec{\omega}_i)$$

Substitute the rotational inertia and the angular velocities.

$$\Delta\vec{L} = (2.934 \times 10^{-5}\,\text{kg} \cdot \text{m}^2)[-20.9\hat{k}\,\text{rad/s} - (-52.4\hat{k}\,\text{rad/s})]$$
$$\Delta\vec{L} = (9.24 \times 10^{-4})\hat{k}\,\text{kg} \cdot \text{m}^2/\text{s}$$

CHECK and THINK

Because the CD slows down and because its angular velocity is in the negative z direction, the change in its angular momentum is in the positive z direction. In Problem 12.82, you should find that the angular acceleration is also in the positive z direction. Again, it is *not* a coincidence.

13-7 Conservation of Angular Momentum

NEWTON'S SECOND LAW

Underlying Principle

We found in Chapter 10 that when no net force acts on a system, the system's *translational* momentum is conserved, and this principle gave us a way to analyze complicated situations such as the decay of a nucleus. In this section, we will learn when and how to apply the principle of conservation of *angular* momentum. However, to better understand conservation of angular momentum, we consider a system whose angular momentum is changing and derive another expression for Newton's second law.

DERIVATION Another Expression of Newton's Second Law

Another way to express Newton's second law for a system is

$$\frac{d\vec{L}_{tot}}{dt} = \sum \vec{\tau}_{ext} \tag{13.27}$$

where $\vec{\tau}_{ext}$ is the total external torque exerted and $\vec{L}_{tot}$ is the system's total angular momentum. Equation 13.27 holds for a single particle, a collection of particles, or an object.

Our derivation is made simpler if we consider a single particle moving along a straight path (Fig. 13.28). A net force $\vec{F}_{tot}$ acts on the particle so that it speeds up along that path. We will show that for this particle,

$$\frac{d\vec{L}}{dt} = \vec{\tau}_{tot} \tag{13.28}$$

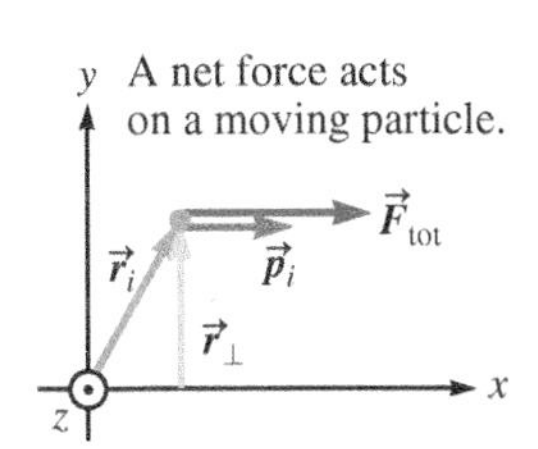

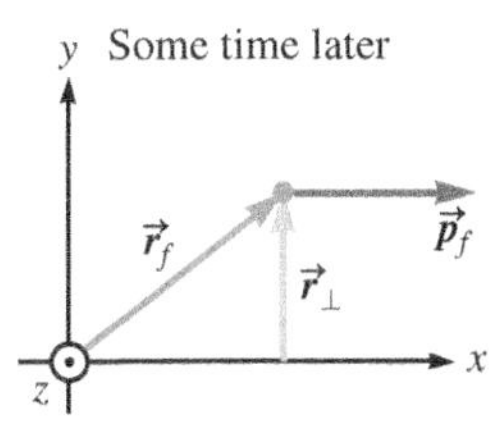

FIGURE 13.28 The particle moves in a straight line, and so $\vec{r}_\perp$ is constant.

Because the particle's speed increases, its translational momentum also increases. The particle's path is along the straight line shown, so $r_\perp$ does not change, but because the translational momentum p increases, the particle's angular momentum L around the z axis increases also.

Take the time derivative of Equation 13.24 to derive an expression for the rate of change of angular momentum.	$\vec{L} = \vec{r} \times \vec{p}$ (13.24) $\frac{d\vec{L}}{dt} = \frac{d(\vec{r} \times \vec{p})}{dt}$
Expanding the right side using the chain rule, we obtain a cross product in each term.	$\frac{d(\vec{r} \times \vec{p})}{dt} = \left(\frac{d\vec{r}}{dt} \times \vec{p}\right) + \left(\vec{r} \times \frac{d\vec{p}}{dt}\right)$
Rewrite the derivative in the first term using Equation 4.8 (definition of translational velocity). Rewrite the derivative in the second term using Equation 10.2 (a general form of Newton's second law).	$\vec{v} = \frac{d\vec{r}}{dt}$ (4.8) $\vec{F}_{tot} = \frac{d\vec{p}}{dt}$ (10.2) $\frac{d(\vec{r} \times \vec{p})}{dt} = (\vec{v} \times \vec{p}) + (\vec{r} \times \vec{F}_{tot})$ $\frac{d(\vec{r} \times \vec{p})}{dt} = [\vec{v} \times (m\vec{v})] + (\vec{r} \times \vec{F}_{tot})$
The first term is zero because $\vec{v}$ and $(m\vec{v})$ are parallel, and the cross product of parallel vectors is zero. The second term is the torque $\vec{\tau}_{tot}$ on the particle due to the net force $\vec{F}_{tot}$ (Eq. 12.23).	$\frac{d(\vec{r} \times \vec{p})}{dt} = 0 + \vec{\tau}_{tot}$

Derivation continues on page 380 ▶

So, a net torque on a particle changes its angular momentum.	$\frac{d\vec{L}}{dt} = \vec{\tau}_{tot}$ ✓ (13.28)

COMMENTS

Equation 13.28 is analogous to $\vec{F}_{tot} = d\vec{p}/dt$ (Eq. 10.2) and is another way to express Newton's second law.

We derived Equation 13.28 for a particle. With two modifications, it also holds for a system of particles or for a rotating object, where $\vec{L}_{tot}$ is the total angular momentum and $\vec{\tau}_{ext}$ is the total external torque acting on the system.	$\frac{d\vec{L}_{tot}}{dt} = \sum \vec{\tau}_{ext}$ (13.27)

Our result $d\vec{L}_{tot}/dt = \sum\vec{\tau}_{ext}$ is analogous to $d\vec{p}_{tot}/dt = \sum\vec{F}_{ext}$ (Eq. 10.10). Also, $d\vec{L}_{tot}/dt = \sum\vec{\tau}_{ext}$ may be applied to a particle. In that case of a single particle, the total angular momentum is $\vec{L}_{tot} = \vec{L}$, and all the torques must be external. So, $\sum\vec{\tau}_{ext} = \vec{\tau}_{tot}$.

CONSERVATION OF ANGULAR MOMENTUM

Underlying Principle

According to $d\vec{L}_{tot}/dt = \sum\vec{\tau}_{ext}$ (Eq. 13.27), if $\sum\vec{\tau}_{ext} = 0$, then $\vec{L}_{tot}$ is constant. Putting this expression into words, **if the net external torque acting on a system is zero, the system's total angular momentum is conserved.**

Recall from Chapters 8 through 11 that a conservation approach works well when we would like to consider a particle or system at two different times, t_i and t_f. If the angular momentum of a particle is conserved, then

$$\vec{r}_i \times \vec{p}_i = \vec{r}_f \times \vec{p}_f \quad (13.29)$$

If the angular momentum of a rotating object (or collection of particles) is conserved, it is convenient to write

$$I_i\omega_i = I_f\omega_f \quad (13.30)$$

by making use of Equation 13.26.

CONCEPT EXERCISE 13.3

In Problem 12.82, (page 359) we found a CD's angular acceleration, and in Example 13.12, we found its change in angular momentum. Use $d\vec{L}_{tot}/dt = \sum\vec{\tau}_{ext}$ (Eq. 13.27) and $\vec{\tau}_{tot} = I\vec{\alpha}$ (Eq. 12.24) to explain why the CD's angular acceleration $\vec{\alpha}$ and change in angular momentum $\Delta\vec{L}$ point in the same direction.

FIGURE 13.29 The rotors on a helicopter rotate in opposite directions. The best way to understand the motion of these vehicles is by considering their angular momentum.

CONCEPT EXERCISE 13.4

Use the principle of conservation of angular to momentum to explain why the rotors on a helicopter such as the one in Figure 13.29 rotate in opposite directions. What would happen if one rotor stopped operating?

CONCEPT EXERCISE 13.5

A simple gyroscope is a spinning disk. Why are gyroscopes used on space telescopes?

EXAMPLE 13.13 Kepler's Second Law Revisited

According to Kepler's second law (Section 7-2), a line joining a planet to the Sun sweeps out equal areas in equal time intervals. Use the principle of conservation of angular momentum to prove Kepler's second law.

INTERPRET and ANTICIPATE

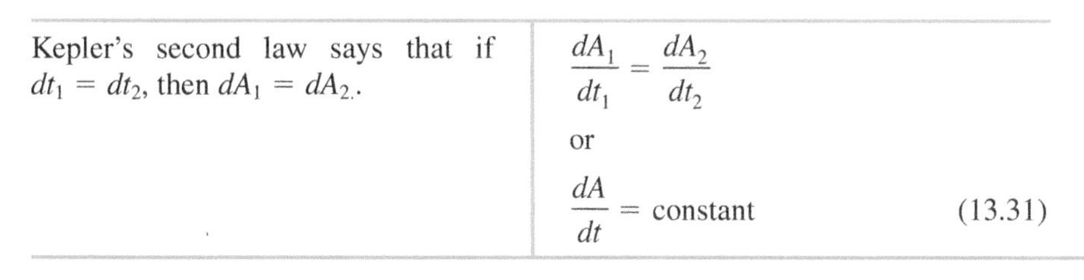

To express this law mathematically, imagine an infinitesimal area dA swept out in a very short time interval dt (Fig. 13.30).

Kepler's second law says that if $dt_1 = dt_2$, then $dA_1 = dA_2$.	$\frac{dA_1}{dt_1} = \frac{dA_2}{dt_2}$ or $\frac{dA}{dt} = \text{constant}$ (13.31)

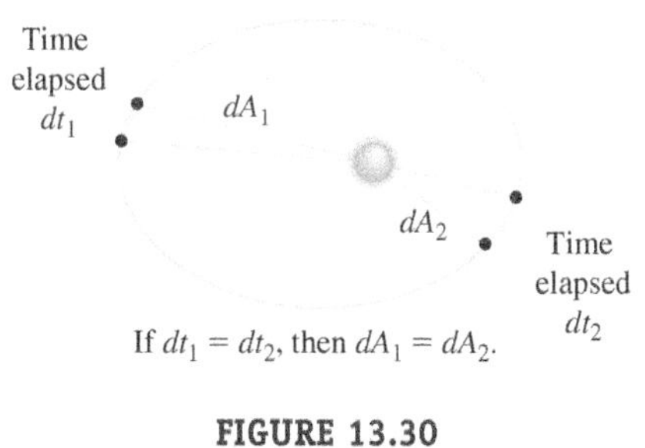

FIGURE 13.30

SOLVE

Now that we have a mathematical expression for Kepler's second law, we can use physical concepts and mathematics to prove it. Our results in Example 13.11 showed that the angular momentum of Comet Halley is nearly constant as it orbits the Sun. Let us see if a particle such as a planet in orbit around the Sun meets the condition for conserved angular momentum.

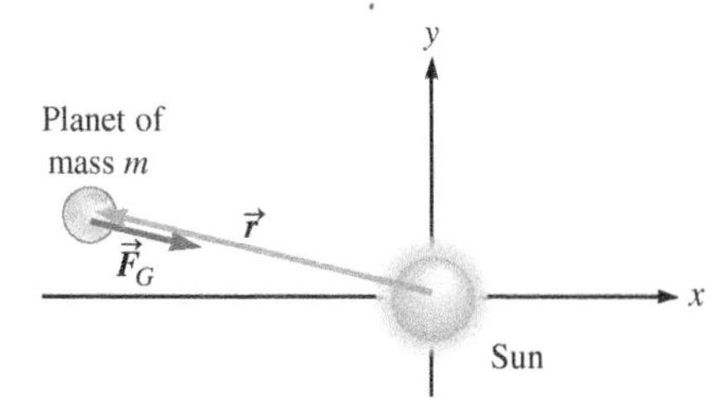

FIGURE 13.31

We assume the other planets and objects in the solar system exert a negligible force on the planet. The only force then acting on the planet is $\vec{F}_G$ due to the Sun's gravity (Fig. 13.31).

Because the angle between $\vec{r}$ and $\vec{F}_G$ is 180° at all times, there is no torque acting on the planet, and its angular momentum around the Sun is conserved.	$\vec{\tau} = \vec{r} \times \vec{F}_G$ $\tau = rF_G \sin 180° = 0$
Write an expression for the planet's constant angular momentum.	$\vec{L}_{\text{const}} = \vec{r} \times \vec{p}$ $\vec{L}_{\text{const}} = \vec{r} \times m\vec{v}$ $\vec{L}_{\text{const}} = m(\vec{r} \times \vec{v})$ (13.32)

The infinitesimal area dA is half the area of the parallelogram in Figure 13.32.

From Example 12.5 (page 347), the area of a parallelogram may be written in terms of a cross product. The area of the parallelogram is the magnitude of $\vec{r} \times d\vec{r}$.	$dA = \frac{1}{2}\lvert\vec{r} \times d\vec{r}\rvert$ (13.33)

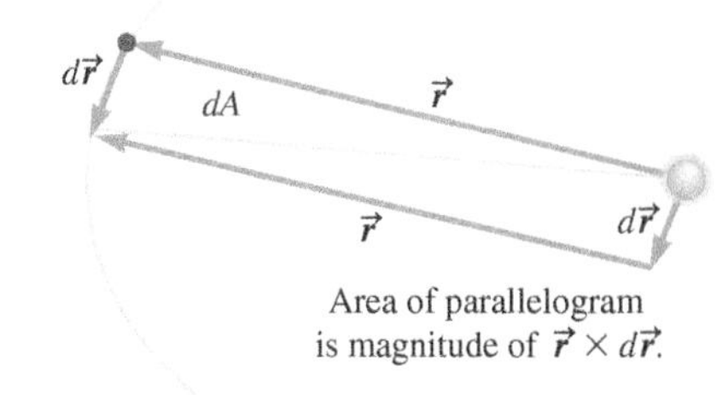

FIGURE 13.32

Write $d\vec{r}$ in terms of the velocity $\vec{v}$ (Eq. 4.8).	$\vec{v} = \frac{d\vec{r}}{dt}$ (4.8) $d\vec{r} = \vec{v}dt$
Substitute $d\vec{r}$ into Equation 13.33.	$dA = \frac{1}{2}\lvert\vec{r} \times \vec{v}dt\rvert$ $dA = \frac{1}{2}\lvert\vec{r} \times \vec{v}\rvert dt$ $\frac{dA}{dt} = \frac{1}{2}\lvert\vec{r} \times \vec{v}\rvert$ (13.34)
Finally, substitute Equation 13.32 into Equation 13.34.	$\frac{dA}{dt} = \frac{1}{2}\left\lvert\frac{\vec{L}_{\text{const}}}{m}\right\rvert$ $\frac{dA}{dt} = \frac{L_{\text{const}}}{2m}$ ✓ (13.35)

CHECK and THINK

Because m and L_{const} are constant for a particular planet, dA/dt is a constant, exactly as Kepler's second law says in Equation 13.31.

EXAMPLE 13.14 More on Spinning Pulsars

Typical neutron stars (the remains of dead stars) have a radius of 10 km. A pulsar is a neutron star that we observe as pulses or flashes of light. Pulsars rotate rapidly with periods roughly between 1 millisecond and 1 second (Example 12.2, page 339). The period of the pulses equals the period of the pulsar's rotation. A pulsar may form when a massive star is unable to produce energy in its core through nuclear fusion. At that time, the core collapses and becomes a pulsar while the outer layers are blown out in a powerful supernova explosion (Fig. 13.33). Before it collapses, the core's radius is roughly the same as that of the Earth (5000 km). Assume a newly formed neutron star's period is 5 ms and none of the core's mass is lost when it collapses. Model the core and the pulsar as solid spheres of uniform density and estimate the rotational period of the core before it collapses.

FIGURE 13.33 The Crab Nebula, the remains of a star that exploded over 1000 years ago.

INTERPRET and ANTICIPATE

Assume no external forces act on the core during its collapse. Because no net external torque acts on the core, its angular momentum is conserved.

SOLVE

Using Equation 13.30, apply the principle of conservation of angular momentum.

$$I_i \omega_i = I_f \omega_f \tag{13.30}$$

Write the angular speeds in terms of the period before and after collapse.

$$I_i\left(\frac{2\pi}{T_i}\right) = I_f\left(\frac{2\pi}{T_f}\right)$$

$$T_i = T_f\left(\frac{I_i}{I_f}\right)$$

Use Table 13.1 to find the rotational inertia of a sphere.

$$T_i = T_f\left(\frac{\frac{2}{5}Mr_i^2}{\frac{2}{5}Mr_f^2}\right) = T_f\left(\frac{r_i}{r_f}\right)^2$$

Substitute the given final period ($T_f = 5$ ms) and radius ($r_f = 10$ km) and initial core radius ($r_i = 5000$ km). We do not need to convert from kilometers to meters in this case.

$$T_i = (5 \times 10^{-3}\ \text{s})\left(\frac{5000\ \text{km}}{10\ \text{km}}\right)^2$$

$$T_i = 1.25 \times 10^3\ \text{s}$$

$$T_i \approx 20\ \text{min}$$

CHECK and THINK

Because the rotational inertia is reduced by the collapse of the core, the core's angular speed must increase after collapse to compensate, in accordance with conservation of angular momentum.

EXAMPLE 13.15 Bicycle Tires and Unicycles

At the Smithsonian Air and Space Museum in Washington, D.C., visitors are invited to sit on a stool and hold a rapidly rotating bicycle tire (Fig. 13.34). Initially, the visitor and the stool are at rest, and the bicycle tire is rotating (Fig. 13.34A). The visitor flips the rapidly rotating tire upside down (Fig. 13.34B), which causes the stool and visitor to rotate. In which direction does the visitor rotate? What happens if the visitor flips the tire right side up again?

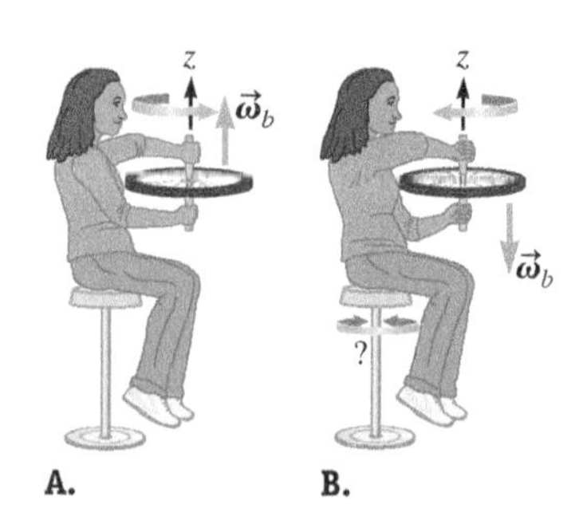

FIGURE 13.34

INTERPRET and ANTICIPATE

Let's include the visitor, the stool, and the bicycle tire in the system. There is no net external force on the system, so there is no net external torque on it; thus, angular momentum is conserved. Before the visitor flips the bicycle tire, the system's angular momentum is upward in the positive z direction. After she flips the tire, the system's angular momentum must still have the same magnitude and direction.

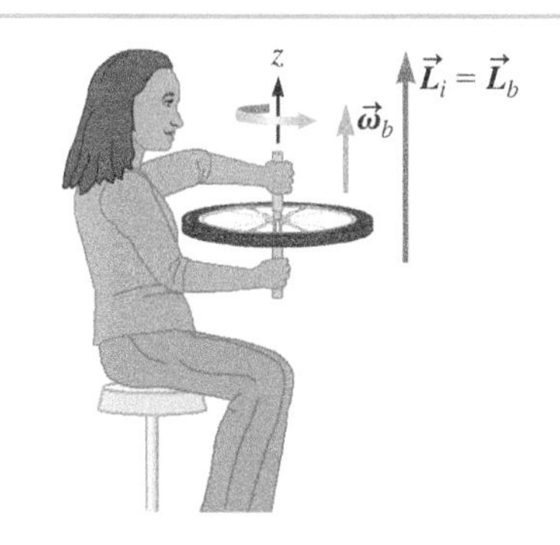

FIGURE 13.35

SOLVE

Initially (Fig. 13.35), the angular momentum of the system equals the angular momentum of the *b*icycle tire $\vec{L}_i = L_b\hat{k}$.

After the visitor flips the bicycle tire upside down, its angular momentum points in the negative z direction (Fig. 13.36). Because angular momentum is conserved ($\vec{L}_f = \vec{L}_i$), the angular momentum of the whole system must still be in the positive z direction. So, the visitor and the stool must rotate, and their angular momentum must be in the positive z direction and given by $\vec{L}_f = L_v\hat{k} - L_b\hat{k}$.

According to $\vec{L}_v = I\vec{\omega}_v$ (Eq. 13.26), the visitor's angular velocity is in the positive z direction. She rotates counterclockwise when viewed from above.

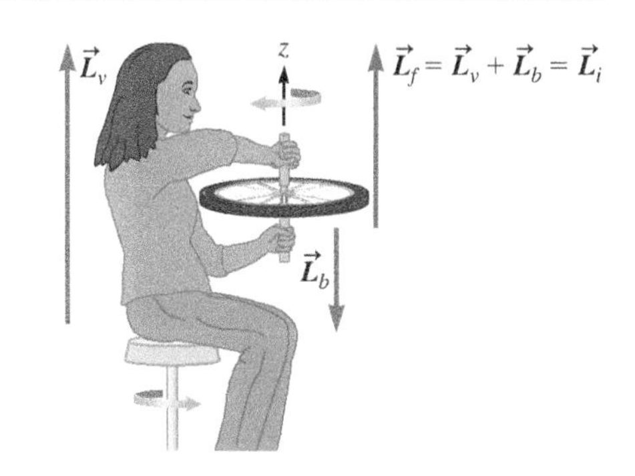

FIGURE 13.36

If the visitor flips the tire right side up again, angular momentum is still conserved. The bicycle tire's angular momentum points in the positive z direction again. All the angular momentum of the system is in the bicycle tire, so the visitor must stop rotating.

CHECK and THINK

Our results give us a way to think about how a person might steer a unicycle. Consider the person, the unicycle, and the Earth as the system. No net torque acts on the system. Ignoring the motion of the Earth, Figure 13.37A shows the initial angular momentum of the system with the unicycle in motion: $\vec{L}_i = \vec{L}_{\text{tire}}$. The person leans to his left so that the angular momentum of the tire rotates downward. The total angular momentum must still point to the right (Fig. 13.37B), so the angular momentum of the person must be upward to compensate. The person's angular velocity therefore points upward (parallel to his own angular momentum), and he is able to make a turn. Another way to analyze this situation is to exclude the Earth from the system and calculate the torque done by gravity. This approach is left as a homework problem.

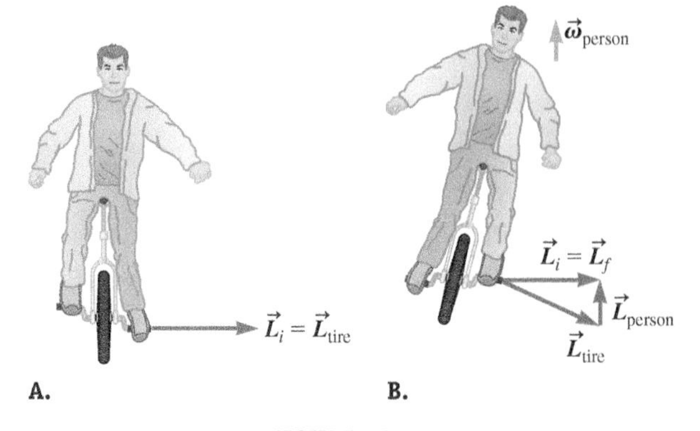

FIGURE 13.37

SUMMARY

Underlying Principles

1. The **conservation of energy** principle is slightly modified to include rotational motion:

$$K_i + K_{ri} + U_i + W = K_f + K_{rf} + U_f + \Delta E_{\text{th}}$$

2. **Newton's second law** can be written in terms of torque and angular momentum:

$$\frac{d\vec{L}_{\text{tot}}}{dt} = \sum \vec{\tau}_{\text{ext}} \tag{13.27}$$

3. A system's **angular momentum is conserved** if the total external torque on the system is zero.

Major Concepts

1. **Rotational inertia** is a measure of an object's tendency to maintain its angular velocity. The rotational inertia of an object that consists of n particles may be found mathematically from
$$I = \sum_{i=1}^{n} m_i r_i^2 \quad (13.4)$$
Rotational inertia of a continuous object is found from
$$I = \int r^2 dm \quad (13.8)$$
Table 13.1 (page 365) provides the result of such integration for a number of simple-shaped objects.
2. According to the **parallel-axis theorem**, if h is the perpendicular distance between a new rotation axis and the axis through the center of mass, the rotational inertia through the new axis is
$$I = I_{\text{CM}} + Mh^2 \quad (13.9)$$
where M is the mass of the object.
3. **Rotational kinetic energy** is most easily found in terms of the object's rotational inertia and angular speed:
$$K_r = \tfrac{1}{2}I\omega^2 \quad (13.10)$$
4. **Work** done by an external torque is
$$W = \int_{\theta_i}^{\theta_f} \tau d\theta \quad (13.18)$$
5. **Power** supplied by an external torque is
$$P = \tau\omega \quad (13.19)$$
6. **Angular momentum** $\vec{L}$ is defined for a particle of mass m moving with translational momentum $\vec{p} = m\vec{v}$:
$$\vec{L} \equiv \vec{r} \times \vec{p} \quad (13.24)$$
In general, a system's angular momentum may be written as
$$\vec{L} = I\vec{\omega} \quad (13.26)$$

Special Case

Pure rolling is a special combination of rotational and translation motion in which the object rolls along a surface without slipping. In this special case, the center-of-mass position is given by
$$x_{\text{CM}} = s = \theta R \quad (13.11)$$
The center-of-mass speed is given by
$$v_{\text{CM}} = \omega R \quad (13.12)$$
The center-of-mass acceleration is given by
$$a_{\text{CM}} = \alpha R \quad (13.13)$$

PROBLEMS AND QUESTIONS

A = algebraic C = conceptual E = estimation G = graphical N = numerical

13-1 Conservation Approach

1. C In the following four situations, you do positive work on a system. In other words, energy is transferred from you to a system. In each case, describe what happens to the energy after it enters the system. If there is kinetic energy in the system after the work is performed, is it possible for the speed to be constant in each case? **a.** You pull a sled across a frozen lake. **b.** You pull a sled up a snowy hill. **c.** You pull a string wrapped around a pulley with a fixed axle. **d.** You pull a cart with large wheels across the flat ground.
2. C CASE STUDY When Chris discovers that the ball does not complete the loop-the-loop challenge, he estimates the increase in thermal energy due to rolling friction. He finds that he needs to release the ball from a greater initial height. **a.** Why does it make sense that the ball must be released from a greater height? **b.** Is Chris's value of the increase in thermal energy an overestimate or an underestimate? Explain.
3. C A Frisbee flies across a field. Determine if the system has translational kinetic energy, rotational kinetic energy, neither, or both as determined by the observer in each of the following cases. **a.** The observer watches the flight of the Frisbee across the field from a park bench. **b.** The observer is a dog that runs directly beneath the Frisbee. **c.** The observer is an ant at rest on the Frisbee.
4. C Does a car moving along a highway have rotational kinetic energy? (You may ignore motions within the car.) Explain.

13-2 Rotational Inertia

Problems 5 and 6 are paired.

5. N A system consists of four boxes modeled as particles connected by very lightweight, stiff rods (Fig. P13.5). The system rotates around the z axis, which points out of the page. Each particle has a mass of 5.00 kg. The distances from the z axis to each particle are r_1 = 32.0 cm, r_2 = 16.0 cm, r_3 = 17.0 cm, and r_4 = 34.0 cm. Find the rotational inertia of the system around the z axis.

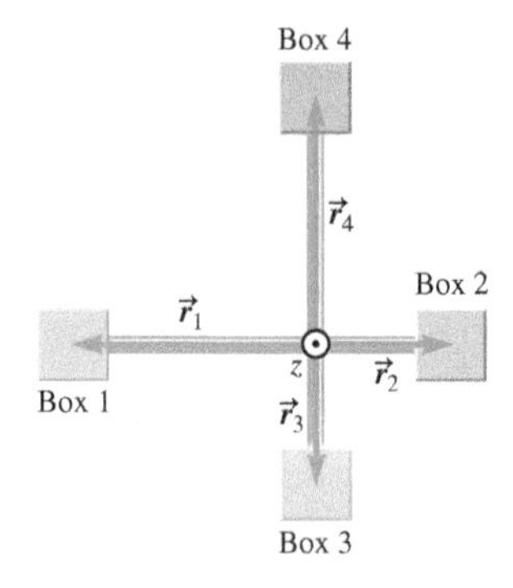

FIGURE P13.5 Problems 5 and 6.

6. N Use the information in Problem 5 to find the rotational inertia of the system around particle 1.
7. N A 12.0-kg solid sphere of radius 1.50 m is being rotated by applying a constant tangential force of 10.0 N at a perpendicular distance of 1.50 m from the rotation axis through the center

of the sphere. If the sphere is initially at rest, how many revolutions must the sphere go through while this force is applied before it reaches an angular speed of 30.0 rad/s?

8. E A figure skater clasps her hands above her head as she begins to spin around a vertical axis that passes through her hands, through the top of her head, and through the bottom of her skates. What is the order of magnitude of her rotational inertia around this axis?

9. A A solid sphere of mass M and radius R is rotating around an axis that is tangent to the sphere (Fig. P13.9). What is the rotational inertia of the sphere in this scenario in terms of M and R?

FIGURE P13.9

10. A Suppose a disk having mass M_{tot} and radius R is broken into four equal parts (Fig. P13.10). What is the rotational inertia of one-fourth of the disk around the z axis shown?

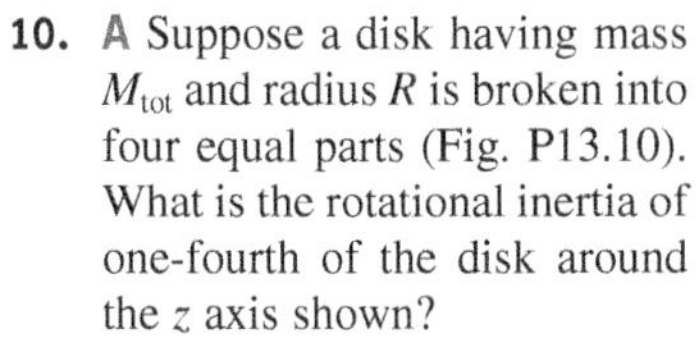

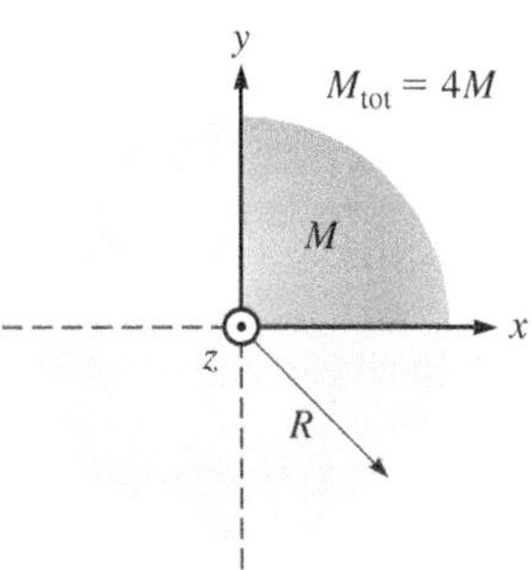

FIGURE P13.10

Problems 11 and 12 are paired.

11. A A thin disk of radius R has a nonuniform density $\sigma = 4.5r^2$, where σ has units of kg/m² when r is in meters. Derive an expression for the rotational inertia of this disk around an axis through its center and perpendicular to the disk's surface, assuming R is given in meters.

12. A Given the disk and density in Problem 11, derive an expression for the rotational inertia of this disk around an axis at the edge of the disk and perpendicular to the disk's surface.

13. N A large stone disk is viewed from above and is initially at rest as seen in Figure P13.13. The disk has a mass of 150.0 kg and a radius of 2.000 m. A constant force of 40.0 N is applied tangent to the edge of the disk for 60.0 s, causing the disk to spin around the z axis. **a.** Calculate the angular acceleration of the stone, finding both the direction and magnitude. **b.** What is the final angular velocity of the stone? **c.** Calculate the translational speed for a point on the edge of the stone after 60.0 s.

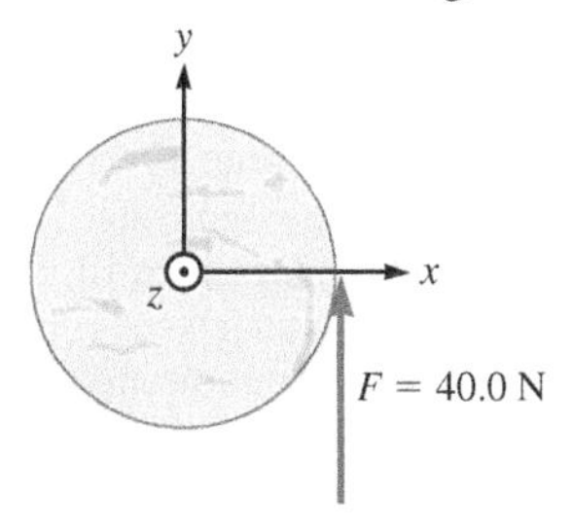

FIGURE P13.13

Problems 14 and 17 are paired.

14. A bike tire has a diameter of 0.775 m and a mass of 3.58 kg. Its mass is concentrated in its rim.

a. N Find its rotational inertia around its axle.

b. N Find its rotational inertia around its diameter.

c. C Which is greater, the rotational inertia around its axle or around its diameter, and why is it greater?

15. N A uniform disk of mass $M = 3.00$ kg and radius $r = 22.0$ cm is mounted on a motor through its center. The motor accelerates the disk uniformly from rest by exerting a constant torque of 1.00 N · m. **a.** What is the time required for the disk to reach an angular speed of 8.00×10^2 rpm? **b.** What is the number of revolutions through which the disk spins before reaching this angular speed?

16. N The net total torque of 50.0 N · m on a wheel rotating around an axis through its center is due to an applied force and a frictional torque at the axle. Starting from rest, the wheel reaches an angular speed of 12.0 rad/s in 5.00 s. At $t = 5.00$ s, the applied force is removed, and the frictional torque brings the wheel to a stop in 30.0 s. **a.** What is the rotational inertia of the wheel? **b.** What is the magnitude of the frictional torque acting on the wheel? **c.** What is the total number of revolutions the wheel undergoes during this 35.0-s interval?

13-3 Rotational Kinetic Energy

17. N A bike tire has a diameter of 0.775 m and a mass of 3.58 kg. Its mass is concentrated in its rim. Assuming the tire rotates with an angular speed of 5.34 rev/s, what is its rotational kinetic energy if it is rotating around its axle?

18. N The system shown in Figure P13.18 consisting of four particles connected by massless, rigid rods is rotating around the x axis with an angular speed of 2.50 rad/s. The particle masses are $m_1 = 1.00$ kg, $m_2 = 4.00$ kg, $m_3 = 2.00$ kg, and $m_4 = 3.00$ kg. **a.** What is the rotational inertia of the system around the x axis? **b.** Using $K_r = \frac{1}{2}I\omega^2$ (Eq. 13.10), what is the total rotational kinetic energy of the system? **c.** What is the tangential speed of each of the four particles? **d.** Considering the system as four particles in motion and using $K = \sum_i \frac{1}{2}mv_i^2$, what is the total kinetic energy of the system? How does this value compare with the result obtained in part (b)?

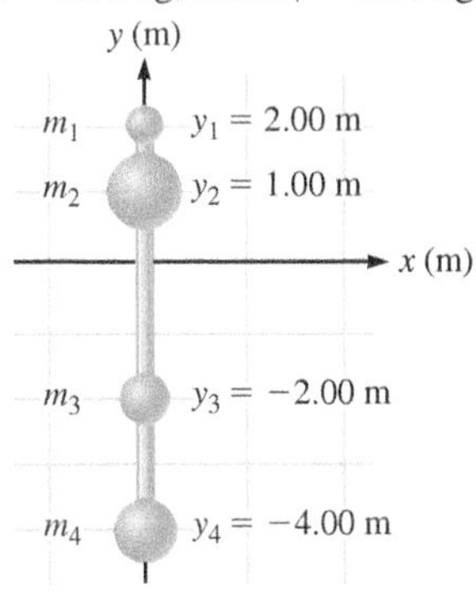

FIGURE P13.18

19. N A 10.0-kg disk of radius 2.0 m rotates from rest as a result of a 20.0-N tangential force applied at the edge of the disk. What is the kinetic energy of the disk 4.00 s after the force is applied?

20. N Model the Earth as a solid sphere rotating around an axis through its center of mass. Find the rotational kinetic energy of the Earth.

Problems 21 and 22 are paired.

21. N A thin, hollow sphere of mass 2.00 kg and radius 0.600 m is rolling on a horizontal surface with a constant angular speed of 60.0 rpm. Find the total kinetic energy of the sphere.

22. N In Problem 21, what fraction of the kinetic energy is translational kinetic energy, and what fraction of the kinetic energy is rotational kinetic energy?

23. N The 2.00-m-long hour hand of a large clock has a mass of 25.0 kg, and the 3.00-m-long minute hand has a mass of 15.0 kg. Assuming the clock hands can be modeled as thin rods and the clock hands rotate at a constant rate, what is the total rotational kinetic energy of the clock hands around their axis of rotation?

24. N When a pitcher throws a baseball or softball, the rotation speed is important when it comes to making the ball "break," or curve away from its expected motion. A baseball that has a mass of 0.143 kg and a radius of 3.65 cm can rotate at a rate of 1.80×10^3 rpm on its way to home plate, causing the ball to curve! Suppose this baseball is rotating around an axis through its center at this extreme rate. What is the rotational kinetic energy of the baseball as it travels toward home plate?

25. N In the fall of 2003, Sanyo announced that it could make 10 compact disks (CDs) from one ear of corn. Each CD has a mass of 16.3 g and a radius $R = 6.00$ cm. From Problem 12.82 (page 359), when information is being read off the innermost ring of a CD, its angular speed is $\omega_0 = 52.4$ rad/s. The CD slows down within the player so that when information is read off the outermost ring, $\omega = 20.9$ rad/s. Assume a Sanyo corn CD can be

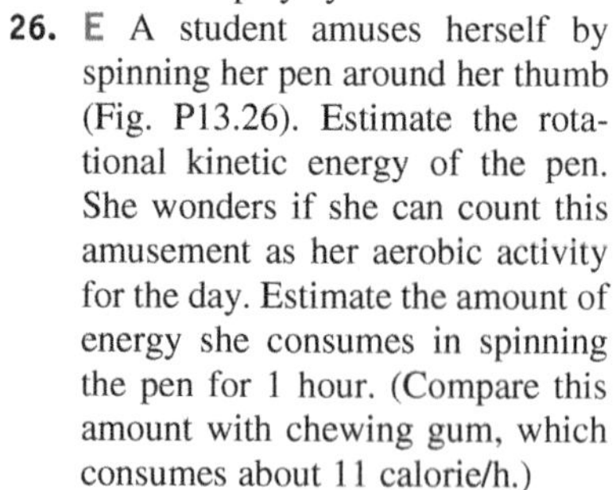

modeled as a solid disk rotating around its central axis. Find the change in its rotational kinetic energy between the beginning and end of its play cycle.

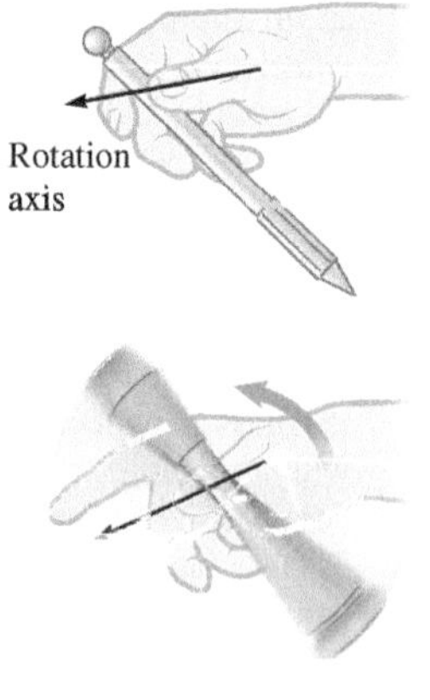

FIGURE P13.26

26. E A student amuses herself by spinning her pen around her thumb (Fig. P13.26). Estimate the rotational kinetic energy of the pen. She wonders if she can count this amusement as her aerobic activity for the day. Estimate the amount of energy she consumes in spinning the pen for 1 hour. (Compare this amount with chewing gum, which consumes about 11 calorie/h.)

27. N The motion of spinning a hula hoop around one's hips can be modeled as a hoop rotating around an axis not through the center, but offset from the center by an amount h, where h is less than R, the radius of the hoop. Suppose Maria spins a hula hoop with a mass of 0.75 kg and a radius of 0.62 m around her waist. The rotation axis is perpendicular to the plane of the hoop, but approximately 0.40 m from the center of the hoop. **a.** What is the rotational inertia of the hoop in this case? **b.** If the hula hoop is rotating with an angular speed of 13.7 rad/s, what is its rotational kinetic energy?

13-4 Special Case of Rolling Motion

28. C A person riding a bicycle on a level road accelerates. Consider the person and the bicycle to make up the system and identify the external force that accelerates the system.

29. A uniform hoop and a uniform solid disk are released from rest from a height h on an incline and roll without slipping.
    a. C Which object reaches the bottom of the incline first? Explain your reasoning.
    b. A Find expressions for the speed with which each object reaches the bottom of the incline to confirm your answer to part (a).

30. A bicycle and its rider have a mass of 85.6 kg. The bike's tires have a radius of 0.382 m. The mass of each tire is 0.980 kg and is concentrated near each tire's rim. The rider is riding at a constant speed of 5.40 m/s.
    a. N Find the kinetic energy of the center of mass.
    b. N Find the rotational kinetic energy of the tires.
    c. C Compare and comment on your results.

Problems 31 and 32 are paired.

31. N Sophia is playing with a set of wooden toys, rolling them off the table and onto the floor. One of the toys is a small sphere with a mass of 0.024 kg and a radius of 0.020 m, and another is a small cylinder that also has a mass of 0.024 kg but a radius of 0.013 m. She rolls each toy so that it has the same translational speed of 0.40 m/s. How much greater is the kinetic energy of the cylinder than the kinetic energy of the sphere?

32. N Consider again the two toys in Problem 31 and their translational speed. **a.** Determine the ratio of the rotational kinetic energy to the translational kinetic energy for each toy. **b.** If the sphere and the cylinder had the same angular speed instead of the same translational speed, how would their translational speeds compare?

Problems 33 and 34 are paired.

33. N A spring with spring constant 25 N/m is compressed a distance of 7.0 cm by a ball with a mass of 202.5 g (Fig. P13.33). The ball is then released and rolls without slipping along a horizontal surface, leaving the spring at point A. The process is repeated, using a block instead, with a mass identical to that of the ball. The block compresses the spring by 7.0 cm and is also released, leaving the spring at point A. Assume the ball rolls, but ignore other effects of friction. **a.** What is the speed of the ball at point B? **b.** What is the speed of the block at point B?

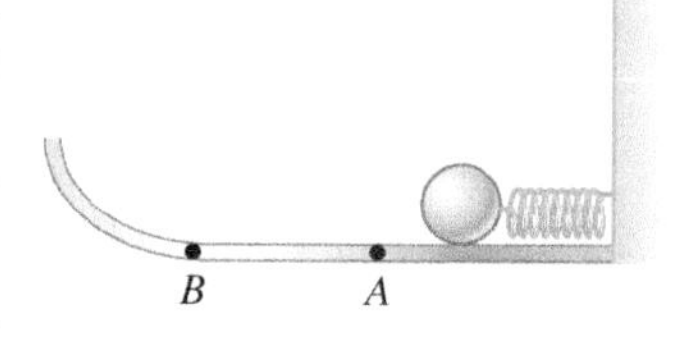

FIGURE P13.33 Problems 33 and 34.

34. N Consider the situation described in Problem 33. Each object travels up the incline shown in Figure P13.33 before coming to rest momentarily. Assume the ball rolls, but ignore other effects of friction. **a.** How high above the starting position is the ball when it comes to rest momentarily? **b.** How high above the starting position is the block when it comes to rest momentarily?

35. C CASE STUDY In the laboratory, you have probably used a small cart on a track. The cart's mass is approximately 0.5 kg, and it rolls on four small wheels. Usually, we ignore the rotational kinetic energy of those four wheels. When Chris did not take the rotational kinetic energy of the ball on the loop-the-loop challenge into account, however, he ran into significant errors. Why is it safe to ignore the kinetic energy of rotation in the case of the cart, but not in case of the ball?

13-5 Work and Power

36. C Suppose you wish to get more power out of a waterwheel (Example 13.10, page 375). How could you improve the wheel?

37. N A tangential force of 1.5 N acts on an object of mass 2.0 kg that is moving in a circular path of radius 3.0 m. The force causes the object's speed to increase. Find the work performed during one complete revolution of the object.

Problems 38 and 39 are paired.

38. N A merry-go-round at a park is subject to a constant torque with a magnitude of 645 N · m as a parent pushes the ride. **a.** How much work is performed by the torque as the merry-go-round rotates through 1.75 revolutions? **b.** If it takes the merry-go-round 4.51 s to go through the 1.75 revolutions, what is the power transferred by the parent to the ride?

39. N A parent exerts a torque on a merry-go-round at a park. The torque has a magnitude given by $\tau = 32\theta^2 + 120\theta - 145$, where the torque has units of newton meters when θ is in units of radians. **a.** How much work is performed by the torque as the merry-go-round rotates through 1.75 revolutions, beginning at $\theta_i = 0$? **b.** If it takes the merry-go-round 4.51 s to go through the 1.75 revolutions, what is the power transferred by the parent to the ride?

40. E Most people know that Benjamin Franklin studied electricity by flying a kite during a storm. He also studied electricity indoors using a hand-cranked electrical generator (Fig. P13.40). For now, we are interested in the work done by the person in turning the crank. The entire apparatus is about 4.5 ft tall. Assume a person can apply a constant force on the crank's handle. Estimate the work done by a person in turning the crank through one revolution. We'll

The Library Company of Philadelphia

FIGURE P13.40

study this device in detail in Chapter 27. The work done by the person goes into electrifying the ball shown on the top of the generator.

41. N Today, waterwheels are not often used to grind food. Instead, we have electrical devices such as blenders, choppers, and mixers. The electric motors in these devices are similar to a waterwheel, but instead of falling water causing the wheel to spin, electricity causes a shaft to spin. The specifications on a particular electric motor reports that at 1.75×10^3 rpm, it puts out 5 hp. What is the corresponding torque in N · m?

13-6 Angular Momentum

42. Model Jupiter's orbit around the Sun as uniform circular motion.
    a. N Find the angular momentum of Jupiter's orbital motion.
    b. N Although Jupiter is a gas giant, model it as a solid uniform sphere and find the angular momentum of Jupiter's rotational motion.
    c. C Compare your results and comment.

43. N A buzzard ($m = 9.29$ kg) is flying in circular motion with a speed of 8.44 m/s while viewing its meal below. If the radius of the buzzard's circular motion is 8.00 m, what is the angular momentum of the buzzard around the center of its motion?

44. A An object of mass M is thrown with a velocity v_0 at an angle θ with respect to the horizontal (Fig. P13.44). Find the angular momentum of the object around the origin when the object is at the highest point of its trajectory.

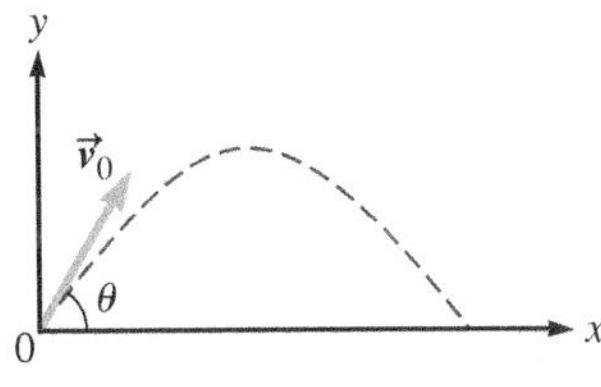

FIGURE P13.44

Problems 45, 46, and 47 are grouped.

45. N A thin rod of length 2.65 m and mass 13.7 kg is rotated at an angular speed of 3.89 rad/s around an axis perpendicular to the rod and through its center of mass. Find the magnitude of the rod's angular momentum.

46. N A thin rod of length 2.65 m and mass 13.7 kg is rotated at an angular speed of 3.89 rad/s around an axis perpendicular to the rod and through one of its ends. Find the magnitude of the rod's angular momentum.

47. N A cylinder of length 2.65 m, radius 0.350 m, and mass 13.7 kg is rotated at an angular speed of 3.89 rad/s around an axis parallel to the length of the cylinder and through its center. Find the magnitude of the cylinder's angular momentum.

48. N Two particles of mass $m_1 = 2.00$ kg and $m_2 = 5.00$ kg are joined by a uniform massless rod of length $\ell = 2.00$ m (Fig. P13.48). The system rotates in the xy plane about an axis through the midpoint of the rod in such a way that the particles are moving with a speed of 3.00 m/s. What is the angular momentum of the system?

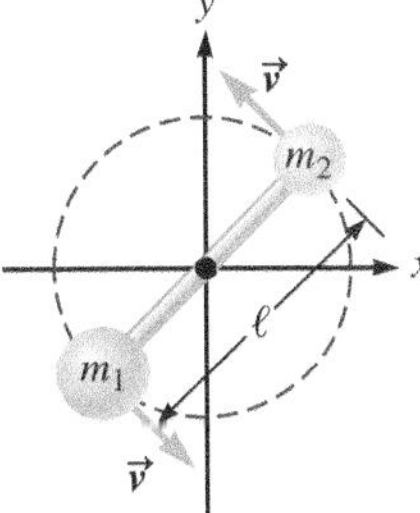

FIGURE P13.48

13-7 Conservation of Angular Momentum

49. N A turntable (disk) of radius $r = 26.0$ cm and rotational inertia 0.400 kg · m² rotates with an angular speed of 3.00 rad/s around a frictionless, vertical axle. A wad of clay of mass $m = 0.250$ kg drops onto and sticks to the edge of the turntable. What is the new angular speed of the turntable?

50. C Reanalyze the unicycle's motion in Example 13.15 (page 382). This time, leave the Earth out of the system and explain how the torque exerted by gravity causes the unicycle to turn. Your explanation should involve a diagram.

Problems 51 and 52 are paired.

51. C While at a party, a friend proposes that if you sit and rotate freely in an office chair holding two heavy barbells close to your chest and then extend your arms, you will rotate so much faster that you will fall out of the chair. Is your friend correct? If yes, explain. If no, correct your friend as to how the outcome might be achieved in this setting. What can you do or change that would cause the increased rotation to be magnified further?

52. N While at a party, a friend challenges you to a game of sorts, where the challenge is to remain sitting throughout. He suggests you hold two 10-kg iron barbells as you sit on a barstool that is free to rotate. Your friend has you extend your arms outward so that the barbells are held away from your body. He then starts the stool rotating. Upon reaching an angular speed of 5.00 rad/s, your friend tells you to pull the barbells back toward your chest. Assuming your body has a rotational inertia of about 8.59 kg · m² without the barbells and your arms are about 0.75 m long, what is your angular speed after you pull the barbells so they are at the center of your chest?

53. N Two children ($m = 30.0$ kg each) stand opposite each other on the edge of a merry-go-round. The merry-go-round, which has a mass of 1.80×10^2 kg and a radius of 1.5 m, is spinning at a constant rate of 0.50 rev/s. Treat the two children and the merry-go-round as a system. **a.** Calculate the angular momentum of the system, treating each child as a particle. **b.** Calculate the total kinetic energy of the system. **c.** Both children walk half the distance toward the center of the merry-go-round. Calculate the final angular speed of the system.

54. A A disk of mass m_1 is rotating freely with constant angular speed ω. Another disk of mass m_2 that has the same radius is gently placed on the first disk. If the surfaces in contact are rough so that there is no slipping between the disks, what is the fractional decrease in the kinetic energy of the system?

Problems 55, 56, 60, and 61 are grouped.

55. N A *month* is the time for the Moon to orbit the Earth. Currently, a month is about 28 days. The same side of the Moon always faces the Earth, so the period of the Moon's rotation equals the period of its orbit. Take into account the orbital motion of the Moon around the Earth and the rotational motion of the Earth and the Moon to find the total angular momentum of the Earth–Moon system. Assume the Moon orbits the Earth in uniform circular motion and the Earth and the Moon are solid, uniform spheres. All the motion is counterclockwise as seen from above the North Pole.

56. In Problem 55, the external torques on the Earth–Moon system are negligible so that the angular momentum of the system is nearly constant. The tidal force between the Earth and the Moon is causing the Earth's rotation to slow down, however. In the distant future, the period of the Earth's rotation will equal the period of the Moon's orbit and rotation. All will be equal to about 50 (24-hour) days.
    a. C What must happen to the Moon's orbit for angular momentum to be conserved? (This question may be answered in the **INTERPRET and ANTICIPATE** step.)
    b. N What will be the radius of the Moon's orbit at that time?

57. A The angular momentum of a sphere is given by $\vec{L} = (-4.59t^3)\hat{\imath} + (6.01 - 1.19t^2)\hat{\jmath} + (6.26t)\hat{k}$, where L has units of kg · m²/s when t is in seconds. What is the net torque on the sphere as a function of time?

General Problems

58. A A ball of mass M is connected to a second ball of mass m by a massless, rigid rod of length L (Fig. P13.58). **a.** Show that for an axis perpendicular to the rod, the minimum rotational inertia of the system is for an axis passing through the center of mass. **b.** What is the minimum rotational inertia of the system for an axis perpendicular to the rod?

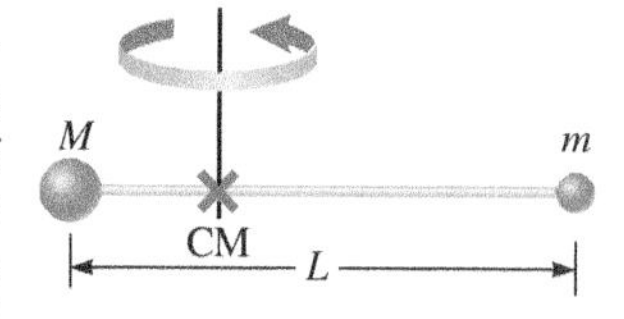

FIGURE P13.58

59. N An 80.0-kg disk with a radius of 2.50 m that can rotate around a frictionless axle through its center is initially at rest. A force of 200.0 N can be applied to the disk at any radial distance from the axis of rotation, perpendicular to the radial direction in the plane of the disk. At what radial distance should the force be applied to accelerate the disk so that it completes 3.00 revolutions in 6.00 s?

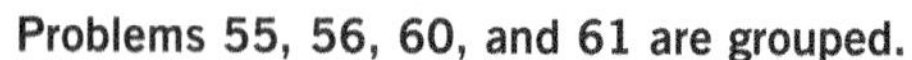

Problems 55, 56, 60, and 61 are grouped.

60. N Find the rotational inertia of the Earth–Moon system around its center of mass.

61. N The Earth and the Moon orbit their system's center of mass roughly every 28 days. Find the system's rotational kinetic energy.

62. A A rigid rod of mass M and length L is pivoted around point P at one of its ends (Fig. P13.62). Find the speed of the center of mass of the rod when it is vertical if it is released from its horizontal position.

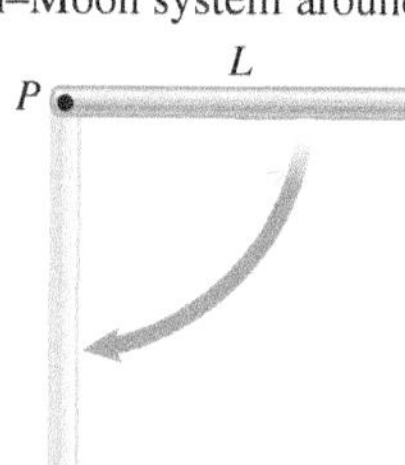

FIGURE P13.62

63. N A uniform cylinder of radius r = 10.0 cm and mass m = 2.00 kg is rolling without slipping on a horizontal tabletop. The cylinder's center of mass is observed to have a speed of 5.00 m/s at a given instant. **a.** What is the translational kinetic energy of the cylinder at that instant? **b.** What is the rotational kinetic energy of the cylinder around its center of mass at that instant? **c.** What is the total kinetic energy of the cylinder at that instant?

64. C Why does a twirling baton (Fig. P13.64) have a fairly massive rubber tip on its ends (aside from it being a safety feature)?

© sshepard/iStockphoto.com

FIGURE P13.64

65. A A thin, spherical shell of mass m and radius R rolls down a parabolic path PQR from height H without slipping (assume $R \ll H$) as shown in Figure P13.65. Path PQ is rough (and so the shell will roll on that path), whereas path QR is smooth, or frictionless (so the shell will only slide, not roll, in this region). Determine the height h above point Q reached by the shell on path QR.

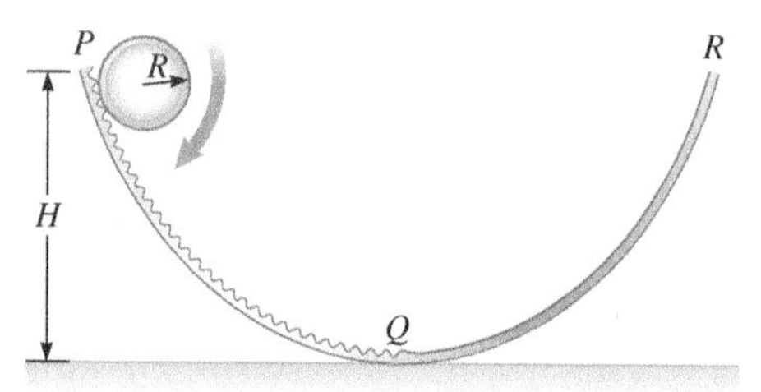

FIGURE P13.65

Problems 66 and 67 are paired.

66. E To give a pet hamster exercise, some people put the hamster in a ventilated ball and allow it roam around the house (Fig. P13.66). When a hamster is in such a ball, it can cross a typical room in a few minutes. Estimate the total kinetic energy in the ball–hamster system.

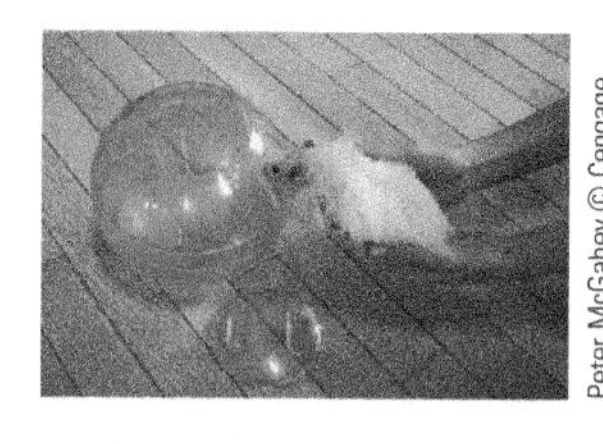

Peter McGahey © Cengage Learning

FIGURE P13.66 Problems 66 and 67.

67. E **Review** Estimate the power consumed by a hamster exercising in a ventilated ball (Fig. P13.66). *Hint*: Read Problem 13.66.

68. Some potters prefer to keep their wheel rotating at a constant speed while they work the clay. Suppose you are going to try your hand at some pottery and want to make a tall, cylindrical vase. Should you be worried about the angular speed changing due to a change in shape of the clay? Assume the wheel–clay system is brought to a modest angular speed (120 rpm) while the clay is in the approximate shape of a sphere in the center of the wheel. Estimate and find appropriate values for the masses of the clay and the wheel, the radius of the wheel, and the density of the clay.

a. E Given your estimates, what would be the total angular momentum of the initial wheel–clay system?

b. C When you shape the clay into a tall, cylindrical vase (one-fourth the radius of the sphere), does the system need to be sped up or slowed down, or should nothing be done to maintain a constant angular speed? (Assume no net external torque is applied as you carefully shape the clay.) Explain.

69. N The velocity of a particle of mass m = 2.00 kg is given by $\vec{v} = -5.10\hat{\imath} + 2.40\hat{\jmath}$ m/s. What is the angular momentum of the particle around the origin when it is located at $\vec{r} = -8.60\hat{\imath} - 3.70\hat{\jmath}$ m?

70. N A ball of mass M = 5.00 kg and radius r = 5.00 cm is attached to one end of a thin, cylindrical rod of length L = 15.0 cm and mass m = 0.600 kg. The ball and rod, initially at rest in a vertical position and free to rotate around the axis shown in Figure P13.70, are nudged into motion. **a.** What is the rotational kinetic energy of the system when the ball and rod reach a horizontal position? **b.** What is the angular speed of the ball and rod when they reach a horizontal position? **c.** What is the linear speed of the center of mass of the ball when the ball and rod reach a horizontal position? **d.** What is the ratio of the speed found in part (c) to the speed of a ball that falls freely through the same distance?

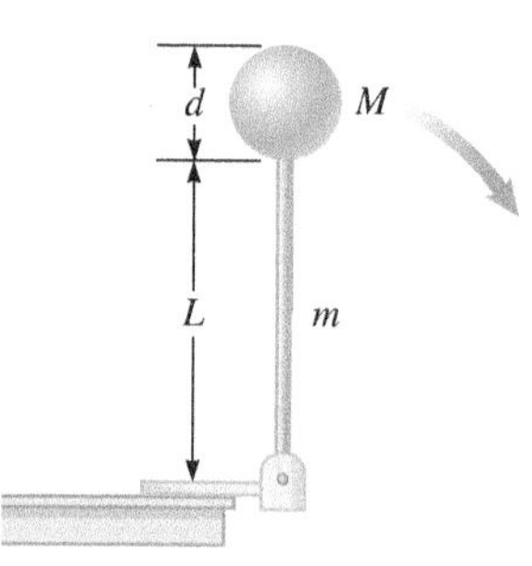

FIGURE P13.70

71. N A long, thin rod of mass m = 5.00 kg and length ℓ = 1.20 m rotates around an axis perpendicular to the rod with an angular speed of 3.00 rad/s. **a.** What is the angular momentum of the rod if the axis passes through the rod's midpoint? **b.** What is the angular momentum of the rod if the axis passes through a point halfway between its midpoint and its end?

72. A A solid sphere and a hollow cylinder of the same mass and radius have a rolling race down an incline as in Example 13.9 (page 372). They start at rest on an incline at a height h above a horizontal plane. The race then continues along the horizontal plane. The coefficient of rolling friction between each rolling object and the surface is the same. Which object rolls the farthest? (Justify your answer with an algebraic expression.)

73. N A uniform disk of mass $m = 10.0$ kg and radius $r = 34.0$ cm mounted on a frictionless axle through its center, and initially at rest, is acted upon by two tangential forces of equal magnitude F, acting on opposite sides of its rim until a point on the rim experiences a centripetal acceleration of 4.00 m/s² (Fig. P13.73). **a.** What is the angular momentum of the disk at this time? **b.** If $F = 2.00$ N, how long do the forces have to be applied to the disk to achieve this centripetal acceleration?

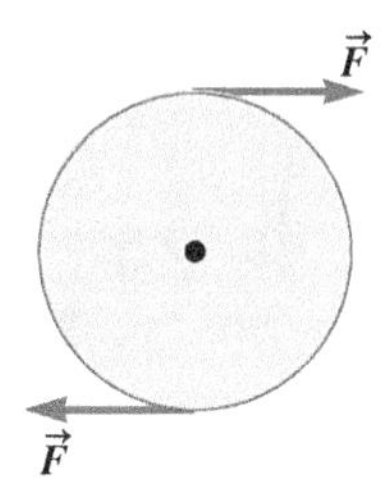

FIGURE P13.73

74. C When a person jumps off a diving platform, she imparts some amount of angular momentum to her body. (To see why, consider a person who jumps straight up: She falls straight down. When a person jumps off a diving platform, she must jump at an angle to the vertical, or she would land on the board.) While she is in the air, angular momentum is conserved. Draw a free-body diagram of the diver in the air and use it to explain why angular momentum is conserved during flight. It may be useful to include the axes of rotation for the two common types of spinning moves that divers perform: twists and flips (Fig. P13.74).

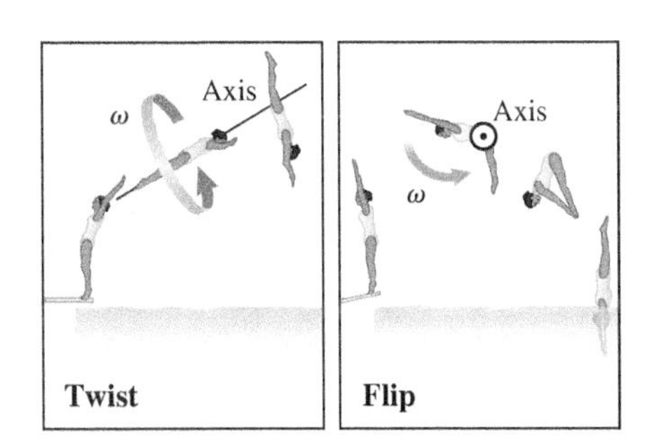

FIGURE P13.74

75. A One end of a massless rigid rod of length ℓ is attached to a wooden block of mass M resting on a frictionless, horizontal tabletop, and the other end is attached to the table through a pivot (Fig. P13.75). A bullet of mass m traveling with a speed v in a direction perpendicular to the rod and parallel to the table impacts the block and embeds itself inside. **a.** What is the angular momentum of this system around a vertical axis through the pivot after the collision? **b.** What is the fraction of the bullet's initial kinetic energy that is lost to internal energy during the collision?

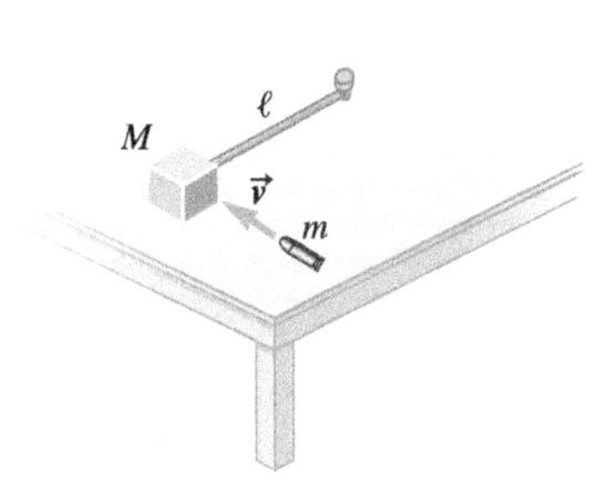

FIGURE P13.75

76. A A uniform solid sphere of mass m and radius r is released from rest and rolls without slipping on a semicircular ramp of radius $R \gg r$ (Fig. P13.76). If the initial position of the sphere is at an angle θ to the vertical, what is its speed at the bottom of the ramp?

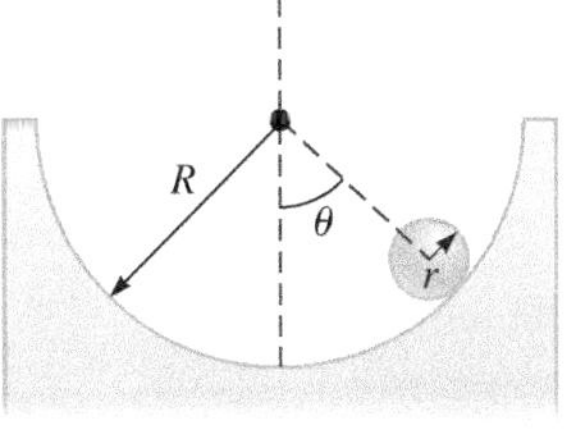

FIGURE P13.76

77. N A rod of length 2.0 m and of negligible mass is constrained to move in a vertical plane (Fig. P13.77). The ends of the rod always touch the x and y axes, respectively, as the rod slides downward. Find the coordinates of the instantaneous axis of rotation of the rod when it makes an angle of exactly 60° with the horizontal.

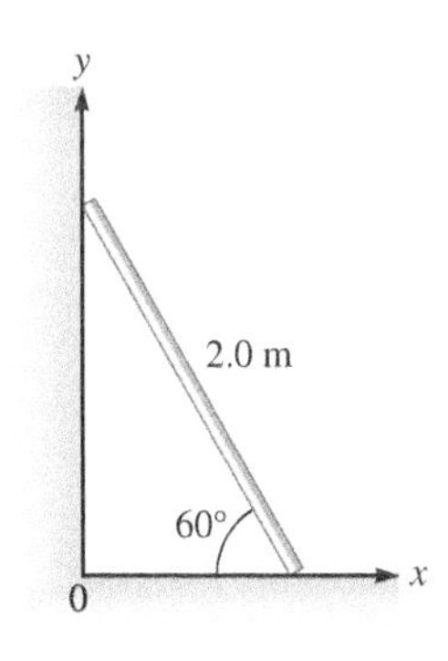

FIGURE P13.77

78. A A cam of mass M is in the shape of a circular disk of diameter $2R$ with an off-center circular hole of diameter R is mounted on a uniform cylindrical shaft whose diameter matches that of the hole (Fig. P13.78). **a.** What is the rotational inertia of the cam and shaft around the axis of the shaft? **b.** What is the rotational kinetic energy of the cam and shaft if the system rotates with angular speed ω around this axis?

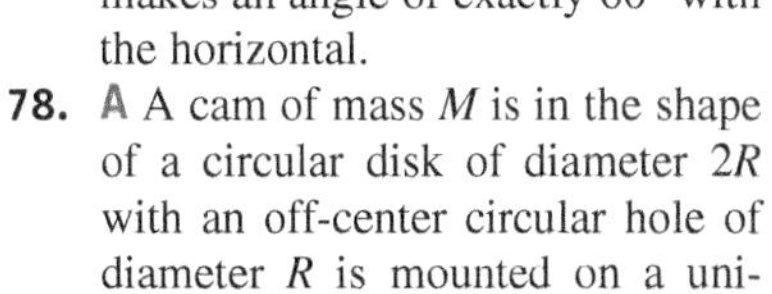

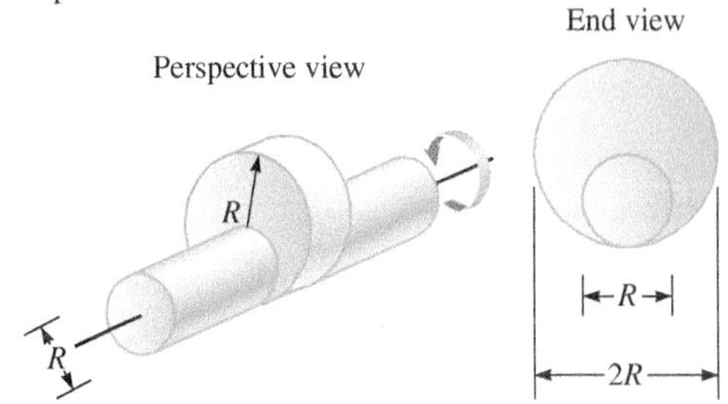

FIGURE P13.78

79. A A uniform solid sphere of radius R initially at rest is subjected to an applied force $\vec{F}$ and begins to roll without slipping (Fig. P13.79). **a.** What is the acceleration of the center of mass of the sphere? **b.** What are the magnitude and direction of the force of friction on the sphere? **c.** What is the speed of the center of mass of the sphere after it has traversed a distance d?

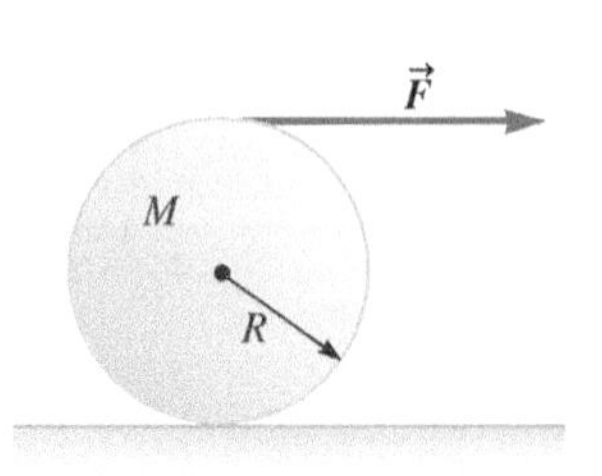

FIGURE P13.79

80. A Consider the downhill race in Example 13.9 (page 372). The acceleration of a particle down an incline is $a_{\text{particle}} = g \sin \beta$, where β is the angle the incline makes with the horizontal. Show that the acceleration of a rolling object down an incline is given by

$$a_{\text{CM}} = \frac{a_{\text{particle}}}{1 + f}$$

where f is the unitless fraction described in Example 13.9.

81. A A disk of mass M and radius R is released from rest along a smooth (frictionless) inclined plane having an angle of inclination θ as shown in Figure P13.81. The disk will slide instead of roll. Find the magnitude of the angular momentum of the disk around the instantaneous point of contact after a time t from the instant of release. Assume dissipative forces are negligible.

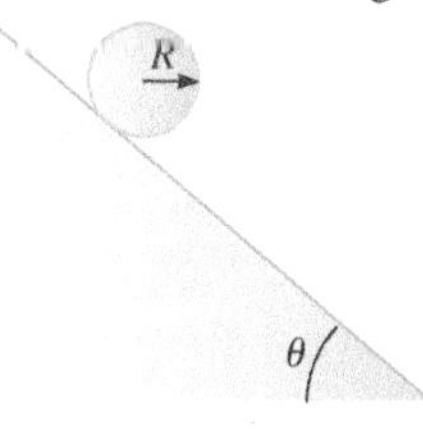

FIGURE P13.81

14 Static Equilibrium, Elasticity, and Fracture

Underlying Principle

Newton's second law applied in the special case of equilibrium

Major Concepts

1. Static equilibrium
   a. Stable static equilibrium
   b. Unstable static equilibrium
   c. Neutral static equilibrium
2. Equilibrium conditions
   a. Force balance condition
   b. Torque balance condition
3. Stress
4. Strain
5. Young's modulus
6. Elastic object and elastic limit
7. Tensile strength
8. Compressive strength
9. Shear modulus

Tools

Cross products

Key Questions

What are the conditions for equilibrium?

How do static objects respond when forces are applied?

14-1 What is static equilibrium? 391

14-2 Conditions for equilibrium 392

14-3 Examples of static equilibrium 394

14-4 Elasticity and fracture 404

Imagine that an engineer is asked to design a roadway to cross a body of water. The engineer has many choices to make. The engineer may choose a suspension bridge like the Golden Gate Bridge in San Francisco or an arch bridge like the Rialto Bridge in Venice (Fig. 14.1) or perhaps a longer and more complex structure such as the Chesapeake Bay Bridge-Tunnel that connects Virginia's eastern shore with the mainland at Virginia Beach near Norfolk. In addition to choosing a type of structure, the engineer has many materials from which to choose: stone, concrete, and steel, just to name a few.

Many structures such as bridges, buildings, and furniture are designed to be stationary, that is, at rest relative to an observer. In this chapter, we take a closer look at the conditions that are necessary for an object to be at rest.

We'll also see how the choice of material affects the structure when forces are applied. Take a classroom chair, for example. When a student sits on the chair, the chair deforms. Bare wooden chairs are less comfortable than cushioned chairs because the wood does not deform as much as does a fabric cushion when compressed.

A.

B.

FIGURE 14.1 **A.** Golden Gate Bridge, San Francisco, California. **B.** Rialto Bridge, Venice, Italy.

14-1 What Is Static Equilibrium?

So far, we have studied objects in translational motion, objects in rotational motion, and even objects in combined translational and rotational motion. In this chapter, we study objects that are at rest, that is, not moving relative to the observer. An object that is at rest is said to be in **static equilibrium**. Why do we need a chapter dedicated to objects at rest? The answer is that objects at rest are very common and important; many engineers and architects spend their entire careers designing structures that must maintain static equilibrium despite the application of large forces.

Static equilibrium is a special case of motion. Start by thinking about a ball rolling (without slipping) down an inclined plane and assume rolling friction is negligible. In Section 13-4, we found it convenient to think about rolling as a combination of translational motion and rotational motion. In this case, the ball speeds up. So, both the translational acceleration $\vec{a}$ of its center of mass and its angular acceleration $\vec{\alpha}$ have some nonzero value. Put another way, the ball's translational momentum $\vec{p} = m\vec{v}_{CM}$ and angular momentum $\vec{L} = I\vec{\omega}$ continually change while it is rolling down the incline.

Now consider a special case of motion in which the ball's translational acceleration and angular acceleration are both zero: $\vec{a} = 0$ and $\vec{\alpha} = 0$. For example, if the ball rolls along a level floor instead of down an incline, the ball will not speed up, slow down, or change direction (as long as dissipative forces are negligible). Therefore, the ball's translational momentum and angular momentum are both constant, and the ball is said to be in *equilibrium*. **Equilibrium** is a special case of motion in which an object's translational momentum $\vec{p}$ and angular momentum $\vec{L}$ are both constant.

EQUILIBRIUM
Underlying Principle

Static equilibrium is a special case of equilibrium in which the object's translational and angular momenta equal a particular constant, zero: $\vec{p} = 0$ and $\vec{L} = 0$. Static equilibrium depends on the inertial reference frame from which the object is observed. For example, if you were sitting on a bridge, you would observe that the bridge is in static equilibrium. If you are drifting with a constant velocity on a raft in the water below the bridge, however, you would say that the bridge has a constant, nonzero translational momentum. You would conclude that the bridge was in equilibrium but not in static equilibrium. Throughout this chapter, we will assume the observer is in the reference frame that allows him to observe static equilibrium.

Types of Static Equilibrium

Static equilibrium is classified into three types: stable, unstable, and neutral. To distinguish between these types, you must imagine what would happen if the object in equilibrium were displaced slightly and then released from its new resting position. If, after being released, the forces acting on the object return it to its equilibrium position, the object is in **stable static equilibrium**. For example, a marble ball at rest in the bottom of a bowl is in stable static equilibrium. If the marble is displaced and then released from its new position, gravity returns the marble to its equilibrium position (Fig. 14.2A). In this case, the marble would roll back and forth through its equilibrium position, and rolling friction would cause it to eventually come to rest at the bottom of the bowl. If, after being released, the object moves farther away from its equilibrium position, the object is in **unstable static equilibrium**. For example, a marble at rest on top of an inverted bowl is in unstable static equilibrium. If the marble is displaced and then released, gravity causes it to roll off the bowl, and the marble will not return

STABLE, UNSTABLE, AND NEUTRAL STATIC EQUILIBRIUM
Major Concept

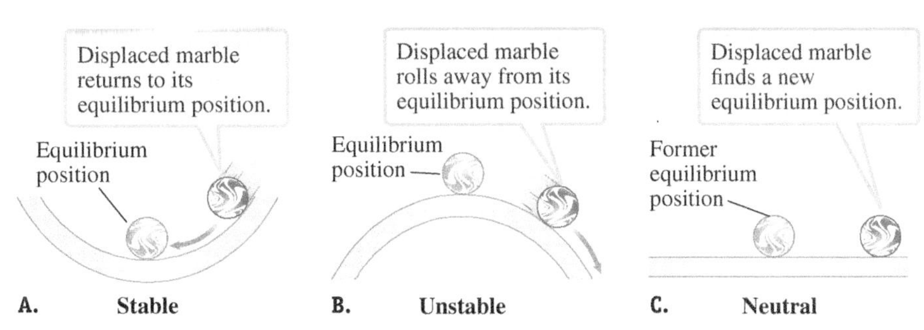

FIGURE 14.2 Types of static equilibrium.

to its equilibrium position (Fig.14.2B). If an object is moved and released from a new position and does not move toward or away from its equilibrium position, the object is in **neutral static equilibrium**. In this case, the forces exerted on the object do not maintain its equilibrium; so, if the object is displaced, it is not restored to its equilibrium position, and it is not forced farther away. For example, a marble at rest on a level surface is in neutral static equilibrium (Fig. 14.2C). If you displace the marble to the right and then release your hold, gravity neither moves the ball back to equilibrium nor farther away. Although the marble is in a new position, there has been no change in the Earth–marble system's gravitational potential energy.

CASE STUDY Designing an Air and Space Exhibit

Besides making sure that a particular design meets the project goals, an engineer must also make sure that the design is safe from collapse, explosion, or other hazards. The designers of the Udvar-Hazy Center—a companion facility to the Smithsonian Air and Space Museum in Washington, D.C.—faced the extraordinary challenge of planning a structure to support 73 hanging aircraft. Their design had to protect the public, the aircraft, and the building. The aircraft must be low enough to be easily seen from the ground and must be accessible by high-lift cherry pickers. To convey a sense of movement and excitement, the aircraft are hung at various angles (Fig. 14.3). To make this design possible, the architects and engineers incorporated important structural elements into the museum. In this case, we explore one such element: the steel cables used to support a single plane. Our goal is to calculate the maximum airplane weight that can be hung from two steel cables without breaking the cables.

FIGURE 14.3 Aircraft at the Udvar-Hazy Center.

CONCEPT EXERCISE 14.1

A rubber duck floats in a bathtub. Imagine moving the duck by a small amount in different directions such as up and down or side to side. Is the floating duck in stable, unstable, or neutral equilibrium?

CONCEPT EXERCISE 14.2

If the rubber duck in Concept Exercise 14.1 is replaced by a small toy boat with an open hull such as a rowboat or canoe, is the boat in stable, unstable, or neutral equilibrium? *Hint*: What if you displace the boat downward or rock it so that water enters the hull?

14-2 Conditions for Equilibrium

For an object to be in equilibrium, two conditions must be met: (1) the object's translational momentum must be constant, and (2) its angular momentum must be constant. We can express these two conditions for equilibrium in mathematical terms. We start with Newton's second law written in terms of momentum:

$$\sum \vec{F} = \vec{F}_{\text{tot}} = \frac{d\vec{p}}{dt} \tag{10.2}$$

In equilibrium, $\vec{p}$ = constant, so the time derivative of momentum is zero:

$$\frac{d\vec{p}}{dt} = 0$$

Then, according to Equation 10.2, the first condition for equilibrium is that the total force exerted on an object is zero:

FORCE BALANCE CONDITION FOR EQUILIBRIUM

Major Concept

$$\vec{F}_{\text{tot}} = 0 \tag{14.1}$$

Conceptually, Equation 14.1 says that for an object to be in equilibrium, the forces must "balance out," so we will refer to it as the **force balance condition**.

For the second equilibrium condition (angular momentum is constant), we start with Newton's second law written in terms of torque and angular momentum:

$$\sum \vec{\tau} = \vec{\tau}_{\text{tot}} = \frac{d\vec{L}}{dt} \quad (13.27)$$

An object in equilibrium has a constant angular momentum $\vec{L}$, so the time derivative of angular momentum is zero:

$$\frac{d\vec{L}}{dt} = 0$$

Then, according to Equation 13.27, the second condition for equilibrium is that the total torque acting on an object is zero:

$$\vec{\tau}_{\text{tot}} = 0 \quad (14.2)$$

TORQUE BALANCE CONDITION FOR EQUILIBRIUM

Major Concept

Conceptually, this second condition for equilibrium says that the torques on the object must "balance out," so we will refer to it as the **torque balance condition**.

These conditions (Eqs. 14.1 and 14.2) are true for all equilibrium situations, static or not. In the next section, we will apply the equilibrium conditions to objects that are in static equilibrium.

Equations 14.1 and 14.2 and are vector equations. It is often practical to apply these equations in their component forms. The condition that the forces must balance may be written in terms of three scalar component equations:

$$\sum F_x = 0 \qquad \sum F_y = 0 \qquad \sum F_z = 0 \quad (14.3)$$

Similarly, the condition that the torques must balance may be written in terms of three scalar component equations:

$$\sum \tau_x = 0 \qquad \sum \tau_y = 0 \qquad \sum \tau_z = 0 \quad (14.4)$$

As with any vector, the subscript indicates the Cartesian component of the torque. For example, $\sum \tau_x$ is the sum of the x components of all the torques.

Cross Product Revisited

CROSS PRODUCTS Tool

Torque is the result of a cross product (Eq. 12.23, $\vec{\tau} = \vec{r} \times \vec{F}$, where $\vec{r}$ is the position vector from the rotation axis to the point of application of the force $\vec{F}$). In Section 12-6, we found the magnitude and direction of a vector resulting from a cross product. Now we need to write the result of a cross product in component form because we will need the components of the torque to use Equation 14.4.

Consider vector $\vec{R}$, the cross product of vectors $\vec{A}$ and $\vec{B}$:

$$\vec{R} = \vec{A} \times \vec{B}$$

To write $\vec{R}$ in component form, we start by writing $\vec{A}$ and $\vec{B}$ in component form:

$$\vec{R} = (A_x\hat{\imath} + A_y\hat{\jmath} + A_z\hat{k}) \times (B_x\hat{\imath} + B_y\hat{\jmath} + B_z\hat{k})$$

The cross product is distributive, meaning that

$$\begin{aligned}\vec{R} = &(A_x\hat{\imath} \times B_x\hat{\imath}) + (A_x\hat{\imath} \times B_y\hat{\jmath}) + (A_x\hat{\imath} \times B_z\hat{k}) \\ &+ (A_y\hat{\jmath} \times B_x\hat{\imath}) + (A_y\hat{\jmath} \times B_y\hat{\jmath}) + (A_y\hat{\jmath} \times B_z\hat{k}) \\ &+ (A_z\hat{k} \times B_x\hat{\imath}) + (A_z\hat{k} \times B_y\hat{\jmath}) + (A_z\hat{k} \times B_z\hat{k})\end{aligned}$$

The cross product of parallel vectors is zero (Eq. 12.22 with $\varphi = 0$). Therefore, the three of the terms containing $(\hat{\imath} \times \hat{\imath})$, $(\hat{\jmath} \times \hat{\jmath})$, and $(\hat{k} \times \hat{k})$ are zero:

$$\begin{aligned}\vec{R} = &\,0 + (A_x\hat{\imath} \times B_y\hat{\jmath}) + (A_x\hat{\imath} \times B_z\hat{k}) \\ &+ (A_y\hat{\jmath} \times B_x\hat{\imath}) + 0 + (A_y\hat{\jmath} \times B_z\hat{k}) \\ &+ (A_z\hat{k} \times B_x\hat{\imath}) + (A_z\hat{k} \times B_y\hat{\jmath}) + 0\end{aligned}$$

The angle between other pairs of unit vectors is $\varphi = 90°$, so the magnitude of their cross products is $|1||1| \sin 90° = 1$. The directions of their cross products are found by using the right-hand rule:

$$\vec{R} = A_xB_y\hat{k} - A_xB_z\hat{\jmath} - A_yB_x\hat{k} + A_yB_z\hat{\imath} + A_zB_x\hat{\jmath} - A_zB_y\hat{\imath}$$

Regrouping to collect all scalar factors of each unit vector,

$$\vec{R} = (A_yB_z - A_zB_y)\hat{\imath} + (A_zB_x - A_xB_z)\hat{\jmath} + (A_xB_y - A_yB_x)\hat{k} \quad (14.5)$$

Equation 14.5 gives us a way to find the components of $\vec{R}$ directly from the components of $\vec{A}$ and $\vec{B}$:

$$R_x = A_yB_z - A_zB_y \qquad R_y = A_zB_x - A_xB_z \qquad R_z = A_xB_y - A_yB_x \quad (14.6)$$

Applying Equations 14.5 and 14.6 to the equation for torque (Eq. 12.23), we have

$$\vec{\tau} = \vec{r} \times \vec{F} = (r_yF_z - r_zF_y)\hat{\imath} + (r_zF_x - r_xF_z)\hat{\jmath} + (r_xF_y - r_yF_x)\hat{k} \quad (14.7)$$

and

$$\tau_x = r_yF_z - r_zF_y \qquad \tau_y = r_zF_x - r_xF_z \qquad \tau_z = r_xF_y - r_yF_x \quad (14.8)$$

Equation 14.8 allows us to apply the torque balance condition (Eq. 14.4) in component form. For many situations, applying the equilibrium conditions in component form is the most direct way to solve a problem.

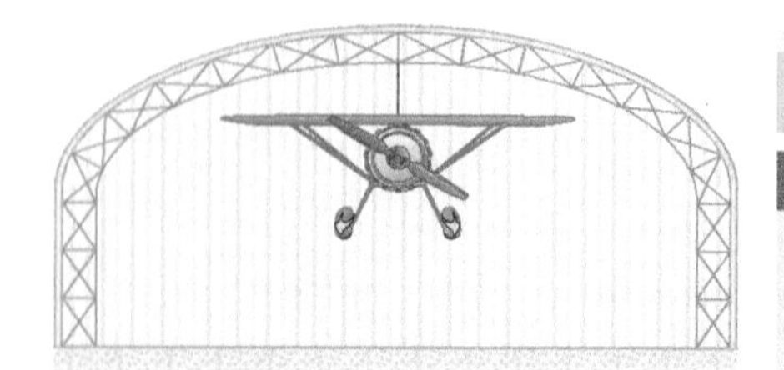

A.

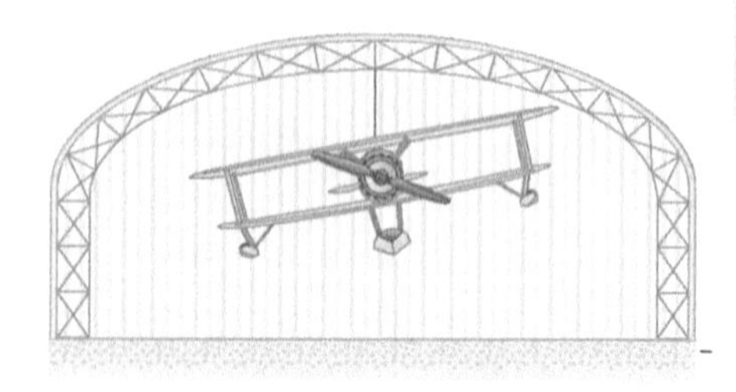

B.

FIGURE 14.4

CONCEPT EXERCISE 14.3

CASE STUDY Hanging a Plane from a Single Point

In the Air and Space Museum, airplanes are usually hung from two or more cables.

a. Is it possible to hang an airplane from a cable attached at a single point such that its wings and fuselage are parallel to the ground as in Figure 14.4A? If so, describe how to do it. If not, why not?

b. Is it possible to hang an airplane from a cable attached at a single point such that its wings are tilted as in Figure 14.4B? If so, describe how to do it. If not, why not?

CONCEPT EXERCISE 14.4

Two people want to sit on opposite ends of an adjustable seesaw.

a. If the two people have equal masses, where should they put the pivot so as to have balanced horizontal seesaw?

b. If one person has more mass than the other person, where should they put the pivot? In the middle? Closer to the heavier person? Closer to the lighter person?

14-3 Examples of Static Equilibrium

In this section, we consider several examples involving objects that are in static equilibrium.

PROBLEM-SOLVING STRATEGY

Static Equilibrium

The strategy we follow is similar to that established in previous chapters (primarily Chapters 5, 6, and 12) when applying Newton's second law.

A modified free-body diagram is helpful for visualizing the problem in the **INTERPRET and ANTICIPATE** step. Because the object is in equilibrium, its acceleration is zero. In Section 5-8, we asked for information about acceleration on the diagram. Here, however, you may wish to keep the diagram clean, so leave the acceleration off the diagram. The modified free-body diagram has five elements.

Element 1. Draw the object as a simple shape. Because we must calculate the torque acting on the object, the object cannot be represented by the particle model. So, represent the object by a simple shape rather than by a dot.

Element 2. As always, your free-body diagram must include all forces acting on the object. You must also **draw the forces at their point of contact.** Recall from Section 9-6 that gravity acts at the object's center of mass, so $\vec{F}_g$ should always be attached to the object's center of mass and point toward the center of the Earth.

Element 3. Choose a rotation axis to calculate torque. Because an object in static equilibrium is at rest (not actually rotating), you may choose any convenient rotation axis. Often, a rotation axis that passes though the point of contact of one of the forces is convenient because that force does not exert a torque around your rotation axis and you will therefore have one torque fewer to calculate. In a situation in which the geometry is complicated—such as when the forces and the points of contact lie in different planes or when the situation has little or no symmetry—you may need to choose more than one rotation axis.

Element 4. It is often convenient to **choose your coordinate system** so that one of the coordinate axes is the same as the rotation axis.

Element 5. Draw the position vector $\vec{r}$ for each force in your free-body diagram. Show each position vector from the rotation axis to the point where the force is exerted on the object. (It is not necessary to draw $\vec{r}$ if the force is on the rotation axis. In that case, $\vec{r} = 0$.) Usually, position vectors are not drawn on a free-body diagram, but they are included on the modified free-body diagram to help calculate torques.

There are three steps in the **SOLVE** procedure:

Step 1 The equilibrium conditions are often best applied in component form. If you calculate the torques using Equation 14.7 or 14.8, it is often helpful to **write every $\vec{F}$ and $\vec{r}$ vector in component form.** Alternatively, you may wish to calculate the torques using $\tau = rF \sin \varphi$ (Eq. 12.16) and the right-hand rule, in which case it may not be necessary to write the vectors in component form. Instead it is helpful to list the magnitude of these vectors if they are known either numerically or algebraically.

Step 2 Finally, **apply the equilibrium condition equations.**

a. Apply the force balance condition.

b. Find the torques and apply the torque balance condition. You will need to find the torques by using either Equation 12.16 ($\tau = rF \sin \varphi$) or Equation 14.7 or 14.8 (in component form) before you apply the torque balance condition. You can use Equations 14.1 (force balance) and 14.2 (torque balance) or apply the conditions in component form Equations 14.3 and 14.4.

Step 3 **Do algebra** as needed to isolate and solve for the unknown quantity you seek before substituting values.

Here is an example of a simple geometry with perpendicular forces and position vectors. The right-hand rule and $\tau = rF \sin \varphi$ (Eq. 12.16) suffice for finding torques.

EXAMPLE 14.1 A Day at the Playground

A granddaughter of mass m_D and her grandfather of mass m_F are on an adjustable seesaw of mass m_s. The seesaw is a board of length ℓ with notches on its underside so that the board can be moved relative to the pivot. For the seesaw to be in equilibrium when the granddaughter and grandfather sit on opposite ends, the pivot must be placed closer to the grandfather than to the granddaughter (Fig. 14.5). Find an expression for the placement of the pivot relative to the seesaw's center of mass.

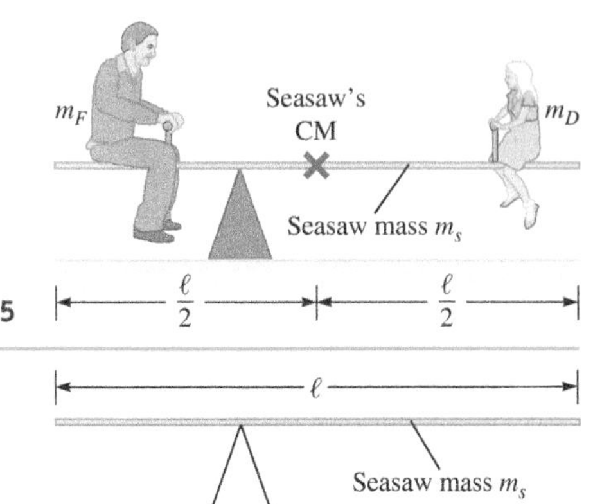

FIGURE 14.5

FIGURE 14.6

INTERPRET and ANTICIPATE

Start by creating a modified free-body diagram containing all five elements.

Element 1 **Simple shape.** The seesaw is the object, represented as a simple thick line (Fig. 14.6). The pivot has been included to help visualize its placement.

Element 2 **Forces at points of application.** Four forces are exerted on the seesaw. Three of them are normal forces: $\vec{F}_F$ due to the grand*f*ather, $\vec{F}_P$ due to the *p*ivot, and $\vec{F}_D$ due to the grand*d*aughter. The fourth force is the gravitational force $\vec{F}_g$ acting at the seesaw's center of mass, which, in this case, is at the seesaw's geometric center. Each force has been drawn at its point of application (Fig. 14.7). At this stage, you do not need to know the relative size of the force vectors.

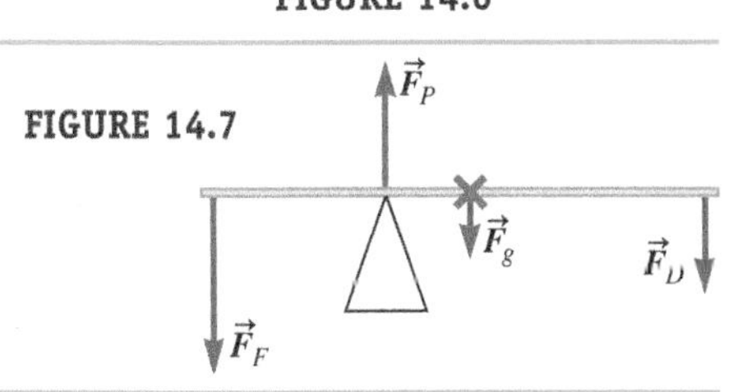

FIGURE 14.7

Element 3 **A rotation axis.** We decided to put the rotation axis through the center of mass so that gravity $\vec{F}_g$ does not exert a torque on the seesaw.

Element 4 **Coordinate system.** It is best to align one of the axes with the rotation axis. Here we choose a coordinate system with its origin at the center of mass and the z axis aligned with the rotation axis (Fig. 14.8).

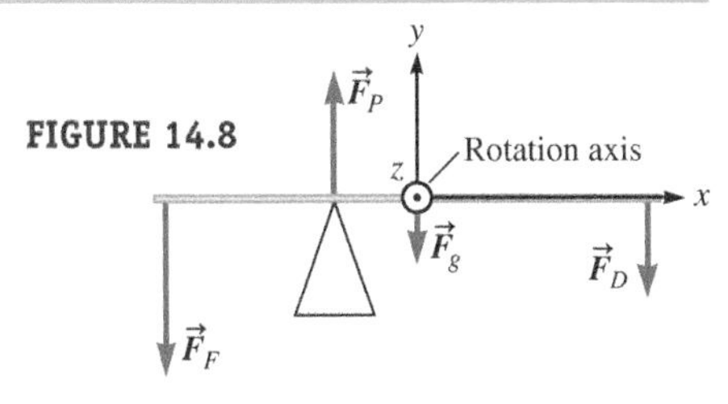

FIGURE 14.8

Example continues on page 396 ▶

Element 5 **Position vectors.** Three position vectors are drawn ($\vec{r}_F$, $\vec{r}_P$, and $\vec{r}_D$) for the three forces ($\vec{F}_F$, $\vec{F}_P$, and $\vec{F}_D$) that exert nonzero torques on the seesaw. The example requires us to find the placement of the pivot; that is, find $\vec{r}_P$ (Fig. 14.9).

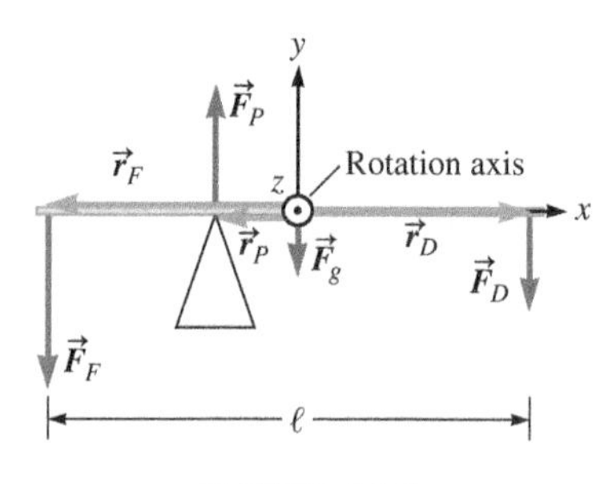

FIGURE 14.9

SOLVE

Step 1 Write each force and position vector in component form. Although the geometry is simple here, we demonstrate the use of unit vectors. Because the seesaw is level, the normal force exerted by each person on the seesaw is equal to the person's weight.

$$\vec{F}_F = -F_F\hat{\jmath} = -m_F g\hat{\jmath} \quad (1)$$

$$\vec{F}_D = -F_D\hat{\jmath} = -m_D g\hat{\jmath} \quad (2)$$

$$\vec{F}_g = -F_g\hat{\jmath} = -m_s g\hat{\jmath} \quad (3)$$

$$\vec{F}_P = +F_P\hat{\jmath} \quad (4)$$

Use Figure 14.9 to write the position vectors in component form. The origin is at the seesaw's center of mass, and each person is at an end, so $r_F = r_D = \ell/2$. We are solving for the pivot's position vector $\vec{r}_P$.

$$\vec{r}_F = -r_F\hat{\imath} = -\frac{\ell}{2}\hat{\imath} \quad (5)$$

$$\vec{r}_D = r_D\hat{\imath} = \frac{\ell}{2}\hat{\imath} \quad (6)$$

$$\vec{r}_{CM} = 0 \quad (7)$$

$$\vec{r}_P = -r_P\hat{\imath} \quad (8)$$

Step 2a Apply the force balance condition. The forces exerted on the seesaw have only y components (Eqs. 1 through 4), so we only need $\sum F_y = 0$ in Equation 14.3.

$$\sum F_y = -F_F - F_D - F_g + F_P = 0 \quad (14.3)$$

$$F_P = F_F + F_D + F_g$$

$$F_P = (m_F + m_D + m_s)g \quad (9)$$

Step 2b Find the torques and apply the torque balance condition. In this case we can use $\tau = rF\sin\varphi$ to find the magnitude of the torque and the right-hand rule to find the direction. Each force is perpendicular to its corresponding position vector (Fig. 14.9), so $\sin\varphi = \sin 90° = 1$ in each case. The magnitude of the torque is just the magnitude of the position multiplied by the magnitude of the force (Eq. 10).

$$\tau = rF\sin\varphi \quad (12.16)$$

$$\tau = rF \quad (10)$$

To find the torque exerted by the grandfather, substitute the magnitude of the force (Eq. 1) and the magnitude of the position (Eq. 5) into Equation (10). The magnitude of a vector is always positive. The right-hand rule shows that the torque is out of the page in the positive z direction.

$$\tau_F = r_F F_F = \left(\frac{\ell}{2}\right)(m_F g)$$

$$\vec{\tau}_F = \left(\frac{\ell}{2}\right)m_F g\hat{k} \quad (11)$$

Next use Equation (10) to find the torque exerted by the granddaughter. Again, we need the magnitude of the force (Eq. 2) and the magnitude of the position (Eq. 6). The right-hand rule shows the torque is into the page, in the negative z direction.

$$\tau_D = r_D F_D = \left(\frac{\ell}{2}\right)(m_D g)$$

$$\vec{\tau}_D = -\left(\frac{\ell}{2}\right)m_D g\hat{k} \quad (12)$$

Now use Equation (10) to find the pivot's torque. The magnitude of the torque τ_P due to the pivot is the magnitude of its position (Eq. 8) multiplied by the magnitude of the force it exerts (Eq. 4). Using the right-hand rule, we find that this torque points into the page in the negative z direction.

$$\tau_P = r_P F_P$$

$$\vec{\tau}_P = -r_P F_P\hat{k} \quad (13)$$

Finally, substitute Equation (9) for F_P in Equation (13).

$$\vec{\tau}_P = -r_P(m_F + m_D + m_s)g\hat{k} \quad (14)$$

The gravitational torque is zero due to our choice of rotation axis.

$$\tau_g = 0$$

because $r_{CM} = 0$ in Equation (7).

We are finally ready to **apply the torque balance condition**. This step is especially easy because all the torques have only z components.

$$\sum\vec{\tau} = \sum\tau_z = 0$$

Step 3 Do algebra. Set the sum of the torques (Eqs. 11, 12, and 14) equal to zero and solve for r_P. The unit vector $\hat{k}$ cancels out.

$$\left(\frac{\ell}{2}\right)m_F g\hat{k} - \left(\frac{\ell}{2}\right)m_D g\hat{k} - r_P(m_F + m_D + m_s)g\hat{k} = 0$$

$$r_P(m_F + m_D + m_s)g = \left(\frac{\ell}{2}\right)m_F g - \left(\frac{\ell}{2}\right)m_D g$$

$$r_P = \left(\frac{\ell}{2}\right)\frac{m_F - m_D}{m_F + m_D + m_s}$$

From Figure 14.9, we see that $\vec{r}_P$ is in the negative x direction. So, the numerator becomes $m_D - m_F$ when we express the vector in component form.

$$\vec{r}_P = \left(\frac{\ell}{2}\right)\frac{m_D - m_F}{m_F + m_D + m_s}\hat{\imath} \quad (15)$$

CHECK and THINK

Our answer tells us where to put the pivot relative to the center of mass so that when the grandfather and granddaughter are on the seesaw, the system is in equilibrium. The dimensions are length, as we would expect for position.

To further check our expression, imagine that the grandfather and granddaughter have the same mass. In that case, we would expect that the pivot would be placed in the middle of the seesaw. Because our coordinate system is at the center of the seesaw, we find $r_P = 0$ if $m_D = m_F$.

However, based on Figure 14.5 we assume that the grandfather is heavier than the granddaughter. So, we expect that he would be closer to the pivot than she is. In other words, we expect $\vec{r}_P$ to point in the negative x direction if $m_F > m_D$ (Fig. 14.9), exactly as we found from Equation (15).

Here is another example showing the use of $\tau = rF \sin \varphi$ and the right-hand rule to find the torques. This method may be especially useful when the unknown quantity is an angle.

EXAMPLE 14.2 Don't Let the Rake Fall Down

A rake is placed against a smooth wall so that it makes an angle θ with the floor of the garage (Fig. 14.10). Assume friction between the wall and the rake is negligible, but the rake's tines are made from a hard rubber that rests on the concrete floor. The coefficient of static friction μ_s for rubber on dry concrete is 1.0. Find the minimum value of θ (other than zero) so that the rake does not slide. The rake's center of mass is one-fourth of the way up from the tines as shown.

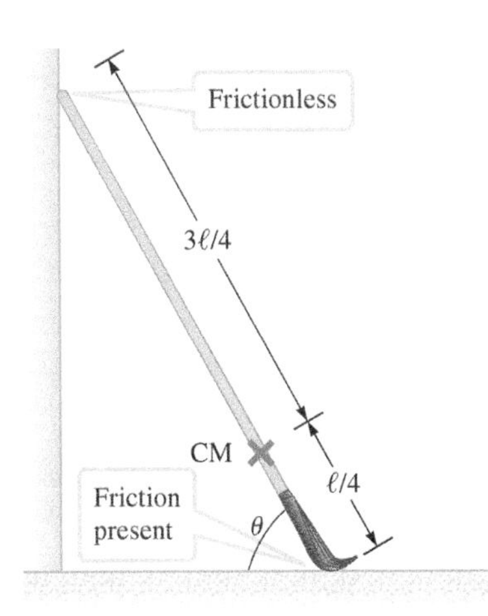

FIGURE 14.10 For a rough floor, what is the smallest angle θ such that the rake won't fall?

INTERPRET and ANTICIPATE

Again, start with a modified free-body diagram that includes all five elements.

Element 1 Simple shape. Represent the rake as a simple thick line.

Element 2 Forces at points of application. Four forces are exerted on the rake: the normal force $\vec{f}_N$ exerted by the wall, the gravitational force $\vec{F}_g$ acting at the center of mass, the normal force $\vec{F}_N$ exerted by the floor, and static friction $\vec{F}_s$ exerted by the floor. Each force has been drawn at its point of contact (Fig. 14.11). But at this stage, you do not need to know the relative size of the force vectors.

Element 3 A rotation axis. We chose a rotation axis through the point of contact between the wall and the rake so that the wall (more specifically, the normal force $\vec{f}_N$) cannot exert a torque on the rake.

Element 4 Coordinate system. We placed the origin at the point of contact with the wall and aligned the z axis with the rotation axis.

Example continues on page 398 ▶

Element 5 **Position vectors.** Draw position vectors $\vec{r}_{CM}$ and $\vec{r}_F$ for the three forces $\vec{F}_g$, $\vec{F}_N$, and $\vec{F}_s$ that exert nonzero torques around the rotation axis (Fig. 14.11). The forces exerted by the floor—$\vec{F}_N$ and $\vec{F}_s$—act at the same point of contact, so the position vector $\vec{r}_F$ is the same for these two forces.

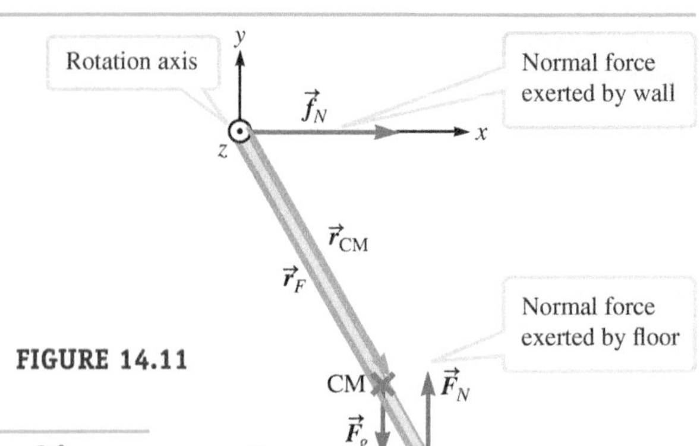

FIGURE 14.11

SOLVE

Step 1 **Write each force in component form.** Because we plan to use $\tau = rF \sin \varphi$ (Eq. 12.16), we don't need to write the forces in component form.

List the position vectors. Writing the position vectors would be complicated and is not necessary in this example. Instead, we list their magnitudes. There are only two positions: the center-of-mass position and the position of the floor.

$$r_{CM} = \frac{3\ell}{4}$$
$$r_F = \ell$$

Step 2a **Apply the force balance condition.** Because there are forces in the x and y directions, we must use Equation 14.3 separately for each of them.

$$\sum F_x = f_N - F_s = 0 \quad \text{(14.3 for } x\text{)}$$
$$f_N = F_s$$
$$\sum F_y = F_N - F_g = 0 \quad \text{(14.3 for } y\text{)}$$
$$F_N = F_g \quad (1)$$

Step 2b **Find the torques and apply the torque balance condition.** When you use $\tau = rF \sin \varphi$, it is helpful to sketch the force and position vectors tail to tail as shown in Figure 14.12. Before calculating the torques, use the right-hand rule and Figure 14.11 to find their directions. Here the torques have only z components.

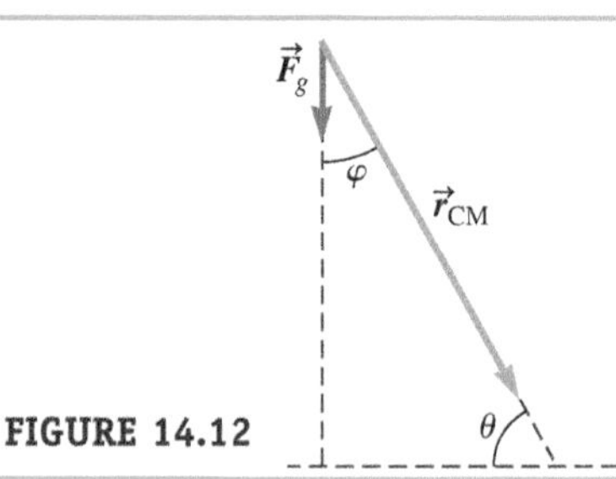

FIGURE 14.12

Now calculate each torque. The angle between $\vec{F}_g$ and $\vec{r}_{CM}$ is φ, and from Figure 14.12, $\varphi = 90° - \theta$.

$$\vec{\tau}_g = -F_g r_{CM} \sin \varphi \hat{k}$$
$$\vec{\tau}_g = -F_g r_{CM} \sin(90° - \theta)\hat{k}$$

Substitute $r_{CM} = 3\ell/4$ and use the trigonometric identity $\sin(90° - \theta) = \cos \theta$.

$$\vec{\tau}_g = -\left(\frac{3\ell}{4}\right)F_g \cos \theta \hat{k} \quad (2)$$

Again, to find the angle φ between $\vec{r}_F$ and $\vec{F}_N$, slide one vector or the other so that their tails are together (Fig. 14.13).

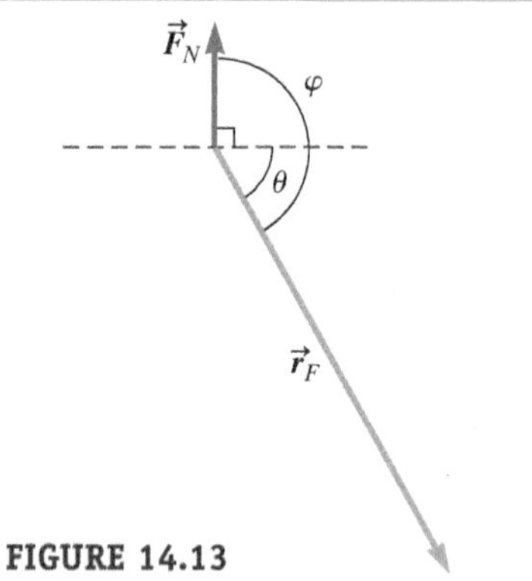

FIGURE 14.13

Find the torque due to the normal force exerted by the floor in terms of θ.

$$\vec{\tau}_N = r_F F_N \sin \varphi \hat{k}$$
$$\vec{\tau}_N = r_F F_N \sin(90° + \theta)\hat{k}$$

Substitute $r_F = \ell$ and use the trigonometric identity $\sin(90° + \theta) = \cos \theta$.

$$\vec{\tau}_N = \ell F_N \cos \theta \hat{k} \quad (3)$$

The angle between $\vec{F}_s$ and $\vec{r}_F$ is θ, so no separate sketch is needed.

$$\vec{\tau}_s = -\ell F_s \sin \theta \hat{k} \quad (4)$$

Now we are ready to **apply the torque balance condition**. The torques in this case only have z components.

$$\vec{\tau}_{tot} = \sum \tau_z = 0$$
$$\vec{\tau}_{tot} = \vec{\tau}_g + \vec{\tau}_N + \vec{\tau}_s$$

Step 3 **Do algebra.** Substitute the expressions for torque (Eqs. 2–4), and solve for θ, the unknown quantity.

$$-\left(\frac{3\ell}{4}\right)F_g \cos \theta + \ell F_N \cos \theta - \ell F_s \sin \theta = 0$$

$$F_s \sin \theta = \left(F_N - \frac{3F_g}{4}\right)\cos \theta$$

By Equation (1), $F_g = F_N$. The minimum angle θ_{min} is possible when static friction is maximum, so $F_s = \mu_s F_N$ (Eq. 6.1) when $\theta = \theta_{min}$.

$$\mu_s F_N \sin \theta_{min} = \left(F_N - \frac{3F_N}{4}\right)\cos \theta_{min} = \frac{F_N}{4}\cos \theta_{min}$$

Solve for $\theta_{\min}$.	$\dfrac{\sin\theta_{\min}}{\cos\theta_{\min}} = \tan\theta_{\min} = \dfrac{1}{4\mu_s}$ $\theta_{\min} = \tan^{-1}\left(\dfrac{1}{4\mu_s}\right)$
Substitute the given value of μ_s.	$\theta_{\min} = \tan^{-1}\left[\dfrac{1}{4(1.0)}\right] = 14°$

CHECK and THINK

To check our result, imagine that the floor is wet so that the coefficient of static friction between the rake's tines and the floor is reduced. We expect the minimum angle to increase. When the floor is slippery, suppose $\mu_s = 0.5$ instead of 1.0. According to our results, $\theta_{\min}$ would be $\tan^{-1}(0.5) = 27°$, a larger minimum angle, as expected.

Here is an example using a complex geometry. In such a case, Equations 14.7 and 14.8 should be used to find the torques.

EXAMPLE 14.3 CASE STUDY Hanging an Airplane

We consider a somewhat simplified version of the sort of problems solved by engineers and architects who designed the Udvar-Hazy Center. To minimize damage to an airplane, it should be hung from as few cables as possible, and the cables should be connected to support elements in the plane. An airplane has mass $m = 948$ kg and is hung from two steel cables so that the wings make an angle $\theta = 10.0°$ with the horizontal. Cable 1 is attached near the middle of the upper wing, and cable 2 is attached to a wing support near the edge (parts A and B of Fig. 14.14). The placement of cable 2 is determined by the location of the support structure. Cable 1, however, can be attached anywhere along another support structure running across the short length of the wing, so the x component of its position is determined by the desired equilibrium configuration of the plane. Use the coordinate system and data provided in Figure 14.14 to find r_{1x} (the x coordinate of cable 1's point of contact) and the tension in the two cables.

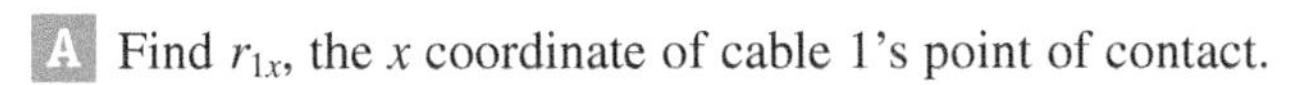

A Find r_{1x}, the x coordinate of cable 1's point of contact.

A. Front view

$\vec{r}_1 = r_{1x}\hat{\imath} + (1.25\text{ m})\hat{\jmath} + (0.22\text{ m})\hat{k}$

$\vec{r}_2 = (-0.52\text{ m})\hat{\imath} + (1.72\text{ m})\hat{\jmath} + (-3.49\text{ m})\hat{k}$

INTERPRET and ANTICIPATE

Because the plane is in equilibrium, the sum of the forces and the sum of the torques acting on it are zero (Eqs. 14.1 and 14.2). While Figure 14.14 is somewhat more detailed than you may draw on your own, it provides the free-body diagram with all the elements. The forces are drawn at points of application. There is a rotation axis and coordinate system. The position vectors have been drawn. Three forces act on the plane: gravity $\vec{F}_g$ and the tension forces $\vec{F}_1$ and $\vec{F}_2$ exerted by each cable. Because the origin of the coordinate system is at the center of mass, it is relatively easy to calculate the torques around an axis passing through the center of mass.

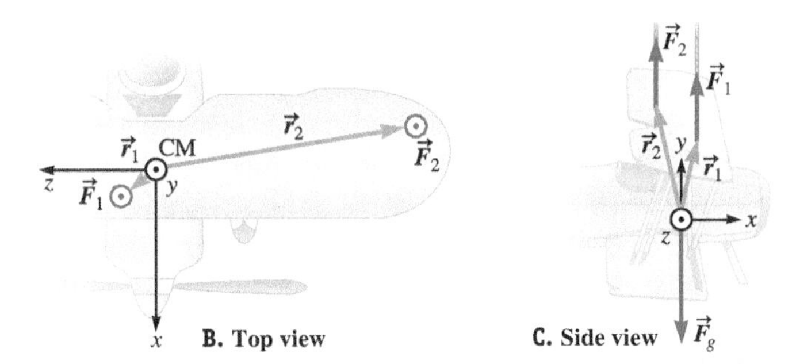

FIGURE 14.14 **A.** Front view of airplane showing a coordinate system with the origin at the center of mass, the x axis out of the page, and the z axis to the left. **B.** Top view looking down from the $+y$ axis. **C.** Side view looking at the fuselage from the $+z$ axis.

SOLVE

Step 1 **Write each force in component form.** As shown in Figure 14.14, the forces on the airplane have only y components.	$\vec{F}_1 = F_1\hat{\jmath}$ $\quad \vec{F}_2 = F_2\hat{\jmath}$ $\quad F_1$ and F_2 are unknown. $\vec{F}_g = -mg\hat{\jmath}$

Example continues on page 400 ▶

We are given the position vectors $\vec{r}_1$ and $\vec{r}_2$ in component form (Fig. 14.14).	$\vec{r}_1 = r_{1x}\hat{\imath} + (1.25\text{ m})\hat{\jmath} + (0.22\text{ m})\hat{k}$ $\vec{r}_2 = (-0.52\text{ m})\hat{\imath} + (1.72\text{ m})\hat{\jmath} + (-3.49\text{ m})\hat{k}$ r_{1x} is unknown.
Step 2a **Apply the force balance condition.** We only need to apply Equation 14.3 along the y axis.	$\sum F_y = 0$ (14.3) $F_1 + F_2 - mg = 0$ $F_1 + F_2 = mg$ (1)
Step 2b **Find the torques and apply the torque balance condition.** Find the total torque around the center of mass. Gravity cannot exert a torque around the center of mass in our chosen coordinate system, so the total torque comes from adding torques $\vec{\tau}_1$ and $\vec{\tau}_2$ exerted by cable 1 and cable 2, respectively.	$\vec{\tau}_{tot} = \sum \vec{\tau} = 0$ $\sum \vec{\tau} = \vec{\tau}_1 + \vec{\tau}_2 = 0$ (2)

Find the torque due to each cable using Equation 14.7. The tensions $\vec{F}_1$ and $\vec{F}_2$ have only y components, so the F_x and F_z terms disappear.

Find the torque due to each cable.	$\vec{\tau}_1 = \vec{r}_1 \times \vec{F}_1 = -r_{1z}F_1\hat{\imath} + r_{1x}F_1\hat{k}$ $\vec{\tau}_2 = \vec{r}_2 \times \vec{F}_2 = -r_{2z}F_2\hat{\imath} + r_{2x}F_2\hat{k}$
Add the torques and set their sum to zero (Eq. 2).	$(-r_{1z}F_1\hat{\imath} + r_{1x}F_1\hat{k}) + (-r_{2z}F_2\hat{\imath} + r_{2x}F_2\hat{k}) = 0$
Step 3 **Do algebra.** Collect terms in $\hat{\imath}$ and $\hat{k}$.	$-(r_{1z}F_1 + r_{2z}F_2)\hat{\imath} + (r_{1x}F_1 + r_{2x}F_2)\hat{k} = 0$
In equilibrium, each component of the total torque is zero.	$\tau_x = r_{1z}F_1 + r_{2z}F_2 = 0$ $\quad\tau_z = r_{1x}F_1 + r_{2x}F_2 = 0$
Solve these two equations for F_1.	$F_1 = -\dfrac{r_{2z}}{r_{1z}}F_2$ and $F_1 = -\dfrac{r_{2x}}{r_{1x}}F_2$ (3)
The x coordinate of cable 1's point of contact r_{1x} comes from equating expressions for F_1 in Equation (3). (The values of the position components are given in Fig. 14.14.)	$\dfrac{r_{2z}}{r_{1z}} = \dfrac{r_{2x}}{r_{1x}}$ $r_{1x} = \dfrac{r_{2x}r_{1z}}{r_{2z}}$ $r_{1x} = \dfrac{(-0.52\text{ m})(0.22\text{ m})}{-3.49\text{ m}} = 3.3 \times 10^{-2}\text{ m}$

CHECK and THINK

This result is the answer to part A. Imagine for a moment that the plane were suspended from only one cable. In that case, the cable would be attached directly above the center of mass, and the tension in the cable would equal the weight of the plane. We now expect that because cable 1 is attached almost directly above the center of mass, the tension in cable 1 should be nearly equal to the weight of the plane.

B Find the tension in the two cables.

SOLVE Our work in part A allows us to jump straight into solving this part. The tension in cable 2 comes from substituting Equation (3) into Equation (1).	$-\dfrac{r_{2z}}{r_{1z}}F_2 + F_2 = mg$ $F_2 = \dfrac{mg}{1 - r_{2z}/r_{1z}} = \dfrac{(948\text{ kg})(9.81\text{ m/s}^2)}{1 - (-3.49\text{ m}/0.22\text{ m})}$ $F_2 = 5.50 \times 10^2\text{ N}$ (4)
Finally, the tension in cable 1 comes from substituting Equation (4) into Equation (3).	$F_1 = -\dfrac{r_{2z}}{r_{1z}}F_2 = -\dfrac{(-3.49\text{ m})}{(0.22\text{ m})}(5.50 \times 10^2\text{ N})$ $F_1 = 8.7 \times 10^3\text{ N}$

CHECK and THINK

The plane's weight is $mg = (948\text{ kg})(9.81\text{ m/s}^2) = 9.30 \times 10^3$ N. So, the tension F_1 in cable 1 is about 94% of the plane's weight, and the tension in cable 2 is about 6% of the plane's weight. As expected, the tension in cable 1 is nearly equal to the plane's weight.

Here is an example with a complicated geometry. The example has symmetry, however, which helps us predict the outcome and makes the calculation a little easier than usual.

EXAMPLE 14.4 Making a Three-Legged Table

Children often hear interesting stories about a parent's childhood. According to the author's grandmother, when the author's father was about eight years old, he decided to see if their table would stand up on just three of its four legs, so he sawed off one of the legs. Assume the tabletop was a square of side a. Find expressions for the force exerted by the remaining three legs on the tabletop just after the fourth leg was removed. Did the table remain standing up? Model the tabletop without the legs as the system.

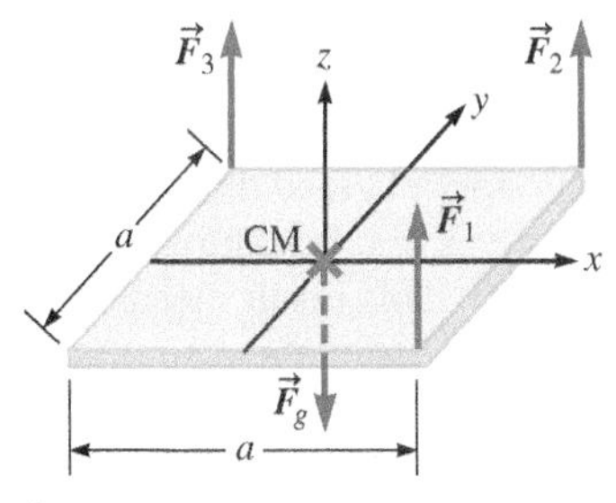

A. Perspective view

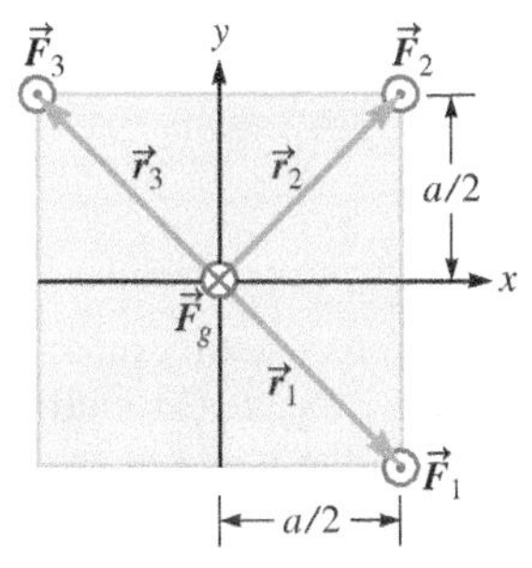

B. Overhead view

FIGURE 14.15 Representation of a tabletop. One leg of the table has been removed.

INTERPRET and ANTICIPATE

The modified free-body diagram in Figure 14.15 is shown from two perspectives. The **tabletop is represented by a square**. Four forces act on the tabletop: gravity and the three normal forces $\vec{F}_1$, $\vec{F}_2$, and $\vec{F}_3$ exerted by the three remaining legs. These forces are drawn at the **point of application** *(but not to scale)*. The origin of the **coordinate system** is located at the center of mass. We will calculate torques around a **rotation axis** passing through the center of mass and aligned with the z axis. The **position vectors** are drawn. Because of the symmetry in this example, we expect that the expressions for $\vec{F}_1$ and $\vec{F}_3$ are identical.

SOLVE

Step 1 **Write each force in component form.** The four forces have only z components.

$$\vec{F}_1 = F_1\hat{k}$$
$$\vec{F}_2 = F_2\hat{k}$$
$$\vec{F}_3 = F_3\hat{k}$$
$$\vec{F}_g = -F_g\hat{k}$$

Write the position vectors in component form. The position vectors $\vec{r}_1$, $\vec{r}_2$, and $\vec{r}_3$ have x and y components (Fig. 14.15B). Because the origin of the coordinate system was placed at the center of mass, $\vec{r}_{CM} = 0$.

$$\vec{r}_1 = \tfrac{a}{2}\hat{\imath} - \tfrac{a}{2}\hat{\jmath}$$
$$\vec{r}_2 = \tfrac{a}{2}\hat{\imath} + \tfrac{a}{2}\hat{\jmath}$$
$$\vec{r}_3 = -\tfrac{a}{2}\hat{\imath} + \tfrac{a}{2}\hat{\jmath}$$

Step 2a **Apply the force balance condition.** We only need the z component of Equation 14.3.

$$\sum F_z = F_1 + F_2 + F_3 - F_g = 0 \quad (14.3\text{ for }z)$$
$$F_1 + F_2 + F_3 = F_g \quad (1)$$

Step 2b **Find the torques.** Many of the terms disappear from Equation 14.7 in this example because $F_x = F_y = 0$ and $r_z = 0$ for each torque.

$$\vec{\tau} = (r_yF_z - r_zF_y)\hat{\imath} + (r_zF_x - r_xF_z)\hat{\jmath} + (r_xF_y - r_yF_x)\hat{k} \quad (14.7)$$
$$\vec{\tau} = r_yF_z\hat{\imath} - r_xF_z\hat{\jmath} \quad (2)$$

Use Equation (2) to find the torque exerted by each of the normal forces. The torque exerted by gravity is zero because $\vec{r}_{CM} = 0$.

$$\vec{\tau}_1 = -\tfrac{a}{2}F_1\hat{\imath} - \tfrac{a}{2}F_1\hat{\jmath}$$
$$\vec{\tau}_2 = \tfrac{a}{2}F_2\hat{\imath} - \tfrac{a}{2}F_2\hat{\jmath}$$
$$\vec{\tau}_3 = \tfrac{a}{2}F_3\hat{\imath} + \tfrac{a}{2}F_3\hat{\jmath}$$

Example continues on page 402 ▶

Apply the torque balance condition using Equation 14.2.	$\vec{\tau}_{tot} = \vec{\tau}_1 + \vec{\tau}_2 + \vec{\tau}_3 = 0$ (14.2) $\vec{\tau}_{tot} = \frac{a}{2}(-F_1 + F_2 + F_3)\hat{\imath} + \frac{a}{2}(-F_1 - F_2 + F_3)\hat{\jmath} = 0$
Step 3 **Do algebra.** Each component of torque must be zero, so we can derive two equations.	$-F_1 + F_2 + F_3 = 0$ $F_1 = F_2 + F_3$ (3) $-F_1 - F_2 + F_3 = 0$ $F_1 + F_2 = F_3$ (4)
Solve Equations (1), (3), and (4) simultaneously for the three normal forces. These forces point upward in the positive z direction (Fig. 14.15A).	$\vec{F}_1 = \vec{F}_3 = \frac{F_g}{2}\hat{k}$ $\vec{F}_2 = 0$

CHECK and THINK

As expected, $\vec{F}_1 = \vec{F}_3$. Because we were able to find real solutions for all the normal forces, it seems that the table should remain standing even after removal of one of its legs. In fact, the father in this story claimed that table remained standing until his mother tried to set the table for dinner.

It seems counterintuitive that the leg across from the missing leg exerts no force on tabletop ($\vec{F}_2 = 0$). This result means that the second leg may also be removed with the same observable effect: The table should remain standing. Our experience, however, tells us that such a two-legged table would be *unstable*. Imagine setting a dinner plate on the three-legged table. As long as the plate is placed on the triangular half of the table that contains all three legs, the table will not tip over. If the plate were placed on the other half of the table, the table would, of course, tip over. Now imagine placing the same plate on a two-legged table. Unless the plate is placed precisely along the line joining the two remaining legs, the table will tip over.

When the forces acting on an object do not lie in a single plane, it may be necessary to consider torques around two separate rotation axes. In the next example, we study a situation in which there is more than one rotation axis.

EXAMPLE 14.5 Designing a Large Floor Fan

Fans are more portable, more reliable, and less expensive to buy and to run than air conditioners, so military bases often use large floor fans to circulate air in a large space such as a gym. (The fan's blades rotate at a constant angular velocity, so there is no net torque on the blades. In this example, consider the force the blades exert on the air.) Because the fans must exert a great force on the air, the air must exert a great force on the fan, similar to the force exerted by the exhaust on a rocket. This force is known as *thrust* (Chapter 10)..If the thrust is too great, the fan will topple over. The large fan in Figure 14.16 weighs 20 lb; it has four legs of length $b = 12$ in., and its height (measured from the floor to the center of the fan blades) is $h = 36$ in. Each leg has a small rubber foot at the end so that the floor is only in contact with the end of each leg. Find the maximum thrust that may be exerted without toppling the fan. Give your answer in pounds and in newtons. At the maximum thrust, the front two feet are just lifting off the floor (Problem 58).

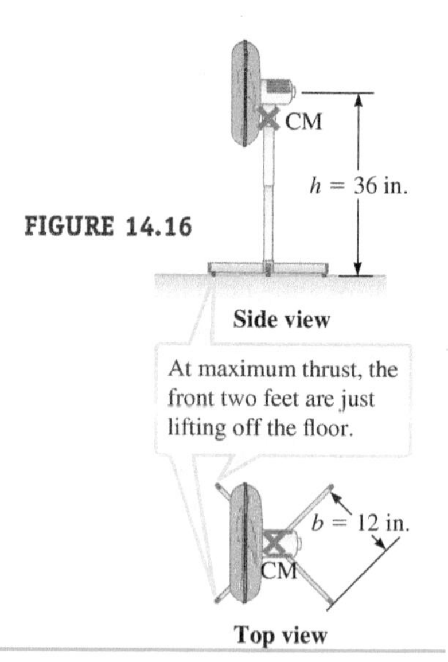

FIGURE 14.16

INTERPRET and ANTICIPATE

The modified free-body diagram (Fig. 14.17) shows two views. The fan is **represented by a simple rectangle** with an X-shaped base.

A total of 10 forces act on the fan and are **drawn at their points of application.** Gravity $\vec{F}_g$ acts on the center of mass. The thrust $\vec{F}_{\text{thrust}}$ is a horizontal force whose point of contact we take to be the center of the fan's blades. The eight other forces are due to the four points of contact between the floor and the fan's feet. At each point of contact, the floor exerts a normal force upward ($\vec{F}_{N1}, \vec{F}_{N2}, \vec{F}_{N3}$, and $\vec{F}_{N4}$) and static friction horizontally ($\vec{F}_{S1}, \vec{F}_{S2}, \vec{F}_{S3}$, and $\vec{F}_{S4}$). Static friction must point in the opposite direction to the thrust; otherwise, the fan would glide across the floor.

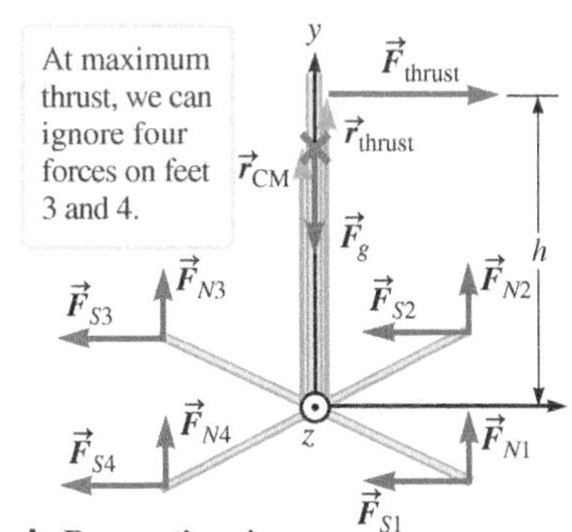

A. Perspective view

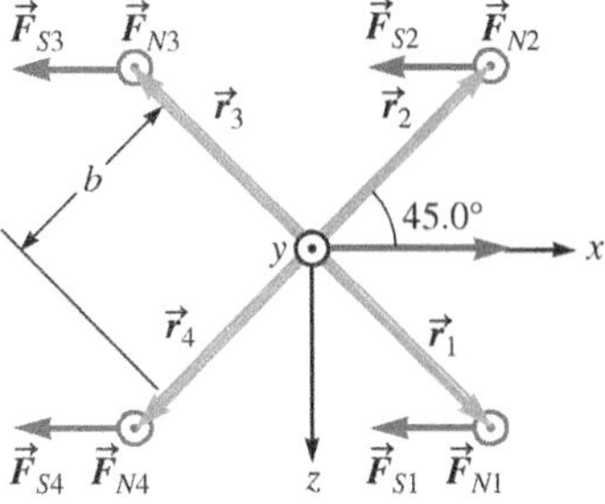

FIGURE 14.17 B. Overhead view

The origin of the **coordinate system** is at the center of the fan's base. We consider **three rotation axes** aligned with the coordinate axes. The forces act at six points of contact, so we **draw six position vectors.** The position of the center of mass is $\vec{r}_{\text{CM}}$. The thrust acts at $\vec{r}_{\text{thrust}}$. The four points of contact between the floor and the fan's feet are $\vec{r}_1, \vec{r}_2, \vec{r}_3$, and $\vec{r}_4$.

The front two feet are not in contact with the floor, however, so there is no normal force or friction exerted on these feet ($\vec{F}_{N3} = \vec{F}_{N4} = \vec{F}_{S3} = \vec{F}_{S4} = 0$), and there is no need to find $\vec{r}_3$ and $\vec{r}_4$.

SOLVE

Step 1 Write forces in component form. Gravity is in the negative y direction, and thrust is in the positive x direction.

$$\vec{F}_g = -F_g\hat{\jmath} \quad (1)$$

$$\vec{F}_{\text{thrust}} = F_{\text{thrust}}\hat{\imath} \quad (2)$$

For each foot, the normal force is in the positive y direction, and static friction is in the negative x direction. For example, consider foot 1. The total force $\vec{F}_1$ exerted by the floor is due to the force of static friction $\vec{F}_{S1}$ and the normal force $\vec{F}_{N1}$.

$$\vec{F}_{N1} = F_{N1}\hat{\jmath}$$

$$\vec{F}_{S1} = -F_{S1}\hat{\imath}$$

$$\vec{F}_1 = -F_{S1}\hat{\imath} + F_{N1}\hat{\jmath} \quad (3)$$

Write a similar equation for $\vec{F}_2$.

$$\vec{F}_2 = -F_{S2}\hat{\imath} + F_{N2}\hat{\jmath} \quad (4)$$

Write the position vectors in component form. Gravity is applied at the center of mass, and thrust is applied at a height h above the center of the fan's X-shaped base.

$$\vec{r}_{\text{CM}} = r_{\text{CM}}\hat{\jmath} \quad (5)$$

$$\vec{r}_{\text{thrust}} = h\hat{\jmath} \quad (6)$$

Express the point of contact for the forces acting on each foot in terms of the leg length b. Refer to the coordinate system shown in Figure 14.17B.

$$\vec{r}_1 = b\sin 45°\hat{\imath} + b\cos 45°\hat{k}$$

$$\vec{r}_1 = \frac{b\sqrt{2}}{2}\hat{\imath} + \frac{b\sqrt{2}}{2}\hat{k} \quad (7)$$

$$\vec{r}_2 = \frac{b\sqrt{2}}{2}\hat{\imath} - \frac{b\sqrt{2}}{2}\hat{k} \quad (8)$$

Step 2a Apply the force balance condition in component form, which gives us two equations.

$$\sum F_x = F_{\text{thrust}} - F_{S1} - F_{S2} = 0 \quad (9)$$

$$\sum F_y = F_{N1} + F_{N2} - F_g = 0 \quad (10)$$

Step 2b Find the torques. With our coordinate system, the position vectors measured from the origin are the ones we need for calculating torque using Equation 14.7.

The torque due to gravity is zero because both $\vec{F}_g$ and $\vec{r}_{\text{CM}}$ only have y components (Eqs. 1 and 5). (Whenever two vectors are parallel or antiparallel, their cross product is zero.)

$$\vec{\tau}_g = \vec{r}_{\text{CM}} \times \vec{F}_g = 0$$

Find the torque due to the thrust using the position vector (Eq. 6) and the force (Eq. 2). All but the last term in Equation 14.7 are zero for the torque due to thrust.

$$\vec{\tau}_{\text{thrust}} = \vec{r}_{\text{thrust}} \times \vec{F}_{\text{thrust}} = -(r_y F_x)\hat{k} = -hF_{\text{thrust}}\hat{k} \quad (11)$$

Example continues on page 404 ▶

Find the torque exerted by forces on foot 1 and foot 2 using Equation 14.7. The force and position vectors for foot 1 are given by Equations (3) and (7).	$\vec{\tau}_1 = \vec{r}_1 \times \vec{F}_1 = \left(\frac{b\sqrt{2}}{2}\hat{\imath} + \frac{b\sqrt{2}}{2}\hat{k}\right) \times (-F_{S1}\hat{\imath} + F_{N1}\hat{\jmath})$ $\vec{\tau}_1 = -\frac{b\sqrt{2}}{2}F_{N1}\hat{\imath} - \frac{b\sqrt{2}}{2}F_{S1}\hat{\jmath} + \frac{b\sqrt{2}}{2}F_{N1}\hat{k}$
The force and position vectors for foot 2 are given by Equations (4) and (8). Notice that the torque exerted on each foot has x, y, and z components.	$\vec{\tau}_2 = \frac{b\sqrt{2}}{2}F_{N2}\hat{\imath} + \frac{b\sqrt{2}}{2}F_{S2}\hat{\jmath} + \frac{b\sqrt{2}}{2}F_{N2}\hat{k}$
Apply the torque balance condition in component form, which gives us three equations. The thrust only contributes to the z component of the torque (Eq. 11).	$\sum \tau_x = \frac{b\sqrt{2}}{2}(-F_{N1} + F_{N2}) = 0$ (12) $\sum \tau_y = \frac{b\sqrt{2}}{2}(-F_{S1} + F_{S2}) = 0$ (13) $\sum \tau_z = \frac{b\sqrt{2}}{2}(F_{N1} + F_{N2}) - hF_{\text{thrust}} = 0$ (14)
Step 3: Algebra. We have five equations (Eqs. 9, 10, 12, 13, and 14), and we are interested in finding the maximum thrust F_{thrust}. Use Equation (10) to find an expression for the total normal force.	$F_{N1} + F_{N2} = F_g$ (15)
Substitute Equation (15) into Equation (14) and solve for thrust.	$\frac{b\sqrt{2}}{2}(F_g) - hF_{\text{thrust}} = 0$ $F_{\text{thrust}} = \frac{\sqrt{2}b}{2h}F_g = \frac{\sqrt{2}(12\text{ in.})}{2(36\text{ in.})}(20\text{ lb})$ $F_{\text{thrust}} = 4.7\text{ lb} = 21\text{ N}$

CHECK and THINK

We found five equations by applying the equilibrium conditions, but we only needed two of them to solve for the thrust. In solving a complicated equilibrium problem such as this one, you may write down more equations than you need. You may find those unused equations useful if you need to solve for some other quantity.

14-4 Elasticity and Fracture

Robert Hooke (1635–1703) and Isaac Newton (1642–1727) were contemporaries, but they each took a very different approach to science. Newton was interested in astronomy, cosmology, and mathematics. Hooke was interested in more practical, mundane matters such as buildings, ships, and fleas. Today we would probably consider Hooke to be primarily an *applied* physicist and Newton primarily a *theoretical* physicist.

Hooke was interested in how objects in static equilibrium responded to being pulled and pushed. That is, he wanted to know how stationary objects reacted to being *loaded.* Rather than trying to come up with a theory, his approach was to collect data first. For example, he hung loads—objects of known weight—on rods made of different materials and of different sizes (Fig. 14.18). He measured the length of the rods before and after the load was attached and found that the rods stretched when loaded and returned to their unstretched length when the load was removed. He concluded that there is no such thing as a rigid object. In other words, all objects are stretched or compressed when forces are applied, even if only slightly.

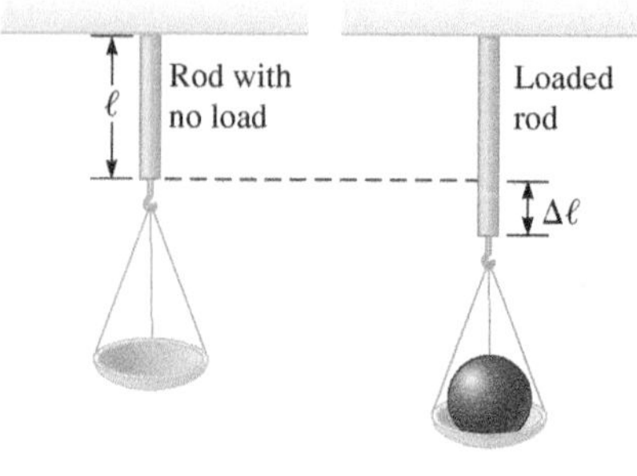

Rod with no load

Loaded rod

ℓ

$\Delta\ell$

FIGURE 14.18 A load is hung from a rod of length ℓ, resulting in a new length $\ell + \Delta\ell$. The strain ε of the rod is $\varepsilon = \Delta\ell/\ell$.

Stress

STRESS Major Concept

Augustin Cauchy (1789–1857), an engineer and academic, continued to study the behavior of loaded objects. He defined **stress** σ as the magnitude of the force applied to an object per unit area (Fig. 14.19):

$$\sigma \equiv \frac{F}{A} \quad (14.9)$$

The SI units of stress are N/m^2 and $1 N/m^2 = 1$ Pa, where Pa stands for *pascals*. There are four types of stress: *tensile stress*, *compression stress*, *shear stress*, and *volume stress*, depending on where and how the force is applied to the object. **Tensile stress** on an object occurs when a force pulls perpendicular to a surface of the object (Fig. 14.19). Tensile stress, compression stress, and shear stress are discussed in this chapter. Volume stress is discussed in Chapter 15.

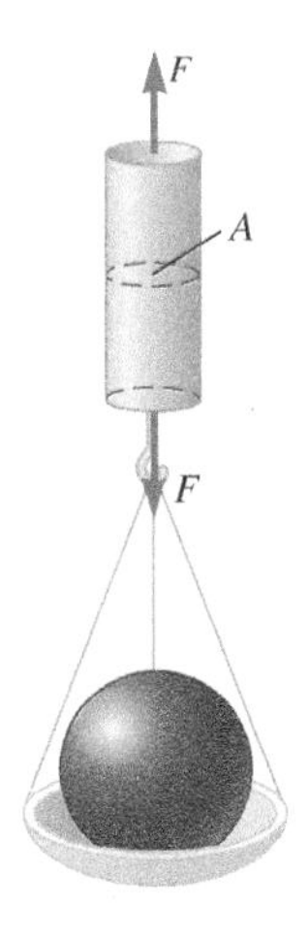

FIGURE 14.19 Stress on a rod is the force applied per unit cross-sectional area. Here is an example of *tensile* stress.

Strain

Imagine performing experiments similar to Hooke's (Fig. 14.18) in which you measure the rod's length as you load objects of known weight into the pan. From these data, you could calculate the stress σ on the rod by dividing the force applied to the rod by the rod's cross-sectional area. You could also calculate the fractional increase in the rod's length by dividing the increase $\Delta\ell$ in the rod's length by its original length ℓ. The fractional increase $\Delta\ell/\ell$ is known as the **strain**. The symbol for strain is the lowercase Greek letter epsilon, ε. Because strain is the fractional increase in the rod's length

STRAIN ✪ Major Concept

$$\varepsilon = \Delta\ell/\ell \tag{14.10}$$

it is dimensionless. The words *stress* and *strain* sound alike and are often confused. It might help to remember the phrase, "A stress on a rod causes a strain."

Tensile Deformation

Figure 14.20 shows a graph of stress as a function of tensile strain for the rod from experimental data. This graph reveals three phenomena.

First, as long as the stress σ is small, the strain is directly proportional to the stress. Over the linear response range on the graph in Figure 14.20,

$$\sigma = Y\varepsilon \tag{14.11}$$

where Y is known as **Young's modulus** and is the slope of the line. Because strain ε is dimensionless, Young's modulus must have the same dimensions as stress. Young's modulus is a measure of the rod's stiffness and depends on its composition. Table 14.1 lists Y for various substances. As long as the stress put on the rod is in the linear response range, the rod returns to its original length when the load is removed. An **elastic object** is one that returns to its original length after the load has been removed.

YOUNG'S MODULUS ✪ Major Concept

ELASTIC OBJECT ✪ Major Concept

Second, as a greater load is placed on the rod past the linear response range, the rod responds by stretching an even greater amount. Now, even when the load is removed, the rod remains somewhat stretched rather than returning to its original size.

TABLE 14.1 Approximate values for moduli and strength of materials.

Material	Young's modulus Y (10^9 N/m²)	Shear modulus G (10^9 N/m²)	Tensile strength σ_{ten} and compression strength σ_{cmp} (10^6 N/m²)
Aluminum	73	30	110
Bone	21 tensile		
	18 compressive		170 (compression only)
Cement	17		40 (compression only)
Diamond	1200		
Glass	70	20	50-1000
Rubber	7×10^{-3}		
Steel	210	80	380 (tensile)
			210 (compression)
Wood (Douglas fir)	11	4.1	130 (tensile only)

Note: These values are reasonable for classroom use, but if your work requires more precision, you should do your own research or experimentation.

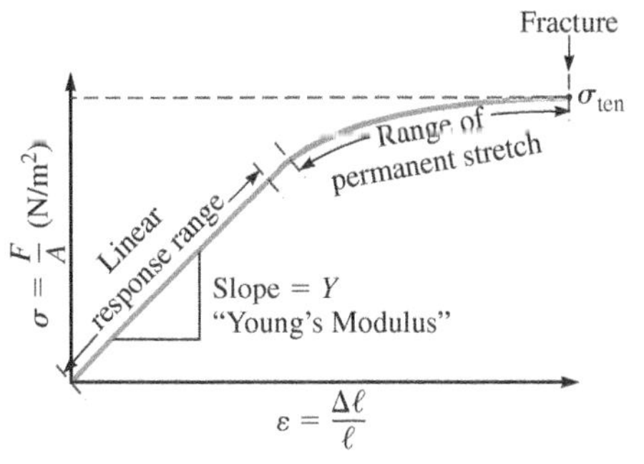

FIGURE 14.20 A graph of stress σ (vertical axis) as a function of strain ε (horizontal axis).

The **elastic limit** is the maximum stress that can be placed on an object without permanently deforming it. The elastic limit of an object is *not* necessarily equal to the maximum stress of the linear response range.

TENSILE STRENGTH

Major Concept

Third, the rod fractures at some maximum tensile stress known as the **tensile strength** σ_{ten}. Experiments show that tensile strength depends on the material of the rod. Table 14.1 includes the tensile strength of many materials. Let's take a microscopic look at what happens to the rod as it is stretched. As in Section 5-7, we model the rod's molecules as particles and the bonds between the molecules as stiff springs (Fig. 14.21). When a rod is loaded, the vertical bonds stretch, but the horizontal bonds do not. As the load is increased, the bonds obey Hooke's law and stretch in proportion to the force applied. When the load is removed, the bonds shrink back to their original size. When the load is too great, however, the molecular bonds no longer obey Hooke's law, and even if the load were to be removed, the bonds would no longer shrink to their original size. On the macroscopic level, the rod has been permanently stretched. When an even greater load is placed on the rod, the molecular bonds break, and the rod fractures.

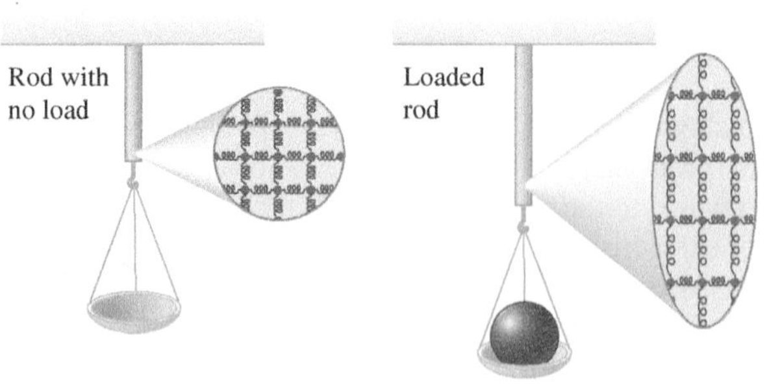

FIGURE 14.21 A microscopic model illustrates the strain on a rod as it reacts to a stress exerted on it by a hanging load. Each molecular bond stretches vertically, obeying Hooke's law.

Compressive Deformation

Engineers must also consider structures that are compressed by a load and structures that are not vertical. Whether the object is compressed or stretched, vertical or not, $\sigma = Y\varepsilon$ (Eq. 14.11) applies (Fig 14.22). If the object is compressed, however, $\Delta\ell$ is the *decrease* in the length of the object. Also, Young's *compression* modulus may not necessarily equal Young's *tensile* modulus (see, for example, the entries for bone in Table 14.1).

COMPRESSIVE STRENGTH

Major Concept

Like a tensile load, a compressive load may cause an object to fracture (Fig. 14.23). The **compressive strength** σ_{cmp} of a material—the stress at which the object fractures under compression—does not necessarily equal the tensile strength σ_{ten} of the material (Table 14.1).

Shear Deformation

Figure 14.24 shows another way a stationary object may be deformed. In this case, a *shear stress* has skewed the house. A **shear stress** τ results when a force is applied tangentially to one surface of a stationary object (Fig. 14.25). Mathematically, shear stress is given by

The Greek letter τ is used for both torque and shear stress.

$$\tau = \frac{F}{A} \tag{14.12}$$

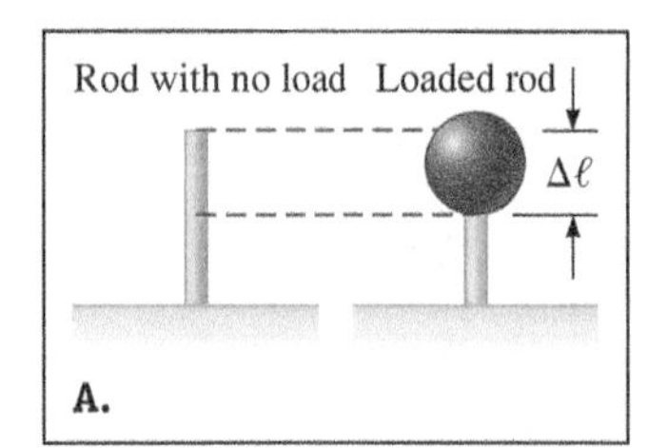

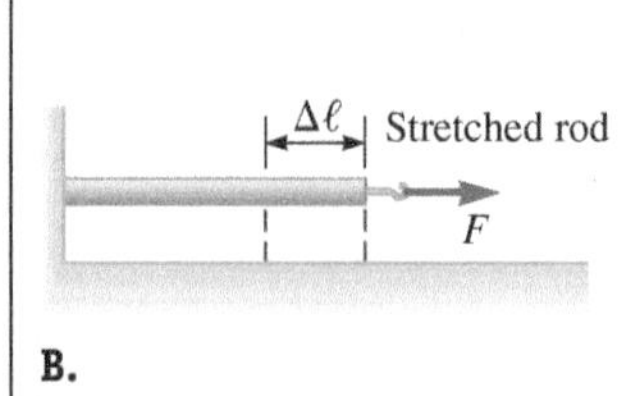

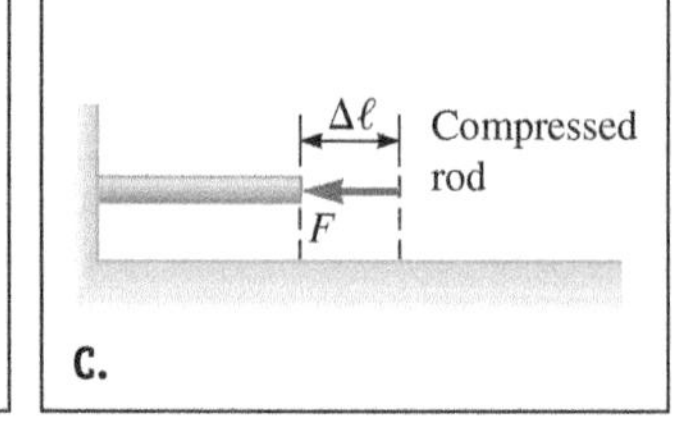

FIGURE 14.22 A. A rod is compressed by placing a load on top of it. **B.** A rod is stretched by a horizontal force. **C.** A rod is compressed by a horizontal force.

FIGURE 14.23 A column that fractured under a compressive load.

FIGURE 14.24 A house that suffered shear deformation in an earthquake.

where F is the magnitude of the force applied tangentially to the surface of area A. The **shear strain** γ is given by

$$\gamma = \frac{\Delta x}{h} \tag{14.13}$$

where Δx is the amount the sheared face moves and h is the height of that face (Fig. 14.25). For small shear stresses τ, the shear strain γ is proportional to the stress:

$$\tau = G\gamma \tag{14.14}$$

where G is called the **shear modulus**, also known as the torsion modulus. Like Young's modulus, G depends on the material making up the object (Table 14.1).

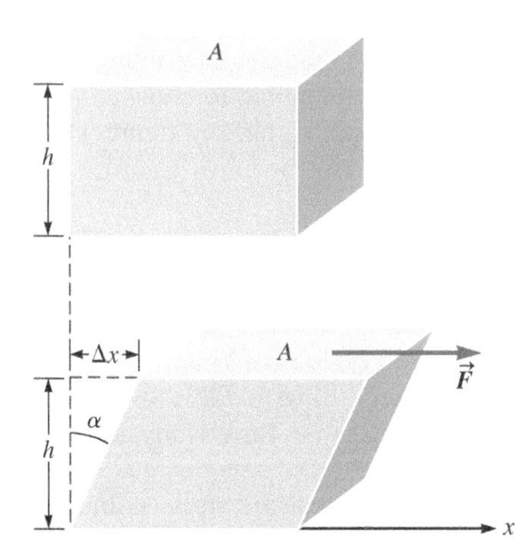

FIGURE 14.25 When a force is applied tangentially to one surface of a stationary object, a shear stress τ is exerted on the object, and a shear strain γ results.

SHEAR MODULUS

Major Concept

CONCEPT EXERCISE 14.5

Imagine two vertical rods initially of equal length. One rod is made of steel, and the other is made of rubber. Suppose you hang a load of the same mass from each rod. You are just able to eyeball the stretch in the rubber rod. Do you expect to notice any stretch in the steel rod? Explain.

EXAMPLE 14.6 Seeing Isn't Always Believing

In Figure 14.26, a car seems to be hanging from some gooey substance. Estimate Young's modulus for this unidentified material.

FIGURE 14.26 A car appears to hang from gooey supports.

© Tim Hornikel

INTERPRET and ANTICIPATE

Young's modulus can be found from $\sigma = Y\varepsilon$ (Eq. 14.11), so we need to estimate the strain ε of the substance and the stress σ exerted on it to find Y. The gooey substance appears to be very stretchy, so we expect Young's modulus to be very low.

SOLVE

Estimate the stress by assuming each gooey support experiences a force F equal to one-fourth of the car's weight F_g. A sedan typically weighs about 1.5 tons. Because this car is hanging, it seems likely that many of the heavy components (the engine, perhaps) may have been removed. So, let's assume the car weighs about 1 ton.

$$1 \text{ ton} = 2000 \text{ lb} \times \frac{1 \text{ N}}{0.2248 \text{ lb}}$$

$$1 \text{ ton} \approx 9000 \text{ N}$$

$$F = \frac{F_g}{4} = \frac{9000 \text{ N}}{4} \approx 2200 \text{ N}$$

The cross-sectional area of each gooey attachment is roughly equal to half the area of a tire's outer surface (the part that would be available to touch the road). Estimate that area by a rectangle of length $L \approx 1$ m and width $W \approx 0.3$ m.

$$A = LW \approx (1\,\text{m})(0.3\,\text{m})$$

$$A \approx 0.3\,\text{m}^2$$

Estimate the stress by dividing the force F by the cross-sectional area A (Eq. 14.9).

$$\sigma = \frac{F}{A} \tag{14.9}$$

$$\sigma = \frac{2200 \text{ N}}{0.3 \text{ m}^2} = 7300 \text{ N/m}^2$$

To estimate the strain, imagine raising the car upward until the gooey substance is compressed into a tight rectangular solid. Perhaps the substance would form a rectangular solid with a height of 5 cm. Its stretched length is a bit longer than the height of the car; let's estimate it to be 1.5 m. Find the strain by dividing the change in length by the original length of the substance.

$$\varepsilon = \frac{\Delta \ell}{\ell}$$

$$\varepsilon = \frac{1.5 \text{ m} - 5 \times 10^{-2} \text{m}}{5 \times 10^{-2} \text{m}} = 29$$

Example continues on page 408 ▶

Find Young's modulus by solving Equation 14.11 for Y. Because we estimated the values from the photo to one significant figure, we report our estimate of Young's modulus to one significant figure. Don't worry if your estimate is off by a factor of a few.

$$\sigma = Y\varepsilon \quad (14.11)$$

$$Y = \frac{\sigma}{\varepsilon} = \frac{7300\,\text{N/m}^2}{29}$$

$$Y \approx 300\,\text{N/m}^2$$

CHECK and THINK

As expected, Young's modulus for this substance is very low. The numbers in column 2 of Table 14.1 (values of Young's modulus) are to be multiplied by 10^9 N/m². Is our answer unrealistic? According to Table 14.1, Young's modulus for rubber (7×10^6 N/m²) is about 20,000 times greater than for this mysterious substance. To increase our estimate of Young's modulus, our estimate of the stress σ must be increased or our estimate of the strain ε must be decreased. It is unlikely that the stress is much greater than our estimate because the car cannot weigh much more than 1 ton (in fact, it probably weighs much less), and, from the photo, the cross-sectional area of the gooey substance cannot be much smaller than the corresponding surface area of the tire. Could our estimate of ε be off by four orders of magnitude? From the photo, the stretched length of the substance is clearly between 1 and 2 m, so our estimate of that quantity is quite good. The original length of the substance was an assumption. Perhaps the substance is not actually that gooey; it may be carved wood or molded metal that has been artistically created to look gooey. In that case, the unstretched length may be very close to the stretched length, which would make ε smaller and Y larger. It seems likely that Figure 14.26 is a photo of a clever illusion and that the substance is not four orders of magnitude more stretchable than rubber.

EXAMPLE 14.7 CASE STUDY Margin of Safety

The yellow airplane in the case study hangs from two steel cables (Fig. 14.14, page 399). Cable 1 has a diameter of 1.2 cm, and cable 2 has a diameter of 0.3 cm. Find the maximum weight of a plane that can be hung from these two cables and compare your answer with the airplane's actual weight (9.3×10^3 N), which we found in Example 14.3.

INTERPRET and ANTICIPATE

When the tension per unit area in one of the cables equals the tensile strength of steel, that cable will break. So, we should find the maximum tension in each cable from the tensile strength of steel and the cross-sectional area of the cable. We can then use $F_{1,\,\text{max}} + F_{2,\,\text{max}} = m_{\text{max}}g$ (Eq. 1 in Example 14.3) to find the maximum weight of a plane that could be hung from these cables. We expect the maximum weight to be several times greater than the actual weight of the airplane.

SOLVE

The tensile strength is the maximum tensile stress. Use $\sigma = F/A$ (Eq. 14.9) to find the maximum tension in the cables.

$$F_{1,\,\text{max}} = \sigma_{\text{ten}} A_1$$

$$F_{2,\,\text{max}} = \sigma_{\text{ten}} A_2$$

Write the cross-sectional area in terms of each cable's diameter, $A = \pi(d/2)^2$.

$$F_{1,\,\text{max}} = \sigma_{\text{ten}}\pi\left(\frac{d_1}{2}\right)^2$$

$$F_{2,\,\text{max}} = \sigma_{\text{ten}}\pi\left(\frac{d_2}{2}\right)^2$$

Consult Table 14.1 to find the tensile strength of steel.

$$F_{1,\,\text{max}} = (380 \times 10^6\,\text{N/m}^2)\pi\left(\frac{1.2 \times 10^{-2}\,\text{m}}{2}\right)^2 = 4.3 \times 10^4\,\text{N}$$

$$F_{2,\,\text{max}} = (380 \times 10^6\,\text{N/m}^2)\pi\left(\frac{0.3 \times 10^{-2}\,\text{m}}{2}\right)^2 = 2.7 \times 10^3\,\text{N}$$

The maximum weight of the plane equals the maximum tension in the cables.	$F_{g,\max} = F_{1,\max} + F_{2,\max}$ $F_{g,\max} = (4.3 \times 10^4\ \text{N}) + (2.7 \times 10^3\ \text{N}) = 4.57 \times 10^4\ \text{N}$
For the required comparison, divide the maximum weight that can be supported by the airplane's actual weight.	$\dfrac{F_{g,\max}}{F_g} = \dfrac{4.57 \times 10^4\ \text{N}}{9.30 \times 10^3\ \text{N}} = 4.91$

CHECK and THINK

As expected, the maximum weight that the cables can support is several times the weight of the hanging airplane. It is common for designers to include this sort of safety margin in their plans.

You might be wondering if each cable is designed to support roughly five times what it actually supports. To answer that question, divide the maximum tension in each cable found in this example by the actual tension in each cable found in Example 14.3. We find that both cables have the same margin of safety.	$\dfrac{F_{1,\max}}{F_1} = \dfrac{4.3 \times 10^4\ \text{N}}{8.7 \times 10^3\ \text{N}} = 4.9$ $\dfrac{F_{2,\max}}{F_2} = \dfrac{2.7 \times 10^3\ \text{N}}{5.50 \times 10^2\ \text{N}} = 4.9$

In this chapter, we looked at a special case of motion, that is, no motion at all. At first, it may seem like a boring case when you compare static equilibrium to motion of stars as they are devoured by a giant black hole or to the accurate dive made by a hawk pursuing its prey or to the flight of the yellow plane before it was hung on display in the Udvar-Hazy Center (Fig. 14.14). When we are standing underneath a plane suspended from the ceiling by a few steel cables, however, static equilibrium is exactly what we want. Structural engineers have an important job in ensuring static equilibrium. In fact, every day we are surrounded by structures that are designed to be in static equilibrium, and if they fail to remain in static equilibrium, the consequences can be tragic. The job of a structural engineer is made more complicated because structures are never truly in static equilibrium. Structures are stretched, compressed, and sheared during the course of normal use and during extraordinary situations such earthquakes and major storms. In Chapter 19, we return to these ideas when we see how changes in temperature can further complicate an engineer's work.

SUMMARY

Underlying Principle

Equilibrium is a special case of Newton's second law, and **static equilibrium** is a special case of equilibrium. An object is in equilibrium if both its translational and angular accelerations are zero. An object in *static* equilibrium is at rest relative to an observer.

Major Concepts

1. There are three types of static equilibrium:
   a. If an object is in **stable static equilibrium** and it is displaced by a small amount, the forces acting on it will return it to its equilibrium position.
   b. If an object is in **unstable static equilibrium** and it is displaced by a small amount, the forces acting on it will move it away from its equilibrium position.
   c. The forces exerted on an object in **neutral static equilibrium** do not maintain its equilibrium, so if such an object is displaced, it is neither restored to its equilibrium position nor is it forced farther away.

2. There are two equilibrium conditions:
   a. According to the **force balance condition**, the net force acting on the object must be zero:

$$\vec{F}_{tot} = 0 \quad (14.1)$$

   b. According to the **torque balance condition**, the net torque acting on the object must be zero:

$$\vec{\tau}_{tot} = 0 \quad (14.2)$$

3. **Stress** is the magnitude of the force applied to an object per unit area.
   a. **Tensile stress** σ results when a force pulls perpendicular to a side of a stationary object.
   b. **Compression stress** σ results when a force pushes perpendicular to a side of a stationary object. For a tensile or compression stress,

$$\sigma \equiv \frac{F}{A} \quad (14.9)$$

   c. **Shear stress** τ results when a force is applied tangentially to one surface of a stationary object:

$$\tau = \frac{F}{A} \quad (14.12)$$

4. **Strain** is a dimensionless measure of the object's deformation due to shear. The fractional increase (or decrease) in an object's length $\Delta\ell/\ell$ due to a tensile (or compression) shear is known simply as the strain,

$$\varepsilon = \Delta\ell/\ell \quad (14.10)$$

The **shear strain** γ that results from a shear stress is given by

$$\gamma = \frac{\Delta x}{h} \quad (14.13)$$

where Δx is the amount that the sheared face moves and h is the height of that face.

5. As long as the tensile (or compressive) stress σ is small, the strain is directly proportional to the stress:

$$\sigma = Y\varepsilon \quad (14.11)$$

where Y is known as **Young's modulus**.

6. An **elastic** object returns to its original length after a load has been removed. The **elastic limit** is the maximum stress that can be placed on an object without permanently deforming it.
7. The **tensile strength** σ_{ten} of a material is the maximum tensile stress that may be applied to an object made of that material without fracturing it. The **compressive strength** σ_{cmp} of a material—the stress at which the object fractures under compression—does not necessarily equal the tensile strength σ_{ten}.
8. For small shear stresses τ, the shear strain γ is proportional to the stress:

$$\tau = G\gamma \quad (14.14)$$

where G is called the **shear modulus**.

Tools

Cross Products

The cross product of two vectors $\vec{R} = \vec{A} \times \vec{B}$ may be expressed in component form as

$$\vec{R} = (A_yB_z - A_zB_y)\hat{\imath} + (A_zB_x - A_xB_z)\hat{\jmath} + (A_xB_y - A_yB_x)\hat{k} \quad (14.5)$$

Using Equation 14.5 to express the torque is often helpful:

$$\vec{\tau} = \vec{r} \times \vec{F} = (r_yF_z - r_zF_y)\hat{\imath} + (r_zF_x - r_xF_z)\hat{\jmath} + (r_xF_y - r_yF_x)\hat{k} \quad (14.7)$$

PROBLEM-SOLVING STRATEGY Static Equilibrium

INTERPRET and ANTICIPATE

Draw a modified free-body diagram.

1. Represent the object as a **simple shape**.
2. Draw all the **forces** exerted on the object at each **point of application**.
3. Choose a **rotation axis**.
4. Choose a **coordinate system**.
5. Draw **position vectors**.

SOLVE

1. Write the **forces and position vectors in component form** (or list magnitudes).
2. Apply **equilibrium conditions**.
   a. Apply the force balance condition.
   b. Find the torques and **apply the torque balance condition**.
3. **Do algebra** before substituting values.

PROBLEMS AND QUESTIONS

A = algebraic C = conceptual E = estimation G = graphical N = numerical

14-1 What Is Static Equilibrium?

Problems 1–3 are grouped.

1. C A ball is attached to a strong, lightweight rod (Fig. P14.1). The rod is supported by a horizontal pin near the top. The ball is at rest. Is the ball in static equilibrium? If not, why not? If so, which type of equilibrium is it—stable, unstable, or neutral? *Hint*: What would happen if you displaced the ball slightly?

FIGURE P14.1

2. C A ball is attached to a strong, lightweight rod (Fig. P14.2). The rod is supported by a horizontal pin near the bottom. The ball is at rest. Is the ball in static equilibrium? If not, why not? If so, which type of equilibrium is it—stable, unstable, or neutral? *Hint*: What would happen if you displaced the ball slightly?

FIGURE P14.2

3. C Two identical balls are attached to a strong, lightweight rod (Fig. P14.3). The rod is supported by a horizontal pin at the system's center of mass. The balls are at rest. Is the system in static equilibrium? If not, why not? If so, which type of equilibrium is it—stable, unstable, or neutral? *Hint*: What would happen if you displaced the balls slightly?

FIGURE P14.3

4. C While working on homework together, your friend stands her pencil on its sharpened tip. You say, "Nice trick. How did you do it?" Why do you think some trick had to be involved? Would you ask the same question if the pencil were lying on its side or if the pencil were unsharpened?
5. C Consider the sketch of a portion of a roller-coaster track seen in Figure P14.5. Identify places on the track that could be considered possible locations of static equilibrium for a roller-coaster car were the car to be placed at any spot on the track. Which places are candidate locations for stable, unstable, and neutral static equilibrium?

FIGURE P14.5

14-2 Conditions for Equilibrium

6. N **Review** Where is the center of mass of the L-shaped uniform metal object shown in Figure P14.6?

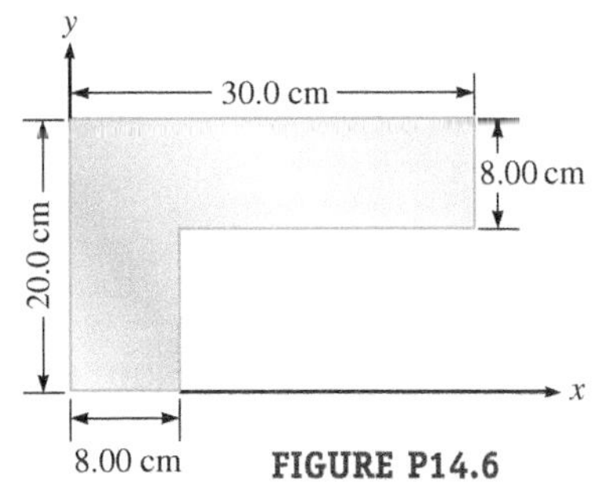

FIGURE P14.6

7. N A heavy iron gate is stuck closed. Opening the gate requires pushing hard along the y axis and lifting at the same time along the z axis. A force of $(920\hat{\jmath} + 110\hat{k})$ N is applied to the top right corner of the gate such that the position vector from the axis of rotation to the top right corner is given by $\vec{r} = 3.0\hat{\imath}$ m. What is the torque on the gate due to this force?
8. Suppose you have two vectors, $\vec{A} = 7.95\hat{\imath}$ and $\vec{B} = 5.13\hat{\jmath}$.
   a. N What is the vector cross product of these two vectors, $\vec{R} = \vec{A} \times \vec{B}$?
   b. C Given the direction of the resultant vector and the directions of the two original vectors, what can you say in general about the relationships between the three vectors involved in the cross product?
   c. N Now suppose the two vectors are $\vec{A} = 7.95\hat{\imath}$ and $\vec{B} = 3.00\hat{\imath} + 5.13\hat{\jmath}$. What is their cross product in this case?
   d. C Given the direction of the resultant vector and the directions of the two original vectors, can you further generalize the relationship between the three vectors involved in the cross product? How are three vectors necessarily oriented relative to one another?
9. A The keystone of an arch is the stone at the top (Fig. P14.9). It is supported by forces from its two neighbors, blocks A and B. Each block has mass m and approximate length L. What can you conclude about the force exerted by each block, $\vec{F}_A$ and $\vec{F}_B$, for the keystone to remain in static equilibrium? That is, show that the equilibrium conditions are satisfied by the components of the forces F_{Ax}, F_{Ay}, F_{Bx}, and F_{By}. Assume the arch is symmetric.

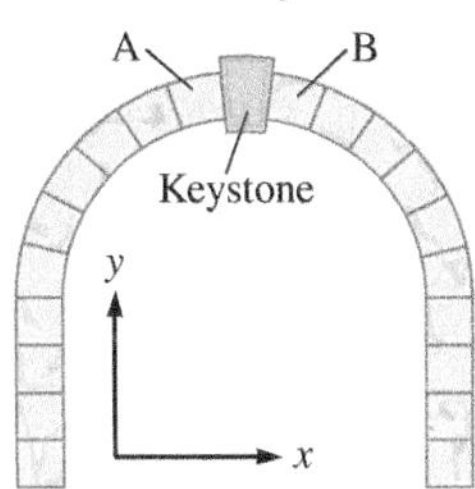

FIGURE P14.9

Problems 10 and 11 are paired.

10. C Stand straight and comfortably with your feet about shoulders' width apart. Your center of mass is located somewhere in your abdomen directly above the point between your feet. Now bend forward, hinging from your hips. What do you notice about the motion of your lower body? Explain why it moves. If you have trouble noticing any motion, hold a dumbbell or other heavy object in each hand. As you bend forward, allow your hands to move toward the ground.
11. C Stand straight and comfortably with your feet about shoulders' width apart. Now lift one of your legs so that you are balanced on only one foot. Notice how your body moves. Explain why it moves.
12. N **Review** Three uniform spheres are placed in the xy plane as follows: The center of mass of an 8.00-kg sphere is located at (2.00, 2.00) m, the center of mass of a 5.00-kg sphere is located at (0, 5.00) m, and the center of mass of a 2.00-kg sphere is located at (3.00, 0) m. Where should a 6.00-kg sphere be placed if the center of mass of the four-sphere system is to be located at the origin?

Problems 13 and 14 are paired.

13. A The uniform block of width W and height H in Figure P14.13 rests on an inclined plane in static equilibrium. The plane is slowly raised until the block begins to tip over. Assume the tipping occurs before the block starts to slide. Find an expression for the angle θ at which the block is on the verge of tipping.

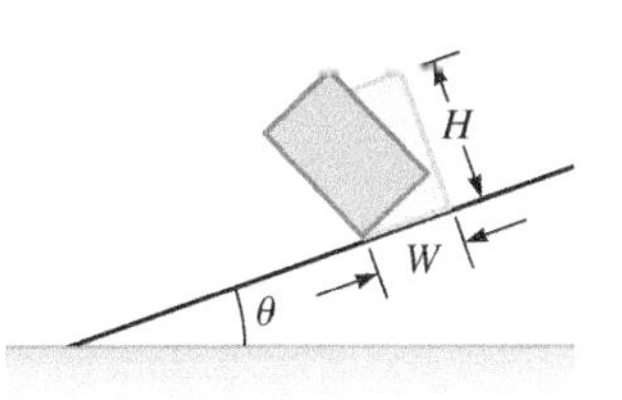

FIGURE P14.13 Problems 13 and 14.

14. A If the block in Problem 13 tips before it slides, determine the minimum coefficient of static friction.

15. N Find the cross product $\vec{R} = \vec{A} \times \vec{B}$, where

$$\vec{A} = (23.45\hat{\imath} + 106.3\hat{\jmath} - 92.81\hat{k}) \text{ and}$$
$$\vec{B} = (-5.010\hat{\imath} + 2.146\hat{\jmath} - 3.114\hat{k}).$$

Give your answer in component form and also state the magnitude of the result.

16. A Is the vector cross product commutative? In other words, does $\vec{A} \times \vec{B} = \vec{B} \times \vec{A}$? Expound on your answer with a mathematical proof or prove that it is not true by example.

Problems 17–19 are grouped.

17. A force $\vec{F} = (65.8\hat{\imath} + 96.3\hat{\jmath})$ N is exerted on an object at a point whose position from the axis of rotation is given by $\vec{r} = -5.80\hat{\imath}$ m.
   a. N Find the torque exerted by this force.
   b. C What happens to your answer if the force's x component increases while the y component remains constant?
   c. C What happens to your answer if the force's y component increases while the x component remains constant?

18. A force $\vec{F} = (65.8\hat{\imath} + 96.3\hat{\jmath})$ N is exerted on an object at a point whose position from the axis of rotation is given by $\vec{r} = -5.80\hat{k}$ m.
   a. N Find the torque exerted by this force.
   b. C What happens to your answer if the force's x component increases while the y component remains constant?
   c. C What happens to your answer if the force's y component increases while the x component remains constant?

19. N A force $\vec{F} = (65.8\hat{\imath} + 96.3\hat{k})$ N is exerted on an object, and the torque that results is given by $\vec{\tau} = (57.1\hat{\jmath})$ N · m. Find $\vec{r} = r_z\hat{k}$.

Problems 20 and 21 are paired.

20. N A man poses for a photo while holding a dumbbell. His arm is bent as shown in Figure P14.20. The torque exerted by the dumbbell around his elbow is 38.9 N · m. What is the mass of the dumbbell?

21. N If the torque exerted by the dumbbell around the man's elbow in Figure P14.20 is 38.9 N · m, what is the torque exerted by the dumbbell around his shoulder?

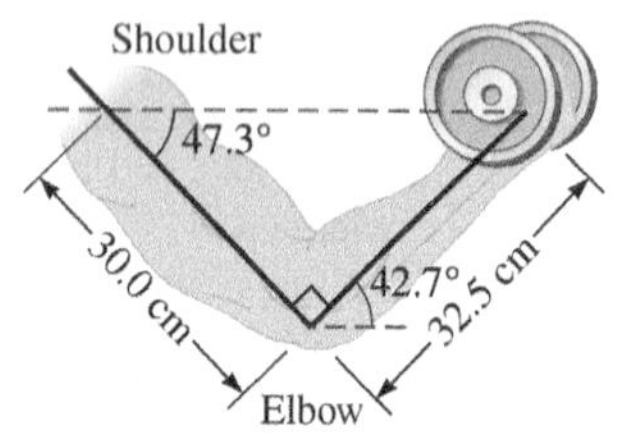

FIGURE P14.20 Problems 20 and 21.

14-3 Examples of Static Equilibrium

22. N The inner planets of our solar system are represented on a mobile constructed from drinking straws and light strings for a school project (Fig. P14.22). The mass of the piece representing the Earth is 25.0 g, and the mass of the straws can be ignored. (The lengths shown are not proportional to actual distances in the solar system.) What is the mass of the piece representing **a.** Mars, **b.** Venus, and **c.** Mercury?

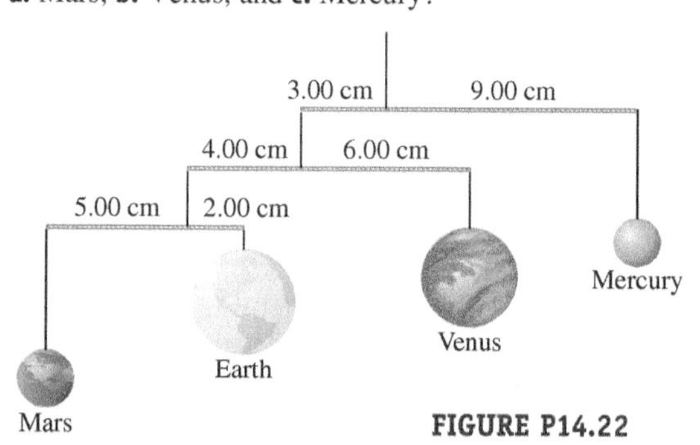

FIGURE P14.22

23. N Two Boy Scouts, Bobby and Jimmy, are carrying a uniform wooden tent pole 4.50 m in length and having mass $m = 23.0$ kg. Bobby is holding the pole 60.0 cm from the back, and Jimmy is holding the pole 40.0 cm from the front. If the pole is held horizontally by the two boys, what is the force Bobby and Jimmy each exert on the beam?

24. G Draw a modified free-body diagram for the bottle-holder system shown in Figure P14.24. Notice that the wine bottle is horizontal. *Hint*: Draw a line that passes through the system's center of mass.

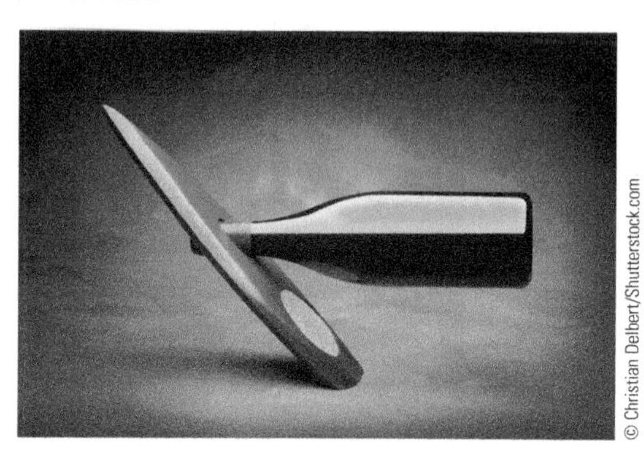

FIGURE P14.24

Problems 25 and 26 are paired.

25. A painter of mass 87.8 kg is 1.45 m from the top of a 6.67-m ladder. The ladder rests against a wall. Friction between the ladder and the wall is negligible. The ladder's mass is 5.50 kg. Assume the ladder is set up according to Occupational Safety and Health Administration standards so that it makes a 75.5° angle with the floor.
   a. N What is the minimum coefficient of static friction required for the painter's safety?
   b. C Would rubber ladder tips on dry concrete be safe?

26. Consider the situation in Problem 25. Tests have shown that when people try to eyeball the proper angle for the ladder, they are often off by as much as 9°. Suppose the painter sets up the ladder at too small an angle.
   a. N At what maximum angle would rubber ladder tips on dry concrete be unsafe?
   b. C How likely is it for a person to make such a dangerous mistake?

27. N Children playing pirates have suspended a uniform wooden plank with mass 15.0 kg and length 2.50 m as shown in Figure P14.27. What is the tension in each of the three ropes when Sophia, with a mass of 23.0 kg, is made to "walk the plank" and is 1.50 m from reaching the end of the plank?

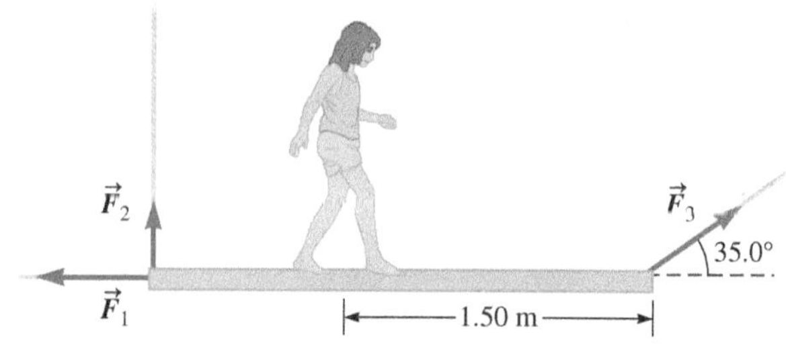

FIGURE P14.27

28. E CASE STUDY After having worked through the case study, you decide to visit the Udvar-Hazy Center at the National Air and Space Museum. While there, you spend more time

thinking about how the exhibits are displayed than you did on prior visits. You see the aircraft in Figure P14.28A, and you think it would be much nicer if the front supports were replaced by a single cable. You imagine that you wouldn't need to drill a hole into the top of the aircraft. Instead, you could attach one end of the cable by slipping a small yellow ring over the nose, and you could attach the other end to a vertical steel column. Your plan is shown in Figure P14.28B. Estimate the tension in the cable by considering the dimensions and weight of the plane in the case study. Keep in mind that the aircraft in Figure P14.28 is smaller than the one in the case study.

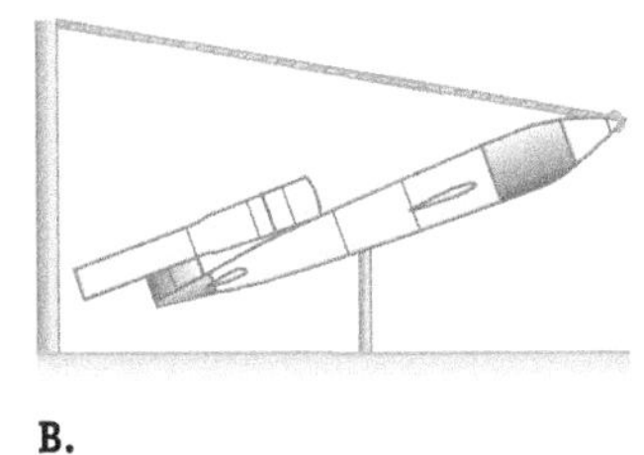

A. **B.**

FIGURE P14.28

29. N A uniform beam rests on a cart with a portion of one end hanging out (Fig. P14.29). The overhanging portion is one-fourth of the total length L of the beam. A gradually increasing force F acts on the end of the beam at point P. The opposite side of the beam begins to rise when the magnitude of the force reaches 2.0×10^2 N. What is the beam's weight?

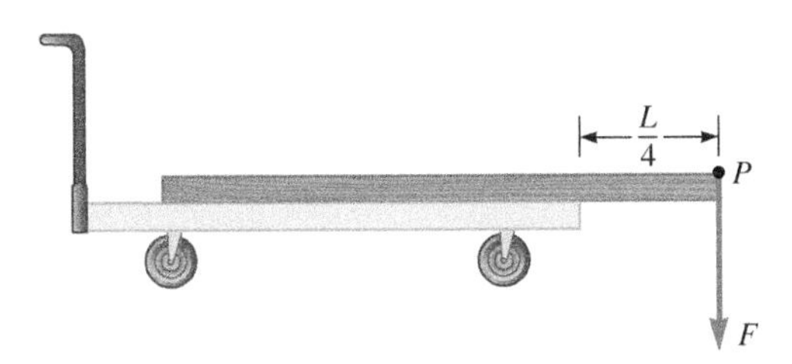

FIGURE P14.29

30. N A 5.45-N beam of uniform density is 1.60 m long. The beam is supported at an angle of 35.0° by a cable attached to one end. There is a pin through the other end of the beam (Fig. P14.30). Use the values given in the figure to find the tension in the cable.

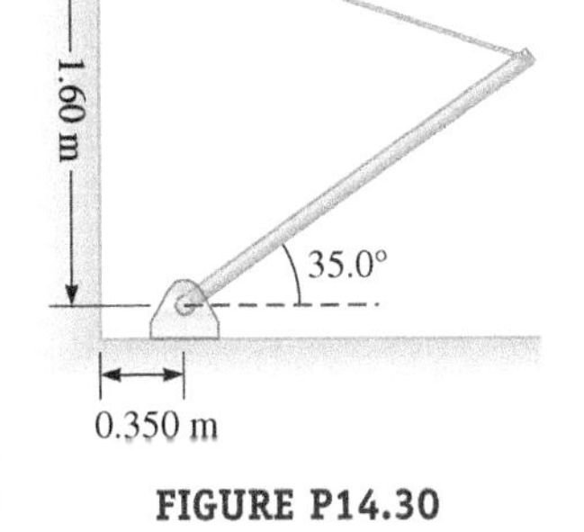

FIGURE P14.30

31. N A wooden door 2.1 m high and 0.90 m wide is hung by two hinges 1.8 m apart. The lower hinge is 15 cm above the bottom of the door. The center of mass of the door is at its geometric center, and the weight of the door is 260 N, which is supported equally by both hinges. Find the horizontal force exerted by each hinge on the door.
32. N A 215-kg robotic arm at an assembly plant is extended horizontally (Fig. P14.32). The massless support rope attached at point B makes an angle of 15.0° with the horizontal, and the center of mass of the arm is at point C. **a.** What is the tension in the support rope? **b.** What are the magnitude and direction of the force exerted by the hinge A on the robotic arm to keep the arm in the horizontal position?

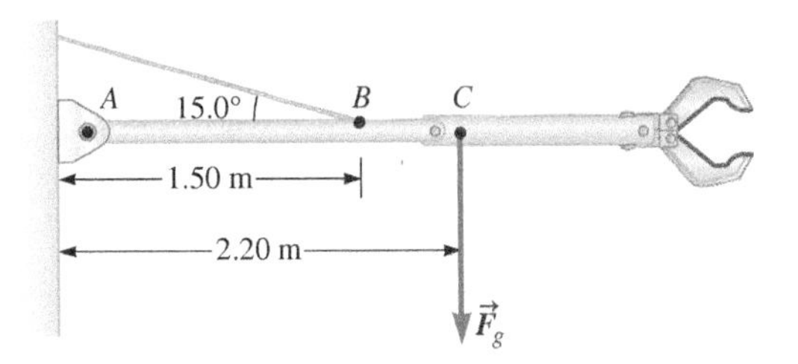

FIGURE P14.32

Problems 33 and 34 are paired.

33. N One end of a uniform beam that weighs 2.80×10^2 N is attached to a wall with a hinge pin. The other end is supported by a cable making the angles shown in Figure P14.33. Find the tension in the cable.
34. N For the uniform beam in Problem 33, find the horizontal and vertical components of the force exerted by the hinge pin on the beam.

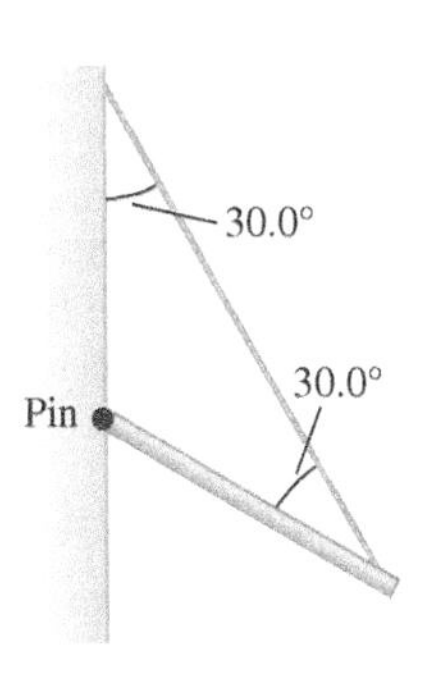

FIGURE P14.33 Problems 33 and 34.

35. N The owner of the Galaxy Café wishes to suspend a fictional spacecraft outside the front door (Fig. P14.35). The spacecraft weighs 1000 lb. The triangular structure is much lighter than the spacecraft, and its weight can be neglected. Use the values for length given in the figure and find the normal forces exerted by the pins at points A and B.

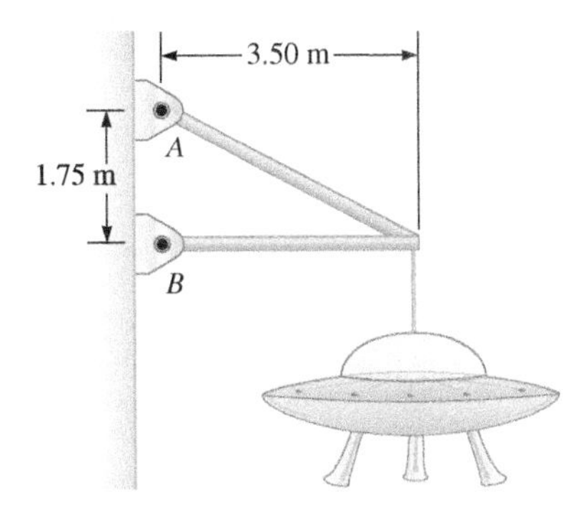

FIGURE P14.35

36. N A square plate with sides of length 4.0 m can rotate about an axle passing through its center of mass and perpendicular to the plate as shown in Figure P14.36. There are four forces acting on the plate at different points. The rotational inertia of the plate is 24 kg · m². Is the plate in equilibrium?

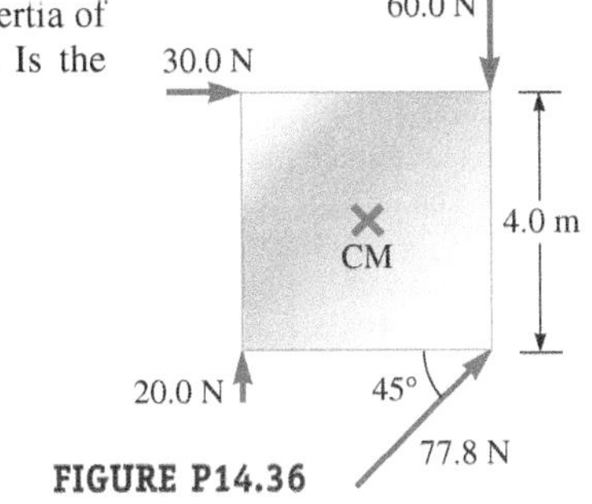

FIGURE P14.36

37. N The circular sign in Figure P14.37 has a mass of 16.7 kg. Use the angle shown in the figure to find the tension in the cable.

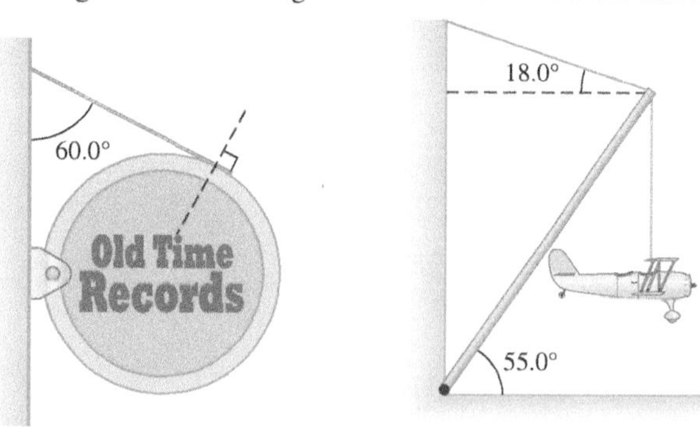

FIGURE P14.37 FIGURE P14.38

38. N At a museum, a 1300-kg model aircraft is hung from a lightweight beam of length 12.0 m that is free to pivot about its base and is supported by a massless cable (Fig. P14.38). Ignore the mass of the beam. **a.** What is the tension in the section of the cable between the beam and the wall? **b.** What are the horizontal and vertical forces that the pivot exerts on the beam?

14-4 Elasticity and Fracture

39. N A uniform wire ($Y = 2.0 \times 10^{11}$ N/m^2) is subjected to a longitudinal tensile stress of 4.0×10^7 N/m^2. What is the fractional change in the length of the wire?

40. C A brass wire and a steel wire, both of the same length, are extended by 1.0 mm under the same force. Is the cross-sectional radius of the brass wire more, less, or equal to the cross-sectional radius of the steel wire? Explain. Young's moduli for brass and steel are 1.0×10^{10} N/m^2 and 2.0×10^{11} N/m^2, respectively.

41. N CASE STUDY In Example 14.3, we found that one of the steel cables supporting an airplane at the Udvar-Hazy Center was under a tension of 9.30×10^3 N. Assume the cable has a diameter of 2.30 cm and an initial length of 8.00 m before the plane is suspended on the cable. How much longer is the cable when the plane is suspended on it?

Problems 42–44 are grouped.

42. N A carbon nanotube is a nanometer-scale cylindrical tube composed of carbon atoms. One of its interesting properties is a very large Young's modulus, measured to be more than 1.000×10^{12} N/m^2. A tensile stress of 5.3×10^{10} N/m^2 is exerted on a particular nanotube with a Young's modulus measured at 2.130×10^{12} N/m^2. Assuming the nanotube obeys Hooke's law, by what percentage does the atomic spacing increase?

43. N A nanotube with a Young's modulus of 1.000×10^{12} Pa is subjected to a stress of 3.14×10^{11} Pa by being pulled at its ends. Assuming the tube had an initial length of 8.12×10^{-6} m, what is the new length of the nanotube?

44. N Consider a nanotube with a Young's modulus of 2.130×10^{12} N/m^2 that experiences a tensile stress of 5.3×10^{10} N/m^2. Steel has a Young's modulus of about 2.000×10^{11} Pa. How much stress would cause a piece of steel to experience the same strain as the nanotube?

45. N The spring in Figure P14.45 has a spring constant of 7.50×10^2 N/m, and it is relaxed when the rod is vertical. The rod is 2.50 m long, with uniform density. The horizontal distance between the pin in the spring and the pin in the rod is 6.75 m. **a.** When the bar is tilted by 41.6° from the vertical as shown, what is the length of the spring? **b.** If the system in Figure P14.45 is in equilibrium, what is the weight of the bar?

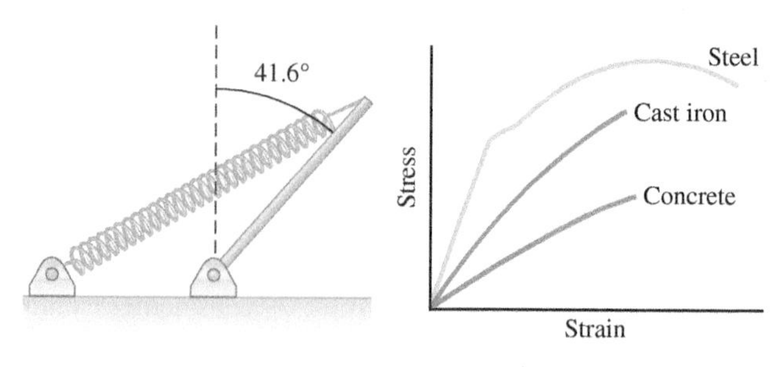

FIGURE P14.45 FIGURE P14.46

46. C Use the graph in Figure P14.46 to list the three materials from greatest Young's modulus to smallest. Explain your reasoning.

47. N A 5.00-m-long copper wire with a diameter of 5.00 mm and Young's modulus 11.0×10^{10} N/m^2 is used to suspend a large bucket of mass 355 kg. What is the increase in the wire's length due to this load?

48. G A company is testing a new material made of recycled plastic for use in construction. Figure P14.48 is a graph of the stress as a function of strain for this material. Find Young's modulus for this material and compare your result with the data for conventional building materials listed in Table 14.1.

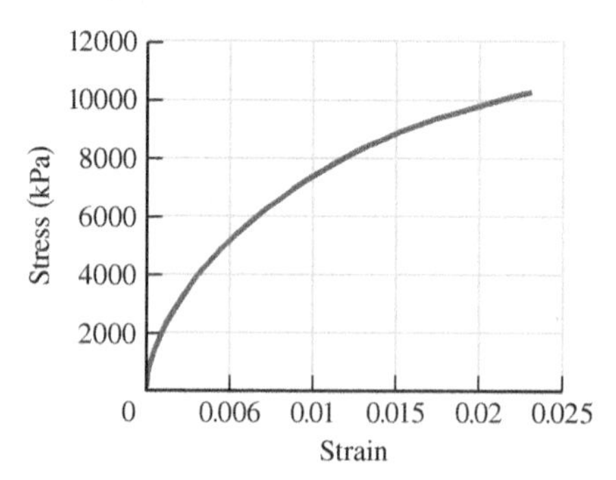

FIGURE P14.48

49. N A tubular steel support is 3.0 m tall and shortens 0.60 mm under a compressive force of 70 kN. If the inner radius of the tube is 0.80 times the outer radius, what is the outer radius? Young's modulus for steel is 2.0×10^{11} Pa.

Problems 50–53 are grouped.

50. N A steel rod has a radius of 15.0 cm and a length of 1.25 m. The rod is held firmly in place, and a 32.5-kN force is applied perpendicular to one face of the rod so that it compresses. What is the strain of the rod?

51. N An aluminum rod has a radius of 15.0 cm and a length of 1.25 m. The rod is held firmly in place, and a 32.5-kN force is applied perpendicular to one face of the rod so that it compresses. What is the strain of the rod? (If you worked Problem 50, compare your answers.)

52. N A steel rod has a radius of 15.0 cm and a length of 1.25 m. The rod is held firmly in place, and a 32.5-kN force is applied perpendicular to one face of the rod so that it stretches. What is the strain of the rod? (If you worked Problem 50, compare your answers.)

53. N A steel rod has a radius of 15.0 cm and a length of 1.25 m. The rod is held firmly in place, and a 32.5-kN force is applied parallel to one face of the rod so that it is sheared. What is the (shear) strain of the rod? (If you worked Problem 50, compare your answers.)

54. N A structural piece of material consists of a solid cylindrical rod 2.00 m in length with one half having a diameter of 1.00 cm and the other half having a diameter of 0.750 cm. It is made from an alloy with a Young's modulus of 1.30×10^{11} N/m^2. If the piece is put under a tensile force of 15 kN, by how much does it elongate?

55. N While trying to catch a badly thrown basketball, Kobe Bryant briefly slides across the basketball court. The court exerts a frictional force of 25.0 N on each foot. The footprint of his size 14 shoes has an area of 420 cm^2, and the rubber soles (shear modulus 6.00×10^5 N/m^2) are 19.0 mm tall. What is the horizontal offset of the upper and lower surfaces of the shoe soles?

56. A A metal wire of length L is stretched an amount ΔL by a weight F_g. What is the fractional change in the wire's volume, $\Delta V/V$?

57. A A copper rod with length 1.4 m and cross-sectional area 2.0 cm^2 is fastened to a steel rod of length L and cross-sectional area 1.0 cm^2. The compound structure is pulled on each side by two forces of equal magnitude 6.00×10^4 N (Fig. P14.57). Find the length L of the steel rod if the elongations (ΔL) of the two rods are equal. Use the values $Y_{steel} = 2.0 \times 10^{11}$ Pa and $Y_{Cu} = 1.1 \times 10^{11}$ Pa.

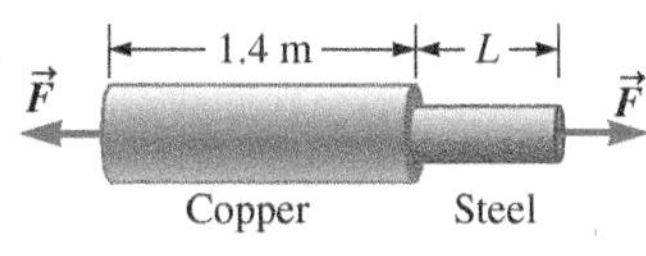

FIGURE P14.57

General Problems

58. A, C In Example 14.5, it was stated that at the maximum thrust the front two feet of the fan are just lifting off the floor; in this problem, you'll show that this statement is true. Return to Figure 14.17 (page 403), which shows all 10 forces exerted on the fan. Write down the force balance and the torque balance equations. Then, reason that as F_{thrust} approaches its maximum, F_{N3} and F_{N4} both approach zero.

59. N **Review** A uniform circle of mass $m_1 = 3.00$ kg, a uniform triangle of mass $m_2 = 1.00$ kg, and a uniform square of mass $m_3 = 6.00$ kg are placed in the xy plane as shown in Figure P14.59. What are the x and y coordinates of the center of mass of the three-object system?

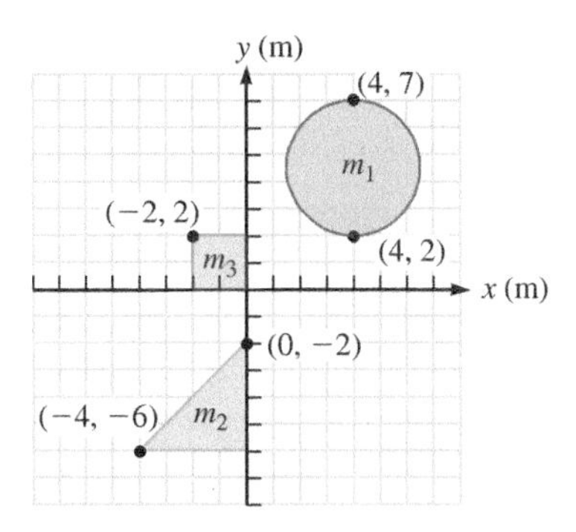

FIGURE P14.59

60. N Bruce Lee was famous for breaking concrete blocks with a single karate chop. From slow-motion video, the speed of his 1.50-kg hand descending on a block was estimated to be 15.0 m/s, which decreased to a speed of 0.500 m/s in the 2.50×10^{-3} s during which his hand made contact with and broke through the block. The maximum shear stress a concrete block can be subjected to before breaking is 9.50×10^5 N/m^2. **a.** What was the force exerted by Lee's hand on the block? **b.** If a typical concrete block broken by Lee was 3.00 cm thick and 15.2 cm wide, what is the shear stress experienced by the concrete block? **c.** Will the concrete block succumb to Lee's karate chop?

Problems 61 and 62 are paired.

61. A A disk of radius R and mass M is rotating and experiencing an angular acceleration $\vec{\alpha} = \alpha\hat{k}$. Define the additional torque that must be applied for the disk to be in equilibrium. Write your answer as a vector in terms of α, R, and M.

62. A disk of radius R and mass M is rotating and experiencing an angular acceleration $\vec{\alpha} = \alpha\hat{k}$. There is actually a range of possible values for the magnitude of a single force that could be applied for the disk to be in equilibrium.

a. N What is the minimum magnitude of a single force that, when applied, causes the disk to be in equilibrium?

b. C Explain why there is a range of possible values for this force and what bounds you can place on its magnitude.

63. A Show that when an object is suspended by a single rope, the object's center of mass must be someplace along the line that extends through the length of the rope.

64. A One end of a metal rod of weight F_g and length L presses against a corner between a wall and the floor (Fig. P14.64). A rope is attached to the other end of the rod. Find the magnitude of the tension in the rope if the angle β between the rod and the rope is 90°.

65. N A 150-kg pile driver starts from rest and impacts a steel beam 15.0 cm in radius that is 7.60 m below its original position. The pile driver rebounds with half its speed before impact after 0.065 s. What is the average strain in the beam during the impact?

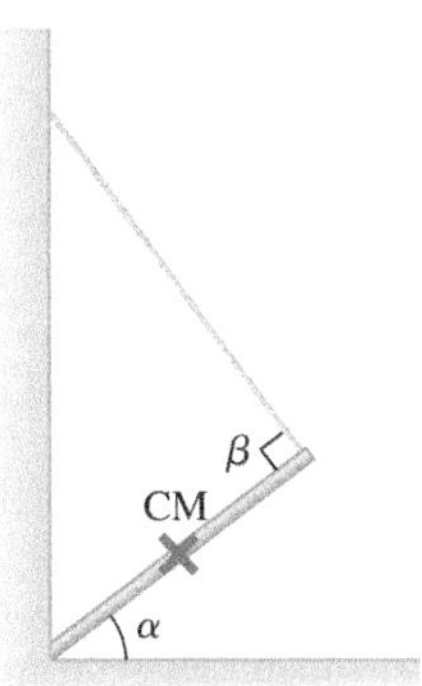

FIGURE P14.64

66. N A steel cable 2.00 m in length and with cross-sectional radius 0.350 mm is used to suspend from the ceiling a 10.0-kg model aircraft that is flying in a horizontal circle with an angular speed of 6.00 rad/s. What is the strain produced in the cable?

Problems 67 and 68 are paired.

67. N While working out at the gym, a Marine finds that the fan (Fig. 14.16, page 402) is too low to keep her cool. She gets the idea to hang the fan from two cables. She removes its pedestal, and the fan's remaining mass is 10.3 kg. When the fan is off, the cables are vertical (Fig. P14.67A). When the fan is on, the cables make an angle θ with the vertical (Fig. P14.67B). She attaches the cables to the fan using pins that allow the fan to remain in the same orientation when it is operating. Assume the fan is exerting the maximum thrust (21.0 N). Use any of the necessary values given and derived in Example 14.5 to find θ.

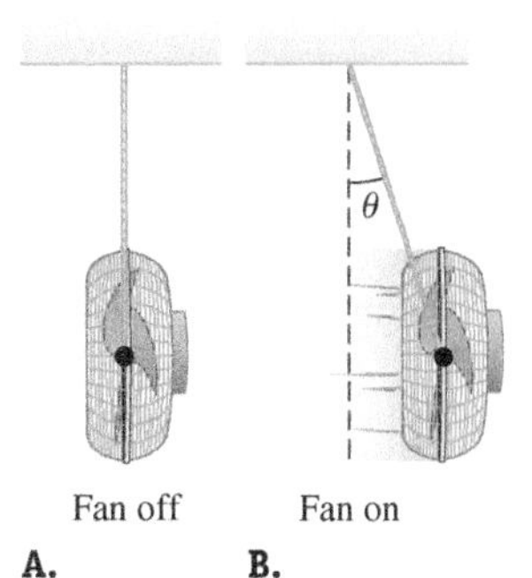

FIGURE P14.67 Problems 67 and 68.

68. N Return to the scenario described in Problem 14.67 and use any of the necessary values given and derived in Example 14.5 (page 402) to find the tension in each cable.

69. A A dresser of mass m is pushed with force F_P at height h above the ground (Fig. P14.69). The dresser has width W and total height H, and the center of mass is exactly in the middle of the dresser. If the

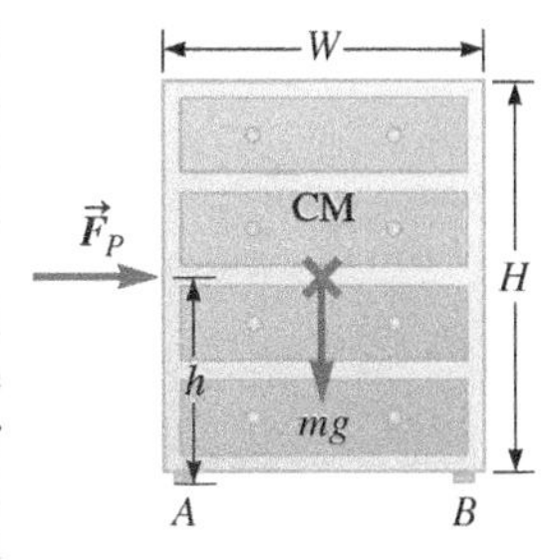

FIGURE P14.69

force is increased, the dresser can be tipped over. At the largest force we can apply before the dresser tips, the dresser is in static equilibrium, and the force between the back legs, labeled A, and the ground is zero. Using this fact, determine an equation for the force required to tip over the dresser in terms of the variables defined above.

70. N Two metal wires of the same material—lengths L and $3L$, respectively, and diameters $3D$ and D, respectively—are under different tensions that stretch each by the same amount ΔL. What is the ratio of the two tensions in the wires?

71. N Brianna is bored while at lunch and begins to play with her food. She has a block of lime gelatin dessert and pushes on the block parallel to the top surface. She notices that the strain experienced by the gelatin was about 0.150 and that the applied stress was about 100.0 Pa. What is the shear modulus of lime gelatin?

Problems 72 and 73 are paired.

72. N A steel rod has a radius of 15.0 cm and a length of 1.25 m. The rod is held firmly in place. What is the maximum **a.** compressive force and **b.** tensile force that can be applied to one face?

73. N An aluminum rod has a radius of 15.0 cm and a length of 1.25 m. The rod is held firmly in place. What is the maximum **a.** compressive force and **b.** tensile force that can be applied to one face? (If you worked Problem 72, compare your answers.)

74. N We know from studying friction forces that static friction increases with increasing normal force between the surfaces, which becomes important for vehicles traveling on icy or snowy roads that have coefficients of static friction much smaller than those of dry pavement. In particular, the greater the normal force on the drive wheels (those coupled to the engine), the better the traction. The horizontal position of the center of mass of a typical compact automobile is located 1.1 m toward the rear as measured from the front wheel axle. The wheelbase (distance from the front wheel axle to the rear wheel axle) is 2.7 m. Assume the car is stationary on level ground and has a weight of 12,000 N. Determine the total normal force on the two front tires and on the two rear tires. Which do you suppose are the drive wheels in this case?

75. N Ruby, with mass 55.0 kg, is trying to reach a box on a high shelf by standing on her tiptoes. In this position, half her weight is supported by the normal force exerted by the floor on the toes of each foot as shown in Figure P14.75A. This situation can be modeled mechanically by representing the force on Ruby's Achilles tendon with $\vec{F}_A$ and the force on her tibia as $\vec{F}_T$ as shown in Figure P14.75B. What is the value of the angle θ and the magnitudes of the forces $\vec{F}_A$ and $\vec{F}_T$?

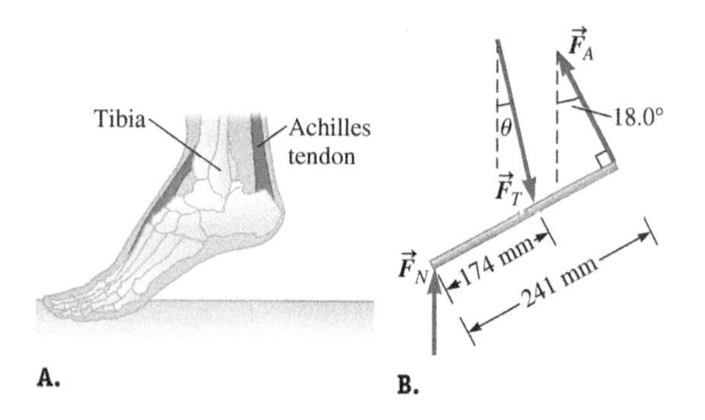

FIGURE P14.75

76. N An object is being weighed using an unequal-arm balance (Fig. P14.76). When the object is in the left pan, a downward force of 3.0 N must be exerted on the right pan to balance the gravitational force on the object. When the object is in the right pan, a downward force of 2.0 N must be exerted on the left pan to balance the gravitational force on the object. Determine the magnitude of the gravitational force on the object.

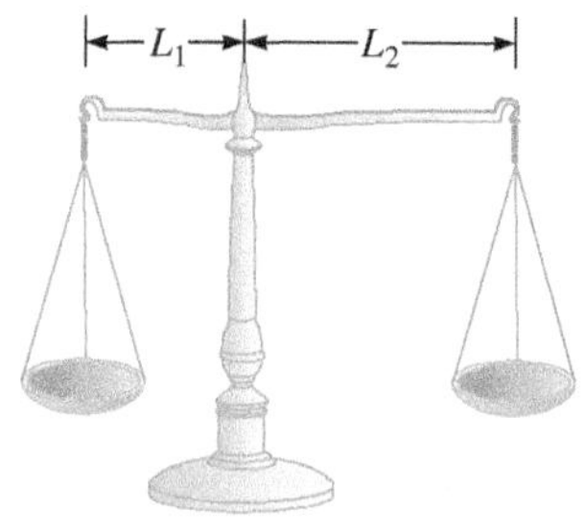

FIGURE P14.76

77. N The normal force exerted on an object by a surface is distributed over the area of contact between the object and the surface. The effective point of application of the normal force must then be located to produce a torque equal to the net torques of the distributed force (much like how we locate the center of mass when determining the torque due to the weight of an object). Consider a filing cabinet of height 1.3 m and width 0.40 m resting on a level, horizontal floor. The cabinet has a weight of 7.0×10^2 N and is loaded so that its center of mass is at its geometric center. Now suppose someone leans against it, pushing horizontally with a force of 50.0 N at the top edge, and the cabinet remains in static equilibrium (Fig. P14.77). Locate the effective point of application of the normal force due to the floor as measured from the left side of the cabinet.

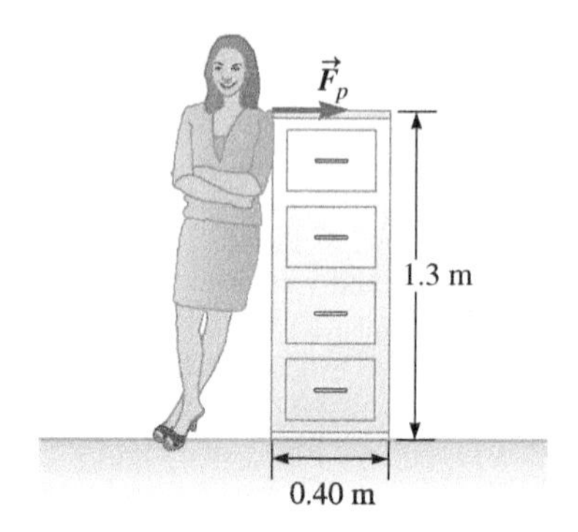

FIGURE P14.77

78. A A massless, horizontal beam and a massless rope support a sign of mass m (Fig. P14.78). **a.** What is the tension in the rope? **b.** In terms of m, g, d, and θ, what are the components of the force exerted by the beam on the wall?

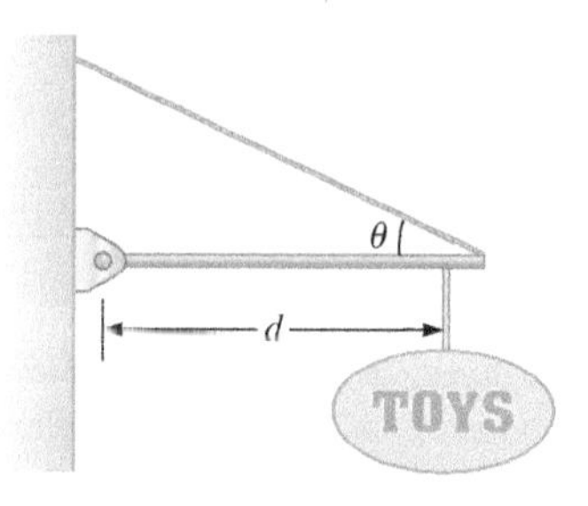

FIGURE P14.78

Problems 79 and 80 are paired.

79. N A rod of length 4.00 m with negligible mass is hinged to a wall. A rope attached to the end of the rod runs up to the wall at an angle of exactly 45°, helping support the rod, while a sign of weight 10.0 N is hanging by two ropes attached to the bottom of the rod. The ropes make an angle of exactly 30° with the rod as shown in Figure P14.79. Another sign with a weight of 10.0 N is attached to the top of the rod with its center of mass at the midpoint of the rod. The entire system is in equilibrium. Find the magnitude of the tension in the rope above the rod that is also attached to the wall.

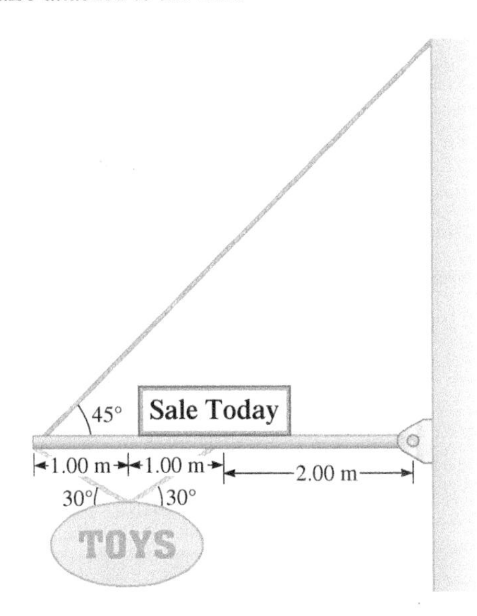

FIGURE P14.79 Problems 79 and 80.

80. N In Problem 79, find the magnitude of the vertical and horizontal components of the force that the hinge must exert on the rod to keep the system in equilibrium.

81. N A horizontal, rigid bar of negligible weight is fixed against a vertical wall at one end and supported by a vertical string at the other end. The bar has a length of 50.0 cm and is used to support a hanging block of weight 400.0 N from a point 30.0 cm from the wall as shown in Figure P14.81. The string is made from a material with a tensile strength of 1.2×10^8 N/m^2. Determine the largest diameter of the string for which it would still break.

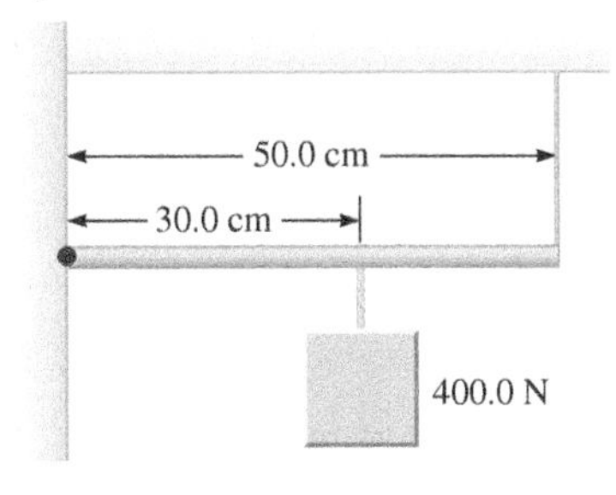

FIGURE P14.81

82. A rope is used to connect a uniform ladder with length L and mass M to a wall (Fig. P14.82). The ladder rests on a rough, horizontal surface with a coefficient of static friction μ_s, and the angle θ is such that the ladder is on the verge of slipping. Assume the ladder is in rotational and translational equilibrium.

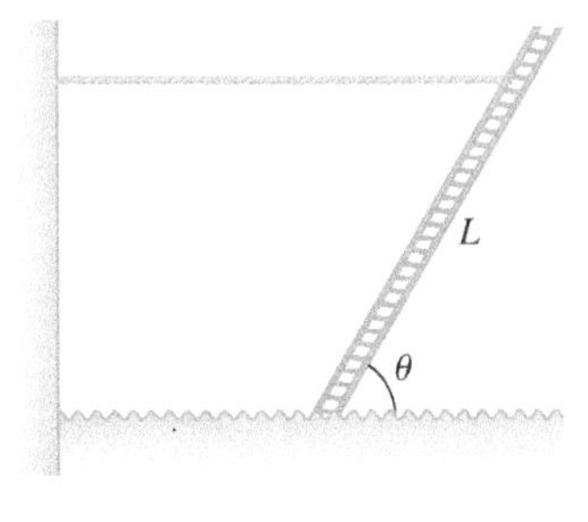

FIGURE P14.82

a. G Draw a free-body diagram of the forces acting on the ladder.

b. A If the ladder is in rotational equilibrium, what is the tension in the rope in terms of M, g, and θ?

c. A Obtain a second expression for the tension in the rope in terms of μ_s, M, and g by considering the ladder in translational equilibrium.

d. A What is the coefficient of static friction μ_s in terms of the angle θ?

e. C What would occur if the ladder were moved slightly so as to reduce the angle θ?

83. N A 185-kg uniform steel beam 8.00 m in length rests against a frictionless vertical wall, making an angle of 55.0° with the horizontal. A 75.0-kg construction worker begins walking up the beam. The coefficient of static friction between the beam and the ground is 0.750. What are the horizontal and vertical forces exerted on the beam by the ground when the worker has walked 3.00 m along the beam?

84. Two smooth planes support a uniform wooden beam of length L and mass M (Fig. P14.84).

a. C Show that for the beam to be in equilibrium, its center of mass must coincide vertically with point P.

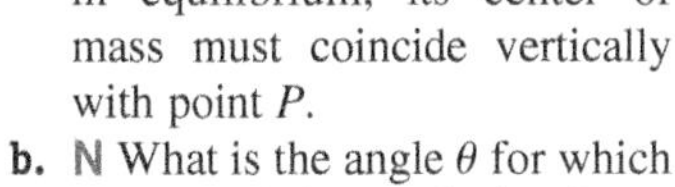

b. N What is the angle θ for which the beam is in equilibrium?

c. C Is the equilibrium found in part (b) stable or unstable? Explain.

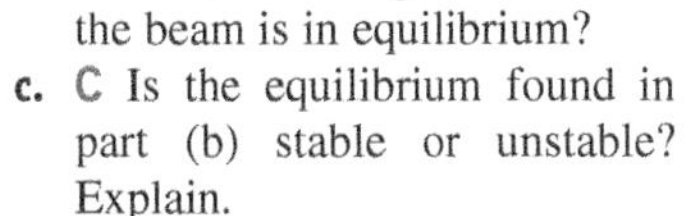

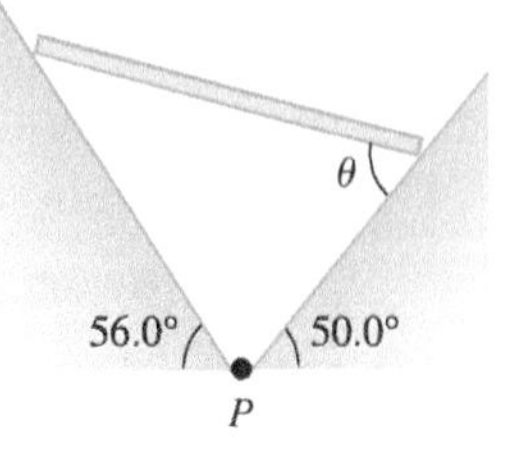

FIGURE P14.84

85. A The length of a metal wire is L_1 when the tension in it is F_{T1}, and it is L_2 when the tension in it is F_{T2}. What is the actual length of the wire under no tension in terms of these lengths and tensions?

15 Fluids

Key Questions

How can Newton's second law and the work–energy theorem be used to predict the behavior of fluids?

When a solid interacts with a fluid, how do we describe the force exerted by the fluid?

15-1 What is a fluid? 418

15-2 Static fluid on the Earth 420

15-3 Pressure 420

15-4 Archimedes's principle 423

15-5 Measuring pressure 428

15-6 Ideal fluid flow 432

15-7 The continuity equation 433

15-8 Bernoulli's equation 436

Underlying Principles

No new *fundamental* physical principles are introduced in this chapter.

Major Concepts

1. Ideal fluid
2. Static fluid
3. Pressure
4. Archimedes's principle
5. Pascal's principle
6. Ideal fluid flow
7. Continuity equation
8. Bernoulli's equation

Special Cases

A fluid is a special case (state) of matter that can flow. Liquids and gases are fluids.

Our lives depend on fluids such as the air we breathe and the water we drink. In fact, the human body consists mostly of water. Engineers design airplanes to fly in air and boats to float in water. Meteorologists observe fluids (the Earth's atmosphere and oceans) to predict the weather. Architects must understand fluids in order to design plumbing and heating systems. Chefs know that fat floats to the top of a sauce and that the ravioli is done when it rises to the top of the water.

In this chapter, we focus our attention on fluids and how they interact with solid objects. We will introduce many new concepts and apply Newton's second law and the work–energy theorem. No new fundamental principles are necessary.

15-1 What Is a Fluid?

FLUID Special Case

The term **fluid** (a substance that can flow) applies to both liquids and gases, but not to solids. Previous chapters dealt with the properties and state of motion of solid objects such as hockey pucks, freight trains, and ancient monuments. How would we describe or predict the properties and motion of liquids and gases? In the following case study, three students discuss the interaction between air, a fluid, and an airplane, a solid object. Their discussion continues throughout the chapter as they attempt to build an understanding or model of fluid behavior and fluid interactions.

CASE STUDY Part 1: Flying Airplanes

Avi: I was just thinking about an air show I saw. They had these planes flying upside down (Fig. 15.1), and I can't figure out why that works. It makes me think that I don't understand how planes fly.

Shannon: Airplanes fly because air exerts a force on their wings.

Cameron: I know you have to be right about the air, but there has to be something about that plane's motion, too. I mean the plane has to keep moving, and it can't just fly any old way and stay up there. Imagine a plane pointing straight down. There will be a little air drag on the plane, but basically it is going to fall out of the sky.

Avi: I think you're kind of getting at the problem. Planes shouldn't be able to fly any old way. The net force of the air has to counteract gravity or else the plane will fall. So, how can you possibly design a plane to fly both upside down *and* right side up?

FIGURE 15.1 How can an airplane fly upside down?

CONCEPT EXERCISE 15.1

Imagine an airplane flying at constant velocity. What must be the net force on the plane? Draw a free-body diagram for the plane.

Fluid Model

To understand why some substances can flow and others cannot, let's see how a microscopic model of a solid differs from microscopic models for liquids and gases. A solid is made up of molecules that are strongly bonded together. We modeled the molecules as particles and the bonds as stiff springs (Fig. 14.21, page 406). Because of the strong bonds, the molecules in a solid have limited motion. They can move slightly as the bonds are stretched, compressed, or sheared, but they cannot slide past one another.

The molecules in a gas are not bonded to one another; the force between them is very weak. Gas molecules move freely throughout their container. The gas fills its container, and the molecules interact with one another only when they collide. We model them as very small, deformable objects similar to tiny, very springy rubber balls that collide elastically with one another and with the walls of their container (Fig. 15.2A). The gas molecules whiz past one another and barely affect the other's motion. Gas is a fluid because the gas molecules are free to flow.

Liquid is the state of matter between solid and gas. The molecules in a liquid are bound together, but the bonds are weak compared with those in a solid. Liquid molecules can easily slide past one another; so, like a gas, a liquid can freely flow. The molecules in a liquid are much closer together than are the molecules in a gas, however. We model the liquid molecules as particles that are tightly packed together and weakly connected (Fig. 15.2B).

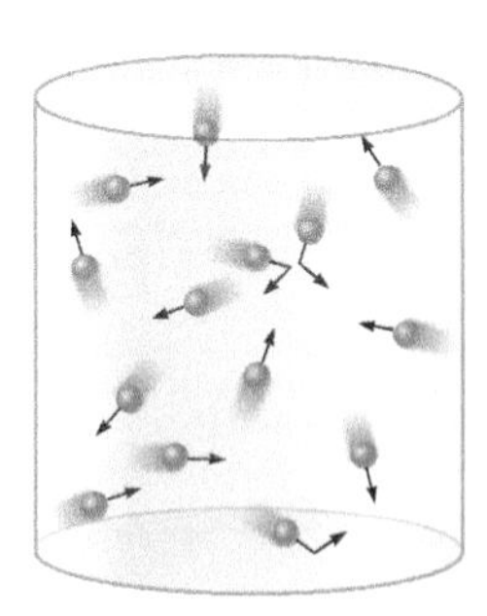

A. Gas

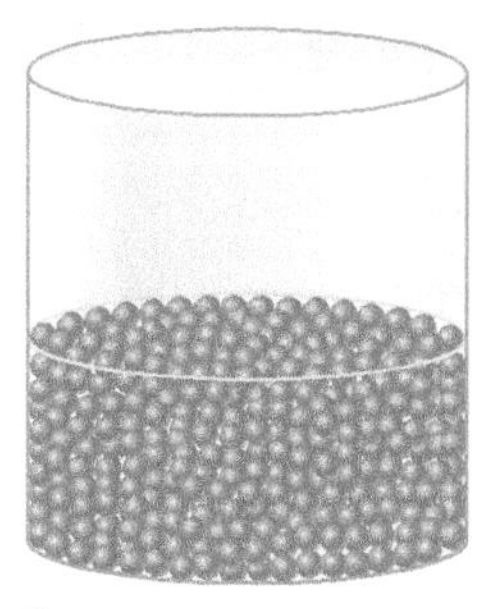

B. Liquid

FIGURE 15.2 A. Gas molecules are far apart. They collide elastically with one another and the walls of their container. **B.** Molecules in a liquid are closely packed, but they are free to slip past one another.

Because the particles in a liquid are much closer together than are the particles in a gas, gases can be compressed easily, whereas liquids cannot. We say that a liquid is nearly *incompressible*. **Incompressible** describes a substance whose volume *cannot* change. On the other hand, gas is *compressible*. The volume of a **compressible** substance is changeable.

In this chapter, we will limit our study to incompressible fluids. Incompressibility is one of the properties of an **ideal fluid**. Therefore, much of what we study is a very good approximation to the behavior of real liquids. Because gases are, in fact, compressible, there are many gas behaviors that we cannot discuss here. Gas compression, however, is discussed in Chapters 19 through 22.

IDEAL FLUID ✪ Major Concept

15-2 Static Fluid on the Earth

STATIC FLUID ✪ Major Concept

In this and the next three sections, we focus on *static fluids*, as in a cup of coffee, the water in a bathtub, or the (still) air in your room. A **static fluid** is able to flow but is not flowing at the time we study it. The individual particles in a static fluid are still in motion, but there is no macroscopic motion of the fluid as a whole. In Section 15-6, we turn our attention to fluids in motion.

If we limit our study to incompressible fluids, how does that limit our study of gases? We will study fluids on or near the surface of the Earth. Therefore, at least one force is exerted on the fluid—the Earth's gravity. Imagine a tall cylinder of air that stretches from sea level to the top of Mount Everest, approximate elevation 8850 m (Fig. 15.3). Gravity exerts a downward force on the air molecules, so you might expect to find all the molecules packed closely together on the Earth's surface like tennis balls in a bucket. Unlike the tennis balls in a bucket, however, air molecules are constantly in motion, colliding elastically with one another and with the walls of their container. All that motion means that the gas particles spread out to fill their container. So, there is a continuous decrease in density as a function of altitude. Recall that density is mass per unit volume (Eq. 1.1). The density of dry air at sea level is $\rho_0 = 1.2929\ \text{kg/m}^3$. The density of air at the summit of Mount Everest is 30% of the density near sea level, or $\rho_{\text{top}} \approx 0.4\,\text{kg/m}^3$. (Climbers must take bottled oxygen when they ascend tall mountains such as Everest.) This variation in density means that we cannot approximate such a large cylinder of air as an incompressible fluid. In this chapter, we are limited to studying much smaller containers of air in which the density is nearly uniform.

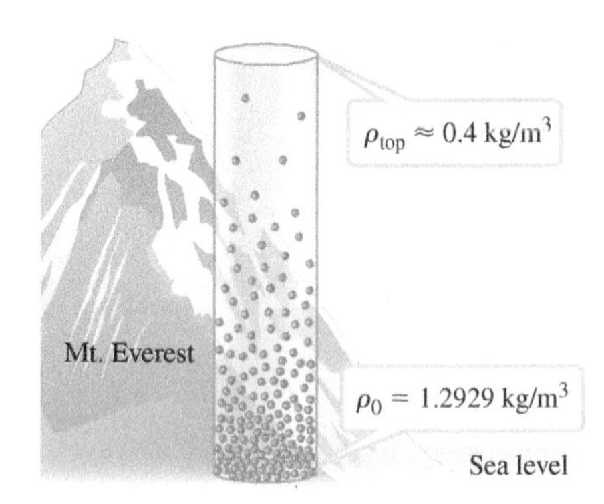

FIGURE 15.3 The density of air molecules is greater near sea level than it is near the top of Mount Everest.

Exactly what size container are we talking about? If we imagine a cylinder as tall as the Willis (Sears) Tower in Chicago (Fig. 15.4), the density at the top is $\rho_{\text{top}} \approx 0.94\rho_0 \approx 1.22\ \text{kg/m}^3$. Depending on what we wish to calculate, such a slight variation in the density may not be ignorable. In this chapter, we usually consider containers that can fit in a laboratory. Even in a container as large as a typical college building, we may safely assume air density is essentially uniform. Unless otherwise specified, we will assume the density of dry air is uniform and given by $\rho = 1.29\ \text{kg/m}^3$.

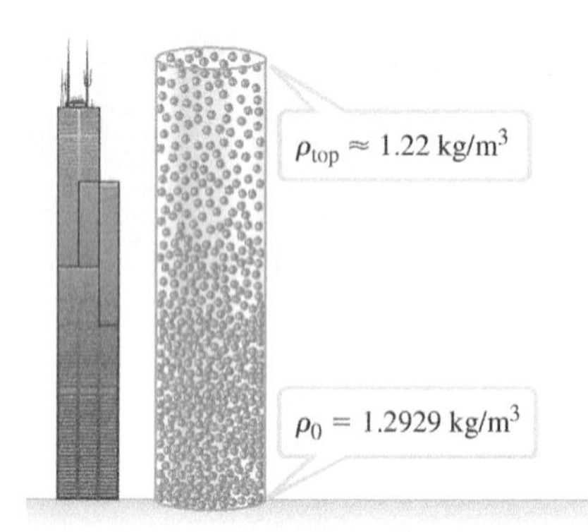

FIGURE 15.4 The air density at the top of the Sears (Willis) Tower is 94% of the density at the bottom of the tower.

Just as the Earth's gravity increases the density of gas molecules near the bottom of a container, we might expect to find an increase in the density of liquid molecules near the bottom of a similar container. Because the molecules in a liquid are much closer together than in air, however, it is difficult to compress them further. The density of seawater near the surface of the Earth is $1.025 \times 10^3\ \text{kg/m}^3$, and even 1000 m below the surface, the density of seawater increases only slightly to $1.028 \times 10^3\ \text{kg/m}^3$. In this chapter, assume liquids are incompressible with a uniform density. Some fluid density values are given in Table 15.1.

TABLE 15.1 Typical density of common fluids.

Fluid	Density ρ (kg/m^3)[a]
Acetone	784.58
Air	1.29
Ethyl alcohol	785.06
Gasoline	737.22
Helium gas	0.18
Hydrogen gas	0.089
Mercury	13,600
Sunflower oil	920
Seawater	1025.18
Water (at 4°C)	1000.00
Water	998.2

[a]Densities are at room temperature (15°–25°C) except where indicated.

15-3 Pressure

Although we are usually unaware of the fluid—air—that surrounds us, a few common experiences may remind us that it is present. For example, if you quickly change elevation by flying in an airplane, driving down a mountain, or even taking an elevator in a very tall skyscraper, you may experience discomfort in your ears. Usually, when it is quiet, the eardrum is bombarded on both sides by air molecules; so, there is no net force on your eardrum, and you don't notice the air pounding on it. If your eardrum gets bombarded more often on one side, you notice a net force on the eardrum. A tube on the inside of the eardrum connects to the throat, allowing air to get to the inside of the eardrum (Fig. 15.5). Rapid elevation changes cause discomfort because the air takes a moment to equalize on both sides. The discomfort may be worse if you have a head cold that closes the tube connecting the inside of the eardrum to the throat. Many people yawn or swallow in such situations in an effort to make their ears "pop" and equalize the air.

Definition and Units of Pressure

Because eardrums and other surfaces are bombarded by a huge number of air molecules, it is not practical to measure the force exerted by each microscopic particle. Instead, we measure a macroscopic quantity called *pressure* that results from the force of the large number of microscopic particles. Consider the simple system shown in Figure 15.6, a piston in a tube with no friction between the piston and the tube's walls. On the left side of the piston, there is a gas, but there is nothing (a vacuum) on the other side. The gas molecules collide with the piston, and because there is no friction or other forces acting on the piston, the piston accelerates to the right. Now imagine holding the piston at rest by applying a force, $\vec{F}$. The force you would need to apply must be equal in magnitude to the force exerted by gas molecules on the other side of the piston. Now imagine the same experiment, but with a much larger piston. More gas molecules would collide with the larger piston, so we would need to apply a larger force to keep the piston at rest. In fact, the force we must apply is proportional to the area of the piston: $F \propto A$. The force exerted per unit area is a constant, however. The **pressure** (P) exerted by the gas on the piston equals the magnitude of the force we must exert per unit area:

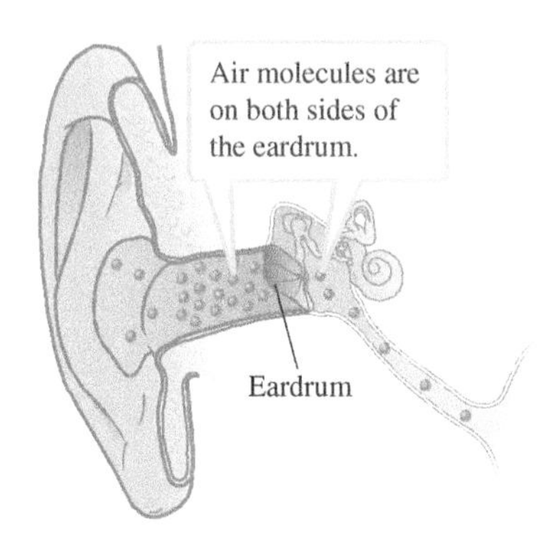

FIGURE 15.5 Usually, the air pressure on the outside of the eardrum equals the air pressure on the inside because a tube connects inside of the eardrum to the throat.

$$P = \frac{F}{A} \tag{15.1}$$

PRESSURE ✪ Major Concept

Pressure is a scalar quantity with the dimensions of force per area. The SI unit of pressure is the pascal (Pa):

$$1 \text{ Pa} = 1 \text{ N/m}^2$$

We use an uppercase P for pressure and a lower case $\vec{p}$ for momentum (Eq. 10.1).

There are other common units for pressure. For example, one *atmosphere* (atm) is the approximate pressure exerted by the Earth's atmosphere at sea level:

$$1 \text{ atm} = 1.01325 \times 10^5 \text{ Pa}$$

Another commonly used pressure unit is the *millimeter of mercury* (mm Hg), also called the *torricelli* (torr) in honor of the inventor of the mercury barometer, Evangelista Torricelli (Section 15-5):

$$1 \text{ Pa} = 7.501 \times 10^{-3} \text{ mm Hg} = 7.501 \text{ mtorr}$$

The U.S. customary pressure unit that we see written on our car and bike tires is *pounds per square inch* (psi or lb/in.^{2}):

$$1 \text{ Pa} = 1.45 \times 10^{-4} \text{ psi}$$

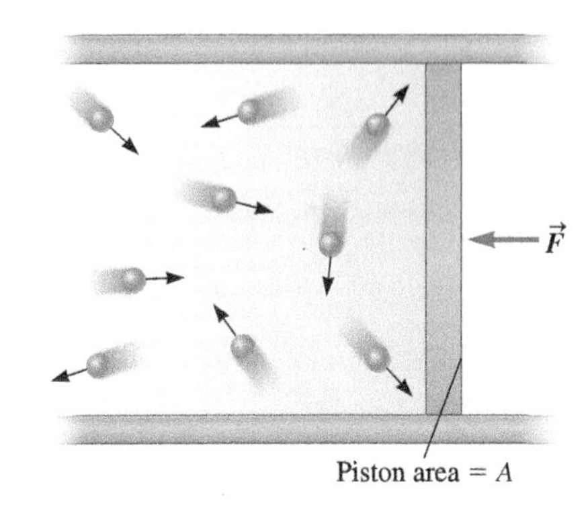

FIGURE 15.6 A piston of cross-sectional area A is in a frictionless tube. Air pressure on one side of the piston is balanced by a force on the other side of the piston. The magnitude of the force exerted must equal $F = PA$.

Pressure Variation with Depth in a Static Fluid

The Earth's gravity affects the pressure in a static fluid such as the water in a swimming pool. Let's look at the forces that act on a portion of the water. As shown in Figure 15.7A, we imagine a cylinder-shaped column of water with a cross-sectional area A. We have chosen a downward-pointing y axis so that the column of water extends from the surface to a depth y. Three vertical forces are exerted on this column of water, shown in Figure 15.7B: (1) the Earth's gravity exerts a force $\vec{F}_g$ in the positive y direction, (2) the air above the column of water also exerts a force $\vec{F}_{air}$ in

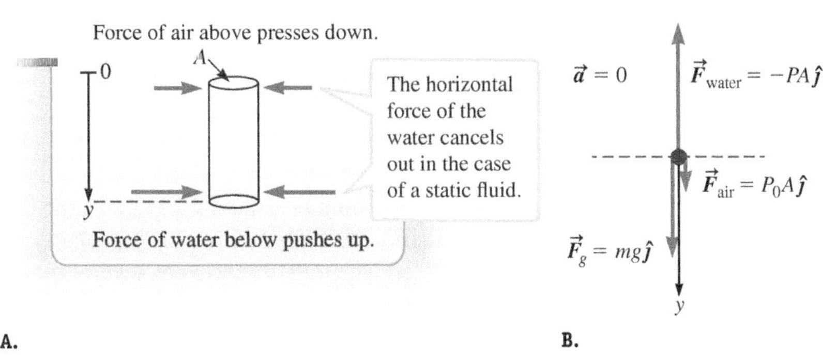

FIGURE 15.7 A. We imagine selecting a cylinder-shaped portion of pool water. **B.** A free-body diagram for the cylinder-shaped portion of water, showing forces acting in the vertical direction.

the positive y direction, and (3) the water below the column exerts an upward force $\vec{F}_{water}$ in the negative y direction. Because the fluid is static, we may apply Newton's second law (in component form) with the acceleration set equal to zero (Eq. 14.3):

$$\sum F_y = F_g + F_{air} - F_{water} = 0 \tag{15.2}$$

If the air pressure is P_0 and the cross-sectional area of the column is A, the force exerted by the air on the column of water is (Eq. 15.1):

$$F_{air} = P_0 A \tag{15.3}$$

Similarly, the force exerted by the water below the column is

$$F_{water} = PA \tag{15.4}$$

where P is the water pressure just below the column at depth y. The weight of the column of water is given by the usual expression $F_g = mg$, where m is the mass of the water column. Because we are modeling the water as an incompressible fluid, we assume the density ρ of water is uniform throughout the column. Therefore, we can write the mass m of the column in terms of the water density ρ and the volume of the column, $V = Ay$. The weight of the column can be expressed as

$$F_g = mg = (\rho V)g = (\rho Ay)g \tag{15.5}$$

Substitute Equations 15.3 through 15.5 into Equation 15.2 and rearrange the terms:

$$\rho Ayg + P_0 A - PA = 0$$

Pressure P of an incompressible, static fluid on the Earth as a function of depth y below the surface

$$P = \rho gy + P_0 \tag{15.6}$$

Equation 15.6 says that the pressure of a fluid increases with depth due to the weight of the column of fluid above.

Pressure and weight are both scalars. When you are swimming under water, the water pressure along the surface of your body is roughly uniform, and the force exerted by the water on you is perpendicular to your surface at every point. If you swim to the bottom of a deep pool, you experience some discomfort in your ears because you are holding your breath with air at the atmosphere's pressure on the inside of your eardrum. The water outside your eardrum has greater pressure near the bottom of the pool than it does near the surface due to the weight of all the water above you. The discomfort does not change if you turn your head because pressure is a scalar, independent of direction. All that matters is your depth. The deeper you swim, the more water is pressing down on you, and the greater your ear discomfort.

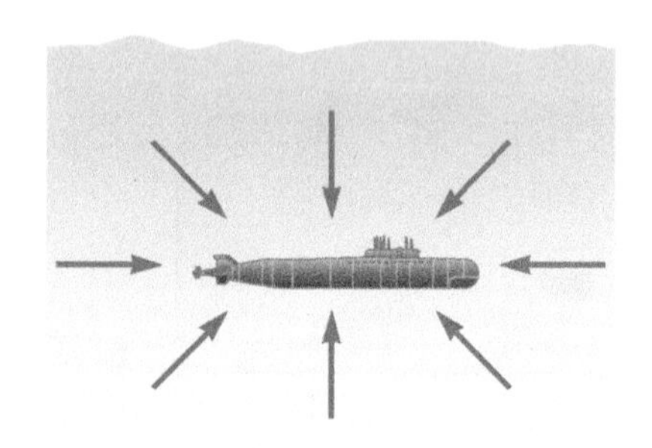

FIGURE 15.8 When an object is submerged in a fluid, the fluid presses the object equally in all directions. The result is the object is compressed, but its shape does not change.

Change in an Object's Volume with Pressure

In this chapter, we model fluids that are *un*compressed, but an object (whether a solid, liquid, or gas) immersed in the fluid may be compressed. For example, if you immerse a balloon or even a submarine in seawater, the fluid presses uniformly all over the object's surface. The result is that the object is compressed, but its shape does not change (Fig. 15.8). The total pressure exerted by the fluid on the object is called the **volume stress**. The **volume strain** is a measure of the object's compression, expressed as the change in the object's volume divided by its original volume, $\Delta V/V_i$. As discussed in Section 14-4, a stress on an object may produce a linear response so that as the stress increases the strain increases proportionally. Often, the volume strain is proportional to the volume stress:

$$P = B\frac{\Delta V}{V_i} \tag{15.7}$$

The constant of proportionality B is called the **bulk modulus**; it has the same dimensions as pressure, so its SI units are pascals. The bulk modulus is a property of the material making up the object. Table 15.2 lists the bulk modulus of several materials. Materials with a low bulk modulus (such as mercury) are more easily compressed than materials with a high bulk modulus (such as steel).

TABLE 15.2 Bulk modulus for several materials.

Material	Bulk modulus B ($\times 10^9$ Pa)
Aluminum	76
Brass	61
Mercury	27
Silicon	98
Steel	140
Tungsten	200
Water	2.2

CONCEPT EXERCISE 15.2

We often hear the word *pressure* in everyday conversations. Give examples from everyday use of the word *pressure* that are consistent with the physics definition and give examples that are inconsistent. Explain.

CONCEPT EXERCISE 15.3

Estimate the water pressure near the bottom of a swimming pool located near sea level. What is the difference in pressure between the surface and the bottom of the pool?

CONCEPT EXERCISE 15.4

Imagine a swimming pool empty of water, but full of air. Estimate the air pressure near the bottom of this swimming pool. Again, assume the pool is located near sea level. Is it reasonable to assume the air pressure is constant over a container of such volume?

15-4 Archimedes's Principle

When you think about the weight of the Earth's atmosphere pressing down on you and everything else, it may seem amazing that you are not crushed flat like a pancake against the surface of the Earth. Three physics students discuss this apparent paradox.

CASE STUDY Part 2: Airplanes and Helium Balloons

Avi: I know that somehow the air must exert an upward force on an airplane, but I can't figure out how. In fact, it seems to me that air should exert a downward force. If you have a plane on the runway, the plane has the weight of the whole atmosphere pressing it down. I can't figure out how it gets up at all.

Shannon: I think I have a way to explain it. Imagine that you have a helium balloon. You know it wants to float.

Cameron: But that is because it is lighter than air. Planes aren't lighter than air. Planes fly because they are moving.

Shannon: Yes. I agree that planes are more complicated. But I think that if we just talk about a helium balloon, it will help us understand more about atmospheric pressure. The balloon doesn't just have the weight of the atmosphere pressing it down. It also has air below pushing it up. The air pressure below the balloon must be greater than the pressure above the balloon, so the net force of the air is upward.

Shannon is correct. The weight of the atmosphere doesn't just press *down* on you. The air is all around you, and the atmospheric pressure pushes you in all directions, even upward. Normally, the air does not exert a net force on you. However, if you evacuate the air (or just reduce the amount of air) on one side of an object, the Earth's atmosphere exerts a net force on the object. For example, you may get a suction cup to hang from the underside of a horizontal surface by burping the suction cup and reducing the amount of air inside. We often use vacuum cleaners to reduce the air on one side of tiny objects, say spilled coffee grounds. Then the air on the other side forces the dirt into the vacuum cleaner's bag (Fig. 15.9).

FIGURE 15.9 The pressure inside the vacuum cleaner is lower than the pressure on the other side of the coffee grounds. The higher pressure forces the grounds into the vacuum cleaner.

Our experience with devices such as vacuum cleaners sometimes makes it more difficult to think about how fluid pressure affects objects in the fluid. For example, when we want to buy a new vacuum cleaner, the salesperson explains that brand X has more *suction* than brand Y. The word *suction* may give you the idea that the vacuum *pulls* the dirt, but air molecules don't have hooks. Instead, the air molecules push the dirt as they move from a region of high pressure to a region of lower pressure.

If the fluid pressure exerted on an object varies across its surface, the fluid exerts a net force on the object. The net force points from high pressure to low pressure. For a static fluid near the Earth's surface, $P = \rho g y + P_0$ (Eq. 15.6), so the pressure below the object is greater than the pressure above it. Therefore, the fluid exerts a net upward force on the object. Imagine carrying your friend across a room; now imagine carrying the same person across a swimming pool in which both of you are mostly submerged. It is much easier to carry your friend in the swimming pool because of the net upward force of the water. Why doesn't the net upward force of the air help when you try to carry your friend across the room? Remember that the air pressure in a container the size of a small building is nearly constant, so the net upward force of the air is very small and doesn't make a noticeable difference.

The Buoyant Force

The net upward force exerted by a fluid on an object is called the **buoyant force** $\vec{F}_B$. The magnitude of the buoyant force was discovered by Archimedes of Syracuse (c. 287–212 BCE), a Greek mathematician and scientist who reportedly was taking a bath when he was inspired.[1] According to **Archimedes's principle**, *the magnitude of the buoyant force is equal to the weight of the fluid displaced by the object.*

Archimedes's principle can be deduced by looking at the forces exerted within a fluid. Figure 15.10A shows an imaginary box enclosing part of the fluid in the container. Two forces are exerted on the box of fluid: the downward force $\vec{F}_g$ due to the Earth's gravity and the upward buoyant force $\vec{F}_B$ from the remaining fluid. The box of fluid is in static equilibrium, so there is no net force on it:

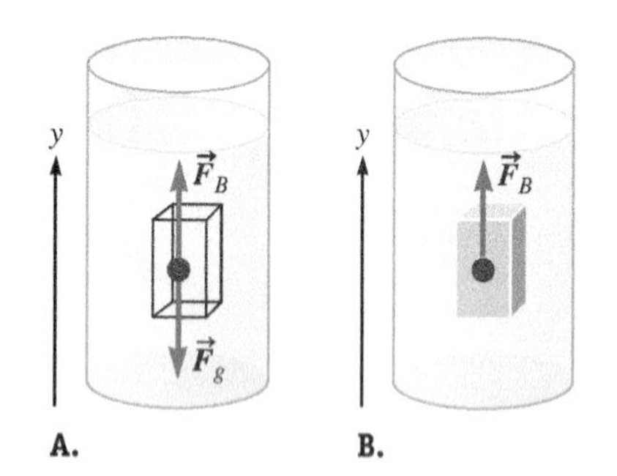

FIGURE 15.10 A. The magnitude of the buoyant force exerted on an imaginary "box" of fluid equals the weight of fluid in the box. **B.** According to Archimedes's principle, the magnitude of the buoyant force exerted by a fluid is equal to the weight of the fluid displaced by the object.

$$\sum F_y = F_B - F_g = 0$$

Therefore, the magnitude of the buoyant force equals the weight of the fluid in the box: $F_B = F_g$. What would happen to the buoyant force if we had an actual object that exactly filled the space of the imaginary box (Fig. 15.10B)? The answer is that nothing would happen; the buoyant force would not change because the buoyant force is the result of the net force exerted by the *surrounding* fluid. Replacing the box of fluid with an object of the same volume does not affect the surrounding fluid. Therefore, it doesn't matter whether we are considering a real box-shaped object or just a boxed-off portion of the fluid: The buoyant force is the same in either case. *The buoyant force equals the weight of the fluid that would fit in the volume of the submerged object.*

The argument illustrated in Figure 15.10 may leave you feeling like you've just watched a magic act as an object is swapped in to replace a boxed-off portion of fluid. In order to fit the box-shaped object into the fluid without changing the level of the fluid, a volume of fluid V_{disp} (for "*displaced*") equal to the object's volume V_{obj} must be removed. The argument helps us arrive at Archimedes's principle, but it is outside our normal experience because usually when we place an object in a fluid, we don't first remove a volume of fluid equal to the object's volume. (We don't fill the bathtub to the top before we get in, however, because we know the water will rise when we enter!)

In Figure 15.11, we start with a container of fluid and then place an object in the fluid so that the object is totally submerged. The volume of the object V_{obj} equals the

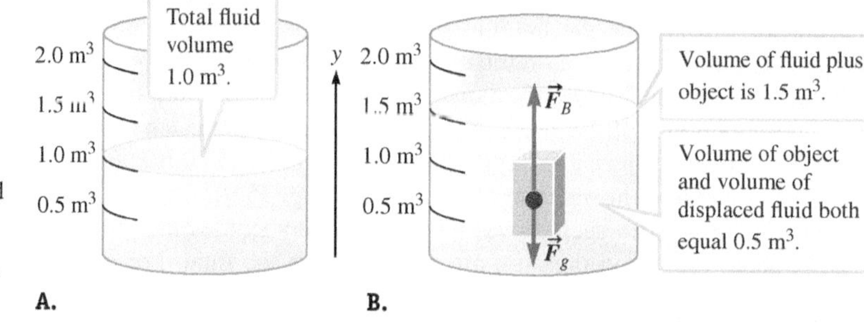

FIGURE 15.11 The volume of a submerged object equals the volume of displaced fluid. If the fluid in the figure is water, the buoyant force exerted on the object is $\vec{F}_B = (1000 \text{ kg/m}^3)(0.5 \text{ m}^3)(9.81 \text{ m/s}^2)\hat{\jmath} = 4.9 \times 10^3 \text{ N}$.

[1] Archimedes's story was reported by ancient Roman writer Vitruvius in Book 9 of *De architectura*.

volume of the displaced fluid, $V_{obj} = V_{disp}$. The magnitude of the buoyant force equals the weight of this volume of displaced fluid:

$$F_B = F_g = m_f g$$

where m_f denotes the mass of *displaced fluid*. It is convenient to write the mass m_f of the displaced fluid in terms of its volume V_{disp} and density ρ_f. The buoyant force is then

$$F_B = \rho_f V_{disp} g \tag{15.8}$$

ARCHIMEDES'S PRINCIPLE

Major Concept

EXAMPLE 15.1 A Totally Submerged Object

A totally submerged object released from rest may rise up, sink farther down, or just rest in place, depending on the relative densities of the object and the fluid. Show that after an object submerged in a fluid is released from rest, its acceleration is given by

$$a_y = \left(\frac{\rho_f - \rho_{obj}}{\rho_{obj}}\right)g \tag{15.9}$$

In the **CHECK and THINK** step, discuss the three cases of remaining at rest, sinking, and rising.

INTERPRET and ANTICIPATE

This example is much like those in Chapters 5 and 6. Use a free-body diagram (Fig. 15.11B) and apply Newton's second law to find the acceleration.

SOLVE

As shown in Figure 15.11B, two forces act on the object: gravity and the buoyant force. Apply Newton's second law in component form along the y direction.	$\sum F_y = F_B - F_g = m_{obj} a_y$ (1)
The object's mass m_{obj} can be expressed in terms of the object's volume and density.	$m_{obj} = \rho_{obj} V_{obj}$ (2)
Write the buoyant force and the gravitational force in terms of the density of the fluid and the density of the object.	$F_B = \rho_f V_{disp} g$ (3) $F_g = \rho_{obj} V_{obj} g$ (4)
Substitute Equations (2) through (4) into Equation (1).	$\rho_f V_{disp} g - \rho_{obj} V_{obj} g = \rho_{obj} V_{obj} a_y$ (5)
Because the object is totally submerged, the volume of the displaced fluid equals the object's volume, $V_{disp} = V_{obj}$. We rewrite Equation (5) in terms of V_{obj}.	$\rho_f V_{obj} g - \rho_{obj} V_{obj} g = \rho_{obj} V_{obj} a_y$
Cancel V_{obj} from both sides of the equation.	$\rho_f g - \rho_{obj} g = \rho_{obj} a_y$
Isolate the acceleration on the left side of the equation.	$a_y = \left(\frac{\rho_f - \rho_{obj}}{\rho_{obj}}\right)g$ ✓ (15.9)

CHECK and THINK

If the object's density equals the fluid's density, the buoyant force equals the object's weight. (Compare Eqs. 3 and 4.) In such a case, $\rho_{obj} = \rho_f$, and the object's acceleration is zero. Because the object was initially at rest, it remains at rest.

If the object is denser than the fluid ($\rho_{obj} > \rho_f$), the acceleration is negative (relative to the upward-pointing y axis in Fig. 15.11). In this case, the gravitational force exceeds the buoyant force.

Finally, if the object is less dense than the fluid ($\rho_{obj} < \rho_f$), the acceleration is positive. In this case, the buoyant force exceeds the gravitational force. According to Equation 15.9, if the fluid vanishes (that is, if $\rho_f \rightarrow 0$), the object is in free fall with $a_y = g$ as expected.

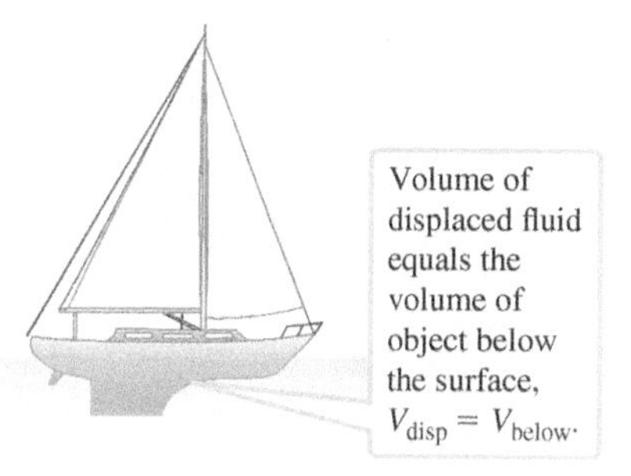

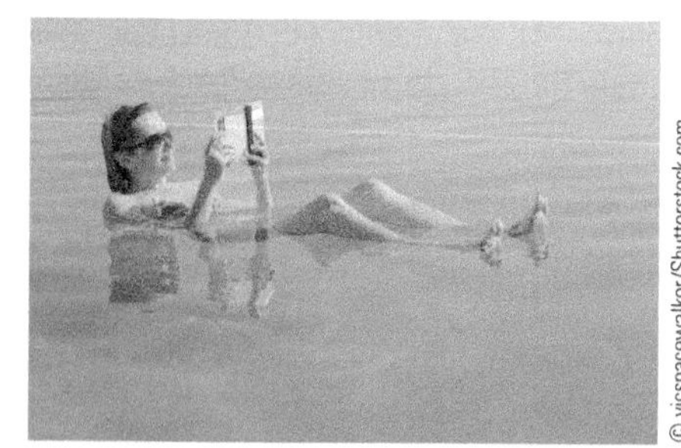

FIGURE 15.12 A. For a floating, partially submerged object, the volume of the displaced fluid equals the volume of the object that is below the surface. **B.** Bathers float in salty water because their weight is balanced by the buoyant force. This woman floats easily in the Dead Sea.

Equation 15.9 only works for a totally submerged object such as a fish in the ocean because only then is the volume of the displaced fluid equal to the volume of the object. Many objects—such as a boat on the water or a person on the Dead Sea (Fig. 15.12)—float partially submerged, however. In these cases, the volume of the displaced fluid does not equal the object's total volume: $V_{\text{disp}} \neq V_{\text{obj}}$. Instead, the volume of the displaced fluid equals the volume of the portion of the object *below* the surface of the fluid: $V_{\text{disp}} = V_{\text{below}}$.

EXAMPLE 15.2 An Object Floating Near the Surface

Consider the sunbather in Figure 15.12B floating with much of her body above the surface of the water. Show that the fraction of the object below the surface of the fluid is determined by the relative densities of the object and the fluid such that

$$\frac{V_{\text{below}}}{V_{\text{obj}}} = \frac{\rho_{\text{obj}}}{\rho_f} \tag{15.10}$$

INTERPRET and ANTICIPATE

This example is much like Example 15.1 for the submerged object. The free-body diagram is similar, with gravity pointing down and the buoyant force pointing up (Fig 15.11B).

SOLVE

Apply Newton's second law in component form along the y direction. Because the sunbather is in equilibrium, her acceleration is zero, and the two forces must be equal in magnitude.	$\sum F_y = F_B - F_g = m_{\text{obj}}a_y$ $F_B = F_g$	(1)
The sunbather displaces a volume of water V_{below}. The weight of the displaced water equals the magnitude of the buoyant force on her.	$F_B = \rho_f V_{\text{below}} g$	(2)
Write her weight in terms of her density ρ_{obj} and her volume V_{obj}.	$F_g = \rho_{\text{obj}} V_{\text{obj}} g$	(3)
Substitute Equations (2) and (3) into Equation (1). The free-fall acceleration g cancels.	$\rho_f V_{\text{below}} g = \rho_{\text{obj}} V_{\text{obj}} g$ $\rho_f V_{\text{below}} = \rho_{\text{obj}} V_{\text{obj}}$	
Solve for the fraction of her body below the surface of the water.	$\dfrac{V_{\text{below}}}{V_{\text{obj}}} = \dfrac{\rho_{\text{obj}}}{\rho_f}$ ✓	(15.10)

CHECK and THINK

It makes sense that each side of the equation is dimensionless. Also, Equation 15.10 shows that the fraction of the object below the surface of the fluid is determined by the relative densities of the object and the fluid. The more dense the object (large ρ_{obj}), the more of its volume is below the surface.

CONCEPT EXERCISE 15.5

Estimate the fraction of the sunbather's body (Fig 15.12B) below the surface of the water in two ways.

a. The human body is made mostly of water; therefore, its density is close to that of water. Very lean people are somewhat denser than water, whereas overweight people are somewhat less dense. So, let's assume the density of the sunbather is about $\rho_{obj} \approx 1000 \text{ kg/m}^3$. The water in the Dead Sea is much denser than seawater: $\rho_f \approx 1250 \text{ kg/m}^3$. Use $V_{below}/V_{obj} = \rho_{obj}/\rho_f$ (Eq. 15.10) to estimate the fraction (and percent) of the sunbather's body that is below the water.

b. Next use Figure 15.12 to check your estimate.

CONCEPT EXERCISE 15.6

CASE STUDY "Lighter than Air"

Cameron says that a helium balloon rises because it is lighter than air. In what way is this statement correct? How can you make Cameron's statement more precise?

EXAMPLE 15.3 Wearable Lead

Apparent weight and *weight* are different. In Section 5-7, apparent weight was defined as the reading on a scale, and weight was defined as the magnitude of the gravitational force. (See Eq. 5.7.)

If an object is submerged in a fluid such as water, its apparent weight may be less than its weight. Sometimes a scuba (for *s*elf-*c*ontained *u*nderwater *b*reathing *a*pparatus) diver is less dense than seawater. To descend, a diver must wear 20.0 lb of lead on his weight belt (Fig. 15.13). What is the apparent weight of this much lead when it is submerged in seawater? The density of lead is 11,350 kg/m³.

FIGURE 15.13 Lead weights on a diver's belt.

INTERPRET and ANTICIPATE

Imagine using a spring scale to weigh the lead as a single piece when it is submerged (Fig. 15.14). The three forces on the piece of lead are the spring force $\vec{F}_H$, gravity $\vec{F}_g$, and the buoyant force $\vec{F}_B$. By Newton's third law, the magnitude of the force exerted by the spring on the lead is equal to the magnitude of the force exerted by the lead on the spring. Therefore, the reading on the scale is the magnitude of $\vec{F}_H$. We expect the magnitude of the spring force to be less than the weight: $F_H < F_g$.

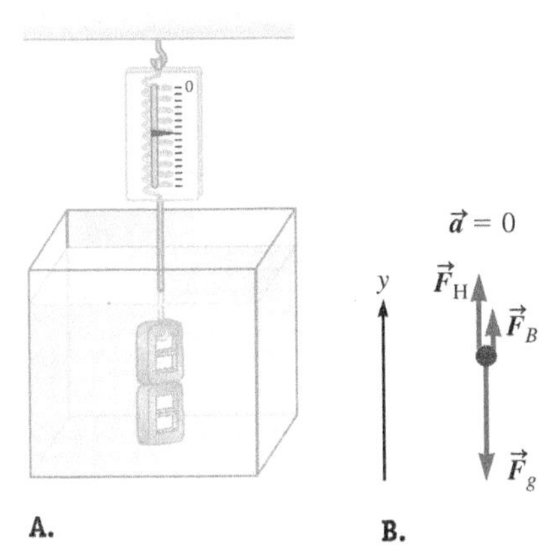

FIGURE 15.14 Finding the apparent weight of lead used by a diver.

SOLVE

Use the free-body diagram to apply Newton's second law in the y direction with the acceleration set equal to zero.	$\sum F_y = F_H + F_B - F_g = 0$ $F_H = F_g - F_B$
Substitute Equation 15.8 with $V_{disp} = V_{lead}$ for F_B.	$F_H = m_{lead}g - \rho_f V_{lead} g$
Express the volume of the lead V_{lead} in terms of its mass m_{lead} and density: $\rho_{lead} = m_{lead}/V_{lead}$.	$F_H = m_{lead}g - \rho_f g\left(\frac{m_{lead}}{\rho_{lead}}\right) = m_{lead}g\left(1 - \frac{\rho_f}{\rho_{lead}}\right)$
Substitute values, starting with the given 20.0 lb of lead. The density of seawater is listed in Table 15.1. (We could have, of course, converted the 20.0 lb directly into newtons.)	$m_{lead} = 20.0 \text{ lb}(1 \text{ kg}/2.20 \text{ lb}) = 9.09 \text{ kg}$ $F_H = (9.09 \text{ kg})(9.81 \text{ m/s}^2)\left(1 - \frac{1025.18 \text{ kg/m}^3}{11{,}350 \text{ kg/m}^3}\right)$ $F_H = 81.1 \text{ N}$

Example continues on page 428 ▶

CHECK and THINK

As expected, the apparent weight (about 18 lb) is less than the weight of the lead (20 lb). The apparent weight depends not just on the object's density and mass, but also on the density of the surrounding fluid. If the scuba diver were in freshwater ($\rho = 1000$ kg/m^3) instead of seawater ($\rho = 1025$ kg/m^3), the lead's apparent weight would be greater.

15-5 Measuring Pressure

Measuring fluid pressure is common in many engineering, scientific, and everyday applications. If you wish to go scuba diving, for instance, there are many safety checks you should perform before jumping into the ocean. One of the most important things to check before you dive is the pressure in your scuba tank. There are several types of devices for measuring pressure.

In this section, we explore two standard pressure-measuring devices: the manometer and the barometer. A manometer is used to measure the pressure of some fluid relative to atmospheric pressure. A barometer is used to measure atmospheric pressure P_0. (To measure the pressure in a scuba tank, you would use a practical device such as a Bourdon-tube pressure gauge.) Studying the operation of a manometer or barometer will give us a deeper understanding of fluid pressure (Section 15-3) and, in particular, of the variation of pressure with depth in a fluid (Eq. 15.6).

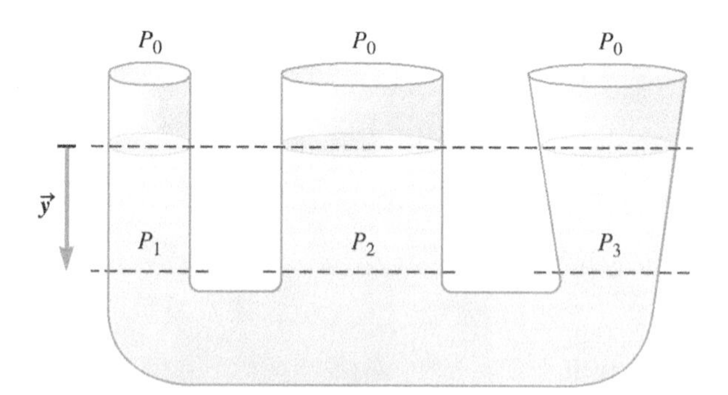

FIGURE 15.15 It may seem surprising, but the fluid comes to the same height in all columns of this complex tube, regardless of column shape or cross-sectional area. Also, the fluid pressure is the same at all points on a horizontal line through the static, incompressible fluid, so $P_1 = P_2 = P_3$.

To find pressure, manometers and barometers measure the height of a fluid. If you had a fluid in a symmetrical U-shaped tube open at both ends, your intuition would probably tell you that the fluid comes to the same height in both sections of the tube. But what about the more complex tube shown in Figure 15.15? Think about the pressure at the same depth y below the surface in each column for any such tube. If the fluid were not at the same height in all columns, the pressures would not be equal at equal depths. The fluid would be forced to move from a high-pressure region to a low-pressure region until it came to the same height in each column. We can make two general statements. First, a static, incompressible fluid comes to the same height in all open regions of a container. Second, the fluid pressure is the same at all points on a horizontal plane through the static, incompressible fluid.

Pascal's Principle

PASCAL'S PRINCIPLE

Major Concept

Next we need to know about pressure changes in a fluid. According to **Pascal's principle,** *a change in the pressure at one point in an incompressible fluid appears undiminished at all points in the fluid and on the walls of its container.* In other words, any *change* in fluid pressure has exactly the same value everywhere in the fluid. As an example, if we start with a fluid in an open container (Fig. 15.16A), the pressure at the surface of the fluid is atmospheric pressure P_0. The pressure at all points below the surface is given by $P_i = \rho g y + P_0$ (Eq. 15.6). Now suppose that we change the surface pressure by attaching this container to another container full of gas at pressure P_{gas} (Fig. 15.16B). Now the pressure at the surface of the fluid is P_{gas}, and the pressure at any point below the surface is $P_f = \rho g y + P_{gas}$. We find the change in pressure ΔP:

$$\Delta P = P_f - P_i$$

$$\Delta P = (\rho g y + P_{gas}) - (\rho g y + P_0)$$

$$\Delta P = P_{gas} - P_0 \tag{15.11}$$

Equation 15.11 shows that every point in the fluid experiences the same change in pressure independent of depth y.

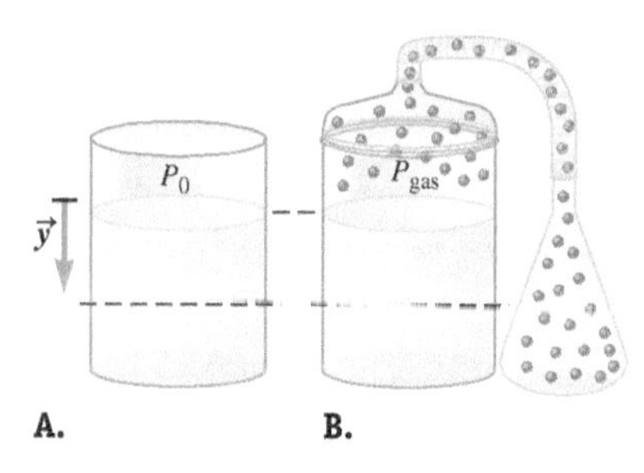

FIGURE 15.16 A. Originally, the pressure at the surface of the fluid is atmospheric pressure P_0. **B.** The pressure at the surface of the fluid is now P_{gas}. According to Pascal's principle, a change in pressure is transmitted undiminished to the entire fluid and the walls of the container.

EXAMPLE 15.4 Hydraulic Lever

One of the most useful applications of Pascal's principle is the hydraulic automobile lift. The basic design is shown in Figure 15.17. A U-shaped tube is filled with a fluid. The ends of the tube are capped off by pistons. A vehicle is placed on the larger piston of area A_L, and a force is applied to the smaller piston of area A_S to raise the vehicle for service.

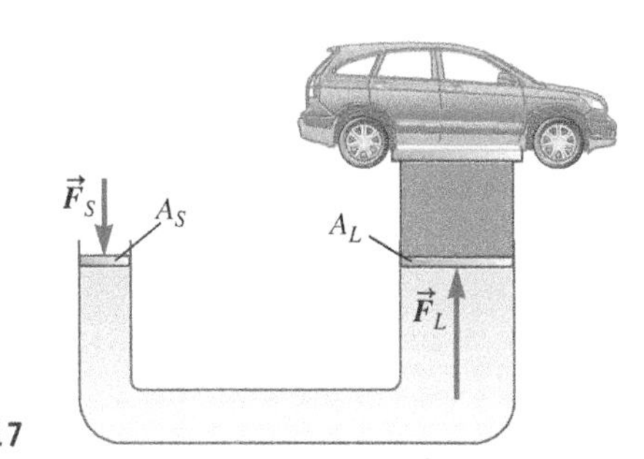

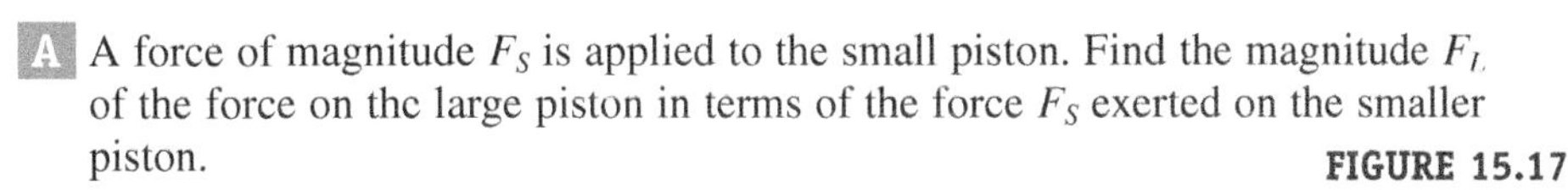

A A force of magnitude F_S is applied to the small piston. Find the magnitude F_L of the force on the large piston in terms of the force F_S exerted on the smaller piston.

FIGURE 15.17

INTERPRET and ANTICIPATE

The force on the small piston increases the pressure in the hydraulic fluid. According to Pascal's principle, that pressure is transmitted undiminished to all parts of the fluid and to the walls of the container that includes the large piston. So, the pressure change on the small piston equals the pressure change on the large piston: $\Delta P_S = \Delta P_L$. We expect to find that the force exerted on the small piston is less than the force on the larger piston ($F_S < F_L$); otherwise, no one would use a hydraulic lift.

SOLVE

Start with Pascal's principle. Use Equation 15.1 and solve for F_L.

$$\Delta P_S = \Delta P_L$$

$$\frac{F_S}{A_S} = \frac{F_L}{A_L}$$

$$F_L = \frac{A_L}{A_S} F_S \qquad (1)$$

CHECK and THINK

Since $A_L > A_S$, we find that $F_S < F_L$, which is what makes devices like hydraulic lifts useful. A small force is magnified by the factor A_L/A_S.

$$\frac{A_L}{A_S} > 1$$

$$F_L > F_S$$

B A large SUV of mass $m = 2.67 \times 10^3$ kg is at rest on the car lift in Figure 15.17. Both pistons have a circular cross-sectional area. The small piston has radius $R_S = 3.81$ cm, and the large piston has radius $R_L = 20.3$ cm. Find the magnitude of the force F_S exerted on the small piston. (Don't worry that we have simplified the hydraulic lift.)

INTERPRET and ANTICIPATE

In this part, start with Equation (1). The two forces on the vehicle are the upward force due to the lift $\vec{F}_L$ and the downward gravitational force $\vec{F}_g$. Because the vehicle is at rest, the magnitude of these two forces must be equal. We expect to find that F_S is less than the weight of the vehicle.

SOLVE

Apply Equation (1) with $F_L = F_g$. Write the weight of the vehicle in terms of its mass.

$$F_L = \frac{A_L}{A_S} F_S = F_g = mg$$

Make substitutions and solve for F_S. Use the familiar formula for the area of a circle.

$$mg = \frac{A_L}{A_S} F_S$$

$$F_S = \frac{A_S}{A_L} mg = \frac{\pi R_S^2}{\pi R_L^2} mg = \left(\frac{R_S}{R_L}\right)^2 mg$$

$$F_S = \left(\frac{3.81 \text{ cm}}{20.3 \text{ cm}}\right)^2 (2.67 \times 10^3 \text{ kg})(9.81 \text{ m/s}^2)$$

$$F_S = 9.23 \times 10^2 \text{ N}$$

CHECK and THINK

As expected, F_S is less than the weight of the vehicle ($mg = 2.62 \times 10^4$ N). In fact, the force required to hold up this SUV is about the same as a typical man's weight.

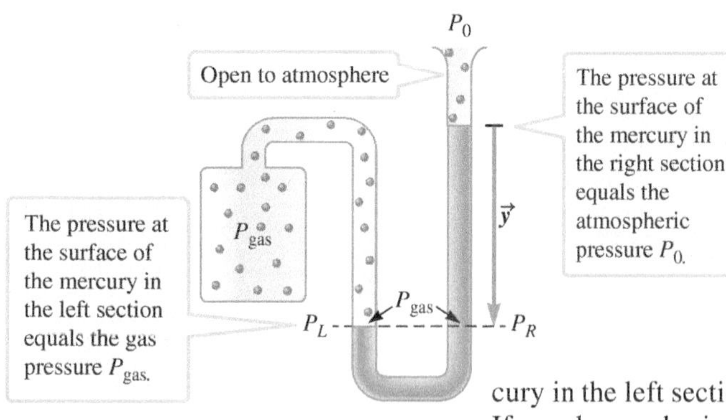

FIGURE 15.18 An open-tube manometer is used to measure gas pressure. The gauge pressure is found by measuring the height y: $P_{gauge} = \rho g y$.

Manometers, Gauge Pressure, and Absolute Pressure

An open-tube **manometer** is a U-shaped tube with one end open to the atmosphere and the other end attached to a container full of a fluid (usually a gas) whose pressure P_{gas} you would like to measure (Fig. 15.18). The U-shaped tube contains an incompressible fluid of known density ρ such as mercury or water; just to be specific, let's assume it is mercury. As long as the gas pressure does not equal atmospheric pressure—$P_{gas} \neq P_0$—the height of the mercury in the left section does not equal the height of the mercury in the right section. If we draw a horizontal plane that passes through a point on the surface of the mercury in the left section and a point at a depth y in the right section, the pressure at those two points must be equal:

$$P_L = P_R \tag{15.12}$$

The pressure on the left equals the gas pressure, so

$$P_L = P_{gas} = P_R \tag{15.13}$$

We use $P_R = \rho g y + P_0$ (Eq. 15.6) to find the pressure on the right and write the gas pressure as

$$P_{gas} = \rho g y + P_0 \tag{15.14}$$

Equation 15.14 gives the **absolute pressure** of the gas: $P_{abs} = P_{gas} = \rho g y + P_0$. As you can see, the absolute pressure depends on the atmospheric pressure P_0. When measuring the pressure in our tires or scuba tanks, we are more interested in the **gauge pressure** P_{gauge} *relative* to atmospheric pressure:

$$P_{gauge} \equiv P_{abs} - P_0 \tag{15.15}$$

The gauge pressure just depends on the height of the mercury in the manometer:

$$P_{gauge} = P_{gas} - P_0$$
$$P_{gauge} = \rho g y + P_0 - P_0$$
$$P_{gauge} = \rho g y \tag{15.16}$$

In terms of practical measurements, if you want the gauge pressure, you only need to measure the height y of the mercury in the manometer. If you need to know the absolute pressure, you must measure y and know the atmospheric pressure P_0.

FIGURE 15.19 Evangelista Torricelli invented the mercury barometer in 1644.

Barometers

Often it is okay to assume $P_0 = 1\text{ atm} = 1.01 \times 10^5$ Pa, but atmospheric pressure depends on altitude and weather conditions, so when a more precise value is needed, atmospheric pressure should be measured. A **barometer** is a device used to measure atmospheric pressure. Italian physicist Evangelista Torricelli (1608–1647) invented the mercury barometer in 1644 (Fig. 15.19). A simple way for Torricelli to build a mercury barometer would have been for him to pour mercury into a dish and also into a long, glass tube. The tube is permanently sealed by the glass on one end, and he may have temporarily sealed the other end with his finger (Fig. 15.20A). He would then invert the tube into a dish of mercury and remove his finger. When done, some (not all) of the mercury in the tube flows into the dish. Because the tube is initially filled with mercury and nothing is allowed to enter the tube, there is essentially nothing—that is, a vacuum—above the mercury in the tube (Fig. 15.20B). Therefore, the pressure at the top surface of the mercury in the tube is essentially zero: $P_{top} = 0$. The atmosphere presses down on the mercury in the dish, so the pressure at that surface is atmospheric pressure P_0. We find P_0 by measuring the height of the mercury in the tube. According to Equation 15.6, the pressure P_{tube} at the depth y is

$$P_{tube} = \rho g y + P_{top}$$

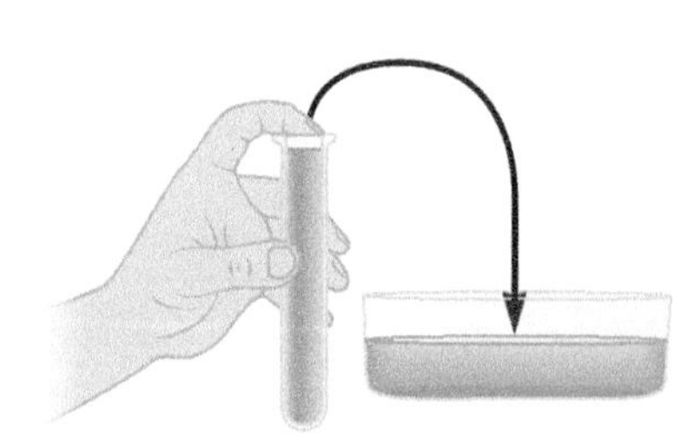

A.

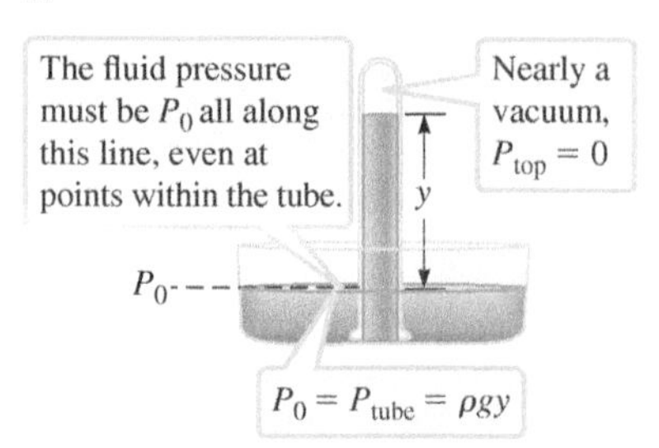

B.

FIGURE 15.20 How a mercury barometer is made. **A.** A glass tube is filled with mercury and sealed off. **B.** The tube is inverted in a dish of mercury.

Because there is a vacuum above the mercury in the tube, $P_{top} = 0$ and

$$P_{tube} = \rho g y + 0 = \rho g y$$

Finally, because the pressure must be equal all along the horizontal plane, $P_{tube} = P_0$, and

$$P_0 = \rho g y \qquad (15.17)$$

If we know the density of the fluid (usually mercury) in the barometer and the acceleration due to gravity g, by measuring y we can find the atmospheric pressure P_0, which is why millimeters of mercury (mm Hg) may be used as a pressure unit.

EXAMPLE 15.5 Building a Barometer

A If you wish to build your own mercury barometer, what is the minimum possible length of your tube?

INTERPRET and ANTICIPATE

Atmospheric pressure will push the mercury up in the tube. The greater the atmospheric pressure, the higher the mercury will go. If the tube is too short, the mercury will be forced to the very top, and you won't be able to measure the pressure. Assume the atmospheric pressure to be measured is roughly the pressure near sea level. Of course, if the pressure increases due to weather changes or because you take the barometer to a lower elevation such as the Dead Sea, you might find that the "sea-level" barometer is too short. From Figure 15.19, we estimate the barometer to be between 2 ft and 3 ft tall.

SOLVE

Solve Equation 15.17 for y and substitute the appropriate values. The density of mercury is given in Table 15.1.

$$P_0 = \rho g y \qquad (15.17)$$

$$y = \frac{P_0}{\rho g} = \frac{1.01325 \times 10^5\,\text{Pa}}{(13{,}600\,\text{kg/m}^3)(9.81\,\text{m/s}^2)}$$

$$y = 0.760\,\text{m}$$

CHECK and THINK

The barometer's height ($y \approx 2.5$ ft) is in the expected range. You can also check the answer from the conversion 1 atm = 760 mm Hg (or torr); it says that 1 atmosphere of pressure raises mercury to a height of 760 mm, or 0.760 m. You may also have heard a weather forecaster report the pressure as 29.92 inches. That is because 29.92 in. (about 2.5 ft) is about 760 mm, so the pressure is 1 atm.

B Mercury vapor is particularly harmful to human health. To avoid the health hazards associated with handling mercury, you decide to build a water barometer. How long would your tube need to be? Why did Torricelli use mercury?

INTERPRET and ANTICIPATE

In this part, use the density of water instead of the density of mercury. We expect the water barometer to be larger than the mercury barometer.

SOLVE

Substitute the appropriate values, including the density of water from Table 15.1.

$$y = \frac{P_0}{\rho g} = \frac{1.01325 \times 10^5\,\text{Pa}}{(1000\,\text{kg/m}^3)(9.81\,\text{m/s}^2)}$$

$$y = 10.3\,\text{m}$$

CHECK and THINK

As expected, the water barometer is larger than the mercury barometer. In fact, a water barometer would need to be roughly three stories tall. It would be difficult—if not impossible—for Torricelli to have a glass tube of this size. However, as of 2013, the tallest barometer in the world (located at Portland State Universiy in Oregon) is made of glass tubes and is more than 14 m tall.

A. B.

FIGURE 15.21 A. Fluid flow can be complex and turbulent as in these river rapids. **B.** A calmly flowing river is an example of ideal fluid flow.

15-6 Ideal Fluid Flow

IDEAL FLUID FLOW

Major Concept

We now shift our attention from fluids at rest to moving fluids such as the water flowing in a river. When water passes through rapids (Fig. 15.21A) or runs close to the shore, fluid motion can be complicated, but the motion of water in a calm river (Fig. 15.21B) is much simpler. In this book, we study the simple motion of **incompressible** fluids, known as **ideal fluid flow.**

There are three characteristics of ideal fluid flow: It is *steady*, it is *irrotational*, and it is *nonviscous*. To understand steady flow, imagine watching cars on an amusement park ride and consider just one small portion of the track near the top of a hill (Fig. 15.22). Each car has the same velocity as it passes over that part of the track. In **steady flow** (also known as **laminar flow**), the fluid's velocity at any fixed point is constant, but a fluid in steady flow may accelerate as it moves along its path. Again imagine cars on the amusement park ride. If you consider another part of the track near the bottom of the hill, you find that each car passing over that part of the track has a greater velocity than what it had near the top. Therefore, each car accelerates as it moves along the track.

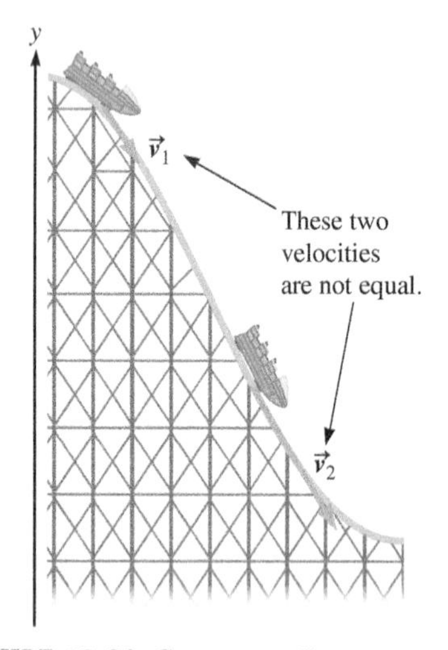

FIGURE 15.22 Cars on a roller coaster are like fluid molecules in steady flow.

An ideal flow is also irrotational; if a test object such as a leaf were placed in a fluid such as a river it would not rotate. The river may have a curved path, but the leaf would always face the same direction, just like the passengers on a Ferris wheel (Fig. 12.2A, page 332).

The third characteristic of ideal flow—the nonviscous part—is analogous to the motion of an object without resistive forces such as kinetic friction or drag. Just as kinetic friction resists motion and causes solid objects to stop, viscosity resists flow and causes fluids to come to rest. Imagine a race between a puddle of honey and puddle of water. Honey has a higher viscosity than water, so the water flows more easily than the honey. More precisely, **viscosity** is a measure of a fluid's resistance to deformation when a shear stress is applied to the fluid.

Although kinetic friction cannot be completely eliminated from any real system, we have seen that it is very useful to study ideal, frictionless systems. Likewise, viscosity cannot be completely eliminated from any fluid. Both kinetic friction and viscosity can often be ignored, however. Some fluids such as water, mercury, and alcohol have low viscosities. We say that these fluids are "thin." "Thick" fluids like honey, oil, and hot lava have higher viscosities. The viscosity of a fluid usually depends on its temperature such that viscosity is lower at higher temperatures. You probably know that the oil in your car is thicker and more viscous when the engine is cold than it is after you warm up the engine, and chocolate sauce kept in the refrigerator is thicker than when it is in a hot fondue pot. Viscosity also tends to be lower far from any walls or obstructions. If you imagine a lava flow, the viscosity is higher near the banks of the flow than near the center. Ideal fluid flow is assumed to be **nonviscous**, meaning that the viscosity is zero; in other words, we assume there is no resistance to flow.

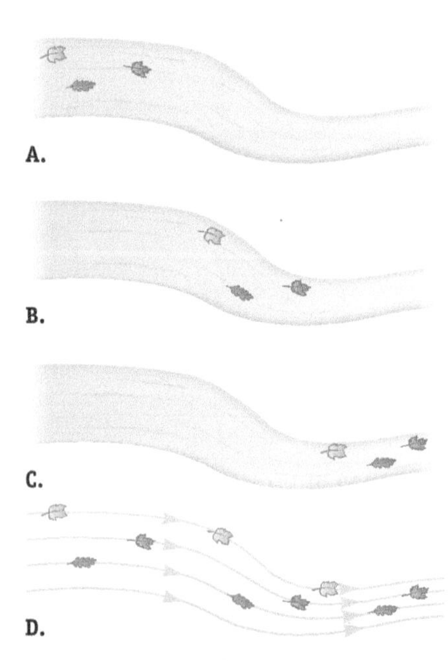

FIGURE 15.23 A–C. Three snapshots of a river show the motion of several leaves. **D.** Streamlines show the path of several leaves flowing in the river. Each leaf follows a streamline.

The study of moving fluids requires a few special tools. When studying the motion of solid objects, it is helpful to make motion diagrams. In Section 2-2, we imagined making a motion diagram of Mars by taking photos of the planet every few nights. Motion diagrams are not helpful for studying flowing fluids. Imagine taking a picture of a river at regular time intervals. You would not be able to detect any changes in the river that would help you study the water's flow. If a few leaves were floating on the water when you took your photos, however, you would be able to detect their motion (parts A–C of Fig. 15.23). Each leaf is attached to one small part of the water known as a **fluid element**. So, the motion diagram for the leaves shows the path of the fluid elements. A fluid element's path is called a **streamline**, and several streamlines are usually drawn to illustrate a flow (Fig. 15.23D). In practice, the leaves shown in Figure 15.23 are too spread out to trace streamlines adequately. Often, dyes or smoke particles (Fig. 15.24) are added to a fluid to better trace the streamlines.

Streamlines are a visualization tool, and interpreting a sketch of a fluid's streamlines is much like interpreting a sketch of a particle's path. A particle's velocity is tangent to its path. Similarly, the velocity of a fluid element is tangent to its streamline (Fig. 15.25), which leads to a few important facts about streamlines. (1) Because a fluid element must have only one velocity at any instant, streamlines cannot cross. If they did cross, at the intersection the fluid element would have two velocities simultaneously. (2) Streamlines for ideal fluid flow are constant in time. That is, the sketch of the streamlines is frozen in time, but each fluid element is in motion along its streamline. (3) The distance between streamlines is related to the speed of the fluid in that region. Where the streamlines are close together, the fluid is moving quickly. Widely separated streamlines characterize slow flow. We will discuss this idea more fully in the next section.

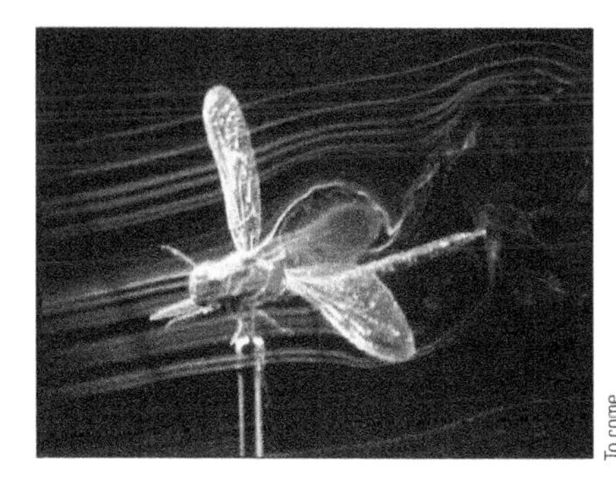

FIGURE 15.24 Smoke particles in a wind tunnel show the streamlines of air as it flows past a dragonfly.

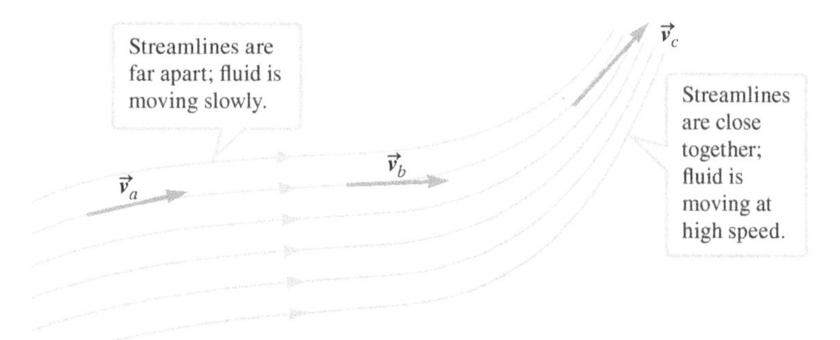

FIGURE 15.25 Streamlines show the path of several fluid elements. Fluid velocity vectors are tangent to the streamlines.

15-7 The Continuity Equation

Zak creates his own shower by partially blocking the opening of the hose with his thumb (Fig. 15.26A). When he removes his thumb, the water trickles out of the end of the hose (Fig. 15.26B). These two situations differ in the size of the hole that the water passes through. Water is essentially incompressible, so all the water that enters the hose in a certain time interval—say1 second—must leave the hose in the same time interval, whether through a partially open hole or a completely open hole. Because the partially open hole is smaller than the completely open hole, the water must pass through it faster to get the same amount of water through the hose in the 1-second time interval.

To quantify this common experience, imagine a fluid flowing through a tube that is wide on one end and narrow on the other (Fig. 15.27). Let's derive an expression for the speed v of the fluid in terms of the cross-sectional area A of the tube. During a time interval Δt, a volume V of fluid enters the left region. Because the fluid is incompressible, an equal volume V of fluid must leave the right region in the

A. B.

Photos © Dr. Jenny Spinner/St. Joseph's University

FIGURE 15.26 A. Zak creates a shower by reducing the cross-sectional area of the hose. When the cross-sectional area is small, the flow is fast. **B.** When the cross-sectional area is large, the flow is slow.

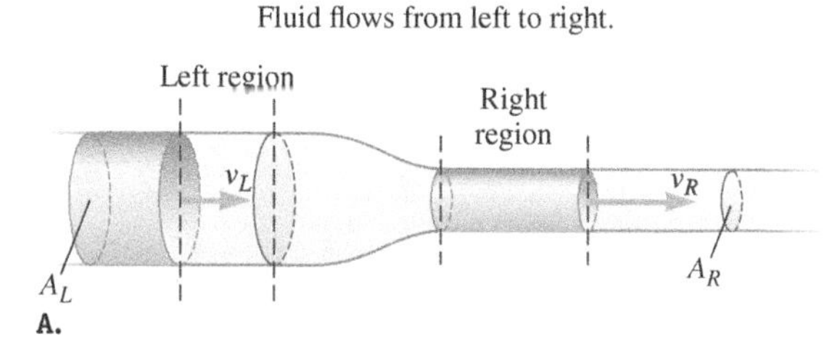

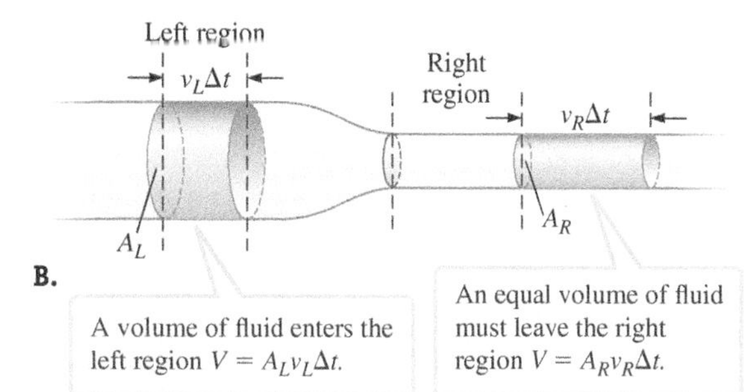

FIGURE 15.27 Fluid flows through a tube that narrows. The volume of fluid that enters the left region is equal to the volume of the fluid that leaves the right region.

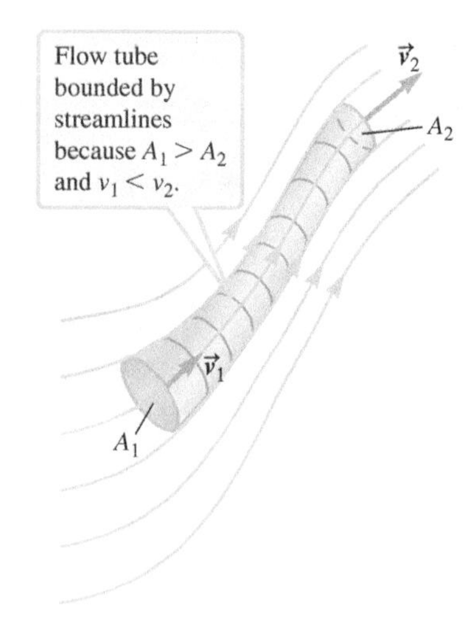

FIGURE 15.28 An imaginary tube known as a *flow tube* whose boundaries consist of streamlines.

same time Δt. We can write this volume in terms of the fluid's speed. On the left, the fluid's speed is v_L, so in the time interval Δt, the fluid's displacement is $v_L\,\Delta t$ along the axis of the tube. The volume of fluid that enters the left region is the cross-sectional area of the left side of the tube multiplied by the displacement of the fluid:

$$V = A_L v_L\,\Delta t \tag{15.18}$$

Similar reasoning can be used to find the volume of fluid exiting the right region:

$$V = A_R v_R\,\Delta t \tag{15.19}$$

Because the volume of fluid entering the right region equals the volume of fluid leaving the left region, we set Equation 15.18 equal to Equation 15.19:

$$A_R v_R\,\Delta t = A_L v_L\,\Delta t$$

$$A_R v_R = A_L v_L \tag{15.20}$$

Although Equation 15.20 had an actual tube (Fig. 15.27) in mind, Equation 15.20 also holds for an imaginary tube known as a **flow tube** whose boundaries consist of streamlines (Fig. 15.28). Because streamlines cannot cross, the fluid moves in the flow tube just as it would in a real tube. If the cross-sectional area is A_1 on one end of the flow tube and A_2 on the other end, the fluid speed at these two ends is given by Equation 15.20:

CONTINUITY EQUATION

Major Concept

$$A_1 v_1 = A_2 v_2 \tag{15.21}$$

Equation 15.21 is called the **continuity equation.**

The quantity Av is known as the **volume flow rate** $R = Av$. Volume flow rate has the dimensions of volume per time, so the SI units are cubic meters per second, or m^3/s. We can restate the continuity equation as follows: *In ideal fluid flow, the volume flow rate is uniform* (that is, it has the same value over the length of the flow tube).

One consequence of the continuity equation is that when a fluid flows from a tube of given cross-sectional area through a region with a smaller cross-sectional area, the fluid speed must increase. Another version of this statement is as follows: *In regions where the streamlines are closely crowded, the fluid flows quickly* (Fig. 15.28).

EXAMPLE 15.6 Keep Your Pets Alive

Koi are a fish known to produce a lot of waste. If you wish to keep a koi fishpond (Fig. 15.29), you will need a very good water filter. Professional-grade filters are very large and may be buried in the ground near the pond. The effectiveness of the filter is determined by the length of time the polluted water is held in it. Ideally, you would like the polluted water to remain in the filter for at least 10 min. A large pond may hold 6500 gallons, and the entire pond should be processed through the filter every 3 hours. What is the required volume flow rate (in cubic meters per second and gallons per minute)? What size filter (in cubic meters and in gallons) is required?

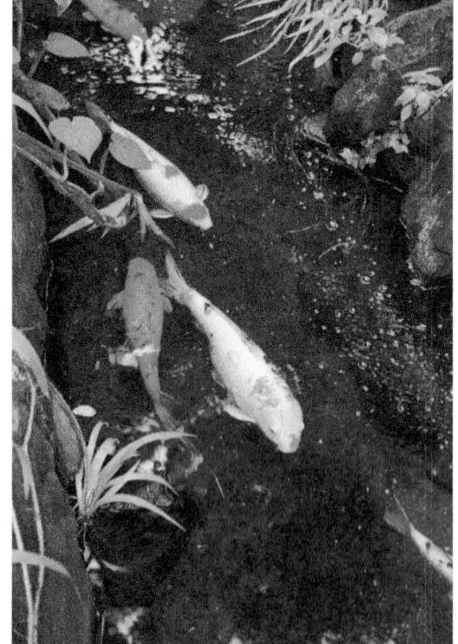

FIGURE 15.29 Koi fish.

INTERPRET and ANTICIPATE

Use the fact that the entire pond must be processed in 3 hours to find the required volume flow rate. Once that value is known, use the fact that the water should remain in the filter for 10 min to find the filter's size.

SOLVE

First, convert the volume of the pond, filter time, and processing time to SI units.

$$V_{\text{pond}} = 6500\text{ gal}\left(\frac{1\text{ m}^3}{264\text{ gal}}\right) = 24.6\text{ m}^3$$

$$t_{\text{process}} = 3\text{ h} = 180\text{ min} = 1.08 \times 10^4\text{ s}$$

$$t_{\text{filter}} = 10\text{ min} = 600\text{ s}$$

The required volume flow rate R comes from the whole volume V_{pond} having to be filtered in time $t_{process}$.	$R = \frac{V_{pond}}{t_{process}} = \frac{24.6\text{ m}^3}{1.08 \times 10^4\text{ s}} = 2.28 \times 10^{-3}\text{ m}^3/\text{s}$ $R = \frac{V_{pond}}{t_{process}} = \frac{6500\text{ gal}}{180\text{ min}} = 36.1\text{ gal/min}$
Use the volume flow rate and filter time to find the volume of the filter.	$V_{filter} = Rt_{filter} = (2.28 \times 10^{-3}\text{m}^3/\text{s})(600\text{ s}) = 1.4\text{ m}^3$ $V_{filter} = Rt_{filter} = (36.1\text{ gal/min})(10\text{ min}) = 360\text{ gal}$

CHECK and THINK

Let's take a moment to appreciate the size of this filtering system. First, a typical 10-minute shower uses 30 to 50 gal of water, so the shower volume flow rate is 3 to 5 gal/min. The koi fishpond system has about 10 times the volume flow rate of a typical shower. A fairly large kitchen refrigerator has an interior volume of about 200 gal, so the required koi filter is nearly the size of two refrigerators. Such large filters are usually installed underground when the pond is constructed.

EXAMPLE 15.7 A Most Educational Demonstration

You have probably seen science demonstrations using common household items. In one such experiment, 13 Mentos mints are dropped quickly into a 2-L bottle of cola. The cola is forced rapidly out of the bottle (Fig. 15.30). Assuming it takes 0.6 s to empty the 2-L bottle, estimate the maximum height of the cola spray.

FIGURE 15.30

INTERPRET and ANTICIPATE

We have enough information (bottle volume and emptying time) to estimate the volume flow rate. From that estimate, use the equation of continuity to find the speed of the cola as it exits the bottle. (We must estimate the area of the bottle opening.) To find its maximum height, model the cola as a projectile launched straight up and assume gravity is the only force acting on the cola after it leaves the bottle.

SOLVE Find the volume flow rate using the fact that the 2 L are emptied in 0.6 s.	$V_{bottle} = 2\text{ L} \times \frac{1\text{ m}^3}{1000\text{ L}} = 2 \times 10^{-3}\text{ m}^3$ $R = \frac{V_{bottle}}{t} = \frac{2 \times 10^{-3}\text{ m}^3}{0.6\text{ s}}$ $R = 3.3 \times 10^{-3}\text{ m}^3/\text{s}$
Estimate the area of the bottle opening. The diameter of the opening is about 1 in., which means that the radius is about 1.3 cm.	$A_{opening} = \pi r^2 \approx \pi(1.3 \times 10^{-2}\text{ m})^2$ $A_{opening} \approx 5.3 \times 10^{-4}\text{ m}^2$
According to the continuity equation, the volume flow rate is a constant.	$R = A_{opening}v_{cola}$ $v_{cola} = \frac{R}{A_{opening}} = \frac{3.3 \times 10^{-3}\text{ m}^3/\text{s}}{5.3 \times 10^{-4}\text{ m}^2} = 6.3\text{ m/s}$
There are many ways to find the correct expression for the maximum height. Here, we have chosen to use kinematics (Eq. 2.13). The initial speed v_{0y} is the speed v_{cola} found above, and the final speed $v_y = 0$ when $\Delta y = y_{max} - 0 = y_{max}$.	$v_y^2 = v_{0y}^2 + 2a_y\,\Delta y$ (2.13) $0 = v_{cola}^2 - 2gy_{max}$ $y_{max} = \frac{v_{cola}^2}{2g} = \frac{(6.3\text{ m/s})^2}{2(9.81\text{ m/s}^2)} \approx 2\text{ m}$

CHECK and THINK

The exit speed of the cola may seem a bit high at roughly 14 mph, but the maximum height (2 m ≈ 6 ft) seems to be about right as seen in Figure 15.31.

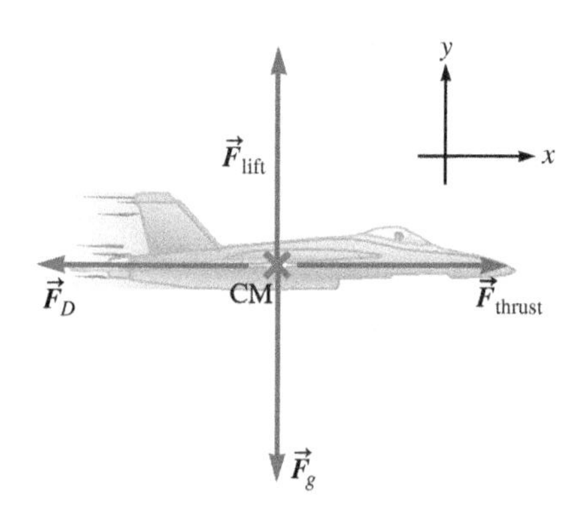

FIGURE 15.31 Free-body diagram for an airplane flying at constant velocity in the positive x direction.

15-8 Bernoulli's Equation

In this section, we return to the issues raised in our case study, namely how can an airplane fly upside down as well as right side up; indeed, how can it fly at all given that it is heavier than an equal volume of air? In Section 15-4, Cameron said that "planes fly because they are moving." It is true that there must be relative motion between the airplane wings and the surrounding air. In this section, we'll see why motion is necessary.

Let's start by drawing a free-body diagram for a jet plane flying at constant velocity in the positive x direction (Fig. 15.31). Gravity acts in the negative y direction (downward in the figure). Air drag (Eq. 6.5) and thrust (Eq. 10.16) act in opposite directions along the x axis. Because the plane is flying at constant velocity, there is no net force acting on it. Because the only forces acting in the x direction are thrust and drag, the magnitude of the thrust must equal the magnitude of the drag force: $F_{\text{thrust}} = F_D$. There must also be an upward force exerted on the plane; otherwise, the plane would accelerate downward. This upward force is known as **lift** ($\vec{F}_{\text{lift}}$ in Fig. 15.31). Lift is exerted by air primarily on the wings of the plane.

For air to exert lift on a wing, there must be relative motion between the air and the wing. As a first approximation, imagine removing a simple, flat door from its hinges to use as a wing. Figure 15.32 shows the short edge of the door (a simple board). Imagine the board tilted with its right edge upward in a wind tunnel (Fig. 15.32A). The air flows from right to left. The streamlines shown on the right are parallel and equally spaced, showing that the air is moving uniformly in a straight path toward the board. The streamlines near the board, however, are bent downward, showing that the board deflects the air downward. So, the board exerts a downward force on the air. According to Newton's third law, the air must exert an upward force on the board. This upward force is lift. If the board is tilted with its right edge downward (Fig. 15.32B), the air is deflected upward. In this case, the air must exert a downward force on the board, which we still call "lift" even though that terminology seems odd. If the board is level, the air is not deflected up or down, and there is no lift exerted on the board (Fig. 15.32C). So, although you might think that a plane, or at least its wing, must be tilted as in Figure 15.32A to have upward lift, wings are designed so that the air is deflected downward even when the plane is flying level (Fig. 15.33).

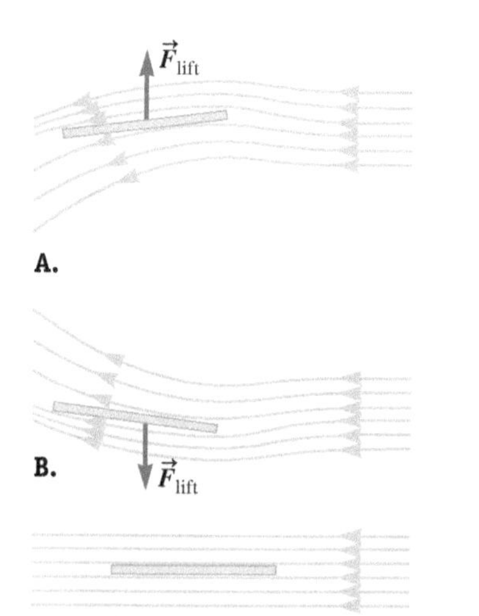

FIGURE 15.32 A flat door is in a wind tunnel. **A.** If the door's right edge is tilted upward, air moving past the wind is deflected downward, and the air exerts an upward force on the board known as *lift*, $\vec{F}_{\text{lift}}$. **B.** If the door's right edge is tilted downward, the air is deflected upward, and the lift is downward. **C.** If the door is not tilted, there is no net deflection of the air, and there is no lift exerted on the board.

CASE STUDY Part 3: A Moving Explanation

Avi: I think I get the basic idea. The wing of the plane is designed so that when the plane moves through the air, the wing forces the air downward. That means that the air must force the plane upward. I just can't see how this is related to air pressure and helium balloons.

Cameron: Helium balloons rise because they are less dense than air. Sometimes you even see helium balloons that just hover in the middle of the room. That's because they've lost some helium and now are just as dense as the air around them.

Shannon: The helium balloon rises in the air because the air exerts a buoyant force on the balloon. And that buoyant force is greater than the balloon's weight.

Avi: That idea works for a balloon, but a plane weighs a lot. There is no way the buoyant force is greater than its weight!

Cameron: That's right. Because the plane is denser than air, the buoyant force cannot get it off the ground.

Shannon: You're both right. Look at it this way. If you blow a bubble under water, it rises to the surface. That's because the water pressure below the bubble is greater than the water pressure above the bubble. When the plane is just sitting on the runway, the air pressure above the wing is about the same as the air pressure below the wing. When the plane moves, the air pressure below the wing is greater than the air pressure above the wing.

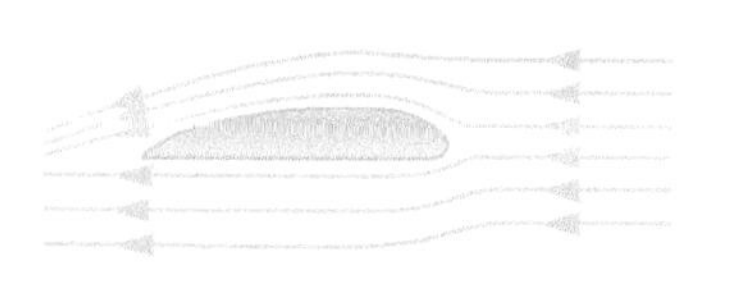

FIGURE 15.33 Cross section of a wing shows that it has a curved upper surface and flat lower surface. This design ensures that when the plane is flying level, air moving past the wing is deflected downward. Flaps on the wings allow pilots to further adjust the shape of the wing to control the lift.

Shannon is correct again. When the plane is stationary, the air pressure is nearly uniform over its entire surface, so the air does not exert a net force on the plane. When the air moves relative to the wings, however, the air pressure above the wings is lower than the air pressure below them, and the air exerts a net force—lift—on the wings.

Pressure in a Moving Fluid

We have just discovered that the motion of a fluid may change its pressure. To derive an expression for the pressure in a moving fluid, we need to take a conservation of energy approach (Chapter 8). Consider a fluid moving in a bent pipe (Fig. 15.34). In this scenario, a piston moves to the right, pushing fluid to the right and upward.

Our system consists of the Earth and the portion of fluid shown in orange in Figure 15.34. The piston and the rest of the fluid (shown in blue) are external to the system and may do work on it. The piston applies a constant force $\vec{F}_{\text{piston}} = F_{\text{piston}}\hat{\imath}$, displacing the point of contact between the fluid and the piston by the amount Δx_1. Therefore, the work done by the piston is positive and is given by

$$W_{\text{piston}} = F_{\text{piston}}\,\Delta x_1 \tag{15.22}$$

According to Newton's third law, the fluid in the system exerts a force on the piston equal in magnitude to F_{piston}. If the piston has an area A_1, we can write F_{piston} in terms of the fluid pressure P_1 in region 1: $F_{\text{piston}} = P_1A_1$. Equation 15.22 now becomes

$$W_{\text{piston}} = P_1A_1\,\Delta x_1 \tag{15.23}$$

As shown in Figure 15.34B, the fluid in the system (shown in orange) pushes the external fluid (shown in blue) in the positive x direction by applying a force $\vec{F}_{\text{system}} = F_{\text{system}}\hat{\imath}$. The area of the pipe in region 2 is A_2, so we can write $\vec{F}_{\text{system}}$ in terms of the pressure P_2 in region 2: $\vec{F}_{\text{system}} = P_2A_2\hat{\imath}$. According to Newton's third law, the external fluid exerts a force on the system equal in magnitude to $\vec{F}_{\text{system}}$, but in the negative x direction: $\vec{F}_{\text{external}} = -P_2A_2\hat{\imath}$. The point of contact between the system and the external fluid is displaced by amount Δx_2 in the positive x direction. Therefore, the external fluid does negative work on the system, given by

$$W_{\text{external}} = -P_2A_2\,\Delta x_2 \tag{15.24}$$

There are no other external forces exerted on the system, so the total work done on the system is just the sum of Equations 15.23 and 15.24:

$$W_{\text{tot}} = P_1A_1\,\Delta x_1 - P_2A_2\,\Delta x_2 \tag{15.25}$$

Comparing parts A and B of Figure 15.34, you see that the system fluid that was initially in region 1 moves out of region 1 and that the system fluid that was just to the left of region 2 moves into region 2. Because the fluid is incompressible, the volume of region 1 must equal the volume of region 2:

$$V_1 = V_2 \equiv V$$

$$A_1\,\Delta x_1 = A_2\,\Delta x_2 = V \tag{15.26}$$

Substitute Equation 15.26 into Equation 15.25:

$$W_{\text{tot}} = P_1V - P_2V$$

$$W_{\text{tot}} = (P_1 - P_2)V \tag{15.27}$$

We now have an expression for the work done on a system that includes fluid moving between two regions.

Next, we need to calculate the changes in the system's kinetic and potential energies. Consider the change in kinetic energy first. Because we assume ideal fluid flow, the speed of a fluid element (and therefore its kinetic energy) depends only on its location in the pipe. Because there is system fluid in the middle region at both the initial and final times, fluid in the middle region does not come into the calculation of the *change* in kinetic energy. The change in kinetic energy ΔK only depends on the fluid that moves out of region 1 and the fluid that moves into region 2. The mass m of the fluid in region 1 equals the mass m of the fluid in region 2 because these two regions have equal volumes and the fluid is incompressible. Therefore,

$$\Delta K = \tfrac{1}{2}mv_2^2 - \tfrac{1}{2}mv_1^2 \tag{15.28}$$

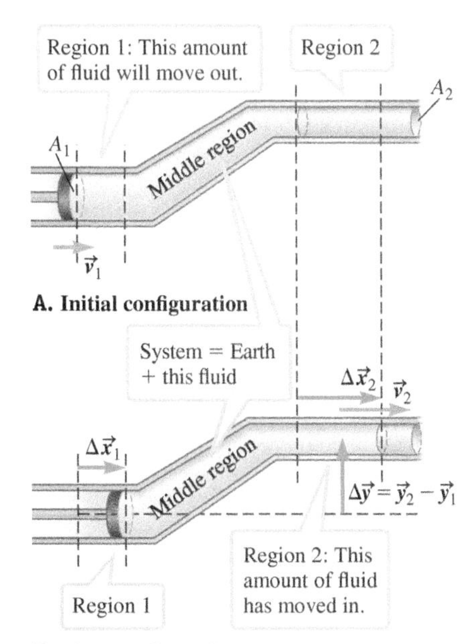

FIGURE 15.34 A piston forces fluid in a pipe to move up and to the right. The fluid in orange is part of our Earth–fluid system. The rest of the fluid, shown in blue, and the piston are outside the system and do work on the system. Fluid moves out of region 1, and fluid moves into region 2. There is no net change in the fluid in the middle region.

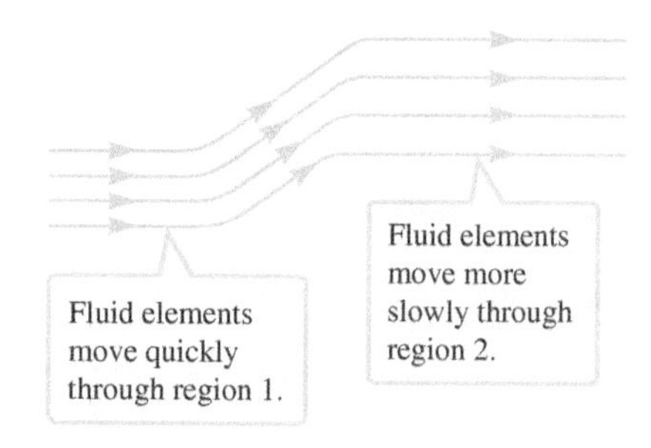

FIGURE 15.35 Fluid elements move into and out of regions 1 and 2, but the overall fluid flow represented by the streamlines does not change in time.

The same sort of reasoning can be applied to the change in gravitational potential energy. Because the *change* in gravitational potential energy is determined by the *change* in vertical position (along the y axis), we only need to consider the fluid that moves out of region 1 and the fluid that moves into region 2:

$$\Delta U_g = mg\,\Delta y = mg(y_2 - y_1) \tag{15.29}$$

In ideal fluid flow, there are no viscous (resistive) forces, so there is no change in the fluid's internal energy. Therefore, we can express the conservation of energy using the work–mechanical energy theorem (Eq. 9.26):

$$W_{\text{tot}} = \Delta K + \Delta U$$

Substitute Equations 15.27, 15.28, and 15.29 for total work, change in kinetic energy, and change in potential energy, respectively:

$$(P_1 - P_2)V = (\tfrac{1}{2}mv_2^2 - \tfrac{1}{2}mv_1^2) + mg(y_2 - y_1)$$

Divide both sides by the volume V:

$$P_1 - P_2 = \frac{1}{V}\left(\frac{1}{2}mv_2^2 - \frac{1}{2}mv_1^2\right) + \frac{m}{V}g(y_2 - y_1)$$

Eliminate m and V by using $\rho = m/V$ (Eq. 1.1):

$$P_1 - P_2 = (\tfrac{1}{2}\rho v_2^2 - \tfrac{1}{2}\rho v_1^2) + \rho g(y_2 - y_1)$$

Finally, we group terms for region 1 on one side of the equation and terms for region 2 on the other side:

BERNOULLI'S EQUATION

Major Concept

$$P_1 + \tfrac{1}{2}\rho v_1^2 + \rho g y_1 = P_2 + \tfrac{1}{2}\rho v_2^2 + \rho g y_2 \tag{15.30}$$

Equation 15.30 is known as **Bernoulli's equation.** It is a statement of the work–energy theorem for ideal fluid flow. Bernoulli's equation looks much like the work–energy theorem (Eq. 9.31) with ΔE_{int} set to zero, but there is an important notational difference between Bernoulli's equation and the work–energy theorem. In the work–energy theorem, the subscripts i and f stand for *i*nitial and *f*inal times. Between these times, the configuration and motion of the system may change. In Bernoulli's equation, the subscripts 1 and 2 represent two different regions in an ideal fluid flow. Fluid elements move in and out of these regions, but the overall fluid flow represented by the streamlines does not change in time (Fig. 15.35).

With this difference in mind, we use Bernoulli's equation much as we use the conservation of energy principle. We used the conservation of energy principle when we had information about a system's energy at one particular time and needed information about another time. Likewise, we use Bernoulli's equation when we have information (such as pressure, speed, height) about a flowing fluid in one region and need information for another region.

PROBLEM-SOLVING STRATEGY Applying Bernoulli's Equation

:• INTERPRET and ANTICIPATE

Step 1 Draw a simple **sketch indicating your coordinate system.** The y axis is most important because vertical position coordinates are needed in Bernoulli's equation. Include streamlines if they help you visualize the problem.

Step 2 Choose region 1 and region 2. Usually, you know a lot of information about one region and need to know something about the other region. Label these regions on your sketch.

:• SOLVE

Step 3 Write **Bernoulli's equation.**

Step 4 Compare the parameters P, v, and y in the two regions. Often, one parameter is equal in the two regions, or you may consider a parameter to be approximately equal in the two regions.

Step 5 The **continuity equation** $A_1v_1 = A_2v_2$ (Eq. 15.21) is often helpful in finding or relating the speed in the two regions. If one cross-sectional area is very large compared with the other area, sometimes the fluid speed in the large area is approximately zero.

Step 6 Do algebra combining information in steps 3 through 5.

EXAMPLE 15.8 Doing Your Own Plumbing Job

At the restaurant where you work at your summer job, a large kitchen sink (1.20 m long, 0.46 m wide, and 0.36 m deep) is clogged and needs to be bailed out. Once there is about 5.0 cm of water left in the bottom, it becomes nearly impossible to continue bailing out the sink by scooping water into containers. Instead, you decide to siphon the remaining water with a hose of radius 1.2 cm into a cooking pot on the floor 0.79 m below the sink bottom.

A Find the speed of water emerging from the siphon hose.

B Find the volume flow rate from the hose.

C How long will it take you to empty the sink of that last 5.0 cm of water?

INTERPRET and ANTICIPATE

Apply Bernoulli's equation for part A, using the steps listed above. For part B, calculate the volume V of water in the sink from the dimensions given in the example opening. Then find the volume flow rate $R = Av$ by finding the cross-sectional area A of the hose and using the speed v of the water found in part A. The time needed to empty the sink comes from dividing the volume of water by the volume flow rate: $t = V/R$.

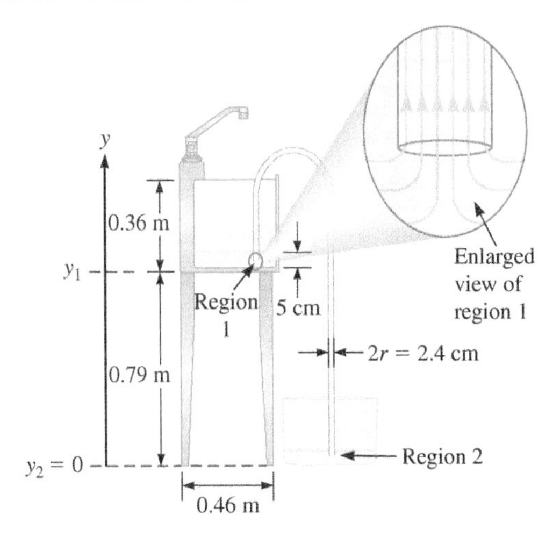

FIGURE 15.36

Steps 1 and 2 Sketch, add coordinate system, and choose regions. In our sketch (Fig. 15.36), we show the sink from the side. This perspective allows us to easily sketch the hose and pot. We indicate our coordinate system and our two regions. Region 1 is at the surface of the water near the entrance to the hose, and region 2 is at the exit. We have set $y_2 = 0$. The figure also shows an enlarged view of region 1 with streamlines in the sink spread out and then bending into the entrance of the hose and becoming closer together. The spacing of the streamlines indicates that the water moves slowly when it is in the sink, but speeds up in the hose.

A Speed of water, v_2

SOLVE

Step 3 Use **Bernoulli's equation** to find v_2 (the speed of the water as it emerges from the hose).

$$P_1 + \tfrac{1}{2}\rho v_1^2 + \rho g y_1 = P_2 + \tfrac{1}{2}\rho v_2^2 + \rho g y_2 \quad (15.30)$$

Step 4 Compare *P*, *v*, and *y* in the two regions. The air pressure is assumed to be constant in the kitchen, so the pressure in regions 1 and 2 is roughly equal. The other parameters—speed v and vertical position y—are different in the two regions.

$$P_1 = P_2$$

Step 5 Relate speeds using the **continuity equation.** The streamlines in Figure 15.36 indicate that the water in the sink moves slowly and water in the hose moves quickly. You can confirm these speeds with the continuity equation $A_1 v_1 = A_2 v_2$ (Eq. 15.21): The area of the sink is large, and the cross-sectional area of the hose is small; so, the water in region 1 moves much slower than the water in region 2. We approximate the speed in region 1 as zero.

$$v_1 \approx 0$$

Step 6 Do algebra. Use the approximations $P_1 = P_2$ and $v_1 \approx 0$ found above to simplify Bernoulli's equation.

$$\cancel{P_1} + \tfrac{1}{2}\rho v_1^2 + \rho g y_1 = \cancel{P_2} + \tfrac{1}{2}\rho v_2^2 + \rho g y_2$$

$$\tfrac{1}{2}\rho(0)^2 + \rho g y_1 = \tfrac{1}{2}\rho v_2^2 + \rho g y_2$$

$$\rho g y_1 = \tfrac{1}{2}\rho v_2^2 + \rho g y_2$$

Solve for v_2.

$$v_2 = \sqrt{2g(y_1 - y_2)} \quad (1)$$

Example continues on page 440 ▶

CHECK and THINK

Equation (1) is the same free-fall expression we would find for a particle dropped from rest (Section 2-10), which makes sense because gravity is the only force exerted on the water.

SOLVE, continued Substitute values into Equation (1) to find v_2.	$v_2 = \sqrt{2(9.81\,\text{m/s}^2)(0.79\,\text{m} - 0)}$ $v_2 = 3.9\,\text{m/s}$
B Volume flow rate, R **SOLVE** To find the volume flow rate, we need v_2 and the cross-sectional area of the hose.	$A = \pi r^2 = \pi(1.2 \times 10^{-2}\text{m})^2 = 4.5 \times 10^{-4}\text{m}^2$ $R = Av = (4.5 \times 10^{-4}\text{m}^2)(3.9\,\text{m/s})$ $R = 1.8 \times 10^{-3}\,\text{m}^3/\text{s}$
C Time to empty sink **SOLVE** When the water is in the sink, it forms a rectangular box. Find the volume of that box.	$V = lwd = (1.20\,\text{m})(0.46\,\text{m})(0.050\,\text{m}) = 2.8 \times 10^{-2}\,\text{m}^3$
Finally, the time to empty the sink is the volume divided by the volume flow rate.	$t = \dfrac{V}{R} = \dfrac{2.8 \times 10^{-2}\,\text{m}^3}{1.8 \times 10^{-3}\,\text{m}^3/\text{s}} = 1.6\,\text{s}$

CHECK and THINK

Of course, this time seems short. In practice, it is difficult to get the water to flow. Some people suck on the opposite end as you would a straw (which is not a great idea if the fluid you need to siphon is toxic). There are also handheld pumps that may be used to start the fluid flowing. In practice, the more water there is in the sink, the easier it is to start the siphon and keep it going. So, instead of scooping out water, it is better to siphon the whole amount. In fact, people with large fish tanks often use a siphon to empty it.

EXAMPLE 15.9 A Venturi Tube

Figure 15.37 shows a horizontal tube in which the middle section—known as the *neck*—is narrower than the rest. This tube is known as a *Venturi* tube, and it may be used to model blood flow through capillaries (small blood vessels) or air flowing between buildings in a large city. Venturi tubes are also used to measure fluid speeds and to mix gasoline with air in automobile carburetors. The cross-sectional area of the wide part of the Venturi tube is A_1 and that of the neck is A_2, where $A_1 > A_2$.

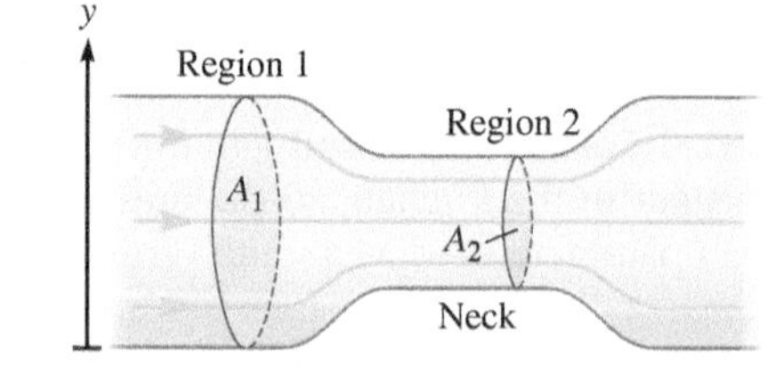

FIGURE 15.37 A Venturi tube.

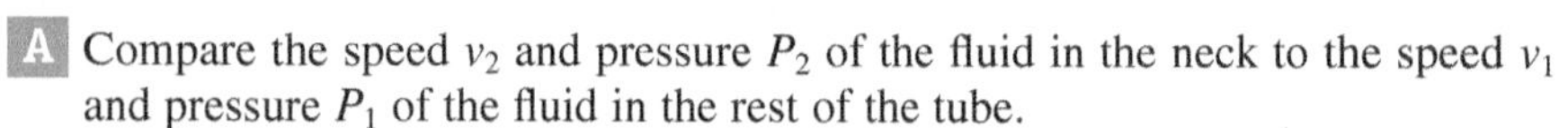

A Compare the speed v_2 and pressure P_2 of the fluid in the neck to the speed v_1 and pressure P_1 of the fluid in the rest of the tube.

INTERPRET and ANTICIPATE

Because we are asked to compare quantities in two regions, we expect to give two answers of the form $Q_1 > Q_2$, $Q_1 < Q_2$, or $Q_1 = Q_2$. We can use the continuity equation to make the comparison for speed. Use the streamlines in Figure 15.37 to anticipate the results. In region 1, the streamlines are far apart, whereas in region 2, they are closer together, so we expect $v_2 > v_1$. The comparison for pressure requires Bernoulli's equation.

SOLVE Begin with the continuity equation.	$A_1 v_1 = A_2 v_2$

Solve for v_2 and use $A_1 > A_2$.	$v_2 = \frac{A_1}{A_2}v_1$ $v_2 > v_1$	(1)

CHECK and THINK
As expected, the fluid speed is higher in region 2 than in region 1.

SOLVE, continued Use Bernoulli's equation to find the pressure in each region. Steps 1 **(sketch)** and 2 **(region selection)** are completed in Figure 15.37. So, we move on to step 3: Write **Bernoulli's equation**.	$P_1 + \frac{1}{2}\rho v_1^2 + \rho g y_1 = P_2 + \frac{1}{2}\rho v_2^2 + \rho g y_2$	(15.30)
In step 4, we **compare parameters** P, v, and y in the two regions. Figure 15.37 shows that y is the same all along the horizontal tube. We expect the pressure to be different in the two regions, and we have already found that $v_2 > v_1$.	$y_1 = y_2$	
We've already applied the **continuity equation** (step 5), and so we are on to step 6: **Do algebra.**	$P_1 + \frac{1}{2}\rho v_1^2 + \cancel{\rho g y_1} = P_2 + \frac{1}{2}\rho v_2^2 + \cancel{\rho g y_2}$ $P_1 + \frac{1}{2}\rho v_1^2 = P_2 + \frac{1}{2}\rho v_2^2$ $P_1 - P_2 = \frac{1}{2}\rho(v_2^2 - v_1^2)$	(2)
Use the previous result, $v_2 > v_1$.	$(v_2^2 - v_1^2) > 0 \qquad P_1 - P_2 > 0$ $P_1 > P_2$	

CHECK and THINK
Our result that the pressure in region 1 is greater than the pressure in region 2 confirms the Venturi effect: In region 2, the speed is higher and the pressure is less, whereas in region 1, the speed is lower and the pressure is greater.

B Derive an expression for the speed v_2 of the fluid in the neck in terms of P_1, P_2, A_1, A_2, and ρ.

INTERPRET and ANTICIPATE
This part of the example is a continuation of part A. Pick up step 5 (apply **continuity equation**) to eliminate v_1 from Equation (2).

SOLVE Step 5 Substitute Equation (1), which came from the continuity equation, into Equation (2).	$v_1 = \frac{A_2}{A_1}v_2$ $P_1 - P_2 = \frac{1}{2}\rho\left[v_2^2 - \left(\frac{A_2}{A_1}v_2\right)^2\right] = \frac{1}{2}\rho v_2^2\left(1 - \frac{A_2^2}{A_1^2}\right)$	
Solve for v_2.	$v_2^2 = \frac{2(P_1 - P_2)}{\rho\left(1 - \frac{A_2^2}{A_1^2}\right)} = \frac{2A_1^2(P_1 - P_2)}{\rho(A_1^2 - A_2^2)}$ $v_2 = \sqrt{\frac{2A_1^2(P_1 - P_2)}{\rho(A_1^2 - A_2^2)}}$	(3)
CHECK and THINK There are two things to check about Equation (3). First, make sure that it gives the square root of a positive number.	$\rho > 0 \qquad A_1^2 > 0 \qquad P_1 > P_2 \qquad A_1^2 > A_2^2$ $\frac{2A_1^2(P_1 - P_2)}{\rho(A_1^2 - A_2^2)} > 0$	

Example continues on page 442 ▶

Second, make sure we have the dimensions (L/T) of speed.	$\sqrt{\frac{\cancel{[A_1^2]}[(P_1 - P_2)]}{[\rho]\cancel{[(A_1^2 - A_2^2)]}}} = \sqrt{\frac{[(P_1 - P_2)]}{[\rho]}} = \sqrt{\frac{F/L^2}{M/L^3}}$ $\sqrt{\frac{(M \times L/T^2)/\cancel{L^2}}{M/L^{\cancel{3}}}} = \sqrt{\frac{\cancel{M} \times L/T^2}{\cancel{M}/L}} = \sqrt{\frac{L^2}{T^2}}$ $[v_2] = \frac{L}{T}$

Equation (3) gives us a way to design a device—known as a *Venturi meter*—to measure the speed in a fluid of known density ρ. If you use a tube of known cross-sectional areas A_1 and A_2 and measure the pressure in regions 1 and 2 using, for example, a manometer (Section 15-5), you can solve for v_2. Of course, once you know v_2, you can use the continuity equation to solve for v_1.

EXAMPLE 15.10 CASE STUDY Flying Planes

Let's return to the case study one more time. Consider a small plane flying at constant velocity. The plane has mass $m = 1353$ kg, including its payload. The area of both wings combined is $A_{\text{wings}} = 16.3 \text{ m}^2$.

A If the air speed below the wing is 153 m/s, what is the air speed above the wing?

INTERPRET and ANTICIPATE

We know from the free-body diagram in Figure 15.31 that the magnitude of the lift exerted by air on the wings must equal the weight of the plane. Because lift is upward (the air exerts an upward force on the wings), we expect the air pressure below the wings to be greater than the pressure above the wings.

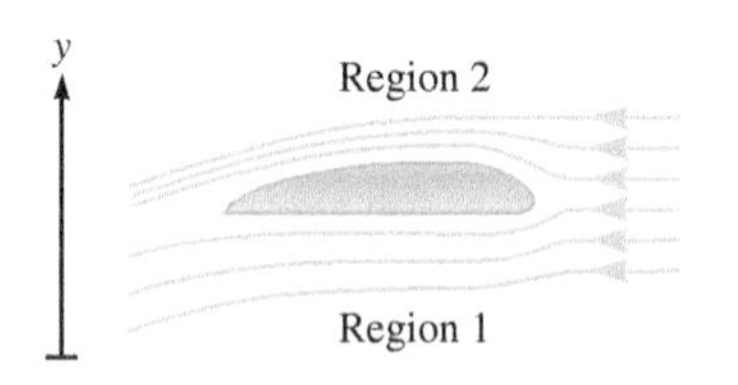

FIGURE 15.38

Relate air pressure to lift by applying Bernoulli's equation. **Steps 1 (sketch)** and **2 (choose regions)**: We have chosen an upward pointing y axis with region 2 above the wing and region 1 below the wing (Fig. 15.38).

The streamlines in region 2 are closer together than the streamlines in region 1, so air speed in region 2 is greater: $v_2 > v_1$. From Example 15.9, the pressure is lower in the region with the higher speed, so $P_2 < P_1$ as expected.

SOLVE

Step 3 Write **Bernoulli's equation.**	$P_1 + \frac{1}{2}\rho v_1^2 + \rho g y_1 = P_2 + \frac{1}{2}\rho v_2^2 + \rho g y_2$ (15.30)
Step 4 **Compare parameters.** The only remaining parameter to compare in the two regions is the y position. The top of the wing is only slightly above the bottom, so these regions are essentially at the same y position.	$y_1 \approx y_2$
Simplify Bernoulli's equation as in Example 15.9.	$P_1 - P_2 = \frac{1}{2}\rho(v_2^2 - v_1^2)$ (1)
We do not need **Step 5** in this part of the example (relating speeds using the continuity equation). Instead, we relate the pressure difference $P_1 - P_2$ to the lift F_{lift} on the wings. The lift is a result of the net pressure on the wings multiplied by the wing area.	$F_{\text{lift}} = (P_1 - P_2)A_{\text{wings}}$
The magnitude of the lift must equal the weight of the plane because the plane is not accelerating.	$F_g = F_{\text{lift}} = (P_1 - P_2)A_{\text{wings}}$ $P_1 - P_2 = \frac{F_g}{A_{\text{wings}}}$ (2)

Substitute Equation (1) into Equation (2).	$\frac{1}{2}\rho(v_2^2 - v_1^2) = \frac{F_g}{A_{\text{wings}}}$
Solve for v_2.	$(v_2^2 - v_1^2) = \frac{2F_g}{\rho A_{\text{wings}}}$ $v_2 = \sqrt{\frac{2F_g}{\rho A_{\text{wings}}} + v_1^2} = \sqrt{\frac{2mg}{\rho A_{\text{wings}}} + v_1^2}$
Substitute given values. The density of air is listed in Table 15.1.	$v_2 = \sqrt{\frac{2(1353\,\text{kg})(9.81\,\text{m/s}^2)}{(1.29\,\text{kg/m}^3)(16.3\,\text{m}^3)} + (153\,\text{m/s})^2} = 157\,\text{m/s}$

CHECK and THINK

This result is a very small difference in air speed, 4 m/s or roughly 3% of the given air speed below the wing. It might seem incredible that it is a great enough difference to create the required lift, but because the wings are large, the lift force is great enough to balance the weight of the plane.

B Answer Avi's concern about airplanes that can fly upside down. Do planes truly fly upside down?

SOLVE

For a plane to have an upward lift, its wings must force the air downward as in Figure 15.38. If you compare the streamlines on the far right of the wing with those to the far left of the wing, you can see that the ones on the left are lower than the ones on the right.

Now, imagine that plane rolled by exactly 180° around its long axis. The result would be Figure 15.39A, in which the wings would force the air upward. According to Newton's third law, the air must push the wings, and therefore the plane, downward. In practice, airplanes are not flown truly upside down. Instead, the plane is tilted slightly upward as in Figure 15.39B. For many airplanes, this upward tilt is just a few degrees, not noticeable to spectators watching an air show.

Exactly upside-down

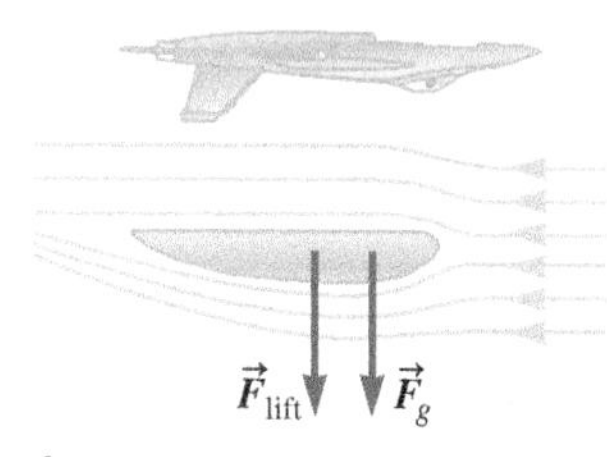

A.

Tilted slightly upward

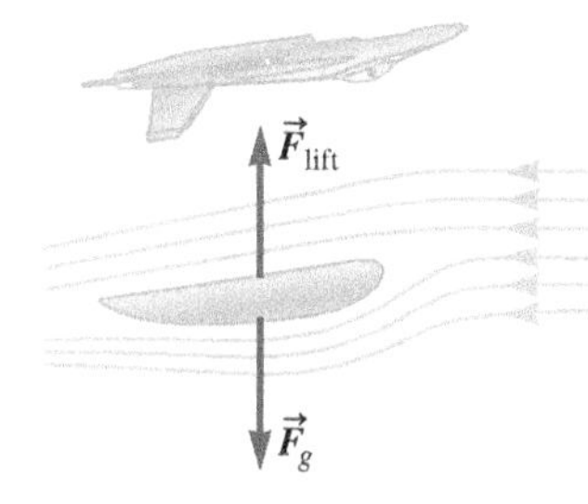

B.

FIGURE 15.39

A Final Note

In this chapter, we restricted our study to ideal fluid flow, in which air is assumed to be incompressible, and airflow around airplane wings is assumed to be laminar (steady). This restriction allowed us to get a good idea of how planes fly and to make calculations as we did in Example 15.10. However, air is actually compressible, and the airflow around a wing is not ideal. The details of a plane's design must be examined in a wind tunnel and are beyond the scope of this book.

SUMMARY

Underlying Principles

Newton's second law is used to derive Archimedes's principle, and the work–energy theorem is used to derive Bernoulli's equation.

Major Concepts

1. An **ideal fluid** is incompressible.
2. The molecules in a **static fluid** move, but overall the fluid does not flow.
3. **Pressure** is a scalar given by

$$P = \frac{F}{A} \quad (15.1)$$

and pressure as a function of depth is

$$P = \rho g y + P_0 \quad (15.6)$$

4. The buoyant force is the net upward force exerted by a static fluid. According to **Archimedes's principle**, the magnitude of the buoyant force is equal to the weight of the fluid displaced by the object.
5. According to **Pascal's principle**, a *change* in the pressure at one point in an incompressible fluid appears undiminished at all points in the fluid and on the walls of its container.
6. The **ideal fluid flow** of an incompressible fluid is *steady* and *nonviscous*. In **steady flow**, also known as **laminar flow**, the fluid's velocity at any fixed point is a constant. In **nonviscous flow,** there is no resistance.
7. The **continuity equation** is

$$A_1 v_1 = A_2 v_2 \quad (15.21)$$

The continuity equation says that in ideal fluid flow the volume flow rate ($R = Av$) is uniform.

8. **Bernoulli's equation** is a statement of the work–energy theorem for ideal fluid flow:

$$P_1 + \tfrac{1}{2}\rho v_1^2 + \rho g y_1 = P_2 + \tfrac{1}{2}\rho v_2^2 + \rho g y_2 \quad (15.30)$$

Special Cases

A **fluid** is a substance that can flow. Therefore, liquids and gases are fluids, but solids are not.

PROBLEM-SOLVING STRATEGY Bernoulli's Equation

INTERPRET and ANTICIPATE

1. Draw a simple **sketch**, including your **coordinate system**.
2. **Pick region 1 and region 2** and indicate these regions on your sketch.

SOLVE

3. Write **Bernoulli's equation.**
4. **Compare the parameters** P, v, and y in the two regions. Make approximations if appropriate.
5. If necessary, use the **continuity equation** (Eq. 15.21). (Keep in mind that the speed in the large area is approximated as zero.)
6. **Do algebra** to solve for the required quantity.

PROBLEMS AND QUESTIONS

A = algebraic C = conceptual E = estimation G = graphical N = numerical

15-2 Static Fluid on the Earth

1. An alpha particle is the nucleus of a helium atom with two protons ($m_p = 1.673 \times 10^{-27}$ kg) and two neutrons ($m_n = 1.675 \times 10^{-27}$ kg) tightly bound together. Each of the particles has a radius of about 1.00 fm (femtometer).
   a. N What is the density of an alpha particle's nucleus?
   b. C How does this density compare with that of osmium, the densest natural element with a density of 2.26×10^4 kg/m^3? What does this comparison say about the space between nuclei in liquids and solids?
2. E What is the approximate weight of the air in your physics classroom?
3. N Dry air is primarily composed of nitrogen. In a classroom demonstration, a physics instructor pours 2.0 L of liquid nitrogen into a beaker. After the nitrogen evaporates, how much volume does it occupy if its density is equal to that of the dry air at sea level? Liquid nitrogen has a density of 808 kg/m^3.
4. C Why is the Earth's atmosphere denser near sea level than it is at a high altitude? Be sure to explain why the atmosphere's density is not uniform and why the air isn't all in contact with the Earth's surface.

15-3 Pressure

5. N Crater Lake in Oregon is the deepest lake in the United States, with a maximum depth of 655 m. The density of freshwater is 1.00×10^3 kg/m^3. **a.** If the air pressure above the lake, which has a surface altitude of 1800 m, is 8.00×10^4 Pa, what is the absolute pressure at the bottom of the lake? **b.** What is the force of the lake's water on each window of a submarine near the lake bottom if each of its circular windows has an area of 100 cm^2?
6. C At a crowded party, a woman wearing flat shoes accidentally steps on your foot. Now imagine that the same woman is wearing high-heeled shoes when she steps on your foot. Which hurts more? Why?

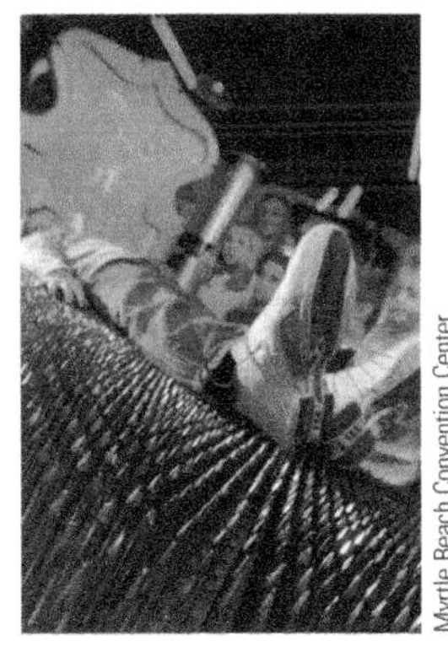

FIGURE P15.7

7. C A child lies on a bed of nails (Fig. P15.7). Explain why he is unharmed.
8. N One study found that the dives of emperor penguins ranged from 45 m to 265 m below the surface of the ocean. **a.** Using the density of seawater, what is the range of pressures experienced by the penguins between these two depths? **b.** At what depth below the surface is the pressure 10 times the normal atmospheric pressure?
9. Exploring the deep ocean is a challenge due to the extremely high pressures encountered. The deepest part of the ocean is at Challenger Deep in the Mariana Trench, with a depth of 10,924 m.
   a. N What is the pressure at this depth?
   b. N What is the volume strain of water compressed at this pressure?
   c. C Comment on our assumption that water is incompressible.
10. N The dimensions of a room's floor are 6 m × 5 m, and its height is 4 m. What is the force on each wall of the room due to the air in the room?
11. N Suppose you are at the top of Mount Everest and you fill a water balloon. The air pressure at the top of Mount Everest is 58 kPa. **a.** What is the fractional change in the balloon's volume $\Delta V/V_i$ when you take it to sea level? **b.** If instead you take it 100 m below the surface of the ocean, what is the fractional change in its volume?
12. E In a Hollywood movie, a salty old sailor on a submarine wishes to frighten a newcomer. When the submarine is at sea level, the old sailor ties a rope to the walls, which are about 3.5 m apart, so the rope is taut. The rope is tied at the height of a man's shoulder. When the submarine is at depth, that rope is slack. Suppose the center of the rope may be pulled down to the sailor's waist while the ends remain at shoulder height. Assume the submarine is made of steel and estimate its depth. Comment on the validity of the old sailor's demonstration.
13. N Imagine a planet that has the same sea-level atmospheric pressure as the Earth, but whose gravitational acceleration is only 0.5 *g*. What is the pressure difference between the bottom and the top of a 2.10-m deep swimming pool of water on this planet?
14. N You may have had the experience of feeling your ears pop as you ascend or descend in the elevator of a tall building. Assume the density of air is constant, and find the difference in pressure from the bottom to the top of a 100-story building whose total height is roughly 333 m.

Problems 15 and 16 are paired.

15. N A 20.0-kg child sits on a four-legged stool. The radius of each of the stool's feet is 0.022 m. **a.** Ignoring the mass of the stool, what is the force that is exerted on the floor? **b.** What is the pressure applied to the floor by the stool?
16. A 20.0-kg child sits on a pillar with a circular base of radius 0.250 m.
    a. N Ignoring the mass of the pillar, what is the force that is exerted on the floor?
    b. N What is the pressure applied to the floor by the pillar?
    c. C Evaluate the answers for all parts of this problem and Problem 15 (if you solved it) and explain any similarities or differences you find.
17. N The dolphin tank at an amusement park is rectangular in shape with a length of 40.0 m, a width of 15.0 m, and a depth of 7.50 m. The tank is filled to the brim to provide maximum splash during dolphin shows. What is the total amount of force exerted by the water on **a.** the bottom of the tank, **b.** the longer wall of the tank, and **c.** the shorter wall of the tank?
18. N In December 1995, a probe from the NASA spacecraft *Galileo* entered Jupiter's atmosphere. It traveled about 150 km into the atmosphere, collecting data for about 1 hour before it was vaporized. This problem is based on that probe's mission. At the top of its flight, the atmospheric pressure on the probe was about 10^{-2} Pa. The pressure exerted on the probe at the bottom of its flight was about 2.2×10^6 Pa. Find the average density of Jupiter's atmosphere over this depth. Compare your answer to the average density of Jupiter, which is about 1300 kg/m³. *Hint*: Jupiter's gravitational field near the top of its atmosphere, $R \approx 70{,}000$ km, is not the same as the Earth's gravitational field near its surface.

15-4 Archimedes's Principle

19. N A block of an unknown material floats in water with 75% of it below the surface. What is the density of the material?
20. C A battleship is an incredibly heavy object—much heavier than, say, a small pebble—yet when thrown into water, a pebble sinks, but a battleship can float. Explain how the battleship can float but the stone sinks.
21. N A block of density 1250.0 kg/m³ floats in an unknown fluid with two-thirds of its volume below the surface. What is the density of the fluid?
22. N A spherical submersible 2.00 m in radius, armed with multiple cameras, descends under water in a region of the Atlantic Ocean known for shipwrecks and finds its first shipwreck at a depth of 1.75×10^3 m. Seawater has density 1.03×10^3 kg/m³, and the air pressure at the ocean's surface is 1.013×10^5 Pa. **a.** What is the absolute pressure at the depth of the shipwreck? **b.** What is the buoyant force on the submersible at the depth of the shipwreck?
23. N What fraction of an iceberg floating in the ocean is above sea level? Assume the density of the iceberg is 917 kg/m³.
24. A An object in a fluid has a density $\rho_{obj} = 1755$ kg/m³. It is submerged in a fluid of density ρ_f. Its acceleration is upward with a magnitude $g/3$. What is the density of the fluid? Ignore any drag force on the object.
25. N A hollow copper ($\rho_{Cu} = 8.92 \times 10^3$ kg/m³) spherical shell of mass $m = 0.950$ kg floats on water with its entire volume below the surface. **a.** What is the radius of the sphere? **b.** What is the thickness of the shell wall?
26. N When a wooden box is placed in a pail of water, it floats with 40% of its height above the waterline. When this box is then placed in a pail full of olive oil, it floats with 30% of its height above the oil. What is the density of **a.** olive oil and **b.** the box?
27. N You have probably noticed that carrying a person in a pool of water is much easier than carrying a person through air. To understand why, find the buoyant force exerted by air and by water on the person. Assume the average volume of a person is 0.45 m³, and that the person is submerged in air and water respectively.
28. C A straw is in a glass of juice. Peter puts his finger over the top of the straw and carefully removes the straw. Juice remains in the straw (Fig. P15.28). **a.** Explain why the juice remains in the straw. **b.** What happens if Peter removes his finger?

FIGURE P15.28 A. B.

29. An object of mass 3.0 kg and volume 0.0020 m^3 is placed into a fluid of density 1.33×10^3 kg/m^3. The fluid is at rest.
    a. N, C Will the object float or will it sink in the fluid?
    b. N If it floats, what fraction of volume is immersed in the fluid? If it sinks, find its acceleration, ignoring any drag force on the object.
30. N Imagine a planet that has an ocean with the same seawater density as that of the Earth, but whose gravitational acceleration is only $0.5g$. A 67.5-kg person is submerged in the planet's ocean. Assume the person's volume is the same as it is on the Earth, 0.57 m^3. **a.** What is the person's weight on the alien planet? **b.** What is the buoyant force exerted on the person? **c.** What is the person's acceleration, ignoring any drag forces on the person?

Problems 31 and 32 are paired.

31. N A ball with a volume of 0.65 m^3 is floating on the surface of a pool of water. (The density of water is 1.00×10^3 kg/m^3.) If 5.41% of the ball's volume is below the surface, what is the density of the ball?
32. N In Problem 31, what is the magnitude of the buoyant force on the ball?
33. N A rectangular block of Styrofoam 25.0 cm in length, 15.0 cm in width, and 12.0 cm in height is placed in a large tub of water. Assume the density of Styrofoam is 3.00×10^2 kg/m^3. **a.** What volume of the block is submerged? **b.** A copper block is now placed atop the Styrofoam block so that the top of the Styrofoam block is level with the surface of the water. What is the mass of the copper block?
34. In Example 1.3 (page 15), we estimated the density of a grape and found that it had a density of 900 kg/m^3. So, if we put a grape in water, we expect that it would float, but a simple kitchen experiment shows that the grape sinks. The grape will float, though, if 12.5 ml of salt is dissolved in 225 ml of water. The dry density of salt is 1233 kg/m^3. The five grapes shown in Figure P15.34A displace 12.5 ml of water.
    a. N Find the density of the salt water.
    b. N From these experimental data, calculate the density and mass of a grape.
    c. C In Example 1.4 (page 16), we found that a raisin is denser than a grape. Does Figure P15.34B support our previous conclusion? Explain.

A.

B.

Peter McGahey © Cengage Learning

FIGURE P15.34

15-5 Measuring Pressure

35. N Imagine a planet that has the same sea-level atmospheric pressure as on the Earth, but whose gravitational acceleration is only $0.5g$. What is the minimum required height of a mercury barometer on this planet?
36. C A manometer is shown in Figure P15.36. Rank the pressures at the five locations indicated from highest to lowest. Indicate equal pressures, if any.

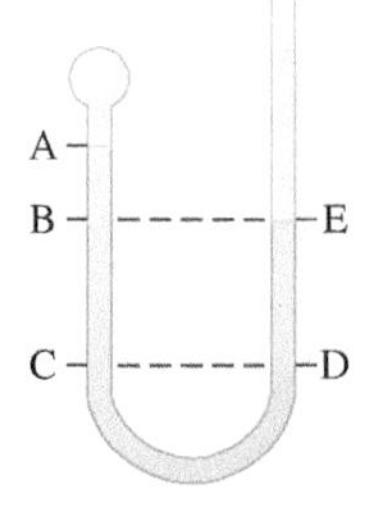

FIGURE P15.36

37. N The gauge pressure measured on a car's tire is 35 psi. What is the absolute pressure? Comment on your answer.
38. The gauge pressure of an *empty* scuba tank is 500 psi, whereas a full tank is at 3000 psi. The tank is connected to an open-tube manometer.
    a. N What is the height of the mercury if the tank is empty?
    b. N What is the height of the mercury if the tank is full?
    c. C **CHECK and THINK** step: Is it practical to use a manometer or a barometer on a scuba tank? Explain.
39. N Calculate the atmospheric pressure in pascals if the height of a mercury column in a barometer is 760 mm. The density of mercury is 13.6×10^3 kg/m^3.
40. N To allow a car to slow down or stop, hydraulic brakes transmit forces from a master cylinder to the brake pads through a fluid. Imagine this system as a tube filled with an incompressible fluid and a piston on each end. A force of 95.0 N is applied to a piston 2.65 cm in diameter on one end of the tube. **a.** What is the magnitude of the force that is exerted on the piston 5.15 cm in diameter on the other side? **b.** If the 2.65-cm piston is displaced by 1.00 cm, by how much is the 5.15-cm piston displaced?
41. N A hurricane cannot develop unless the height of the column of mercury in a barometer falls below 749 mm. During Hurricane Katrina in 2005, drops in the mercury level of barometers as large as 75.0 mm from the normal level of 760 mm were recorded. If normal atmospheric pressure is 1.013×10^5 Pa, what was the lowest atmospheric pressure recorded during this hurricane?
42. N In a hydraulic lever such as that seen in Example 15.4, the cross-sectional areas of the pistons are 0.250 m^2 and 1.00 m^2. What is the minimum force that should be applied to the narrower piston to support a car with a mass of 1.20×10^3 kg that is resting on the larger piston?
43. N A schematic of a homemade hydraulic jack is shown in Figure P15.43, with cylindrical tube ends of different radii, $r_1 = 0.0500$ m and $r_2 = 0.100$ m, respectively. **a.** If a downward force of $F_1 = 15.0$ N is applied to the end with the smaller radius, how much force F_2 would the jack exert on a car if it were sitting above the other end? **b.** If the smaller piston is pushed down a distance $d_1 = 0.200$ m, through what distance d_2 would the other end move?

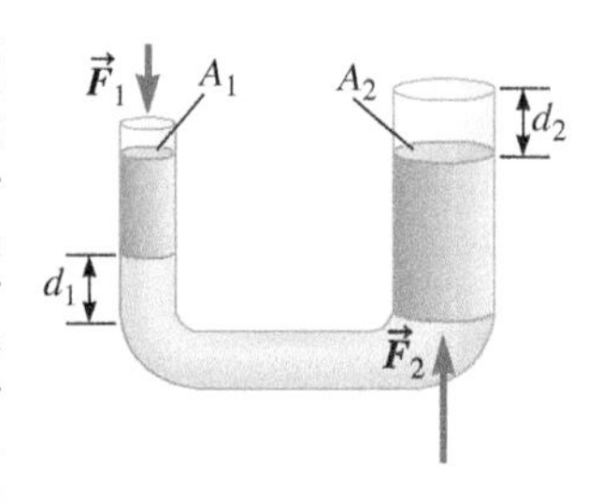

FIGURE P15.43

15-7 The Continuity Equation

44. N Water enters a smooth, horizontal tube with a speed of 2.0 m/s and emerges out of the tube with a speed of 8.0 m/s. Each end of the tube has a different cross-sectional radius. Find the ratio of the entrance radius to the exit radius.

Problems 45 and 46 are paired.

45. A fluid flows through a horizontal pipe (Fig. P15.45). The fluid enters the pipe with speed v_0 from the left.
    a. G Sketch streamlines for the fluid flow in this pipe and describe how the speed varies as it flows through the pipe.
    b. A The pipe has a circular cross section with a radius given by the expression $r(x) = B + Cx^2$, where the units of the constants B and C are such that r has units of centimeters. Derive an expression for the speed $v(x)$ of the fluid as a function of position for $0 < x < 1$.

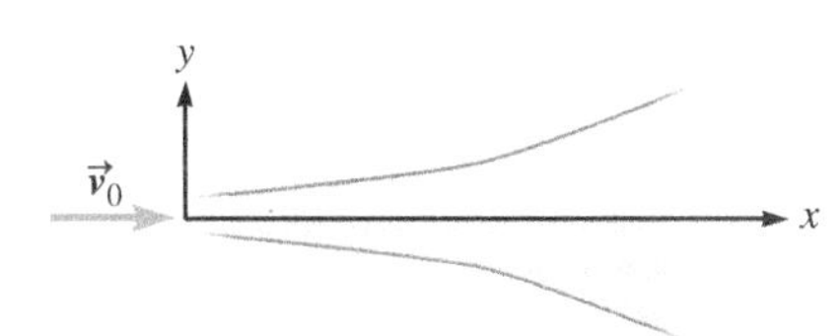

FIGURE P15.45 Problems 45 and 46.

46. Consider the pipe of variable radius described in Problem 45.

a. G If $v_0 = 1.35$ m/s, $B = 4.5$ cm, $C = 0.025$ cm^{-1}, and the pipe is 25 cm long, plot $v(x)$ versus x for $0 < x < 25$ cm.

b. N Water enters the pipe at $x = 0$ and leaves the pipe at $x = 25$ cm. Given the information in part (a), what is the volume flow rate for the water entering the pipe?

c. N What is the volume flow rate for water leaving the pipe?

47. N When connected to a standpipe, a fire hose 5.08 cm in diameter must deliver a minimum flow rate of 1.00×10^3 L/min. **a.** At this rate, with what speed does water leave the fire hose? **b.** A standard fire hose nozzle has an opening 0.950 cm in radius. What is the speed of the water leaving the fire hose with the nozzle attached?

48. A fluid flows through a horizontal pipe that widens, making a 45° angle with the y axis (Fig. P15.48). The thin part of the pipe has radius R, and the fluid's speed in the thin part of the pipe is v_0. The origin of the coordinate system is at the point where the pipe begins to widen. The pipe's cross section is circular.

FIGURE P15.48

a. A Find an expression for the speed $v(x)$ of the fluid as a function of position for $x > 0$.

b. G Plot your result: $v(x)$ versus x.

49. N Water is flowing through a pipe that has a constriction opening into a region with a wider cross-sectional area. If the pipe regions are cylindrical with radii of 0.10 m and 0.35 m, respectively, and the water is moving with a speed of 1.50 m/s in the wider section, what is the speed of the water in the constricted section?

50. The stream of water flowing from a faucet is wider near the top as shown in Figure P15.50. (It is easier to observe if the faucet does not have an aerator.)

a. C Explain this phenomenon.

b. A The cross-sectional area of the stream in a region near the top is A_1, and the cross sectional area lower region is A_2. If the distance between the two regions is h, find an expression for the speed of the water in top region in term of A_1, A_2, and h.

c. N If $A_1 = 1.4$ cm^2, $A_2 = 0.25$ cm^2, and $h = 6.5$ cm, what is the speed at the top of the flow?

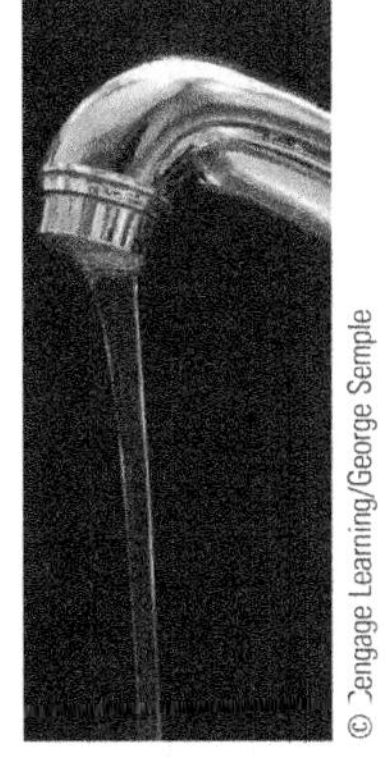

FIGURE P15.50

15-8 Bernoulli's Equation

51. N A water cannon 2.25 m long and with a cross-sectional area of 10.0 cm^2 is aimed straight up and ejects water out of the top at a speed of 9.5 m/s. It is fed by a pump at the bottom of the cannon, which has a cross-sectional area of 45.0 cm^2. What is the difference in pressure of the water between the pump and the length of the cannon? Assume the density of the water is 998.2 kg/m^3.

52. Figure P15.52 shows a *Venturi meter*, which may be used to measure the speed of a fluid. It consists of a Venturi tube through which the fluid moves and a manometer used to measure the pressure difference between regions 1 and 2. The fluid of density ρ_{tube} moves from left to right in the Venturi tube. Its speed in region 1 is v_1, and its speed in region 2 is v_2. The neck's cross-sectional area is A_2, and the cross-sectional area of the rest of the tube is A_1. The manometer contains a fluid of density ρ_{mano}.

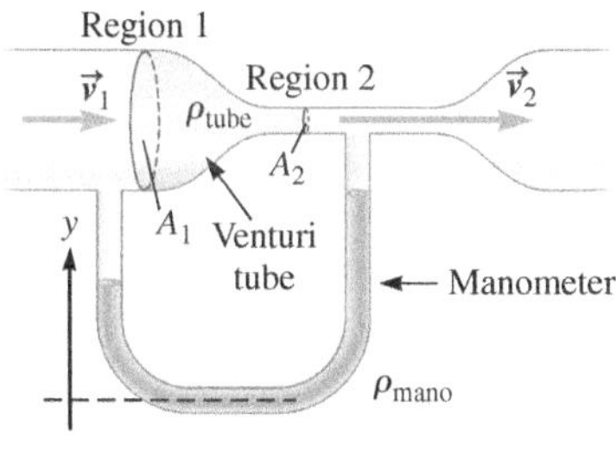

FIGURE P15.52

a. C Do you expect the fluid to be higher on the left side or the right side of the manometer?

b. A The speed v_2 of the fluid in the neck comes from measuring the difference between the heights $(y_R - y_L)$ of the fluid on the two sides of manometer. Derive an expression for v_2 in terms of $(y_R - y_L)$, A_1, A_2, ρ_{tube}, and ρ_{mano}.

53. N At a fraternity party, drinking straws have been joined together to make a giant straw that will be used to drink punch placed in a bowl on the ground from atop the fraternity house building. What is the maximum allowable height of the building if the partygoers are successful in drinking the punch? Assume the density of the punch is the same as the density of water.

54. N Liquid toxic waste with a density of 1752 kg/m^3 is flowing through a section of pipe with a radius of 0.312 m at a velocity of 1.64 m/s. **a.** What is the velocity of the waste after it goes through a constriction and enters a second section of pipe with a radius of 0.222 m? **b.** If the waste is under a pressure of 850,000 Pa in the first section of pipe, what is the pressure in the second (constricted) section of pipe?

55. N Water is flowing in the pipe shown in Figure P15.55, with the 8.00-cm diameter at point 1 tapering to 3.50 cm at point 2, located $y = 12.0$ cm below point 1. The water pressure at point 1 is 3.20×10^4 Pa and decreases by 50% at point 2. Assume steady, ideal flow. What is the speed of the water **a.** at point 1 and **b.** at point 2?

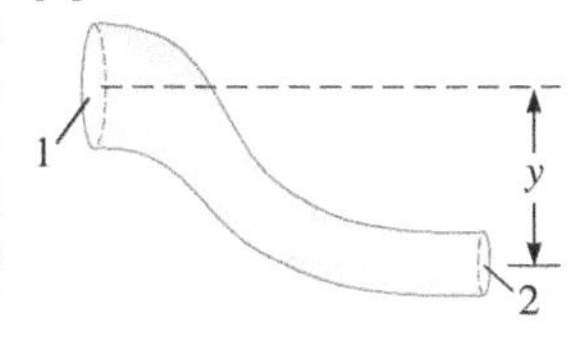

FIGURE P15.55

56. C A baseball is thrown through still air. In the reference frame of the ball, the air is moving around the ball. In Figure P15.56A, streamlines are shown around a ball that is *not* spinning. Figure P15.56B shows streamlines around a ball that is spinning counterclockwise as seen from above. **a.** Compare the airspeed in region L with the airspeed in region R in each case. **b.** Compare the air pressure in region L with the air pressure in region R in each case. **c.** Describe the motion of the ball in each case.

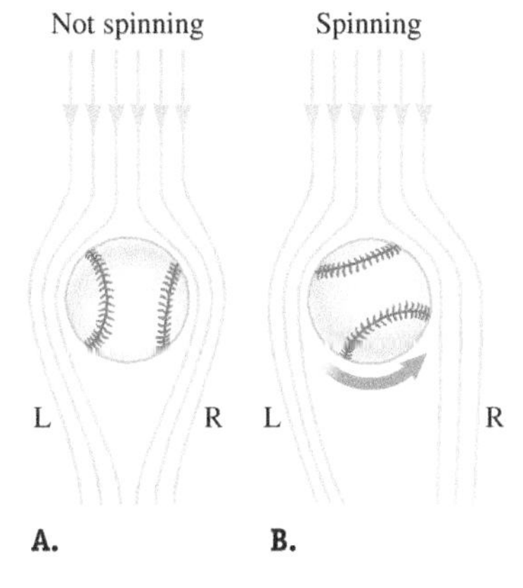

FIGURE P15.56

57. N Water flows through a pipe that gradually descends from a height of 6.78 m to the ground. Near the top, the cross-sectional area is 0.400 m^2, and the pipe gradually widens so that its area near the ground is 0.800 m^2. Water leaves the pipe at a speed of 16.8 m/s. What is the difference in the water pressure between the top and bottom of the pipe?

58. N Air flows horizontally with a speed of 108 km/h over a house that has a flat roof of area 20.0 m^2. Find the magnitude of the net force on the roof due to the air inside and outside the house. The density of air is 1.30 kg/m^3, and the thickness of the roof is negligible.

59. N A cylindrical tank of height 0.40 m is open at the top and has a diameter of 0.16 m. It is filled with water up to a height of 0.16 m. Find the time it takes to empty the tank through a hole of radius 5.0×10^{-3} m in its bottom.

General Problems

60. N A 75.0-kg man is sitting on a three-legged 15.0-kg barstool with circular legs of diameter 2.50 cm. What is the pressure exerted by each leg of the barstool on the floor?

61. N A rural water tank filled to capacity is subjected to subzero temperatures overnight, freezing the water inside the tank. Water expands by about 9.00% as it freezes, and the bulk modulus of ice is 2.00×10^9 N/m^2. What is the increase in pressure inside the tank?

62. N A spherical beach ball is 7.00 cm in radius and has a mass of 50.0 g. What is the force that must be exerted on the ball to keep it completely submerged in a swimming pool?

Problems 63–65 are grouped.

63. C CASE STUDY **Before airplanes, people were able to fly using hot-air balloons.** In June 1783, two brothers, Joseph and Jacques Montgolfier, demonstrated that a balloon full of hot air rises. They got the idea by observing that smoke rises and that a paper bag placed over a fire expanded and rose as well. They believed that such bags were filled with a unique gas, later called Montgolfier gas. Why do you suppose their hot-air balloon rose?

64. E CASE STUDY **Before airplanes, people were able to fly using hot-air balloons.** The Montgolfier brothers (Problem 63) experimented with hot-air balloons. The first flight of live animals (a sheep, a rooster, and a duck) was launched from Versailles. France, on September 19, 1783, and the first human passengers were launched about a month later, on October 15. These hot-air balloon demonstrations drew the attention of more than 100,000 spectators, including Louis XVI, Marie Antoinette, and Benjamin Franklin. The Montgolfier brothers believed that the balloons were full of a unique gas later called Montgolfier gas, but today we know that it was just hot air. Use information from the first human flight to estimate the density of Montgolfier gas. The balloon's capacity was 60,000 cubic feet. There may have been one or two adults on the ride. The balloon was tethered, and it gently rose 25 m, staying aloft for about 4 min.

65. N CASE STUDY **Before airplanes, people were able to fly using hot-air balloons.** The Montgolfiers' hot-air balloon (Problems 63 and 64) proved that human flight was possible, but their balloon was impractical. As the air cooled, the balloon descended, and it was dangerous to try to maintain a fire under a balloon made of cloth and paper. Jacques Alexandre Cesar Charles replaced the hot-air balloon with a silk balloon full of hydrogen gas. He launched his balloon from the Tuileries Gardens in Paris on December 1, 1783. This balloon carried a barometer and thermometer, making it the first scientific balloon flight. Assume Charles's balloon was a sphere 4.0 m in diameter full of hydrogen gas. Find the buoyant force exerted by the air on his balloon.

66. N Imagine a planet that has a sea with the same water density (1250 kg/m^3) as the Dead Sea on the Earth, but whose gravitational acceleration is only $0.5g$. A 67.5-kg person floats in the planet's sea. What fraction of the person is above the water's surface? Think about your answer by comparing it with what you would find if the person were on the Earth.

67. N At a processing plant, olive oil of density 875 kg/m^3 flows in a horizontal section of hose that constricts from a diameter of 3.00 cm to a diameter of 1.00 cm. Assume steady, ideal flow. **a.** What is the volume flow rate if the change in pressure between the two sections of hose is 5.00 kPa? **b.** What is the volume flow rate if the change in pressure between the two sections of hose is 12.5 kPa?

68. N A device designed for undersea exploration is a solid copper sphere ($B = 120.0 \times 10^9$ Pa) with a radius of 2.50 m. Find the change in volume of the copper sphere when the device is at a depth of 100.0 m below the surface of the ocean.

69. N A student is designing a piece of equipment that holds a camera and remains neutrally buoyant in fresh water. That is, the device will remain at the same height at which it is placed under water. This device has a mass of 25 kg and a volume of 0.025 m^3. If it is instead brought to the ocean and released under water, what would be the acceleration experienced by the device? Would it accelerate up or down?

70. N A vessel contains oil of density 8.00×10^2 kg/m^3 resting on top of mercury of density 13.6×10^3 kg/m^3. A solid, homogeneous sphere floats with half its volume immersed in the mercury and the other half in the oil. What is the density of the sphere?

71. N The density of air in the Earth's atmosphere decreases according to the function $\rho = \rho_0 e^{-h/h_0}$, where $\rho_0 = 1.20$ kg/m^3 is the density of air at sea level and h_0 is the scale height of the atmosphere, with an average value of 7640 m. What is the maximum payload that a balloon filled with 2.50×10^3 m^3 of helium ($\rho_{He} = 0.179$ kg/m^3) can lift to an altitude of 10.0 km?

72. N A manometer containing water with one end connected to a container of gas has a column height difference of 0.60 m (Fig. P15.72). If the atmospheric pressure on the right column is 1.01×10^5 Pa, find the absolute pressure of the gas in the container. The density of water is 1.0×10^3 kg/m^3.

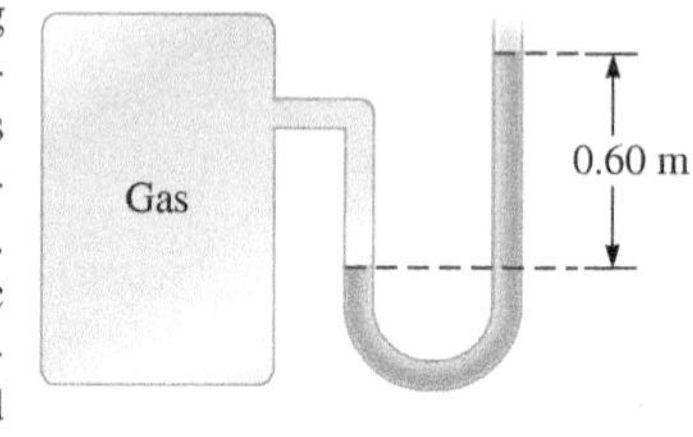

FIGURE P15.72

73. N In the novel *20,000 Leagues Under the Sea*, Jules Verne's *Nautilus* dove to the bottom of the sea by taking on seawater ($\rho_{seawater} = 1.03 \times 10^3$ kg/m^3) into three floodable tanks. Captain Nemo wishes to dive at a constant speed of 2.00 m/s. Model the submarine as a cylinder 70.0 m in length and 8.00 m in diameter. If the resistive force on the 1.50×10^6 kg submarine is 1.20×10^5 N in the upward direction, what is the mass of seawater the *Nautilus* must take on for this descent?

Problems 74 and 75 are paired.

74. Ancient Romans built long aqueducts to bring water from mountain springs into their cities. In imperial times, when Rome had more than one million inhabitants, the aqueducts delivered more than 1 m^3 of water per person daily.

a. E Place a lower limit on the volume rate of flow into Rome during this time.

b. E Roughly how much water do you consume each day?

c. N Assuming your water usage is typical, what is the volume rate of flow into your city?

75. Ancient Romans built long aqueducts to bring water from mountain springs into their cities. The longest was 59 miles. These aqueducts passed through variable terrain, and the

Romans used several different types of structures to keep the water flowing. One of the most costly and difficult structures to build was the inverted siphon (Fig. P15.75), which was used to cross valleys. It was expensive because it required lead pipes, and the lead had to be imported. The pipes were run down one side of the valley and up the other side to a slightly lower elevation. A major problem facing the ancient engineers was that it was hard to make joints strong enough to withstand the great pressure. The inverted siphon in Alatri, Italy, is 3000 m long and 101 m deep. Assume the pipe's diameter is constant.

a. C Where do you expect the maximum pressure?

b. N Calculate the change in the water's pressure between the top and bottom of the siphon. Express your answer in pascals and atmospheres.

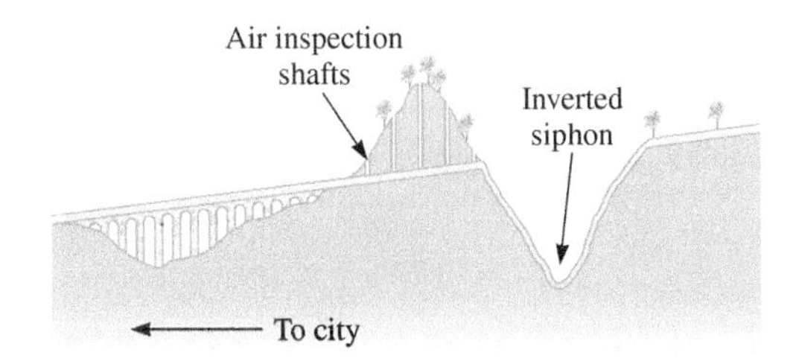

FIGURE P15.75

76. N A mother is administering cold medicine with the density of water using an oral syringe to her child. The medicine is pushed by a plunger from the barrel with a cross-sectional area of 1.25 cm^2 into the opening with a cross-sectional radius of 1.20 mm. The syringe is held horizontally, and it is a distance of 7.00 cm in the horizontal direction and 5.00 cm in the vertical direction from the child's mouth. Assume atmospheric pressure is 1.00 atm and neglect air resistance. **a.** What is the time of flight of the medicine from the syringe into the child's mouth? **b.** With what speed must the medicine leave the opening of the syringe to reach the child's mouth? **c.** What is the speed with which the mother must push the plunger for the medicine to reach the child's mouth? **d.** What is the pressure of the medicine in the opening of the syringe? **e.** What is the pressure in the barrel of the syringe?

77. N Strato-lab was a 1950s project in which large balloons were used to carry astronauts in training to the Earth's stratosphere. In one flight, a balloon 39.0 m in diameter was used to lift a total payload of 595 kg. Assume the volume of the balloon was filled with helium gas of density 0.179 kg/m^3. The air density was 1.20 kg/m^3. **a.** What was the buoyant force acting on the Strato-lab balloon? **b.** What was the net force on the Strato-lab? **c.** How much more mass could the Strato-lab have carried had it ascended at a constant velocity?

78. N CASE STUDY Shannon uses the example of a helium balloon to explain the buoyant force. Large helium "blimp" balloons are sometimes used as an advertisement (Fig. P15.78). The blimp balloon has a volume of 42.8 m^3, and the mass of the empty blimp is 13.6 kg. It is held down by either a large-link steel chain or a large-link aluminum chain. Each link of steel has a mass of 2.6 kg, and each link of aluminum has a mass of 0.87 kg. The chain rests on the ground but is not attached to it. The density of helium gas is 0.180 kg/m^3. **a.** How many links hang from the blimp if the steel chain is used? **b.** Compare your answer with the number of links that would hang if the aluminum chain were used instead.

FIGURE P15.78

79. Ships crossing the Atlantic bound for Canada often cross the Laurentian Abyss, which has a maximum depth of 6.00 km, corresponding to a pressure of 6.00×10^7 N/m^2. The density of seawater is 1.03×10^3 kg/m^3.

a. N What is the change in volume of 1.00 L of seawater if a ship sinks from the surface into the deepest point of the Laurentian Abyss?

b. N What is the density of seawater at the deepest point of the Laurentian Abyss?

c. C Is the assumption that water is an incompressible fluid justified?

80. N Carolyn, living on the fourth floor of an apartment building, opens a faucet tap 1.25 cm in radius. The tap is 12.0 m above the main water supply pipe to the building. The main water pipe is 4.00 cm in radius and has a volume flow rate of 1.50 L per second. **a.** What is the speed of the water exiting Carolyn's faucet? **b.** If no other taps are open in the building, what is the gauge pressure in the main water supply pipe?

81. A A uniform wooden board of length L and mass M is hinged at the top of a vertical wall of a container partially filled with a certain liquid (Fig. P15.81). (If there were no liquid in the container, the board would hang straight down.) Three-fifths of the length of the board is submerged in the liquid when the board is in equilibrium. Find the ratio of the densities of the liquid and the board.

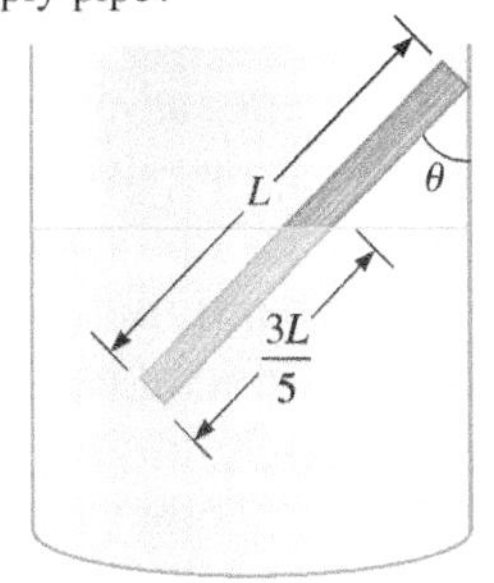

FIGURE P15.81

16 Oscillations

Key Questions

What is an oscillator?

How can we describe the kinematics, dynamics, and energy of an oscillator mathematically?

16-1 Picturing harmonic motion 451
16-2 Kinematic equations of simple harmonic motion 453
16-3 Connection with circular motion 456
16-4 Dynamics of simple harmonic motion 459
16-5 Special case: Object–spring oscillator 461
16-6 Special case: Simple pendulum 463
16-7 Special case: Physical pendulum 466
16-8 Special case: Torsion pendulum 468
16-9 Energy in simple harmonic motion 470
16-10 Damped harmonic motion 473
16-11 Driven oscillators 476

Underlying Principles

No new *fundamental* physical principles are introduced in this chapter.

Major Concepts

1. Simple harmonic motion (SHM) and simple harmonic oscillator (SHO)
2. Angular frequency
3. Amplitude
4. Phase and initial phase
5. Damped harmonic motion
6. Time constant
7. Critically damped, underdamped, and overdamped oscillators
8. Driven oscillator
9. Resonance

Special Cases

1. Object–spring oscillator
2. Simple pendulum
3. Physical pendulum
4. Torsion pendulum

We are surrounded by time-keeping devices, all of which are based on periodic motion. There are clocks on the classroom wall, next to our beds, and displayed on our phones. It doesn't matter whether you spent a lot of money on your wristwatch or found it in your box of cereal; the same physics makes it run. Something has to move around and around, back and forth, or in and out *periodically*, repeating at regular intervals.

The earliest timekeepers were based on the rotation of the Earth, but the Earth's rotational period varies slightly. (See Problem 12.66; the Earth's rotation is slowing down.) Also, you know from your own experience that we would like to measure time to the nearest minute, not day. In hopes of building a more accurate and precise device, inventors looked for objects that either oscillated back and forth or vibrated in and out at regular time intervals. Until the early part of the 20th century, the most accurate clocks were based on the oscillations of a pendulum (Fig. 16.1A). A pendulum clock can be designed so that the pendulum bob oscillates back and forth with a very regular period, but it must be vertically oriented to work. Wind-up wristwatches are based on the back-and-

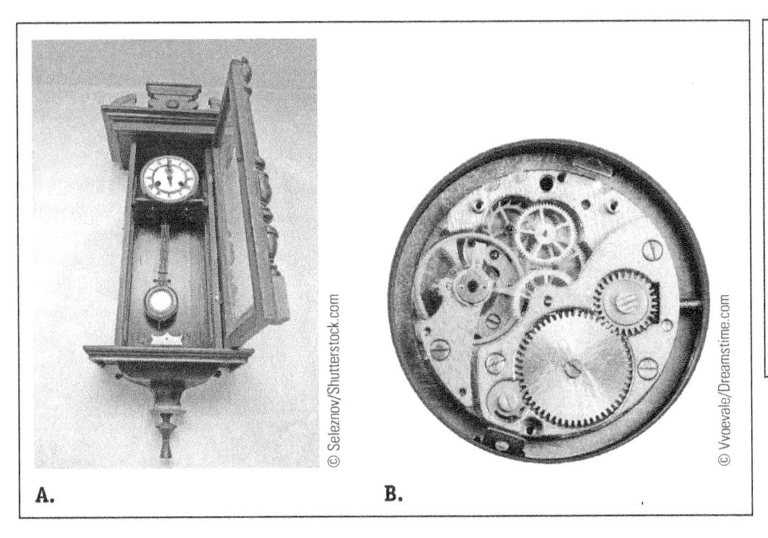

FIGURE 16.1 Timekeeping devices employ periodic motion. **A.** Pendulum clock. **B.** Wind-up watch.

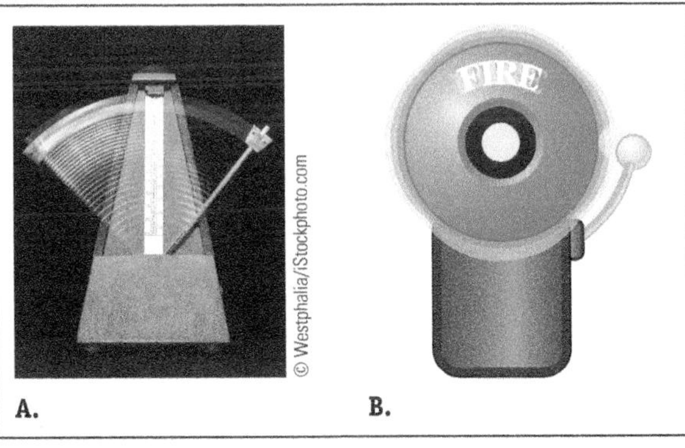

FIGURE 16.2 Oscillation versus vibration. **A.** The back-and-forth motion of a metronome is an oscillation. **B.** The ringing of a bell is a vibration.

forth rotation of a small wheel (Fig. 16.1B). Around the 1930s, quartz crystals were introduced into watches and clocks. When a quartz crystal is placed in an electrical circuit, the crystal expands and contracts with a regular period. In 1949, the first atomic clock, based on the vibration of ammonium, was built. Today the most accurate timepieces are based on the vibration of the cesium atom.

In this chapter, we study oscillations such as the motion of a pendulum bob or a particle attached to a spring. Oscillations are at the heart of wave motion (Chapters 17 and 18), common electrical circuits (Chapter 33), light (Chapters 35–38), and many aspects of modern physics (Chapters 40–43).

16-1 Picturing Harmonic Motion

Your beating heart, orbiting planets, and a child playing on a swing are all examples of objects whose motion repeats. Any motion that repeats at regular time intervals is **periodic**. We have studied two examples of periodic motion: circular motion (Chapter 4) and rotation (Chapter 12). In both of these cases, the object continues moving in just one angular direction; it does not reverse direction. We now consider motion that reverses direction periodically.

Many physicists use the words *oscillation* and *vibration* interchangeably to refer to repetitive motion. In this book, we make a slight distinction between the two terms. An **oscillation** is the back-and-forth motion of a particle or object over the same path; examples are the back-and-forth motion of a balance wheel in a watch and the motion of a metronome (Fig. 16.2A). A **vibration** is the expansion and contraction of an object or the back-and-forth motion of parts of the object. A particle such as the bob at the end of a pendulum may oscillate, but because a particle has no spatial extent, it cannot vibrate. Objects such as molded gelatin, a fire alarm bell (Fig. 16.2B), and a water molecule may change in shape or size and so can vibrate. An object's vibration may be modeled as the oscillation of particles making up the object.

The term ***harmonic motion*** describes periodic oscillations or vibrations. The word *harmonic* shows the important role periodic oscillations play in making sound and music. We've seen harmonic motion in previous chapters when studying the motion of a particle attached to a spring. In this chapter, we take a closer look at harmonic motion by considering an experiment involving a relatively simple example of a harmonic oscillator.

CASE STUDY Experiment, Part 1: Bouncy

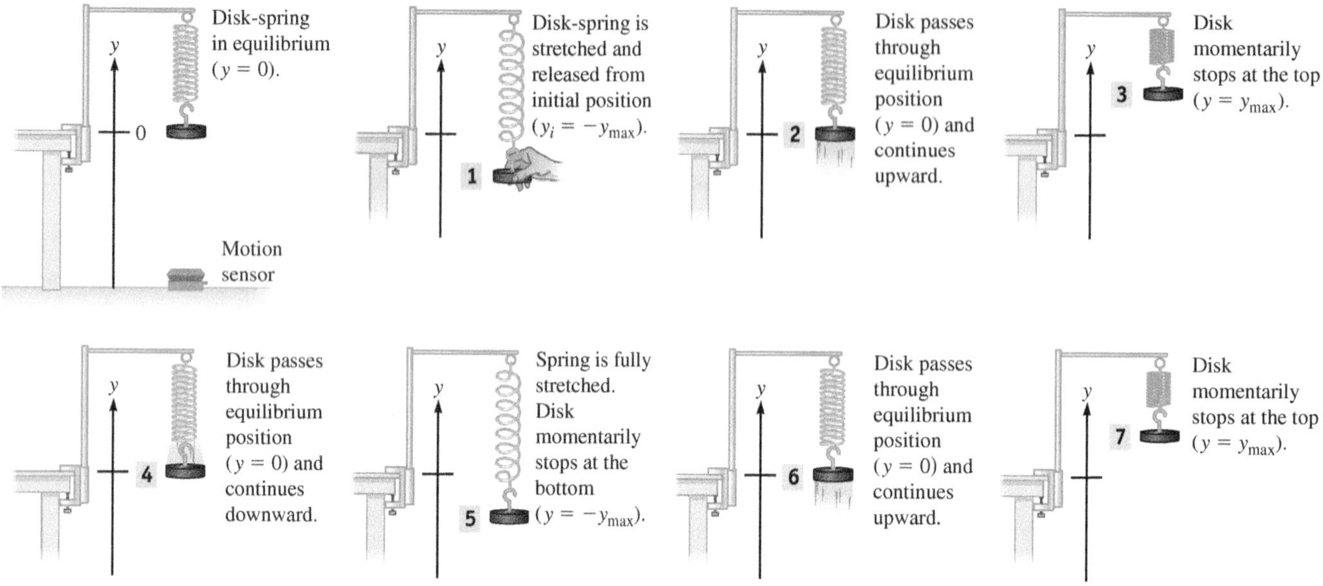

FIGURE 16.3 Crall and Whipple's experiment consists of a disk oscillating at the end of a vertical spring. A motion sensor on the floor measures the disk's position, velocity, and acceleration.

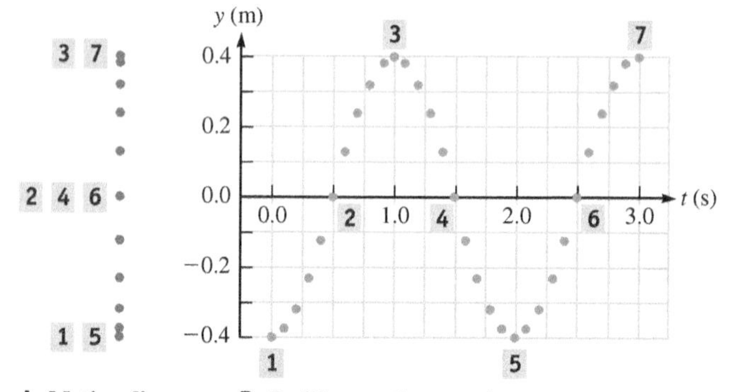

FIGURE 16.4 Data from the experiment in Figure 16.3.

Crall and Whipple hang a thick rubber disk of mass $m = 0.100$ kg from a vertical spring with spring constant $k = 0.987$ N/m. They choose an upward-pointing y axis with its origin set to the equilibrium position of the disk (Fig. 16.3). Whipple stretches the spring so that the disk is initially at $y_i = -0.40$ m and releases it from rest (1). The disk oscillates, first moving upward and passing through its equilibrium position (2) until it reaches its maximum height (3). It stops momentarily at this maximum height and then falls back down, again passing through its equilibrium position (4). The disk continues to fall until it reaches its initial position again (5). Now the motion repeats from the beginning: The disk moves upward, passes through its equilibrium position (6), continues to its maximum height (7), and so on.

Crall and Whipple made a motion diagram for the rubber disk, labeling the points 1 through 7 as described above (Fig. 16.4A). The points near the equilibrium position (2, 4, 6) are more spread out than the points near the top (3, 7) or near the bottom (1, 5). From Section 2-2, the disk must be moving faster when it is near equilibrium than when it is near either the top or bottom of its range.

To study the disk's kinematics, Crall and Whipple made a position-versus-time graph (Fig. 16.4B) that looks like a stretched-out version of the motion diagram (Section 2-4). Compare the labeled points 1 through 7 in Figures 16.3 and 16.4. Points 2, 4, and 6 are at $y = 0$ because the disk is at the equilibrium position at those times ($t_2 = 0.5$ s, $t_4 = 1.5$ s, $t_6 = 2.5$ s). Points 1 and 5 are at the bottom of the graph because at these times ($t_1 = 0$, $t_5 = 2.0$ s) the disk is at its lowest position ($y = -0.4$ m). Points 3 and 7 are at the top of the graph because at these times ($t_3 = 1.0$ s, $t_7 = 3.0$ s) the disk is at its highest position ($y = 0.4$ m). The disk's position-versus-time graph (Fig. 16.4B) is consistent with the motion diagram (Fig. 16.4A) and the sketch (Fig. 16.3).

Crall and Whipple then connected the points in the position-versus-time graph (Fig. 16.5A) with a smooth curve and computed the slopes of lines tangent to that curve to find velocities at various times (Fig. 16.5B). Compare the four tangent lines at 2, 3, 4, and 5 in the position-versus-time graph to the velocities at these points in Figure 16.5B. At points 3 and 5, the disk is momentarily at rest ($v = 0$). At point 2, the disk is moving quickly upward. Finally, at point 4, the disk is moving quickly downward. These results are exactly what we expected from the motion diagram (Fig. 16.4A) and the sketch (Fig. 16.3).

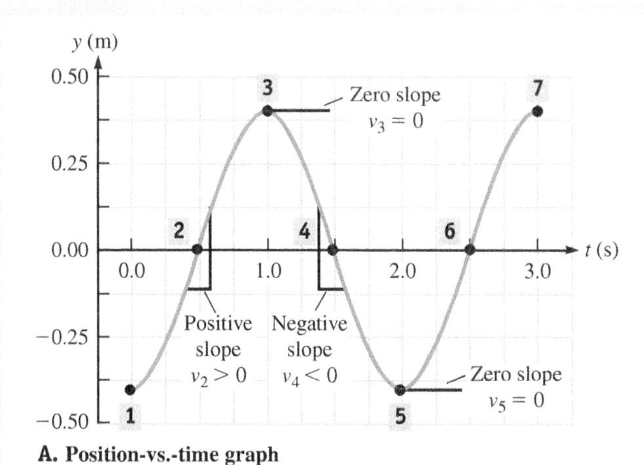

A. Position-vs.-time graph

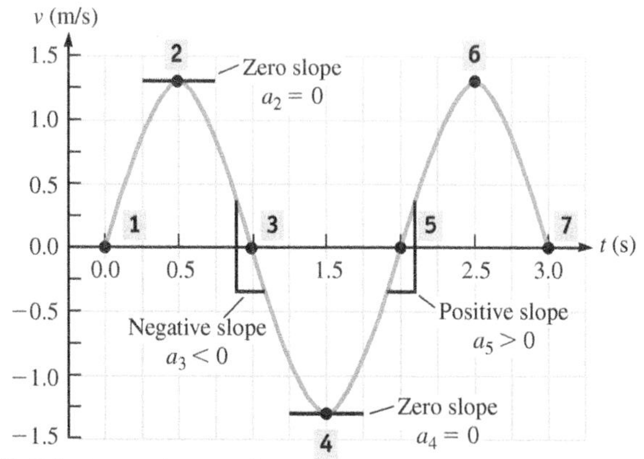

B. Velocity-vs.-time graph

FIGURE 16.5 Graphical analysis of the motion experiment data in Figure 16.3.

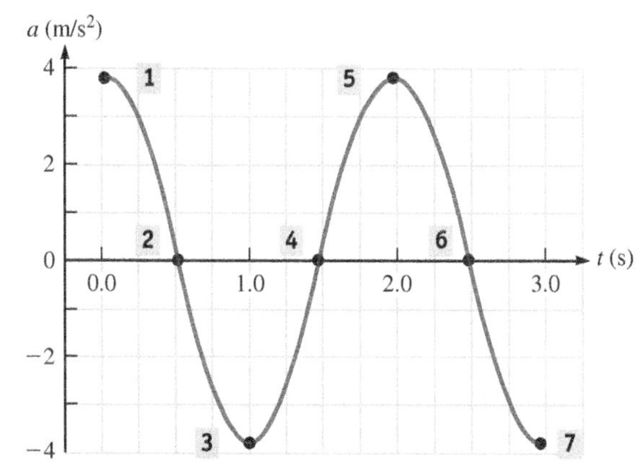

C. Acceleration-vs.-time graph

Crall and Whipple made an acceleration-versus-time graph (Fig. 16.5C) following a similar procedure. They connected the points in the velocity-versus-time graph with a smooth curve and then found the slope of that curve at various times. As before, compare the four tangent lines shown in the velocity-versus-time graph (Fig. 16.5B) with the accelerations in Figure 16.5C. At points 2 and 4, the acceleration is zero. At point 3, the acceleration is relatively great and pointing downward. At point 5, the acceleration is relatively great and pointing upward.

CONCEPT EXERCISE 16.1

CASE STUDY Disk Velocity and Acceleration

For each position listed, state the velocity and acceleration of the disk in Crall and Whipple's experiment (Figs. 16.3 through 16.5). There may be more than one possible answer for each given position.

a. $y = 0$ **b.** $y = 0.4$ m **c.** $y = -0.4$ m

16-2 Kinematic Equations of Simple Harmonic Motion

Crall and Whipple's rubber disk displays a relatively uncomplicated periodic oscillation known as **simple harmonic motion (SHM)**. In this section, we work through the kinematic equations that describe SHM and the mathematical signature of SHM. A **simple harmonic oscillator (SHO)** is a particle, object, or system displaying simple harmonic motion such as Crall and Whipple's disk.

SIMPLE HARMONIC MOTION (SHM); SIMPLE HARMONIC OSCILLATOR (SHO)

Major Concepts

Any single repetition of periodic motion is called a **cycle**. In Figure 16.3 or Figure 16.4, the disk's motion from 1 to 5, from 2 to 6, or from 3 to 7 is one cycle. Do not mistake half a cycle for a full cycle; for example, the motion from 1 to 3, from 2 to 4, or from 3 to 5 is *half* of a cycle.

The time to complete one cycle is called the **period**, symbolized by an uppercase T. From Figure 16.4B, the period of Crall and Whipple's oscillator is 2.0 s. The number of cycles per unit time is known as the **frequency**. The frequency f is the inverse of the period:

$$f = \frac{1}{T} \tag{16.1}$$

The SI unit of frequency is the *hertz* (Hz), and

$$1\,\text{Hz} = 1\ \text{cycle/s} = 1\ \text{s}^{-1}$$

In the mathematical description of harmonic motion, frequency f is often multiplied by 2π to define a new quantity known as the **angular frequency** ω such that

ANGULAR FREQUENCY

Major Concept

$$\omega \equiv 2\pi f = \frac{2\pi}{T} \tag{16.2}$$

Angular frequency is measured in radians per second. For Crall and Whipple's data, the angular frequency is

$$\omega = \frac{2\pi}{T} = \frac{2\pi}{2.0\ \text{s}} = 3.1\ \text{rad/s}$$

SHM has a single, constant angular frequency.

In Crall and Whipple's experiment, the disk's motion is one-dimensional along the y axis. Using their experimental data, we will write expressions for the y component of position, velocity, and acceleration. Similar equations could be written for oscillations along the x or z direction if needed.

Position Versus Time in Simple Harmonic Motion

The curve in the position-versus-time graph (Fig. 16.5A) is described by either a sine function or a cosine function. In fact, **the position versus time of any simple harmonic oscillator can be described by a single sine or cosine.** In this book, we have arbitrarily decided to use the cosine function for this time-varying position:

A similar equation can be written for SHM along the x or z axis.

$$y(t) = y_{max} \cos(\omega t + \varphi) \tag{16.3}$$

The maximum displacement of the oscillator *from the equilibrium position* (here, $y = 0$) is called the **amplitude** y_{max}. The amplitude is always positive, but the position may take on positive, zero, or negative values. In Figure 16.5A, the amplitude is $y_{max} = 0.40$ m. Do not mistake the maximum displacement for the amplitude. For example, the disk's displacement from 1 to 3 is 0.80 m, which is twice the amplitude. **In SHM, the amplitude is constant**, so the particle always moves between $-y_{max}$ and y_{max}.

AMPLITUDE; PHASE; INITIAL PHASE

Major Concepts

The argument $(\omega t + \varphi)$ of the cosine is called the **phase**. The **initial phase** φ is the phase when $t = 0$ as determined by the initial position $y_i \equiv y(0)$:

$$y(0) = y_{max} \cos(0 + \varphi)$$

$$y_i = y_{max} \cos \varphi$$

$$\varphi = \cos^{-1}\left(\frac{y_i}{y_{max}}\right) \tag{16.4}$$

Both the phase and the initial phase are expressed in radians. Usually, the initial phase is between $-\pi$ and π rad.

The sign of the initial phase cannot be determined from Equation 16.4 because the cosine function is symmetric about the y axis: $\cos \pi = \cos(-\pi)$. Here is a rule for choosing that sign: If the oscillator's position is increasing at $t = 0$—that is, if the oscillator is initially moving in the *positive* direction—you should choose the *negative* initial phase.

Of course, the converse is true: If the oscillator's position is decreasing at $t = 0$, you should choose a *positive* initial phase.

The initial phase for Crall and Whipple's data is

$$\varphi = \cos^{-1}\left(\frac{-0.4\ \text{m}}{0.4\ \text{m}}\right) = \cos^{-1}(-1) = \pm\pi$$

$$\varphi = -\pi$$

We chose the negative sign because the disk initially moves upward in the positive y direction. For Crall and Whipple's data, Equation 16.3 becomes

$$y(t) = (0.40\ \text{m}) \cos(3.1t - \pi) \tag{16.5}$$

Velocity Versus Time in Simple Harmonic Motion

We can find an expression for the simple harmonic oscillator's velocity by taking the time derivative of its position function (Eq. 16.3):

$$v_y(t) = \frac{dy(t)}{dt} = \frac{d[y_{max}\cos(\omega t + \varphi)]}{dt} = y_{max}\frac{d\cos(\omega t + \varphi)}{dt}$$

Use the chain rule (Appendix A):

$$v_y(t) = -y_{max}\omega\sin(\omega t + \varphi) \qquad (16.6)$$

A similar equation can be written for SHM along the *x* or *z* axis.

The term $y_{max}\omega$ is the amplitude of the velocity curve and also the maximum speed (Fig. 16.5B):

$$v_{max} = y_{max}\omega \qquad (16.7)$$

We can write Equation 16.6 in terms of the maximum speed:

$$v_y(t) = -v_{max}\sin(\omega t + \varphi) \qquad (16.8)$$

The maximum speed is always positive: $v_{max} > 0$. We can read v_{max} from Crall and Whipple's graph (Fig. 16.5B) or use Equation 16.7. Either way, we find

$$v_{max} = y_{max}\omega = (0.40\ \text{m})(3.1\ \text{rad/s}) \approx 1.3\ \text{m/s}$$

So, for Crall and Whipple's data, Equation 16.8 becomes

$$v_y(t) = -(1.3\ \text{m/s})\sin(3.1t - \pi)$$

Acceleration Versus Time in Simple Harmonic Motion

Finding an expression for the acceleration of a simple harmonic oscillator requires taking the time derivative of the velocity (Eq. 16.6):

$$a_y(t) = \frac{d[-y_{max}\omega\sin(\omega t + \varphi)]}{dt} = -y_{max}\omega\frac{d\sin(\omega t + \varphi)}{dt}$$

Again, use the chain rule:

$$a_y(t) = -y_{max}\omega^2\cos(\omega t + \varphi) \qquad (16.9)$$

The term $y_{max}\omega^2$ is the amplitude of the acceleration curve (Fig. 16.5C) and also the maximum acceleration:

$$a_{max} = y_{max}\omega^2 \qquad (16.10)$$

We can write Equation 16.9 in terms of the maximum acceleration:

$$a_y(t) = -a_{max}\cos(\omega t + \varphi) \qquad (16.11)$$

A similar equation can be written for SHM along the *x* or *z* axis.

The maximum acceleration is always positive: $a_{max} > 0$.

Compare parts A and C of Figure 16.5 and notice that the acceleration curve looks like an upside-down version of the position curve. In fact, the acceleration (Eq. 16.9) can be written in terms of the position (Eq. 16.3):

$$a_y(t) = -\omega^2[y_{max}\cos(\omega t + \varphi)]$$

$$a_y(t) = -\omega^2 y(t) \qquad (16.12)$$

Equation 16.12 is the chief feature or "hallmark" of simple harmonic motion: The periodic oscillator's acceleration is proportional to its position and is directed opposite to its displacement from equilibrium. In Section 16-4, we will derive this feature by considering the forces that act on the oscillator.

The acceleration of Crall and Whipple's oscillator is given by either Equation 16.9 or Equation 16.12.

$$a_y(t) = -(3.1\ \text{rad/s})^2[(0.40\ \text{m})\cos(3.1t - \pi)] = -(3.8\ \text{m/s}^2)\cos(3.1t - \pi) \qquad (16.13)$$

CONCEPT EXERCISE 16.2

CASE STUDY Choosing the Starting Time

Suppose Crall sets the initial time $t = 0$ to the moment when the disk is at point 2 in Figure 16.4 and Whipple sets the initial time to the moment when the disk is at point 3.

a. Write expressions for $y(t)$, $v_y(t)$, and $a_y(t)$ for each researcher's choice of the initial time.

b. Compare your expressions and comment on the significance of their differences and similarities.

CONCEPT EXERCISE 16.3

For each expression, identify the angular frequency ω, period T, initial phase φ and amplitude y_{max} of the oscillation. All values are in SI units.

a. $y(t) = 0.75 \cos (14.5t)$ **b.** $v_y(t) = -0.75 \sin (14.5t + \pi/2)$

c. $a_y(t) = 14.5 \cos (0.75t + \pi/2)$

16-3 Connection with Circular Motion

The equations that describe simple harmonic motion might remind you of uniform circular motion (UCM; Section 4-6). Understanding the close connection between SHM and UCM will help you digest the mathematics in the previous section.

Figure 16.6 shows a motion diagram of a ball moving in a vertical circle at a constant angular speed ω. The origin of our chosen coordinate system is at the center of the circle. The ball's initial position is at point 1, and it then moves counterclockwise. The ball completes one revolution as it passes through 1 again.

Uniform circular motion is periodic; that is, the motion repeats with a regular time interval. The ball, however, does not move back and forth along a single path. Instead, it continues to move in one direction along the same path. Therefore, the ball is not oscillating, so you probably wouldn't expect the SHM equations to describe the ball's kinematics.

Actually there are two simple harmonic oscillators hidden in the ball's circular motion. To see these two oscillators, think of lights above the circle showing the ball's shadow on the floor below and lights on the left showing the ball's shadow on the right wall. Figure 16.6 shows motion diagrams (in gray) for both the shadow on the floor and the one on the wall. The shadow on the floor moves back and forth along a line parallel to the x axis, whereas the shadow on the wall moves up and down along a line parallel to the y axis.

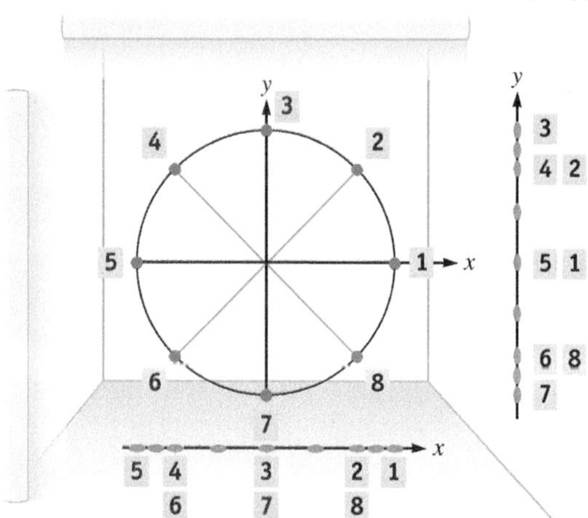

FIGURE 16.6 A ball travels in uniform circular motion. Lights on the ceiling show the ball's shadow on the floor, and lights on the left wall show the ball's shadow on the right wall.

The motion diagrams of both shadows look like Crall and Whipple's motion diagram (Fig. 16.4A). The points are close together near the ends and far apart near the middle, so the shadow moves more slowly near the ends of its path than in the middle. To confirm that the shadow's motion is simple harmonic, we must verify that the kinematic equations describe the motion we see.

Let's start with position. (You may be asked to find the velocity and acceleration for homework; see Problem 17.) Figure 16.7 shows the ball at some arbitrary time t at position $\vec{r}(t) = \vec{x}(t) + \vec{y}(t)$ or angular position $\theta(t) = \omega t$, where ω is the ball's angular speed. As usual, we can use trigonometry to find the x component of the ball's position:

$$x(t) = r \cos \omega t \tag{16.14}$$

Because the shadow on the floor is always directly below the ball, Equation 16.14 also gives this shadow's position along the x axis. The magnitude r is a constant equal to the radius of the circle. From Figure 16.7, that radius is also the maximum x or y value of:

$$r = x_{max} = y_{max} \tag{16.15}$$

We can use this result to rewrite the shadow's position (Eq. 16.14):

$$x(t) = x_{max} \cos \omega t \qquad (16.16)$$

Compare Equation 16.16 with the position of a simple harmonic oscillator, $x(t) = x_{max} \cos(\omega t + \varphi)$ (Eq. 16.3). Equation 16.16 gives the position of a simple harmonic oscillator along the x axis with the initial phase φ set to zero.

We used the same symbol ω for both the angular speed of the ball and the angular frequency of the shadow because these two quantities are equal in magnitude; the ball and the shadow have the same period. Be careful not to confuse angular frequency and angular speed, however. The distinction between these two quantities is important when studying pendulums (Section 16-6).

Now let's find the y component of the ball's position. From Figure 16.7, we find

$$y(t) = y_{max} \sin(\omega t) \qquad (16.17)$$

Use a trigonometric identity (Appendix A) to write Equation 16.17 in terms of a cosine:

$$y(t) = y_{max} \cos\left(\omega t - \frac{\pi}{2}\right) \qquad (16.18)$$

Comparing Equation 16.18 with Equation 16.3, we have obtained the position of a simple harmonic oscillator with an initial phase $\varphi = -\pi/2$.

We just found another way to think about UCM: as the combination of two simple harmonic oscillators along perpendicular axes. In Figure 16.6, one oscillator is along the x axis and the other along the y axis. These two oscillators have equal amplitudes $r = x_{max} = y_{max}$ and equal angular frequencies ω, but their initial phases differ by $\pi/2$. In other words, these oscillators are $\pi/2$ *out of phase*.

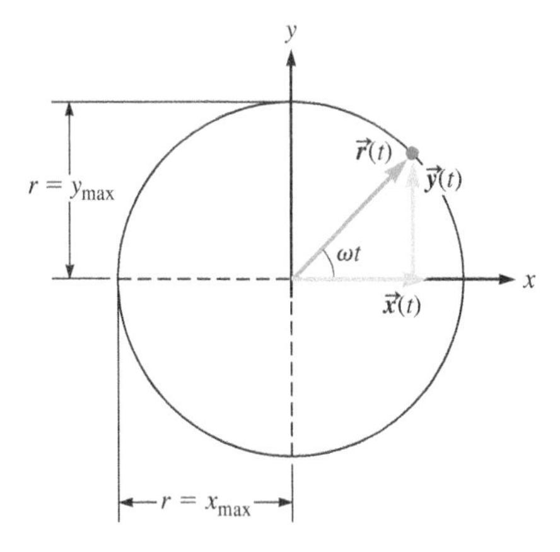

FIGURE 16.7 A ball in uniform circular motion. At an arbitrary time t, the ball's position is $\vec{r}(t)$, and its angular position is ωt.

EXAMPLE 16.1 Earth Satellite

Model the orbit of a satellite as UCM around the Earth with an orbital period of 90.0 min, ignoring the Earth's motion. Use the coordinate system shown in Figure 16.8. The Sun is at some distant position on the positive y axis, and the satellite moves clockwise as shown. Initially, the satellite is in opposition, on the opposite side of the Earth from the Sun. Find expressions for the x component of the satellite's position, velocity, and acceleration. *Hint*: Think of the shadow that would be cast by the satellite due to sunlight.

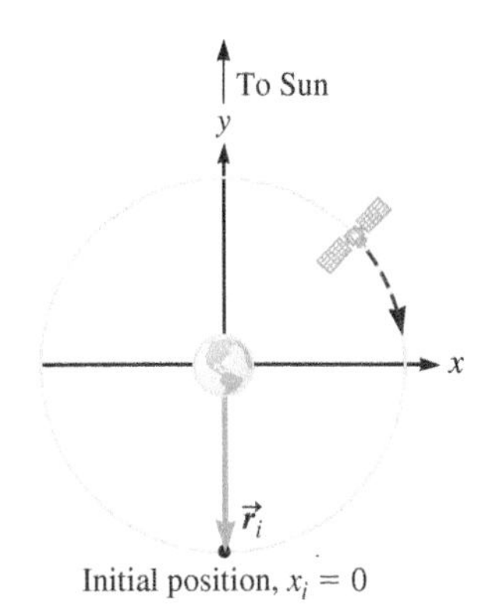

FIGURE 16.8

INTERPRET and ANTICIPATE

Once we have the position equation, it is relatively simple to find velocity and acceleration equations. The hint suggests that we compare Figure 16.8 with Figure 16.6. The satellite's solar shadow, like the ball's shadow on the floor, is along the x axis, but the initial positions and directions of motion differ in the two figures. The ball's initial position is $\vec{r}_i = x_{max}\hat{\imath}$, and it moves counterclockwise. The satellite's initial position is $\vec{r}_i = -y_{max}\hat{\jmath}$, and it moves clockwise. We expect our results to be like the x components of position, velocity, and acceleration for the ball, but with specific values for the satellite's amplitude x_{max}, angular speed ω, and initial phase φ.

SOLVE

A To find the x component of the satellite's position, start with Equation 16.3 written in terms of x. We need to find x_{max}, ω, and φ.

$$x(t) = x_{max} \cos(\omega t + \varphi) \qquad (16.3)$$

We can find the satellite's angular speed from its period using Equation 16.2. This angular speed is equal in magnitude to the angular frequency of the shadow's motion.

$$T = (90 \text{ min})(60 \text{ s/min}) = 5.40 \times 10^3 \text{ s}$$

$$\omega = \frac{2\pi}{T} = \frac{2\pi}{5.4 \times 10^3 \text{ s}}$$

$$\omega = 1.16 \times 10^{-3} \text{ rad/s}$$

Example continues on page 458 ▶

Use Kepler's third law (Eq. 7.5) to find x_{max}. As in the case of the ball in Figures 16.6 and 16.7, $r = x_{max}$.

$$T^2 = \left(\frac{4\pi^2}{GM_\oplus}\right)x^3_{max} \tag{7.5}$$

$$x_{max} = \left(\frac{GM_\oplus T^2}{4\pi^2}\right)^{1/3}$$

$$x_{max} = \left[\frac{(6.67 \times 10^{-11}\ \text{N}\cdot\text{m}^2/\text{kg}^2)(5.98 \times 10^{24}\ \text{kg})(5.40 \times 10^3\ \text{s})^2}{4\pi^2}\right]^{1/3}$$

$$x_{max} = 6.65 \times 10^6\ \text{m}$$

The initial phase comes from Equation 16.4, using $x_i = 0$. As the satellite moves clockwise from its initial position, x decreases, so we should choose the positive sign.

$$\varphi = \cos^{-1}\left(\frac{x_i}{x_{max}}\right) = \cos^{-1} 0 = \pm\frac{\pi}{2}$$

$$\varphi = \frac{\pi}{2}$$

Substitute the results for x_{max}, ω, and φ into Equation 16.3.

$$x(t) = (6.65 \times 10^6\ \text{m}) \cos\left(1.16 \times 10^{-3} t + \frac{\pi}{2}\right) \tag{1}$$

CHECK and THINK

Equation (1) has the form we expected. It is the same equation we would find for a satellite whose initial position is on the same side as the Sun, $\vec{r}_i = +y_{max}\hat{\jmath}$, and whose orbit is counterclockwise.

INTERPRET and ANTICIPATE

There are at least two different ways to find (B) velocity and (C) acceleration. We could take the time derivative of position (Eq. 1) to find velocity and then take the time derivative of velocity to find acceleration, or we could use Equation 16.6 for velocity and Equation 16.12 for acceleration. Let's use the second approach here.

SOLVE

B Write Equation 16.6 for velocity but in the x direction and substitute values found above.

$$v_x(t) = -x_{max}\omega \sin(\omega t + \varphi) \tag{16.6}$$

$$v_x(t) = -(6.65 \times 10^6\ \text{m})(1.16 \times 10^{-3}\ \text{rad/s}) \sin\left(1.16 \times 10^{-3} t + \frac{\pi}{2}\right)$$

$$v_x(t) = -(7.71 \times 10^3\ \text{m/s}) \sin\left(1.16 \times 10^{-3} t + \frac{\pi}{2}\right)$$

C Write Equation (16.12) for acceleration but in the x direction and substitute values found above.

$$a_x(t) = -\omega^2 x(t)$$

$$a_x(t) = -(1.16 \times 10^{-3}\ \text{rad/s})^2(6.65 \times 10^6\ \text{m}) \cos\left(1.16 \times 10^{-3} t + \frac{\pi}{2}\right)$$

$$a_x(t) = -(8.95\ \text{m/s}^2) \cos\left(1.16 \times 10^{-3} t + \frac{\pi}{2}\right)$$

CHECK and THINK

The solutions for velocity and acceleration have the form we expected, and we can further check the acceleration equation. The maximum acceleration of the shadow along x should equal the centripetal acceleration of the satellite, $a_{max} = a_c = 8.95\ \text{m/s}^2$. In this case, the centripetal acceleration should equal the acceleration due to gravity at the satellite's position, $a_{max} = a_c = g(x_{max})$, where $g(x_{max})$ is given by Equation 7.11. Aside from a little rounding error, the numbers are in agreement.

$$g(x_{max}) = \frac{GM_\oplus}{x^2_{max}} \tag{7.11}$$

$$g(x_{max}) = \frac{(6.67 \times 10^{-11}\ \text{N}\cdot\text{m}^2/\text{kg}^2)(5.98 \times 10^{24}\ \text{kg})}{(6.65 \times 10^6\ \text{m})^2}$$

$$g(x_{max}) = 9.02\ \text{m/s}^2 \approx a_{max}$$

16-4 Dynamics of Simple Harmonic Motion

In Sections 16-1 and 16-2, we *described* the motion of the disk in Crall and Whipple's experiment. In this section, we examine the *cause* or dynamics of that motion. In particular, we find two ways to express the net force on the disk.

First, according to Newton's second law, all we need to do is multiply the mass ($m = 0.100$ kg, page 452) by its acceleration:

$$\sum F_y(t) = ma_y(t) = -ma_{max} \cos(\omega t + \varphi) \tag{16.19}$$

For Crall and Whipple's data (Eq. 16.13), we find

$$\sum F_y(t) = (0.100 \text{ kg})(-3.8 \text{ m/s}^2) \cos(3.1t - \pi)$$

$$\sum F_y(t) = -(0.38 \text{ N}) \cos(3.1t - \pi) \tag{16.20}$$

So, the net force depends on time. A graph of force versus time looks similar to the acceleration-versus-time graph with each value on the vertical axis multiplied by mass (Fig. 16.9).

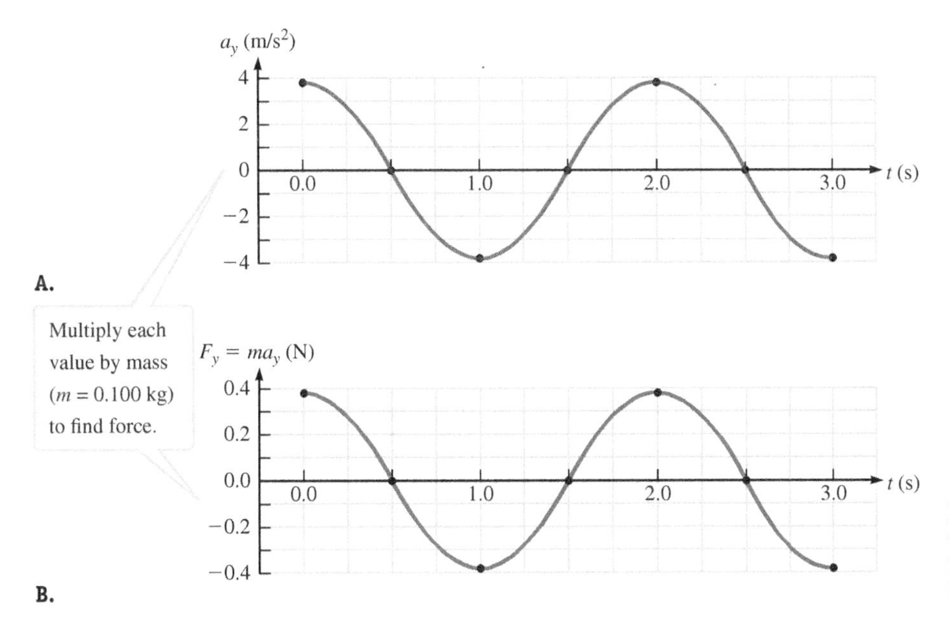

FIGURE 16.9 A. Acceleration data for Crall and Whipple's experiment, repeated from Figure 16.5C. **B.** To find force versus time, multiply by mass.

Equation 16.19 is one expression for the net force on a simple harmonic oscillator; we found it without considering the specific forces—gravity and the spring force—exerted on the disk. Let's consider these forces to find a second expression for the net force by taking a closer look at how Crall and Whipple set up their experiment. First, they hung a spring of length ℓ from its support (Fig. 16.10A). Next, they hung the disk from the spring (Fig. 16.10B). The spring then stretched to its new length $\ell + \Delta\ell$. They set the origin ($y = 0$) of the coordinate system to the disk's position. The disk was at rest and two forces—gravity and the spring force obeying Hooke's law—were exerted on it as shown in the free-body diagram in Figure 16.10B. Because the acceleration was zero when the disk was at rest, these forces were equal in magnitude at that time:

$$F_H = F_g$$

$$k\,\Delta\ell = mg \tag{16.21}$$

Finally, Whipple set the disk into motion. Figure 16.10C shows it at some arbitrary time when the disk is at $\vec{y}(t)$, measured from the origin at $y = 0$. Two forces are exerted on the disk: the downward force of gravity and the upward force exerted by the spring. According to Hooke's law, the upward force exerted by the spring

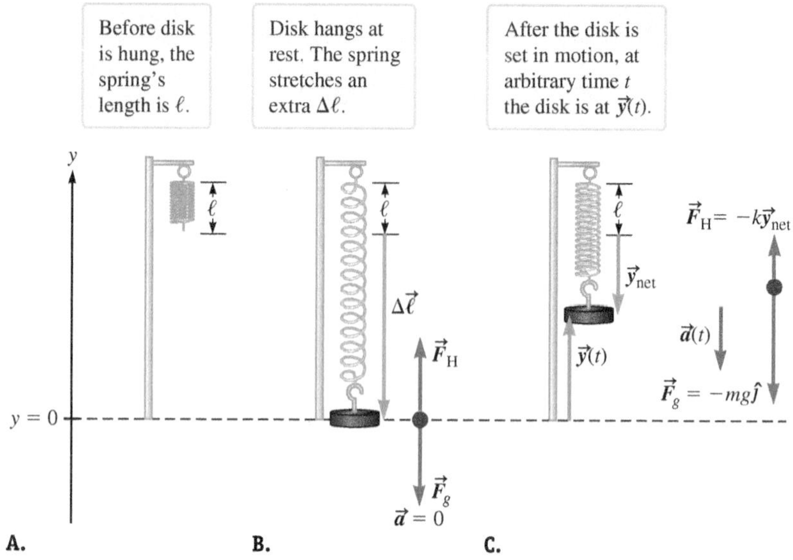

FIGURE 16.10 **A.** Vertical spring of length ℓ. **B.** Crall and Whipple then hung a disk from the spring, stretching it by an amount $\Delta\ell$. Two forces are exerted on the disk: the downward force of gravity and the upward force exerted by the spring. With the disk at rest, $F_H = F_g$. **C.** The disk oscillates up and down. At an arbitrary time t, the disk is at $y(t)$.

depends on its net stretch, which is $\vec{y}_{net}(t) = -[\Delta\ell - y(t)]\hat{\jmath}$ in the coordinate system shown. Therefore,

$$\vec{F}_H = -k\vec{y}_{net}$$

$$\vec{F}_H = k[\Delta\ell - y(t)]\hat{\jmath} \tag{16.22}$$

The total force exerted on the disk is

$$\sum \vec{F} = \vec{F}_H + \vec{F}_g = m\vec{a}$$

$$\sum \vec{F} = k[\Delta\ell - y(t)]\hat{\jmath} - mg\hat{\jmath} = ma_y(t)\hat{\jmath}$$

Regrouping terms,

$$\sum \vec{F} = (k\,\Delta\ell - mg)\hat{\jmath} - ky(t)\hat{\jmath} = ma_y(t)\hat{\jmath}$$

Simplify by substituting $k\,\Delta\ell = mg$ (Eq. 16.21):

$$\sum \vec{F} = (mg - mg)\hat{\jmath} - ky(t)\hat{\jmath} = ma_y(t)\hat{\jmath}$$

$$\sum \vec{F} = -ky(t)\hat{\jmath} = ma_y(t)\hat{\jmath} \tag{16.23}$$

So, another way to express the net force on a simple harmonic oscillator is in terms of its position $y(t)$:

$$\sum F_y = -ky(t) = -ky_{max}\cos(\omega t + \varphi) \tag{16.24}$$

where we have substituted Equation 16.3, $y(t) = y_{max}\cos(\omega t + \varphi)$.

Let's apply Equation 16.24 to Crall and Whipple's data by multiplying Equation 16.5 for the disk's position, $y(t) = (0.40\text{ m})\cos(3.1t - \pi)$, by minus one times the spring constant $k = 0.987$ N/m (case study, page 452):

$$\sum F_y = -(0.987\text{ N/m})(0.40\text{ m})\cos(3.1t - \pi)$$

$$\sum F_y = -(0.39\text{ N})\cos(3.1t - \pi)$$

which (except for a slight error) is what we found before (Eq. 16.20). So, we have just found a second way to make a force-versus-time graph: Start with a position-versus-time graph and multiply the values on the vertical axis by $-k$ (Fig. 16.11).

Moreover, by writing Equation 16.23 as

$$a_y(t) = -\frac{k}{m}y(t) \tag{16.25}$$

we gain a deeper insight into Equation 16.12, the hallmark of SHM $[a_y(t) = -\omega^2 y(t)]$. Equation 16.25 expresses the hallmark of SHM for the special case of an object–spring oscillator. The negative sign indicates that acceleration and position vectors point in opposite directions. By using Newton's laws as we did in this section, we

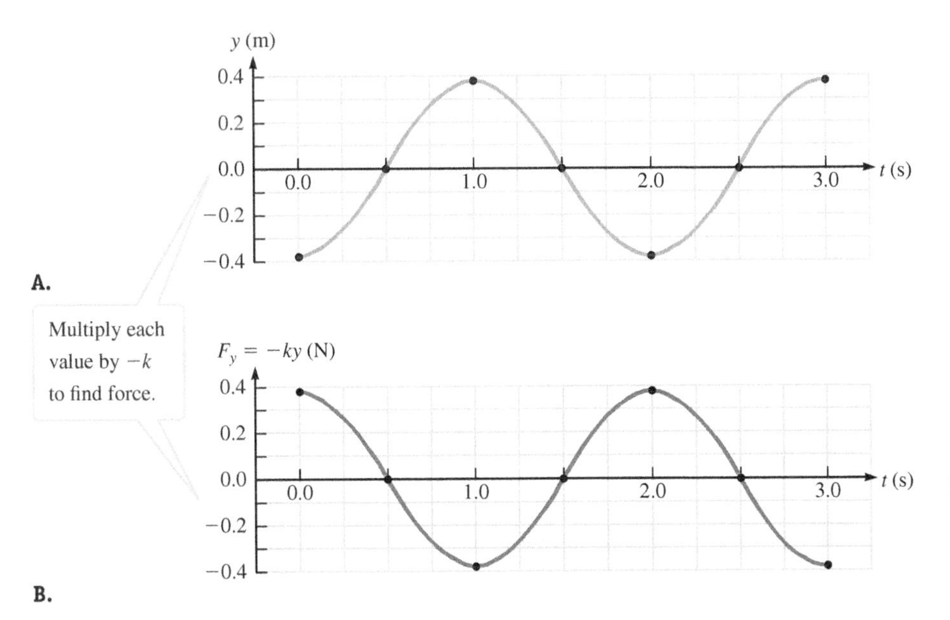

FIGURE 16.11 A. Crall and Whipple's position-versus-time graph. **B.** A force-versus-time graph is made by multiplying the position-versus-time graph by $-k$. Compare with Figure 16.9.

see why they are in opposite directions. The net force exerted on the oscillator is a restoring force (Eq. 16.24), which always points toward the equilibrium position ($y = 0$) in the opposite direction of the displacement.

16-5 Special Case: Object–Spring Oscillator

OBJECT–SPRING OSCILLATOR

Special Case

In this section, we continue to study the special case of a simple harmonic oscillator consisting of an object attached to a spring. We just found that for this specific oscillator, the hallmark of SHM is

$$a_y(t) = -\omega^2 y(t) = -\frac{k}{m}y(t)$$

where we have combined Equations 16.12 and 16.25. By canceling $-y(t)$, we find

$$\omega = \omega_s \equiv \sqrt{\frac{k}{m}} \tag{16.26}$$

Equation 16.26 gives the angular frequency of an oscillating object with mass m attached to a spring with spring constant k, so we have added the subscript s to denote the spring. The angular frequency is related to the frequency and period by Equation 16.2 ($\omega \equiv 2\pi f = 2\pi/T$) so once you know one of these parameters, you know all three. It might seem amazing, but **the frequency of SHM does not depend on the amplitude.** In other words, once Crall and Whipple chose their disk (mass m) and spring (spring constant k), the frequency was already set, and it did not matter how far Whipple initially displaced the disk.

The spring does not need to be vertical like the one Crall and Whipple used. Figure 16.12 shows a block attached to a horizontal spring that oscillates on a frictionless table. Three forces are exerted on the block: the normal force, gravity, and the spring force. The block does not accelerate in the y direction, so gravity is balanced by the normal force. The spring force accelerates the block in the x direction. When the block is at $x = 0$, the spring is relaxed. At some arbitrary time t, the block is at $\vec{x}(t)$. For $x(t) > 0$, the spring is stretched and exerts a force in the negative x direction, accelerating the block in that direction. From Hooke's law and Newton's second law,

$$\vec{F}_H(t) = m\vec{a}(t)$$
$$-k\vec{x}(t) = -kx(t)\hat{\imath} = ma_x(t)\hat{\imath}$$

$$a_x(t) = -\frac{k}{m}x(t) \tag{16.27}$$

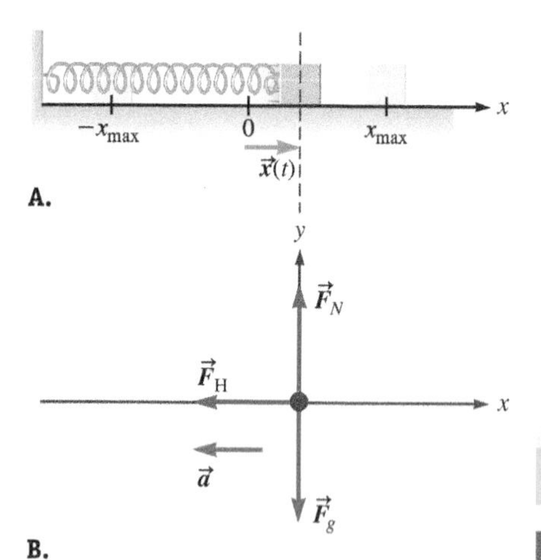

FIGURE 16.12 A. A block is connected to a horizontal spring. **B.** Free-body diagram for the block.

Equation 16.27 also looks like the hallmark of SHM (Eq. 16.12), so the block's angular frequency is given by Equation 16.26. As in the case of the vertical spring, the frequency does not depend on the amplitude of the oscillation.

We can use our derivation of Equation 16.26 to come up with a general procedure for finding the angular frequency of any simple harmonic oscillator. First, use Newton's laws to find an expression for the object's acceleration. Next, compare the object's acceleration with $a_y(t) = -\omega^2 y(t)$ (Eq. 16.12) to see if the acceleration obeys this hallmark of SHM. If it does, identify factors that determine ω, the angular frequency. We will use this procedure in the next three sections to find the angular frequency of several common oscillators.

CONCEPT EXERCISE 16.4

CASE STUDY Angular Frequency

Use Equation 16.26 to predict the angular frequency and period of Crall and Whipple's oscillator. How does your prediction compare with their actual data (Fig. 16.5)?

EXAMPLE 16.2 CASE STUDY A Stronger Spring

Crall and Whipple repeat their experiment using a stiffer spring of spring constant 65.3 N/m. If everything else is exactly the same, find new expressions for position, velocity, acceleration, and force.

INTERPRET and ANTICIPATE

Our results should be similar to the previous expressions for Crall and Whipple's data, but a stiffer spring means that the restoring force is greater, so we expect a greater acceleration and speed. The stiffer spring also means a higher angular frequency.

SOLVE

Equation 16.26 gives the new angular frequency.

$$\omega_s = \sqrt{\frac{k}{m}} = \sqrt{\frac{65.3\ \text{N/m}}{0.100\ \text{kg}}}$$

$$\omega_s = 25.6\ \text{rad/s}$$

The amplitude and initial phase are unchanged, so the position equation is the same as Equation 16.5 except for the new angular frequency.

$$y(t) = (0.40\ \text{m})\cos(25.6t - \pi)$$

The velocity comes from Equation 16.6.

$$v_y(t) = -y_{max}\omega \sin(\omega t + \varphi) \quad (16.6)$$

$$v_y(t) = -(0.40\ \text{m})(25.6\ \text{rad/s})\sin(25.6t - \pi)$$

$$v_y(t) = -(10.2\ \text{m/s})\sin(25.6t - \pi)$$

The acceleration comes from Equation 16.9.

$$a_y(t) = -y_{max}\omega^2 \cos(\omega t + \varphi) \quad (16.9)$$

$$a_y(t) = -(0.40\ \text{m})(25.6\ \text{rad/s})^2\cos(25.6t - \pi)$$

$$a_y(t) = -(262\ \text{m/s}^2)\cos(25.6t - \pi)$$

Finally, find the force by multiplying the acceleration by the mass.

$$F_y = ma_y = (0.100\ \text{kg})[-(262\ \text{m/s}^2)\cos(25.6t - \pi)]$$

$$F_y = -(26.2\ \text{N})\cos(25.6t - \pi)$$

CHECK and THINK

As expected, the maximum speed $v_{max} = 10.2$ m/s, the maximum acceleration $a_{max} = 262$ m/s^2, and the maximum force $F_{max} = 26.2$ N are all greater in the case of the stiffer spring than with the original spring. The amplitude y_{max} is unchanged, however.

EXAMPLE 16.3 Measuring Mass and Force

A block of unknown mass is attached to a horizontal spring ($k = 75$ N/m) as in Figure 16.12. The block oscillates on a frictionless table between $x = \pm 1.2$ m. The block's period is measured to be 2.7 s.

What is the mass of the block?

INTERPRET and ANTICIPATE

The mass is related to the period (and angular frequency) of the block.

SOLVE

Solve Equation 16.26 for mass.

$$\omega_s = \sqrt{\frac{k}{m}} \tag{16.26}$$

$$m = \frac{k}{\omega_s^2}$$

Use Equation 16.2 to write mass in terms of period.

$$T = \frac{2\pi}{\omega_s} \tag{16.2}$$

$$T^2 = \frac{4\pi^2}{\omega_s^2}$$

$$m = \frac{k}{4\pi^2}T^2 = \frac{75 \text{ N/m}}{4\pi^2}(2.7 \text{ s})^2 = 13.8 \text{ kg} = 14 \text{ kg}$$

CHECK and THINK

Verify for yourself that the results for mass have the expected dimensions. This example describes a way to find the mass of an object without using a gravitational field.

16-6 Special Case: Simple Pendulum

Today when we think of Galileo, we tend to focus on his work in astronomy. Galileo's initial fame, however, was based on a much more down-to-Earth discovery. When he was 20 years old, he used his pulse to time the period of a lamp swinging overhead in a cathedral. Galileo discovered that the period did *not* depend on the amplitude of the lamp's oscillation. The cathedral lamp is an example of a simple pendulum. A **simple pendulum** consists of a particle—known as the bob—hanging from a massless, unstretchable string or rod. The pendulum bob oscillates back and forth along a portion of a circular path. Of course, no real pendulum is such an idealized simple pendulum, but many real pendulums can be modeled in this way. Galileo's discovery had important practical significance. Because of its regular period, the pendulum was the basis of the most accurate clocks for nearly 300 years (Fig. 16.1A). Clock pendulums are usually approximated as simple pendulums.

Let's find the angular frequency of a simple pendulum. Figure 16.13 shows a simple pendulum consisting of a bob at the end of a string of length ℓ. The origin of the coordinate system is placed at the point where the string is attached to the support, so the bob's position $\vec{r}$ is measured from that point. As the bob swings along its circular arc, the magnitude of $\vec{r}$ is constant and equal to the length of the string, $r = \ell$. The x axis has been chosen to point straight down so that when the bob hangs vertically, its angular position is $\theta = 0$. When the bob is to the right of the vertical, its angular position is positive. Its angular position is negative to the left of the vertical.

At some arbitrary time t, the bob's angular position is $\theta(t)$ as shown in Figure 16.13A. Figure 16.13B shows a free-body diagram for the bob at that moment. Two forces are exerted on the bob: gravity and the tension force. In our model of a simple pendulum, the string is unstretchable. Therefore, the bob cannot move radially (along the radius of the circular arc), only tangentially. The forces in Figure 16.13B are broken into components that are either parallel (or antiparallel) or perpendicular

SIMPLE PENDULUM Special Case

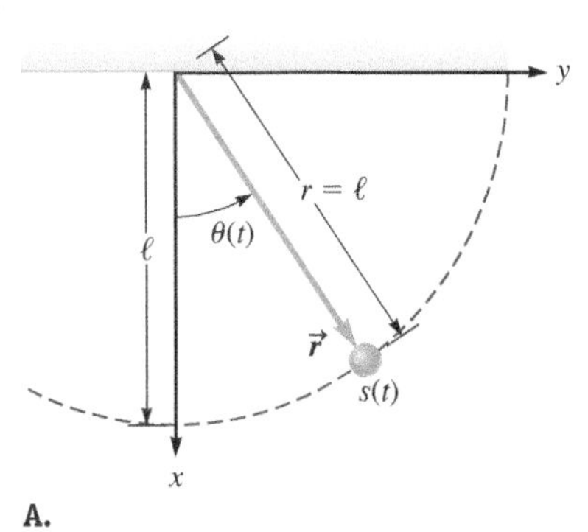

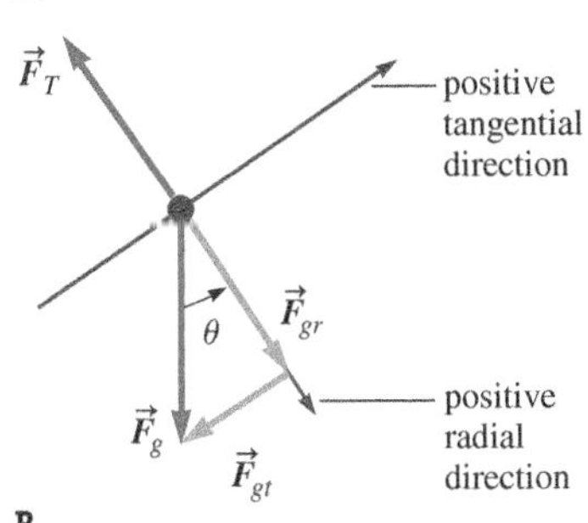

FIGURE 16.13 **A.** A simple pendulum consists of a particle at the end of a string of length ℓ. **B.** Free-body diagram for the pendulum bob.

to the bob's velocity. In this case, those two directions are radial and tangential to the circular arc. The tension force $\vec{F}_T$ has only a radial component, but gravity $\vec{F}_g$ has a *r*adial component $\vec{F}_{gr}$ and a *t*angential component $\vec{F}_{gt}$. Because the bob does not move radially, we only need to consider the tangential component to find expressions for the bob's SHM.

The tangential component of gravity $F_{gt} = -mg\sin\theta(t)$ provides a restoring force just like the spring provides a restoring force on Crall and Whipple's disk. In the case of a simple pendulum, gravity causes the bob to oscillate. Newton's second law gives an expression for the tangential acceleration $a_t(t)$:

$$\sum F_t = ma_t$$
$$-mg\sin\theta(t) = ma_t(t)$$
$$-g\sin\theta(t) = a_t(t)$$

We have a mixture of angular parameters $\theta(t)$ and translational parameters $a_t(t)$. Equation 12.14, $a_t = \ell\alpha$, gets rid of the translational parameter, so we can write $a_t(t)$ in terms of the bob's angular acceleration $\alpha(t)$:

$$-g\sin\theta(t) = \ell\alpha(t)$$
$$\alpha(t) = -\frac{g}{\ell}\sin\theta(t)$$

We have just found the angular acceleration of the bob as a function of its angular position, but the result does *not* fit the form of Equation 16.12 that characterizes SHM, $\alpha(t) = -\omega^2\theta(t)$. For SHM, $\alpha(t)$ must be proportional to $\theta(t)$, but we found that $\alpha(t)$ is proportional to the *sine* of $\theta(t)$. We are forced to conclude that a simple pendulum oscillates, but its motion is *not* simple harmonic.

If, however, we limit the bob's angular position $\theta(t)$ to small values—say no more than 15°—its motion is very nearly simple harmonic. From Appendix A, the sine of a small angle measured in radians approximately equals that angle:

$$\sin\theta \approx \theta \quad \text{for } \theta \ll 1 \text{ rad}$$

If the bob's angular displacement is small, we can write the angular acceleration as

$$\alpha(t) \approx -\frac{g}{\ell}\theta(t) \tag{16.28}$$

Equation 16.28 has the same mathematical form as the hallmark of SHM if the angular frequency is given by

$$\omega_{\text{smp}} = \sqrt{\frac{g}{\ell}} \tag{16.29}$$

where the subscript "smp" stands for simple pendulum. *For a small angular displacement, a simple pendulum's motion is simple harmonic with an angular frequency that depends only on the length of the string and the local gravitational acceleration.*

The kinematic equations for position (Eq. 16.3), velocity (Eq. 16.8), and acceleration (Eq. 16.11) may be written in terms of the *angular* position, *angular* speed, and *angular* acceleration of a simple pendulum by using the same analogy we used in Chapter 12:

Be careful; the symbol ω is used both for angular frequency and angular speed. You must distinguish between the two quantities from the context. For instance, when used inside the argument of a sine or cosine function, ω is the angular frequency, so ω_{smp} in Equation 16.31 represents angular frequency.

$$y \to \theta \qquad v \to \omega \qquad a \to \alpha \tag{16.30}$$

For example, Equation 16.3 becomes

$$\theta(t) = \theta_{\max}\cos(\omega_{\text{smp}}t + \varphi) \tag{16.31}$$

where $\theta_{\max}$ is the maximum angular position or amplitude of the bob.

CONCEPT EXERCISE 16.5

Calculate the fractional difference $|(\sin\theta - \theta)/\sin\theta|$ for $\theta = 0, \pm5°, \pm10°$, and $\pm15°$. *Hint*: Work in radians using two significant figures.

CONCEPT EXERCISE 16.6

Using the analogy in Equation 16.30,

a. Write an expression for angular acceleration from Equation 16.11, $a_y(t) = -a_{max} \cos(\omega t + \varphi)$.
b. Write an expression for the maximum angular speed from Equation 16.7, $v_{max} = y_{max}\omega$.
c. Write an expression for the maximum angular acceleration from Equation 16.10, $a_{max} = y_{max}\omega^2$.
d. What does each ω represent in your expressions?

EXAMPLE 16.4 Pendulum on the Moon

A simple pendulum in a clock is displaced to $\theta_i = 5.00°$ (8.73×10^{-2} rad) and released from rest. When the pendulum passes through its equilibrium position, the bob's speed is 0.300 m/s.

A What are the length, angular frequency, and period of the pendulum?

INTERPRET and ANTICIPATE

We must interpret two important pieces of information from the example statement. The amplitude of the oscillation is $\theta_{max} = 8.73 \times 10^{-2}$ rad, and the bob's maximum *translational* speed is $v_{max} = 0.300$ m/s. The maximum translational speed is *not* the same as the maximum *angular* speed ω_{max}.

SOLVE

Find an expression for the maximum *angular* speed from the maximum *translational* speed (Eq. 12.13). In this case, the radius of the circular arc is the length of the pendulum's rod.

$$v_{max} = \ell\omega_{max} \quad (12.13)$$

$$\omega_{max} = \frac{v_{max}}{\ell} \quad (1)$$

By analogy with Equation 16.7, the bob's maximum angular speed ω_{max} depends on the angular frequency ω_{smp} and the amplitude of the oscillation. (Also see Concept Exercise 16.6b.)

$$\omega_{max} = \theta_{max}\omega_{smp} \quad (2)$$

Substitute $\omega_{smp} = \sqrt{g/\ell}$ (Eq. 16.29) for the angular frequency ω_{smp} of a simple pendulum into Equation (2).

$$\omega_{max} = \theta_{max}\sqrt{\frac{g}{\ell}} \quad (3)$$

Substitute Equation (1) into Equation (3) and solve for ℓ.

$$\frac{v_{max}}{\ell} = \theta_{max}\sqrt{\frac{g}{\ell}} \qquad v_{max} = \ell\theta_{max}\sqrt{\frac{g}{\ell}} = \theta_{max}\sqrt{g\ell}$$

$$v^2_{max} = \theta^2_{max}g\ell$$

$$\ell = \frac{v^2_{max}}{g\theta^2_{max}} = \frac{(0.300 \text{ m/s})^2}{(9.81 \text{ m/s}^2)(8.73 \times 10^{-2} \text{ rad})^2}$$

$$\ell = 1.20 \text{ m}$$

Now that we know the pendulum's length, use $\omega_{smp} = \sqrt{g/\ell}$ (Eq. 16.29) to find angular frequency.

$$\omega_{smp} = \sqrt{\frac{g}{\ell}} = \sqrt{\frac{9.81 \text{ m/s}^2}{1.20 \text{ m}}}$$

$$\omega_{smp} = 2.86 \text{ rad/s}$$

Use Equation 16.2 to find the period.

$$T = \frac{2\pi}{\omega_{smp}} = \frac{2\pi}{2.86 \text{ rad/s}}$$

$$T = 2.20 \text{ s}$$

Example continues on page 466 ▶

CHECK and THINK

The dimensions are correct, and the length of the pendulum seems to fit a pendulum clock (Fig. 16.1A). Because angular frequency and angular speed are both represented by ω, one of the trickiest parts of this problem is keeping clear in your mind the distinction between the two; carefully labeled subscripts are your best aid in doing so.

B If the bob is displaced to $\theta_i = 10.0°$ instead, what are the angular frequency and period of the pendulum and the maximum speed of the bob?

INTERPRET and ANTICIPATE

The initial displacement of the bob is the amplitude. So, displacing the bob to 10.0° doubles the pendulum's amplitude. The angular frequency and period do not depend on amplitude and so remain unchanged—$\omega_{smp} = 2.87$ rad/s and $T = 2.20$ s—but the maximum speed of the bob increases.

SOLVE

The new maximum speed comes from combing Equations (1) and (2). Convert the new θ_{max} to radians and substitute values.

$$\omega_{max} = \frac{v_{max}}{\ell} = \theta_{max}\omega_{smp}$$

$$v_{max} = \ell\theta_{max}\omega_{smp}$$

$$v_{max} = (1.20\text{ m})(2)(8.73 \times 10^{-2}\text{ rad})(2.86\text{ rad/s}) = 0.600\text{ m/s}$$

CHECK and THINK

Because the amplitude θ_{max} has doubled while the length and angular frequency are constant, the new maximum speed is double the previous value of 0.300 m/s.

$$(v_{max})_{new} = \theta_{max}(\ell\omega_{smp}) = \theta_{max}(v_{max})_{old}$$

$$(v_{max})_{new} = 2(0.300\text{ m/s}) = 0.600\text{ m/s}$$ ✓

C If the same pendulum clock were on the Moon ($g = 1.62$ m/s^2), what would its period be?

INTERPRET and ANTICIPATE

The gravitational acceleration on the surface of the Moon is lower than that on the Earth, so the restoring force on the pendulum bob is weaker. We expect the bob to have a lower acceleration, giving the pendulum a longer period.

SOLVE

Combine Equations 16.2 and 16.29 to find the period.

$$T = \frac{2\pi}{\omega_{smp}} \quad \text{and} \quad \omega_{smp} = \sqrt{\frac{g}{\ell}}$$

$$T = 2\pi\sqrt{\frac{\ell}{g}} = 2\pi\sqrt{\frac{1.19\text{ m}}{1.62\text{ m/s}^2}}$$

$$T = 5.39\text{ s}$$

CHECK and THINK

As expected, the period is longer on the Moon than on the Earth.

PHYSICAL PENDULUM

Special Case

16-7 Special Case: Physical Pendulum

A **physical pendulum** is any real pendulum that *cannot* be modeled as having its mass concentrated in a small bob at the end of a relatively massless string. Isaac in Figure 16.14A is a physical pendulum because his mass is well distributed over his entire body. If his amplitude is not far from his equilibrium position, however, his motion can still be modeled as SHM.

Isaac rotates as he swings back and forth, so we apply the concepts and mathematical tools of rotation (Chapter 12) to find an expression for his angular acceleration α.

We set the z axis along Isaac's rotation axis so that his motion is in the xy plane (Fig. 16.14A). We model Isaac as a rigid object and draw a free-body diagram by representing him as a simple shape (Fig. 16.14B). At some arbitrary time t, his center of mass is at $\vec{r}_{CM}$, which has an angular position $\theta(t)$. Two forces are exerted on him: the normal force of the bar on the back of his knees and gravity exerted on his center of mass. To find his angular acceleration, we need to find the net torque exerted on him. Because the bar runs along the rotation axis, the normal force cannot exert a torque on Isaac, so we have omitted the normal force from our free-body diagram. The only torque exerted on Isaac is due to gravity. From the definition of torque and from Newton's second law in rotational form (Eqs. 12.23 and 12.24), Isaac's angular acceleration is

$$\vec{\tau} = \vec{r} \times \vec{F} = I\vec{\alpha}$$

Substituting the parameters from Figure 16.14B and solving for α gives

$$\vec{\tau} = -r_{CM}F_g \sin\theta(t)\hat{k} = I\alpha(t)\hat{k}$$

$$-r_{CM}mg \sin\theta(t)\hat{k} = I\alpha(t)\hat{k}$$

$$\alpha(t) = -\frac{mgr_{CM}}{I}\sin\theta(t)$$

As in the case of a simple pendulum, a physical pendulum is *not* a simple harmonic oscillator because $\alpha \propto \sin\theta$ instead of $\alpha \propto \theta$. If the maximum angular displacement is small ($\theta_{max} \leq 15°$), however, the physical pendulum (like the simple pendulum) may also be modeled as a simple harmonic oscillator:

$$\alpha(t) = -\frac{mgr_{CM}}{I}\theta(t) \quad (16.32)$$

Comparing Equation 16.32 with Equation 16.12, we find that the angular frequency of a physical pendulum is

$$\omega_{phy} = \sqrt{\frac{mgr_{CM}}{I}} \quad (16.33)$$

where the subscript "phy" stands for *phy*sical pendulum.

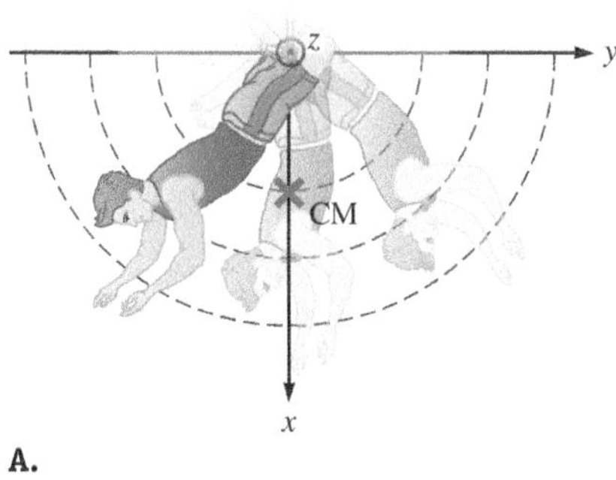

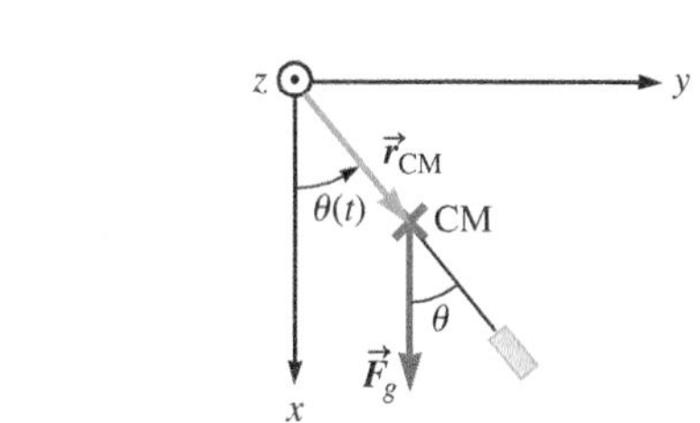

FIGURE 16.14 A. Isaac hangs by his knees and swings from a horizontal bar. **B.** Isaac is modeled as a physical pendulum.

EXAMPLE 16.5 A Comfortable Stroll

When you walk, your legs may be modeled as physical pendulums (Fig. 16.15). Walking is most comfortable and consumes the least amount of energy when you swing your legs at their natural angular frequency given by Equation 16.33. For the typical leg of a tall man (2 m in height), the pendulum period is 0.9 s. The length of his leg is roughly $L \approx 1$ m, and its mass is roughly one-seventh of the man's total mass, which is $m = 95$ kg. Assume his leg's center of mass is roughly $L/3$ from his hip socket and use these data to estimate the rotational inertia of a tall man's leg.

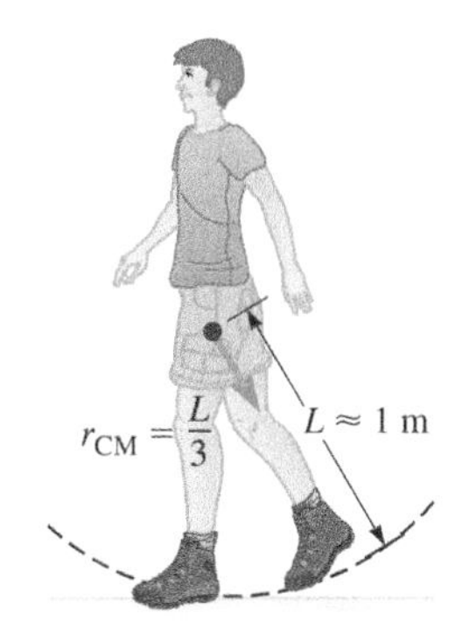

FIGURE 16.15

INTERPRET and ANTICIPATE

The man's leg is modeled as a physical pendulum, so its angular frequency depends on its rotational inertia. In Example 13.6 (page 367), we found the rotational inertia of a thin rod rotated about an axis through one end: $I = \frac{1}{3}mL^2$. If we treat the leg as a uniform rod pivoted at the hip, its rotational inertia would be $I = \frac{1}{3}[(95/7)\text{ kg}](1\text{ m})^2 = 4.5\text{ kg}\cdot\text{m}^2$. We should obtain a result less than this one with our physical pendulum model because we are taking the center of mass to be closer to the pivot (that is, center of mass at $L/3$, not at $L/2$).

SOLVE

Begin with Equations 16.2 and 16.33.

$$T = \frac{2\pi}{\omega_{phy}} \quad (16.2)$$

$$\omega_{phy} = \sqrt{\frac{mgr_{CM}}{I}} \quad (16.33)$$

Example continues on page 468 ▶

Substitute to obtain an expression for the period that includes the rotational inertia.	$T = 2\pi\sqrt{\dfrac{I}{mgr_{CM}}}$
Solve for I and substitute the given information.	$I = \dfrac{mgr_{CM}T^2}{4\pi^2} = \dfrac{[(95/7)\text{ kg}](9.81\text{ m/s}^2)(1/3\text{ m})(0.9\text{ s})^2}{4\pi^2}$ $I = 0.91\text{ kg}\cdot\text{m}^2$

CHECK and THINK

As expected, our result is lower than if we had modeled the leg as a thin, uniform rod rotated around an axis through one end.

TORSION PENDULUM

Special Case

16-8 Special Case: Torsion Pendulum

Computer scientist Danny Hillis published an essay titled "The Millennium Clock" in *Wired* magazine. In his article, Hillis said that he wanted

> to build a clock that ticks once a year. The century hand advances once every one hundred years, and the cuckoo comes out on the millennium. I want the cuckoo to come out every millennium for the next 10,000 years.[1]

FIGURE 16.16 The prototype of Hillis's 10,000-year "millennium clock" is in the Science Museum in London.

A prototype for Hillis's clock was built (Fig. 16.16) and bonged twice at midnight on January 1, 2000. Unlike the clocks in Figure 16.1A, Hillis's clock is based on a *torsion pendulum* (Fig. 16.17).

A **torsion pendulum** consists of an object such as the three-armed wheel in Figure 16.17A or the tire in Figure 16.17B connected to a *torsion spring*. A **torsion spring** resists being twisted just as a regular spring resists being stretched or compressed. For example, imagine a parent rotating and then releasing the tire (Fig. 16.17B), the torsion spring (the chains) provides a restoring force, twisting the tire and the child back toward the equilibrium position. In a torsion pendulum, the object rotates back and forth, passing the equilibrium position much like the disk moves up and down in Crall and Whipple's experiment.

In this book, we only study springs that obey Hooke's law ($\vec{F}_H = -k\vec{x}$). Similarly, we only study torsion springs that obey an analogous relation for torque:

$$\tau = -\kappa\theta \tag{16.34}$$

where θ is the object's angular position measured from equilibrium and κ (lowercase Greek letter kappa) is the **torsion spring constant**. The torsion spring constant depends on physical properties of the spring and has the SI units of newton meters per radian.

[1]http://www.longnow.org/essays/millennium-clock/.

A.

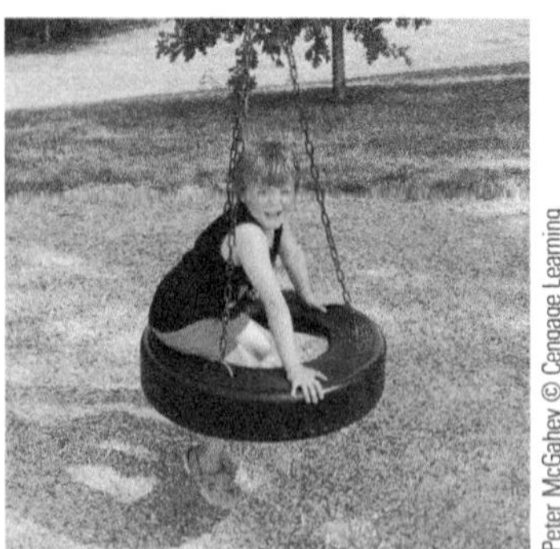

B.

FIGURE 16.17 A. The interior of the millennium clock shows that a metal ribbon is the torsion spring that causes the balls to rotate back and forth. **B.** A child on a tire swing rotates back and forth due to a torsion spring (the chains).

The negative sign in Equation 16.34 means that the torque's direction always accelerates the torsion pendulum toward its equilibrium position (Fig. 16.18). Newton's second law for rotation (Eq. 12.24) leads to an expression for the angular acceleration of the object having rotational inertia I:

$$\tau = I\alpha(t) = -\kappa\theta(t)$$

$$\alpha(t) = -\frac{\kappa}{I}\theta(t) \tag{16.35}$$

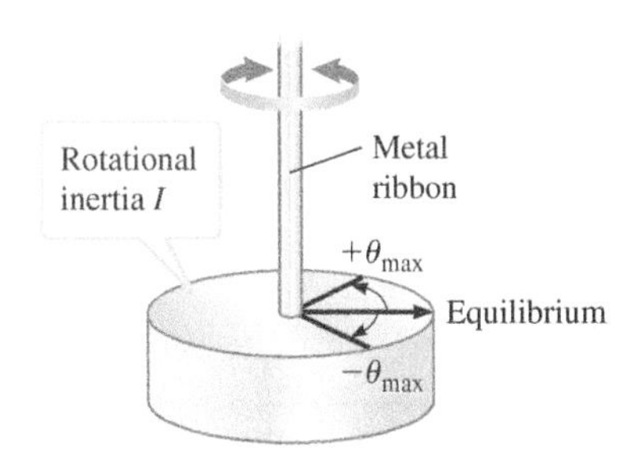

FIGURE 16.18 A basic torsion pendulum.

Comparing Equation 16.35 with Equation 16.12 shows that the torsion pendulum is a simple harmonic oscillator with an angular frequency of

$$\omega_{tor} = \sqrt{\frac{\kappa}{I}} \tag{16.36}$$

There is no need to use the small-angle approximation as we did for the other pendulums.

EXAMPLE 16.6 Hillis's Prototype Clock

The period of the three-armed torsion pendulum in Hillis's prototype clock is 60 s. Each ball (Fig. 16.17A) is made of tungsten ($\rho = 1.935 \times 10^4$ kg/m^3) and weighs about 7 lb. Estimate the torsion spring constant used in the clock. For comparison, the balance wheel in a wind-up watch has a rotational inertia $I \sim 10^{-6}$ kg · m^2 and a torsion spring constant $\kappa \sim 10^{-4}$ N · m/rad. Also, the period of the prototype clock's tick is about two orders of magnitude longer than that of a wind-up watch.

INTERPRET and ANTICIPATE

Our estimate of κ will be based on the angular frequency of a torsion pendulum (Eq. 16.36), which we can obtain from the given period. First estimate the rotational inertia of the three-armed pendulum in Figure 16.17A.

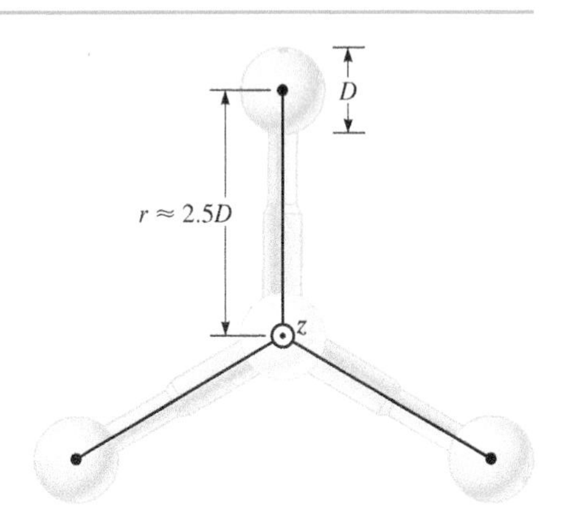

FIGURE 16.19

SOLVE

From Figure 16.17A, it looks like the arms are about twice as long as a ball's diameter D. We estimate the distance from the rotation axis labeled z in Figure 16.19 to the ball's center of mass to be about two and a half times the diameter: $r \approx 2.5D$.

Estimate the ball's diameter from its mass and density. A 7-lb ball has a mass of about 3 kg. The volume of a sphere is $V = \frac{4}{3}\pi R^3$.	$m = \rho V$ $V = \frac{4}{3}\pi R^3 = \frac{4}{3}\pi\left(\frac{D}{2}\right)^3$ $m = \rho\frac{4}{3}\pi\left(\frac{D}{2}\right)^3 = \rho\frac{\pi}{6}D^3$
Solve for D. Because this result is an estimate, keep just one significant figure.	$D = \left(\frac{6}{\pi}\frac{m}{\rho}\right)^{1/3} = \left(\frac{6}{\pi}\cdot\frac{3\text{ kg}}{1.935 \times 10^4\text{ kg/m}^3}\right)^{1/3}$ $D = 6.67 \times 10^{-2}\text{ m} \approx 7\text{ cm}$
Find the rotational inertia of the three-armed pendulum from Equation 13.4. Because $r \approx 2.5D$ and $D \approx 7$ cm, $r \approx 18$ cm. (We are modeling the three balls as three particles and ignoring the mass of the arms.)	$I = \sum_{i=1}^{n} m_i r_i^2$ (13.4) $I = 3mr^2 = 3(3\text{ kg})(18 \times 10^{-2}\text{ m})^2 = 0.29\text{ kg} \cdot \text{m}^2$ $I \approx 0.3\text{ kg} \cdot \text{m}^2$

Example continues on page 470 ▶

Finally, to estimate the torsion spring constant, use the period and $\omega_{tor} = \sqrt{\frac{\kappa}{I}}$ (Eq. 16.36).	$\omega_{tor} = \sqrt{\frac{\kappa}{I}} = \frac{2\pi}{T}$ (16.36) $\frac{\kappa}{I} = \frac{4\pi^2}{T^2}$
Solve for κ and substitute values, keeping just one significant figure in the answer.	$\kappa = \frac{4\pi^2}{T^2}I = \frac{4\pi^2}{(60\text{ s})^2}(0.29\text{ kg}\cdot\text{m}^2)$ $\kappa = 3.20 \times 10^{-3}\text{ kg}\cdot\text{m}^2/\text{s}^2 \approx 0.003\text{ N}\cdot\text{m/rad}$
CHECK and THINK Perhaps the greatest sources of error in our answer are that we did not include the mass of the three arms in our estimate of rotational inertia and we modeled the balls as particles. So, we have underestimated I, and κ should be somewhat greater than our value. To check our answer, let's compare the period of the prototype clock with the period of a wind-up watch. As expected, the period of the clock is about two orders of magnitude greater than for a watch.	$\frac{T_{clock}}{T_{watch}} = \sqrt{\frac{I_{clock}}{\kappa_{clock}}\frac{\kappa_{watch}}{I_{watch}}}$ $\frac{T_{clock}}{T_{watch}} \sim \sqrt{\frac{0.3\text{ kg}\cdot\text{m}^2}{0.003\text{ N}\cdot\text{m/rad}}\cdot\frac{10^{-4}\text{ N}\cdot\text{m/rad}}{10^{-6}\text{ kg}\cdot\text{m}^2}}$ $\frac{T_{clock}}{T_{watch}} \sim 10^2$ ✓

16-9 Energy in Simple Harmonic Motion

So far, we have analyzed simple harmonic motion based on the force approach. For example, in Crall and Whipple's experiment (case study, page 452), the disk is at rest in its equilibrium position before Whipple displaces it. Afterward, the restoring force causes the disk to oscillate in SHM up and down while passing above and below equilibrium.

We can gain a greater insight into SHM by taking a conservation of energy approach as in Chapter 8. First, we must define the system to include the oscillating object and the source of the restoring force. In the case of a block attached to a horizontal spring (Fig. 16.12), we would include the block and the spring in the system. If the system is isolated and there are no dissipative forces like friction, we expect the mechanical energy E of the system (Eq. 8.11) to be constant:

$$E = K + U = \text{constant}$$

Because the system is oscillating, both the kinetic and potential energies are continually changing, but the sum of these two time-varying quantities is constant.

DERIVATION Mechanical Energy of a Simple Harmonic Oscillator

We show that in the absence of dissipative forces the mechanical energy of an isolated simple harmonic oscillator is a constant given by

$$E = \tfrac{1}{2}my_{max}^2\omega^2 \tag{16.37}$$

To show formally that mechanical energy is conserved, we use the kinematic equations in Section 16-2 to find expressions for the kinetic energy $K(t)$ and the potential energy $U(t)$. We then add these expressions together to find an expression for the mechanical energy.

To find an expression for the kinetic energy $K(t)$ of a simple harmonic oscillator as a function of time, we substitute the speed of a simple harmonic oscillator (Eq. 16.6) into the definition of kinetic energy (Eq. 8.1).	$K = \frac{1}{2}mv^2 = \frac{1}{2}mv_y^2(t)$ (16.38) $K(t) = \frac{1}{2}m[-y_{max}\omega\sin(\omega t + \varphi)]^2$ $K(t) = \frac{1}{2}my_{max}^2\omega^2\sin^2(\omega t + \varphi)$ (16.39)

Finding an expression for the potential energy as a function of time $U(t)$ takes a few more steps. First, we must find an expression for the net force F_{tot} acting on the object by substituting the hallmark of SHM (Eq. 16.12) into Newton's second law.	$F_{tot}(y) = ma_y(t) = -m\omega^2 y(t)$	(16.40)
Second, we must integrate the force to find an expression for the potential energy as a function of position $U(y)$.	$U(y_f) - U(y_i) = -\int_{y_i}^{y_f} F_{tot}(y)dy$	(8.3)
Substitute Equation 16.40 into Equation 8.3 and integrate.	$U(y_f) - U(y_i) = -\int_{y_i}^{y_f} -m\omega^2 y\, dy$ $U(y_f) - U(y_i) = \frac{1}{2}m\omega^2(y_f^2 - y_i^2)$	
The usual convention is to choose $y_i = 0$ as the equilibrium position, set $U(y_i) = 0$, and drop the subscript f.	$U(y) = \frac{1}{2}m\omega^2 y^2$	(16.41)
We now have a simple form for $U(y)$, and we know how y varies with time. To find $U(t)$, substitute the time-dependent position of a simple harmonic oscillator (Eq. 16.3) into Equation 16.41.	$U(t) = \frac{1}{2}m\omega^2[y_{max}\cos(\omega t + \varphi)]^2$ $U(t) = \frac{1}{2}my_{max}^2\omega^2\cos^2(\omega t + \varphi)$	(16.42)
Both the kinetic energy (Eq. 16.39) and the potential energy (Eq. 16.42) increase and decrease as the oscillator moves back and forth. Add them together to find the mechanical energy.	$E = K(t) + U(t)$ $E = \frac{1}{2}my_{max}^2\omega^2\sin^2(\omega t + \varphi) + \frac{1}{2}my_{max}^2\omega^2\cos^2(\omega t + \varphi)$ $E = \frac{1}{2}my_{max}^2\omega^2[\sin^2(\omega t + \varphi) + \cos^2(\omega t + \varphi)]$	
The term in brackets is the trigonometric identity $\sin^2\alpha + \cos^2\alpha = 1$ (Appendix A).	$E = \frac{1}{2}my_{max}^2\omega^2$ ✓	(16.37)

COMMENTS

The mass m, the amplitude y_{max}, and the angular frequency ω are all constants for SHM, so the total mechanical energy E is indeed a constant for an isolated system with no dissipative forces. Equation 16.37 holds for any simple harmonic oscillator. Our next step is to consider the special case of an object–spring oscillator such as Crall and Whipple's system.

Energy of an Object–Spring Oscillator

For an object–spring oscillator, the angular frequency is $\omega^2 = k/m$. So, the system's potential energy (Eq. 16.41) becomes

$$U(y) = \frac{1}{2}m\omega^2 y^2 = \frac{1}{2}ky^2$$

as we found for a spring in Equation 8.10. The total mechanical energy of the object–spring oscillator is

$$E = \frac{1}{2}my_{max}^2\omega^2 = \frac{1}{2}ky_{max}^2 \quad (16.43)$$

which is the maximum value of the system's potential energy:

$$E = U_{max} = \frac{1}{2}ky_{max}^2 \quad (16.44)$$

The potential energy is a maximum when the object momentarily stops at either end of its path ($y = \pm y_{max}$).

The maximum kinetic energy (Eq. 16.38) occurs when the object moves though its equilibrium position ($y = 0$) and its speed is a maximum ($v = v_{max}$):

$$K_{max} = \frac{1}{2}mv_{max}^2 \quad (16.45)$$

When the object is at the equilibrium position, the system's potential energy is zero, so the mechanical energy equals the maximum kinetic energy:

$$E = K_{max} = \frac{1}{2}mv_{max}^2$$

Because the mechanical energy is a constant, we write

$$E = K_{max} = U_{max} \quad (16.46)$$

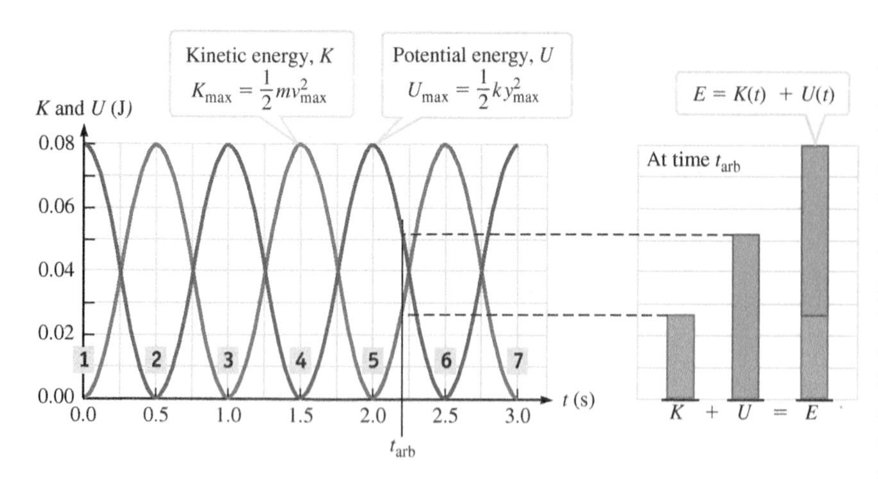

FIGURE 16.20 Kinetic energy (red) and potential energy (blue) versus time for Crall and Whipple's data.

Let's visualize a typical oscillating system's energy using Crall and Whipple's data (case study, page 452). Figure 16.20 shows the kinetic and potential energies versus time on the same graph. To make the kinetic energy graph, square the velocity data in Figure 16.5B and multiply by $\frac{1}{2}m$. To make the potential energy graph, square the position data in Figure 16.5A and multiply by $\frac{1}{2}k$. The times labeled 1 through 7 in Figure 16.20 correspond to the same configurations and times in Figures 16.3 through 16.5. For example, when Whipple first releases the disk at time 1, the system's kinetic energy is zero, and its potential energy is at a maximum ($U_{max} \approx 0.08$ J). When the disk passes through its equilibrium position at time 2, the kinetic energy is at its maximum ($K_{max} \approx 0.08$ J), and the potential energy is zero.

To show that the sum of the kinetic and potential energies in Crall and Whipple's experiment is constant, we draw energy bars in Figure 16.20. We have drawn a kinetic energy bar and a potential energy bar at an arbitrary time t_{arb}. When we stack those energy bars on top of each other, their combined height reaches the total mechanical energy ($E \approx 0.08$ J) of the system. Adding the kinetic and potential energy bars at any arbitrary time gives the same result: $E \approx 0.08$ J. Figure 16.20 shows that the mechanical energy $E \approx 0.08$ J also equals the maximum of either the kinetic or the potential energy.

The conservation of energy approach gives an additional description of simple harmonic motion that the force approach cannot supply. As the object oscillates back and forth, potential energy is converted into kinetic energy and back again repeatedly. The kinetic energy is at its maximum when the object whizzes through its equilibrium position. The potential energy is at its maximum when the object stops momentarily at either end of its path. The mechanical energy is constant. No energy is lost or gained by the system, and oscillation continues indefinitely, again provided that the system is isolated and that no dissipative forces are present.

EXAMPLE 16.7 Today's Lab: A Horizontal Spring

In laboratories all over the world, physics students observe the motion of objects attached to springs. In one such laboratory, a horizontal spring of spring constant $k = 170$ N/m is attached to an object of mass $m = 0.85$ kg. As in Figure 16.12, friction between the object and the table is negligible. A student stretches the spring so that the object's initial position is $x_i = 0.20$ m and releases it at $t = 0$. Find the maximum speed.

INTERPRET and ANTICIPATE

There are at least two ways to approach this problem. (1) We'll calculate the mechanical energy and then find the maximum speed based on the maximum kinetic energy. (2) As a check, we'll use the angular frequency and amplitude to find the maximum speed.

SOLVE

We can find the maximum potential energy from the given amplitude using $U_{max} = \frac{1}{2}ky^2_{max}$ (Eq. 16.44) expressed in terms of horizontal position x.

$$U_{max} = \tfrac{1}{2}kx^2_{max} = \tfrac{1}{2}(170 \text{ N/m})(0.20 \text{ m})^2$$
$$U_{max} = 3.4 \text{ J}$$

The maximum potential energy equals the total mechanical energy and also equals the maximum kinetic energy (Eq. 16.46). The maximum speed then comes from Equation 16.45.

$$E = K_{max} = U_{max} = 3.4 \text{ J}$$
$$K_{max} = \tfrac{1}{2}mv^2_{max} = 3.4 \text{ J}$$
$$v_{max} = \sqrt{\frac{2K_{max}}{m}} = \sqrt{\frac{2(3.4 \text{ J})}{0.85 \text{ kg}}} = 2.8 \text{ m/s}$$

CHECK and THINK
To check the result, use Equation 16.26 to calculate the angular frequency and use Equation 16.7 (in terms of x) to find v_{max}. We find the same answer using this approach.

$$\omega_s = \sqrt{\frac{k}{m}} = \sqrt{\frac{170\ \text{N/m}}{0.85\ \text{kg}}} = 14\ \text{rad/s}$$

$$v_{max} = x_{max}\omega_s \qquad (16.7)$$

$$v_{max} = (0.20\ \text{m})(14\ \text{rad/s}) = 2.8\ \text{m/s}$$

16-10 Damped Harmonic Motion

A simple harmonic oscillator is an *ideal* oscillator that never stops. A real oscillator such as Crall and Whipple's rubber disk on a spring will not continue to move up and down forever. Many real oscillators can be well approximated as ideal simple harmonic oscillators if their motion continues for much longer than one period. For example, Crall and Whipple's oscillator took several minutes to stop on its own—much longer than the measured period ($T = 2.0$ s), so their system is well modeled as a simple harmonic oscillator. Crall and Whipple then returned to the laboratory and tried another experiment with slightly modified equipment.

DAMPED HARMONIC MOTION
Major Concept

CASE STUDY Experiment, Part 2: Not Very Bouncy

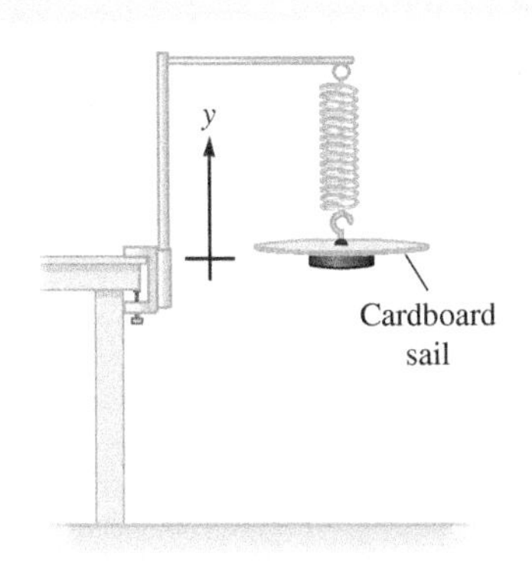

FIGURE 16.21 Crall and Whipple modified their experimental setup by adding a cardboard sail just above the disk.

This time, our experimenters added a cardboard sail just above the disk (Fig. 16.21), displaced and released the disk as in the first experiment, and collected new position and velocity data (Fig. 16.22). Because of air drag, their new oscillator stopped in a few seconds after completing just a few cycles. Their new system cannot be modeled as a simple harmonic oscillator because its position and velocity data in Figure 16.22 do not fit the cosine and sine function of Equations 16.3 and 16.6.

Crall and Whipple's sail experiment must be modeled as **damped harmonic motion**. Dissipative forces such as friction and drag cause a **damped oscillator** to lose mechanical energy and stop. We can use Crall and Whipple's position and velocity data to calculate and plot the energy of their damped oscillator at various times. Figure 16.23 shows that $K(t)$ and $U(t)$ are converted back and forth as in simple harmonic motion, but over a short time both decrease. The energies $K(t)$ and $U(t)$ still add up to the total mechanical energy, but as the mechanical energy decreases to zero, the oscillator stops.

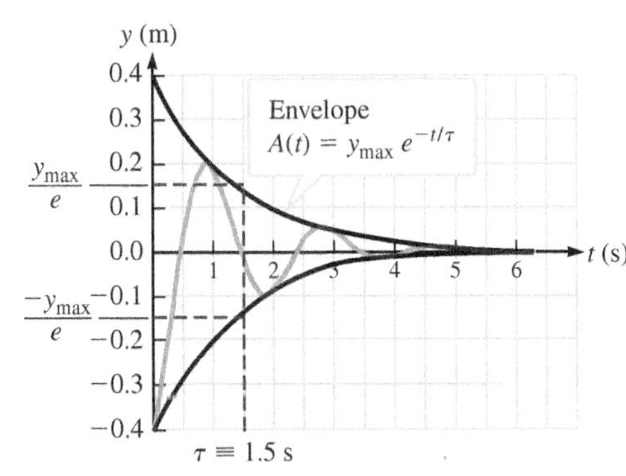

A. Position-versus-time

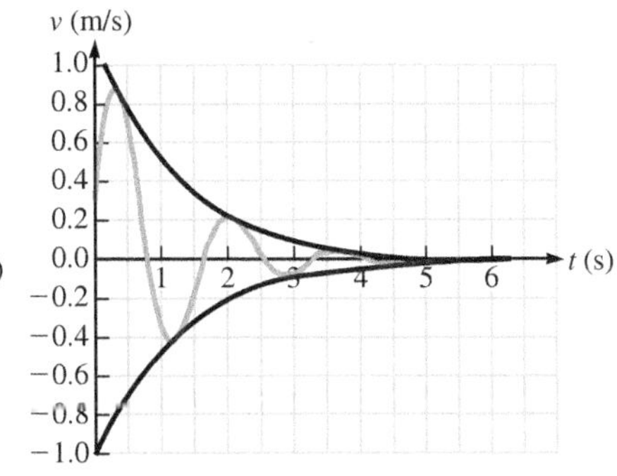

B. Acceleration-versus-time

FIGURE 16.22 A. In damped harmonic motion, oscillations decrease until the oscillator comes to rest in its equilibrium configuration. **B.** The speed and energy of a damped harmonic oscillator decrease to zero.

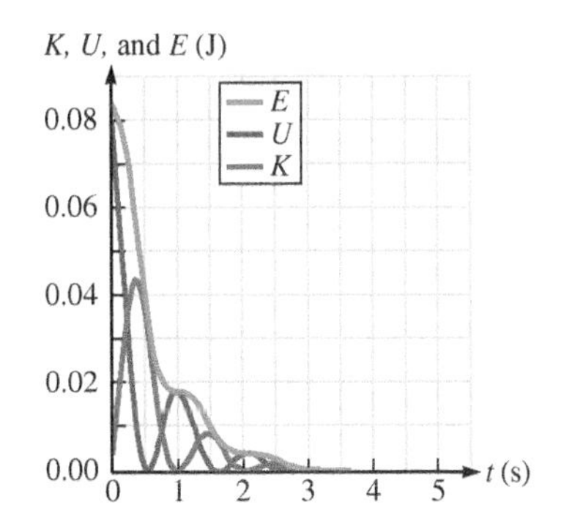

FIGURE 16.23 The mechanical energy (green), potential energy (blue), and kinetic energy (red) of a damped harmonic oscillator decrease to zero.

Describing Damped Oscillations: The Time Constant

Because the damped oscillator's motion cannot be described by the kinematic equations in Section 16-2, we must derive new equations. Using dynamics in Section 16-4 led us to Equation 16.23 for the *undamped* SHO:

$$-ky(t) = ma_y(t) \tag{16.23}$$

Substitute $a_y(t) = d^2y(t)/dt^2$ (Eq. 4.17):

$$-\frac{k}{m}y(t) = \frac{d^2y(t)}{dt^2} \tag{16.47}$$

Equation 16.47 involves derivatives of the variable $y(t)$ and is therefore a differential equation. It says that if we take two time derivatives of $y(t)$, we get back $y(t)$ multiplied by a constant $(-k/m)$. Equation 16.47 cannot be solved for $y(t)$ by direct integration; the techniques for solving such differential equations are beyond the scope of this book. One solution is Equation 16.3:

$$y(t) = y_{max} \cos(\omega t + \varphi) \tag{16.3}$$

As an exercise, you should verify that Equation 16.3 satisfies Equation 16.47 for the undamped oscillator (Problem 52).

We can arrive at a relation for position versus time for a damped oscillator in a similar way. The sail on Crall and Whipple's oscillator exerts a drag force $\vec{F}_D = -b\vec{v}$ (Eq. 6.4) on the oscillator, so Equation 16.23 becomes

$$-ky(t) - bv_y(t) = ma_y(t) \tag{16.48}$$

By the definitions of velocity and acceleration, Equation 16.48 becomes a differential equation for $y(t)$:

$$-ky(t) - b\frac{dy(t)}{dt} = m\frac{d^2y(t)}{dt^2} \tag{16.49}$$

Equation 16.49 for damped harmonic motion looks similar to Equation 16.47 for SHM except for the extra term involving the first derivative of $y(t)$. The solution to Equation 16.49 is

$$y(t) = A(t) \cos(\omega_D t + \varphi) \tag{16.50}$$

which is similar to the solution for SHM (Eq. 16.3) except that the amplitude $A(t)$ is time-dependent, *not* constant. This amplitude is given by

$$A(t) = y_{max} e^{-t/\tau} \tag{16.51}$$

Equation 16.51 describes a decreasing amplitude $A(t)$. The **time constant** τ is a measure of how quickly the amplitude decreases and is defined as

TIME CONSTANT ✪ **Major Concept**

$$\tau \equiv \frac{2m}{b} \tag{16.52}$$

where b is sometimes referred to as the *damping constant* or *damping coefficient*. Problem 54 asks you to verify that the time constant has dimensions of time.

Figure 16.22A illustrates the decreasing amplitude $A(t)$ and the time constant τ. The cosine part of Equation 16.50 says that Crall and Whipple's disk is still oscillating, corresponding to the peaks and valleys on the y-versus-t graph. The amplitude is decreasing (Eq. 16.51), however, so the peaks and valleys get smaller as t increases. Equation 16.51 describes an exponentially decreasing envelope, and the peaks and valleys in Figure 16.22A fit inside the envelope.

The time constant τ determines how quickly the envelope closes in, as is best seen by finding the amplitude $A(\tau)$, when $t = \tau$:

$$A(\tau) = y_{max} e^{-\tau/\tau} = y_{max} e^{-1}$$

$$A(\tau) = \frac{y_{max}}{e} \approx \frac{y_{max}}{2.718} \approx 0.3679 y_{max}$$

The time constant τ is the time it takes the amplitude envelope A to drop from y_{max} to y_{max}/e, or just more than one-third of its original value. In Figure 16.22A, $y_{max} \approx 0.4$ m, so $y_{max}/e \approx 0.15$ m. We can read the time constant off this y-versus-t graph as $\tau \approx 1.5$ s.

By substituting Equations 16.51 and 16.52 into Equation 16.50, the time-varying position of a damped oscillator can be written as

$$y(t) = y_{max} e^{-bt/2m} \cos(\omega_D t + \varphi) \tag{16.53}$$

Frequency of Damped Harmonic Motion

The angular frequency ω_D of damped harmonic motion is not the same as the angular frequency ω of simple harmonic motion:

$$\omega_D = \sqrt{\omega^2 - \frac{1}{\tau^2}} = \sqrt{\omega^2 - \frac{b^2}{4m^2}} \tag{16.54}$$

where ω is the angular frequency of the corresponding undamped system. If the dissipative force disappears, $b \rightarrow 0$ and $\omega_D \rightarrow \omega$. Equation 16.54 says that the angular frequency of a damped oscillator is less than the angular frequency of the simple harmonic oscillator that would result if the dissipative force were removed. Therefore, the damped oscillator has a longer period than the corresponding simple harmonic oscillator. The angular frequency of Crall and Whipple's damped oscillator is given by

$$\omega_D = \sqrt{\omega^2 - \frac{1}{\tau^2}} = \sqrt{(\pi \text{ rad/s})^2 - \frac{1}{(1.5 \text{ s})^2}} = 3.07 \text{ rad/s}$$

which is less than $\omega = 3.14$ rad/s of their simple harmonic oscillator.

The strength of the dissipative force depends on b. If a strong dissipative force (large b) is exerted on an oscillator, the oscillator will have a short time constant τ and a low angular frequency ω_D. Such an oscillator will stop quickly without completing very many cycles. If the dissipative force is very strong such that $b^2/4m^2 > \omega^2$, then ω_D according to Equation 16.54 is imaginary, and the system does not oscillate.

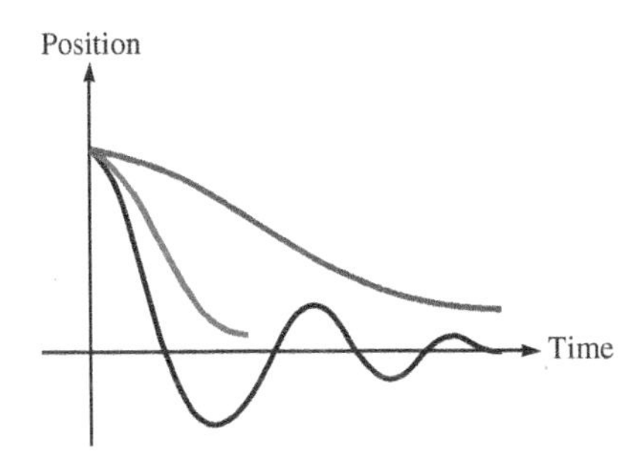

FIGURE 16.24 Position versus time for a critically damped oscillator (red), underdamped oscillator (black), and overdamped oscillator (blue).

Equation 16.54 is used to classify damped oscillators into three types, depending on the relative strength of the dissipative force (Fig. 16.24). A **critically damped** oscillator's angular frequency is zero, so

CRITICALLY DAMPED, UNDERDAMPED, AND OVERDAMPED OSCILLATORS

Major Concept

$$\omega_D = \sqrt{\omega^2 - \frac{b^2}{4m^2}} = 0$$

$$\omega = \frac{b}{2m}$$

$$b = 2m\omega$$

After a critically damped oscillator is displaced from its equilibrium position, it returns to equilibrium in the shortest possible time without passing through to the other side of equilibrium. A weaker dissipative force ($b < 2m\omega$) is exerted on an **underdamped** oscillator, which oscillates briefly before stopping at its equilibrium position. Finally, an **overdamped** oscillator results when a stronger-than-critical dissipative force ($b > 2m\omega$) is exerted. When an overdamped oscillator is displaced from its equilibrium position, it returns directly to equilibrium, but more slowly than does a critically damped oscillator.

If you are designing a clock, you should avoid damping so that the pendulum, watch spring, crystal, or whatever else is oscillating does so for as long as possible. Other systems such as the shock absorbers on your car or bicycle are deliberately damped. A shock absorber consists of a spring attached to a piston (Fig. 16.25). The piston moves through a viscous fluid that exerts a dissipative force. When you hit a pothole, a properly working shock absorber will quickly bring your vehicle back to equilibrium without oscillating. Therefore, the shock absorber should be critically damped. If the shock absorber is underdamped, your vehicle will bounce up and down several times after you hit the pothole. If the shock absorber is overdamped, your vehicle will take a longer time to return to equilibrium.

FIGURE 16.25 The shock absorber attaches the vehicle's frame to the wheel; it is designed to act as a critically damped oscillator, reducing oscillations and making the ride more comfortable.

EXAMPLE 16.8 My Cat Lives with a Physicist

A cat toy consists of a lightweight, hollow ball ($m = 2.0$ g) hanging on the end of a thin string of length $\ell = 32$ cm. A pet owner displaces the ball by 6° and releases it. The cat enthusiastically watches the ball swing back and forth, but the ball's amplitude is reduced to 1° in 24 s. Both the cat and the owner are disappointed. What is the time constant?

INTERPRET and ANTICIPATE

You might be tempted to use Equation 16.52, which defines the time constant τ, but you have not been given enough information to use that equation. Instead, use Equation 16.51, noting that the amplitude drops from 6° to 1° in 24 s. We expect that because the amplitude has dropped by more than one-third, the 24-s observation interval is greater than the time constant.

SOLVE

Write Equation 16.51 for an angular amplitude envelope $A_\theta(t)$. The maximum angular amplitude is θ_{max}.

$$A_\theta(t) = \theta_{max} e^{-t/\tau}$$

To solve for τ, take the natural logarithm of both sides (Appendix A).

$$\frac{A_\theta(t)}{\theta_{max}} = e^{-t/\tau}$$

$$\ln\left(\frac{A_\theta(t)}{\theta_{max}}\right) = \ln e^{-t/\tau} = -\frac{t}{\tau}$$

$$\tau = -\frac{t}{\ln\left(A_\theta(t)/\theta_{max}\right)}$$

Substitute the values $t = 24$ s, $A_\theta(t) = 1°$, and $\theta_{max} = 6°$. There is no need to convert the angles to radians because the ratio of the angles in the logarithm is unitless.

$$\tau = -\frac{24\text{ s}}{\ln(1°/6°)} = 13\text{ s}$$

CHECK and THINK

As expected, the time constant is less than 24 s. The length of the string was extraneous information. In Problem 55, we explore ways to improve this toy.

16-11 Driven Oscillators

DRIVEN OSCILLATOR

Major Concept

In 1851, French physicist Jean Bernard Leon Foucault used a 67.5-m-long piano wire to hang a 28-kg pendulum bob from the dome of the Pantheon in Paris. His pendulum provided the first nonastronomical demonstration of the Earth's rotation. A restored version of Foucault's pendulum still oscillates in the Pantheon today (Fig. 16.26). Although Foucault's pendulum oscillates for many cycles, like all real pendulums it is damped and so would lose energy and eventually stop. There is a person whose job is to push or *drive* the pendulum's bob several times during the workday, however. The worker's push does work on the pendulum, transferring energy from the environment (the employee) to the system. The force exerted by the employee is called a **driving force**. Such an oscillator is damped and driven, so we refer it as a **damped-driven oscillator** or just a **driven oscillator**.

FIGURE 16.26 Foucault's pendulum inside the Pantheon in Paris, France. The circular scale gives the time.

Although you have never been employed at the Pantheon to drive Foucault's pendulum, you have probably pushed a child on a swing, which is like driving a pendulum. You stand on one side of the pendulum and push on the bob just as it moves away from you (Fig. 16.27A). The force-versus-time graph in Figure 16.27A illustrates two important properties of the driving force. (1) The driving force is not constant; you do not push on the bob continuously. (2) The driving force is periodic with roughly the same period as the oscillator; you push on the bob every time it returns to the same position near you.

You could be more efficient at transferring energy to the pendulum if you had a friend stand on the opposite side of the pendulum (Fig. 16.27B). He would push on the bob when the bob moves away from him (in the negative x direction), and you would continue to push on the bob as it moves away from you (in the positive x direction). The new force-versus-time graph in Figure 16.27B shows your force as the positive spikes and his force as the negative spikes. There are still gaps in the force-versus-time graph, however. A motor can be designed to exert a still more efficient driving force F_{drv}, represented by a sine or cosine function. (If you have ever seen a contemporary version of Foucault's pendulum, it may have been driven by such a motor.) We arbitrarily choose the cosine function (Fig. 16.27C):

$$F_{drv} = F_{max} \cos \omega_{drv} t \tag{16.55}$$

The driving force has an angular frequency ω_{drv}.

If an oscillator is damped and driven, (at least) three forces are exerted on it:

1. A restoring force $\vec{F}_{restore}$. In the case of Foucault's pendulum, the restoring force is exerted by gravity. In the case of Crall and Whipple's experiments, the restoring force is exerted by the spring: $\vec{F}_{restore} = -k\vec{y}$.
2. A dissipative force. The only dissipative force we study here has the form $\vec{F}_D = -b\vec{v}$, which is the result of air drag.
3. A a driving force of the form $F_{drv} = F_{max} \cos \omega_{drv} t$.

When we apply Newton's second law to a damped-driven oscillator, we have to add all three of these forces:

$$\sum \vec{F} = \vec{F}_{restore} + \vec{F}_D + \vec{F}_{drv} = m\vec{a} \tag{16.56}$$

Imagine that a driving force is applied to Crall and Whipple's damped oscillator (Fig. 16.22) and apply Equation 16.56:

$$-ky - bv_y + F_{max} \cos \omega_{drv} t = ma_y \tag{16.57}$$

Except for the additional driving term ($F_{max} \cos \omega_{drv} t$), Equation 16.57 looks like Equation 16.48, $-ky(t) - bv_y(t) = ma_y(t)$. As we did for that equation, we rewrite Equation 16.57 as a differential equation for $y(t)$ using the definitions of velocity and acceleration:

$$-ky(t) - b\frac{dy(t)}{dt} + F_{max} \cos \omega_{drv} t = m\frac{d^2y(t)}{dt^2}$$

which is traditionally written as

$$F_{max} \cos \omega_{drv} t = m\frac{d^2y(t)}{dt^2} + b\frac{dy(t)}{dt} + ky(t) \tag{16.58}$$

As before, we will provide a solution to this differential equation without carrying out the solution method. When the driving force is first exerted on the oscillator, the system's mechanical energy increases and the amplitude increases. After a while, however, the amplitude A_{drv} is constant, and the solution to Equation 16.58 is

$$y(t) = A_{drv} \cos (\omega_{drv} t + \varphi_{drv}) \tag{16.59}$$

as you may be asked to verify for homework (Problem 82). Equation 16.59 looks much like the time-varying position of a simple harmonic oscillator (Eq. 16.3). The most significant difference is that for a damped-driven oscillator, the amplitude A_{drv} depends on the driving force's angular frequency ω_{drv}:

$$A_{drv} = \frac{F_{max}}{\sqrt{m^2(\omega^2 - \omega_{drv}^2)^2 + b^2\omega_{drv}^2}} \tag{16.60}$$

The angular frequency ω with no subscript is the angular frequency that the oscillator would have in the absence of damping or driving. In other words, ω is the angular frequency of the equivalent simple harmonic oscillator and is called the **natural angular frequency**. The natural angular frequency is given by Equation 16.26, 16.29, 16.33, or 16.36, depending on the type of oscillator.

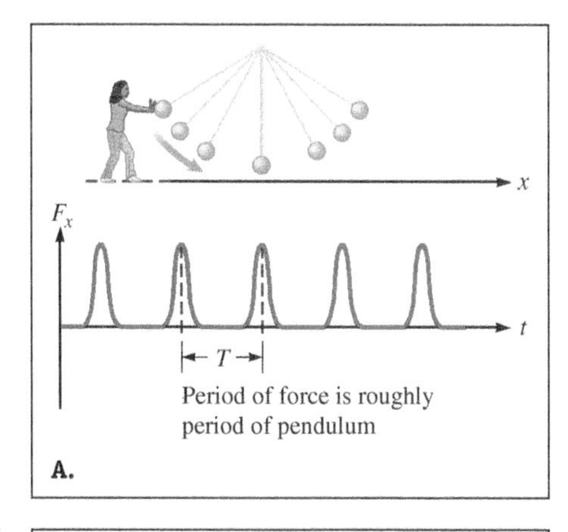

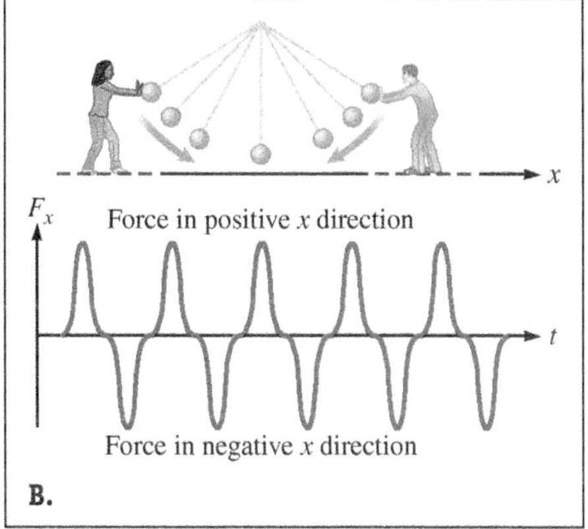

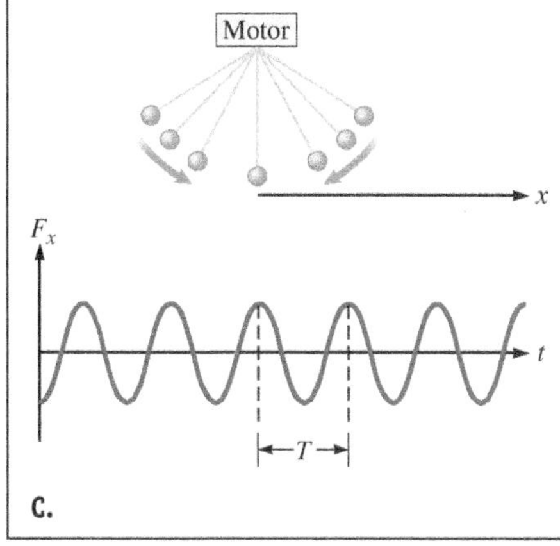

FIGURE 16.27 **A.** A pendulum is driven by a person who pushes in the positive x direction. **B.** The pendulum is driven more efficiently when a second person additionally pushes in the negative x direction. **C.** A pendulum is driven by a motor so that the driving force is $F_{drv} = F_{max} \cos \omega_{drv} t$.

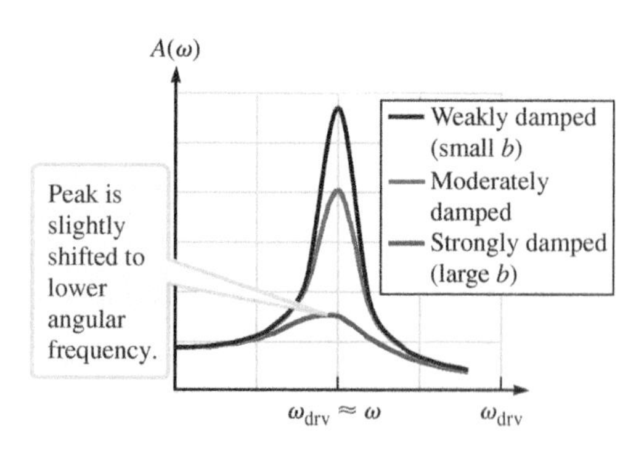

FIGURE 16.28 The amplitude A of a damped-driven oscillator peaks when the driving angular frequency equals the oscillator's natural angular frequency $\omega_{drv} = \omega$.

Because Equation 16.60 is a messy function of ω_{drv}, it is helpful to sketch a graph of A_{drv} versus ω_{drv} (Fig. 16.28). The amplitude of a damped-driven oscillator peaks when the driving angular frequency *approximately* equals the oscillator's natural angular frequency, $\omega_{drv} \approx \omega$. The driving force transfers energy to the system most effectively when its angular frequency matches the natural angular frequency of the oscillator. The term **resonance** describes the condition of an oscillator that is driven nearly at its natural angular frequency.

Often the terms *natural frequency f* and *driving frequency* f_{drv} are used instead of natural *angular* frequency and driving *angular* frequency. The natural frequency is given by Equation 16.2:

$$f = \frac{\omega}{2\pi}$$

and the driving frequency is

$$f_{drv} = \frac{\omega_{drv}}{2\pi}$$

Finally, the natural frequency is also called the **resonance frequency**.

RESONANCE ✪ **Major Concept**

Imagine using a motor to drive Foucault's original pendulum ($\ell = 67.5$ m). The natural angular frequency of the pendulum is given by Equation 16.29:

$$\omega = \omega_{smp} = \sqrt{\frac{g}{\ell}} = \sqrt{\frac{9.81 \text{ m/s}^2}{67.5 \text{ m}}}$$

$$\omega = 0.381 \text{ rad/s}$$

and its natural frequency is

$$f = \frac{0.381 \text{ rad/s}}{2\pi} = 6.06 \times 10^{-2} \text{ Hz}$$

Suppose you start with the pendulum at rest and the motor's frequency set at zero. You slowly increase the motor's frequency f_{drv}. As you do, you find that the pendulum does not respond until the motor's frequency is close to the natural frequency of the pendulum. The pendulum's response is most pronounced when $f_{drv} = 6.06 \times 10^{-2}$ Hz, and we say that the motor matches the pendulum's resonance. As you continue to *increase* the motor's frequency past resonance, the pendulum's amplitude *decreases*, and its motion actually dies out. The motor is still running, but it cannot transfer energy to the pendulum because its driving frequency is not close to the pendulum's resonance frequency.

EXAMPLE 16.9 CASE STUDY Driving Crall and Whipple's Oscillator

Suppose Crall and Whipple use a driving force of the form $F_{max} \cos \omega_{drv} t$ to compensate for the mechanical energy lost by their damped harmonic oscillator in Figure 16.21. They set the driving frequency equal to the natural frequency of their oscillator. Recall that the disk's mass is 0.100 kg and that the time constant of their damped oscillator is 1.5 s. If the amplitude A_{drv} of the motion is the same as that of their simple harmonic oscillator ($y_{max} = 0.40$ m), find F_{max}.

INTERPRET and ANTICIPATE

The maximum force exerted by the driving force is related to the amplitude of the driven oscillator.

SOLVE

If the driving frequency equals the natural frequency, the driving *angular* frequency equals the *angular* frequency of the equivalent simple harmonic oscillator, $\omega_{drv} = \omega = 3.1$ rad/s. Simplify Equation 16.60 by setting the driving frequency equal to the natural frequency.

$$A_{drv} = \frac{F_{max}}{\sqrt{m^2(\omega^2 - \omega_{drv}^2)^2 + b^2\omega_{drv}^2}} \quad (16.60)$$

$$A_{drv} = \frac{F_{max}}{\sqrt{0 + b^2\omega^2}} = \frac{F_{max}}{b\omega}$$

$$F_{max} = b\omega A_{drv} \quad (1)$$

Equation 16.52 relates the given time constant τ to the parameter b.	$\tau \equiv \frac{2m}{b}$ (16.52) $b = \frac{2m}{\tau} = \frac{2(0.100\text{ kg})}{1.5\text{ s}} = 0.13\text{ N}\cdot\text{s/m}$
In this case, the amplitude of the driven oscillator equals the amplitude of the undamped oscillator: $A_{drv} = y_{max}$. Substitute into Equation (1) to find F_{max}.	$F_{max} = b\omega A_{drv} = b\omega y_{max}$ $F_{max} = (0.13\text{ N}\cdot\text{s/m})(3.1\text{ rad/s})(0.40\text{ m}) = 0.16\text{ N}$

CHECK and THINK

Because the driving frequency equals the resonance frequency, even a small driving force is effective at producing a large amplitude. The maximum force is considerably less than the disk's weight, $w = 0.98$ N. If the driving frequency is set much higher or lower, the required driving force F_{max} must increase in order to achieve the same amplitude.

SUMMARY

Underlying Principles

We applied kinematics, dynamics (Newton's laws of motion), and conservation of energy to **oscillations**, a special case of periodic motion in which a particle moves back and forth over a path.

Major Concepts

1. **Simple harmonic motion (SHM)** is oscillatory motion that is relatively easy to describe mathematically. The hallmark of SHM is

$$a_y(t) = -\omega^2 y(t) \qquad (16.12)$$

The kinematic equations describing a **simple harmonic oscillator (SHO)** are as follows:

$$\text{position: } y(t) = y_{max}\cos(\omega t + \varphi) \qquad (16.3)$$

$$\text{velocity: } v_y(t) = -y_{max}\omega\sin(\omega t + \varphi) \qquad (16.6)$$

$$\text{acceleration } a_y(t) = -y_{max}\omega^2\cos(\omega t + \varphi) \qquad (16.9)$$

2. **Angular frequency** is written in terms of frequency or period as

$$\omega \equiv 2\pi f = \frac{2\pi}{T} \qquad (16.2)$$

3. The maximum displacement from the origin is called the **amplitude** y_{max}.
4. The argument $(\omega t + \varphi)$ of the cosine or sine function is called the **phase**. The **initial phase** φ is determined by the initial position $y_i \equiv y(0)$ such that

$$\varphi = \cos^{-1}\left(\frac{y_i}{y_{max}}\right) \qquad (16.4)$$

5. **Damped harmonic motion** occurs when dissipative forces cause an oscillator to lose mechanical energy. The position of a damped harmonic oscillator is given by

$$y(t) = y_{max}e^{-bt/2m}\cos(\omega_D t + \varphi) \qquad (16.53)$$

Its angular frequency is given by

$$\omega_D = \sqrt{\omega^2 - \frac{1}{\tau^2}} = \sqrt{\omega^2 - \frac{b^2}{4m^2}} \qquad (16.54)$$

6. The **time constant** τ is a measure of how quickly the amplitude of a damped oscillator is decreasing:

$$\tau \equiv \frac{2m}{b} \qquad (16.52)$$

7. An oscillator is **critically damped** if $b = 2m\omega$. If $b < 2m\omega$ the oscillator is **underdamped,** and if $b > 2m\omega$ the oscillator is **overdamped**.
8. An oscillator is a **driven oscillator** if a driving force transfers energy from the environment to the oscillator system.
9. **Resonance** describes the condition of an oscillator that is driven at its natural angular frequency.

Special Cases

Angular frequency of an **object–spring oscillator**:

$$\omega = \omega_s \equiv \sqrt{\frac{k}{m}} \tag{16.26}$$

Angular frequency of a **simple pendulum**:

$$\omega_{\text{smp}} = \sqrt{\frac{g}{\ell}} \tag{16.29}$$

Angular frequency of a **physical pendulum**:

$$\omega_{\text{phy}} = \sqrt{\frac{mgr_{\text{CM}}}{I}} \tag{16.33}$$

Angular frequency of a **torsion pendulum**:

$$\omega_{\text{tor}} = \sqrt{\frac{\kappa}{I}} \tag{16.36}$$

PROBLEMS AND QUESTIONS

A = algebraic C = conceptual E = estimation G = graphical N = numerical

16-1 Picturing Harmonic Motion

1. G CASE STUDY For each velocity listed, state the position and acceleration of the rubber disk in Crall and Whipple's experiment (Figs. 16.3–16.5). There may be more than one possible answer for each given velocity. **a.** $v_y = 1.3$ m/s **b.** $v_y = -1.3$ m/s **c.** $v_y = 0$
2. G CASE STUDY For each acceleration listed, state the position and velocity of the disk in Crall and Whipple's experiment (Figs. 16.3–16.5). There may be more than one possible answer for each given acceleration. **a.** $a_y = 3.8$ m/s^2 **b.** $a_y = -3.8$ m/s^2 **c.** $a_y = 0$

16-2 Kinematic Equations of Simple Harmonic Motion

3. N A particle in simple harmonic motion has a period of 4.3 s. What is the particle's acceleration when it is at $\vec{r} = -1.65\hat{\jmath}$ m?
4. N A simple harmonic oscillator's position is given by $y(t) = (0.850\text{ m})\cos(10.4t - 5.20)$. Find the oscillator's position, velocity, and acceleration at each of the following times. **a.** $t = 0$ **b.** $t = 0.500$ s **c.** $t = 2.00$ s
5. N A simple harmonic oscillator's velocity is given by $v_y(t) = (0.850\text{ m/s})\sin(10.4t - 5.20)$. Find the oscillator's position, velocity, and acceleration at each of the following times. **a.** $t = 0$ **b.** $t = 0.500$ s **c.** $t = 2.00$ s
6. N A simple harmonic oscillator's acceleration is given by $a_y(t) = (0.850\text{ m/s}^2)\cos(10.4t - 5.20)$. Find the oscillator's position, velocity, and acceleration at each of the following times. **a.** $t = 0$ **b.** $t = 0.500$ s **c.** $t = 2.00$ s
7. N The equation of motion of a simple harmonic oscillator is given by $x(t) = (8.0\text{ cm})\cos(10\pi t) - (6.0\text{ cm})\sin(10\pi t)$, where t is in seconds. **a.** Find the amplitude. **b.** Determine the period. **c.** Determine the initial phase.
8. N The expression $x = 8.50\cos(2.40\omega t + \pi/2)$ describes the position of an object as a function of time, with x in centimeters and t in seconds. What are the **a.** frequency, **b.** period, **c.** amplitude, and **d.** initial phase of the object's motion? **e.** What is the position of the particle at $t = 1.45$ s?
9. A A simple harmonic oscillator has amplitude A and period T. Find the minimum time required for its position to change from $x = A$ to $x = A/2$ in terms of the period T.
10. N A particle is moving in simple harmonic motion with an amplitude of 4.0 cm and a maximum velocity of 10 cm/s. Assume the initial phase is 0. **a.** At what position is its velocity 4.0 cm/s? **b.** What is its velocity when its position is 2.0 cm?
11. A 1.50-kg mass is attached to a spring with spring constant 33.0 N/m on a frictionless, horizontal table. The spring–mass system is stretched to 4.00 cm beyond the equilibrium position of the spring and is released from rest at $t = 0$.
    a. N What is the maximum speed of the 1.50-kg mass?
    b. N What is the maximum acceleration of the 1.50-kg mass?
    c. A What are the position, velocity, and acceleration of the 1.50-kg mass as functions of time?

16-3 Connection with Circular Motion

12. C A bicycle pedal has reflective stripes on the front and back (Fig. P16.12) You are in a parked car with its headlights on, and a cyclist is riding toward you. It is very dark, and the car's headlights only illuminate the front reflective stripe. Describe the motion you see.

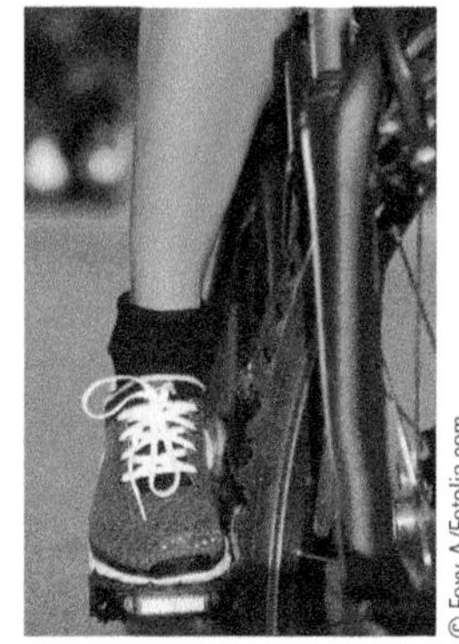

© Foxy-A/Fotolia.com

FIGURE P16.12

13. C A conical pendulum consists of a small ball of mass m attached to a string of length L. It executes uniform circular motion in a horizontal plane such that the string makes a fixed angle θ with the vertical, sweeping out the surface of a cone. The angular frequency of this motion is given by

$$\omega = \sqrt{\frac{g}{L\cos\theta}}$$

Suppose a shadow of the moving mass is projected onto a vertical screen adjacent to the pendulum by a light source in the plane of motion. Describe the motion of the shadow.

14. C When the Earth passes a planet such as Mars, the planet appears to move backward for a time, a phenomenon known as **retrograde motion**. Ancient astronomers believed that the Earth did not move and that the planets moved around the Earth. They also believed that uniform circular motion was perfect and that heavenly objects such as planets exhibited this perfect motion. How do you suppose ancient astronomers accounted for retrograde motion? Include a sketch with your explanation.

Problems 15 and 16 are paired.

15. N A point on the edge of a child's pinwheel is in uniform circular motion as the wheel spins counterclockwise with a frequency of 1.53 Hz. The point is at the location $x = 30.00$ cm and $y = 0$

when a stopwatch is started to track the motion (Fig. P16.15). **a.** What is the period of the circular motion? **b.** What is the velocity of the point at the instant described? **c.** What is the acceleration of the point at the instant described?

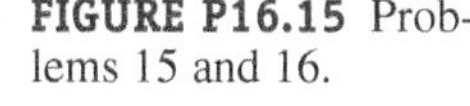

FIGURE P16.15 Problems 15 and 16.

16. Consider the scenario in Problem 15 (Fig. P16.15).

a. C, N How do the maximum values of the x and y components of the velocity compare? Discuss the validity of your answer.

b. N When is the earliest time at which the x and y components of the velocity are the same?

c. G Sketch the path of the point and mark the location of two points where the velocity components are equal.

17. A Find expressions for the y component of the ball's velocity and acceleration in Figure 16.6 (page 456).

16-4 Dynamics of Simple Harmonic Motion

18. A jack-in-the-box undergoes simple harmonic motion after it pops out of its box with a frequency of 3.4 Hz and an amplitude of 15 cm.

a. N What is the maximum acceleration experienced by the jack-in-the-box?

b. N If the mass of the jack-in-the-box is 0.210 kg, what is the maximum net force it experiences during the motion?

c. A If the initial phase angle is $\pi/3$ rad, write an equation that describes the net force as a function of time.

19. C, N A uniform plank of length L and mass M is balanced on a fixed, semicircular bowl of radius R (Fig. P16.19). If the plank is tilted slightly from its equilibrium position and released, will it execute simple harmonic motion? If so, obtain the period of its oscillation.

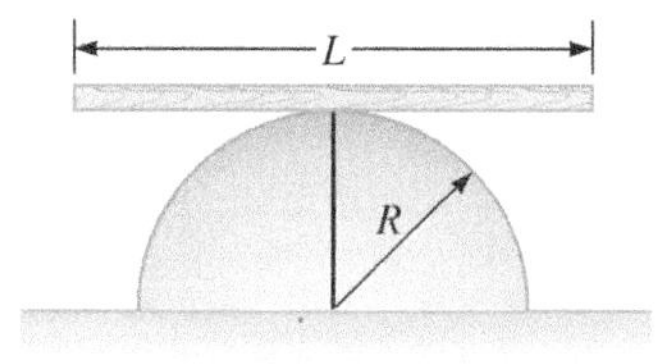

FIGURE P16.19

16-5 Special Case: Object–Spring Oscillator

20. C If Crall and Whipple performed their experiment (Fig. 16.3) in a laboratory on a high mountain where $g = 9.72$ m/s^2, what changes would there be to $y(t)$, $v_y(t)$, $a_y(t)$, and $F_y(t)$?

21. N A block of mass $m = 5.94$ kg is attached to a spring with spring constant $k = 1592$ N/m and rests on a frictionless surface. The block is pulled, stretching the spring a distance of 0.150 m, and is held still. The block is then released and moves in simple harmonic motion about the equilibrium position. **a.** What is the frequency of this oscillation? **b.** Where is the block located 3.24 s after it is released? **c.** What is the velocity of the mass at that time?

22. A A block of mass m rests on a frictionless, horizontal surface and is attached to two springs with spring constants k_1 and k_2 (Fig. P16.22). It is displaced to the right and released. Find an expression for the angular frequency of oscillation of the resulting simple harmonic motion.

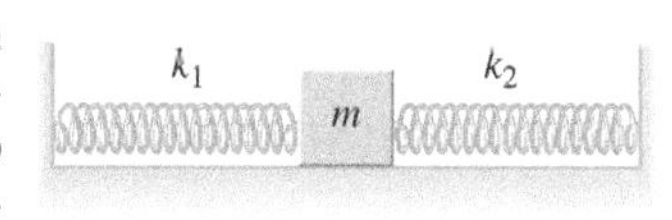

FIGURE P16.22 Problems 22 and 81.

23. N It is important for astronauts in space to monitor their body weight. In Earth orbit, a simple scale only reads an apparent weight of zero, so another method is needed. NASA developed the body mass measuring device (BMMD) for Skylab astronauts. The BMMD is a spring-mounted chair that oscillates in simple harmonic motion (Fig. P16.23). From the period of the motion, the mass of the astronaut can be calculated. In a typical system, the chair has a period of oscillation of 0.901 s when empty. The spring constant is 606 N/m. When a certain astronaut sits in the chair, the period of oscillation increases to 2.37 s. Determine the mass of the astronaut.

FIGURE P16.23

24. N A popular item for infants consists of a seat that hangs from a spring attached to the top of a door frame. Most infants quickly learn to bounce up and down in simple harmonic motion. Suppose the total mass of an infant and the chair is 11.0 kg and the spring constant is 1.20×10^3 N/m. Find the amplitude of her oscillation such that she just loses contact with the seat when she reaches maximum height above her equilibrium position.

25. A A spring of mass m_s and spring constant k is attached to an object of mass M and set into simple harmonic motion on a frictionless, horizontal table. All portions of the spring are assumed to oscillate in phase, and the velocity of each segment dx of the spring with mass dm can be assumed to be proportional to the distance x of that segment from point A in Figure P16.25. **a.** What is the kinetic energy of the system at the instant the object is moving with speed v? **b.** What is the frequency of oscillation of the system?

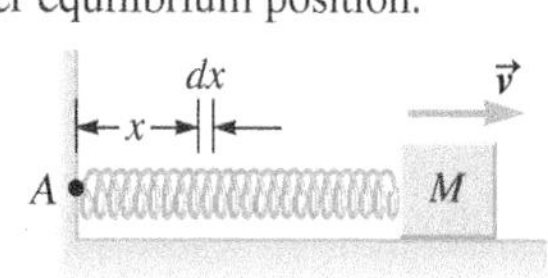

FIGURE P16.25

16-6 Special Case: Simple Pendulum

26. N In an undergraduate physics lab, a simple pendulum is observed to swing through 75 complete oscillations in a time period of 2.25 min. What are the **a.** period and **b.** length of the pendulum?

27. C A simple pendulum of length L hangs from the ceiling of an elevator. **a.** While the elevator is moving up with constant acceleration a, is the period of the pendulum affected? If so, how? **b.** Now suppose we hang a particle of mass m on a spring of spring constant k and attach it to the ceiling of the same elevator. How does an upward acceleration a affect the period of this simple harmonic oscillator?

28. A We do not need the analogy in Equation 16.30 to write expressions for the *translational* displacement of a pendulum bob along the circular arc $s(t)$, *translational* speed $v(t)$, and *translational* acceleration $a(t)$. Show that they are given by

$$s(t) = s_{\max} \cos(\omega_{\text{smp}} t + \varphi)$$

$$v(t) = -v_{\max} \sin(\omega_{\text{smp}} t + \varphi)$$

$$a(t) = -a_{\max} \cos(\omega_{\text{smp}} t + \varphi)$$

respectively, where $s_{\max} = \ell\theta_{\max}$ with ℓ being the length of the pendulum, $v_{\max} = s_{\max}\omega_{\text{smp}}$, and $a_{\max} = s_{\max}\omega_{\text{smp}}^2$.

29. N Dr. Chaos uses his watch to hypnotize unsuspecting citizens in attempting to rob them. His watch is a simple pendulum that he sets into simple harmonic motion. The length of the pendulum is 0.150 m, and the mass at the end is 3.25 kg. Use the equations described in Problem 28 to model the motion of the watch when it is pulled back to a maximum angle of 5° and released. Assume the initial phase is zero. What are the **a.** period of motion, **b.** maximum speed of, and **c.** maximum acceleration of the watch?

30. N A simple pendulum near sea level ($g = 9.81$ m/s^2) has a period of 1.50 s. If you observe the motion of the pendulum on a high mountain ($g = 9.72$ m/s^2), what is the new period?

31. A simple pendulum comprised of a bob of mass $m = 122$ g and a lightweight string of length 75.0 cm is released from rest from an initial angle of 23.0° from the vertical.
   a. N Using the approach of simple harmonic motion, what is the maximum speed of the pendulum bob?
   b. N Using the approach of simple harmonic motion, what is the maximum angular acceleration of the bob?
   c. C How does the speed found in part (a) compare with that found by using a conservation of energy approach?

32. N A simple pendulum is constructed from a bob of mass $m =$ 150 g and a lightweight string of length $\ell = 1.50$ m. What are the periods of oscillation for this pendulum **a.** in a physics lab at sea level, **b.** in an elevator accelerating upward at 2.00 m/s^2, **c.** in an elevator accelerating downward at 2.00 m/s^2, and **d.** in a school bus accelerating horizontally at 2.00 m/s^2?

16-7 Special Case: Physical Pendulum

33. C Three scenarios are listed, each involving a physical pendulum. In each case, determine what happens to the period. Does the period increase, decrease, or stay the same? **a.** A young man is crouched on a moving tire swing and then stands up. **b.** A gymnast swings by her arms from a high bar and then swings by her knees. **c.** A trapeze artist swings alone, standing on top of a bar. Another person then grabs the bottom of the bar so that the two people swing together.

34. A Show that angular frequency of a physical pendulum $\omega_{\text{phy}} = \sqrt{mgr_{\text{CM}}/I}$ (Eq. 16.33) equals the angular frequency of a simple pendulum $\omega_{\text{smp}} = \sqrt{g/\ell}$ (Eq. 16.29) in the case of a particle at the end of a string of length ℓ.

35. A A uniform annular ring of mass m and inner and outer radii a and b, respectively, is pivoted around an axis perpendicular to the plane of the ring at point P (Fig. P16.35). Determine its period of oscillation.

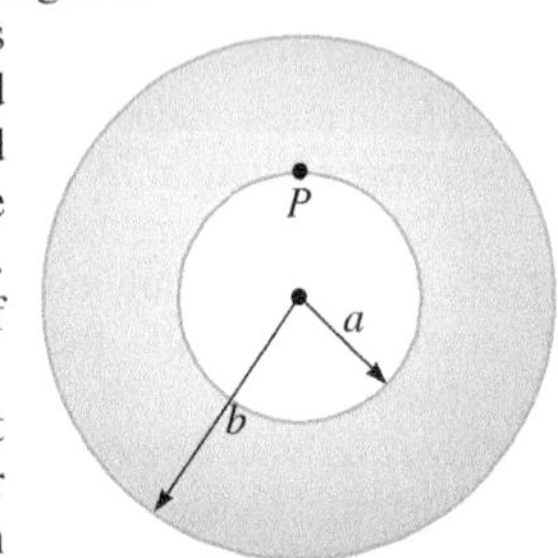

FIGURE P16.35

36. N A child works on a project in art class and uses an outline of her hand on a sheet of construction paper to draw a turkey (Fig. P16.36). The teacher pins the turkey to the bulletin board in the front of the classroom by using a thumbtack. The student notices that if she flicks her finger on the end of the turkey, it oscillates back and forth with a frequency of about 1.65 Hz. If the rotational inertia of the paper turkey is 1.25×10^{-5} kg · m^2 and its mass is 0.005 kg, what is the distance between the thumbtack and the center of mass of the turkey?

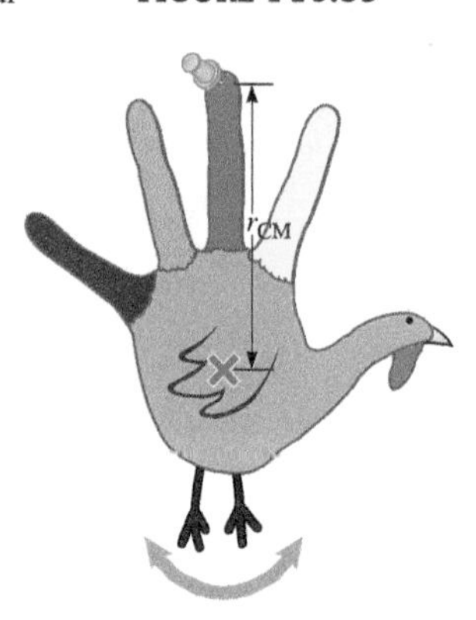

FIGURE P16.36

37. N Three thin sticks of equal mass, each of length 20.0 cm, are connected at the ends to form an equilateral triangle. When the triangle is pivoted around an axis perpendicular to the plane of the triangle and passing through one of the vertices, what is the period of the triangle's oscillation as a physical pendulum?

38. N A physical pendulum in the form of a thin rod of length 1.00 m is pivoted at point P, a distance x from the pendulum's center of mass. Find the value of x for which the period of the pendulum will be a minimum.

39. E In the short story *The Pit and the Pendulum* by 19th-century American horror writer Edgar Allen Poe, a man is tied to a table directly below a swinging pendulum that is slowly lowered toward him. The "bob" of the pendulum is a 1-ft steel scythe connected to a 30-ft brass rod. When the man first sees the pendulum, the pivot is roughly 1 ft above the scythe so that a 29-ft length of the brass rod oscillates above the pivot (Fig. P16.39A). The man escapes when the pivot is near the end of the brass rod (Fig. P16.39B). **a.** Model the pendulum as a particle of mass $m_s = 2$ kg attached to a rod of mass $m_r = 160$ kg. Find the pendulum's center of mass and rotational inertia around an axis through its center of mass. (Check your answers by finding the center of mass and rotational inertia of just the brass rod.) **b.** What is the initial period of the pendulum? **c.** The man saves himself by smearing food on his ropes so that rats chew through them. He does so when he has no more than 12 cycles before the pendulum will make contact with him. How much time does it take the rats to chew through the ropes?

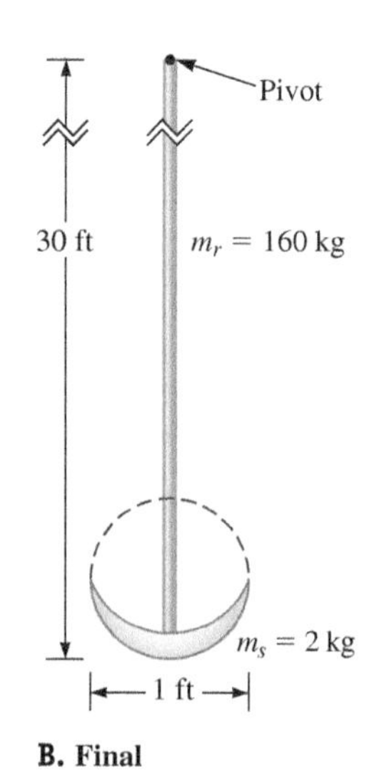

FIGURE P16.39

16-8 Special Case: Torsion Pendulum

40. C A pendulum clock that keeps time on the Earth such as that in Figure 16.1A cannot be used on the Moon because the pendulum's period would be different there. Can Hillis's millennium clock based on a torsion pendulum (Fig. 16.17, page 468) be used on the Moon to keep accurate time? Explain.

41. N A restaurant manager has decorated his retro diner by hanging (scratched) vinyl LP records from thin wires. The records have a mass of 180 g, a diameter of 12 in., and negligible thickness. The records oscillate as torsion pendulums. **a.** Records hung from a small hole near their rims have a period of roughly 3.5 s (Fig. P16.41A). What is the torsion spring constant of the wire? **b.** If a record is hung from its center hole using a wire of the same torsion spring constant (Fig. P16.41B), what is its period of oscillation?

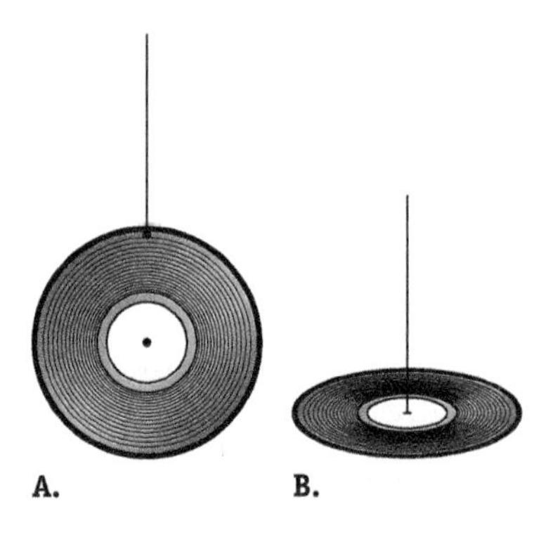

FIGURE P16.41

16-9 Energy in Simple Harmonic Motion

42. N A simple harmonic oscillator consists of a 695-g block attached to a lightweight spring. The total energy of the system is 8.50 J, and its period of oscillation is 0.330 s. **a.** What is the maximum speed of the block? **b.** What is the force constant of the spring? **c.** What is the amplitude of the motion of the block?

43. N A wooden block ($m = 0.600$ kg) is connected to a spring and undergoes simple harmonic motion with an amplitude of oscillation of 0.075 m. The frequency of the motion is 12.50 Hz. **a.** What is the spring constant? **b.** What is the maximum speed of the block? **c.** What is the speed of the block when it is 0.015 m away from the equilibrium position?

44. N A box of mass 0.900 kg is attached to a spring with $k = 125$ N/m and set into simple harmonic motion on a frictionless, horizontal table. The amplitude of motion is 5.00 cm. **a.** What is the total energy of the box–spring system? **b.** What is the speed of the box when the spring is compressed by 2.00 cm? **c.** What is the kinetic energy of the box at this position? **d.** What is the potential energy of the box–spring system at this position?

Problems 45 and 46 are paired.

45. N A block of mass $m = 1.23$ kg is attached to the end of a spring with a spring constant of 565 N/m. The block rests on a frictionless surface, is pulled to the right, and is held there. When released, the block undergoes simple harmonic motion. **a.** If the maximum speed of the block is 7.12 m/s, what is the amplitude of the motion? **b.** What is the speed of the block when it is halfway between the equilibrium point and the maximum displacement ($x = \frac{1}{2}A$)?

46. N In Problem 45, the block's maximum speed is 7.12 m/s. There is an instant in the motion when the potential energy of the system is equal to the kinetic energy of the block. At what position x does this situation occur?

47. N This problem is a follow-up to Example 16.7 (page 472). A horizontal spring of spring constant $k = 170$ N/m is attached to an object of mass $m = 0.85$ kg. Friction between the object and the table is negligible. A student stretches the spring so that the object's initial position is $x_i = 0.20$ m and releases the system at $t = 0$. Find the kinetic and potential energies at $t = 2.5$ s.

48. N An object of mass 0.20 kg executes simple harmonic motion along the x axis with a frequency $f = 25/\pi$ Hz. At a particular moment, the object has 0.50 J of kinetic energy and 0.40 J of potential energy. Find the amplitude of oscillation.

16-10 Damped Harmonic Motion

49. N A 2000-kg car is lowered by 1.50 cm when four passengers, each of mass 70 kg, sit down in it. **a.** Determine the damping constant b of the shock absorbers that will provide critical damping. **b.** Suppose for the same car the shock absorbers are so worn that they provide almost no damping. Find the period of up-and-down oscillation of the car after hitting a bump in the road.

50. N The bob of a simple pendulum is displaced by an initial angle of 22.0° and released from rest. Because of friction, the amplitude of oscillation is observed to be half the initial value after 765 s. What is the value of the time constant τ for damping for this pendulum?

51. C A object–spring oscillator is critically damped. The damping force exerted on the object is of the form $F_D = -bv$. If the object's mass is increased, is the oscillator still critically damped? If not, is it underdamped or overdamped? Explain.

52. A Show that for an object–spring oscillator, Equation 16.3

$$y(t) = y_{\max} \cos(\omega t + \varphi)$$

is a solution to Equation 16.47,

$$-\frac{k}{m}y(t) = \frac{d^2y(t)}{dt^2}.$$

53. A Show that Equation 16.50,

$$y(t) = A(t)\cos(\omega_D t + \varphi)$$

is a solution to Equation 16.49,

$$-ky(t) - b\,dy(t)/dt = m\,d^2y(t)/dt^2$$

where $A(t) = y_{\max}e^{-t/\tau}$ (Eq. 16.51), $\tau \equiv 2m/b$, and $\omega_D = [\omega^2 - (b^2/4m^2)]^{1/2}$.

54. A Verify that the time constant $\tau \equiv 2m/b$ has the dimensions of time.

55. C A cat toy consists of a lightweight, hollow ball ($m = 2.0$ g) hanging on the end of a thin string of length $\ell = 32$ cm. A pet owner displaces the ball by 6° and releases it. The cat enthusiastically watches the ball swing back and forth, but the ball's amplitude is reduced to 1° in 24 s. Both the cat and owner are disappointed. Would it help the cat if the hollow ball were filled with BBs, increasing its mass but not changing its aerodynamics? Explain your answer.

16-11 Driven Oscillators

56. C A university has a nonoperating Foucault pendulum. Senior students at the university decided to donate a new motor. Avi, Cameron, and Shannon have been asked to install the motor and bring the pendulum back to life. The pendulum looks much like the one in the Pantheon (Fig. 16.27, page 477) with a bob at the bottom of a steel cable of length 40 ft. It should have a 7.00-s period. The students have connected the motor and turned it on, but the pendulum is not responding. The motor has two dials, one for frequency and one for amplitude. It has one display for the frequency. Read the three students' discussion and answer the questions.

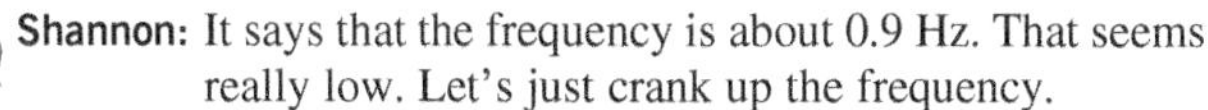

Shannon: It says that the frequency is about 0.9 Hz. That seems really low. Let's just crank up the frequency.

Avi: I calculated the frequency for a simple pendulum with a 40-ft cable, and that's about right. I think we just need to turn up the amplitude.

Shannon: It's already maxed out.

Avi: Really? Then I would say that the motor is broken. Or maybe the seniors didn't pay for a strong enough motor. Let's get our money back.

Cameron: Avi, you used the formula for a *simple* pendulum. No real pendulum is a simple pendulum. You need to model it as a physical pendulum.

Shannon: Cameron, you can calculate that if you want. I'm just going to turn up the frequency.

a. What happens as Shannon slowly turns up the frequency?
b. Do you agree with Avi's calculation?
c. Is Cameron correct that no real pendulum can be modeled as a simple pendulum? If so, calculate the correct frequency.
d. The amplitude is maxed out, and the pendulum is not oscillating. Do those problems mean that the motor is defective? In other words, does the pendulum just need a bigger push than this motor is delivering?

57. N To demonstrate the concept of resonance to your son and his friends, you suspend your smartphone by a lightweight string of length L and set the phone on vibrate. Using your landline phone, you call the cell phone, which vibrates with a frequency of 0.900 Hz, causing your makeshift pendulum to oscillate at a very large amplitude. What is the length L of the string you used in this experiment?

58. N An ideal simple harmonic oscillator comprises a 255-g ball hanging from a lightweight, vertical spring with spring constant $k = 8.50$ N/m. The system is driven by a sinusoidal force of amplitude 3.00 N. If the ball vibrates with an amplitude of 75.0 cm, what is the frequency of the driving force?

General Problems

Problems 59–65 are grouped.

59. G Table P16.59 gives the position of a block connected to a horizontal spring at several times. Sketch a motion diagram for the block.

TABLE P16.59

Time of measurement, t (s)	Position of block, x (m)
0	5.00
0.25	3.50
0.50	0
0.75	−3.50
1.00	−5.00
1.25	−3.50
1.50	0
1.75	3.50
2.00	5.00
2.25	3.50
2.50	0

60. G Use the position data for the block given in Table P16.59. Sketch a graph of the block's **a.** position versus time, **b.** velocity versus time and **c.** acceleration versus time. There is no need to label the values of velocity or acceleration on those graphs.

61. C Consider the position data for the block given in Table P16.59. What are the signs of the block's velocity and acceleration at the first five times listed?

62. A Use the data in Table P16.59. Write an expression for the magnitude of the block's **a.** position, **b.** velocity, and **c.** acceleration. Assume the maximum position observed is the amplitude.

63. Use the data in Table P16.59 for a block of mass $m = 0.250$ kg and assume friction is negligible.

a. N What is the spring constant k?

b. N What is the maximum force exerted on the block?

c. C, N If you replace the block with a new block of mass $m =$ 0.125 kg, which parameters (x_{max}, v_{max}, a_{max}, F_{max}) change? For each parameter that changes, give its new value.

64. Use the data in Table P16.59 for a block of mass $m = 0.250$ kg and assume friction is negligible.

a. N Write an expression for the force F_H exerted by the spring on the block.

b. G Sketch F_H versus t.

65. Consider the data for a block of mass $m = 0.250$ kg given in Table P16.59. Friction is negligible.

a. N What is the mechanical energy of the block–spring system?

b. N Write expressions for the kinetic and potential energies as functions of time.

c. G Plot the kinetic energy, potential energy, and mechanical energy as functions of time on the same set of axes.

66. N A mass on a spring undergoing simple harmonic motion completes 4.00 cycles in 14.0 s. **a.** What is the period of motion for this system? **b.** What is the frequency, in hertz, of this system? **c.** What is the angular frequency of this system?

67. A particle initially located at the origin undergoes simple harmonic motion, moving first in the positive z direction, with a frequency of 3.20 Hz and an amplitude of 1.40 m. The particle oscillates between $z = 1.40$ m and $z = -1.40$ m.

a. A What is the equation describing the particle's position as a function of time?

b. N What is the maximum speed of the particle?

c. N What is the maximum acceleration of the particle?

d. N What is the total distance covered by the particle in the first 2.50 s of this motion?

68. A Consider the system shown in Figure P16.68 as viewed from above. A block of mass m rests on a frictionless, horizontal surface and is attached to two elastic cords, each of length L. At the equilibrium configuration, shown by the dashed line, the cords both have tension F_T. The mass is displaced a small amount as shown in the figure and released. Show that the net force on the mass is similar to the spring-restoring force and find the angular frequency of oscillation, assuming the mass behaves as a simple harmonic oscillator. You can assume the displacement is small enough to produce negligible change in the tension and length of the cords.

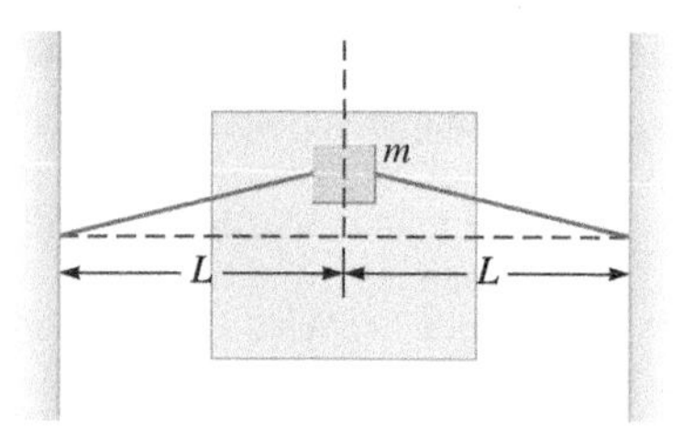

FIGURE P16.68

69. N A simple pendulum comprising a bob of mass $m = 2.50$ kg and a massless string of length 1.25 m is observed to have a speed of 3.24 m/s at its equilibrium position. **a.** What is the period of this pendulum? **b.** What is the total energy of the pendulum? **c.** What is the maximum displacement of the pendulum from its equilibrium position?

70. C An object–spring oscillator and a simple pendulum are each set into simple harmonic motion on the Earth and are each found to oscillate with a frequency of 4.80 Hz. If they were instead set into simple harmonic motion on the Moon, how would their frequencies compare? Explain any similarities or differences.

71. C A hollow, metal sphere is filled with water, and a small hole is made at the bottom. The sphere hangs by a long thread and is made to oscillate. How will the period of oscillation change over time if water is allowed to flow through the hole until the sphere is empty?

72. N A system exists in which a block of mass $m_1 = 0.500$ kg connected to a spring is extended to a displacement of 5.00 cm and released from rest. The block then oscillates with simple harmonic motion on a frictionless, horizontal tabletop. What is the maximum frequency of motion for which a second block, of mass $m_2 = 0.125$ kg, placed atop the first block, will not slide off? Assume the coefficient of static friction between the two blocks is $\mu_s = 0.650$.

73. A Determine the period of oscillation of a simple pendulum of length L suspended from the ceiling of a car that rolls down an inclined plane of angle α (Fig. P16.73). Dissipative forces between the car and the plane are negligible.

FIGURE P16.73

74. The total energy of a simple harmonic oscillator with amplitude 3.00 cm is 0.500 J.

a. N What is the kinetic energy of the system when the position of the oscillator is 0.750 cm?

b. N What is the potential energy of the system at this position?

c. N What is the position for which the potential energy of the system is equal to its kinetic energy?

d. C For a simple harmonic oscillator, what, if any, are the positions for which the kinetic energy of the system exceeds the maximum potential energy of the system? Explain your answer.

75. **A** A spherical bob of mass m and radius R is suspended from a fixed point by a rigid rod of negligible mass whose length from the point of support to the center of the bob is L (Fig. P16.75). Find the period of small oscillation.

FIGURE P16.75

76. **N** The frequency of a physical pendulum comprising a nonuniform rod of mass 1.25 kg pivoted at one end is observed to be 0.667 Hz. The center of mass of the rod is 40.0 cm below the pivot point. What is the rotational inertia of the pendulum around its pivot point?

77. A lightweight spring with spring constant k = 225 N/m is attached to a block of mass m_1 = 4.50 kg on a frictionless, horizontal table. The block–spring system is initially in the equilibrium configuration. A second block of mass m_2 = 3.00 kg is then pushed against the first block, compressing the spring by x = 15.0 cm as in Figure P16.77A. When the force on the second block is removed, the spring pushes both blocks to the right. The block m_2 loses contact with the spring–block 1 system when the blocks reach the equilibrium configuration of the spring (Fig. P16.77B).

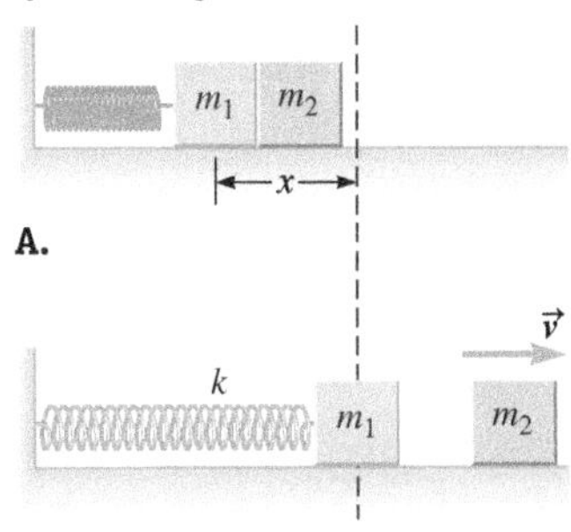

FIGURE P16.77

a. **N** What is the subsequent speed of block 2?

b. **C** Compare the speed of block 1 when it again passes through the equilibrium position with the speed of block 2 found in part (a).

78. **A** Determine the angular frequency of oscillation of a thin, uniform, vertical rod of mass m and length L pivoted at the point O and connected to two springs (Fig. P16.78). The combined spring constant of the springs is k ($k = k_1 + k_2$), and the masses of the springs are negligible. Use the small-angle approximation ($\sin\theta \approx \theta$).

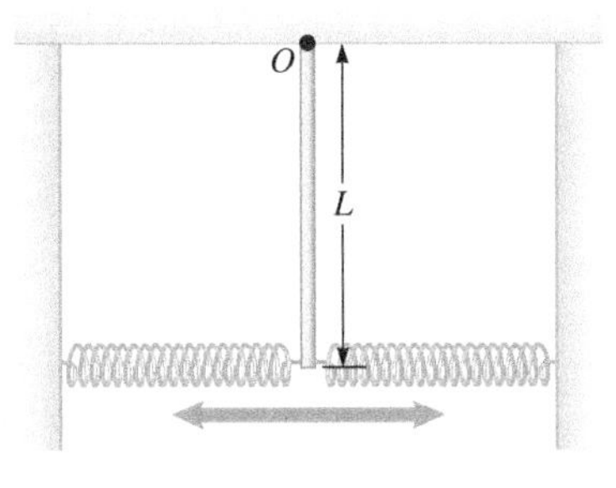

FIGURE P16.78

79. **N** Air resistance in a lab causes the motion of a 5.00-kg disk attached to a vertical spring with spring constant $k = 5.000 \times 10^3$ N/m to be damped at a rate given by a damping coefficient $b = 4.50$ N · s/m. **a.** What is the frequency of the damped oscillation of the system? **b.** What is the percentage by which the amplitude of motion decreases after each cycle?

Problems 80 and 81 are paired.

80. **A** Two springs, with spring constants k_1 and k_2, are connected to a block of mass m on a frictionless, horizontal table (Fig. P16.80). The block is extended a distance x from equilibrium and released from rest. Show that the block executes simple harmonic motion with a period given by

$$T = 2\pi\sqrt{\frac{m(k_1 + k_2)}{k_1 k_2}}$$

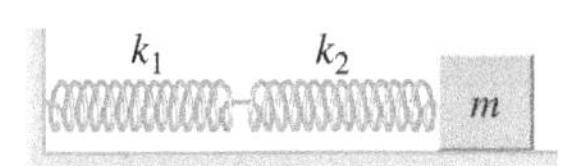

FIGURE P16.80

81. **A** Figure P16.22 shows a system in which two springs, with spring constants k_1 and k_2, are connected to a block of mass m on a frictionless, horizontal table. The block is moved a distance x to the right and released from rest. Show that the block executes simple harmonic motion with a period given by

$$T = 2\pi\sqrt{\frac{m}{(k_1 + k_2)}}$$

82. **A** Show that Equation 16.59,

$$y(t) = A_{drv}\cos(\omega_{drv}t + \varphi_{drv})$$

is a solution to Equation 16.58,

$$F_{max}\cos\omega_{drv}t = m\frac{d^2y(t)}{dt^2} + b\frac{dy(t)}{dt} + ky(t)$$

where $A_{drv} = \dfrac{F_{max}}{\sqrt{m^2(\omega^2 - \omega_{drv}^2)^2 + b^2\omega_{drv}^2}}$ (Eq. 16.60). *Hint*: Use $e^{i\alpha} = \cos\alpha + i\sin\alpha$.

17 Traveling Waves

Key Questions

How can we describe a traveling wave mathematically (that is, its kinematics)?

What is carried or transferred by a wave?

17-1 Introducing mechanical waves 487

17-2 Pulses 487

17-3 Harmonic waves 490

17-4 Special case: Transverse wave on a rope 494

17-5 Sound: Special case of a traveling longitudinal wave 497

17-6 Energy transport in waves 500

17-7 Two- and three-dimensional waves 502

17-8 Refraction and diffraction 506

17-9 The Doppler shift 507

17-10 The wave equation 513

Underlying Principles

One-dimensional wave equation

Major Concepts

1. Longitudinal wave
2. Transverse wave
3. Harmonic wave
4. Wave function of transverse harmonic wave
5. Wave function of longitudinal harmonic wave
6. Intensity
7. Law of refraction
8. Diffraction
9. Doppler shift equation
10. Shock wave

Special Cases

1. Speed of transverse wave on a rope 2. Sound

Sometimes, a small group of friends in a large stadium or music hall will start a "wave" going around the audience. The friends all stand up in unison, lifting their arms above their heads and then sitting back down together (Fig. 17.1). They hope that a group of people—say to their immediate left—will imitate them soon afterward. If the people to the left of those people imitate them and so on, the wave propagates around the audience. If the stadium seating is arranged in a loop, the wave can propagate around and around the loop until people grow tired of the activity.

A stadium wave has the features of traveling waves we study in this chapter. A disturbance—people standing and then sitting—travels around the audience. Each person rises and falls in place; no one walks around the stadium. Enthusiasm or energy, however, flows around the stadium.

FIGURE 17.1 Fans in a stadium create a wave by standing and lifting their arms. The enthusiasm travels around the stadium.

17-1 Introducing Mechanical Waves

Suppose you would like to ask one of your classmates to join you for a movie. You could write a note and toss it on your (soon-to-be) friend's desk. A second—perhaps more effective—method is to simply ask. In the first case, you have used matter to transport your message; paper was displaced from your hand to your friend's desk. In the second case, no matter was transported. Your vocal cords created a disturbance in the air, and that disturbance—not the air molecules—traveled to your friend's eardrums. The disturbance is a sound wave. If you don't want to drop off a note or ask in person, you could send a text message. The text message also uses a wave in the form of an electromagnetic signal traveling from your cell phone to a relay tower and then to your friend's phone.

A **wave** is a disturbance in a medium (or field). For a sound wave, the medium is air as disturbed by a moving object such as your vocal cords. Waves are distinguished in part by the medium through which they travel. In this chapter and the next, we study **mechanical waves** (waves in a material medium such as a solid or fluid). In Chapters 34 through 38, we study light and radiation (electric and magnetic field waves requiring no material medium for propagation).

Traveling waves disturb a medium as they move from one region to another region. For example, when you ask your friend a question, your vocal cords make air molecules oscillate. The resulting disturbance in the air moves away from you. When this disturbance reaches your friend's eardrums, they oscillate. No matter what type of medium a wave travels in, it transports energy and momentum, but the medium itself is not carried along with the wave.

CASE STUDY Rolling Stones and Kidney Stones

College students discuss the warning label that came with Shannon's new earbuds.

Shannon: Check out all the warning labels on my new earbuds. Do you really think that sound can damage your hearing? Everyone is just worried about getting sued.

Avi: I think it's true. In high school health class, they made us read this article in *Rolling Stone* about how all these old rock stars were losing their hearing. Now they are trying to make people more aware of the dangers of loud music.

Shannon: But those guys didn't even wear earbuds. I just don't believe that music can damage your ears. Maybe an explosion can bust your eardrums because it blows particles right through your body.

Cameron: Look, it doesn't have to be that dramatic. When I volunteered at the hospital last summer, they broke up kidney stones with ultrasound.

Shannon: Well, isn't using *ultrasound* on a kidney stone like jackhammering a boulder? It's not like I'm jamming a pencil in my ear canal.

Avi: Yeah, but sound is a wave. It sends air molecules down your ear canal. It's like when you use a sandblaster to take the paint off a car. The air molecules coming out of an earbud just blast your eardrums.

Are any of these students correct? In this case study, we'll explore the damage that sound can do to your hearing as well as the use of sound to break up kidney stones. In the process, we'll learn more about waves, what they transmit, and how we hear.

17-2 Pulses

To see how a disturbance travels in a medium, imagine squeezing together several coils of a stretched-out horizontal spring that has a small bell dangling from the far end (Fig. 17.2). Your hands have created a disturbance in the medium (the spring). When you release the coils, the disturbance travels to the right. Beads on the coils of the spring highlight the wave. Notice that the coils never move very much, and, after the wave has passed

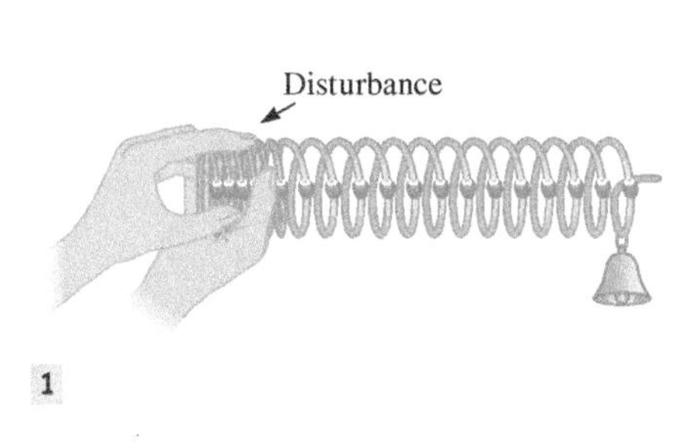

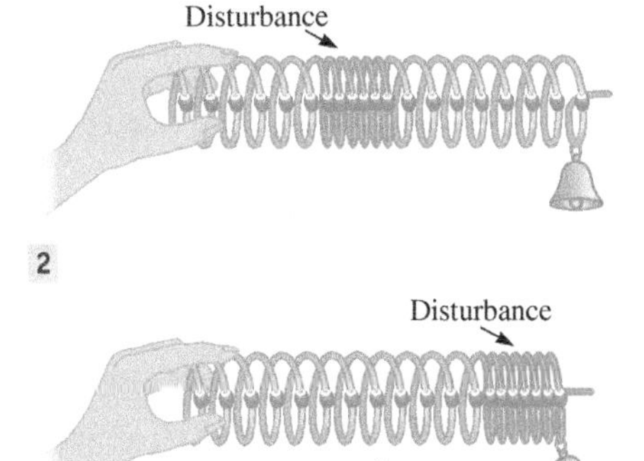

FIGURE 17.2 1 You create a longitudinal pulse by compressing a few coils together and releasing them. 2 The pulse travels to the right as the coils oscillate. 3 The pulse reaches the bell, causing it to ring.

LONGITUDINAL WAVE; TRANSVERSE WAVE

Major Concepts

a particular part of the spring, the coils return to their original positions. Contrast the coils' motion to the motion of the disturbance. The disturbance propagates the full length of the spring from your hands to the opposite end, causing the hanging bell to oscillate.

We will model a wave as the motion of a collection of particles represented by beads on the coils in Figure 17.2. As the wave propagates along the spring, each particle's motion forward and back is parallel to the spring's long axis. The wave in Figure 17.2 is an example of a **longitudinal wave**, one in which the motion of each particle is parallel to the direction of the wave's propagation. It might help to remember that in a *long*itudinal wave, the particles in the medium jiggle a*long* the direction of propagation.

In a **transverse wave**, particles in the medium move perpendicular to the direction of propagation. Figure 17.3 shows how you could set up a transverse wave in the horizontal spring. By flicking your hand up and then back down, you create a disturbance that is perpendicular to the spring's long axis. Each bead moves up and then back down while the disturbance moves to the right.

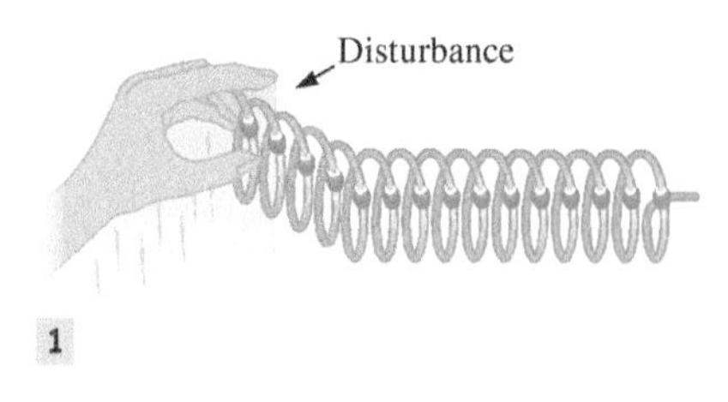

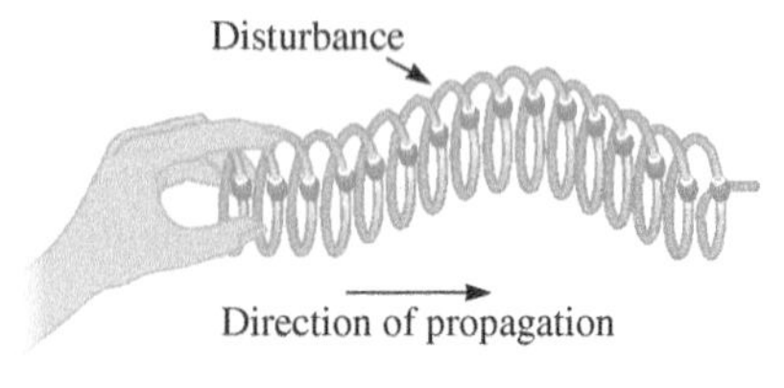

FIGURE 17.3 1 You create a transverse pulse by shaking the spring up and down. 2 The pulse travels to the right.

Wave Function for a Particular Pulse

The waves in Figures 17.2 and 17.3 are both **pulses** because they are discrete disturbances that do not repeat. Let's take a closer look at a single transverse pulse traveling in the positive x direction along a long, straight medium such as a stretched rope (Fig. 17.4). Again we model the rope as a collection of particles and use beads to help picture those particles. Our goal is to come up with a **wave function**, a mathematical description of the wave's kinematics.

Before there is a pulse, the rope is at rest; all the beads lie along the x axis (Fig. 17.4A). Mathematically, we can write $y(x) = 0$, meaning that $y = 0$ for all values of x. If no one ever plucks the rope so that there is never a pulse on it, we can write $y(x, t) = 0$, which means that $y = 0$ for all values of x at all times t. The notation (x, t) means that y (the wave function) depends on x and t.

If the rope is plucked at $t = 0$, a pulse is created. Imagine taking a photo at $t = 0$ (Fig 17.4B) so that you could easily see the shape of the pulse. To make our discussion specific, we have chosen a pulse with a simple shape. At $t = 0$, the rope's shape is mathematically described by

$$y(x) = \frac{1}{x^2 + 1} \tag{17.1}$$

where the constants are in appropriate SI units. Equation 17.1 is unique to this particular pulse shape. If the rope is plucked differently, the shape of the pulse will be different, and Equation 17.1 must be changed. The mathematical description $y(x)$ is known as the **profile** of the wave. The profile is only a function of the position x. To find an expression for the wave function $y(x, t)$ that is a function of both position x and time t, we consider the pulse at later times.

Let's start by considering one specific later time, say $t = 1$ s. At this time, the profile (shape) is unchanged, but the pulse has moved to the right by 2 m as you can see by examining the peak (Fig. 17.4C). The wave function must produce the same profile, but shifted to the right. At $t = 1$ s, the mathematical description is

$$y(x) = \frac{1}{(x - 2)^2 + 1}$$

where the 2 in $(x - 2)$ is needed because the pulse has shifted by 2 m. If you are wondering why there is a minus sign in $(x - 2)$ instead of a plus sign, notice that the peak of the pulse is $y_{peak} = 1$ m. If you substitute $x = 2$ m, you will find that the peak of the pulse is still 1 m, but if there had been a plus sign instead, as in

$$y(x) = \frac{1}{(x + 2)^2 + 1}$$

the peak would have been 1/17 m. So, only the negative sign is consistent with the profile of the wave. (We'll explore a leftward-moving pulse in part B of Example 17.1.)

We seek a way to describe the pulse mathematically at any arbitrary time t. Think of parts B through D of Figure 17.4 as a series of photos taken every second from $t = 0$ to $t = 2$ s. We see that the peak moves to the right at speed $v_x = 2$ m/s. So, the

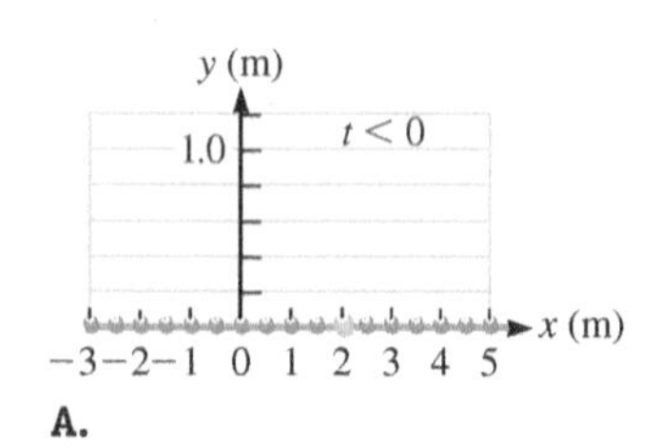

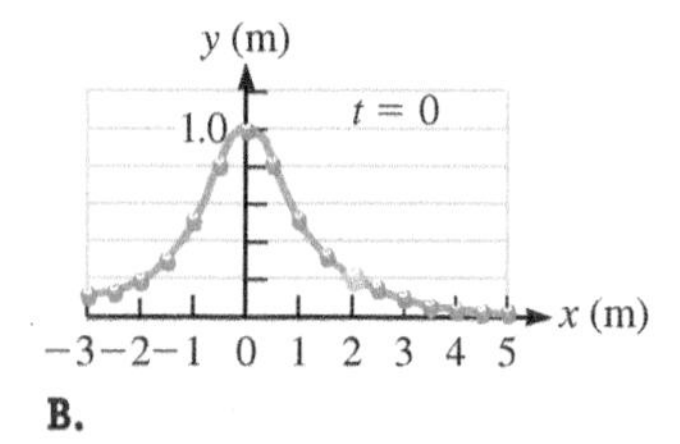

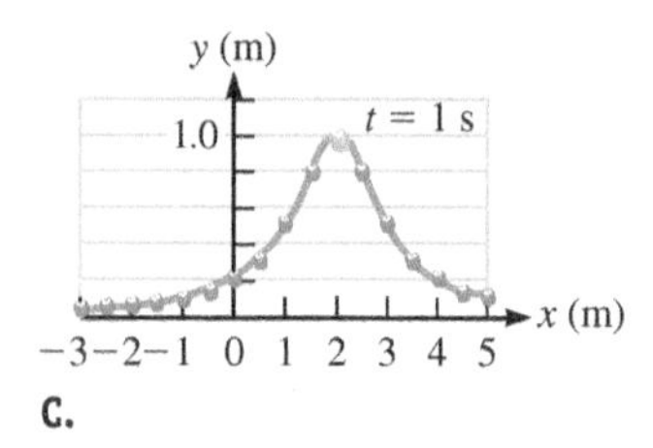

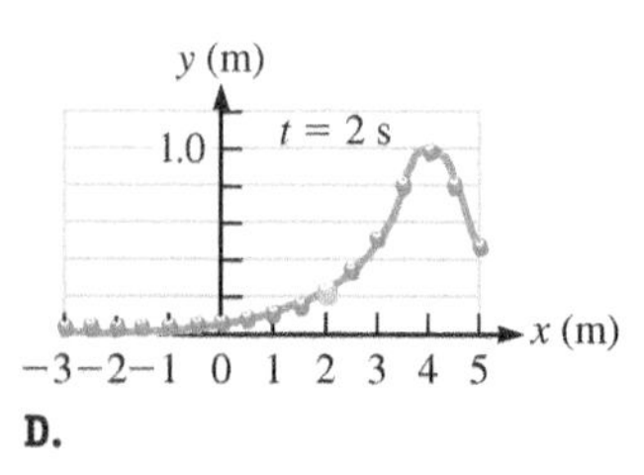

FIGURE 17.4 A transverse pulse on a rope moves to the right.

pulse shifts to the right by $v_x t$ relative to its initial position shown in Figure 17.4B. The wave function of this particular pulse is

$$y(x, t) = \frac{1}{(x - v_x t)^2 + 1} \quad (17.2)$$

where $v_x = 2$ m/s.

Equation 17.2 is not a general expression for a wave function; it is only the wave function of this particular pulse. Like all wave functions, however, this one provides a full description of the wave on the rope, giving the vertical position y of a rope particle (a bead) as a function of that particle's horizontal position x and time t. Wave functions are complicated because they depend on two variables x and t. Let's think about each variable individually.

First, if we substitute a specific time in Equation 17.2, it is like freezing the pulse in time by taking a photo. The wave function at a specific time is a mathematical description of the profile (the rope's shape).

If we instead substitute a specific position x into Equation 17.2, we then have a description of the up-and-down motion of one particle (one bead). Imagine watching the gold bead at $x = 2$ in Figure 17.4 through a narrow slit so that you cannot see the rest of the rope. The up-and-down motion you would see through the slit is best represented by a motion diagram (Fig. 17.5A). As usual, its position-versus-time graph (Fig. 17.5B) looks like a stretched-out version of its motion diagram; this graph also looks similar to the pulse's profile (Fig. 17.4), but don't confuse the two. The mathematical description of this gold bead's motion comes from substituting $x = 2$ m into Equation 17.2:

$$y(t) = \frac{1}{(2 - v_x t)^2 + 1} \quad (17.3)$$

where $v_x = 2$ m/s as before. Equation 17.3 gives the vertical position y as a function of time of only the gold bead.

Now that we've thought about the wave function at one particular time (by taking a photo) and at one particular position (by looking though a narrow slit at one particular bead), we see how the wave function describes the rightward motion of the pulse. According to the wave function, each bead only moves vertically along the y axis (as seen through the slit). No bead moves horizontally along the x axis. At any instant (as seen in a photo), the beads form a shape that is the profile of the pulse. According to the wave function, as the beads move vertically, that profile moves along the x axis.

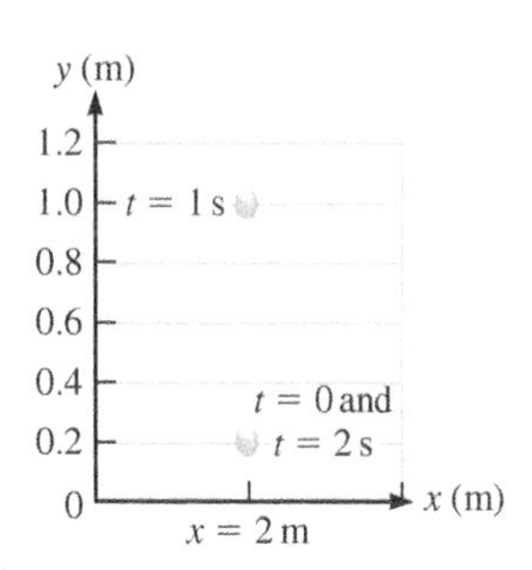

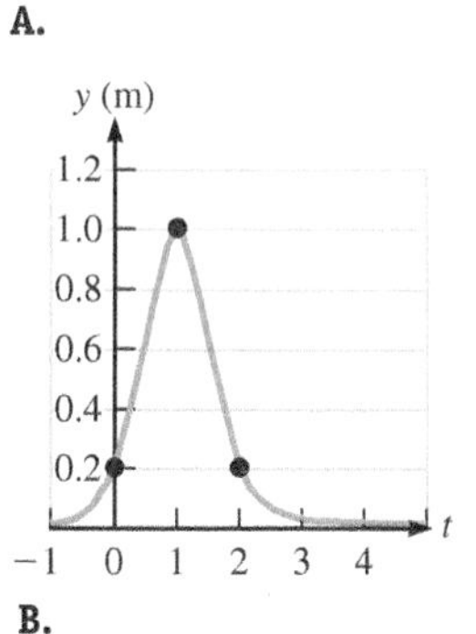

FIGURE 17.5 **A.** Motion diagram for the gold bead located at $x = 2$ m (Fig. 17.4). **B.** Position-versus-time graph for this gold bead.

CONCEPT EXERCISE 17.1

As we've seen before, terms used in physics often differ in meaning from the same terms used in everyday language. How does the use of the word *pulse* in one instance, *Be sure to check your pulse to ensure a productive workout*, differ from the physics definition of *pulse*?

CONCEPT EXERCISE 17.2

A graph of a pulse's profile and a position-versus-time graph for one particle in the medium look similar. For a transverse pulse traveling in the x direction, on both graphs the vertical axis is y. What is plotted on the horizontal axis in each graph?

EXAMPLE 17.1 Test It for Yourself

A Verify that Equation 17.3 with $v_x = 2$ m/s gives the position of the gold bead in Figures 17.4 and 17.5 by substituting $t = 0$, 1, 2, and 3 s.

Example continues on page 490 ▶

INTERPRET and ANTICIPATE
If Equation 17.3 describes the motion of the gold bead, we expect to find values for y that are consistent with those shown in Figure 17.5B.

SOLVE
Substitute $v_x = 2$ m/s and the times into Equation 17.3.

$$y(t) = \frac{1}{(2 - v_x t)^2 + 1} \quad (17.3)$$

$$y(t) = \frac{1 \text{ m}^3}{[2 \text{ m} - (2 \text{ m/s})t]^2 + 1 \text{ m}^2}$$

$$y(0) = 0.2 \text{ m} \qquad y(1) = 1 \text{ m}$$

$$y(2) = 0.2 \text{ m} \qquad y(3) = 0.06 \text{ m}$$

CHECK and THINK
Our results are consistent with the positions in Figure 17.5B.

B If the pulse in Figure 17.4 has the same profile but moves to the left with the same speed, what changes are needed in Equations 17.2 and 17.3?

INTERPRET and ANTICIPATE
If the pulse moves to the left instead of to the right, the peak of the pulse at $t = 1$ s will be at $x = -2$ m. Our revised equation should give $y(-2, 1) = 1$ m. We also expect that the gold bead at $x = 2$ m should only move slightly because the pulse moves away from that bead and to the left. We'll check that the gold bead's vertical position y is never above 0.2 m for all times. (That condition is the "mirror image" of what happens to the bead at $x = -2$ m in Fig. 17.4.)

SOLVE
Because the pulse moves to the left, replace $v_x = 2$ m/s with $v_x = -2$ m/s in Equations 17.2 and 17.3 to find the new wave function.

$$y(x, t) = \frac{1}{[x + (2 \text{ m/s})t]^2 + 1}$$

CHECK and THINK
As expected, at $t = 1$ s, the wave function gives $y(-2, 1) = 1$ m, showing that the pulse moves to the left.

$$y(-2, 1) = \frac{1 \text{ m}^3}{[(-2 \text{ m}) + (2 \text{ m/s})(1 \text{ s})]^2 + 1 \text{ m}^2}$$

$$y(-2, 1) = 1 \text{ m}$$

We can then determine the vertical position y of the gold bead by substituting $x = 2$ m.

$$y(2, t) = \frac{1 \text{ m}^3}{[2 \text{ m} + (2 \text{ m/s})t]^2 + 1 \text{ m}^2}$$

The maximum vertical position of the gold bead occurs when $t = 0$. At any later time, its vertical position is closer to zero because the denominator is larger. We have confirmed that $y(2, t) < 0.2$ m for the gold bead in the case of a leftward-moving pulse.

$$y(2, 0) = \frac{1 \text{ m}^3}{[2 \text{ m} + (2 \text{ m/s})(0)]^2 + 1 \text{ m}^2} = 0.2 \text{ m}$$

17-3 Harmonic Waves

A wave's profile is determined by how the disturbance was created. For example, the gentle up-and-down motion of the hand in Figure 17.3 creates a nearly triangular pulse. If a rope is jiggled up and down many times, the wave's profile is much longer (Fig. 17.6). Any real wave has a beginning and an end determined by the time the disturbance began and ended. A **periodic wave** is an ideal wave that repeats endlessly. Of course, you cannot produce a periodic wave by jiggling the end of a rope because that result would require you to jiggle an infinitely long rope in the same way forever.

Many real waves, however, can be modeled as periodic waves. Figure 17.7 displays a note held by an opera singer using a microphone whose signal was fed to an

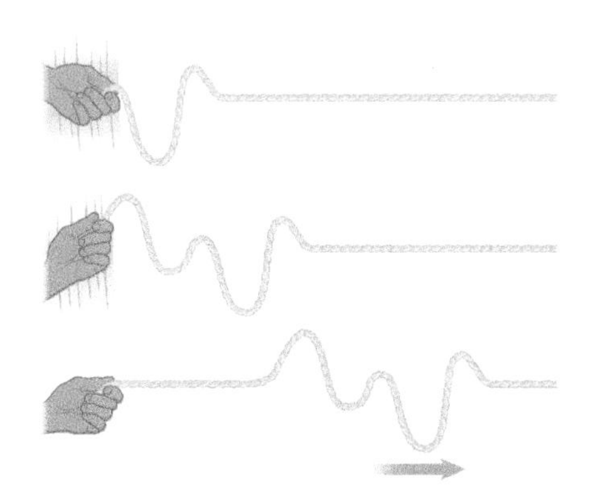

FIGURE 17.6 By jiggling the rope several times, you can generate a complicated wave profile. The resulting wave is neither periodic nor harmonic.

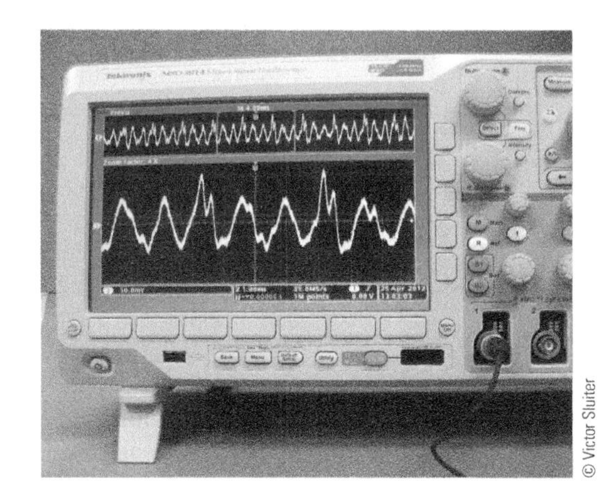

© Victor Sluiter

FIGURE 17.7 Oscilloscope trace of a single note held by opera singer.

oscilloscope. Although the oscilloscope trace has a mathematically complicated wave function, it can be modeled as a periodic wave. In this chapter, we focus on simpler periodic waves with wave functions in the form of a single sine or cosine. Such waves are called **harmonic waves**.

HARMONIC WAVE ✪ Major Concept

Transverse Harmonic Waves

The easiest way to generate a harmonic wave is to use a simple harmonic oscillator (SHO) to create the disturbance. In Figure 17.8A, a vertical spring is attached to one bead at the end of a rope. Think of Figure 17.8A as a snapshot that cannot show motion. The end bead oscillates up and down in simple harmonic motion (SHM) similar to the disk in Crall and Whipple's experiment (Fig. 16.3, page 452). Because the spring in Figure 17.8A oscillates in SHM consistently for a long time, the resulting wave can be modeled as a harmonic wave.

The wave moves to the right, but no bead in Figure 17.8A moves to the right. Instead, the wave causes each bead to oscillate up and down in SHM. Imagine using two thin slits to observe the motion of the gold and silver beads. Figure 17.8B shows motion diagrams for these two beads, similar to the motion diagram for Crall and Whipple's disk (Fig. 16.4, page 452). The beads move faster near the centers of their paths and slower near the ends. The main difference between the gold and silver bead's motion is that when one is moving upward, the other is moving downward. For example, as the gold bead moves downward from point 1 to point 4 (left), the silver bead moves upward from its own point 1 to point 4 (right).

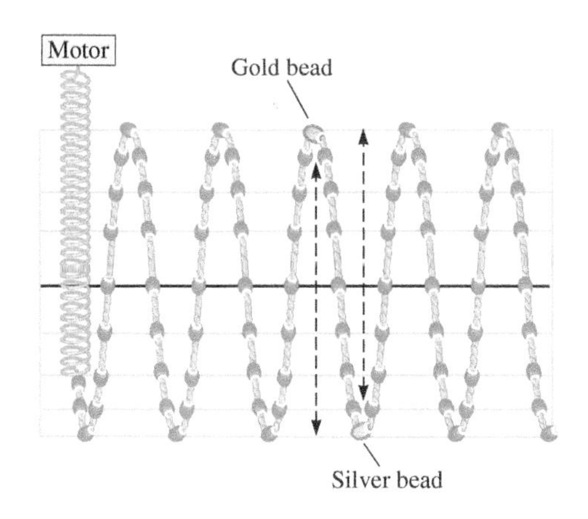

A.

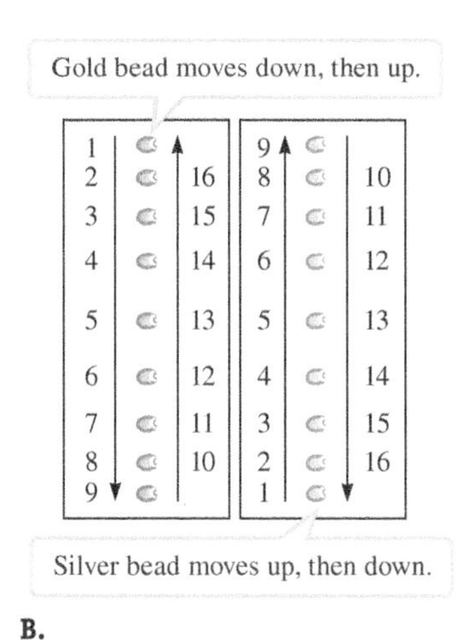

B.

FIGURE 17.8 **A.** A transverse harmonic wave is created on a beaded rope by a vertical spring driven by a motor. The resulting wave is harmonic and (nearly) periodic. **B.** Each bead oscillates up and down in simple harmonic motion.

The vertical position y of each bead can be described by the equation for SHM (Eq. 16.3), but with two different initial phases:

$$y_{gold}(t) = y_{max}\cos(\omega t + \varphi_{gold})$$
$$y_{silver}(t) = y_{max}\cos(\omega t + \varphi_{silver})$$

The angular frequency and amplitude are the same for both beads. In fact, each bead on the rope obeys Equation 16.3 with an initial phase that depends on its horizontal position x along the rope, and each bead moves with the same angular frequency ω and amplitude y_{max}.

Our observation of the gold and silver beads provides important information about the wave function $y(x, t)$ of a harmonic wave: The wave function gives the vertical position y of a series of simple harmonic oscillators whose initial phase depends on their horizontal position x. In the coordinate system in Figure 17.9, the x axis passes through the equilibrium position of the rope, and y is perpendicular to x. A sine or cosine may be used for a harmonic wave function, and we arbitrarily choose the sine:

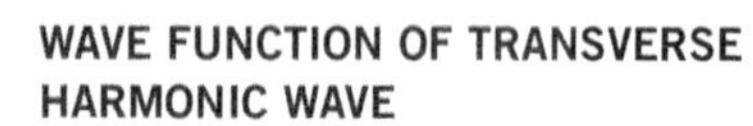

Major Concept

$$y(x, t) = y_{max}\sin(kx - \omega t) \tag{17.4}$$

where k is the **angular wave number**. The angular wave number is constant for a particular wave, closely related to the **wavelength** λ of the wave:

$$k = \frac{2\pi}{\lambda} \tag{17.5}$$

The wavelength is the length of a single repetition of the wave pattern as found from a photo (with a scale) or profile of the wave (Fig. 17.9). The SI unit of wavelength is the meter, so the angular wave number is measured in radians per meter.

At a single horizontal position x, Equation 17.4 becomes the position equation for SHM (Eq. 16.3). Let's check Equation 17.4 for a particular bead in Figure 17.8A, say the one at $x = \pi/2k$:

$$y\left(\frac{\pi}{2k}, t\right) = y_{max}\sin\left[k\left(\frac{\pi}{2k}\right) - \omega t\right] = y_{max}\sin\left[\left(\frac{\pi}{2}\right) - \omega t\right]$$

Using the trigonometric identity $\sin(\pi/2 - \theta) = \cos\theta$ (Appendix A),

$$y\left(\frac{\pi}{2k}, t\right) = y_{max}\cos\omega t$$

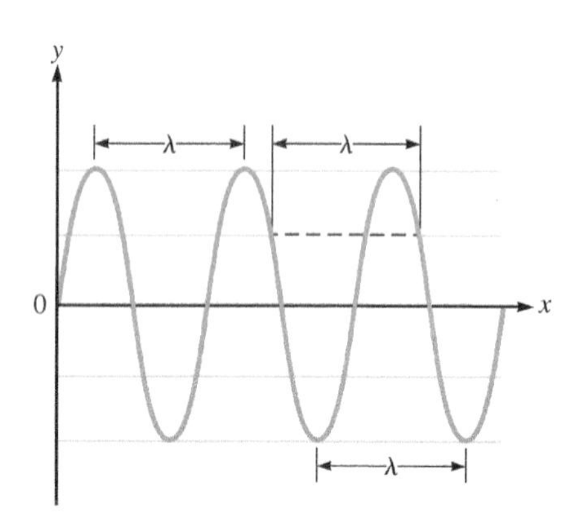

FIGURE 17.9 The profile of a harmonic wave. The wavelength may be found from the peak-to-peak distance, the valley-to-valley distance, or any other distance representing one complete wave cycle.

which is the equation for the vertical position y of an SHO (Eq. 16.3) with the initial phase equal to zero. The angular frequency ω is related to the period T and frequency f of each bead's oscillation by Equation 16.2:

$$\omega \equiv 2\pi f = \frac{2\pi}{T} \tag{16.2}$$

EXAMPLE 17.2 A Harmonic Wave Is Really a Collection of Oscillators

Show that Equation 17.4 can be rewritten in the form of Equation 16.3, $y(t) = y_{max}\cos(\omega t + \varphi)$. Find an expression for the initial phase in terms of each particle's horizontal position x.

INTERPRET and ANTICIPATE

Sine and cosine functions are connected through trigonometric identities. This example is asking us to use an identity to rewrite the sine function in terms of a cosine function. In doing so, we will discover an expression for the initial phase $\varphi(x)$. We expect φ to be in radians.

SOLVE

Equation (1) relates the sine and cosine functions (Appendix A).

$$\sin\theta = \cos\left(\frac{\pi}{2} - \theta\right) \tag{1}$$

Apply this trigonometric identity to Equation 17.4 by setting $\theta = kx - \omega t$.	$y(x, t) = y_{max} \sin(kx - \omega t) = y_{max} \cos\left[\frac{\pi}{2} - (kx - \omega t)\right]$ $y(x, t) = y_{max} \cos\left[\omega t + \left(\frac{\pi}{2} - kx\right)\right]$ (2)
Find an expression for $\varphi(x)$ by comparing Equation (2) with Equation 16.3.	$y(t) = y_{max} \cos(\omega t + \varphi)$ (16.3) $\varphi(x) = \frac{\pi}{2} - kx$ (3)

CHECK and THINK

Equation (2) is the equation of an SHO with the phase given by Equation (3). Because the angular wave number k is measured in radians per meter and the position x is in meters, the phase is in radians as expected. We can interpret our results this way: A harmonic wave is a collection of simple harmonic oscillators (Eq. 2) each of whose phase depends on its horizontal position (Eq. 3).

Longitudinal Harmonic Waves

Equation 17.4 is the wave function for a *transverse* harmonic wave. The wave moves along the x axis, and each particle in the medium (each "bead") oscillates along the y axis, perpendicular to the wave's motion. With just a slight modification, we can write Equation 17.4 for a *longitudinal* harmonic wave. First, imagine how such a wave might be established. Figure 17.10 shows a long, horizontal spring attached to a simple pendulum. Beads on the spring's coils help visualize the particle model. (Each bead represents a piece of the medium, which in this case is the spring.) The pendulum is displaced slightly so that the motion of the bob is roughly along the x axis. As the pendulum swings back and forth, the beads oscillate in SHM. All the beads oscillate with the same angular frequency and amplitude, but with differing initial phase constants φ that depend on the coil's equilibrium position.

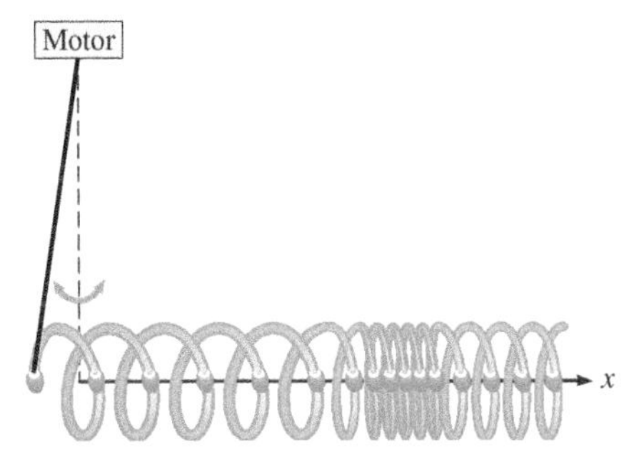

FIGURE 17.10 A motor drives a simple pendulum that is used to create a harmonic longitudinal wave on a spring.

The beads on the spring (Fig. 17.10) oscillate much like the beads on the rope (Fig. 17.8). The difference is that each bead on the spring is displaced back and forth in the $\pm x$ direction instead of in the $\pm y$ direction. We can modify Equation 17.4 to come up with the wave function of this longitudinal wave on the spring. To avoid confusion with the x that appears in Equation 17.4 and to keep the notation compact, we will write horizontal displacement as S without a Δ. The wave function for a longitudinal wave is written as

WAVE FUNCTION OF LONGITUDINAL HARMONIC WAVE

Major Concept

$$S(x, t) = S_{max} \sin(kx - \omega t) \quad (17.6)$$

where S_{max} is the amplitude or maximum displacement of any bead from its equilibrium position.

Figure 17.11 connects the longitudinal wave function (Eq. 17.6) with the motion of particles in the medium. Figure 17.11A shows the particles in their equilibrium positions, corresponding to beads on the spring before the pendulum began its motion. After the pendulum is set in motion, particles will be displaced from equilibrium. Figure 17.11B shows a graph of displacement S versus horizontal position x for time t long after the pendulum is set in motion. To find a bead's displacement, draw a line that extends from its position in Figure 17.11A to Figure 17.11B and then read the displacement off the vertical S axis. The displacement may then be used to find the bead's position at time t. Figure 17.11C shows the beads at time t. We will use Figure 17.11 in Section 17-5 to model sound waves.

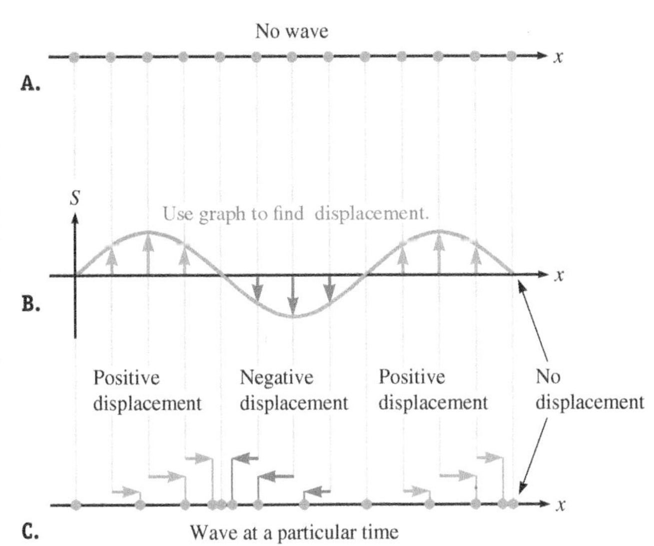

FIGURE 17.11 **A.** Initially, the particles are uniformly distributed. **B.** The graph of S versus x gives the displacement of each particle. **C.** The new positions of the particles are obtained by applying the displacements from part B to the particle positions from part A.

Speed of a Harmonic Wave

To find the propagation speed of a harmonic wave, imagine riding along with one of the peaks in Figure 17.8A. Because you are at a peak and traveling at the propagation speed of the wave, you always see the bead at its maximum displacement y_{max}. When we substitute $y = y_{max}$ into the wave function Equation 17.4, we find

$$y(x, t) = y_{max} = y_{max} \sin(kx - \omega t)$$

$$\sin(kx - \omega t) = 1$$

For $\sin(kx - \omega t) = 1$, the argument $kx - \omega t$ must equal a constant of the form

$$kx - \omega t = \frac{\pi}{2}, \frac{5\pi}{2}, \frac{9\pi}{2}, \ldots$$

For the purpose of finding the propagation speed, we can choose $kx - \omega t = \pi/2$. Solve for horizontal position x:

$$x = \frac{\pi}{2k} + \frac{\omega}{k}t$$

Take the time derivative to find the propagation velocity in the x direction:

$$v_x = \frac{dx}{dt} = \frac{d}{dt}\left(\frac{\pi}{2k} + \frac{\omega}{k}t\right)$$

$$v_x = \frac{\omega}{k} \tag{17.7}$$

Equation 17.7 is the propagation speed of a harmonic wave; in this case, the wave is traveling in the positive x direction. Using Equations 17.5 and 16.2, the propagation speed can also be written as

$$v_x = \frac{\lambda}{T} = \lambda f \tag{17.8}$$

If the wave is traveling in the negative x direction instead, the wave function is given by

$$y(x, t) = y_{max} \sin(kx + \omega t) \tag{17.9}$$

and (as you may show in Problem 10) the velocity is given by

$$v_x = -\frac{\omega}{k} \tag{17.10}$$

CONCEPT EXERCISE 17.3

A longitudinal wave function is given by $S(x, t) = 0.75 \sin(0.30x - 655t)$ in SI units.

a. What are the amplitude, angular wave number, and angular frequency?
b. What is the propagation velocity?

17-4 Special Case: Transverse Wave on a Rope

The profile of a wave is determined by how the disturbance was created, but the propagation speed is determined by the properties of the medium. For a transverse wave on a rope, the two properties affecting wave speed are the tension and mass per unit length. For a given rope, a higher tension means a faster wave. For a given tension, a thinner rope (smaller mass per unit length) results in a faster wave. Mathematically, the propagation speed of a wave on a rope of linear mass density μ and under tension F_T is

SPEED OF TRANSVERSE WAVE ON A ROPE

Special Case

$$v_x = \sqrt{\frac{F_T}{\mu}} \tag{17.11}$$

DERIVATION Propagation Speed of a Transverse Wave on a Rope

We will derive Equation 17.11. The profile of the wave doesn't matter, so we choose a simple pulse traveling to the right. Imagine that you are traveling along with the pulse so it appears to be stationary. From your point of view, however, the rope appears to be pulled to the left at a constant speed through a clear tube shaped like the pulse. Figure 17.12 illustrates three snapshots that you might take of the rope from your moving reference frame. The pulse appears stationary, but the colored beads move to the left.

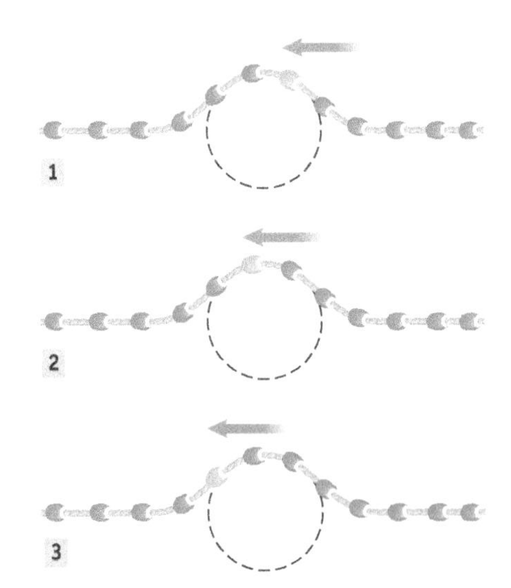

FIGURE 17.12 As seen from a reference frame that moves with the pulse, the rope moves to the left from time 1 to time 3.

The three beads near the top of the pulse momentarily move in a circle. No matter what the pulse's shape, you can always find a small enough portion of the rope to approximate its motion as circular as seen from the reference frame that moves along with the wave.

Start with a free-body diagram for a small segment of the rope marked by a single bead (Fig. 17.13). The bead momentarily moves in a circle of radius r. From the center of the circle, the bead subtends a small angle $\Delta\theta$, so the bead's diameter is $r\,\Delta\theta$. The segment of rope has mass Δm. (The bead is just there to mark that portion of the rope, so we treat it as massless.) Tension in the rope pulls the segment in two directions, both tangent to the rope and making an angle $\Delta\theta/2$ with the x axis.

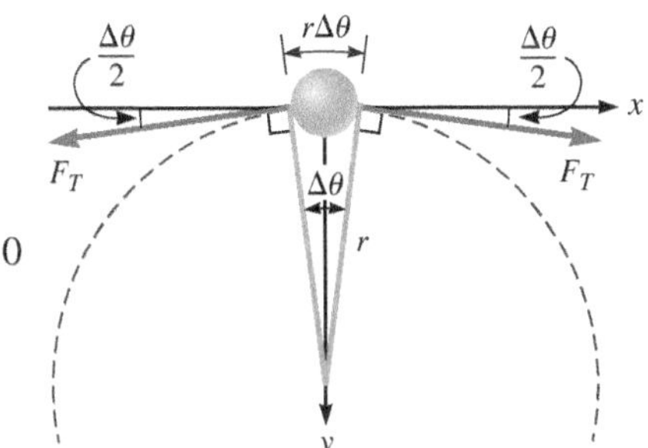

FIGURE 17.13 A rope particle momentarily moves in a circular path.

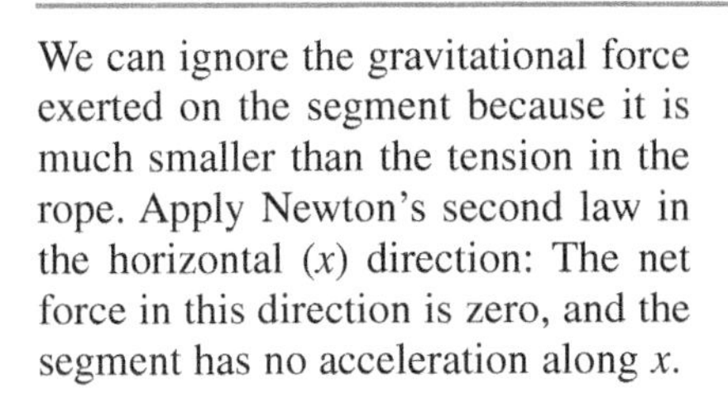

We can ignore the gravitational force exerted on the segment because it is much smaller than the tension in the rope. Apply Newton's second law in the horizontal (x) direction: The net force in this direction is zero, and the segment has no acceleration along x.

$$\sum F_x = F_T \cos\left(\frac{\Delta\theta}{2}\right) - F_T \cos\left(\frac{\Delta\theta}{2}\right) = 0$$

There is a net force in the y direction (chosen as positive downward), pointing to the center of the circle.

$$\sum F_y = F_T \sin\left(\frac{\Delta\theta}{2}\right) + F_T \sin\left(\frac{\Delta\theta}{2}\right) = \Delta m a_y$$

$$\sum F_y = 2F_T \sin\left(\frac{\Delta\theta}{2}\right) = \Delta m a_y \quad (1)$$

This net force results in centripetal acceleration (Eq. 4.36).

$$a_y = a_c = \frac{v_x^2}{r} \quad (2)$$

Substitute Equation (2) into Equation (1).

$$2F_t \sin\left(\frac{\Delta\theta}{2}\right) = \Delta m \frac{v_x^2}{r}$$

The segment of rope is small so that the angle $\Delta\theta$ is small; therefore, we use the small-angle approximation $\sin(\Delta\theta/2) \approx \Delta\theta/2$ and isolate the speed on one side.

$$2F_T\left(\frac{\Delta\theta}{2}\right) \approx \Delta m \frac{v_x^2}{r}$$

$$F_T \Delta\theta \approx \Delta m \frac{v_x^2}{r}$$

$$v_x^2 \approx F_T \frac{r\,\Delta\theta}{\Delta m}$$

Derivation continues on page 496 ▶

The rope's linear mass density μ is defined to be its mass per unit length. The mass of the small segment is Δm, and its length is the diameter of the bead $r\Delta\theta$. The resulting mass per unit length is $\mu = \Delta m/r\Delta\theta$.

$$v_x^2 = \frac{F_T}{\mu}$$

$$v_x = \sqrt{\frac{F_T}{\mu}} \checkmark \quad (17.11)$$

COMMENTS

Equation 17.7 ($v_x = \omega/k$) holds for any harmonic wave, but our result (Eq. 17.11) holds only for a transverse wave on a rope. Nonetheless, our derivation leads to some insight into the speed of waves traveling in other media. When the pulse passes through some portion of the rope, it deforms the rope. If the pulse passes through quickly, the rope quickly returns to its straight configuration. How could the two factors—tension and linear mass density—in Equation 17.11 cause the pulse to move quickly? First, consider the free-body diagram. If there is a lot of tension in the rope, the bead at the top is quickly pulled down, and the rope is quickly restored to its original configuration. Second, if the mass per unit length is very great, the portion of the bead displaced has a lot of inertia. It resists being moved back to its original position. So, roughly speaking, the speed of a wave in a medium depends on the force that restores its original configuration and on the medium's ability to resist returning to equilibrium (inertia), which we can express as

$$\text{speed} = \sqrt{\frac{\text{restoring ability}}{\text{inertia resisting the return to equilibrium}}} \quad (17.12)$$

We'll use this conceptual relationship to find the speed of waves in other media.

EXAMPLE 17.3 Piano String Factory

Piano wire must be both strong and flexible. The wire is placed under high tension and then subjected to repeated blows by the piano's hammer (Fig. 17.14), so modern piano strings are made from steel. Consider a thick piano wire of diameter d = 4.8 mm and linear mass density μ = 64.9 g/m. If the propagation speed of a wave on this wire is 129 m/s, what is the tension in the wire? The tensile strength of steel is 3.8×10^8 N/m² (Table 14.1). **CHECK and THINK:** Will the wire break?

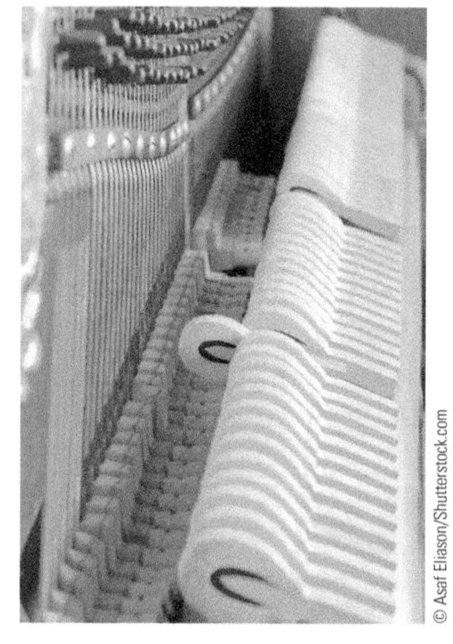

FIGURE 17.14 Inner workings of a piano.

INTERPRET and ANTICIPATE

We can find the tension in the piano wire using Equation 17.11, and we expect that the string does not break because if it did break, it wouldn't be useful in a piano.

SOLVE

Solve Equation 17.11 for tension.

$$v_x = \sqrt{\frac{F_T}{\mu}} \quad (17.11)$$

$$v_x^2 = \frac{F_T}{\mu}$$

$$F_T = \mu v_x^2 = (64.9 \times 10^{-3}\ \text{kg/m})(129\ \text{m/s})^2$$

$$F_T = 1.08 \times 10^3\ \text{N}$$

CHECK and THINK

To determine if the wire will break, find the tensile stress $\sigma = F_T/A$ (Eq. 14.9) resulting from this force.

$$\sigma = \frac{F_T}{A} = \frac{F_T}{\pi r^2} = \frac{F_T}{\pi (d/2)^2}$$

$$\sigma = \frac{(1.08 \times 10^3\ \text{N})}{\pi (4.8 \times 10^{-3}\ \text{m}/2)^2}$$

$$\sigma = 6.0 \times 10^7\ \text{N/m}^2$$

Because the tensile strength of steel (3.8×10^8 N/m²) is greater than the tensile stress on the wire, the piano wire does not break.

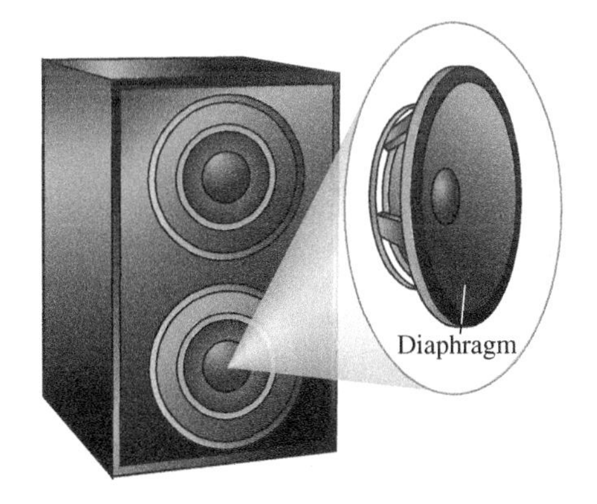

FIGURE 17.15 A stereo speaker creates longitudinal waves in air.

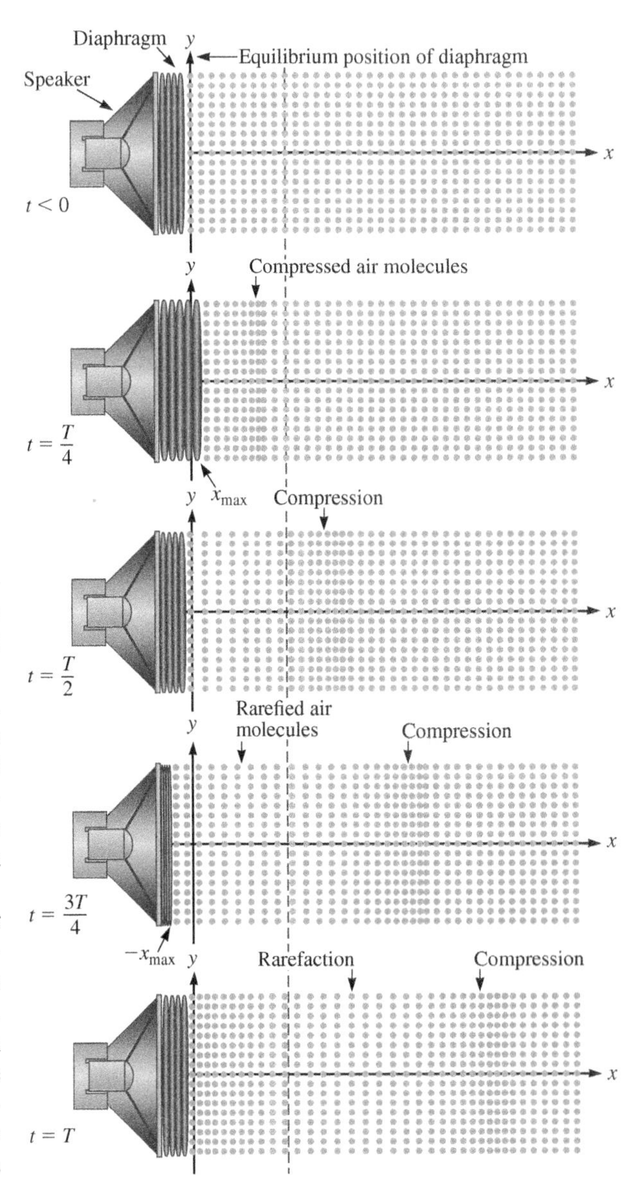

FIGURE 17.16 Longitudinal wave in a column of air near a speaker diaphragm. The gold molecules move back and forth over a short distance along the x axis. (The relative shape and size of the diaphragm are exaggerated.)

17-5 Sound: Special Case of a Traveling Longitudinal Wave

A basic speaker is a relatively simple device (Fig. 17.15). A diaphragm made of a lightweight material such as paper oscillates back and forth, creating longitudinal waves in air known as **sound waves** or **sound**.

Imagine attaching a speaker to the end of a tube filled with air. This scenario could be a model for one of Shannon's earbuds (case study, page 487) at the end of an ear canal. To keep our model simple, we'll ignore the diaphragm's curvature. Before the diaphragm begins to oscillate (that is, at $t < 0$), air molecules are uniformly distributed throughout the tube. The air is modeled as a collection of particles (Fig. 17.16).

The diaphragm begins by moving to the right in the positive x direction, so at $t = T/4$, the diaphragm is at x_{max}, and the air molecules close to the diaphragm are compressed. The air molecules farther downstream are not yet affected by the diaphragm's motion.

The air molecules oscillate back and forth along the x axis much as the coils in Figure 17.10. No molecule moves down the entire length of the tube. The pattern of compression moves in the positive x direction, but each individual molecule oscillates only around its own equilibrium position. Examine the gold molecules in Figure 17.16; they move slightly to the right and left, but never are very far from their original positions.

At $t = T/2$, the diaphragm passes through its equilibrium position as it moves in the negative x direction. At $t = 3T/4$, the diaphragm is at $-x_{max}$, and a region of low density (rarefied) air molecules has formed to the right of the diaphragm. The compression has also moved farther to the right. By $t = T$, one cycle is complete. Both a compression and a rarefaction are propagating to the right, and a new compression is just forming near the diaphragm.

If the diaphragm continues to move back and forth in SHM, the resulting sound wave is harmonic, and we would hear a single tone or note. Usually, a speaker is used to transmit spoken words or music, so the motion of a diaphragm is usually much more complicated than simple harmonic. A detailed study of such complicated motion is beyond the scope of this textbook.

CONCEPT EXERCISE 17.4

CASE STUDY **Sound from Earbuds**

In the case study (page 487), Avi models the sound from an earbud as sand from a sandblaster. Is this model a good model for sound? Explain.

Speed of Sound

Instead of a formal derivation for the speed of sound v_s, we use our derivation of the speed of a transverse wave on a rope as a guide. For simplicity, imagine a compression pulse traveling along the positive x axis through a column of fluid such as air (Fig. 17.16). Like the rope in Figure 17.12, the air molecules are distorted by the pulse. In this case, as the pulse passes, the volume occupied by a particular group of molecules is reduced. The decrease in volume means an increase in pressure. This pressure works to restore the medium to its undisturbed state. The fractional change in volume is proportional to the pressure. Recall that the bulk modulus is the constant of proportionality in Equation 15.7: $P = B(\Delta V/V_i)$. For our work here, we can think of the bulk modulus B as a measure of the medium's ability to restore its original configuration.

To find an expression for the speed of sound, we also need a measure of the fluid's "inertia." Here, that "inertia" is its density (mass per unit volume) ρ.

Using the conceptual relationship we found in Equation 17.12 when deriving the speed of a wave on rope, the speed of sound in a fluid depends on the fluid's density ρ (inertia) and bulk modulus B (restoring ability):

$$v_s = \sqrt{\frac{\text{restoring ability}}{\text{inertia resisting the return to equilibrium}}} = \sqrt{\frac{B}{\rho}} \quad (17.13)$$

According to Equation 17.13, sound travels faster in a fluid with lower density than one with higher density. The speed of sound is also higher in a less compressible fluid (one with greater B). Although water is denser than air, it is less compressible than air; as a result, the speed of sound is higher in water than in air.

For an ideal gas, the ratio B/ρ depends on the gas temperature, so the speed of sound depends on temperature. For a sound wave in room-temperature air, $v = 343$ m/s. At other air temperatures T_C (measured in Celsius), the speed of sound is given by

$$v = 331\,\text{m/s}\sqrt{1 + \frac{T_C}{273°\text{C}}} \quad (17.14)$$

When sound waves propagate in a solid rod, the longitudinal compression causes a slight transverse expansion. For a fluid, there is no such transverse motion. The speed of sound in a solid rod is given by a relation similar to Equation 17.13 except that the bulk modulus is replaced by Young's modulus (Section 14-4):

$$v_s = \sqrt{\frac{Y}{\rho}} \quad (17.15)$$

Table 17.1 provides the speed of sound in several media.

TABLE 17.1 Speed of sound in various media.

Medium	Speed of sound, v_s (m/s)
Air (0°C)	331.45
Air	343.37
Aluminum (rolled)	6420
Brass	4700
Helium	965
Human tissues at body temperature	1540
Hydrogen	1284
Glass	5640
Mercury	1450
Seawater	1531
Steel	5960
Water (0°C)	1402
Water	1496.7

Speed of sound is at room temperature except where indicated. Air is assumed to be dry.

CONCEPT EXERCISE 17.5

The bulk modulus of water is 2.2×10^9 Pa (Table 15.2). The density of water is 10^3 kg/m^3 (Table 15.1). Find the speed of sound in water and compare your answer with the value given in Table 17.1.

Pressure Waves

So far, we've been thinking of a harmonic sound wave in a tube (Fig. 17.16) like the longitudinal harmonic wave on the spring in Figure 17.10. When studying sound, though, it is not convenient or practical to account for the displacement of the many molecules involved. Instead, we measure changes in the fluid's density or pressure.

So, let's think of sound as a pressure wave passing through the medium. Before the wave passes through the region, the pressure equals the equilibrium pressure P_0 of the fluid. We denote a change in the pressure relative to the equilibrium pressure as $\Delta P = P - P_0$. In a sound wave, the pressure oscillates between $P_0 + \Delta P_{\max}$ and $P_0 - \Delta P_{\max}$.

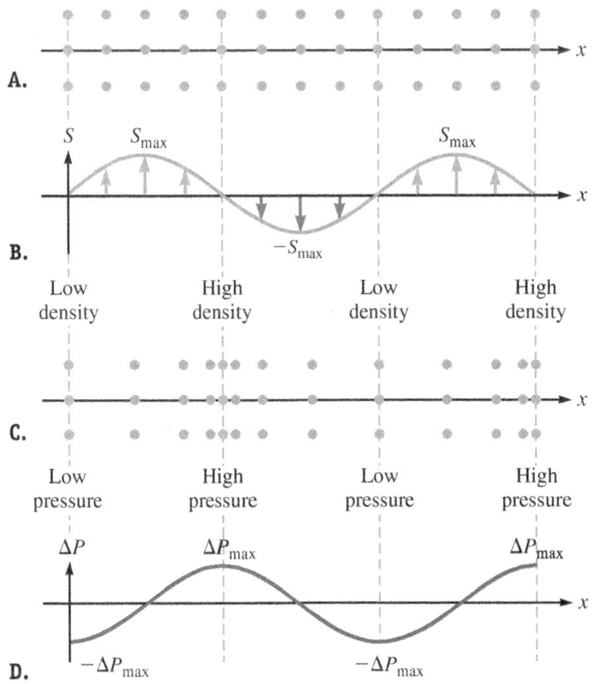

FIGURE 17.17 **A.** Initially, the particles are uniformly distributed. **B.** The graph of S versus x gives the displacement of the particles. **C.** The particles have oscillated into a new arrangement. **D.** The density and pressure are highest where $S = 0$ and are lowest when S is at a maximum or minimum.

Now, to come up with a mathematical expression for the pressure wave, let's return to the speaker causing air molecules to oscillate in a tube. Before the speaker is turned on, air molecules are evenly distributed (Fig. 17.17A). After the speaker has been on for some time, a sound wave travels in the positive x direction, and we can use Equation 17.6 to find the horizontal displacement $S(x, t)$ of the molecules:

$$S(x, t) = S_{max} \sin (kx - \omega t) \quad (17.6)$$

Figure 17.17B shows a graph of S versus x at a particular time t. Figure 17.17C shows the corresponding position of air molecules at that time. Figure 17.17 is very similar to Figure 17.11 except that instead of a single row of particles to model the spring, the air molecules are modeled by a three-dimensional grid of particles.

Figure 17.17C indicates regions of high and low density, and pressure is directly proportional to density. High-density regions are also high-pressure regions, and low-density regions are low-pressure regions.

Figure 17.17D shows ΔP versus x. The relative pressure ΔP is zero when the displacement S is a maximum or minimum. When the displacement is zero, the relative pressure is a maximum or minimum. The relative pressure is 90° or $\pi/2$ radians out of phase with the displacement, so the pressure wave corresponding to Equation 17.6 is

$$\Delta P = \Delta P_{max} \sin\left(kx - \omega t - \frac{\pi}{2}\right)$$

$$\Delta P = \Delta P_{max} \cos (kx - \omega t) \quad (17.16)$$

PRESSURE WAVE Special Case

where maximum relative pressure is given by

$$\Delta P_{max} = \rho \omega v_s S_{max} \quad (17.17)$$

This maximum relative pressure (also called "pressure amplitude") depends on properties of the medium (density and speed of sound) and on properties of the wave (amplitude and angular frequency). Another way to think about it is that the change in pressure—a macroscopic quantity—caused by the wave depends on the microscopic motion (amplitude and angular frequency) of the molecules.

EXAMPLE 17.4 CASE STUDY No Whispering

Hair cells inside your ear convert sound waves into electrical signals that are sent to your brain. Each hair cell has about 100 closely packed tufts of hair. Even an extremely small displacement of one of these tufts, roughly 10^{-11} m, can send a signal to the brain. Suppose you have just barely heard a whisper at $f = 1000$ Hz. Assume the displacement amplitude of the sound wave is $S_{max} = 1 \times 10^{-11}$ m, $v_s = 343$ m/s (speed of sound in air), and the density of air is 1.3 kg/m^3 (Table 15.1). Find the pressure amplitude ΔP_{max}. Compare your answer to atmospheric pressure at sea level.

INTERPRET and ANTICIPATE

We expect to find a numerical answer of the form $\Delta P_{max} =$ ______ Pa.

SOLVE

Use Equation 16.2 to find the angular frequency.	$\omega = 2\pi f = 2\pi(1000 \text{ Hz}) = 6280$ rad/s

Example continues on page 500 ▶

Substitute numerical values into Equation 17.17 to find ΔP_{max}.	$\Delta P_{max} = \rho \omega v_s S_{max} = (1.3\ \text{kg/m}^3)(6.3 \times 10^3\ \text{rad/s})(343\ \text{m/s})(1 \times 10^{-11}\ \text{m})$ $\Delta P_{max} = 3 \times 10^{-5}\ \text{Pa}$

CHECK and THINK

Our result is in the expected form. Atmospheric pressure at sea level is 1 atm $\approx 1 \times 10^5$ Pa (Section 15-3). The quietest whisper you can hear has a pressure amplitude 10 orders of magnitude lower than the normal air pressure, making the ear a very sensitive pressure detector.

EXAMPLE 17.5 CASE STUDY No Shouting

A loud sound can permanently damage your ears. The maximum pressure amplitude ΔP_{max} the human ear can withstand is roughly 30 Pa at $f = 1000$ Hz. What is the corresponding displacement of the hair tufts? Use the same speed of sound and air density as in Example 17.4 and compare your answer with the value of S_{max} given there.

INTERPRET and ANTICIPATE

This example is similar to Example 17.4, but here we are given ΔP_{max} and need to find S_{max}. We expect S_{max} to be greater than 10^{-11} m.

SOLVE Solve Equation 17.17 for S_{max}.	$\Delta P_{max} = \rho \omega v_s S_{max}$ (17.17) $S_{max} = \dfrac{\Delta P_{max}}{\rho \omega v_s}$
Substitute numerical values.	$S_{max} = \dfrac{30\ \text{Pa}}{(1.3\ \text{kg/m}^3)(6.3 \times 10^3\ \text{rad/s})(343\ \text{m/s})}$ $S_{max} = 1.1 \times 10^{-5}\ \text{m}$

CHECK and THINK

A displacement of 10^{-5} m is about the thickness of a red blood cell. It may be hard to believe that such a small displacement of a hair can cause permanent damage to your hearing, but this displacement is six orders of magnitude larger than the minimum displacement required for you to hear a whisper. Not only is the ear a very sensitive pressure detector, it has an amazing range. It is worth protecting, so Shannon should heed the warning labels on the earbuds.

17-6 Energy Transport in Waves

A sound wave such as one from your earbuds carries energy and momentum from the speaker, but no air molecules travel from the speaker to your ears while that happens. Waves transport energy without transporting matter, and this statement is true of all waves. For example, in Figure 17.2, each coil oscillates around its equilibrium position, but no matter is transported along the spring. The hand, though, does work on a few coils, increasing their mechanical energy. This energy is transported along the spring, causing the bell at the end to oscillate when the wave arrives. In this section, we find expressions for mechanical energy and power transmitted by a wave.

Energy and Power of a Transverse Harmonic Wave

Particles in a medium oscillate as a harmonic wave passes through. Because the particles are in motion, we can associate kinetic energy with their motion. Also, as it passes, a wave distorts the medium (the system), changing its configuration. A change in configuration is associated with a change in potential energy. So, when a wave exists in a medium, the medium (the system) has mechanical energy (kinetic energy plus potential energy).

Consider a harmonic transverse wave on rope. We can find the energy E_λ associated with a portion of the rope that is one wavelength long. Each particle of the rope behaves as an SHO, and from Equation 16.37, the mechanical energy of an SHO is $E = \frac{1}{2}my_{max}^2\omega^2$. It doesn't make sense to consider the mass of each particle that makes up a rope. Instead, we work with the linear mass density μ (mass per unit length). The mass in one wavelength's portion of the rope is then found by multiplying the linear mass density by the wavelength: $m = \mu\lambda$. So, the average mechanical energy in one wavelength of a transverse harmonic wave is

$$E_\lambda = \tfrac{1}{2}\mu\omega^2 y_{max}^2\lambda \tag{17.18}$$

Often, we are not interested in the energy associated with a wave, but in the rate at which a wave *transfers* energy. Think of one particle (say the gold bead) as a transverse wave on a rope passes (Fig. 17.8A). The spring at the end of the rope does work on the end bead, increasing its mechanical energy. This mechanical energy travels along the rope. As one wavelength of the wave passes the gold bead, mechanical energy is transferred to that bead. At first, the bead is at rest with no kinetic energy. As the wave passes, the bead moves, gaining kinetic energy. The bead momentarily stops at the top, so its kinetic energy is momentarily zero. The bead then picks up speed and kinetic energy. After passing through equilibrium, it slows down again, losing kinetic energy. The bead completes one full cycle with the passage of each wavelength of the wave. So, E_λ is also the amount of mechanical energy transferred (to the gold bead) in one period T.

To find the average power P transferred by the wave, we divide the energy transferred in one period by one period:

$$P_{av} = \frac{\Delta E}{\Delta t} = \frac{E_\lambda}{T} = \frac{\frac{1}{2}\mu\omega^2 y_{max}^2\lambda}{T}$$

The symbol P is used for both power and pressure. Be sure to get the proper meaning from the context.

Substitute Equation 17.8 to express the power transferred by a transverse harmonic wave in terms of its propagation speed v_x:

$$P_{av} = \tfrac{1}{2}\mu\omega^2 y_{max}^2 v_x \tag{17.19}$$

Energy and Power of Sound

We can use our results for the transverse harmonic wave to find the energy and power in a harmonic sound wave. Again, start with Equation 16.37 for the energy of an SHO, $E = \frac{1}{2}my_{max}^2\omega^2$. Just as the beads on the rope oscillate between $+y_{max}$ and $-y_{max}$ (Fig. 17.8A), the particles (molecules) in the tube of air oscillate between $+S_{max}$ and $-S_{max}$ (Figs 17.16 and 17.17). Clearly, we need to replace y_{max} with S_{max}.

Next, we need the mass m of a small volume of air of density ρ. Consider a sound wave in a column of cross-sectional area A (Fig. 17.18). The volume of the box of length λ is $V = A\lambda$. So, the mass in the box is

$$m = \rho V = \rho A\lambda$$

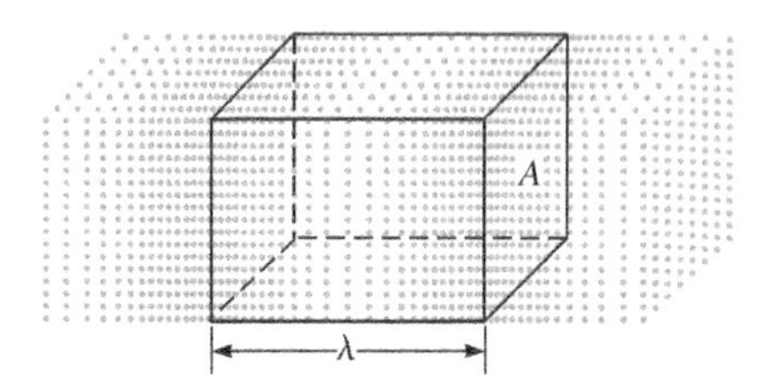

FIGURE 17.18 A sound wave in a column of air with cross-sectional area A. The volume of the box of length λ is $V = A\lambda$.

The mechanical energy in one wavelength of a harmonic sound wave is

$$E_\lambda = \tfrac{1}{2}\rho A\omega^2 S_{max}^2\lambda \tag{17.20}$$

As before, we find the average rate at which energy is transferred by dividing by the period. So, the average power transmitted by sound is given by

$$P_{av} = \frac{E_\lambda}{T} = \frac{\frac{1}{2}\rho A\omega^2 S_{max}^2\lambda}{T}$$

or

$$P_{av} = \tfrac{1}{2}\rho A\omega^2 S_{max}^2 v_s \tag{17.21}$$

where v_s is the speed of sound.

Comparing the average power transmitted by a harmonic wave on a rope (Eq. 17.19) with that transmitted by a harmonic sound wave (Eq. 17.21), shows that they have much in common. In both cases, the average power transmitted depends on properties of the medium (mass density and speed) as well as properties of the wave itself (amplitude and angular frequency). Perhaps it is no surprise that a dense medium with a high wave speed or a wave with a large amplitude and high frequency transmits power effectively.

A.

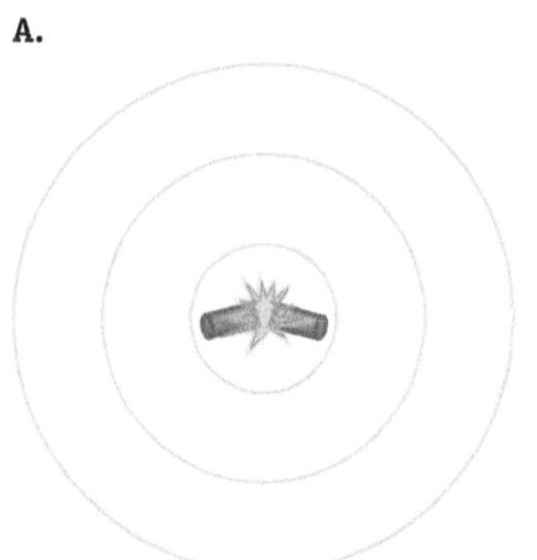

B.

FIGURE 17.19 A. A two-dimensional water wave on the surface of a still pond. **B.** A three-dimensional sound wave traveling outward from a source at its center.

CONCEPT EXERCISE 17.6

CASE STUDY Sound from Earbuds Revisited

In the case study (page 487), Avi models the sound from an earbud as sand from a sandblaster. In Concept Exercise 17.4, we argued that no air molecules are shot out of a speaker. Sound does not transmit particles, but it can damage your hearing. What does sound transmit and how can it damage your hearing? *Hint*: Examples 17.4 and 17.5 briefly explain human hearing.

17-7 Two- and Three-Dimensional Waves

So far, we have studied one-dimensional waves (those that travel along a single, straight axis). When a drop of water falls into a still pond, it creates circular ripples that travel along the two-dimensional plane of the water's surface (Fig. 17.19A). If a firecracker explodes, it creates spherical ripples of sound traveling outward in three dimensions (Fig. 17.19B).

Figure 17.20 represents the circular waves in Figure 17.19A or a cross section of the spherical waves in Figure 17.19B. The source of the disturbance is shown as a dot at the center of the wave. Such a source is known as a **point source**. Circles in Figure 17.20 represent the wave crests, also called the **wave fronts**. The distance between wave fronts is one wavelength.

Like a photo, Figure 17.20 shows the traveling wave at just one particular instant. The wave fronts move outward and expand as the wave travels. **Rays** are lines perpendicular to the wave fronts. The arrowhead on a ray indicates the direction of the wave's motion. Rays are not vectors; to distinguish them from vectors, we draw the arrowhead on the line, but not at the end.

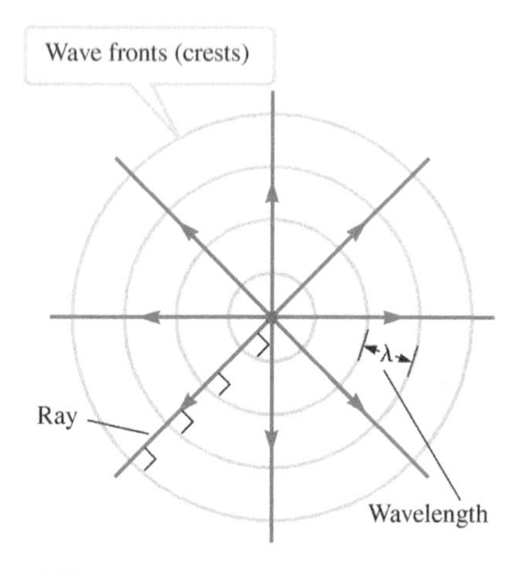

FIGURE 17.20 The wave fronts of a spherical or circular wave. The rays are perpendicular to the wave fronts.

In a small region far from the sources in Figures 17.19 and 17.20, the wave fronts are nearly parallel. **Plane waves** have straight, flat wave fronts as shown in Figure 17.21A. Waves far from a point source can be modeled as plane waves. You could also make a plane wave by using a large, straight source such as a rod oscillating up and down with its length parallel to the surface of the water as in Figure 17.21B.

Intensity and Loudness

If you hear fireworks exploding in midair or a clap of thunder, the sound is louder the closer you are to the source. A thunderclap can be modeled as a source that emits sound uniformly in all directions. The energy and power transmitted by the sound wave is uniformly distributed over any sphere centered on the source (Fig. 17.22). **Intensity** I is defined as the average power P_{av} per unit area A perpendicular to the direction of propagation:

INTENSITY Major Concept

$$I \equiv \frac{P_{av}}{A} \tag{17.22}$$

Intensity is a scalar with SI units of watts per square meter.

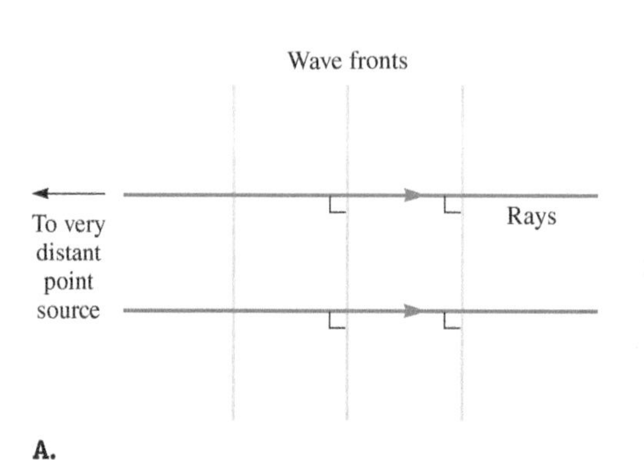

A.

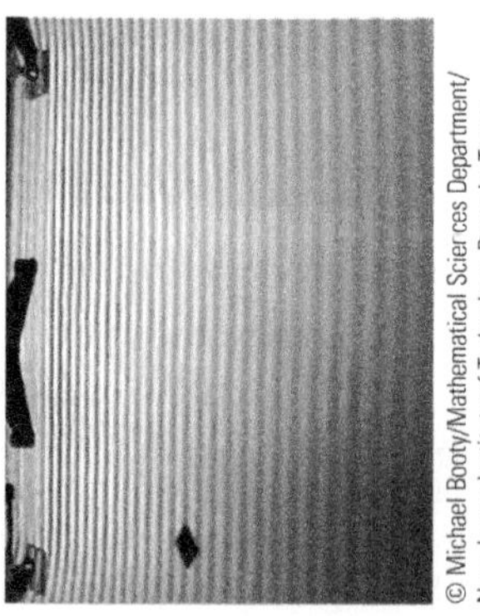

B.

FIGURE 17.21 A. The wave fronts of a plane wave are straight lines or planes. **B.** Plane waves can be generated on a water surface by oscillating a horizontal rod.

The same power passes through each sphere shown in Figure 17.22. The area of the sphere is greater for a more distant observer ($A_2 > A_1$), however, so the intensity of the sound wave is lower, $I_2 < I_1$. The intensity of a point source decreases as distance r from the source increases:

$$I = \frac{P_{av}}{4\pi r^2} \tag{17.23}$$

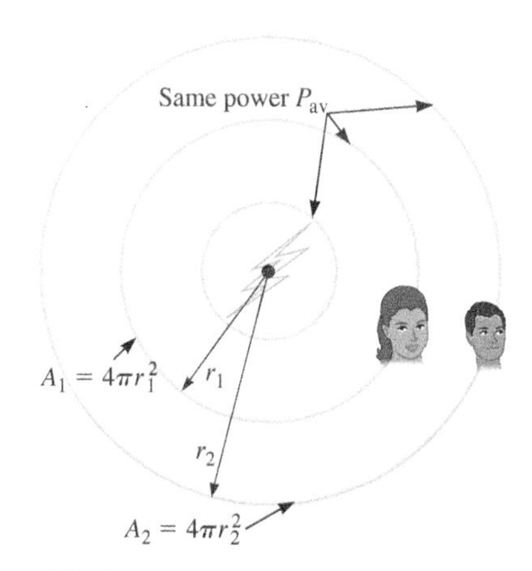

FIGURE 17.22 Intensity decreases as the distance from the source increases.

The average power of a sound wave is given by Equation 17.21, so the intensity can be written as

$$I = \frac{P_{av}}{A} = \frac{\frac{1}{2}\rho A\omega^2 S_{max}^2 v_s}{A} = \frac{1}{2}\rho\omega^2 S_{max}^2 v_s$$

We can eliminate the displacement amplitude S_{max} in favor of the pressure amplitude by using Equation 17.17:

$$I = \frac{1}{2}\rho\omega^2\left(\frac{\Delta P_{max}}{\rho\omega v_s}\right)^2 v_s$$

Simplifying, the intensity of a sound wave is given by

$$I = \frac{1}{2}\frac{\Delta P_{max}^2}{\rho v_s} \tag{17.24}$$

Remember that P_{av} is average **power** and ΔP_{max} is the **pressure amplitude**.

The intensity range of human hearing spans 12 orders of magnitude, from just barely audible ($I_0 = 10^{-12}$ W/m^2) to painful ($I = 1$ W/m^2). Human response to or perception of sound depends logarithmically on the intensity. American inventor Alexander Graham Bell defined **sound level** β in terms of base 10 logarithms as

$$\beta \equiv 10 \log\left(\frac{I}{I_0}\right) \tag{17.25}$$

where $I_0 = 10^{-12}$ W/m^2. Sound level is measured in **decibels** (dB), although the quantity defined in Equation 17.25 is actually dimensionless. The threshold of human hearing is

$$\beta = 10 \log\left(\frac{I_0}{I_0}\right) = 0 \text{ dB}$$

and sound is painful at

$$\beta = 10 \log\left(\frac{1 \text{ W/m}^2}{10^{-12} \text{ W/m}^2}\right) = 120 \text{ dB}$$

Table 17.2 provides the sound level of a number of sources.

TABLE 17.2 Typical sound levels.

Intensity I (W/m^2)	Sound level β (dB)	Examples	Time required for onset of hearing damage
10^{-12}	0	Barely audible	
10^{-9}	30	Soft whisper	
10^{-8}	40	Quiet bedroom; mosquito buzzing	
10^{-7}	50	Gentle breeze; refrigerator	
10^{-6}	60	Normal conversation	
10^{-5}	70	Noisy restaurant; vacuum cleaner	Some damage if continuous
10^{-4}	80	City traffic; factory noise	More than 8 hours
10^{-3}	90	Truck traffic; power lawn mower	Less than 8 hours
10^{-2}	100	Chain saw; subway	2 hours
1	120	Rock concert; nearby thunderclap	Immediate danger
10^2	140	Machine gun; nearby jet plane	Immediate danger
10^4	160	Rocket launchpad	Hearing loss inevitable

People can hear sounds in the frequency range from 20 Hz to 20,000 Hz. This range varies from person to person and generally decreases with increasing age. Sound with frequency higher than the upper limit of human hearing ($f > 20{,}000$ Hz) is known as **ultrasound**, and sound with a frequency below the lower limit ($f < 20$ Hz) is known as **infrasound**. Seismographs that measure earthquakes operate in the infrasound, and large animals such as elephants and whales communicate with infrasound. Ultrasound is used in medical applications to probe inside the human body and to deliver energy inside the body without surgery. Some animals such as dogs and bats can hear ultrasound.

EXAMPLE 17.6 CASE STUDY Ultrasound to the Rescue

Mineral deposits such as calcium can build up in the kidneys to form kidney stones. Although some kidney stones are as small as a grain of sand, others have been the size of golf balls. Often, the patient does not notice kidney stones until they move into the ureter (Fig. 17.23), causing a great deal of pain. Before the mid-1980s, kidney stones were surgically removed, but now physicians use lithotripsy to break up the stones into smaller pieces that easily pass through the ureter with very little pain. In lithotripsy, ultrasound pulses deliver energy to the kidney stones, as Cameron mentioned (page 487). Typically, a patient is exposed to thousands of pulses during a 45-minute procedure.

Assume the average intensity of a pulse near the lithotripter is $I_0 = 6.5 \times 10^8\ \text{W/m}^2$ and the face area of the pulse is $A = 1.9 \times 10^{-4}\ \text{m}^2$. In a single medical procedure, 2000 such pulses are used to break up a kidney stone 15.5 mm in diameter that is 10 cm from the lithotripter.

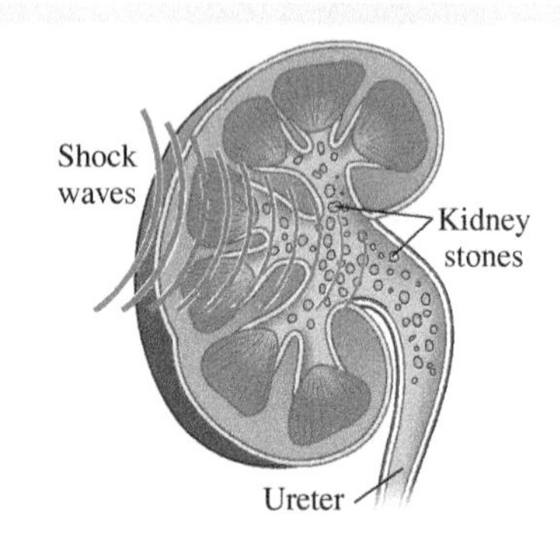

FIGURE 17.23 Kidney stones form in the kidney and pass into the ureter.

A What is the average power output of a single pulse?

INTERPRET and ANTICIPATE

Ultrasound is just high-frequency sound, so all the equations developed for sound can be applied to ultrasound. We expect a numerical result with the SI units of power (watts).

SOLVE

Rearrange Equation 17.22 to find P_{av}.

$$P_{av} = IA = (6.5 \times 10^8\ \text{W/m}^2)(1.9 \times 10^{-4}\ \text{m}^2)$$

$$P_{av} = 1.2 \times 10^5\ \text{W}$$

CHECK and THINK

Our result has the expected units. For comparison, a single incandescent lightbulb in your room might put out 100 W. A single pulse of the lithotripter is approximately equivalent to 1200 such lightbulbs, or roughly the number of lightbulbs in a small public building.

B What is the intensity of the ultrasound wave near the kidney stone?

INTERPRET and ANTICIPATE

Intensity decreases with distance, so we expect a numerical value that is smaller than the intensity near the lithotripter ($I < I_0$).

SOLVE

Use Equation 17.23 and the result from part A to find the intensity of the ultrasound wave at the kidney stone.

$$I = \frac{P_{av}}{4\pi r^2} = \frac{1.2 \times 10^5\ \text{W}}{4\pi(0.10\ \text{m})^2}$$

$$I = 9.5 \times 10^5\ \text{W/m}^2$$

CHECK and THINK

As expected, $I < 6.5 \times 10^8\ \text{W/m}^2$.

C What is the average power delivered to the kidney stone in one pulse?

INTEPRET and ANTICIPATE
We need the area of the kidney stone and the intensity of the pulse near the kidney stone from part B.

SOLVE Solve Equation 17.22 for P_{av} and use the kidney stone's diameter $d = 15.5$ mm.	$P_{av} = IA = I(\pi d^2/4)$ $P_{av} = (9.5 \times 10^5\ \text{W/m}^2)\dfrac{\pi(15.5 \times 10^{-3}\ \text{m})^2}{4}$ $P_{av} = 1.8 \times 10^2\ \text{W}$

CHECK and THINK
Because the intensity drops off dramatically with increasing distance, the power delivered to the kidney stone is roughly equivalent to just two 100-W incandescent lightbulbs. The remaining power is delivered to tissues in the body. According to some studies, this power can damage healthy tissues.

D If the duration of a single pulse is 5.2 μs, what is the total energy delivered to the kidney stone by 2000 pulses?

INTERPRET and ANTICIPATE
First, find the energy delivered by a single pulse. Then, multiply by 2000 to find the energy delivered during the entire medical procedure.

SOLVE The energy E delivered by a single pulse is the product of the power delivered to the kidney stone and the duration Δt of the pulse.	$E = P_{av}\Delta t = (1.8 \times 10^2\ \text{W})(5.2 \times 10^{-6}\ \text{s})$ $E = 9.4 \times 10^{-4}\ \text{J}$
Multiply by 2000 to find the total energy delivered to the kidney stone.	$E_{tot} = 2000E = (2000)(9.4 \times 10^{-4}\ \text{J}) = 1.9\ \text{J}$

CHECK and THINK
Although 2 J may not seem like a lot of energy, laboratory experiments have shown that even 0.5 J can break up artificial kidney stones.

EXAMPLE 17.7 Buying the Right Speakers

A certain manufacturer's middle-of-the-line speakers each have an input power $P_{in} = 60$ W and reach a maximum sound level $\beta = 95$ dB at 1 m in front of each speaker. Use this information to estimate the power output P_{out} in the form of sound and the efficiency (P_{in}/P_{out}) of this manufacturer's product.

INTERPRET and ANTICIPATE
Use the sound level to find intensity (Eq. 17.25). To find the power, assume the intensity is uniform over a sphere of radius 1 m. Speakers are designed to send sound waves out primarily in the forward direction, so our assumption means that we will overestimate P_{out}. We expect that some energy is lost in producing sound, so $P_{out} < P_{in}$, and the efficiency should be less than 1.

Example continues on page 506 ▶

SOLVE

Use Equation 17.25 to relate the intensity at 1 m from the speaker to the sound level at that distance. The reference intensity I_0 is always 10^{-12} W/m^2. (The inverse of the log function is exponential base 10; see Appendix A.)	$\beta = 10 \log\left(\frac{I}{I_0}\right)$ $I = I_0 10^{\beta/10} = (10^{-12}\ \mathrm{W/m^2})10^{95/10}$ $I = 3.2 \times 10^{-3}\ \mathrm{W/m^2}$
Intensity is power per unit area. Assume the speaker emits sound uniformly in all directions and use Equation 17.23 to find the power output.	$I = \frac{P_{out}}{4\pi r^2}$ (17.23) $P_{out} = 4\pi r^2 I = 4\pi(1\ \mathrm{m})^2(3.2 \times 10^{-3}\ \mathrm{W/m^2})$ $P_{out} = 4.0 \times 10^{-2}\ \mathrm{W}$
Find the efficiency by dividing the output power by the input power.	$\frac{P_{out}}{P_{in}} = \frac{4.0 \times 10^{-2}\ \mathrm{W}}{60\ \mathrm{W}} = 7 \times 10^{-4} = 0.07\%$

CHECK and THINK

As expected, the efficiency is less than 1 (and perhaps smaller than you expected). Basic home speakers have an efficiency of around 0.5 to 4%, so this speaker is much less efficient than the average. You may wish to purchase a more efficient speaker than this one.

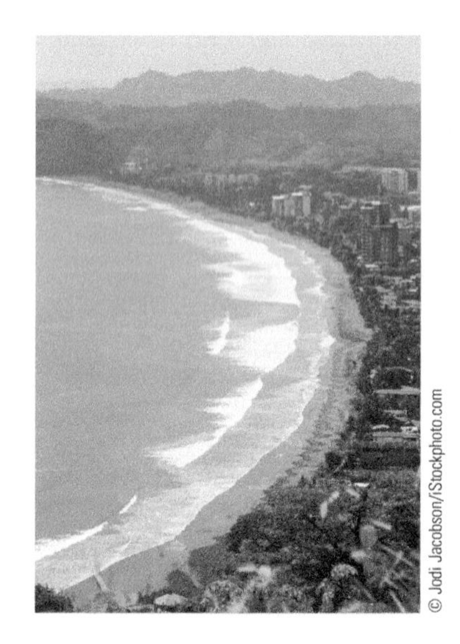

FIGURE 17.24 Ocean waves are parallel to the shoreline.

17-8 Refraction and Diffraction

Imagine that you wish to share a letter with a person who is on the other side of a heavy, closed door. The only way to get the letter to the person is through the mail slot. Now imagine that you wish to talk to the person on the other side of the door. You would probably be able to carry on a conversation even if neither of you put your ears or mouths near the mail slot because sound waves are able to bend around obstacles and through holes. Particles do not bend in the same way, so the letter must be slipped through the mail slot. In this section, we explore this unique feature, the bending of waves.

Refraction

Have you ever wondered why water waves near a beach are nearly parallel to the shore (Fig. 17.24), but waves in the middle of the sea may be seemingly random? The reason is that as the waves move through shallow water near the shore, they are bent or **refracted**. No matter what their original orientation is, when they get near the shore, they are refracted so that the wave fronts become roughly parallel to the shoreline.

A two- or three-dimensional wave is refracted when it is transmitted from one medium to another medium having a different wave propagation speed. Because the speed of water waves is lower in the shallow water near the shore than in deep water, waves are refracted near the shore. You can picture this phenomenon by imagining a marching band as it encounters a change in medium from pavement to mud (Fig. 17.25A). The band models a plane wave, each row of band members represents a wave front, and the distance between adjacent rows is a wavelength. All the musicians obediently march at the same pace, but the length of their stride is shortened by the mud. As the first few musicians step into the mud, they slow down (Fig. 17.25B). The first few rows are bent. After the entire band has entered the mud (Fig. 17.25C), all the rows are nearly parallel

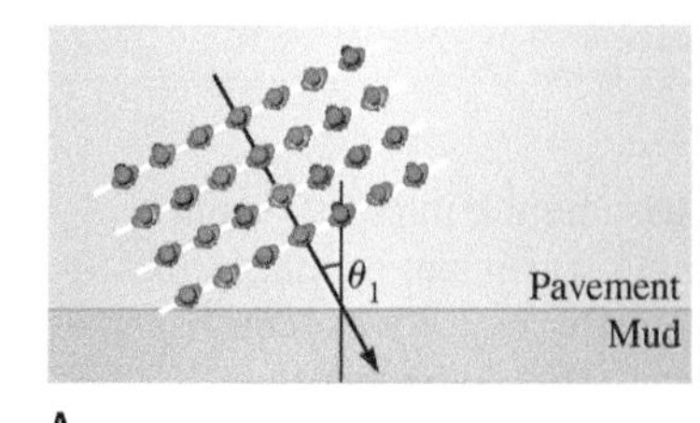

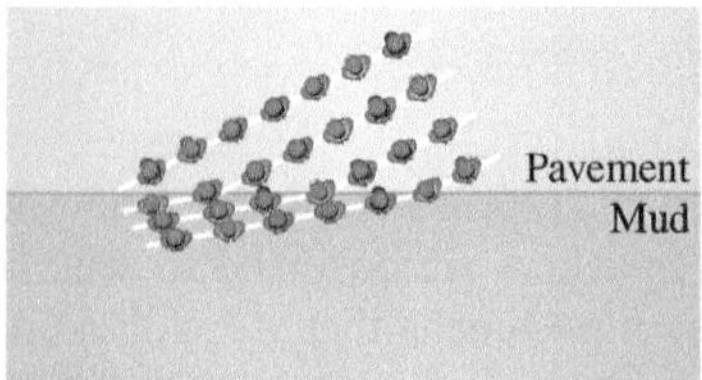

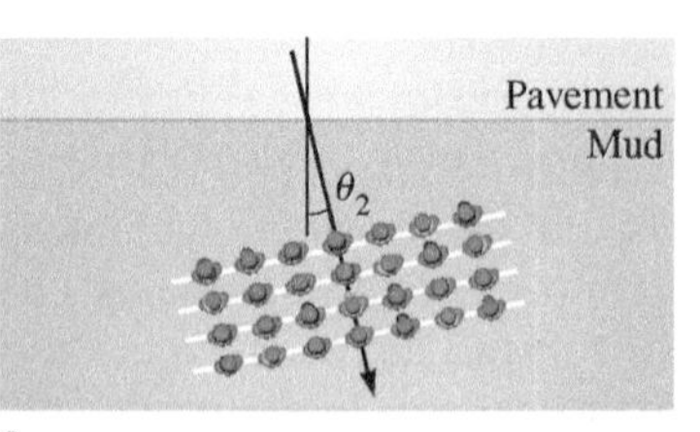

FIGURE 17.25 A band marches from pavement to mud. Although this analogy is contrived, a real marching band (on a single surface) does turn by this method. The analogy illustrates that as a wave moves from one medium to another, it bends or refracts.

to the line that separates the mud from the pavement. In a natural setting such as a seashore, the wave propagation speed drops off slowly and the refraction of the water waves is gradual, but the wave fronts still end up roughly parallel to the shore.

The ray in Figure 17.25A makes an angle θ_1 with the line perpendicular to the pavement–mud boundary. The ray in Figure 17.25C makes an angle θ_2 with the perpendicular, and $\theta_2 < \theta_1$. We can find the mathematical relationship between these two angles by taking a closer look at two of the musicians as they enter the mud (Fig. 17.26). Musician 2 is at the boundary between pavement and mud, and musician 1 must march a distance d_1 before entering the mud. Musician 1 marches on the pavement at speed v_1 parallel to ray 1. In a time interval Δt, he marches the distance $d_1 = v_1\Delta t$ to the boundary. During this same time interval, musician 2 marches in the mud on a path parallel to ray 2. Musician 2's speed in the mud is v_2, so she marches a distance $d_2 = v_2\Delta t$. Figure 17.26 shows two right triangles that share a common hypotenuse h. Using the definition of the sine function,

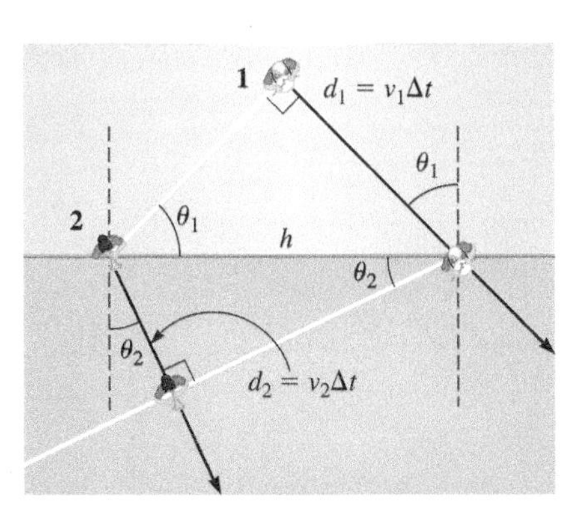

FIGURE 17.26 Two band members approach the mud.

$$h = \frac{d_1}{\sin\theta_1} = \frac{d_2}{\sin\theta_2}$$

Rewriting the distances d_1 and d_2 in terms of the speeds v_1 and v_2,

$$\frac{v_1\Delta t}{\sin\theta_1} = \frac{v_2\Delta t}{\sin\theta_2}$$

After canceling Δt and rearranging, we are left with the **law of refraction**:

LAW OF REFRACTION Major Concept

$$\sin\theta_2 = \frac{v_2}{v_1}\sin\theta_1 \tag{17.26}$$

where θ_1 and θ_2 are measured with respect to the **normal**, the line perpendicular to the boundary between medium 1 and medium 2. If the propagation speed in medium 2 is lower than in medium 1 (that is, if $v_2 < v_1$), the ray bends toward the perpendicular, $\theta_2 < \theta_1$. In this case, the wave fronts tend to align with the boundary (see Figs. 17.25 and 17.26). On the other hand, if the propagation speed in medium 2 is higher than in medium 1 (that is, if $v_2 > v_1$) then the ray bends away from the perpendicular, $\theta_2 > \theta_1$.

Diffraction

We just showed that waves bend at an interface between media in which they have different speeds. Waves also bend when they encounter an obstacle that partially blocks the wave fronts. The bending of waves passing through a hole or around an obstacle is known as **diffraction**. The amount of diffraction depends of the relative size of the obstacle or hole compared to the wavelength. If the hole or obstacle is much larger than the wavelength, diffraction is small. If the obstacle or hole is comparable in size to the wavelength, diffraction becomes significant.

DIFFRACTION Major Concept

Diffraction allows waves to transmit a disturbance and its energy around obstacles. Again imagine communicating with a friend on the other side of a closed door: Energy is transferred from your mouth to your friend's ears through sound waves that bend around the door and through the mail slot.

17-9 The Doppler Shift

If you have ever been to an auto race, you have probably noticed that as the cars move toward you, you hear a high-frequency hum that sounds like a swarm of bees. Just after the cars pass, you hear a low-frequency roar. You may also have heard a change in frequency if you have ever waited by a railroad track as a train passed. The pitch of the approaching train whistle sounds high, and the pitch of the receding whistle sounds low.

The change in frequency you perceive in these cases is known as the **Doppler**[1] **shift** and is due to the relative motion between you and the sound source. When the

[1]Christian Doppler, a 19th-century Austrian physicist, predicted that relative motion causes a frequency shift for both sound and light waves.

source moves toward you, you hear a higher frequency than you would were the source stationary with respect to air. When the source moves away from you, you hear a lower frequency than you would were the source stationary. In this section, we focus on sound, but there is a Doppler shift for other waves as well, including water waves and light.

The relative motion is what matters. If you were moving and the source were stationary with respect to the medium, you would still hear a Doppler shift. For example, if you are approaching a parked emergency vehicle with its siren blaring, you hear the siren at a higher frequency than what was originally emitted. After you pass the vehicle, you hear the siren at a lower frequency.

Stationary Source, Moving Observer

Figure 17.27 shows a stationary siren. Hannah is approaching the siren from the right, and Aaron is moving to the left away from the siren. The four parts of the figure show four instants from $t = 0$ to $t = 3T$. The siren creates spherical wave fronts that expand outward. You can see their motion by following the history of one wave front such as the innermost wave front shown in green in Figure 17.27A. In Figure 17.27B, the green wave front has expanded by one wavelength λ. By $t = 3T$, the green wave front has expanded to the edge of the panel in Figure 17.27D.

Initially, Hannah is at the red wave front. If she were not moving, she would have encountered three other wave fronts—orange, yellow, and green—in a time interval of $3T$. So, if Hannah were stationary, she would hear a frequency

$$f = \frac{3\text{ cycles}}{3T} = \frac{1}{T}$$

During that time interval of $3T$, however, suppose she moves toward the siren a distance 2λ. Her speed is given by

$$v_{\text{obs}} = \frac{2\lambda}{3T}$$

The subscript "obs" stands for observer. We use $v_s = \lambda/T$ (Eq. 17.8) to write her speed in terms of the speed of sound v_s:

$$v_{\text{obs}} = \tfrac{2}{3}v_s$$

which is unrealistically fast, but that won't change our results. Because she is traveling toward the siren, she encounters five wave fronts—orange, yellow, green, blue, and violet—instead of just three. Hannah hears a frequency given by

$$f_{\text{obs}} = \frac{5\text{ cycles}}{3T} = \frac{5}{3}\frac{1}{T} = \frac{5}{3}f$$

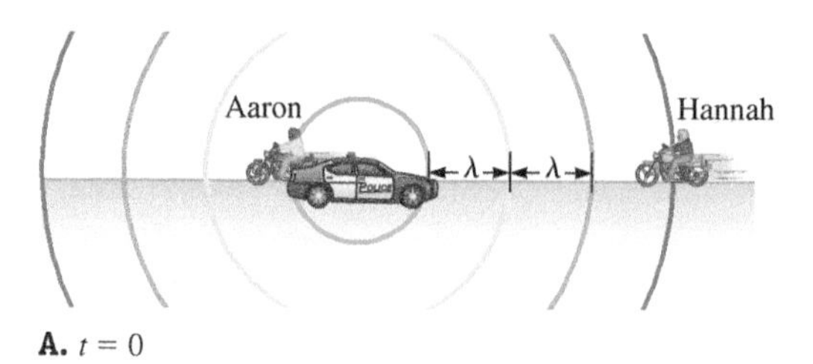

A. $t = 0$

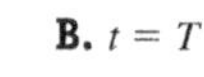
B. $t = T$

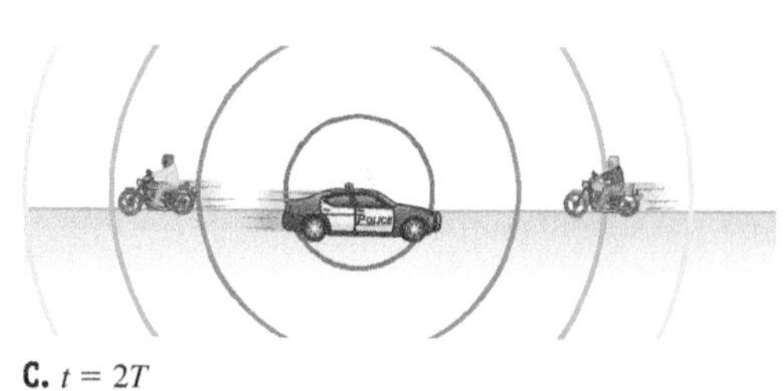
C. $t = 2T$

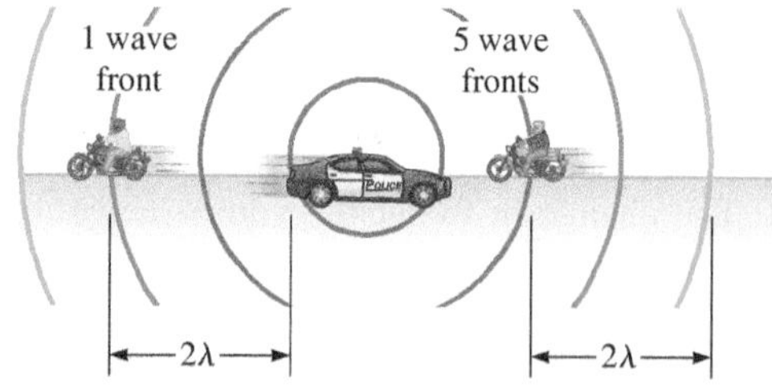

D. $t = 3T$

FIGURE 17.27 An observer moving toward a source hears a higher frequency than when standing still, and an observer moving away from a source hears a lower frequency.

Aaron is moving away from the source at the same speed as Hannah, $v_{obs} = \frac{2}{3}v_s$. Initially, he is at the green wave front (Fig. 17.27A). During the time interval of $3T$, he only encounters one other wave front, the blue one. Thus, the frequency Aaron hears is

$$f_{obs} = \frac{1 \text{ cycle}}{3T} = \frac{1}{3}f$$

DERIVATION Frequency Observed by Moving Observer

We'd like a general expression for the frequency heard by an observer in terms of the observer's speed and the source's speed. The frequency heard by a moving observer when the source is stationary is:

$$f_{obs} = \left(\frac{v_s \pm v_{obs}}{v_s}\right)f \qquad (17.27)$$

where we choose the plus sign for an approaching observer and the minus sign for a receding observer. We will show that this expression is consistent with the frequency heard by the people in Figure 17.27.

Because Hannah moves toward the siren at v_{obs}, the relative speed between her and the sound wave is $v_s + v_{obs}$. She encounters the wave fronts more rapidly than she would were she stationary. She hears a frequency given by Equation 17.8, where v_x is replaced by her relative speed.

$$v_x = \lambda f$$

$$f = \frac{v_x}{\lambda} \qquad (17.8)$$

$$f_{obs} = \frac{v_s + v_{obs}}{\lambda} \qquad (1)$$

Equation (1) can easily be modified for an observer moving away from the source. In this case, the relative speed is given by $v_s - v_{obs}$, and only the numerator is modified.

$$f_{obs} = \frac{v_s - v_{obs}}{\lambda} \qquad (2)$$

Combine Equations (1) and (2) to include the possibility of an approaching observer or a receding observer.

$$f_{obs} = \frac{v_s \pm v_{obs}}{\lambda} \qquad (3)$$

The source's wavelength λ is not affected by the observer's motion, so it is still given by Equation 17.8, where v_x is the speed of the sound wave v_s.

$$\lambda = \frac{v_s}{f} \qquad (4)$$

Substitute Equation (4) into Equation (3) to find the frequency heard by an observer moving toward or away from a source.

$$f_{obs} = \frac{v_s \pm v_{obs}}{(v_s/f)}$$

$$f_{obs} = \left(\frac{v_s \pm v_{obs}}{v_s}\right)f \checkmark$$

COMMENTS

Check this result for Hannah in Figure 17.27, who is approaching the source at $v_{obs} = \frac{2}{3}v_s$.

$$f_{obs} = \left(\frac{v_s + \frac{2}{3}v_s}{v_s}\right)f = \left(1 + \frac{2}{3}\right)f$$

$$f_{obs} = \frac{5}{3}f \checkmark$$

Now check this result for Aaron, who is moving away from the source at the same speed.

$$f_{obs} = \left(\frac{v_s - \frac{2}{3}v_s}{v_s}\right)f = \left(1 - \frac{2}{3}\right)f$$

$$f_{obs} = \frac{1}{3}f \checkmark$$

Moving Source, Stationary Observer

If an observer is stationary with respect to the medium but the source is moving, the observer still hears a higher or lower frequency, depending on whether the source is approaching or receding. Figure 17.28 shows a source moving to the right. At time

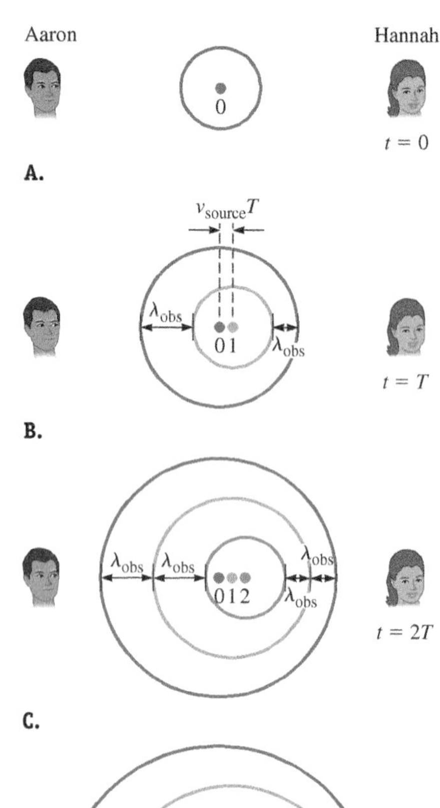

FIGURE 17.28 When the source is moving, an observer in front hears a higher frequency as the source approaches than she would if the source were stationary, and an observer in back hears a lower frequency as the source recedes.

$t = 0$ (Fig. 17.28A), the source is at position 0, and a single wave front shown in red is centered on this position. At $t = T$, the source has moved to position 1. The original red wave front has expanded, and a new wave front shown in orange and centered on position 1 has appeared. Parts C and D of Figure 17.28 show that as the source continues to move to the right, each wave front expands in a sphere centered on its own initial position. The result is that the wave fronts ahead of the source are crowded together and those behind the source are spread apart. Therefore, the wavelength is shorter in front of the source and longer behind the source.

Imagine that Hannah stands directly in front of the source. At $t = 0$, the source emits a wave front that travels outward toward her (Fig. 17.28A). One period later, at $t = T$, the source emits another wave front. If the source did not move, Hannah would find that the distance between wave fronts is λ and the frequency is $f = v_s/\lambda$. The source moves to the right, however, so the wave front emitted at $t = T$ is shifted toward her. She observes a wavelength λ_{obs} as the distance between the two wave fronts, but shortened by the distance that the source moved. The source's speed is v_{source}, so it moves a distance $v_{source}T$ in one period. The wavelength Hannah observes is then given by

$$\lambda_{obs} = \lambda - v_{source}T$$

The frequency she hears is

$$f_{obs} = \frac{v_s}{\lambda_{obs}} = \frac{v_s}{\lambda - v_{source}T}$$

Now use $f = 1/T$ and $v_x = \lambda/T$ (Eqs. 16.1 and 17.8) to write the observed frequency f_{obs} of a source moving at speed v_{source} toward a stationary observer,

$$f_{obs} = \frac{v_s}{(v_s/f) - (v_{source}/f)}$$

$$f_{obs} = \frac{v_s}{v_s - v_{source}}f \tag{17.28}$$

where v_s is the speed of sound and f is the frequency the source would emit if it were stationary.

If the source is moving away from the observer, the observed wavelength is lengthened by the amount $v_{source}T$: $\lambda_{obs} = \lambda + v_{source}T$. The observed frequency is lowered and is given by

$$f_{obs} = \frac{v_s}{v_s + v_{source}}f \tag{17.29}$$

Combining Equations 17.28 and 17.29, we have one expression for a moving source and a stationary observer:

$$f_{obs} = \frac{v_s}{v_s \mp v_{source}}f \tag{17.30}$$

where we choose the minus sign for a source moving toward the observer and the plus sign for a source moving away from the observer.

Source and Observer Both Moving

It is possible that both the source and the observer are moving with respect to the medium. For example, imagine driving along the highway and hearing a siren on a moving police car. Both you (the observer) and the source (the police siren) are moving with respect to the medium (the air). We can combine Equations 17.27 and 17.30 to cover all the possible combinations of source and observer motion. The all-purpose equation is

DOPPLER SHIFT EQUATION

Major Concept

$$f_{obs} = \left(\frac{v_s \pm v_{obs}}{v_s \mp v_{source}}\right)f \tag{17.31}$$

where f_{obs} is the observed frequency, v_s is the speed of sound, v_{obs} is the observer's speed relative to the medium, and v_{source} is the source's speed. The signs of these speeds depend on the relative direction of motion.

1. If the observer moves toward the source, choose the top sign (+) for v_{obs} in the numerator of Equation 17.31. If the source moves toward the observer, choose the top sign (−) for v_{source} in the denominator. Either choice (or both) increases the observed frequency as source and observer approach each other. (It helps to remember that both *t*oward and *t*op begin with *t*.)
2. If the observer moves away from the source, choose the bottom sign (−) in the numerator of Equation 17.31. If the source moves away from the observer, choose the bottom sign (+) in the denominator. Either choice (or both) decreases the observed frequency as source and observer recede from each other.

If the source and the observer are moving at the same speed in the same direction, there is no Doppler shift according to Equation 17.31. For example, if the source is followed by the observer and $v_{obs} = v_{source}$, then

$$f_{obs} = \left(\frac{v_s + v_{obs}}{v_s + v_{source}}\right)f = \left(\frac{v_s + v_{obs}}{v_s + v_{obs}}\right)f$$

$$f_{obs} = f$$

where we have chosen the top sign in the numerator because the observer is heading toward the source and the bottom sign in the denominator because the source is heading away from the observer. So, if the observer is moving along with the source, the observed frequency equals the frequency of a stationary source. (No matter how fast an ambulance is going, the ambulance driver always hears the same frequency from the siren.)

EXAMPLE 17.8 Where's the Fire?

A fire truck's siren has a frequency of 875 Hz. You maintain your speed at 65 mph (29.0 m/s) on a multilane highway as the fire truck comes up from behind at 80 mph (35.7 m/s). Assume the speed of sound in air is 343.4 m/s.

A What frequency do you hear as the fire truck approaches? What frequency do you hear after the truck passes you?

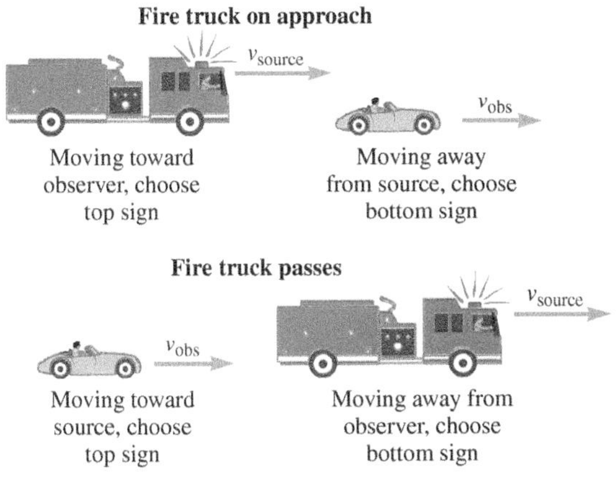

FIGURE 17.29

INTERPRET and ANTICIPATE

Use Equation 17.31 twice with the same speed values, changing signs as appropriate to each situation (fire truck approaches car and after fire truck passes car). A sketch helps in picking out the correct signs (Fig. 17.29). We expect to hear a higher frequency when the fire truck is approaching and a lower frequency after it passes.

SOLVE

Substitute speed values into Equation 17.31 and choose signs, first for when the truck is approaching the car. The source moves toward the observer and that the observer moves away from the source *at the same time*. We need the top sign in the denominator and the bottom sign in the numerator.

$$f_{obs} = \left(\frac{v_s \pm v_{obs}}{v_s \mp v_{source}}\right)f$$

$$f_{obs} = \left(\frac{v_s - v_{obs}}{v_s - v_{source}}\right)f \text{ after choosing signs}$$

$$f_{obs} = \frac{(343.4\text{ m/s}) - (29.0\text{ m/s})}{(343.4\text{ m/s}) - (35.7\text{ m/s})}(875\text{ Hz})$$

$$f_{obs} = 894\text{ Hz as fire truck approaches}$$

Example continues on page 512 ▶

CHECK and THINK
As expected, the siren frequency is shifted higher when the fire truck is approaching the observer.

SOLVE
After the truck passes, only the signs change.

$$f_{obs} = \left(\frac{v_s + v_{obs}}{v_s + v_{source}}\right) f$$

$$f_{obs} = \frac{(343.4 \text{ m/s}) + (29.0 \text{ m/s})}{(343.4 \text{ m/s}) + (35.7 \text{ m/s})}(875 \text{ Hz})$$

$$f_{obs} = 860 \text{ Hz} \text{ after truck passes}$$

CHECK and THINK
As expected, the siren frequency is shifted lower when the fire truck is receding from the observer.

B If you are driving on the other side of the highway in the opposite direction as the fire truck, what frequency do you hear as the fire truck approaches? What frequency do you hear after the fire truck passes?

INTERPRET and ANTICIPATE
The solution resembles part A, but we need a new sketch to get the signs correct (Fig. 17.30). We expect an even higher frequency upon approach than in part A because the relative speed of approach is greater. Similarly, we expect an even lower frequency after the truck passes than in part A due to the greater relative speed of recession.

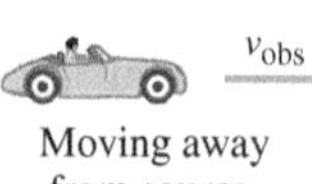

FIGURE 17.30

SOLVE
When the observer and source both move toward each other, choose the top sign for each speed in Equation 17.31.

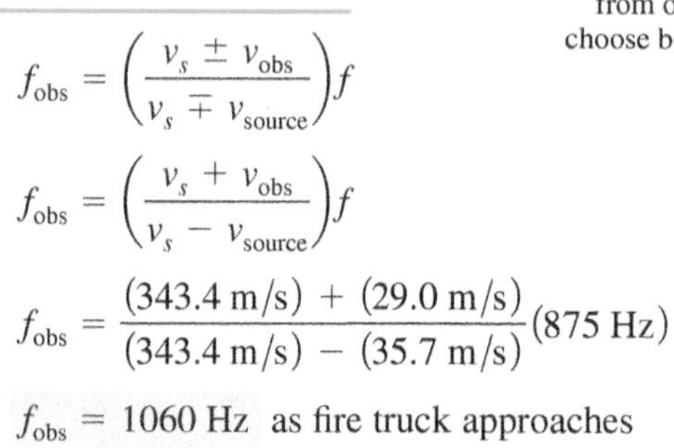

$$f_{obs} = \left(\frac{v_s \pm v_{obs}}{v_s \mp v_{source}}\right) f$$

$$f_{obs} = \left(\frac{v_s + v_{obs}}{v_s - v_{source}}\right) f$$

$$f_{obs} = \frac{(343.4 \text{ m/s}) + (29.0 \text{ m/s})}{(343.4 \text{ m/s}) - (35.7 \text{ m/s})}(875 \text{ Hz})$$

$$f_{obs} = 1060 \text{ Hz} \text{ as fire truck approaches}$$

CHECK and THINK
As expected, this result is higher than the "approach frequency" in part A. In fact, this result corresponds to the highest frequency that can be heard from this particular siren at these speeds.

SOLVE
When the observer and source both move away from each other, choose the bottom sign for each speed.

$$f_{obs} = \left(\frac{v_s - v_{obs}}{v_s + v_{source}}\right) f$$

$$f_{obs} = \frac{(343.4 \text{ m/s}) - (29.0 \text{ m/s})}{(343.4 \text{ m/s}) + (35.7 \text{ m/s})}(875 \text{ Hz})$$

$$f_{obs} = 726 \text{ Hz} \text{ after truck passes}$$

CHECK and THINK
As expected, this result is lower than the "receding frequency" in part A and corresponds to the lowest frequency that can be heard from this siren at these speeds.

FIGURE 17.31 A swimming duck moves faster in the water than waves do and creates a wake.

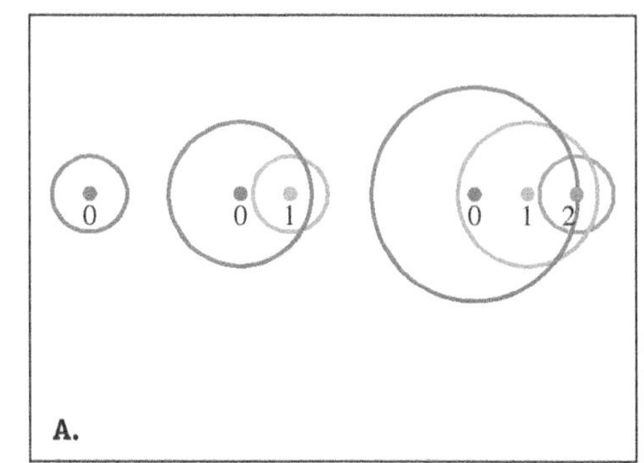

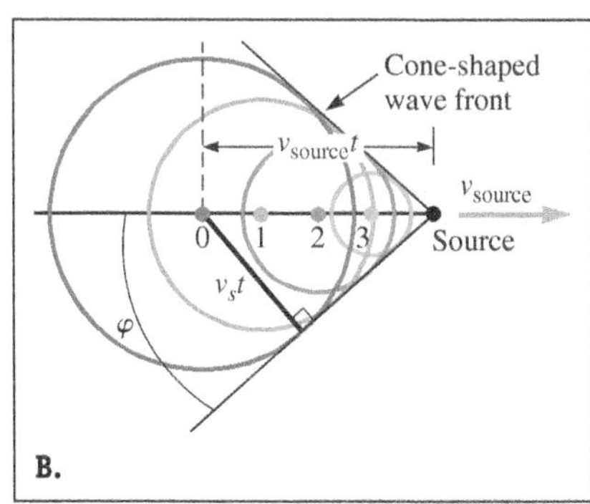

FIGURE 17.32 A shock wave forms when the source is moving faster than the speed of the wave. Compare with Figure 17.28.

Shock Waves

The source in Figure 17.28 is moving at a speed that is less than the speed of the wave it produces. What happens if the source is moving faster than waves travel in that medium? Some jet airplanes and certain race cars exceed the speed of sound in air, and even the most modest boats and good swimmers can exceed the speed of the water waves they produce. When the source moves faster than the wave it emits, a cone-shaped wave front known as a **shock wave** forms (Fig. 17.31).

SHOCK WAVE Major Concept

Figure 17.32 illustrates how the shock wave forms for a source emitting spherical sound waves in air. The source moves to the right at a speed v_{source} that is greater than the sound speed v_s. The leading edge of each subsequent spherical wave front is to the right of the previous wave front. For example, the leading edge of the blue wave front is to the right of the orange wave front in Figure 17.32A. The combination of the offset spherical waves produces a cone-shaped shock wave.

The half-angle of the cone is known as the **Mach angle** φ and can be found with trigonometry. In Figure 17.32B, the source is at the tip of the cone, having moved a distance $v_{source}t$ from position 0. During that same time interval t, the spherical sound wave that was emitted from position 0 has a radius $v_s t$. As shown, $v_{source}t$ is the hypotenuse of a right triangle, and $v_s t$ is the leg opposite the Mach angle φ. So,

$$\sin \varphi = \frac{v_s t}{v_{source}t} = \frac{v_s}{v_{source}} \tag{17.32}$$

The ratio v_{source}/v_s is known as the **Mach number** M. If an object moves faster than the speed of sound, then $M > 1$, and the object is said to be *supersonic*.

If you have ever have been near a moving supersonic jet airplane, you have probably heard a loud *sonic boom*. You hear the boom when the shock wave reaches you, some time after the airplane has passed (Fig. 17.33). Because the shock wave is the sum of many spherical waves, it has a very large amplitude and carries a lot of energy. The shock wave and sonic boom are continuously produced, but you can only hear the boom for the moment that shock wave makes contact with you. When you are in front of the cone-shaped shock wave, you cannot hear any sound from the plane because the sound waves are behind the cone. If you are inside the cone, you hear the normal Doppler-shifted sound from the plane.

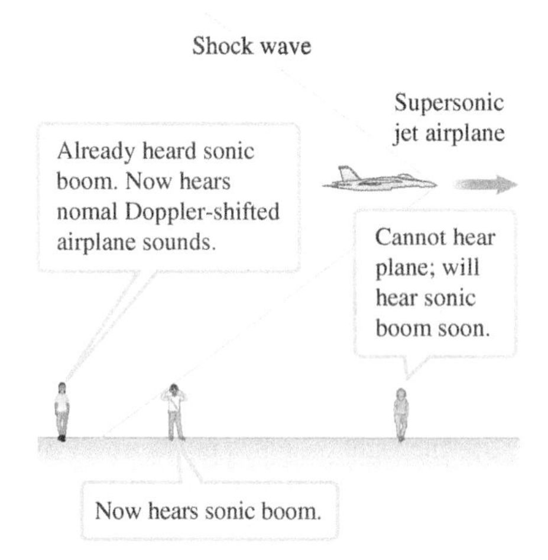

FIGURE 17.33 A sonic boom is only heard when a shock wave is in contact with the listener.

17-10 The Wave Equation

We began this chapter by thinking about fans standing and sitting back down, forming a "wave" around the stadium. Of course, people in a stadium stand up and sit back down all the time, but a wave forms only when the choreography is just right. So, when is a disturbance a wave, and when is it just a disturbance? The answers are given by a (partial) differential equation known as the **wave equation**. If a disturbance satisfies the wave equation, it is a wave. In this section, we derive a somewhat simple version of this equation.

ONE-DIMENSIONAL WAVE EQUATION Underlying Principle

DERIVATION One-Dimensional Wave Equation

Figure 17.34 shows a small segment of rope at a particular moment as a transverse wave passes. We will apply Newton's second law to that small segment and show that it leads to the wave equation:

$$\frac{\partial^2 y(x,t)}{\partial x^2} = \frac{1}{v_x^2}\frac{\partial^2 y(x,t)}{\partial t^2} \quad (17.33)$$

In this differential equation, the symbol ∂ denotes a *partial* derivative. The wave function y depends on x and t. When you take a partial derivative, you treat one of these variables as a fixed constant and take the derivative with respect to the other variable. (This derivation is much like the derivation of the propagation speed of a transverse wave on page 495.)

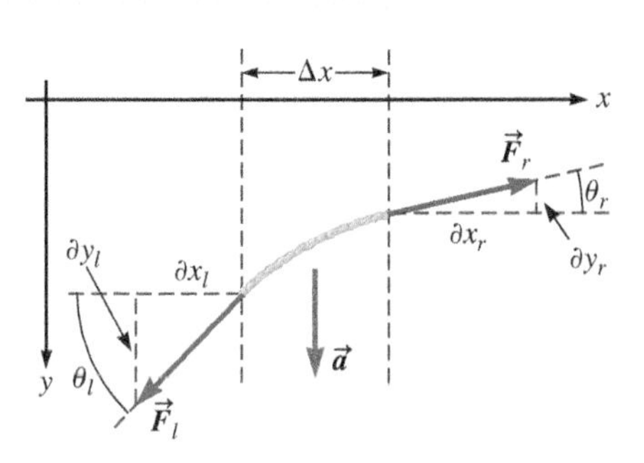

FIGURE 17.34 A small portion of a rope at one instant as a transverse wave passes. Here the notation ∂ denotes small changes. Angles are exaggerated for clarity.

There are two forces exerted on the rope segment. (We can ignore gravity because this segment is very small with very little mass.) Each force has a magnitude equal to the tension F_T in the rope. The force $\vec{F}_r$ on the right pulls up and to the right at an angle θ_r with the x axis, whereas $\vec{F}_l$ pulls down and to the left at an angle θ_l. Both angles are small. The segment is accelerated downward.

Apply Newton's second law in component form starting with the x direction. The angles are small, so $\cos\theta \approx 1$. Because the magnitude of the forces are equal, there is no net force in the x direction, confirming that there is no acceleration in the x direction.	$\sum F_x = F_r\cos\theta_r - F_l\cos\theta_l = ma_x$ $\sum F_x \approx F_r - F_l = ma_x$ $\sum F_x \approx F_T - F_T = 0$ $a_x = 0$
Now apply Newton's second law in the y direction. The magnitude of the force is the tension F_T.	$\sum F_y = F_l\sin\theta_l - F_r\sin\theta_r = ma_y$ $F_T(\sin\theta_l - \sin\theta_r) = ma_y$
Because the angles are small, $\sin\theta \approx \tan\theta$.	$F_T(\tan\theta_l - \tan\theta_r) = ma_y$
Use the right triangles in Figure 17.34 to write the tangent of each angle in terms of the differential distances ∂x and ∂y. (Although y is a function of x and t, here we are only concerned with the segment at one fixed time shown in Fig. 17.34).	$F_T\left[\left(\frac{\partial y}{\partial x}\right)_l - \left(\frac{\partial y}{\partial x}\right)_r\right] = ma_y$
As in our derivation of the propagation speed, we write the mass of the segment in terms of the rope's mass per unit length μ and the length Δx of the segment. (Don't worry about the slight upward bend of the segment; the segment's length is very close to Δx.)	$F_T\left[\left(\frac{\partial y}{\partial x}\right)_l - \left(\frac{\partial y}{\partial x}\right)_r\right] = (\mu\Delta x)a_y$
The acceleration is the second (partial) time derivative of position y.	$F_T\left[\left(\frac{\partial y}{\partial x}\right)_l - \left(\frac{\partial y}{\partial x}\right)_r\right] = (\mu\Delta x)\frac{\partial^2 y}{\partial t^2}$
Rearrange terms.	$\frac{(\partial y/\partial x)_l - (\partial y/\partial x)_r}{\Delta x} = \left(\frac{\mu}{F_T}\right)\frac{\partial^2 y}{\partial t^2}$ (1)
As the segment's length goes to zero, the term on the left becomes a (partial) derivative.	$\lim_{\Delta x \to 0}\frac{(\partial y/\partial x)_l - (\partial y/\partial x)_r}{\Delta x} = \frac{\partial}{\partial x}\left(\frac{\partial y}{\partial x}\right) = \frac{\partial^2 y}{\partial x^2}$
Rewrite Equation (1) in terms of this second (partial) derivative.	$\frac{\partial^2 y}{\partial x^2} = \left(\frac{\mu}{F_T}\right)\frac{\partial^2 y}{\partial t^2}$
Finally, use $v_x = \sqrt{F_T/\mu}$ (Eq. 17.11) to eliminate the term in parentheses.	$\frac{\partial^2 y(x,t)}{\partial x^2} = \frac{1}{v_x^2}\frac{\partial^2 y(x,t)}{\partial t^2}$ ✓ (17.33)

COMMENTS

We derived the wave equation for a transverse one-dimensional wave rather than for a more general case, but the form of the equation we derived is fundamental enough for us to appreciate its significance.

The wave equation tells us that a disturbance is a wave if the medium's curvature $\partial^2 y(x, t)/\partial x^2$ is proportional to the medium's acceleration $\partial^2 y(x, t)/\partial t^2$, where the constant of proportionality depends on the wave's propagation speed. One of the amazing things about nature is that the wave equation describes so many diverse phenomena. The wave equation applies to the mechanical waves in this chapter, including transverse and longitudinal waves, waves on water, and sound waves in air. Moreover, the wave equation describes the propagation of light through a vacuum and of the gravitational force through the Universe. According to quantum mechanics, the wave equation is a fundamental description of matter. The overarching goal of physics is to discover the fundamental laws of nature. The wave equation is a great achievement of physics, illustrating the simple beauty of nature behind its apparent diversity.

SUMMARY

Underlying Principles

One-dimensional wave equation:

$$\frac{\partial^2 y(x, t)}{\partial x^2} = \frac{1}{v_x^2}\frac{\partial^2 y(x, t)}{\partial t^2} \quad (17.33)$$

Major Concepts

1. A **longitudinal wave** is a mechanical wave in which the back-and-forth motion of each particle is parallel to the direction of wave propagation.
2. A **transverse wave** is a mechanical wave in which the particles in the medium move perpendicular to the direction of wave propagation.
3. A **harmonic wave** is a periodic wave in which each particle of the medium oscillates in simple harmonic motion (SHM).
4. The **wave function** of a **transverse** harmonic wave is
$$y(x, t) = y_{max} \sin (kx - \omega t) \quad (17.4)$$
5. The **wave function** of a **longitudinal** harmonic wave is
$$S(x, t) = S_{max} \sin (kx - \omega t) \quad (17.6)$$
In Equations 17.4 and 17.6, the angular wave number is $k = 2\pi/\lambda$ (Eq. 17.5) and the angular frequency is $\omega = 2\pi f$ (Eq. 16.2); the speed of the wave is $v = \omega/k$ (17.7).
6. **Intensity** I is defined as the average power P_{av} per unit area A perpendicular to the direction of propagation:
$$I \equiv \frac{P_{av}}{A} \quad (17.22)$$
The intensity of a point source decreases as the distance r from the source increases:
$$I = \frac{P_{av}}{4\pi r^2} \quad (17.23)$$
7. **Refraction** is the bending of a wave as it travels between media having different wave propagation speeds v_1 and v_2. The **law of refraction** states that
$$\sin \theta_2 = \frac{v_2}{v_1}\sin \theta_1 \quad (17.26)$$
where θ is the angle between the direction of wave propagation and the normal to the boundary between media.
8. **Diffraction** is the bending of a wave due to an obstacle or hole. If the obstacle or hole is comparable in size to the wavelength, diffraction becomes significant.
9. The **Doppler shift** indicates the change in frequency due to relative motion between the wave source and the observer. For sound waves, the observed frequency is
$$f_{obs} = \left(\frac{v_s \pm v_{obs}}{v_s \mp v_{source}}\right)f \quad (17.31)$$
where v_s is the speed of sound. Speeds v_{obs} and v_{source} are measured with respect to the medium (usually air). See page 511 for rules on choosing the signs.
10. A **shock wave** is a cone-shaped wave front created when a wave source moves faster than waves travel in the medium.

Special Cases

1. **Transverse waves on a rope** travel at speed

$$v_x = \sqrt{\frac{F_T}{\mu}} \quad (17.11)$$

where F_T is the tension in the rope and μ is the linear mass density.
The power transported is

$$P_{av} = \tfrac{1}{2}\mu\omega^2 y_{max}^2 v_x \quad (17.19)$$

2. **Sound** is a longitudinal wave in a medium with bulk modulus B and density ρ. The speed of sound is

$$v_s = \sqrt{\frac{B}{\rho}} \quad (17.13)$$

A harmonic sound wave can be written as a **pressure wave**:

$$\Delta P = \Delta P_{max} \cos(kx - \omega t) \quad (17.16)$$

Sound intensity:

$$I = \frac{1}{2}\frac{\Delta P_{max}^2}{\rho v_s} \quad (17.24)$$

Sound level:

$$\beta \equiv 10 \log\left(\frac{I}{I_0}\right) \quad (17.25)$$

where $I_0 = 10^{-12}$ W/m^2.

PROBLEMS AND QUESTIONS

A = algebraic C = conceptual E = estimation G = graphical N = numerical

17-1 Introducing Mechanical Waves

1. C A dog swims from one end of a pool to the opposite end. Is the dog's motion described as a wave? Explain.
2. C A boat crosses a still pond, causing a person in an inner tube floating on the pond to bob up and down. Use the terms *particle motion*, *wave*, and *oscillates* to describe this situation.

17-2 Pulses

3. C If the pulse in Figure 17.4 is twice as tall so that $y(0, 0) = 2$ m, what changes are needed in the wave function (Eq. 17.2)

$$y(x, t) = \frac{1}{(x - v_x t)^2 + 1}$$

4. A pulse on a rope has a profile given by

$$y(x) = \frac{1}{4x^2 + 1}$$

The pulse moves to the right at $v_x = 2$ m/s.
   a. A Write an expression for the wave function.
   b. G How does this new pulse compare with the one in Figure 17.4? *Hint*: This problem is best approached graphically.
5. N The wave function for a traveling pulse, similar to the one discussed in Section 17-2, is

$$y(x, t) = \frac{8}{1 + (x + 4t)^2}$$

where x and y are in meters and the time t is in seconds. In which direction and with what speed is the pulse propagating?

Problems 6 and 7 are paired.

6. G The wave function of a pulse is

$$y(x, t) = \frac{7.5}{(x + 3.5t)^2 + 0.5}$$

where the values are in the appropriate SI units. **a.** Consider a particle at $x = 5.0$ m and sketch a graph of its vertical position y versus time. **b.** Repeat part (a) for a particle at $x = -5.0$ m. Compare your results in the **CHECK and THINK** step.
7. The wave function of a pulse is

$$y(x, t) = \frac{7.5}{(x + 3.5t)^2 + 0.5}$$

where the values are in the appropriate SI units.
   a. N What are the speed and direction of the pulse? *Hint*: Sketch the wave pulse at $t = 2$, 4, and 6 s as in Figure 17.4.
   b. A Write the wave function for a pulse with the same profile and speed but moving in the opposite direction.
8. G The wave function of a transverse pulse on a rope is

$$y(x, t) = \frac{2}{(3x - 4.0t)^2 + 1}$$

where the values are in the appropriate SI units. Sketch a graph of the wave's profile.

17-3 Harmonic Waves

9. N A sinusoidal traveling wave is generated on a string by an oscillating source that completes 120 cycles per minute. What is the wavelength of the wave if individual crests are observed to travel 15.0 m in 1.00 min?
10. A If a harmonic wave is traveling in the negative x direction and its wave function is $y(x, t) = y_{max} \sin(kx + \omega t)$, show that $v_x = -\omega/k$.
11. C Suppose a simple harmonic oscillator creates a transverse wave on a rope as in Figure 17.8A (page 491). If the angular frequency of the oscillator is increased, what properties of the wave must change? Explain your answers.
12. The equation of a harmonic wave propagating along a stretched string is represented by $y(x, t) = 4.0 \sin(1.5x - 45t)$, where x and y are in meters and the time t is in seconds.
   a. C In what direction is the wave propagating?
   b–e. N What are the **b.** amplitude, **c.** wavelength, **d.** frequency, and **e.** propagation speed of the wave?

13. **N** Neonatal ultrasound machines typically operate in the 7.00-Mhz to 18.0-MHz range. What is the wavelength of 7.00-MHz ultrasound traveling through human tissue at 1497 m/s?

14. As in Figure 17.8A (page 491), a simple harmonic oscillator is attached to a rope, creating a transverse wave. The oscillator has an angular frequency of 2.34 rad/s and an amplitude of 34.6 cm.
    a. **N** If the propagation speed of the wave is 2.67 m/s, what is its wavelength?
    b. **A** Write an expression for the wave function.
    c. **C, A** If the oscillator's angular frequency doubles and the propagation speed remains the same, what changes in the wave function? Write the new wave function.

15. **N** The wave function

$$y(x, t) = (0.0500 \text{ m}) \sin\left(3\pi x + \frac{\pi}{4}t\right)$$

describes a transverse wave on a rope, with x and y in meters and t in seconds.
What are the **a.** wavelength, **b.** period, and **c.** speed of the wave? **d.** At $t = 1.00$ s, what is the transverse velocity of a rope element located at $x = 0.200$ m? **e.** At $t = 1.00$ s, what is the transverse acceleration of a rope element located at $x = 0.200$ m?

16. A harmonic transverse wave function is given by

$$y(x, t) = (0.850 \text{ m}) \sin(15.3x + 10.4t)$$

where all values are in the appropriate SI units.
    a. **N** What are the propagation speed and direction of the wave's travel?
    b. **N** What are the wave's period and wavelength?
    c. **N** What is the amplitude?
    d. **C** If the amplitude is doubled, what happens to the speed of the wave?

Problems 17 and 18 are paired.

17. A graph (profile) of a traveling wave at a moment in time is shown in Figure P17.17. The wave is traveling in the negative x direction with a propagation speed of 56.0 cm/s.
    a. **N** What is the amplitude of the wave?
    b. **N** What is the frequency of the wave?
    c. **A** Write the transverse harmonic wave function for this wave.

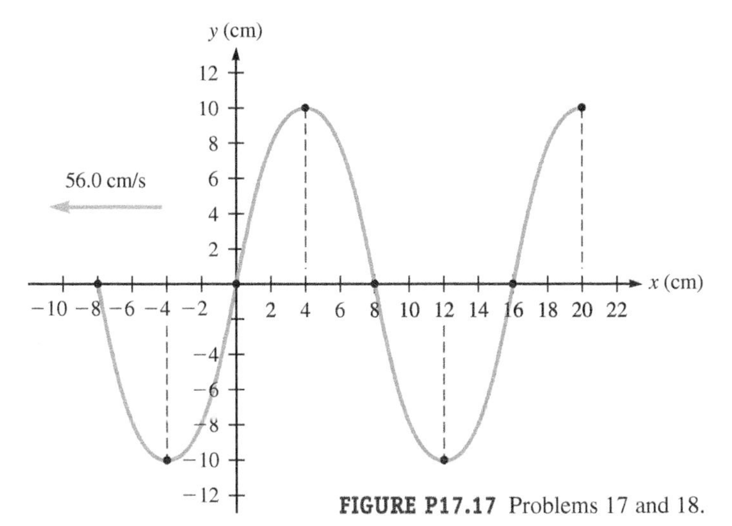

FIGURE P17.17 Problems 17 and 18.

18. In Problem 17,
    a. **N** What is the period of the wave?
    b. **G** Plot the wave over the same range of x values shown when $t = T/2$ (Assume the graph shown is for $t = 0$.)

19. **N** A longitudinal harmonic wave is given by

$$S(x, t) = (0.850) \sin(10.4x - 5.20t)$$

where all values are in the appropriate SI units. **a.** What are the period and wavelength of this wave? **b.** Consider a particle whose equilibrium position is $x = \lambda/2$. Find the displacement S of this particle at $t = 0$, $T/4$, $T/2$, $3T/4$, and T. **c.** What is the position of the particle at these times?

17-4 Special Case: Transverse Wave on a Rope

20. **C** Your laboratory instructor gives your team a long, very thin thread. Your challenge is to find its mass precisely. You have a scale, several cylinders of different weights each with a hook, a meterstick, and a timer. Your lab has familiar equipment such as pulleys and vertical and horizontal supports. The thread is very light and strong, similar to spider's silk. When you place it on the scale, you find that the scale is not sensitive enough to report a meaningful mass. Come up with a better procedure to measure the thread's mass. Describe your procedure and your plan to interpret your data.

21. **N** A copper wire with linear mass density 6.30×10^{-2} kg/m is placed under 2.25×10^3 N of tension. What is the speed of traveling waves on this wire?

22. **A** A string of length L and mass M hangs freely from a fixed point on the ceiling. **a.** Determine the propagation speed of a transverse wave traveling along the string at any position y. *Hint*: The tension is not constant throughout the string. The tension at any point along the string is equal to the weight $g\,dm$ of the string below that point. **b.** Determine the time required for a transverse pulse to travel the entire length of the string.

23. **N** A wave on a string with linear mass density 5.00×10^{-3} kg/m travels at 105.0 m/s. What is the tension in the string?

24. **N** A traveling wave on a thin wire is given by the equation $y = (0.112 \text{ m}) \sin(4.35x + 46.7t)$, where all values are in the appropriate SI units. **a.** What is the propagation speed of this wave? **b.** If the linear mass density of the string is 3.20×10^{-3} kg/m, what is the tension in the string?

25. A sinusoidal wave with amplitude 1.00×10^{-2} cm and frequency 325 Hz is observed to be traveling with a speed of 55.0 m/s on a wire held under 195 N of tension.
    a. **A** What is the wave function for this wave in SI units?
    b. **N** What is the mass per unit length for the wire?

26. As in Figure 17.8A, a simple harmonic oscillator is attached to a rope, creating a transverse wave of wavelength 7.47 cm. At the other end of the rope is a pulley and a hanging block of mass 6.30 kg. The oscillator has an angular frequency of 2340 rad/s and an amplitude of 34.6 cm.
    a. **N** What is the linear mass density of the rope?
    b. **C** If the angular frequency of the oscillator doubles, what properties (speed, angular wave number, amplitude) of the wave must change? Give the new values of these properties.
    c. **C** If the frequency remains at its original value but the mass of the hanging block is doubled, what properties of the wave must change? Give the new values of these properties.

27. **N** An aluminum wire with a radius of 1.00 mm is held under tension. If the speed of traveling waves on the wire is 180 m/s and the density of aluminum is 2.70 g/cm³, what is the tension in the wire?

28. **C** People of the Incan empire built a long series of roads and bridges in the highlands of Peru. The bridges were made of rope and had to be rebuilt yearly by nearby villagers (Fig. P17.28).

In the construction of a new bridge, two long ropes were secured across the gorge. A worker then scooted across the two ropes while attaching smaller lengths of rope to form the footpath. Suppose it is your job to scoot out over the gorge on just two ropes. You may want to check the tension in both ropes to make sure it is the same in each rope. Explain how you could use a traveling pulse to check the tension.

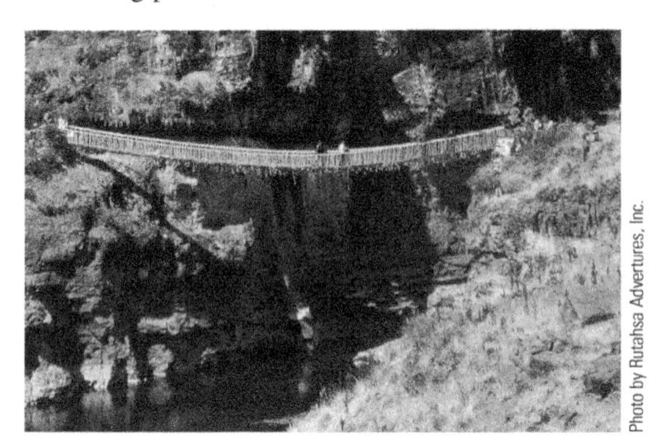

FIGURE P17.28

17-5 Sound: Special Case of a Traveling Longitudinal Wave

29. N Assume the smallest amplitude detectable by your ear is $S_{min} = 1 \times 10^{-11}$ m. If $f = 1000$ Hz, what is the smallest-pressure amplitude you can hear under the ocean? Compare your result with that in Example 17.4 (quietest whisper in air, $\Delta P_{max} = 3 \times 10^{-5}$ Pa). The speed of sound in seawater is 1531 m/s.

30. N Dolphins have been trained to understand human voice commands. Imagine a trainer standing on a platform with her head and shoulders 5.20 m above the surface of a pool, with a dolphin waiting 10.4 m directly below the trainer (Fig. P17.30). Assume the speed of sound in air is 343 m/s and in water is 1497 m/s.

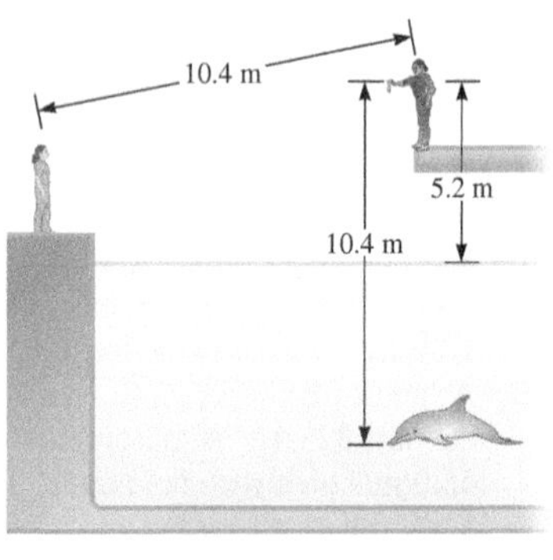

FIGURE P17.30

a. N If the trainer gives the command to jump, how long will it take the dolphin to hear the command?

b. C Would the dolphin hear the command before or after another person who is 10.4 m from the trainer? Justify your answer.

31. A whistle is often used in dog-training exercises and is an integral part of field-marking competitions, where signals and commands are relayed to the dog via whistle. Suppose the whistle produces a sound wave with a frequency of 25,100 Hz (outside the range of human hearing) and the propagation speed of sound in air is 343 m/s.

a. N What is the wavelength of this sound from the whistle?

b. N What is the wave number k corresponding to the wavelength you found?

c. A Write the wave function for the longitudinal sound wave described above, assuming the amplitude of the sound wave is $S_{max} = 1.57 \times 10^{-6}$ m and it is moving to the right (in the positive x direction).

Problems 32 and 33 are paired.

32. N Seismic waves travel outward from the epicenter of an earthquake. A single earthquake produces both longitudinal seismic waves known as P waves and transverse waves known as S waves. Both transverse and longitudinal waves can travel through solids such as rock. Longitudinal waves can travel through fluids, whereas transverse waves can only be sustained near the surface of a fluid, not inside the fluid. When seismic waves encounter a fluid medium such as the liquid outer core of the Earth, only the longitudinal P wave can propagate through. Geophysicists can model the interior of the Earth by knowing where and when S and P waves were detected by seismographs after an earthquake (Fig. P17.32). Assume the average speed of an S wave through the Earth's mantle is 5.4 km/s and the average speed of a P wave is 9.3 km/s. After an earthquake, a seismograph finds that the P wave arrives 1.5 min before the S wave. How far is the epicenter from the detector?

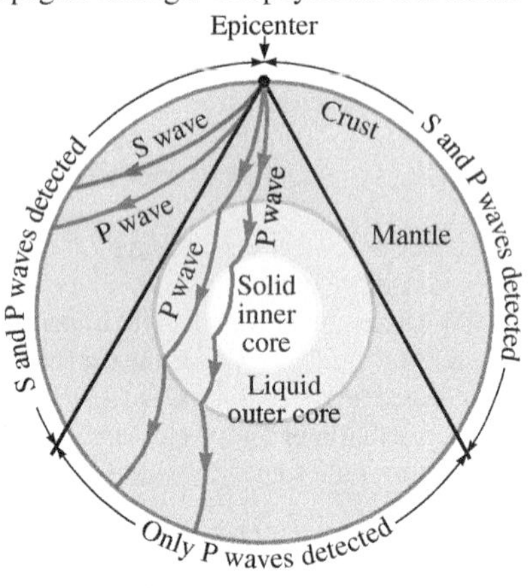

FIGURE P17.32

33. N On July 17, 2006, an earthquake hit just south of Java, Indonesia (between Australia and Malaysia). The time of the earthquake was 08:19:25.02 UTC (Coordinated Universal Time). The strong earthquake was detected by more than 400 seismograph stations worldwide. Table P17.33 provides the arrival time of the P wave and the distance to three such stations. Assume the P wave traveled through the crust and the crust's average density is 3300 kg/m^3. Estimate the bulk modulus of the Earth's crust. *Hint*: See Problem 32.

TABLE P17.33 Java earthquake data.

Arrival time at seismograph (UTC)	Distance from epicenter to station (km)
08:21:55.58	1190
08:22:05.09	1230
08:23:29.89	1970

17-6 Energy Transport in Waves

34. N A harmonic wave travels on a string with a linear mass density of 4.51×10^{-3} kg/m. It travels with a speed of 112 m/s and has a frequency of 65.0 Hz. If the wave has an amplitude of 0.118 m, what is the amount of energy in three wavelengths of the wave?

Problems 35 and 36 are paired.

35. A Show that the kinetic energy in one wavelength of a harmonic transverse wave on a rope (Fig. 17.8A) is given by $K_\lambda = (1/4)\mu\omega^2 y_{max}^2 \lambda$.

36. A Use the result of Problem 35 to show that the potential energy in one wavelength of a harmonic transverse wave on a rope (Fig. 17.8A) is given by $U_\lambda = (1/4)\mu\omega^2 y_{max}^2 \lambda$.

37. N A 5.00-m rope with a mass of 0.650 kg is held under tension. If 1.00 kW of power is supplied to the rope, what is the amplitude of sinusoidal waves that will be generated with a wavelength of $\pi/4$ m and speed 50.0 m/s?

38. C Suppose you have a transverse harmonic wave on a rope. If you wish to increase the power transferred by the wave, what should you do to the rope?

39. N Sound with a wavelength of 6.59 m is produced in air where it travels with a propagation speed of 343 m/s. The average power of the sound wave is 40.0 W. If the density of the air in this region of space happens to be 1.20 kg/m^3, what is the energy contained in one wavelength of the wave?

40. N An oscillating lever delivers a maximum power of 0.500 kW to a string with linear mass density of 1.50×10^{-2} kg/m that is under 125 N of tension. What is the maximum angular frequency of the lever that will result in the generation of sinusoidal waves with an amplitude of 3.00 cm?

17-7 Two- and Three-Dimensional Waves

41. N A sound produces a pressure amplitude of $\Delta P_{max} = 32$ Pa. Determine the intensity of this sound wave. Assume the density of air is 1.3 kg/m^3 and the speed of sound is 343 m/s.

42. CASE STUDY Besides music, other loud sounds can damage your hearing. When standing 1 m from an operating chainsaw, the sound level is 105 dB.

a. G Plot the sound level versus distance from the chainsaw.

b. N At what distance from the chainsaw is the sound level equivalent to a whisper at the threshold of human hearing? (Read this distance off your graph.)

Problems 43 and 44 are paired.

43. N Sunlight is an example of a wave that, like sound, can be thought of as being emitted in all directions with its power uniformly distributed over the surface of any sphere centered on the source (the Sun). Therefore, the intensity of sunlight will be different when measured at the locations of the various planets (Fig. P17.43). The intensity of sunlight that reaches the Earth's atmosphere is about 1400 W/m^2. The Sun-Earth distance is 1 AU (1.496×10^{11} m). **a.** What is the intensity that reaches Mars if Mars is 1.5 times farther from the Sun than the Earth? **b.** Saturn is about 10 times farther from the Sun than the Earth. What is the intensity of sunlight at Saturn?

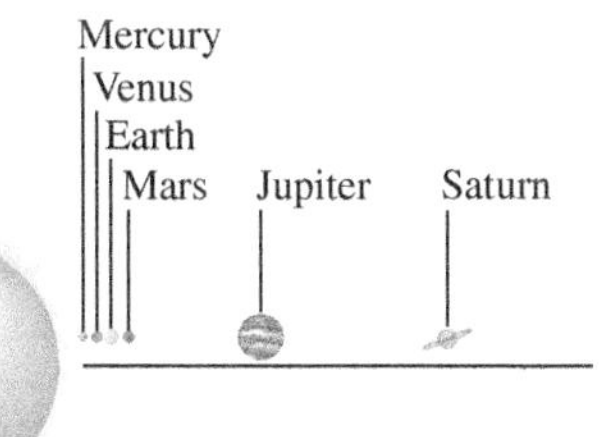

FIGURE P17.43 Problems 43 and 44.

44. Consider the discussion of light waves in Problem 43. Solar cells absorb energy from light waves and convert it to electrical energy. A certain solar cell has an area of 2.00 m^2 and converts 10% of the absorbed energy into usable electric energy.

a. N What is the power output of the solar cell if placed in space near the Earth?

b. N What is the power output of the solar cell if placed in space near Mars?

c. C Consider your answers to parts (a) and (b). Describe an issue that must be considered when scientists and engineers plan research missions to Mars and beyond. In what practical ways can researchers address this issue? Why might we choose to use nuclear power rather than solar power when planning a mission to Mars?

45. N A violin is played with an initial intensity I_i increasing to a final intensity I_f. If the final intensity is 10 times the initial intensity, what is the difference in sound level between these two extremes?

46. N What is the sound level of a sound wave with intensity 6.40×10^{-5} W/m^2?

47. N A listener finds that the sound level of a flute is 3 dB higher than the sound level of a cello. How does the intensity of the flute compare with the intensity of the cello?

48. N The speaker system at an open-air rock concert forms a ring around the entire circular stage and delivers 50,000 W of power output. Assume the sound radiates in all directions equally as if it were generated by an isotropic point source and assume the sound energy is not absorbed by air. **a.** At what distance is the sound from the speakers barely audible? **b.** What is the closest distance audience members can be to the speakers if the sound is not to be painful to their ears?

17-8 Refraction and Diffraction

Problems 49 and 50 are paired.

49. N The Earth's mantle is denser near the core than it is near the crust, so the speed of seismic waves varies throughout the interior. Therefore, seismic waves are refracted as they travel deep within the Earth. This refraction is used to help model the density of interior layers. A seismic longitudinal P wave traveling at 8.5 km/s encounters a region in which the propagation speed is 12.5 km/s (Fig. P17.49). If the P wave strikes the boundary between regions at an angle of 65° with respect to the boundary, what is the angle of refraction θ in the dense region?

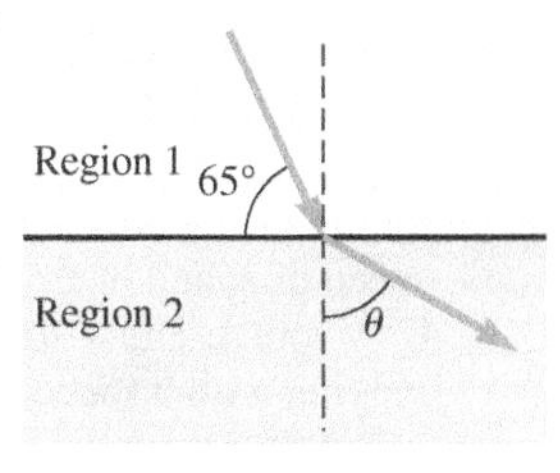

FIGURE P17.49

50. N A seismic P wave strikes a boundary between two types of material having the same bulk modulus. At the boundary, the density increases abruptly from 3400 kg/m^3 to 4100 kg/m^3 (Fig. P17.50). If the wave front of the P wave is planar and makes a 25° angle with the boundary, what is the angle of refraction θ?

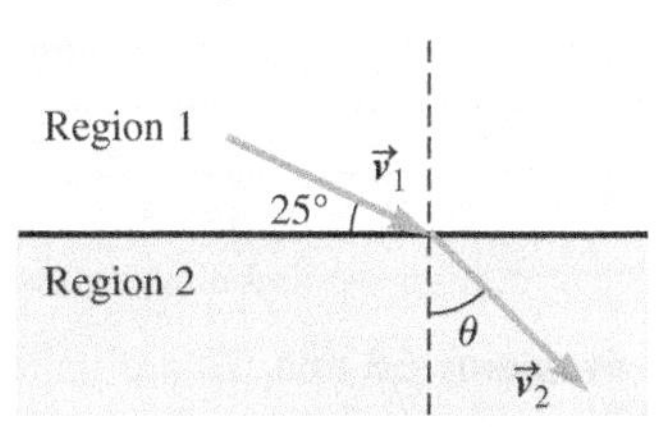

FIGURE P17.50

Problems 51 and 52 are paired.

51. N Light can be modeled as a wave. You can see because your eye collects the light waves that bounce off objects (or light produced by objects). When viewing a fish in the water, Wanda's eye collects light that has bounced off the fish, such as the ray indicated in Figure P17.51. The direction of wave propagation is altered due to refraction as the light wave leaves the water. The speed of light in air is 3.00×10^8 m/s and in water is 2.25×10^8 m/s. Given these speeds and the information in the figure, find the angle θ measured from the vertical.

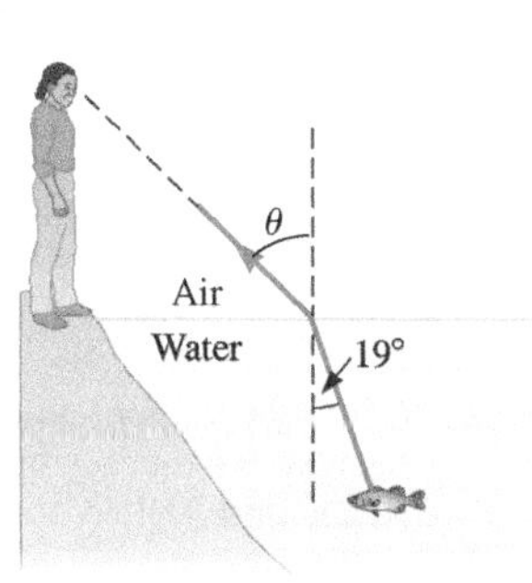

FIGURE P17.51 Problems 51 and 52.

52. C Spearfishing from shore or a dock is a practical means by which to catch fish and is still practiced in parts of the world. Use Figure P17.51 (Problem 51) to think about Wanda's aim when she attempts to spear a fish. Should she aim at a location farther away or closer than that at which she perceives the fish?

53. C Sometimes when you hear an airplane, it is difficult to locate its position in the sky immediately. Assume the density of air at the plane's altitude is less than the density of air near the Earth's surface. How can refraction make it difficult to locate the source of the sound?

54. C Using the concept of diffraction, discuss how the sound from a concert stage is affected if you sit behind a large pillar (greater than 1 m wide) near the middle of the concert hall. Is the answer different for treble (shorter-wavelength) or bass (longer-wavelength) sounds? Explain your answer.

17-9 The Doppler Shift

55. N For an ecology project, Maria is recording some ambient sounds in the woods. A hawk emits a cry with a frequency of 8.00×10^2 Hz. If the hawk is flying toward Maria's microphone at a speed of 18 m/s as it sustains this sound frequency, what is the frequency that is recorded by the microphone? Assume the speed of sound in air is 343 m/s.

56. C A source emits sound with a frequency of 440 Hz and is moving with a speed of 24 m/s. An observer driving a car at a speed of 24 m/s is chasing the source. Assume the speed of sound is 343 m/s. Will the frequency heard by the observer be higher than, lower than, or equal to 440 Hz? Explain your answer.

57. N An ambulance traveling eastbound at 140.0 km/h with sirens blaring at a frequency of 7.00×10^2 Hz passes cars traveling in both the eastbound and westbound directions at 55.0 km/h. What is the frequency observed by the eastbound drivers **a.** as the ambulance approaches from behind and **b.** after the ambulance passes them? What is the frequency observed by the westbound drivers **c.** as the ambulance approaches them and **d.** after the ambulance passes them?

Problems 58 and 59 are paired.

58. N Joe, a skateboarder, wants to measure his downhill speed, thinking that he may reach 45 mph. While riding down the hill, he holds a panic alarm that operates at a frequency of 1250 Hz. Rochelle, who has perfect pitch, claims that she can identify any note on a piano from 27.5 Hz to 4186.0 Hz.

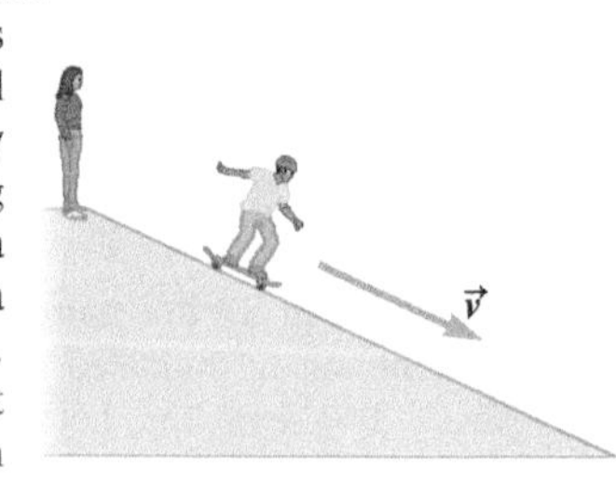

FIGURE P17.58

a. Rochelle listens from the top of the hill as Joe rolls away (Fig. P17.58). If Joe really reaches 45 mph, what frequency will Rochelle hear? **b.** From the top of the hill, Rochelle hears a C6, which has a frequency of 1046.5 Hz. What is Joe's speed? In the **CHECK and THINK** step, describe whether or not your answer seems reasonable.

59. N Consider the scenario in Problem 58. This time, Rochelle listens from the bottom of the hill as Joe rolls toward her (Fig. P17.59). If the skateboarder really reaches 45 mph, what frequency will Rochelle hear?

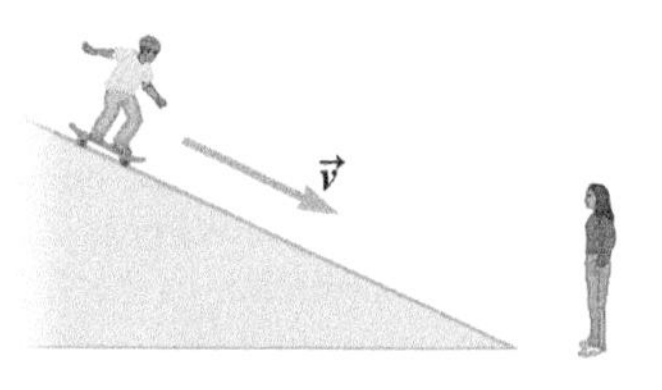

FIGURE P17.59

60. N The Mach number of a supersonic jet is 3.5. What is the Mach angle?

Problems 61 and 62 are paired.

61. C You are sitting a certain distance away from a speaker that is playing a 450-Hz sound at a constant power. Assume the speed of sound in air is 343 m/s. **a.** Describe what happens to the frequency and wavelength of the sound you hear if the speaker is moved toward your location. **b.** Describe what happens to the frequency and wavelength of the sound you hear after the speaker has moved past you and continues receding.

62. (G, C) In Problem 61, **a.** Sketch an image of the wave fronts being emitted by the speaker as the speaker approaches you. In what way does the sketch illustrate the change in the wavelength you observe as the speaker is moving toward your location? **b.** Sketch an image of the wave fronts being emitted by the speaker as it recedes away from you. In what way does the sketch illustrate the change in the wavelength you observe as the speaker is moving away from your location?

63. N A woman sees a supersonic plane directly overhead and then hears a sonic boom 3.7 s later. If the plane's altitude is 1.75×10^3 m, what is the Mach number of the plane? Assume the speed of sound is 331 m/s.

General Problems

64. A Begin with the wave function of a one-dimensional transverse wave on a string and show that it leads to the wave equation

$$\frac{\partial^2 y(x, t)}{\partial x^2} = \frac{1}{v_x^2} \frac{\partial^2 y(x, t)}{\partial t^2}$$

In this differential equation, the symbol ∂ denotes a *partial* derivative. The wave function y depends on x and t. When you take a partial derivative, treat one of these variables as constant and take the derivative with respect to the other variable.

65. How far are you located from a lightning strike if the thunderclap arrives 7.50 s after you see the lightning flash? Assume the speed of light in air is 3.0×10^8 m/s and the speed of sound at this location is 3.40×10^2 m/s.

66. N A traveling wave on a thin wire is given by the equation $y = (0.405 \text{ m}) \sin[0.354x - 1150t]$, where all values are in the appropriate SI units. What are the **a.** propagation speed and **b.** period of the wave? **c.** How far horizontally does the wave travel during one period?

67. N A traveling sinusoidal wave is described by the wave function

$$y(x, t) = (0.600 \text{ m}) \sin\left(7.50\pi t - \pi x + \frac{\pi}{2}\right)$$

where x and y are in meters and t is in seconds. What are the **a.** wavelength, **b.** frequency, **c.** amplitude, and **d.** speed of propagation of the wave?

68. A A sinusoidal wave with amplitude 4.50 cm, wavelength 10.0 cm, and frequency 1.50 Hz is observed to travel in the negative direction. **a.** What is the wave function for this wave if $y(x, t) = 4.50$ cm at $x = 0$ and $t = 0$? **b.** What is the wave function for this wave if $y(x, t) = 4.50$ cm at $x = 5.00$ cm and $t = 0$?

69. N If the intensity of a sound wave increases by 1%, what is the change in the sound level of the wave?

70. N The equation of a traveling wave is given as

$$y(x, t) = 0.50 \sin^2(2.0x - 4.0t)$$

where x and y are in meters and the time t is in seconds. Find the amplitude, wavelength, and angular frequency of the wave. *Hint*: You will need to use the trigonometric identity $\cos(2\theta) = 1 - 2\sin^2\theta$.

71. N A block of mass $m = 5.00$ kg is suspended from a wire that passes over a pulley and is attached to a wall (Fig. P17.71). Traveling waves are observed to have a speed of 33.0 m/s on the wire. **a.** What is the mass per unit length of the wire? **b.** What would the speed of waves on the wire be if the suspended mass were decreased to 2.50 kg?

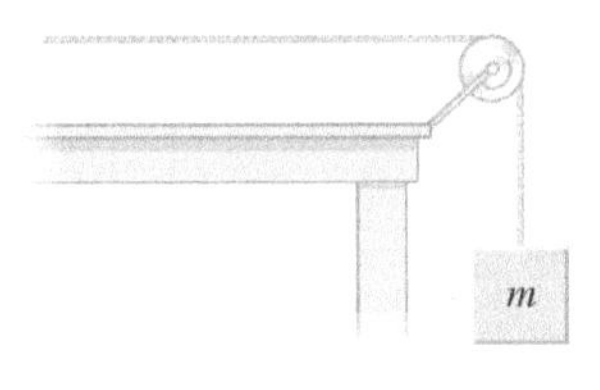

FIGURE P17.71

72. A The equation of a harmonic wave propagating along a stretched string is given as

$$y(x, t) = y_{\max} \sin\left[2\pi\left(\frac{x}{a} - ft\right)\right]$$

where a is a constant and f is the frequency. If the magnitude of the maximum vertical speed

$$v_{y,\max} = \left.\frac{\partial y}{\partial t}\right|_{\max}$$

is equal to three times the propagation speed, what is the value of a? Express your answer in terms of $y_{\max}$.

73. N A wave on a string with mass density 0.450 g/cm under tension is described by

$$y = (0.650\text{ m}) \sin\left(4\pi x + 5\pi t - \frac{3\pi}{4}\right)$$

where x and y are in meters and t is in seconds. **a.** What is the average rate of energy transfer along the wave on the string? **b.** What is the mechanical energy content of each cycle of this wave?

74. N Sound waves leaving a speaker deliver a total energy of 275 J in 4.00 s to a location where the energy of the sound waves is spread over an area of 5.00 m^2. **a.** What is the intensity of the sound at this location? **b.** What is the sound level at this location? **c.** What is the intensity of the sound at a point farther away where the energy is spread over an area of 20.00 m^2? **d.** What is the sound level at this new location?

75. N The wave function $y(x, t) = (0.400\text{ m}) \sin(4.00x + 25.0t)$ describes the motion of a transverse wave on a string, with x and y in meters and t in seconds. **a.** For an element of the string located at $x = 0.100$ m, what is the time interval that elapses between the first and second times that the element is located at $y = -0.300$ m? **b.** In the time interval found in part (a), what is the distance covered by the wave?

76. N A stationary observer receives sound waves from two tuning forks, each oscillating at a frequency of 512 Hz. One of the tuning forks is approaching the observer, and the other tuning fork is receding from the stationary observer with the same speed as the first one. The observer hears a frequency difference (beat frequency; see Chapter 18) of 4 Hz when comparing the sound from each tuning fork. Find the speed of each tuning fork. Assume the speed of sound in air is 343 m/s.

77. N A siren emits a sound of frequency 1.44×10^3 Hz when it is stationary with respect to an observer. The siren is moving away from a person and toward a cliff at a speed of 15 m/s. Both the cliff and the observer are at rest. Assume the speed of sound in air is 343 m/s. What is the frequency of the sound that the person will hear **a.** coming directly from the siren and **b.** reflected from the cliff?

78. C, N Female *Aedes aegypti* mosquitoes emit a buzz at about 4.00×10^2 Hz, whereas male *A. aegypti* mosquitoes typically emit a buzz at about 6.00×10^2 Hz. As a female mosquito is approaching a stationary male mosquito, is it possible that he mistakes the female for a male because of the Doppler shift of the sound she emits? How fast would the female have to be traveling relative to the male for him to make this mistake? Assume the speed of sound in the air is 343 m/s.

79. N A careless child accidentally drops a tuning fork vibrating at 450 Hz from a window of a high-rise building. How far below the window is the tuning fork when the child hears sound waves with frequency 425 Hz? Remember to account for the time required for the sound to reach the child.

80. Tsunamis, waves that displace large volumes of water, are generated by underwater disturbances, including earthquakes, volcanic eruptions, and meteorite impacts. The speed of the wave in a tsunami depends on the depth of the water d and is given by $v = \sqrt{gd}$. A meteorite 50.0 m in diameter impacts the Atlantic Ocean near Greenland, generating a tsunami that can be modeled as a series of circular waves emanating from the point of impact. Satellites measure the speed of the waves to be 300 m/s and their amplitude to be 4.00 m, a distance 1.00 km from the point of impact.

a. C Why does the amplitude of the tsunami wave increase as the wave approaches the Greenland shore?

b. N If the waves emanating from the point of impact are measured to have an amplitude of 0.500 m and a speed of 400 m/s, what will be their amplitude when the depth of the water decreases to 5.00 m near the Greenland coast? Assume the wave power does not diminish between the source and the coast.

81. A A wire with a tapered cross-sectional area is placed under a tension force F_T. The radius of the wire is given by $r = 2.50 - 0.200x$. If the density of the wire is ρ, what is the speed of a wave as a function of position along this wire?

18 Superposition and Standing Waves

Key Questions

What happens when more than one wave exists in a single medium?

How do traveling waves and standing waves differ?

18-1 Superposition 522

18-2 Reflection 524

18-3 Interference 526

18-4 Standing waves 531

18-5 Guitar: Resonance on a string fixed at both ends 534

18-6 Flute: Resonance in a tube open at both ends 537

18-7 Clarinet: Resonance in a tube closed at one end and open at the other end 540

18-8 Beats 541

18-9 Fourier's theorem 543

Underlying Principles

1. Superposition
2. Fourier's theorem

Major Concepts

1. Boundary conditions
   a. Fixed-end reflection
   b. Free-end reflection
2. Law of reflection
3. Constructive and destructive interference
4. Standing wave
5. Beats

Special Case: Standing Waves

1. String fixed at both ends
2. Tube open at both ends
3. Tube open at one end and closed at the other end

When you talk to a friend, watch ripples on a pond, or play a guitar, you are experiencing a combination of many waves. In Chapter 17, we focused on the physics and mathematics of a single harmonic wave traveling in a medium, but we rarely encounter a single harmonic wave. More often, we find a combination of harmonic waves. In this chapter, we develop the mathematics used to describe the combination of waves, known as **superposition**. We will explore many interesting physical phenomena such as reflection, interference, standing waves, and beats.

SUPERPOSITION

Underlying Principle

18-1 Superposition

In Chapter 17, we studied one wave at a time traveling in some medium such as a transverse pulse on a rope (Fig. 17.4). Now let's see what happens when two (or more) waves travel in the same medium. Suppose you work with a friend to send two pulses on a rope. The rope is taut; you stand at one end of the rope, and your friend stands at the other end. You pluck the rope on your end, and your friend plucks the rope on her end; the result is two pulses traveling in opposite directions toward each other (Fig. 18.1). What happens when the two pulses meet?

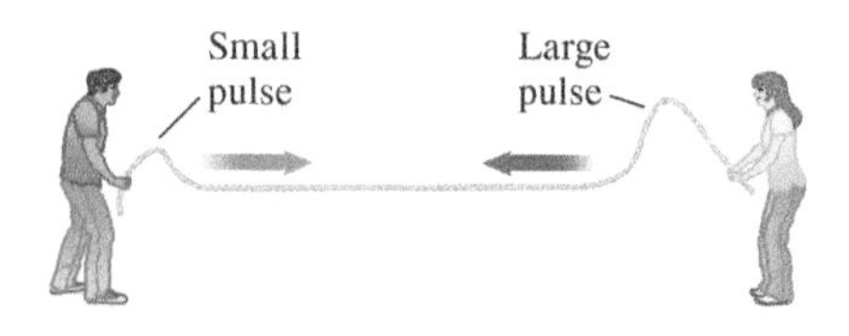

FIGURE 18.1 Two pulses on the same rope are made by people standing on opposite ends.

Each pulse is a disturbance on the rope. Both disturbances exist on the rope simultaneously, and the result is the sum of the two disturbances (Fig. 18.2). The resulting wave function $y(x, t)$ on the rope is found by adding (**superimposing**) the wave functions $y_1(x, t)$ and $y_2(x, t)$ of the individual pulses:

$$y(x, t) = y_1(x, t) + y_2(x, t)$$

At any particular time, the wave on the rope is found by adding the contribution of the pulses at each point x. The resultant wave is the **superposition** of the individual pulses. Notice this quirk in our language: The verb is to super*im*pose, but the letters *im* are dropped from the noun: superposition.

You might have expected that the result of two waves on a rope is the superposition of the individual waves, but this result leads to quite an amazing consequence: The two pulses pass through each other (Fig. 18.2). Each individual pulse continues to travel undisturbed with its original velocity and amplitude.

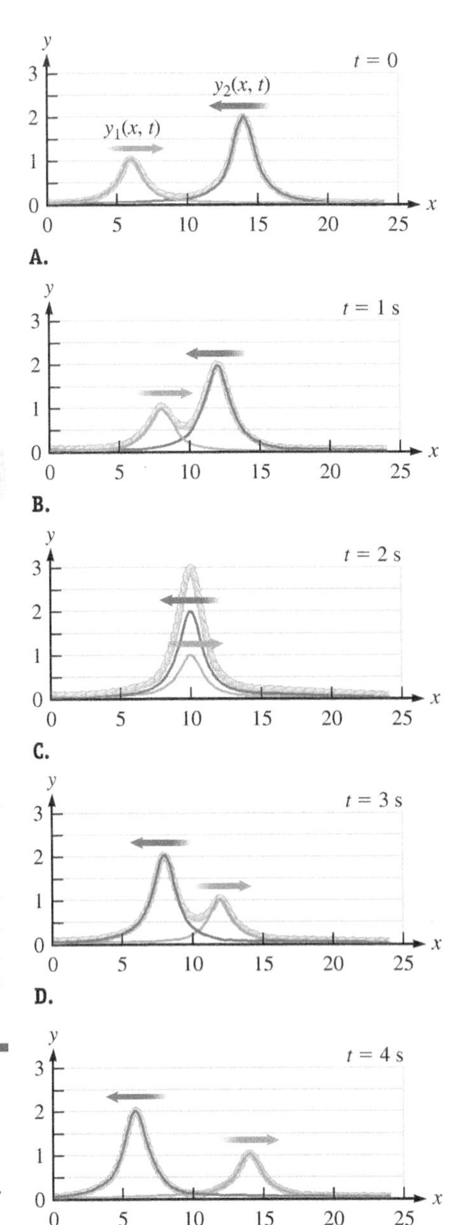

FIGURE 18.2 Two transverse pulses travel on the same rope. The individual pulse wave functions $y_1(x, t)$ in blue and $y_2(x, t)$ in red are plotted at five times. The resulting wave function $y(x, t)$ on the rope may be found by superimposing the individual pulse wave functions: $y(x, t) = y_1(x, t) + y_2(x, t)$.

CASE STUDY Part 1: A Big-Band Concert

Superposition is important whenever two or more waves exist in the same medium. In a band concert, the instruments emit sound waves through several speakers; those sound waves are reflected off the walls, the ceiling, the floor, and even the people in the room. All those sound waves travel in the same medium, the air. The music that reaches your ear is the superposition of all those waves. In addition, the sound produced by the instruments is a superposition of sound waves in the body of the instrument itself.

William Basie (1904–1984), better known as "Count" Basie, led a big-band jazz ensemble with about 20 members playing more than a dozen different instruments. Big-band music was extremely popular in the 1930s and 1940s, and there are still big bands today, including the Count Basie Orchestra. That orchestra performs concerts all over the world, including at festivals in Tokyo and Toronto, at universities, and in formal concert halls. In this case study, we'll explore how a concert hall or an outdoor venue affects the music. Of course, the richness of a big band comes from the great number of instruments, each adding its own voice to the music. So, we'll also explore how three of the big band's instruments (the flute, clarinet, and guitar) produce their unique sounds.

CONCEPT EXERCISE 18.1

As shown in Figure 18.3, two pulses traveling along the same rope toward each other have the same amplitude. Draw a sketch of the rope at the instant the pulses completely overlap.

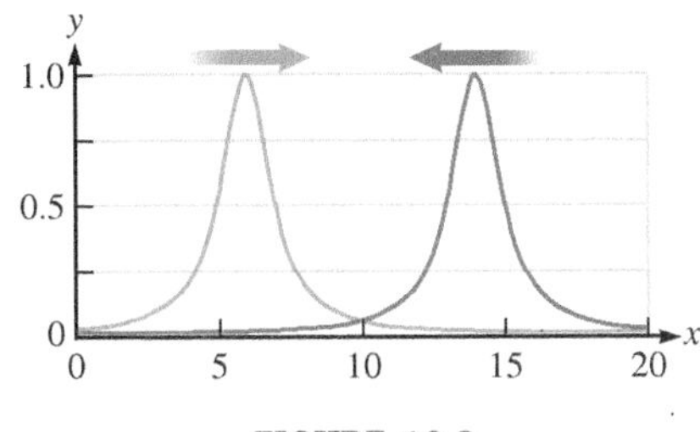

FIGURE 18.3

CONCEPT EXERCISE 18.2

As shown in Figure 18.4, two pulses traveling along the same rope toward each other have the same amplitude, but one is upside down. Draw a sketch of the rope at the instant the pulses completely overlap.

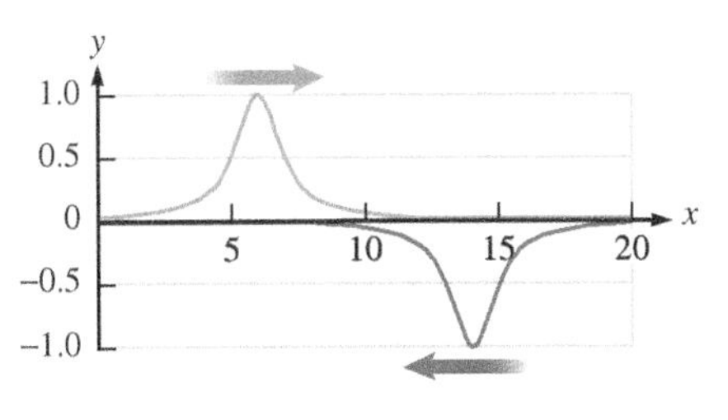

FIGURE 18.4

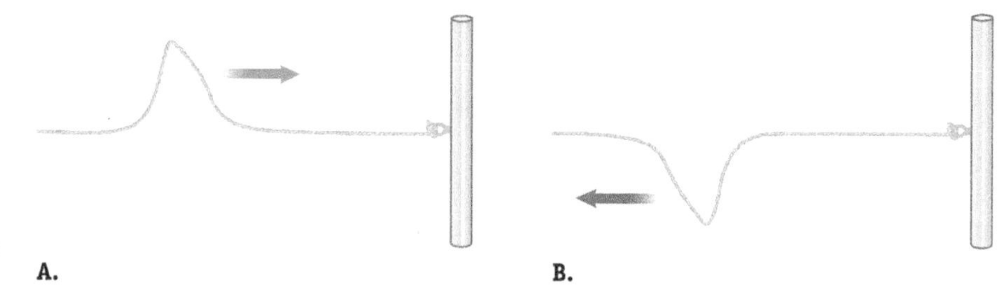

FIGURE 18.5 When a pulse reflects from a fixed end, the reflected pulse is inverted.

18-2 Reflection

BOUNDARY CONDITIONS: FIXED-END REFLECTION AND FREE-END REFLECTION

Major Concept

When you shout into a gorge or at a large wall, you hear your voice echo because when a wave travels to the end of a medium—in this case, the air—the wave is reflected back into the medium. The reflection requires that there be a **boundary**, or an end to the medium. The nature of the reflected wave depends on the state of the boundary known as the **boundary conditions**. Here we consider two important boundary conditions.

Fixed-End Reflection

To study reflection, let's look at a pulse on a rope. A pulse shaped something like a shark's fin travels to the right along a horizontal rope (Fig. 18.5A). The rope is tied to a rigid ring on the right, so the right end is *fixed* and cannot move. As the pulse—called the *incident wave*—approaches the fixed end, it exerts a force on the rigid ring. According to Newton's third law, the ring exerts a force on the rope; as a result, there is a *reflected* pulse that looks like an inverted shark's fin traveling to the left (Fig. 18.5B).

Imagine that instead of having a rope fixed on the right end, we had two shark-fin pulses traveling toward each other on a long rope (Fig. 18.6). One pulse looks like a shark's fin traveling to the right, and the other is an inverted shark's fin traveling to the left. The pulses pass through each other as expected. Now cover up the right-hand portion of the rope shown in the gray region and labeled "imaginary part." The remaining rope on the left looks like the incident pulse and the reflected pulse in Figure 18.5. So, the formation of the reflected pulse is similar to the superposition of two pulses on a rope. To find the reflected pulse, just *imagine* a pulse traveling along a similar rope toward the boundary at the same speed as the incident pulse. The imaginary pulse must be inverted and flipped so that when the two meet at the fixed boundary, they cancel each other.

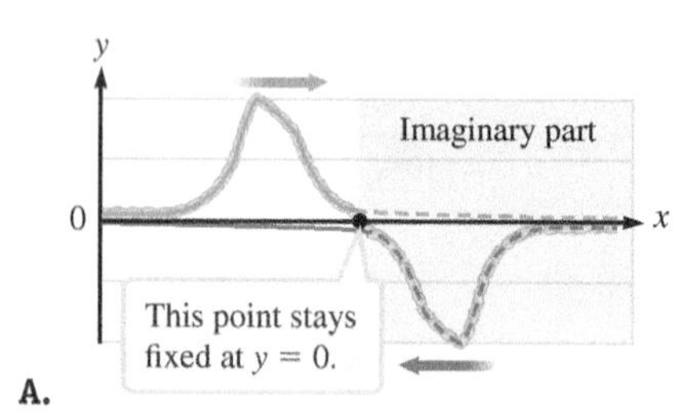

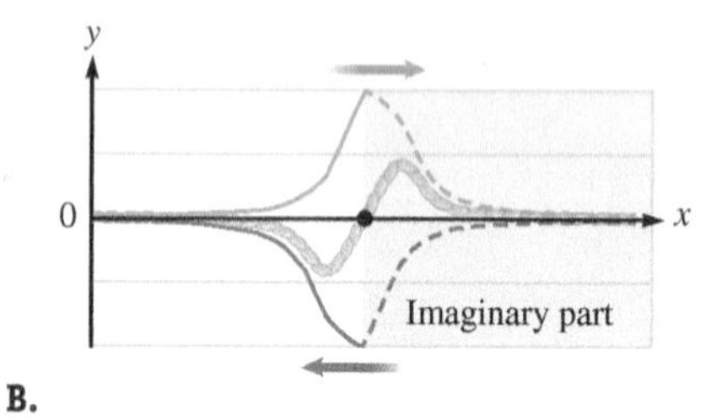

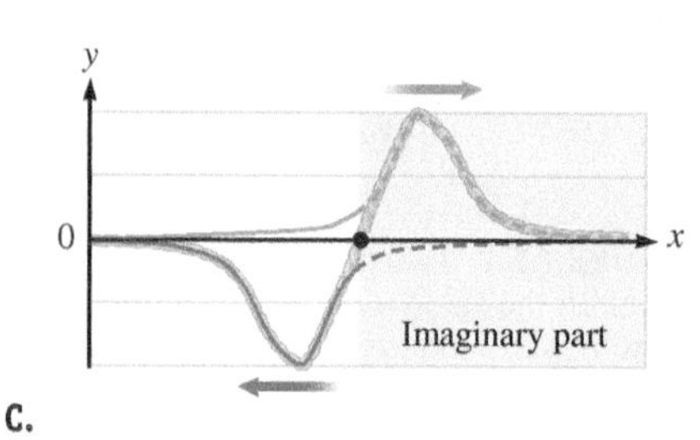

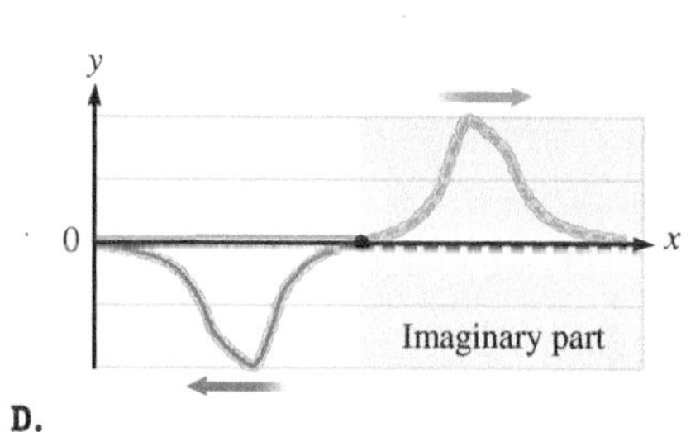

FIGURE 18.6 A pulse travels to the right along a rope that is fixed at one end. Imagine a second pulse traveling to the left. The two pulses must cancel at the fixed end of the rope, so the imaginary pulse must be inverted.

Free-End Reflection

What happens if the right-hand end of the rope is attached to a ring that is *free* to slide along a pole (Fig. 18.7)? As the pulse approaches the free end, the ring slides upward along the pole, rising to a height that is twice that of the pulse (Fig. 18.7B). The ring is pulled back down by the tension in the rope, and a reflected pulse travels to the left. In the case of a free end, the reflected pulse is upright and looks like a shark's fin traveling to the left.

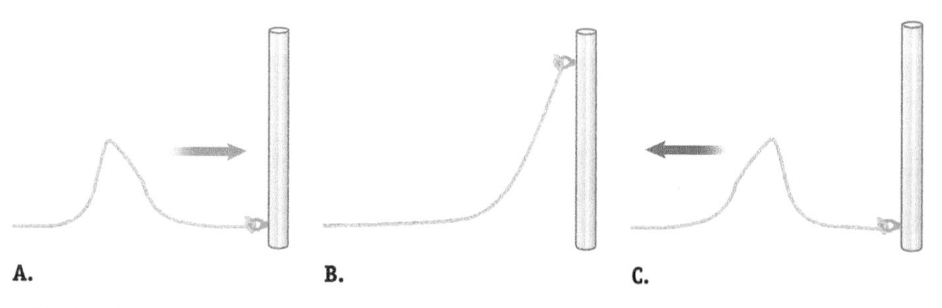

FIGURE 18.7 When a pulse reflects from a free end, the reflected pulse is upright.

As in the case with a fixed right end, it is helpful to imagine free-end reflection as the superposition of two pulses along a rope. In Figure 18.8, two pulses shaped like shark's fins travel toward each other and pass through each other. Again cover up the right-hand side of the rope shown in gray and compare the remaining portion of the rope with the incident and reflected pulses in Figure 18.7. As before, the formation of the reflected pulse is similar to the superposition of two pulses on the rope. In this case, however, the imaginary pulse must be flipped but not inverted so that when the two pulses meet at the free boundary, they add together.

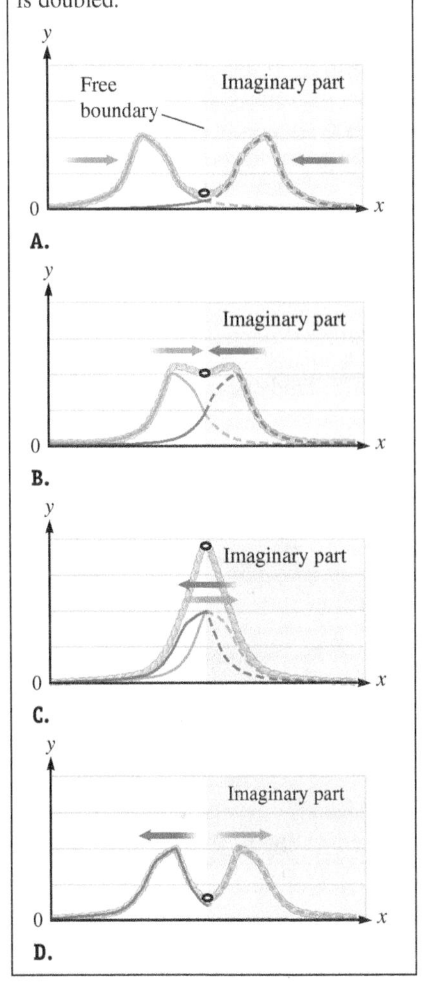

FIGURE 18.8 A pulse travels to the right along a rope that is free at the right-hand end, imagine a second pulse traveling to the left. The two pulses add together, so when the two pulses "meet" at the free end of the rope, the amplitude is doubled.

CONCEPT EXERCISE 18.3

A wave pulse travels to the left on a rope as shown in Figure 18.9. Draw a sketch of the reflected pulse if the left end is

a. fixed. **b.** free.

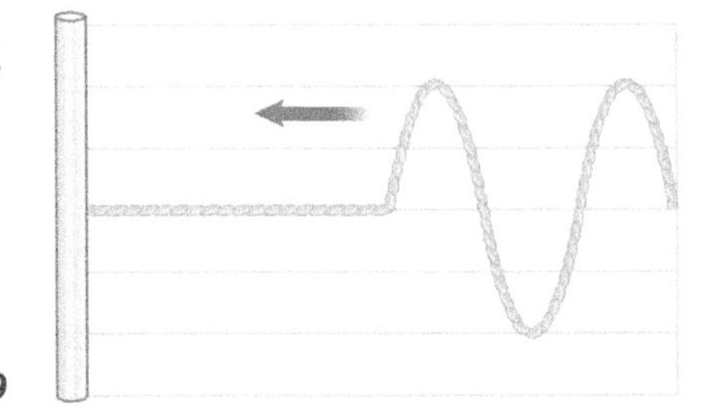

FIGURE 18.9

Law of Reflection

A wave on a rope is one-dimensional, so a reflected wave must simply travel in the opposite direction as the incident wave. A two- or three-dimensional wave does not necessarily reflect along the path of the incoming wave. Figure 18.10 shows an incident ray that reflects from a boundary. The incident ray makes an angle θ_i with the line that is perpendicular to the boundary (called the **normal**), and the reflected ray makes an angle θ_r with that perpendicular line. According to the **law of reflection**, the incident angle equals the reflected angle:

$$\theta_i = \theta_r \quad (18.1)$$

LAW OF REFLECTION

✪ **Major Concept**

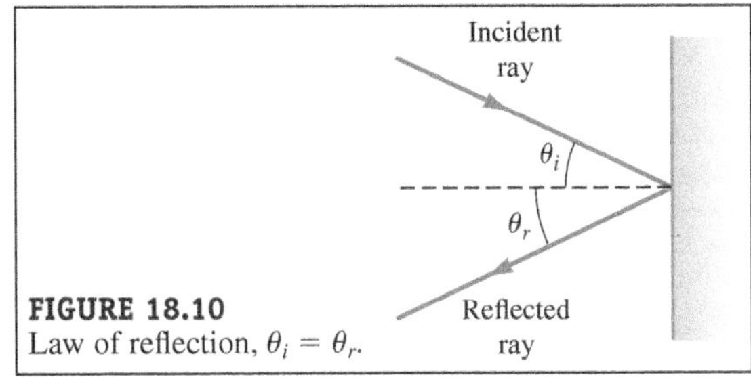

FIGURE 18.10 Law of reflection, $\theta_i = \theta_r$.

EXAMPLE 18.1 A Whisper Dish

Sophia, standing roughly 30 ft from her father, whispers. He can clearly understand her because he is standing in front of a whisper dish (Fig. 18.11A). The dish is designed so that the sound waves from a distant source such as Sophia are reflected toward a single point. The father's ears are near that point, so he hears the sum of these many reflected waves.

The sound waves near Sophia are roughly hemispherical. Far from Sophia, the waves are nearly plane waves, and the rays are a series of parallel lines (Section 17-7). Figure 18.11B shows three such rays near the surface of a spherically shaped whisper dish. Sketch the three corresponding reflected rays.

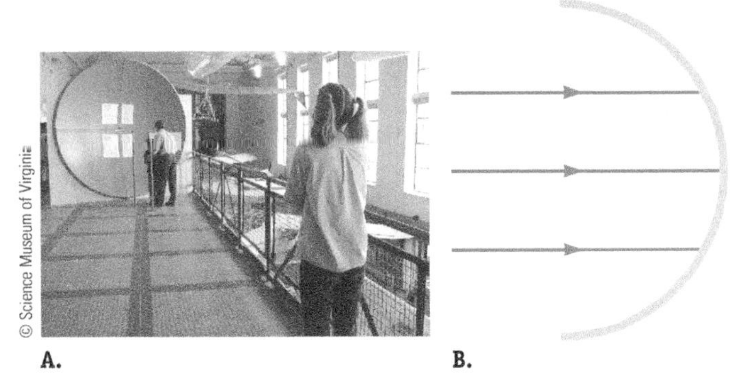

FIGURE 18.11 **A.** A whisper dish in use. **B.** Parallel rays strike the whisper dish.

Example continues on page 526 ▶

INTERPRET and ANTICIPATE

The law of reflection tells us how to draw the reflected rays. We expect from the description of the whisper dish that the reflected rays intersect.

SOLVE

According to the law of reflection (Eq. 18.1), the incident angle equals the reflected angle. Both angles are measured from the line that is perpendicular to the surface. For a sphere, radii are perpendicular to the surface (along the dashed lines in Fig. 18.12). In the figure, we have labeled the incident rays A, B, and C, and the corresponding reflected rays a, b, and c.

CHECK and THINK

As expected, the three reflected rays (a, b, and c) intersect, and even a soft whisper from a distant source can easily be heard if a listener stands near the dish close to the point of intersection.

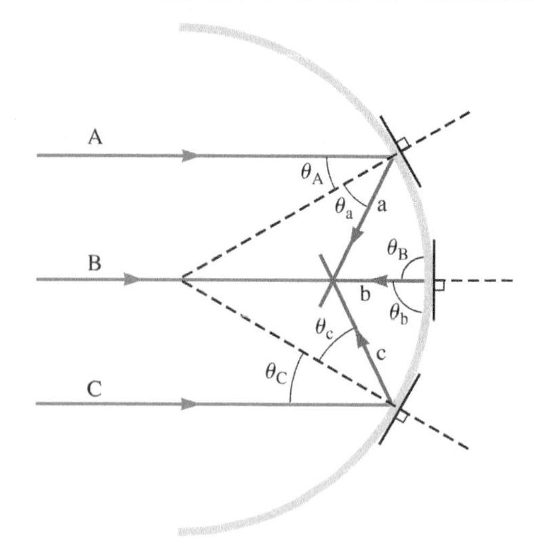

FIGURE 18.12 Reflected rays are shown in red.

CASE STUDY Part 2: Designing Concert Halls and Auditoriums

Musicians who go on tour play in many different venues, and they know that some rooms make better concert halls than others. A major factor that determines a good hall from a bad one is reflection of the sound.

When sound waves are reflected off a surface, we may hear either an echo or a reverberation. A sound affects the human brain for about 0.1 s. If the reflected wave reaches your ears in less than 0.1 s, you perceive that the sound is prolonged and hear a reverberation. Because the reflection must reach you in a short time, you generally experience reverberation in a small space (such as when you sing in the shower). If, on the other hand, the reflected wave takes longer than 0.1 s to reach your ears, the reflection is heard as a second separate sound (an echo). You probably have heard an echo by shouting at the wall of a canyon or a building.

A concert hall designer does not want the hall to have an obvious echo, but does want it to have just the right amount of reverberation. So, the hall should be designed so that reflections come to each person from many different directions. The walls and ceiling of a well-designed concert hall are not made of smooth, hard materials such as concrete, which is why school gyms and cafeterias don't make good concert halls (Fig. 18.13A). During a concert in a school's gym, a listener will hear reflections off the walls (and perhaps the ceiling). Because the walls are smooth and hard, the reflected sound heard by any individual in the audience must have come from one or just a few places. Musicians say that such a room is "dead."

The walls of a well-designed concert hall are made of a rough material and are broken into several panels oriented at various angles so that the reflected sound heard by any one individual comes from many places in the room (Fig. 18.13B). Such a room with the correct amount of reverberation is said to be "live." Because there are no people in the audience reflecting the sound waves, musicians report that even a well-designed concert hall can sound dead when they rehearse.

A.

B.

FIGURE 18.13 A. Music sounds dull in a school gym or cafeteria because the walls and ceiling are smooth. **B.** Walls in concert halls are made of many rough and curved panels.

18-3 Interference

Contrast the two pulses in Figure 18.2 with the motion of two particles on a collision path. The waves pass through each other, but the particles don't. So, we need to think about the encounter of two (or more waves) differently than the way we think about

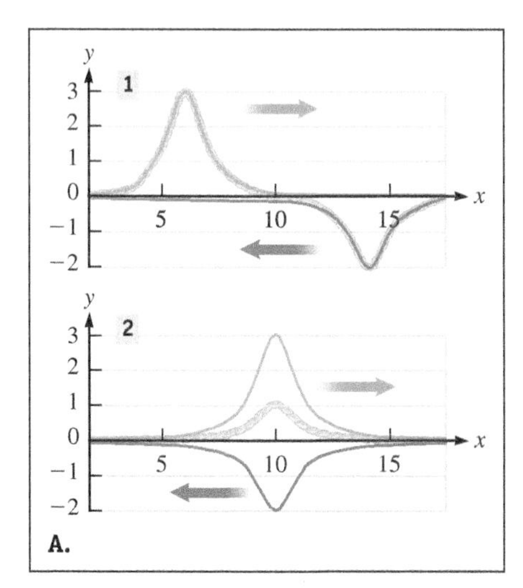

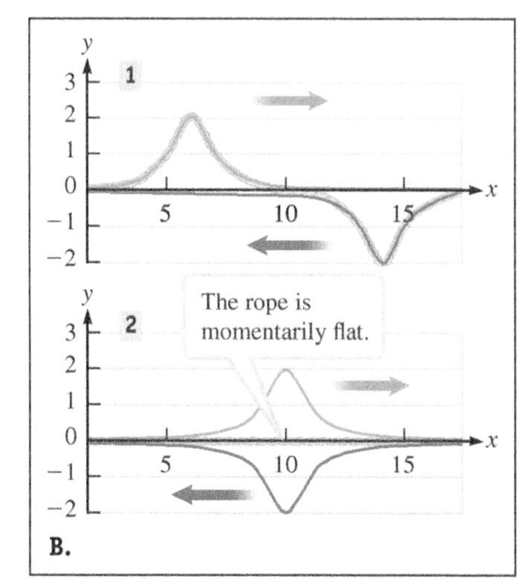

FIGURE 18.14 A. **1** Two pulses (red and blue) move toward each other. **2** The resulting pulse exhibits partial destructive interference. **B.** **1** Two pulses (red and blue) move toward each other. **2** The resulting pulse is completely flat, so it shows fully destructive interference.

colliding particles. When two waves exist in the same medium, they don't collide; they *interfere* with each other. Their interference creates a new wave in the medium.

Interference in Pulses and One-Dimensional Waves

The wave pattern in Figure 18.2 results from the interference of the two traveling wave pulses. When the resulting pulse's height is greater than that of both original waves, the result is *constructive interference* as in Figure 18.2C.

When the resulting pulse's height is smaller than that of each original wave as in Figure 18.14A, the result is *destructive interference*. If the two pulses have the same amplitude but one is inverted with respect to the other, the resulting wave profile is momentarily flat as in Figure 18.14B.

Now suppose there are two harmonic waves traveling in one dimension in the same medium. For example, imagine two speakers sending out harmonic sound waves in a column of air as in Figure 18.15. Each wave is represented by a single cosine function as in Equation 17.16, $\Delta P = \Delta P_{max} \cos(kx - \omega t)$. Such a sound wave wouldn't sound like music; it is just a single tone that you might have heard such as the tone broadcast by an emergency alarm system. Because the waves are traveling in the same medium, they have the same speed. In addition, suppose these speakers produce the same amplitude for each wave.

Now let's further assume the speakers are arranged so that these two waves are in phase, which means that all the peaks of one wave line up with all the peaks of the other wave (Fig. 18.15A). The superposition of two harmonic waves that are identical produces a harmonic wave that has twice the amplitude of either one of the origi-

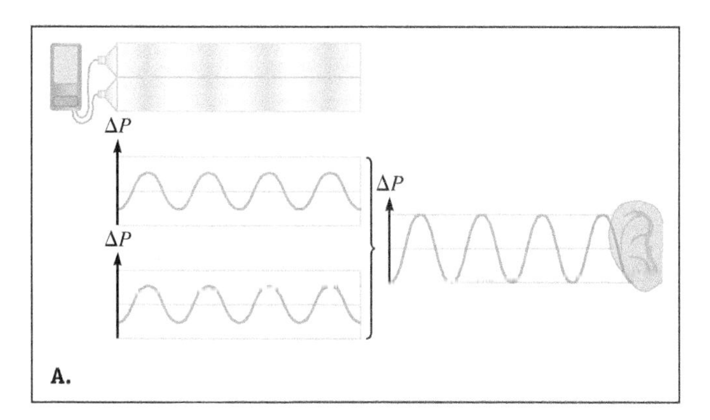

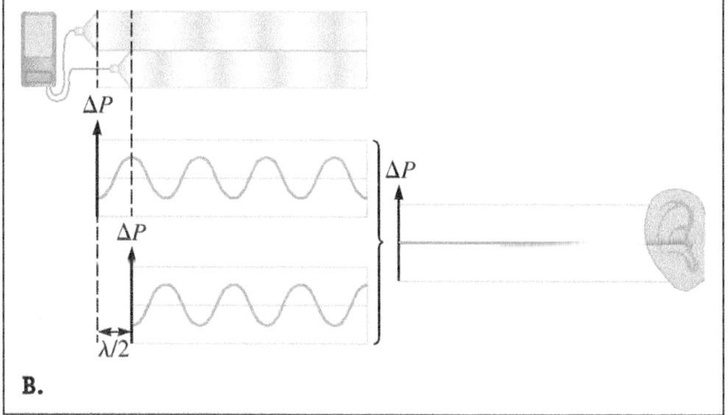

FIGURE 18.15 A. The two speakers create two harmonic sound waves that are in phase. The amplitude of the wave heard by the person is two times greater than the sound heard from either speaker alone. **B.** The two speakers create two harmonic sound waves that are 180° or $\lambda/2$ out of phase. The two waves cancel, and the person cannot hear a sound.

nal waves. **Constructive interference** results from the superposition of two waves that are **in phase**. If you were to place your ear in front of the two speakers in Figure 18.15A, the pressure changes on your eardrum would be twice as great with two speakers than it would be if there were only one speaker.

CONSTRUCTIVE AND DESTRUCTIVE INTERFERENCE

Major Concept

You might have guessed that two speakers always produce twice as much pressure variation on your eardrum as one speaker, but it is not true in all cases. In Figure 18.15B, the speakers have been rearranged so that one speaker is half of a wavelength closer to the person's ear. Now the two waves are out of phase by 180°, meaning that the peaks of one wave line up with the valleys of the other wave. The superposition of two waves that are 180° out of phase results in **destructive interference**. If you place your ear in front of the two speakers in Figure 18.15B, the pressure change on your eardrum would be zero; the pressure would equal atmospheric pressure, and you wouldn't hear a sound.

What would happen if the speaker were moved another $(1/2)\lambda$ closer to the person's ear so that the two speakers were separated by λ? The two waves would once again be in phase, and the result would be constructive interference. If the difference in the distance traveled by the two waves is an integer multiple of the wavelength, the result is constructive interference. Mathematically, we write the condition for constructive interference as

$$\Delta d = n\lambda \qquad (n = 0, 1, 2, 3, \ldots) \tag{18.2}$$

CONDITION FOR CONSTRUCTIVE AND DESTRUCTIVE INTERFERENCE

Major Concept

where Δd is the difference in the distance traveled by the two waves. If the difference in the distance traveled by the two waves is a half-integer number of wavelengths, the result is destructive interference:

$$\Delta d = \frac{n}{2}\lambda \qquad (n = 1, 3, 5, \ldots) \tag{18.3}$$

Two- and Three-Dimensional Interference

We usually listen to sound waves that travel in three dimensions instead of one-dimensional sound waves in a column of air. Two- and three-dimensional waves also interfere constructively and destructively according to the conditions given in Equations 18.2 and 18.3. For example, circular waves on the surface of the water interfere constructively and destructively. In Figure 18.16, the interference pattern of two circular waves on the surface of water is best seen in the lower center of the photo, where you can see a grid of light and dark spots.

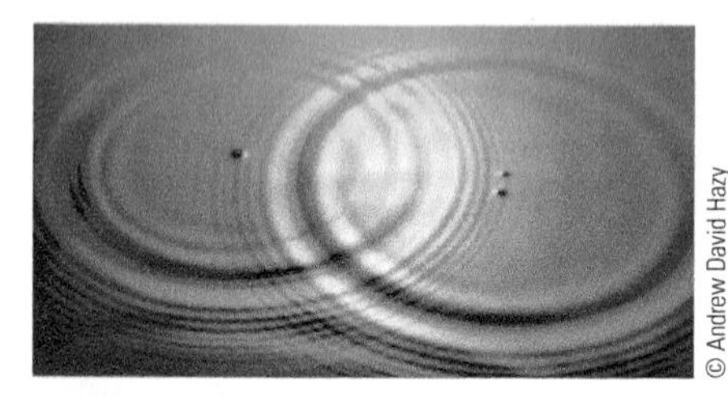

© Andrew David Hazy

FIGURE 18.16 Two circular waves in the water interfere constructively and destructively.

Figure 18.17 models the interference of two identical circular harmonic waves. The solid circles represent the wave's peaks, and the dashed circles represent the valleys; so, the distance between any two solid circles or between any two dashed circles is one wavelength λ. As shown in Figure 18.17A, constructive interference occurs at any point where two peaks or two valleys intersect. Destructive interference occurs at any point where one peak meets one valley.

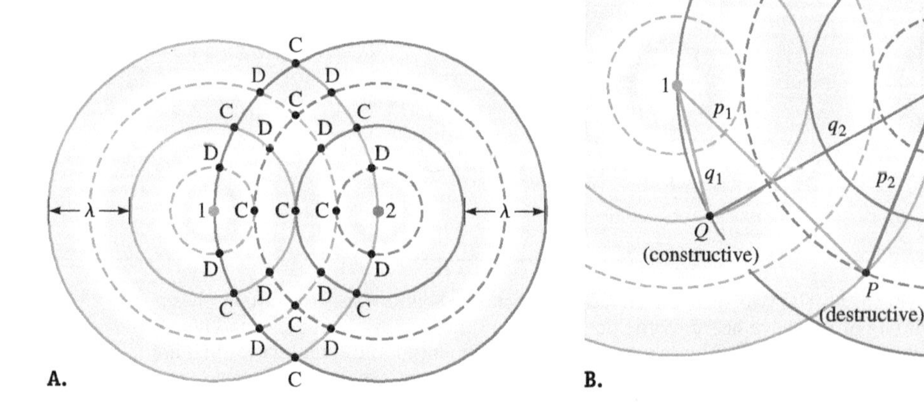

FIGURE 18.17 A. Constructive interference occurs where two valleys or two peaks meet, as indicated by "C." Destructive interference occurs where a valley and a peak meet, as indicated by "D." **B.** Close-up of part A.

At point Q in Figure 18.17B, there is constructive interference. The distance from source 1 to point Q is q_1, and the distance from source 2 to Q is q_2. From the figure, we find that

$$q_1 = \lambda$$

and

$$q_2 = 2\lambda$$

The difference in the distance to point Q traveled by the two waves is

$$\Delta d = |q_2 - q_1| = |2\lambda - \lambda| = \lambda$$

This equation fits Equation 18.2, which says that for constructive interference, Δd must be an integer multiple of λ. For point Q, that integer is $n = 1$.

At point P in Figure 18.17B, there is destructive interference. The distance from source 1 to point P is p_1, and distance from source 2 to P is p_2. From the figure, we find that

$$p_1 = 2\lambda$$

and

$$p_2 = \frac{3}{2}\lambda$$

The difference in the distance to point P traveled by the two waves is

$$\Delta d = |p_2 - p_1| = \left|\frac{3}{2}\lambda - 2\lambda\right| = \frac{1}{2}\lambda$$

This equation fits Equation 18.3 which says that for destructive interference, Δd must be a half-integer multiple of λ. For point P, the integer $n = 1$.

CONCEPT EXERCISE 18.4

Noise cancellation headphones use a microphone to pick up background noise and a speaker to produce a sound wave canceling the noise. These headphones are used on planes or in noisy dormitory rooms to reduce distracting sounds, allowing the wearer to study a physics textbook. Figure 18.18 is a graph of ΔP versus t for the noise produced by a human voice (perhaps a roommate). Sketch a corresponding graph of ΔP versus t that must be generated by the speaker so as to cancel the noise.

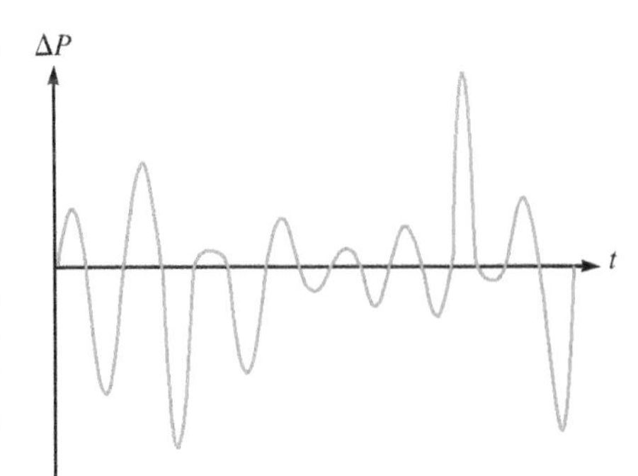

FIGURE 18.18

EXAMPLE 18.2 CASE STUDY Choosing Your Seats

When buying tickets to a concert, you are asked to choose your seats. Usually, seat prices are based on the view, but you are interested in the sound. Suppose you are offered either seats D8 and D9 or E4 and E5. (Figure 18.19 includes distances that are not usually provided on the seating chart.) You know that before a performance, instruments may be tuned with an electronic tuner that can produce a harmonic sound wave with a frequency of 440 Hz. So, you assume it is important to hear the 440-Hz note clearly. Also assume each speaker produces hemispherical waves and the speed of sound in this (hot, humid) auditorium is 350 m/s.

A Should you choose the seats in row D or E?

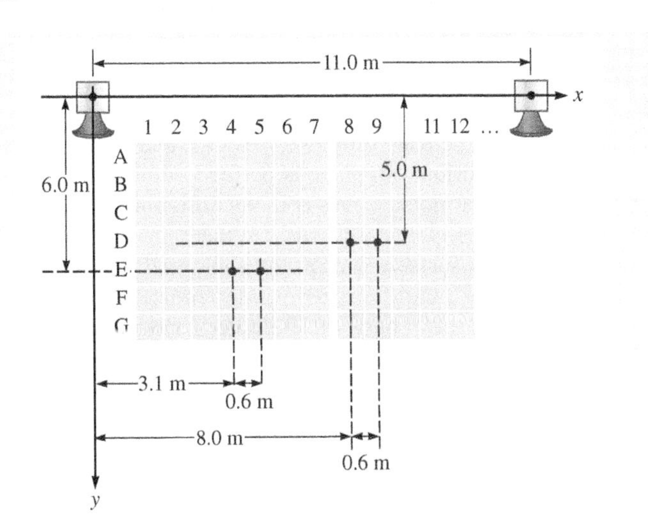

FIGURE 18.19 Not drawn to scale; for example, the aisle (between 7 and 8) is not narrower than the width of a seat.

INTERPRET and ANTICIPATE

We'd like to hear the 440-Hz note. So, we do not want to choose seats if there is destructive interference between the speakers at this frequency.

Example continues on page 530 ▶

SOLVE

First, find the corresponding wavelength of sound in air (Eq. 17.8).

$$v = \lambda f \quad (17.8)$$

$$\lambda = \frac{v}{f} = \frac{350\text{ m/s}}{440\text{ Hz}} = 0.80\text{ m}$$

Next, find the distance between each speaker and each seat. Let's start with speaker 1 and the seats in row D. (We keep an extra significant figure in this intermediate step.)

$$d_1 = \sqrt{x^2 + y^2} = \sqrt{x^2 + (5.0\text{ m})^2}$$

$$\text{D8: } d_1 = \sqrt{(8.0\text{ m})^2 + (5.0\text{ m})^2} = 9.43\text{ m}$$

$$\text{D9: } d_1 = \sqrt{(8.6\text{ m})^2 + (5.0\text{ m})^2} = 9.95\text{ m}$$

Now, do the same for speaker 2 and the seats in row D.

$$d_2 = \sqrt{(11\text{ m} - x)^2 + y^2} = \sqrt{(11\text{ m} - x)^2 + (5.0\text{ m})^2}$$

$$\text{D8: } d_2 = \sqrt{(11\text{ m} - 8.0\text{ m})^2 + (5.0\text{ m})^2} = 5.83\text{ m}$$

$$\text{D9: } d_2 = \sqrt{(11\text{ m} - 8.6\text{ m})^2 + (5.0\text{ m})^2} = 5.55\text{ m}$$

Next, find the distance between speaker 1 and the seats in row E.

$$d_1 = \sqrt{x^2 + y^2} = \sqrt{x^2 + (6.0\text{ m})^2}$$

$$\text{E4: } d_1 = \sqrt{(3.1\text{ m})^2 + (6.0\text{ m})^2} = 6.75\text{ m}$$

$$\text{E5: } d_1 = \sqrt{(3.7\text{ m})^2 + (6.0\text{ m})^2} = 7.05\text{ m}$$

Find the distance between speaker 2 and the seats in row E.

$$d_2 = \sqrt{(11\text{ m} - x)^2 + y^2} = \sqrt{(11\text{ m} - x)^2 + (6.0\text{ m})^2}$$

$$\text{E4: } d_2 = \sqrt{(11\text{ m} - 3.1\text{ m})^2 + (6.0\text{ m})^2} = 9.92\text{ m}$$

$$\text{E5: } d_2 = \sqrt{(11\text{ m} - 3.7\text{ m})^2 + (6.0\text{ m})^2} = 9.45\text{ m}$$

Find the absolute value of the difference in the distance Δd between a seat and each speaker.

$$\text{D8: } |d_2 - d_1| = |\Delta d| = |5.83\text{ m} - 9.43\text{ m}| = 3.60\text{ m}$$

$$\text{D9: } |\Delta d| = 4.40\text{ m}$$

$$\text{E4: } |\Delta d| = 3.17\text{ m}$$

$$\text{E5: } |\Delta d| = 2.40\text{ m}$$

Divide by 0.80 m to write each difference in terms of wavelength.

$$\text{D8: } |\Delta d|/\lambda = 3.60\text{ m}/0.80\text{ m} = 4.5$$

$$\text{D8: } |\Delta d| = 4.5\lambda \qquad \text{D9: } |\Delta d| = 5.5\lambda$$

$$\text{E4: } |\Delta d| = 4.0\lambda \qquad \text{E5: } |\Delta d| = 3.0\lambda$$

The best seats are the ones where there is constructive interference, which are the ones where Δd is an integer multiple of the wavelength.

Choose seats E4 and E5.

CHECK and THINK

Destructive interference at seats D8 and D9 implies that a 440-Hz note cancels at these two seats. Does it mean that there is silence there? No. The situation described in this example is *ideal* (as opposed to real). There are only two speakers, and they only emit two identical harmonic waves. In an actual auditorium, no place is perfectly quiet because (1) musical instruments produce a superposition of harmonic waves at many different frequencies (Section 18-9); (2) speakers usually do not produce identical waves; (3) there are often more than two speakers, and the sound reflects from many surfaces such as the walls, ceiling, and seats; and (4) the number and location of the constructive (and destructive) interference points is continually changing as the waves from each speaker travel outward. Because of destructive interference, music aficionados nevertheless claim that the sound is richer at some seats than at others.

B How would your answer to part A change if the tuner were to produce a 220-Hz wave instead of a 440-Hz wave?

INTERPRET and ANTICIPATE

It isn't necessary to redo all the work in part A. Just find how changing the frequency of the tone played through the speakers affects the answers.

SOLVE Because the frequency is lower, the corresponding wavelength is longer.	$\lambda = \frac{v}{f} = \frac{350\ \text{m/s}}{220\ \text{Hz}} = 1.6\ \text{m}$	
Divide the absolute value of each difference Δd by 1.6 m to write each difference in terms of the new wavelength.	D8: $\lvert\Delta d\rvert = 2.3\lambda$ E4: $\lvert\Delta d\rvert = 2.0\lambda$	D9: $\lvert\Delta d\rvert = 2.8\lambda$ E5: $\lvert\Delta d\rvert = 1.5\lambda$
In this case, the only seat where the note cannot be heard is E5. Because both seats in row D can hear the note, they are the better choice.	Choose seats D8 and D9.	

CHECK and THINK

Each musical instrument emits many harmonic waves. Because the interference pattern depends on the wavelength (or, equivalently, frequency), only a few of the instrument's harmonic waves interfere at any one point in the auditorium.

18-4 Standing Waves

A guitar produces a sound wave in the air that travels outward from the guitar. The wave on the guitar string is *not* a traveling wave, however. Each part of the string oscillates back and forth as it would for any wave on a string, but the wave pattern (the envelope) does not move along the string's length (Fig. 18.20). Such a wave is known as a **standing wave**.

STANDING WAVE ✪ Major Concept

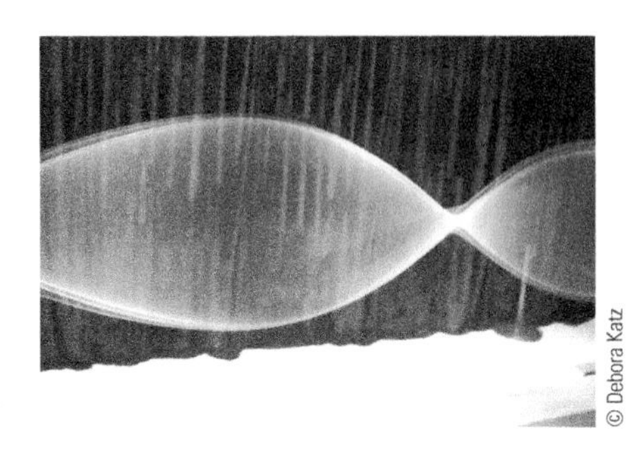

FIGURE 18.20 For a standing wave, each part of the string oscillates back and forth, but the wave pattern does not move along the string's length.

Producing a Standing Wave

You and a friend might produce a standing wave on a stretched string by repeatedly shaking opposite ends in a certain way. Suppose the string's length is L. You and your friend agree to create identical harmonic waves of wavelength $\lambda = 2L/3$. The topmost panels of Figure 18.21 show these two waves traveling on the same string at four different times. The waves have the same amplitude and wavelength. The next row of panels shows that the superposition of these two waves is a standing wave that does not move to the right or the left.

We can represent pieces of the string with beads. Any bead on the string—for example, the bead shown in black in Figure 18.21 (middle)—oscillates up and down in simple harmonic motion. The wave pattern is fixed, so each bead's amplitude is determined by its position along the string. Figure 18.22 shows the string at five different times. The black bead has a relatively small amplitude. Some beads do not oscillate at all and are said to be at the **nodes** of the standing wave. Beads with the greatest amplitude are at the **antinodes**.

Wave Function of a Standing Wave

Because a standing wave on a string may be produced by the superposition of two traveling waves, we can find the wave function for a standing wave by adding together the wave functions for two identical harmonic waves traveling in the opposite direction. A harmonic wave traveling in the positive x direction is given by Equation 17.4:

$$y_1(x, t) = y_{\max} \sin(kx - \omega t) \tag{18.4}$$

and a harmonic wave traveling in the negative x direction is given by

$$y_2(x, t) = y_{\max} \sin(kx + \omega t) \tag{18.5}$$

The resulting wave is the superposition of the two traveling waves:

$$y(x, t) = y_1(x, t) + y_2(x, t)$$
$$y(x, t) = y_{\max} \sin(kx - \omega t) + y_{\max} \sin(kx + \omega t)$$
$$y(x, t) = y_{\max}[\sin(kx - \omega t) + \sin(kx + \omega t)]$$

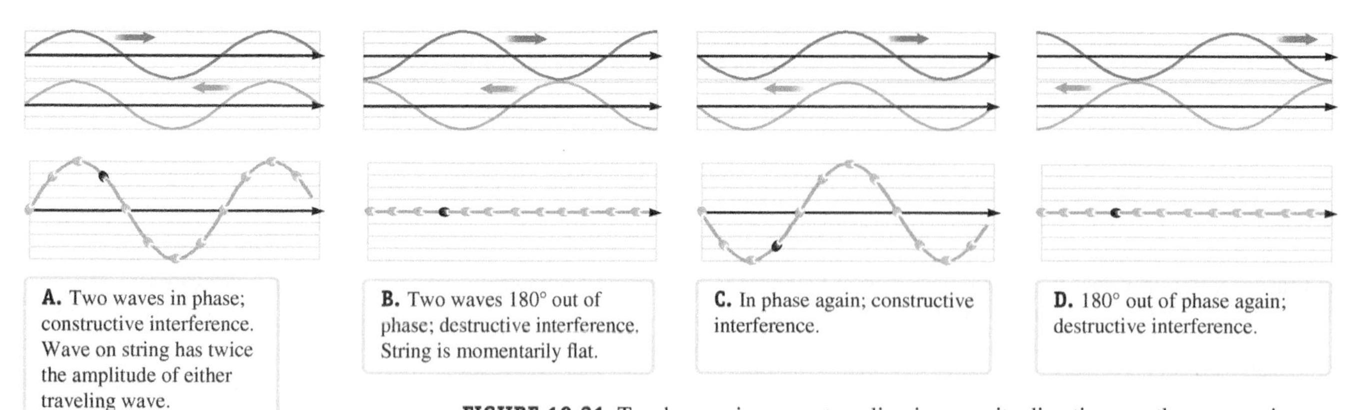

A. Two waves in phase; constructive interference. Wave on string has twice the amplitude of either traveling wave.

B. Two waves 180° out of phase; destructive interference. String is momentarily flat.

C. In phase again; constructive interference.

D. 180° out of phase again; destructive interference.

FIGURE 18.21 Two harmonic waves traveling in opposite directions on the same string. The superposition of the two waves results in a standing wave. The wave pattern does not move along the string. Each bead on the rope moves up and down in simple harmonic motion, and its amplitude is determined by its position along the rope.

We can use a trigonometric identity $\sin\alpha + \sin\beta = 2\sin\left[\frac{1}{2}(\alpha + \beta)\right]\cos\left[\frac{1}{2}(\alpha - \beta)\right]$ (Appendix A) with $\alpha = (kx - \omega t)$ and $\beta = (kx + \omega t)$ to simplify this expression:

$$y(x, t) = y_{max}\{2\sin\tfrac{1}{2}[(kx - \omega t) + (kx + \omega t)]\cos\tfrac{1}{2}[(kx - \omega t) - (kx + \omega t)]\}$$

$$y(x, t) = 2y_{max}\sin(kx)\cos(-\omega t)$$

$$y(x, t) = [2y_{max}\sin(kx)]\cos(\omega t) \qquad (18.6)$$

since $\cos(-\omega t) = \cos\omega t$.

Equation 18.6 is the wave function for a standing transverse wave, which appears to be very different mathematically from Equation 18.4 for a traveling wave. However, comparing Equation 18.6 with Equation 16.3 for a simple harmonic oscillator, $y(t) = y_{max}\cos(\omega t + \varphi)$, shows that those two equations look mathematically similar. Equation 18.6 describes a simple harmonic oscillator with a phase constant φ of zero and an amplitude given by the absolute value of the term in square brackets, $|2y_{max}\sin(kx)|$. Equation 18.6 supports what we find graphically in Figures 18.21 and 18.22: *A standing harmonic wave is made of many simple harmonic oscillators, each of whose amplitude is determined by its position x.*

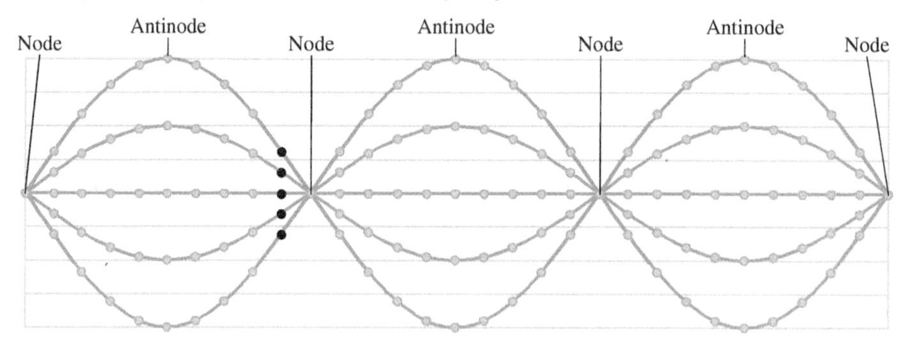

FIGURE 18.22 The amplitude at the nodes of a standing wave is zero; at the nodes, the string remains at rest. The amplitude at the antinodes is a maximum.

Position of Nodes and Antinodes

Because the nodes of a standing wave have zero amplitude, we can find the position of the nodes by setting $|2y_{max}\sin(kx)|$ to zero and solving for x:

$$|2y_{max}\sin(kx)| = 0$$

$$\sin(kx) = 0$$

$$kx = n\pi \qquad (n = 0, 1, 2, 3, \ldots)$$

Substitute $k = 2\pi/\lambda$ (Eq. 17.5) for the wave number k:

$$x = \frac{n\pi}{k} = \frac{n\pi}{(2\pi/\lambda)}$$

$$x = n\frac{\lambda}{2} \qquad (n = 0, 1, 2, 3, \ldots) \qquad (18.7)$$

Equation 18.7 gives the position of the nodes, showing that the distance between adjacent nodes (Fig. 18.22) is $\lambda/2$.

We can find the position of the antinodes by finding the locations of the maximum amplitude. Set $|2y_{max}\sin(kx)|$ equal to $2y_{max}$ and solve for x:

$$|2y_{max}\sin(kx)| = 2y_{max}$$
$$\sin(kx) = \pm 1$$
$$kx = \frac{1}{2}\pi, \frac{3}{2}\pi, \frac{5}{2}\pi$$
$$kx = (n + \tfrac{1}{2})\pi \quad (n = 0, 1, 2, 3, \ldots)$$

Again, substitute $k = \dfrac{2\pi}{\lambda}$ for wave number k and solve for the positions of the antinodes:

$$x = \left(n + \frac{1}{2}\right)\frac{\lambda}{2} \quad (n = 0, 1, 2, 3, \ldots) \tag{18.8}$$

The distance between adjacent antinodes is one-half of a wavelength ($\lambda/2$), and the distance between a node and an adjacent antinode is a one-fourth of a wavelength ($\lambda/4$).

Standing Waves in Musical Instruments

When a musician plays an instrument, a standing wave is established in the instrument. How can a solo musician do so when, according to Figure 18.21, we need two traveling waves to make a standing wave? Imagine that you attach one end of a string to a fixed ring as in Figure 18.5, but instead of creating a pulse shaped like a shark's fin you oscillate your end of the string in simple harmonic motion. A harmonic wave travels from you toward the ring, reflects from the ring, and creates a second wave traveling toward you. The reflected wave has the same amplitude and wavelength as the incident wave you created. The superposition of these two waves creates a standing wave on the rope just as if you had a friend at the other end helping you. The basis of musical instruments is the formation of standing waves from the superposition of reflected waves. In the next three sections, we focus on standing waves that form in three instruments in a big-band ensemble: the guitar, the flute, and the clarinet.

EXAMPLE 18.3 Making Standing Waves

In a large laboratory, two waves with the same amplitude and wavelength travel in opposite directions along the same rope. The two waves interfere, and the result is a standing wave on the rope. One traveling wave is described by

$$y_1(x, t) = 0.150\sin(2.50x - 6.35t) \tag{1}$$

where numerical values have the appropriate SI units.

A Write an expression for the standing wave and find the amplitude of the antinodes.

INTERPRET and ANTICIPATE

Equation (1) is the wave function of a wave traveling in the positive x direction. The identical wave traveling in the negative x direction must by given by $y_2(x, t) = 0.150\sin(2.50x + 6.35t)$. The sum of the two waves $y_1 + y_2$ is a standing wave.

SOLVE

Identify y_{max}, k, and ω from Equation (1) and substitute these values into Equation 18.6 to find an expression for the standing wave $y(x, t)$.

$$y_{max} = 0.150 \text{ m}$$
$$k = 2.50 \text{ rad/m}$$
$$\omega = 6.35 \text{ rad/s}$$
$$y(x, t) = 2(0.150)\sin(2.50x)\cos(6.35t)$$
$$y(x, t) = 0.300\sin(2.50x)\cos(6.35t)$$

Example continues on page 534 ▶

The amplitude A of an antinode is the maximum amplitude of the standing wave, and it is twice the amplitude of either of the traveling waves.	$A = 2y_{\text{max}} = 0.300 \text{ m}$

B What is the separation between adjacent nodes? What is the separation between a node an adjacent antinode?

SOLVE

The distance between adjacent nodes is half a wavelength (Eq. 18.7). Use Equation 17.5 to find λ from k.

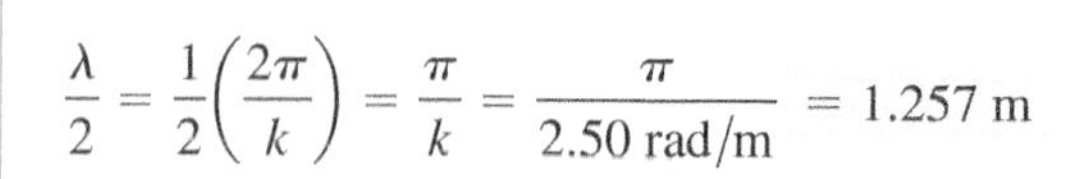

$$\frac{\lambda}{2} = \frac{1}{2}\left(\frac{2\pi}{k}\right) = \frac{\pi}{k} = \frac{\pi}{2.50 \text{ rad/m}} = 1.257 \text{ m}$$

The distance between a node and an adjacent antinode is one-fourth of a wavelength.

$$\frac{\lambda}{4} = \frac{1}{4}\left(\frac{2\pi}{k}\right) = \frac{\pi}{2k} = \frac{\pi}{2(2.50 \text{ rad/m})} = 0.6283 \text{ m}$$

C If there are seven nodes, including a node on one end, and if there is an antinode on the other end, find the length of the rope. (The antinode could be generated by attaching one end of the rope to a motor.)

INTERPRET and ANTICIPATE

It is helpful to sketch a standing wave with seven nodes as described (Fig. 18.23). We can then use our results from part B to find the length of the rope by adding up the distance between adjacent nodes and one node adjacent to an antinode.

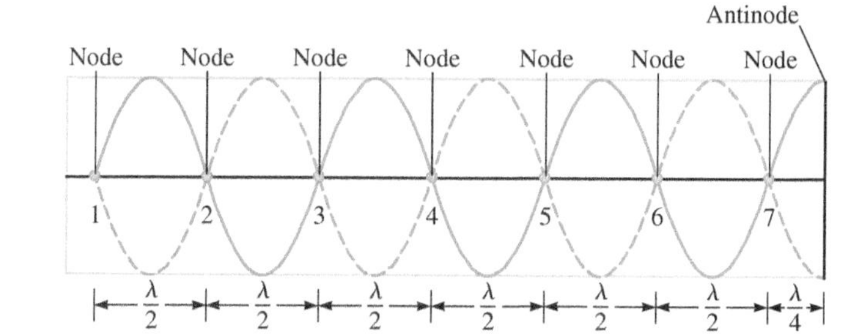

FIGURE 18.23

SOLVE

The length L of the rope is the sum of the six distances between adjacent nodes $(6\lambda/2)$ and the distance between the last node and adjacent antinode $(\lambda/4)$. (Use an extra significant figure from results of the intermediate steps.)

$$L = 6\frac{\lambda}{2} + \frac{\lambda}{4} = 6(1.257 \text{ m}) + (0.6283 \text{ m}) = 8.17 \text{ m}$$

CHECK and THINK

The rope may seem a bit long (nearly 30 ft). According to the problem statement, the rope is in a *large* laboratory, so 30 ft is not unreasonable.

18-5 Guitar: Resonance on a String Fixed at Both Ends

FIGURE 18.24 A guitarist plays several different notes on a single string by holding the string down against different frets on the neck of the guitar.

A guitar string is fixed at both ends. One end is held by the guitarist, and the other is attached to the guitar's bridge. A guitarist plays many different notes on a single guitar string by using one hand to hold down the end of the string at different points along the neck and strumming or plucking the string with the other hand (Fig. 18.24). Waves reflect from both ends of the string, and the resulting superposition creates a standing wave. Sound from the waving string is then enhanced or amplified in a way that depends on the type of guitar.

Because the ends of the string are fixed, they cannot oscillate; so, the standing wave that forms on the string must have a node on each end. Therefore, only certain wavelengths will fit on the string. Figure 18.25 shows the family of such waves on a string of fixed length L. The wavelength of the standing waves that fit on the string depends on the number of antinodes n such that

$$L = n\frac{\lambda_n}{2}$$

or

$$\lambda_n = 2\frac{L}{n} \quad (n = 1, 2, 3, \ldots) \tag{18.9}$$

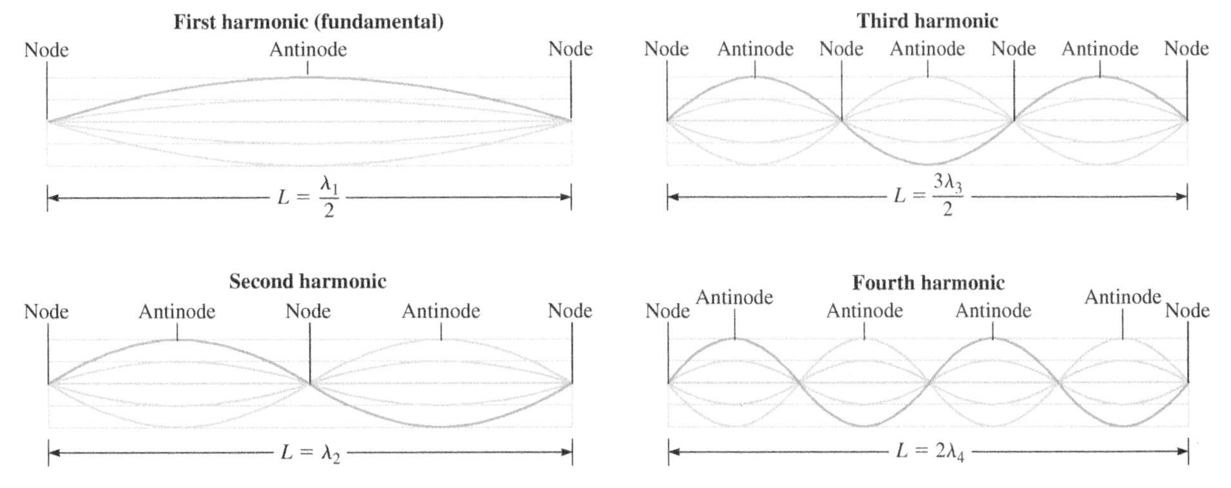

FIGURE 18.25 When a string is fixed at both ends, only standing waves that have nodes on each end can oscillate on the string. The wavelength of harmonic waves that can be sustained is given by $\lambda_n = 2L/n$, where n is the number of antinodes and L is the length of the string.

The integer subscript n is called the **harmonic number**, and each possible standing wave is referred to as the nth **harmonic**. For example, the last wave shown in Figure 18.25 has four antinodes; it is the fourth harmonic, and $\lambda_4 = L/2$.

It is often more convenient to work in terms of the wave's frequency instead of its wavelength. We find frequency by substituting Equation 18.9 into $v = \lambda f$ (Eq. 17.8):

$$f_n = \frac{v}{\lambda_n}$$

$$f_n = n\left(\frac{v}{2L}\right) \qquad (n = 1, 2, 3, \ldots) \qquad (18.10)$$

where v is the speed of the traveling waves on the string. The first harmonic ($n = 1$) is referred to as the **fundamental**, and $f_1 = v/2L$ is the **fundamental frequency**. The frequency of the higher-number harmonics is easily written in terms of the fundamental frequency:

STANDING WAVE; STRING FIXED AT BOTH ENDS

Special Case

$$f_n = nf_1 \qquad (18.11)$$

For a particular string, the distance between the two fixed ends determines the fundamental frequency (Eq. 18.10). So, when the guitarist moves his hand along the neck of the guitar, he changes the distance between the fixed ends and the note sounded by the string.

The speed of a wave on a string is given by $v = \sqrt{F_T/\mu}$ (Eq. 17.11), so the fundamental frequency can be written as

$$f_1 = \frac{1}{2L}\sqrt{\frac{F_T}{\mu}}$$

where F_T is the tension in the string and μ is the string's mass per length. Each guitar string has a different μ and F_T, and the guitarist tunes the instrument by adjusting the tension in the strings.

Another way to model a standing wave on a string such as the ones in Figure 18.25 comes from Equation 18.6, which shows that a standing wave is made of many oscillating particles. An extended object made of many oscillating particles is **vibrating**. In Section 16-11, we found that a particle—such as the bob at the end of a simple pendulum—oscillates only at a particular frequency known as the natural or resonance frequency, and when you try to drive the pendulum, the maximum amplitude occurs when the driving frequency equals the resonance frequency. The same sort of thing is true for a vibrating object except that such an object has a family of resonance frequencies. The object must be driven at one of its resonance frequencies; if not it won't vibrate very much. In the case of a string fixed at both ends, the resonance frequencies are given by Equation 18.10.

EXAMPLE 18.4 CASE STUDY Making Your Own Guitar

A guitar usually has six strings that are kept at roughly the same tension. Each string has a unique mass per unit length. In principle, a guitarist can play any note on any string by changing the length of the string that is free to vibrate. If he does not press on the string, the length of string that vibrates extends from the "0" fret to the bridge on the other end of the guitar (Fig. 18.26). When a guitarist wishes to play a higher note on the same string, he presses the string against one of the frets on the neck. Each fret marks the position for a particular musical note. Notice that the frets are not evenly spaced along the neck of the guitar.

Suppose you wish to make your own guitar. One particular string is 0.650 m long and sounds an A at $f_A = 440$ Hz when it is played open, without being pressed down. Find the location of the frets starting from "0" so that the next 12 notes, A-sharp, B, C, C-sharp, D, D-sharp, E, F, F-sharp, G, G-sharp and (octave) A, may be played. The fundamental frequency of these higher notes is given by

$$f_m = 2^{m/12} f_A \quad (1)$$

where m is an integer that corresponds to each note such that $m = 1$ for A-sharp and $m = 12$ for (octave) A.

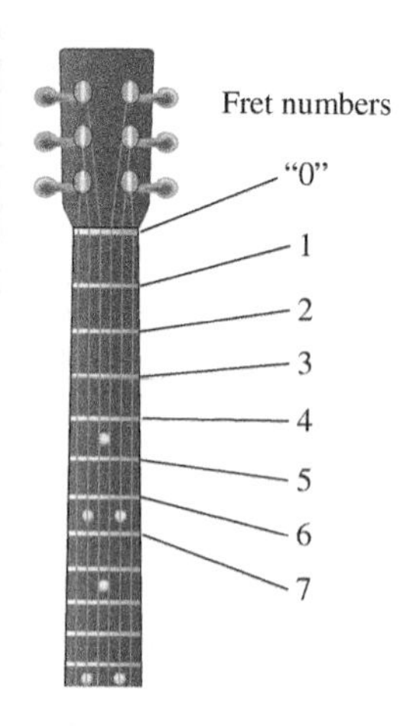

FIGURE 18.26 Frets on a guitar.

INTERPRET and ANTICIPATE

We only need the fundamental harmonic f_m to find the length L of the string for each note m, and we can derive an equation for L_m as a function of m. The position of each fret is then given by $L_A - L_m$, where $L_A = 0.650$ m. To check the results, we compare them to the location of the frets on an actual guitar (Fig. 18.26).

SOLVE

Set $n = 1$ in Equation 18.10 to write an expression for the length L_A of the string sounding an A at $f_A = 440$ Hz.

$$f_A = \frac{v}{2L_A}$$

$$L_A = \frac{v}{2f_A} \quad (2)$$

To find L for a higher-frequency note, substitute Equation (1) for f_m into Equation 18.10 with $n = 1$.

$$f_m = \frac{v}{2L_m} \text{ from Equation 18.10}$$

$$L_m = \frac{v}{2f_m} = \frac{v}{2(2^{m/12})f_A}$$

Use Equation (2) to rewrite L_m in terms of L_A.

$$L_m = \frac{L_A}{2^{m/12}}$$

The position of each fret measured from "0" is the difference between L_A and L_m.

$$L_A - L_m = L_A - \frac{L_A}{2^{m/12}}$$

$$L_A - L_m = L_A(1 - 2^{-m/12}) \quad (3)$$

Table 18.1 provides the frequency of each note (Eq. 1) and the position of each fret (Eq. 3).

TABLE 18.1

Note	m	f (Hz)	L_m (m)	$L_A - L_m$ (m)
A	0	440.00	0.650	0.000
A-sharp	1	466.16	0.614	0.036
B	2	493.88	0.579	0.071
C	3	523.25	0.547	0.103
C-sharp	4	554.37	0.516	0.134
D	5	587.33	0.487	0.163
D-sharp	6	622.25	0.460	0.190
E	7	659.26	0.434	0.216
F	8	698.46	0.409	0.241
F-sharp	9	739.99	0.386	0.264
G	10	783.99	0.365	0.285
G-sharp	11	830.61	0.344	0.306
Octave A	12	880.00	0.325	0.325

CHECK and THINK

If you examine the positions in Table 18.1, you will see that the frets are not evenly spaced. (The space between A and A-sharp is 0.036 m, and the space between is G-sharp and octave A is 0.019 m.) The frets get closer together as they get closer to the bridge, just as in the real guitar shown in Figure 18.26.

18-6 Flute: Resonance in a Tube Open at Both Ends

In a wind instrument such as the flute (Fig. 18.27), a standing wave forms in a tube of air. As with standing waves on a string, waves in a tube of air reflect from the ends, resulting in superposition of the reflected waves. If the tube is closed at one end, there is a node at that (closed) end. If the tube is open at one end, there is an antinode at that (open) end. The resonance frequencies depend on whether one or both ends of the tube are open.

A flute is best modeled as a tube open at both ends. In our model, we consider a straight tube of length L. Figure 18.28 shows the motion of air molecules vibrating at the fundamental frequency in a tube open at both ends. The sound wave is longitudinal, with the air molecules oscillating back and forth parallel to the tube. When the fundamental harmonic standing wave is established, there are antinodes at each end because both ends of the tube are open, and there is a single node at the tube's center. Molecules near the end of the tube are at the antinode of the wave and move with the greatest amplitude, bringing them beyond the end of the tube as well as deep inside it. Molecules near the center of the tube are at the node and do not move at all.

Figure 18.29 gives us a simpler way to represent the position of air molecules as they oscillate in a tube. In Figure 18.29A, a thin metal rod is held at its center by a clamp. A hammer strikes one end of the rod parallel to its long axis. Molecules in the rod oscillate back and forth along its length, creating a longitudinal wave. The wave reflects from the ends, resulting in a standing longitudinal wave. Molecules near the ends are free to oscillate, so they are at antinodes. Molecules near the center are fixed in place by the clamp, so there is a node at the center. The standing wave in the metal rod is similar to the standing wave in the tube of air illustrated in Figure 18.28. Rather than representing the positions of the numerous metal molecules, it is much simpler to plot their displacement $S(x, t)$ versus their position x along the rod's length. Figure 18.29B shows S versus x at two different times separated by half a period ($T/2$). The result looks much like a transverse wave. In fact, it might help to picture such a transverse wave in the rod. Instead of using the hammer to strike the rod along its axis, imagine striking the rod perpendicular to its long axis (Fig. 18.29C). Because the ends are free to oscillate, antinodes are established at the ends. Because the center is held fixed by the clamp, a node is established at the center. Figure 18.29C shows the shape of the rod at two different times separated by half a period, a shape that resembles the graph of S versus x.

Now we are ready to represent the harmonic series of standing longitudinal waves in a tube open at both ends. Figure 18.30 shows a graph of S versus x like Figure 18.29B for each harmonic. There is an antinode at each end and a node at any point that never moves. As in Figure 18.29C, we can imagine that there are "clamps" at the nodes that prevent the molecules from moving. The first harmonic has one node in the middle, and half a wavelength fits between the two ends of the tube: $L = \lambda_1/2$.

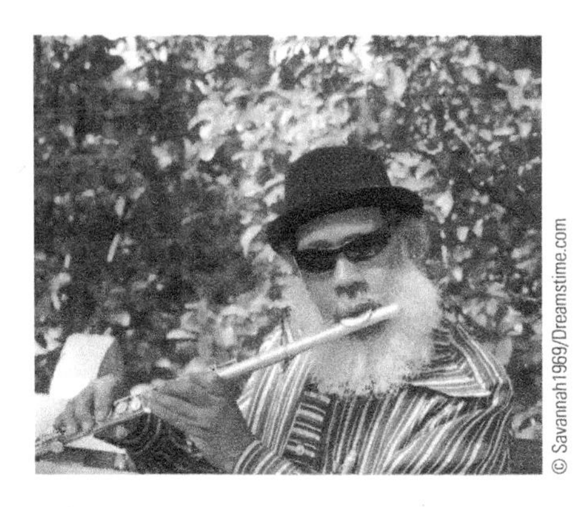

FIGURE 18.27 A musician plays a flute.

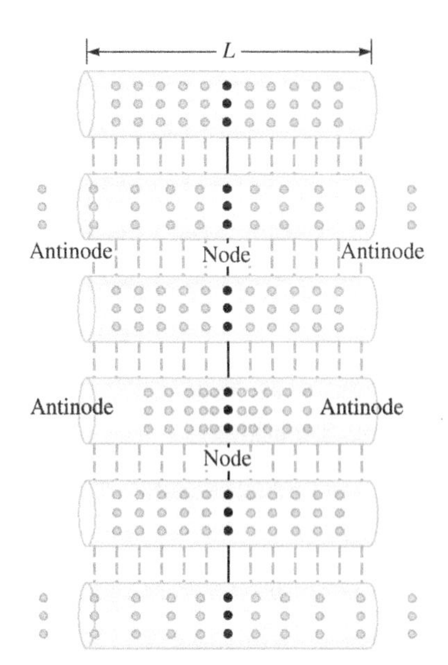

FIGURE 18.28 The molecules in a tube open at both ends move back and forth as part of a longitudinal wave.

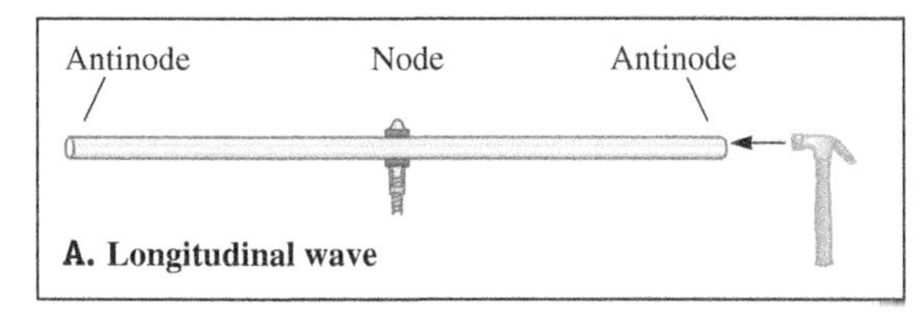

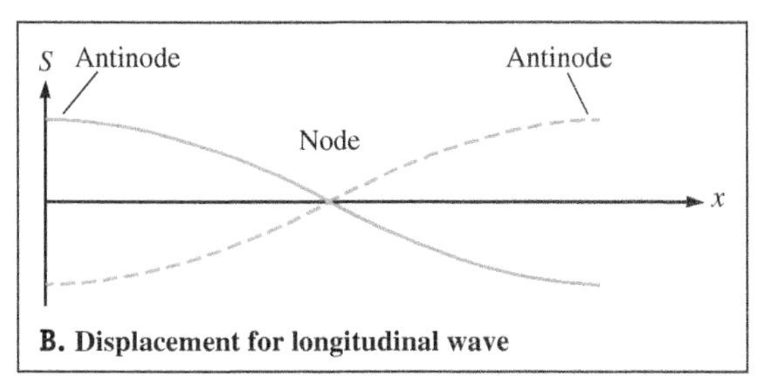

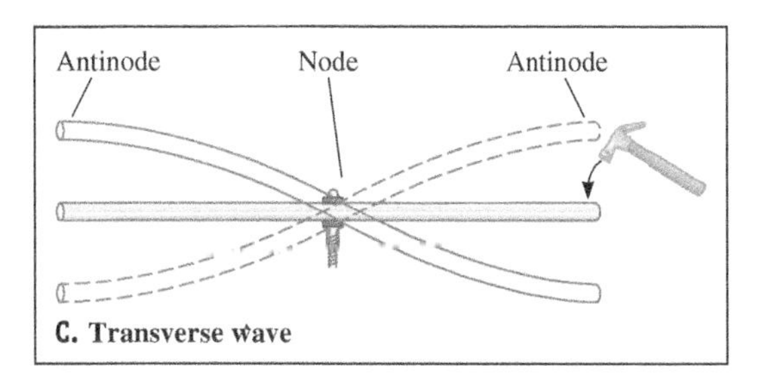

FIGURE 18.29 **A.** A longitudinal wave in a metal rod is set up by striking it horizontally. **B.** A graph of S versus x for a longitudinal wave. **C.** A transverse wave in the same rod is set up by striking it vertically.

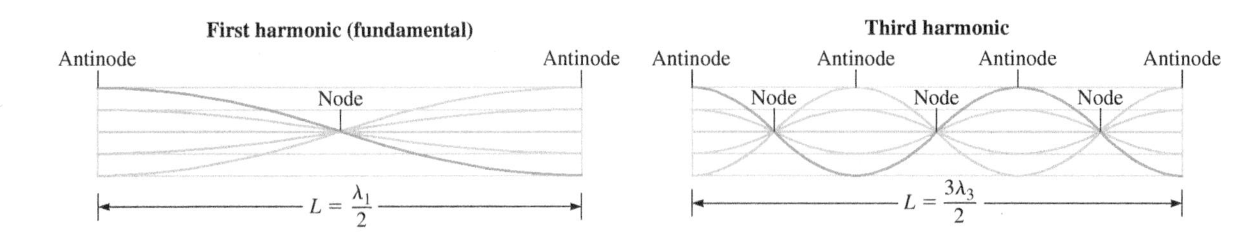

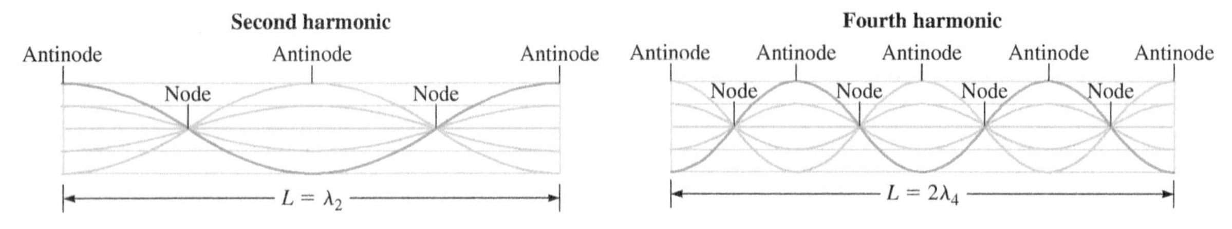

FIGURE 18.30 First four harmonics in a tube open at both ends. There must be an antinode on each end, so only those waves whose wavelength is given by $\lambda_n = 2L/n$, where n is the number of nodes, can fit in the tube. Compare with Figure 18.25.

Examining the higher harmonics in Figure 18.30 shows that the wavelength of the harmonic depends on the number of nodes n such that

$$\lambda_n = 2\frac{L}{n} \qquad (n = 1, 2, 3, \ldots) \tag{18.12}$$

The frequency of the harmonics comes from substituting Equation 18.12 into $f_n = v/\lambda_n$ (Eq. 17.8), where v is the speed of the longitudinal wave:

STANDING WAVE; TUBE OPEN AT BOTH ENDS

Special Case

$$f_n = n(v/2L) = nf_1 \qquad (n = 1, 2, 3, \ldots) \tag{18.13}$$

where $f_1 = \dfrac{v}{2L}$ is the fundamental frequency.

Equations for the wavelength (Eq. 18.12) and frequency (Eq. 18.13) of the harmonics in a tube open at both ends are similar to those for the wavelength (Eq. 18.9) and frequency (Eq. 18.10) of the harmonics on a string fixed at both ends. The integer n refers to the harmonic number in both cases, but in the case of a string fixed at both ends, n is the number of *antinodes*, and in the case of a tube open at both ends, n is the number of *nodes*.

EXAMPLE 18.5 CASE STUDY An Outdoor Concert

Many musicians hate to play outdoor concerts because the air temperature is not regulated and it is therefore difficult to keep their instruments in tune. Musicians often tune their instruments indoors before the performance. Suppose a flute is tuned such that the note A has a fundamental frequency of 440 Hz indoors where the temperature is $T_{\text{in}} = 24.0°\text{C}$ (75°F). Recall that the speed of sound in air as a function of temperature is given by Equation 17.14:

$$v = (331\ \text{m/s})\sqrt{1 + \frac{T_C}{273°\text{C}}}$$

A What is the fundamental frequency of the flute's A played outdoors in the summer where the temperature is $T_{\text{out}} = 35.0°\text{C}$ (95°F)? (Treat 440 Hz as three significant figures.)

INTERPRET and ANTICIPATE

Because the temperature outdoors is higher than it is indoors, the speed of sound is greater outdoors. The length of the flute doesn't change significantly. So, by Equation 18.13, the higher speed of sound means that the frequency of all notes will be higher. (The instrument will sound sharp.)

SOLVE
Use Equation 17.14 to determine v_{out}, the speed of sound outdoors at $T_{out} = 35.0°C$, and v_{in}, the speed of sound indoors at $T_{in} = 24.0°C$.

$$v_{out} = (331 \text{ m/s})\sqrt{1 + \frac{35.0°C}{273°C}} = 351.6 \text{ m/s}$$

$$v_{in} = (331 \text{ m/s})\sqrt{1 + \frac{24.0°C}{273°C}} = 345.2 \text{ m/s}$$

CHECK
As expected, the speed of sound is greater outdoors than it is indoors.

SOLVE
Use Equation 18.13 to solve for the fundamental frequency f_{out} when the flute is played outdoors in terms of the fundamental frequency f_{in} when the flute is played indoors.

$$\frac{f_{out}}{f_{in}} = \frac{v_{out}/2L}{v_{in}/2L} = \frac{v_{out}}{v_{in}}$$

$$f_{out} = \left(\frac{v_{out}}{v_{in}}\right)f_{in} = \left(\frac{351.6 \text{ m/s}}{345.2 \text{ m/s}}\right)(440 \text{ Hz}) = 448 \text{ Hz}$$

CHECK and THINK
As expected, the outdoor frequency is higher than the indoor frequency. The 8-Hz difference would certainly be noticeable to listeners.

B What is the fundamental frequency of the flute's A played outdoors in the fall where the temperature is $T_{out} = 10°C$ (50°F)? How could the flutist adjust his instrument so that it plays an A at the fundamental frequency of 440 Hz? (Round your answer to the nearest millimeter.)

INTERPRET and ANTICIPATE
In this part, the outdoor temperature is lower than that indoors, so we expect the speed of sound outdoors to be less and the flute's frequency to be lower than that indoors. According to $f_n = n(v/2L)$, the musician can correct for this difference by making his instrument shorter.

SOLVE
Following the procedure in part A, find the speed of sound and the frequency outdoors.

$$v_{out} = (331 \text{ m/s})\sqrt{1 + \frac{10.0°C}{273°C}} = 337.0 \text{ m/s}$$

$$f_{out} = \left(\frac{v_{out}}{v_{in}}\right)f_{in} = \left(\frac{337.0 \text{ m/s}}{345.2 \text{ m/s}}\right)(440 \text{ Hz}) = 430 \text{ Hz}$$

CHECK
As expected, the speed of sound and the frequency are both lower outdoors than they are indoors.

SOLVE
Use Equation 18.13 to find ΔL, the amount by which the flute needs to be shortened. Set the desired ("final") frequency to 440 Hz when the speed of sound outdoors is $v_{out} = 337.0$ m/s.

$$f_i = \frac{v_{out}}{2L_i} = 430 \text{ Hz and } f_f = \frac{v_{out}}{2L_f} = 440 \text{ Hz}$$

$$L_i = \frac{v_{out}}{2f_i} \text{ and } L_f = \frac{v_{out}}{2f_f}$$

$$\Delta L = L_f - L_i = \frac{v_{out}}{2f_f} - \frac{v_{out}}{2f_i} = \frac{v_{out}}{2}\left(\frac{1}{f_f} - \frac{1}{f_i}\right)$$

$$\Delta L = \left(\frac{337.0 \text{ m/s}}{2}\right)\left(\frac{1}{440 \text{ Hz}} - \frac{1}{430 \text{ Hz}}\right)$$

$$|\Delta L| = 8.91 \times 10^{-3} \text{ m} \approx 9 \text{ mm}$$

CHECK and THINK
In principle, the flutist can compensate for the cold weather by making his flute about 9 mm shorter than its indoor length. Because there is a physical limit to how short any instrument can be made, however, musicians may adjust how they blow or may use alternate fingering. Of course, they try to keep their instruments warm, perhaps by using their own body heat. At least one flutist is known to put his flute into his pants leg between songs.

FIGURE 18.31 A musician plays the clarinet.

18-7 Clarinet: Resonance in a Tube Closed at One End and Open at the Other End

Some wind instruments such as a clarinet are best modeled by a tube of air open at one end and closed at the other end. A clarinet has an antinode at the open end as does the flute, but it differs in having a node at the closed end. The "closed" end of the clarinet has a reed that the musician holds in his mouth (Fig. 18.31), whereas the flute has a hole that the musician blows across without closing it off (Fig. 18.27).

As in the case of the flute, we represent a series of harmonic waves in the tube by a set of S versus x sketches (Fig. 18.32). The first harmonic has just one node at the closed end and a single antinode at the open end. One-fourth of a wavelength ($\lambda/4$) fits in the tube of length L. The next harmonic has an additional node and antinode in the tube, so three-fourths of a wavelength ($3\lambda/4$) fit there.

Looking at the higher harmonics in Figure 18.32, we find the wavelength of the nth harmonic is given by

STANDING WAVE; TUBE CLOSED AT ONE END Special Case

$$\lambda_n = 4\frac{L}{n} \qquad (n = 1, 3, 5, \ldots) \tag{18.14}$$

Only the odd harmonics exist in a tube that is open at just one end. We still use the number n to refer to the harmonics. For example, the last pattern shown in Figure 18.32 is referred to as the seventh harmonic (not the fourth). The harmonic number is *not* equal to either the number of nodes or antinodes. Instead, n is one less than the number of nodes plus antinodes:

$$n = \begin{pmatrix}\text{number} \\ \text{of nodes}\end{pmatrix} + \begin{pmatrix}\text{number} \\ \text{of antinodes}\end{pmatrix} - 1$$

The frequency of the harmonics is found by substituting Equation 18.14 into $f_n = v/\lambda_n$ (Eq. 17.8), where v is the speed of the longitudinal sound wave:

$$f_n = n\frac{v}{4L} = nf_1 \qquad (n = 1, 3, 5, \ldots) \tag{18.15}$$

where $f_1 = v/4L$ is the fundamental frequency.

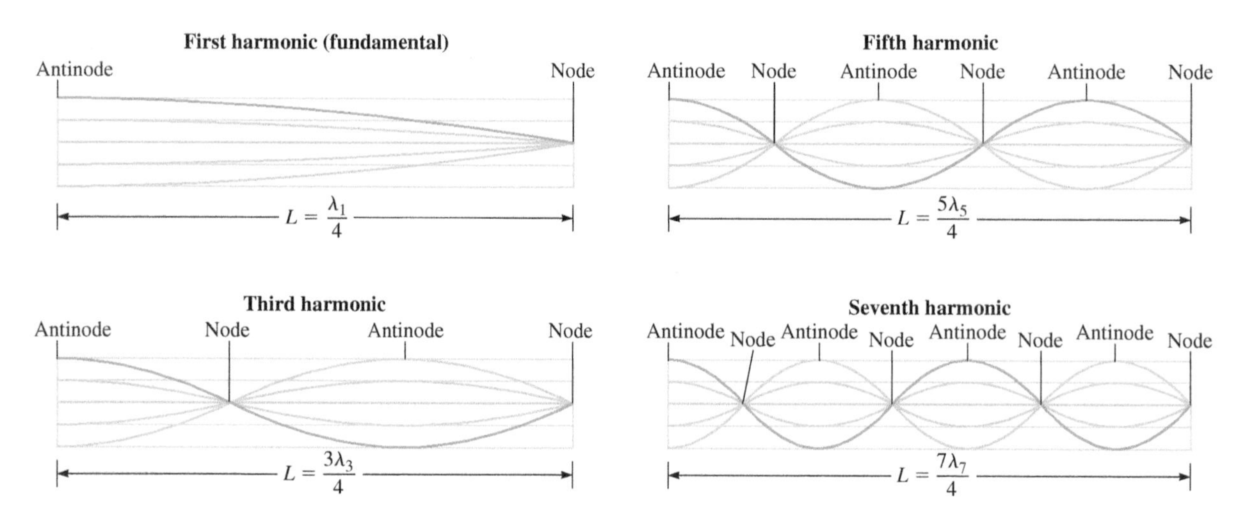

FIGURE 18.32 First four harmonics in a tube open at one end and closed at the other end. There must be an antinode at the open end and a node at the closed end, so only those waves whose wavelength is given by $\lambda_n = 4L/n$, where n is an odd integer, can fit in the tube. A tube that is only open at one end has only odd harmonics. Compare with Figures 18.25 and 18.30.

EXAMPLE 18.6 Aluminum Rod

In a physics demonstration, an aluminum rod is used to create sound waves. The rod of length $L = 1.50$ m is fixed at one end and free at the other end. By striking the free end, a standing longitudinal wave is set up in the rod. What audible frequencies are produced in the rod? (As discussed in Section 17-7, people can hear sounds in the frequency range from 20 Hz to 20,000 Hz.) The speed of sound in aluminum is 6420 m/s.

INTERPRET and ANTICIPATE

Because the rod is fixed at one end and free at the other end, it has a node at the fixed end and an antinode at the free end. The harmonics look like those in Figure 18.32. We need to find all the harmonic frequencies set up in the rod in the range from 20 Hz to 20,000 Hz.

SOLVE

Find the fundamental frequency by setting $n = 1$ in Equation 18.15.

$$f_1 = \frac{v}{4L} = \frac{6420\ \text{m/s}}{4(1.50\ \text{m})} = 1070\ \text{Hz}$$

Find the higher frequencies by multiplying the fundamental frequency by odd integers n beginning with 3. Most people cannot hear the $n = 19$ harmonic, so the audible frequencies are f_1 through f_{17}.

$$f_3 = 3f_1 = 3210\ \text{Hz} \qquad f_{13} = 13f_1 = 13{,}910\ \text{Hz}$$
$$f_5 = 5f_1 = 5350\ \text{Hz} \qquad f_{15} = 15f_1 = 16{,}050\ \text{Hz}$$
$$f_7 = 7f_1 = 7490\ \text{Hz} \qquad f_{17} = 17f_1 = 18{,}190\ \text{Hz}$$
$$f_9 = 9f_1 = 9630\ \text{Hz} \qquad f_{19} = 19f_1 = 20{,}330\ \text{Hz}$$
$$f_{11} = 11f_1 = 11{,}770\ \text{Hz}$$

CHECK and THINK

It might seem surprising that so many audible frequencies are produced at once, but the amplitude of the harmonic waves can vary greatly. Many such waves have very small amplitudes and are barely noticeable.

18-8 Beats

BEATS Major Concept

Before a performance, members of a jazz ensemble tune their instruments. If a piano is part of the ensemble, the pianist plays an A at 440 Hz. The other musicians tune their instruments to match the piano. The guitarist, for example, will sound an A on her instrument while the pianist plays an A. If the guitar is out of tune, the musicians will hear a modulated sound wave known as a *beat*, in which the sound alternates between louder and quieter (greater and smaller intensity). A **beat** is the interference of two waves that are at slightly different frequencies. The guitarist adjusts the tension of the string until the intensity of the sound is stable (in other words, until the beats disappear).

Figure 18.33 shows how beats are generated. When two nearly identical harmonic waves in a medium—such as sound waves in air—are superimposed, beats result if their frequencies are just slightly different. To see how, consider two time instants t_c and t_d. At time t_d, the two waves are nearly out of phase and nearly cancel each other at that instant (Fig. 18.33C). So, in the case of sound waves, it is momentarily very quiet at time t_d. At time t_c, the two waves are nearly in phase; the result is constructive interference and—in the case of a sound wave—a loud sound is heard at time t_c.

To better understand the beats in Figure 18.33C, let's add two harmonic waves that are identical except that one has angular frequency ω_1 and the other has a slightly different angular frequency ω_2. Because we are thinking about sound waves, we'll use Equation 17.6, $S(x, t) = S_{max} \sin(kx \pm \omega t)$, for longitudinal wave displacement. We hear the beats at a fixed position x, so we only need the time portion of this equation. The blue and red waves in Figure 18.33 are described by

$$S_1(x, t) = S_{max} \sin(\omega_1 t)$$
$$S_2(x, t) = S_{max} \sin(\omega_2 t)$$

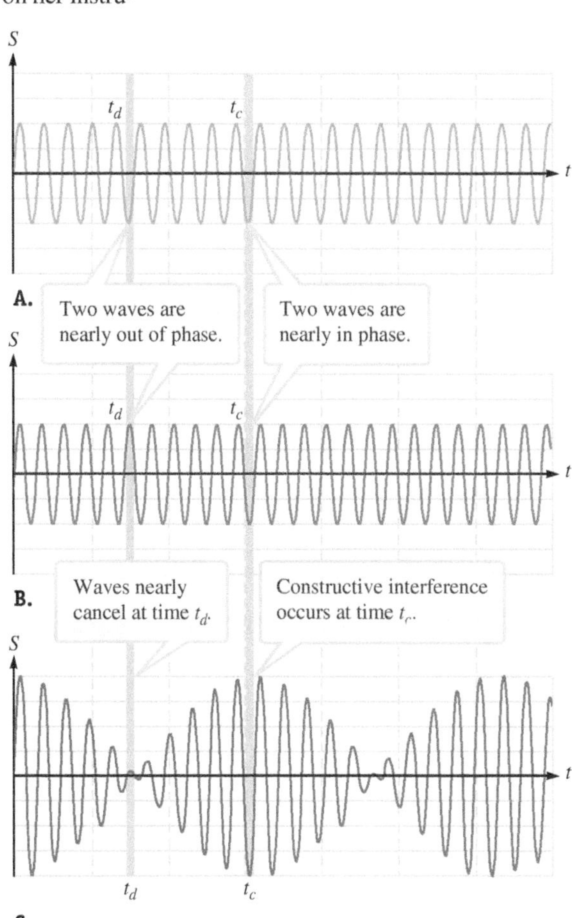

FIGURE 18.33 Beats result when two harmonic waves that are nearly identical except for a slight difference in frequency are superimposed. Adding the waves in parts **A.** and **B.** together produces the wave in part **C.**

respectively. The superposition of these two waves is the sum

$$S = S_1 + S_2 = S_{max}[\sin(\omega_1 t) + \sin(\omega_2 t)]$$

Use the trigonometric identity $\sin\alpha + \sin\beta = 2\cos[\frac{1}{2}(\alpha - \beta)]\sin[\frac{1}{2}(\alpha + \beta)]$ from Appendix A to rewrite S:

$$S = 2S_{max}\cos[\tfrac{1}{2}(\omega_1 - \omega_2)t]\sin\left[\frac{\omega_1 + \omega_2}{2}t\right] \quad (18.16)$$

Finally, we define two new angular frequencies, the average angular frequency,

$$\omega_{av} = \frac{\omega_1 + \omega_2}{2} \quad (18.17)$$

and the angular beat frequency,

$$\omega_{beat} = |\omega_1 - \omega_2| \quad (18.18)$$

(Recall that $\cos(-\theta) = \cos\theta$, and an angular frequency should be positive.) Rewriting Equation 18.16 in terms of these angular frequencies gives

$$S = 2S_{max}\cos(\tfrac{1}{2}\omega_{beat}t)\sin(\omega_{av}t) \quad (18.19)$$

Equation 18.19 describes the graph in Figure 18.33C. The superposition of two harmonic waves with slightly different frequencies results in a wave with angular frequency ω_{av}. The amplitude depends on time according to $S = 2S_{max}\cos(\frac{1}{2}\omega_{beat}t)$, oscillating with angular frequency $\frac{1}{2}\omega_{beat}$.

In most applications such as music, it is customary to talk about the beat frequency,

$$f_{beat} = \frac{\omega_{beat}}{2\pi} = \left|\frac{\omega_1}{2\pi} - \frac{\omega_2}{2\pi}\right| = |f_1 - f_2| \quad (18.20)$$

and the average frequency,

$$f_{av} = \frac{\omega_{av}}{2\pi} = \frac{f_1 + f_2}{2} \quad (18.21)$$

Figure 18.34 shows S versus t for three sets of beats. The beat period is the time between successive amplitude maxima (Fig. 18.34A) or successive amplitude minima (parts B and C of Fig. 18.34). The beat period is related to the beat frequency and angular frequency by

$$\omega_{beat} = 2\pi f_{beat} = \frac{2\pi}{T_{beat}} \quad (18.22)$$

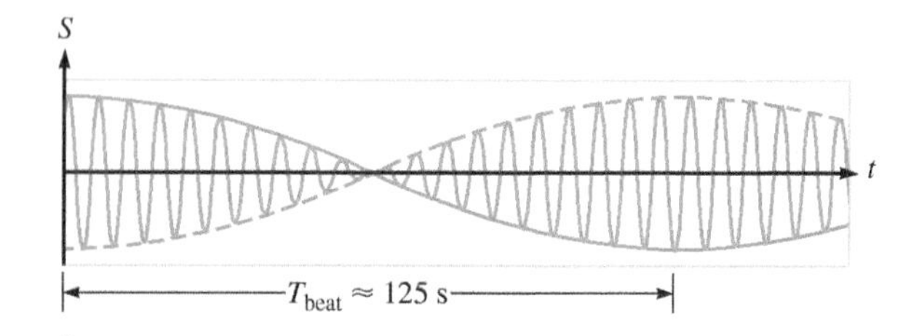

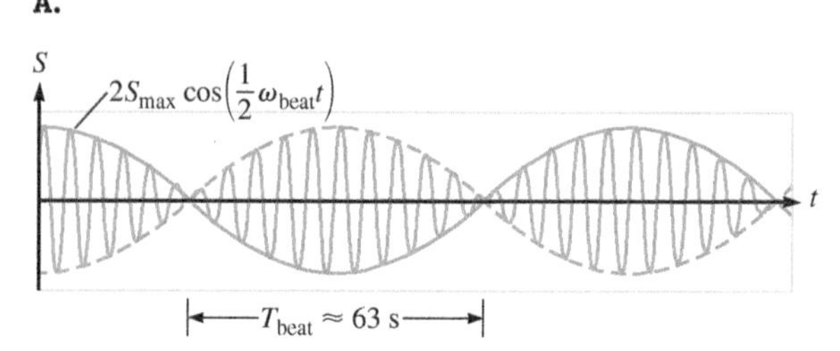

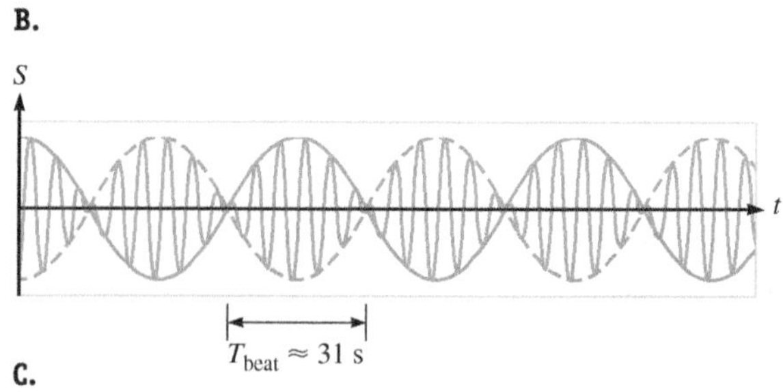

FIGURE 18.34 **A.** Beats for $\omega_{beat} = 0.05$ rad/s ($f_{beat} \approx 8$ mHz). **B.** Beats for $\omega_{beat} = 0.1$ rad/s ($f_{beat} \approx 16$ mHz). **C.** Beats for $\omega_{beat} = 0.2$ rad/s ($f_{beat} \approx 32$ mHz).

Figure 18.34 also shows three different beat patterns. Figure 18.34A results from two waves with the smallest difference in angular frequency $|\omega_1 - \omega_2| = 0.05$ rad/s, so this pattern has the lowest beat frequency ($f_{beat} \approx 8$ mHz) and the longest beat period ($T_{beat} \approx 125$ s). So, the more beats per unit time (greater f_{beat}) that a musician hears, the more out of tune her instrument is.

CONCEPT EXERCISE 18.5

CASE STUDY Tuning the Guitar

Before a performance, a piano is tuned so that the middle A string emits sound at 440 Hz. The guitarist sounds an A on her instrument while that note is played simultaneously on the piano and hears beats at a frequency $f_{beat} = 2$ Hz. As she tightens the string of her guitar, the beat frequency decreases. The beats completely disappear when her guitar is in tune with the piano. What was the fundamental frequency of her guitar before it was tuned?

EXAMPLE 18.7 Speed of a Submarine

Sonar (*so*und *n*avigation *a*nd *r*anging) uses sound waves in water to detect other vessels. Two submarines are under water, each emitting identical sonar waves at 990 Hz. Submarine A is at rest, and submarine B is moving toward sub A. On sub A, beats are detected with a beat frequency of 11 Hz. What is the speed of sub B? The speed of sound waves in seawater is 1531 m/s (Table 17.1).

INTERPRET and ANTICIPATE

Sub B is the source of sonar waves that are detected by sub A. According to the Doppler effect, because the source (sub B) is moving toward the detector (sub A), the observed sonar frequency f_{obs} is higher than the emitted frequency $f_{emit} = 990$ Hz. Sub A produces its own sonar wave at 990 Hz that is superimposed on the sonar wave produced by sub B. Because the waves are identical except for the slight difference in frequency due to the Doppler effect, beats are created. Typical submarine speeds are between 10 and 100 km/h, so we should get an answer in this range.

SOLVE

Use Equation 18.20 to find the frequency f_{obs} of the sonar wave detected by sub A. Because sub B is moving toward the detector, assign it the higher of two possible frequencies.

$$f_{beat} = |f_1 - f_2| = |f_{obs} - f_{emit}|$$

$$11 \text{ Hz} = |f_{obs} - 990 \text{ Hz}|$$

$$f_{obs} = 979 \text{ Hz or } 1001 \text{ Hz; choose } 1001 \text{ Hz}$$

The speed v_{source} of the source (sub A) can be found from the all-purpose Doppler effect relation (Eq. 17.31). In the case of a stationary observer, we have $v_{obs} = 0$. Because the source is moving toward the observer, we choose the minus sign (top sign) in the denominator. Solve for v_{source}.

$$f_{obs} = \left(\frac{v_s \pm v_{obs}}{v_s \mp v_{source}}\right) f_{emit} \quad (17.31)$$

$$f_{obs} = \left(\frac{v_s \pm 0}{v_s - v_{source}}\right) f_{emit} \text{ after choosing signs}$$

$$f_{obs}(v_s - v_{source}) = v_s f_{emit}$$

$$v_{source} = \frac{v_s(f_{obs} - f_{emit})}{f_{obs}}$$

$$v_{source} = \frac{(1531 \text{ m/s})(1001 \text{ Hz} - 990 \text{ Hz})}{1001 \text{ Hz}}$$

$$v_{source} = 16.82 \text{ m/s} = 60.57 \text{ km/h}$$

CHECK and THINK

Sub B's speed is in the expected range. Sonar is important on a submarine because there are no windows (and it is hard to see through water). Without sonar, a sub may not discover the location and velocity of other subs (or undersea creatures) until there is a collision. In February 2009, a British sub and a French sub collided in the Atlantic Ocean. Both subs were equipped with active and passive sonar. Active sonar sends a sound wave out and then "listens" for the echo from other objects, whereas passive sonar just listens. When on patrol, subs generally use just passive sonar because active sonar makes their presence known. Passive sonar tells you the direction (bearing) of the other sub, but not its distance. The colliding subs were probably using only their passive systems.

18-9 Fourier's Theorem

Any wind instrument can be modeled as either a tube open at both ends or closed at one end and open at the other, and strings on any instrument are fixed at both ends. It would seem at first glance, then, that various musical instruments should make very similar sounds. When two different instruments play the same note, however, it is possible to identify the type of instrument being played because an instrument does not produce a pure, single harmonic wave. In this section, we take a brief look at more complicated waves.

French physicist Jean Baptiste Joseph Fourier discovered that any wave can be mathematically represented by the superposition of harmonic (sine and cosine) wave functions; this statement is known as **Fourier's theorem**. For example, the sound wave produced by a musical instrument is a complex wave that cannot be represented by

FOURIER'S THEOREM

Underlying Principle

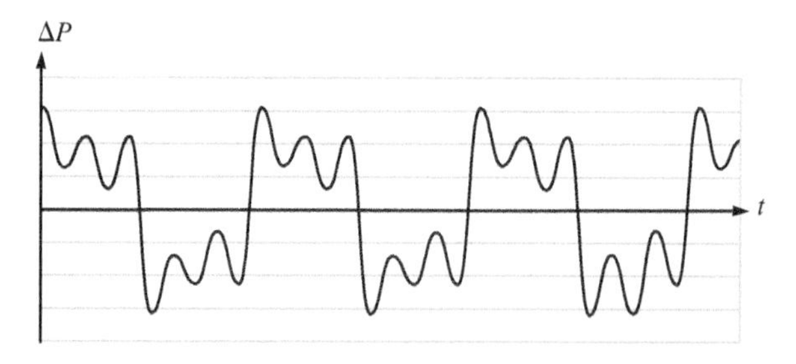

FIGURE 18.35 A clarinet produces a complex pressure wave. The wave is periodic with a frequency called the *fundamental frequency* f_1.

a *single* sine or cosine function; instead, it is represented by a sum of sine and cosine functions.

Figure 18.35 represents a complex pressure wave that may be produced by an instrument such as a clarinet. This sound wave is periodic with a frequency f_1 called the **fundamental frequency**. According to Fourier's theorem, this periodic wave can be represented by a sum of harmonic waves. The first term in the sum is a wave with this same frequency, so f_1 is also called the **first harmonic frequency**. Subsequent harmonic waves in the sum have frequencies that are integer multiples of the fundamental frequency called the **second harmonic frequency** $2f_1$, the **third harmonic frequency** $3f_1$, and so on. The six harmonic waves shown in Figure 18.36 add together to produce the clarinet's sound wave in Figure 18.35. The amplitude of each harmonic wave indicates the relative importance of that harmonic frequency in the complex wave. For example, the first harmonic sine wave (Fig. 18.36A) is much stronger than the fifth harmonic sine wave (Fig. 18.36C). Even more striking, the clarinet's sound wave does not contain any even harmonics (second, fourth, sixth). Different instruments playing at the same fundamental frequency sound different from one another because of the relative strength of the harmonics that make up the complex sound waves.

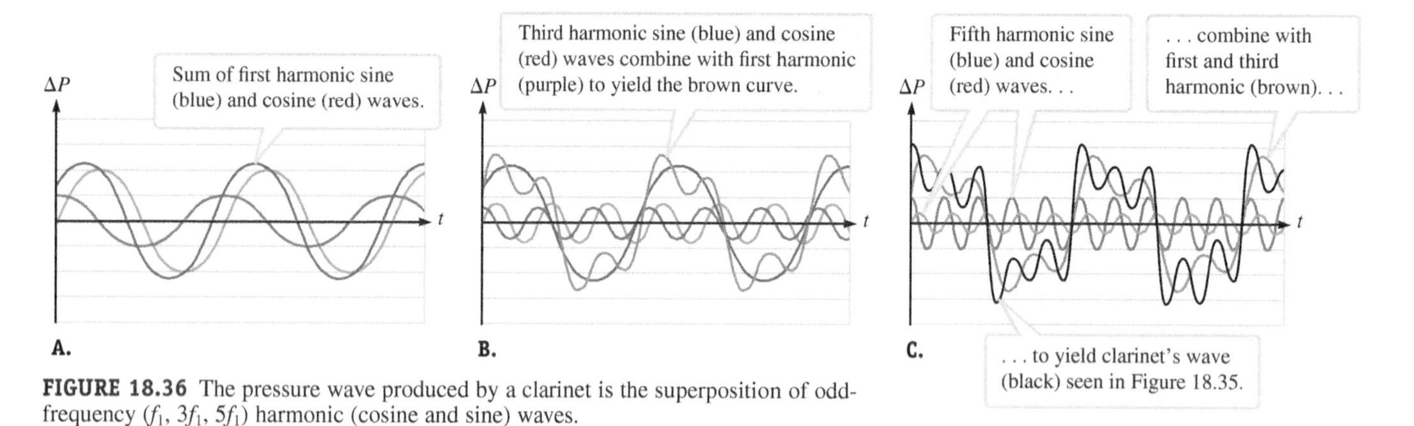

FIGURE 18.36 The pressure wave produced by a clarinet is the superposition of odd-frequency (f_1, $3f_1$, $5f_1$) harmonic (cosine and sine) waves.

CASE STUDY Part 3: Guitar String

When a guitarist strums or plucks a string, he distorts the shape of the string as shown in Figure 18.37A. According to Fourier's theorem, the pulse on the string is made of a number of harmonic waves (Fig. 18.37B). Only the harmonic waves with frequencies that resonate with the string will cause significant vibration. The resulting standing wave is the superposition of many harmonic waves. The fundamental harmonic is the strongest, but the higher harmonics give the instrument a rich sound. Tones without the higher harmonics don't sound like music, but instead sound like an emergency alarm. If the same note were to be played on a different instrument—say a violin—the same fundamental frequency would be heard, but the relative strength of the higher harmonics would be different. A characteristic set of higher harmonics gives each instrument its unique sound, and many instruments give a big band its rich sound.

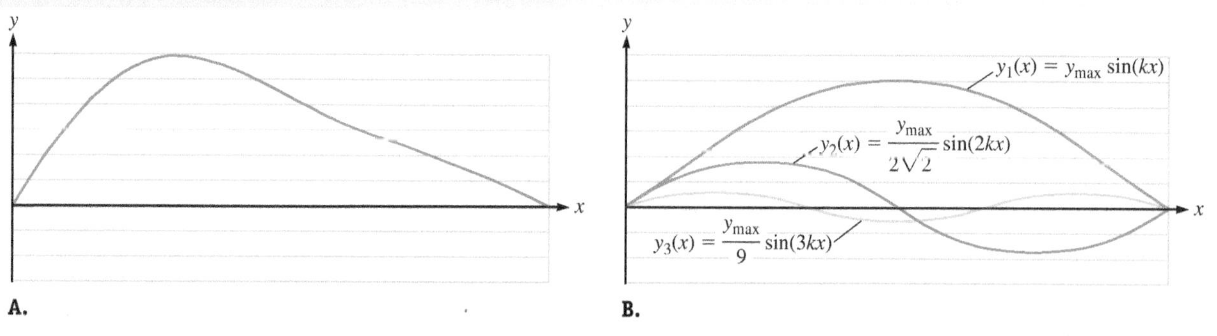

FIGURE 18.37 **A.** A plucked guitar string has a roughly triangular profile. **B.** The guitar string's profile is made of many harmonic waves. These three harmonic waves nearly reproduce the profile in part A.

SUMMARY

Underlying Principles

1. When more than one wave exists in a medium, the resulting wave function is found by **superimposing** (adding) the individual wave functions. For two such wave functions $y_1(x, t)$ and $y_2(x, t)$,
$$y(x, t) = y_1(x, t) + y_2(x, t)$$
2. According to **Fourier's theorem**, any wave can be mathematically represented by the superposition of harmonic (sine and cosine) wave functions.

Major Concepts

1. A **boundary condition** is the state of the medium's end.
   a. The profile of a wave reflected from a **fixed end** is inverted (Fig. 18.5).
   b. The profile of a wave reflected from a **free end** is upright (Fig. 18.7).
2. **Law of reflection:** For reflected waves, the incident angle equals the reflected angle:
$$\theta_i = \theta_r \quad (18.1)$$
where angles are measured between the direction of propagation and the line normal (perpendicular) to the reflecting surface (Fig. 18.10).
3. The condition for **constructive interference** is
$$\Delta d = n\lambda \quad (n = 0, 1, 2, 3, \ldots) \quad (18.2)$$
The condition for **destructive interference** is
$$\Delta d = \frac{n}{2}\lambda \quad (n = 1, 3, 5, \ldots) \quad (18.3)$$
where Δd is the difference in distance traveled by the two interfering waves.
4. In a **standing wave**, each particle of the medium oscillates back and forth, but unlike in a traveling wave, the wave pattern does not propagate. The wave function for a standing harmonic wave is
$$y(x, t) = [2y_{max} \sin(kx)]\cos(\omega t) \quad (18.6)$$
Particles that do not oscillate at all are at the **nodes** of the standing wave. Nodes are at
$$x = n\frac{\lambda}{2} \quad (n = 0, 1, 2, 3, \ldots) \quad (18.7)$$
Particles with the greatest amplitude are at the **antinodes**. Antinodes are at
$$x = \left(n + \frac{1}{2}\right)\frac{\lambda}{2} \quad (n = 0, 1, 2, 3, \ldots) \quad (18.8)$$
5. The interference of two waves that are at slightly different frequencies is a **beat**, given by
$$S = 2S_{max} \cos(\tfrac{1}{2}\omega_{beat}t)\sin(\omega_{av}t) \quad (18.19)$$
where $\omega_{beat} = |\omega_1 - \omega_2|$ and $\omega_{av} = (\omega_1 + \omega_2)/2$. The quantities f_{beat} and f_{av} are used in music and other applications:
$$f_{beat} = \frac{1}{T_{beat}} = \frac{\omega_{beat}}{2\pi} = |f_1 - f_2| \quad (18.20)$$
and
$$f_{av} = \frac{1}{T_{av}} = \frac{\omega_{av}}{2\pi} = \frac{f_1 + f_2}{2} \quad (18.21)$$

Special Case: Standing Waves

1. Harmonic frequencies for a standing wave on a **string fixed at both ends** (Fig. 18.25) are given by
$$f_n = n\left(\frac{v}{2L}\right) = nf_1 \quad (18.10 \;\&\; 18.11)$$
where $n = 1, 2, 3, \ldots$ and v is the speed of waves traveling on the string. All harmonics are present.
2. Harmonic frequencies for a standing wave in a **tube open at both ends** (Fig. 18.30) are given by
$$f_n = n\frac{v}{2L} = nf_1 \quad (n = 1, 2, 3, \ldots) \quad (18.13)$$
All harmonics are present.
3. Harmonic frequencies for a standing wave in a **tube open at one end and closed at the other** (Fig. 18.32) are given by
$$f_n = n\frac{v}{4L} = nf_1 \quad (n = 1, 3, 5, \ldots) \quad (18.15)$$
Only the odd harmonics are present.

PROBLEMS AND QUESTIONS

A = algebraic C = conceptual E = estimation G = graphical N = numerical

18-1 Superposition

1. There are two waves on a string given by $y_1(x, t) = 2\cos(3t - 10x)$ and $y_2(x, t) = 2\cos(3t + 10x)$.
   a. C In what ways do the two waves differ?
   b. N Find the wave that results on this string. *Hint*: Recall that $\cos\alpha + \cos\beta = 2\cos\left[\frac{1}{2}(\alpha + \beta)\right]\cos\left[\frac{1}{2}(\alpha - \beta)\right]$.
   c. C What is the resultant wave's amplitude?
2. G Two pulses travel in opposite directions along a string. One pulse is traveling to the right. The other pulse is inverted, with half the width and twice the amplitude of the first pulse, moving to the left as shown in Figure P18.2. Sketch the shape of the string at the moment in time when the centers of each pulse are at the same location.

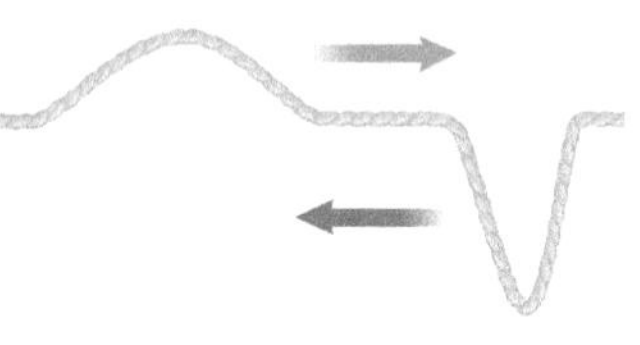

FIGURE P18.2

3. Two waves in the same medium are given by $y_1(x, t) = 2.5\sin(8x - 3.2t)$ and $y_2(x, t) = 2.5\sin(8x - 3.4t)$.
   a. C In what ways do the two waves differ?
   b. N Find the wave that results on this string. *Hint*: Recall that $\sin\alpha + \sin\beta = 2\sin\left[\frac{1}{2}(\alpha + \beta)\right]\cos\left[\frac{1}{2}(\alpha - \beta)\right]$.
   c. C What is the resultant wave's amplitude?
4. G Two individual waves exist in a medium as shown in Figure P18.4. Graphically find the superposition of these waves.

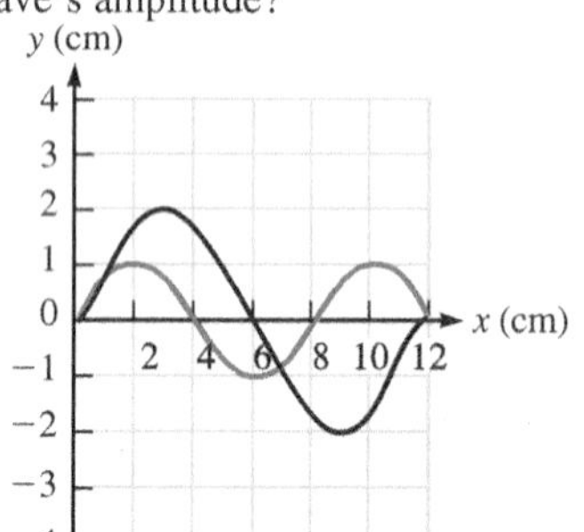

FIGURE P18.4

5. N The wave functions $y_1(x, t) = (0.150 \text{ m})\sin(3.00x - 1.50t)$ and $y_2(x, t) = (0.250 \text{ m})\cos(6.00x - 3.00t)$ describe two waves superimposed on a string, with x and y in meters and t in seconds. What is the displacement y of the resultant wave at **a.** $x = 0.100$ m and $t = 0$, **b.** $x = 1.00$ m and $t = 1.00$ s, and **c.** $x = 3.00$ m and $t = 3.00$ s?

18-2 Reflection

Problems 6 and 7 are paired.

6. G The wave function for a pulse on a rope is given by

$$y(x, t) = \frac{0.43}{(x - 13.6t)^2 + 1}$$

where all constants are in the appropriate SI units. Sketch the wave profile for **a.** the incident pulse, **b.** the reflected pulse if the end is free, and **c.** the reflected pulse if the end is fixed.

7. A The wave function for a pulse on a rope is given by

$$y(x, t) = \frac{0.43}{(x - 13.6t)^2 + 1}$$

where all constants are in the appropriate SI units. What is the wave function for the reflected pulse if the end is **a.** free or **b.** fixed?

8. G Jeff and Zak are sound tourists, exploring amazing sound effects around the world. At St. Paul's Basilica in Vatican City, they test out the whispering gallery inside the circular wall of the dome. If Jeff faces Zak who stands on the opposite side of the gallery, Zak cannot hear him, but if Jeff cups his mouth and whispers toward the wall while Zak puts his ears near the wall on the opposite side, Jeff's message is easily heard. Figure P18.8 shows two rays emitted by Jeff after they have reflected from the wall he faces. Draw the rays' paths as they reflect to Zak.

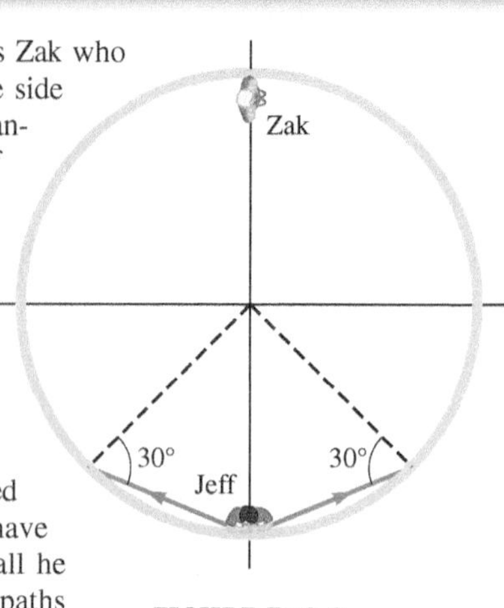

FIGURE P18.8

9. CASE STUDY Sounds that reach your ear within 0.1 s of the initial sound are perceived as *reverberation* rather than as a distinct echo. In a music hall, a little reverberation can be a good thing. Too much can be distracting. Because sound reflects off the many surfaces in the hall several times, a listener may hear many reflections of the same sound. Each reflection absorbs some of the wave's energy, and the reflected waves get quieter. So, after a short time, the reflected waves are unheard. If the surfaces don't absorb enough energy, however, the reflected waves may be perceived for too long, and an echo is heard. Often, panels are placed in the interior of a music hall to control the reflected sounds. Let's consider a simple example. (We won't worry about speakers, reflections off of people, furniture, or other objects.) A single source of sound is on stage, and a listener sits directly in front of the source (Fig. P18.9). There are six reflecting panels set inside the hall. Suppose each panel absorbs sound so that each reflected wave has 10% of the intensity of the incident wave. The range of human hearing is quite large, so let's say that if the intensity is reduced by a factor of one million, the sound is just barely audible.
   a. N How many reflections can the sound wave undergo before becoming inaudible?
   b. A Use the distances shown in Figure P18.9 to write an expression for the duration Δt of a sound heard due to its many reflections in terms of the speed of sound v_s in the room.

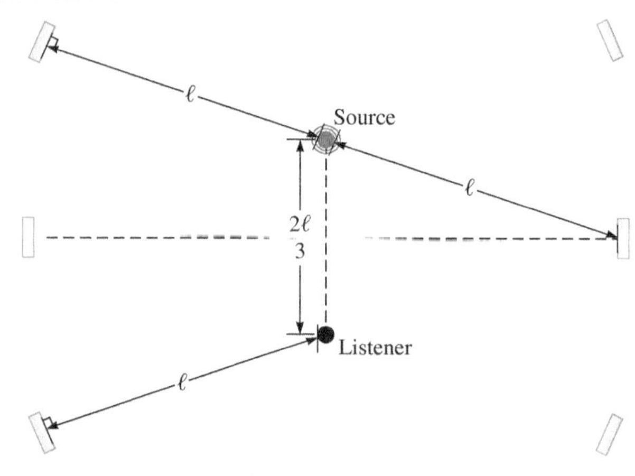

FIGURE P18.9

c. N The duration can be longer than 0.1 s because reflected waves are quieter than the original sound and because the music being played overwhelms the quieter reflections. So, a duration Δt of about 1 s is quite good. Find the distance ℓ that gives a duration of 1 s.

10. At many broadcast sporting events, you might notice that some crew members use microphones surrounded by a parabolic dish to record the sounds on the field (Fig. P18.10).

a. C What advantage does the parabolic dish provide? Why is it necessary at large events?

b. G Draw a sketch of a side view of a parabolic dish and show how sound would be reflected by the dish. Also indicate where the microphone should be placed for the best recording of the sound.

© Michael King/Cornell University

FIGURE P18.10

11. C Some animals use echolocation to navigate. Although the process may be complicated, describe in principle how emitting a sound pulse can allow a creature to locate other objects.

18-3 Interference

12. Two speakers, facing each other and separated by a distance d, each emit a pure tone of the same amplitude A with frequency f. The speed of each of the sound waves is v_s. A listener stands between the speakers, a distance x from one of the speakers.

a. A What frequencies would cause a dead spot (complete destructive interference) at the listener's position?

b. N If the speakers are separated by 5.00 m with the listener 2.00 m from one of the speakers, what is the lowest frequency for which there is a dead spot? The speed of sound in air is 343 m/s.

13. Two waves traveling on the same string are given by $y_1(x, t) = y_{max}\sin(kx - \omega t)$ and $y_2(x, t) = y_{max}\sin(kx - \omega t + \varphi)$. The two waves have the same amplitude y_{max}, angular wave number k, and angular frequency ω, but are out of phase by φ.

a. A Show that the resulting wave on the string is given by

$$y(x, t) = \left[2y_{max}\cos\frac{\varphi}{2}\right]\sin\left(kx - \omega t + \frac{\varphi}{2}\right)$$

where the absolute value of the term in brackets is the amplitude $A = |2y_{max}\cos(\varphi/2)|$. *Hint:* Recall that $\sin\alpha + \sin\beta = 2\cos[\frac{1}{2}(\alpha - \beta)]\sin[\frac{1}{2}(\alpha + \beta)]$.

b. N What is the amplitude if $\varphi = 0$? If $\varphi = \pi/2$? If $\varphi = \pi$?

14. N Two speakers generate harmonic sound waves that are in phase at 686 Hz. The speakers face the same direction. The speed of sound is 343 m/s. Find the phase difference between the two waves received at **a.** 4.00 m from each speaker, **b.** 3.00 m from one speaker and 3.50 m from the other, **c.** 3.00 m from one speaker and 4.00 m from the other, and **d.** 3.00 m from one speaker and 3.25 m from the other.

15. C It is thought by some that ancient monuments like Stonehenge were constructed with the knowledge that the sounds created by drummers drumming would become augmented or diminished at certain locations, which we now know is due to the interference of sound waves at these locations in the monument. Figure P18.15 shows an image of the Maryhill World War I Memorial Monument in Maryhill, Washington, which is a circular structure similar to Stonehenge. If a drum is struck repeatedly at the center such that the sound reflects off the outer wall, describe what will determine the locations where the sound is louder than expected and locations where the sound is diminished. Discuss these locations in terms of the radius of the circle, the speed of sound in air, and the rate at which the drum is struck.

© Michael D. Martin

FIGURE P18.15

18-4 Standing Waves

16. N As in Figure P18.16, a simple harmonic oscillator is attached to a rope of linear mass density 5.4×10^{-2} kg/m, creating a standing transverse wave. There is a 3.6-kg block hanging from the other end of the rope over a pulley. The oscillator has an angular frequency of 43.2 rad/s and an amplitude of 24.6 cm. **a.** What is the distance between adjacent nodes? **b.** If the angular frequency of the oscillator doubles, what happens to the distance between adjacent nodes? **c.** If the mass of the block is doubled instead, what happens to the distance between adjacent nodes? **d.** If the amplitude of the oscillator is doubled, what happens to the distance between adjacent nodes?

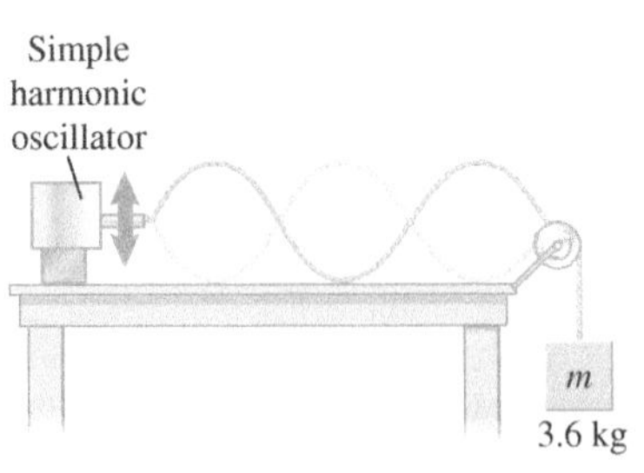

FIGURE P18.16

17. N A standing wave on a string is described by the equation $y(x, t) = 1.25\sin(0.0350x)\cos(1450t)$, where x is in centimeters, t is in seconds, and the resulting amplitude is in millimeters. **a.** What is the length of the string if this standing wave represents the first harmonic vibration of the string? **b.** What is the speed of the wave on this string?

18. N The resultant wave from the interference of two identical waves traveling in opposite directions is described by the wave function $y(x, t) = (2.40\text{ m})\sin(0.0450x)\cos(8.00t)$, where x and y are in meters and t is in seconds. What are the **a.** frequency, **b.** wavelength, and **c.** speed of the two interfering waves?

19. N A standing transverse wave on a string of length 60 cm is represented by the equation $y(x, t) = 4.0\sin(\pi x/15)\cos(96\pi t)$, where x and y are in centimeters and t is in seconds. **a.** What is the maximum value of the standing wave at the point $x = 5.0$ cm? **b.** Where are the nodes located along the string for this particular standing wave? **c.** What is the vertical velocity v_y of the string at $x = 7.5$ cm when $t = 0.25$ s?

20. C People of the Inca Empire built a long series of roads and rope bridges in the highlands of Peru (Fig. P18.20). Imagine a couple of brave kids playing on such a bridge. By jumping up and down at a frequency of 0.55 Hz, they set up a standing wave with a node on each end and one in the middle. If the bridge is 40.7 m long, what is the speed of the wave?

Photo by Rutahsa Adventures, Inc.

FIGURE P18.20

21. N A standing wave is described by $y(x, t) = [34.0 \sin(4.15x)]\cos(215t)$, where all constants are in appropriate SI units. What are the **a.** period, **b.** wavelength, and **c.** maximum y displacement of this wave?

22. N A standing wave is described by $y(x, t) = [34.0 \sin(4.15x)]\cos(215t)$, where all constants are in appropriate SI units. What are the wave functions for the two traveling waves that make up this standing wave?

23. N Two harmonic waves travel in opposite directions in the same medium. They have the same period (34.2 s), wavelength (6.23 m), and amplitude (3.45 m). What is the wave function of the resulting standing wave?

18-5 Guitar: Resonance on a String Fixed at Both Ends

24. N A violin string vibrates at 294 Hz when its full length is allowed to vibrate. If it is now held so that two-thirds of it is allowed to vibrate in the same harmonic, what frequency will it produce?

25. N Two successive harmonics on a string fixed at both ends are 66 Hz and 88 Hz. What is the fundamental frequency of the string?

26. N Two strings on a musical instrument are tuned to play at the fundamental frequencies 622.25 Hz (D-sharp) and 554.37 Hz (C-sharp). **a.** If the strings have the same length and are held under the same tension, what is the ratio (D-sharp to C-sharp) of their masses? (This situation describes a guitar or violin.) **b.** If the strings are held under the same tension and have the same mass per unit length, what is the ratio (D-sharp to C-sharp) of their lengths? (This situation describes a piano or harp.)

27. N When a string fixed at both ends resonates in its fourth harmonic, the wavelength is 0.345 m. How long is the string?

28. N A 50.0-cm-long copper wire with radius 0.100 cm and density 8.96 g/cm^3 is placed under 45.0 N of tension. **a.** What is the fundamental frequency of vibration for the wire? **b.** What are the next two harmonic frequencies for standing waves on this wire?

29. N As Toki tunes his guitar, he observes that by playing with the tension on a string he is able to change the string's fundamental frequency. He changes the fundamental frequency from 520.0 Hz to 550.0 Hz. The length of the string is 0.600 m. **a.** What is the wavelength of the fundamental harmonic in each case? **b.** What is the range of velocities of traveling waves on the string as he performs the change?

30. A A string fixed at both ends resonates in its fourth harmonic. The tension in the string is quadrupled. What happens to **a.** the speed of the wave, **b.** its wavelength, and **c.** its frequency?

31. A Two strings fixed at both ends oscillate in the fourth harmonic. String A has twice the mass per unit length of string B. The length of the strings and the tension in each are identical. Find the ratio (A to B) of their **a.** wavelengths and **b.** frequencies.

32. N A wire 1.50 m in length and with a mass of 25.0 g is placed under a tension of 33.0 N. **a.** What are the first two harmonic frequencies of vibration for this wire? **b.** The wire is observed to have a node at 30.0 cm from one end. What are the possible harmonic modes of vibration of the wire?

Problems 33 and 34 are paired.

33. C A standing wave exists on a string fixed at both ends. If you touch the string at a node, what happens to the standing wave?

34. C If you touch the string in Problem 33 at an antinode, what happens to the standing wave?

35. N A 0.530-g nylon guitar string 58.5 cm in length vibrates with a fundamental frequency of 196 Hz. **a.** What is the tension in the guitar string? **b.** The string is later observed to vibrate with two antinodes. What is the frequency of vibration?

36. N A sonometer is a single-stringed musical instrument on a wooden box with two movable bridges (Fig. P18.36). The string is fastened at one end, and the other end is attached to a hanging weight holder that is extended over a pulley. The distance between the end of the pulley and the fastened end define the length of the string. By adding weights to the weight holder, you can change the tension in the string. A particular sonometer wire has a length of 114 cm. Where should the two bridges be placed along this length so as to divide the wire into three segments whose fundamental frequencies are in the ratio 1:3:4? Assume the tension is the same throughout the wire.

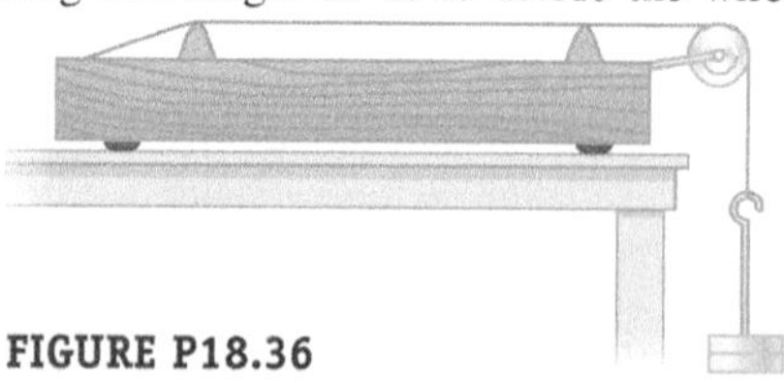

FIGURE P18.36

37. N A 2.20-g guitar string 60.0 cm in length is placed under a tension of 53.0 N and plucked so that it vibrates with its third harmonic frequency. **a.** Which is longer, the wavelength of the wave on the guitar string or the wavelength of the sound it emits in the room-temperature air surrounding the string? **b.** What is the ratio of the wavelength of the wave on the string to the wavelength of the sound it produces in air?

18-6 Flute: Resonance in a Tube Open at Both Ends

38. E A barrel organ is shown in Figure P18.38. Such organs are much smaller than traditional organs, allowing them to fit in smaller spaces and even allowing them to be portable. Use the photo to estimate the range in fundamental frequencies produced by the organ pipes in such an instrument. Assume the pipes are open at both ends. How does that range compare to a piano whose strings range in fundamental frequency from 21.7 Hz to 4186.0 Hz?

© Andrzej Barabasz (Chepry)

FIGURE P18.38

39. N A very large organ may have pipes that are as short as about 1 ft and as long as 16 ft. Imagine such an organ kept in a church whose temperature in the middle of winter is 55°F and in the middle of summer is 90°F. Find the seasonal change in both the shortest and the longest pipe's fundamental frequency. Assume the pipes are open at both ends.

40. C A standing wave exists in a pipe open at both ends. If you were to close off one end with your hand, what happens to the wave?

41. N The Channel Tunnel, or Chunnel, stretches 37.9 km undersea across the Strait of Dover in the English Channel between France and the United Kingdom. What are the resonant frequencies along this direction for the air in the Chunnel?

42. N A dog whistle emits frequencies above the range of human hearing, but within the range that dogs can hear. If a dog whistle is modeled as a tube open at both ends, how long would the whistle need to be if the fundamental frequency of the whistle is at the upper limit of human hearing, assumed to be 20,000 Hz?

Problems 43 and 44 are paired.

43. N A musical artist fits one piece of PVC (plastic) piping around another to create a variable-length tube in which harmonics can be excited by causing the tube to vibrate. She places a microphone on the inside of the tube to record the sound and send it to an amplifier. The tube has a minimum length of 1.2 m and a maximum length of 2.00 m, and the speed of sound in air in the tube is 343 m/s. Suppose she wants to create a sound that has a harmonic frequency of 540.0 Hz. Find each possible length of the tube such that a harmonic frequency of 540.0 Hz is created.

44. C In Problem 43, where should the musical artist place the microphone to best sample the sound? Consider that she would like all existing harmonic frequencies to be recorded by the microphone.

18-7 Clarinet: Resonance in a Tube Closed at One End and Open at the Other End

45. N If the aluminum rod in Example 18.6 were free at both ends, what audible frequencies would be heard? Compare your results with the results of Example 18.6 and explain the difference.

46. At the end of a party, Ezra finds that he can play the beginning notes (B, A, and G) to the children's song *Three Blind Mice* by blowing across the top of three identical soft-drink bottles (Fig. P18.46). One is completely empty and sounds in the fundamental harmonic, but the other two still have some liquid inside. Assume the sound is reflected from the surface of the liquid and the bottles are modeled as pipes closed at one end and open at the other end. Each bottle resonates at one of the three notes $f_B = 493.88$, $f_A = 440.00$ Hz, and $f_G = 392.00$ Hz.

FIGURE P18.46

a. C Are the three frequencies the fundamental frequency of each of the respective bottles? Explain your answer.

b. C Which bottle is empty? How do you know? (Assume the bottles are opaque so you cannot see the liquid level.)

c. N How tall is each bottle?

d. N What is the height of the liquid in the partially filled bottles?

47. N The third harmonic of an organ pipe that is closed at one end is in resonance with the first harmonic of an organ pipe that is open at both ends. Find the ratio of the lengths of the two pipes (L_{closed}/L_{open}).

48. E If you blow across the top of a plastic bottle, it's possible to hear an audible tone. Estimate the frequency for a 2-liter soda bottle if the frequency heard is the fundamental frequency.

Problems 49 and 50 are paired.

49. C A tube that is closed at one end and open at the other is filled with room-temperature air, where the speed of sound is 343 m/s. Striking the tube produces a characteristic set of harmonic wavelengths and frequencies. **a.** What happens to the harmonic wavelengths if the tube is suddenly filled with water, where the speed of sound in water is 1482 m/s? **b.** What happens to the characteristic frequencies when the water is added?

50. A tube that is closed at one end and open at the other is filled with room-temperature air, where the speed of sound is 343 m/s. Then water is used to fill the tube. The speed of sound in water is 1482 m/s.

a. N Calculate the wavelength of the fundamental and the third harmonic both before and after the water is added.

b. N Calculate the fundamental and third harmonic frequencies both before and after the water is added.

c. C If someone were striking the tube and the sound inside were amplified through a speaker, would a listener hear a higher-pitched tone or a lower-pitched tone after the water is added? Explain your answer.

51. N An auto accident with a deer can be fatal for both the people and the deer. One way to prevent such an accident is with a deer whistle. Online directions tell you how to make your own deer whistle out of PVC pipes. According to the directions, you need a 16-in. length of pipe. One end is closed. The pipe is attached vertically to the front of the vehicle. Determine the fundamental frequency produced by such a pipe. Would it be audible to you?

52. N A cylindrical pipe of length 1.0 m is fitted with a thin, flexible diaphragm in the middle of the pipe (Fig. P18.52). Each half of the pipe is of equal length and represents a pipe that is closed at one end and open at the other. The portions A and B contain hydrogen and oxygen respectively. Each half of the pipe is set into vibration. What is the minimum harmonic number in each of the portions A and B for which they have the same frequency? Assume the speed of sound in hydrogen is 1300 m/s and in oxygen is 300 m/s.

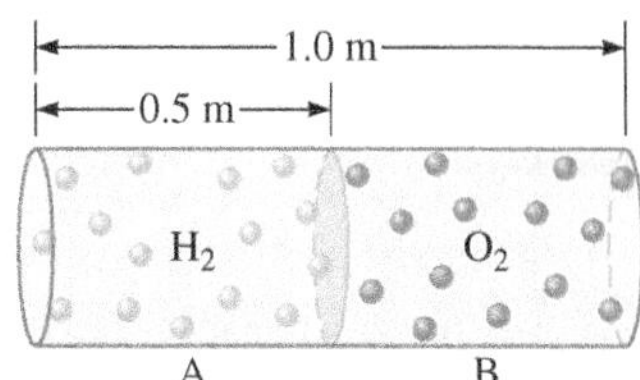

FIGURE P18.52

18-8 Beats

53. N CASE STUDY Before a band plays at an outdoor concert, the instruments are tuned indoors. As the concert goes on for a while, the temperature drops. One of the clarinetists was unable to keep her clarinet warm while she was required to play her second instrument, but the other clarinetist used his body heat to keep his clarinet warm. What beat frequency will be heard if the two identical clarinets attempt to play D (293.66 Hz) while one perfectly tuned clarinet is at 20°C and the other is at 3°C?

54. C Dog whistles operate at frequencies above the range of human hearing. Explain how two such whistles operating at slightly different frequencies may be used to make a sound audible to a person.

55. N A violinist and a pianist each play an A note. The piano was recently tuned to 440.0 Hz, but when the musicians play together, there are clear beats heard at a rate of about three per second. The violinist tightens the violin string, and the number of beats increases to about five beats per second. What is the frequency of the violin string after it is tightened?

56. N Two identical wires under tension are initially vibrating at 200.0 Hz. The tension in one string is increased slightly, resulting in 3.00 beats per second from the vibration of the two strings. What is the frequency with which the higher tension wire is vibrating?

57. N Find the beat frequency between the two waves given by $y_1 = A\sin(0.250x - 420\pi t)$ and $y_2 = A\sin(0.250x - 426\pi t)$.

58. N In a classroom demonstration, two pipes of length 1.25 m open at both ends oscillate in their fundamental harmonic. One pipe has a collar allowing it to be lengthened by 15.0 cm. As one pipe is slowly lengthened, beats are heard. **a.** Find the maximum beat frequency. **b.** What is the length of the longer pipe when the beat frequency heard is half of its maximum?

59. N The air in an organ pipe is set into vibration and compared with the sound from a tuning fork of frequency 256 Hz. When the air in the pipe is at a temperature of 16°C and both objects are vibrating, 24 beats are heard in 12 s. The experiment is repeated with another tuning fork that has a slightly lower frequency than the previous one, and the beat frequency is not observed. Given that information, what change in temperature of the air in the pipe is necessary to bring the pipe and the original tuning fork into resonance, where no beat frequency is observed?

18-9 Fourier's Theorem

60. At $t = 0$, three waves are given by

$$y_1(x, 0) = \frac{4}{\pi}\sin\left(\frac{\pi x}{2}\right)$$

$$y_2(x, 0) = \frac{4}{3\pi}\sin\left(\frac{3\pi x}{2}\right)$$

$$y_3(x, 0) = \frac{4}{5\pi}\sin\left(\frac{5\pi x}{2}\right)$$

a. G Sketch these three wave functions from $x = -4.0$ m to 4.0 m.
b. G Sketch $y(x, 0) = y_1(x, 0) + y_2(x, 0) + y_3(x, 0)$ in the interval $x = -4.0$ m to $x = 4.0$ m.
c. C Are $y_1(x, 0)$, $y_2(x, 0)$, and $y_3(x, 0)$ the harmonic components of a square wave, a triangular wave (a series of equilateral triangle shapes), or a sawtooth wave (a series of right triangular shapes)? Explain.
A spreadsheet program may be helpful.

General Problems

61. N What is the amplitude of the resultant wave for two sinusoidal waves with equal amplitudes of 6.50 cm traveling in the same direction on a taut string, but 90.0° out of phase?

62. G A pulse on a string travels to the right (Fig. P18.62). It moves one box to the right per millisecond. The right end of the string is tied to a post so that it cannot move up and down at that location. Sketch the shape of the string at a time 6 milliseconds later.

FIGURE P18.62

63. The functions

$$y_1 = \frac{2}{(2x + 5t)^2 + 4} \quad \text{and} \quad y_2 = \frac{-2}{(2x - 5t - 3)^2 + 4}$$

describe two pulses traveling along a stretched string.
a. C What is the direction of travel for each of the waves?
b. N At what position x do the two pulses always cancel each other?
c. N At what time t do the two pulses cancel each other everywhere?

64. N The low E string and the high E string on a guitar are tuned to have fundamental frequencies of 82.4 Hz and 329.6 Hz, respectively. The difference in the fundamental frequency is primarily due to the different thickness, and therefore different mass per unit length, of the two strings. We could, however, try to design a guitar using the same type of strings for both notes by varying either the length of the string or its tension.
a. If the mass per unit length and the length were kept the same for both strings, what would be the ratio of the tension in the high E string to the tension in the low E string?
b. If the mass per unit length and the tension were kept the same for both strings and we found a way to alter the length of the strings, what would be the ratio of the length of the high E string to the length of the low E string?

65. N The wave functions

$$y_1 = (3.00 \text{ m})\sin(2.00\pi x - 15.0\pi t)$$

$$y_2 = (3.00 \text{ m})\sin\left(2.00\pi x - 15.0\pi t + \frac{\pi}{2}\right)$$

describe two sinusoidal waves traveling on the same string, where x, y_1, and y_2 are in meters and t is in seconds. What are the **a.** amplitude, **b.** frequency, and **c.** wavelength of the resultant wave?

66. N The wave functions

$$y_1(x, t) = (1.20 \text{ m})\sin\left(0.350\pi x - \frac{\pi}{2}t\right)$$

$$y_2(x, t) = (1.20 \text{ m})\sin\left(0.350\pi x + \frac{\pi}{2}t\right)$$

describe two transverse sinusoidal waves traveling along a rope, where x and y are in meters and t is in seconds. What is the maximum transverse displacement of a rope element located at **a.** $x = 1.00$ m and **b.** $x = 3.00$ m? **c.** Where are the first three antinodes located along the rope?

67. N An air column in a variable-length pipe, which is closed at one end, is in resonance with a vibrating tuning fork of frequency 264 Hz. If the pipe has a maximum length of 1.0 m and a minimum length of 0.20 m, what is (are) the possible length(s) of the pipe? Assume the speed of sound in air is 334 m/s.

68. N A standing wave with three antinodes is formed on a 0.890-m-long wire under tension. **a.** Which harmonic mode of vibration does this wave represent? **b.** What is the wavelength of this wave? **c.** What is the number of nodes in this standing wave?

69. Two successive harmonic frequencies of vibration in a pipe are 532 Hz and 684 Hz.
a. C Is the pipe open at both ends, or is it closed at one end and open at the other?
b. N What is the fundamental frequency of the pipe?

70. N In Oshin's home theater, an oscillator operating at 425 Hz drives a pair of identical speakers placed 4.00 m apart and facing each other in phase. Where are the relative minima of sound pressure located on the line joining the two speakers?

71. N A block of mass $m = 2.50$ kg is suspended by a 10.0-g wire that is 2.25 m long and is draped over a massless pulley. The other end of the wire is attached to a wall (Fig. P18.71). If the distance between the wall and the pulley is 1.50 m, what is the fundamental frequency of vibration for the horizontal portion of the wire?

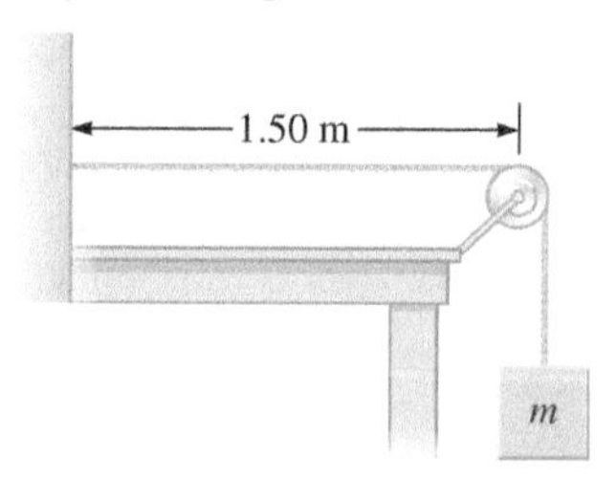

FIGURE P18.71

72. G The example of beats shown in Figure 18.34 (page 542) is the result of two waves with similar frequencies and equal amplitudes. If the two waves have different amplitudes instead, is there still a beat frequency? To answer, consider two waves that are incident at the same point in space such that $S_1(t) = \sin(10t)$ and $S_2(t) = 3\sin(10.5t)$. Plot the superposition of these waves versus time and comment on the effect on the beats. A spreadsheet program may be helpful.

73. N A pipe is observed to have a fundamental frequency of 345 Hz. Assume the pipe is filled with air ($v = 343$ m/s). What is the length of the pipe if the pipe is **a.** closed at one end and **b.** open at both ends?

74. N The wave function for a standing wave on a 2.00-m-long string is given by $y = 0.0500\sin(2\pi x)\cos(76\pi t)$, where x and y are in meters and t is in seconds. **a.** How many antinodes are present? **b.** What is the fundamental frequency of this wave?

75. N A cylindrical glass tube is partially filled with water to create an air column with a length of 0.20 m. The air column is vibrating at its fundamental frequency. After the water column is slowly lowered, the same frequency is heard. What should be the new length of the air column so that the third harmonic has the same frequency as the previous fundamental frequency? Treat the air column as a case of resonance with one open end and one closed end.

76. C A common technique of tuning a guitar is to use a string that is currently in tune to create a sound with the proper frequency that should also be the fundamental frequency of the next string. When the next string is close but not exactly in tune and both are plucked at the same time, a beat frequency can be heard. If this test is performed, explain how the guitarist knows whether to increase or decrease the tension in the string being tuned. Try to construct a logic flow chart (step-by-step instructions) for tuning a string in this manner. Consider that the second string's tension might be too high or too low initially.

77. N A pipe open at both ends has a fundamental frequency of 145 Hz. A second pipe, which is closed at one end, vibrates at this frequency in its fourth harmonic mode. What is the length of the pipe **a.** open at both ends and **b.** closed at one end?

78. N After a successful business meeting, a salesman steps into his hotel room's shower stall and closes the door, creating a rectangular pipe closed at both ends with dimensions 1.20 m $\times$ 1.20 m $\times$ 2.43 m. He begins to sing, covering frequencies from 80.0 Hz to 1100.0 Hz. If nodes form at the opposite sides of the shower stall, which frequencies will be in resonance and therefore sound the richest? Assume the speed of sound in the shower stall is 360 m/s.

79. N In attempting to tune the G string of a guitar to 196 Hz, a musician hears 4.00 beats/s between the guitar string and a reference tone of 196 Hz, indicating that the current tuning of the string is not correct. **a.** What are the possible frequencies to which the guitar string is currently tuned? **b.** The musician now loosens the guitar string and hears 2.00 beats/s. To what frequency is the guitar's G string now tuned? **c.** By what percentage must the tension in the guitar string change for the string to be tuned to 196 Hz?

80. N An accurate electronic tuner with an adjustable frequency is used to determine the actual frequency of a tuning fork marked as 512 Hz. When the tuner is set to 514 Hz and the tuning fork is sounded, a beat frequency of 2 Hz is heard. When the tuner is set to 510 Hz and the tuning fork is sounded, the beat frequency is 6 Hz. What is the frequency of the tuning fork?

19 Temperature, Thermal Expansion, and Gas Laws

Key Questions

How is temperature measured?

What is the connection between temperature and other macroscopic properties?

19-1 Thermodynamics and temperature 553

19-2 Zeroth law of thermodynamics 555

19-3 Thermal expansion 555

19-4 Thermal stress 560

19-5 Gas laws 564

19-6 Ideal gas law 566

19-7 Temperature standards 570

Underlying Principles

1. Thermodynamics
2. Zeroth law of thermodynamics

Major Concepts

1. Temperature
2. Thermal contact
3. Thermal equilibrium
4. Thermal linear expansion
5. Thermal volume expansion
6. Equation of state
7. Equilibrium state
8. Ideal gas
9. Mole
10. Avogadro's number
11. Kelvin scale
12. Absolute zero

Special Cases: Equations of State

1. Charles's law
2. Gay-Lussac's law
3. Boyle's law
4. Avogadro's law
5. Ideal gas law

Many fads—love beads, bell-bottom pants, pet rocks—took hold in the United States in the 1970s. Although some fads fade away forever, others, such as the mood ring (Fig. 19.1), reappear. The claim is that the "stone" of the ring changes color according to the mood of the wearer. The stone is actually a glass shell filled with liquid crystals whose molecular structure depends on their temperature. The molecular structure determines which colors of light are absorbed and which are reflected. The stone's color actually depends on the temperature of the liquid crystals, not on the mood of the wearer.

The color of mood rings is an example of a property that depends on temperature. The color of other objects also depends on temperature. Blue stars are hotter than red stars, for instance, and the heating coil on a stove top turns bright red when it is hot. Matter has other temperature-dependent properties. For example, most materials expand when they are hot and contract when they are cold. This chapter is about temperature: What is it, and how does it affect an object or system?

FIGURE 19.1 Mood rings change color depending on the temperature of the wearer.

19-1 Thermodynamics and Temperature

When you slide your textbook across your desk, the work you do on the book–desk system goes into the internal *thermal energy* of the system. You observe that the book and the desktop get warmer. **Thermal energy** is associated with the kinetic energy of the particles that make up the objects in the system (in this case, the book and the desk).

Thermodynamics is the study of thermal energy and the transfer of energy through *heat* and work. To understand thermodynamics, we connect macroscopic observations to microscopic phenomena. For example, when the book slides on the desktop, we cannot directly observe the increase in the molecules' kinetic energy (that is, we cannot see the molecules moving faster), but we can sense the increase in thermal energy by placing our hand against each surface and feeling the increased warmth. Our experience is macroscopic, or large scale. Much of this chapter focuses on the macroscopic property of *temperature* and its connection with other macroscopic properties such as pressure, volume, and density. **Temperature** is a measure of the average kinetic energy of the particles in a system. For instance, the molecules in a cup of ice-cold tea have on average less kinetic energy than the molecules in a cup of hot tea.

THERMODYNAMICS
Underlying Principle

TEMPERATURE **Major Concept**

Units of Temperature

Temperature is one of the seven base quantities that make up the SI (Système International d'Unités) set of units (see Appendix B and Section 1-5). The SI unit for temperature is the **kelvin**, abbreviated simply as K.

We do *not* say "degrees Kelvin" or write "°K."

For 18 chapters, we have been working with three of those seven SI base units: kilogram, meter, and second. In this chapter, we add two more to our list: *kelvin* and *mole*. Here we focus on kelvin; we'll return to mole later in the chapter. Each fundamental unit must have a precise definition or standard. The SI standard for the kilogram is a specific platinum–iridium alloy cylinder kept at the International Bureau of Weights and Measures in Sèvres, France.

Some standards are given in terms of a definition or procedure that can be reproduced in a laboratory. The SI standard for temperature is a prescription or recipe for measuring temperature that can be done in any well-equipped laboratory as discussed in Section 19-7. For now, it is sufficient to know how to convert from temperature scales in common use—Fahrenheit and Celsius—to the Kelvin scale (also known as the *absolute scale*). A **temperature scale** is defined by assigning values to two well-separated reference temperatures.

Until the 1960s and 1970s, most English-speaking countries used the Fahrenheit scale, but now the only industrialized country in the world to continue using it is the United States. The Fahrenheit scale was developed by Dutch physicist Daniel Gabriel Fahrenheit (1686–1736). Today on the Fahrenheit scale, the freezing point of water is 32°F, and the boiling point of water is 212°F. Fahrenheit did not originally use water to set the reference temperatures for his scale. Instead, he set the zero of his scale to the temperature of a mixture of water, ice, and sea salt, and he set the other reference temperature to healthy human body temperature at 96°F. It is not clear why he intended to set human body temperature to 96°F instead of some round number like 100°F. On the contemporary Fahrenheit scale, healthy human body temperature is 98.6°F.

The Celsius scale (sometimes known as the centigrade scale) is in common use worldwide. It was developed by Swedish astronomer Anders Celsius (1701–1744) about 30 years after the Fahrenheit scale. On the Celsius scale, the freezing point of water is 0°C, and the boiling point of water is 100°C.

In the United States, the Celsius scale is used in laboratories, and the Fahrenheit scale is used in medicine, in homes, and in weather reports. Figure 19.2 shows the Kelvin, Fahrenheit, and Celsius scales, allowing easy comparison. As shown in Figure 19.2, water freezes at

$$T_f = 273.15\,\text{K} = 0°\text{C} = 32°\text{F}$$

Water boils at

$$T_b = 373.15\,\text{K} = 100°\text{C} = 212°\text{F}$$

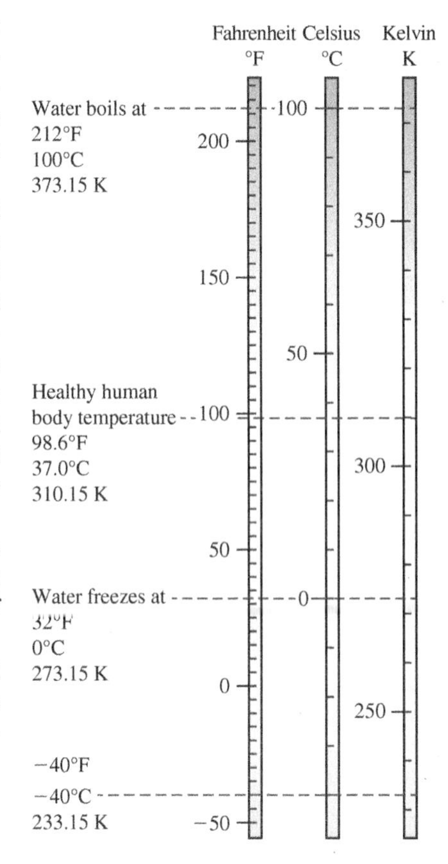

FIGURE 19.2 Three commonly used scales for measuring temperature.

So, the difference in temperature between the boiling point and the freezing point of water is

$$T_b - T_f = 100 \text{ K} = 100°\text{C} = 180°\text{F}$$

In other words, a 1 K increase in temperature equals a 1°C increase; the separation between increments of 1 K on the Kelvin scale equals the separation between increments of 1° on the Celsius scale. The only difference between the Kelvin scale and the Celsius scale is the zero point; zero on the Celsius scale is 273.15 K. To convert from the Celsius scale to the Kelvin scale, you only need to add 273.15:

$$T(\text{K}) = T(°\text{C}) + 273.15 \qquad (19.1)$$

The relationship between the Fahrenheit scale and the Celsius scale is a little more complicated, however, because an increment of 1° on the Fahrenheit scale is smaller than an increment of 1° on the Celsius scale. Based on the difference between the boiling point and the freezing point of water, an increase of 1°C equals an increase of 1.8°F. So, the separation between degree marks on the Fahrenheit scale is 1/1.8 = 5/9 of the separation between degree marks on the Celsius scale. In addition, the zero points are not the same; zero on the Celsius scale is 32°F. The conversion from Fahrenheit to Celsius is given by

$$T(°\text{C}) = \tfrac{5}{9}[T(°\text{F}) - 32] \qquad (19.2)$$

Equation 19.2 can be inverted to find the conversion from Celsius to Fahrenheit:

$$T(°\text{F}) = \tfrac{9}{5}T(°\text{C}) + 32 \qquad (19.3)$$

FIGURE 19.3 Thermometers play an important role in our daily lives including in industrial applications.

CONCEPT EXERCISE 19.1

The Fahrenheit scale remains useful in part due to personal experience with temperature. For example, the average temperature range in a typical midlatitude city such as Gdansk, Poland (formerly Dansk, where Fahrenheit was born), is 25°F to 78°F over a year.

a. Find the temperature range in Gdansk, Poland, in degrees Celsius and in kelvins.

b. Many people say that 0°F is a very cold day and 100°F is an unpleasantly hot day. Convert these temperatures to the Celsius and Kelvin scales.

c. When someone has an infection, her body may respond with a fever. A high fever ($T > 105°$F) can damage the brain. Convert this temperature to degrees Celsius.

CASE STUDY Part 1: Inventing the Thermometer

You know that water boils at 212°F, but three centuries ago scientists were studying the properties of boiling water. Without a thermometer, you wouldn't know that all pots of water boil at the same temperature, and you might speculate that some other property is more important. When Fahrenheit learned that a French scientist, Guillaume Amontons, found that water boils at a particular temperature, he was inspired to design and build his own thermometer.

Having been surrounded by thermometers and accurate temperature measurements our whole lives (Fig. 19.3), it is hard to imagine living without them, but for most of human history people didn't have thermometers. Since ancient times, however, many human endeavors—such as pottery, metallurgy, and cooking—required at least an estimate of temperature. In many cases, temperature was estimated by touch, but that is not possible if the object is very hot or very cold. The temperature of a hot metal was estimated from the color of its glow, and a kiln or oven's temperature may be estimated from its effects on other objects: Does it melt wax? Boil water? Melt lead? Accurate thermometers are necessary in thermodynamics because the laws of thermodynamics involve temperature. The case study in this chapter describes the story and physics of thermometers.

Although some types of thermometers we use today are based on recent technology, many thermometers can trace their roots to the end of the 16th century, especially to Galileo Galilei. The basic idea behind his thermometer and most of the thermometers we use today is that many macroscopic properties of matter depend on its temperature. These properties can be measured and exploited. For example, a mercury thermometer determines temperature based on expansion of the mercury as it warms.

19-2 Zeroth Law of Thermodynamics

Imagine working in a laboratory and needing to measure the temperature of some liquid in a flask. You put a mercury thermometer in the liquid and watch the mercury rise or fall (Fig. 19.4). You don't read the thermometer until the mercury level stops because the level of the mercury depends on temperature. If the level is changing, the temperature of the mercury is changing. The mercury's temperature will continue to change until it has the same temperature as the liquid in the flask. Then, by reading the temperature of the mercury from the thermometer scale, you have measured the temperature of the liquid.

This simple laboratory technique for measuring temperature illustrates the concepts of thermal contact and thermal equilibrium. When two objects are in **thermal contact**, thermal energy can pass from one object to the other. If the objects remain in thermal contact for a sufficient period of time, they will eventually reach thermal equilibrium and have the same temperature. In other words, two objects are in **thermal equilibrium** if they are in thermal contact with each other and their temperatures do not change. In the simple laboratory experiment (Fig. 19.4), the mercury thermometer is brought into thermal contact with the liquid. Thermal energy passes either from the thermometer to the liquid or from the liquid to the thermometer until they reach thermal equilibrium. When they are in thermal equilibrium, the thermometer's temperature equals the temperature of the liquid.

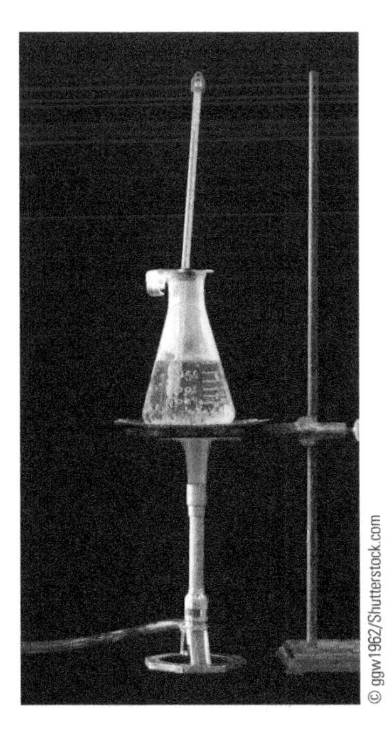

FIGURE 19.4 When using a mercury thermometer, you must wait until the mercury stops moving before reading the temperature.

Some processes require objects to be in thermal equilibrium (in other words, to have the same temperature) before they are brought into thermal contact. Suppose you are following a recipe for homemade mayonnaise that requires you to mix egg yolks (usually stored in the refrigerator) and oil (usually stored at room temperature) only after they have reached thermal equilibrium. Because the eggs are initially colder than the oil, you should put the yolks in a small container and put the container in a warm-water bath.

THERMAL CONTACT; THERMAL EQUILIBRIUM

Major Concept

Without putting them in contact with each other, how can you be sure that the egg yolks and oil are in thermal equilibrium? You could use a thermometer to measure their temperatures. Suppose you put the thermometer in the oil and, after a short time, the oil and thermometer are in thermal equilibrium. Their temperature—read off the thermometer—is 22°C. You then place the thermometer in the egg yolks, and after a while the yolks and the thermometer are in thermal equilibrium with a temperature of 22°C. You conclude that because both the oil and the yolks are in thermal equilibrium with the thermometer at 22°C, they must be in thermal equilibrium with each other. That is correct and, in fact, is an example of the **zeroth law of thermodynamics**:

ZEROTH LAW OF THERMODYNAMICS

Underlying Principle

If two objects A and B are in thermal equilibrium with a third object C, then A and B they are in thermal equilibrium with each other.

We can write the zeroth law of thermodynamics mathematically:

$$\text{If } T_A = T_C \text{ and } T_B = T_C, \text{ then } T_A = T_B.$$

There are four laws of thermodynamics. The other laws were stated before the zeroth law. You might think that it makes more sense to designate the zeroth law as the next-highest-number law, but because the zeroth law is more fundamental than the other laws, it was given a lower number. The zeroth law is considered more fundamental because it contributes an additional way to define temperature. **Temperature** is the macroscopic property of a system that determines whether that system is in thermal equilibrium with another system.

CONCEPT EXERCISE 19.2

Explain this statement: *A thermometer reports its own temperature.* How is it possible to use a thermometer to measure another object's temperature?

19-3 Thermal Expansion

When an engineer or architect designs a bridge, a large building, or even a sewer system, she must leave a gap between pieces (Fig. 19.5). The gap is necessary because most building materials (concrete and metals) expand as their temperature increases and contract as their temperature decreases. Without these gaps, the structures could buckle and fracture.

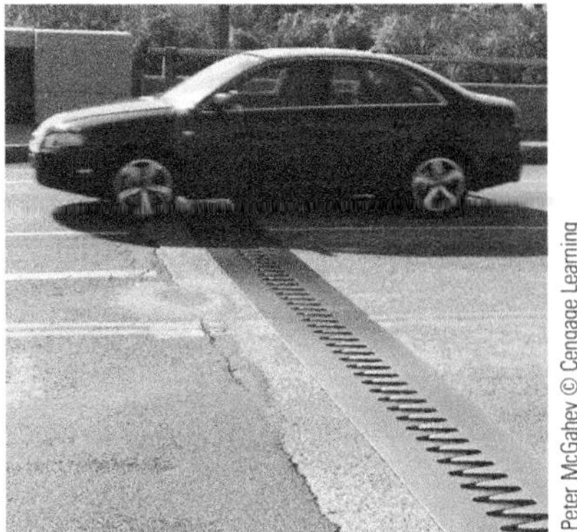

FIGURE 19.5 A gap in the road prevents the concrete from cracking when the road expands or contracts.

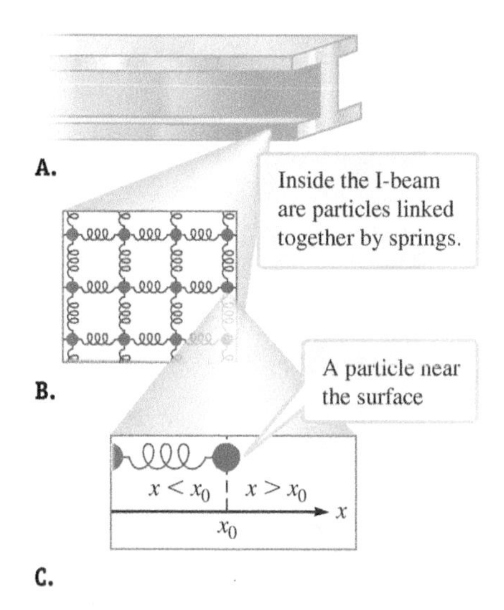

FIGURE 19.6 **A.** A steel beam. **B.** A model of the molecules that make up the steel beam. Particles are attached to one another by springs. **C.** A close-up of one particle near the surface.

Microscopic Model of Thermal Expansion

To see how temperature changes can cause a steel beam or similar object to expand, let's model the molecules in the object as particles and the molecular bonds as springs (parts A and B of Fig. 19.6) as we did in Section 14-4 when studying tensile and compressive stresses. Consider one such molecule at the surface of the beam with its horizontal bond (Fig. 19.6C). The potential energy curve for the molecule–bond system is shown in Figure 19.7. The system's potential energy is plotted on the vertical axis, and the molecule's position is plotted on the horizontal axis. When the molecule is at $x = x_0$, the molecular bond is relaxed, and the system's potential energy is at its minimum. When the molecule moves to the left ($x < x_0$), the bond is compressed, and when the molecule moves to the right ($x > x_0$), the bond is stretched.

Unlike the curve for the particle–spring system in Figure 8.32C (page 233), the potential energy curve in Figure 19.7 for a molecule–bond system is not a parabola. Instead, it is asymmetric, rising rapidly as $x \to 0$ and approaching zero as x gets very large. This curve makes sense: When the molecule is near $x = x_0$, the bond acts like a spring obeying Hooke's law, and the potential energy curve is nearly parabolic. As the molecule moves to the left, it runs up against other molecules that repel it, and the potential energy curve shoots upward. However, as the molecule moves to the right, there is nothing but the bond to keep it attached, and the molecule may overstretch and even break that bond.

As the temperature of the steel beam increases, so does its thermal energy. The molecules in a hot beam have more kinetic energy on average than do the molecules in a cool beam. According to $E = K + U$ (Eq. 8.11), the mechanical energy E of the molecules must increase as their kinetic energy K increases. Energy graphs for the molecule–bond system (near the surface) in a cool beam and a hot beam are shown in Figure 19.8. A molecule in the cool beam moves about as far to the right of its relaxed position x_0 as it moves to the left (Fig. 19.8A). On the other hand, because the asymmetry of the potential energy curve is more pronounced at higher energy, a molecule in the hot beam moves much farther to the right (outward) than it moves to the left (Fig. 19.8B). Once the surface bonds lengthen, interior bonds are able to lengthen and the beam expands.

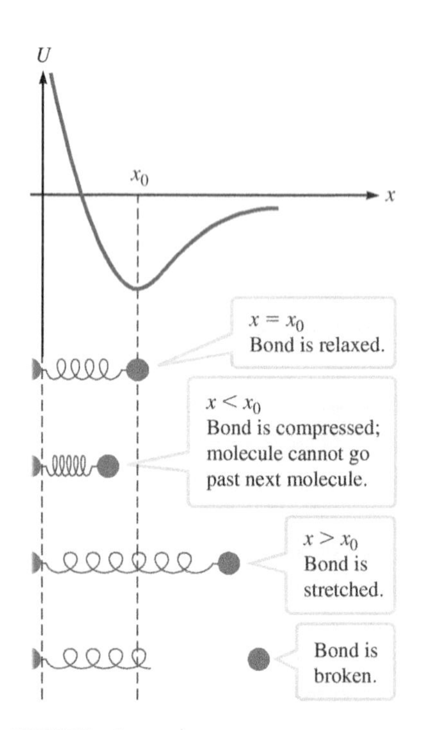

FIGURE 19.7 For a particle near the surface, the potential energy curve for a particle–bond system is not symmetric, so the particle moves farther to the right than it does to the left.

CONCEPT EXERCISE 19.3

Explain this statement: *If the potential energy curve were symmetric (as in Fig. 8.32C), hot materials would not expand.*

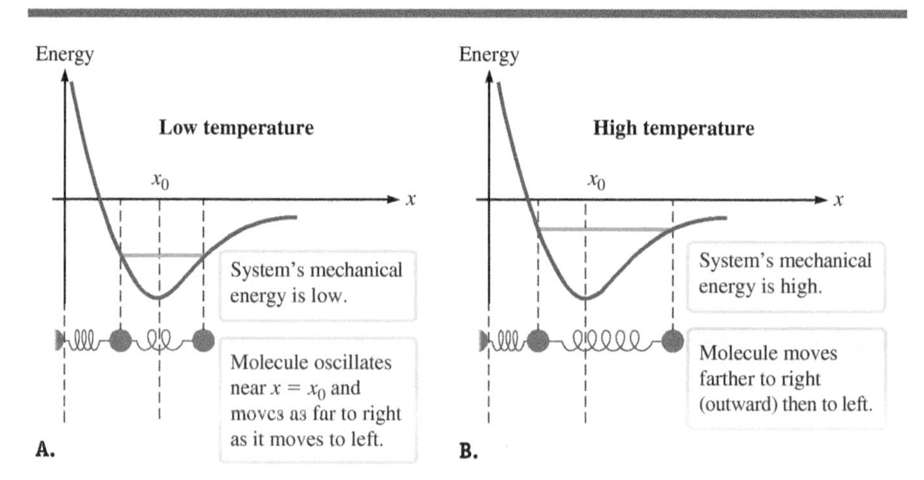

FIGURE 19.8 Due to energy conservation, mechanical energy E is represented by a straight horizontal line on an energy graph (Section 8-8). **A.** The energy graph for a cool beam shows that the mechanical energy E is low, and each surface molecule oscillates back and forth very close to its equilibrium position x_0. The asymmetry of the potential energy curve is not noticeable, and the bonds act much like springs obeying Hooke's law. **B.** The energy graph for a hot beam shows that the mechanical energy E is high, and each surface molecule's oscillation takes it far from its equilibrium position. The asymmetric nature of potential energy is noticeable, and the molecules move much farther outward than they move inward.

Macroscopic Observation of Thermal Expansion

On the macroscopic scale, when a solid rod of length L_0 experiences a temperature change ΔT, we observe a **thermal linear expansion** given by

THERMAL LINEAR EXPANSION
Major Concept

$$\Delta L = \alpha L_0 \Delta T \quad (19.4)$$

where ΔL is the change in the rod's length and α is called the **coefficient of linear expansion**. The coefficient of linear expansion has the dimensions of inverse temperature and depends on the type of material. Values of α for some common materials are given in Table 19.1.

Equation 19.4 is consistent with the microscopic model illustrated in Figure 19.8. If ΔT is positive, the molecule's mechanical energy increases, and its average distance to its neighbor increases. The initial length L_0 of the rod is proportional to its number of molecules and bonds; the more molecules and bonds in the rod, the greater its change in length. The coefficient of linear expansion α is related to the shape of the potential energy curve; a material with a large α must have a very asymmetric potential energy curve.

TABLE 19.1 Coefficients of linear expansion near room temperature.

Material	α ($\times 10^{-6}$ K^{-1} or °C^{-1})	Material	α ($\times 10^{-6}$ K^{-1} or °C^{-1})
Aluminum	22–24	Glass	4.0–9.0
Brass	18.7	Iron	10.4–12.0
Brick	5.5	Marble	12
Bronze	18.0	Mortar	7.3–13.5
Cement	10.0	Steel	13.0
Concrete	14.5	Wood (oak, parallel to grain)	4.9
Copper	16.8	Wood (oak, across grain)	5.4

These values are approximate and can be used for problems in this book if no other values are given.

DERIVATION Thermal Volume Expansion

Equation 19.4 works well for long, thin objects such as metal rods. For other shapes such as cubes or spheres, each dimension expands according to Equation 19.4, resulting in a change in the object's volume. Consider a rectangular solid whose original dimensions are L_0, W_0, and H_0 (Fig. 19.9) so that the original volume is

$$V_0 = L_0 W_0 H_0$$

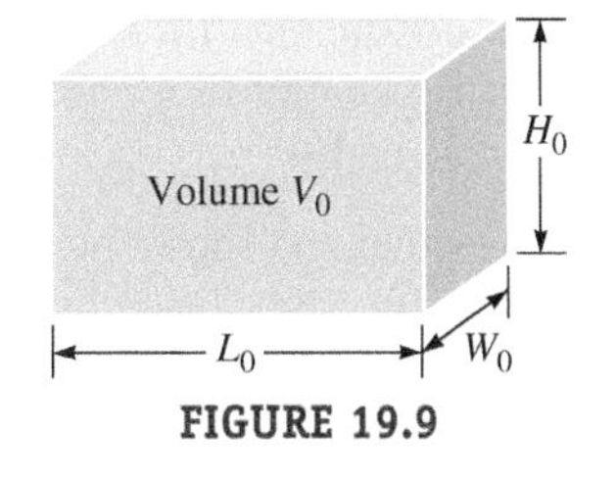

FIGURE 19.9

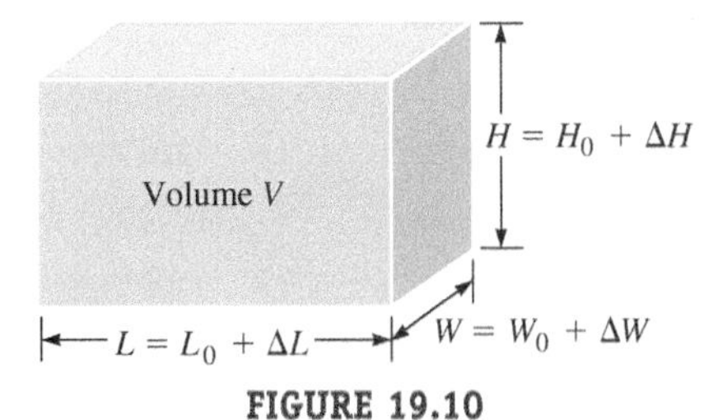

FIGURE 19.10

If the object experiences a temperature change ΔT, each dimension changes according to Equation 19.4. We will show that the change in the object's volume, known as **thermal volume expansion**, is given (approximately) by

THERMAL VOLUME EXPANSION
Major Concept

$$\Delta V \approx \beta V_0 \Delta T \quad (19.5)$$

where β is the coefficient of volume expansion: $\beta \approx 3\alpha$.

When the temperature increases, the object expands. Our sketch (Fig. 19.10) shows what happens when we apply $\Delta L = \alpha L_0 \Delta T$ (Eq. 19.4) to each dimension of the box.

Find the new length L, width W, and height H algebraically for thermal expansion.	$L = L_0 + \Delta L = L_0 + \alpha L_0 \Delta T = L_0(1 + \alpha\,\Delta T)$ $W = W_0 + \Delta W = W_0 + \alpha W_0 \Delta T = W_0(1 + \alpha\,\Delta T)$ $H = H_0 + \Delta H = H_0 + \alpha H_0 \Delta T = H_0(1 + \alpha\,\Delta T)$
The new volume V is the new length multiplied by the new height multiplied by the new width.	$V = LHW = L_0 W_0 H_0 (1 + \alpha\,\Delta T)^3 = V_0(1 + \alpha\,\Delta T)^3$

Derivation continues on page 558 ▶

Expand the cubed term.	$V = V_0[1 + 3\alpha\,\Delta T + 3(\alpha\,\Delta T)^2 + (\alpha\,\Delta T)^3]$
The increase in volume is represented as the difference between the expanded box and the original-size box (Fig. 19.11).	$\Delta V = V - V_0$ FIGURE 19.11
The change in the object's volume is the new volume V minus the original volume V_0.	$\Delta V = V - V_0 = V_0[1 + 3\alpha\,\Delta T + 3(\alpha\,\Delta T)^2 + (\alpha\,\Delta T)^3] - V_0$ $\Delta V = V_0[3\alpha\,\Delta T + 3(\alpha\,\Delta T)^2 + (\alpha\,\Delta T)^3]$
As long as the expansion is much smaller than the object's original size, $\alpha\Delta T \ll 1$, so the terms $(\alpha\,\Delta T)^2$ and $(\alpha\,\Delta T)^3$ are even smaller than $\alpha\,\Delta T$ and can be ignored.	$\Delta V \approx V_0 3\alpha\,\Delta T$
Use the definition of the coefficient of volume expansion, $\beta \approx 3\alpha$.	$\Delta V \approx \beta V_0 \Delta T$ ✓

COMMENT

The coefficient β also depends on the type of material (and temperature); values for some common materials are given in Table 19.2.

TABLE 19.2 Coefficients of volume expansion near room temperature.

Material	β ($\times 10^{-6}$ K^{-1} or °C^{-1})
Aluminum	66–74
Brass	60
Copper	51
Ethanol (liquid)	750
Ethyl and isopropyl alcohol (liquid)	1100
Gasoline (liquid)	950
Glass	12–27
Mercury (liquid)	180
Steel	36
Water (liquid)	210

These values are approximate and can be used for problems in this book if no other values are given.

Several points need to be made about Equations 19.4 and 19.5:

1. If the object cools off, its temperature decreases ($\Delta T < 0$). According to Equations 19.4 and 19.5, the object contracts ($\Delta L < 0$ or $\Delta V < 0$).
2. The coefficients of linear expansion α and volume expansion β depend on the temperature of the object. We ignore that complication in this textbook and simply note that the values in Tables 19.1 and 19.2 are valid when the object is at approximately room temperature.
3. Because liquids and gases conform to the shape of their containers, the linear expansion coefficient has no meaning for liquids or gases. Volume expansion coefficients for several liquids are given in Table 19.2.
4. Equations 19.4 and 19.5 are only valid if ΔL or ΔV are small compared with L_0 or V_0.

The second and fourth points make it difficult to model gases with Equation 19.5 because β for gases has a strong dependence on temperature and because ΔV can be very large for a gas. In Sections 19-5 and 19-6, we will develop a better model for gases.

CONCEPT EXERCISE 19.4

No physical object is truly one-dimensional. Imagine a steel rod with a circular cross section whose temperature increases by $\Delta T = 1$ K. If the length of the rod is 1 m and its radius is 1 cm, find its change in length and radius. Is it reasonable to model such a rod as a one-dimensional object?

EXAMPLE 19.1 CASE STUDY Building an Alcohol Thermometer

In 1709, Fahrenheit invented the alcohol thermometer using "spirit of wine" (also known as ethyl alcohol, ethanol, or grain alcohol) as the expanding fluid. Consider making a simple version of his thermometer using dyed isopropyl alcohol in a container with a straw poking through the top (Fig. 19.12). Fahrenheit put his thermometer into a mixture of water, ice, and salt and marked the level of the alcohol as zero on his scale. If you wish to use the Celsius scale, you would place your thermometer in a bath of water and ice, leaving out the salt. This mixture corresponds to 0°C.

Alcohol boils at 78°C, so if you put your thermometer in boiling hot water, the alcohol will expand up the straw as its temperature rises until it reaches its boiling point. (Such a thermometer cannot operate above 78°C.)

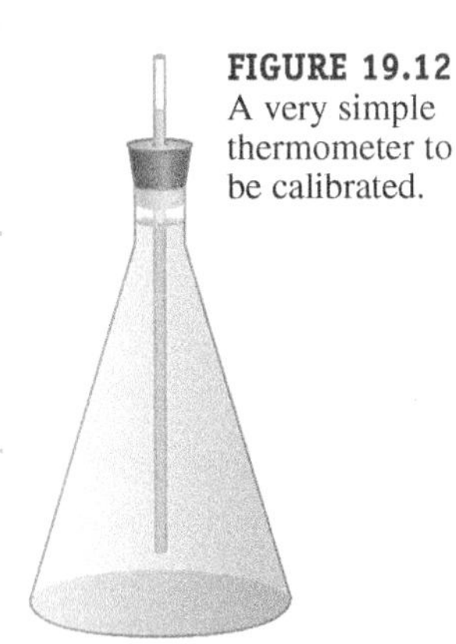

FIGURE 19.12 A very simple thermometer to be calibrated.

A If your thermometer contains an entire bottle of isopropyl alcohol (volume 473 mL at room temperature, 20°C) and the diameter of your straw is 5 mm, find the length of straw required to mark off the boiling point of alcohol. Assume only the alcohol expands appreciably.

INTERPRET and ANTICIPATE

As the alcohol's temperature increases, its volume increases accordingly. Because a liquid's shape is determined by its container, the alcohol will rise up the straw. The thinner the straw, the greater the alcohol's height.

SOLVE

The change in alcohol volume ΔV is proportional to the change in alcohol height Δh in a straw of radius r.	$\Delta V = \pi r^2 \Delta h$
Equation 19.5 relates ΔV to ΔT.	$\Delta V = \beta V_0 \Delta T = \pi r^2 \Delta h$
Solve for Δh. The coefficient of volume expansion is found in Table 19.2. The initial volume is 473 mL = 0.473×10^{-3} m^3. The alcohol's temperature changes from room temperature (T = 20°C) to its boiling point (T = 78°C), so ΔT = 58°C. The radius of the straw is 2.5 mm.	$\Delta h = \dfrac{\beta V_0 \Delta T}{\pi r^2}$ $\Delta h = \dfrac{(1100 \times 10^{-6}{}^\circ\text{C}^{-1})(0.473 \times 10^{-3}\text{m}^3)(58^\circ\text{C})}{\pi(2.5 \times 10^{-3}\text{m})^2}$ $\Delta h = 1.5$ m

CHECK and THINK

Even if the alcohol in the straw is level with the rest of the alcohol at room temperature, the straw would need to be 1.5 m or 5 ft tall to measure the required temperature! Clearly, such a large thermometer is not practical. There are a number of ways to make the thermometer smaller, such as using less alcohol (smaller V_0) and a wider straw (larger r). Fahrenheit did not try to measure such high temperatures with his early thermometer. If the maximum temperature you wished to measure was of a healthy person (only 37°C), our simple thermometer's straw would only need to be about 0.5 m or about 1.5 ft tall.

B Why do you suppose Fahrenheit invented the mercury thermometer 5 years later in 1714? *Hint*: The boiling point of mercury is 357°C.

INTERPRET and ANTICIPATE

Think about the problems you would encounter using an alcohol thermometer: (1) It cannot measure very high temperatures, and (2) it must be very large.

SOLVE

(1) Because the boiling point of mercury is so high, a mercury thermometer can easily measure the temperature of other hot substances such as boiling water.

(2) The volume expansion coefficient for mercury is about six times smaller than that for alcohol (Table 19.2). So, if the alcohol in your thermometer (Fig. 19.12) were replaced with mercury, the straw could be six times shorter. It would only need to be about 0.25 m or about 10 in. tall to measure the required temperature.

CHECK and THINK

Today, alcohol thermometers such as the one in the back of your refrigerator are used to measure relatively cool objects over a relatively small range of temperatures.

Example continues on page 560 ▶

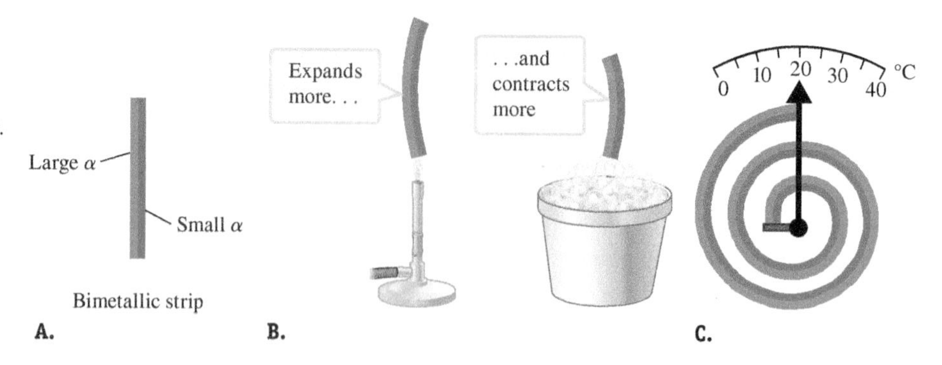

FIGURE 19.13 A. A bimetallic strip consists of two different metals bonded together. **B.** When the strip is heated or cooled, the strip bends because each metal expands or contracts by a different amount. **C.** Bimetallic strips are commonly used in thermometers and thermostats.

CASE STUDY Part 2: Bimetallic Strip Thermometer

In many contemporary applications, a thermometer made of glass and liquid is not practical. One alternative is the bimetallic strip thermometer. The basic physics behind this device is shown in Figure 19.13. Two strips of different metals are bonded together (Fig. 19.13A). Each metal has a different coefficient of linear expansion, so when the bimetallic strip's temperature changes, each metal expands or contracts by a different amount, causing the strip to bend (Fig. 19.13B). In a thermometer, the strip is formed into a coil and attached to a pointer (Fig. 19.13C). A scale is calibrated so that the pointer's position indicates temperature. As the temperature increases, the coil becomes tighter, and the pointer is pushed to the right. As the temperature decreases, the coil becomes looser, and the pointer is pulled to the left.

19-4 Thermal Stress

Figure 19.5 shows an example of the gaps architects and engineers put into their designs to allow structures to expand and contract. What would happen if a designer left out such a gap? Imagine, for example, a steel beam whose ends are free. If the beam's temperature rises, the beam will expand; similarly, if its temperature drops, the beam will contract. Now imagine that the ends of the steel beam are clamped and that the clamps that keep the beam from changing length apply compressive or tensile stress to the beam (Section 14-4) when its temperature changes. Because this stress is a result of the beam's temperature change, it is called **thermal stress.** If the thermal stress is large enough, the structure may crack or burst.

Thermal stress is quantified as the stress needed to keep the beam's length constant when its temperature changes. To find the thermal stress on a beam clamped at both ends, let's first use $\Delta L = \alpha L_0 \Delta T$ (Eq. 19.4) to find the fractional change ε in the beam's length if it were allowed to expand when its temperature increases by ΔT:

$$\varepsilon = \frac{\Delta L}{L_0} = \alpha \, \Delta T \tag{19.6}$$

To keep the beam's length constant, the clamps must apply a compressive stress that produces a strain equal in magnitude to ε. This compressive stress σ is given by $\sigma = Y\varepsilon$ (Eq. 14.11), where Y is Young's modulus. The thermal stress comes from substituting Equation 19.6 into Equation 14.11:

$$\sigma = Y\alpha|\Delta T| \tag{19.7}$$

We have taken the absolute value of ΔT because stress is a magnitude of force per area (Eq. 14.9), but the sign of ΔT determines whether the thermal stress is compressive or tensile. If the temperature increases ($\Delta T > 0$), the thermal stress must be compressive. In this case, if the thermal stress meets or exceeds the compressive strength σ_{cmp} of the beam, it will fracture. A decrease in temperature can also damage a structure. If the temperature decreases ($\Delta T < 0$), the thermal stress must be tensile. In this case, if the thermal stress meets or exceeds the tensile strength σ_{ten} of the beam, it will fracture. The tensile and compressive strengths of various materials are given in Table 14.1 (page 405).

EXAMPLE 19.2 Car Talk

On August 12, 2006, Gerald from Portland, Oregon, called the Tappet brothers on the National Public Radio program *Car Talk*.[1] Gerald had loaned his car to his girlfriend, who had parked it in the direct sunshine on a very hot day ($T \approx 100°F$). The next day, Gerald discovered a crack in his windshield. He told his girlfriend that she was at fault because she had turned on the air conditioner. He asked the Tappet brothers if turning on the air conditioner could have caused the crack. The Tappet brothers said that it was very unlikely for two reasons:

1. The air conditioner *gradually* cools the glass so that "it does not immediately go from 150°F to 60°F."
2. They said, "Windshields are designed to take those kinds of temperature differentials. . . . Most windshields have a bunch of dots around the perimeter or some kind of a transition zone so that where the window does touch the rest of the car there is allowance for expansion and movement of the glass so that it doesn't crack."

A third reason was provided by Gerald's girlfriend:

3. Windshields are made of tempered glass.

The Tappet brothers agreed with Gerald's girlfriend. Tempered glass has a tensile strength up to six times that of regular glass. In addition, tempered glass has a low coefficient of linear expansion α. The Tappet brothers recommended that Gerald apologize to his girlfriend and take her out to dinner.

Assume the temperature of the windshield decreased from 150°F to 60°F and the car manufacturer erred by omitting the transition zone between the glass and the rest of the car.

A Find the thermal stress on the windshield and determine if the glass will break. Use the maximum value of α listed in Table 19.1 for glass. Young's modulus for glass is 70×10^9 N/m^2 (Table 14.1).

INTERPRET and ANTICIPATE

By choosing the maximum value for α and considering the compressive strength of normal glass, we are checking to see if such a temperature change could possibly crack normal glass if the glass were improperly installed. If the thermal tensile stress on the glass due to the air conditioning is less than the tensile strength of the glass, the glass shouldn't break. If normal glass wouldn't crack, tempered glass, which is stronger and has a much lower α, shouldn't crack either.

SOLVE

We need to find the temperature change in kelvins, which is equivalent to finding the temperature change in Celsius degrees. An expression for the change in temperature is derived from Equation 19.2.

$$\Delta T(°C) = T_f(°C) - T_i(°C)$$

$$\Delta T(°C) = \tfrac{5}{9}[T_f(°F) - 32] - \tfrac{5}{9}[T_i(°F) - 32]$$

$$\Delta T(°C) = \Delta T(K) = \tfrac{5}{9}\Delta T(°F) \tag{19.8}$$

$$\Delta T(K) = \tfrac{5}{9}(60°F - 150°F) = -50\ K$$

Calculate the thermal stress from Equation 19.7 and use the given value for Young's modulus. The maximum value of α for glass from Table 19.1 is $9.0 \times 10^{-6}\ K^{-1}$.

$$\sigma = Y\alpha|\Delta T|$$

$$\sigma = (70 \times 10^9\ N/m^2)(9.0 \times 10^{-6}\ K^{-1})|-50\ K|$$

$$\sigma = 3.2 \times 10^7\ N/m^2$$

Because $\Delta T < 0$, the thermal stress on the glass is tensile. Compare the thermal stress to the lowest tensile strength of normal glass found in Table 14.1 (50×10^6 N/m^2). We find that the tensile strength of glass is stronger than the thermal stress. So, the glass should survive the change in temperature.

$$\sigma_{ten} = 50 \times 10^6\ N/m^2 = 5.0 \times 10^7\ N/m^2$$

$$\sigma_{ten} \approx 1.5\sigma$$

Example continues on page 562 ▶

[1]National Public Radio *Car Talk* show number 200632 originally aired on August 12th, 2006; http://www.cartalk.com/ct/review/show.jsp?showid=200632.

CHECK and THINK

The Tappet brothers were correct in thinking that the temperature change would not cause the glass to break. Even if the glass were installed improperly, the thermal stress is small compared to the tensile strength of glass. Taking into account that tempered glass is even stronger and has a lower α, it becomes even less likely that turning on the air conditioner cracked the glass.

B Why did the Tappet brothers give their first reason that using the air conditioner did not crack the windshield (gradual cooling)?

SOLVE

If you have dropped ice cubes into a cup of hot chocolate to cool it off quickly, you have probably noticed that the ice cubes crack. Temperature differences within an object can set up thermal stresses in different parts of the object. As another example, if you pour hot tea into a cold glass, the glass may break because the thermal stress between the cold and hot parts of the glass exceeds its strength. The Tappet brothers' first reason is a statement that the windshield's temperature remains relatively uniform as it cools, so they don't expect significant thermal stress within the glass itself. In the program transcript, one of the brothers says that turning on the air conditioner is "not like someone is throwing dry ice on the windshield from the inside."

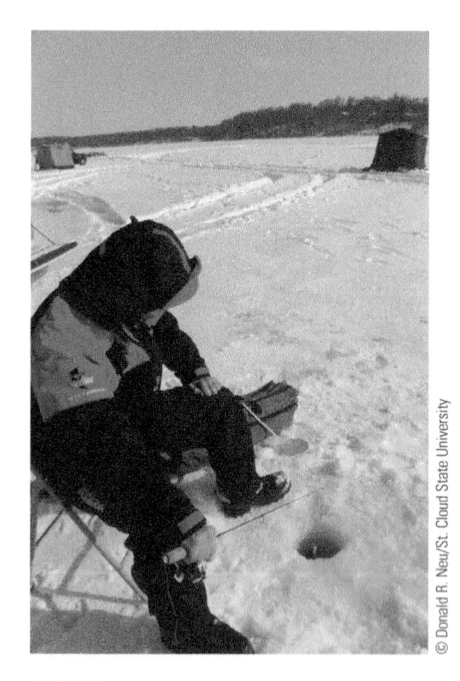

FIGURE 19.14 The water on the surface of a lake freezes and insulates the water below, so the water below remains liquid. Life is safe below the ice, except for the fish who take the bait.

Thermal Expansion of Water

Most materials contract as they are cooled. That is, most coefficients of volume expansion β in $\Delta V = \beta V_0 \Delta T$ (Eq. 19.5) are positive numbers, so if the material is cooled, ΔT is negative and so is ΔV. Water—perhaps the most importance substance for life on the Earth—does not always contract as its temperature drops, however.

When water is cooled from 4°C to its freezing point at 0°C, it expands rather than contracts. When water is between 0°C and 4°C, β is negative, so a decrease in temperature results in an increase in volume. When the water's temperature is above 4°C, β is positive, and a decrease in temperature means a decrease in volume. (Water ice contracts as its temperature decreases.) Therefore, the density of liquid water is greatest at 4°C.

This somewhat odd property of water has important implications for life (Fig. 19.14). Imagine a lake in the winter when the air temperature drops below 0°C. The water at the surface of lake is in contact with the cold air, so it cools first. When the water below the surface is above 4°C, the colder water at the surface sinks because it is denser than the water below. The warmer water from below rises up to the surface, where it in turn cools and sinks. This process continues until the entire lake is at 4°C and of uniform density. As the water at the surface continues to cool, it expands and becomes less dense. Now this colder water remains on the surface and continues to cool. When the lake's surface temperature reaches 0°C, the water at that surface turns to ice, which is less dense than water and therefore floats at the top. The ice on the surface helps insulate the water below, which remains liquid with a temperature of about 4°C. Some plants and animals survive the winter in this cold liquid water.

If water were like other substances and continually contracted as it cooled, the lake would freeze from the bottom up, and the entire lake would probably turn to ice. Plants and animals could not survive the winter in solid ice.

CONCEPT EXERCISE 19.5

Suppose you wish to make sure there is a very tight fit between a metal peg and a hole. The hole is just a little too small for the peg to fit. Should you try to heat the peg or cool it before inserting it?

EXAMPLE 19.3 Holey Sheet Metal

Three holes are drilled into a rectangular sheet of aluminum when its temperature is 95°C. The dimensions of the holes and the aluminum are given in Figure 19.15. The aluminum is cooled, and its new length is $L_f = 4.990$ m. (Assume $\alpha = 22.2 \times 10^{-6}\ \text{K}^{-1}$.)

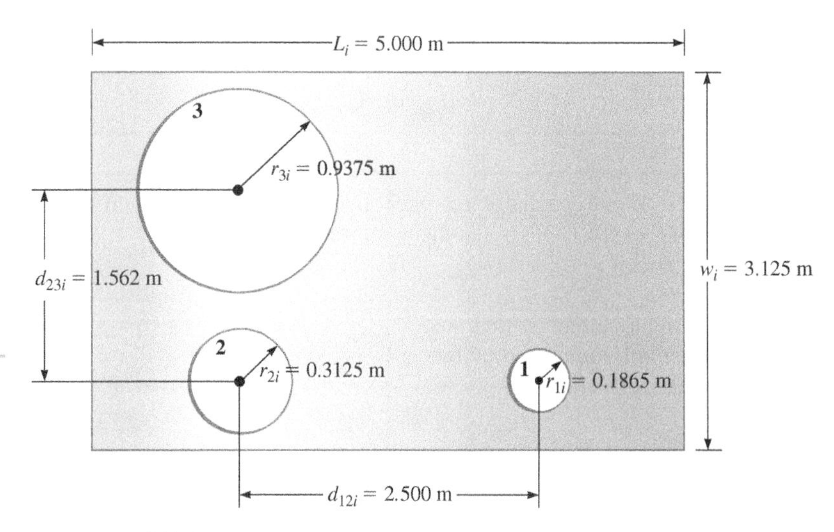

FIGURE 19.15 Three holes in a sheet of aluminum at 95°C.

A Find the final (cooler) temperature of the aluminum.

INTERPRET and ANTICIPATE

The aluminum sheet contracts as it cools. Its new dimensions are determined by the coefficient of linear expansion for aluminum and the change in temperature, and we can use this information to find the new temperature.

SOLVE

Start by solving Equation 19.4 for the new temperature T_f.

$$\Delta L = \alpha L_i \Delta T$$

$$\Delta T = \frac{\Delta L}{\alpha L_i} = \frac{L_f - L_i}{\alpha L_i}$$

$$T_f = \frac{L_f - L_i}{\alpha L_i} + T_i$$

Substitute values.

$$T_f = \frac{(4.990\ \text{m}) - (5.000\ \text{m})}{(22.2 \times 10^{-6}\,°\text{C}^{-1})(5.000\,\text{m})} + 95.0°\text{C} = 4.9°\text{C}$$

CHECK and THINK

The temperature of the aluminum must drop by 90.1°C for its length to decrease by 1 cm. This temperature change is nearly the same as that required to go from boiling water to ice. Does such a great change in temperature make sense? Think about baking with an aluminum cookie sheet. Initially, the sheet is at room temperature, about 20°C. The oven temperature is about 400°F or about 200°C, so the temperature change is about 180°C. Have you ever noticed that the sheet is larger when you take it out of the oven? You probably haven't because the change in size of the sheet is proportional to its size, which is small (only about 0.5 m long), and the change is too small to notice. For the much larger aluminum sheet in this example, the temperature change is great enough to get a noticeable change in size.

B Find the change in width Δw of the rectangular sheet; the change in radius Δr_1, Δr_2, and Δr_3 of each circular hole; and the change in hole separations Δd_{12} and Δd_{23} between the warmer and cooler temperatures.

INTERPRET and ANTICIPATE

The entire aluminum sheet with its holes experiences the same change in temperature ($\Delta T = -90.1$°C). According to Equation 19.4, the change in size ΔL depends on the original length L_0, so we don't expect the width of the rectangular aluminum sheet to change by as much as the length: $|\Delta w| < |\Delta L|$. Likewise, we expect that the distance between holes 1 and 2 should change more than the distance between holes 2 and 3: $|\Delta d_{12}| > |\Delta d_{23}|$.

Perhaps the most difficult prediction to make is whether the holes contract or expand when the temperature decreases. Think about the aluminum disks that were punched out to make the holes. At the high temperature T_i, each aluminum disk fits perfectly into its respective hole. If the holes were not punched out and the aluminum sheet were cooled to T_f, those disks would still be part

Example continues on page 564 ▶

of the aluminum sheet and would still fit perfectly into their holes. Now imagine punching out the holes when the aluminum is hot at T_i and then cooling all the aluminum, including the disks, to temperature T_f. The aluminum disks must still fit perfectly inside their respective holes. Because the disks must contract when they are cooled, the holes must also contract by the same amount. Finally, we expect that the largest hole should experience the greatest change, so we predict that $|\Delta r_3| > |\Delta r_2| > |\Delta r_1|$.

SOLVE

Use Equation 19.4 repeatedly for each linear dimension. The fractional length change ε is the same for each part of the sheet, so find that value first.

$$\Delta L = \alpha L_i \Delta T \qquad \varepsilon = \frac{\Delta L}{L_i} = \alpha\,\Delta T$$

$$\varepsilon = (22.2 \times 10^{-6}\,{}^\circ\mathrm{C}^{-1})(-90.1^\circ\mathrm{C}) = -2.00 \times 10^{-3}$$

The change in any length is then just the original length multiplied by the fractional change.

$$\Delta w = w_i\varepsilon = (3.125\ \mathrm{m})(-2.00 \times 10^{-3}) = -6.25 \times 10^{-3}\ \mathrm{m} = -6.25\ \mathrm{mm}$$

$$\Delta d_{12} = d_{12,i}\varepsilon = (2.500\ \mathrm{m})(-2.00 \times 10^{-3}) = -5.00 \times 10^{-3}\ \mathrm{m} = -5.00\ \mathrm{mm}$$

$$\Delta d_{23} = d_{23,i}\varepsilon = -3.12\ \mathrm{mm}$$

$$\Delta r_1 = r_{1i}\varepsilon = (0.1865\ \mathrm{m})(-2.00 \times 10^{-3}) = -3.73 \times 10^{-4}\ \mathrm{m} = -0.373\ \mathrm{mm}$$

$$\Delta r_2 = r_{2i}\varepsilon = -0.625\ \mathrm{mm}$$

$$\Delta r_3 = r_{3i}\varepsilon = -1.88\ \mathrm{mm}$$

CHECK and THINK

Check the predictions we made for the relative changes.

$\lvert\Delta w\rvert < \lvert\Delta L\rvert$	0.625 mm < 1 cm ✓
$\lvert\Delta d_{12}\rvert > \lvert\Delta d_{23}\rvert$	5.00 mm > 3.12 mm ✓
$\lvert\Delta r_3\rvert > \lvert\Delta r_2\rvert > \lvert\Delta r_1\rvert$	1.88 mm > 0.625 mm > 0.373 mm ✓

19-5 Gas Laws

In the previous sections, we focused on liquids and solids; gases require a different approach. Usually, a gas—such as the air in your room—fills its container. Under the right conditions, a gas expands when its temperature rises and contracts when its temperature drops. If you place a balloon filled with air in liquid nitrogen, which has a temperature of 77 K (−196°C), the balloon collapses, and if you then allow the balloon to return to room temperature, it returns to its previous size (Fig. 19.16). As the air temperature changed its volume could change because its container—a balloon—is flexible. (A well-sealed, inflexible container may break when the air inside expands or contracts.)

So, $\Delta V = \beta V_0 \Delta T$ (Eq. 19.5) is not a good model for gas volume changes because it does not take into account the interrelationships between temperature, pressure,

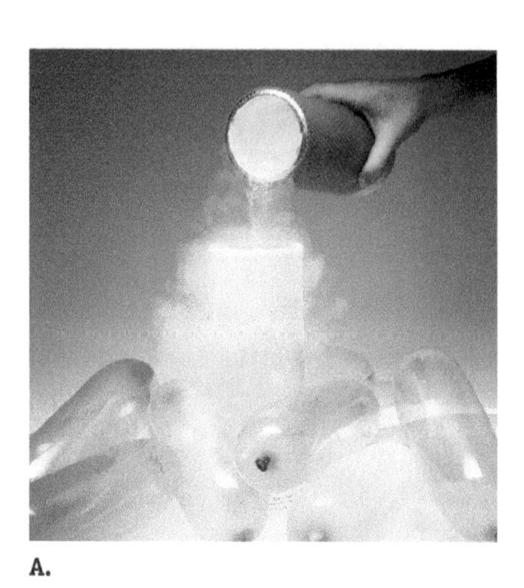

A.

B.

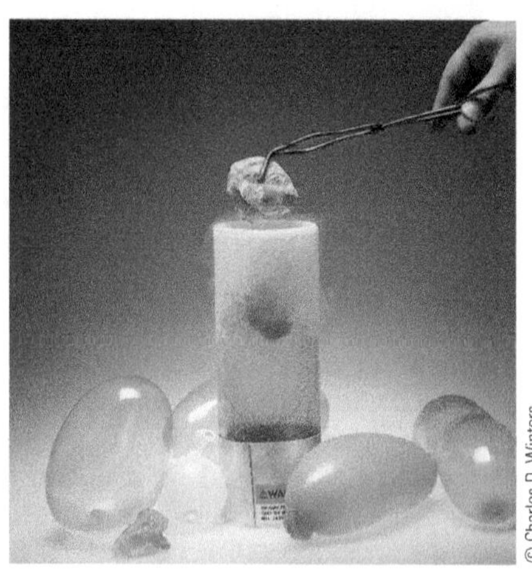

C.

FIGURE 19.16 When balloons are placed in a flask of liquid nitrogen, their temperature drops and so does their volume. When the balloons are removed from the liquid nitrogen, their temperature and volume are restored.

volume, and the amount of a gas. We need to find an equation or series of equations that relate these macroscopic physical variables; such an equation is called an **equation of state**. The state of a gas is a list of its physical properties (for example, temperature, pressure, and volume) that can vary. We study gases that are in an **equilibrium state**, meaning that the physical properties are at least temporarily constant and uniform throughout the gas. The equations of state described in this section were found experimentally and are valid for gases that are not very dense and are at approximately atmospheric pressure.

EQUATION OF STATE
Major Concept

EQUILIBRIUM STATE
Major Concept

Consider again the balloons in Figure 19.16. When the air temperature in a balloon is low, the balloon's volume is small. The air pressure is constant, approximately atmospheric pressure. In the late 1700s, French scientist Jacques Charles carefully carried out experiments on gases held at constant pressure. He found that gas volume under such conditions is proportional to its temperature. **Charles's law** is written as

$$V \propto T \quad \text{(if } P \text{ is constant)} \tag{19.9}$$

CHARLES'S LAW
Special Case

or

$$\frac{V_f}{V_i} = \frac{T_f}{T_i} \quad \text{(if } P \text{ is constant)} \tag{19.10}$$

Charles never published his results. They were published by another French scientist, Joseph Louis Gay-Lussac, in the early 1900s. Gay-Lussac conducted his own experiments and found that if the gas volume is held constant using a container with rigid walls such as a scuba tank, pressure is proportional to temperature expressed in kelvins. **Gay-Lussac's law** is expressed as

$$P \propto T \quad \text{(if } V \text{ is constant)} \tag{19.11}$$

GAY-LUSSAC'S LAW
Special Case

or

$$\frac{P_f}{P_i} = \frac{T_f}{T_i} \quad \text{(if } V \text{ is constant)} \tag{19.12}$$

Charles's law is valid for a gas held at constant pressure, and Gay-Lussac's law is valid for a gas held at constant volume, but what happens if the temperature is held constant and the pressure and volume are allowed to change? To answer, imagine a squeezing a balloon full of air. As you decrease the volume, the air pressure goes up. The absolute pressure (Eq. 15.14) and volume are inverses of each other, as found by Irish chemist Robert Boyle through careful experimentation in the mid-1600s. **Boyle's law** may be written as

$$P \propto \frac{1}{V} \quad \text{(if } T \text{ is constant)} \tag{19.13}$$

BOYLE'S LAW
Special Case

or

$$P_f V_f = P_i V_i \quad \text{(if } T \text{ is constant)} \tag{19.14}$$

Equations 19.9 through 19.14 are all called *laws*, but they are really *empirical fits* to experimental data. Empirical fits model the data but do not attempt to explain the causes (as a theory does). Kepler's *laws* of orbital motion are really empirical fits as well (Chapter 7). Kepler's laws fit Tycho's celestial data, but they do not provide a theoretical explanation for the planets' motion. Such explanation comes from Newton's laws of motion and gravity. In Chapter 20, we will provide a theoretical basis for the gas laws. In the next section, we combine the empirical fits in Equations 19.9 through 19.14 into a single equation.

CONCEPT EXERCISE 19.6

Sketch the following graphs:

a. V as a function of T for Charles's law
b. P as a function of T for Gay-Lussac's law
c. P as a function of V for Boyle's law

19-6 Ideal Gas Law

Each gas law from the previous section (due to Charles, Gay-Lussac, and Boyle) was discovered by holding one of three variables (pressure, volume, or temperature) constant while allowing the other two to change. In many practical situations, none of those three variables is constant. For example, if you fill a balloon with helium gas and let it go, it will rise high into the atmosphere where it will experience lower pressure, larger volume, and colder temperatures. All three variables change, and we must use a combination of Charles's, Gay-Lussac's, and Boyle's laws to arrive at an equation of state. Mathematically, the combination is

$$PV \propto T \tag{19.15}$$

Simply combining these three gas laws doesn't describe the state of all gases because Equation 19.15 is only valid for a fixed amount of a particular gas; it doesn't tell you anything about the state of the gas if the *amount* of gas changes or what the state would be for a different *type* of gas. For instance, what happens to the state of helium in a balloon if some of it leaks out? And how does the state of a helium-filled balloon compare with the state of an oxygen-filled balloon? The goal in this section is to come up with one gas law that takes into account the amount and type of gas.

How do we measure the *amount* of gas? There are several ways, but one particularly important measure is a count of the number of particles N in the gas. By *number of particles*, we mean either the number of atoms or the number of molecules, depending on whether the gas is monatomic (like He) or polyatomic (like water vapor, H_2O).

FIGURE 19.17 If you remove much of the air from this storage bag, the bag's volume decreases.

Your experience tells you the number of particles in a gas is proportional to its volume. For example, before sealing a storage bag, you remove as much of the air as possible (Fig. 19.17). Removing air reduces the volume. This simple procedure illustrates that a gas held at constant temperature and pressure (room temperature and atmospheric pressure in this case) has a volume proportional to the amount of gas.

Now, imagine two balloons filled to the same volume at the same temperature and pressure, but one balloon is filled with oxygen and the other with helium. Will these two balloons have the same number of particles? In 1811, Amedeo Avogadro, an Italian scientist, came up with an insightful hypothesis that deals with the *type* of gas. According to **Avogadro's hypothesis** (now also known as **Avogadro's law**), *equal volumes of gas at the same temperature and pressure contain the same number of particles independent of the type of gas.* Mathematically, we can write Avogadro's law as

AVOGADRO'S LAW ▶ Special Case

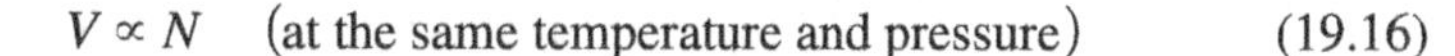

$$V \propto N \quad \text{(at the same temperature and pressure)} \tag{19.16}$$

So, because the oxygen balloon and the helium balloon have the same volume (temperature and pressure), they contain equal numbers of gas particles.

We now combine Charles's, Gay-Lussac's, and Boyle's laws with Avogadro's law to arrive at the **ideal gas law**:

IDEAL GAS LAW ▶ Special Case

$$PV = Nk_BT \tag{19.17}$$

where k_B is **Boltzmann's constant**, P is absolute pressure, and T is absolute temperature. Boltzmann's constant has the same value for all gases:

$$k_B = 1.38 \times 10^{-23}\,\text{J/K}$$

IDEAL GAS ✪ Major Concept

An **ideal gas** is a hypothetical gas made up of identical particles that do not interact with one another. In practice, a real gas may be modeled as an ideal gas if its density is low, its pressure is less than about one atmosphere, and the gas temperature is not near its liquefaction point (boiling point of the liquid). The ideal gas law (Eq. 19.17) is also known as the **equation of state for an ideal gas** because it relates P, V, and T. In Section 20-6, we present slightly modified versions of Equation 19.17 that models the equation of state for a real gas under nonideal conditions. For the rest of this chapter, assume all gases can be modeled by the ideal gas law.

CONCEPT EXERCISE 19.7

The mass of an oxygen molecule is 5.3×10^{-26} kg, and that of a helium atom is 6.6×10^{-27} kg. One balloon is full of oxygen gas (O_2), and another is full of helium gas (He). If the balloons are at the same temperature and pressure and have the same volume, find the ratio of their masses and densities.

Avogadro's Number

When an administrative assistant of a large company orders office supplies, he orders pencils by the *gross*. One gross of pencils equals 12 dozen, or 144 pencils. The gross is a convenient unit for business because it is the correct size for what a company needs, and it is far more convenient than specifying some number of individual pencils and figuring out the cost of that special order.

MOLE Major Concept

Counting individual molecules in a gas is far more inconvenient than counting pencils, so it is more convenient to count in *moles*. The **mole** (abbreviated "mol") is the SI unit for the number of particles in a substance. One mole is defined as the amount of substance containing as many particles as there are atoms in 12 g of carbon-12. Exactly 12 g of carbon-12 contain **Avogadro's number** N_A of atoms, where

Avogadro never measured N_A. The quantity was measured in the middle of the twentieth century and is named to honor him.

$$N_A = 6.022 \times 10^{23}$$

AVOGADRO'S NUMBER Major Concept

A mole is like a dozen or a gross in that a dozen means 12 of something, a gross means 144 of something, and a mole means 6.022×10^{23} of something. If there is a mole of oxygen gas in a balloon, there are 6.022×10^{23} oxygen molecules in that balloon.

The ideal gas law can be written in terms of the number of moles n instead of the number of molecules N. The number of moles comes from dividing the number of molecules by Avogadro's number:

$$n = \frac{N}{N_A} \quad (19.18)$$

Solving for N and substituting into the ideal gas law (Eq. 19.17) gives

$$PV = nN_A k_B T \quad (19.19)$$

Avogadro's number multiplied by Boltzmann's constant, $N_A k_B$, is usually combined to form another constant R known as the **universal gas constant**:

$$R = N_A k_B$$
$$R = 8.315 \text{ J/(mol} \cdot \text{K)} \quad (19.20)$$

Substituting Equation 19.20 into Equation 19.19, we can write the ideal gas law in terms of the universal gas constant and the number of moles:

$$PV = nRT \quad (19.21)$$

In practice, the number of moles n of a gas is often determined from its mass. To determine this number, you must first look up the **molar mass** M_{mol} for that type of gas. The molar mass is the mass of 1 mole of particles. Although the SI units of molar mass are kilograms per mole, we usually work in grams per mole for convenience.

The molar mass in grams per mole can be found on the Periodic Table of Elements in Appendix B. Usually, the key on a periodic table gives each element's **atomic mass**—the mass of one atom—in atomic mass units; $1 \text{ u} = 1.6605402 \times 10^{-27}$ kg. Molar mass in grams per mole is numerically equivalent to the atomic mass in atomic mass units, u.

EXAMPLE 19.4 How Much Is in the Balloon?

By weighing a balloon both before and after it is filled with molecular oxygen gas (O_2), a student determines that the mass of the oxygen is 18.3 g. According to the periodic table (Appendix B), the atomic mass of oxygen is 15.999 u. Find the number of moles n of oxygen molecules and the number of molecules N in the balloon.

Example continues on page 568 ▶

INTERPRET and ANTICIPATE

The molar mass of molecular oxygen (O_2) is twice the molar mass of atomic oxygen (O) found in the periodic table. Divide the mass of oxygen in the balloon by the molar mass of molecular oxygen to find the number of moles. The number of molecules comes from multiplying the number of moles by Avogadro's number, the number of molecules per mole.

SOLVE

First, find the molar mass of molecular oxygen by multiplying the molar mass of atomic oxygen by 2.	$M_{\text{mol}} = 2(15.999 \text{ g/mol}) = 31.998 \text{ g/mol}$
Find the number of moles of oxygen gas in the balloon by dividing the mass of the gas M_{oxygen} by its molar mass M_{mol}.	$n = \frac{M_{\text{oxygen}}}{M_{\text{mol}}} = \frac{18.3 \text{ g}}{31.998 \text{ g/mol}} = 0.572 \text{ mol}$
Use Equation 19.18 to find the number of molecules.	$n = \frac{N}{N_A} \qquad N = N_A n$ $N = (6.022 \times 10^{23} \text{ molecules/mol})(0.572 \text{ mol})$ $N = 3.44 \times 10^{23} \text{ molecules}$

CHECK and THINK

Our result makes sense. The molar mass of molecular oxygen is about 32 g. Therefore, if the mass of oxygen in the balloon were 32 g, there would be 1 mole of oxygen in the balloon, and the number of molecules would equal Avogadro's number. The mass of oxygen in the balloon is just over half the molar mass of molecular oxygen, however, so the number of molecules N is just over half of Avogadro's number.

EXAMPLE 19.5 Standard Temperature and Pressure (STP)

Standard temperature and pressure (STP) is defined to be 0°C and 1 atm. Find the volume of 1 mol of an ideal gas at STP. Give your answer in cubic meters (m^3) and liters (L).

INTERPRET and ANTICIPATE

Because this example involves an ideal gas, we can use either expression of the ideal gas law (Eq. 19.17 or 19.21). The expression $PV = nRT$ is easier to use here because it is written in terms of moles, one of the parameters given in the example.

SOLVE

Set $V = V_{\text{STP}}$ and solve Equation 19.21 for volume.	$PV_{\text{STP}} = nRT \qquad V_{\text{STP}} = \frac{nRT}{P}$
Convert the given pressure and temperature to SI units and find V_{STP}. Then, convert V_{STP} to liters using $1000 \text{ L} = 1 \text{ m}^3$.	$P = 1 \text{ atm} = 1.0133 \times 10^5 \text{ Pa}$ $T = 0°\text{C} = 273.15 \text{ K}$ $V_{\text{STP}} = \frac{(1 \text{ mol})(8.315 \text{ J/(mol} \cdot \text{K}))(273.15 \text{ K})}{1.0133 \times 10^5 \text{ Pa}}$ $V_{\text{STP}} = 2.241 \times 10^{-2} \text{ m}^3 = 22.41 \text{ L}$

CHECK and THINK

Our result says that 1 mol of any gas at STP fills a volume equal to about the size of ten 2-L soda bottles. This handy result makes future problem solving easier. If a gas is at STP and you know the number of moles of that gas, you can find its volume by multiplying the number of moles n by V_{STP}. For example, if the balloon in Example 19.4 is at STP, its volume is $0.572 \times 22.41 \text{ L} = 12.8 \text{ L}$ or about six soda bottles.

EXAMPLE 19.6 Tire Blowouts

To help prevent blowouts, manufacturers of car and bicycle tires often recommend keeping the tires at a slightly lower pressure during the hottest part of the summer. When the tire is in motion, friction between it and the road can raise the temperature of the air in the tire. The increased temperature can change the air volume and pressure.

Racing slicks are large, smooth tires used on certain types of race cars. Unlike the tires on an ordinary car, the air density in a racing slick is meant to be low. Often, wrinkles appear in the racing slick when the car is near the starting line (Fig. 19.18). Soon after the car speeds along the track, the tires appear firm and round. When the car is racing, the air in the slicks increases in temperature, volume, and pressure.

FIGURE 19.18 At the starting line, a racing slick shows wrinkles. When the tire warms up, it inflates.

A Assume the air temperature in the slick before the race begins is 15°C and soon after the race begins rises to 45°C. What is the fractional increase in the quantity PV for air in the tire?

INTEPRET AND ANTICIPATE

We model the air in the slick as an ideal gas, which is probably valid because the density is low. Because the temperature increases, we expect that the quantity PV will also increase.

SOLVE

Use the subscript f for the parameters during the race and the subscript i for the parameters before the race. Write down the ideal gas law (Eq. 19.17 or 19.21) once for each set of parameters. (We have arbitrarily used Eq. 19.17.) The amount of air in the slick does not change as long as there are no leaks. So, when the quantities during the race are divided by the quantities before the race, N cancels.

$$(PV)_f = Nk_BT_f \quad (19.17)$$

$$(PV)_i = Nk_BT_i$$

$$\frac{(PV)_f}{(PV)_i} = \frac{T_f}{T_i} \quad (1)$$

The fractional change in a quantity is the quantity's final value minus its initial value divided by its initial value.

$$\frac{(PV)_f - (PV)_i}{(PV)_i} = \frac{(PV)_f}{(PV)_i} - 1 \quad (2)$$

Substitute Equation (1) into Equation (2) and calculate the fractional change in PV. Because the ideal gas law is written in SI units, we must convert all temperatures to kelvins.

$$\frac{(PV)_f - (PV)_i}{(PV)_i} = \frac{T_f}{T_i} - 1$$

$$\frac{(PV)_f - (PV)_i}{(PV)_i} = \frac{(45 + 273.15)\text{K}}{(15 + 273.15)\text{K}} - 1 = 0.104$$

CHECK and THINK

As expected, the fractional change in PV is positive (meaning that the quantity has increased).

B For the cars that most of us drive, there is usually only a small change, if any, in the volume of air in the tires, and an increase in temperature mostly means a change in pressure. In racing slicks, the change in volume is small but not negligible. If the tire volume increases by 5% and the initial gauge pressure in the tire is 55.2 kPa (about 8 psi), what is the gauge pressure during the race?

INTERPRET and ANTICIPATE

A 5% increase in volume can be interpreted as a fractional increase of 0.05. In part A, we found the fractional increase in the quantity PV. We can use that result to find the new pressure. If the volume did not change at all, we would expect the fractional increase in pressure to equal the fractional increase in temperature, which is 0.104. Because the volume increases, we expect the fractional increase in pressure to be less than 0.104. The ideal gas law is written in terms of absolute pressure and not gauge pressure, so we must find the gauge pressure from the absolute pressure.

Example continues on page 570 ▶

SOLVE Use the result from part A to find an expression for P_f. The result will be an absolute pressure.	$\frac{(PV)_f - (PV)_i}{(PV)_i} = \frac{(PV)_f}{(PV)_i} - 1 = 0.104$ $\frac{P_f V_f}{P_i V_i} = 1.104$ $P_f = 1.104\left(\frac{V_i}{V_f}\right)P_i$ (3)
Use the fractional change in V to find a value for (V_i/V_f).	$\frac{V_f - V_i}{V_i} = \frac{V_f}{V_i} - 1 = 0.05$ $\frac{V_i}{V_f} = \frac{1}{1.05}$
From Equation 15.15, absolute pressure is the sum of the gauge pressure and atmospheric pressure. Calculate the initial absolute pressure and substitute values into Equation (3).	$P_i = 55.2 \times 10^3\ \text{Pa} + 1.013 \times 10^5\ \text{Pa} = 1.565 \times 10^5\ \text{Pa}$ $P_f = (1.104)\left(\frac{1}{1.05}\right)(1.565 \times 10^5\ \text{Pa}) = 1.65 \times 10^5\ \text{Pa}$
Find the gauge pressure during the race by subtracting the atmospheric pressure from P_f.	$P_{\text{gauge}} = P_f - P_0$ from Equation 15.15 $P_{\text{gauge}} = 1.65 \times 10^5\ \text{Pa} - 1.013 \times 10^5\ \text{Pa}$ $P_{\text{gauge}} = 6.325 \times 10^4\ \text{Pa}$
CHECK and THINK To check our result, let's calculate the fractional increase in the absolute pressure.	$\frac{P_f - P_i}{P_i} = \frac{1.65 \times 10^5\ \text{Pa} - 1.565 \times 10^5\ \text{Pa}}{1.565 \times 10^5\ \text{Pa}} = 0.0543$

As expected, the fractional increase in absolute pressure is less than 0.104. The gauge pressure increases from about 8 psi to about 9 psi, an increase of roughly 13%. In the cars we normally drive, such an increase would mean that a cool gauge pressure of about 30 psi would increase to nearly 34 psi after driving. The recommended pressure given by the tire manufacturer is meant to be measured when the tire is cool, so you should not drive your car for more than about 1 mile before measuring its gauge pressure.

19-7 Temperature Standards

The Kelvin temperature scale (Section 19-1) is named for Lord Kelvin, born in 1824 in Belfast, Ireland, and originally named William Thomson. Kelvin, an impressive scientist and engineer, was given his title of nobility for his success in laying a submarine telegraph cable from England to France in 1866. (Kelvin is the name of the river that runs past the university in Glasgow, Scotland, where he worked.) In 1848, Kelvin published an article outlining the need for an absolute temperature scale.

To understand the problems Kelvin raised in his article, imagine two different laboratories, one using an alcohol thermometer and the other using a mercury thermometer. Each thermometer is calibrated according to the Celsius scale so that the freezing point of water is at 0°C and the boiling point of water is at 100°C. The two thermometers do not necessarily agree at intermediate temperatures, however. The mercury thermometer may measure a person's temperature as 36.5°C, whereas the alcohol thermometer may measure the same temperature as 37.8°C. The researchers are left wondering which temperature, if either, is correct. Kelvin wrote that researchers in different laboratories must be able to make independent temperature measurements that agree with one another and that an absolute temperature scale must not depend on the material used in the thermometer. Today, these difficulties have been overcome by (1) the absolute temperature scale in use by agreement of the international scientific community and (2) the use of constant-volume ideal gas thermometers.

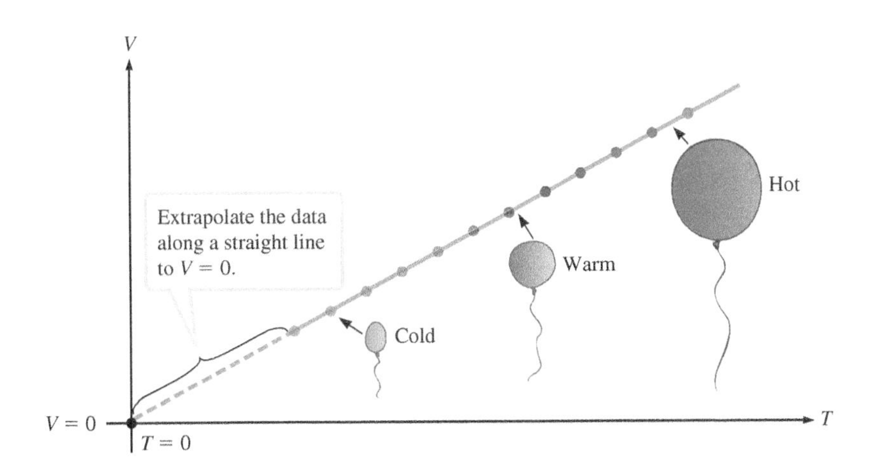

FIGURE 19.19 A graph of volume as a function of temperature shows that as the temperature of a balloon decreases, so does its volume. These data follow a straight line, and *absolute zero* is found by extrapolating the line back to zero volume.

KELVIN SCALE ✪ Major Concept

The **Kelvin scale** (also known as the **absolute scale**) in today's SI system of units is based on the principles laid out by Lord Kelvin in the mid-1800s. Kelvin used Charles's law (Eq. 19.9) to define zero on the absolute scale. According to Charles's law, as gas temperature decreases, its volume decreases. Kelvin reasoned that there must be a limit to how cold a gas could be because the gas volume could not be negative. He took zero to be the theoretical limit of any gas's volume. For a balloon filled with gas at constant pressure, a graph of the volume as a function of temperature would fit a line according to Charles's law (Fig. 19.19). Kelvin's idea was to extrapolate that line back to zero volume and set the corresponding temperature to zero on the absolute scale ($T = 0$ K). This temperature is known as **absolute zero**, equal to –273.15°C. No matter what type of gas you use, your data for volume as a function of temperature would fit a line extrapolating back to absolute zero ($T = -273.15$°C).

ABSOLUTE ZERO ✪ Major Concept

This definition of absolute zero does not require that any actual gas or object be at absolute zero. In fact, no measurement of absolute zero has been made. Three researchers—William D. Phillips, Steven Chu, and Claude Cohen-Tannoudji—won the Nobel Prize in Physics in 1997 for cooling atoms to about 900 nK, however, and in 2003, a team of researchers from the Massachusetts Institute of Technology reported that they had lowered a gas's temperature to 500 pK.[2] We will return to the topic of absolute zero in later chapters, but for now it is sufficient to know that absolute zero is the lowest temperature that is theoretically possible.

Because the lowest possible temperature is zero on the absolute temperature scale, only one reference temperature is needed to calibrate an actual thermometer. The reference temperature used for the Kelvin scale is the temperature of the **triple point of water,** the pressure and the temperature at which water exists in three phases: solid, liquid, and gas. The triple point of water is a condition that is more precisely reproducible than either the boiling point or freezing point of water. For example, because water boils at a lower temperature when it is at a lower pressure, people who live at high altitude (greater than 3000 ft) must modify the cooking instructions when they boil a pot of rice or pasta. Likewise, researchers in a laboratory located in the mountains would have a hard time calibrating a thermometer based on the boiling point of water because it would boil off before reaching 100°C.

Water can only exist in all three phases at once when the pressure and the temperature are at particular values. The triple point of water is defined to have a temperature $T = 273.16$ K $= 0.01$°C and requires that the water vapor pressure be 610 Pa, roughly 6×10^{-3} atm. So, when you see water in all three phases, you know the temperature (and pressure) precisely.

The triple point requires a fairly low pressure. To maintain and establish the required temperature and pressure, laboratories use a **triple-point cell** as shown in Figure 19.20. A triple-point cell is a glass tube partially filled with a mixture of water and ice, above which is water vapor. The tube has a large depression so that a thermometer bulb may be brought into thermal contact with the cell.

A.

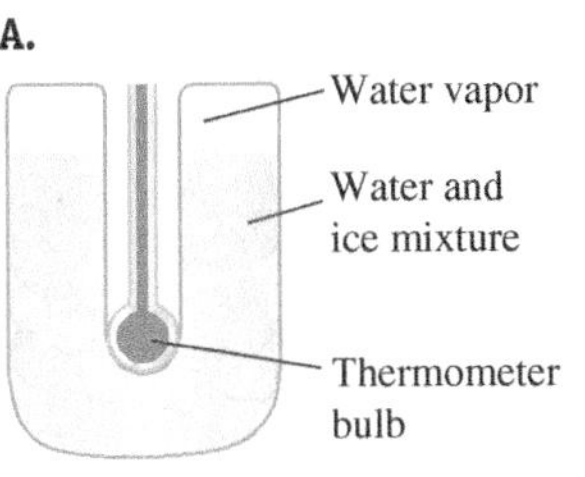

B.

FIGURE 19.20 Triple-point cells are used to keep water at its triple point: a mixture of solid, liquid, and gas.

---

[2]See A. E. Leanhardt et al., "Cooling Bose-Einstein Condensates Below 500 Picokelvin," *Science* 301, no. 5639 (2003): 1513–1515.

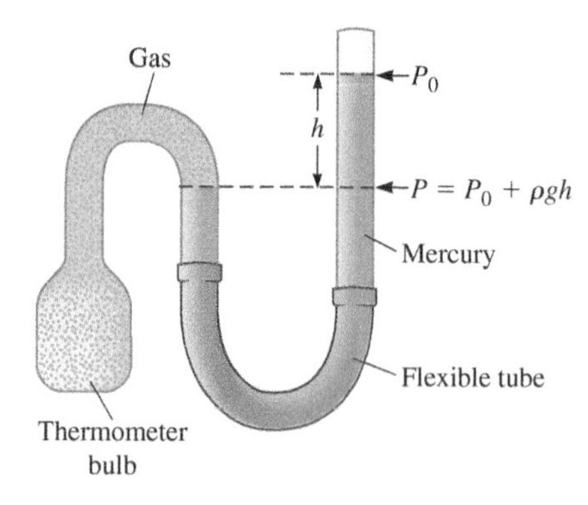

FIGURE 19.21 In a constant-volume gas thermometer, a gas is held at a constant volume so that a change in temperature means that only its pressure can change. The change in pressure is measured by a mercury manometer that may be calibrated in terms of gas temperature.

CASE STUDY Constant-Volume Gas Thermometer

In a laboratory where temperature must be measured carefully, a constant-volume ideal gas thermometer is used to calibrate the Kelvin scale. This thermometer is the final device in our case study on the construction of thermometers. A **constant-volume ideal gas thermometer** is shown in Figure 19.21. The bulb contains a low-density gas and is connected to a mercury manometer (Section 15-5). We'll assume such a gas is well modeled as an ideal gas. The bulb is placed in thermal contact with an object whose temperature is to be measured. No gas can escape from the thermometer, so the amount of gas N or n in the ideal gas law (Eqs. 19.17 and 19.21) is constant.

In general, as the gas temperature changes, both its volume and pressure also change; the constant-volume thermometer, however, is designed so that the experimenter can keep the gas volume constant. The mercury is in a U-shaped tube with the middle part of the tube made of a flexible material so that raising or lowering the right-hand tube of the mercury manometer (Fig. 19.21) keeps the gas volume at its original value, depending on the temperature of the object in thermal contact with the gas.

Figure 19.22 shows how to operate a constant-volume gas thermometer. Initially, the gas temperature is T_i, and the mercury in the right-hand tube is at height h_i above the gas–mercury boundary in the left tube (step 1). The gas temperature T_i is read by measuring the height h_i of the mercury. Now if you increase the gas temperature (step 2), the gas volume increases. To return the gas to its original volume, you must raise the right-hand tube of mercury (step 3). The volume of gas in step 3 is the same as the original volume of gas in step 1. The final temperature T_f of the gas is read by measuring the new height $h_i + \Delta h$ of the mercury.

Now let's take a mathematical look at the constant-volume thermometer. According to $P = P_0 + \rho gy$ (Eq. 15.6), the initial gas pressure P_i is given by the height of the mercury h_i:

$$P_i = P_0 + \rho g h_i \tag{19.22}$$

where P_0 is the atmospheric pressure and ρ is the density of mercury. According to the ideal gas law (Eq. 19.17), the absolute pressure P_i is proportional to the initial gas temperature T_i:

$$P_i = \frac{Nk_B}{V}T_i \tag{19.23}$$

With the volume held fixed, a change in the gas's temperature means a change in its pressure and nothing more, so N and V are unchanged and do not require a subscript to indicate initial and final values. The height of the mercury in the right-hand tube

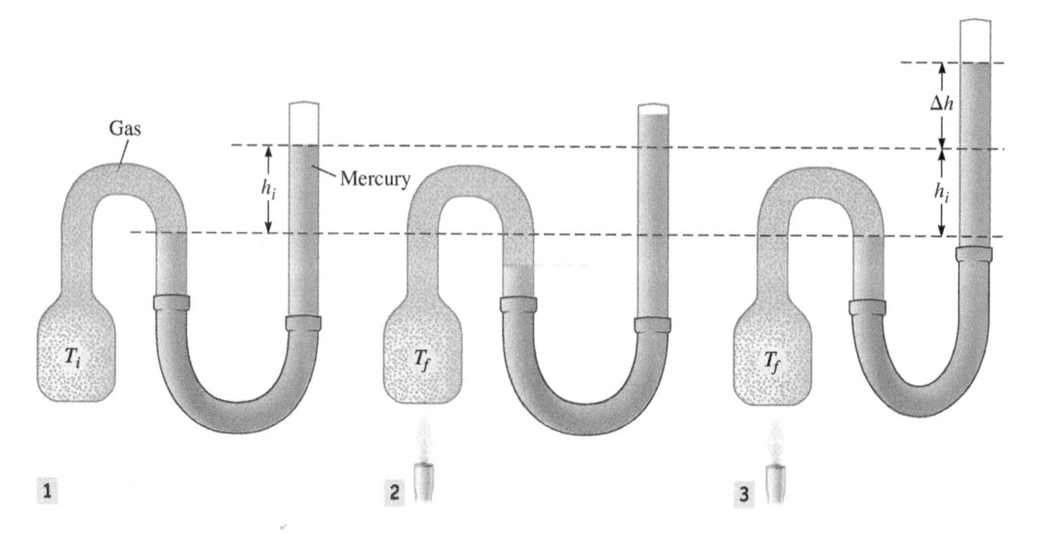

FIGURE 19.22 **1** Gas in the constant-volume gas thermometer is at T_i, and the mercury in the manometer is at height h_i above the gas–mercury boundary. **2** The thermometer is placed in thermal contact with an object, and its temperature increases. The increased temperature increases the gas volume. **3** By raising the right-hand tube of the mercury manometer, the gas volume is restored to its original size. The mercury is now at $h_i + \Delta h$ above the gas–mercury boundary. This increase in height is a result of the increase in gas temperature and pressure.

in Figure 19.21 is used to find the pressure of the gas and therefore the gas temperature according to Equation 19.23.

Similar equations hold when the gas is at its final temperature T_f as in Figure 19.22, far right (step 3). When the gas is at this higher temperature, its pressure P_f can be written as

$$P_f = P_0 + \rho g h_f$$

$$P_f = P_0 + \rho g(h_i + \Delta h) = (P_0 + \rho g h_i) + \rho g \,\Delta h$$

$$P_f = P_i + \rho g \,\Delta h \qquad (19.24)$$

The final temperature T_f of the gas comes from the ideal gas law:

$$P_f = \frac{Nk_B}{V} T_f \qquad (19.25)$$

For calibration, the thermometer is initially put in contact with a triple-point cell so that its initial temperature is $T_3 = 273.16$ K and the gas pressure in the thermometer (not the water vapor pressure in the triple-point cell) has some value P_3. We can find the temperature T of other objects in terms of the triple point temperature T_3 by dividing Equation 19.25 by Equation 19.23:

$$\frac{P}{P_3} = \left(\frac{Nk_B/V}{Nk_B/V}\right)\left(\frac{T}{T_3}\right)$$

$$T = \frac{P}{P_3} T_3 \qquad (19.26)$$

where P is the gas pressure when the thermometer is in thermal equilibrium with an object at temperature T.

Equation 19.26 is a prescription for how to use a constant-volume gas thermometer. First, put the thermometer in thermal contact with a triple-point cell and measure the height h_i of the mercury. From this height, use $P_3 = P_0 + \rho g h_i$ (Eq. 19.22) to calculate the gas pressure P_3 in the thermometer. Now put the thermometer in thermal contact with an object whose temperature you wish to measure. As in Figure 19.22, adjust the height of the mercury until the gas volume equals its original volume. Measure the height of the mercury and use Equation 19.24 to find the gas pressure P at this new temperature T. The temperature T is found from Equation 19.26 with $T_3 = 273.16$ K. In Example 19.7, we see how to turn this prescription into tick marks made on the thermometer.

EXAMPLE 19.7 CASE STUDY Making a Constant-Volume Ideal Gas Thermometer

You are asked to design your own constant-volume gas thermometer (Fig. 19.23). Your thermometer uses alcohol ($\rho = 785.06$ kg/m^3) instead of mercury. The thermometer maintains 0.500 L of O_2 gas in its bulb. The total mass of the O_2 gas in the bulb is 0.500 g. Your job is to set the tick marks on the scale shown in Figure 19.23 so that each tick mark is 0.5 K. What is the separation in centimeters between adjacent tick marks?

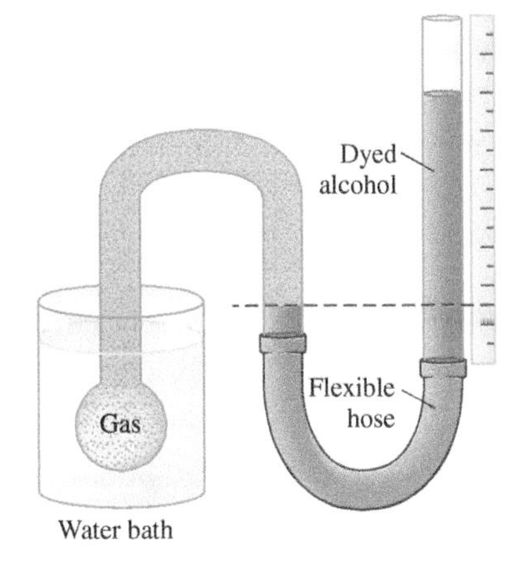

FIGURE 19.23 How far apart are the tick marks?

INTERPRET and ANTICIPATE

Figure 19.23 illustrates your task. For each change in temperature of 0.5 K, your job is to find the change Δh in the alcohol's height. The amount and volume of the O_2 gas are held constant. Only its pressure changes, resulting in a change in the height of the alcohol.

Example continues on page 574 ▶

SOLVE

Start with Equation 19.26 written in terms of an initial temperature T_i and pressure P_i and a final temperature T_f and pressure P_f.	$T_f = \frac{P_f}{P_i} T_i$
Substitute $P_f = P_i + \rho g \, \Delta h$ (Eq. 19.24) for P_f.	$T_f = \frac{P_i + \rho g \, \Delta h}{P_i} T_i$
Solve for ΔT.	$T_f = \frac{\cancel{P_i} T_i}{\cancel{P_i}} + \rho g \, \Delta h \frac{T_i}{P_i}$ $T_f - T_i = \Delta T = \rho g \, \Delta h \frac{T_i}{P_i}$
Use $P_i = (Nk_B/V)T_i$ (Eq. 19.23) to eliminate T_i/P_i.	$\Delta T = \rho g \, \Delta h \left(\frac{V}{Nk_B} \right)$
Solve for Δh. Remember that ρg has to do with the fluid in the manometer (in this case, alcohol), whereas Nk_B/V has to do with the gas in the bulb (in this case, oxygen).	$\Delta h = \frac{1}{\rho g} \left(\frac{Nk_B}{V} \right) \Delta T$ (1)
Find N for the oxygen gas in the bulb. For molecular oxygen (O_2), multiply the molar mass of oxygen found in the periodic table by 2.	number of moles $n = \frac{\text{mass of oxygen present}}{2M_{\text{mol}}}$ $n = \frac{0.500 \text{ g}}{2(15.999 \text{ g/mol})} = 1.56 \times 10^{-2} \text{ mol}$ number of molecules $N = nN_A$ $N = (1.56 \times 10^{-2} \text{ mol})(6.022 \times 10^{23} \text{ molecules/mol})$ $N = 9.41 \times 10^{21}$ molecules of O_2
Substitute numerical values into Equation (1). The density of alcohol is given in the opening statement. Convert the bulb volume into SI units: $V = 0.500 \text{L} = 0.500 \times 10^{-3} \text{ m}^3$.	$\Delta h = \frac{(9.41 \times 10^{21})(1.38 \times 10^{-23} \text{ J/K})(0.5 \text{ K})}{(785.06 \text{ kg/m}^3)(9.81 \text{ m/s}^2)(0.500 \times 10^{-3} \text{ m}^3)}$ $\Delta h = 1.69 \times 10^{-2} \text{ m} = 1.69 \text{ cm}$

CHECK and THINK

So, the distance between each tick mark is 1.69 cm. To check our answer, imagine that we replace the alcohol with mercury. Because mercury is denser than alcohol, we would expect that it would not rise as high as the alcohol and the tick marks would therefore be closer together, due to increased ρ in Equation (1). As one further check, imagine that we put more gas in the bulb, increasing the gas pressure (if its volume is unchanged). A greater gas pressure would mean that the alcohol rises higher and the tick marks would be farther apart, due to increased N in Equation (1).

No matter what type of gas is used in constant-volume gas thermometers, the temperatures they measure agree. These thermometers are bulky and awkward to use, however. In practice, constant-volume gas thermometers are used to calibrate other thermometers that are easier to manage.

SUMMARY

Underlying Principles

1. **Thermodynamics** is the study of thermal energy and the transfer of energy through heat and work.
2. **Zeroth law of thermodynamics:** If two objects A and B are in thermal equilibrium with a third object C, they are in thermal equilibrium with each other. Mathematically,

$$\text{If } T_A = T_C \text{ and } T_B = T_C, \text{ then } T_A = T_B.$$

Major Concepts

1. **Temperature** is a measure of the average kinetic energy of the particles in a system. It is the macroscopic property of a system that determines whether that system will be in thermal equilibrium with another system.
2. When two objects are in **thermal contact**, thermal energy can pass from one object to the other.
3. Two objects are in **thermal equilibrium** if they are in thermal contact with each other and their temperatures do not change.
4. Most building materials (concrete and metals) expand as their temperature increases and contract as their temperature decreases. **Thermal linear expansion** is given by

$$\Delta L = \alpha L_0 \Delta T \qquad (19.4)$$

5. **Thermal volume expansion** is approximately given by

$$\Delta V \approx \beta V_0 \Delta T \qquad (19.5)$$

6. An **equation of state** is an equation or series of equations that relate the macroscopic physical variables (pressure, temperature, volume, and amount) of a gas.
7. When a gas is in an **equilibrium state**, its physical properties are constant and uniform throughout the gas.
8. An **ideal gas** is a hypothetical gas made up of identical particles that do not interact with one another. A real gas may be modeled as an ideal gas if its density is low, its pressure is less than about 1 atm, and the gas temperature is not near its liquefaction point.
9. The SI unit for the number of particles in a substance is the **mole**. One mole is the amount of substance containing as many particles as there are atoms in 12 g of carbon-12.
10. Exactly 12 g of carbon-12 contain **Avogadro's number** N_A of atoms, where $N_A = 6.022 \times 10^{23}$.
11. The **Kelvin scale (absolute scale)** is the temperature scale in the SI system.
12. On the Kelvin scale, $T = 0$ K is known as **absolute zero** and is found by extrapolating the temperature an ideal gas would have when its volume is zero. The triple point of water is defined as 273.16 K.

Special Cases: Equations of State

1. **Charles's law:**

$$V \propto T \quad (\text{if } P \text{ is constant}) \qquad (19.9)$$

2. **Gay-Lussac's law:**

$$P \propto T \quad (\text{if } V \text{ is constant}) \qquad (19.11)$$

3. **Boyle's law:**

$$P \propto \frac{1}{V} \quad (\text{if } T \text{ is constant}) \qquad (19.13)$$

4. **Avogadro's law:** Equal volumes of gas at the same temperature and pressure contain the same number of particles independent of the type of gas. Mathematically,

$$V \propto N \ (\text{at the same } T \text{ and } P) \qquad (19.16)$$

5. The **ideal gas law** is a combination of Charles's, Gay-Lussac's, Boyle's, and Avogadro's laws:

$$PV = Nk_BT \qquad (19.17)$$

PROBLEMS AND QUESTIONS

A = algebraic C = conceptual E = estimation G = graphical N = numerical

19-1 Thermodynamics and Temperature

1. N Convert the following temperatures from the Fahrenheit scale to the Celsius and Kelvin scales: **a.** The seawater temperature near Hawaii is about 76°F. **b.** McDonald's was sued for serving dangerously hot coffee. Their coffee was served between 180°F and 190°F. **c.** The coldest day recorded on Earth was −129°F in Vostok, Antarctica.
2. C Correct the following sentences by changing the temperature units. **a.** It is best to cook a turkey slowly at 250 K. **b.** A healthy dog's temperature is 374°C. **c.** A comfortable room temperature is 22°F. **d.** A crisp fall day in New England is about 50°C.
3. N Stars (like the Sun) that are actively converting hydrogen into helium in their cores are known as main sequence stars. Their surface temperatures range from about 1700 K to 53,000 K. Express this temperature range on the Fahrenheit scale.

Problems 4 and 5 are paired.

4. C Figure P19.4 shows a graph of the two temperature scales—Fahrenheit and Celsius—versus the Kelvin scale. Which line represents which scale? Justify your answer.

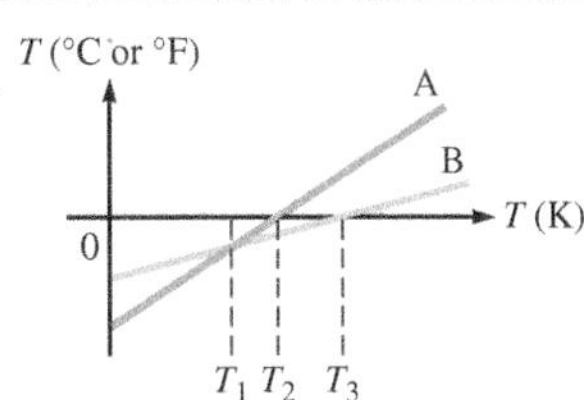

FIGURE P19.4 Problems 4 and 5.

5. N In Figure P19.4, what are the values of the temperatures T_1, T_2, and T_3 on all three scales?
6. N Hypothermia occurs when the human body temperature falls below 35.0°C, and hyperpyrexia occurs when the human body temperature exceeds 41.5°C. What are the equivalent temperatures on the **a.** Fahrenheit scale and **b.** Kelvin scale?
7. N The lowest temperature conceivable is absolute zero (0 K). What Fahrenheit temperature corresponds to 0 K?
8. N Liquid oxygen, a primary fuel component of rocket boosters, boils at a temperature of 90.2 K. What is the boiling point of liquid oxygen on the **a.** Celsius scale and **b.** Fahrenheit scale?

19-2 Zeroth Law of Thermodynamics

9. C Object A is placed in thermal contact with a very large object B of unknown temperature. Objects A and B are allowed to reach thermal equilibrium; object B's temperature does not change due to its comparative size. Object A is removed from thermal contact with B and placed in thermal contact with another object C at a temperature of 40°C. Objects A and C are of comparable size. The temperature of C is observed to be unchanged. What is the temperature of object B?

10. C A healthy boy wishes to miss school. His father says that the boy can stay home if he has a fever. The father leaves the boy with an alcohol thermometer in his mouth. (There are no lights or heaters near the boy's bed.) What is one thing the boy can do to increase the thermometer's temperature without leaving the bed?

11. C Answer each of the following questions and explain your reasoning. **a.** After you hold an ice cube for several minutes, are your hand and the ice cube in thermal equilibrium? **b.** You store ice cream and ice pops in your freezer for several days. Are they in thermal equilibrium? **c.** You swim in the ocean near Hawaii for several hours. Is your bathing suit in thermal equilibrium with the ocean? **d.** You toss your bathing suit in a sink full of cool water and let it soak overnight. Is your bathing suit in thermal equilibrium with the water?

19-3 Thermal Expansion and 19-4 Thermal Stress

12. C In the past, the coolant from a car's radiator would flow onto the ground; cars today have an overflow container to collect the excess. Why is the overflow container necessary? *Hint*: Imagine that you fill the car's radiator on a cold day.

13. C If a metal lid is stuck on a glass jar, a trick for opening it is to hold the lid and neck of the jar under hot water. **a.** Why does this method help? **b.** What would happen if the jar and lid were both made of the same type of metal? Explain.

14. N The tallest building in Chicago is the Willis Building (formerly the Sears Tower), which reaches 527 m including the antenna. The lowest and highest temperatures recorded in Chicago so far were −33.0°C (−27°F) and 44.0°C (111°F), respectively. If we assume the building is a steel structure, what is the difference in the height of the building at these two extreme temperatures? The coefficient of linear expansion for steel is $13.0 \times 10^{-6}\,°C^{-1}$.

15. N A bridge spanning a river gorge is made up of concrete sections 15.24 m in length, poured and cured at 15.0°C. What is the minimum spacing between bridge sections required to prevent buckling if the summertime temperatures at the bridge reach 43.0°C?

16. N A thin strip of copper 0.250000 m long is stored at 68.0°F. **a.** If the strip of copper is heated to 90.0°F, find the new length of the copper strip. Keep six significant figures as you work and report your answer to the same precision as the initial length. The coefficient of linear expansion for copper is $16.8 \times 10^{-6}\,K^{-1}$. **b.** What is the thermal stress on the copper?

17. N At 22.0°C, the radius of a solid aluminum sphere is 7.00 cm. **a.** At what temperature will the volume of the sphere have increased by 3.00%? **b.** What is the increase in the sphere's radius if it is heated to 250°C? Assume $\alpha = 22.2 \times 10^{-6}\,K^{-1}$ and $\beta = 66.6 \times 10^{-6}\,K^{-1}$.

18. C CASE STUDY Suppose you wish to replace a mercury thermometer, which is often used to measure a person's temperature, with a water thermometer. Estimate the dimensions of such a thermometer and comment on the pros and cons of using water as the fluid.

19. N A sphere of diameter 7.0 cm and mass 266.5 g floats in a liquid that has a temperature of 0°C. As the temperature is raised, the sphere begins to sink as the liquid reaches a temperature of 35°C. If the density of the liquid is 1.527 g/cm³ at 0°C, find the coefficient of volume expansion of the liquid. Ignore the volume expansion of the sphere.

20. An exterior wooden door fits comfortably in a steel frame on a day when the outdoor temperature is 50°F. The wood has been cut so that the grain runs vertically. The door is in a climate with a yearly temperature range from 0°F to 100°F. Keep four significant figures as you work and report final answers to one significant figure. Use values from Table 19.1.

a. C Would you expect the door to stick on a hot day or on a cold day? Explain.

b. N On a day when the temperature is 50°F, the door is 36 in. wide and 80.5 in. tall. What are its maximum and minimum width and height, given the range of temperatures?

c. N On a day when the temperature is 50°F, there is a 0.50-cm gap between the door and the frame. The steel frame has a uniform width of 2.5 in. What are the maximum and minimum gaps between the door and the frame, given the range of temperatures?

d. C Will the door ever get stuck in its frame? Explain.

21. N The distance between telephone poles is 30.50 m in a neighborhood where the temperature ranges from −35°C to 40°C. If you hang a copper cable between two adjacent poles on a day when the temperature is 22.30°C, what is the minimum length of the copper cable you must use for the cable to remain connected to the poles all year? Consider α to have four significant figures and report your answer to four significant figures.

22. N A copper cube with density 8.96×10^3 kg/m³ and mass 10.0 kg at 20.0°C is heated to 130°C. What is **a.** the density and **b.** the mass of the copper cube at this new temperature?

23. N A section of the Alaska railroad needs urgent repairs during a cold winter day with a temperature of −10.0°C, when 11.89-m-long steel beams are delivered to the repair site. **a.** What is the length of each beam in the summertime, when the temperature can reach 27.0°C? **b.** What is the thermal stress experienced by one of the beams between these two temperatures?

24. A clock with an iron pendulum is made so as to keep the correct time when at a temperature of 20°C. Assume the coefficient of linear expansion of iron is $12 \times 10^{-6}\,°C^{-1}$ and the pendulum is a simple pendulum.

a. C Will the clock be ahead or behind the correct time if the temperature rises to 30°C?

b. N What is the amount of time lost or gained each day?

25. N You notice that a concrete sidewalk is cracked and wonder if it is a result of thermal stresses. Assume the lowest temperature in the winter is −20°C, the highest temperature in summer is 40°C, and the edges of the 1-m-wide concrete sidewalk are fixed in place in the winter. Comment on whether thermal stresses might be responsible for the cracks in the sidewalk as summer temperatures are reached. Assume Young's modulus for concrete is 17×10^9 N/m² and the coefficient of linear expansion is $10 \times 10^{-6}\,K^{-1}$. The compressive strength of concrete is 40×10^6 N/m².

19-5 Gas Laws

26. C **Review** The gas laws are empirical, like Kepler's laws. Let's use Kepler's first law to think about the difference between an empirical model and a theory. According to Kepler's first law, planetary orbits are elliptical. Newton's laws of motion and of universal gravity provide the theoretical bases for Kepler's third law. If a planet is discovered that has a nonelliptical orbit (such as the rosette orbit of Mercury), does that discovery pose a challenge to Kepler's first law, Newton's laws, neither, or both? Explain your reasoning.

27. N A plastic bottle is closed outdoors on a cold day when the temperature is −13.0°C and is later brought inside where the temperature is 21.0°C. What is the pressure of the air in the bottle after it reaches room temperature, assuming the air in the bottle was at a pressure of 1.00 atm upon reaching thermal equilibrium with the outdoor temperature?

Problems 28 and 29 are paired.

28. N An 80-ft^3 scuba tank is filled with air at room temperature (22°C) so that the gauge pressure is 2987 psi. The tank is forgotten in the Sun for hours, and the tank's temperature reaches 125°F. Ignore the slight increase in the tank's size. What is the gauge pressure at this higher temperature?

29. N A scuba tank is filled as described in Problem 28. The tank is forgotten on the shore in winter for hours, and the tank's temperature drops to −45°F. Ignore the slight decrease in the tank's size. What is the gauge pressure at this lower temperature?

30. A 0.025-m^3 tank filled with helium had been kept in a refrigerated room at a temperature of 3.15°C and a pressure of 2.00 atm. A careless researcher moves the tank into the hallway while working in the room and forgets to put it back. The air in the hallway is at standard atmospheric pressure (1.00 atm) and room temperature (20.0 °C).

a. C What happens to the pressure in the tank as it remains in the hallway for a significant amount of time?

b. A Write an equation for the pressure of the helium as a function of its temperature.

31. N A balloon is filled with air outdoors on a cold day when the temperature is −13.0°C so that it occupies a volume of 3.00 L. It is then brought inside where the temperature is 21.0°C. Assuming the pressure in the balloon remains constant, what is the volume of the balloon indoors?

32. E Paul gets a 1.50-ft-diameter helium-filled balloon as a birthday present. He wants to preserve the balloon, so he stores it in the freezer (along with his leftover cake). Assume no helium leaks out. Estimate **a.** the volume and **b.** the diameter of the balloon when Paul finds it in the freezer the next day. Report your answers in US customary units. Consider Figure 19.16 and comment on your results.

19-6 Ideal Gas Law

33. N The ideal gas in a container is under a pressure of 20 atm at a temperature of 27°C. If half of the gas is released from the container and the temperature is increased by 50°C, what is the final pressure of the gas?

34. N Dexter has developed a special knockout gas that is kept in a 0.0650-m^3 cylinder at a temperature of 17.25°C. **a.** If the container is holding 10 mol of gas, what is the gas pressure? **b.** If the radius of the top of the cylinder is 0.30 m, what is the force the gas exerts on the top of the cylinder?

Problems 35 and 36 are paired.

35. G Figure P19.35 shows a graph of pressure P as a function of volume V for an ideal gas in a balloon with a tight seal so that no gas leaks out. Describe what is happening to the gas. What is changing, and what is constant?

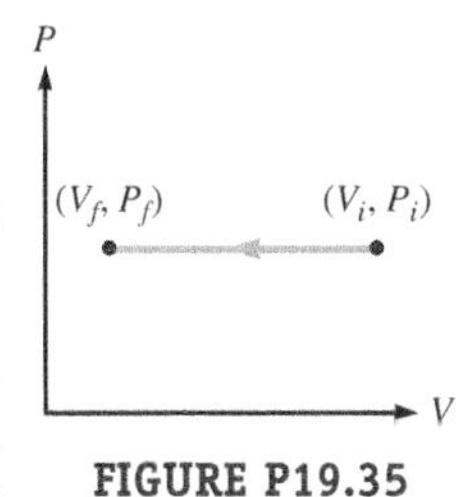

FIGURE P19.35

36. G Figure P19.36 shows a graph of pressure P as a function of volume V for an ideal gas in a balloon. The temperature of the balloon remains constant. Describe what is happening to the gas. What is changing and what is constant?

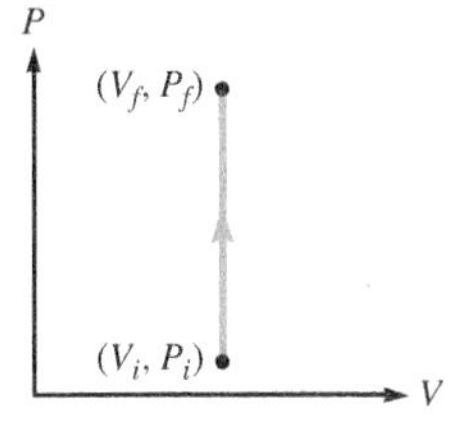

FIGURE P19.36

37. N An oxygen tank is filled at $T = 20.3$°C to an absolute pressure of 1.00 atm. The tank is then stored in the trunk of a car where the temperature reaches 46.8°C. What is the pressure in atmospheres?

38. N A very good laboratory vacuum can evacuate the gas in a chamber so that its pressure is about 10^{-12} Pa. Assuming the gas is at room temperature ($T \approx$ 20°C), estimate the number density (number per unit volume) of gas molecules in the chamber.

39. N What is the ratio of the number of molecules per cubic meter on top of Mount Everest at a pressure of 30.0 kPa and temperature of 2.00×10^2 K (around −100°F) compared with that at sea level (assuming standard temperature and pressure)?

40. On a hot summer day, the density of air at atmospheric pressure at 35.0°C is 1.1455 kg/m^3.

a. N What is the number of moles contained in 1.00 m^3 of an ideal gas at this temperature and pressure?

b. N Avogadro's number of air molecules has a mass of 2.85×10^{-2} kg. What is the mass of 1.00 m^3 of air?

c. C Does the value calculated in part (b) agree with the stated density of air at this temperature?

41. N A thermos is constructed such that there is an evacuated region between two walls that acts as a lining, leading to good thermal insulation. **a.** If the pressure in the lining is 1.32×10^{-6} atm at room temperature (293 K) and the volume of the lining is 30.0 cm^3, how many molecules are in the evacuated lining? **b.** How many moles are in the thermos lining?

42. N To decrease the pressure within a tank containing 2.00 mol of an ideal gas from 50.0 atm to 8.00 atm, a fraction of the gas is removed. If the volume and the temperature of the gas in the container remain constant, what is the number of moles of the gas that must be removed for the decrease in pressure to occur?

43. N Most of the Universe is made up of hydrogen, often found in huge clouds. The Orion Nebula is an example. The number density of hydrogen molecules (H_2) in such a cloud is estimated to be between 10^9 and 10^{12} molecules per cubic meter. The cloud temperature is typically between 10 K and 30 K. What is the range in such a cloud's gas pressure? Give your answer in pascals and atmospheres and comment on your results.

44. N The gauge pressure of a truck tire is 2.72 atm at 22.0°C. As the truck is laden with deliveries and driven around town, the temperature of its tires increases to 52.0°C. What is the gauge pressure in the tire at 52.0°C if the volume of the tires is assumed not to change?

45. N The 100-m^3 European JET tokamak fusion device achieves a near vacuum pressure of 5.00×10^{-6} Pa. If the temperature within the device is 22.0°C while it is not operating, what is the number of molecules in the device at this pressure?

46. A transmission electron microscope (TEM) is a very sensitive instrument that uses high-speed electrons to view materials on the nanometer-size scale. The sample you wish to observe must be kept in an ultrahigh vacuum chamber, where the pressure in the chamber is on the order of 100 nanopascals. Suppose the volume of a TEM vacuum chamber is 0.520 m^3 and there are about 3.00×10^{11} molecules/m^3 in the air inside the chamber. Treating the air as an ideal gas, what is the temperature of the air inside the chamber if the pressure is 100 nanopascals?

19-7 Temperature Standards

Problems 47 and 48 are paired.

47. C CASE STUDY A constant-volume ideal gas thermometer such as the one shown in Figure 19.21 (page 572) is constructed and calibrated at sea level. It is then carefully shipped to Denver, which is 1 mile above sea level. Do the researchers in the Denver laboratory need to recalibrate the thermometer? Explain.

48. C A triple-point cell such as the one shown in Figure 19.20 (page 571) is constructed and calibrated at sea level. It is then shipped to Denver, which is 1 mi above sea level. Can the researchers in the Denver lab use the triple-point cell just as in a laboratory at sea level, or will special adjustments be needed? Explain.

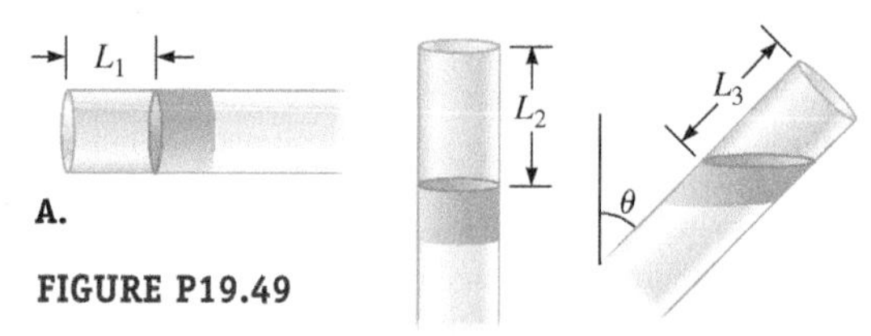

FIGURE P19.49

49. A An ideal gas is trapped inside a tube of uniform cross-sectional area sealed at one end as shown in Figure P19.49. A column of mercury separates the gas from the outside. The tube can be turned in a vertical plane. In Figure P19.49A, the column of air in the tube has length L_1, whereas in Figure P19.49B, the column of air has length L_2. Find an expression (in terms of the parameters given) for the length L_3 of the column of air in Figure P19.49C, when the tube is inclined at an angle θ with respect to the vertical.

50. C CASE STUDY When Lord Kelvin defined absolute zero, why couldn't he choose the zero point to be the temperature of some measurable phenomenon such as the temperature of liquid helium? *Hint*: Does the Kelvin scale include negative temperature?

51. C CASE STUDY A researcher misuses a constant-volume thermometer by failing to lift the right tube in Figure 19.22, when the bulb is in contact with an object that is hotter than the triple point of water. Does the researcher measure a temperature that is too high or too low? Explain.

52. C CASE STUDY When a constant-volume thermometer is in thermal contact with a substance whose temperature is lower than the triple point of water, how does the right tube in Figure 19.22 need to be moved? Explain.

53. N An air bubble starts rising from the bottom of a lake. Its diameter is 3.6 mm at the bottom and 4.0 mm at the surface. The depth of the lake is 2.50 m, and the temperature at the surface is 40°C. What is the temperature at the bottom of the lake? Consider the atmospheric pressure to be 1.01×10^5 Pa and the density of water to be 1000 kg/m^3. Model the air as an ideal gas.

54. C Rather than use the triple point of water as the basis for precisely determining a temperature scale, you would like a temperature reference that is easier to produce. You're aware that the transition between freezing and melting depends on pressure, but you consider defining a specific temperature standard as the "freezing point of pure water at atmospheric pressure." What advantage does the triple point of water have over this definition?

General Problems

55. N To facilitate transport by tanker trucks, natural gas is often liquefied, which is achieved by decreasing the temperature of the gas below −162°C. What is the temperature on **a.** the Kelvin scale and **b.** the Fahrenheit scale at which natural gas liquefies?

56. N Jerome's physics lab on heat expansion involves a copper ring with a diameter of 3.000 cm and a stainless steel ball with a diameter of 3.015 cm. The linear expansion coefficient for copper is 17×10^{-6} °C^{-1} and that for stainless steel is 11×10^{-6} °C^{-1}, and both the copper ring and the steel ball are at 22.0°C. **a.** In the first phase of the experiment, Jerome warms the copper ring so that the stainless steel ball will just slip through. How hot is the copper ring when that occurs? **b.** In the second phase of the experiment, both the copper ring and the stainless steel ball are warmed together. What is the temperature at which the ball slips through the ring?

57. N A royal clock pendulum is one that has a period of precisely 1.000 s at a temperature of 20.0°C. One such clock pendulum, made of copper, is placed in a room with temperature 5.00°C. What is the change in the period of the pendulum?

58. N A university lecture hall is 12.0 m wide, 15.0 m deep, and 4.00 m high. The temperature in the hall is 22.0°C, and the pressure is 1.00 atm. What is the number of molecules of air in the lecture hall?

Problems 59 and 60 are paired.

59. N A solid aluminum sphere has a radius of 0.5700 m when it is at a temperature of 90.0°F. What is the change in volume of the sphere when it is heated to a temperature of 110.0°F?

60. C A solid aluminum sphere is heated from an initial temperature to a higher temperature. If, instead, the aluminum sphere were hollow but started with the same radius at the same initial temperature and was then heated to the same final temperature, how would the new volume of the hollow sphere compare with that of the solid aluminum sphere? Explain your reasoning.

61. N A rigid tank contains 3.00×10^{-2} m^3 of an ideal gas at a pressure of 5.15×10^5 Pa and a temperature of 300.0 K. How many **a.** moles and **b.** molecules of the gas are contained in the tank?

62. N In December 2006, two Missouri representatives filed legislation requiring that the price of motor fuel be adjusted for volume changes caused by temperature fluctuations. The industry standard assumes the temperature is 60°F. In the summer, when the temperature easily exceeds 60°F in Missouri, gasoline expands and its density decreases. Because gasoline is sold by volume, you buy fewer gasoline molecules (less energy) per gallon during the hot months. This change in volume is good for consumers in the winter months, when cold temperatures increase gasoline density and consumers get more molecules of gasoline per gallon. The industry has accepted temperature-based price adjustments in cold places like Canada, but not in warm places. In 2006, it was estimated that Missouri consumers paid $12 million extra due to the expansion of hot fuel. **a.** The density of gasoline at 60°F is 737.22 kg/m^3. What is the density of gasoline on a very hot day (T = 95°F)? **b.** If you buy 12.0 gal of gasoline at 60°F, what mass of gasoline have you purchased? **c.** If you buy 12.0 gal of gasoline at 95°F, what mass of gasoline have you purchased? **d.** If gasoline costs $2.50 per gallon (in 2006), how much money did a consumer lose by buying gasoline on a very hot day?

63. A A glass container of volume V_0 is completely filled with a liquid, and its temperature is then increased by ΔT. Find an expression for the volume of the liquid that will overflow. Assume the coefficient of linear expansion of glass is α_g and the coefficient of volume expansion of the liquid is β_ℓ, where $\beta_\ell > 3\alpha_g$.

64. G Some substances (such as water between 0°C and 4°C) expand when they are cooled and contract when they are heated. Sketch a potential energy curve for molecules in such a substance. Explain your sketch.

65. N A 30-quart (28.4-L) pressure cooker initially at 22.0°C is filled with 1.00 L of water and brought to a temperature of 450°C. What is the pressure within the pressure cooker?

66. An aluminum canister is filled with oxygen for use by hikers. The canister holds 20.0 L at a gauge pressure of 2200 psi at 20.0°C. The hiker empties the canister by taking 75 breaths from it. Assume each breath fills the same volume and the air is an ideal gas at constant temperature.
  a. G Plot the number density ($\rho = N/V$) of oxygen in the canister versus (absolute) pressure.
  b. C Does the hiker get the same number of oxygen molecules for each breath? If not, does the hiker get more oxygen molecules on the first breath or on the last breath?

67. N A tank contains gas at 15.0°C pressurized to 14.0 atm. The temperature of the gas is increased to 90.0°C, and half the gas is removed from the tank. What is the pressure of the remaining gas in the tank?

68. C Consider making a bimetallic strip from two metals listed in Table 19.1. Which two metals would create the most curvature (tightest circle) when the strip's temperature is changed? Explain.

69. N Chemical vapor deposition (CVD) is a chemical process used to manufacture several materials, including thin films and carbon nanotubes. Consider the manufacture of silicon dioxide (SiO_2) thin films (which are used in electronic devices) achieved using silane, SiH_4. Suppose the silane is drawn into a CVD chamber with a volume of 3.00 m^3 and it can be modeled as an ideal gas. The pressure of the silane gas is 0.050 atm, and it is at 825 K. **a.** Find the number of moles of silane in the chamber. **b.** What is the total mass of the silane in the chamber?

Problems 70 and 71 are paired.

70. C While studying a colony of bacteria, you find that the number N of bacteria grows exponentially, $N = N_0 e^{t/\tau}$, where N_0 is the initial number of bacteria and τ is the time constant. You further find that at $t = 54\tau$, the colony reaches a maximum and does not grow any further. Have you discovered an empirical fit or a theoretical model? Explain.

71. C You believe that a colony of bacteria should grow exponentially, following $N = N_0 e^{t/\tau}$, where N_0 is the initial number of bacteria and τ is the time constant until the colony reaches the maximum size. This is because as long as resources are available, each individual bacterium cell will divide into two new daughter cells, which will divide into two more cells and so on. Is your idea an empirical fit or a theoretical model? Explain.

72. A A steel plate has a circular hole drilled in its center. If the diameter of the hole varies according to thermal linear expansion, show that the area of the original circle A_0 changes with an increase in temperature ΔT, following the approximate relation $\Delta A \approx CA_0\,\Delta T$, where $C = 2\alpha$. *Hint*: $(\alpha\,\Delta T)^2 \ll \alpha\,\Delta T$.

73. N Before climbing a tough hill, a bicycle tire is inflated with air at 1.00 atm and 20.0°C. **a.** What is the change in the volume of the air in the tire in terms of the initial volume V_i if the pressure in the tire increases to 7.00×10^5 Pa and its temperature doubles? **b.** As the rider crests the hill, the tire temperature increases to 70.0°C, and the tire expands its volume by 3.50%. What is the absolute pressure in the bicycle tire?

74. A A gas is in a container of volume V_0 at pressure P_0. It is being pumped out of the container by a piston pump. Each stroke of the piston removes a volume V_s through valve A and then pushes the air out through valve B as shown in Figure P19.74. Derive an expression that relates the pressure P_n of the remaining gas to the number of strokes n that have been applied to the container.

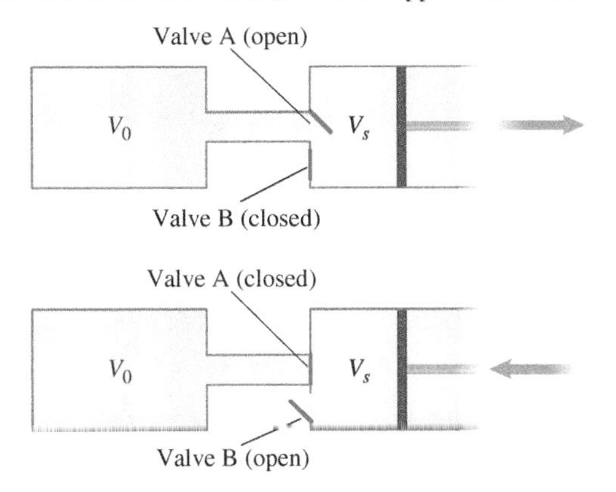

FIGURE P19.74

75. N CASE STUDY A constant-volume thermometer uses mercury and oxygen. In the thermometer, 0.750 g of oxygen is kept at a volume of 1.000 L. When the thermometer is placed in contact with a triple-point cell, the mercury rises to 1.674 cm. When the thermometer is placed in contact with an unknown substance, the mercury rises to 7.896 cm. What is the temperature (in kelvins) of the unknown substance?

76. A gas is confined to a cylindrical volume with a movable piston on one end. Initially, the volume of the container is 0.175 m^3, and the gas is at 305 K. The gas is heated, but the piston is moved such that the temperature of the gas remains constant.
    a. A If there are 2.00 mol of gas in the container, write an equation for the pressure of the gas as a function of the volume of the container.
    b. G Display your equation from part (a) on a graph, from the initial volume to five times the initial volume.

77. N The boiling point and the freezing point of water are marked as 80° and 10°, respectively, on an arbitrary temperature scale using a homemade thermometer. What is the Celsius temperature when this thermometer reads 59° on its arbitrary temperature scale? Assume there is a linear relationship between the Celsius scale and this arbitrary temperature scale.

78. N Children playing in the neighborhood park on a cold day find a section of PVC pipe 50.0 cm long and open at both ends, and they begin blowing through it to produce trumpet-like sounds. The pipe is initially at 2.00°C and warms to 25.0°C when warm air is blown through it. What is the change in the fundamental frequency of the pipe after it warms? The coefficient of thermal expansion for PVC is $54 \times 10^{-6}\ °C^{-1}$.

79. N The legendary Fender Stratocaster electric guitar has a scale length of 65.78 cm over which the steel strings of the guitar are stretched. One such string, 0.0254 cm in diameter, has a fundamental frequency of oscillation of 196 Hz at a temperature of 20.0°C. The density of steel in the string is 7.80 g/cm^3, its Young's modulus is 200 GPa, and its coefficient of linear expansion is $11 \times 10^{-6}\ °C^{-1}$. **a.** If the density of the steel in the string is 7.80 g/cm^3, what is the mass per unit length of the guitar string? **b.** What is the tension in the guitar string? **c.** What is the tension in the guitar string if its temperature increases to 45.0°C during an intense guitar solo? **d.** What is the fundamental frequency of the guitar string at 45.0°?

80. G A cylinder filled with an ideal gas is sealed by a piston on one end as shown in Figure P19.80. The piston can freely slide up and down the cylinder without any loss of gas. **a.** The cylinder is held over a burner so that the gas temperature rises (Fig. P19.80, middle). If the pressure is constant, sketch a graph of pressure as a function of volume and a graph of temperature as a function of volume. Describe what you would observe as the gas is heated. **b.** Now the temperature of the gas remains constant as you press the piston downward (Fig. P19.80, right). Sketch graphs of pressure and temperature as functions of volume. Compare your results with Charles's, Gay-Lussac's, and Boyle's laws and comment.

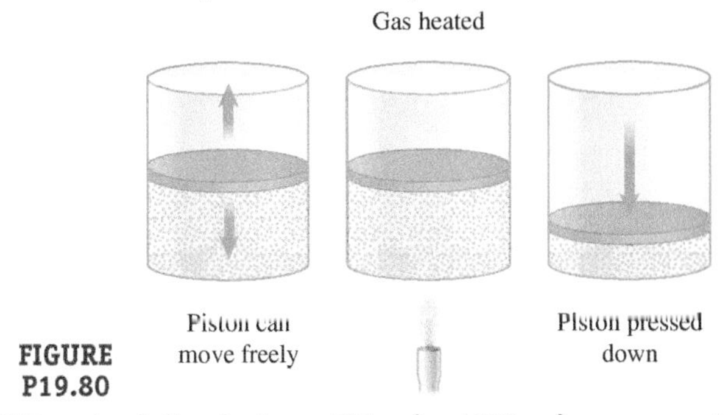

FIGURE P19.80

81. N Two glass bulbs of volumes 500 cm^3 and 200 cm^3 are connected by a narrow tube of negligible volume. The apparatus is filled with air and sealed. The initial pressure of the air is 1.0×10^5 Pa, and the initial temperature is 17°C. The smaller bulb is immersed in a large amount of ice at 0°C, and the larger bulb is immersed in a large pot of boiling water at a temperature of 100°C for a long period of time. Determine the final pressure of the air in the bulbs. Ignore any change in the volume of the glass. Model the air as an ideal gas.

20 Kinetic Theory of Gases

Key Question

How does the kinetic theory—a microscopic model—account for the macroscopic properties of ideal and real gases?

20-1 What is the kinetic theory? 581

20-2 Average and root-mean-square quantities 582

20-3 The kinetic theory applied to gas temperature and pressure 584

20-4 Maxwell-Boltzmann distribution function 588

20-5 Mean free path 591

20-6 Real Gases: The Van der Waals equation of state 596

20-7 Phase changes 599

20-8 Evaporation 601

Underlying Principles

Kinetic theory

Major Concepts

1. Connection between average kinetic energy and temperature
2. Connection between rms velocity and temperature
3. Maxwell-Boltzmann speed distribution
4. Most probable speed
5. Average speed
6. Mean free path
7. Van der Waals equation of state
8. Critical temperature; critical point; critical pressure
9. Saturated vapor pressure
10. Relative humidity

Tools

1. Root mean square
2. Pressure–volume (*PV*) diagram
3. Phase diagram

As you read this sentence, air molecules are flying into you at an average speed of roughly 500 m/s or about 1250 mph. At that speed, you could get from Baltimore to Minneapolis or from Buenos Aires to Rio de Janeiro in about an hour, but the air molecules undergo many collisions, so it would take a molecule much longer to cross that distance. When the molecules hit your skin, you experience those high-speed collisions as warmth. If the air temperature drops, the molecules slow down, and you feel cooler. On a hot day, the molecules speed up, and you feel warmer. If you get too hot, your body produces perspiration (sweat). Water molecules on your skin evaporate, removing energy as they leave and cooling you off. In this chapter, we explore a microscopic model for gases known as the kinetic theory, which successfully explains their macroscopic properties and behavior.

20-1 What Is the Kinetic Theory?

KINETIC THEORY

Underlying Principle

The **kinetic theory** is a model for matter stating that all matter is made up of particles in motion. The particles may be either atoms or molecules. In a solid, the particles oscillate back and forth around some equilibrium position (Fig. 19.6). In a solid, there is no net motion of molecules through the body of the sample. In a liquid, however, the particles may oscillate and slide past one another as they travel throughout the sample (Fig. 15.2B). The particles in a gas move randomly and freely throughout its container (Fig. 15.2A).

The kinetic theory can be applied to an ideal gas or a real gas. Until we reach Section 20-6, we focus on gases that are well described by the ideal gas law, $PV = Nk_BT$ (Eq. 19.17). A real gas may be modeled as an ideal gas if it is at low pressure, has a low density, and is not close to its liquefaction point. In applying the kinetic theory to gases that follow the ideal gas law, we start with the following five assumptions.

1. **A gas consists of a large number of molecules or atoms, and they can be modeled as particles.** That is, the molecules or atoms are assumed to have no physical extent, so they do not vibrate or rotate. The total volume occupied by the particles is negligible compared to the volume occupied by the gas as a whole, so the particles are small compared to the average distance between them.
2. **The particles in an ideal gas do not interact with one another except when they collide.** When the particles are far apart, they do not exert forces on one another. Thus, only contact forces (not field forces) are exerted.
3. **The particles make elastic collisions with the walls, and the duration of each such collision is short.**
4. **The particles are free to move in any direction at any speed.**
5. **The gas is made up of identical particles.** Therefore, only one type of gas is present, such as pure oxygen.

In this chapter, we apply the kinetic theory to derive the ideal gas law and other properties of ideal gases. When we apply the theory to some real gases, we may need to change one or more of the five assumptions. These assumptions will be referred to by number in subsequent sections.

CASE STUDY Part 1: The Earth's Atmosphere

Life on the Earth depends on its atmosphere (Fig. 20.1). Not only do we breathe the atmosphere, it also regulates the Earth's temperature and weather. The Earth's temperature is typically between 250 K and 300 K, making it suitable for life. A different balance of gases in the atmosphere could cause major climate changes and make the planet uninhabitable.

You only need to consider our Moon to imagine what our planet might be like if it didn't have an atmosphere. The sunlit side of the Moon reaches temperatures of about 380 K (above the boiling point of water), and the dark side may be as cold as 110 K (about −170°C, or about 100°C colder than Antarctica).

© NASA

FIGURE 20.1 The atmosphere of the Earth is contained by the Earth's gravity. If the temperature of the Earth were higher, more components of the atmosphere would escape.

Even a planet with an atmosphere can be unlivable if its atmosphere has the wrong composition. For example, Venus is the second planet from the Sun, but it is the hottest planet in the solar system. Venus's temperature is about 700 K, or about 250 K higher than it would be without an atmosphere. Venus has this high temperature because 96% of its atmosphere is carbon dioxide (CO_2). Carbon dioxide is called a greenhouse gas because it traps infrared radiation, causing the atmosphere to retain solar energy that would otherwise escape into space. The number of greenhouse gases in our atmosphere has been increasing; as a result, the temperature of the Earth has been increasing, a phenomenon known as **global warming**. As you are no doubt aware, the recent increases in carbon dioxide and other greenhouse gases in our atmosphere are most likely due to human activities. If the percentage of greenhouse gases continues to rise, the Earth may become unlivable. In this case study, we will study the Earth's atmosphere and our interaction with it.

20-2 Average and Root-Mean-Square Quantities v_j

When we apply the kinetic theory to a gas, we must consider a large number of particles (assumption 1, Section 20-1). As an analogy, imagine studying the entire human population. Some insights come from focusing on one particular person, but to understand the population as a whole, we often measure properties that have been averaged over all people. Likewise, macroscopic observations of gases depend on properties that are averaged over all the particles making up the gas. In this section, we describe two ways to find the *mean* or average. The second—the root mean square—is useful in the kinetic theory of gases. To better understand the root mean square, we start with a common way to find an average.

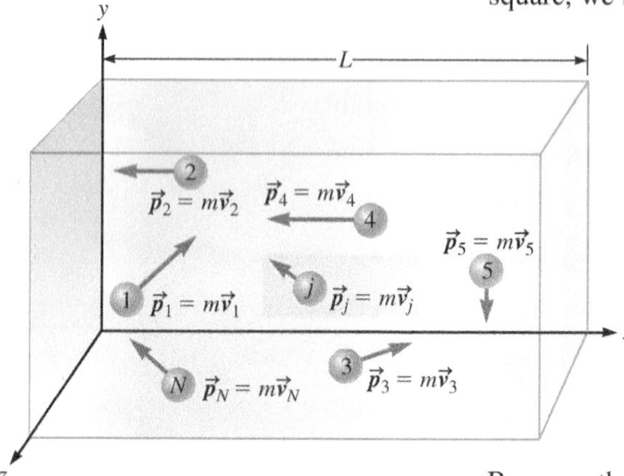

FIGURE 20.2 A gas in a box is made up of many moving, identical particles. For convenience, we number each particle from 1 to N and refer to an arbitrary particle as the jth particle.

Suppose we want to find the average kinetic energy of a set of gas particles. The gas consists of a large number N of identical particles of mass m (assumptions 1 and 5) contained in a box (Fig. 20.2). For convenience, we number each particle from 1 to N and refer to an arbitrary particle as the jth particle. The particles move randomly (assumption 4), so the jth particle has an arbitrary momentum $\vec{p}_j = m\vec{v}_j$. The particles cannot rotate or vibrate (assumption 1), so the kinetic energy of the jth particle is purely due to its translational motion and is given by $K_j = \frac{1}{2}mv_j^2$ (Eq. 8.1). The average kinetic energy is found by adding the kinetic energy of each particle and dividing by the number of particles:

$$K_{\text{av}} = \frac{1}{N}\sum_{j=1}^{N} K_j = \frac{1}{N}\sum_{j=1}^{N}\frac{1}{2}mv_j^2$$

Because the particles are identical (assumption 5), we can bring m outside the summation.

$$K_{\text{av}} = \frac{1}{2}m\left(\frac{1}{N}\sum_{j=1}^{N} v_j^2\right) = \frac{1}{2}mv_{\text{av}}^2 \tag{20.1}$$

where the average of the speed squared is

$$v_{\text{av}}^2 = \frac{1}{N}\sum_{j=1}^{N} v_j^2 \tag{20.2}$$

ROOT MEAN SQUARE Tool

Equation 20.2 leads to another kind of average known as the **root mean square (rms)** that is useful when we apply the kinetic theory to a gas. For example, the root-mean-square *speed* is found by (1) *squaring* the speed of each particle, (2) calculating the average (also known as the *mean*) of the speed squared, and (3) taking the square *root* of the average of the speed squared. Equation 20.2 includes the first two steps: The speed of each particle is squared, and the squares are averaged. To find the **rms speed**, take the square root of Equation 20.2:

$$v_{\text{rms}} = \sqrt{v_{\text{av}}^2} = \sqrt{\frac{1}{N}\sum_{j=1}^{N} v_j^2} \tag{20.3}$$

The root mean square of a quantity does not necessarily equal the mean of that quantity. For example, imagine some stretch of highway where there are about as many cars headed northbound as there are cars headed southbound. If the cars travel at roughly the same speed and if you choose a coordinate system so that north is positive and south is negative, the average velocity of all the cars traveling north and south would be near zero. The rms velocity would *not* be zero because squaring all the velocities eliminates negative values before taking the average. In fact, an rms quantity is never negative, so rms speed is the same as rms velocity.

In Section 20-3, we'll show that the temperature of a gas is closely related to the rms speed of its particles. To do so, we need to derive an expression for the particles' average momentum, and to do that, it is helpful to write the average velocity squared v_{av}^2 in terms of one scalar component v_x. First consider just the jth particle and write v_j^2 in terms of the x, y, and z components of $\vec{v}_j$ (Eq. 3.13):

$$v_j^2 = v_{jx}^2 + v_{jy}^2 + v_{jz}^2 \tag{20.4}$$

When we substitute Equation 20.4 into Equation 20.2, we find that v_{av}^2 can be written as

$$v_{av}^2 = \frac{1}{N}\sum_{j=1}^{N}(v_{jx}^2 + v_{jy}^2 + v_{jz}^2)$$

$$v_{av}^2 = \frac{1}{N}\sum_{j=1}^{N}v_{jx}^2 + \frac{1}{N}\sum_{j=1}^{N}v_{jy}^2 + \frac{1}{N}\sum_{j=1}^{N}v_{jz}^2$$

Each term on the right side of the equal sign is the average velocity squared in that particular direction:

$$v_{av}^2 = v_{av,x}^2 + v_{av,y}^2 + v_{av,z}^2 \quad (20.5)$$

In Figure 20.2, the particles' motion is random (assumption 4). So, there is no preferred direction for the motion of all the particles. Therefore, the average velocity squared should be the same in all three dimensions x, y, and z:

$$v_{av,x}^2 = v_{av,y}^2 = v_{av,z}^2$$

We arbitrarily replace $v_{av,y}^2$ and $v_{av,z}^2$ in Equation 20.5 with $v_{av,x}^2$:

$$v_{av}^2 = v_{av,x}^2 + v_{av,x}^2 + v_{av,x}^2$$

$$v_{av}^2 = 3v_{av,x}^2 \quad (20.6)$$

So, the average velocity squared is three times the average velocity in any one single dimension.

The last average property we need to derive is the average of the x component of momentum squared:

$$p_{av,x}^2 = \frac{1}{N}\sum_{j=1}^{N}p_{jx}^2 = \frac{1}{N}\sum_{j=1}^{N}m^2v_{jx}^2 = m^2\left(\frac{1}{N}\sum_{j=1}^{N}v_{jx}^2\right) \quad (20.7)$$

The term in parentheses in Equation 20.7 is $v_{av,x}^2$:

$$p_{av,x}^2 = m^2v_{av,x}^2 = \tfrac{1}{3}m^2v_{av}^2 \quad (20.8)$$

where we have used Equation 20.6.

EXAMPLE 20.1 Highway Velocities

The velocity of vehicles on a straight stretch of highway is given in Table 20.1. A coordinate system was chosen so that north is positive. Find the average velocity and the rms velocity.

TABLE 20.1 Velocities on a straight stretch of highway.

Number	Velocity (mph)	Number	Velocity (mph)	Number	Velocity (mph)	Number	Velocity (mph)
1 southbound	−67	10 southbound	−70	19 northbound	62	28 northbound	74
2 southbound	−92	11 southbound	−100	20 northbound	76	29 northbound	56
3 southbound	−88	12 southbound	−80	21 northbound	67	30 northbound	51
4 southbound	−82	13 southbound	−74	22 northbound	66	31 northbound	80
5 southbound	−76	14 southbound	−63	23 northbound	70	32 northbound	63
6 southbound	−78	15 southbound	−80	24 northbound	79	33 northbound	69
7 southbound	−55	16 northbound	55	25 northbound	65	34 northbound	62
8 southbound	−81	17 northbound	73	26 northbound	67	35 northbound	79
9 southbound	−77	18 northbound	65	27 northbound	82		

INTERPRET and ANTICIPATE

We expect the average and rms velocities to be very different because nearly half the velocities are negative.

Example continues on page 584 ▶

SOLVE The sum of the velocities (v_{tot}) is fairly small because nearly half the velocities are negative.	$\vec{v}_{tot} = \sum_{j=1}^{N} \vec{v}_j = (-67 - 92 + \cdots + 79)$ mph north $\vec{v}_{tot} = 198$ mph north
Find the average velocity v_{av} by dividing by the total number of vehicles.	$\vec{v}_{av} = \dfrac{\vec{v}_{tot}}{N} = \dfrac{198\,\text{mph}}{35}$ north $= 5.7$ mph north
To find the rms velocity v_{rms}, square each velocity and add the results.	$v_{tot}^2 = \sum_{j=1}^{N} v_j^2 = (-67\text{ mph})^2 + (-92\text{ mph})^2 + \cdots + (79\text{ mph})^2$ $v_{tot}^2 = 185{,}992\,(\text{mph})^2$
Next, divide by the number of vehicles to find the average of the squared velocities.	$v_{av}^2 = \dfrac{v_{tot}^2}{N} = \dfrac{185{,}992(\text{mph})^2}{35} = 5314\,(\text{mph})^2$
Finally, take the square root to find the rms velocity.	$v_{rms} = \sqrt{v_{av}^2} = \sqrt{5314(\text{mph})^2} = 73$ mph

CHECK and THINK

As predicted, the average velocity and the rms velocity are very different. Interpreting an average velocity of nearly zero is somewhat ambiguous. It may be that roughly equal numbers of vehicles are traveling in opposite directions at roughly the same speeds, or it may be that a few very fast vehicles are moving in one direction and many very slow vehicles are moving in the other direction. Depending on what you would like to know about traffic patterns, the average velocity may not be helpful. For example, if you want to know if drivers are exceeding the speed limit, the average velocity isn't useful. The rms velocity may be more helpful; because the velocities are squared first, the rms calculation does not depend on the sign (direction) of the velocity. The rms velocity is slightly higher than the average speed because squaring the values first means that higher speeds have a slightly greater weight when taking the average.

CONCEPT EXERCISE 20.1

In Example 20.1, we found that the rms value of a quantity was greater than the average of that quantity. Is it possible for the rms value to equal the average? If so, give an example of six numbers whose rms value equals their average. If not, explain why not.

20-3 The Kinetic Theory Applied to Gas Temperature and Pressure

One of the major goals of the kinetic theory is to connect the microscopic model of a gas to our macroscopic observations. In this section, we connect the particles' average kinetic energy and rms speed to the temperature of an ideal gas.

The pressure we observe in a gas is due to the total force per unit area exerted by the particles on the container walls. Once again, consider a gas inside a box. To find the total force, start by focusing on the jth particle as it collides with the shaded wall at the left in Figure 20.3. The particle collides elastically with the wall (assumption 3), and the wall exerts a force in the positive x direction on the particle. According to Newton's second law $\vec{F} = d\vec{p}/dt$ (Eq. 10.2), only the x component of the particle's momentum is subject to change as a result of the collision. Because the wall has a much greater mass than the particle, the particle reverses direction, and its velocity after the collision is $\vec{v}_{1f} \approx -\vec{v}_{1i}$ (Fig. 11.13B). So, the change in the particle's momentum is

$$\Delta p_{jx} = (p_{jx})_f - (p_{jx})_i = 2(p_{jx})_i = 2p_{jx} \tag{20.9}$$

where we have dropped the subscript i in the last quantity for simplicity. The symbol p_{jx} now denotes the particle's initial momentum.

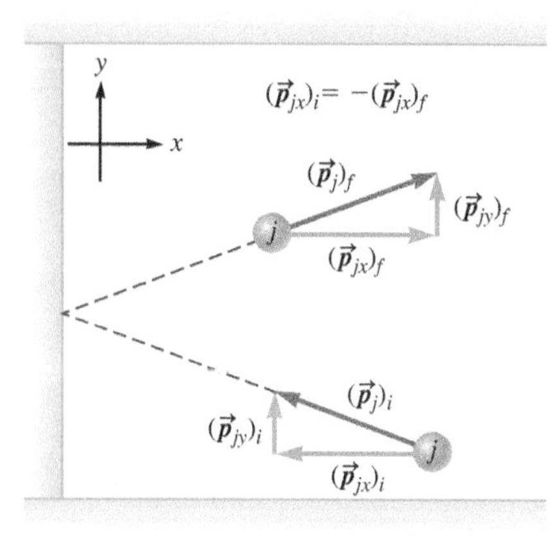

FIGURE 20.3 When the jth particle collides with the wall on the left, only the x component of its momentum is changed. The y component is identical before and after the collision.

We use the impulse–momentum theorem $\vec{I}_{tot} = \Delta\vec{p} = \vec{F}_{av}\Delta t$ (Eqs. 11.3 and 11.4) to find the average force $(F_{jx})_{av}$ exerted by the left wall on the jth particle:

$$I_x = \Delta p_x = (F_{jx})_{av}\,\Delta t_{\text{collision}}$$

$$(F_{jx})_{av} = \frac{\Delta p_x}{\Delta t_{\text{collision}}} \tag{20.10}$$

The impulse Δp_x is the area under the force-versus-time curve (Fig. 20.4). We have an expression for Δp_x (Eq. 20.9), so once we find an expression for $\Delta t_{\text{collision}}$, we will have an expression for $(F_{jx})_{av}$. Finding $\Delta t_{\text{collision}}$ can be tricky because the duration of the collision is so short. In this situation, however, there is a clever solution to this problem. Imagine that at $t = 0$, the jth particle is at the *right* wall, and after some time interval Δt, the particle is back at the right wall, having collided with the left wall. (During that time interval, the particle may bounce off other walls, but because those collisions cannot change the x component of its momentum, we may ignore such collisions in our analysis.) The only force in the x direction that is exerted on the particle during this time interval is due to the left wall, so the area under the force-versus-time curve is equal to that of the thin rectangular box shown in Figure 20.4. The height of this box is the average x component of force exerted on the particle during the time Δt.

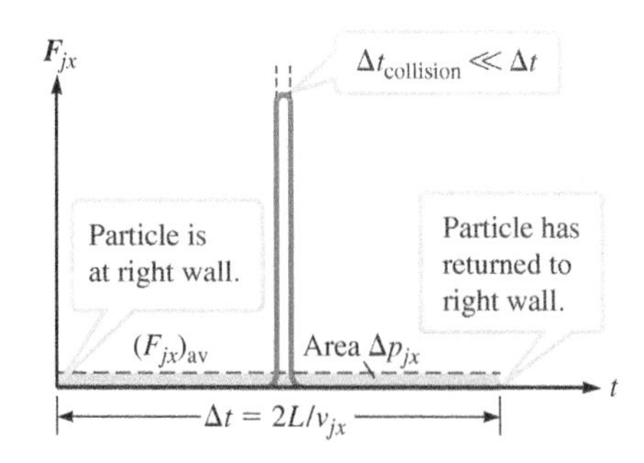

FIGURE 20.4 The change in the jth particle's momentum is the area under the force-versus-time curve (red), which also equals the area of the very thin shaded rectangular box $\Delta p_{jx} = (F_{jx})_{av}\,\Delta t$.

All we need to do is find an expression for Δt, the time the particle takes to go from the right wall to the left wall and back again. The collision with the left wall only changes the direction of the particle's momentum. So, the jth particle's velocity parallel to the x axis reverses direction, but its magnitude v_{jx} remains unchanged. The time for the particle to return to the right wall is given by the total distance traveled ($2L$ for a box of length L) divided by that magnitude v_{jx}:

$$\Delta t = \frac{2L}{v_{jx}} \tag{20.11}$$

We find an expression for the average force exerted by the left wall on the jth particle by substituting Equations 20.11 and 20.9 into Equation 20.10:

$$(F_{jx})_{av} = \frac{2p_{jx}}{2L/v_{jx}}$$

$$(F_{jx})_{av} = \frac{p_{jx}v_{jx}}{L} = \frac{p_{jx}^2}{mL} \tag{20.12}$$

where we have used $p_{jx} = mv_{jx}$.

By Newton's third law, the average force exerted by the jth particle on the left wall $(f_{jx})_{av}$ is equal in magnitude and opposite in direction to the force exerted by the wall on the particle $(F_{jx})_{av}$. So, from Equation 20.12,

$$(f_{jx})_{av} = -\frac{p_{jx}^2}{mL} \tag{20.13}$$

Equation 20.13 is the average force exerted by the jth particle on the left wall. If there were only a few particles in the box, the wall would experience a relatively strong force when it was hit by a particle and no force for some period of time when it was not being hit. According to assumption 1, however, there are many particles in the box. So, the wall is continually bombarded by particles, and the force of the particles on the wall is a constant that we can find by adding the average force exerted by each particle:

$$f_{av,x} = \sum_{j=1}^{N}(f_{jx})_{av} = -\sum_{j=1}^{N}\frac{p_{jx}^2}{mL} = -\frac{1}{mL}\sum_{j=1}^{N}p_{jx}^2$$

Multiply the right side by N/N, where N is the number of particles:

$$f_{av,x} = -\frac{N}{mL}\left(\frac{1}{N}\sum_{j=1}^{N}p_{jx}^2\right)$$

The term in parentheses is the average of the momentum's x component squared:

$$f_{av,x} = -\frac{N}{mL}p_{av,x}^2$$

Now substitute $p^2_{\text{av},x} = \frac{1}{3}m^2v^2_{\text{av}}$ (Eq. 20.8):

$$f_{\text{av},x} = -\frac{N}{mL}\left(\frac{1}{3}m^2v^2_{\text{av}}\right) = -\frac{N}{3L}mv^2_{\text{av}}$$

The pressure on the left wall is the absolute value of the force exerted by the particles divided by the area of the wall:

Uppercase P stands for pressure, and lowercase p stands for momentum.

$$P = \frac{|f_{\text{av},x}|}{A} = \frac{N}{3LA}mv^2_{\text{av}} \tag{20.14}$$

The volume V of the box is LA, so

$$P = \frac{N}{3V}mv^2_{\text{av}}$$

$$PV = \frac{N}{3}mv^2_{\text{av}} \tag{20.15}$$

Uppercase V stands for volume, and lowercase v stands for velocity (speed).

Compare Equation 20.15 with the ideal gas law $PV = Nk_BT$ (Eq. 19.17) to find

$$PV = Nk_BT = \frac{N}{3}mv^2_{\text{av}}$$

$$k_BT = \tfrac{1}{3}mv^2_{\text{av}}$$

$$v^2_{\text{av}} = \frac{3k_BT}{m} \tag{20.16}$$

Substitute into Equation 20.1 for average kinetic energy:

CONNECTION BETWEEN AVERAGE KINETIC ENERGY AND TEMPERATURE

Major Concept

$$K_{\text{av}} = \tfrac{1}{2}mv^2_{\text{av}} = \tfrac{3}{2}k_BT \tag{20.17}$$

Equation 20.17 confirms what we stated in Section 19-1: **The absolute temperature is a measure of the average translational kinetic energy of (ideal) gas particles.** Specifically, temperature is directly proportional to the average translational kinetic energy.

The absolute temperature is also a measure of the rms velocity of the particles. To find the rms velocity, take the square root of each side of Equation 20.16 and use Equation 20.3:

CONNECTION BETWEEN RMS VELOCITY AND TEMPERATURE

Major Concept

$$v_{\text{rms}} = \sqrt{v^2_{\text{av}}} = \sqrt{\frac{3k_BT}{m}} \tag{20.18}$$

Equations 20.17 and 20.18 connect the microscopic motion of gas particles to macroscopic observations of the gas's absolute temperature. According to these equations, a higher temperature means that the particles have more kinetic energy and are moving, on average, faster, which gives us another way to interpret absolute zero (Section 19-7). No one has observed absolute zero, but if a system were at absolute zero, the average kinetic energy and rms velocity of its particles would be zero. So, the particles would not be moving. According to quantum mechanics (Chapters 40–42), however, this situation is not possible; particles actually have some kinetic energy even at absolute zero.

CONCEPT EXERCISE 20.2

If the temperature of a gas is doubled, what happens to the average kinetic energy and rms speed of its particles?

CONCEPT EXERCISE 20.3

Gas 1 and gas 2 are at the same temperature. Gas 1 is made up of molecules whose mass is four times that of gas 2's molecules: $m_1 = 4m_2$. Compare the average kinetic energy of the molecules in the two gases. Compare the rms speed of the molecules in the two gases.

EXAMPLE 20.2 CASE STUDY The Earth's Atmosphere

Hydrogen and helium are the most abundant elements in the Universe, together making up almost 100% of it, whereas all other elements make up just a fraction of a percent. The Sun and the four giant planets Jupiter, Saturn, Uranus, and Neptune have a lot of pure hydrogen and helium in their atmospheres. The inner planets Mercury, Venus, the Earth, and Mars have essentially no pure atmospheric hydrogen or helium. (Hydrogen compounds such as water are found in the Earth's atmosphere.)

A planet's atmosphere is not contained by the walls of a box but instead is bound by gravity. Molecules and atoms that are moving at the escape speed

$$v_{esc} = \sqrt{\frac{2GM}{R}} \tag{8.17}$$

or faster will leave the atmosphere, where M is the mass of the planet and R is its radius.

The rms speed of gas particles is determined by their temperature and mass (Eq. 20.18), so different planets have different gases in their atmosphere depending on the planet's temperature and mass. If $v_{rms} = v_{esc}$ for a given type of particle, that gas will leave the atmosphere in only a few days. The condition for retaining a certain gas in the atmosphere for several billion years (the age of the solar system) is $10v_{rms} \le v_{esc}$. Find the mass (in kilograms and atomic mass units) of the lightest molecule or atom that can be retained in the Earth's atmosphere. Use the periodic table (Appendix B) to determine why there is no hydrogen or helium gas in the Earth's atmosphere. Assume the atmospheric temperature is a uniform 277 K.

INTERPRET and ANTICIPATE

According to $v_{rms} = \sqrt{3k_BT/m}$, lighter particles have a greater rms velocity. By setting $10v_{rms} = v_{esc}$, we can solve for the mass m_{min} of the lightest particles in the atmosphere. We expect to find that major components of the atmosphere such as nitrogen gas (N_2) and oxygen gas (O_2) are easily retained by the Earth's gravity.

SOLVE

Set $10v_{rms} = v_{esc}$. Use $v_{rms} = \sqrt{3k_BT/m}$ (Eq. 20.18) and $v_{esc} = \sqrt{2GM/R}$ (Eq. 8.17).

$$10v_{rms} = v_{esc}$$

$$10\sqrt{\frac{3k_BT_\oplus}{m_{min}}} = \sqrt{\frac{2GM_\oplus}{R_\oplus}}$$

Square each side and solve for m_{min}.

$$100\left(\frac{3k_BT_\oplus}{m_{min}}\right) = \frac{2GM_\oplus}{R_\oplus}$$

$$m_{min} = \frac{150k_BR_\oplus T_\oplus}{GM_\oplus}$$

Find m_{min} in kilograms and then convert to atomic mass units (u).

$$m_{min} = \frac{150(1.38 \times 10^{-23}\,\text{J/K})(6.387 \times 10^6\,\text{m})(277\,\text{K})}{(6.67 \times 10^{-11}\,\text{N}\cdot\text{m}^2/\text{kg}^2)(5.97 \times 10^{24}\,\text{kg})}$$

$$m_{min} = 9.2 \times 10^{-27}\,\text{kg}$$

$$m_{min} = 9.2 \times 10^{-27}\left(\frac{1\,\text{u}}{1.66 \times 10^{-27}\,\text{kg}}\right) = 5.5\,\text{u}$$

CHECK and THINK

Our result means that the Earth's gravity can retain gas particles whose mass is at least 5.5 u. According to the periodic table, atomic hydrogen and helium have atomic masses of 1 and 4 u, respectively. Both are too light to be retained in the atmosphere. Even molecular hydrogen gas (H_2) is too light to be retained. We expected molecular nitrogen and oxygen gases to be easily retained, however. The molecular masses of N_2 and O_2 are 28 and 32 u, respectively, both much heavier than the minimum retainable mass.

20-4 Maxwell-Boltzmann Distribution Function

Because the kinetic theory of gases has been so successful, it might be hard to imagine that there were once alternative theories. One alternative model was a *static* model, according to which gas atoms (such as those in the surrounding air) were in constant contact with one another, normally at rest and not moving around randomly as we assumed in Figure 20.2. Instead, this model attributed the gas pressure we experience to the many atoms pressed against us. Proponents of the static model believed that atoms were small, fundamental objects whose size could change. Further, if there were a net motion of the gas, such as when you breathe air into or out of your lungs, all the gas atoms would move together at the same velocity.

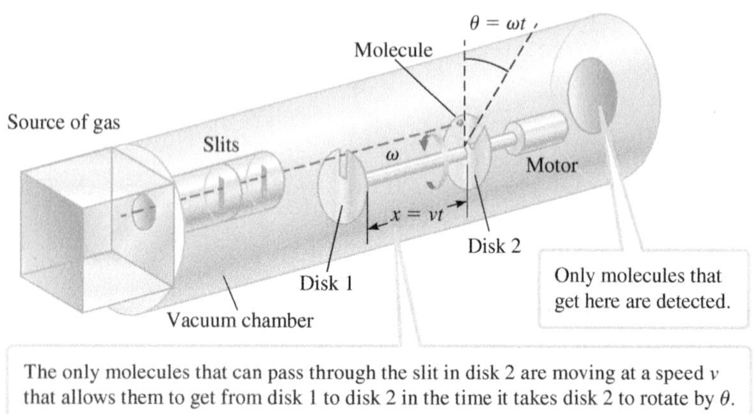

FIGURE 20.5 Molecules escape through the slits and travel in a beam toward the spinning disks. Only molecules within a certain range of speeds will make it all the way to the detector.

Figure 20.5 shows an apparatus that may be used to collect evidence supporting either the static model or the kinetic theory. On the left is a source of gas particles. The particles escape through a series of slits to the right of the source, creating a narrow beam of particles in a vacuum chamber. The beam is aimed at two subsequent disks rotating at angular speed ω. Each disk has a narrow slit allowing molecules to pass. The distance between the rotating disks is x, and the slit in disk 2 is offset by an angle θ from the slit in disk 1. The time it takes the slit in disk 2 to line up with the beam (after the beam has passed through the slit in disk 1) is $t = \theta/\omega$, so particles traveling at speed $v = x/t = x(\omega/\theta)$ will pass through the slit in disk 2, and the other particles will be blocked. Actually, because the slits have some finite width, a range of speeds can get through disk 2. The particles passing through disk 2 are counted by the detector on the right end of the vacuum chamber. By changing the angular speed ω of the disks, particles of different speed ranges are counted.

To find the number of particles in a particular speed range, ω is set for that range. Each bar in Figure 20.6 comes from counting the number of particles at a particular ω. The height of each bar is determined by dividing the number of particles in each particular range by the total number of particles detected in all the speed ranges. So, the height of each bar represents the fraction of particles that are in that particular speed range. If the static model were correct, all the particles would be in contact with one another and moving at the same speed. Therefore, the static model predicts that there should only be one bar of height 1. In Figure 20.6, however, we find a variety of ranges of possible speeds, which supports the kinetic theory.

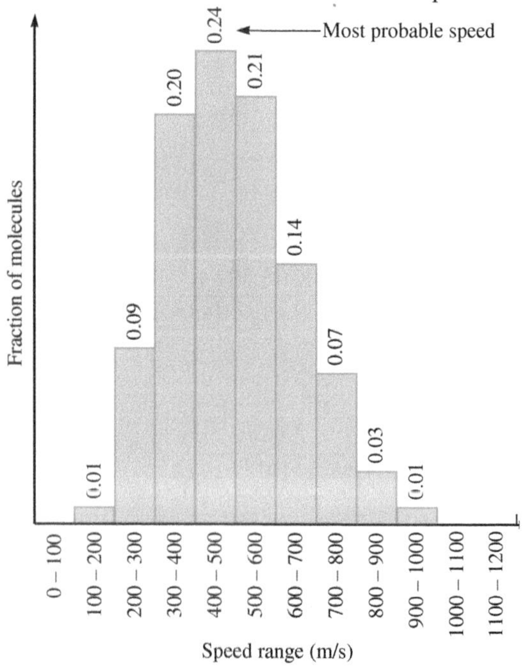

FIGURE 20.6 The height of each bar represents the fraction of particles in the sample that are in the particular speed range.

Figure 20.6 tells us more about the motion of the particles in the gas. The height of each bar is the fraction of particles in that particular speed range and also the probability that a randomly selected particle's speed is in that range. In this case, the most probable speed is somewhere in the range of 400 to 500 m/s.

We see that the most probable speed is not in the very center of the distribution, but is somewhat lower than the median (middle) speed. In other words, there are more particles with speeds greater than this most probable speed than particles with lower speeds. By adding the fractions of particles that are below the most probable range, we find that $(0.01 + 0.09 + 0.20) = 0.30$ of the particles are at speeds lower than the most probable speed. The fraction of particles with speeds higher than the most probable speed is $(0.21 + 0.14 + 0.07 + 0.03 + 0.01) = 0.46$.

If the temperature of the source of particles in Figure 20.5 increases, the most probable speed increases as shown in Figure 20.7. Compare the speed distribution for the cool gas (blue bars) with the speed distribution for the hot gas (red bars). The most probable speed for the hot gas is in the range of 900 to 1000 m/s, or about 500 m/s greater than that for the cool gas. The hot gas also has a greater overall range of speeds. The high-

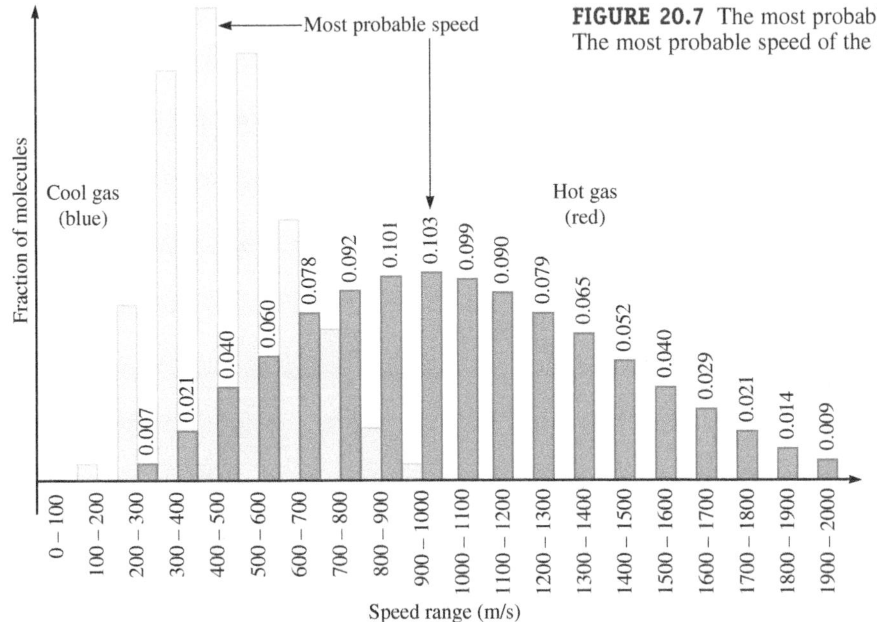

FIGURE 20.7 The most probable speed depends on the temperature of the gas. The most probable speed of the hotter gas is greater than that of cooler gases.

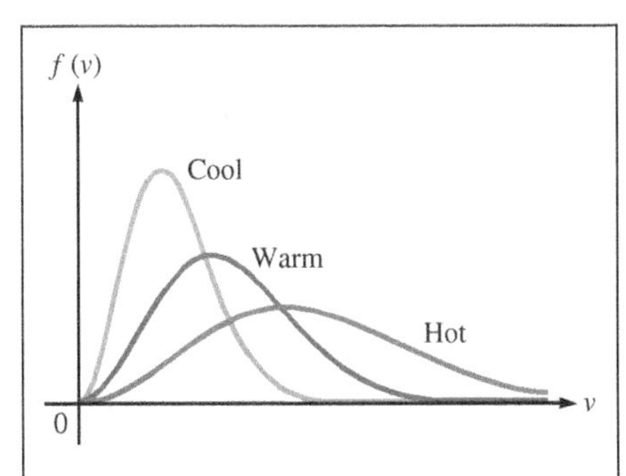

FIGURE 20.8 Maxwell-Boltzmann distribution function for a gas at three different temperatures.

est range of speeds in the hot gas is 1900 to 2000 m/s or about 1000 m/s faster than the top speeds in the cool gas.

Maxwell-Boltzmann Distribution

Imagine reducing the slit width in the rotating disks in Figure 20.5. As the widths are reduced, the range of speeds represented by each bar in Figures 20.6 and 20.7 gets smaller and smaller. As the width approaches zero, the distribution becomes continuous (Fig. 20.8) and is given by

$$f(v) = 4\pi\left(\frac{m}{2\pi k_B T}\right)^{3/2} v^2 e^{-(mv^2/2k_B T)} \qquad (20.19)$$

MAXWELL-BOLTZMANN SPEED DISTRIBUTION ✪ Major Concept

Equation 20.19 is known as the **Maxwell-Boltzmann speed distribution**. The function $f(v)$ is called a distribution function, and it gives the *probability of a particle's speed per unit speed interval*. The probability that a randomly chosen particle has a speed between v and $(v + dv)$ is the probability per unit speed interval $f(v)$ multiplied by the width dv of the speed interval: $f(v)dv$. Because probability does not have dimensions, the probability per unit speed $f(v)$ must have the SI units of seconds per meter.

DERIVATION Most Probable Speed

Figure 20.8 shows the Maxwell-Boltzmann speed distribution for a hot, a warm, and a cool gas. As we saw before, the distribution is more spread out for a hot gas than it is for a cool gas. Figure 20.8 also shows that the most probable speed in a hot gas is greater than the most probable speed in a cool gas. We will show that the most probable speed is a function of temperature given by

$$v_{mp} = \sqrt{\frac{2k_B T}{m}} \qquad (20.20)$$

MOST PROBABLE SPEED ✪ Major Concept

The most probable speed v_{mp} occurs at the peak of the distribution, so we can find an expression for the most probable speed by setting the derivative of Equation 20.19 equal to zero.	$\frac{df(v)}{dv} = 0$ at $v = v_{mp}$ $\frac{d}{dv}\left[4\pi\left(\frac{m}{2\pi k_B T}\right)^{3/2} v^2 e^{-(mv^2/2k_B T)}\right] = 0$
Pull the constant terms outside the derivative.	$4\pi\left(\frac{m}{2\pi k_B T}\right)^{3/2} \frac{d}{dv}\left[v^2 e^{-(mv^2/2k_B T)}\right] = 0$

Derivation continues on page 590 ▶

To complete the derivative, use the chain rule (Appendix A).	$4\pi\left(\frac{m}{2\pi k_B T}\right)^{3/2}\left[2ve^{-(mv^2/2k_BT)} - v^2\left(\frac{2mv}{2k_BT}\right)e^{-(mv^2/2k_BT)}\right] = 0$ $4\pi\left(\frac{m}{2\pi k_B T}\right)^{3/2}\left[2ve^{-(mv^2/2k_BT)}\right]\left[1 - v^2\frac{m}{2k_BT}\right] = 0$
At the most probable speed, the last term must equal zero.	$\left[1 - v_{mp}^2\frac{m}{2k_BT}\right] = 0$
Solve for the most probable speed.	$v_{mp} = \sqrt{\frac{2k_BT}{m}}$ (20.20)

DERIVATION Average Speed

To find the average speed, we add the speed of each particle and divide by the number of particles. We can extend that procedure by integrating instead of adding. So, we will integrate over a continuous range of speeds using the Maxwell-Boltzmann speed distribution to show that the average speed of the particles in a gas is given by

$$v_{av} = \sqrt{\frac{8k_BT}{\pi m}} \tag{20.21}$$

AVERAGE SPEED
Major Concept

Because there are more particles at speeds above the most probable speed v_{mp} than below (Fig. 20.6), we expect the average speed v_{av} to be greater than the most probable speed: $v_{av} > v_{mp}$.

Let's call the total number of gas particles N and the number of particles in a particular speed range dN. Write an expression for dN.	$dN(v) = \left(\text{number of particles}\right)\left(\text{fraction of particles per unit speed range}\right)\left(\text{width of speed range}\right) = Nf(v)dv$ (1)
Find the average speed by adding the speed of each particle (integrating in this case for a continuous distribution) and dividing by the number of particles.	$v_{av} = \frac{\int v\,dN(v)}{N}$
Substitute Equation (1) for dN. Now the integral is over dv, whose range is from 0 to ∞.	$v_{av} = \frac{\int vNf(v)dv}{N}$ $v_{av} = \int_0^\infty vf(v)dv$
Substitute the Maxwell-Boltzmann distribution (Eq. 20.19) for $f(v)$.	$v_{av} = 4\pi\left(\frac{m}{2\pi k_BT}\right)^{3/2}\int_0^\infty v^3 e^{-(mv^2/2k_BT)}dv$
Integrate by parts (Appendix A) or look up the integral.	$v_{av} = 4\pi\left(\frac{m}{2\pi k_BT}\right)^{3/2}\left(\frac{2k_B^2T^2}{m^2}\right)$
Simplify this expression.	$v_{av} = \sqrt{\frac{8k_BT}{\pi m}}$ (20.21)
COMMENTS To check our results, compare the most probable speed (Eq. 20.20) with the average speed (Eq. 20.21). As expected, the average speed is greater than the most probable speed.	$v_{mp} = \sqrt{\frac{2k_BT}{m}} \approx 1.4\sqrt{\frac{k_BT}{m}}$ $v_{av} = \sqrt{\frac{8k_BT}{\pi m}} \approx 1.6\sqrt{\frac{k_BT}{m}}$ $v_{av} > v_{mp}$

CASE STUDY Part 2: Smog

Smog is often seen as a brownish-yellow haze around an urban center (Fig. 20.9). Spending even a couple of hours in smog can irritate your eyes, nose, and throat. Repeated or extended exposure to smog can cause respiratory illnesses and may prematurely age the lungs. High levels of smog can damage crops and vegetation. Various government agencies issue smog warnings, and you might have noticed that smog warnings tend to be worse on hot days. In this part of the case study, we use the Maxwell-Boltzmann distribution to explain why smog depends on temperature.

© Hampelmann/Dreamstime.com

FIGURE 20.9 Smog over Los Angeles harbor.

The term *smog* originally referred to a mixture of *smo*ke and f*og*. Today, it refers to a noxious mixture of many air pollutants (including gases and particulate matter) suspended in the atmosphere near the surface of the Earth, and it is more precisely called *photochemical smog*, a result of sunlight (*photons*) and *chemical* reactions in the atmosphere.

Many human activities produce hydrocarbons (molecules composed of hydrogen and carbon) and nitrogen oxides (molecules composed of nitrogen and oxygen) that react in the atmosphere to become smog. Industrial societies are generally making some effort to reduce the production of such pollutants. For example, cars produce both hydrocarbons and nitrogen oxides, but catalytic converters on cars destroy these pollutants before they become part of the atmosphere. Even with such measures in place, many cities such as Los Angeles, California, still have serious smog problems and are looking for other ways to reduce smog. One way is to reduce the city's temperature.

From your experience in the kitchen, you know that the rate at which a chemical reaction takes place depends on temperature. To preserve food, you put it in the refrigerator or freezer, and when you want to cook something (change its chemical composition), you put that food in the oven or on the stovetop.

Chemical reactions in the atmosphere also depend on temperature. A chemical reaction is a result of a collision between two or more molecules, forming one or more new molecules. For a chemical reaction to occur, the two colliding molecules must have a certain minimum amount of energy: The molecules must exceed a certain minimum relative speed when they collide. Atmospheric hydrocarbons, nitrogen oxides, hydrogen oxides, and oxygen must react with one another to make smog. These various types of molecules have Maxwell-Boltzmann speed distributions. (In the case of a mixture of gases such as our atmosphere, the Maxwell-Boltzmann distribution holds for each type of molecule separately.) When the atmosphere is at a high temperature, there are many fast particles as shown in Figure 20.8 for the hot gas, so when two molecules collide, it is likely that they will be going fast enough to undergo a chemical reaction. In a much cooler gas, very few molecules may have enough energy to undergo a chemical reaction. For this reason, more smog is produced on hot days than on cool days. In fact, smog is not a problem on days that are cooler than about 70°F.

Smog in urban centers is made worse because cities are warmer than the surrounding countryside and warmer than they used to be. For example, in the 1930s, Los Angeles was full of orchards. In 1934 (one of the hottest years on record), the highest temperature in Los Angeles was 97°F. Today, those orchards have been replaced by buildings, roads, and parking lots. The high temperature of Los Angeles has risen by more than 8°F; in 2010, the hottest temperature in Los Angeles was 113°F. Some solutions to reducing smog focus on lowering the city's temperature by making the pavement and rooftops reflective and by planting trees. Lowering the temperature means that fewer atmospheric molecules will undergo the chemical reactions that produce smog.

20-5 Mean Free Path

Imagine waking up to the sight and scent of fresh hot coffee. Aromatic molecules from the coffee travel to your nose. You probably don't notice much of a time delay between when you see the coffee and when you smell it because your nose isn't very far from the cup. In Annapolis, Maryland, a small coffee company roasts coffee beans in a residential neighborhood (Fig. 20.10A). The people in the shop are the first to smell the coffee. On a calm day without wind, residents cannot smell the coffee for several minutes or more, depending on how far they are from the roaster. If the molecules were truly point particles, those molecules would not collide with one another but would travel in straight lines out from the roaster at hundreds of meters per second. Residents who live within a few blocks of the roaster would smell the coffee in

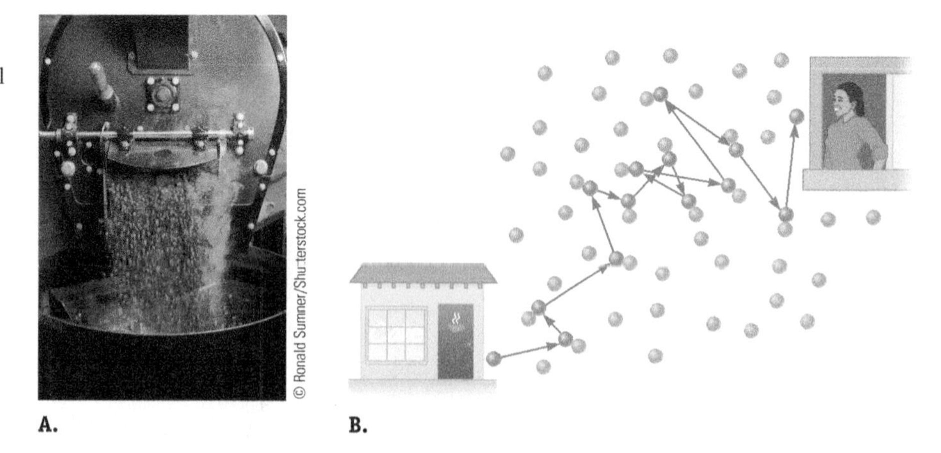

FIGURE 20.10 A. A coffee roaster in a small residential neighborhood. **B.** On a calm day, aromatic coffee molecules travel in a zigzag path throughout the neighborhood.

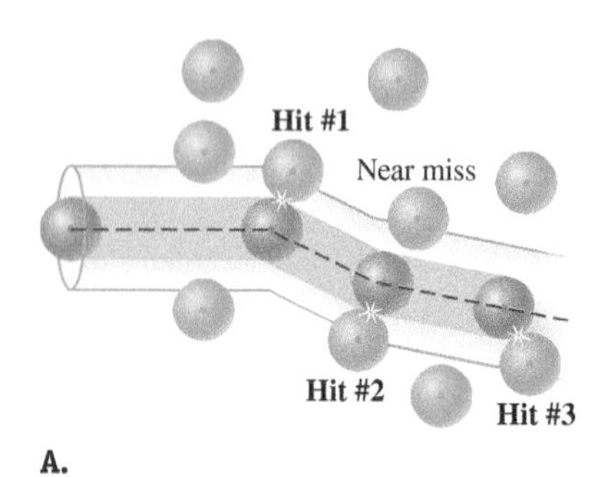

A.

Near miss
Hit #1
$2d = 4r$
$d = 2r$
Hit #2
Hit #3
$v_{av}\Delta t$

B.

FIGURE 20.11 A. The red particle undergoes three collisions along this short path. **B.** For convenience, this scenario is drawn as a straight path.

a fraction of a second. The molecules are not collisionless point particles, though. Instead, they collide as they move in the air, and with each collision their directions of motion are altered so that the molecules travel along a zigzag path (Fig. 20.10B).

Between each collision, a molecule's trajectory is a straight line. If there are many molecules or if the molecules are very large, the straight line parts of their paths will be fairly short. The **mean free path** is the average length of the straight portion of a molecule's path. When two populations of particles encounter each other, such as when two gases mix or when two galaxies of stars merge, the mean free path is a measure of how likely the particles are to collide. A long mean free path means that collisions are rare.

In this section, we relax assumption 1 (that the molecules are particles with no physical extent) to find an expression for a molecule's mean free path. Let's trace the path of one molecule shown in red in Figure 20.11A; the molecule undergoes three collisions along this short path in a time Δt. We model all the molecules as hard spheres of radius r, and for now we assume the red molecule is moving at speed v_{av} and the remaining molecules are at rest. It is somewhat easier to derive the red molecule's mean free path if we unbend its path as shown in Figure 20.11B. Now we can see that the red molecule sweeps out a cylindrical space of diameter $d = 2r$. Figure 20.11B also shows a cylinder of diameter $4r$. The red molecule will collide with any other molecule whose center is within this larger cylinder. The number of such molecules is given by the density of molecules N/V multiplied by the volume of the cylinder. The length of this large cylinder is $v_{av}\,\Delta t$, so its volume is $\pi(2r)^2 v_{av}\,\Delta t$. The number of collisions that occur in the time Δt is given by the number of molecules whose center is within this cylinder:

$$\text{number of collisions} = \frac{N}{V}\pi(2r)^2 v_{av}\,\Delta t \tag{20.22}$$

The mean free path λ is the total distance traveled in the time Δt divided by the number of collisions in that time:

$$\lambda \approx \frac{v_{av}\,\Delta t}{(N/V)\pi(2r)^2 v_{av}\,\Delta t}$$

$$\lambda \approx \frac{1}{4\pi r^2(N/V)} \quad \text{(approximate)} \tag{20.23}$$

Equation 20.23 is approximate because it was derived under the assumption that only the red molecule moves. Of course, that assumption is not valid because the other molecules must also be moving. The number of collisions occurring in the time Δt depends on the *relative* speed of the colliding molecules; so, to find a more accurate expression, we replace v_{av} in Equation 20.22 with v_{rel}. For a Maxwell-Boltzmann distribution, the relative speed is $v_{rel} = \sqrt{2}v_{av}$, so the mean free path is given by

MEAN FREE PATH ✪ **Major Concept**

$$\lambda = \frac{1}{4\sqrt{2}\pi r^2(N/V)} \tag{20.24}$$

The average time t_{mf} between collisions is called the **mean free time**, found by dividing the mean free path by the average speed of the molecules:

$$t_{mf} = \frac{\lambda}{v_{av}} \tag{20.25}$$

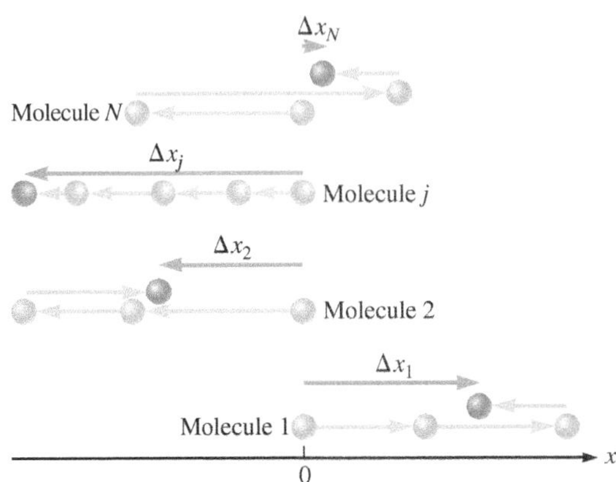

FIGURE 20.12 After each collision, a molecule is just as likely to move to the left as to the right. The average displacement of a collection of molecules depends on their mean free path and the mean free time.

Diffusion

Diffusion is a process that mixes the particles of two substances (such as the molecules of two gases) as a result of the particles' kinetic energy. For example, the aromatic molecules of coffee mix with air molecules. The zigzag path of the aromatic molecules (Fig. 20.10B) is called a **random walk**.

In Example 20.4, we will estimate the time it takes aromatic coffee molecules traveling in air to make their way throughout the neighborhood by diffusion. To do so, we would like to know how far (on average) an aromatic molecule can travel in some time Δt. To start, imagine an aromatic molecule that is constrained to move in just one dimension along the x axis. After each collision (with a particle in the air), the molecule is just as likely to move to the left as to the right, and after some time Δt, the magnitude of its displacement is Δx. Now consider N molecules that start at the origin and walk randomly along the x axis (Fig. 20.12). The average number of steps each molecule takes in a time Δt is

$$\text{average number of steps} = \frac{\Delta t}{\text{average time between collisions}} = \frac{\Delta t}{t_{mf}}$$

Some molecules have a positive displacement and some a negative one, but we are *not* interested in their direction; we are only interested in the average distance the molecules have traveled from the origin. So, it is convenient to find an expression for the square of the average displacement Δx_{av}^2. Because the average size of a molecule's step is its mean free path λ,

$$\Delta x_{av}^2 = (\lambda^2) \times (\text{average number of steps}) = \lambda^2(\Delta t/t_{mf})$$

The mean free path and the mean free time both depend on size and density of the two types of molecules. These two factors are usually combined as $D = \lambda^2/2t_{mf}$, where D is known as the **diffusion constant**. So, in one dimension, the average magnitude of the displacement is

$$\Delta x_{av} = \sqrt{2D\,\Delta t}$$

The diffusion constant has the dimensions of length squared per time; it depends on the properties of both gases, so it is reported for diffusion of one type of molecule in a particular medium at a given temperature. Table 20.2 gives D for several gases in air. Usually, the molecules may move in all three dimensions, and the average displacement is given by

$$\Delta r_{av} = \sqrt{6D\,\Delta t} \tag{20.26}$$

TABLE 20.2 Diffusion constant for gases in air.

Gas or vapor	D (m²/s)	T (°C)
Alcohol vapor	1.37×10^{-5}	40.4
Carbon dioxide	1.39×10^{-5}	0.0
Carbon disulfide	1.02×10^{-5}	19.9
Ether vapor	8.9×10^{-6}	19.9
Hydrogen, H_2	6.34×10^{-5}	0.0
Oxygen, O_2	1.78×10^{-5}	0.0
Water vapor	2.39×10^{-5}	8.0

Adapted from *The Handbook of Chemistry and Physics*, 68th ed. (Boca Raton, FL: CRC Press, 1987–1988), p. F-47.

EXAMPLE 20.3 It's in the Air

The air in your room is made up mostly of nitrogen and oxygen. Approximate these molecules as spheres of radius $r = 2.0 \times 10^{-10}$ m.

A Estimate the average distance between the air molecules in your room. Compare your answer to the size of a molecule.

INTERPRET and ANTICIPATE

We can find the number density (N/V) of air molecules from the ideal gas law. Number density is molecules per unit volume with the dimensions (L)$^{-3}$, so to find a distance L per molecule, we take the reciprocal of the cube root of the number density. We expect—based on assumption 1—that the average distance between molecules is greater than the size of the molecules.

Example continues on page 594 ▶

SOLVE

Use the ideal gas law (Eq. 19.17) to find N/V. Take room temperature to be 20°C (293 K) and the air pressure to be 1 atm (1.01×10^5 Pa). Keep an extra significant figure in the calculations for now.

$$PV = Nk_BT \tag{19.17}$$

$$\frac{N}{V} = \frac{P}{k_BT} = \frac{1.01 \times 10^5 \text{ Pa}}{(1.38 \times 10^{-23}\text{ J/K})(293\text{ K})}$$

$$\frac{N}{V} = 2.5 \times 10^{25}\text{ molecules/m}^3$$

To find the average distance d_{av} between molecules, take the reciprocal of the cube root of the number density.

$$d_{av} = \left(\frac{N}{V}\right)^{-1/3} = [2.5 \times 10^{25}\text{ m}^{-3}]^{-1/3}$$

$$d_{av} = 3.4 \times 10^{-9}\text{ m}$$

Compare the average distance between molecules to the radius of a molecule by taking a ratio.

$$\frac{d_{av}}{r} = \frac{3.4 \times 10^{-9}\text{ m}}{2.0 \times 10^{-10}\text{ m}} = 17$$

$$d_{av} \approx 20r$$

CHECK and THINK

As expected, the average distance between molecules is greater than the size of an individual molecule. To help visualize such a separation, imagine sitting in a classroom auditorium. If the whole room is filled, the separation between you and the next person is perhaps half the size of your diameter. You could easily hand that person a pencil. Now imagine having about 20 times more space between you and the next person. You might share the row with just one other person, and there might be no one sitting in front of you for 10 rows. Now imagine trying to hand a pencil to someone. The molecules in Figure 20.10B are drawn too close together to represent air molecules at STP.

B Estimate the mean free path of the air molecules in your room. Compare your answer to the average distance between molecules found in part A.

INTERPRET and ANTICIPATE

Use the number density found in part A to find the mean free path. From Figures 20.10B and 20.11, we expect the mean free path to be greater than the average separation between molecules.

SOLVE

Substitute values into Equation 20.24.

$$\lambda = \frac{1}{4\sqrt{2}\pi r^2(N/V)} \tag{20.24}$$

$$\lambda = \frac{1}{4\sqrt{2}\pi(2.0 \times 10^{-10}\text{ m})^2(2.5 \times 10^{25}\text{ m}^{-3})}$$

$$\lambda = 5.6 \times 10^{-8}\text{ m}$$

Take a ratio to compare the mean free path to the average separation between molecules.

$$\frac{\lambda}{d_{av}} = \frac{5.6 \times 10^{-8}\text{ m}}{3.4 \times 10^{-9}\text{ m}} = 16$$

$$\lambda \approx 20d_{av}$$

CHECK and THINK

Because the mean free path is greater than the average separation between molecules, the molecules do not necessarily collide with their nearest neighbors as shown in Figure 20.11B.

C What is the mean free time of the air molecules, and with what frequency are they undergoing collisions? Assume air is about 70% N_2 and 30% O_2 by mass. *Hint*: Use Equation 20.21.

INTERPRET and ANTICIPATE

The mean free time depends on the mean free path and the average speed of the molecules. We can calculate the average speed from the temperature of the gas. The frequency of collisions is the number of collisions per unit time, which we can find from the reciprocal of the mean free time.

SOLVE

Equation 20.21 gives the average speed of the molecules.

$$v_{av} = \sqrt{\frac{8k_BT}{\pi m}} \quad (20.21)$$

We also need the molecule's mass. The mass of N_2 is 28.0 u, and the mass of O_2 is 32.0 u. Find the average mass of air molecules from the assumption that air is 70% N_2 and 30% O_2.

$$m_{av} = 0.7(28.0\,u) + 0.3(32.0\,u)$$

$$m_{av} = 29.2\text{ u} \times \left(\frac{1.66 \times 10^{-27}\text{ kg}}{1\text{ u}}\right) = 4.85 \times 10^{-26}\text{ kg}$$

Substitute numerical values. As in part A, we can assume the temperature is 20°C = 293 K.

$$v_{av} = \sqrt{\frac{8(1.38 \times 10^{-23}\text{ J/K})(293\text{ K})}{\pi(4.85 \times 10^{-26}\text{ kg})}} = 4.61 \times 10^2\text{ m/s}$$

Now use Equation 20.25 to find the mean free time. Substitute the mean free path λ from part B.

$$t_{mf} = \frac{\lambda}{v_{av}} = \frac{5.6 \times 10^{-8}\text{ m}}{4.61 \times 10^2\text{ m/s}} = 1.2 \times 10^{-10}\text{ s}$$

The frequency f of the collisions is the reciprocal of the mean free time.

$$f = \frac{1}{t_{mf}} = \frac{1}{1.2 \times 10^{-10}\text{ s}} = 8.2 \times 10^9\text{ collisions/s}$$

CHECK and THINK

The mean free time depends on the speed of the molecules; slower molecules have a greater mean free time. The high frequency of collisions explains why molecules take such a long time to mix by diffusion.

EXAMPLE 20.4 The Whole Neighborhood Stays Awake

If the aromatic molecules from the coffee roaster (Fig. 20.10A) did not undergo any collisions, they would pass through the neighborhood in just a second or so. Many molecular collisions occur (Fig. 20.10B), though, and if the air were calm, it would take years for the aroma to fill the neighborhood. Assuming the diffusion coefficient is the same as for water vapor, estimate to one significant figure the time it would take to fill a neighborhood with the scent of coffee by diffusion alone.

INTERPRET and ANTICIPATE

Our primary job is to estimate the size of the neighborhood and then find Δt, the time for the aromatic molecules to diffuse into the air in the neighborhood. From the opening statement, we expect that Δt is several years.

SOLVE

Let's assume that the coffee roaster sits in the middle of a neighborhood that is about 0.5 km in radius. The average displacement of the molecules must then be about 500 m for people in the whole neighborhood to smell the coffee.

The aromatic molecules travel in three dimensions, so we can use Equation 20.26, which we solve for Δt. Use D for water vapor from Table 20.2 and report the final answer to one significant figure. (D in Table 20.2 is for water vapor at a chilly 8°C, but the value of D is good enough for our one-significant-figure estimate.)

$$\Delta r_{av} = \sqrt{6D\,\Delta t} \quad (20.26)$$

$$\Delta t = \frac{\Delta r_{av}^2}{6D} = \frac{(500\text{ m})^2}{6(2.39 \times 10^{-5}\text{ m}^2/\text{s})}$$

$$\Delta t = 1.7 \times 10^9\text{ s} \times \left(\frac{1\text{ yr}}{3.15 \times 10^7\text{ s}}\right)$$

$$\Delta t = 50\text{ yr}$$

CHECK and THINK

As we expected, it would take about half a century for the coffee aroma to diffuse throughout the neighborhood. Neighborhood residents, however, know that they can smell the coffee soon after the shop starts roasting it. The aroma is actually transported by currents in the air (Chapter 21).

CONCEPT EXERCISE 20.4

When people smell something unpleasant such as cigarette smoke, they sometimes fan the air. Does fanning the air help get rid of the smell? Explain your answer.

20-6 Real Gases: The Van der Waals Equation of State

The ideal gas law works (that is, it does a good job of describing a real gas) if the gas density is low enough. For many real gases, however, the ideal gas law is not a good approximation. In those cases, we must use the **Van der Waals equation of state**, which frees us from two of the assumptions of the kinetic theory. As in the last section, we can disregard assumption 1 (that the molecules are particles with no physical extent). Instead, we model gas molecules as spheres of radius r. Also, we disregard assumption 2 (that the molecules only exert contact forces on one another). The molecules actually exert an attractive field force on one another due to the electromagnetic force that we study in Parts III and IV.

Think about the motion of a single gas molecule. It cannot move through the entire volume V of the container because the other molecules take up some of that volume. Instead, the volume $V_{\text{available}}$ that the molecule can move through is reduced by the volume occupied by the other molecules:

$$V_{\text{available}} = V - V_{\text{molecules}} \quad (20.27)$$

It is common convention to write the volume occupied by the molecules in terms of b, the volume of molecules per mole. The volume occupied by the molecules is then the number of moles n multiplied by b, and Equation 20.27 is written as

$$V_{\text{available}} = V - nb \quad (20.28)$$

Because molecules take up some volume, we correct the ideal gas law by replacing V in the ideal gas law $PV = nRT$ (Eq. 19.21) by $V_{\text{available}}$ in Equation 20.28:

$$P_{\text{Clausius}}(V - nb) = nRT \quad (20.29)$$

Equation 20.29 is called the **Clausius equation of state**. It models the molecules as spheres of radius r, but it does not take into account the field force they exert on one another.

We can modify the Clausius equation of state to include that attractive field force. Imagine the motion of one molecule as it approaches one of the container's walls, where it is pulled back by the attractive force of the other molecules. The force the molecule exerts on the wall is smaller than it would be if there were no field forces. Because all the molecules experience such a force, the pressure exerted on the walls is reduced:

$$P = P_{\text{Clausius}} - P_{\text{reduction}} \quad (20.30)$$

We can find an expression for the reduction in pressure $P_{\text{reduction}}$. The force experienced by each molecule as it approaches the wall is proportional to the molar density of the molecules n/V. Further, the pressure (force per unit area) on the walls is also proportional to the density of the molecules, so the reduction in pressure is proportional to the density squared. We write the reduction in pressure as $P_{\text{reduction}} = a(n/V)^2$ and substitute this equation and Equation 20.29, $P_{\text{Clausius}} = nRT/(V - nb)$, into Equation 20.30:

$$P = \frac{nRT}{V - nb} - a\left(\frac{n}{V}\right)^2 \quad (20.31)$$

Equation 20.31 is the Van der Waals equation of state. It is often written as

VAN DER WAALS EQUATION OF STATE
Major Concept

$$\left(P + a\frac{n^2}{V^2}\right)(V - nb) = nRT \quad (20.32)$$

so that the right side follows the same form as the ideal gas law. The constants a and b depend on the type of gas and are determined experimentally (Table 20.3).

TABLE 20.3 Constants *a* and *b* in the Van der Waals equation of state.

Gas	a (Pa · m^6/mol^2)	b (m^3/mol)
Carbon dioxide, CO_2	0.359	4.27×10^{-5}
Carbon monoxide, CO	0.149	3.99×10^{-5}
Helium, He	3.46×10^{-3}	2.37×10^{-5}
Nitrogen, N_2	0.139	3.91×10^{-5}
Oxygen, O_2	0.136	3.18×10^{-5}
Water vapor	0.546	3.05×10^{-5}

Adapted from *The Handbook of Chemistry and Physics*, 68th ed. (Boca Raton, FL: CRC Press, 1987–1988), p. D-188.

Comparing Van der Waals and the Ideal Gas Equations of State

We expect that the Van der Waals equation of state should reduce to the ideal gas law when the density of the gas is very low or when the volume is very large. When the volume is very large,

$$P \gg a\frac{n^2}{V^2}$$

and

$$V \gg nb$$

so that the Van der Waals equation (Eq. 20.32) is identical to the ideal gas law in that case:

$$\left(P + a\frac{n^2}{V^2}\right)(V - nb) \approx (P + 0)(V - 0) = nRT$$

A graph of pressure as a function of volume—a ***PV* diagram** for short—is an important visualization tool. We can use a *PV* diagram to compare the ideal gas law to the Van der Waals equation of state (Fig. 20.13). Figure 20.13A is the *PV* diagram for a gas at constant temperature. Comparing the two laws for a gas at the same pressure, shows that the Van der Waals gas has a smaller volume because attractive forces between molecules tend to pull them together. (In an ideal gas, we assume there are no such long-range forces between molecules, so the volume is larger.) The two curves are nearly the same, however, when the gas volume is large, which is consistent with modeling a real gas as an ideal gas when the volume is large.

PRESSURE–VOLUME (*PV*) DIAGRAM

Tool

Figure 20.13B shows *PV* diagrams for a gas at two different temperatures, showing that a warm gas behaves much more like an ideal gas even when its volume is small. At higher temperatures, the molecules are moving faster, and the attractive force between the molecules has less influence on their motion. We will return to *PV* diagrams in the next section.

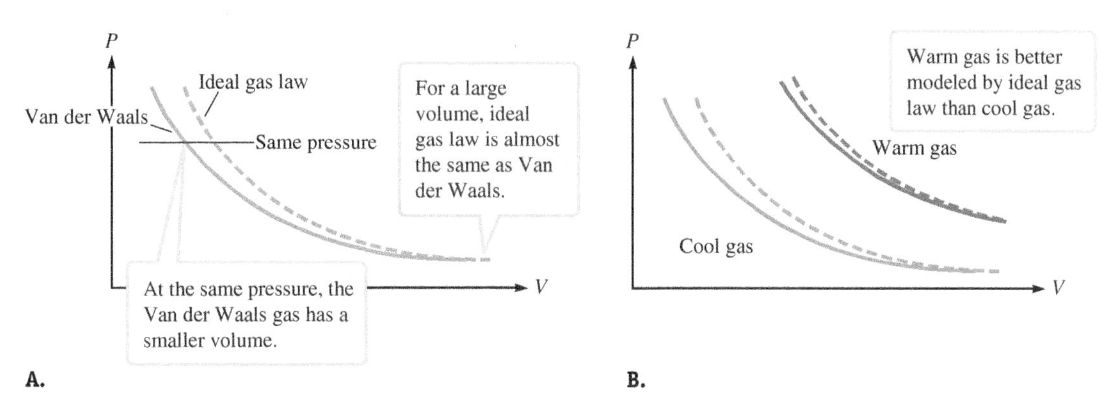

FIGURE 20.13 **A.** *PV* diagram for a gas using the ideal gas law (dashed curve) and Van der Waals equation of state (solid curve). **B.** *PV* diagrams for gases at two different temperatures using the ideal gas law (dashed curves) and Van der Waals equation of state (solid curves).

EXAMPLE 20.5 Real Gas Versus Ideal Gas

In this example, we compare the pressure found by using the ideal gas law with the pressure found by using the Van der Waals equation of state. In all cases, there are 1000 mol of oxygen gas (O_2) at a warm room temperature of 300 K. Give your answers to three significant figures.

A Use the ideal gas law to calculate the pressure exerted by the oxygen gas in a container approximately the size of a laboratory with a volume of 100 m^3.

INTERPRET and ANTICIPATE

This example is a straightforward problem that we could have tackled in Section 19-6. The density (1000 mol/100 m^3) is less than the density of air in your room at atmospheric pressure, so we expect our answer to be less than atmospheric pressure.

Example continues on page 598 ▶

SOLVE

Solve the ideal gas law (Eq. 19.21) for pressure.

$$PV = nRT$$

$$P = \frac{nRT}{V} = \frac{(1000\,\text{mol})[8.314\,\text{J/(mol}\cdot\text{K)}](300\,\text{K})}{100\,\text{m}^3}$$

$$P = 2.49 \times 10^4\,\text{Pa}$$

CHECK and THINK

As expected, the pressure is less than atmospheric pressure (1.01×10^5 Pa).

B Now use the Van der Waals equation of state to calculate the pressure exerted by the oxygen gas in a container with a volume of 100 m^3.

INTERPRET and ANTICIPATE

We need to use the Van der Waals equation for pressure with the constants a and b for oxygen given in Table 20.3. Because the density is very low, we expect that the pressure should be close to the pressure we found in part A.

SOLVE

Substitute appropriate values into Equation 20.31.

$$P = \frac{nRT}{V - nb} - a\left(\frac{n}{V}\right)^2 \tag{20.31}$$

$$P = \frac{(1000\,\text{mol})[8.314\,\text{J/(mol}\cdot\text{K)}](300\,\text{K})}{100\,\text{m}^3 - (1000\,\text{mol})(3.18 \times 10^{-5}\,\text{m}^3/\text{mol})} - \left(\frac{0.136\,\text{Pa}\cdot\text{m}^6}{\text{mol}^2}\right)\left(\frac{1000\,\text{mol}}{100\,\text{m}^3}\right)^2$$

$$P = 2.49 \times 10^4\,\text{Pa}$$

CHECK and THINK

The pressure found using the Van der Waals equation of state matches the pressure we found using the ideal gas law to three significant figures. We can see that it is okay to approximate such a low-density gas as ideal.

C Use the ideal gas law to calculate the pressure exerted by the same amount of oxygen gas at the same temperature in a container about the size of a large duffel bag (volume 1 m^3).

INTERPRET and ANTICIPATE

Use the same process as in part A. This time, the density (1000 mol/m^3) is greater than the density of the air in your room at atmospheric pressure, so we expect to find that the pressure is greater than atmospheric pressure.

SOLVE

Substitute values into the ideal gas law.

$$P = \frac{nRT}{V} = \frac{(1000\,\text{mol})[8.314\,\text{J/(mol}\cdot\text{K)}](300\,\text{K})}{1\,\text{m}^3}$$

$$P = 2.49 \times 10^6\,\text{Pa}$$

CHECK and THINK

As expected, the pressure is about 25 times greater than atmospheric pressure.

D Now use the Van der Waals equation of state to calculate the pressure exerted by that oxygen gas in a container with a volume of 1 m^3.

INTERPRET and ANTICIPATE

Repeat the process from part B. Because the density is great, we expect that the pressure should differ from the pressure we found in part C.

SOLVE

Substitute appropriate values into Van der Waals equation of state in the form of Equation 20.31.

$$P = \frac{nRT}{V - nb} - a\left(\frac{n}{V}\right)^2 \quad (20.31)$$

$$P = \frac{(1000\text{ mol})[8.314\text{ J}/(\text{mol}\cdot\text{K})](300\text{ K})}{1\text{ m}^3 - (1000\text{ mol})(3.18 \times 10^{-5}\text{ m}^3/\text{mol})} - \left(\frac{0.136\text{ Pa}\cdot\text{m}^6}{\text{mol}^2}\right)\left(\frac{1000\text{ mol}}{1\text{ m}^3}\right)^2$$

$$P = 2.44 \times 10^6\text{ Pa}$$

CHECK and THINK

The pressure found using the Van der Waals equation of state is 5×10^4 Pa (about 0.5 atm) lower than the pressure we found using the ideal gas law. In this low-volume case, if we model the gas as an ideal gas, our calculations are incorrect to three significant figures. We can easily measure such a discrepancy.

20-7 Phase Changes

The molecular forces in a real gas can cause it to change phase, that is, become liquid or solid. (In an ideal gas, the particles do not exert an attractive field force on one another, so an ideal gas would not change phase.) In this section, we look at the conditions required for phase changes. Pressure–volume (PV) diagrams are one way to visualize these conditions.

Figure 20.13A shows that at the same pressure, a real gas has a smaller volume than an ideal gas, and we account for this smaller volume in terms of the attractive field force between molecules. At a sufficiently low temperature (or at a high pressure), these attractive forces can cause the gas to become a liquid. Follow the PV diagram (Fig. 20.14) for a liquefying gas from point A (on the right) to point D (on the left). As the gas liquefies, its pressure remains constant and its volume decreases; then, as the pressure of the liquid increases, its volume is nearly constant.

Under the pressure and volume conditions on the portion of the curve between points B and C in Figure 20.14, the substance exists as both gas and liquid. In fact, there is a range of pressure, temperature, and volume conditions in which the sub-

CRITICAL TEMPERATURE; CRITICAL POINT; CRITICAL PRESSURE

Major Concepts

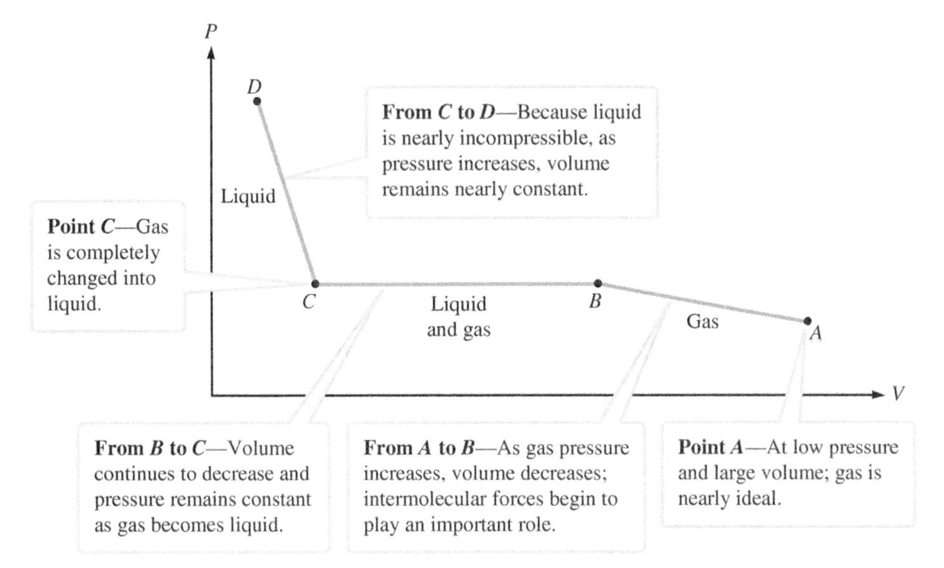

FIGURE 20.14 As a gas liquefies, its pressure remains constant, and its volume decreases as shown by the horizontal line between points B and C.

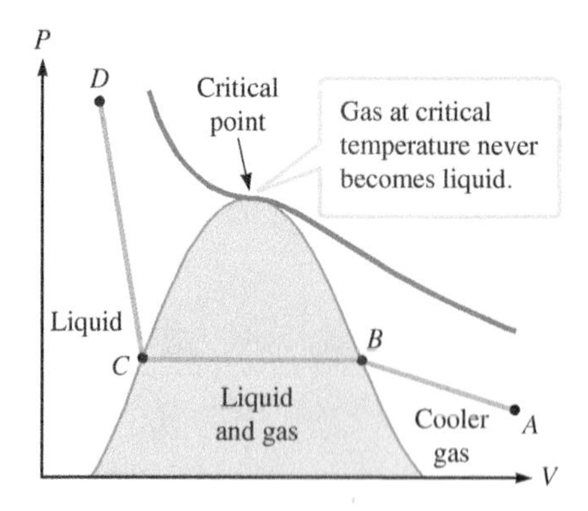

FIGURE 20.15 *PV* diagram for a hot gas (red) and cooler gas (orange). The green area represents the range where the substance exists in equilibrium between the liquid and gas phases.

stance exists in equilibrium between the liquid and gas phases. The green area in Figure 20.15 represents the range of liquid–gas equilibrium conditions.

If a gas is warmer than the **critical temperature**, it cannot become a liquid no matter how high the pressure. The red curve in Figure 20.15 shows a gas at its critical temperature. When a gas is at its critical temperature, its *PV* curve does not pass through the green region; instead, it touches the green region at the **critical point** where the curve is horizontal. The pressure at the critical point is known as the **critical pressure**. As more pressure is applied to the gas, it becomes denser. It acquires properties similar to those of a liquid, but no liquid forms. Cooler gases pass through the wide part of the green region, becoming liquid. Critical temperatures and pressures of five common gases are found in Table 20.4.

The term **vapor** describes a substance that is in the gaseous phase and below its critical temperature, and the term **gas** is reserved for a substance above its critical temperature. So, the substance represented by the orange curve in Figure 20.15 should be properly referred to as a *vapor* because its temperature is lower than the critical temperature.

PHASE DIAGRAM ⊙ Tool

A **phase diagram** (another way to visualize phase changes) is a graph of pressure as a function of temperature. Curves on the phase diagram show the conditions under which a substance may change between two of the three states: solid, liquid, and gas (or vapor). The phase diagram in Figure 20.16 holds for a substance such as carbon dioxide. We often think of carbon dioxide as a gas that we exhale; it is found naturally as a liquid venting under the sea (right inset). Carbon dioxide in its solid form is known as dry ice, and its vapor is used to create an eerie atmosphere (left inset). The three curves in Figure 20.16 represent the equilibrium conditions between two phases. The orange curve represents the pressure and temperature combinations that allow the substance to exist as a solid and as a vapor. The green curve defines the equilibrium between the solid and liquid phases, and the purple curve does the same for the liquid and vapor phases.

TABLE 20.4 Critical temperatures and pressures.

Substance	Critical temperature (°C)	Critical pressure (Pa)
Carbon dioxide, CO_2	31	7.38×10^6
Hydrogen, H_2	−239.9	1.30×10^6
Nitrogen, N_2	−147	3.39×10^6
Oxygen, O_2	−118.4	5.08×10^6
Water, H_2O	374.1	2.21×10^7

Adapted from *The Handbook of Chemistry and Physics*, 68th ed. (Boca Raton, FL: CRC Press, 1987–1988), p. F-66.

A block of dry ice is a solid, so its pressure and temperature are represented by a point labeled *A* in the yellow region in Figure 20.16. If the temperature of the dry ice increases sufficiently and the pressure remains constant, the new pressure and temperature are represented by point *B*. Because point *B* is in the red region, the carbon dioxide has become a vapor. The process of changing from a solid to a vapor without becoming a liquid is called **sublimation**. Sublimation only occurs at low pressure. If the dry ice were at a higher pressure such as at point *C*, as the temperature increases the carbon dioxide first becomes a liquid (point *D*). As the temperature increases further, the substance becomes a vapor (point *E*). Of course, if the temperature increases above the critical temperature, the substance is properly referred to as a gas.

Figure 20.16 shows the following two other important points on the phase diagram for any substance.

1. The **critical point**, which also appeared on the *PV* diagram (Fig. 20.15), represents the critical temperature and pressure (Table 20.4). A gas at a pressure greater than the critical pressure does not undergo a distinct phase transition to a liquid as its temperature decreases. That is, for a substance above the critical pressure, there is no distinction between liquid and gas phases. For substances we are familiar with, the critical pressure is much greater than atmospheric pressure, so we don't usually observe the blending of the gas and liquid phases.
2. The **triple point** is a unique combination of temperature and pressure that allows the substance to exist in all three phases simultaneously. The triple point of water is used as a reference point for the absolute temperature scale (Section 19-7).

Figure 20.17 shows the phase diagram for water. As indicated on the diagram, at 1 atm the freezing point of water is at 0°C or 273.15 K and the boiling point is at 100°C or 373.15 K. The triple point of water occurs at 6.0×10^{-3} atm or 0.61 kPa and 0.01°C or 273.16 K. (The triple point of carbon dioxide occurs at 5.1 atm or 520 kPa and −56.4°C or 217 K.) The most important difference between the phase diagram for water (Fig. 20.17) and the phase diagram for a substance like carbon

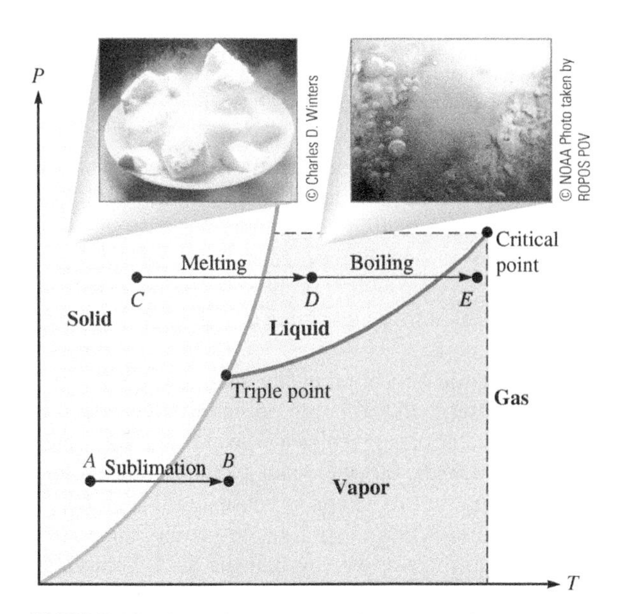

FIGURE 20.16 A phase diagram is a graph of pressure as a function of temperature. Carbon dioxide is shown in liquid form escaping from vents deep in the ocean and in solid form known as dry ice.

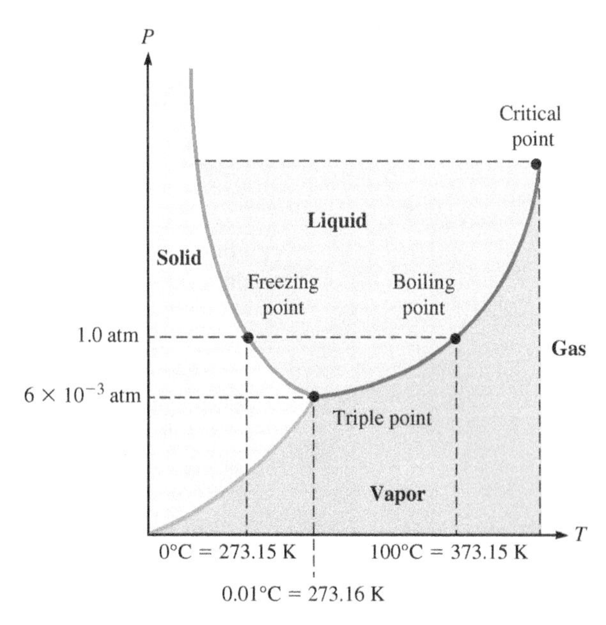

FIGURE 20.17 A phase diagram for water shows that the curve (green) between the solid and liquid phases slopes up and to the left from the triple point. Note that the temperature axis is not linear.

dioxide (Fig. 20.16) is in the shape of the curve between the solid and liquid phases (shown in green). For carbon dioxide, this curve slopes upward and to the right from the triple point, as is true for all substances that contract when freezing. For water and other substances that expand when freezing, the curve slopes upward and to the left from the triple point. For such a substance, a lower temperature is required to freeze it at higher pressure.

CONCEPT EXERCISE 20.5

a. In what state (phase) is water when it is at 100°C and its pressure is above its critical pressure?
b. In what state is water when it is at 100°C and its pressure is between 1 atm and its critical pressure?
c. In what state is water when it is at 100°C and its pressure is below 1 atm?

20-8 Evaporation

In Section 20-7, we considered phase changes between three states of matter. In this section, we look at the interplay of the liquid and gaseous states. Although much of what follows can be applied to any substance, our focus is primarily on water and its transition between a liquid and a vapor.

Evaporation is the process whereby atoms or molecules in a liquid gain enough kinetic energy to leave the liquid and enter the gaseous (vapor) phase. **Condensation** is the opposite process, in which atoms or molecules in a gas become liquid. A substance is in equilibrium when the rate of evaporation equals the rate of condensation and the relative amount of liquid and vapor remains constant. If the temperature is increased after equilibrium is reached, the evaporation rate exceeds the condensation rate, and the relative amount of liquid decreases.

We can use the kinetic theory to explain how temperature can change the rate of evaporation. The molecules in a liquid are strongly attracted to one another, but a molecule with sufficient kinetic energy near the liquid surface may leave. What happens next depends on the molecule's kinetic energy. If the kinetic energy is below a certain value, the molecule will fall back into the liquid (much like a ball thrown upward and then falling back toward the ground). A molecule with a great enough kinetic energy will escape the liquid and become part of the gas above the surface (much like an object with a speed greater than the escape speed of the Earth). So the fastest molecules escape the liquid and become a gas.

The molecules in a liquid have a speed distribution similar to the Maxwell-Boltzmann distribution (Fig. 20.8). An increase in the liquid's temperature increases the relative number of high-speed molecules, which in turn means an increase in the rate of evaporation. Therefore, evaporation is a cooling process. When you perspire on a very hot day, the water evaporates from your skin, taking away the fastest molecules. Because the average speed of the remaining molecules is lower, the temperature of the water is lower, and you feel cooler.

When you set your pet's water dish outside on a hot day, the water evaporates and disappears from the dish, but what if you set a closed bottle of water in the hot sunshine? If the bottle has been moved from a cooler place to a warmer one, the water will evaporate at a faster rate than it condenses, but the water vapor cannot escape from the sealed bottle. For a while, the amount of vapor increases until the rate of evaporation equals the rate of condensation. When those rates are equal, the liquid and vapor are then in equilibrium. At equilibrium, we say that the air is **saturated**, and the resulting pressure of the vapor is called the **saturated vapor pressure**.

SATURATED VAPOR PRESSURE

Major Concept

The saturated vapor pressure depends only on temperature and not on the volume of the container. Suppose the volume of the bottle was suddenly reduced (perhaps by squeezing the bottle). The volume of the liquid would not change because liquids are incompressible, but the volume of the vapor would decrease, so the vapor density would increase. More vapor molecules would collide with the liquid's surface each second. The condensation rate would exceed the evaporation rate until a new equilibrium was reached at the same saturated vapor pressure.

Both the evaporation rate and the saturated vapor pressure increase as temperature increases. Table 20.5 gives the saturated vapor pressure for water at several temperatures from $-50°C$ to $150°C$, and Figure 20.18 is a graph of saturated vapor pressure for water in the temperature range from 20°C to 40°C.

Evaporating is not the same as boiling. A liquid **boils** when the saturation vapor pressure equals the external pressure. Consider a pot of water on a hot stove at sea level. As the water temperature increases, tiny vapor bubbles begin to form. If the vapor pressure inside the bubbles is less than the external pressure, the bubbles collapse. As the water temperature continues to increase, the vapor pressure in the bubbles increases until it equals the external pressure. Bubbles rise to the surface, and the vapor escapes. At sea level, the external pressure is 1 atm $= 1.013 \times 10^5$ Pa, and according to Table 20.5, that value corresponds to a water tempera-

TABLE 20.5 Saturated vapor pressure for water.

Temperature (°C)	Temperature (K)	Saturated vapor pressure (Pa)
Ice, –50	223	3.939
Ice, –25	248	7.011×10^1
Freezing point, 0	273	6.103×10^2
10	283	1.227×10^3
15	288	1.705×10^3
20	293	2.337×10^3
25	298	3.166×10^3
30	303	4.242×10^3
35	308	5.621×10^3
40	313	7.452×10^3
50	323	1.233×10^4
60	333	1.991×10^4
70	343	3.115×10^4
80	353	4.733×10^4
90	363	7.008×10^4
Boiling point, 100	373	1.013×10^5
125	398	2.320×10^5
150	423	4.759×10^5

Adapted from *The Handbook of Chemistry and Physics*, 68th ed. (Boca Raton, FL: CRC Press, 1987–1988), pp. D-189–D-191.

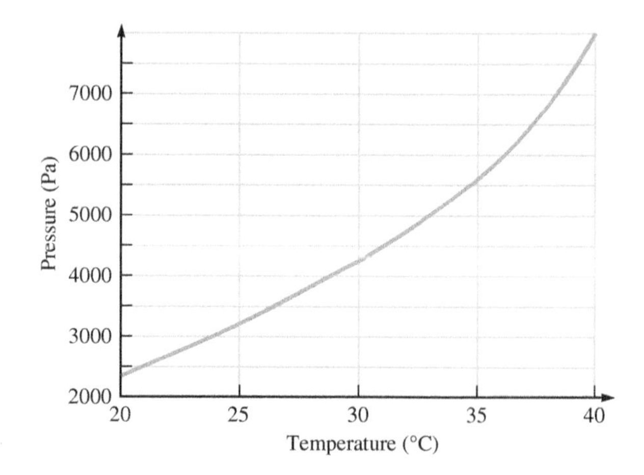

FIGURE 20.18 Saturated vapor pressure of water.

ture of 100°C, as you would expect. If the pot of water is at a higher elevation, however, the external pressure is less than 1 atm, and the water will boil at a lower temperature. Thus, cooking instructions must be modified for people who live at high altitudes.

Humidity

Air is a mixture of several gases: about 78% nitrogen by volume, 21% oxygen, and less than 1% argon and carbon dioxide. The pressure that would be exerted by each gas if the other gases were not present is called the **partial pressure** of that gas. The total air pressure is the sum of the partial pressures due to each gas. If no water vapor is in the air, the partial pressure of water is zero. The partial pressure of water cannot exceed the saturated vapor pressure at that temperature if water is to remain in the vapor phase. On a hot day, there may be more water vapor in the air than on a cooler day, and we say that it is "muggy."

Relative humidity (RH) is the ratio of the partial pressure of water P_{water} to the saturated vapor pressure P_{svp} at some temperature, expressed as a percentage:

$$RH = \frac{P_{\text{water}}}{P_{\text{svp}}} \times 100\% \qquad (20.33)$$

RELATIVE HUMIDITY

Major Concept

The maximum partial pressure of water is the saturated vapor pressure, giving a relative humidity of 100%. In this case, we say that the air is *saturated*.

It is possible for the partial pressure of water to exceed the saturated vapor pressure temporarily, which can happen if the temperature drops as it does at night. When the temperature is high during the day, the relative humidity may get quite high. When the temperature drops, the saturated vapor pressure decreases. If the partial pressure of water is greater than the saturated vapor pressure at that lower nighttime temperature, water condenses into a liquid in the form of rain, fog, or dew, which is why we often find dew in the early morning.

In fact, dew can be used to measure the relative humidity. As the air cools, it reaches a temperature T_{DP} at which the partial pressure of water equals the saturated vapor pressure, known as the **dew point**. You can measure the dew point with a metal can and a thermometer (Fig. 20.19). Pour some water into the can and slowly add ice to lower the temperature. You need to measure the temperature of the water, so it is best to use the thermometer to mix the water as you add ice. The metal can must be in thermal contact with the air as you cool the can. Watch for dew to form on the outside of the can; the temperature at which dew forms is the dew point. Because the partial pressure of water equals the saturated vapor pressure at the temperature of the dew point, once you know the dew point you can look up the partial pressure of water in either Table 20.5 or Figure 20.18.

FIGURE 20.19 To measure the dew point, cool an object that is in thermal contact with the air and measure the temperature at which dew forms on the object.

EXAMPLE 20.6 CASE STUDY Would You Rather Be in Phoenix?

Some people like to live in a humid climate, and others prefer a dry one. Houston, Texas, is known for it high humidity. On a typical August day, the high temperature in Houston may reach 93°F, and the low temperature is about 73°F. The average dew point in August is close to the low temperature: $T_{\text{DP}} = 72$°F. Phoenix, Arizona, is known for its dry climate. In February, its relative humidity is 28% in the middle of the afternoon when the temperature reaches its high.

A Assume the dew point is constant throughout the day. What is the relative humidity in Houston when the temperature is at its typical August high point? At its low point?

INTERPRET and ANTICIPATE

At the dew point, the partial pressure of water equals the saturated vapor pressure. To find the relative humidity, find the saturated vapor pressure at the high and low temperatures. Because Houston is known for high humidity, we expect its relative humidity to be higher than that in Phoenix, and because Houston's low temperature is just above the dew point, we expect the relative humidity at that temperature to be nearly 100%.

Example continues on page 604 ▶

SOLVE

First, convert all the Fahrenheit temperatures to Celsius (Eq. 19.2).	$T_{hi}(°C) = \frac{5}{9}[T_{hi}(°F) - 32]$ $T_{hi}(°C) = \frac{5}{9}[93°F - 32] = 34°C$ $T_{lo} = 73°F = 23°C$ $T_{DP} = 72°F = 22°C$
The partial pressure of water equals the saturated vapor pressure at the dew point. Use Figure 20.18 to find P_{svp} at 22°C. From the graph, P_{svp} is about 2.7×10^3 Pa.	$P_{water} = P_{svp}(22°C) \approx 2.7 \times 10^3$ Pa
Find the relative humidity from Equation 20.33. Now we need to find the saturated vapor pressure at the high and low temperatures. Again, use Figure 20.18 to read off P_{svp} at the high temperature $T_{hi} = 34°C$. We estimate P_{svp} at 34°C as 5.3×10^3 Pa.	$RH = \frac{P_{water}}{P_{svp}} \times 100\%$ (20.33) $RH_{hi} = \frac{2.7 \times 10^3 \text{ Pa}}{5.3 \times 10^3 \text{ Pa}} \times 100\%$ $RH_{hi} = 51\%$
At the low temperature $T_{lo} = 23°C$, P_{svp} is about 2.8×10^3 Pa.	$RH = \frac{2.7 \times 10^3 \text{ Pa}}{2.8 \times 10^3 \text{ Pa}} \times 100\%$ $RH_{lo} = 96\%$

CHECK and THINK

As expected, the relative humidity is greater than in Phoenix and nearly 100% at the low temperature. Dew does not form, however, because the actual temperature does not get down to the dew point.

B In Phoenix, the average high temperature in February is 70°F, and the average low temperature is 44°F. Roughly what is the typical February dew point in Phoenix? Does dew form there on a typical February day?

INTERPRET and ANTICIPATE

Use the relative humidity at the high temperature (28%, given in the problem statement) to find the partial pressure of water. The dew point occurs at the temperature where the partial pressure of water equals the saturated vapor pressure. We expect the dew point for Phoenix to be lower than the dew point for Houston. For dew to form, the temperature in Phoenix must be lower than the temperature in Houston.

SOLVE

As in part A, convert all temperatures to Celsius.	$T_{hi} = 70°F = 21°C$ $T_{lo} = 44°F = 6.7°C$
The relative humidity is 28% when the temperature is at its high. Solve Equation 20.33 for the partial pressure of water, P_{water}.	$RH = \frac{P_{water}}{P_{svp}} \times 100\% = 28\%$ (20.33) $\frac{P_{water}}{P_{svp}} = 0.28$ $P_{water} = 0.28P_{svp}$
Read the saturated vapor pressure at 21°C from Figure 20.19. We get 2.5×10^3 Pa.	$P_{water} = 0.28(2.5 \times 10^3 \text{ Pa}) = 7.0 \times 10^2 \text{ Pa}$

At the dew point, the saturated vapor pressure equals the partial pressure of water, so we need to find the temperature that corresponds to $P_{svp} = 7.0 \times 10^2$ Pa. That pressure is below the range in Figure 20.18, but it can be estimated from Table 20.5. According to the table, the dew point is between 0°C and 10°C. In fact, it is much closer to 0°C, so let's estimate it to be about 1°C. Because the low temperature in Phoenix on a typical February day is nearly 7°C, no dew forms.

CHECK and THINK

As expected, the dew point in Phoenix is much lower than it is in Houston. For dew to form in Phoenix, the temperature must be near freezing, whereas in Houston, it would still be balmy.

CONCEPT EXERCISE 20.6

CASE STUDY **Global Warming and Evaporation**

An increase in the amount of greenhouse gases is causing the temperature of the Earth to increase. As mentioned in Section 20-1, this change is known as global warming. Would you expect the evaporation rate to increase or decrease as a result of global warming?

CASE STUDY Part 3: Water Cycle and Global Dimming

The **water cycle** is the continuous circulation of the Earth's water stored in oceans, in other bodies of water such as lakes, and in the soil. Water evaporates from these sources and condenses in the atmosphere as clouds, and the cycle continues. Clouds transport water, and **precipitation** (such as rain, snow, and hail) occurs when the condensed water vapor in the atmosphere falls to the Earth. The water cycle is critical to life.

© Illinois State Water Survey

FIGURE 20.20 An evaporation pan is used to monitor the Earth's water cycle.

One way to monitor the Earth's water cycle is with an **evaporation pan**, an open metal container with carefully measured dimensions (Fig. 20.20). The pan is usually set on a palette, and the water level is measured daily. Measurements going back at least 50 years show an overall decrease in evaporation rate (the volume of water evaporating per day) until the 1990s. A decrease in evaporation rate indicates a slowdown in the water cycle, which could have serious consequences for all life on the Earth.

The measured decrease in the evaporation rate from the 1950s to the 1990s is believed to be an indication of **global dimming**, a long-term decrease in the amount of solar radiation that reaches the Earth's surface. One explanation for global dimming is an increase in particulate air pollution such as sulfates, nitrates, and soot. Such particles in the atmosphere block sunlight from reaching the Earth's surface, so the surface becomes cooler. The temperature decrease causes a decrease in the evaporation rate. The global dimming trend seems to have reversed since the 1990s, perhaps due to legislation such as the Clean Air Act, which has reduced the amount of pollution going into the air in the United States.

It may seem ironic that human activities produce pollutants that may cause both global warming and global dimming. Global warming mainly affects the troposphere (up to 20 km), whereas global dimming mainly affects the Earth's surface; in any event, global dimming has probably masked the effects of global warming. In fact, a strange solution to the problem of global warming proposes burning sulfur in the stratosphere to block out some sunlight. This solution was proposed several decades ago and has been recently revived by Paul Crutzen, winner of the 1995 Nobel Prize in Chemistry, as an "escape plan" because greenhouse gas reduction efforts have not been sufficient. Climatologist Mikhail Budyko—who is credited with first proposing the idea—warned, however, that such action would cause different climate changes in different regions in the world. For instance, some countries might become deserts, whereas others might become jungles. It may be better to increase our efforts to cut back on greenhouse gas emissions.

SUMMARY

Underlying Principles

The **kinetic theory** states that all matter is made up of particles in motion. When the kinetic theory is applied to an ideal gas, the following five assumptions are made:

1. A gas consists of a large number of molecules or atoms that can be modeled as particles.
2. The particles in an ideal gas do not interact with one another.
3. The particles make elastic collisions with the walls, and the duration of each such collision is short.
4. The particles are free to move in any direction at any speed.
5. The gas is made up of identical particles.

Major Concepts

1. The **average kinetic energy** of gas particles is directly proportional to the gas **temperature**:
$$K_{av} = \tfrac{1}{2}mv_{av}^2 = \tfrac{3}{2}k_B T \quad (20.17)$$
2. The **root-mean-square velocity** of gas particles is proportional to the square root of the **temperature**:
$$v_{rms} = \sqrt{v_{av}^2} = \sqrt{\frac{3k_B T}{m}} \quad (20.18)$$
3. **Maxwell-Boltzmann speed distribution:** The probability of a particle's speed per unit speed interval (Fig. 20.8) is
$$f(v) = 4\pi\left(\frac{m}{2\pi k_B T}\right)^{3/2} v^2 e^{-(mv^2/2k_B T)} \quad (20.19)$$

For Maxwell-Boltzmann speed distribution,

4. the **most probable speed** is given by
$$v_{mp} = \sqrt{\frac{2k_B T}{m}} \quad (20.20)$$
5. and the **average speed** is given by
$$v_{av} = \sqrt{\frac{8k_B T}{\pi m}} \quad (20.21)$$
6. **Mean free path:** The average length of the straight portion of a molecule's zigzag path (between collisions) is
$$\lambda = \frac{1}{4\sqrt{2}\pi r^2 (N/V)} \quad (20.24)$$
7. **Van der Waals equation of state:** The relation between pressure, volume, and temperature in a real gas, in which assumptions 1 and 2 for an ideal gas are not valid, is
$$\left(P + a\frac{n^2}{V^2}\right)(V - nb) = nRT \quad (20.32)$$
where a and b are experimentally determined constants for each gas.
8. **Critical temperature** is the temperature of the **critical point**. At $T > T_{crit}$, a gas cannot become a liquid no matter how high the pressure. The pressure at the critical point is known as the **critical pressure.**
9. **Saturated vapor pressure** is the pressure of a vapor when the rate of evaporation equals the rate of condensation so that the liquid and vapor are in equilibrium.
10. **Relative humidity (RH)** is the ratio of the partial pressure of water P_{water} to the saturated vapor pressure P_{svp} at some temperature, expressed as a percentage:
$$RH = \frac{P_{water}}{P_{svp}} \times 100\% \quad (20.33)$$

Tools

1. The **root mean square (rms)** of some quantity can be found as follows:
   a. Square each value.
   b. Take the average (mean) of those squared values, and
   c. Take the square root of the averaged squared values.

   The rms velocity is given by
$$v_{rms} = \sqrt{v_{av}^2} = \sqrt{\frac{1}{N}\sum_{j=1}^{N} v_j^2} \quad (20.3)$$
2. A **pressure–volume (*PV*) diagram** is a graph of pressure as a function of volume (Fig. 20.13).
3. A **phase diagram** is a graph of pressure as a function of temperature. Curves on the phase diagram show the conditions under which a substance may change between two of the three states: solid, liquid, and gas (or vapor); see Figures 20.16 and 20.17.

PROBLEMS AND QUESTIONS

A = algebraic C = conceptual E = estimation G = graphical N = numerical

20-1 What Is the Kinetic Theory?

1. C Use the kinetic theory to explain why a gas fills its container.
2. C One of the five assumptions for an ideal gas is worse for a molecular gas than for an atomic gas. Which one? Explain your answer.
3. C CASE STUDY Is the Earth's atmosphere contained by walls? Why doesn't the atmosphere escape? Why is the Earth's atmospheric pressure greater at sea level than it is on the tops of mountains?

20-2 Average and Root-Mean-Square Quantities

4. A The speeds of three molecules are v, $4v$, and $8v$, respectively. What are the average and rms speeds for the system consisting of these molecules?
5. N A system consists of 10 identical particles with the following speeds: two particles with speeds of 4.00 m/s, one particle with a speed of 8.00 m/s, four particles with speeds of 9.00 m/s, and three particles with speeds of 2.00 m/s. What is the **a.** average speed and **b.** rms speed of the particles in this system?

6. Five particles have velocities given by

$$\vec{v}_1 = (3.4\hat{\imath} + 5.6\hat{\jmath} + 6.7\hat{k})\ \text{m/s}$$
$$\vec{v}_2 = (2.4\hat{\imath} - 6.6\hat{\jmath} - 3.7\hat{k})\ \text{m/s}$$
$$\vec{v}_3 = (-0.4\hat{\imath} + 1.6\hat{\jmath} - 10.7\hat{k})\ \text{m/s}$$
$$\vec{v}_4 = (9.2\hat{\imath} - 9.8\hat{\jmath} + 3.6\hat{k})\ \text{m/s}$$
$$\vec{v}_5 = (0.9\hat{\imath} + 3.6\hat{\jmath} + 4.0\hat{k})\ \text{m/s}$$

N Find their **a.** average velocity, **b.** average speed, and **c.** rms speed.
d. C Compare your results and comment.

7. A For two particles with velocities $2v_0\hat{\imath}$ and $-v_0\hat{\imath}$, where $v_0 > 0$, consider the magnitude of the average velocity, the average of the two speeds, and the root-mean-square velocity. Order the magnitude of these three quantities from lowest to highest.

8. N CASE STUDY Nitrous oxide (N_2O) is a gas commonly used for its anesthetic effects in dentistry or surgery. When it interacts with oxygen, nitrous oxide gives rise to nitric oxide (NO), which will interact with ozone. Thus, nitrous oxide is a greenhouse gas that affects the ozone layer in our stratosphere. Suppose a container contains 3 mol of nitrous oxide, where the rms speed of the molecules in the container is 411.1 m/s. What is the average kinetic energy of the molecules inside the gas?

9. N Particles in an ideal gas of molecular oxygen (O_2) have an average momentum in the x direction of 2.726×10^{-23} kg·m/s. What is **a.** their average speed and **b.** the average of their x component of velocity?

10. N A cylinder has a nearly perfect vacuum within its interior except for 12 identical atoms. The atoms are moving with speeds 1.50 km/s, 5.00 km/s, 8.00 km/s, 1.25 km/s, 2.95 km/s, 4.40 km/s, 5.60 km/s, 3.50 km/s, 7.80 km/s, 2.70 km/s, 2.90 km/s, and 6.00 km/s. What is the **a.** average speed and **b.** rms speed of the 12 atoms in the cylinder?

11. N Table P20.11 gives the hourly temperature for Sudbury, Ontario, for November 1, 2006. Find the average temperature and rms temperature for that day.

TABLE P20.11

Time	Temperature (°C)	Time	Temperature (°C)
00:00	0.5	12:00	2.1
01:00	−0.7	13:00	2.2
02:00	−0.9	14:00	2.3
03:00	−1.3	15:00	1.3
04:00	−1.4	16:00	1.1
05:00	−1.4	17:00	0.4
06:00	−1.7	18:00	−0.4
07:00	−1.9	19:00	−0.5
08:00	−1.4	20:00	−0.7
09:00	−0.5	21:00	−0.8
10:00	0.5	22:00	−0.9
11:00	1.1	23:00	−1.2

20-3 The Kinetic Theory Applied to Gas Temperature and Pressure

12. N Find the rms speed of hydrogen molecules (H_2) at a temperature of 27°C.

13. N Oxygen gas (O_2) at 673 K is confined in a cylinder. What is the rms speed of oxygen molecules in the container?

14. C Three physics students are discussing the concept of temperature and have the following conversation.

Avi: The temperature is basically the average kinetic energy of the molecules. You have to find the average speed of the molecules and then use $k_B T = \frac{1}{2}mv^2$.

Shannon: I think you mean *velocity*. I think it's probably safer to use the average velocity of the molecules.

Cameron: I'm not sure. What was the rms velocity? Maybe we need to find that instead? Does it matter?

Who is on the right track? Explain your reasoning.

15. N The mass of a single hydrogen molecule is approximately 3.32×10^{-27} kg. There are 5.64×10^{23} hydrogen molecules in a box with square walls of area 49.0 cm^2. If the rms speed of the molecules is 2.72×10^3 m/s, calculate the pressure exerted by the gas.

16. C What is $\Delta t_{\text{collision}}$ in Equation 20.10? Why was a "trick" needed to find $\Delta t_{\text{collision}}$? Describe the trick. In particular, what is Δt in Equation 20.11?

17. N The noble gases neon (atomic mass 20.1797 u) and krypton (atomic mass 83.798 u) are accidentally mixed in a vessel that has a temperature of 90.0°C. What are the **a.** average kinetic energies and **b.** rms speeds of neon and krypton molecules in the vessel?

18. C In what way is $K_{\text{av}} = \frac{1}{2}mv_{\text{av}}^2 = \frac{3}{2}k_B T$ (Eq. 20.17) a connection between the microscopic and the macroscopic properties of an ideal gas?

19. N An 8.00×10^{-3} m^3 cylinder is filled with 3.00 mol of hydrogen gas. If the pressure within the cylinder is 5.00×10^5 Pa, what is the average translational kinetic energy of the hydrogen molecules in the cylinder?

20. E Estimate the rms speed of the oxygen molecules in your room. Estimate the rms speed of the oxygen molecules on a very cold day.

21. N In a cubical container with sides 4.00 cm in length, 2.00×10^{24} helium atoms, each with mass 6.64×10^{-27} kg, move with an rms speed of 1.50 km/s. What is the magnitude of the average force exerted on a wall by the helium gas in the container?

Problems 22 and 23 are paired.

22. G A particular roulette wheel is made up of 26 boxes, each labeled by one of the letters of the alphabet such that each letter is used. On each spin of the wheel, a ball lands in one of the 26 boxes. Plot the letters versus the frequency of the ball landing in a particular square. What letter is most probable?

23. G A particular roulette wheel is made up of 26 boxes, each labeled by only some of the letters of the alphabet (A–H). On each spin of the wheel, a ball lands in one of the 26 boxes. Figure P20.23 is a bar graph of the letters versus the frequency of the ball landing in a particular square. Determine the number of boxes labeled by each letter. What letter is most probable?

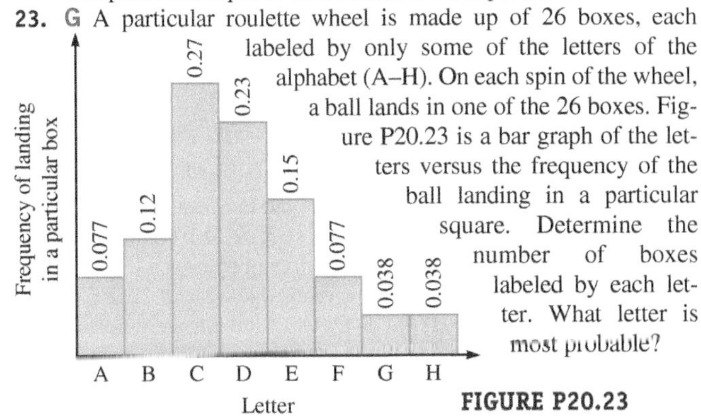

FIGURE P20.23

24. C CASE STUDY New stars form out of gaseous material (mostly molecular hydrogen) in a galaxy. Spiral galaxies like our own Milky Way have plenty of gas, so new stars are still forming. Some galaxies do not have enough gas for stars to form because the gas has escaped. A dwarf elliptical galaxy has an escape speed of 42 km/s. If the gas temperature is around 10^6 K, will the gas be retained in the galaxy? What does your answer tell you about star formation in the dwarf elliptical galaxy?

25. N CASE STUDY Because the Moon is about the same distance from the Sun as is the Earth, assume the Moon is at the same temperature as the Earth. Find the mass (in kilograms and atomic mass units) of the lightest molecule or atom that can be retained in the Moon's atmosphere. Comment on your results.

20-4 Maxwell-Boltzmann Distribution Function

26. G For the Maxwell-Boltzmann speed distribution, the most probable speed is less than the average speed. Sketch a speed distribution for which the most probable speed **a.** equals the average speed and **b.** is greater than the average speed.
27. N Consider a container of oxygen gas (O_2) at room temperature, 22°C. **a.** Find the most probable speed, average speed, and rms speed for the oxygen gas. **b.** How do you expect your results to change for nitrogen (N_2) instead of oxygen? Try it and compare your results.

Problems 28 and 29 are paired.

28. G Plot the Maxwell-Boltzmann distribution function for a gas composed of nitrogen molecules (N_2) at a temperature of 295 K. Identify the points on the curve that have a value of half the maximum value. Estimate these speeds, which represent the range of speeds most of the molecules are likely to have. The mass of a nitrogen molecule is 4.68×10^{-26} kg.
29. N Consider the Maxwell-Boltzmann distribution function plotted in Problem 28. For those parameters, determine the rms velocity and the most probable speed, as well as the values of $f(v)$ for each of these values. Compare these values with the graph in Problem 28.
30. A container of helium is at 20°C.
    a. N If you increase the temperature to 25°C, by what amount does the most probable speed increase?
    b. N If you increase the temperature to 100°C, by what amount does the most probable speed increase?
    c. C Compare your results and comment.
31. A A gas is cooled and trapped in a two-dimensional region (known as an optical lattice) by laser beams and magnets. The speed distribution function of the gas molecules obeys the two-dimensional Maxwell-Boltzmann distribution law,

$$f(v) = 2\pi\left(\frac{m}{2\pi k_B T}\right)ve^{-(mv^2/2k_BT)}$$

where m is the mass of a gas molecule, T is the temperature, and v is the speed. What is the **a.** most probable speed and **b.** average speed of the gas molecules?

20-5 Mean Free Path

32. N The Sun's surface is mainly composed of hydrogen atoms, with density $\rho = 2.1 \times 10^{-4}$ kg/m^3 and temperature 5.75×10^3 K. The radius of a hydrogen atom is 5.29×10^{-11} m. Find the **a.** mean free path and **b.** mean free time of hydrogen.
33. N A 5.00-L cylinder is filled with nitrogen gas at 22.0°C and a pressure of 75.0 atm. If the diameter of the nitrogen molecule is 1.08×10^{-10} m, what is the mean free path for nitrogen molecules in this cylinder?
34. N Galaxies are usually found in clusters. There are so many galaxies that it is common for them to pass through one another, but the stars rarely collide. A typical star has a radius of $0.63R_\odot$. The density of stars is about 0.098 star per cubic parsec (1 pc = 3.086×10^{16} m). **a.** What fraction of the galaxy's volume is occupied by stars? Use this result to help you think about your answer to part (c). **b.** Find the mean free path of a star through a galaxy. **c.** If an intruder star travels 1000 pc through a galaxy, what is the probability of a collision?
35. N Gaseous water condenses at a temperature of 100.0°C and a pressure of 1.0 atm. What is the mean free path of gaseous water molecules at this temperature? (Assume water molecules have a radius of 0.15 nm.)
36. N The mean free path of the molecules of an ideal gas under pressure P and temperature T is λ. By what factor will the mean free path increase or decrease when **a.** the pressure changes to $P/6$ while remaining at a temperature T and **b.** the temperature changes to $2T$ while the pressure changes to $2P$?
37. N A very good laboratory vacuum can evacuate the air in a chamber so that its pressure is about 10^{-12} Pa. Assume air is an ideal gas at room temperature ($T \approx 20°$C) with molecules of radius $r = 2.0 \times 10^{-10}$ m. **a.** Estimate the number density (number per unit volume) of gas molecules in the chamber. **b.** Estimate the mean free path and compare it with the mean free path at atmospheric pressure in Example 20.3.
38. E In Example 20.4, we found that it would take about 50 years to fill a neighborhood with the scent of coffee purely by diffusion. Because we know that the time is actually much shorter than 50 years, we conclude that currents help spread the aromatic molecules quickly. Estimate the time it would take for the scent of coffee to fill a home kitchen purely by diffusion. What do you conclude?
39. Monica and Kennedy are going to have a race of sorts. They measure the distance across a room as 10.0 m, declaring one end the starting line and the other the finish line. They each have a container filled with a different gas and plan to release the gases simultaneously, timing the travel of the gases across the room. (Don't worry about the details such as how they detect the gases and don't worry about chemical reactions.)
    a. C If Monica has a container of hydrogen and Kennedy has a container of oxygen, who wins the race? Explain your answer.
    b. N Assume the room's temperature is 0°C. By what factor is the one gas faster than the other?

20-6 Real Gases: The Van der Waals Equation of State

40. C You find carbon dioxide at its triple point in all three phases in a single container. Describe the phases from the bottom of the container to the top. Compare these phases to what you would find for water at its triple point.
41. E Use the value of b for oxygen from Table 20.3 (3.18×10^{-5} m^3/mol) to estimate the radius of an oxygen molecule.
42. N There are 1789 moles of carbon dioxide in a container of volume 15.75 m^3 at a temperature of 45.6°C. Can you model this situation as an ideal gas? Show your work and explain your conclusion.
43. N At very high pressures, the volume of a real gas is larger than that predicted by the ideal gas law because real molecules occupy space, which the ideal gas law neglects. At what pressure would the Van der Waals equation of state result in a volume that was 2% larger than that predicted by the ideal gas law for 1 mol of nitrogen at a temperature of 0°C? Assume $a = 0.139$ Pa·m^6/mol^2 and $b = 3.91 \times 10^{-5}$ m^3/mol.
44. C Examine the Van der Waals equation of state (Eq. 20.32):

$$\left(P + a\frac{n^2}{V^2}\right)(V - nb) = nRT$$

When the density of the confined gas is severely reduced, the Van der Waals equation is approximately equivalent to the ideal gas law ($PV = nRT$) as described in Section 20-6. Explain why that makes sense. In other words, qualitatively explain why the pressure and volume would no longer need to be modified as they are in the Van der Waals equation.

20-7 Phase Changes

45. E Figure P20.45 shows a phase diagram of carbon dioxide in terms of pressure and temperature. **a.** Use the phase diagram to explain why dry ice (solid carbon dioxide) sublimates into vapor at atmospheric pressure rather than melting into a liquid. At what temperature does the dry ice sublimate when at atmospheric pressure? **b.** Estimate what pressure would be needed to liquefy carbon dioxide at room temperature.

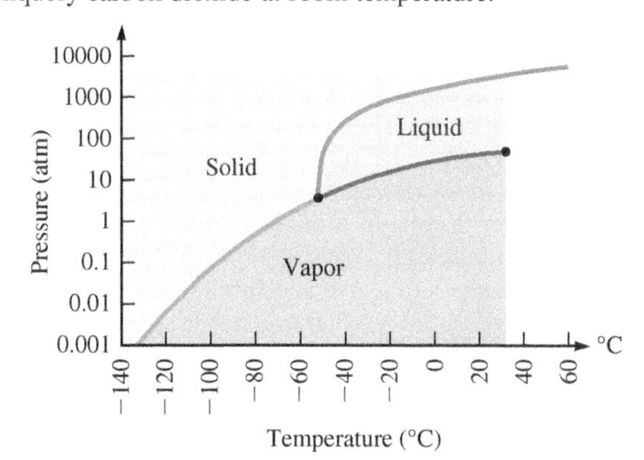

FIGURE P20.45

46. Initially, a substance has a pressure and temperature P_i and T_i.

a. G The substance undergoes a process at constant pressure. At the end of the process, its new pressure and temperature are given by $P_f = P_i$ and $T_f > T_i$. Draw a curve on a phase diagram that represents this process.

b. G Instead, the substance undergoes a process at constant temperature. At the end of the process, its new pressure and temperature are given by $P_f > P_i$ and $T_f = T_i$. Draw a curve on the same phase diagram that represents this process.

c. C Come up with a general statement about how these two different processes can be recognized on a phase diagram.

Problems 47 and 48 are paired.

47. C Consider water at 1.0 atm and initially at some temperature below 0°C. Its temperature is then raised to a value above the critical temperature while the pressure is held constant. *Hint*: Consult Figure 20.17. **a.** Describe the phases that water passes through during this process. **b.** If the curve between the solid phase and the liquid phase sloped upward and to the right as it does for carbon dioxide, how would your answer to part (a) change?

48. C Consider water at 0°C and initially at some pressure just below the critical pressure. Its pressure is then lowered to a value near zero while the temperature is held constant. *Hint*: Consult Figure 20.17. **a.** Describe the phases that water passes through during this process. **b.** If the curve between the solid phase and the liquid phase sloped upward and to the right as it does for carbon dioxide, how would your answer to part (a) change?

20-8 Evaporation

49. N At high altitude, water boils at 95°C. What is the external pressure at that altitude?

50. C On a humid day, would you expect the dew point to be particularly high or particularly low? Explain.

51. N In a pressure cooker, water boils at 120°C. What is the pressure inside the cooker?

52. Minneapolis, Minnesota, is famous for its cold winters. On a typical January day, the high temperature is 21°F, and the low temperature is 3°F. The dew point is 5°F.

a. N Calculate the high relative humidity and the low relative humidity.

b. C Does dew form? Explain your answer.

53. N On a certain day in Orlando, Florida, the temperature of the air and dew point were 16°C and 7.6°C, respectively. Find the relative humidity of the air on that day. Assume the saturated vapor pressure of water at 7°C, 8°C, and 16°C are 7.5 mm Hg, 8.0 mm Hg, and 12.0 mm Hg, respectively.

54. N On a summer day, the temperature is 34.5°C, and the partial pressure of water vapor in air is 2.75×10^3 Pa. What is the relative humidity?

General Problems

55. N A container is filled with a gas at a temperature of 300 K such that its pressure is 2.02×10^5 Pa. If the average kinetic energy of a molecule in the gas is to be **a.** tripled and **b.** halved, by what factor must the rms speed decrease or increase? In each case, state whether the rms speed decreases or increases.

Problems 56 and 57 are paired.

56. A A box with volume V confines 18 particles with identical mass m and the following speeds: three particles with speed v, one particle with speed $2v$, four particles with speed $3v$, three particles with speed $4v$, five particles with speed $5v$, and two particles with speed $6v$. What are the **a.** most probable speed, **b.** average speed, and **c.** rms speed of the particles in this box?

57. A Consider again the box and particles with the speed distribution described in Problem 56. **a.** What is the average pressure exerted by the particles on the walls of the box? **b.** What is the average kinetic energy per particle in this box?

58. N The average translational kinetic energy and the rms speed of molecules in a container of oxygen at a particular temperature are 6.20×10^{-21} J and 434 m/s, respectively. What are the corresponding values when the temperature is doubled?

59. N The average kinetic energy of an argon atom in a spherical container filled with argon gas and with a volume of 1.00×10^{-2} m^3 is 4.50×10^{-21} J. If the pressure inside the container is 2.00 atm, how many moles of argon are in the container?

60. N For the exam scores given in Table P20.60, find the average score and the rms score.

TABLE P20.60

Number	Score	Number	Score
1	67	19	72
2	92	20	86
3	88	21	77
4	82	22	76
5	76	23	80
6	78	24	89
7	22	25	75
8	81	26	77
9	77	27	92
10	70	28	84
11	100	29	66
12	89	30	61
13	74	31	90
14	63	32	73
15	94	33	79
16	55	34	72
17	83	35	89
18	75		

61. N The total translational kinetic energy of hydrogen molecules contained in a sealed 10.0-L container is 7.5×10^3 J. Determine the pressure of the gas molecules. *Hint*: See Equation 21.2, $NK_{av} = \frac{3}{2} N k_B T$.

Problems 62 and 63 are paired.

62. N A 0.500-m^3 container is filled with 30.0 mol of carbon dioxide gas (CO_2) such that its pressure is 2.00 atm. What is the rms speed of the molecules in the gas?

63. N Consider the carbon dioxide gas in Problem 62. Suppose the temperature of the gas is tripled. **a.** What is the new rms speed of the molecules in the gas? **b.** If one end of the container has an area of 0.0325 m^2, what will be the average force exerted on that end by the carbon dioxide?

64. N A spherical balloon with a radius of 10.0 cm is filled with oxygen gas at atmospheric pressure and at a temperature of 15.0°C. What are the **a.** number of molecules, **b.** average kinetic energy, and **c.** rms speed of oxygen in the balloon?

65. N A lifeguard pours chlorine into a pool and assumes it will eventually diffuse throughout the pool. Assuming diffusion alone is acting and the diffusion constant of chlorine in water is 2×10^{-9} m^2/s, how long would it take for the chlorine to diffuse from one end of the pool across the 100-m length to the other side?

66. N At what temperature will the mean free path of the molecules of an ideal gas be twice that at 27°C, if the pressure is kept constant but the volume changes?

67. N Determine the rms speed of an atom in a helium balloon at standard temperature and pressure ($T = 273.15$ K and $P = 1.00 \times 10^5$ Pa). Is it lower or higher than the rms speed of a nitrogen molecule in the atmosphere?

Problems 68 and 69 are paired.

68. C Consider a gas filling two connected chambers that are separated by a removable barrier (Fig. P20.68). The gas molecules on the left (red) are initially at a higher temperature than the ones on the right (blue). When the barrier between the two chambers is removed, the molecules begin to mix and move from one chamber to the other. **a.** Describe what happens to the temperature in the left chamber and in the right chamber as time goes on, once the barrier is open. Discuss in terms of the mixing of the molecules from each gas. **b.** Describe what happens to the most probable speed and average speed in the left chamber and in the right chamber as time goes on, once the barrier is open. Do they increase or decrease by the same factor? Explain.

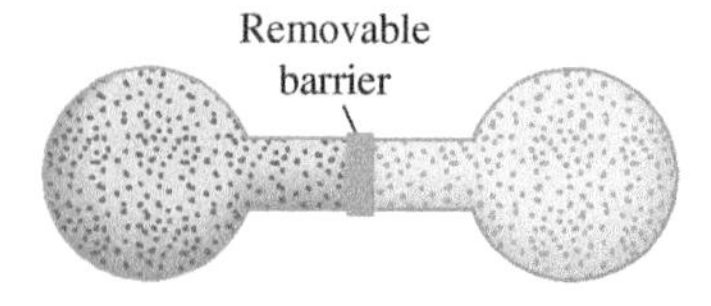

FIGURE P20.68 Problems 68 and 69.

69. Consider the situation described in Problem 68 and focus on the right chamber shown in Figure P20.68. Suppose the gas filling the chambers is O_2. Assume the initial average speed is 900.0 m/s.

a. A Derive an expression for the rate of change of temperature in terms of the rate of change of the average speed.

b. N If the temperature rises at a rate of 2.00 K/s, what is the rate of change of the average speed?

70. N The relative humidity of air in a closed room at 20°C is 50%. What will be the relative humidity of air if the temperature is increased to 25°C? The saturated vapor pressure of water at 20°C is 17.4 mm of mercury (Hg) and at 25°C is 23.6 mm of Hg.

Problems 71 and 72 are paired.

71. N A 0.500-m^3 container holding 3.00 mol of ozone (O_3) is kept at a temperature of 250 K. Assume the molecules have radius $r = 2.50 \times 10^{-10}$ m. What are the **a.** mean free path and **b.** mean free time between collisions for an ozone molecule in the container?

72. Consider the gas described in Problem 71.

a. N What is the average distance between molecules?

b. C Compare the mean free path found in Problem 71 with your answer to part (a) of this problem. Comment and explain the differences you find in comparing these quantities.

73. N CASE STUDY Although the nebula from which our solar system formed was comprised mostly of hydrogen and helium, the terrestrial planets including the Earth and Mars quickly lost these gases from their atmospheres. The mass of a helium atom is 6.64×10^{-27} kg. What is the temperature at which the average speed of helium atoms exceeds the escape speed from **a.** the Earth (11.2 km/s) and **b.** Mars (5.00 km/s)? **c.** If the average temperature of air at sea level is 20.0°C, how is it possible for helium to escape from the Earth's atmosphere?

Heat and the First Law of Thermodynamics

21

Underlying Principles

First law of thermodynamics

Major Concepts

1. Heat
2. Thermal energy
3. Heat capacity
4. Specific heat
5. Thermodynamic process
6. Degrees of freedom
7. Principle of equipartition of energy

Special Cases

Six specific thermodynamic processes:

1. Adiabatic
2. Isothermal
3. Constant volume
4. Constant pressure
5. Cyclic
6. Free expansion

Three specific ways for heat to flow:

1. Conduction
2. Convection
3. Radiation

Tools

Energy bar charts

Key Questions

What is heat?

How can heat be incorporated into the principle of conservation of energy?

What are some examples of processes that involve heat?

21-1 What is heat? 612

21-2 How does heat fit into the conservation of energy? 613

21-3 The first law of thermodynamics 615

21-4 Heat capacity and specific heat 616

21-5 Latent heat 620

21-6 Work in thermodynamic processes 625

21-7 Specific thermodynamic processes 628

21-8 Equipartition of energy 631

21-9 Adiabatic processes revisited 637

21-10 Conduction, convection, and radiation 638

During spring break, you go to the beach with your friends. One afternoon, you take your physics textbook and a cold drink so that you can study in the Sun. You forget your cooler, though, so your cold drink doesn't stay cold very long. The drink's increased temperature is a clear indication that its thermal energy increased, but where did that energy come from? Of course, the answer is shining down on you: the Sun. How could the Sun, which is more than 93 million miles away, transfer energy to your drink? From Chapter 9, energy can be transferred to a system if the environment does work on that system. The drink was not displaced, so clearly nothing did any work on it. Therefore, the energy must have been transferred by other means. This other means of transferring energy is known as *heat*. The Sun transfers energy to the rest of the solar system through heat. If you had put your drink in the cooler, your drink would have been fairly well insulated and unable to absorb very much energy. The next time you forget your cooler, try wrapping your drink in a towel.

21-1 What Is Heat?

HEAT ✪ Major Concept

When you hold your cold hands in front of a nice hot campfire (Fig. 21.1), they warm up because energy is transferred from the environment (the fire) to the system (your hands). **Heat** is the energy transferred from the environment to the system (or from the system to the environment) due to their temperature difference. The term *heat* is often misused and misunderstood.

FIGURE 21.1 Holding your hands near a campfire is one way to warm them.

Historical theories about heat contribute to the misunderstanding of that term. In the late 1700s, many scientists believed that heat was a fluid called "caloric." Caloric was believed to flow from hot objects to cooler objects, much like lemonade flowing from a pitcher into a glass. An 18th-century scientist, for example, might have said that when you put a cold potato in a hot oven, caloric flowed from the oven to the potato. The oven would have less caloric in it, and the potato would gain caloric. Today, we often use the same sort of language for heat. We might say that heat flows from the oven to the potato, but this wording gives the wrong impression because *heat is not a substance*, and it does not flow from the oven to the potato in the same way that liquids flow from one container into another.

By our contemporary definitions, neither heat nor work describes the energy contained *in* a system (or in the environment). Instead heat and work only describe energy that is transferred between a system and its environment. Work describes the transfer of energy due to forces between macroscopic objects. Heat is work on the microscopic scale. Again, imagine a cool potato surrounded by hot air molecules in the oven. Because the air in the oven is hotter than the potato, the air molecules have a greater average kinetic energy than the potato molecules. Therefore, during a collision, it is more likely that an air molecule will lose kinetic energy and a potato molecule will gain energy. So, on the whole, thermal energy is transferred from the hot air to the potato, and we say that *heat flows from the oven to the potato*. The wording is the same as above, but now the phrase means that the hot air in the oven environment did work on the potato at the microscopic level, transferring energy.

James Joule described his experiments in a paper, *On the Mechanical Equivalent of Heat*, but the Royal Society of London did not allow him to include his conclusion about heat. Joule worked on the establishment of units involved in thermodynamics, and today's SI unit for energy is named in his honor.

If the potato's temperature is equal to the temperature of the air in the oven, equal numbers of air molecules lose energy and gain energy when they collide with the potato molecules. So, when the potato is in thermal equilibrium with the oven, there is no heat flow, meaning that there is no net work done on the microscopic level.

Caloric fluid was never detected, and by the 19th century, other models for heat were pursued. James Joule (1818–1889), an English scientist and brewer, conducted experiments leading to the idea that heat is work on the microscopic level. Usually, we cannot directly observe the kinetic and potential energies of microscopic particles. Instead, we observe macroscopic properties such as temperature. If the temperature of a system rises, we can infer that energy was transferred to the system from the environment either by heat or by work.

Properties such as temperature that describe the condition or state of the system (and the environment) are called **state variables**. Energy may be transferred through work or through heat. If energy is transferred between the environment and the system, one or more of these state variables will change, and we say that "the state of the system changes."

Joule focused on changing one state variable—temperature—of a system consisting of a can of water. He found that the water temperature could be raised by two different methods. As every cook knows, if Joule placed the water on a hot stove, the water's temperature increased due to heat from the environment. What is very interesting is that he could also raise the temperature of water by allowing gravity to do macroscopic work on the system. Joule's system included a can of water and a paddle wheel attached to a hanging disk, but not the Earth (Fig. 21.2). The Earth exerts a downward force on the hanging disk, pulling it downward. The Earth does positive work on the system, transferring energy to it and ultimately causing the water temperature to rise. Joule's experiment showed that heat and work are both ways to transfer energy between a system and its environment.

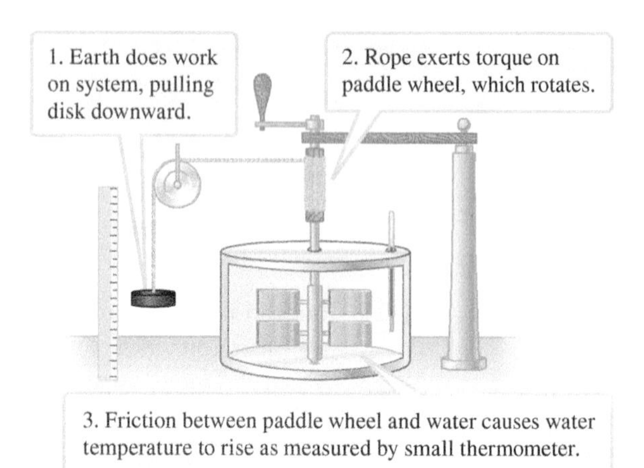

FIGURE 21.2 Joule's apparatus. The system consists of the device, including the water in the can. The Earth—which is not in the system—does work on the system as the disk falls. The falling disk is connected by a rope to a paddle wheel. The paddle wheel rotates in the can filled with water, and the water's temperature rises.

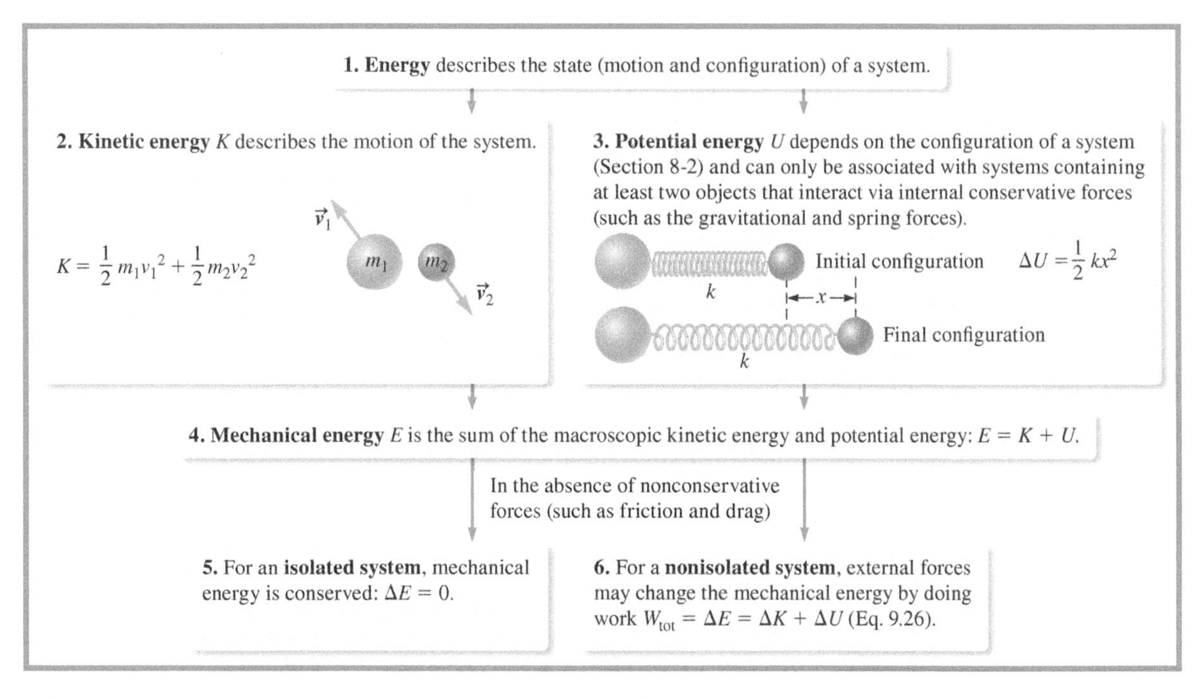

FIGURE 21.3 This concept map is a review of energy concepts from Chapters 8 and 9.

21-2 How Does Heat Fit into the Conservation of Energy?

In this chapter, we extend the concepts of energy conservation and the work–energy theorem from Chapters 8 and 9 to include heat. Review this material by studying Figure 21.3 to set heat and the first law of thermodynamics (Section 21-3) in context. Pay particular attention to when energy is transferred between a nonisolated system and its environment (panel 6 in Fig. 21.3). The total work W_{tot} done on a system by external forces in the absence of nonconservative forces (such as friction and drag) changes the system's mechanical energy by the work–mechanical energy theorem:

$$W_{tot} = \Delta E = \Delta K + \Delta U \tag{9.26}$$

Thermal Energy, Work, and Heat

The concepts of work and energy (Fig. 21.3) are convenient when studying a macroscopic system consisting of a small number of objects that are well modeled as particles (such as a tossed ball and the Earth). When the objects in the system are not well modeled as particles or when there are many particles in the system, though, we must modify and add to these concepts.

Figure 21.4 maps how we think about energy in the case of a system consisting of deformable objects or a large number of particles. The following are key points corresponding to Figure 21.4.

1. Energy still describes the state of the system.
2. Kinetic energy and potential energy are defined as in Figure 21.3, but those terms are reserved for the macroscopic state of the system. For example, in a system consisting of a hamburger at rest in a frying pan, we would say that the hamburger's kinetic energy is zero because the hamburger is at rest (in our frame). Microscopically, however, molecules within the hamburger are in motion.
3. Molecular motion is accounted for by **thermal energy** (Section 9-7), the sum of each microscopic particle's kinetic energy plus the total potential energy stored in the forces between microscopic particles. In the absence of phase changes, thermal energy depends on the temperature of the system.

THERMAL ENERGY
Major Concept

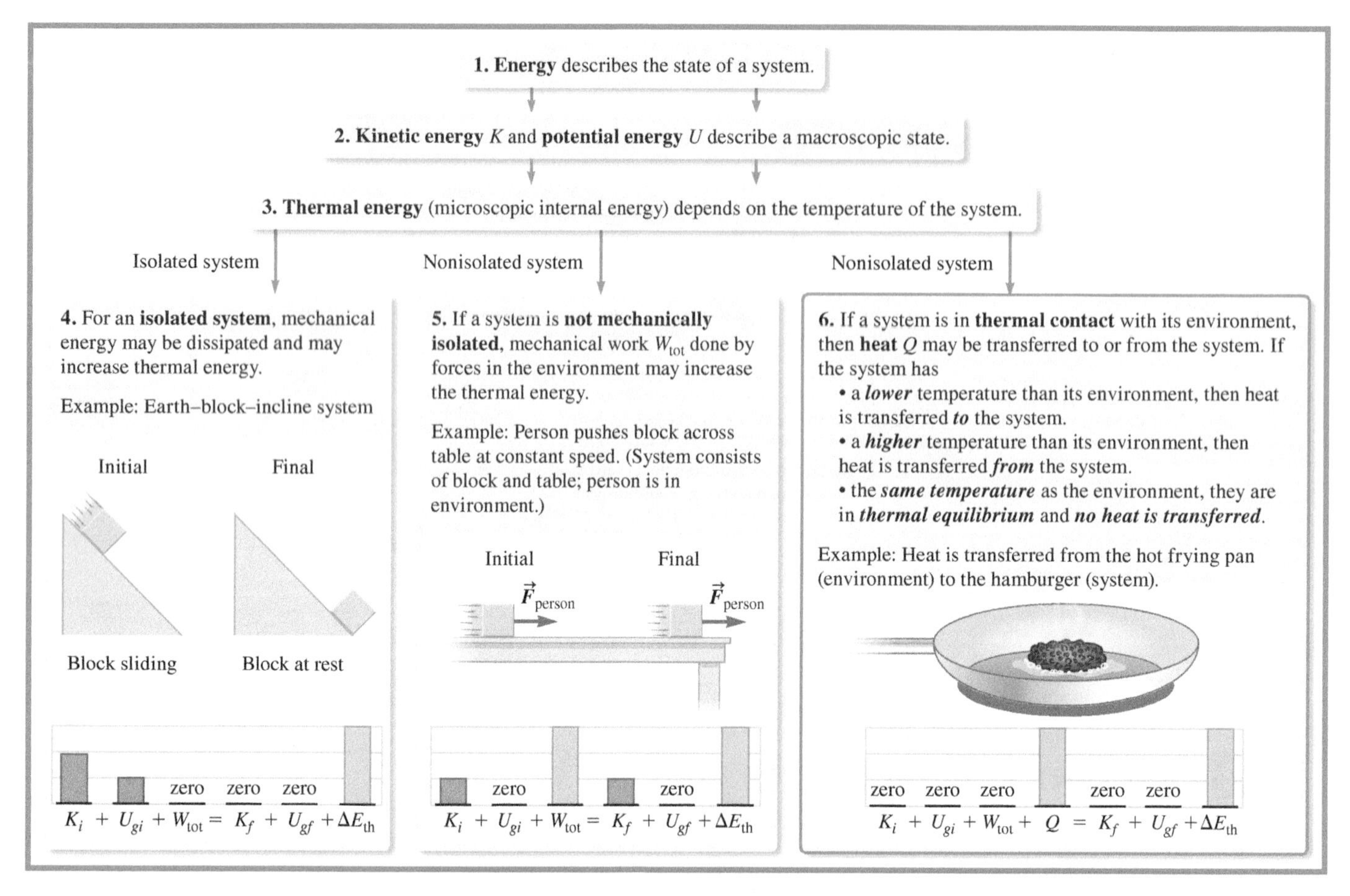

FIGURE 21.4 This concept map shows how heat (panel 6) fits into the concepts of energy.

Any changes in thermal energy depend on what contact the system has with its environment.

4. If a system is **isolated** from its environment, no energy may be exchanged. In this case, the system's mechanical energy may be transformed into thermal energy if dissipative forces such as kinetic friction are present. When a block slides down a rough, inclined plane, mechanical energy is transformed into thermal energy, and the block and incline are both warmer.
5. If the system is not mechanically isolated from its environment, the environment may do work on the system, and that work can alter the system's thermal energy. When a person (who is in the environment) pushes the block across a table at constant velocity, the positive work done by the person increases the block–table system's thermal energy. The block and the table are warmer.

Methods 4 and 5 of changing the thermal energy were discussed in Chapter 9. We now formally introduce a third way for thermal energy to change as described in the blue outlined box (panel 6) in Figure 21.4.

HEAT ✪ Major Concept

6. If the system is in **thermal contact** with its environment, energy may be transferred by heat. **Heat** Q is the amount of energy transferred between a system and its environment due to their difference in temperature. Consider a system consisting of a cool hamburger at rest in a hot frying pan. The frying pan is in the environment. The hamburger remains at rest, so its kinetic energy remains zero. The system consists only of the hamburger, so on the *macroscopic* scale there are no internal forces. No external forces do work on the hamburger. Energy (heat) is transferred from the hot frying pan to the hamburger, increasing the hamburger's thermal energy. We need to add a bar for heat Q to the bar chart (panel 6). Because heat is work on the *microscopic* level, we place Q next to W_{tot}. In the case of a hamburger in a hot frying pan, the bar for heat Q is the

same size as the bar for the change in thermal energy ΔE_{th}. In a sense, the next section is about how adding a bar for heat changes the conservation of energy principle.

CONCEPT EXERCISE 21.1

Decide which of the following statements are incorrect according to the definition of *heat* used in physics. Explain your answers and provide corrections where possible.

a. "The heat in Texas will make you tired."
b. "When you slide a book across a table, kinetic friction converts kinetic energy into heat."
c. From a science textbook: "The steam engine marked the first time in human history that heat was used to do work."

21-3 The First Law of Thermodynamics

FIRST LAW OF THERMODYNAMICS
Underlying Principle

The first law of thermodynamics is both an extension and a special case of the work–energy theorem. In Equation 9.30, we wrote the work–energy theorem as

$$\begin{pmatrix}\text{initial}\\ \text{mechanical}\\ \text{energy}\end{pmatrix} + \begin{pmatrix}\text{energy}\\ \text{transferred}\\ \text{(total work)}\end{pmatrix} = \begin{pmatrix}\text{final}\\ \text{mechanical}\\ \text{energy}\end{pmatrix} + \begin{pmatrix}\text{change in}\\ \text{internal}\\ \text{energy}\end{pmatrix}$$

First, we broaden the work–energy theorem to include the idea that energy may be transferred by heat:

$$\begin{pmatrix}\text{initial}\\ \text{mechanical}\\ \text{energy}\end{pmatrix} + \begin{pmatrix}\text{energy}\\ \text{transferred}\\ \text{(total work)}\end{pmatrix} + \begin{pmatrix}\text{energy}\\ \text{transferred}\\ \text{(heat)}\end{pmatrix} = \begin{pmatrix}\text{final}\\ \text{mechanical}\\ \text{energy}\end{pmatrix} + \begin{pmatrix}\text{change in}\\ \text{internal}\\ \text{energy}\end{pmatrix}$$

This equation is a very general expression for the conservation of energy in a system because it accounts for the energy transferred *by work* and *through heat.*

Our next step is to write this general expression for a special case. Let's start by taking a close look at the last term, the change in internal energy. Of course, mechanical energy is "internal" to the system, but the term ***internal energy*** refers to *microscopic* internal energy. Internal energy is not a new form of energy; it is potential energy and kinetic energy on the microscopic level. For convenience, internal energy is further divided into other categories such as thermal energy, chemical energy, and nuclear energy. Throughout this textbook, our focus is primarily on thermal energy. So, as we did in Section 9-8, we write this last term for the special case in which the only category of internal energy we deal with is thermal energy:

$$\begin{pmatrix}\text{initial}\\ \text{mechanical}\\ \text{energy}\end{pmatrix} + \begin{pmatrix}\text{energy}\\ \text{transferred}\\ \text{(total work)}\end{pmatrix} + \begin{pmatrix}\text{energy}\\ \text{transferred}\\ \text{(heat)}\end{pmatrix} = \begin{pmatrix}\text{final}\\ \text{mechanical}\\ \text{energy}\end{pmatrix} + \begin{pmatrix}\text{change in}\\ \text{thermal}\\ \text{energy}\end{pmatrix}.$$

Our final step is to write the conservation of energy for the special case in which the mechanical energy of the system is constant. In this case, the initial mechanical energy equals the final mechanical energy, and we have:

$$\begin{pmatrix}\text{energy}\\ \text{transferred}\\ \text{(total work)}\end{pmatrix} + \begin{pmatrix}\text{energy}\\ \text{transferred}\\ \text{(heat)}\end{pmatrix} = \begin{pmatrix}\text{change in}\\ \text{thermal}\\ \text{energy}\end{pmatrix}$$

Using symbols, we have the **first law of thermodynamics**:

$$W_{tot} + Q = \Delta E_{th} \tag{21.1}$$

The sign convention used throughout this book is that *if energy is transferred from the environment to the system, the sign of W_{tot} or Q is positive.*

The first law is a statement of the conservation of energy for a system that may exchange energy with its environment by work or through heat, and the only form of internal energy that may change is thermal energy. For the rest of this chapter and throughout Chapter 22, we consider systems whose energy changes are completely described by the first law of thermodynamics.

If a system is **mechanically isolated** (or simply **isolated**), energy cannot be transferred between the system and its environment by work. Likewise, if a system is **thermally insulated** (or simply **insulated**), energy cannot be transferred between the system and its environment through heat.

Blend Images/Stockphoto.com

FIGURE 21.5 This case study is about you, how you exchange energy with your environment and how you maintain your healthy body temperature. Of course, it all begins with you getting energy from the food you eat.

CASE STUDY Part 1: You

Every day, you eat food that supplies your body with chemical energy (Fig. 21.5). If you are a typical adult, you eat about 2400 Cal (1.00×10^7 J) per day. Some of this energy is transformed into macroscopic kinetic energy that allows you to walk, allows you to scan your eyes across this page, and keeps your heart beating.

The *calorie*, a commonly used unit of energy, comes from the word *caloric*. In the United States, the food energy used by the human body is reported in Calories (with an uppercase C), where 1 Calorie = 1000 calories. In European countries, the energy content of food is reported in kilocalories (kcal), where 1 kcal = 1 Calorie.

Most of the time, your body maintains its healthy temperature of 37°C. To regulate your temperature, your body must exchange heat with the environment. Throughout this chapter, we consider the human body in terms of the first law of thermodynamics. We'll see how forensic experts can use the temperature of a corpse to estimate the time of death, how much work you do just by breathing, and how you maintain your body temperature in different environments such as deserts, oceans, and space.

CONCEPT EXERCISE 21.2

In each situation listed, an object's temperature increases. For each situation, decide if energy was transferred to the object through heat or work. Explain your answers.

a. A pitcher of water is removed from the refrigerator and left on the kitchen counter.
b. A rubber ball is bounced (repeatedly) against the floor.
c. A meteor falls through the Earth's atmosphere.

21-4 Heat Capacity and Specific Heat

There is a close connection between the temperature of a system and its thermal energy. To see that connection, consider the special case of an ideal gas whose molecules may be modeled as particles that do not exert any forces on one another (except during a collision). On the microscopic level, the gas has kinetic energy but no potential energy. Therefore, the thermal energy of an ideal gas is just the sum of the kinetic energies of its particles. The average kinetic energy of the particles is given by $K_{av} = \frac{3}{2} k_B T$ (Eq. 20.17). The thermal energy of an ideal gas (total kinetic energy of the particles) is found by multiplying the average kinetic energy by the number N of particles:

$$E_{th} = NK_{av} = \frac{3}{2} N k_B T = \frac{3}{2} nRT \tag{21.2}$$

where $Nk_B = nR$ (Eqs. 19.18 and 19.20).

According to Equation 21.2, we can observe a change in thermal energy by just measuring the system's temperature change. (This technique holds not just for ideal gases, but for other systems as well.) As long as a system is not undergoing a phase change, a change in the system's thermal energy is proportional to a change in the system's temperature:

The term *heat capacity* suggests that it has the same dimensions as heat and that a system has a capacity to "hold" heat. Neither of these impressions is true, so it is best to think of heat capacity as a name for the constant of proportionality in Equation 21.3.

$$\Delta E_{th} = \mathcal{C} \Delta T \tag{21.3}$$

The constant $\mathcal{C}$ has the dimensions of energy per temperature (as does Boltzmann's constant k_B in Eq. 21.2). Unfortunately, $\mathcal{C}$ was named when people still thought of heat as a substance, and it is called **heat capacity**.

HEAT CAPACITY Major Concept

Let's consider situations in which heat is exchanged but no work is done (such as a hamburger in a hot frying pan). We start with systems that are solids or liquids and will return to gases in Section 21-8. When no work is done on the system, the first law of thermodynamics becomes

$$\Delta E_{th} = Q$$

So, we can write Equation 21.3 as

$$Q = \mathcal{C}\Delta T \tag{21.4}$$

Heat capacity $\mathcal{C}$ depends on the *type* and the *amount* of material making up the system.

For convenience, we separate heat capacity's dependence on the *amount* of material from its dependence on the *type* of material, so we write Equation 21.4 as

$$Q = mc\Delta T \tag{21.5}$$

where $\mathcal{C} = mc$. The amount of material is expressed in terms of the system's mass m, and the constant c is known as the **mass specific heat capacity** or simply the **specific heat**. The specific heat is the heat capacity of a material per unit mass. The dimensions of specific heat are energy per mass per temperature, so its SI units are $J/(kg \cdot K)$.

SPECIFIC HEAT ✪ Major Concept

As in Section 19-6, it is often convenient to express the amount of a substance in terms of the number of moles, so we define the **molar specific heat capacity** or just **molar specific heat** as the heat capacity of a material per unit mole. It has the SI units $J/(mol \cdot K)$. We write Equation 21.4 in terms of the molar specific heat C:

$$Q = nC\Delta T \tag{21.6}$$

where the amount of substance is expressed as the number of moles n. The heat capacity $\mathcal{C} = nC$. Table 21.1 provides the specific heat and molar specific heat for a number of different solids and liquids.

In this textbook, heat capacity is denoted by an uppercase script $\mathcal{C}$, specific heat by a lowercase c, and molar specific heat by an uppercase C.

The first law of thermodynamics (Eq. 21.1) states that when energy is transferred to or from a system (by work or heat), a state variable (thermal energy) changes. Heat capacity and specific heat allow us to rewrite the first law of thermodynamics in terms of a change in temperature. First, substitute heat capacity (Eq. 21.3):

$$W_{tot} + Q = \Delta E_{th} = \mathcal{C}\Delta T$$

Then, eliminate heat capacity $\mathcal{C}$ and rewrite this equation in terms of the specific heat c ($\mathcal{C} = mc$):

$$\Delta E_{th} = W_{tot} + Q = mc\Delta T \tag{21.7}$$

Because a change in kelvin temperature is numerically the same on the Celsius scale, specific heat expressed as $J/(kg \cdot °C)$ is numerically equivalent to $J/(kg \cdot K)$.

Let's see how Equation 21.7 fits in with Joule's experiment (Fig. 21.2). In Joule's experiment, no heat was transferred to the system (a container of water), so $Q = 0$.

TABLE 21.1 Specific heat and molar specific heat for solids and liquids near room temperature and at atmospheric pressure (except where noted).

Substance	Specific heat c [$J/(kg \cdot K)$]	Molar specific heat C [$J/(mol \cdot K)$]	Substance	Specific heat c [$J/(kg \cdot K)$]	Molar specific heat C [$J/(mol \cdot K)$]
Aluminum	900	24.4	Lead	128	26.5
Beef, flank	2340		Marble	860	
Brass	380		Mercury	140	28.1
Cabbage	3940		Milk	3930	
Copper	386	24.5	Seawater	3900	
Ethyl alcohol	2430	110.4	Silver	230	25.5
Glass	840		Water (ice at −5°C)	2100	37.6
Gold	128	25.6	Water (liquid)	4190	75.4
Granite	790		Wood	1700	
Iron or steel	450	25.1			

Note: Specific heat and molar specific heat depend on temperature, but for small changes in temperature, they may be considered constant.

The Earth did work on the system, however; so, for Joule's experiment, Equation 21.7 is

$$W_{tot} = mc\Delta T \quad (21.8)$$

When Joule placed the water on a hot stove, no work was done ($W_{tot} = 0$), so $Q = mc\,\Delta T$. Joule's experiment showed that energy transferred by work (Eq. 21.8) or by heat (Eq. 21.5) raises the temperature of the water, leading to the idea that heat is work on the microscopic level.

CONCEPT EXERCISE 21.3

Come up with your own terms to replace *heat capacity*, *specific heat*, and *molar specific heat*.

CONCEPT EXERCISE 21.4

Suppose you double the number of particles in an ideal gas without changing the gas temperature. What happens to the average kinetic energy of the gas particles? What happens to the thermal energy?

EXAMPLE 21.1 Joule's Experiment, or Another Way to Make a Pot of Tea

Suppose you wish to make a liter of hot tea by using Joule's equipment (Fig. 21.2). The density of water is 1 kg/L, so 1 L of water has a mass of 1 kg. The specific heat of water is listed in Table 21.1.

A Start with a more modest goal of raising the water temperature just 1 K. If the hanging disk has a mass of 5.00 kg, how far must it fall to raise the temperature of 1.00 L of water 1.00 K? If you imagine building a tower from which to drop the disk, this distance would be the minimum required height for the tower.

INTERPRET and ANTICIPATE

The system consists of Joule's apparatus, including 1.00 kg of water in the can. The Earth is in the environment, and it does work on the system as the hanging disk falls. From the amount of work required to raise the water 1.00 K, find the distance through which the disk must fall.

SOLVE

Using the specific heat of liquid water from Table 21.1, find the work required.

$$W_{req} = mc\Delta T \quad (21.8)$$

$$W_{req} = (1.00\text{ kg})\left(4190\frac{\text{J}}{\text{kg}\cdot\text{K}}\right)(1\text{ K}) = 4190\text{J}$$

The work done by gravity on the falling disk is the weight of the disk multiplied by its displacement (Eq. 9.15). There are no other external forces doing work on the system.

$$W_{tot} = \vec{F}\cdot\Delta\vec{y} = mg\,\Delta y$$

$$\Delta y = \frac{W_{tot}}{mg}$$

If all the work done by gravity goes into heating the water, we can set the required work W_{req} equal to the work done by gravity W_{tot}. Joule's system may not be 100% efficient, and some of the work done by gravity may go into other parts of the system. (For example, kinetic friction may raise the temperature of the rope, pulley, and paddle wheel.) So, our result is a lower limit to how far the disk must fall.

$$\Delta y = \frac{W_{tot}}{mg} = \frac{4190\text{ J}}{(5.00\text{ kg})(9.81\text{ m/s}^2)}$$

$$\Delta y = 85.4\text{ m}$$

CHECK and THINK

If Joule used a 5-kg weight, he would have needed to drop the weight off a tower roughly 30 stories tall to see a rise in temperature of 1 K (= 1°C). Of course, he could reduce the height needed by increasing the mass of the disk. Perhaps by using a strong rope, he could replace the 5-kg disk with a 25-kg disk; the tower would then only need to be 17 m or about 6 stories tall.

B Replace the 5-kg disk with a 25-kg disk. If the water is initially at room temperature (22°C), how far must the disk fall to raise the water temperature to the boiling point? (Tea should be made with boiling water.)

INTERPRET and ANTICIPATE

We need only two modifications to our work in part A. First, calculate the work required to raise the water temperature from 22°C to 100°C. Then recalculate the work done by gravity on the more massive disk.

SOLVE

The difference in temperature on the Celsius scale is the same as that on the Kelvin scale. Find the work required to raise the 1 kg of water to the new temperature.

$$W_{req} = mc\Delta T \quad (21.8)$$

$$W_{req} = (1\text{ kg})\left(4190\frac{\text{J}}{\text{kg}\cdot\text{K}}\right)(78\text{ K}) = 3.3 \times 10^5\text{ J}$$

As before, assume all the work done by gravity goes into heating the water.

$$\Delta y = \frac{W_{tot}}{mg} = \frac{3.3 \times 10^5\text{ J}}{(25\text{ kg})(9.81\text{ m/s}^2)} = 1.3 \times 10^3\text{ m}$$

CHECK and THINK

To make a pot of tea using Joule's equipment, we would need a tower nearly 450 stories tall. Clearly, it is much more convenient to use a kitchen stove.

EXAMPLE 21.2 A Calorimeter

A water calorimeter is a device used to measure the specific heat of a substance (Fig. 21.6). A well-insulated container of mass m_c is filled with water of mass m_w. The specific heats of water c_w and the container c_c are known. A thermometer fits through the lid of the container, and very little (if any) heat can pass through the container. Suppose a hot sample of the substance is placed in the water. Eventually, the sample and the water are in thermal equilibrium. The mass of the sample is m_s. By measuring the change in the sample's temperature ΔT_s and the change in the water's temperature ΔT_w, the sample's specific heat c_s can be calculated. Find an expression for the sample's specific heat.

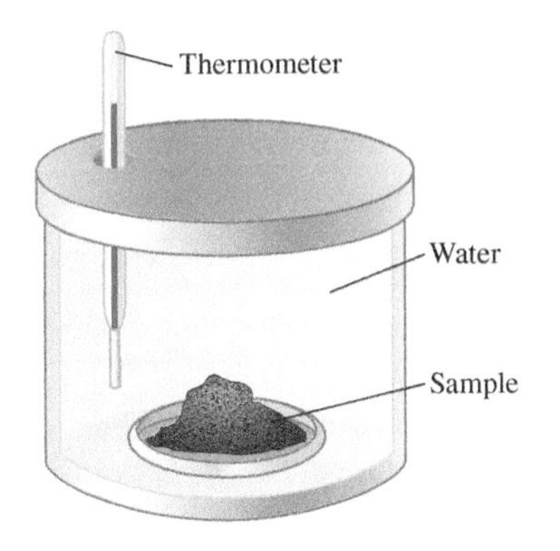

FIGURE 21.6 A water calorimeter is used to find the specific heat of a sample of some material.

INTERPRET and ANTICIPATE

Because the container is insulated, any energy lost by the sample must be gained by the water and its container. Mathematically, we write $-\Delta E_s = \Delta E_w + \Delta E_c$, where *s*, *w*, and *c* stand for *s*ample, *w*ater, and *c*alorimeter, respectively. We expect that if the sample has a relatively large change in temperature ΔT_s, it must have a low specific heat.

SOLVE

Use the first law of thermodynamics (Eq. 21.7) to express changes in thermal energy in terms of changes in temperature. The water is in thermal equilibrium with its container, so they both have the same change in temperature ΔT_w.

$$\Delta E_{th} = mc\,\Delta T \quad (21.7)$$

$$-\Delta E_s = \Delta E_w + \Delta E_c$$

$$-m_s c_s\,\Delta T_s = m_w c_w\,\Delta T_w + m_c c_c\,\Delta T_w$$

Solve for c_s.

$$c_s = -\frac{(m_w c_w + m_c c_c)\,\Delta T_w}{m_s\,\Delta T_s} \quad (1)$$

CHECK and THINK

First, because the dimensions of mass and temperature cancel, we see that our expression has the correct dimensions for specific heat. Second, our expression also has the correct sign. We expect the water temperature to increase and the sample's temperature to decrease. The minus sign in front of the expression ensures that $c_s > 0$, as expected. Finally, as predicted, if ΔT_s is large, then c_s is small.

EXAMPLE 21.3 CASE STUDY Time of Death

You have probably seen a movie where someone is found dead and the "lab report" gives the time of death. Have you ever wondered how forensic scientists can really determine the time of death? There are many methods, but one simple way is based on the core temperature of the corpse.

A living male adult consumes about 2400 Cal (1.00×10^7 J) per day. So a man extracts on average about 116 J/s from his food, which, while alive, he must lose to maintain his temperature. To estimate the time of death from a corpse's temperature, we make several assumptions: (1) at the time of death, the person's temperature was normal; (2) the corpse's temperature is above the environment's temperature; (3) the corpse continues to lose energy at a constant rate of 116 J/s; and (4) the environment's temperature has been constant.

Suppose a 65-kg male corpse with a core temperature of 35°C is found at 3:25 PM. The room's temperature has been regulated for days at 21°C. The average specific heat of the human body is $3470\ \text{J}/(\text{kg}\cdot\text{K})$. Estimate the time of death.

INTERPRET and ANTICIPATE

We assumed that the corpse loses thermal energy at a constant rate of 116 J/s. Because a change in temperature is proportional to a change in thermal energy, we can use the drop in temperature relative to normal (living) body temperature, 37°C, to find the time of death.

SOLVE

Start with the first law of thermodynamics to write an expression for ΔE_{th} in terms of ΔT.

$$\Delta E_{\text{th}} = mc\Delta T \quad (21.7)$$

Divide both sides by Δt to find an expression for the rate at which thermal energy and temperature change with time.

$$\frac{\Delta E_{\text{th}}}{\Delta t} = mc\frac{\Delta T}{\Delta t}$$

Solve for Δt. We were given that $\Delta E_{\text{th}}/\Delta t = 116\ \text{J/s}$. The temperature change is the same on the Kelvin scale as it is on the Celsius scale.

$$\Delta t = mc\frac{\Delta T}{\Delta E_{\text{th}}/\Delta t}$$

$$\Delta t = (65\ \text{kg})\left(3470\frac{\text{J}}{\text{kg}\cdot\text{K}}\right)\frac{(2\ \text{K})}{116\ \text{J/s}}$$

$$\Delta t = 3.9 \times 10^3\ \text{s} \times \left(\frac{1\ \text{hr}}{3600\ \text{s}}\right)$$

$$\Delta t = 1{:}05$$

This result is the elapsed time; we find the time of death t_{death} by subtracting the elapsed time from the time the body was found.

$$t_{\text{death}} = 3{:}25\ \text{PM} - 1{:}05$$

$$t_{\text{death}} = 2{:}20\ \text{PM}$$

CHECK and THINK

Our assumption that a dead body loses energy at the same rate as a living body is too simple. From your own experience, you know that the rate at which your thermal energy changes depends on your level of activity. In fact, a dead body loses energy at a somewhat lower rate than we assumed. The rule of thumb is that a dead body's temperature decreases 0.8 K per hour, so a drop in temperature of 2 K means that the time elapsed is 2:30 or the time of death was approximately 12:55 PM. (Of course, even this rule of thumb is too simple. The corpse cannot continue to lose energy at a constant rate forever.)

21-5 Latent Heat

Consider a simple experiment you could do by placing a block of ice in a hot container. The system is the ice, everything else is in the environment, and a thermometer is kept in contact with the ice. Because there are no external forces doing work on the system and as long as the ice remains below its melting point, the total heat transferred to it is directly proportional to its temperature change $Q = mc\,\Delta T$ (Eq. 21.5). A graph of temperature versus total heat transferred (Fig. 21.7) is a straight line whose slope is $1/mc$.

On the microscopic level, as heat is transferred to the ice, the water molecules oscillate more vigorously. Once it reaches its melting point (0°C at atmospheric pressure), the ice goes from its solid to its liquid phase. Although the ice melts, there is *no* change in the system's temperature (Fig. 21.7). During the phase change, the first law of thermodynamics still holds: $Q = \Delta E_{th}$, but $\Delta T = 0$. Heat is still flowing from the environment to the system, so the thermal energy of the system continues to increase. On the microscopic level, the increase in the system's thermal energy loosens the bonds between the molecules, so in the liquid phase the molecules can slip around one another.

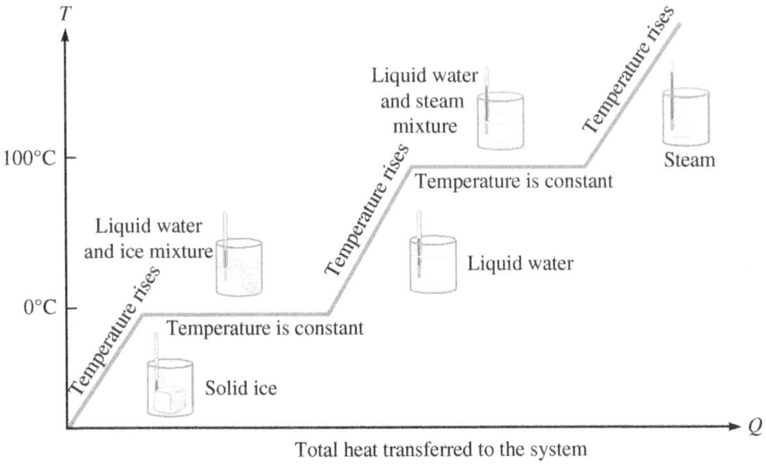

FIGURE 21.7 Temperature as a function of heat transferred to water (beginning as ice). When heat is added to the system and its temperature remains constant, the system is undergoing a phase change.

Latent heat L (also known as the **heat of transformation**) is the energy per unit mass that must be transferred to the system for the system to transform completely from one phase to the next. The heat required to melt a solid is found by multiplying the latent heat by the mass of the substance:

$$Q = mL_F \tag{21.9}$$

where L_F is the **heat of fusion**. The heat of fusion is the latent heat required to melt or freeze a substance.

Sign convention: If the substance is melting, heat must flow from the environment to the system, so Q is positive. If the substance is freezing, heat must flow from the system to the environment, so Q is negative.

Once the system has completely transformed into a liquid, a change in water temperature is directly proportional to the total heat added to the system (Fig. 21.7). On the microscopic level, the molecules move more rapidly as energy is transferred to the system.

When the water reaches its boiling point (100°C at atmospheric pressure), the liquid begins to transform into a gas. During this phase change, there is also no change in the system's temperature (Fig. 21.7). Heat is still flowing from the environment to the system, so the thermal energy of the system continues to increase. On the microscopic level, when a system changes from a liquid to a gas, the increase in the system's thermal energy breaks the bonds between molecules; in the gas phase, the molecules have little interaction with one another.

The **heat of vaporization** L_V is the latent heat required to transform a substance from a liquid to a gas or from a gas to a liquid. The heat required for such a transformation is

$$Q = mL_V \tag{21.10}$$

Again by our sign convention, if the substance is boiling, heat is transferred to the system, so Q is positive. If the substance is condensing, heat is transferred from the system, so Q is negative.

The heats of fusion and of vaporization for many substances are given in Table 21.2. The heat of vaporization is several times greater than the heat of fusion because

TABLE 21.2 Melting temperature, boiling temperature, and latent heat at 1 atm.

Substance	Melting temperature (K)	Heat of fusion L_F (J/kg)	Boiling temperature (K)	Heat of vaporization L_V (J/kg)
Ammonia	195.2	3.3×10^4	239.6	1.37×10^5
Copper	1356	2.07×10^5	2868	4.73×10^6
Ethyl alcohol	159	1.04×10^5	477	8.5×10^5
Hydrogen	14.0	5.8×10^4	20.3	4.55×10^5
Iron	2081	2.89×10^5	3296	6.34×10^6
Lead	600	2.5×10^4	2017	8.58×10^5
Mercury	234	1.14×10^4	630	2.96×10^5
Nitrogen	63	2.6×10^4	77.2	2.00×10^5
Oxygen	54.2	1.39×10^5	90	2.13×10^5
Silver	1235	1.05×10^5	2323	2.336×10^6
Tungsten	3410	1.84×10^5	6173	4.8×10^6
Water	273	3.33×10^5	373	2.256×10^6

the molecules in a liquid are loosely bound, whereas the molecules in a gas are essentially free. When a solid is melting, the latent heat required is the amount of energy per mass needed to loosen the bonds. More latent heat is required to break the bonds when a substance transforms from a liquid to a gas.

Evaporating is not the same as boiling. Recall from Section 20-8 that evaporation is a cooling process whereby the fastest particles escape from the liquid. During evaporation, the liquid's thermal energy and temperature decrease.

EXAMPLE 21.4 "Mom, How Long Should I Put This Food in the Microwave?"

Microwave ovens are very efficient at transferring heat to water. Because the major component of food is water, microwave ovens can speed up cooking so much that it is possible to take food from the freezer and eat it just minutes later. Imagine preparing a meal that has been in a cold freezer at −20°C. Model the food as 0.500 kg of water (initially ice). People generally like to eat hot food at a temperature of about 75°C. If the power transferred by the microwave oven to the food is 700 W, what time should you set on the oven timer?

INTERPRET and ANTICIPATE

This example is described by Figure 21.7. Heat Q_1 transferred to the ice will first raise its temperature. When the ice reaches its melting point, additional heat Q_2 will melt the ice without changing its temperature. After the ice completely melts, additional heat Q_3 will raise the temperature of the water. Calculate the total heat Q_{tot} that must be transferred to the ice to turn it into hot water. Then use the power rating of the microwave oven to find the cooking time.

SOLVE

Find the amount of heat Q_1 required to raise the ice temperature from −20°C to 0°C, where ΔT is the same on the Kelvin scale as it is on the Celsius scale. The specific heat of ice is given in Table 21.1.

$$Q_1 = mc\,\Delta T \qquad (21.5)$$

$$Q_1 = (0.500\text{ kg})[2.100 \times 10^3\text{ J/(kg}\cdot\text{K)}](20\text{ K})$$

$$Q_1 = 2.10 \times 10^4\text{ J}$$

Once the ice reaches 0°C, it begins to melt. Find the heat Q_2 required to melt all the ice. The latent heat of fusion is found in Table 21.2.

$$Q_2 = mL_F \qquad (21.9)$$

$$Q_2 = (0.500\text{ kg})(3.33 \times 10^5\text{ J/kg}) = 1.67 \times 10^5\text{ J}$$

After the ice has been changed into water, additional heat Q_3 increases the water's temperature. Find Q_3, this time with the specific heat of liquid water (Table 21.1) and not ice.

$$Q_3 = mc\,\Delta T \qquad (21.5)$$

$$Q_3 = (0.500\text{ kg})[4.19 \times 10^3\text{ J/(kg}\cdot\text{K)}](75\text{ K})$$

$$Q_3 = 1.57 \times 10^5\text{ J}$$

The total heat required is the sum of the heat required for each stage.

$$Q_{\text{tot}} = Q_1 + Q_2 + Q_3$$

$$Q_{\text{tot}} = (2.10 \times 10^4\text{ J}) + (1.67 \times 10^5\text{ J}) + (1.57 \times 10^5\text{ J})$$

$$Q_{\text{tot}} = 3.45 \times 10^5\text{ J}$$

Power is the rate of change of energy (Section 9-9). To find the time required, divide the total heat required by the power transferred.

$$\frac{\Delta Q_{\text{tot}}}{\Delta t} = 700\text{ W}$$

$$\Delta t = \frac{\Delta Q_{\text{tot}}}{700\text{ W}} = \frac{3.45 \times 10^5\text{ J}}{700\text{ J/s}} = 4.93 \times 10^2\text{ s} \times \left(\frac{1\text{ min}}{60\text{ s}}\right)$$

$$\Delta t = 8.22\text{ min} = 8\text{ min }13\text{ s}$$

CHECK and THINK

Our result seems about right. A cook's rule of thumb for microwave cooking is 6 min per pound. The food's mass is 0.500 kg, so it weighs slightly more than a pound, and our result is a little more than 6 min.

21-6 Work in Thermodynamic Processes

The state of a system is described by state variables such as temperature, pressure, volume, and thermal energy. Often, a *PV* diagram (Section 20-6) is used to visualize the state of a gaseous system (Fig. 21.8). A **thermodynamic process** is a way for a system to change from some initial state to some final state. On a *PV* diagram, a thermodynamic process is represented by a curve or *path* connecting the initial and final states of the system, and there are many possible paths that can connect two endpoints. The word *path* here means a curve on the *PV* diagram, not through space.

THERMODYNAMIC PROCESS

Major Concept

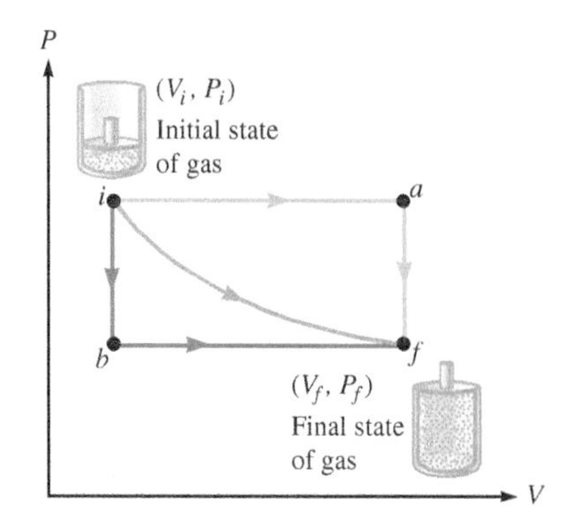

FIGURE 21.8 The state of a system is represented by a point on a *PV* diagram. Points *i* and *f* here represent an initial state and a final state of an ideal gas that expands. Three curves or paths connecting the initial state to the final state represent three different thermodynamic processes.

To record a thermodynamic process as a path, the gas must expand (or contract) slowly so that all the gas stays in equilibrium at the same pressure and temperature and so that its volume is well-defined. Such a slow process is called **quasistatic** (meaning "almost static"). Throughout this section and the next, we primarily study quasistatic thermodynamic processes of an ideal gas.

Consider an ideal gas in a container with a movable, frictionless piston on one end (Fig. 21.9). The gas is the system, and everything else (including the container and piston) is in the environment. When the gas does a small amount of work dW on the piston, the piston moves a short distance dy (Fig. 21.9A). The area of the piston is A, and the force exerted by the gas on the piston is PA. When the force exerted by the gas and the displacement of the piston are both upward, the work done by the gas to raise the piston by dy is positive and given by

$$dW = PA\,dy = P\,dV$$

where dV is the small change in the gas's volume. If the gas continues to do work on the piston, raising it to a new height and increasing the volume of the container (Fig. 21.9B), the total work done by the gas is

$$W = \int dW = \int_{V_i}^{V_f} P\,dV = \text{work done } by \text{ the system} \quad (21.11)$$
$$(\text{energy } leaving \text{ the system})$$

where V_i is the initial volume of the gas and V_f is the final volume.

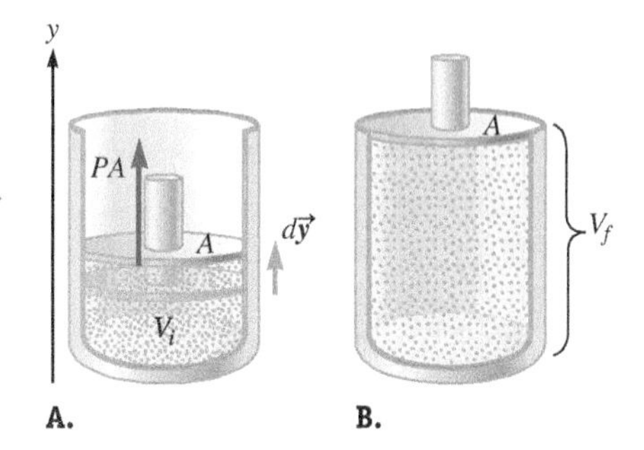

FIGURE 21.9 A. An ideal gas in a container with a movable piston. **B.** The gas raises the piston, doing positive work on the piston.

According to Equation 21.11, the work done by the gas on its environment is the area under the curve connecting the initial state of the gas to its final state on a *PV* diagram. Because there are many possible paths connecting the endpoints, however, the work done by the gas depends on the particular thermodynamic process (or path). For example, both path *A* and path *B* (Fig. 21.10A) represent **expansion processes** (final volume V_L is greater than the initial volume V_S). The area under the curve for process *A* is greater, however, so the gas does more work on its environment during process *A*.

In Figure 21.10A, $W_A > W_B$ because during process *A*, the gas pressure is higher than it is in process *B*. So, the force ($F = PA$) exerted by the gas on the piston is greater during process *A*. Because the force is greater and the piston is lifted by the

FIGURE 21.10 A. Two thermodynamic expansion processes. **B.** A compression process.

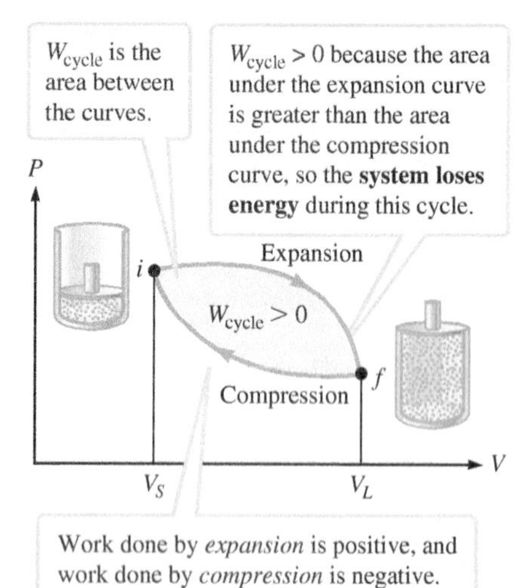

A.

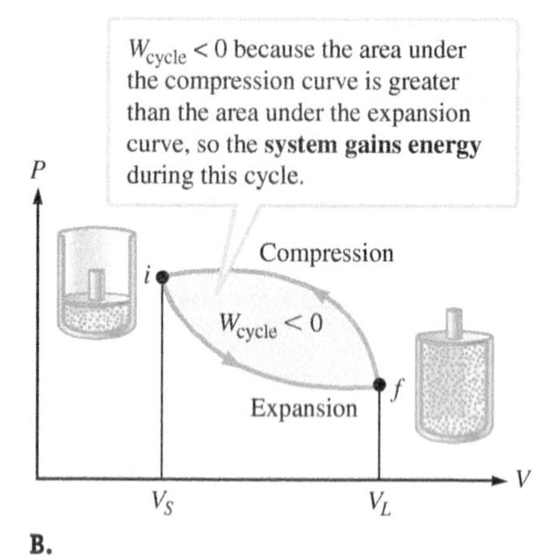

B.

FIGURE 21.11 Two thermodynamic cycles.

same amount during either process, the gas does more work during process A than during process B (that is, more energy leaves the system during process A).

What happens if the piston (Fig. 21.10B) were lowered so that gas volume is reduced from a large volume V_L to a small volume V_S? In this **compression process**, the gas pressure would still push upward on the piston, but the piston's displacement would be downward in the opposite direction. During the compression, the work done by the gas is negative. On a *PV* diagram, the work done by the gas during compression is the negative of the area under the curve. You can interpret the negative work done by the gas during a compression as the piston doing *positive* work on the gas. So, during a compression, energy is transferred from the environment (piston) to the system (gas).

Now imagine that the gas undergoes an expansion and then a compression process that returns it to its original volume. A process that returns the system to its original state is known as a **thermodynamic cycle** (Fig. 21.11). The work done by the gas is the area under the curve, which is positive during expansion and negative during compression. The net work done by the gas during a cycle is represented by the area between the two curves.

The area under a path in a *PV* diagram is the work done *by the system* on its environment (Eq. 21.11), but that is *not* the same as the work in the first law of thermodynamics $W_{tot} + Q = \Delta E_{th}$ (Eq. 21.1). The work W_{tot} in the first law of thermodynamics is the work exerted *by the environment* on the system. By Newton's third law, the work done by the environment on the system has the opposite sign:

$$\left(\begin{array}{c}\text{work done}\\ \textit{by the environment}\\ \text{on the system}\end{array}\right) = -\left(\begin{array}{c}\text{work done}\\ \textit{by the system}\\ \text{on the environment}\end{array}\right) \quad (21.12)$$

To match the notation in the first law of thermodynamics, we will use the subscript "tot" for the total work W_{tot} exerted *by the environment* on the system. By inserting a negative sign into Equation 21.11, we find an expression for the work done by the environment:

$$W_{tot} = -\int_{V_i}^{V_f} P\,dV = \text{work done by the environment} \quad (21.13)$$
$$(\text{energy } \textit{entering} \text{ the system})$$

EXAMPLE 21.5 CASE STUDY Work You Do When Breathing

Since the moment you were born, you have been breathing air. During each cycle of inhalation and exhalation, your body does work on the air. With each inhalation, your chest expands your lungs, and with each exhalation, your lungs contract.

Figure 21.12 shows a *PV* diagram for the air in your lungs. The cycle shown in green is for a single breath taken while you are resting. The cycle shown in red is for a single breath taken while you are exerting yourself, perhaps while exercising. Use the figure to estimate the *work you do* on the air while you are resting and how much more work you do while you are exercising.

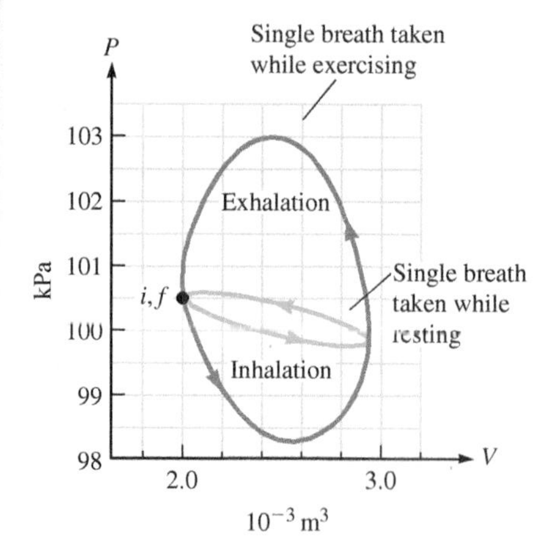

FIGURE 21.12

INTERPRET and ANTICIPATE

The system is the air, and the environment is your body. The net work done by the air is negative because the area under the inhalation (expansion) curve is smaller than the area under the exhalation curve. (Compare with Fig. 21.11B.) In the case of breathing, the environment (your chest) does positive work on the air (Eq. 21.12). Find the *positive* area enclosed by the inhalation and exhalation cycle.

SOLVE

Find the area A enclosed by one rectangle on the grid. The width of one rectangle is 0.2×10^{-3} m^3, and its height is 0.5×10^3 Pa. The area has the dimension of energy.

$$A = hw = (0.2 \times 10^{-3}\ \text{m}^3)(0.5 \times 10^3\ \text{Pa}) = 0.10\ \text{m}^3\ \text{Pa}$$
$$1\ \text{m}^3\ \text{Pa} = 1\ \text{m}^3\frac{\text{N}}{\text{m}^2} = 1\ \text{m}^3\frac{\text{N}}{\text{m}^2} = 1\ \text{N}\cdot\text{m} = 1\ \text{J}$$
$$A = 0.10\ \text{J}$$

Estimate the number of rectangles enclosed by the green cycle for the breaths taken while resting. You can probably estimate to the nearest one-half of a rectangle. Multiply the number of rectangles by the area of each rectangle to find the work you exert on the air when resting, W_{rest}.

3.5 rectangles
$$W_{rest} = 3.5A = 3.5(0.10\ \text{J})$$
$$W_{rest} = 0.35\ \text{J}$$

CHECK and THINK

When you are asleep, your body uses 4800 J per minute. You breathe about 10 times per minute. Our results show that you need only 3.5 J per minute for breathing when you sleep, less than 0.1% of the sleeping energy consumption. Most of the energy consumed when sleeping is used by your liver, spleen, brain, skeletal muscles, kidney, and heart.

SOLVE

Estimate the number of rectangles enclosed by the red cycle and multiply by A to find the work W_{exer} you exert to breathe while you are exercising.

34.5 rectangles
$$W_{exer} = 34.5A = 34.5(0.10\ \text{J})$$
$$W_{exer} = 3.45\ \text{J}$$

Subtract the work you do while you are at rest from the work you do while you are exercising to find the extra work required, W_{xtra}.

$$W_{xtra} = W_{exer} - W_{rest}$$
$$W_{xtra} = 3.45\ \text{J} - 0.35\ \text{J} = 3.1\ \text{J}$$

CHECK and THINK

When you exercise, your breathing rate is 150 breaths per minute, which means that you need an extra 460 J per minute when you undergo this very strenuous exercise. For some people, this difference is noticeable, and they might find such rapid breathing exhausting.

21-7 Specific Thermodynamic Processes

In Section 21-6, we used a *PV* diagram and Equation 21.13 to find the work done by the environment on a gaseous system. In this section, we apply the first law of thermodynamics to six specific thermodynamic processes for gaseous systems, illustrated using *PV* diagrams and energy bar charts.

Because the first law of thermodynamics does not involve mechanical energy, an energy bar chart (Fig. 21.13) is somewhat simplified. To see what is required, let's rewrite the first law of thermodynamics (Eq. 21.1):

$$W_{tot} + Q = \Delta E_{th} = E_{th,f} - E_{th,i}$$

where $E_{th,i}$ is the initial thermal energy and $E_{th,f}$ is the final thermal energy. Bringing the initial thermal energy to the left side of the equation, we have

$$E_{th,i} + W_{tot} + Q = E_{th,f} \tag{21.14}$$

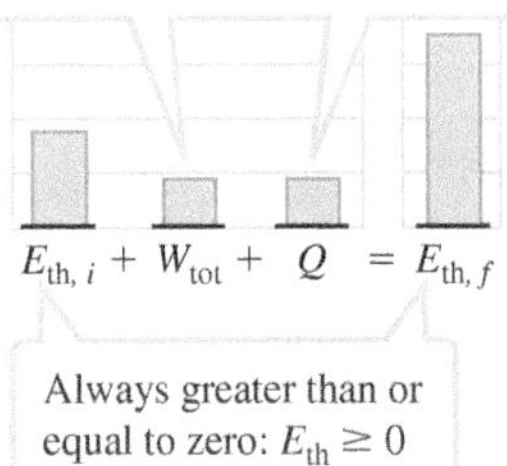

FIGURE 21.13 A bar chart helps when using the first law of thermodynamics.

ENERGY BAR CHARTS Tool

Adiabatic Process ($Q = 0$)

ADIABATIC PROCESS Special Case

For the first specific thermodynamic process, consider the expansion of an ideal gas in a cylindrical container with a movable piston on top (Fig. 21.9). During an **adiabatic process**, no heat flows into or out of the system. You might guess that the system must be insulated to ensure that there is no heat flow. Although that is one possibility, if a thermodynamic process is fast enough, it is possible that little or no heat flows during the process even if a system is not insulated. For example, the rapid

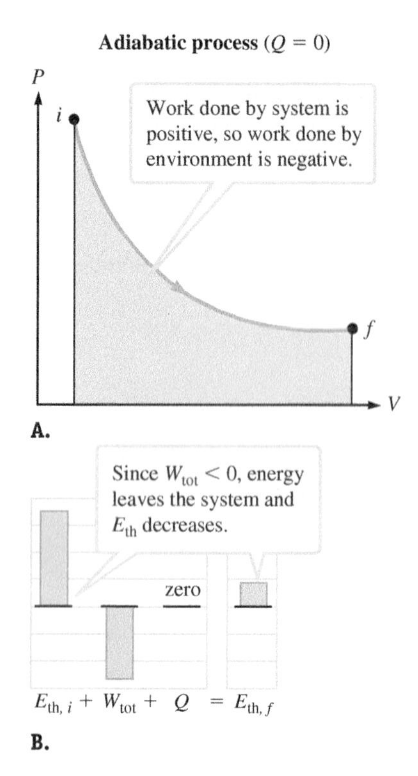

FIGURE 21.14 A. *PV* diagram for an adiabatic expansion. **B.** Bar chart for an adiabatic expansion.

expansion of the gas in your car's engine may be approximated as an adiabatic process. Such a process can be fast enough to be adiabatic, but slow enough to be considered quasistatic.

As the gas expands adiabatically, its pressure decreases as shown in the *PV* diagram (Fig. 21.14A). The work done *by the gas* in lifting the piston is the area under the curve (positive for an expanding gas). The work done *by the environment*, W_{tot}, has the opposite sign (in this case, negative). So, W_{tot} is represented by a bar below the line (Fig. 21.14B).

Initially, as long as the gas is not at absolute zero, it has some thermal energy as represented by a bar above the line (Fig. 21.14B). Because the process is adiabatic, no heat is exchanged with the environment as indicated by the "zero" written on the line for heat. According to the first law of thermodynamics (Eq. 21.14), with $Q = 0$, the final thermal energy is

$$E_{th,f} = E_{th,i} + W_{tot}$$

and because the work done by the environment is negative, the final thermal energy is less than the initial thermal energy. We return to adiabatic expansion of a gas in Section 21-9.

CONCEPT EXERCISE 21.5

Figure 21.14 is drawn for a gas that expands adiabatically. Draw a *PV* diagram and an energy bar chart for a gas that is compressed adiabatically.

ISOTHERMAL PROCESS

Special Case

Isothermal Process ($\Delta T = 0$)

Isothermal means "same temperature" in Greek, so in an **isothermal process**, the temperature of the system is unchanged: $\Delta T = 0$. Because the temperature does not change, the system's thermal energy does not change: $\Delta E_{th} = 0$.

Let's add some details to our apparatus so that we can consider the isothermal expansion of an ideal gas. The walls and piston are insulated so that no heat can flow through them (Fig. 21.15). The bottom of the container is in thermal contact with a **heat reservoir**, an object so large that when heat is exchanged with the system, the reservoir's temperature does not change significantly. In practice, a heat reservoir may be a stove top with a control that allows the researcher to set its temperature.

Because the system is modeled as an ideal gas,

$$PV = Nk_BT = \text{constant}$$

$$P = \frac{Nk_BT}{V} = \frac{\text{constant}}{V}$$

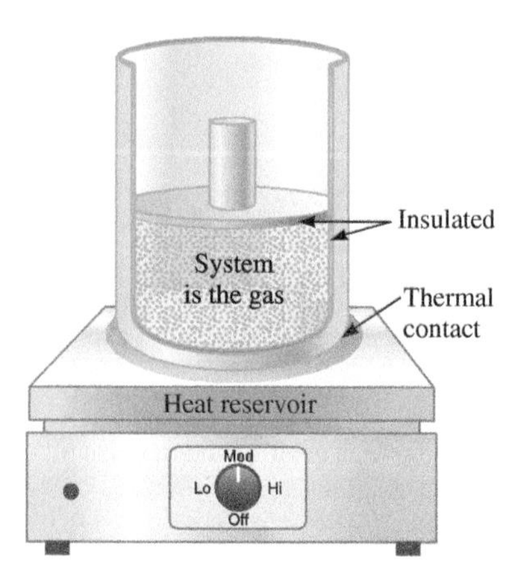

FIGURE 21.15 The container is in thermal contact with a heat reservoir. The temperature of the reservoir can be set by turning the control knob. The container's walls and piston are insulated. The piston is free to move.

which is plotted on the *PV* diagram (Fig. 21.16A) and labeled "Isothermal." Let's compare an isothermal expansion of an ideal gas to an adiabatic expansion. Suppose the gas expands from V_i to V_f. In both processes, the gas pressure drops, but the temperature drops only for the adiabatic expansion. Because temperature is proportional to pressure, the decrease in pressure is greater in the adiabatic process than in the isothermal process. So, the path in the *PV* diagram for the adiabatic process falls below the path for the isothermal process (Fig. 21.16A). Therefore, the work done *by the expanding gas* (area under the *PV* curve) in raising the piston is greater in the isothermal process.

The gas lifts a piston as it expands, which means that the environment does negative work on the system: $W_{tot} < 0$. In an isothermal process, the thermal energy remains constant. Therefore, the energy lost by the system as a result of work must be balanced by heat transferred to the system from the heat reservoir:

$$W_{tot} = -Q \text{ for an isothermal process}$$

The bar for Q is the same height as that for W_{tot} except that Q is positive (Fig. 21.16B).

We find an expression for the work done *by the environment* on an ideal gas in an isothermal process by substituting the ideal gas law into Equation 21.13:

$$W_{tot} = -\int_{V_i}^{V_f} P\,dV = -\int_{V_i}^{V_f} \frac{Nk_BT}{V}\,dV$$

For an isothermal process, T is constant, and as long as the system is closed (no gas escapes or enters), N is constant. So,

$$W_{tot} = -Nk_BT\int_{V_i}^{V_f} \frac{dV}{V}$$

Using Appendix A, we find

$$W_{tot} = -Nk_BT(\ln V_f - \ln V_i) = Nk_BT(\ln V_i - \ln V_f)$$

$$W_{tot} = Nk_BT\ln\frac{V_i}{V_f} \quad \text{for an isothermal process on an ideal gas} \qquad (21.15)$$

FIGURE 21.16 A. PV diagram for an isothermal expansion. B. Bar chart for an isothermal expansion.

CONCEPT EXERCISE 21.6

Figure 21.16 is drawn for a gas that expands isothermally. Draw a PV diagram and an energy bar chart for a gas that is compressed isothermally.

CONCEPT EXERCISE 21.7

The PV diagram in Figure 21.16 shows that an ideal gas must do more work on a piston in an isothermal expansion than in an adiabatic expansion. Draw a PV diagram for the compression of an ideal gas from V_f to V_i both isothermally and adiabatically. Compare the work done by the piston on the gas for these two compression processes.

Constant-Volume Process ($\Delta V = 0$)

CONSTANT VOLUME (ISOCHORIC) PROCESS
Special Case

In a constant-volume process, the system's volume is fixed. In practice, we must add a mechanism (such as a screw top, a clamp, or wedges) for holding the piston in place when needed.

Because the piston cannot move, it cannot do work on the gas, and the gas cannot do work on the piston:

$$W_{tot} = 0 \quad \text{for a constant-volume process}$$

Energy can still be exchanged through heat. As an example, suppose you turn down the temperature on the heat reservoir so that heat flows from the hotter system to the cooler reservoir and the gas cools. According to the ideal gas law $PV = Nk_BT$, a reduction in temperature (at constant volume) means a reduction in pressure, which is shown on the PV diagram by a vertical, downward path (Fig. 21.17A). The vertical path has no area under it, so no work is done by the system as expected.

Because heat flows from the gas to the reservoir, energy leaves the system as represented by a bar for Q below the zero line in the bar chart (Fig. 21.17B). According to the first law of thermodynamics, the system's thermal energy must be reduced by the amount of heat that flows out:

$$\Delta E_{th} = Q \quad \text{for a constant-volume process}$$

In this case, $Q < 0$, so the initial thermal energy bar is higher than the final thermal energy bar (Fig. 21.17B)

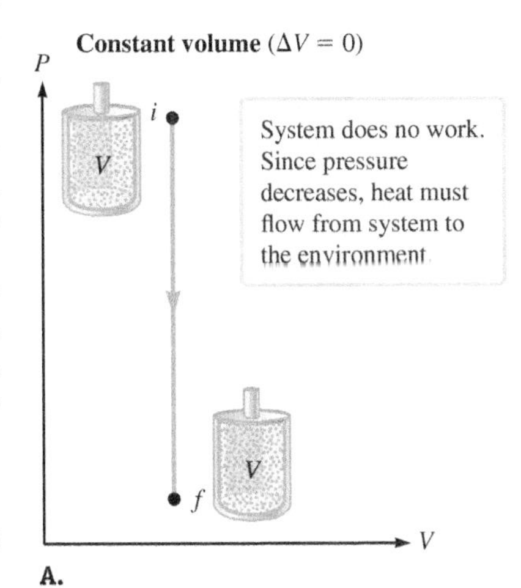

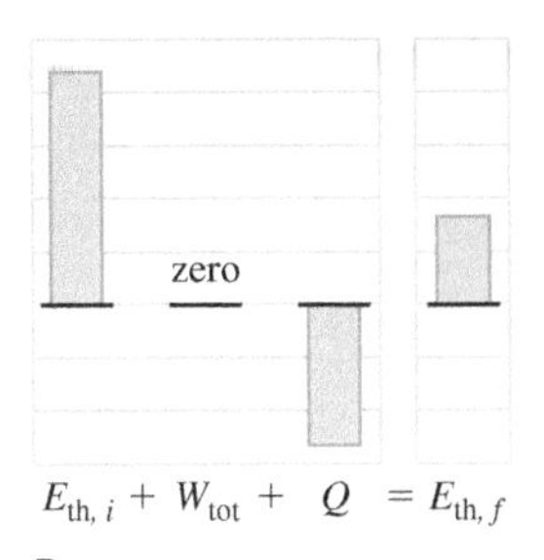

FIGURE 21.17 A. PV diagram for an isochoric (constant-volume) process. B. Bar chart for a constant-volume process.

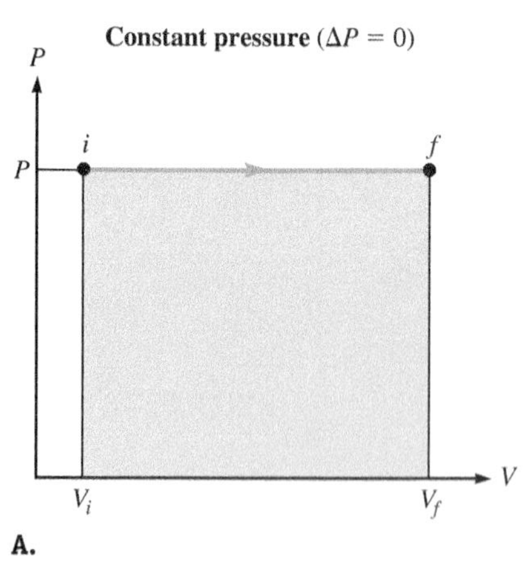

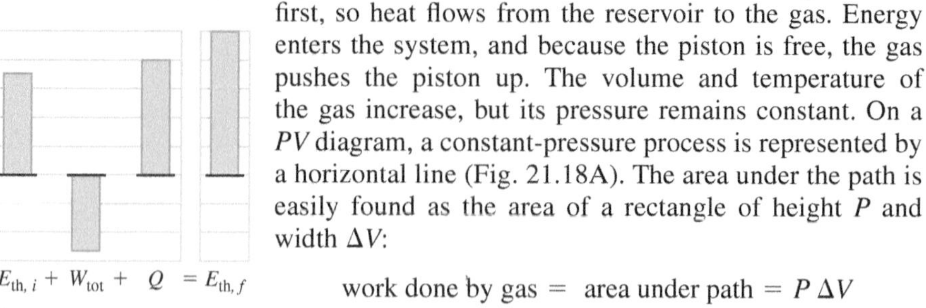

A. **B.**

FIGURE 21.18 A. *PV* diagram for an isobaric (constant-pressure) process. **B.** Bar chart for a constant-pressure process.

Constant-Pressure Process ($\Delta P = 0$)

Return to the apparatus shown in Figure 21.15 once again with the piston free to move. Suppose the gas temperature is lower than the heat reservoir's temperature at first, so heat flows from the reservoir to the gas. Energy enters the system, and because the piston is free, the gas pushes the piston up. The volume and temperature of the gas increase, but its pressure remains constant. On a *PV* diagram, a constant-pressure process is represented by a horizontal line (Fig. 21.18A). The area under the path is easily found as the area of a rectangle of height P and width ΔV:

$$\text{work done by gas} = \text{area under path} = P\,\Delta V$$

The work done by the gas in this case is positive. Energy leaves the system, and W_{tot} (work done by environment) is negative as represented by the bar in Figure 21.18B:

CONSTANT-PRESSURE (ISOBARIC) PROCESS

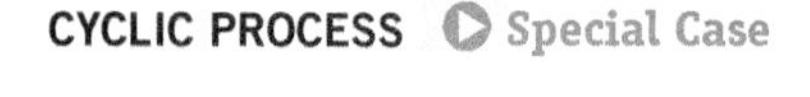

$$W_{tot} = -P\Delta V \text{ for a constant-pressure process} \tag{21.16}$$

If the system is an ideal gas, we can substitute the ideal gas law to find the work done on the system,

$$W_{tot} = -\frac{Nk_BT_i}{V_i}(V_f - V_i) = -\frac{Nk_BT_f}{V_f}(V_f - V_i)$$

$$W_{tot} = Nk_BT_i\left(1 - \frac{V_f}{V_i}\right) = Nk_BT_f\left(\frac{V_i}{V_f} - 1\right) \tag{21.17}$$

for a constant-pressure ideal gas.

Other than the first law of thermodynamics, there is no particular constraint on the heat or change in the system's thermal energy. In Figure 21.19, we have arbitrarily shown a system that has heat flowing into the system and an increase in the system's thermal energy.

Cyclic Process ($\Delta E_{th} = 0$)

CYCLIC PROCESS Special Case

During a **cyclic process**, energy may be exchanged as a result of heat and work, but the system returns to its original state. Therefore, all the system's state variables—including its thermal energy—must return to their original values, and there is no change in the system's thermal energy: $\Delta E_{th} = 0$.

We already know that a cyclic process makes a closed path on a *PV* diagram (Fig. 21.19A) and that the area between the expansion and compression curves is the net

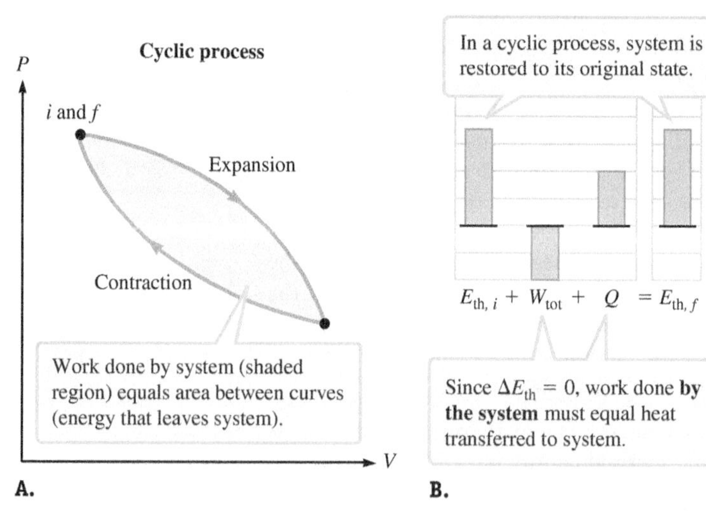

A. **B.**

FIGURE 21.19 A. *PV* diagram for a cyclic process. **B.** Bar chart for a cyclic process.

work exerted *by the gas*. So, the net work done *by the environment* W_{tot} is the negative of the area between the two curves (Eq. 21.11):

$$W_{tot} = -(\text{area between curves on } PV \text{ diagram})$$

In the cyclic process in Figure 21.19, W_{tot} is negative; the system loses energy through work.

Because the thermal energy returns to its original value after one cycle, the system must gain energy through heat. The amount of heat that flows into the system must equal the amount of energy the system loses through work:

$$W_{tot} = -Q \quad \text{for a cyclic process}$$

The energy transferred out of the system by work W_{tot} is shown as a bar below the line representing zero, and the heat that flows into the system is shown as a bar of equal height but above that line (Fig. 21.19B). We will look at cyclic processes in more detail in Chapter 22.

CONCEPT EXERCISE 21.8

Suppose that the cyclic process in Figure 21.19 is reversed so that the gas starts with a large volume and is compressed by the piston (upper path), and then the gas lifts the piston, returning the gas to its original volume (lower path). Draw a *PV* diagram.

a. Is the total work done by the gas on the piston (energy leaving the system) during the cycle positive, negative, or zero?

b. Is the total work done by the piston on the gas (energy entering the system) during the cycle positive, negative, or zero?

c. Is the net flow of heat from the environment to the system or from the system to the environment?

Free Expansion ($Q = W = 0$)

A gas fills the left side of a container in Figure 21.20. The container is well insulated so that heat cannot be exchanged with the environment. When the valve is opened, the gas expands to fill both sides of the container. During this expansion, the gas pressure is not the same at all points, and its volume is not clearly defined. So, this expansion process is not quasistatic and cannot be plotted on a *PV* diagram (Fig. 21.21A).

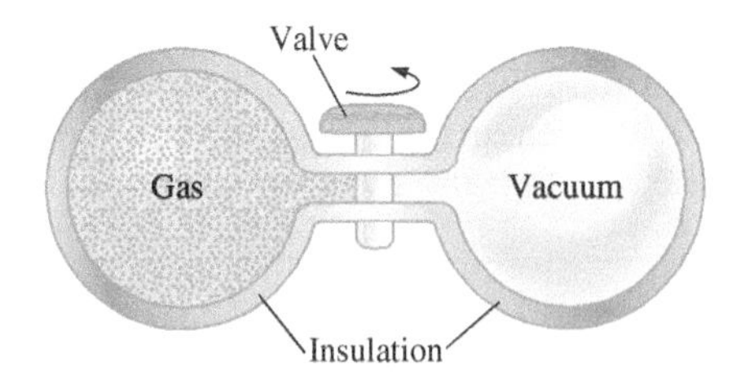

FIGURE 21.20 This apparatus allows a gas to expand freely.

Because there is no piston for the gas to lift, the gas does no work. An expansion that involves no work or heat is called a **free expansion**. Because $W_{tot} = 0$ and $Q = 0$, there is no change in the system's thermal energy, and $\Delta E_{th} = 0$. The bar chart for a free expansion is shown in Figure 21.21B.

FREE EXPANSION Special Case

CONCEPT EXERCISE 21.9

Which of the following quantities depend on the path taken in a *PV* diagram, and which quantities only depend on the endpoints?

a. Change in thermal energy
b. Change in temperature
c. Change in volume
d. Change in pressure
e. Heat transferred
f. Work done by the environment
g. Work done by the system

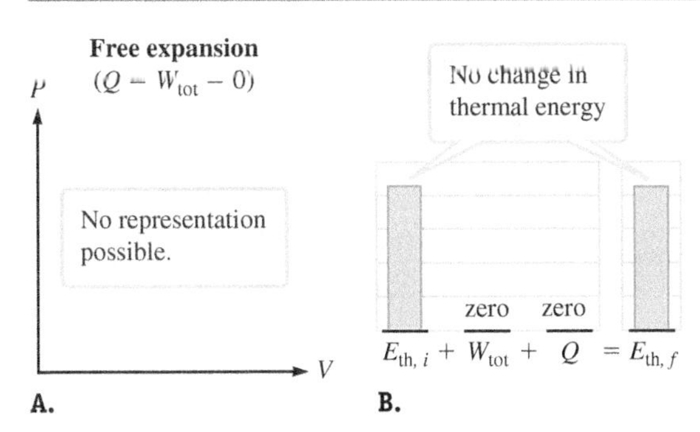

FIGURE 21.21 A. Free expansion cannot be plotted on a *PV* diagram. **B.** Bar chart for free expansion. A free expansion is adiabatic ($Q = 0$) and no work is done ($W = 0$).

EXAMPLE 21.6 Comparing Processes

As shown on the PV diagram in Figure 21.22, a gas may be compressed either isothermally or by a two-part process consisting of a constant-pressure stage followed by a constant-volume stage. Find the work done by the environment and the heat transferred to the system in each process. Assume it is an ideal gas. The initial position on the PV diagram is at (19.00 m³, 1.29 × 10⁴ Pa) and the final position is at (5.00 m³, 4.89 × 10⁴ Pa).

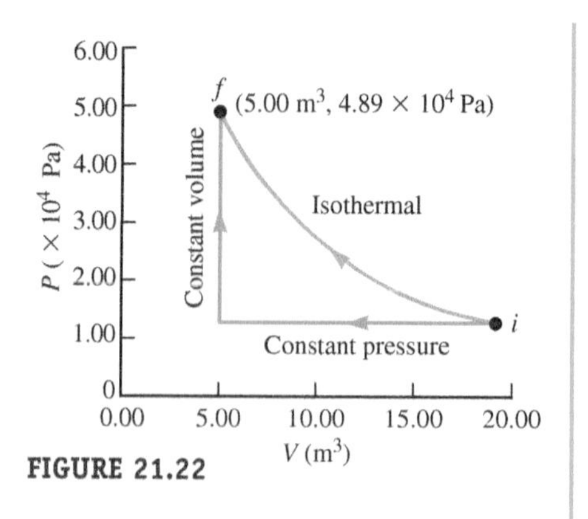

FIGURE 21.22

Heat flowing out of system must equal work done by environment.

Isothermal process

$E_{th,i} + W_{tot} + Q = E_{th,f}$

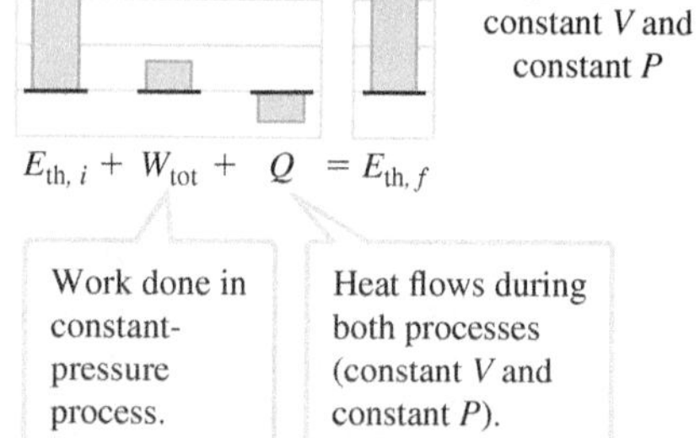

FIGURE 21.23

INTERPRET and ANTICIPATE

By inspection of the PV diagram, the area under the isothermal curve is greater than the area under the combined constant-pressure–constant-volume curve, so we expect that the environment must do more work in the isothermal process. Constructing a bar chart (Fig. 21.23) helps provide an idea of the amount of heat transferred during each process. The change in thermal energy is the same whether the gas is compressed isothermally or by the two-step process. Furthermore, $\Delta E_{th} = 0$ and $W_{tot} = -Q$ for any isothermal process. Therefore, the thermal energy bars are all identical in the two bar charts. In the isothermal process, more work is done by the environment, and more heat must flow out of the system.

SOLVE

Step 1 Isothermal process. Because we know the initial pressure but not the initial temperature, we can use the ideal gas law and Equation 21.15 to find an expression for the work W_1 done by the environment in terms of P_i. (We could have also used the final pressure; our choice is arbitrary.)

$$W_1 = Nk_BT \ln \frac{V_i}{V_f} \quad (21.15)$$

$$Nk_BT_i = P_iV_i$$

$$W_1 = P_iV_i \ln \frac{V_i}{V_f}$$

Substitute values found on the PV diagram.

$$W_1 = (1.29 \times 10^4\ \text{Pa})(19.0\ \text{m}^3) \ln \left(\frac{19.0\ \text{m}^3}{5.00\ \text{m}^3}\right)$$

$$W_1 = 3.27 \times 10^5\ \text{J (isothermal)}$$

Because there is no change in thermal energy, the heat Q_1 that flows out of the system must equal the work that the environment does on the system.

$$\Delta E_{th} = W_1 + Q_1 = 0$$

$$Q_1 = -W_1$$

$$Q_1 = -3.27 \times 10^5\ \text{J (isothermal)}$$

Step 2 Two-part process (constant-pressure process followed by **constant-volume process)**. No work is done during the constant-volume process, so the only work done by the environment during this two-part process is given by Equation 21.16. During the constant-pressure part, the pressure is P_i.

$$W_2 = -P\,\Delta V = -P_i(V_f - V_i) \quad (21.16)$$

$$W_2 = -(1.29 \times 10^4\ \text{Pa})(5.00\ \text{m}^3 - 19.0\ \text{m}^3)$$

$$W_2 = 1.81 \times 10^5\ \text{J (two-part process)}$$

The heat that leaves the system must equal the work done by the environment. Heat flows during both parts of the two-part process. Heat flows out of the system during the constant-pressure stage, and heat flows into the system during the constant-volume stage. The heat we find here is the net heat that flows out of the system.

$$\Delta E_{th} = W_2 + Q_2 = 0$$

$$Q_2 = -W_2$$

$$Q_2 = -1.81 \times 10^5\ \text{J (two-part process)}$$

CHECK and THINK

As expected, the environment does more work during the isothermal process than during the two-step process ($W_1 > W_2$). During the two-step process, compression occurs in the constant-pressure stage when the gas is at a lower pressure P_i. Therefore, the force exerted by the environment is smaller for the same change in volume than it is for the isothermal process in which pressure rises continuously.

21-8 Equipartition of Energy

Molar Specific Heat of Gases

In Section 21-4, we skipped over the specific heat of gases because the specific heat of a gas depends on the thermodynamic process. When a solid or liquid is in thermal contact with a heat reservoir, heat is exchanged, but the work done on the system is typically zero: $W_{\text{tot}} = 0$. Thus, for a solid or a liquid, $\Delta E_{\text{th}} = Q$. Because a change in temperature is directly proportional to a change in thermal energy, heat entering (or leaving) the system is directly proportional to the change in the system's temperature. The molar specific heat C of a substance in the solid or liquid phase is the constant of proportionality in

$$Q = nC\,\Delta T \tag{21.6}$$

The situation for a gas is more complicated because work is often done on (or by) the system, so $W_{\text{tot}} \neq 0$. Because the work done depends on the path between endpoints, the amount of heat required to raise the temperature of a gas by a certain amount ΔT depends on the thermodynamic process. Figure 21.24A shows the PV diagram for an ideal gas at two different temperatures. The curve in red is for the hot gas, $P = Nk_BT_{\text{hot}}/V$, and the blue curve is for the cool gas, $P = Nk_BT_{\text{cool}}/V$. Figure 21.24A also shows two possible paths (constant volume and constant pressure) for raising the gas temperature. Both processes result in the same temperature change ΔT, and because thermal energy is proportional to temperature, both processes result in the same thermal energy change ΔE_{th}. Both work and heat depend on the particular process, however. We define the amount of heat Q_V required during a constant volume process to be

$$Q_V \equiv nC_V\Delta T \tag{21.18}$$

where C_V is the molar specific heat of the gas at constant volume. We also define the amount of heat Q_P required during a constant pressure process to be

$$Q_P \equiv nC_P\Delta T \tag{21.19}$$

where C_P is the molar specific heat of the gas at constant pressure.

Values for C_V and C_P for various gases are provided in Table 21.3. From the data, we see that (1) $C_P > C_V$, and (2) their difference is approximately constant: $C_P - C_V = R$, where $R = 8.315\ \text{J/(mol}\cdot\text{K)}$.

To explain these experimental results, let's take a close look at each process for raising the gas temperature. If the gas goes through a constant-volume process, no work is involved. Using Equation 21.18 and $W_{\text{tot}} = 0$, the first law of thermodynamics (Eq. 21.1) becomes

$$Q_V = \Delta E_{\text{th}} = nC_V\Delta T \tag{21.20}$$

If the gas goes through a constant-pressure process, the gas lifts the piston (energy leaves the gas), so the environment does negative work on the system (Eq. 21.16), and the amount of heat that flows from the environment is given by Equation 21.19. So, the first law of thermodynamics becomes

$$\Delta E_{\text{th}} = W_{\text{tot}} + Q_P$$

$$\Delta E_{\text{th}} = -P\Delta V + nC_P\Delta T \tag{21.21}$$

The change in thermal energy and temperature must be the same no matter which path is taken, so Equation 21.20 must equal Equation 21.21. Before looking at this

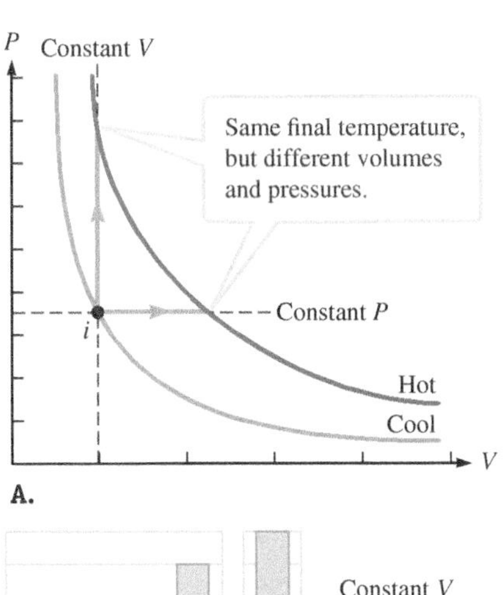

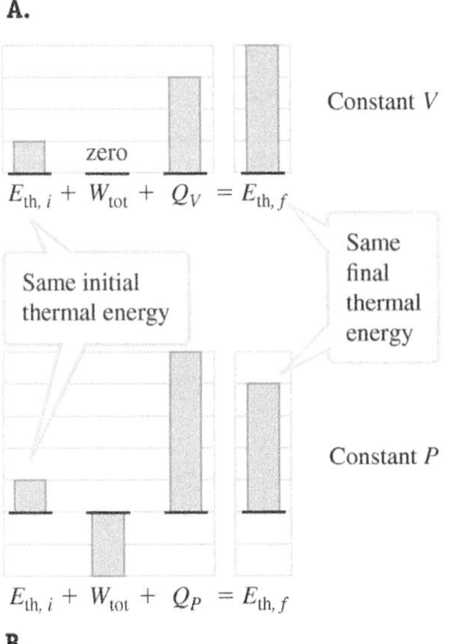

FIGURE 21.24 A. PV diagram for an ideal gas at two different temperatures with two possible paths (constant volume and constant pressure) for raising the gas temperature. **B.** Bar charts for constant-volume and constant-pressure processes.

TABLE 21.3 Molar specific heat for gases at room temperature except where noted.

Gas	C_P [J/(mol· K)]	C_V [J/(mol· K)]	$C_P - C_V$ [J/(mol· K)]	$\gamma = C_P/C_V$
Monatomic				
Ar	20.8	12.5	8.3	1.67
He	20.8	12.5	8.3	1.67
Ne	20.8	12.7	8.1	1.64
Diatomic				
CO	29.3	21.0	8.3	1.40
H_2	28.8	20.4	8.3	1.41
N_2	29.1	20.8	8.3	1.40
O_2	29.2	21.1	8.3	1.40
Triatomic				
CH_4	35.5	27.1	8.4	1.31
CO_2	37.0	28.5	8.5	1.30
H_2O (at 100°C)	35.4	27.0	8.4	1.30
Polyatomic				
C_2H_6	51.7	43.2	8.5	1.20

situation algebraically, consider the energy bar charts (Fig. 21.24B). The bars for $E_{\text{th},\,i}$ are the same in both processes, as are those for $E_{\text{th},\,f}$. The difference is that in the constant-pressure process, negative work is done, and in the constant-volume process, no work is done by the environment. As a result, the heat entering the gas is greater in the constant-pressure case:

$$Q_P > Q_V$$
$$nC_P\Delta T > nC_V\Delta T$$
$$C_P > C_V$$

To find exactly how much greater, let's first use the ideal gas law to find an expression for $P\,\Delta V$ in Equation 21.21:

$$\Delta(PV) = \Delta(nRT)$$

Because Equation 21.21 is for the constant-pressure process, the only quantity that changes on the left is V. Because no gas escapes or is added, n is constant, and the only quantity that changes on the right is T. We find that

$$P\Delta V = nR\Delta T \tag{21.22}$$

which we can substitute into Equation 21.21:

$$\Delta E_{\text{th}} = -nR\Delta T + nC_P\Delta T \tag{21.23}$$

Set Equation 21.23 equal to Equation 21.20:

$$-\,nR\Delta T + nC_P\Delta T = nC_V\Delta T$$

Solve for $C_P - C_V$:

$$C_P - C_V = R \tag{21.24}$$

This result is nearly what is found experimentally as shown in the fourth column in Table 21.3.

EXAMPLE 21.7 Raising Gas Temperature: Constant Pressure Versus Constant Volume

The initial pressure, volume, and temperature of a sample of He gas are $P_i = 2.45 \times 10^5$ Pa, $V_i = 0.75$ m^3, and $T_i = 235$ K, respectively. The final temperature of the gas is $T_f = 354$ K. The gas is well modeled by the ideal gas law.

A For a constant-volume process, what are the final pressure and volume of the gas? Also, find the work done by the environment, the heat transferred to the gas, and the change in thermal energy.

INTERPRET and ANTICIPATE

The final volume equals the initial volume. Given the initial and final temperatures, use the ideal gas law to find the final pressure. From Figure 21.24A, we expect the final pressure to be greater than the initial pressure. We can also use the ideal gas law to find the number of moles, which we need to find the heat absorbed. Because the temperature increases, we expect that the heat entering the system is positive.

SOLVE

First, use the ideal gas law to find the number of moles n. No gas escapes or is added, so n is a constant. (We'll need this result for part B.)

$$P_i V_i = nRT_i$$

$$n = \frac{P_i V_i}{RT_i} = \frac{(2.45 \times 10^5 \text{ Pa})(0.75 \text{ m}^3)}{\left(8.315 \frac{\text{J}}{\text{mol} \cdot \text{K}}\right)(235 \text{ K})}$$

$$n = 94.0 \text{ mol}$$

The final volume equals the initial volume.

$$V_f = V_i = 0.75 \text{ m}^3$$

Because there is no change in volume, no work is done.

$$W_{\text{tot}} = 0$$

Find the final pressure from the ideal gas law.

$$V_f = V_i = 0.75 \text{ m}^3$$

$$P_f = \frac{nRT_f}{V_f} = \frac{(94.0 \text{ mol})\left(8.315 \frac{\text{J}}{\text{mol} \cdot \text{K}}\right)(354 \text{ K})}{0.75 \text{ m}^3}$$

$$P_f = 3.69 \times 10^5 \text{ Pa}$$

CHECK and THINK

As expected, the final pressure is greater than the initial pressure.

SOLVE

For a constant-volume process, heat transferred to the gas is given by Equation 21.20. The molar specific heat C_V of helium is listed in Table 21.3.

$$Q_V = nC_V \Delta T \qquad (21.20)$$

$$Q_V = (94.0 \text{ mol})\left(12.5 \frac{\text{J}}{\text{mol} \cdot \text{K}}\right)(354 \text{ K} - 235 \text{ K})$$

$$Q_V = \Delta E_{\text{th}} = 1.40 \times 10^5 \text{ J}$$

CHECK and THINK

As expected, the heat entering the gas is positive, and the thermal energy increases.

B If the process is instead at constant pressure, what are the final pressure and volume of the gas? Find the work done by the environment, the heat transferred to the gas, and the change in thermal energy.

INTERPRET and ANTICIPATE

The final pressure equals the initial pressure. The number of moles is the same as in part A, so use the ideal gas law to find the final volume. From Figure 21.24A, we expect the final volume to be greater than the initial volume. During a constant-pressure process, energy is transferred by both work and heat (Fig. 21.24B). Find the work done on the gas using $W_{\text{tot}} = -P\,\Delta V$ (Eq. 21.16) and the heat transferred to the gas using $Q_P = nC_P\,\Delta T$ (Eq. 21.19). From Figure 21.24B, we expect the work done on the gas to be negative (energy leaving the system), the heat transferred to the gas to be positive, and the change in thermal energy to be the same as it is in part A. We also expect that more heat is transferred in the constant-pressure process than in the constant-volume process.

SOLVE

The final pressure equals the initial pressure.

$$P_f = P_i = 2.45 \times 10^5 \text{ Pa}$$

Example continues on page 634 ▶

Find the final volume from the ideal gas law.	$V_f = \dfrac{nRT_f}{P_f} = \dfrac{(94.0\text{ mol})\left(8.315\dfrac{\text{J}}{\text{mol}\cdot\text{K}}\right)(354\text{ K})}{2.45\times10^5\text{ Pa}}$ $V_f = 1.13\text{ m}^3$

CHECK and THINK

As expected, the final volume is greater than the initial volume.

SOLVE For a constant-pressure process, heat transferred to the gas is given by Equation 21.19. The molar specific heat C_P of helium is listed in Table 21.3.	$Q_P = nC_P\Delta T$ (21.19) $Q_P = (94.0\text{ mol})\left(20.8\dfrac{\text{J}}{\text{mol}\cdot\text{K}}\right)(354\text{ K} - 235\text{ K})$ $Q_P = 2.33\times10^5\text{ J}$
The work done by the environment during a constant-pressure process is given by Equation 21.16.	$W_{tot} = -P\Delta V$ (21.16) $W_{tot} = -(2.45\times10^5\text{ Pa})(1.13\text{ m}^3 - 0.75\text{ m}^3)$ $W_{tot} = -9.31\times10^4\text{ J}$
The change in thermal energy is the sum of the energies entering the system through work and heat.	$\Delta E_{th} = W_{tot} + Q_P$ $\Delta E_{th} = (-9.31\times10^4\text{ J}) + (2.33\times10^5\text{ J}) = 1.40\times10^5\text{ J}$

CHECK and THINK

As expected, energy leaves the system through work ($W_{tot} < 0$), energy enters the system through heat ($Q > 0$), and the change in thermal energy is the same in the constant-pressure process (part B) as it is in the constant-volume process (part A). The constant-pressure process requires more heat than the constant-volume process because energy is lost by the system through work. Thus, C_P is greater than C_V.

Degrees of Freedom

The values of C_V or C_P in Table 21.3 depend on the structure of the gas molecules. Monatomic molecules, those made up of just one atom, are well modeled as particles with no internal structure. Diatomic, triatomic, and polyatomic molecules are not well modeled as particles; instead, we must account for their internal structure.

To find C_V for a monatomic ideal gas, start with $E_{th} = \frac{3}{2}nRT$ (Eq. 21.2) to find an expression for ΔE_{th}.

$$\Delta E_{th} = \tfrac{3}{2}nR\Delta T \quad (21.25)$$

Set $\Delta E_{th} = nC_V\Delta T$ (Eq. 21.20) equal to Equation 21.25:

$$\Delta E_{th} = nC_V\Delta T = \tfrac{3}{2}nR\Delta T \quad (21.26)$$

Solve for C_V for a monatomic ideal gas:

$$C_V = \tfrac{3}{2}R \text{ monatomic} \quad (21.27)$$

Equation 21.27 is in good agreement with experimental results: $C_V = 12.5$ J/(mol · K) for a monatomic ideal gas, which is close to the values in Table 21.3.

More complicated molecules can rotate and vibrate, so they cannot be modeled as particles. Equation 21.25, $\Delta E_{th} = \frac{3}{2}nR\,\Delta T$, does not hold for complicated molecules. So, our next step is to find a more general expression for ΔE_{th}. We start by considering the ways molecules may move.

DEGREES OF FREEDOM

Major Concept

First, a particle (such as a monatomic molecule) can undergo translational motion in up to three independent directions—x, y, and z—and we say that it has three *degrees of freedom*. In thermodynamics, **degrees of freedom** are the number of independent ways a molecule can have energy. (It is possible for the particle to change its motion in one dimension without changing its motion in the other two dimensions.)

Now, consider a diatomic molecule (Fig. 21.25) modeled as two particles connected by a massless spring. The molecule can rotate around two independent axes (Fig. 21.25A), and the particles can also vibrate in and out from the center of mass (Fig. 21.25B). If the molecule vibrates, it has two forms of energy: kinetic energy due the motion of the particles and potential energy due to the compression and extension of the "spring." In addition, the entire molecule can have translational motion in three dimensions. A diatomic molecule therefore can have up to seven degrees of freedom:

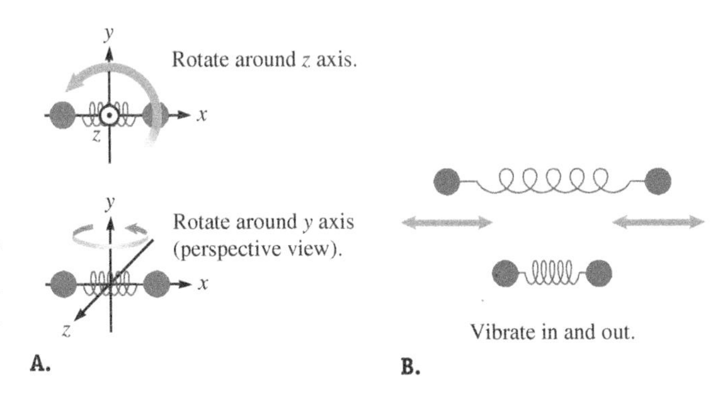

FIGURE 21.25 A diatomic molecule can **A.** rotate around two different axes and **B.** vibrate in and out.

- Three degrees of freedom for its translational kinetic energy in three dimensions
- Two degrees of freedom for its rotational kinetic energy around two different axes
- Two degrees of freedom for its kinetic energy and potential energy of vibration

Not all these seven degrees of freedom must be exhibited; the ones that are exhibited are known as the **active degrees of freedom**.

In Section 20-3, we found that the average kinetic energy of the particles in an ideal gas is $K_{avg} = \frac{3}{2}k_B T$. The "3" in the average kinetic energy is due to the individual particles' three degrees of freedom (Eq. 20.6). The **principle of equipartition of energy** states that energy is shared equally between the active degrees of freedom, each one having on average $\frac{1}{2}k_B T$ of energy.

PRINCIPLE OF EQUIPARTITION OF ENERGY ✪ Major Concept

How do the principle of equipartition of energy and a molecule's active degrees of freedom determine the molar specific heat of a gas? To answer, we need to generalize our derivation of C_V. We replace $\Delta E_{th} = \frac{3}{2}nR\Delta T$ (Eq. 21.25), which is good for monatomic molecules, with a more general expression for ΔE_{th} for polyatomic molecules:

$$\Delta E_{th} = \frac{(\text{d.o.f.})}{2} nR\Delta T \tag{21.28}$$

where "d.o.f." is the number of active *d*egrees *o*f *f*reedom. We set Equation 21.28 equal to $\Delta E_{th} = nC_V\Delta T$ (Eq. 21.20):

$$\Delta E_{th} = nC_V\Delta T = \frac{(\text{d.o.f.})}{2} nR\,\Delta T$$

and solve for C_V:

$$C_V = \frac{(\text{d.o.f.})}{2} R \tag{21.29}$$

A more complicated molecule may have more active degrees of freedom, and according to Equation 21.29, the corresponding gas has a greater molar specific heat. The molar specific heat of diatomic and more complicated molecules depends on the gas temperature because more degrees of freedom become active at higher temperatures.

At room temperature, the diatomic molecules in Table 21.3 have five active degrees of freedom: three for translational motion and two for rotational motion. According to Equation 21.29, such a diatomic gas should have a molar specific heat of

$$C_V = \frac{(\text{d.o.f.})}{2} R = \frac{5}{2} R = \frac{5}{2}[8.315\,\text{J}/(\text{mol}\cdot\text{K})] = 20.8\,\text{J}/(\text{mol}\cdot\text{K})$$

which is confirmed experimentally (Table 21.3).

Although the molar specific heats in Table 21.3 are for gases at room temperature, many of those values are, in fact, fairly good for a broad range of temperatures (150 K to 750 K). At much cooler temperatures, however, diatomic molecules tend to stop rotating, so they have only three active degrees of freedom, and their molar specific heat is $C_V = \frac{3}{2}R$. At much higher temperatures, diatomic molecules tend to rotate and vibrate, and their molar specific heat is $C_V = \frac{7}{2}R$.

EXAMPLE 21.8 Monatomic Versus Diatomic Gas

We have two containers; one holds 1 mol of a monatomic gas, and the other holds 1 mol of a diatomic gas. Both gases start with an initial temperature $T_i = 245$ K and reach a final temperature $T_f = 295$ K by a constant-volume process. Find the initial thermal energy $E_{th,\,i}$ and heat Q_V absorbed by each gas.

INTERPRET and ANTICIPATE

There is 1 mol of each gas, but the diatomic gas has more degrees of freedom than the monatomic gas. Therefore, at the same temperature, we expect the thermal energy will be greater for the diatomic gas. Both gases experience the same change in temperature, so the thermal energy increases more for the diatomic gas than the monatomic gas. Therefore, because the process is constant-volume (no work), more heat must be required in the diatomic case.

SOLVE

Equipartition of energy allows $\frac{1}{2}k_B T$ of energy for each active degree of freedom.

$$K_{avg} = (\text{d.o.f.})(\tfrac{1}{2}k_B T)$$

For a gas, thermal energy is the number of molecules N multiplied by their average kinetic energy (Eq. 21.2). Write N in terms of n (number of moles) and N_A (Avogadro's number).

$$E_{th} = NK_{avg} \qquad N = nN_A$$

$$E_{th} = nN_A K_{avg}$$

$$E_{th} = \frac{(\text{d.o.f.})}{2} nN_A k_B T$$

Use $N_A k_B = R$ (Eq. 19.20).

$$E_{th} = \frac{(\text{d.o.f.})}{2} nRT$$

A monatomic gas has three degrees of freedom.

$$E_{th,\,i} = \frac{3}{2} nRT_i$$

$$E_{th,\,i} = \frac{3}{2}(1 \text{ mol})\left(8.315 \frac{\text{J}}{\text{mol}\cdot\text{K}}\right)(245 \text{ K})$$

$$E_{th,\,i} = 3.06 \times 10^3 \text{ J} \quad (\text{monatomic})$$

A diatomic gas has five active degrees of freedom in the given temperature range.

$$E_{th,\,i} = \frac{5}{2} nRT_i$$

$$E_{th,\,i} = \frac{5}{2}(1 \text{ mol})\left(8.315 \frac{\text{J}}{\text{mol}\cdot\text{K}}\right)(245 \text{ K})$$

$$E_{th,\,i} = 5.09 \times 10^3 \text{ J} \quad (\text{diatomic})$$

Because the process is constant volume (page 627), the change in thermal energy equals the heat absorbed. The change in thermal energy is given by Equation 21.28, where the change in temperature is the same (50 K) for both gases.

$$\Delta E_{th} = Q_V = \frac{(\text{d.o.f.})}{2} nR\Delta T \tag{21.28}$$

$$\Delta T = 50 \text{ K}$$

First solve for a monatomic gas.

$$Q_V = \frac{3}{2}(1 \text{ mol})\left(8.315 \frac{\text{J}}{\text{mol}\cdot\text{K}}\right)(50 \text{ K})$$

$$Q_V = 6.2 \times 10^2 \text{ J} \quad (\text{monatomic})$$

Now solve for a diatomic gas.

$$Q_V = \frac{5}{2}(1 \text{ mol})\left(8.315 \frac{\text{J}}{\text{mol}\cdot\text{K}}\right)(50 \text{ K})$$

$$Q_V = 1.0 \times 10^3 \text{ J} \quad (\text{diatomic})$$

CHECK and THINK

As expected, the diatomic gas has more thermal energy than the monatomic gas, and more heat is required to raise the temperature of the diatomic gas by the same amount.

EXAMPLE 21.9 Solids

The equipartition of energy principle holds for solids as well as gases. Model a solid (a crystal) as a collection of particles connected by springs (Fig. 21.26). Each particle can vibrate around its equilibrium position as if it were connected by springs to its neighbors. How many degrees of freedom are there per particle? What is the molar specific heat of a solid? Assume all degrees of freedom are active and remember that a solid's volume is nearly constant.

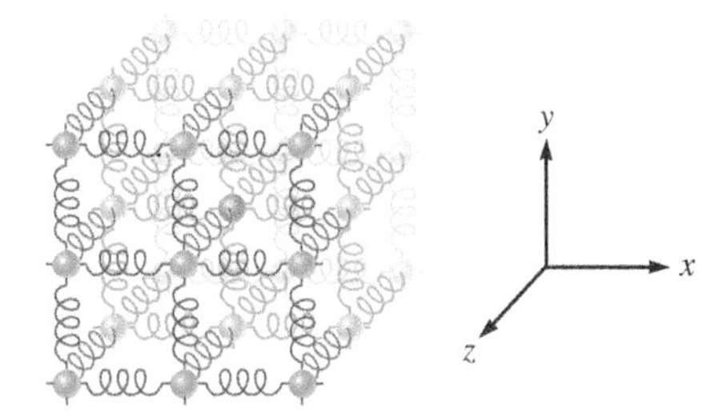

FIGURE 21.26 Particle-spring model of a solid.

INTERPRET and ANTICIPATE

Once we determine the number of active degrees of freedom, the equipartition of energy principle allows us to find the molar specific heat.

SOLVE

In Figure 21.26, the particle shown in red near the center of the solid is connected by springs to its neighbors so that it can oscillate in the x, y, and z directions independently. The particle has one degree of freedom for potential energy and kinetic energy in each of these directions.

6 degrees of freedom

The molar specific heat depends on the number of active degrees of freedom.

$$C_V = \frac{(\text{d.o.f.})}{2}R \tag{21.29}$$

$$C_V = \frac{6}{2}R = 3R = 3 \times 8.315 \frac{\text{J}}{\text{mol} \cdot \text{K}}$$

$$C_V = 24.9 \frac{\text{J}}{\text{mol} \cdot \text{K}}$$

CHECK and THINK

Our result agrees very well with the values for molar specific heat given in Table 21.1 for solids.

21-9 Adiabatic Processes Revisited

Adiabatic processes play an important role whenever a gas expands or contracts rapidly enough to experience a change in temperature without any exchange of heat. An example is the engine, which played a major role in the development of thermodynamics and is a major topic of Chapter 22. So, in this section, we take another look at the adiabatic process. Our goal is to derive equations that connect the state variables.

DERIVATION Adiabatic Path

Figure 21.27 shows a PV diagram and bar chart for an ideal gas that is initially at some cool temperature and then is adiabatically compressed (by a piston) until it reaches a higher temperature. Show that the adiabatic path on the PV diagram is given by

$$P_f V_f^\gamma = P_i V_i^\gamma = \text{constant} \tag{21.30}$$

where γ is the ratio of the specific heats:

$$\gamma \equiv \frac{C_P}{C_V} \tag{21.31}$$

(The ratio γ is given in Table 21.3.)

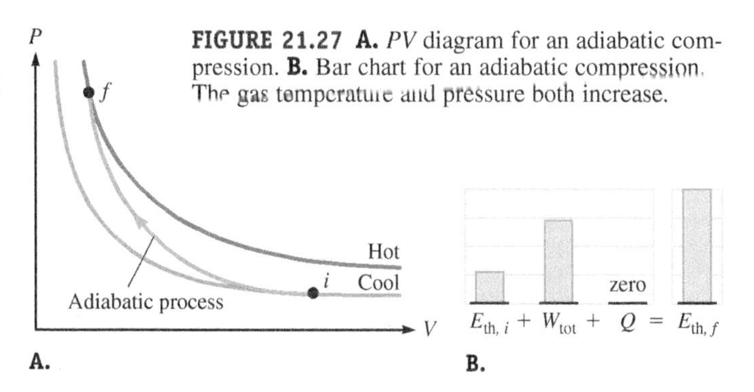

FIGURE 21.27 A. PV diagram for an adiabatic compression. **B.** Bar chart for an adiabatic compression. The gas temperature and pressure both increase.

Derivation continues on page 638 ▶

Consider a small increase in the gas's thermal energy dE_{th} as a result of a small amount of work dW done by the piston on the gas. No heat is exchanged. So, according to the first law of thermodynamics, the work done by the piston equals the change in thermal energy.	$dE_{th} = dW$	(1)
The piston compresses the gas, so the gas volume decreases by a small amount. The work dW done by the piston is positive, and the change in volume dV is negative.	$dW = -P\,dV$	
Eliminate P using the ideal gas law $PV = nRT$.	$dW = -\frac{nRT}{V}dV$	(2)
The gas temperature rises slightly, and we can use $\Delta E_{th} = nC_V \Delta T$ (Eq. 21.26) to write an expression for dE_{th} in terms of dT.	$dE_{th} = nC_V\,dT$	(3)
Substitute Equations (2) and (3) into Equation (1) and simplify.	$nC_V\,dT = -\frac{nRT}{V}dV$ $\frac{dT}{T} = -\frac{R}{C_V}\frac{dV}{V}$	(4)
Using Equation 21.24 and $\gamma \equiv C_P/C_V$ (Eq. 21.31), we can rewrite the constant R/C_V in terms of γ.	$C_P - C_V = R$ (21.24) $\frac{C_P}{C_V} - \frac{C_V}{C_V} = \frac{R}{C_V}$ $1 - \gamma = -\frac{R}{C_V}$	(5)
Substitute Equation (5) into Equation (4) and integrate. The limits on the left are from T_i to T_f, and the limits on the right are from V_i to V_f. The gas temperature rises ($T_i < T_f$) and the volume decreases ($V_i > V_f$).	$\frac{dT}{T} = (1-\gamma)\frac{dV}{V}$ $\int_{T_i}^{T_f}\frac{dT}{T} = (1-\gamma)\int_{V_i}^{V_f}\frac{dV}{V}$ $\ln T_f - \ln T_i = (1-\gamma)(\ln V_f - \ln V_i)$	
Combine the natural logarithm functions (Appendix A). Equation 21.32 is important because it connects two state variables, temperature and volume.	$\ln\frac{T_f}{T_i} = (1-\gamma)\ln\frac{V_f}{V_i} = \ln\left(\frac{V_f}{V_i}\right)^{(1-\gamma)}$ $\frac{T_f}{T_i} = \left(\frac{V_f}{V_i}\right)^{(1-\gamma)}$	(21.32)
Use the ideal gas law to eliminate temperatures and simplify.	$\frac{P_fV_f}{P_iV_i} = \left(\frac{V_f}{V_i}\right)^{(1-\gamma)} = \frac{V_f}{V_i}\left(\frac{V_f}{V_i}\right)^{-\gamma}$ $\frac{P_f}{P_i} = \left(\frac{V_f}{V_i}\right)^{-\gamma} = \left(\frac{V_i}{V_f}\right)^{\gamma}$ $P_fV_f^{\gamma} = P_iV_i^{\gamma} = \text{constant}$ ✓	(21.30)

COMMENTS

Equations 21.30 and 21.32 along with the ideal gas law hold whenever an ideal gas expands or contracts adiabatically. We will return to these equations when we study engines in Chapter 22.

21-10 Conduction, Convection, and Radiation

When there is a temperature difference between a system and its environment, heat flows from the hotter to the colder of these two. In this section, we consider three mechanisms for heat transfer: conduction, convection, and radiation. These mechanisms do not change the first law of thermodynamics; instead, they are three specific ways to exchange energy by heat. In a given situation, one or more of them may be operating.

Conduction

CONDUCTION Special Case

Thermal conduction (or simply **conduction**) is a process for transporting thermal energy requiring a physical connection between the system and its environment through an object called a **conductor.**

When you hold the dry end of a metal spoon whose other end is submerged in a hot mug of coffee, your fingers may feel warm or even burn. That is because heat is transported from the hot coffee (the environment) to your fingers (the system) by conduction through the spoon. On the microscopic level, the spoon's molecules that are in contact with the hot coffee begin to vibrate vigorously due to collisions with the hot coffee molecules. Thermal energy is transported throughout the spoon until thermal equilibrium is reached.

The power (rate at which heat is transported) by conduction depends on the temperature difference between the system and the environment. Figure 21.28 shows a cylindrical conductor connecting a hot reservoir with temperature T_{hot} to a cooler reservoir with temperature T_{cool}. The conductor is made out of a single substance of length ℓ and cross-sectional area A. The amount of heat Q that flows per time interval Δt has dimensions of power and is given by

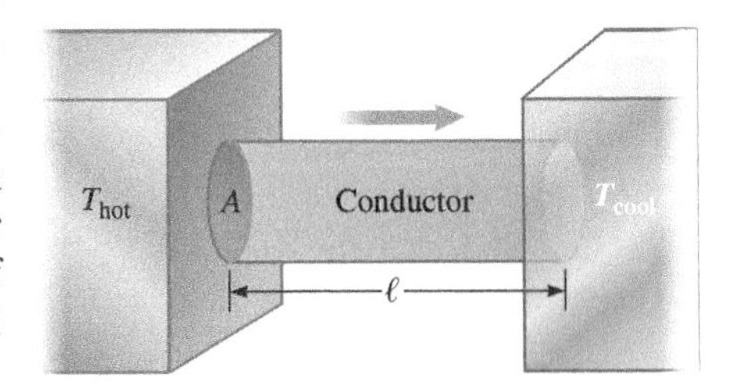

FIGURE 21.28 A conductor transports heat from the hot reservoir at T_{hot} to the cool reservoir at T_{cool}.

$$\frac{Q}{\Delta t} = k\frac{A}{\ell}(T_{\text{hot}} - T_{\text{cool}}) \quad (21.33)$$

where k is the **thermal conductivity** of the substance. Thermal conductivity has the dimensions of power per length per temperature, with SI units W/m · K. Thermal conductivity depends on the type of material, and selected values are provided in Table 21.4. A substance such as polyurethane with a low thermal conductivity is called a good **thermal insulator** or simply an **insulator**. A substance such as a metal with a high thermal conductivity is called a good **thermal conductor** or simply a **conductor.** Many metal objects conduct heat effectively because metals have free electrons that may travel for great distances throughout the sample. These free electrons collide with one another and with the metal ion cores, speeding up the transportation of thermal energy.

TABLE 21.4 Thermal conductivity.

Substance	$k\left(\frac{\text{W}}{\text{m}\cdot\text{K}}\right)$	Conductor or insulator?
Air	0.026	Insulator
Aluminum	235	Conductor
Brick	0.84	Insulator
Concrete	0.84	Insulator
Copper	401	Conductor
Cork	0.042	Insulator
Diamond[a]	2300	Conductor
Glass	1.0	
Goose down	0.025	Insulator
Fiberglass	0.048	Insulator
Helium	0.15	Insulator
Human tissue	0.2	Insulator
Hydrogen	0.18	Insulator
Ice	2	
Iron	79.5	Conductor
Lead	35	Conductor
Polyurethane	0.024	Insulator
Silver	428	Conductor
Stainless steel	14	
Water	0.56	Insulator
Wood	0.11	Insulator
Wool	0.040	Insulator

[a]Diamond is not a metal, but it is a good thermal conductor. It is not a good electrical conductor, however.

CASE STUDY Part 2: Why You Wear Clothes

From Table 21.4, air has low conductivity and is therefore a good insulator. You probably feel most comfortable when your skin temperature is 34°C. How can you maintain such a high skin temperature when you are in a room with an air temperature of 22°C? As heat is transferred from your body to the surrounding air, the air around you would eventually reach 34°C, and heat would no longer be transferred from you to the air. This temperature might keep you comfortable even without clothing if you remained stationary with a warm bubble of surrounding air. People usually need to move around, however, and air also moves, so each of us is continually surrounded with fresh, cool air. Clothing traps air near the skin, and this trapped air moves around with the wearer so that there is much less conduction of heat than without clothing. Cold-weather clothing made of wool or down feathers contains many pockets for trapping air. Many animals stay warm by trapping air in their fur or feathers. When they are particularly cold, their fur or feathers stand up, creating large air pockets. People do the same thing when they get "goose bumps," raising the hair on their arms and legs. This hair is mostly vestigial, however, so goose bumps don't help people trap very much air.

Water is more than 21 times more conductive than air (Table 21.4), so you feel cold when you are in water at 85°F (30°C), but warm when you are in air at that temperature. If you wish to swim in much colder water, you would need to wear a wetsuit or a drysuit. Wearing a wetsuit in water is much like wearing clothes in air. You heat the water near your skin, which the wetsuit traps, so as you swim you are surrounded by a bubble of warm water. Of course, no wetsuit is perfect, and you will continue to lose energy slowly. If you want to swim in much colder water, you must use a drysuit, which keeps a layer of warm air near your skin. Because air is a better insulator than water, you lose less energy by conduction. In Example 21.10, we explore the effects of a drysuit.

EXAMPLE 21.10 CASE STUDY Your Body in Cold Water

You want to swim in a lake at 15°C while wearing a drysuit that surrounds you with a layer of air 0.75 cm thick. When you enter the water, your skin temperature is a comfortable 34°C. You can estimate your surface area A in square meters from your mass m in kilograms and your height h in meters from the following empirical relationship (Problem 1.60):

$$A = 0.202 m^{0.425} h^{0.725}$$

For this example, assume $m = 84$ kg and $h = 1.8$ m. (You may wish to recalculate with your own specifications.)

A Find the power you lose due to conduction.

INTERPRET and ANTICIPATE

The system is your body, the conductor is the layer of air, and the environment is the cold lake.

SOLVE

First, find your surface area A.

$$A = 0.202 m^{0.425} h^{0.725}$$

$$A = (0.202)(84\,\text{kg})^{0.425}(1.8\text{ m})^{0.725} = 2.0\text{ m}^2$$

Find the power you lose through conduction from Equation 21.33 using the conductivity of air listed in Table 21.4. The temperature difference can be calculated in Celsius degrees or kelvin.

$$\frac{Q}{\Delta t} = k\frac{A}{\ell}(T_{\text{hot}} - T_{\text{cool}})$$

$$\frac{Q}{\Delta t} = \left(0.026\frac{\text{W}}{\text{m}\cdot\text{K}}\right)\left(\frac{2.0\,\text{m}^2}{7.5 \times 10^{-3}\,\text{m}}\right)(34°\text{C} - 15°\text{C})$$

$$\frac{Q}{\Delta t} = 130\,\text{W}$$

CHECK and THINK

This result is comparable to the 116 W (= 116 J/s) we know that a person loses under normal conditions (Example 21.3).

B If you stay in the water for half an hour, how much energy do you lose, and what is your skin temperature? Assume the specific heat of skin is the value given in Example 21.3 (3470 J/kg · K).

INTERPRET and ANTICIPATE

Because we know the power lost, we can find the energy lost in half an hour by multiplying by the time. This energy is lost as heat, which is related to change in temperature. We expect that your skin temperature will be lower than 34°, but—assuming half an hour is a safe amount of time to spend in the lake—we expect that your temperature will still be much warmer than the water temperature.

SOLVE

The total energy lost by conduction is the power lost multiplied by the time you are in the lake.

$$\frac{Q}{\Delta t} = 130\text{ J/s}$$

$$Q = (130\text{ J/s})(0.5 \times 3600\text{ s}) = 2.3 \times 10^5\text{ J}$$

CHECK and THINK

This result is about the amount of energy you get from eating an apple.

SOLVE

Find the change in your skin temperature from Equation 21.5 with the given mass $m = 84$ kg and specific heat $c = 3470$ J/kg · K. Because heat flows out of the system, we must include a negative sign.

$$Q = -mc\,\Delta T \tag{21.5}$$

$$\Delta T = -\frac{Q}{mc} = -\frac{2.3 \times 10^5\text{ J}}{(84\text{ kg})\left(3470\frac{\text{J}}{\text{kg}\cdot\text{K}}\right)} = -0.79\text{ K}$$

Solve for your skin temperature when you leave the lake. The temperature difference is the same in Celsius degrees as it is in kelvin.	$T_f - T_i = -0.79°C$ $T_f = T_i - 0.79°C = 34°C - 0.79°C$ $T_f = 33.2°C$

CHECK and THINK

As expected, your skin temperature is slightly cooler than when you first went in the lake. In Problem 62, you will compare this result to what you would find if you didn't use a drysuit.

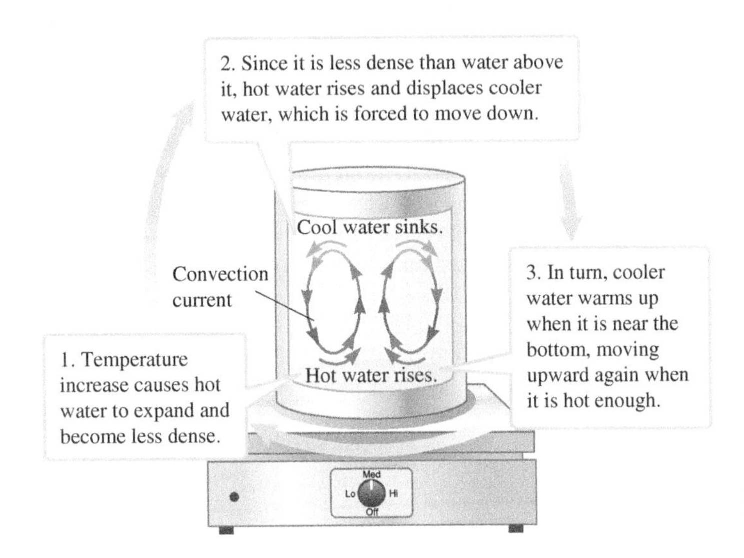

FIGURE 21.29 Convection cells bring hot water upward and cold water downward.

Convection

On a cold day, you might warm your hands by holding them near a warm fire (Fig. 21.1). The only substance between your hands and the fire is air, and according to Table 21.4, air is not a good conductor. So, how can heat get to you?

CONVECTION ▶ Special Case

In fluids (gases and liquids), the **convection** process transports heat by the large-scale motion of molecules through the fluid. If you place a pot of cool water on a hot stove, a cycle of rising warm fluid and sinking cool fluid sets up convection currents (Fig. 21.29).

An important difference between convection and conduction is that in conduction, the molecules vibrate near their equilibrium positions but do not travel throughout the conductor. During convection, molecules travel throughout the fluid, so convection transports heat and matter. For example, when you make a pot of coffee, aromatic coffee molecules are carried by convection currents in the air (Example 20.4).

Radiation

RADIATION ▶ Special Case

Life on the Earth depends on energy transferred from the Sun. There is essentially no medium between the Earth and the Sun, however, so there can be no conduction and no convection. Instead, heat is transferred by **thermal radiation**, electromagnetic waves that can travel through empty space and carry energy (Chapter 34). One of the primary examples of electromagnetic radiation is visible light. Electromagnetic waves are characterized by their wavelength. Ultraviolet (UV) radiation has a shorter wavelength than light. You cannot see UV radiation, but when your skin is exposed to it, the skin may tan or even burn. Infrared (IR) radiation has a longer wavelength than light, and you feel its effect when you stand in front of a hot fire. In fact, when you are near a fire, most of the heat transferred to you is by radiation; convection plays only a minor role.

The power emitted by an object through radiation depends on its surface area A, temperature T, and composition. The amount of heat Q radiated away per time interval Δt is given by the **Stefan-Boltzmann equation**:

$$\frac{Q}{\Delta t} = \sigma \varepsilon A T^4 = \text{power lost through radiation} \quad (21.34)$$

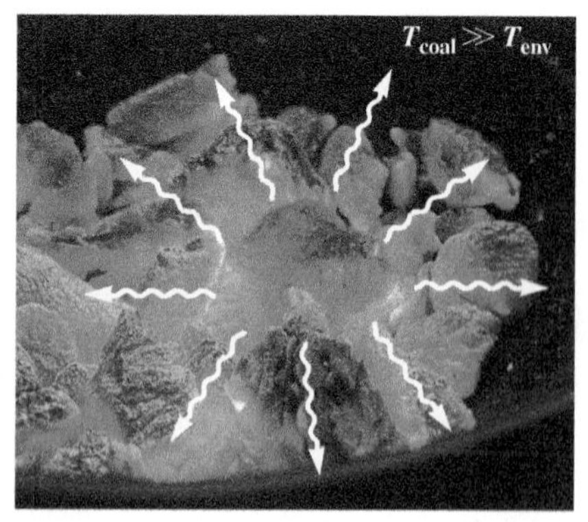

A. Glowing coal

B. Black coal

FIGURE 21.30 A. When the temperature of an object is warmer than its surroundings, it gives off more radiation than it absorbs. **B.** When the temperature of an object is cooler than its surroundings, it absorbs more radiation than it gives off.

TABLE 21.5 Emissivity.

Substance	ε
Aluminum	0.02
Brass	0.03
Brick (red)	0.93
Carbon	0.95
Coal	0.95
Cotton cloth	0.77
Cement	0.96
Glass	0.80
Gold (polished)	0.02
Ice	0.97
Iron (dull)	0.94
Iron (polished)	0.28
Platinum	0.05
Soil	0.38
Snow	0.82–0.89
Silver (polished)	0.01
Water	0.67
Wood	0.80–0.90

where ε (Greek letter epsilon) is the **emissivity**, a number between 0 and 1 that depends on the type of material (Table 21.5). Emissivity is a measure of a substance's ability to emit or absorb radiation. A good emitter ($\varepsilon \approx 1$) is also a good absorber. For example, a very black object such as coal is usually a good emitter, with $\varepsilon = 0.95$. When coal is much hotter than its surroundings, it gives off considerable radiation and glows brightly (Fig. 21.30A). When coal is cooler than its surroundings, it absorbs much radiation and appears black (Fig. 21.30B). On the other hand, a shiny object is a poor emitter and poor absorber ($\varepsilon \approx 0$); most of the radiation is reflected off a shiny object such as when light reflects off a mirror. In SI units, the **Stefan-Boltzmann constant** σ is

$$\sigma = 5.6703 \times 10^{-8} \frac{\text{W}}{\text{m}^2 \cdot \text{K}^4} \tag{21.35}$$

If a system has a temperature T_{sys} above absolute zero and is in an environment with a temperature T_{env}, the system will absorb radiation from the environment and emit radiation at the same time. The net power absorbed is the power absorbed minus the power emitted:

$$\left.\frac{Q}{\Delta t}\right|_{net} = \left.\frac{Q}{\Delta t}\right|_{abs} - \left.\frac{Q}{\Delta t}\right|_{emit} = \sigma\varepsilon A T_{env}^4 - \sigma\varepsilon A T_{sys}^4$$

$$\left.\frac{Q}{\Delta t}\right|_{net} = \sigma\varepsilon A(T_{env}^4 - T_{sys}^4) = \text{net power absorbed} \tag{21.36}$$

Power Absorbed from Sunlight

A star such as our Sun has an emissivity of essentially 1, so for a star we write Equation 21.34 as

$$\frac{Q}{\Delta t} = 4\pi R^2 \sigma T^4 \tag{21.37}$$

where R is the radius of the star. The power radiated by the Sun is

$$\left.\frac{Q}{\Delta t}\right|_{\odot} = 4\pi R_{\odot}^2 \sigma T_{\odot}^4 = 4\pi(6.96 \times 10^8 \text{ m})^2\left(5.67 \times 10^{-8} \frac{\text{W}}{\text{m}^2 \cdot \text{K}^4}\right)(5770 \text{ K})^4$$

$$\left.\frac{Q}{\Delta t}\right|_{\odot} = 3.83 \times 10^{26} \text{ W}$$

In the same way that a spherical sound wave's intensity I (power per unit area) decreases with distance (Section 17-7), the intensity of the Sun's radiation also decreases. The Sun's intensity at the Earth must be calculated from Equation 17.23:

$$I = \frac{1}{4\pi r^2}\left(\left.\frac{Q}{\Delta t}\right|_{\odot}\right)$$

where r is the distance between the Sun and the Earth. The intensity of the Sun's radiation at the top of the Earth's atmosphere is called the **solar constant** and is

$$I = \frac{1}{4\pi(1.50 \times 10^{11}\ \text{m})^2}(3.83 \times 10^{26}\text{W}) = 1350\ \text{W/m}^2$$

According to NASA, the averaged yearly solar constant measured by satellites is 1368 W/m^2. The Earth's atmosphere absorbs some of this radiation, and the intensity near the surface of the Earth on a clear day is about 1000 W/m^2.

Let's come up with an expression for the power absorbed by an object such as bowl of water lying in the Sun (Fig. 21.31). First, we need to multiply the radiation intensity by the effective area of the object. The effective area is the area perpendicular to the Sun's rays. We must also multiply by the emissivity ε of the object to account for the object's ability to absorb radiation. The amount of heat Q absorbed by an object per time interval Δt is given by

$$\frac{Q}{\Delta t} = (1000\ \text{W/m}^2)\varepsilon A\cos\theta = \text{power absorbed from Sun's radiation} \quad (21.38)$$

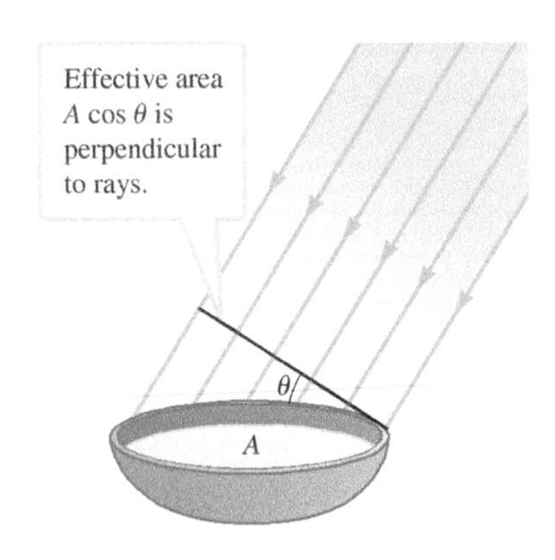

FIGURE 21.31 The power absorbed by a bowl of water exposed to sunlight depends on its cross-sectional area and the angle θ.

CASE STUDY Part 3: What Should You Wear in the Desert?

You probably like to wear white clothes in hot weather because white objects generally absorb less radiation than do dark or black objects. Bedouins who dwell in deserts such as the Sahara, Negev, and Sinai traditionally wear black robes, including a black hood (Fig. 21.32). If black robes absorb more radiation than white robes, why do Bedouins wear black? This question was answered by four researchers in an article published in *Nature* in 1980.[1] The researchers found that a black robe can be as much as 6°C warmer than a white robe, but the skin temperature of the wearer does not depend on the color of the robe. The hot black robe warms the air inside, which escapes through the porous fabric while fresh cooler air enters through the bottom of the robe. Thus, the wearer has his own convection cell with a breeze that makes the wearer more comfortable, if not cooler. Also, because the white robe reflects a lot of sunlight, looking at a white robe in the desert may be as uncomfortable to the eyes as sunlight reflected off snow, which can be blinding.

[1]A. Shkolnik, C. R. Taylor, V. Finch, and A. Borut, "Why Do Bedouins Wear Black Robes in Hot Deserts?" *Nature* 283: 373–374 (1980).

FIGURE 21.32 Bedouins wear black robes in the desert. The black robes allow for stronger convection currents, keeping the wearer cool and comfortable.

EXAMPLE 21.11 CASE STUDY What Should You Wear in Orbit?

When we think of outer space, we often think of a very cold environment and suppose a spacesuit must be designed to keep astronauts warm. It is true that objects with little exposure to the Sun's radiation get very cold. The dark side of the Moon has a low temperature of about 110 K (–163°C). If you were to come into contact with such a cold object, you would lose energy by conduction, so spacesuits are designed to keep the feet and hands warm. The interior of a spacesuit can become very hot, however, especially if the suit is worn in sunlight. For example, on the sunny side of the Moon, the temperature is 387 K (114°C).

An object such as an astronaut in space is oriented so that one side points toward the Sun (Fig. 21.33). The object maintains this orientation for a long time so that the side oriented toward the Sun comes into equilibrium. Find the equilibrium temperature of the sunny side of an object near the Earth (1 AU from the Sun).

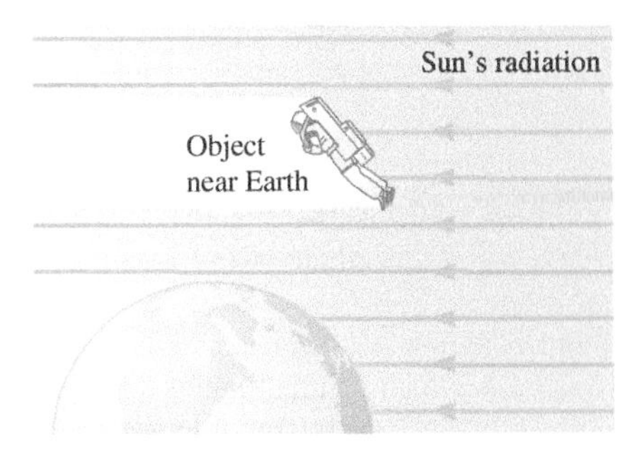

FIGURE 21.33

Example continues on page 644 ▶

INTERPRET and ANTICIPATE

The object absorbs radiation from the Sun and also emits radiation. Before the sunny side reaches equilibrium, it absorbs radiation at a higher rate than it emits. When the object reaches its equilibrium temperature, the power absorbed equals the power emitted, so the net power absorbed is zero. Because the Moon is near the Earth, has a very thin atmosphere, and has a slow rotational period, we can check our results against the high temperature of the Moon.

SOLVE

The object is near the Earth and above the atmosphere, so the intensity of sunlight intercepting the object is given by the solar constant. To find the power absorbed, modify Equation 21.38 by using the solar constant.

$$\left.\frac{Q}{\Delta t}\right|_{\text{abs}} = (1350\ \text{W/m}^2)\varepsilon A \cos\theta$$

The sunlight is direct as shown in Figure 21.33, so we set $\theta = 0$.

$$\left.\frac{Q}{\Delta t}\right|_{\text{abs}} = (1350\ \text{W/m}^2)\varepsilon A$$

In equilibrium, the net power absorbed is zero (power absorbed equals power emitted). The power emitted is given by Equation 21.34, where T_{obj} is the equilibrium temperature of the object.

$$\left.\frac{Q}{\Delta t}\right|_{\text{abs}} = \left.\frac{Q}{\Delta t}\right|_{\text{emit}}$$

$$(1350\ \text{W/m}^2)\varepsilon A = \sigma\varepsilon A T_{\text{obj}}^4$$

Solve for T_{obj}. The Stefan-Boltzmann constant is given in Equation 21.35.

$$T_{\text{obj}} = [(1350\ \text{W/m}^2)/\sigma]^{1/4} = \left[\frac{1350\ \text{W/m}^2}{5.670 \times 10^{-8}\ (\text{W/m}^2 \cdot \text{K}^4)}\right]^{1/4}$$

$$T_{\text{obj}} = 393\ \text{K} = 120^\circ\text{C}$$

CHECK and THINK

As expected, our result is very close to the high temperature on the sunny side of the Moon and shows that objects in direct sunlight near the Earth get very hot. Spacesuits are designed to keep an astronaut at a comfortable temperature. Spacesuits are well insulated with Mylar and other materials that help prevent the absorption of radiation, but the astronaut and the electrical components inside the suit warm up just as they would on the Earth. On the Earth, people and machines give off heat through conduction, convection, and radiation. In space, there are generally no conductors or air, so heat cannot be transferred in space through conduction or convection. Radiation is too slow to keep the astronaut at a constant comfortable temperature. In fact, if the temperature inside the suit becomes warm enough, the astronaut will sweat, making the inside of the suit uncomfortably humid. The solution is a liquid cooling garment (Fig. 21.34) worn against the astronaut's skin and laced with tubes of water. Water tubes are also used to cool the electronics. Heat is transmitted to the water through conduction, and thermal energy in the water is then ejected into space either through evaporation or sublimation. Each pound of water evaporated removes about 1000 J of thermal energy.

FIGURE 21.34 A liquid cooling garment for astronauts.

One final note: The astronaut shown in Figure 21.33 is not exposed to radiation on the side facing away from the Sun. In the absence of heat transfer from the sunny side of the object, the dark side of the object would radiate energy into space until it reaches the temperature of its environment (roughly 3 K in deep space). Temperatures in the solar system are more moderate than in deep space. Objects such as planets or moons rotate so that both sides of the object have some time in the Sun. Planets and moons also have an atmosphere (if very thin sometimes), enabling heat transfer by convection. The Earth's Moon has a rotational period of roughly one month, so any particular spot on the Moon spends roughly two weeks per month in darkness and two weeks in sunshine. This rotation plus the Moon's thin atmosphere maintain its surface low temperature around 110 K, well above the temperature of deep space.

SUMMARY

Underlying Principles

The **first law of thermodynamics** is a statement of conservation of energy for a system that may exchange energy with its environment by work or through heat:

$$W_{tot} + Q = \Delta E_{th} \quad (21.1)$$

where W_{tot} and Q are positive if energy is transferred from the environment to the system.

Major Concepts

1. **Heat** is defined as energy Q transferred between a system and its environment due to their temperature difference.
2. **Thermal energy** is the sum of each microscopic particle's kinetic energy plus the total potential energy stored in the forces between microscopic particles in a system.
3. **Heat capacity** is the constant of proportionality in Equation 21.3:
$$\Delta E_{th} = \mathcal{C}\,\Delta T \quad (21.3)$$
4. A **thermodynamic process** is a way for system to change from an initial state to a final state.
5. The mass specific heat capacity or simply the **specific heat** c is related to the heat capacity by $\mathcal{C} = mc$. When no work is done on the system, the heat transferred is then given by
$$Q = mc\Delta T \quad (21.5)$$
6. **Degrees of freedom** indicate the number of independent ways a molecule can have energy. Not all degrees of freedom are active at a given temperature.
7. The **principle of equipartition of energy** states that energy is shared equally between the active degrees of freedom, each one having on average $\frac{1}{2}k_BT$ of energy.

Special Cases

Six specific thermodynamic processes:

Process	Heat entering system	Work done by environment (energy entering system)	Change in thermal energy
Adiabatic ($Q = 0$)	$Q = 0$	—	—
Isothermal ($\Delta T = 0$)	$Q = -W_{tot}$	For an ideal gas, $W_{tot} = Nk_BT\ln\frac{V_i}{V_f}$ (21.15)	$\Delta E_{th} = 0$
Cyclic ($\Delta E_{th} = 0$)		$Q = -W_{tot}$	$\Delta E_{th} = 0$
Free expansion	$Q = 0$	$W_{tot} = 0$	$\Delta E_{th} = 0$
Constant volume ($\Delta V = 0$)	$Q = nC_V\Delta T$	$W_{tot} = 0$	$\Delta E_{th} = Q$
Constant pressure ($\Delta P = 0$)	$Q = nC_P\Delta T$	$W_{tot} = -P\Delta V$ (21.16) For an ideal gas, $W_{tot} = Nk_BT_i\left(1 - \frac{V_f}{V_i}\right)$ (21.17)	—

For an **adiabatic** expansion or compression of an ideal gas, the state variables P, V and T are related by

$$\frac{T_f}{T_i} = \left(\frac{V_f}{V_i}\right)^{(1-\gamma)} \quad (21.32)$$

and

$$P_fV_f^\gamma = P_iV_i^\gamma \quad (21.30)$$

There are three ways to transfer heat.

1. **Conduction** is a process for transporting thermal energy requiring a physical connection between the system and its environment through an object (a *conductor*). The heat Q flowing per time interval Δt as a result of conduction is
$$\frac{Q}{\Delta t} = k\frac{A}{\ell}(T_{hot} - T_{cool}) \quad (21.33)$$
where k is the **thermal conductivity**.
2. **Convection** transports heat by the large-scale motion of molecules through a fluid.
3. (Thermal) **radiation** transfers heat via electromagnetic waves. The heat Q radiated away per time interval Δt is given by the **Stefan-Boltzmann equation**:
$$\frac{Q}{\Delta t} = \sigma\varepsilon AT^4 \quad (21.34)$$
where ε is the **emissivity** and ε is the **Stefan-Boltzmann constant.**

Tools

We modify the **energy bar charts** from Chapters 8 and 9 for the first law of thermodynamics by adding a bar for heat (Q) and removing the mechanical energy bars (K and U). See Figure 21.13, page 625.

PROBLEMS AND QUESTIONS

A = algebraic C = conceptual E = estimation G = graphical N = numerical

21-1 What Is Heat?

1. C Come up with a word or phrase to replace *heat*. Explain how your new term fits the concept of heat.
2. C To keep the contents of a house thermally isolated from the external environment, the walls and roof of the building are insulated. **a.** Is it fair to say that *insulation keeps heat in*? **b.** Is insulation only important in the cold months? Explain.

Problems 3 and 56 are paired.

3. C You extend an impromptu invitation to a friend for dinner. The only food you have is a couple of frozen steaks. You wish to defrost the steaks before grilling them. You defrost one by using your microwave oven, and you defrost the other by placing it in a bowl of very warm water. In each case, decide whether energy is transferred by heat or by work. Explain your answers.
4. C If a system is in thermal equilibrium with its environment, is it possible to transfer energy from the system to the environment through heat, work, both, or neither? Is it possible to transfer energy from the environment to the system through heat, work, both, or neither? Explain your answers.

21-2 How Does Heat Fit into the Conservation of Energy?

5. C The words *isolated* and *insulated* sound similar. In our everyday language, we use these words metaphorically; we may say that "Bobby is isolated" or "Bobby is insulated." What subtle difference exists between these two phrases? Explain how each word is used in physics. What is the major difference between these two terms in physics?
6. C A can of paint is vigorously shaken by a paint mixer. Would you expect the temperature of the paint to increase, decrease, or remain unchanged? Explain your answer.
7. N Late for his morning workout, a 75.0-kg man eats a 200.0-Calorie energy bar for breakfast. **a.** What is the energy equivalent of this energy bar in joules? **b.** On a stair-climbing exercise machine, each "step" can be thought of as increasing the gravitational potential energy of the man–Earth system. If the equivalent height of each step is 12.0 cm, how many steps must the man take to work off the energy of the energy bar? Assume the efficiency of the human body in converting chemical energy to mechanical energy is 20.0%.
8. C A wooden block slides along a horizontal aluminum plane, eventually coming to rest. Explain what happened to the kinetic energy of the block–plane system.
9. C After driving your car for an hour, you place your hand on the hood of the car and discover that it is very hot. Explain how the temperature increased. Would your answer change if you were driving at night? Would your answer change if you were driving on a very cold winter night?

21-3 The First Law of Thermodynamics

Problems 10 through 13 are grouped.

10. C Is it possible for heat to be transferred to a system but for the system's thermal energy to remain unchanged? Explain.
11. N A system experiences no change in thermal energy when the environment does 15.0 J of work on it. According to the first law of thermodynamics, how much heat is transferred to the system? Does energy enter or leave the system by work? Does energy enter or leave the system by heat?
12. C According to the first law of thermodynamics, is it possible for the thermal energy of an object to decrease if heat is entering the object from its surroundings? Justify your answer.
13. N A system's thermal energy increases by 15.0 J when the environment does 15.0 J of work on it. According to the first law of thermodynamics, how much heat is transferred to the system? Does energy enter or leave the system by work? Does energy enter or leave the system by heat?
14. C CASE STUDY To maintain your current weight, suppose you must eat 2100 Cal per day. If you only eat 1900 Cal per day, you find that you lose weight. Explain why consuming less energy per day results in a mass loss. Where did the mass go?

21-4 Heat Capacity and Specific Heat

15. N If 30.0 g of milk at a temperature of 4.00°C is added to a 255-g cup of coffee at a temperature 90.0°C, what is the final temperature of the mixture? Assume coffee has a specific heat of 4.19×10^3 J/(kg · K) and milk has a specific heat of 3.93×10^3 J/(kg · K).
16. N A block of metal of mass 0.250 kg is heated to 150.0°C and dropped in a copper calorimeter of mass 0.250 kg that contains 0.160 kg of water at 30°C. The calorimeter and its contents are insulated from the environment and have a final temperature of 40.0°C upon reaching thermal equilibrium. Find the specific heat of the metal. Assume the specific heat of water is 4.190×10^3 J/(kg · K) and the specific heat of copper is 386 J/(kg · K).

Problems 17 and 18 are paired.

17. C Is it possible for a cool bathtub of water to have more thermal energy than a hot cup of water? Explain.
18. C Suppose a bathtub of cool water has more thermal energy than a hot potato. If the hot potato is placed in the water, will heat flow from the potato to the water or from the water to the potato? Explain.
19. N If you wish to keep a cup of coffee warm, is it better to use a glass cup or a steel cup? Before you decide, find the final temperature of 0.125 kg water poured into **a.** a glass cup and **b.** a steel cup. In each case the water has an initial temperature of 95°C. Each cup has a mass of 0.130 kg and an initial temperature of 22°C. Comment on your results.

20. N From Table 21.1, the specific heat of milk is 3.93×10^3 J/(kg · K), and the specific heat of water is 4.19×10^3 J/(kg · K). Suppose you wish to make a large mug (0.500 L) of hot chocolate. Each liquid is initially at 5.00°C, and you need to raise their temperature to 80.0°C. The density of milk is about 1.03×10^3 kg/m^3, and the density of water is 1.00×10^3 kg/m^3. **a.** How much heat must be transferred in each case? **b.** If you use a small electric hot plate that puts out 455 W, how long would it take to heat each liquid?

21. N A 125-g sample of lead with specific heat c_{Pb} = 128 J/(kg · °C) at 85.0°C and a 320-g sample of silver with c_{Ag} = 230.0 J/(kg · °C) at 34.0°C are added to 0.500 kg of water at 22.0°C where c_W = 4190 J/(kg · °C) in an insulated container. Assuming the system is thermally isolated, what is its final equilibrium temperature?

22. E You bought a new electric teakettle. The manufacturer claims that the teakettle's power output is 5.2 kW. You decide to test it by heating 1 L of water. The water is initially at room temperature, and in about 3 minutes it reaches its boiling point. What do you make of the manufacturer's claim?

23. N An ideal gas is confined to a cylindrical container with a movable piston on one end. The 3.57 mol of gas undergo a temperature change from 300.0 K to 350.0 K. If the total work done on the gas during this process is 1.00×10^4 J, what is the energy transferred as heat during this process? Is the heat flow into or out of the system?

24. N Suppose 5.15 kg of helium gas is at a temperature of 30.0°C. The specific heat of helium is 5193 J/kg · K. **a.** Find the heat exchanged between this system and the surroundings when the temperature of the gas is lowered to 10.0°C. **b.** What is the heat exchanged if the helium is heated from 30.0°C to 50.0°C? **c.** What is the heat capacity of helium?

21-5 Latent Heat

25. N You place frozen soup ($T = -17$°C) in a microwave oven for 3.5 min. The oven transfers 650 W to the soup. Model the soup as 0.35 kg of water (initially ice). What are the temperature and state of the soup when the oven stops?

26. N A 25-g ice cube at 0.0°C is heated. After it first melts, the temperature increases to the boiling point of water (100.0°C), and the water then boils to form 25 g of water vapor at 100.0°C. How much energy in total is added to the ice/water? Which process (melting, increasing temperature, or boiling) requires the most energy? Water has a latent heat of vaporization of 2.256×10^6 J/kg, a latent heat of fusion of 3.33×10^5 J/kg, and specific heat of 4190 J/(kg · K).

27. N What is the energy required to transform an ice cube of mass m = 45.0 g and temperature −5.00°C to steam at 120.0°C?

Problems 28 and 29 are paired.

28. N A jeweler must melt 45.75 g of silver. Assume the silver starts at room temperature and reaches a temperature just above its melting point (T = 1241 K). Use the values found in Tables 21.1 and 21.2. How much heat is transferred **a.** in order for the whole process to take place, **b.** in order to raise the silver to its melting point and, **c.** during the phase change? In what way are your results approximations?

29. N A jeweler must melt 45.75 g of copper. Assume the copper starts at room temperature (T = 22.3°C) and reaches a temperature just above its melting point (T = 1361 K). Use the values found in Tables 21.1 and 21.2. How much heat is transferred **a.** in order for the whole process to take place, **b.** in order to raise the copper to its melting point and, **c.** during the phase change? In what way are your results approximations? If you solved Problem 28, compare your results and comment.

30. N Two 40.0-g ice cubes initially at 0°C are added to 450 g of water at 22.0°C. Assuming this system is insulated and ignoring heat transfer with the glass, what is the equilibrium temperature of the mixture?

Problems 31 and 32 are paired.

31. N Consider the latent heat of fusion and the latent heat of vaporization for H_2O, 3.33×10^5 J/kg and 2.256×10^6 J/kg, respectively. How much heat is needed to **a.** melt 2.00 kg of ice and **b.** vaporize 2.00 kg of water? Assume the temperatures of the ice and steam are at the melting point and vaporization point, respectively.

32. C Consider the latent heat of fusion and the latent heat of vaporization for H_2O, 3.33×10^5 J/kg and 2.256×10^6 J/kg, respectively. Why is the latent heat of vaporization so much larger than the latent heat of fusion? Think about what is happening to the molecules as the phase changes occur in each case. Why is more energy needed to change the same amount of material to vapor than is necessary to melt it?

21-6 Work in Thermodynamic Processes

33. N How much energy is transferred from a system by heat during a thermodynamic process in which the work done on the system is 4.00×10^2 kJ and the system's thermal energy decreases by 2.50×10^2 kJ?

34. N A thermodynamic cycle is shown in Figure P21.34 for a gas in a piston. The system changes states along the path $ABCA$. **a.** What is the total work done by the gas during this cycle? **b.** How much heat is transferred? Does heat flow into or out of the system?

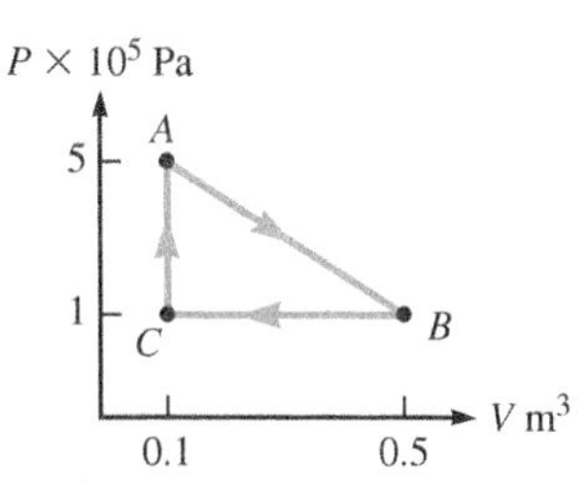

FIGURE P21.34

35. A During a process in which the pressure and volume of a gas are directly proportional, an ideal gas is taken from the initial state ($4P_0$, $4V_0$) to the final state (P_0, V_0). Assuming no gas is lost, what is the work done on the gas during this process?

36. N Figure P21.36 shows a cyclic thermodynamic process $ABCA$ for an ideal gas. **a.** What is the net energy transferred into the system by heat during each cycle? **b.** What would be the net energy transferred into the system by heat if the cycle followed the path $ACBA$ instead?

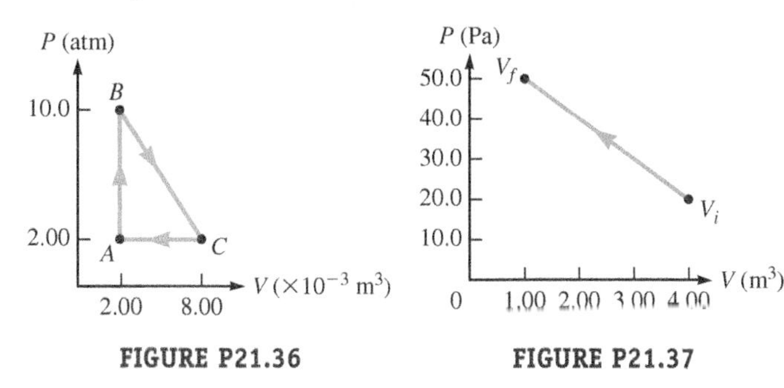

FIGURE P21.36 **FIGURE P21.37**

37. N Figure P21.37 shows a PV diagram for a gas that is compressed from V_i to V_f. Find the work done by the **a.** gas and **b.** environment during this process. Does energy enter the system or leave the system as a result of work?

38. C CASE STUDY You have injured an ankle and need to soak it in ice water. Can you model yourself as a heat reservoir if **a.** you put your foot in a bucket of ice water or **b.** you soak your whole body in a lake of ice water? Explain.

21-7 Specific Thermodynamic Processes

39. N A gas is at a constant pressure of 2.4×10^5 N/m^2. Its volume increases by 1.2×10^{-2} m^3 as 3200 J of energy is transferred to the gas in the form of heat. Find the change in thermal energy of the gas.

40. C You observe the state of a gas at two different times, measuring P, V, and T, and discover that its temperature is unchanged. How would comparing the initial and final values of the volume and pressure help you determine if the gas has undergone an isothermal expansion, a cyclic process, or some other process?

41. N During an isothermal process, 0.500 mol of an ideal gas expands to a final volume of 8.00 L and a pressure of 2.03×10^5 Pa while doing 2.50 kJ of work on its surroundings. **a.** What is the initial volume of the gas? **b.** What is the temperature of the gas?

42. An ideal gas at $P = 2.50 \times 10^5$ Pa and $T = 295$ K expands isothermally from 1.25 m^3 to 2.75 m^3. The gas then returns to its original state through a two-part process: constant-pressure followed by constant-volume.

a. G Draw a PV diagram for this gas.
b. C What is the change in thermal energy?
c. N Find the work done by the environment on the gas.
d. N Find the heat that flows into the gas.

43. N Suppose 3.67 mol of a monatomic gas is at a temperature of 300.0 K and undergoes a constant-pressure expansion from an initial volume of 0.025 m^3 to a final volume of 0.065 m^3. **a.** What is the final temperature of the gas? **b.** What is the change in thermal energy of the gas as it undergoes this process? **c.** What is the work done on the gas?

44. In Figure P21.44, an ideal gas undergoes a change in state from A to C by two different paths: ABC and AC.

a. C Along which path is the least amount of work done by the gas?
b. N If the thermal energy of the gas at A is 12 J and the heat transferred to the gas along the path AC is 220 J, find the thermal energy of the gas at C.
c. N If the thermal energy of the gas at state B is 25 J, find the amount of heat added to the gas to change its state from A to B.

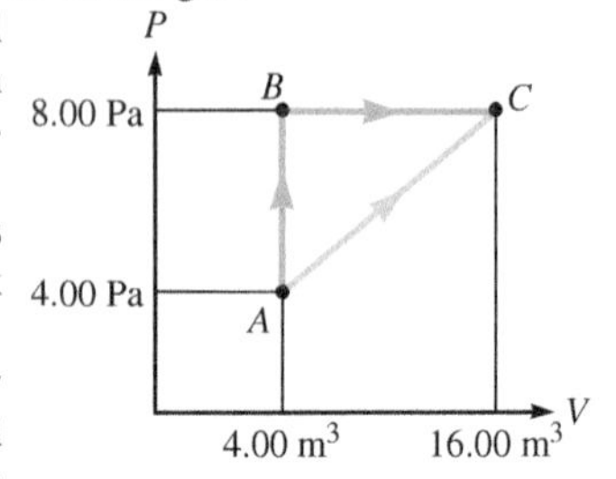

FIGURE P21.44

45. N Figure P21.45 shows a cyclic process $ABCDA$ for 1.00 mol of an ideal gas. The gas is initially at $P_i = 1.50 \times 10^5$ Pa, $V_i = 1.00 \times 10^{-3}$ m^3 (point A in Fig. P21.45). **a.** What is the net work done on the gas during the cycle? **b.** What is the net amount of energy added by heat to this gas during the cycle?

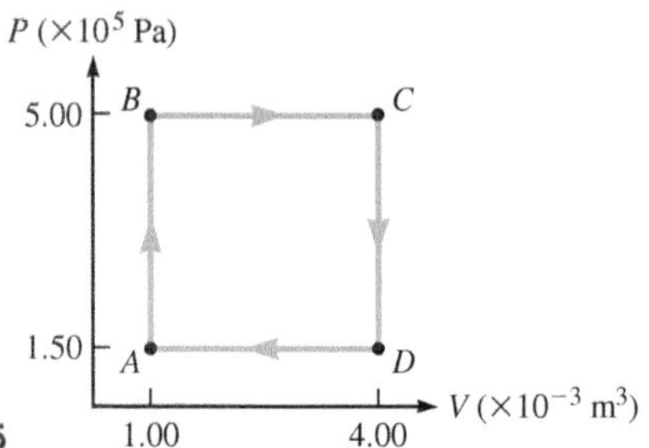

FIGURE P21.45

46. An ideal gas expands such that no heat is exchanged with the surroundings. It then undergoes a process for which no work is done to return to the original pressure. Finally, it undergoes a constant-pressure process to return to the initial state of the gas.

a. C What thermodynamic process is occurring in each of the three steps described? Explain your reasoning.
b. G Sketch this cycle on a PV diagram.

47. N A monatomic gas undergoes the set of thermodynamic processes shown in Figure P21.47, where $P_A = 2.00 \times 10^5$ Pa, $P_B = 1.00 \times 10^5$ Pa, $V_A = 0.0550$ m^3, and $V_C = 0.0750$ m^3. (The subscripts refer to the labeled points on the PV diagram.) There are 5.50 mol of gas. **a.** Find the temperature at B and C. **b.** What is the total work done on the gas during one cycle? **c.** How much heat is transferred when the gas goes from B to C?

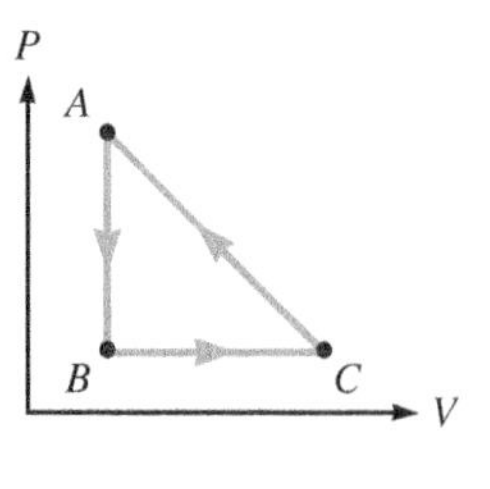

FIGURE P21.47

21-8 Equipartition of Energy

48. N Use the data in Table 21.3 to find the active number of degrees of freedom in a triatomic gas at room temperature.

49. N In a constant-pressure expansion, a diatomic gas performs 84 J of work on the environment at room temperature. Calculate the heat transferred to the gas during this process.

50. N A sample of a monatomic gas is in a container with a movable piston such that it is maintained at constant pressure. If 25.0 J of heat were transferred into the gas, the temperature would increase by 75.0°C. **a.** If, instead, the piston was initially locked in place such that the *volume* remains fixed, by how much would the temperature increase if 25.0 J of heat were transferred into the container? **b.** If, instead, the same number of moles of a *diatomic* gas were contained in the box and 25.0 J of heat were transferred in at constant pressure, by how much would the temperature increase? (Assume there are five active degrees of freedom for the diatomic gas.)

51. N A gas mixture contains 3.0 moles of oxygen (O_2) and 9.0 moles of argon (Ar) at a temperature of 300.0 K. What is the total thermal energy of the system consisting of both gases? Ignore any vibrational degrees of freedom.

52. N A diatomic gas is in a container with a fixed volume of 0.250 m^3. Assume the diatomic gas has five degrees of freedom. **a.** What is the temperature of the gas if there are 6.00 mol present and the gas is under a pressure of 1.50 atm? **b.** If the gas is to be heated by 225 K, how much heat is necessary to accomplish this task? **c.** Repeat parts (a) and (b) for a monatomic gas. Assume the monatomic gas has three degrees of freedom.

21-9 Adiabatic Processes Revisited

53. N An ideal gas is compressed adiabatically so that its volume is cut in half. **a.** If the gas is monatomic, find the ratio of its final pressure to its initial pressure, P_f/P_i. **b.** If the gas is diatomic at room temperature, find the ratio of its final pressure to its initial pressure, P_f/P_i. Explain the difference you find between the two gases in terms of the number of degrees of freedom.

54. A monatomic ideal gas is initially at a pressure of 1.00 atm in a 1.00-L cylindrical container with a piston on one side. It is compressed to a volume of 0.250 L.

a. G Plot the pressure versus volume during this process, assuming the process is either (i) adiabatic or (ii) isothermal.
b. N Determine the final pressure, assuming the process is either (i) adiabatic or (ii) isothermal.

55. A diatomic ideal gas at room temperature undergoes an adiabatic process such that its final pressure is 2.75 times its initial pressure.

a. C Did the gas expand or contract?
b. N What is the ratio of its final volume to its initial volume?

21-10 Conduction, Convection, and Radiation

56. C You extend an impromptu invitation to a friend for dinner. The only food you have is a couple of frozen steaks. You wish to defrost the steaks before grilling them. You defrost one by using your microwave oven, and you defrost the other by placing it in a bowl of very warm water. In each case, decide whether heat is transferred by conduction, convection, radiation, or some combination of these mechanisms. Explain your answers.

57. N Most of the energy emitted by an incandescent lightbulb is in the form of radiation. Assume all the power of a 75-W lightbulb is released from a tungsten filament at a temperature of 2800 K with an emissivity of 0.35. What is the effective area of the filament?

58. **a.** E Estimate the rate at which heat is transmitted through a typical window of a house by thermal conduction when the outside temperature is 0°C.
b. C Why do you suppose many thermally efficient windows are double-paned, with two pieces of glass separated by a small air gap?

59. N A lake is covered with ice that is 2.0 cm thick. The temperature of the ambient air is −20°C. Find the rate of thickening of ice. Assume the thermal conductivity of ice is 200.0 W/(m · K), the density of ice is 9.0×10^2 kg/m^3, and the latent heat of fusion is 3.33×10^5 J/kg.

60. N A concerned mother is dressing her child for play in the snow. The child's skin temperature is 36.0°C, and the outside air temperature is 2.00°C. If the emissivity of the child's skin is 0.790 and he has 1.10×10^{-2} m^2 of exposed skin area, what is the amount of energy transferred from his body to the surroundings in 1.00 h?

61. N On a hot summer day, a section of the interstate highway near Los Angeles receives 985 W/m^2 of sunlight. If the ground below the highway can be considered an insulator and the highway surface transfers energy by radiation only, what is the equilibrium temperature of the highway surface?

62. N **CASE STUDY** In Example 21.10, we calculated your skin temperature if you used a drysuit while swimming in a lake at 15°C. If you used a normal bathing suit, what would your body temperature be in half an hour? Assume your hair and the suit maintain a thin layer of water with an average thickness of 2 mm near your skin.

63. Imagine 1 kg of water in a container that does not allow heat transfer from the surrounding air to the water by conduction or convection. This container allows radiation to pass freely into it, but little radiation is allowed to escape.
a. E Estimate the time it would take to make a pot of tea if the container were left in the Sun on a clear day. Tea is properly made with boiling water.
b. C People often make *sun tea* by leaving water with tea leaves in a clear glass jar in the Sun. They claim that the tea tastes smoother because, in contradiction to part (a), the water does not boil! Explain why the water doesn't boil.

64. C Figure P21.64 shows a silver teapot. Explain why the handle has two small pieces of a ceramic material.

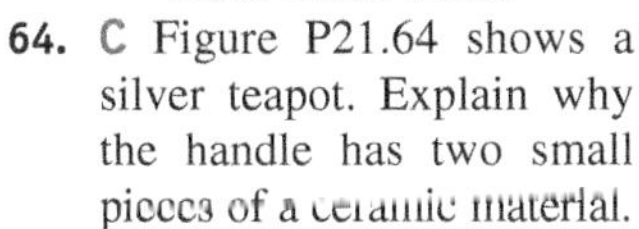

FIGURE P21.64

General Problems

65. N A water calorimeter is a device used to measure the specific heat of a substance (Fig. 21.6). A 0.125-kg sample has a temperature of 1030 K when it is placed into 1.00 kg of water at a temperature of 280 K. The container is made of steel and has a mass of 0.250 kg. After the sample, the water, and its container reach thermal equilibrium, their common temperature is 293 K. What is the specific heat of the substance?

66. C An ideal gas is in the apparatus shown in Figure 21.9. Process A and process B (Fig. P21.66) represent two thermodynamic processes that expand the system from an initial state i to a final state f. **a.** If you wish to *minimize* the work done by the gas during the process, describe how to use process A or B to do so. Include a description of the apparatus during each part of the process. **b.** Repeat part (a) assuming you wish to *maximize* the work done by the gas.

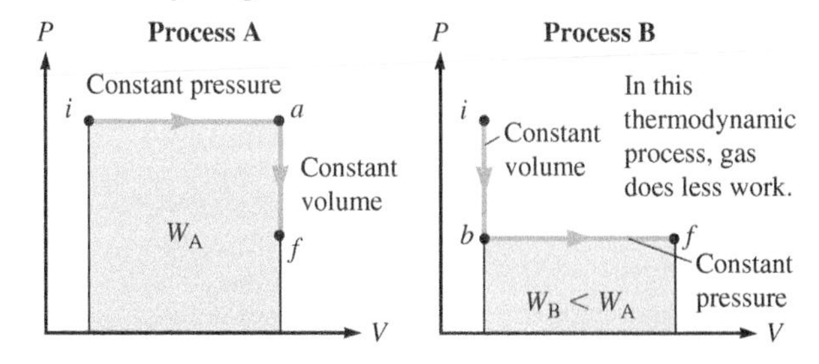

FIGURE P21.66

67. N Many scientists are concerned about the effects of melting icebergs and polar ice on the Earth's climate and its evolving patterns. **a.** Given a large amount of ice with a mass of 2540.0 kg at a temperature of −20.0°C, find the heat necessary to completely melt the ice so that it becomes water at 0°C. **b.** What energy flow into or out of the water is necessary to refreeze the water at 0°C?

68. A 350-g sample of an unknown (nongaseous) material experiences a 20.0°C increase in temperature after absorbing 5.53×10^3 J of energy.
a. N What is the specific heat of this material?
b. C Which substance listed in Table 21.1 best matches the specific heat of this unknown material?

69. N Three 100.0-g ice cubes initially at 0°C are added to 0.850 kg of water initially at 22.0°C in an insulated container. **a.** What is the equilibrium temperature of the system? **b.** What is the mass of unmelted ice, if any, when the system is at equilibrium?

70. N In a constant-pressure process, an ideal gas at a pressure of 800 Pa and initial temperature of 325 K increases in volume from 2.00 m^3 to 5.50 m^3. During the process, 3.00 kJ of energy is transferred into the gas by heat. **a.** What is the change in the internal energy of the gas? **b.** What is the final temperature of the gas?

71. N An ice cube of mass 0.200 kg at 0°C is placed in an insulated container that is at 227°C. The specific heat c of the container varies linearly with temperature as $c = A + BT$, where $A =$ 419 J/(kg · K) and $B = 8.38 \times 10^{-3}$ J/(kg · K^2). If the final temperature of the container is 27.0°C upon reaching thermal equilibrium with the ice cube, determine the mass of the container.

72. C In Chapter 20, we related temperature to the average kinetic energy of the molecules in an ideal gas. We wouldn't say that a baseball flying through the air is at a higher temperature than a baseball sitting on the ground, however, even though the kinetic energy of the moving ball is larger than that of the stationary ball. Explain how these two ideas are consistent with each other.

73. N A blacksmith forging a 950.0-g knife blade drops the steel blade, initially at 700.0°C, into a water trough containing 40.0 kg of water at 22.0°C. Assuming none of the water boils away or is lost from the trough and no energy is lost to the surrounding air, what is the final temperature of the water–knife blade system? Assume ($c_{steel} = 450$ J/kg · °C).

74. A In Figure 21.27, we saw that the gas initially has a large volume, low pressure, and low temperature and finally it has smaller volume, higher pressure, and higher temperature. Check our derivation of the adiabatic path by making sure these changes are correctly predicted by the equations $P_f V_f^\gamma = P_i V_i^\gamma =$ constant and $T_f/T_i = (V_f/V_i)^{(1-\gamma)}$.

75. A A container contains equal moles of two ideal gases A and B at temperature T. Gas A is monatomic, and gas B is diatomic. Find the average thermal energy per molecule at temperature T. Assume the diatomic molecule has five active degrees of freedom.

76. A vessel with a movable piston contains 1.00 mol of an ideal gas with initial pressure $P_i = 2.03 \times 10^5$ Pa, initial volume $V_i = 1.00 \times 10^{-2}$ m³, and initial temperature $T_i = 305$ K.

a. N What is the work done on the gas during a constant-pressure compression, after which the final volume of the gas is 3.00 L?

b. N What is the work done on the gas during an isothermal compression, after which the final pressure of the gas is 5.00 atm?

c. N What is the work done on the gas during a constant-volume process, after which the final pressure of the gas is 5.00 atm?

d. G Sketch each of the processes in parts (a) through (c) on a *PV* diagram.

77. N The temperature of a 10.0-kg block of lead is increased from 18.0°C to 55.0°C at a constant pressure of 1.00 atm. The coefficient of linear expansion of lead is 28.0°C⁻¹, its density is 11.3×10^3 kg/m³, and its specific heat is 128 J/(kg · °C). **a.** What is the work done on the lead block during this process? **b.** How much energy is added to the block by heat during this process? **c.** What is the change in the internal energy of the block during this process?

78. C All main-sequence ("living") stars convert hydrogen nuclei into helium nuclei in their hot cores. When the supply of hydrogen in the core is used up, the star dies. In high-mass stars, the energy produced in the core makes its way out primarily through radiation (Fig. P21.78A). In low-mass stars, however, energy is transferred mainly through convection (Fig. P21.78B). Low-mass stars live longer than high-mass stars. The outer layers in both types of stars are primarily hydrogen. Explain how convection helps extend the lifetime of low-mass stars.

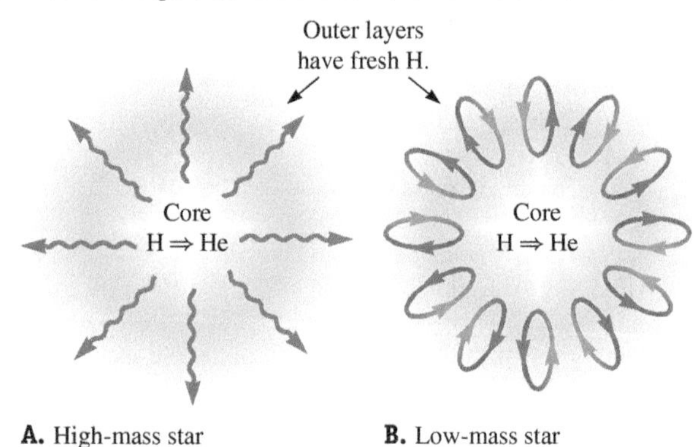

FIGURE P21.78

79. N How much faster does a cup of tea cool by 1°C when at 373 K than when at 303 K? Consider the tea to be a blackbody (an ideal object whose emissivity is 1) and assume the room temperature is 293 K.

80. N The *PV* diagram in Figure P21.80 shows a set of thermodynamic processes that make up a cycle *ABCDA* for a monatomic gas, where *AB* is an isothermal expansion occurring at a temperature of 375 K. There are 2.00 mol of gas undergoing the cycle with $P_A = 1.01 \times 10^6$ Pa, $P_B = 5.05 \times 10^5$ Pa, and $P_C = 2.02 \times 10^5$ Pa. **a.** Find the volumes V_A and V_B. **b.** Find the work done in each process of the cycle and the total work done for the whole cycle. **c.** Find the change in thermal energy during the constant-volume process *BC*.

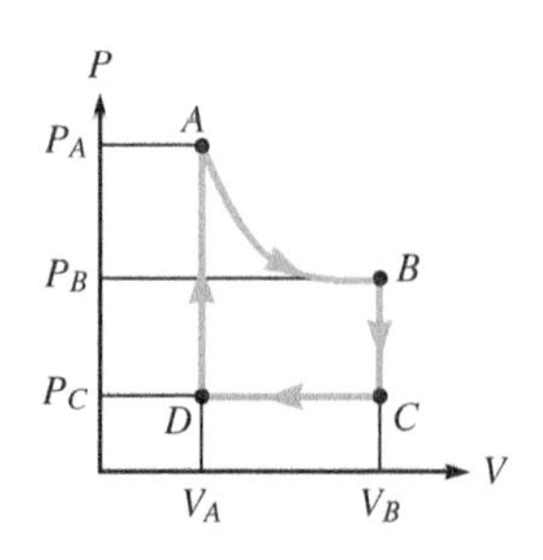

FIGURE P21.80

81. N An insulated canister contains 4.00×10^{-2} m³ of liquid argon at 87.3 K, which is the boiling point for argon. A small, 7.00-W heat lamp within the canister is switched on. If the latent heat of vaporization for argon is 1.61×10^5 J/kg, what is the mass of the argon that evaporates in the first 1.00 h?

82. A In an adiabatic process, the relationship between pressure and volume is $PV^\gamma =$ constant, where γ is the adiabatic exponent that depends on the properties of the gas. Two different adiabatic processes *AD* and *BC* intersect two isothermal processes *AB* at temperature T_1 and *DC* at temperature T_2 as shown in Figure P21.82. Is the ratio V_A/V_D greater than, less than, or equal to the ratio V_B/V_C?

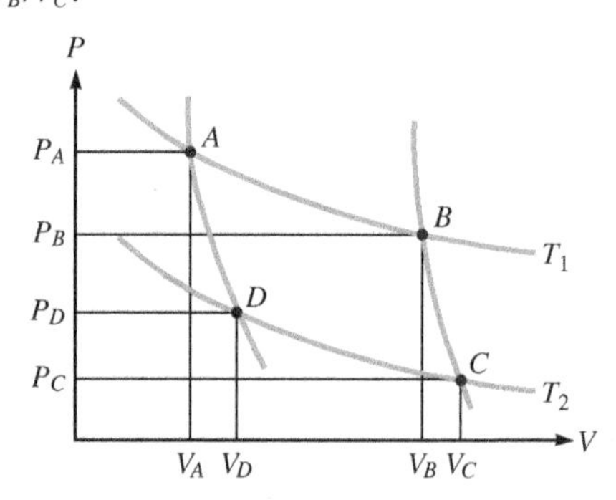

FIGURE P21.82

83. N Between 1793 and 1837, the U.S. penny was made entirely of copper and had a mass of 3.20 g. A coin collector accidentally drops one such penny from the balcony of his high-rise apartment, 38.0 m above the street. If 40.0% of the change in gravitational energy of the penny–Earth system is converted into internal energy of the penny, what is the increase in the temperature of the coin just before it impacts with the sidewalk?

84. N In the 1970s cartoon *The Super Friends*, the Wonder Twins helped Superman and others fight crime. One of the Wonder Twins was Zan, who was able to take on the form of H_2O in its various phases (ice, liquid water, water vapor). After taking a physics class, he decides to conduct a little experiment by turning into an ice igloo in the park on a hot day. **a.** If his total mass as an igloo is 65.88 kg and he starts with a temperature of −10.00°C, how much heat must flow into him to completely melt his body? **b.** He continues absorbing energy up to the point where he would start to vaporize (100.0°C). How much more heat is required to raise his temperature to the vaporization point? **c.** How much more heat is required to turn him into water vapor at 100.0°C? **d.** If we model the water vapor molecules making up his body as having only six degrees of freedom, what is the average kinetic energy of one of the water vapor molecules that make up Zan?

85. N On a cold winter day when the outside temperature is −5.00°C, the interior of a house is heated to a comfortable 22.0°C. The dining room window of the house is 1.50 m tall by 2.50 m long, and the glass windowpane has a thickness of 0.450 cm. **a.** What is the rate of energy transfer through the windowpane? **b.** If the interior and exterior temperatures remain constant for 6.00 h, how much energy is transferred through the windowpane during this time period?

Entropy and the Second Law of Thermodynamics

22

Underlying Principles

Second law of thermodynamics

Major Concepts

1. Efficiency
2. Reversible and irreversible processes
3. Entropy

Special Cases: Three Specific Engines

1. Carnot
2. Otto and internal combustion
3. Refrigerator

Tools

Energy transfer diagram

Key Questions

Why are some processes that are allowed by the first law of thermodynamics never seen in nature?

Why is it impossible to build a 100% efficient engine?

22-1 Second law of thermodynamics, Clausius statement 652

22-2 Heat engines 652

22-3 Second law of thermodynamics, Kelvin-Planck statement 656

22-4 The most efficient engine 656

22-5 Case study: Refrigerators 664

22-6 Entropy 665

22-7 Second law of thermodynamics, general statements 670

22-8 Order and disorder 672

22-9 Entropy, probability, and the second law 673

If you want to make a group of kids laugh, play a movie backward. For example, imagine a scene in which someone fries an egg. In the backward version, a completely cooked egg floats up from the plate into a tilted frying pan, the pan is set on the stove, and heat is transferred away from the egg as it becomes *un*cooked. When the egg is completely raw, it flies straight upward into a cracked shell, and finally the parts of the shell come together. Kids laugh at this movie because they can see that something is wrong—eggs don't spontaneously fly up into cracked shells—but why not?

Why can't an egg use the heat transferred from the stove to gain macroscopic kinetic energy and then turn that kinetic energy into potential energy by flying upward? There is nothing in the work–energy theorem or the first law of thermodynamics that would prevent such a conversion of energy. In this chapter, we introduce the second law of thermodynamics, which would be violated if an egg used heat from the stove to launch itself upward. The first law of thermodynamics is an extension and a special case of the work–energy theorem, but the second law of thermodynamics is a completely separate and independent principle, involving a new concept: entropy.

22-1 Second Law of Thermodynamics, Clausius Statement

Imagine making a potato salad. After cooking the potatoes, you place them in a very large sink full of cool water. The water is cooler than the potatoes:

$$T_{\text{water}} < T_{\text{potato}}$$

Assume the potatoes are small compared to the amount of water, so there are many more water molecules than potato molecules:

$$N_{\text{water}} \gg N_{\text{potato}}$$

Thermal energy E_{th} is proportional to the number of molecules and the temperature (Eq. 21.2), and because there are many more water molecules than potato molecules, there is more thermal energy in the cool water than in the hot potatoes:

$$N_{\text{water}}T_{\text{water}} > N_{\text{potato}}T_{\text{potato}}$$
$$E_{\text{th, water}} > E_{\text{th, potato}}$$

It might seem reasonable for heat to flow from the water to the potatoes because there is more thermal energy in the water. If that happened, the potatoes would get hotter by sitting in cool water. Experience, however, tells us that the hot potatoes cool off because heat flows from the potatoes to the cool water until they are at the same temperature.

Nothing about the first law of thermodynamics or the conservation of energy principle tells us which way heat will be transferred. Another law—the *second law of thermodynamics*—tells us which way heat flows. There are many versions of the second law; we start with the one about heat transfer. This statement is attributed to Rudolf Julius Emanuel Clausius (1822–1888), a German physicist and mathematician. The **Clausius statement** of the **second law of thermodynamics** is as follows:

SECOND LAW OF THERMODYNAMICS (CLAUSIUS STATEMENT)

Underlying Principle

Heat flows naturally from a hot object to a cooler object; heat does not naturally flow from a cold object to a hotter object.

CASE STUDY Part 1: Engines and Refrigerators

Why is the word *naturally* included in this statement of the second law of thermodynamics? *Naturally* is included because it is possible for heat to be transferred from a cooler system to a hotter environment if work is done on the system. For example, a refrigerator does work on the cool air inside it so as to transfer heat from that air to the hotter room. So, the word *naturally* is included in Clausius's wording to indicate that in the absence of work done, heat flows from a hot object to a cooler object.

The word *naturally* is also a reminder that the second law of thermodynamics was discovered because people were trying to use the principles of thermodynamics to design and build new machines, the most important of which is the *heat engine*. Examples of heat engines are the steam engines that were once used to power trains, internal combustion engines that make most of today's automobiles run, and many power plants that supply households with electricity.

You might hear the phrase *a heat engine turns heat into work*. Such a phrase is actually shorthand for the idea that energy enters the engine through heat and leaves the engine through work. It is okay to use this phrase as long as you keep in mind what it really means.

A **heat engine** is a device that absorbs energy from its environment through heat and then does work on its environment. Our case study throughout this chapter involves both heat engines and refrigerators. A refrigerator is the opposite of a heat engine; it takes in energy as work and gives off energy as heat.

22-2 Heat Engines

We have already studied a crude heat engine such as the simple device shown in Figure 21.15 and reproduced here in Figure 22.1. At first in part A, the piston is held in place, and the heat reservoir is warmer than the gas. As a result, heat Q_h is transferred to the gas, and its pressure rises. The piston is then freed in part B, and thermal energy is transferred from the gas to the piston as the gas does positive work W_{out} in lifting the piston. This device is a heat engine because the system took in energy as heat and released energy as work. You can imagine that this device may be useful if you wish to do something like lift an elephant.

We are only interested in engines that run continuously in a thermodynamic cycle, returning repeatedly to their initial state. Figure 22.2 is a *PV* diagram for a system that undergoes an adiabatic and constant-volume cycle. As before, we consider the gas to be the system and everything else (the heat reservoir, the piston, and the Earth) to be in the environment. The cylindrical walls and piston are insulated, so the system can only exchange heat with the heat reservoir under its base.

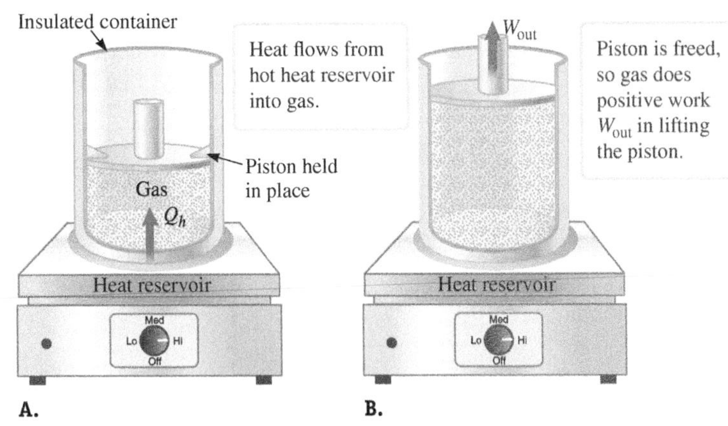

FIGURE 22.1 The basic principles of a heat engine are illustrated with this simple apparatus. **A.** The temperature of the heat reservoir may be adjusted by turning a knob. The piston may be held in place by wedges. The system is the gas inside the cylinder. Heat flows from the reservoir into the system. **B.** The system does work on the piston, raising it to a new height.

Figure 22.2 also introduces a new visualization tool known as an **energy transfer diagram**. The system (the gas) is represented as a rectangle in the middle of the diagram. A hot reservoir (with a temperature higher than that of the system) is always drawn above the rectangle, and a cold reservoir (with a temperature lower than that of the system) is always drawn below the rectangle. Thick arrows represent energy transferred into or out of the system as either heat or work. There is no need to draw a symbol for thermal energy because we only study heat engines that operate in a cycle, and in a cycle, the thermal energy returns to its original state.

Developers of practical heat engines were interested in the net work W_{eng} done by the engine, which is the difference between the energy transferred out of the system by work and the energy transferred in:

$$W_{eng} \equiv W_{out} - W_{in} \tag{22.1}$$

Let's find an expression for W_{eng} by applying the first law of thermodynamics (Eq. 21.1),

$$\Delta E_{th} = W_{tot} + Q$$

to one cycle of the heat engine.

ENERGY TRANSFER DIAGRAM

Tool

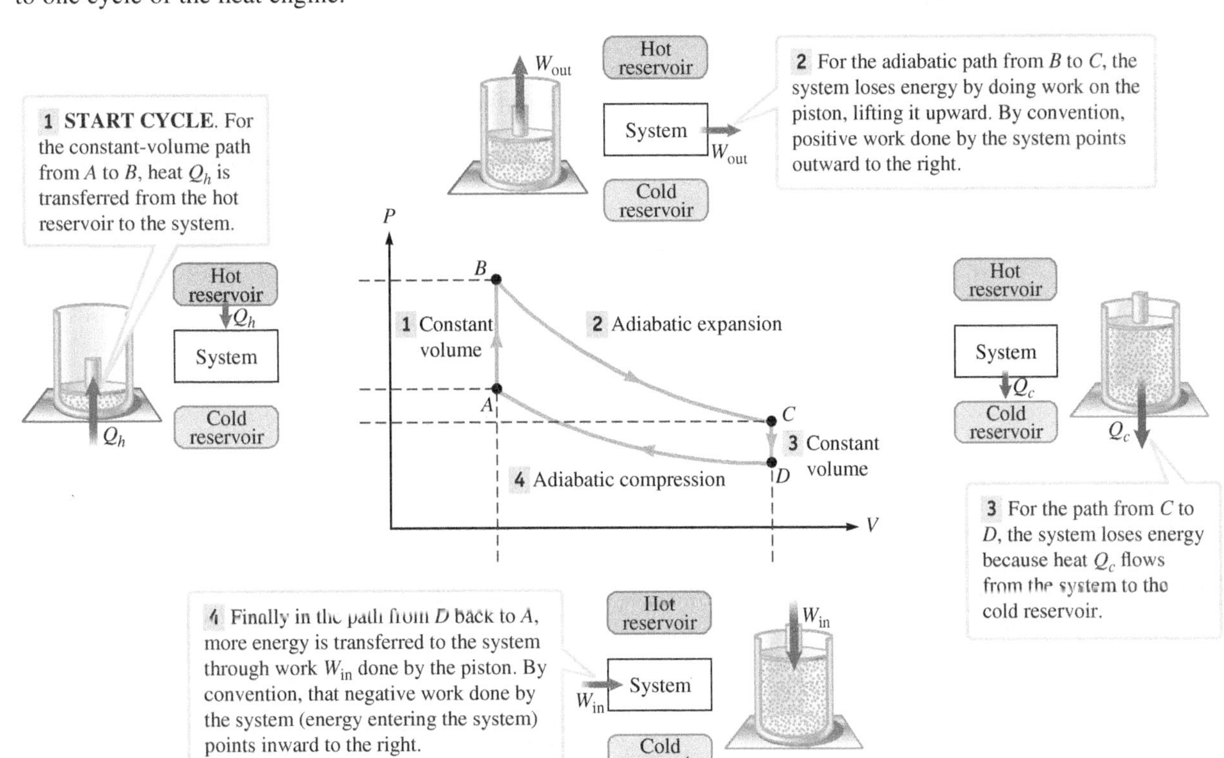

FIGURE 22.2 The system undergoes a cyclic thermodynamic process from state A to B to C to D and back to A. This figure shows the connection between a series of energy transfer diagrams and the *PV* diagram.

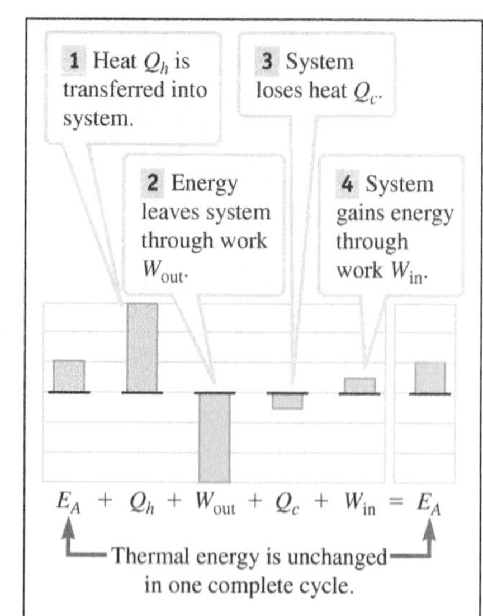

FIGURE 22.3 The bar chart for the thermodynamic cycle is made by considering the energy transferred during each of the four processes indicated in Figure 22.2. Notice that the thermal energy is unchanged.

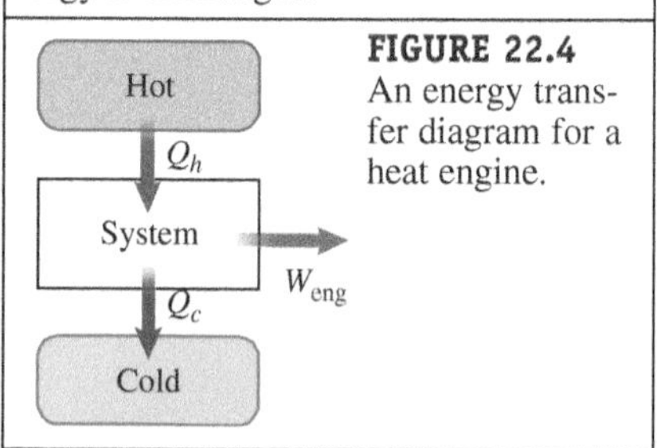

FIGURE 22.4 An energy transfer diagram for a heat engine.

Whenever we apply the first law, a bar chart is helpful. Figure 22.3 is the bar chart that results from the four-part cycle in Figure 22.2. Using this bar chart and the convention that energy entering the system either by work or by heat is positive, we have

$$\Delta E_{th} = E_A - E_A = \underset{\text{(process 1)}}{(0 + Q_h)} + \underset{\text{(process 2)}}{(-W_{out} + 0)} + \underset{\text{(process 3)}}{(0 - Q_c)} + \underset{\text{(process 4)}}{(0 + W_{in})}$$

The thermal energy is unchanged at the end of the cycle. So, the net result of one cycle is $(Q_h - Q_c) - (W_{out} - W_{in}) = 0$, and

$$(Q_h - Q_c) = (W_{out} - W_{in}) \tag{22.2}$$

By substituting $W_{eng} = W_{out} - W_{in}$ (Eq. 22.1) into Equation 22.2, we find that the work done by the engine W_{eng} on its environment equals the net heat that enters the engine:

$$W_{eng} = (Q_h - |Q_c|) \tag{22.3}$$

Equation 22.3 may be represented as an energy transfer diagram as shown in Figure 22.4. The negative sign in Equation 22.3 explicitly takes care of the heat Q_c being transferred out of the system. To ensure that we don't substitute a negative number into Q_c, we use an absolute value sign.

CONCEPT EXERCISE 22.1

Imagine a perfect engine in which all the heat that enters is completely converted to work. Estimate the heat required to lift an elephant from the ground into a truck using the heat engine in Figure 22.1. Would a real engine require more heat or less heat to do the job?

CASE STUDY Part 2: Steam Engines and Internal Combustion Engines

The fundamental ideas behind the steam engine are ancient, dating to around 200 BCE in Greece, but the steam engine didn't change the world until the Industrial Revolution in the 18th and 19th centuries (Fig. 22.5A). Today we power cities with giant steam engines located inside power plants (Fig. 22.5B).

Figure 22.6 is a schematic diagram of a steam engine. Heat is transferred from a hot reservoir to water in the boiler shown on the left. The hot reservoir needs some sort of fuel, such as a fossil fuel like coal or gas, or even a nuclear fuel like uranium. The system that is heated and cooled in an engine is called the **working substance**; in a steam engine, the working substance is water in its liquid and vapor form. In a typical steam engine, the water is kept at very high pressure so that it vaporizes at a very high temperature. High-temperature steam passes through the intake valve as shown in middle of Figure 22.6 and is then in contact with a piston, doing work on it. The piston may be connected to a wheel known as a crank. The up-and-down motion of the piston rotates the crank, which may be attached to any number of devices such as the wheels of a train. In Figure 22.6, the intake valve is open and the exhaust valve is closed, as is the case whenever the steam is expanding and the piston is moving downward. When the steam is being compressed and the piston is moving upward, the intake valve is closed and the exhaust valve is open. The

Courtesy Muskegon Heritage Museum

A.

JoLin/Shutterstock.com

B.

FIGURE 22.5 A. Steam engines such as the one shown changed the world. **B.** This nuclear power plant is located in Bay City, Texas.

exhausted steam moves through the condenser, which is in thermal contact with a cold reservoir. Heat is transferred from the hot steam to the cold reservoir, and the steam condenses back into liquid water. That water is then recycled into the boiler, and the whole process continues.

The cold reservoir may be a river or another natural body of water. Heat transferred from the hot steam to the body of water is called **thermal pollution**. Natural bodies of water are becoming warmer, which is a problem for life in the water. Unfortunately, the second law of thermodynamics tells us that all engines must give off some heat (Section 22-3).

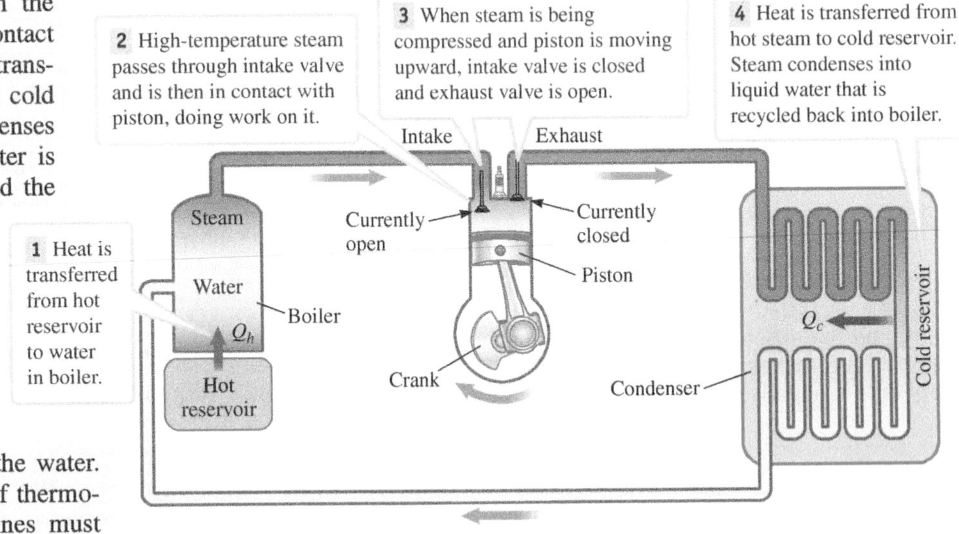

FIGURE 22.6 Components of a simplified steam engine.

Steam engines used to provide electricity are similar to the one shown in Figure 22.6, but with the piston replaced by a rotating turbine as shown in Figure 22.7. High-pressure steam from the boiler pushes the paddle wheel of the turbine around, and low-pressure steam is exhausted through a pipe on the other side of the turbine. We return to electricity generation in Chapter 32 when we describe how the motion of the turbine is used to produce an electric current.

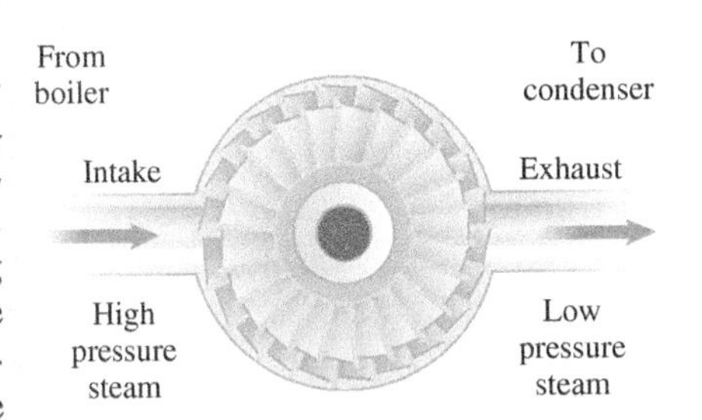

FIGURE 22.7 A steam engine may be used to rotate a turbine.

The engine in most cars is an **internal combustion engine** in which the working substance is a mixture of air and gasoline vapor. The main difference between the steam engine (Fig. 22.6) and the internal combustion engine is that the high temperature in an internal combustion engine is achieved by igniting the air–gasoline mixture *in* the cylinder instead of maintaining a hot reservoir *outside* the engine. The internal combustion engine is described as having four *strokes* or movements as shown in Figure 22.8.

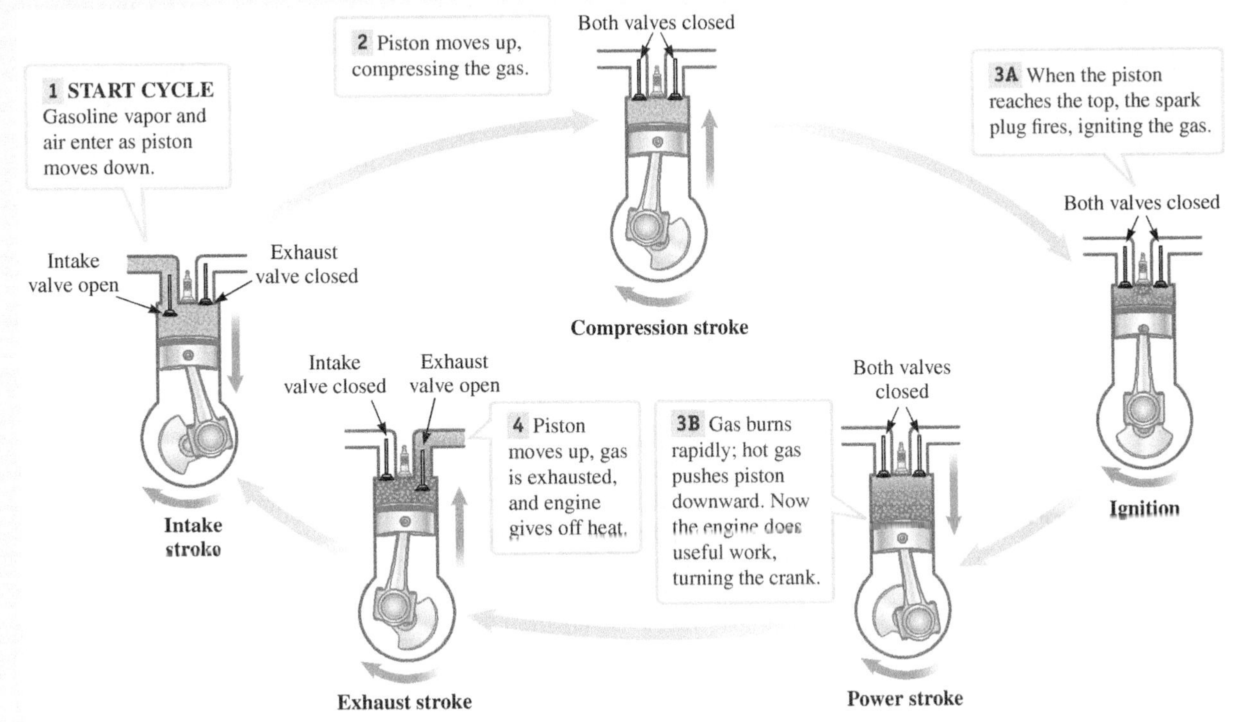

FIGURE 22.8 The processes of a four-stroke internal combustion engine such as a car engine. (For each cycle, the piston moves up twice and down twice for a total of four *strokes*.)

22-3 Second Law of Thermodynamics, Kelvin-Planck Statement

The purpose of a heat engine is to do useful work by taking in a net amount of heat. The **efficiency** e of an engine is defined as

EFFICIENCY Major Concept

$$e \equiv \frac{W_{eng}}{Q_h} = \frac{\text{what you get from the engine}}{\text{what you put into the engine}} \tag{22.4}$$

By substituting Equation 22.3, $W_{eng} = (Q_h - |Q_c|)$, into Equation 22.4, we can rewrite efficiency as

The negative sign takes into account that Q_c is energy that *leaves* the system, and the absolute value in Equation 22.5 makes sure we don't eliminate this negative sign.

$$e = \frac{(Q_h - |Q_c|)}{Q_h} = 1 - \frac{|Q_c|}{Q_h} \tag{22.5}$$

where Q_c is the heat expelled by the engine. It is waste, so the smaller $|Q_c|$, the more efficient the engine is.

A perfect engine would expel no heat. If $|Q_c|$ were zero, then e would equal 1, and we would have a 100% efficient engine. A 100% efficient engine cannot be built, though, because it would violate the second law of thermodynamics. In fact, another way to state the second law of thermodynamics is as follows:

SECOND LAW OF THERMODYNAMICS (KELVIN-PLANCK STATEMENT) Underlying Principle

There can be no perfect (100% efficient) heat engine.

The **Kelvin-Planck statement** of the second law of thermodynamics says the same thing:

It is impossible to construct a cyclic heat engine that takes in heat and produces only work without also giving off waste heat.

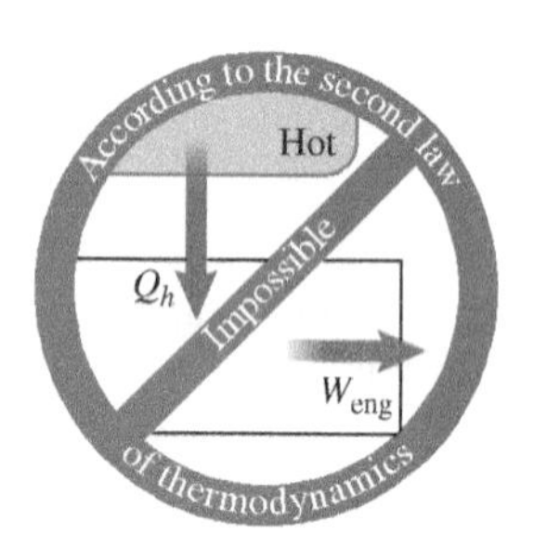

FIGURE 22.9 According to the second law of thermodynamics, all heat engines must expel heat. Compare with Figure 22.4, where waste heat and a cold reservoir must be present.

Figure 22.9 illustrates the Kelvin-Planck statement of the second law of thermodynamics. Although no perfect engine can be built, it is possible to maximize an engine's efficiency, as discussed in the next section.

22-4 The Most Efficient Engine

Sadi Carnot (1796–1832), a French physicist, was interested in maximizing the efficiency of heat engines. In 1824, he published *Reflections on the Motive Power of Fire* in which he described an ideal engine that has maximum efficiency. According to Carnot, the most efficient engine is *reversible*, meaning that the working substance returns to its original state after each cycle. Today we call this ideal engine the **Carnot engine**. No real engine can be made reversible, so no real engine is a Carnot engine. Before looking at the Carnot engine, we must first consider reversible and irreversible processes.

Reversible and Irreversible Processes

REVERSIBLE AND IRREVERSIBLE PROCESSES Major Concept

A **reversible process** is an *ideal* thermodynamic process in which a system that goes from state i to state f can go back again to state i, exchanging the same amount of heat and work so that the system and the environment have been returned to their original conditions. Other processes are known as **irreversible processes**.

Three general conditions determine whether a process is reversible or irreversible. First, according to the second law of thermodynamics (Clausius statement), heat flows from hotter objects to cooler objects but not the other way around. So, if energy is spontaneously transferred (that is, with no work required) through heat from a hot object to a cooler object, the process is irreversible. It would take work to transfer the energy back from the cold object to the hot one. Second, when a block slides down an inclined plane, the block and the plane warm up due to friction between the surfaces, so mechanical energy is transformed into thermal energy. You can never use friction to transform thermal energy into mechanical energy, however, so processes that involve friction are irreversible. Third, if the system changes states too quickly (as in an explosion), the process is irreversible. Put more generally, a process is reversible if the system can pass back through a series of equilibrium states in reverse order. (Explosions don't happen in reverse.) A reversible process is an ideal process because it must be carried out infinitely slowly so that the process can be a considered a series of equilibrium states taking the system from state i to state f. All real processes are irreversible, but a real process can be approximated as a reversible process if the following conditions hold:

1. The system's temperature is close to that of the environment. The environment's temperature may be changed, but this change must be slow.
2. Friction and other dissipative forces are very small so that little mechanical energy is transformed into thermal energy.
3. The process must be done very slowly—quasistatically—so that the system is always in (or very close to) an equilibrium state.

FIGURE 22.10

CONCEPT EXERCISE 22.2

You suspend a pot of water above a campfire (Fig. 22.10). Assume the pot is tightly sealed. In the evening, water boils and becomes a vapor trapped in the pot, and then the fire dies out. During the night, the water condenses back into liquid.

a. Has the water undergone an approximately reversible cycle?
b. Has the wood undergone an approximately reversible cycle?

Carnot Engine

CARNOT ENGINE ▶ Special Case

Let's imagine a Carnot engine made out of an ideal apparatus as in Figure 22.1. For this ideal apparatus, there is no friction between the piston and the walls of the container, the system temperature is very close to that of the heat reservoir, and all changes are made slowly. Furthermore, assume the working substance is an ideal gas.

The working substance in a Carnot engine goes through four reversible processes in each cycle. Any thermodynamic cycle that is made up of reversible processes is known as a **reversible cycle**. For the Carnot engine, the four reversible processes are an isothermal ($\Delta T = 0$) expansion, an adiabatic ($Q = 0$) expansion, an isothermal compression, and an adiabatic compression. Together, these four processes are known as the **Carnot cycle**. Figure 22.11 illustrates the Carnot cycle on a PV diagram.

FIGURE 22.11 The Carnot engine is an ideal engine; no real engine is as efficient as the Carnot engine. This engine undergoes a cyclic process made up of four parts. For each of these parts be sure you can identify the illustration, bar chart, and curve on the PV diagram.

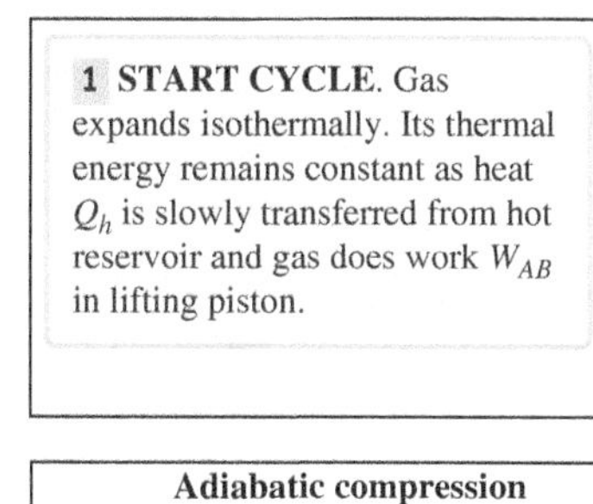

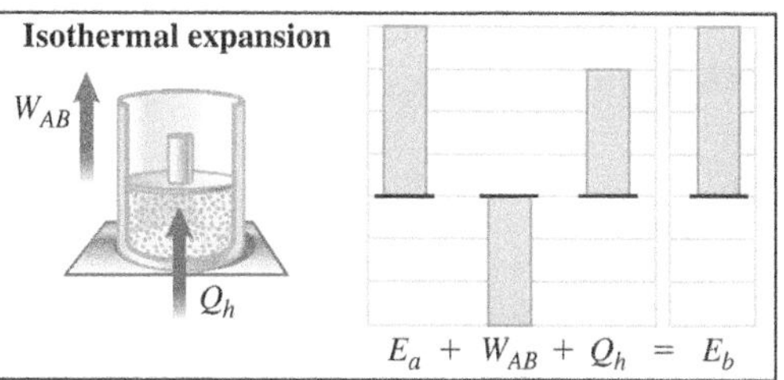

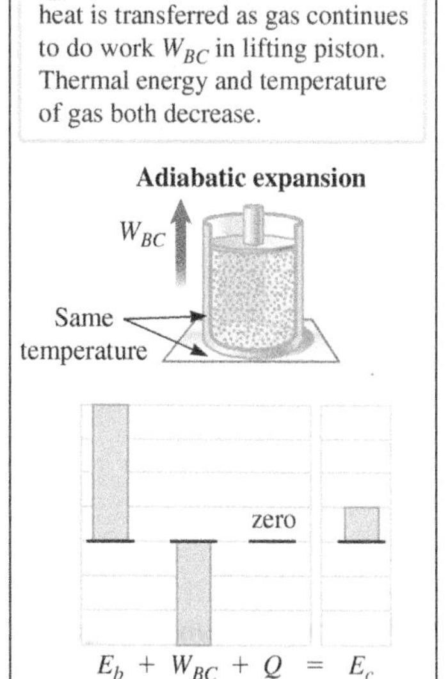

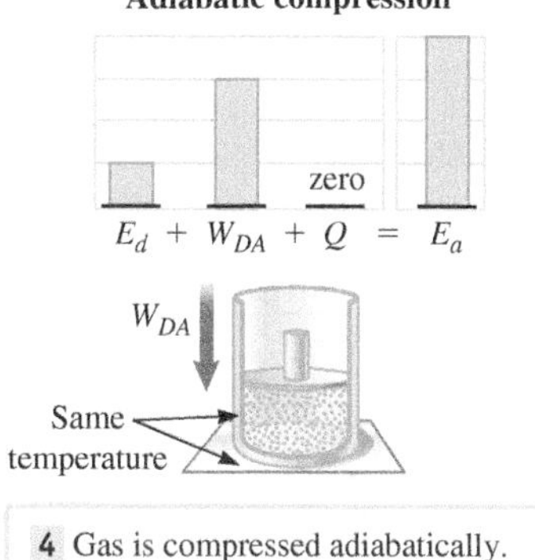

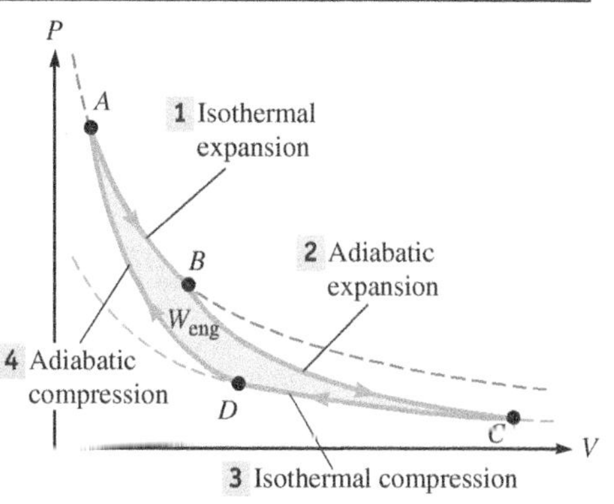

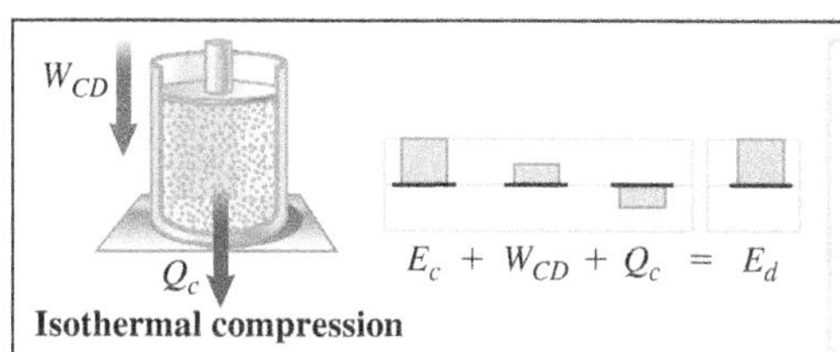

The total work done by the engine is the sum of the work done by the gas as the piston is raised in the isothermal and adiabatic expansions minus the work done by the piston as gas is compressed in the isothermal and adiabatic compressions:

$$W_{\text{eng}} = W_{AB} + W_{BC} - W_{CD} - W_{DA}$$

The work W_{eng} done by the engine can also be found from the area enclosed by the four paths in the *PV* diagram. This area is positive, so the total work done by the engine is positive, and the engine does useful work just as we would expect.

DERIVATION Carnot Efficiency

The Carnot engine is an ideal engine that has the best possible efficiency. We will derive **Carnot's theorem**, which states that the efficiency of this ideal, reversible engine is

$$e_{\text{Carnot}} = 1 - \frac{T_c}{T_h} \qquad (22.6)$$

We can find an expression for its efficiency using either $e = W_{\text{eng}}/Q_h$ (Eq. 22.4) or $e = 1 - |Q_c|/Q_h$ (Eq. 22.5). Our job is somewhat simpler if we use Equation 22.5. So, we need to find the heat Q_h absorbed by the engine during the isothermal expansion and the heat Q_c lost during the isothermal compression.

From the bar charts for the two isothermal processes in Figure 22.11, the work done by the gas (negative of the work done by the environment) equals the heat absorbed by the gas.	$W_{AB} = Q_h$ $W_{CD} = Q_c$
Equation 21.15 gives the work done *by the piston* (environment) on an ideal gas that expands isothermally. To find the work done *by the gas*, multiply by -1. (The "-1" flips the numerator and denominator in the natural log.)	$W_{\text{by piston}} = Nk_BT\ln\frac{V_i}{V_f}$ (21.15) $W_{\text{eng}} = -Nk_BT\ln\frac{V_i}{V_f}$ $W_{\text{eng}} = Nk_BT\ln\frac{V_f}{V_i}$ (22.7)
Use Equation 22.7 to write an expression for Q_h. The temperature of the hot reservoir is T_h.	$Q_h = W_{AB} = Nk_BT_h\ln\frac{V_B}{V_A}$
Now find $\|Q_c\|$, where the temperature of the cold reservoir is T_c. Equation 22.7 is positive when the gas expands; when the gas contracts, we need to multiply by -1 to find the absolute value of the heat transferred.	$\|Q_c\| = -W_{CD} = -Nk_BT_c\ln\frac{V_D}{V_C}$ $\|Q_c\| = Nk_BT_c\ln\frac{V_C}{V_D}$
Find the ratio of the heat lost ($\|Q_c\|$) to the heat gained (Q_h) by dividing these two expressions.	$\frac{\|Q_c\|}{Q_h} = \frac{T_c\ln(V_C/V_D)}{T_h\ln(V_B/V_A)}$ (22.8)
Adiabatic processes connect states *B* and *C* and states *D* and *A*. Use Equation 21.32 (page 638) for the adiabatic expansion or compression of an ideal gas to find the ratio of the volumes at each set of endpoints.	$\frac{T_f}{T_i} = \left(\frac{V_f}{V_i}\right)^{(1-\gamma)}$ (21.32)
For the adiabatic expansion from a small V_B to a larger V_C, the initial gas temperature is T_h, and the final temperature is T_c.	$\frac{T_c}{T_h} = \left(\frac{V_C}{V_B}\right)^{(1-\gamma)}$ (22.9)
For the adiabatic compression from a large V_D to a smaller V_A, the initial gas temperature is T_c, and the final temperature is T_h.	$\frac{T_h}{T_c} = \left(\frac{V_A}{V_D}\right)^{(1-\gamma)}$ (22.10)

By substituting Equation 22.10 into Equation 22.9, we find that the ratio of volumes are equal.	$\frac{T_c}{T_h} = \left(\frac{V_C}{V_B}\right)^{(1-\gamma)} = \left(\frac{V_D}{V_A}\right)^{(1-\gamma)}$ $\frac{V_C}{V_B} = \frac{V_D}{V_A}$ and $\frac{V_C}{V_D} = \frac{V_B}{V_A}$
The terms involving the natural logarithms cancel out of Equation 22.8.	$\frac{\lvert Q_c \rvert}{Q_h} = \frac{T_c \ln(V_C/V_D)}{T_h \ln(V_C/V_D)}$ $\frac{\lvert Q_c \rvert}{Q_h} = \frac{T_c}{T_h}$ (22.11)
Finally, substitute this result into Equation 22.5 to find the efficiency of a Carnot engine in terms of the temperatures of the hot and cold reservoirs.	$e = 1 - \frac{\lvert Q_c \rvert}{Q_h}$ (22.5) $e_{\text{Carnot}} = 1 - \frac{T_c}{T_h}$ ✓ (22.6)

COMMENTS

A Carnot engine is an ideal engine. All real, irreversible engines are less efficient than a Carnot engine. In practice, a well-designed irreversible engine may reach about 80% of the Carnot efficiency.

Third Law of Thermodynamics

The temperatures in Equation 22.6 must be expressed on the Kelvin scale. The efficiency of the Carnot engine only depends on the temperature of the two reservoirs. If the temperature of the cold reservoir is lowered or if the temperature of the hot reservoir is raised, the engine becomes more efficient. You might imagine that a perfect (100% efficient) engine could be made if the hot reservoir were infinitely hot, but it would take an infinite amount of energy to achieve an "infinite" temperature. Having dismissed the idea of an infinitely hot reservoir, you might imagine that a perfect engine could be made if the cold reservoir were at absolute zero ($T_c = 0$). Experiments show, however, that the closer the temperature gets to absolute zero, the harder it is to reduce the temperature further. In fact, the **third law of thermodynamics** says that it is not possible to lower the temperature to zero by any finite process. The third law of thermodynamics means that there are no perfect engines.

EXAMPLE 22.1 Lifting an Elephant

You wish to use a steam engine to lift an elephant into a truck as in Concept Exercise 22.1. The steam engine operates using a hot reservoir at $T_h = 250°\text{C}$, and the cold reservoir is the surrounding air at $T_c = 32°\text{C}$.

A What is the maximum efficiency of the steam engine?

INTERPRET and ANTICIPATE

The maximum efficiency is the efficiency of a Carnot engine. No real engine can operate more efficiently than this ideal engine.

SOLVE Substitute the given temperatures into Equation 22.6 after converting them to the Kelvin scale.	$e_{\text{Carnot}} = 1 - \frac{T_c}{T_h} = 1 - \frac{305\text{ K}}{523\text{ K}} = 0.42 = 42\%$

CHECK and THINK

Even an ideal engine is only 42% efficient at these operating temperatures, meaning that only 42% of the heat absorbed from the hot reservoir goes into useful work. More than half the energy absorbed is simply expelled as heat into the cold reservoir.

Example continues on page 660 ▶

B Estimate the heat required to lift a single elephant into a truck using this engine. *Hint*: See Concept Exercise 22.1.

INTEPRET and ANTICIPATE

In Concept Exercise 22.1, we estimated the amount of work (as heat) that must be done by an engine to lift the elephant into the truck, assuming a 100% efficient engine. (All the heat absorbed by such an engine goes into work done.) No engine operates at 100% efficiency. Even an ideal Carnot engine is not 100% efficient. Use the efficiency of the Carnot engine from part A to estimate the minimum heat required.

SOLVE

The work required to lift the elephant is about 6×10^4 J from Concept Exercise 22.1. We found the efficiency of the engine in part A. Use Equation 22.4 to find the heat absorbed, Q_h.

$$e = \frac{W_{eng}}{Q_h} \quad (22.4)$$

$$Q_h = \frac{W_{eng}}{e} = \frac{6 \times 10^4 \text{ J}}{0.42} \approx 1 \times 10^5 \text{ J}$$

CHECK and THINK

Of course, a real engine would require an even greater input of heat. While this isn't a lot of energy (about the same as a few grapes), it may be better to just guide the elephant up a ramp.

CASE STUDY Part 3: Otto Cycle

The internal combustion engine in your car can be modeled by the **Otto cycle** (Fig. 22.12). The Otto cycle consists of two constant-volume processes and two adiabatic processes. The Otto cycle is ideal and reversible. A real car engine (Fig. 22.8) can only be approximated as an Otto cycle because a real car engine cycle is *ir*reversible.

Compare the *PV* diagrams of the Carnot cycle (Fig. 22.11) and the Otto cycle (Fig. 22.12). In the Carnot cycle, heat is transferred into or out of the engine isothermally, but when heat is transferred during the Otto cycle, the temperature is changing. Even the ideal Otto cycle is not as efficient as the Carnot cycle.

FIGURE 22.12 The Otto engine is an ideal engine. The engine "stroke" positions have been moved relative to their positions in Figure 22.8 to help you identify where each stroke occurs on the *PV* diagram.

OTTO ENGINE

Special Case

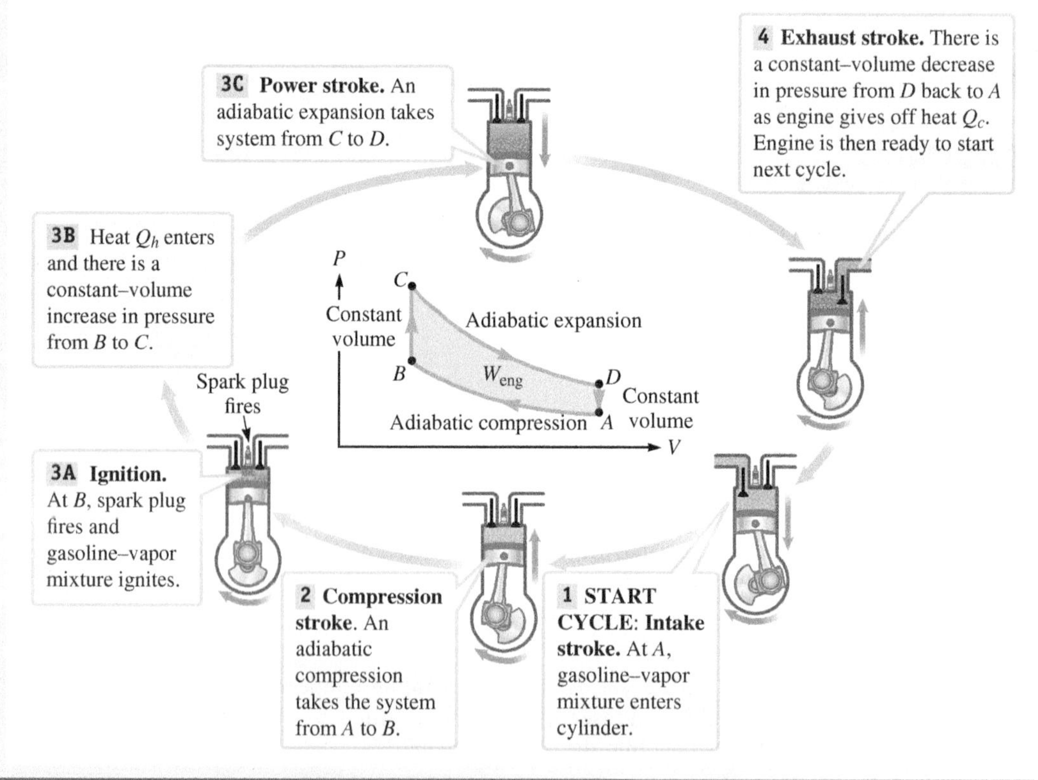

EXAMPLE 22.2 CASE STUDY Otto Engine Efficiency

A Find an expression for the efficiency of an Otto engine in terms of the maximum volume V_{max} and minimum volume V_{min} of the system.

INTERPRET and ANTICIPATE

Follow a procedure similar to our derivation of the efficiency of a Carnot engine (Eq. 22.6). The main difference is that in the Otto engine, heat is transferred in a constant-volume process instead of in an isothermal process. From Figure 22.12, a great change in volume means a large area enclosed on the *PV* diagram. So, we expect that the greater the change in volume, the more efficient an Otto engine will be.

SOLVE

Heat is transferred in a constant-volume process from *B* to *C*, so we can use Equation 21.20 to write an expression for the heat transferred in terms of the molar specific heat at constant volume, C_V. Because $T_C > T_B$, heat Q_h is positive, meaning that energy enters the engine.

$$Q_h = nC_V\,\Delta T \qquad (21.20)$$

$$Q_h = nC_V(T_C - T_B)$$

Heat Q_c is transferred out of the engine during the constant-volume process from *D* to *A*. Because $T_D > T_A$, heat Q_c is negative, meaning that energy leaves the engine. Equation 22.5 requires the absolute value of Q_c.

$$Q_c = nC_V\,\Delta T = nC_V(T_A - T_D)$$

$$|Q_c| = -nC_V(T_A - T_D) = nC_V(T_D - T_A)$$

To find the efficiency, substitute Q_h and Q_c into Equation 22.5.

$$e = 1 - \frac{|Q_c|}{Q_h} \qquad (22.5)$$

$$e_{\text{Otto}} = 1 - \frac{nC_V(T_D - T_A)}{nC_V(T_C - T_B)} = 1 - \frac{(T_D - T_A)}{(T_C - T_B)} \qquad (1)$$

Adiabatic processes connect states *C* and *D* and states *A* and *B*. Use Equation 21.32 for the adiabatic expansion or compression of an ideal gas to eliminate *T* in favor of *V* at each set of endpoints.

$$\frac{T_f}{T_i} = \left(\frac{V_f}{V_i}\right)^{1-\gamma} \qquad (21.32)$$

$$\frac{T_D}{T_C} = \left(\frac{V_D}{V_C}\right)^{1-\gamma} \qquad \frac{T_B}{T_A} = \left(\frac{V_B}{V_A}\right)^{1-\gamma}$$

Because the processes from *D* to *A* and from *B* to *C* are constant-volume processes, we can eliminate two of the four volumes to get expressions in terms of the maximum and minimum volumes.

$$V_A = V_D = V_{max}$$

$$V_B = V_C = V_{min}$$

$$\frac{T_B}{T_A} = \left(\frac{V_B}{V_A}\right)^{1-\gamma} = \left(\frac{V_{min}}{V_{max}}\right)^{1-\gamma} \qquad (22.12)$$

$$\frac{T_D}{T_C} = \left(\frac{V_D}{V_C}\right)^{1-\gamma} = \left(\frac{V_{max}}{V_{min}}\right)^{1-\gamma} \qquad (22.13)$$

Find T_D from Equation 22.13.

$$T_D = T_C\left(\frac{V_{max}}{V_{min}}\right)^{1-\gamma} \qquad (2)$$

Find T_A from Equation 22.12.

$$T_A = \frac{T_B}{(V_{min}/V_{max})^{1-\gamma}} = T_B\left(\frac{V_{max}}{V_{min}}\right)^{1-\gamma} \qquad (3)$$

Substitute Equations (2) and (3) into Equation (1).

$$e_{\text{Otto}} = 1 - \frac{T_C(V_{max}/V_{min})^{1-\gamma} - T_B(V_{max}/V_{min})^{1-\gamma}}{T_C - T_B}$$

$$e_{\text{Otto}} = 1 - \frac{(T_C - T_B)(V_{max}/V_{max})^{1-\gamma}}{T_C - T_B}$$

$$e_{\text{Otto}} = 1 - \left(\frac{V_{max}}{V_{min}}\right)^{1-\gamma} \qquad (22.14)$$

Example continues on page 662 ▶

CHECK and THINK

From Table 21.3, $\gamma > 1$, so the exponent in Equation 22.14 is negative. The greater the ratio V_{max}/V_{min}, the smaller the quantity $(V_{max}/V_{min})^{1-\gamma}$ and the greater the efficiency, as expected.

B Typically, $V_{max}/V_{min} \approx 8$ and $\gamma = 1.4$ for air. Find the typical efficiency of an Otto engine.

INTERPRET and ANTICIPATE

In this part of the example, we are given a chance to find the efficiency of an Otto engine numerically by substituting into Equation 22.14 from part A. We expect that the Otto engine is more efficient than an actual engine.

SOLVE Substitute values into Equation 22.14.	$e = 1 - \left(\frac{V_{max}}{V_{min}}\right)^{1-\gamma} = 1 - (8)^{1-1.4} = 1 - (8)^{-0.4}$ $e = 1 - 0.44 = 0.56$

CHECK and THINK

The efficiency of an Otto engine is about 56%. A typical car engine has an efficiency between 15% and 35%, so the Otto engine is more efficient than a real car engine, as expected. An Otto engine is an ideal engine whose model does not include friction; also, the gasoline vapor–air mixture is treated as an ideal gas that changes slowly through a series of equilibrium steps.

EXAMPLE 22.3 CASE STUDY Gas Mileage

When checking the specs on a car you are interested in buying, you might see that the car gets 32 miles per gallon on the highway at 65 mph and its engine is rated at about 200 hp (horsepower). In this example, we estimate the work done by the engine per second and compare our result to the car's specs. The heat of combustion for gasoline (the energy released per unit mass) is about $L_{com} = 44$ MJ/kg. The density of gasoline is about $\rho = 740$ kg/m^3.

A How much heat does the engine absorb per second on the highway? This result is the power input P_{in}. Give your answer in watts and horsepower.

INTERPRET and ANTICIPATE

Find the gasoline consumed in 1 hour of highway driving from the given specs. Then, by using the heat of combustion, find the heat absorbed by the engine in 1 hour. Dividing by the number of seconds in 1 hour will give the power input to the engine.

SOLVE If the car travels at 65 mph for 1 hour and its gas mileage is 32 mi/gal, find the volume of gasoline consumed in 1 hour. Convert that volume to SI units.	$V = 65\,\text{mi}\left(\frac{1\text{ gal}}{32\text{ mi}}\right) = 2.03\text{ gal}$ $V = 2.03\text{ gal} \times \frac{3.79\text{ L}}{1\text{ gal}} \times \frac{10^{-3}\text{ m}^3}{1\text{ L}}$ $V = 7.69 \times 10^{-3}\text{ m}^3$ in 1 h
The mass of gasoline consumed in 1 hour follows from its density.	$m = \rho V = (740\text{ kg/m}^3)(7.69 \times 10^{-3}\text{ m}^3)$ $m = 5.69$ kg in 1 h
The heat of combustion is similar to heat of transformation (Section 21-5), so we can find the heat absorbed in 1 hour as a result of combustion by multiplying the mass of gasoline consumed by the heat of combustion.	$Q_h = mL_{com} = (5.69\text{ kg})(44 \times 10^6\text{ J/kg})$ $Q_h = 2.5 \times 10^8$ J in 1 h

The power input is the heat absorbed per second, so divide Q_h by the number of seconds in 1 hour. To find power in horsepower, use the conversion factor 1 hp = 746 W.	$P_{in} = \frac{Q_h}{\Delta t} = \frac{2.5 \times 10^8 \text{ J}}{3600 \text{ s}} = 7.0 \times 10^4 \text{ W}$ $P_{in} = 93 \text{ hp}$

CHECK and THINK

This result may seem low. The car's specs say that the output power is more like 200 hp. Because the output power must be less than the input power, we might expect the input power to be several hundred or even a thousand horsepower. We will address this issue in parts B and C.

B If the engine has an efficiency of 30% (a very efficient real engine), what is the power output P_{out} of the engine? Give your answer in horsepower.

INTERPRET and ANTICIPATE

Equation 22.4 is written in terms of energy input Q_h and energy output W_{eng}. We must modify this equation so that we can work with power input and power output. We expect the output power to be less than the input power.

SOLVE Power is energy per unit time interval, so rewrite Equation 22.4 in terms of power.	$e = \frac{W_{eng}/\Delta t}{Q_h/\Delta t} = \frac{P_{out}}{P_{in}}$
Rearrange to express the power output as the efficiency multiplied by the power input.	$P_{out} = eP_{in} = (0.30)(93 \text{ hp})$ $P_{out} = 28 \text{ hp}$

CHECK and THINK

As expected, the output power is less than the input power. Our result is about one order of magnitude below the specs given by the car manufacturer because the manufacturer's reported number is not the power output during highway driving. Instead, the engine is tested in a laboratory off the road. The engine is loaded and run at different rates; the power output reported is the maximum output of the engine during the test. When you drive, you will probably never cause the engine to put out as much power.

C When the engine is being tested, its maximum power output is found to be 250 hp. At what rate is gasoline consumed by the engine if its efficiency is 20% at its maximum output? Give your answer in kilograms per second and gallons per hour.

INTERPRET and ANTICIPATE

Reverse the procedure from parts A and B to find the amount of gasoline consumed per hour. We expect a higher rate of gasoline consumption during the test than is actually consumed during highway driving.

SOLVE Reversing the process in part B, find the power input by dividing the power output by the efficiency of the engine.	$P_{in} = \frac{P_{out}}{e} = \frac{250 \text{ hp}}{0.20} = 1250 \text{ hp} \times \frac{746 \text{ W}}{1 \text{ hp}}$ $P_{in} = 9.33 \times 10^5 \text{ W}$
The power input is the heat absorbed mL_{com} per time interval Δt.	$P_{in} = \frac{mL_{com}}{\Delta t}$
Solve for the rate at which gasoline is consumed.	$\frac{m}{\Delta t} = \frac{P_{in}}{L_{com}} = \frac{9.33 \times 10^5 \text{ W}}{44 \times 10^6 \text{ J/kg}}$ $\frac{m}{\Delta t} = 2.1 \times 10^{-2} \text{ kg/s}$

Example continues on page 664 ▶

Use the density of gasoline to convert to gallons per hour.	$\frac{2.1 \times 10^{-2}\text{ kg}}{\text{s}} \times \frac{1\text{ m}^3}{740\text{ kg}} \times \frac{1\text{ L}}{10^{-3}\text{ m}^3} \times \frac{1\text{ gal}}{3.79\text{ L}} \times \frac{3600\text{ s}}{1\text{ hr}}$ $= 27\text{ gal/hr}$

CHECK and THINK

As expected, the rate of gasoline consumption during the test is greater—by about 13 times—than the rate during normal highway driving. At this rate of consumption, most cars would be out of gas in just about half an hour.

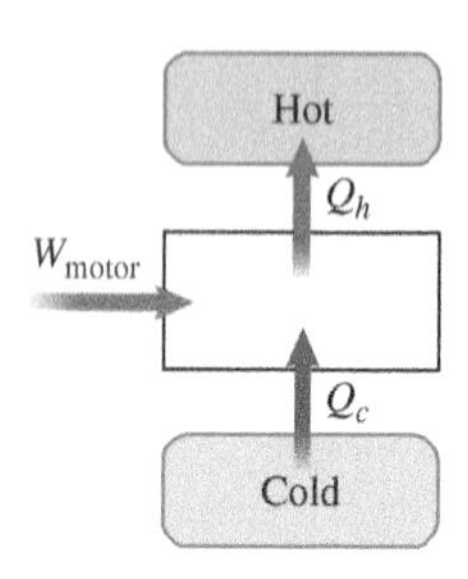

FIGURE 22.13 The energy transfer diagram for a refrigerator looks like that for a heat engine running in reverse. Compare with Figure 22.4.

22-5 Case Study: Refrigerators

A refrigerator absorbs heat from the cold food compartment and gives off heat to the warm room. It might seem like this process is a violation of the second law of thermodynamics, which says that heat flows naturally from a hot object to a colder object, not from a cold object to a hotter one (Clausius statement, Section 22-1). The key to the Clausius statement is the word *naturally* because heat can flow from a cold object to a hotter object if work is involved. A refrigerator does not violate the second law of thermodynamics. As shown in Figure 22.13, a refrigerator is like a heat engine running in reverse: The refrigerator motor does work so that heat is absorbed from the low-temperature compartment inside the fridge and expelled into the warm room. You may have felt a rush of warm air coming from behind or beneath your refrigerator.

Figure 22.14A identifies the major components of a refrigerator. A fluid known as a refrigerant is the working substance. As the refrigerant flows through a tube, it undergoes transitions between its liquid and vapor phases. The boiling point of the refrigerant is lower than the temperature you would like inside the freezer. For example, ammonia boils at −27°F, so it may be used as a refrigerant.[1]

Figure 22.14B is a schematic showing the function of each component of the refrigerator. The refrigerant on the left side of the circuit is at low temperature and low pressure; on the right side of the circuit, it is at high temperature and high pressure. The sum of the heat transferred to the refrigerant from the freezer and food compartment is Q_c. When the refrigerant enters the condenser it is a vapor, which gives off heat Q_h and becomes a liquid. The cycle then begins again as the cold, low-pressure liquid enters the freezer compartment.

A measure of the refrigerator's performance known as the **coefficient of performance** κ is given by

REFRIGERATOR Special Case

$$\kappa = \frac{|Q_c|}{|W_{\text{motor}}|} = \frac{\text{what you get from the refrigerator}}{\text{what you put into it}} \tag{22.15}$$

According to the second law of thermodynamics, the motor must do work to transfer heat from the cold interior of the refrigerator to the hot exterior. Because the work done by the motor cannot be zero, the coefficient of performance κ cannot be infinite.

As in the case of an engine, the work done by the motor can be written in terms of the net heat transferred:

$$|W_{\text{motor}}| = |Q_h| - |Q_c| \tag{22.16}$$

The coefficient of performance may be written as

$$\kappa = \frac{|Q_c|}{|Q_h| - |Q_c|} \tag{22.17}$$

[1]In the past, a common refrigerant was chlorofluorocarbon (CCl_2F_2), a member of the Freon family, but because it and similar substances deplete the stratospheric ozone layer, they are no longer used as refrigerants.

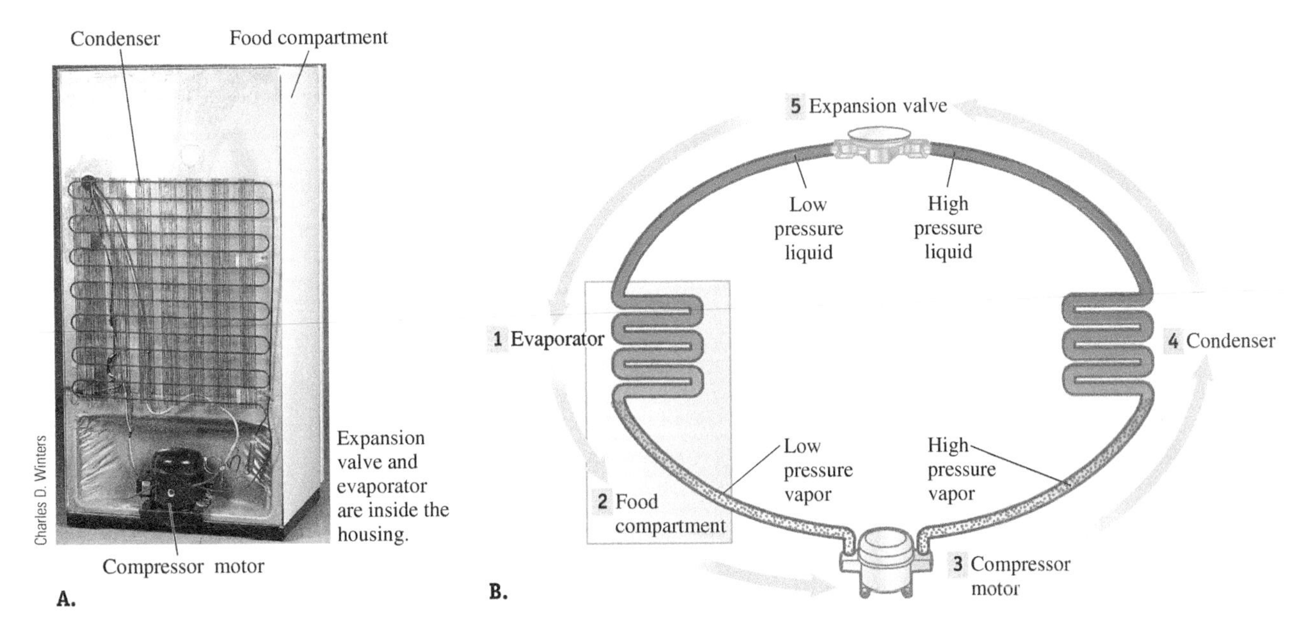

FIGURE 22.14 A. Components of a refrigerator. **B.** A schematic of the processes in a refrigerator shows refrigerant passing counterclockwise through a closed tube. **1** Liquid refrigerant passing through the evaporator absorbs heat from the air and becomes a vapor. **2** Vapor entering the food compartment is colder than the air in the compartment, so heat is transferred to the refrigerant. **3** Next the vapor enters the compressor, whose piston compresses it adiabatically. Positive work W_{motor} is done on the vapor and it leaves the compressor at a higher pressure and higher temperature. **4** Vapor entering the condenser gives off heat and becomes a hot, high-pressure liquid. **5** This liquid expands adiabatically at a rate controlled by the expansion valve so that the liquid is greatly cooled.

If you imagine an ideal refrigerator as a Carnot engine running in reverse, the best possible coefficient of performance may be written in terms of the cold temperature T_c inside and the hot temperature T_h outside (Problem 33):

$$\kappa = \frac{T_c}{T_h - T_c} \tag{22.18}$$

CONCEPT EXERCISE 22.3

On a hot summer day, can you cool your room by opening the door to your refrigerator?

22-6 Entropy

ENTROPY Major Concept

Neither the Clausius nor the Kelvin-Planck statement of the second law of thermodynamics explains how we know when we are watching a movie played backward. A more general and mathematical statement of the second law of thermodynamics requires another state variable known as *entropy*. By using the concept of entropy in the second law, we can explain why some processes are not reversible.

Entropy is like potential energy in that only changes in entropy are physically meaningful. A mathematical definition of the **change in entropy** ΔS for an isothermal process at temperature T is

$$\Delta S \equiv \frac{Q}{T} \tag{22.19}$$

where ΔS is the change in the system's entropy and Q is the amount of heated added to the system. Entropy has the dimensions of energy per temperature, so its SI units are joules per kelvin.

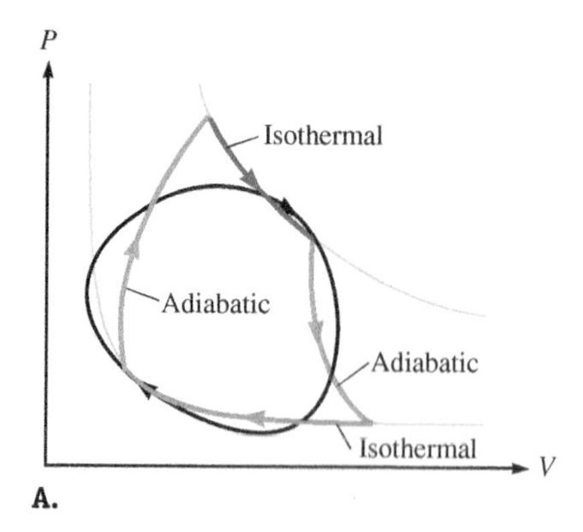

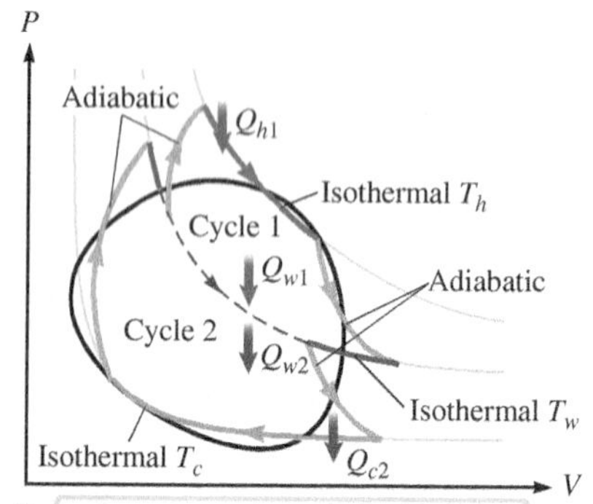

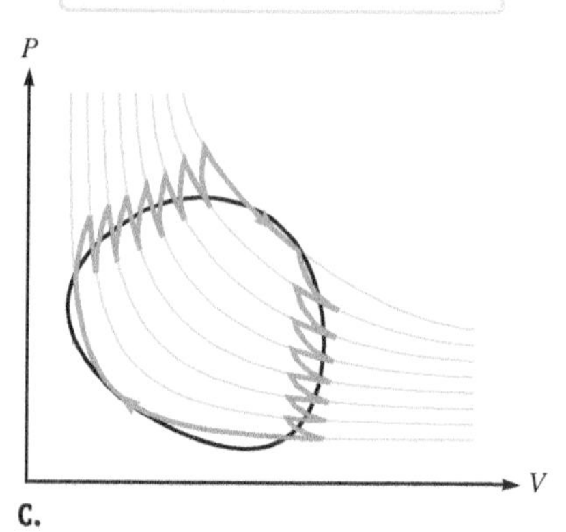

FIGURE 22.15 **A.** A reversible thermodynamic cycle is approximated by a single Carnot cycle. **B.** Two Carnot cycles approximate the thermodynamic cycle. **C.** More Carnot cycles are a better model for this arbitrary cycle.

Let's build our intuition of entropy by taking another look at the Carnot cycle. In deriving the efficiency of the Carnot engine, we found a simple relationship between the absorbed heat Q_h and emitted heat Q_c and the temperature of the hot and cold reservoirs (Eq. 22.11):

$$\frac{|Q_c|}{Q_h} = \frac{T_c}{T_h}$$

We can remove the absolute value signs if we remember that the emitted heat Q_c is negative:

$$\frac{-Q_c}{Q_h} = \frac{T_c}{T_h}$$

$$\frac{-Q_c}{T_c} = \frac{Q_h}{T_h}$$

$$\frac{Q_c}{T_c} + \frac{Q_h}{T_h} = 0 \quad (22.20)$$

Equation 22.20 can be rewritten in terms of entropy. In the *PV* diagram in Figure 22.11 (page 657), label the change in entropy from state *A* to state *B* as ΔS_{AB} and the change in entropy from state *B* to state *A* as ΔS_{BA}. So, Equation 22.20 for the two isothermal processes is

$$\Delta S_{AB} + \Delta S_{BA} = 0$$

There is no heat exchanged in the adiabatic processes, so according to Equation 22.19, there is no change in entropy during these two processes ($\Delta S_{BC} = \Delta S_{DA} = 0$). Thus the total change in entropy around the Carnot cycle is zero: $\Delta S_{\text{cycle}} = 0$.

Recall that state variables such as pressure, volume, temperature, and thermal energy depend only on the endpoints in a *PV* diagram and not on the path. If a system goes through a thermodynamic cycle, it must return to the same point in the *PV* diagram, and the state variables must also return to their original values. Thus, the total change in a state variable during a cyclic process is zero. The change in entropy around a Carnot cycle is zero, supporting the idea that entropy is a state variable.

In fact, the change in entropy is zero around any reversible cyclic path. Figure 22.15 shows an arbitrary cycle on a *PV* diagram. Any reversible cyclic process can be approximated by a series of Carnot cycles. In Figure 22.15A, the cyclic path is approximated by a single Carnot cycle. The approximation improves if more Carnot cycles are used.

In Figure 22.15B, the cycle is approximated with portions of two Carnot cycles (cycle 1 and cycle 2) indicated by the solid curves. Heat is only transferred during the isothermal processes. For the complete solid-curve cycle shown in Figure 22.15B, most of the heat is transferred during the isothermal processes at T_h and at T_c. Little heat is transferred for the portion of the cycle at the intermediate warm temperature T_w because the process along that isotherm is very short. So, the change in entropy for this complete cycle is approximately

$$\Delta S_{\text{complete}} \approx \frac{Q_{h1}}{T_h} + \frac{Q_{c2}}{T_c}$$

To see this, consider the entropy changes for Carnot cycles 1 and 2:

$$\Delta S_1 = \frac{Q_{h1}}{T_h} + \frac{Q_{w1}}{T_w} = 0 \quad \text{and} \quad \Delta S_2 = \frac{Q_{w2}}{T_w} + \frac{Q_{c2}}{T_c} = 0$$

The heat Q_{w2} absorbed during cycle 2 is approximately equal to the negative of the output heat Q_{w1} from cycle 1. When we add the changes in entropy for both cycles, we find

$$\Delta S_1 + \Delta S_2 = \frac{Q_{h1}}{T_h} + \frac{Q_{w1}}{T_w} + \frac{Q_{w2}}{T_w} + \frac{Q_{c2}}{T_c} = 0$$

$$\Delta S_1 + \Delta S_2 \approx \frac{Q_{h1}}{T_h} + \frac{Q_{w1}}{T_w} - \frac{Q_{w1}}{T_w} + \frac{Q_{c2}}{T_c} = 0$$

$$\Delta S_1 + \Delta S_2 \approx \frac{Q_{h1}}{T_h} + \frac{Q_{c2}}{T_c} = 0$$

which is the same as $\Delta S_{\text{complete}}$ for the complete solid cycle shown in Figure 22.15B. So, the change in entropy around the cycle is approximately zero: $\Delta S_{\text{complete}} \approx 0$.

Figure 22.15C shows that the approximation becomes better if more Carnot cycles are used. The same argument we made for two Carnot cycles may be used to show that the change in entropy around any reversible cyclic path is zero. In the limit of infinitely many Carnot cycles, we write the change in entropy around a reversible cycle as

$$\Delta S = \oint \frac{dQ}{T} = 0 \tag{22.21}$$

where the circle on the integral sign means that the integration is over a closed path on a PV diagram and dQ is an infinitesimal amount of heat.

The integral in Equation 22.21 can be calculated for either a clockwise or a counterclockwise path, and you can begin the calculation at any point on the cycle. For example, in Figure 22.16, you could integrate along path A from point i clockwise through point f and continue along path B clockwise back to point i:

$$\Delta S = \oint \frac{dQ}{T} = \left[\int_i^f \frac{dQ}{T}\right]_{\text{path A}} + \left[\int_f^i \frac{dQ}{T}\right]_{\text{path B}} = 0$$

By switching the limits on the integral for path B, we find

$$\Delta S = \left[\int_i^f \frac{dQ}{T}\right]_{\text{path A}} - \left[\int_i^f \frac{dQ}{T}\right]_{\text{path B}} = 0$$

$$\left[\int_i^f \frac{dQ}{T}\right]_{\text{path A}} = \left[\int_i^f \frac{dQ}{T}\right]_{\text{path B}}$$

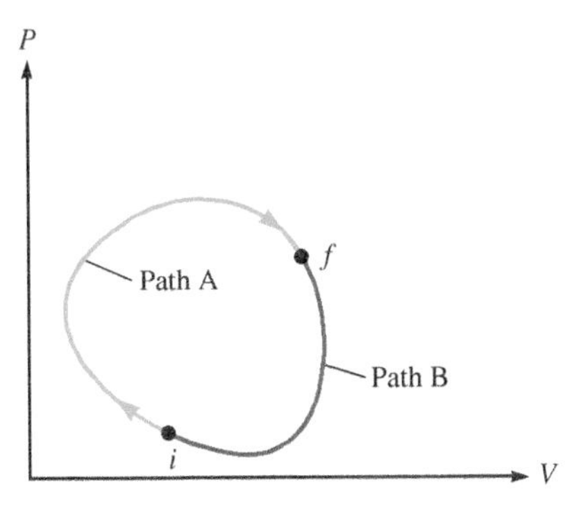

FIGURE 22.16 The change in entropy does not depend on path.

Just as you would expect for a state variable, the difference in entropy ΔS does not depend on the path:

$$\Delta S = S_f - S_i = \left[\int_i^f \frac{dQ}{T}\right]_{\text{any reversible path}} \tag{22.22}$$

Equation 22.22 is for any *reversible* process, but all real processes are *ir*reversible. To find the change in entropy ΔS for a real process, we must imagine some ideal, reversible process that would take the system from state i to state f and then calculate ΔS using Equation 22.22. Because ΔS does not depend on the path, ΔS calculated for the ideal process equals ΔS for the real process.

DERIVATION Entropy Change in an Isothermal Process

We will show that the entropy change during a reversible isothermal process is given by

$$\Delta S = \frac{Q}{T} \text{ (isothermal)} \tag{22.19}$$

In an isothermal process, temperature is constant. We will use Equation 22.22 with T constant. Our result can then be used for any isothermal process.

Because T is constant, we pull it out of the integral.	$\Delta S = \int_i^f \frac{dQ}{T}$ $\Delta S = \frac{1}{T}\int_i^f dQ$	(22.22)
The integral is just the total amount of heat Q transferred during the process from state i to state f.	$\Delta S = \frac{Q}{T}$ ✓	(22.19)

COMMENTS

Equation 22.19 is valid for any real process that can be modeled as reversible and isothermal.

EXAMPLE 22.4 Put the Ice Tray Away!

Imagine leaving a tray of 14 ice cubes out of the freezer. As ice cubes melt, their temperature remains constant at 0°C. Each cube is 4.00 cm on a side. Find the entropy change when the ice melts. The density of ice is 0.917×10^3 kg/m³, and its latent heat of fusion is 3.33×10^5 J/kg.

INTERPRET and ANTICIPATE

The ice melts slowly, so we can model the process as reversible and isothermal and thus can use Equation 22.19. To find the heat required to melt the ice, use $Q = mL_F$ (Eq. 21.9). First find the mass of the 14 ice cubes.

SOLVE	
To find the mass of the ice cubes, calculate their volume and use the density of ice.	$V_{\text{cube}} = \ell^3 = (4.00 \times 10^{-2}\text{ m})^3 = 6.40 \times 10^{-5}\text{ m}^3$ $m_{\text{cube}} = \rho V_{\text{cube}} = (0.917 \times 10^3\text{ kg/m}^3)(6.40 \times 10^{-5}\text{ m}^3) = 5.87 \times 10^{-2}\text{ kg}$ $m = (14\text{ cubes})m_{\text{cube}} = 14(5.87 \times 10^{-2}\text{ kg}) = 0.822\text{ kg}$
Find the heat required to melt all the ice using Equation 21.9 and the latent heat of fusion (Table 21.2).	$Q = mL_F = (0.822\text{ kg})(3.33 \times 10^5\text{ J/kg}) = 2.74 \times 10^5\text{ J}$
Finally, find the change in entropy. The temperature must be expressed in kelvins.	$\Delta S = \frac{Q}{T} = \frac{2.74 \times 10^5\text{ J}}{273\text{ K}}$ (22.19) $\Delta S = 1.00 \times 10^3\text{ J/K}$

CHECK and THINK

Entropy is a new and abstract concept. This example helps build intuition about entropy. When a tray of ice cubes melts, the change in entropy is about 1000 J/K.

DERIVATION Entropy Change in a Constant-Pressure Process

We show that the entropy change during a reversible constant-pressure process is given by

$$\Delta S = mc_P \ln\left(\frac{T_f}{T_i}\right) \tag{22.23}$$

where c_P is the mass specific heat capacity at constant pressure.

This derivation is much like the previous one (entropy of an isothermal process) except now the process is constant pressure instead of isothermal. Because the temperature is not constant, we cannot pull T outside the integral. Instead, we find an expression for dQ when pressure is constant.

Equation 21.5 gives the heat required to change the temperature of a substance. We can modify this equation for a small change in temperature dT, including a subscript P for specific heat at constant pressure.	$Q = mc\Delta T$ (21.5) $dQ = mc_P\,dT$
Substitute dQ into Equation 22.22 and integrate.	$\Delta S = \int_i^f \frac{dQ}{T} = \int_i^f \frac{mc_P\,dT}{T} = mc_P \int_i^f \frac{dT}{T}$ $\Delta S = mc_P \ln\left(\frac{T_f}{T_i}\right)$ ✓ (22.23)

COMMENTS

Equation 22.23 is valid for any process in which the temperature changes from T_i to T_f as long as the final pressure equals the initial pressure. If the substance is cooled so that $T_f < T_i$, the change in entropy is negative (entropy decreases).

EXAMPLE 22.5 Mixing Barrels of Water

You have two barrels of water. Barrel 1 contains $m_1 = 35$ kg of water at temperature $T_1 = 30.0°C$, and barrel 2 contains $m_2 = 75$ kg of water at temperature $T_2 = 90.0°C$. If you mix the two barrels of water in an isolated tub of negligible heat capacity so that heat is only exchanged between the hot and cold water, find the temperature of the completely mixed water. Find the change in entropy of the water from barrel 1, the change in entropy of the water from barrel 2, and the change in entropy of the Universe. *Hint*: The specific heat of water is listed in Table 21.1.

INTERPRET and ANTICIPATE

Mixing the water is a constant-pressure process. Because heat is only exchanged between the water, the heat lost by the hot water must equal heat gained by the cold water. We expect the temperature of the mixed water to be between 30.0°C and 90.0°C. We also expect that the cold water's entropy increases and the hot water's entropy decreases. Because the only heat exchanged is between the water, we can add the two entropy changes to find the entropy change for the Universe.

SOLVE

Set the heat lost by the hot water equal to the heat gained by the cold water and use $Q = mc\Delta T$ (Eq. 21.5) to find the temperature of the mixture. Because hot water loses energy, we must remember Q_{lost} is negative. The final temperature of the mixture is T_f for both the hot water and the cold water. It is okay to work in Celsius for this part of the example, but we'll need this final temperature in kelvins for the next step.

$$Q_{lost} = Q_{gained}$$

$$-m_2 c\,\Delta T_{hot} = m_1 c\,\Delta T_{cold}$$

$$-m_2(T_f - T_2) = m_1(T_f - T_1)$$

$$(m_1 + m_2)T_f = m_1 T_1 + m_2 T_2$$

$$T_f = \frac{m_1 T_1 + m_2 T_2}{m_1 + m_2}$$

$$T_f = \frac{(35.0\,\text{kg})(30.0°\text{C}) + (75.0\,\text{kg})(90.0°\text{C})}{35.0\,\text{kg} + 75.0\,\text{kg}}$$

$$T_f = 70.9°\text{C} = 343.9\ \text{K}$$

CHECK and THINK

As expected, the temperature of the mixture is between the two original temperatures. Because there is initially more hot water, the final temperature is higher than the average of the two original temperatures.

SOLVE

Find the entropy change for cold water from barrel 1 and for hot water from barrel 2 using Equation 22.23. The temperatures must be in kelvins. The specific heat of water from Table 21.1 is valid near room temperature and atmospheric pressure.

$$\Delta S = mc_P \ln\frac{T_f}{T_1} \tag{22.23}$$

$$\Delta S_1 = (35.0\,\text{kg})(4190\,\text{J/kg}\cdot\text{K})\ln\left(\frac{343.9\ \text{K}}{303\ \text{K}}\right)$$

$$\Delta S_1 = 1.86 \times 10^4\ \text{J/K}$$

$$\Delta S_2 = (75.0\,\text{kg})(4190\,\text{J/kg}\cdot\text{K})\ln\left(\frac{343.9\ \text{K}}{363\ \text{K}}\right)$$

$$\Delta S_2 = -1.70 \times 10^4\ \text{J/K}$$

CHECK and THINK

As expected, the entropy of the cold water from barrel 1 increased, and the entropy of the hot water from barrel 2 decreased.

SOLVE

Because heat is only exchanged between the hot water and the cold water, the only entropy changes are in the water. The entropy change in the Universe as a result of mixing the two barrels of water is the sum of ΔS_1 and ΔS_2.

$$\Delta S_{Universe} = \Delta S_1 + \Delta S_2$$

$$\Delta S_{Universe} = (1.86 \times 10^4\ \text{J/K}) - (1.70 \times 10^4\ \text{J/K})$$

$$\Delta S_{Universe} = 1.6 \times 10^3\ \text{J/K}$$

Example continues on page 670 ▶

CHECK and THINK

The entropy of the Universe increases as a result of mixing the two barrels of water. This example is another illustration of the second law of thermodynamics. (See Section 22-7.)

EXAMPLE 22.6 An Ideal Gas Doubles in Volume

The free expansion of an ideal gas is described in Section 21-7. It can be shown (Problem 70) that the entropy change of an ideal gas during a free expansion is given by

$$\Delta S = nR \ln\left(\frac{V_f}{V_i}\right) \tag{22.24}$$

Use this expression to find the entropy change when 1 mol of an ideal gas undergoes a free expansion that doubles its volume as in Figure 21.20.

INTERPRET and ANTICIPATE

We will develop some intuition about entropy by substituting values into Equation 22.24. Because the gas expands, we expect that the entropy increases.

SOLVE

Substitute given values into Equation 22.24.	$\Delta S = (1 \text{ mol})(8.31 \text{ J/mol} \cdot \text{K}) \ln\left(\frac{2V_i}{V_i}\right) = 5.76 \text{ J/K}$

CHECK and THINK

As expected, the entropy increased. Compare with Example 22.4: When a tray of ice cubes melts, the entropy of the water increases by about 1000 J/K. In this example, we found that when a gas doubles in volume—picture a balloon of gas doubling in size—the change in entropy is only about 6 J/K.

SECOND LAW OF THERMODYNAMICS (GENERAL STATEMENTS)

Underlying Principle

22-7 Second Law of Thermodynamics, General Statements

The concept of entropy as a state variable allows us to make several general statements of the second law of thermodynamics. One of them is as follows:

The entropy of an isolated system never decreases: $\Delta S \geq 0$.

The equal sign holds for a reversible process, $\Delta S = 0$. Entropy always increases in the case of an irreversible process, $\Delta S > 0$.

The Universe is defined to be everything, so the ultimate isolated system is the Universe. In addition, all real processes are irreversible. Therefore, according to the second law of thermodynamics:

The entropy of the Universe must be increasing:

$$\Delta S_{\text{Universe}} > 0$$

In thermodynamics, we often choose part of the Universe as a system and then take everything else to be in the environment. Another way to state the second law of thermodynamics for the Universe is as follows:

The total entropy of a system plus its environment must increase as a result of any real process:

$$\Delta S_{\text{sys}} + \Delta S_{\text{env}} > 0$$

It is thus possible that the entropy of one part of the Universe decreases during some process, but, if so, the entropy of another part of the Universe increases so that the overall entropy of the Universe increases.

The second law of thermodynamics is very different from other underlying principles we've studied. For example, compare the second law of thermodynamics to the conservation of energy principle. On the one hand, conservation of energy states that the energy of the Universe remains constant. If energy increases in one part of the Universe, energy must decrease in another part of the Universe by the same amount. On the other hand, the second law of thermodynamics says that the entropy of the Universe is always increasing.

To understand the importance of this distinction, imagine mixing 1 liter of hot water with 1 liter of cool water. Energy is transferred in the form of heat from the hot water to the cool water, resulting in 2 liters of warm water with a temperature between the two original temperatures. Such a transfer of energy obeys the principle of energy conservation and the second law of thermodynamics. Consider the entire 2 liters of water to be an isolated system. The total energy of the water remains constant, but the entropy increases.

Now imagine that you have 2 liters of warm water, and later you find that the water has spontaneously separated itself into two parts: 1 liter is hot, and the other is cold. According to the principle of energy conservation, such a process is possible—the sum of the water's thermal energy remains constant—but according to the second law of thermodynamics, such a process is impossible because the entropy of the water would decrease. Recall Example 22.5, in which we mixed hot water and cold water and found that the entropy of the Universe increased by $\Delta S_{\text{Universe}} = 1.6 \times 10^3$ J/K. If the warm water mixture were to separate spontaneously into hot water and cold water, the entropy of the Universe would decrease by $\Delta S_{\text{Universe}} = -1.6 \times 10^3$ J/K.

The Arrow of Time

The second law of thermodynamics has broad consequences. It says that processes that decrease the entropy of the Universe are not possible. At the beginning of this chapter, we thought about how we know when we are watching a movie in reverse. We know that an egg will not spontaneously fly off a plate and into a hot frying pan, become uncooked, and then fly up into the open eggshell that closes up without a trace. How do we know that this order of events is impossible? It is impossible because the entropy of the Universe would have to decrease. When we watch a movie, we can tell whether the movie is being played forward or in reverse by observing changes in entropy, so entropy is *time's arrow* because it tells us which way time is going. In Sections 22-8 and 22-9, we develop a broader understanding of entropy and the consequences of the second law of thermodynamics.

CONCEPT EXERCISE 22.4

You have considerable intuition about whether some process increases or decreases the entropy of the Universe. Just imagine watching the process in reverse. If it doesn't make sense in reverse, entropy must have increased. This test isn't always foolproof; some process may *seem* reversible, but the entropy of the Universe still increases. Use your intuition to decide whether the entropy of the Universe increases during the following processes:

a. A car crashes into a brick wall.
b. An ice cube floating in a large cup of water melts.
c. A puddle of water freezes to ice slowly.
d. The frozen puddle of water melts slowly.

EXAMPLE 22.7 Clausius Statement

Heat is transferred from a hot reservoir at T_h to a cold reservoir at T_c. Recall that the temperature of a heat reservoir is assumed to be constant. Find an expression for the net change in the entropy of the Universe $\Delta S_{\text{Universe}}$ and then show that heat cannot flow from the cold reservoir to the hot reservoir.

Example continues on page 672 ▶

INTERPRET and ANTICIPATE

A heat reservoir is so large that when heat flows into or out of the reservoir, its temperature stays approximately constant. Therefore, we can use $\Delta S = Q/T$ (Eq. 22.19) to find the change in entropy of each heat reservoir.

SOLVE

Heat flows out of the hot reservoir, so heat is negative ($-Q$), and entropy decreases.	$\Delta S_{\text{hot}} = -\dfrac{Q}{T_h}$
Heat flows into the cold reservoir, so heat is positive (Q), and entropy increases.	$\Delta S_{\text{cold}} = \dfrac{Q}{T_c}$
Find the net entropy change of the Universe by adding the entropy changes for the hot and cold reservoirs.	$\Delta S_{\text{Universe}} = \Delta S_{\text{hot}} + \Delta S_{\text{cold}} = -\dfrac{Q}{T_h} + \dfrac{Q}{T_c}$ $\Delta S_{\text{Universe}} = Q\left(\dfrac{1}{T_c} - \dfrac{1}{T_h}\right)$ (1)
We have used the convention that when heat flows from hot to cold, Q is positive. The high temperature is greater than the low temperature, so the Universe's entropy increases.	$T_h > T_c$ $\dfrac{1}{T_h} < \dfrac{1}{T_c}$ $\left(\dfrac{1}{T_c} - \dfrac{1}{T_h}\right) > 0$ $\Delta S_{\text{universe}} > 0$
If heat flows from the cold to the hot reservoir, we must modify Equation (1) by inserting a negative sign because $Q \rightarrow -Q$, and we find that the entropy of the Universe decreases. The second law of thermodynamics does not allow the entropy of the Universe to decrease, so we conclude that heat cannot spontaneously flow from cold to hot.	$\Delta S_{\text{Universe}} < 0$

CHECK and THINK

We have just found that the general statement of the second law of thermodynamics (entropy of the Universe does not decrease) includes the Clausius statement (heat does not naturally flow from cold to hot).

22-8 Order and Disorder

ENTROPY Major Concept

As a state variable, entropy must tell us something about the condition of a system. Just as kinetic energy—another state variable—tells us about the motion in a system, **entropy** is a measure of the *disorder* in a system.

We see order and disorder in our lives everyday. A cup of black coffee and a carton of cream have more order than a blended cup of coffee with cream. When the coffee and cream are separated, all the coffee molecules are in one container, and all the cream molecules are in a different container. When you stir the cream into your coffee, disorder increases because the coffee and cream molecules are randomly distributed. As another example, a drawer full of socks rolled together by type has more order than a laundry basket full of clothes thrown in arbitrarily.

When the entropy of a system increases, the system becomes more disordered. In Example 22.4, we found that when ice melts, the entropy of the water increases. On the microscopic level, water molecules in cubes of ice are very orderly, with each molecule holding a position in a three-dimensional crystalline structure (Fig. 22.17A). When the ice melts, the bonds weaken; the structure then falls apart, becoming more disordered (Fig. 22.17B).

This statement is true in general; when heat flows into a system, the system's disorder increases because of an increase in molecular motion. From Equation 22.19, $\Delta S = Q/T$. When heat is added to a relatively cold system, the ratio Q/T is large, so there is a great change in entropy. When the system is cold, it has very little molecular motion and is relatively ordered. Adding heat causes a relatively large increase in the particles' motion and therefore a relatively large increase in the system's disorder. A system that is initially hot has considerable motion and therefore much disorder. When heat is added, it does little to increase the motion, and the increase in entropy Q/T is therefore relatively small.

Consider another example. When hot water is mixed with cold water, the entropy of the water increases (Example 22.5). Initially, the water is ordered by its temperature; afterward, that order is lost because the molecules are mixed together randomly. As molecules collide with one another, energy is redistributed among them until they become a single population of randomly moving particles at some temperature between the original temperatures.

The idea that mixing hot and cold parts of a system increases the system's entropy has an important practical implication. Suppose the hot and cold parts are the hot and cold reservoirs in a heat engine or refrigerator. For example, the hot part could be the air in the kitchen, and the cold part could be the air inside the refrigerator. Without two separate heat reservoirs, heat engines and refrigerators won't operate. Put more generally, when entropy increases, disorder increases, and the ability of the system to do useful work decreases.

In fact, in any natural process, energy is *degraded.* Of course, energy is conserved, but some of the energy goes into a more disordered, useless form. For example, when a block slides down an inclined plane, potential energy goes into kinetic energy and thermal energy. Kinetic energy is an ordered and useful form of energy that may be transformed back into potential energy, but thermal energy is disordered. That energy cannot be transformed back into mechanical energy, and we say that friction has "dissipated" some energy.

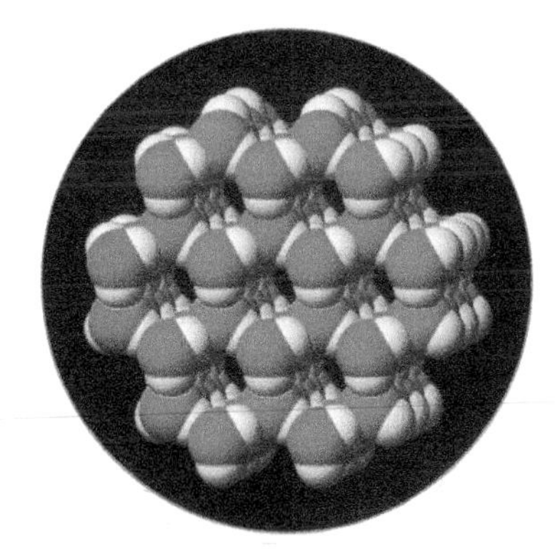

A.

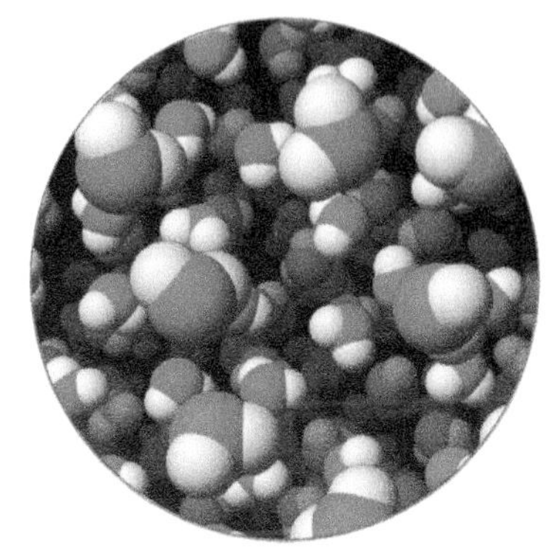

B.

FIGURE 22.17 Frozen water (**A**) is more ordered than liquid water (**B**).

The idea that entropy is a measure of disorder has an important implication for the entire Universe. According to the second law of thermodynamics, the entropy of the Universe is increasing. So, another way to state the second law of thermodynamics is as follows:

The Universe is becoming more disordered.

SECOND LAW OF THERMODYNAMICS (GENERAL STATEMENT)

Underlying Principle

The idea that the Universe is becoming more disordered may not sit well with you; somehow, you came to exist, and you are clearly a highly ordered part of the Universe. In fact, evolution is a process that increases order and reduces entropy. According to the second law of thermodynamics, any process such as evolution that decreases entropy in one part of the Universe must increase entropy in another part of the Universe. Evolution—and all biological processes—produces waste molecules. These waste molecules are very disordered, and the overall entropy of the Universe increases as a result.

22-9 Entropy, Probability, and the Second Law

Entropy is a *quantitative* measure of the disorder of a system. To illustrate how to calculate the disorder in a system, let's look at a simple system consisting of four coins: a penny, a nickel, a dime, and a quarter. Each coin can show either a head or a tail. If you observe that all four coins show the same face (either four tails or four heads), you would say the system had a high degree of order. If the system shows a mixture of heads and tails, however, you would conclude the system had more disorder.

The state of the system as a whole (in this case, the tally of heads and tails) is called the **macroscopic state**. The term *macroscopic state* is equivalent to the term *state* in previous sections. State variables such as pressure, temperature, volume, and entropy describe the macroscopic state of a system. The **microscopic state** of a system is the condition of each particle that makes up the system. The microscopic state of the four-coin system specifies the condition (head or tail showing) of each coin.

Macroscopic state	Microscopic state of each coin	Number of microscopic states
4 heads		1
3 heads 1 tail		4
2 heads 2 tails		6
1 head 3 tails		4
4 tails		1

FIGURE 22.18 For this four-coin system, a microscopic state indicates the condition of each coin. The macroscopic state is the total number of heads and the total number of tails.

If the system is the air in your room, its microscopic state is the velocity and position of each air molecule. Ordinarily, we do not know the microscopic state of a thermodynamic system because there are far too many particles to have information about each one of them at every given moment. For illustration purposes, we have chosen the four-coin system because it is so simple that we can observe its microscopic state.

Imagine arranging the coins in all possible microscopic states as shown in Figure 22.18, indicating the condition of each coin. For example, in the first microscopic state in the second row, the penny is tail up and the other coins are heads up. The macroscopic state gives the total number of heads and tails, so all the microscopic states in the second row have the same macroscopic state: three heads and one tail. From Figure 22.18, there are 16 possible microscopic states that are categorized into just five macroscopic states: (1) all heads, (2) three heads and one tail, (3) two heads and two tails, (4) one head and three tails, and (5) all tails.

Entropy S is a macroscopic state variable, and it is directly related to the number $\mathcal{W}$ of possible microscopic states associated with a particular macroscopic state:

$$S = k_B \ln \mathcal{W} \tag{22.25}$$

where k_B is Boltzmann's constant. Like potential energy, only differences in entropy are physically significant. By convention associated with Equation 22.25 if $\mathcal{W} = 1$, then $S = 0$, but it is possible to add an arbitrary constant S_0 so that if $\mathcal{W} = 1$, then $S = S_0$. Throughout this book, we use the convention that $S_0 = 0$.

Taking the four-coin example as an illustration, the macroscopic state consisting of all heads or all tails has the most order and the lowest entropy. In fact, because $\mathcal{W} = 1$ for either of these macroscopic states, the entropy of either of these states is zero:

$$S = k_B \ln \mathcal{W} = k_B \ln(1) = 0$$

The macroscopic state consisting of two heads and two tails has the greatest number of microscopic states ($\mathcal{W} = 6$) and thus also has the greatest entropy.

So, the macroscopic state with the greatest number of microstates is the most disordered state. Think of it this way: The description *two heads and two tails* gives very little information about each coin. You don't know if the penny is head up or tail up, and if you guess, you have only a 50% chance of being correct. Compare that situation to the macroscopic state of *all heads up*. That statement contains all the information about each coin; you know that the penny is head up. In general, the more information provided by the macroscopic state about the microscopic state of each particle, the more ordered that state is.

This four-coin example gives us another way to think about the second law of thermodynamics. Imagine tossing the four coins on a table. What macroscopic state are you most likely to find? You wouldn't count on finding, say, all heads or all tails because there is only a 1 in 16 chance of getting one of those states. The most probable outcome is two heads and two tails because there is a 6 in 16 chance of getting that macroscopic state. So, the most probable outcome is the macroscopic state with the greatest entropy.

Another general way to state the second law of thermodynamics is as follows:

SECOND LAW OF THERMODYNAMICS (GENERAL STATEMENT)

Underlying Principle

The Universe is moving from a state of low probability to a state of higher probability.

Stating the second law in terms of probability means that it is not *impossible* for heat to flow spontaneously from cold to hot or for entropy to decrease; it just means that is very *improbable*. In terms of the four-coin example, if you toss the coins enough times, once in a while you will get all heads. That this outcome is unlikely is not the same as saying it is forbidden. We have never observed heat spontaneously flowing from cold to hot or all the air molecules in a room contracting into a corner. In both of these examples, the number of particles is very large, and it is extremely unlikely to find a large number of particles in an ordered state.

To illustrate, imagine tossing 100 coins instead of just four. The number of microstates is found using

$$\mathcal{W} = \frac{N_{\text{tot}}!}{N_h! N_t!} \quad (22.26)$$

where N_h is the number of heads, N_t is the number of tails, and N_{tot} is the total number of coins, $N_{\text{tot}} = N_h + N_t$. (The symbol "!" stands for factorial. For example, $3! = 3 \times 2 \times 1 = 6$.) Table 22.1 lists a few macroscopic states for these 100 coins. There are a total of about 1.27×10^{30} microstates, so if you toss 100 coins, you have about a 1 in 1.3×10^{30} chance of throwing all heads. Still, 100 is a small number of objects; consider that 1 mole of particles is more than 10^{23} particles. Imagine finding 10^{23} coins, all heads up. It is very unlikely that such a great number of particles would be found in an ordered state. Put another way, you may measure state variables such as the pressure, temperature, and volume of the air in your room, but those measurements would tell you almost nothing about the microstate of a particular oxygen molecule that you inhaled half an hour ago.

TABLE 22.1 Macroscopic states and microscopic states for a collection of 100 coins.

Macroscopic state		Number of microscopic states $\mathcal{W}$
Heads	Tails	
100	0	1
99	1	100
90	10	1.73×10^{13}
80	20	5.36×10^{20}
60	40	1.37×10^{28}
55	45	6.14×10^{28}
54	46	7.35×10^{28}
53	47	8.44×10^{28}
52	48	9.32×10^{28}
51	49	9.89×10^{28}
50	50	1.01×10^{29}
49	51	9.89×10^{28}
48	52	9.32×10^{28}
47	53	8.44×10^{28}
46	54	7.35×10^{28}
45	55	6.14×10^{28}
40	60	1.37×10^{28}
20	80	5.36×10^{20}
10	90	1.73×10^{13}
1	99	100
0	100	1

CONCEPT EXERCISE 22.5

Use Table 22.1 to find the following:

a. The probability of finding 50 heads and 50 tails in a toss of 100 coins
b. The probability of finding between 45 and 55 heads

EXAMPLE 22.8 Entropy of Free Expansion Revisited

In Example 22.6, we calculated the change in entropy when 1 mole of gas undergoes a free expansion and doubles in volume (Fig. 21.20). Repeat that calculation using Equation 22.25.

INTERPRET and ANTICIPATE

We expect to find the same result, 5.76 J/K (Example 22.6). Initially, the gas occupies a volume V_i and has a number of microscopic states $\mathcal{W}_i$. When the valve is open, no work is done, so the velocity of the molecules is unchanged. The number of microscopic states increases, however, because each molecule has twice as much volume in which to move. The set of possible positions of each molecule has doubled, so the number of microstates for all N molecules has increased by 2^N. This increase in microstates is directly related to the increase in entropy.

SOLVE

The number $\mathcal{W}_f$ of microscopic states when the volume is $V_f = 2V_i$ is 2^N multiplied by the initial number of microscopic states.

$$\mathcal{W}_f = 2^N \mathcal{W}_i \quad (1)$$

Use $S = k_B \ln \mathcal{W}$ (Eq. 22.25) to find an expression for the change in entropy.

$$S_i = k_B \ln \mathcal{W}_i \qquad S_f = k_B \ln \mathcal{W}_f$$

$$\Delta S = S_f - S_i = k_B \ln \mathcal{W}_f - k_B \ln \mathcal{W}_i$$

$$\Delta S = k_B \ln \frac{\mathcal{W}_f}{\mathcal{W}_i}$$

Example continues on page 676 ▶

Eliminate $\mathcal{W}_f$ with Equation (1).	$\Delta S = k_B \ln \frac{2^N \mathcal{W}_i}{\mathcal{W}_i} = k_B \ln 2^N$ $\Delta S = N k_B \ln 2$
Substitute values. For 1 mole of molecules, $N = N_A$.	$\Delta S = N_A k_B \ln 2$ $\Delta S = (6.02 \times 10^{23})(1.38 \times 10^{-23}\ \text{J/K}) \ln 2$ $\Delta S = 5.76\ \text{J/K}$

CHECK and THINK

As expected, this result is exactly what we found in Example 22.6. This time, we did not need to think about a specific thermodynamics process and only needed to consider the increase in the number of microscopic states.

EXAMPLE 22.9 Who Stole My Air?

You and your roommate have split your room in half (Fig. 22.19). Estimate the probability of finding all the air molecules in your half of the room, leaving your roommate with no air.

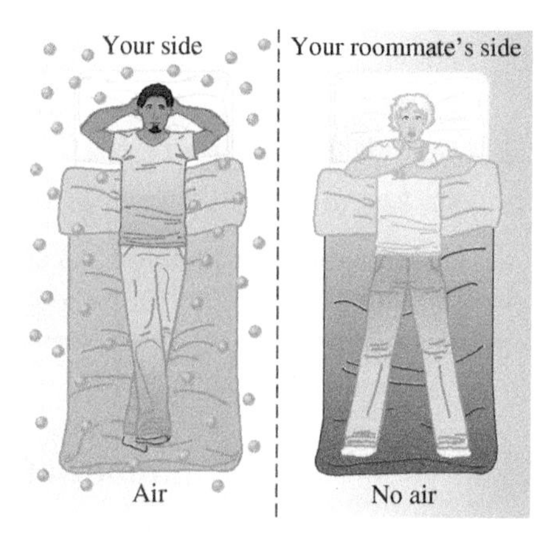

FIGURE 22.19

INTERPRET and ANTICIPATE

This example involves the opposite of a free expansion, and we know that such a process does not naturally occur. We expect the probability of finding all the air on one side of your room to be extremely low, nearly zero.

SOLVE

Suppose there are N molecules filling the entire room. The chance of finding one particular molecule on your side of the room is 50% or $\frac{1}{2}$, the same as the probability of flipping a single coin and having it land head up. The chance of finding two particular molecules on your side of the room is $(\frac{1}{2})(\frac{1}{2}) = (\frac{1}{2})^2 = \frac{1}{4}$, the same as the probability of tossing two coins and having them both land heads up. The probability of finding three particular molecules on your side of the room or of throwing three coins heads up is $(\frac{1}{2})(\frac{1}{2})(\frac{1}{2}) = (\frac{1}{2})^3 = \frac{1}{8}$.

Find a general expression for the probability $\mathcal{P}$ of finding N particular molecules on your side of the room or of throwing N coins heads up.	$\mathcal{P} = (\frac{1}{2})^N$ $\mathcal{P} = 2^{-N}$
There are probably a few moles of air in your room. For the purpose of our estimate, we can assume there is 1 mole of air so that $N \approx N_A$.	$\mathcal{P} \approx 2^{-N_A}$ $\mathcal{P} \approx 2^{-(6 \times 10^{23})}$

CHECK and THINK

This probability is so close to zero that if you try to enter it into your calculator (which probably displays 21 digits), you will find all zeros. To get a sense of how small this probability truly is, try entering 2^{-600} into your calculator; you will find that it is approximately 10^{-181}. So, even if you only had 600 molecules in your room, it is *very* unlikely that you will find them all on just your side. The probability of finding an entire mole of molecules in just half the available space is so unlikely that is essentially impossible. So, it is no surprise that the second law of thermodynamics is often stated so strongly (heat does not spontaneously flow from cold to hot) instead of in probabilistic terms (heat is unlikely to flow from cold to hot).

SUMMARY

Underlying Principles

The **second law of thermodynamics** can be expressed in many different ways.

1. **Clausius statement:**

   Heat flows naturally from a hot object to a cooler object; heat does not naturally flow from a cold object to a hotter object.

2. **Kelvin-Planck statement:**

   It is impossible to construct a cyclic heat engine that takes in heat and produces only work without also giving off waste heat.

3. The concept of entropy allows for several **general and mathematical statements** of the second law of thermodynamics:

   a. *The entropy of an isolated system never decreases:* $\Delta S \geq 0$.

   b. *The entropy of the Universe must be increasing:* $\Delta S_{\text{Universe}} > 0$.

   c. *The total entropy of a system plus its environment must increase as a result of any real process:* $\Delta S_{\text{sys}} + \Delta S_{\text{env}} > 0$.

   d. *The Universe is becoming more disordered.*

   e. *The Universe is moving from a state of low probability to a state of higher probability.*

Major Concepts

1. **Efficiency** e of an engine:

$$e \equiv \frac{W_{\text{eng}}}{Q_h} = 1 - \frac{|Q_c|}{Q_h} \quad \text{(22.4 and 22.5)}$$

2. **Reversible process** is an ideal thermodynamic process in which a system that goes from state i to state f can go back again from state f to state i, exchanging the same amount of heat and work so that the system and the environment have been returned to their original conditions. Other processes are **irreversible processes**.

3. **Entropy** is a state variable measuring a system's disorder. Only changes in entropy are physically meaningful.

   a. The change in entropy ΔS for an isothermal process at temperature T is

$$\Delta S \equiv \frac{Q}{T} \quad (22.19)$$

   b. The change in entropy around a reversible cycle is

$$\Delta S = \oint \frac{dQ}{T} = 0 \quad (22.21)$$

   c. The difference in entropy ΔS does not depend on the path:

$$\Delta S = S_f - S_i = \left[\int_i^f \frac{dQ}{T}\right]_{\text{any reversible path}} \quad (22.22)$$

   d. The entropy of a system depends on the number of microscopic states $\mathcal{W}$:

$$S = k_B \ln \mathcal{W} \quad (22.25)$$

Special Cases: Three Specific Engines

1. A **Carnot** engine goes through four reversible processes known as the **Carnot cycle:** (1) isothermal expansion ($\Delta T = 0$), (2) adiabatic expansion ($\Delta Q = 0$), (3) isothermal compression, and (4) adiabatic compression. The Carnot engine is an ideal engine with efficiency

$$e_{\text{Carnot}} = 1 - \frac{T_c}{T_h} \quad (22.6)$$

   No real engine can deliver this efficiency.

2. The **Otto** engine is a good model for the **internal combustion** engine. The ideal, reversible **Otto cycle** consists of four processes: two constant-volume processes and two adiabatic processes. The efficiency of the Otto cycle is

$$e_{\text{Otto}} = 1 - \left(\frac{V_{\text{max}}}{V_{\text{min}}}\right)^{1-\gamma} \quad (22.14)$$

   where $\gamma = C_P/C_V$ (Eq. 21.31).

3. A **refrigerator** is a heat engine running in reverse; its motor does work, and heat is absorbed from the low-temperature compartment inside and is expelled into the warm room. The **coefficient of performance** κ is given by

$$\kappa = \frac{|Q_c|}{|W_{\text{motor}}|} \quad (22.15)$$

Tools

Energy transfer diagrams are useful for studying heat engines. The system is represented as a rectangle in the middle of the diagram, the hot reservoir is drawn above the system, and the cold reservoir is drawn below the system. Thick arrows represent energy transferred into or out of the system.

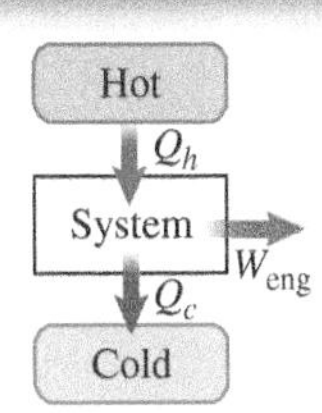

PROBLEMS AND QUESTIONS

A = algebraic C = conceptual E = estimation G = graphical N = numerical

22-1 Second Law of Thermodynamics, Clausius Statement

1. C If heat naturally flowed from cold objects to hotter ones, what would happen if you put a hot potato into a large cold body of water such as a lake? What can you say about the final temperature of the potato and of the lake?

22-2 Heat Engines

2. G Figure P22.2 shows a Carnot cycle. The system expands isothermally in path A, expands adiabatically in path B, is compressed isothermally in path C, and is compressed adiabatically in path D, returning to the initial state. Draw an energy transfer diagram, similar to Figure 22.4, for this cycle.

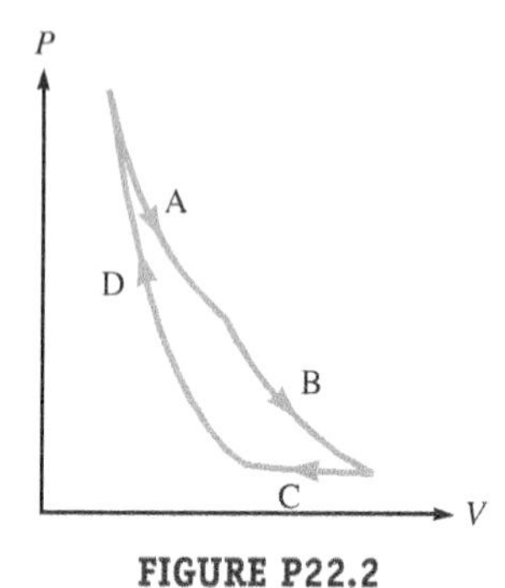

FIGURE P22.2

3. G Use a *PV* diagram such as the one in Figure 22.2 (page 653) to figure out how you could modify an engine to increase the work done.

22-3 Second Law of Thermodynamics, Kelvin-Planck Statement

4. N During each cycle, a heat engine does 50.0 J of work and absorbs 425 J of energy from a hot reservoir. **a.** What is the efficiency of this engine? **b.** What is the amount of energy that is exhausted to the cold reservoir during one cycle?

5. N An engine is able to convert 30.0% of the energy it absorbs from a reservoir into work. **a.** What is the thermal efficiency of this engine? **b.** What is the percentage of the energy absorbed that is exhausted into the cold reservoir?

6. C According to the Kelvin-Planck statement of the second law, it is impossible to construct a cyclic heat engine that takes in heat and produces only work without also giving off waste heat. Is the opposite possible? In other words, can a device operate in a cycle by having work done on it by the environment and give off heat without doing any work? If so, give an example. If not, explain why not.

7. N An engine with an efficiency of 0.36 can supply a power of 140 hp. At what rate does it exhaust heat to the environment? Express your answer in watts.

8. C An environmentally conscious driver asked Click and Clack (the Tappet Brothers), who host the National Public Radio show *Car Talk*, if it was environmentally friendly to drive a convertible with its top down and run the heater. Click and Clack assured the driver that such a practice does not consume any additional energy than driving the car with the heater off, so it is as environmentally friendly as driving with the heater off. (They did, however, warn that driving the car with the top down and the air conditioner running did consume more energy and was therefore worse for the environment.) Why did Click and Clack say that it was okay to drive with the top down and the heater on, but not okay to turn the air conditioner on with the top down?

9. N A heat engine operates in contact with a cold reservoir and a hot reservoir such that 1.620×10^4 J of heat is transferred into the cold reservoir and 7.481×10^4 J of heat is absorbed from the hot reservoir. **a.** What is the work done by the engine on its environment? **b.** What is the efficiency of the engine?

10. N A heat engine with 40.0% efficiency produces 2.50×10^4 W of power while exhausting 1.00×10^4 J to the cold reservoir during each cycle. **a.** How much energy does the engine absorb from the hot reservoir during each cycle? **b.** How long does each cycle take?

22-4 The Most Efficient Engine

11. N What is the maximum efficiency for a heat engine operating between a 22.0°C reservoir and a 535°C reservoir?

12. C What are the similarities and differences between a Carnot engine and a perfect engine?

13. N An internal combustion engine ignites a mixture of fuel and air that reaches a temperature of 1800 K and exhausts gas at 440 K. In the process, the volume of the gas expands such that its maximum volume is nine times larger than the minimum volume. The adiabatic coefficient for air is $\gamma = 1.4$. **a.** What is the maximum efficiency possible for the engine, assuming gas in the piston follows the Otto cycle? **b.** What is the ideal efficiency for a Carnot cycle operating between the same high and low temperatures?

14. C CASE STUDY If a wealthy car enthusiast says that she wants to buy an Otto engine, what would you tell her?

15. N What is the efficiency of a Carnot engine operating between a hot reservoir at 800.0 K and a cold reservoir at 400.0 K?

16. C Does the type of working substance affect the efficiency of a Carnot engine? Does the type of working substance affect the efficiency of real engines? Explain.

17. N A Carnot heat engine operates in contact with a cold reservoir and a hot reservoir such that 4.310×10^5 J of heat is transferred into the cold reservoir and 9.678×10^5 J of heat is absorbed from the hot reservoir. **a.** If the temperature of the cold reservoir is 315.0 K, what is the temperature of the hot reservoir? **b.** What ratio of volumes for an Otto cycle would cause it to have the same efficiency as the above engine? Assume $\gamma = 1.4$.

18. C Why is the size of a car engine important?

19. N A Carnot engine operating between a reservoir at 25.0°C and a reservoir at 340.0°C has a power output of 1.10×10^4 W. **a.** How much energy is extracted from the hot reservoir during each hour of operation? **b.** How much energy is exhausted to the cold reservoir during each hour of operation?

20. CASE STUDY Model a car engine as an Otto engine.

a. N If the engine has an efficiency of 25%, what is the compression ratio $r \equiv V_{max}/V_{min}$ of the Otto engine? Assume $\gamma = 1.4$.

b. C How does that ratio compare with a real engine's compression ratio, assuming the real engine's efficiency is 25%?

21. N A car engine has an efficiency of 22%. If it produces 1.20×10^4 J of work, how much heat does it expel?

22. In 1816, Robert Stirling, a Scottish minister, patented an engine now known as the **Stirling engine**. The Stirling engine follows a four-process cycle: two isothermal processes and two constant-volume processes as shown in Figure P22.22. The working substance used in the Stirling engine never leaves the engine. There are no exhaust valves and no internal combustion; instead, the Stirling engine uses an external hot reservoir. Thus, the Stirling engine is very quiet and is used in applications where a quiet engine is important, such as on

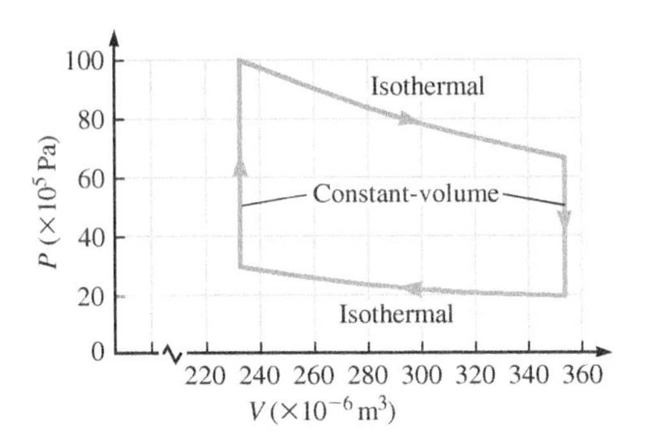

FIGURE P22.22

submarines. Today, the engine is being developed for a wider range of applications.

a. **E** Use Figure P22.22 to estimate the work done by a Stirling engine in one cycle.

b. **N** If the engine delivers 500 hp, how many cycles does it make per second?

c. **N** If the engine has an efficiency of 20%, how much input power does the engine require?

23. **A** **CASE STUDY** Consider the Otto engine as depicted in Figure 22.12. **a.** Show that the efficiency of an Otto engine can be expressed as

$$e_{\text{Otto}} = 1 - \frac{T_A}{T_B} = 1 - \frac{T_D}{T_C}$$

where the subscripts A through D refer to the states indicated in Figure 22.12.

b. The lowest temperature during an Otto cycle is T_A and the highest temperature is T_C. Find the difference ($e_{\text{Carnot}} - e_{\text{Otto}}$) for Carnot and Otto engines operating between these high and low temperatures. Does your answer to this part make sense? Explain.

Problems 24 and 25 are paired.

24. In the late 1800s, Rudolf Diesel invented an internal combustion engine that does not require a spark plug. Figure P22.24 is the *PV* diagram for an ideal **diesel engine**. Air is taken into the cylinder (at A) and compressed adiabatically to a high pressure (A to B). Then, diesel fuel (petroleum) is injected into the cylinder (at B). The fuel is ignited by the high pressure and high temperature of the air, and no spark is needed. Heat enters the system at constant pressure (B to C), and the gas then expands

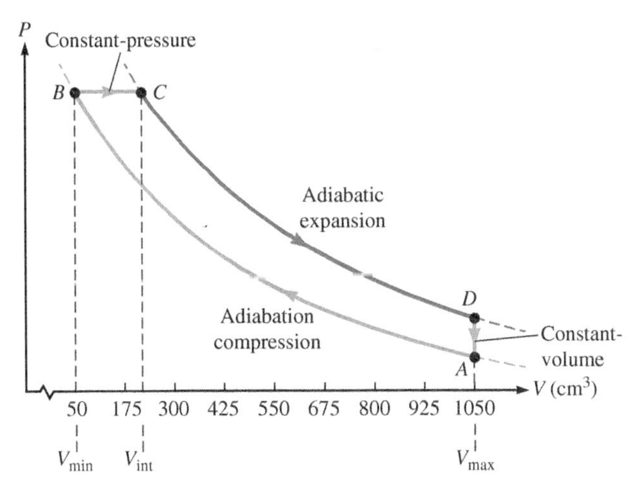

FIGURE P22.24 Problems 24 and 25.

adiabatically (C to D). Finally, the pressure drops at constant volume from D back to A. As shown in the figure, the gas has its maximum volume V_{max} during the constant-volume process from D to A, its minimum volume V_{min} is at point B, and at point C its volume is V_{int}, an intermediate value.

a. **A** Show that the efficiency of a diesel engine is given by

$$e_{\text{diesel}} = 1 - \frac{1}{\gamma}\left(\frac{r^{-\gamma} - R^{-\gamma}}{r^{-1} - R^{-1}}\right)$$

where $r = V_{\text{max}}/V_{\text{int}}$, $R = V_{\text{max}}/V_{\text{min}}$, and γ is the ratio of specific heats, $\gamma = C_P/C_V$.

b. **E** Usually, air is used, so $\gamma = 1.4$. Estimate the efficiency of the diesel engine represented by Figure P22.24.

25. A diesel engine (Problem 24) has the volume ratios $r = 4.57$ and $R = 21.0$. When air is taken into the cylinder, the pressure is 1.00 atm, and the temperature is 22.5°C.

a. **N** What is the temperature of the air when the fuel is injected?

b. **C** The air–fuel mixture must be at least 483 K for ignition to take place. Is ignition achieved? What minimum intake temperature would achieve ignition?

22-5 Case Study: Refrigerators

26. **N** What is the maximum possible coefficient of performance for an ideal refrigerator operating between a high-temperature reservoir at 2.0 K and a low-temperature reservoir at 0.020 K?

27. **N** A refrigerator is a heat engine running in reverse. What is the coefficient of performance κ of a Carnot engine with an efficiency of 43.0% that is run in reverse?

28. **C** In Concept Exercise 22.3, we found that you could not cool your house by opening your refrigerator door. Any baker, however, will tell you that you can heat a bakery by turning on the oven. Why can you heat a (small) house by opening the oven, but you cannot cool it by opening the refrigerator? Be sure your explanation illustrates the fundamental difference between refrigerators and ovens.

29. **N** What is the best possible coefficient of performance for a household refrigerator in which an internal temperature of 3.0°C is desired? Assume room temperature is 20.0°C.

30. **C** An air conditioner is similar to a refrigerator. By doing work, the air conditioner takes heat Q_c from the cool building and expels heat Q_h out into the hot environment. Why is a room air conditioner placed in a window or through a hole in the wall so that part of the unit is outside the room?

31. **N** The coefficient of performance of a refrigerator that takes 210.0 J of energy from the cold reservoir during each cycle is $\kappa = 4.00$. **a.** What is the work done during each cycle? **b.** How much energy is exhausted into the hot reservoir during each cycle?

32. **C** Consider the coefficient of performance of an ideal refrigerator $\kappa = T_c/(T_h - T_c)$ as in Equation 22.18. Notice that the performance depends on the temperatures of the hot and cold reservoirs. Suppose you plug in a refrigerator and the temperatures of the hot and cold reservoirs are initially close in value. Model your refrigerator as an ideal one and describe what happens to the coefficient of performance as time goes on and you allow the refrigerator to run continuously.

33. **A** Using the derivation of the Carnot engine efficiency as a guide (Section 22-4), derive Equation 22.18, the coefficient of performance for an ideal refrigerator: $\kappa = T_c/(T_h - T_c)$.

34. A *heat pump* is a device that can heat a building in winter by taking heat Q_c from outside at low temperature and giving off heat Q_h to the warm interior by doing work. The operation of a heat pump is much like that of a refrigerator. Heat exchangers inside and outside the building resemble the coils inside the refrigerator and the condenser outside it (Fig. 22.14).

a. C Draw an energy transfer diagram for a heat pump.

b. A Because the goal of a heat pump is to deliver heat Q_h into the building, the coefficient of performance is defined as $K \equiv |Q_h|/W$. Assume an ideal heat pump can be modeled as a Carnot cycle and derive an expression for the ideal heat pump's coefficient of performance in terms of the indoor and outdoor temperatures.

c. N What is the coefficient of performance for an ideal heat pump if the indoor temperature is 75.0°F and the outdoor temperature is 20.0°F?

22-6 Entropy

35. N What is the change in entropy of 100.0 g of water when heated so that its temperature changes from 20.0°C to 100.0°C? The specific heat capacity of water is 4186 J/(kg · K).

36. E Estimate the change in entropy of the Universe if you pop a balloon.

37. N An ideal gas, initially at atmospheric pressure and a temperature of 300.0 K, is compressed from a volume of 1.00 L to 0.50 L. What is the change of entropy of the gas if this process is performed **a.** adiabatically and **b.** isothermally?

38. E Estimate the change in entropy of a pond when it freezes.

39. N What is the change in entropy of a 125-g sample of liquid water at 0°C that slowly freezes at this temperature?

40. N An ideal gas undergoes a free expansion, doubling in volume. If there are 15 mol of the gas, what is the change in the gas's entropy? Use Example 22.6 to check your results.

41. N A 30.0-mol monatomic carbon gas undergoes a constant-pressure process at a pressure of 2.35 atm from an initial volume of 0.525 m³ to a final volume of 0.340 m³. What is **a.** the work done on the gas, **b.** the heat transferred and **c.** the change in entropy during this process?

42. N A cold reservoir at T_c = 15.0°C and a hot reservoir at T_h = 450.0°C are connected via a steel bar. The energy transferred from the hot reservoir to the cold reservoir in this irreversible process is 1.85×10^4 J. What is the change in the entropy of **a.** the cold reservoir and **b.** the hot reservoir during this process? **c.** Ignoring the change in entropy of the steel bar, what is the change in the entropy of the Universe during this process?

43. N A cold block of copper is placed in contact with a hot block of copper. The blocks have the same volume, and each has a mass of 0.345 kg. The original temperature of the cold copper is 6.25°C, and the original temperature of the hot copper is 45.2°C. **a.** What is the final equilibrium temperature? **b.** What is the change in entropy for each block? **c.** What is the change in entropy of the Universe?

44. N A railway car with mass $m = 5.50 \times 10^3$ kg moving at 8.00 m/s has a totally inelastic head-on collision with a second identical railway car also moving at 8.00 m/s. The air temperature at the collision site is initially 15.0°C. Consider the system to consist of the two railway cars and the surrounding air. If the kinetic energy lost in the collision goes into the air surrounding the cars, what is the change in entropy of the air at the collision site as a result of the collision?

45. N A nonporous membrane divides a vessel in the form of two identical 5.00-L chambers that are welded together (Fig. P22.45). Chamber A is filled with 1.22 mol of argon gas, and chamber B is filled with 1.22 mol of krypton gas. What is the change in entropy of the system if the membrane is removed and the gases mix together?

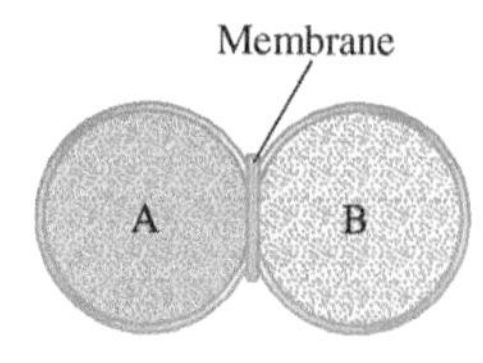

FIGURE P22.45

22-7 Second Law of Thermodynamics, General Statements

46. C If you stretch a rubber band, does the entropy of the rubber band change? If so, does it increase or decrease? How do you know?

47. A spring obeys Hooke's law and has a spring constant of 125 N/m. You stretch the spring from 1.60 m to 1.85 m at 22.2°C.

a. N What is the change in the spring's entropy?

b. C Is the process reversible? Explain.

c. N What is the change in the Universe's entropy?

48. C Two students discuss entropy on a very cold day.

Avi: So, the water in the birdbath that froze last night is more ordered now that it's ice, right? And the entropy has gone down?

Cameron: Well, I think entropy can't decrease in natural processes, so it seems like it can't decrease. Maybe it stays the same?

Avi: I thought it just had to do with heat coming out of the water to make ice. Is entropy something else?

Cameron: Entropy is unchanged for a reversible process and increases for an irreversible process.

What part of this conversation do you agree with? Which claims do you disagree with?

22-8 Order and Disorder

49. C Consider this analogy: A neatly arranged sock drawer is like a cool system, and a messy basket of laundry is like a hot system. Use this analogy to explain why adding the same amount of heat to either system causes the cool system's entropy to increase much more than the hot system's entropy.

50. C Describe three examples from your everyday experience that are representative of the second law of thermodynamics expressed as "the Universe is becoming more disordered."

51. C A physics student tells her dad that she cannot fold her laundry because it would violate the second law of thermodynamics. How does her dad (also a physicist) convince her that folding laundry does not violate any laws of physics?

52. C A watch is clearly a very ordered system but, the process of producing a watch cannot violate the second law of thermodynamics. If making a watch produces an ordered system, how can the overall entropy of the Universe increase? Name as many specific parts of the Universe as possible whose entropy must have increased in the process.

53. C The principles of biological evolution and the development of life on planet Earth seem to require a decrease in entropy to occur. Does that violate the second law of thermodynamics? Justify your answer by explaining why it is a violation or by proposing what could account for a necessary decrease in entropy, allowing life to develop.

22-9 Entropy, Probability, and the Second Law

54. E You and your roommate have split your room in two; your side of the room is one-third the volume of the whole room. Estimate the probability of finding all the air molecules in your part of the room, leaving your roommate with no air.

55. N If you roll two identical six-sided dice, what is the probability that you roll "snake eyes" (both dice show a single dot on the top surface)?

56. N If you toss a coin 100 times, what is the probability of finding between 0 and 10 heads? How does your answer compare with your answer to Concept Exercise 22.5? Use this comparison to make sense of your answer to this problem.

57. N What is the entropy of a freshly shuffled deck of 52 playing cards?

58. Feeling lucky, George is rolling a pair of dice at a craps table.
  a. N His first roll of the dice is a 2. How many microscopic states are there for this roll?
  b. N His second roll of the dice is an 8. How many microscopic states are there for this roll?
  c. C Which of the two rolls corresponds to a state with lower entropy?

Problems 59 and 60 are paired.

59. G Using Table 22.1 (page 675) and given that there are 1.27×10^{30} microstates in total, plot the probability for each outcome versus the number of heads for values listed in the table. In words, describe the shape of the curve and what it says about the likely outcome of flipping 100 coins.

60. **a.** N Using Equation 22.26, create a table similar to Table 22.1, but do so for the case of flipping six coins. Include the outcomes of 0, 1, 2, 3, 4, 5, and 6 heads as possible microscopic states. Calculate the number of microscopic states, the probability, and the entropy for each macroscopic state.
  b. G Plot the entropy of each macroscopic state versus number of heads.

General Problems

61. N The Carnot efficiency of a heat engine operating between two reservoirs is 54.0%. If the temperature of the hot reservoir is 355°C, what is the temperature of the cold reservoir?

62. C As you have seen in the discussion of the Carnot engine, it is impractical to think that we could have a cold reservoir at absolute zero or an infinitely energetic hot reservoir, which means that we cannot not have a 100% efficient engine. What other limitations do we have when it comes choosing two real heat reservoirs? Consider what reservoirs we might have readily available and the energy required to create a hot or cold reservoir.

Problems 63 and 64 are paired.

63. C Consider two different scenarios: (1) An ice cube melts to form water. (2) The same amount of water boils to form steam. Which of these processes would lead to a larger change in entropy? Justify your answer.

64. **a.** N What is the entropy change for 5.0 g of ice that melts at 0.0°C to form water? The heat of fusion of water is 333 J/g. **b.** What is the entropy change for 5.0 g of water that boils at 100°C to form steam? The heat of vaporization for water is 2260 J/g.

65. N A Carnot engine operates between a cold reservoir with T_c = 22.0°C and a hot reservoir with T_h = 300.0°C. **a.** Consider keeping the temperature of the hot reservoir fixed at 300.0°C while the temperature of the cold reservoir is variable. What is the rate of change in efficiency of the engine per change in T_c (what is de_{Carnot}/dT_c)? **b.** Consider keeping the temperature of the cold reservoir fixed at 22.0°C while the temperature of the hot reservoir is variable. What is the rate of change in efficiency of the engine per change in T_h (what is de_{Carnot}/dT_h), when T_h = 300.0°C?

66. N A 3.00-mol diatomic ideal gas undergoes a free expansion. The constant temperature during the process is 350 K, and the gas begins with a pressure of 1.5 atm and finishes with a pressure of 1.0 atm. What is the change in entropy of the gas as it undergoes this free expansion?

Problems 67 and 68 are paired.

67. N A 1.55-kg sample of liquid water at 50°C is heated and boiled away, and eventually it is converted entirely to water vapor. The resulting water vapor is then heated until it reaches a final temperature of 115°C. Assume the specific heat of the liquid water and the water vapor are 4190 J/(kg · K) and 2010 J/(kg · K), respectively, and the latent heat of vaporization is 2.256×10^6 J/kg. No work is performed on or by the water or the water vapor. **a.** What is the change in entropy of the liquid water as it is heated to the vaporization point? **b.** What is the change in entropy of the water vapor as it is heated from the vaporization point to the final temperature of 115°C? **c.** What is the change in entropy of the water as it undergoes the phase transformation?

68. C Consider heating water as described in Problem 67. **a.** Describe what is happening in the water and water vapor as they are heated that would indicate that the entropy is increasing. What might we observe happening that leads us to say that the "entropy has increased" in each case? **b.** What is happening during the phase change from liquid to gas that indicates that the entropy is increasing?

69. N One mole of an ideal gas is taken through a Carnot cycle in which the gas extracts 1.55 kJ of energy from the hot reservoir during each cycle. The isothermal expansion for this cycle occurs at 300.0°C, and the isothermal compression occurs at 85.0°C. **a.** How much energy is exhausted to the cold reservoir during each cycle? **b.** What is the net work done by the gas during each cycle?

70. A The free expansion of an ideal gas is described in Section 21-7. Show that the entropy change of an ideal gas during a free expansion (Fig. 21.20) is given by Equation 22.24,

$$\Delta S = nR \ln\left(\frac{V_f}{V_i}\right)$$

Hint: A free expansion is not a reversible process, so we cannot use Equation 22.22,

$$\Delta S = S_f - S_i = \left[\int_i^f \frac{dQ}{T}\right]_{\text{any reversible path}}$$

Instead, we must consider an equivalent reversible process that takes the gas from its initial state to its final state. Like a free expansion, a reversible isothermal process can take the ideal gas from state i to state f with no change in thermal energy. Start by applying the first law of thermodynamics to an isothermal expansion.

71. N A system consisting of 10.0 g of water at a temperature of 20.0°C is converted into ice at −10.0°C at constant atmospheric pressure. Calculate the total change in entropy of the system. Assume the specific heat of water is 4.19×10^3 J/(kg · K), the specific heat of ice is 2.10×10^3 J/(kg · K), and the latent heat of fusion is 3.33×10^5 J/kg.

72. C When a gas undergoes a free expansion, its entropy increases according to

$$\Delta S = nR \ln\left(\frac{V_f}{V_i}\right)$$

Explain the increase in entropy in terms of the change in order (or disorder).

73. N Figure P22.73 illustrates the cycle *ABCA* for a 2.00-mol sample of an ideal diatomic gas, where the process *CA* is a reversible isothermal expansion. What is **a.** the net work done by the gas during one cycle? **b.** How much energy is added to the gas by heat during one cycle? **c.** How much energy is exhausted from the gas by heat during one cycle? **d.** What is the efficiency of the cycle? **e.** What would be the efficiency of a Carnot engine operated between the temperatures at points *A* and *B* during each cycle?

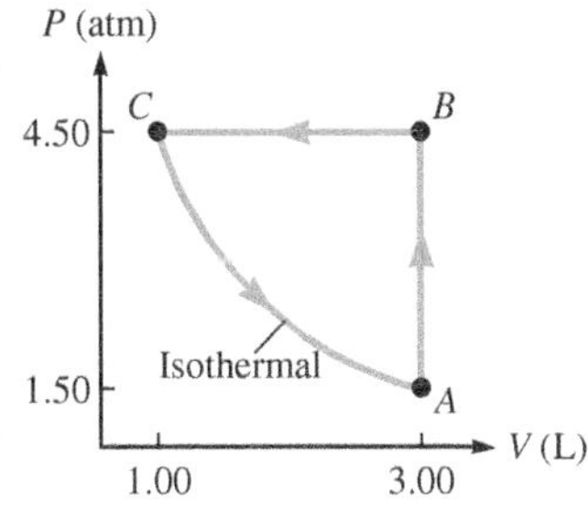

FIGURE P22.73

74. N Experimental measurements of heat capacities of graphite at constant pressure show that over a wide range of temperatures, the molar specific heat varies with temperature as

$$C_p = a + bT - \frac{c}{T^2}$$

where $a = 16.86$ J/K, $b = 4.77 \times 10^{-3}$ J/K^2, and $c = 8.54 \times 10^5$ J · K. Calculate the change in entropy when one mole of graphite is heated at constant pressure from 250°C to 45°C.

75. N A cylinder with a movable piston contains 2.00 mol of a diatomic ideal gas at atmospheric pressure and a temperature of 325.0 K. The air within the cylinder is cooled slowly to 250.0 K. **a.** If the volume of the cylinder does not change, what is the change in entropy of the gas? **b.** If the pressure inside the cylinder does not change, what is the change in entropy of the gas?

76. C You pop a DVD of your favorite movie into your DVD player. How might the example of watching the playback of a movie be thought of as being similar to the Clausius statement of the second law of thermodynamics? Consider the ability to watch the movie as it is playing or rewinding. In which direction do things make sense?

77. N During each cycle, a heat engine extracts 3.25×10^4 J from a hot reservoir at $T_h = 400$°C and does 4.00 kJ of work. The cold reservoir is at $T_c = 22.0$°C. **a.** What is the change in entropy of the Universe for each cycle? **b.** How much more work would a Carnot engine operating between these reservoirs have performed?

78. N The coefficient of performance of a household heater that consumes 850.0 W of power is $\kappa = 2.95$, where $\kappa = |Q_h|/W$ represents the coefficient of performance for a heater. **a.** If the heater is in operation for 6.00 h each night, how much energy does it deliver? **b.** What is the energy extracted by the heater from the outside air (the cold reservoir) during this time?

79. N A sample of 1.00 mol of a monatomic ideal gas is initially at atmospheric pressure and occupies a volume of 5.00 L. It then undergoes a reversible isothermal expansion followed by a reversible adiabatic contraction such that its final pressure is 1.00 atm and its final volume is 20.0 L. What is the change in entropy of the gas after the two processes?

80. N Consider a chemical reaction A + B → D. The three substances have molar specific heat capacities at constant pressure given by $C_A = 2\sqrt{T}$, $C_B = 4\sqrt{T}$, and $C_D = 3\sqrt{T}$ where T is in kelvin. If this reaction occurs at a temperature of 300.0 K, use Eq. 22.22 to determine how much heat is absorbed or released if 1.0 mol of substance D is produced. To determine the relative entropy of substances A, B, and D, respectively, assume the heat capacities of the reactants A and B and the product D are zero at $T = 0$ K.

81. N Consider a system A with two subsystems A_1 and A_2. The total number of microstates of A_1 and A_2, respectively, are 10^{10} and 2×10^{10}. **a.** What is the number of microstates available to the combined system, A? **b.** What are the entropies of the system A and the subsystems A_1 and A_2?

APPENDIX

A

Mathematics

This appendix is not meant to serve as a tutorial or review. It is a short list of useful formulas. A more comprehensive list may be found in various handbooks such as the *Handbook of Chemistry and Physics* (Boca Raton, FL: CRC Press, published annually). These handbooks are available in hardcovcr and as e-books, and they make great birthday presents for science and engineering students.

A-1 Algebra and geometry 683

A-2 Trigonometry 684

A-3 Calculus 685

A-4 Propagation of uncertainty 687

A-1 Algebra and Geometry

Quadratic formula: If $ax^2 + bx + c = 0$, then $x = \dfrac{-b \pm \sqrt{b^2 - 4ac}}{2a}$

Factorial notation: $n! = n(n-1)\cdots 2 \cdot 1$

Binomial theorem: $(1+x)^n = 1 + \dfrac{nx}{1!} + \dfrac{n(n-1)x^2}{2!} + \cdots \quad (x < 1)$

Commonly used approximation: $(1+x)^n \approx 1 + nx \; (x \ll 1)$

Exponential expansion: $e^x = 1 + x + \dfrac{x^2}{2!} + \dfrac{x^3}{3!} + \cdots$

Logarithms (any base):

$$\log(x)^n = n\log x$$
$$\log(AB) = \log A + \log B$$
$$\log(A/B) = \log A - \log B$$

Logarithms (base 10):

$$10^{\log x} = x$$
$$\log 10^x = x$$

Natural logarithms (base *e*):

$$e^{\ln x} = x$$
$$\ln e^x = x$$

Area of common shapes:

Rectangle $A = \ell w$		Circle $A = \pi r^2$	
Parallelogram $A = bh$	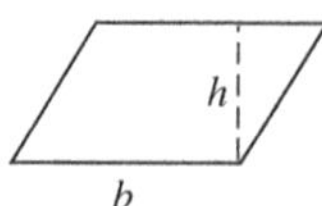	Ellipse $A = \pi ab$	
Triangle $A = \frac{1}{2}bh$	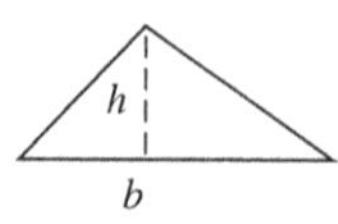		

Equation of a straight line of slope *m* and *y* intercept *b*: $y = mx + b$

Equation of a parabola: $y = ax^2 + bx + c$

Equation of a circle: $x^2 + y^2 = r^2$

Equation of an ellipse: $\left(\frac{x}{a}\right)^2 + \left(\frac{y}{b}\right)^2 = 1$

Circumference of a circle: $c = 2\pi r$

Volume and surface area of common solids:

Rectangular box	$V = \ell wh$	$A = 2(\ell w + \ell h + wh)$	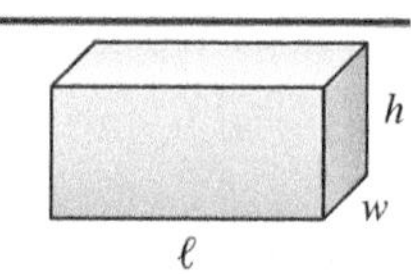
Right circular cylinder	$V = \pi r^2 h$	$A = 2\pi r^2 + 2\pi rh$	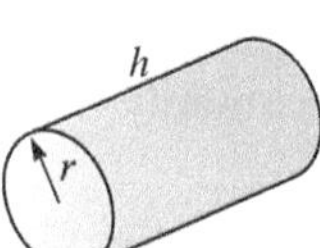
Sphere	$V = \frac{4}{3}\pi r^3$	$A = 4\pi r^2$	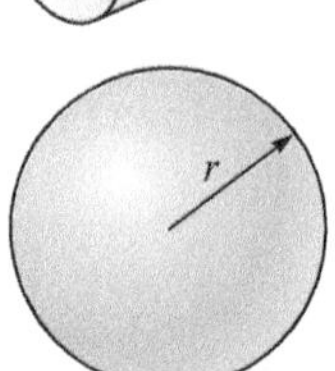

A-2 Trigonometry

Pythagorean theorem (applied to a right triangle): $x^2 + y^2 = r^2$

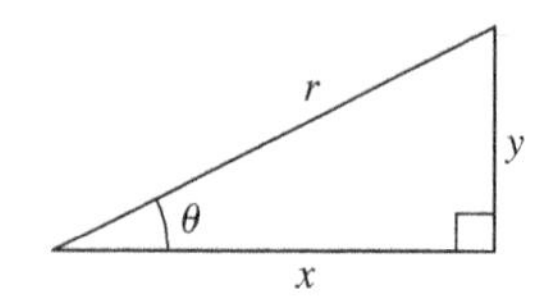

Trigonometric functions (applied to a right triangle):

$\sin\theta = \frac{y}{r}$	$\csc\theta = \frac{1}{\sin\theta} = \frac{r}{y}$
$\cos\theta = \frac{x}{r}$	$\sec\theta = \frac{1}{\cos\theta} = \frac{r}{x}$
$\tan\theta = \frac{y}{x}$	$\cot\theta = \frac{1}{\tan\theta} = \frac{x}{y}$

Commonly used approximations:

$$\sin\theta \approx \tan\theta \approx \theta \text{ for small } \theta \text{ in radians}$$

$$\cos\theta \approx 1 \text{ for small } \theta$$

Trigonometric identities:

$\sin(-\theta) = -\sin\theta$	$\sin 2\theta = 2\sin\theta\cos\theta$
$\cos(-\theta) = \cos\theta$	$\cos 2\theta = \cos^2\theta - \sin^2\theta$
$\tan(-\theta) = -\tan\theta$	
$\sin(90° - \theta) = \sin\left(\frac{\pi}{2} - \theta\right) = \cos\theta$	$\sin(\alpha \pm \beta) = \sin\alpha\cos\beta \pm \cos\alpha\sin\beta$ $\cos(\alpha \pm \beta) = \cos\alpha\cos\beta \mp \sin\alpha\sin\beta$
$\cos(90° - \theta) = \cos\left(\frac{\pi}{2} - \theta\right) = \sin\theta$	
$\sin(90° + \theta) = \sin\left(\frac{\pi}{2} + \theta\right) = \cos\theta$	
$\cos(90° + \theta) = \cos\left(\frac{\pi}{2} + \theta\right) = -\sin\theta$	
$\tan\theta = \frac{\sin\theta}{\cos\theta}$	$\sin\alpha \pm \sin\beta = 2\sin\frac{1}{2}(\alpha \pm \beta)\cos\frac{1}{2}(\alpha \mp \beta)$ $\cos\alpha + \cos\beta = 2\cos\frac{1}{2}(\alpha + \beta)\cos\frac{1}{2}(\alpha - \beta)$
$\sin^2\theta + \cos^2\theta = 1$	$\cos\alpha - \cos\beta = -2\sin\frac{1}{2}(\alpha + \beta)\sin\frac{1}{2}(\alpha - \beta)$

A-3 Calculus

Derivatives

In this section, f, g, u, and v are functions of x; a, b, C, and n are constants.

Derivative of $f(x)$: $\dfrac{df}{dx} = \lim\limits_{\delta x \to 0} \dfrac{f(x + \delta x) + f(x)}{\delta x}$

Derivative of a constant: $\dfrac{dC}{dx} = 0$

The power rule: $\dfrac{dx^n}{dx} = nx^{n-1}$

The derivative of a sum: $\dfrac{d}{dx}[f(x) + g(x)] = \dfrac{df}{dx} + \dfrac{dg}{dx}$

The product rule: $\dfrac{d}{dx}[f(x)g(x)] = g\dfrac{df}{dx} + f\dfrac{dg}{dx}$

Special case of the product rule: $\dfrac{d}{dx}[Cf(x)] = C\dfrac{df}{dx}$

The chain rule: $\dfrac{df}{dx} = \dfrac{df}{du}\dfrac{du}{dx}$

Second derivative: $\dfrac{d^2f}{dx^2} = \dfrac{d}{dx}\left(\dfrac{df}{dx}\right)$

Derivatives of special functions:

$\frac{d}{dx}\ln x = \frac{1}{x}$	$\frac{d}{dx}\sin x = \cos x$
$\frac{d}{dx}e^x = e^x$	$\frac{d}{dx}\cos x = -\sin x$
$\frac{d}{dx}e^u = e^u\frac{du}{dx}$	$\frac{d}{dx}\tan x = \sec^2 x$

L'Hôpital's rule: If the limit of the numerator $f(x)$ and of the denominator $g(x)$ of a fraction both approach zero or both approach infinity, the limit of the fraction is indeterminate in the form of type $0/0$ or ∞/∞. In such cases, it may be possible to find the limit using L'Hôpital's rule:

$$\lim\frac{f(x)}{g(x)} = \lim\frac{df/dx}{dg/dx}$$

Essentially, L'Hôpital's rule replaces the limit of a fraction with the limit of a new fraction, where the numerator and denominators are the derivatives of their original counterparts.

Integrals

Indefinite integral or antiderivative: $\int df(x) = f(x) + C$

Indefinite integral of a constant: $\int a dx = ax + C$

The power rule: $\int x^n dx = \frac{x^{n+1}}{n+1} + C\ (n \neq -1)$

Indefinite integral of a sum: $\int [f(x) + g(x)]dx = \int f(x)\,dx + \int g(x)\,dx$

Indefinite integrals of particular functions:

$\int \frac{1}{x}dx = \ln\lvert x\rvert + C$	$\int (\sin ax)dx = -\frac{1}{a}\cos ax + C$	$\int \frac{dx}{\sqrt{x^2 + a^2}} = \ln\lvert x + \sqrt{x^2 + a^2}\rvert + C$
$\int e^x dx = e^x + C$	$\int (\cos ax)\,dx = \frac{1}{a}\sin ax + C$	$\int \frac{dx}{\sqrt{a^2 - x^2}} = \sin^{-1}\frac{x}{\lvert a\rvert} + C$
$\int e^{-ax}dx = -\frac{1}{a}e^{-ax} + C$	$\int (\tan ax)\,dx = -\frac{1}{a}\ln(\cos ax) + C$	$\int \frac{dx}{x^2 + a^2} = \frac{1}{a}\tan^{-1}\frac{x}{a} + C$
	$\int \sin^2 ax\,dx = \frac{x}{2} - \frac{\sin 2ax}{4a} + C$	$\int \frac{dx}{(x^2 + a^2)^{3/2}} = \frac{x}{a^2\sqrt{x^2 + a^2}} + C$
	$\int \cos^2 ax\,dx = \frac{x}{2} + \frac{\sin 2ax}{4a} + C$	$\int \frac{x\,dx}{(x^2 + a^2)^{3/2}} = -\frac{1}{\sqrt{x^2 + a^2}} + C$

U substitution integration method: $\int f(x)\frac{df}{dx}dx = \int u\,du$, where $u \equiv f(x)$ and $du = \frac{df}{dx}dx$

Integration by parts: $\int udv = uv - \int vdu$

Definite integrals and the Fundamental Theorem of Calculus:
If $F(x)$ is continuous on the interval from $x = a$ to b, then

$$\int_a^b F(x)\,dx = f(x)\rvert_a^b = f(b) - f(a)$$

where $f(x)$ is the antiderivative of $F(x)$ and a and b are known as the ***limits***.

Average value of a function in the interval from $x = a$ to $x = b$: $f_{av} = \dfrac{1}{b-a}\displaystyle\int_a^b f(x)\,dx$

Definite integrals of particular functions:

$\int_0^\infty x^n e^{-ax}\,dx = \dfrac{n!}{a^{n+1}}$	$\int_0^\pi (\sin^2 ax)\,dx = \dfrac{\pi}{2}$
$\int_0^\infty e^{-ax^2}\,dx = \dfrac{1}{2}\sqrt{\dfrac{\pi}{a}}$	$\int_0^\pi (\cos^2 ax)\,dx = \dfrac{\pi}{2}$
$\int_0^\infty xe^{-ax^2}\,dx = \dfrac{2}{a}$	$\int_0^\infty e^{-ax}(\sin nx)\,dx = \dfrac{n}{a^2+n^2}\ (a > 0)$
$\int_0^\infty x^2e^{-ax^2}\,dx = \dfrac{1}{4}\sqrt{\dfrac{\pi}{a^3}}$	$\int_0^\infty e^{-ax}(\cos nx)\,dx = \dfrac{a}{a^2+n^2}\ (a > 0)$

A-4 Propagation of Uncertainty

Physics is an experimental science that relies on measurements. All measurements have uncertainty or error. We often seek quantities that result from the combination of uncertain measurements, so the resulting quantity is also uncertain. One way to estimate the uncertainty δQ in a quantity q calculated from uncertain measurements is to find the best estimate Q and then the extreme possible values of q.

For example, if $q = 2(a+b)/c$, $a = 5.2 \pm 0.2$, $b = 7.5 \pm 0.3$, and $c = 57.6 \pm 0.5$, the best estimate of q is

$$Q = \frac{2(5.2 + 7.5)}{57.6} = 0.44$$

The maximum value of q is

$$Q_{max} = \frac{2(5.4 + 7.8)}{57.1} = 0.46$$

where we chose the maximum possible values in the numerator and the minimum possible value in the denominator. Similarly, we find that the minimum value of q is

$$Q_{min} = \frac{2(5.0 + 7.2)}{58.1} = 0.42$$

The quantity must fall between its minimum and maximum values: $0.42 \leq q \leq 0.46$. So,

$$q = Q \pm \delta Q = 0.44 \pm 0.02$$

where $\delta Q = Q_{max} - Q = Q - Q_{min}$.

We apply this technique to come up with three rules for propagating uncertainty.

Sums and Differences

If $q = a + b$, $a = A \pm \delta A$, and $b = B \pm \delta B$, then

$$q = Q \pm \delta Q = (A + B) \pm (\delta A + \delta B)$$

If $q = a - b$, $a = A \pm \delta A$, and $b = B \pm \delta B$, then

$$q = Q \pm \delta Q = (A - B) \pm (\delta A + \delta B)$$

When measured quantities are added or subtracted, their errors add.[1]

[1]If the original uncertainties δA and δB are random and independent, then $\delta A + \delta B$ is an overestimate of the propagated error.

Products, Quotients, and Powers

If $a = A \pm \delta A$, the *fractional uncertainty* in a is $\delta A/|A|$. If $q = ab$, $a = A \pm \delta A$, and $b = B \pm \delta B$, then $Q = AB$, and the fractional uncertainty in q is approximately

$$\frac{\delta Q}{|Q|} \approx \frac{\delta A}{|A|} + \frac{\delta B}{|B|} \tag{1}$$

and

$$q = Q \pm \delta Q \approx AB \pm |AB|\left(\frac{\delta A}{|A|} + \frac{\delta B}{|B|}\right)$$

If $q = a/b$, $a = A \pm \delta A$, and $b = B \pm \delta B$, then $Q = A/B$. The fractional uncertainty in q is given by Equation (1), and

$$q = Q \pm \delta Q \approx \frac{A}{B} \pm \left|\frac{A}{B}\right|\left(\frac{\delta A}{|A|} + \frac{\delta B}{|B|}\right)$$

If $q = a^n$ and $a = A \pm \delta A$, then $Q = A^n$, and the fractional uncertainty in q is

$$\frac{\delta Q}{|Q|} \approx n\frac{\delta A}{|A|}$$

and

$$q = Q \pm \delta Q \approx A^n \pm |A^n|\left(n\frac{\delta A}{|A|}\right)$$

When measured quantities are multiplied or divided, their fractional errors add.

Multiplication by an Exact Number

If $q = ab$, a is exact, and $b = B \pm \delta B$, then

$$q = Q \pm \delta Q = aB \pm |a|\delta B$$

APPENDIX

B

Reference Tables

B-1 Symbols and units 689

B-2 Conversion factors 691

B-3 Some astronomical data 692

B-4 Rough magnitudes and scales 693

Periodic table of the elements 696

B-1 Symbols and Units

Prefixes for powers of 10

Name	Abbreviation	Value
yocto	y	10^{-24}
zepto	z	10^{-21}
atto	a	10^{-18}
femto	f	10^{-15}
pico	p	10^{-12}
nano	n	10^{-9}
micro	μ (Greek letter "mu")	10^{-6}
milli	m	10^{-3}
centi	c	10^{-2}
deci	d	10^{-1}
deka	da	10^{1}
hecto	h	10^{2}
kilo	k	10^{3}
mega	M	10^{6}
giga	G	10^{9}
tera	T	10^{12}
peta	P	10^{15}
exa	E	10^{18}
zetta	Z	10^{21}
yotta	Y	10^{24}

Greek alphabet

Name	Uppercase	Lowercase	Name	Uppercase	Lowercase	Name	Uppercase	Lowercase
Alpha	A	α	Iota	I	ι	Rho	P	ρ
Beta	B	β	Kappa	K	κ	Sigma	Σ	σ
Gamma	Γ	γ	Lambda	Λ	λ	Tau	T	τ
Delta	Δ	δ	Mu	M	μ	Upsilon	Υ	υ
Epsilon	E	ε	Nu	N	ν	Phi	Φ	φ
Zeta	Z	ζ	Xi	Ξ	ξ	Chi	X	χ
Eta	H	η	Omicron	O	o	Psi	Ψ	ψ
Theta	Θ	θ	Pi	Π	π	Omega	Ω	ω

SI base units

Dimension	SI unit	Symbol	Definition
Time	second	s	1 second is the duration of 9,192,631,770 periods of the radiation (corresponding to the transition between hyperfine levels of the ground state) of the cesium-133 atom.
Length	meter	m	1 meter is the distance light travels through empty space in 1/299,729,458 second.
Mass	kilogram	kg	1 kilogram is the mass of a prototype (a particular platinum-iridium cylinder).
Thermodynamic temperature	kelvin	K	1 kelvin is the fraction 1/273.16 of the thermodynamic temperature of the triple point of water.
Amount of substance	mole	mol	1 mole is the amount of a substance of a system that contains as many elementary entities as there are atoms in 0.012 kg of carbon-12.
Electrical current	ampere	A	1 ampere is the constant current that, if maintained in two straight parallel conductors of infinite length, of negligible circular cross section, and placed 1 m apart in a vacuum, would produce between these conductors a force per unit length equal to 2×10^{-7} N/m.
Luminous intensity	candela	cd	1 candela is the luminous intensity, in a given direction, of a source that emits monochromatic radiation of frequency 540×10^{12} Hz and that has a radiant intensity in that direction of 1/683 watt per steradian.

Source: Adapted from "Definitions of the SI base units," National Institute of Standards and Technology. See http://physics.nist.gov/cuu/Units/current.html.

Symbols and abbreviations for units

Unit	Symbol	Unit	Symbol
ampere	A	light-year	ly
atmosphere	atm	liter	L
atomic mass unit	U	meter	m
British thermal unit	Btu	mile	mi
calorie	cal	miles per hour	mph
coulomb	C	millimeter of mercury (torricelli)	mm Hg (torr)
day	d	minute	min
degree Celsius	°C	mole	mol
degree Fahrenheit	°F	newton	N
electron volt	eV	ohm	Ω
farad	F	pascal	Pa
foot	ft	pound	lb
gallon	gal	pounds per square inch	psi
gauss	G	radian	rad
gram	g	revolution	rev
henry	H	revolutions per minute	rpm
hertz	Hz	second	s
horsepower	hp	tesla	T
inch	in.	volt	V
joule	J	watt	W
kelvin	K	weber	Wb
kilocalorie	Cal	yard	yd
kilogram	kg	year	yr
kilowatt-hour	kWh		

B-2 Conversion Factors

Length

	meter	cm	km	in.	ft	mi
1 meter	1	10^2	10^{-3}	39.37	3.281	6.214×10^{-4}
1 centimeter	10^{-2}	1	10^{-5}	0.3937	3.281×10^{-2}	6.214×10^{-6}
1 kilometer	10^3	10^5	1	3.937×10^4	3281	0.6214
1 inch	2.540×10^{-2}	2.540	2.540×10^{-5}	1	8.333×10^{-2}	1.578×10^{-5}
1 foot	0.3048	30.48	3.048×10^{-4}	12	1	1.894×10^{-4}
1 mile	1609	1.609×10^5	1.609	6.336×10^4	5280	1
1 angstrom	10^{-10}	10^{-8}	10^{-13}	3.937×10^{-9}	3.281×10^{-10}	6.214×10^{-14}
1 AU	1.496×10^{11}	1.496×10^{13}	1.496×10^8	5.890×10^{12}	4.908×10^{11}	9.296×10^7
1 nautical mile	1852	1.852×10^5	1.852	7.291×10^4	6076	1.151
1 light-year	9.461×10^{15}	9.461×10^{17}	9.461×10^{12}	3.725×10^{17}	3.104×10^{16}	5.878×10^{12}
1 parsec	3.086×10^{16}	3.086×10^{18}	3.086×10^{13}	1.215×10^{18}	1.012×10^{17}	1.917×10^{13}
1 yard	0.9144	91.44	9.144×10^{-4}	36	3	5.682×10^{-4}

Mass

	kilogram	g	slug	u
1 kilogram	1	10^3	6.852×10^{-2}	6.022×10^{26}
1 gram	10^{-3}	1	6.852×10^{-5}	6.022×10^{23}
1 slug	14.59	1.459×10^4	1	8.786×10^{27}
1 atomic mass unit	1.661×10^{-27}	1.661×10^{-24}	1.138×10^{-28}	1

Force

	newton	dyne	lb	oz	ton
1 newton	1	10^5	0.2248	3.597	1.124×10^{-4}
1 dyne	10^{-5}	1	2.248×10^{-6}	3.597×10^{-5}	1.124×10^{-9}
1 pound	4.448	4.448×10^5	1	16	5×10^{-4}
1 ounce	0.2780	2.780×10^4	6.250×10^{-2}	1	3.125×10^{-5}
1 ton	8.896×10^3	8.896×10^8	2000	3.2×10^4	1

Pressure

	pascal	atm	Torr (mm Hg)	psi	dyne/cm^2
1 pascal	1	9.869×10^{-6}	7.501×10^{-3}	1.450×10^{-4}	10
1 atm	1.013×10^5	1	760	14.70	1.013×10^6
1 Torr	1333	1.316×10^{-2}	1	0.1934	1.333×10^4
1 psi	6.895×10^3	6.805×10^{-2}	51.71	1	6.895×10^4
1 dyne/cm^2	0.1	9.869×10^{-7}	7.501×10^{-4}	1.405×10^{-5}	1

Energy

	joule	erg	ft·lb	cal	eV
1 joule	1	10^7	0.7376	0.2389	6.242×10^{18}
1 erg	10^{-7}	1	7.376×10^{-8}	2.389×10^{-8}	6.242×10^{11}
1 ft·lb	1.356	1.356×10^7	1	0.3238	8.464×10^{18}
1 cal	4.184	4.184×10^7	3.088	1	2.612×10^{19}
1 eV	1.602×10^{-19}	1.602×10^{-19}	1.182×10^{-19}	3.827×10^{-20}	1

B-3 Some Astronomical Data

Object	Symbol	Rotation period (hh:mm:ss.s or days)	Mass (× 10^{24} kg)	Equatorial radius (× 10^6 m)	Free-fall acceleration near surface (m/s^2)	Escape speed (km/s)	Blackbody temperature (K)
Sun	☉	≈ 25 to 36 days[1]	1.9891×10^6	695.51	274	618	5777
Mercury	☿	58.65 days	0.3302	2.4397	3.7	4.3	440.1
Venus	♀	243 days	4.87	6.052	8.9	10.36	184.2
Earth	⊕	23:56:4.1	5.9736	6.3871	9.81	11.186	254.3
Moon	☾	27.3 days	0.07	1.738	1.6	2.38	270.7
Mars	♂	24:37:22.6	0.64	3.397	3.7	5.03	210.1
Ceres	⚳	09:04:19	9.6×10^{-4}	0.48			239
Jupiter	♃	9:50:30	1900	71.493	24.8	59.5	110.0
Saturn	♄	10:14:00	569	60.268	10.4	35.5	81.1
Uranus	⛢	17:14:00	87	25.559	8.87	21.3	58.2
Neptune	♆	16:03:00	103	24.764	11.2	23.5	46.6
Pluto	♇	6.387 days	0.01	1.135	0.58	1.2	37.5
Eris			$\approx 10^{-2}$	1.2			30

[1]The Sun is gaseous and does not rotate as a solid body; its period near the equator is shorter than at the poles.

Orbital parameters for objects that orbit the Sun

Object	Orbital period (days or years)	Semimajor axis (AU)	Eccentricity
Mercury	87.969 days	0.387	0.2056
Venus	224.701 days	0.723	0.0067
Earth	365.26 days	1.000	0.0167
Mars	1.524 years	1.524	0.0935
Ceres	4.603 years	2.767	0.097
Jupiter	5.203 years	5.204	0.0489
Saturn	9.54 years	9.5482	0.0565
Uranus	19.18 years	19.201	0.0457
Neptune	30.06 years	30.047	0.0113
Pluto	39.44 years	39.482	0.2488
Eris	559 years	67.89	0.4378

Some natural satellites

Satellite	Planet	Orbital period (days)	Semimajor axis (10^6 m)	Mass (10^{22} kg)	Radius (10^6 m)
Moon	Earth	27.322	384.4	7.349	1.7371
Io	Jupiter	1.769	421.6	8.932	1.8216
Europa	Jupiter	3.551	670.9	4.800	1.5608
Ganymede	Jupiter	7.155	1070.4	14.819	2.6312
Callisto	Jupiter	16.689	1882.7	10.759	2.4103
Titan	Saturn	15.945	1221.8	13.455	2.575
Triton	Neptune	5.877	354.8	2.14	1.3534

B-4 Rough Magnitudes and Scales

Numbers

Quantity	Approximate value or order of magnitude
Number of atoms in the Earth	10^{50}
Number of atoms in a 70-kg person	7×10^{27}
Number of cells in a person	5×10^{13}
Number of mobile phones in U.S.	330 million
Number of mobile phones worldwide	5 billion
Number of dogs in U.S.	78 million
Number of dogs in Italy	8 million
Population of the Earth	7 billion
Population of students at University of CA	220,000
Population of U.S.	300 million
Population of China	1.3 billion
Population of New York City	8 million
Population of Annapolis, MD	38,000
Population of Chesterton, IN	13,000
Population of Morris, MN	5,000
Veterans in U.S.	23 million
Percentage of people in U.S. under age of 18	24%
Money spent in film investments in U.S.	$15 billion
Money spend in film investments in India	$200 million

Sizes: Lengths, diameters, areas, and volumes

Quantity	SI or metric units	U.S. customary units
Area of a $1 bill	100 cm^2	17 $in.^2$
Area of a typical college campus	15 km^2	6.5 mi^2
Area of a cell phone	30 cm^2	5 $in.^2$
Area of continents	1.5×10^{14} m^2	6×10^7 mi^2
Area of oceans	3.6×10^{14} m^2	1.4×10^8 mi^2
Area of palm	40 cm^2	6 $in.^2$
Area of U.S. land	9×10^{12} m^2	3.5×10^6 sq miles
Average human stride	1 m	1 yd
Diameter of a hydrogen atom	10^{-10} m	
Diameter of a pollen grain	10–100 μm	
Diameter of a proton	10^{-15} m	
Diameter of a U.S. nickel	2.121 cm	0.835 in.
Diameter of the Milky Way galaxy	10^{21} m	10^5 ly
Height of a typical adult human	2 m	5–6 ft
Height of a typical story	3 m	10 ft
Length of a house fly	0.5 cm	0.2 in.
Length of a human thumb	5 cm	2 in.
Length of a match stick	5 cm	2 in.
Size of a living cell	10 μm	
Size of the smallest visible dust particle	0.1 μm	
Thickness of a human hair	50 μm	
Thickness of a U.S. nickel	1.95 mm	

Speeds

	SI or metric units	U.S. customary units
Top speed of a car	200 km/h	120 mph
Top speed of a *typical* bicycle	50 km/h	30 mph
Walking	1.3 m/s	3 mph
Running or jogging	1.5 km/h	6-min. mile (10 mph)
Commercial airplane cruising speed	250 m/s	550 mph
Speed of a snail	1 mm/s	2-3 inch/min
Speed of a cheetah	28 m/s	62 mph
Speed of a rifle bullet	700 m/s	1600 mph

Weights and masses

	SI or metric units	U.S. customary units
Mass of a U.S. nickel	5.000 g	3×10^{-4} slug (approx)
Weight of a car	10000-20000 N	1-2 tons
Mass of a car	1000-2000 kg	70-140 slug
Weight of a physics book	50 N	10 lb
Mass of a physics book	5 kg	0.4 slug
Weight of a U.S. quarter	6×10^{-2} N	0.2 oz
Mass of a U.S. quarter	6 g	4×10^{-4} slug
Mass of the Milky Way galaxy	10^{42} kg	10^{41} slug
Mass of an elephant	5×10^{3} kg	340 slug
Mass of a frog	100 g	7×10^{-3} slug
Mass of a house fly	8-20 mg	$(5\text{-}14) \times 10^{-4}$ slug
Weight of an adult human	500-1000 N	110-200 lbs
Mass of an adult human	50-100 kg	4-7 slug

Times, ages, periods, frequencies, and angular momentum

Quantity	Convenient units
Resting heart beat	60–80 per min
Age of the Universe	14 billion years
Age of human written history	10^4 years
Age of the Earth	4.5 billion years
Age of oldest fossil	2.7 billion years
Time for light to travel from the Sun to the Earth	10 min
Time for light to cross the diameter of a proton	3.3×10^{-24} s
Period of Halley's comet	2.4×10^9 s
Period of a typical x-ray	10^{-19} s
Time for light to travel from nearest star	4.3 years
Angular speed of record turntable	33 rpm
Angular momentum of record (33 rpm)	6 mJ · s
Angular momentum of electric fan	1 J · s
Angular momentum of Frisbee	0.1 J · s
Angular momentum of helicopter rotor (320 rpm)	5×10^4 J · s

Periodic Table of the Elements

Group I	Group II	Transition elements						
H 1 1.007 9 $1s$								
Li 3 6.941 $2s^1$	**Be** 4 9.0122 $2s^2$							
Na 11 22.990 $3s^1$	**Mg** 12 24.305 $3s^2$							
K 19 39.098 $4s^1$	**Ca** 20 40.078 $4s^2$	**Sc** 21 44.956 $3d^14s^2$	**Ti** 22 47.867 $3d^24s^2$	**V** 23 50.942 $3d^34s^2$	**Cr** 24 51.996 $3d^54s^1$	**Mn** 25 54.938 $3d^54s^2$	**Fe** 26 55.845 $3d^64s^2$	**Co** 27 58.933 $3d^74s^2$
Rb 37 85.468 $5s^1$	**Sr** 38 87.62 $5s^2$	**Y** 39 88.906 $4d^15s^2$	**Zr** 40 91.224 $4d^25s^2$	**Nb** 41 92.906 $4d^45s^1$	**Mo** 42 95.94 $4d^55s^1$	**Tc** 43 (98) $4d^55s^2$	**Ru** 44 101.07 $4d^75s^1$	**Rh** 45 102.91 $4d^85s^1$
Cs 55 132.91 $6s^1$	**Ba** 56 137.33 $6s^2$	57–71*	**Hf** 72 178.49 $5d^26s^2$	**Ta** 73 180.95 $5d^36s^2$	**W** 74 183.84 $5d^46s^2$	**Re** 75 186.21 $5d^56s^2$	**Os** 76 190.23 $5d^66s^2$	**Ir** 77 192.2 $5d^76s^2$
Fr 87 (223) $7s^1$	**Ra** 88 (226) $7s^2$	89–103**	**Rf** 104 (261) $6d^27s^2$	**Db** 105 (262) $6d^37s^2$	**Sg** 106 (266)	**Bh** 107 (264)	**Hs** 108 (277)	**Mt** 109 (268)

Key:

Symbol — **Ca** 20 — Atomic number
Atomic mass† — 40.078
$4s^2$ — Electron configuration

*Lanthanide series	**La** 57 138.91 $5d^16s^2$	**Ce** 58 140.12 $5d^14f^16s^2$	**Pr** 59 140.91 $4f^36s^2$	**Nd** 60 144.24 $4f^46s^2$	**Pm** 61 (145) $4f^56s^2$	**Sm** 62 150.36 $4f^66s^2$
**Actinide series	**Ac** 89 (227) $6d^17s^2$	**Th** 90 232.04 $6d^27s^2$	**Pa** 91 231.04 $5f^26d^17s^2$	**U** 92 238.03 $5f^36d^17s^2$	**Np** 93 (237) $5f^46d^17s^2$	**Pu** 94 (244) $5f^67s^2$

Note: Atomic mass values given are averaged over isotopes in the percentages in which they exist in nature.
† For an unstable element, mass number of the most stable known isotope is given in parentheses.

			Group III	Group IV	Group V	Group VI	Group VII	Group 0
							H 1 1.007 9 $1s^1$	**He** 2 4.002 6 $1s^2$
			B 5 10.811 $2p^1$	**C** 6 12.011 $2p^2$	**N** 7 14.007 $2p^3$	**O** 8 15.999 $2p^4$	**F** 9 18.998 $2p^5$	**Ne** 10 20.180 $2p^6$
			Al 13 26.982 $3p^1$	**Si** 14 28.086 $3p^2$	**P** 15 30.974 $3p^3$	**S** 16 32.066 $3p^4$	**Cl** 17 35.453 $3p^5$	**Ar** 18 39.948 $3p^6$
Ni 28 58.693 $3d^84s^2$	**Cu** 29 63.546 $3d^{10}4s^1$	**Zn** 30 65.41 $3d^{10}4s^2$	**Ga** 31 69.723 $4p^1$	**Ge** 32 72.64 $4p^2$	**As** 33 74.922 $4p^3$	**Se** 34 78.96 $4p^4$	**Br** 35 79.904 $4p^5$	**Kr** 36 83.80 $4p^6$
Pd 46 106.42 $4d^{10}$	**Ag** 47 107.87 $4d^{10}5s^1$	**Cd** 48 112.41 $4d^{10}5s^2$	**In** 49 114.82 $5p^1$	**Sn** 50 118.71 $5p^2$	**Sb** 51 121.76 $5p^3$	**Te** 52 127.60 $5p^4$	**I** 53 126.90 $5p^5$	**Xe** 54 131.29 $5p^6$
Pt 78 195.08 $5d^96s^1$	**Au** 79 196.97 $5d^{10}6s^1$	**Hg** 80 200.59 $5d^{10}6s^2$	**Tl** 81 204.38 $6p^1$	**Pb** 82 207.2 $6p^2$	**Bi** 83 208.98 $6p^3$	**Po** 84 (209) $6p^4$	**At** 85 (210) $6p^5$	**Rn** 86 (222) $6p^6$
Ds 110 (271)	**Rg** 111 (272)	**Cn** 112 (285)	113†† (284)	**Fl** 114 (289)	115†† (288)	**Lv** 116 (293)	117†† (294)	118†† (294)

Eu 63 151.96 $4f^76s^2$	**Gd** 64 157.25 $4f^75d^16s^2$	**Tb** 65 158.93 $4f^85d^16s^2$	**Dy** 66 162.50 $4f^{10}6s^2$	**Ho** 67 164.93 $4f^{11}6s^2$	**Er** 68 167.26 $4f^{12}6s^2$	**Tm** 69 168.93 $4f^{13}6s^2$	**Yb** 70 173.04 $4f^{14}6s^2$	**Lu** 71 174.97 $4f^{14}5d^16s^2$
Am 95 (243) $5f^77s^2$	**Cm** 96 (247) $5f^76d^17s^2$	**Bk** 97 (247) $5f^86d^17s^2$	**Cf** 98 (251) $5f^{10}7s^2$	**Es** 99 (252) $5f^{11}7s^2$	**Fm** 100 (257) $5f^{12}7s^2$	**Md** 101 (258) $5f^{13}7s^2$	**No** 102 (259) $5f^{14}7s^2$	**Lr** 103 (262) $5f^{14}6d^17s^2$

††Elements 113, 115, 117, and 118 have not yet been officially named. Only small numbers of atoms of these elements have been observed.
Note: For a description of the atomic data, visit *physics.nist.gov/PhysRefData/Elements/per_text.html.*

Answers to Concept Exercises and Odd-Numbered Problems

CHAPTER 1: Concept Exercises

1.1 The puns are **a.** a megaphone **b.** a piccolo **c.** 2 kilomockingbirds **d.** a microfiche **e.** a terrapin

1.2 To find the mass of the raisins in kilograms,

$$42.5\ \text{g}\left(\frac{1\ \text{kg}}{1000\ \text{g}}\right) = 4.25 \times 10^{-2}\ \text{kg}$$

1.3 To find the dimensions of energy,

$$[E] = [m][c^2] = \text{M}\left(\frac{\text{L}}{\text{T}}\right)^2$$

1.4 Only (**b**) may represent energy. Use dimensional analysis to find the dimensions of each quantity.

a. $[Q] = \text{M}\frac{\text{L}}{\text{T}}$

b. $[Q] = \text{M}\left(\frac{\text{L}}{\text{T}}\right)^2$, which may represent energy because it has the correct dimensions.

c. $[Q] = \text{M}\frac{\text{L}}{\text{T}^2}$ **d.** $[Q] = \text{M}\frac{\text{L}^2}{\text{T}}$

1.5 $[\rho] = \frac{\text{M}}{\text{L}^3}$, and the SI units are kg/m^3.

1.6 a. Exact. Numbers that are not measured but are instead derived mathematically are exact **b.** 3 **c.** Ambiguous: 2 or 3. The final zero may be significant, or it may just be a placeholder. Scientific notation avoids such ambiguity. **d.** 3 **e.** 2 **f.** 2

1.7 a. 9.95×10^1 **b.** 3.6×10^4 **c.** -7.3 **d.** -3.90

CHAPTER 1: Problems and Questions

1. 2.6×10^9 s
3. 1.4×10^{17} s
5. a. 5.3×10^{-3} m **b.** 0.12892 kg **c.** $3.57 \times 10^{-5}\ \text{m}^3$ **d.** $6.57 \times 10^4\ \text{kg/m}^3$
7. $1.93 \times 10^4\ \text{kg/m}^3$
9. 1.5×10^{11} m
11. $2 \times 10^2\ \text{bloobits/bot}^3$
13. 2.8×10^{-3} m/s
15. a. 1.10×10^3 gal/min **b.** 69.4 L/s **c.** 576 s
17. $2.12 \times 10^4\ \text{kg/m}^3$
19. a. $\frac{\text{M}\cdot\text{L}}{\text{T}^2}$ **b.** $\frac{\text{M}\cdot\text{L}}{\text{T}^2}$ **c.** $\frac{\text{M}\cdot\text{L}}{\text{T}}$ **d.** $\frac{\text{M}\cdot\text{L}^2}{\text{T}}$
21. Both quantities have dimensions $\text{M}\,(\text{L/T})^2$
23. $f \propto \sqrt{k/m}$
25. a. $\text{M}\cdot(\text{L/T}^2)$ **b.** $\frac{\text{kg}\cdot\text{m}}{\text{s}^2}$ **c.** kg/s^2
27. 3
29. a. 4 **b.** 3 **c.** 2 **d.** 3
31. a. 8.65 **b.** 177 **c.** 25.891
33. −6.71
35. a. $1.0868 \times 10^{21}\ \text{m}^3$ **b.** $5.50 \times 10^3\ \text{kg/m}^3$
37. a. The result should be reported as 1.2 ± 0.1 g instead. **b.** The reported result makes sense. **c.** The reported result makes sense. **d.** The reported result contains only one digit to the right of the decimal, whereas the reported uncertainty would suggest knowledge of the value out to two digits to the right of the decimal. Either the uncertainty has been underestimated, or the measured value has been reported erroneously.
39. Assuming 10 hours per week for 12 weeks, then you will study 120 h. If you are a typical student and there are 40 students in your class, then that is about 5000 h total. (Answers will vary.)
41. about 2 m
43. $5 \times 10^{-3}\ \text{m}^3$ (lungs) and $2 \times 10^{-3}\ \text{m}^3$ (stomach). Answers may vary by a factor of two or so.
45. 10^{14} cells
47. 15 days. Answers may vary by a factor of two or so.
49. 1.83×10^3
51. a. $3.58 \times 10^3\ \text{m}^3$ **b.** 4.24×10^4 N
53. 9.13 in.
55. Both quantities have dimensions $\text{M}\,(\text{L/T}^2)$.
57. 1500 ft
59. a. 205.9 g and 204.7 g **b.** (205.3 ± 0.6) g
61. 4.2 ly, 4.0×10^{16} m

CHAPTER 2: Concept Exercises

2.1 The spacing between the dots reveals how the particle is moving.

Case 1. Even spacing between dots: Particle is moving at constant speed.
Case 2. Dots get farther apart: Particle is speeding up.
Case 3. Dots get closer together: Particle is slowing down.
Case 4. Dots get farther apart: Particle is speeding up.
Case 5. All dots lie on top of one another: Particle is at rest.

2.2 All four vectors have the same magnitude, 2.4 m. To find the components, look for the unit vector. For example, if $\hat{\imath}$, the vector component is $\vec{x} = x\hat{\imath}$, where x is the scalar component. The vector components and scalar components are as follows.

a. $\vec{x} = -2.4\hat{\imath}$ m; $x = -2.4$ m
b. $\vec{z} = -2.4\hat{k}$ m; $z = -2.4$ m
c. In this case, the vector component $\vec{z}$ is given; $\vec{z} = -2.4\hat{k}$ m; $z = -2.4$ m
d. In this case, the scalar component x is given; $\vec{x} = -2.4\hat{\imath}$ m; $x = -2.4$ m

2.3 If the displacement of every point on the object is the same, the object is undergoing purely translational motion and may be modeled as a particle.

Case 1. A person on a Ferris wheel may be modeled as a particle.
Case 2. A person on a loop-the-loop roller coaster may not be modeled as a particle because the person flips upside down as he moves along the track.
Case 3. The tire on a bicycle may not be modeled as a particle.
Case 4. The person on a bicycle may be modeled as a particle.

2.4 For each course, divide the displacement by 2 hours to get the magnitude of the runner's average velocity. For the course in Figure 2.14A, we find 10 mph, and for that in Figure 2.14B, we find 6 mph. In both cases, the magnitude of the average velocity is less than the average speed of 13.1 mph because the magnitude of the displacement is shorter than the distance traveled by the runner. Average velocity is not a particularly useful quantity for describing the motion of a marathon runner.

2.5 It refers to magnitude of the displacement.

2.6 Because the tiger's displacement is zero, his average velocity is zero. The tiger's average speed is the total distance (100×20 m) traveled divided by 2 hours: $\frac{2000 \text{ m}}{2 \text{ h}} \times \frac{1 \text{ h}}{60 \text{ min}} \times \frac{1 \text{ min}}{60 \text{ s}} = 0.3$ m/s. If the tiger remains in one place, both his average speed and average velocity are zero.

2.7 **a.** Plot acceleration on the vertical axis and time on the horizontal one. **b.** The slope of the tangent line for each point on the position-versus-time graph gives the velocity, and the slope on the velocity-versus-time graph gives acceleration. **c.** Position-, velocity- and acceleration-versus-time graphs are a concave-up parabola, an upward-sloping line and a horizontal line above the time axis, respectively.

CHAPTER 2: Problems and Questions

1. No; the path is nearly circular when viewed from the north celestial pole, indicating two-dimensional motion.

3. a. $x_A = -6$ m, $x_B = 0$, $x_C = 4$ m **b.** $z_A = 0$, $z_B = 6$ m, $z_C = 10$ m

5. a. $\vec{v}_y = 35.0\hat{\jmath}$ m/s, $v_y = 35.0$ m/s, $v = 35.0$ m/s **b.** $\vec{v}_x = 53.0\hat{\imath}$ m/s, $v_x = 53.0$ m/s, $v = 53.0$ m/s **c.** $\vec{v}_z = -3.50\hat{k}$ m/s, $v_z = -3.50$ m/s, $v = 3.50$ m/s **d.** $\vec{v}_x = -5.30\hat{\imath}$ m/s, $v_x = -5.30$ m/s, $v = 5.30$ m/s

7. a. Less than **b.** Speeding up **c.** Less than because the slope of the line drawn from $t = 0$ to 60 s for B is less than the slope of the similar line for A.

9. a.

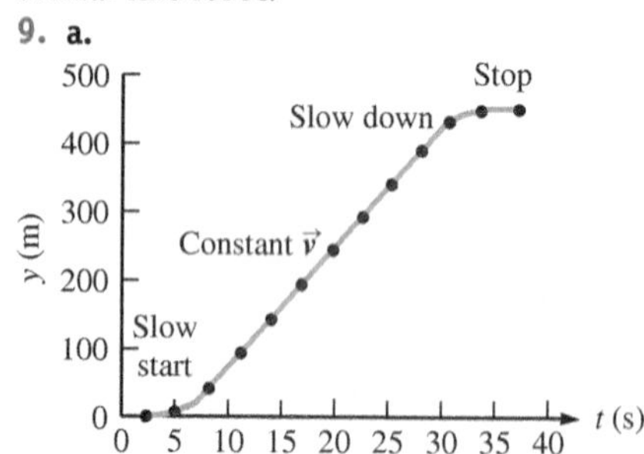

b. The elevator's speed gradually increases for the first 5 seconds, is nearly constant for about 30 seconds, and gradually decreases over the last 5 seconds. **c.** 17 m/s, from about 5 s to about 30 s **d.** Make sure that the acceleration is gradual at the beginning and end.

11. Never

13. 1.00×10^2

15. a. 6.44×10^5 m **b.** $2.99 \times 10^5\hat{\imath}$ m

17. a. $14.5\hat{\jmath}$ cm **b.** $14.5\hat{\jmath}$ cm

19. a. Zero; the initial and final positions are equal. **b.** 1.56 m; distance is the whole path length, but displacement only depends on the initial and final positions. **c.** There are many answers; here are two: the atomic bonds in a solid and the muscles in an animal are sometimes modeled by springs.

21. a. $9.0\hat{\jmath}$ m/s **b.** 0

23. a. The estimate depends on the distance between you and the lamp, but you should find an answer on the order of nanoseconds. **b.** 5.00×10^2 s **c.** The photon moves a very great distance in a very short amount of time. The jogger's speed is low enough to be measured with ordinary metersticks (or measuring tape) and stopwatches.

25. a. About 3×10^3 m **b.** About 1×10^{-5} s. Yes; the light travel time is much less than the sound travel time. **c.** The maximum difference in measurements would be approximately 200 m, or about one-tenth of a mile. Depending on the needed level of accuracy, it can probably be neglected.

27. 5.1×10^6 yr

29. a. $+L/t_1\hat{\jmath}$, **b.** $-L/t_2\hat{\jmath}$ **c.** 0 **d.** $2L/(t_1 + t_2)$

31. a. It is speeding up. **b.** 0, $-7.58\hat{\jmath}$ m/s, $-2.27 \times 10^2\hat{\jmath}$ m/s **c.** 0, 7.58 m/s, 2.27×10^2 m/s

33. a. $4.2\hat{\jmath}$ m/s **b.** $6.0\hat{\jmath}$ m/s **c.** 3.7 s

35. a. $-22.5\hat{k}$ m/s, $-52.5\hat{k}$ m/s **b.** $-37.5\hat{k}$ m/s

37. 1.33×10^4 m/s²

39. a. No, the acceleration is not necessarily zero when the velocity is zero. **b.** A ball thrown vertically upward comes to a stop momentarily at the top of its path, but it moves back down again. So, its acceleration at the top cannot be zero.

41. It must be positive.

43. a. The derivative of v is negative, and so it is slowing down. **c.** No and no **d.** A motor must compensate for deceleration.

45. a. $-(492.6 \text{ m} \cdot \text{s})\left(\frac{1}{t + 2.0 \text{ s}}\right)^3\hat{k}$ **b.** Slowing down. The velocity and the acceleration have opposite signs. **c.** No; it is always moving forward, although its speed becomes vanishingly small as time goes on.

47. 27.1 m/s²

49. a. $31.2\hat{\imath}$ m/s **b.** $-5.55\hat{\imath}$ m/s²

51. a. 30.4 m/s **b.** 2.56 m/s²

53. a. $36.8\hat{\imath}$ m **b.** 36.8 m **c.** $36.8\hat{\imath}$ m **d.** 38.2 m

55. a. $307\hat{\imath}$ ft **b.** $116\hat{\imath}$ ft/s **c.** $24.0\hat{\imath}$ ft/s²

57. 2.31 s

59. a. 44.1 m/s, upward **b.** 99.3 m

61. a. $-23\hat{\jmath}$ m/s **b.** $-9.2\hat{\jmath}$ m/s²

63. 1.2 s and 3.7 s

65. a. 46.3 s **b.** 2.16 km **c.** 206 m/s, downward

67. About 4 m/s²

69. 16 m

71. About 0.008 s (Answers within a factor of two are probably okay.)

73. a. $-3.04\hat{\jmath}$ m **b.** $5.00\hat{\jmath}$ m/s

75. 455 m east

77. a. 13.0 AU/yr **b.** 5.00 AU/yr and 21.0 AU/yr **c.** 0.375 yr

79. a. $34.0\hat{k}$ m/s **b.** $39.5\hat{k}$ m/s

81. There are many good choices; here are some tips for choosing good coordinate systems. (1) It is best to place an axis along the direction of motion. (2) It is sometimes helpful to choose an origin so that the position of the particle is always positive. (3) Choose positive as the direction in which the moving particle speeds up. (4) When the motion is caused by a spring, it is common to put the origin at the spring's relaxed position; because the motion switches direction, it doesn't matter which direction is chosen to be positive. Use these tips to check your coordinate system and see if you can make improvements.

83. $\vec{v}_y = \lim_{\Delta t \to 0} \frac{\Delta \vec{y}}{\Delta t} = \frac{d\vec{y}}{dt} = \frac{dy}{dt}\hat{\jmath}$, $\vec{v}_z = \lim_{\Delta t \to 0} \frac{\Delta \vec{z}}{\Delta t} = \frac{d\vec{z}}{dt} = \frac{dz}{dt}\hat{k}$

85. $\vec{a}_y = \frac{dv_y}{dt}\hat{\jmath}$, $\vec{a}_z = \frac{dv_z}{dt}\hat{k}$

87. **a.** $\vec{a}_x = 0.0157\hat{\imath}$ m/s² **b.** 92.6 s **c.** $\vec{a}_x = -0.0157\hat{\imath}$ m/s². ThrustSSC's acceleration is nearly 400 times greater than Jeantaud's, and Jeantaud's timed mile took about 20 times longer than ThrustSSC's.

CHAPTER 3: Concept Exercises

3.1 Using any graphical method, we find
a. $\vec{A} - \vec{B} = 0$ **b.** $\vec{B} - \vec{A} = 0$ **c.** $\vec{A} - \vec{C} \neq 0$
d. $\vec{C} - \vec{A} \neq 0$ **e.** $\vec{A} + \vec{C} = 0$

3.2 **a.** Vector addition is commutative, so $A + C = C + A$ for any two vectors; in this case $\vec{A} + \vec{C} = \vec{C} + \vec{A} = 0$. **b.** The resultant of $\vec{A} - \vec{C}$ points to the right, and the resultant of $\vec{C} - \vec{A}$ points to the left, so they are not equal. **c.** Part (b) shows that vector subtraction is not commutative.

3.3 **a.** 9.81 cm **b.** (9.81 cm)/2 = 4.90 cm

3.4 Only case 2 shows a right-handed coordinate system.

3.5 **a.** $\alpha + \theta = 270°$
$\alpha = 270° - \theta = 270° - 215°$
$\alpha = 55°$

b. $\vec{B} = B\cos\theta\hat{\imath} + B\sin\theta\hat{\jmath}$
$\vec{B} = B\cos 215°\hat{\imath} + B\sin 215°\hat{\jmath}$
$\vec{B} = -0.82B\hat{\imath} - 0.57B\hat{\jmath}$

$\vec{B} = -B\sin\alpha\hat{\imath} - B\cos\alpha\hat{\jmath}$
$\vec{B} = -B\sin 55°\hat{\imath} - B\cos 55°\hat{\jmath}$
$\vec{B} = -0.82B\hat{\imath} - 0.57B\hat{\jmath}$ as before.

CHAPTER 3: Problems and Questions

1. **a.** 7.20 cm **b.** 1 cm → 200 m/s **c.** No; they both seem to fit comfortably on the page and are large enough.
3. $A\sqrt{2}$ or $B\sqrt{2}$
5. 24 units in the negative x direction
7. The magnitude is correct, but the direction is reversed.
9. The direction is correct, but the magnitude is half as long as it should be.
11. $\frac{1}{3}\vec{A} - \vec{B}$
13. B, A, and then C
15. **a.** 4.5×10^{-5} m/s (Answers may vary slightly due to measurement error.) **b.** The direction is the same as the displacement vector.
17. **a.** He is farthest from the starting point in the final drill. He is closest to his starting point in the second drill. **b.** He is farther than d in all three cases. **c.** 120°
19. The painting won't fit in the gallery on any wall.
21. 9.02 km
23. **a.** 4.24 blocks at 45.0° north of east **b.** 10.0 blocks
25. 3.47 mi south and 4.14 mi west
27. 36.9 at 63.4° clockwise from the x axis
29. $B_x = -11.79$, $B_y = 15.47$
31. 60°
33. **a.** $\vec{A} = -4.00\hat{\imath} - 2.00\hat{\jmath}$ **b.** 4.47 at 207° **c.** $\vec{B} = -2.00\hat{\imath} + 2.00\hat{\jmath}$
35. $(v_{1x}, v_{1y}) = (20.90, 12.07)$ m/s, $(v_{2x}, v_{2y}) = (0, 8.96)$ m/s, $(v_{3x}, v_{3y}) = (-13.25, 0)$ m/s, $(v_{4x}, v_{4y}) = (-7.648, -21.01)$ m/s
37. $(-35.76\hat{\imath} + 61.94\hat{\jmath})$ m
39. **a.** $\vec{A} = (15.0\hat{\imath} - 6.00\hat{\jmath} - 3.00\hat{k})$ m **b.** $\vec{B} = (5.00\hat{\imath} - 2.00\hat{\jmath} - 1.00\hat{k})$ m **c.** $\vec{C} = (-45.0\hat{\imath} + 18.0\hat{\jmath} + 9.00\hat{k})$ m
41. **a.** 19° below the horizontal direction **b.** 14° below the horizontal direction; she is safe.
43. 65.6° north of east (or 24.4° east of north)
45. −0.740 m
47. $-18.5\hat{\imath} + 36.0\hat{\jmath} - 4.00\hat{k}$
49. $\sqrt{d_1^2 + d_2^2 + 2d_1d_2\cos\theta}$
51. $C_x = 6.10$ m, $C_y = -10.1$ m
53. 147°
55. 334 m
57. The spider's displacement is between d and $2d$. Add the horizontal displacements to find $1.7d$.
59. $(3.4\hat{\imath} + 1.3\hat{\jmath})$ m
61. $R_x = 6.58$ units, $R_y = -21.5$ units
63. 55.8 at 234°
65. Greater than 5, and $90° > \theta > 0$. The x component of each vector must be equal to 5.
67. The angle between $\vec{A}$ and $\vec{B}$ must be 60°.
69. **a.** The vectors must point in opposite directions. **b.** $s < 0$
71. **a.** 1.7 **b.** Each of the components must be proportional with the same proportionality constant m (given the function in the problem), but that is not the case here.
73. 72°
75. The direction of the instantaneous velocity at both times is up and to the right (clockwise). The direction of the average velocity is down and to the left (counterclockwise).

CHAPTER 4: Concept Exercises

4.1 **a.** Two dimensions **b.** Two dimensions
c. Two dimensions **d.** One dimension

4.2 **a.** Two dimensions
b. The particle slows down from A to G and then speeds up to L.

4.3 Because the initial and final positions are the same for the ball as they are for the cart, the displacement of the cart is equal to the displacement of the ball, $\Delta\vec{r} = 0.75\hat{\imath}$ m.

4.4 To answer these questions, we need to know the difference between range and displacement.
a. Because the balloon returns to the ground, its vertical displacement is zero, which means that its displacement equals the range. In this case, $\Delta r = R = 300$ yd.
b. The vertical displacement is not zero in this case. From the figure, we see that the horizontal displacement in this case is greater than the range.

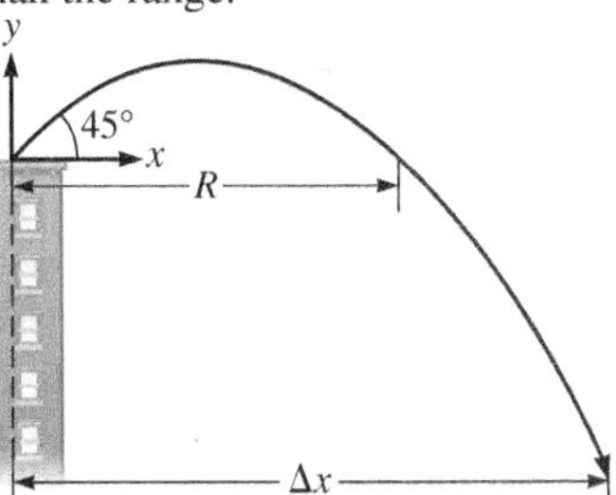

4.5 From $R = (v_0^2/g)\sin 2\theta$ (Eq. 4.28), we see that the range of a device depends on g. The free-fall acceleration is lower on the Moon than it is on the Earth. So, the range of the device on the Moon is longer that it is on the Earth.

4.6 a. This solution is a straightforward application of $v = 2\pi r/T$ (Eq. 4.30):

$$v = \frac{2\pi(0.5\ \text{m})}{12\ \text{s}} = 0.26\ \text{m/s}$$

b. This solution is a straightforward application of $\omega = 2\pi/T$ (Eq. 4.33):

$$\omega = \frac{2\pi\ \text{rad}}{12\ \text{s}} = 0.52\ \text{rad/s}$$

c. In 4 s, the particle has completed only $\frac{4}{12} = \frac{1}{3}$ of its circular path. That means that it has swept out only one third of 2π rad: $\theta = \frac{1}{3}2\pi = \frac{2}{3}\pi$.
Because we are looking at a time that is one third of a revolution, the particle has traveled one third of the circle's circumference: $s = \frac{1}{3}(2\pi r) = \frac{2}{3}\pi r = \frac{2}{3}\pi(0.5\ \text{m}) = 1.0\ \text{m}$ (to 2 significant figures). It is the same as simply using $s = \theta r$ with $\theta = \frac{2}{3}\pi$.

4.7 Because we know the distance between the two cities as measured in the frame of the ground, we need the groundspeed to find the flight time.

CHAPTER 4: Problems and Questions

1. a. One dimension **b.** One dimension **c.** One dimension because the train moves in a single straight line in all cases
3. Neither player has an advantage as long as the ship moves with a constant velocity.
5. a. Two dimensions **b.** Points near the bottom of each arc (DEF and JKL) **c.** Points near the top of each arc (AB, GHI, and MN)
7. a. 5.00 m

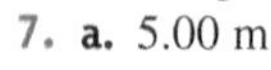

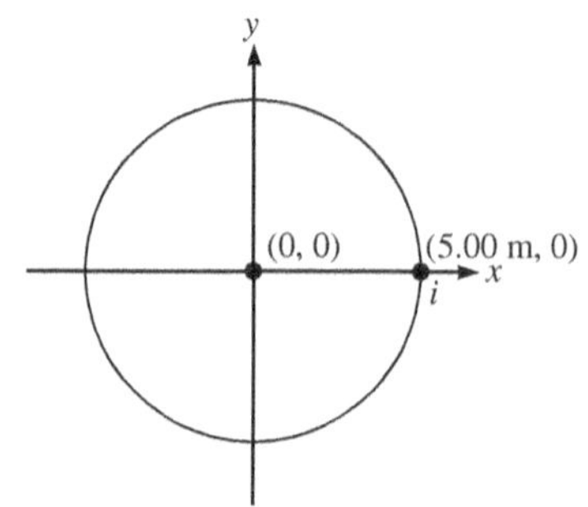

b. $x = (5.00\ \text{m})\cos[(\pi\ \text{rad/s})t]$
c.

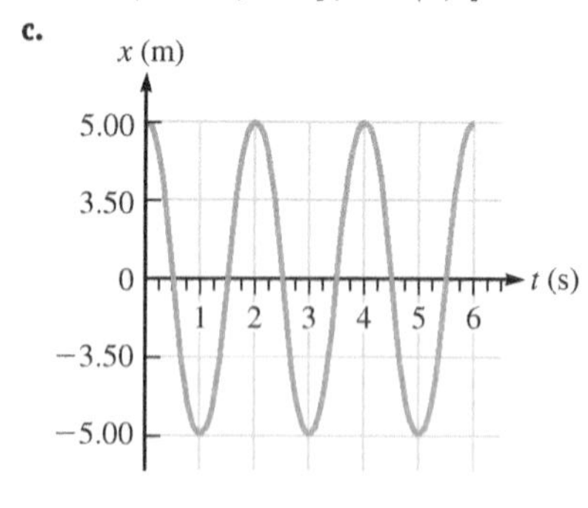

d. The graph also repeats such that x is between -5.00 m and $+5.00$ m.
9. a. 3.59 m **b.** No change
11. a. $(1.25t^2\hat{\imath} + 3.00t\hat{\jmath})$ m **b.** $(2.50t\hat{\imath} + 3.00\hat{\jmath})$ m/s **c.** $(11.3\hat{\imath} + 9.00\hat{\jmath})$ m **d.** 8.08 m/s
13. $90° < \theta \leq 180°$
15. a. $\vec{v}_i = (15.72\hat{\imath} - 7.88\hat{\jmath})$ m/s **b.** $\vec{a} = (-7.88\hat{\jmath} + 11\hat{k})$ m/s^2
17. $2b/3c$
19. $\vec{v} = A\omega\{[\cos(\omega t) - \omega t\sin(\omega t)]\hat{\imath} + [\sin(\omega t) + \omega t\cos(\omega t)]\hat{\jmath}\}$
21. a. 2.50 m/s **b.** −56.7°
23. 2.4×10^2 m
25. $x = 72.5$ m, $y = 2.95$ m
27. a. 23 m above the ground **b.** No
29. 15.1 m/s
31. a. 26.5 m **b.** 87.0 m **c.** 2.05 s
33. a. 7.8 m **b.** $(8.2\hat{\imath} - 5.6\hat{\jmath})$ m/s **c.** 1.58 m
35. 323.7 m/s^2
37. a. 7.40×10^2 m/s^2 **b.** Greater with the longer string **c.** 6.32×10^2 m/s^2
39. 2:1
41. 15.8 m/s^2
43. a. 2.6×10^{-6} rad/s **b.** 2.6×10^{-3} m/s^2
45. a. 27.0 km/h (west) **b.** 27.0 km/h (east) **c.** 43.3 s
47. a. While riding the train, pitch the ball in the same direction as the train's velocity. **b.** Reference frame fixed to the Earth; no
49. 5 min
51. 90°
53. 15.6° south of east
55. a. 1.34 h **b.** 1.15 h **c.** 1.24 h
57. a. 10.1 m/s^2 at 13.4° west from the vertical **b.** 9.81 m/s^2 vertically downward
59. 90°
61. $\sqrt{2gh}/\tan\theta$
63. 2.4 m from the end of the track
65. A parabola
67. $\vec{v} = A\hat{\imath} + (B - 2Ct)\hat{\jmath}$; $\vec{a} = -2C\hat{\jmath}$
69. 82.9°
71. 566 m/s
73. 3.34 m above the floor
75. Because the relative motion between the Earth and Mars is zero at those times
77.
a. $\left(2.00 + \frac{2\sqrt{3}}{3}t^{3/2}\right)\hat{\imath} + \left(7.00 - \frac{t^2}{2}\right)\hat{\jmath}$
b. $\left(2.00t + \frac{4\sqrt{3}}{15}t^{5/2}\right)\hat{\imath} + \left(7.00t - \frac{t^3}{6}\right)\hat{\jmath}$
79. a. 5.5 m **b.** $(3.2\hat{\imath} - 15\hat{\jmath})$ m/s
81. a. Circular **b.** $(\vec{v}_p)_M = -R\omega\sin(\omega t)\hat{\imath} + R\omega\cos(\omega t)\hat{\jmath}$ **c.** $(\vec{a}_p)_L = (\vec{a}_p)_M = -R\omega^2\cos(\omega t)\hat{\imath} - R\omega^2\sin(\omega t)\hat{\jmath}$ **d.** The acceleration should be equal as found in part (c).

CHAPTER 5: Concept Exercises

5.1 a. Because friction cannot be completely eliminated, any sliding object has at least one force acting on it. So friction made it appear that rest was a natural state and that it takes a force to maintain motion. **b.** Newton's first law implies rest is **not** a natural state and replaces the idea of a "natural state of rest" with "constant velocity." **c.** No force is required to maintain constant velocity.

5.2 Avi's first statement is false. No force throws you through the windshield. Cameron's first statement is true. The passengers are already in motion, and (according to Newton's first law) it would take a force to stop them (when the vehicle stops suddenly). Avi's second statement is true. Cameron's second statement is true. Shannon's underlined statement is false. A seat can exert a force on you.

5.3 Case 1. Source: glove. Direction: up. Contact force.
Case 2. Source: freight train. Direction: right. Contact force.
Case 3. Source: Earth. Direction: toward Earth. Field force.

5.4 Identify which sources are outside the system in each case, and then decide whether these sources exert an external force on the system. **a.** The spring scale, and the Earth. The elevator car and the cable are external, but are not in contact with the system, so they cannot exert a force on the system. **b.** The elevator car, and the Earth.

The cable is not in contact with the system, so it cannot exert a force on the system. **c.** The Earth and the cable are outside the system, and they both exert forces on it.

5.5 a. A person with "a lot of inertia" can't seem to get moving on tasks, analogous to an object with a lot of mass that is difficult to accelerate from rest. **b.** It is just as difficult to stop an object with a lot of mass as it is to get it to start moving, but we don't say that a person that is working on task has a lot of inertia. **c.** In our everyday language, *massive* means heavy or large, such as "I have a massive amount of homework."

5.6 The only inertial frame is **a**, an airplane cruising in a straight path at constant speed. The others are accelerating frames.

5.7 a. Newton's first law states that acceleration requires at least one force; and is used to identify inertial frames. Newton's second law mathematically connects net force with inertia (mass) and acceleration. **b.** According to the second law, if the total force is zero, the object does not accelerate. This statement is consistent with the first law. It is likely that Newton stated the first law separately from the second law to address the notion of a *natural state*, which had been discussed for centuries before he published the *Principia*.

5.8 The apparent weight of the bunch of bananas does not change when it is hung from scale 2, so $\Delta y_2 = \frac{1}{3}\Delta y_1$.

5.9 Because the magnitude of the tension is the same all along the rope, in all three cases the magnitude of the tension force from each rope on Rochelle is 15 N and the rope pulls Rochelle. **a.** Its direction is toward Buddy. **b.** Its direction is toward the pole. **c.** Rochelle's left hand is pulled toward Joe, and her right hand is pulled toward Buddy.

5.10

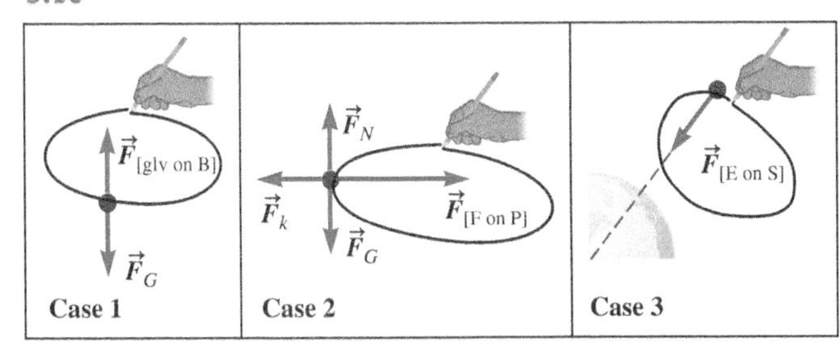

5.11 The magnitude of the force exerted on the child is the same as the magnitude of the force exerted on the Earth:

$$m_E a_E = m_c a_c$$

$$\frac{a_E}{a_c} = \frac{m_c}{m_E}$$

The Earth accelerates, but because its mass is so very much larger than the mass of the child, the Earth's acceleration is negligible.

5.12 Shannon is correct. According to Newton's third law, each train exerts a force on the other, with both forces of the same magnitude. (According to Newton's second law, the passenger train was accelerated more because it is lighter than the freight train.)

CHAPTER 5: Problems and Questions

1. It is easier to lift the beach ball because it is lighter. Other properties such as its size and shape aren't important.

3. The light block has less mass and so is easier to accelerate than the heavy block. The same push will make the light block go faster initially and take longer to stop.

5. Choice (b) is correct. The hero's inertia keeps him moving at a constant velocity with little force affecting him as he travels.

7. a. No force is necessary. **b.** A force must cause the deceleration. **c.** A force is needed to cause the change in direction.

9. 17.98 N

11. $(-2.32\hat{\imath} + 23.57\hat{\jmath})$N

13. The fish exerts a contact force. It pushes on the water, and the water pushes back. The fish would also exert a gravitational force on the surrounding water, but the magnitude would be much less than the contact force.

15. The ball has the same mass in each instance and therefore the same inertia whether it is at rest or rolling.

17. The heavy man is more massive and thus has more inertia.

19. The wind causes the deflection. The cart is moving at a constant velocity and is thus an inertial frame.

21. 3.13 kg

23. 769 N

25. a. $\vec{F}_{\text{tot}} = (-1.63\times10^6\hat{\imath} + 7.59\times10^6\hat{\jmath})$ N **b.** $\vec{a} = (-641\hat{\imath} + 2.98 \times 10^3\hat{\jmath})$m/s^2

27. a. 0.294 **b.** 0.966 m/s^2

29. a. $\vec{a} = (6.089\hat{\imath} - 6.595\hat{\jmath})$ m/s^2 **b.** 8.976 m/s^2 at $-47.28°$

31. a. $F_x = 0.0800$ N, $F_y = 0.480$ N **b.** 0.487 N

33. At $t = b/3c$

35. 1.5×10^{-2} N

37. 1.7 lb/in. and 3.0×10^2 N/m

39. 1.20×10^3 N, upward

41. 124 N and 2.00 m/s^2

43. $$\frac{\mu_k mg}{\mu_k \sin\theta + \cos\theta}$$

45. 0.400

47. 38.6 N, 53.9 N, 3.07 m/s^2

49. 6.89 m/s^2

51. a.

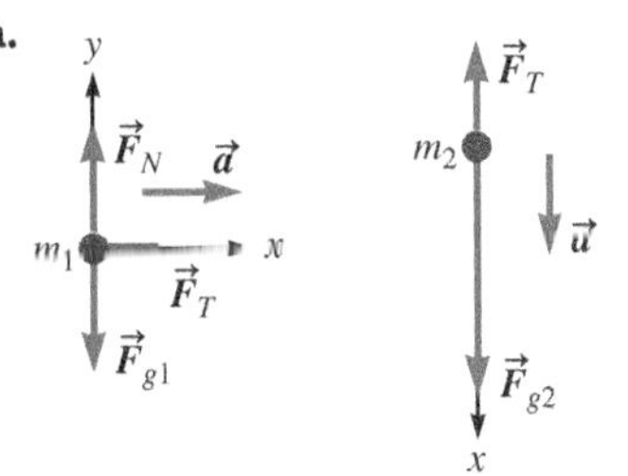

b. 21.8 N **c.** 7.25 m/s^2

53. By Newton's third law, you push on the ground, and the ground pushes back with a force equal in magnitude.

55. a. and **b.**

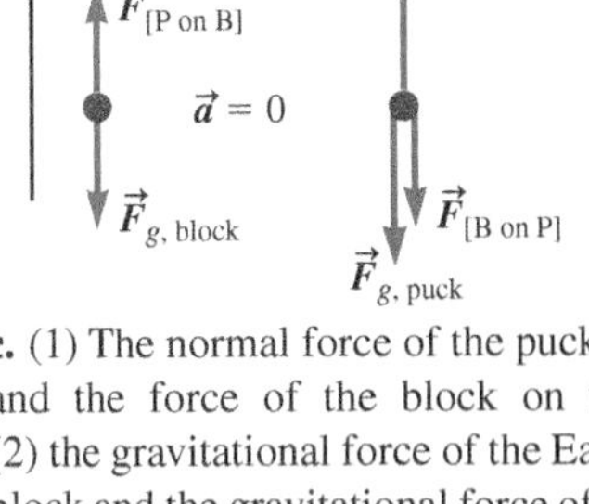

c. (1) The normal force of the puck on block and the force of the block on the puck; (2) the gravitational force of the Earth on the block and the gravitational force of the block on the Earth, (3) the gravitational force of the Earth on the puck and the gravitational force of the puck on the Earth, and (4) the normal force of the frozen lake on the puck and the force of the puck on the frozen lake

57. a. When she pushes the backpack, the backpack exerts an equal and opposite force on her according to Newton's third law. So, when she forces the backpack away from the craft, the backpack forces her toward the craft. **b.** 7.04 m/s^2 **c.** 2.78 m/s^2

59. The force is directed toward you, or pointing upward from the ground toward you. The Earth is so massive that the effect of the gravitational force you exert on the Earth is not observed.

61. By Newton's third law, each train exerts a force of equal magnitude on the other.
63. 3.63×10^3 N to the right
65. 47.9 N
67. **a.** 3.13×10^{-15} N **b.** The force is 1.70×10^{12} greater than the muon's weight.
69. North
71. **a.** $F_{pull} = 95.0$ N, $F_g = 147$ N, $F_N = 147$ N **b.** 9.75 m/s
73. $\tan^{-1}\left[\dfrac{mg - F_N}{\mu_k F_N}\right]$

75. **a.**

y, $\vec{F}_T$, α, α, $\vec{F}_T$, x, $\vec{a} = 0$, $\vec{F}_g$

b. 552 N

77. $F_{T1} = 549$ N, $F_{T2} = 1000$ N, $F_{T3} = 1230$ N

79. **a.**

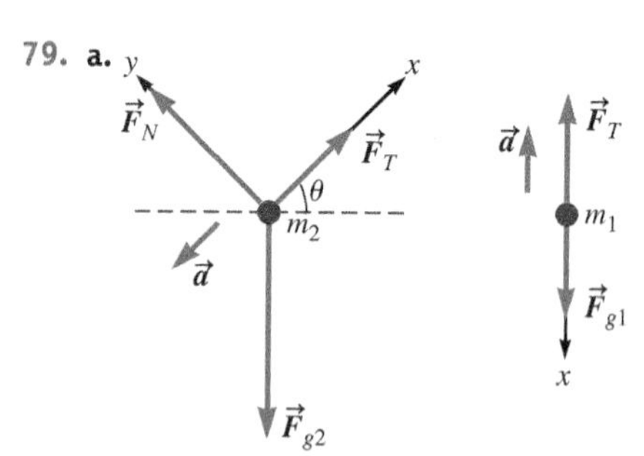

b. 0.701 m/s² **c.** 42.0 N **d.** 2.10 m/s
81. **a.** 46.7° **b.** 1.01 N
83. **a.** 61.3 N **b.** 150 N

CHAPTER 6: Concept Exercises

6.1 **a.** Avi is correct. Air does exert a force (the drag force) on the skydivers. It may be helpful to think of an extreme example of two skydivers, one in free fall and the other with an open parachute. The skydiver with the open parachute falls much more slowly than the skydiver in free fall. If the first skydiver to leave the plane is the one with the open parachute, the skydiver in free fall would be able to catch up. The same is true of all the other skydivers in the photograph. **b.** Shannon suggests that the plane does not fly along a level path but instead points downward so that each diver leaves the plane at a lower altitude than the previous diver. The problem here is that the skydivers are (nearly) in free fall. If the plane were to keep near them, it would also be in free fall. People on a free-falling plane would feel weightless. **c.** Cameron is correct; we often study physics under ideal conditions. Doing so helps make the problems simpler, but physical laws and principles govern more complicated problems as well. It is true that in some situations, people cannot be modeled as particles, but that is not an issue in this Case Study. The skydivers' motion is nearly translational. The reason this case is more complicated is due to the air. In many situations, we can ignore the medium in which a particle moves, but not in the case of skydivers.

6.2 Panel 1. Aluminum on steel has a higher coefficient of static friction than Teflon on steel (Table 6.1), so it requires a greater tension force to move the box with the aluminum side down.

Panel 2. The area of the object does not change the maximum static friction. Therefore, this comparison is the same as the one made in Panel 1.

Panel 3. The coefficient of static friction is the same in both cases, but a greater normal force is exerted on the box with extra weight so a greater tension force is required for motion to occur.

6.3 **a.** Because the particle is at rest and no force is applied horizontally, $F_s = 0$. **b.** The tension is less than the maximum static friction force. Because the object is at rest, $F_s = 5$ N. **c.** The tension equals the maximum static friction force $F_s = 15$ N. **d.** The object moves when the tension exceeds $F_{s,\max} = 15$ N.

6.4 When we push a heavy sofa, we increase the force we apply until the sofa begins to move. The force we apply must be at least equal to $F_{s,\max}$. Once the sofa begins to slide, kinetic friction takes over. Kinetic friction is weaker than $F_{s,\max}$ and therefore weaker than the force we are applying at the moment the sofa begins to move. Therefore the sofa accelerates. We usually reduce our applied force in order to match kinetic friction. The sofa then moves at a constant speed.

6.5 We lubricate engines, bicycle chains, and pocket knives to reduce friction because kinetic friction causes abrasions and wears out the equipment.

6.6 a. Parked:

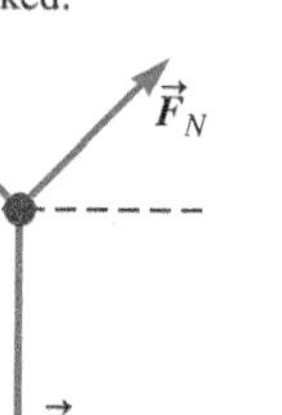

b. Rolling:

$\vec{F}_r$, $\vec{F}_N$, θ, $\vec{F}_g$

c. Sliding:

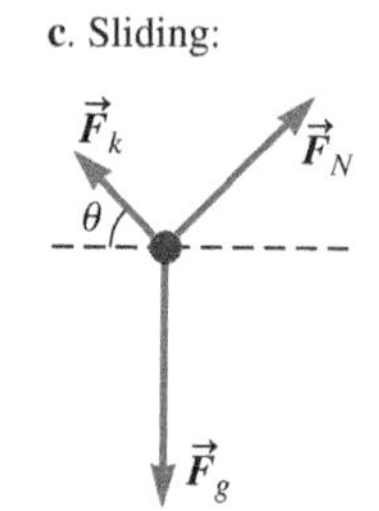

6.7 Three forces act on you as shown in the free-body diagram: gravity, the normal force, and the force of static friction. This friction force, which is exerted on you by the floor, is the force that moves you. Newton's third law says that the harder you press on the floor, the harder the floor presses back on you. To walk, you must apply a horizontal component of force to the floor. If the floor is slippery, static friction is smaller. If the horizontal force you apply to the floor exceeds the maximum value of the static friction force, your foot slips. On icy days, you might notice that people's steps are more vertical.

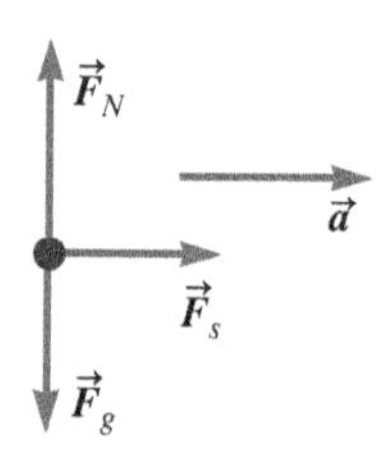

6.8 The cross-sectional areas are
Panel 1: $A = wd$, Panel 2: $A = \pi R^2$, Panel 3: $A = 2Rh$, and Panel 4: $A = \pi R^2$

6.9 Not all objects reach their terminal speed before they land. For example, if you drop a ball from a sufficiently low height, it will not reach terminal speed before it hits the ground.

6.10 **a.** An object falling through a vacuum is in free fall. Its acceleration is constant, which makes Graph 3 the best representation. Its speed continues to increase as best represented by Graph 4. **b.** An object falling through air starts in free fall, and then drag reduces its acceleration. Graph 2 best represents its acceleration. Initially, the object is at rest and its speed increases until it reaches its terminal speed, a condition best represented by Graph 6.

6.11 The force responsible for the centripetal force is circled in each free-body diagram.

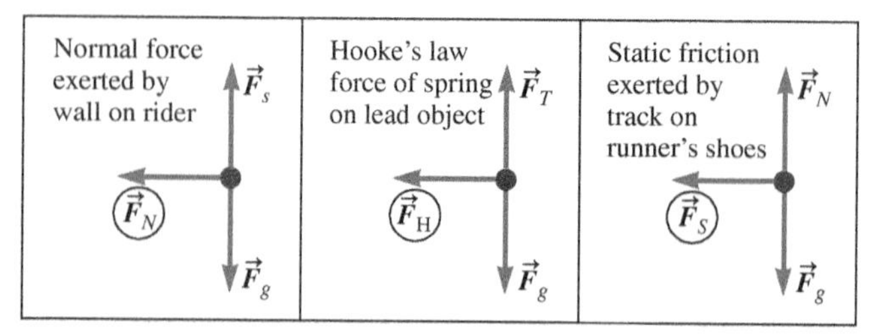

CHAPTER 6: Problems and Questions

1. Consider one example. In Chapter 5, Problem 3 we are asked to consider two blocks on ice. Of course, there is friction between the ice and the blocks, but it is fairly weak. So the problem is asking us to assume that there is no friction.
3. No; air still exerts a force on the skydivers.
5. a. They are equal but in opposite directions. **b.** The normal force will also be reduced. **c.** The static friction force will also be reduced because the normal force is reduced.
7. Static friction between the rod and each wall is strong enough to counteract the gravitational force on the rod. No tension is involved because the rod presses on the wall and the wall presses back. Perhaps the rod should be called a compression rod instead.
9. a. 6.52 N **b.** 8.4 N (The box won't slip because its weight is less than the maximum value of static friction.)
11. a. 249 N **b.** 374 N
13. a. 385 m **b.** 48.9 m
15. a. $\mu_S g$ **b.** It will not change.
17. 1.7×10^3 N
19. $\tan^{-1} \mu_S$, independent of the mass of the rider
21. No; the maximum static friction force is not enough to keep them from sliding given the angle of the hill because the downhill component of the gravitational force (186 N) is greater than the maximum value of static friction (118 N).
23. The block will remain motionless because $F_{s,\,max} = 10.2$ N is greater than $mg \sin \theta = 9.81$ N.
25. a. 0.387 s **b.** 56.4°
27. a. $\vec{a} = -\mu_k g \hat{\imath}$ **b.** $\dfrac{v_i^2}{2\mu_k g}$
29. 1.47 m/s². The result will be the same if the rider's mass is increased because acceleration doesn't depend on mass.
31. 32°
33. 2.8 m/s²
35. a. 118 m/s **b.** 52.7 m
37. 24 m/s
39. a. $v_{A,\,min} = 32.2$ m/s, $v_{A,\,max} = 103$ m/s, $v_{B,\,min} = 30.1$ m/s, $v_{B,\,max} = 78.2$ m/s, $v_{C,\,min} = 29.0$ m/s, $v_{C,\,max} = 70.0$ m/s **b.** Skydiver C leaves first at minimum speed, then skydiver B jumps headfirst until catching up with skydiver C, and skydiver A jumps headfirst last. Both skydivers B and A should flatten out as they each catch up with skydiver C. **c.** 84 s. If we take the acceleration time into account, the wait is longer because terminal speed is not reached immediately.
41. 4.46 m/s
43. a. 0° **b.** 8.18°
45. The sphere, like an aerodynamic car, does not have sharp edges and would have a lower drag coefficient.
47. 11.6 N upward
49. 4.90 m/s²
51. a. 9.05 m/s **b.** 0 and 136 m/s² **c.** 3.02 m/s²
53. $v_{ucm} = \sqrt{F_T R/m}$
55. 340 N toward the center of the circular motion
57. 4.7×10^{30} N
59. a. Between 2.26 m/s and 37.6 m/s **b.** 0.404
61. 0.148
63. a. 0.296 **b.** 0.217
65. $\tan^{-1}(1/\mu_s)$
67. $\dfrac{2\pi d \cos\theta}{\sqrt{gd \sin\theta}}$
69. a. 1.89 N **b.** 4.48 m/s
71. $\dfrac{2\pi d \cos\theta \sqrt{\cos\theta + \mu_s \sin\theta}}{\sqrt{gd \cos\theta(\sin\theta - \mu_s \cos\theta)}}$
73. a. 38 m/s **b.** 4.4×10^3 N
75. a. 1.75 m/s **b.** $F_1 = 3.70 \times 10^2$ N and $F_2 = 455$ N
77. $\dfrac{v_i}{1 + v_i kt}$
79. a. 9.06×10^{22} m/s² inward **b.** 8.26×10^{-8} N inward **c.** 5.16×10^{-9} N inward
81. a. 3.70×10^3 m/s **b.** 1.75 h

CHAPTER 7: Concept Exercises

7.1 Tycho worked to eliminate procedural errors from his observations. We should evaluate our laboratory procedures. Often, redesigning an experiment leads to better results.
7.2 The orbits A and C are possible. Orbit A is circular with the Sun at the center. A circle is a special ellipse in which the two foci are the same point. Orbit C is an ellipse with the Sun at one focus. Orbit B is not possible because the Sun is not at a focus.
7.3 a. According to Kepler's third law, the periods are equal because the semimajor axis of the orbits are equal. (For a circle, the radius becomes the semimajor axis.) **b.** The speed of planet A is constant because its orbit is circular. The speed of planet B varies, being fastest near perihelion. Planet B's speed is higher than planet A's speed whenever planet B is closer to the Sun than planet A.
7.4 a. The semimajor axis is found from Kepler's third law $T^2 = a^3$:

$$a = T^{\frac{2}{3}} = 500^{\frac{2}{3}} = 63.0\,\text{AU}$$

b. The aphelion distance comes from $a = \dfrac{r_P + r_A}{2}$ (Eq. 7.1):

$$r_A = 2a - r_P = 2(63.0\,\text{AU}) - 0.5\,\text{AU} = 125.5\,\text{AU}$$

c. The comet moves very slowly when it is near aphelion. It is faster near perihelion. Therefore, the comet spends most of its time in the outer solar system.
7.5 Kepler's contemporaries maintained that the planets moved in uniform circular motion. Kepler hypothesized that planetary motion was nonuniform and elliptical.

CHAPTER 7: Problems and Questions

1. These terms give the impression that the Sun moves around the Earth. However, we know that heliocentric model has been verified.
3. 280 yr
5. Anywhere between 0 and 10 cm. The length of the semi-major axis does not depend on the distance between the foci.
7. a. 3.84×10^8 m **b.** 2.1×10^7 m **c.** A small version of the sketch should look similar to this drawing. A larger scale shows the Earth's offset better.

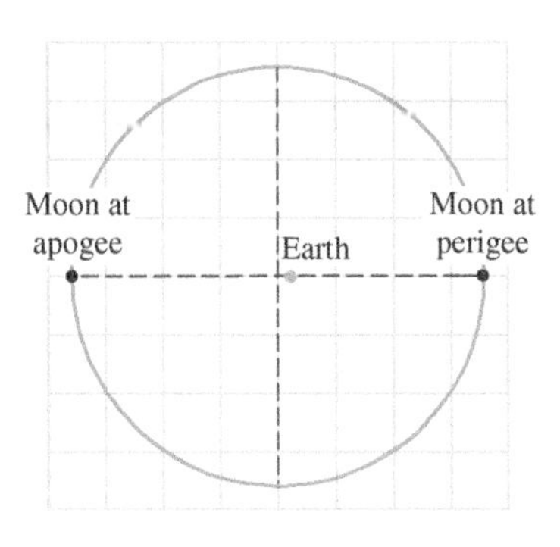

9. a. 1.5 **b.** 1.2 AU and 1.8 AU **c.** 0.3 AU
11. 4.97 days
13. a. 7.5×10^{36} kg **b.** 1.5×10^6 m/s. The Earth's speed is about 2% of this star's speed. This star's speed is about 0.5% of the speed of light.
15. a. 480 AU **b.** 884 AU
17. a. 1.99×10^{20} N **b.** 40 N
19. 1.96×10^{-8} N
21. a. 4.27×10^{-8} N toward the 750-kg sphere **b.** 2.97 m from the 750-kg sphere

23. a. 2.83×10^3 N toward the Moon **b.** 2.83×10^3 N toward the LRO
25. 4.66×10^4 m/s
27. a. $T_{\text{outer}}/T_{\text{inner}} = 1.8$ **b.** 7.9 h **c.** 4.74×10^8 m
29. 3.62×10^{22} N at perihelion, 3.44×10^{22} N at aphelion
31. 1.18 m/s² toward the Earth
33. $\dfrac{Gm}{L^2}\left(\sqrt{2} + \dfrac{1}{2}\right)$ at an angle of 45° to the horizontal
35. a. No, unless you and your dog have the same mass (which is not likely) **b.** The field is still the same. The source of the field (the Earth) has not changed, nor has the location under consideration.
37. a. 5.90×10^{-3} m/s²
b.

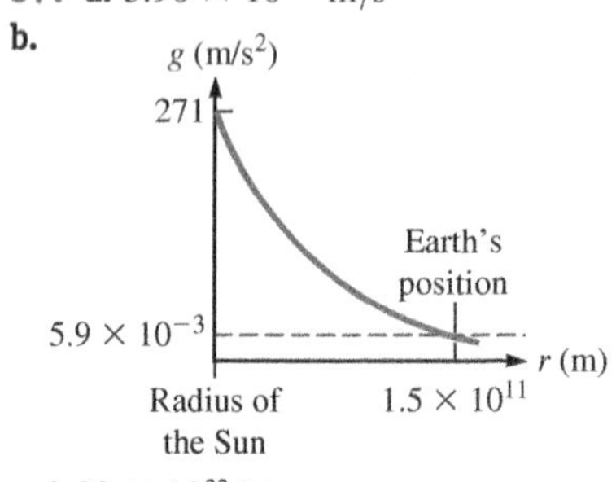

c. 3.53×10^{22} N
39. The mass of the Sun is much larger than the mass of the Earth. When calculating the gravitational force on each object, you are comparing a weaker field strength acting on a large mass and a stronger field acting on a smaller mass.
41. $(-4 \times 10^{-12}\,\hat{\imath} + 1 \times 10^{-11}\,\hat{\jmath})$ m/s²
43. No. To remain above the same point, the satellite must orbit in the same direction as the Earth, which is only possible at the equator.
45. a. $-9.24\hat{r}$ m/s², which is about 94% of $\vec{g}$ at the surface of the Earth. **b.** In essence, astronauts in orbit are in a continued state of free fall.
47. a.

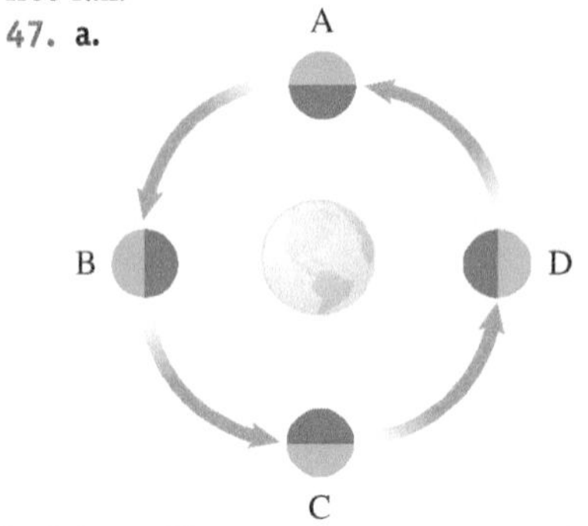

b. 2.4×10^6 s **c.** $w_{\text{pole}} - w_{\text{equator}} = 1.2 \times 10^{-5}$ N **d.** 3.4×10^{-2} N is the difference on the Earth, so the difference on the Moon is about 2800 times less than that on the Earth.
49. a. 360° **b.** 0°
51. 1.52 h
53. 1.00×10^{30} kg and 3.00×10^{30} kg
55. a. 3.44×10^{-5} m/s² **b.** 3.22×10^{-5} m/s² **c.** 3.32×10^{-5} m/s²
d.

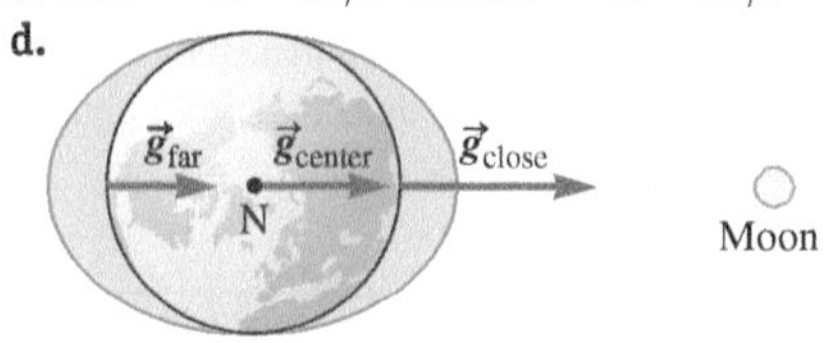

e. The two bulges in the figure show that there are two high tides a day. These bulges form as a result of the gradient in Moon's gravitational field across the Earth. The side closest to the Moon feels the greatest force and is pulled the most in the general direction of the Moon. The side farthest from the Moon feels the weakest force and is not pulled nearly as much as the center of the Earth or the close side.
57. a. 1.07×10^{-2} m/s² toward the Earth **b.** 7.15×10^{13} N
59. 1.9 rpm, 0.20 rad/s
61. No; the satellite must be in orbit around the center of the Earth.
63. a. 2.3 **b.** 1.2 **c.** With a 20% difference in speed, it would not be proper to approximate the orbit as circular.
65. a. 3.748×10^8 s (approximate) and 3.746×10^8 s (proper). So, the difference is 2×10^5 s, or about 0.05%. **b.** Somewhat valid. Equation 7.6 is most valid when the masses are very different in magnitude. The mass of the Moon is about 1.2% that of the Earth. We might expect a 1% difference between the two calculations. **c.** Not valid. Assuming the masses of the asteroids are quite similar, the difference between these two calculations might be on the order of 30%.
69. b. $-400\hat{r}$ m/s² **c.** This difference is 40 times greater than the gravitational acceleration on the Earth. You are being "spaghettified."

CHAPTER 8: Concept Exercises

8.1 a. Use Kepler's third law to find Comet Halley's semimajor axis from its period given in the case study:

$$T^2_{[\text{yr}]} = a^3_{[\text{AU}]} \tag{7.2}$$

$$a = t^{2/3} = (76\,\text{yr})^{2/3} = 17.9 \approx 18\,\text{AU}$$

b. Comet Halley's aphelion distance can be found using

$$a = \frac{r_P + r_A}{2} \tag{7.1}$$

$$r_A = 2a - r_p = 35.2\,\text{AU}$$

which is quite close to the measured aphelion distance.

8.2 At the top of its one-dimensional flight, the ball momentarily stops. Therefore, its kinetic energy is momentarily zero.
8.3 Scalars can be negative; an example is a negative temperature. Kinetic energy cannot be negative because neither mass nor speed (squared) can be negative.
8.4 The change in gravitational potential energy only depends on the vertical displacement. The ball has the same vertical displacement in both cases 1 and 2, so $\Delta U_1 = \Delta U_2$, and because that vertical displacement is negative, $\Delta U_1 = \Delta U_2 < 0$. In case 3, the ball's vertical displacement is zero, so $\Delta U_3 = 0$, which means that the system's potential energy did not change.
8.5

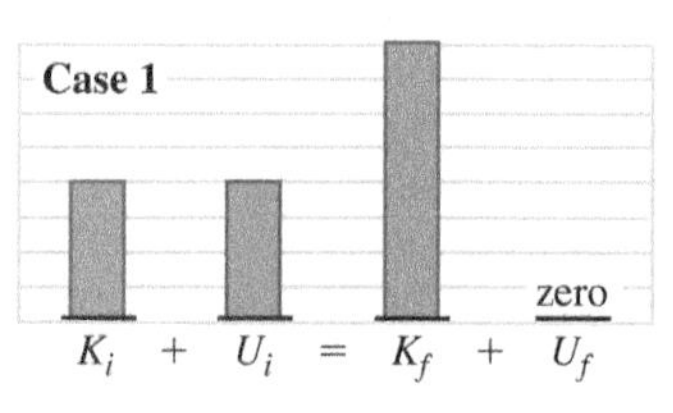

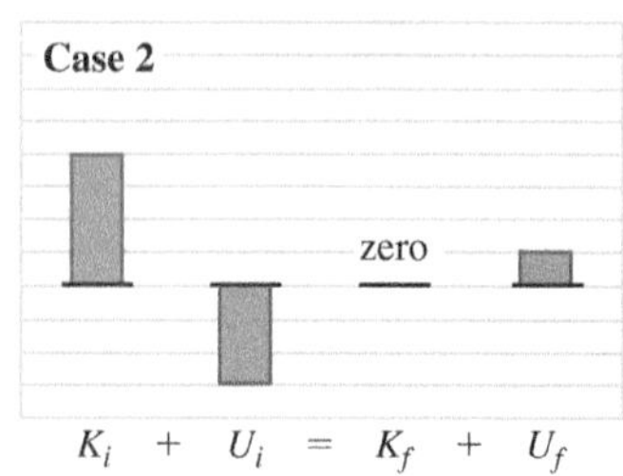

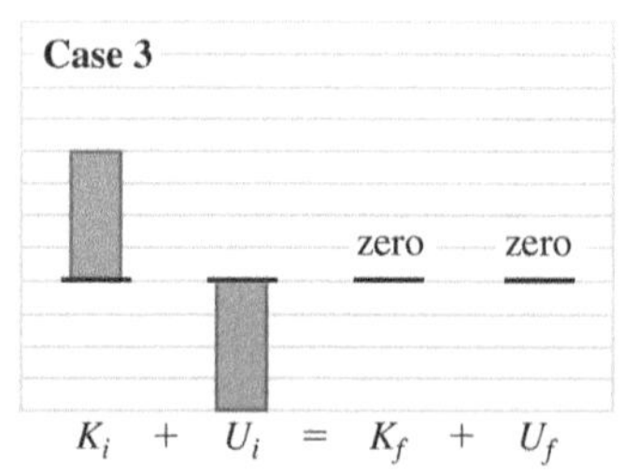

CHAPTER 8: Problems and Questions

1. The comet is slowing down just before it reaches aphelion because the angle between the gravitational force and the velocity will be greater than 90°. It is speeding up just after passing aphelion because the angle between the gravitational force and the velocity will be less than 90°.
3. **a.** 55 J **b.** 43 J
5. 167 J
7. **a.** 48.5 J **b.** 17.6 J
9. 2.26 m/s
11. 280 J
13. Case 1: $-mgh$; Case 2: $-\frac{1}{2}mgR$; Case 3: 0
15. −25.8 J
17. **a.** $U_T = 717$ J, $U_B = 0$, $\Delta U = -717$ J **b.** $U_T = 0$, $U_B = -717$ J, $\Delta U = -717$ J
19. 0.34 J
21. **a.** -5.28×10^{33} J **b.** -1.11×10^{27} J **c.** There is significantly more negative potential energy in the Earth–Sun system than in the Earth–Mars system. Thus, the force on the Earth due to the Sun is greater than that due to Mars. The Sun's mass is the dominant factor.
23. **a.** 0.0300 J **b.** 0.0675 J **c.** Elastic potential energy becomes kinetic energy when the spring relaxes. Because more energy is stored when the car is pulled back farther, the final speed of the car is greater in that case.
25. 0.29 J
27. $-mg\ell \sin\theta + \frac{1}{2}k\ell^2$
29. 30.8 m/s
31. 1.16×10^6 m
33. 4.15×10^4 m/s
35. Mechanical energy is conserved for the watermelon–Earth system. The initial and final total energies are the same regardless of which direction the watermelon is tossed.
37. **a.** 7.25 m **b.** 21.0 m/s
39. **a.** 20 J and 5 J, respectively **b.** 4.0 m/s and 2.0 m/s, respectively
41. 1.09×10^4 m/s
43. 4.5 m/s
45. 4.9 m/s
47. 15.5 m
49. 25.0 m
51.

U
mgh_i
x
d

53.

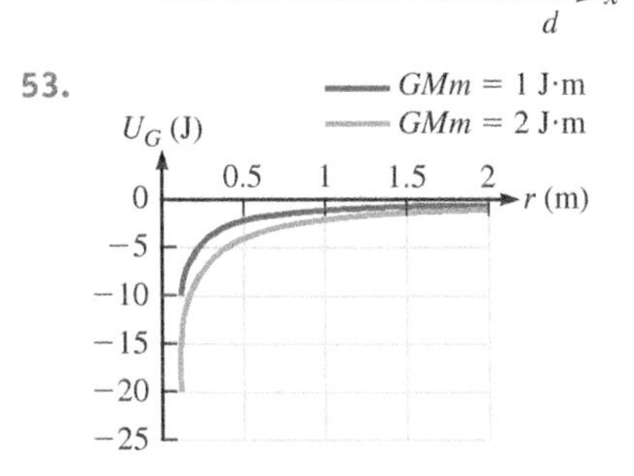

55. **a.**

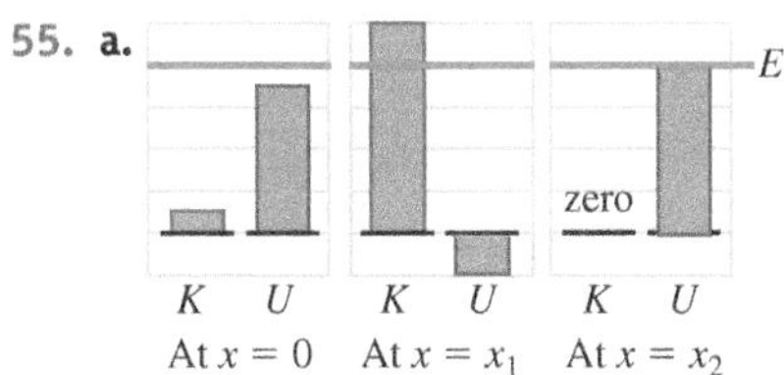

b. x_1 **c.** The particle cannot get there because the total energy is not sufficient.
57. **a.** 13.4 J **b.** 5.9 J **c.** The particles will not touch because the system does not have enough energy.
61. **a.** 8.4×10^4 m/s **b.** 3.0×10^8 m/s **c.** His speed is a factor of 3000 or so less than the speed of light in part (a) and approximately equal to the speed of light in part (b). He would have to go faster than light if he were to orbit any closer (which is theoretically impossible).
63. **a.** 0.989 **b.** 2540 yr **c.** -3.11×10^{22} J
65. **a.** $2\sqrt{gR}$ **b.** 1.47 N downward
67. -1.41×10^{32} J and 1.41×10^{32} J, respectively
69. $\sqrt{\dfrac{2gH}{1 + m_1/m_2}}$
71. **a.** 9.13 m/s **b.** 3.9 m
73. **a.** 238 km **b.** −500 MJ **c.** 1000 MJ
75. **a.** 110 m **b.** 29.4 m/s²
77. 1.95 m/s
79. 1.54 kg
81. We are looking for locations where the derivative of U as a function of position is equal to 0. This would indicate locations where the net force on the object could be 0, and could thus be in equilibrium. The object could be in equilibrium where the potential energy curve is at its peak. A small displacement in either direction would destroy the equilibrium, however, because there will be a force on the object that increases in magnitude as the object moves. The force drives the object away from the equilibrium point in this case. This makes the equilibrium an unstable one because small displacements from the equilibrium point do not drive the object back toward the location where there is no net force.
83. **a.** -3.07×10^{34} J **b.** -2.71×10^{34} J

CHAPTER 9: Concept Exercises

9.1 Work (area under curve) is positive in the first two cases and negative in the last case. Work is positive when the force is parallel to the displacement. So, in the first two cases, the force is parallel to the displacement, and in the last case, the force is antiparallel to the displacement.
9.2 **a.** The normal force on her feet accelerates her upward.
b. Because the point of application is not displaced (when the normal force is being applied), the force does zero work on her.
9.3 **a.** The normal force on his feet accelerates him.
b. This force does positive work on the man because the point of application is displaced in the same direction as the force is applied.
9.4 Gravity is a conservative force, so the change in the gravitational potential energy depends only on the initial and final configurations. Because the Earth–book system is restored to its initial configuration when the book is returned to Avi, there is no change in gravitational potential energy.
9.5 Thermal energy depends on the total path length. In other words, each time the book slides across the table, the thermal energy of the book–tabletop system increases. This answer is in contrast to Concept Exercise 9.4, which concerns gravitational potential energy that is unchanged when the system is restored to its original configuration.
9.6 There are 720 hours in 30 days, and 100 W is 0.1 kW. So, in 30 days, her lightbulb requires a total of (720 hr)(0.1 kW) = 72 kWh, and her cost is (72 kWh)($0.0472/kWh) = $3.40.

CHAPTER 9: Problems and Questions

1. **a.** Earth and satellite **b.** Earth, atmosphere, and plane **c.** Earth, truck, road, and air **d.** Earth, person, and floor
3. No. The force is always perpendicular to the direction of a rider's motion.
5. −12 J
7. 1.60×10^4 J
9. Both objects require forces that point to the west. Assuming the objects are restricted

to move in the east–west direction, the first object must have work performed on it to bring it to a stop. This amount of work will be negative. Once that occurs, both objects require the same magnitude of work to get them moving to the west at 25 m/s. Because it first needs to be stopped, the first object requires a net amount of work equal to zero, whereas the second object experiences a positive amount of net work.

11. 7.19

15. a. 0° **b.**

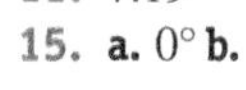

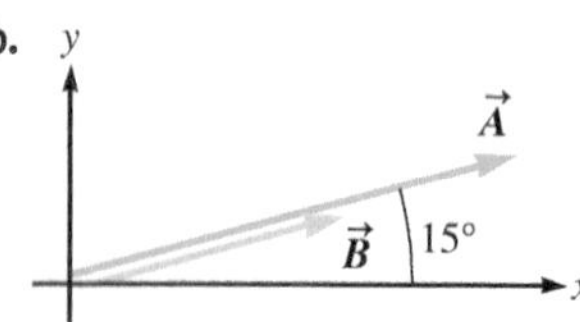

17. 7.9 J in all cases

21. −23

23. 1.3 J

25. 2.7×10^3 J

27. a. 75.0 J **b.** −75.0 J **c.** 0

29. 1.166×10^7 J

31. $-(\frac{1}{2})(k_1 + k_2)x_f^2 + (k_1 + k_2)[\ell\sqrt{x_f^2 + \ell^2} - \ell^2]$

35. The sled is speeding up (accelerating). $\vec{F}_P$ in the figure represents the force applied by Paul, and $\vec{F}_S$ represents the static friction force.

37. a.

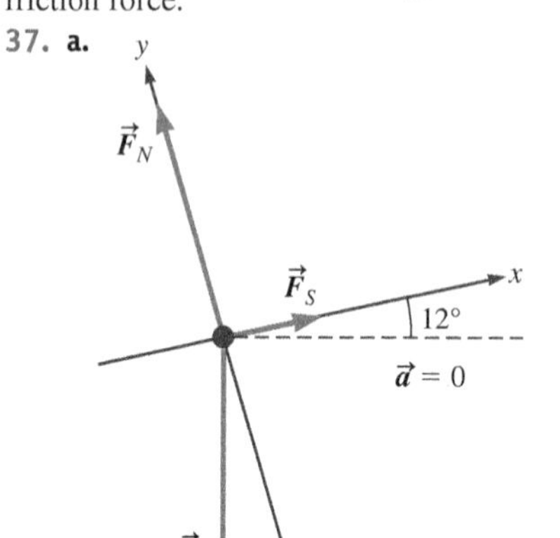

b. 14 J

39. a. 5.00 cm **b.** 2.40 J

41. a. 11.3° **b.** Both surfaces are warmed as the thermal energy increases, so it is best to include both the block and the incline in the system. **c.** 20.3 J

43. 1.9×10^6 J

45. $\sqrt{\dfrac{m_b v_b^2 - 2E_{th}}{m_b + m_{wb}}}$

47. a. 39.8 m/s **b.** 5.94×10^5 J **c.** 0.350

51. a. To account for the increase in thermal energy and the changes in gravitational potential energy, the system of choice is the Earth–box–hill–spring surface. **b.** 8.6 m **c.** The block slides a shorter distance on this surface because friction acts throughout the entire motion.

53. 0.623 m

55. 190 N

57. 3.75 m

59. 7.3 N

61. 8.3×10^4 W

63. 6.28×10^4 W

65. 978 W

67. a. 90° **b.** 120° **c.** 86°

69. a. 1.60×10^3 J **b.** 491 J **c.** 1.11×10^3 J **d.** 9.42 m/s

71. 1.50×10^8 J

73. a. 37.9 J **b.** 44.6 N

75. a. 100 J **b.** −100 J **c.** 102 J

77. a. 3.3 m/s **b.** 1.8 N

79. a. 1.21×10^5 J **b.** 24.1 m/s

81. 99.4 s

83. 7.84 m

85. a. $K = 36t^2 + 72t^3 + 36t^4$ **b.** $a = 6.00 + 12.0t$ and $F = 12.0 + 24.0t$ **c.** $P = 72.0t + 216t^2 + 144t^3$ **d.** 5040 J

CHAPTER 10: Concept Exercises

10.1 Thrusters are not just used to launch spacecraft; they are used to maneuver spacecraft in space. (Check out NASA's website.)

10.2 Set the origin of the coordinate system to the center of the Earth. Then Equation 10.5 becomes

$$\vec{r}_{CM} = \frac{1}{M}\sum_{j=1}^{n} m_j\vec{r}_j = \frac{1}{M_\oplus + M_{Moon}}(M_{Moon}r_{Moon})\hat{r} = 4.67 \times 10^6\ \hat{r}\text{ m}$$

The center of mass of the Earth–Moon system is inside the Earth, $r_{CM} < R_\oplus = 6.37 \times 10^6$ m. For most problems, assuming the center of mass of the Earth–Moon system is at the center of the Earth is a good approximation.

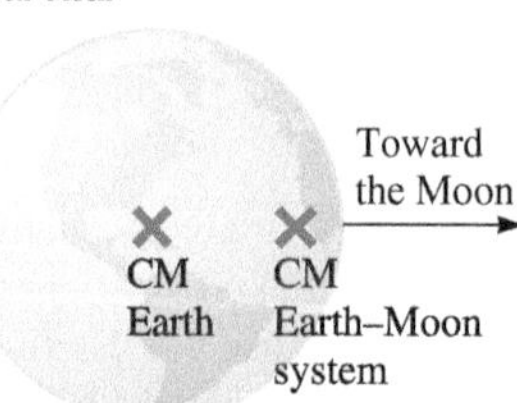

10.3 The center of mass moves closer to the trunk.

10.4 The center of mass is in the center of the can, so $(x_{CM}, y_{CM}, z_{CM}) = (0, 7.5, 0)$ cm.

10.5 When a cannonball is fired, the cannon must move backward for the momentum of the ball–cannon system to be conserved. The backward motion of the cannon is called *recoil.* The ropes are used to restrain recoil and therefore prevent damage to the ship or crew.

CHAPTER 10: Problems and Questions

1. Neither (b) nor (c) can be answered using the conservation of energy principle because both ask about direction. Energy is a scalar and therefore does not contain directional information.

3. a. $78\hat{\imath}$ kg · m/s **b.** $-1.7\hat{\imath}$ kg · m/s

5. $(4.64\hat{\imath} - 0.640\hat{\jmath})$ kg · m/s

9. 1.8×10^{29} kg · m/s

11. a. 2.35 kg **b.** 17.0 m/s

13. 13.7 N

15. (0.67 m, 0.29 m)

17. a. 3.76 m **b.** 4.02 m. The two answers do not agree because there is more mass to the right of the average position of the students.

19. a. 0.250 m from the center in the direction of the sister. **b.** She moved 0.429 m closer to the center of the see-saw.

21. $(0, 6.51 \times 10^{-12}$ m$)$

23. 4.56×10^5 m from the center of the Sun

25. 2.0×10^4 m from the center of Jupiter

27. 60.0 kg · m/s to the left

29. 1.0 m/s^2

31. Equal. The net change in momentum of the system is zero.

33. $v/2$

35. $x = 1.00$ m

37. 3.73×10^{-23} m/s

39. $(5.30\hat{\imath} - 2.52\hat{\jmath})$ m/s

41. a. There is no net external force on the system. **b.** $-1.01\hat{\imath}$ m/s

43. a. Yes; 2.56×10^{-2} m/s toward the shore. **b.** The forces acting on the child are a gravitational force between the child and the

Earth, a normal force on the child from the icy surface, and the force of the ball pushing on the child as he throws it. From Newton's third law, as the child exerts a force and throws the ball, the ball exerts an equal and opposite force on the child.
45. $(4.04\hat{\imath} - 2.67\hat{\jmath})$ m/s
47. 2.9 m/s
49. 13.4 m/s
51. a. 20.2% **b.** 3.73×10^5 kg
53. When Shannon throws the ball, the ground exerts static friction on Shannon. Friction creates a net external force, so momentum is not conserved. When analyzing the raft rocket in Section 10-6, we ignored the drag of the water on the boat. If we take that into account, then momentum is not conserved. Because the drag on a boat is not very large, the boat will still accelerate in response to the water being launched, but its acceleration will be less than in a frictionless case.
55. 4.64 m/s
57. 362 kg/s
59. 20.4 N perpendicular to the wall
61. $(1.2 \times 10^{-2}\hat{\imath} + 1.2 \times 10^{-2}\hat{\jmath})$ kg · m/s
63. 2.7 m/s
65. $(-1.26 \times 10^3\hat{\imath} + 4.30 \times 10^2\hat{\jmath})$ N
67. a. $(6.00\hat{\imath} - 11.6\hat{\jmath})$ m
b. $(5.57\hat{\imath} - 8.14\hat{\jmath})$ m/s
c. $(19.5\hat{\imath} - 28.5\hat{\jmath})$ kg · m/s
69. 5.55×10^{-3} m/s
71. $(2/3)L$
73. $(4/7)L$
75. a. $v_{\text{sled}} = 3.36$ m/s **b.** $v_{\text{pack}} = 1.64$ m/s in the opposite direction to the sled's velocity relative to the ground
77. Gravity pulls straight down at the center of mass. Your body shifts so that the normal force exerted on the bottom of your foot is directly below your center of mass.
79. 0.107 m/s and 0.214 m/s in opposite directions

CHAPTER 11: Concept Exercises

11.1 The car leaves skid marks when the brakes have locked and kinetic friction decelerates the car. If the car comes to a stop, we know its final speed (zero). If it collides with something before coming to a stop, we may be able to use information about the collision to find the speed of the car at the end of the skid marks (that is, the speed just before impact). If we know the car's speed at impact, we can use the length of the skid marks and kinematics (Chapter 2) to find the speed of the car just as the brakes were locked.
11.2 By bending her legs as she lands on the ground, the gymnast increases the time Δt over which her momentum is reduced to zero. Increasing Δt reduces the average force she experiences from the ground.
11.3 Newton's third law
11.4 The two trains are stuck together after the collision; therefore, the collision is completely inelastic. Equations 11.5 through 11.9 apply (conservation of momentum and Newton's third law).
11.5 From Example 11.6, we know that $\Delta v_S = 2v_P$. We find Venus's speed from its orbital information, assuming its orbit is nearly circular:

$$v_P = \frac{2\pi r}{T} = \frac{2\pi(1.08 \times 10^{11}\text{ m})}{1.94 \times 10^7\text{ s}} = 3.50 \times 10^4\text{ m/s}$$

$$\Delta v_S = 2v_P = 7.00 \times 10^4\text{ m/s} = 70\text{ km/s}$$

11.6 a. The eight-ball will sink if

$$\theta_2 = \tan^{-1}\left(\frac{0.50\text{ m}}{0.75\text{ m}}\right) = 34°$$

which is the angle shown. So, the eight-ball is sunk.
b. The cue ball will sink if

$$\theta_1 = \tan^{-1}\left(\frac{0.50\text{ m}}{0.75\text{ m}}\right) = 34°$$

For an elastic collision between pool balls, we have (from Eq. 11.30):

$$\theta_1 = 90° - \theta_2 = 90° - 34° = 56°$$

so the player does not lose the game.

CHAPTER 11: Problems and Questions

1. When the force between two interacting objects is a field force, the objects do not need to touch to collide. The force exerted between a spacecraft and a planet is gravity (a field force), but the truck and the car interact through the normal force and friction.
3. 3.9×10^2 N. Adding padding does not change the impulse, but it does increase the time needed for the man's head to stop. A longer stopping time means a weaker average force.
5. a. 8.11 kg · m/s upward **b.** 54.1 N upward
7. a. 11 kg · m/s to the left **b.** 0.23 s
9. a. $-748\hat{\imath}$ kg · m/s **b.** $-4.98 \times 10^3\hat{\imath}$ N
11. 17.2 N
13. Equal. Both objects experience forces of the same magnitude for the same amount of time.
15. a. 0 **b.** 0 **c.** 0.900 m
17. a. 5.55×10^{-3} m/s **b.** 5.55×10^{-3} m/s
19. $\dfrac{M^2v^2}{2m^2\mu_k g}$
21. 3.3 m/s
23. a. 2.50 m/s **b.** 0.319 m
25. 1.3×10^3 N
27. a. 2.5 m/s to the left **b.** 2.96×10^5 J
29. $\dfrac{(m + M)}{m}\sqrt{2\mu_k gD}$
31. 207 m/s
33. b. 0
35. $v_1 = 6.80$ m/s and $v_2 = 2.80$ m/s, both to the right
37. Particle 1 must be much more massive than particle 2. Particle 2's velocity is in the same direction as particle 1's initial velocity, and its speed is about twice that of particle 1.
39. $v_1 = 11.7$ m/s to the left and $v_2 = 6.26$ m/s to the right
41. $v'_{1f} = -6.2\hat{\imath}$ m/s and $\vec{v}_{2f} = -7.6 \times 10^{-2}\hat{\imath}$ m/s
43. a. $\dfrac{4m_1^2}{(m_1 + m_2)^2}h_1$ **b.** 1.78 m
45. $\vec{v}_{1i} = 0.647\hat{\imath}$ m/s and $\vec{v}_{2i} = 0$; $\vec{v}_{1f} = 1.22\hat{\imath}$ m/s and $\vec{v}_{2f} = 0.793\hat{\imath}$ m/s
47. 25.6 m
49. $h_1 = 2.94$ m and $h_2 = 5.22$ m
51. $6.93\hat{\jmath}$ m/s
53. $(1.5\hat{\imath} + 2.0\hat{\jmath})$ m/s
55. a. 4.43 m/s at 2.38° west of south **b.** 2.87×10^4 J **c.** More energy is lost in this collision than in Problem 54. In a completely inelastic collision, the energy lost is maximized.
57. 50.0 m/s at an angle of 143° relative to the direction of A
59. 0
61. $(2.10\hat{\imath} + 1.35\hat{\jmath})$ m/s
63. 141 m/s
65. a. 2.72 m/s at 34.1° north of east (b) 677 J
67. a. $(v_{1i}/5)\hat{\imath} + (4v_{1i}/15)\hat{\jmath}$ **b.** $\frac{5}{13}$
69. 1.3 kg
71. 1.1×10^2 N
73. a. 1.37 m/s to the left **b.** No. As long as there is no net external force on the system, the total momentum is conserved regardless of the order of the collision events.
75. 5.8×10^5 m/s
77. 9.8 m
79. 0.059 m/s in the direction the ball was traveling initially
81. a. $v_m = \sqrt{2}v_i$ and $v_{3m} = (\sqrt{\tfrac{2}{3}})v_i$ **b.** 215°

CHAPTER 12: Concept Exercises

12.1 The rotation axis for each object is shown in the figure.

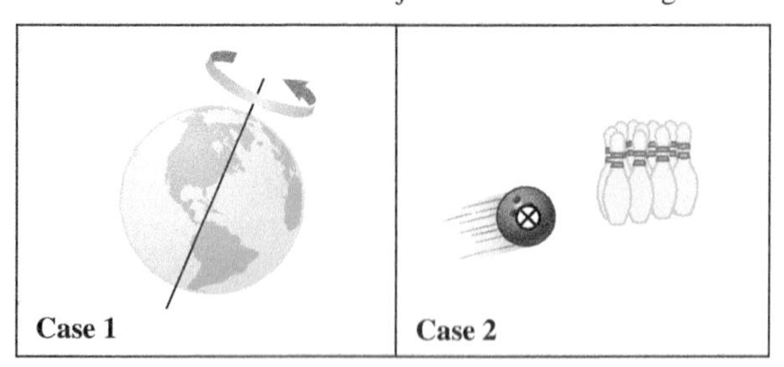

Case 1. The rotation axis of the Earth is fixed in the Earth's center-of-mass frame.

Case 2. The bowling ball's rotation axis is fixed in a frame that translates to the right at the same speed as the ball's center of mass.

12.2 Solve Equation 12.1 for s:

$$s = r\theta_{[\text{rad}]}$$

In Equation 12.1, θ is the ratio of two lengths, so θ has no units. (The "radian" is dimensionless.) We are given θ in degrees, however. To convert from an angle measured in degrees to one measured in radians, recall that there are 360° in a circle and that the ratio of the circumference c of a circle to its radius r is 2π:

$$\frac{c}{r} = 2\pi \text{ rad} = 360°$$

So, if θ is measured in degrees instead of radians, we can find s from

$$s = r\theta_{[\text{deg}]}\left(\frac{2\pi \text{ rad}}{360°}\right)$$

$$s = (0.25\,\text{m})(15.5°)\left(\frac{2\pi \text{ rad}}{360°}\right) = 6.8 \times 10^{-2}\text{m}$$

If θ is measured in radians, Equation 12.1 takes its simplest form without the factor $(2\pi \text{ rad}/360°)$. We say that θ is most naturally measured in radians.

12.3 Use the values of θ from Table 12.1 to mark off the angular position of the reference line for each given time.

a.

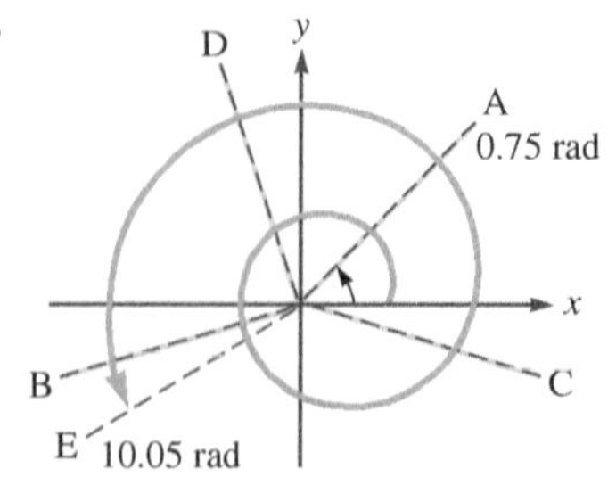

b. The angular displacement from A to E is given by Equation 12.2:

$$\Delta\theta = \theta_E - \theta_A = 10.05 \text{ rad} - 0.75 \text{ rad}$$
$$\Delta\theta = 9.30 \text{ rad}$$

The positive angular displacement indicates that the bottle's rotation is counterclockwise.

12.4 Use Equation 12.3 and the angular displacement found in Concept Exercise 12.3 to find the average angular speed of the bottle:

$$\omega_{av} = \frac{\Delta\theta}{\Delta t} = \frac{9.3 \text{ rad}}{8\,\text{s}} = 1.2 \text{ rad/s}$$

Notice that we kept an extra significant figure because the first digit is 1.

12.5 The rotation is counterclockwise when viewed from $+z$, so $\vec{\omega}$ points in the positive z direction. We know that the bottle is slowing down because the reference lines get closer together for later time intervals. Because the bottle is slowing down, the angular acceleration must point in the opposite direction of the angular velocity. In this case, $\vec{\alpha}$ points in the negative z direction.

12.6 Imagine the force exerted on each dumbbell. The moment arm is the component of $\vec{r}$ that is perpendicular to the force $\vec{F}$.

Case 1. The moment arm is the forearm from the elbow to the hand.

Case 2. The moment arm is the whole arm from the shoulder to the hand.

Case 3. There is (almost) no moment arm. The force on the dumbbell is nearly vertical along the person's arm. So, there is no perpendicular component of $\vec{r}$.

CHAPTER 12: Problems and Questions

1.

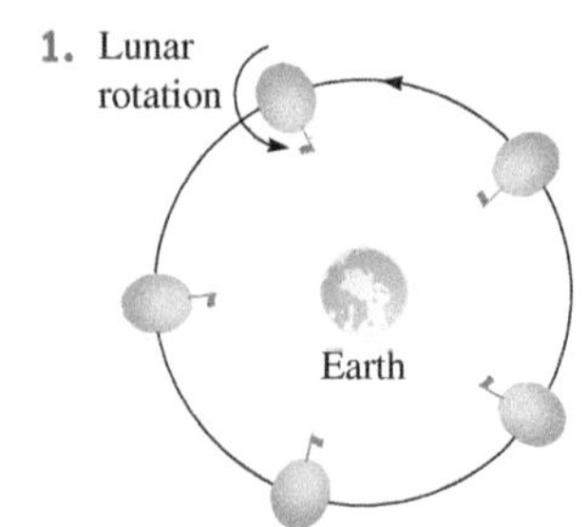

The same side of the Moon always faces the Earth; so, as the Moon orbits the Earth, it rotates. Our sketch shows the Moon as an exaggerated oval to emphasize the rotation. A rotating object cannot be modeled as a simple particle in translational motion.

3. The ice skater is rotating. The axis of rotation is vertical, passing through his foot. Each off-axis point on the skater's body moves in a circle around this axis.

5. a. 13 rad/s **b.** 25 rad **c.** 4.0 rev

7. a. 3.13 rad/s **b.** 3.08 rad/s^2

9. a. 1.76×10^{-4} rad/s **b.** Because of its angular speed, Jupiter bulges at the equator.

11. a. $\Delta\vec{\theta} = -\pi/2\hat{k}$ rad **b.** $\vec{\omega} = -1.7 \times 10^{-3}\hat{k}$ rad/s **c.** 0

13. $\sim 10^7$ rev/yr

15. a. $(189 + 4.54 \times 10^{-6})$ rad/s **b.** 2.40×10^{-9} rad/s^2 **c.** 1.9×10^3 s

17. a. $\theta(0) = 2.00$ rad, $\theta(1.50) = 5.45$ rad **b.** $\omega(0) = 0.800$ rad/s, $\omega(1.50) = 3.80$ rad/s **c.** $\alpha = 2.00$ rad/s^2. It is the same at both times and is constant.

19. $\vec{\alpha} = 1.62 \times 10^{-8}\hat{k}$ rad/s^2, $\vec{\omega} = -1.69 \times 10^{-3}\hat{k}$ rad/s

21. a. 17.0 rad/s^2 **b.** 1.63 rev

23. a. $\vec{\alpha} = -2.22\hat{k}$ rad/s^2 **b.** 10.9 s

25. a. No. The derivative of the angular velocity will be a function of time. **b.** $\alpha(t) = (9.34 \text{ rad/s}^2) + (4.48 \text{ rad/s}^3)t$

27. $v_{large}/v_{small} = 2.40$

29. a. 0.453 rad/s **b.** 10.9 m/s^2 toward the center of the circular track

31. 9.2 m/s^2

33. a. All points on the 40-m blade meet this condition. Both blades have the same

angular speed, so any point on the 40-m blade has the same angular speed as the point at the end of the 20-m blade. **b.** The midpoint at 20.0 m would have the same translational speed as the point at the end of the 20-m blade. The translational speed at a point depends on both the angular speed and the distance from the center of rotation.
35. a. 12.0 rad/s **b.** 2.40 m/s **c.** 28.8 m/s^2 at 1.19° with respect to the radial direction **d.** 24.6 rad
37. 52.1 m/s^2
39. The axis of rotation is vertical and passes through the hinges. By putting the doorknob on the opposite edge, the torque we exert has the longest moment arm. Assuming we exert a perpendicular force, the force we exert is the minimum needed to rotate the door open or closed.
41. $\tau_1 = 4.53 \times 10^2$ N · m, $\tau_2 = 2.60 \times 10^2$ N · m, $\tau_3 = 0$
43. 0.891 N · m
45. $\vec{F}_3 = \vec{F}_5, \vec{F}_2, \vec{F}_1$, and $\vec{F}_4$
47. a. 0 **b.** $225\hat{k}$ **c.** $-225\hat{k}$
49. 30.0°
51. $(5\hat{\imath} - 10\hat{\jmath} + 5\hat{k})$N · m
53. a. $-53\hat{k}$ N · m **b.** $-2.2\hat{k}$ rad/s^2
55. a. 450 N · m **b.** 420 N · m **c.** $\alpha_a = 2.7$ rad/s^2, $\alpha_b = 2.6$ rad/s^2 **d.** Case 1 is more effective because the resulting angular acceleration is greater and will cause the disk's angular speed to increase more rapidly.
57. a. $\omega_0 = 4.00$ rad/s, $b = 0.277$ s^{-1} **b.** 0.277 rad/s^2 **c.** 1.72 rev
59. a. 2.54 s **b.** 11.3 rad
61. a. 26.2 rad **b.** increase by a factor of nine
63. a. They are the same. **b.** Assuming there is nonzero angular acceleration, the tangential acceleration at H will be half that at R because $a_T = r\alpha$. **c.** Assuming H is not the same distance as R from the center, the answer to (a) will remain unchanged, but we can only say that the tangential acceleration is less at H than at R for (b). **d.** In this case, both the angular and tangential accelerations would be the same and are equal to zero.
65. a. 61.4 rev **b.** 24.6 rev/s
67. $a_{ES} = 3.4 \times 10^{-15}$ m/s^2, $a_{NO} = 2.9 \times 10^{-15}$ m/s^2
69. 248 rad
71. $\tau_1 = F_1R$, $\tau_2 = (1/2)F_1R$, $\tau_3 = 2F_1R$
73. $-(5/2)F_1R\hat{k}$
75. 5.25 N
77. 0.982
79. 39.1 N · m
81. 44.6 N
83. a. 0.625 rad/s **b.** 0.125 m/s **c.** 0.833 s **d.** 0.0553 rev

CHAPTER 13: Concept Exercises

13.1 Mechanical energy is conserved under two conditions: (1) No energy is lost or gained by the system, and (2) there are no changes to the system's internal energy. If the system is isolated, its energy (mechanical and internal) is conserved. When Chris defined his system to include the ball, the track, and the Earth, he ensured that the system is isolated. There is nothing left in the environment that could do work on the system. Chris has every reason to believe that energy is conserved. In fact, it is conserved.

13.2 Use the right-hand rule to find the direction of $\vec{L}$.

Case 1. The angular momentum is into the page, in the negative z direction: $\vec{L} = -(rp\sin\varphi)\hat{k}$.

Case 2. The angular momentum is zero because $\vec{r}$ is parallel to $\vec{p}$.

Case 3. The angular momentum is out of the page, in the positive z direction: $\vec{L} = (rp\sin\varphi)\hat{k}$.

13.3 Set $\frac{d\vec{L}_{tot}}{dt} = \sum\vec{\tau}_{ext}$ (Eq. 13.27) and $\vec{\tau}_{tot} = I\vec{\alpha}$ (Eq. 12.24) equal to each other:

$$\frac{d\vec{L}_{tot}}{dt} = \sum\vec{\tau}_{ext} = I\vec{\alpha}$$

So, a change in angular momentum $d\vec{L}_{tot}$ is proportional $\vec{\alpha}$, and the two vectors must point in the same direction.

13.4 When the helicopter is on the ground with its rotors off, its total angular momentum is zero. The rotors rotate in opposite directions so that the total angular momentum of the system remains constant at zero. If one of the rotors stopped operating, the body of the helicopter would have to rotate (in the opposite sense as the functioning rotor) for the total angular momentum of the system to remain constant.

13.5 If no net external torque acts on the disk, its angular momentum (a vector) is conserved, always pointing in the same direction. Imagine arranging the gyroscope so that the angular momentum vector points at Polaris, the North Star. The computers on a space telescope can then use the direction of Polaris as a reference direction to find other astronomical objects. To precisely aim the telescope at particular objects, space telescopes use several gyroscopes with angular momentum vectors pointing in different directions.

CHAPTER 13: Problems and Questions

1. a. The kinetic energy of the sled increases. Because the system is not completely frictionless, the thermal energy of the ice and sled increases. So, it is possible to pull the sled at constant speed. **b.** Same as part (a), but the energy also goes into increasing the gravitational potential energy of the sled–Earth system. In fact, it may be possible to pull the sled at constant speed even if there were no friction between the sled and the snow. **c.** The rotational kinetic energy of the pulley increases. If there is friction or another dissipative force, the work you do may also increase the thermal energy of the system, and it is possible to rotate the pulley at constant angular speed. **d.** The translational kinetic energy of the center of mass and the rotational kinetic energy of the wheels both increase. If there are dissipative forces, it is possible to move the cart with constant angular speed and constant center-of-mass speed.
3. a. The observer sees the center of mass translate and the rotation of the Frisbee. So, the system has both translational and rotational kinetic energy. **b.** The dog runs under the Frisbee and so does not observe its translational motion. The dog would only observe the system's rotational kinetic energy. **c.** The ant is at rest on the Frisbee and so does not observe its motion. According to the ant, the system has no kinetic energy.
5. 1.36 kg · m^2
7. 51.6 rev
9. $\frac{7}{5}MR^2$
11. $(3\pi/2)R^6$
13. a. $0.270\hat{k}$ rad/s^2 **b.** $16.0\hat{k}$ rad/s **c.** 32.0 m/s
15. a. 6.08 s **b.** 40.5 rev
17. 303 J

19. 6.4×10^2 J
21. 23.7 J
23. 6.89×10^{-5} J
25. -3.39×10^{-2}J
27. **a.** 0.41 kg · m² **b.** 38 J
29. **a.** The disk reaches the bottom first because the ratio of its rotational inertia to its mass is smaller than for the hoop; this result is independent of the radius. **b.** $v_{disk} = \sqrt{4gh/3}$, $v_{hoop} = \sqrt{gh}$
31. 1.92×10^{-4} J
33. **a.** 0.66 m/s **b.** 0.78 m/s
35. In the case of a rolling ball, the entire system rotates. No part of it is in pure translational motion. In the case of a cart on a track, however, most of the system's motion is translational; only the wheels, which have very low rotational inertia, are rotating.
37. 28 J
39. **a.** 1.98×10^4 J **b.** 4.40×10^3 W
41. 20 N · m
43. 627 kg · m²/s
45. 31.2 kg · m²/s
47. 3.26 kg · m²/s
49. 2.88 rad/s
51. Your friend is incorrect. Your rotational inertia will increase when you extend your arms, and your angular speed will slow due to conservation of angular momentum. If the experiment were repeated but with the motions reversed (arms begin extended and are later drawn in), you would observe an increase in angular speed after you change your rotational inertia. The effect would be magnified by increasing the mass of the objects held in your hands.
53. **a.** 1.1×10^3 kg · m²/s **b.** 1.7×10^3 J **c.** 4.5 rad/s
55. 3.53×10^{34} kg · m²/s
57. $(-13.8t^2)\hat{\imath} + (-2.38t)\hat{\jmath} + (6.26)\hat{k}$
59. 1.31 m
61. 3.59×10^{28} J
63. **a.** 25.0 J **b.** 12.5 J **c.** 37.5 J
65. $0.6H$
67. 1.6×10^{-3} W
69. $-3.54\hat{k}$ kg · m²/s
71. **a.** 1.80 kg · m²/s **b.** 3.15 kg · m²/s
73. **a.** 1.98 kg · m²/s **b.** 1.46 s
75. **a.** $mv\ell$ **b.** $M/(M + m)$
77. (1.0 m, 1.7 m)
79. **a.** $10F/7M$ **b.** $3F/7$ to the right **c.** $\sqrt{20Fd/7M}$
81. $MRgt \sin \theta$

CHAPTER 14: Concept Exercises

14.1 If you displace the duck up or down slightly, it will return to its original position. So, it is in stable equilibrium in the vertical direction. If you move the duck from side to side, however, it will not return to its original position, illustrating neutral equilibrium in the horizontal direction.

14.2 Like the duck, if you move the boat from side to side, you will see that it is in neutral equilibrium horizontally. If you press the boat downward far enough that water gets in, however, the boat will sink; so, the boat is unstable in the vertical direction.

14.3 **a.** You just attach one cable to a point directly above the center of mass in the middle of the plane.
b. The center of mass will be directly below the cable. By picking the point of attachment to be off the midline, you can cause the plane to be tilted.

14.4 **a.** The pivot should be in the middle of the seesaw.
b. The pivot should be closer to the heavier person.

14.5 Because Young's modulus is much greater for steel, if you can just barely eyeball the stretch in the rubber rod, your eyes will not be able to detect the stretch in the steel cable.

CHAPTER 14: Problems and Questions

1. Yes, the ball is in a stable, static equilibrium because if it were displaced by a small amount, it would return or pass through the equilibrium position.
3. Yes, the system is in a neutral, static equilibrium because if you displaced the system from equilibrium, it would stay in the new position.
5. A and G would be locations of neutral equilibrium; C and E would be locations of stable equilibrium; and B, D, and F would be locations of unstable equilibrium.
7. $(-3.3 \times 10^2\hat{\jmath} + 2.8 \times 10^3\hat{k})$ N · m
9. $F_{Ay} = F_{By} = \frac{1}{2}mg$ and $F_{Ax} = -F_{Bx}$. Each neighboring block supports half the weight of the keystone.
11. You lean toward the side of your supporting foot to keep from falling. This leaning is necessary to ensure your center of mass is directly above your foot so that the net torque on your body is zero.
13. $\tan^{-1}(W/H)$
15. $\vec{R} = (-131.8\hat{\imath} + 538.0\hat{\jmath} + 582.9\hat{k})$ and $R = 804.1$
17. **a.** $-559\hat{k}$ N · m **b.** The torque is unaffected. **c.** The torque will increase if the y component of the force is increased.
19. $\vec{r} = 0.868\hat{k}$ m
21. 72.0 N · m
23. $F_B = 119$ N and $F_J = 106$ N
25. **a.** 0.198 **b.** The coefficient of static friction between rubber and concrete is approximately 1.0, so rubber ladder tips on dry concrete would be safe.
27. $F_1 = 234$ N, $F_2 = 209$ N, and $F_3 = 286$ N
29. 2.0×10^2 N
31. $F_{upper} = F_{lower} = 65$ N, but in opposite directions
33. 242 N
35. $F_A = 9.95 \times 10^3$ N and $F_B = 8.90 \times 10^3$ N
37. 1.09×10^2 N
39. 2.0×10^{-4}
41. 8.53×10^{-4} m
43. 1.07×10^{-5} m
45. **a.** 8.62 m **b.** 1.88×10^3 N
47. 8.06×10^{-3} m
49. 0.039 m
51. 6.30×10^{-6}
53. 5.75×10^{-6}
55. 1.88×10^{-5} m
57. 1.3 m
59. (0.467 m, 1.48 m)
61. $(-MR^2\alpha/2)\hat{k}$
63. $|\vec{\tau}_{F_g}| = |\vec{F}_g||\Delta\vec{r}| \sin\theta = 0$, so $\sin\theta = 0$
65. 2.85×10^{-6}
67. 11.7°
69. $F_P = Wmg/2h$
71. 667 Pa
73. **a.** 7.78×10^6 N **b.** 7.78×10^6 N
75. 12.8°, $F_A = 666$ N, $F_T = 927$ N
77. 0.29 m from the left side of the cabinet
79. 17.7 N
81. 1.6×10^{-3} m
83. $F_{vertical} = 3.49 \times 10^3$ N and $F_{horizontal} = 2.61 \times 10^3$ N
85. $(F_{T1}L_2 - F_{T2}L_1)/(F_{T1} - F_{T2})$

CHAPTER 15: Concept Exercises

15.1 The net force on the airplane must be zero because the airplane is not accelerating. Air drag must be balanced by the thrust of the plane's engines, and gravity must be balanced by an upward force exerted by the air. The free-body diagram for an airplane flying at constant velocity is shown here.

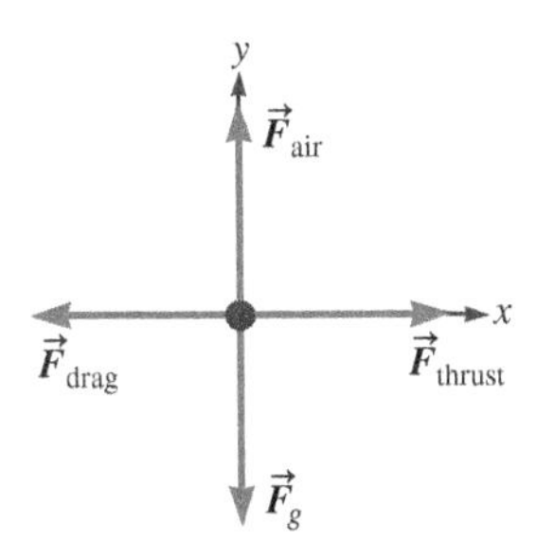

15.2 Consistent with the physics use of the word *pressure*: *His mom and dad are pushing him, and he is feeling pressure from both parents.* Pressure is a scalar, so pressure applied does not cancel out. Instead, it adds! So, the son feels squeezed like a balloon under water.

Inconsistent with physics use of the word *pressure*: *Peer pressure is just one form of negative pressure on him.* The word *negative* here seems to imply that peer pressure pushes in particular direction (the wrong way?). In physics, pressure is a scalar. A better analogy may be with the word *force*: *His peers force him down* is analogous to *gravity exerts a downward force on the apple.* In both *peer force* and *gravity*, that force is always in the same direction.

15.3 Estimate the depth of the pool to be about 2 m. The pressure at the bottom of the pool is given by Equation 15.6:

$$P = \rho g y + P_0 = (1000\ \text{kg/m}^3)(9.81\ \text{m/s}^2)(2\ \text{m}) + 1.01 \times 10^5\ \text{Pa}$$

$$P = 1.21 \times 10^5\ \text{Pa}$$

The difference in pressure is

$$P - P_0 = \rho g y = (1000\ \text{kg/m}^3)(9.81\ \text{m/s}^2)(2\ \text{m})$$

$$P - P_0 = 1.96 \times 10^4\ \text{Pa}$$

15.4 If the same pool only contains air, the pressure at the bottom is

$$P = \rho g y + P_0 = (1.29\ \text{kg/m}^3)(9.81\ \text{m/s}^2)(2\ \text{m}) + 1.01 \times 10^5\ \text{Pa}$$

$$P = 25.3\ \text{Pa} + 1.01 \times 10^5\ \text{Pa} = 1.01 \times 10^5\ \text{Pa}$$

There is only a very slight difference in air pressure between the bottom and top of the "empty" pool, so it is safe to ignore this pressure difference and assume the pressure is constant.

15.5 a. We simply substitute the densities into Equation 15.10:

$$\frac{V_{\text{below}}}{V_{\text{obj}}} = \frac{1000\ \text{kg/m}^3}{1250\ \text{kg/m}^3} = 0.8$$

which means that about 80% of the sunbather's body is below the surface of the water and about 20% of her body is above the surface.

b. We see from Figure 15.12B that the sunbather's head, shoulders, forearms, hands, knees, and toes are above the water. The question is: Does that amount to 20%, or one-fifth, of the sunbather's volume? It is helpful to sketch a woman and break her volume into roughly five pieces. As shown in the sketch, it seems about right that one-fifth of the sunbather's body is above water.

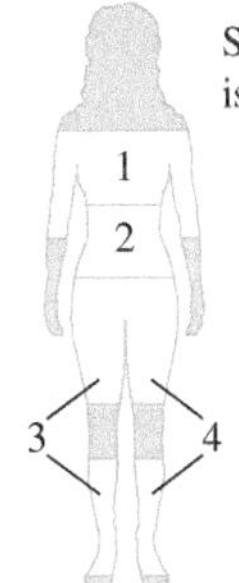

15.6 Many people use the words *light* and *heavy* to describe density. So, the phrase *It is lighter than air* should be translated as *It is less dense than air.* It isn't clear if Cameron's statement is correct because we don't know if the volumes being compared are the same.

CHAPTER 15: Problems and Questions

1. a. $\sim 4 \times 10^{17}$ kg/m^3 **b.** The density of an atomic nucleus is about 10^{13} times greater than the density of osmium, which shows that an atom is mostly empty space. Solids and liquids, as well as gases, are mostly empty space.
3. 1.25 m^3
5. a. 6.51×10^6 Pa **b.** 6.51×10^4 N
7. Because there are many nails in a bed of nails, the total area of their contact is fairly large. If half the person's cross-sectional area is in contact with the nails, the pressure exerted by the nails isn't very great. In fact, when the person stands on one foot, the floor probably exerts a greater pressure because the area of one foot is smaller than the area covered by the nails.
9. a. 1.10×10^8 Pa **b.** 0.050 **c.** Water is compressible, but the strain near atmospheric pressures can be ignored. At about 1000 times atmospheric pressure, the volume change of water is only about 5%.
11. a. 2.0×10^{-5} **b.** 4.8×10^{-4}
13. 1.03×10^4 Pa
15. a. 196 N **b.** 3.23×10^4 Pa
17. a. 1.05×10^8 N **b.** 4.14×10^7 N **c.** 1.55×10^7 N
19. 750 kg/m^3
21. 1875.0 kg/m^3
23. 0.105
25. a. 0.0610 m **b.** 2.37×10^{-3} m
27. $(F_B)_{\text{air}} = 5.7$ N and $(F_B)_{\text{water}} = 4.4 \times 10^3$ N
29. a. The object will sink ($F_g > F_B$). **b.** 1.1 m/s^2
31. 54.1 kg/m^3
33. a. 1.35×10^{-3} m^3 **b.** 3.15 kg
35. 1.52 m
37. 3.4×10^5 Pa. The absolute pressure in the tire is only about three times more than atmospheric pressure, so gauge pressure is very convenient when you want to know the pressure inside your tire.
39. 1.01×10^5 Pa
41. 9.13×10^4 Pa
43. a. 60.0 N **b.** 0.0500 m
45. a. The fluid slows down as it moves in the positive x direction.

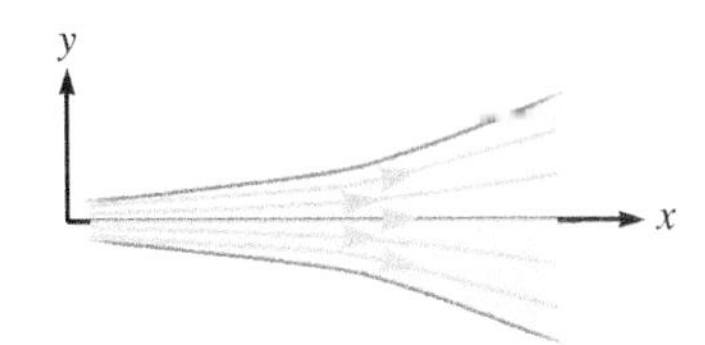

b. $v_0/[1 + (C/B)x^2]^2$
47. a. 8.22 m/s **b.** 58.8 m/s
49. 18 m/s
51. 6.5×10^4 Pa
53. 10.3 m

55. **a.** 1.14 m/s **b.** 5.97 m/s
57. 4.90 × 10^5 Pa
59. 46 s
61. 1.80 × 10^8 Pa
63. The surrounding air exerts pressure on all sides of the balloon, but the pressure on the bottom of the balloon is slightly greater than at the top of the balloon. So, the air exerts a net upward (buoyant) force. Because the balloon is full of low-density hot air, the downward gravitational force on it is relatively small. Thus, the net force (buoyant force minus gravity) on the balloon is upward.
65. $\vec{F}_B = 4.2 \times 10^2 \hat{\jmath}$ N
67. **a.** 2.67 × 10^{-4} m^3/s **b.** 4.22 × 10^{-4} m^3/s
69. 0.25 m/s^2 upward
71. 363 kg
73. 2.14 × 10^6 kg
75. **a.** The pressure should be highest at the bottom of the siphon. **b.** 9.91 × 10^5 Pa or 9.78 atm
77. **a.** 3.66 × 10^5 N **b.** 3.05 × 10^5 N upward **c.** 3.11 × 10^4 kg
79. **a.** −0.027 L **b.** 1.06 × 10^3 kg/m^3 **c.** Even at these great depths where the pressure is very high compared to atmospheric pressure, the volume of the water is only changed by about 2.7%. It is acceptable in most circumstances to consider water to be incompressible.
81. 25:21

CHAPTER 16: Concept Exercises

16.1 Read the times for each given position off the position-versus-time graph (Fig. 16.5A). Then read the corresponding velocity and acceleration off parts B and C of Figure 16.5 for those times.

Position y	Times (s)	Velocity v_y (m/s)	Acceleration a_y (m/s^2)
0	0.5, 1.5, and 2.5	1.3, −1.3, and 1.3	0 for all three times
0.4 m	1.0, 3.0	0 for both times	−3.8 m/s^2 for both times
−0.4 m	0, 2.0	0 for both times	+3.8 m/s^2 for both times

16.2

a. Crall has set the initial time to the moment when the disk is at $y_i = 0$. The initial phase comes from Equation 16.4:

$$\varphi = \cos^{-1} 0 = \pm\frac{\pi}{2}$$

Because the disk's position is increasing at point 2, we choose the negative sign. The disk's position, velocity, and acceleration according to Crall are

$$y(t) = (0.4 \text{ m}) \cos\left(3.1t - \frac{\pi}{2}\right)$$

$$v_y(t) = -(1.3 \text{ m/s}) \sin\left(3.1t - \frac{\pi}{2}\right)$$

$$a_y(t) = -(3.8 \text{ m/s}^2) \cos\left(3.1t - \frac{\pi}{2}\right)$$

Whipple has set the initial time to the moment when the disk is at $y_i = y_{max}$. The initial phase comes from Equation 16.4:

$$\varphi = \cos^{-1}\left(\frac{y_{max}}{y_{max}}\right) = \cos^{-1} 1 = 0$$

According to Whipple, the disk's position, velocity, and acceleration are given by

$$y(t) = (0.4 \text{ m}) \cos(3.1t)$$
$$v_y(t) = -(1.3 \text{ m/s}) \sin(3.1t)$$
$$a_y(t) = -(3.8 \text{ m/s}^2) \cos(3.1t)$$

b. The only difference is the initial phase φ, which reflects the observers' different choices of "when to start the clock," giving a different position of the disk at the chosen initial time.

16.3

	Angular frequency ω (rad/s)	Period T (s)	Initial phase φ	Amplitude y_{max} (m)
a.	14.5	0.433	0	0.75
b.	14.5	0.433	$\pi/2$	5.17 × 10^{-2}
c.	0.75	8.4	$\pi/2$	26

16.4 Using Equation 16.26,

$$\omega_s = \sqrt{\frac{k}{m}} = \sqrt{\frac{0.987 \text{ N/m}}{0.100 \text{ kg}}} = 3.1 \text{ rad/s}$$

$$T = \frac{2\pi}{\omega_s} = \frac{2\pi \text{ rad}}{3.1 \text{ rad/s}} = 2.0 \text{ s}$$

These results fit Crall and Whipple's data.

16.5

θ (degrees)	θ (rad)	$\left\lvert\frac{\sin\theta - \theta}{\sin\theta}\right\rvert$
0	0	0
±5°	±8.7 × 10^{-2}	1.3 × 10^{-3}
±10°	±0.17	4.8 × 10^{-3}
±15°	±0.26	1.1 × 10^{-2}

When the angle is between −15° and +15°, the difference between sine of the angle and the angle itself is about 1% or less.

16.6

a. $\alpha(t) = -\alpha_{max} \cos(\omega t + \varphi)$
b. $\omega_{max} = \theta_{max}\omega$
c. $\alpha_{max} = \theta_{max}\omega^2$
d. In all three equations for parts (a) through (c), ω without a subscript is angular frequency. In part (b), ω_{max} is the maximum angular speed.

CHAPTER 16: Problems and Questions

1. **a.** At 0.5 s and 2.5 s, $y = 0$ and $a = 0$ **b.** At 1.5 s, $y = 0$ and $a = 0$ **c.** At 0, 1.0, 2.0 and 3.0 s, $y = -0.4$, 0.4, −0.4, and 0.4 m, respectively, and $a = -3.8$, 3.8, −3.8 and 3.8 m/s^2, respectively.
3. 3.5$\hat{\jmath}$ m/s^2
5. **a.** −0.0383 m, 0.751 m/s, 4.14 m/s^2 **b.** −0.0817 m, 0 m/s, 8.84 m/s^2 **c.** 0.0813 m, 0.0916 m/s, −8.79 m/s^2

7. **a.** 0.10 m **b.** 0.20 s **c.** 37°
9. $T/6$
11. **a.** 0.188 m/s **b.** 0.880 m/s^2
c. $x = (4.00 \text{ cm}) \cos(4.69t)$,
$v = (18.8 \text{ cm/s}) \sin(4.69t)$,
$a = (88.0 \text{ cm/s}^2) \cos(4.69t)$
13. The shadow of the ball will oscillate back and forth horizontally in simple harmonic motion with the frequency given in the problem statement. In general, the position of the object as a function of time would be given by $x(t) = (L \sin\theta) \cos(\omega t + \varphi)$.
15. **a.** 0.654 s **b.** $\vec{v} = 2.88\hat{j}$ m/s
c. $\vec{a} = -27.7\hat{\imath}$ m/s^2
17. $v_y(t) = y_{\max}\omega \cos(\omega t)$,
$a_y(t) = -y_{\max}\omega^2 \sin(\omega t)$
19. Yes; $\pi L/\sqrt{3gR}$
21. **a.** 2.61 Hz **b.** 0.140 m on the opposite side of the equilibrium position **c.** 0.876 m/s away from the point where the block was released
23. 73.8 kg
25. **a.** $\frac{1}{2}\left(M + \frac{m}{3}\right)v^2$ **b.** $\frac{1}{2\pi}\sqrt{\frac{k}{M + m/3}}$
27. **a.** When the elevator is accelerating upward, the apparent weight of the pendulum bob is increased to $m(g + a)$, which means that there is a stronger restoring force and a shorter period. **b.** The period of a mass hanging by a spring only depends on m and k, so the acceleration of the elevator has no effect on the period.
29. **a.** 0.777 s **b.** 0.106 m/s **c.** 0.856 m/s^2
31. **a.** 1.09 m/s **b.** 5.25 rad/s^2 **c.** It is 1.08 m/s using the conservation of energy approach, so the answers match very well.
33. **a.** The center of mass moves towards the axis of rotation, so the period decreases. **b.** The center of mass moves toward the axis of rotation, so the period decreases. **c.** The center of mass moves away from the axis of rotation, so the period increases.
35. $2\pi\sqrt{\frac{3a^2 + b^2}{2ga}}$
37. 1.45 s
39. **a.** 4.6 m and 1.2×10^3 kg · m^2 **b.** 4.9 s **c.** 60 s
41. **a.** 3.4×10^{-3} N × m **b.** 4.9 s
43. **a.** 3.70×10^3 N/m **b.** 5.89 m/s **c.** 5.77 m/s
45. **a.** 0.332 m **b.** 6.17 m/s
47. $K = 1.7$ J and $U = 1.7$ J
49. **a.** 4.09×10^4 N·s/m **b.** 0.701 s
51. The oscillator is no longer critically damped. If the mass is increased the term k/m decreases. The term $(b/2m)^2$decreases even more. So $\omega_D = \sqrt{\frac{k}{m} - \frac{b^2}{4m^2}}$ is real and the oscillator is underdamped.
55. If the ball were filled with BBs, increasing its mass but not changing the drag force (which depends on the ball's shape and cross-sectional area), the time constant would increase. Yes, the ball would swing longer and presumably hold the cat's attention.
57. 0.307 m
59. The block starts at the right and oscillates back and forth. Numbers below the line give the block's position in meters.

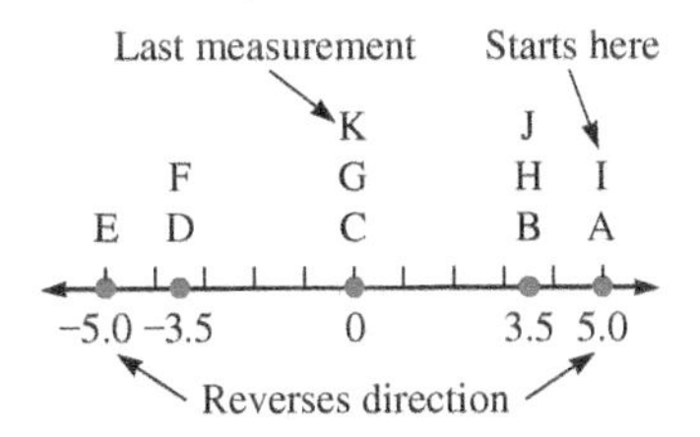

61. The block is moving in the *negative x* direction, so the velocity is negative. For the first two times the acceleration is in the negative x direction and for the last two times the acceleration is the positive x direction. There is no acceleration at the 0.50 s.
63. **a.** 2.47 N/m **b.** 12.3 N
c. $v_{\max} = 22.2$ m/s and $a_{\max} = 98.7$ m/s^2
65. **a.** 30.9 J **b.** $K = 30.9 \sin^2(3.14t)$ and $U = 30.9 \cos^2(3.14t)$
c.

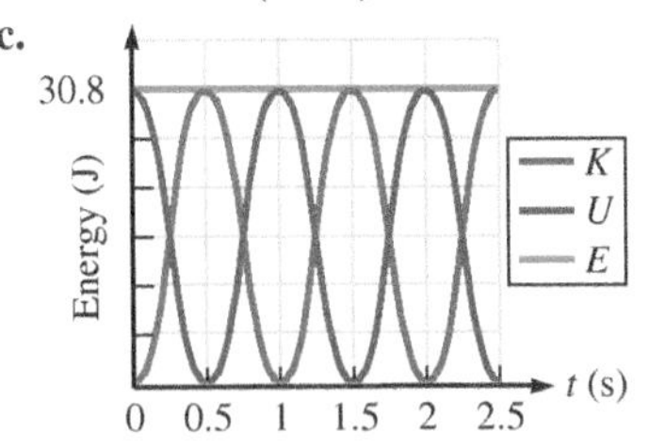

67. **a.** $z(t) = 1.40 \cos(6.40\pi t - \pi/2)$
b. 28.1 m/s **c.** 566 m/s^2 **d.** 44.8 m
69. **a.** 2.24 s **b.** 13.1 J **c.** 1.16 m
71. The period of a simple pendulum is proportional to the square root of its effective length (from the top to the center of mass of the sphere). Initially, the center of mass of the water-filled sphere is at its center. As the water flows out through the hole at the bottom of the sphere, the center of mass of the sphere moves downward. Thus, the effective length of the pendulum increases, so its period increases, reaching a maximum value. The period goes back to the original period when the sphere becomes empty.
73. $2\pi\sqrt{L/(g \cos\alpha)}$
75. $2\pi\sqrt{\frac{2R^2 + 5L^2}{5gL}}$
77. **a.** 0.822 m/s **b.** The speed found in part (a) is that of block A and block B at the equilibrium point. Beyond this point, block B moves with the constant speed of 0.822 m/s, whereas block A starts to slow down due to the restoring force of the spring. When block A returns to the equilibrium point, its speed will be equal to that of block B.
79. **a.** 5.03 Hz **b.** 8.55%

CHAPTER 17: Concept Exercises

17.1 When you check your pulse, you are measuring the period of your heart's vibrational motion. In physics, a pulse is a discrete disturbance in a medium.

17.2 The horizontal axis is x on the graph of the profile and t on the position-versus-time graph.

17.3 a. $S_{\max} = 0.75$m, $k = 0.30$ rad/m, $\omega = 655$ rad/s

b. $v_x = \frac{\omega}{k} = \frac{655 \text{ rad/s}}{0.30 \text{ rad/m}} = 2.2 \times 10^3$ m/s

17.4 No. Avi's model is not a good model for sound. Sand is shot out of a sandblaster. Particles move outward. No air molecules are shot out of a speaker, however. In fact, the air molecules oscillate back and forth, staying relatively close to their original positions. Only the wave pattern moves outward, not the air molecules.

17.5

$$v_s = \sqrt{\frac{B}{\rho}} = \sqrt{\frac{2.2 \times 10^9 \text{ Pa}}{10^3 \text{kg/m}^3}} = 1.5 \times 10^3 \text{ m/s}$$

which is consistent with the speed of sound (1496.7m/s) in water given in Table 17.1.

17.6 Like any wave, sound transmits energy. So, the speaker's energy is transmitted to the hair cells in your ear through sound, and that energy is converted into motion of those hairs. The hairs oscillate back and forth. Imagine wiggling a toothpick back and forth. You can do it for a while, but if you do it for too long or if you wiggle too much, the toothpick breaks.

CHAPTER 17: Problems and Questions

1. No. The dog is modeled as a particle moving from one end to the other end of the pool.
3. We need only to multiply the original wave function by 2, so the new wave function is $y(x, t) = 2/[(x - v_x t)^2 + 1]$.
5. 4 m/s along the negative x axis
7. a. $\vec{v} = -3.5\hat{\imath}$ m/s
b. $y(x, t) = 7.5/[(x - 3.5t)^2 + 0.5]$
9. 0.125 m
11. The frequency of the wave increases, while the wavelength and angular wave number could possibly remain unchanged. Therefore, the speed of the wave $v_x = \omega/k$ must also increase.
13. 2.14×10^{-4} m
15. a. 0.667 m **b.** 8.00 s **c.** 0.0833 m/s **d.** $-0.0350\hat{\jmath}$ m/s **e.** $-0.0140\hat{\jmath}$ m/s^2
17. a. 10 cm **b.** 3.5 Hz **c.** $y(x, t) = (10 \text{ cm}) \sin(0.39x + 22t)$
19. a. 1.21 s and 0.604 m **b.** $S(0) = 0$, $S(T/4) = 0.850$ m, $S(T/2) = 0$, $S(3T/4) = -0.850$ m, $S(T) = 0$ **c.** $x(0) = 0.302$ m, $x(T/4) = 1.15$ m, $x(T/2) = 0.302$ m, $x(3T/4) = -0.548$ m, $x(T) = 0.302$ m
21. 189 m/s
23. 55.1 N
25. a. $y = (1.00 \times 10^{-4} \text{ m}) \sin[37.1x - (2.04 \times 10^3)t]$ **b.** 0.0645 kg/m
27. 275 N
29. 9.9×10^{-2} Pa; about 3000 times greater than the pressure amplitude in air
31. a. 0.014 m **b.** 4.60×10^2 rad/m **c.** $S(x, t) = (1.57 \times 10^{-6}) \sin[(4.60 \times 10^2)x - (1.58 \times 10^5)t]$
33. 2.0×10^{11} Pa
37. 0.0439 m
39. 0.769 J
41. 1.1 W/m^2
43. a. 6.20×10^2 W/m^2 **b.** 14 W/m^2
45. 10 dB
47. The flute is about twice as intense as the cello.
49. 38°
51. 26°
53. The speed of sound in the thinner air above is different from the speed of sound in denser air below. As the sound wave travels, it refracts (changes direction). To an observer, the sound will seem to have come from a different direction.
55. 8.4×10^2 Hz
57. a. 754 Hz **b.** 657 Hz **c.** 825 Hz **d.** 601 Hz
59. 1.3×10^3 Hz
61. a. The frequency increases and the wavelength decreases as the source approaches. **b.** The frequency decreases and the wavelength increases as the source moves away.
63. 1.4
65. 2.55 km
67. a. 2.00 m **b.** 3.75 Hz **c.** 0.600 m **d.** 7.50 m/s
69. 0.043 dB
71. a. 0.0450 kg/m **b.** 23.3 m/s
73. a. 2.93 W **b.** 1.17 J
75. a. 0.0578 s **b.** 0.361 m
77. a. 1.38×10^3 Hz **b.** 1.51×10^3 Hz
79. 22.0 m
81. $\sqrt{F_T/[\rho\pi[(2.50 \times 10^{-6})x - 2.00 \times 10^{-7}]^2]}$

CHAPTER 18: Concept Exercises

18.1

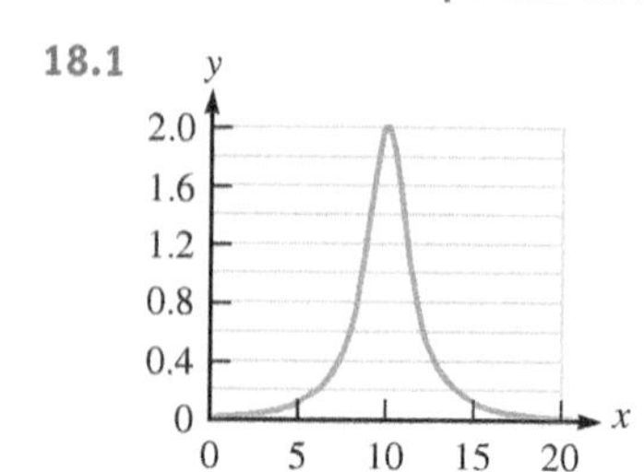

18.2

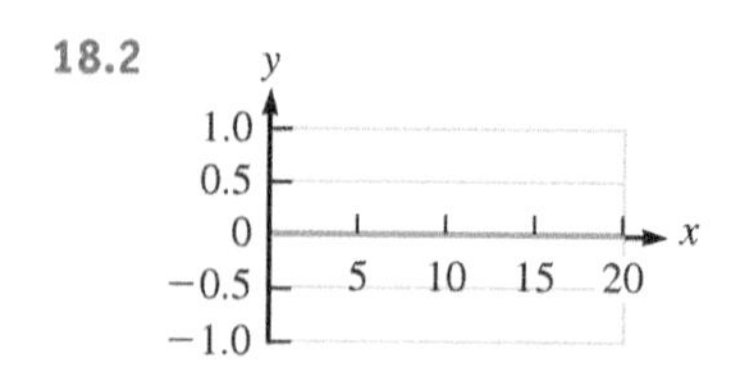

18.3

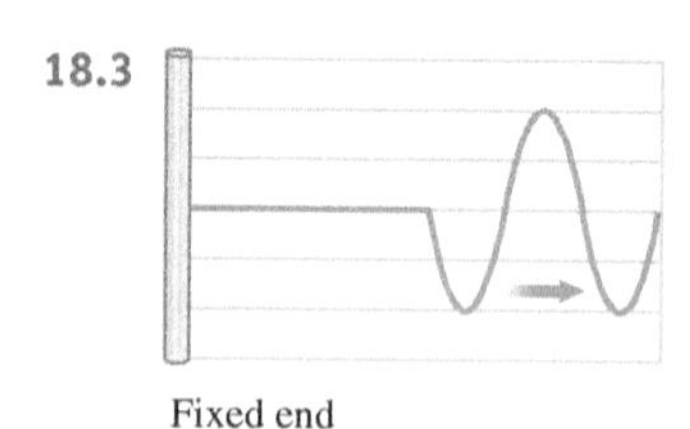

Fixed end

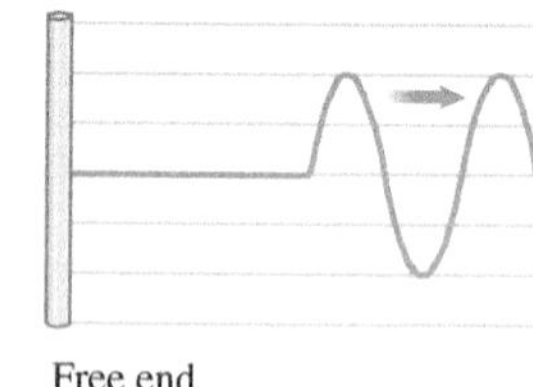

Free end

18.4

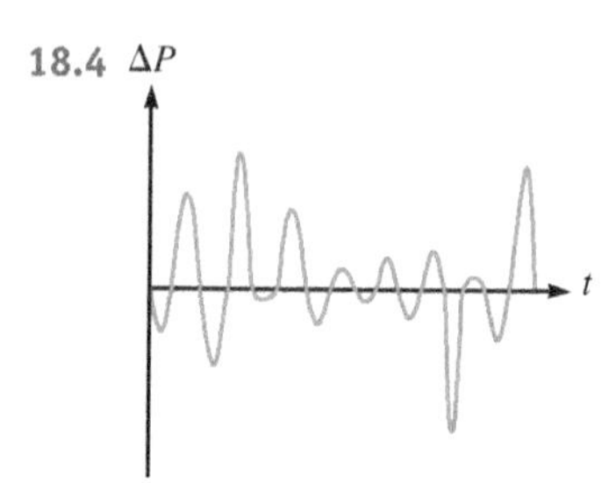

18.5 Because she hears a beat frequency of 2 Hz, we know from $f_{beat} = |f_2 - f_1|$ (Eq. 18.20) that her guitar either sounded at 440 Hz + 2 Hz = 442 Hz or at 440 Hz − 2 Hz = 438 Hz. Tightening the string made the beats disappear, and, according $v = \sqrt{F_T/\mu}$ (Eq. 17.11), an increase in the tension increases the speed of the wave. Finally, because $f_n = n(v/2L)$ (Eq. 18.10), the frequency increases as the speed increases. So, the guitar must have been at too low a frequency before it was tuned. The guitar was initially sounding an A at 438 Hz, and tightening the string increased the note's frequency to 440 Hz.

CHAPTER 18: Problems and Questions

1. a. Wave 1 propagates in the positive x direction, whereas wave 2 propagates in the negative x direction.
b. $y(x, t) = 4 \cos 3t \cos 10x$ **c.** $4 \cos 10x$
3. a. Wave 1 has an angular frequency of $\omega_1 = 3.2$ rad/s, whereas wave 2's angular frequency is $\omega_2 = 3.4$ rad/s.
b. $y(x, t) = 5.0 \cos(0.1t)\sin(8x - 3.3t)$ **c.** $5.0 \cos(0.1t)$
5. a. 0.251 m **b.** −0.0979 m **c.** −0.374 m
7. a. $y(x, t) = \dfrac{0.43}{(x + 13.6t)^2 + 1}$
b. $y(x, t) = -\dfrac{0.43}{(x + 13.6t)^2 + 1}$
9. a. 6 **b.** $11\ell/v_s$ **c.** 30 m
11. An animal could emit a sound and listen for the reflected sound, or echo. After learning to interpret the time between emission and echo as a relative distance in its environment, the animal can determine the location of nearby objects.

13. b. $A(0) = 2y_{max}$, $A(\pi/2) = y_{max}\sqrt{2}$, $A(\pi) = 0$
15. The sound will be louder where constructive interference occurs. If we assume the sound from the drum radiates equally in all directions, waves spread outward from the drum with two-dimensional cylindrical symmetry, traveling at the speed of sound in air v_s. The waves reflect off the wall and encounter the next outbound sound wave. If the time between drum strikes is equal to the time it takes the first wave to reach the outer edge of the monument, the reflected wave and the next generated sound wave will meet at a point on a circle with radius equal to half the monument radius, which will be a location of constructive interference. Other rates of striking could also be considered, resulting in different locations where constructive interference would occur.
17. a. 0.898 m **b.** 414 m/s
19. a. 3.5 cm **b.** Starting from the left end ($x = 0$) of the string, the nodes are at $x = 0$, 15 cm, 30 cm, 45 cm, and 60 cm. **c.** 0
21. a. 2.92×10^{-2} s **b.** 1.51 m **c.** 34.0 m
23. $y(x, t) = [6.90 \sin(1.01x)]\cos(0.184t)$
25. 22 Hz
27. 0.690 m
29. a. 1.20 m **b.** 624 m/s to 6.60×10^2 m/s
31. a. 1 **b.** $1/\sqrt{2}$
33. There is no change; the string is not moving at a node.
35. a. 47.6 N **b.** 392 Hz
37. a. The wavelength in air of the sound produced by the string is longer because the wave speed is greater. **b.** 0.351
39. $\Delta f_{short} = 19$ Hz and $\Delta f_{long} = 1.2$ Hz
41. $n(4.53 \times 10^{-3})$ Hz, where $n = 1, 2, 3, \ldots$
43. 1.27 m, 1.59 m, and 1.91 m
45.

Free at both ends	Fixed at one end (from Example 18.6)
$f_1 = 2140$ Hz	$f_1 = 1070$ Hz
$f_2 = 4280$ Hz	$f_3 = 3210$ Hz
$f_3 = 6420$ Hz	$f_5 = 5350$ Hz
$f_4 = 8560$ Hz	$f_7 = 7490$ Hz
$f_5 = 10{,}700$ Hz	$f_9 = 9630$ Hz
$f_6 = 12{,}840$ Hz	$f_{11} = 11{,}770$ Hz
$f_7 = 14{,}980$ Hz	$f_{13} = 13{,}910$ Hz
$f_8 = 17{,}120$ Hz	$f_{15} = 16{,}050$ Hz
$f_9 = 19{,}260$ Hz	$f_{17} = 18{,}190$ Hz
f_{10} is inaudible	$f_{19} = 20{,}330$ Hz

The rod that is free at both ends has a slightly smaller range, but the difference between adjacent resonant frequencies in both cases is 2140 Hz.
47. 3/2
49. a. The harmonic wavelengths will remain unchanged. **b.** The frequencies will increase when the speed of sound in the medium increases.
51. 210 Hz, which is an audible low tone
53. 8.65 Hz
55. 445 Hz
57. 3.0 Hz
59. 4.5°C
61. 9.19 cm
63. a. Wave 1 travels in the negative x direction, and wave 2 travels in the positive x direction. **b.** 0.75 m **c.** 0.300 s
65. a. 4.24 m **b.** 7.50 Hz **c.** 1.00 m
67. 0.32 m and 0.95 m
69. a. The pipe is closed at one end. **b.** 76 Hz
71. 24.7 Hz
73. a. 0.249 m **b.** 0.497 m
75. 0.60 m
77. a. 1.18 m **b.** 4.14 m
79. a. 192 Hz or 200 Hz **b.** 198 Hz **c.** The tension must be reduced by 2.01%.

CHAPTER 19: Concept Exercises

19.1 Use Equation 19.2 to convert from Fahrenheit to Celsius degrees and then Equation 19.1 to convert from Celsius degrees to kelvins. You should become comfortable thinking about temperature using all three scales.

a. The temperature range in Gdansk is –3.9°C to 26°C or 269 K to 299 K.

b. A very cold day (0°F) is –18°C or 255.15 K, and a very hot day (100°F) is 38°C or 311.15 K.

c. A high fever (105°F) is 41°C.

19.2 A thermometer reading indicates the temperature of the thermometer. If the thermometer is in thermal equilibrium with some object, the object and the thermometer are at the same temperature, and the reading on the thermometer reports the temperature of the object.

19.3 If the potential energy curve is symmetric (as in Fig. 8.36C), the molecules move just as far outward as they move inward when the temperature increases, and the average distance between molecules remains unchanged. Therefore, the material would not expand at higher temperatures.

19.4 Find the changes in length and radius from Equation 19.4, using the coefficient of linear expansion for steel in Table 19.1. The length of the steel rod increases by

$$\Delta L = L_0 \alpha\, \Delta T = (1 \text{ m})(13.0 \times 10^{-6} \text{ K}^{-1})(1 \text{ K})$$
$$\Delta L = 13.0 \times 10^{-6} \text{ m} = 13.0\ \mu\text{m}$$

The factor $\alpha \Delta T$ is the same for the radius of the rod, but because the radius is initially 1 cm, it changes by only 0.130 μm:

$$\Delta r = r_0 \alpha\, \Delta T = (1 \text{ cm})(13.0 \times 10^{-6} \text{ K}^{-1})(1 \text{ K})$$
$$\Delta r = 13.0 \times 10^{-6} \text{ cm} = 0.130\ \mu\text{m}$$

Because the change in radius is 100 times smaller than the change in length, the change in the cross-sectional area is 10,000 times smaller than the change in length. It is reasonable to model the expansion of the rod as a one-dimensional expansion.

19.5 You should cool the peg, perhaps with liquid nitrogen. The peg will shrink and therefore fit into the hole. When the peg warms up again, there will be a very tight fit.

19.6

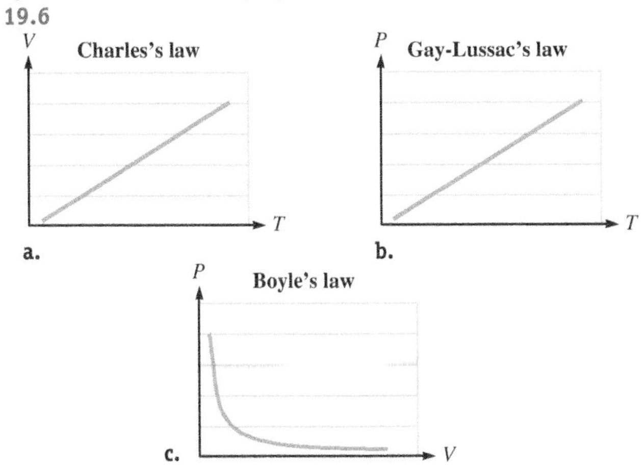

19.7 Label the mass of one molecule of O_2 as m_{O_2} and that of one He atom as m_{He}. The entire mass of the oxygen in the balloon is the number of oxygen molecules multiplied by the mass of each molecule:

$$M_{O_2} = N_{O_2} m_{O_2}$$

and the entire mass of the helium in the balloon is the number of helium atoms multiplied by the mass of each atom:

$$M_{He} = N_{He} m_{He}$$

According to Avogadro's law, because the gases have the same volume and are at the same temperature and pressure, they must contain the same number of particles:

$$N_{O_2} = N_{He}$$

So, the ratio of the balloon's masses is the same as the ratio of the particle's masses:

$$\frac{M_{O_2}}{M_{He}} = \frac{m_{O_2}}{m_{He}} = \frac{5.3 \times 10^{-26}}{6.6 \times 10^{-27}} = 8.0$$

Because the balloons have the equal volumes, the ratio of their densities is also just the ratio of the particle masses:

$$\frac{\rho_{O_2}}{\rho_{He}} = \frac{M_{O_2}/V}{M_{He}/V} = \frac{m_{O_2}}{m_{He}} = 8.0$$

CHAPTER 19: Problems and Questions

1. **a.** 24°C, 297 K **b.** 82°C to 88°C, 355 K to 361 K **c.** −89°C, 184 K
3. 2.6×10^3 °F to 9.5×10^4 °F
5. $T_1 = -40$°F $= -40$°C $= 233.15$ K
$T_2 = 0$°F $= -17.78$°C $= 255.37$
$T_3 = 32$°F $= 0$°C $= 273.15$ K
7. −460°F
9. 40°C
11. **a.** No. Your hand is at 98°F, and the ice is at 32°F. **b.** Yes. They are likely in thermal equilibrium with the freezer and thus with each other by the zeroth law of thermodynamics. **c.** No. Your bathing suit is in contact with your body, which is not in thermal equilibrium with the ocean. **d.** Yes. Your bathing suit and the water have had a sufficient amount of time to reach the same temperature.
13. **a.** The metal expands more than the glass because its thermal coefficient of expansion is greater. **b.** If they are made of the same metal, they will expand by the same amount, and the lid will not loosen.
15. 6.19×10^{-3} m
17. **a.** 439°C **b.** 3.81×10^{-4} m
19. 8.3×10^{-4} °C^{-1}
21. 30.53 m
23. **a.** 11.896 m **b.** 1.20×10^9 Pa
25. $\sigma_{com} \approx 4\sigma$. The compressive strength of cement is greater than the thermal stress by a factor of four. This result suggests that thermal stress is not the cause of the cracking.
27. 1.14×10^5 Pa
29. 1.603×10^7 Pa
31. 3.39 L
33. 1.1×10^6 Pa
35. The pressure and number of particles remains constant, whereas the temperature and the volume must be decreasing.
37. 1.09 atm
39. 0.405
41. **a.** 9.92×10^{14} **b.** 1.65×10^{-9} mol
43. 1×10^{-13} Pa to 4×10^{-10} Pa, 1×10^{-18} atm to 4×10^{-15} atm. Pressure in this cloud is about the same a good laboratory vacuum.
45. 1.23×10^{17}
47. Yes. As in any lab, the bulb should be placed in contact with a triple-point cell and adjusted until the temperature reads 273.15 K.
49. $\dfrac{L_1 L_2}{L_2 - (L_2 - L_1)\cos\theta}$
51. The temperature measured is too low because the height of the mercury is too low. To use a constant-volume thermometer, the volume of the sample must remain constant. By failing to lift the right tube, the researcher has allowed the sample to expand.
53. 284 K
55. **a.** 111 K **b.** −260°F
57. 1.26×10^{-4} s
59. 6.21×10^{-4} m^3
61. **a.** 6.19 mol **b.** 3.73×10^{24} molecules
63. $V_0 \Delta T[\beta_\ell - 3\alpha_g]$
65. 12.0 MPa
67. 8.82 atm
69. **a.** 2.2 mol **b.** 0.071 kg
71. Your idea is a theoretical model because you have a *theory* or *explanation* for the colony's growth; that is, each cell divides into two daughter cells and so on.
73. **a.** $-0.714V_i$ **b.** 1.13×10^5 Pa
75. 315 K
77. 70°C
79. **a.** 3.95×10^{-4} kg/m **b.** 26.3 N **c.** 22.8 N **d.** 182 Hz
81. 1.2×10^5 Pa

CHAPTER 20 Concept Exercises

20.1 If all the numbers are equal, the rms value and the average are equal. For example, we can find the average and rms values of the set of six numbers {3, 3, 3, 3, 3, 3}:

$$\text{average} = \frac{3+3+3+3+3+3}{6} = 3$$

$$\text{RMS} = \sqrt{\frac{3^2+3^2+3^2+3^2+3^2+3^2}{6}} = \sqrt{\frac{54}{6}} = 3$$

If the numbers are not all equal, the rms weights the larger numbers; in that case, the rms is greater than the average value.
20.2 The average kinetic energy is proportional to temperature; so, if the temperature doubles, that kinetic energy also doubles. The rms speed is proportional to the square root of the temperature; so, if the temperature doubles, the rms speed goes up by $\sqrt{2}$.
20.3 Because the temperatures of the gases are equal, the average kinetic energies of their particles are equal: $K_{av,1} = K_{av,2}$. The rms speed depends on the mass of the particles, and we find

$$\frac{v_{rms,2}}{v_{rms,1}} = \frac{\sqrt{3k_BT/m_2}}{\sqrt{3k_BT/m_1}} = \sqrt{\frac{m_1}{m_2}} = \sqrt{\frac{4m_2}{m_2}} = 2$$

So, the rms speed of particles in gas *B* is twice that of particles in gas *A*.
20.4 Yes, fanning helps. The smoke or other noxious particles take a long time to mix into the air by diffusion. Usually, convection currents speed up that process. Fanning the air also helps by creating a breeze. Once the smoke is mixed into the air, it is less noticeable than before.
20.5 **a.** Above the critical pressure, water must be a blended state in which there is no distinction between liquid and gas. **b.** Between 1 atm and the critical pressure, water at 100°C is a liquid. **c.** Below 1 atm, water at 100°C is a vapor.
20.6 Because the Earth's temperature increases due to global warming, we would expect the evaporation rate to increase.

CHAPTER 20: Problems and Questions

1. The gas particles are in motion in random directions, and they do not interact with one another. Therefore, their motion is only altered by the walls of the container and the gas must fill the container.
3. The Earth's atmosphere is not contained by walls, but rather by the Earth's gravity. The pressure is greater at sea level than on the tops of mountains because gravity causes a higher density of particles near the surface, and therefore a higher pressure.
5. a. 5.80 m/s **b.** 6.57 m/s
7. The magnitude of the average velocity is less than the average speed, which is less than the rms speed: $v_0/2 < 3v_0/2 < 1.58v_0$.
9. a. 883.5 m/s **b.** 510.1 m/s
11. $T_{avg} = -0.18°C$, $T_{rms} = 1.3°C$
13. 722 m/s
15. 1.35×10^7 Pa
17. a. 7.51×10^{-21} J **b.** $v_{rms, Ne} = 668$ m/s and $v_{rms, Kr} = 328$ m/s
19. 3.32×10^{-21} J
21. 2.49×10^5 N
23. C is the most probable.

Letter	Number of boxes	Letter	Number of boxes
A	2	E	4
B	3	F	2
C	7	G	1
D	6	H	1

25. 2.03×10^{-25} kg, 122 u. This is heavier than any single atom on the periodic table, so only molecules of at least this mass would be retained.
27. a. $v_{mp} = 390$ m/s, $v_{avg} = 440$ m/s, and $v_{rms} = 478$ m/s **b.** We expect N_2 molecules to be faster because their mass is lower: $v_{mp} = 417$ m/s, $v_{avg} = 471$ m/s, and $v_{rms} = 511$ m/s.
29. $v_{rms} = 511$ m/s and $f(v_{rms}) = 0.00181$, $v_{mp} = 417$ m/s and $f(v_{mp}) = 0.00199$. The most probable speed is indeed at the peak of the distribution function. Because the function is not symmetric, the rms velocity is somewhat higher than the most probable speed.
31. a. $\sqrt{k_B T/m}$ **b.** $\sqrt{(\pi k_B T)/(2m)}$
33. 1.04×10^{-8} m
35. 1.3×10^{-7} m
37. a. 2.47×10^8 m^{-3} **b.** 5.69×10^9 m in the vacuum chamber, about 10^{17} times longer than in air.
39. a. Monica's hydrogen will win. Assuming the gases are at the same temperature, the molecules have the same average kinetic energy, but hydrogen molecules have less mass. Thus, the hydrogen will move faster, resulting in a higher diffusion rate. **b.** The diffusion rates differ by a factor of $\sqrt{32}$.
41. 2.33×10^{-10} m
43. 2.26×10^7 Pa
45. a. Imagine drawing a horizontal line at a pressure of 1 atm; this line does not cross the liquid region at any temperature. As the temperature of solid carbon dioxide (dry ice) is increased, the line crosses into the gas region without ever becoming liquid. The temperature at which this transition occurs appears to be around −75°C. **b.** Imagine drawing a vertical line at a temperature of 20°C to intersect the gas/liquid line. Tracing from this point to the pressure axis shows that the transition occurs at a pressure of around 50 atm.
47. a. The water would go from ice to liquid water to vapor and then to gas. **b.** There would be no change.
49. 8.6×10^4 Pa
51. 2.1×10^5 Pa
53. 65%
55. a. Increase by a factor of $\sqrt{3}$ **b.** Decrease by a factor of $1/\sqrt{2}$
57. a. $(96mv^2)/V$ **b.** $8mv^2$
59. 1.12 mol
61. 5.0×10^5 Pa
63. a. 827 m/s **b.** 1.97×10^4 N
65. 8×10^{11} s
67. 1300 m/s, higher than the rms speed of a nitrogen molecule in the atmosphere
69. a. $\frac{dT}{dt} = \left(\frac{v_{av}\pi m}{4k_B}\right)\left(\frac{dv_{av}}{dt}\right)$ **b.** 0.731 m/s²
71. a. 2.49×10^{-7} m **b.** 7.53×10^{-10} s
73. a. 2.37×10^4 K **b.** 4.72×10^3 K **c.** Although the sea-level temperature of the atmosphere is low, solar energy heats the upper atmosphere such that the thermosphere, at an altitude of 180 km, has a temperature exceeding 2200 K. Thus, a small fraction of the helium gas in the upper atmosphere always has a velocity that exceeds Earth's escape speed and is lost to space.

CHAPTER 21 Concept Exercises

21.1 a. Incorrect. Heat is not contained in Texas. The high temperature in Texas means that the air has a lot of thermal energy. **b.** Incorrect. Kinetic energy is converted into thermal energy due to friction between the table and the book. **c.** Incorrect. Heat and work are both methods of transferring energy. One does not become the other.
21.2 a. *Heat* is transferred from the air in the room to the water. **b.** If the Earth is not in the system, it does *work* on the ball. When the ball hits the floor, the normal force distorts the ball, which can be modeled as a spring. Kinetic energy is converted into elastic potential energy and thermal energy. **c.** As in part (b), the Earth does *work* on the system. Friction between the atmosphere and the meteor converts mechanical energy into thermal energy.
21.3 Many possible terms would be more helpful than the traditional terms *heat capacity*, *specific heat*, and *molar specific heat*. For example, the constant of proportionality in the equation for static friction (Eq. 6.1) is called the *coefficient of static friction*. How about replacing *heat capacity* with *heat coefficient*? Then, specific heat could be called the *mass heat coefficient*, and molar specific heat could be called the *molar heat coefficient*. Unfortunately, the traditional terms are in standard use.
21.4 Average kinetic energy only depends on the gas temperature, so the average kinetic energy does not change. Thermal energy depends on temperature and the amount of gas. Doubling the number of particles doubles the thermal energy.
21.5

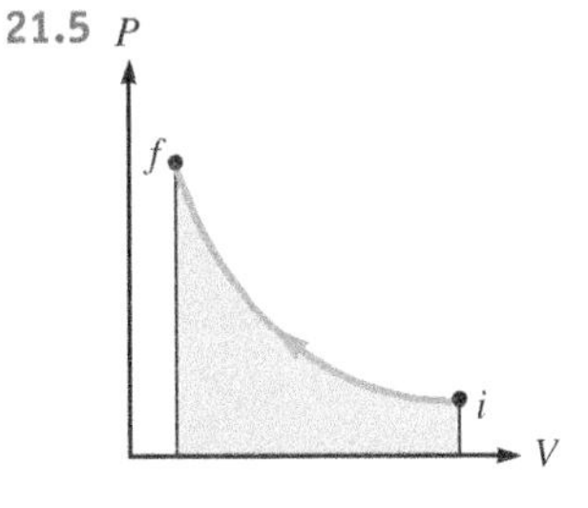

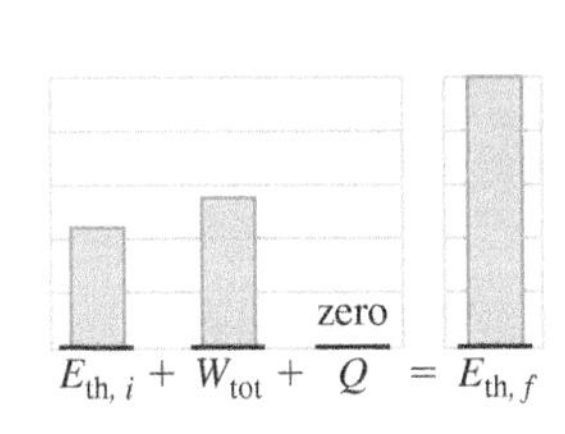

21.6

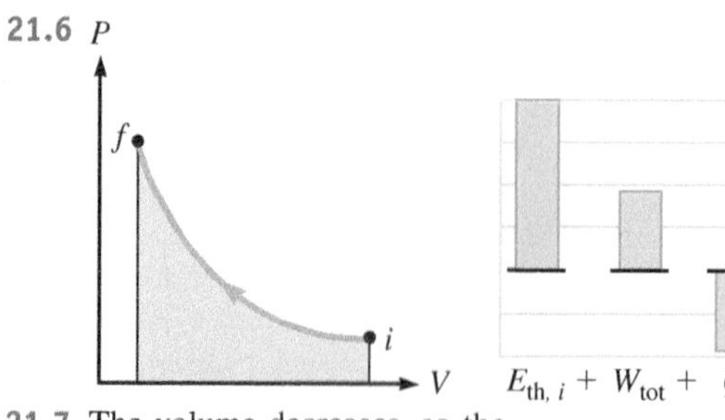

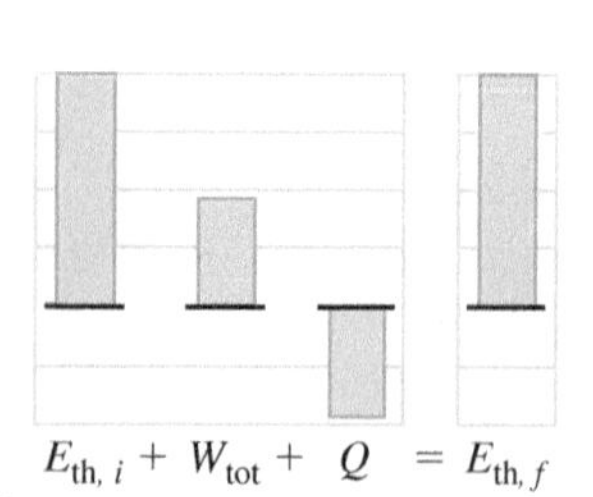

21.7 The volume decreases, so the work done by the gas is negative in both adiabatic compression and isothermal compression. The environment (the piston) does positive work on the gas. Because the area under the adiabatic path in the figure here is greater than the area under the isothermal path, the work done by the piston is greater in the adiabatic process.

P, f, f, Adiabatic, i, Isothermal, V

21.8 a. According to the *PV* diagram, the area under the expansion path is less than the area under the contraction path. Because the gas does positive work during expansion and negative work during contraction, the net work done by the gas is *negative* for this cycle. **b.** Because the gas does negative work during this cycle, we can conclude that the piston does *positive* work on the gas. Energy enters the system through work. **c.** Because the thermal energy is constant in a cyclic process, the amount of heat that flows into the system must equal the amount of energy the system loses through work. So, the net flow of heat is from the system to the environment.

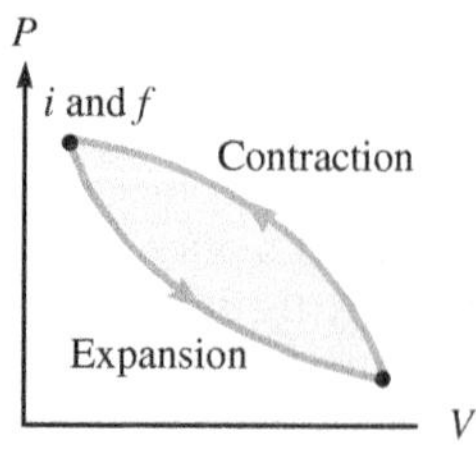

21.9 Changes in state variables only depend on the endpoints in a *PV* diagram. The method of energy transfer depends on the thermodynamic process and therefore depends on the path. So, (e) through (g) depend on the path, and (a) through (d) depend on the endpoints.

CHAPTER 21: Problems and Questions

1. Heat is the energy transferred from a system to the environment or vice versa due to a temperature difference between the two. Perhaps *thermotransportation energy* is a good term because it describes the temperature difference between the system and environment and the concept of energy transfer.
3. Energy is transferred by heat in both cases. The energy is transferred from the water or oven to the steaks because the steaks are at the lower temperature in each case.
5. The word *isolated* suggests that Bobby is cut off, remote, or secluded. The word *insulated* suggests that he is protected or shielded. *Isolated* means that the system cannot exchange energy through work, whereas *insulated* means that the system cannot exchange energy in the form of heat.
7. a. 8.37×10^5 J **b.** 1.90×10^3 steps
9. Energy is transferred by heat from the engine to the hood when the engine is hotter than the hood. During the day, sunlight will also transfer energy by heat from the Sun to the hood. The air around the hood will transfer energy to it by heat if the air is at a higher temperature than the hood. So if you are traveling on a cold winter night, you would conclude that the energy was only transferred from the engine to the hood.
11. -15.0 J. The energy enters by work and leaves by heat.
13. 0 J. The energy enters by work, but no energy is exchanged by heat.
15. 81.5°C
17. Yes. Thermal energy depends on n and T. So, if n is larger in one case (as it is for the bathtub), then T can be lower and the system can still have more thermal energy.
19. a. 82°C **b.** 88°C It would be better to use a steel cup than a glass cup.
21. 22.9°C
23. -7.77×10^3 J. The energy transferred by heat is leaving the system.
25. 5.1°C; cold liquid
27. 1.38×10^5 J
29. a. 2.83×10^4 J **b.** 1.87×10^4 J **c.** 9.47×10^3 J. Results are approximate because the value for c is for *solid* copper, and we used it for *liquid* copper in part (a). It takes more energy to melt the copper than it does to melt the silver.
31. a. 6.66×10^5 J **b.** 4.51×10^6 J
33. -6.50×10^2 kJ
35. $7.50 P_0 V_0$
37. a. -105.0 J **b.** 105.0 J **c.** Energy enters the system because the work done by the gas is negative.
39. 3.2×10^2 J
41. a. 0.00172 m^3 **b.** 391 K
43. a. 7.8×10^2 K **b.** 2.2×10^4 J **c.** -1.5×10^4 J
45. a. -1.05×10^3 J **b.** 1.05×10^3 J
47. a. $T_B = 1.20 \times 10^2$ K and $T_C = 164$ K **b.** 1.00×10^3 J **c.** 5.02×10^3 J
49. 2.9×10^2 J
51. 5.2×10^4 J
53. a. 3.2 **b.** 2.7. Because some energy goes into internal degrees of freedom for nonmonatomic gases, the final pressure of the diatomic gas is lower.
55. a. The gas contracted. **b.** 0.486
57. 6.1×10^{-5} m^2
59. 6.67×10^{-6} m/s
61. 363 K
63. a. Approximately 9 h **b.** The water will not get much hotter than the air temperature due to convection and thus will never boil.
65. 6.07×10^2 J/(kg · K)
67. a. 9.53×10^8 J **b.** -8.46×10^8 J
69. a. 0°C **b.** 0.0647 kg
71. 1.06 kg
73. 297 K
75. $2k_BT$
77. a. -0.279 J **b.** 47.4 kJ **c.** 47.4 kJ
79. 11.3 times faster
81. 0.157 kg
83. 0.386°C
85. a. 2.25×10^4 W **b.** 4.86×10^8 J

CHAPTER 22: Concept Exercises

22.1 The back of the truck is at a height $h \approx 1$ m, and the mass of an elephant is about 6000 kg. The work required to lift the elephant into the back of the truck is approximately

$$W_{\text{required}} = mgh = (6000\text{ kg})(9.81\text{ m/s}^2)(1\text{ m})$$
$$W_{\text{required}} \approx 6 \times 10^4\text{ J}$$

A perfect engine would require the same amount of heat, $Q_h = 60{,}000$ J. A real engine would require more heat because it would lose energy (for example, due to friction between the piston and the cylinder).

22.2 **a.** It is possible that the water has undergone a nearly reversible cycle. The temperature could be kept near the boiling point. No work is involved, so no friction dissipates energy. Finally, the process could be done very slowly. **b.** The wood has not undergone a reversible cycle. The chemical reactions that burn the wood cannot be reversed.

22.3 No. Opening the refrigerator door allows the hot and cold reservoirs to come into thermal contact. As they approach the same temperature $T_c \to T_h$, the coefficient of performance approaches infinity, $\kappa \to \infty$. The second law of thermodynamics does not allow the coefficient of performance to reach infinity.

Here's another way to think about it. Usually, a refrigerator removes heat from the cold interior and deposits heat outside. If the door is open, there is no distinction between the inside and the outside. So, where does the extracted heat go?

22.4 **a.** You know that a smashed car does not spontaneously roll back from a brick wall and return to perfect condition. Thus, the entropy of the Universe must increase when a car crashes into a wall. **b.** Ice cubes don't spontaneously form in cups of water, so the entropy of the Universe must increase when ice melts in a cup of water. **c** and **d.** You can imagine watching a puddle of water freeze or melt in reverse. It is possible that the entropy of the Universe does not change during these processes, but any change in the water's entropy must be balanced by a change in the heat reservoir's entropy. When the ice melts, its entropy increases. If the heat reservoir—the Earth and surrounding air—has an equal decrease in entropy, it is possible that the entropy of the Universe is unchanged.

22.5 **a.** The probability is found by dividing the number of microscopic states for the 50 heads/50 tails macroscopic state by the total number of microscopic states:

$$\frac{1.0 \times 10^{29}}{1.27 \times 10^{30}} = 0.079$$

So, there is a 7.9% chance of finding 50 heads and 50 tails.

b. Now we need to find the total number of microscopic states that are possible for all the macroscopic states that produce between 45 and 55 heads:

$$[2(6.14) + 2(7.35) + 2(8.44) + 2(9.32) + 2(9.89) + 10.1] \times 10^{28} = 9.24 \times 10^{29}$$

As in part (a), the probability is found by dividing this result by the total number of possible microscopic states:

$$\frac{9.24 \times 10^{29}}{1.27 \times 10^{30}} = 0.728$$

So, there is about a 73% chance of finding between 45 and 55 heads.

CHAPTER 22: Problems and Questions

1. The potato would get hotter, and the lake would cool. The temperature of each would get farther and farther apart, and equilibrium would not be reached if that were true.

3. The engine would do more work in the adiabatic process, *B* to *C* in Figure 22.2, if the process occurred at a higher pressure. Likewise, if the work done by the environment in the process from *D* to *A* occurred at a lower pressure, the net amount of work done by the engine would also increase. The area enclosed by the cycle on the *PV* diagram is the net work performed by the engine. These changes would increase that area.

5. **a.** 0.300 **b.** 70.0%

7. 1.9×10^5 W

9. **a.** 5.861×10^4 J **b.** 0.7835

11. 0.635

13. **a.** 0.58 **b.** 0.76

15. 0.5000

17. **a.** 707.3 K **b.** 7.556

19. **a.** 7.71×10^7 J **b.** 3.75×10^7 J

21. 4.3×10^4 J

23. **b.** The difference in efficiencies is $1 - (T_B/T_C)$. The ratio of temperatures at *B* and *C* will result in $(T_D/T_C) < 1$, which means that the Carnot efficiency is greater than the Otto efficiency, as expected.

25. **a.** 999 K **b.** Yes. The air must initially be at a temperature of at least 483 K.

27. 1.33

29. 16.2

31. **a.** 52.5 J **b.** 262.5 J

35. 101 J/K

37. **a.** 0 **b.** 0.23 J/K

39. −153 J/K

41. **a.** 4.40×10^4 J **b.** -1.10×10^5 J **c.** −271 J/K

43. **a.** 25.7°C **b.** $\Delta S_{\text{cold}} = 8.98$ J/K, $\Delta S_{\text{hot}} = -8.40$ J/K **c.** 0.58 J/K

45. 14.1 J/K

47. **a.** −0.183 J/K **b.** Yes. You can watch this process in reverse without it appearing odd or improper. **c.** 0.183 J/K

49. The neatly arranged drawer is like a cool system because entropy is low. The messy basket is like a hot system because entropy is high. When you add one more piece of laundry to the basket, the disorder does not increase much compared to the amount of disorder that was already present. When you add a messy item to the drawer, the relative disorder increases significantly. Likewise, when you add a little heat to a cool system, the relative disorder increases significantly as opposed to when you add the same amount of heat to an already hot system.

51. She can decrease the entropy of the clothes as long as there is a corresponding increase in entropy of greater magnitude in the surrounding environment. When she uses her muscles to fold the laundry, she will produce excess heat that increases entropy.

53. A local decrease in entropy is possible as long as there is a corresponding increase in entropy somewhere else that is equal or greater in magnitude. Evolution does not violate the second law because the Earth is not a closed system, nor is any localized system on Earth a closed system. The Earth radiates energy back into space. This is one example of a process where there can be a nearby increase in entropy.

55. 1/36

57. 2.16×10^{-21} J/K

59. It is very likely that the number of heads will be near 50 because the curve has a broad hump near 50 but falls off quickly for values greater than 60 and less than 40.

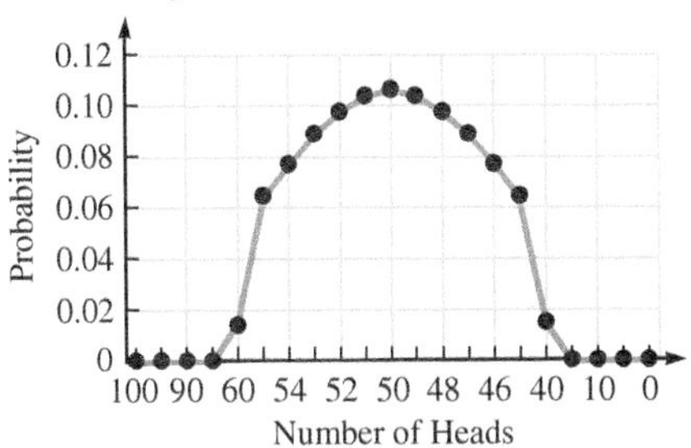

61. 289 K

63. Boiling would lead to a larger change in entropy. Each process occurs at a constant temperature such that $\Delta S \equiv Q/T$. The boiling temperature is higher than the melting temperature (373 K versus 273 K). The heat added depends on the latent heat of fusion for melting and the latent heat of vaporization for boiling, however, and the latter is much higher (2260 J/g versus 333 J/g). The latent heat is almost seven times larger for boiling, whereas the boiling temperature is not even twice as large as the melting temperature; therefore, heat exchange is the dominant factor and the change of entropy must be higher for boiling, even though boiling occurs at a higher temperature.

65. a. $-1.75 \times 10^{-3}\ K^{-1}$ **b.** $8.98 \times 10^{-4}\ K^{-1}$
67. a. 934 J/K **b.** 123 J/K **c.** 9.37×10^{3} J/K
69. a. 968 J **b.** 582 J
71. −15.9 J/K
73. a. −410 J **b.** 2.77×10^{3} J **c.** 3.18×10^{3} J **d.** 0.148 **e.** 0.667
75. a. −10.9 J/K **b.** −15.3 J/K
77. a. 48.9 J/K **b.** 1.43×10^{4} J
79. 28.8 J/K
81. a. 2×10^{20} **b.** $S_{A_1} = 3.18 \times 10^{-22}$ J/K, $S_{A_2} = 3.27 \times 10^{-22}$ J/K, $S_A = 6.45 \times 10^{-22}$ J/K

Index

A

Abbreviations and symbols, 689–690
Absolute pressure, 430, 570
Absolute scale, 553, 571, 575
Absolute temperature, 586
Absolute zero, 571, 575, 586, 659
Accelerating reference frames, 125
Acceleration
  angular (*See* Angular acceleration)
  average (*See* Average acceleration)
  centripetal (*See* Centripetal acceleration)
  constant (*See* Constant acceleration)
  instantaneous, 37–38
    in circular motion, 93–94
    defined
      one-dimensional motion, 37, 51
      three-dimensional motion, 111
    due to gravity, 130–131
    free-body diagrams, representation in, 134
    free-fall acceleration, 103
    in multidimensional motion, 91
    relationship to mass, 124
    in relative motion, 104, 107
    for simple harmonic oscillator, 455, 479
  tangential, 341
Acceleration-versus-time graph, 38, 51
  for simple harmonic motion, 453, 455
Active degrees of freedom, 635
Adiabatic process, 625–626, 637–638, 645
  in Carnot cycle, 657–658
  in Otto cycle, 660–661
Air density, 420
Airplanes
  atmospheric pressure and flying, 419, 423, 436–437, 442–443
  hanging in exhibit, 392, 394, 399–401, 408–409, 412–413
  speed of, 107–108
Air resistance, 110
Airspeed, 107–108
Alcohol thermometer, 558–559, 570
Aldrin, Edwin E. (Buzz), 131, 297
Amplitude
  defined, in simple harmonic motion, 454, 479
  for object-spring oscillator, 461–462
Angular acceleration, 336–337, 354
  average, 336
  constant, 338–339
  instantaneous, 336
  rotational compared to translational, 336, 341
  and torque, 345
Angular beat frequency, 542
Angular displacement, 334–335, 341, 353
Angular frequency
  average, 542
  of damped harmonic motion, 475, 480
  defined, in simple harmonic motion, 454, 457, 480
  natural, 477–478
  for object-spring oscillator, 461–462, 480
  of physical pendulum, 467, 480
  of simple pendulum, 463–464, 480
  of torsion pendulum, 469, 480
Angular momentum, 376–379, 384
  orbital, of comet, 377–378
  of particle, 376
  of rotating rigid object, 377
  See also Conservation of angular momentum
Angular position, 334–335, 353
Angular speed, 100–101, 102
  defined, 100, 111, 335, 354
Angular velocity, 335–336, 354
  average, 335
  instantaneous, 335
  rotational compared to translational, 336, 341
Angular wave number, 492
Antinodes, 531, 545
  position of, 532–533
  resonance in musical instruments, 534–535, 537–538, 540–541
Antiparallel, 68
Aphelion, 188, 377
Apparent weight, 132
  and spring scale, 139
  submerged in fluid, 427–428
Aqueducts in ancient Rome, 448–449
Archimedes of Syracuse, 351, 424
Archimedes's principle, 423–428, 444
Area under the curve, 36
Armstrong, Neil, 297
Arrow of time, 671
Astronomical data, 692
  See also Universe
Astronomical unit (AU), 188
Atmosphere, unit of pressure, 421
Atmospheric pressure, 420–421, 423
  and absolute pressure, 570
  measuring, 430–431
Atomic clocks, 6–7
Atomic mass, 567
Atwood machine, 368–369
Average acceleration, 37–38
  defined
    one-dimensional motion, 37–38, 51
    three-dimensional motion, 111
  finding direction geometrically, 64
  in multidimensional motion, 91
Average angular acceleration, 336
Average angular frequency, 542
Average angular velocity, 335

Average kinetic energy of gas, 582, 586, 606
Average speed, 31
for block and spring, 35
defined, one-dimensional motion, 31, 50
of gas particles, 590, 606
Average velocity, 30–34
for block and spring, 35
for constant acceleration, 41
defined
one-dimensional motion, 31, 50
three-dimensional motion, 111
of laboratory cart, 32–34
in multidimensional motion, 90
of particle, 32
from position-versus-time graph, 32–34
from vector components of motion variables, 77
of vehicles on highway, 583–584
Avogadro's hypothesis (law), 566, 575
Avogadro's number, 567–568, 575
Axes of ellipse, 187
Axis (coordinate axis) in Cartesian coordinate system, 24, 51, 65
Axis of rotation, 363–364

B

Ballistic pendulum, 304
Bar charts, 228
for conservation of energy, 361–362, 369, 371, 372
for heat engine, 654, 657
for mechanical energy, 228–232, 237, 239
for rotational kinetic energy, 368–369
for thermal energy, 625–631, 637, 646
for work-energy theorem, 264, 266–269
Barometers, 428, 430–431
Basie, William "Count," 523
Beat frequency, 542, 543
Beat period, 542
Beats, 541–543, 545
Bell, Alexander Graham, 503
Bernoulli's equation, 436–443
defined, 438, 444
pressure in moving fluid, 437–438
problem-solving strategy, 438
siphoning water, 439–440, 449
Venturi tube, 440–442
Bicycle
gears, 343–344
tires, 366, 370, 382–383, 569
Billiards, 321–322
Bimetallic strip thermometer, 560
Black holes, 208, 242
Block and spring. *See* Spring-block system
Boiling liquid, 602
Boiling temperatures, 621
Boltzmann's constant, 566, 567, 616, 674
Boundary, 524
Boundary conditions, 524, 545
Boyle's law, 565, 575
Bulk modulus, 422, 498
Buoyant force, 424–428, 444

C

Calculus, 41–42, 685–687
Calorie, 616
Calorimeter, water, 619
Carbon dioxide
as greenhouse gas, 581
phase changes, 600–601
Cardinal compass system (on maps), 72, 108
Carnot, Sadi, 656
Carnot cycle, 657–658, 677
Carnot engine, 656–659, 677
Carnot's theorem, 658
Car Talk, 561–562, 678
Cartesian coordinate system, 65–68
axes and coordinates, 65
coordinates (*See* Coordinates, Cartesian)
defined, 65, 80
and Newton's second law, 126
reading coordinates, 66–67
right-handed systems, 67–68
unit vectors, 67
Cauchy, Augustin, 404
Cavendish, Henry, 192
Celsius scale, 553–554
Center of gravity, 260–261
Center of mass, 260–261, 274
locating, 284–287, 299
in rolling motion, 369–373
and rotational inertia, 365–366, 384
velocity during collision, 311
Center-of-mass frame, 288
Centrifuge, 103
Centripetal acceleration, 101–103
in Milky Way galaxy, 186, 201–202
of Moon, 197–198
of rotating rigid object, 341
Centripetal component, 175
Centripetal force, 168–177
nonuniform circular motion, 174–177
problem-solving strategy, 169
uniform circular motion, 168–174
as zero-work force, 259
Cesium clocks, 6–7
Change in entropy, 665–667, 677
around reversible cycle, 667
in constant-pressure process, 668–670
depends on number of microscopic states, 674
does not depend on path, 667
free expansion, 670, 675–676
in isothermal process, 667–668
Change in position. *See* Displacement
Charles's law, 565, 571, 575
Chemical energy, 265, 274
Chemical vapor deposition (CVD), 579
Circular motion
acceleration and velocity in, 93–94
connection to rotation, 340–344
connection to simple harmonic motion, 456–458
mathematical description of circular path, 88–90
mechanical energy for orbit system, 235–236, 239
velocity and speed in, 92
See also Uniform circular motion
Classical mechanics, 2, 127
Clausius, Rudolf Julius Emanuel, 652
Clausius equation of state, 596
Clausius statement of second law of thermodynamics, 652, 671–672, 677
Clocks
atomic, 6–7
history of timekeepers, 450–451
Millennium Clock, 468, 469
pendulum clocks, 450–451, 463, 465–466
Clockwise, 334
Closed system, 290, 299
Coefficient
of drag, 165, 177, 181
of kinetic friction, 134, 158, 161, 177
of linear expansion, 557, 558
of performance, 664–665, 677
of rolling friction, 158, 163, 177
of static friction, 157–158, 159, 177
of volume expansion, 557–558, 562

Cold-weld, 156
Collins, Michael, 297
Collisions, 306–330
  conservation of kinetic energy during, 311–312
  conservation of momentum during, 310–311, 312
  defined, 307, 324
  impulse, 307–310
  one-dimensional elastic, 315–320
    (*See also* One-dimensional elastic collisions)
  one-dimensional inelastic, 312–315
  two-dimensional, 320–323
  See also Completely inelastic collision
Comet Hale-Bopp, 245
Comet Halley, 214, 215–216, 237–238, 377–378
Comets, 235–238, 306
Completely inelastic collision, 311, 324
  conservation of momentum, 311–312
  one-dimensional, 312–315, 324
  two-dimensional two-particle, 322–323
Component form, of vectors, 69, 79
  for multidimensional vectors, 88
  translating magnitude and direction into, 75
  See also Vector components
Compressible, 419
Compression processes, 624, 625–627, 637
Compression stress, 405, 410
Compressive deformation, 406
Compressive strength, 406, 410
  and thermal stress, 560, 561
Conan wheels, 373–374
Condensation, 601–602
Conduction, heat transfer, 639–641, 645
Conductor, thermal, 639
Conservation of angular momentum, 379–383
  and Kepler's second law, 380–381
  and Newton's second law, 379–380
Conservation of energy, 213–246
  conservation of mechanical energy, 225–228
    applications of, 228–232
  elastic potential energy, 223–225
  energy defined, 215
  energy graphs, 232–235
  gravitational potential energy near Earth, 218–221
  heat and, 613–615
  for ideal fluid flow, 437–438
  including rotational motion, 361–362, 368, 383
  kinetic energy, 215–217
  orbital energies, 235–238
  potential energy, 217–218
  and second law of thermodynamics, 671
  universal gravitational potential energy, 221–223
Conservation of energy approach
  compared to Newton's force approach, 214, 215, 228, 234–235, 282
  in simple harmonic motion, 470–472
Conservation of kinetic energy during collision, 311–312, 324
Conservation of mechanical energy, 225–228
  applications of, 228–232
Conservation of momentum, 289–292, 299
  during collision, 310–311, 312, 324
  nuclear decay, 291–292
Conservative forces, 218, 256–259
  defined, 257, 274
  gravity as, 257–258
Constant acceleration, 40–45
  defined, one-dimensional motion, 40, 51
  of electrons, 44–45
  free fall as special case of, 45–50
    (*See also* Free fall)
  kinematic equations for, 40–41, 338
  problem-solving strategy, 42
  for projectile motion, 94–95
  sound speed barrier, breaking, 42–44
Constant angular acceleration, 338–339
  kinematic equations for, 338
  problem-solving strategy, 338
Constant-pressure (isobaric) process, 628, 630, 631, 632–634, 645
  change in entropy, 668–670
Constant-volume ideal gas thermometer, 572–574
Constant-volume (isochoric) process, 627, 630, 631, 632–634, 645
  in Otto cycle, 660–661
Constructive interference, 527–530
  beats, 541
  condition for, 528, 545
Contact forces, 123
  normal force as component of, 132
Continuity equation, 433–435, 444
Convection, heat transfer, 641, 643, 645
Conversion factors, 7–8, 18, 691
Coordinate axis, 24, 51
Coordinates, Cartesian, 65
  conversion to polar coordinates, 100
  reading, 66–67
Coordinate systems, 24–25
  Cartesian (*See* Cartesian coordinate system)
  defined, 24, 51
  in free-body diagrams, 134
  map compass (*See* Cardinal compass system)
  for multidimensional motion, 87
  polar, 100
  reference frame, 104
  for rotational kinematics, 333, 334–335
Copernican theory, 2–3
Copernicus, Nicolaus, 2, 185
Cosmic rays, 118, 151
Coulomb's constant, 245
Counterclockwise, 100, 334
Critically damped oscillator, 475, 479
Critical point, 600, 606
Critical pressure, 600, 606
Critical temperature, 600, 606
Cross product, 346–347, 354, 376
  torque balance condition for equilibrium, 393–394
Cycle, 453
Cyclic process, 628–629, 645
  thermodynamic cycle, 624, 653
Cyclotron, 170

D

Damped-driven oscillator, 476–478
Damped harmonic motion, 473–476
Damped oscillator, 473
Damping coefficient, 474
Damping constant, 474
Dark matter, 186–187, 201–203
Deceleration, 37
Decibels, 503
Deformable object, 259
Deformable systems, 259–261
Degrees of freedom, 634–637, 645
Density
  of air and common fluids, 420
  Cavendish apparatus, 192
  drag dependence on, 163–164
  estimating, 15–17
  formula for, 9
Derived quantities, 9
Descartes, René, 213
Destructive interference, 527–530, 545
Dew point, 603

Diagrams
free-body, 134–141, 147
phase, 600–601, 606
See also Graphs; Motion diagrams
Diatomic gas molecules, 632, 634–636
Diesel engine, 679
Diffraction, 507, 515
Diffusion, 593, 595
Diffusion constant for gases in air, 593
Dimensional analysis, 9–10, 18
Direction of vectors, 72, 79
and inverse trigonometric functions, 72–73
translating unit-vector information to, 73–75
Disorder of Universe, 672–673
Displacement, 27–30
angular, 334–335, 341, 353
defined
multidimensional motion, 88, 110
one-dimensional motion, 28, 50
as function of velocity and time, 40–41
in multidimensional motion, 88–90
rotational compared to translational, 335, 341
and translation and particle model, 28–29
from velocity-versus-time graph, 36–37
Displacement vector, 77
Dissipative forces, 263, 264, 274, 473, 475
Distance traveled, 29–30
of block and spring, 29–30
defined, one-dimensional motion, 29, 50
by point on rotating object, 340
Doppler shift, 507–513
defined, 507, 515
frequency observed by moving observer, 509
moving source, stationary observer, 509–510
shock waves, 513
sonar waves, 543
source and observer both moving, 510–512
stationary source, moving observer, 508–509
Dot product, 252–254, 274
Drag coefficient, 165, 177, 181
Drag force (drag), 163–165, 177
as nonconservative force, 257, 263
and rocket thrust, 296–297
Driven oscillators, 476–479
Driving angular frequency, 478
Driving force, 476–479
Dry ice, 600
Dynamics, 120–121, 147
conservation approach, 214, 215, 228, 234–235, 282
Newtonian, 120
rotational, 331
of simple harmonic motion, 459–461
See also Rotational dynamics

E

Earth
astronomical data, 692
atmosphere of, 581, 587, 591
gravitational field of, 196–199, 203–206
as inertial reference frame, 126
kinetic energy of, 216
and Moon system, center of mass, 285
as noninertial reference frame, 204–205
orbit of, 188
rotational kinematics of, 337
rotational parameters compared to translational parameters, 342–343
satellite orbit and circular motion, 457–458
water cycle and global dimming, 605
See also Universe
Easter Island statues, 332–333, 353
Effective cross-sectional area, 167, 177
Effective gravitational field, 204–206
Efficiency of engines, 272, 656, 658–659, 661–662, 677
Egyptian obelisks, 332–333, 350
Einstein, Albert, 185, 195
Elastic collisions, 311, 324
conservation of kinetic energy, 311–312
one-dimensional, 315–320
stationary target, 316–317
two-dimensional two-particle, 321–322
See also One-dimensional elastic collisions
Elasticity, 404–409
compressive deformation, 406
shear deformation, 406–407
strain, 405
stress, 404–405
tensile deformation, 405–406
Elastic limit, 406
Elastic object, 405, 410
Elastic potential energy, 223–225, 230–231, 239
Electric force, 123, 127–128
Electromagnetic force, 146, 155
Electromagnetic waves, thermal radiation, 641–644
Elements, periodic table of, 696–697
Elevators, 53, 138–140, 271
Elliptical orbits
mechanical energy for, 236–238, 239
in planetary motion, 187–188, 207
Emissivity, 642–643, 645
Empirical fits to experimental data, 565
Empirical laws, 190, 206
Energy, 215, 239
common fuels and energy content, 265
exchange of heat, 638–644
in simple harmonic motion, 470–473
transfer to and from the environment, 248
transport in harmonic waves, 500–501
See also Conservation of energy; Elastic potential energy; Kinetic energy; Mechanical energy; Potential energy; Thermal energy
Energy bar charts. *See* Bar charts
Energy graphs, 232–235, 239
Energy in nonisolated systems, 247–280
conservative and nonconservative forces, 256–259
dot product, 252–254
energy transfer to and from environment, 248
particles, objects, and systems, 259–261
power, 269–273
thermal energy, 261–263
work done by constant force, 248–251
work done by nonconstant force, 254–256
work-energy theorem, 264–269
Energy transfer diagram, 653, 677
Engines
Carnot, 656–659
diesel, 679
heat, 652–655
Otto, 660–662
Stirling, 678
See also Pistons
Entropy, 665–670
defined, 672, 677
as measure of disorder, 672–673
and probability, 673–676
of Universe, 670–671, 673
See also Change in entropy
Equation of state, 565, 575
Clausius, 596
for ideal gas, 565, 566, 575
Van der Waals, 597–599

Equilibrium, 391, 409
conditions for, 392–394
stable compared to unstable, 246
See also Static equilibrium
Equilibrium state of gases, 565, 575
Equipartition of energy, 631, 635–637, 645
Error, 10, 17
Escape speed, 229–230, 587
Estimate, 12
order-of-magnitude estimates, 12–17, 18
Evaporation, 601–605
humidity, 603–604
water cycle and global dimming, 605
Evaporation pan, 605
Expansion processes, 623
Experimental science, 3
External forces, 123
total, 290–291

F

Fahrenheit, Daniel Gabriel, 553, 554, 558–559
Fahrenheit scale, 553–554
Ferris wheel, 102–103
Feynman, Richard, 262
Field, 196
Field forces, 123
First harmonic. *See* Fundamental harmonic
First harmonic frequency, 544
First law of thermodynamics. *See* Thermodynamics, first law of
Fixed axis, 332, 353
Flow tube, 434
Fluid element, 433
Fluid model, 419
Fluids, 418–449
Archimedes's principle, 423–428
Bernoulli's equation, 436–443
continuity equation, 433–435
defined, 418–419, 444
fluid model, 419
ideal fluid flow, 432–433
pressure, 420–423
pressure, measuring, 428–431
static fluid on Earth, 420
Foci of ellipse, 187
Force, 122–124
compared to torque, 345
contact versus field forces, 123
defined, 123, 147
internal versus external forces, 123
newton as unit of, 127
point of application of, 260
See also Centripetal force; Drag force; Gravity; Kinetic friction; Normal force; Spring force; Tension force; Thrust
Force approach
compared to conservation approach, 214, 215, 228, 234–235, 282
in simple harmonic motion, 459–461
Force balance condition, 392, 409
used in applications, 396, 398, 400, 401, 403
Force pairs in Newton's third law, 141–142
Force-versus-position graph, 254, 255
Force-versus-time graph, 459–460, 477, 585
Forensic science, 307, 620
Foucault, Jean Bernard Leon, 476
Foucault pendulum, 210, 476, 478
Fourier, Jean Baptiste Joseph, 543
Fourier's theorem, 543–544, 545
Fracture, 404–409
Frame, reference, 104, 111
Free-body diagrams, 134–141, 147
Free expansion, 629, 645
change in entropy, 670, 675–676
Free fall, 45–50
defined, 45, 51
and gravitational force, 130
model rocket launch, 46
problem-solving strategy, 46
Free-fall acceleration, 45, 103
Earth's gravitational field, 197
Frequency
in harmonic motion, 454, 461
range of sound, 504
See also Angular frequency
Friction
and Newton's first law, 121
See also Kinetic friction; Rolling friction; Static friction
Fundamental forces, 146
Fundamental frequency, 535, 536, 544
Fundamental harmonic, 535, 538, 540

G

Galaxies
globular cluster, 262
gravitational force among, 193–194
Milky Way, and dark matter, 186, 201–202
spiral galaxies and expanding Universe, 22
See also Universe
Galileo Galilei, 6, 23, 122, 190, 203, 463, 554
Gases
in fluid model, 419
measuring pressure, 430
molar specific heat of, 631–636
in phase change, 599–601
See also Ideal gas; Kinetic theory of gases; Thermodynamic processes
Gas laws, 564–570
Avogadro's number, 567–568
constant volume, temperature, pressure relationships, 564–565
ideal gas law, 566–570, 575
standard temperature and pressure (STP), 568
Gauge pressure, 430, 570
Gay-Lussac's law, 565, 575
Gears, 343–344
Geocentric model, 185, 190, 203, 206
Geometric calculations of vectors, 60–65
adding vectors, 61, 62–63
drawing vectors, 60–61
finding direction of acceleration, 64
multiplying vector by scalar, 61–62
subtracting vectors, 62, 63
using a scale, 64–65
Geometry formulas, 684
Geosynchronous satellite, 198–199
Global dimming, 605
Global warming, 581
Goddard, Robert, 282
Graphical representation
in free-body diagrams, 134
solution for free fall problems, 49–50
using vector information, 71–72
Graphs
acceleration-versus-time graph (*See* Acceleration-versus-time graph)
energy, 232–235, 239
force-versus-position graph, 254, 255
force-versus-time graph, 459–460, 477, 585
position-versus-time graph (*See* Position-versus-time graph)
pressure-volume diagram, 597, 599, 606
velocity-versus-time graph (*See* Velocity-versus-time graph)
Gravitational field, 196–203, 206
of Earth, 196–199, 203–206
of a particle, 199
Gravitational force, 130

Gravitational mass, 195–196, 206
Gravitational potential energy
  near Earth, 218–221, 239
  universal, 221–223, 239
  and waterwheels, 374–376
Gravitational slingshot, 319–320
Gravity, 184–212
  center of, 260–261
  as conservative force, 257–258
  and escape speed of gas particles, 587
  as field force, 123, 130–131
  as fundamental force, 146
  gravitational field, 196–203
  gravitational field, variations in Earth's, 203–206
  historical theories of the Universe, 185–186
  Kepler's laws of planetary motion, 187–190, 194–195
  Newton's law of universal gravity, 190–196
Greek alphabet, 689
Greenhouse gas, 581
Groundspeed, 107–108

H

Harmonic frequency, 544
  for standing waves, 535, 538, 540, 545
Harmonic motion, 451–453
Harmonic number, 535, 540
Harmonic oscillator, simple, 452–453
Harmonic waves, 490–494, 515
  collection of oscillators, 492–493
  longitudinal waves, 488, 493
  speed of, 494
  transverse waves, 488, 491–492
Hawking, Stephen, 212
Head-to-tail vector addition, 61, 62–63, 79
Heat, 611–650
  of combustion, 662
  and conservation of energy, 613–615
  defined, 612, 614, 645
  degrees of freedom, 634–637
  equipartition of energy, 635–637
  first law of thermodynamics, 615–616
  of fusion, 621
  heat capacity and specific heat, 616–620
  latent heat, 620–622
  mechanisms for heat transfer, 638–644
  molar specific heat, 631–637
  thermodynamic processes, 625–631, 637–638
  of transformation, 621, 662
  of vaporization, 621
Heat capacity, 616–620, 645
Heat engines, 652–655, 656
Heat pump, 679–680
Heat reservoir, 626
Heat transfer, and second law of thermodynamics, 652
Heat transfer mechanisms, 638–644, 645
  conduction, 639–641
  convection, 641, 643
  radiation, 641–644
Helical motion, 91–92
Heliocentric model, 185, 190, 203, 206
Hertz, SI unit of frequency, 454
Hooke, Robert, 404
Hooke's law
  and elastic potential energy, 223–224
  mathematical expression of, 132
  and tensile strength, 406
Horizontal spring, 234
Horsepower, 270, 271
Hot-air balloons, Montgolfiers', 448
Human body
  body mass measuring device (BMMD), 481
  breathing as thermodynamic cycle, 624–625
  eardrums and air pressure, 420–421, 422
  ears and sound waves, 499–500
  food energy, 616
  heat transfer in what we wear, 639–641, 643–644
  specific heat to estimate time of death, 620
  typical values for estimates, 12, 693
Humidity, 603–604
Hydraulic levers, 429

I

Ideal fluid, 419, 444
Ideal fluid flow, 432–433, 444
Ideal gas, 566, 575
  entropy change during free expansion, 670, 675–676
  equation of state, 565, 566, 575
    compared to Van der Waals equation of state, 597–599
  thermal energy of, 616
  See also Gases
Ideal gas law, 566–570, 575
  Avogadro's number, 567–568
  standard temperature and pressure (STP), 568
  See also Kinetic theory of gases
Impulse, 307–310, 324
Impulse approximation, 307, 324
Impulse-momentum theorem, 309, 324, 585
Incident wave, 524
Incompressible fluids, 419, 432
Inelastic collisions, 311, 324
  conservation of momentum, 311–312
  one-dimensional, 312–315
Inertia, 124, 147
  See also Rotational inertia
Inertial mass. *See* Mass
Inertial reference frames, 124–126, 147
  and center of mass, 288
  Earth as noninertial reference frame, 204–205
Infrared (IR) radiation, 641
Infrasound, 504
Initial phase, in simple harmonic motion, 454, 479
Instantaneous acceleration. *See* Acceleration
Instantaneous angular acceleration. *See* Angular acceleration
Instantaneous angular velocity. *See* Angular velocity
Instantaneous power. *See* Power
Instantaneous speed. *See* Speed
Instantaneous velocity. *See* Velocity
Insulated system, 616
Insulator, thermal, 639
Integral calculus. *See* Calculus
Intensity of waves, 502–506, 515
  sound intensity, 503, 515
Interference, 526–531
  in pulses and one-dimensional waves, 527–528
  two- and three-dimensional, 528–531
Internal combustion engine, 655, 660, 662–664
Internal energy, 262, 265, 274, 615
Internal forces, 123
International Bureau of Weights and Measures, 7
International Space Station, 138
Inverse-square law, 191, 207
Inverse trigonometric functions and vector directions, 72–73
Irreversible processes, 656–657, 665, 667, 677
Irrotational ideal flow, 432
Isobaric (constant-pressure) process, 628, 630, 631, 632–634, 645
  change in entropy, 668–670
Isochoric (constant-volume) process, 627, 630, 631, 632–634, 645
  in Otto cycle, 660–661
Isolated system, 226, 239, 614, 616
Isothermal process, 626–627, 630–631, 645
  in Carnot cycle, 657–658
  change in entropy, 667–668

J

Joule, James, 215, 612, 617–619
Joule, SI unit of energy, 215
Jupiter
  atmospheric pressure, 445
  Kepler's third law and, 189
  tidal forces on, 211

K

Kelvin, Lord, 570, 571
Kelvin-Planck statement of second law of thermodynamics, 656, 677
Kelvin scale, 553–554, 570–571, 575
Kepler, Johannes, 2, 186, 187–190
Kepler's laws of planetary motion, 187–190
  empirical fits to data, 565
  first law, 187–188, 206
  second law, 188, 206
    and conservation of angular momentum, 380–381
  third law, 188–189, 194–195, 206
    and comet speed, 215–216
Kilogram, 7, 124, 131
Kinematic equations
  rotational, for constant angular acceleration, 338
  for simple harmonic motion, 453–456, 464, 479
  translational motion, for constant acceleration, 40–41, 95
Kinematics, 23, 50
  one-dimensional translational, 23
  rotational, 331
  rotational compared to translational, 331–338, 341–344, 347
  See also Rotational kinematics
Kinetic energy, 215–217, 239
  of Comet Halley, 214, 215–216
  energy graphs, 232–233
  and potential energy bar charts, 228–232
  See also Rotational kinetic energy
Kinetic friction, 133–134
  coefficient of, 134, 158, 161, 177
  model for, 160–162, 177
  as nonconservative force, 257
  See also Moving friction
Kinetic theory of gases, 580–610
  applied to gas temperature and pressure, 584–587
  assumptions, 581, 605
  average kinetic energy of gas, 582
  defined, 581, 605
  diffusion, 593, 595
  evaporation, 601–605
  humidity, 603–604
  ideal gas equations of state compared to Van der Waals, 597–599
  Maxwell-Boltzmann speed distribution, 589–591
  mean free path, 591–595
  phase changes, 599–601
  root-mean-square, 582–584
  and static model of gases, 588
  Van der Waals equation of state, 596–599

L

Laboratory cart, 27, 32–33
  kinetic energy of, 217
  work done on, 251, 253–254
Lagrange point, 199–201
Laminar flow, 432, 444
Latent heat, 620–622
Launch angle, 96, 109, 117
Law, 2
Law of inertia, 124
Law of reflection, 525–526, 545
Law of refraction, 507, 515
Law of universal gravity. *See* Newton's law of universal gravity
Laws, gas. *See* Gas laws
Laws of motion. *See* Newton's laws of motion
Laws of planetary motion. *See* Kepler's laws of planetary motion
Length, units of, 7
Levers, 351–353
Lift, 436–437
Light-year, 8–9
Linear expansion coefficient, 557, 558
Linear momentum, 283
Linear speed, 99, 100–101, 102
Line of action, 345
Liquids, in fluid model, 419
Local acceleration due to gravity, 130
Longitudinal harmonic waves, 488, 515
  sound waves, 497–500
    energy and power, 501–502
  wave function, 493, 515
Loop-the-loop challenge, 361–362, 371–372
Loudness, 502–506
Lowell, Percival, 207
Luminous matter, 186, 201

M

Mach angle, 513
Mach number, 513
Macroscopic state, 673–675
Magnitude of vectors, 72, 79
  and direction, 72, 73–75
Major axis of ellipse, 187, 207
Manometers, 428, 430, 572
Mars, motion of, 23
Mass density, 9, 17
Mass (inertial mass), 124, 147, 206
  center of, 260–261
    (*See also* Center of mass)
  compared to rotational inertia, 348–350
  and gravitational mass, 195–196
  in harmonic motion, 463
  reference table of typical masses, 694
  units of, 7, 124
Mass specific heat capacity, 617
Mathematical formulas, 683–688
Maximum range, 96, 111
Maxwell-Boltzmann speed distribution, 589–591, 606
Mean free path, 591–595, 606
  and diffusion, 593
Mean free time, 593, 595
Measurement errors, 10
Mechanical energy, 226
  for circular orbit system, 235–236
  for elliptical orbits, 236–238
  energy graphs, 232–233
  in harmonic waves, 500–501
  of simple harmonic oscillator, 470–471
  and work, 260
  and work-energy theorem, 265
  See also Conservation of mechanical energy
Mechanically isolated system, 616
Mechanical waves, 487
Mechanics
  conservation approach, 214, 215, 228, 234–235, 282
  Newtonian laws of motion, 120
Megalithic monuments, ancient, 332–333, 350, 352, 353
Melting temperatures, 621
Mercury, and Kepler's third law, 189–190
Mercury barometer, 430–431
Mercury thermometer, 559, 570
Meteors
  meteor shower, 248
  size of, 272–273
  work done by gravity on, 255–256
Meter, SI unit of length, 7

Microscopic state, 673–676
Microwave ovens, and heat transfer, 622
Milgrom, Mordehai, 187, 202
Millennium Clock, 468, 469
Millimeters of mercury, 421, 431
Minor axis of ellipse, 187, 207
Modeling, 156
Model rockets, 46
Models
  for kinetic friction, 160–162
  rotational compared to translational, 335
  of solar system, historical, 185
  for static friction, 156–160
Molar mass, 567–568
Molar specific heat, 617
  of gases, 631–636
  of solid, 637
Mole, 567, 568, 575
Molecular bonds, 133
Molecular collision, 323
Moment arm, 345
Momentum, 281–305
  angular (*See* Angular momentum)
  and center of mass, 284–287
  conservation of, 289–292
    (*See also* Conservation of momentum)
  defined, 281, 283, 299
  of a particle, 283–284
  and rockets, 282, 292–296
  and rocket thrust, 296–298
  of systems of particles, 287–289
Monatomic gas molecules, 632, 634, 636
MOND (modified Newtonian dynamics), 187, 202–203
Montgolfiers' hot-air balloon, 448
Moon
  astronomical data, 692
  and Earth system, center of mass, 285
  landing of *Eagle,* 297–298
  pendulum clock on, 466
  period of orbit and centripetal acceleration, 197–198
  temperature and radiation from Sun, 643–644
Most probable speed, 589–590, 606
Motion
  helical, 91–92
  rolling, 369–373
  vector components of motion variables, 77
  See also Circular motion; Multidimensional motion; Newton's laws; One-dimensional motion; Projectile motion; Relative motion in one dimension; Relative motion in two dimensions; Rotational motion; Simple harmonic motion; Uniform circular motion
Motion diagrams, 23–24, 51
  for fluid flow, 432–433
  for multidimensional motion, 87–88
  for projectile motion, 94
  for simple harmonic oscillator, 452
  for wave pulse, 489
Moving friction, 163
  change in thermal energy due to, 262–263
  as nonconservative force, 257, 263
Multidimensional motion, 85–118
  defined, 86–87
  motion diagrams, 87–88
  position and displacement, 88–90
  projectile motion, special case, 94–99
  relative motion in two dimensions, 106–110
  skateboarding, 86–87, 90
  uniform circular motion, special case, 99–103
  velocity and acceleration, 90–94
Multiplication of vectors
  cross product, 346
  dot product, 252
  multiplying vectors by scalars, 61–62
Music
  beats, 541–542
  concert halls and sound, 523, 526, 529–531
  musical notes, wavelength and frequency, 20, 536
  See also Standing waves
Musical instruments
  clarinet, 540–541, 544
  flute, 537–539
  and Fourier's theorem, 543–544
  guitar, 534–536, 542, 544, 579
  standing waves in, 533
  and temperature, 538–539

N

Nanotubes, carbon, 414
Natural angular frequency, 477–478
Natural (resonance) frequency, 478, 535
Net force, 126
Neutral static equilibrium, 392, 409
Newton, Isaac, 120, 122, 190–191, 404
Newton, SI unit of force, 127
  conversion between pound and newton, 130
Newton's law of universal gravity, 190–196
  Cavendish's experiment, 192–193
  gravitational mass and inertial mass, 195–196
  historical beginnings of, 185, 190–191
  and Kepler's third law, 194–195
Newton's laws of motion
  applications of, 153–183
    centripetal force, 168–177
    drag force, 163–165
    kinetic friction, 160–162
    normal force, 155–156
    rolling friction, 163
    static friction, 155–160
    terminal speed, 166–168
  first law of motion, 121–126
    1st statement of, 121–122, 146
    2nd statement of, 122–123
    3rd statement of, 126
    4th statement of, 283
    and train collision, 122
  force approach compared to conservation approach, 214, 215, 228, 234–235, 282
  second law of motion, 126–130, 147
    and angular momentum, 379–380, 383
    applied to systems of particles, 287–289, 290, 299
    compared to third law, 141–142, 144
    free fall and gravitational force, 130
    general form of, stated with momentum, 283, 299
    impulse-momentum theorem, 309
    in rotational form, 348, 351
    solving problems with free-body diagrams, 134–140, 143–144
  third law of motion, 141–146, 147
    compared to second law, 141–142, 144
    mathematical expression of, 141
    and rockets, 282
Nodes, 531, 545
  position of, 532–533
  resonance in musical instruments, 534–535, 537–538, 540–541
Nonconservative forces, 256–259
  defined, 257, 274
  dissipative, 263, 264, 274, 473, 475
Noninertial reference frames, 125, 147
  astronaut in elevator, 139
  Earth as, 204–205
Nonisolated systems. *See* Energy in nonisolated systems
Nonuniform circular motion, and centripetal force, 174–177
Nonuniform motion, 87
Nonviscous flow, 432, 444
Normal, 507, 525

Normal force, 132–133, 155–156
*n*th harmonic, 535, 540
Nuclear decay and conservation of momentum, 291–292

O

Object-spring oscillator, 461–463, 471–473, 480
One-dimensional elastic collisions, 315–320, 324
  gravitational slingshot, 319–320
  in two reference frames, 317–318
One-dimensional inelastic collisions, 312–315
One-dimensional motion, 22–23, 50
  compared to two- and three-, 86
  relative motion, 104–106
One-dimensional translational kinematics, 23
One-dimensional waves, 527–528
Open system, 296, 299
Orbital energies, 235–238
Order and disorder, 672–673
Order of magnitude, 14
Order-of-magnitude calculation, 14
Order-of-magnitude estimates, 12–17, 18
Origin, 24
Oscillations, 450–485
  connection with circular motion, 456–458
  damped harmonic motion, 473–476
  defined, 451
  driven oscillators, 476–479
  dynamics of simple harmonic motion, 459–461
  energy in simple harmonic motion, 470–473
  harmonic motion, 451–453
  and harmonic waves, 492–493
  kinematic equations of simple harmonic motion, 453–456
  object-spring oscillator, 461–463
  physical pendulum, 466–468
  simple pendulum, 463–466
  torsion pendulum, 468–470
  See also Simple harmonic oscillator
Otto cycle, 660, 677
Otto engine efficiency, 661–662, 677
Overdamped oscillator, 475, 479

P

Parabolic path, 96
Parallel-axis theorem, 366–367, 384
Parallelogram
  and cross products, 347
  vector addition, 61, 79
Partial pressure of gas, 603–604
Particle accelerator, 170
Particle model, and displacement and translation, 28–29
Particles, 23, 50, 259
  colliding, 313, 315–316
  and deformable objects, 259
  and degrees of freedom, 634–637
  law of universal gravity and, 192
  momentum of, 283–284
Pascal, SI unit of pressure, 421
Pascal's principle, 428–429, 444
Path independence
  change in potential energy, 219–220
  work done by conservative force, 257
Path integral, 254
Pendulums
  ballistic, 304
  clocks, 450–451, 463, 465–466
  conical, 480
  Foucault, 210, 476, 478
  physical, 466–468
  simple, 463–466
  torsion, 468–470
Performance coefficient, 664–665, 677
Perihelion, 188, 377
Period
  in circular motion, 99, 111
  of planet, 188, 189
  of satellite in low orbit, 212
  in simple harmonic motion, 453
Periodic motion, 451
Periodic table of elements, 696–697
Periodic wave, 490
Phase, in simple harmonic motion, 454, 479
Phase changes, 599–601, 621
Phase diagram, 600–601, 606
Physical pendulum, 466–468, 480
Physical principle, 2
Physics, 1–6
Pistons, 421, 437, 623–629, 637–638, 652, 654–655, 657–658
Planetary motion
  Copernican theory, 2–3
  Kepler's laws of (*See* Kepler's laws of planetary motion)
Plane waves, 502
Point of force application, 260
Point source (of wave), 502
Polar coordinates, 100, 334
Polar coordinate system, 100
Polyatomic gas molecules, 632, 634
Population of selected places, 693
Position, 24–25
  angular, 334–335, 353
  defined
    multidimensional motion, 88, 110
    one-dimensional motion, 24, 50
  displacement as change in, 28
  in multidimensional motion, 88–90
  in projectile motion, 94–95
  rotational compared to translational, 335
  for simple harmonic oscillator, 454, 479
  in uniform circular motion, 88, 99
Position vector, 77
Position-versus-time graph, 25–27, 51
  average velocity from, 32–34
  features of, 26
  for laboratory cart, 27
  for simple harmonic motion, 452–453, 454
  for wave function, 489
Potential energy, 217–218, 239
  bar charts for, 228–232
  elastic, 223–225, 230–231, 239
  general expression for change in, 257, 274
  gravitational, near Earth, 218–221
  near surface of planet, 218–219
  path independence, 219–220
  relationship to work, 256
  universal gravitational, 221–223
Potential energy curves, 232–233
Pound, 7, 127, 130, 131
Power, 270, 274
  absorbed from sunlight, 642–644
  instantaneous, 269–273
  in rotational motion, 373–376, 384
  transport in harmonic waves, 500–501
Power-of-10 prefixes, 7, 689
Precipitation, 605
Pressure, 420–423
  defined and units of, 421, 444
  in gas laws, 564–566
  kinetic theory applied to, 584–586
  measuring, 428–431
  in moving fluid, 437–438
  object's volume change, 422
  Pascal's principle, 428–429
  variation with depth in static fluid, 421–422
  See also Constant-pressure (isobaric) process

Pressure amplitude, 499–500
Pressure-volume diagram, 597, 599, 606
Pressure waves, 498–500, 516
Principia (Newton), 120, 122
Principle of conservation of mechanical energy, 226, 227–228
Principle of equipartition of energy, 631, 635–637, 645
Probability, entropy, and second law of thermodynamics, 673–676
Problem-solving strategies, 5–6
 Bernoulli's equation, 438
 when centripetal force is present, 169
 constant acceleration, 42
 constant angular acceleration, 338
 free fall, 46
 Newton's second law with free-body diagrams, 135
 principle of conservation of mechanical energy, 229
 projectile motion, two-dimensional, 97
 static equilibrium, 394–395
 work-energy theorem, 266
Profile of wave, 488
Projectile, 94
 in collision, 312
Projectile motion, 94–99
 and air resistance, 110
 defined, 94, 111
 equations for, 94–95
 parabolic path, 96
 problem-solving strategy, 97
 range equation, 96
Ptolemy, 185
Pulleys, free-body diagrams and Newton's second law, 140–141
Pulsars, 339, 355, 382
Pulses, 487–490
 defined, 488
 interference in, 527–528
 and superposition, 522–523, 524–526
 wave function for particular pulse, 488–490
Pure rolling, 371

Q

Quasistatic thermodynamic processes, 623

R

Radian, 100, 334
Radiation, heat transfer, 641–644, 645
Random walk, 593
Range, 96, 111
Range equation, 96
Rays (in wave fronts), 502
Real gases
 modeled as ideal gas, 581
 Van der Waals equation of state, 596–599
Reese, Cassidi, 52, 60, 63, 74, 78, 167
 See also Skydiving
Reference configuration, 219
 for gravitational potential energy, 219
 for spring potential energy, 224
 for universal gravity, 222–223
Reference frames, 104, 111, 124–126
 one-dimensional collision in, 317–318
Reference line, 334
Reference tables, 689–697
Reflection, 524–526
 fixed-end, 524, 545
 free-end, 524–525, 545
Refraction, 506–507, 515
Refrigerators and second law of thermodynamics, 652, 664–665, 677
Relative humidity, 603–604, 606
Relative motion in one dimension, 104–106
 defined, 104, 111
 and reference frames, 124
Relative motion in two dimensions, 106–110
 air resistance, 110
 groundspeed, 107–108
 launching ball from cart, 109–110
Resistive forces, 163
Resonance, 478, 479
 on string fixed at both ends: guitar, 534–536
 in tube open at both ends: flute, 537–539
 in tube open at one end: clarinet, 540–541
Resonance frequency, 478, 535
Restoring force, 132, 464
Resultant (resultant vector), 61
Retrograde motion, 480
Reverberation, 526, 546
Reversible cycle, 657, 666–667
Reversible process, 656–657, 677
Right-handed coordinate systems, 67–68, 80, 333
Right-hand rule convention
 for angular momentum, 376
 for angular velocity, 336, 354
 for cross products, 346–347
 for translational motion, 68, 80
Rigid object, 332, 353, 377
rms speed, 582–584, 587
rms velocity, 582–584, 586
Roche limit, 211
Rockets
 first rocket equation, 293–294, 299
 launching of, 38–39, 46
 and momentum, 282, 292–296
 second rocket equation, 296, 298, 299
Rocket thrust, 296–298
 and drag, 296–297
Roller coaster, 231–232
Rolling friction, 134, 163, 177
 coefficient of, 158, 163
 as nonconservative force, 257
 See also Moving friction
Rolling motion, 369–373
Root-mean-square (rms) speed and velocity, 582–584, 606
Rope, unstretchable, 133
Rotating rigid object, 332
Rotational dynamics, 347–353
 defined, 331
 levers, 351–353
 Newton's second law in rotational form, 348, 351
 rotational inertia, 348–350
Rotational inertia, 348–350, 354, 362–367, 384
 of bicycle tire, 366
 of continuous objects, 365–367
 and parallel-axis theorem, 366–367
Rotational kinematics, 333–337
 angular acceleration, 336–337
 angular acceleration, constant, 338–339
 angular displacement, 334–335
 angular position, 334–335
 angular velocity, 335–336
 compared to translational, 331–338, 341–344, 347
 defined, 331
Rotational kinetic energy, 367–369, 384
 Atwood machine, 368–369
 gravitational potential energy and waterwheels, 374–376
Rotational motion, 23
 and conservation of motion, 360–362
 cross product, 346–347
 relationship to circular motion, 340–344
 rolling motion, 369–373
 torque, 334–346
 work and power, 373–376

Rotation axis, 332, 353
Rutherford, Ernest, 329–330

S

Saturated air, 602
Saturated vapor pressure, 602, 603, 604, 606
Scalar components, 25, 68–69, 79
Scalar product, 252–254, 274
Scalars (scalar quantities), 23, 51
 multiplication of vectors by, 61–62
Scales, using with vectors, 64–65
Schwarzchild radius, 242
Scientific evidence, 2–3, 17
Scientific notation, 7, 11
Second law of thermodynamics. *See* Thermodynamics, second law of
Second, SI unit of time, 6
Seismic waves, 518, 519
Semimajor axis, 187–189
Shear deformation, 406–407
Shear modulus, 405, 407, 410
Shear strain, 407, 410
Shear stress, 405, 406, 410
Shock absorber, 475
Shock waves, 513, 515
Significant figures, 10–11, 14, 17
Simple harmonic motion, 453, 479
 dynamics of, 459–461
 energy in, 470–473
 kinematic equations of, 453–456
  acceleration versus time, 455
  position versus time, 454
  velocity versus time, 455
Simple harmonic oscillator, 452–453
 defined, 453, 479
 harmonic wave generation, 491
 mechanical energy of, 470–471
 and standing harmonic wave, 532
Simple pendulum, 463–466, 480
SI system, 6, 17
SI units
 common prefixes for, 7
 estimates using, 12–13, 693–694
 standards for fundamental units, 553
Skateboarding, 86–87, 90
Skydiving
 and Newton's laws of motion, 154–155, 164, 167–168
 and vectors, 60, 63–64, 74–75, 78–79
Slug, 124
Smog, 591
SOHO (solar and heliospheric satellite), 199–201
Solar constant, 643
Solar system historical models, 185–186
Sonar waves, 543
Sonic boom, 513
Sound intensity, 503, 515
Sound level, 503, 516
Sound speed barrier, 42–44
Sound waves (sound), 497–500, 516
 speed of sound, 498
 See also Superposition
Source of force, 123
Spacecraft
 escape speed, 229–230
 gravitational slingshot, 319–320
 Moon landing, 297–298
Spacesuits, 131, 643–644
Speakers, 499, 505–506
Specific heat, 616–620, 645
Speed, 35, 50
 angular (*See* Angular speed)
 average (*See* Average speed)
 instantaneous, 35–36
  in circular motion, 92
  of harmonic wave, 494
  linear or translational speed, 100
  of point on rotating object, 340–341
  of sound, 498
  sound speed barrier, breaking, 42–44
  of transverse wave, 494–496
 most probable, 589–590, 606
 reference table of typical speeds, 694
Spring-block system
 displacement, velocity, and speed, 29–30, 35–36
 elastic potential energy, 223–224
 work and mechanical energy, 260
Spring constant, 132
Spring force, 132
Spring-loaded dart gun, 230–231, 267–268
Spring potential energy, 223–225
Springs
 horizontal, 234
 torsion, 468
 vertical, 224–225
Spring scales, 132, 138
Stable equilibrium, 246
Stable static equilibrium, 391, 409
Standard temperature and pressure (STP), 568
Standard unit, 6, 17
Standing waves, 531–541
 beats, 541–543
 defined, 531, 545
 making, 533–534
 in musical instruments, 533
 resonance
  in clarinet, 540–541
  in flute, 537–539
  in guitar, 534–536
 wave function of, 531–532
 See also Superposition
Stars
 pulsars, 339, 355, 382
 Sirius, 186
 See also Sun; Universe
State of matter, 419
State variables, 612, 665, 670, 673
Static equilibrium, 391–392, 409
 conditions for equilibrium, 392–394
 examples of, 394–404
 problem-solving strategy, 394–395
Static fluid, on Earth, 420, 444
 pressure variation with depth in, 421–422
Static friction, 134
 and centripetal force, 171
 coefficient of, 157–158, 159
 model for, 156–160
 and normal force, 155–156, 177
 as zero-work force, 261
Static model of gases, 588–589
Steady flow, 432, 444
Steam engines, 654–655, 659–660
Stefan-Boltzmann constant, 642, 644, 645
Stefan-Boltzmann equation, 641, 645
Stirling engine, 678–679
Stonehenge trilithons, 332–333, 352, 359
Strain, 405, 410
Streamline, 433, 434
Strength of materials, 405
Stress, 404–405, 410

Strong force, as fundamental force, 146
Structural engineering, 390, 409, 555
Subject of force, 123
Sublimation, 600
Submarine, 104–105, 106–107
Sun
  intensity of sunlight, 519
  as nonisolated system, 247
  nuclear fusion of, 244
  power absorbed from sunlight, 642–644
Superimposing, 523, 545
Superposition, 522–531
  beats, 541–543
  defined, 523
  Fourier's theorem, 543–544
  interference, 526–531
  reflection, 524–526
  See also Standing waves
Supersonic, 513
Symbols and abbreviations, 689–690
System, 123, 147, 215
  isolated, 226, 239, 614, 616
Systems of particles
  momentum of, 288–289, 299
  Newton's second law applied to, 287–289, 290
Systems of units, 6–9

T

Tail-to-tail vector subtraction, 62, 80
Tangential acceleration, 341, 464
Tangential component, 175
Target in collision, 312, 316
Temperature, 553, 555, 575
  in gas laws, 564–566
  kinetic theory applied to, 584–586, 606
  relation to thermal energy, 261
  standards, 570–574
  and thermodynamics, 552–554
  units of, 553–554
  See also Heat; Heat transfer mechanisms; Thermodynamic processes
Temperature scales, 553–554
Tempered glass, 561–562
Tensile deformation, 405–406
Tensile strength, 405, 406, 410
Tensile stress, 405, 410
Tension, 133
Tension force, 133
Terminal speed, 166–168, 177
Theory, 2, 17
Thermal conduction, 639
Thermal conductivity, 639, 645
Thermal contact, 555, 575, 614
Thermal energy, 261–263
  change due to moving friction, 262–263
  defined, 261, 274, 553, 645
  work and heat, 613–615
  See also Heat
Thermal equilibrium, 555, 575
Thermal expansion, 555–560
  of aluminum sheet, 563–564
  macroscopic observation of, 557–560
  microscopic model of, 556
  thermal volume expansion, 557–558
  of water, 562
Thermal insulator, 639
Thermal linear expansion, 557, 558, 575
Thermally insulated system, 616
Thermal pollution, 655
Thermal radiation, 641
Thermal stress, 560–562
Thermal volume expansion, 557–558
Thermodynamic cycle, 624, 628–629, 653
Thermodynamic processes, 625–631
  adiabatic process, 625–626, 637–638
  constant-pressure (isobaric) process, 628, 630, 631, 632–634
  constant-volume (isochoric) process, 627, 630, 631, 632–634
  cyclic process, 628–629
  defined, 623, 645
  free expansion, 629
  isothermal process, 626–627, 630–631
  quasistatic, 623
  work in, 623–625
Thermodynamics, 2, 553, 574
  first law of, 615–616, 617, 625, 645
  second law of, 651–682
    Carnot engine, 656–659
    entropy, 665–670
    entropy and probability, 673–676
    heat engines, 652–655
    order and disorder, 672–673
    Otto cycle, 660–662
    refrigerators, 652, 664–665
    statements of
      Clausius, 652, 671–672
      disorder, 673
      entropy, 670–672
      Kelvin-Planck, 656
      probability, 674
  and temperature, 552–554
  third law of, 659
  zeroth law of, 555, 574
Thermometers
  alcohol, 558–559, 570
  bimetallic strip, 560
  constant-volume ideal gas, 572–574
  invention of, 554, 558
  mercury, 559, 570
  temperature scales, 553–554
Third law of thermodynamics, 659
Three-dimensional Cartesian coordinate system, 67–68
Three-dimensional interference, 528–531
Three-dimensional motion
  compared to one- and two-, 86
  helical, 91–92
  See also Multidimensional motion
Three-dimensional waves, 502–504, 506
Thrust, 296–298, 402
ThrustSuperSonicCar (ThrustSSC), 42–44
Time
  in simple harmonic motion, 454–456
  units of, 6–7, 695
  and velocity
    and constant acceleration, 40
    and displacement, 40–41
  See also Position-versus-time graph; Velocity-versus-time graph
Time constant in damped oscillations, 474–475, 479
Timekeeping devices. *See* Clocks
Time's arrow, 671
Torque, 344–346, 354
  compared to force, 345
  and conservation of angular momentum, 379–380
  work in terms of, 374
Torque balance condition, 393–394, 409
  used in applications, 396, 398, 400, 402, 404
Torricelli, Evangelista, 421, 430
Torsion balance, 192
Torsion modulus, 407
Torsion pendulum, 468–470, 480
Torsion spring, 468

Torsion spring constant, 468
Total external forces, 290–291
Total force, 126
Train collision, 120–121, 122, 137–138, 146, 307, 314–315
Translation, and displacement and particle model, 28–29
Translational momentum, 283
Translational motion, 23, 50
  compared to rotational motion, 331–338, 341–344, 347
Translational speed, 99, 100
  of point on rotating object, 340–341
Transmission electron microscope (TEM), 577
Transverse harmonic waves, 491–492
  defined, 488, 515
  energy and power, 500–501
  piano wires, 496
  propagation speed of wave on rope, 494–496, 516
  wave equation for, 514
Traveling waves, 486–521
  defined, 487
  diffraction, 507
  Doppler shift, 507–513
  energy transport in waves, 500–502
  harmonic waves, 490–494
  longitudinal waves: sound, 497–500
    (*See also* Longitudinal harmonic waves)
  mechanical waves, 487
  pulses, 487–490
  refraction, 506–507
  shock waves, 513
  transverse waves on a rope, 494–496
    (*See also* Transverse harmonic waves)
  two- and three-dimensional waves, 502–506
  wave equation, 513–515
Triatomic gas molecules, 632, 634
Trigonometry, 684–685
  inverse functions and vector directions, 72–73
Triple point, 571, 600–601
Triple-point cell, 571, 573
Two-dimensional collisions, 320–323, 324
Two-dimensional interference, 528–531
Two-dimensional motion, 86
  See also Linear speed; Multidimensional motion; Projectile motion
Two-dimensional waves, 502–504, 506
Tycho Brahe, 185–186

U

Ultrasound, 487, 504–505
Ultraviolet (UV) radiation, 641
Uncertainty, 10, 17, 687–688
Underdamped oscillator, 475, 479
Unequal-arm balance, 416
Uniform circular motion, 99–103
  acceleration and velocity in, 93–94
  centrifuge, 103
  centripetal acceleration, 101–103
  centripetal force, 168–174
  defined, 99, 111
  linear and angular speed, 100–101, 102
  mathematical description of circular path, 88–90
  polar coordinate system, 100
  velocity and speed in, 92
  See also Circular motion
Uniform motion, 87
Units, 6–8, 689–690
  See also SI units; U.S. customary units
Unit vectors, 25, 67, 73
Universal gas constant, 567
Universal gravitational constant, 191–192
Universal gravitational potential energy, 221–223, 239
Universe
  astronomical data, 692
  dimensions in, 85
  entropy of, 670–671, 673
  historical theories of, 185–186
  and laws of physics, 1–2
  spiral galaxies and expanding Universe, 22
  See also Comets; Dark matter; Galaxies; Meteors; Stars
Unstable equilibrium, 246
Unstable static equilibrium, 391, 409
U.S. customary units, 12–13, 693–694

V

Van der Waals equation of state, 596–599, 606
  compared to ideal gas equations of state, 597–599
Vapor, 600
Vector components, 68–75
  combining vectors by components, 76–78
  defined, 25, 68, 79
  expressing vectors in component form, 70–71
  graphical representation of, 71–72
  inverse trigonometric functions, 72–75
  magnitude and direction, 72, 73–75
  of motion variables, 77
  resolving vectors into components, 69–71, 79
Vector field, 196
Vector product, 346–347, 354, 376
Vectors (vector quantities), 59–84
  adding by combining components, 76–78, 79
  adding geometrically, 61, 62–63
  angular velocity, average and instantaneous, 335
  basics of, 24–25
  components of (*See* Vector components)
  coordinate system (*See* Cartesian coordinate system)
  defined, 23, 51
  drawing, 60–61
  force as, 123
  free-body diagrams, representations in, 134
  geometric calculations, 60–65
  momentum as, 281
  multiplying of (*See* Multiplication of vectors)
  skydiving example, 60, 63–64, 74–75, 78–79
  subtracting by combining components, 76–78, 79, 80
  subtracting geometrically, 62, 63
  torque, 345–347
  unit, 25, 67
  using a scale, 64–65
Velocity
  angular (*See* Angular velocity)
  average (*See* Average velocity)
  instantaneous, 34–37
    calculus derivation, 41–42, 90–91
    in circular motion, 92, 93–94
    defined
      one-dimensional motion, 34–35, 50
      three-dimensional motion, 111
    as function of time for constant acceleration, 40
    in multidimensional motion, 90–91, 92
    of point on rotating object, 340–341
    in projectile motion, 94–95
    in relative motion, 104, 107
    for simple harmonic oscillator, 455, 479
Velocity-versus-time graph, 36, 51
  displacement from, 36–37, 38
  for simple harmonic motion, 453, 455
Venturi meter, 442, 447
Venturi tube, 440–442
Venus, 190, 203, 581
Vertical spring, 224–225
Vibrating object, 535

Vibrational motion, 23
Vibrations, 451
Viscosity, 432
Volume
  changed by pressure, 422
  in gas laws, 564–566
  pressure-volume diagram, 597, 599, 606
  See also Constant-volume (isochoric) process
Volume expansion coefficient, 557–558, 562
Volume flow rate, 434–435
Volume strain, 422
Volume stress, 422

W

Water
  cycle of Earth, 605
  molecular collision, 323
  thermal conductivity of, 639–641
  triple point of, 571, 600–601
Water calorimeter, 619
Waterwheels, 374–376
Watt, unit of power, 270
Wave equation, 513–515
Wave fronts, 502
Wave functions, 488, 515
  of longitudinal harmonic wave, 493, 515
  of standing waves, 531–532
  of transverse harmonic wave, 492, 514
Wavelength, 492
Waves, 487
  See also Harmonic waves; Standing waves; Superposition; Traveling waves
Weak force, as fundamental force, 146
Weight, 124, 130
  compared to apparent weight submerged in fluid, 427–428
  reference table of typical weights, 694
  See also Apparent weight
Weighted average of position, as center of mass, 285
Weightlessness of astronaut, 138–140
Whisper dish/whispering gallery, 525–526, 546
Work, 248, 274
  done by constant force, 248–251
    in same direction as displacement, 249–250
  done by general force, 255
  done by gravity on meteor, 255–256
  done by nonconstant force, 254–256
  done by piston, 437
  general expression for work done, 254, 274
  and mechanical energy, 260
  relationship to potential energy, 256
  in terms of torque in rotational motion, 373–376, 384
  in thermodynamic processes, 623–625
Work-energy theorem, 264–269, 274
  and Bernoulli's equation, 438
  and first law of thermodynamics, 615
  problem-solving strategy, 266
Working substance, 654–655, 656
Work-kinetic energy theorem, 250, 274
Work-mechanical energy theorem, 260, 262, 274

Y

Young's compression modulus, 406
Young's modulus, 405, 407–408, 410
  in speed of sound, 498
  in thermal stress, 560, 561
Young's tensile modulus, 406

Z

Zero significant figures, 14
Zeroth law of thermodynamics, 555, 574
Zero-work force, 259, 261, 274